D1447457

SPACE SCIENCE SERIES

Tom Gehrels, General Editor

Planets, Stars and Nebulae, Studied with Photopolarimetry
Tom Gehrels, editor, 1974, 1133 pages

Jupiter
Tom Gehrels, editor, 1976, 1254 pages

Planetary Satellites
Joseph A. Burns, editor, 1977, 598 pages

Protostars and Planets
Tom Gehrels, editor, 1978, 756 pages

Asteroids
Tom Gehrels, editor, 1979, 1181 pages

Comets
Laurel L. Wilkening, editor, 1982, 766 pages

Satellites of Jupiter
David Morrison, editor, 1982, 972 pages

Venus
D.M. Hunten, L. Colin, T.M. Donahue and V.I. Moroz, editors,
1983, 1143 pages

Saturn
Tom Gehrels and Mildred S. Matthews, editors, 1984, 968 pages

Planetary Rings
Richard Greenberg and André Brahic, editors, 1984, 784 pages

Protostars and Planets II
David C. Black and Mildred S. Matthews, editors, 1985, 1293 pages

Satellites
Joseph A. Burns and Mildred S. Matthews, editors, 1986, 1021 pages

The Galaxy and the Solar System
Roman Smoluchowski, John N. Bahcall and Mildred S. Matthews, editors,
1986, 485 pages

Meteorites and the Early Solar System
John F. Kerridge and Mildred S. Matthews, editors, 1988, 1269 pages

Mercury
Faith Vilas, Clark R. Chapman and Mildred S. Matthews, editors,
1988, 794 pages

Origin and Evolution of Planetary and Satellite Atmospheres
S.K. Atreya, J.B. Pollack and M.S. Matthews, editors, 1989, 881 pages

Asteroids II
Richard P. Binzel, Tom Gehrels and Mildred S. Matthews, editors,
1989, 1258 pages

Uranus
Jay T. Bergstralh, Ellis D. Miner and Mildred S. Matthews, editors,
1991, 1076 pages

The Sun in Time
C.P. Sonett, M.S. Giampapa and M.S. Matthews, editors, 1991, 996 pages

Solar Interior and Atmosphere
A.N. Cox, W.C. Livingston and M.S. Matthews, editors, 1991, 1414 pages

Mars
H.H. Kieffer, B.M. Jakosky, C.W. Snyder and M.S. Matthews, editors,
1992, in press

SOLAR INTERIOR
AND
ATMOSPHERE

SOLAR INTERIOR AND ATMOSPHERE

A.N. Cox
W.C. Livingston
M.S. Matthews

Editors

With 101 collaborating authors

THE UNIVERSITY OF ARIZONA PRESS
TUCSON

About the cover:

Photograph of the total eclipse of 11 July 1991 from Mauna Kea, Hawaii. Taken as part of the International Multistation Coronal Experiment, this picture represents the first of a series of recordings on coordination with other telescopes in Baja and Brazil. The objective was to measure temporal changes in coronal structure during passage of the lunar shadow across the Earth's surface; and such changes were detected. This is a maximum activity corona.

Proceeding clockwise from the top of the image, the N-W limb, we find near 2 o'clock a fan streamer with embedded H-alpha prominence and coronal rays; at 4 o'clock a helmet streamer over the south heliographic pole; at 8 o'clock another prominence; and at 10 o'clock another helmet streamer. The team leader was Serge Koutchmy, Institut d'Astrophysique, Paris. The instrument was a 20-cm aperture, 3-meter focal length telescope with radial graded filter in front of Ektachrome 64 film. Artifacts as a result of the filter produced the halo and other background effects.

The University of Arizona Press

Copyright © 1991
The Arizona Board of Regents
All Rights Reserved

⊖ This book is printed on acid-free, archival-quality paper.
Manufactured in the United States of America.

96 95 94 93 92 91   6 5 4 3 2 1

Library of Congress Cataloging-in-Publication Data

Solar interior and astmosphere / A.N. Cox, W.C. Livingston, M.S.
    Matthews, editors ; with 102 collaborating authors.
        p.   cm. — (Space science series)
    Includes bibliographical references and index.
    ISBN 0-8165-1229-9
    1. Sun—Internal structure.   2. Solar atmosphere.   I. Cox, Arthur
N.   II. Livingston, W. C. (William Charles), 1927–
III. Matthews, Mildred Shapley.   IV. Series.
OB539.I5S65   1991
523.7'6—dc 20   91-34154
                                                          CIP

British Library Cataloguing in Publication data are available.

To
Leo Goldberg
and
Paul Ledoux

CONTENTS

COLLABORATING AUTHORS

Preface

This textbook for graduate students and solar and stellar astrophysicists studying the Sun is the most comprehensive of many recent books on this subject due to the contributions of 102 authors, all with considerably different interests in solar research. Readers may notice that during the more than two years that these 36 chapters and 3 appendices (some with as many as eight authors) were written, significant progress has been made, because some of these chapters are more up-to-date than those that were finished earlier. Such variations are inevitable in a grand project like this.

The usual procedures for writing the books in the University of Arizona Space Science Series were followed for this one. The first editor was contacted by Tom Gehrels, General Editor of the Space Science Series, early in 1986. Agreement to attempt this book was consummated in January 1987 when Livingston agreed to become a coeditor using his knowledge of solar observations to assemble a team of writers for those subjects. An initial letter announcing our proposal was mailed to about 800 potential authors in March 1987. A rough outline of over 30 chapters using about 180 responses was made at our only planning meeting in June 1987. In August an announcement letter was sent to over two hundred stellar astronomers and astrophysicists seeking their specific interests. The three editors and 22 other members of the Scientific Organizing Committee organized the structure of the book and selected as authors those who were particularly enthusiastic about participating. When most of the authors were committed, a meeting was held November 15–18, 1988 in Tucson to hear scientific presentations of the chapter material represented by abstracts, many generated in conversations during the earlier Baltimore International Astronomical Union General Assembly. In Tucson, as can be expected from any meeting with aggressive scientists, we not only found necessary revisions to the book plan, but we also inspired new research projects. Many of these research results are found in the chapters of this book.

The actual writing of these chapters was very time consuming because the experts assembled all had many other duties and interests. Even getting two or three referee reports promptly was a challenge for the editors. Some issues are addressed in several chapters, and sometimes different conclusions are reached. Such inconsistencies have been retained because, obviously, consensus has not been reached and the flavor and focusing of controversies is what makes science so interesting.

Financial support for the editorial staff at the University of Arizona has been provided by NASA and by the Institute for Geophysics and Planetary Physics of the Los Alamos National Laboratory. In addition, appreciation is due to the many voluntary efforts of the authors and their own support teams.

The editors and authors dedicate this book to two outstanding stellar astrophysicists who passed away while this book was being planned and written: Leo Goldberg and Paul Ledoux. Their personal influences on our interests in solar physics and our research results were greater than most will ever realize.

Arthur N. Cox
William C. Livingston
Mildred S. Matthews
July 1991

PART I
The Solar Interior

THE GLOBAL SUN

JEAN-CLAUDE PECKER
Collège de France

After a definition of the various terms used to identify the solar layers, from the center to the exterior, and a physical description of these layers, we show that various couplings are controlling the physics in the core and the outer layers, and even the planets. One of these couplings is between convection, rotation and magnetism (the dynamo), and another coupling is between solar activity and planetary physics. These couplings allow us to use observed data (oscillations, neutrinos, emergence of active regions, and of course their evolution) to infer properties of the solar interior. The theoretical knowledge of the Sun must take into account the existence of these couplings, as well as the existence of another type of coupling, the one that links the past and the present states of our star the Sun.

The Sun presents us with a thousand-fold face. Depending on the ways we observe it, whether it be through a groundbased telescope, during an eclipse, or from a space observatory, we see a different Sun. One can distinguish the wavelengths in which it is observed: ultraviolet, radio, visible, or X ray. One can distinguish the time when it is observed: near its maximum activity, or near its minimum activity. Still, this multifaced Sun of ours is a single celestial object, and it is this idea of the Sun as a whole that we wish to develop in this chapter.

I. THE SUN'S LAYERS

Operational Definition

In this book, the various regions of the Sun will be discussed. Starting from the outside and going inward, we have first the solar wind, then the solar corona, the chromosphere, the photosphere, the interior and finally the deep interior or core where nuclear reactions take place, and where the observed neutrinos originate. The distinction between some of these layers is fuzzy and somewhat artificial. One should therefore remember the definition and etymology of these names, for the benefit of clarity.

The *photosphere* is the part of the Sun from which come the observed visible photons, the largest part of the radiated energy, or light, of the Sun. Hence its name, from the Greek: *photos* (in English: *light*). Its outer boundary (top) is the layer at which the tangential optical depth is equal to unity (Fig. 1). This operational definition is, of course, a function of wavelength, but as the opacity does not vary by more than a factor of 2 between the part of the visible spectrum where it is most transparent ($\sim$ 1700 nm) and the part of the visible spectrum where the matter is more opaque ($\sim$ 800 nm), the difference in height is of the order of 100 km between the layers responsible for the two extremes. This is very small, when we take into consideration that this upper layer is the one that appears in the sky as the visible limb of the Sun, defined, observationally, with an accuracy not better than perhaps 0″.2, i.e. $\sim$ 150 km.

The inner boundary of the photosphere (bottom) is less easy to define. We can say that it is the layer from which only 1% (or even 0.1%) is directly observed without being absorbed. With 1%, it means that the optical depth has τ_2 such that $e^{-\tau_2} = 0.01$; (or with 0.1%, it means that the optical depth is such that $e^{-\tau_2} = 0.001$). This leads to $\tau_2 \simeq 4.5$ (or, respectively, $\tau_2 \simeq 6.9$). In geometrical depths, $\Delta h \simeq 370$ km (or respectively, $\Delta h \simeq 385$ km), when Δh is the difference in height between top and bottom of the photosphere. These borders of course are artificial. Below the photosphere, is the "interior," which is not directly observable.

Above the photosphere, one defines the *chromosphere* in the same way as the photosphere, except that the opacity is taken at the center of the Hα line (i.e., orders of magnitude larger) instead of being that of the continuum spectrum. This defines the upper boundary of the chromosphere. The lower border is of course defined as the upper border of the photosphere. The suffix

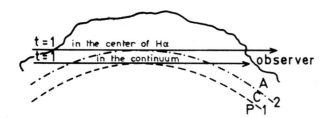

Fig. 1. The operational definition of the solar limb. The optical depth equal to unity, in the continuum, defines the solar limb, and the surface (or outer boundary or top) of the photosphere. The optical depth equal to unity in the center of H α line defines the top of the chromosphere.

sphere reflects a general feature of the observed layers. Spherical symmetry appears to be a realistic description, at least in first approximation, of the observations, as they were conducted in the 19th century, by Janssen, Lockyer, Hale or Deslandres. It emphasizes the dominant role of gravity in these layers. Needless to say, these excellent observers did notice, during eclipses, or when using the large-slit spectra of the solar limb, that the chromosphere, defined by the Hα emission, has a somewhat irregular shape; hence the notion of *prominences*, or local departures from the globally spherical appearance of the chromosphere. Later, in the early 1940s, Roberts (1945) discovered the *spicules* as constituents of the upper layers of the chromosphere.

Coming to the *corona,* initially observed only during eclipses (and then sometimes accredited to a lunar atmosphere), the suffix sphere cannot be employed. The irregular shape of the corona, easily visible during total eclipse, is very characteristic. One sees jets, streamers, loops, etc.; modern techniques completely confirm this view. It is obvious from the first that there is no trace of sphericity; one cannot speak of a coronosphere. This essentially means that in the corona we are very far from hydrostatic equilibrium.

The corona has been divided into three components: the E corona, responsible for emitting emission lines; the K corona, marked by a continuum without spectral features; and the F corona, where the lightly broadened Fraunhofer lines are the identifying signatures. Their dependence as a function of radial distance from Sun's center are each different; they can be disentangled, but they are thoroughly interpenetrating.

The corona displays large streamers and jets. It was shown at the beginning of this century, that the *zodiacal light* is the extension of the F corona. This implies that matter flows out from the Sun to the solar system; indeed, it has been demonstrated, first by the analysis of comet tails, and then by *in situ* data, that the solar system, as a whole, is permeated by a *solar wind,* which is the external part of the corona.

In the physical sense, these operational definitions we have given above meet with difficulties. Obviously, the passage of some optical depth through some critical value is a function of wavelength, and greatly differs from one part of the spectrum to another, even if well defined in the visible. That the average observed photon comes from the photosphere and not from the interior is not a precise physical definition. True, the rise in temperature as one goes outwards in the chromosphere is linked with the fact that its optical depth is very small in the continuum, leading to strong departures from local thermodynamical equilibrium (LTE), the local properties of matter being decoupled from the radiation field. But this rise does not occur exactly at the top of the photosphere, as defined above; nor does it coincide with the place where departures from LTE begin to become noticeable.

For the corona and chromosphere, the situation is just as unclear. Non-sphericity obviously affects the corona, as already noted, because hydrostatic equilibrium is not shaping matter when gravitational forces are superseded by thermal or MHD velocities. The chromosphere is also not really spherical. Departures from hydrostatic equilibrium exist there as well, as exemplified by spicules, local phenomena that belong to a nonspherical chromosphere in motion. Interior to the photosphere, we find the same fuzzy situation. True enough, energy transport in the photosphere is mostly radiative, but it cannot be defined as such. Granular patterns of all sizes, are directly observable there, because they are at the top of the convective zone.

In the interior, the picture is perhaps less ambiguous. It can be divided at least into the *convective zone* and the *deeper radiative zone*. Still this division is very primitive, as differential rotation and magnetism distribution may suggest other subdivisions, both from empirical and theoretical points of view. The *deep interior* or core, is defined theoretically as the region that produces energy through thermonuclear reactions. It also produces neutrinos. Hence, we find no contradiction here, the operational definition being inspired only by theory, and not by the observations. Figure 2 shows one of several possible qualitative models of the Sun. The main regions (lettered) are indicative as well as various (numbered) subdivisions. Whenever layer 1 (top of photosphere) is well defined, layers 2 and 3 (limits of the convective zone according Schwarzschild's [1948] criteria) are theoretically well defined, but strongly model dependent. Layer 4, the lower limit for penetrative convection, is poorly known. The rotation velocity is affected by differential rotation, in latitude (not represented in this figure), and in depth. At layer 5, there seems to be a clear slowing down, going inward; layer 6 is rotating slowly. On the contrary, layer 7 and the core appear (but it is a disputed fact) to rotate twice as fast as the photospheric layers. The distances indicated for each layer to the center, in term of solar radius, should still be considered as only approximate (Fig. 2).

In layer 8, we have displayed the magnetic field, often assumed to be concentrated under the classical convection zone. Tubes of force may be dragged from there, and appear eventually as active regions at the Sun's surface, as shown in Fig. 7 below.

Physical Description of the Solar Layers

It is no longer adequate to describe the various solar layers for which we have given operational definitions by simply assigning a series of numbers to the thermodynamical quantities: temperature, pressure, density. A few decades ago, such a simple approach was called a solar model. We must now ask questions concerning the distribution of all scales of hydrodynamical motion (rotation and differential rotation, convection, turbulent motions, etc.); questions concerning the distribution of magnetic fields; and even questions concerning the distribution of energy sources in the core. In some cases, we must also take into consideration the local nonspherical distribution of the thermodynamical quantities.

Typical present-day models are given in Appendix B. The core of the

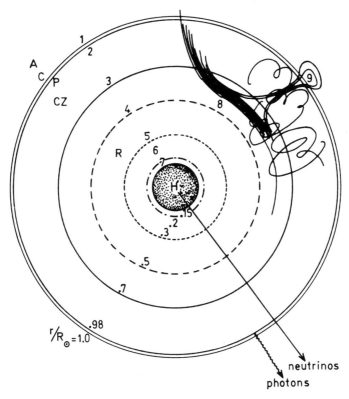

Fig. 2, A qualitative model of the Sun. The main regions from inside to outside are *H:* heart, or core; *R:* radiative inner zone; *CZ:* convective zone; *P:* photosphere; *C:* chromosphere; *A:* corona. Subdivisions to these layers may be introduced as shown. See text for explanation.

Sun, where thermonuclear reactions are taking place, is the source of high-energy photons and of neutrinos. Whatever magnetic field is inside the core, it cannot be observed nor even inferred from observations. It may be a fossil field dating back to early solar evolution, as suggested by Weiss (Rosner and Weiss 1985). The rotation may be nearly that of a rigid body, with little latitudinal differential rotation. Some indications deduced from surface acoustic oscillations of low l modes seem to indicate a fast rotation for the core, perhaps more than twice that of the superficial layers. Modeling the core is basically a theoretical task of course, but it must use observed global data such as photon and neutrino luminosities, and also perhaps the mass flux in solar wind (although this flux implies a very small production of radiant energy transferred into mass motion). Also core modeling must take into account global properties such as solar mass, solar radius and the chemical composition of the Sun.

The interior of the Sun is mainly radiative. However, turbulent diffusion

induced by differential rotation, as suggested by Schatzman et al. (1981),
may take place. This will affect the composition and structure of regions in
the core. It seems likely from several studies concerned with emerging active
phenomena and with acoustic oscillations of the photospheric layers, that the
interior outside of the core rotates nearly as a rigid body at a velocity similar
to but possibly faster than the photosphere. It also appears that the upper part
of the interior is the seat of a strong magnetic field, rather concentrated, of
the order of 10^4 Gauss, or more; much further study is needed to clarify the
dynamics in this region.

The interior, being bounded by the convective zone, is affected in its
outer parts by the penetrative convection (or overshooting) associated with
the inertial motions of convective elements. The role of these motions may
be essential in dragging out part of the magnetic field, in the form of tubes
of magnetic force. These tubes emerge as active regions in the directly observable parts of the photosphere (see Fig. 2). Again the mechanisms of these
processes are not well understood as to how one could get an efficient dynamo
in the radiative zone.

The convective zone itself is a very complex machine. Convective instability arises from the ionization of hydrogen and helium that gives a low
specific heat ratio γ and a high opacity. It is probably affected by differential
rotation in latitude and depth, animated by a complex system of motions at
all scales, and divided into convective cells that may be a function of the
magnetic activity. The structure of the convective zone may be of paramount
importance in understanding the phenomena of the solar cycle including the
migrations of localized active phenomena. It can be sounded by moderately
high modes of acoustic oscillations, or even more importantly, through the
analysis of granulation, mesogranulation and supergranulation. The properties of these cellular patterns are probably cycle dependent.

The photosphere is the visible locus of almost all indices concerning the
Sun. The propagation properties of waves in it creates systems of stationary
waves; their analysis is one of the studies in solar physics which has evolved
most rapidly in recent years. Photospheric magnetic fields occur also at all
scales, from very small features, like filigrees, to the impressive developments of active regions. These fields emerge from below, to form a basis for
coronal structure. Because the photosphere links the outer and inner layers,
it is the center of focus of all solar studies. The oscillations are characteristic
of properties of the resonant cavities that can be identified with photospheric
layers, but they may be triggered from above or from below. Solar matter can
be compared to a flute, the trigger being like the blowing by the flutist!

As one proceeds from the interior of the Sun to the photosphere, magnetic fields and gas motions do not materially affect the main equilibria existing there. These are *local thermodynamical equilibrium* (LTE or strong coupling between radiation fields and local properties of matter), *radiative
equilibrium* (energy being transported largely by radiation), and *hydrostatic*

equilibrium. In photospheric or higher regions that are traversed by the flux of photons without being strongly affected by them, and where the density is so low as to allow the forces involved in the motions to overcome the constraints of gravity, we encounter a situation where the situation progresses toward larger and larger departures from equilibria. These departures from LTE register only slightly in the photosphere, at least in its spectrum, but in the chromosphere and corona LTE computations are not even valid first approximations. The radiative equilibrium, similarly, does not allow us to predict the distribution of temperatures in these regions; there, the dissipation of magnetohydrodynamic (MHD) waves, and mechanical motions are dominant processes in commanding the thermodynamical conditions. We have already noted that the hydrostatic equilibrium hypothesis does not apply. Homogeneous, and plane parallel (or spherical) models are possibly good scholarly exercises, but they do not approach physical reality in the chromosphere and corona. There is only one physical concept which remains more or less valid in all regions from the interior to the exterior of the Sun: namely, the classical equation of state which can be applied to matter (see Fig. 3).

So far as the solar wind is concerned, it is essentially a dynamical phenomenon, which does not resemble, in any way, what one would expect when treating stellar structure. Still, it is a part of the global Sun, as are also the whole of the interplanetary medium, the planets and other objects of the solar system. This progressive departure from thermodynamical equilibrium to decoupled many-parameter physics, from inside to outside, can be used to define, in stellar cases, such features known as emission line shells, outer layers, circumstellar regions, etc. These concepts overlap. They all affect the

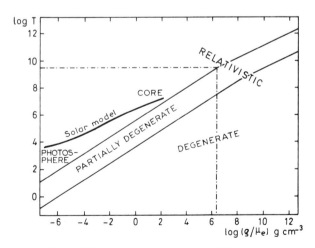

Fig. 3. Illustration of the validity of the nondegenerate, nonrelativistic equation of state. In this classical T, ρ diagram, the solar model is entirely located in the zone affected neither by degeneracy nor by relativistic effects.

outer layers above photosphere; but departures from local thermodynamic equilibrium, from radiative and hydrostatic equilibrium have different consequences that, according to the values of luminosity, mass and radius, dominate more or less the structures observed in stars. Although the Sun is only one star, we have displayed in Fig. 4, in a schematic way, the dominant effects in some of the identified stellar-circumstellar features.

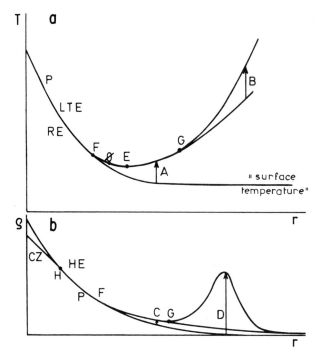

Fig. 4. Illustration of nonequilibrium in the outer layers of stars. (a) the temperature T is displayed as a function of an arbitrary height scale. P is the classical photosphere, in LTE and in radiative, as well as in hydrostatic equilibrium. Departures from LTE induce an increase of the temperature, from point F outwards. Extra heating by dissipation mechanisms starts to be effective from point G. Accordingly, arrows A and B, respectively, represent the effects of departures from LTE and from radiative equilibrium. Although they may depend on the density behavior (coupled effects), these are the primary effects affecting the T distribution. Point O and E represent, respectively, the top of the photosphere, as defined in Fig. 1, and the temperature minimum. (b) The density ρ is a function of an arbitrary height scale. Again, P is the photosphere. The convection zone CZ starts at deeper layers than the point H. Above point F, departures from hydrostatic equilibrium (coupled with the temperature behavior described in (a) lead to an extended envelope. If the variation of ρ undergoes a maximum above G, then we have a shell, or better a circumstellar shell. Whether this shell will give a shell spectrum depends mainly on its location, extent and depth. Arrows C and D represent, respectively, the effects of departures from local thermodynamic plus those from radiative equilibrium, and from hydrostatic equilibrium.

II. COUPLINGS IN THE SUN

Core and Solar System

The Sun has now reached the main sequence, and is an adult star. In the earlier stages of its evolution, the Earth and planets barely existed as such. In later stages, they will have strongly evolved. The intimately connected behavior of the central condensation, the star and the planetary system as a whole, is a permanent feature of the evolution of solar-type stars. This concept is clearly beyond the scope of this chapter, although it well illustrates the fact that the solar system should be considered as a whole. While in the present epoch, one may distinguish and identify as almost independent bodies the various planets, asteroids and comets, even now they are not entirely independent, always being subjected to Newton's universal attraction.

Due to the high pressures necessarily present, the solar core is the hottest part of the Sun, where nuclear reactions take place. The *p-p* reactions are present and actively producing photons and neutrinos. Nuclear energy is released as photons of high energy. These photons are progressively absorbed by matter and re-emitted, progressively degraded into many more photons of lower and lower energy, until they leave the photosphere. The transit time of radiant energy is of the order of 10^6 yr, based on rough estimates of their average mean free path. This suggests that whatever coupling might appear between the core of the Sun and the solar system, it must be a retarded coupling. Neutrinos do not play any appreciable part in this coupling. They essentially pass through the solar system and its matter, without affecting it.

Long-term variations of solar luminosity might affect the climatological phenomena on planets. Moreover, the Sun is not a true sphere. It has a rotational axis, and a magnetic axis; and as the solar system is markedly shaped in the form of an equatorial disk, any latitudinal variation with time of the distribution of brightness on the solar surface might affect the planetary atmospheres. This may happen even if the link does not involve an appreciable variation of global luminosity.

One should certainly not forget the effects of solar-emitted particles on the planetary atmospheres: the ionospheric disturbances, the geomagnetism and the aurorae. These are all linked with the solar wind, and with its fluctuations associated with solar activity. But, indirectly, wind and activity, and their evolution are perhaps associated with properties of the core.

A very interesting field of study, so far hardly touched by astrophysicists, is the study of long-term relations between planetary physics and solar behavior: these are surely of the same nature as solar-terrestrial effects. By taking into account, for example, Jovian reactions to solar activity (as done by Focas in the 1950s; see Focas and Banos 1964), one might hope to distinguish solar effects from other effects, such as effects due to the geometry of the Sun-planet system, or from effects linked with those due to a change in the solar output. Altogether, solar-terrestrial relations are good symptoms of the solar activity, but it is difficult, in view of the retarded effects we have mentioned, to deduce from their study any idea about the variation of properties in the deep interior of the Sun. This point must be kept in mind when discussing the evolution of the solar core.

The delay needed for central energy to reach out to the solar system certainly imposes some spread in time between couplings, and explains the fuzzy behavior of possible phenomena linked with a secular variation of the solar core. To the best knowledge of the author, there is but one single phenomenon that may be linked with variations of the core: the secular variation of the shape of the solar cycle as perhaps detected in geologic data by Sonett and Williams (1987). Phenomena such as glaciations are likely to be due to the variation of the geometry of the Earth-Sun system (Milankovitch theory; Berger et al. 1984).

There is also the possibility that phenomena due to the astrophysical environment at large (the Sun crossing the spiral arms of the Galaxy, for example) may affect the properties of planetary atmospheres, either directly or indirectly. By the returning to the Sun some of its own radiation, the outer properties of the Sun might then be changed by a type of blanketing effect.

Coupling Between Convection and Rotation

The Sun rotates at the rate of about one turn in 27 days, ignoring differential rotation in the first approximation. Convective motions mix the layers located between 0.7 and 0.99 solar radii from the center, at velocities of the order of a few km s^{-1}. These phenomena are coupled, obviously, through the Coriolis forces. One must estimate these in order to allow a discussion of their influence (see the chapter by DeLuca and Gilman).

Let us assume a locally rotational velocity expressed by the vectorial quantity Ω, implying a local velocity of $r \times \Omega$ m s^{-1}. This must be with respect to an absolute frame of reference A, the one defined by fixed stars. Next we assume a convective velocity field V in a frame of reference K, corotating within the system A with the Sun. The velocity of any atom with respect to A is

$$V' = V + \Omega \times r. \tag{1}$$

This expression must be introduced into the Lagrangian function. The expression of the latter in an external field U is

$$L = m|V|^2/2 - U(r,t). \tag{2}$$

The equation of motion is

$$m\frac{d|\mathbf{V}|}{dt} = -\frac{\partial U}{\partial r} \tag{3}$$

or, more generally, the Lagrangian equation is given by

$$\frac{d}{dt}\frac{\partial L}{\partial V} = \frac{\partial L}{\partial r}. \tag{4}$$

Putting Eq. (1) into (Eq. 2), gives

$$L' = |\mathbf{V}'|^2/2 - U = m|\mathbf{V}|^2 - U + m\mathbf{V}\,(\mathbf{\Omega} \times \mathbf{r}) + (m/2)(\mathbf{\Omega} \times r)^2. \tag{5}$$

The two differentiations of Eq.(5) will allow us to rewrite the Lagrangian Eq. (4), and the analog of Eq. (3) will then become

$$m\frac{d\mathbf{V}}{dt} = -\frac{\partial U}{\partial r} + \underbrace{m\mathbf{r} \times \mathbf{\Omega}}_{A} + \underbrace{2m\mathbf{V} \times \mathbf{\Omega}}_{B} + \underbrace{m\mathbf{\Omega} \times (\mathbf{r} \times \mathbf{\Omega})}_{C}. \tag{6}$$

Inertial forces have thus three components, **A, B** and **C**, which are, respectively, the second, third and fourth terms of the right-hand side of Eq. (6). The force **A** is linked with the nonuniformity of rotation; it essentially vanishes in the solar case. The force **C** is the well-known centrifugal force, which is perpendicular to the solar sphere. The force **B**, the Coriolis force has a horizontal component. It depends on the proper velocity **V**, unlike the centrifugal force (Fig. 5).

Values of $|\mathbf{\Omega}| = 3 \times 10^{-6}$ rad s^{-1}, $|\mathbf{r}| = R_\odot \sim 7 \times 10^8$ m, and $|\mathbf{V}| = 1$ km s^{-1} are typical for the solar photosphere. One sees that Coriolis force per kilogram is of the order of 3×10^{-3} newton, whenever the centrifugal force per kilogram is equal to $7 \times 10^8 \times 9 \times 10^{-12} \sim 6 \times 10^{-3}$ newton and the local gravitational force of the order of 3×10^2 newton. The Coriolis force is by no means a small quantity, and accounts for many of the horizontal flows at the solar surface.

The physics of the Coriolis forces seems straightforward; they are described by simple equations and they have, like centrifugal forces, the character of inertial forces, acting to transfer the momentum. However, the concept of a fundamental system of reference is involved, and this, by no means a clear concept, will not be pursued here. One should only note that cyclonic motions often observed in the Sun, even at distances away from active regions, are in part linked with the Coriolis forces.

As expected, Coriolis motions are of opposite direction in the two hemispheres. They are also of opposite direction whenever linked with descending or ascending convections. In many cases a detailed treatment may appear

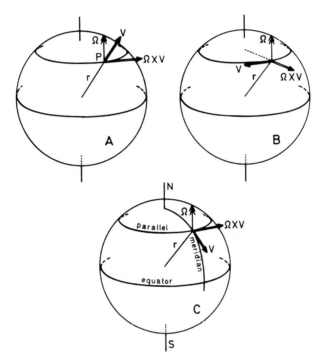

Fig. 5. The Coriolis forces. As a consequence of vertical motions A, they appear as oriented
along the latitude parallel. It is similar when the motions are oriented along the meridian C.
When motions are horizontal, along a latitude parallel B, Coriolis forces are oriented as the
centrifugal force, normal to the rotation axis of the Sun. By combining the three effects, one
can get the Coriolis force, in any circumstance. The figure assumes motions are corotating
with the Sun.

necessary. When the Sun is quiet, the Coriolis forces have a tendency to
dominate over the magnetic forces. In that case, they influence the shape of
convective cells, that have a tendency to appear in banana-shaped form (see
Fig. 6). Without rotation, the cells would be quite different, and the general
pattern of convection, through the effect of Coriolis force, is a clear consequence of the rotation-convection coupling. All surface phenomena are controlled by these patterns. Hence this coupling is essential to the understanding
of the solar machinery. Equations of motion of course do imply other terms.
Of particular importance to account for are the forces due to turbulent viscosity.

Coupling Between Magnetism and Convection

In a similar way, the magnetism of the Sun can be described in the corotating reference frame K. Concerning the motions of matter with respect to
frame K (denoted by their velocity V, and the matter assumed to be partially

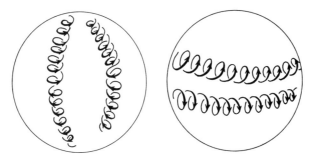

Fig. 6. Convective cells. At the left, banana shaped cells. Compare the two hemispheres: the sense of the spiraling is reversed. At the right, roll-shaped cells. Of course, the figure is simplified. Both banana-shaped cells and roll-shaped cells can coexist with each other, producing alternating spiraling motions.

or completely ionized), the Laplace-Lorentz force (LL force, hereafter) due to the action of the field $\mathbf{B}$ upon the current $\mathbf{j}$ can be equated to

$$\mathbf{F} = \mu \mathbf{j} \times \mathbf{B}/c \qquad (7)$$

with μ being the permeability.

The current $\mathbf{j}$ is linked with the nonuniformity of the field. If the current is in the plane of the solar surface, as for the field $\mathbf{B}$ (to a first approximation), the LL force $\mathbf{F}$ is normal to the surface and depends strongly on the magnitude of the field. The evaluation of the current density $\mathbf{j}$ is linked with Ohm's law

$$\mathbf{j} = \rho(\mathbf{E} + \mathbf{V} \times \mathbf{B}) \qquad (8)$$

where ρ is the conductivity, which is strong at large depths.

Without entering into too much detail, the order of magnitude of $|\mathbf{F}|$ could thus be approximately $|\mathbf{V}|\,|\mathbf{B}|^2$ per unit of mass. For $|\mathbf{B}|$ about 10^2 gauss, it leads to $|\mathbf{F}| = 10^{-8}$ newton; for $|\mathbf{B}|$ of the order of 10^4 gauss, it is 10^4 larger, hence $\sim 10^{-4}$ newton. We can therefore consider, by reference to the estimations given above, that, in periods of solar activity, the role of LL forces is certainly quite important relative to that of the Coriolis force. This will be more marked at large depths, where conductivity may be high, which means in turn, that computations of the internal structure cannot neglect the influence of the LL forces in the equation of motion when the magnetic field is large. When rotation is superimposed on the LL forces, we find that motions due to the LL forces can be introduced just in the term in $\mathbf{V}$. These can modify the effect of the Coriolis inertial component.

Altogether, magnetism, convection and rotation are therefore intimately coupled. This is particularly the case for the large scale, commanding the shape of convective cells. We have seen that Coriolis forces lead to banana-shaped cells elongated along meridional structures. LL forces, on the other hand, impose rolls, i.e., convective cells of a toroidal form, aligned with the solar axis. These cells are formed at times of high magnetic activity by toroidal magnetic structures. A dipole field is more likely representative of the quiet magnetic Sun near or before sunspot minimum (Fig. 7).

The importance of the horizontal component is clear. So is the difference between the three types of granulation so far identified. Not only are they different in size but also in velocity. However, the peaks are not so well defined, and one could conceive of a continuous variation.

Coupling also acts locally, in active regions, where strong magnetic fields impose cyclonic convection, and where coronal loops are manipulated by magnetic field evolution. Detailed computations can be made, but we must not forget the importance of the coupling forces, whether they be inertial Coriolis forces, or LL forces. Both impose difficult numerical problems as their introduction delinearizes the equations of motion, making it necessary to solve the magnetic-field equations simultaneously. These numerical difficulties are so great that, when one wants to achieve realistic solutions, some

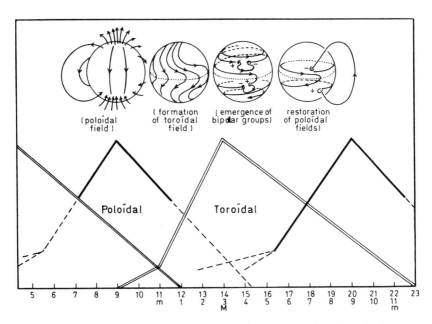

Fig. 7. The succession of magnetic structures in the solar cycle. A clear alternation between poloidal behavior and toroidal behavior is observed. It can be understood on the basis of the old Babcock (1961) theory (bottom after Legrand and Simon 1980; top after various publications inspired by Babcock's theories).

linearization must be found in another way. The result is possibly no more realistic than the solutions avoiding the coupling effects. Realizing the importance of these couplings does not mean that we can yet readily master the equations and solve them. In many cases, very simple models from the numerical point of view, may still be more suggestive than elaborate archetype computations that do not fit the basic nonlinear aspects of the problems. Order-of-magnitude models such as Babcock's (1961), although purely phenomenological, may still have an important role in the discussion of solar phenomena. In other words, we are still looking for good working hypotheses, i.e., good from the operational point of view.

Another point concerns the long duration for ascent of solar phenomena through the convective layers. At velocities between 1 m s^{-1} and 1 km s^{-1}, the time of ascent can vary between 2×10^{10} s and 2×10^7 s, this latter period of the order of 1 yr. Convection is essentially slow, and one should be careful, when modeling the mutual effects of convection, rotation and magnetism, to take into account the fact that, during the length of time covered by the computation, the field is changing its structures, and this change may be different in the deep layers than in the outer layers. The complexity of the problems involved forces us to recognize that theoretical efforts are often rather inadequate to describe the astrophysical phenomena. However, diverse ways of solving the basic equations, using different type of approximations, may still give physical insight to the understanding of the basic physics at work in the solar machine. More on these problems are found in the two chapters on convection by Chan et al. and by Gilman and DeLuca.

Coupling Between the Active Phenomena and Planetary Physics

Assuming the position of a given planet (e.g., Earth) to be continually in the solar energy flow, which is assumed to be constant, then one may think that some climato-meteorological state of equilibrium would be achieved, modified only by purely terrestrial causes, whether they be man's existence, man-made activities, or only subsuperficial phenomena such as seismic or volcanic. As a matter of fact, meteorologists completely ignore the variation of the solar energy flow. Climatology is still in such an elementary state that it recognizes only the secular variation of the Earth's orbit (as in Milankovitch theories; see Berger et al. 1984). However, a few scientists have regularly put forward the idea that even a low-energy perturbation of solar energy flow could trigger some modification in the Earth's upper atmosphere, and even of more consequence, its tropospheric layers (see Sonett et al. 1990).

In recent years, more and more convincing arguments have come to light favoring these ideas. At the same time, some warnings have been given against abuses of this type of thinking. For example, Fourier-type analysis have led Pecker (1982) to show that equatorial meteorological phenomena (rains) do not follow any of the known solar active-cycle periodicities. Reconciling these types of points of view may be possible, by some proper consideration of the way the solar activity may influence the near-surface phenomena on Earth, but for now we are still far from this reconciliation.

Without going into the methodology of discovering valid correlations between solar and terrestrial phenomena, we should indeed pay a good deal more attention to the way the energy is carried, to the mechanisms responsible for a minute change in the incoming energy (compared to the solar constant), and to the average energy of the wind, sufficient to trigger meteorological changes and even, as an integrated consequence, climatological changes. As for triggering mechanisms, we should certainly consider that, on Earth as in the interior of the Sun, climate processes are in very near equilibrium. These large-scale phenomena imply therefore efficient triggers, but little, if anything, has been really developed along these lines. We would note, however, that variations in the solar flux (of light or particles) may be large without affecting the Sun itself. The stored energy is some 10^{14} greater than the outflowing energy per year.

Obviously the solar wind and its long-term variations is a candidate for this trigger. Another source is the relatively short-term variations of the wind, linked with the duration of some important center of activity favorably located on the Sun, or with the duration of the coronal structures, or coronal holes, associated with some complex chromospheric-photospheric state. Shorter-term variations, linked with the explosive release of energy previously stored in the magnetic structures, such as flares of various strengths, is yet another source. However, we still lack any convincing theory for this wind, and for most of its modulations. If the mechanisms for the acceleration of the flow are relatively well known from Parker's (1958; 1963) theory, one still encounters difficulties in accounting for the deep origin of the wind, which must satisfy the equations of the continuity of mass. Its distribution at a large distance from the Sun is no doubt influenced by the shape of the Sun's magnetic field. We know little about where the components of the solar wind originated (from the polar regions, or elsewhere). In the low corona, one wishes to know how much of the visible mass is outflowing and, how much is inflowing. Observational discussions of the mass-flow balance is certainly progressing, but there is not enough information to enable us to put the theory even of the average solar wind on a firm basis. As to the influence of solar activity, although we observe the obvious difference in latitude distribution between a minimum corona and a maximum corona, we have no good quantitative theory to explain it.

We have only vague ideas relating the coronal holes and their low-altitude structures with a network of activity at the photospheric levels. The coronal holes are "avoided," as suggested by Pecker and Roberts (1955), by the slower particles of the wind. Magnetic energy stored in flare regions is, no doubt, liberated through some quick reorganization of unstable magnetic structures. Still some very different detailed pictures can be made of these violent events and of their coronal counterparts. Much progress has already been made in this domain.

The lack of understanding of the solar wind phenomena does not suppress the importance of it (the study of solar-terrestrial relationships may be of an essential importance). We may not understand these phenomena, either on our side of the wind, or on the solar side. The weakness of the theoretical approach may explain why it seems that solar physicists are more interested in these relationships than are the climatologists or the meteorologists (except the specialists of the upper atmosphere, who are very conscious of the problems).

Solar activity influences the planetary climate and meteorology, but is the reverse true? Although this question has scarcely been tackled, and is still highly controversial, some authors (see, e.g., Trellis 1966a,b) have suggested that the location of planets, through perhaps their tidal effects, small as they may be, might trigger the buoyancy emergence of magnetic tubes of force, kept in the convective zone in a quasi-equilibrium state. Obviously at this time, we cannot say more about this very daring and still unconvincing hypothesis.

III. PROBING THE DEEP SUN

Granulation, Mesogranulation and Supergranulation as Indicators of Convective Layers

Granulation has now been known for more than a century; until recently, its nature has been the subject of many controversies. Are we dealing with a convective cell of the Bénard type? This was extensively discussed by Schwarzschild (1959), after a period when granulation was more or less considered as the tail of a turbulent distribution. Without entering here into details, we must now face a much more complicated situation (see the chapter by von der Lühe et al.).

On the one hand, the size distribution of granules is not simple, as well exemplified by the Pic-du-Midi observers (Roudier and Muller 1986). Smaller-size granules follow a turbulent behavior, in the physical sense; the large-size granules do not. The slope of the size distribution of eddies is a function of the solar activity. It is different at solar maximum than at solar minimum. At least, the larger granules seem to be well described by hydrodynamical models implying little turbulent dissipation. Exploding granules, for example, can be adequately modeled. But the variation of properties of the granules during the activity cycle, although probably linked with concomitant variations of the structure of the convective zone, is unaccounted for.

The vertical extent of the granules is unknown. The mixing-length type of theory for the convective zone seems to predict a continuous distribution of eddy sizes, but this appears not to be the case. Supergranulation (cells of

30,000 km), mesogranulation (7000 km), and granulation ($\sim$ 700 km), seem to define three distinct peaks of the distribution, to which should correspond three vertical scales in the convective zone. The least that can be said is that we see no trace of this in the various current theoretical models of the convection zone with their limited spatial resolution.

One obvious reason for the inadequacy of the models of the convective zone to explain the photospheric and chromospheric observations, is that they use extremely simplified boundary conditions. Often they take little account of the dominant role of radiation transport in controlling the scales of convective motions. The result of this inadequacy is that the emergence of active regions is practically the only indicator of the structure of the convective cell. There, the indications in favor of toroidal rolls are stronger than those in favor of banana-shaped cells. But, again, the variation across the cycle of the structure of convective cells may be important, due to the evolution from a Coriolis-force-dominated hydrodynamics to an MHD situation where the LL forces may play a dominant role.

The size of eddies is obviously not the only characteristic of granular features that needs to be introduced as an indicator of deep convection. Clearly, the shapes of granules, their number, their evolution, their contrast and their spectrophotometric properties, should be introduced. Also, one should not forget that the so-called unresolved motions may play an important part in the diagnosis of the hydrodynamics of the photosphere, and in the understanding of the deep convection zone. It is now admitted generally that horizontal components of turbulent motions are stronger near the surface (see Fig. 9 below). But what meaning does such a remark have for the structural properties of convection cells? This question has not necessarily been answered.

The features of real phenomena are quite complex; they are oversimplified in Fig. 8. There is little doubt that this is a field in which further observational data, through a continuous survey of the solar features, should considerably progress.

The Emergence of Active Centers as Indicators of Convective Layers and of the Depth-Dependent Differential Rotation

This problem may actually be tackled in several ways such as through the observations of young spot groups or through the behavior of coronal holes. We have expressed earlier the idea that magnetic tubes of force are extracted from the upper layers of the radiative inner zone, by convective motions. However, we should keep in mind that this is by no ways proven. It may well be that the emergence of flux tubes is due to buoyancy forces applied to floating tubes located in the upper part of the convective zone. The author of this review feels more convinced by the empirical arguments favoring the deep origin of sunspots. It is then clear that the emergent spots may reveal the properties of the convective motions that bring them to the surface.

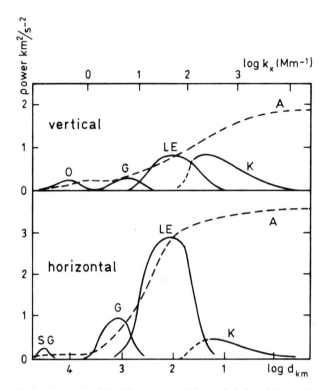

Fig. 8. Distribution of velocity fields. The meaning of the symbols is as follows: *O:* oscillations; *G:* granulation; *SG:* supergranulation; *LE:* deduced from limb effects (unresolved features?); *K:* turbulence at very small scale (Kolmogorov tail); *A:* integrated power up to a horizontal wave number. The figure is drawn at $\tau \sim 0.1$. The velocities are smaller at smaller depths.

Whatever happens later to these active regions is dominated by differential rotation, which elongates the spot groups along lines tending to be parallel to the equator, and which acts on filaments in a similar way. The regions of emergence have been noted by Maunder and others (see Kiepenheuer 1953, p. 338) as appearing at some selected Carrington longitudes. This leads to the idea that the regions where the activity originates are likely to be deep (even deeper than the bottom of the convective zone), and rigidly rotating. This idea has been reinforced by the study of the evolution of filaments of which the pivot points reveal a deep solid-like rotation.

A similar argument came from the consideration of M regions (the regions that scientists some decades ago associated with the magnetic peaks of activity), or with the coronal holes (which, like the M regions, are regions "avoided" by the flow of matter, due to the magnetic guiding of ionized matter in the corona). M regions, defined from geomagnetic data, and coronal holes as observed in X rays, indeed rotate rigidly at about the Carrington velocity.

This can be seen directly on images, that show clearly the latitudinal elongation of coronal holes, almost from pole to pole at successive rotations.

Spots and associated active regions, after they have emerged, are affected by proper motions on the solar surface. Setting aside the effects of differential rotation in the superficial photospheric layers, this leads to a residual motion either polewards or equatorwards often affecting both of two spots near each other. The analysis of data at any given period of the solar cycle, has shown in a rather clear way (Ribes et al. 1985a) that the proper motion of newly born spots favors roll-shaped convective cells, more or less controlled by the toroidal magnetic fields. This tendency is not as obvious at the time of activity maximum when the distribution of field polarity seems to indicate that the field is closer to a poloidal structure.

However, some aspects of this behavior have not yet been firmly accepted. During the course of the magnetic evolution of the Sun, do the successive convective cells move towards the poles, as suggested by Ribes and Laclare (1988)? If so, why do the emerging spot groups seem to appear from the emergent regions of successive convective rolls at lower and lower latitudes? Should the origin of this be looked for in the evolution of the deep-rooted magnetic field, below the convection zone or in the evolution with the cycle, of the depth of penetrating convection, or elsewhere? Moreover, toroidal rolls imply motions that favor, in any Fourier analysis of photospheric motions, the modes $m \simeq 0$. On the other hand, observations of tachyfouriermeters, such as Brown's (1988b), seem to indicate a dominance of the motions with spherical harmonic indices $m \simeq \pm l$. But this may be a function of the phase of the solar cycle. Clearly, we need observations covering the whole duration of either one single migration, or one single activity cycle.

Another problem relates to the consideration of high-latitude phenomena. Are they related to deep layers of both the convective zone and radiative interior? This relation rests on controversial facts. Some authors favor a continuous migration equatorwards of the active phenomena, which shows up in the form of bright X-ray points at high latitude, and spots and active regions in lower latitudes. This migration would last from 18 to 22 yr. Two successive migrations occur at intervals (cycles) of about 11 yr. Others do not agree with this view. They see activity markers emerging from some intermediate latitude and moving together, but with opposite polarities, towards the pole and towards the equator in successive periods of 11 yr. This latter view, suggested by some aspects of the data by Leroy and Noëns (1983), is at present advocated by Gilman (see the chapter by DeLuca and Gilman). The former description, in different form, is that of Legrand and Simon (1981), that of Howard and Labonte (1980,1983), or that of Wilson (1988) (Fig. 9). A combination of two processes is obviously not excluded.

Although solar activity is not a central treatment in this book (see, however, the chapters by Rabin et al. and Semel et al.), it is impossible not to take into due consideration evolution and migration of active phenomena dur-

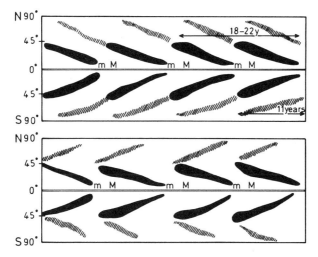

Fig. 9. Equatorwards migrations of 18 to 22 years. Above, the possible schematical migration description; below, the double undecennial migration, also schematical. The darker area represent the usual butterfly diagram. Other parts of the diagrams are inspired (above) from the behavior of torsional waves, bright facular spots, etc., and (below) from the partial behavior of coronal data.

ing the evolution of solar activity in order to understand the behavior of the convective layers, and the underlying interior. Coupling between magnetism and convection is still poorly understood, or at least poorly modeled.

Seismology as an Indicator of Deeper Layers

In recent years, much has been accomplished in this field (see the chapters in Part II). The distribution of energy has been measured in the various modes, with indices l and m, of the photospheric (and chromospheric) velocity fields. A not always obvious problem is the disentanglement between those motions that are due to the existence of some resonant quasi-cavities, and those associated with convection, and its dissipative effects. Another unclear problem is that any oscillation must be triggered; the flute does not produce music unless one blows in it, so to speak. Therefore one is led to the question: who is blowing the pipe? As the outer atmosphere is animated by active (and sometimes violent) phenomena, these are possible sources to trigger the oscillations. Oscillations may also be triggered from below by the convection and the dissipative phenomena associated with them. Whatever the solution to these physical problems as discussed in the Cox et al. chapter, one can consider the oscillations, as they are measured, as a signature of the deep regions of the Sun. Just as the spectrum, in the ordinary optical sense of the word, is an image of the photospheric layers, one can say that the frequency spectrum of the oscillations is an image of the interior structures of the Sun.

In order to transform this observed image into some logical description, one needs to have an algebraic relation that can be inverted; hence a theory must be introduced at this stage. However, theories of first approximation seem to show without ambiguity that oscillations of about a five-min period (the first discovered and the best known) are definitely acoustic oscillations. Theory can then be refined, and the modes of low l are influenced more or less by all layers. Modes of high l, as can be intuitively demonstrated, are influenced mostly by the superficial layers. Next, the m-splitting relation can be inverted and it leads to a distribution of rotational velocity with depth. The results are quite striking in that they indicate little or no variation of rotational velocity outside a layer located at 0.3 solar radii from the Sun's center. Below that layer, the rotation seems slower than above and at the surface. The core seems, on the contrary, to rotate twice as fast as the outer layers. However, the data concerning deep layers are so strongly affected by large errors, that even qualitatively, we cannot be sure that the above described mechanism corresponds well to the situation.

Nevertheless, some remarks may be made. First, the numerical problem is not easy. Second, one starts from data pertaining to a given period, and the m splitting, as suggested in the preceding paragraph, may well be cycle dependent: it can be quite different when the field is toroidal and when it is poloidal. Finally, the propagation time of mechanical phenomena is controlled by dissipative phenomena. Although it takes months, or years, for signals to ascend from the center of the Sun, the analysis assumes a quasi stationarity of the deep interior. Little is known about this except that data concerning oscillations and convection do not cover adequately the evolution of the overall structure of the Sun with its activity cycle. One might just as well suggest that the behavior of the distribution of rotation with depth is a function of the cycle. If this were so, conservation of angular momentum must imply that outwards transport of momentum as well as energy play an important role in the Sun's life. It also means that some downward transport of momentum must take place as well, to insure the stability of the Sun over billions of years. Large-scale oscillatory behavior of the deep differential rotation may well be the consequence of this mechanism.

Therefore, we should perhaps pay a good deal more attention to the direct measurements of the solar diameter. Is it changing with a quasi period of about 1000 days, as suggested by Laclare (1983) and others? Does it have a variable amplitude, implying beats of some kinds, or is it fixed, as suggested by Brown's (1985) data? If it is changing, are these changes correlated with some changes of the solar luminosity? Surely, measuring the solar constant is not a good way to sense this, as the distribution of emerging flux in photospheric layers has no reason to be spherically symmetric. Out-of-the-ecliptic polar missions covering a solar cycle, or, better, a solar migration of

18 to 22 yr, would give the only appropriate answer to this question. These changes might not be linked with any change in luminosity, but with changes in efficiency of convective transport during the solar magnetic activity cycle (Schatzman and Ribes 1987), hence of the depth of the convective zone. One sees how far-reaching that type of data can be.

Oscillations of about 160.01-min (note the precision) have been observed and confirmed by several observers. Their identification as solar gravity waves is advocated. If so, they must reveal information about the deep layers of the Sun. Little has been done recently on this question, so it is difficult to comment on it. The evidence of the observed facts is large, but progress is not rapid. We are facing a source of important information that we are not able to use completely. It should be said with reference to this 160-min oscillations that some observers, such as Kotov and Lyuty (1988a,b, 1989) have observed it (although with much less precision) in the spectrum of several nonsolar sources, some of them extragalactic; hence it would not reveal much knowledge concerning the Sun. Although the findings of Kotov are severely criticized by many, who suggest that some resonance with terrestrial atmospheric phenomena is possibly influencing the data (160 min = 1/9 day), one should perhaps leave the door open to some far-reaching, but highly speculative, conclusions such as perhaps the cosmological nature of the reported oscillations.

The Neutrino Flux, as an Indicator of the Central Layers

Neutrinos are produced in the core of the Sun by the well-known reactions of the p-p chain. In spite of the difficulty of the measurement, they have been detected when reaching the Earth, notably by the epoch-making observations of Davis (see the chapter by Davis).

The observation of solar neutrinos is now the preoccupation of many groups. This should allow us soon to detect solar neutrinos of several energies, whereas at the present time, the Davis data measures only those neutrinos that are able to transform ^{37}Cl into ^{37}A, i.e., those with energy > 0.814 MeV. The other reactions of the p-p chain are not considered because their neutrinos escape detection.

Nevertheless, the Davis data has focused on at least two important problems. First of all, the neutrino fluxes are about 3 or more times smaller than predicted for that energy range by standard solar models of the core. In order to bring the standard solar core models in agreement with the data, one must either decrease the abundance of heavy elements, or decrease the central temperature. Both operations would modify the global properties of the Sun too greatly to be readily acceptable. True enough, some progress remains to be made with respect to the physical constants in use in the computation of the energy output of the solar core and of the opacities; therefore there remains some uncertainty in the solar model. Still, the main phenomenon needs an explanation.

Should we look into the physics of the neutrino, or into some solar phenomenon? Although suggestions have been made in both directions, the matter evidently is not completely settled. Whatever the physicist may say (and they may well have the last word), astrophysicists, such as Schatzman and Maeder (1981) and others, may be right in introducing the turbulent diffusion as a process that brings additional hydrogen into the Sun's center. Even if they are wrong as to the magnitude of the effects (which, in order to account for neutrino flux, are incompatible with the helioseismological data), turbulent diffusion must exist. One must now evaluate, physically, the importance of turbulent diffusion in the various layers of the Sun, and its various effects, strongly linked with the depth dependence of thermodynamical quantities, of turbulence and of other dissipative mechanisms.

A second problem brought to attention by Davis's data is the variation with time of the neutrino flux. Since the time lag between the departure of neutrinos from the core, and the observation of active phenomena at the solar surface must be counted in months and years, we should be very cautious in drawing any conclusion, at this time, from any correlation that could be found. Only some sort of Fourier-type analysis could possibly push us towards some association between core and surface activity. One could ask if convective cells or the behavior of magnetic field are linked with the core phenomena. The question is evidently wide open. It may be more interesting to look for relations between the neutrino flux and the variations of the solar diameter already mentioned.

Measurements of the whole energy spectrum of neutrinos will no doubt help to clarify the processes at work; measurements of the variations of this whole spectrum with time will also bring out interesting problems, though this cannot be expected before the end of this century.

The Evolution of Active Phenomena after Emergence

The evolution of active regions, from their first appearance to their disappearance, has been studied by a number of techniques and in great detail, over the years. If we consider 1891 as the year of the construction of spectroheliographs at Mount Wilson (Hale) and at Meudon (Deslandres), the study has been in progress for a century (see Abetti [1929] for a detailed account of this historical period). We do consider active regions as "epiphenomena," secondary to the basic processes leading to their emergence. They play an accessory role in the study of solar physics, but their study may give hints about these basic processes.

We know that active regions are affected by differential rotation, and that their evolution, under this influence, elongates their structures. The rotation of these structures around some pivot point is an indicator of their deep roots. The evolution of an active area is influenced by differential rotation, and by conductivity of the photospheric medium, leading to complex magnetic structures that tend to convert (and store) to magnetic energy, some of the primarily mechanical energy from the deformation of the region. This storing of magnetic energy, and its subsequent release are well observed, if not always well described by the theory. However, it clearly poses a difficult problem; i.e., how the stored energy in magnetic flux tubes can lead to complicated structures that affect the dissipation mechanisms at the top of the convection zone.

The overall physics of active regions, from far below the observed surface to the coronal structures at high altitudes above it, is not only that of an independent structure that emerges, evolves and vanishes. At both ends of its evolution it is closely linked with the overall structure of the Sun, and even during its lifetime it interacts with other active regions, i.e., with the quiet part of the Sun, with the unipolar large regions that are known to exist, and with the polar regions from which emerge the polar plumes.

We know that at different times during the activity cycle, the structure of active regions is different, and so is their behavior with time. This may be an indication, by the effect it has on the behavior of flux tubes, of the interactive processes that are acting between convection and magnetism. The properties of the emerging tubes of force (with which we are mostly concerned) may well be closely linked with the properties of the time evolution of the active features. So little is known that we can hardly go beyond suggesting the importance of this point. And again, this point may be crucial to understanding how the solar wind is modified by the development of solar activity across the cycle.

Whatever may be the answer to these questions, it is quite clear that they should be studied carefully. Data are numerous, proper theories are many, but they often contradict one another (see the chapters in Part IV). These are good problems for future work.

IV. THEORETICAL KNOWLEDGE OF THE SUN

Principles of Solar Modeling

At the beginning of the century, the trend was to follow the early works by Ritter (1882) and Homer Lane (1869), and to study, like Emden (1907), the equilibrium of gaseous spheres posed in very general terms. Later, it became necessary to introduce a source of energy at the center. Standard models of the Eddington type were discussed and studied extensively. They were indeed adequate to represent the main solar parameters of mass, luminosity and radius. And they were also relatively adequate in regard to the chemical composition as deduced from analysis of the spectrum, at the accuracy that one was able to reach after Russell's (1929) analysis. But the obvious phenomena of rotation, convection and magnetism, displayed by the observations, were not described. Moreover, boundary conditions of the early models were extremely simple: zero pressure at surface, and one single value of the boundary temperature. Actually, the problem could not have been solved even with complicated boundary conditions, and spherical symmetry was necessarily imposed.

Then came the era of large computers. Theoreticians now divide the Sun into the various layers described above. In each case, they compute the best of the possible physical parameters to be introduced in the physical equations. But, now difficulties arise, not only because it is hard to get excellent physical quantities, but also because the astrophysical data (such as chemical abundances) are not always the best possible ones. Although the present situation seems rather satisfactory, one should possibly still discuss these last points.

Basic Needs for Solar Modeling

It is necessary, first, in order to compute properly the rate of energy production, to use the best possible values of the reaction rates, for each of the thermonuclear reactions that enter the picture; one must know them as a function of the physical parameters. The astronomers therefore depend heavily upon the data obtained by high-energy physics.

The opacities come from both the physical data pertaining to a given atomic species, and from determinations of the absolute abundances of elements. If one assumes that the Sun is homogeneous, then atmospheric abundances are the same as abundances in the core. One may use them to model the core and the solar interior. But are these abundances well determined? Since the early estimations by Russell (1929), several dramatic revisions have occurred. The use of correct curve-of-growth methodology, the discovery by Wildt (1939) that H^-, not atomic hydrogen, is the main contributor to continuous opacity (at least in the visible), the discovery of the importance of departures from LTE (hence the use of weak lines and lines from highly excited levels), all have led to a consistent set of abundances. But it is quite possible that the last word has not been spoken. Simplifying hypotheses are still underlying the analysis, and we know little about their effect. For example, the "roughening effect" (introduced by Redman [1943]) may affect differently the formation of metallic lines, and that of the continuous spectrum. Another example is the possible influence of a multitemperature model: different lines of different elements do result from the averaging process in which the respective weights of hot (granules?) and cold (intergranular spaces?) elements vary one from another. A single-temperature spherical (or plane parallel) atmosphere is simply not an adequate representation. Hence, further study of the quantitative influence of these effects is required, for the benefit of stellar interior modelers.

Solar Modeling at Present

As it stands, the present methods of modeling the Sun are rather satisfactory, at least in the core; there, the convergence of solutions is relatively good. But how can we explain the differential rotation of the core with respect to the interior and how does this affect the models? In the radiative interior, this is an easy problem to solve. But again, one encounters questions about the role of turbulent diffusion, about the localization of the magnetic fields, and their magnitude. If the magnetic fields are located in the radiative interior, what is then the motor behind the solar activity cycle, and the migrations? If they are not located there, what is the process of interaction between the convective zone and radiative interior? Penetrative convection is one thing, another may be the structure itself. The interior may not necessarily be spherically symmetric in the sense of the boundary between the two regions, convective and radiative, especially if the whole Sun is seismically active.

When we pass to the convective zone, we have at our disposal the mixing-length theory, a purely phenomenological approach. We know that it is not entirely adequate, although it fits well in practice the overall requirements of the modeling (see the chapters by Sofia et al. and Ulrich and Cox). However, it does not account for the various scales of the granulation. In addition to this evident weakness of the theory, the boundary conditions are poorly known at the lower boundary of the convection zone. At the upper boundary, conditions are better known (such as granulations at all scales and magnetic photosphere), but rarely taken properly into account. Observations give us a wealth of detail about the photosphere, chromosphere and the corona. Yet we have difficulty in matching the observations with a theory, e.g., to mention only one, how quantitatively the corona is heated.

Thus it appears that the only way to treat the problem is indeed to proceed as before, but in this process take better into account along with our own computations the findings of others. In a sense, the photospheric physicists are the ones who should provide boundary conditions for theorists of the convective cells. The knowledge of each layer must be used as a way to provide better boundary conditions for other layers. This reflects what we wish especially to express in this introductory chapter, i.e., the feeling that the Sun is one, that the theoretical approach, as well as the empirical approach, has to face a global Sun, even if equations to solve problems may differ from layer to layer.

An important remark to emphasize, at this point, is that diffusive processes, at the boundaries, play an important role in carrying mass, energy and momentum through these boundaries. Diffusive processes are an important coupling mechanism (see below and the chapter by Michaud and Vauclair concerning their influence on element separation).

The Study of the Sun's Past Evolution: the Time Coupling

Another question emerges from the preceding ones given above, notably from the recognized need for better opacities in the core: the Sun is, at present, a rather unmixed star. Except in the convection zone, which is relatively far from the core where thermonuclear processes do modify the composition (see Fig. 2), there is actually no mixing, or at least none that can be derived from the usual Schwarzschild stability condition; although in the past, some convection may have mixed the core. In these conditions, what kind of abundances must we use for deriving the present composition (hence the present opacity) of the solar core? We cannot use the atmospheric abundances, which are more representative of the primitive Sun than the present-day Sun. We must use the present-day atmospheric abundances, and make them, together with the Sun, evolve from the primitive Sun. Some convective models may mix part of the Sun, or even the whole Sun. This means that using the atmospheric present-day abundance might still be misleading. By using successive choices of the "initial" central chemical composition, and then computing the evolution of the Sun from this initial step to the present stage (with proper account of mixing processes whenever and wherever they occur) and finally making a comparison between the computed atmospheric abundances and the observed ones, we may be able to select the best choice from the initial composition. In any case, it is necessary to compute the past evolution of the Sun.

A mixing process which is quite effective is the convection. But we should not forget in modeling the evolution that overshooting has also a mixing role and is poorly known. We should also introduce all types of diffusive processes (mentioned above in another context; see the chapter by Michaud and Vauclair). Thermal diffusion and turbulent diffusion may indeed play a role in mixing, and not all of these processes are well known: for example, the turbulent diffusion has, at present, received only a phenomenological treatment. In addition, if these diffusive processes also affect the neutrino production rate, or the rate of lithium destruction, it is clear that they must be taken into account at all phases of the modeling. Much progress must still be made in this respect (see the chapter by Michaud and Vauclair).

Other interesting developments are occurring in this evolution modeling. For example, the fact that the present rate of solar rotation results from an earlier fast rotation, modified by the mass loss (hence the loss of momentum), which may have been in some early stages of solar life much more important than it is now, must be considered.

V. CONCLUSIONS

We have given several arguments in favor of a global treatment of the Sun. No matter what layer we choose, they are all physically interconnected and, accordingly, the main phenomena in the interior can be diagnosed only through observations of the outer layers. In this sense the whole planetary system, of which the Sun is a part, is influenced by solar particles and solar radiation, and may help towards its diagnosis.

This review has only alluded to the many facts that form the basis of my argument. They are described at length in this book. However, I have not limited myself to the description of the state of the art as it now appears to solar physicists. I have tried to define the difficulties, notably with regards to many important unsolved problems. And I have tried to outline, in a very schematical way, what, in my opinion, should be the philosophy of the solar physicist. I have, as well, emphasized that no one solar physicist can ignore the observations or computations of another solar physicist.

However, there is another point, which I have not touched, and which I must mention at least at the end of this introductory chapter. It is an indisputable fact that the Sun is a star. Many other stars exist, and display, to a smaller or larger extent, some of the phenomena observed on and in the Sun. We know about stellar rotation and stellar activity. We begin to know about stellar seismology. We know as yet little about stellar granulation, but may know more in the coming years through interferometric techniques and image treatment. We know at least something about the nonthermal motions in photospheres. Perhaps the only field in which we shall know little for a very long time (except in the cases of stars quite different from the Sun such as supernovae) is that of the neutrino production. Although it is difficult to define now a methodology, I would like to suggest that, although we have used the Sun to help us to understand the processes at work in the stars, we have still done little to reach a better knowledge of the Sun through a careful examination of its stellar properties (see the chapter by Sofia et al.). In particular, much is to be gained through the examination of early stellar evolution, since the present state of our star is a result of its past history.

Finally, I wish to comment on a bibliography for further reading. Obviously, this introductory review covers far too large a field of solar physics for a proper bibliography to be given therein. As this book is mainly oriented towards graduate students, I give a list in chronological order (however, quite incomplete) of reference books dealing with the Sun, and often used since the early 1950s.

Among *introductory* readings, one should certainly list: Menzel 1959; Brandt 1966; Tandberg-Hanssen 1967; White, ed. 1977; Eddy J., ed. 1978; Noyes 1982; Pecker (in French) 1984; Zirin 1988; Wentzel 1989.

More *advanced* introductions, *symposium volumes* or *monographs* dealing with selected topics are: Kuiper, ed. 1953; Thomas and Athay 1961; Thomas, ed. 1961; de Jager, ed. 1962; Bray and Loughead 1964,1967; Athay and Newkirk 1969; Macris, ed. 1971; Bray and Loughead 1974; Tandberg-Hanssen 1974; Athay, 1976, Svestka 1976; Bruzek and Durrant eds. 1977; Bonnet and Delache, eds. 1977; Dumont and Rösch, ed. 1978; Muller ed. 1984; "Giovanelli Commemorative Symposium" 1985 (Aust. J. Phys, *38*, no. 5); Marsden 1986; Schroter, ed. 1987; Durney and Sofia, eds. 1987; Tandberg-Hanssen 1988; Stix 1989.

Basic treaties, full of references, are: Cox and Giuli 1978; Jordan ed. 1981; Priest 1982; Sturrock et al., ed. 1986.

The author would also like to mention two of his own recent review papers on the subject (namely, Pecker 1988,1989) as well as the regular reports of various IAU commissions published every three years (in particular: commissions 10, 12, 13—now discontinued—, 35, 36, 40).

Acknowledgments: The author wishes to thank A. Cox, W. Livingston, E. Schatzman for their detailed comments on his text, as well as to thank various authors of this book for their suggestions. He gratefully acknowledges the bibliographical help kindly provided by S. Laloë.

NUCLEAR ENERGY GENERATION IN THE SOLAR INTERIOR

PETER D. MacD. PARKER
Yale University

and

CLAUS E. ROLFS
Westfälische Wilhelms-Universität Münster

The present status of our knowledge of nuclear energy generation in the solar interior is reviewed, together with a discussion of the role of the solar neutrino measurements as tests of our understanding of the solar interior. The theory of nuclear reaction rates is reviewed, including the effects of screening. Recent energy production rates are given for the many reactions of the p-p chains as well as the CNO cycles. The current disagreement between observations and predictions for the solar neutrino output does not seem to be any fault of nuclear reaction uncertainties.

On the basis of his comment in Cardiff, Wales, that "what is possible in the Cavendish Laboratory may not be too difficult in the Sun," it is clear that as early as 1920 Eddington was already making the connection between nuclear reactions and energy generation in the solar interior. The recognition of this connection was not based on any direct observation of these reactions but was instead arrived at more on the basis of default. The realization that the Earth was at least a few billion years old, coupled to the known mass, radius and luminosity of the Sun, demonstrated the inadequacy of classical (gravitational and chemical) energy sources for the Sun; the enormous energy available from nuclear reactions could, however, readily supply the required luminosity. This connection was further developed by Eddington during the

1920s (see, e.g., Eddington 1926) and by Gamow's (1928) recognition of the role of tunneling under the Coulomb barrier at low energies which then made it possible for these reactions to occur in the solar interior where kT was most probably equivalent to only $\leq$ few keV, well below the Coulomb barrier (Atkinson and Houtermans 1929). Following this early work which laid the foundations for nuclear astrophysics, by the end of the 1930s Weizsäcker (1937,1938), Bethe and Critchfield (1938) and Bethe (1939) had laid out the energetics and many of the theoretical details for the specific reactions in the CN cycle and the p-p chain, as processes for carrying out the conversion of hydrogen to helium.

By the end of the 1950s, initial laboratory measurements had been made of the cross sections and/or decay rates for each of the important nuclear reactions in the CN cycle (Fig. 1) and the p-p chain (Fig. 2) [except, of course, for the $^1H(p,e^+\nu)D$ reaction] and then extrapolated/corrected for conditions in the solar interior. Of equal significance during this time was the recognition (Davis 1955; Cameron 1958; Fowler 1958) of the possibility and the significance of detecting the neutrinos from the radioactive decays associated with some of these reactions as a way of probing the solar interior for direct evidence of the role of these thermonuclear reactions. In parallel with the design and implementation of the ^{37}Cl solar-neutrino detector (Davis 1964; Davis et al. 1968) and motivated by the obvious significance of the solar-neutrino measurements, during the 1960s, 1970s and 1980s, substantial effort has been devoted to the refinement of experimental measurements of the important p-p chain and CN-cycle reactions. For detailed discussions of these measurements and their results, see Bahcall et al. (1982), Parker (1986) or Rolfs and Rodney (1988).

The net result of all of this effort is that, averaged over the period 1970–

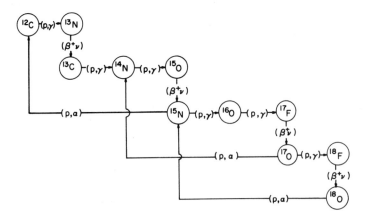

Fig. 1. The CNO tri-cycle.

$4{}^1H \rightarrow {}^4He + 2e^+ + 2\nu$ $Q = +26.731$ MeV

$^1H + e^- + {}^1H \rightarrow {}^2D + \nu$ 1.442 MeV $E_\nu = 1.442$ MeV

$^1H + {}^1H \rightarrow {}^2D + e^+ + \nu$ 1.442 MeV $E_{\nu\,max} = 0.420$ MeV

(a) $^2D + {}^1H \rightarrow {}^3He + \gamma$ 5.493 MeV

$^3He + {}^3He \rightarrow {}^4He + 2{}^1H$ 12.859 MeV

(b) $^3He + {}^4He \rightarrow {}^7Be + \gamma$ 1.587 MeV

(b$_1$) $^7Be + e^- \rightarrow {}^7Li + \nu$ 0.862 MeV $E_\nu = \begin{matrix} 0.862 \text{ MeV } 89.5\% \\ 0.384 \text{ MeV } 10.5\% \end{matrix}$

$^7Li + {}^1H \rightarrow 2{}^4He$ 17.347 MeV

(b$_2$) $^7Be + {}^1H \rightarrow {}^8B + \gamma$ 0.135 MeV

$^8B \rightarrow {}^8Be^* + e^+ + \nu$ 15.079 MeV $E_{\nu\,max} = 14.02$ MeV

$^8Be^* \rightarrow 2{}^4He$ 2.995 MeV

Fig. 2 The proton-proton chain, showing the three most important terminations: a, b$_1$ and b$_2$.

1985, there is now (1) a measured (see, e.g., Davis et al. 1987) yield of ^{37}Ar from the ^{37}C1 detector corresponding to 2.1 ± 0.3 SNU (1 SNU $= 10^{-36}$ solar-neutrino captures per ^{37}C1 atom s^{-1}) which is in disagreement with the current theoretical predictions of 7.9 ± 0.9 SNU (Bahcall and Ulrich 1988) as calculated using standard solar-model codes incorporating the best values for the various physical input parameters such as the nuclear reaction rates, opacities, etc.; and (2) a measured $\nu_e + e^-$ scattering result from the Kamiokande-II H$_2$O detector (Hirata et al. 1987a) which corresponds to an upper limit on the ^{8}B neutrino flux that is a factor of 2 below the standard solar-model predictions. For a more detailed discussion of these measurements and of possible correlations with solar-activity cycles, see the chapter by Davis. The observation of the solar-neutrino spectrum is a crucial test of our understanding of the solar interior, and until we can understand the disagreement between the current model predictions and the current experimental results, we may be forced to conclude that after more than 60 yr, we still have only qualitative evidence for thermonuclear reactions in the solar interior.

I. DETERMINATION OF STELLAR-REACTION RATES

At the temperature $T_6 = 15.5$ (in units of 10^6 K) and density $\rho = 156$ g cm^{-3} in the solar interior, the interacting nuclides reach a Maxwellian equilibrium distribution in a time that is infinitesimal compared to the mean lifetime of a nuclear reaction. Thus, the thermally averaged reaction rate per particle pair for nuclear reactions between nuclides of type 1 and 2 with an energy-dependent cross section $\sigma(E)$ is obtained by weighting $\sigma(E)$ by the Maxwell-Boltzmann distribution (see, e.g., Burbidge et al. 1957; Parker et al. 1964; Fowler et al. 1967; Fowler et al. 1975; Barnes 1971; Parker 1986; also see two textbooks which deal extensively with this matter [Clayton 1968 and Rolfs and Rodney 1988])

$$<\sigma v> = \left[\frac{8}{\pi\mu(kT)^3}\right]^{1/2} \int_0^\infty \sigma(E)\exp(-E/kT)EdE \qquad (1)$$

where μ is the reduced mass of particles 1 and 2, E their relative (center-of-mass) kinetic energy and T the temperature. If n_1 and n_2 are the number densities of particles 1 and 2, the total reaction rate R is determined by the relation

$$R = n_1 n_2 <\sigma v>/(1+\delta_{12}) \qquad (2)$$

where the Kronecker symbol δ_{12} prevents double counting in the case of identical particles. The mean lifetime τ of, say, particles 1 against destruction by particles 2 is given by

$$\tau_2(1) = (<\sigma v>n_2)^{-1} \qquad (3)$$

(see also Eqs. 7, 9, and 16) and the rate of nuclear energy generation ϵ is given by

$$\epsilon = QR/\rho \qquad (4)$$

where the Q value of the nuclear reaction is based on atomic masses. Thus, an understanding of most of the critical solar features, such as time scales, energy generation and nucleosynthesis of elements, hinges directly on the magnitude of $<\sigma v>$. The approach for arriving at an analytic expression for $<\sigma v>$ in terms of temperature T is determined by the energy dependence of $\sigma(E)$. Two commonly occurring cases are considered below.

For nonresonant charged-particle-induced reactions at low energies, the steepest energy dependence in $\sigma(E)$ is contained in the penetration factor for the Coulomb and angular momentum barriers. For low incident energies and for s waves (zero relative angular momentum), this factor is approximately proportional to $\exp(-2\pi\eta)$ (called the Gamow penetration factor), where η is the Sommerfeld parameter given by

$$2\pi\eta = 2\pi Z_1 Z_2 e^2\hbar v = (E_G/E)^{1/2} = 31.290\ Z_1 Z_2(\mu/E)^{1/2}. \qquad (5)$$

In this expression Z_1 and Z_2 are the integral nuclear charges of the two particles, v is their relative velocity, and (in numerical units) μ is given in units of amu and E in keV. It is convenient to factor out this energy dependence, as well as an additional factor of $1/E$ arising from the squared DeBroglie wavelength λ^2, which always appears in cross sections (geometrical factor):

$$\sigma(E) = S(E)E^{-1}\exp(-2\pi\eta). \qquad (6)$$

The function $S(E)$ defined by this equation contains all the strictly nuclear effects and is therefore referred to as the nuclear or astrophysical $S(E)$ factor.

For nonresonant reactions, such as ${}^3\text{He}({}^3\text{He},2p){}^4\text{He}$ (Fig. 3), the $S(E)$ factor is a smoothly varying function of energy, that varies much less rapidly with beam energy than $\sigma(E)$. Thus, $S(E)$ is much more useful in extrapolating measurements of $\sigma(E)$ to the very low energies of astrophysical interest, essentially to zero energy. Inserting Eq. (6) into Eq. (1) yields

$$<\sigma v> = \left(\frac{8}{\pi\mu(kT)^3}\right)^{1/2} \int_0^\infty S(E)\exp[-(E/kT)-(E_G/E)^{1/2}]dE. \qquad (7)$$

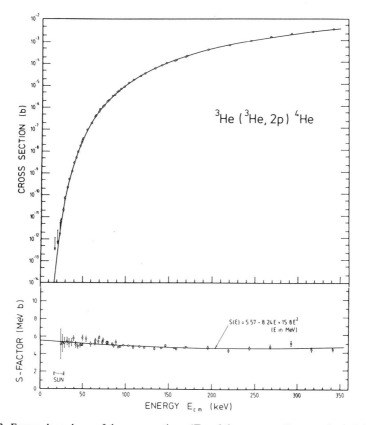

Fig. 3. Energy dependence of the cross section $\sigma(E)$ and the corresponding astrophysical $S(E)$ factor for the ${}^3\text{He}({}^3\text{He},2p){}^4\text{He}$ reaction (Krauss et al. 1987). The line through the data points is the results of a polynomial fit to the $S(E)$-factor data (described by the relation given) and is used in the extrapolation of the data to zero energy. Also indicated (hatched area) is the thermal energy region (Gamow peak) in the Sun ($E_o \pm \Delta E_o/2 = 21.9 \pm 6.2$ keV).

The energy dependence of the integrand is governed primarily by the product of the Maxwell-Boltzmann factor and the penetration factor, leading to a peak of the integrand (called the Gamow peak) with its maximum value at energy E_0 and a full width ΔE_0 (at $1/e$ of its maximum value). The nuclear reactions take place in stars in this relatively narrow energy region of $E_0 \pm (\Delta E_0/2)$. Expressing $S(E)$ as a MacLaurin series around zero energy

$$S(E) = S(0) + S'(0)E + S''(0)E^2/2 + \ldots \ldots \ldots \tag{8}$$

[with $S'(0) = dS/dE(0)$ and $S''(0) = d^2S/dE^2(0)$], leads to a simple approximation of the integral (see, e.g., Burbidge et al. 1957; Parker et al. 1964; Fowler et al. 1967; Fowler et al. 1975; Barnes 1971; Clayton 1968; Parker 1986; Rolfs and Rodney 1988)

$$<\sigma v> = (2/\mu)^{1/2}[\Delta E_0/(kT)^{3/2}]S_{eff}\exp(-\tau) \tag{9}$$

with

$$E_0 = [\pi e^2 Z_1 Z_2 kT(\mu c^2/2)^{1/2}\hbar c]^{2/3} \tag{10}$$

$$\Delta E_0 = 4(E_0 kT/3)^{1/2} \tag{11}$$

$$S_{eff} = S(0)[1 + (5kT/36E_0)] + S'(0)[E_0 + (35kT/36)]$$
$$+ S''(0)[E_0^2 + (89E_0 kT/36)]/2 + \ldots \ldots \tag{12}$$

$$\tau = 3E_0/kT \tag{13}$$

$$\mu e^2 = [A_1 A_2/(A_1 + A_2)]931.494 \text{ MeV} \tag{14}$$

For example, for the $^3\text{He}(^3\text{He},2p)^4\text{He}$ reaction at $T_6 = 15.5$, one finds $E_0 = 21.9$ keV and $\Delta E_0 = 12.4$ keV. For the determination of $<\sigma v>$, the experimentalist must determine the coefficients of the MacLaurin expansion [e.g., Fig. 3].

For nuclear reactions, in which $\sigma(E)$ exhibits narrow resonances, $\sigma(E)$ is described by the Breit-Wigner expression, in the usual notation (see, e.g., Clayton 1968; Rolfs and Rodney 1988),

$$\sigma(E) = \pi\lambda^2\omega\Gamma/_{12}\Gamma_{34}/[(E - E_R)^2 + (\Gamma/2)^2]. \tag{15}$$

For the frequently occurring case $\Gamma << \Delta E_0$, the resonance acts as a delta function and Eq. (1) can be integrated immediately to yield

$$<\sigma v> = (2\pi/\mu kT)^{3/2}\hbar^2(\omega\gamma)\exp(-E_R/kT). \tag{16}$$

The quantity $\omega\gamma = \omega\Gamma_{12}\Gamma_{34}/\Gamma$ is called the resonance strength and can be determined experimentally (see, e.g., Rolfs and Rodney 1988). If several narrow resonances occur within ΔE_0, the total reaction rate is the sum of terms like those in Eq. (15). In order to calculate these rates, the experimenter has to determine the strength $\omega\gamma$ and the location E_R of the resonance. (For other cases, such as broad resonances, subthreshold resonances, etc., see, e.g., Rolfs and Rodney 1988, and references therein.)

All of the nuclear reactions, relevant for nuclear energy generation in the solar interior, involve charged particles in the entrance channel, and thus their rates are dominated by Coulomb barriers. The thermal energy of particles in the solar interior is $kT = 1.3$ keV, which is much smaller than the height E_c of Coulomb barriers among light nuclides, e.g., $E_c = 1.7$ MeV for $^3\text{He} + ^3\text{He}$. As one can see in Fig. 3, the Gamow penetration factor is responsible for the decrease of $\sigma(E)$ by many orders of magnitude between energies near the Coulomb barrier and solar energies near E_o. With improved experimental techniques, measurements of $\sigma(E)$ can be extended towards lower energies, but in practice for most nuclear reactions, one barely approaches the relevant solar (stellar) energy regions. Such experiments are usually very difficult and frequently push the techniques of nuclear physics to extreme limits. The determination of $\sigma(E)$ for solar (stellar) fusion reactions usually requires measuring $\sigma(E)$ at as low an energy as feasible, and then extrapolating $\sigma(E)$, or equivalently $S(E)$, to still lower energies (essentially to zero energy), with the guidance of theory and other arguments. Sometimes the shape of the $S(E)$ curve is well understood beforehand, and it is then a matter of determining the normalization of the curve at enough energies to confirm the assumed theoretical shape. Of course, the basis for extrapolation is improved if extremely low-energy data with high accuracy are available (see also below). However, low-energy resonances that are just above or just below the particle threshold and that are narrow enough so that they do not extend into the measurable range can completely dominate the reaction rate for low stellar temperatures and yet not be directly measurable. In such cases one must use indirect coincidence experiments to measure the $\omega\gamma$'s for these resonances; if that is not practical, then the extrapolated data discussed above represent only lower limits of the rates. While this situation prevails in some reactions involved in the CNO cycles, all of the reactions involved in the p-p chain are of the nonresonant type. Numerous laboratory measurements have been carried out to determine the coefficients $S(0)$, $S'(0)$, and $S''(0)$ for these reactions. The present status of the measurements and extrapolations for these reactions is discussed in Sec. II.

In the above treatments, it is assumed that the Coulomb barrier of the interacting particles results from bare nuclei and that it thus would extend to infinity. However, for nuclear reactions studied in the laboratory the target nuclei are usually in the form of atoms or molecules, and the projectiles are usually in the form of positively charged ions (reaching, after a few atomic target layers, an equilibrium charge-state distribution). Thus, in reality the atomic (or molecular) electron clouds surrounding the target and projectile nuclides act as a screening potential (Assenbaum et al. 1987); an incoming projectile does not see the repulsive Coulomb force until it penetrates beyond the atomic (molecular) radius R_a, and thus it effectively sees a reduced Coulomb barrier. At low projectile energies, when the classical turning point R_c of an incoming projectile for the bare nucleus is near or outside the atomic radius, the magnitude of this shielding effect becomes significant. The condition $R_c \geq R_a$ corresponds to energies $E \leq U_e = Z_1 Z_2 e^2 / R_a$. Setting R_a equal to the radius of the innermost electrons of the target (or projectile) atoms, the resulting energies U_e are quite low [e.g., $U_e = 0.22$ keV for ^{3}He(^{3}He,2p)^{4}He], and thus the shielding effects might appear to be effectively unimportant. However, due to the acceleration of the projectiles by the electron clouds, the penetration through a shielded Coulomb barrier at energy E is equivalent to that of bare nuclei at energy $E_{\text{eff}} = E + U_e$ (Assenbaum et al. 1987). Thus, the shielding effect increases the cross section $\sigma_b(E)$ [or, equivalently $S_b(E)$] for bare nuclei to the value $\sigma_s(E)$ [or $S_s(E)$] for shielded nuclei

$$\sigma_s(E) = f_{\text{lab}}(E) \sigma_b(E) \qquad (17)$$

with the enhancement factor $f_{\text{lab}}(E)$ given by

$$f_{\text{lab}}(E) = \exp(\pi \eta U_e / E) \geq 1, \text{ for } U_e << E \qquad (18)$$

i.e., the factor $f_{\text{lab}}(E)$ increases exponentially with decreasing energy. For energies $E/U_e \geq 1000$ the shielding effects are negligible [e.g., $f_{\text{lab}} \simeq 1.006$ for ^{3}He(^{3}He,2p)^{4}He], and laboratory experiments can be regarded as measuring essentially $\sigma_b(E)$. However, at energies $E/U_e \leq 100$, the shielding effects cannot be disregarded [e.g., $f_{\text{lab}} \simeq 1.18$ and 131 at $E/U_e = 100$ and 10, respectively for ^{3}He(^{3}He,2p)^{4}He] and become important for the understanding of the low-energy data; relatively small enhancements due to electron screening at energies near $E/U_e = 100$ could cause significant errors in the extrapolation to lower energies, if the cross-section curve is forced to follow the trend of the enhanced cross section $\sigma_s(E)$ without correcting for screening. Recent experimental measurements of the ^{3}He(d,p)^{4}He reaction (Engstler et al. 1988) at energies as low as $E_{cm} = 5.9$ keV have provided proof of the existence of the effects of electron screening on cross-section measurements of low-energy fusion and have also shown that the effects (i.e., the magnitude of the screening potential U_e) depend on the aggregate state of the target (i.e., atomic or molecular form) Electron screening effects also have to be considered for the ^{3}He(^{3}He,2p)^{4}He and ^{7}Li(p,α)^{4}He reactions, where data have been measured as low as $E/U_e \simeq 72$ and 42, respectively (Secs. II.C and II.F). A systematic study of the screening effects for nuclear reactions between light nuclides is continuing at the present time.

The effects of electron screening also occur in a stellar plasma, as pointed out first by Salpeter (1954). In the plasma, the cross section for bare nuclei $\sigma_b(E)$ is enhanced to $\sigma_p(E)$ with the enhancement factor $f_0(E)$:

$$\sigma_p(E) = f_o(E)\sigma_b(E). \tag{19}$$

The factor $f_o(E)$, (by which the Eqs. (1), (7), (9) and (16) have to be multiplied) depends on the details of the stellar plasma (Salpeter 1954), such as temperature, density and chemical composition. For the solar interior, a weak screening approximation appears to be appropriate (Parker 1986) leading to an enhancement factor $f_o(E)$ of at most a few percent. Clearly, the calculation of $\sigma_p(E)$ requires a knowledge of $\sigma_b(E)$, which (as discussed above) is, however, at low energies not the quantity measured in the laboratory. An improved understanding of the laboratory enhancement factor $f_{lab}(E)$ is needed in order to correct the measured $\sigma_s(E)$ for screening effects in order to determine $\sigma_b(E)$ and thus $\sigma_p(E)$. At the same time, an improved understanding of these screening effects might also provide an improved basis for the calculation of $f_o(E)$ in a stellar plasma, such as the inclusion of dynamic effects (see, e.g., Carraro et al. 1988).

II. HYDROGEN BURNING IN THE SOLAR INTERIOR

According to current standard (traditional) solar models (see, e.g., Bahcall et al. 1982), the CN cycle (Fig. 1) is expected to play only a very minor role ($\sim 1.5\%$) in the generation of energy in the Sun. It has also been argued (see, e.g., Bahcall et al. 1968; Bahcall 1979) that the present results of the solar-neutrino experiment place an observational limit ($<10\%$) on the role of the CN cycle in the Sun. However, we should keep an open mind on this situation until we understand better the disagreement between the solar-model calculations and the present experimental data and, for example, the possible role that neutrino properties may be playing in these measurements.

In the p-p chain (Fig. 2), $^1H(p,e^+\nu)D$ is by far the slowest reaction, and therefore its cross section determines the overall rate of this sequence and hence the temperature of the solar interior. Two important branch points occur in the p-p chain, one involving the burning of ^3He through either the $^3He(^3He,2p)^4He$ or $^3He(\alpha,\gamma)^7Be$ reactions, and the other involving the subsequent burning of ^7Be through either the $^7Be(e^-, \nu)^7Li$ or $^7Be(p,\gamma)^8B$ reactions. Although the $^7Be(e^-,\nu)^7Li$ and $^7Be(p,\gamma)^8B$ terminations play only relatively minor roles in the generation of energy in the Sun [$\approx 12.4\%$ for the $^7Be(e^-,\nu)^7Li$ termination and only $\approx 0.02\%$ for the $^7Be(p,\gamma)^8B$ termination, compared to 86.4% for the $^3He(^3He,2p)^4He$ termination], the neutrinos from these two side chains—especially the $^7Be(p,\gamma)^8B$ branch—play extremely important roles in the ^{37}Cl solar-neutrino experiment and in other solar-neutrino detectors (e.g., Kamiokande-II and Sudbury-SNO) which are almost exclusively sensitive to high-energy neutrinos. The present status of our knowledge of the cross sections and decay rates for the important nuclear reactions in the p-p chain are discussed in the following subsections.

A. The ¹H(p,e⁺ν)D Reaction

Because of its very small cross section, at the present time there is no practical way to measure directly the rate of this reaction in the laboratory. [It can be estimated that with a total cross section of 10^{-47} cm² at $E_p(\text{lab}) = 1$ MeV, for a proton beam of 1 mA incident on a thick hydrogen target, there would be only one ¹H(p,e⁺ν)D reaction in 10^6 yr.] Instead, its rate must be calculated on the basis of our knowledge of the weak interaction. Bahcall and May (1969) have derived a formula for $S(E)$ for this reaction (their Eq. 12) which is proportional to the inverse of the free-neutron lifetime. There have been 8 published measurements of the free-neutron lifetime in the past 20 yr, each with an uncertainty of $< \pm 2.5\%$. These are listed in Table I and plotted in Fig. 4. Four of these are direct measurements [D] of the free neutron lifetime, while the other four are indirect measurements in which one measures the ratio $|g_A/g_V|$ that can then be related to the free-neutron lifetime (see Wilkinson 1982). In these indirect experiments, one can measure either the angular correlation between the direction of the electron and the direction of the spin of the neutron (the beta-decay asymmetry) [A] or the angular correlation between the electron and the anti-neutrino (inferred from the momentum distribution of the recoil proton) [C].

The direct and indirect measurements of $t_{1/2}(\text{neutron})$ can be viewed as forming two separate, but consistent, data sets which are characterized by $t_{1/2}(\text{direct}) = 617.7 \pm 4.7$ s and $t_{1/2}(\text{indirect}) = 623.5 \pm 1.4$ s. The combined data sets determine $t_{1/2}(\text{neutron}) = 619.9 \pm 2.9$ s. Using this value, together with the most recent calculations (Gari 1978) of the square of the overlap matrix element between the deuteron and the p-p initial state $[\Lambda^2(0) = 6.91]$

TABLE I
Neutron Half-Life

Neutron Half-Life (s)	Experiment[a]	Reference
637 ± 10	D	Christensen et al. (1972)
629 ± 12.5	A	Krohn and Ringo (1975)
608 ± 11	D	Bondarenko et al. (1978)
625 ± 14	C	Stratawa et al. (1978)
627 ± 10	A	Erozolimskii et al. (1979)
649 ± 12	D	Byrne et al. (1980)
622 ± 4	A	Bopp et al. (1986)
607 ± 15	D	Last et al. (1988)
615 ± 3	D	Mampe et al. (1989)
619.9 ± 2.9	Weighted mean	

[a] A = Measurement of beta-decay asymmetry.
C = Measurement of $e - \nu$ angular correlation.
D = Direct Measurement of half-life.

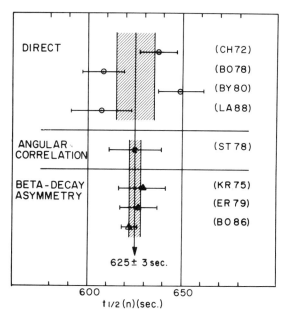

Fig. 4. Comparison of the most recent determinations of the half-life of the free neutron (see Table I).

and of the mesonic-exchange correction term $[(1+0.02)^2]$, for low energies $S_{11}(E)$ can be evaluated using $S_{11}(0) = 4.10 \times 10^{-25}$ MeV-b and $S'_{11}(0) = 4.59 \times 10^{-24}$ b with an estimated uncertainty in $S_{11}(0)$ of $\pm 1.2\%$, due to the uncertainty in the neutron lifetime ($\pm 0.5\%$), the uncertainty in the meson-exchange correction ($\pm 0.7\%$), and the uncertainty in $\Lambda^2(0)$ ($\pm 0.8\%$).

A variation of this reaction is the $p + e^- + p \rightarrow D + \nu$ reaction. Its rate is even slower (typically ~ 500 times smaller) than the p-p reaction (Bahcall and May 1969)

$$R_{pep} \approx 5.51 \times 10^{-5} \, \rho \, (1+X_H)T_6^{-1/2}(1+.02 \, T_6) \, (1+.0088 \, T_6^{2/3})^{-1}R_{pp} \quad (20)$$

so that it does not play a significant role in solar hydrogen burning. It does, however, play an important role in the ^{37}Cl solar-neutrino experiment because its monoenergetic neutrino ($E_\nu = 1.44$ MeV) is above the ^{37}Cl(ν,e^-)^{37}Ar threshold and provides a lower limit of 0.25 SNU (Bahcall and May 1969) for the theoretical predictions for this detector, regardless of the uncertainties in the opacities or in the relative roles of the various p-p chain terminations, assuming only that the Sun is currently generating sufficient thermonuclear energy via the p-p chain to balance its external luminosity.

B. The $D(p,\gamma)^3$He Reaction

This reaction has the lowest Coulomb barrier of all the nuclear reactions in the p-p chain, and measurements have been extended to energies as low as $E_{cm} = 16$ keV (Griffiths et al. 1963; Bailey et al. 1970). Since the experimental data over a wide range of energies ($E/U_e \geq 590$; Sec. I) are well described by the direct capture (DC) model, there is considerable confidence in the validity of its extrapolation to zero energy: $S_{12}(0) = 0.25 \times 10^{-3}$ keV-b; $S'_{12}(0) = 0.75 \times 10^{-5}$ b.

Other D-Burning Reactions: The $D(d,n)^3$He and $D(d,p)^3$H reactions have cross-section factors, $S(0)$, which are $\sim 10^5$ times larger than for the $D(p,\gamma)^3$He reaction, but, because the relative deuterium-to-proton abundance is $<10^{-17}$ in the solar interior, these reactions do not have any significant role there (see, e.g., Parker et al. 1964). The $D(^3$He,p$)^4$He reaction has a cross-section factor which is more than 10^6 times larger than for the $D(p,\gamma)^3$He reaction, but this is counterbalanced by the smaller ^{3}He abundance and the larger $D + ^3$He Coulomb barrier so that this reaction accounts for $<10^{-3}$ of the deuterium burning in the solar interior (see, e.g., Parker et al. 1964).

C. The ^{3}He(^{3}He,2p$)^4$He Reaction

Studies of this reaction have been carried out over a wide range of energies, down to as low as $E_{cm} \simeq 90$ keV (Bacher and Tombrello 1965; Neng-Ming et al. 1966; Tombrello 1967; Dwarakanath and Winkler 1971). The $S(E)$ factor data have been fitted to the polynomial function given above in Eq. (8) with $S(0) = 5.5 \pm 0.5$ MeV-b, $S'(0) = -3.1$ b, and $S''(0) = 2.8$ MeV^{-1} (Dwarakanath and Winkler 1971). Theoretical models (May and Clayton 1968) describe the observed energy dependence of the data very well, thus providing confidence in the extrapolation via the polynomial function. Based partially on theoretical arguments, it has been suggested (Fowler 1972; Fetisov and Kopysov 1972, 1975) that a low-energy resonance might exist in this reaction. If it were sufficiently low and narrow in reaction energy, it might have been unobserved in the measurements. With such a resonance, the discrepancy between predicted and observed solar-neutrino fluxes might be accounted for or at least decreased. Such a resonance would correspond to an excited state in ^{6}Be near the ^{3}He + ^{3}He threshold ($E_x \simeq 11.6$ MeV). However, searches for this state, using a variety of other nuclear reactions, have not been successful (McDonald et al. 1977, and references therein). Dwarakanath (1974) carried out a direct search for this hypothetical resonance state by extending the ^{3}He(^{3}He,2p$)^4$He reaction studies down to $E_{cm} = 30$ keV. Although the data might allow an increase in the $S(E)$ factor at the lowest energies, the large uncertainties in the data points below 40 keV (about 200% at the lowest energies) precluded any conclusion about the existence or nonexistence of this resonance below $E_{cm} = 40$ keV.

In the absence of such a resonance, Fowler et al. (1975) and Harris et al. (1983) recommended $S(0) = 5.5$ MeV-b. Kavanagh (1982) pointed out that the absolute $S(E)$ values reported by Neng-Ming et al. (1966) and by Dwarakanath (1974) are systematically about 20% lower than the combined results of Bacher and Tombrello (1965) and Dwarakanath and Winkler (1971), and therefore recommended a lower value of $S(0) = 5.0 \pm 0.5$ MeV-b. From similar considerations, Bahcall et al. (1982) used $S(0) = 4.7 \pm 0.25$ MeV-b for the calculation of the solar-neutrino flux.

In view of these discussions a renewed investigation of this reaction has recently been carried out down to low energies approaching the solar-energy region (Krauss et al. 1987). The experimental setup included equipment such as high-current accelerators, windowless gas-target systems, detectors placed in close geometry to the target, and beam calorimeters. As a first step in these investigations, several background problems had to be examined, such as deuterium contamination of order 10^{-6} in the ^{3}He target and in the ^{3}He ion beam (as HD^+), cosmic-ray background (of order 1 event per hr), and background from high-frequency noise in the electronics (of order 1 event per 10 hr). When these problems were solved, it was possible to study this reaction down to energies as low as $E_{cm} = 24$ keV (Fig. 3). This lowest data point, with $\sigma = 7 \pm 2$ pb, required a total running time of 6 days, or an accumulated charge of about 180 Coulomb. Upper limits for the reaction yields were also obtained over similar running times down to $E_{cm} = 16$ keV ($\sigma \leq 1.5$ pb). The data give no evidence for the suggested resonance and thus show that the solar-neutrino problem cannot be solved by the suggested resonance, unless it is located near or below $E_{cm} = 16$ keV. The deduced $S(0)$ value for these data is 5.57 ± 0.31 MeV-b (Fig. 3); if the data shown in Fig. 3 are corrected for the predicted enhancement using $U_e = 0.22$ keV, a fit to these modified data for bare nuclei leads to $S_b(0) = 5.24$ MeV-b (Assenbaum et al. 1987). Combining this result with the average of the previous, higher-energy measurements, $S(0) = 4.7 \pm 0.3$ MeV-b (Bahcall et al. 1982), yields a value of $S_{33}(0) = 5.0 \pm 0.3$ MeV-b.

D. The ^{3}He(α,γ)^{7}Be Reaction

Parker and Kavanagh (1963) studied this reaction via γ-ray spectroscopy over a wide range of energies, $E_{cm} = 180$ to 2500 keV, using a gas cell with a conventional entrance foil. The resulting $S(E)$ factor indicated a somewhat steeper rise with decreasing energy than was suggested by the theoretical predictions based on the DC model (Tombrello and Parker 1963; for more recent theoretical work, see Kajino and Arima [1984, and references therein]). Normalizing the DC model prediction to the data set, determined $S(0) = 0.47 \pm 0.05$ keV-b. Subsequently, measurements were made (Nagatani et al. 1969) at lower energies ($E_{cm} = 164$ to 245 keV) using a windowless-gas target to minimize the energy loss and energy straggling of the beam. The combined results were fitted by a polynomial, leading to $S(0)$

$= 0.61 \pm 0.07$ keV-b (Nagatani et al. 1969). However, a re-analysis of both data sets separately, guided by the DC model predictions, led to a mean value of $S(0) = 0.52 \pm 0.05$ keV-b (Parker 1986; Kavanagh 1982). More recently, data were measured by Kräwinkel et al. (1982) as low as $E_{cm} = 107$ keV ($E/U_e \geq 490$; Sec. I) using high-resolution Ge(Li) detectors in combination with windowless-gas targets. The observed angular distributions for the capture γ-ray transitions, the energy dependence of their branching ratio, and the energy dependence of the total $S(E)$ factor data were all in excellent agreement with the DC model calculations (Tombrello and Parker 1963), supporting its use in extrapolation of these data to zero energy, determining $S(0) = 0.32 \pm 0.04$ keV-b (Kräwinkel et al. 1982). Other recent measurements (Osborne et al. 1982, 1984) using similar experimental techniques, confirmed the energy dependence of the branching ratio and of the $S(E)$ factor but determined an $S(0)$ value of 0.52 ± 0.03 keV-b. The higher value (Osborne et al. 1982, 1984) for $S(0)$ was subsequently confirmed by other measurements (Robertson et al. 1983; Volk et al. 1983; Alexander et al. 1984), involving both prompt γ-ray measurements as well as measurements of the delayed ^{7}Be radioactivity induced by the bombardment (Table II).

The inconsistent absolute cross sections reported by Kräwinkel et al. (1982) were derived predominantly from measurements using a quasi-point

TABLE II
Summary of $S_{34}(0)$ for ^{3}He(α,γ)^{7}Be

$S^{34}(0)$[a](keV-b)	Reference
Method: prompt capture γ-ray transitions	
0.47 ± 0.05	Parker and Kavanagh (1963)
0.58 ± 0.07[b]	Nagatani et al. (1969)
0.45 ± 0.06[c]	Kräwinkel et al. (1982)
0.52 ± 0.03	Osborne et al. (1982)
0.47 ± 0.04	Alexander et al. (1984)
0.53 ± 0.03	Hilgemeier et al. (1988)
Method: ^{7}Be activity	
0.55 ± 0.05	Osborne et al. (1982)
0.63 ± 0.04	Robertson et al. (1983)
0.56 ± 0.03	Volk et al. (1983)
<0.69[d]	Hilgemeier et al. (1988)
0.553 ± 0.017	Weighted mean

[a]The DC-model prediction for the energy dependence of $S(E)$ (Tombrello and Parker 1963) is used in the extrapolation of the data to zero energy.
[b]Results of re-analysis (Parker 1986; Kavanagh 1982); $S(0) = 0.61 \pm 0.07$ keV-b was reported originally.
[c]Results based only on jet density as measured via elastic scattering yields (Kräwinkel et al. 1982; Hilgemeier et al. 1988); $S(0) = 0.32 \pm 0.04$ keV-b was reported originally.
[d]Not included in the weighted average.

supersonic jet-gas target. A crucial quantity in these measurements was the number of target atoms in the jet-target zone. Due to the point-like nature of these jet-gas targets (about 2 to 3 mm diameter), the overlap of the ion-beam profile with the gas-density profile is extremely critical for such measurements and depends on the characteristics of the incident ion beam, the precision in alignment, and the size of the beam-defining collimators. *In situ* measurements of this overlap are thus of utmost importance in absolute cross-section measurements. For example, in the work of Kräwinkel et al. (1982), the number of target atoms in the He jet N_t was measured via elastic-scattering yields and via energy-loss data; the first method led to N_t values that were a factor of 1.4 smaller than the values adopted by Kräwinkel et al. (1982). On the basis of these smaller N_t values, the absolute $S(0)$ value would increase to 0.45 ± 0.06 keV-b, consistent with the $S(0)$ values determined by the other experiments (Table II). Thus, it was suggested by Hilgemeier et al. (1988) that the He gas density of the supersonic jet was actually lower than the values used in the previous analyses (Kräwinkel et al. 1982), and that correcting the gas density to the value indicated by the elastic scattering essentially removes the discrepancy.

This suggestion was tested recently (Hilgemeier et al. 1988) using an improved experimental setup to allow for *in situ* measurement of the overlap between the ion-beam profile and the jet-density profile, as well as allowing for a higher precision in the determination of other relevant quantities. These new measurements of the prompt-capture γ-ray transitions gave $S(0) = 0.53 \pm 0.03$ keV-b, in good agreement with the other measurements (Table II), leading to a weighted average of $S_{34}(0) = 0.533 \pm 0.017$ keV-b and $S'_{34}(0) = -3.1 \times 10^{-4}$ b, where the energy derivative is determined from the observed energy dependence of $S(E)$ (Krawinkel et al. 1982; Osborne et al. 1982, 1984) together with DC model calculations (see, e.g., Tombrello and Parker 1963).

Other ^{3}He-Burning Reactions. The ^{3}He(d,p)^{4}He reaction (see also Sec. II.B, above) and the ^{3}He(^{3}He,2p)^{4}He reaction have comparable cross-section factors, $S(0)$; however, the much larger abundance of ^{3}He relative to deuterium more than compensates for the larger ^{3}He + ^{3}He Coulomb barrier so that the ^{3}He(d,p)^{4}He reaction only accounts for $<10^{-3}$ of ^{3}He burning in the solar interior (see, e.g., Parker et al. 1964). The ^{3}He(p,$\beta^+\nu$)^{4}He reaction is a potential source of high-energy neutrinos ($E_{\nu\,max} = 18.77$ MeV). However, on the basis of calculations by Werntz and Brennan (1973) and Tegner and Bargholtz (1983), and measurement of the ^{3}He(n,γ)^{4}He cross section (see, e.g., Wolfs et al. 1989), this reaction plays an even smaller role ($\sim 5 \times 10^{-7}$) in the termination of the p-p chain in the solar interior than the ^{7}Be(p,γ)^{8}B reaction ($\approx 2 \times 10^{-4}$), and these "hep" neutrinos are therefore expected to account for $< 1\%$ of the predicted capture rates for any of the solar neutrino detectors currently under discussion (Bahcall and Ulrich 1988) but at some point may be important in providing a more model-independent lower limit on the flux of high-energy solar neutrinos.

E. The $^7Be(e^-,\nu)^7Li$ and $^7Be(p,\gamma)^8B$ Reactions

The interpretation of data from solar neutrino detectors that are primarily sensitive to the high-energy neutrinos from the positron decay of ^8B (such as the Homestake-^{37}Cl tank, Kamiokande-II and the proposed Sudbury-D_2O facility is particularly dependent on an accurate determination of the relative rates of the various ^7Be-burning reactions in order to determine what fraction of the ^7Be produced in the $^3He(\alpha,\gamma)^7Be$ reaction will be subsequently converted into ^8B.

$^7Be(e^-,\nu)^7Li$.   The electron-capture decay of ^7Be goes to both the ground state and the 478-keV first excited state of ^7Li. (The distinction between these two decays is important to the solar-neutrino experiment because of the differences in the associated neutrino energies.) In 1983–84, there were 9 new measurements of the fraction of these decays that go through the first excited state, and the weighted mean of these modern results is $(10.52 \pm 0.06)\%$ (Ajzenberg-Selove 1988). The measured laboratory half-life for the decay of ^7Be is (53.29 ± 0.07) days; however, in the solar interior the half-life of ^7Be will be substantially longer due to the ionization of the ^7Be. The relative contributions of electron capture (a) from residual and transient bound states, and (b) directly from the continuum are functions of the temperature and density. By averaging over the appropriate regions of the solar interior, Iben et al. (1967) and Bahcall and Moeller (1969) have shown that in almost all solar-model calculations it is sufficient simply to increase the continuum capture rate by 20%. The continuum capture rate is proportional to the electron density, and over the temperature range $10 \leq T_6 \leq 16$ this decay rate can be written as (Bahcall and Moeller 1969)

$$\lambda_c = 1/\tau_c = 4.62 \times 10^{-9}(\rho/2)\ (1+X_H)\ T_6^{-1/2}[1+.004(T_6-16)]\ s^{-1}. \quad (21)$$

Therefore the total capture rate is given by

$$\lambda \approx 1.2 \times \lambda_c = 5.5 \times 10^{-9}\ (\rho/2)\ (1+X_H)\ T_6^{-1/2}[1+.004(T_6-16)]\ s^{-1}. \quad (22)$$

$^7Be(p,\gamma)^8B$.   Six independent experimental studies of this reaction have been reported (Kavanagh 1960; Parker 1966,1968; Kavanagh et al. 1969; Kavanagh 1972; Vaughn et al. 1970; Wiezorek et al. 1977; Filippone et al. 1983). All these measurements (except that of Wiezorek et al.) determine the amount of ^7Be in their targets by measuring the rate at which ^7Li is being produced in the target from the decay of the ^7Be. The ^7Li density is measured using the 770-keV resonance in the $^7\text{Li}(d,p)^8\text{Li}$ reaction in which the ^8Li and the ^8B are detected using the same geometry and the same detector (i.e., either by detecting the high-energy β^- or the delayed α's), thereby cancelling out a number of geometrical-efficiency calibrations. The absolute cross section at the peak of the 770-keV resonance has been determined from a large number of independent measurements to be 157 ± 10 mb (Elwyn et al. 1982; Filippone et al. 1982). An independent check on this number is provided in a comparison performed by Filippone et al. (1983) in which they determined the ^7Be areal density in their target using both the $^7\text{Li}(d,p)^8\text{Li}$ resonance and a mapping of the ^7Be activity distribution on the target using a tightly collimated γ-ray detector; the two methods agreed to within 7%, well within their individual uncertainties.

The results of these 6 measurements of $S_{17}(0)$ are listed in Table III. A direct-capture model calculation (see, e.g., Tombrello 1965; Williams and Koonin 1981) has been used to extrapolate the measured $S(E)$ to $E = 0$. The weighted mean of these results gives $S_{17}(0) = 0.0243 \pm 0.0018$ keV-b with an energy derivative of $S'_{17}(0) = -3 \times 10^{-5}$b determined from the direct-capture model.

Other ^7Be-Burning Reactions. A number of alternative ways to destroy ^7Be and ^8B in the solar interior have been examined in order to investigate whether the solar-neutrino problem (i.e., the apparent lack of detectable ^8B neutrinos) could be explained on the basis of (a) a reduced production rate for ^8B caused by the burning of ^7Be through other nuclear reactions, or (b) the destruction of ^8B via nuclear reactions with time scales that are shorter than (or comparable to) the ^8B beta-decay lifetime (see, e.g., Parker et al. 1964; Wagoner 1969; Parker 1972; Hardie et al. 1984). Specifically, these studies examined the

$$^7\text{Be} + d \rightarrow p + \alpha + \alpha$$
$$^7\text{Be} + {}^3\text{He} \rightarrow p + p + \alpha + \alpha$$
$$^7\text{Be} + \alpha \rightarrow {}^{11}\text{C} + \gamma \qquad (23)$$

TABLE III
Summary of $S_{17}(0)$ for $^7\text{Be}(p,\gamma)^8\text{B}$

$S_{17}(0)$ (keV-b)	Reference
0.016 ± 0.006	Kavanagh (1960)
0.028 ± 0.003	Parker (1966, 1968)
0.0273 ± 0.0024	Kavanagh et al. (1969); Kavanagh (1972)
0.0214 ± 0.0022	Vaughn et al. (1970)
0.045 ± 0.011	Wiezorek et al. (1977)
0.0221 ± 0.0028	Filippone et al. (1983)
0.0243 ± 0.0018	Weighted Mean

reactions as alternatives to the $^7Be(e^-\nu)^7Li$ and $^7Be(p,\gamma)^8Be$ reactions and the

$$^8B + d \rightarrow p + p + \alpha + \alpha$$
$$^8B + {}^3He \rightarrow p + p + p + \alpha + \alpha$$
$$^8B + \alpha \rightarrow {}^{11}C + p \qquad (24)$$

reactions as alternatives to ^8B beta decay. The results of these studies indicate that in the solar interior, even under the most favorable circumstances [with a (currently unknown) s-wave resonance with $\theta_i^2 = 1$ located at the effective thermal energy (E_o; Eq. 10)], these reactions are at least 10^{10} times less important than the nonresonant $^7Be(p,\gamma)^8B$ reaction or the beta decay of ^8B, respectively.

F. The $^7Li(p,\alpha)^4He$ Reaction

The work of Rolfs and Kavanagh (1986) confirmed the absolute cross sections reported by Spinka et al. (1971) and thus removed the apparent uncertainties in the reaction rate for $^7Li(p,\alpha)^4He$. The data for $E_{cm} \leq 100$ keV (Fig. 5b) suggested an extrapolated $S(0)$ factor of 40 to 65 keV-b with a probable value of $S(0) = 52 \pm 8$ keV-b (Rolfs and Kavanagh 1986). In the search for electron screening effects (Sec. I), the $^7Li(p,\alpha)^4He$ and $^6Li(p,\alpha)^3He$ reactions have been studied recently to as low as $E_{cm} \simeq 10$ keV (Neldner 1988). The results for $^6Li(p,\alpha)^3He$ (Fig. 5a) suggest a screening potential of $U_e \simeq 300$ eV, slightly higher than expected ($U_e = 240$ eV; Assenbaum et al. 1987), and for $^7Li(p,\alpha)^4He$ $U_e \simeq 210$ eV, slightly below the expected value. As both reactions should have identical screening potentials, further investigations are desirable. The present analysis of the $^7Li(p,\alpha)^4He$ data suggest $S(0) \simeq 57$ keV-b.

G. The CNO Cycles

Due to the high Coulomb barriers of the reactions involved in the CNO cycles (compared to those in the p-p chain) and the relatively low temperature in the solar interior (compared to more massive stars), the CNO cycles (Parker 1986; Rolfs and Rodney 1988, and references therein) constitute only a small contribution to the total solar luminosity, of order 1.5% in the standard solar model (Bahcall et al. 1982). However, the CNO cycles play an important role in the understanding of the solar-neutrino spectrum (Parker 1986; Bahcall et al. 1982).

The $^{14}N(p,\gamma)^{15}O$ reaction is the slowest reaction in the main CN cycle and thus controls the energy generation in the CNO cycles. Various investigators (Schröder et al. 1987, and references therein) had obtained data for this reaction to energies as low as $E_{cm} \simeq 100$ keV. However, where data of different investigators overlapped, the absolute cross sections $\sigma(E)$ differed by about a factor of 2. Furthermore, the available data allowed no clear picture of the capture mechanisms involved and thus extrapolated $S(0)$ values

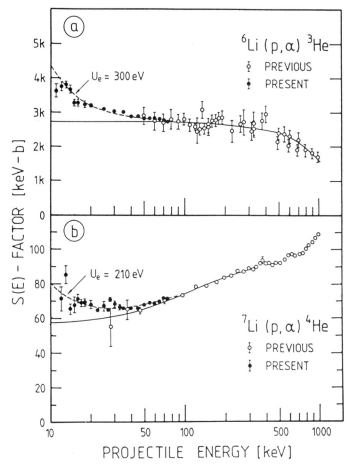

Fig. 5. Energy dependence of the $S(E)$ factor for the reactions $^7\text{Li}(p,\alpha)^4\text{He}$ and $^6\text{Li}(p,\alpha)^3\text{He}$ (Spinka et al. 1971; Rolfs and Kavanagh 1986; Neldner 1988, and references therein). The solid curves are obtained from a fit to the previous data at higher energies and are assumed to represent the case of bare nuclei. The dashed curves are calculated enhancements due to the effects of electron screening, with the fitted potentials of $U_e = 300$ and 210 eV.

were uncertain. Fowler et al. (1975) recommended the value of $S(0) = 3.32$ keV-b. Recent work (Schröder et al. 1987) has removed most of the apparent discrepancies and has shown that the capture process into the 6.79-MeV state of ^{15}O and the capture process into the high-energy wing of this subthreshold state account for 93% of the total $S(0)$ value. Although, this work improved the understanding of the reaction mechanisms involved and thus of the ex-trapolated $S(0)$ value for this astrophysically important reaction, the effects of two major changes in the previously suggested reaction mechanisms (i.e., a negligible contribution of the DC $\rightarrow$ 6.18 MeV process and a significant contribution of the $E_R = -504$ keV subthreshold resonance, the 6.79-MeV state) cancelled each other so that the new total $S(0)$ value of 3.20 ± 0.54 keV-b is essentially identical with the previously recommended value of 3.32 keV-b.

III. CONCLUSIONS

As noted at the beginning of this chapter, the observation of the solar-neutrino spectrum is a crucial test of our understanding of the solar interior. The current disagreement between the solar-model calculations and the experimental measurements would appear to be well outside the uncertainties associated with any of the nuclear reaction rates, including the $\pm$ 15% uncertainty in S_{17}. In order to investigate further this disagreement and the many speculative questions that have been raised concerning the possible consequences of as yet unmeasured neutrino properties, the next logical step would appear to be the development and implementation of the next generation of solar-neutrino detectors which will measure additional, complementary information about the solar-neutrino flux. Such detectors need to provide either information about other parts of the neutrino energy spectrum (e.g., the p-p neutrinos) or, more importantly, measurements of the time, energy, direction and type (ν_e vs ν_μ) of the interacting neutrinos (e.g., a real-time detector such as the proposed SNO-D_2O facility [see, e.g., Aardsma et al. 1987]).

SOLAR NEUTRINO EXPERIMENTS

RAYMOND DAVIS, JR.
University of Pennsylvania

and

ARTHUR N. COX
Los Alamos National Laboratory

Observational methods for detecting solar neutrinos are reviewed, and the current results presented. Four operating systems, the Homestake Mine ^{37}Cl radiochemical method, the real-time neutrino scattering detector at Kamiokande II, in Japan, and two radiochemical Gallium-Germanium experiments are discussed in some detail. The results from these operating detectors are compared to solar model calculations and various neutrino oscillation processes that could alter the primary electron neutrino flux from the Sun. It is of great interest to study the solar neutrino particle spectrum with a number of detectors that have a potential for exploring the solar energy processes and the interactions of neutrinos. A few old and new solar neutrino detectors are reviewed that have promise for revealing new information about the neutrinos from the Sun.

I. INTRODUCTION

This is a review of the experimental results and the methods that have been used in solar neutrino research. At the present time two experimental observations of the solar neutrino flux have been performed by two totally different experimental techniques. The Homestake chlorine detector uses a simple radiochemical method, and has been operating for over 20 yr. The Kamiokande II detector has been observing for 3 yr and uses a real-time directional electronic technique based on neutrino-electron elastic scattering.

The two experiments are complementary in that the former is sensitive to neutrinos with energies > 0.814 MeV, and observes only the integrated flux from all solar sources above the threshold energy. The Kamiokande II detector observes neutrinos > 7.5 to 9 MeV and has the capability of observing the direction of the neutrino, and uses this directional information to distinguish solar neutrinos from background events. The results of these two experiments, interpreted in terms of our understanding of the solar energy production processes, are in reasonable agreement. According to solar model calculations, the Homestake experiment is expected to be predominantly sensitive to the relatively high-energy neutrinos from ^{8}B decay in the Sun, accounting for approximately 77% of the rate. From this viewpoint both of the detectors are measuring the same source. However, the measurements find that present solar neutrino flux is lower than predicted by standard models of the Sun. The Homestake experiment reports that there is an apparent variation in the observed rate that anti-correlates with the solar activity cycle. Clearly one needs continuing observations with the present detectors and a new generation of solar neutrino detectors. Two new radiochemical experiments began observations in 1990–1991. These experiments use gallium as a target element and will be able to observe the low-energy neutrinos from the p-p reaction. Now is an appropriate time to review solar neutrino experimental techniques, to serve as a guide to earlier work and stimulate new efforts to study the solar energy processes.

In this chapter we will discuss in more detail the experiments that are currently operating, and those that will begin observations in the very near future. Various proposed experiments will be discussed more briefly. Solar neutrino research has had a long and difficult history that has greatly retarded its development as an experimental science. Almost every experiment that is now being proposed was suggested many years ago. In the course of this chapter, we will only mention some historical matters and give appropriate historical references. In presenting the results of experiments, various theoretical matters of direct interest to the topic being discussed will be brought out.

More discussions of solar models and concepts in neutrino physics are discussed in other chapters by Parker and Rolfs, by Rosen, by Ulrich and Cox and by Demarque and Guenther.

II. RADIOCHEMICAL EXPERIMENTS

A. General Considerations

The Sun produces energy by the proton-proton chain of reactions and the carbon-nitrogen cycle. The principal nuclear processes are shown in Table I along with the neutrino energies and fluxes. The Sun has a very broad energy spectrum of neutrinos, from 0 to over 14 MeV, and the individual

TABLE I
The Principal Nuclear Fusion Processes in the Sun

Reaction	Neutrino Energy (MeV)	Neutrino Flux (cm^{-2} s^{-1})
The Proton-Proton Chain		
PP-I $\begin{cases} H + H \to D + e^- + \nu_c \text{ (99.75\%)} \\ \text{or} \\ H + H + e^+ \to D + \nu_c \text{ (0.75\%)} \\ D + H \to {}^3He + \gamma \\ {}^3He + {}^3He \to 2H + {}^4He \text{ (87\%)} \end{cases}$	0–0.420 spectrum 1.44 line	6.0 × 10^{10} 1.4 × 10^8
PP-II $\begin{cases} {}^3He + {}^4He \to {}^7Be + \gamma \text{ (13\%)} \\ {}^7Be + e^- \to {}^7Li + \nu_e \\ {}^7Li + H \to \gamma + {}^8Be \to 2{}^4He \end{cases}$	0.861 (90%) line $\Big\}$ 0.383 (10%) line	4.7 × 10^9
PP-III $\begin{cases} {}^7Be + H \to {}^8B + \gamma \text{ (0.017\%)} \\ {}^8B \to {}^9Be^* + e^+ + \nu_e \\ \quad\quad \llcorner\to 2{}^4He \end{cases}$	0–14.1 spectrum	5.8 × 10^6
The Carbon-Nitrogen Cycle		
$H + {}^{12}C \to {}^{13}N + \gamma$ ${}^{13}N \to {}^{13}C + e^+ + \nu_c$ $H + {}^{13}C \to {}^{14}N + \gamma$	0–1.20 spectrum	6.1 × 10^8
$H + {}^{14}N \to {}^{15}O + \gamma$ ${}^{15}O \to {}^{15}N + e^+ + \nu_c$ $H + {}^{15}N \to {}^{12}C + {}^4He$	0–1.73 spectrum	5.2 × 10^8

[a]Neutrino fluxes from Bahcall and Ulrich (1988).

components overlap one another. The spectrum is shown in Fig. 1. All of these components are of great interest for understanding the nuclear processes responsible for the solar energy, and exploring concepts in neutrino physics.

Radiochemical detectors depend on the production of a specific product that can be removed from a large mass of the target isotope and measured in a small radiation detector or other device capable of measuring the product isotope. These detectors are especially useful for solar neutrino detection because the neutrino capture rate is extremely low, often in the range of a few atoms per day in hundreds of tons of target material.

The nuclear processes for neutrino or anti-neutrino detectors can be illustrated by the examples,

$$\nu_e + {}^{37}Cl \rightleftarrows e^- + {}^{37}Ar, \text{ half life } = 35 \text{ days} \tag{1}$$

$$\bar{\nu}_e + {}^{35}Cl \rightleftarrows e^+ + {}^{35}S, \text{ half life } = 87 \text{ days} \tag{2}$$

where the double arrows indicate neutrino absorption and decay of the product. The element chlorine not only is useful here to illustrate neutrino and

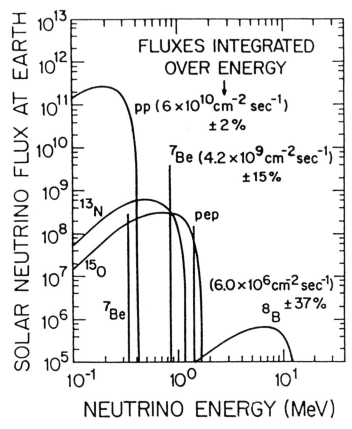

Fig. 1. The solar neutrino spectrum.

anti-neutrino processes, but chlorine is as well a practical element for detecting both neutrinos and anti-neutrinos. In this chapter we will only be concerned with neutrino detectors, though there is interest in observing anti-neutrinos from the Earth to measure its radioactive heat production (Krauss et al. 1984). The above neutrino or anti-neutrino capture reaction is the inverse of a beta-decay process. The cross section for these processes can be calculated accurately from the principle of detailed balancing and the known beta-decay process (Bahcall 1964, 1978). In general, only the cross section to produce the product nucleus in the ground state can be calculated in this way. There are a few examples of interest to solar neutrino detection in which the excited state lives sufficiently long for its beta-decay branch to be measured. The cross section to produce the product nucleus in all bound states is extremely important because of the broad energy spectrum of the neutrinos from the Sun. It is essential to determine the sensitivity of any detector to the entire solar neutrino spectrum. Cross sections for neutrino capture to produce the product nucleus in excited states are difficult to determine with the required accuracy. However, in some cases reasonably good values have been determined empirically by (p,n) reaction studies (Goodman 1985), nuclear structure modeling (Lanford and Wildenthal 1972), or the nuclear properties of the product of neighboring nuclei (Bahcall 1978). These matters are of course important for all neutrino detectors that use complex nuclei as targets. In a few radiochemical detectors of special interest, considerable experimental and theoretical effort has been devoted to this subject. Examples of the neutrino capture cross-section studies will be given when specific experiments are discussed.

The radiochemical method of neutrino detection can be designed to have a very high sensitivity, low-energy threshold, high efficiency, and a low-background signal rate from cosmic rays, neutrons and gamma radiation. The high sensitivity is achieved by choosing an abundant target isotope with a favorable inverse beta transition, allowed or superallowed, to yield a product nucleus with a convenient lifetime and decay scheme that will allow a clear separation of its radiations from background radiations. The efficiency of the chemical isolation of the product can be very high, in excess of 90%. The recovery yield is easily measured by using an enriched or natural isotopic carrier. It is essential that the chemical process be simple, well understood and tested.

Identifying the neutrino source is usually difficult with a detector based on an inverse-beta process, because this process has an unfavorable angular distribution between the incoming neutrino and the emitted low-energy electron. Such a detector responds to the flux-cross section product for all neutrino sources. An electronic detector using the inverse-beta process can measure the neutrino energy, and thereby give information useful for identifying the source of the neutrinos. However, a radiochemical detector only measures the total rate of all processes that form the radioactive product that is measured. In applying the radiochemical method to studying solar neutrinos, it is presumed that the Sun provides the highest flux of neutrinos. However, background processes can give a false signal that could be attributed to neutrinos. It is therefore important to understand and test for all background processes that could lead to the product being observed. The major background arises from cosmic-ray muons. Energetic muons usually in the range of hundreds of GeV, interact with nuclei by electromagnetic interactions to produce secondary particles, pions and evaporation products. These active nuclear particles produce the radioactive product being observed from the target isotope principally by (p,n) reaction. To reduce these background processes to acceptable levels requires that the detector be located deep underground. Auxiliary experiments are needed to evaluate these cosmic-ray muon processes. Backgrounds also come from fast neutrons that are produced in the surrounding rock by spontaneous fission of ^{238}U and (α,n) reactions in the rock followed by (n,p) and (p,n) reactions in the detector. This secondary process is usually small. Fast neutrons can be readily eliminated by a modest water or paraffin shield.

The most insidious background arises from internal radioactive contaminants, usually uranium and thorium, that can produce the product nucleus by alpha-particle reactions. These ever present contaminants must be reduced to an insignificant level. However, to insure that the detector is free of a built-in background necessitates careful monitoring of the target material, the walls of the containment vessel, and any reagents used in the chemical processing. If a geological deposit is used as a target material, alpha background effects are in general relatively high. In a few cases, the minerals containing target isotope can be separated from the host rock, and thus greatly reduce the production of the product nucleus in the sample being measured. One must choose a deposit that is accessible, sufficiently deep underground and of a suitable mineral and chemical composition as discussed by Rowley et al. (1980) and Cowan and Haxton (1982).

At the present stage of development of experimental neutrino detection, radiochemical detectors are the only means of observing the low-energy neutrino sources in the Sun, those from the p-p reaction and the decay of ^{7}Be, ^{13}N and ^{15}O. We will now discuss in some detail the ^{37}Cl and ^{71}Ga radiochemical detectors that are operating. In the course of the last twenty years, there have been a number of excellent suggestions for other radiochemical detector systems. A brief discussion will be given of the merits and effort that have been devoted to developing these experiments. There have been several earlier reviews on solar neutrino research (see, e.g., Bahcall, 1978, 1979,1980,1989; Bahcall and Sears 1972; Kuchowitz 1976; Bahcall and Davis 1976,1982; Bahcall et al. 1982, 1988; Weneser and Friedlander 1987; Friedlander and Weneser 1987; Bahcall and Ulrich 1988; Davis et al. 1989,1990; Wolfsberg and Kocharov 1991; Davis et al. 1989a,b). In addition, three conferences devoted specifically to this topic that are of special interest are reported by Reines and Trimble (1973a,b), Friedlander (1978), and Cherry et al. (1985). A list of all inverse-beta processes with an energy threshold $<$ 1 MeV that could be used conceivably as neutrino (ν_e) detectors are given in Reines and Trimble (1973a).

B. The ^{37}Cl Experiment

1. Early Development. The chlorine solar neutrino experiment was built in the Homestake Gold Mine at Lead, South Dakota by Brookhaven National Laboratory during 1965–1967. The construction of the detector was stimulated by the realization that the ^{3}He(^{4}He,γ)^{7}Be and the ^{7}Be(p,γ)^{8}B reactions were of importance in the proton-proton chain (Bahcall and Davis 1982; Pinch 1986; Davis 1989). Prior to 1958 it was generally believed that the proton-proton chain ended with the ^{3}He + ^{3}He → ^{4}He + 2H reaction (Burbidge et al. 1957), a reaction suggested by Schatzman (1951). Furthermore, the carbon-nitrogen cycle produced $<$ 2% of the Sun's energy (Burbidge et al. 1957), a conclusion that still holds. Therefore the principal source of neutrinos from the Sun was believed to be the low-energy neutrinos from the *p-p* reaction. The energy of these neutrinos was well below the 0.814 MeV threshold of the chlorine radiochemical detector. When it was realized that the Sun could be the source of more energetic neutrinos, those from ^{7}Be and ^{8}B decay, there was great interest in measuring the neutrino flux.

Over 40 years ago, Pontecorvo (1946) pointed out that the reaction $^{37}Cl + \nu_e \rightarrow {}^{37}Ar + e^-$, producing the 35 day ^{37}Ar electron capturing isotope, could be used for detecting neutrinos by radiochemical method. Alvarez (1949) amplified this suggestion in the form of a proposal to carry out an experiment at a nuclear reactor. (Interestingly, Alvarez's proposal contained a calculation of the neutrino capture cross section, attributed to Leonard Schiff.) The importance of including captures that produce ^{37}Ar in excited states was pointed out, though at that time it was not possible to determine quantitatively the increase in neutrino capture cross section contributed by excited states. The method Cl-Ar was developed by Davis (1955) and applied in two experiments at the Savannah River reactors, as reported by Davis (1957) and Davis and Harmer (1959) to test lepton conservation (1956–1962). A theoretical analysis of this experiment is given by Primakoff (1978) and by Bahcall and Primakoff (1978). A pilot solar neutrino experiment was carried out in a mine in Ohio (1960–1962), and reported by Davis (1964).

2. The Homestake Experiment.   The experiment is located at a depth of 4200 ± 100 hectograms cm^{-2} standard rock (1478 m or 4850 feet) in a cavity designed to house the facility, provide containment for safety and serve as a water shield for fast neutrons (see Fig. 2). The detector contains 615 metric tons of liquid perchloroethylene ($CCl_2 = CCl_2$) corresponding to 2.19×10^{30} atoms of ^{37}Cl. The apparatus displayed in Fig. 2a consists of a single horizontal steel tank 14.6 m long and 6.1 m in diameter filled with liquid perchloroethylene and helium gas. The extraction of argon is accomplished by a helium gas purging system driven by two liquid circulation pumps that force the helium simultaneously through the entire liquid volume by means of a double set of 20 eductor nozzles shown in Fig. 2b. To extract the ^{37}Ar, helium is circulated in series through a condenser, a molecular sieve adsorber to remove perchloroethylene vapors, and through liquid nitrogen cooled charcoal to collect the argon. Finally, the helium is then returned to the tank. The extraction efficiency is determined by the volume of helium circulated through the trapping system, and also measured with ^{36}Ar or ^{38}Ar carrier gas introduced and recovered in each experiment. Davis et al. (1968) describe the experiment in more detail and present the initial results. The procedures followed and various tests performed are described in several articles and reports (see Bahcall and Davis 1976; Reines and Trimble 1973a; Davis 1978; Rowley et al. 1985).

The entire sample of argon collected is purified and placed in a small

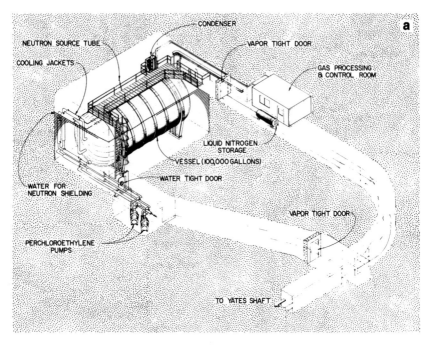

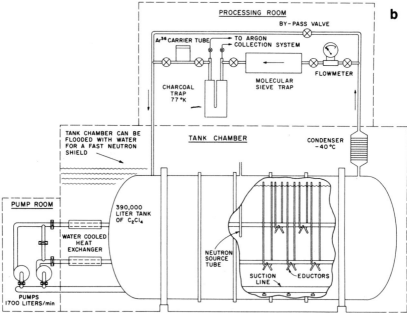

Fig. 2. (a and b) Schematic diagrams of the Homestake radiochemical detector tank and pumping system.

proportional counter to observe ^{37}Ar decays. Initially only the pulse height and the time distribution were recorded. An upper limit of 3 SNU was given in 1968 for the neutrino capture rate by Davis et al. (1968). (Here SNU stands for a solar neutrino unit, defined as 10^{-36} captures per second per target atom.) In 1970, a pulse rise time system was introduced that greatly increased the selectivity of ^{37}Ar decay events, and thereby increased the sensitivity of the chlorine detector (Davis et al. 1972). The time, pulse height, and pulse rise time are continuously recorded over a period of 200 to 300 days. Events that have the correct pulse height and rise time are selected, and the time distribution of these ^{37}Ar-like events are resolved into a decaying component with a 35-day half life, and a constant background component using a maximum likelihood method discussed by Cleveland (1983).

From the earliest observations, the chlorine experimental rates have been lower than predicted by standard theory. Many special experiments were performed to test the Homestake experiment for efficiency of collecting and recording ^{37}Ar atoms to answer various criticisms and suggestions. The tests that were performed are aired in many reports, primarily in the three solar neutrino conferences referred to above (Bahcall and Davis 1976), and a sociological study of solar neutrino research by Pinch (1986). The experimental apparatus has not been tested with a neutrino source, although in the original design of the apparatus this possibility was considered and a re-entrant source tube designed. Alvarez (1973) proposed that a megacurie source of ^{65}Zn could be prepared in a reactor that would be excellent for this purpose. The matter was carefully considered and a test irradiation was performed. However, a source of this magnitude was difficult to prepare except in a production reactor, and the experiment would be exceedingly expensive. Another approach to this problem was to use a μ^+ decay source at the beam stop at the Los Alamos meson facility. This source (0 to 53 MeV) would feed all excited states, and be a far more interesting experiment. Considerable effort was devoted to building the experiment and making background studies (Davis et al. 1973), but the experiment was not given a high enough priority to be carried out. Needless to say this experiment is entirely feasible with the present intense beam at this facility, provided that a thicker overhead shield is added to the room.

The chemical aspects of the recovery of a few atoms of ^{37}Ar from perchloroethylene was tested in several ways. A 7 millicurie source of perchloroethylene labeled with ^{36}Cl was prepared to test the chemical fate of ^{36}Ar that produced the beta decay of ^{36}Cl (half life 308,000 yr, maximum energy 0.709 MeV). When ^{36}Cl decays to ^{36}Ar, the dynamics of the recoiling argon atom and its chemical behavior is essentially identical to that of an ^{37}Ar atom produced in perchloroethylene by neutrino absorption. In this experiment the produced ^{36}Ar was collected by a helium purge and measured by neutron activation analysis. The experiment was carried out in a 20-liter iron tank to correspond to the conditions in the Homestake experiment, though on a

smaller scale. It was found that the ^{36}Ar observed by this procedure corresponded to the quantity of ^{36}Ar expected from ^{36}Cl decay. Two experiments
were performed on the Homestake tank to test that a small number of ^{37}Ar
atoms could be recovered quantitatively. A neutron source was placed in the
center of the tank to generate ^{37}Ar radioactivity. In the second test, a measured
quantity of ^{37}Ar was introduced directly into the detector. These quantities of
^{37}Ar were recovered by the procedures described above.

The Homestake experiment has operated continuously since 1967 except
for an 18 month period from May 1985 to October 1986 when the liquid
circulation pumps suffered electrical failures (October 1984 and May 1985,
respectively). The new pumps have a 10 to 20% lower pumping speed requiring a correspondingly longer time to collect argon. The Homestake facility was operated by Brookhaven National Laboratory until 1986, and subsequently by the University of Pennsylvania.

The Homestake solar neutrino experiment is a unique facility and an
effort was made to continue observations for a number of reasons. This detector served to monitor the Sun's neutrino flux for short- and long-period
changes that may be associated with solar flares and the solar activity cycle.
New solar neutrino experiments were continually being proposed, and it was
important to overlap the observations of the new experiments. The chlorine
experiment is a specific detector of electron neutrinos and could be of critical
importance in the event of a stellar collapse within 10 kpc of our solar system
(Burrows 1984). Unfortunately the supernova 1987A in the Large Magellanic
Cloud was beyond this range. Finally, the chlorine experiment could be converted to another radiochemical detector, for example, one using a bromine-
or iodine-containing liquid (see Secs. II.E 2 and 3).

3. Results.   The data from the chlorine experiment for each individual
experimental run is plotted in Fig. 3. Over the period 1970.3 to 1990.3 the
average ^{37}Ar production rate in the 615-ton detector was 0.500 ± 0.033 ^{37}Ar
day^{-1} (Davis et al. 1989*a*). There is a cosmic-ray background of 0.08 ±
0.03 atoms day^{-1} as discussed by Wolfendale et al. (1972); by Cassidy
(1973); by Fireman et al. (1985); and by Zatsepin et al. (1981) that must be
subtracted. The total ^{37}Ar production rate ascribed to solar neutrinos is
0.420 ± 0.045 atoms day^{-1} or 2.23 ± 0.26 SNU. This result is usually
compared to solar model calculations.

However, before making a direct comparison of this result with solar
model calculations, it is useful to discuss several matters: the cross section
for the ^{37}Cl$(\nu,e^-)^{37}$Ar reaction, whether the results are influenced by short
periodic increases in neutrino fluxes, and whether the solar neutrino signal is
constant with time.

4. The ^{37}Cl$(\nu,e^-)^{37}$Ar Cross Section.   We now discuss briefly the cross-
section values of ^{37}Cl for the various neutrino-emitting sources in the Sun and

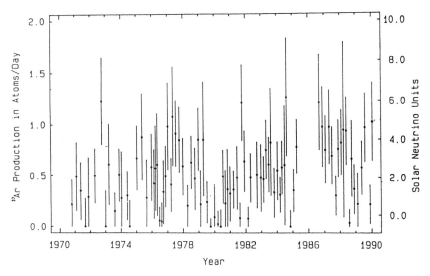

Fig. 3. A plot of the data from the Homestake chlorine experiment.

energetic neutrino sources of interest. For the low-energy sources, ^{7}Be, ^{15}O, ^{13}N, and the *pep* reaction, the cross sections are determined from the ground-state transition. However, the energetic neutrinos from ^{8}B will feed various excited states in ^{37}Ar. Fortunately, these transitions can also be accurately evaluated from the proton decay of ^{37}Ca to ^{37}K. The latter nucleus is the mirror nucleus to ^{37}Ar. Studies of the beta-decay branches of ^{37}Ca allow one to evaluate empirically the neutrino capture cross section of ^{37}Cl (for details see Bahcall 1964,1978; Haxton and Donnelly 1977; Bahcall and Ulrich 1988; Bahcall and Holstein 1986; Adelberger and Haxton 1990, and references therein). Calculations using nuclear models by Lanford and Wildenthal (1972), Itoh et al. (1977), and Itoh and Kohyama (1978), (p,n) reactions by Rapaport (1981), and for energetic neutrinos by Schramm (1989) and Fukugita et al. (1989) have also been carried out. The neutrino capture rates in ^{37}Cl calculated for the model of Bahcall and Ulrich (1988) are given in Table II. Note that the major contribution to the capture rate is from the ^{8}B flux, approximately 77% of the total rate. This fact is important in interpreting the experimental results. The production of ^{8}B is extremely sensitive to the central temperatures in the Sun, and therefore a measurement of this rate is a most critical test of the details of the solar model calculations.

We mentioned earlier that the chlorine detector has not been tested with a neutrino source. The possibility is being explored of measuring the cross section for μ^+ decay neutrinos (energy 0 to 53 MeV) using the neutrino facility at the Los Alamos Meson Factory. The cross section for these ener-

TABLE II
A Comparison of the Predicted Rates for the Chlorine Detector

Neutrino Source	Neutrino Flux at Earth cm^{-2} s^{-1}		Neutrino Capture Rate in SNU
	Bahcall & Ulrich[a]	Turck-Chièze et al.[b]	^{37}Cl[a]
$pp \rightarrow De^+v_e$	$6.0 \pm 0.12 \times 10^{10}$	$5.98 \pm 0.18 \times 10^{10}$	0
$pep \rightarrow Dv_e$	$1.4 \pm 0.07 \times 10^8$	1.30×10^7	0.2
$^3hep \rightarrow {}^4He\ e^+v$	7.6×10^3	—	0.03
$^7Be \rightarrow {}^7Li\ v_e$	$4.7 \pm 0.7 \times 10^9$	4.18×10^9	1.1
$^8B \rightarrow {}^8Be^*e^+v_e$	$5.8 \pm 2.1 \times 10^6$	$3.8 \pm 1.1 \times 10^6$	6.1
$^{13}N \rightarrow {}^{13}C\ e^+v$	$6.1 \pm 3.1 \times 10^8$	—	0.1
$^{15}O \rightarrow {}^{15}N\ e^+\ v$	$5.2 \pm 3.0 \times 10^8$	—	0.3
$^{17}F \rightarrow {}^{17}O\ e^+v$	$5.2 \pm 2.4 \times 10^6$	—	0.004
			7.9 ± 2.6

[a] Bahcall and Ulrich 1988; similar results in Sackman et al. 1990.
[b] Turck-Chiéze et al. 1988.

getic neutrinos is $1.0 \pm 0.1 \times 10^{-40}$ cm^2. The plans include measuring the $^{127}I(v,e^-)^{127}Xe$ cross section as well.

5. Solar Flares and Gamma Bursts.   During the last 20 yr, the Homestake chlorine experiment was the only neutrino detector with a low threshold energy and high sensitivity that was continuously monitoring the flux of neutrinos over all space. It is natural to employ the data recorded by this purely v_e detector toward searching for astrophysical sources of neutrinos and by noting increases in the signal rate. Increases could conceivably result from collapsing stars, neutron star quakes, gamma bursters, solar flares, etc. The data has been studied by many investigators who have reached various conclusions on the question of whether or not there are any changes or trends in the data beyond those expected by normal statistical fluctuations. In this section, we will discuss these matters with special emphasis on the questions: is there evidence for short-period increases associated with solar flares or other astronomical phenomena, and is there a variation in the ^{37}Ar production with time?

One could consider the highest individual experimental runs as candidates for events of this nature. Three experiments recorded an ^{37}Ar production rate of 1.25 ^{37}Ar atoms day^{-1} that could be regarded as possible candidates for a source. During the exposure periods of these experiments there were intense solar flares, and an intense gamma burst observed by the Solar Maximum Mission (SMM) satellite. These events were discussed by Davis and Evans (1973), Evans et al. (1974), Bazilevskaya et al. (1984), Rowley et al. (1985) and Davis (1986), but unfortunately there were no other neutrino detectors operating with similar sensitivities. There does appear to be a reasonable correlation of high experimental values with energetic cosmic-ray increases measured at the top of the Earth's atmosphere with daily balloon flights (Bazilevskaya et al. 1984). They identified two high experimental runs (numbers 27 and 71) in which large solar flares occurred during their exposure interval. These were the great flares of 4 and 7 August 1972 and the large flares of 7 and 12 October 1981. Ever since these occurrences, we have been alerted to the possibility of neutrino production associated with solar flares. Extremely large flares were not observed in solar cycle 21. But in the early development of solar cycle 22 a few intense flares were observed that were as intense as the flare in 1972. A Monte Carlo simulation of the chlorine data estimated that 1 run as high as 1.25 ^{37}Ar atoms day^{-1} should be observed, whereas 3 were observed. Without confirmation one cannot regard a few high experiments as significant evidence for attributing such events to neutrinos unless they are extraordinarily high, >1.5 atoms day.$^{-1}$ Searches in other detectors with much higher energy thresholds did not reveal an increase during these periods as reported by Alexeyev et al. (1987) and Hirata et al. (1988a,b). It should be noted that if the time of occurrence of the event is precisely known, as in the case of a gamma burst, a very careful search is required, as even a single event may be significant. Experiment number 86 was high (Davis et al. 1987) with a 1.26 ^{37}Ar atoms day^{-1} average rate, and this experiment included the highest gamma burst observed by the SMM on August 5,23:48:00 UT (Share 1986), a beautiful pulse with a periodic gamma ray spectrum. To our knowledge, a careful search has not been made for this event in other neutrino detectors, e.g., the 300-ton scintillation detector in Baksan Valley or the 90-ton scintillator in the Mt. Blanc tunnel. In the case of solar flares, the source of the neutrinos may be a deep thermal event not observable by high-threshold detectors. It is important to continue the search, noting high results when they occur.

Lande (1974) and his associates observed a most unusual series of 4 precise millisec pulses on 4 January 1974, each with a microsec duration using 7 modular Cherenkov detectors in the Homestake mine. A search by Evans et al. (1974) was made for an increase in the chlorine detector, but no increase was observed. No event was reported from Vela satellites, or in any other detector. Possibly this unique event could have been an anti-neutrino pulse. We might point out here that the chlorine detector did not observe the ν_e pulse from SN 1987A, because this source was beyond the range of this detector.

6. An Apparent Variation in the Solar Neutrino Flux.   The Sun's activity cycle has been studied for hundreds of years revealing a periodic 11-yr sunspot cycle and the 22-yr magnetic cycle. It is informative to examine the data from the chlorine experiment for a possible relation to solar activity (Zatsepin and Kuzmin 1964; Sheldon 1969). The first study of this possibility was made by Bazilevskaya et al. (1984). Their analysis showed that there was an apparent correlation of the Homestake data with their measurement of the cosmic-ray intensity at the top of the Earth's atmosphere. It is well known that the galactic cosmic-ray intensity varies inversely with the sunspot cycle as a result of the varying magnetic fields associated with the solar wind.

To search for these effects in the solar neutrino data, it is helpful to apply a smoothing of the data because of the large statistical error in the individual measurements. Smoothing may be accomplished by combining the data as yearly averages (Rowley et al. 1985), or by using a running average (Davis 1986; Davis et al. 1989a,1990). Figure 4 shows a plot of a 5-point running average of the Homestake data compared to sunspot occurrences in solar cycles 20, 21, and the beginning of cycle 22. The ^{37}Ar production rate anticorrelates with the sunspot occurrences, and correlates with the solar diameter measurements reported by Delache et al. (1985). The most significant change in the ^{37}Ar production rate occurs at a time when the solar activity cycle begins. The rate was highest at solar minimum and lowest at solar maximum. The magnitude in the effect is summarized in Table III. The changes in the rates observed are large at the extremes and are apparently well correlated in time with the onset of the solar activity in cycles 21 and 22. It is interesting to note that the solar maximum rate in cycle 21 is not distinguishable from the cosmic-ray muon background, approximately 0.4 SNU in 1977.

Two possible explanations have been suggested: it results from a modulation of the neutrino flux by the solar magnetic fields, through neutrino spin

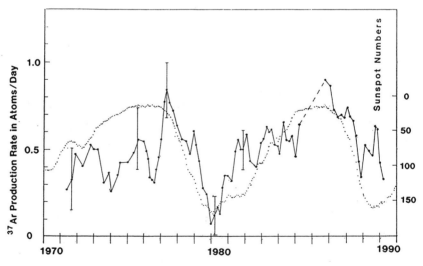

Fig. 4. A plot of the 5-point running average rates from the Homestake chlorine detector. The individual data points used are shown in Fig. 3. The small dots are the monthly average sunspot numbers.

TABLE III
Periodic Changes in the Observed Solar Neutrino Flux

Solar Activity	Dates	Rate in SNU	Expt. Nos.
Solar Minimum Cycle 20	1977	4.1 ± 0.9	47–52
Solar Maximum Cycle 21	1979.5–1980.7	0.4 ± 0.1	59–65
Solar Minimum Cycle 21	1986.8–1988.3	4.2 ± 0.7	90–100
Onset of Cycle 22	1988.4–1990.3	2.5 ± 0.5	101–110

rotation, or by a Mikheyev-Smirnov-Wolfenstein (MSW) effect in the convective zone of the Sun (see Secs. II.B 8 and 9 below). Assuming the former is correct, one could use the flux during solar minimum as an approximation to the neutrino flux generated in the solar core.

It is well known that the low-energy ($\leq$ 10 GeV) end of the galactic cosmic-ray spectrum anti-correlates with the solar activity cycle. One may ask, is the anti-correlation of the ^{37}Ar production rate with solar activity also related to a cosmic-ray background effect. As discussed earlier, there is a small background effect arising from energetic cosmic-ray muons (300 GeV) at the great depth of the Homestake experiment. This background effect is well understood, is observed to be relatively small (0.08 ^{37}Ar atoms day^{-1} or 0.4 SNU), and is expected to be constant. The production of ^{37}Ar by cosmic-ray-produced neutrinos, ν_e or ν_μ in the Earth's atmosphere is far below detection by the Homestake experiment (0.002 SNU) according to Domogatskii and Eramzhyan (1977). There would be an even lower ^{37}Ar production from positron emitters produced in the Earth's atmosphere by galactic cosmic rays (de la Zerda-Lerner and O'Brien 1987). One might conclude that the conventional background muon and neutrino sources cannot account for the ^{37}Ar production rate observed or its variation.

Numerous authors have examined the data from the Homestake experiment as the data evolved using various statistical methods to decide whether the data set is consistent with a constant flux of neutrinos. They have also examined whether there is evidence for a variation in the observed ^{37}Ar production rates and whether there are unique correlations with solar or astronomical phenomena.

Below, we refer to these analyses and indicate the statistical treatment and conclusions. Ehrlich (1982) found no periodicity in the range 0.4 to 5.0 yr. Basu (1982) reported a correlation with the high-level Ap index. Haubold and Gerth (1985, 1990) made a harmonic analysis of data from 1970 to 1989 and found periods of 8.33, 5.00, 2.13, 1.61, 0.83, 0.61, 0.54 and 0.51 yr. Lanzerotti and Raghaven (1981) found no correlation with the Ap solar index. Bazilevskaya et al. (1984) found a correlation with cosmic-ray fluxes and intense solar flares. Gavrin and Kopylov (1984) used Kolomogorov-Smirnov and chi-squared tests to find no variations. Bahcall et al. (1987*a*) noticed that the correlations depend on just a few experimental points. Subramanian and Lal (1987) and Subramanian et al. (1988) found that the correlations were suggestive of neutral-penetrating particles. Wilson (1987) found no evidence for variations from 1971 to 1981. Raychaudhuri (1986, 1989) published many papers using various statistical tests that favored variations. Attolini et al. (1988) using a superposed epochs method found support for a 2.1-yr period. Davis et al. (1990), Davis (1986), and Rowley et al. (1985), discussed a correlation analysis with sunspots and solar diameter data. Baltz et al. (1989) did an analysis of correlations and streaks of high and low results. Nunokawa and Minakata (1991) found that any anti-correlation with sunspots depends on the treatment of errors and is not definitely established. Veselov et al. (1987) pointed out that there is a half-year correlation with observed solar latitude in 1979 and 1980. Bahcall and Press (1991) showed that there is an anti-correlation with sunspots in 1977–1989, but not in 1970–1977. Krauss (1990) found a correlation with variations in low-order acoustic solar oscillations in 1977–1988. Bieber et al. (1990) showed that the data is strongly correlated with monthly sunspot numbers, and the strength of this correlation rises when the semi-annual variation is included. Filippone and Vogel (1990) found that a time dependence with sunspot numbers is more probable than a constant rate. Dorman and Wolfendale (1990) suggested that there is an apparent correlation with solar magnetic fields. Fiorentini and Mezzorani (1991) note that a fit to sunspot occurrences is slightly better than a fit to a constant flux. Sakurai (1990) showed that a sequential analysis supports quasi-biennial variations. Finally Gavryuseva et al. (1990) found that a power spectrum and correlations analysis reveal 3 significant peaks that are unlikely to be of random origin.

Most authors that have analyzed the full set of data from 1970 to 1990 find a reasonably significant correlation between the ^{37}Ar production rate and sunspot occurrences, solar oscillation frequency changes and solar magnetic fields. Obviously one needs a solar neutrino detector with a higher signal rate to make a definitive test of this phenomenon.

7. Comparison with Theoretical Solar Models.   A great many solar model calculations have been made since the first detailed solar model by Sears (1964) and by Pachoda and Reeves (1964). We will only refer to the most recently published models, and let the interested reader use the extensive literature cited in these papers. The models of Bahcall and Ulrich (1988), Turck-Chieze et al. (1988; Turck-Chieze 1990) and Sackman et al. (1990) use, in the authors' view, the best nuclear reaction cross sections, solar compositions, opacity values, and solar data. In making the calculations, the usual assumptions of the standard model are incorporated. Table II summarizes the standard solar model neutrino capture rates from these most recent calculations. The total neutrino capture rate in ^{37}Cl of Bahcall and Ulrich is 7.9 ± 1.1 SNU (1σ, for comparison, though the authors prefer to use a 3σ

error, ± 3.3 SNU), while Turk-Chieze et al. obtained 5.8 ± 1.3 SNU and Sackman et al. obtain 7.7 SNU (errors discussed). The results of the Homestake experiment given above are about a factor of 2 to 3 lower than these predictions. The results of the Homestake experiment appear to be confirmed by the recent results from the Kamiokande II imaging water Cherenkov detector. See Sec. III.B.

Stellar model calculations have been remarkably successful in explaining the evolution of main-sequence stars. These models are generally used to determine the main-sequence lifetime of stars, determine the age of star clusters, and serve as a basis for advanced phases of evolution. Of great importance to solar neutrino research, they provide a valuable quantitative prediction of the solar neutrino spectrum essential for the design and interpretation of experiments, and for interpreting solar seismology data as discussed in the chapters of Ulrich and Cox and of Christensen-Dalsgaard and Berthomieu. The term "standard solar model" is a relatively recent one that refers to specific calculations with a set of basic assumptions that are accepted as valid. Some of the assumptions are: the Sun was initially homogeneous in composition identical to that observed in the present convection zone, and had no differential rotation, mixing, or mass loss during its evolution.

We must recognize these assumptions when comparing standard solar models (see chapter by Ulrich and Cox) with results from the chlorine and Kamiokande II (see Sec. III) experiments. The chlorine experiment only measures a rate equal to the total neutrino flux-cross section product above its 0.814 MeV threshold. This rate is dominated by the neutrino flux from ^{8}B decay as listed in Table II. The production of ^{8}B is very temperature sensitive, depending on the central temperature to the 18th power. The solar internal temperatures are critically dependent on the calculated opacity, the internal composition, and various dynamic effects. Therefore, a comparison of the results of the chlorine and Kamiokande II experiments with the standard model is a very critical test of the model, perhaps revealing its defects. It is not surprising that the results are lower than expected by a factor of 3.

Many solar models have been put forward to explore the effects of chemical composition, internal rotation, mass loss and mixing, and the presence of weakly interacting massive particles (WIMPs); available reviews are by Rood (1978), Newman (1986), Bahcall and Ulrich (1988) and Gough (1990). Recent models are discussed by Cox et al. (1989,1990; considering element diffusion and WIMPs), by Bahcall and Loeb (1990; considering diffusion), by Guzik et al. (1987; considering significant mass loss), by Spergel and Faulkner (1988; considering WIMPs), by Sienkiewics et al. (1990; considering mixing). In general, the nonstandard models give reduced total solar neutrino capture rates for the chlorine experiment in agreement with the observed average rate.

One should recognize that there are various known physical processes that occur in the deep interior of the Sun, that are not included in the standard model, e.g., mass loss in the early Sun, differential rotation, and diffusion. These and other effects are discussed by Bahcall and Ulrich (1988), Rood (1978,1989), Schatzman and Maeder (1981), and Newman (1986).

Since the initial finding that the total rate in the chlorine experiment was a factor of 3 lower than the standard model predictions, a search was made for an explanation in terms of various modifications of the standard solar model or radiochemical effects in the Homestake detector itself that would lead to an erroneous result. These led to the development of the radiochemical gallium experiment to test whether the primary p-p reaction was operating as expected. The mean flux over the past million years could possibly be measured with the molybdenum and thalium radiochemical experiment. There was also concern about the neutrino itself: was it oscillating (Pontecorvo 1968a), decaying (Bahcall et al. 1972), or did it have a magnetic moment (Cisneros 1971)? In the last 6 yr, there has been a growing interest in neutrino physics, and in the possibility that the neutrino flux is modulated by resonance neutrino oscillations and a neutrino spin flip by solar magnetic fields. These topics are discussed only briefly in the next two sections, as a more complete discussion is given in the chapter by Rosen.

8. *The Mikheyev-Smirnov-Wolfenstein Effect.*   It was suggested by Pontecorvo (1968a,1983) in 1964 that vacuum oscillations of neutrinos could explain a deficit in the solar neutrino flux. Wolfenstein (1978a) suggested that matter oscillations within the Sun may also play an important role. However, large mixing angles were regarded as necessary to obtain a reduction in the flux as large as a factor of 3. In 1984, Mikheyev and Smirnov (1985) pointed out that in the Sun resonant enhancement of matter oscillations could occur if the squared neutrino mass differences between various neutrino flavors were in the broad range 10^{-4} to 10^{-8} eV^2, even if mixing angles were as small as 10^{-3}. This concept was immediately adopted as a suitable explanation of the apparent disagreement between solar model calculations and the results of the chlorine experiment by Rosen and Gelb (1986a) and by Bethe (1986,1989). Many calculations of the effects have been given in the literature. These are summarized in Mikheyev and Smirnov's recent review articles (1987,1989), by Kuo and Pentaleone (1989), by Barger et al. (1991) and by Bilenky and Petcov (1987). A possible verification of the Mikheyev-Smirnov-Wolfenstein (MSW) affect can be obtained in several ways: from studying the day/night or seasonal difference in the solar neutrino flux (Mikheyev and Smirnov 1987, 1989; Baltz and Weneser 1988; Cherry and Lande 1988), by observing a distortion in the shape of the ^{8}B flux spectrum, by observing neutral-current interactions of ν_μ or ν_τ from the Sun, by measuring the response of a radiochemical detector like the gallium-germanium one that has a dramatically different response to the solar neutrino spectrum, or by measuring an up-down effect in the intensity of cosmic-ray neutrinos (Losecco 1985,1986; Berger et al. 1990).

In analyzing and comparing the results of various neutrino detectors, it is important to recognize that most of the calculations on the MSW effect consider only oscillations between two neutrino flavors. Three neutrino flavors can behave quite differently, as mentioned by Rosen in his chapter. Also there are effects resulting from errors in the solar model as discussed by Dearborn and Fuller (1989) and by Bahcall and Haxton (1989). This is an active field of investigation and is the primary motivation for a number of newly proposed solar neutrino detectors. A more complete discussion of the MSW effect may be found in the chapter by Rosen.

9. The Effects of a Neutrino with a Magnetic Moment.   Voloshin et al. (1986*b*) pointed out that the anti-correlation with solar activity could be explained by the precession of the spin of a left-handed electron-neutrino $v_e L$ into a right-handed electron-neutrino $v_e R$ in the magnetic fields in the convective zone of the Sun. Only the left-handed neutrinos would be observable in the chlorine or gallium experiments. If the transverse magnetic fields in the Sun are a few thousand Gauss and extend over a distance of order 10^5km, then a neutrino magnetic moment of $(1 - 10) \times 10^{-11} \mu_B$ (Bohr magnetons) is needed. It is generally believed that the solar magnetic fields are toroidal in nature and have intensities of this magnitude. The fields increase as the solar cycle develops, and decrease as the solar cycle declines explaining a solar cycle variation of the ^{37}Ar production rate. Magnetic fields of this intensity are observed in sunspots. Although the internal solar fields, their intensity, location, and extent are very poorly known, it is generally thought that considerably larger fields may exist at the base of the convective zone. Further, Voloshin et al. (1986*a,b*) suggested that a half-year solar latitude effect could arise in the observed solar neutrino flux. The solar rotational axis is inclined by 7.25 deg with respect to the plane of the ecliptic, and the magnetic fields exist only at higher solar latitudes, the solar equator being essentially free of magnetic fields. It follows then that the neutrino flux may exhibit a biennial sinusoidal variation at the Earth. Observing this feature of the neutrino flux allows one to test for a neutrino magnetic moment. The effect was apparently present in the experimental data during the period 1979–1982 when the fields were well developed during this period of solar maximum (Veselov et al. 1987).

The spin precession mechanism was considered unlikely by theorists when first suggested, because a neutrino with a small mass would have a correspondingly small magnetic or electric moment, around 10^{-19} Bohr magnetons. However, Barr et al. (1990) point out that this simple relationship between the neutrino's mass and its magnetic moment is not necessarily valid, and that the neutrino could well have an enhanced magnetic moment relative to its mass. The experimental limit on the electron neutrino magnetic moment is $10^{-10} \mu_B$; limits are set by various astronomical observations (Fujikawa and Shrock 1980). More recently it was implied that the neutrino must have a magnetic moment $< 10^{-12} \mu_B$, based on the fact that neutrinos were observed from the SN 1987A event (Barbieri and Mohapatra 1989; Lattimer and Cooperstein 1988; Nötzold 1988). However, there is the possibility that the neutrino has a transition magnetic moment sufficiently large to allow a spin rotation change in the Sun; see Voloshin et al. (1986a,b), Fukugita and Yanagida (1987), Barbieri and Mohapatra (1989) and Babu and Mohapatra (1989) for a discussion of the theoretical question of the neutrino magnetic moment. Another consequence of a neutrino with a transition magnetic moment was pointed out by Lim and Marciano (1988), and by Akhmedov (1988a,b, 1989; Akhmedov and Khlopov 1988). They showed that there is a resonant conversion of a $\nu_e(L)$ into a $\nu_\mu(R)$ or $\bar{\nu}_\mu(R)$ in the presence of matter and a magnetic field. Flavor-changing neutrino spin rotation (FCνSR) is analogous to the MSW affect discussed earlier, and will effect the solar neutrino flux if the neutrino is a Majorana particle with a transition magnetic moment. Calculations have been made for the combined MSW and FCνSR for the solar flux by Akhmedov and Bychuk (1989) and by Minakata and Nunokawa (1989). One interesting conclusion is that spin rotation for the low-energy p-p neutrinos may not occur because to be efficient Δm^2 must be $< 2 \times 10^{-9}$ eV^2, and therefore this flux may not show a variation. There is considerable interest in this subject. The FCνSR mechanism is an attractive explanation for a solar cycle variation of the solar neutrino flux (Pulido 1990; Fukugita et al. 1989).

C. The Two Gallium Experiments

1. Early Developments. The gallium solar neutrino experiment was suggested by Kuzmin in 1964. He pointed out that the neutrino capture reaction, $\nu_e + {}^{71}\text{Ga} \rightarrow {}^{71}\text{Ge} + e^-$, has a threshold (0.233 MeV) low enough to observe the neutrinos from the basic p-p reaction in the Sun. Moreover, the decay of ^{71}Ge, 11.4 day half life, has an unusually low ft value (log $ft = 4.35$ s, spin change $3/2 \rightarrow 1/2$), and therefore this reaction would have a very favorable cross section (Kuzmin 1966). Approximately 20 tons of gallium would be needed to obtain one neutrino capture per day, at that time an impossibly large amount of gallium. However, in the early 1970s, the electronics industry developed a need for gallium to produce light-emitting diodes. By 1974 it was clear that tens of tons could become available, but the cost would be very high.

A detector to observe the p-p reaction neutrinos was considered an important next step in solar neutrino research. Two other neutrino capture reactions were under discussion at the time that could be used to observe the low-energy neutrinos from the p-p reaction. These were: ^{55}Mn$(\nu,e^-)^{55}$Fe with a half life of 2.6 yr and an energy threshold of 231 keV, and ^{87}Rb$(\nu,e^-)^{87}$Sr*, with this decaying to ^{87}Sr with a half life of 2.8 hr and an energy threshold of 115 keV. However these possibilities were not favorable from an experimental viewpoint. A radiochemical gallium detector seemed an excellent prospect. Work began as a summer project in 1974 at Brookhaven with K. Lande and W. Frati of the University of Pennsylvania. By the end of the summer two methods were devised for removing trace quantities of germanium from gallium target material. One was an extraction of germanium from gallium metal (melting point 30° C), the usual industrial product, by an acid solution of hydrogen peroxide. The other method was based on removing volatile germanium chloride from a concentrated aqueous gallium chloride solution containing hydrochloric acid. The metal method was suggested by J. Evans at the Brookhaven National Laboratory, and the gallium chloride method was tried following a suggestion by A. Pomanskii at the Institute of Nuclear Research (INR), Moscow. Subsequently, these techniques were refined, and background processes from natural alpha processes and high-energy muons were studied. The results from these studies showed that the gallium experiment could be carried out by either of the two methods (Dostrovsky 1978; Bahcall et al. 1978). The Soviet group at INR independently developed and adopted the gallium metal process. Because there are no deep mines in the USSR, it was necessary for them to build an underground facility to house their solar neutrino experiments, as discussed by Markov (1977) and Zatsepin (1982).

2. Development of the Full-Scale Experiments. A collaboration was formed in 1978 between Brookhaven National Laboratory, the Max-Planck Institut für Kernphysik, Heidelberg, the Weizman Institute, and the University of Pennsylvania to develop and build a full-scale experiment. They ultimately chose the gallium chloride solution method, and prepared a pilot experiment with a volume of 2500 liters of solution containing 1.3 tons of gallium. The solution method is considerably easier to carry out and uses a chemical procedure that is well understood. The acidity of the solution is chosen to optimize the evolution of germanium chloride ($GeCl_4$). The procedure is simply to pass gas, usually air, through the solution and the germanium chloride vapors are recovered from the gas stream by a water-spray scrubber. The transfer of germanium chloride to the scrubber can be carried out with 100% efficiency. The recovered germanium chloride is purified and concentrated further by solvent extraction to a volume of 50 to 100 cm^3. Finally the germanium chloride is reduced to germane (GeH_4) gas, a suitable gas for counting. The ^{71}Ge decays are recorded in a small proportional counter. These developments are described in a Brookhaven proposal to carry out a full-scale experiment using 50 tons of gallium (Davis et al. 1981). The full-scale experiment was not approved for funding in the United States, and the collaboration was dissolved. A second attempt to secure funding in the U.S. in 1985 again failed to receive support. Subsequently, the Max-Planck-Institut, Heidelberg formed a new collaboration in Europe, and obtained funding for a 30-metric-ton gallium detector. Their experiment will be performed in the Gran Sasso tunnel in Italy. The experiment (GALLEX) is now built, and it is currently prepared to begin solar neutrino observations in June 1991 (Kirsten 1990; Vignaud 1991).

The Soviet scientists built a 7-ton gallium metal pilot experiment, and demonstrated that germanium can be recovered from this quantity of gallium with a high efficiency of approximately 80%. Various background processes were studied, and appropriate counting techniques were developed (Barabanov et al. 1985). The metal process uses large volumes of hydrochloric acid. To avoid possibly contaminating their detector with germanium from reagent hydrochloric acid, a recovery system was built so that the acid could be reused. The underground laboratory in Baksan valley in the North Caucasus mountains is now completed, and a 60-ton gallium experiment is installed. At the time of writing, one-half of the detector is operational, and measurements with the full detector are expected during 1991 (Gavrin et al. 1990). At a late stage (1986), collaborators from the University of Pennsylvania and Los Alamos National Laboratory joined the effort. A preliminary report of the initial results was presented at the Neutrino 90 conference. They reported that the total rate was <70 SNU (1σ confidence level). The p-p reaction alone is expected to provide a rate of 71 SNU (see Table IV). If this result is confirmed by subsequent measurements, it would strongly indicate a modulation of the solar neutrino flux by the MSW, FCνSR, or other effects (Bahcall and Bethe 1990; Baltz and Weneser 1991).

3. The Determination of the Cross Section. The gallium detector was expected to be primarily sensitive to the low-energy neutrinos from the p-p reaction. The next most intense neutrino source, ^{7}Be, would populate 2 low-lying states at 0.175, and 0.500 MeV in ^{71}Ge, and of course ^{8}B decay neutrinos would feed many excited states. The isobaric analog state of ^{71}Ge is

TABLE IV
Total Neutrino Capture Rates for Various Radiochemical Detectors[a]

Target	p-p	pep	hep	^{7}Be	^{8}B	^{13}N	^{15}O	^{17}F	Total	Rate d^{-1} ton^{-1}
										0.36
^{7}Li	0.0	9.2	0.06	4.5	22.5	2.6	12.8	0.1	51.8	
										0.0028
^{37}Cl	0.0	0.2	0.03	1.1	6.1	0.1	0.3	0.004	7.9	
										0.097
^{71}Ga	70.8	3.0	0.06	34.3	14.0	3.8	6.1	0.06	132	
										0.0089
^{81}Br	0.0	1.1	0.07	8.6	15.3	0.9	1.9	0.02	27.8	
										0.0023
^{98}Mo	0.0	0.0	0.08	0.0	17.3	0.0	0.0	0.0	17.4	

[a]From Bahcall and Ulrich 1988; rates in body of table are in SNU.

unbound and therefore would not contribute to the cross section. It was of great importance to evaluate the contribution of excited states to determine the sensitivity of the gallium experiment of the neutrino spectrum. This was accomplished by (p,n) reaction studies and from nuclear models. The (p,n) reaction experiments at 120, and 200 MeV indicated a large contribution from highly excited states (Krofcheck et al. 1987). Cross sections derived from shell-model studies of the $^{71}Ga(\nu,e^-)^{71}Ge$ reaction are in essential agreement with the (p,n) reaction values (Mathews et al. 1985; Grotz et al. 1986). As matters stand now, the model of Bahcall and Ulrich predicts a total neutrino capture rate in ^{71}Ga of 132 ± 18 SNU, out of which the p-p reaction contributes 71 SNU, 54% of the total according to Bahcall and Ulrich (1988). The model of Turck-Chieze et al. (1988) and Turck-Chieze (1990) predicts 125 ± 5 SNU.

Since ^{69}Ga is also present in the detector, neutrino capture by that isotope can also produce ^{69}Ge. In this case, the analog state decays primarily by gamma emission, and thus enhances the production of this product by energetic neutrinos (Champagne et al. 1988a,b). This results from the fact that alpha-particle emission from the analog state is isospin forbidden. Since the neutrino capture in ^{69}Ga has an energy threshold of 2.22 MeV, this isotope (60% abundant) would only serve as a detector of solar neutrinos from ^8B decay. A gallium detector could therefore be used to measure independently the low-energy solar neutrino sources by measuring ^{71}Ge, and only ^8B decay neutrinos by measuring ^{69}Ge. The two radioactivities could be easily separated by their radiations and lifetime. However, with only 60 tons of gallium, it would be well to devise a counting system capable of clearly distinguishing these products. The gallium chloride method is amenable to rapid recovery of germanium, and it would be possible to carry out a successive set of experiments directed toward measuring the 39 hr ^{69}Ge production rate.

4. The Source Calibration.   The GALLEX (Kirsten 1990) and Soviet collaborations intend to test their detectors with a ^{51}Cr neutrino source prepared by activation of enriched ^{50}Cr metal in the nuclear reactor (Barabanov et al. 1985; Cribier et al. 1988). This source emits neutrinos with energy 751 MeV (90%) and 431 keV (10%). A chromium source only feeds the ground-state transition and the first excited state (175 keV) of germanium: the expected cross section is $59.2(1 \pm 0.1) \times 10^{42}$ cm^2, (total)/(ground) = 1.05 (Bahcall and Ulrich 1988). The preparation of this source and carrying out the experiment is a very large effort. It is anticipated that a 0.6 megacurrie source can be prepared, and a measurement of the cross section to 20% can be achieved. Using four sources, by repeated irradiations, the error can be reduced to 10%. This effort is motivated by the prospect of calibrating the detectors with a known neutrino source, though only the ground-state cross section is tested by the ^{51}Cr source. The calculated ground-state cross section is unlikely to be in error.

Two other neutrino-emitting sources have been considered. Alvarez (1973) suggested that ^{65}Zn could be prepared in a nuclear reactor by bombarding natural ^{64}Zn (48.6%) with neutrons. ^{65}Zn emits mono-energetic neutrinos with energies of 1.343 (48%) and 0.227 (51%) MeV. The more energetic neutrinos are particularly important for evaluating the cross sections for the excited states in ^{71}Ge at 0.411 and 0.736 MeV, and for testing the chlorine and iodine experiments. The zinc source has the decided advantage in that ^{65}Zn has a half life of 244 days, but is more difficult to shield because of the 1.112 MeV gamma ray (51%). Recently Haxton (1988b) pointed out that ^{37}Ar would be a suitable source that could be prepared by neutron capture in ^{36}Ar or by an (n,γ) reaction with fast neutrons. This source has the advantage of requiring very modest shielding and produces 0.814 MeV neutrinos (Slansky 1989). This source would be particularly attractive if an emanating calcium target could be found that would withstand an intense fast-neutron flux. This preparation method would produce a highly enriched ^{37}Ar source that could be generated repeatedly in a suitable facility.

D. A Geochemical Experiment

It has often been said that the Sun's energy production could conceivably vary over periods of millions of years or more. This thought has brought forward suggestions to measure the solar neutrino flux in the past (Scott 1976; Cowan and Haxton 1982; Haxton and Johnson 1988). It is most unlikely that the Sun's luminosity changes with time any more than is expected from the normal evolutionary increase of 5% per billion years. Even so, geochemical experiments do have the potential of exploring the neutrino flux millions of years ago, and this subject should be explored. One exceptionally good experiment is very close to yielding a result. The number of favorable cases is quite limited because background processes from cosmic rays and natural radioactivities usually overwhelm the weak signal from neutrinos (Rowley et al. 1980).

Cowan and Haxton (1982) noted that the neutrino capture cross sections of ^{97}Mo and ^{98}Mo to form the long-lived technetium isotopes ^{97}Tc ($t_{1/2}$ = 2.6 Myr) and ^{98}Tc ($t_{1/2}$ = 4.2 Myr) could be high enough to be useful for monitoring the solar neutrino flux during the past few Myr. In this case, one takes advantage of the fact that these products would be contained in molybdenum ores (MoS_2) 25 to 30 Myr old being mined at great depth in the Henderson Mine in Colorado. In addition, the processing of the minerals by flotation not only concentrates the technetium, but also serves to reduce the background production of technetium isotopes by uranium and other isotopes, because this particular mineral is low in uranium and thorium. Another advantage of this method is that the element technetium does not have any stable isotopes. The result of these considerations is that the molybdenum experiment appears to be capable of measuring the ^{8}B decay solar neutrino flux in the past few Myr. Another interesting application of the experiment is that it could give information on the galactic neutrino flux in the past from collapsing stars (Haxton and Johnson 1988).

Such an experiment is now in progress at Los Alamos National Laboratory (Wolfsberg and Kocharov 1991). They have devised a method of collecting the relatively volatile oxides of technetium and rhenium from the ore roasting stage of the process. These products will be purified and analyzed by surface ionization mass spectroscopy. A result from this interesting experiment is expected soon.

The only means presently available for determining the neutrino capture cross sections for the molybdenum isotopes is by nuclear systematics and (p,n) reaction studies. Studies have been made with 200 MeV protons using the (p,n) method and enriched ^{98}Mo. These studies give a value of 3×10^{-42} cm^2 for the ^{8}B neutrino cross section (Rapaport et al. 1985). Bahcall argues that the combined errors in the cross sections and the standard model, both around 30%, limit the sensitivity of a test for a long-period solar neutrino variation.

^{99}Tc ($t_{1/2} = 2.1 \times 10^5$ yr) is a fission product, and the amount observed should correspond to that from spontaneous fission and neutron-induced fission of uranium. ^{99}Tc will also be produced from ^{100}Mo by energetic muons in electromagnetic interactions at a sufficiently high rate to be observed in this experiment. This conclusion is based on the studies of energetic muons underground by Fireman et al. (1985) and his associates. A proper analysis of the muon production rate may allow one to determine the average depth of the molybdenum during the last few 100 Myr.

E. Proposed Radiochemical Experiments

A number of solar neutrino detectors have been considered over the last 20 yrs. Many of these experiments were suggested as a means of studying the low-energy solar neutrino spectrum by a set of radiochemical detectors each having a different sensitivity to the principal components of the spectrum: the p-p reaction, ^{7}Be, and ^{8}B decay neutrinos (see, e.g., the analysis of Bahcall 1978). If the solar neutrino flux is affected by the MSW or FCνSR mechanisms, radiochemical detectors can study the spectral changes needed to understand the basic mechanisms. Radiochemical detectors have the required sensitivities and low backgrounds needed to observe the solar neutrino spectrum below 1 MeV.

A considerable effort has been devoted to a few unusually promising cases, but at the present time none of these are sufficiently well developed to build a full scale detector. Usually there are some difficulties that must be overcome such as background effects, reaction cross sections not sufficiently well known, or the product nucleus not able to be measured and characterized with sufficient sensitivity. Another factor that has retarded progress is that very few laboratories are willing to devote a major effort in radiochemical solar neutrino research, and there has been little funding available. We will review some of the prospective radiochemical techniques that appear to be especially promising in the hope of stimulating further work in this field. Table IV, with data from Bahcall and Ulrich (1988), summarizes the expected solar neutrino capture rate from their standard solar model for the radiochemical solar neutrino detectors that will be discussed. A lengthy list of possible neutrino capture reactions with an energy threshold of <1 MeV that could be the basis of a radiochemical detector is included in the Irvine solar neutrino conference (Reines and Trimble 1973b).

1. Lithium. The lithium experiment, based upon the neutrino capture reaction ^{7}Li $(\nu,e^-)^7$Be reaction, has a low threshold energy (0.862 MeV), and a very high neutrino capture cross section. These mass 7 nuclei are mirror nuclei and therefore the neutrino capture cross sections to produce ^{7}Be in the ground state, and its first excited state at 0.48 MeV have very high values (Bahcall 1964,1978). As a result, even the low neutrino flux from the *pep* reaction and ^{7}Be decays in the Sun, that are barely over threshold contribute a significant rate (see Table IV). Lithium is a relatively inexpensive target element, the ^{7}Li isotope is 93% abundant, and the half life of ^{7}Be is 53 days. Because of these considerations, a lithium detector was regarded in the early 1970s as the most promising one to follow the chlorine experiment.

At Brookhaven a chemical solvent extraction procedure was developed for removing ^{7}Be from concentrated aqueous solutions of lithium chloride (Rowley 1978). Two background problems are associated with this approach. Commercial lithium chloride solutions contains trace quantities of radium and other alpha emitters that lead to the production of ^{7}Be by (α,p) scattering followed by (p,n) reaction (Zahkarov 1977). This background process requires that the radium concentration be less than 0.03 picograms per liter. This problem in chemical purification has not been solved. The other problem with the aqueous solution method is the fast-muon production of ^{7}Be by oxygen spallation. Because of the muon background effect, the experiment must be built deep underground, greater than 5500 hectograms cm^{-2}, a depth available in the Sudbury mine in Canada. The Soviet group at INR has developed a method of extracting ^{7}Be from metallic lithium that avoids these background problems (Veretenkin et al. 1985). Their technique involves filtering lithium metal through a stainless steel gauze. Apparently this simple procedure efficiently collects beryllium on the surface of the gauze. This approach requires a modest temperature to melt the lithium, over 180° C. However, the cost of producing the metal greatly increases the cost of the target material.

A major problem with the lithium detector is that an efficient means for counting and characterizing ^{7}Be decays has not been developed. One can observe the 480 keV gamma radiation that occurs in 10% of the decays with a low background germanium crystal, but this method is rather inefficient. Other detection methods have been suggested in various reports, but no serious effort has been made to develop these techniques. Some of the methods suggested are: observe the 57 eV recoiling ^{7}Li ion from ^{7}Be decay electrostatically or by thermal evaporation (Rowley 1978; Davis 1952), counting the recoiling ion with a cryogenic solid-state counter suggested by Lowry (personal communication), laser excitation of the resulting lithium atom in a gas proportional counter (Kramer et al. 1979; Hurst et al. 1979), and detecting ^{7}Be by accelerator mass spectroscopy (Litherland et al. 1987).

With a solution to the counting problem, the lithium radiochemical detector could become a solar neutrino detector of major importance. It has the largest rate per unit target weight of all radiochemical solar neutrino detectors (see Table IV), and it has a well-known cross section for the solar neutrino spectrum. According to the standard solar model of Bahcall and Ulrich, the total neutrino capture rate is 52 SNU and this rate corresponds to an event rate of 0.36 ^{7}Be atoms day^{-1} ton^{-1} of lithium. The major contribution to this rate arises from the ^{8}B flux (43%).

2. Bromine. The bromine experiment is based on the ^{81}Br$(\nu,e^-)^{81}$Kr* reaction producing ^{81}Kr in the 190 keV isomeric state. This state has a 13 s half life, sufficiently long to measure its beta-decay branch. The neutrino threshold energy for this process is 0.471 MeV. Because of the long lifetime of the product, the bromine method was originally suggested as a geochemical experiment in which ^{81}Kr (half life of 2.1 × 10^6 yr) would be extracted from a salt deposit (Scott 1976). Unfortunately, this approach is impractical because salt deposits are impure and have low concentrations of bromine. As a consequence, there are relatively large ^{81}Kr productions by fast neutrons and alpha processes in natural salt deposits (Rowley et al. 1980). In addition there is an enormous technical problem of isolating krypton from salt deposits. However, Hurst suggested that a real-time ^{81}Br experiment could be made practical by single-atom counting of ^{81}Kr by a laser-mass spectroscopic method (Hurst et al. 1980). This method of measuring ^{81}Kr was developed at Oak Ridge National Laboratory that had the capability of measuring a few hundred atoms of ^{81}Kr (Chen et al. 1984). Using this method of ^{81}Kr counting, a bromine detector could be built using a suitable bromine liquid like ethylene-dibromide. The same extraction procedure used in the chlorine experiment could be used (Hurst et al. 1984). At this stage, the bromine detector was regarded as one with a high sensitivity for the ^{7}Be neutrinos from the Sun.

The cross section depends on knowing the beta-decay branch of the 13 s isomeric state in ^{81}Kr, a very difficult measurement to carry out. This decay rate was measured by Lowry et al. (1987) and Davids et al. (1987). In addition, it is necessary to know the production rate to excited states that would be fed by ^{8}B neutrinos. The (p,n) measurements that were carried out showed a strong contribution from excited states (Krofcheck et al. 1987). The predictions from the standard model of Bahcall and Ulrich give a total rate of 28 SNU, with a 31 and 55% contribution, respectively, to the rate from ^{7}Be and ^{8}B fluxes (see Table IV). The interest in the bromine experiment was its sensitivity to the ^{7}Be flux, and the fact that the Homestake chlorine detector could easily be converted into a bromine detector. At present the chlorine detector is the only detector with a long observational record, and for the immediate future is being usefully used to test for variations in the solar neutrino flux. It could be converted easily to a bromine detector by filling the tank with ethylene dibromide or other suitable liquid. Tests of the extraction efficiency for krypton were carried out with the Homestake detector to demonstrate that the procedures were valid (Hurst et al. 1984).

3. Iodine. The iodine experiment was recently proposed by Haxton (1988*a*) as a high-rate radiochemical detector useful for observing ^{7}Be and ^{8}B decay neutrinos. At present the value of the cross section can only be estimated, but Haxton argues that it is relatively high, based on evaluations of the Gamov-Teller transitions by (p,n) reaction studies and estimates of transition ^{127}I (ground state $5/2+$) to a 125 keV state in ^{127}Xe ($3/2+$). The expected neutrino capture rate in an equal volume of an organic iodine compound or an iodine solution in an aqueous potassium iodide will be approximately 10 to 20 times higher than in perchloroethylene. The extraction of xenon and the measurement of ^{127}Xe decays (half life 36.4 days) would follow the methods already developed for the Homestake experiment.

There is great interest in having a high-sensitivity radiochemical experiment with a low neutrino energy threshold for monitoring and testing for variation in the neutrino flux from the Sun. Of particular interest is the effect of solar magnetic fields on the flux. The iodine experiment could have considerable promise for this purpose.

An experiment to measure the cross section for μ^+ decay electron neutrinos is underway at the beam stop at the LAMPF 800 MeV accelerator at Los Alamos. This experiment will test the theoretical calculations of the cross section.

III. DIRECT NEUTRINO COUNTING DETECTORS

A. Introduction

Direct observation of neutrino interaction by electronic methods serve as an effective means of determining the time, energy, neutrino flavor and the direction to the neutrino source. The observation of the anti-neutrinos from the supernova 1987A is a singular and outstanding example (Hirata et al. 1987*b*,1989; Bionta et al. 1987; Aglietta et al. 1989). The detection of this event was successful because the neutrinos were modestly energetic anti-neutrinos (10 to 40 MeV) that have a very favorable capture cross section, and the pulses occurred in a period of only 11 s. For this event the neutrino flavor and direction of the source could not be determined because of the unfavorable angular distribution of the emitted positron produced by the $p(\bar{\nu}_e, e^+)n$ reaction. The recording of this single event by the Kamiokande and Irvine-Michigan-Brookhaven detector was a remarkable confirmation of the theory of neutrino cooling of a collapsed star. In addition, because of the narrow interval of the pulse and the long travel time, it was possible to derive useful limits on the mass of the electron neutrino and other neutrino properties.

Note that these direct counting experiments did not observe the electron neutrinos from the initial collapse of SN 1987A. The chlorine radiochemical detector observes only electron neutrinos, and an observation of this pure ν_e pulse would be an added check on the theory, in particular, whether the MSW effect was operating. A special experiment was made to recover ^{37}Ar from this event, and only a single count was observed in 40 days of counting. That count could be attributed to counter background. The chlorine experiment is capable of observing a neutrino from as far away as the center of our Galaxy (Burrows 1984), but SN 1987A was a factor of 5 more distant.

Observing a steady source of low-energy neutrinos from the Sun is a far more difficult undertaking because of the low value of the cross section for the scattering of neutrinos with electrons, and presence of various background processes. However, one can take advantage of the fact that for the modestly energetic neutrinos the recoil electron is in the direction of the neutrino. A detector that measures the direction of the recoil electron can therefore resolve events from the direction from the Sun. Since 1987, the Kamiokande II imaging water Cerenkov detector has measured the flux of ^{8}B neutrinos from the Sun. Below we will discuss this experiment, and other electronic techniques for observing solar neutrinos. Many proposals have been made for measuring the solar neutrino flux by electronic methods, but we will confine our review to the experiments that are being planned, and a few experiments that use unusual experimental techniques that are being developed.

The basic solar neutrino spectrum consists of only electron neutrinos with energies corresponding to the *p-p* chain and the CNO cycle. However, because of possible MSW effects and neutrino spin rotation, the observable spectrum could be distorted and contain muonic and tauonic neutrinos, their anti-particles and sterile neutrinos. If this is indeed the case, one can in principle employ various electronic detectors to search for these neutrinos. This is a matter of great interest that must be explored. The original goal of solar neutrino research, studying directly the solar energy processes, and comparing the experimental observations with theoretical solar models, could become a very complex matter. Electronic detectors and low-temperature instrumental detectors have great promise in studying these questions that have been put forward in the last five years.

B. The Kamiokande II Experiment

The imaging water Cerenkov detector near Kamioka, Japan was originally built to observe the decay of the proton, a process predicted by theory to be in the range 10^{30} to 10^{32} yr. It was designed to observe energetic decay products, μ^+, e^+, π^0, π^+ etc. by measuring the cone of Cerenkov light on the walls of a very large tank of water (16 m diameter and 16 m tall) by means of 2500–50 cm diameter photomultiplier tubes shown in Fig. 5. The facility was located at a depth of 2700 m of water equivalent, sufficiently deep to reduce the intensity of cosmic-ray muons to a tolerable level. It was realized in 1985 that this facility could be used to observe the energetic neutrinos from ^8B decay in the Sun. These neutrinos could be observed by elastic scattering from electrons, a process that produces a recoil electron with sufficient energy to emit Cerenkov light (> 0.26 MeV). The elastic scattering process does not have a favorable cross section, but it does have the important advantage that the recoil electron is favorably directed along the path of the neutrino allowing one to determine the neutrino direction. The resolution for observing the neutrino direction is broadened by scattering of the recoil electron in the water to an opening angle of around 30 deg.

The recoil energy is measured by the number of photoelectrons recorded by the photomultiplier tubes. However, the electron recoil energy does not reflect the energy of the neutrino, because, for a given neutrino energy, the recoil energy spectrum is flat, extending from zero to the full energy of the neutrino.

Some modifications of the Kamiokande proton-decay detector were necessary to identify single low-energy electrons from neutrino interactions and

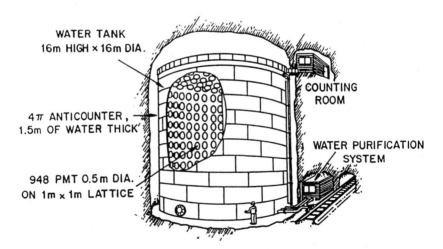

Fig. 5. The Kamiokande II experiment located at a depth of 2700 m of water equivalent in the Kamioka zinc mine, Gifu prefecture, Japan.

determine their direction. The detector was improved in 1986 to permit observing solar neutrino events. Fast-timing circuitry was introduced to allow distinguishing low-energy events. An anti-coincidence shield was formed around the central regions of the detector by adding an external water Cerenkov counter and re-arranging the photomultiplier tubes at the top and bottom of the tank. This arrangement provided a monitor for cosmic-ray muons. Fast muons produce short-lived energetic beta emitters in the water by photonuclear reactions with oxygen. The delayed decay of the beta emitters would simulate a neutrino event. Care was taken to purify the water from uranium, thorium and radium radioactivities. These radioisotopes produce a general background level of light that limits the detection of low-energy events. The Kamiokande II detector, as it is now called, is described by Hirata et al. (1988a,b,1989,1990b,1991).

Observing solar neutrinos requires reducing the data recorded in 2140 tons of water to eliminate events from various background effects. The reduction of the data was carried out in the following sequence: (1) only events in an internal fiducial volume of 680 tons were considered that had an energy above a chosen threshold (7.5 to 10.5 MeV); (2) events were discarded that occurred 30 s following the passage of a muon; and (3) the events were ordered with respect to the direction of the Sun. The resulting events were then compared to a Monte Carlo calculation of the signal expected from the standard solar model. (The model of Bahcall and Ulrich [1988] was used.) The result in Fig. 6 assures us that the neutrino signal is indeed coming from the Sun.

The result for 1040 days of observations (Jan. 1987 to April 1990) are given in Fig. 7 for two energy thresholds, 9.3 MeV (for 450 days) and 7.5 MeV (for 590 days). The final result is expressed as the ratio of the total observed rate in the solar direction to the standard model prediction; Kamiokande II data/standard model = 0.46 ± 0.05 (stat.) ± 0.06 (syst). The average rate in the nonsolar direction was used as the background. This result compares favorably with the rate from the Homestake chlorine experiment obtained for the same period, that is 4.2 ± 0.12 SNU compared to the same model that predicts 7.9 ± 2.6 SNU (3σ error, all experimental errors are 1σ). The comparison of the two experiments is made under the presumption that the solar energy is produced principally by the p-p chain, and therefore the chlorine experimental rate is dominated by the flux of ^{8}B decay neutrinos. The Kamiokande II collaboration also reports a comparison of the observed spectrum to that expected for ^{8}B decay events. The statistical uncertainties are large but the conclusion is that the shape of the spectrum is unaltered by matter oscillations. This is an important result (Bahcall and Bethe 1990). The experimenters also note that a day-night effect was not observed, although again the statistical uncertainties are too large to draw definite conclusions (Hirata et al. 1990b,1991). Finally, no measurable time variation was observed. A search for solar flare correlations was made with Kamiokande data

R. DAVIS AND A. N. COX

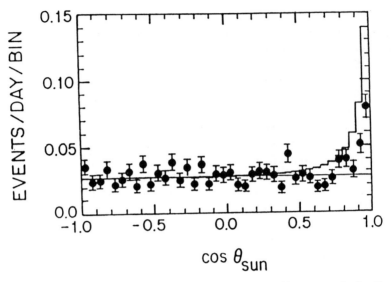

Fig. 6. A plot of the cosine of the direction of the recoil electron with respect to the direction of the Sun. The rise in rate approaching $\cos\theta_{Sun} = 1$ distinguishes the events from solar neutrinos. The histogram represents the rate expected from the model of Bahcall and Ulrich (1988). The period of observation was 1040 live days from January 1987 to April 1990.

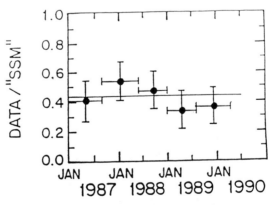

Fig. 7. A plot of the observed rate by the Kamiokande II detector for the 1040-day period January 1987 to June 1990. Figure from Hirata et al. 1991, energy threshold 9.3 Mev.

from July 1983 to July 1988 at neutrino energies 50–100 Mev; no correlations were observed (Hirata et al. 1988*a*).

The Kamiokande II observations correspond in time with a period in which the Homestake experiment observed a higher than average rate. It is interesting to make a comparison between the two experiments although this is difficult because the Kamiokande II experiment presumably observes only the ^{8}B flux, whereas the Homestake experiment observes an average rate for all electron neutrinos above 0.814 MeV. A rough comparison was made by assuming the standard model solar neutrino spectrum with all its components lower, e.g., by the same fraction. With this assumption, the chlorine experimental rate includes 77% ^{8}B decay neutrinos and 23% from all other components. Then one can calculate the ^{8}B rate for the individual experimental runs and average them. In Fig. 8 from Hirata et al. (1991), the data, using a threshold of 9.3 MeV, was binned in 5 time intervals in an effort to search for time variations. The chlorine data points in each of the bins was averaged and the ^{8}B decay neutrino flux calculated using the model of Bahcall and Ulrich (1988). The Kamiokande points in Fig. 7 were also scaled to the ^{8}B flux, and the two results compared in Fig. 8. Note that there is a general agreement between the two experiments with only the one Homestake point during the period June 1988 and March 1989 being well below the Kamiokande II point.

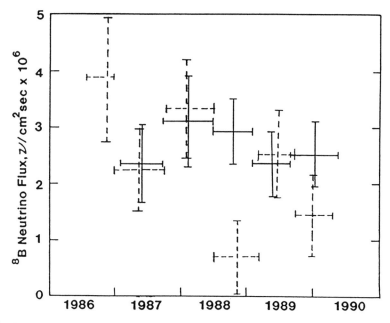

Fig. 8. A comparison of the calculated ^{8}B neutrino flux from the Kamiokande II and Homestake chlorine detectors (dashed symbols). See the text for the assumptions and method of analysis.

The remarkable fact is that these two experiments using entirely different techniques are in essential agreement. It would be interesting to examine the data from the two experiments in more detail.

Following these observations, the detector was shut down to replace failed photomultiplier tubes and to replace the electronic recording system. Presumably observations will continue. One looks forward to new results, particularly on whether the rate will decrease as solar cycle 22 continues.

A very much larger and improved version of the detector is proposed (Kajita 1989) for the same location. This Super-Kamiokande detector will use 50,000 tons of water resulting in a 22,000-ton fiducial mass, with a doubling of the fraction of the area covered by photomultiplier tubes, 11,200 tubes covering 40% of the surface. It is expected to have a threshold for recoil electrons from neutrino scattering of around 5 MeV, and a greatly increased accuracy for determining the vertex position of the event. The facility will serve to monitor the ^{8}B decay solar neutrino flux and measure its spectral shape above 5 MeV, monitor for supernova events in the Galaxy, determine the v_e and v_μ flux produced by cosmic rays, and provide increased sensitivity to proton-decay events.

C. The Sudbury Heavy-Water Detector

A considerable effort has been devoted to the design of a heavy-water imaging Cerenkov detector for studying the ^{8}B solar neutrino flux. The effort was stimulated by the possibility of using 1000 tons of heavy water from the Atomic Energy of Canada Limited and a very deep chamber in the Creighton mine of the International Nickel company near Sudbury, Canada (Chen 1985, 1988). Neutrino detection depends upon the two processes, the inverse-beta reaction

$$v_e + D \rightarrow p + p + e^- \tag{3}$$

and the neutral-current process

$$v_e, v_\mu \text{ or } v_\tau + D \rightarrow n + p + v_e, v_\mu, \text{ or } v_\tau. \tag{4}$$

The first reaction responds only to electron neutrinos, whereas the second neutral-current reaction has the identical cross section for all neutrino types or flavors. By distinguishing the solar neutrino rate from these two processes, one can test in principle whether the solar neutrino flux is altered by vacuum or matter oscillations. The charge-current reaction rate and the neutrino energy spectrum can be measured by observing the energy of the electron emitted. However, the angular distribution of the emitted electron with respect to the neutrino direction is not very favorable for an inverse-beta process. The neutral-current process can be observed by measuring the neutron capture gamma radiation from the neutron produced, a process that can be enhanced by introducing a small amount of an element with a high neutron capture cross section, e.g., chlorine. The heavy-water detector responds as well to the elastic scattering of neutrinos. These events can be distinguished by their favorable angular relation to the direction of the Sun. Of course a clean signal from elastic scattering could be obtained by filling the detector with light water. The basic cross sections for all these processes can be calculated with relatively small errors, approximately 10% (Bahcall et al. 1989). However, as with all multidetection processes, the detector must be designed to allow these numerous processes to be resolved.

The 1000 tons of heavy water in the Sudbury detector will be contained in a transparent acrylic plastic cylinder 5 m in diameter that is supported in the center of a closed cylindrical tank of water 20 m in diameter and 30 m high. The entire system is viewed by photomultiplier tubes chosen in design to have a low background and the best possible photon sensitivity and pulse timing characteristics. Approximately 40% of the surface will be covered with photomultiplier tubes. The arrangement allows the sensitive region to be filled with heavy or light water. The walls of the cavity and the base will be covered with sulfurcrete blocks to provide a shield against gamma radiation and neutrons from the rock. The central cylinder will be shielded with at least 4 m of water. Great care is needed to maintain an extremely high water and D_2O purity that is free of uranium and thorium radioactivities, less than parts per trillion. These precautions are essential for the neutral-current processes to be resolved from the detector background. The design of the detector is described in the proposal (Ewan 1987). This unique facility has been approved and should be built in four years at an estimated cost of 35 million dollars. It would be operated as a Canadian national facility.

THEORY OF NEUTRINO OSCILLATIONS AND THE SUN

S. P. ROSEN

The University of Texas at Arlington

This chapter describes the enhancement by matter of neutrino oscillations with small mixing angles and indicates how this phenomenon can be used to solve the solar neutrino problem. There are three possible solutions to the ^{37}Cl experiment and they yield different predictions for the Kamiokande II solar neutrino-electron scattering experiment and for the ^{71}Ga experiment. Present data tend to favor the nonadiabatic solution but do not exclude the large-angle one. They do, however, definitely disfavor the adiabatic solution.

I. INTRODUCTION

In recent years there has been a great deal of excitement about neutrino oscillations and the solar neutrino problem. Although the problem has been with us for a long time and oscillations have always appeared to be a potential solution, they did not draw the attention they deserved because the oscillation parameters needed to solve the solar problem seemed to fall outside the range endorsed by the prejudices of particle theorists. In particular, the solar neutrino problem required large mixing angles for the standard *in vacuo* oscillations, whereas our only experience of mixing angles, namely in the quark sector of particle physics, pointed towards small angles. Then, in July 1985, Mikheyev and Smirnov (1986,1987; Rosen and Gelb 1986*a,b;* Kuo and Panteleone 1989) found that matter oscillations, originally invented by Wolfenstein (1978*a;* Barger et al. 1980) for neutrinos passing through the Earth, could greatly enhance the oscillations of neutrinos traveling from the core of the Sun to Earth; in a most elegant paper, they showed that because of matter oscillations, small *in vacuo* mixing angles became large effective angles and hence they brought the solution of the solar neutrino problem into a domain acceptable to particle theory.

This beautiful result has revitalized the field of solar neutrino physics. Several major experiments designed to test whether the MSW (Mikheyev Smirnov Wolfenstein) effect, as matter oscillations have come to be known, really are the explanation of the solar neutrino puzzle are running and others are being planned. Within a decade we will indeed know whether neutrinos oscillate or whether the physics of the Sun itself differs from the expectations of the standard solar model. If neutrinos oscillate, then we will have made a discovery that would be impossible with purely terrestrial experiments, and that would reveal new physics beyond the standard model of Glashow, Weinberg and Salam (Bilenky and Petcov 1987).

II. HISTORICAL SURVEY

The idea of neutrino oscillations was inspired by an inaccurate rumor about one experiment performed by Davis in the mid 1950s and established as a serious notion by accurate accounts of another experiment begun by Davis in the 1960s. In both cases, Davis was trying to detect neutrinos using the famous reaction in which the neutral particle is captured by a ^{37}Cl nucleus to yield a ^{37}Ar nucleus plus an electron; but in the first experiment he used a reactor, which is a source of electron-type anti-neutrinos, while in the second he used the Sun, which is expected to be a source of electron-type neutrinos.

Distinguishing between neutrinos and anti-neutrinos is important because of a conservation law for leptons, which is satisfied to a high degree of accuracy. According to this law, anti-neutrinos, which are created in the decay of neutrons into protons, will always produce positrons when they interact with nuclei. But neutrinos, which are created in the conversion of protons into neutrons, will always produce electrons. Therefore in the first Davis experiment (Davis 1955,1956), we would expect to produce no ^{37}Ar whatsoever, and in the second one we would expect to produce a number of ^{37}Ar nuclei, which can be calculated from the flux and energy spectrum of solar neutrinos together with the relevant reaction cross sections.

Sometime in the summer of 1957 Pontecorvo (1958a,b), working in the Soviet Union, heard a rumor that Davis was seeing a signal in the reactor experiment. Using an analogy with K-meson physics, he proposed the oscillation hypothesis for reactor anti-neutrinos: as they travel from the reactor core to the ^{37}Cl target some of them could transform into neutrinos, which are capable of interacting with the target to produce an electron plus ^{37}Ar.

Unfortunately this proposal ran into difficulties because the rumor proved to be inaccurate (Davis 1955,1956). It did not die, but merely lay dormant until Davis performed his second experiment a decade later, this time with neutrinos produced in the fusion reactions that are believed to be the source of solar energy. He saw a signal but it was much smaller than the one predicted by Bahcall and his colleagues on the basis of the standard model of the Sun (Davis et al. 1968; Bahcall et al. 1968). In fact, the signal was at most one-half of the prediction and could be as little as one-quarter.

On hearing the news of the solar neutrino experiment, Pontecorvo (1968a,b) revived the idea of oscillations and this time he included the possibility of oscillations between muon-type and electron-type neutrinos as well as those between particle and anti-particle. In both cases, he argued that if the distance over which a typical oscillation takes place is much smaller than the distance between the Earth and the Sun, then by the time the neutrinos reach the Earth, one half of them will have transformed into a kind that cannot stimulate the ^{37}Cl to ^{37}Ar transition. Thus, the expected signal should be reduced by a factor of 2 in comparison with the original prediction.

Today we believe that oscillations between different families of neutrino are more credible than those between particle and anti-particle, and so it is only fair to point out that family oscillations were discussed independently by Sakata, Maki and Nakagawa several years before Pontecorvo. Sakata had evidently realized shortly after the discovery of the muon that the electrically neutral particle associated with it in the decay of the pi-meson might be different from the electrically neutral particle associated with the electron in beta decay. Over the years, he and his colleagues developed a theory of the two neutral particles, and in 1962 they published a paper in which they predicted oscillations between them (Maki et al. 1962). Unfortunately this paper did not become known in the West for many years.

An attractive feature of family oscillations is that, with enough families, it becomes possible to reduce the expected signal by more than a factor of 2. As we shall discuss below, the phenomenon of oscillations arises from a mismatch between the flavor eigenstates, so-called because they are the neutrinos associated with the electron, the muon, and the tau-lepton in charged-current weak interactions, and the eigenstates of the neutrino mass matrix. The flavor eigenstates are specific linear combinations of the mass eigenstates with definite phase relations between them; as time evolves, these phase relations change because of the differing mass eigenvalues and hence the neutrino becomes an admixture of different flavors. With n independent flavors, it is possible for what was initially an electron neutrino to become an equal admixture of all n flavors; thus the expected signal is reduced by a factor n. We know that there are three families of light neutrinos and so we expect a signal of one-third that predicted theoretically.

The problem with this argument is that it requires large mixings of mass eigenstates, whereas in the one mixing phenomenon with which we are familiar, namely that among quarks, we encounter small mixings. In 1985, Mikheyev and Smirnov (1986) solved it when they realized that matter oscillations occurring in the passage of neutrinos through the material of the Sun could, under the right conditions, greatly amplify even the smallest of mixing angles. Wolfenstein (1978a) had invented the notion of matter oscillations several years earlier when he investigated the behavior of neutrinos traveling through the Earth. He realized that because electron neutrinos can scatter from electrons through the exchange of charged W-bosons as well as neutral Z-bosons, their refractive indices in matter will be different from the refractive indices of other types of neutrino. As the refractive index affects the phase of the neutrino, and as neutrinos oscillations are themselves a matter of phase, it follows that the passage through matter can have a significant influence on the manner in which neutrinos may oscillate.

Mikheyev and Smirnov were the first to recognize that the Sun itself was the perfect place for this resonance enhancement effect, and that it would yield a most elegant solution of the solar neutrino problem with small *in vacuo* mixing angles. Their work stimulated many investigations, of both a numerical and analytic nature (Messiah 1986; Barger et al. 1986; Haxton 1986,1987; Parke 1986; Parke and Walker 1986; Notzold 1987; Pizzochero 1987; Toshev 1987a,b; Petcov 1988; Kim et al. 1987; Dar et al. 1987; Balantekin et al. 1988), in which the MSW effect has been applied to the chlorine experiment, to the gallium experiment and to solar neutrino-electron scattering. The outcome of these investigations will be discussed in subsequent sections.

III. *IN VACUO* OSCILLATIONS

The phenomenon of neutrino oscillations is a pure quantum mechanical effect involving two almost degenerate levels. Consider two states v_1 and v_2 with masses m_1 and m_2, respectively, and with a common momentum p, that is much greater than both masses. The energies of the two states are then given by the approximate formula

$$E_i = p + m_i^2/2p \quad (i = 1,2).$$
(1)

As these states, or any linear combinations of them, evolve in time, they acquire the appropriate phase factors

$$\exp(-iE_j t) \quad (j = 1,2)$$
(2)

and hence the phase difference between them oscillates with time. Therefore, whenever we have new states defined as coherent combinations of these two states with definite phase relations between them, the character of the new states will oscillate in time along with the phase.

Let us define the electron neutrino v_e as one combination of the mass eigenstates:

$$v_e = v_1 \cos \theta + v_2 \sin \theta$$
(3)

and the muon neutrino v_m as the orthogonal combination:

$$v_m = v_2 \cos \theta - v_1 \sin \theta. \tag{4}$$

By electron neutrino and muon neutrino we mean those flavor eigenstates that take part in weak interactions with the electron and muon, respectively. As these states evolve in time, the relative phase between the states with definite mass, namely v_1 and v_2, will change, and what was initially a pure electron neutrino or a pure muon neutrino will become an admixture of the two flavor states. In other words, an electron neutrino with sufficient energy can, some time after its birth, interact with nuclei and produce muons, and vice versa.

We can calculate the probability that an electron neutrino gives rise to muon-type interactions in terms of the mixing angle θ and the time t that it takes to traverse the distance from the point of birth of the neutrino to the target, or detector. Using the prescription of Eq. (2), we can write the state at time t of a neutrino that began life as an electron-type object at $t = 0$ as

$$v_e(t) = v_1 \cos \theta \exp(-iE_1 t) + v_2 \sin \theta \exp(-iE_2 t). \tag{5}$$

Taking the inner product of this state with the flavor states in Eqs. (3) and (4), we obtain the probability amplitudes for finding neutrinos of definite flavor at time t:

$$[v_e, v_e(t)] = \cos^2 \theta \exp(-iE_1 t) + \sin^2 \theta \exp(-iE_2 t) \tag{6}$$

and

$$[v_m, v_e(t)] = \cos \theta \sin \theta [\exp(-iE_2 t) - \exp(-iE_1 t)]. \tag{7}$$

The square moduli of these expressions then give us the probabilities for v_e to remain v_e at time t and to become v_m at time t:

$$P(v_e, v_e; t) = 1 - \sin^2 2\theta \sin^2(E_2 - E_1)t/2 \tag{8}$$

and

$$P(v_m, v_e; t) = \sin^2 2\theta \sin^2(E_2 - E_1)t/2. \tag{9}$$

It is usually the case that the momentum of the neutrino is much greater than the neutrino mass eigenvalues, and so we can use Eq. (1) to express the energy difference in terms of the mass difference and thence to define an oscillation length:

$$L = 2\pi/(E_2 - E_1) = 4\pi p/(m_2^2 - m_1^2)$$
$$= 4\pi p/\Delta m^2$$
$$= 2.5(p/\text{MeV}/(\Delta m^2/\text{eV}^2)$$

(10)

expressed in meters. The neutrino masses are taken to be so small that the particles travel with the speed of light; in units where $c = 1$ we can then equate the time t to the distance R traveled. The probabilities for the survival of the original flavor and the appearance of a new flavor can now be written in their usual forms:

$$P(v_e, v_e; R) = 1 - \sin^2 2\theta \sin^2(\pi R/L)$$
$$P(v_m, v_e; R) = \sin^2 2\theta \sin^2(\pi R/L).$$

(11)

The survival probability varies between 1 and $\cos^2 2\theta$ as R ranges between an integer multiple of the oscillation length L and the next half-integer multiple. Thus, depending on the angle θ, the probability can lie between 1 and 0. When R is very much larger than L, which is often the case in practice, we replace the distance-dependent factor by its average value of $1/2$, and find that the survival probability is always greater than or equal to $1/2$. It reaches its minimum value in the case of maximal mixing when $\theta = \pi/4$.

Were we to consider the mixing of three neutrinos, we would find a minimum of the survival probability of $1/3$ when R is much greater than the oscillation lengths in the problem; and again we would need maximal mixing to achieve this minimum. An instructive example can be constructed by using the (complex) cube roots of unity as the elements of the mixing matrix for flavor eigenstates in terms of mass eigenstates (Wolfenstein 1978b). For n neutrinos, the corresponding minimum probability would be $1/n$.

Let us now examine the experimental situation from the point of view of neutrino oscillations. The essential fact is that in the ^{37}Cl experiment, Davis and colleagues (Davis et al. 1989a) see an average signal of (2.0 ± 0.3) SNU, whereas the comprehensive theoretical analysis of Bahcall and Ulrich (1988; Bahcall 1989) predicts a capture rate of $7.9(1 \pm 0.33)$ SNU and another analysis based upon a smaller value for the ^{7}Be(p, γ) ^{8}B cross section predicts a rate approximately 25% smaller, 5.8 ± 1.3 SNU (Turck-Chieze et al. 1988; Turck-Chieze 1990). The solar neutrino unit (abbreviated to SNU) is defined to be one neutrino capture per 10^{36} atoms of ^{37}Ar per second; in terms of the amount of chlorine used by Davis, the observed capture rate corresponds to approximately 0.4 atoms of ^{37}Ar per day, and the predicted rate to between 1 and 1.5 atoms per day.

In order to reduce the predicted rate to the one observed by means of *in vacuo* oscillations, we need large mixings between the electron neutrino and the other flavor eigenstates. For oscillation lengths, which are much smaller than the astronomical unit, the mixing angles must be close to the values for maximal mixing. If the oscillation lengths are of the order of the astronomical unit, then we can use somewhat smaller angles because the distance-dependent factors in Eqs. (10) and (11) can achieve their maximal values of 1 instead of their average values of 1/2; however, even in this case, the angles cannot become too small. Thus an *in vacuo* oscillation solution for the solar neutrino problem requires large mixing angles—an unattractive possibility from the viewpoint of particle physics, as explained in Sec. II above. To resolve this problem one must turn to the phenomenon of matter oscillations.

IV. MATTER OSCILLATIONS

The neutrinos that Davis and his colleagues are trying to detect are born in the central region of the Sun where the density varies from $\sim$150 to 100 g cm^{-3}, $\sim$10 times greater than the density at the core of the Earth. We must therefore ask whether passage through such dense material can affect neutrino oscillations (Wolfenstein 1978a; Barger et al. 1980).

This question arises because all types of neutrinos undergo coherent forward scattering from ordinary matter (i.e., neutrons, protons and electrons) and they thereby acquire refractive indices that influence the phases of their wave functions. For the most part, the scattering occurs through the neutral-current interaction and is the same for all neutrino flavors. There is, however, one charged-current scattering that contributes to the electron neutrino alone. Consequently, the electron neutrino will have a different refractive index from the other neutrinos, and will acquire a different phase in its passage through matter. As oscillations themselves are a phenomenon of phase, we anticipate that matter can exert a nontrivial influence on them.

In all neutral-current processes, the neutrino scatters from the target through the exchange of a neutral Z^0 vector boson. The universality of the standard electro-weak model ensures that, for a given target, the amplitude is the same for all flavors of neutrino to lowest order of perturbation theory. Higher-order corrections can change from one flavor to another because the corresponding charged-leptons have different masses, but they are too small to be of any importance for our purposes.

The one charged-current process that comes into play is essentially a charge-exchange reaction involving the electron-type neutrino and the electron. It occurs through the exchange of a charged W^+ boson between the incident neutrino and the electron; this causes the neutrino to transform into an electron and vice versa (see Fig. 1). In the standard model the effective Hamiltonian for the scattering of electron neutrinos is given by

$$H_{\text{eff}} = G_F/2^{1/2}[e^+ \ \gamma_4\gamma_\mu(1 + \gamma_5)\nu_e] \times [\nu_e^+ \ \gamma_4\gamma_\mu(1 + \gamma_5)e]. \quad (12)$$

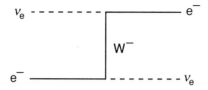

Fig. 1. The charged-current, or W-exchange diagram for the scattering of electron neutrinos by electrons.

After a Fierz re-arrangement in which we bring the electron fields into the same bilinear bracket, we obtain

$$H_{\text{eff}} = (-1)^2 G_F/2^{1/2}[e^+ \, \gamma_4\gamma_\mu(1 + \gamma_5)e] \times [\nu_e^+ \, \gamma_4\gamma_\mu(1 + \gamma_5)\nu_e]. \quad (13)$$

The two factors of (-1) arise from the Fierz transformation of the Dirac γ matrices and the anti-commuting Fermion fields, respectively. In the rest frame of the electron only the fourth component of the electron bilinear survives, and the expression reduces to

$$H_{\text{eff}} = (+1)G_F/2^{1/2}(e^+e) \times [\nu_e^+(1 + \gamma_5)\nu_e]. \quad (14)$$

We identify the electron factor as the number density of electrons in the matter traversed by the neutrino, and we evaluate the neutrino factor by noting that the incident and scattered neutrinos are left handed, i.e., eigenstates of the projection operator $(1 + \gamma_5)/2$. We therefore gain a factor 2 from the neutrino bilinear and obtain the final expression for the effective Hamiltonian for the coherent forward scattering of electron neutrinos by unpolarized electrons at rest:

$$H_{\text{eff}} = +G_F 2^{1/2}N_e \qquad \text{(neutrinos)} \qquad (15)$$

where N_e is the density of electrons. It must be emphasized that this effective Hamiltonian represents the difference between the coherent forward scatterings of electron neutrinos and other types of neutrinos, and that the difference is the important quantity for the MSW effect.

The same formalism can be used to calculate the coherent forward amplitude for the scattering of anti-neutrinos of the electron variety. In this case, however, the roles of the neutrino fields in Eq. (12) are interchanged: ν_e creates the scattered anti-neutrino instead of destroying the incident neutrino, and ν_e^+ destroys the incident anti-neutrino instead of creating the scattered neutrino. As a result, the effective Hamiltonian changes sign:

$$H'_{\text{eff}} = -G_F 2^{1/2} N_e \qquad \text{(anti-neutrinos)}. \qquad (16)$$

This change of sign has important consequences for matter oscillations.

In order to incorporate the effect of matter on neutrino oscillations, we recast the equations for the time dependence of the amplitudes for the survival of electron neutrinos and the appearance of muon neutrinos in the form of a time-dependent Schroedinger equation. We then add the effective matter Hamiltonian to the *in vacuo* one. This procedure to combine the two effects as used by Wolfenstein in his original paper, and the rigorous justification for it in terms of relativistic wave equations has been discussed by Halprin (1986) and by Mannheim (1988). From Eqs. (6) and (7), we can construct the Schroedinger equation for *in vacuo* oscillations as

$$i\frac{dA}{dt} = HA \qquad (17)$$

where A represents a column vector of the two probability amplitudes a_e and a_m, respectively,

$$A = \begin{pmatrix} a_e \\ a_m \end{pmatrix} \qquad (18)$$

and the *in vacuo* Hamiltonian H is given by

$$H = \begin{pmatrix} X & Y \\ Y & Z \end{pmatrix} \qquad (19)$$

with

$$
\begin{aligned}
X &= \frac{m_1^2 c^2 + m_2^2 s^2}{2p} \\
Z &= \frac{m_1^2 s^2 + m_2^2 c^2}{2p} \\
Y &= \frac{m_2^2 - m_1^2}{2p} cs = \frac{\Delta m^2}{2p} cs \qquad (20) \\
c &= \cos \theta \\
s &= \sin \theta.
\end{aligned}
$$

For neutrinos propagating through the Sun, we incorporate matter oscillations by replacing the Hamiltonian in Eq. (17) by (Wolfenstein 1978a)

$$i\frac{dA}{dt} = (H + H_M)A \qquad (21)$$

where the additional matrix is given by

$$H_M = \begin{pmatrix} H_{eff} & 0 \\ 0 & 0 \end{pmatrix}. \tag{22}$$

From this we see that the only modification to the *in vacuo* equations of motion occurs in the top left-hand element of the Hamiltonian matrix (the *ee* element) where X is replaced by

$$X + H_{eff} = \frac{m_1^2 c^2 + m_2^2 s^2}{2p} + 2^{1/2} G_F N_e. \tag{23}$$

This seemingly innocent modification can have powerful consequences: for, under the right conditions, it gives us a Hamiltonian matrix that is not only symmetric (see Eqs. 19–22), but also one that has equal elements down the diagonal. The eigenvectors of such a matrix are equal admixtures of electron and muon neutrino, and for a given off-diagonal element, the separation between the eigenvectors is minimal. In other words, the extra term in the *ee* element gives us a chance to progress from nonmaximal mixing *in vacuo* to maximal mixing in matter.

The condition for equal diagonal elements, $X + H_{eff} = Y$ (see Eqs. 20 and 23) can be written as

$$2^{1/2} G_F N_e = \frac{\Delta m^2 \cos 2\theta}{2p}. \tag{24}$$

Now the electron density N_e is inherently positive, and the Fermi constant G_F, because it arises from the exchange of a gauge boson, is also positive: therefore the product of Δm^2 and $\cos 2\theta$ must also be positive, if the condition is to be satisfied.

If the mixing angle is close to zero, then $\cos 2\theta$ is close to $(+1)$ and the squared mass difference

$$\Delta m^2 = m_2^2 - m_1^2 \tag{25}$$

must be positive; that is

$$m_2 > m_1. \tag{26}$$

If the mixing angle is close to $\pi/2$, then $\cos 2\theta$ is close to (-1) and Δm^2 must be negative; that is,

$$m_2 < m_1. \tag{27}$$

In both cases, the electron neutrino is dominantly composed of the lighter of the two mass eigenstates, and the muon neutrino of the heavier. Therefore, a necessary condition for obtaining effective maximal mixing when an electron neutrino propagates through matter is that the electron neutrino be lighter than the muon neutrino. Had we considered anti-neutrinos instead of neutrinos, the matter contribution to our Schroedinger-like equation would change sign by virtue of Eq. (16), and this necessary condition would be reversed. For the possibility of maximal mixing when anti-neutrinos propagate through matter, the electron neutrino must be the heavier of the two. Therefore, the effective enhancement of oscillations can occur either for the particle or for the anti-particle but not both. It is most fortunate that the condition for neutrinos is consistent with the general prejudice that the electron neutrino should be lighter than the muon neutrino, just as the electron is lighter than the muon.

V. CONSTANT DENSITY

When the density of electrons is constant, we can describe the oscillation problem in the same way as in the *in vacuo* case except that the mixing angle and the oscillation length are both modified. The survival probability for an electron neutrino is

$$P(v_e, v_e; R) = 1 - \sin^2 2\theta_M \sin^2(\pi R/L_M) \tag{28}$$

and the new oscillation parameters are obtained by diagonalizing the equations of motion in Sec. IV above:

$$
\begin{aligned}
\sin^2 2\theta_M &= \sin^2 2\theta/[\sin^2 2\theta + (L/L_0 - \cos 2\theta)^2] \\
L_M &= L/[\sin^2 2\theta + (L/L_0 - \cos 2\theta)^2]^{1/2} \\
L &= 4\pi p/\Delta m^2 \\
L_0 &= 2\pi/2^{1/2}G_F N_e.
\end{aligned}
\tag{29}
$$

The formula for the new mixing angle in matter has two important properties: first, no matter how small the *in vacuo* angle θ may be, the matter angle θ_M reaches its maximal value when

$$L/L_0 = \cos 2\theta \tag{30}$$

which is just another way of writing the equal diagonal element condition of Eq. (24). Therefore, as long as θ is different from zero, there is always a density for which the neutrino will oscillate with maximal mixing.

This amplification, or enhancement, of the oscillations does not come without a price. We see from Eq. (29) above, that when the enhancement condition is satisfied, the oscillation length in matter becomes (Mikheyev and Smirnov 1986)

$$L_M = L/\sin \theta \tag{31}$$

which is much longer than the *in vacuo* oscillation length itself, especially when θ is small. This means that in order for the enhanced oscillations to take place, we must not only satisfy the enhancement condition, but also have a sufficiently long path length for the neutrinos.

It is apparent from Eq. (29) that we can regard $\sin^2 2\theta_M$ as a function of the ratio L/L_0, which we denote by l. When the electron density is low and l is very small, the matter mixing angle is almost the same as the *in vacuo* angle and the effects of matter on oscillations are negligible. As the density and l grow, so the matter angle grows until it reaches its maximal value at $l = \cos 2\theta$. For larger values of l, the matter angle then falls until it reaches zero for the very largest l; in this case, the high density of matter has the effect of wiping out the oscillations altogether.

The general shape of the curve for $\sin^2 2\theta$ is similar to that of a resonance (Mikheyev and Smirnov 1986,1987), and its full width at half maximum is $2 \sin 2\theta$:

$$2\Delta l = 2\Delta(L/L_0) = 2[L/L_0 - \cos 2\theta] = 2 \sin 2\theta. \tag{32}$$

Thus, the smaller the *in vacuo* angle θ, the narrower the peak; for very small angles, the peak becomes a very sharp spike. Outside the spike, the matter angle θ_M is essentially zero.

To gain some insight into the densities and oscillation lengths for which we might find the enhancement or resonant effect, we express the electron density in units of Avogadro's number,

$$N_e = 6 \times 10^{23} \rho_e \tag{33}$$

and then calculate L_0 in meters

$$L_0 = 1.77 \times 10^7/\rho_e. \tag{34}$$

Using this and the last part of Eq. (10) in the enhancement condition of Eq. (30), we find that the condition becomes

$$\frac{p}{\Delta m^2} = \frac{7 \times 10^6 \cos 2\theta}{\rho_e}. \tag{35}$$

The values of ρ_e encountered in the Sun vary from ~ 150 in the core to close to zero at the edge; for neutrino momenta of 1 to 10 MeV, typical of the ^{8}B neutrinos to which the Davis experiment is sensitive, and a small mixing angle, this condition corresponds to Δm^2 in the range of 10^{-7} to 10^{-3} eV^2. In the Earth, ρ_e varies from ~3 at the surface to 13 at the center, and so the range of values for Δm^2 is 10^{-8} to 10^{-6} eV^2. As there is a band of overlap between these ranges, we may anticipate that for certain sets of neutrino parameters, neutrinos can undergo enhanced oscillations in the Earth as well as the Sun; as we shall see later (Sec. IX), this can give rise to a "day-night" effect in which electron neutrinos are rejuvenated in their passage through the Earth (Carlson 1986; Bouchez et al. 1986).

Another interesting feature of the enhancement condition is an inverse scaling between the neutrino momentum and the electron density. For given Δm^2 and cos 2θ, the product of momentum times density has a fixed value, and any pair of parameters p and ρ_e with this value will satisfy the enhancement condition. Thus, if a 10 MeV neutrino undergoes enhancement at a density of 100, then a 100 MeV neutrino will undergo enhancement at a density of 10, and a 1000 MeV neutrino at 1. By this scaling we can, at least in principle, test the matter oscillations of solar neutrinos by studying the more energetic atmospheric neutrinos in their passage through the Earth.

VI. VARYING DENSITY IN THE SUN

Because the density of the Sun varies from a large value at the core to a very small value at the edge, the enhancement condition of Eq. (35) can be satisfied by neutrinos with momenta or equivalently oscillation lengths, in a wide band. The band covers several orders of magnitude,

$$10^4 \leq \frac{p}{\Delta m^2} \leq 10^8 \tag{36}$$

and so it is possible for all, or a significant part of the solar neutrinos produced in the core region to undergo large oscillations.

The history of a neutrino within this parameter range as it travels through the Sun can be divided into three parts (Mikheyev and Smirnov 1986,1987; Rosen and Gelb 1986a,b; Kolb et al. 1986). Initially, it is in a region of high density for which the parameter l is very much greater than 1; the effective mixing angle is much smaller than the *in vacuo* angle and oscillations are suppressed. The neutrino then moves into a region of densities corresponding to $l \approx 1$ and oscillations are enhanced. Finally, it moves into a low-density region where l is much smaller than 1 and *in vacuo* oscillations set in.

These three stages are well illustrated by Figs. 2 and 3 in which we follow the probability for an electron neutrino to remain an electron neutrino as it travels through the Sun. In both cases, the mean value of the probability when the neutrino leaves the Sun is much less than it would have been without matter oscillations, and the change in the mean value occurs over a relatively

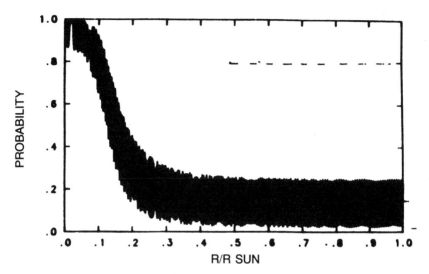

Fig. 2. Probability for an electron neutrino to retain its flavor as it travels through the Sun. This is an example of the adiabatic transition with $\sin^2 2\theta = 0.4$ and $p/\Delta m^2 = 1.5 \times 10^5$ (figure from Rosen and Gelb 1986a).

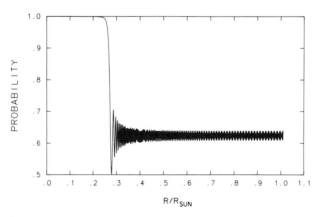

Fig. 3. Probability for an electron neutrino to retain its flavor as it travels through the Sun. This is an example of the nonadiabatic transition with $\sin^2 2\theta = 0.001$ and $p/\Delta m^2 = 6 \times 10^5$ (figure from Rosen and Gelb 1986a).

short distance. The size of this region, ~0.2 of a solar radius for Fig. 2 and much less for Fig. 3, serves to indicate the basic type of approximation one can make in solving the equations of motion. In the former case, it is the adiabatic approximation and in the latter it is the nonadiabatic, or sudden, or slab approximation (Rosen and Gelb 1986a,b).

The adiabatic approximation holds when the eigenvalues of the equations of motion (Eqs. 17–23) change slowly as the neutrino travels through the Sun (Messiah 1986); and the nonadiabatic approximation must be used when the eigenvalues change rapidly in a small region (Rosen and Gelb 1986a,b). To distinguish between these two cases, we compare the physical size $2\Delta x$ of the region in which enhanced oscillations take place with the oscillation length in matter L_M (Eq. 31): when $2\Delta x$ is much larger than L_M the adiabatic approximation holds, and when it is much smaller, the nonadiabatic one is valid.

To calculate the size of the enhancement region, we make use of the fact that as a function of the parameter l above, the width of the resonance is

$$2\Delta l = 2 \sin 2\theta. \tag{37}$$

Because l is equivalent to the electron density for fixed $p/\Delta m^2$, we can determine the physical extent of the region in terms of the change in the density:

$$2\Delta x = 2(\tan 2\theta/h_0) \tag{38}$$

where h_0 is the logarithmic derivative of the density at the point of enhancement:

$$h_0 = \left| \frac{1}{\rho} \frac{d\rho}{dx} \right| \text{ enhancement.} \tag{39}$$

Except for the first 5% of the solar radius, the electron density in the Sun can be approximated by an exponential (Bahcall 1989):

$$\begin{aligned} \rho_e &= \rho_0 \exp(-x/R_0) \\ \rho_0 &= \approx 150 \\ R_0 &= \approx R_\odot/10 \approx 7 \times 10^7 \text{ meters.} \end{aligned} \tag{40}$$

Consequently, the scale height h_0 is roughly a constant:

$$h_0 = 1/R_0 \approx [7 \times 10^7]^{-1} \tag{41}$$

and the enhancement region is given by

$$2\Delta x = 2 \tan 2\theta \, R_0. \tag{42}$$

For small mixing angles this is a small fraction of a solar radius. We now use this expression to write down the conditions for the validity of the two approximations we have discussed.

The adiabatic approximation is valid when the enhancement region in the Sun is much larger than the effective oscillation length at the point of enhancement (Messiah 1986). We can translate this into a bound on the parameter $p/\Delta m^2$:

$$\frac{p}{\Delta m^2} << \frac{\sin 2\theta \tan 2\theta}{2\pi h_0}. \tag{43}$$

The essential feature of this approximation is that the eigenvectors of the equations of motion (Eqs. 17–23) change so slowly that for all practical purposes the neutrino remains in the same eigenstate as it crosses the enhancement region, but the meaning of the eigenstate in terms of neutrino flavor changes. An electron born at the core of the Sun is dominantly in the "heavier" of the two eigenstates by virtue of the effective mass (Eq. 15) it acquires from the charged-current scattering diagram; but when it emerges from the Sun, the heavier state is now the muon neutrino. So, by remaining in the same eigenstate, the neutrino undergoes a change of flavor.

In the nonadiabatic regime (Rosen and Gelb 1986a,b), the criterion of Eq. (43) is reversed,

$$\frac{p}{\Delta m^2} >> \frac{\sin 2\theta \tan 2\theta}{2\pi h_0} \tag{44}$$

and the probability grows that the neutrino will make a sudden jump from one eigenstate to the other, thereby preserving its flavor. A naive model for this is to assume, especially for small mixing angles, that the neutrino does not oscillate in the high- and the low-density regions of the Sun, where l is either much greater than 1 or much smaller, and that its only oscillations take place in a small slab around the point of enhancement, where l is close to 1.

The probabilities for ν_e to remain ν_e at Earth in both approximations have been calculated by several authors. For the adiabatic approximation, the probability is (Messiah 1986)

$$P^{ad} = \cos^2 \phi_0 \sin^2 \theta + \sin^2 \phi_0 \cos^2 \theta \tag{45}$$

where $(\cos \phi_0, -\sin \phi_0)$ is the "heavier" eigenvector of the Hamiltonian operator in Eqs. (19–23) at the point of birth of the neutrino. If the density at this point is high, or if $p/\Delta m^2$ is large, ϕ_0 approaches zero; and if the density is low it approaches $\pi/2 + \theta$. For the nonadiabatic case, Haxton and Parke (Haxton 1987; Parke 1986) have used the Landau-Zener formula to obtain the probability:

$$P^{non} = \exp\left[- \frac{\pi \Delta m^2 \sin 2\theta \tan 2\theta}{4ph_0} \right]. \tag{46}$$

As functions of $p/\Delta m^2$ for fixed θ, these two expressions join together as illustrated in Fig. 4. P^{ad} remains close to 1 in the vicinity of $p/\Delta m^2 = 10^{4-5}$ and then falls rapidly to its asymptotic value of $\sin^2 \theta$ for larger values. It remains at the asymptotic value until the adiabatic approximation breaks down and the nonadiabatic one takes over (cf. Eqs. [43] and [44]). P^{non} has the property that it rises from a small value and tends towards 1 as $p/\Delta m^2$ increases. The net effect of this behavior is to yield a "suppression gap" in the middle range of the momentum over mass parameter in which the probability for ν_e to remain ν_e is small; the gap begins at $\sim 10^5$ and its width increases with the angle θ, covering more than an order of magnitude in most

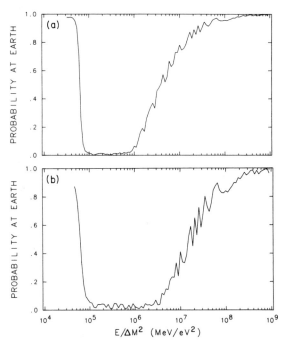

Fig. 4. Probability for an electron neutrino to arrive at Earth as an electron neutrino as a function of $p/\Delta m^2$. The upper curve (a) corresponds to $\sin^2 2\theta = 0.01$ and the lower one (b) to 0.04. Notice that the suppression gap grows as the *in vacuo* mixing angle increases (figure from Rosen and Gelb 1986a).

cases. The types of solution we find for the solar neutrino problem depend upon how the spectrum of solar neutrinos overlaps with the suppression gap.

VII. THE SOLAR NEUTRINO PROBLEM

We believe that the Sun generates its radiant energy by means of nuclear fusion, and that the most important reactions are the chains that fuse protons into ^{4}He. Neutrinos are by-products of these reactions and detecting them would help to confirm our views.

The three most important sources of neutrinos are (Bahcall 1989): the fusion of two protons to form a deuteron; electron capture by ^{7}Be; and the beta decay of ^{8}B. Of these, the p-p neutrinos are the most copious in numbers, but the lowest in energy; the flux of p-p neutrinos is directly correlated with the solar luminosity and is essentially independent of solar models.

$$p + p \rightarrow {}^2H + e^+ + \nu_e$$
$$E_\nu(\text{max}) = 420 \text{ keV} \qquad (47)$$
$$\text{Flux} = 6 \times 10^{10} \text{cm}^{-2}\text{s}^{-1}$$

at Earth. Next come the monoenergetic ^{7}Be neutrinos that are fewer, but higher in energy:

$$e^- + {}^7Be \rightarrow {}^7Li + \nu_e$$
$$E_\nu = 861 \text{ keV } (90\%)$$
$$= 383 \text{ keV } (10\%) \qquad (48)$$
$$\text{Flux} = 4 \times 10^9 \text{cm}^{-2}\text{s}^{-1}$$

at Earth. Finally there are the ^{8}B neutrinos that are very much less abundant but very much higher in energy:

$$^8B \rightarrow {}^8Be^* + e^+ + \nu_e$$
$$^8Be^* \rightarrow 2{}^4He$$
$$E_\nu(\text{max}) = 14.06 \text{ MeV} \qquad (49)$$
$$\text{Flux} = 6 \times 10^6 \text{cm}^{-2}\text{s}^{-1}$$

at Earth. It is because the cross sections for neutrino reactions on nuclei vary as $(E_\nu)^2$ that the ^{8}B neutrinos are so important.

For many years now, Davis and his colleagues have been trying to detect these neutrinos by means of the reaction (Davis et al. 1989a)

$$\nu_{\text{solar}} + {}^{37}Cl \rightarrow e^- + {}^{37}Ar. \qquad (50)$$

The threshold for the transition to the ground state of ^{37}Ar is 810 keV, and so the *p-p* neutrinos cannot contribute to the reaction; by far the strongest transition is to an excited state, the so-called isobaric-analog state, which is 5.1 MeV above the ^{37}Cl ground state and can only be reached in the reactions of ^{8}B neutrinos. Thus the ^{8}B neutrinos make the dominant contribution (~80%) to the reaction.

The theoretical prediction for the capture rate is (7.9 ± 2) SNU (we use the analysis of Bahcall and Ulrich [1988] rather than the 25% smaller value of Turck-Chieze et al. [1988]), which is considerably larger than the experimental value of (2.1 ± 0.3) SNU. This experimental value is the average of almost 20 yr of observations and it demonstrates clearly that there are too few ^{8}B neutrinos.

This deficiency has very recently been confirmed by another experiment, the Kamiokande II detector (Hirata et al. 1989,1990*a,b*) which also observed the supernova neutrinos. It detects the ^{8}B neutrinos through the neutrino-electron scattering process, and it observes less than half of the number of neutrinos expected on the basis of the standard solar model of Bahcall and Ulrich (1988); in the analysis of Turck-Chieze et al. (1988), the deficiency amounts to 30 ± 3%.

Given the lack of solar neutrinos, we may argue that either the standard solar model is incorrect or that the neutrino has some special property that inhibits its interactions with nuclei. As far as the solar model is concerned, the number of ^{8}B neutrinos is very sensitive to the temperature of the core of the Sun, and a reduction of the temperature by 7 to 10% can reduce the number of neutrinos to the observed level. Alternatively, the solar model may be correct and the neutrino has the special property of oscillating into other flavors. Because of the energy spectrum of solar neutrinos, these other flavors interact with nuclei only through neutral currents and give no ^{37}Ar signal. *In vacuo* oscillations require large mixing angles and we have already expressed a prejudice against them. We shall therefore examine the matter oscillation solutions, which allow the possibility of small mixing angles. We shall also argue that certain experiments will enable us to distinguish between various solutions, including the option of an incorrect solar model (Bahcall 1989).

VIII. MSW SOLUTIONS FOR ^{37}Cl

We would like now to determine those values of the oscillation parameters $\sin^2 2\theta$ and Δm^2 which, when combined with matter oscillations, yield the observed reduction in the ^{37}Cl experimental signal. By direct computation and by analytical means, we find that there are three classes of solution the qualitative properties of which can be inferred by laying the solar neutrino spectrum shown in Fig. 5 over probability curves of the kind shown in Fig. 4.

The line-up of the two figures is determined by the value of Δm^2: for

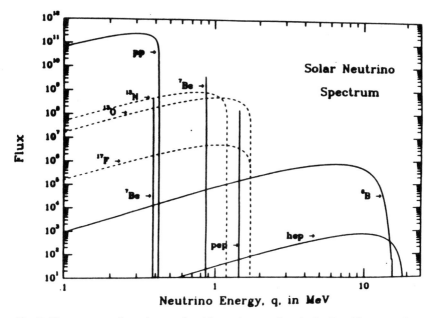

Fig. 5. The spectrum of neutrinos produced by nuclear reactions in the Sun. The appropriate reaction is indicated with each curve (figure from Bahcall and Ulrich 1988).

example, if $\Delta m^2 = 10^{-5}$, then the point corresponding to $E_\nu = 1$ MeV on the spectrum lines up with the point corresponding to $E/\Delta m^2 = 10^5$ on the probability curve. Consequently, most of the neutrinos with energies below 1 MeV will fall in a region of large probability and hence will arrive at Earth as electron neutrinos. By contrast, most of the neutrinos with energies > 1 MeV fall into the suppression gap and arrive at Earth as muon (or tau) neutrinos. In other words, the p-p and ^{7}Be neutrinos preserve their flavor while the bulk of the ^{8}B neutrinos lose theirs.

Solutions of this type correspond to the adiabatic approximation (Messiah 1986; Haxton 1986) and were discussed by Bethe (1986). For small mixing angles, the location of the rapid fall in probability is effectively independent of mixing angle and so Δm^2 remains fairly constant, in the vicinity of a few times 10^{-4}, for all small angles. In terms of neutrino energy, the rapid fall occurs in the vicinity of 5 to 7 MeV.

As a second example, suppose that $\Delta m^2 = 10^{-6}$. The spectrum must be moved to the right so that $E_\nu = 1$ MeV now coincides with $E/\Delta m^2 = 10^6$. Neutrinos with energies below 1 MeV now have almost zero probability of arriving at Earth as electron neutrinos, while those with energies > 1 MeV have increasing probabilities of doing so. As we make Δm^2 smaller, the spectrum must be moved further to the right and a lower electron energy coincides with the 10^6 point; as we make it larger, a larger energy coincides with 10^6.

Solutions of this second type were first emphasised by Rosen and Gelb (1986a,b), and they correspond to the nonadiabatic approximation. In contrast to the adiabatic solutions, they have the property that lower-energy neutrinos are much more readily converted to nonelectron flavors than the higher energy ones; thus the p-p and ^{7}Be neutrinos are largely converted to other flavors while a much higher proportion of the ^{8}B neutrinos remain as electron neutrinos (see Fig. 6). Another difference between this type of solution and the adiabatic one is that the rising part of the probability curve moves to the right as $\sin^2 2\theta$ increases; therefore, in order to keep the same proportion of the neutrino spectrum overlapping the rising probability, the value of Δm^2 must decrease as the mixing angle increases. The characteristic feature of these two types of solution is that one part of the spectrum falls within the suppression gap while the other does not.

There is a third type of solution, first discussed by Parke (1986) and by Parke and Walker (1986), in which the entire solar neutrino spectrum falls inside the suppression gap. This means that the probability for electron neutrinos to preserve their flavor at Earth is independent of energy; it must lie in the range of 1/4 to 1/2 to fit the observations of the Davis experiment. As the probability in the suppression gap is given by the asymptotic value of the adiabatic approximation, namely $\sin^2 \theta$, it follows that the angle θ must be large, lying between 30 and 45°. Because the gaps for such large angles span several decades in $E/\Delta m^2$, it follows that the values of Δm^2 for this type of solution will also span several decades for a given angle. In this way we close the triangle of solutions shown in Fig. 7.

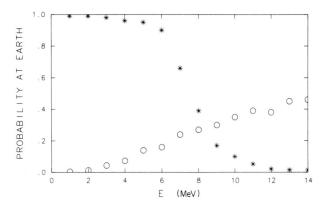

Fig. 6. Probability for an electron neutrino to retain its flavor at Earth as a function of its energy for the adiabatic solution (stars) and the nonadiabatic solution (circles). The mixing angle $\sin^2 2\theta$ is 0.01 in both cases, but Δm^2 is 10^{-4} in the adiabatic case and 3.6×10^{-6} in the nonadiabatic one (figure from Rosen and Gelb 1986a).

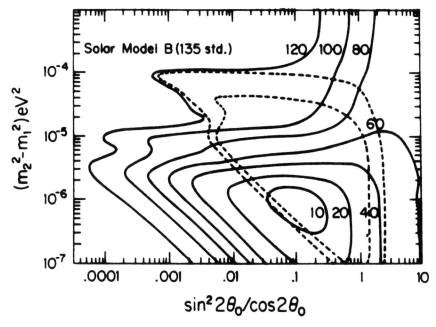

Fig. 7. Consequences of the different MSW solutions of the ^{37}Cl experiment for the ^{71}Ga solar neutrino experiment. The dashed triangle represents the ^{37}Cl solutions and the solid lines are iso-SNU curves for ^{71}Ga, the corresponding values of the capture rate being beside each curve (figure from Parke and Walker 1986).

IX. CONSEQUENCES FOR OTHER EXPERIMENTS

Before turning to other specific experiments on solar neutrinos, we consider general properties of solutions of the ^{37}Cl anomaly based upon neutrino oscillations. It is generally believed that, if neutrinos oscillate, they transform from one known flavor into another. Now, in the standard Glashow-Weinberg-Salam[3] model of electro-weak interactions, neutrinos of all known flavors undergo neutral-current interactions with quarks with equal strength in lowest order. Therefore the neutral current signal from solar neutrinos should be exactly as predicted by the standard solar model (SSM) if conventional oscillations are the proper solution.

It is always possible that in some nonstandard electro-weak model, neutrinos oscillate into "sterile" partners that do not interact with quarks, or do so extremely weakly. In this case, the neutral-current signal from solar neutrinos will be suppressed to the same extent as the charged-current one. This is exactly what is expected in solutions that change the solar physics, and so to distinguish between these possibilities, we must measure other quantities.

From a theoretical point of view, the best branch of the solar spectrum to measure is that of the p-p neutrinos. As the most copious branch, its flux is closely correlated with the luminosity of the Sun and does not depend sensitively on the solar model (Bahcall 1989). Therefore a measurement of these neutrinos can provide us with direct information about oscillations: if the signal is significantly lower than expected, then oscillations must be taking place. Whether these oscillations are of the *in vacuo* type or involve the MSW effect will depend on a comparison between the reduction factors for the ^{8}B neutrinos and the *p-p* ones. Unequal factors would point towards the nonadiabatic MSW solution (Rosen and Gelb 1986*a,b*), while equal factors would indicate a large-angle solution (Parke and Walker 1986). In the large-angle case, it may be difficult to distinguish between the MSW and the *in vacuo* solutions; one needs to measure Δm^2 in order to do so.

There is one special case in which we might be able to make this distinction. Should it happen that Δm^2 falls in the range of $10^{-5,6}$, then ^{8}B neutrinos can undergo a significant "day-night" effect (Baltz and Weneser 1987,1988). The original electron neutrinos are converted to muon neutrinos in the Sun and give a reduced signal during the day; at night, however they must travel through the Earth to reach the detector, and for the mass range they are reconverted to electron neutrinos, thereby giving a larger signal. Thus for neutrinos the Sun could shine at night, and in the process it would demonstrate the MSW effect.

So far we have assumed a *p-p* signal much smaller than expected; suppose now that the signal is essentially equal to the expectation. In this case, we must choose between the adiabatic MSW solution and a modification of the solar model as the proper resolution of the ^{37}Cl anomaly, and the way to do this is to measure the spectrum of electron neutrinos arriving at Earth. For the adiabatic MSW case, low-energy neutrinos do not change flavor, but high-energy ones do (Bethe 1986); consequently, the shape of the arriving electron-neutrino spectrum will differ from the known shape of the ^{8}B spectrum. For the modified solar model option, the normalization of the spectrum is reduced, but its shape remains the same. In general, the observation of a spectral shape different from the known one is an unambiguous signal for the MSW effect and neutrino oscillations.

Experiments to detect the *p-p* neutrinos are based on the reaction (Bahcall 1989):

$$\nu_e + {}^{71}\text{Ga} \rightarrow e^- + {}^{71}\text{Ge}. \tag{51}$$

The threshold for the ^{71}Ga reaction is 236 keV and so it can be initiated by *p-p* neutrinos. Interpretation of the experiment is, however, rendered more difficult by the fact that other branches of the solar neutrino spectrum can also initiate the reaction. The signal based on the standard model is expected to be 132 SNU, of which 71 come from *p-p* neutrinos, 34 from ^{7}Be neutrinos, 14 from ^{8}B neutrinos and the remainder from less important branches. If the signal turns out to be well below the 71 SNU associated with the *p-p* neutrinos, then we can be sure that oscillations are taking place—indeed, some nonadiabatic MSW solutions predict a signal of 10 SNU or less. If it is at the level of 70 SNU, we can still be reasonably confident of oscillations because few nonstandard models can eliminate the ^{7}Be neutrinos as effectively as they do the ^{8}B ones. On the other hand, a signal of about 100 SNU could be explained in various ways, and in this case we would need more information, such as the spectrum of electron neutrinos arriving at Earth (Rosen and Gelb 1986a,b).

Solar neutrino scattering by electrons provides another means of distinguishing between the three types of MSW solutions (Bahcall et al. 1987b). Neutrino-electron scattering involves a coherent mixture of charged and neutral currents for electron neutrinos, but only neutral currents for the other flavors. Now the cross section for the combined charged- plus neutral-current reaction is $\sim$7 times larger than that for the purely neutral-current process. Thus, the oscillation from electron flavor to the other flavors leads to a significant reduction in the observed cross section. Interpretation of this reduction in terms of the MSW effect is not as direct as for charged-current reactions, but one can make some general statements regarding the MSW solutions of the existing ^{37}Cl anomaly.

If the observed cross section is less than 1/3 of the one expected in the standard solar model (SSM), then the adiabatic solution is correct. Cross sections between 1/3 and 3/5 of SSM are consistent with large-angle solutions, and values close to 1/2 are also consistent with the nonadiabatic solution (Rosen and Gelb 1989; Rosen 1990). Calculations of the cross section for solar neutrinos as a function of the mass difference and mixing angle are shown in Fig. 8.

X. SUMMARY AND OUTLOOK

Searches for neutrino oscillations have been carried out at reactors and at accelerators covering a wide range of energies, but so far they have yielded no substantial evidence for the phenomenon. The bounds on oscillation parameters obtained from these experiments are (Vuilleumier 1990)

$$\Delta m^2 \leq 0.02 \text{ eV}^2 \text{ for large } \sin^2 2\theta$$
$$\sin^2 2\theta \leq 0.002 \text{ for large } \Delta m^2. \tag{52}$$

Solar neutrino experiments probe much smaller mass differences and a comparable range of mixing angles, and they provide the only indication that oscillations may be occurring.

The fact that the signals in both the ^{37}Cl and Kamiokande II experiments (Davis et al. 1989a; Hirata et al. 1989,1990a,b) are significantly below the predictions of the standard solar model (SSM) demonstrates that the number of ^{8}B neutrinos arriving at Earth as electron-type neutrinos is well below the

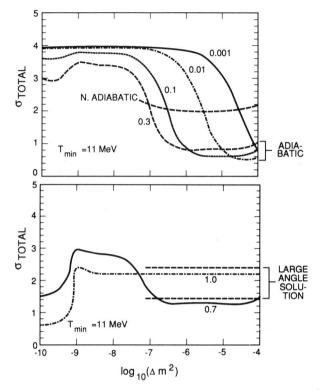

Fig. 8. Cross sections for solar-neutrino electron scattering as a function of Δm^2 for the values of $\sin^2 2\theta$ indicated beside each curve. The experimental cutoff energy for recoil electrons is 11 MeV in this case, but the general shape of the curves is much the same for all cutoff energies; the normalization increases as the cutoff energy decreases. Also shown are the predictions of the three solutions to the chlorine anomaly (figure from Rosen and Gelb 1989).

expectations of the SSM. One explanation for this is that the SSM is wrong and the Sun is actually producing fewer ^{8}B neutrinos. Another is the matter-enhanced MSW oscillation hypothesis.

An important test between these explanations is the ^{71}Ga experiment. Should its signal also turn out to be well below the predictions of the SSM, then we will have to conclude that oscillations are taking place. Within the context of the MSW effect, such a result would support either the nonadiabatic solution or the large-angle one; the choice between them would depend on the degree of suppression of the ^{71}Ga signal, the large-angle solution being able to yield about the same as in ^{37}Cl and the nonadiabatic one much greater.

Preliminary indications at the time of this writing point to the nonadiabatic MSW solution as being the most likely one to solve the solar neutrino problem. The latest value for the ratio of the actual Kamiokande II signal to the SSM prediction is very close to the nonadiabatic prediction based upon the Bahcall and Ulrich (1988) predictions for ^{8}B neutrinos (Hirata et al. 1989,1990a,b):

$$R = 0.46 \pm 0.05 \text{ (statistical)} \pm 0.06 \text{ (systematic)} \tag{53}$$

and preliminary results from the gallium experiment (Gavrin 1991) suggest that the signal is very much smaller than expected. Taken together, these two results support the general notion of matter-enhanced oscillations and tend to exclude both the adiabatic and large-angle solutions. The likely oscillation parameters are in the neighborhood of (Rosen and Gelb 1986a,b; Kolb et al. 1986; Bahcall and Haxton 1989)

$$\Delta m^2 = 10^{-6} \text{eV}^2$$
$$\sin^2 2\theta = 0.02 - 0.05. \tag{54}$$

We look forward to more specific data from both the SAGE and GALLEX experiments on the precise signal in the gallium detector. The next generation of experiments, such as SNO (Chen 1985; Aardsma et al. 1987) and BOREX (Raghavan and Pakvasa 1988) will provide better statistics on charged-current interactions of solar neutrinos, and will also make measurements of neutral-current interactions which are crucial for the flavor oscillation interpretation of all solar neutrino experiments.

Whether the masses of neutrinos will turn out to be much larger than, or comparable to the differences in Eq. (54) might eventually be learned from tritium beta-decay measurements (Robertson 1990) and the search for no-neutrino double beta decay (Avignone and Brodzinski 1988; Doi et al. 1985; Haxton and Stephenson 1984; Rosen 1989).

We have not considered oscillations amongst three flavors (Kuo and Pantaleone 1987; Krastev and Petcov 1988) because no new fundamental physics is introduced and the number of parameters is much larger than in the two-flavor case. Eventually we will have to analyze it in terms of three flavors. We have also not considered "just-so" oscillations in which the mass difference is finely tuned to give an oscillation length of the order of the astronomical unit (Glashow and Krauss 1987; Barger et al. 1990; Acker et al. 1990); such oscillations tend to reduce the gallium signal by a similar factor as in chlorine and predict a semiannual variation in the signal due to the variation in the distance between Earth and Sun.

EQUATION OF STATE AND OPACITY

WERNER DÄPPEN
European Space Research and Technology Centre

JOHN KEADY
Los Alamos National Laboratory

and

FORREST ROGERS
Lawrence Livermore National Laboratory

The three most important physical ingredients of stellar models are the nuclear-energy generation rates (discussed in the chapter by Parker et al.), the equation of state and the opacity. Additionally, the initial chemical composition must be known, either from observation, or by assumption. The equation of state is crucial for the internal structure of the star, and the opacity determines the flow of radiation energy through the star. We begin this chapter with a discussion of the equation of state, not only because the equation of state plays an important role in stellar structure, but also because it is by itself a fundamental part of any opacity calculation. Furthermore, the computation of the equation of state is the simpler problem. There are two reasons for this. First, already at lower densities, where atoms exist (i.e. where many-body effects can be neglected), it suffices for the equation of state to know the energy levels of atoms (and their occupation); for the opacity, however, transition matrix elements between all sorts of atomic states have to be known as well. Second, at higher densities, where many-body effects become important (and where one cannot speak of atoms), there are at least roads to a correct treatment of the equation of state (we will show some of them in this chapter). Extending these techniques to opacity calculations faces the difficulty that again more detailed information about the quantum-mechanical many-body states is required.

I. THE EQUATION OF STATE

For many astrophysical applications, crude recipes for the equation of state are quite adequate (e.g. Saha's equation with ground-state partition functions, together with an imposed total ionization above certain temperatures and densities). More elaborate partition functions have been included in calculations of stellar opacities, for which ionization fractions and occupation numbers are crucial (Cox 1965; Huebner 1986). But even these more accurate formalisms do not always satisfy thermodynamic consistency expressed by the condition that the equation of state and the thermodynamic quantities must be derived from a single thermodynamic potential. The Saha equation (even with more detailed partition functions) cannot predict the physically expected ionization at high pressures (called pressure or density ionization; see below). As a cure, total ionization is often postulated for the conditions in stellar cores. However, the resulting two-zone formalism obviously violates thermodynamic consistency.

Such inconsistencies can cause problems in, for instance, finer helioseismological applications. First, because in the usual formulation of the stellar pulsation equations thermodynamic identities are used several times. Therefore, the equation of state and the thermodynamic quantities have to be at least formally consistent. Second, helioseismology has attained such a level of precision that a better description of the physics of stellar interiors is needed. In Sec. I.F., we discuss two recent examples for the need of an accurate, not just formally correct equation of state. The first is the influence of the hydrogen and helium ionization zones on high-degree p-mode frequencies, the second is the seismological determination of the solar helium abundance by a thermodynamic method, which is based on the signature of the second helium ionization zone on local sound speed.

We first discuss the requirements demanded from an equation of state, before going to some typical equations of state (for extensive reviews of the more recent literature see, e.g., Eliezer et al. (1986), Hummer and Mihalas (1988), Kraeft et al. 1986 and Däppen et al. (1987). We begin with a simple but consistent and astrophysically useful equation of state. Then, on the way to more realistic formalisms, we define the terms of the "chemical picture" and the "physical picture", and show the most recent realizations of each picture. We end Sec. I with applications from solar physics.

A. Requirements of an Equation of State

We wish to emphasize the importance of a consistent equation of state for stellar applications. Without exaggeration, one can say that consistency is even more important than accuracy of the physical description. The reason is that in many conventional applications formal consistency of thermodynamic quantities is used several times during the manipulations of the hydrodynamical equations. Consider, as an example, how the adiabatic temperature gradient enters the derivation of the adiabatic pulsation equations after the substitution of pressure perturbations by density perturbations (which are further transformed using mass conservation). The adiabatic gradient thus inserted must be consistent with the equation of state used in the equation of hydrostatic support.

From these considerations it is clear that the degree of formal accuracy required of an equation of state strongly depends on the chosen purpose. Equilibrium models and adiabatic pulsations need second-order thermodynamic quantities (the terminology refers to derivatives of the free energy or any other equivalent thermodynamic potential). Nonadiabatic pulsation calculations have to go one level deeper: third-order quantities such as derivatives of the adiabatic gradient or specific heat are also required. In a complicated equation of state that comprises many nonideal effects, it is a highly nontrivial matter to achieve accurate third-order quantities.

Formal accuracy is of course not enough, and perhaps the most transparent solar physics application that demonstrates the need for *absolute* accuracy of the physical description is the thermodynamic method to determine the helium abundance of the solar convection zone (see Sec. I.F).

B. Simple Equations of State

Disregarding the even simpler polytropic relations for the equation of state, the simplest prescriptions available consist in mixtures of ideal gases with ionization (and molecular dissociation) reactions. The ionization equilibrium is found by the mass-action law, which in the language of astrophysicists is the Saha equation. For a simple hydrogen plasma (without molecules) it is

$$\frac{N_H}{N_{H^+}N_{e^-}} = \frac{\lambda^3}{V} \sum_i g_i \exp(-E_i/kT). \tag{1}$$

Here, N_H, N_{H^+} and N_{e^-} are the numbers of hydrogen atoms, protons and electrons in the volume V, respectively, $\lambda \equiv h/\sqrt{2\pi m_e kT}$ is the thermal wavelength of electrons, and E_i and g_i are statistical weight and (negative) energy of the internal states of hydrogen atoms, labeled by the index i, respectively. The g_i do not contain the factor 2 due to the spin of the electron, because it cancels in the ratio N_H / N_{e^-}. The divergence of the sum in Eq. (1) has always plagued the statistical mechanics of partially ionized plasmas, but the simplest equations of state avoid the problem by including only the ground state. The Saha equation thus becomes

$$\frac{N_H}{N_{H^+}N_{e^-}} = \frac{\lambda^3}{V} \exp(I/kT) \tag{2}$$

where $I = -E_o$ is the (positive) ionization potential of hydrogen. More realistic internal partition functions with elaborate cutoff procedures have been developed, e.g., for opacity calculations (Cox 1965; Huebner 1986). They allow calculations of ionization fractions and occupation numbers, at least at not too high densities. Nevertheless, the Saha equation suffers from its inability to describe pressure ionization. From physical arguments it is clear that atoms must be ionized at very high densities, but the Saha equation predicts just the contrary, i.e., an unphysical recombination of atoms. In the solar center, for instance, it predicts as much as 30% neutral hydrogen. [To understand this, note that the thermal wavelength λ is of the order of a Bohr radius for a temperature $T = I/k$ ($\sim 1.6 \times 10^5$ K). Also note that the Boltzmann factor exp (I/kT) is essentially 1 at high temperatures (i.e. $kT \gg I$).]

This Saha recombination is clearly at variance with elementary volume considerations, from which one concludes that at densities of 150 g cm^{-3} (of the solar center) there is no room for neutral hydrogen atoms, which have (in tightly packed configurations) densities of the order of 1 g cm^{-3}. Since the aforementioned simple equations of state know nothing about the radius of the hydrogen atom, their predicted recombination is, in this approximation, a legitimate quantum-mechanical effect, reflecting the fact that at higher densities (despite simultaneously higher temperatures) the continuum states of the electrons become less accessible. This is due to a smaller density of electronic states per energy interval at higher densities (think of the problem of electrons in a box, and note how the discrete energy spectrum gets wider-spaced the smaller the box is). The rest is done by the Pauli principle, which causes, at high densities, a piling up to such high continuum energies that the system finally responds with favoring atomic recombination as the lesser evil. Only more realistic equations of state can get rid of this unphysical recombination.

Eggleton, Faulkner and Flannery (1973) developed a simple equation of state (EFF) that is formally consistent and includes an *ad hoc* pressure ionization device that works correctly at least qualitatively. The device is not based on a physical model (e.g., a description of an atom and its surrounding particles), but is imposed by forcing the anticipated result, i.e., full ionization at high densities. In addition, the EFF equation of state incorporates a correct treatment of the partially degenerate electrons according to Fermi-Dirac statistics.

Despite the lack of a physical foundation of its pressure-ionization device, the EFF equation of state is nonetheless useful because of its thermodynamic consistency. Compared to simple prescriptions, like those imposing full ionization above an empirically determined temperature (such procedures can still be found in some of the current programs of stellar evolution), the EFF equation of state is better suited for stellar pulsation applications, and has been, e.g., successfully employed in the solar models of Christensen-Dalsgaard (1982). In addition, the relative simplicity of the EFF formalism allows accurate numerical computation of higher-order thermodynamical derivatives in a way that can be put directly into the programs of stellar evolution, without interpolation of pretabulated results. Furthermore, the manifestation of pressure ionization in stellar structure can be discussed with the help of the parameters that fix the actual location of the EFF pressure-ionization zone. By varying these parameters, we might get some indication about the sensitivity of stellar and solar models on the precise nature of pressure ionization, as has been suggested by Bahcall and Ulrich (1988).

On the negative side of the EFF equation of state, we mention its principal limitations, given by the absence: (1) of a physical mechanism for pressure ionization; (2) of excited states in the bound systems; (3) of a treatment of hydrogen molecules (important for low-mass stars; see Lebreton and Däppen [1988]); and (4) of the Coulomb-pressure correction.

C. Realistic Equations of State

Realistic statistical mechanical models of plasmas can essentially be classified in terms of the "chemical picture" or the "physical picture" (Krasnikov 1977). While in the more conventional chemical picture, bound configurations such as hydrogen atoms are introduced and treated as new and independent species, only *fundamental* particles (electrons and nuclei) appear in the physical picture. In the chemical picture, reactions between the various species occur, and thus the thermodynamic equilibrium must be sought among the stoichiometrically allowed set of concentration variables by means of a maximum entropy (or minimum-free-energy) principle. In contrast, the physical picture has the aesthetic advantage that there is no need for a minimax principle; the question of bound states is dealt with implicitly through the Hamiltonian describing the interaction between the fundamental particles. The different physical assumptions of the two pictures have been the source of considerable difficulties in recent comparisons and interpretations (see, e.g., Rouse 1983; Ebeling et al. 1985; Däppen et al. 1987).

D. The Chemical Picture

1. The free-energy minimization method. Most realistic equations of state that have appeared in the last 30 yr are based on the free-energy minimization method. The free-energy minimization method uses approximate statistical mechanical models (for example, the nonrelativistic electron gas, Debye-Hückel theory for ionic species, hard-core atoms to simulate pressure ionization via configurational terms, quantum mechanical models of atoms in perturbed fields, etc.). From these models, a macroscopic free energy is constructed as a function of temperature T, volume V and the concentrations $N_1, \ldots . N_k$ of the k components of the plasma.

There is an intuitive simplicity in the chemical picture (we usually take the existence of atoms in plasmas for granted, at least at not too high densities), but this simplicity has to be paid for by additional minimization procedures in the multidimensional space of abundances of each species, restricted by the appropriate stoichiometrical relations and by mass and charge conservation. The physical idea behind this minimization is simple: the "internal" degrees of freedom (like ionization degrees) are not adjustable by the experimentor, who can only control "external" parameters (like temperature, density and mass fractions of each chemical element). The thermodynamic equilibrium is then determined as the one configuration (compared to those having different internal parameters) that minimizes the free energy or equivalently, maximizes entropy. Once this minimum is found, the model free energy delivers all thermodynamic quantities in a straightforward way by differentiation.

2. Tables and approximative analytic versions.   Fontaine, Graboske and van Horn (1977) (hereafter FGH) published extensive tabular material based on the free-energy minimization method at lower and intermediate densities, and on Thomas-Fermi theory at high densities. Between the two zones, a thermodynamically consistent interpolation procedure was employed. However, the relative coarseness of the tables rendered them difficult for use in stellar models. In the first place, only second-order quantities are tabulated, and the spacing in temperature and density chosen does not allow numerical differentiation within the tables. Therefore, it is not possible to obtain third-order quantities. In the second place, practical applications of the tables are restricted by the lack of choice in the chemical composition: only pure-hydrogen and pure-helium mixtures were presented, together with a prescription for a volume-weighted interpolation for stellar H-He mixtures.

Interpolations of that kind do, however, violate thermodynamic consistency, because each thermodynamic quantity has a different nonlinear dependence on the chemical composition. Däppen (1980) has therefore proposed a simplified model, based on the confined-atom model (hereafter CAM), which can produce the results of these tables approximately, and which is simple enough to be programed within stellar evolution codes. In the confined-atom model, the Coulomb potential outside a sphere of radius R is replaced by an infinitely high potential wall (this is equivalent to a zero boundary condition of the wave function at R). The value of R is chosen as a function of the volume available for a given bound species. For $R < \infty$, all bound-state energies are lifted from their unperturbed values; with decreasing R, the higher states are gradually spilled over into the continuum. While physically, this is certainly not a very realistic procedure, it has some formal advantages (like being able to ionize "from the cold", i.e., without "seed electrons" or other starters (usually assumed available from alkali-type metals).

Compared with the simple equations of state discussed above, both FGH and CAM represent major physical improvements: realistic pressure-ionization devices, Coulomb pressure and polynomial fits for excited states.

3. The MHD Equation of State.   Mihalas, Hummer and Däppen (Hummer and Mihalas 1988; Mihalas et al. 1988a; Däppen et al. 1988a) have recently developed a new treatment of the equation of state (hereafter MHD), which is part of the "Opacity Project" (see Sec. II; Seaton 1987). While the basic concept of the MHD equation of state is conventional (it is based on the free energy minimization method), it is characterized by detailed internal partition functions of a large number of atomic, ionic and molecular species. Full thermodynamic consistency is assured by analytical expressions of the free energy and its first- and second-order derivatives. This not only allows an efficient Newton-Raphson minimization, but, in addition, the ensuing thermodynamic quantities are of analytical precision and can therefore be differentiated once more, this time numerically. Reliable third-order thermodynamic quantities are thus calculated.

Apart from formal thermodynamic consistency, the MHD equation of state also achieves some degree of statistical mechanical consistency (Hummer and Mihalas 1988). Statistical mechanical consistency refers to the more subtle requirement that each time a bound configuration (like an atom) is modified by its surroundings in the plasma, then the relevant force has to be described in the physical description of the surroundings as well. To be more specific consider, as an example, the EFF equation of state, which is thermodynamically consistent, but not statistical mechanically consistent, because atoms are pressure ionized by an *ad hoc* mechanism which does not have its counterpart in the free particles.

In the MHD equation of state, perturbations by charged and neutral particles are taken into account. These perturbations are described by an occupation probability w_i for each energy level in the sum over internal states in Eq. (1), which is generalized to the multicomponent plasma and renormalized so that the energies of the excited states are measured with respect to the ground state. The resulting weighted internal partition function Z_s^{internal} is (*is* labeling the state i of species s)

$$Z_s^{\text{internal}} = \sum_i w_{is} g_{is} \exp\left[-\frac{E_{is} - E_1 s}{kT}\right].$$ (3)

The coefficients w_{is} take into account charged and neutral surrounding particles. In physical terms, w_{is} gives the fraction of all particles of species s that can exist in state i with an electron bound to the atom or ion, and $1 - w_{is}$ gives the fraction of those that are so heavily perturbed by nearby neighbors that the state is effectively destroyed. Perturbations by neutral particles are based on an excluded volume treatment and perturbations by charges are calculated from a fit to a quantum-mechanical Stark-ionization theory. Hummer and Mihalas's (1988) choice has been

$$\ln w_{is} = -\left(\frac{4\pi}{3V}\right)\left\{\sum_\nu N_\nu (r_{is} + r_{1\nu})^3 + 16\left[\frac{(Z_s + 1)e^2}{\chi_{is} k_{is}^{1/2}}\right]^3 \sum_{\alpha \neq e} N_\alpha Z_\alpha^{3/2}\right\}.$$ (4)

Here, the index ν runs over neutral particles, the index α runs over charged ions (except electrons), r_{is} is the radius assigned to a particle in state i of species s, χ_{is} is the (positive) binding energy of such a particle, k_{is} is a quantum-mechanical correction, and Z_s is the net charge of a particle of species s. Note that $\ln w_{is} \simeq n^6$ for large principal quantum numbers n (of state i), and hence provides a (density-dependent) cutoff for Z_{is}^{internal}. A first comparison of these occupation probabilities with experiment has been made: Däppen et al. (1987) have used them to simulate the radiation from a precision plasma experiment (Wiese et al. 1972). Though the agreement is good, it serves only as a necessary condition (see Däppen et al. 1987; Ruzdjak and Vuinovic 1977; Dyachkov et al. 1988).

The MHD equation of state is a computational heavy weight. Unless drastically simplified to a small number of atomic and ionic species, it has to be used in the form of tables. A tape with first results is already available (Mihalas et al. 1988b). Furthermore, Lebreton and Däppen (1988) have developed programs that can automatically create tables that are centered around the temperatures and densities of stellar interiors. These table-creating programs can also handle the changes in chemical composition during the main-sequence evolution of stars. As we have mentioned above, the smoothness of the MHD formalism allows tabulation of third-order thermodynamic quantities. Furthermore, by tabulating the results from EFF in the same way as those of MHD, and by comparing the models that use the interpolated EFF results with the models that call EFF directly (Christensen-Dalsgaard et al. 1988c), we can control the accuracy of the interpolation process and adjust the necessary fineness of the tables.

Before leaving the MHD equation of state we wish to add one general remark. Though the MHD formalism was developed for stellar envelopes, there is no problem in applying it to interiors. Though one is leaving the domain for which MHD was originally conceived, the physical ingredients are, even at these high densities, quite the same as the ones conventionally used in models of stellar interiors. Only theories that treat higher-order correlations in plasmas seriously would distinctly go beyond our assumptions. Such theories still wait to be applied in the context of stellar interiors, and, in order to extend them to the low-density regime, special care will have to be given to overall consistency.

E. The Physical Picture

While the chemical picture has proven to be practical for constructing astrophysical equations of state, from a fundamental point of view, its major drawback is the necessity to assert what effect the plasma has on the bound energy states of ions and atoms. The strength of the physical picture is that it views the partially ionized plasma in terms of its fundamental constituents, such that the plasma effects on bound states arise naturally.

1. The Livermore Equation of State.   In order to determine the ramifications of this physics difference, the Livermore group has chosen to pursue the physical picture in their new equation of state and opacity effort (Rogers 1986; Iglesias et al. 1987). The Livermore group uses a many-body activity expansion of the grand canonical partition function (Rogers 1981). To explain the advantages of this approach for partially ionized plasmas, it is instructive to discuss first the activity expansion for gaseous hydrogen. The interactions in this case are all short ranged and the pressure is determined from a self-consistent solution of the equations (Hill 1960)

$$\frac{p}{kT} = z + z^2 b_2 + z^3 b_3 + \ldots \tag{5}$$

$$\rho = \frac{z}{kT}\left(\frac{\partial p}{\partial z}\right) \tag{6}$$

where $z = \lambda^{-3}\exp(\mu/kT)$ is the activity, λ is the thermal (de Broglie) wavelength of electrons, defined in Eq. (1), μ is the chemical potential and T is the temperature. The b_n are cluster coefficients such that b_2 includes all two-particle states, b_3 includes all three-particle states, etc. The second cluster coefficient for hydrogen includes the formation of H_2 molecules as well as scattering states in the $^1\Sigma_g$ potential. It also includes scattering states in the $^3\Sigma_u$ potential and all excited electronic state potentials. The third cluster coefficient includes H_3 bound states, $H - H_2$ and $H - H - H$ scattering states. Equation (5) demonstrates that the equation of state for associating gases can be obtained without an explicit knowledge of the occupation numbers of associate pairs.

For low-density gases, the bound state contributions to the b_n can be important at low temperature while the scattering contributions are too small to matter. Strict application of Eq. (5) would contain a large amount of unimportant information which is very hard to calculate. Consequently, it is necessary to reorganize Eq. (5) such that the bound-state terms from each b_n are treated as being of the same order as the ideal-gas term, i.e., of order z. Terms of order z in the physical picture are roughly equivalent to what in the chemical picture is called the Saha equation. Similar reorganization of terms involving scattering from composite particles is also required. Assuming that H_2 molecules are the only bound complex to form, Eq. (5) becomes

$$\frac{p}{kT} = z_H + z_{H_2} + z_H^2 b_2^s + 2 z_H z_{H_2} b_3^* + z_{H_2} b_4^* + \ldots \tag{7}$$

and Eq. 6 is replaced with the conditions

$$\rho_H = \frac{z_H}{kT}\left(\frac{\partial p}{\partial z_H}\right) \qquad \rho_{H_2} = \frac{z_{H2}}{kT}\left(\frac{\partial p}{\partial z_{H_2}}\right) \tag{8}$$

where $z_{H_2} = z_H^2 b_2^b$ is the activity for molecules and the superscripts b and s refer to bound and scattering parts of b_2, b_3^* is the part of b_3 involving scattering of H from H_2, and b_4^* is the part of b_4 involving scattering of H_2 from H_2.

In the case of partially ionized plasmas, very similar steps are required except that now even Eq. (5) must involve at least two species (nuclei and electrons) to assure electrical neutrality. In addition, due to the long range of the Coulomb potential each, of the b_n is composed of a number of divergent terms, some of which are fictitious and some of which are real.

An example of real divergence is afforded by the classical ring diagrams occurring in each b_n (Mayer 1950). They are individually divergent but the many-body correlations introduced by summing over the b_n yields the well-known Debye-Hückel correction. This type of divergence occurs even for an electron gas in a neutralizing background for which there are no bound states. Although the original equations involve only entire powers in the activity (see Eq. 7), as a result of many-body Coulomb correlations, the Debye-Hückel term is $z^{3/2}$ in the activity (also in density).

An important example of a fictitious divergence is that associated with the atomic partition function. This divergence is fictitious in the sense that the bound-state part of b_2 is divergent, but the scattering-state part, which is omitted in the Saha approach, has a compensating divergence. Consequently, the total b_2 does not contain a divergence of this type (Ebeling et al. 1976; Rogers 1977). A major advantage of the physical picture is that it incorporates this compensation at the outset. As a result, the Boltzmann sum appearing in the atomic (ionic) free energy is replaced with the so-called Planck-Larkin partition function (PLPF), given by (Ebeling et al. 1976)

$$\text{PLPF} = \sum_{nl} (2l + 1)\left[\exp\left(-\frac{E_{nl}}{kT}\right) - 1 + \frac{E_{nl}}{kT}\right]. \qquad (9)$$

The PLPF is convergent without additional cutoff criteria as are required in the chemical picture.

The power of the activity-expansion method, arising in the physical picture, lies in the fact that it produces expressions for thermodynamic quantities that systematically take account of density corrections without the introduction of models or cutoffs. This approach clearly shows that if the leading correction to thermodynamic quantities is assumed to be the Debye-Hückel term, then the internal sum over bound states that must be used to calculate ideal free energies for ions and atoms is the PLPF (Rogers 1986). This is because many-body effects on high-lying states cause most of the contribution normally included in a multilevel Saha equation to be smeared out in energy such that it acts like a continuum. This contribution is included in the many-body sum that produces the Debye-Hückel correction; i.e., adding the Debye-Hückel correction to a multilevel Saha equation double-counts the effect of high-lying states on thermodynamic properties.

An important feature of the PLPF is that it picks out the states that are highly occupied. In a perturbation sense it corresponds to that part of the occupation of the allowed states that is large enough to be treated as a new variable when the plasma reorganization equivalent to Eq. (7) is carried out. Some residual effects of the discreteness of high-lying states appear in correction terms of order z^2 that are the next higher corrections beyond Debye-Hückel. This additional occupation of high-lying states is important for frequency-dependent opacities and must be added to the PLPF occupation numbers when opacities are calculated (Rogers 1986).

The chemical picture being pursued by the Opacity Project and the physical picture being pursued by the Livermore group are very different attempts to remove some of the arbitrariness of previous calculations of astrophysical equations of state. Quantitative agreement between these approaches would greatly increase confidence in the adequacy of the equation of state. A comparison of the results is just now being initiated.

F. Implications for Solar Physics

The influence of the equation of state for solar structure and oscillations has been discussed in many places (see, e.g., Ulrich 1982; Ulrich and Rhodes 1984; Shibahashi et al. 1983; Noels et al. 1984; Stix and Knölker 1987; Bahcall and Ulrich 1988; Christensen-Dalsgaard et al. 1988). Here, we would merely like to give two typical examples to illustrate the importance of the equation of state.

1. High-degree p-*modes.*   The most convincing demonstration so far of the sensitivity of an observable astrophysical quantity on details of the equation of state is the influence of hydrogen and helium partition functions on high-degree p-mode frequencies. Observed oscillation frequencies have been compared with theoretical results, based in one case on the EFF and in the other case on the MHD equation of state. The results of this comparison are discussed in Christensen-Dalsgaard et al. (1988c) and also in Christensen-Dalsgaard (1988c). To demonstrate the link between the equation of state and observable p-mode frequencies, we will here give the physical explanation for the different behavior of EFF and MHD. First, however, we will summarize the situation.

It is convenient to plot the difference between observed and theoretically predicted frequencies of p-modes (with order n and angular degree l) against their frequency v_{nl} (see Fig. 1). Actually, *scaled* frequency differences are shown, i.e., $Q_{nl}v_{nl}$, where the scale factor Q_{nl} compensates the l dependence of the frequency differences v_{nl} (being about 10 times higher for modes with $l = 1000$ than for radial modes due to smaller inertia (see Christensen-Dalsgaard 1987; Christensen-Dalsgaard et al. 1988b). In conventional standard models, represented for our purposes by the EFF calculation, the frequency differences δv_{nl} between observation and theory (Fig. 1a) depend both on the frequency v_{nl} and the angular degree l. The frequency dependence (which is similar for all degrees l) is thought to be due to an inadequate (because adiabatic) treatment of the pulsations in the uppermost part of the convection zone and the solar atmosphere. This is clear from the nonadiabatic treatments of Cox et al. (1989), Delache and Fossat (1988) and Stein et al. (1988). But there is also a distinct dependence on the angular degree l which suggests inadequacies of the model in deeper layers, coming from the position of the lower turning point. Apart from these two features, one notes a substantial frequency error at low frequencies, where one would expect no problems arising from the usual assumptions (like adiabatic oscillations). This low-frequency discrepancy also suggests inadequacies of the model in the interior.

The same comparison between observations and a model using the MHD equation of state (Fig. 1b) shows: (1) little systematic variation with angular degree l; (2) a significant reduction of the low-frequency discrepancy; and (3) no change in the overall frequency dependence of the frequency differences. Since the first two effects are related to the interior, it appears that much of the inadequacy of the theory of the interior has been eliminated by using the MHD equation of state. Furthermore, the fact that the overall frequency dependence is unaffected by changing the equation of state supports the interpretation that it is caused by errors in the treatment of the surface layers.

Figure 1c and d again compare EFF and MHD models with observations, but the opacities of the models are changed; the Huebner et al. (1977) (LAAOL) opacities are used instead of Cox and Tabor's (1976) opacities (see the chapter by Christensen-Dalsgaard et al.). Now, the overall trend in the frequency dependence of the frequency differences has virtually disappeared. This is somewhat surprising, since nonadiabatic effects are still neglected. It could well be that another inadequacy of the model just happens to compensate the error of the adiabatic treatment.

Finally, Fig. 1e shows the frequency differences between two *model* frequencies. The models are identical except that one uses the EFF and the other the MHD equation of state. The models use old opacities, but in this comparison between two models little would change if the new opacities were used in both. The absence of observational noise and systematic errors in this figure allows us to demonstrate most clearly the influence of the equation of state on p-mode frequencies.

Let us now come to the physical explanation. First, we note that the effect of the MHD equation of state on oscillation frequencies can be understood in terms of the local sound speed (Christensen-Dalsgaard et al. 1988b). Figure 2 shows that the relative sound-speed difference $\delta c/c$ ($c^2 = p\Gamma_1/p$) alternates between ± 1 and 2% in the subphotospheric layers (with $0.95 < r/R < 1.0$). Second, it can be shown that the sound-speed difference is qualitatively dominated by the influence of Γ_1, thus roughly $\delta c/c = 0.5 \, \delta\Gamma_1/\Gamma_1$. This relatively simple relation is by no means evident, since changing the

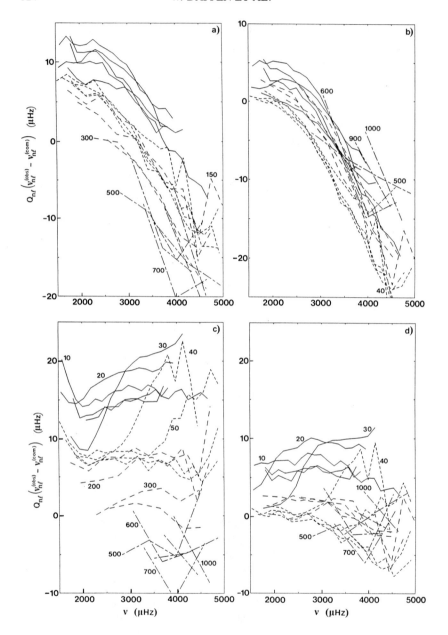

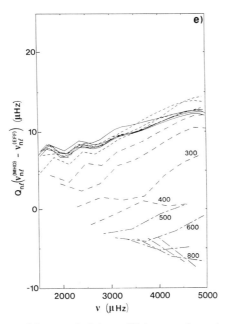

Fig. 1. Scaled frequency differences Q_{nl} ($\nu_{nl}^{obs} - \nu_{nl}^{EFF}$) between observed and computed frequencies for the EFF model (a), and the corresponding scaled differences Q_{nl} ($\nu_{nl}^{obs} - \nu_{nl}^{MHD}$) for the MHD model (b), for selected values of l. The scaling Q_{nl} (according to mode inertia) is explained in the text. Points corresponding to a given value of l have been connected as follows: l = 0–30 (solid line); l = 40–100 (short-dashed line); l = 150–400 (long-dashed line); and l = 500–1000 (alternate short and long dashes). A few selected values of l are shown in the figures. Plots (c) and (d) are as (a) and (b), respectively, except that the newer (LAAOL) opacities are used (see text). To demonstrate the effect of the equation of state clearly, plot (e) shows the difference Q_{nl} ($\nu_{nl}^{MHD} - \nu_{nl}^{EFF}$) between the frequencies of the two models (with old opacities) that are identical except for their equation of state. (This set of figures provided by J. Christensen-Dalsgaard.)

equation of state also causes changing the equilibrium structure. Nevertheless, Christensen-Dalsgaard et al. (1988*b*) show that the influence of Γ_1 is indeed the most important one. Third, Γ_1 is connected to the internal partition functions. In this part of the Sun ($0.95 < r/R < 1.0$), the principal effect of the MHD equation of state is caused by the larger than usual internal partition functions of H, He and He$^+$, which increase the statistical weight of these species. Another effect is caused by Coulomb pressure, included in MHD, but not in EFF.

    Why are the internal partition functions of the MHD equation of state significantly larger? It is true that normally the weight of the ground state (with [negative] energy E_o) dominates the one of each excited state (with [negative] energy E_j) by a factor of $\exp[-(E_j - E_o)/kT]$. For the temperatures encountered in the H and He ionization zones, this domination can easily be

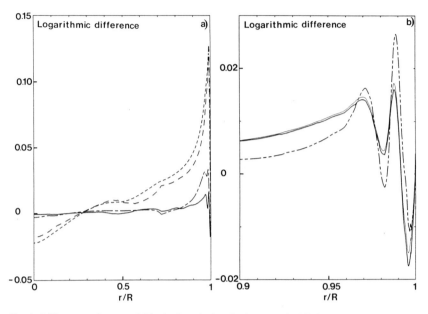

Fig. 2. Differences of state variables in the solar interior between the MHD and the EFF models, in the sense MHD − EFF, at fixed radius r. Illustrated are $\delta \ln p$ (short-dashed line), $\delta \ln \rho$ (long-dashed line), $\delta \ln c$ (heavy solid line), and $\delta \ln \Gamma_1$ (alternate short and long dashes). An expanded section of plot a is shown in plot b, for the region close to the solar surface. The thin solid line in plot b shows $\delta \ln c$ resulting from an approximate calculation, which supposes that $\delta \ln c$ is solely caused by $\delta \ln \Gamma_1$. The closeness of the thin solid line to the heavy solid line confirms this approximation (see Christensen-Dalsgaard et al. 1988b).

by a factor of 10^2 to 10^4. However, the sheer number of excited states explicitly dealt with in MHD can lead to a sizeable correction of the total statistical weight of the H and He atoms or the He$^+$ ion. More specifically, this increase of the statistical weight can become as high as 10 to 30%. This increase is then translated into a decrease of the H, He and He$^+$ ionization fractions, as is directly seen from the Saha equation, Eq. (1). The decrease is of the same order as the increase of the statistical weights. MHD thus pushes deeper down the location of the ionization zones. All this, however, does not yet explain the change of sign in $\delta c/c$ as a function of depth (Fig. 2). The change of sign is explained by considering Γ_1. We know that in a partially ionized plasma Γ_1 is lowered from its ideal value of 5/3 (at full ionization or recombination); the minimum values are about 1.20 and 1.55 in the H-ionization and the second He-ionization zone, respectively. Consequently, the MHD equation of state also pushes down the zones where this lowering of Γ_1 takes place. The $\delta \Gamma_1$ that results from this translation of the Γ_1 "dip" has the opposite sign on the upper flank than on the lower flank of the Γ_1 dip, thus explaining the alternation of positive and negative $\delta c/c$.

2. Seismological Helium Abundance Determination.   It has been argued (Gough 1984c; Däppen and Gough 1984,1986) that the helium abundance in the solar convection zone could be determined directly from frequencies of high-degree solar oscillations by measuring the depression in the bulk modulus of the gas brought about by the second ionization stage of helium. In the adiabatically stratified region of the convection zone (which contains the He II ionization zone) the first adiabatic exponent $\Gamma_1 = (\partial \ln p / \partial \ln \rho)_s$ and its thermodynamic derivatives are related to the sound speed c according to

$$\Theta \equiv \frac{1 - \Gamma_{1,\rho} - \Gamma_1}{1 - \Gamma_{1,c^2}} = \frac{r^2}{Gm} \frac{dc^2}{dr} \equiv W \qquad (10)$$

where r is a radial coordinate, $\Gamma_{1,\rho}$ and Γ_{1,c^2} are the partial logarithmic derivatives of Γ_1 with respect to ρ and c^2 at constant c^2 and ρ, respectively, m is the mass in the sphere r = constant, and G is the gravitational constant. The quantity Θ is a function solely of thermodynamic state variables; it takes the value $-2/3$ for a perfect monoatomic gas with $\Gamma_1 = 5/3$, and it exceeds that value when the gas is in a state of partial ionization. The quantity W is determined entirely by a knowledge of $c^2(r)$, the equations of hydrostatic support and a boundary condition relating p and ρ in the photosphere. Thus, a seismological determination of $c^2(r)$ would permit one to infer $\Theta(r)$, and in particular its deviation from $-2/3$ in the He II ionization zone. This deviation (the "helium hump") is shown in Fig. 3 for a typical solar model. Evidently the magnitude of the deviation depends on the amount of helium present.

An attempt to quantify this intuitive idea and to develop a procedure for estimating the helium abundance Y was discussed by Däppen and Gough

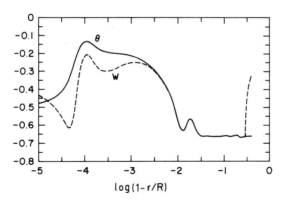

Fig. 3. The functions W and Θ defined by Eq. (10) plotted against $\log_{10}(1 - r/R)$ for a model of the solar envelope with a mixing-length parameter $\alpha = 1.9$, $Y = 0.22$ and a heavy element abundance $Z = 0.02$ (figure from Däppen and Gough 1986).

(1986). Sound speed was estimated from oscillation frequencies using an asymptotic procedure discussed by Gough (1986a) (cf. chapter by Gough et al.). Unfortunately, by the time c^2 had been differentiated to obtain W, the result was so noisy that progress was temporarily halted. Däppen, et al. (1988b) have resumed the study by replacing the previously used *absolute* inversion for sound speed by the *differential* method discussed by Christensen-Dalsgaard et al. (1988b). They have found, by using artificial data, that the method is potentially useful. However, a reliable estimate of the helium abundance of the Sun has not yet been possible, partly because the real solar data are apparently too noisy, and partly because of the uncertainty in the equation of state (see Däppen 1987). Table I summarizes the results obtained so far.

The method to determine the solar helium abundance using the helium hump in the sound-speed derivative is no more than a sophisticated calibration of envelope models, carried out by matching oscillation frequencies. Therefore, to some degree, it rests upon the validity of the theory used to calculate those models and their eigenfrequencies. Nevertheless, by comparing only that combination of frequencies that we believe is dependent predominantly on the region in which helium is undergoing second ionization, we hope to be able to eliminate the possibly larger discrepancies between the absolute frequencies of any of the theoretical models and those of the Sun that arise from errors in the theory that have little or no bearing on the helium abundance. Thus, the calibration should be more robust than the early naive comparison of $k - \omega$ diagrams (Gough 1982). Moreover, it is certainly independent of all of the uncertain assumptions of the theory of stellar evolution that are required for determining the structure of the radiative interior, and upon which the calibration with low-degree modes (Christensen-Dalsgaard

TABLE I

Helium Abundance Y and Mixing-Length Parameter α of 4 Test Models

Model[a]	Grid[b]	Actual		Inferred		
		Y	α	Y	α	E_{min}[c]
1	1	0.235	2.00	0.233	2.02	1.2×10^{-3}
2	1	0.254	1.60	0.265	1.79	1.5×10^{-2}
3	1	0.254	1.60	0.366	3.22	3.8×10^{-2}
4	2	0.237	2.25	0.233	2.21	3.7×10^{-3}

[a] Actual values were inferred by fitting the function W (defined by Eq. 10) in a grid of functions W belonging to 9 other different models ($\alpha = 1.4, 1.9, 2.4$; $Y = 0.23, 0.26, 0.29$ in all combinations).
[b] The models of grid 1 differ from those of grid 2 by their equation of state: The former use the Saha equation with all species in their ground state, the latter use the MHD equations of state.
[c] A gain function E was introduced to measure the quality of the fit; E_{min} is the minimum value found for the inferred parameters ($Y\alpha$).

and Gough 1980,1981; Gabriel et al. 1982; Guenther and Sarajedini 1988) depends.

In the calibration tests to date, it seems that the errors are of a greater magnitude than the spread in values one finds in the literature from the calibration of evolved solar models. That is not to say that the helium-hump method will be less reliable. Indeed, its strength will lie in the fact that it rests on fewer unconfirmed assumptions of the theory of stellar structure, and therefore is less susceptible to serious systematic flaws. Evidently the method will require accurate knowledge of the equation of state, and for this reason Däppen et al. (1988*b*) have generated a grid of models. Our hope is that we shall refine the calibration procedure to the point where the equation of state is the most uncertain step. Then at least we shall know exactly on what that calibration depends. Moreover, since we generate a function (such as W), and Y is but a number, the data contain much more information than we set out to seek. Therefore there is hope to reverse the process to make a serious seismic investigation of the equation of state.

II. THE OPACITY

Most modern stellar models use the Los Alamos Astrophysical Opacity Library (Huebner et al. 1977); a detailed description of the underlying physics is found in Huebner (1986). The Los Alamos Astrophysical Opacity Library (hereafter LAAOL) can be used to calculate opacities for an arbitrary stellar mixture. The method is based on earlier work by Cox (1965) that resulted in the publication of extensive tables (see, e.g., Cox and Stewart 1969,1970*a,b;* Cox and Tabor 1976), which are still widely used. The only other source of stellar opacities that has been published is the table by Carson et al. (1981). However, there are two current efforts to compute stellar opacities, one being the so-called Opacity Project, described in Seaton (1987), and the other pursued at Livermore (see Sec. II.C below). The Opacity Project is an international collaborative effort of which the aforementioned MHD equation of state (Sec. I.D.3) is a part. It restricts itself to the case of stellar *envelopes,* where density is sufficiently low that the concept of atoms makes sense. Therefore, the Opacity Project is mainly an atomic-physics effort: its aim is to compute up-to-date atomic data (energy levels, oscillator strengths, photoionization cross sections and parameters for pressure broadening of spectral lines), and use these in opacities; plasma effects on occupation numbers are secondary.

We begin this opacity part with a typical illustration of opacity computations. Then we discuss recent Los Alamos and Livermore efforts, as well as a first preliminary comparison between their results. Finally, we mention possible consequences of some of the remaining opacity uncertainties for solar physics.

A. Methods of Opacity Calculations

In the solar interior the photon mean free path is sufficiently small so that radiation transport occurs via diffusion. Use of the Rosseland mean opacity

$$\kappa_R^{-1} = \left(\frac{\pi}{4\sigma T^3}\right) \int_0^\infty \frac{1}{\kappa_\nu} \frac{\partial (T)_\nu}{\partial T} d\nu \tag{11}$$

guarantees the correct integrated flux transport when the *diffusion approximation* is valid.

1. Atomic Models. The atomic data central to the calculation of the opacity can be generated in a number of fashions. These data are in general influenced by the plasma interactions which modify the interatomic potential and wavefunctions, and hence energy levels, level widths, transition moments, spectral line shapes, etc. In the most general formulation, the system is the atom or ion and the plasma. This many-body problem is a very formidable one. In the lowest density regime, many of the effects are perturbative, so that for increasing density a useful approach is to utilize an isolated atom model with plasma "corrections" to the energy levels, level widths, etc. It is important to note that these plasma perturbations effectively limit the number of bound states which would otherwise exist.

Some remarks related to the overall characterization of the energy levels arising from atomic configurations are in order. In the *central field approximation,* the pair of indices nl completely specifies the electron configuration, while the inclusion of electrostatic effects between the electrons and the spin-orbit interaction splits the energy levels into a number of sublevels. When the electrostatic interaction dominates over the spin-orbit, we have the familiar case of LS coupling. In the opposite extreme, we have the case of jj coupling. Although the spin-orbit interaction increases in importance with increasing atomic charge Z, the occurrence of pure jj coupling is relatively rare, with the intermediate case (intermediate coupling) being more common. Highly excited states in noble gas atoms often experience intermediate coupling. However, because of its historical importance in the analysis of light-element spectra, many analyses assume LS coupling.

The atomic models may be divided into two broad categories; the average atom model and detailed configuration-accounting models. In the average atom model, a mean field approximation is used to calculate the (fictitious) fractional occupancy of the one-electron energy levels. This provides an immediate gross indication of plasma ionization and excitation. An application of first-order perturbation theory allows a deconvolution to the detailed ionic species and excited configurations. Historically, this method has been used at high temperatures ($kT > 100$ eV). The method of detailed configuration accounting uses either experimental data or *ab-initio* data (or a mixture of both) for the individual energy levels for all ions. A more extended discussion of these approaches can be found in Cox (1965) and Huebner (1986).

Contributions of the opacity arise from several distinct physical processes; bound-bound transitions, bound-free (photoionization), free-free (*bremsstrahlung*) and scattering. For large enough densities (not achieved through most of the Sun), electron conduction can also contribute. The following subsections only review the principal methods and results. More details can be found in the references cited, and, e.g., in Peach (1965, 1967a,b,c, 1970).

2. Bound-Bound Transitions.   The spectrum arising from the totality of all spectral line absorption has a profound effect on the radiative opacity. Since the Rosseland mean opacity is a harmonic mean, it is especially sensitive to the windows that arise from the distribution of transition arrays (the totality of transitions between two configurations), and to windows that arise within the transition arrays. While the distribution of transition arrays and line patterns within the arrays is determined by the atomic model, the line-profile functions are also crucial in determining the overall intensity distribution.

In addition to the natural broadening, arising from the finite lifetime of a radiating atom, which results in a Lorentz profile

$$\phi \nu_2 = \frac{\left(\dfrac{\Gamma}{2\pi}\right)^2}{(\nu - \nu_0)^2 + \left(\dfrac{\Gamma}{2\pi}\right)^2} \tag{12}$$

and Doppler broadening, resulting from a thermal ensemble of radiating atoms, which has a gaussian distribution

$$\phi = \sqrt{\frac{\pi m}{2kT}} \exp\left(-\frac{m\nu^2}{2kT}\right) \tag{13}$$

there is also collision broadening. The nature of this broadening depends explicitly on the nature of the target and perturber.

In the hot ionized gas of the solar interior, electron impact broadening is an important contributor. Classical impact theories of Lindholm (1946) and Foley (1946) result in shifted Lorentz profiles, the evaluation of whose damping constants depends on the nature of the collision partners. The neglect of internal excitation during the collision process (the so-called adiabatic approximation) underestimates the damping constants and overestimates the frequency shift (Griem et al. 1962). Electron impact broadening of non-hydrogenic ions (quadratic Stark effect) is usually treated in the impact approximation.

For a well-ionized gas, broadening by ionic perturbations as well as electron impacts can be important, and a Stark profile (linear Stark effect for hydrogen and hydrogenic ions) must be folded with a Lorentz profile. A comprehensive discussion of line broadening in the quantum mechanical context can be found in Griem (1974).

To the extent that these broadening mechanisms are physically uncorrelated, the natural and electron impact Lorentzians can be convoluted to yield a Lorentzian whose damping width is the sum of the individual damping constants. A subsequent convolution with the gaussian yields a Voigt profile

$$H(a,v) = \frac{a}{\sqrt{\pi}} \int_{-\infty}^{\infty} \frac{\exp(-y^2)}{(v-y)^2 + a^2} dy. \tag{14}$$

In an absorption spectrum containing millions of lines, the evaluation of the Voigt profile is tremendously time consuming, and it pays to seek out an efficient numerical algorithm (see Hui et al. 1978). For lines with a damping parameter a less than some typical value (for instance $a \leq 0.1$), resorting to the use of a direct table can be quite efficient (R. Kurucz, personal communication).

3. Bound-free absorption.   In a partially ionized gas, photoionization is an important component of the absorption. The earliest result for photoionization is the semi-classical result of Kramers

$$\sigma_{bf} = \frac{2^6 \pi^4 (Z^*)^4 m e^{10}}{3\sqrt{3} \ ch^6 n^5 v^3} g_{bf} \ (cm^2) \tag{15}$$

where Z^* is an effective charge, which has a characteristic frequency dependence. Quantum mechanical corrections (i.e., Gaunt factors) that introduce the multiplicative correction g_{bf} have been calculated by Karzas and Latter (1958,1961; also see Peach 1965,1967a,b,c,1970). These have been extensively used for photoionization out of excited ionic states. In some cases, reasonable cross sections for ionic ground states can be obtained from scaling one-electron self-consistent-field (SCF) calculations of the neutral atom (Huebner et al. 1978).

Atomic models, such as the Hartree-Fock or close-coupling methods now being used to generate energy levels and oscillator strengths, also permit the direct calculation of the photoionization spectra, allowing for the interaction of the bound states with the continua. When properly done, these calculations ensure continuity in the absorption strength across the Rydberg series into the continuum.

4. Free-free absorption. A free electron in the field of an atom or ion can absorb a photon (inverse *bremsstrahlung*) or emit one (*bremsstrahlung*). A frequently used form is

$$\sigma_{\text{ff}} = \frac{2^4 \pi^2}{3\sqrt{3}} \frac{(Z^*)^2 e^6}{hc(2\pi m)^{3/2}} \frac{N_e}{\sqrt{kT}} \frac{g_{\text{ff}}}{\nu^3} \equiv \sigma_k g_{\text{ff}} \tag{16}$$

where σ_k is the Kramers cross section and g_{ff} is the free-free Gaunt factor (Green 1960; Nakagawa et al. 1987).

Modern atomic-structures calculations permit a direct calculation of this cross section. Excitation during the scattering process can introduce structures into the cross sections not accounted for in the simpler treatment. In contrast to the photoionization process, these analogous effects in ionic *bremsstrahlung* have received scant attention. For a partially ionized plasma, the free-free process seldom is the dominant component of the Rosseland mean opacity, although it will dominate the monochromatic absorption at long enough wavelengths. At temperatures above about 1 keV in astrophysical mixtures, where the gas is dense and nearly completely ionized, the free-free processes with hydrogen ions is important and the above treatment should be accurate.

5. Scattering. The Thomson formula

$$\sigma_T = \frac{8\pi e^4}{3m^2 c^4} \tag{17}$$

describes the behavior of photons scattering off both bound and free electrons. For conditions appropriate to the center of the Sun, high-energy effects (the Compton regime) can be incorporated via the Klein-Nishina formula (Jackson 1962), which represents the scattering process in the electron rest frame. For the high temperatures encountered, electron motion in the observer's (laboratory) frame should not be neglected (Sampson 1959). Electron correlation effects (see Diesendorf and Ninham [1969], Watson [1969] and Huebner [1986, especially Eq. 85]) must also be considered; they are important when the product of the photon wavenumber and Debye radius is approximately equal to or less than unity.

For a plasma of pure hydrogen, at a temperature $kT = 1.5$ keV and electron density of 4.5×10^{25} cm^{-3}, Fig. 4 shows the high-energy correction factor (ratio of effective Compton to Thomson cross section) due to high-temperature electrons (curve labeled thermal motion), and the correction factor due to electron correlation (curve labeled correlation). Note that they both reduce the scattering cross section. Incorporating these effects into a calculation of a Rosseland mean opacity leads to about a 20% reduction in the

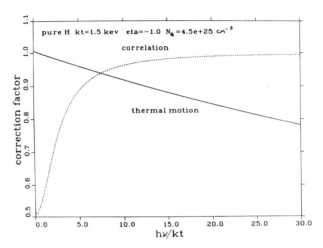

Fig. 4. Correction factors to Thomson scattering due to collective behavior of electrons (correlation), and to high-energy effects (Klein-Nishina formula) of high-temperature electrons (thermal motion), for conditions at the solar center. The Thomson cross section must be multiplied by these correction factors.

scattering part of opacity, with most of the effect arising from electron correlation. For a discussion of further nuances in the scattering process see Boercker (1987).

B. Recent Los Alamos Efforts

Recent efforts at Los Alamos have centered around the generation and application of internally consistent atomic data sets to the calculation of group mean-absorption coefficients and opacities (as opposed to spectrum-synthesis calculations).

The Cowan atomic structures codes (cf. Cowan 1981) have been restructured (Abdallah et al. 1988) to allow large-scale production runs, and easy and quick retrieval of the voluminous data generated.

To make the data handling more tractable, some of the concepts of quantum defect theory have been applied (Clark and Merts 1987) to model the behavior of the (fine structure averaged) LS term energies along a Rydberg sequence. This relies upon the fact that electrons sufficiently far from the core experiences an approximately Coulombic field, with the existence of the core introducing a shift. The form of the energy fit is

$$E = \frac{(Z - N + 1)^2}{\left[n - (c_o + c_1 n^{-1} + c_2 n^{-2})\right]^2} \ (\text{Ry}) \qquad (18)$$

where Z is the atomic charge, N is the number of bound electrons and n is the principal quantum number. A linear least-squares routine fits the *LS* energies along a Rydberg series. It might be expected that the worst case would be low-lying levels in neutral atoms. Figure 5 shows a comparison of some energy-level fits for neutral oxygen with National Bureau of Standards data (Moore 1976). The worst case is an error of several percent. In general (particularly for ions), the fits are an order of magnitude better. So far, a similar procedure applied to the dipole matrix elements is also successful, especially at tracking effects such as Cooper minima (where the transition matrix element changes sign), which occasionally appear along the Rydberg series.

As an example for oxygen, approximately 4600 fit coefficients and angular factors allow the generation of 70,000 lines (with the fine structure averaged). For equation-of-state purposes in a dilute plasma, the approximately 2300 *LS* energy-level fit coefficients allow the calculation of some 150,000 energy levels (from extrapolation up the Rydberg sequences). The atomic structures codes also provide the necessary photoionization cross sections. The calculations of the new opacities using these data are presently being performed by N. H. Magee.

C. Recent Livermore Efforts

In response to the need for improved astrophysical opacities described in Sec. I.C., the Livermore group has developed a new opacity code called

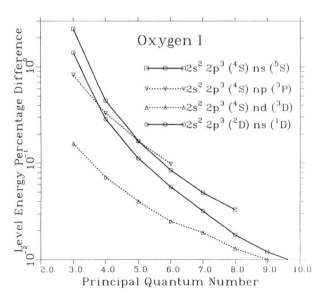

Fig. 5. Percentage difference between level energies fitted according to Equation (18) and the corresponding NBS table values for 4 Rydberg series of neutral oxygen.

OPAL. They use the physics developed in the calculation of the equation of state as theoretical bases for this code. This introduces important differences with existing calculations with regard to occupation numbers and plasma (density) effects on transition rates.

The OPAL code uses the method of detailed configuration accounting and requires detailed atomic data. This data is calculated on-line using prefitted parametric potentials (Rogers et al. 1988) of the form

$$V_{ei} = -\frac{2(Z - \nu)}{r} - \frac{2}{r} \sum_{n=1}^{n^*} N_n \exp(-\alpha_n r) \tag{19}$$

where Z is the atomic number, ν is the number of core electrons, N_n is the occupation of shell n and n^* is the principal quantum number of the largest occupied shell. The parameters α_n are obtained by iteratively solving the Dirac equation to match experimental valence electron-binding energies. The potential V_{ei} is also used to calculate all excited states having a single electron promoted out of the valence shell. Potentials for excited parent configurations are obtained by scaling the α_n from the solutions of the Dirac equation for the potential V_{ei}.

Using the parametric potentials, as just described, photoionization and bound-bound absorption cross sections are calculated for every subshell in every electron configuration of the various ion stages. For high-lying bound states and for scattering states, plasma effects on the wavefunctions are obtained by exponential screening of the long-range component of V_{ei}. Oscillator strengths are generally calculated in the LS coupling scheme but can also be calculated in intermediate coupling. Transitions to states with $n > 10$ are not explicitly included, but instead are approximately treated by extending the photoionization edge to lower energies. In comparisons with available experimental data, the parametric potential oscillator strengths and line energies are comparable in accuracy to self-consistent field results.

At present, the line shapes are assumed to be Voigt profiles with the gaussian width given by Doppler and the Lorentz width by a hydrogenic-scaled electron-impact formula identical to that in the LAAOL (Huebner 1986). However, further work is necessary on line shapes. For astrophysical conditions, ions are usually highly stripped and many lines experience linear Stark broadening that is familiar in hydrogen-like systems for which Voigt profiles are invalid. In addition, for hydrogen-rich mixtures at low densities, the protons may contribute significantly to the impact widths.

A small set of results from the OPAL code have been reported (Iglesias et al. 1987) for a seven-element mixture of H, He, C, O, Mg, Si and Fe with $X = 0.7$, $Y = 0.28$ and $Z = 0.02$. At a density of 10^{-5} g cm^{-3} and a temperature of 6.9×10^5 K, the OPAL Rosseland mean is 1% higher than the corresponding LAAOL value. At the same density and a temperature of

2.3×10^5 K, the OPAL Rosseland mean is 37% above LAAOL. This increase is mainly due to improved atomic physics and is attributed to a detailed treatment of 3–3 transitions. Recent more extensive calculations with a correspondingly detailed treatment of 3–4 and 3–5 transitions does not significantly change the results at 6.9×10^5 K but at the lower temperature the Rosseland mean has increased to 2.2 times the value of LAAOL.

D. Opacity Uncertainties in Solar and Stellar Models

A detailed description of opacity-related uncertainties of the standard solar model can be found in the chapter by Christensen-Dalsgaard et al. There is further material in the reviews Bahcall et al. (1982) and Bahcall and Ulrich (1988) and in the references therein. Among the recent articles on the solar standard model that deal with opacity-related changes are those by Lebreton and Maeder (1986), Turck-Chièze et al. (1988) and Cox et al. (1989). Here, we discuss a few issues that have been evolving in the last 20 yr. First, we mention that in the bulk of the convection zone, opacity plays no role; the temperature gradient is uniquely fixed by thermodynamics, because the real temperature gradient is very close to the adiabatic temperature gradient. This is not the case close to the photosphere, where convection is inefficient and the interplay between radiation and convective motion is important, and it is not the case beneath the convection zone.

An important manifestation of the progress in opacity calculations, from Cox and Stewart (1969,1970a,b) to the LAAOL, is the increase of the calibrated value of the Sun's initial helium abundance, which rose from a typical 0.25 in mass fraction (Bahcall et al. 1982; Christensen-Dalsgaard 1982) to 0.27 (Bahcall and Ulrich 1988; Turck-Chièze et al. 1988), 0.28 (Lebreton and Maeder 1986) and 0.29 (Cox et al. 1989). This is a consequence of the considerable higher LAAOL opacities in the solar center.

Note that uncertainties in the equation of state also affect the calibrated helium abundance by as much as 5% (Lebreton and Däppen 1988; Cox et al. 1989). This uncertainty mainly comes from the Coulomb pressure correction in the solar center (which is compensated with a change in the helium abundance to satisfy the constraint of the solar luminosity). It should not be confused with the uncertainty of the seismological helium-abundance determination (see Sec. I.D.) due to unknown details of the hydrogen and helium partition functions. Speaking of helium, we also should mention the role of the isotope ^{3}He. As far as its atomic structure is concerned, it is very similar to the usual ^{4}He (the small difference is due to a slightly different reduced mass of the atom). Therefore ^{3}He and ^{4}He atoms are practically indistinguishable in equation-of-state and opacity calculations. Nevertheless, the *number* of ^{3}He nuclei per unit mass is larger than that of ^{4}He (by a factor of 4/3), which directly influences the equation of state. In formalisms that do not explicitly treat the two isotopes, one obtains a sufficiently accurate description by computing, for given mass fractions of ^{3}He and ^{4}He, the appropriate number of ^{3}He and ^{4}He nuclei and using their sum as the usual ^{4}He abundance (by number).

Another opacity issue that has bothered the constructors of standard models is the correction due to the collective behavior of partially degenerate electrons (for details see Sec. II.A). Diesendorf (1970) found that for conditions near the center of the Sun, this effect reduced the Thomson cross section by about 35%. Huebner (1986) pointed out that Diesendorf's correction to the opacity was overestimated. Boercker (1987) performed a detailed calculation and found a correction of about 28%, which lies between the values of Diesendorf (1970) and Huebner (1986). In standard models of the Sun, this correction to the opacity can influence the predicted ^{8}B neutrino flux by up to 10% (Bahcall and Ulrich 1988; Turck-Chièze et al. 1988).

Furthermore, it was suggested by Christensen-Dalsgaard et al. (1985b), and by Korzennik and Ulrich (1989) that *larger* opacities (about 15 to 20% above the LAAOL values) below the solar convection zone (at temperatures between 2 and 7 $\times$ 10^6 K) would explain some of the discrepancies between theoretical and observed p-mode frequencies. Cox et al. (1989) confirm this explicitly by deliberately adjusting the opacity in this range in order to obtain good agreement for the p-mode frequencies. They justify this adjustment by the fact that such an increase is within the theoretical uncertainty of the opacity at these temperatures and densities (A.L. Merts, personal communication).

Finally, we mention an opacity-related issue concerning δ Scuti and Cepheid stars. For an astrophysical mixture of chemical elements with $\log T = 5.36$ and $\rho = 10^{-5}$ g cm^{-3}, recent Livermore calculations have resulted in Rosseland mean-opacity values that are a factor 2.2 larger than those of the LAAOL. This difference is mostly attributed to inclusion of $\Delta n = 0$ transitions ignored in the older Los Alamos calculations (see Sec. II.C).

A numerical experiment by Andreasen (1988) finds that arbitrarily increasing the opacity, by a factor of 2.5, in the domain $10^{-8} < \rho < 10^{-3}$ g cm^{-3} and $1.5 \times 10^5 < T < 8 \times 10^5$ K, substantially resolves the period ratio discrepancies for both double-mode δ Scutis and Cepheids. However, at $\rho = 10^{-5}$ g cm^{-3} and $\log T = 5.84$, the Livermore and LAAOL results are essentially identical. The incorporation of the $\Delta n = 0$ transitions into the new Los Alamos calculations introduces an effective oscillator strength redistribution, decreasing the effective strength of the previously used $\Delta n \neq 0$ transitions. To the extent that the $\Delta n \neq 0$ transitions losing appreciable strength to the $\Delta n = 0$ transitions occur under the peak of the Rosseland weighting function, the new Los Alamos opacities could conceivably decrease somewhat in places relative to the older Opacity Library results.

A definitive answer to the question of how improved opacities modify solar and stellar structure simulations awaits the recalculation, over the entire relevant domain of temperature and density, of realistic mixes involving the twenty or so astrophysically important elements.

Acknowledgments. We are grateful to A. Cox for his constant help at all stages of development of this chapter. We thank J. Christensen-Dalsgaard, D. Mihalas, and S. Turck-Chièze for stimulating discussions and critical comments on the manuscript. We are grateful to J. Christensen-Dalsgaard also for Fig. 1 a–e and their underlying calculations of various solar models and oscillation frequencies.

PRE-MAIN-SEQUENCE EVOLUTION

S. SOFIA, S. KAWALER, R. LARSON

and

M. PINSONNEAULT
Yale University

The pre-main-sequence (PMS) stage of a star's evolution begins when a sub-condensation forms in a dense interstellar cloud, and it ends when the stellar core becomes sufficiently hot to begin nuclear processing of the hydrogen fuel. The PMS stage is very short lived in terms of evolution time scales. In the earlier stages, when the gravitational forces overcome internal pressure, the object is essentially in free fall. This dynamical stage comes to an end when hydrostatic equilibrium is re-established with the possible involvement of magnetic and rotational forces. At this point, evolution slows down, and it is regulated by the ability to shed internal energy and/or by the pace of dissipation of angular momentum (perhaps in an accretion disk). Meanwhile the structure of the object changes drastically from a radiative to a fully (or nearly fully) convective object. The details of the evolution process are still very sketchy, especially in the dynamic stage. The main uncertainty in the hydrodynamic stage involves the possible role of accretion. In this chapter, we describe the various possible scenarios envisioned in the evolution, state the uncertainties involved, and refer to existing or potential observational data which may help unravel the details of the PMS evolution. Because the degree of uncertainty decreases for the latest stages of the PMS, we emphasize them in this chapter. In particular, we follow recent evolution calculations that include rotation, and conclude that this important feature of the PMS can best be tested in terms of the chemical abundance of trace elements in the solar system, on the Sun, and on solar analogs of different ages.

I. INTRODUCTION

The brief but interesting pre-main-sequence (PMS) phase of stellar evolution is the link between interstellar clouds and young stars. Consequently, a complete history of the Sun from its conception to the present day requires us to understand the PMS phase. In fact, the PMS is usually ignored in modeling stellar evolution. Two reasons have been invoked to justify this procedure. First, the physical conditions (densities, temperatures, sources and sinks of energy) prevailing in the early phases of the PMS domain are very different from their stellar descendents; standard stellar evolution codes cannot properly handle these early phases (i.e., dynamical collapse) without extensive modifications. Second, the main effects of the PMS phase on subsequent evolution are on the rotational state and on the abundances of trace species; both are at least second-order effects, and as such are not usually treated in (or important for) most modeling exercises. The usual context of pre-main-sequence evolution studies is in star formation, where the tools and emphasis are very different from the detailed structure studies of later evolution.

In the case of the Sun, the above excuses for neglecting PMS evolution are not justifiable on several accounts. To begin with, the Sun is surrounded by the only known planetary system, whose formation must be understood in the context of the pre-main-sequence stage of the Sun. Second, there are solar phenomena, such as its magnetic activity, which are of fundamental importance to us even though they are second-order effects from the stellar viewpoint. For example, it is currently believed that the solar magnetic activity is produced by a dynamo process. A dynamo requires a differentially rotating magnetized plasma to operate. Although the current rotation state of the Sun may eventually be derived from oscillation data, it is still necessary to find out what the rotation state was like in the past (and what it will be in the future) to forecast variation of the dynamo process. This requires us to treat rotation as an initial value problem, rooted in the pre-main-sequence stage. Finally, because of the Sun's nearness to us, and our vital interest in its behavior, it has been subjected to much closer scrutiny than any other star. For example, we can determine current trace element abundances by observing the Sun, *and* the abundances of the forming Sun by measuring the abundances of pristine comets. To explain these data requires a comprehensive analysis of the nuclear burning that took place in the young Sun as well as internal mixing (by convection, rotation, etc.). Again, we must know the rotational history of the Sun from its earliest infancy to perform such an analysis.

A significant question to address is how early the PMS calculation must be started. Clearly, since the Sun's ultimate origin was some phase of the interstellar medium, one might consider the possibility of going all the way back to that point. However, as stated earlier, this procedure is difficult, and probably unnecessary. The difficulty arises because the physical processes, the time scales and the parameter range of nearly all variables are extremely different between a young star and a cloud; no tools exist that are equally capable and efficient to deal with both situations. The effort may be unnecessary, because of the properties of a phase of evolution just prior to the main-sequence stage reached by most, if not all, of the protostar scenarios. Here, the protostellar object is fully (or nearly fully) convective, and vigorous convection is very efficient at obliterating the details of the previous evolution of the object. Only global properties such as average chemical composition and total angular momentum are carried over. Obviously, a state like this is the appropriate starting point for an initial value problem. However, we need to understand the earlier evolution in order to know precisely how good an approximation such a fully convective model is. This requires that we deal with the very early phases of star formation even though they will not be part of the subsequent initial value problem. Here we can use tools appropriate to the diffuse regime rather than to the stellar one.

The very early dynamical phase of protostellar evolution is described in Sec. II. Section III describes the PMS evolution in the hydrostatic contraction phase by means of detailed evolutionary models. In all sections, reference is made to observations to guide the selection of appropriate model parameters and to test the consequences.

II. DYNAMICAL PHASE

A. Initial Conditions

The extensive evidence now available leaves little doubt that stars of solar mass are currently forming in the dense cores of dark molecular clouds such as those in Taurus and Auriga (see, e.g., Beichman et al. 1986; Myers et al. 1987; Myers 1987; Fuller and Myers 1987). These cloud cores typically have masses of the order of 1 $M_\odot$ and diameters of order 0.1 pc as defined by the half-maximum contours of maps of the NH_3 molecule, and they are sometimes closely associated with infrared sources that are believed to be very young stellar objects. Some of these infrared sources coincide with visible T Tauri stars, which are variable stars with emission lines and strong surface activity that represent the earliest visible stages of the pre-main-sequence evolution of solar-type stars (Herbig 1962; Bertout 1989). In the observed cloud cores, pressure and gravity are roughly in balance and conditions correspond approximately to those required for cloud fragmentation (Larson 1985). Turbulent motions (Myers 1983) and rotational motions (Heyer 1988) are both too small to contribute importantly to the support of these cores or to affect strongly the early stages of their collapse. Magnetic fields may play a more important role, but there is evidence that the initial magnetic field may already have largely decoupled from the gas in the observed star-forming cores (Heyer 1988; Myers and Goodman 1988).

The available evidence thus suggests that the simple idealization of a nearly spherical, quasi-equilibrium configuration not dominated by turbulence, rotation, or magnetic fields may be a reasonable zero-order representation of the initial state of a collapsing solar-mass protostar. Magnetic fields and turbulence may continue to decay as the collapse of such a protostar proceeds; on the other hand, rotation must become progressively more important, and the effect of rotation on the later stages of the collapse can almost certainly not be neglected. The detailed outcome of collapse with rotation is very sensitive to the initial distribution of mass and angular momentum in the collapsing cloud. The simplest assumption is that the initial configuration has uniform density and uniform rotation (see, e.g., Larson 1969a,1972a), but it is also possible that before the onset of dynamical collapse, a cloud core develops a centrally condensed structure similar to that of a singular isothermal sphere with $\rho \propto r^{-2}$ (Shu 1977; Shu et al. 1987). In the former case, the outcome of the collapse may sometimes be fragmentation into a binary or multiple system of stars (Larson 1972a,1978; Bodenheimer et al. 1980; Boss 1988); however, with a centrally condensed initial configuration, fragmentation is strongly suppressed (Boss 1987) and the innermost part of the cloud may collapse almost spherically to form a single central star, around which a disk later forms (Terebey et al. 1984). Since fragmentation into a binary or multiple stellar system evidently did not occur during the formation of the Sun and most of the mass ended up in a single star, the Sun probably formed from a cloud core that was somewhat centrally condensed before beginning to collapse dynamically.

B. Spherical Collapse

The only case in which collapse has been followed all the way through to the formation of a hydrostatic pre-main-sequence star is that of spherically symmetric collapse. As will be seen, even this very idealized case yields predictions that are in good agreement with the observed radii and luminosities of the youngest T Tauri stars. It may therefore be possible to neglect the effects of rotation in an initial crude treatment of the formation of a solar-type star, and to consider them later on as a perturbation to the spherical case.

The essential features of spherical collapse from initial conditions like those seen in typical dark cloud cores are now well established. The physical conditions expected in collapsing protostars were first systematically analyzed by Hayashi and Nakano (1965), and fully dynamical calculations starting from appropriate low-density initial conditions have been made by Larson (1969a,1972b,1973), Appenzeller and Tscharnuter (1975) and Winkler and Newman (1980a,b; see also Newman and Winkler 1980). The results show that after an initial phase during which the density distribution becomes increasingly sharply peaked at the center, a small stellar core forms and proceeds to grow in mass by accreting infalling gas; the later development of this "embryo star" is then almost independent of the details of how it forms. Stahler et al. (1980a,b,1981) made use of this fact to study in greater detail the growth of such a forming star by accretion from an infalling envelope, taking pains to treat carefully the physics of the accretion shock at the surface of the star. The results are quite similar to those of the dynamical calculations cited above except for a somewhat larger final stellar radius, as will be discussed below.

The early stages of the collapse are approximately isothermal and lead to a density distribution of the form $\rho \propto r^{-2}$, even if the initial configuration is not centrally condensed (Bodenheimer and Sweigart 1968; Larson 1969a,1973). Collapse is first halted in a small central region when the optical depth in this region becomes large and the temperature begins to rise, but dissociation of hydrogen molecules soon causes the collapse to resume and it is not permanently halted at the center until a very small core region becomes ionized and attains nearly stellar density. Thereafter the evolution of the system consists essentially of the growth in mass of this stellar core as matter continues to fall onto it. The mass inflow rate remains roughly constant with time while the infall velocity steadily increases, producing an ever stronger accretion shock at the surface of the star. The part of the infalling envelope just outside the shock is optically thin except for a brief initial period, and this allows most of the kinetic energy thermalized in the shock to be radiated away during the accretion process; eventually all of this energy escapes from the infalling envelope as infrared radiation (Larson 1969b). As a result of this continuing loss of energy by radiation, the star finally formed at the end of the accretion process has a modest radius of only several solar radii which is much smaller than the maximum radius that is possible in principle for a hydrostatic pre-main-sequence object. At this stage, the star is essentially a conventional pre-main-sequence star of the type first studied by Hayashi (1961) and Hayashi et al. (1962), and it falls on the lower part of the nearly vertical "Hayashi track" in the HR diagram appropriate for fully convective stars still contracting toward the main sequence, before they become mostly radiative and follow nearly horizontal "Henyey" tracks (Bodenheimer 1972; Stahler 1988b).

The final protostellar radius (or initial stellar radius) determines how far above the main sequence a newly formed star will appear in the HR diagram when it first becomes visible, and this provides the main point of contact with observations of T Tauri stars. Both Larson (1972b) and Winkler and Newman (1980a) predicted a relatively small initial radius of about 2.1 $R_\odot$ for a star of one solar mass, while Stahler et al. (1980a) predicted an initial stellar radius of about 4.7 $R_\odot$, which is in better agreement with observations of the youngest T Tauri stars. Stahler (1988a) has shown that the principal reason for this difference is the inclusion of deuterium burning in the detailed evolutionary calculations of Stahler et al. but not in the dynamical calculations of Larson or of Winkler and Newman. Deuterium burning acts as a "thermostat" that keeps the central temperature close to 10^6K during the later stages of the accretion process, and the result is that a newly formed star of one solar mass is predicted to have a radius of approximately 4 to 5 $R_\odot$ for a significant range of accretion rates. For other masses, newly formed stars are predicted to appear along a well-defined "birthline" in the HR diagram which coincides well with the upper envelope of the observed distribution of T Tauri stars (Stahler 1983). Acceptable agreement with the observed properties of T Tauri stars is obtained as long as the accretion rate is within about a factor of 3 or 4 of the theoretically expected value of 10^{-5} $M_\odot yr^{-1}$ (Stahler 1988a), or equivalently as long as the time scale for building a solar-mass star by accretion is not greatly different from 10^5 yr, which is comparable to the expected free-fall time of about 2×10^5 yr in typical dark cloud cores.

If a newly formed star of one solar mass has an initial radius of at least 4 $R_\odot$ as a result of deuterium burning, it will be essentially fully convective when accretion ceases and it first appears on the "birthline." However, the possibility cannot presently be excluded that a small central region of very low entropy surviving from early stages of the collapse, as was found in the calculations of Larson and of Winkler and Newman, never becomes convective at any time, even though the associated temperature inversion disappears before the star reaches the main sequence.

C. Effects of Rotation

A recent study of cloud cores in Taurus by Heyer (1988) finds a median core velocity gradient, presumed to be due to rotation, of only about 0.4 km $s^{-1}pc^{-1}$. If a typical cloud core with this angular velocity of rotation and with a radius of 0.05 pc collapses with conservation of angular momentum to form a rotationally supported configuration such as a disk, the radius of the disk will be about 50 AU. Many newly formed stars appear to possess remnant circumstellar disks with sizes of this order (see, e.g., Strom et al. 1988). Our planetary system, which is believed to be a fossilized protostellar disk, also has a size of this order, and so may have originated from the collapse of a typical slowly rotating cloud core. In this case, it is likely that the Sun actually acquired part of its final mass from the pre-planetary disk or "solar nebula" that initially surrounded it, rather than by direct infall. The fraction of the Sun's mass that may have been acquired from a disk cannot be predicted without a detailed knowledge of the initial conditions, but strong dynamical coupling between the Sun and the solar nebula, possibly via accretion, is at least suggested by the near alignment between the angular momentum vector of the Sun and that of the planetary system.

The evolution of a forming star growing by accretion from a disk has been studied by Mercer-Smith et al. (1984), using a schematic treatment which assumes that a fixed fraction of the rotational energy of the accreted material is radiated from the surface of the star during the accretion process. For an assumed accretion rate of $10^{-5} M_\odot yr^{-1}$, the results are very similar to those previously obtained by Stahler et al. (1980a) for spherical accretion, except for a slightly smaller final radius of about $3.4 R_\odot$. The stellar radius is smaller in this case because half of the binding energy of the incoming material has already been radiated from the disk before the material reaches the star, so that both the stellar surface luminosity and the entropy of the newly accreted material are smaller than in the spherical case. These results suggest that as long as disk accretion can supply matter to a growing embryo star at a rate similar to spherical accretion, the essential features of the evolution of the star should not be greatly different from the spherical case.

The rate at which matter can be accreted from a disk depends on the rate at which angular momentum can be redistributed within or removed from the disk. At present, the transfer of angular momentum in accretion disks remains a poorly understood subject. The possible mechanisms that may act to transfer angular momentum in protostellar disks have been reviewed by Larson (1989). Neither turbulence nor magnetic fields can clearly be justified as being of major importance, but both gravitational torques (Larson 1984) and acoustic-wave motions may play a significant role in transporting angular momentum at different stages in the evolution of such a disk. If, at an early stage, the disk has a mass comparable to that of the central star, then gravitational torques associated with trailing spiral density enhancements of any origin can transport angular momentum outward rapidly enough to permit accretion on a time scale of 10^5 yr or less (Larson 1989), as required to produce a star with a radius large enough to agree with the observations of T Tauri stars. Later, when the disk mass becomes small, or in a disk that never attains a very large mass, wave motions produced by effects such as tidal perturbations by Jupiter-like giant planets may help to drive accretion and/or dispersal of the disk on a time scale of 10^6 yr (Larson 1989); this could account for the typical disk lifetime of order 10^6 yr that has been inferred from the observations. However, the rate of residual disk accretion in the latter case is probably too small to have any further important consequences for the evolution of the central star.

If disk accretion is an important part of the star-formation process, it will tend to spin up a forming star to the maximum possible rotation rate, unless angular momentum is somehow removed from the star as fast as it is added. The great majority of T Tauri stars are actually observed to rotate much more slowly than the maximum possible rate (Vogel and Kuhi 1981; Bouvier et al. 1986; Hartmann et al. 1986; Hartmann and Stauffer 1989), indicating either that rotation plays no significant role in their formation or, more likely, that rotational braking acts very effectively during even the earliest stages of stellar evolution, and probably during the accretion process itself. Possibly the braking mechanism is related to the vigorous bipolar outflows that are frequently observed in association with the youngest stellar objects; magnetic fields probably play a role in both the braking and the bipolar flows, but the actual mechanism of these outflows is not yet understood (Shu et al. 1987).

It may be significant that among the few pre-main-sequence objects to show clear evidence for rapid rotation are the FU Orionis variables, which have exhibited dramatic and prolonged flare ups in luminosity (Herbig 1977, 1989). It has been widely suspected that the FU Orionis flare ups are in some way caused by an episode of rapid accretion of material from a disk. The flare up may result from the accretional heating and expansion of the outer layers of the central pre-main-sequence star (Larson 1983), or the enhanced luminosity may originate in the disk itself as a result of strong viscous heating (Kenyon and Hartmann 1988). In either case, a substantial amount of mass must be accreted, so that some spin up of the central star would be predicted. The observed rapid rotation of the FU Ori objects, if they are in fact stellar in nature rather than disks, might then actually constitute evidence for episodic spin up by accretion. It has also been suggested that some of the spectroscopic peculiarities of the extreme T Tauri stars can be explained by disk accretion (Bertout et al. 1988; Hartmann and Kenyon 1988), but even if this is the case, the inferred accretion rates are only of order $10^{-7} M_\odot$ yr^{-1} or less and are probably too small in most cases to alter importantly either the mass or the angular momentum of the accreting star. Thus, apart from possible sporadic FU Ori accretion events, it seems safe to assume that even the extreme T Tauri stars are essentially fully formed stars no longer gaining significant mass or angular momentum from their surroundings.

III. THE HYDROSTATIC PHASE

The classical theory of pre-main-sequence evolution developed by such workers as Hayashi (1961), Hayashi et al. (1962), Iben (1965) and Bodenheimer (1972) used as its starting point the hydrostatic phase. This phase, which temporally follows the dynamic phase, is characterized by the development of sufficient internal pressure support to produce a much slower contraction than the earlier phase; a hydrostatic protostellar model begins in a fully convective, contracting (not collapsing) configuration. The early work on the hydrostatic phase did not include rotation. Current work includes rotation and takes into account the recent advances in our understanding of the dynamic phase (reviewed in Sec. II). To describe the properties of this phase, we will use the calculations recently carried out in Pinsonneault, Kawaler, Sofia and Demarque (1989b—hereafter PKSD). We begin by following a typical hydrostatic model of the proto-Sun from an initial fully convective state to the main sequence; because rotation is a perturbation on the structure, the global features of the models (such as the effective temperature and luminosity as a function of time) will be essentially the same as found in the nonrotating case.

Where is the natural starting point for a hydrostatic model? Hayashi first noted that for a star of a given mass and composition there exists a minimum temperature at which the star becomes fully convective; models with cooler temperatures are unstable against collapse. The Hayashi limit is nearly a vertical line in the HR diagram; a protostar will arrive at the Hayashi limit with some initial luminosity and will remain close to that limit as it evolves down in luminosity towards the main sequence. This phase of evolution is referred to as the Hayashi track.

The chosen initial luminosity in the classical theory is essentially arbitrary. Fortunately, the time scale of evolution decreases rapidly as the radius of the initial hydrostatic model is increased. Therefore the properties of the later evolution are minimally influenced as long as the initial model is luminous enough still to be fully convective. From a high initial luminosity a protosolar model will remain fully convective for approximately 1.5 Myr. The time scale of evolution prior to this age is too short to alter the surface lithium abundance, or to allow significant angular momentum loss by a magnetic stellar wind. Convection quickly obliterates the details of the earlier internal angular momentum distribution by enforcing some relationship between angular velocity and radius (probably solid-body rotation); the internal angular-momentum distribution is thus insensitive to the prior evolution. We therefore begin our detailed calculation, (i.e., the initial value) of hydrostatic pre-main-sequence evolution at an age of 1.5 Myr. At this point, the protosolar model has a radius of 2.0 $R_\odot$ and a central temperature of 3.8 $\times$ 10^6K.

As the model evolves towards the main sequence, it undergoes drastic structural changes. The evolution in an HR diagram is shown in Fig. 1; the evolution of the central temperature and depth of the surface convection zone as a function of time is shown in Fig. 2. The model initially contracts rapidly at almost constant surface temperature. This contraction causes a rapid decrease in luminosity. The energy released in the contraction is accompanied by an increase in the central temperature which lowers the opacity. Once the opacity has become low enough, a radiative core is established. This core grows rapidly as the luminosity drops. By an age of 10 Myr, the surface convection zone has retreated to a depth of approximately 40% by mass and the central temperature has risen to 6.5 $\times$ 10^6 K. The radius has decreased to 1.1 $R_\odot$.

As the radiative core grows, the properties of the model undergo a transition from a convection-dominated to a radiation-dominated regime. The star departs from the convective Hayashi track and evolves on the radiative "Henyey" track towards the main sequence. Because the opacity in radiative regions drops as the temperature increases, the model can maintain almost constant luminosity as it contracts; the surface temperature therefore increases. The global contraction of the star and the retreat of the surface convection

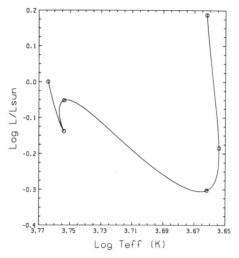

Fig. 1. Evolutionary track of a rotating solar model, X = 0.736, Y = 0.244, Z = 0.020, ratio of mixing length to pressure scale height = 1.37. Circles indicate the position of the model at ages of 10^6, 3×10^6, 10^7, 3×10^7, 5×10^7, and 4.7×10^9 yr.

zone both continue at a reduced rate on the Henyey track. By an age of 30 Myr, the star is essentially on the main sequence and hydrogen burning is well established in the core. There is a transient convective core as the initial relative ^{12}C, ^{13}C and ^{14}N abundances are re-adjusted by CNO cycle burning to the appropriate equilibrium values. The central temperature is not high enough to process efficiently ^{16}O, so it remains essentially unburned until much later on the main sequence. The model reaches its minimum radius (0.89 $R_\odot$ at an age of 50 Myr, and has a luminosity of 0.73 $L_\odot$. We can define the main sequence evolution as beginning at this point. Its central temperature is 13×10^6 K and the depth of the surface convection zone is 3% by mass; by comparison, the current solar model central temperature is 15×10^6 K and the surface convection zone depth is slightly less than 2% by mass.

Dynamical models should be used as a guide to determining the initial luminosity (or equivalently radius) of the hydrostatic phase. They give starting radii for solar-mass models ranging from 2.1 $R_\odot$ (Larson 1972b; Winkler and Newman 1980a) without deuterium burning to 4.7 $R_\odot$ (Stahler et al. 1980a) when deuterium burning is included (Sec. II.B). Our starting radius is comparable to those of Larson or Winkler and Newman. A larger initial radius does not significantly affect the properties of later evolution.

The evolution could also be affected by the possible existence of a small core which is never convective and the uncertain role of late accretion. As noted in PKSD, the small size of such a core coupled with rotational instabilities (and the absence of any effective mechanism for impeding them in

S. SOFIA ET AL.

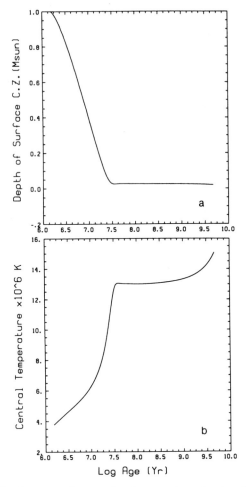

Fig. 2. (a) Depth of surface convection zone (in solar masses) as a function of time for the model shown in Fig. 1. (b) Central temperature (in 10⁶K) as a function of time for the model shown in Fig. 1.

this phase of evolution) ensure that the angular momentum distribution in the core will not depart greatly from that of a fully convective model. The potential role of late accretion is more difficult to gauge; it is possible, for example, that the energy input from the accreted material will hold up the evolution near the deuterium-burning "birthline" until further accretion is negligible. In this case, the "initial" model will have some error in its age; however, most other properties (including some assumed initial angular momentum) will be similar to a traditional hydrostatic model. If stars evolve down the Hayashi track and simultaneously accrete a substantial quantity of matter, however, the evolutionary effects will be considerable. These possibilities can be tested observationally by studying patterns of light-element depletion (Pinsonneault et al. 1989*a*). For the purpose of this review, we will assume that late accretion is negligible.

The above scenario is relatively insensitive to the angular-momentum evolution which occurs simultaneously. However, the rotational history is determined to a large extent by the structural changes described above. Of course, we need to know the initial angular momentum, the angular-momentum loss rates, and the extent of angular-momentum redistribution to describe effectively the rotational history of the Sun. Understanding the rotational history is crucial to explaining other features of the Sun such as the trace-element abundances, and eventually the magnetic-field properties.

A. Initial Angular Momentum

The properties of the eventual collapse of a protostellar cloud into a main-sequence star are affected by the history of the removal and redistribution of the primordial angular momentum. The initial angular momentum of the star in its protostellar, diffuse state is several orders of magnitude too large to allow collapse to stellar dimensions. The star's solution to this problem may involve fragmentation, magnetic fields, disk formation, accretion, and planetary system formation and evolution, as described in Sec. II. In any case, by the time it appears as a true stellar object near the Hayashi track, it retains only a tiny fraction of the primordial angular momentum. Yet this residual spin manifests itself in the observed solar surface rotation, the solar magnetic-field dynamics, and other key properties of the solar interior. In addition, the final adjustments of the solar interior to the demands of rotational stability produce important changes in the internal angular-momentum distribution and the surface abundances of several elements.

We begin with a discussion of the appropriate "initial" solar angular momentum. By initial, we mean the angular momentum the Sun retained when it reached the Hayashi track and emerged as a star, following the initial dynamical collapse phase. It is well known that stars less massive than about 1.5 $M_\odot$ lose angular momentum with time. The best evidence for this comes from the run of stellar angular momentum with mass, as first determined by Kraft (1970) from observed rotation velocities. Below this mass, even young stars are observed to rotate systematically more slowly than more massive stars and this discrepancy becomes much larger for older stars. Because this change in behavior occurs in coincidence with the onset of deep surface convection zones in stars, it is likely that the current rotation of low-mass stars (including the Sun) was affected by angular momentum loss, presumably from a magnetic wind (Schatzman 1962). This suggests two possible techniques for deducing the *initial value* of the solar angular momentum.

Initial Solar Rotation Rate from Observations of Young Low-Mass Stars.   Young stellar objects of approximately solar mass allow us to look backwards in time at the rotation rate of the early Sun. Rotation velocities have been measured for very young pre-main-sequence T Tauri stars (Bouvier et al. 1986; Hartmann et al. 1986; Hartmann and Stauffer 1989). We turn to the T Tauri star observations reluctantly because for these stars it is extremely difficult to assign accurate masses, and the rotation velocities themselves can only be determined if greater than 5 to 10 km s^{-1}. In addition, T Tauri stars display vigorous winds, and these winds carry angular momentum away from the star. Thus the rotation of T Tauri stars only provides a broad range of possible values for the initial solar angular momentum.

With these uncertainties in mind, we examine the study of 50 T Tauri stars in the Taurus-Auriga and Orion complexes by Hartmann et al. (1986). They found 10 objects with rotation velocities below their detection limit of 10 km s^{-1}, almost half with velocities below 20 km s^{-1}, and only 7 with velocities greater than 30 km s^{-1}. On average, these low-mass T Tauri stars rotate at about 10 km s^{-1}, or about 5 times the current solar equatorial rotation rate. This would imply that at the base of the Hayashi track, the Sun's angular momentum was about 10 to 25 times greater than its current value because the moments of inertia of T Tarui stars are roughly 2 to 5 times greater than used for stars on the main sequence. Extrapolation back in time from the T Tauri stars suggests an initial solar angular momentum of about 2 to 5 $\times$ 10^{49} g cm^2 s^{-1}. However, because some T Tauri stars rotate at much higher velocities, a higher initial angular momentum for the Sun is not excluded.

Initial Solar Rotation Rate from Observations of High-Mass Stars.
Stars of higher masses (M $>$ 1.5 M$_\odot$) have extremely shallow surface convection zones and weak magnetic fields, which leads to little angular momentum loss near the main sequence (Schatzman 1962). Because they also evolve quite rapidly through the convective phases of pre-main-sequence evolution, their current angular momentum probably represents the distribution of initial angular momentum for these massive stars.

For stars of a given mass, the observed range in rotation velocities covers at least an order of magnitude. However, as shown by Kraft (1970; see also Kawaler 1987), the *mean* rotation velocities follow a power law in stellar mass; equivalently, the homologous character of the main sequence allows us to say that the mean main-sequence rotation velocities are approximately ⅓ breakup over the range from 1.5 to 6 solar masses. Assuming solid-body rotation, this implies that the mean angular momentum scales as the mass squared on the main sequence. Extrapolation of these relationships to low-mass objects provides an estimate of the mean initial angular momentum of solar mass stars as 1 to 2 $\times$ 10^{50} g cm^2 s^{-1}, with a total range of at least an order of magnitude. This value is 2 to 10 times larger than the mean value obtained from observations of T Tauri stars. The initial solar angular momentum probably lies within the range defined by these two estimates.

Once we have chosen the initial angular momentum, it remains to specify the initial angular momentum distribution within the protosolar model. Upon reaching the Hayashi track, the solar model was essentially fully convective. Because the turnover time scale for a convection zone (months to years) is short compared to the evolutionary time scale, convection quickly obliterates any previous rotation history and establishes its own rotation law. However, it is not possible from first principles to specify what the resulting rotation law will be. In particular, one can argue for anything from solid-body rotation for heavily interacting convective flows, to constant specific angular momentum ($\Omega \propto r^{-2}$) if convective elements do not exchange angular momentum with their surroundings. Solid-body rotation has been assumed to be the rule in convection zones for most calculations. This assumption is based in part on laboratory experimental results, and in part on initial estimates from helioseismology. Thus PKSD began with models that rotate as a solid body on the Hayashi track, with angular momenta of order several times 10^{49} g cm^2 s^{-1} to a few times 10^{50} g cm^2 s^{-1}.

B. Angular Momentum Loss by Magnetic Braking

As the radiative core grows in the evolving proto-Sun, the differentially contracting core spins up with respect to the convective envelope, and differential rotation develops. When this contrast in rotation rate between the convective envelope and radiative core develops, dynamo-generated magnetic-field modulation begins, and modifies (and organizes) the primordial solar magnetic field. This intimate relationship between rotation and magnetic fields extends out into the solar wind; the global solar magnetic field constrains the wind to co-rotate with the Sun out to very large distances. This greatly enhances the efficiency of the wind in removing angular momentum, as initially suggested by Schatzman (1962). Strong observational evidence for angular momentum loss can be found in stars, as stated above. For example, we estimated the Sun's initial angular momentum to be of order 10^{50} g cm^2 s^{-1}; whereas currently, the solid-body angular momentum is 10^{48} g cm^2 s^{-1}, and the total cannot be as much as an order of magnitude larger. Thus, the Sun has lost all but a small fraction of its angular momentum through the solar wind over its lifetime.

With some simple assumptions about the velocity structure of the solar wind, and using the observed relationship between the rotation rate and global magnetic field strength, Kawaler (1988a) has rewritten the Mestel (1983) prescription for angular-momentum loss via a magnetic stellar wind as:

$$\frac{dJ}{dt} = K\Omega^3 \left[\frac{R}{R_\odot}\right]^{1/2} \tag{1}$$

for stars of one solar mass. Skumanich (1972) and Soderblom (1983) have determined the rotation rate for solar-mass stars in clusters of known age; their empirical determinations of the spin-down rate for solar-mass stars shows that the rotation velocity decreases as $t^{-1/2}$. On the main sequence, where the stellar radius and moment of inertia change very slowly with time, the above angular-momentum loss law can be integrated to yield

$$v_{eq} \propto t^{-1/2} \tag{2}$$

which matches the observed spin-down rate for solar-type stars.

This model of angular-momentum loss extends backwards in time to allow us to trace the early angular-momentum history of the Sun. Converting the loss rate to a time scale in yr for angular-momentum loss, one finds

$$\tau_J \approx 2.3 \times 10^{10} \left[\frac{I}{I_\odot}\right]\left[\frac{\Omega}{\Omega_\odot}\right]^{-2} \tag{3}$$

It is easy to see that angular-momentum loss occurs only when I is small and Ω is large. In the T Tauri phase, the evolutionary time scale is of order 10^6 to 10^7 yr, and Ω and I are larger, by factors of about 4, than the current Sun. Thus, the angular-momentum loss time scale of several times 10^9 yr is far longer than the evolutionary time scale in the T Tauri phase, and the proto-Sun continues to spin up as it contracts. As the Sun approaches the main sequence, the evolutionary time scale increases and the braking time scale shortens (since Ω increases and I decreases). These factors result in the increased importance of angular-momentum loss for the Sun when it reaches the main sequence; most of the angular-momentum loss of the Sun after the protostellar collapse phase occurred just as it settled on the main sequence. Indeed, with the above angular-momentum loss model it is possible for solar models to reach the current rotation rate at the current age, even if the Sun first arrived on the main sequence rotating (at near breakup) with a surface rate almost 200 times the current value.

C. Angular-Momentum Redistribution

As the Sun evolved, and the radiative core grew to its current size, loss of angular momentum from the surface resulted in the spin down of the surface convection zone. In the absence of angular-momentum redistribution, a discontinuity in rotation velocity would have developed between the outer convection zone and the radiative layers that lie below. In addition, the differential contraction of the radiative core leads to differential rotation with depth. Thus, even if the Sun started out rotating as a solid body, by the time it reached the main sequence, it could conceivably have steep gradients of angular velocity with depth, and a very rapidly rotating core.

Hydrodynamical instabilities are triggered when the gradient in Ω becomes steeper than some critical values (see Zahn 1987). These instabilities cause mixing of stellar material, and result in the transport of angular momentum, which tend to reduce the gradients. The resulting rotation curve will tend to be one of marginal stability; however, the time scale over which the angular-momentum redistribution processes act may be longer than the evolutionary time scale. Hence the internal rotation curve of the Sun is not quite one of marginal stability; one must compute numerical models of the evolving solar interior including the transport of angular momentum to find the correct state of its internal rotation starting at the initial value point.

Because the approach required to solve meaningfully the PMS problem is an initial value problem, and because not all initial values are directly observable, it is necessary to explore the consequences of different values of the unknown parameters for models of the Sun at different ages, as well as for other stars of different masses, ages or chemical composition. This can only be done by using evolution codes which allow the calculation of main-sequence as well as post-main-sequence stages, both of which are discussed in detail in two other chapters of this book (see, e.g., Ulrich and Cox, and Demarque and Guenther). The current discussion will be based on the code, and the work reported in a recent paper by PKSD. This code follows the evolution of rotating stars as an initial value problem. It includes angular-momentum loss through a stellar wind, angular-momentum redistribution, and material mixing. Since this is the only recent calculation which explicitly treats the transport of angular momentum in evolving solar models beginning during the pre main sequence, we limit the discussion to the redistribution mechanisms employed in their work, which are the same as those treated in Endal and Sofia (1978). It does not explicitly consider magnetic fields, although some of its effects can be approximately mimicked by varying one of the code parameters. The best solar model is that which most closely agrees with observed properties of the present Sun, and which does not violate observed behavior in other stars. Of course, the best solar model is not independent of the code used to evolve the model, and consequently, it is a tentative (thus current) model subject to revisions and improvements.

Hydrodynamical instabilities of relevance to the solar interior come in two varieties that are classified in terms of the time scales over which they operate. Those that operate on time scales that are short compared with the evolutionary time scales, such as convection or dynamical shear, we call "dynamical" instabilities. When triggered, these processes can be regarded as acting instantaneously to redistribute angular momentum and maintain a rotation curve that represents rotation at marginal stability. Other processes, such as Eddington-Sweet circulation, secular shear, and the Goldreich-Schubert-Fricke (GSF) instabilities, act on longer time scales; we refer to them as "secular" instabilities. They do not usually act fast enough to attain marginal stability before the solar interior evolves "out from under" them, so they can be modeled as transporting angular momentum through a diffusive process. Because the time scales for these processes are not precisely known, we must develop independent means of estimating them.

We can constrain the time scale for angular-momentum transport from observations of young stars (Stauffer and Hartmann 1986). In the Alpha Perseus cluster (age ~30 Myr), G and K stars rotate rather rapidly. By the age of the Pleiades (~70 Myr), the G stars have spun down, while the K stars have not. The K stars rotate more slowly than the G stars by the age of the Hyades (500 Myr). If the time scale of transport from the interior to the surface is longer than the age of the Pleiades, then initially the surface convection zones of the G stars spin down while the deeper convection zones of the K stars take longer to spin down. By the age of the Hyades, the interior is fully coupled with the surface so the G stars, with their larger internal reservoir of angular momentum, rotate more rapidly than the K stars.

By modeling angular momentum transport as a diffusive process, the whole problem reduces to the calculation of the appropriate diffusion coefficients resulting from each mechanism. Because of this, PKSD only considered mechanisms for which theoretical estimates for the appropriate diffusion coefficients existed. However, because they considered the effect of altering the diffusion coefficients, their results could apply to redistribution mechanisms yet to be proposed. In addition, their work included the possibility of angular-momentum transport more efficient than the corresponding material mixing. This procedure can mimic the transport of some angular momentum by magnetic fields or waves.

In Fig. 3, we show the internal rotation curve for the solar model, beginning on the Hayashi track and continuing on to the present. Initially, the outer convection zone spins down independently from the radiative core. However, upon reaching the main sequence, the secular shear instability begins to funnel angular momentum out from the core and into the envelope. This mechanism flattens out the rotation curve below the convection zone, and makes the angular momentum stored in the interior during the pre-main-sequence phase available to be depleted by magnetic braking. As time passes on the main sequence, the curve approaches one of marginal stability for secular shear near the base of the convection zone, and for GSF in the inner regions. Near a fractional radius of 0.2, the gradient in mean molecular weight caused by nuclear burning in the core provides a very effective barrier against angular-momentum transport; thus the inner core of the solar model continues to rotate rapidly even at the age of the current Sun. We point out that even with this rapid rotation, this inner core contains only 1/3 of the total angular momentum.

It is interesting to note that the "final" rotation curve for the solar models is almost independent of the initial angular momentum outside the central region where mean molecular-weight gradients are significant. Even in the core, the dependence on J_0 is weak for a relatively large variation of the parameters. The strong dependence of the angular-momentum loss rate on

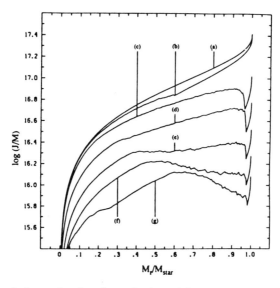

Fig. 3. Angular velocity as a function of mass fraction and time in the core of the solar model in Fig. 1. Curves (a-g) represent models at ages of $(3 \times 10^6, 10^7, 3 \times 10^7, 1.2 \times 10^8, 4.4 \times 10^8, 1.2 \times 10^9, 4.7 \times 10^9)$ yr.

the rotation rate feeds back into the rotational history, resulting in a surface rotation velocity that reflects only the properties of the angular-momentum loss law. Since the coupling time scale for the secular instabilities is less than the age of the Sun, most of the solar model participates in the spin down.

D. Mixing and the Effects on Surface Abundances

To reach this state of quasi-solid-body rotation the interior must have undergone angular momentum transport. If at least some of this transport was caused by hydrodynamic instabilities (e.g. PKSD) then some material mixing will have occurred. This mixing will have brought up material to the surface that spent some time at depths well below the surface convection zone. As a result, the surface abundances of certain elements would reflect nuclear processing at very high temperatures relative to the surface convection zone. The element of primary interest to solar physicists as a probe of internal mixing is ^{7}Li. This element is destroyed by the ^{7}Li$(p,\alpha)^4$He reaction, which is extremely temperature sensitive. For densities of interest, the destruction of ^{7}Li occurs when the temperature rises above about 2.5×10^6K, and the time scale drops rapidly with increasing temperature. It is one of the great fortunate accidents of astrophysics that the temperature at which lithium burns is almost precisely the temperature at the base of the solar convection zone throughout its entire pre-main-sequence evolution. Thus, the degree to which ^{7}Li is depleted in the current Sun tells us the history of the retreat of the surface convection zone. If the convection zone was only slightly deeper in the Sun than in the models, then the models should over-estimate the lithium abundance; if the convection zone were shallower, then the models should be depleted with respect to the Sun.

The rapid early evolution of models of the proto-Sun results in negligible lithium depletion prior to 1.5 Myr. As the time scale of evolution increases, lithium is rapidly depleted by nuclear burning in the surface convection zone; the lithium abundance drops to half its original value by 5 Myr. The retreat of the surface convection zone, and the resulting slow decrease in the temperature at its base, eventually shuts down nuclear destruction of lithium. In standard (i.e., nonrotating) solar models, by an age of 13 Myr the surface lithium abundance reached its final value of about one quarter of the original abundance (PKSD; Proffitt and Michaud 1989). The observed solar depletion of a factor of 200 (Müller et al. 1975) is thus an embarrassment for nonrotating solar models. Two general categories of solutions to the solar lithium problem have been proposed. One solution invokes an increase of the nuclear burning by altering the structural properties of the models; the others invoke a mixing mechanism not present in standard solar models. As we will see, solutions of the first type face severe difficulties.

A proposed solution of the first type invokes "convective overshoot" to arbitrarily mix the surface convection zone to a greater depth (and thus higher temperature material, where lithium burns) (D'Antona and Mazzitelli 1984). Any desired degree of lithium depletion can then be arrived at by adjusting the depth of such an overshoot layer. Changes in the opacity can also cause a deeper surface convection zone and increase pre-main-sequence lithium burning (Stringfellow et al. 1987). However, the bulk of the lithium burning in models without extra mixing occurs during the pre-main-sequence phase of evolution regardless of the means by which the surface convection zone is made deeper. If the solar lithium depletion is explained by convective overshoot or opacity changes, then young G stars should be severely depleted in lithium by pre-main-sequence nuclear burning. In fact, young cluster G stars show only modest lithium depletion in the amount predicted by models without overshoot (Balachandran et al. 1988). As we look at successively older stars, the surface lithium abundance steadily decreases (Hobbs and Pilachowski 1988); this calls for mixing mechanisms, such as rotationally induced instabilities, which can operate on the main sequence without drastic consequences for the pre-main-sequence depletion. The observed solar ^{9}Be depletion (Reeves and Meyer 1978; Ross and Aller 1974) cannot be explained by the same degree of convective overshoot required in standard models to explain the solar ^{7}Li abundance. If the surface convection zones in the models were deepened enough to burn the required amount of ^{9}Be, it would totally destroy lithium. Rotational mixing, on the other hand, occurs to varying degrees throughout the interior; this provides a natural explanation for the abundances of both elements in the Sun. The process of rotational mixing is also relevant to study the depletion of trace elements in other stars. The history of trace-element depletion can be inferred from observations of open cluster stars with different ages. This provides an excellent example of how stellar observations serve to constrain the properties of the Sun.

Rotating models show lower surface lithium abundances because lithium-depleted material from below the convection zone is mixed to the surface, diluting the surface lithium abundance. The models of PKSD use the observed lithium abundance to determine how much material is transported during the transport of angular momentum. For natural choices of parameters, their models reproduce the solar lithium abundance when the diffusion of matter occurs at 3% to 10% the rate of diffusion of angular momentum. The same models are very successful in explaining the observed lithium abundances in stars of various masses in clusters of various ages, thus giving strong support to the rotational mixing hypothesis. Also, since those models which rotated faster initially lose more angular momentum on the main sequence, they transport more lithium-depleted material to the surface. Therefore, the observed lithium abundance in all stars should be inversely proportional to their initial angular momentum. A range of initial angular momenta will produce a spread in lithium abundance which is seen in the Hyades and in older clusters (Pinsonneault et al. 1989*a,* and references therein), as shown in Fig. 4.

Other trace elements probe rotational mixing to different depths than lithium. In nonrotating models, the abundance of ^{3}He shows a peak at a temperature of about 10^7 K. Therefore, if rotational mixing does indeed take place, the surface ^{3}He abundance should be enhanced relative to the cosmic abundance. Unfortunately, the solar helium abundance (^{4}He) is difficult to measure directly, so this test of rotational mixing is not too useful. In later phases of evolution, the abundances of the CNO elements and isotopes, which have different distributions with depth in rotating models, will provide important probes of internal mixing.

E. Observational Constraints

Currently, many of the observational constraints on the evolution of the solar rotation come from observations of other stars. The initial angular momentum is constrained by the observed rotation rate of higher-mass main-sequence stars, and younger solar-mass stars (Sec. III.A). The Skumanich (1972) spin-down law constrains the parameters describing angular-momentum loss by magnetic braking. Observations of the lithium abundance in solar analogs in young clusters provide tight constraints on the degree of rotational mixing just below the surface convection zone. The rotation velocities of stars in young clusters, and T Tauri stars, provide glimpses into the past rotational history of the Sun.

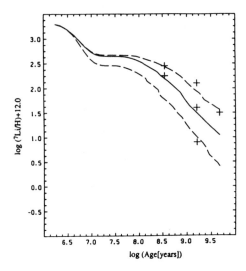

Fig. 4. Spread in surface lithium abundance as a function of time due only to a range in the initial angular momentum J_o. The solid line is a solar model with $J_o = 1.6 \times 10^{50}$ g cm^2 s^{-1}. The crosses are the observed solar analogs from clusters of known age. The upper dashed line is a model with initial angular momentum 3 times smaller than J_o. The lower dashed line is a model with initial angular momentum 3 times larger than J_o.

There are four basic solar observations which are relevant to study of the internal rotation. The first and most obvious is the surface rotation velocity field. Models that accurately describe younger or older stars must also satisfy the challenge of matching the solar surface rotation rate at the solar age. The mean solar rotation rate then provides the calibration point for models of stellar rotation and angular momentum loss.

Another basic observation is the solar lithium abundance (Sec. III.D). The third observational constraint comes from the solar oblateness and quadrupole moment. The current upper limit on the solar oblateness is 10^{-5} (Hill and Stebbins 1975a). This surface deformation caused by rotational support of the equatorial regions reflects the basic structural effects that are characteristic of the solar quadrupole moment. The precession of the perihelion of Mercury provides a firm upper limit to the Sun's quadrupole moment of 3×10^{-5} (that is, if you believe in General Relativity). This value effectively rules out rapid rotation within the Sun between the base of the convection zone and the composition gradient; the moments of models such as PKSD are well within the constraints imposed by the quadrupole moment and oblateness.

The most promising probe of the solar internal rotation, and therefore of the evolutionary history of the solar rotation, is the rotational inversion of helioseismology data. In principle, this technique can determine the rotation rate of the solar interior as a function of depth and latitude. In practice, however, the data are extremely difficult to analyze from both theoretical and practical considerations. For example, to obtain adequate detail in the inner half of the Sun requires observation of rotational splittings of the lowest l-modes, modes which currently have the largest uncertainties in the measured splittings. From a theoretical standpoint, any departure from spherical symmetry in the solar interior causes splittings that can mimic the effects of rotation; hence magnetic fields or turbulence confuse the issue of determining rotation rates below the convection zone. Current results indicate that the Sun rotates essentially as a solid body in the convection zone, and somewhat below, with uncertainties increasing dramatically in the radiative interior. For a thorough review of observational determination of rotation via helioseismology see the chapter by Libbrecht and Morrow, and for the theory see the chapters by Christensen-Dalsgaard and Berthomieu and by Dziembowski and Goode.

THE COMPUTATION OF STANDARD SOLAR MODELS

ROGER K. ULRICH
University of California, Los Angeles

and

ARTHUR N. COX
Los Alamos National Laboratory

Procedures for calculating standard solar models with the usual simplifying approximations of spherical symmetry, no mixing except in the surface convection zone, no mass loss or gain during the solar lifetime, and no separation of elements by diffusion are described. The standard network of nuclear reactions among the light elements is discussed including rates, energy production and abundance changes. Several of the equation of state and opacity formulations required for the basic equations of mass, momentum and energy conservation are presented. The usual mixing-length convection theory is used for these results. Numerical procedures for calculating the solar evolution, and current evolution and oscillation frequency results for the present Sun by some recent authors are given.

I. INTRODUCTION

A. Why Compute Models of the Sun?

Solar models have served as a test case for stellar structure and evolution theorists throughout most of the history of the field. The Sun is a particularly simple type of star: it is in the middle of its main sequence lifetime; it has a shallow convective envelope, has modest mass loss, has no convective core (at least according to the standard theory), has no companion; and, probably most importantly, it has well-known properties including mass, radius, luminosity, age and composition. The detailed thermal structure of the outer solar atmosphere is available. Given this simple problem and all this information, it is clear that stellar structure and evolution theorists should be able to compute a good solar model. For this reason, solar models are commonly used by individual investigators to verify the correct operation of new stellar structure codes and by the community generally to verify the correctness of the basic theories. Codes that correctly reproduce the solar properties are then applied to the calculation of similar stars and the results can be used in a differential analysis with much greater confidence than would be the case without calibration on the Sun.

During the early years of stellar structure theory, solar models were used to test the validity of the fundamental ideas through the ability of the models to reproduce the correct solar radius and luminosity. More recently, the study of solar models has progressed beyond this limited test because of the additional information beyond the mass, radius and luminosity that is now available. Observations of the neutrino capture rate onto ^{37}Cl extend back nearly 20 yr. These observations are in disagreement with the results of theoretical calculations, i.e., observations according to Davis have a count rate of 2.1 $\pm$ 0.3 SNU vs a predicted rate for 3.5 times that level.

At one level the observed frequencies of solar oscillations have been reasonably well accounted for by theoretical calculations. However, statistically significant discrepancies remain and as the observations improve in precision we anticipate that further improvements in the models will be required. The helioseismology database is rich—about 10^5 individual mode frequencies have been measured. The deduction of properties of the solar interior from these frequencies utilizes the fact that the frequencies depend on the thermodynamic properties of the gas. Owing to the similarity of many oscillation eigenfunctions, the number of constraints on solar models is much smaller than the number of measured frequencies. The redundancy contained in the data depends on the precision of the frequency measurement. After appropriately scaled linear dependencies have been removed, a smaller number of new and independent parameters constraining the solar interior structure and dynamics have been deduced. The actual number of independent free parameters is difficult to quantify precisely but, as an example, the analysis of the opacity by Korzennik and Ulrich (1989) found that about 20 independent parameters were available from the observational data. Some of the frequencies have been observed on a regular basis for $\sim$ 10 yr. Because of this wide range of information, it is possible to probe the solar structure at a very detailed level and test a variety of physical hypotheses.

Interest in the solar interior will be maintained by the development of several new experiments for the measurement of additional neutrino interactions that will add constraints to the fluxes of solar neutrinos from individual nuclear sources and by several major international collaborations to improve the helioseismology observations. Davis in his chapter has described all the neutrino experiments in detail. We summarize their status briefly here. In Japan, the Kamioka collaboration has put into operation a new experiment called Kamiokande II that detects neutrinos through their scattering interaction with electrons. This experiment is sensitive to the high-energy electron neutrinos only and thus can test primarily the results of the Davis experiment. At present two experiments are in operation based on detection by the isotope ^{71}Ga being conducted by a European group (GALLEX) and by a USSR/US group (SAGE). Initial operations are concentrating on understanding and removing background contributions to the signal. These experiments are sensitive to the low-energy $p-p$ neutrinos and if the rate of detection is too low, the ^{71}Ga experiment can require the solution to be in the physics of the neutrino rather than in the calculation of the solar models. Other projects further from completion include the Sudbury Neutrino Observatory, which uses D_2O and resembles the Kamiokande II experiment in its sensitivity, and the ICA-RUS experiment, which uses liquid Ar and is also sensitive to the high-energy neutrinos. These last two experiments, as well as Kamiokande II, yield their signal immediately and can provide directional information about the interacting neutrinos. In addition to these four, there are about 10 other proposed neutrino experiments in various stages of study and development.

Four major international helioseismology experiments, the Global Oscillation Network Group (GONG) organized by the US National Solar Observatory, and three space experiments on the joint NASA/ESA Solar and Heliospheric Observatory (SOHO) mission: the Solar Oscillation Investigation (SOI) experiment, the Global Oscillations at Low Frequency (GOLF) experiment and the Variability of Iradiance Global Oscillations (VIRGO) experiment are in progress. These efforts all have the goal of overcoming the detrimental effects of day/night sidebands in the power spectra. Each has a special emphasis: the GONG experiment may be able to operate over a more extended period than the space experiments because it will not be subject to the fundamental resource limitations inherent in a space-borne experiment; the SOI experiment will achieve the highest possible spatial resolution that will allow the most complete analysis of the surface structure and dynamics; GOLF will provide the most sensitive measurement of solar velocity oscillations while VIRGO will provide the ultimate in sensitivity to luminosity oscillations. An important objective of all the helioseismology experiments is the clear detection of solar g-modes. These will allow a considerable extension and refinement of the deduced structure of the solar interior. With the data that may come from g-mode analysis, we would be able to deduce the solar structure with great precision and provide unambiguous predictions of solar neutrino fluxes.

B. A Brief History

The initial models that tested the basic theory of stellar structure required the correct identification of the nuclear fuel that powers the Sun. Although models in radiative equilibrium obey a unique mass-luminosity relation that could be deduced without reference to the solar energy source, such models could not reproduce the solar radius without arbitrary assumptions. One of the earliest complete solar models following the identification of the CNO cycle as a possible nuclear energy source for the Sun was that by Schwarzschild (1946). This model was completely radiative, did not include the effects of the conversion of hydrogen to helium and omitted the $p-p$ chain as an energy source. It had a convective core and a central temperature of 19.8×10^6 K—a rather poor approximation to modern models.

Shortly thereafter Schwarzschild, et al. (1956) presented a model based on the $p-p$ chain that included a convective envelope and the effects of nuclear conversion. Their models depended on the assumed heavy-element abundance Z just as do modern models and had the initial helium abundance Y adjusted to yield the solar luminosity. For (Z,Y) values of $(0.04,0.26)$ and $(0.015,0.185)$ the models had central temperatures and densities of 15.8 and 14.8×10^6 K and 127 and 132 g cm^{-3}, respectively. Although the required value of Z is higher than is typical now, these results are a good approximation to modern models especially when consideration is given for the computational resources and physical data in the form of opacities and nuclear cross sections available at that time. These discussions were motivated by a desire to provide some constraint on the solar helium content as well as the verification of the validity of stellar structure theory by computing a model for a star that has a complete and accurate set of constraining observations. The numerical techniques for model calculation are well described by the classic work of Schwarzschild (1958) where reference is also made to solar models of the period.

Following the development by Davis (1955) of methods of using ^{37}Cl to detect neutrinos, Fowler (1958) discussed the possibility of detecting neutrinos produced by the Sun. During the 1960s while the experimental program to follow through on Fowler's suggestion was being implemented, theoretical interest in the calculation of solar models was sharpened. Beginning with Bahcall et al. (1963) and Sears (1964), models of the Sun have nearly always included a prediction of the rates for the fluxes of the neutrinos from the source reactions of the $p-p$ chain. At that time Sears (1964) began his discussion with the remark "Theoretical models of the Sun are no longer at the frontier of the theory of stellar structure and evolution." This comment suggesting that the computation of solar models was a solved problem is in marked contrast to the present level of activity in the field. Sears' comment can be understood by realizing that it referred primarily to the question of constructing a solar model, that satisfies the constraints of radius and luminosity at the correct age. The great vitality in the field of solar model calculation today results from the addition of many more constraints into the theory beyond those of the radius and luminosity.

Between these early days and the present time there has been a gradual evolution in the solar models as defined by the input physics and assumptions made in the computation. The main changes in the input physics have come in the adopted values of the nuclear cross-section factors and in the adopted opacities. During the period beginning in 1970, most of these improvements were stimulated by the persistent neutrino flux discrepancy. Models computed prior to this period were presented by Demarque and Percy (1964), Salpeter (1968), Iben (1969), Bahcall and Shaviv (1968) and Torres-Peimbert et al. (1969). This period also saw the first calculations that explored the problem of the sensitivity of derived conclusions to the uncertainties in the input parameters. Discussions of this type were given by Iben (1969) and by Bahcall et al. (1969) and these results continue to be helpful in understanding where it is advantageous to search for errors in the modeling parameters. A similar approach of sensitivity studies is now being followed in the field of helioseismology in the papers by Christensen-Dalsgaard (1982), Ulrich et al. (1983), Ulrich and Rhodes (1983), Christensen-Dalsgaard (1988c), Guenther and Sarajedini (1988), Guenther et al. (1989), Cox et al. (1989,1990), and Guenther (1991).

The development of the standard solar model theory during the early part of the 1970s is well presented in the review by Bahcall and Sears (1972). Improvements to the opacities and nuclear parameters led to a series of solar models reviewed by Bahcall et al. (1982) and Bahcall and Ulrich (1988). This first paper reviews the evolution in our understanding of the neutrino flux from the Sun and how it never was even close to the very low-flux observations of Davis. The most recent studies are discussed below in Sec. IV.

C. Other Reviews

The topic of the calculation of solar models itself has not been reviewed in isolation. A recent general discussion of solar models has been given by Roxburgh (1985). There have been a number of reviews of solar models in the context of helioseismology (Gough 1985b; Toomre 1986; Christensen-Dalsgaard 1988c) and the solar neutrino problem in the past few years (see above). In addition to these well-known topics, there are two other interesting aspects of the solar models, the depletion of lithium necessary to explain the observed abundance of lithium on the solar surface, which is 2% of the lithium abundance observed in the interstellar medium, and the low solar luminosity of models of the young Sun, which would have caused the Earth to freeze over if it were like the present Earth when the Sun first formed. These two problems have been recognized since the Schwarzschild et al. (1956)

paper but have received little attention because they do not provide as quantitative a constraint on the structure as do the neutrino fluxes and the oscillation frequencies.

Helioseismology reviews have concentrated more on the theory of the solar oscillations than on the theory of the solar models. The works by Ulrich and Rhodes (1983), Christensen-Dalsgaard (1988c), and the Yale papers by Guenther and colleagues referenced above have discussed a variety of uncertainties in the computation of the solar models in the context of the effects on the derived frequencies. Similarly, the reviews focusing on the solar neutrinos have emphasized the physics needed for the computation of the neutrino fluxes. Volume 1 of the *Physics of the Sun* series edited by Sturrock et al. (1986) provides a comprehensive review of all aspects of the solar interior. In particular, Chapter 7 by Brown et al. (1986) reviews all aspects of helioseismology. Chapters 2 and 3 by Parker (1986) and Huebner (1986) provide detailed reviews of the physics of the nuclear reaction rates and the radiative opacities, respectively. There have been no extensive reviews of the equation of state as applied to the solar interior partly because this aspect of the theory has been assumed to be well established.

D. What is a Standard Solar Model?

The term "standard solar model" might seem to imply that this model has been adopted by the community as a standard against which other models are to be compared. This has not been the intent although in principle the availability of such a standard would be desirable. Instead, this designation has been used to distinguish those models computed using the simplest possible physical assumptions and the best available physical input from the nonstandard solar models that make assumptions based on one or another novel hypothesis. Several such nonstandard model hypotheses have been described by Newman (1986). It is a hope that a "nonstandard model" might become the "standard solar model" in which the solar neutrino discrepancy is resolved. Thus far none of the nonstandard models has succeeded in this fashion.

The use of the best available physics is an important characteristic of the standard solar model. This aspect requires some judgment, and decisions as to which algorithms or data to use cannot be made in an unambiguous fashion. The papers by Bahcall et al. (1982) and Bahcall and Ulrich (1988) contain the description of an effort to produce standard models according to this definition.

E. Scope of this Chapter

Except for the discussion of the treatment of the nuclear physics, this chapter includes a brief description of the input physics with references to sources where details can be obtained. For the case of the nuclear physics, there is not a single reference where a complete implementation of the reaction rates and abundance change system has been presented. As this part of the calculation is essential to the calculation of the neutrino fluxes, we give here a set of formulae and procedures that can be used to set up such a calculation. Other details can be found in the Parker and Rolfs chapter. One implementation of this approach is available as the code SUNEV that is in the public domain through the GONG project at the National Solar Observatory. Additional topics discussed in this chapter include a critique of equation of state approaches, a summary of available opacity results, convection theory as applied to solar structure calculations, numerical methods with an emphasis on the less popular fitting technique, and the chapter concludes with a summary of recent results including derived helium abundances and oscillation frequencies.

II. THE INPUT PHYSICS

A. Nuclear Reactions

1. Rates. The nuclear reaction rates are crucially important to the predictions of the neutrino fluxes. Because of the long-standing problem in accounting for the observed rate of neutrino capture in the Davis experiment, the adopted values for the cross-section factors S_{eff} have been well studied and reported on by a variety of authors. Particular reference should be made to the works by Parker (1986), Bahcall et al. (1982) and Bahcall and Ulrich (1988). These papers contain very extensive discussion of the experimental data.

The numerical implementation of the reaction rate calculation from the cross-section factors is a straightforward task. Although there are discussions of some aspects of the numerical formulation necessary to treat the nuclear reactions and the nuclear transformations, there is not at present a single place where all the parts are collected together. The cross sections are given by Bahcall et al. (1982) and Bahcall and Ulrich (1988). Formulations giving the rates as a function of the cross sections are found in Fowler et al. (1967,1975) and Parker (1986). Conservation equations giving the time dependence of the abundances have been given by Bodenheimer et al. (1965), Iben (1965), Christensen-Dalsgaard et al. (1974), Maeder (1983), Roxburgh (1985) and others. Numerical techniques for solving these equations were discussed by the first two of the above works and also by Arnett and Truran (1969) and Wagoner (1969). In order to assist investigators in implementing studies of the solar interior, we provide here a summary of the nuclear reaction theory.

We discuss the archetype reaction between species

$$0 + 1 \rightarrow 2 + 3 \tag{1}$$

where the reactants **0** and **1** are converted into the products **2** and **3**. The number of reactions per gram per second depends on the product of the densities of the reacting species and an average of the cross section and thermal velocity of the reactants. Parker (1986; see also chapter by Parker and Rolfs) gives a concise summary of the relevant formulae. The rates depend strongly on the relative kinetic energy between the reacting species due to the barrier penetration factor. It is usual to express the reaction cross sections in terms of a slowly varying cross-section factor $S(E)$ where $S(E) = E\sigma(E) \exp(2\pi\eta)$ and η is the barrier penetration factor: $\eta = Z_0 Z_1 e^2/\hbar v$ for particles moving with a relative velocity v. The experimental data on the measured cross sections is normally provided in the form of coefficients in an expansion of $S(E)$:

$$S(E) = S(0) + E\,S'(0) + \frac{1}{2} E^2\, S''(0) + \ldots \tag{2}$$

The average of the barrier penetration factor over the thermal energy distribution using the representation of Eq. (2) for $S(E)$ has been carried out by Bahcall (1966) to yield:

$$\langle\sigma v\rangle = \left(\frac{2}{\mu(kT)^3}\right)^{1/2} (\Delta E_0)(f_0\, S_{\text{eff}}) \exp(-\,3E_0/kT). \tag{3}$$

The f_0, the screening factor from Salpeter (1954), is

$$f_0 = \exp(Z_0\, Z_1\, \delta) \tag{4}$$

with

$$\delta = 0.188\, \rho^{1/2}\, T^{-3/2}\, \zeta \tag{5}$$

and

$$\zeta^2 = \sum_i \frac{Z_i^2 + Z_i}{A_i} X_i \tag{6}$$

and other quantities in Eq. (3) are given in Parker (1986). The effective cross-section factor, S_{eff}, involves all the $S(0)$, $S'(0)$ and S'' but is not itself needed. The working formula is written in terms of a sum over powers of $T^{1/3}$ where the coefficients depend on the values of the S factors. Because these S factors change with time as the experimental data improves, there is a need to be able to change them easily. The code used by Bahcall and Ulrich has been written in such a way that the cross-section factors are read into the program in the measured units of MeV-barns and then converted into the appropriate internal units by the code. This procedure reduces the danger of bookkeeping errors and makes the current status of the cross-section data readily apparent.

The values of the S factors used by Bahcall and Ulrich (1988) are given in Table I. This table also gives the reference from which the values have been adopted. These numbers have not been updated since the middle of 1987 and may be slightly out of date at this time. The most recent data are in the Parker and Rolfs chapter. Table I contains no information beyond that in the Bahcall et al. (1982) and Bahcall and Ulrich (1988) works but may nonetheless provide a convenient reference. The format of the presentation of S'/S and S''/S is consistent with the numerical implementation of Eq. (3) which is described below.

To make the formulation explicit, consider a reaction of the type given in Eq. (1). The description starts with the a reaction rate parameter R_i which is related to the quantity $[01]_i = \rho N_A \langle \sigma v \rangle_i$ defined by Fowler et al. (1967) by

$$R_i = \frac{1}{1 + \delta_{01}} [01]_i \frac{1}{A_0} \frac{1}{A_1}. \tag{7}$$

The subscript i denotes the reaction in Table I and the A's are the atomic masses in atomic mass units. The R_i parameter when multiplied by the product of mass fractions X_0 and X_1 is the number of moles of reaction per gram

TABLE I
Reaction Rate Cross-Section Factors

i	Reaction	$S(0)$ (MeV − barn)	$S'(0)/S$ (MeV^{-1})	$\frac{1}{2}S''(0)/S$ (MeV^{-2})	Reference
1.	^{1}H$(p,e^+v_e)^2$D	4.07×10^{-25}	11.2	0.0	Bahcall and Ulrich (1988)
2.	^{3}He$(^3$He,$2p)^4$He	5.15	−0.19	0.0	Bahcall et al. (1982) Krauss et al. (1987)
3.	^{4}He$(^3$He,$\gamma)^7$Be	5.4×10^{-4}	−0.574	0.0	Alexander et al. (1984) Parker (1986)
4.	^{3}He$(p,e^+v_e)^4$He	8.1×10^{-23}	0.0	0.0	Werntz and Brennan (1973) Tegner and Bargholtz (1983)
5.	^{7}Li$(p,\alpha)^4$He	5.2×10^{-2}	0.0	0.0	Rolfs and Kavanagh (1986)
6.	^{7}Be$(p,\gamma)^8$B	2.43×10^{-5}	−1.0	0.0	Filippone et al. (1983) Parker (1986)
7.	^{12}C$(p,\gamma)^{13}$N	1.45×10^{-3}	1.69	23.4	Rolfs and Azuma (1974)
8.	^{14}N$(p,\gamma)^{15}$O	3.32×10^{-3}	−1.78	1.36	Fowler et al. (1975)
9.	^{15}N$(p,\alpha)^{12}$C	78.0	4.50	71.1	Zyskind and Parker (1979)
10.	^{15}N$(p,\gamma)^{16}$O	6.4×10^{-2}	0.50	31.1	Rolfs and Rodney (1974)
11.	^{16}O$(p,\gamma)^{17}$F	9.4×10^{-3}	−2.45	6.4	Rolfs (1973)

per second of stellar material[a]. To apply this formulation to a particular reaction of the type in Eq. (1), replace the subscripts 0 and 1 with the chemical symbol for the species. Thus for the $p-p$ reaction, the number of moles of reactions per second per gram is $R_{^1H(p,e^+\nu_e)^2D}\, X_H^2$. Each mole of reactions consumes $1 + \delta_{01}$ moles of reactants and produces 1 or 2 moles of products. Substituting into Eqs. (51) to (59) of Fowler et al. (1967) yields a result that can be written:

$$R_i = C_{0i}\, f_0 \rho (T_9^{1/3})^{-2} \exp(-C_{ei}/T_9^{1/3}) \left(1 + \sum_{j=1}^{5} C_{ji}(T_9^{1/3})^j\right) \qquad (8)$$

per second where T_9 is the temperature in 10^9 K and the result is in reactions per second when the numerical data and units used here are adopted. After defining A as the reduced mass between species 0 and 1: $A_{01} = A_0 A_1/(A_0 + A_1)$ and defining $W_i = Z_0^2 Z_i^2 A_{01}$, the coefficients $C_{0i} \ldots C_{5i}$, C_{ei} are given by:

$$C_{0i} = 7.8327 \times 10^9 \left(\frac{Z_{0i} Z_{1i}}{A_{01i}}\right)^{1/3} S_{0i} \qquad (9)$$

$$C_{1i} = 0.09897\, W_i^{-1/3} \qquad (10)$$

$$C_{2i} = 0.122\, W_i^{1/3} \left(\frac{S'}{S_0}\right)_i \qquad (11)$$

$$C_{3i} = 0.08378 \left(\frac{S'}{S_0}\right)_i \qquad (12)$$

$$C_{4i} = 0.01489\, W_i^{2/3} \left(\frac{1}{2}\frac{S''}{S_0}\right)_i \qquad (13)$$

$$C_{5i} = 0.026\, W_i^{1/3} \left(\frac{1}{2}\frac{S''}{S_0}\right)_i \qquad (14)$$

$$C_{ei} = 4.2487\, W_i^{1/3}\,. \qquad (15)$$

The coefficients can be calculated once per run from the adopted values of the S parameters.

In addition to the above charged-particle nonresonant reactions, there are two other weak interaction reactions that are part of the $p-p$ chain: $^7Be(e^-, \nu)^7Li$ and $^1H(e^-p,\nu)^2D$. The reaction rates for these two processes

---

[a] Recall that 1 mole is $N_A = 6.0225 \times 10^{23}$ of nuclei or reactions. This number is the inverse of the mass, which is the basis of the atomic masses used to calculate the number of moles in a gram of matter. Since these mass units have several possible definitions, care should be exercised to ensure that all are consistent. The calculations by Bahcall and Ulrich have generally followed the recommendations by Fowler et al. (1967) and adopted the $A_{^{12}C} = 12$ scale for these units.

are not in the same form as above and can be written (Bahcall and Moeller 1969; Bahcall et al. 1982):

$$R_{^7\mathrm{Be}(e^-,\nu)^7\mathrm{Li}} = 8.76 \times 10^{-11} \, \rho T9^{-1/2} \, [1 + 4(T9 - 0.016)] \tag{16}$$

and (Bahcall and May 1969; Bahcall et al. 1982)

$$R_{^1\mathrm{H}(e^-p,\nu)^2\mathrm{D}} = 1.74 \times 10^{-6} \, \rho T_9^{-1/2} \, (1 + 20T_9) \, R_{^1\mathrm{H}(p,e^+\nu e)^2\mathrm{D}}. \tag{17}$$

For $^7\mathrm{Be}(e^-,\nu)^7\mathrm{Li}$ the number of moles of reactions per gram of matter per second is given by $R_{^7\mathrm{Be}(e^-,\nu)^7\mathrm{Li}} X_{\mathrm{Be}} (1 + X_{\mathrm{H}})$ and for $^1\mathrm{H}(e^-p,\nu)^2\mathrm{D}$, the number of moles of reactions per gram of matter per second is given by $R_{^1\mathrm{H}(e^-p,\nu)^2\mathrm{D}} X_{\mathrm{H}}^2 (1 + X_{\mathrm{H}})$.

The notation above is cumbersome and we will designate the R_i and X_i rate coefficients and mass fractions by the atomic masses of the species rounded to the nearest integer. In this way, the rate factor $R_{^1\mathrm{H}(p,e^+\nu e)^2\mathrm{D}}$ becomes $R_{1,1}$ and the mass fraction of hydrogen becomes X_1. This notation only breaks down for the case of mass 7 where both Li and Be would receive the same designation. We use subscripts Li to denote those reactions and abundances that refer to Li and retain the subscript 7 to denote the cases referring to Be. The electron capture reaction onto $^7\mathrm{Be}$ is indicated by $R_{0,7}$. The two reactions between $^{15}\mathrm{N}$ and $^1\mathrm{H}$ are distinguished by adding a superscript C or O to denote the product. Finally, we use the R and X notation of Bodenheimer et al. (1965) instead of the [01] and Y notation of Arnett and Truran (1969). The latter is more compact but does not refer directly to the mass fractions. The two formulations are equivalent at this stage.

2. Energy Generation. The energy generation for each reaction is the product of the energy released per reaction times the rate of the reaction occurrence. These energy values Q have been given by Bahcall and Ulrich (1988) for the $p-p$ chain and by Fowler et al. (1967) for the CNO cycle. The Bahcall and Ulrich (1988) implementation combines the energy released from all the reactions that follow immediately after the slow initiating step. For example, following every $^1\mathrm{H}(e^-p,\nu)^2\mathrm{D}$ we assume that a $^2\mathrm{D}(p,\gamma)^3\mathrm{He}$ reaction follows essentially instantaneously. The reaction sequences could be grouped differently with little effect on the results. For more massive stars than the Sun or later stages of evolution, the CNO cycle should be broken into more parts. The treatment of the Li as a separate step is necessary only to provide information about the depletion of the primordial solar Li. Otherwise, sequences 5 and 12 of Table II would have been combined. Also, the rare *pep* and *hep* reactions are not included in the energy generation. The Q values in MeV per reaction are converted to ergs per mole of reactions by multiplying by 9.649×10^{18}. Table II gives the reaction sequences and energy equivalents that were used by Bahcall and Ulrich (1988). Denoting the Q_i values by the index of Table II, the resulting expression for the nuclear energy generation rate becomes in ergs per gram per second:

$$\epsilon_{\text{nuc}} = 9.649 \times 10^{18}[Q_1R_{1,1}X_1^2 + Q_2R_{3,3}X_3^2 + Q_3R_{3,4}X_3X_4$$
$$+ Q_5R_{1,Li}X_1X_{Li} + Q_6R_{1,7}X_1X_7 + Q_7R_{1,12}X_1X_{12} + Q_8R_{1,14}X_1X_{14}$$
$$+ (Q_9R_{1,15}^C + Q_{10}R_{1,15}^O)X_1X_{15} + Q_{11}R_{1,16}X_1X_{16} + Q_{12}R_{0,7}(1 + X_1)X_7]. \quad (18)$$

3. The Abundance Change Equations. The reaction sequences given in Table II cause the abundances of the reactants and products to change. Each sequence is initiated by the first reaction and then followed immediately by the subsequent reactions. The abundance changes thus need to include all reactions in each sequence. The rate, however, is governed only by the initiating reaction. The sequences can be broken up into additional individually followed reactions in a straightforward manner. Each intermediate species is in a steady state between production by the previous reaction in the sequence and destruction by the next. Note that one should avoid the use of the term "equilibrium" to describe the abundance of these species because equilibrium should be reserved for those cases where a true thermodynamic equilibrium is achieved as in the *e*-process of Burbidge et al. (1957). In fact, the case of Li is a good example as it is appropriate to treat X_{Li} as if it is in a steady state throughout most of the solar interior. The equation that expresses the steady

TABLE II
Reaction Sequence Energies

i	Reaction Sequence	Q(MeV per rx)
1.	$^1H(p,e^+\nu_e)^2D(p,\gamma)^3He$	6.664
2.	$^3He(^3He,2p)^4He$	12.860
3.	$^4He(^3He,\gamma)^7Be$	1.586
4.	$^3He(p,e^+\nu_e)^4He$	10.16
5.	$^7Li(p,\alpha)^4He$	17.347
6.	$^7Be(p,\gamma)^8B(e^+\nu)^8Be^*(\alpha)^4He$	11.499
7.	$^{12}C(p,\alpha)^{13}N(e^+\nu)^{13}C(p,\lambda)^{14}N$	11.008
8.	$^{14}N(p,\gamma)^{15}O(e^+\nu)^{15}N$	9.054
9.	$^{15}N(p,\alpha)^{12}C$	4.966
10.	$^{15}N(p,\gamma)^{16}O$	12.128
11.	$^{16}O(p,\gamma)^{17}F(e^+\nu)^{17}O(p,\gamma)^{18}F$	3.553
	$^{18}F(e^+\nu)^{18}O(p,\alpha)^{15}N$	
12.	$^7Be(e^-,\nu)^7Li$	0.047

state will always involve the abundance of the steady-state species and can be used to eliminate that abundance from all other equations.

Using the notation defined in the preceding sections and neglecting the two extremely rare reactions *pep* and *hep*, the abundance equations can be written:

$$\frac{1}{A_1}\frac{dX_1}{dt} = -3\,R_{1,1}X_1^2 + 2\,R_{3,3}\,X_3^2 - R_{1,Li}\,X_1X_{Li} - R_{1,7}X_1X_7$$

$$- 2\,R_{1,12}X_1X_{12} - R_{1,14}X_1X_{14} - (R_{1,15}^C + R_{1,15}^O)X_1X_{15} - 3R_{1,16}X_1X_{16} \quad (19)$$

$$\frac{1}{A_4}\frac{dX_4}{dt} = R_{3,3}X_3^2 - R_{3,4}\,X_3X_4 + 2\,R_{1,Li}X_1X_{Li} + 2\,R_{1,7}X_1X_7$$

$$+ R_{1,12}\,X_1X_{12} + R_{1,15}^C\,X_1X_{15} + R_{1,16}X_1X_{16} \quad (20)$$

$$\frac{1}{A_3}\frac{dX_3}{dt} = R_{1,1}\,X_1^2 - 2\,R_{3,3}\,X_3^2 - R_{3,4}X_3X_4 \quad (21)$$

$$\frac{1}{A_{Li}}\frac{dX_{Li}}{dt} = R_{0,7}\,X_7(1 + X_1) - R_{1,Li}\,X_1X_{Li} \quad (22)$$

$$\frac{1}{A_7}\frac{dX_7}{dt} = R_{3,4}\,X_3X_4 - R_{0,7}\,X_7(1 + X_1) - R_{1,7}\,X_1X_7 \quad (23)$$

$$\frac{1}{A_{12}}\frac{dX_{12}}{dt} = R_{1,15}^C\,X_1X_{15} - R_{1,12}\,X_1X_{12} \quad (24)$$

$$\frac{1}{A_{14}}\frac{dX_{14}}{dt} = R_{1,12}\,X_1X_{12} - R_{1,14}\,X_1X_{14} \quad (25)$$

$$\frac{1}{A_{15}}\frac{dX_{15}}{dt} = R_{1,14}\,X_1X_{14} - (R_{1,15}^C + R_{1,15}^O)\,X_1X_{15} \quad (26)$$

$$\frac{1}{A_{16}}\frac{dX_{16}}{dt} = R_{1,15}^O\,X_1X_{15} - R_{1,16}\,X_1X_{16}. \quad (27)$$

As a check on the precision of the numerical calculation it is desireable to augment Eq. (20) with an alternate determination of the abundance of ^{4}He from

$$X_4 = 1 - X_1 - X_3 - X_7 - X_{Li} - X_{12}$$

$$- X_{14} - X_{15} - X_{16} - \sum_{heavy} X_{heavy} \quad (28)$$

which measures the round off error and when adequately controlled ensures that the stellar mass is properly accounted for. The abundance for ^{15}N is normally taken to be in steady state so that X_{15} can be eliminated from the system. This elimination introduces a branching ratio $B_C = R^C_{1,15}/(R^C_{1,15} + R^O_{1,15})$ that is introduced as a multiplier for the two source terms in the CNO cycle, which otherwise would have depended on the ^{15}N reactions. Thus $R^O_{1,15} X_1 X_{15} \to (1 - B_C)R_{1,14} X_1 X_{14}$ and $R^C_{1,15} X_1 X_{15} \to B_C R_{1,14} X_1 X_{14}$.

B. Equation of State

Throughout most of the solar interior the equation of state has not been considered as a source of significant theoretical uncertainty. Typically the most difficult regimes for the equation of state occur at higher densities than are found in the Sun. However, the requirements imposed by the precision and accuracy of the oscillation frequency measurements now bring the equation of state uncertainties to a point where they must be treated along with those of the opacities and nuclear reaction rates. The equation of state is primarily that of a perfect gas modified by the effects of radiation, partial electron degeneracy, multiparticle Coulomb interactions and incomplete ionization. All the corrections are very small so that each effect can be treated in isolation with the combined effects added linearly. For example, the correction to the electron pressure due to their partial degeneracy can be calculated from the electron density without considering partial ionization at the same time. However, the relationship between the free electron density and the density of nuclei would be affected by partial ionization. The correction to the electron pressure due to their partial degeneracy is well known and not subject to any substantial uncertainties. There is no known uncertainty in the application of the perfect gas law with radiation pressure. The most uncertain aspect of the equation of state is in the effect of the Coulomb interactions and partial ionization on both the particle pressure and on the internal energy of the gas. Because the derivatives of both these quantities are needed to determine the temperature distribution within the convection zone as well as to determine the sound speed, small effects that are restricted to limited range of temperature or density could have a big effect within that range.

Equation of state treatments that have been applied to solar conditions have been described by Bodenheimer et al. (1965, their appendix A); Eggleton et al. (1973); Däppen (1980); Ulrich (1982); Noels et al. (1984); and Mihalas et al. (1988a). The treatments prior to 1980 do not include the corrections due to the Coulomb interaction whereas all treatments from 1980 to the present do. All these treatments have recognized the need to maintain thermodynamic consistency.

Another aspect of the equation of state that has only recently received much attention is the question of pressure ionization and the elimination of atomic states close to the continuum. The importance of an appropriate resolution of this question is easily demonstrated by applying the Saha equation to the solar center where 20% of the hydrogen is found to be neutral. Normally, partial ionization of hydrogen is ignored for temperatures above 10^6 K so the recombination is never produced in theoretical models. This effect should not be present in thermodynamically consistent theory. A closely related part of the problem is the need to truncate the partition function sum which nominally includes an infinite number of states and will diverge. Evidently an appropriate theory should prevent the occurrence of neutral hydrogen at the solar center by restricting the range of the sum for this simple species to no states at all.

The truncation of the partition function for the high-lying states is the most troublesome aspect of the equation of state. Eggleton et al. (1973) have lowered the ionization energy as a function of the density in order to assure complete ionization. Däppen (1980) has used a confined atom model to modify the energy levels so that at high densities no bound state is present. Ulrich (1982) used the Planck-Larkin partition function (Larkin 1960) which includes the compensating effects of high-lying bound states and scattering states of free electrons and does not exhibit divergences. Noels et al. (1984) have used a modified form of the Graboske et al. (1969) theory in which a Yukawa potential is used in place of the Coulomb potential to describe the atomic wave functions. In this approach bound states disappear discontinuously when their binding energy vanishes. Noels et al. (1984) smoothed over these discontinuities. The chapter by Däppen et al. discusses the equation of state problem in the context of the most recent available approaches.

The issues concerning the truncation of the partition function have not been fully resolved at present. They are important to helioseismology but not to the solar neutrino problem. Because the outer layers of the Sun are most influenced by these uncertainties, it may be difficult to establish the correct approach based on solar data alone. The outer layers contain other uncertainties such as inhomogeneities and magnetic fields that could mask the influence of the equation of state.

Numerical implementations of the more complex equations of state have been explicit and based on analytic representations of the functional dependences of the actual thermodynamic quantities. Previously, the equation of state theories have been simple enough that their analytic formulae were compact and did not require significant computational resources. The most recent work by Däppen et al. (1988a) differs from the previous theories applied to solar models in that the results are sufficiently complex that they have to be represented by fitting formulae. This was the approach used by Christensen-Dalsgaard (1988c).

C. Opacity

Opacities have been obtained from the Los Alamos Opacity Library. This library was described by Huebner et al. (1977) and has been made generally available to the community. Theoretical background to the opacity library was presented by Huebner (1986). The opacity library consists of a database with the monochromatic opacities tabulated as functions of T and η^*, an effective electron degeneracy parameter that is the (electron chemical potential)/kT. A table of this sort exists for each elemental species. The mixture is formed by choosing the values of η^* to be the same for each constituent and multiplying the monochromatic opacities by the relative abundance assigned to the species. Each table entry includes 2000 monochromatic opacities distributed over λ appropriate to that temperature. The full Los Alamos Opacity Library consists of a large number of magnetic tapes, one for each species. The number of tapes required to compute the opacity of a mixture is a function of the number of species that need to be included. The process of combining a full set of opacities to produce the mixture opacity is sufficiently computationally intensive and the cost of distributing a full set of data tapes sufficiently great that few installations received such a set. Most opacities for mixtures were computed in a two-step process where the mixing of the heavy elements that go to make up a stellar Z distribution was carried out at Los Alamos. Huebner and collaborators have then distributed 3 tapes per mixture—one each for H and He and a third for Z. The user then has the option of scaling Z up or down but not of changing the relative abundances that define Z.

The Los Alamos Opacity Library routines compute the Rosseland mean opacities at the grid points for which the monochromatic data is available. Then in a two-step process, the opacity is interpolated to the user-selected densities and temperatures using a cubic spline procedure. Because the distribution of points in the library database is fixed, it is not possible for the user to define the opacity on an arbitrarily fine grid without performing some interpolation. Figures 1 and 2 show the Rosseland mean opacities for two values of Z as determined by the Los Alamos Opacity Library routines. The points for which the monochromatic opacity database is available are shown by the + signs. At constant values of T, the solid lines show the dependency of the opacity as determined by the cubic spline interpolation routine distributed with the library. The places where solar models with the two values of Z cross the opacity lines are indicated by the $\odot$ symbols. The value of X_1 is constant in the two figures so that the portions of the diagrams where conditions are appropriate to the center of the Sun do not have the correct value of X_1. The relative accuracy of the interpolation is similar for the correct case of X_1 to that shown in Figs. 1 and 2. It is evident that the cubic spline interpolation procedure cannot be responsible for errors in the derived opacity at a level close to the quoted uncertainty of 20%.

One uncertainty in the Los Alamos Opacity Library as commonly applied is the conversion of the initial C into N by the CNO cycle. Bahcall and Ulrich (1988) applied a correction to the library opacities that they felt would account for this conversion. Cox (1990b) has pointed out to Ulrich that his attempts to reproduce this result were unsuccessful when carried out directly with the full set of elemental monochromatic opacities. Based on Cox's suggestion, Ulrich has checked the CNO correction using a restricted set of opacities that had been received outside the normal Los Alamos Opacity Library

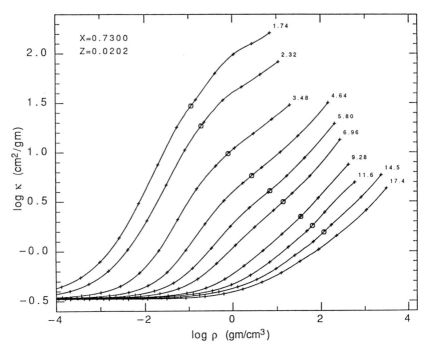

Fig. 1. The opacity of the solar material plotted vs density for lines of constant temperature in
10^6 K. Here the hydrogen mass fraction of the mixture is 0.7300, and the metals have $Z =$
0.0202.

distribution and has verified that the CNO correction should not be applied.
The change found by Bahcall and Ulrich (1988) was the result of using Huebner's interpolation formulae to calculate derivatives. The error of the fit in
Heubner's interpolation formulae then dominated the calculation of the
changes due to the alteration in CNO abundance.

D. Convection Theory

The outer layers of stars are usually unstable against thermal convection.
In the hotter stars the presence of a convection zone does little to modify the
structure from what would have prevailed without convection. However, for
stars cooler than $\sim$ 9000 K, the convective motions are able to carry most of
the stellar flux and prevent the occurrence of a highly superadiabatic temperature gradient. The Sun has an outer envelope that is convective with a temperature gradient very close to the adiabatic gradient. Where the gradient is
nearly adiabatic, the only theory required is a good equation of state. However, where the gradient is different from the adiabatic gradient, a theory must
be available to permit the calculation of the actual gradient from the physical
properties of the gas. Although progress has been made in the numerical

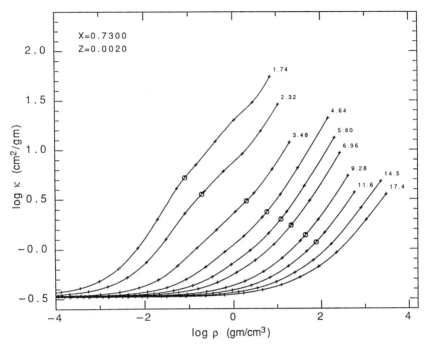

Fig. 2. The opacity of the solar material plotted vs density for lines of constant temperature in 10^6 K. Here the hydrogen mass fraction of the mixture is 0.7300, and the metals have $Z = 0.0020$.

modeling of convection through a direct integration of the hydrodynamic equations of motion, such calculations are too numerically intensive for use in the calculation of solar interior structure. Instead, the mixing-length theory of convection in essentially the form first derived by Vitense (1953) is used in calculating the structure of the superadiabatic regions of stellar envelopes. This formulation yields the temperature gradient from the total flux of energy and the physical properties of the medium. Its most important feature is an adjustable parameter called the mixing length that governs the relationship between energy flux and superadiabatic temperature gradient. A detailed discussion of this theory together with numerical results has been given by Cox and Giuli (1968) and the reader who needs a full exposition of the mixing-length theory should consult that reference.

We will not present the equations in full here because there is little new to add. However, there is one general point we wish to make that requires reference to the equations. The structure of the atmosphere/envelope region is described in terms of the logarithmic temperature gradients ∇ which is the actual value of $d \log T/d \log P$, and ∇_{ad} which is the adiabatic logarithmic temperature gradient $(\partial \log T/\partial \log P)_{ad}$ calculated from the thermodynamic description of the stellar material. The mixing-length equations are complicated substantially by the modifications due to radiative interactions the Vitense (1953) paper introduced. The zone in which this effect is important in the Sun is also the zone near the boundary where the mixing-length description is least likely to be valid. In the interior where the radiative interaction and boundary effects can be neglected, the superadiabatic gradient $\nabla - \nabla_{ad}$ is given by:

$$(\nabla - \nabla_{ad})^{3/2} = \frac{2}{\alpha^2} \frac{F_{Tot}}{F_{c0}} \tag{29}$$

where F_{Tot} is the total flux $L/(4\pi r^2)$, $\alpha = \ell/H_P$, ℓ is the mixing length, and H_P is the pressure scale height. The quantity F_{c0} has dimensions of an energy flux and represents twice the flux that would be carried by convection with $\nabla - \nabla_{ad} = 1$ and $\alpha = 1$. The quantity F_{c0} is given by:

$$F_{c0} = (gH_P\delta/\nu)^{1/2} \, \rho C_P T. \tag{30}$$

In Eq. (30) the quantities not previously defined are g, the acceleration of gravity; ν, a parameter describing the energetics of the mixing-length eddy; and the thermodynamic parameters, $\delta = -(\partial\log\rho/\partial\log T)_P$ and $C_P = T(\partial S/\partial T)_P$, the heat capacity at constant pressure.

Because of the increase in ρT toward the interior of the Sun, the superadiabatic gradient $\nabla - \nabla_{ad}$ rapidly becomes very small and the actual temperature gradient may be approximated by ∇_{ad} for most of the convective zone. Within a large range of possible values of α, this result remains valid and the uncertainties of the mixing-length theory have no influence on the actual structure. For the bulk of the convection zone, the calculation may proceed by simply replacing the energy transport equation with the requirement that the entropy should remain constant. The region beginning in the photosphere where this replacement is inadequate is also where all aspects of convection theory are poorly described by the mixing-length theory and is only ~ 1000 km thick at the top of the convection zone. This layer determines the adiabat the remainder of the convection zone follows and very little else. For the purposes of calculating the flux of neutrinos, the details of the thin superadiabatic layer have no effect. However, some of the solar oscillations do depend on these details, but at present no analysis of this layer has been carried out based on oscillation data due to the large measurement errors in the frequencies at high ℓ. The work by Guenther (1991) shows how low ℓ-modes near the acoustic cutoff frequency can be used to probe the structure of this zone when more precise observed frequencies are available.

III. NUMERICAL METHODS

A. The Structure Equations

The equations of stellar structure including boundary conditions are standard and will not be repeated here. Good presentations are found in a variety of standard texts such as Clayton (1968, chapter 6) and Cox and Giuli (1968). There are, however, several issues related to the numerical integrations that require reference to the individual equations, which are: (1) the mass conservation condition; (2) the hydrostatic equilibrium equation; (3) the equation of energy conservation; and (4) an energy transport equation. The energy conservation equation includes terms that depend on derivatives with respect to time while the other three include only derivatives with respect to the mass or distance. The time derivatives can cause numerical difficulties when small time steps are used, and in codes that must be able to handle all phases of stellar evolution, a method that is invulnerable to these difficulties must be used. Typically, following Henyey et al. (1959), numerical relaxation methods based on first-order difference forms of the differential equations are used.

For solar model computations the temporal derivatives in the energy equation are small corrections and small time steps must be used for only a restricted class of solar models. This special aspect of solar model computation permits the use of direct numerical integration of the structure equations from the center toward the surface and from the surface toward the center with the solutions joined at an arbitrarily selected midpoint. This model-fitting approach is a throwback to the earliest studies of stellar structure and is not commonly in use today. However, there is an important advantage to the "fitting" approach that has led to its continued use by Bahcall and Ulrich (1988) in their studies of the solar neutrino problem. Both ways of computing stellar models begin with trial values of the parameters defining the structure, and then they are improved by use of equations of condition that express the stellar structure equations numerically. In a fitting approach, only 4 parameters are needed to define the model (2 at the surface and 2 at the center) and the equations of stellar structure are used to calculate 8 parameters (4 from the inner zone and 4 from the outer zone) at a fitting point through numerical integration. The equations of condition are then simply the requirement that corresponding variables from the inner and outer zones match at the fitting point. In a Henyey approach, 4 parameters are needed for each mass zone of the model. The direct numerical integrations of the fitting approach can be written in such a way that the details of the definition of all aspects of the structure equations are hidden from the algorithm that carries out the relaxation of the surface and central parameters to achieve the fit at the midpoint. In contrast, codes based on a Henyey approach explicitly depend on details of the structure equations to derive corrections to the stellar structure variables throughout the model. Because of the wide variety of hypotheses that have been considered as potential solutions to the solar neutrino problem, the separation of the structure equations from the relaxation process of the fitting approach has been a great convenience.

In order to carry out the relaxation calculation, it is necessary to evaluate the derivatives of equations of condition with respect to the variable parameters. The use of backward time differences in the calculation of the nuclear abundances as discussed below introduces a complication in the evaluation of these derivatives. Nominally the Henyey relaxation method solves the spatial dimension alone. However, the backward time differences in the abundance equations spoils this separation because the abundances at the advanced time step are dependent on the temperature and density through the nuclear reaction rates. This dependence is especially important in the calculation of the derivative of ϵ_{nuc}, the nuclear energy generation rate. If this dependence is neglected, the rate of convergence of the model structure can be slowed and, in extreme cases, the algorithm can fail. To avoid such problems, the abundance variation should be included in the evaluation of $\partial \epsilon_{nuc}/\partial T$ by recomputing the abundances as well as the nuclear reaction rates at an altered temperature. In the case of a model-fitting algorithm, the derivatives of the parameters at the fitting point are found by numerical integrations with each of the central and surface parameters altered one at a time. As long as the abundances are computed by the normal procedure for each trial integration, the indirect effect of the abundance variation is automatically included in all derivatives.

Two disadvantages of the fitting method are the inability to treat transient phenomena such as may be caused by solar activity or mixing episodes and the serial nature of the algorithm that prevents effective use of parallel processors. These problems have not proven to be serious for the solar model calculations carried out to date. However, for the study of transient effects or for the extension of the calculation to post-main-sequence stages, the more commonly used relaxation methods are preferred. This approach was initially applied to stellar evolution calculations by Henyey et al. (1959) and extended by many authors to the study of all stages of stellar evolution. A concise description of the method can be found in the paper by Eggleton (1971) in combination with a description of the Gaussian elimination process in Press et al. (1986, Sec. 16.3).

There are a number of other important aspects of generating standard solar models. Many of these, such as boundary conditions, have been discussed by Ulrich and Rhodes (1983). The very large number of required mass zones for helioseismic studies has been discussed in the Christensen-Dalsgaard and Berthomieu chapter. Initial models for evolution studies are discussed in the chapter by Sofia et al.

B. The Nuclear Reactions

The treatment of the abundance changes arising out of the nuclear reactions is straightforward for the conversion of hydrogen into helium. The fractional changes in the abundances of these species is small for each time step and almost any algorithm will suffice. Usually no more than 10 time steps are required to being a solar model from initial main sequence to its current age with enough precision to ensure independence from time-step size. However, the treatment of the abundances of the minor species in the nuclear reaction network is a more difficult problem. Typically at the outer parts of the nuclear burning core, such species make a transition from their primordial abundance to a steady-state abundance that is achieved as an intermediary of a sequence of nuclear reactions. Either of the limiting cases can be treated in a simple way but the intermediate zone requires a more general algorithm. Because deviations from a simple steady state are potentially useful diagnostics of the solar interior, it is desirable to have such a general algorithm.

The archetypical reaction sequence for an intermediary species is:

$$\mathbf{j + k \rightarrow i + m} \tag{31}$$

and

$$\mathbf{i + l \rightarrow p + q} \tag{32}$$

which represents the formation of **i** through the reaction of **j** and **k** followed by the destruction of **i** through its reaction with **l**. Species **m, p** and **q** do not influence the abundance of **i**. For the species discussed in Sec. II.A, only ^{1}H and ^{4}He do not have this character in the reactions governing their abundance. We will assume that the evolution of the model is being followed by a series of models at specific time points $t = t^{(0)}, t^{(1)}, \ldots, t^{(N)}, t^{(N+1)}$. In the discussion below we shall denote the time step by a parenthetical superscript so that $t^{(N)}$ represents the time at step N. Although most actual species are governed by equations that are more complicated, we discuss here the simplified case:

$$\frac{1}{A_i} \frac{dX_i}{dt} = R_{j,k} X_j X_k - R_{i,l} X_i X_l. \tag{33}$$

As long as $R_{j,k} X_j X_k$ is approximately constant, this equation has a solution:

$$X_i(t^{(N+1)}) = X_i(t^{(N)}) \exp(-\Delta t / \tau_i) + X_i^{ss} [1 - \exp(-\Delta t / \tau_i)] \tag{34}$$

where X_i^{ss} is the steady-state abundance and τ_i is the lifetime of species **i**. These quantities are given by:

$$X_i^{ss} = \frac{R_{j,k} X_j X_k}{R_{i,l} X_l} \tag{35}$$

$$\tau_i = \frac{1}{A_i R_{i,l} X_l} \tag{36}$$

and

$$\Delta t = t^{(N+1)} - t^{(N)}. \tag{37}$$

As long as the time steps can be chosen so that $\Delta t/\tau_i$ is of order unity or smaller for all species **i**, no special procedures need to be followed. However, for the short-lived species, this requirement would force unacceptably short time steps.

The difference form of Eq. (33) is readily obtained by replacing dX_i/dt by $(X_i^{(N+1)} - X_i^{(N)})\,/\Delta t$ where now the parenthetical superscript indicates the time step of all quantities, not just the time. It is now necessary to specify the time step to be used for the evaluation of the quantities on the right side of Eq. (33). As is well known (Richtmeyer and Morton 1967), backward differences using the advanced time step on the right side are required for numerical stability. We use backward differences and rearrange Eq. (33) into:

$$X_i^{(N+1)} = \frac{X_i^{(N)} + A_i R_{j,k}^{(N+1)} X_j^{(N+1)} X_k^{(N+1)} \Delta t}{1 + A_i R_{i,l}^{(N+1)} X_l^{(N+1)} \Delta t} = \frac{X_i^{(N)} + X_i^{ss} \Delta t/\tau_i}{1 + \Delta t/\tau_i}. \tag{38}$$

There are several things to note about Eq. (38). First, the values of X_i^{ss} and τ_i in the second form of the equation are to be evaluated at the advanced time step. Second, when changes in the conditions in the model imply changes in the steady-state abundance and $\Delta t/\tau_i$ is large, the numerical solution will not track the steady-state value nearly as closely as it should. For example, with $\Delta t/\tau_i = 10$, Eq. (34) gives a 5×10^{-5} deviation from X_i^{ss} while Eq. (38) gives a 9% deviation. Third, as long as the backward differences are evaluated in a fully self-consistent fashion, the system of equations formed out of all those like Eq. (38) describes a set of abundances that depend on the conditions at the advanced time step. To account fully for this dependence, the abundances should be updated during the iterative cycle that defines the model at the advanced time step. If the abundance variation is ignored in taking the derivative of ϵ_{nuc} with respect to temperature as well as other less sensitive derivatives this dependence can cause convergence difficulty.

One method of dealing with the difficulties described in the preceding paragraph might be to use the rates from the previous time step instead of the advanced time step to derive the abundances. For numerical stability it is necessary to use the abundances from the advanced time step so that one is required to use products of the form $R_{j,k}^{(N)} X_j^{(N+1)} X_k^{(N+1)}$ in computing the abundance. However, the energy generation will involve products of the form $R_{j,k}^{(N+1)} X_j^{(N+1)} X_k^{(N+1)}$ and a fundamental inconsistency will be present in the model. Because the rates generally increase during the evolution of solar models, this approach will overestimate the energy production or underestimate the consumption of hydrogen. In some cases spurious convection zones can occur as a result of using a mixed algorithm of this type.

The proper approach to the treatment of the abundance is to use a fully consistent backward difference system of the sort indicated by Eq. (38). The derivatives of quantities with respect to temperature should be taken numerically and should include a redetermination of the abundances at the perturbed temperature. Application of this approach in the context of the fitting method is particularly simple because the temperature and density are always known in the trial integration models. Convection zones where they appear in some nonstandard solar models cannot be treated in a fully self-consistent fashion because the zone abundance depends on the conditions throughout the unstable region and these conditions cannot be known until the zone is computed. A proper treatment has to include the abundances as iterated parameters like the temperature and density. Models including diffusion suffer some of the problems of a convection zone but to a degree that depends on the strength of the diffusion.

Equations (19) to (27) can be written in difference form following the expansion of Eq. (33) into Eq. (38). The resulting system of algebraic equations is nonlinear in the $X_i^{(N+1)}$ but can be solved straightforwardly by a variety of techniques. The approach described by Bodenheimer et al. (1965) for a similar set of equations has proven effective in solving the abundance network defined here. Maeder (1983) has described a method in which the composition is always slightly out of phase with the rates during the iterative cycle of the Henyey stellar structure calculation. By taking small time steps he finds that the composition changes are negligible after 3 iterations. Lebreton and Maeder (1986,1987) and Lebreton et al. (1988) have used this approach in their solar model calculations.

IV. RECENT RESULTS

A. Evolution Calculations

Recent standard solar models by several groups are presented in Table III. Most of this activity has been inspired by the continuing solar neutrino problem discussed in the chapters by Davis and by Rosen. Some investigations by Turck-Chieze et al. (1988) and by Guenther et al. (1989) were made to see why there are still important differences in the helium contents, the neutrino outputs and the mixing lengths, but more work is needed. Even though all these papers match the observed solar mass, radius, luminosity and age to reasonable precision, most do not make any detailed comparisons with the observed oscillation frequencies. These later data may be the most constraining of all observations.

It appears that using the most recent Los Alamos opacities from the Opacity Library, the helium mass fraction is about 0.28 ± 0.01. Guenther and his associates have shown that the older Cox-Stewart (1970a,b) or Cox-Tabor (1976) opacities produce Y near 0.24 instead of currently favored 0.28. That is why the Guenther and Sarajedini (1988) and the Christensen-Dalsgaard et al. (1988b) papers and another one by Christensen-Dalsgaard (1990a), based on older model calculations and not in the table because Y, SNU and l/H_p data are not given, report Y values much lower.

The SNU values for the Homestake mine Davis chlorine detector all cluster within one SNU of 8 except for the Turck-Chieze one, when the pressure systematic error (due to only approximate atomic weights of hydrogen and helium) of Cox et al. (1989) was corrected by Cox et al. (1990) and Guzik and Cox (1991). It appears that most of the reason why the 5.8 SNU for the Turck-Chieze group is lower than the others is the lower ^7Be proton-capture cross section used. Another reason this group gets a smaller neutrino output is because they inadvertently applied two opacity corrections for the collective effects on the free-electron scattering. This effect was already included in the Los Alamos opacity tables, and the double correction reduces the central temperature even more to give lower neutrino flux. Since there may still be a small question about the ^7Be proton capture to produce ^8B, SNU values in Table III may be overestimated. But no predictions are as small

TABLE III
Some Recent Solar Models

Authors	Y	SNU	ℓ/H_p
Bahcall, Ulrich, 1988	0.271	7.9	—
Lebreton, Däppen, 1988	0.278	7.6	2.16
Guenther, Sarajedini, 1988	0.240	—	1.35
Lebreton, Berthomieu, Provost, Schatzman, 1988	0.287	8.0	2.18
Lebreton, Berthomieu, Provost, Schatzman, 1988	0.291	8.4	2.11
Turck-Chieze, Cahen, Cassé, Doom, 1988	0.276	5.8	1.55
Christensen-Dalsgaard, Däppen, Lebreton, 1988b	0.237	—	—
Korzennik, Ulrich, 1989	0.271	8.2	—
Guenther, 1989	0.282	—	1.25
Guenther, Jaffe, Demarque, 1989	0.28	—	1.24
Cox, Guzik, Kidman, 1989	0.291	11.4	1.89
Cox, Guzik, Raby, 1990	0.28	8.0	1.89
Sackman, Boothroyd, Fowler, 1990	0.278	7.7	2.1
Sienkiewicz, Bahcall, Paczynski, 1990	0.280	7.7	1.62
Kim, Demarque, Guenther, 1991	0.274	—	2.18
Guzik, Cox, 1991	0.270	8.7	2.29

as near 2.1 ± 0.3 SNU, the mean value observed by Davis over the last 20 years.

The variations in the mixing length to pressure scale height ratios in the last column of Table III are not something to worry about. It could be that different teams (such as Guenther and Turck-Chieze) use slightly different definitions for this quantity. Most teams use the Vitense (1953) formulation for convection as described in Cox and Giuli (1968). Atmospheric opacities can also influence the derived value of the mixing-length parameter as discussed by Kim et al. (1991). Models without using the most modern solar atmosphere and molecular opacities have considerably lower mixing lengths. The main point is that this mixing-length ratio is adjusted so that the model radius at the solar age is exactly that observed.

B. Oscillation Frequencies

Adiabatic and nonadiabatic solar p-mode frequencies have been calculated most recently by Christensen-Dalsgaard (1990a) and by Cox et al. (1989). Both these studies have emphasized that the highest-precision opacities and equations of state are necessary to predict frequencies that are close to those observed. These results have been confirmed independently by Kim et al. (1991). Opacity increases below the convection zone were suggested by Christensen-Dalsgaard et al. (1985b) and by Korzennik and Ulrich (1989), and they have been confirmed by Cox et al. (1989) with much better agreement between the observed and predicted p-mode frequencies. The latest comparisons with observations (Libbrecht et al. 1990) using nonadiabatic theoretical frequencies of Guzik and Cox (1991) are given in Figs. 3, 4 and 5. The first of these compares the observed minus the calculated low-degree p-mode oscillation frequencies. For these deeply penetrating modes, there seems little improvement found with the MHD equation of state that mainly considers more accurately the details of the hydrogen and helium ionizations in the convection zone of solar models.

Figure 4 shows that the more accurate MHD equation of state begins to be important for modes that are completely contained in the convection zone, and sample much of the helium ionization region. In Fig. 5 for very high-degree modes, the observational uncertainties are so large that it is difficult to confirm the superiority of the MHD equation of state, even though it certainly has better physics.

Only the Los Alamos predictions for solar oscillation frequencies consider nonadiabatic effects. These effects reduce the frequencies by typically 5 to 10 μHz, and make agreement with observations much better. However, at frequencies approaching 4000 μHz, the pulsation damping begins to occur in the surface regions at or even above the photosphere. In that case, both the model structure and the time-dependent energy conservation equation in the pulsation equations are inaccurate because they assume photon diffusion

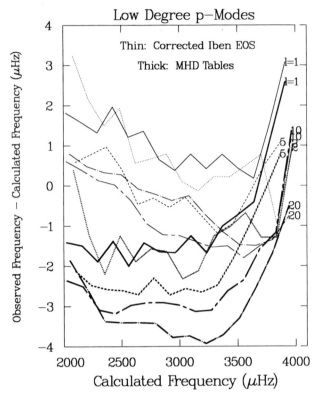

Fig. 3. The observed minus calculated nonadiabatic frequencies plotted vs frequency for low-degree *p*-modes. Both the corrected Iben and the new MHD equations of state are used. Individual mode frequencies are connected with lines to identify clearly the ℓ values.

rather than transport. Modern predictions cannot be made accurately for the highest observed solar oscillation frequencies (Guenther 1991).

C. Nonstandard Models

We here mention three nonstandard features of solar models: individual element diffusion and the floating of hydrogen, large-scale mixing during the solar life in addition to that in the surface convection zone, and low central temperature models caused by the efficient conductivity of weakly interacting massive particles (WIMPs). The reader is directed to learn about other nonstandard solar models discussed by Maeder (1990) and by Sienkiewicz et al. (1990).

Element diffusion has been discussed by Michaud and Vauclair in their chapter. The latest results are discussed by Proffitt and Michaud (1991). The situation with lithium as observed at the surface is covered in detail by Schatzman in his chapter, as he discusses both diffusion and large-scale mix-

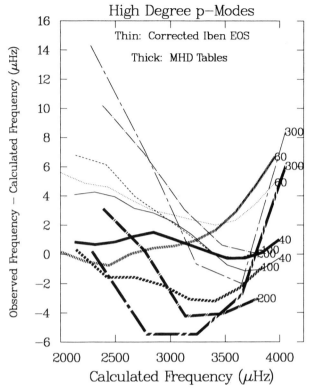

Fig. 4. The observed minus calculated nonadiabatic frequencies plotted vs frequency for high-degree *p*-modes. Both the corrected Iben and the new MHD equations of state are used. Individual mode frequencies are connected with lines to identify clearly the ℓ values.

ing. The comparison of observed with predicted oscillation frequencies seem to rule out any major departures from the standard solar models. Cox et al. (1989) and Bahcall and Pinsonneault (1991) have given recent diffusion results.

Several authors (Gilliland et al. 1986; Gilliland and Däppen 1988) recently have considered in detail the possibility that the solar neutrino problem (and the missing mass of the Universe) can be solved with 5 proton mass particles named cosmions. These WIMPs could orbit the solar center so that energy acquired by rare collisions at the center deposits 10% of the mass and radius farther out. This very efficient conduction could cool the central solar temperature from 15.6×10^6 K to about 13×10^6 K. Recent models by Cox et al. (1990), using many particle physics considerations, indicate that such low temperatures are very unlikely, but earlier Gilliland and Däppen (1988) using earlier opacities from Cox and Stewart (1970a) came to the opposite conclusion. Differences between the low radial order oscillation frequencies

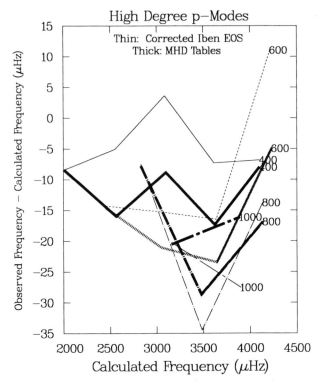

Fig. 5. The observed minus calculated nonadiabatic frequencies plotted vs frequency for very high-degree *p*-modes. Both the corrected Iben and the new MHD equations of state are used. Individual mode frequencies are connected with lines to identify clearly the ℓ values.

for the radial and quadrupole modes are very sensitive to the central structure of the Sun. This is because these modes reach to the center with significant amplitude, and they are very similar in their eigenvector structure for the model regions away from the very center. Subtraction of the frequencies cancels much of the sensitivity for all parts of the model except at the center where these modes do have some different structures. Standard models predict the observed differences, but WIMP models seem to produce significantly smaller ones.

This lack of confirmation of WIMPs in the Sun is also discussed by Christensen-Dalsgaard and Berthomieu and by Gough and Thompson in their chapters. The first of these chapters uses models calculated with the latest equation of state and opacity, while the second uses inversion techniques that can reveal the internal sonic velocity of the Sun. Both suggest the standard solar model for the Sun.

It has been recognized for a long time that detection of low-degree and radial-order *g*-modes would be an excellent probe of the solar center. The controversial detections of these modes (see the H. Hill et al. chapter on *g*-modes) have been discussed by Cox (1990*a*) in terms of the central temperature and the predicted *g*-mode frequencies. Again standard, nonmixed and non-WIMP models fit the reported observations best.

MIXING IN THE RADIATIVE ZONE

EVRY SCHATZMAN
Observatoire de Meudon

This is a brief review of the present situation concerning the process of mixing in the radiative zone. After an introduction in Sec. I, Sec. II presents the problem of penetration of the convective motion, and concludes that the penetration is likely to be small. Section III is an analysis of the problem of turbulent mixing, with a discussion of the different mechanisms, which are usually supposed to be at the origin of the turbulence. In Sec. IV is discussed the effect of rotation and time dependent mixing. The fact that these mechanisms are very unlikely to be active in the present Sun, and the difficulty of explaining the abundances of lithium as a function of mass and age suggest that some other mechanism should be present and efficient. Section V concerns the possible role of internal waves, which can as well carry angular momentum and create a random walk, equivalent to a diffusion effect, which can transport lithium to the burning level.

I. PRESENTATION OF THE PROBLEM

Mixing in the radiative zone is a physical process that cannot be rejected. Mixing takes place in many stars. The evidence comes from a number of facts such as the difference between the chemical composition of the atmosphere of Am and normal A stars, between Am and δ Scuti stars (Baglin 1972,1975), the lithium burning in main-sequence stars, the dependence on spectral type and the remarkable age effect on lithium abundance (Soderblom 1983), the ($^{13}C/^{12}C$) abundance ratio in giants of the ascending branch (Bienaymé et al. 1984), and the bewildering quasi-solid rotation inside the Sun (Brown et al. 1989; Dziembowski et al. 1989; Thompson 1990).

The major methodological problem of dealing with the question of the solar internal mixing, consists in the proper application to the Sun of physical models, which are applicable to other stars. There is the difficulty that there is much information on the Sun that has no stellar counterpart (such as the solar neutrino flux or the internal solar rotation).

Mixing in the solar interior concerns either the transport of passive contaminants (such as ^{3}He, ^{7}Li, ^{9}Be, etc.) or the transport of angular momentum. Physically speaking, it can be due to overshooting from a convective zone to the neighboring stable radiative zone, to turbulent flows generated by various instabilities, meridional circulation, or to any kind of random motion, for example the flow of internal gravity waves induced by the turbulent motions in the convective zone (Press 1981; Press and Rybicki 1981).

II. PENETRATIVE CONVECTION

A. Transport Process in the Convective Zone

From observations it is easy to derive an order of magnitude of the turbulent diffusion coefficient D_T in the outer part of the convective zone, $D_T \cong 10^{12}$ to 10^{13} cm^2 s^{-1}. Near the bottom of the convective zone, where the motion of the convective cells is quasi adiabatic, the turbulent velocity is approximately

$$v_T = \left(\frac{F}{10\rho}\right)^{1/3} \tag{1}$$

where F is the total energy flux of the star. The factor 10 comes from the value of the heat capacity at constant pressure $c_p = (5/2) \, \mathfrak{R}$ and from the numerical coefficients coming from the classical mixing-length theory of the convective zone. With a mixing length of the order of the pressure scale height, this gives at the bottom of the solar convective zone

$$D_T \cong (1/3)lv_b \cong 6 \times 10^{12} \text{ cm}^2\text{s}^{-1} \tag{2}$$

for a density $\rho = 0.16$ g cm^{-3} and a temperature $T = 2 \times 10^6$ K.

It is important to notice the large value of D_T. The time scale of the turbulent diffusion mixing (TDM) over one scale height is of the order of 46 days, thus an incredibly short interval of time, and when discussing the problem of overshooting, this has to be carefully kept in mind.

B. Constraints on Penetrative Convection in the Radiative Zone

Surface abundances of lithium in the Sun and in solar-like stars put constraints on the properties of penetrative convection. In the case of the Sun, it is possible to derive from helioseismological data the depth of the convective zone. The distance h from the bottom of the convective zone (defined by the condition that the radiative gradient (d $\log T$/d $\log P$)$_{rad}$ is superadiabatic) to the level of lithium burning depends on the model of the convective zone. Helioseismology (chapter by Christensen-Dalsgaard and Gough; Cox et al. 1989; Merryfield et al. 1991) places in the present Sun the bottom of the adiabatic region around 2.3×10^6. With a burning temperature of lithium around 2.7×10^6 K, this gives h of the order of 40,000 km. With a time scale of lithium destruction of the order of $t \cong 10^9$ yr, we can derive the constraint on D_T in two ways: (a) the average $< D_T >$ should be $\sim (h^2/t) \cong 500$; (b) the turbulent diffusion coefficient D_T decreases exponentially down from the bottom r_l of the convective zone. It is then possible to estimate the vertical scale height from the relation giving in order of magnitude the diffusion time t_D:

$$\left(\int \frac{dr}{\sqrt{D}} \right)^{-2} = t_D. \tag{3}$$

With $D = \exp[-k(r_l - r)]$, this gives the scale height

$$1/k = 0.035 \, h = 1400 \text{ km.} \tag{4}$$

We can comment these results in the following way. If we consider hypothesis (a), the low value of D_T in the radiative zone underlying the convective zone suggests the presence of a discontinuity; with hypothesis (b), there is a very sharp decrease of the diffusion coefficient, with a very small scale height $(1/k) \cong 1400$ km, suggesting a very abrupt change in the hydrodynamics of the Sun.

C. Overshooting: State of the Art

There are presently three approaches to the problem of overshooting: the phenomenological theory, which is based on the concept of convective eddies and their penetration in the radiative zone, the nonlinear modal analysis, and the numerical simulation. In the picture of Saslaw and Schwarzschild (1965), the stellar gradient

$$\nabla^* = (\text{d} \log T/\text{d} \log P)^* \tag{5}$$

in the radiative zone is supposed to be identical to the local radiative gradient $\nabla^* = \nabla_R$. The difference $(\nabla_{ad} - \nabla^*) \equiv (\nabla_{ad} - \nabla_R)$ grows rapidly away from the boundary of the convective zone as given by the Schwarzschild condition (Schwarzschild 1906). The consequence is a great efficiency of the negative buoyancy and the distance of overshooting is very small. However, when a nonlocal description of the mixing-length theory is used (Shaviv and Salpeter 1973; Cogan 1975; Maeder 1975), the stellar gradient is different from the local radiative gradient. The sign of $(\nabla_{ad} - \nabla^*)$ has indeed changed across the Schwarzschild boundary, but the negative buoyancy is smaller, and the distance of overshooting is larger, by about one order of magnitude (let us say, from 1% of the pressure scale height to 10% of the pressure scale height). Roxburgh (1978,1989) has clearly shown that the convective zone extends beyond the boundary defined by the Schwarzschild condition, into a subadiabatic region, where $L_{rad} > L^*$. However, from the preceding paragraph, a thin boundary layer must exist between the fully mixed region and the almost quiet radiative zone.

The nonlinear modal analysis as carried by Massaguer et al. (1984), following former developments by Zahn et al. (1982), based on the introduction of hexagonal cells, lead to different results, depending on whether axial motions are going upward or downward. Massaguer et al. (1984) use a simplified description of a convective zone bounded above and below by a radiative zone. The polytropic index is equal to 1 in the instable zone and is equal to 3 in the radiative zones. They give, as the depth of penetration, the distance Δ to the first node of the vertical velocity. With the units that they have chosen they obtain for the penetration below the convective zone $\Delta_D \cong 0.80$ (downward motion), and $\Delta_U \cong 0.1$ (upward motion), corresponding, respectively, to 2.9 and 0.36 scale heights. The behavior of the mesogranulation and of the supergranulation suggests that the axial plume is going upwards. With the same value of the Rayleigh number, $Ra(\cong 10^9)$, this would correspond to a scale of vertical penetration of the order of 18,000 km.

However, the dependence on the Rayleigh number, Ra (Zahn et al. 1982; Massaguer et al. 1984) shows that Δ_U decreases when Ra increases. Considering the asymptotic equation describing the vertical motion

$$(D^2 - a^2)^3 W = - RaWa^2 \tag{6}$$

where a is the horizontal wave number, and $D = \partial/\partial z$, it is easily seen that the constant k, describing the asymptotic exponential decrease $\exp(-kz)$, is proportional to $Ra^{1/6}$. With the assumption made by Massaguer et al. (1984), that in the definition of the Rayleigh number it is legitimate to introduce the turbulent viscosity and the turbulent thermal diffusivity instead of the usual microscopic quantities, and extrapolating their results to higher values of the Rayleigh number ($Ra/Ra_{crit} >> 10^6$), it is possible to obtain a penetration depth of the order of a few thousand kilometers, which is compatible with the constraints derived from the lithium abundance.

The numerical simulation (Graham 1977; Nordlund 1985b; Glatzmaier 1985a; Hurlburt et al. 1986) possesses strong downflows for the two-dimensional as well as for the three-dimensional models. In the model of Hurlburt et al. (1986), with a relatively shallow convective zone (1.32 H_P), the downward penetration depth is 0.92 pressure scale height. The authors in their conclusions raise some doubts about whether "the behavior revealed in [their] two-dimensional simulation is actually representative of stellar convection." Comparing the results of the nonlinear modal analysis to the numerical simulation, the conclusion is certainly that there is a great need of studying further the problem.

III. TURBULENT DIFFUSION MIXING

A. Overview and Observational Tests

The destruction of lithium in solar-like main-sequence stars is a relatively slow process; for the Sun, the time scale is $\sim 10^9$ yr with similar orders of magnitude for solar-like stars. For stars with a smaller mass, having a deeper convective zone, the time scale is shorter. In a first approach (Schatzman 1981), this seemed related to the fact that the distance from the bottom of the convective zone to the level of lithium burning is smaller. As we shall see in Secs. IV and V, this is not the physical reason.

The time dependence of the lithium abundance, like those obtained by Boesgaard (1987), Boesgaard et al. (1988a,b), Boesgaard and Tripico (1986a,b), Hobbs and Pilachowski (1986a,b), is a strong constraint, as it cannot be explained by a pre-main-sequence destruction of lithium as suggested by Bodenheimer (1965) and discussed by D'Antonia and Mazziteli (1984), by Michaud (1986), Proffitt and Michaud (1989) and by Swenson et al. (1990). It cannot be explained either by convective penetration (overshooting), as it would need a very fine adjustment of the convective zone and of the convective penetration in order for the stars to burn lithium exactly at the proper rate. This would imply that the physical parameters defining the properties of the convective zone and the depth of overshooting are mass dependent. Finally, as mentioned in Sec. II, due to high Rayleigh numbers in deep convective zones, the depth of convective penetration, is probably small.

It has been noticed, for example, by Soderblom (1983) that there is a close connection between rotation and age and between surface abundance of lithium and age. It is perhaps daring to use the surface abundance of lithium as a precise age indicator, but it should be noticed that age calibration by the surface abundance of lithium leads to a remarkably smooth relation between age and rotation, which suggests a universal time-dependent and mass-dependent law of rotation for main-sequence stars. The existence of a few misfits, as noticed by Boesgaard and Tripico (1986b), by Balachandran et al. (1988), and by Balachandran (1990), suggests possible differences in angular momentum history (see Sec. IV.B).

All these considerations have led to the hypothesis that there is one physical process that is active in radiatively stable regions: macroscopic diffusion. The close relation between rotation, lithium and age has led now for a long time to the idea that turbulent diffusion depends on rotation and governs both the abundance of nuclear processed elements and the distribution of angular momentum (Endal and Sofia 1978; Pinsonneault et al. 1989; Kawaler 1988*a*; Deliyannis et al. 1990). It is important to present a brief review of this theory, as it gives us the opportunity of emphasizing its weak points.

B. Phenomenological Approach

The interest of the phenomenological approach is to provide orders of magnitude of the turbulent diffusion coefficient D_T. Several models have been studied by Schatzman and Maeder (1981) and by Lebreton and Maeder (1987). Schatzman and Maeder (1981) have chosen a turbulent diffusion coefficient $D_T = Re^*v$, where Re^*, a pseudo-Reynolds number, is a constant, space and time independent; Lebreton and Maeder (1986,1987) have chosen a variety of space dependent functions for D_T. Whereas the main aim of these models was to provide an explanation for the neutrino flux deficiency, the application to the mixing of ^{3}He, ^{9}Be and ^{7}Li provides constraints on the numerical values of D_T.

Consider how the constraints come out when traveling from the inside of the Sun to the bottom of the convective zone. The ^{3}He concentration presents a strong maximum in the neighborhood of 0.5 $M_\odot$. This corresponds to a steep gradient of the concentration around 0.4 $M_\odot$. In the presence of turbulent diffusion, ^{3}He would be carried by diffusion to the surface of the Sun. The estimate of the amount of ^{3}He carried by diffusion to the convective zone depends on the assumptions concerning the primordial abundance of ^{2}D in the Sun. On the other hand, the comparison by Geiss (1973*a,b*) and Bochsler and Geiss (1975) of the (^{3}He/^{4}He) abundance ratio at the surface of the Moon, in the present solar wind and in the Moon ashes, suggest a slow increase of the ^{3}He abundance during the last 1 Gyr, with an overall overabundance of about 15%. Altogether, it is possible to derive from the results of Schatzman and Maeder (1981) an upper limit to the turbulent diffusion coefficient $D_T = Re^*v$ with $Re^* < 35$ (Schatzman 1984). With $v \cong 5$ this gives an upper limit to D_T of the order of 150 at $R \cong 0.4$ $R_\odot$.

In the same way, a simple model of lithium burning is sufficient to get an idea of the value of the time average $< D_T >$ in the region just below the convective zone. The diffusion equation can be solved with the following boundary conditions (Schatzman and Maeder 1981): $c(Li) = 0$ at the level of lithium burning; continuity of the flux of lithium at the boundary of the convective zone, $-D_T (\partial c/\partial z) = -\lambda H_p c$, where λ is the rate of destruction of lithium, with an exponential decay of lithium, $\exp(-\lambda t)$. λ is given by the relation

$$(\lambda h^2/D_T)^{1/2} \tan (\lambda h^2/D_T)^{1/2} = (h/H_P) \qquad (7)$$

where h is the distance from the bottom of the convective zone to the level of lithium burning, and H_P is the pressure scale height at the bottom of the convective zone. The solution of Eq. (7) for $h = 0.43 \times 10^{10}$ cm, $H_P = 0.443 \times 10^{10}$ cm is

$$(\lambda h^2/D_T) = 0.851. \tag{8}$$

The present lithium deficiency is $[c(t)/c(0)] \cong 10^{-2}$ is obtained with $\lambda t = 4.605$ and then $< D_T > \cong 860$.

A similar result can be obtained with a model of ^{9}Be burning, that takes place at a greater depth than ^{7}Li burning (Schatzman and Maeder 1981). With a beryllium depletion factor of 1/2 one obtains an average diffusion coefficient $< D_T > \sim 300$. It is clear that the diffusion coefficient decreases with depth below the convective zone. The upper limit derived from the constraint on ^{3}He abundance is compatible with the results obtained from ^{7}Li and ^{9}Be.

We can judge from these numbers that the turbulence present in the radiative zone is of a very moderate strength, compared to the turbulence in the convective zone. Any physical theory of the turbulent diffusion mixing should provide values of the turbulent diffusion coefficient comparable to the values derived from the phenomenological theory.

C. Differential Rotation and Turbulence

It is well known that meridional circulation and differential rotation generate a variety of instabilities (see, e.g., Tassoul 1978; Spruit 1984b). It is generally agreed that these instabilities generate some sort of turbulence. Kippenhahn and Thomas (1981), for example, give an estimate of the characteristics of the turbulent flow generated by the Goldreich-Schubert-Fricke instability (Goldreich and Schubert 1967; Fricke 1968).

1. Meridional Circulation. The meridional circulation (Zahn 1983,1987) induced by rotation generates a differential rotation that feeds a two-dimensional turbulence on equipotentials. At first look, it seems that it can decay into a three-dimensional turbulence at small scales, when the inertial effects overcome the Coriolis force. l being the correlation length and u the *rms* of the corresponding velocity, this takes place when the Rossby number $Ro = (u/|\Omega|)$ becomes > 0.2 (from the data of Hopfinger et al. [1982]). The turbulent diffusion coefficient D_z given by Zahn (1987), when $\lambda = (\Omega^2 R^3/GM)$ is small, is

$$D_z = \frac{4}{5} \frac{L\Omega^2 r^6}{G^2 M_r^3} \frac{<\rho>}{\rho} \left| \left(1 - \frac{\Omega^2}{2\pi G\rho}\right) \right|. \tag{9}$$

This can eventually be adjusted by writing $D_{\text{turb}} = fD_z$, where f is an efficiency parameter.

However, it is necessary to check whether the Richardson-Townsend condition is fulfilled with describing the generation of the three-dimensional turbulence by the decay of the two-dimensional turbulence on equipotentials. Zahn gives an estimate of the energy available for the feeding of the three-dimensional turbulence.

$$\varepsilon_t = \eta D_{th} \Omega^2 \qquad (10)$$

where D_{th} is the radiative diffusivity. η is the efficiency parameter

$$\eta = \frac{\Omega^2 r}{g} \Delta\nabla^{-1}. \qquad (11)$$

The energy ε_t must exceed the work that is necessary to carry the turbulent cell vertically over the mixing distance l. If there were no radiative exchange of heat with the surrounding medium, the work done per unit of mass and per second would be

$$W = glv\Delta\nabla \frac{1}{H_P} \qquad (12)$$

where the product lv will be considered later as representing the diffusion coefficient. This should be multiplied by the ratio of the cooling time

$$t_{cool} = (l^2/D_{th}) \qquad (13)$$

to the characteristic time of the turbulence (l/v), which is the way of deriving the flux Richardson number (Townsend 1958). We then obtain the condition

$$\frac{L}{4\pi GM_r\rho} \frac{\nabla_{ad}}{\nabla_{rad}} \frac{\Omega^2 H_P^{1/2} \, r^{5/2}}{GM_r\Delta\nabla} > lv. \qquad (14)$$

Near the bottom of the convective zone of the present Sun, this gives $D_{turb} < 250$. When compared to the expression of the diffusion coefficient D_Z given by Zahn, this gives the condition

$$\left(\frac{H_P}{r}\right) > 1 \qquad (15)$$

where the number 1 on the right-hand side of the inequality is just an order of magnitude. Physically, it can be said that the power generating the turbulence is proportional to $(1/r)$, whereas the power necessary to overcome the gravity, which depends on the difference $(\nabla_{rad}\rho - \nabla_{ad}\rho)$, is proportional to $(1/H_P)$. The exact value depends: (a) on the efficiency factor f (if we assume that the characteristic Rossby number is $Ro = 0.2$, $f = 6.25$); and (b) on the way of estimating the time scale of the heat exchange of a turbulent element of size l by radiative diffusivity. The condition (Eq. 15) is fulfilled inside the Sun, for $r < 0.11\ R_\odot$. In other words, we have to consider, as a possibility, that the energy condition, necessary for the vertical motions to take place, will not be fulfilled in the presence of a density gradient, except in regions close of the center of the star. We shall conclude that there are doubts about the efficiency of the process described by Zahn (1983). This applies also to the description of the turbulent motions by Pinsonneault et al. (1989). Christensen-Dalsgaard (1990b) underlines the inconsistency between the frequency splitting derived from the model of Pinsonneault et al. (1989) and helioseismological data.

2. *Shear Flow.* The vertical gradient $(\mathrm{d}\ \Omega/\mathrm{d}\ \ln r)$ of the angular velocity can generate a turbulent flow if the Richardson-Townsend instability condition is satisfied (Zahn 1974). The Richardson number Ri, written here

$$Ri = \frac{N^2}{(\mathrm{d}\ \Omega/\mathrm{d}\ \ln r)^2} \tag{16}$$

with the usual definition of the Brunt-Väisälä is frequency N:

$$N^2 = (g\delta/H_P)\Delta\nabla \tag{17}$$

$$\delta = -\left(\frac{\partial\ln\rho}{\partial\ln T}\right)_P \tag{18}$$

and

$$\Delta\nabla = \left(\frac{\partial\ln T}{\partial\ln P}\right)_{\mathrm{ad}} - \left(\frac{\partial\ln T}{\partial\ln P}\right)_{\mathrm{rad}} \tag{19}$$

compares the buoyancy frequency with the vorticity. In a real fluid, the diffusion of heat weakens the restoring force. When the cooling time becomes shorter than the time $(\partial\Omega/\partial\ln r)^{-1}$ the instability criterion becomes

$$\frac{N^2 t_{\mathrm{cool}}}{(\mathrm{d}\Omega/\mathrm{d}\ \ln r)} < 1. \tag{20}$$

This condition was established by Townsend (1958) on heuristic arguments, in order to explain the dynamical behavior of the stratosphere.

If we wish to apply this criterion to predict the instability of a stratified shear flow, it remains to evaluate the cooling time t_{cool}. In a medium of thermal diffusivity D_{th}, $t_{\mathrm{cool}} \cong (l^2/D_{th})$, l being the characteristic length of the instability. In the present case, a reasonable conjecture is to identify the length l with the smallest scale that is unstable in the shear flow, namely verifying

$$\frac{l^2}{\nu} \frac{d\Omega}{d \ln r} \cong Re_{crit}. \tag{21}$$

Combining Eq. (20) and the expression of t_{cool}, one obtains the following instability criterion

$$\frac{N^2}{(d\Omega/d \ln r)^2} \frac{\nu}{D_{th}} < 1. \tag{22}$$

If this criterion is valid, we can consider that the Reynolds number associated with a given shear flow is

$$Re = \frac{D_{th}}{\nu} \frac{(d\Omega/d \ln r)^2}{N^2} > Re_{crit}. \tag{23}$$

With the expression of the thermal diffusivity

$$D_{th} = \frac{L_r}{4\pi r^2} \frac{\mu}{C_p \rho T} \frac{H_P}{\nabla_{rad}} = \frac{L_r}{4\pi GM_r \rho} \frac{\nabla_{ad}}{\nabla_{rad}} \tag{24}$$

we can write

$$D_{shear} = \frac{L_r r^4 (\mathcal{R}T/\mu)}{4\pi G^2 M^3} \frac{\nabla_{ad}}{\nabla_{rad}} \frac{(d\Omega/d \ln r)^2}{G\rho} \frac{1}{\delta\Delta\nabla}. \tag{25}$$

It is important to consider orders of magnitude of the constraint on the shear. If $(d\Omega/d \ln r)$ is of the order of Ω (within a factor of a few units), we can write

$$\Omega^2 \geq N^2 \left(\frac{\nu}{D_{th}}\right) Re_{crit}. \tag{26}$$

With $Re_{crit} = 10^3$, $\Delta\nabla \cong 0.1$, $g = 10^5$, $H_P = 10^{10}$, $(\nu/K) \cong 10^{-6}$, we obtain, respectively, $N^2 \cong 10^{-6}$ and $\Omega^2 \cong 10^{-9}$, which means at the surface

$$V_{equ} \cong 20 \text{ km s}^{-1}. \tag{27}$$

This is 10 times the present equatorial velocity of the Sun, but it is clear that turbulence due to vertical shear flow may have been important at an early stage of the evolution of the Sun.

3. Gradient of the Mean Molecular Weight.   If the μ-gradient is too large, there is a complete inhibition of the instabilities. With the definition of the μ-buoyancy

$$N_\mu^2 = \frac{g\delta}{H_P} \nabla\mu \qquad (28)$$

with

$$\nabla\mu = \frac{d\,\ln\mu}{d\,\ln P} \qquad (29)$$

it is possible to write the conditions of stabilization of the turbulent flow (Boesgaard and Tripico 1986a).

The three-dimensional turbulent flow induced by the differential rotation is inhibited by the μ gradient (Zahn 1983) if the nonthermal part of the Brunt-Väisälä frequency (which will persist even in the presence of strong thermal diffusion) is larger than the turnover frequency of the smallest eddies present in the turbulent spectrum, those of the Kolmogoroff scale. This condition can be written:

$$\nabla\mu(\Delta\nabla + \nabla\mu) > \frac{4}{5}\frac{1}{\nu}\frac{L_r r^8 \Omega^4 H_P}{G^3 M_r^4}. \qquad (30)$$

Similarly, the three-dimensional turbulent flow induced by the vertical shear can be inhibited when a condition similar to the Richardson condition is fulfilled. This is due to the fact that the transfer of momentum and the diffusion of chemical contaminants are both governed by the microscopic diffusivity. We have then the stability condition

$$\nabla\mu > \frac{H_P}{4g\delta}\left(\frac{d\Omega}{d\,\ln r}\right)^2. \qquad (31)$$

It is readily seen that these stability conditions are very easily met in the central region, and this is due to the very large μ gradient associated with hydrogen burning. More surprising is the way in which the μ gradient, in the outer half of the Sun can suppress the three-dimensional turbulence. In the present Sun, there is a strong gradient of ^{3}He in the neighborhood of $M = 0.7\ M_\odot$. With X_3 of the order of 10^{-3}, the μ gradient due to ^{3}He turns out to be $\sim 10^{-4}$ in the present Sun. At that level, the stability condition (Eq. 30) gives $\nabla\mu > 3 \times 10^{-6}$ in the present Sun. If there had been no mixing, it turns out that the μ gradient due to ^{3}He might have stabilized the three-dimensional turbulence produced by the meridional circulation when the Sun was $\sim$ 2 Gyr old. The rise of μ due to the production of ^{3}He cannot inhibit the meridional circulation. With a maximum concentration Y_3 = 0.003 reached at t = 4.6 Gyr, the μ excess is $\Delta\mu$ > 0.0011. In order to stop the meridional circulation, we should fulfull condition (31) and this is not the case.

A similar problem arises when considering the gravitational separation of ^{4}He, such as estimated by Noerdlinger (1977) and by Cox et al. (1989). In a nonrotating Sun, the drift velocity at 0.1 of the mass from the center is of the order of a few times 10^{-10} cm s^{-1}, which corresponds to a time scale of the diffusion process of 10^{19} s. The surface decrease of the helium concentration over the solar life time is approximately -0.03. If we apply the Mestel criterion on the μ-currents (Mestel 1953), we find that gravitational separation can inhibit the meridional circulation if Ω is smaller than $\sim 10^{-5}$, which corresponds to an equatorial circulation velocity of $\sim$ 10 km s^{-1}. In the Sun, meridional circulation could presently be inhibited by gravitational separation of helium and hydrogen. As a consequence, there would be no turbulence induced by meridional circulation.

The instability condition for the vertical shear flow, in presence of a ^{3}He gradient is fulfilled if the Ω gradient, defined by k = (d ln Ω/d ln$r$), is very steep. With $\nabla \mu \cong 0.004$, this gives k > 180. We shall therefore conclude that the ^{3}He gradient does not inhibit the instability of the vertical shear flow, but can stop the turbulence generated by the meridional circulation (if it exists).

It is readily seen that these stability conditions are very easily met in the central region. But the vertical shear flow instability can be present in the outer layers of a fast rotating Sun and has to be included when considering the history of the solar rotation.

IV. EVOLUTION WITH TURBULENT DIFFUSION MIXING

A. Basic Assumptions

We are concerned first by the initial conditions. A standard view is to consider the approach to the main sequence of a solar-like star along the Hayashi track, with an initial rotational velocity close to the rotational breakup. The loss of angular momentum is such that the equatorial velocity hardly increases or even decreases, depending on the model of angular-momentum losses and transport inside the star during the contraction (Schatzman 1962,1989; Pinsonneault et al. 1989), and the star reaches the main sequence with an equatorial velocity of the order of 20 km s^{-1}, as it appears to be the case for the slow rotators in the Pleiades (van Leeuwen and Alphenaar 1982; Soderblom et al. 1983; Stauffer et al. 1984; van Leeuwen et al. 1987) and in α Per (Butler et al. 1987; Balachandran et al. 1988; Stauffer et al. 1984).

However, both the Pleiades and α Per contain fast rotators (up to 200 km s^{-1}) which are as well on the main sequence. It seems possible to compare the situations to the existence of two kinds of T Tau stars (Cohen and Kuhi 1979), the wide-line T Tau surrounded by a disk, detectable by its infrared emission (Mendoza 1968; Cohen and Kuhi 1979; Bertout et al. 1988; Bertout 1989), and the narrow-line T Tau stars that do not have a disk. When there is a disk, it appears to be in strong interaction with the central star. The proportion of T Tau stars with a disk seems to be approximately 10% of all T Tau stars. It seems possible to imagine that the late presence of a disk has the effect of feeding the central star with a large amount of angular momentum. This could explain the presence simultaneously in the Pleiades and in α Per of slow and fast rotators. Stars having lost their disk at an early phase of their evolution would have reached the main sequence at a moderate equatorial velocity. However, Bouvier (1990) noticed that there is no indication of the presence of a disk around the fast rotators of α Per (Balachandran et al. 1988).

Anyhow, the existence of the planetary system around the Sun strongly suggests that the Sun had a disk and that it can possibly have been a fast rotator. It seems therefore necessary to consider two scenarios for the history of the rotating Sun, and to study the consequences both for the transfer of angular momentum and for the transport of passive contaminants.

B. Standard Model

The loss of angular momentum depends on several effects: the geometry of the magnetic field, which determines the fraction of the stellar surface that is occupied by open lines of force; the surface value of the magnetic field, which is related to the dynamo mechanism, and the rate of mass loss. The problem of the geometry of the magnetic field has been considered by Roxburgh (1983) and more recently by Mestel and Spruit (1987). The result depends on the choice of the structure of the magnetic field (dipole, quadrupole or a distribution of bipolar magnetic spots). The value of the magnetic field at the Alfvénic distance is then related to the surface value in a complicated way.

A parametric expression of the loss of angular momentum has been given by Kawaler (1988a). We shall follow here, as Durney and Latour (1978), the simple assumption of flux conservation, $B \propto r^{-2}$. Similarly, the simplest assumption concerning the nonlinear dynamo is to assume that the maximum rate of growth of the magnetic field is compensated by the losses due to buoyancy (Schatzman 1989). Finally, following Durney and Latour (1978), we shall assume, as shown by Parker (1975a,b) that the velocity of escape of the stellar wind for solar-like stars (Linsky 1985) is proportional to the velocity of escape from the surface of the star $(GM/R)^{1/2}$.

We shall assume that the rate of loss of angular momentum is given by

$$I\frac{d\Omega}{dt} = - K_F \Omega^{7/3}. \tag{32}$$

The model of Schatzman (1989), does not give the exact value of K_F, but, without any parametric adjustment, it is very close to the value needed to explain the equatorial velocity of the Sun. We have the asymptotic expression

$$\Omega = \left(\frac{4}{3}\frac{K_F t}{I}\right)^{-3/4}. \tag{33}$$

For the Sun, $(4/3)K_F = 1.16 \times 10^{44}$. It should be noticed that the $t^{-3/4}$ law, which differs from the Skumanich relation (1972), has been obtained by Bohugas et al. (1986).

Let us discuss the validity of the assumption of transport of angular momentum by diffusion. It is easy to obtain the order of magnitude of the angular velocity gradient, which, at the boundary of the convective zone, is sufficient to carry away the flux of angular momentum. With a turbulent diffusion coefficient of 1000, the turbulence being generated by some instability, and the present rate of loss of angular momentum, it turns out that an angular velocity gradient

$$(d\Omega/dr) \cong 5 \times 10^{-16} \tag{34}$$

is sufficient to carry away the angular momentum from the bottom of the convective zone. This corresponds however to a logarithmic gradient

$$(d \ln\Omega/d \ln r) = 8.6. \tag{35}$$

The Richardson-Townsend condition of instability of the turbulent shear flow leads to a critical value

$$(d \ln\Omega/d \ln r)_{crit} \cong 30. \tag{36}$$

The vertical shear flow gradient does not allow the generation of turbulence, but it should be noticed that, compared to the results of helioseismology (Dziembowski et al. 1989; Brown et al. 1989; Thompson 1990) this is very large. It would correspond to a change of Ω by a factor of 2 over 5% of the solar radius. We are meeting here one of the major problems concerning mixing inside the Sun: *the turbulent diffusion coefficient for the transport of angular momentum is much larger than the turbulent diffusion coefficient for the transport of passive contaminants.* When considering the quasi-solid body rotation from 0.7 $R_\odot$ to 0.3 $R_\odot$, we have to introduce a turbulent diffusion coefficient at least of the order of

$$D_T = (0.16 \, R_\odot 2/t_\odot) \cong 6000 \qquad (37)$$

which is 6 to 10 times what is needed in order to explain lithium burning. Tassoul and Tassoul (1989) introduced a turbulent diffusion coefficient for the angular momentum, $D_{T\Omega}$, which is 10 to 20 times the diffusion coefficient for passive contaminants. Pinsonneault et al. (1989) met the same problem. It is clear that the vertical shear flow, at the boundary of the convective zone, is at most $(d \ln \Omega/d \ln r) \cong 0.3$. We can suspect here that the assumption of the transport of angular momentum by turbulent diffusion induced by rotation is not valid. The physical question is then the following: what is the physical origin of this efficient transport process? A magnetic field is a possible answer, as it is sufficient, in principle, to have a magnetic field such that the propagation of a perturbation at the Alfvén velocity from the center to the surface takes < 4.6 Gyr. This corresponds to a magnetic field of the order of 10^{-5} Gauss. However, the exact nature of the magnetic field in the radiative zone is not known, and the recent discussions of Spruit (1987) and Mestel and Weiss (1987) have just shown the difficulty of the problem. In fact, the turbulent flow that is needed for explaining the lithium depletion corresponds to a dynamo number < 1. The dissipation of the magnetic field by the turbulent flow is not compensated by a dynamo effect. It seems difficult to attribute the transport of angular momentum to an MHD process.

With a larger angular velocity, the turbulent diffusion coefficient due to differential rotation varies as Ω^2, but the rate of loss of angular momentum varies as $\Omega^{7/3}$. If angular momentum is carried by the dissipative process of turbulent diffusion, the gradient of angular momentum at the boundary of the convective zone varies as $\Omega^{1/3}$. A factor of 100 on the angular velocity corresponds to an increase of the angular momentum gradient by a factor of only 4.6, and this is still below the critical value for the vertical shear flow instability. In such conditions, the major effect seems to come from the differential rotation. Numerical solutions have been given by Pinsonneault et al. (1989) but we raise doubts here about the validity of their assumptions concerning the generation of the turbulent flow that they need. As we shall see in Secs. IV.D and E, a unique set of parameters describing the turbulent flow cannot explain the mass dependence of the lithium abundance in the Hyades.

C. The Fast Rotating Stars

The presence, among fast rotators in α Per (Stauffer 1987), of lithium-rich stars (Balachandran et al. 1988) (they are either not or only slightly deficient) is quite remarkable. The implications for the solar history are so important that we must discuss the meaning of these facts.

Coming back to the expression of the turbulent diffusion coefficient, it can be seen that this coefficient does vanish at a certain level, given by

$$\Omega^2 = 2\pi G\rho. \qquad (38)$$

The effect is a diffusion barrier. It is important only when it is located below the bottom of the convective zone. However, even in that case, the barrier is not opaque. Microscopic diffusion allows the connection between the two circulation zones. The transport of the elements is not suppressed but only diminished. It is easy to obtain, in a simple model, an analytical solution which involves Bessel functions of the first and second kind, and to obtain the order of magnitude of the eigenvalues for the rate of lithium depletion. The analytical results have been confirmed by the numerical results of Charbonneau and Michaud (1990) and Charbonneau et al. (1989). It is therefore necessary to look for another explanation than the presence of a singularity with a vanishing diffusion coefficient.

The hypothesis of the presence of a disk provides simultaneously the explanation of the high equatorial velocity and of the presence of lithium. The disk provides angular momentum and lithium. The abundance of lithium is then the result of a balance between the lithium brought by accretion and the fast destruction inside the star. If p is the rate of lithium destruction, X_7 the concentration in the disk, and X_7^* the concentration in the convective zone, we have

$$X_7^* = X_7 \frac{K_F \Omega^{4/3}}{4\pi R^4 \left(\frac{r}{R}\right)^2 \rho H_P \lambda + K_F \Omega^{4/3}}. \tag{39}$$

For a 0.9 $M_\odot$ star, with $T_{\text{eff}} = 5140$ K, $H_P = 4.21 \times 10^9$, $T = 2.33 \times 10^6 K$, $\rho = 0.41$ g cm^{-3}, one finds $\lambda = 4.63 \times 10^{-14}$ s^{-1} for an equatorial velocity of 200 km s^{-1} or $\Omega = 3.71 \times 10^{-4}$, and $(X_7^*/X_7) = 0.36$, which is quite a reasonable value.

After the disappearance of the disk, the spin-down process can start. There must be a very rapid spin-down of the layers immediately below the convective zone, in such a way that the turbulent diffusion coefficient drops quickly, otherwise there would be no lithium left in these fast rotators. In initially fast rotating stars, there should be a fast transfer of angular momentum from the outer regions of the star to circumstellar space. In the case of the Sun, this suggests a possible memory of a high initial angular velocity (Vigneron et al. 1990). However, as shown by Bouvier (1990), the absence of spectroscopic features due to the presence of such a disk, raises some doubts about this picture. The difficulty is confirmed by a theoretical argument: a great efficiency of the transfer of angular momentum by internal waves (see Sec. V) and a transport of lithium by nonrotation-dependent mixing process would make the whole picture inconsistent.

D. Difficulties

We shall now show that a completely consistent theory of lithium depletion, assuming that the only agent of the transport process is a turbulent flow due to rotation, meets difficulties that are impossible to overcome in the present state of the physical representation of the stellar interiors. Baglin and Lebreton (1990) and Schatzman and Baglin (1990) have used a sequence of models of $M = 0.9$, 1.0, and 1.1 $M_\odot$ obtained by a full treatment of the evolution from the initial homogeneous state to the age of the Hyades. The chemical composition is $X = 0.70$, $Y = 0.28$, $Z = 0.02$. The opacities come from the Los Alamos Opacity Tables (Huebner et al. 1977), supplemented by the contribution of the molecular absorbents at low temperature. The convective zone has been calculated with a mixing-length parameter $\alpha = 2$, close to the solar value. The method of computation of the diffusion equation, as well as the models has been described elsewhere (Baglin et al. 1985).

In order to show the importance of the difficulties, they have first computed depletions due to a diffusion mechanism which is space and time independent. The properties of the eigenvalues allow us to estimate, in a semi-analytical way, the effect of a time-dependent diffusion coefficient. We recall here only the main points. Writing $c \propto e^{-\lambda t}$ they compute the largest eigenvalues of the diffusion equation as proposed by Baglin et al. (1985).

1. Constant Diffusion Coefficient: Depletion as a Function of D. A first approach consists in assuming a constant diffusion coefficient: it depends neither on the radius nor on the stellar age. It is then possible to derive the values of a constant "effective" diffusion coefficient (space and time independent) which would explain the lithium depletion of the Hyades. The smaller the mass, the shorter the distance over which the diffusive process takes place, and nevertheless, as mentioned above, the larger are the values needed for values of D.

2. r-Dependent Diffusion Coefficient. It is assumed that the Zahn coefficient of diffusion is the dominant one. The eigenvalues λ are proportional to D^p, and the exponent p is ~ 0.73. If we introduce a factor f, $D = fD_z$, we need to multiply the values of λ by f^p. Adjusting the efficiency parameter f allows us to adjust the diffusion coefficient to fit the observations. This is just the kind of artificial manipulation that must be rejected. If the diffusion coefficient is adjusted with a value of the factor f, such as fitting the observed abundance of the 1.1$M_\odot$ stars, then there is a large overabundance of lithium in the 0.9 $M_\odot$ stars. However, if the factor f is adjusted in such a way that the solution fits the abundances of the 0.9 $M_\odot$ stars, there is no lithium left in the 1.1 $M_\odot$ stars (Baglin and Morel 1989).

3. Time-Dependent Diffusion Coefficient. The diffusion coefficient depends on the angular velocity through a term Ω^2. As Ω decreases with time, due to spin-down, D decreases also with time. In principle, in order to obtain the evolution of the lithium abundance at the surface as a function of time, it should be necessary to solve the diffusion equation, with a diffusion coefficient $D(r,t)$ depending both on space and time. In order to show the nature of the physical problems which come up, an approximate solution of the diffusion equation is sufficient. Assuming a concentration $c \propto \exp(-\lambda t)$, we can obtain at any time t and for the value of the diffusion coefficient at that time, the eigenvalue λ of the diffusion equation. If we take into account the time dependence of the diffusion coefficient, we have now a time-dependent eigenvalue, $\lambda(t)$. The depletion will then be described by $c \propto \exp(-\int \lambda dt)$. It can be shown in the following way that the approximation is valid.

In the WKB approximation, the solution of the differential equation is either trigonometric or exponential. The change takes place at the transition point, where the rate of depletion λ is equal to the local rate K of lithium burning. It is easy to study the analytical solution of the simplified problem, where the right-hand side (rhs) is replaced by the following approximations:

$$\text{if } -\lambda c + K(r)c < 0, \text{ then, rhs} \equiv -\lambda c \qquad (40)$$
$$\text{if } -\lambda c + K(r)c > 0, \text{ then } c = 0.$$

The validity of the method is checked in two steps. First, it is possible to compare the solution to the exact eigenvalues obtained in the case of time-independent diffusion (Baglin et al. 1985; Baglin and Morel 1989). Second, assuming that D varies slowly with time, it is easy to show that the radius r defined by $\lambda = K(r)$ varies slowly with time. If it is assumed (Schatzman 1990a,b) that $\Omega \propto (1 + (t/t_0))^{-3/4}$ and if the depth dependence of lithium burning is written $K(r) \propto (R - r)^{-q}$ with $q \cong 24$, the transition point moves slowly, the time scale of the motion being of the order of 25 times the spin-down time scale. This verifies the approximation $\exp(-\int \lambda \, dt)$.

With a time-dependent diffusion coefficient, it is possible with a proper choice of the parameters (the initial velocity of rotation V_0, the spin-down time scale t_0, and the efficiency parameter f) to fit the observed distribution of lithium abundances. We shall show that this fitting has to be rejected, as it disagrees with other observational data.

We know the depletion factor and the final equatorial velocity at the age of the Hyades and we can neglect the pre-main-sequence destruction of lithium, which is a well-established observational fact. With a proper choice of the three parameters V_0, t_0 and f, it is possible to fit the observed abundances, but, as we shall show in the following subsections, at the cost of physically unacceptable values of the parameters.

(a) Adjustment of the Diffusion Coefficient.   Assuming that the angular velocity as a function of time is given by the relation obtained by Schatzman (1989):

$$\Omega = \Omega_0 (1 + (t/t_0))^{-3/4} \qquad (41)$$

we can consider the effect of the values of the two parameters Ω_0 and t_0 on the depletion factor. A numerical solution has been obtained for the eigenvalue λ for different values of the equatorial velocity (from 5 to 50 km s^{-1}), and with an efficiency factor $f = 6.25$. It is possible to write an interpolation formula, $\lambda \propto D^p$ with $D \propto \Omega^2$, and then by integration, to obtain the depletion factor $\delta = \int \lambda dt$:

$$\delta = \lambda_0(50) \left(\frac{V_0}{50}\right)^{2p} \frac{t_0}{3/2p - 1} \left\{ 1 - \left(1 + \frac{t}{t_0}\right)^{-(3/2\ p\ -\ 1)} \right\} \qquad (42)$$

where λ_0 (50) is the reference eigenvalue for an equatorial velocity of 50 km s^{-1}. We can also eliminate the time scale t_0 of the spin-down between Eqs. (41) and (42) and derive the depletion measure δ, for any initial value of the equatorial velocity:

$$\delta = \lambda_0(50) \left(\frac{V_0}{50}\right)^{2p} \frac{t}{(V_0/V)^{4/3} - 1} \frac{1}{3/2\ p\ - 1} \left\{ 1 - \left(\frac{V_0}{V}\right)^{-4/3\ (3/2\ p\ -1)} \right\}. \qquad (43)$$

Any change in the parameter D_0 by a factor f gives a value of δ multiplied by a factor f^p. On the whole, it appears that if the factor f is adjusted in such a way that the 1.1 M$_\odot$ stars of the Hyades have the correct depletion, the 0.9M$_\odot$ stars keep too much lithium. This remains true for all values of the initial velocity V_0, being given the values of the equatorial velocities V of the stars in the Hyades.

(b) Evaluation of the Time Scale of the Spin-Down. We have two known quantities, the present velocity of rotation V and the depletion factor δ. We do not know the initial velocity of rotation V_0, and we do not know the efficiency factor f. For a given value of f, it is possible to derive the initial equatorial velocity from the present value of the equatorial velocity and the known value of the depletion measure δ. We obtain then the time scale of the spin-down. Among other results, it appears:

(i) That the time scale of the spin-down decreases for increasing values of the stellar mass. This is exactly the opposite of what is derived from the observations (Kraft 1969) and is contradictory with the theory of the spin-down effect (Schatzman 1987,1990b; Durney and Latour 1978; Mestel and Spruit 1987).

(ii) That the values of the time scale of the spin-down obtained from this adjustment has nothing to do with the observational data.

The physical reason for this last result can be found in two effects acting in opposite directions. When the stellar mass decreases, the convective zone becomes deeper and the rate of diffusion towards the region of lithium burning has a tendency to decrease. But the turbulent diffusion coefficient varies as r^{-6}, and the convective zone, which is the reservoir of lithium is larger. These two effects overtake the influence of the deepening of the convective zone, giving more lithium burning.

Altogether, it appears that it is possible to fit this model to the measured velocities of rotation by choosing a mass-dependent parameter f of the diffusion coefficient $D = fD_z$. This provides unacceptable properties of the spin-down time scales and rests on the assumption that the efficiency of the process that converts the two-dimensional turbulence into a three-dimensional turbulence is mass dependent. The efficiency would increase with decreasing stellar masses. As there is no evident physical reason for such a property, Baglin and Schatzman consider that it has to be rejected.

E. Lithium Depletion and Rotational Velocities in the Lithium Gap

Assuming that transfer of angular momentum and lithium depletion have as a common origin turbulence generated by rotation leads to difficulties. This can be emphasized by considering the earlier spectral types, for which the velocity dispersion does not seem to correspond to the $V \sin i$ distribution. This is especially true of the region of the gap (Boesgaard and Tripico 1986a,b). It should be noted here that the integrated distribution function (Fig. 1) obtained here with the data collected by Kraft (1965)

$$\int f(V \sin i) \, d \cos i \tag{44}$$

looks more like representing $(1/C) \int \int dV \, d \cos i$ than $\int \int \delta(C - V) \, dV$ $d \cos i$, where C is the maximum velocity of rotation and $\delta(C - V)$ is the delta function. The exact distribution in normalized variables is

$$\int_0^v f(v)dv = \left(\frac{v}{C}\right) \arctan \sqrt{\left(\frac{C}{v}\right)^2 - 1} + 1 - \sqrt{\left(\frac{v}{C}\right)^2 - 1}. \tag{45}$$

It thus appears that there is an almost uniform distribution of the equatorial velocities between 0 and the maximum equatorial velocity C rather than a pure $\sin i$ effect. This, in itself, suggest that something other than rotation is acting in the depletion mechanism of lithium (Spruit 1984b,1990; Garcia-Lopez and Spruit 1990). If the effect were due to rotation only, it would not be possible to explain the lithium gap by the envelope of the velocity distribution function, when the observed velocity dispersion is due to an intrinsic distribution of the velocities. The same question comes up when discussing the mutual exclusion in the same area of the Hertzsprung-Russell diagram of the Am and δ Scuti stars (Baglin 1975) where it appears that there is a need for some other physical process besides rotation.

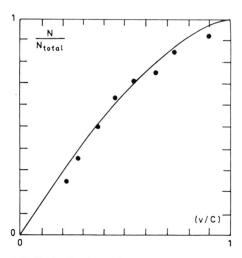

Fig. 1. The integrated distribution function of the observed velocity of rotation $v = V \sin i \int f(v)$ dv for stars of the lithium gap (Boesgaard and Tripico 1986a). The points represent, after normalization, the data from R. Kraft (1965). The curve represents Eq. 45.

V. MIXING BY INTERNAL GRAVITY WAVES

The difficulty of explaining lithium depletion by turbulent diffusion mixing, induced by rotation, leads to the consideration of another physical process. It appears that this has to be connected with the problem of interpreting other observational results. The problems of transport and redistribution of angular momentum and the spin-down problem have taken on a new aspect with the recent results coming from the interpretation of the pressure oscillation modes splitting (Brown et al. 1989; Thompson 1990). At the equator, where the main fraction of the stellar angular momentum exists, the angular velocity ω in the radiative zone is lower than the angular velocity ω_E in the convective zone, and decreases with the radius down to about 0.4 $R_\odot$. In the radiative zone, the angular velocity seems to be only slightly dependent on latitude.

The decrease of ω when going towards the center of the Sun is contrary to what would be expected if the angular momentum were carried outwards by a diffusive process only (Endal and Sofia 1978; Schatzman 1987; Pinsonneault et al. 1989); these properties of the angular velocity cannot be explained by a flux of angular momentum proportional to the gradient of the angular velocity. In order to explain this contradiction, it seems necessary to take into account another physical process, the transport of angular momentum by internal waves (Spruit 1990), generated by the turbulent motion in the convective zone. It seems that transport of angular momentum by internal waves is a very efficient process (see also Vigneron et al. 1990).

Press (1981) has estimated the flux of mechanical energy carried in the radiative zone by gravity waves excited at the boundary of the convective zone and has also given an estimate of their radiative damping $A(r)$. Goldreich and Nicholson (1989a,b), following the results derived from the theory of the mean Lagrangian (Dewar 1970; Grimshaw 1984), have shown how gravity waves carry angular momentum. In fact, as gravity waves generate a random-walk motion of the fluid elements (Press 1981), they can also generate a diffusive process that will transport chemical elements. It is necessary to study together the transport of angular momentum and the transport of chemical contaminants, as two characteristic quantities—the macroscopic diffusivity D_M and the flux of angular momentum—depend on the same parameters. In the following, we summarize the recent results of Schatzman (1990b).

A. Angular Momentum Transport

The review paper by Grimshaw (1984) is a good introduction to the question of wave action and wave-mean flow interaction. The general idea is that when a wave propagates along a mean flow, there is an exchange of momentum between the wave and the mean flow. In the case of a rotating system, there is an exchange of angular momentum. As shown clearly by Grimshaw (1984) and by Goldreich and Nicholson (1989a), dissipative processes play an important role, as they provide the transfer of momentum from the waves to the mean flow. In the absence of dissipation, we just have conservation of angular momentum. In the case of the 5 $M_\odot$ considered by Goldreich and Nicholson (1989b), they take into account the dissipative effects only when the internal waves enter in the central convective zone and carry the angular momentum to the core.

The treatment of the problem by Goldreich and Nicholson can be generalized to the case where internal waves are not generated by tidal effects, but are produced by the motions in the surface convective zone. We show first the properties of monochromatic internal waves, and next take into account the frequency spectrum of the internal waves generated by the turbulent motion present in the convection zone. The frequency of the waves, at the level of production, and seen in the local frame of reference, will be called ω_c. In the inertial system of reference, the local rotation frequency will be called ω.

In this oversimplified treatment, it is assumed that the waves are produced in a system rotating like a solid body, with a circular frequency ω_E, despite the fact that the convective zone does not present solid-body rotation, showing a latitude dependence of the angular velocity. However, as the largest fraction of the angular momentum is concentrated near the equator, it can be considered as an acceptable approximation: this is like dealing with a cylindrical Sun. The choice of the circular frequency ω_E will be part of the boundary conditions. The gravity waves are propagating in a radiative zone rotating with a local value of the circular frequency $\omega(r)$, and are seen with a Doppler shift $\omega_E - \omega(r)$. There is a superimposition of waves with the frequencies $+\omega_c$ and $-\omega_c$, and the matter of the radiative zone does see waves with the frequencies $\omega_c + \omega_E - \omega$ and $-\omega_c + \omega_E - \omega$. It is clear that the angular momentum carried by these waves has an opposite sign and a different absolute value. The transport of angular momentum results from the effect of the superimposition of these two kinds of waves. Calling σ_+ and σ_- the frequencies of these two waves, we follow Goldreich and Nicholson (1989a,b), showing that the mechanical energy carried by the internal waves can be written

$$L_E = K\sigma^3 \tag{46}$$

where K is a constant determined by the mechanism of production of the internal waves. The corresponding flux of angular momentum, neglecting any damping effect is

$$L_H = \frac{L_E}{\omega_E}. \tag{47}$$

We need now to obtain an estimate of L_E. The flux of mechanical energy is proportional to the product of the square of the horizontal velocity of the waves, multiplied by the radial group velocity:

$$L_E = 4\pi r^2 \, \rho u_H^2 v_g. \tag{48}$$

The group velocity v_g is negative,

$$v_g = -\frac{\omega_c}{k_r} \tag{49}$$

and does not depend on the amplitude of the waves. If we take into account the relation between the vertical and the horizontal velocity,

$$\frac{u_r}{u_H} = \frac{k_H}{k_r} = \left(\frac{\omega_c^2}{N^2 - \omega_c^2}\right)^{1/2} \cong \frac{\omega_c}{N} \tag{50}$$

where N is the Brunt-Väisälä frequency, we can write for the flux of mechanical energy

$$L_E = -4\pi r^2 \rho \, u_H^2 \frac{\omega_c^2}{k_H N}. \tag{51}$$

B. Spectral Effects

The turbulent flow generates a spectrum of scales k and frequencies ω_c. If we assume a Kolmogoroff spectrum with a cutoff at a certain wave number k_M and at a certain frequency ω_M, we have the following relations:

$$
\begin{aligned}
k \text{ spectrum } &= f(k)\mathrm{d}k = \frac{3}{2} \frac{k_M^{2/3}}{k^{5/3}} \frac{\mathrm{d}k}{} \\
\text{velocities } v &= v_M \left(\frac{k_M}{k}\right)^{1/3} \\
\text{frequencies } \omega &= v_M \, k_M^{1/3} \, k^{2/3}.
\end{aligned}
\tag{52}
$$

As the dispersion relation implies both k_H and k_r we shall assume with Press (1981) that the horizontal wave number is determined by the large-scale flow pattern of the convective zone, but that the frequency spectrum and the velocity spectrum are determined by the effect of the pressure fluctuations in the convective zone. If the large-scale flow pattern of the convective zone is so important, we have to take into account that it has the geometry of two-dimensional convective cells. The waves then should reflect this geometry with horizontal wave numbers of the same order of magnitude in two horizontal directions (Garcia-Lopez and Spruit 1990).

It is the mean square horizontal velocity that contributes to the transport of angular momentum. If we call δu_{ij} the contribution to the velocity due to the pressure exerted on the boundary by the turbulent vortices 1, 2, ...i,... V, in the wave number interval δk_i, we have to consider the quantity

$$
< \left(\frac{1}{V} \sum_{ij} \delta u_{ij}\right)^2 >.
\tag{53}
$$

Due to the fact that the action of the vortices are not in phase, we must take into account the fact that over a surface k_M^{-2}, there are $(k/k_M)^2$ convective cells of a scale k^{-1}. Consequently, the amplitude of the motion that is generated at the frequency ω_c must be divided by the number of convective cells to the power $1/2$ and the square of the velocity must be divided by $(k/k_M)^2$.

In expression (51) we have to replace u_H^2 by the contribution of the spectral interval $\mathrm{d}k$, $v_M^2 f(k)\, \mathrm{d}k$, and include the reduction factor $(k_M/k)^2$. This gives the contribution to the flux of mechanical energy. As emphasized by Press (1981), in the WKB approximation, the fitting of the wave motion in the convective zone to the motion in the radiative zone must take into account the behavior of the Brunt-Väisälä frequency as a function of depth. If the frequency N increases quickly from zero to to a finite value, it is necessary to write the fitting condition a little below the boundary of the convective zone, at a place where the Brunt-Väisälä frequency has a small positive value, which we write here as N_I. The index will be affected by all quantities "close" to or at the boundary of the convection zone:

$$dL_E = -4\pi r_I^2 \rho_I v_M^2 \frac{\omega_c^2}{k_M N_I} \left(\frac{k_M}{k}\right)^2 f(k)dk \qquad (54)$$

or

$$dL_E = -4\pi r_I^2 \rho_I v_M^4 \frac{k_M}{N_I} \frac{2}{3} \left(\frac{k_M}{k}\right)^{7/3} d\left(\frac{k}{k_M}\right). \qquad (55)$$

Due to the negative value of the group velocity, the energy luminosity is negative. This is important for what follows.

It is suggested by Press to take for v_M the velocity u of the turbulent flow in the convective zone, given by the asymptotic formula of the quasi-adiabatic flow (Cox and Giuli 1968) and to assume that

$$4\pi r^2 \rho u^3 = (1/10)L_{star} = \phi L_{star}. \qquad (56)$$

C. Radiative Damping

It is then necessary to take into account in the propagation the effect of the damping factor. Calling $A(r)$ the square of the radiative damping factor obtained by Press (1981) for the vertical velocity of the waves, we write

$$L_H = \frac{K\sigma^3 A}{\omega_E} \qquad (57)$$

where σ is either σ_+ or σ_-. In fact, we have to consider the superimposition of waves with the frequencies $+\omega_c$ and $-\omega_c$. Following Press, we just introduce the factor A in Eq. (54):

$$dL_E = -4\pi r_I^2 \rho_I v_M^2 \frac{\omega_c^2}{k_M N_I} \left(\frac{k_M}{k}\right)^2 Af(k)dk \qquad (58)$$

where N_I is the Brunt-Väisälä frequency close to the boundary of the convective zone (N_I goes from zero to an almost constant value over a distance of $0.01 \, R_\odot$); the term $4\pi r^2 \, \rho_I v_c^3$ is an estimate of the mechanical energy flux from the convective zone to the radiative zone, taken near its boundary, ρ_I being the density at the boundary, given in order of magnitude (Eq. 56). We assume, for the time being, as mentioned above, that $v_M = u$, but this assumption, as we shall see in Sec. V.D needs further discussion.

The calibration of the constant K is made by considering the energy luminosity in a nonrotating star. We can then proceed to the superposition of waves of the same amplitude, but with opposite circular frequencies $\pm \omega_c$, and obtain finally the contribution of the spectral interval dk to the flow of angular momentum:

$$L_H = -4\pi r_I^2 \, \rho_I v_c^3 \, \frac{\omega_c}{N_I} \frac{1}{2} \frac{(\omega_c + \omega_E - \omega)^3 + (-\omega_c + \omega_E - \omega)^3}{\omega_c^3} \frac{A}{\omega_E}. \quad (59)$$

It is clear, with this expression, that the flux of angular momentum vanishes when $\omega = \omega_E$. If the frequency difference $\omega_E - \omega$ is small compared to ω_c, we can write

$$dL_H = -4\pi r_I^2 \rho_I v_c^3 \left(\frac{k_M}{k}\right)^2 \frac{\omega_c}{N_I} \, 3 \, \frac{(\omega_E - \omega)}{\omega_c} \, Af(k)dk. \quad (60)$$

As the flux is proportional to the square of the velocity of the internal wave, the damping factor A derived from the result of Press will be

$$A = \exp\left\{-\int_r^{r_I} \frac{D_{th}k_H^2}{N} \left(\frac{N}{\omega_c}\right)^4 k_H dr\right\} \quad (61)$$

where the thermal diffusivity is given by Eq. 24.

If we take into account the fact that, at the border of the convective zone, we can write with Press $v_M = \omega_M k_M$, we obtain for the damping factor the expression

$$A = \exp\left\{-\int_r^{r_I} \frac{1}{\phi} \frac{\nabla_{ad}}{\nabla_{rad}} \frac{N^3}{g_r \omega_M} \left(\frac{r_I}{r}\right)^3 \left(\frac{\omega_M}{\omega}\right)^4 dr\right\} \quad (62)$$

where we have made the same choice as Press (1981) for the wave number k_H and write

$$k_H = \frac{2\pi}{(H_P)_I} \frac{r_I}{r} \quad (63)$$

where $(H_P)_I$ is the scale height at the boundary of the convective zone.

We introduce here the spectrum of velocities and frequencies, given by Eq. (52). We can then write

$$dL_H = -4\pi r_I^2 \rho_I v_M^3 \frac{1}{N_I} \frac{\omega_E - \omega}{\omega_E} d \left(\frac{k_M}{k}\right)^3 A. \quad (64)$$

In order to carry the integration we need only to express the damping factor A, which we write

$$A = \exp\left\{-\left(\frac{\omega_M}{\omega}\right)^4 F\right\} \tag{65}$$

with

$$F = \int_r^{r_I} \frac{\nabla_{ad}}{\nabla_{rad}} \frac{1}{\phi} \frac{N^3}{g\omega_M} \left(\frac{r_I}{r}\right)^3 dr. \tag{66}$$

The integration of Eq. (64) gives, when F is large (and this is true very close to the bottom of the convective zone),

$$L_H = -\phi L \frac{1}{N_I} \frac{9}{2} \frac{3/4}{F^{3/4}} \frac{\omega_E - \omega}{\omega_E}. \tag{67}$$

In stationary conditions, we can write $L_H = -(I\omega_E/\tau)$, where τ is the time scale of the spin-down. L_H is negative (there is a loss of angular momentum), and so is the flow of angular momentum given by Eq. (67).

The first aim being to obtain an order of magnitude, we shall write

$$F = \frac{1}{\phi} \frac{\nabla_{ad}}{\nabla_{rad}} \frac{N_I^3}{g_I\omega_M} r_I \ln\left(\frac{r_I}{r}\right). \tag{68}$$

We finally obtain for the angular velocity, including a factor $(2/3)$ in the expression of L_H, which takes into account the distance to the axis of rotation in the evaluation of the flux of angular momentum:

$$\omega_E - \omega = \frac{IN_I\omega_E^2}{\tau} \frac{F^{3/4}}{(3/4)!} \frac{(3/2)}{\phi L} \tag{69}$$

where τ is the characteristic time scale of the solar spin-down, of the order of the age of the Sun.

With the following values of the different quantities, $r_I = 5 \times 10^{10}$, $k_H = (\pi/H_P)$ with $H_P = 5.4 \times 10^9$, $v_M = 3.8 \times 10^3$, $\omega_M = 2.21 \times 10^{-6}$, $N_I = 1.26 \times 10^{-3}$, $g_I = 6 \times 10^4$ and the moment of inertia $I = 5.88 \times 10^{53}$, the angular velocity $\omega_E = 2.85 \times 10^{-6}$, and the characteristic time $\tau = 5$ Gyr, we obtain

$$(\omega_E - \omega) = 312 \times 10^{-9} \, [\ln(r_I/r)]^{3/4} \tag{70}$$

which gives for $(r_I/r) = (0.72/0.567)$,

$$(\omega_E - \omega) = 74.5 \times 10^{-9} \text{ and } (1/2\pi)(\omega_E - \omega) = 11.9 \times 10^{-9}. \quad (71)$$

This numerical value shows that transport of angular momentum by internal waves is a very efficient process. Before comparing it to the results of helioseismology, it is necessary to make two comments:

(a) This result has been obtained with a description of the boundary conditions appropriate for a cylindrical Sun. The latitudinal dependence of the angular velocity should be taken into account. As the polar regions rotate more slowly than the solar interior, their contribution to the whole Sun must be of the opposite sign to the contribution of the equatorial regions.

(b) The parameter ϕ (Eq. 56), from which is derived the parameter v_M of the turbulence spectrum, has to be adjusted in order to describe the inertial part of the spectrum. In Eq. (56), when taking $\phi = 0.1$ one obtains the velocity of large-scale elements (of the order on 1 or 2 scales heights), and this does *not* describe the inertial part of the spectrum for which a larger value of ϕ must be chosen.

When we compare this angular momentum transport rotation to the result given by the analysis of the helioseismological data, $\delta\omega/2\pi = 16.6 \times 10^{-9}$, as given by Brown et al. (1989), we can be very satisifed that we have obtained the same order of magnitude, despite the oversimplifications that we have introduced. We shall conclude that internal waves carry angular momentum very efficiently. This is essentially due to the high value of the mechanical energy luminosity (Eq. 51) of the order of $\phi(\omega/N_l)L^*$. The waves definitely have to be taken into account when describing the internal rotation of the radiative part of the Sun.

D. Dissipative Process

Garcia-Lopez and Spruit (1990), following Press (1981) consider the turbulence induced by the shear flow present in the field of internal waves, but this process does not seem strong enough. Press (1981) has shown that, due to nonlinear dissipative effects, the *rms* of the displacement of a piece of fluid is finite; this implies the existence of a diffusion coefficient, which turns out to be proportional to ε^4, where ε is the ratio of the horizontal displacement to the horizontal wavelength, the condition for applying the linear theory being that $\varepsilon < 1$.

We have now to express the diffusion coefficient D_{mix}. The diffusion coefficient can be written as the product of a length by a velocity. The length is the average distance l reached by a fluid element in an irreversible process after an overturn ω^{-1}, ω being now the frequency of the internal waves generated by the motions in the convective zone. We have dropped the index c, as the inertial frame of reference does not play any role in the diffusion problem. The velocity v is the product of the distance l by the circular frequency ω, $v = l\omega$. For a gravity wave, with a vertical wave number k_v and a vertical velocity v_v, Press (1981) gives as an order of magnitude

$$l = \frac{k_v u_v^2}{\omega^2} \frac{D_{th} k_v^2}{\omega^2} \tag{72}$$

where D_{th} is the thermal diffusivity, $(D_{th} k_v^2/\omega)$ the fraction of the entropy memory which makes the process irreversible. It is assumed again that the horizontal wave number k_H is defined by the geometry of the convection and is approximately $k_M = (2\pi/\alpha H_p)$ where α is the analog of the mixing-length parameter and H_p the pressure scale height in the neighborhood of the boundary of the convective zone. We have now to take into account the way in which the gravity waves are generated. For this purpose, we express k_v and u_v in terms of the horizontal components k_H and u_H,

$$k_H = \frac{\omega}{N} k_v \qquad u_H = \frac{N}{\omega} u_v \tag{73}$$

and write

$$l = \frac{k_H^3 u_H^2 N D_{th}}{\omega^4} \tag{74}$$

$$v = \frac{k_H^3 u_H^2 N D_{th}}{\omega^3} . \tag{75}$$

The horizontal velocity is generated by the turbulent flow, the frequency ω corresponding to a wave number k and u_H being identified with the turbulent velocity u. The summation of the contributions of the wave number intervals dk must include the distribution function $f(k)\,dk$ and the number (k/k_M) of cells of size k which contribute to the horizontal displacement generating the wave (ω, k_H). We must also take into account the expression given by Press of the horizontal velocity as a function of depth,

$$u_v \propto k_H^{3/2} \rho^{-1/2} \left(\frac{\Omega}{N}\right)^{1/2} A^{1/2}. \tag{76}$$

We also write $\omega = k_M u_M (k/k_M)^{2/3}$ and obtain with the index I for the boundary of the convective zone,

$$l = \int_{k_M}^{\infty} \left(\frac{r_I}{r}\right)^6 \frac{\rho_I}{\rho} D_{th} \frac{N_I}{k_M v_M^2} \frac{2}{3} \frac{3}{7} A \, d\left(\frac{k_M}{k}\right)^{7/3} \tag{77}$$

$$v = \int_{k_M}^{\infty} \left(\frac{r_I}{r}\right)^6 \frac{\rho_I}{\rho} D_{th} \frac{N_I}{v_M} \frac{2}{3} \frac{3}{5} A \, d\left(\frac{k_M}{k}\right)^{5/3}. \tag{78}$$

The factor F in the damping factor A becomes large very quickly below the boundary of the convective zone and the same approximation, used for the flux of angular momentum, gives the coefficient of diffusion D_M for the elements of matter. It is useful to give here the expressions of D_M and F:

$$D_M = \frac{\alpha}{2\pi} \frac{4}{35} \frac{7}{8}! \frac{5}{8}!(\Delta\nabla)_I \frac{L_r}{\phi L_I} \left(\frac{M_I}{M_r}\right)^2 \frac{L_r}{4\pi GM_I\rho_I} \left(\frac{\rho_I}{\rho_r}\right)^4 \left(\frac{r_I}{r}\right)^{12} \left(\frac{\nabla_{ad}}{\nabla_{rad}}\right)_r^2 \frac{1}{F^{3/2}} \quad (79)$$

with

$$F = \int_r^{r_I} \frac{\alpha}{2\pi} \frac{GM_I}{r_I^2} r_I \frac{1}{v_M} \frac{L_r}{\phi L_I} \frac{M_I}{M_r} \left(\frac{\rho_I}{\rho_r}\right) \left(\frac{r_I}{r}\right)^7 \left(\frac{\mu}{\mathcal{R}T}\right)^{1/2} \left(\frac{\nabla_{ad}}{\nabla_{rad}}\right)_r^2 (\Delta\nabla_r)^{3/2} \frac{dr}{r_I}. \quad (80)$$

The WKBJ method is used here in the following way: the local rate of lithium burning q is neglected, down to the point where it becomes equal to the eigenvalue s of the differential equation $q = s$. If we call t_{diff} the diffusion time from the boundary r_I of the convective zone to the radius r,

$$t_{diff} = \left(\int_r^{r_I} \frac{dr}{\sqrt{D_M}}\right)^{-2} \quad (81)$$

we have to solve the system:

$$\begin{cases} \sqrt{st_{diff}} \, tan \, \sqrt{st_{diff}} = \dfrac{r_I - r}{H_P} \\ s(r) = q(r). \end{cases} \quad (82)$$

An important parameter is the velocity v_M. It has been introduced in the definition of the energy spectrum. But the Kolmogoroff spectrum is valid only in the inertial range. The main contribution to the integrals (77) and (78) comes from a narrow peak around $(k/k_M) \cong F^{3/8}$ which falls in the inertial range. Then, the parameter v_M^2 in the Kolmogoroff spectrum should be taken as the parameter that describes the inertial range, and not as the velocity coming from the definition (Eq. 56) of the energy flow.

When applied to the problem of lithium depletion in the Hyades, the conclusions are the following:

(i) The rate of lithium burning in the 1 $M_\odot$ stars is definitely smaller than the rate of lithium burning in the 0.9 $M_\odot$ stars. The boundary of the convective zone is higher in the 1 $M_\odot$ stars than in the 0.9 $M_\odot$ stars. As the diffusion coefficient D_M decreases very quickly with depth, the lithium nuclei do not reach the burning level when the boundary of the convective zone is a little too high.

(ii) In stellar evolution, the boundary of the convective zone is receding upwards. This means that the rate of lithium burning can become very small compared to what it is during the early phases of stellar evolution.

The present rate of lithium burning in the Sun is then probably very small. This is an important conclusion, as an extrapolation of the lithium abundance in the 1 $M_\odot$ stars of the Hyades to the present Sun would give a very great deficiency, with $\lambda t = -6$, instead of -4.6, even when including the effect of the spin-down. A drop of the rate of lithium burning some time around 1 Gyr would ease the difficulty.

(iii) With $\alpha = 2$, $\phi = 0.1$, and according to the discussion above, $v_M \cong 2.5 \times 10^4$, the results are the following:

$$0.9 \ M_\odot \ \text{stars,} \ \lambda t_{\text{diff}} = 5.04, \ (\lambda t)_{\text{observed}} = 7 \tag{83}$$
$$1.0 \ M_\odot \ \text{stars,} \ \lambda t_{\text{diff}} = 1.26, \ (\lambda t)_{\text{observed}} = 3.1.$$

The theoretical values are a little too small compared to observations. This can be brought to the observed values by taking a mixing-length parameter $\alpha = 2.2$, which provides a greater depth of the convective zone.

As a conclusion, we state that the lithium abundance is a critical test of the depth of the convection zone and of the mixing process in the radiative zone.

RECENT DEVELOPMENT IN SOLAR CONVECTION THEORY

KWING L. CHAN
Applied Research Corporation

Å. NORDLUND
Copenhagen University Observatory

MATTHIAS STEFFEN
Universitat Kiel

and

R. F. STEIN
Michigan State University

In recent years, the theory of solar (and stellar) convection has made funda-
mental advances due to the increasing cost effectiveness of supercomputers and
the constant improvement of numerical techniques. It is expected that the nu-
merical approach will become a dominant trend for the future. Here, we report
on these new advances. First, we provide a brief review of the subject. In Sec.
I, references to theoretical studies on phenomena related to solar convection
are compiled. The objective is to provide a view of the breadth, not to be ex-
haustive. The next three sections then discuss three numerical studies of solar
convection in greater detail, so as to provide the readers with some general
understanding of the numerical techniques being used and the results obtained.
The discussion starts, in Sec. II, with a two-dimensional study of the spectro-
scopic properties of solar granules. While the two-dimensional limitation is
severely detrimental to some important hydrodynamical processes, it is both
economical and able to provide some initial understanding of the gross features
of solar convection. Section III discusses the testing of the well-known mixing-

length theory with three-dimensional numerical experiments. It also gives an example of applying the numerically gained knowledge to analytical study, in this case the behavior of compressible convection as a heat engine. Section IV describes a realistic, three-dimensional simulation of solar granulation; many observational features of solar granules are faithfully reproduced. It is the most sophisticated numerical calculation of this sort today.

I. INTRODUCTION

As illustrated by discussions in various other chapters, it is difficult to find a solar phenomenon that is not somewhat affected by the solar convection zone. The need for accurate modeling of convective processes is great. In recent years, with the aide of supercomputers, the theoretical handling of many complicated, highly nonlinear phenomena associated with solar convection have made remarkable progress. In this chapter, we discuss some new developments in this very active area of solar research.

As solar convection touches upon many related areas and the length of discussion must be limited, we arrange this chapter in the following manner: In this section, a compilation of references to the theoretical studies of solar convection problems is given. It is intended to provide a bird's-eye-view of the subject and therefore the bibliography is not exhaustive. However, by putting the many related phenomena together, we hope to present an integrated picture of the solar convection system. In later sections, in depth discussions are presented for several numerical studies that illustrate the techniques and results of present-day efforts in the theory of solar convection.

Table I gives a list of problems closely associated with the solar convection zone. The other columns show different types of theoretical approaches. For each entry, examples of a few of the earliest works are given. The theoretical approaches are roughly divided into three groups: the analytical/one-dimensional group, the modal group and the multi-dimensional simulation group. The analytical and one-dimensional approaches are grouped together as the final steps of most analytical studies result in differential equations of one spatial dimension. The modal approach cannot be strictly separated from the multi-dimensional group as the modes are multi-dimensional. However, the number of modes used in actual calculations are usually very limited and the amplitudes of the modes are described by one-dimensional differential equations that are relatively simple to solve. Multi-dimensional simulations usually require much more computing resources than the other approaches and their results display much more complexity.

Approximations are sometimes used in multi-dimensional simulations to reduce the demand on computing resources. The most popular are the Boussinesq (1903; Spiegel and Veronis 1960) and anelastic approximations (Ogura and Phillips 1962; Gough 1969). The main reason for using these approximations is to eliminate the acoustic modes which generally limit the time steps of a calculation to very small values (the CFL condition; see Richtmyer

TABLE I
Theoretical Studies of Problems Related to Solar Convection

Problem	Analytical/1-D Approach	Modal Approach	Multi-Dimensional Simulation
Dynamics of stratified convection	Bierman (1932) Vitense (1953) Spiegel (1963)	Gough et al. (1975b); Latour et al. (1976)	Graham (1975) Deupree (1977) Hurlburt et al. (1984)
Overshooting	Unno (1957) Shaviv and Salpeter (1973)	Veronis (1963) Latour et al. (1981)	Sofia and Chan (1984); Hurlburt et al. (1986)
Granulation	Musman et al. (1976)	Nelson and Musman (1977) Van der Borght and Fox (1983)	Cloutman (1979) Nordlund (1980)
Photospheric effects	Voight (1956) Schroter (1957)	Beckers and Nelson (1978)	Dravins et al. (1981)
Interaction with pulsation	Unno (1967) Stein (1967)	Gabriel et al. (1975); Goldreich and Kumar (1977)	Steffen (1988) Chan and Sofia (1988); Stein et al. (1988)
Interaction with rotation	Wasiutynski (1946) Kippenhahn (1963)	Durney (1970) Busse (1970)	Gilman (1972) Young (1974) Glatzmaier (1984)
Interaction with small-scale magnetic field	Parker (1963) Weiss (1964)	Weiss (1966); Clark and Johnson (1967)	Schussler (1979) Galloway and Moore (1979) Nordlund (1985b)
Interaction with large-scale magnetic field	Parker (1955b) Babcock (1961) Steenbeck et al. (1966)	Yoshimura (1972) Stix (1973)	Gilman et al. (1981); Glatzmaier (1985a)

and Morton 1968) and therefore make the total number of steps prohibitively large. The Boussinesq approximation is a straightforward extension of the technique used in studying liquids; it eliminates all the effects of compressibility except the idealized buoyancy term. In recent years, it has generally been abandoned for computations involving large density stratifications. The anelastic approximation eliminates sound waves by neglecting the Eulerian variation of density in the continuity equation, but the Lagrangian variation of density is kept; it is applicable to situations with some density stratification. However, it is not valid unless the Mach number is much less than 1 (not always true in the upper region of the solar convection zone) and when the acoustic waves are unimportant to the dynamics. More and more researchers now chose to solve the fully compressible fluid equations. For conciseness, most of the references listed below will not be differentiated according to their levels of approximation as discussed here.

A. Convection Dynamics

The most direct problem of solar convection concerns the hydrodynamics of the process (1st row of Table I). The following questions need to be addressed: what are the patterns of the flows; how can the convective turbulence be described; and how should the flows be related to the structure of the convection zone (i.e., depth distributions of the mean temperature, pressure, etc.)? The last question is particularly important for understanding the internal structure and evolution of the Sun. Up to now, almost all such calculations use the mixing-length theory (Biermann 1932; Vitense 1953). This theory has the advantage of being simple to manipulate analytically and easy to implement computationally. However, it is based on a hypothetical picture of heat carrying bubbles which is not rigorously derived from the fluid equations. Therefore, this theory is only applied with some reluctance, and the results are viewed with caution.

Numerous attempts to improve on the mixing-length theory have been made. The most significant improvement is the generalization to nonlocal theories (Spiegel 1963; Unno 1969; Ulrich 1970; Travis and Matsushima 1973; Nordlund 1976; Xiong 1981; Kuhfuss 1986). Furthermore, modification of the theory to account for the opacity difference in upward and downward flows has been considered (Deupree 1979). Variable mixing-length ratios have also been proposed (Deupree and Varner 1980; Chan et al. 1981; Cloutman 1987).

To study the problem on a more fundamental level, attempts have been made to reconstruct a uniform convective flux by a linear superposition of unstable modes generated by the superadiabatic structure (Hart 1973; Bogart et al. 1980; Narashima and Antia 1982; Antia et al. 1983). However, effects of the neglected nonlinear interactions cannot be properly assessed.

The most reliable way to study convection theory is to solve the fluid dynamical equations. To reduce the computational load, the modal approach uses a very limited number (usually 1 or 2) of planforms to represent the horizontal patterns of the convective motions; the vertical variation of the amplitudes of the modes can be computed in high resolution (Gough et al. 1975b; Van der Borght 1975; Latour et al. 1976,1983; Toomre et al. 1976; Fox and Van der Borght 1985; Fox 1985; Legait 1986).

Multi-dimensional solutions of the Navier Stokes equations for convective layers traversing several pressure scale heights were first obtained by Graham (1975,1977). Since then, the simulation approach flourishes in the study of convection dynamics (Deupree 1977; Marcus 1979,1980; Nordlund 1980,1982; Chan et al. 1982,1987; Hurlburt et al. 1984,1986; Yamagushi 1984,1985; Woodward and Porter 1987; Gigas 1989; Cataneo et al. 1990; Nordlund and Dravins 1990). However, the stratification of the solar convection zone is much too large for any single computation; so far, numerical computations can only study different regions of the solar convection zone.

B. Overshooting

The study of convective overshooting into neighboring stable regions is a natural extension of the study of dynamics. It has significant implication in the elemental distribution and evolution of the Sun (Bohm 1963; Weymann 1965; Straus et al. 1976; Chiosi 1986). Most of the theories are based on some extension of the mixing-length theory (Unno 1957; Shaviv and Salpeter 1973; Cogan 1975; Maeder 1975; Roxburgh 1978; Cloutman and Whitaker 1980; Van Ballegooijen 1982; Schmitt et al. 1984; Pidatella et al. 1986). Enlightening results have been obtained by modal calculations (Veronis 1963; Latour et al. 1981; Massaguer et al. 1984) and multi-dimensional simulations (Sofia and Chan 1984; Hurlburt et al. 1986).

C. Granulation

Granulation is the most prominent convective feature on the solar surface. A one-dimensional model based on energy balance has been considered by Musman and Nelson (1976), and two-dimensional single-mode models have been computed by Nelson and Musman (1977) and Van der Borght and Fox (1983). By solving the Navier Stokes equations in two dimensions, Cloutman's model (1979) included a treatment of the thermodynamics of the partially ionized gas. Nordlund's (1980,1982,1985b) anelastic model solved the Navier Stokes equations in three dimensions and included a sophisticated treatment of radiative transfer. Detailed behavior of axisymmetric two-dimensional models have been studied by Steffen and Muchmore (1988) and Steffen et al. (1989); also see Sec. II. Recently, Stein and Nordlund extended Nordlund's code to compute fully compressible three-dimensional models (see Sec. IV).

D. Effects on the Photosphere

The effects of convection on the photosphere is closely related to the granulation, the most important aspect being the influence on the radiation spectrum, because that is the fundamental diagnostic tool for observation. The significance of the convective turbulence on the radiative transfer in the photosphere was recognized very early (Voight 1956; Schroter 1957). Using a sinusoidal model to represent the motions of the granules, Beckers and Nelson (1978) presented a theory for the solar limb effect based solely on the effects of convective motions. Using the kinematic information from Nordlund's granulation model, Dravin et al. (1981,1986,1990a,b) have computed details of line asymmetries and wavelength shifts of photospheric spectral lines. Steffen (1987) made a similar study on the spectroscopic properties of solar granulation with his axisymmetric stationary models.

E. Interaction with Pulsation

The study of pulsations or waves in the Sun has grown into a very large field. The 5-min oscillations discovered by Leighton et al. (1962) and Noyes and Leighton (1963) have become a very important diagnostic tool for the internal condition of the Sun (Ulrich 1970; Leibacher and Stein 1971; Wolff 1972; Unno et al. 1979; Cox 1980; Duval and Harvey 1983; Ulrich and Rhodes 1983; Deubner and Gough 1984; Gough 1985b; Christensen-Dalsgaard et al. 1985a; Christensen-Dalsgaard 1986). However, studies on the connection of solar convection and waves were originally developed for the heating of the chromosphere and corona (Whitaker 1963; Athay 1966; Schatzman and Souffrin 1967; Ulmschneider 1971). The convective turbulence was proposed to be the source of waves that supply mechanical energy to the upper atmosphere of the Sun (Unno 1967; Stein 1967,1968). This branch of development is merging with the study of convective effects on nonradial pulsations (Gabril et al. 1975; Goldreich and Keeley 1977b; Gonczi and Osaki 1980; Stellingwerf 1982; Goldreich and Kumer 1988). The generation of p-mode oscillations by convection was recently simulated in multidimensional calculations (Steffen 1988; Chan and Sofia 1988; Stein et al. 1988,1989). Excitation of gravity waves by penetrative convection was illustrated by the calculation of Hurlburt et al. (1986).

F. Interaction with Rotation

Like most astrophysical objects, the Sun is rotating. Rotation introduces much complication in the theory of convection. One important problem is the generation of the latitudinal differential rotation (Newton and Nunn 1951; Howard and Harvey 1970). Many theories on the origin of solar differential rotation have been proposed; most of them were based on the anisotropic action of the convective turbulence or on the latitudinal variation of convective heat transfer (Wasiutynski 1946; Kippenhahn 1963; Weiss 1965; Durney 1970; Busse 1970; Durney and Roxburgh 1971; Belvedere and Paterno 1977; Durney and Spruit 1979; Rudiger 1980,1989; Stix 1981; Pidatella et al. 1986; Tuominen and Rudiger 1989; Durney 1989). It also has been proposed that the differential rotation is the zonal velocity field of an axisymmetric convective mode (Chan et al. 1987). Numerical simulation of solar differential rotation was pioneered by Gilman (1972,1977a); elaborations were made in later years (Young 1974; Gilman and Glatzmaier 1981; Glatzmaier and Gilman 1981; Glatzmaier 1984). However, recent results from helioseismology (Brown et al. 1989; Libbrecht 1989; Dziembowski et al. 1989) indicate that the distribution of angular velocity inside the Sun is substantially different from those predicted by the numerical models. Thus a reconsideration of the numerical problem is being undertaken (Gilman et al. 1989; Chan and Mayr 1989).

G. Interaction with the Global Magnetic Field: the Dynamo Model

The magnetic field adds another dimension of difficulty to the problem of solar convection. The origin of the 11-yr sunspot cycle is a problem that has occupied the minds of many generations of solar physicists (Hale 1908; Cowling 1934; Alfvén 1950). The prediction of the magnitude of a cycle has very practical application in space aeronautics (Schatten and Sofia 1987). In most solar dynamo models, the azimuthal field is generated by the nonuniform rotation stretching the poloidal field, and the poloidal field is produced by the cyclonic convective motions twisting the azimuthal field (Parker 1955b; Babcock 1961; Steenbeck et al. 1966; Steenbeck and Krause 1969a,b; Leighton 1969; Kohler 1970,1973; Roberts 1972). Global kinematic dynamo models compute the evolution of the global magnetic field with assumed patterns of the global scale convection (Yoshimura 1971,1972,1975a; Stix 1973,1976a). Dynamically self-consistent dynamo models further take account of the feedback of magnetic actions on the fluid motions (Gilman and Miller 1981,1986; Gilman 1983; Glatzmaier 1985b; Brandenburg et al. 1990). So far, such numerical models have not been successful in generating solutions compatible with the essential features of the solar magnetic cycle. Recently, attentions have been brought to locating the dynamo region in a thin layer near the bottom of the convection zone (DeLuca and Gilman 1986,1988; Gilman et al. 1989).

H. Interaction with the Magnetic Field in Smaller Scales

Studying the interaction of convection with the magnetic field in the small scales is fundamental to the understanding of sunspots, active regions and the fibrils that are the constituents of the global field (Parker 1955b,1975a,1979a,1982a,b; Weiss 1964; Leighton 1964; Gurm and Wentzel 1967; Mullan 1974; Piddington 1975; Schussler 1979,1980; Moffatt 1978; Spruit and Zweibel 1979b; Schatten et al. 1986). Many numerical studies have been made on the different aspects of the problem (Weiss 1966,1981a; Clark and Johnson 1967; Proctor and Weiss 1982; Galloway and Moore 1979; Schussler 1979; Nordlund 1983,1985b,1986; Deinzer et al. 1984a,b; Cattaneo and Hughes 1986; Hurlburt and Toomre 1988; Knolker et al. 1988; Nordlund and Stein 1989; Brandenburg et al. 1989).

II. TWO-DIMENSIONAL NUMERICAL SIMULATIONS AND SPECTROSCOPIC PROPERTIES OF SOLAR GRANULES

A. Introduction

In this section we describe two-dimensional (cylindrical) hydrodynamical models of solar granular convection cells, obtained from numerical simulations of turbulent compressible convection in a stratified medium, including realistic thermodynamics and a detailed treatment of radiative energy transfer. Based on these models, we discuss the dynamical and thermal structure of the solar photosphere. For a direct comparison with various spectroscopic observations, spatially resolved and horizontally averaged synthetic spectra have been computed from the models. We conclude that basic properties of the solar granulation can be reproduced and understood by our two-dimensional simulations.

The theoretical understanding of the phenomenon of solar granulation is one of the topics of present-day solar physics. Despite extensive observational and theoretical efforts, our knowledge about the velocity and temperature fluctuations associated with granulation in the solar photosphere is still incomplete. Commonly, granulation is interpreted as a pattern of surface convection cells at the boundary between the hydrogen convection zone and the photosphere, but from time to time even the convective character of granulation has been questioned, as in a recent study by Roudier and Muller (1986) who suggest that granules smaller than about 1000 km might actually be turbulent eddies, owing their existence to larger-scale flows of which they are just decay products, in contrast to convection cells, that are autonomous structures driven by buoyancy.

It has become evident over the years that it is impossible to obtain a consistent quantitative picture of the conditions prevailing in the solar granulation layers from a purely empirical analysis of existing observational material (see, e.g., Bray et al. 1984). To make progress, theoretical granulation models having the potential to provide data that can be compared directly to solar observations are indispensable.

Notoriously, the calculation of convection in stellar atmospheres is a complex problem, requiring the application of time-dependent, nonlinear hydrodynamics to a highly turbulent flow that, to complicate the situation, strongly interacts with the photospheric radiation field. Although in principle the problem is well defined by a few differential equations, an analytical solution is impossible when realistic background physics is to be included. Relevant results can only be obtained from numerical simulations on powerful computers.

The purpose of this section is to demonstrate what kind of results can be obtained from two-dimensional models of granular convection cells. In Sec. II.B we give some motivation for doing two-dimensional calculations. Section II.C briefly describes the basic foundations of the numerical simulations, while in Secs. II.D and E some of the results are presented along with corresponding observations.

B. Why Two-Dimensional Hydrodynamics?

Turbulent convection is intrinsically a three-dimensional process. Strictly two-dimensional flows are nowhere found in nature; they are an idealization applied to flows in computer simulations. It is well known that two-dimensional turbulence has basically different cascade properties compared to three-dimensional turbulence (cf. review by Kraichnan and Montgomery 1980). While in three-dimensional systems the energy cascade is from larger to smaller spatial scales, it is from higher to lower wavenumber modes in two-dimensional flows, i.e., the largest possible spatial scales are preferred here. But it is not necessarily true that all the energy in two-dimensional flows resides at the largest wavelengths. If the flow is strongly driven at smaller wavelengths, there can be significant energy also at those smaller wavelengths. However, it is clear that some interesting phenomena are missed even qualitatively in two dimensions. For example, vertical vorticity cannot be modeled and there is no vortex stretching.

Nevertheless, there are good reasons to perform two-dimensional simulations of stellar convection. One important aspect is that in the case of the solar granulation the preferred horizontal scales are known observationally. Although granules down to very small sizes may exist, the overwhelming contribution to the surface area (and hence to the emergent spectrum) comes from granules measuring between 700 and 1500 km in diameter (Roudier and Muller 1986). Furthermore, the visual impression of the granulation pattern suggests that the structure of a single granule is roughly axially symmetrical, especially for undisturbed granules. On this basis, it seems reasonable to model isolated granular convection cells in cylindrical symmetry, with model diameters corresponding to scales that dominate the appearance of the solar granulation.

There are in fact examples of two-dimensional convection calculations that have been successfully applied to describe actual convective phenomena accessible to experimental verification. For example, the behavior of fireballs could be realistically predicted by two-dimensional models for a large range of events (Ruppel and Norton 1975). For a reasonable agreement between simulation and nature it seems important that the appropriate spatial scales are imposed on the models and that the actual mean flow is essentially two-dimensional.

Further justification for two-dimensional calculations comes from a comparison between otherwise identical simulations in two and three dimensions, as carried out, e.g., by Deupree (1984a,b) and by Chan and Sofia (1986). These authors find that the corresponding results are phenomenologically similar and many basic properties found in two-dimensional flows persist in three dimensions.

Finally, it is important to note that restricting the problem to two dimensions is an enormous advantage in terms of computer requirements. The calculation of three-dimensional models is much more severely restricted by available computer capacities than two-dimensional calculations, in particular if detailed radiative transfer is to be taken into account. This has important consequences: the spatial resolution achievable with two-dimensional models is at least an order of magnitude better than with three-dimensional models; or, alternatively, the simulated volume can be correspondingly larger in two dimensions. Moreover, it is often necessary to repeat a simulation with a variety of different parameters to understand the physical (or numerical) cause for a certain phenomenon found in the calculations. It is simply not feasible to do a large number of test runs with fully three-dimensional models. In practice, this is not merely a quantitative difference, but constitutes a qualitative advantage of two-dimensional calculations. For example, the three-dimensional compressible simulations described by Stein et al. (1989) took 600 CPU hr on a vector machine to cover 3 solar hr. In contrast, the two-dimensional models discussed here need approximately 5 to 10 hr of CPU time on a CRAY X-MP to simulate 3 hr of real time.

Apart from the problem of properly displaying three-dimensional information, the added complexity makes the numerical results more difficult to understand. In contrast, the two-dimensional (cylindrical) models described in the following are comparatively simple, especially if the steady-state solutions are considered. Under these circumstances it is more readily possible to study in some detail the physical mechanisms governing granular convection.

Summarizing, modeling granular convection cells in two dimensions is well motivated. It is clear that two-dimensional simulations cannot replace fully three-dimensional calculations. Rather, they must be understood as a complementary approach, which may prove advantageous for certain applications. Our aim is to investigate to what extent such an idealized description is useful to explain the observations. The least we can expect is that two-dimensional models including detailed thermodynamics and realistic radiative transfer will be able to give a much better representation of the dynamical and thermal structure of the solar photosphere than the commonly used one-dimensional solar atmospheric models based on mixing-length concepts.

C. The Numerical Simulations

1. Foundations of the Numerical Simulations. The framework of the model calculations is given by the time-dependent, nonlinear equations of hydrodynamics prescribing the conservation of mass, momentum and energy in a stratified compressible fluid. To account for the highly turbulent character expected for the solar granular flow (Reynolds number $\approx 10^9$) viscosity terms are included to model roughly the turbulent exchange of momentum and energy on subgrid scales. The corresponding subgrid scale eddy viscosity is calculated according to the scheme given by Deardorff (1970,1971). In cylindrical coordinates (r,φ,z), assuming axial symmetry but not permitting a φ-component of the flow velocity, the conservation equations for mass and momentum read:

$$\frac{D\rho}{Dt} + \rho \text{ div } \mathbf{v} = 0 \tag{1}$$

$$\frac{Du}{Dt} + \frac{1}{\rho}\frac{\partial p}{\partial r} - g_{r,vis} = 0 \tag{2}$$

$$\frac{Dv}{Dt} + \frac{1}{\rho}\frac{\partial p}{\partial z} - g_{z,vis} + g = 0 \tag{3}$$

where $\dfrac{D}{Dt} \equiv \dfrac{\partial}{\partial t} + u\dfrac{\partial}{\partial r} + v\dfrac{\partial}{\partial z}$ is the Lagrangean (substantial) derivative along the particle path, t denotes time, r and z represent the horizontal and vertical spatial coordinates (z increasing upwards), ρ is the mass density, p the gas pressure; u and v are the horizontal and vertical velocity components, respectively, div $\mathbf{v} \equiv \partial u/\partial r + u/r + \partial v/\partial z$; g is the acceleration of gravity (directed downwards) while $g_{r,vis}$ and $g_{z,vis}$ stand for the viscous acceleration in r and z direction, respectively, representing functions of the spatial derivatives of the velocity field.

The energy equation may be written as

$$\frac{Ds}{Dt} = \left.\frac{Ds}{Dt}\right|^{rad} + \left.\frac{Ds}{Dt}\right|^{dis} + \left.\frac{Ds}{Dt}\right|^{dif} \tag{4}$$

stating that the specific entropy s of a moving fluid element is in general not constant (as in the adiabatic case), but changes due to: (1) exchange of radiation, (2) viscous dissipation of kinetic energy, (3) turbulent diffusion of heat. The radiative damping term is given by

$$\left.\frac{Ds}{Dt}\right|^{rad} = \frac{4\pi}{\rho T}\int_0^\infty k_\nu (J_\nu - S_\nu)d\nu \tag{5}$$

where T is temperature, k_ν is the monochromatic total absorption coefficient (cm^{-1}), J_ν is the angle-averaged monochromatic intensity and S_ν is the source function which in LTE is equal to the Planck function B_ν. The dissipation term reads

$$\left.\frac{Ds}{Dt}\right|^{dis} = \frac{\Phi}{\rho T} \tag{6}$$

with

$$\Phi = \eta \left[2\left(\frac{\partial u}{\partial r}\right)^2 + 2\left(\frac{\partial v}{\partial z}\right)^2 + 2\left(\frac{u}{r}\right)^2 + \left(\frac{\partial u}{\partial z} + \frac{\partial v}{\partial r}\right)^2 - \frac{2}{3}(\text{div } \mathbf{v})^2 \right]. \tag{7}$$

η is the dynamical viscosity which itself is a complicated function of the velocity field. The turbulent heat transfer is modeled as

$$\left.\frac{Ds}{Dt}\right|^{dif} = - \frac{\text{div } \mathbf{f}_{di\,f}}{\rho T} \tag{8}$$

where

$$\mathbf{f}_{di\,f} = - \frac{2}{Pr} T\eta \text{ grad } s. \tag{9}$$

The Prandtl number Pr is taken to be 1. Note that here the heat diffusion is proportional to the local entropy gradient, which is substantially different from the commonly used proportionality to the temperature gradient. This concept is more compatible with the idea that turbulence has the tendency to produce isentropic conditions, as opposed to the diffusive action based on temperature (see also Chan and Sofia 1986). It is worth pointing out that the convective energy flux according to mixing-length theory may be written, without restriction to the case of an ideal gas, as

$$F_c = - \frac{1}{2} T \rho v l \frac{ds}{dz}. \tag{10}$$

This is essentially the same expression as Eq. (9). The main quantitative difference is that while the mixing length l in Eq. (10) is of the order of one pressure scale height, it is of the order of the grid resolution in Eq. (9). The calculations show that in the solar photosphere, the effect of dissipation is about one order of magnitude smaller than that of turbulent heat transfer, which, in turn, is roughly one order of magnitude smaller than radiative damping, the dominant mode of energy exchange, at least in the layers around optical depth unity.

   The thermodynamical relations entering the computations explicitly allow for temperature and pressure dependent ionization of hydrogen that has a critical influence on the specific heats, the adiabatic temperature gradient, and hence on the strength of convection in the upper layers of the solar convection zone. Helium ionization is negligible in the temperature/pressure domain represented by the current models.

Equally important, a realistic modeling of granular convection requires a reasonable description of the interaction of hydrodynamics and radiative transfer. Because the simulation of photospheric convection has to include the optically thick and optically thin (visible) layers at the same time, it is not appropriate to calculate the radiation field from the local temperature gradient by applying the often-used diffusion approximation; this is only acceptable in the deep, optically thick layers where, however, radiative transfer is of minor importance for the gas dynamics. In the transition region from optically thick to optically thin conditions, radiative cooling becomes a very important factor. Here and in the higher layers, the radiation field has a nonlocal character and the diffusion approximation is no longer valid. More accurate (but also more costly) methods are needed to derive the radiation field at the critical boundary between convective and radiative layers. Notably, the thermal and dynamical structure of this so-called overshoot region are of considerable interest, because it is here that most of the emergent spectrum originates.

Our approach is to solve the equation of radiative transfer.

$$dI_\nu = -k_\nu(I_\nu - S_\nu)dl \qquad (11)$$

along a large number of rays crossing the model in various directions and with different inclinations. Angle-averaging of the intensities resulting from Eq. (11) yields J_ν needed in Eq. (5). In this way, nonlocal radiative exchange is taken into account both vertically and horizontally. Using a realistic Rosseland mean opacity as a function of pressure and temperature, the gray approximation in LTE has been adopted so far.

The equations of radiative transfer and hydrodynamics are solved simultaneously without introducing simplifications such as linearizations or the anelastic approximation. The numerical scheme uses an iterative procedure based on the method of bi-characteristics. The code was derived from that of Stefanik et al. (1984). Several extensions were applied to adapt it for the simulation of solar granulation, including the introduction of turbulent viscosity, hydrogen ionization and two-dimensional radiative transfer.

2. Model Parameters, Boundary and Initial Conditions. Typically, the models extend vertically from 250 km below to 600 km above the $\tau_{\rm ross} = 1$ level, i.e., they span several pressure scale heights. The vertical grid distance ranges from about 20 km in the lower, convective part (more than 10 grid points per pressure scale height) to about 40 km in the upper, radiative region.

The diameter of the cylindrical model is a free parameter which has been varied between 260 and 2600 km. Depending on the model diameter, the horizontal resolution of the grid lies between 10 and 40 km.

The time step is typically less than 1 s by the requirement that it must not exceed the sound-travel time between any two adjacent grid points (Courant condition).

Boundary conditions appropriate to the situation in the solar atmosphere must be imposed before the differential equations can be solved numerically. At the axis of symmetry ($r = 0$) the horizontal component of the velocity must be zero, while for all remaining variables the horizontal derivative must vanish. Similar conditions are imposed at the lateral boundary ($r = R$). For the upper boundary we have two options. Either we use a stress-free closed top ($v = \partial u/\partial z = 0$) or the upper boundary is made transmitting for simple acoustic waves as described by Stefanik et al. (1984), to permit (initial) pressure disturbances to leave the computational domain instead of being reflected.

For our purpose, the stratification of the solar convection zone is much too large to be included in a single simulation. We model only the very top of the convection zone. In this situation, the lower boundary is most critical because there is no way to place it at a position where conditions naturally allow a simple boundary condition to be used; considering that at the location of the lower boundary nearly all the energy is carried by convection, a closed bottom seems unreasonable. Much effort was needed to devise an open lower boundary condition, allowing a free flow of gas out of and into the model. The basic idea in the formulation of this boundary condition is to assume a spatially constant pressure p^* along those parts of the bottom where the flow is directed upward. Two principal versions have been tested. In the first one (a), p^* is constant also in time. Then the value of p^* fixes the depth of the lower boundary within the atmosphere. Alternatively (version b) p^* is adjusted from time step to time step such that the total mass within the model volume is conserved. In this case, the depth of the model is determined by its (initial) total mass. The entropy of the gas entering the model from below is automatically adjusted in such a way as to make the radiative flux through the *upper* boundary correspond to the specified effective temperature. The time constant for this entropy adjustment is chosen to be of the order of 1 turnover time. Both versions of the lower boundary condition are physically consistent and flexible enough to allow the flow itself to choose the horizontal positions of rising and sinking regions at any time during the simulation. While versions (a) and (b) are the same for steady-state situations, condition (b) seems in general to give more reasonable results.

For radiative transfer we assume no incident radiation at the top, while at the bottom we can safely apply the diffusion approximation as a boundary condition. The lateral boundary is chosen to be reflective in order to mimic closely conditions in the solar granulation where each convection cell is surrounded by several similar ones. This choice also guarantees that the net energy flow through the side walls is zero, and no assumptions about the surroundings of the model need to be made.

In principle, any arbitrary configuration can be taken as an initial condition. In practice, however, the initial state must not be too far from the mean relaxed state that develops during the subsequent time evolution. Otherwise numerical problems can arise if the flow velocities go supersonic and shocks are generated in a too violent initial phase of relaxation.

3. Computation of the Emergent Spectrum. It is essential to have the possibility to calculate synthetic spectra from the numerical models in order to investigate their observational implications. Although in the numerical simulation itself radiative transfer is treated in the gray approximation, it is possible (somewhat inconsistently) to use the two-dimensional hydrodynamical model atmospheres for detailed line-formation calculations to derive basic spectroscopic properties of the models. Employing a modified version of the LTE package ATMOS/LINFOR, developed by the Kiel Group for the analysis of stellar spectra, the emergent spectrum at arbitrary wavelengths can be obtained as a function of the inclination of the line of sight against the vertical axis μ. Such spectrum synthesis calculations are performed only at selected instants of time for diagnostic purposes, e.g., to evaluate the continuum intensity contrast or the asymmetry of spectral line profiles.

D. Resulting Granular Flows

1. Steady-State Solutions. Starting with appropriate initial conditions, we can follow the time evolution of the flow. From a first series of models, we found that the flow developed towards a steady state if the chosen diameter of the cell was less than a critical upper limit of roughly 2000 km. Subsequently, the calculation of the radiation field was improved to give a better angular resolution, using more than 10 times as many rays as before. Furthermore, the new scheme was designed to assure a much better numerical conservation of energy. The resulting series of second-generation models shows no steady-state solutions down to cell diameters of about 1000 km. It is presently not known whether the new models become stationary if the cell size is further reduced; the corresponding runs have not yet been carried out.

Although the new, nonstationary models are certainly more realistic, we discuss the first-generation steady-state models here, because they reveal more clearly some of the basic physics governing granular convection. A typical steady-state solution is displayed in Fig. 1, where the model diameter is 1750 km. All stationary models show similar characteristic flow patterns in the lower, convectively unstable part of the model. A strong downdraft at the axis of symmetry with maximum velocities of the order of 6 km s^{-1} is surrounded by a broader ring-like upflow of hot gas with lower velocities (≤ 2.5 km s^{-1}) that again turns into a narrow downflow near the side walls. The convective velocity field extends considerably into the stable layers, a

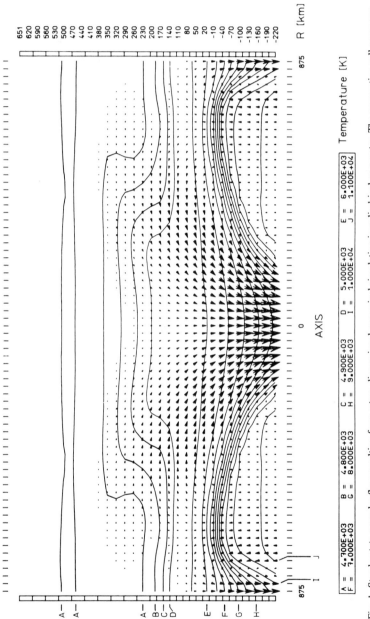

Fig. 1. Steady-state granular flow resulting from a two-dimensional numerical simulation in cylindrical symmetry. The convection cell measures 1750 km in diameter and extends from 220 km below to 650 km above $\tau = 1$ ($z = 0$). $\tau = 0.1$ corresponds to $z = 135$ km, $\tau = 0.01$ to $z = 280$ km. Arrows indicate direction and magnitude of the local velocity where $|v|$ exceeds 300 m s^{-1}; their lengths scale in proportion to $|v|$. The maximum velocity is 6.3 km s^{-1} in the central downdraft, while upward velocities are < 2.5 km s^{-1}. The thermal structure is represented by lines of constant temperature in steps of 500 K in the lower atmosphere, and in steps of 100 K where temperature is below 5000 K.

result that cannot be obtained within the framework of local mixing-length theories. The amount of overshooting is clearly a function of the horizontal size of the convection cells. While the absolute value of the *rms* vertical velocity v_{rms} is typically 1 km s^{-1} at a height of approximately 30 km above $\tau = 1$, quite independent of cell size, we find that the scale height of v_{rms} in the overshooting region H_v depends on the model diameter D roughly as $H_v \sim D^{0.7}$, i.e., vertical motions decline more rapidly with height above smaller granules as compared to the larger ones. Similarly, the magnitude of the maximum *rms* horizontal velocity, $u_{rms,max}$ depends on D approximately as $u_{rms,max} \sim D^{0.5}$. This means that the horizontal return flows occurring in the layers around $\tau = 1$, where the vertical motions are decelerated and turned into a horizontal direction, are more pronounced in large granules.

The corresponding temperature structure is also indicated in Fig. 1. The lines of constant temperature illustrate that granulation generates large horizontal temperature differences, typically more than 4000 K just 100 km below the visible surface ($\tau = 1$). At first sight, this value seems exceedingly large. However, it is easily verified that temperature differences of this magnitude are necessary to carry the solar energy flux by convection with flow velocities of a few km s^{-1}.

Equally remarkable, the calculations produce a very steep temperature gradient at the top of the ascending part of the flow ($\gtrsim 40$ K km^{-1}) where the hot gas reaches the optically thin layers and loses its excess energy within a short time by efficient radiative cooling. The steep temperature gradient in concert with the recombination of hydrogen produces a local density inversion about 50 km below $\tau = 1$; i.e., a layer of relatively higher density lies on top of gas with lower density. In contrast, density increases monotonically with depth in the cool, intergranular regions.

In the overshooting layers, the temperature fluctuations change sign: these layers are relatively cooler above the ascending granular parts and relatively warmer above the descending intergranular regions. This behavior is a consequence of the penetration of the convective motions into stably stratified atmospheric layers.

The uppermost part of the model is essentially in radiative equilibrium. Here the temperature is nearly constant with height as expected for a gray radiative atmosphere. In these layers, horizontal temperature fluctuations become insignificant for the steady-state models. For a detailed study of the calculated steady-state velocity and temperature fields as a function of horizontal cell size see Steffen et al. (1989).

Test calculations have shown that details of the initial conditions are unimportant in the sense that the final steady state seems to be largely independent of the initial configuration. In particular, an initial model suggesting the flow to ascend at the axis of symmetry and descend in a ring surrounding the central part, resulted in a reversal of the motions after a short time. This behavior seems to be a consequence of the symmetry conditions required at the axis, and not due to a problem with the code or the lower boundary condition. This notion is supported by the numerical simulation study of compressible convection by Chan and Sofia (1986). They found that downdrafts are attracted and enhanced by impenetrable lateral boundaries (centers of symmetry). Thus, the ring-shaped granules emerging from our calculations may well be related to the so-called "exploding granules," a common phenomenon in quiet granulation (Title et al. 1987a).

Finally, it is worthwhile mentioning that oscillations, superimposed on the convective flow, seem to be ubiquitous. Typically, we find periods of the order of 250 s (Steffen 1988). Current evidence suggests that the oscillation frequency is related to the acoustic cutoff frequency in the layers around $\tau = 1$. Test runs have confirmed the frequency to increase in proportion to g, the acceleration of gravity.

2. Time-Dependent Flows.   Simulations with model diameters exceeding the critical upper limit mentioned above never reach a steady state, not even asymptotically; they are truly nonstationary. For diameters close to the critical limit, we find long quiet periods of time (about 20 to 30 min) where the topology of the flow is basically like that of the steady-state solutions with smaller horizontal size. After some time, however, oscillations with increasing amplitude develop and finally lead to a violent reorganization of the flow structure: the extended rising regions become separated by a downflow, implying that large convection cells temporarily split into smaller fragments (rings). This configuration typically lives for 5 to 10 min, after which adjacent regions of hot rising gas have the tendency to merge again. When the initial topology is restored, another quiet-time interval begins. The downdrafts at the axis of symmetry and at the side walls persist throughout the simulated time interval of approximately 2 hr of real time.

A similar behavior is found from the more recent simulations carried out with the improved version of the code (see above). For all cases studied so far, the resulting flows exhibit a distinctly nonstationary character.

The time evolution of the flow can be characterized as stochastic. Significant changes occur on time scales of the order 10 min (corresponding to approximately 1 turnover time), comparable to typical granular life times. Alterations of the flow topology may be interpreted as a continuous splitting and merging of granules under the constraints imposed by the cylindrical symmetry. Sometimes these changes are strong enough to cause supersonic flow velocities in the higher layers, leading to the formation of upward traveling shocks above the central downdraft.

A snapshot from a nonstationary simulation, using the more advanced version of the code, is shown in Fig. 2. Here the center of the rising part of the flow has collapsed, resulting in the formation of two "granules" separated by a downflow. In the subphotospheric layers, velocity and temperature fluctuations are comparable in magnitude to those found in the steady-state mod-

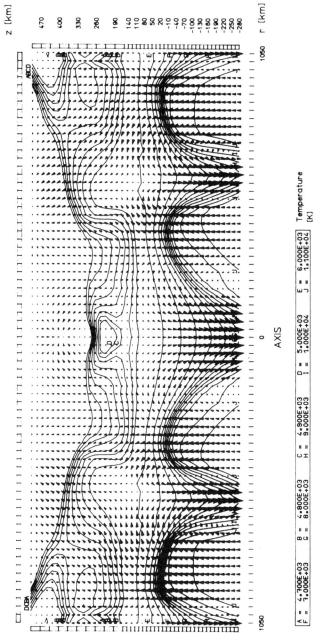

Fig. 2. Snapshot from a nonstationary numerical simulation of granular convection in a cell measuring 2100 km in diameter. The velocity and temperature field is indicated as in Fig. 1. The maximum velocity is $|v| = 7.6$ km s^{-1}.

els. However, the layers above $\tau = 1$ are much more strongly affected by granular convection. In contrast to the steady-state models, we have to conclude that the thermal structure of the higher layers (where the spectral lines are formed) is not exclusively determined by the condition of radiative equilibrium. Rather, the velocity field due to overshooting motions from the convection zone seems to have an important impact on the temperature stratification of the upper solar photosphere.

The development of more than one ring of rising gas may be an indication that the assumption of cylindrical symmetry is no longer a good choice. Perhaps a more reasonable alternative would be to use two-dimensional rectangular coordinates, where the granules are rolls of infinite length instead of rings. Of course, the real topology of granulation can only be modeled in three dimensions (see Sec. III and IV).

E. Spectroscopic Properties of the Two-Dimensional Model Granules

1. Continuum. From radiative transfer calculations along rays parallel to the axis of symmetry, we can simulate spectroscopic observations at disk center. Viewed in continuum light, the steady-state models show a dark center surrounded by a bright ring that is bounded by an outer dark lane (resembling an "exploding granule"). For steady-state models with diameters between 1000 and 2000 km the *rms* intensity contrast of the two-dimensional intensity pattern ranges between 14 and 16% in the continuum at 5000 Å, which seems to be in reasonable agreement with observational evidence (see, e.g., Wittmann 1979; Bray et al. 1984). For the larger, time-dependent models we obtain a somewhat higher value of the *rms* intensity contrast. Averaged over time, 20% at 5000 Å is a typical value.

Towards smaller granular scales the amplitude of the horizontal intensity fluctuations declines strongly. As discussed in detail by Steffen et al. (1989), the main reason is that horizontal radiative exchange becomes increasingly more efficient with decreasing cell size, reducing horizontal temperature fluctuations particularly in the continuum-forming layers. Observations and model calculations indicate that the continuum intensity contrast δI_{rms} depends on wavelength roughly as $\delta I_{rms} \sim 1/\lambda$, essentially reflecting the wavelength dependence of the Planck function.

Spectrum synthesis for different disk positions requires more extensive calculations with inclined rays ($\mu = \cos \theta < 1$). Table II gives $\delta I_{rms}(\mu)/\delta I_{rms}(1)$ at λ 5380 Å as a function of μ for a steady-state model measuring 1050 km in diameter (column A) and for the larger model shown in Fig. 2 (column B), representing a typical phase during the time evolution of a nonstationary simulation (no time average).

We note a monotonic decline of the granular contrast towards the limb for both models. It is obvious that the nonstationary model (column B) produces a slower decrease of the granular contrast towards the limb, because it shows larger temperature inhomogeneities in the higher layers (see above).

TABLE II
Simulated and Observed Center-to-Limb Variation
of *rms* Intensity Contrast at 5380 Å

μ	A	$\delta I_{rms}(\mu)/\delta I_{rms}(1)$ B	C
1.0	1.00	1.00	1.00
0.8	0.88	0.89	0.93
0.6	0.70	0.76	0.75
0.4	0.44	0.61	0.52
0.2	0.14	0.52	0.25

The original observational results found by Schmidt et al. (1979) are given in column C of Table II showing the same general trend as derived from the simulations. It must be mentioned, however, that according to Wiesmeier and Durrant (1981), the values given in column C should be corrected by a factor $1/\sqrt{\mu}$. Accordingly, the observed center-to-limb variation of the granular intensity contrast is even smaller than indicated by the simulated results listed in column B.

2. Line Spectrum. The profiles of spectral lines can be calculated with a spatial resolution that corresponds to the horizontal grid distance of the hydrodynamical models (typically between 10 and 40 km). Although currently no spectroscopic observation is capable of such an extremely high spatial resolution, it is nevertheless instructive to look at the predicted individual line profiles. The general picture emerging from the numerical simulations may be summarized as follows.

The cores of absorption lines originating from the bright granular regions are blue shifted (relative to the laboratory wavelength), and the blue wing of the line profile is depressed relative to the red wing, resulting in a considerable line asymmetry. Lines formed in the dark intergranular lanes exhibit an even stronger asymmetry, but in the opposite direction, their cores being red shifted (Fig. 3). The horizontally averaged line profile, obtained as a superposition of the spatially resolved profiles, turns out to be much less asymmetrical than most of the line profiles seen at high spatial resolution.

The residual intensity in the cores of weak spectral lines, which are formed near the continuum-forming layers, varies across the granulation pattern in accordance with the continuum intensity. In contrast, the intensity in the cores of the stronger lines tends to be anti-correlated with the continuum intensity: the cores of these lines are darker in the granules and brighter in the intergranular lanes (Fig. 3). This behavior is due to the change of sign of the temperature fluctuations in the overshooting layers mentioned in Sec. II.D.1.

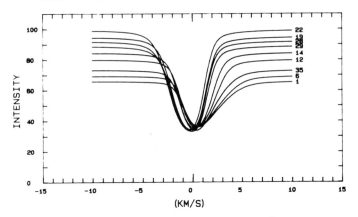

Fig. 3. Spatially resolved profiles of an artificial Fe I line at λ 6000 Å at disk center, calculated from a steady-state model measuring 1050 km in diameter. In granular regions with high continuum intensity, the line profiles are blue-shifted and exhibit a blue asymmetry, whereas the lines are red-shifted and show a strong red asymmetry in the intergranular regions with low continuum intensity. Numbers indicate the horizontal position within the cylindrical model seen from above (1 = near axis, 35 = near lateral boundary). The horizontally averaged line profile has an equivalent width of 53.5 mÅ. Referred to the local continuum, the line is weakest in the dark regions with an equivalent width of 42.5 mÅ and strongest in the bright granular parts, equivalent width 59.5 mÅ. From this model we obtain a continuum intensity contrast of δI_{rms} (6000 Å) = 11.5%.

It is encouraging to see that recent spectroscopic observations with high spatial resolution carried out at the Observatorio del Teide in the Canary Islands (Wiehr and Kneer 1988; Holweger and Kneer 1989) indeed confirm these spectroscopic characteristics predicted by our hydrodynamical models.

Figure 4 illustrates the situation in terms of the line bisectors. In the bright parts of the granulation, the corresponding line bisectors are inclined to the blue (the top portion near the continuum being blue shifted relative to the line core); in the dark intergranular regions, the line bisectors are inclined even stronger, but to the red. The different slopes of the spatially resolved bisectors reflect the different depth dependence of temperature and convective velocity at the various horizontal positions within a granular convection cell. Remarkably, the bisector of the horizontally averaged line profile exhibits the typical C-shape in close agreement with observation. Clearly, opposite asymmetries of the spatially resolved profiles cancel to a large degree when the spectrum is averaged over the granulation pattern. Spectroscopic evidence for such a behavior was found by Mattig et al. (1989).

A series of synthetic line spectra has been obtained from a nonstationary simulation for Fe II, λ 5197.6 Å. The bisector of the horizontally averaged

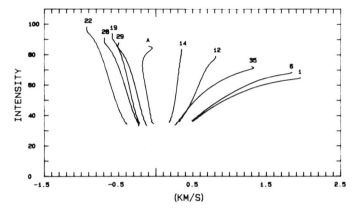

Fig. 4. Bisectors corresponding to the line profiles shown in Fig. 3. They are given on an absolute velocity scale (solar reference frame). Numbers indicate the horizontal position within the cylindrical model seen from above (1 = near axis, 35 = near lateral boundary). A denotes the bisector of the horizontally averaged line profile (C-shape).

line profile is shifted back and forth (typically ± 1 km s^{-1}) and heavily distorted as the flow evolves in time. However, the shape of a "C" is more or less retained; inverted "C's" were not found. Spatial and temporal averaging results in a line profile can be compared directly to standard solar spectra. In Fig. 5 we compare the observed line bisector of Fe II, λ 5197.6 Å with that of the mean synthetic line profile obtained by averaging 70 computed spectra (each one again an average of 35 spatially resolved spectra) separated in time by 100 s, covering a total interval of almost 2 hr. The agreement between observed and calculated bisector shape is excellent. However, the absolute convective blueshift of about 0.2 km s^{-1} (line core) seems too small if the value of $\Delta V = -0.8$ km s^{-1}, given by Dravins et al. (1986), is taken as a reference. On the other hand, the line broadening provided by the simulated photospheric velocity field is sufficient to account fully for the observed line width without invoking the usual *ad hoc* parameters micro- and macroturbulence.

Finally, an example of how the bisector of a synthetic spectral line varies across the solar disk is shown in Fig. 6, based on the snapshot model displayed in Fig. 2. Note that the large blueshift indicated by the bisectors is due to the fact that the whole upper atmosphere is moving upward at this instant. A time-averaged spectrum will result in a substantially smaller net blueshift (cf. Fig. 5). In qualitative agreement with observation, the bisector near the limb at $\mu = 0.2$ has the shape of an inverted "C".

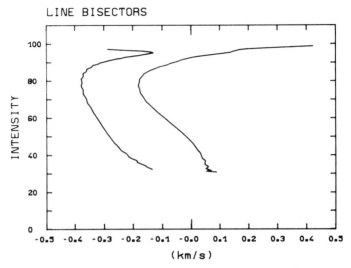

Fig. 5. Comparison of calculated and observed bisector of Fe II, λ 5197.6 Å at solar disk center, which has an equivalent width of roughly 90 mÅ. The observed bisector (right) is arbitrarily displaced relative to the calculated bisector (left, on an absolute velocity scale). Based on a nonstationary simulation in a cell measuring 2100 km in diameter, the synthetic bisector was obtained as a combined spatial and temporal average of 2450 individual line profiles. The differences near the continuum are due to a weak blend in the observed spectrum. (FTS observation courtesy of W. Livingston, National Solar Observatory, Tucson, U.S.A.)

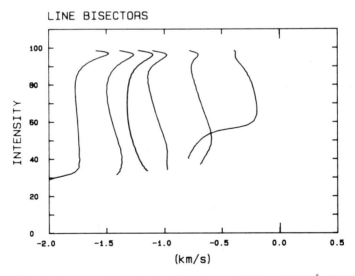

Fig. 6. Calculated center-to-limb variation of the bisector of Ni I, λ 6767.8 Å (GONG line). From left to right $\cos \theta = \mu$ is 1.0, 0.8, fd, 0.6, 0.4 and 0.2, respectively (fd = bisector of disk-integrated line profile). Spectrum synthesis is based on the model granule shown in Fig. 2 (no time average).

F. Conclusions and Future Prospects

We have calculated detailed hydrodynamical models of cylindrical photospheric convection cells. This two-dimensional problem requires considerably less computational effort than the full three-dimensional case. Based on the two-dimensional models, we have discussed the dynamical and thermal structure of the solar photosphere. The nonstationary simulations suggest that even the higher photosphere is not in radiative equilibrium but exhibits substantial temperature inhomogeneities induced by overshooting motions from the convection zone. Using spectrum synthesis techniques, observable quantities have been derived from our models. The results obtained so far indicate that basic properties of solar granular convection can be reproduced and understood by the two-dimensional model calculations. For example, the computed continuum intensity contrast and asymmetry of spectral lines compare well with observations.

We plan to perform further simulations of solar convection using two-dimensional rectangular geometry instead of axial symmetry, allowing the use of periodic boundary conditions in the horizontal direction. In this way, the development of exceedingly strong downdrafts at the lateral boundaries will be avoided. Another improvement under way is the implementation of nongray radiative transfer. Simulations with the future code should be extended to somewhat deeper layers than the present models.

Apart from studying solar granulation, future applications of the two-dimensional simulations will include the exploration of the hydrodynamical conditions in the atmospheres of F- and A-type stars.

III. BEHAVIOR OF DEEP, EFFICIENT CONVECTION

A. The Search for Hydrodynamical Principles of Convection

In the central region of the solar convection zone (from 1 to 16 pressure scale heights below the photosphere), radiation is ineffective for energy transport. Over 98% of the outward energy flux is carried by convection. This is a region where convection is mainly controlled by purely hydrodynamical effects, little complicated by radiation; we call this *efficient convection*. It would be easier to extract the hydrodynamical principles (if any) of deeply stratified convection by studying this region. These principles are not only important for understanding the structure and evolution of the Sun, but similarly important for other stars.

The mixing-length theory of convection is a set of assumed hydrodynamical principles. It supposes that energy transport is performed by heat-carrying bubbles which travel and then dissolve in about 1 to 2 pressure or density scale heights (the mixing length). The extent of the bubbles is scaled by the mixing length. This picture is mainly based on considering the effects of stratification (Schwarzschild 1961). As neighboring rising bubbles travel across some scale heights, their volumes and cross sections have to expand by a large factor. There would be no room for them to continue, and the pushing and squeezing would generate turbulent (twisting) motions which destroy the vertical coherence of the velocity (Chan et al. 1981). However, all numerical simulations of convection before 1980 did not reveal any scaling effects or dynamical significance associated with the scale heights. The only important vertical length scale found was the total depth of the convective region.

B. Controversy over Effects of Stratification

In 1982, Chan, Sofia, and Wolff reported that convection cells with sizes ranging from the total depth of the convection zone to the smallest scale height at the top of the zone co-exist in their two-dimensional simulation. Furthermore, the longitudinal correlation function of the vertical velocity was found to be scaled by the local scale heights. This report has generated controversies because for quite a few years after that, no other group obtained similar results. In particular, the flows of the other calculations would produce vertical auto-correlation coefficients close to unity throughout the total depths of the convection zones. Conclusive comparison could not be made between the calculations of the different groups because the model problems were very different. The study made by Chan et al. differed from the other studies in two important aspects. First, they studied efficient and turbulent convection. Effects of diffusion which tend to smother small-scale motions were minimized. Second, they paid special attention to resolving the scale heights at all depths. This of course is a necessary requirement for any study interested in the effects of scale heights.

Did some kind of numerical approximation or parameterization generate the reported behavior artificially? Specifically, the following questions have been raised: (1) Chan et al. used an alternating direction implicit method to increase the numerical speed of solving the Navier Stokes equations (Chan and Wolff 1982). The temporal truncation error of this method is relatively large due to the use of an operator-splitting technique (Chan 1983). Are the smaller cells simply numerical noise? (2) To parameterize the effects of subgrid-scale turbulence, Chan et al. used a nonlinear viscosity (Smagorinsky 1963) in their calculation. Can this added nonlinearity generate the smaller cells? These are valid concerns.

To examine these possibilities, Chan and Sofia (1986) made several three-dimensional calculations with different numerical techniques and viscosity models to intercompare. The scaling effects of the local scale heights persisted, even when a standard, explicit, time-marching scheme was used, as well as when a constant viscosity was used. Therefore, the behavior is robust.

The three-dimensional flows do not behave in a way identical to that conceived by the mixing-length picture (which is not surprising however).

Particularly, the upflows and downflows are highly asymmetric, and the stronger, more concentrated downflows can penetrate many scale heights, in agreement with those found by modal and two-dimensional calculations (see the references quoted in Sec. I). Recently, Stein and Nordlund reported that effects associated with stratification and scale heights appeared in their simulation of solar granules (see Sec. IV for a vivid description of such phenomena).

C. Dynamics of Efficient Convection

The demonstration of the effects of stratification, although important, is only one of the many aspects of efficient convection studied in the series of numerical work by Chan et al. To present an overall view on the behavior of this kind of convection, the following results are collected here to make a concise summary:

1. The energy-containing eddies of the convective turbulence decrease in size toward the top region where the lengths of the scale heights drop. The amplitudes and time scales of the fluctuation of the convective flux also decrease toward the top (Chan et al. 1982). The implication of this is that while the brightness fluctuation at the solar surface is only moderate, the absolute fluctuation can be much larger in the interior.

2. The convective velocity, temperature fluctuation and the enthalpy flux can be approximately computed from the mean superadiabatic gradient, with mixing-length type formulae (Chan and Sofia 1987). The result thus gives support to the mixing-length theory not only qualitatively, but also quantitatively.

3. The vertical correlation lengths of the vertical velocity and the temperature fluctuation are both scaled by the pressure scale height, not by the density scale height. This provides evidence that the mixing length in the mixing-length theory is indeed scaled by the pressure scale height, as most stellar evolution codes use.

4. In the upper part of the convection zone, vorticity in the vertical direction tends to associate with funnel-like downflows (Chan and Sofia 1986). This is similar to the behavior of the flows obtained by Nordlund (1985b). The correlation coefficient of the downward velocity and the absolute vertical vorticity is, however, found to be small, only ~ 0.1. This is consistent with the rare occurrence of fully developed vortices in the photosphere (Brandt et al. 1989).

5. Contrary to an implicit assumption of the mixing-length theory, the kinetic energy generated by the buoyancy can work form a large energy flux with amplitude approaching the size of the total flux. The direction of the mechanical flux is downward in most of the convection zone. This confirms the modal result of Massaguer and Zahn (1980) and the two-dimensional result of Hurlburt et al. (1984). The distribution of this flux is poorly

modeled by the diffusion approximation, but is found to be scaled by F/C_p where F is the total flux and C_p is the specific heat under constant pressure (Chan and Sofia 1989).

6. The production and dissipation of the kinetic energy do not parallel each other. Production is scaled by the total flux, and the local production rate is essentially a function of the local mean variables (see Sec.III.D). The dissipation is clearly nonlocal; a significant amount of kinetic energy is carried away from the production region, to be dissipated in lower regions.

7. The effective viscosity generated by the convective turbulence can be roughly estimated as $V_z'' H_p/3$ where V_z'' is the root-mean-square vertical velocity and H_p is the pressure scale height (Chan et al. 1987).

8. Long-lived oscillations co-exist with the convective turbulence (Chan and Sofia 1988). Their frequencies are shown to be almost identical to the eigenfrequencies of acoustic modes.

Chan and Sofia (1989) have compiled a list of numeral-empirical formulae which describe the quantitative relationships among the mean variables, root-mean-square fluctuations and correlations of the convective turbulence. However, the variables used in that paper are dimensionless. For the convenience of application, the list is recast in Table III in terms of dimensional variables. These formulae contain interesting information about the thermodynamical behavior of efficient convection. Below, two of them will be used to show that efficient convection is a very peculiar heat engine.

D. Efficient Convection as a Heat Engine

If no motions were allowed in the solar convection zone, the outward transfer of energy would have been very difficult. With convection, the process is much easier, but a price must be paid for moving the fluid against the turbulence (eventually the gas viscosity). This supply of energy comes from the heat entering at the bottom with a higher temperature T_b and leaving the top with a lower temperature T_t (see Fig. 7); considered in this way, the convecting gas is a heat engine. The rate of work W is the production rate of kinetic energy, which is eventually dissipated back into heat; therefore, the mean heat flux F going through the upper level must be the same as that going through the lower level.

At first glance, one may think that W is restricted by the Carnot limit: $(1 - T_t/T_b)F$. However, we now show that this limit can apparently be exceeded. Applying the formulae 15 and 20 in Table III to relate the local production rate of kinetic energy w to the flux F, one obtains

$$w = - \langle V_z \rho' \rangle g = \langle V_z \rangle \langle \rho \rangle g \sim 0.8 \, (F/C_p) \, (g/T) \qquad (12)$$

where V_z, ρ', g are the vertical velocity, density fluctuation and the gravitational acceleration, respectively; the brackets $\langle \ \rangle$ denote averaging. At the

TABLE III

A List of Approximate Relationships Obtained from fitting Three-Dimensional Numerical Results

Identifier	Approximate Formula
1	V_x'' (or V_y'') $\approx 0.61\ V_z''$
2	$\rho''/\langle\rho\rangle \approx 0.89\ T''/\langle T\rangle$
3	$p''/\langle p\rangle \approx 0.57\ T''/\langle T\rangle$
4	$S'' \approx 0.94\ C_p T''/\langle T\rangle$
5	$p'' \approx 0.26\ \langle\rho\rangle\ V''^2$
6	$p'' \approx 0.51\ \langle\rho\rangle\ V_z''^2$
7	$T''/\langle T\rangle \approx 0.90\ \langle\rho\rangle\ V_z''^2/\langle p\rangle$
8	$C[T',S'] \approx 0.99$
9	$C[\rho',S'] \approx -0.89$
10	$C[\rho',T'] \approx -0.82$
11	$C[p',T'] \approx 0.49$
12	$C[V_z,T'] \approx 0.81$
13	$C[V_z,S'] \approx 0.81$
14	$C[V_z,\rho'] \approx -0.74$
15	$\langle V_z\rho'\rangle \approx -\langle V_z\rangle\langle\rho\rangle$
16	$\langle V_z p\rangle \approx 1.24\ \langle V_z\rangle\langle p\rangle$
17	$\langle V_z T'\rangle \approx 1.26\ \langle V_z\rangle\langle T\rangle$
18	$\langle V_z S'\rangle \approx 1.20\ C_p\langle V_z\rangle$
19	$\langle V_z\rangle \approx 0.58\ \langle\rho\rangle V_z''^3/\langle p\rangle$
20	$F_{ep} \approx 1.25\ (C_p/R)\ \langle p\rangle\langle V_z\rangle$
21	$F_{ep} \approx 0.72\ (C_p/R)\ \langle\rho\rangle V_z''^3$
22	$T''/\langle T\rangle \approx 1.05\ \Delta\nabla + 0.0027$
23	$\langle\rho\rangle V_z''^2/\langle p\rangle \approx 1.17\ \Delta\nabla + 0.0032$
24	$\Delta\nabla \approx 1.04\ [(R/C_p)F]^{2/3}\ \langle\rho\rangle^{1/3}\langle p\rangle^{-1} - 0.002$
25	$\langle V_z\rho V^2\rangle \approx 1.03\ \langle\rho\rangle\langle V_z V^2\rangle$
26	$\langle V_z V^2\rangle \approx 1.13\ \langle V_z^3\rangle$
27	$\langle p'\nabla\cdot V\rangle \approx \langle V_z p'\rangle/H_\rho$

Notations

$'$	fluctuation with respect to the mean
$''$	root-mean-square (*rms*) deviation
$\langle\ \rangle$	the mean value at a certain depth
$C[,]$	correlation coefficient
C_p	specific heat under constant pressure
F	total flux
F_{ep}	enthalpy flux
H_ρ	density scale height
p	pressure
ρ	density
R	gas constant
T	temperature
S	entropy
$V_{x,y}$	horizontal velocities
V_z	vertical velocity
V''	*rms* velocity

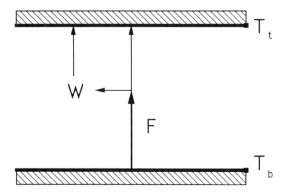

Fig. 7. The convection zone considered as a heat engine operating between two temperatures. The temperature at the bottom of the zone T_b is higher than the temperature at the top T_t. When a heat flux F flows through the system, a certain amount of mechanical power W is generated to drive the fluid motions.

moment, without affecting the validity of our argument, we ignore the difference between the total flux F and the enthalpy flux here. Integrating w over the depth z of the convection zone, one obtains

$$W = \int w \, dz = 0.8 \, (F/C_p) \int (g/T) \, dz \sim 0.8 \, F \, (R/C_p) \int d(\ln p)$$
$$\sim 0.8 \, F \, \ln(T_b/T_t) \qquad (13)$$

for which the hydrostatic approximation $d(\ln p)/dz \sim g/RT$ and the almost-adiabatic approximation $\ln(p_b/p_t) \sim (R/C_p) \ln (T_b/T_t)$ have been used. Equation (13) shows that W is proportional to the total number of temperature scale heights in the convection zone; it can make W greater than F. Does this mean that the laws of thermodynamics are being violated?

The Carnot limit given above is based on taking the whole convection zone as a single engine, but how about considering the zone as composed of a series of heat engines stacking one over another as shown in Fig. 8. For the convenience of analysis, let us suppose that the ratios of temperature drops that the engines operate on are the same; namely, $T_{i-1}/T_i = \lambda$, where $\lambda > 1$ is a constant. To satisfy the boundary conditions, $\lambda^N = T_b/T_t$ where N is the total number of engines. Now the total allowed power is

$$W = \Sigma \, W_i = F \, N \, (1 - 1/\lambda) = F \, \ln(T_b/T_t) \, (1 - 1/\lambda)/\ln(\lambda). \qquad (14)$$

The optimal value of this sum is obtained as $\lambda \to 1$; the upper limit is now

$$W \leq F \, \ln(T_b/T_t). \qquad (15)$$

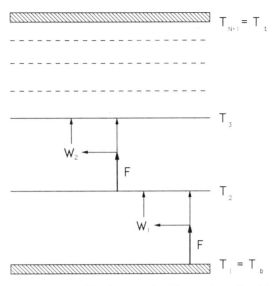

Fig. 8. The convection zone considered as a stack of heat engines. The i-th engine operates between temperatures T_i and T_{i+1}; it can generate mechanical work at the rate W_i.

Therefore, Eq. (13) is only at the 80% level of this limit and is in line with the laws of thermodynamics. One may consider such nonoptimal efficiency to be due to the existence of nonadiabatic heat diffusion. Alternatively, one could assume that the engines have 100% efficiency, then Eq(14) can be used to deduce a value of 1.6 for λ, implying that the engines operate between levels separated by about half a temperature scale height. The real situation should be somewhere between these two extremes.

The story does not stop here. When the buoyancy work integrals are evaluated directly with numerical data from the computations, some cases (with low C_p) show that W can even exceed slightly the limit given by Eq.(15) (on the order of 1%).

This turns out to be caused by the negative flux of kinetic energy that feeds energy to the lower portion of the convection zone. In the above derivation, we have ignored the difference between the enthalpy flux and the total flux. In fact, the enthalpy flux is not uniform and is larger than F in most of the convective region because it has to balance the substantial negative flux of kinetic energy. From another point of view, one can say that the back-feeding of mechanical energy to the lower portion of the convective region enhances the energy supply to the enthalpy flux. The sum over W_i in Eq.(14) can actually be larger.

The example shown in Fig. 9 illustrates the operation of the energy loop-back process. For simplicity, a single heat engine is considered here. The

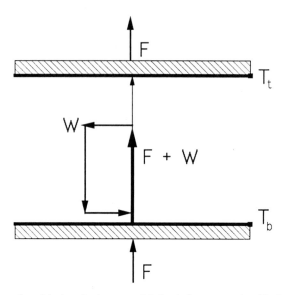

Fig. 9. An illustration of the loop-back process. Mechanical energy produced in the upper region is transported to the lower region, dissipated back to heat, and fed into the thermal flux inside the system.

mechanical energy, instead of being dissipated and fed to the low-temperature reservoir (top) as assumed previously, is now transferred to the bottom, dissipated there, and looped into the thermal flux. The Carnot limit for such a situation can be written as:

$$W < (1 - T_t/T_b)\,(F + W) \tag{16}$$

which can be rearranged as

$$W < (T_b/T_t)\,(1 - T_t/T_b)\,F. \tag{17}$$

This limit is much more liberal than (Eq. 15). The loop-back effect is more significant for cases with smaller C_p because the magnitudes of the mechanical fluxes are relatively larger (see point 5 in the previous section—III.C).

In all of the above arrangements, the real trick is that W does not deliver energy to outside the system. It does not generate real work for the external world. But taken at face value, the convection zone can indeed generate an amazing amount of mechanical power.

IV. REALISTIC, 3D SIMULATIONS OF SOLAR CONVECTION

A. Introduction

The nature of the upper solar convection zone can be explored using numerical simulations, and we have made such a model of the upper 2.5 Mm of the convection zone and the photosphere. By using a realistic equation of state and by including radiative energy transfer, we have been able to compare the results directly with observations. The simulation shows that the topology of solar convection is dominated by effects of stratification, and that convection consists of broad, gentle, structureless, warm, adiabatic, diverging upflows, with embedded filamentary, cool, fast, twisting, converging downdrafts. The flow topology is hierarchical, with downflows around many small cells close to the surface merging into fewer, filamentary downdrafts at greater depths. This merging of downdrafts into fewer, more widely separated plumes may persist through the entire depth of the convection zone. Radiative cooling at the surface provides the entropy deficient plasma that drives the circulation. A comparison of observable features from the simulation with recent granulation and mesogranulation observations shows that they are in accord.

B. Simulation

The upper solar convection zone is modeled by solving the equations of hydrodynamics, i.e., the equation of mass conservation

$$\frac{\partial \ln \rho}{\partial t} = -\mathbf{u} \cdot \nabla \ln \rho - \nabla \cdot \mathbf{u} \tag{18}$$

the equation of momentum conservation

$$\frac{\partial \mathbf{u}}{\partial t} = -\mathbf{u} \cdot \nabla \mathbf{u} + \mathbf{g} - \frac{P}{\rho}\nabla \ln P \tag{19}$$

and the equation of energy conservation

$$\frac{\partial e}{\partial t} = -\mathbf{u} \cdot \nabla e - \frac{P}{\rho}\nabla \cdot \mathbf{u} + Q_{\text{rad}} + Q_{\text{visc}}. \tag{20}$$

These are rewritten from Eqs. 1–4. Here ρ is the mass density, $\mathbf{u}$ is the velocity, e is the internal energy, P is the pressure, and Q_{rad} and Q_{visc} are the radiative heating and viscous dissipation, respectively. In order to model actual solar convection as realistically as possible, we use a tabular equation of state $P = P(\rho,T)$, which includes ionization and excitation of hydrogen and other abundant atoms, and formation of H_2, CO and other molecules (Gustafsson 1973).

This form of the equations, in terms of per-unit-mass quantities rather than the conventional per-unit-volume quantities, is advantageous when dealing with a strongly stratified atmosphere. The per-unit-volume quantities (density, momentum and energy per unit volume), vary exponentially with depth, and we are presently performing simulations with density ratios bottom/top of the order of 10^5. The logarithm of the density and the energy per unit mass vary nearly linearly with depth, both in the optically thin atmosphere and in the subsurface layers. Also, the velocities vary much less with depth than the linear momenta. The per-unit-mass quantities therefore allow a more accurate finite difference representation of derivatives. Although the per-unit-volume variables offer the possibility of exact conservation of mass, momentum and energy, small errors in mass and energy conservation are harmless in a stratified model, because the hydrostatic and energy equilibria enforce a very well-defined mean state. These equations were tested for one-dimensional shock tubes and found to give a good representation of the solution for pressure jumps up to one million.

We evaluate the radiative energy exchange

$$Q_{rad} = \int_\lambda \int_\Omega \kappa_\lambda (I_{\lambda,\Omega} - S_\lambda) d\Omega d\lambda \qquad (21)$$

by solving the transfer equation for the specific intensity $I_{\lambda,\Omega}$

$$\frac{dI_{\lambda,\Omega}}{d\tau_\lambda} = I_{\lambda,\Omega} - S_\lambda \qquad (22)$$

along inclined rays. $d\tau_\lambda$ is the optical depth increment along the ray, and S_λ is the source function. We assume that the source function is equal to the Planck function, and approximate the detailed wavelength integral with a sum over 4 bins. Absorption coefficients are calculated with a standard stellar atmosphere code (Gustafsson et al. 1975), and are sorted into bins representing continua, weak lines, intermediate lines and strong lines (cf. Nordlund 1982; Nordlund and Dravins 1990). This treatment is sufficiently elaborate to describe the sudden release of radiation by ascending hot gas in a thin layer at the solar surface, and the subsequent re-absorption of a small but energetically significant fraction of this energy in the upper layers of the photosphere.

Nonlinear time-stepping methods, such as van Leer's monotonic second-order upwind method (Van Leer 1977), and Colella and Woodward's piecewise parabolic method (Colella and Woodward 1984) are difficult to combine with radiative energy transfer and magnetic fields. As we are primarily interested in using our code as a test bed for understanding the interaction of convection, radiation and magnetic fields near stellar surfaces, conceptual simplicity and ease of implementation are important factors in our choice of numerical methods.

We advance the numerical solution in time with the Adams-Bashforth method, chosen for its high accuracy and modest memory requirements (cf. Gear 1971; Gazdag 1976). Because of a truncation error which corresponds to a weak but *negative* diffusion, Adams-Bashforth time stepping is weakly unstable and requires a small compensating positive diffusion to be added to all equations for stability. Stable methods were tested on model problems and found to offer no advantage, but to require more memory or input/output. Most stable methods (except those with large truncation errors) still need numerical diffusion near shocks. Independent of whether the diffusivity is provided explicitly (through spatial numerical diffusion) or implicitly (through truncation errors in the time stepping), the required net diffusivity is similar, and so are the actual results. As the magnitude of the numerical dissipation is decreased, the effective resolution of a code increases. If the dissipation is too small, structures smaller than the code can resolve and develop, and eventually the time evolution becomes unstable.

We use a diffusion coefficient with three types of contributions: (1) a term proportional to the sound speed, to stabilize the Adams-Bashforth time stepping; (2) a term proportional to the fluid velocity, to prevent ringing at sharp changes in advected quantities; and (3) a term proportional to the finite difference velocity convergence (where positive) to stabilize shock fronts.

One of the main difficulties with applying these equations to a specific simulation problem is the treatment of the "virtual boundaries"; i.e., boundaries of the computational domain that do not correspond to real boundaries, but just delimit the volume we choose (or can afford) to simulate. We deal with this problem by using periodic horizontal boundary conditions, and by constructing top and bottom boundary conditions which are as transmitting as we can easily make them while still preserving stability. At the top, we take an extra large boundary zone ($\geq$ scale height). In this zone, we impose the conditions that the amplitude of the velocity and the density fluctuations remain constant, while the energy density at the boundary is fixed at its initial average value. At the bottom, we impose constant pressure by adjusting the density, and we require $\partial \mathbf{u}/\partial z = 0$. The vertical heat flux is kept from drifting by specifying the internal energy of inflowing material at the bottom boundary. For additional discussion of our numerical methods see Nordlund and Stein (1990).

We simulated a region 6×6 Mm horizontally, in order to cover scales at least marginally larger than granulation, and extending vertically from the temperature minimum (-0.5 Mm) to a depth of 2.5 Mm (Stein and Nordlund 1989). We used a grid of $63 \times 63 \times 63$ points, which gave a resolution of

95 km horizontally and 50 km vertically. The calculations were performed on an Alliant FX/80 at the University of Colorado.

C. Results

1. Surface Topology: Granulation.   The granulation pattern visible at the solar surface consists of hot upflowing plasma in disconnected cells, surrounded by cold downflowing plasma in narrow topologically connected intergranular lanes (Fig. 10). The primary process that occurs in the granulation is the radiation of energy from the hot upwelling fluid, which cools it and reduces its entropy. Higher pressure in the hot, ascending granule centers pushes the cooling fluid toward the intergranular lanes. The entropy deficient fluid then starts to sink under the pull of gravity.

The granulation pattern is asymmetrical with respect to the direction of time, as well as with respect to the connectivity of ascending and descending gas and, indeed, the two are closely related. The ascending fluid, which expands horizontally because of the density stratification, meets expanding fluid from neighboring cells along common borders. At these cell borders, the flow is deflected horizontally along the border and vertically downward. The flows along borders eventually converge at the common corners of three or more cells. The "arrow of time" is provided by the dissipative processes in conjunction with the asymmetry between the expansion of the ascending flow and the convergence of descending flow.

The horizontal flow is driven by horizontal pressure fluctuations, which in turn are caused by temperature fluctuations and the Bernoulli effect. In the anelastic approximation ($\nabla \cdot (\rho \mathbf{u}) = 0$), the pressure is determined by

$$\nabla^2 P = \nabla \cdot [\rho(P,T)\mathbf{g} - \rho\mathbf{u}\,\nabla \cdot \mathbf{u}] \qquad (23)$$

which shows that the instantaneous pressure field is determined by the divergence of the forces. The first term on the right-hand side is the gravitational force, and the second term is the inertial "force".

Qualitatively, the resulting pattern of horizontal pressure fluctuations is easy to understand. For small velocities and sufficiently large horizontal scales the vertical derivatives dominate; the vertical pressure gradient force and gravity almost balance, i.e., large-scale fluctuations are close to hydrostatic equilibrium. In hydrostatic equilibrium, the change of pressure with height is determined by the local pressure scale height, which is proportional to temperature. As temperature fluctuations decrease rapidly with depth in a stratified convection zone, the pattern of horizontal pressure fluctuations at any given depth is dominated by the temperature fluctuations in the next few scale heights below that depth. Thus, the pressure excess in a granule (which provides the driving for the horizontal flow from the granule centers to the surrounding intergranular lanes), is a consequence of the temperature excess of the ascending granular flow just below the surface.

GRANULATION

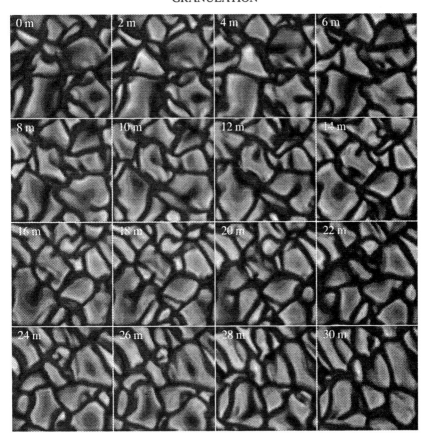

Fig. 10. Sixteen simulated snapshots of integrated radiation intensity at the solar surface. These approximate monochromatic continuum intensity snapshots. Each snapshot is 6 × 6 Mm (8″ × 8″). The sequence spans 30 min.

In the center of intergranular lanes, the inertial forces associated with the convergence of flows from neighboring cells also causes local pressure enhancements. This is obvious both from direct inspection of Eq. (23) above, and from physical grounds; the horizontal flow towards the intergranular lanes must be decelerated by a local pressure excess there. Hurlburt et al. (1984) have pointed out that this may be viewed as a Bernoulli effect along approximately horizontal streamlines near the solar surface—the pressure is a maximum where the horizontal velocity is a minimum, at the cell centers and in the intergranular lanes.

What determines the size of granules? When attempting to answer this question, one should keep in mind that there is no one size of granules, but rather a continuous distribution of sizes (and shapes), with a relatively well-defined upper limit to the size of single bright patches. This upper limit may be understood in terms of the constraints from mass conservation, pressure and energy balance.

The larger the granule, the larger the horizontal velocities needed to carry the increasing amounts of overturning gas. The amount of ascending fluid is proportional to the horizontal surface area which scales quadratically with the linear size of the granule, whereas the granule boundary, through which it must flow, only scales linearly with the size. The pressure fluctuations required to accelerate this material horizontally scale as the square of the velocities, and hence as the square of the linear size of the granule. This scaling has two important, and in some sense, opposite consequences. On the one hand, the larger pressures achieved by a large granule will tend to force the common border with neighboring cells to expand, thus further increasing the size of the larger granule. On the other hand, the increasing pressure in the interior of the granule decreases the buoyancy and eventually leads to buoyancy braking of the ascending gas in the center of the cell. This happens when the pressure excess in the granule is sufficient to cancel the buoyancy due to the temperature excess. For the Sun, the temperature excess is of the order of 2 (about 11,000 K in granules relative to about 6000 K on the average), and hence the excess pressure can support close to sonic horizontal velocities. The temperature excess can be supported and the granule continues to grow, as long as the rate of energy advection to the surface

$$\rho(e + P/\rho)u_z + \frac{1}{2}\rho\mathbf{u}^2 u_z \qquad (24)$$

exceeds the rate of radiation, σT^4. Once the vertical velocity has been reduced to

$$u_z \leq \frac{\sigma T^4}{\text{several} \times nkT} \qquad (25)$$

the vertical advection of excess entropy to the surface is no longer able to supply the entropy lost through radiation, and the granule center begins to cool, thus strangling the granule through a lack of heat input (cf. Nordlund 1985b). The critical ascent velocity is of the order of 2 km s^{-1}. A large granule, which is permitted to grow with little influence from neighboring granules, may develop into a ring of hot, ascending material, surrounding a cool, dark and eventually descending granule center, produced by the buoyancy braking (cf. Fig. 10; Nordlund 1985b). This phenomenon has been called exploding granules in the literature.

For a rough estimate of the ratio of ascent vertical to horizontal velocity as a function of granule size, assume that the ascending flow has cylindrical symmetry and a scale height of the vertical mass flux equal to $H_{\rho u_z}$. Then, if the medium is anelastic and the ascending and expanding fluid is overturning at a radius of r, continuity requires

$$\pi r^2 \, \rho u_z / H_{\rho u_z} = 2\pi r \rho u_r \qquad (26)$$

or

$$d \equiv 2r = 4H_{\rho u_z} \, (u_r/u_z) \qquad (27)$$

From the pressure equation, Eq. (23), it can be shown that large-scale pressure fluctuations (and hence velocity amplitudes) decrease relatively slowly with height as compared to the pressure itself (cf. Nordlund 1982), so the density factor dominates the mass flux scale height.

Collecting the constraints from the continuity equation (the ratio of ascent velocity to horizontal velocity), the momentum equation (the maximum horizontal velocity), and the energy equation (the minimum ascent velocity), we obtain an estimate of the maximum granule size by inserting the sound speed (8 km s^{-1}) for u_r, the density scale height (0.2 Mm) for $H_{\rho u_z}$, and 2 km s^{-1} for the minimum ascent velocity. The result is a diameter of about 3 Mm, in good agreement with observations of solar granulation (Bray et al. 1984).

Using an earlier, anelastic version of the present code, Nordlund and Dravins (1990) found that, for stars in the vicinity of the Sun in the Hertzsprung-Russell diagram, the size of granules scales roughly as the density scale height in the photosphere.

Apart from this type of "self-inflicted" death of relatively large granules, the simulations show granules breaking up because of influence from neighboring granules, whose expanding flows and pressure fluctuations constitute a highly time-dependent environment. The external perturbations experienced by an individual granule are not arranged in a nicely symmetric pattern around the granule. Rather, the neighbors surrounding any particular granule are likely to be of different strengths; some strong, some weak. Neighboring granules with large expansion velocities inhibit granule growth in that direction. Thus, the external influence from neighboring granules results in distortion of the shape, or even break up, of a granule.

From the perspective of an individual granule, the influence from surrounding granules is a pseudo-random function of space and time; although the surrounding granules may undergo similar evolutions, the evolutions are not in phase. As a consequence, a pattern consisting of many granules evolves chaotically. The evolution appears subjectively chaotic, and is indeed most likely mathematically chaotic in the sense that two neighboring points in phase space depart exponentially with time. A pattern of granules evolves through many "points of decision," where the evolution may go one way or the other, and a slight perturbation may tip the balance in favor of one granule or the other. Two slightly different configurations (phase-space points) will take different paths through such "points of decision," and thus rapidly diverge. Such "points of decision" occur, for example, when granules of nearly equal size compete for space. They also occur when a granule is breaking up, and details of the external velocity field may have a decisive influence. For instance, tongues of protruding cool material from surrounding granules aid in the break up of a granule. An increase of the small-scale structure in granules increases the interaction between them; highly structured flows have a harder time arranging a "peaceful coexistence."

Initially, we used overly conservative estimates of the necessary numerical diffusion (away from shocks), which caused a lack of small-scale details in granules (Lites et al. 1989). Tests showed that the coefficients that control viscosity in nonshocking parts of the flow could be reduced by about a factor of 3, while still retaining stability. As expected, the reduction of the viscosity resulted in increased small-scale structure, and increased the number of small granules in the simulation. However, we found that the size of even the largest granules depended slightly on the amount of numerical diffusion. Diffusion increases the smoothness of the simulated granules, and hence delays the break up of granules.

2. Subsurface Topology: Mesogranulation.   Beneath the surface, the connected intergranular lane downflow converges into topologically disconnected, finger-like structures (Fig. 11). The flow topology becomes large-scale, slow, diverging, structureless upflow of warm plasma, with embedded, twisting, narrow, isolated, fast, converging downdrafts of cool plasma (cf. Graham 1975; Nordlund 1985*b;* Chan and Sofia 1986). This change of topology takes place over a vertical distance ($\simeq 0.5$ Mm) which is only a fraction of the typical horizontal cell size. To understand this remarkable change of topology, we proceed to discuss qualitatively the properties of convection below the solar surface, basing the discussion partly on the numerical results from our simulations, and partly on inspection of the governing equations.

The flow below the visible surface may be characterized as almost pure advection. Mass is advected with only small Eulerian changes of the density (i.e., almost anelastically), and entropy is advected with negligible influence from radiation and dissipation (i.e., almost adiabatically). Given the negligible energy exchange, the flow may be understood in terms of *fluid parcel trajectories;* the properties of a fluid parcel at a certain time is the result of the histories of its constituent parcels.

The flow topology is primarily controlled by the density stratification. The continuity equation, Eq. (18), may also be written

TEMPERATURE   VERTICAL VELOCITY   LOG PRESSURE   HELICITY

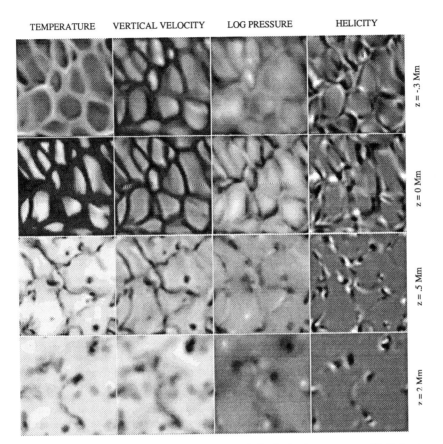

Fig. 11. Horizontal slices, showing temperature, vertical velocity, log pressure and helicity at depths -0.35, 0 (visible surface), 0.5 and 2.0 Mm. The flow topology changes radically below the surface, from isolated cells of warm upflowing plasma surrounded by connected lanes of cool downflowing plasma, to isolated downdrafts of cool plasma embedded in a broad gentle upflow of warm plasma. The pressure at the surface is high in the granule centers, low at the granule boundaries, and often has a secondary maximum in the intergranule lanes. In addition, the pressure shows a larger-scale variation corresponding to the meso-scale subsurface topology.

$$\text{div}(\mathbf{u}) = \frac{\text{D}}{\text{D}t}(-\ln\rho) \qquad (28)$$

which simply states that the expansion of the fluid flow is given by the relative rate of decrease of the density, following the motion. To be specific, a fluid parcel ascending from the bottom of our computational box (at $z = 2.5$ Mm) to the solar surface (at $z = 0$ Mm) expands by a factor of about 90; e.g., its linear size would increase by about a factor of 4.5 if the expansion were isotropic. In principle, one could imagine the expansion to be entirely in the vertical, but this would not be consistent with the energy equation (constant total energy flux is obtained with a much smaller variation of vertical velocity with height). The ascent velocity actually varies with depth approximately as $\rho^{-0.5}$, and thus the linear horizontal size of a fluid parcel increases with about a factor of 3 over this 2.5 Mm depth interval. Because of the horizontal expansion, only a small fraction (about 10%) of the ascending fluid at the lower boundary ever makes it to the surface. Conversely, descending fluid at the surface contracts as it descends. Descending fluid parcels have a strong tendency to be sheared and stretched out into long, twisting filaments. Thus the horizontal size of a descending fluid parcel decreases even more rapidly than the horizontal size of an ascending fluid parcel increases. Consequently, the entropy deficient fluid from the surface rapidly becomes a smaller and smaller fraction of the descending fluid. The rest of the descending flow is made up of overturning fluid that never made it to the surface (and hence never lost any entropy).

Thus, the topology of the flow is primarily controlled by the continuity equation. The role of the energy equation is subsidiary; below a thin layer near the surface, the flow is very nearly adiabatic, and the energy equation simply traces the path of the entropy-deficient fluid from the surface as it descends, and the (mostly numerical) spatial diffusion of the entropy fluctuations. The tendency for the entropy fluctuation to spread horizontally is counteracted by the horizontal convergence of the descending flow. This keeps the descending cool filaments well defined.

Since the entropy-deficient fluid from the surface becomes a smaller and smaller fraction of the descending flow, the relative temperature fluctuations decrease rapidly with depth. These relative temperature fluctuations determine the pressure fluctuations (via the pressure scale height; cf. earlier discussion), which in turn determine the velocity amplitudes, which closes the causal loop. A selfconsistent (average) state is characterized by a vanishing average mass flux (pressure equilibrium), and a constant total energy flux (energy equilibrium; sum of radiative, convective and kinetic energy flux is constant).

Figure 12 shows the origin and destination of fluid parcels ascending through the visible surface (plane $z = 0$) at time $t = 0$. Most of these parcels were also ascending at 9 solar min earlier. They ascend slowly, with nearly the same speed, and all originate from a small source volume, both vertically (because of the nearly uniform vertical velocity) and horizontally (because of the diverging topology of the upflow). Going back further in time, we find the particles in progressively smaller source volumes. Only a small subset of all the ascending fluid at depth reaches the surface. Most of the fluid that does reach the surface, turns around and descends rapidly, concentrating into a few

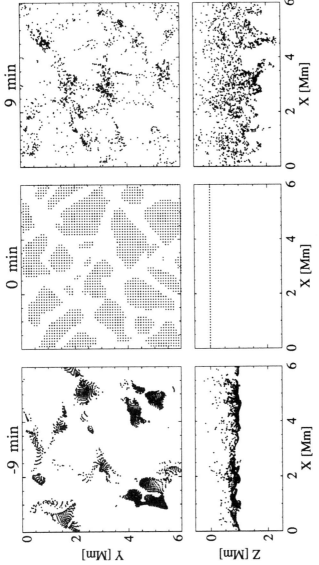

Fig. 12. Location of fluid parcels ascending through the solar surface at time $t = 0$ are shown 9 solar min earlier and later. Note the slow, diverging, nearly structureless nature of the ascending flow, and the fast descending flow, that converges into downdrafts that merge into a few widely separated filaments at large depth.

narrow downdrafts. Nine solar min later, most of the originally ascending fluid has descended a substantial distance and outlines these filamentary downdrafts.

The amplitude of entropy fluctuations decreases rapidly with depth in the convection zone. The mapping of very small source volumes of these nearly isentropic deeper layers onto all of the ascending flow close to the surface explains why the ascending flow is so nearly isentropic, and also justifies the use of isentropic inflow at the lower boundary of the computational domain.

We now have the necessary conceptual building blocks to understand the topology of convection qualitatively. Ascending flow is gentle and featureless because it is expanding; small-scale features are constantly being washed out by the expansion of the ascending flow. Overturning flows from neighboring updrafts collide along common borders, and are deflected towards corners between several updrafts. Fluid that reaches the surface, and *only fluid that reaches the surface,* looses entropy before overturning. Thus, at the surface, the characteristic cell topology is made clearly visible, with bright disconnected islands of ascending fluid, and connected dark lanes of descending fluid. Below the surface, the entropy-deficient fluid from the surface rapidly contracts, and is engulfed in overturning isentropic fluid that never reached the surface. Thus the entropy contrast between ascending and descending fluid decreases rapidly with depth, and the connectivity of the intergranular lanes is lost because entropy-deficient fluid in the lanes is replaced by overturning entropy neutral fluid.

Continuity allows fluid overturning below the surface to do so on increasingly larger scales at larger depths. At a depth of 2.5 Mm, the density scale height is significantly larger than at the surface (about 1 Mm), because of the ($\simeq$ 3 times) larger temperature, and also because of the ($\simeq$ 2 times) smaller mean molecular weight. Hence, Eq. (27) indicates that cell diameters of the order of 8 to 12 Mm are permitted, if the ratio of horizontal to vertical velocities remain of order 2 to 3. Our box was not large enough to allow this, but we do see the scale changing from many downdrafts at vertices near the surface to one or a few near the bottom. The scale change is caused by the advection of the small-scale downdrafts by the horizontal velocity field associated with the expansion of ascending fluid and is, of course, limited by the horizontal extent of our periodic box. In the Sun, no such restrictions occur and we expect that the scale of the horizontal flows continues to increase with depth.

Experiments where the fluid ascending through the lower boundary was not given a uniform entropy did not yield qualitatively different results. We have recently started a much larger and deeper (12 × 12 × 9 Mm) simulation. Similar topology occurs in this deeper simulation.

The vertical extent of the convective flows is greater than the depth of our computational box. The granulation pattern does *not* correspond to a closed circulation pattern on the scale of the granulation. In fact, there is no evidence for multiple cells in the vertical direction. The flows do not close within our domain, and the merging downdrafts of cool material may possibly extend the entire depth of the convection zone (Fig. 13). However, the solar plasma is much less viscous than our simulated plasma, and so has a huge Reynolds number. Hence, although our upflows are smooth, and our downflows are only moderately turbulent, in the Sun they are likely to be strongly turbulent with significant generation of small-scale vorticity and eddies.

3. Convective Driving.   The flow topology has important consequences also for the *driving* of flows on different scales. Driving is provided by the buoyancy fluctuations associated with entropy fluctuations. But the ever-expanding, gentle upflow is almost isentropic, so the main source of entropy fluctuations is the entropy loss at the surface. The surface acts as a source of cool, relatively dense material, which descends into a nearly isentropic interior. From the discussion of the topology of granulation at the surface, we know the initial topology of the entropy-deficient material: a connected network of intergranular lanes, with accumulation of particularly cool and dense fluid at the corners between several granules. The situation is similar to the Rayleigh-Taylor instability of a dense fluid on top of a less dense fluid, and a similar evolution ensues: filaments of dense material are formed, with denser filaments descending faster; additional dense fluid is pulled down in the wakes of the descending filaments, and the denser filaments gain an additional advantage.

The accumulation of cool, dense fluid at the vertices is self-amplifying; the flow is driven by the pressure deficiency caused by the smaller pressure scale height of cooler material at the vertices, and causes further draining of cool material from the intergranular lanes into the vertices. The overturning material that replaces the intergranular material below the surface is not entropy deficient, and hence the entropy deficiency at the intergranular lanes vanishes rapidly below the surface. The topology thus changes from one with cool material in connected lanes to one with cool material in narrow, nearly vertical and descending filaments.

We showed earlier that larger-scale flows are *allowed* at larger depths by the continuity equation. The descending cool filamentary material also provides *driving* of these larger-scale flows. The overturning fluid of larger-scale flows advect cool descending filaments of smaller-scale flows towards the boundaries and vertices of these larger-scale flows. Thus, the small-scale downdrafts close to the surface merge into large-scale, more widely separated downdrafts further from the surface, and provide the supply of entropy-deficient material that is necessary to drive the large-scale flows. By this process, the horizontal distribution of entropy deficient material changes gradually with depth, from one with many closely spaced filaments near the surface to one with fewer, less closely spaced filaments at depth.

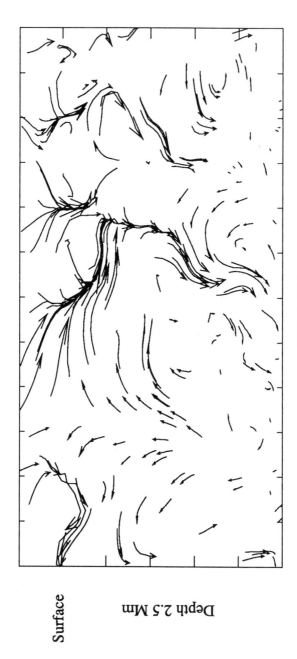

Width 6 Mm

Surface

Depth 2.5 Mm

Fig. 13. Fluid velocity in the x-z plane. The ascending flow is broad and diverging; the descending flow is filamentary and converging. The cells span the entire vertical extent of the computational domain. There are no multiple cells in the vertical direction.

This can be clearly seen in the vertical kinetic energy flux, $\rho\mathbf{u}^2 u_z$ (Fig. 14). Only the downflow is visible, because the upflow is slow and structureless. The merging of the granular downflows into the isolated, filamentary, mesogranulation downflow is clearly revealed. Notice also, that in our simulation, with its transmitting lower boundary, there is no sign of the downflow turning around into upflow within our simulation domain.

At first glance, the situation may seem somewhat absurd, with the low-density surface layers apparently providing the driving for large-scale flows at great depth, where the density is orders of magnitude larger ("tail wagging the dog"). However, consider the distribution and evolution of kinetic energy density in the flow. Most of the kinetic energy is concentrated in the strong downflows (cf. Fig. 14), and its evolution is controlled by a balance of buoy-

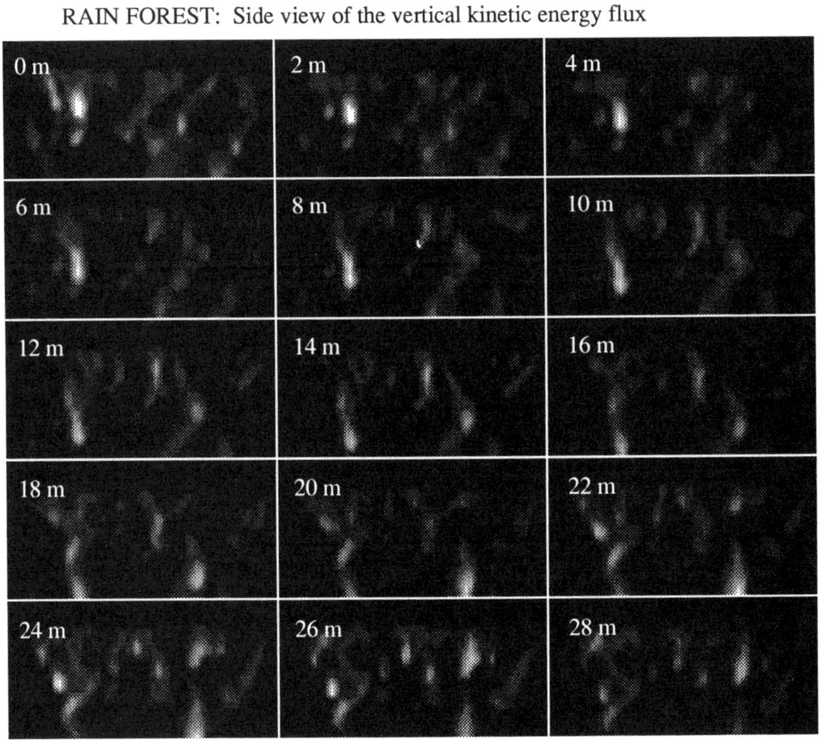

RAIN FOREST: Side view of the vertical kinetic energy flux

Fig. 14. Vertical kinetic energy flux, $\rho\mathbf{u}u_z$. Transparent side views through the computational domain, along the x-axis, are shown at 2-min intervals. Only the downward flux is visible, because the upflow is much slower, broader and featureless. The merging of downflows into isolated filaments, and the increasing horizontal separation of fewer filaments at greater depths is clearly revealed. The downdrafts penetrate through the transmitting lower boundary with little, if any, return flow.

ancy work, kinetic energy flux divergence and dissipation. Kinetic energy flux is even more concentrated into the downflows than the kinetic energy itself. The net kinetic energy flux is directed downwards, and almost entirely represents advection of kinetic energy density by the downflows. Buoyancy work (which is proportional to the product of density fluctuation and velocity) is positive in both the up and down flows, but is mainly performed on the downflows, where the density fluctuations and velocity are much larger. Hence, the energy input to the convective flow occurs primarily in the cool material from the surface layers.

This situation, where the "flow of information" is directed downwards (even though the total energy flux is directed outwards) is fortunate for the numerical simulation of solar surface layers, because it diminishes the consequences of having to introduce a computational lower boundary in the midst of the convection zone. Conversely, it necessitates a special treatment at a computational *upper* boundary in simulations of global solar convection (where these same surface layers have much too short characteristic time scales to be included).

4. Photospheric Phenomena.   Our simulation results can be analyzed for other phenomena associated with the upper solar convection zone. We briefly indicate two of these.

Shocks developing from vertically propagating acoustic waves are thought to contribute to heating the chromosphere. Our simulation did not extend above the temperature minimum, so we can not investigate chromospheric heating. However, we do see shocks developing in the photosphere. These shocks are not due to vertically propagating acoustic waves, but rather to quasi-steady horizontal flows associated with the granulation pattern. Some of the ascending gas penetrates into the stable photosphere before being deflected sideways. The horizontal flow produces nearly stationary vertical shocks around bright granules which are being squeezed out of existence by their neighbors.

Helicity in the convection zone is thought to be essential in the dynamo process that produces the solar magnetic field (Parker 1955*b;* Gilman 1983). Although our simulation does not include rotation, we find that significant helicity occurs in the intergranule lanes and especially in the downdrafts (cf Fig. 11). When a magnetic field is present, this helicity will produce copious Alfvén waves, that will propagate along the field, and may contribute to chromospheric and coronal heating (Osterbrock 1961; Wentzel 1976; Ionson 1978; Holweg 1984).

D. Comparison with Observations

1. Granulation.   Lites et al. (1989) used a sequence of simultaneous spectra in the Fe I 6302.5 Å line and narrowband slit-jaw images from the Swedish Vacuum Solar Telescope on La Palma, Canary Islands, Spain to determine granular intensity and velocity fluctuations, and the granule size distribution. They compared these with the results of our simulation. The original simulations had a granule size distribution with significantly fewer small granules than observed. Because the amount of artificial viscosity needed for stabilization had initially been overestimated, a reduction of the magnitude of the terms that are proportional to the flow and sound speeds by a factor of 2.5 was possible, while still maintaining stability. This increased the effective spatial resolution of the code sufficiently to bring the sizes of the large simulated and observed granules into agreement.

The radiative energy transfer calculation, which is performed at each time step, yields, as a side effect, images in the total radiation intensity at the surface of the model. These images may be used as a convenient approximation of monochromatic continuum images. This has been verified by direct comparisons of images in total radiation intensity with monochromatic continuum images calculated with a separate program. Figure 10 shows a sequence of 16 such images. The time separation between images is 2 min. The size of each image is 6 × 6 Mm, corresponding to approximately 8″ × 8″. Because most features in an image change only moderately in a few minutes, there is some continuity from frame to frame in the sequence, and a single strip is in some sense analogous to an 8″ × 32″ image. More direct comparisons may be made, by smearing the synthetic images with point spread functions representing the telescope and atmospheric transmission functions (cf. Lites et al. 1989).

Figure 15 shows an 8″ × 24″ strip from an observed continuum image on the same scale as Fig. 10 (Lites et al. 1989). It is evident that the simulations do indeed produce surface patterns that qualitatively resemble the observed solar granulation. However, even with a reduced viscosity (the minimum allowed that maintains stability of the simulations), there are more

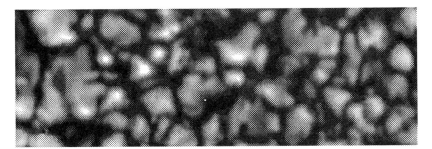

Fig. 15. Monochromatic continuum image of solar granulation, obtained at the Swedish Vacuum Solar Telescope on La Palma (Lites et al. 1989). The image is of a region 6 × 18 Mm (8″ × 24″). Compare this image with Fig. 10, which has the same scale. Note the similarity in large-scale features between the observation and simulation, but the presence of much more small-scale structure in the observation.

small-scale structures in the observations than in the simulations. Thus, the observations currently have a higher resolution than the simulations. A further increase of the spatial resolution of the simulations can only be obtained by decreasing the horizontal mesh spacing. Such simulations are under way.

Lites et al. also compared the amplitudes of the intensity and velocity fluctuations in the observed slit-jaw spectra to synthetic slit-jaw spectra. The magnitude of the observed intensity and velocity fluctuations depend on the (unknown) modulation transfer function (MTF) of the atmosphere (seeing) and telescope, so a direct comparison could not be made. However, by applying trial MTF's to the synthetic images, it was found that the *ratio* of intensity to velocity fluctuations is insensitive to the amount of smearing by the MTF, and that the ratio obtained from the simulations agrees with the observed ratio. On the other hand, the width of the (spatially unresolved) Fe I line provides a seeing independent check on the magnitude of the velocity fluctuations. From the average line width of the slit-jaw spectrum, Lites et al. conclude that the *rms* amplitude of the vertical velocities in the Fe I formation layers are consistent with the observations, and that consequently the intensity fluctuations are likely to be consistent too.

At the visible surface, the granule centers are warm and the intergranule lanes are cool. Our simulation shows that in the photosphere, due to adiabatic expansion, the diverging upflow above the bright granules becomes cooler than the converging downflow above the intergranule lanes (cf. Fig. 11). This reversal of bright and dark regions is observed (Evans and Catalano 1972; Canfield and Mehltretter 1973; Altrock and Musman 1976; Keil and Canfield 1978; Balthasar et al. 1990).

2. *Mesogranulation.* Granulation is the only scale for which the cell structure is directly visible at the surface, with cell boundaries delineated by cool, dark material. However, the larger scales are *indirectly* visible on the surface. The pressure fields induced by the larger-scale flows at depth extend to the surface. From the solutions of the anelastic pressure equation (Eq. 23) for large horizontal scales, Nordlund (1982, Eq. 46) has shown that the *relative* pressure fluctuations for horizontal wavenumber k in a medium with pressure scale height H have an $\exp(-k^2Hh)$ dependence on height h; i.e., cells with a horizontal size large relative to the scale height induce their relative pressure fluctuations over a height range comparable to or larger than their horizontal size. These relative pressure fluctuations drive the horizontal flows associated with the cell. Hence, the horizontal velocity fields associated with the larger-scale subsurface cells extend to the surface. The presence of large-scale subsurface cells thus becomes visible at the surface through the horizontal advection of smaller-scale cells.

The large (mesogranule) scale flow manifests itself at the visible solar surface by its effect on the growth as well as the motion of granules. Examination of a sequence of surface intensity images (Fig. 10) reveals areas where granule growth is enhanced and granules are large, and other areas where granule growth is suppressed and granules are small.

This pattern of granule motion from regions of enhanced growth towards regions of suppressed growth is just what is observed as mesogranular flows by November and Simon (1988) and Title et al. (1989). We find that the areas of horizontal convergence and small granules correspond to the persistent cool downdrafts and the areas of horizontally diverging motion and large granules correspond to the regions of warm upflows.

E. Conclusions

Only the small-scale cells at the surface (granulation) are made clearly visible by a large temperature contrast, with bright disconnected islands of ascending fluid, and connected dark lanes of descending fluid. The larger-scale flows are visible at the solar surface only by the advection of smaller-scale flows by their horizontal velocity fields. Our simulations reproduce the observed large end of granule scale structures, but are missing the smaller-scale structures. The simulated vertical velocities near the surface agree with those inferred from observations, and the ratio of intensity to velocity fluc-tuations in synthetic slit-jaw spectra agree with observed ones. The advection of granules and variation of granule sizes over larger regions, referred to as mesogranulation, are also reproduced.

The usual picture of turbulent convection has been of a hierarchy of eddies, or in the mixing length picture of bubbles that move some distance and then mix with their surroundings. Our simulations suggest a very differ-ent picture. The dominant topology of the outer solar convection zone appears to be one of broad, gentle, structureless, warm, adiabatic, diverging, up-flows, with intermixed narrow, filamentary, cool, fast, twisting, converging downdrafts. The horizontal velocity field has a hierarchical structure, with small-scale cells at the surface, and successively larger-scale flows at larger depths, driven by the merging of the filamentary downdrafts of the smaller-scale cells closer to the surface. Our computational box, which is only 2.5 Mm deep, supports meso-scale flows with a horizontal extent comparable to the horizontal size of our box (6 Mm). Stronger turbulence in the downdrafts will increase the rate of mixing between ascending and descending fluid, but will not change the overall contraction of descending, entropy-deficient fluid. Thus, the filamentary downdrafts may possibly persist through the entire depth of the convection zone, merging into fewer, more widely separated plumes as they descend. Presumably, successively larger (supergranular) hor-izontal flows at the surface, with sizes ranging up to at least 50 Mm (Simon and Leighton 1964) reflect the successively increasing separation of the merg-ing, descending plumes.

Although the descending vertical flows extend over many scale heights, estimates based on the mixing length concept may still be relevant, as the overturning of most of the ascending flow within a scale height, and the associated dilution of descending fluid, in effect resembles a mixing.

Acknowledgments.   KLC thanks the National Science Foundation for support. RFS and AN wish to express their appreciation for the support of the National Science Foundation, the Carlsberg Foundation and the Danish Natural Science Research Council. MS thanks the Deutsche Forschungsgemeinschaft for support. The authors wish to express their appreciation for this support.

THE SOLAR DYNAMO

E. E. DeLUCA

University of Chicago

and

P. A. GILMAN

National Center for Atmospheric Research

In this chapter we discuss the present state of our understanding of the origin of the Sun's magnetic field. We begin with an introduction to the theory of magnetic field generation in rotating, conducting fluids. Next we consider a dilemma that has persisted for some 15 yr, namely the inconsistency of the kinematic and dynamic convection zone dynamo models. A resolution of this dilemma has been suggested by the recent helioseismology observations on the rotation rate as a function of latitude and depth. These observations, together with other observational and theoretical constraints, suggest that the solar dynamo operates not in the convection zone, but rather in a thin layer between the convection zone and the radiative interior. A model of such a dynamo is presented and discussed. We conclude by discussing the problems posed by the placement of the dynamo below the convection zone.

I. INTRODUCTION

To maintain the Sun's complex magnetic field requires a system of electric currents somewhere beneath the photosphere. To maintain these currents against dissipation by turbulent cascades and magnetic reconnection almost certainly requires hydromagnetic dynamo action. The main questions are where the dynamo is located, and what fluid motions are of primary importance in it.

The Sun's magnetic field evolves with time in a quasi-cycle pattern, known as the solar cycle. Some Sun-like stars (stars with convective envelopes) are also known to have magnetic cycles (see, e.g., Vaughan 1984), while others show large amplitude but essentially chaotic magnetic activity. Thus, hydromagnetic dynamo action is a common phenomenon in stars, and understanding how it drives the solar magnetic cycle is of great importance for understanding the physics of stars in general.

Modern hydromagnetic dynamo theory for the Sun has its origin principally in a now classic paper by Parker (1955b). In this chapter we sketch out solar dynamo theory as it has evolved from that work. Much of the early theory was kinematic, in that the driving-motion fields were assumed rather than predicted. Application of this kinematic theory to the Sun, in some cases with *ad hoc* nonlinear additions, has led to apparently good agreement with such well-known features as the Maunder "butterfly diagram" (see the chapter by Rabin et al. and Semel et al.). However, the motions assumed in the induction equation to produce this agreement have not been confirmed by either deductive theory or by recent inferences from measurements of acoustic modes on the Sun. In particular, kinematic theory needs an angular velocity increasing with depth in the bulk of the convection zone, while global convection theory predicts that it will decrease with depth, and solar acoustic modes indicate it is essentially independent of depth within the convection zone. Thus, both dynamo theory and global convection theory applied to the Sun need to be modified.

For a variety of reasons, including the above, the focus of dynamo modeling for the Sun has now shifted to the interface between the base of the convection zone and the radiative interior. In this chapter we describe the formulation of one such model, by the authors, as well as some early solutions. Finally, we outline the many problems that remain to be addressed, if we are to understand thoroughly how the solar dynamo works.

II. ESSENTIAL ELEMENTS OF DYNAMO THEORY

In this section we introduce the physical processes that are necessary for the maintenance of a magnetic field in a fluid with finite conductivity. The velocity fields will be assumed to be known. Emphasis will be placed on the physical discussion of the processes involved rather than the details of the derivations. The reader is referred to Roberts (1967) for an in depth introduction to magnetohydrodynamics (MHD), and to Moffatt (1978) for more details on the dynamo problem and mean field equations.

We begin by writing the basic equations of MHD in a reference frame rotating with angular velocity Ω. The momentum equation is

$$\rho\frac{\partial \mathbf{U}}{\partial t} + \rho(\mathbf{U} \cdot \nabla)\mathbf{U} = -2\rho\mathbf{\Omega} \times \mathbf{U} - \nabla P$$

$$- \rho\mathbf{g} + \mathbf{j} \times \mathbf{B} + \mu(\nabla^2\mathbf{U} + \frac{1}{3}\nabla(\nabla \cdot \mathbf{U})) \quad (1)$$

where ρ is the density, P is the gas pressure, $\mathbf{U}$ is the velocity in the rotating frame, $\mathbf{\Omega}$ is the angular velocity, $\mathbf{g}$ is the gravitational acceleration, $\mathbf{j}$ is the current density, $\mathbf{B}$ is the magnetic field and μ is the dynamic coefficient of viscosity (assumed to be constant). The continuity equation is

$$\frac{\partial \rho}{\partial t} + \nabla \cdot (\rho\mathbf{U}) = 0. \quad (2)$$

The reader is referred to any introductory text in fluid dynamics for a derivation and discussion of these equations (i.e., Batchelor 1967). The induction equation governs the evolution of the magnetic field in the MHD approximation,

$$\frac{\partial \mathbf{B}}{\partial t} = \nabla \times (\mathbf{U} \times \mathbf{B}) - \eta\nabla \times (\nabla \times \mathbf{B}) \quad (3)$$

where η is the magnetic diffusivity and is taken to be constant. In addition the magnetic field must be solenoidal,

$$\nabla \cdot \mathbf{B} = 0. \quad (4)$$

From Eq. (3) it is easy to see that if the magnetic field is initially solenoidal the induction equation will keep it solenoidal.

The above equations together with a suitable equation of state, and boundary conditions, describe the evolution of a magneto-fluid. There are two parameters, $\mu = \rho\nu$ and η, which must be specified before approximate solutions to the above equations can be found. The simplest assumption is that μ and η are constants. The justification for the assumption is based on the small molecular values of μ and η; this implies that the length scales on which momentum and magnetic fields are dissipated are well separated from the scales on which the flow is being driven (for our problem the flow is driven on a global scale comparable to the radius of the Sun). Between the global scale forcing and the very small dissipation lies a large "inertial" range whose role is to transport energy from larger scales to smaller scales. If all of the above statements about the inertial range are valid, and if one is only interested in the dynamics of the global-scale features, then one can choose "eddy" diffusivities that are much larger than the molecular values. These might be assumed to be constant, or they might be taken to depend on the local shear in the velocity. In the following discussion, we assume that the viscosity and magnetic diffusivity are known and that the concept of an eddy viscosity in a compressible, rotating, magneto-fluid is valid. The reader should be aware that these are simplifying assumptions.

We proceed with an investigation of the induction equation with the assumption that the velocity fields are specified and for simplicity are divergence free (i.e., the flow is incompressible). There are two limiting forms of the induction equation: (1) The high-conductivity limit in which the magnetic Reynolds number, $R_m = UL/\eta$ is large (U is a characteristic velocity amplitude and L is a characteristic length scale for the problem under consideration); and (2) the low-conductivity limit R_m is small. In the former case, the induction equation can be written,

$$\frac{\partial \mathbf{B}}{\partial t} + (\mathbf{U} \cdot \nabla)\mathbf{B} = (\mathbf{B} \cdot \nabla)\mathbf{U}. \qquad (5)$$

On the left-hand side of Eq. (5) is the Lagrangian derivative of the magnetic field, on the right-hand side is a term that represents the stretching and twisting of magnetic field lines by gradients in the velocity field. In this limit the field is "frozen" to the fluid. Notice that the field can be amplified indefinitely by a flow with constant shear, e.g., if $\mathbf{B} = (a(t),0,c), \mathbf{U} = (wz,0,0)$ then $\partial a/\partial t = cw$ implies that $a(t) = cwt + a(0)$. This stretching process is a crucial part of the dynamo, but, as argued below, it is insufficient by itself; diffusion is the other key ingredient. In the limit of small magnetic Reynolds number (Eq. 3) becomes

$$\frac{\partial \mathbf{B}}{\partial t} = \eta \nabla^2 \mathbf{B}. \qquad (6)$$

Simple scale analysis gives the decay time of the field as $\tau_d = L^2/\eta$, where L is the characteristic scale over which the field varies. While it is clear that Eq. (6) alone cannot generate magnetic fields, one of the fascinating aspects of dynamo theory is the role played by diffusion.

We can satisfy the solenoidal condition, Eq. (4), by expressing the magnetic field in terms of the sum of two vector potentials (a toroidal and poloidal part),

$$\mathbf{B} = \nabla \times [\mathbf{r}T(\mathbf{r})] + \nabla \times \nabla \times [\mathbf{r}P(\mathbf{r})]. \qquad (7)$$

The current density $\mathbf{j}$ is determined from the curl of the magnetic field from Ampere's law. Taking the curl of Eq. (7) and using the vector identities $\nabla \times \nabla \times \mathbf{f} = \nabla(\nabla \cdot \mathbf{f}) - \nabla^2\mathbf{f}$, and $\nabla^2[\nabla \times (\mathbf{r}h)] = \nabla \times [\mathbf{r}\nabla^2 h]$ yields

$$\mathbf{j} = \nabla \times \mathbf{B} = -\nabla \times (\mathbf{r}\nabla^2 P(\mathbf{r})) + \nabla \times \nabla \times [\mathbf{r}T(\mathbf{r})]. \qquad (8)$$

The current can also be expressed in terms of a toroidal and poloidal part. The toroidal magnetic field is generated by a poloidal current, and the poloidal magnetic field is generated by the toroidal current. The problem of regeneration of the magnetic field can then be reduced to the regeneration of the poloidal part and the toroidal part, separately.

To illustrate the above points and to illustrate an important result known as Cowling's theorem, we will now consider the Cartesian analog of an axisymmetric field in a spherical shell. Let $\hat{\mathbf{y}}$ be the northward, $\hat{\mathbf{z}}$ be the radial, and $\hat{\mathbf{x}}$ be the eastward unit vectors. The variables will depend only on y and z. The magnetic field can be written

$$\mathbf{B} = \mathbf{B}_T + \mathbf{B}_P = \nabla \times [\hat{\mathbf{z}}T(y,z)] + \nabla \times \nabla \times [\hat{\mathbf{z}}P(y,z)]$$

or

$$\mathbf{B} = \nabla \times (A(y,z)\hat{\mathbf{x}}) + \mathbf{B}_T(y,z)\hat{\mathbf{x}} \qquad (9)$$

where $A \equiv \partial P(y,z)/\partial y$, and $B_T \equiv \partial T(y,z)/\partial y$. The toroidal field is directed in the $\hat{\mathbf{x}}$ direction (azimuthal direction), and the poloidal field has components in both the $\hat{\mathbf{y}}$ and $\hat{\mathbf{z}}$ directions (in the meridional plane). If the velocity field is taken to be incompressible, then we can also express it in terms of toroidal and poloidal parts. The nonlinear term in the induction Eq. (3) can then be separated into a poloidal and toroidal part,

$$\mathbf{U} \times \mathbf{B} = (\mathbf{U}_T \times \mathbf{B}_P + \mathbf{U}_P \times \mathbf{B}_T) + \mathbf{U}_P \times \mathbf{B}_P. \qquad (10)$$

Now the curl of the poloidal part of Eq. (10) is in the toroidal direction and the curl of the toroidal part is the poloidal direction. The poloidal part of the induction Eq. (3) is then

$$\frac{\partial \mathbf{B}_P}{\partial t} = \nabla \times (\mathbf{U}_P \times \mathbf{B}_P) + \eta\nabla^2 B_P \qquad (11)$$

or, using Eq. (9) with a suitably defined gauge,

$$\frac{\partial A}{\partial t} + (\mathbf{U}_P \cdot \nabla)A = \eta A. \qquad (12)$$

The toroidal induction equation is

$$\frac{\partial B_T}{\partial t} + (\mathbf{U}_P \cdot \nabla)B_T = (\mathbf{B}_P \cdot \nabla)U_T + \eta\nabla^2 B_T. \qquad (13)$$

Comparing Eqs. (12) and (13), we see that the vector potential A has no source term to balance the loss due to ohmic dissipation. In contrast, the toroidal field can be maintained temporarily by the shearing of poloidal field lines. As the poloidal field decays this source term will decrease, and eventually the toroidal field will also decay. These two equations illustrate Cowling's theorem which states that axisymmetric magnetic fields cannot be maintained against ohmic dissipation by axisymmetric velocity fields. The maintenance of the poloidal field by a three-dimensional velocity field is discussed next.

The seminal work on the regeneration of the poloidal field, and on the solar cycle in general, is the subject of Parker's (1955) paper. He showed that a mean toroidal current can be maintained by the action of small-scale, short-lived, cyclonic eddies on the toroidal field. The lifting and twisting of a toroidal field line to create a toroidal current and poloidal field is shown in Fig. 1. In Parker's model, diffusion allows the field to relax after twisting. Parker used what he called the "short sudden" approximation in this paper, by which he meant that the magnetic field was first twisted by the cyclonic eddies under the assumption that the field is frozen in (infinite conductivity) and that resistive diffusion then acts to reconnect the field lines and re-establish an axisymmetric field. A second ingredient in Parker's model is large-scale differential rotation. When coupled with the small-scale cyclonic motions, mentioned earlier, propagating magnetic field solutions can be found, the so-called dynamo waves. It is the presence of just these solutions that makes Parker's model so compelling. The mechanics of how the fields propagate are shown in Fig. 2. The shear shown at the left in the figure acts on the vertical component of the poloidal fields to produce new toroidal field of the same sign on the right-hand edge, and of opposite sign on the left-hand edge, of each concentration of toroidal field. Thus, the whole pattern propagates to the right. In kinematic dynamo models, the effect of the cyclonic motions is often

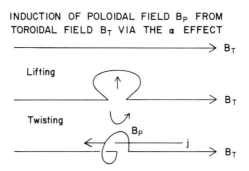

Fig. 1. A schematic drawing of lifting and twisting of a toroidal field B_T into a poloidal field B_p due to helicity represented by the α-effect.

PRODUCTION OF TOROIDAL FIELD B   FROM
POLOIDAL FIELD B$_P$ BY DIFFERENTIAL ROTATION

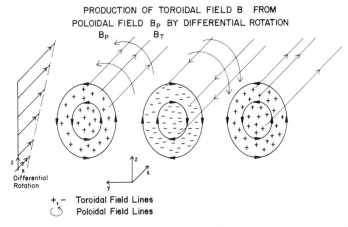

+, −   Toroidal Field Lines
⟲   Poloidal Field Lines

Fig. 2. A schematic drawing of the production of toroidal field (+ and − symbols in cross section) from poloidal field due to the shearing by the radial differential rotation shown at the left.

represented by a parameter α and called the α-effect (after Steenbeck et al. 1966). Similarly, the shearing of the toroidal field by the differential rotation is often represented by ω, and called the ω-effect. The combinations of poloidal and toroidal field shown in Fig. 2 are for $\alpha < 0$. Note that the direction of propagation depends both on the sign of α ($\alpha < 0$ implies that the poloidal field direction is given by the left-hand rule) and by the sign of the shear ω. If the sign of either is changed, the fields propagate leftward in the figure; if both signs are changed, the fields continue to propagate toward the right.

The cyclonic eddies, when acting on the toroidal magnetic field, create a toroidal current. This process can be parameterized by a constant α times the toroidal field B_T in the poloidal induction equation

$$\frac{\partial A}{\partial t} + (\mathbf{U}_P \cdot \nabla)A = \alpha B_T + \eta A. \tag{14}$$

Equations (13) and (14) form the basis of kinematic α-ω dynamo theory (so called because α maintains the poloidal field and ω maintains the toroidal field). They are linear in the magnetic field and have been extensively used and discussed for more than twenty years (see, e.g., Stix 1981b). A more formal derivation of the kinematic dynamo equations, from the point of view of a mean field theory, is presented at the end of this section. The cyclonic eddies also act on the poloidal field, generating a poloidal current, that can contribute to the maintenance of the toroidal field, resulting in an $\alpha^2\omega$ dynamo. This α term is generally neglected because it is small compared with the differential rotation term ω. In the overshoot region, the additional α term in the toroidal field equation can have important consequences for the period of the dynamo waves (see Gilman et al. [1989] for a detailed discussion of these effects).

Equations (13) and (14) can be easily solved in a periodic domain. Let $\mathbf{u} = (U_T(z),0,0)$, and $\mathbf{B} = [B_T(y,z), -\partial A/\partial z, \partial A/\partial y]$, and let α be positive. Then Eq. (13) becomes

$$\frac{\partial \mathbf{B}_T}{\partial t} = -\omega \frac{\partial A}{\partial y} + \eta \nabla^2 B_T \tag{15}$$

where $\omega \equiv \partial U_T/\partial z$, the radial shear of the azimuthal velocity. Equation (14) becomes

$$\frac{\partial A}{\partial t} = \alpha B_T + \eta \nabla^2 A. \tag{16}$$

Expanding the toroidal field and vector potential in plane waves $e^{i(\omega t + ky)}$ yields the dispersion relation

$$(i\omega + k^2\eta)^2 = -ik\omega\alpha. \tag{17}$$

The condition for growing solutions is $k^2\eta < \sqrt{k\omega\alpha}/2$ and the frequency of the oscillations is given by $2\pi\sqrt{k\omega\alpha}$. (The condition for growing solutions can also be expressed in terms of a critical magnetic Reynolds number: $R_{mc} = UL/\eta = 2$ where $\alpha \sim U$, $\omega \sim U/L$, and $k \sim 1/L$.) These are exponentially growing or decaying solutions that propagate in the $\hat{y}$ (north-south) direction. The sign of the product $\alpha\omega$ determines the direction of propagation (see Fig. 2).

We conclude this section with a sketch of the derivation of the kinematic dynamo Eqs. (13) and (14) from mean field theory. The reader is referred to Moffatt (1978), Roberts (1971) and Roberts and Stix (1971) for a more thorough treatment. The velocity and magnetic fields are separated into mean and fluctuating parts

$$\mathbf{B} = \bar{B} + b'; \mathbf{u} = \bar{U} + u'. \tag{18}$$

where the averaging operator $\langle \ldots \rangle$ averages over the entire azimuthal directions and over some characteristic "large" scale in the radial and latitudinal directions. The mean part is then axisymmetric and "large" scale, with $\langle \mathbf{B} \rangle = \bar{B}$, while the fluctuating part contains all the nonaxisymmetric and small-scale components, with $\langle b' \rangle = 0$. Under the kinematic approximation, we assume that the mean velocity fields are completely specified, and the statistical properties of the fluctuating velocity fields are known. Note that the average of products of fluctuating terms do not necessarily vanish, $\langle u' \times b' \rangle$ $\neq 0$, as the fluctuations may be correlated. Substituting Eq. (18) into Eq. (3) yields

$$\frac{\partial}{\partial t}(\bar{B} + b') = \nabla \times \{\bar{U} \times \bar{B} + u' \times b'$$
$$+ \bar{U} \times b' + u' \times \bar{B}\} + \eta \nabla^2 (\bar{B} + b'). \quad (19)$$

Applying the averaging operator to Eq. (19) yields an equation for the mean magnetic field

$$\frac{\partial \bar{B}}{\partial t} = \nabla \times \{\bar{U} \times \bar{B} + \overline{u' \times b'}\} + \eta \nabla^2 \bar{B}. \quad (20)$$

The difference of Eqs. (19) and (20) yields an equation for the fluctuating magnetic field

$$\frac{\partial b'}{\partial t} = \nabla \times \{\bar{U} \times b' + u' \times \bar{B} + u' \times b' - \overline{u' \times b'}\} + \eta \nabla^2 b'. \quad (21)$$

So far, no approximations have been made; we have merely replaced a single equation for the magnetic field for two equations for separate components of the field. We want to solve Eq. (20), but the evolution of the mean field depends on the fluctuating field through the quadratic quantity $\overline{u' \times b'}$. If we can express this term as a function of $\bar{U}$ and $\bar{B}$, then Eq. (20) can be simply integrated. From Eq. (21) we can see that b' depends linearly on $\bar{B}$, and the quadratic term $\overline{u' \times b'}$ is also linear in $\bar{B}$. If b' and u' also vary over much smaller spatial scale than $\bar{B}$, then the relationship can be approximated by a Taylor series in $\bar{B}$

$$\overline{u' \times b'} \approx a_{ij} \bar{B}_j + b_{ijk} \frac{\partial \bar{B}_j}{\partial x_k} \quad (22)$$

where the tensor coefficients a_{ij} and b_{ijk} can be expressed in terms of the statistical properties of the fluctuating (turbulent) velocity field (see Roberts [1971] or Moffatt [1978] for a more detailed discussion).

When specific assumptions are made about the nature of the turbulent field, closed form expressions can be found for $\overline{u' \times b'}$. One of the simplest arises when the term $G \equiv u' \times b' - \overline{u' \times b'}$ in Eq. (21) is sufficiently small to be neglected. Then if the turbulence is steady, homogeneous and isotropic $\overline{u' \times b'}$ can be shown to depend on the helicity $\mathbf{u} \cdot \nabla \times \mathbf{u}$ and on the magnetic diffusivity such that $\overline{u' \times b'} \to 0$ as $\eta \to 0$. It can further be shown that if the velocity field lacks parity-invariance (which often means that the helicity is not zero, but see Gilbert et al. [1988] for a counter example), then the coefficient a_{ij} in Eq. (22) can maintain the magnetic field against dissipation. If the turbulence is steady, homogeneous and isotropic then a_{ik} and b_{ijk} are constants, the only appropriate isotropic tensors of rank 2 and 3 are $\alpha\delta_{ij}$ and $\beta\varepsilon_{ijk}$ (Roberts 1971), and the mean field Eq. (20) becomes

$$\frac{\partial \bar{B}}{\partial t} = \nabla \times \{\bar{U} \times \bar{B} + \alpha\bar{B}\} + \eta_T\nabla^2\bar{B} \tag{23}$$

where $\eta_T = \eta + \beta$. There is a good deal of algebra between Eqs. (20, 21) and (23). The point we wish to make here is that the arguments of Parker under the short sudden approximation can be substantiated by a more formal analysis. The crucial approximation is not the smallness of the fluctuating components u' or b' but rather the smallness of the uncorrelated part of the current G and the separation of scales between the fluctuating and mean components. Moffatt (1978, §7.5) identifies two distinct circumstances under which the assumption of small G may be justified: (1) if the magnetic Reynolds number based on the small scale-motions is small, or (2) if the velocity fluctuations are due to random waves, rather than conventional turbulence. Unfortunately neither of these conditions is likely to be appropriate for the Sun. Thus, although we possess an appealing physical model of how magnetic fields may be regenerated in the presence of small-scale cyclonic eddies, it cannot be rigorously justified for the Sun. It suffices to say that something like the α-effect is likely to occur in the Sun. Perhaps simulations of MHD turbulence will give us a fuller understanding of the small-scale processes of field regeneration in the near future.

The discussion in this section has centered on dynamo action at modest magnetic Reynolds numbers. The growth rate for mean field dynamos decreases as the magnetic Reynolds number increases (Zeldovich et al. 1983). The concern is that the true magnetic Reynolds number in the Sun is large, $\geq 10^8$ ($U \approx 10^5$ cm s^{-1}, $L \approx 2.0 \times 10^5$ cm and $\eta \approx 10^7$ cm^{2}s^{-1}). Recently there has been a good deal of effort to find a way for dynamo action to persist in the limit of large magnetic Reynolds number. Both the mean-field dynamo models and the dynamical-convective dynamo models rely on turbulent eddys to enhance the diffusion of magnetic flux and momentum. In contrast, so-called "fast" dynamos have been shown to exist for a few specific velocity fields in the limit of infinitely high magnetic Reynolds number, with no parameterized eddy diffusivity (see Strauss 1986; Soward 1987; Finn and Ott 1988). There are no specific models of the solar dynamo based on this approach, so it is too early to judge whether a competitive solution to the solar cycle will emerge from these efforts.

III. DYNAMO THEORY APPLIED TO THE SUN: A DILEMMA

A. Application of Kinematic Dynamo Theory to the Sun

Kinematic dynamo theory has played (and continues to play) an important role in developing our intuition about how magnetic field may be generated in rotating turbulent fluids. These models also provide constraints on the form the velocity fields may take and still satisfy the basic observational constraints of the solar cycle. By definition, the kinematic theories cannot tell us whether a particular flow is realizable. Unfortunately, many of the early dynamo models were concerned with only matching the observations, and did not address the question of whether the particular form of the differential rotation and the helicity can be maintained by the convective motions.

There exists an extensive literature, mostly in the 1970s, that discusses the application of kinematic dynamo theory, as represented by Eq. (23), to the solar dynamo problem. In particular, Eq. (23) written in spherical geometry is solved for growing dynamo waves using assumed profiles for the differential rotation, meridional circulation (contained in U) and assumed profiles of the helicity parameter α and eddy magnetic diffusivity η. At the time these solutions were studied, there was almost no guidance from observations for appropriate values and profiles for these parameters. The differential rotation of the Sun with latitude at the surface was well known, but not the radial gradient within the convection zone. From theoretical considerations, it was easy to argue that α should have opposite signs in north and south hemispheres, and with somewhat less certainty, that α should be > 0 in the north, < 0 in the south. But the amplitude of α cannot be estimated to even within a factor of 10. Values of η are also quite uncertain, given the uncertainty even in the concept of an eddy diffusivity. So what was typically done was to "tune" the model solutions to behave qualitatively like the Sun. In particular, profiles of radial differential rotation and magnitudes of α were chosen to produce migration of toroidal and poloidal field patterns toward the equator, at the right rate to simulate the butterfly diagram. To accomplish this required that the rotation rate increase significantly with depth within the convection zone, perhaps by as much as 40% compared to surface values, and that α be extremely small, of order a few cm s^{-1}. These small values of α would mean that only a small fraction of a typical velocity in the bulk of the convection zone, as estimated by mixing-length arguments, would contribute to the helicity available for induction of magnetic fields. Given that the typical turnover time for convection in the lower reaches of the convection zone should be a month or more, Coriolis forces would have a strong influence on the flow and generate a lot of helicity; this seems unlikely. A possible alternative is that, even at depth in the convection zone, the magnetic field is concentrated into small regions of space, and does not feel the full effect of the bulk α (Childress 1979).

But in any case, the values of U and α were chosen virtually without regard to the physics that would generate them, as represented by Eqs. (1) and (2) and the thermodynamic energy equation. In particular, U and α are not created independently but have a common origin in the influence of rotation upon convection. Therefore choosing them independently and arbitrarily is bound to lead to violation of the governing laws of physics and render agreement with the Sun's behavior fortuitous at best.

The alternative to kinematic theory, of course, is to solve the full equations of MHD, including the induction equation, the momentum equation, the equation of mass continuity and the thermodynamic energy equation. This is obviously a much more difficult task, requiring major computing resources and has been attempted by only a few.

B. Magnetohydrodynamic Dynamo Models.

To calculate a full MHD dynamo model for the Sun first requires a theoretical model for global convection and differential rotation of the Sun. Historically, two large classes of theories for solar differential rotation have been developed. Gilman (1986) gives a review of these different approaches. One of the two classes involves representing the global effects of convection in a small number of parameters, that describe either anisotropies in the angular momentum transfer by convection, or latitudinal variations in radial heat flux produced by the influence of rotation upon convection. A recent typical example of differential rotation models of this type is given in Pidatella et al. (1986). We do not attempt to review this class of models further here, because they do not lead, in and of themselves, to MHD dynamos of the Sun. This is because all of the effects of convection are parameterized rather than calculated explicitly.

The second class of models, that we will focus on, involves full three-dimensional calculations of nonlinear convection in a deep rotating spherical shell. In such models, the full electromagnetic induction effects of the calculated convection are included in the dynamo, as are the nonlinear feedbacks on the convection and differential rotation through the electromagnetic body force. Work on these models prior to 1982 is reviewed by Gilman (1986). More recent calculations which we discuss in somewhat greater detail here include Gilman (1983), Glatzmaier (1984,1985a,b), Gilman and Miller (1986). Dynamo models included only incompressible convection until Glatzmaier's work, which represents the only calculations reported in the scientific literature that include both compressible global convection and dynamo action.

Let us consider first the theory of differential rotation driven by global convection in the absence of magnetic fields. The basic formulation of the model equations that has been used in these calculations is given in Gilman and Glatzmaier (1981), with some subsequent modifications. Nonlinear model results from this formulation, using quite different solution techniques, are given in Glatzmaier (1984,1985a) and in Gilman and Miller (1986).

Many results found in these calculations were also seen in earlier incompressible ones carried out by Gilman in the 1970s. Relevant qualitative results are perhaps best summarized in Gilman (1980a).

In all of these calculations the nonlinear convection model is formulated as a finite amplitude perturbation system, perturbed about a reference state of no motion relative to a rotating reference frame. This state is not one of radiative equilibrium, but rather one, particularly in the compressible case, determined by mixing-length arguments to be close to an adiabatic stratification. Small-scale diffusion of momentum and entropy is assumed to occur. The diffusivities assumed in this formulation are identified with small-scale turbulent transport processes on the Sun. But, unlike the first class of models alluded to above, these diffusivities are passive, in the sense that, in the absence of other transport processes, the fluid shell assumes a state of solid rotation and spherically symmetric heat flux and temperature distribution. In the compressible case, the heat flux is matched to the observed solar luminosity.

The crucial parameter in these calculations is the ratio of the rotation time to the turnover time for the convection which results when the reference state is unstable to global motions. If the influence of rotation is strong, so that the rotation time is about the same or shorter than the turnover time, the convection generates an equatorial acceleration, by transporting angular momentum from high latitudes to low. The deeper the convection zone is, the broader the equatorial acceleration. When the models produce an equatorial acceleration, they also produce an angular velocity that decreases with depth, in such a way that the angular velocity is roughly constant on cylinders concentric with the rotation axis.

Although compressible calculations can be matched much more closely to the real Sun than incompressible calculations, they nevertheless contain a number of simplifying assumptions. But, in any case, these calculations show that the models predict a surface differential rotation similar in magnitude and profile to that of the Sun when they have as input the solar luminosity, an average solar rotation rate and plausible eddy diffusivities. In Glatzmaier's models, eddy diffusivities are not assumed, but rather calculated from the motions themselves.

Thus, it would seem we are close to a deductive theory for the solar general circulation. However, the reality is not so clear. For example, the models predict amplitudes for global or "giant cell" convection a factor of 2 or 3 larger than the upper limits determined from observations (LaBonte et al. 1981). The models do not actually include the outermost several percent of the solar radius, because the spatial and time resolution requirements are much too great for a global model to handle, but intuitively we should expect such motions to penetrate through the uppermost layers to the surface. Van Ballegooijen (1986) has a possible mechanism for hiding these giant cells below the surface, but it remains to be tested in a full nonlinear model. A perhaps more serious challenge to these results is that the dynamo action predicted from the models gives very different results from the real Sun.

The nonlinear dynamo calculations to test the dynamo action of the differential rotation models described above are reported principally in Gilman and Miller (1981), Gilman (1983) and Glatzmaier (1985a). Gilman and Miller's calculations are for the incompressible case, Glatzmaier's for the compressible case. Since there is a high degree of similarity in the solutions for the motion fields in the compressible and incompressible models, there should be a high degree of similarity in the dynamos too. In Gilman and Miller (1981), all of the dynamo solutions found were for essentially random magnetic fields with no cyclic magnetic reversals. The reason was that, for the particular values of eddy diffusivities chosen, the convection driving the equatorial acceleration contained about 2/3 of the total kinetic energy of the system. Convection dominated over differential rotation in the dynamo process. In the language of mean-field kinematic dynamo theory, the model acted more like an "α^2" dynamo. But with a reduction of a factor of 10 in the diffusivities, still within a plausible range from mixing-length theory, the convection was reduced to about 20% of the total kinetic energy, with differential rotation the other 80%. Under these conditions, cyclic dynamos were produced.

In all the cyclic dynamo solutions, the toroidal and poloidal magnetic fields migrate toward the poles with increasing time, which, unfortunately, is opposite to the observed behavior of the magnetic field. Also, the magnetic field migrated quite fast, with magnetic cycle periods about 1 order of magnitude too short for the Sun. Given the magnitude and sign of the helicity of the convection calculated, and the profile of differential rotation, the dynamo behaved about as would be expected from mean-field kinematic dynamo theory. In fact, the model results indicated that the mean-field theory probably applies beyond the range of its assumptions, which, strictly speaking, require averaging over eddies and magnetic structures small compared to the dimensions of the convective shell. The convection and magnetic features actually calculated in the full MHD dynamos just described are comparable in scale to the shell dimensions.

The compressible MHD dynamo calculations in Glatzmaier (1985a) confirmed the basic result of poleward propagation of the fields with time, although he found dynamo periods only a factor of 2 shorter than the real solar cycle.

C. The Dynamo Paradox and Its Possible Resolution

We have seen above that the differential rotation and helicity required by kinematic mean-field dynamo theory are not predicted by global convection models. The predicted radial gradient of rotation has the wrong sign, and the helicity is 2 to 3 orders of magnitude too large. Thus, although global convection models successfully predict the observed surface differential rotation, they predict fundamentally wrong dynamo behavior. What is the resolution of this paradox? Is the global convection theory wrong, or the kinematic dynamo theory (or both)? This paradox has been known at least since 1975, and was discussed at IAU Symposium 71 in Prague (see Durney et al. 1976 for a summary). At the time, most scientists concluded (P. A. Gilman excepted) that it must be the global convection theory that was wrong, mainly because kinematic mean-field dynamo theory seemed to work so well. But until recently there was no crucial test of either theory, namely an observational estimate of the radial gradient of angular velocity within the convection zone. As discussed in the chapter by Libbrecht and Morrow, helioseismology has now provided such a test, and it indicates that the angular velocity is not increasing substantially with depth within the convection zone. At present (Morrow 1988a; Brown et al. 1989), the most likely possibility is that there is essentially no radial gradient within the convection zone, at least in low and mid-latitudes where the butterfly diagram is formed. This result, if correct, rules out the kinematic mean-field dynamo theory as it has been applied to the bulk of the convection zone. But, the lack of radial gradient also requires that the global convection theory for the differential rotation be modified. This represents a much more modest change than is needed to arrive at a satisfactory dynamo.

Kinematic dynamo theory can certainly be applied to any subregion of the Sun. If the bulk of the convection zone is ruled out by observations, then the lower part of the convection zone and the overshoot region can be investigated. In fact, the first spherical $\alpha\omega$-dynamo model had a radial shear concentrated at the base of the convection zone (Steenbeck and Krause 1969a). The signs of α and gradient of the angular velocity are wrong for this model, but substantially the same results would be obtained if both signs were changed. This was recognised by Durney (1976); he argued that the lower part of the convection zone would be the "most appealing" region for the dynamo. As stated above, the kinematic models are useful for demonstrating that the solar dynamo could exist in a certain region, but dynamical models are needed to demonstrate that the assumed flows are realizable.

The recent results from helioseismology also indicate that the deep interior below the convection zone is nearly in solid rotation, at a rate intermediate between the maximum and minimum of the convection zone. This means that there must be a transition layer that matches these two rotation profiles. This intermediate layer must contain substantial radial gradients, of opposite sign in low latitudes and high. Gilman et al. (1989) discuss the implications of the helioseismological results for both angular momentum balance and dynamo action on the Sun; they point out that these strong radial gradients in rotation suggest that the transition layer is a very likely location for the solar dynamo. If the helicity in this transition layer were of the same sign as predicted by the global convection models for the bulk of the convection zone, which agrees in sign with the requirements of kinematic mean-field dynamo theory, then these radial gradients would imply magnetic field migration toward mid-latitudes from both low and high latitudes. But the global convection models of Gilman and Miller (1981) and Glatzmaier (1985b) predict the opposite sign of helicity at the base of the convection zone compared to the bulk of the zone above so that the magnetic field migrations should be toward the equator in low latitudes, and toward the poles in higher latitudes. This is in qualitative agreement with observations, though only the equatorward branch produces sunspots.

The concept of placing the solar dynamo at the base of the convection zone rather than in the bulk of the layer above predates the recent helioseismological evidence by several years. However, the arguments in its favor seem strengthened considerably by this evidence. Gilman and collaborators began developing quantitative models for dynamos at the convection zone base in 1981. Glatzmaier (1985b) attempted a calculation of such a model using his global convection and dynamo models achieving limited success, although some of his assumptions were rather unrealistic and he was unable to run the calculation long enough to verify that a full magnetic cycle with long-term field migration was produced.

In the next two sections, we discuss in detail our own modeling efforts (DeLuca 1986).

IV. A NEW APPROACH: DYNAMOS AT THE BASE OF THE CONVECTION ZONE

We have seen that placing the dynamo in the bulk of the convection zone leads to a paradox. One way of resolving the paradox is to place the dynamo in a thin layer at the base of the solar convection zone where overshooting convective plumes stir up the stably stratified radiative zone. In fact, such a placement has been advanced by a number of individuals to solve other problems associated with having the magnetic fields seen in sunspots generated in the bulk of the convection zone. We examine these problems below and then discuss the physical properties of the overshoot region in Sec. IV.A.

One of the major problems with locating the dynamo in the bulk of the convection zone is the buoyant rise of field in a convectively unstable environment. Amplification by the dynamo requires that the magnetic field remain in the convection zone for times comparable to the solar cycle. Buoyancy effects then limit the amplitude of the magnetic field to about 200 gauss (Parker 1979a; §8.7). In contrast, large-amplitude fields ($\sim 10^4$ gauss) can be maintained in a stably stratified region beneath the solar convection zone (Van Ballegooijen 1982). The flux-tube calculations of Moreno-Insertis (1986) show how part of a tube can buoyantly rise to the surface while the rest is firmly anchored in a stably stratified region. Nonlinear calculations of magnetic Rayleigh-Taylor instabilities (Cattaneo and Hughes 1988) show that the growth of disturbances in a layer of concentrated magnetic field is strongly influenced by vortex interactions between adjacent buoyant concentrations of fields. The result is a rapid mixing and dispersal of the field, rather than a rapid rise of flux. These calculations indicate that the motions in the convection zone will play a large role in the confinement of the field to the overshoot layer.

The interaction of convection and diffuse magnetic fields has been the subject of a number of studies. Galloway and Weiss (1981) showed that closed stream lines quickly amplify the field near the cell boundaries and eventually (within several turnover times) expel the flux from the center of the cell (see also Weiss 1981a,b; Proctor and Weiss 1982). As the field is concentrated into the intercell regions, convection in these regions is reduced. The result is regions of highly concentrated field separated by convecting regions. These solutions hold for two-dimensional convection. The compressible calculations of Hurlburt and Toomre (1988) show similar behavior with the additional feature that the magnetic fields are concentrated most strongly in the downward plumes. These results may be dependent on the relatively simple cellular structure present in two-dimensional calculations. In three-dimensional compressible convection, the flow is no longer cellular (Cattaneo and Hurlburt, personal communication); near the upper boundary, the upflows are distinct from each other and the downflows are connected, while near the lower boundary the upflows are connected and the downflows are distinct. The connectivity of the upwellings and downwellings changes with height. It is not yet clear what will happen to a magnetic field injected into one of these flows. If turbulent convection is efficient at concentrating the magnetic field into small regions at the cell boundaries, then the convective flows may help contain a strong magnetic field at the base of the convection zone. An important question is what happens to compressible flows in the presence of strong rotation? The pattern of convection may be very different when the Coriolis force is important.

If the magnetic field is excluded from a significant fraction of the volume of the convection zone, then the problem of a maximum field strength of 200 gauss becomes more acute. Galloway and Weiss (1981) estimate that about 10^{24} Maxwell emerge from the Sun during the course of the solar cycle. The dynamo region must then be large enough and stable enough to contain 10^{24} Maxwell worth of flux. If the field strength is limited to 200 gauss, the full volume of the convection zone would be needed to contain the necessary flux. It is possible that most of the magnetic flux seen at the surface is pulled back under the surface and only a small fraction escapes (Parker 1984b). However, the toroidal field must change sign during the course of the solar cycle, so either the surface flux is dissipated beneath the solar convection zone or it is reprocessed into toroidal flux of the opposite sign. The former possibility still requires 10^{24} Maxwell of new flux to be generated; the latter requires a mechanism for changing the sign of the toroidal field without massive reconnection. Such a mechanism has yet to be proposed.

The surface magnetic fields are characterized by regions of intense magnetic field, separated by nearly field-free regions. In contrast, the dynamo models of Secs. II and III are derived from continuum fields. The situation is further complicated by the fact that Hale's Polarity law is almost never violated. The small-scale intense surface magnetic fields must "know" about the large-scale field in their hemisphere. If the magnetic fields are generated in the convection zone, one must either use the mean-field equations to generate the large-scale field and then appeal to local processes near the surface of the Sun to form flux tubes and sunspots; or alternatively, use flux tubes as the basic quantity and try to create a dynamo model based on twisting of flux tubes (Schüssler 1980). A major difficulty of the latter approach is how one treats diffusion and reconnection of field lines. If the dynamo is located in a thin region below the convection zone, then one might naturally think of a portion of the field becoming unstable and rising to the surface with its foot points attached fixed in the overshoot layer (Moreno-Insertis 1986).

Finally, recent observational evidence supports the argument that only at the base of the convection zone is there the correct sign of the differential rotation and kinetic helicity. The angular velocity of the Sun as a function of radius and latitude, can (in principle) be found by measuring the rotational splitting of p-modes at the surface of the Sun (see the chapter by Libbrecht and Morrow). The problem is that the determination is extremely ill posed. All that one can say is that certain models of the solar angular velocity profile are consistent with the observations, while others are inconsistent. One consistent model (Morrow 1988a; Gilman, et al. 1989) has an angular velocity that is constant on radial spokes, and decreases with latitude in the convection zone, and is constant on spherical shells in the radiative zone. The angular velocity must vary smoothly in the transition zone between the convection zone and the radiative zone. This smooth connection results in a radial gradient of angular velocity in the overshoot region (which is just a portion of the transition zone). The sign of the gradient depends on how fast the interior shells are rotating. In the Morrow model, the interior rotation rate is equal to the convection zone rate at about $\pm 30°$ latitude. Interestingly, this latitude is consistent with a balance of the torques between spinning down the equatorial regions of the convection zone and spinning up the polar regions. The same model, with $\alpha < 0$ in the northern hemisphere, results in equatorward propagation of the dynamo waves at low latitudes and polarward propagation at high latitudes. (Note that if α is due to magnetostrophic waves and changes sign with latitude, as suggested by D. Schmitt [1987], then all the dynamo waves will propagate toward the equator.)

A. Overshoot Region

We now describe the physical processes that are important in the overshoot region, i.e., that region at the top of the stably stratified radiative interior into which convective plumes penetrate. The most important property of the overshoot region is its thickness. This determines the strength of the magnetic field needed to supply the surface with 10^{24} Maxwell of flux over the course of a solar cycle. It also determines whether the fluid in the overshoot region can be treated as incompressible. Unfortunately this depth is not well determined either from theoretical and computational work or from observational constraints.

Standard mixing-length models are unable to treat convective overshooting because they are local in nature. However, Shaviv and Salpeter (1973) extended the local mixing-length theory to include the past history of a parcel of fluid. A parcel that starts in an unstable region and enters a stable region will be decelerated and eventually will stop. These models estimate the depth of the overshoot region to be on the order of 0.2 of a pressure scale height, and to be nearly adiabatically stratified (Pidatella and Stix 1986).

Schmitt et al. derive a set of phenomenological plume equations for convective overshoot in the solar interior. This model includes equations for mass flux, momentum flux, and buoyancy flux in a downward-directed plume together with an equation of state and the condition that the total energy flux through the overshoot region is constant. When the downward velocity and filling factor are known, the model determines both the depth of the overshoot region and the stratification. The concentration of the downward plume (the filling factor) and the downward velocity are not well known, but recent results from three-dimensional simulations of compressible convection (Cattaneo, personal communication) indicate that the downflows are strongly concentrated and have large velocities. For a downward velocity v of 10^4 cm s^{-1} and a filling factor f of 0.1, Schmitt et al.'s model gives an overshoot region about 0.3 pressure scale heights thick, with the depth of penetration varying as $v^{3/2}$ and $f^{1/2}$.

Dynamical models of penetrative convection give different results. Hurlburt (1983) solves the equations of two-dimensional compressible convection with stable layers both above and below the unstable one. The solutions are characterized by large downward plumes with penetration into the stable layer on the order of a pressure scale height, and mixing throughout the layer (about two pressure scale heights). The stratification in these models was not chosen to mimic solar conditions, but his calculations do suggest that the overshoot layer may be as deep as one scale height.

We will consider a model of the overshoot region that is a few tenths of a pressure scale height in depth. The next question that needs to be addressed is the nature of the velocity fields in the overshoot layer. It is convenient to consider separately the axisymmetric and fully three-dimensional velocity fields. This is a natural separation to make if, as we shall argue below, the flows in the overshoot region are strongly modified by rotation. For a mean-field dynamo to operate in the overshoot region, the convective plumes that enter the layer must deposit a net helicity, and must generate a sufficient amount of turbulence so that the viscosity and magnetic diffusivity are much larger than their molecular values. The sign of the mean helicity, $\alpha \propto -\mathbf{u} \times (\nabla \times \mathbf{u})$, in the overshoot layer should be the same as in the lower part of the convection zone. As discussed in Sec. III.C, three-dimensional convection calculations by Glatzmaier (1985b) and Gilman (1983) predict that the helicity will be positive in the lower part of the convection zone, $\alpha < 0$. These small-scale turbulent flows are asymmetric; the only axisymmetric flow imposed on the overshoot region is the differential rotation with radial and latitudinal shear that is suggested by the Morrow model. The combination of the differential rotation, the net helicity, and the enhanced diffusivities, provide all the ingredients needed for a $\alpha^2\omega$ kinematic dynamo.

In the following section we discuss a particular model of magnetic field generation in a thin layer. The solutions to this simple model highlight the role played by the velocity field driven by the Lorentz force in determining both the structure of the magnetic fields, and the period of the dynamo wave solutions.

V. QUANTITATIVE MODELS FOR INTERFACE DYNAMOS

A. Model Formulation

In this section we present a simple Cartesian model of a dynamo that operates in a thin layer between the convection zone and the radiative zone. We view the overshoot layer as a region where overshooting convective plumes are sufficiently vigorous so that the diffusivities are enhanced from their molecular values. The convective plumes also deposit a net helicity in the layer. The layer has a differential rotation imposed upon it by its location between a region where the angular velocity is constant on cones (the convection zone) and a region where the angular velocity is constant on spherical shells (the radiative zone). We parameterize the effect of the convective plumes in the overshoot layer by an α term in both the toroidal and poloidal induction equation; and by enhanced (constant) turbulent diffusivities.

The magnetic field that evolves in such a region will have dynamo wave solutions with the direction of propagation determined by the sign of the α and ω (the imposed differential rotation). As the field grows, it will drive a flow through the Lorentz force. The nature of this flow is determined by the momentum equation. If the advection time is long compared with the rotation time, the Rossby number will be small, and the inertial terms scale out of the momentum equation. Further, if the layer is smaller or on the order of a pressure scale height, then the well-mixed fluid may be taken to be homogeneous, incompressible and in hydrostatic balance.

The magnetic fields present in the overshoot layer must reach the surface, i.e., they have to escape. If, as we have argued, the overshoot region is stably stratified, then the means of escape is most likely magnetic buoyancy. We parameterize the loss of magnetic flux by a simple loss term with a coefficient proportional to the magnetic energy (our loss term is the same as that used by Yoshimura [1975*b*]).

One feature that distinguishes our model equations from much of the other work is that we derive the model equations by a detailed scale analysis of the Navier-Stokes equation, with rotation, and the MHD induction equation. We include the α-effect, turbulent diffusivities, a parameterization of magnetic flux loss due to magnetic buoyancy and an imposed shear across the layer. The derivation is presented in DeLuca and Gilman (1986) and proceeds as follows: we choose a Cartesian geometry and admit only axisymmetric field (there is no variation in the east-west direction); we take the rotation axis anti-parallel to gravity, and we assume that the fluid is homogeneous and incompressible (the thickness of the layer is small compared with any scale height, and the flow is very subsonic). We define a basic state with no magnetic fields and only an azimuthal flow with linear height dependence (see DeLuca and Gilman 1988); this state defines the reference state equations which are then subtracted from the full equations. The resulting perturbations are then made dimensionless using solar values of the rotation rate, the local gravitational acceleration and the mean density. The vertical length scale is taken to be 10^9 cm (consistent with Schmitt et al. 1984); the horizontal length scale is taken to be 10^{10} cm (a fraction of the pole to equator distance). The magnetic field strength is 10^4 gauss, sufficiently large for the layer to contain enough flux from large active region complexes (Golub et al. 1981). The velocity field is scaled by the size of α; we take $\alpha = 10^3$ cm s^{-1}. The result of this scale analysis is a set of equations in which the magnetic field is evolved by the induction equation with additional terms for the α-effect, the azimuthal shear and the enhanced flux loss due to magnetic buoyancy. The poloidal magnetic field is maintained against dissipation by the α-effect and the toroidal field is maintained by both the α-effect and differential rotation, a $\alpha^2\omega$ dynamo. The large magnetic fields drive a significant flow through the Lorentz force in the momentum equations; this flow feeds back nonlinearly on the magnetic fields in the induction equation. The momentum equations reduce to a set of balance conditions: viscous and Coriolis forces balancing Lorentz and pressure forces. The inertial terms scale out of the equations because the fluid velocity is much smaller than the Alfvén speed (by a factor of 3×10^{-2}).

The derivation outlined above results in a set of equations for the velocity and magnetic fields in three directions. The momentum equation is replaced by balance conditions in the $\hat{x}$ (east-west) and $\hat{y}$ (north-south) directions. The dimensionless form of these equations are

$$v(\partial^2 u/\partial z^2 + \delta^2 \partial^2 u/\partial y^2) + v = -b\partial a/\partial y - c\partial a/\partial z \qquad (24)$$

and

$$v(\partial^2 v/\partial z^2 + \delta^2 \partial^2 v/\partial y^2) - u = -\gamma \partial \eta/\partial y - b\partial b/\partial y - c\partial b/\partial z. \quad (25)$$

Here $v \equiv R_o/(R_e \delta^2)$ is the Ekman number, R_o is the Rossby number and R_e is the Reynolds number. v is a scaled diffusivity; for larger values of v, the flow is more viscous. δ is the aspect ratio of our layer, the ratio of vertical to horizontal dimensions. As we are considering a thin layer δ is small. We take advantage of the thinness of the layer by scaling the vertical components of the velocity and magnetic field by δ; the dimensionless vertical velocity is $w = w_d/(|\alpha|\delta)$, where $|\alpha|$ is the amplitude of the α-effect, and w_d is the dimensional vertical velocity. The only powers of δ that remain in our dimensionless equations are those associated with the horizontal diffusion terms. It is imperative that these "small" terms be retained because the nonlinear interactions in the equations are very effective at generating small-scale structures in both the horizontal and the vertical directions. Without horizontal diffusion all the energy in the magnetic and velocity field quickly piles up in the highest wavenumbers retained in the calculation. The term $\gamma\partial\eta/\partial y$ is a pressure gradiant term due to the presence of a free surface in our calculation. The pressure has been eliminated from the equations by integrating the vertical momentum equation. Taken together, Eqs. (24) and (25) represent the balance of the Lorentz force by Coriolis and viscous forces. We shall see in the solutions that follow that very different feedbacks result from balances by the Coriolis force and by the viscous force.

The vertical velocity is found from the incompressibility condition

$$\partial v/\partial y + \partial w/\partial z = 0. \quad (26)$$

The vertical component of the momentum equation reduces, under our approximations, to a statement of magnetostatic balance. The pressure is then determined by the magnetic fields and the height of the free surface η.

The induction equations are essentially nonlinear generalizations of Eqs. (13) and (14),

$$Da/Dt = b\partial u/\partial y + c(\partial u/\partial z + \partial\omega/\partial z) - S_\alpha\partial b/\partial z + \eta(\partial^2 a/\partial z^2 + \delta^2\partial a/\partial y^2) - \kappa a \quad (27)$$

$$Db/Dt = b\partial v/\partial y + c\partial v/\partial z + S_\alpha\partial a/\partial z + \eta(\partial^2 b/\partial z^2 + \delta^2\partial^2 b/\partial y^2) - \kappa b \quad (28)$$

and

$$\delta(Dc/Dt) = (b\partial w/\partial y + c\partial w/\partial z - S_\alpha\partial a/\partial y + \eta(\partial^2 c/\partial z^2 + \delta^2\partial^2 c/\partial y^2) - \kappa c)\delta. \quad (29)$$

Notice again the role played by the aspect ratio δ in reducing the effective diffusivity in the horizontal direction. In Eq. (29) the coefficients δ are retained on both sides of the equations to remind us that the vertical components of the magnetic field is also small compared with the horizontal components. S_α is ± 1 depending on whether α is positive or negative (recall that the velocities are scaled by the magnitude of α). The parameter κ is used to simulate the loss of magnetic flux from the layer due to magnetic buoyancy. κ is taken to be spatially constant with an amplitude dependent on the total magnetic energy in the layer. When the magnetic energy is less than some critical value E_c, κ is zero; as the magnetic energy exceeds E_c, κ grows and eventually limits the amplitude of the magnetic field.

The magnetic fields must also satisfy the divergence free condition

$$\partial b/\partial y + \partial c/\partial z = 0. \qquad (30)$$

In practice we determine a and c from Eqs. (27) and (29) and b from (30). Equation (27) is then used as a check on the calculation.

We choose boundary conditions that isolate our thin region from the rest of the Sun. We recognize that this is unrealistic, but it has the advantage of allowing us to concentrate on the physical processes in the layer. The horizontal ($\hat{y}$) boundary conditions are taken to be periodic. The bottom boundary is assumed to be a rigid, stress-free, infinitely conducting surface. The top boundary is assumed to be a stress-free, infinitely conducting, material surface. There are no external magnetic fields or electric currents. The height of the free surface is then determined by the vertical velocity at the top of the layer.

The equations are solved by expanding the variables in a fourier series in the horizontal direction and truncating after n modes. The variables are mapped onto a grid in the vertical direction, and then solved by centered finite differences. The magnetic fields are advanced explicitly in time with the diffusion terms lagged.

B. Model Results

We present a few examples of solutions to these nonlinear model equations. The solutions are for an imposed shear across the interface of 10 m s^{-1}, a plausible value given the helioseismological results. We also take $\alpha = 10$ m s^{-1}, consistent with mixing-length estimates of convective velocities at the base of the convection zone. The magnetic diffusivity is chosen so that the magnetic Reynolds number is 1.7 times the critical magnetic Reynolds number from linear theory (the magnetic Reynolds number at which linearized equations become unstable to growing magnetic fields is called the critical magnetic Reynolds number).

Figures 3a and b show the kinetic and magnetic energy spectra as func-

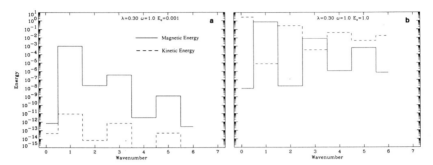

Fig. 3. (a and b) Dimensionless kinetic and magnetic energy spectra in y-wavenumber produced by dynamo action with a low-energy cutoff $E_c = 0.001$ (see text for definition) and large cutoff $E_c = 1.0$.

tions of y wavenumber. These solutions differ only in the magnetic energy level E_c at which magnetic flux begins to be expelled from the layer by magnetic buoyancy. These spectra represent equilibrium solutions to the model equations. As a point of reference the total kinetic energy associated with the imposed differential rotation has the value 0.2. Magnetic energies are given by the solid lines, kinetic energies by the dashed lines.

For both solutions, the dominant y wavenumber in the magnetic field is 1, in agreement with extrapolation from linear theory. With E_c at 0.001, the spectrum falls off rapidly (wavenumber 3 has 10^{-3} of the energy of wavenumber 1), and there is very little induced flow. In this case, the flux expulsion cuts in at low enough magnetic energies that we have essentially a single mode, linear solution. By contrast, the $E_c = 1.0$ solution shows that induced velocities have grown to quite large amplitude. Magnetic fields are essentially confined to odd wavenumbers, velocity to even wavenumbers. The tendency for both spectra to flatten out at high wavenumber is undoubtedly due to the aliasing which arises from truncation of the horizontal expansion at 7 wavenumbers, indicating that more modes are needed if we wish to go to higher-energy cutoffs.

The alternation in wavenumber space of magnetic and kinetic energy peaks can be easily explained. Self-interactions of the wavenumber 1 magnetic field produce a $j \times B$ force of wavenumbers 0 and 2, leading to strong velocities in these wavenumbers. Then the velocity/magnetic field interactions in the induction equations between wavenumber 1 in the magnetic field and 2 in the velocity field excite wavenumber 3 in the magnetic field, and so on. The large amplitude of wavenumber zero velocity fields is of particular interest, since it represents in part the feedback of the induced magnetic fields on the imposed differential rotation. Because the differential rotation is responsible for the propagation of dynamo waves, the dynamo period can be significantly altered by this feedback. We illustrate this below.

The typical magnetic field and velocity patterns for the $E_c = 1$ solution are shown in Fig. 4. The phase relationship of poloidal and toroidal fields is familiar from classical α-ω dynamo theory. The poloidal field encircles the toroidal field with the direction determined by the sign of α, determined by the left-hand for $\alpha < 0$ (same as in Fig. 2). The induced meridional flow nearly coincides in pattern with the poloidal field, so that there is very little induction of magnetic field by this interaction. The shading for differential rotation indicates the presence of "torsional oscillations," whose profile is only weakly dependent on depth, but with strong latitudinal shear near the boundaries between adjacent toroidal fields of opposite signs (corresponding to adjacent "sunspot cycles"). The amplitude of this torsional oscillation is similar to the average imposed differential rotation, leading to the net profile seen, which has largely vertical shading contours. If the imposed differential rotation dominated, then the contours would be strictly horizontal.

The patterns shown in Fig. 4 all propagate to the right, essentially without change of shape. Figure 5 shows a succession of patterns equally spaced in time, for both the $E_c = 0.001$ and 1.0 solution. Of particular interest is the difference in slopes between these two cases, showing that the induced differential rotation, that largely opposes the imposed one, in this case slows down the propagation by a factor of 2 or so. In Fig. 6 we show how the nonlinear feedbacks reduce the differential rotation in the center of the layer where most of the induction takes place. The total differential rotation (solid) and the imposed-differential rotation (dotted) are shown for the large-amplitude solution, $E_c = 1$. Notice that the magnitude of the shear is reduced by a factor of about 3 in the center of the layer. A similar plot for the nearly

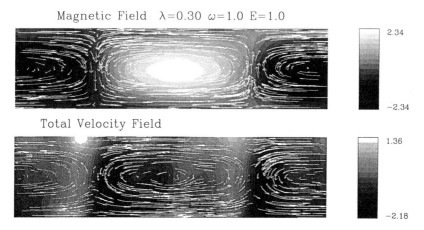

Fig. 4. Typical induced magnetic and total velocity solutions for the large energy cutoff case. Shading represents magnitude of toroidal field (upper) and differential rotation (lower). Arrows represent poloidal field (upper) and meridional flow (lower).

Toroidal Magnetic Field   λ=0.30 ω=1.0 E=0.001

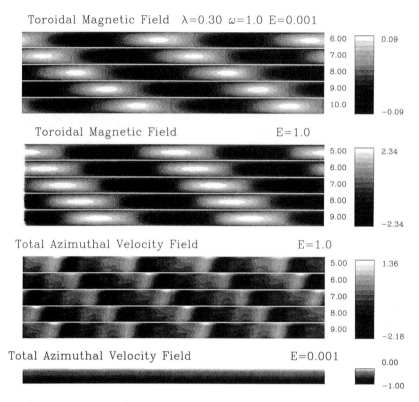

Fig. 5. Toroidal fields and differential rotation for both energy cutoffs for a succession of time intervals (two y-wavelengths shown) illustrating the propagation of the patterns in latitude y with time.

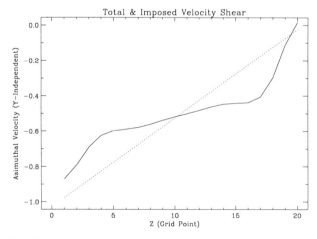

Fig. 6. Profiles of imposed and total radial velocity shear (y independent), for the E_c = 1.0 case.

linear solution, $E_c = 0.001$, would show very little change between the total differential rotation and the imposed differential rotation.

C. Discussion

It should be evident from these few examples that the nonlinear feedback by the induced magnetic fields is quite important in determining the final amplitudes and periods of the dynamo wave. And this feedback is controlled largely by the efficiency of magnetic buoyancy or other process that is responsible for removing flux from the interface layer. The induced azimuthal flow that opposes the imposed shear must arise from a balance of the Lorentz force and the viscous force in the x-momentum equation. It is an example of Lenz's law; the Lorentz force drives a current that reduces the time rate of change of the magnetic fields. It is likely there are solutions in other parts of the parameter space in which the feedbacks are even stronger than this. These remain to be explored.

Our model is highly simplified compared to the Sun, but a few comparisons seem appropriate. First, torsional oscillations are a natural consequence of the feedback, and take a rather simple form. The torsional oscillations arise in our calculations from a balance between the Coriolis force, the pressure gradient due to the free surface and the Lorentz force in the y-momentum equation. The two significant parts of the induced azimuthal velocity result from different balance conditions. However, the phase relation to the toroidal field is different than inferred from observation (Howard and LaBonte 1980). Note, however, that we have calculated the torsional oscillation at the base of the convection zone, while it is observed at the surface.

The poloidal fields and induced meridional circulation in these particular solutions are similar in amplitude to the toroidal fields and induced differential rotation. On the Sun, it appears the toroidal field is much larger than the poloidal field (which is what keeps the orientation of sunspot groups nearly east-west). It seems likely that there are other relevant dynamo solutions in which the imposed differential rotation is large compared to the α-effect, for which the toroidal field should be large compared to the poloidal field. What little observational evidence exists suggests that meridional flow associated with torsional oscillations does have similar energy to them (Snodgrass, personal communication). Finally, in our model there is a meridional flow independent of y-wavenumber that is allowed because there are no lateral walls or spherical shell to confine the flow by allowing latitudinal pressure gradients to build up. Only a spherical model can tell us whether this flow is real or an artifact.

VI. PROBLEMS RAISED BY A DYNAMO AT THE BASE OF THE SOLAR CONVECTION ZONE

In Sec. IV, we discussed the various problems associated with making a dynamo work at the interface of the solar convection zone with the interior.

In this section we suppose all those issues are resolvable, and that such a dynamo will work there. There is, then, an additional set of problems associated with getting the magnetic flux maintained by this dynamo into the convection zone and up to the surface, distributed in the form we actually see. We list several of these problems below. In most of these areas some work has been done, but there are no definitive treatments.

1. How does magnetic flux get injected into the convection zone? Magnetic buoyancy and entrainment by convection may both be important. Modeling of some aspects of this problem has been done by Cattaneo and Hughes (1988). From Sec. V's results, it is clear that the rate of this injection has profound influence on field amplitude and dynamo period.

2. Once in the convection zone, flux tubes may be driven toward the surface by magnetic buoyancy, or carried by convection. We know from modeling by Moreno Insertis (1986) that parts of a flux tube may rise to the surface while other parts remain anchored in the interface layer. Detailed interactions between rising flux tubes and convection have not been calculated, but they must be important. If magnetic buoyancy is the only driver, flux tubes will rise through the convection zone along a trajectory parallel to the rotation axis (Choudhuri and Gilman 1987) rather than radially (unless they traverse the zone extremely fast, $\sim$ a few days or less). This is due to the influence of Coriolis forces. Such tubes would emerge at latitudes poleward of where sunspots are found.

3. While rising, the flux tubes are subject to many stresses and distortions due to turbulent interactions with the ambient flow. These interactions should fragment tubes, cause some to coalesce and contribute to the observed distribution of spot sizes seen at the surface (Bogdan et al. 1987). Quantitative theories of these processes for realistic conditions are extremely difficult. Some relatively idealized calculations have been done by Bogdan and Lerche (1985), but much further work is needed. Interactions of tubes with small-scale convection (granulation and supergranulation) near the Sun's surface may be particularly important for the small end of the flux-tube size spectrum, including tubes too small in cross section to be identified as sunspots.

4. Differential rotation with latitude within the bulk of the convection zone may have important induction effects if magnetic field lines are not purely aligned with latitude circles. In fact, such latitudinal shears may force the field lines to stay in constant latitude.

5. Once at the surface, field lines may recoalesce into sunspots, or may be further fragmented into smaller tubes, and undergo random walk and meridional drift across the solar surface. Much modeling of these processes has been done, but how the field lines are connected below the surface is unknown from observations and has received very little theoretical attention. Deduction of subsurface magnetic structures from their scattering of acoustic modes has promise (Bogdan and Cattaneo 1989).

6. Observations of differential rotation at the solar surface suggest the existence of small-amplitude torsional oscillations (Howard and LaBonte 1980) that are associated with the solar cycle. Our dynamo calculations discussed in Sec. V indicate that perturbations in differential rotation of this general type are produced as a result of feedbacks from the electromagnetic body force. But such perturbations in differential rotation must be transmitted to the solar surface in a coherent form. Torsional oscillations have been cited along with other solar features as evidence of a greatly extended solar cycle (see, e.g., Wilson et al. 1988), perhaps as long as 20 yr for a single half magnetic cycle, with sunspots occurring only in the later stages. Gilman (1989) has given arguments why the evidence for such an extended cycle seems very weak. Our view is that there is no present justification for theoretical calculations specific to that possibility.

7. Sunspots are seen at the surface only in latitudes equatorward of 40°. Yet we expect dynamo action to occur at the base of the convection zone at both low and high latitudes. Is there some fundamental difference between the convection zone at the low and high latitudes that prevents sunspot emergence above 40°? Or is there diminished penetration of helicity into the interface layer at high latitudes? These questions are completely unanswered.

8. Stix (1987b) has noted that from the observed phase relation of the poloidal and toroidal fields on the surface of the Sun, the sign of the both α and $\partial\omega/\partial r$ can be independently inferred. Unfortunately the inferred signs are opposite to those used by the overshoot layer dynamos. It is not clear whether this is an indication of a problem in the basic theory of α-ω dynamos or whether the surface observations are misleading.

In closing, we remark that, after many years through which the prevailing opinion was that the problem of the solar dynamo was "solved" by mean field electro-dynamics applied to the bulk of the solar convection zone, new observational and theoretical results have now overturned that belief, leading to a stimulating new period of proliferation of solar dynamo theories. Graduate students embarking on theses and new postdoctoral students should find this a particularly stimulating time to explore the solar dynamo problem.

Acknowledgment.   The National Center for Atmospheric Research is sponsored by the National Science Foundation.

ELEMENT SEPARATION BY ATOMIC DIFFUSION

G. MICHAUD
Université de Montréal

and

S. VAUCLAIR
Observatoire du Pic-du-Midi et de Toulouse

Atomic diffusion is a basic physical process and it should, in principle, be included in the standard solar model. In order to understand its possible importance in evolutionary calculations, the various physical processes that cause atomic diffusion are discussed in turn. Uncertainties in the atomic diffusion coefficients are considered. It is, in particular, shown that shielding effects reduce considerably the contribution of thermal diffusion. For most elements, the gravitational settling time scale from the convection zone is shown to be about 1.5×10^{10} yr. The diffusion velocity is smaller close to the solar center. At the Sun's age, the abundances in the convection zone could be reduced by some 16% while the abundance of He could be increased at the solar center by about 2%. These results are in better agreement with the recent numerical calculations of Cox, Guzik and Kidman than with the earlier results of Noerdlinger or those of Wambsganss. A detailed comparison of the various diffusion coefficients used by these authors and their effect in the same evolutionary code is instructive. Atomic diffusion also appears to play a role in the outer layers of the Sun. The correlation between the first ionization potential and abundance anomalies observed in the solar wind and corona suggests that atomic diffusion is involved. Because oxygen and helium must be neutral, the separation must occur at low temperature. In spite of the large-scale motions observed in the chromosphere and in the transition region, it appears that chemical separation does occur there. Models that suggest chemical separation in the solar wind are also discussed, especially for helium.

I. THE ASTROPHYSICAL CONTEXT

It was thought some sixty or seventy years ago that the solar chromosphere was largely supported by radiation pressure on calcium (Milne 1924,1925; Eddington 1926,§252–254). Radiation pressure was the mechanism then imagined to lead to the observed extension of the solar chromosphere. However, it should then also lead to considerable element separation while observationally it was soon evident that the chromosphere was not made up of pure calcium. The model was modified to minimize the selective effects of radiation pressure (Chandrasekhar 1934a,b) while keeping its essential aspects. It is now easy to verify that radiative acceleration cannot support the whole mass of the chromosphere if it has a normal calcium/hydrogen ratio. Alternates involving hydrodynamical motions were soon considered (McCrea 1929,1934).

At the same time, Eddington strongly argued that chemical separation could not play a major role in modifying photospheric solar abundances (Eddington 1926,§199). He claimed (§193) that the presence of the light and heavy elements in stellar spectra showed that the extreme separation of the elements, that diffusion can lead to, had not occurred. He went on to suggest that this could be explained both by the time it takes for chemical separation to arrive at equilibrium and by the presence of meridional circulation opposing separation. His arguments against chemical separation in the solar interior, and in stellar interiors in general, were so largely accepted that little work on the subject was done until the 1960s. Some research, however, was done on gravitational separation in outer solar layers (Biermann 1937; Wasiutynski 1958).

It was shown by Chapman (1958) that the thermal-diffusion coefficient for ionized elements could be much larger than that for neutral elements originally discussed by Chapman (1917). Aller and Chapman (1960) then suggested that this could add significantly to the contribution of gravitational settling and they re-evaluated the effect of diffusion below the solar convection zone. They introduced the thermal-diffusion term into a diffusion velocity equation and evaluated the appropriate values of the diffusion coefficients. They showed how additional fields of forces could be introduced into the diffusion equation but, surprisingly, they did not discuss the possible role of differential radiative pressure (Michaud 1970). They concluded that diffusion could reduce the superficial abundance of heavy elements by 10 to 25% while, close to the center, their abundance could increase by about 1%. Delcroix and Grevesse (1968) extended the work of Aller and Chapman (1960) by evaluating the diffusion velocity below the solar convection zone for essentially all chemical elements.

In this chapter, we review the properties of the diffusion equation and in particular the properties of the diffusion coefficients, as well as the time scale for abundance variations (Sec.II), in order to understand the effect of atomic diffusion in solar evolutionary models (Sec.III). We then review the evidence for abundance separation in outer solar regions and the models that have been suggested to explain this separation (Sec.IV). A more complete review on many aspects of diffusion calculations may be found in Vauclair and Vauclair (1982).

II. THE PHYSICS OF SOLAR DIFFUSION

A. Diffusion Velocity Equation

The diffusion velocity may be written for a trace element

$$v_D = -D_{12}\left\{ \frac{\partial \ln c}{\partial r} + \left[\left(A - \frac{Z}{2} - \frac{1}{2}\right)g - Ag_R\right]\frac{m_p}{kT} - k_T\frac{\partial \ln T}{\partial r}\right\} \quad (1)$$

where D_{12} and k_T are, respectively, the atomic and thermal diffusion coefficients and may be found in Paquette et al (1986a). When the element of interest of atomic mass A and charge Z is a trace element, the concentration $c = n(A)/n(H)$. The dominant element is assumed to be hydrogen. The first term appearing on the right-hand side is the "classical" diffusion term. It is usually unimportant in stars since concentration gradients remain relatively small and equilibrium abundances are rarely reached. The second term represents gravitational settling while the third term is due to the radiative acceleration caused by the photon flux and the last term to thermal diffusion.

A more general equation, valid when the diffusing element is not trace and the dominant element not necessarily hydrogen, is given by Pelletier et al. (1986). This equation can also be used in the presence of electron degeneracy. An alternate treatment, based on Burgers (1969), has been used by some authors (Noerdlinger 1977,1978; Iben and MacDonald 1985; Cox et al. 1989). It is easily included in evolutionary calculation codes. It allows treating elements in any proportion but complicates the discussion of the physical processes involved. The accuracy gained is probably illusory since the main uncertainty in these calculations comes from the diffusion coefficients. Indeed, Noerdlinger (1978) carried out a detailed comparison with Montmerle and Michaud (1976) for He III and obtained essentially the same result as they did for gravitational settling.

B. Diffusion Coefficients

The "classical" evaluation of diffusion coefficients (Chapman 1958; Aller and Chapman 1960; Chapman and Cowling 1970) assumes that the density is small enough that the natural logarithm of Λ (the ratio of the Debye length, r_D, to the impact parameter for a deflection by 90°; see, e.g., Spitzer 1962) is large and can be taken out of integrals over energy. Assuming asymptotically small densities overestimates the thermal diffusion coefficient by

20% even at solar corona densities (see Table 10 of Paquette et al. 1986a). The effect is much larger below the solar convection zone. In a solar model calculated with $\alpha = 1.1$ (the ratio of the mixing length to the pressure scale height) and the Cox and Stewart (1970a) opacities, thermal diffusion dominates atomic diffusion if one uses the classical evaluation. Taking shielding into account reduces thermal diffusion by a factor of up to 10 on certain elements (see Table I), making it less important than gravitational settling. This effect reduces the total diffusion time scale by a factor of 1.3 (Table 1). Taking shielding effects into account leads to larger values of D_{12} but smaller values of k_T. Consequently, thermal diffusion is generally not the dominant process in stellar interiors. To obtain accurate diffusion velocities, it is *essential* to use the results of Paquette et al. (1986a) for both the atomic-diffusion coefficient and the thermal-diffusion contribution.

While a detailed description of the physics of diffusion coefficients is outside the scope of this chapter, it is worth describing briefly the origin of the effect of using shielded potentials. In the classical approximation, pure Coulomb potentials are used but the integrations to obtain collision frequencies are stopped at a distance r_D, in order to avoid the divergence that $1/r$ potentials lead to. This causes an energy dependence for Λ, but to simplify calculations the value of Λ is assumed independent of the energy of the interacting particles in the classical treatment. This leads to an overestimate of the contribution of the low-energy collisions thereby reducing D_{12} but increasing k_T. This is the main cause of the difference between the results of Paquette et al. and the classical results in main-sequence stars. The proper energy dependence was included in the calculations of Paquette et al. The use of shielded potentials has other effects but these dominate only at densities higher than encountered in main-sequence stars.

It can be argued that the Paquette et al. (1986a) calculations are highly uncertain in the solar interior because the Debye-Huckel potential they use may be a poor representation of the ion-ion interaction at those densities. In Fig. 1, the helium self-diffusion coefficient obtained using their calculations

TABLE I
Diffusion Time Scales θ Below the Solar Convection Zone

Element	Paquette et al. (1986a)		Low-Density Limit	
	θ(Gyr)	f[a]	θ(Gyr)	f
He	26	0.27	25	0.80
Li	25	0.30	24	0.92
Ca	34	0.50	25	3.4
Fe	31	0.43	23	3.2
Hg	16	0.20	16	2.2

[a] The quantity f is the ratio of the thermal diffusion to the gravitational settling velocity.

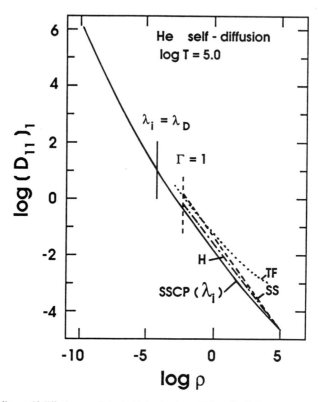

Fig. 1. Helium self-diffusion coefficient obtained using a Debye-Huckel potential (full line) and a Thomas-Fermi potential (dotted line) by Paquette et al. (1986a), compared to the Monte Carlo studies of Hansen (1973; dashed line), and the metallic helium model of Stevenson and Salpeter (1977; point-dashed line). In the shielding potential, Paquette et al. used the larger of the Debye length λ_D and the inter-nucleon distance λ_i. The density where they are equal is indicated by a vertical line. The line $\Gamma = 1$ identifies where the classical distance of closest approach equals the average inter-particle distance.

is compared to those obtained in the Monte Carlo studies of Hansen (1973) and the metallic-helium model of Stevenson and Salpeter (1977). The agreement is very good even at the high-density limit, which is much higher than solar densities. There exists no similar high-density calculation of thermal diffusion to compare with the results of Paquette et al. Since the corrections to the low-density coefficient are larger for thermal diffusion than for D_{12}, this is worrisome and it would be crucial to have further research on this point, especially for applications to white dwarfs.

It had been noted by Noerdlinger (1977) that $\ln \Lambda$ is small in the solar interior and that this lead to considerable uncertainty in the diffusion velocity. He obtained that $\ln \Lambda$ was not much larger than 1 in most parts of the solar model. He correctly pointed out the need for more precise evaluations of the atomic-diffusion coefficient, D_{12}. He did not realize, however, that the effect on the thermal-diffusion coefficient is even larger. He used the low-density limit of the thermal-diffusion coefficient and this led to his overestimating considerably the thermal diffusion. He noted that his value is a factor of 1.5 larger than the low-density Aller and Chapman (1960) limit which is larger than values obtained when shielding effects are taken into account. He also noted that ln Λ is so small that one could argue that it be reduced by a further factor of 5 and he evaluated what effect this would have. From our results with the shielded potentials, it does not, however, appear that diffusion velocities increase so dramatically when ln Λ is reduced and the uncertainty is probably smaller than suggested by Noerdlinger.

There has been some confusion in the literature as to the proper value of the low-density limit of the thermal-diffusion coefficient. While Aller and Chapman (1960) obtained 2.65 for the constant multiplying the temperature gradient, Burgers (1960, 1969) obtained 3.45. Roussel-Dupré (1981) explains this discrepancy by Burgers properly adding the electron contribution to the constant in the ternary mixture while Aller and Chapman did not properly compensate for the presence of the electrons in their binary gas equation. Other corrections, such as the effect of the finite mass of the electron and the inclusion of other terms in series developments, however, reduce the constant to approximately the value of Aller and Chapman (1960; see also Roussel-Dupré 1982). The effect of shielding described above reduces the value of the constant by a much larger factor in the solar interior. The results of Paquette et al. may be used with either the Burgers or the Chapman and Cowling formalism.

Monchick and Mason (1985) calculate "dynamic shielding" and quantum corrections on the atomic and thermal diffusion coefficients. For the atomic-diffusion coefficient, one should first note that only the ion-ion diffusion is important since the electron-ion diffusion coefficient is always large enough for electrons not to play a role in slowing down diffusion. However, both the electron-ion and ion-ion thermal-diffusion coefficients are important. The interesting results are contained in Monchick and Mason (1985; their Table 2) where it appears that, in the solar interior, the corrections they calculate could be important. However, they note that, where these corrections are important (last value of n_e in their Table 2), the theory they use does not apply. Therefore, they only establish that such corrections are not excluded and further work is needed to establish that they are really important.

Thermal diffusion can dominate at the base of the solar corona, as will be seen below, because of the large temperature gradient there. Furthermore, Geiss and Burgi (1986a,b,c) have shown that partial ionization, as present at the base of the corona, can considerably enhance the effect of thermal diffusion.

C. Radiative Acceleration

As they stream out of the star, photons interact preferentially with certain elements that have large absorption cross sections where the photon flux is maximum, thereby creating a differential radiative acceleration. When the star arrives on the main sequence, it is assumed to be thoroughly mixed and that the concentration gradient is equal to zero in the outer envelope. If the radiative acceleration on an element with atomic mass number A is greater than the combined effect of gravity and thermal diffusion, diffusion is upward and the concentraiton of element A increases in superficial layers; otherwise, its concentration decreases. To estimate the potential effects of radiative acceleration, it is useful to evaluate it under the most favorable assumptions. We assume that element A is concentrated in states that all have lines with f ~ 1 (f is the oscillator strength) and that these lines have frequencies close to the frequency of the maximum of the photon flux. We further assume that those lines are not saturated, so that the flux at the line frequency is not significantly reduced from that in the continuum. Under those very favorable conditions (Michaud et al. 1976) the radiative acceleration is

$$g_R = 1.7 \times 10^8 \frac{T_{\text{eff4}}^4 R^2}{A T_4 r^2} \tag{2}$$

in cm s^{-2} where R is the stellar radius and r the radius where the evaluation is made. The temperatures (T_4) are in units of 10^4 K, so that T_{eff4} is the effective temperature in units of 10^4 K. At the solar surface, g_R would be 40 times larger than gravity for an iron peak element ($A = 50$) if its lines were not saturated. It is 10 times smaller than gravity below the convection zone (at $T_4 = 200$). In the Sun, the radiative acceleration could play a role at the surface where it can be larger than gravity (see also Sec. IV below) but not below the convection zone, where $T_4 > 100$. This is consistent with the rough evaluation of Sienkiewicz et al. (1988) described below (Sec.III) and this explains why the effect sought by these authors is negligible. Note that Eq. (2) is sometimes a fairly good approximation at very high temperatures ($T_4 > 200$).

D. Settling Time Scale

Below the solar convection zone, diffusion is downward and progressively empties the surface of its heavy element ($A > 1$) content. Applying element conservation one obtains

$$\frac{d}{dt} (c \, \Delta M_{CZ}) = c v_D \rho \tag{3}$$

where ΔM_{CZ} is the mass in the convection zone (CZ) in g cm^{-2}, and the diffusion velocity v_D is positive upward. Equation (3) leads to

$$c = c_0 \exp(-t/\theta) \text{ with } \theta = -\Delta M_{CZ}/\rho v_D. \tag{4}$$

The time scales are given in Table I. The solar model used here was computed with $\alpha = 1.2$, and the Cox and Stewart (1970a) opacities. Its convection zone has a mass of 0.017 $M_\odot$ and the temperature at its bottom is 2×10^6 K. Settling time scales calculated using both the calculations of Paquette et al. (1986a) and the low-density limit of Aller and Chapman (1960) are shown. The settling time depends rather sensitively on the mass in the convection zone. It scales as (Michaud 1977)

$$\theta = 2.6 \times 10^{10} (\Delta M/0.017)^{0.545} \tag{5}$$

in yr where ΔM is the mass in the convection zone in solar mass units. The settling time scale varies little from element to element except for elements heavier than the iron peak. In this solar model, assuming no mixing below the convection zone, most elements are underabundant by a factor of about 0.8 (compared to their original abundance) at an age of 4.6 Gyr. However, this result depends rather sensitively on the mass in the convection zone.

More precisely, it is the mass in the mixed zone that matters. If there is any turbulence below the convection zone, the mixed mass is considerably increased. For instance, the fact that Li is underabundant by a factor of 100 while Be has about normal abundance suggests that the mixed mass extends to temperatures in excess of 2.5×10^6 K but less than 3.5×10^6 K. Assuming complete mixing, this increases the mixed mass by a factor of about 3 and the diffusion time scale by a factor of about 1.8. This in turn leads to underabundance factors of only 0.9. In reality, the Li underabundance is probably caused by the turbulent transport of the Li in the convection zone to depths where the temperature is high enough for Li to burn. The turbulent particle-transport coefficient probably varies with depth (Cayrel et al. 1984; Baglin et al. 1985; Zahn 1987; chapter by Sofia et al.). To what extent increasing the convective zone mass, as done above, reproduces the effect of turbulence on both the gravitational settling of Li and its burning has never been studied in detail. The most thorough discussion may be found in Vauclair et al. (1978).

Opacities have a large effect on the mass of the convection zone. If one adjusts the value of α so that the bottom of the convection zone is always at the same temperature, increasing the opacity by a factor of 1.5 (Magee et al. 1975) reduces the mass in the convection zone by a factor of about 1.2 and reduces all diffusion time scales by the same factor.

At the bottom of the solar convection zone, our two evaluations of the time scale give about the same result. This comes from a cancellation of two rather large effects. As can be seen in Table I, the value of f, the ratio of the thermal diffusion to the gravitational settling velocity, can be up to 10 times smaller in the Paquette et al. results compared to the low-density limit; this, however, is canceled by an increase of the atomic-diffusion coefficient. That the cancellation of the two effects should be so close is a coincidence here and is not generally the case. For instance, the total diffusion time scale below the convection zone of white dwarfs calculated using the Paquette et al. (1986a) results is smaller by orders of magnitude than the low-density limit (see Paquette et al. 1986b).

E. Transport Velocities

To understand whether or not atomic diffusion should be expected to have large effects on solar abundances, it is useful to compare atomic-diffusion velocities to other transport velocities in stellar envelopes. Only a very brief description of the origin of the other velocities is given here.

The estimates of Tassoul and Tassoul (1984a,b) will be used for meridional circulation. They took into account the presence of a μ-gradient. Their results confirm those of Mestel (1965) that the μ-gradient reduces the circulation. Results shown in Table II assume that the initial rotation period of the Sun was 10 days.

Turbulent transport is discussed in the chapters by Pecker, by Sofia et al. and by Schatzman. To allow a comparison with other types of particle transport, the turbulent-diffusion velocity can be represented by

$$v_T = D_T \left(- \frac{\partial \ln c}{\partial r} \right) = P_* D_{12} \left(- \frac{\partial \ln c}{\partial r} \right) \qquad (6)$$

where D_T is the turbulent-diffusion coefficient taken to be P_* times larger than the atomic-diffusion coefficient D_{12}. Particle transport by turbulent diffusion, as calculated here, is more uncertain in so far as D_T is only assumed. Since D_T is the same for all elements but D_{12} varies from element to element, P_* also varies from element to element. From the discussion presented in Michaud (1985), it appears that $P_* > 400$ is excluded for He. For the numerical

TABLE II
Transport Velocities (cm s^{-1})

$r/R_\odot$	$v_D(\text{He})$	$v_D(\text{Li})$	v_T[c]	v_w[d]	v_{mc}[e]
0.05[a]	2×10^{-10}	—	-10^{-9}		10^{-14}
0.125[b]	-10^{-9}	—	-10^{-8}		10^{-13}
0.74	-10^{-8}	-10^{-8}	-10^{-7}	10^{-10}	6×10^{-10}

[a]For ^{4}He. [c]Turbulent velocity. [e]meridional circulation velocity.
[b]For ^{3}He. [d]Wind velocity.

examples presented here, it is assumed that P_* = 20 for He (this corresponds to Re_* = 5 in Michaud 1985) but 40 for Li. The abundance gradients will be read from solar models except for Li whose abundance will be assumed to vary by a factor of 10 over a pressure scale height.

The mass loss rate, dM/dt, is assumed small enough to have negligible effect on the solar structure. Its main effect is to cause a global outward velocity of matter v_w through a static model

$$\frac{dM}{dt} = -4\pi r^2 \rho v_w. \tag{7}$$

It expresses the mass conservation of the main constituent. A trace element diffusing in the presence of mass loss must satisfy the conservation equation

$$-N_H \frac{\partial c}{\partial t} = \frac{1}{r^2} \frac{\partial}{\partial r} [r^2 c N_H (v_w + v_d)] \tag{8}$$

where it is assumed that the only other transport process is atomic diffusion and where the diffusion velocity is given by Eq. (1). In the quantitative estimates, the measured solar wind value ($10^{-14} M_\odot \, yr^{-1}$) is used.

Estimates of the various transport velocities at a few places within the Sun are given in Table II. No value is given within the convection zone itself, where the turbulent velocity related to the convection is assumed to be very large and maintains very nearly uniform abundances. Velocities were evaluated immediately below the convection zone and where the ^{3}He ($r/R_\odot$ = 0.125) and ^{4}He ($r/R_\odot$ = 0.05) diffusion could be most important. It should first be noted that turbulence dominates all other velocities by at least one order of magnitude. This is assuming that the turbulent-diffusion coefficient is 20 times the atomic-diffusion coefficient for He and is consistent with values for D_T evaluated elsewhere (Zahn 1987; chapter by Sofia et al.). It is 10 times smaller than the turbulence that significantly affects evolutionary models. If present, turbulence dominates over all other transport processes; it should be present at some level. [See the estimates of Baglin et al. (1985) and Cayrel et al. (1984) based on their models for Li destruction. The values they obtain depend rather sensitively on the depth of the convection zones.] Meridional circulation and mass-loss velocities are always much smaller. Atomic diffusion is the most important of the transport processes that we can calculate from first principles. Given that, to have large effects, elements must be transported over a significant fraction of a scale height, atomic diffusion can have effects by only a few percent in the solar interior ($r/R_\odot$ = 0.05 or 0.125).

III. CHEMICAL SEPARATION IN EVOLUTIONARY
SOLAR MODELS

The standard solar model should in principle incorporate all basic physical phenomena including atomic diffusion. Most evolutionary calculations have been carried out without atomic diffusion because it is assumed to play a minor role in stellar evolution. While this is true, it has some effect on the internal structure of the Sun, on its pulsations and on the neutrino flux; so atomic diffusion cannot arbitrarily be neglected without either realizing the error made or without introducing competing processes that are assumed to eliminate its effect. In this section the influence of atomic diffusion on the evolution of a solar model is discussed, assuming that it is unimpeded by macroscopic motion.

Solar models calculated without diffusion differ from each other because of different input physics (Turk-Chièze et al. 1988). In order to investigate the effect of diffusion, we will compare the helium abundance in two models calculated by the same author which only differ by the inclusion of diffusion.

Three solar evolutionary models will be discussed, those of Noerdlinger (1977), of Wambsganss (1988) and of Cox, Guzik and Kidman (1989; hereafter CGK). In comparing evolutionary models, it must first be realized that Noerdlinger compared a solar model calculated without any effect of diffusion and using the radius and luminosity of the Sun at an age of 4.5 Gyr to an evolutionary sequence calculated with exactly the same parameters and abundances but including diffusion effects. The model with diffusion has a luminosity that is 1.5% larger than the solar luminosity at the age of the Sun; its radius is 1.7% larger. The diffusion model of Noerdlinger is not a properly converged solar model with diffusion but it permits evaluating the effects of diffusion in a solar model. On the other hand, CGK and Wambsganss (1988) compare two properly converged solar models, one without diffusion and the other with diffusion. In order to have the solar radius and luminosity at an age of about 4.5 Gyr forces a change in α (from 1.894 to 1.951 for CGK, and from 1.64 to 1.75 for Wambsganss) and original He abundance Y (from 0.291 to 0.289 for CGK, and from 0.2699 to 0.2679 for Wambsganss). Much of the input physics is also different. In particular, different opacities were used. The two CGK models have slightly different solar ages. That without diffusion had an age of 4.66 Gyr and that with diffusion an age of 4.54 Gyr; this is probably close enough not to affect the comparison.

All series of calculations overestimate the effect of thermal diffusion since all use the low-density limit of the thermal-diffusion coefficient. Since CGK have used the shielded values of the atomic-diffusion coefficient D_{12} while keeping the low-density limit of the thermal-diffusion coefficient, they overestimate the diffusion velocity by a factor of about 1.5 below the convection zone. They use the larger coefficient both for k_T and D_{12}, while partial cancellation occurs when the two coefficients are calculated with either the low-density limit or the shielded potentials (see Sec.II.B).

The most direct effect of diffusion is on the helium abundance in the convection zone and at the center. They are compared for the three models in Table III. The models of CGK and of Noerdlinger agree that the He abundance in the convection zone is reduced by about 13%. This is also in reasonable agreement with the estimate of Sec.II.E where a 16% reduction was obtained. However, Wambsganss obtains only a 4% reduction of the He abundance in the convection zone and this is in contradiction with all other results. CGK also calculate a settling of heavy elements from the convection zone by 10% (their Z is reduced from 0.0200 to 0.0179) which is a somewhat smaller change than the 16% estimated in Sec. II.D. Further contradictions appear at the solar center. While Noerdlinger obtains a change of $\Delta Y = 0.04$, CGK obtain a change of $\Delta Y = 0.0044$, and Wambsganss obtains $\Delta Y = 0.0116$ if one takes into account the difference in the original He abundance. The results of CGK and Wambsganss appear in better agreement with our discussion in Sec.II.E since there the diffusion velocity at the solar center was calculated to be more than 10 times smaller than below the convection zone. Since the distances that have to be traveled within the solar model are of the same order in both cases, it would be difficult to understand how the effect in ΔY could be the same, as obtained by Noerdlinger. Noerdlinger may have overestimated the effect of diffusion at the solar center.

Given the small final effect of diffusion on the central He abundance, it is not surprising that it has a small effect on the original He abundance required and on the B^8 and Be^7 neutrinos. According to CGK, the original He abundance only needs to be changed from 0.291 to 0.289 to obtain a solar model at about the age of the Sun. In SNUs, the B^8 neutrinos increase from 12.6 to 13 and the Be^7 neutrinos from 1.9 to 2.0. These increases are small compared to other uncertainties. The changes are small because the solar

TABLE III
Comparison of Helium Abundances, Y

	Noerd.[a] Without diffusion	Noerd.[a] With diffusion	CGK[b] Without diffusion	CGK[b] With diffusion	Wamb.[c] Without diffusion	Wamb.[c] With diffusion
Original	0.23	0.23	0.291	0.289	0.2699	0.2679
Final (CZ)	0.23	0.20	0.2912	0.2556	0.2699	0.2563
Final (Center)	0.70	0.74	0.6556	0.6580	0.6169	0.6265

[a] Noerdlinger (1977) model.
[b] Cox, Guzik and Kidman (1989) model.
[c] Wambsganse (1988) model.

luminosity at a given age was imposed onto the model. As evolution continues, the effects would probably become larger but CGK did not present results for later periods. The changes that Noerdlinger obtained were much
larger.

It was noted by Noerdlinger (1977) that the diffusion in the central parts
of the Sun was speeded up as time passed because of the increasing temperature gradient in the solar interior. In Sec.II.B, it was discussed that thermal
diffusion is overestimated by the classical diffusion coefficients used by
Noerdlinger. It was noted that while more exact calculations lead to a large
decrease of the thermal-diffusion coefficient, the effect was largely canceled
under the convection zone by the increase of the atomic-diffusion coefficient.
However, deep in the Sun there is also an abundance gradient term. The He
abundance is increasing inward due to nuclear burning. This opposes the
settling of He. The efficiency of this abundance gradient term depends on the
size of the driving terms in the diffusion equation. If the thermal diffusion
term is larger, the abundance-gradient term is less efficient in opposing downward diffusion. Below the convection zone, the comparison of settling time
scales was made assuming negligible contributions of the abundance-gradient
term. If, however, the abundance gradient should be nearly able to stop diffusion with the thermal-diffusion coefficients as calculated by Pelletier et al.
(1986), this may not be true if thermal diffusion were larger, as in the low-
density limit approximation. Consequently, when the low-density limit of the
diffusion coefficients is used, diffusion might be considerably overestimated
in the central solar regions, where there are large gradients in the He abundance.

It has been imagined by Sienkiewicz et al. (1988) that radiation pressure
could solve the solar neutrino problem by depleting the deep solar interior of
its metals, thereby reducing the opacity there. Because of the lower opacity,
the central temperature would be reduced and so the neutrino flux would be
reduced. These authors argued that, for metals, the ratio of the radiative acceleration to gravity could be much larger in the central solar regions than at
the surface because, in the energy generation region, the L/M ratio is 10 times
larger than at the surface. Since the dominant opacity in the center is due to
metals, all that radiative acceleration would go to them and might be sufficient to counter gravity on metals. However, in their detailed calculations,
they find that the radiative acceleration on metals is always more than 10
times smaller than gravity, in agreement with our discussion in Sec.II.C,
above. Metals then cannot be pushed out of the center. Furthermore, their
analysis was only based on the value of the radiative acceleration and neglected the time scale for the abundance variations to materialize. Since the
radiative acceleration would at most have been of the same order as gravity,
the diffusion time scale would at best have been about the same as the settling
time scale. As discussed above, this is too long for the heavy-element abundances to be modified by more than a few percent at the solar age. Even if

the radiative acceleration had been larger than gravity, this model was condemned not to reduce significantly the metal abundance and so the opacity.

In principle, it is straightforward to test observationally the direct effect of diffusion on superficial abundances. In practice, however, the effects are small enough (10–20%, see above) that it is not possible to develop a meaningful test. The neutrino problem would be enhanced by the settling of He toward the center. As to the pulsation modes, the comparison of CGK is inconclusive because the disagreements with standard models are still large and the agreement is not significantly improved by diffusion. Only when both the diffusion calculations and the pulsation calculations have substantially improved will a meaningful comparison be possible.

IV. CHEMICAL SEPARATION IN THE OUTER SOLAR LAYERS

A. Observational Evidence

Before precise chemical abundances were measured in the outer solar layers (corona, solar wind, solar energetic particles) and compared to the photospheric chemical composition, it seemed unlikely that chemical separation could occur in regions where large macroscopic motions were present. Delache (1965,1967), Jokipii (1965,1966) and Nakada (1969) studied element segregation in the transition region, as caused by thermal diffusion in the steep thermal gradient. They showed that it could lead to large overabundances of heavy ions in the corona (see also Dupree 1972). As such overabundances were not observed, it was later concluded that macroscopic motions prevented at least partly this diffusion.

The first real evidence of element separation in outer layers of the Sun came from helium measurements in the solar wind. While variations of helium abundance by > 2 orders of magnitude are observed, the average ^{4}He/H $\simeq 4 \times 10^{-2}$, i.e., 2 to 3 times smaller than the standard value (a good discussion of the first observations of helium abundance in the solar wind can be found in Hundhausen [1972]). Space measurements have made available more complete and continuous records of the ^{4}He/H and ^{3}He/^{4}He values. Summaries of these measurements may be found in Gosling et al. (1981) and Borrini et al. (1982b). Values of He/H between 0.001 and 0.8 have been reported. The helium abundance appears strongly related to the large-scale structure of the plasma. Large He values (He/H > 0.15) are generally associated with interplanetary shocks and large solar flares (Bame et al. 1979, 1981; Gosling et al. 1980; Borrini et al. 1982a,b). Very low helium abundances appear to accompany the polarity reversal in the interplanetary magnetic field (Borrini et al. 1981). Helium abundances in coronal holes associated with high-speed streams show a nearly constant value of 0.048, while interstream values vary by a factor 50 with an average of 0.038 (Bame et al. 1977). The overall He/H ratio may show a long-term variation with solar activity (Ogilvie and Hirshberg 1974; Feldman et al. 1978).

The ^{3}He abundance, first measured in the solar wind by Bame et al. (1968), was thoroughly studied by Coplan et al. (1984) using four years of continuous data from the ISEE-3 spacecraft. They found an average ^{4}He/^{3}He = 2050 ± 200, confirming previous estimates (Geiss et al. 1972; Hall 1975). The ^{4}He/^{3}He ratio is larger at low ($\simeq$3000) than at high velocities ($\simeq$ 2000). It is also more nearly constant for high than for low velocities which suggests that the formation of the slow solar wind is more complex than that of high-speed streams. Very large ^{3}He abundances are observed in some solar energetic-particle events (Anglin 1975; Hurford et al. 1975; Kocharov and Kocharov 1984; Reames and Lin 1985; Reames et al. 1985; Kahler et al. 1987) but not in the solar wind (Coplan et al. 1984).

In spite of the large helium abundance variations in the outer layers of the Sun, the solar wind and the solar energetic particles, the average ^{4}He depletion and ^{3}He/^{4}He enhancement compared to the photosphere suggest that a separation process occurs in the wind, as discussed extensively by Geiss (1982). Helium suffers fewer Coulomb collisions than heavier ions due to its smaller charge. (More precisely, for most heavy elements the Z^2/A ratio is > 1, its value for He III). Heavy ions are more efficiently dragged by the wind than helium so that helium can be depleted in the wind even when heavy ions have normal abundances. The abundances of heavier ions also vary significantly in the solar wind even under quiet conditions. Some correlations have been determined, for example, between Fe/H and Si/H (Bame et al. 1975), Fe/H and He/H (Mitchell et al. 1983), and Ne/H and ^{3}He/^{4}He (Geiss et al. 1972). They are systematically enhanced in ^{3}He-rich solar energetic-particles events (Kahler et al. 1987).

An extensive review of observed average element abundances in the solar photosphere, the solar corona, the solar wind, the solar energetic particles and the galactic cosmic rays has been given by Meyer (1985a,b). Although the composition of the solar energetic particles is highly variable, it has been known for some time that it is influenced by the first ionization potential (FIP) of the elements: low-FIP elements are systematically more abundant than high-FIP ones (Geiss 1982; Geiss and Bochsler 1985; Meyer 1985a,b; Fan et al. 1984; Bochsler et al. 1986; Schmid et al. 1987). Meyer (1985a) presented a review of the observational results (see also Appendix by Grevesse and Anders). A detailed study of the available information enabled him to stress the existence of a constant baseline for the chemical composition of the solar outer layers, the solar wind, the solar energetic particles and galactic cosmic rays. His study confirmed: (1) that the average ionic abundances are the same in the corona, solar wind and solar energetic particles; and (2) that a correlation exists with the first ionization potential: the elements with a FIP $\gtrsim$ 9 eV are, with respect to the solar photosphere, systematically depleted by the same factor ($\sim$ 4) in all the observed sites (Fig. 2). This observational result first suggests that galactic cosmic rays originate in the winds of solar type stars: they have similar compositions. Second, it confirms the suggestion of

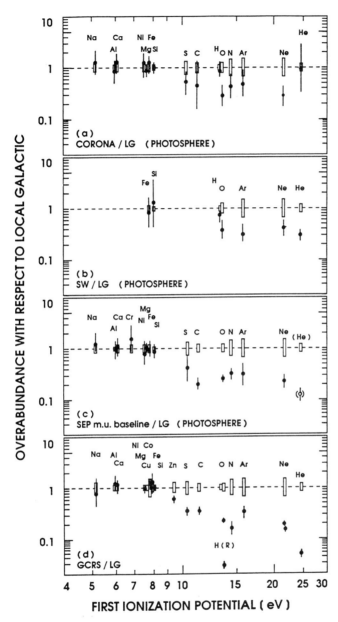

Fig. 2. Relative abundances of the elements in the solar corona, solar wind, solar energetic particles (SEP) and galactic cosmic rays vs their first ionization potentials. Each plot is normalized to the weighted average of the elements with FIP < 9 eV (figure after Meyer 1985a).

Geiss (1982), that some separation between neutral and ionized elements occurs where the elements with FIP < 9 eV are ionized while those with > FIP are still neutral, namely at a temperature of around 8000 K. Such a temperature is found in the chromospheric plateau, below the transition zone. Computations of element ionization in a homogeneous chromospheric model by Vernazza, et al. (1981; their model C and tables 17 to 24) show indeed that, in the chromospheric plateau, where hydrogen is neutral (between 1000 and 2000 km above the photosphere) carbon (FIP $\simeq$ 11 eV) is still at least 50% neutral while silicon (FIP $\simeq$ 8 eV) is mostly ionized (see also Geiss and Bochsler 1985). The separation in the solar chromosphere is discussed in Sec.IV.B, while the element separation in the wind is discussed in Sec.IV.C.

B. Diffusion in the Solar Chromosphere

Several models have been proposed to account for the physical separation between neutral and ionized atoms in the solar chromosphere (Geiss 1982; Geiss and Bochsler, 1985, 1986; Vauclair and Meyer 1985).

The large temperature gradient at the base of the corona suggests that thermal diffusion might be involved. It is known (Burgers 1960,1969; Chapman 1917,1958; Chapman and Cowling 1970; Ferziger and Kaper 1972) that heavy ions in an ionized gas of smaller atomic mass diffuse towards the hotter region. The thermal-diffusion coefficient nearly vanishes when collisions between ions and neutral atoms are involved (ions diffusing in a neutral medium or neutrals diffusing in a plasma). It is thus expected that, in the presence of a large thermal gradient in a partially ionized medium, ions will diffuse towards the hotter region while neutrals will suffer a negligible effect. Furthermore, the thermal-diffusion coefficient for charged ions in a plasma, k_T (see Eq. 1), is nearly proportional to Z^2 in low-density plasmas (Chapman 1958; Burgers 1960; Paquette et al. 1986a). As the atomic diffusion coefficient D_{12} varies proportionally with Z^{-2}, the thermal-diffusion time scale is nearly the same for all ions. Thermal diffusion could then, perhaps, account for the enhancement of low-FIP elements in the corona compared to the photosphere: ions would diffuse upward relatively to neutrals. However, a strong thermal gradient rapidly ionizes even high-FIP elements so that the separation can occur only in a very thin layer. An equilibrium state is reached for the relative abundances of the elements when the diffusion velocity vanishes, which occurs for (see Eq. 1)

$$\frac{\partial \ln c}{\partial r} = k_T \frac{\partial \ln T}{\partial r}. \tag{9}$$

The separation must take place between about 7000 and 9000 K above which even high-FIP elements are ionized. With $k_T \simeq 2.6$, this leads to a relative concentration difference of $d \ln c \simeq 0.7$, a factor of 2 too small to account for the observations. This argument applies whatever the details be of the model, so long as it is based on thermal diffusion.

Another argument could be used against thermal diffusion. Where oxygen is neutral, so also is hydrogen; therefore the high-FIP elements must diffuse with respect to a largely neutral hydrogen gas. The thermal-diffusion coefficient is then very small. However, Geiss and Burgi (1986a) made a detailed study of thermal diffusion in a partially ionized medium and showed that, for specific relative concentrations of ions and neutrals, the thermal-diffusion coefficient of partially ionized elements could be strongly enhanced. Although this effect increases thermal diffusion, it has not yet been used in a model involving diffusion between low- and high-FIP elements. It clearly must be used in any quantitative model but it is not clear how it would increase the relative separation of high- and low-FIP elements.

More recent models are based on the magnetic structure of the chromospheric regions. Von Steiger and Geiss (1985, unpublished, as quoted in Geiss and Bochsler 1986) proposed that neutrals could diffuse horizontally across the vertical magnetic field lines surrounding the spicule while matter goes up inside them. In such a scheme, the spicule would inject into the corona a plasma depleted in high-FIP elements. However, diffusion, in this case, is only due to the concentration gradient between the interior and the exterior of the spicule, and the diffusion time scale is of the order of one day, according to Meyer (1987a), while the life time of spicules is only 5 min. According to that estimate, no observational effect would be expected.

Vauclair and Meyer (1985) proposed another scenario based on the fact that most of the solar surface is covered by horizontal magnetic fields of 5 to 25 gauss, at altitudes of $\simeq$ 700 to 2000 km, corresponding to the chromospheric plateau (see also Meyer and Vauclair 1989, in preparation). As shown in Fig. 3, this magnetic field, which emerges vertically from the Sun in the chromospheric network, extends horizontally above the supergranules, and becomes vertical again in the upper coronal regions, creating the so-called "canopy" configuration (Whithbroe and Noyes 1977; Giovanelli and Janes 1982; Jones and Giovanelli 1983; Mein et al. 1982,1985). The Vauclair and Meyer model supposes that most of the matter which is carried up in the spicules falls back down either in the spicules themselves, or along the vertical magnetic lines above the chromosphere (Fig. 3). The chromospheric plateau, with its horizontal magnetic lines, acts as a diffusion zone, in which a leakage of neutral elements occurs due to gravitational settling, while ions are confined by the magnetic field. Meanwhile, some of this processed matter (about 0.7% only) is carried up into the higher corona, refilling it with a plasma depleted in high-FIP elements, as observed. This plasma later leaves the Sun, giving rise to the solar wind.

The computation of the involved time scales must take partial ionization into account. When the atoms are partially ionized, they spend part of their time in each stage of ionization, and the resulting velocity may be written

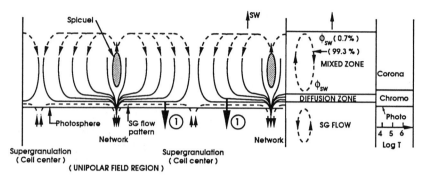

Fig. 3. Diffusion model for partially ionized elements in the chromospheric plateau (figure after Vauclair and Meyer 1985). The horizontal magnetic field in the canopies acts as a filter for neutral particles while matter is mixed in the transition region and low corona.

$$v_D = \sum_i x_i v_{Di} \tag{10}$$

where x_i is the fraction of atoms in the ionization stage i and v_{Di} the corresponding diffusion velocity (Montmerle and Michaud 1976). Across a magnetic field, the transverse diffusion velocities of the ions are reduced by a factor $(1 + \omega_i^2\tau^2)$ where ω is the cyclotron frequency and τ the characteristic time between two collisions (Spitzer 1962; Chapman and Cowling 1970; Vauclair et al. 1979; Alecian and Vauclair 1981). In this case the diffusion velocity becomes approximately

$$v_D = x_1 v_{D1} + \sum_{i \neq 1} x_i v_{Di} \frac{1}{(1 + \omega_i^2 \tau_i^2)}. \tag{11}$$

There have been suggestions, based on laboratory measurements, that the magnetic field would be less efficient than this in reducing diffusion perpendicular to magnetic field lines; see Michaud (1986) for a brief discussion. In the model of Vernazza et al. (1981), at a chromospheric temperature of about 6000 K, with a magnetic field of 5 to 25 gauss, the factor $\omega^2\tau^2$ becomes equal to 1 at an altitude of about 950 km, reaching a value of 100 at about 1300 km, so that the ions are effectively prevented from settling while neutrals may diffuse freely. For high-FIP elements, $x_1 \simeq 1$ (x_1 is the fraction of the element which is neutral) in this region, except for carbon for which x_1 may go down to 0.4. The diffusion time scales for the neutrals vary as the collision cross section with the hydrogen atoms and protons; these differ from atom to atom by a factor 3 at most. The computed time scales vary from about 20 days at 900 km, to 0.2 day at 2000 km except for carbon, for which the time scales are longer due to its larger ionization. These time scales have to be compared to the lifetime of the network, which is of the order of 1 day. Although more precise computations are needed, this model seems promising to account for the observed depletion of high-FIP elements in the solar corona and solar wind.

Given the uncertainties, there remains considerable room for controversy with these models and it is appropriate to consider alternatives. Roussel-Dupré and Beerman (1981) developed a model involving diffusion in the corona-chromosphere transition region. Both thermal diffusion in the transition region and gravitational settling in the corona are involved in their model. They explain the observed profile of EUV lines that show evidence of mass motion. No calculations using this model have yet been made to explain the relative depletion of chemical elements with high FIP and, as discussed above, it is not clear how thermal diffusion could explain this relation.

C. Diffusion in the Solar Corona and Solar Wind

The observations of helium depletion in the solar wind (see above) have led many authors (Jokipii 1966; Delache 1967; Nakada 1969,1970; Yeh 1970; Geiss et al. 1970; Cuperman and Metzler 1975; Hollweg 1981b; Geiss 1982) to conclude that helium is left behind in the solar wind, due to insufficient Coulomb drag. This model is similar to a model proposed by Vauclair (1975), Montmerle and Michaud (1976) and Michaud et al. (1987) to account for helium-rich stars. In a Lagrangian frame linked to the wind, helium and heavier elements diffuse downward with respect to hydrogen by gravitational settling, and the equation for element separation is similar to that given in Sec.II.D (see, for instance, Michaud et al. 1987). For ionized elements, the coupling constant varies as Z^2/A where Z is the charge number of the considered element and A its mass number. Geiss (1982) pointed out that this ratio, equal to 1 for ^{4}He III, becomes much larger for heavier elements in the corona because of the much larger degree of ionization (the ratio is 3 for O VIII). Thus, gravitational settling can be important for helium, and negligible for heavier elements.

Detailed computations of helium diffusion in the solar wind are complicated. As already discussed in Sec.IV.A, the solar wind configuration is quite complex. It shows various components which leave the Sun with different speeds and chemical compositions. The low-speed solar wind is highly variable, and its source region and acceleration process are not yet well known. The steady-state approximation is a crude one in this case and, as discussed by Geiss (1982), the number of parameters is large and does not allow one to reach a definitive conclusion about the importance of element separation from first principles. Any ^{4}He/H ratio can be reproduced with a good choice of model and parameters.

The situation is better for high-speed streams which originate from coronal holes and which, as already mentioned, are more stable. Burgi and Geiss (1986) have recently done detailed calculations of a model for chemical separation in the solar wind including constraints coming from the electron density observed in the corona, from the fluxes of proton and helium, and from the observed charge distribution of a few elements at 1 AU. They used the observed electron-density profile to infer a temperature gradient in the corona. This temperature gradient is affected by the presence of He and it must be obtained iteratively with the solution for chemical separation in the wind.

In the model they prefer, *model 4,* Burgi and Geiss (1986) find that the matter arrives at 1 AU with the same $n(\text{He})/n(\text{H})$ ratio as at the base of the wind model. No net separation occurs in the wind itself. This is in contrast to the results of a number of earlier studies (see above). The difference is caused by the different geometry and temperature structure assumed. These are not obtained from first principles but are partly obtained from detailed solar observations and partly treated as arbitrary parameters. In particular, they assume an appropriate geometry for the magnetic flux tubes. By reducing the surface from which the wind originates close to the photosphere, flux tubes increase the particle density there and so reduce the particle separation (see also Joselyn and Holzer 1978). The temperature structure used by Geiss and Burgi (1986) also reduces the separation, mainly through its effect on the friction coefficient. The friction coefficient is proportional to $T^{-1.5}$ and so increases as the temperature decreases close to the star in the Burgi and Geiss (1986) model. A constant temperature model does not have this lower-temperature region close to the surface and so maximizes the separation. [The temperature is yet lower in the chromosphere but separation still occurs there because the elements are mainly neutral. The collision cross section is much larger for ionized elements which interact through the Coulomb potential.] The only chemical separation in the Burgi and Geiss model occurs close to 0.5 $R_\odot$ above the surface where there is a local He accumulation. This modifies the coronal structure there but leads to no separation in the flux carried by the wind. While the work of Burgi and Geiss (1986) is the most comprehensive to date, it still requires a number of assumptions about the electron density as a function of radius, flux tube geometry, etc. and it does not consider the heating mechanism, which is still unknown. It illustrates our limited understanding of such winds.

It then appears that no real separation may occur in the wind itself. Instead, differential diffusion induced by incomplete ionization of the elements in the chromosphere, as proposed by Geiss (1982), may be the solution to the problem of underabundance of high-FIP elements (see Sec.IV.B). According to the Burgi and Geiss model, some accumulation of He would, however, occur in the corona while the Vauclair and Meyer (1985) model gives a reduced flux of all high-FIP elements through the corona, including helium. The combination of the two processes could lead to a normal helium abundance in the corona. Note that, although the observations of helium in the corona are very uncertain (Fig. 2), helium might be less depleted there than in the solar wind, solar energetic particles and galactic cosmic rays. For this model to be tenable, however, the other FIP elements must have nearly the same abundance in the solar wind and corona. This is currently within observational uncertainties. It could be that the two processes, in the chromospheric plateau and in the wind, work together to lead to the observed features.

CONCLUSIONS

In summary, the observations of element abundances in the photosphere, corona, solar wind and solar energetic particles suggest that separation occurs in the chromospheric plateau, where elements with a first ionization potential $\gtrsim 9$ eV become depleted in the outgoing matter. It is possible that there is no enhancement of any element in the wind itself. While the He velocity may be smaller than the hydrogen velocity ~ 0.5 $R_\odot$ above the solar surface, this is compensated for by a local accumulation of He, and the He flux at the Earth would be the same as that which enters at the base of the wind. Diffusion in the transition region is probably negligible due to the large macroscopic motions which take place there. In any case, as discussed by Geiss and Bochsler (1986), the situation is quite complex, and more observations from space and from the ground are needed to understand with greater accuracy the motions and the chemical composition of the Sun's outer layers.

Acknowledgments.   G.M. was partially supported by Conseil de Recherche en Sciences Naturelles et en Génie and Fonds pour la Formation de Chercheurs et l'Aide à la Recherche.

PART II
What Surface Oscillations Reveal About the Inside

OSCILLATION OBSERVATIONS

FRANK HILL
National Solar Observatory

FRANZ-LUDWIG DEUBNER
Universitat Würzburg

and

GEORGE ISAAK
The University of Birmingham

This chapter reviews recent observations of solar oscillations. The oscillations discussed are global and local 5-min p-modes, the 160-min oscillation and oscillations in the solar atmosphere. Experimental and data reduction methods are described. Summaries of recent results are provided, including measurements of frequencies, amplitudes, line widths and splittings. Other topics include active-region tomography, solar cycle changes, the chromospheric cavity and diameter measurements.

I. INTRODUCTION

Solar oscillations can be divided into two classes: oscillations that exist in the atmosphere of the Sun, and oscillations that propagate through the solar interior. The oscillations that exist inside the Sun have led to the development of helioseismology, the study of both global and local modes of solar oscillations that are analogous in many respects to the waves that travel through the Earth after an earthquake. Just as terrestrial seismology has provided information about the internal structure of the Earth, so too has helioseismol-

[329]

ogy produced new insights about the solar interior. Helioseismology has produced estimates of the depth of the solar convection zone (Berthomieu et al. 1980; Christensen-Dalsgaard et al. 1991), the internal solar rotation rate (Duvall et al. 1984; Brown et al. 1989; Rhodes et al. 1990; Goode et al. 1990), sound speed (Christensen-Dalsgaard et al. 1985a), and opacity (Korzennik and Ulrich 1989). It promises to provide information on the flows in the convection zone (F. Hill 1990b), the evolution of active regions before they are visible at the surface (Braun et al. 1990), and the helium abundance (Däppen et al. 1988b). It constrains solar models (see, e.g., Christensen-Dalsgaard 1988c,1990b), and thereby may also shed light on the solar neutrino problem (Bahcall and Ulrich 1988; Dziembowski et al. 1990).

Waves in the solar atmosphere also probe their environment. Atmospheric seismology provides information on the wave transport of energy that probably plays a role in the heating of the chromosphere and corona; the interaction of convection with the stable atmosphere; the conditions in the chromosphere; and the conditions in magnetic structures.

Solar oscillations can also be classified according to the restoring force that controls them. Pressure is the restoring force for acoustic waves, buoyancy for internal gravity waves, and magnetic fields for Alfvén waves. More than one restoring force can be present, resulting in a rich variety of physical wave behavior. In the context of helioseismology, acoustic waves are known as p-modes, and internal gravity waves are called g-modes. There is also a third type of global oscillation called the fundamental or f-mode. This mode is related to the surface waves that exist on the Earth's oceans.

Because the oscillations are so useful for probing the solar interior, much work has been done to obtain and understand measurements of their properties. The high level of effort in this area is reflected in the large portion of this book that is devoted to the topic. In addition to this chapter, the theory of solar oscillations is discussed in the chapter by Christensen-Dalsgaard and Berthomieu, while the chapter by Libbrecht and Morrow discusses helioseismic estimates of internal rotation. Solar g-modes are reviewed in the chapter by H. Hill et al., the inversion of helioseismic data to yield information about the solar interior is discussed in the chapter by Gough and Thompson, the coupling of the modes with other processes is reviewed in the chapter by Dziembowski and Goode, while the chapter by Cox et al. presents the excitation of the oscillations. An Appendix by Pallé contains the observed frequencies of the oscillations. Other recent reviews of and introductions to helioseismology can be found in the following papers (Andersen 1989; Brown et al. 1986; Christensen-Dalsgaard et al. 1985a; Deubner and Gough 1984; Duvall 1990; Harvey 1988; Leibacher et al. 1985; Libbrecht 1988b; Unno et al. 1989; Vorontsov and Zharkov 1989). In addition, much of the literature of helioseismology is contained in conference proceedings; see, e.g., some recent ones listed below. (An extensive bibliography of helioseismology and related topics can be found in F. Hill [1989a].) For compactness, these proceedings are identified by the location of the conference in the list. The full references are:

Aarhus: *Advances in Helio- and Asteroseismology, IAU Symp. 123,* ed. J. Christensen-Dalsgaard and S. Frandsen (Dordrecht: Reidel), 1988.

Cambridge: *Seismology of the Sun and the Distant Stars,* ed. D. O. Gough (Dordrecht: Reidel), 1986.

Hakone: *Oji International Seminar on Progress of Seismology of the Sun and Stars,* ed. H. Shibahashi and Y. Osaki (Berlin: Springer), 1990.

Santa Barbara: *Challenges to Theories of Moderate Mass Stars,* ed. D. Gough and J. Toomre (Berlin: Springer), 1991.

Snowmass: *Solar Seismology From Space,* ed. R. K. Ulrich, J. Harvey, E. J. Rhodes, Jr., and J. Toomre (Pasadena: NASA JPL 84–84), 1984.

Tenerife: *Seismology of the Sun and the Sun-Like Stars,* ed. E. J. Rolfe (Paris: ESA SP-286), 1988.

Versailles: *Inside the Sun, IAU Colloq. 121,* ed. G. Berthomieu and M. Cribier (Dordrecht: Kluwer), 1990.

In this chapter, a review of recent observations of solar oscillations is presented. We begin with a brief history of the subject, then describe the methods of measuring the properties of the modes. We summarize some of the more recent results in the major areas of unresolved observations, resolved measurements, and atmospheric oscillations. It should be noted that helioseismology is a rapidly expanding field; the reader is advised to read the recently appearing literature to keep abreast in this field's development.

II. A BRIEF HISTORY

In the summers of 1960 and 1961, one of the most important sets of observations in the history of solar physics was obtained at Mount Wilson Observatory. The data comprised spectroheliograms made in the red and blue wings of a number of spectral lines from which were produced line-of-sight Doppler images of the Sun. These images immediately led to the discovery of two fundamental solar phenomena: the existence of large convective cells called supergranulation, and the presence of oscillations in the solar atmosphere (Leighton 1961; Leighton et al. 1962). Almost simultaneously, work was underway at Sacramento Peak to measure fluctuations in the brightness and wavelength of spectral lines observed with a Littrow spectrograph. These observations (Evans and Michard 1962) confirmed the existence of the oscillations and the field of helioseismology was born, if not yet conceived.

Over the next 15 years, numerous studies were done and papers published both on the observed characteristics of the oscillations, and on the cause of the phenomenon. A review of the early observational evidence can be found in Noyes (1967). The motions were seen to be primarily vertical and to have a range of periods of ~240 to 360 s, with a dominant period of

~300 s. The power in the oscillations was observed to shift towards higher frequencies with increased height in the atmosphere. The phase relationships between velocities at different heights showed a lag at the onset of oscillations which quickly vanished as the motion continued. The onset was also seen to be preceded by a brightening in the continuum, suggesting that the oscillations were initiated by an impulse presumably caused by a rising granule acting as a piston. The phase relationship between the velocity and the intensity at a single height was observed to be 90°, implying a standing wave. The oscillations were observed to be reduced in amplitude in regions of strong magnetic fields.

Early observations of both the spatial scale and the lifetime of the oscillations gave contradictory results. Published estimates for the spatial scale ranged from 1700 to 100,000 km and, for the lifetime, 6 to 100 min. These wide ranges reflect the difficulty of understanding the superposition of what are now known to be 10^7 modes with closely spaced frequencies and wavenumbers by looking at the beat pattern; this was essentially what early workers were doing by obtaining one-dimensional (usually temporal) observations and then performing one-dimensional power spectral analyses. The widely varying measured spatial and temporal scales actually reflected the observer's choice of spatial and temporal instrumental parameters, rather than the solar conditions. A major step towards the understanding of the oscillations came in 1966 when the two-dimensional power spectrum was first applied to the problem (Mein 1966; Frazier 1968). In these spectra, also known as the diagnostic diagram, the observed power is plotted as function of the horizontal spatial wavelength and the temporal frequency of the wave. Application of these diagrams to the solar oscillations conclusively demonstrated that the oscillations were evanescent in the solar atmosphere; this was also indicated by the nearly zero phase difference observed between spectral lines formed at different heights.

The early diagnostic diagrams also showed hints of structure in the region where the oscillations have the most power. The resolution of this structure was later to prove essential in establishing the correct physical model of the oscillations. The earliest model was of a rising granule acting like a piston and providing an impulse to the atmosphere that then continued to oscillate. However, the acoustic waves emitted by this process should be isotropically propagating both horizontally and vertically, whereas the observations showed that the solar 5-min oscillations were primarily vertical motions, as conclusively demonstrated by Stix and Wöhl (1974). Other models consisted of either gravity or acoustic waves trapped in various regions of the Sun, ranging downward in height from the transition region to the convection zone below the photosphere. Each of these models predicted a different pattern of power in the diagnostic diagram. The early diagrams immediately ruled out the gravity waves, and also showed that there were at least two patches of power in the evanescent acoustic wave region of the diagram (Frazier 1968).

As observers increased the resolution and decreased the noise in the diagnostic diagram by observing larger areas for longer time spans, these patches began to take on the character of slanted ridges of power (Tanenbaum et al. 1969; Deubner 1972). Finally, in 1975, a diagram was obtained that unmistakably showed the presence of several well-separated ridges of power (Deubner 1975). An example of a modern-day high-resolution diagnostic diagram is shown in Fig. 1.

These ridges were predicted by the models of acoustic waves trapped below the photosphere (Ulrich 1970; Leibacher and Stein 1971). They arise from the interference of acoustic waves trapped in the temperature structure inside the Sun. At deeper depths, waves whose path of propagation is not oriented precisely vertically are refracted upwards by the increasing sound speed with temperature and depth. Just below the photosphere, the pressure scale height rapidly decreases outwards to a value less than the wavelength of the waves, reflecting them back downward. These two reflection points define a cavity within which the waves are trapped. Only waves that have two of their vertical nodes at the reflection points will survive destructive interference. The position of the lower reflection point depends on the number of vertical nodes of the wave, resulting in a series of ridges in the diagnostic diagram. An individual ridge contains power from modes with the same number of vertical nodes. As it was then clear that the oscillations were trapped

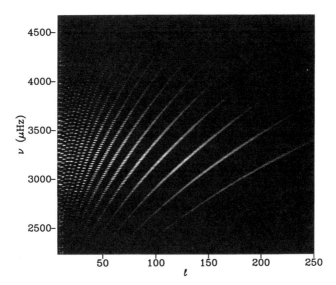

Fig. 1. A diagnostic, k-ω or l-ν diagram of the solar p-mode oscillations. The power in this two-dimensional spectrum is distributed in individual peaks that are aligned along curved ridges. Each peak corresponds to an individual mode with a specific l and n value, where l is the spherical harmonic degree and n is the radial order of the mode. Each ridge contains modes with equal values of n (figure courtesy J. Harvey, from Duvall et al. 1987).

in the solar interior, it was quickly realized that they could be used to infer the physical conditions inside the Sun (see the remark by McIntyre following Deubner [1976a]). Thus, helioseismology was finally conceived, 15 yr after having been born.

The early studies of solar oscillations focused on a plane-wave, or local description of the modes. However, a parallel description of the oscillations as a global phenomenon involving the entire Sun as a star also developed (Wolff 1972). Luminosity variations of large amplitude in a small fraction of stars have been known since fluctuations in the brightness of Mira were observed by Fabricius in 1596, and in δ Cephei and η Aquilae by Goodricke and Pigott in 1784. It took more than a century and the efforts of Kelvin, Ritter, Belopolsky, Emden, Shapley, Eddington and others to interpret these variations in terms of the global oscillations of gaseous spheres. It was not until the 1970s that it was realized that the Sun was also an intrinsically pulsating star.

The global description starts from the premise that all stars, and the Sun in particular, may be oscillating but with such minute amplitudes that the fluctuations escape detection. It is surmised that any disturbance or noise excites the fundamental and low-order harmonics of the radial and nonradial normal modes of a sphere and that these oscillations could show up in the case of the Sun, when viewed as a star without spatial resolution, at an observable level. A search for periodicities on the order of 1 hr in the Doppler shift of the potassium and sodium Fraunhofer absorption lines using atomic resonance scattering methods originally devised many years earlier (Isaak 1961; Fossat and Roddier 1971) resulted in the discovery of a 160-min oscillation (Brookes et al. 1976) at a 3 m s^{-1} level. Simultaneous measurements by Severny et al. (1976) using a modified magnetograph showed similar oscillations which, on subsequent comparison, were found to be in phase with the published (Birmingham) data.

It had been noticed in earlier spatially integrated full-disk observations that 5-min oscillations were readily visible with *rms* amplitudes of 1 m s^{-1} (Fossat and Ricort 1973,1975). The discovery of structure in the global 5-min spectrum came at the end of the decade (Claverie et al. 1979). This study showed that the spatial extent, the long coherence time of the oscillations and the agreement with theoretical models (Iben and Mahaffy 1976; Christensen-Dalsgaard et al. 1979) proved these oscillations to be nonradial normal modes of the whole Sun. This discovery of the detectable presence of these modes was unexpected in the context of contemporary theoretical ideas (see, e.g., Gough 1978; Unno et al. 1979). The observation of global velocity oscillations immediately implied the existence of detectable luminosity oscillations of the Sun. It also held out the promise that similar techniques of spectroscopy and photometry could be extended to stellar seismology (Isaak 1980; Fossat 1981).

The detection of the global modes was soon followed by a study using some 120 hr of continuous velocity data from the South Pole (Grec et al. 1980) which, with improved resolution and without the aliases of one day, resolved fine structure due to incomplete degeneracy between neighboring modes. The following year the Birmingham group managed to obtain 88 days of data of up to 22 hr per day operating from two stations, Tenerife and Hawaii, separated by over 9 hr in longitude. The resulting high-resolution power spectrum (Claverie et al. 1984) and the success of this project paved the way for the current flourishing of networks to observe the solar oscillations. In 1983, data from the Active Cavity Radiometer (ACRIM) experiment (Willson 1981) on the Solar Maximum Mission (SMM) confirmed the existence of luminosity oscillations at a level of a few parts in 10^6 (Woodard and Hudson 1983) and established that photometry can be used to study global solar and stellar oscillations, at least above the atmosphere.

III. OBSERVATIONAL TECHNIQUES

In this section, we discuss the general techniques used to obtain oscillation observations. We begin by presenting the spherical harmonic and plane wave descriptions of the oscillations. We then turn to issues of sampling, observing strategies, velocity vs intensity measurements, instrumentation, atmospheric effects and data reduction including methods for analyzing atmospheric waves. More information can be found in Harvey (1985a, 1990b), Brown (1988c) and Appourchaux (1988,1989).

A. Descriptions of the Oscillations

Much of the discussion of solar oscillations centers on the description of the waves as either spherical harmonics or as plane waves. The spherical harmonic description is appropriate for global modes which sense the spherical geometry of the Sun. In this representation, the radial component of a single mode of oscillation is described by an eigenfunction E in polar spherical coordinates (r, θ, ϕ) where r is the radius, θ is the colatitude, and ϕ is the longitude:

$$E(l, m, n) \equiv K_{ln}(r)Y_l^m (\theta, \phi)e^{i\omega_{tmn}t} \tag{1}$$

where $K_{ln}(r)$ is the radial dependence of the eigenfunction, and n is the radial order, or number of radial nodes. A spherical harmonic is represented by $Y_l^m(\theta, \phi) = P_l^m(\cos \theta)e^{im\phi}$, where $P_l^m(\cos \theta)$ is the associated Legendre function. The spherical harmonic degree l is the total number of nodal lines on the solar surface, or the number of nodes along a great circle at an angle $\cos^{-1}[m/\sqrt{l(l + 1)}]$ to the equator. The azimuthal order m is the number of nodal lines around the equator of the coordinate system, usually chosen as the solar rotational equator. The value of m is restricted to be between $+l$ and $-l$; thus each mode with a given l has $2l + 1$ values of m associated with it. If $m = 0$, then the mode is known as a *zonal* mode; if $m = l$, the mode is called a *sectoral* mode, otherwise the mode is *tesseral*. Sectoral modes are limited in their extent in θ. Examples of spherical harmonic patterns are shown in Fig. 2. The cyclic temporal frequency, ω_{lmn}, depends on the mode which is labeled by the values of l, m and n. It is often written as $\nu = \omega/2\pi = 1/P$, where P is the period of the mode. Measurements of ν are the basic data of helioseismology.

It has become customary to refer to different ranges of l in terms of their relative values, as shown in Table I.

In the plane wave description, the modes are considered to be local waves that are not affected by the spherical solar shape. This approximation is only valid for modes with short wavelengths, or high values of l. In this representation, the eigenfunction is

$$E(\mathbf{k}_h, n) \equiv K_{ln}(r)e^{i(\omega t - \mathbf{k}_h \cdot \mathbf{x})} \tag{2}$$

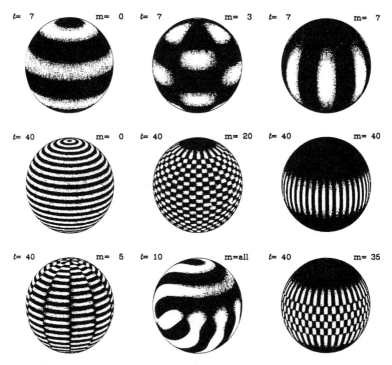

Fig. 2. Some examples of spherical harmonic patterns. *Top row:* zonal, tesseral and sectoral patterns with $l = 7$. *Middle row:* zonal, tesseral and sectoral patterns with $l = 40$. *Bottom row:* tesseral patterns with $l = 40$, and the sum of 11 spherical harmonics with $l = 10$ and m ranging from 0 to 10 inclusive.

TABLE I
Common ℓ-ranges

Low	$0 \leq l \leq 3$
Intermediate	$3 \leq l \leq 200$
High	$200 \leq l \leq 1500$
Very High	$1500 \leq l$

where $\mathbf{k}_h$ is the horizontal wavenumber vector, and $\mathbf{x}$ is the displacement vector in a horizontal coordinate system on the solar surface. The magnitude of the wavenumber vector k_h is related to λ, the horizontal wavelength of the wave: $k_h = 2\pi/\lambda$, and to the spherical harmonic degree:

$$k_h = \sqrt{l(l + 1)}/R_\odot \tag{3}$$

where $R_\odot$ is the solar radius. The components k_x and k_y of $\mathbf{k}_h$ are approximately related to m and l by $k_x \approx m/R_\odot$ and $k_y \approx (l^2 - m^2)^{1/2}/R_\odot$.

B. Spatial and Temporal Sampling and Extent

For a given set of observations, the maximum value of l obtainable is determined by the spatial resolution of the detector. In the classical Fourier transform sense, the maximum spatial wavenumber accessible in an observation with a spatial sampling of Δx is the Nyquist wavenumber k_{Ny} given by

$$k_{Ny} = \frac{\pi}{\Delta x}. \tag{4}$$

An expression for L_{Ny}, the Nyquist value of $L \equiv \sqrt{l(l + 1)}$, can be derived by combining Eqs. (3) and (4):

$$L_{Ny} = \frac{\pi N}{2} \tag{5}$$

where N is the number of samples across the solar diameter. However, this is only true in plane geometry. In a spherical geometry, foreshortening must be taken into account as a sampling interval in the plane of the sky transforms into a larger sampling interval on the solar surface towards the limb. Taking this effect into account results in the following approximate expression for the maximum L visible at a radius vector ρ:

$$L_{Ny} \approx \sqrt{1 - \rho^2} \, \frac{\pi N}{2}. \tag{6}$$

Thus, at the limb where $\rho = 1$, $L_{Ny} = 0$ because the solar surface is parallel to the line of sight, and no spatial variation can be detected. This gives a somewhat false impression of the range of L available in the data, because one might then draw the wrong conclusion that only radial harmonics can be measured from spatially resolved full-disk data. A perhaps more appropriate measure of L_{Ny} for spherical geometry may be obtained by computing its average from disk center to a radius vector ρ. The result is:

$$\langle L_{Ny} \rangle = \frac{1}{2\rho}\left[\rho\sqrt{1 - \rho^2} + \sin^{-1}\rho\right]\frac{\pi N}{2}. \tag{7}$$

If $\rho = 1$, then $\langle L_{Ny} \rangle = 0.785\ \pi N/2$. Thus, the spherical shape of the Sun reduces the highest value of l available in the data. For a much more sophisticated discussion, see Appourchaux (1988).

The preceding discussion is applicable to spatially resolved observations, but the global view of a spherical object without spatial resolution acts as a spatial filter. Such observations average to nearly zero the contributions of the high-l modes because of the cancellation of positive and negative areas of velocity or intensity. For these unimaged observations, the use of Eq. (7) with $N = 1$ and $\rho = 1$ suggests that $L_{Ny} = 1.23$. In fact, more sophisticated calculations of the spatial response function of these observations show that the residual signal consists of the radial and low-l nonradial modes with $l = 0, 1, 2, 3$ and possibly 4. These calculations show that the sensitivity to radial velocity and luminosity oscillations varies as l^{-3} and as $l^{-1.5}$, respectively, assuming equal weight for all parts of the visible disk (H. Hill 1978; Christensen-Dalsgaard and Gough 1982). It has been pointed out that this assumption is not valid as a monochromatic view of the rotating Sun (or star) weights different portions of the stellar disk unequally (Brookes et al. 1978b; cf. also the techniques of Doppler imaging). A recent treatment discusses this in more detail (Christensen-Dalsgaard 1989). The rotational weighting may actually be of some benefit, as it could enhance the signal of the $l = 4$ mode (Christensen-Dalsgaard and Gough 1989). The spatial response function for low-resolution annular observations such as those obtained in the Crimea (Kotov et al. 1982) and at Stanford (Scherrer et al. 1982,1983) is more complicated, with an oscillatory response extending as far as $l = 20$ (Balandin et al. 1987; Pallé et al. 1989).

The resolution in l obtainable from the observations can also be calculated. In plane geometry, the spatial wavenumber resolution, k_R, is given by

$$k_R = \frac{2\pi}{X} = \frac{2\pi}{N\,\Delta x} \tag{8}$$

where X is the total spatial extent of the observations, N is the number of spatial samples, and Δx is the spatial sampling. Using Eq. (3), and assuming that N is the number of samples across the solar diameter so that $N \, \Delta x = 2R_\odot$, results in

$$L_R = \pi. \tag{9}$$

In the spherical case, this becomes

$$L_R = \frac{\pi}{\sin^{-1}\rho}. \tag{10}$$

In observations covering the entire disk, $\rho = 1$ and $L_R = 2$, showing that we will never be able to completely resolve individual values of l and m because only one solar hemisphere is visible. Spherical geometry improves the resolution in l, while lowering the highest value of l obtainable.

Observations of intermediate- and high-l modes must naturally be spatially resolved. If the images are full disk, then the amount of data is very large, creating practical problems in the reduction process. A commonly used alternative has been to collect data from only a restricted portion of the disk. This reduces the data glut at the expense of loss of information about the spatial scale of the observations, which is most accurately determined by measuring the solar diameter in pixels. An accurate spatial scale is essential in helioseismology because it is the ultimate source of the identification of the l-value of the modes. The spatial scale of partial-disk observations has been calibrated by examining the f-mode. This mode has a particularly simple theoretical dispersion relationship: $\omega = \sqrt{gk}$, or $\nu = 3159(l/1000)^{1/2}$ μHz. A spatial scale can be determined by adjusting it until this relation is satisfied. However, care must be taken with this procedure because the simple dispersion relation is only valid for a range of l, and it may be affected by the presence of magnetic fields (Campbell and Roberts 1989; Evans and Roberts 1990). In addition, the f-mode provides information about the density structure of the Sun, and is heavily weighted in velocity inversions (Haber et al. 1991b), so use of the mode to calibrate the spatial scale compromises its use as a probe. It is thus far better to calibrate the data spatially by other means.

Aliasing occurs in the power spectrum when the observed process contains frequencies or wavenumbers that are higher than the Nyquist value. These components are then undersampled and appear in the spectra folded about the Nyquist point, masquerading (or under aliases) as incorrect lower frequency components. This can be a problem if the aliased components have substantial amplitudes. For helioseismic experiments with low spatial resolution, anti-aliasing techniques must be applied to attenuate the modes with $L > L_{Ny}$. These techniques are generally optical, and consist either of defocusing the image slightly, or inserting a carefully constructed spatial filter in the optical train. In addition, the typically small aperture of helioseismic telescopes effectively filters out very high spatial wavenumber components.

Temporal aliasing is also of concern. The oscillations have been observed to have periods as short as 137 s (Jefferies et al. 1988c), requiring a temporal sampling rate of no longer than 60 s to insure that the Nyquist frequency is comfortably above the highest-frequency component of the oscillations. In practice, observations have been obtained at cadences of 60 to 90 s, with the slower cadence observations showing temporal aliasing (Jefferies et al. 1988c; Woodard and Libbrecht 1988). The temporal Nyquist frequency v_{Ny} is given by

$$v_{Ny} = \frac{1}{2\Delta t} \tag{11}$$

where Δt is the sampling rate in seconds. A period of 137 s corresponds to $v = 7300$ µHz, so a sampling rate of 60 s ($v_{Ny} = 8333$ µHz) is sufficient. To obtain high resolution in v, the observations must be made for a long time. The temporal frequency resolution attained in the data string v_R is given by

$$v_R = \frac{1}{T} \tag{12}$$

where T is the total length of the data set in seconds. Table II provides some useful examples of the frequency resolution associated with typical observing run lengths.

C. Observing Strategies for Continuous Data

The precise determination of solar oscillation frequencies requires long unbroken temporal sequences of data. For example, in order to barely resolve the splitting of modes due to the solar rotation, the oscillations must be observed for at least two rotation periods or about 60 days. Even longer strings of data are essential to improve the frequency resolution of the observations and allow us to distinguish between different models of the solar internal rotation.

The pattern of ones or zeroes that denotes usable (ones) or unusable

TABLE II
Typical Frequency Resolution

T	v_R, µHz
8 hours	34.72
1 day	11.57
30 days	0.3858
1 year	0.03169

(zeroes) observation times is called the window function. The observed data string is the product of the window function times the actual unbroken string of solar data. Taking the Fourier transform of the product of two functions results in the convolution of the transforms of the functions. If the window function is broken by periodic gaps, such as those caused by the diurnal rising and setting of the Sun, then its spectrum consists of a series of peaks at multiples of $\pm$ 11.57 μHz (1/day). The convolution of this spectrum with the complex line spectrum of the oscillations results in a set of daily sidelobes surrounding every solar spectral line. As the oscillations are on average more closely spaced than 11.57 μHz, the diurnal sidelobes frequently overlap the solar lines, impeding mode identification and frequency measurement as shown in Fig. 3. Randomly spaced gaps do not produce a line window spectrum, but instead increase the background noise (F. Hill 1984a).

It is thus of considerable importance to remove the daily cycle from the observations. Several schemes to fill in the diurnal gaps have been investigated. The CLEAN algorithm developed in radio astronomy has been used in searches for g-modes (Delache and Scherrer 1983; Scherrer 1984), and in the production of p-mode spectra (Duvall and Harvey 1984; Henning and Scherrer 1986; Pallé et al. 1989). Other schemes include rearrangement of the data to fill in the gaps (Kuhn 1982,1984); prediction of data into the gaps using maximum entropy methods (Fahlman and Ulrych 1982; Brown and Christensen-Dalsgaard 1990); and both linear and nonlinear spectral deconvolution techniques (Connes and Connes 1984; Scherrer 1986). All known schemes can either misidentify solar modes or degrade the solar spectrum, and there appears to be no substitute for using an observing strategy to obtain nearly continuous long-data sequences.

Three main strategies have been developed to produce the desired sequences of data: observe from the polar regions; observe from space; or observe with a network. Observations have been obtained at the South Pole where the Sun stays continuously above the horizon during the Austral summer (Grec et al. 1980; Harvey et al. 1982; Stebbins and Wilson 1983; Pomerantz 1986; Jefferies et al. 1988c; Harvey 1989; Fossat et al. 1989). The experience of these observers has shown that the practical upper limit of the duration of South Pole data is ~10 days, far less than the 60 days required for rotational splitting studies. Spacecraft can provide observations free of terrestrial atmospheric effects, as well as continuous observations if they are not in low Earth orbits (Noyes and Rhodes 1984). Nonimaged observations of the oscillations have been obtained from space by the ACRIM instrument on SMM (Solar Maximum Mission) (Woodard and Hudson 1983) and by the IPHIR (Interplanetary Helioseismology by Irradiance) instrument on the PHOBOS mission (Fröhlich et al. 1988b,1990). Both nonimaged and imaged observations will be provided by the SOHO spacecraft that will be in a halo orbit around the L_1 Lagrangian point (Domingo 1988a,b; Domingo and Poland 1988; Bonnet 1990). Irradiance observations similar to the IPHIR mea-

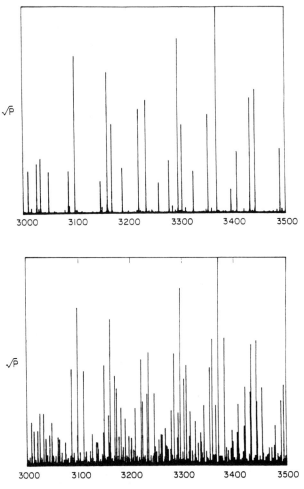

Fig. 3. The effect of the diurnal setting of the Sun on solar oscillation power spectra. *Top panel:* the 3000 to 3500 μHz portion of a synthetic low-degree solar oscillation spectrum that might be obtained from a one-year unbroken sequence of data. *Bottom panel:* the same spectrum that might be obtained from a single midlatitude site with data gaps from weather and the day-night cycle (figure from F. Hill and Newkirk 1985).

surements will be obtained by the VIRGO (Variability of Solar Irradiance and Gravity Oscillation) instrument (Fröhlich et al. 1988*a*). This instrument will also obtain low-resolution luminosity observations with a custom-built detector optimized to detect modes with *l*-values between 0 and 7. A nonimaging Doppler instrument known as GOLF (Global Oscillation of Low Frequencies) will observe the low-*l* global modes (Gabriel et al. 1989). The spatially resolved instrument, called SOI (Solar Oscillations Investigation), will be able to observe oscillations with l-values of up to 1200 (Scherrer et al. 1988). Spatially resolved intensity observations of the solar oscillations will also be made by the Japanese SOLAR-A satellite (Sakurai 1990).

A network of observing sites placed at suitably chosen longitudes around the Earth will greatly increase the continuity of observations and reduce the impact of weather and instrumental failures (Aindow et al. 1988; F. Hill 1990a; F. Hill and Leibacher 1991). A model of the performance of networks predicts that data is obtainable nearly 94% of the time from a 6-site network (F. Hill and Newkirk 1985). This prediction has been confirmed by the site survey being carried out by the Global Oscillation Network Group (GONG) project (Harvey et al. 1987; Harvey et al. 1988a; F. Hill et al. 1988a). The first network designed to obtain helioseismic observations is the Birmingham network, which began operations in 1978. This network provides nonimaging observations from 4 sites, with 2 more to be added in the future (Elsworth et al. 1988; Aindow et al. 1988, New 1990). Another nonimaging network, known as International Research on the Interior of the Sun or IRIS, has also recently begun operations (Fossat 1988a,1990). Founded by the Université de Nice, this network will have 7 sites; 5 are currently operating. Power spectra from the Birmingham network have the best signal-to-noise ratio yet attained with daily sidelobe amplitudes below 3% of the signal (Aindow et al. 1988). The results of the GONG site survey indicate that daily sidelobes with amplitudes below 0.1% should be easily achievable with a 6-site network (F. Hill et al. 1988a).

D. Velocity vs Intensity Measurements

Solar oscillations are typically detected by observing either Doppler shifts of spectral lines formed in the solar atmosphere, or intensity variations in the emergent radiation in various wavelength bands (Fossat 1984). The amplitude of a single mode of oscillation in velocity is typically $\sim$5 to 10 cm s^{-1}, but the incoherent addition of 10^7 modes results in an oscillatory signal with an amplitude of 300 to 500 m s^{-1}. The intensity contrast signal of a single mode is $\sim$1 to 2 parts in 10^6, with an incoherent sum of $\sim$5 parts in 10^3. There is evidence that intensity oscillations in some infrared wavelengths may have amplitudes as high as 2 parts in 10^2 (Leifsen and Maltby 1988,1990). The solar convective velocity background noise has been estimated theoretically (Harvey 1985a), and confirmed observationally (Jiménez et al. 1988b). In the 3-mHz range of the 5-min oscillations, the dominant contribution to the solar background noise is from granulation, which has an *rms* radial velocity of $\sim$1 km s^{-1}. Granulation is also the dominant source of noise for intensity measurements, having an *rms* contrast of 15%.

The oscillation signal to granulation noise ratio is thus about an order of magnitude lower for intensity than for velocity. This makes it more difficult to observe the oscillations in intensity, but there are some advantages to using this method. The oscillations are predominantly vertical motions, so their amplitude as seen in velocity decreases towards the limb where the line of sight becomes more horizontal in the solar atmosphere. In addition, the strong supergranular motions are predominately horizontal, and thus dominate the velocity signal near the limb. These effects dictate a limit to the utility of helioseismic data near the limb. There are no such limits to intensity data, because supergranulation has a weak intensity signal, and the oscillatory temperature fluctuations are isotropic. Thus, intensity observations are able to make use of a larger fraction of the solar disk, improving separation and identification of modes. Further, it is easier to make precise photometric measurements than Doppler measurements. Wavelength stability is crucial for Doppler measurements, and difficult to achieve. Doppler observations can be easily compromised by line-shape changes that masquerade as line shifts; intensity measurements do not face this problem.

Excellent results have been obtained with intensity observations, but because of the much better signal-to-noise ratio, most studies have utilized spectral-line Doppler-shift measurements. These have been performed in one of three ways: centroiding or convolution of the entire line profile; comparison of the relative intensity of two or more segments of the profile; or determination of the phase shift between the spectral line and an interferometric zero point. All of these techniques will interpret line shape changes as line shifts, even though the shape changes may be due to effects other than bulk motion of the emitting plasma. The two-point method is the technique most commonly used for Doppler measurements. In this method, the intensity difference between images obtained in the red and blue wings of a spectral line are related to the Doppler shift of the line. It is very sensitive to the line shifts because the solar spectral lines are quite narrow. It is also a simple method, well suited for large detector arrays. However, the relationship between the velocities and the intensity differences is not linear, resulting in complications due to the large rotational velocity gradient across the Sun. The two-point method is also affected by variations of the relative line depth.

Consideration must be given to the choice of spectral line in all Doppler techniques. The line should be narrow to maximize the sensitivity of the velocity determination. It should be free of both blends and contamination by terrestrial atmospheric lines, and have a clean surrounding continuum. Since all Doppler techniques are contaminated by spurious apparent line shifts arising from line-shape changes, the chosen line should be as insensitive as possible to the processes that cause shape changes. The two main processes that cause shape changes are magnetic fields and granulation. The presence of a magnetic field will split the majority of solar spectral lines through the Zeeman effect. It is thus best to use a spectral line with a low (or zero) Landé g-factor. Magnetic fields can also produce apparent red shifts (Robillot et al. 1984; Ulrich et al. 1988a) that may compromise nonimaged full-disk helioseismology measurements and result in apparent solar cycle changes in the oscillation frequencies. The presence of granulation results in asymmetrical line profiles, as shown by the so-called C-shaped bisector of the profiles. The asymmetry arises because the relatively hotter central portions of the granules are also rising, resulting in an enhanced intensity in the blue wing of the line. Because the motions are vertical, observations of spectral lines at the limb have a reduced asymmetry, resulting in the apparent limb red shift, which is actually an absence of a blue shift. The result is that the asymmetry of the spectral lines changes as a function of ρ, the radius vector (for recent discussions see Dravins et al. 1986; Cavallini et al. 1987a; Gomez et al. 1987; Andersen 1987; Bertello and Caccin 1990). Local changes in the convective efficiency, such as suppression by magnetic fields, may also affect the apparent Doppler shift of a spectral line (Miller et al. 1984; Alamanni et al. 1990b). These changes of line shape appear in helioseismic measurements as noise which has unknown spatial and temporal characteristics. Lines formed at heights above the region of strong granular temperature fluctuations may still be affected by the granulation through nonlocal thermodynamic effects (Rutten et al. 1988). However, theoretical calculations suggest that the Fourier tachometer technique of determining Doppler shifts is remarkably linear, even in the face of competing velocity fields (Jones 1989).

E. Instrumentation

In this section we discuss the instrumentation that has been used to observe the solar oscillations. As the four broad classes of helioseismic observations (nonimaged whole-disk, differential whole-disk, imaged and diameter) have developed rather different observing devices, we discuss each of them separately. A general discussion of instrumental issues for helioseismology can be found in Appourchaux (1988,1989).

1. Nonimaged Whole-Disk Observations.   The predominant technique used in unimaged velocity measurements is that of resonance spectroscopy (Isaak 1961; Fossat and Roddier 1971; Grec and Fossat 1977; Brookes et al. 1978a; Hoyng 1989; Grec et al. 1990). The method is based on an atomic frequency standard, is differential in content, uses switching techniques and has direct continuous calibration. In this class of instrument, sunlight first passes through a prefilter that isolates the desired spectral line, then typically through a linear polarizer and a quarter-wave modulator. The resulting alternately right and left circularly polarized light next enters an atomic resonant scattering cell containing either potassium or sodium vapor in a magnetic field parallel to the incident beam. In some instruments, the order of the major optical components is reversed, with the light passing through the resonance cell before the polarizers (Grec et al. 1990). In either case, by alternating the polarization between left and right, the σ^+ and σ^- Zeeman components of the spectral line (769.9 nm for K, 589.0 and 589.6 nm for Na) are alternately resonantly scattered. From the measurements of the intensities I_B and I_R scattered from the blue and the red wings of the solar Fraunhofer absorption line, the ratio $R = (I_B - I_R)/(I_B + I_R)$ is evaluated. If the magnetic field is chosen so that the Zeeman components are near the point of inflection of the solar line, then the ratio is nearly linearly related to velocity displacements which are small compared to the line width. Small nonlinearities can be corrected for by adding higher-order terms to the observed R using a Taylor expansion about the $R = 0$ point. The optical depth of the vapor can also be chosen to optimize the sensitivity of the measurement. Among the advantages of the resonant-scattering technique are high stability, high sensitivity, and relative simplicity. Among its drawbacks are a small nonlinear response and the restricted number and type of spectral lines that can be used.

Another whole-disk nonimaging technique used to observe the global low-l oscillations has been to observe either the total solar irradiance or the luminosity in certain wavelength bands. The active cavity radiometer irradiance monitor (ACRIM) on the Solar Maximum Mission satellite (SMM) measures the solar flux over a very wide spectral region. Sunlight alternately strikes, or is blocked off, an absorbing cavity while a servosystem maintains the temperature of the cavity at its illuminated value. The electrical power required to do this is a direct measure of the solar input power. ACRIM was not designed for searching for solar oscillations and has a high digitization noise and a rather low Nyquist frequency. In addition, it is believed that the background solar noise level in luminosity is higher. Nevertheless, careful analysis of the ACRIM data led to the first successful measurements of the solar luminosity oscillations (Woodard and Hudson 1983). Previous ground-based whole-disk photometric observations (Claverie et al. 1981c) suffered from excessive noise due to the Earth's atmosphere, as did attempts using reflections from the planets Uranus and Neptune (Deubner 1981) and using pinhole cameras (Schmidt-Kaler and Winkler 1984).

Full-disk Sun photometers, using photodiodes and appropriate means to digitize the photocurrent, have also been used to measure the solar flux at a number of different wavelengths. Initially only measurements from balloons (Fröhlich 1984) were successful, but more recent extended groundbased observations from Tenerife and Mexico have shown positive results (Andersen et al. 1988; Jiménez et al. 1987,1988d,1990). These 4-channel measurements, made at 680, 517.8, 516.2 and 770.2 nm, have made it possible to correlate the magnitudes and phases of intensity and velocity fluctuations (Jiménez et al. 1988a,d,1990). The IPHIR experiment on the mission to Mars is a 3-channel photometer operating in the 335, 500 and 865 nm bands with half-power bandwidth of 5 nm (Fröhlich et al. 1988b). This instrument has produced a very clean low-l spectrum despite problems with the pointing of the spacecraft (Fröhlich et al. 1990). The SOHO mission will be carrying two instruments that will study the low-l modes. These are the GOLF experiment, which is essentially a resonant scattering spectrometer (Damé 1988; Gabriel et al. 1989), and the VIRGO instrument, an updated IPHIR-type irradiance monitor (Fröhlich et al. 1988a).

2. Nonimaged Differential Observations.   This class of observations compares Doppler measurements made in two complementary regions that together include the entire solar disk. The measurements are made by projecting the solar image onto a circular mask that covers a portion of the center of the disk, while leaving an outer annular portion uncovered. The mask is composed of a polarizer and a quarter-wave plate, so the light transmitted through the central portion of the image is circularly polarized. Additional polarizing optics (typically a KDP crystal plus another polarizer) after the mask permit high-frequency chopping between sunlight from the two regions. The Doppler shift of the light is then measured using either a spectrograph (Kotov et al. 1978, 1982; Dittmer et al. 1978) or a resonant spectrometer (Pallé et al. 1989). Differences in the line-of-sight velocity can be determined almost independently of long-period instrumental drifts and with high precision by rapidly comparing the two regions.

3. Imaged Observations.   Three main classes of instrumentation have been used to obtain imaged helioseismic Doppler observations: grating spectrometers, narrowband filters and interferometers. Grating spectrographs provide the best line profiles, but the one-dimensional slit must be rastered to produce a two-dimensional image, and this introduces complications in the temporal sampling of the oscillations, as well as noise. Narrowband filters can produce instantaneous two-dimensional images. Three types of filters have been used: tunable birefringent Lyöt filters (Libbrecht and Zirin 1986; F. Hill et al. 1986; Tarbell et al. 1988), magneto-optical filters (Cacciani and Rhodes 1984; Cacciani et al. 1988; Appourchaux 1987) and tunable Fabry-Perot etalons (Rust and Appourchaux 1988). All of the filters must be stabilized by comparison with a stable external wavelength reference, and, while commonly used in conjunction with the two-point method, can also be used to provide line profile observations.

Interferometers have become the current instrument of choice in imaged Doppler helioseismology in the form of glass-cube Michelson interferometers. They have the advantage of being linear, easily stabilized, and less sensitive to line-shape changes. However, they have intrinsically low signal-to-noise ratios, and the influence of the solar continuum results in the presence of so-called spurious modulation, a systematic effect that is difficult to calibrate. The first of these instruments is the Fourier tachometer, which has been in operation since 1985 (Brown 1984a). A similar instrument is under development by the Global Oscillation Network Group (GONG) project (Harvey et al. 1988b). A two-interferometer version will be used for the Solar Oscillations Imager (SOI) on board the SOHO spacecraft (Hoeksema et al. 1988; Scherrer et al. 1988).

As mentioned, intensity observations have also yielded excellent information for helioseismology. The most successful results have been obtained using narrowband observations in the Ca II K line (see, e.g., Duvall et al.

1986,1987,1988b; Jefferies et al. 1988c,1990a,1990a,b). These observations use an interference filter with a 6 Å-wide bandpass, along with a two-dimensional detector. A similar instrument is currently under development as the High-Degree Helioseismometer at the National Solar Observatory. Intensity observations have also been carried out with a narrowband filter in the CN bandhead (Kneer et al. 1982). The same group has performed studies of the oscillations in the chromosphere using observations of Hα intensity (see, e.g., von Uexküll et al. 1989). Observations of continuum intensity oscillations have been performed by Brown and Harrison (1980) and Didkovskii et al. (1988); in white light by Nishikawa et al. (1986); and from space using the SOUP (Solar Optical Universal Polarimager) instrument.

The recent development of infrared detectors has opened up a new avenue for observational helioseismology. Heterodyne spectroscopy in the infrared has been used to observe the chromospheric observations (Glenar et al. 1986,1988a; Deming et al. 1986,1988a). Broadband photometric observations in several channels have also yielded observations of the oscillations in intensity (Leifsen and Maltby 1988,1990; Lindsey et al. 1990). The amplitudes of the oscillations observed in the 2.23-μm band are surprisingly high, being some 5 times stronger than typical intensity oscillations observed in the visible. If confirmed, this may open up a fruitful new observing technique for helioseismology.

4. Diameter Measurements. Observational studies of the oscillations have also been performed using measurements of the solar diameter. This technique has its roots in studies of the solar oblateness, a quantity of great importance to theories of general relativity. Observations of the solar diameter have been made with an astrometric telescope originally designed to measure the gravitational deflection of starlight grazing the Sun (Oleson et al. 1974), at the Santa Catalina Laboratory for Experimental Relativity and Astrometry (SCLERA). As the solar diameter is monitored as a length reference, this instrument is able to provide observations of the diameter. Various detector configurations have been placed at diametrically opposed positions on the solar limb, typically at the equator. The experiment has also been attempted at the South Pole with the image rotating past the detectors, improving the mode identification capability (Stebbins and Wilson 1983). A Fourier transform definition is used to locate the edge of the Sun (H. Hill et al. 1975a), although several other possible definitions exist (Brown 1982). The diameter fluctuations can be interpreted as variations in the limb-darkening function, and hence as changes in the temperature of the solar atmosphere. The Earth's atmosphere is a major source of both random and systematic noise for these observations. An experiment that produces two overlapping solar images, the Solar Disk Sextant, has been developed (Sofia et al. 1984). This experiment is readily adaptable for space flight, which will remove the terrestrial noise source and provide the best determination yet of the solar diameter and its possible oscillations.

F. Atmospheric Effects

The Earth's atmosphere produces many sources of noise for helioseismic observations. Seeing, scattering, refraction and transparency variations all degrade the quality of helioseismic measurements in different ways. The obvious strategy for dealing with these effects is to obtain the observations in the best possible conditions, but this is incompatible with the need for long temporal sequences of data. Thus, some effort has been expended on understanding and correcting for terrestrial atmospheric effects on the oscillation observations. The most ambitious effort has been in the context of the GONG project, which has initiated a program to produce artificial helioseismology data that is degraded by simulations of the Earth's atmosphere (Global Oscillation Network Group 1987; Hathaway 1988).

The effects of turbulence, or seeing, include blurring, differential stretching, and image jitter. The effects of seeing on the power spectra of the oscillations have been modeled (F. Hill et al. 1984a,d,1987a,1991). In this model, seeing is simulated by moving a pixel to some random position, partitioning the data and adding it into the surrounding pixels. The new position is computed from a two-dimensional spatial power spectrum of the seeing. The results show that the dominant effect of seeing on the power spectra is to redistribute the power due to nearly steady background solar velocity fields, such as supergranulation or rotation, into the region of the oscillations. This is confirmed by observations (Ulrich et al. 1984). The results also reproduce the attenuation of the high spatial wavenumber components of the oscillations, or blurring. The power redistribution can be greatly reduced by integrating the observations over several seeing correlation times. Seeing noise can also severely impact phase and coherence studies of solar atmospheric waves, especially at very high temporal and spatial frequencies.

The presence of dust in the Earth's atmosphere scatters light. The redistribution of light from other parts of the solar disk with different Doppler shifts effectively smears the large-scale steady flow fields into the oscillation velocity signal. This will again redistribute the power in the spectra and degrade the signal-to-noise ratio of the oscillations. These effects can be computed and corrected for, as long as a measurement of the scattering function is made (Andersen 1985,1986).

Refraction distorts the shape of the solar image, especially near sunrise and sunset. A common observing strategy is to halt observations within 1 hr of sunrise and sunset. It is also common practice to fit an ellipse to the solar image, and then interpolate the data to a circular image, thereby removing much of the residual refraction distortion. However, refraction does not produce an elliptical Sun, so data treated in this way still contains an unknown residual distortion. Refraction fluctuations are a source of noise for diameter observations (Fossat et al. 1981).

Transparency variations can appear as spurious velocities. A spatial transparency gradient across a resolution element results in an apparent velocity variation for instruments that are sensitive to an intensity-weighted mean of the velocity. The effect is negligible for imaged observations, but very important for nonimaged whole-disk observations (Belmonte et al. 1988). Temporal transparency variations can also produce apparent velocities but, because these variations are typically low-frequency, rapid modulation of the observations is an effective filter. Both the temporal and spatial transparency variations are a major source of noise for diameter observations (Clarke 1978; Fossat et al. 1977; H. Hill et al. 1983; Yerle 1986).

G. Reduction Techniques

Here again, a division of topics is provided by the different requirements for nonimaging and imaging observations. We will also discuss data reduction for the study of waves in the solar atmosphere.

1. Nonimaged Observations. The calibration of the observed velocity V measured by nonimaging observations is provided by the accurately known and varying velocity changes of the observer relative to the Sun (see, e.g., Brookes et al. 1976; Pallé et al. 1986a; Ehgamberdiev et al. 1990). Figure 4 illustrates a typical result. The velocity V is given by

$$V = V_{\text{orbit}} + V_{\text{observer}} + V_{\text{grs}} + V_{\text{conv}} + V_{\text{osc}} \qquad (13)$$

where V_{orbit} is the radial part of the Earth's orbital velocity, varying about zero by ± 500 m s^{-1}; $V_{\text{observer}} = V_{\text{spin}} \cos(\text{declination}) \cos(\text{latitude}) \sin(\text{hour angle})$ is the radial velocity of the observer relative to the Sun providing a diurnal variation of precisely known value at any given time; V_{grs} is the contribution due to the gravitational potential difference between the solar photosphere and the Earth's surface and additional second-order Doppler shifts; V_{conv} is due to the convective blue shift (Schröter 1957) and possible isotope effects (for potassium only); and finally, V_{osc} is the Doppler shift due to oscillations. The diurnal variation provides a continuous calibration which, however, vanishes at the South Pole. An additional calibration, relying on rapid modulation of the magnetic field and thereby of the positions of the laboratory Zeeman components is currently being developed (Isaak and Jones 1988).

In Eq. (13), it has been tacitly assumed that the Sun's rotation and the atmospheric extinction gradient across the solar disk average out to zero. The first of these assumptions is only approximately valid. Thus, various authors (Durrant and Schröter 1983; Andersen and Maltby 1983; Edmunds and Gough 1983; Duvall et al. 1983) have pointed out that sunspots and plages contribute to a modulation effect at twice the solar rotation frequency and its

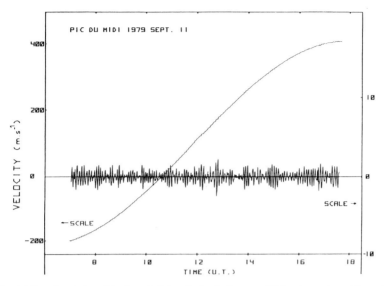

Fig. 4. Diurnal variation of the line-of-sight velocity for integrated-light observations. The residuals from a fit to the diurnal variation are also shown; these residuals are the V_{osc} term in Eq. (13) (figure from Calverie et al. 1981).

harmonics. A modulation of the correct order of magnitude was first seen by Claverie et al. (1982). Atmospheric extinction contributes to the noise, particularly at low frequencies, and great care has to be taken in extracting information relating to oscillations with very low frequencies such as g-modes. Partially successful attempts have been made in the past to reduce these effects by appropriate modeling of the atmospheric extinction in terms of observed intensity variations (Isaak 1986; Belmonte et al. 1988). However, it is better to measure the gradient of extinction across the solar disk (McLeod and Isaak 1988).

After the calibration of the raw data and the application of the transparency correction, the data are fitted using Eq. (13), and daily time series of the residual V_{osc} are formed. These daily time series must be combined to obtain high resolution in temporal frequency. If the time series are available from different stations in a network, the daily observations must also be merged together. For the unimaged whole-disk or differential observations, this merging is relatively simple, consisting of a weighted average of at most four samples per time step (Gelly 1990). However, even this is not an easy task as discontinuities in the slope or offset of the signal can introduce high-frequency noise. The merging problem is many orders of magnitude more difficult for imaged observations from a network such as GONG.

Gaps in the time series will remain even after the merging. Provided that the data has a high signal-to-noise ratio, and that substantial lengths of data surround the gaps, then short gaps of a few minutes can easily be filled either by simple interpolation techniques, or by more sophisticated maximum entropy methods (Fahlman and Ulrych 1982; Brown and Christensen-Dalsgaard 1990). Recently, application of linear deconvolution techniques to data obtained at the South Pole has also yielded good results in the removal of the effects of the observing window on the solar power spectrum (Duvall, personal communication, 1989).

After merging and gap filling, the data can be tapered with one of several possible functions to further reduce the sidelobes of the window function. Next, the power spectrum of the data is computed. This can be done using a fast Fourier transform, however the relatively small amount of data also allows the use of interactive sinusoid fitting methods which can provide more precise frequencies. An example of the power spectrum that results from the whole-disk nonimaged type of observations is shown in Fig. 5. This spectrum was obtained in 1981 using data from two stations (Tenerife and Haleakala) of the Birmingham network. Identification of the modes in this diagram can be accomplished by first re-arranging the spectrum into an echelle diagram. In this diagram, an example of which is shown in Fig. 6, the spectrum is cut into equal-length sections and then are displayed the sections above each other (Grec et al. 1983). For the solar spectrum, the length of the sections is about 135 μHz. Comparison of the observed echelle diagram with theoretical predictions allows the identification of the l-values of the modes.

The differential nonimaged observations are reduced in a manner similar to the whole-disk nonimaged data. A major difference between the two types of data lies in the spatial response functions. As discussed in Sec. III.B, nonimaged observations act as spatial filters, restricting the range of l-values present in the power spectrum. Figure 7 shows the spatial response functions for both the whole-disk and the differential observations. The difference in the responses is reflected in different patterns of peaks present in the power

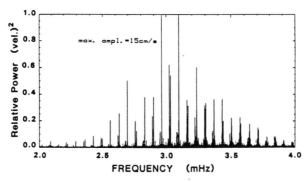

Fig. 5. Power spectrum of low-l modes from Birmingham integrated-light observations obtained at Tenerife and Hawaii over 3 mon in 1981 (figure from Deubner and Gough 1984).

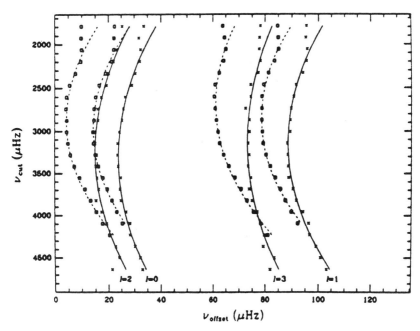

Fig. 6. Echelle diagram of the observed and theoretical frequencies of the solar oscillation spectrum for low l-modes. To produce this figure, a representation of the frequencies of the modes in spectra such as those seen in Figs. 5 and 10 is cut into equal-length sections, which are then displayed in strips that are displaced vertically from each other. The vertical axis gives the starting values of the strips, ν_{cut}; the horizontal axis shows the frequency offset from the beginning of the strip, ν_{offset}. The x symbols and solid lines show the observed frequencies from Pallé et al. (1986b), while the open squares and dashed lines show the theoretical frequencies from the CNO Cor model of Bahcall and Ulrich (1988) (figure from Bahcall and Ulrich 1988).

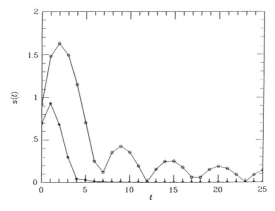

Fig. 7. The spatial response functions of nonimaged whole-disk observations (filled circles) and nonimaged annular observations (open circles) (figure from Pallé et al. 1989).

spectra and echelle diagrams of the two methods. Comparison of spectra obtained by the two methods allows the identification and removal of modes common to both, thereby increasing the confidence of the mode identification (Pallé et al. 1989).

Frequencies, amplitudes and line widths can be measured from the non-imaged spectra by a variety of methods. Since the modes are well separated, the most accurate method of determining the parameters of the oscillation is to fit a functional form to the clean isolated peaks in the spectra. In the past, the function was typically a gaussian (Libbrecht 1986), but a more accurate description of the line shape of a stochastically excited oscillator is provided by a Lorentzian profile. It is not correct to apply standard least-squares fitting techniques to this problem because the noise in the process is not normally distributed, but instead is distributed as χ^2 with two degrees of freedom (Woodard 1984a,b; Duvall and Harvey 1986). Thus, a maximum likelihood fitting method is typically used. The fits provide frequency and amplitude measurements, as well as estimates of the widths of the spectral lines and thus the lifetimes of the modes, an important input to studies of the excitation and damping of the oscillations. In addition to individual peak fitting, cross correlation between spectra obtained at different epochs is used to measure frequency changes, and autocorrelation functions provide information on periodicities present in the power spectrum itself.

2. Imaged Observations.   Imaged helioseismic observations often produce enormous amounts of data that must undergo complex numerical processing. To quantify this, consider the example of the GONG project. This project will collect a set of three, 16-bit, 256×256 pixel images of the Sun every 60 s at six sites around the world for a total of 3 yr. Assuming an average of 10 hr of observing per site per day, the total amount of raw data collected by the project is estimated to be 1200 Gbytes. The data products of the project will more than double the amount to 2900 Gbytes (Kennedy and Pintar 1988; Pintar et al. 1988). Another instructive (or frightening) example is provided by the SOI instrument on SOHO. This instrument will produce two, 16-bit, 1024×1024 pixel images every minute, resulting in 2200 Gbytes per year of raw data. The mission duration may be as long as six years, producing a total raw data base at the detector of 13,200 Gbytes. Telemetry limitations reduces this to 300 GBytes/year of usable raw data at the ground, but subsequent processing will expand the final data set to a size of 6000 to 18,000 Gbytes (Scherrer et al. 1988). Data sets from a single ground-based site are of course much smaller than the GONG and SOI examples, nevertheless they still represent substantial data processing challenges.

These gigantic data sets must flow through a series of processing steps. *The first step* is a preliminary reduction that transforms the raw data into images of either Doppler velocity or relative intensity variations. Several sub-steps must be performed:

a. Gain and dark current corrections must be made, as well as any other instrument-dependent calibrations. These are hopefully well known and straightforward to remove. Problems can arise however, particularly from the difficulty of obtaining a uniform flat field from an extended structured source such as the Sun.

b. The effects of the terrestrial atmosphere should be removed. These effects include scattered light, distortion of the circular image by refraction, differential stretching from atmospheric seeing, and blurring. All of these processes are sources of noise for oscillation observations, typically affecting the power spectra either by reducing the signal-to-noise ratio in certain bands, or by redistributing the power between bands. Each of the effects is computationally difficult and expensive to remove; consequently, all or parts of this step are frequently omitted.

c. The raw data must be transformed into either Doppler velocity or relative intensity measurements. This is usually a straightforward application of a simple arithmetic algorithm.

d. The images must be registered to remove image motion from terrestrial atmospheric seeing, guiding errors and jitter. This is accomplished by fitting a function to the estimated solar limb to obtain an image center and size, and then shifting the images to a common position and size via an interpolation technique.

e. Other velocity fields exist on the Sun that are not oscillatory in nature, and are a source of noise for helioseismology. These fields include the surface differential rotation, the limb red shift, and the line-of-sight components of the Earth's orbital and rotational motions. These effects are not important for intensity measurements of the oscillations, and are well-known velocity effects that can be easily subtracted. However, the convective velocity fields of granulation and supergranulation are continually evolving, making their removal difficult.

f. In the context of a network, an additional and crucial preliminary step must be performed. This is the process of merging two or more solar images obtained simultaneously at different sites. The details of this procedure are currently undefined, but the final merged image must generally consist of a weighted average of the contributing images. The major challenge is the determination of the weights, which involves the development of quality assessment measures for the images. Another challenge is to determine just where in the data flow the merging should be done, as it is possible that inspection of the spatial power spectrum of the images will be a more informative quality assessment procedure. A last challenge is the efficient implementation of the merging scheme, a seemingly complex task that may require artificial intelligence.

The second major step after the preliminary reduction is the spatial filtering of the images. For full-disk-imaged measurements, this most frequently means decomposing the observed velocity or intensity field into its spherical harmonic components. This is a computationally intensive task, requiring extremely efficient algorithms. The simplest method is to multiply the data images directly by a series of masks, which are suitably normalized images of each of the desired spherical harmonics on the same sampling grid and central position as the data. The result is then integrated over the solar disk to produce an estimate of the amplitude of the spherical harmonic contained in the data. This simple method unfortunately is extremely inefficient, with the number of operations required to filter a single image scaling as N^4, where N is the length of one side of the square image in pixels. The computations required to create the masks can be eliminated by calculating them once, storing them, and then reading them when needed, but this then requires an enormous storage capability, and input/output is considerably slower than computation. The practical limitations of the method have led to its use only for studies of small sets of spherical harmonics (see, e.g., Libbrecht and Zirin 1986).

The quest for an efficient spherical harmonic transform has led to the development of a technique that separates the process into two one-dimensional transforms (Brown 1985,1986,1988c). The observed velocity or intensity field is first interpolated onto an equally spaced grid in longitude and sine of the latitude. The interpolation must include compensation for the B angle of the Sun. A fast Fourier transform is applied in the longitude direction, resulting in an image of complex amplitudes as a function of m and sine latitude. Then, a one-dimensional Legendre transform is applied to the sine latitude direction resulting in complex amplitudes as a function of both l and m. The Legendre functions are recursively generated as needed, rather than precomputed and stored. As less than half of the solar surface is available for analysis, the Legendre functions need only be computed for every other value of m. This technique requires N^3 operations per image, a factor of N fewer than the masking process. Thus, the computational savings rapidly increase with N.

Currently, the combination of an FFT and a Legendre transform appears to be the most efficient method available, and consequently it has become almost universal in the analysis of imaged helioseismic data. However, it is still an expensive method, motivating attempts to develop faster methods. Dilts (1985) proposed a method that uses a two-dimensional FFT of the data image, and then constructs the spherical harmonic coefficients from the complex Fourier components. Unfortunately, tests of the method showed that it failed to provide any advantage over the masking technique when applied to the task of computing spherical harmonic coefficients for many values of l and m (Elowitz et al. 1989).

An alternative method of spatially filtering the observations is simply to perform a two-dimensional Fourier transform after the images have been interpolated onto an evenly spaced longitude and latitude grid (F. Hill 1988a).

Typically, this is done on a portion, or subraster, of an image allowing a mosaic of spatial filters to be formed on the solar disk. The oscillations are represented as locally propagating plane waves as a function of the components k_x and k_y of the horizontal wavenumber vector $\mathbf{k}$ in this analysis, rather than global spherical harmonics characterized by l and m. This is one step of the process that leads to oscillation ring diagrams (Morrow 1988a; F. Hill 1988a).

A simple spatial filter that isolates nearly sectoral modes is the averaging of the data along lines perpendicular to the solar equator. A more sophisticated filter averages the data along lines of constant longitude. These techniques can be optically implemented, and have been used extensively in observations of intermediate- and high-l sectoral modes (Duvall and Harvey 1983; Harvey and Duvall 1984; Rhodes et al. 1977; F. Hill et al. 1986; Haber et al. 1990a).

After spatial filtering, *the third major step* in the reduction of helioseismic data is time series analysis, or the computation of temporal power spectra of the spatial filter coefficients. This process typically begins with an inspection of the time series to remove obviously bad data values. The inspection process is nearly impossible to automate, and is thus a tedious job that requires the use of a human data analyst. Next, the time strings of individual spatial filter coefficients must be extracted. This involves the transposition of the three-dimensional spatial filter coefficients array into a useful arrangement. The process is conceptually simple, but again because of the typically large data volumes, it is a fairly difficult practical challenge.

The next step in the time series analysis is the filling of gaps in the temporal strings as discussed previously. After gap filling, the time strings can then be apodized, or multiplied by a smoothing function intended to reduce the sidelobes of the window function of the data. Finally, power spectra for each of the spatial filter coefficients are computed from the time strings via a fast Fourier transform.

The power spectra are an important product of the data reduction process. Displays of the spectra show the striking presence of localized regions of enhanced power, as shown in Fig. 1. These structures are the result of the dispersion relation of the oscillations and the physical conditions of the solar plasma in the region of propagation of the waves. The classical example of these structures is the ridges seen in the two-dimensional spectrum, frequently called the k-ω or l-ν diagram. The existence and the position of the ridges in the spectrum confirmed the description of the oscillations as acoustic modes trapped below the photosphere. In three-dimensional spectra, the structures are either roughly triangular curved sheets for the spherical harmonic decomposition, or trumpet-shaped closed surfaces for the plane-wave Fourier analysis. The three-dimensional spectra are most easily and usefully displayed as slices perpendicular to one of the coordinate axes. For the spherical analysis, the slices are typically made either at constant l, producing m-

ν diagrams, or at constant m, once again producing l-ν diagrams. In the plane analysis, the slices are made at constant ν, producing ring diagrams as a function of k_x and k_y. Figure 8 displays examples of observed power spectra slices, while Fig. 9 shows theoretically generated three-dimensional power spectra for the spherical harmonic and plane-wave cases.

The measurement of the locations, shapes and dimensions of these structures provides the information that is used to infer the physical conditions in the solar interior and to calibrate solar models. The most basic quantities of interest are the frequency and amplitude of the oscillations as a function of l, m and n. The frequency is determined by the position of the structure along the ν-axis of the spectrum. The most accurate frequency measurements come from spectra that clearly isolate a single mode of the oscillations. This is possible using spectra obtained from data free of the diurnal gaps, or in regions of the spectrum where the modes are well separated and the diurnal sidelobes can be clearly avoided. Even in this case, some confusion is generated by the appearance of a single mode in several spectral slices with similar values of l and m. These ghost modes result from incomplete spatial filtering due to the restricted observable area on the solar surface, and can be seen in Fig. 8a. Their presence can make it difficult to identify reliably individual peaks in the power spectrum with the correct l and m. In addition, the presence of differential rotation may result in the distortion of the spherical harmonic pattern of the modes with $l > 100$ (Woodard 1989). This could shift the maximum response of the spatial filtering away from the target mode to other ghost modes in the spectrum, further complicating mode identification. Identification can be done by comparing theoretically predicted frequencies with the spectra. Sophisticated automated mode identification techniques have been developed (Brown 1988a; Morrow and Brown 1988) which can consider diurnal sidelobes, ghost modes, noise peaks, and the overall signal-to-noise ratio of the observations. The techniques have proven to be useful for high-quality data but, as might be expected, they tend to fail for either poor data, or poor guesses about the distribution of noise in the data.

Individual spectral line fitting can only be performed for isolated mode peaks. In practice, this has so far been possible for data with $l < 100$ (Duvall et al. 1988a), but recent data from the South Pole indicate that it is possible to extend this type of analysis up to at least $l = 150$ (J. W. Harvey, personal communication). Automated techniques to determine the frequencies in large numbers of spectra are currently under development, motivated by the GONG project (Anderson et al. 1990). Above some currently unknown l-value, it is likely that the character of an oscillation will change from a global mode to a locally standing acoustic wave. This change will occur at the l-value above which the mode lifetime is shorter than its transit time around the solar circumference. Under these conditions, the wave will not survive long enough to self-interfere constructively in the horizontal direction, and hence it will lose its global nature. The lifetime is reduced by transmission and reflection

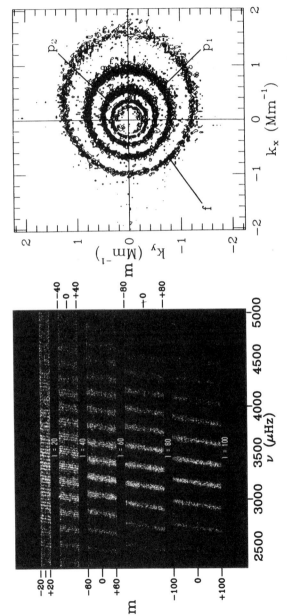

Fig. 8. Examples of slices of observed multidimensional power spectra. *Left Panel*: A set of five *m-ν* diagrams or slices at constant *l*. "Ghost" ridges arising from incomplete spatial filtering can be seen most clearly in the *l* = 40 spectrum. The convention that prograde modes correspond to negative values of *m* results in a positive frequency shift for those modes (figure courtesy J. Harvey from Duvall et al. 1987). *Right Panel*: A ring diagram of the oscillation power as a function of the components of the horizontal wavenumber vector, k_x and k_y, at constant *ν* = 3010 μHz (figure from F. Hill 1989*b*).

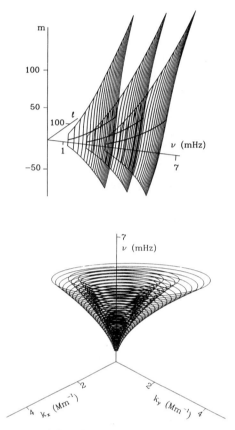

Fig. 9. Theoretically generated three-dimensional power spectra for spherical harmonic and
plane-wave representations of the solar p-mode oscillations. *Upper Panel:* Spherical harmonic
"sail" surfaces for $n = 4$, 12 and 20. The surfaces have been computed with the solar surface
rotation value of a_1, but the values of a_3 and a_5 have been exaggerated by a factor of 50. The
curves have been computed for every tenth value of l with $10 \leq l \leq 200$. The bold curves in
the l-ν plane connect modes with $m = 0$. *Lower Panel:* Plane-wave "trumpet" surfaces for
$n = 1$, 2 and 3. The surfaces have been computed from Eq. (19) with depth-independent
values of $U_x = 1850$ m s^{-1}; $U_y = 100$ m s^{-1}, $p = 0.5$ and the polytropic value of c (figure
from F. Hill 1988a).

of the wave through inhomogeneities such as magnetic or convective velocity
fields. The effect of the inhomogeneities increases as the spatial wavelength
of the mode decreases, or as l increases. Thus, above the l-value at which the
lifetime of the waves becomes short compared with its global transit time, the
structure of power in the spectrum should change from a string of isolated
peaks to a continuous ridge of power, even in spectra obtained with high-
frequency resolution. In this case, individual spectral line fitting is not possible.

In the absence of individually resolved modes, the usual strategy for measuring quantities is to fit a function to the shape of an entire discrete structure in the spectrum. For the ridges in the two-dimensional l-v diagram, Gaussians have been fitted across the ridges in v (Libbrecht and Kaufman 1988), and cubic splines have been fitted to the length of the ridges (F. Hill et al. 1984d). Ellipses have been fitted to ring diagrams (F. Hill 1988a,b). Fitting techniques have been applied to the ridges in the m-v diagrams (Libbrecht 1988c,1989), but in this case cross-correlation techniques have more often been used (Brown and Morrow 1987a; Duvall et al. 1986; Tomczyk et al. 1988) to determine frequency splittings.

Other quantities besides frequency and amplitude are of great interest for helioseismologists. Processes such as the solar internal differential rotation remove the degeneracy of frequencies of modes with the same l and n but different m, and produce ridges with a distinctive functional dependence $v(m)$. The difference δv, defined as

$$\delta v \equiv v(l, m, n) - v_0 \tag{14}$$

is called the frequency splitting. In Eq. (14), v_0 is the frequency of the mode free of splitting effects. In practice, v_0 is taken to be either $v(l, 0, n)$, the frequency of the zonal ($m = 0$) mode, or $\bar{v}(l,n)$, the m-averaged frequency of the multiplet ridge. For differential rotation,

$$\delta v = -m \; \frac{\displaystyle\int_{-1}^{1} \bar{\Omega}/2\pi(\theta) \left[P_l^m(\cos\theta)\right]^2 d\cos\theta}{\displaystyle\int_{-1}^{1} \left[P_l^m(\cos\theta)\right]^2 d\cos\theta} \tag{15}$$

where $\bar{\Omega}/2\pi(\theta)$ is the solar rotation rate averaged over the depth range of the oscillation with given l and n, θ is solar latitude, and P_l^m is the Legendre function for the mode (Cuypers 1980; Brown 1986). As the integrals in this expression are symmetric in m, the frequency splittings due to differential rotation are antisymmetric in m. However, other processes can produce splittings that are symmetric in m, such as changes in the internal temperature or magnetic field.

It has become customary to represent the splitting of a multiplet ridge as coefficients of a series expansion. The first expansion used in this context was powers of m (Brown 1985,1986). The basis functions for this series are not orthogonal, and thus the coefficients were found to depend upon the number of terms retained in the expansion. A nearly orthogonal series is provided by using Legendre polynomials instead of powers of m as the basis functions (Duvall et al. 1986). In this representation, the splitting is given by

$$\delta v = L \sum_{i=0}^{i=N} a_i P_i(-m/L) \tag{16}$$

where P_i is the Legendre polynomial of degree i, $L = \sqrt{l(l+1)}$, and N is number of terms retained in the expansion, typically $N = 5$. The choice of m/L rather than m/l reduces the variation of the coefficients a_i with degree l. However, the coordinate m/l is also used, producing a small difference in the values of the a_i for $l < 20$. The odd coefficients can be added to approximate the equatorial rotation rate:

$$\Omega_{eq}/2\pi \approx a_1 + a_3 + a_5 \equiv a_{135}. \tag{17}$$

An alternative expansion has also been proposed (Durney et al. 1988):

$$\delta v = m \sum_{i=0}^{i=N} b_i P_i(m/L) \tag{18}$$

where the symbols have the same meaning as above. In this expansion, the coefficients b_i are more closely related to a Legendre polynomial expansion of the solar rotation rate as a function of latitude than the a_i, which reduces the cross-talk between the components of the rotation in the expansion. Comparison of the predicted relationship between the b_i and the a_i determined from NSO/South Pole data shows that the expansion is valid (Durney 1990).

As mentioned, cross-correlation techniques are most often used to measure the splittings. In this method, first a reference spectrum, $S_0(v)$, is created for a given l either by removing the splitting effects and averaging over m, or by simply using the zonal spectrum. This reference spectrum contains peaks for each value of n available at the selected l. The reference spectrum is cross correlated with the full spectrum at each value of m, and the offset is then δv. The method reduces the noise in the measured δv at the expense of the loss of resolution in radius and latitude due to the elimination of information about the dependence of the splittings on n. If the signal-to-noise ratio in the data is high, then the multiplets can be fit individually. This preserves all of the available information, and will become a more common practice as the quality of the observations improves.

In the plane-wave decomposition, the oscillations appear as rings in slices of the power spectrum at constant ω, where $\omega = 2\pi v$ (Morrow 1988a; F. Hill 1988a). If a simple power law representation, $\omega_0 = ck^p$, is assumed for the dispersion relation of the waves in the absence of any flows, then in the presence of flows the equation of the rings for a slice at frequency ω_d is given by

$$\omega_d = c(k_x^2 + k_y^2)^{p/2} + k_x U_x + k_y U_y \qquad (19)$$

where k_x and k_y are the horizontal components of the wavenumber vector **k**, and U_x and U_y are the depth weighted average horizontal components of a subsurface velocity field **U** (F. Hill 1988a). The rings can be approximated by ellipses, and then it can be shown that U_x and U_y are proportional to the central coordinates of the ring in k_x and k_y. Furthermore, the semi-major and minor axes of the ellipse can be used to measure the constants c and p in the unperturbed dispersion relation (F. Hill 1988b). The information thus derived from the rings can be used to infer a map of the horizontal flows as a function of depth in the convection zone (F. Hill 1989b,1990b).

3. Atmospheric Waves.   The same techniques used to obtain imaged helioseismic data are also used for observing atmospheric waves, and much of the same preliminary data reduction steps are also used. However, there are differences in the later stages of the processing. For these running waves, a cavity does not exist either because they are small scale, and the reflecting layers are too rough to permit coherent reflection, or because the waves are scattered or dissipated before the reflecting layers are reached. Without a cavity, these waves do not develop coherent standing-wave patterns as the global p-modes do. They have no eigenfrequencies or particular wavenumbers with enhanced power. Their power spectra are continuous, and so are the phase and coherence spectra derived from velocity V and intensity I fluctuations observed in spectral lines (and continuum) chosen to represent certain levels or conditions in the solar atmosphere. These fluctuations are recorded as functions of position x on the solar surface, and time t, and analyzed in much the same way as one studies seismic disturbances in the Earth's crust: differences in travel time between certain locations and for different seismic-wave modes are measured.

For comparison with theoretical predictions, Fourier transforms $T_v(k, \nu)$, and $T_i(k, \nu)$ of the observed time series $V(x, t)$ and $I(x, t)$ are computed and combined to yield spectra of power,

$$P_a(k, \nu) = \frac{\sum_k \sum_\nu T_a T_a^*}{(\Delta k \cdot \Delta \nu)} \qquad (20)$$

phase differences,

$$\Phi_{ab}(k, \nu) = \arg \sum_k \sum_\nu T_a T_b^* \qquad (21)$$

and coherence,

$$C_{ab}(k,\nu) = \frac{\left|\sum_k \sum_\nu T_a T_b^*\right|}{\sum_k \sum_\nu |T_a T_b^*|} \tag{22}$$

as a function of frequency ν or spatial wavenumber k, or both. The summation extends over suitably chosen frequency and wavenumber intervals Δk and $\Delta\nu$; it is optional for power P and phase Φ, where it is often applied to enhance statistical stability.

The amplitudes and phases of atmospheric waves as a function of height, spatial position, and time can also be studied by calculating the analytic signal through application of the Hilbert transform (Stebbins and Goode 1987). This technique then allows the empirical determination of the vertical eigenfunctions of the waves, providing information on the energy transport.

IV. RECENT RESULTS

In this section, we summarize some of the recent results of oscillation observations. We discuss results pertaining to nonimaged observations, imaged observations and atmospheric waves. Topics includes measurements of frequencies, amplitudes and the line widths; fine structure; solar cycle variations; splittings; active-region effects; diameter measurements; the 160-min oscillation; very high-degree and high-frequency oscillations; energy transport; interaction of convection and oscillations; and chromospheric oscillations.

A. Unimaged Observations: Global Oscillations

We first turn to a discussion of the global scale low-l modes obtained from nonimaged observations, both whole-disk and differential. These modes have lifetimes on the order of months, substantially longer than the few days for the higher-degree modes. As these modes are very long lived, their frequencies can be accurately measured, and the relative sparseness of modes simplifies study of their spectrum. The low-l modes are also of great interest to stellar physics because they penetrate to the energy-generating core regions of the Sun and may eventually be useful for probing other stars. Thus, the study of the frequencies of these modes provides information on the structure and dynamics of stellar cores. Observations of the low-l modes may help to resolve the solar neutrino problem, although it now appears that the resolution of this problem lies in the domain of particle physics. Information on the solar cycle is also provided by studying the change of the frequencies with the phase of the cycle.

1. Spectra and Echelle Diagrams. An example of a low-l power spectrum obtained with nonimaged whole-disk observations is shown in Fig. 5.

This spectrum was obtained by the Birmingham group using two stations at Tenerife and Hawaii in 1981. Figure 10 shows two additional examples of low-l spectra, obtained at the South Pole in 1980 (Grec et al. 1983), and with the IPHIR instrument on the PHOBOS spacecraft in 1989 (Fröhlich et al. 1990). All three of these spectra are substantially free of the diurnal side-lobes, and are examples of the three observing strategies used to achieve nearly continuous observations. These results show a discrete spectrum of many modes superimposed on a continuum.

The discrete spectrum, rising clearly above the continuum between 1.8 mHz and 5 mHz, shows a periodicity of approximately 135 μHz. This periodicity can be displayed in an echelle diagram, as seen in Fig. 6. The periodicity arises from the asymptotic behavior of modes with low l and high n (Tassoul 1980,1990). For such modes,

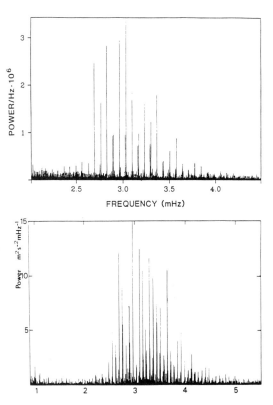

Fig. 10. Examples of low-l power spectra. *Top panel:* Spectrum obtained from the IPHIR experiment on the PHOBOS space craft (figure from Fröhlich et al. 1990). *Bottom panel:* Spectrum obtained by the Nice group at the South pole (figure from Grec et al. 1983). The spectra in this figure and in Fig. 5 illustrate the three observing strategies used to eliminate the day-night cycle.

$$\nu_{nl} = \left[n + \frac{1}{2} \left(l + \frac{1}{2} \right) + \alpha \right] \nu_0 + \varepsilon(n,l) \tag{23}$$

where

$$\nu_0 = \left(2 \int_0^{R_\odot} \frac{dr}{c} \right)^{-1} \tag{24}$$

and c is the sound speed, α is a constant of order unity, and ε is a small correction term that depends on conditions in the solar core. From Eq. 23, it can be seen that $\nu_{nl} \approx \nu_{n+1,l-2}$ and that $\nu_{nl} \approx \nu_{n,l+1} + \nu_0/2$. Thus the periodicity in the low-l spectrum is $\nu_0/2$, and ν_0 measures the sound travel time from the center to the solar surface and back. The oscillations can be identified as acoustic modes of degree $l = 0, 1, 2$ and 3 by comparing the echelle diagram with predictions of solar models (Grec et al. 1983). A definite determination of the number of radial nodes n between the surface and the lower reflection boundary could only be made with certainty when the high-l modes of known (n,l) were traced to near $l = 0$ (Duvall and Harvey 1983). Thus we now know the (n,l) assignment of the various modes; the values of n for these modes range from 10 to 40. Searches for low-l, low-n modes with $0 \le n \le 10$ have been made (Anguera Gubau et al. 1990), but the low amplitudes (less than 2 cm s^{-1}) of modes with $\nu < 2$ μHz have hampered their detection.

Spectra from differential observing techniques have so far only been obtained from single groundbased midlatitude sites; thus they are all degraded by the presence of the daily sidelobes. Nonetheless, comparison of these spectra with the whole-disk spectra and with theory have enabled the identification and measurement of modes with l-values of 4 and 5 (Henning and Scherrer 1986,1989; Pallé et al. 1989).

A measurement of the solar oscillation spectral continuum was published by Jiménez et al. (1988b) but no allowance was made for the noise due to the terrestrial atmosphere, the apparatus nor photon statistics. Thus, that spectrum provides us only with an upper bound; however, it is in general agreement with the theoretical prediction of Harvey (1985a).

2. Eigenfrequencies.   Figure 6 shows a comparison between an observed echelle spectrum (Pallé et al. 1986b) and a theoretical fit (Bahcall and Ulrich 1988). It can be seen that the discrepancies between observation and model are less than some 15 μHz, i.e., 0.5% of the frequencies. Such excellent agreement would generally be regarded as a triumph in general astrophysics. In the case of the Sun, where most of the parameters are better known, we expect agreement to within observational errors which are, on

average, some 2 orders of magnitude smaller. We thus conclude that the physics of the Sun is not fully understood.

This is an area of much current theoretical work, and it will be treated
in detail in the chapter by Ulrich and Cox. In a brief summary, the major
causes of the discrepancy between theoretically calculated and actually observed frequencies are currently thought to be either an inadequate description of the equation of state and the inaccurate computation of the opacity in
the solar interior (see, e.g., Christensen-Dalsgaard et al. 1988; Guenther et
al. 1989; Korzennik and Ulrich 1989; Cox et al. 1989), or a lack of detailed
knowledge about the interaction of the oscillations and the turbulent granulation near the solar surface (Christensen-Dalsgaard 1988c; Stein et al. 1989).
Other theoretical investigations include mixing (Cox et al. 1990) and exotic
particles (see, e.g., Spergel 1990; Cox et al. 1989).

Tables of measured low-l high-n frequencies can be found in many of
the papers reporting on the observations (see, e.g., Grec et al. 1983; Scherrer
et al. 1983; Harvey and Duvall 1984; Woodard 1984a; Henning and Scherrer
1986; Libbrecht and Zirin 1986; Jiménez et al. 1988e; Pallé et al. 1989).
Comprehensive summaries can be found in Duvall et al. 1988. Many of the
published tables of low-l frequencies do not explicitly state errors for the
measurements. Those that do (Jiménez et al. 1988e; Woodard 1984a) typically show errors ranging from 0.1 to 1.3 μHz, with an average of about 0.4
μHz. A table of frequencies for low-l low-n modes can be found in Anguera
Gubau et al. (1990), but these are difficult modes to detect.

3. Variations in Eigenfrequencies.   Variations in the frequencies of the
low-l modes as a function of the solar activity cycle have been searched for
since their discovery. These variations could result from changes in the dimensions and shape of the solar cavity and/or in the speed of sound. Precision
on the order of a few parts in 10^5 is required to measure frequency changes
corresponding to estimated changes in diameter (Gilliland 1981; Delache et
al. 1985), and in luminosity (Willson and Hudson 1988), and if no fortuitous
cancellations occur.

A decrease in the frequencies of ~0.4 μHz between solar maximum in
1980 and minimum in 1984–86 was first measured by Woodard and Noyes
(1985,1986,1988) using ACRIM data, by the Nice group comparing all AC-
RIM data with 1984–85 South Pole velocity data (Fossat et al. 1987; Gelly
et al. 1988b, Fossat 1988b), and by the Stanford group using a long time base
of differential observations (Henning and Scherrer 1988a). On the other hand,
van der Raay (1984); Isaak et al. (1988); Jefferies et al. (1988b); and Pallé et
al. (1988b), using different techniques analyzed various portions of the Bir-
mingham/Izaña velocity data sets between 1977 and 1987, obtaining a variety
of inconsistent results. The situation was similar for the intermediate-l
modes, as discussed in Sec. IV.B.1.

In late 1989, this situation changed dramatically. Re-analysis of the existing data sets with improved techniques, and the discovery of errors in the previous analyses resulted in agreement in both the sign and magnitude of the observed change in frequency from the Birmingham and Izaña data (Isaak et al. 1990; Pallé et al. 1989,1990a,b; Elsworth et al. 1990b) and also with the previous ACRIM and Nice South Pole results (Woodard and Noyes 1985; Gelly et al. 1988b). Similar agreement was reached with the spatially resolved observations as discussed later. It now appears that the frequencies of the low-l modes change over the course of the solar cycle with an amplitude of ∼0.4 μHz, and that the modes have higher frequencies at solar maximum than at minimum. There is some evidence that the variation depends on l, being absent for $l = 0$ (Pallé et al. 1989,1990b). In the future, our knowledge of the variation of the low-l modes over the course of the solar cycle will be greatly improved by the data from the IRIS network, which is scheduled to operate for an entire 11-yr solar activity cycle.

    4. Fine Structure.   In addition to v_0, the dominant periodic spacing of the modes discussed in Sec. IV.A.1, there is fine structure in the spectrum due to the small correction term ε in Eq. 24. According to Tassoul (1980), this term depends on l and v (and thus on n) according to

$$\varepsilon \approx [l(l + 1) + \delta]Av_0^2/v \qquad (25)$$

where A and δ are constants and v_0 is defined in Eq. 24. The parameter A is of interest to helioseismologists because its value depends on the structure of the solar core. By fitting polynomials to the curved shape of the lines corresponding to constant values of l in the echelle diagram, it can be shown that

$$v_{nl} - v_{n-1,l+2} \approx (4l + 6)D_0 \qquad (26)$$

where D_0 is directly proportional to A (Grec et al. 1983; Scherrer et al. 1983; Ulrich 1986; Christensen-Dalsgaard 1988b,c; Gelly et al. 1988a; Gabriel 1989). Thus, for example, $6D_0$ is the frequency separation between the $n = 22$, $l = 0$ and $n = 21$, $l = 2$ modes and $10D_0$ is the frequency separation between the $n = 22$, $l = 1$ and $n = 21$, $l = 3$ modes. These separations can be easily measured, providing values for D_0 and hence A. Furthermore, it may be possible to measure both v_0 and D_0 in other stars, thus providing an independent method for determining stellar masses and ages (Ulrich 1986; Christensen-Dalsgaard 1988b).

    Several studies of the low-l modes have measured the quantity $6D_0$. Claverie et al. (1981c) give 8.3 ± 0.3 μHz as the mean spacing between $l = 0$, $l = 2$ for eigenfrequencies between 2.4 and 3.85 mHz, while Pallé et al. (1986b) found a mean spacing of 9.2 ± 0.7 μHz. From space-borne observations, the ACRIM measurements found a value of 9.30 μHz (Woodard and Hudson 1983), while the IPHIR results provided a measurement of

$6D_0$ = 9.35 μHz (Fröhlich et al. 1990). Another analysis of the Birmingham/Izaña data derived a value of $6D_0$ of 9.12 ± 0.08 μHz (Jiménez et al. 1988e). More recently, Gelly et al. (1988b,c) found a value of D_0 of 1.52 ± 0.03 μHz. This value is in excellent agreement with the value predicted by some so-called standard solar models, but in disagreement with others. One difficulty with comparing the values of $6D_0$ between studies is that choices in the polynomial fit to the echelle diagram can alter the measured separations by a few tenths of a μHz. While this does not seem like a large effect, it can be larger than the theoretical variations between different solar models (Cox 1990b; Cox et al. 1989,1990).

5. Rotational and Magnetic Splittings.   Rotation removes the degeneracy of modes with the same values of n and l but different values of m, giving rise to an even finer splitting of the nonradial modes (Cowling and Newing 1949; Ledoux 1951) which should be measurable (Brookes et al. 1976). The measurement of the rotational splitting for the low-l, and especially the $l = 1$ modes is extremely difficult. As the total observed splitting is proportional to m, and as the maximum value of m for a mode is l, the rotational splitting for the $l = 1$ mode is not amplified, and is nearly a direct measure of the rotation rate in the core regions of the Sun. The surface rotation rate is ~0.4 μHz, so the time span required to achieve this frequency resolution is the surface rotation period of ~28 days. Data strings of this length are currently degraded by noise created by the finite lifetime of the modes, by background solar and instrumental sources, and by the window function of the observations. The small number of modes hinders the reduction of these problems by averaging. Thus, in order to measure the low-l rotational splittings reliably, observers will have to gather data nearly continuously for many mode lifetimes.

In spite of these difficulties, there have been several attempts to measure the $l = 1$ rotational splitting. The measured values have basically fallen into two classes. In one class, a value of ~0.75 μHz has been obtained, predominately from the Birmingham data (Claverie et al. 1981a,b; Isaak 1986; van der Raay et al. 1986; Pallé et al. 1988c). Another analysis of the data using a different technique gave a value of 0.68 μHz (Jefferies et al. 1988a). Comparison of the line width of the $l = 1$ mode to that of the $l = 0$ mode in the Nice South Pole data gave an upper limit of 0.75 μHz (Grec et al. 1983). A more recent unpublished measurement by the Nice group has lowered this upper limit to 0.6 μHz. These values are almost twice as high as the surface rotation rate, and, if correct, suggest a rapidly rotating solar core. However, the second class of splitting results have provided values consistent with the surface rate. Woodard (1984b) using ACRIM data placed an upper limit on the $l = 1$ splitting of 0.5 μHz. A line-width comparison of the $l = 1$ and $l = 0$ modes in the IPHIR spectrum provides an $l = 1$ splitting of 0.42 μHz (Fröhlich et al. 1990). These space-borne results are also supported by mea-

surements of low-l splittings by the Stanford differential method, which measured values of 0.4 μHz for the $l = 2$ to 5 modes (Henning and Scherrer
1986). Recently, a spectroscopic scheme has been proposed in an attempt to
improve the separation of the prograde and retrograde $l = 1$ modes (Cacciani
et al. 1990). It remains to be seen if this method will work.

Large magnetic fields are also capable of removing the degeneracy (see,
e.g., Goossens 1976; Gough and Thompson 1990) and it has been suggested
that these effects are seen as a perturbation in the intensities of the rotationally
split multiplet (Isaak 1982). Such a perturbation could show up before it
becomes noticeable as an additional splitting, analogous to the appearance of
the Hanle effect in atomic spectroscopy before the full Zeeman splitting. An
internal magnetic field of 10^7 gauss could result in an additional splitting of
0.05 μHz and, because each rotationally split component will then itself be
split into $2l + 1$ lines, there will be a total of $(2l + 1)^2$ components for each
line in the solar oscillation spectrum (Gough and Thompson 1990). So far,
no data set has both the required frequency resolution and the signal-to-noise
ratio to observe these possible magnetic splittings, but that should change
with the advent of the IRIS, GONG and SOHO projects.

6. Amplitudes.   As shown in Figs. 5 and 10, the solar oscillation spectral lines emerge out of the noise background above 1.6 mHz, reach a peak
at frequencies around 3.0 mHz, and then decrease until the vicinity of the
acoustic cutoff frequency of 5.7 mHz. This frequency dependence is also
obvious at higher values of l. The amplitude envelope is not completely
understood, and its shape provides clues to the damping and driving of the
modes, as discussed in Sec. IV.B on spatially resolved results.

The measured peak amplitude in Doppler velocity depends on the choice
of spectral line, reflecting the changing amplitude of the oscillations with
height in the solar atmosphere. In potassium, the observed maximum amplitude is 15 cm s^{-1}, while in sodium it is 25 cm s^{-1}. In irradiance, the observed
relative amplitude is 2 to 3 × 10^{-6}. Tables of measured Doppler velocity
amplitudes can be found in Grec et al. 1983; Scherrer et al. 1983; Pallé et al.
1989; and Anguera Gubau et al. 1990. A detailed comparison of the amplitudes of the low-l modes in the potassium and sodium lines can be found in
Isaak et al. (1989). A table of intensity oscillation amplitudes measured in
four wavelength bands can be found in Jiménez et al. 1990.

7. Line Widths.   As the frequency of low-l spectra improved, it became
clear that there was a width of the oscillation spectral lines in excess of the
width imposed by the duration of the observations. The line width Γ is presumably a measure of the lifetime of the modes, and provides information on
the driving and damping of the oscillations. It was soon obvious from a qualitative inspection of the early high-resolution spectra that the line width in-

creases with increasing frequency, implying that higher-n modes have shorter lifetimes.

The first attempt to measure the width of the $l = 0$ line could only estimate that it was less than the frequency resolution $\Delta \nu$ of 8.7 μHz (Claverie et al. 1979). The Nice South Pole data with $\Delta \nu = 2.0$ μHz estimated an upper limit for Γ of 5 μHz for modes with $l \leq 3$ (Grec et al. 1980). Analysis of a much longer time series with $\Delta \nu = 0.4$ μHz provided an estimate of $\Gamma \approx 0.5$ μHz for the $l = 0, 1$ modes (Claverie et al. 1981a,b). Re-analysis of the 1980 Nice South Pole data refined the measurement of Γ to 2 μHz, and also provided the first measurement of the increase of Γ with ν (Grec et al. 1983). The ACRIM data have a $\Delta \nu$ of 0.04 μHz; early estimates of the width of the $l = 0$ line gave $\Gamma = 1$ μHz (Woodard and Hudson 1983), which was later revised upward to 1.6 μHz (Woodard 1984a). The preliminary IPHIR results have measured $\Gamma = 1.45$ μHz for the $l = 0$ mode (Fröhlich et al. 1990).

More recent work has focused on the dependence of Γ on ν, l, and the level of solar activity. Isaak (1986) plotted $\Gamma(\nu)$ for the $l = 0$ mode, with the results ranging from $\Gamma = 0.2$ μHz at $\nu = 1.7$ mHz to $\Gamma = 10$ μHz at $\nu = 4$ mHz. Recently, Elsworth et al. (1990a) have published measurements of the line widths as a function of ν for the $l = 0$ and 1 modes. Their results are shown in Fig. 11. The well-established increase of Γ with increasing ν is apparent and there also appears to be an increase of Γ with increasing l. Finally, measurements of Γ during the course of the solar cycle are now becoming available. Pallé et al. (1990a) present evidence that the line width of the low-l modes increases with the level of solar activity. Measurements of the line widths from spatially resolved observations is discussed in Sec. IV.B.3.

8. Correlations and Phase Relations.   There is some interest in comparing observations of the oscillations as obtained in different quantities, such as velocity and intensity, in different spectral lines, or in different photometric colors. Such comparisons of amplitudes and relative phases provide information on the behavior of the oscillations in the solar atmosphere, where they are evanescent waves tunneling their way into the chromosphere and corona. These sorts of studies will be considered in more detail in the Sec. IV.C. on atmospheric waves; here we focus on results for the low-l modes.

Coordinated low-l velocity V and intensity I observations were first performed by Fröhlich and van der Raay (1984). Using balloon-borne three-channel photometric measurements and the groundbased Doppler observations, this study found I/V amplitude ratios of 15 to 50 ppm / m s^{-1}, and $V - I$ phase differences of 120 to 180°. Later, a single groundbased luminosity telescope was installed at Izaña (Jiménez et al. 1987,1988d) and used to extend this type of observations to four channels. More recently, the two-

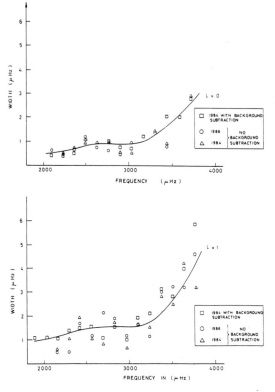

Fig. 11. Line widths of low-*l* modes as a function of *ν. Upper panel: l* = 0. *Lower panel: l* = 1.
The solid curves are meant to be guides for the eye rather than precise fits (figure from Elsworth
et al. 1990*b*).

station SLOT (Solar Luminosity Oscillations Telescope) network (Andersen
et al. 1988) has further improved these observations. This study has examined
the results for 36 low-*l* modes (Jiménez et al. 1988*d;* Jiménez et al. 1990).
They found *I/V* amplitude ratios ranging from 10 to 100 ppm / m s⁻¹, in
agreement with the earlier work. The measured frequency dependence of the
ratios of the amplitudes are in approximate agreement with the theoretical
expectations of Gough (1980). However, a similar comparison using spatially
resolved observations has failed to confirm the frequency dependence, but the
cause of the discrepancy is not known (Libbrecht 1990). The luminosity oscillations for the four channels were observed to be in phase, and the measured *V* − *I* phase difference was found to be ~140°, again in agreement
with the previous work. As the observed phase differences are not 90°, there
must be some radiative damping occurring in the solar atmosphere.
     Another correlation that has been studied is that of the amplitude and

phases of velocity oscillations in the K and Na lines (Isaak et al. 1989; Pallé et al. 1988b). The results of these observations are again consistent with an energy decaying solution for the waves in the solar atmosphere. A study of the amplitudes and phases of nonresolved intensity oscillations in two photometric bands was done using Fourier Transform spectrometer observations at Kitt Peak (Ronan et al. 1990). The bands considered were from 3833 to 4237 Å and 6640 to 6803 Å. The observations covered only a limited area at disk center, and so they contain some signal from modes with $l > 3$. This study found that the relative intensity fluctuations of the p-modes increased with height in the solar atmosphere, and that the relative phase variations were consistent with an adiabatic evanescent description of the waves in the solar atmosphere.

9. 160-Minute Oscillation. The 160-min oscillation was discovered in nonimaged whole-disk measurements (Brookes et al. 1976) and in differential observations (Severny et al. 1976) and at that time had a large amplitude of some 3 m s^{-1} for potassium and over 4 m s^{-1} for sodium. In subsequent years the amplitude was observed to be smaller, variable and possibly even nonexistent (Brookes et al. 1978b). Comparison of the Crimean data and the Stanford data for the period of 1976 through 1979 (Scherrer et al. 1979,1980a) showed that the oscillation maintained phase coherence both in time and between the two sites, suggesting that the oscillation had a solar origin. A comparison of later data from the two sites showed that the phase coherence had been substantially reduced (Henning and Scherrer 1989), casting doubt on the solar origin. However, the Crimean group has continued to see temporal phase coherence in their observations (Kotov et al. 1983; Kotov 1985; Kotov et al. 1988).

Additional doubt comes from the difficulty of theoretically describing the nature of the oscillation. The period is too long for it to be a p-mode and, if it is a g-mode, then why is only a single g-mode present? One possibility is that the 160-min oscillation is the result of nonlinear interactions between invisible g-modes (Guenther and Demarque 1984).

Grec and Fossat (1979), Ryzhikova (1988a,b) and Elsworth et al. (1989) have proposed that the 160-min oscillation may be an artifact due to differential extinction and aliases of the day. In the 1980s the oscillation was small or absent and a very careful analysis is undoubtedly required to reduce the effect of the daily aliases. Such an analysis is yet to be made. Tides in the Earth's atmosphere may also be responsible for the 160-min oscillation. These tides can be remarkably coherent, and have periods at several high harmonics of a day (see, e.g., Grec and Fossat 1979). The 160-min period is 1/9 of a day, and the Crimean and Stanford observatories are almost exactly 4/9 of the Earth's circumference apart. However, the Crimean group has argued that the period is actually 160.010 min and that if a tide were responsible, then the period should depend on the equation of time and this is not observed in the Crimean data (Kotov et al. 1980). It seems clear that conclusions regarding the solar origin of the 160-min oscillation are premature.

B. Imaged Observations

In this section, some recent results of spatially resolved observations of the solar oscillations will be summarized. These results include the basic measurements of frequencies, line widths and amplitudes, and splittings. Additional topics discussed include solar cycle effects, diameter measurements, flare effects, absorption of the oscillations by sunspots, large-scale horizontal flows, very high-l and v modes, and anisotropies.

1. Frequencies. The basic data for helioseismology are the observed frequencies of the oscillations that can then be compared with theoretical frequencies. Extensive tables of frequencies can be found in Duvall et al. (1988*a*), Libbrecht and Kaufman (1988) and Libbrecht et al. (1990). The first two papers list the frequencies for 2237 modes with $0 \le l \le 1320$ and $0 \le n \le 35$. Most of the frequencies for $4 \le l \le 99$ were determined from two data sets obtained at the South pole in 1981 and at Big Bear Solar Observatory in 1985. Three additional data sets from Big Bear were used to measure the frequencies at values of l up to 1320. Previously published tables of frequencies for modes with $l \le 5$ are also included. An Appendix in this book provides approximation formulae for the frequencies as a function of n and l. Recently, Libbrecht et al. (1990) have published an extensive table of frequencies measured from three Big Bear data sets. This table contains frequencies for most of the p- and f-modes with $0 \le l \le 1860$, $0 \le n \le 26$ and $1.0 \le v \le 5.3$ μHz.

The precision with which the frequencies of the oscillations can be measured depends upon what range of l is considered. Table III shows the average and the standard deviation of the error in frequency measurements for various ranges of l as published in Duvall et al. (1988*a*) and Libbrecht and Kaufman (1988*a*). For intermediate l from 20 to 100, where the available spectra show very clean and isolated peaks, frequencies can be determined with accuracies as high as 0.01 μHz (Libbrecht et al. 1990). For a mode with a frequency of 3000 μHz, this represents a proportional error of 0.0003%, one of the more accurate astronomical measurements. This represents an improvement of more than an order of magnitude over previous work, and was achieved primarily through the use of a longer data set (100 days as compared to 12 days). At intermediate l, the peaks and their sidelobes begin to overlap, forcing the adoption of less-accurate measurement techniques and increasing the errors. At high-l, the modes cannot be resolved, further increasing the errors.

Systematic errors are of more concern than random errors, because they cannot be removed by simply averaging, and may remain undetected in the data. Comparison of the South Pole and Big Bear data sets showed systematic

TABLE III
Errors in Frequency Measurements

ℓ-range	Number of Measured Frequencies	Average Error, μHz	Standard Deviation of Error, μHz
0—3	111	0.40	0.25
4—10	105	0.32	0.23
11—20	173	0.29	0.18
21—30	168	0.30	0.23
31—40	160	0.45	0.68
41—50	180	0.77	1.0
51—60	182	0.83	0.92
61—70	183	0.95	1.1
71—80	181	1.1	1.2
81—90	169	1.0	0.93
91—100	145	1.0	1.1
110—200	124	5.5	8.1
210—300	104	11	12
310—500	96	19	13
520—800	63	31	9.1
820—1320	64	39	11

differences of 0.1 μHz at low-l, to 0.6 μHz at $l = 99$. No instrumental or computational sources could be found for the systematic difference but, as the data were acquired at different phases of the solar activity cycle, it is possible that the difference has a solar origin. Systematic errors due to image-scale errors and incomplete spatial filtering were removed from the three Big Bear sets used for the higher-l analysis, but it was estimated that residual errors of 2 to 10 μHz still remained.

Comparison of the observed frequencies with theoretical calculations shows discrepancies of up to 15 μHz (Ulrich and Rhodes 1983; Bahcall and Ulrich 1988; Christensen-Dalsgaard 1988c; Cox et al. 1990). This suggests that the standard solar model is inaccurate, a suggestion further strengthened by the observed solar neutrino deficit. The most likely sources of the discrepancy are the equation of state (Christensen-Dalsgaard et al. 1988b), the calculated opacity in the solar interior (Korzennik and Ulrich 1989), or the surface boundary conditions (Christensen-Dalsgaard and Pérez Hernández 1988). Exotic models, such as WIMPS (Faulkner et al. 1986; Däppen et al. 1986; Gilliland and Däppen 1988; Cox et al. 1990), and high core concentrations of iron (Rouse 1986) have also been studied.

2. Dispersion Relation. It was reasoned by Duvall (1982a) that a simple transformation of the observed *p*-mode frequencies should result in the collapse of all of the ridges onto a single curve as the consequence of the resonant trapping of the modes in cavities. At the bottom of the cavity, the horizontal phase speed of the waves, ω/k_h, is equal to the sound speed. Because the top of the cavity is essentially the same for all modes, modes with identical values of ω/k_h are trapped in the same cavity, and have the same sound travel time across the cavity. The sound travel time is given by the expression $(n + \alpha)\pi/\omega$ where α is a constant. Thus, a plot of $(n + \alpha)\pi/\omega$ vs ω/k_h for different modes that are trapped in the same cavity should result in the collapse of the ridges onto a single curve that provides information about the sound speed dispersion relation within the Sun. Following this procedure showed that this was indeed the case for $\alpha \approx 1.5$, and allowed the determination of the sound speed as a function of depth (Christensen-Dalsgaard et al. 1985b).

   3. Line Widths and Amplitudes.   The measurement of line widths and amplitudes provides clues to the excitation and damping mechanisms of the solar oscillations. The widths and amplitudes are obtained from fitting a line-profile function to individually resolved modes in the power spectrum. The line widths are directly related to the lifetimes of the modes, and the amplitudes can be converted to the power and energy per mode. Early measurements of the line widths of intermediate-l modes were hampered by insufficient resolution in ν (Duvall and Harvey 1983; Libbrecht and Zirin 1986). With the advent of spectra that resolve the individual modes, amplitudes and line widths were more reliably measured first for modes with l-values of around 20 to 100 (see, e.g., Duvall et al. 1988b). The results of that study indicated that the line width increased both with ν and l. Even more thorough measurements of these quantities have been made on the Big Bear data (Libbrecht 1988a,d). The results of these measurements are shown in Fig. 12. The observed dependences of the line widths and energies with frequency can be compared to models of the driving and damping of the oscillations (Goldreich and Kumar 1988; Kumar et al. 1988; Kumar and Goldreich 1989; Christensen-Dalsgaard et al. 1989b). The comparisons suggest that perturbations induced by the oscillations themselves in the convective flux rather than nonlinear interactions among the modes probably limit the growth of the oscillations.

   Recently, the NSO/Bartol South Pole data has been analyzed to study the line widths as a function of ν, l and solar activity (Jefferies et al. 1990a,b). This study found that the line widths increase with all three of these parameters. It also found that the observed line-width variation as a function of l was in excess of the variation predicted from radiative and convective damping, suggesting the existence of additional damping mechanisms. Two additional mechanisms are postulated to act for different ranges of l. For values of $l > 25$, the extra damping may be due to increased absorption of p-mode energy by active regions, as observed by Braun et al. (1987,1988). At values of $l < 40$, the damping could be caused by increased leakage of wave

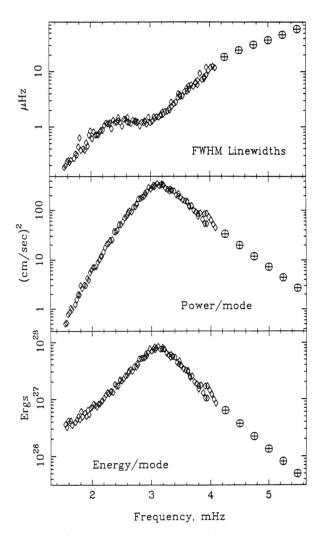

Fig. 12. FWHM line widths, power per mode, and total (kinetic plus potential) energy per mode as a function of frequency for $l \approx 20$. Diamonds represent individual mode fits with $19 \leq l \leq 24$; circles indicate results from an indirect measurement. The power per mode is normalized to represent the *rms* radial surface velocity averaged over both time and the solar surface. This normalization could be in error by 10 to 20%. The energy per mode assumes that the observations were made at $\tau_{5000} = 0.05$ (figure courtesy of K. Libbrecht, from Libbrecht 1988*d*).

energy into the corona from modes tunneling through the temperature minimum region during high solar activity.

4. Splittings.   Several groups have published tables of splittings determined from the power spectra as a function of ν and m for a given l. The results all refer to modes with intermediate $l < 120$, where it is possible to obtain reasonably accurate splittings either via the cross-correlation method or by direct fitting to the observed frequencies. Figure 13 shows the observed splittings for recently published results from Big Bear (Libbrecht 1989), the South Pole (Jefferies et al. 1988c), HAO (Brown and Morrow 1987a), and Mount Wilson (Tomczyk 1988). See also Libbrecht (1988c), Tomczyk et al. (1988), Duvall et al. (1986,1987), Brown (1985), Duvall and Harvey (1984), Rhodes et al. (1987), and Didkovskii (1989) for additional reports on observed intermediate-l splittings.

Figure 13 shows several features. There appear to be systematic differences in the values of a_1 measured by the four groups. For example, the Mount Wilson results are ~ 5 nHz greater than the others. The cause of this disagreement may lie either in various image defects (drifts, distortions, erroneous diameter), or in temporal variations in the solar internal rotation rate. There is agreement in all results that a_1 decreases with decreasing l, suggesting an inward decrease of the average rotation rate. The values of a_3 also display systematic differences, with the South Pole results larger at $l < 40$, and the Mount Wilson measurements smaller at $l > 50$. The results are generally close to the surface value, but decline at low-l, implying an inward decrease of the latitudinal dependence of the differential rotation. The values of a_5 are in general agreement, but this may be more a reflection of the increased noise in the measurements. It appears that a_5 may be greater than the surface value for $l > 40$, and tend towards zero for $l < 40$, again implying an inward decrease in the differential rotation.

An approximate measure of the equatorial rotation rate can be gained by plotting a_{135}. Figure 13 shows the results for the four groups, as well as direct sectoral measurements from Duvall and Harvey (1984). The plot shows only slight systematic differences between the five measurements. The differences are due in part to the systematic differences in the odd a_i noted above, but the discrepancies in a_{135} are smaller than those for a_1. This suggests that either the equatorial rotation rate can be more accurately measured, or it is more stable. There is a decrease of a_{135} with decreasing l, implying that the rotation rate as sampled by the sectoral modes decreases inwardly. However, since the low-l sectoral modes sample a wider range of latitude than the higher-l sectoral modes, this does not necessarily mean that the equatorial rotation rate decreases inwardly.

The odd a_i provide information about the internal solar rotation rate. The chapter by Libbrecht and Morrow is devoted entirely to this subject, and several papers have discussed the topic (see, e.g., Harvey 1988; Rhodes et

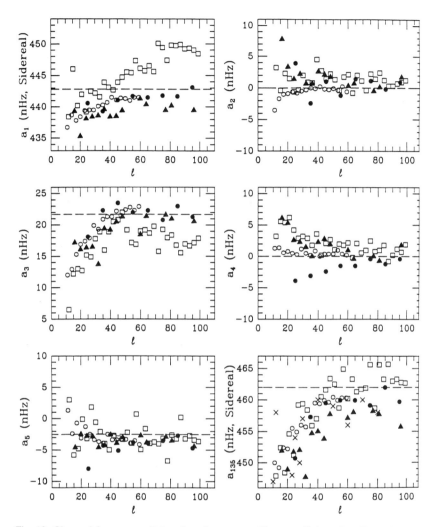

Fig. 13. Observed frequency splittings from four groups. The a_i coefficients from Eq. (16) are displayed for $i = 1$ to 5. The combination a_{135} from Eq. (17) is also displayed. Open circles: Big Bear 1986.4, individual l-values, from Libbrecht (1989). Filled triangles: HAO 1984.8, averaged over four l-values, from Brown and Morrow (1987). Filled circles: South Pole 1987.9, averaged over ten l-values, from Jefferies et al. 1988b). Open squares: Mount Wilson 1984.5, individual l-values, from Tomczyk (1988). Crosses: Kitt Peak 1983.5, individual l-values, sectoral modes, from Duvall and Harvey (1984) in a_{135} panel only.

al. 1988c; Morrow 1988b; Korzennik et al. 1988; Christensen-Dalsgaard and Schou 1988; Brown et al. 1989; Dziembowski et al. 1989; Goode et al. 1990; Christensen-Dalsgaard 1990b). The results to date have ruled out the theoretical model that the rotation rate is constant along cylinders, but there are still several qualitatively different rotation models that cannot yet be ruled out on the basis of the data (Morrow 1988a,b).

High-l splittings have also been measured and used to infer the solar rotation rate in the extreme outer layers of the solar convection zone. These measurements require high spatial resolution and therefore very high data rates if the full disk of the Sun is imaged. If only a portion of the disk is used, then the data flow becomes more manageable, but the exact spatial scale of the image and thus an accurate calibration for l is not possible except indirectly using the f-mode ridge. Thus, high-l splittings have not yet been exploited to the extent of the intermediate-l modes. Nevertheless, the high-l modes were the first to be used to infer the solar rotation (see, e.g., Deubner et al. 1979; F. Hill et al. 1984a,1988b,c; F. Hill 1987b). The results of these studies suggest that the equatorial rotation rate may have a small inward increase in the outer layers, and that it then decreases inwardly. Studies in progress will undoubtedly shed more light on the rotation rate in the shallow outer solar layers in the near future (Woodard and Libbrecht 1988; Rhodes et al. 1988a).

The even splittings, a_2 and a_4, are also displayed in Fig. 13. It can be seen that there are large systematic differences between the independent sets of measurements. It appears that the measured coefficients may be nonzero, suggesting the presence of a solar physical process that is insensitive to the direction of propagation of the oscillations. This process could be an anisotropic sound speed, magnetic field, or a measure of the asphericity or oblateness of the Sun (Kuhn 1989b). The disagreement of the measured values could reflect solar cycle variations, which are discussed in the next section.

The measured frequencies have been plotted without error bars in Fig. 13 to improve the comparison of the results. Typical estimated errors in the observed a_i are given in Table IV.

TABLE IV
Estimated Errors for Splitting Coefficients, nHz

Data Set	a_1	a_2	a_3	a_4	a_5
South Pole	0.7	0.9	1.1	1.2	1.3
HAO	1.2	1.6	1.5	1.3	1.6
Big Bear	0.5	0.6	0.7	0.9	0.9
Mt Wilson	1.2	1.3	1.5	1.7	1.9

5. *Solar Cycle Changes.*   As stated above, the values of a_2 and a_4 provide information on the symmetric component of the solar structure, such as sound speed (and thus temperature) or magnetic fields. Brown and Morrow (1987a) found that the values of a_2 and a_4 varied as l^{-1}, so that the quantity $\alpha_i \equiv a_i l$ could be defined. When the values of α_2 and α_4 are computed from various splitting measurements and then plotted as a function of time, then an apparently consistent temporal variation appears correlated with the solar activity cycle in the sense that both a minimum of α_2 and a sign change of α_4 occur around the time of minimum solar activity (Jefferies et al. 1988c; Libbrecht 1989). Kuhn (1988a,b,1989a) has suggested that the observed variations are due to changes in the internal temperature structure, and are correlated with observations of variations in the surface temperature (Kuhn et al. 1988b; Goode and Kuhn 1990; Gough 1990). There is some question as to the magnitude of this effect (Gough and Thompson 1988b), and an alternative explanation (and prediction for future measurements) has been proposed from the number of sunspots (Gough 1988b; Kuhn and Gough 1989). The accuracy of the predictions remains to be seen, as does the robustness of the measured variations themselves.

There is some evidence for a 1% decrease in the value of a_1 over the declining phase of the solar cycle (Jefferies et al. 1988c), implying that the equatorial rotation rate is slightly slower at solar minimum. Inversions of the splittings suggest that the internal rotation rate does vary with the solar cycle (Goode et al. 1990) but the effect of systematic errors is not easy to dismiss from measurements of variations with magnitudes of a few nHz.

The correlation of the temporal behavior of the solar frequencies with the phase of the solar activity cycle has also been investigated. Such a change may provide information on the magnetic field strength in the solar interior, or on variations of the solar radius and internal sound speed. The results for the low-l modes have been discussed earlier in this chapter, and a similar situation of early confusion and recent agreement exists for the spatially resolved results. Early searches for the effect in intermediate degree modes (Rhodes et al. 1988b; Jefferies et al. 1988c) found no significant variation of the mode frequencies during the declining phase of cycle 21, in contrast with the early low-l results which found both an increase (Isaak et al. 1988), and a decrease (Woodard and Noyes 1985). As with the low-l observations, this situation changed at the Hakone meeting in late 1989. Recent spectra with very high resolution in v from Big Bear showed clear evidence of a solar cycle variation in the frequencies (Libbrecht and Woodard 1990a,b). This variation depends on the frequency and degree of the modes, and is shown in Fig. 14. The shift increases with both v and l and could be caused by changes in the magnetic field at the top of the convection zone. Also apparent in Fig. 14 is a sharp drop of the shift near $v \approx 4$ mHz, which may be a signature of the chromospheric mode discussed later (Sec. IV.C.4). The maximum mag-

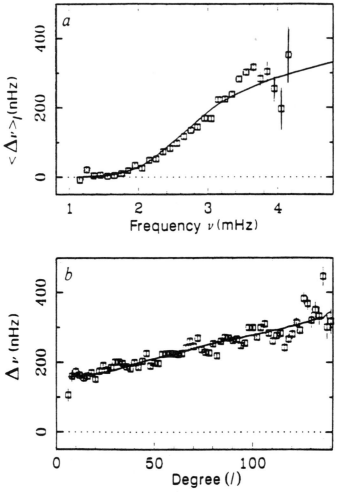

Fig. 14. The difference in solar p-mode oscillation frequencies observed at Big Bear between 1988 (high solar activity) and 1986 (low activity). *Panel a:* The difference $\Delta\nu = \nu_{88} - \nu_{86}$ as a function of frequency after averaging over modes with $5 \le l \le 60$. *Panel b:* The difference as a function of l at $\nu = 3$ mHz (figure from Libbrecht and Woodard 1990a).

nitude and the sign of the bulk change agree with the results obtained from the low-l observations, and a re-analysis of the NSO/Bartol South Pole observations brought further agreement with another data set (Jefferies et al. 1990b). There is thus now virtually unanimous agreement among observers that the frequencies of the solar oscillations increases as the level of solar activity increases. The dominant cause of the frequency change is thought to be changes in the solar surface structure during the course of the solar cycle, but much work remains to be done in this area.

6. *Interaction of* p-*Modes and Active Regions.*   Recent observations have revealed that the amplitude of the p-mode oscillations that are travelling inward towards a sunspot is as much as 50% greater than the amplitude for outgoing modes (Braun et al. 1987,1988). This implies that the sunspot is either absorbing the p-modes, or scattering them into different spectral regions. Observations of photospheric umbral oscillations suggest that sunspots can act as selective acoustic-wave filters (Abdelatif et al. 1986; Abdelatif and Thomas 1987; Lou 1990). It may thus become possible to use observations of the interaction of the sunspots and the p-modes to helioseismically determine the subsurface structure of active regions. A measurement of the frequencies inside the spot could be used to determine the depth dependence of the spot diameter (Thomas et al. 1982). It is possible to build a map of acoustic absorption, both inside and outside active regions (Braun et al. 1990), thereby gaining a picture of the density of scattering fibrils within the region (see, e.g., Bogdan and Zweibel 1985,1987a; Zweibel and Bogdan 1986; Bogdan 1987a,b; Zweibel and Däppen 1989; Bogdan and Cattaneo 1989). It may ultimately be possible to detect active regions on the invisible hemisphere (Lindsey and Braun 1990).

Besides absorption processes, emission of acoustic radiation may also occur in active regions during flares. Observations have shown that a single ridge in the k-ω diagram can become enhanced after a large flare (Haber et al. 1988b), and that the absorption of the p-mode radiation may disappear during a flare (Haber et al. 1988a). The presence of both emission and absorption of radiation suggests a possibility of the future development of active region tomography (see, e.g., Brown 1990a).

In addition to tomography of the magnetic field structure, the development of ring diagrams allows the mapping of flows associated with active regions (F. Hill 1988a,1989b,1990b). Preliminary results suggest the presence of a deeply situated strong meridional flow with a large active region (F. Hill 1990b). These results were inferred from partial-disk observations, and thus suffer from spatial scale uncertainties. In the future, maps of the flows in the convection zone over the entire disk and solar cycle will become available (F. Hill et al. 1990b).

Recently, interest in the effects of active regions on the amplitudes and frequencies of the oscillations has been generated by the suggestion that such effects could contaminate measurements of the properties of the global solar modes (Ulrich et al. 1988a). Early observational work suggested that the magnetic field suppressed the amplitude of the photospheric velocity oscillations and did not substantially alter the frequencies (see Stein and Leibacher [1974] for a summary). Additional work produced evidence that the shape of the oscillation power spectrum (and thus the measured frequencies) was affected by magnetic activity (Woods and Cram 1981; Bonet et al. 1988; Alamanni et al. 1990b), and that the effect seemed to depend on the spectral line used for the observations. Theoretical studies of the interaction of the p-

modes and the magnetic fields in the solar atmosphere have also suggested that the frequencies of the high-l modes can be altered (Bogdan and Zweibel 1985; Zweibel and Bogdan 1986; Bogdan 1987a,b; Bogdan and Cattaneo 1989; Zweibel and Däppen 1989; Campbell and Roberts 1989; Evans and Roberts 1990). The question has been recently addressed by a study that inserts an artificial active region into quiet Sun observations and then compares the global oscillation spectra (Haber and Hill 1990). Preliminary results of this work suggest that the amplitudes of some modes are suppressed, but that the frequencies are unchanged. The inconsistent results between previous observational studies may be due to changes in the beating of unresolved modes with different amplitudes.

7. Very High Degree and Frequency Observations.   The seeing in the Earth's atmosphere places an upper limit on the maximum value of l observable from the ground. The SOUP experiment on Spacelab 2 was free of this limitation, and detected the f-mode ridge up to $l = 3500$ in white light observations (Tarbell et al. 1988). The p-modes, however, were not visible. Subsequent groundbased observations using filtergrams and three-dimensional Fourier transforms were able to detect f-mode oscillations to $l = 2500$, the p_1 ridge to $l = 1800$, and the p_2 ridge to $l = 1200$. As granulation has an average diameter corresponding to $l \approx 3000$, the very high-degree modes should interact strongly with the granulation, thus perhaps providing a solar limit to l.

Spatially resolved observations have also shown that the ridge structure of the diagnostic diagram extends up to 7 mHz, well above the acoustic cutoff frequency at the temperature minimum of 5.5 mHz (Jefferies et al. 1990b). This is surprising because these high-frequency waves are not reflected at the solar surface, are thus not trapped in the solar interior, and should not display a ridge structure. Three explanations have so far been offered. One possibility is that the waves are reflected further up in the solar atmosphere at the transition region (Balmforth and Gough 1990a). However, the predicted spacing in v between adjacent ridges from this theory is smaller than the observed spacing. Another possibility is that the observed ridge structure is due to the spherical harmonic projection of untrapped progressive waves (Kumar et al. 1990). This theory successfully predicts the ridge spacing, but incorrectly predicts the absolute position of the ridges. A third theory postulates that the high-frequency "modes" are emitted by localized turbulence in the convection zone (Brown 1990b).

8. Correlations and Phase Relations.   Comparison of the amplitudes of the low-l oscillations observed in intensity and velocity showed a variation of nearly an order of magnitude in the brightness to velocity power ratio as a function of temporal frequency of the modes (Jiménez et al. 1988a,d,1990). A recent similar comparison using intermediate-l modes with $30 \leq l \leq 60$

found a much smaller variation of this ratio with ν (Libbrecht 1990). Because the qualitative physical features such as the mode mass and the ratio of vertical and horizontal spatial wavelengths of these intermediate-degree modes and the low-degree modes are very similar, many of the mode properties should be independent of l. The two observations should thus show similar frequency variations, making the discrepancy puzzling. One possible source of the difference is in the spectral information content of the intensity observations. The Big Bear observations were performed using a narrowband filter centered on a wavelength of 6687 Å in a region of clean continuum, while the SLOT measurements were broadband and included spectral lines. In addition, the intermediate-degree measurements have a higher signal-to-noise ratio.

9. Sectoral Mode Anisotropies.   It has been proposed that the solar rotation may play a role in the excitation of the oscillations (see, e.g., Ando 1985). If differential rotation does affect the wave generation and propagation, then the amplitude and hence the power should be different for modes propagating east-west along the equator, and north-south over the poles of the Sun. There is thus some interest in comparing the power of sectoral modes obtained from observations oriented north-south and east-west.

The existence of possible anisotropies in the oscillations was first investigated by Deubner (1975), who concluded that there was no difference between spectra of sectoral modes oriented east-west on the Sun and modes oriented north-south. This conclusion was also reached by Rhodes (1977). However, Kuhn and O'Hanlon (1983) found that for $l < 30$, the east-west modes had more power than the north-south modes, and the converse for $l > 60$. Using higher spatial resolution, Deubner at first found a similar behavior, except that the turnover occurred at about $l = 900$. Later re-analysis removed this effect (Deubner, 1986b). F. Hill et al. (1986) again found similar behavior, but this time the turnover was at $l = 250$, and the turnover point appeared to depend on the choice of spatial filtering used in the data reduction. This filtering dependence, and the fact that the turnover l-value appeared to be $\sim 1/2$ of the number of spatial resolution elements across the solar diameter, led those authors to suggest that the apparent anisotropy was due to spatial filtering effects, rather than an actual anisotropy in the amplitude of the oscillations. Using a spherical harmonic filtering, Hill et al. found that the north-south sectoral modes had more power than the east-west modes for $l > 40$. The sensitivity of possible anisotropies to systematic effects was further explored by Kuhn et al. (1988a), who showed that the interaction of finite spatial resolution and the difference in rotational Doppler line asymmetry between east-west and north-south observations could explain the earlier Kuhn and O'Hanlon (1983) results. Recently, Caccin et al. (1990) have analyzed a time series of Doppler images obtained at Mount Wilson in 1984. They found that the north-south sectoral modes had a higher amplitude than the east-west modes for $l > 25$, consistent with the earlier investigations. In the light of the earlier results regarding systematic effects, we can only conclude that further work must be done to carefully separate the systematic effects of spatial filtering and line asymmetries from any possible solar phenomena.

10. Diameter Measurements.   The detection of oscillations in the solar diameter is a controversial topic. As discussed previously, the method is very susceptible to noise sources in the terrestrial atmosphere. Nonetheless, the SCLERA group has claimed the identification of many oscillation multiplets, including *f*-, *g*- and *p*-modes. Independent studies have so far failed to confirm these observations (see, e.g., Kuhn et al. 1986). Rotational frequency splittings of 1.8 μHz have been reported for low-*l* 5-min modes (H. Hill 1985*a*; H. Hill and Rosenwald 1986*a*), and of 0.6 to 1.6 μHz for long-period *p*-modes. These measurements have high internal consistency (H. Hill and Caudell 1985), but have not been confirmed by independent observations and are in disagreement with other observations by factors of 2 to 4. However, disk-center integrated intensity measurements made by SCLERA have confirmed the diameter results (Ogelsby 1987*a*). The diameter measurements have been used to infer the internal rotation rate (H. Hill et al. 1984,1986*b*), with the estimate being in substantial disagreement with all other investigations. Long-period *f*-modes have also been identified in these observations (Rabaey et al. 1988), again with no independent confirmation.

C. Oscillations in the Atmosphere

In addition to the global *p*-modes seen as evanescent waves in the visible atmosphere, there are also various propagating modes present that have been utilized to infer properties of the outer parts of the solar envelope in the range of optical depth from $\tau = 1$ to $\tau = 10^{-6}$. In this section, we discuss the diagnostic and scientific questions of "atmospheric seismology." For details, the reader is referred to reviews on the subject by Frandsen (1988) and Deubner et al. (1984), and the references therein. In the following sections we wish to discuss and emphasize some of the more recent developments in this area.

1. Phase and Coherence Spectra.   With a single exception, the oscillations discussed in this section are running waves of either the acoustic-gravity or magneto-acoustic-gravity type. Depending on the frequency and the ambient medium, the acoustic or the internal gravity wave character dominate. MHD modes with dominant magnetic restoring force will be discussed in the chapter by Semel et al. As mentioned previously, for these waves a cavity does not exist and these waves do not develop coherent standing wave patterns as the global *p*-modes do. Thus, their power, phase and coherence spectra are continuous, and so are the phase and coherence spectra computed as discussed in the section on observational techniques.

The classical example of the application of power spectra to the diagnosis of waves in the atmosphere is Frazier's (1968) work, which demonstrated the nonpropagating or evanescent character of the 5-min p-modes in the solar atmosphere. Later diagnostic diagrams with better resolution in frequency and wavenumber lead to the recognition of the global character of the p-mode wave field (Deubner 1975). Measurements of oscillatory power at given frequency and wavenumber as a function of height in the atmosphere also help to reveal propagation properties (Schmieder 1976).

The nature of propagation is clarified by the use of phase diagrams, which immediately reveal the phase velocity of a wavetrain that has been observed at two different height levels (i.e., spectral lines) simultaneously, as function of frequency. In 1968, Howard and Livingston discussed whether certain rapid ($P \approx 5$s) fluctuations of Doppler shifts found superimposed on the 5-min p-mode signal in recordings of photospheric velocity fields were caused by high-frequency sound waves in the solar atmosphere. A phase diagram constructed from simultaneous measurements in spectral lines of different strength would probably have settled the discussion in favor of a nonsolar origin of the signal at once. High-frequency sound waves are expected to propagate through the atmosphere with a well-defined phase speed; this signature is not shown by seeing noise (Endler and Deubner 1983).

Phase and coherence spectra have been applied by numerous authors with great success in searching for running acoustic and internal gravity waves as potential sources of atmospheric heating, and for defining the vertical structure of the so-called chromospheric cavity where standing waves may be found. Pioneering observational and theoretical research in this field has been published by Mein and Mein (1976), Mein (1977), Schmieder (1977), Cram (1978) and Lites and Chipman (1979). More recent observations have been made by Staiger et al. (1984), Kneer and von Uexküll (1985), Keil and Mossman (1989) and Alamanni et al. (1990a), and recent theoretical studies have been done by Frandsen (1986) and Marmolino and Stebbins (1989), among others.

Running acoustic waves have been identified (see also Deubner 1976b, 1985) up to frequencies as high as 25 mHz, and the existence of internal gravity waves seemed likely in view of the V-V phases observed at low frequencies (Schmieder 1976). However, the range of frequencies below $\nu = 2.5$ mHz is a particularly treacherous terrain, because low-frequency branches of the low-n 5-min p-modes and motions caused by granular convection and overshoot overlap on a large range of spatial scales and interfere with the propagating waves, judging from the low coherence of the observed signals.

Deubner and Fleck (1989) have separated these contributions by computing phase diagrams as function of frequency and wavenumber, and have shown that at high wavenumbers k (or l), and in the range of internal gravity waves, the phase spectra are in qualitative agreement with the theoretical V-V dispersion relations (Fig. 15): The vertical phase observed between two fixed levels increases with horizontal wavenumber and with decreasing frequency, and the downward pointing phase vector indicates upward mechanical energy transfer potentially important for the heating of the higher layers (Mihalas and Toomre 1981).

Finally, in the range of high-frequency acoustic waves with nearly frequency-independent phase velocity, it is possible to establish a geometric height scale simply by relating the gradient of the linear high-frequency phase spectrum to a length scale via the well-known phase velocity of sound,

$$d\phi/d\nu = \Delta h/c_{ph}. \tag{27}$$

This method works well in the range of the atmosphere below 800 km (Deubner et al. 1984). Above this level, however, the acoustic waves are predominantly standing in character and the phase differences cannot be used to infer geometric heights.

   2. Transport of Mechanical Energy by Waves.   Experimental assessment of the contribution of acoustic wave flux to the heating of the lower chromosphere is still one of the outstanding issues in solar physics (cf. Andersen and Athay 1989a), even though observations in the upper chromosphere and transition region seem to indicate that there is not enough acoustic energy available at those layers to account for the losses inferred from model calculations (Athay and White 1978).

All attempts to derive the mechanical energy flux carried by running acoustic waves directly from observations of waves and traveling pulses face several difficulties which until quite recently have caused a large scatter of the mechanical flux values quoted in the literature (Lites and Chipman 1979; Mein and Schmieder 1981; Staiger 1987; Stebbins and Goode 1987; Keil and Mossman 1989). The energy flux cannot be derived from the observed power alone, because this power contains contributions from nonpropagating, noncoherent components of the wave field, and considerable nonsolar noise through image motion. At high frequencies $\nu \geq c/D$ (sound speed over width of the line-forming region), the amplitude of waves believed to be efficiently generated by turbulent convection (Stein 1968; Ulmschneider et al. 1978) is reduced by optical phase averaging.

In this situation the analysis of cross-covariance spectra and their spatial averages extracted from time series of velocity fluctuations observed in pairs of spectral lines formed at different heights in the atmosphere has proven to be very useful for rejecting the noncoherent power at all frequencies, detecting the seeing noise signal (which has a purely real cross-covariance ampli-

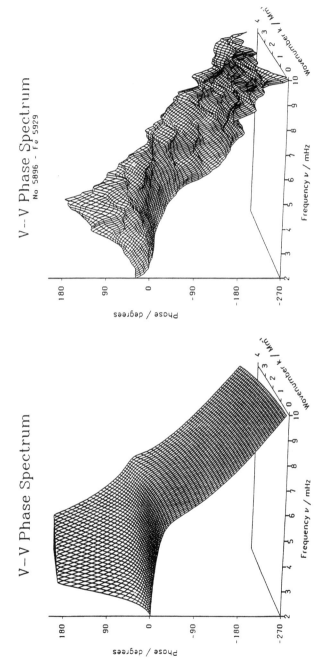

Fig. 15. Theoretical (*left*) and observed (*right*) photospheric velocity phase difference spectra. The steep rise at low frequencies and high wavenumbers, characteristic of internal gravity waves, is somewhat subdued in the observed spectrum due to seeing and to the one-dimensionality in the spatial coordinate (figure from Deubner and Fleck 1989).

tude), and determining the phase velocity of the wave (Deubner et al. 1988). Finally, it is possible with standard line synthesis routines to estimate the filtering effect, or modulation transfer function (MTF), of the line-forming regions on the apparent wave amplitudes.

The mechanical energy flux can then be computed as a function of height as

$$\Phi(z) = \bar{\rho}(z) \int \left| \langle \tilde{V}_1 \cdot \tilde{V}_2 \rangle_x \right|_{\text{corr}} c(\nu)_{\text{gr}} d\nu \qquad (28)$$

where z represents an effective mean height in between z_1 and z_2, the effective heights of the line-forming layers; the exact value of z depends on the flux profile $d\Phi/dz$. In this equation,

$$\bar{\rho}(z) = \sqrt{\rho_1 \rho_2}; \quad \rho_i = \rho(z_i); \qquad (29)$$

$$\left| \langle \tilde{V}_1 \cdot \tilde{V}_2 \rangle_x \right|_{\text{corr}} \qquad (30)$$

is the mean cross-covariance power corrected for seeing noise and the MTF; and

$$c(\nu)_{\text{gr}} = c^2/c(\nu)_{\text{ph}} \qquad (31)$$

is the frequency-dependent group velocity of the wavetrain determined from the observed phase spectra.

Using this scheme, we have derived flux values around 2×10^7 erg $cm^{-2}s^{-1}$ in the low and middle photosphere. The flux drops to 1.2×10^6 erg $cm^{-2}s^{-1}$ at the level of Na I 5896, and to 4.5×10^5 erg $cm^{-2}s^{-1}$ at the Ca II 8542 (core) level. These estimates are lower bounds because the acoustic power beyond the frequency where seeing dominates the cross-covariance spectra was not accounted for.

A substantial uncertainty still remains in the limited accuracy that can be achieved in determining the fraction of power apparent in the low chromosphere due to propagating waves. The flux estimates appear to be in satisfactory agreement with theoretical predictions of the radiative losses in the lower and middle chromosphere. Potential errors are discussed by Deubner et al. (1988).

3. Convection and Waves.   Both granulation and supergranulation are established phenomena, with well-defined spatial and time scales. On the other hand, mesogranulation (November et al. 1981) has remained somewhat elusive until very recently. Neither its location in the atmosphere (photosphere or chromosphere?) (Damé 1985), nor its dynamical character (convection or waves?) have been well assessed. It has been suggested that mesogranulation is of convective origin because its spatial scale ($5''$ to $10''$) agrees well with the depth of the He I ionization layer within the convection zone (Simon and Leighton 1964). In analogy, granulation and supergranulation had been associated with the H and He^+ ionization zones, respectively.

A major step in identifying the nature of mesogranulation was the detection of horizontal flow patterns in white-light time-lapse sequences of photospheric granulation (November et al. 1987, Brandt et al. 1988), using individual granules as a tracer of larger-scale horizontal motions (see Fig. 16). These studies clearly indicate photospheric cellular motions with typical spatial scales of the order of $10''$ and horizontal velocities up to ~ 800 m s^{-1}. Simon and Weiss (1989b) modeled the mesoscale flows and stress their importance in the redistribution of small-scale magnetic flux at the photospheric level.

If mesogranulation is convective in origin, the vertical flow pattern observed by November et al. (1981) should be correlated with a brightness distribution as in the case of granules. Mesoscale brightness fluctuations have been observed as early as 1932 by Strebel and Thüring, and again by Koutchmy and Lebecq (1986) and by Oda (1984). A study of the dynamics of mesogranulation by Deubner (1989a), based on time series of photographic spectra in various photospheric lines, confirmed the convective char-

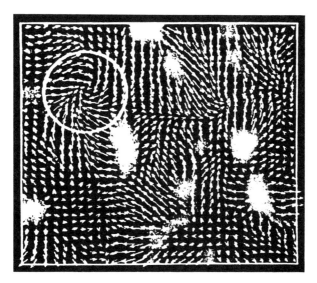

Fig. 16. Horizontal motions observed in white light in the photosphere using individual granules as tracers. The size of the field is $14''.2$ by $12''.2$ (figure from Brandt et al. 1988).

acter of the mesoscale fluctuations, and demonstrated the close relationship of the three scales of granulation observed on the Sun.

In this study, both power and coherence spectra of temporal brightness and velocity fluctuations support the notion of a separate regime of motions in the 5″ to 10″ range, with a typical life time of several hours (F. Hill et al. 1984c). The horizontal velocities inferred from the power spectra agree well with those derived from high-resolution observations of granule proper motions. The observed vertical motions are fairly strong ($\sim$300 m s^{-1}) and quite sufficient to suppress, impede or accelerate the evolution of individual granules, as seen in white-light granulation movies (Brandt et al. 1989a). The observed ratio of the vertical to horizontal flow velocity of convective motions seems to be linearly dependent on the spatial scale, being on the order of 1 for granules, 2.6 for mesogranules, and $\sim$20 for supergranules. Such a linear scaling law is suggested by the continuity equation.

The temporal phase differences observed between velocity fluctuations at different levels of the atmosphere, and between velocity and brightness fluctuations in various spectral lines as function of the spatial scale both testify for the convective origin of the observed quasi-cellular motions at all spatial scales in the photosphere. Deviations of the phase spectra from uniformity can be interpreted in part as a consequence of thermal relaxation varying with height in the atmosphere, and partly by the effects of internal gravity waves, which are detectable at the lowest observable levels in the photosphere. Table V presents some dynamical properties of mesogranulation derived in this study. The theory of convection should reproduce these properties and phase values in its three-dimensional models. However, it will be challenging to determine whether the three different scales observed in solar

TABLE V
Physical Properties of Mesogranules (granules)

		C I 5380		Fe I 5383
Diameter: 4″.5–10″				
Lifetime: $\gtrsim$ 30 min				
Horizontal velocity (*rms*)		750 m s^{-1}		450 m s^{-1}
Vertical velocity (*rms*)		300 m s^{-1}		200 m s^{-1}

	cos θ	*V—I*	*V—V*	*V—I*
Coherence	1.0	0.8 (0.9)	0.8 (0.8)	0.6(0.7)
	0.8	0.75(0.85)	0.85(0.8)	0.4(0.6)
Phases	1.0	+12° ($\pm$0°)	$\pm$0° ($-$12°)	$-$100° ($-$140°)
	0.8	$-$25°($-$15°)	$-$10° ($-$12°)	$-$90° ($-$165°)

convection are actually related to different depths of the relevant source regions, or instead to a topological effect caused by the tendency of the downdrafts visible at the surface to collect into fewer and stronger funnels as deeper layers of the convection zone are reached (Stein et al. 1989).

While granular brightness and velocity fluctuations are tightly correlated at low photospheric levels, the degree of correlation decreases with height and the correlation changes sign at a level of approximately 100 km before it is lost completely. This well-established observation was first described by Evans and Catalano (1972), and has in recent years been discussed extensively by Nesis and coworkers (cf. Nesis et al. 1988). If one compares the velocity fluctuations at increasingly higher levels with those at the bottom of the photosphere, a continuous decrease of their correlation without a reversal of sign is observed.

The relationship between velocity and brightness signals at the spatial scale of granulation has been interpreted using the concept of overshoot, the inertial penetration of previously convective elements into the stably stratified atmosphere where they become cooler and therefore darker than the ambient material due to the continuing adiabatic expansion. The dynamics of this process have been studied using time series of granular fluctuations observed simultaneously in various spectral lines by Keil (1980), and by Altrock et al. (1984). A similar series of observations has yielded phase-difference spectra as a function of frequency (Deubner 1974b; Staiger 1985,1987) for brightness and intensity fluctuations.

The temporal phase spectra seem to show the signature of gravity waves in the frequency regime below 2.5 mHz, namely a downward propagating phase in the V-V signal (Fig. 15), corresponding to an upward energy flux. Also, at high wavenumbers $k > 3.5$ Mm^{-1} we observe phase lags larger than 90° between brightness and (upward) velocity (Fig. 17).

There is, however, a feature in the V-I phase spectra of the lower photospheric lines, which is not accounted for by the dispersion relation of gravity waves, and which needs further analysis: at ~2.2 mHz the phase rapidly changes sign and assumes values of up to 60°, with upward velocity leading (Staiger 1985; Deubner and Fleck 1989).

The significance of this feature has become apparent only recently when two-dimensional phase diagrams (V-I phase as function of frequency and wavenumber) were studied (Deubner and Fleck 1989). These diagrams reveal that the peak of positive phases is strongest at the largest spatial scales, and that its position shifts to higher frequencies at higher wavenumbers (Fig. 17). Also, the lower frequency limit of the feature moves to higher values with increasing height within the lowest 150 km of the photosphere, whereas the upper frequency limit remains stationary up to the temperature minimum (unpublished data from Staiger 1985). At the level of the Ca II infrared lines, however, the feature has almost disappeared (Fleck and Deubner 1989, Figs. 5a,b). A more detailed account of this phenomenon is given in Deubner

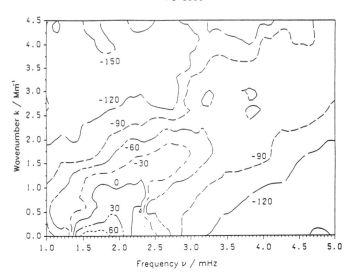

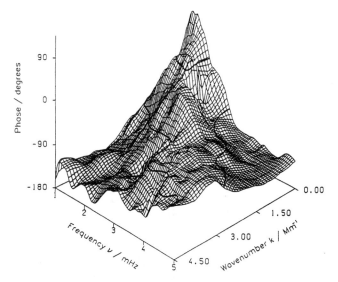

Fig. 17. Phase difference spectrum of photospheric brightness and velocity fluctuations at low frequencies. The positive values at 1.6 mHz < ν < 2.7 mHz are restricted to low wavenumbers (figure from Deubner and Fleck 1989).

(1990), and an interpretation in terms of scattered evanescent waves with downward energy propagation is proposed. Nonadiabatic model calculations in the corresponding region of the k-ω diagram (Marmolino and Severino 1990, Deubner et al. 1990) suggest that the observed fluctuation in the V-I spectra is related to phase discontinuities occurring both near $\omega_L = v_s \cdot k_H$ (Lamb waves) and $\omega_F = \sqrt{g k_h}$ (divergence free waves) in the evanescent regime.

It thus appears possible to understand all of the observed overshoot effects in terms of the interaction between convection and various wave modes excited over a large range of wavelengths and periods in the convectively stable atmosphere. Nevertheless, our understanding of the V-I phase spectra of the evanescent p-mode oscillations close to the photosphere appears still incomplete (see also Sec. IV.C.4 below for chromospheric oscillations). Theoretical reasoning suggests phase values between 90° (upward motion lagging) in the nearly adiabatic upper layers, and values rather closer to 180° near the bottom of the photosphere, where the thermal relaxation time decreases to a few seconds. In striking contrast to this picture, a rapid decrease of the phase from a maximum of ~140° at a geometric height of 250 or 300 km to about 40° in the very low C I 5380 line is observed (Staiger 1985; Deubner 1974b; Fig. 18). A phase of 90° is measured at ~150 km for a frequency of 3.3 mHz (Staiger 1985). It needs to be determined whether opacity effects coupled with convective motions, or an upward propagating wave component compensating the energy losses that occur in the evanescent mode through radiative damping, causes the observed phase anomaly.

4. The Chromospheric Cavity. The theory of solar atmospheric structure (Ando and Osaki 1977; Leibacher and Stein 1981; Deubner and Gough 1984) predicts a shallow chromospheric cavity for p-mode oscillations formed by the temperature minimum at the bottom, and by the steep temperature rise of the transition region above. Simulations of the hydrodynamic wave field show a pattern of standing waves with a period of ~3 min from a height of ~800 km all the way up to the constant pressure boundary assumed for this model (Leibacher et al. 1982). The predicted signature of this chromospheric mode in the diagnostic diagram is one or two ridges at constant ω independent of k.

Early observations of velocity and brightness oscillations in chromospheric lines have shown (see, e.g., Orrall 1966) that there is a separate peak near 5.5 mHz present in the power spectra of the oscillations. This peak was subsequently confirmed by many authors, but attempts to localize one (or two) chromospheric modes corresponding to these oscillations as a distinct power ridge in k-ω diagrams have so far failed. It is conceivable that the development of large-scale coherent patterns is precluded by comparatively small-scale horizontal inhomogeneities of the chromospheric network. An alternative explanation for the observed 5.5 mHz oscillations as the high-l

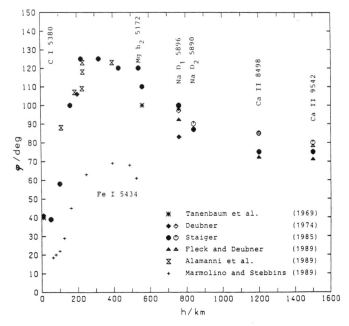

Fig. 18. Phase differences between brightness and velocity (upward) fluctuations for $\nu = 3.3$ mHz as function of height in the atmosphere. For chromospheric lines, values for $\nu = 5.5$ mHz are indicated by open symbols. The measurements of Marmolino and Stebbins (1989) deviate substantially from the others, because the 2.2 mHz feature discussed in the text was included in the integration interval; but they still follow the same trend.

(~3400) branch of the f-mode has been suggested (F. Kneer, personal communication). Recently, analysis of the frequency shifts of the p-modes observed at Big Bear during different phases of the solar cycle has shown a sharp drop in the shifts at a frequency of ~4 mHz, corresponding to one of the predicted chromospheric modes. This drop may be due to a change in the reflection properties of the boundaries of the chromospheric cavity.

Since a clear discrimination among possible candidate modes for these oscillations is crucial for our understanding of the atmospheric structure, phase and coherence spectra were introduced in later observational studies (Mein 1977; Schmieder 1978) to look for the signature of standing waves: 180° phase jumps at certain frequencies connected to the existence of horizontal nodal planes in the atmosphere. Unfortunately, the studies remained largely inconclusive in this respect, probably because of insufficient statistics. Recent investigations by Fleck and Deubner (1989), and Staiger (1987) have attempted to overcome these difficulties by employing observations of considerably longer duration (>3 hr), thereby enhancing the signal with regard to the incoherent noise due to nonuniformities in the solar atmosphere and to terrestrial seeing.

Both studies show clear examples of phase jumps measured in low chromospheric lines (most notably in the core of the Na D_2 line; see Fig. 19). In addition, Fleck and Deubner (1989) note an amplitude depression in the 7 to 9 mHz range of the intensity power and in the V-I coherence spectra of the Na line, suggesting the presence of at least one nodal plane at a certain level in the atmosphere. However, they also point to an inconsistency which appears to arise from such straightforward interpretation of the observations. In order to produce nodal planes, propagating sound waves must exist at this level, and such waves should be reflected at a higher level (presumably the transition region) to interfere with the following, upward propagating wavetrains. If dispersion and dissipation of the waves are sufficiently small to allow significant interference at a comparatively low level in the atmosphere after traversing more than 1000 km in height twice, then why have nodal planes at higher levels, and thus at higher frequencies, not been detected? Perhaps coherence between adjacent spatial elements, and therefore signal amplitude of unresolved areas, is gradually lost with increasing height in the atmosphere, but can it all be restored on the way back after reflection?

Lindsey and Roellig (1987) have recently contributed measurements of solar continuum fluctuations at 350 μm and 800 μm in the far infrared, corresponding to heights of 570 and 820 km, respectively, in the atmosphere. They found phase differences between the two levels of 25° and 35°, with the

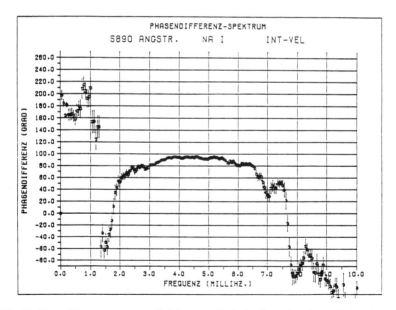

Fig. 19. Phase difference spectrum of brightness and velocity fluctuations in the Na I D_2 line. A 180° phase jump indicating the presence of a pressure nodal plane is clearly visible in this line near $\nu = 7.8$ mHz (figure from Staiger 1985).

upper layer lagging. This range of values is in excellent agreement with phase differences of brightness signals obtained from Fraunhofer lines in the visible bridging the same height interval, such as Mg I 5172 and Na D$_1$ 5896 (Staiger 1985). In these two lines *V-I* phases of 120° and 90°, respectively, have been determined, suggesting that in the lower (Mg I) layer the brightness phase is accelerated by radiative damping, whereas 250 km further up, the signature of almost ideal adiabatic conditions is observed. Still higher up, in the infrared Ca II lines at 8498 Å (1200 km) and 9542 Å (1500 km), the *V-I* phases range from 65° to 85° according to Staiger (1985) and Fleck and Deubner (1989), although one might expect adiabatic conditions in this region as well.

The chromospheric cavity is not well understood theoretically, and it is only poorly observed. But it is here that the magnetic field takes control over most of the plasma, and it is here that the remaining mechanical energy flux emerging from quiet regions couples into the large-scale magnetic chromospheric and coronal structures. Observations with good height discrimination as well as spatial resolution are required to disentangle the dynamical processes in this layer.

5. Magnetic Effects.   The structure and dynamics of the chromospheric network differ considerably from the conditions in the quiet atmosphere. The obvious reason for these changes is the presence of magnetic flux penetrating the photosphere in the form of subarcsecond flux concentrations which spread somewhere above the temperature minimum to form a more uniformly distributed "canopy" or umbrella.

In accordance with the topology of the magnetic field, which occupies a very small fractional area in the photosphere, only small effects are seen in the fluctuations of photospheric lines. The mechanical energy flux carried by acoustic or internal gravity waves should be uniformly distributed in these layers. In the infrared and the H and K lines of Ca II, and also in Na D, however, a number of changes are observed with respect to the quiet atmosphere, not all of which are yet well understood.

In the chromosphere, there is a dominant 150 to 180 s period standing oscillation in the cell interior, compared with the bright network regions where the evanescent 5-min oscillations are strongest and where longer-period motions (internal gravity waves and stochastic motions) are more prominent than in the cell interior (Damé et al. 1984; Deubner and Fleck 1990). It is easy to understand that conditions for trapping of a chromospheric *p*-mode ($P \approx 3$ min) are less favorable in the highly filamentary and irregular network than in the cell interior, and are made even worse by the enhancement of the bottom temperature, and the geometry of the magnetic field. We do not understand, however, why the 3-min mode is not at all apparent in the Doppler measurements in the core of the Ca 8498 line. At the corresponding level of ~1200 km, a pressure node is expected rather than a velocity node (see, e.g., Fleck and Deubner 1989, Fig. 3).

Significant differences between network and cell interior are found in the
V-V phase-difference spectra, both in the gravity-wave regime and for high-
frequency acoustic waves (Lites et al. 1982; Deubner and Fleck 1990). Phases
computed from *V* observations in the cores of Ca 8542 and Ca 8498 are
marginally positive (i.e., downward propagating) in the 1 to 5 mHz region in
the cell interior, but distinctly negative (up to 15°) in the bright network (Fig.
20). We are therefore led to believe that internal gravity waves, supposedly
modified by the presence of magnetic fields, and propagating with *downward
group velocity* into the magnetic funnels are the source of these negative
phase signals (Deubner 1986*a*).

We speculate that mechanical energy is collected within magnetic canopies over extended areas in the low chromosphere from low-frequency vertical motions, and transformed into magneto-gravity waves. These waves "fall"
into the magnetic funnels, and are subsequently dissipated as the amplitudes
steepen through convergence of the field lines. Thus, heating of magnetic
structures in the chromosphere can, at least partially, be affected by magnetically guided waves from above rather than from below, where a very efficient
process is required to feed the necessary energy selectively through the narrow supergranular and mesogranular network lanes.

In the acoustic regime the linear increase of the phase differences measured between velocity fluctuations in the photosphere (Fe I 8496) and the
lower chromosphere (Ca II 8542) depends strongly on brightness in the chro-

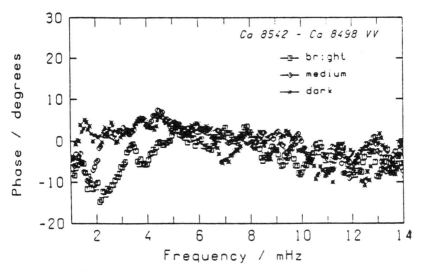

Fig. 20. Phase-difference spectra of velocity fluctuations measured in the core of the Ca II infrared lines at 8542 Å and 9498 Å. In the gravity-wave regime, bright chromospheric network
elements exhibit negative phases corresponding to downward propagation of wave energy (figure from Deubner and Fleck 1990).

mospheric network. The gradient in the phase spectrum is ~30% flatter in the bright network elements, presumably because the transition from running to standing acoustic waves occurs at a lower height in the network and, perhaps more importantly, because the phase velocity increases with temperature or magnetic flux density, or both (Deubner and Fleck 1990).

V. CONCLUSION

It is obvious by the space devoted to the topic in this book that much has been learned about the oscillations of the Sun. It is also obvious that much remains to be done. For example, it is not yet possible to distinguish between several qualitatively different descriptions of internal rotation. The splitting of the $l = 1$ modes has not yet been reliably measured. The existence of magnetic splittings remains to be observationally investigated. The possibility of equatorial and meridional anisotropies in the properties of the sectoral modes needs further clarification. There are inconsistencies in the measurements of the relative amplitudes of velocity and intensity oscillations. The existence of a chromospheric mode remains to be either verified or disproven and, if it does exist, study of its properties will provide information on the structure of the chromosphere. There are questions about the effects of solar magnetic activity on the oscillation properties. Infrared observations are in their infancy and may provide a valuable new tool for studying the oscillations. There is still much to learn about the amplitudes, line widths, and line shapes of the modes; these studies will provide information on the excitation mechanism of the oscillations. The interaction of the oscillations and the surface convection is still poorly understood. More statistics on the behavior of the oscillations through the course of the solar cycle need to be accumulated, and correlative studies would also be of benefit. The mapping of the flows in the convection zone is just beginning, as is the study of active-region tomography.

While helioseismic measurements of the frequencies of the global oscillations are already by far the most precise of all astronomical observations, higher-precision measurements are essential if the promise of helioseismology is to be completely fulfilled. The forthcoming data from the IRIS network, the GONG project and the SOLAR-A and SOHO spacecraft should increase our knowledge of the inside of the Sun greatly. With this new information, perhaps we will eventually know as much about why the Sun acts as it does as we know about how it acts.

THEORY OF SOLAR OSCILLATIONS

JØRGEN CHRISTENSEN-DALSGAARD
Aarhus Universitet

and

GABRIELLE BERTHOMIEU
Observatoire de Nice

The last decade has witnessed a revolution in our knowledge about the interior of the Sun, through the development of helioseismology. By analyzing observed frequencies of solar oscillation, it has been possible to obtain detailed information about many aspects of the properties of the Sun, and of the physics of matter in the solar interior. Major efforts are currently under way to improve the observations, and hence helioseismic investigations will probably remain a very active field of research for the foreseeable future. This chapter provides theoretical background for understanding the properties of solar oscillations and their relation to the structure of the Sun. Although accurate numerical computations are required to match the precision of the observed frequencies, a great deal of insight can be obtained from considering simple asymptotic approximations. In particular, the wide range of modes observed allows a separation of the effects of different parts of the Sun; as discussed in detail in the chapter by Gough and Thompson, this allows inversion of the observed frequencies to determine localized properties of the solar interior. Also, changes in the assumed physics of solar models, within the range of uncertainty in the physics, lead to frequency changes that are substantial compared with the intrinsic precision of the observations. Thus as probes of the solar interior, the frequencies are both sensitive and straightforward to interpret.

I. INTRODUCTION

The observed frequencies of solar oscillation have opened a window on the solar interior. To use the information provided by the observations, however, it is necessary to consider the relationship between the behavior of the oscillations and the underlying properties of the Sun. This requires the ability to compute sufficiently accurate oscillation frequencies, and if possible other aspects of the oscillations, for a given solar model; however, equally important is a well-developed understanding of how different modes of oscillation respond to particular features of the solar interior.

In principle all properties of the oscillations provide information about the structure and dynamics of the solar interior. In practice, however, only those aspects of the physics of the oscillations that determine their frequencies are sufficiently well understood to enable us to get reliable information about the interior of the Sun. This chapter explores the properties of the modes of oscillation, and the relation between their frequencies and the structure of the solar interior. As a simplification, we assume that the oscillations are adiabatic, and hence neglect the processes that excite or damp the modes; these processes, which also have a significant effect on the oscillation frequencies, are treated in the chapter by Cox et al. Furthermore, we consider only the dependence of the frequencies on the *structure* of the solar interior, and neglect effects of rotation, as well as possible effects of a large-scale magnetic field. These effects form the subject of the chapters by Libbrecht and Morrow and by Dziembowski and Goode.

To provide a background for, and motivate, the development of the theory, it is useful to summarize the basic properties of the oscillations of the Sun. We assume that the oscillations can be described with sufficient accuracy in terms of linear perturbation theory. As discussed in the chapter by F. Hill et al., and justified in Sec.II.B of this chapter, for a single mode the dependence on colatitude θ and longitude ϕ is then given by a *spherical harmonic* $Y_l^m (\theta, \phi)$. Here the *degree* l determines the overall complexity of the perturbation over spherical surfaces. More precisely, the horizontal wavelength λ_h and the length k_h of the horizontal component of the wavenumber at a distance r from the center of the Sun are given by

$$k_h = \frac{2\pi}{\lambda_h} = \frac{\sqrt{l(l+1)}}{r} \tag{1}$$

(cf. Sec.II.B). The *azimuthal order* m determines the number of nodes around the equator. For a spherically symmetric star, the orientation of the coordinate system used to describe the oscillation can have no effect on the properties of the modes, including their frequencies. As the definition of m is related to the location of the equator, it follows that frequencies of a spherical star must be independent of m.

For each l and m there is a spectrum of modes, distinguished by the dependence of the perturbation on r, and characterized by the *radial order n,* which measures the number of zeros in the radial direction in, say, the radial component of the displacement associated with the perturbation.

Because we neglect the excitation and damping of the modes, as a function of time t they behave as purely harmonic functions $\cos\omega t$, where ω is the *angular frequency* of oscillation. For a theoretical description, the frequency is most simply characterized by ω; on the other hand, observed frequencies are normally given in terms of the *cyclic frequency*

$$v = \frac{\omega}{2\pi} = \frac{1}{P} \tag{2}$$

where P is the oscillation period.

To illustrate the basic properties of the spectrum of possible oscillations, Fig. 1 shows computed frequencies for a typical solar model. It is immediately obvious that there are two distinct classes of modes, which are conventionally distinguished by the sign of n. Those with $n > 0$, which are normally called p-modes, have frequencies that for large degree increase with l roughly as $l^{1/2}$, whereas for modes with $n < 0$, normally called g-modes, the frequencies tend to a limit as l increases. The modes with $n = 0$ form an intermediate case, despite the similarity of the behavior of their frequencies to that of the p-modes; these modes are called f-modes. The description of the physical nature of these modes forms an important part of this chapter.

II. EQUATIONS AND BOUNDARY CONDITIONS
OF SOLAR OSCILLATIONS

A. General Properties of the Motion

Although the main purpose of this section is to present the equations and boundary conditions governing adiabatic stellar oscillations of small amplitude, we begin by considering briefly the more general equations of fluid motion. We neglect viscosity, and assume that gravity is the only body force; thus, effects of, e.g., magnetic fields are neglected. Then the equations of motion can be written

$$\rho\frac{D\mathbf{v}}{Dt} = -\nabla p + \rho\nabla\Phi \tag{3}$$

where $\mathbf{v}$ is the velocity, p is pressure, ρ is density, and Φ is the gravitational potential. As usual D/Dt denotes the material time derivative,

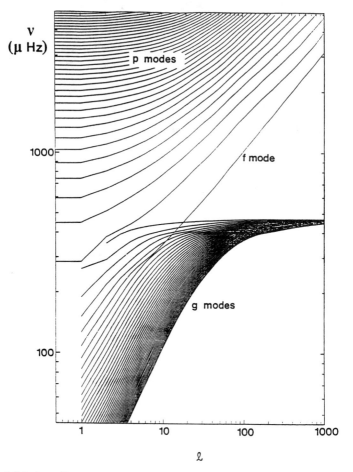

Fig. 1. Adiabatic oscillation frequencies for a normal model of the present Sun, as functions of the degree l. For clarity, points corresponding to modes with a given radial order have been connected by straight lines. Only g-modes with radial order < 40 have been included.

$$\frac{Df}{Dt} = \frac{\partial f}{\partial t} + \mathbf{v} \cdot \nabla f \tag{4}$$

for any quantity f. The density satisfies the equation of continuity,

$$\frac{D\rho}{Dt} = -\rho \operatorname{div} \mathbf{v} \tag{5}$$

and Φ is determined by Poisson's equation,

$$\nabla^2\Phi = -4\pi G\rho \qquad (6)$$

where G is the gravitational constant.

The relation between p and ρ is determined by the energy equation, which may be written

$$\frac{1}{p}\frac{Dp}{Dt} - \Gamma_1\frac{1}{\rho}\frac{D\rho}{Dt} = \frac{\Gamma_3 - 1}{p}(\rho\epsilon - \mathrm{div}\mathbf{F}). \qquad (7)$$

Here Γ_1 and Γ_3 are adiabatic indices, defined by

$$\Gamma_1 = \left(\frac{\partial\ln p}{\partial\ln\rho}\right)_s \qquad \Gamma_3 - 1 = \left(\frac{\partial\ln T}{\partial\ln\rho}\right)_s \qquad (8)$$

where T is temperature, and the derivatives are at constant specific entropy s. Also ϵ is the rate of energy generation (e.g., from nuclear reactions) per unit mass, and $\mathbf{F}$ is the energy flux. Equation (7) is valid for radiative transfer in the equilibrium diffusion approximation, provided the thermodynamic quantities, including Γ_1 and Γ_3, are based on an equation of state where the contribution from radiation, in the diffusion approximation, is included consistently in the pressure and internal energy (Mihalas 1984). This is an excellent approximation in the solar interior, but fails in the solar atmosphere; in this case, the contributions from radiative pressure and energy density must be treated separately. However, radiation pressure makes a very small contribution to the dynamics of the motion throughout the Sun.

Equations (3)–(7) provide a description of the general behavior of the fluid. However, in the case of solar oscillations the situation is vastly complicated by the instability of the outer layers to convection, which gives rise to convective motion. Thus, for example, the velocity field $\mathbf{v}$ in Eq. (3) contains a convective component, in addition to the oscillations which we wish to describe. To eliminate the convective fluctuations, one has to carry out some suitable averaging procedure (see, Gough 1977a; Unno et al. 1979). This introduces second moments of the fluctuations in the equations describing the average motion. The most significant of these is the convective flux, which in most of the convection zone dominates the energy transport in the average energy equation corresponding to Eq. (7); however, the average momentum equation should also contain a term in the turbulent Reynolds stress tensor, often approximated by a turbulent pressure $p_t \sim \rho u_t^2$, where u_t is a typical convective velocity. Very near the solar surface, where the convective velocities are comparable with the sound speed, p_t may make a significant contribution to the pressure balance.

No satisfactory procedure exists for dealing with the convection in a static star, let alone in the presence of pulsations. As discussed in more detail in the chapter by Cox et al., this introduces severe uncertainty in the calculation of the stability of solar oscillations. Furthermore, the turbulent Reynolds stress is likely to have a significant effect on the oscillation frequencies. In this chapter, apart from the inclusion of the convective flux in the calculation of the equilibrium model, we shall generally ignore effects of convection; this introduces errors in the properties of the oscillations, in particular the frequencies, which must be kept in mind when computed frequencies are compared with the observations.

In most of the Sun, the right-hand side of Eq. (7) is negligible compared with the magnitude of the terms on the left-hand side. If we approximate div $\mathbf{F}$ by F/d, where d is a typical length scale, and ignore ϵ, we obtain to order of magnitude

$$\frac{\Gamma_3 - 1}{p} (\rho\epsilon - \mathrm{div}\mathbf{F}) \sim \tau_{th}^{-1} = \frac{F}{pd} \approx \frac{1 \text{ dyn/cm}^2}{p} \frac{R}{d} \mathrm{s}^{-1} \qquad (9)$$

where R is the solar radius, if the solar surface flux and radius are assumed. When averaged over the entire Sun, and taking $d \approx R$, the thermal time scale τ_{th} is approximately the Kelvin-Helmholz time of the Sun, which is of order 10^7 yr. In the solar envelope we may take $d \sim H_p$, where

$$H_p = \frac{p}{g\rho} \qquad (10)$$

is the pressure scale height, g being the gravitational acceleration. Very near the solar surface, τ_{th} is comparable with the typical oscillation periods, and the heating term in the energy equation has a substantial effect on the motion. However, due to the rapid increase of pressure with increasing depth (p increases by ~ 4 orders of magnitude in the outer 1% of the solar radius), τ_{th} increases rapidly, and in most of the Sun the heating term is very small compared with the rate of change of p and ρ. Thus it is reasonable to treat the motion as being *adiabatic*, i.e., occurring without heat loss or gain, so that

$$\frac{1}{p} \frac{Dp}{Dt} = \Gamma_1 \frac{1}{\rho} \frac{D\rho}{Dt}. \qquad (11)$$

The adiabatic approximation is evidently highly restrictive. In particular, the motion is conservative, and hence the total energy is constant. Thus, when considering a single mode of oscillation in the adiabatic approximation, the amplitude of the mode is constant, and therefore it is not possible to treat its excitation or damping. Also, close to the surface, and in the atmosphere, the thermal time scale is comparable with, or smaller than, the periods, and here there are large departures from adiabaticity. These effects are discussed in the chapter by Cox et al.; they are crucial for the calculation of the stability of the modes, as well as for the interpretation of the phase and amplitude relations observed in the solar atmosphere. The nonadiabatic effects also modify the dynamical properties of the motion and hence affect the oscillation frequencies. However, by far the largest fraction of the mass involved in the oscillations is in regions where τ_{th} is large and the dynamics is barely affected by nonadiabaticity. Thus, although the nonadiabatic effects on the frequencies are not negligible, they are generally small.

The adiabatic approximation very considerably simplifies the discussion of the properties of the oscillations. Thus, in most of this chapter we shall assume that the oscillations are adiabatic, keeping in mind that we thereby introduce errors in the computed frequencies.

B. Linearized Equations

The observed amplitudes of solar oscillations are very small. Thus, in principle the modes can be treated in *linear perturbation theory*. We consider a static, spherically symmetric equilibrium model, with pressure $p_0(r)$, density $\rho_0(r)$, etc. These quantities satisfy the usual equations of stellar structure. Thus the gravitational acceleration is $\nabla\Phi_0 = -g_0\mathbf{a}_r$, where $\mathbf{a}_r$ is a unit vector in the radial direction, and

$$g_0 = \frac{Gm_0}{r^2}. \tag{12}$$

Here r is the distance to the center, and m_0 is the mass of the sphere interior to r, which satisfies

$$\frac{dm_0}{dr} = 4\pi r^2 \rho_0. \tag{13}$$

The equation of hydrostatic support becomes

$$\frac{dp_0}{dr} = -g_0\rho_0. \tag{14}$$

The time-dependent quantities are written as, e.g., $p(r,t) = p_0(r) + p_1'(r,t)$, where the prime denotes the *Eulerian perturbation*, i.e., the perturbation at a given point. Also we introduce the displacement $\delta\mathbf{r}$ by $\mathbf{v} = \partial\delta\mathbf{r}/$

∂t. Dropping subscripts 0 on equilibrium quantities we obtain the following linearized equations:

$$\text{Momentum: } \rho\frac{\partial^2 \delta r}{\partial t^2} = -\nabla p' + \frac{\rho'}{\rho}\nabla p + \rho\nabla\Phi' \tag{15}$$

$$\text{Continuity: } \rho' + \text{div}(\rho\,\delta r) = 0 \tag{16}$$

$$\text{Adiabaticity: } \frac{\rho'}{\rho} + \frac{1}{\rho}\delta r\cdot\nabla\rho = \frac{1}{\Gamma_1}\left(\frac{p'}{p} + \frac{1}{p}\delta r\cdot\nabla p\right) \tag{17}$$

$$\text{Poisson's equation: } \nabla^2\Phi' = -4\pi G\rho'. \tag{18}$$

Because of the spherical symmetry and time independence of the equilibrium mode, we can separate the perturbation quantities in spherical polar coordinates (r, θ, ϕ) and time. The equations can be combined in such a way that all derivatives with respect to θ and ϕ appear in the form of the tangential Laplace operator

$$\nabla_t^2 \equiv \frac{1}{r^2\sin\theta}\frac{\partial}{\partial\theta}\left(\sin\theta\frac{\partial}{\partial\theta}\right) + \frac{1}{r^2\sin^2\theta}\frac{\partial^2}{\partial\phi^2} \tag{19}$$

Consequently the variation of a scalar quantity with θ and ϕ can be written in terms of an eigenfunction of $r^2\nabla_t^2$. This may be chosen as a spherical harmonic Y_l^m. Similarly, the time dependence can be expressed in terms of a harmonic function. Thus, e.g., the pressure perturbation is written as

$$p'(r,\theta,\phi,t) = \sqrt{4\pi}Re[p'(r)Y_l^m(\theta,\phi)e^{-i\omega t}] \tag{20}$$

where for simplicity we use p' to denote the full perturbation and the amplitude function. Y_l^m is defined by

$$Y_l^m(\theta,\phi) = (-1)^m c_{lm}P_l^m(\cos\theta)e^{im\phi} \tag{21}$$

where P_l^m is a Legendre function, and the normalization constant c_{lm} is determined by

$$c_{lm}^2 = \frac{2l+1}{4\pi}\frac{(l-m)!}{(l+m)!} \tag{22}$$

such that the integral of $|Y_l^m|^2$ over the unit sphere is unity. It follows from the momentum Eq. (15) that the displacement vector can be written as

$$\delta r(r,\theta,\phi,t) = \sqrt{4\pi}Re\left\{\left[\chi_r(r)Y_l^m a_r + \xi_h(r)\left(\frac{\partial Y_l^m}{\partial\theta}a_\theta + \frac{1}{\sin\theta}\frac{\partial Y_l^m}{\partial\phi}a_\phi\right)\right]e^{-i\omega t}\right\} \tag{23}$$

where $\mathbf{a}_\theta$ and $\mathbf{a}_\phi$ are unit vectors in the θ and ϕ directions. It may be noted that with this separation of variables the effect of ∇_t^2 on any scalar variable corresponds to multiplication by $-k_h^2$, where

$$k_h = \frac{\sqrt{l(l+1)}}{r} \equiv \frac{L}{r}. \tag{24}$$

This justifies the identification, already made in Eq. (1), of k_h with the local horizontal wave number of the mode. The root mean squares, over a spherical surface at radius r and time, of the vertical and horizontal components of the displacement, are given by

$$<\delta r>_{rms} = \frac{1}{\sqrt{2}}\,\xi_r(r), \qquad <\delta h>_{rms} = \frac{L}{\sqrt{2}}\xi_h(r). \tag{25}$$

The separation given in Eqs. (20) and (23) describes the so-called *spheroidal modes*, which is the only class of modes considered in the following. It should be mentioned, however, that there exists a second class of modes, the *toroidal modes* (see e.g., Aizenman and Smeyers 1977). In a nonrotating star these have purely tangential displacement, zero Eulerian perturbations of scalar quantities and zero frequency. Physically, they correspond to infinitely slow rotation of spherical shells relative to each other. In a rotating star, the toroidal modes give rise to oscillations whose frequencies are of order the rotation frequency (see e.g., Papaloizou and Pringle 1978; Provost et al. 1981).

By substituting the spherical harmonic representations into Eqs. (15)–(18), the following relations between the amplitude functions are obtained:

$$-\omega^2\rho\xi_r = -\frac{dp'}{dr} - \rho'g + \rho\,\frac{d\Phi'}{dr} \tag{26}$$

$$-\omega^2\rho\xi_h = -\frac{1}{r}\,(p' - \rho\Phi') \tag{27}$$

$$\frac{1}{r^2}\frac{d}{dr}\left(r^2\,\frac{d\Phi'}{dr}\right) - \frac{L^2}{r^2}\Phi' = -4\pi G\rho' \tag{28}$$

$$\rho' = -\frac{1}{r^2}\frac{d}{dr}\,(r^2\rho\xi_r) + \frac{L^2}{r}\rho\xi_h \tag{29}$$

$$p' = p\left[\Gamma_1\frac{\rho'}{\rho} - \left(\frac{d\ln p}{dr} - \Gamma_1\frac{d\ln\rho}{dr}\right)\xi_r\right]. \tag{30}$$

Equations (26)–(30) can be rearranged into the following fourth-order set of ordinary differential equations for the amplitude functions $(\xi_r(r),\xi_h(r),\Phi'(r))$:

$$\frac{d\xi_r}{dr} = -\left(\frac{2}{r} + \frac{1}{\Gamma_1}\frac{d\ln p}{dr}\right)\xi_r + \frac{r\omega^2}{c^2}\left(\frac{S_l^2}{\omega^2} - 1\right)\xi_h - \frac{1}{c^2}\Phi' \tag{31}$$

$$\frac{d\xi_h}{dr} = \frac{1}{r}\left(1 - \frac{N^2}{\omega^2}\right)\xi_r + \left(\frac{1}{\Gamma_1}\frac{d\ln p}{dr} - \frac{d\ln\rho}{dr} - \frac{1}{r}\right)\xi_h + \frac{N^2}{rg\omega^2}\Phi' \tag{32}$$

$$\frac{d^2\Phi'}{dr^2} = -\frac{2}{r}\frac{d\Phi'}{dr} - 4\pi G\rho\left(\frac{N^2}{g}\xi_r + \frac{r\omega^2}{c^2}\xi_h\right) + \left(\frac{L^2}{r^2} - \frac{4\pi G\rho}{c^2}\right)\Phi'. \tag{33}$$

Here c is the adiabatic sound speed,

$$c^2 = \frac{\Gamma_1 p}{\rho} \approx \frac{\Gamma_1 k_B T}{\mu m_u} \tag{34}$$

where the last approximation is valid for an ideal gas; k_B is Boltzmann's constant, m_u is the atomic mass unit and μ is the mean molecular weight. Also we have introduced the Lamb frequency

$$S_l = \frac{Lc}{r} \tag{35}$$

and the buoyancy frequency N, defined by

$$N^2 = g\left(\frac{1}{\Gamma_1}\frac{d\ln p}{dr} - \frac{d\ln\rho}{dr}\right). \tag{36}$$

For *radial* oscillations, with $l = 0$, it follows immediately from Eqs. (28) and (29) that

$$\frac{d\Phi'}{dr} = 4\pi G\rho\xi_r \tag{37}$$

if we require $d\Phi'/dr$ to be finite at the center. If the differential equations are written in terms of ξ_r and p', it is then possible to eliminate the terms in Φ', to obtain a second-order system of differential equations. With a little further manipulation, this reduces to

$$\frac{1}{r^3}\frac{d}{dr}\left(r^4\Gamma_1 p\frac{d\zeta}{dr}\right) + \frac{d}{dr}[(3\Gamma_1 - 4)p]\zeta + \rho\omega^2 r\zeta = 0 \tag{38}$$

where $\zeta = \xi_r/r$. This equation is very convenient for discussing the properties of radial oscillations (see, e.g., Ledoux and Walraven 1958; Cox 1980). In the solar case, however, little is gained by distinguishing between the radial and the general nonradial case.

C. Boundary Conditions

The amplitude functions must satisfy boundary conditions at the center and surface of the Sun. The center is a regular singular point of Eqs. (31)–(33). The behavior of the solution can be analyzed by expansion in r around $r = 0$, using the expansion of the equilibrium quantities. For the regular solution, $\xi_r \propto r^{l-1}$ (except for $l = 0$, where $\xi_r \propto r$), and p', ρ', $\Phi' \propto r^l$. Furthermore one finds that

$$\xi_r \simeq l\xi_h, \qquad \text{for } r \to 0. \tag{39}$$

From the expressions for the coefficients in the higher-order terms more precise relations can be established among $(\xi_r, \xi_h, \Phi', d\Phi'/dr)$ at small, but nonzero r (Christensen-Dalsgaard et al. 1974). When applied at the innermost meshpoint $r = r_1$ used in the computation, these relations determine two inner boundary conditions.

As discussed in Sec. III, p-modes of high degree have substantial amplitudes only near the solar surface, and decrease exponentially with depth at greater depths. For such modes it may be sufficient to solve the oscillation equations in a model of the solar envelope, rather than in a complete solar model. Furthermore, as illustrated in Sec.IV.D, the perturbation in the gravitational potential may be neglected. Thus one has to specify only a single boundary condition at the bottom of the envelope model. Given the small amplitude of the modes in the deep interior, it may be reasonable to take the displacement to be zero at this point (see, e.g., Ando and Osaki 1975). However, it is straightforward to derive a more accurate condition from the asymptotic behavior of the eigenfunction, discussed in Sec.IV.

At the surface, Φ' and its first derivative must match continuously onto the decreasing solution to Laplace's equation outside the Sun. If the density in the solar atmosphere is neglected, this leads to the condition

$$\frac{d\Phi'}{dr} + \frac{l+1}{r}\Phi' = 0, \qquad \text{at } r = R \tag{40}$$

(see also Christensen-Dalsgaard et al. 1974). The dynamical surface boundary condition should in principle take into account the detailed behavior of the oscillation in the solar atmosphere. In the idealized case where the model is assumed to have a free outer surface, the absence of forces on the surface corresponds to demanding that the pressure perturbation vanish on the perturbed surface.

$$\delta p = p' + \frac{dp}{dr}\xi_r = 0, \qquad \text{at } r = R \tag{41}$$

where δp is the *Lagrangian* pressure perturbation, i.e., the perturbation following the motion. In practice the outer boundary condition is applied at a suitable point within the solar atmosphere. As the equilibrium temperature varies slowly in the region around the temperature minimum, a reasonable condition can be obtained by matching the solution continuously to the analytically known solution to the adiabatic wave equation in an isothermal atmosphere (see, e.g., Unno et al. 1979). Also, as discussed in Sec. III, at the frequencies where most of the oscillations are observed, the modes are evanescent in the atmosphere. Hence the solution is relatively insensitive to the details of the boundary condition, as long as it is applied sufficiently high in the atmosphere (see also Ulrich and Rhodes 1983; Noels et al. 1984).

By using Eqs. (27), (14) and (12), Eq. (41) can also be written as

$$\xi_h = \frac{GM}{R^3\omega^2}\xi_r - \frac{1}{R\omega^2}\Phi' \qquad (42)$$

where M is the total mass of the Sun. As argued below, Φ' is often small. Hence Eqs. (42) and (25) show that the ratio between the root mean squares of the horizontal and vertical components of the surface velocity is approximately given by

$$\frac{<V_h>_{rms}}{<V_r>_{rms}} \approx \frac{GM}{R^3\omega^2}L \qquad (43)$$

and hence is a function of the frequency and the degree, but not of the detailed nature of the mode. Also it is evident that for high-frequency modes of low degree, the surface velocity is predominantly in the radial direction.

D. The Modes of Oscillation

A nonzero solution to Eqs. (31)–(33), with the boundary conditions (39)–(41), is only possible for selected values of ω^2, which is therefore an *eigenvalue* of the problem. For each value of l, one obtains a discrete spectrum of eigenfrequencies ω_{nl}, labeled by the radial order n. It is convenient to let n take both positive and negative values, such that ω_{nl} increases with increasing n. The definition of n may be arranged in such a way that when large $|n|$ gives the number of zeros in ξ_r, both for positive and negative n, excluding (except for $l = 0$) a possible zero at $r = 0$. It can be shown that $\omega_{nl} \rightarrow 0$ for $n \rightarrow -\infty$. The behavior at large positive n depends on the assumed surface properties of the model and the oscillations. For simplified models where the temperature tends to zero at the surface, or if the boundary condition (Eq. 41) is used, $\omega_{nl} \rightarrow \infty$ for $n \rightarrow \infty$; as discussed in Sec.III.B, with more realistic models of the solar atmosphere, the modes cease to be trapped at frequencies exceeding a critical frequency, and in this case, at each value of l there is only a finite number of trapped modes with positive n.

The definition of n can be given a precise mathematical meaning, at least in the Cowling approximation (see below), where Φ' is neglected (Eckart 1960; Scuflaire 1974; Osaki 1975). It may be shown (see, e.g., Gabriel and Scuflaire 1979; Christensen-Dalsgaard 1980) that this definition is invariant under a continuous change of the equilibrium model. From a mathematical point of view, it is customary to classify modes with $n < 0$, $n = 0$ and $n > 0$ as g-, f- and p-modes, respectively. The relation of this classification to the physical nature of the modes is discussed in more detail in Sec.IV.A.

III. ASYMPTOTIC THEORY OF THE OSCILLATIONS

A numerical solution of the equations of adiabatic oscillations is straightforward. However, the understanding of the numerical results and the interpretation of the observed oscillations have been greatly assisted by asymptotic analyses of the oscillation equations. The usefulness of asymptotics is to a large extent due to the fact that the observed acoustic modes in the five-min region have high radial order or high degree; similarly gravity modes corresponding to the oscillations that may have been detected at long periods have high radial order. For this reason asymptotics can provide relations of acceptable accuracy between the properties of the Sun and the properties of the oscillation frequencies.

A. Asymptotic Properties of the Oscillation Equations

The asymptotic analysis is considerably simplified by noting that at high degree or radial order the perturbation Φ' in the gravitational acceleration can be neglected. This approximation, known as the *Cowling approximation*, was first suggested by Cowling (1941). It may be justified by noting that at high degree or order, a mode gives rise to many regions of alternating sign in the density perturbation, the effects of which largely cancel in the perturbation of the gravitational potential. As a very rough approximation we may take

$$\frac{d^2}{dr^2} \sim -\frac{n^2}{r^2} \tag{44}$$

in Poisson's equation (Eq. 33), where $|n|$ is the number of zeros in the radial direction. This results in the following estimate of the Φ' term in Eq. (31):

$$\frac{1}{c^2}\Phi' \sim \frac{4\pi G \rho r^2}{c^2}\frac{1}{n^2 + l^2}\frac{d\ln p}{dr}\xi_r. \tag{45}$$

Here the factor $4\pi G \rho r^2/c^2$ is smaller than ~ 3 in a normal solar model, and hence for large $|n|$ or l the term is small compared with the other term in ξ_r. In Eq. (32) a similar estimate may be made. Thus when the order or the degree is large, we may be able to neglect Φ'. The effects of the Cowling approximation are discussed in more detail in Sec.IV.E, below.

When Φ' is neglected, the oscillation equations reduce to a second-order system consisting of Eqs. (31) and (32) without the terms in Φ', with one boundary condition (the first of Eqs. [39]) at the center and one (e.g., Eq. [41]) at the surface. It was shown by Gough (cf. Deubner and Gough 1984) that a particularly convenient approximate representation of these equations can be obtained by generalizing an analysis by Lamb (1932) for the plane-parallel case. We introduce the quantity

$$\Psi(r) = c^2\rho^{1/2}\text{div }\delta\mathbf{r}. \tag{46}$$

Neglecting derivatives of g and of r, one finds that Ψ satisfies

$$\frac{d^2\Psi}{dr^2} + \frac{1}{c^2}\left[\omega^2 - \omega_{co}^2 - S_l^2\left(1 - \frac{N^2}{\omega^2}\right)\right]\Psi = 0. \tag{47}$$

Here ω_{co} is a generalization of Lamb's (1909) acoustical cut-off frequency; it is defined by

$$\omega_{co}^2 = \frac{c^2}{4H^2}\left(1 - 2\frac{dH}{dr}\right) \tag{48}$$

where $H = -(\text{d}\ln\rho/\text{d}r)^{-1}$ is the density scale height.

In addition to the modes described by Eq. (47), the approximate analysis shows the existence of modes whose frequencies are given by

$$\omega^2 = \frac{g_s L}{R} \tag{49}$$

where g_s is the surface gravity. For these modes div $\delta\mathbf{r}$ vanishes, and ξ_r decreases with increasing depth as $\exp[-k_h(R - r)]$; they are entirely equivalent to surface gravity waves at a free surface. Corrections to Eq. (49) caused by curvature and the variation in g were discussed by Gough (1984a).

From Eq. (47) it follows that the behavior of the mode is determined by the variation of the characteristic frequencies S_l, N and ω_{co} with r. They are illustrated in Fig. 2, for a normal model of the present Sun. N has a maximum, where $N/2\pi \approx 450$ μHz, near the center of the Sun; in the convection zone N is imaginary and, except very near the surface, of small absolute value. S_l decreases monotonically from the center towards the surface. Finally ω_{co} is large predominantly near the surface. It is illustrated in more detail in Fig. 2b, together with the acoustical cut-off frequency according to Lamb's (1909) original definition,

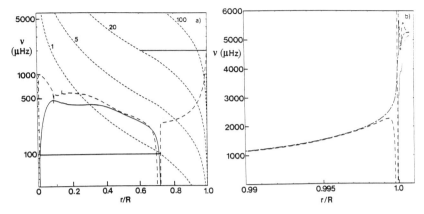

Fig. 2. (a) Characteristic acoustic frequency S_l (cf. Eq. (35); (-------, labeled by the values of l), bouyancy frequency N (cf. Eq. (36); ———————), and acoustical cut-off frequency ω_{co} (cf. Eq. (48); – – – – –), shown in terms of the corresponding cyclic frequencies, against fractional radius r/R. The heavy horizontal lines indicate the trapping regions for a g-mode with frequency $\nu = 100$ μHz, and for a p-mode with degree 20 and $\nu = 2000$ μHz. (b) Acoustical cut-off frequency in the region where the p-modes are reflected. Illustrated are ω_{co} (– – – – –), as well as the Lamb acoustical cut-off frequency $\omega_{co,p}$ for an isothermal atmosphere (cf. Eq. (50); — · — · — · — · —). The thin continuous line shows $N/2\pi$ in the solar atmosphere.

$$\omega_{co,p} = \frac{c}{2H_p} \qquad (50)$$

where H_p is the pressure scale height (cf. Eq. [10]); ω_{co} reduces to $\omega_{co,p}$ for an isothermal layer. The rapid variation in ω_{co} just below $r/R = 1$ is associated with the region where convection is substantially superadiabatic. In the solar atmosphere, $\omega_{co}/2\pi$ and $\omega_{co,p}/2\pi$ vary relatively little, around a value of ~ 5300 μHz.

A mode is an oscillating function of r in the regions where

$$c^2 \mathcal{K} \equiv \omega^2 - \omega_{co}^2 - S_l^2\left(1 - \frac{N^2}{\omega^2}\right) > 0 \qquad (51)$$

and behaves approximately exponentially outside them. A region where Eq. (51) is satisfied is known as a *trapping region* for the mode. From the behavior of the characteristic frequencies shown in Fig. 2 it follows that Eq. (51) is satisfied, approximately, where

$$\omega^2 > S_l^2 \qquad \text{and} \qquad \omega^2 > \omega_{co}^2 \qquad (52a)$$

or

$$\omega^2 < N^2. \qquad (52b)$$

In general a given mode of oscillation has large amplitude in one of the trapping regions, and decreases exponentially outside it. Thus the frequency of the mode is mainly determined by the structure of the Sun within that region.

As shown in Fig. 2, there are two distinct types of trapping regions:

(p) a region corresponding to Eq. (52a), extending from a point in the interior to just below the surface;
(g) a region in the interior corresponding to Eq. (52b), limited by N.

The p type of region typically occurs at high frequencies; modes corresponding to this type of trapping are called p-modes. The g type typically occurs at low frequencies; the corresponding modes are called g-modes.

Physically the terms in N^2 and S_l^2 in Eq. (47) correspond to the two dominant restoring forces for the oscillations, *viz.* buoyancy and pressure fluctuations, respectively. Modes at low frequency where buoyancy dominates (i.e., g-modes) correspond to standing gravity waves, whereas modes at high frequency where pressure fluctuations dominate (i.e., p-modes) correspond to standing acoustic waves.

By applying JWKB analysis to Eq. (47) one obtains the following approximate relation for the eigenfrequencies:

$$\omega \int_{r_1}^{r_2} \left[1 - \frac{\omega_{co}^2}{\omega^2} - \frac{S_l^2}{\omega^2} \left(1 - \frac{N^2}{\omega^2} \right) \right]^{1/2} \frac{dr}{c} \sim \pi(n' + \epsilon). \tag{53}$$

Here r_1 and r_2 are consecutive *turning points* or zeros of $\mathcal{K}$. The phase constant ϵ is $-1/2$, if $\mathcal{K}$ varies approximately linearly with r in the neighborhood of both turning points. [Note that we use the same symbol as for the rate of energy generation (cf. Eq. 7). As the latter quantity is not used below, this should cause little confusion.] The actual behavior may be more complicated; as discussed in Sec.III.C, this is particularly true for g-modes, due to the fact that $|N|$ is very nearly zero in the convection zone. (For p-modes, on the other hand, the computed frequencies can be represented with reasonable accuracy by Eq. [53] with $\epsilon = -1/2$, provided ω_{co} is replaced by $\omega_{co,p}$; cf. Sec.III.B.) Finally, n' is here defined such that the number of zeros in the eigenfunction Ψ in the trapping region is $n' - 1$; it may be shown that for high-order modes this definition corresponds to $n' = |n|$, where n is the formal labeling of computed modes introduced in Sec.II.D.

B. p-Mode Asymptotics

Here we assume that $N^2/\omega^2 \ll 1$. This is certainly true for the solar 5-min oscillations, whose frequencies are typically around 3000 μHz. Furthermore, since in general ω is much larger than ω_{co} near the lower turning point, this is approximately at the point $r = r_t$ where $\omega = S_l$, or

$$\frac{r_t}{c(r_t)} = \frac{L}{\omega}. \tag{54}$$

Thus the location of the lower turning point is a function of ω/L alone. Figure 3 shows how r_t varies with l and the frequency, for a normal solar model. Low-degree modes extend almost to the center of the Sun, whereas modes at the highest degrees observed are confined to the outer fraction of a percent of the solar radius. The upper turning point is approximately at the point where $\omega = \omega_{co}$. Within the present approximation, waves whose frequencies exceed the maximum of ω_{co} in the solar atmosphere are not reflected at the photosphere, but propagate out into the atmosphere. Therefore the concept of trapped modes becomes questionable at high frequencies. In practice the modes may be partially reflected at the transition between the chromosphere and the corona, where the temperature increases steeply. Nonadiabatic calculations show strongly enhanced damping in the atmosphere at frequencies exceeding the adiabatic cut-off frequency (see, e.g., Christensen-Dalsgaard and Frandsen 1983). This, possibly together with damping by nonlinear effects, may account for the decrease in the observed amplitudes at higher frequencies.

We may obtain an approximate asymptotic expression for the p-mode oscillation frequencies from Eq. (53). We neglect the term in N^2/ω^2, and note that the term in ω_{co} is only significant near the surface. By approximating the variation of ω_{co} and c near the surface by the appropriate expressions for a

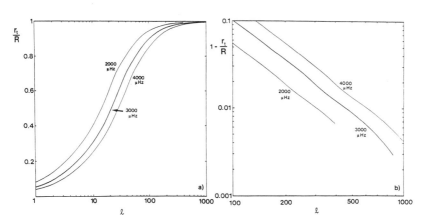

Fig. 3. The turning point radius r_t (a) and the penetration depth $R - r_t$ (b), in units of the solar radius R, as a function of degree l for three values of the frequency ν. The curves have been calculated from Eq. (54) for a normal model of the present Sun. In (b) the curves terminate at the degree where the frequency equals the f-mode frequency; for p-modes the degree is below this value at a given frequency (see also Fig. 1).

polytropic model, it is possible to expand the dependence of the integral on ω_{co}, to obtain

$$\omega \int_{r_t}^{R} \left(1 - \frac{L^2 c^2}{\omega^2 r^2}\right)^{1/2} \frac{dr}{c} \sim \pi(n + \alpha). \tag{55}$$

Here α contains ϵ, as well as a contribution resulting from the expansion of ω_{co} and c; hence, roughly speaking, α describes the phase change at the inner turning point and at the surface. Since $S_l \ll \omega$ near the surface, the l-dependent term in Eq. (53) can be neglected there; consequently α does not depend on l. In the polytropic case α is a constant, which is related to the polytropic index; for realistic models one finds that α is a function of frequency.

From JWKB analysis of Eq. (47), neglecting the terms in N^2 and ω^2_{co}, one obtains the following approximate solution for ξ_r, valid in the trapping region except near the turning point r_t:

$$\xi_r(r) \approx A_p(r) \cos\left[\omega \int_r^R \left(1 - \frac{L^2 c^2}{\omega^2 r'^2}\right)^{1/2} \frac{dr'}{c} - (1/4 + \alpha)\pi\right]. \tag{56}$$

Here $A_p(r)$ is, apart from a normalization constant, given by

$$A_p(r) \approx r^{-1}(\rho c)^{-1/2} \tag{57}$$

at least when r is not close to r_t. As ρ decreases by $\sim$ 9 orders of magnitude from the center to the surface of the Sun, Eq. (57) shows that *the displacement is strongly concentrated towards the surface*. The sensitivity of the frequencies to the structure of the model is approximately determined by the *energy density* in the mode, which is proportional to $\rho r^2 \xi_r^2 \propto c^{-1} \propto T^{-1/2}$, (cf. Eq. 34). This varies by less than a factor 100 through the model. Thus energetically, the p-modes have significant amplitude throughout the trapping region. In particular, low-degree modes are sensitive to the structure of the entire Sun.

The asymptotic behavior of p-modes can be understood very simply in terms of the propagation of sound waves (Christensen-Dalsgaard et al. 1985b). Because the wavelength of high-order modes is small compared with the typical scale over which equilibrium structure changes, the modes can be approximated locally by plane sound waves. These have the dispersion relation

$$k^2 \equiv k_r^2 + k_h^2 = \frac{\omega^2}{c^2} \tag{58}$$

where k_r and k_h are are the radial and horizontal components of the wave vector. For a wave corresponding to a mode of oscillation, k_h is given by Eq. (24). From Eq. (58) we then obtain

$$k_r^2 = \frac{\omega^2}{c^2} - \frac{L^2}{r^2}. \tag{59}$$

Close to the surface, c is small and hence k_r is large. Here the wave propagates almost vertically. With increasing depth, c increases and k_r decreases (see Fig. 4), until the point is reached where $k_r = 0$ and the wave propagates horizontally. It corresponds to the turning point $r = r_t$ defined previously; the condition in Eq. (54) is evidently equivalent to the vanishing of k_r. This is a point of total internal reflection; below r_t, $k_r^2 < 0$, and the mode decays exponentially. The reflection of the wave at the surface is not immediately contained in this simple description; however, it is at least plausible that it results from the steep density gradient. Thus the wave propagates in a series of "bounces" between the surface and the turning point (cf. Fig. 4). A mode of oscillation is formed as the interference pattern between such bouncing waves.

A more direct derivation of this behavior, which avoids explicit separation of the spherical harmonics, can be carried out on the basis of ray theory (Gough 1984b, 1986b).

This description of the p-modes also yields the asymptotic dispersion relation for their frequencies. In fact, it follows from Eq. (59) that the con-

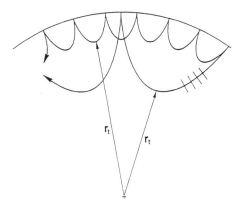

Fig. 4. Schematic illustration of the propagation of sound waves in a star. Due to the increase of the sound speed with depth, the deeper parts of the wave fronts move faster. This causes the refraction of the wave described by Eq. (59). Notice that waves with a smaller wavelength, corresponding to a higher value of the degree l, penetrate less deeply.

dition in Eq. (55) is equivalent to requiring that the change in radial phase between the lower turning point and the surface be an integral multiple of π, apart from α which describes the phase change at the turning points.

As r_t is a function of ω/L, Eq. (55) may be written as

$$\frac{\pi(n + \alpha)}{\omega} = F\left(\frac{\omega}{L}\right) \tag{60}$$

where

$$F(w) = \int_{r_t}^{R}\left[1 - \left(\frac{c}{rw}\right)^2\right]^{1/2}\frac{dr}{c}. \tag{61}$$

A relation like Eq. (60) was first found by Duvall (1982a) for observed frequencies. In fact, both observed and computed frequencies satisfy the Duvall relation quite accurately. Figure 5 shows a plot in this form of observed fre-

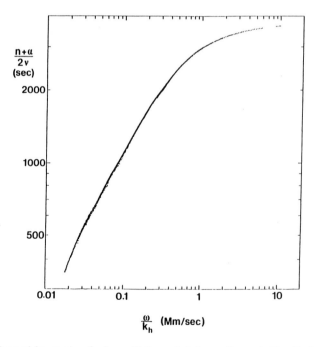

Fig. 5. Observed frequencies of solar oscillations, plotted according to the Duvall relation (60). (Note that $\omega/k_h = \omega R/L$ has been used as abscissa instead of ω/L). The value 1.58 of the constant α was obtained by minimizing the spread of the points around the mean curve (Christensen-Dalsgaard et al. 1985b).

quencies; α was assumed to be constant and was determined in such a way that the scatter in the relation was minimized.

It is of interest to use Eqs. (60) and (61) to estimate the effects on the frequencies of *changes* in the equilibrium model. If δc denotes the difference in c between two solar models, at fixed r, and $\delta\alpha$ denotes the difference in α at fixed frequency, it follows by linearization, assuming δc and $\delta\alpha$ to be small, that (Christensen-Dalsgaard et al. 1988c)

$$S\frac{\delta\omega}{\omega} \approx \int_{r_t}^{R} \left(1 - \frac{L^2 c^2}{r^2 \omega^2}\right)^{-1/2} \frac{\delta c}{c} \frac{dr}{c} + \pi \frac{\delta\alpha}{\omega} \tag{62}$$

where

$$S = \int_{r_t}^{R} \left(1 - \frac{L^2 c^2}{r^2 \omega^2}\right)^{-1/2} \frac{dr}{c} - \pi \frac{d\alpha}{d\omega}. \tag{63}$$

As α is assumed to be a function of ω alone, and r_t is determined by ω/L, the scaled frequency difference $S\delta\omega/\omega$ is predicted to be of the form

$$S\frac{\delta\omega}{\omega} \approx H_1\left(\frac{\omega}{L}\right) + H_2(\omega) \tag{64}$$

where the two functions H_1 and H_2 are determined by Eq. (62). Furthermore, the last term on the right-hand side of Eq. (63) is in general relatively small compared with the first, and hence S is approximately a function of ω/L. Equation (64) evidently describes a very special dependence of $\delta\omega$ on ω and L.

Because c/r decreases quite rapidly with increasing r, $(Lc/r\omega)^2 \ll 1$ except near the turning point r_t; hence as a rough approximation, $1 - L^2c^2/r^2\omega^2$ may be replaced by 1 in the integrals in Eqs. (62) and (63). If, furthermore, the term in $\delta\alpha$ can be neglected, the result is the very simple relation between the changes in sound speed and frequency:

$$\frac{\delta\omega}{\omega} \approx \frac{\int_{r_t}^{R} \frac{\delta c}{c} \frac{dr}{c}}{\int_{r_t}^{R} \frac{dr}{c}}. \tag{65}$$

This shows that the change in sound speed in a region of the Sun affects the frequency with a weight determined by the time the mode, regarded as a superposition of traveling waves, spends in that region. Thus changes near the surface, where the sound speed is low, have relatively large effects on the frequencies.

As discussed in the chapter by Gough and Thompson, the asymptotic relations (55) and (62) permit an almost direct, and surprisingly precise, inversion for the solar sound speed or the sound-speed difference between the Sun and a model.

Due to the importance of the asymptotic relations for the frequency, it is of obvious interest to test their accuracy. One way of doing this is to compute effective phases by demanding that Eqs. (53) and (55) be exactly satisfied for computed frequencies of oscillation. Thus we define ϵ_{nl} and α_{nl} by

$$\epsilon_{nl} \equiv \frac{\omega_{nl}}{\pi} \int_{r_1}^{r_2} \left[1 - \frac{\omega_{co}^2}{\omega_{nl}^2} - \frac{S_l^2}{\omega_{nl}^2}\left(1 - \frac{N^2}{\omega_{nl}^2}\right)\right]^{1/2}\frac{dr}{c} - n \qquad (66)$$

and

$$\alpha_{nl} \equiv \frac{\omega_{nl}}{\pi} \int_{r_t}^{R} \left(1 - \frac{L^2c^2}{\omega_{nl}^2 r^2}\right)^{1/2}\frac{dr}{c} - n. \qquad (67)$$

Here ω_{nl} is the frequency of a mode of order n and degree l. Christensen-Dalsgaard (1984a) found that the rapid variation in ω_{co} near the surface (cf. Fig. 2b) causes a jump in ϵ_{nl} at the frequency corresponding to the local maximum in ω_{co} where the upper turning point r_2 jumps discontinuously. This effect is evidently due to a failure of the JWKB analysis when the coefficients in the differential equation vary rapidly. It might be noted that if ω_{co} is replaced by $\omega_{co,p}$ in Eq. (66), this effect disappears, and ϵ_{nl} is close to $-1/2$, as predicted by simple JWKB theory.

Possibly more interesting is the behavior of α_{nl}, which is more directly related to inversion by means of the asymptotic expressions. It has been found that if L as given in Eq. (24) is replaced by

$$L_0 = l + \frac{1}{2} \qquad (68)$$

in Eq. (67), α_{nl} for radial modes, with $l = 0$, shows a behavior that is very similar to that obtained for nonradial modes; a corresponding improvement of the asymptotic description of the low-degree modes was noted by Brodsky and Vorontsov (1987,1988). This property, that has a close analogy in the asymptotic theory of the Schrödinger equation (Kemble 1937), is related to the asymptotic behavior of the solution near the singular point at $r = 0$ (see Vandakurov 1967; Tassoul 1980). Thus in the subsequent discussion of the asymptotic behavior of the p-modes, we use the definition (68). Computed values of α_{nl}, based on frequencies of a normal solar model, are shown in

Fig. 6. At low degree α_{nl} shows some dependence on l. This, however, is almost entirely caused by effects of the perturbation in the gravitational potential; in a corresponding plot based on frequencies computed in the Cowling approximation, the curves at all degrees essentially coincide with the upper envelope in Fig. 6. Otherwise α_{nl} is largely a function of frequency alone. Hence the computed frequencies can be represented quite accurately by an expression of the form given in Eq. (55), α being a function of frequency. Kosovichev (1988) made a careful analysis of the departures of computed frequencies from the asymptotic expression.

The dependence of α on frequency is determined by the structure of the outermost layers of the Sun. Thus it is possible to express the difference in α between two models in terms of integrals of appropriate kernels multiplying changes in the structure of the model (Christensen-Dalsgaard and Pérez Hernández 1988). This may eventually permit a study of the outer layers of the

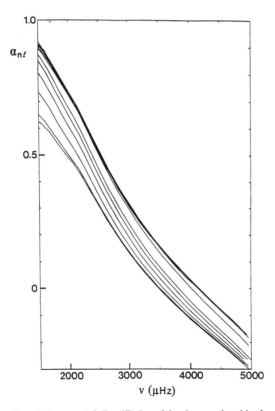

Fig. 6. Effective Duvall phase α_{nl} (cf. Eq. 67); here L has been replaced by $L_0 = l + (1/2)$ for modes of a normal solar model. Each curve corresponds to a fixed value of l; the curves are for $l = 0 - 5$, 10, 20, 30, 40, 50, 70 and 100, with α_{nl} increasing with l at fixed frequency.

Sun from inversion of α as determined from the observed frequencies. It should be noted, however, that the computed α is undoubtedly affected by the uncertainties in our description of the physics of the outer layers of the Sun.

When l is small, $(L_0 c/\omega r)^2 \ll 1$ except close to the center of the Sun. Then it is possible to expand Eq. (55), with L replaced by L_0, to take out the dependence on L_0, in a manner similar to the derivation of Eq. (55) from Eq. (53). To lowest order in the expansion one obtains

$$\nu_{nl} = \frac{\omega_{nl}}{2\pi} \sim (n + \frac{l}{2} + \frac{1}{4} + \alpha)\Delta\nu \qquad (69)$$

where

$$\Delta\nu = \left[2 \int_0^R \frac{dr}{c}\right]^{-1} \qquad (70)$$

is the inverse of twice the sound travel time between the center and the surface (see also Vandakurov 1967; Tassoul 1980). Thus to this asymptotic order, there is a uniform spacing $\Delta\nu$ between modes of same degree but different order. In addition, Eq. (69) predicts the approximate equality

$$\nu_{nl} \approx \nu_{n-1 l+2}. \qquad (71)$$

This frequency pattern has been observed for the solar 5-min modes of low degree (cf. the chapter by F. Hill et al.), and may be used in the search for stellar oscillations of solar type.

The *deviations* from the simple relation (69) have considerable diagnostic potential. By taking the expansion of Eq. (53) to the next order (Gough 1986b), or from a direct JWKB analysis of the oscillation equations (Tassoul 1980), one obtains

$$\delta\nu_{nl} \equiv \nu_{nl} - \nu_{n-1 l+2} \approx -(4l + 6)\frac{\Delta\nu}{4\pi^2\nu_{nl}} \int_0^R \frac{dc}{dr}\frac{dr}{r}. \qquad (72)$$

It is often convenient to represent observed or computed frequencies in terms of a limited set of parameters associated with the asymptotic description of the modes. This may be accomplished by fitting the asymptotic expression to the frequencies. By carrying out a polynomial fit in the quantity $x - x_0$, where $x = n + l/2$ and x_0 is a suitable reference value (Scherrer et al. 1983; Christensen-Dalsgaard 1988a) one obtains an average over n of $\delta\nu_{nl}$ as

$$<\delta\nu_{nl}>_n \approx (4l + 6)D_0 \qquad (73)$$

where

$$D_0 \approx -\frac{1}{4\pi^2 x_0}\int_0^R \frac{dc}{dr}\frac{dr}{r}. \tag{74}$$

Thus $\delta\nu_{nl}$ is predominantly determined by conditions in the solar core. Physically this may be understood from the fact that only near the center is k_h comparable with k_r. Elsewhere the wave vector is almost vertical, and the dynamics of the oscillations is largely independent of their horizontal structure, i.e., of l; therefore at given frequency, the contributions of these layers to the frequency are nearly the same, and hence almost cancel in the difference in Eq. (72).

It should be noted that the accuracy of expressions (73) and (74) is questionable; they appear to agree fortuitously with frequencies computed for models of the present Sun, whereas they are less successful for models of different ages or masses (Christensen-Dalsgaard 1988a). Indeed, it is shown in Sec.IV.E that the neglect of the perturbation of the gravitational potential causes frequency changes that are comparable with $\delta\nu_{nl}$ and depend strongly on l. Thus it is not surprising that the asymptotic description, assuming the Cowling approximation, cannot reproduce exactly the details of the frequencies computed from the full equations. However, the general form of the dependence of $\langle\delta\nu_{nl}\rangle$ on l shown in Eq. (73), as well as the argument that this quantity is most sensitive to conditions in stellar cores, probably have a broader range of validity.

The solar convection zone is essentially adiabatically stratified, except for a thin boundary layer near the surface (see, e.g., Gough and Weiss 1976). Furthermore, except in the outer 2% of the radius, Γ_1 is approximately constant. Thus we can approximate the structure of the convection zone by a simple adiabatically stratified polytropic layer. In this case, Eq. (47) can be solved in terms of confluent hypergeometric functions (Lamb 1932; Spiegel and Unno 1962; Gough 1977b; Christensen-Dalsgaard 1980); the eigenfrequencies are given by

$$\omega_{nl}^2 = \frac{g_s L}{R}\left(1 + \frac{n}{\mu}\right) \tag{75}$$

where μ is an appropriately averaged polytropic index. Note that for $n = 0$ this expression fortuitously reduces to Eq. (49) for surface gravity waves.

C. g-Mode Asymptotics

For g-modes, we assume that $S_l^2/\omega^2 \gg 1$. This is marginally true for a 160-min oscillation with $l = 1$ (cf. Fig. 2), but a good approximation otherwise. The g-modes have frequencies such that $\omega < N$. They are evanescent in the convective regions ($N^2 < 0$); hence their trapping regions are confined to the radiative regions, the limits being approximately given by the zeros r_1 and r_2 of $N^2/\omega^2 - 1$ (cf. Eq. 52b). For the model of the present Sun, the buoyancy frequency has a single maximum N_{max} in the inner radiative zone and increases to very high values in the outer radiative zone (see Fig. 2); this defines two trapping regions. To each one corresponds a set of gravity modes. If we neglect the trapping surface layer, the internal gravity modes are confined in the inner radiative zone, with $\omega < N < N_{max}$, and strongly decaying in the convection zone except for very low degree. In a typical normal solar model, $N_{max}/2\pi$ is around 450 μHz. From Eq. (31), where Φ' is neglected and $S_l^2/\omega^2 \gg 1$, and assuming $d/dr \sim |n|/r$, we have

$$\xi_r \sim \frac{L^2}{|n|}\xi_h \qquad (76)$$

which indicates that low-degree high-order modes have a predominantly horizontal motion while for modes with $l \gg n$ the motion is mostly in the vertical direction.

When l becomes large, low-order modes are split into modes trapped near the surface and modes trapped in the radiative zone, below the local maximum of the buoyancy frequency N_{max}. The frequencies of the latter modes tend to this maximum as l^{-1}. From Eq. (53), using a parabolic approximation to the buoyancy frequency around its maximum: $N^2 = N_{max}^2 [1 - \beta^2(x/x_m - 1)^2]$ and assuming that S_l^2 is much greater than ω^2, the periods of the modes are approximately given by (Christensen-Dalsgaard 1980):

$$P = \frac{2\pi}{N_{max}}\left[1 + \frac{\beta N_{max}^2(2|n| + 1)}{L}\right]. \qquad (77)$$

For low-degree, high-order modes, the theory developed up to the second order for the nonradial stellar pulsations by Tassoul (1980), taking into account the possible movable singularities of the coefficients, can be used. The solution is approximated by Bessel functions around the transition points which are the zeros of N^2, the center and the surface, and it depends on the behavior of the buoyancy frequency close to these points. However, in the major part of the convection zone, $|N|/2\pi$ is of the order of 1μHz or less and the asymptotic approximation fails because $\omega > |N|$ in a large part of the low-frequency range. Thus the behavior of the oscillation in the convection zone must be described either by the analytical solutions of the asymptotic equations with $N = 0$ (Zahn 1970), or by numerically integrated solutions (Provost and Berthomieu 1986; Gabriel 1986b; Ellis 1986,1987).

The matching of the solutions, subject to the boundary conditions, gives the asymptotic relation for the periods:

$$P = \frac{P_0}{L}\left(|n| + \frac{l}{2} - \frac{1}{4} - \theta\right) + \frac{P_0^2}{P}\frac{L^2 V_1 + V_2}{L^2} \tag{78}$$

with

$$P_0 = \frac{2\pi^2}{\int_0^{x_1} \frac{N}{x}dx} \tag{79}$$

and

$$V_1 = \frac{1}{P_0}\lim_{\delta\to 0}\left(\int_\delta^{x_1}\frac{dx}{Nx} - \frac{1}{N(\delta)}\right). \tag{80}$$

The coefficient V_2 is given by a more intricate formula (Tassoul 1980). The phase factor θ is determined by

$$\theta = \frac{\upsilon}{2} - \frac{1}{\pi}\arctan\left(\frac{\sin\upsilon\pi}{K + \cos\upsilon\pi}\right). \tag{81}$$

Here N^2 is assumed to vanish as $(x_1 - x)^\zeta$ at the base of the convection zone, and $\upsilon = 1/(\zeta + 2)$. For a standard solar model, N^2 behaves almost linearly below the convection zone; hence $\zeta = 1$ and $\upsilon = 1/3$.

The value of K is obtained by matching the eigenfunctions on both sides of the lower boundary of the convection zone, assuming continuity of the displacement and of the Lagrangian pressure perturbation (or the horizontal displacement). It is related to the ratio of these two quantities at that point and depends on the derivatives of the buoyancy frequency just below the convection zone. For $\upsilon = 1/3$,

$$K = (3N_0)^{1/3}\frac{\Gamma(2/3)}{\Gamma(1/3)}r_c^{-1}\left[L^2\left(\frac{\xi_h}{\xi_r}\right) - \frac{1}{2\Gamma_1}\frac{d\ln p}{d\ln r} - \frac{N_1}{5}\right]_{rc}^{-1}(LP)^{2/3}$$

with

$$N_0 = \lim_{r\to r_c^-}\frac{N^2}{r_c - r}, \qquad N_1 = -\frac{1}{2}\frac{d}{dr}\ln\left(\frac{d}{dr}\frac{N^2}{r^2}\right)\bigg|_{r_c^-}. \tag{82}$$

N_1 is related to the curvature of N^2. Thus θ is a function of the frequency and of the degree of the modes. In the limit of very low frequency, $K \sim K_0(LP)^{2/3}$ and

$$\theta \sim \frac{1}{6} - \frac{\sqrt{3}}{(2\pi)^{1/3}K_0} (LP)^{-2/3} \tag{83}$$

where K_0 is independent of l and P.

For large l, θ tends to 1/6 and the observational determination of this asymptotic limit would provide a constraint on the stratification below the convection zone. However, for a standard model, θ varies significantly as a function of ω and l and approaches its asymptotic behavior only for very large values of the radial order.

The asymptotic solution in terms of Bessel functions does not exist for all values of ζ (Olver 1956). Some restrictive conditions on the coefficients of the initial equations are required for $\zeta \geq 1$. In the case of g-mode oscillations, these conditions are not satisfied for $\zeta \geq 2$, leading to divergent integrals in the second-order terms. For $\zeta = 2$, i.e., $\upsilon = 1/4$, a development in Weber functions just below the convection zone leads to an expression for P that has the same form as in Eq. (78) with modified expressions for V_1 and V_2 which include a dependence on the logarithm of the period; a similar expression is also obtained for θ, giving the asymptotic limit $\theta = 1/8$ (Ellis 1987). An estimate from observed periods of the limiting behavior of θ could therefore provide information about the behavior of N^2 just below the convection zone.

In the case $\upsilon = 1/3$, Gabriel (1986b) noted that θ depends on both the slope of N^2 and the depth of the convection zone through the term between brackets in Eq. (82). This last dependence is dominant so that if the slope of N^2 is not too poorly known, it is possible to obtain an estimate of the depth of the convection zone from the determination of θ as a function of P. From theoretical periods of modes of radial order n from 30 to 100, for which the second-order terms can be neglected in the asymptotic formula, Gabriel gave a fitting procedure for the variation of θ with frequency in terms of two parameters which depend on the depth of the convection zone.

The main feature which results from the asymptotic analysis is that, to first order, the periods of the modes of a given degree are equally spaced, with a period spacing that depends on the degree. Given a set of observed periods, this property can be used to identify sets of equally spaced periods, each corresponding to a given degree, and hence ideally can lead to an identification of the degree of the modes as well as a determination of P_0 (Delache and Scherrer 1983; van der Raay 1988; see also Kawaler 1988b, who used this technique on white dwarf oscillations).

The way in which the periods P are approximated by the expression (78) is illustrated in Fig. 7, where the quantities LP are plotted for $l = 1$ and 4 as functions of $(|n| + l/2 - 1/4 - \theta)$, using the asymptotic limit $\theta = 1/6$. The first- and second-order asymptotic approximation are given, respectively, by the straight line with slope P_0 and the two hyperbolas. The computed

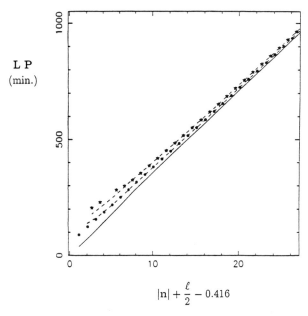

$$|n| + \frac{\ell}{2} - 0.416$$

Fig. 7. Plot of the g-mode periods (multiplied by L) of a normal solar model as a function of $|n| + l/2 - 1/4 - \theta$, with $\theta = 1/6$, for $l = 1$ (●) and $l = 4$ (★). The straight line depicts the linear approximation. The two hyperbolic curves indicate the asymptotic approximation up to the second order.

periods are very close to the asymptotic predictions from Eq. (78). A more careful analysis of computed periods in terms of the asymptotic expression, for a number of different models, is given in Sec. VI.C.

Except in the vicinity of the turning points, an approximate expression for the radial variation of the displacement in the trapping region may be derived from the asymptotic expansion of the Bessel functions:

$$\xi_r(r) \sim A_g(r)\cos\left(\frac{L}{\omega} \int^r \frac{N}{r'} \, dr'\right)$$

$$\xi_h(r) \sim \frac{N}{L\omega} A_g(r)\sin\left(\frac{L}{\omega} \int^r \frac{N}{r'} \, dr'\right)$$

(84)

with

$$A_g(r) \sim r^{-3/2}(\rho N)^{-1/2}.$$

(85)

In a normal solar model, the buoyancy frequency N is proportional to r at the center; the ratio N/r is maximum there (the dimensionless quantity $\tilde{N}/x \equiv$

$N/r\sqrt{R/GM} \sim 90$) and decreases rapidly to zero at the bottom of the convection zone. As a consequence, the amplitude of the g-modes is very large in the core of the Sun with a small secondary maximum near the boundary of the convection zone. The energy density of the modes is proportional to $\rho r^2 \xi_h^2 \propto N/r$. Equation (84) also shows that the radial wavelength around the point r, which can be approximated by $\Lambda \sim 2\pi\omega r/LN$, gets smaller toward the center so that the radial nodes are concentrated in the inner layers.

In the convection zone, the g-modes are evanescent and their amplitude decreases rapidly toward the surface following the law (Christensen-Dalsgaard et al. 1980):

$$\log\left(\frac{\xi_x}{\xi_x}\right) \sim \log\left(\frac{r_c}{r_s}\right) \sqrt{L} + \frac{1}{2}\log\left(\frac{3_{r,c}\rho_c}{3_{r,s}\rho_s}\right). \tag{86}$$

This means that when their degree increases, g-modes are strongly attenuated by the effect of the convection zone.

IV. PROPERTIES OF ADIABATIC OSCILLATIONS

The dependence of the oscillation frequencies on the equilibrium structure is determined by the coefficients in Eqs. (31)–(33). They can be calculated from $p(r)$, $\rho(r)$, $g(r)$ and $\Gamma_1(r)$. However, assuming that the Sun is in hydrostatic equilibrium, these quantities are related by Eqs. (12)–(14). Also, the mass $m(r)$ must satisfy the boundary condition $m(0) = 0$. Thus, for example, if $\rho(r)$ is known, $m(r)$ may be calculated from Eq. (13); given $m(r)$, $g(r)$ is determined by Eq. (12), and finally $p(r)$ can be found by integrating Eq. (14), with the boundary condition $p(R) = 0$. Hence if we know $\rho(r)$, we can obtain $m(r)$, $g(r)$ and $p(r)$ by integration. Thus all coefficients in Eqs. (31)–(33), and hence ω, are determined completely by $\rho(r)$ and $\Gamma_1(r)$. Conversely, if the oscillations can be adequately represented as adiabatic modes of a spherical star in hydrostatic equilibrium, observations of their frequencies *only* give information about $\rho(r)$ and $\Gamma_1(r)$, or equivalently about any other pair of independent quantities from which the coefficients in Eqs. (31)–(33) can be derived. In fact, it follows from the asymptotic properties of p-modes that their frequencies predominantly provide information about the adiabatic sound speed $c(r)$. As discussed in more detail in the chapter by Gough and Thompson, a consequence of this dependence of the frequencies on solar structure is that inversion for the structure can be organized in order to determine a pair among the variables $(\rho(r), \Gamma_1(r), p(r), c(r))$ or any other variable, such as the buoyancy frequency N, which may be derived from them. To get information about other properties of the Sun, such as the temperature or the composition, one must use the equations of stellar structure.

The equations and boundary conditions do not contain the azimuthal order m. Thus the solution, in particular the eigenfrequencies, are independent of m. This is a consequence of the assumption of spherical symmetry. Physically, it follows from the fact that in this case there is no preferred axis in the Sun; as m depends on the choice of coordinate axis, the physics of the oscillations, and hence their frequencies, must be independent of m. For slow rotation, a modal description as in Eqs. (20) and (23) is still possible, provided that the rotation axis is chosen as coordinate axis. As discussed in the chapter by Libbrecht and Morrow, the dependence of ω on m can then be found from a perturbation analysis: rotation introduces a splitting in ω of order $m\Omega$, where Ω is an average of the rotation frequency.

A. Oscillation Frequencies

It is instructive to consider various properties of oscillations computed for a normal solar model. These can, to a large extent, be interpreted by means of the asymptotic results presented in Sec. III.

Computed frequencies for a model of the present Sun were shown in Fig. 1. Here the modes were labeled according to the mathematical classification discussed in Sec. II.D. It is clear, however, that the physical nature of the modes, and the behavior of the frequencies as functions of l, in general corresponds to the asymptotic properties discussed in Sec. III. The p-mode frequencies increase roughly as $l^{1/2}$ for high l, in accordance with the polytropic approximation in Eq. (75). The g-mode frequencies tend towards a constant limit at large l, given by the maximum in the solar interior of the buoyancy frequency N. Finally, the f-modes are intermediate in frequency between the p- and g-modes, and, for degrees exceeding 16, behave approximately as the surface gravity wave described by Eq. (49).

At degrees below about 16 the f-mode frequency apparently intersects the g-mode frequencies. To understand this behavior, it is convenient to follow the frequencies as continuous functions of the degree, which from a mathematical point of view may take nonintegral values; this shows that the interaction between the modes takes place in a sequence of avoided crossings (Christensen-Dalsgaard 1980) where the frequencies approach very closely without actually crossing. This creates a discrepancy between the mathematical classification, which is unchanged as one follows an eigenvalue through the avoided crossing, and the physical nature. Thus for $l < 16$, the modes labeled as f-modes in the figure behave physically as the lowest-order trapped g-mode, as described by Eq. (77); on the other hand, with decreasing l a sequence of "mathematical" g-modes of increasing order take on the physical nature of the surface gravity wave, with most of their energy concentrated near the surface. Similar avoided crossings, and consequent conflicts between mathematical classification and physical nature, are found among modes of evolving massive stars (see, Osaki 1975; Shibahashi and Osaki 1976; Aizenman, et al. 1977; Roth and Weigert 1979; Shibahashi 1979). The study of analogous phenomena has a long history in quantum mechanics; in fact, the analysis of von Neuman and Wigner (1929) has been very useful in understanding the properties of the avoided crossings found among stellar eigen-frequencies (Gabriel 1980; Christensen-Dalsgaard 1981).

B. Oscillation Eigenfunctions

Figure 8 shows eigenfunctions for a selection of p-modes with approximately the same frequency, but varying degree. To show the distribution of kinetic energy in the modes, $r\sqrt{\rho}\xi_r$ is plotted. As also found from the asymptotic analysis, the low-degree modes extend to the solar core, whereas modes with higher l are increasingly confined close to the solar surface. Notice also that except near and below the turning points, the eigenfunctions are very similar. Physically, this may be understood from the fact that only near r_t is the term in L in Eq. (56) significant. Elsewhere the wave vector is almost vertical, and the dynamics of the oscillations is largely independent of their horizontal structure, i.e., of l; the same is therefore true for the eigenfunctions. The location of the asymptotic turning point, shown in Fig. 3, is indicated with arrows in the figure; it is clearly in good agreement with the actual behavior of the eigenfunctions.

To illustrate the reflection of the modes at the surface, Fig. 9 shows the eigenfunctions in the outer few percent of the solar radius of modes of degree 1 with different frequencies. From the behavior of ω_{co} (cf. Fig. 2b) follows that for $\nu \gtrsim 2500$ μHz, the modes are propagating essentially to the photosphere; hence energetically, the amplitude of the mode is as large at the photosphere as in the interior. However, even at 2286 μHz the effect of ω_{co} already causes a decrease in the photospheric amplitude relative to its interior value, and at 1604 μHz there is a substantial evanescent region between the interior and the photosphere, causing the photospheric amplitude to be considerably smaller than the amplitude in the interior. Notice that except when ν is close to the acoustical cut-off frequency in the atmosphere, the pulsation energy decreases rapidly above the photosphere.

Figure 10 shows g-modes with different degrees and periods of approximately 160 min; the energy of such modes is clearly large near the center. Thus observation and identification of g-modes would provide data with high sensitivity to conditions in the solar core.

The envelopes of the eigenfunctions shown in Figs. 8 and 10 closely follow the asymptotic Eqs. (57) and (85) in the trapping regions. Thus for p-modes $r\sqrt{\rho c}\xi_r$, and for g-modes $r^{3/2}\sqrt{\rho N}\xi_r$, are very close to being harmonic functions with a nonuniform phase.

An important global property of a mode of oscillation, besides the frequency, is its integrated kinetic energy or, equivalently, its inertia. Here we consider the dimensionless inertia[a]

---

[a]Note that the quantity considered here is greater by a factor 4π than the quantity plotted in Figs. 7–9 of Christensen-Dalsgaard (1986) and in Fig. 6 of Christensen-Dalsgaard (1988b). The "4π" in the denominator of Eq. (4.1) of the former paper, and Eq. (3.5) of the latter, should be removed.

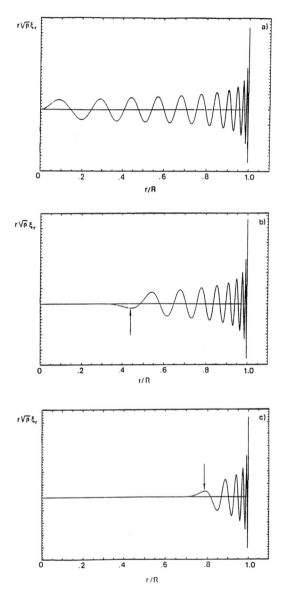

Fig. 8. Eigenfunctions for p-modes, with (a) $l = 0$, $n = 23$, $\nu = 3310$ μHz; (b) $l = 20$, $n = 17$, $\nu = 3373$ μHz; (c) $l = 60$, $n = 10$, $\nu = 3233$ μHz. The arrows mark the asymptotic location of the turning points r_t (cf. Fig. 3).

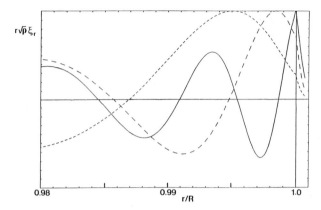

Fig. 9. Eigenfunctions of p-modes with $l = 1$, plotted against fractional radius r/R in the outermost parts of the Sun. The cases shown are: $\nu = 1604$ μHz (------) $\nu = 2286$ μHz (– – – –) $\nu = 3653$ μHz (———).

$$E_{nl} = \frac{4\pi \int_0^R [\xi_r^2 + L^2 \xi_h^2] \rho r \, dr}{M[\xi_r^2(R_{phot}) + L^2 \xi_h^2(R_{phot})]} \qquad (87)$$

where R_{phot} is the photospheric radius. This is related to the modal mass M_{nl} (cf. Goldreich and Keeley 1977a) by $M_{nl} = E_{nl}M$, and defined such that the time-averaged kinetic energy in the oscillation is

$$\frac{1}{2}M_{nl}V_{rms}^2 = \frac{1}{2}E_{nl}\,MV_{rms}^2 \qquad (88)$$

where V_{rms}^2 is the mean, over the solar surface and time, of the squared total photospheric velocity of the mode. Figure 11 shows E_{nl} for selected p-modes in the present Sun. The dominant feature is the strong decrease with increasing frequency up to ~ 3000 μHz. This is caused by the behavior of the eigenfunctions shown in Fig. 9, as determined by the variation of ω_{co}. At low frequencies, the interior amplitude is substantially larger than the surface value, and hence E_{nl} is large. At high frequencies, on the other hand, the mode is oscillatory almost to the surface, and the surface and interior amplitudes are comparable. It is also noticeable that E_{nl} decreases substantially with increasing l at fixed frequency; the increase in l causes the mode to be confined closer to the surface, and hence to involve a smaller fraction of the mass of the Sun.

The normalized inertia of g-modes in the Sun is illustrated in Fig. 12. It shows a much more dramatic variation than the inertia for p-modes, spanning about 30 orders of magnitude for $l \leq 25$. The reason for this is the trapping of the modes at high degrees and frequencies near the maximum in the buoyancy frequency N. It is particularly pronounced for modes whose frequencies

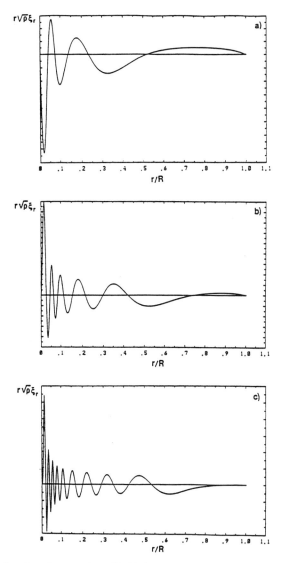

Fig. 10. Eigenfunctions for g-modes with (a) $l = 1$, $n = -5$, $\nu = 108$ μHz; (b) $l = 2$, $n = -10$, $\nu = 101$ μHz; (c) $l = 4$, $n = -18$, $\nu = 103$ μHz.

exceed the secondary maximum in N at 400 μHz. The very low values of E_{nl} found at selected frequencies for higher values of l occur where the g-modes pass through avoided crossings and take on the character of surface gravity waves; this happens when their frequencies are near the extensions of the f-mode frequencies shown in Fig. 1. It is interesting that this phenomenon can be traced in inertia to a degree as low as 3.

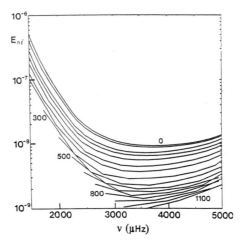

Fig. 11. Normalized inertia E_{nl} (cf. Eq. 87) for f- and p-modes of the present Sun, as function of the cyclic frequency v. For clarity points corresponding to modes with a given degree l have been connected. Selected curves are labeled with l.

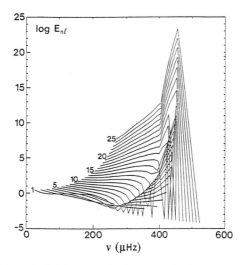

Fig. 12. Normalized inertia E_{nl} (cf. Eq. 87) for f- and g-modes of the present Sun, plotted against frequency v. For clarity points corresponding to modes with a given degree l have been connected. Selected values of l are indicated in the figure.

Whatever the excitation mechanism that may give rise to g-modes in the Sun, it is unlikely that g-modes of high degree can be excited to observable surface amplitudes. In particular, if all modes are arbitrarily assumed to be excited to the same kinetic energy, the predicted velocity amplitudes at high degrees are very small (see, Dziembowski et al. 1985; Provost and Berthomieu 1988; Berthomieu and Provost 1990). On the other hand, it is interesting that the increase in E_{nl} with l is weak for $l \leq 10$. Although all g-modes are evanescent in at least the convection zone, the decrease in amplitude predicted by Eq. (86) is relatively modest at low l.

Predictions of the corresponding amplitudes in intensity require consideration of the nonadiabatic behavior of the modes near the solar surface, and hence may be subject to the uncertainty in the description of convection (cf. the chapter by Cox et al.). Cox et al. (1989) estimated the temperature perturbation at the photosphere, neglecting the perturbation of the convective flux. The observed amplitude of the intensity results from a mean over the surface of the Sun of the intensity perturbation, integrated along the line of sight in the atmospheric layers. With simplifying assumptions concerning the transfer of radiation, and neglecting the perturbation of the convective flux, Provost and Berthomieu (1988) and Berthomieu and Provost (1990) obtained results for the intensity fluctuations, which indicated a zero in the intensity amplitude for dipole modes at a frequency of ~ 60 μHz. However, a full account of the transfer, including the effects of opacity fluctuations, could modify these features (Toutain and Gouttebroze 1988).

C. The Functional Analysis of Adiabatic Oscillations

A great deal of insight into the properties of adiabatic oscillations can be obtained by regarding the equations as an eigenvalue problem in a Hilbert space (Eisenfeld 1969; Dyson and Schutz 1979; Christensen-Dalsgaard 1981). We write Eqs. (26) and (27) as

$$\frac{1}{\rho}\frac{dp'}{dr} + \frac{\rho'}{\rho}g - \frac{d\Phi'}{dr} = \omega^2\xi_r \tag{89}$$

$$\frac{1}{r}\left(\frac{p'}{\rho} - \Phi'\right) = \omega^2\xi_h. \tag{90}$$

Here ρ' is determined in terms of ξ_r and ξ_h by Eq. (29), and, given ρ', p' is determined by Eq. (30). Also Φ' can be found in terms of ρ' from the usual integral solution to Poisson's equation:

$$\Phi' = \frac{4\pi G}{2l+1}\left[r^{-(l+1)}\int_0^r \rho'r'^{l+2}dr' + r^l\int_r^R \rho'r'^{-(l-1)}dr'\right]. \tag{91}$$

We introduce the vector

$$\xi \equiv (\xi_r, \xi_h) \in \mathcal{H} \tag{92}$$

where $\mathcal{H}$ is a subspace of the space of pairs of quadratically integrable functions on the interval $[0, R]$. Then ρ', p' and Φ' are obtained by linear operations on ξ. The same is therefore true of the left-hand sides of Eqs. (89) and (90), and thus these equations can be written as

$$\mathcal{L}\xi = \omega^2 \xi \tag{93}$$

which defines the linear operator $\mathcal{L}$ in $\mathcal{H}$. The boundary conditions on ξ_r and ξ_h can be imposed by restricting the part of $\mathcal{H}$ where $\mathcal{L}$ is defined. For reasons which become clear below, we choose to use the surface condition (41), and define the *domain* of $\mathcal{L}$ by

$$\mathcal{D}(\mathcal{L}) = \{(\xi_r, \xi_h) | \xi_r - l\xi_h \to 0 \text{ for } r \to 0 \text{ and } \delta p(R) = 0\}. \tag{94}$$

The boundary conditions on Φ' are satisfied automatically by the integral expression (91).

We introduce a scalar product on $\mathcal{H}$. Let $\xi = (\xi_r, \xi_h)$ and $\eta = (\eta_r, \eta_h)$ be two vectors in $\mathcal{H}$. Then we define

$$\langle \xi, \eta \rangle \equiv 4\pi \int_0^R [\xi_r^*(r)\eta_r(r) + L^2 \xi_h^*(r)\eta_h(r)] r^2 \rho \, dr, \tag{95}$$

where the star denotes the complex conjugate. It is straightforward, using the explicit expression, to show that the operator $\mathcal{L}$ is symmetric, i.e.

$$\langle \mathcal{L}\xi, \eta \rangle = \langle \xi, \mathcal{L}\eta \rangle \qquad \text{for all } \xi, \eta \in \mathcal{D}(\mathcal{L}). \tag{96}$$

However, for unbounded operators such as $\mathcal{L}$ symmetry is not enough for the completeness of the (possibly generalized) eigenfunctions: we have to show self-adjointness, i.e., that the domain of $\mathcal{L}$ is equal to that of its adjoint. Certain symmetric operators can be successfully extended to self-adjoint operators, by enlarging their domain (which entails a similar reduction of the domain of its adjoint) precisely to the point that both domains become identical. There are sufficient conditions for such extensions: a frequent criterion is given when the quadratic form associated with the operator is bounded from below. This permits the so-called Friedrich extension (Reed and Simon 1975). Dyson and Schutz (1979) have based their completeness proof (valid even for differentially rotating stars) on this idea.

From Eq. (96) follows immediately a number of useful properties of $\mathcal{L}$. The simplest is that the squared eigenfrequencies are real. We introduce the functional Σ on $\mathcal{D}(\mathcal{L})$ by

$$\Sigma(\xi) = \frac{\langle \xi, \mathcal{L}\xi \rangle}{\langle \xi, \xi \rangle}. \tag{97}$$

It follows from Eq. (96) that $\Sigma(\xi)$ is real. If ω_0^2 is an eigenvalue of the problem with eigenvector ξ_0, i.e.

$$\mathcal{L}\xi_0 = \omega_0^2 \xi_0 \tag{98}$$

then

$$\Sigma(\xi_0) = \omega_0^2 \tag{99}$$

and hence ω_0^2 is real. As the coefficients in Eqs. (31)-(33) are then real, it follows that we may also choose the eigenfunctions to be real at all r.

As is well known, a second property of a Hermitian operator is that eigenvectors corresponding to different eigenvalues are orthogonal. Thus if

$$\mathcal{L}\xi_1 = \omega_1^2 \xi_1 \qquad \mathcal{L}\xi_2 = \omega_2^2 \xi_2 \qquad \omega_1^2 \neq \omega_2^2 \tag{100}$$

then

$$\langle \xi_1, \xi_2 \rangle = 0. \tag{101}$$

In addition, ω^2 satisfies a variational principle (cf. Chandrasekhar 1964). From Eq. (96) it is easy to show that if $\delta\xi \in \mathcal{H}$ is a small change to the eigenvector, then

$$\Sigma(\xi_0 + \delta\xi) = \omega_0^2 + O(\|\delta\xi\|^2) \tag{102}$$

where $\| \ldots \|$ is a suitable norm. Thus Σ is stationary at the eigenfrequencies. From a physical point of view this reflects Hamilton's principle for the system consisting of the pulsating star. It is conservative, because of the assumption of adiabaticity, and isolated because of the boundary condition (Eq. 41).

It is evident that the variational property, and equivalently the Hermiticity of the operator $\mathcal{L}$, are not valid for nonadiabatic oscillations, where energy is not conserved. Also it depends on the choice of boundary conditions that isolate the system, such as the vanishing of the Lagrangian pressure perturbation. If a different condition is used, such as the match to a solution in an isothermal atmosphere discussed in Sec. II.C, the variational property is no longer guaranteed. However, regardless of the boundary condition, it is possible to write down expressions analogous to Eq. (99), but possibly containing surface terms; if the effects of the surface properties on the oscillations are small, these expressions may be expected to be approximately variational.

In fact it follows from the equations of linear adiabatic oscillations that for any radius $r*$

$$\omega^2 \int_0^{r*} \rho(\xi_r^2 + L^2\xi_h^2)r^2dr = \int_0^{r*} \left(\Gamma_1 p D_1^2 + 2\frac{dp}{dr}\xi_r D_1\right.$$

$$+ \frac{1}{\rho}\frac{d\rho}{dr}\frac{dp}{dr}\xi_r^2\right)r^2dr - \frac{8\pi G}{2l+1}\int_0^{r*} r^{-(l-1)}D_2(r)\int_0^r r'^{l+2}D_2(r')dr'dr$$

$$+ p'(r*)\xi_r(r*)r*^2 + \frac{4\pi G}{2l+1}\left\{2\rho(r*)\xi_r(r*)r*^{-(l-1)}\int_0^{r*} D_2(r)r^{l+2}dr\right.$$

$$\left. - [\rho(r*)\xi_r(r*)]^2r*^3\right\}.$$

$$(103)$$

Here

$$D_1 = \frac{1}{r^2}\frac{d}{dr}(r^2\xi_r) - \frac{l(l+1)}{r}\xi_h, \quad D_2 = \frac{1}{r^2}\frac{d}{dr}(r^2\rho\xi_r) - \frac{l(l+1)}{r}\rho\xi_h \quad (104)$$

are the amplitudes of div ($\delta\mathbf{r}$) and div ($\rho\delta\mathbf{r}$), respectively. For radial oscillations, equation (103) may be considerably simplified, to give

$$\omega^2 \int_0^{r*} \xi_r^2\rho r^2dr = \int_0^{r*}\left\{\Gamma_1 p r^4\left[\frac{d}{dr}\left(\frac{\xi_r}{r}\right)\right]^2 - r\xi_r^2\frac{d}{dr}[(3\Gamma_1 - 4)p]\right\}dr$$

$$+ r*[3\Gamma_1 p(r*)\xi_r(r*) + r*\delta p(r*)]\xi_r(r*). \quad (105)$$

These equations are valid at any $r*$. If the surface radius is chosen for $r*$, and the surface terms are neglected, they reduce to the expression defined symbolically in Eq. (99). In fact, the surface terms are in general relatively small, and even the complete expressions are approximately variational. As discussed by Christensen-Dalsgaard et al. (1979) and J. Christensen-Dalsgaard (1982), they may be utilized in the computation of accurate oscillation frequencies (see also Sec. V. F., below).

Finally we consider the effects of a perturbation $\delta\mathcal{L}$ in the operator $\mathcal{L}$. This perturbation could be due to changes in the equilibrium model, or to changes in the physics of the oscillations (e.g., the inclusion of nonadiabatic effects; see Christensen-Dalsgaard [1981]). Thus we consider the eigenvalue problem

$$(\mathcal{L} + \delta\mathcal{L})\xi = \omega^2\xi \quad (106)$$

where $\omega^2 = \omega_0^2 + \delta\omega^2$ and, as in Eq. (98), ω_0^2 is an eigenvalue of the unperturbed problem with eigenvector ξ_0. From first-order perturbation theory (see Schiff 1949), it then follows that

$$\delta\omega^2 \approx \frac{\langle \xi_0, \delta\mathcal{L}\xi_0 \rangle}{\langle \xi_0, \xi_0 \rangle}. \tag{107}$$

Thus the frequency change can be computed from the unperturbed eigenvector.

D. Effects on Frequencies of a Change in the Model

As an example of the use of Eq. (107), we consider in more detail the interpretation of changes in the frequencies caused by changes in the equilibrium model. We consider a mode (n,l), with eigenvector $\xi_{nl} = (\xi_{r,nl}, \xi_{h,nl}) \in \mathcal{H}$; without loss of generality we can assume that $\xi_{r,nl}(r)$ and $\xi_{h,nl}(r)$ are real. The relative frequency change caused by $\delta\mathcal{L}$ is then, according to Eq. (107)

$$\frac{\delta\omega_{nl}}{\omega_{nl}} = \frac{1}{2} \frac{\delta\omega_{nl}^2}{\omega_{nl}^2} = \frac{\langle \xi_{nl}, \delta\mathcal{L}\xi_{nl} \rangle}{2\omega_{nl}^2 \langle \xi_{nl}, \xi_{nl} \rangle}. \tag{108}$$

Here the denominator is proportional $\omega_{nl}^2 E_{nl}$, where E_{nl} was defined in Eq. (87). Also, we represent $\delta\mathcal{L}$ on component form as

$$\delta\mathcal{L}\xi_{nl} = (\phi_r[\xi_{nl}], \phi_h[\xi_{nl}]) \tag{109}$$

where $\phi_r[\xi_{nl}](r)$ and $\phi_h[\xi_{nl}](r)$ are functions of r. Then we write Eq. (108) as

$$E_{nl} \frac{\delta\omega_{nl}}{\omega_{nl}} = I_{nl} \tag{110}$$

where

$$I_{nl} = \frac{2\pi \int_0^R [\xi_{r,nl}(r)\phi_r[\xi_{nl}](r) + L^2\xi_{h,nl}(r)\phi_h[\xi_{nl}](r)]\rho r^2 dr}{M\omega_{nl}^2 [\xi_{r,nl}^2(R_{phot}) + L^2\xi_{h,nl}^2(R_{phot})]} \tag{111}$$

Thus I_{nl} gives the integrated effect of the perturbation, normalized to the total photospheric displacement.

There is a close analogy between the exact Eqs. (110) and (111) and the asymptotic expressions (62) and (63). In both cases the factor multiplying $\delta\omega$ represents the fact that modes with larger inertia are more difficult to perturb. The effect on the frequencies of the change in the model is described by the right-hand sides of Eqs. (62) and (111), and depends on the overlap between the change in the model and the eigenfunction.

We now assume that $\delta\mathcal{L}$ is localized near the solar surface, in the sense that

$$\phi_r[\xi](r) \simeq 0 \qquad \phi_h[\xi](r) \simeq 0 \quad \text{for} \quad R - r > \delta \qquad (112)$$

for some small δ. For modes extending substantially more deeply than the region of the perturbation, i.e., with $R - r_t \gg \delta$, the eigenfunctions are nearly independent of l at fixed frequency in that region (see also Fig. 8 and the associated discussion). Hence I_{nl} depends little on l at fixed ω. To get a more convenient representation of this property, we introduce

$$Q_{nl} = \frac{E_{nl}}{\bar{E}_0(\omega_{nl})} \qquad (113)$$

where $\bar{E}_l(\omega)$ is obtained by interpolating to ω in E_{nl} at fixed l. Then

$$Q_{nl}\delta\omega_{nl} \qquad (114)$$

is independent of l, at fixed ω, for modes such that $R - r_t \gg \delta$. Conversely, if $Q_{nl}\delta\omega_{nl}$ is independent of l at fixed ω for a given set of modes, then $\delta\mathcal{L}$ is probably largely localized outside $r = \max(r_t)$ over the set of modes considered.

Q_{nl} has been plotted in Fig. 13, for selected values of l. Its variation with l is largely determined by the change in the penetration depth. Modes with higher degree penetrate less deeply and hence have a smaller inertia at a given photospheric displacement. As a consequence of this, their frequencies are more susceptible to changes in the model.

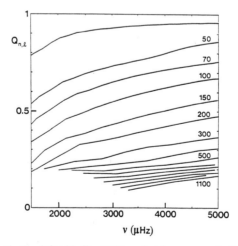

Fig. 13. The inertia ratio Q_{nl}, defined in Eq. (113), against frequency ν, for f- and p-modes in a normal solar model. Each curve corresponds to a given degree l, selected values of which are indicated.

It should be noted that Eqs. (110) and (111) provide a linear relation between the change in the model and the change in the frequency. As discussed in the chapter by Gough and Thompson this may be expressed in terms of kernels relating, for example, changes in ρ and Γ_1 to the changes in the frequencies.

E. The Cowling Approximation

Given the importance of the Cowling approximation for the asymptotic analysis, it is of some interest to consider its implications in more detail. Results for polytropic models were obtained by Robe (1968). The effects in a model of the present Sun may be judged from Fig. 14, which shows differences between frequencies computed in the Cowling approximation and from the full equations. As discussed in Sec. III (cf. Eq. 45), the effect of the perturbation in the gravitational potential decreases rapidly with increasing mode order and degree.

It was pointed out by Cowling (1941) that the effects of Φ' can be estimated from perturbation analysis. This in fact follows from the analysis in

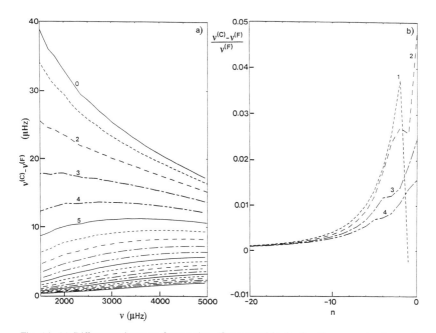

Fig. 14. (a) Differences between frequencies $\nu^{(C)}$ computed in the Cowling approximation and frequencies $\nu^{(F)}$ resulting from the full equations, for p-modes, plotted against frequency. All modes with degree $l = 0 - 20$ are shown. Each curve corresponds to a given value of l, which has been indicated on the first few curves. (b) Relative differences between $\nu^{(C)}$ and $\nu^{(F)}$ for g- and f-modes, plotted against mode order. Each curve corresponds to a given value of l, which has been indicated.

Sec. IV.C, if we regard the terms in Φ' in Eqs. (89) and (90) as small perturbations, to be evaluated from the unperturbed eigenfunctions by means of Eq. (91). Christensen-Dalsgaard (1984a) found that the resulting corrected frequencies differed by $< 0.05\%$ from the frequencies computed without making the Cowling approximation, for p-modes of degree > 0 and frequency > 1700 μHz.

It should be mentioned that in principle the Cowling approximation involves mathematical subtleties which have not been fully resolved. These are related to the fact that the approximation reduces the order of the system by 2, and hence eliminates two independent solutions of the differential equations. A striking manifestation is the fact that in the Cowling approximation there is an $l = 1$ f-mode with no nodes in the radial displacement, intermediate in frequency between the p- and the g-modes; in the full case there is also a solution to the equations with no nodes in the radial displacement, but this has zero frequency, and corresponds to an infinitely slow uniform translation of the model (see also Christensen-Dalsgaard 1976). If the effects of the gravitational potential are gradually introduced, the transition from the dipolar f-mode in the Cowling approximation to the zero-frequency "mode" in the full case takes place through a sequence of avoided crossings with the g-modes (Christensen-Dalsgaard 1978). A related problem concerns the classification of the modes. As discussed in Sec. II.D, a precise mathematical classification is possible in the Cowling approximation. In contrast, a naive application of this classification procedure to the lowest-order dipolar p-modes, computed from the full equations, results in a misidentification of the modes with $n = 1$ and 2 in typical models of the present Sun; for more evolved models, the problem apparently extends to modes of higher order. However, for those modes that have been observed on the Sun, the neglect of Φ' does not change the overall structure of the spectrum.

In the light of the utility of the asymptotic treatment, there is evidently a great need for an asymptotic description that takes the effects of Φ' into account. The leading-order correction to Eq. (55) was presented by Vorontsov (1988). A more complete description has been developed by Dziembowski and Gough (in preparation).

V. NUMERICAL TECHNIQUES

The differential Eqs. (31) - (33), in combination with boundary conditions such as Eqs. (39)-(41), constitute a two-point boundary value problem. Nontrivial solutions to the problem can be obtained only at selected values of the frequency ω, which is therefore an eigenvalue of the problem. Problems of this nature are extremely common in theoretical physics, and hence there exists a variety of techniques for solving them. Nevertheless, the computation of solar adiabatic oscillations possesses special features, which merit discussion. In particular, we typically need to determine a large number of frequencies very accurately, to match the volume and precision of the observed data.

Specific numerical techniques are discussed in considerable detail by, for example, Unno et al. (1979) and Cox (1980). Here we concentrate more on general properties of the solution method, and give examples illustrating typical results on the numerical precision of the solution. The choice of techniques and examples is unavoidably biased by our personal experience, but should at least give an impression of what can be achieved, and how to achieve it.

A. Difference Equations

The numerical problem can be formulated generally as that of solving

$$\frac{dy_i}{dx} = \sum_{j=1}^{I} a_{ij}(x)y_j(x) \quad \text{for} \quad i = 1, \ldots, I \tag{115}$$

with suitable boundary conditions at $x = x_1$ and x_2, say. Here the order I of the system is 4 for the full nonradial case, and 2 for radial oscillations or nonradial oscillations in the Cowling approximation.

To handle these equations numerically, we introduce a mesh $x_1 = x^{(1)} < x^{(2)} < \ldots < x^{(N_{me})} = x_2$ in x, where N_{me} is the total number of mesh points. Similarly, we introduce $y_i^{(n)} \equiv y_i(x^{(n)})$, and $a_{ij}^{(n)} \equiv a_{ij}(x^{(n)})$. A commonly used, very simple representation of the differential equations is in terms of *second-order centered differences*, where the differential equations are replaced by the difference equations

$$\frac{y_i^{(n+1)} - y_i^{(n)}}{x^{(n+1)} - x^{(n)}} = \frac{1}{2} \sum_{j=1}^{I} \left[a_{ij}^{(n)} y_j^{(n)} + a_{ij}^{(n+1)} y_j^{(n+1)} \right], \quad i = 1, \ldots, I. \tag{116}$$

These equations allow the solution at $x = x^{(n+1)}$ to be determined from the solution at $x = x^{(n)}$.

More elaborate and accurate difference schemes (see, e.g., Press et al. 1986; Cash and Moore 1980) can be set up which allow the rapid variation in high-order eigenfunctions to be represented with adequate accuracy on a relatively modest number of mesh points. Alternatively, one may approximate the differential equations on each mesh interval $(x^{(n)}, x^{(n+1)})$ by a set of equations with constant coefficients, given by

$$\frac{d\eta_i^{(n)}}{dx} = \sum_{j=1}^{I} \bar{a}_{ij}^{(n)} \eta_j^{(n)}(x) \quad \text{for} \quad i = 1, \ldots, I \tag{117}$$

where $\bar{a}_{ij}^{(n)} \equiv 1/2(a_{ij}^{(n)} + a_{ij}^{(n+1)})$ (Gabriel and Noels 1976). These equations may be solved analytically on the mesh intervals, and the complete solution is obtained by continuous matching at the mesh points. This technique clearly permits the computation of modes of arbitrarily high order. We have considered its use only for systems of order 2, i.e., for radial oscillations or non-radial oscillations in the Cowling approximation.

B. Shooting Techniques

Perhaps the conceptually simplest technique for handling a boundary value problem is the shooting technique. For simplicity, we consider first the case of a second-order system, such as results from making the Cowling approximation. Then, there is one boundary condition, namely the first of Eqs. (39), at the center, and one condition, Eq. (41), at the surface. For any value of ω, the equations may be integrated numerically, imposing the central boundary condition on ξ_r and ξ_h, and the quantity

$$\Delta(\omega) \equiv \left. \left(p' + \frac{dp}{dr} \xi_r \right) \right|_{r=R} \tag{118}$$

may be evaluated. The eigenfrequencies are obviously the zeros of $\Delta(\omega)$. A convenient method of locating them is to evaluate $\Delta(\omega)$ at a sequence of points $\omega_1, \omega_2, \ldots$; once an interval has been found where Δ changes sign, the zero can be found, for instance, by applying the secant method. An attractive feature of the method is precisely this ability to search automatically for all modes in a given frequency range, particularly when it is combined with a method for determining the order of a given mode, so that a check can be made that no modes have been skipped.

A slight elaboration of this basic technique is required to make it computationally efficient. Due to the rapid decrease of temperature near the solar surface, the equations are almost singular here. Far from the eigenfrequencies, the solution therefore generally increases rapidly towards the surface; this translates into a dramatic variation of Δ with ω, which complicates the determination of the zeros. To avoid this problem, one may compute solutions $(\xi_r^{(i)}, \xi_h^{(i)})$ and $(\xi_r^{(s)}, \xi_h^{(s)})$ satisfying the inner and the surface boundary conditions, respectively. A continuous match of the interior and exterior solutions requires the existence of nonzero constants $C^{(i)}$ and $C^{(s)}$ such that

$$C^{(i)} \, \xi_r^{(i)}(r_f) \; = C^{(s)} \, \xi_r^{(s)}(r_f) \tag{119}$$
$$C^{(i)} \, \xi_h^{(i)} \, (r_f) \; = \; C^{(s)} \, \xi_h^{(s)} \, (r_f)$$

where r_f is an appropriately chosen fitting point. This set of equations has a solution only if the determinant

$$\Delta_f(\omega) \; = \; \xi_r^{(i)}(r_f) \xi_h^{(s)}(r_f) - \xi_h^{(i)}(r_f) \xi_r^{(s)}(r_f) \tag{120}$$

vanishes. Hence the eigenfrequencies are determined as the zeros of Δ_f, as before.

The choice of fitting point should be guided by the expected behavior of the eigenfunction, in such a way that the integration of the differential equations proceeds in a stable fashion. Thus, for instance, when solving for a strongly trapped g-mode of high degree, r_f should be near the maximum in the buoyancy frequency where the mode is trapped; in this way, the integration from both the center and the surface is in the direction where the solution increases. Similarly, when integrating for a p-mode of high degree, r_f should be in the oscillatory region near the surface.

The solution of the full fourth-order problem proceeds in a very similar fashion. Here there are two linearly independent solutions that satisfy the boundary conditions at the center, and two linearly independent solutions that satisfy the conditions at the surface. The condition that these two sets of solutions match continuously at a point r_f leads to a set of equations whose solution requires the vanishing of a 4×4 determinant. It should be noted, however, that problems arise when the effect of the perturbation in the gravitational potential is small. In this case, although the two separate solutions from, e.g., the center are formally linearly independent, they are in practice very close to being linearly dependent, and the zeros of Δ_f are therefore ill determined. This is no major concern in practice, because under these circumstances the Cowling approximation is in general adequate. However, as discussed below, the problem may be avoided through the use of some variant of the relaxation technique.

C. Relaxation Techniques

The relaxation technique considers the set of difference equations, such as Eqs. (116), together with the homogeneous boundary conditions and a normalization condition, as a set of equations for the unknown quantities $[y_i^{(n)}; i = 1, \ldots, I; n = 1, \ldots, N_{me}; \omega]$. Due to the appearance of the eigenfrequency, the equations are nonlinear in the unknowns. They are solved by linearizing around an assumed initial trial solution, and the solution is obtained by iteration. This technique is equivalent to what is commonly known as the Henyey technique in computations of stellar evolution (Henyey et al. 1964; see also Baker et al. 1971).

A disadvantage of this technique is that it requires a reasonably accurate trial solution, both for the eigenfrequency and the eigenfunction, if the iteration is to converge to the desired mode. Also it is not immediately possible to search a given part of the spectrum. These problems may be avoided by dropping one of the boundary conditions, and regarding ω as given (see, e.g., Castor 1971; Osaki and Hansen 1973). The difference equations are then a linear set of equations for the $[y_i^{(n)}]$ which may be solved directly. Given the solution, the remaining boundary condition, now regarded as a function of

ω, is solved to obtain the eigenfrequencies. Thus in this form, the relaxation technique retains the advantages of the shooting method, in that a region of the spectrum can be scanned. Once a sufficiently close approximation to the solution has been found, the rate of convergence can be increased by switching to simultaneous iteration for the eigenfrequency and eigenfunction.

As for the shooting technique, the straight determination of the eigenfrequency through root seeking on one of the boundary conditions is rather ill behaved. This problem may be avoided by imposing all boundary conditions, but permitting for general ω a discontinuity in one component of the eigenfunction at a suitable interior fitting point r_f. The eigenfrequencies are then determined by requiring that the discontinuity vanish. We have found that this technique allows stable solution of the full set of equations for all relevant degrees and frequencies.

D. Formulation as a Matrix Eigenvalue Problem

As discussed in Sec. IV.C, the equations of adiabatic oscillation, written as in Eqs. (89) and (90), constitute a linear eigenvalue problem in function space. If the operator on the right-hand side is discretized, the result is a linear discrete eigenvalue problem. By solving this, one obtains (approximations to) the eigenvalues and eigenfunctions of the continuous problem.

A method of this nature (but generalized to the nonadiabatic case) was used by Keeley (1977) for radial oscillations. Knölker and Stix (1983) used it for adiabatic nonradial oscillations in the Cowling approximation. In these cases, the operator describing the left-hand side of the oscillation equations is a pure differential operator; hence its discrete representation only couples the solution at a few neighboring mesh points and results in an eigenvalue problem where the matrix is banded with only a few off-diagonal elements. Consequently, efficient techniques exist for the determination of the eigenvalues. In contrast, in the full nonradial problem the terms in Φ' couple all parts of the model (see also Eq. 91); then the corresponding matrix is full, although for large l it is diagonally dominated, due to the factors $(r'/r)^{l+1}$ and $(r/r')^l$ occurring, respectively, in the first and second term on the right-hand side of Eq. (91). In this case, it is not evident that sufficiently fast algorithms exist for the determination of the matrix eigenvalues to make the technique competitive with the shooting or relaxation techniques. No attempt has apparently been made to apply it to this problem.

The matrix eigenvalue problem can also be derived directly from the variational principle as expressed in Eq. (103), by means of the so-called Rayleigh-Ritz method (see, e.g., Strang and Fix 1973). To do so, the eigenfunction is expanded on a set of suitable basis functions, and the expansion coefficients are determined by imposing the condition that the expression (103) be stationary. Although this method has proven useful in atomic physics (see, e.g., B. L. Christensen-Dalsgaard 1982), the effects of the gravitational potential once again lead to a full matrix in the resulting eigenvalue problem.

We finally note that Pesnell (1990) has developed an efficient algorithm, based on the method of Castor (1971), for computing nonradial oscillations both in the adiabatic and the nonadiabatic case. This involves formulating the oscillation equations as a generalized linear algebraic eigenvalue problem; in contrast to the techniques discussed above, Poisson's equation is left in differential form, and hence the resulting matrices are sparse. Cox et al. (1989) applied this method to the computation of solar oscillations.

E. Richardson Extrapolation

The difference scheme (116), which is used by at least some versions of the shooting, relaxation and matrix eigenvalue techniques, is of second order. Consequently the truncation errors in the eigenfrequency and eigenfunction scale as N_{me}^{-2}. If $\omega(1/2N_{me})$ and $\omega(N_{me})$ are the eigenfrequencies obtained from solutions with $1/2N_{me}$ and N_{me} mesh points, the leading order error term therefore cancels in

$$\omega^{(Ri)} \equiv \frac{1}{3}\left[4\omega(N_{me}) - \omega\left(\frac{1}{2}N_{me}\right)\right]. \tag{121}$$

The evaluation of $\omega^{(Ri)}$, known as *Richardson extrapolation*, was used by Shibahashi and Osaki (1981) to compute frequencies of solar oscillation. As shown in Sec. V. H., it provides an estimate of the eigenfrequency that is substantially more accurate than $\omega(N_{me})$, although of course at some added computational expense.

F. Variational Frequencies

The variational property discussed in Sec. IV.C can be used to obtain an estimate of the oscillation frequency which is at least formally more accurate than the frequency obtained as eigenvalue of the solution of the oscillation equations (Christensen-Dalsgaard et al. 1979; J. Christensen-Dalsgaard 1982). It follows from Eq. (102) that if the computed eigenfunction is substituted into the functional $\Sigma(\xi)$, the result agrees with the squared eigenfrequency to within an error that is quadratic in the error in the eigenfunction. If the latter error goes as N_{me}^{-2}, the error in $\Sigma(\xi)$ would be expected to vary as N_{me}^{-4}; this assumes that the evaluation of $\Sigma(\xi)$, for given ξ, is sufficiently accurate.

For realistic solar models the complete expressions (103) or (105) must be used, including the surface terms. However, as these terms are in general relatively small, the variational property is still approximately satisfied. Hence the expressions may be used to provide estimates of the frequency which are less sensitive to numerical error than the eigenfrequency. On the other hand, it should be noted that the variational property, and the analysis leading to Eqs. (103) and (105), assume that the solar model satisfies the equations of hydrostatic equilibrium and the mass equation exactly. When the model is itself the result of a numerical solution of the equations of stellar structure, this is evidently not the case; then, even if they were to be evaluated with infinitely high precision for the given model, the variational frequency and the eigenfrequency would not agree. The discrepancy provides an estimate of the effect on the frequencies of the inconsistencies in the model. Examples of this are discussed below.

G. The Determination of the Mesh

Computational efficiency demands that the distribution of mesh points be chosen appropriately. It is immediately obvious from the eigenfunctions (cf. Figs. 8 and 10) that a mesh uniform in r is far from optimal; also the distribution of points should clearly be different for p- and for g-modes.

Procedures exist that determine the optimal mesh as part of the numerical solution of a set of differential equations (Gough et al. 1975a); the use of such techniques in calculations of stellar evolution was discussed, for example, by Eggleton (1971) and J. Christensen-Dalsgaard (1982). In the present case, however, the requirements on the mesh are essentially driven by the behavior of the modes of high radial order, whose eigenfunctions are given, with considerable precision, by the asymptotic expressions (56) and (84). Thus, to have a roughly constant number of mesh points between the nodes in the eigenfunction, the mesh should be approximately uniformly spaced in terms of the integrals in these equations.

To define a flexible method for setting up the mesh we have adopted a simplified version of the procedure developed by Gough et al. (1975a). Thus we introduce a variable z, with a range from 0 to 1, such that the mesh is uniform in z, and determined by

$$\frac{dz}{dr} = \lambda H(r). \tag{122}$$

Here λ is a normalization constant, relating the ranges of z and r, and the function H determines the properties of the mesh. Given H, z is obtained as

$$z(r) = \lambda \int_0^r H(r')dr' \tag{123}$$

with

$$\lambda = \left(\int_0^R H(r)dr \right)^{-1}. \tag{124}$$

The mesh $\{r^{(n)}, n = 1, \ldots, N_{me}\}$ is finally determined by solving the equations

$$z(r^{(n)}) = \frac{n-1}{N_{me}-1} \qquad (125)$$

by interpolating in the computed $z(r)$.

The choice of the function H must be guided by the asymptotic behavior of the modes. We have used

$$H(r)^2 = R^{-2} + c_1 \frac{\omega_a^2}{c^2} + c_2 \frac{|N|^2}{\omega_a^2 r^2} + c_3 \left(\frac{d \ln p}{dr}\right)^2 \qquad 126$$

where $\omega_a^2 \equiv GM/R^3$ is a characteristic squared frequency. Here the terms in c_1 and c_2 are clearly aimed at reproducing the asymptotic behavior of the p- and the g-mode eigenfunctions, respectively. The term in c_3 provides extra mesh points near the surface, where the reflection of the p-modes takes place. Finally, the constant term ensures a reasonable resolution of regions where the other terms are small.

The parameters in this expression can be determined by testing the numerical accuracy of the computed frequencies, as discussed in the following section. For normal solar models reasonable choices, based on a fairly extensive (but far from exhaustive) set of calculations, are given in Table I. In the p-mode case the mesh is predominantly determined by the variation of sound speed, with the term in N giving a significant contribution near the center and the term in $d \ln p/dr$ contributing very near the surface. The g-mode mesh is dominated by the term in N in most of the radiative interior, whereas in the convection zone, where $|N|$ is generally small, the constant term dominates; the term in $d \ln p/dr$ is again important in the surface layers.

H. Numerical Accuracy

Estimates of the numerical accuracy of computed frequencies should in principle regard the physics and assumptions of the underlying model as given, and should therefore take into account the numerical errors introduced in the computation of the models. However, here we restrict the discussion to the calculation of frequencies for a given model. As an indication of the desired accuracy, it might be recalled that the typical errors in the observed

TABLE I
Parameters in Eq. (126) for the Determination of Meshes Suitable for
Computing p- and g-Modes in a Model of the Present Sun

	c_1	c_2	c_3
p-mode mesh	10.0	0.01	0.015
g-mode mesh	0.025	0.1	0.0001

frequencies are around, or < 0.1 μHz over a wide range of modes (cf. the chapter by F. Hill et al.). It is clearly desirable that the errors in the computed frequencies should be no greater.

The only practical way of estimating the numerical error in the computed frequencies is to compare results obtained with different numbers of mesh points. This technique was applied to adiabatic oscillation frequencies of solar models by e.g. Shibahashi and Osaki (1981), J. Christensen-Dalsgaard (1982), Noels et al. (1984) and Guenther and Sarajedini (1988). Mullan and Ulrich (1988) and Mullan (1989) made a careful analysis of the accuracy of frequencies of polytropic models, although restricted to calculations in the Cowling approximation. Here we give examples of such estimates of the errors in calculations of p-mode frequencies by various techniques. In each case, the error is obtained as the difference between the results of a computation with $N_{me} = 600$ points, and a computation with $N_{me} = 2400$ points. Except where otherwise noted, the computations used the centered difference approximation (Eq. 116) to the differential equations. For $l \leq 20$, the full set of equations were solved by means of the relaxation technique with interior fitting discussed in Sec. V.C. For higher degree, the Cowling approximation was employed, the frequency being corrected for the effect of the perturbation in the gravitational potential by means of the perturbation analysis discussed in Sec. IV.E; for modes of degree 20 and frequency below 5000 μHz, the error resulting from this approach is < 0.01 μHz for the Richardson extrapolated frequencies, and < 0.05 μHz for the variational frequencies.

Figure 15 shows the errors in the straight eigenfrequency, in the result of applying Richardson extrapolation according to Eq. (121), and in the variational frequency. In each case, the error generally grows with increasing frequency, and hence mode order, as would be expected. The eigenfrequencies computed with 600 points, using the centered difference Eqs. (116), are evidently totally inadequate, the maximum error within the observed frequency range being as high as ~ 15 μHz. In contrast, the error in the frequencies obtained from Richardson extrapolation, and using the variational method, are modest, although still somewhat higher than our goal of 0.1 μHz.

The variation in error with number of mesh points was estimated by also computing frequencies with $N_{me} = 1200$. For the eigenfrequencies the error varies as N_{me}^{-2}, as expected. Thus to decrease the eigenfrequency error below 0.1 μHz would require about 7000 points. On the other hand, the error in both the Richardson and the variational frequencies decreases much more rapidly with increasing N_{me}. At $N_{me} = 1200$ the errors in the Richardson frequencies are well below 0.05 μHz, and the errors in the variational frequencies are 0.1 μHz. Thus we may estimate that the errors in these quantities for $N_{me} = 2400$ are at or below 0.01 μHz, which should be adequate to match the observational accuracy within the foreseeable future.

For comparison, we also computed eigenfrequencies by using the con-

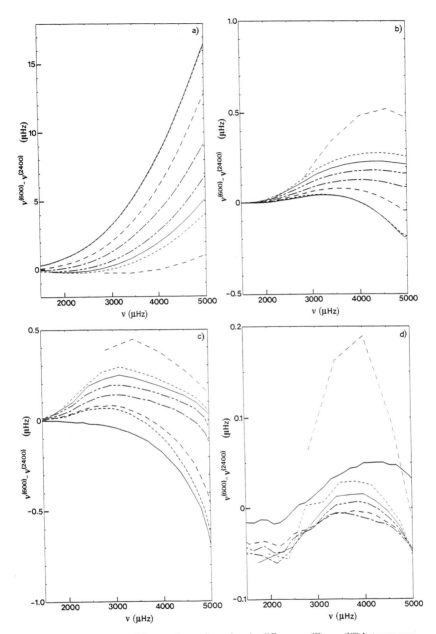

Fig. 15. Errors in computed frequencies, estimated as the difference $\nu^{(600)} - \nu^{(2400)}$ between computations using 600 and 2400 mesh points. The results in panels (a) $-$ (c) were obtained with
the centered difference algorithm in Eq. (116), whereas panel (d) shows results of using
the constant coefficient technique described by Eq. (117). Each curve corresponds to a given
value of l. *Heavy lines:* $l = 0$ (———); $l = 1$ (-------); $l = 20$ (– – – –); $l = 50$
(— · — · —); $\mathbf{l = 100}$ (— ·· — ·· —). *Thin lines:* $l = 150$ (———); $l = 200$ (-------);
$l = 500$ (– – – – –). (a) Straight eigenvalues. (b) Eigenvalues corrected with Richardson
extrapolation (cf. Eq. 121). (c) Variational frequencies. (d) Straight eigenvalues with the constant coefficient technique; modes with $l = 1$ are not included.

stant coefficient method outlined in Eq. (117); only radial modes, and modes of degree $l \geq 20$ where the Cowling approximation is sufficiently accurate, were considered. The resulting errors are shown in Fig. 15d. Here even 600 points are sufficient to achieve errors below 0.1 μHz. As an indication of the consistency of the different integration methods, it might also be mentioned that with $N_{me} = 2400$ the Richardson and the constant coefficient frequencies agree to within 0.005 μHz.

As mentioned above, comparison of the eigenfrequencies and the variational frequencies provides a check of the accuracy of the equilibrium model. Figure 16a shows differences between the Richardson extrapolation and the variational frequencies, computed for Model 1 of J. Christensen-Dalsgaard (1982), interpolated to a mesh with 2400 points. In accordance with the discussion in Sec. IV.C, the differences have been scaled with the energy ratio Q_{nl}. There are evidently discouragingly large differences, indicating that the errors in the model are unacceptably large. Furthermore, the behavior for $l = 0$ and $l > 0$ is different, due to the different formulations in Eqs. (103) and (105), which are equivalent only if the model is consistent. This differ-

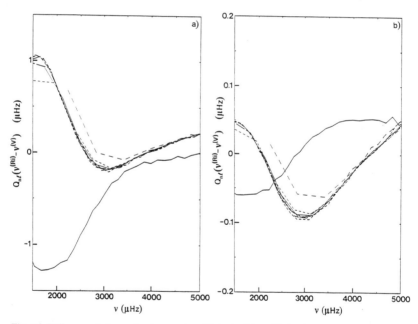

Fig. 16. Differences between the frequencies $\nu^{(Ri)}$ obtained from Richardson extrapolation and the variational frequencies $\nu^{(V)}$; the differences have been scaled by the inertia ratio Q_{nl} (cf. Eq. 113). The frequencies were computed on a mesh with 2400 points. The line styles are the same as in Fig. 15. (a) Results for Model 1 of J. Christensen-Dalsgaard (1982); (b) results for a model computed with higher numerical accuracy.

ence causes problems if variational frequencies are used to compute the low-degree frequency separations in Eq. (72) (Fossat 1985; Christensen-Dalsgaard 1986). In contrast, results for a model computed with substantially higher numerical accuracy are shown in Fig. 16b. Although there are still differences, and a different behavior of the radial and nonradial modes, the effect is now reduced to an acceptable level. It should also be noted that for $l > 0$ the scaled differences depend little on l, indicating that the error in the model is predominantly located near the surface. Further analysis of results such as these is clearly of considerable importance for the optimization of the calculation of solar models.

VI. SENSITIVITY OF THE FREQUENCIES TO THE STRUCTURE OF THE SUN

To analyze observations of solar oscillations, it is common to study the differences between the observed frequencies and those computed for a variety of solar models. Given that significant differences exist, the goal is evidently to identify the likely sources of the errors in the models. To do so, one must consider the sensitivity of the computed frequencies to changes in the model.

As discussed in the chapter by Gough and Thompson, it is possible to obtain kernels that relate, say, changes $\delta\rho$ in density and $\delta\Gamma_1$ in the adiabatic exponent Γ_1 to the changes in the frequencies. Inspection of these kernels then immediately shows how different modes respond to modifications in particular regions of the model. Furthermore, from the differences between observed and computed frequencies, one may in principle obtain the corrections $\delta\rho$ and $\delta\Gamma_1$ to the assumed solar model by means of inverse analysis.

This technique permits the determination of important aspects of the internal structure of the Sun. Of more fundamental interest, however, are the material properties and other features that determine the structure. The properties of matter in the Sun (equation of state, opacity and nuclear reaction rates) are required in computations of solar models, as are assumptions about the features that need to be included in the computation. By studying solar oscillations we may hope to obtain information about matter under solar conditions, and to uncover, or constrain, possible additional features that might be required in the computation. Information of this type is evidently important for the computation of models of other stars; in addition it allows us to use the Sun as a test of the basic physical description of plasmas under conditions that are difficult to achieve in the laboratory (see the chapter by Däppen et al.).

A helioseismic determination of ρ and Γ_1, as discussed above, can evidently be used to constrain solar models in this manner. Nevertheless, it is of obvious interest to consider more directly the relation between the assumptions underlying the model computation and the oscillation frequencies.

Through a suitable parametization of some of these assumptions, keeping the others fixed, it is possible to obtain kernels that relate aspects, say, of the physics to the oscillation frequencies. An interesting example is the analysis by Korzennik and Ulrich (1989) of the effects of the opacity. However, a commonly used, although less general, technique is to compare frequencies of specific models computed under different assumptions. Examples are the calculations by Scuflaire et al. (1975), Iben and Mahaffy (1976), Christensen-Dalsgaard et al. (1979), Ulrich and Rhodes (1983), Shibahashi et al. (1983), Noels et al. (1984) and Christensen-Dalsgaard (1984b, 1986). More recently, Cox et al. (1989) studied the effects of microscopic diffusion, and Christensen-Dalsgaard et al. (1988b) compared frequencies computed with different assumptions about the equation of state.

To be useful, calculations of this nature require considerable care. It is essential that the two models considered differ only in the particular property which is under investigation. Also, given that the radius and luminosity of the Sun are known with high accuracy, the two models should be calibrated (by adjusting the convective efficiency and the initial composition) to have the same radius and luminosity; this of course excepts calculations aimed specifically at testing the effect of radius or luminosity changes. Numerical errors in the calculation of the models and the frequencies should either be negligible, or have the same effect for the two models. Finally, the presentation of the results should provide a reasonably comprehensive overview of the effect of the modification on a representative set of modes, and should illustrate the relationship between the changes in the model and the corresponding frequency changes. A convenient format is provided by plots of relative differences between relevant quantities in the two models, and of the frequency differences. In particular it should be noted from Eqs. (62) and (63) that the change in the frequencies of p-modes, with which we are mainly going to be concerned, is largely determined by the change in the sound speed.

A detailed discussion of the effects of various modifications to the solar model was presented by Christensen-Dalsgaard (1988c). Of major concern are clearly the effects of the uncertainties in the physics. The effect of the treatment of the equation of state on the oscillation frequencies was considered in the chapter by Däppen et al. It may also be noticed that changes in the nuclear reaction rates have very small effects on the frequencies of at least the 5-min oscillations (see, e.g., Christensen-Dalsgaard 1988c). Indeed, it follows from the high temperature sensitivity of the reactions, and the requirement that the model have the correct luminosity, that a change in the overall energy generation rate, as controlled by the cross section of the $p + p$ reaction, can be compensated by a much smaller change in temperature; changes in the rates of the other reactions in the pp chains are evidently entirely insignificant for the frequencies (Bahcall and Ulrich 1988).

Here we concentrate on two aspects of the model:

1. The opacity;
2. The composition profile.

Analysis of these examples illustrates the theoretical properties of the oscillations which have been presented in the preceding sections. Furthermore, it
may be argued that, apart from the very uncertain region near the solar surface, the opacity is the least well determined of the physical properties required to compute a model of the Sun (see the chapter by Däppen et al.). In
Sec. VI.A, we discuss the effects on the structure and the p-mode frequencies
of modifying these two aspects of the model. Section VI.B deals with the
properties of low-degree p-modes, which are particularly important in providing information about the solar core. Finally, Sec. VI.C considers the
properties of g-modes in terms of the asymptotic expression for their frequencies.

A. Effects of Changes in the Model

We have used a reference model which is similar to Model 1 of J. Christensen-Dalsgaard (1982); the equation of state was calculated according to
the Eggleton et al. (1973) prescription, and the opacity was obtained from
spline interpolation in the Cox and Tabor (1976) tables. The heavy element
abundance was $Z = 0.02$. The initial abundance of hydrogen, and the mixing
length parameter, were adjusted to obtain a model at solar age, assumed to
be 4.75 Gyr, with the solar luminosity 3.846×10^{33}erg s^{-1} and the solar
radius 6.96×10^{10} cm. However, the numerical precision was considerably
higher than for Model 1: the calculation used 600 spatial mesh points, and 26
time steps between the zero-age-main-sequence and the present Sun. As was
illustrated in Fig. 16, this leads to a considerable improvement in the consistency of the computed frequencies. The frequencies were computed with
the centered difference technique and Richardson extrapolation, as discussed
in Secs. V.E and V.H.

The modified models were also adjusted to obtain the same radius and
luminosity, to within a relative error of $<10^{-5}$. To study the effects of the
modifications of the opacity, full evolution models were calculated, whereas
the modification in the composition profile was studied by means of a static
model of the present Sun.

1. Opacity Modifications. To study the effect of the opacity on the
structure and frequencies of a solar model, we have computed models where
the opacity was artificially modified. Specifically, the opacity κ as a function
of density ρ and temperature T, was determined from

$$\log \kappa = \log \kappa_0 + Af(\log T) \qquad (127)$$

where the unmodified opacity κ_0 was obtained from the tables of Cox and
Tabor (1976). Here A is a constant determining the magnitude of the opacity

modification, and the function f is defined by $f(x) = \exp[-(y/\Delta \log T)^2]$, where

$$y = \begin{cases} x - \log T_1 & \text{for } x < \log T_1 \\ 0 & \text{for } \log T_1 < x < \log T_2 \\ x - \log T_2 & \text{for } \log T_2 < x \end{cases} \qquad (128)$$

(log is to base 10).

While this definition is evidently entirely arbitrary, it has the advantage of providing an opacity increase over a well-defined region in log T, with a continuous transition to zero on either side.

It should be noted that the structure of the Sun is independent of the local value of the opacity in the adiabatic part of the convection zone, i.e., approximately between $\log T = 4$ and $\log T = 6.3$. Thus modifications to the opacity in this temperature interval have no effect on the model.

Recent opacity calculations (see, e.g., Cox et al. 1989) have shown a very substantial increase, by up to a factor 2, at conditions corresponding to the solar photosphere. To investigate the effects of such a change, we have computed a model with $\log T_1 = 3.5$, $\log T_2 = 3.8$, $\Delta \log T = 0.2$ and $A = 0.3$, thus roughly corresponding in magnitude to the actual increase in the tabulated opacity. Figure 17 shows the resulting differences between the models. In the deeper interior, the model is virtually unchanged. The photospheric pressure, and hence density, is roughly inversely proportional to the opacity, and is therefore drastically changed. However, there is also a significant change in the sound speed, which leads to substantial changes in the frequencies. These are illustrated in Fig. 18. In accordance with the discussion in Sec. IV.D, the differences have been scaled by the ratio Q_{nl} of mode inertias (cf. Eq. 113). It follows from Fig. 13 that for the highest-degree modes the raw differences are larger by a factor of ~ 5 than those shown. Modes with $l \leqslant 500$ all penetrate well beyond the region where the sound speed is affected (cf. Fig. 3), and for these modes the scaled frequency change is mainly a function of frequency, but depends little on the depth of penetration of the mode, and hence on l. At higher degree, the modes sample only part of the negative sound-speed difference, and the frequency change is smaller.

Quite apart from the specific modification considered, this example illustrates the important property that changes to the model near the surface lead to scaled frequency changes that depend only on frequency, and not on degree. Also, the frequency change is very small at low frequency. As discussed in Sec. IV.B, the low-frequency modes are evanescent near the surface, and hence their amplitude is small, compared with the amplitude in the interior (see also Fig. 9); as a result, they are barely affected by changes in the surface region. Qualitatively similar changes result from modifications to the treatment of the upper, significantly superadiabatic part of the convection zone (Christensen-Dalsgaard 1986); more generally, it seems likely that the

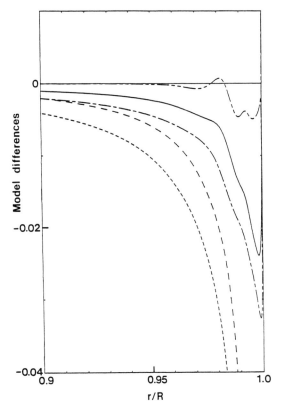

Fig. 17. Differences, at fixed fractional radius r/R, between a model of the present Sun where the opacity has been artificially increased by up to about a factor 2 near the surface and a normal model, in the sense (opacity modified) − (normal). The variables shown are $\delta\ln c$ (————), $\delta\ln p$ (--------), $\delta\ln\rho$ (— — — —), $\delta\ln T$ (— · — · —), and $\delta\ln\Gamma_1$ (— ·· — ·· —).

uncertainties in the treatment of the surface layers (nonadiabaticity of the oscillations, the treatment of convection, possible effects of magnetic fields, etc.) would give rise to a similar behavior of the frequency changes. Thus in analyzing observed frequencies, it is possible to absorb these uncertainties by allowing an undetermined frequency-dependent part of the scaled differences between observed and computed frequencies.

To illustrate the effect of a modification in the opacity near the base of the convection zone, we also computed a model with $\log T_1 = 6$, $\log T_2 = 6.6$, $\Delta\log T = 0.15$ and $A = 0.1$, corresponding to a maximum change in opacity at fixed T and ρ of 26%. This increases the depth of the convection zone in the model by $\sim 0.02\,R$. The resulting changes in the model are shown in Fig. 19. A striking feature is the comparatively small change in the sound speed in much of the convection zone. In fact, it is easy to show that except

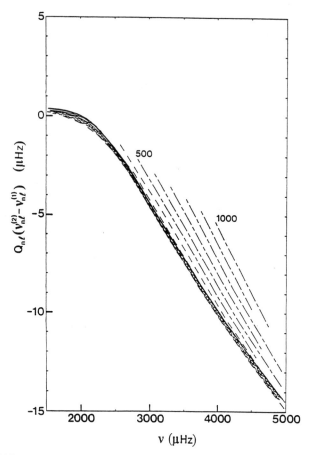

Fig. 18. Differences between the frequencies $\nu_{nl}^{(2)}$ of the model with modified surface opacity and the frequencies $\nu_{nl}^{(1)}$ of a normal model, for selected values of l. The differences have been scaled by the inertia ratio Q_{nl} (cf. Eq. 113). Points corresponding to a given value of l have been connected, according to the following convention: $l = 0, 5, 10, 20, 30$ (————); $l = 40, 50, 70, 100$ (-------); $l = 150, 200, 300, 400$ (– – – –); and $l = 500, 600, 700, 800, 900, 1000$ (— · — · —).

in the outer part of the convection zone, the sound speed is approximately given by

$$c^2 \approx (\Gamma_1 - 1)Rg_s \left(\frac{R}{r} - 1\right) \tag{129}$$

where $\Gamma_1 \approx 5/3$ was assumed to be constant (see, e.g., Christensen-Dalsgaard 1986). Hence the sound speed is essentially given by the total mass and surface radius of the model, which are fixed. Because the convection zone ex-

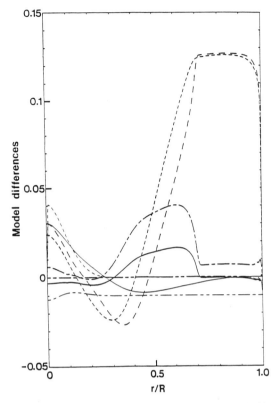

Fig. 19. Differences, at fixed fractional radius r/R, between the model with modified opacity in the interior and the reference model, in the sense (modified model) – (reference model). The lines have the following meaning: *Thin lines:* $\delta \ln q$, where $q \equiv m/M$ is the mass fraction (——————); $\delta \ln L$, where L is the luminosity at r (-------); $\delta \ln r$ (– – – –); δX, where X is the hydrogen abundance (— ·· — — ·· —). *Heavy lines:* $\delta \ln c$ (——————); $\delta \ln p$ (----------), $\delta \ln \rho$ (– – – –); $\delta \ln T$ (— · — · —).

tends more deeply in the modified model, the temperature gradient is higher (being adiabatic) in this model than in the reference model just beneath the bottom of the convection zone of the latter. This is the reason for the increase in $\delta \ln T$ and $\delta \ln c$ with increasing depth just beneath the convection zone. The changes in the deep interior of the model are more difficult to describe in simple terms. However, it should be noted that, due to the high sensitivity of the energy generation rate on temperature, the temperature change in the core has to be small.

Scaled frequency differences between the modified model and the reference model are shown in Fig. 20. These differences can be understood relatively simply in terms of the differences in sound speed shown in Fig. 19, by

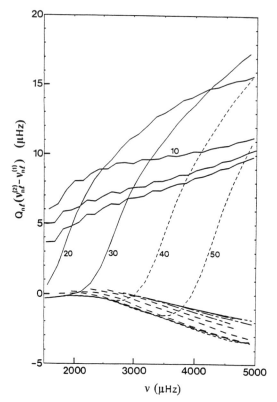

Fig. 20. Scaled frequency differences between the model computed with modified opacity in the interior (2), and the reference model (1), for selected values of l. Points corresponding to a given value of l have been connected, according to the following convention: $l = 0, 5, 10, 20, 30$ (——————); $l = 40, 50, 70, 100$ (--------); $l = 150, 200, 300, 400$ (– – – –); and $l = 500, 600, 700, 800, 900, 1000$ (— · — — — · — ·—). In addition a few values of l have been indicated in the figure.

using Eq. (62) or (65), and the behavior of the turning point illustrated in Fig. 3. At very low degree, the modes penetrate almost to the center, and the frequency change is given by the weighted average in Eq. (65), which is dominated by the region of positive δc beneath the convection zone. As l increases to 10 the turning point moves out through the region of slightly negative δc near the core, and $Q_{nl}\nu_{nl}$ increases. At higher degrees, beginning at low frequency for $l = 20$, and at increasing ν when l increases to 50, the modes become largely confined within the convection zone. Here δc is negative and significant only near the surface; thus the frequency differences are negative, depend little on l, and are furthermore small at low frequency. It should be noted that the negative δc near the surface has a substantial effect on the frequencies, despite its insignificant appearance in Fig. 19. The reason is the weighting with c^{-1} (cf. Eq. 65) which makes the frequencies very sensitive to changes in the model near the surface.

To illustrate the effect of a presumably realistic change in the opacity, we finally consider the difference between the reference model, computed with the Cox and Tabor (CT; 1976) opacity, and a model computed with Los Alamos Opacity Library tables (LAOL; Huebner et al. 1977), combined with a separate opacity calculation for conditions corresponding to the solar atmosphere (Cox et al. 1989). Figure 21 shows the differences between these models. The LAOL opacities are considerably higher in the solar core than the CT values, and as a consequence the initial hydrogen abundance required to obtain the proper luminosity is decreased from 0.732 in the CT model to 0.694 in the LAOL model. On the other hand, the opacity in the LAOL tables at the temperatures corresponding to the base of the convection zone is smaller than the CT values, leading to a shallower convection zone in the

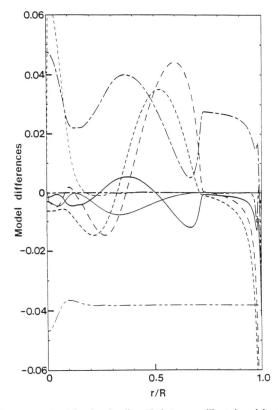

Fig. 21. Differences, at fixed fractional radius r/R, between calibrated models of the present Sun computed using the Los Alamos Opacity Library and the Cox and Tabor tables, in the sense (LAOL) − (CT). The variables plotted and the line styles are as in Fig. 19.

LAOL model, and consequently to the negative $\delta c/c$ in the region just below the convection zone. The scaled frequency differences shown in Fig. 22 reflect the difference in sound speed. At high degree the behavior is similar to that shown in Fig. 18, as it is dominated by the opacity increase in the atmosphere. However, the sound-speed variation at the base of the convection zone causes a substantial variation in $\delta\nu$ with degree at degrees between 20 and 40.

It should be pointed out that from inversion of the frequencies of the 5-minute oscillations, the depth of the solar convection zone has been deter-

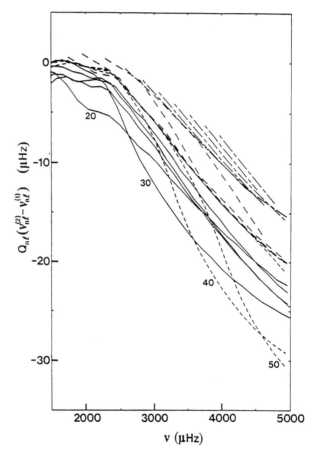

Fig. 22. Scaled frequency differences between the LAOL model (2) and the CT model (1) illustrated in Fig. 21, for selected values of l. Points corresponding to a given value of l have been connected, according to the following convention: $l = 0$ −5, 10, 20, 30 (————); $l = 40$, 50, 70, 100 (---------); $l = 150$, 200, 300, 400 (– – – – –) and $l = 500$, 600, 700 800, 900, 1000 (— · — · — · —).

mined as $0.287\ R$ (Christensen-Dalsgaard et al. 1990*b*; see also the chapter by Gough and Thompson). This is significantly greater than the depth of $0.267\ R$ obtained in the LAOL model. Thus there are indications that the LAOL opacities are actually somewhat too small at temperatures of a few million degrees. A similar conclusion was reached by Cox et al. (1989); Korzennik and Ulrich (1989) also found an opacity increase in this temperature interval in one of their inversions for the opacity.

2. Effects of Partial Mixing.   A strong assumption in normal computations of solar models is the absence of material mixing except in the outer convection zone. As a result, the variation of the hydrogen abundance X with position is solely determined by the rate of nuclear reactions. There is an evident interest in considering the effects on the oscillation frequencies of relaxing this assumption.

Several different instabilities, mostly associated with solar rotation, have been identified that may cause some degree of mixing of the solar core (for reviews, see, e.g., Zahn 1983; Spruit 1984*b*; chapter by Schatzman). However, it appears difficult to achieve substantial mixing across the gradient in molecular weight that arises due to nuclear burning. Without going into such details, Schatzman et al. (1981) considered the effects of an assumed "turbulent diffusion," characterized by a so-called critical Reynolds number Re^* which they took to be constant. Lebreton and Maeder (1987) later considered the, presumably more realistic, case of a nonconstant Re^*, taking into account the stabilizing effect of the molecular weight gradient.

Regardless of their physical plausibility, the Schatzman et al. models are interesting as examples of the consequences of fairly major modifications to the hydrogen abundance profile. To make a careful comparison with the normal model considered here, we estimated the hydrogen profile $X(q)$, q being the mass fraction, for the $Re^* = 100$ model shown by Schatzman et al. (1981, Fig. 2), and recomputed the model by scaling this $X(q)$ to get the proper solar luminosity. The procedure was discussed in more detail by Christensen-Dalsgaard (1986); relative to that work, we have used a slightly modified $X(q)$, taking particular care to obtain reasonable behavior near the solar center.

Differences between the partially mixed and the normal model are shown in Fig. 23. The dominant effects are evidently in the core, where the hydrogen abundance is increased by the mixing near the center, leading to a decrease in the mean molecular weight and consequently to an increase in the sound speed (cf. Eq. 34). This trend is partially reversed in the outer parts of the core. Also, as the convection zone is slightly less deep in the mixed model, there is a jump in the sound-speed difference near the base of the convection zone, similar to that found in Fig. 19, although of opposite sign.

The scaled frequency differences, shown in Fig. 24, correspond closely to the difference in sound speed. Only the lowest-degree modes penetrate into

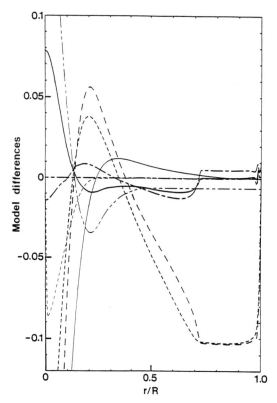

Fig. 23. Differences, at fixed fractional radius r/R, between the model with a partially mixed core and the reference model, in the sense (modified model) − (reference model). The variables plotted and the line styles are as in Fig. 19.

the region of increased sound speed in the core; this effect, however, is compensated by the negative sound-speed difference in the rest of the radiative interior, which dominates for modes of somewhat higher degree. For degrees between 20 and 50, there is again, as in Fig. 20, a transition to modes that are essentially trapped in the convection zone, and for which the frequency differences are dominated by the small region of positive sound-speed difference near the surface.

B. The Frequency Separation Between Low-Degree Modes

It was argued in Sec. III.B that for low-degree modes the frequency difference $\delta\nu_{nl} = \nu_{nl} - \nu_{n-1 l+2}$ is a measure of conditions in the solar core. To illustrate this, we consider the quantity

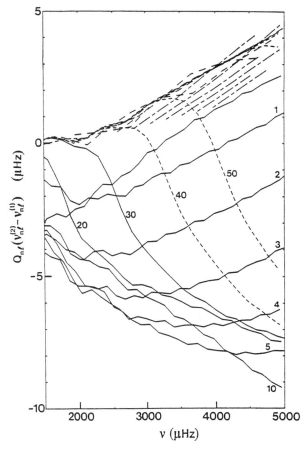

Fig. 24. Scaled frequency differences between the model with partially mixed core (2) and the reference model (1), for selected values of l. Points corresponding to a given value of l have been connected, according to the same convention as in Fig. 20.

$$d_{nl} = \frac{3}{2l+3} \delta \nu_{nl} \qquad (130)$$

which, according to Equation (72), is expected to depend on frequency roughly as ν^{-1} and to be independent of l. Figure 25 shows d_{nl} for $l = 0$ and 1, both for the normal and the partially mixed model, as well as for the observed frequencies compiled by Duvall et al. (1988a). Similar results were presented by Fossat (1985), Pallé et al. (1986b) and Cox et al. (1989). It is evident that, although decreasing with increasing ν, d_{nl} does not precisely

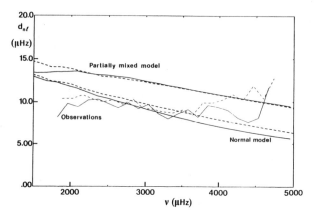

Fig. 25. Normalized separation d_{nl} (cf. Eq. 130) between frequencies of low-degree modes. Curves are shown for $l = 0$ (——————) and $l = 1$ (------). The heavy lines show results for theoretical frequencies. The lower pair is for a normal solar model, and the upper pair for a partially mixed model. The thin lines are based on observed frequencies, from Duvall et al. (1988a).

follow the expected ν^{-1} dependence. Also, there is a significant difference between the values at $l = 0$ and 1 for the normal model. In fact, as argued in Sec. III.B, the asymptotic theory is not sufficiently accurate to represent the finer details of the low-degree frequencies; in particular, the effect of the gravitational potential depends strongly on both frequency and degree, and has a significant effect on d_{nl}. On the other hand, there is a substantial difference between the results for the normal and the partially mixed model. Faulkner et al. (1986) showed that this is entirely consistent with the asymptotic expression Eq. (72): in the normal model, there is a considerable increase in c with r in the core of the model, due to the increase in the hydrogen abundance and the resulting decrease in the mean molecular weight; according to Eq. (72) this gives rise to a *negative* contribution to $\delta\nu_{nl}$; in the partially mixed model, the variation in X is smaller, c is everywhere a decreasing function of r, and consequently $\delta\nu_{nl}$ is higher. Also, it should be noted that for the normal model, d_{nl} is roughly consistent with the observed values (except at the highest frequencies, where the observational errors are rather large); this, however, is certainly not the case for the mixed model. Thus, mixing as severe as that proposed by Schatzman et al. appears to be ruled out by the observed frequencies (see also Cox and Kidman 1984; Provost 1984; Christensen-Dalsgaard 1986; Berthomieu and Provost 1987).

It is evident from Fig. 25 that there is still considerable scatter in the observed values of d_{nl}. For this reason, it is common to carry out comparisons between theory and observations in terms of an average frequency separation, such as the quantity D_0 (cf. Eq. 73), which has been measured with considerable precision (see, e.g., Jiménez et al. 1988e; Gelly et al. 1988b). However, as the asymptotic relation Eq. (72) is not exact, the value of D_0 depends on how it is obtained. Here we use a least-squares fit (cf. Scherrer et al. 1983; Christensen-Dalsgaard 1988a) to the frequencies of modes of degree $0 - 3$, including those for which $17 \leq n + 1/2l \leq 29$, corresponding in frequency to the range between approximately 2500 and 4100 μHz. For each value of l the frequencies are fitted to a polynomial of second order in $n + 1/2l - 23$; the coefficients in this polynomial are then in turn fitted to linear functions of L^2, which leads to a determination of D_0, as well as an average value $\Delta\nu_0$ of the overall frequency spacing $\Delta\nu$ (cf. Eq. 69).

We have applied this analysis to the models discussed in the preceding section, to a model utilizing the so-called MHD equation of state (for Mihalas, Hummer and Däppen; cf. Hummer and Mihalas 1988; Mihalas et al. 1988a; Däppen et al. 1988a; see also the chapter by Däppen et al.), and to a set of observed frequencies. The results are shown in Table II. This immediately confirms that the frequency spacing, as reflected in D_0, is too large in the Schatzman et al. model. For the remaining models there is some scatter in the values of D_0, the general tendency being that the computed values are slightly higher than the value obtained from the observations. It should be noted, however, that this tendency is not evident in the comparison of d_{nl} shown in Fig. 25; thus it may to some extent be caused by effects of the fit, possibly related to systematic errors in the dependence of the observed frequencies on l.

It has been pointed out (see, e.g., Steigman et al. 1978; Spergel and Press 1985; Faulkner and Gilliland 1985) that the presence in the solar core of a very small population of hypothetical "weakly interacting massive particles" (WIMPs) could contribute to the energy transport in the core and hence lower the central temperature. The reduction in the central temperature, which occurs without substantial modifications in the composition profile,

TABLE II
Average Frequency Separations[a] for a Number of Solar Models and the Compilation of Observed Frequencies

Frequencies	$\Delta\nu_0$	D_0
Normal model	136.46 μHz	1.511 μHz
MHD equation of state[b]	136.68 μHz	1.519 μHz
Opacity increase at surface (cf. Fig. 17)	135.70 μHz	1.505 μHz
Opacity increase in interior (cf. Fig. 19)	136.65 μHz	1.481 μHz
LAOL opacity (cf. Fig. 21)	135.24 μHz	1.496 μHz
Schatzman et al. mixed model (cf. Fig. 23)	136.68 μHz	1.976 μHz
Duvall et al. (1988a) observed frequencies	135.15 μHz	1.487 μHz

[a] See Eqs. (69) and (73).
[b] See the chapter by Däppen et al.

would also lead to a reduction of the sound speed in the core, and hence, according to Eq. (74), to a reduction in D_0 (Faulkner et al. 1986; Däppen et al. 1986; Gilliland and Däppen 1988). By choosing the parameters for the WIMPs appropriately, it is possible to construct models for which the predicted neutrino capture rate is consistent with the observed values. In these models, D_0 is typically reduced by 8 to 15% relative to the corresponding normal models. Faulkner et al., Däppen et al. and Gilliland and Däppen argued that this apparently led to an improvement in the agreement between the computed and the observed values of D_0. In fact, the computed values presented in Table II are somewhat smaller than the values for the normal models used by those authors, the difference being probably due to an improvement in the numerical precision of the model and the frequency computations (cf. Sec. V). Although we have not attempted to compute models with WIMPs with comparable precision, it appears likely, based on the changes in D_0 induced by the WIMPs in the earlier calculations, that for such models D_0 would be significantly *lower* than the observations. Certainly there is no evidence in the present results that modifications of the "standard" model are required to bring theory and observations into agreement on D_0. The same conclusion was reached by Cox et al. (1989) on the basis of detailed model calculations.

The computed values of $\Delta\nu_0$ are generally somewhat higher than the observations. Indeed, for these models the differences between the observed and the computed frequencies at all degrees have a general negative slope as functions of frequency (see, e.g., Christensen-Dalsgaard and Gough 1984; Christensen-Dalsgaard et al. 1988b; see also the chapter by Däppen et al.), which is probably associated with errors in the outermost layers of the model; this evidently corresponds to $\Delta\nu_0$ being too large in the models. When the opacity is artificially increased in the surface layers, or when the LAOL opacities are used, the computed $\Delta\nu_0$ is reduced to close to the observed value. As shown in the chapter by Däppen et al., this also reduces the general frequency dependence of the differences between observations and theory; it should be recalled, however, that there are other serious uncertainties in the treatment of the surface layers, which could affect $\Delta\nu_0$ at this level. In any case, it is apparent from these results that $\Delta\nu_0$ is predominantly a measure of conditions in the outer parts of the Sun; this is consistent with the asymptotic expression (70), as much of the contribution to the integral comes from near the surface, where the sound speed is low (see also Gough 1986b).

C. Analysis of *g*-Mode Frequencies

Just as in the case of *p*-modes, it is convenient to analyze computed (or, when available, observed) *g*-mode frequencies in terms of their asymptotic properties, discussed in Sec. III.C. In this section, we consider to what extent the asymptotic relation for the periods approximates computed periods, and how the parameters of this relation can be determined from a set of periods. To do so we analyze a set of computed g-mode periods for a normal model. Furthermore, results are presented for modified models, to illustrate the sensitivity of the g-mode parameters to the properties of the model.

Estimates of the coefficients P_0, V_1, V_2 and θ can be obtained by means of least-squares analysis of sets of computed periods. A least-squares fit of Eq. (78) to periods of gravity modes at a given degree l, with moderate or large radial order, gives values of P_0, θ_l and $V_l = L^2 V_1 + V_2$. Given V_l for several values of l, values of V_1 and V_2 are obtained from a separate fit. In our analysis, we mostly assume θ to be constant; this evidently limits the accuracy of the determination.

The asymptotic formula is valid in principle only for sufficiently large periods; hence there is a lower limit for the period below which the higher-order terms, as well as the frequency dependence of θ, are too large to be neglected. Moreover, to allow the determination of the second-order coefficient $V_l = L^2 V_1 + V_2$ (or V_1 and V_2), the relative departure of P from the first-order linear behavior must be sufficiently large relative to the numerical error (or noise in the case of observed periods). This is well satisfied for the computed modes which we consider.

A final limitation of the asymptotic analysis is that it is valid only in the Cowling approximation. In fact, as shown in Fig. 26, the effect of the perturbation in the gravitational potential on the period is proportional to $1/P$ and is much smaller than the second-order terms. Thus it is reasonable to apply the asymptotic expression to periods computed from the full set of equations, or to observed periods. This is in contrast to the p-mode case, where the effect of the perturbation in the gravitational potential for low-degree modes is comparable with the second-order term in the asymptotic expansion, as measured, for example, by the quantity D_0 (cf. Fig. 14).

Because the asymptotic expression is not exactly satisfied, the values of P_0, V_1, V_2 and θ derived from least-square analysis depend on the set of periods which is used. Thus we need to investigate the sensitivity of the results on the mode selection. We consider sets of modes with degrees $l = 1$, . . . ,4, for various ranges n_{min} to n_{max} in the radial order at each l. This gives a collection of estimates of P_0, V_1, V_2 and θ; their scatter around the mean values gives an indication of the extent to which the asymptotic behavior is satisfied. We have used three different choices of ranges:

1. Mode set (i): vary n_{min} from 10 to 15 and n_{max} from 27 to 35 for all l; the resulting mean values of P_0, V_1, V_2 and θ are given in Table III, in the line labeled "normal model (*);"
2. Mode set (ii): n_{min} between 10 and 15 and n_{max} between 27 and 35 for $l = 1$; n_{min} between 10 and 15 and n_{max} between 51 and 59 for $l = 2,3,4$. In this

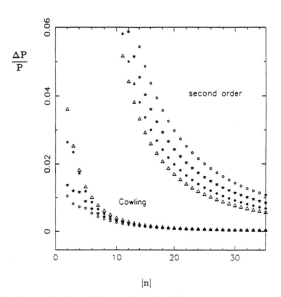

Fig. 26. Relative changes in the periods as a function of the radial order n describing the effect of the perturbation of the gravitational potential (lower set of points), and the importance of the second-order terms in asymptotic formula (78) (upper set of points). The symbols indicate the degrees of the modes: $l = 1 (\triangle)$; $l = 2 (\bullet)$; $l = 3(\star)$; $l = 4(\circ)$.

TABLE III

Asymptotic g-Mode Parameters Obtained from Least-Squares Fitting to Computed Periods for Modes of Degree 1 to 4[a]

Model[a]	P_0	V_1	V_2	θ
Normal model (*)	35.58	0.428	6.18	0.215
Normal model (**)	35.56	0.432	5.71	0.169
Normal model	35.55	0.441	5.68	0.167
Normal model[b]	35.55	0.441	5.25	0.167
MHD equation of state	35.95	0.437	5.56	0.168
Opacity increase in interior	35.54	0.435	5.58	0.161
LAOL opacity	35.45	0.458	6.06	0.162
Schatzman et al.[c]	54.36	0.173	3.78	0.157

[a]The cases normal model (*) and normal model (**) used mode sets (i) and (ii), respectively, whereas for the remaining cases mode set (iii) was used.
[b]With Cowling approximation.
[c]Partially mixed model.

way, the period ranges are more nearly the same for all l. The resulting
mean values of the parameters are labeled "normal model (**)" in Table
III;

3. Mode set (iii): n_{min} between 16 and 20 and n_{max} between 30 and 35 for
   $l = 1$; n_{min} between 30 and 35 and n_{max} between 51 and 59 for $l = 2, 3, 4$.
   The resulting mean values of the parameters are labeled "normal model"
   in Table III.

The quantities P_0 and V_1 are very similar in all three analyses. V_2 and θ are
similar for mode sets (ii) and (iii), θ being close to the expected value of
1/6, whereas they are lower than for mode set (i) by about 8% and 25%,
respectively. Finally, we find that the scatter of the values V_2 and θ are smaller
for mode set (iii) than for mode set (ii). Unless otherwise noted mode set (iii)
is therefore used for all the other computations reported in this section.

The uncertainty in the estimates of the quantities P_0, V_1, V_2 and θ can be
visualized by inspecting the histograms of the relative departure of the results
from their mean values. They are presented in Fig. 27 for a normal model,
for an analysis using mode set (iii). This shows that the two global quantities
P_0 and V_1 are insensitive to the precise set of modes selected. Moreover, the
values obtained from the fit are in reasonable agreement with the values

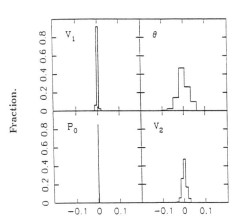

Fig. 27. Determination of P_0, V_1, V_2 and θ by least-squares analysis from computed g-mode
periods. For all these quantities, the ordinate represents the percentage of values falling in a
given range, when performing a least-squares analysis with n_{max} varying from 30 to 35 and n_{min}
from 16 to 20 for $l = 1$ and n_{max} from 50 to 60 and n_{min} from 30 to 35 for $l = 2, 3, 4$, and
different weighting functions. The abscissa represents the relative departure from the mean
values: $P_0 = 35.55$ min, $V_1 = 0.44$, $V_2 = 5.7$ and $\theta = 0.167$.

$P_0 = 35.56$ min and $V_1 = 0.402$ computed from the expressions (79) and (80).

It should be noted that the coefficient V_1 is sensitive to the description of the very center of the model, in particular to the smoothness of quantities like the gradient of molecular weight, which affects the behavior of N. When the physical quantities are well behaved near the center, the value of V_1 obtained by this analysis is insensitive to the precise set of periods considered; otherwise there may be a considerable scatter.

The results obtained for V_2 are scattered by $\sim 4\%$ around the mean value 5.7. For θ the scatter is 7% around 0.167 (which is close to the asymptotic limit of 1/6). As noted above it is substantially smaller than for mode sets including modes of lower radial order. The latter result is due to the variation of θ with frequency and degree. For example, the scatter in V_2 and θ is much larger for mode set (i); in this case the mean value of θ, 0.215, is also quite far from the asymptotic limit. Good results for the fit have been obtained by Ellis (1987, 1988), for modes of lower radial order, but utilizing a simplified model with a polytropic convection zone where $N = 0$. In this case, it can be shown that the variation of θ with ω is not as large (Provost and Berthomieu 1986). Ellis (1988) also studied the effect on the fit of noise in the periods; the conclusion was that the second-order parameters could only be faithfully estimated for a relative noise level below 10^{-4}.

We may also consider, for each value of the degree l, the set of estimates obtained for θ when the range in order is varied, as well as the resulting mean value θ_l which depends on l. This is illustrated in Fig. 28a, for mode set (ii), and in Fig. 28b, for mode set (iii). Contrary to what might have been expected, the scatter around the mean is generally larger than in the case where all degrees were combined. The dispersion around θ_l is due partly to the dependence of θ on ω, although there may also be a contribution from third-order terms. For $l = 1$ the scatter is smaller for mode set (iii), where the radial order is higher; however, somewhat surprisingly, this trend is reversed for higher l. Also we note that, contrary to expectations based on Eq. (83), the values of θ_l generally decrease with increasing l; this behavior is less evident for higher values of n_{min} as shown in Fig. 28b. There is no obvious correlation between this behavior and the scatter in the combined case. It is evident that further work is required to interpret these results. However, they indicate that it is necessary to deal with high radial order modes of different degrees to obtain good determinations of the asymptotic constants.

To test the effect of the Cowling approximation, a least-squares analysis was performed with a set of modes computed in this approximation. As indicated in Table III, this gives essentially the same values of P_0 and V_1 while the values of V_2 is lower by $\sim 9\%$. Hence the perturbation in the gravitational potential has a moderate effect on the asymptotic fit. This is evidently consistent with the comparison made in Fig. 26.

We have considered the effects on the g-mode parameters of the various

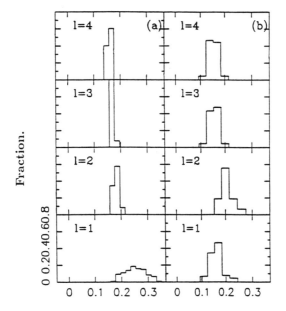

Departure from mean.

Fig. 28. Determination of θ_l by least-squares analysis from computed g-mode periods of given degree l, for the normal solar model; (a) mode set (ii); (b) mode set (iii). The percentage of values falling in a given range, when performing a least-squares analysis varying n_{max}, n_{min} is plotted relative to the value of θ and shows a decrease of the mean value θ_l with l.

modifications to the model which were also considered in the preceding sections. The use of the MHD equation of state (cf. Sec. VI.A) leads to a slightly higher value of P_0, with little change in the remaining parameters. The model with artificially enhanced interior opacity is very close to the normal one, whereas using LAOL opacities leads to a small decrease in P_0, and to slight increases in V_1 and V_2. Thus, it appears that modifications to the physics, within the framework of "standard" models, have little effect on the properties of the g-modes.

On the other hand, P_0 and V_1 are very sensitive to the behavior of the composition, through its effect on the buoyancy frequency, near the center of the Sun (Berthomieu et al. 1984a,b). Thus observational determination of these quantities would provide two constraints on the deep interior of solar models. This is potentially very important in allowing a distinction to be made between the different nonstandard models which have been constructed to satisfy the neutrino constraint. As an example, Table III shows results for the partially mixed model, approximating the $Re^* = 100$ model of Schatzman et al. (1981), which was discussed in Sec. VI.A (cf. Fig. 23). Here P_0 is increased to 54 min, whereas V_1 is very substantially reduced. These results are related to the behavior of the quantity $\tilde{N}/x$ which enters in the determination of P_0 and V_1 in equations (79) and (80); the central value decreases from ~ 90 in the normal model to ~ 28 in the partially mixed one. Moreover, V_2 is much lower in the partially mixed model, while θ, which is predominantly determined by conditions near the base of the convection zone, is barely affected.

The value of V_1 is extremely sensitive to the details of the assumed hydrogen abundance profile near the center. We have found that slight variations in the $X(q)$ which was fitted to the Schatzman et al. curve resulted in substantial changes in V_1. This can in fact be understood from the asymptotic expression in Eq. (80), because V_1 is computed as the difference between two quantities that are singular at the center. Thus an observational determination of V_1 would provide a very sensitive probe of the evolutionary history of the Sun.

Models with WIMPs (cf. Sec. VI.B) have a quasi-isothermal central core; relative to a normal model, this increases N in the core, and hence the value of P_0 is lower. Faulkner et al. (1986) and Däppen et al. (1986) obtained values around 29 min.

Considerable uncertainty surrounds the region just beneath the convection zone. This includes the extent of the overshooting region, and the possible presence of a magnetic field concentration. Thus observational information about this region is badly needed. The asymptotic value of θ is directly a measure of ζ and thus of the variation of the temperature gradient. In all the models considered here, N^2 varies linearly with r just below the base of the convection zone, and hence $\zeta = 1$. Despite the uncertainty in the determination of θ from the computed periods, the estimates are in reasonable agreement with the predicted value of 1/6, whereas they are much further from, for example, the value of 0.125 which would correspond to a parabolic behavior of N^2 at the lower boundary of the convection zone. It would be of obvious interest to consider models with a different behavior of N^2 in this region, due to, for example, the presence of a magnetic field.

VII. CONCLUSION

The principal feature of the theory of adiabatic solar oscillations is its simplicity. For a given solar model, it is straightforward to compute oscillation frequencies with a precision that at least approaches the precision attainable in the observed frequencies. Furthermore, it is possible to relate the behavior of the frequencies, as functions of mode degree and order, to the properties of the solar interior. It is these features that make the oscillation frequencies such powerful probes of the solar interior.

It is true that the theory of solar oscillations, as presented here, is over-simplified. We have neglected effects of nonadiabaticity and other processes that damp or excite the modes; these may change the frequencies by several μHz (see, e.g., Christensen-Dalsgaard and Frandsen 1983; Cox et al. 1989). Furthermore, there are bound to be dynamical interactions between the oscillations and convection in the uppermost layers of the convection zone, where the convective time scale is comparable with the oscillation period, and where the Mach number of convection is not small. Finally, the inhomogeneity of the solar atmosphere, and the presence of magnetic field concentrations, may well have significant effects on the frequencies. However, for the study of the solar interior these effects may to a first approximation be eliminated. They occur in a region whose vertical extent is small compared with the horizontal wavelength of most of the modes observed, and where these modes propagate predominantly vertically. Hence they affect all modes of a given frequency equally, regardless of the degree l, provided that the l dependence of the mode inertia is taken into account. Thus the uncertainties near the solar surface translate into an undetermined frequency-dependent component of the frequencies, which is common to all modes; if allowance is made for this, the dependence of the frequencies on l can be used to study how the properties of the solar interior depend on radius.

Irrespective of this, it is evidently important to improve our understanding of the uppermost layers of the Sun and their influence on the frequencies. Some aspects of this problem are discussed in the chapter by Cox et al. Encouraging progress has been made recently both in the application of simplified models of convection to the study of oscillations (Balmforth and Gough 1988,1990b), and in detailed hydrodynamical simulations of granulation (Stein and Nordlund 1989; Stein et al. 1988a,1989b). Eventually we may hope to use observations of solar oscillation frequencies, line widths, amplitudes and phase relations to test models for the excitation of oscillations and for the interaction between oscillations and convection. This would be of obvious benefit to studies of other types of pulsating stars.

Within the framework of "standard" solar evolution theory, the structure of the model, and hence the oscillation frequencies, are determined by the assumed physics of the solar interior, as well as by the global parameters of the model. Thus in this sense the frequencies can be regarded as providing information about the physics. As such, they are sensitive measures. It was demonstrated in Sec. VI that changes in opacity that are well within the range of uncertainty in current opacity calculations lead to frequency changes that are large compared with the internal precision in the measurement or the calculation of the frequencies, and which have a recognizable signature. Similarly, it has been found that improvements in the equation of state may have substantial effects on the frequencies, generally leading to an improved agreement between theory and observation (Christensen-Dalsgaard et al. 1988b; see also the chapter by Däppen et al.). Thus it appears that, particularly with the vastly improved observational data that can be expected in the coming decade, we may be able to use the Sun as a laboratory for testing our ideas about the properties of matter. This is of obvious importance for the application of these properties to the computation of models of other stars, but it may also provide basic physical insight into the behavior of hot and dense plasmas.

Of course this assumes that the Sun is similar to a normal solar model. It is possible that a combination of nonstandard features could mimic an error in the physics of a standard model, and hence fool us completely. More likely, however, substantial departures from the standard assumptions would be recognizable in the observed frequencies. For instance, departures from a spherically symmetric structure would give rise to a definite signature in the m dependence of the frequencies. Nevertheless, when interpreting the observed data we should keep in mind that the Sun may have surprises in store for us.

Acknowledgments.   We are grateful to W. Däppen, D. O. Gough, Å. Nordlund, J. Provost and M. J. Thompson for useful discussions. We thank A. Kosovichev for pointing out a mistake in some earlier calculations, and Y. Lebreton for providing the Los Alamos Opacity Library opacities. A. N. Cox, W. Däppen, J. A. Guzik, M. J. Thompson and the anonymous referees are thanked for careful reading of earlier versions of the manuscript; their perceptive comments led to a considerable improvement of the text. The computations reported here were partly supported by the Danish Natural Science Research Council and the French C.N.R.S.

THE SOLAR ROTATION

K. G. LIBBRECHT
Big Bear Solar Observatory

and

C. A. MORROW
University of Colorado

We review in this chapter the rotation of the Sun, with emphasis on the obser-vations. While it has been known for centuries that at the solar surface the equator rotates with a higher angular velocity than the poles, different surface measurements suggest that the near-surface rotation is not as simple as one might expect. Recently have we been able to infer the interior rotation rate from helioseismology. These new observations indicate that surface-like rotation ex-tends through the convection zone, changing to solid-body rotation in the upper part of the radiative interior; observations are inconclusive regarding the rota-tion of the deep core of the Sun. The dynamics that produces the observed solar rotation profile remains a mystery.

I. INTRODUCTION

We review our knowledge of the surface and interior rotation of the Sun, with emphasis on current observations. All of the various measurements of the latitudinal differential rotation of the solar surface agree nicely at about the 2 to 3% level, but they provide a confusing picture when viewed more closely. There is little clear agreement among the observations below the 2% level, and no satisfactory explanation exists to explain these discrepancies. Indeed, it is difficult to determine what fraction of the differences between the measurements is of solar origin, and how much is due to unknown systematic errors in the different observing techniques. We note below, however, that application of new measurement techniques may soon considerably reduce the confusion.

Helioseismology measurements have recently begun to provide a picture of the Sun's angular velocity as a function of both depth and latitude. The rate of improvement in this picture has been swift, and is likely to remain so in the near future. Helioseismology has the potential to provide a nearly complete picture of the rotation throughout the solar interior, thus offering very powerful new constraints for modeling the way convection, rotation and magnetic fields interact to produce differential rotation and the solar dynamo. While computer simulations of the solar convection zone have been able to reproduce the Sun's differential rotation at the surface, at present these models fail to generate the helioseismically determined internal rotation. Thus, the detailed physical processes that transport angular momentum to establish and sustain the Sun's differential rotation remain a mystery. Clearly our understanding of turbulent convection and dynamo action in the Sun is incomplete.

We must apologize at the outset for the variety of units used to describe the solar rotation. In the past, degrees per day seems to have been the preferred unit of angular velocity, which was replaced in more modern times by microradians per second (μrad s^{-1}). Helioseismologists, including both the authors, tend to use nanohertz (nHz) as the natural unit of what one might prefer to call a rotational frequency rather than an angular velocity. Where possible we have converted to nHz, but some of the figures taken from the literature retain the other units. To establish the scales: 460 nHz = 2.890 μrad s^{-1} = 14.31 deg day^{-1}.

II. SOLAR SURFACE ROTATION

The rotation of the solar photosphere can be measured using a variety of techniques, including: frequency shifts of short-wavelength solar acoustic oscillations, the Doppler shift of photospheric spectral lines, and the tracking of tracers, such as sunspots, faculae, low-level magnetic features, supergranulation cells, the Ca K network, etc. (see the chapter by Howard et al.). One might expect all the measurements to give the same result, and within a few percent that is indeed the case. Each reveals the Sun's differential rotation as a function of latitude, which is approximately given by a sidereal rotational frequency $\nu(R_\odot, \phi) = 462 - 75\sin^2\phi - 50\sin^4\phi$ nHz, where ϕ is the solar latitude, and $R_\odot$ is the solar radius. This formula implies an equatorial rotational period of 25 days and a polar period of ~34 days. At a precision of better than a few percent, however, different techniques give different rotation rates.

Sunspot Tracers

The use of sunspots to trace the solar surface rotation has a long history, and until recently the classic work of Newton and Nunn (1951), giving a sunspot rotation rate of $\nu(R_\odot,\phi) = 462.3(\pm 0.3) - 95(\pm 2.9)\sin^2\phi$ nHz, was taken as the standard measure of solar rotation. Howard et al. (1984) have since assessed the rotation of sunspots by analyzing Mount Wilson's collection of white-light plates covering the period from 1921 to 1982. Figure 1 shows their resulting rotational profiles, determined as a function of sunspot size.

Note from this figure that the measurements extend only to latitudes equatorward of ± 35 deg because there are no sunspots nearer the solar poles. We also see that small spots rotate approximately 2% faster than large spots, and spot groups rotate at an intermediate rate that agrees well with the recurrent sunspot rate from Newton and Nunn (1951), but depends to some extent on how one defines a spot group (Howard et al 1984). Because very few long time series of solar data like the Mount Wilson collection exist, the results in Fig. 1 have not been confirmed in every detail. However, it is likely that the systematic errors in the measurements are not severe. The rotation of spot

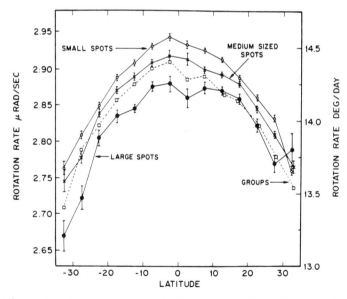

Fig. 1. Sidereal rotational rates for sunspots of various sizes, along with the rate for sunspot groups (from Howard 1984). Small, medium and large refer to spots smaller than 5, between 5 and 15, and larger than 15 millionths of the area of the solar disk, respectively. As discussed in the text, much of the variation in the angular velocity with spot size may be related to the local dynamics and proper motions of sunspots.

groups is in agreement with similar measurements using white-light photographs from the Greenwich Observatory (see Howard 1984).

Unfortunately, it is difficult to interpret the results in Fig. 1 without a
better understanding of the dynamics of sunspots. For example, solar observers know well that when a new bipole, consisting of a leader and follower
sunspot, emerges in an existing sunspot group, the follower spot remains
roughly stationary with respect to the group, while the new leader spot moves
rapidly away from its partner. Indeed, detailed measurements of sunspot motions show that follower spots rotate at roughly the sunspot group rate (which
is essentially the same as that for isolated single spots) while leader spots
show up to a 4% faster rate of rotation (Neidig 1980; Gilman and Howard
1985, 1986). This is clearly a significant dynamical feature of sunspot
groups, which is hidden by the simple separation of spots by size alone. If
one better understood why sunspots exhibit this and other types of internal
proper motion, then one might be able to explain the results in Fig. 1 in
detail. Conversely, without a proper understanding of the details of sunspot
dynamics and proper motions it is likely that the global properties in Fig. 1
will remain enigmatic.

Sunspots have also been used to look for variations in the solar surface
rotation with time, and in particular for a solar cycle dependence. One was
found in the Mount Wilson data by Gilman and Howard (1984b), and is
displayed in Fig. 2. Note from this figure that the peak-to-peak variation in
the rotation rate is about 1%. This result was recently confirmed using the
Greenwich sunspot measurements (Tuominen and Virtanen 1987). Additional
analysis of the Mount Wilson measurements shows a correlation of rotation
rate with sunspot number (Hathaway and Wilson 1990). These observations
are still without an adequate explanation.

Doppler Shift Measurements

Determining the Doppler shift of solar spectral lines across the solar disk
is one of the most obvious ways to try to measure the solar rotation, but it is
also one of the most controversial. The controversy originated with one of
the first modern Doppler measurements made by Howard and Harvey (1970),
who used data taken in 1966–68 with the Mount Wilson magnetograph. They
found the average rotational frequency to be $\nu(R_{\odot},\phi) = 442 - 56\sin^2\phi - 70\sin^4\phi$ nHz, which is approximately 4% slower than the rate for sunspot
groups. This is a significant difference, amounting to about 80 m s^{-1} in rel-
ative velocity at the equator. Day-to-day scatter in their data was as large as
10%, but improvements in the apparatus led to reductions in the daily variations to of order 1%; these fluctuations remain to this day and may be of solar
origin (Howard 1984). Nevertheless, even with the reduction in the scatter,
the angular velocity determined using Doppler shifts remained ~2% slower
than the rotation rate for sunspot groups.

This puzzling result prompted others to repeat the measurements, but

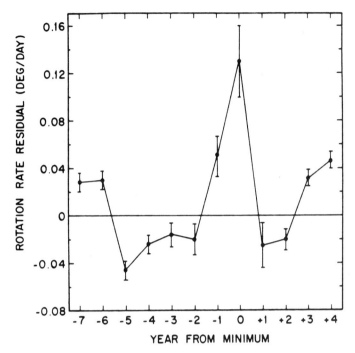

Fig. 2. Superposed epoch plot of residual rotational velocity of sunspots in the latitude interval
$-30 < B < 30$ deg, as a function of years from solar minimum (from Howard 1984). These
data represent an average over the 62-yr period 1921–82.

the results were mixed. Livingston and Duvall (1979) used Doppler data from
Kitt Peak to find that the equatorial rotational rate, $v_{eq} = v(R_{\odot}, \phi = 0)$, increased with time from ~438 nHz in 1966 to ~454 nHz in 1978, a 3.6%
increase in 12 yr. These rates and the temporal increase were in fairly good
agreement with independent Doppler measurements from Mount Wilson covering the period up to 1976, although the noise in both data sets was not
small. However, additional Mount Wilson measurements (see Howard 1984)
revealed that the agreement between these two data sets in 1977–78 was quite
poor, suggesting that at least one of them was somehow being contaminated
with a spurious solar or instrumental signal.

Duvall (1982b), using separate measurements made at Kitt Peak in
1978–80, found $v_{eq} = 454.5 \pm 1.3$ nHz, which supported the notion that
the Doppler-determined rate was indeed 1 to 2% lower than that of sunspot
groups. On the other hand, the Stanford group (Scherrer et al. 1980b) measured a Doppler rate in 1976–79 that did not vary with time and that was
within 1% of the sunspot group rate, even after a dispersion correction (see
Howard 1984).

Foukal (1979) set out to measure spectroscopically the relative velocities of sunspots and the surrounding photospheric gas directly. He found a relative velocity of 148 ± 31 m s^{-1}, an extremely large difference. Here again we must be careful in accepting this result at face value, as high-resolution Doppler images of active regions often display a great deal of velocity structure. Without a good understanding of sunspot dynamics, the meaning of Foukal's result may not be so clear.

Given the large scatter in all the data sets, plus the lack of agreement between observers, it cannot be said with any great certainty that there is a real solar difference between the rotation of the surface gas and the sunspots. If the difference is real, then we have no completely satisfying theoretical explanation for it.

The Mount Wilson Doppler observations have also revealed a torsional oscillation (see Howard 1984; chapter by Howard et al.), consisting of bands of abnormally high rotational velocity situated alongside the active latitudes. The amplitude of this phenomenon is small, only of order 5 m s^{-1} or 0.25% of the underlying rotation. This observation has not been confirmed by any other measurements (however, see Howard 1984), and there is to date no natural theoretical reason to expect it.

Other Techniques

Using full-disk magnetic images provides a good way to determine the solar rotation, as less-concentrated magnetic features can be seen all the way up to the poles, and the small features make excellent tracers. The Mount Wilson magnetograph data have been used in this capacity to generate a new "standard" surface rotational profile (Snodgrass 1983) based on a 2-day autocorrelation lag: $v(R_\odot,\phi) = 461.9(\pm 0.3) - 73.8(\pm 2.9)\sin^2\phi - 52(\pm 5)\sin^4\phi$ nHz.

A particularly interesting technique for measuring the solar surface rotation is that of using supergranular cells as tracers. The measurement is difficult because the individual cells are short lived (with lifetimes of order 1 day), and because the predominantly horizontal velocities of supergranular cells cannot easily be seen at disk center. Nevertheless, because the technique uses a tracer it is not susceptible to the systematic errors that may plague the Doppler method, and the ubiquitous supergranules are not magnetic and thus should not suffer from the unusual proper motions that can confuse sunspot measures of rotation.

Duvall (1980) measured the solar surface rotation using supergranules, and found $v_{eq} = 473 \pm 2.3$ nHz. This value is close to and certainly not less than the sunspot rate of 462 nHz. Because the supergranular cells are tied very closely to the photospheric gases, one is again suspicious of the relatively slow angular velocities determined from Doppler measurements. The supergranule result should agree with that obtained using the Ca K network as a tracer, as the network and the supergranules are closely connected. How-

ever, Ca K measurements have produced results that show a large time dependence (see Howard 1984). Clearly, both of these important measurements
should be repeated, preferably using data that span a longer time base.

It is possible to reconcile all the measurements mentioned above, albeit
with an *ad hoc* model. Consider the following simple explanation offered by
Foukal and Jokipii (1975), and further explored by Gilman and Foukal
(1979). If specific angular momentum were conserved during the radial transport of material in supergranulation, then the rotation rate would go like
$v \sim r^{-2}$, and therefore v at the surface would be a few percent lower than at
the base of the supergranular convection. If sunspots and the supergranular
cells were both somehow tied to a level at the base of the supergranular convection, then the rotation assessed from Doppler shifts, sunspot motions and
supergranule drift would all fit together consistently. Add random motion
from giant cells, and even some of the time dependence would be explained.
Unfortunately this model is very *ad hoc,* and is therefore quite unsatisfactory
as a true explanation of the surface rotation data. One does not know that
specific angular momentum is conserved in supergranular motions—certainly
it is not conserved throughout the convection zone, as the interior measurements below indicate—nor can one say how sunspots should be "tied" to the
base of the supergranular cells (however, see Schussler 1987*a*).

Fortunately there is yet another technique for measuring surface rotation, which has not been fully exploited, and which may shed a great deal of
light on this subject in the near future. The technique involves using the
rotational frequency splittings of solar acoustic oscillations with very short
wavelength (i.e., high spherical harmonic degree l) to determine the interior
angular velocity as a function of depth very near the solar surface (the helioseismological approach will be described in more detail in Sec. III). In this
way, one obtains a true measurement of the gas velocity, as it is the gas that
carries the acoustic waves. The acoustic wave fronts are effectively tracers
that can be observed near the disk center (as one would prefer to do). This is
an ideal technique, and preliminary work by Hill (1988*a*), Rhodes et al.
(1990) and Woodard and Libbrecht (1988) is promising. Unfortunately, the
high-l splittings have proven difficult to measure accurately, and so it is still
too soon to have confidence in the early results.

Summary

The above discussion raises many unanswered questions about the angular velocity observed at the Sun's surface. We know the solar surface rotation at about the 3% level, where all the measurements agree. At the 1%
level, the various sunspot measurements agree well (given our meager understanding of detailed sunspot dynamics), but we also start to see differences
between sunspot and Doppler rates. Below the 1% level, the scene is very
confusing, with time variations in all the measurements, and probably a host
of systematic errors which confuse the results. With only a modest theoretical

understanding of the way in which rotation, convection and magnetic fields interact, it is extremely difficult to make complete sense of the observations. Helioseismic measurements should soon provide the most accurate measurements of the near-surface rotation.

III. SOLAR INTERIOR ROTATION

Until it was realized that frequency shifts in solar acoustic oscillations (p-modes) could be used to measure the rotation rate of the solar interior (Deubner et al 1979), we could do little more than speculate on the differential rotation as a function of both depth and latitude. In the last 10 yr, there has been a great push to produce very accurate measurements of p-mode frequency splittings, and to invert the measurements to infer the Sun's internal angular velocity. Early results were crude, and only recently have such helioseismological approaches advanced to the stage where the interior rotation is determined to an accuracy of a few percent through most of the convection zone, and down to a radius of $\sim 0.4\,R_\odot$. A taste of the latest rotation results is presented here, and a detailed discussion is given in the chapters by Dziembowski and Goode and by Gough and Thompson. However, the reader should be advised that helioseismology is developing very rapidly. The most recent literature should be consulted for the best current opinion on the Sun's internal rotation.

The Concept

The global, acoustic, normal-mode oscillations of the Sun (p-modes) have been reviewed elsewhere (see, e.g., the chapters by Hill et al. and by Christensen-Dalsgaard and Berthomieu; Voronstov and Zharkov 1989; Libbrecht 1988b). The p-mode frequencies ν_{nlm} are characterized by three "quantum numbers": n is the quantum number for the radial part of the wavefunction, and l and m are the degree and azimuthal order which come from the spherical harmonic part of the wavefunction $R_n(r)Y_l^m(\theta,\phi)$. If the Sun were perfectly spherically symmetric, then the oscillation frequencies would be degenerate in azimuthal order m. Rotation is one mechanism that lifts this degeneracy, making the frequencies m-dependent. Multiplets with a fixed n and l are thus said to exhibit a frequency splitting, $\Delta\nu_{nlm} = \nu_{nlm} - \bar{\nu}_{nl}$, somewhat analogous to the Zeeman or Stark splitting of the degenerate energy levels of an atom. The splitting caused by solar rotation is 2 orders of magnitude larger than that from other physical causes, such as magnetic fields (Gough and Thompson 1990) or large-scale temperature fluctuations in the Sun (Kuhn 1988b; Gough 1988a), so for the present we will neglect the frequency splitting not due to rotation.

The p-mode frequency splitting is easily seen directly in solar oscillation power spectra, and a typical example (sometimes referred to as an $m - \nu$ diagram), is shown in Fig. 3. Observers have found it useful, in fitting the

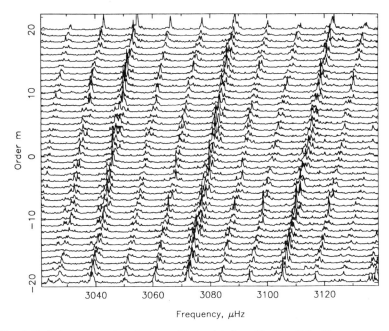

Fig. 3. Typical power spectra of solar oscillation data from Big Bear Solar Observatory; each horizontal trace is a section of a power spectrum for different m with $l = 20$. The peaks in the $m = 0$ spectrum at $\nu = 3047$, 3080 and 3114 μHz are from modes with $(n,l) = (15,19)$, $(15,20)$ and $(15,21)$, respectively. Peaks spaced ± 11.6 μHz around these features are temporal sidelobes arising from the day/night observing window. The shift in the frequency of the peaks as a function of m illustrates the rotational frequency splitting.

peaks in Fig. 3, to express the measured splittings as coefficients a_i of Legendre polynomials in m/L, where $L = \sqrt{l(l + 1)}$:

$$\Delta\nu_{nlm} = L \sum_{i=1}^{N} a_i P_i(m/L) \qquad (1)$$

where usually N $= 5$. Other expansions have been proposed (Durney 1990), but for observational reasons (principally the orthogonality of the Legendre polynomials) the above expansion is the most commonly used.

For each p-mode multiplet (i.e., for fixed n and l), the rotational splitting $\Delta\nu_{nlm}$ depends on the rotation rate of the gas in which the mode propagates. Because each mode propagates through a substantial portion of the solar interior, the splitting of each multiplet gives a weighted average of the solar rotation rate over the region sampled by the mode. If the Sun rotated uniformly $(\nu_\odot(r,\phi) = \nu_\odot)$, then the rotational frequency splitting would depend linearly on m (to first order in $\nu_\odot$):

$$\Delta \nu_{nlm} = \nu_{nlm} - \bar{\nu}_{nl} = m\nu_{\odot} \qquad (2)$$

giving $a_1 = \nu_{\odot}$. We see immediately from Fig. 3 that $\partial \nu_{nlm}/\partial m \approx 430$ nHz, which is roughly the solar (synodic) rotation rate.

If the solar rotational frequency varied with depth alone $\nu_{\odot}$ $(r, \phi) = \nu_{\odot}(r)$, then the frequency splitting would still be linear in m, but a_1 would depend on n and l. For example, at a fixed mode frequency, a_1 at large l would give the rotation close to the surface, while a_1 at lower l would reflect the rotation rate deeper inside the Sun. In general, the frequency splitting is related to the rotational frequency through an integral expression involving a rotational kernel K. For this special case of rotation constant on spheres, the relationship between frequency splitting and rotational frequency may be expressed to good approximation by

$$\Delta \nu_{nlm} = ma_1(n,l) = m \frac{\int_0^{R_{\odot}} \nu_{\odot}(r)K(r; n,l)dr}{\int_0^{R_{\odot}} K(r; n,l)dr}. \qquad (3)$$

The radial kernel K is derived theoretically from the known oscillation eigenfunctions. Gough (1981) provides an explicit expression for the full radial kernel, and Brown et al. (1986) discuss the approximation considered above. Note that higher-order odd terms such as a_3 and a_5 are still zero, which reflects the lack of latitudinal differential rotation when angular velocity is constant on spheres.

For the real solar case, in which the rotation rate depends on both radius and latitude, the frequency splittings are no longer linear in m. Terms that are cubic and of higher odd order in m are needed to describe the advective and Coriolis effects on the frequencies due to latitudinal differential rotation. For example, rotation that is more rapid near the equator than near the poles would result in a greater frequency splitting for the sectoral modes $(m = \pm l)$, and thus the top and bottom of the ridges in Fig. 3 should be slightly flared out into an S-like shape (as is the case). Hansen et al. (1977) offer a good derivation of the radial and colatitudinal rotational kernels. Using a reasonable approximation of their colatitudinal kernel gives a simple expression for the frequency splittings associated with solar angular velocity that varies with both radius and colatitude $\theta = 90° - \phi$ (Brown and Morrow 1987b):

$$\Delta \nu_{nlm} \approx m \frac{\int_{-1}^{1} \bar{\nu}_{\odot}(\theta)[P_l^m(\cos \theta)]^2 \, d \cos \theta}{\int_{-1}^{1} [P_l^m(\cos \theta)]^2 d \cos \theta} \qquad (4)$$

where $\bar{\nu}_\odot$ (θ) is the appropriate radial average of the rotational frequency using the radial kernel K in Eq. (3). As the integral over θ is even in m/L, the frequency splittings are expressible as a product of L and odd powers of m/L. Thus, the higher-order odd coefficients in the Legendre expansion (e.g., a_3 and a_5) are nonzero, while the even coefficients remain zero (to first order in $\nu_\odot$; the solar oblateness produces a nonzero a_2; see the chapter by Dziembowski and Goode).

As another example, if the observed surface rotation were to extend to all depths, $\nu_\odot$ $(r,\theta) = \nu_\odot(R_\odot,\theta)$, then all of the odd coefficients would have nonzero, constant values. For surface rotation expressed as even powers of $\cos\theta$:

$$\nu_\odot (R_\odot,\theta) = A + B \cos^2 \theta + C \cos^4\theta \qquad (5)$$

the constant values are obtained by performing the colatitudinal integration analytically (see, e.g., Morrow 1988a,c) giving:

$$a_1 = A + \frac{1}{5} B + \frac{3}{35} C \qquad (6)$$

$$a_3 = -\left(\frac{1}{5} B + \frac{2}{15} C\right)$$

$$a_5 = \frac{1}{21}C.$$

Using the surface rotation of magnetic features (Snodgrass 1983) gives: $a_1 = 442.8$ nHz, $a_3 = 21.7$ nHz and $a_5 = -2.5$ nHz, and observers have often plotted these lines along with the a_i data for comparison. Note that the equatorial rotation rate is given by setting $m = L$ in the Legendre expansion (Eq. 1), and so is just the sum of the odd coefficients, $a_1 + a_3 + a_5 = A = 462$ nHz. Note also that one could crudely estimate the surface coefficients implied by the a_i data by inverting Eqs. (6) for A, B and C, and then using the values of the a_i at the highest available degrees (Morrow 1988 a,c).

The relationships in Eqs. (6) above have the same form for a rotational frequency that depends on depth as well as latitude. In this case, the surface coefficients would be functions of radius, and the a_i would be related to the radial averages of $A(r)$, $B(r)$ and $C(r)$ weighted by the radial kernel K (as in Eq. 3), instead of to the constant values of A, B and C, as expressed above. Using the more general connections between the observed a_i and the theoretical integrals, Morrow (1988a,c) showed how to make linear combinations of the coefficients a_i that reflect the radial variations of angular velocity at any latitude. This enables one to explore the general character of the solar angular velocity implied by the data, even before proceeding with more sophisticated analyses.

Note that in all of the above, the even-order terms in m $(a_2, a_4,$ etc.) are identically zero to first order in $\nu_\odot$. Centrifugal effects (which are second order in $\nu_\odot$), as well as other causes of acoustic asphericity (e.g., latitudinal variations in magnetic field or temperature) lead to quadratic and other even-order items in m, which are discussed in the chapter by Dziembowski and Goode.

Observational Results

Brown (1985) made the first attempt to estimate the way the Sun's latitudinal differential rotation varied with depth. Many observers have since helped to refine the initially crude rotational picture by obtaining data of steadily improving quality (Duvall et al. 1986; Libbrecht 1986,1989; Brown and Morrow 1987b; Rhodes et al 1987,1990; Tomczyck 1988). The best currently published data are those of Libbrecht (1989), which are based on 100 days of observation. He measured the a_i for each nl multiplet with angular degrees $l =$ 10 to 60, and with n values in the frequency range $\nu =$ 1.5 to 4 mHz. Such data contain information about latitudinal differential rotation in a depth range of $0.4 < r/R_\odot < 0.85$. Libbrecht's odd a_i vary with both l and n due to solar rotation. In practice, the n dependence is small over the measured frequency range, providing little more information than can be gleaned from the l dependence. The n dependence is important for achieving very high-depth resolution, but one can obtain a basic picture of the Sun's internal rotation by considering just the l dependence of the odd a_i, either at constant ν or averaged over n for each l.

Figure 4 displays Libbrecht's odd splitting coefficients interpolated to a constant $\nu =$ 2.5 mHz. Note that the scatter in the a_i is only a few nHz (Libbrecht 1989), making the uncertainty in the estimate of the solar rotation rate quite small. Because these coefficients are based on observations of global oscillation modes, the systematic errors should also be small—less than a few nHz (Libbrecht 1989). Thus, with helioseismic measurements we can quickly do better for the interior rotation than we can for the surface rotation. Note immediately the bend toward smaller absolute values in the coefficients at lower l values, due to the change in $\nu_\odot$ with depth. The behavior of a_1 suggests a slight radial decline in latitudinally averaged angular velocity; a_3 and a_5 reflect a distinct decrease in the degree of latitudinal differential rotation below the convection zone.

Forward Analysis

There are generally two different formal approaches to determining the solar rotation from frequency splittings of solar oscillations: forward and inverse. Both approaches make use of the theory described earlier that connects frequency splittings $\Delta\nu_{nlm}$ to integrals involving kernels and the rotational frequency $\nu_\odot$ (r,ϕ). Inversion methods basically solve the integral equation for $\nu(r,\phi)$ using the observed $\Delta\nu_{nlm}$. Inversion techniques are reviewed in the

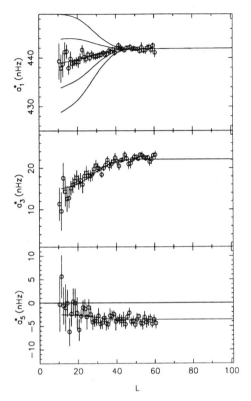

Fig. 4. Data for p-mode splitting from Big Bear Solar Observatory (Libbrecht 1988a), where here a_i^* is the splitting coefficient a_i at a mode frequency of 2.5 mHz. Along with the data, we have plotted the frequency splittings for a family of models of the solar rotation. The models have surface-like rotation throughout the convection zone (best fit with $A = 461$ nHz, $B = -60.5$ nHz, $C = -75.4$ nHz) followed by a transition to constant rotational frequency v_{ri} in the radiative interior. From bottom to top, the a_i curves correspond to rotational models with v_{ri} evenly spaced in 16 nHz intervals, from 404 to 468 nHz. This figure shows that the radiative interior, at least above $r = 0.4$ R$_\odot$ cannot be rotating at a speed much different from the surface rate.

chapters by Gough and Thompson and by Dziembowski and Goode. The forward approach uses physically motivated models for the rotational frequency to calculate frequency splittings which may then be compared to the observed splittings. Neither approach yields a unique answer because many different rotation laws may map into the same set of measured frequency splittings. Nevertheless, by making appropriate assumptions based on physical plausibility and theoretical and observational experience, much can be gained form both approaches. Forward modeling of the kind presented below is described in detail by Morrow (1988a,b). One benefit of this approach is the ability to learn how changes in the magnitude and nature of the solar rotation affect the splitting coefficients a_i.

Figure 4 shows our attempt to fit the observed splitting data using frequency splittings generated by a model that has surface-like rotation throughout the convection zone, and various values of constant angular velocity in the underlying radiative interior. The best fit values for the surface rotation coefficients and the rotational frequency of the radiative interior ν_{ri} were: $A = 461$ nHz, $B = -60.5$ nHz, $C = -75.4$ nHz and $\nu_{ri} = 436$ nHz. This suggests that the radiative interior, at least down to a radius of ~ 0.4 $R_\odot$ (see Libbrecht 1988c, and references therein), is rotating at a speed that is intermediate between the fastest and slowest rates of the surface profile. The result is generally consistent with other p-mode splitting data sets (see below). Below 0.4 solar radii, we will need much better data at very low l values to place very interesting limits on the rotation rate.

Figure 5 displays splittings for rotational models that are again surface-like in the bulk of the convection zone with solid-body rotation in the radiative interior at $\nu_{ri} = 436$ nHz; but this time each model has a thin surface layer in which we conserve specific angular momentum, giving $\nu \sim r^{-2}$ (hereafter referred to as a CAM [constant angular momentum] layer). Here, we see that the decreased surface rotation, owing to the CAM layer, affects the splitting at all l. This is because the amplitudes of p-modes of all degrees are concentrated near the surface. From top to bottom of the a_1 curves, the depth of the CAM layer is 0 to 5% of the solar radius. These results weakly suggest that there cannot be a surface layer rotating more slowly than the interior, as Foukal suggested to explain the discrepancy between the Doppler and sunspot measurements of surface rotation. However, small decrements to the value of A used for the surface-like profile in the bulk of the convection zone would have a similar effect on the a_1 curves, and so a model with some other combination of A and CAM layer depth is not completely ruled out. These data combined with higher l data will be more capable of reaching a satisfactory conclusion on the near-surface rotation.

Figure 6 attempts to illustrate that a model with angular velocity constant on cylinders throughout the convection zone is not consistent with the splitting data. The top a_l curve corresponds to a model with pervasive surface-like rotation in the convection zone; whereas the bottom curve corresponds to pervasive cylindrical rotation. The two intermediate curves represent splittings due to models which have surface-like rotation (i.e., equal to the surface rate and independent of radius) in the outer portion of the convection zone, and angular velocity "constant on cylinders" (i.e., surfaces of constant angular velocity are cylindrical) in the inner portion of the convection zone. Note that a rotation model that is constant on cylinders throughout the convection zone produces a considerable variation of the calculated a_l with degree l, which reflects the positive radial gradients of angular velocity inherent to the cylindrical profile. Such a slope is not seen in the a_l data, and so this

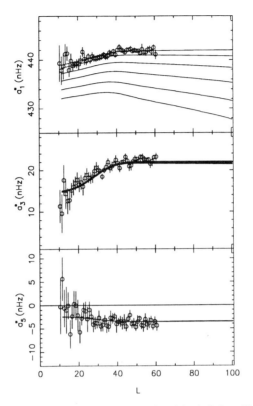

Fig. 5. Same as in Fig. 4, but with another family of model calculations. Here, we have taken the rotation profile to consist of solid-body rotation in the radiative interior at 436 nHz, surface-like rotation throughout most of the convection zone with the surface parameters in Fig. 4, and a thin layer on top where $v \sim r^{-2}$, simulating the conservation of specific angular momentum (CAM layer). Note each model here has a different rate at $r = R_\odot$, which depends on the depth of the CAM layer. From top to bottom the a_1 curves correspond to models with CAM layer thicknesses equal to 0, 0.01 $R_\odot$, 0.02 $R_\odot$, 0.03 $R_\odot$, 0.04 $R_\odot$ and 0.05 $R_\odot$.

model is absolutely ruled out, as was first demonstrated by Morrow (1988a). As we discuss below, this cylindrical model was predicted by state-of-the-art computer simulations of the solar convection zone. The best fit to the data out of this family of rotation profiles is that with surface-like rotation throughout the convection zone.

Currently our best estimate of the real solar interior rotation profile, consistent with the above forward model calculations, is given by an inversion of the BBSO (Big Bear Solar Observatory) data set done by Christensen-Dalsgaard and Schou (1988), shown in Fig. 7. Different inversions of this data set have been performed (Dziembowski et al 1989), and other p-mode splitting data sets have been analyzed (see, e.g., Morrow 1988a,b; Brown et

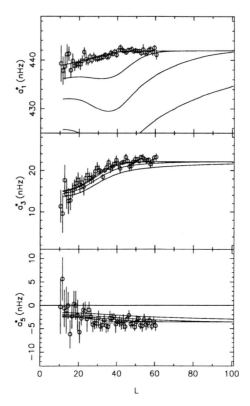

Fig. 6. Same as Fig. 4, but here the models consist of surface-like rotation $v(r, \phi) = v(R_\odot, \phi)$ with $A = 461$ nHz, $B = -60.5$ nHz and $C = -75.4$ nHz in the range $r_a \leq r \leq R_\odot$, where r_a is variable; and angular velocity constant on cylinders between r_a and the base of the convection zone at $r_c = 0.7$ $R_\odot$. This is followed by a transition to solid-body rotation with $v_{ri} = 436$ nHz beneath the convection zone ($r < 0.7$ $R_\odot$). The different curves have $r_a/R_\odot = 1.0, 0.9, 0.8, 0.7$, where the top curve corresponds to pervasive surface-like rotation in the solar convection zone. Note that rotation constant on cylinders throughout the convection zone (bottom a_1 curve) does not fit the data.

al. 1989; Duvall et al 1986; Rhodes et al 1990). In all of these, the following features of the solar interior rotation appear: (1) the angular velocity in the convection zone is roughly independent of radius, in sharp contrast to a profile that has angular velocity constant on cylinders matched to the surface rotation; (2) the rotational frequency of the radiative interior is in the range 425 to 435 nHz, at least for $r > 0.4$ $R_\odot$; (3) there is a relatively sharp transition from the convective to radiative rotation profiles just below the base of the convection zone, although it is impossible now to say just how sharp.

The future of this field is very bright, and we expect that the results here will be confirmed and probably slightly modified as more data are accumulated and the analysis procedures are improved.

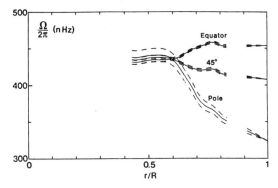

Fig. 7. Solar interior profile, as inferred by an inversion (Christensen-Dalsgaard and Schou 1988) of the Big Bear Solar Observatory data set (Libbrecht 1988c). Note the surface-like rotation profile throughout the convection zone, with a transition to solid-body rotation in the deep interior. This inversion does not extend above 0.8 $R_\odot$; the observed surface rotation is indicated in the plot. The rotation rate below 0.4 $R_\odot$ remains largely unknown. This picture will doubtless be somewhat modified in the near future as better data are accumulated and the inversion techniques are refined.

Rotation of the Deep Interior

Below a radius of 0.4 $R_\odot$, our knowledge of the solar rotation rate is poor, due primarily to the inaccuracy of the current p-mode splitting data at very low l. Observations of other stars indicate that when the Sun was a young star, its surface rotation rate was much larger than observed today (see Skumanich 1972; also chapter by Noyes et al.). A number of authors have suggested that while the rotation of the convection zone would have been reduced in time through the action of magnetic braking, the massive core may have retained its original rapid rotation.

A recent attempt to model the evolution of a rotating Sun by Pinsonneault et al. (1989) suggests that the deep solar interior may indeed be rotating much more rapidly than the surface, due to a lack of strong angular momentum coupling to the surface layers. On the other hand, a weak internal magnetic field could provide sufficient coupling to slow the core (Spruit 1987). While current data confine such a rapidly rotating core to below 0.2 to 0.3 $R_\odot$ (Morrow 1988a), there is some suggestion that the p-mode splittings do turn up at the lowest l-values (Pallé et al. 1988c). Rotational splitting of solar g-modes would give us a very clear picture of the rotation of the deep interior, because these modes are concentrated in the solar core (see the chapter by Hill and Gabriel). Unfortunately, although there have been claimed detections in the past, it is now the opinion of most solar researchers that no g-modes have been seen in the Sun. Clearly the detection of g-modes and the accurate measurement of rotational splitting for low-l p-modes are very important goals for continued research.

Summary

Over the past few years the helioseismic method of determining the solar internal rotation has produced an unexpected result. The Sun's surface latitudinal differential rotation extends through the convection zone, with the angular velocity being roughly independent of radius. Below the convection zone (but above 0.4 $R_\odot$), the Sun rotates roughly as a solid body with a rotation rate of about 436 nHz; the sharpness of the transition between these two states is not known. As the next generation of helioseismology instruments comes on line, we expect this picture to be improved greatly, and to extend more deeply into the solar core. The angular velocity of the innermost core, the width of the transitional layer between the radiative interior and convection zone, the near-surface rotational profile, the torsional oscillation, as well as a slight time-dependent rotation rate, should all be within the grasp of the newer measurements.

IV. THEORETICAL INTERPRETATION

Now that the helioseismic measurements are converging on a picture of the Sun's internal rotation, it is worthwhile to consider more carefully how to explain this picture theoretically. While a great deal has been written on the topic of rotating stars (see, e.g., Tassoul 1978), our understanding of the phenomenon is quite poor. The solar differential rotation remains a challenging theoretical problem, as the transport of angular momentum in the Sun by the many scales of turbulent convection seems to defy any simplifying approximations.

Before helioseismology, our only real knowledge of the solar rotation was from surface measurements. It appeared that supercomputer simulations of the solar convection zone were fairly close to the right answer, because they were able to reproduce the surface profile. Nonlinear dynamical simulations of the interaction between rotation and giant cell convection (see, e.g., Glatzmaier 1987; Gilman and Miller 1986) continue to find a surface differential rotation that agrees fairly well with the observations, but they also produce angular velocity roughly constant on cylinders in the convection zone. This is illustrated in Fig 8. Ironically, while the match with the solar surface rotation had given one confidence in the simulations, the predicted interior rotation turns out to be quite incorrect. It is a significant problem for the models of global convection that the well-established predictions from the simulations were not borne out by the helioseismic data. It remains unclear what deficiency in the simulations is responsible for the discrepancy.

Gilman et al. (1989), Morrow et al. (1988) and Morrow (1988a) discuss implications of the new picture of the solar internal angular velocity for angular momentum transport and dynamo action in the Sun. They consider how transport of angular momentum implied by the helioseismic data contrasts

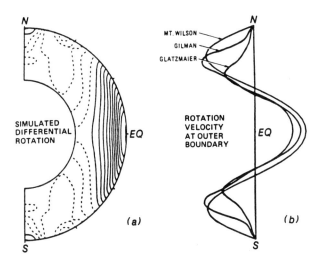

Fig. 8. (a) Computer simulation of the solar convection zone, which shows lines of constant angular velocity relative to a rotating frame of reference (taken from Glatzmaier 1987). (b) Simulated surface rotation, compared with observations. Note that the simulation agrees fairly well with the surface rotation, but gives an incorrect interior rotation profile.

with the transport suggested by computer simulations of the rotational influence of giant cell convection. Figure 9 makes the comparison pictorially. The first difference between the two scenarios for momentum transport is implied by the significantly different profiles for angular velocity within the convection zone: the simulations give angular velocity constant on cylinders, whereas the helioseismic result is angular velocity constant on radial spokes. This means that the processes operating to transport angular momentum within the convection zone are not adequately simulated by the models. The second major difference between the two scenarios for momentum transport involves whether or not angular momentum is exchanged between the convection zone and radiative interior: the computer simulations have not accounted for this possibility, whereas the helioseismic results are indicative of such an exchange, albeit on a time scale that is likely to be much longer than that needed for the transport processes within the convection zone to establish the profile of angular velocity there.

To produce a profile with angular velocity constant on cylinders requires the net radial transport of angular momentum to be outward. This is illustrated by the radial component of the large arrows in Fig. 9b. Also note from the figure that the latitudinal component of momentum transport is equator ward. These large arrows correspond to the transport of angular momentum by the convection under the influence of either Coriolis forces or rotational shearing; such mechanisms can produce differential rotation in both radius

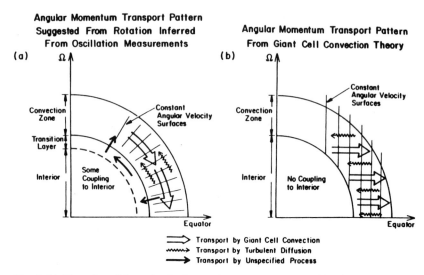

Fig. 9. (a) Illustration of the cycle of solar angular momentum suggested by the measurements of the internal rotation profile. (b) The cycle of solar angular momentum suggested by computer simulations of global convection. Both figures are from Gilman et al. (1989).

and latitude. The squiggly arrows represent the transport of angular momentum by turbulent diffusion, that opposes the production of gradients in the angular velocity.

The discussions of the aforementioned authors point out that to achieve a rotational profile devoid of radial gradients of angular velocity within the convection zone requires that there be no net radial transport within the convection zone. It is not yet clear what crucial piece of physics is lacking in the models that keeps them from producing the required balance of radial transport. There are hints that the problem involves the treatment of the smaller-scale convection, which for computational feasibility has been simply parameterized by an isotropic turbulent viscosity. The possibility of introducing anisotropy in this parameterization of turbulent viscosity has yet to be tried in a fully consistent dynamical calculation. Mean-field models simulate the effect of rotation on convection without explicitly calculating the convection (see Stix 1987a; Durney 1987, and references therein), and they offer insight into transport mechanisms that may be operating in the solar convection zone. Ideally, the smaller-scale convection, as well as the acoustic waves of a fully compressible treatment, could be calculated explicitly along with the large-scale convection; but currently this is not computationally feasible.

Morrow (1988a) and Brown et al. (1989) first noted that the rigid rotation rate of the interior is intermediate between the fastest and slowest rates of the surface rotational profile. This implies that near the base of the convection zone, the angular velocity decreases with depth at lower latitudes,

and increases with depth at higher latitudes. Figure 7 clearly shows this feature of the best-guess rotational model being considered. Current helioseismic data is unable to offer an accurate thickness for the transitional layer,
but the implied existence of fairly steep gradients there highly suggests that
angular momentum is exchanged between the convection zone and the radiative interior, and indeed transported latitudinally within the transitional layer
(Gilman, Morrow and Deluca 1989, hereafter GMD). Figure 9 shows the
directions of transport dictated by sense of the angular velocity gradients.
The possible means of transporting angular momentum in the transitional
layer are many and include magnetic fields, gravity waves, overshooting convection, shear-induced turbulence, diffusive instabilities and meridional circulations (see, e.g., Spruit 1987).

It is natural to ask whether or not the angular momentum leaving the
convection zone in low latitudes is balanced by that transported into the convection zone at high latitudes. Before addressing this question, however, recall that this process of exchange between the two regions is unlikely to occur
fast enough to be relevant to establishing the basic differential nature of the
rotational profile within the convection zone. It is, however, relevant to the
content of angular momentum in the convection zone over the course of time
in much the same way as the solar wind torque is relevant in this respect.

Using an idealized model for the transitional layer, GMD derive a criterion for a balanced exchange of angular momentum between the convection
zone rotating as seen at the surface and the radiative interior rotating at v_{ri}:

$$A + \frac{1}{5}B + \frac{3}{35}C = v_{ri}, \qquad (7)$$

where A, B, and C are defined in Eq. (5) for the surface rotation. The criterion
is essentially expressing the continuity of angular momentum at the boundary
between the base of the convection zone and the top of the radiative interior.
This amounts to equating the $\sin^3\theta$-weighted average of the "surface" differential profile, which is at the base of the convection zone, to the rigid rotation
rate at the top of the radiative interior. Note from Eqs. (6) that this combination of A, B, and C is the same as a_1, indicating that the variation of a_1 with
degree l reflects the radial variation of a $\sin^3\theta$-weighted average of the angular velocity.

We use the values of A, B, and C which produced the best model fit to
the Libbrecht data (see Fig. 4) in Eq. (7) to calculate a value for v_{ri}. With
$A = 461$ nHz, $B = -60.5$ nHz and $C = -75.4$ nHz, the angular velocity
of the radiative interior that implies a balanced exchange with the convection
zone is 442 nHz. This is very close to the value $v_{ri} \approx 436$ nHz determined with
our forward analysis. Thus, using the newer and better Libbrecht data, we
offer the same suggestion as GMD did using the Brown and Morrow (1987b)
data, namely that the torques exerted in low latitudes are approximately bal-

500 K. G. LIBBRECHT AND C. A. MORROW

anced by the torques exerted in the opposite sense in high latitudes, and thus the angular momentum is fairly continuous across the interface between the convection zone and radiative interior.

If indeed there are no radial gradients of angular velocity in the convection zone, this result also implies that the total amount of angular momentum in the differentially rotating convection zone is the same as if it were rotating rigidly at the rate of the radiative interior. This of course suggests that the total angular momentum of the Sun (convection zone plus radiative interior) is the same as for a rigid body rotating at 436 nHz. Note, however, that this comment presumes that the deepest portions of the radiative interior are rotating at the same constant rate as the upper portions where we have measured it; verification of this presumption awaits better observations of low-degree oscillation modes.

GMD also describe a possible new scenario for the solar dynamo, given the newly emerging picture of the Sun's internal angular velocity. We leave the details of this description to the chapter on the solar dynamo by Gilman and DeLuca. However, because this dynamo would be located in the overshoot region just below the convection zone, GMD argue in favor of magnetic stresses playing a significant role in transporting momentum in the transitional region.

In the future, we expect that the derivation of the solar interior rotation will be an obvious problem, perhaps an example to be assigned in a graduate course in fluid dynamics. [A.D. 2020: Problem 1. Using the fractal theory of turbulent convection discussed in class, formulate a model of the rotation of a self-gravitating, convecting plasma sphere, the Sun, on your personal supercomputer. For extra credit, derive the period of the solar magnetic dynamo in terms of fundamental quantities.] For the present, unfortunately, we do not know the solution to this problem. We have, however, seen a dramatic and significant advance in this field in the recent past, namely the measurement of the solar interior rotation profile using helioseismic techniques. We hope that these new measurements will stimulate additional thought directed to this fundamental problem.

Acknowledgment.   This work was supported in part by the National Science Foundation and a Presidential Young Investigator Award for KGL.

THE INTERNAL ROTATION AND MAGNETISM OF THE SUN
FROM ITS OSCILLATIONS

W. A. DZIEMBOWSKI

N. Copernicus Astronomical Center

and

PHILIP R. GOODE

New Jersey Institute of Technology

The study of solar oscillations has revealed knowledge of the internal rotation of the Sun and something of its internal magnetic field. We present the formalism needed to determine the internal rotation from oscillation data. Equations are developed that describe centrifugal distortion and results are given. We sketch the formalism required to treat poloidal and toroidal magnetic fields inside the Sun. Results are presented for a toroidal field concentrated near the base of the convection zone and for assumed relic poloidal and toroidal fields in the deep interior.

I. INTRODUCTION

The 22-yr activity cycle is an overt manifestation of the Sun's internal rotation and magnetism. This cycle has been studied observationally and theoretically for many years; nonetheless, the cycle's dynamo mechanism is not completely understood. In particular, we do not know the size of the dynamo's magnetic field. Furthermore, we do not know where the dynamo is seated except that it is somewhere between the overshoot region, just beneath the convection zone, and the top of the zone. Our information about rotation and magnetism in the Sun's core is even poorer. A global magnetic field in the core tends to be unstable to ohmic diffusion, but this fact alone does not imply that the field in the core is small. The time scale for ohmic diffusivity is comparable to the solar lifetime. There could be a sizeable relic field in the deep interior allowing this region to be a reservoir for both the magnetic field and the angular momentum. Such reservoirs could play roles in the Sun's activity cycle and the transport of angular momentum. We know that the Sun has been losing reservoir angular momentum through its surface to the solar wind. We also know that there is no factor of 2 or so difference in rotation between the convection zone and the radiative interior. Thus, the rotation rates of these two regions are close together without a known dynamical instability to cause the locking. This further suggests a role for an interior magnetic field which may or may not be coupled to the dynamo field. A better understanding of the solar cycle would follow from knowledge of the internal rotation and magnetism of the Sun. If we understood solar activity, we would gain fundamental insight into the activity of young solar-type stars. Many of these stars exhibit rapid rotation and high activity compared to the Sun.

The promise of helioseismology is that we will determine the internal rotation and assess the size and structure of the internal magnetic field of the Sun by studying solar oscillations. The oscillations of interest here are acoustic with periods of about 5 min. The observed oscillations are the surface manifestation of trapped, standing sound waves inside the Sun. These standing waves are the normal modes of the Sun with different modes sampling different regions of the solar interior. The waves have their frequencies modified in prescriptible ways by internal dynamical effects like rotation and magnetism. With a sufficient variety of oscillation data, we can determine the Sun's internal angular velocity and, at least, place limits on the magnetic field. The seismology would be aided by data on gravity modes which sample the deep interior. Unfortunately, there are no compelling observations of these modes. Nonetheless, recent progress in the determination of the internal rotation rate of the Sun has been remarkable.

Duvall et al. (1984) determined that, near its equatorial plane, the Sun rotates essentially rigidly throughout the convection zone with the outer part of the radiative zone rotating somewhat more slowly. Also, there is just the hint of a rapidly rotating core. Duvall et al. (1986) reported that the whole convection zone rotates with the surface differential rate, meaning that there is no sizeable radial gradient in rotation there. This conclusion has been observationally verified by Rhodes et al. (1987), Brown and Morrow (1987a) and Libbrecht (1989). Brown et al. (1989) used the data of Brown and Morrow (1987) to show that there appears to be a gradual transition from latitudinal differential rotation in the convection zone to solid-body rotation in the radiative interior. At low latitudes, the outer part of the radiative region spins more slowly than the inner part of the convection zone; at higher latitudes, the outer part of the radiative region spins faster than the inner part of the polar convection zone. The certainty of this conclusion decreases with increasing latitude. Dziembowski et al. (1989) and Christensen-Dalsgaard and Schou (1988) used the data of Libbrecht (1989) to confirm the basic results of Brown et al. (1989). Further, they concluded that the aforementioned transition is quite abrupt and has its point of inflection at the base of the convection zone. Solar rotation is discussed in detail in the chapter by Libbrecht and Morrow.

The standard $\alpha - \omega$ dynamo theories to describe the 22-yr solar activity cycle are reviewed in the chapter by Gilman and DeLuca. It is generally expected for the postulated dynamos that radial and latitudinal gradients in rotation are required at the seat of the dynamo. If the co-existence of these gradients were the condition for locating the dynamo, then results from the oscillation data imply that it is located near the base of the convection zone. That is, the rotation laws of Brown et al. (1989), Dziembowski et al. (1989) and Christensen-Dalsgaard and Schou (1988) reveal angular and radial gradients in that region. There are other theoretical arguments, as well, for assuming that the dynamo is centered near the base of the convection zone. These arguments are primarily based on the buoyancy of the magnetic field—the difficulty of stabilizing the required toroidal field in the convection zone over the solar cycle. These matters are reviewed in the chapter by Gilman and DeLuca.

Describing the dynamo is a complicated matter. In particular, the interaction between rotation, magnetism and convection involves many arbitrary assumptions. The helioseismic determination of rotation and magnetism has its own difficulties. In the oscillation data there are random errors and finite lifetime effects which, to date, have resulted in data with rather large errors. The bright side is that, unlike the dynamo problem, there is no basic theoretical obstacle in the determination of rotation and magnetism. Nonetheless, the formalism requires major computational and calculational efforts. This is particularly true in the case of the magnetic field.

Efforts to determine the magnetic field inside the Sun from oscillation data have not enjoyed the same success as efforts to determine internal rotation. The problem here is that even if magnetic field and rotational energies were comparable in the Sun, rotation would be easier to determine. This is because the lowest-order effect of rotation of the oscillation spectrum is linear in Ω, whereas the lowest-order magnetic-field perturbation is quadratic in the field. Dziembowski and Goode (1988) and Gough and Thompson (1988) used the data of Duvall et al. (1986) in an effort to learn about the internal toroidal magnetic field. The conclusions of those efforts were only speculative because a greater accuracy in the data was required. Subsequently, Dziembowski and Goode (1989) used the data of Libbrecht (1989) to find evidence for a megagauss toroidal field just beneath the convection zone. The field, if any, in the deeper interior is another matter. Gough and Thompson (1990) have estimated the effect on the oscillations of various aligned, axisymmetric fields with particular configurations. Still much work needs to be done. For instance, what is the effect of an inclined field in the core? Since the problem here is not primarily on the theoretical side, future efforts in the development of the formalism for the determination of the magnetic field inside the Sun from oscillation data is done with the prospect of more accurate data in the future. In particular, we anticipate the data from the Global Oscillations Network Group (GONG), a network of six groundbased observing sites dispersed in longitude.

The spectrum of solar oscillations is characterized by multiplets labelled by n and l, the radial order and angular degree of the oscillation. Rotation and magnetism induce a fine structure in each multiplet labelled by m, the azimuthal order. The lower the degree of the oscillation the deeper it samples inside the Sun. To study the region near the solar surface, oscillations of degree > 100 are required. To study the energy generating region near the center of the Sun, the lowest-degree oscillations are required. Thus, to study rotation and magnetism in the region of the dynamo, oscillations of degree between 10 and 100 are sufficient. The well-sampled region in this case is between 0.6 and 0.8 of the solar radius. Our purpose here is to review the helioseismology used to learn about the internal rotation and magnetism of the Sun.

II. THE LINEAR EFFECT OF ROTATION

The usual description of the fine structure in oscillation data is due to Duvall et al. (1986) and Brown and Morrow (1987a). That is,

$$\nu_{nlm} - \nu_{nl0} = L \sum_{i=1}^{l} a_{i,nl} P_i\left(\frac{m}{L}\right) \tag{1}$$

where $L = l$ or $\sqrt{l(l + 1)}$ depending on the choice of the data analyst and where ν_{nlm} is the frequency of the nlm-oscillation and P is a Legendre polynomial. The odd-a coefficients are associated with the linear effect of rotation. The even-a coefficients are due to the second-order effect of rotation and perhaps, an aligned, axisymmetric magnetic field. The even-a coefficients will be discussed in the next section. The expansion in Eq. (1) is usually cut off at $l = 5$ (cf. Brown et al. (1989). The 3 odd-a coefficients can be directly associated with a rotation law of the form:

$$\Omega(r,\theta) = \Omega_0(r) + \Omega_1(r)\mu^2 + \Omega_2(r)\mu^4 \tag{2}$$

where $\mu = \cos \theta$ and θ is the co-latitude.

If we knew the internal rotation rate of the Sun, we could calculate the frequency splittings defined in Eq. (1) from

$$\nu_{nlm} - \nu_{nl0} = \int \xi^*_{nl} \cdot \hat{K} \, \xi_{nl} \frac{\Omega(r,\theta)}{2\pi} \rho d^3 r \tag{3}$$

where

$$\hat{K} = m - i\hat{e}_z x. \tag{4}$$

and where $\hat{e}_z$ is the unit vector along the rotation axis and i is $\sqrt{-1}$. The $\hat{K}$ operator represents the linear effect of rotation in an inertial frame. The displacement, ξ_{nl}, for the nl oscillation is usually calculated employing a standard model, and

$$\xi_{nlm}(r,\theta,\phi) = r\left[y_{nl}(r), \quad z_{nl}(r)\frac{\partial}{\partial\theta}, \quad z_{nl}(r)\frac{1}{\sin\theta}\frac{\partial}{\partial\phi}\right]Y_l^m(\theta,\phi). \tag{5}$$

For a detailed development of Eq. (3) for a rotation law of the form of Eq. (2), see Cuypers (1980) or Brown et al. (1989). If the rotation were rigid at Ω_R, Eq. (3) reduces to

$$\nu_{nlm} - \nu_{nl0} = \frac{m\Omega_R}{2\pi}(1 - C_{nl}) \tag{6}$$

where C_{nl} is the Ledoux (1951) constant arising from the Coriolis force. Equation (3) gives the general form of the inverse problem to be solved. The observers give us the left-hand side of Eq. (3) and the theorists the kernel of the integrand of the right-hand side.

The specific forms of the inverse problems for $\Omega_0(r)$, $\Omega_1(r)$, and $\Omega_2(r)$ follow from Eqs. (1)–(5). To determine $\Omega(r,\theta)$, first we solve

$$\int\left[K_{nl}(r) - 14k_{nl}(r)\right]\frac{\Omega_2(r)}{2\pi}\,dr = H_{25}a_{5,nl} \xrightarrow[\text{p-modes}]{L^2 \gg 1} 21a_{5,nl} \tag{7}$$

then

$$\int[K_{nl}(r) - 5k_{nl}(r)]\frac{\Omega_1(r)}{2\pi}\,dr = -6\int k_{nl}\frac{\Omega_2(r)}{2\pi}\,dr + H_{13}a_{3,nl}$$
$$+ H_{15}a_{5,nl} \xrightarrow[\text{p-modes}]{L^2 \gg 1} -5a_{3,nl} - 14a_{5,nl} \tag{8}$$

then

$$\int K_{nl}(r)\frac{\Omega_0(r)}{2\pi}\,dr = \int k_{nl}(r)\frac{\Omega_1(r)}{2\pi}\,dr + a_{1,nl} + H_{03}a_{3,nl}$$

$$+ H_{05}a_{5,nl} \xrightarrow[p\text{-modes}]{L^2 \gg 1} a_{1,nl} + a_{3,nl} + a_{5,nl}. \quad (9)$$

This formalism is different than, but consistent with that of Brown et al. (1989). The large and small kernels or sampling functions are given by

$$K_{nl}(r) = [y_{nl}^2(r) + l(l + 1)z_{nl}^2(r) - 2y_{nl}(r)z_{nl}(r) - z_{nl}^2(r)]\rho r^4 \quad (10)$$

and

$$k_{nl}(r) = z_{nl}^2(r)\rho r^4 \quad (11)$$

respectively. The lower the degree of the oscillation, the deeper is the sampling. In the p-mode's high-frequency limit, Eqs. (7)–(9) decouple because the small kernel can be ignored. In fact, the small kernel may be safely ignored in the 5-minute period band at the present level of accuracy in the data, because the large kernel K typically dominates the small kernel k by 3 orders of magnitude. If $L^2 \gg 1$, then Eqs. (7)–(9) take the particularly simple form shown. For $L^2 \gg 1$, one can estimate the rotation rate by inspecting the oscillation data as given, for instance, in the chapter by Libbrecht and Morrow. Goode et al. (1991) have shown that the large L^2 limit is reached at much lower l value if $L = \sqrt{l(l + 1)}$ instead of $L = l$.

The H coefficients follow from connecting the angular integrals from the right-hand side of Eq. (3) to the observational data as given by the Legendre expression in Eq. (1). The angular integrals arising in Eq. (3) are defined by

$$M_{slm} = \int \mu^{2s}|Y_l^m|^2 d\mu d\phi. \quad (12)$$

Once M_{0lm} and M_{1lm} have been calculated, the remaining angular integrals can be determined from the recursion relation

$$M_s = \frac{1}{4l(l + 1) - 4s^2}\frac{2s - 1}{s}\left\{M_{s-1}\left[-2m^2 + \right.\right.$$
$$\left.\left. 2l(l + 1) - (2s - 1)^2\right] + M_{s-2}(s - 1)(2s - 3)\right\} \quad (13)$$

where l and m have been suppressed. This recursion relation is developed elsewhere (see Dziembowski and Goode (1991a). Next the M integrals can be expanded in a power series in m/L

$$M_s = \sum_{j=0}^{s} AM_{sj} \left(\frac{m}{L}\right)^{2j}. \tag{14}$$

The connection to the data is straightforward. It only requires expressing odd powers of m/L in terms of odd polynomials in m/L, so that in the case $L = l$

$$H_{25} = \frac{63}{8} \frac{1}{AM_{22}} \xrightarrow{l^2 \gg l} 21, \tag{15}$$

$$H_{13} = \frac{5}{2} \frac{1}{AM_{11}} \xrightarrow{l^2 \gg l} -5, \tag{16}$$

$$H_{15} = -\frac{35}{4AM_{11}} \left(1 + \frac{9}{10} \frac{AM_{21}}{AM_{22}}\right) \xrightarrow{l^2 \gg l} -14, \tag{17}$$

$$H_{03} = \frac{-3}{2} + \frac{5}{2} \frac{AM_{10}}{AM_{11}} \xrightarrow{l^2 \gg l} 1, \tag{18}$$

$$H_{05} = \frac{63}{8} \left(\frac{5}{21} + \frac{10}{9} \frac{AM_{10}}{AM_{11}} + \frac{AM_{10}}{AM_{11}} \frac{AM_{21}}{AM_{22}} - \frac{AM_{20}}{AM_{22}}\right) \xrightarrow{l^2 \gg l} 1. \tag{19}$$

We must find a suitable inverse method and properly use it to determine $\Omega(r,\theta)$. These are not trivial matters because Eq. (3), with real oscillation data, presents an ill-posed problem. Approaches to the inverse problem are reviewed in the chapter by Gough and Thompson. Solutions to the inverse problem for rotation are shown in the chapter by Libbrecht and Morrow (1989). Knowing the internal rotation is critical for determining the centrifugal distortion of the Sun.

III. QUADRATIC EFFECT OF ROTATION

The second-order effect of rotation is manifest in the symmetric part of the fine structure in the oscillation spectrum. The formal treatment here is a sketch of that given by Dziembowski and Goode (1991b).

The frequency splitting due to the quadratic effect of rotation, in an inertial frame, is given by

$$\omega_2 = \frac{J_D + J_I}{2\omega} + J_T + J_P \tag{20}$$

where ω is the unperturbed oscillation frequency and J_D and J_I represent the lowest-order effects of distortion and inertia. The J_T and J_P terms represent the result of the linear effect of rotation acting on the perturbed eigenfunction split into its toroidal and poloidal components. The second-order effect of rotation may be safely ascribed to distortion. The reasons that the other three terms are negligible will be discussed later in this section.

Distortion, the dominant quadratic effect of rotation, is given by J_D, where

$$J_D = \int \boldsymbol{\xi}^* \cdot \mathbf{L}_D(\boldsymbol{\xi}) d^3r. \tag{21}$$

The distortion operator is

$$\mathbf{L}_D(\boldsymbol{\xi}) = -\omega^2 \rho_D \boldsymbol{\xi} - \nabla[(p\Gamma)_D \nabla \cdot \boldsymbol{\xi}] - \nabla(\boldsymbol{\xi} \cdot \nabla \rho_D) - \nabla \cdot (\rho_D \boldsymbol{\xi}) \nabla \phi$$
$$+ (\rho \nabla \cdot \boldsymbol{\xi} + \boldsymbol{\xi} \cdot \nabla \rho)(\mathbf{F}_c - \nabla \phi_D) + \rho_D \nabla \delta \phi + \rho \nabla \delta_D \phi. \tag{22}$$

The centrifugal force is given by

$$\mathbf{F}_c = \Omega^2 r \sin \theta \, \hat{e}_s = (F_r, F_o, 0) \tag{23}$$

where $\hat{e}_s$ is the radial unit vector in a cylindrical system having the rotation axis as its axis of symmetry. Furthermore, we define the distorted gravitational potential, pressure and density by

$$\delta_D \phi = \int \frac{\nabla \cdot (\boldsymbol{\xi}\rho_D) d^3r}{|\mathbf{r} - \mathbf{r}'|} \tag{24}$$

$$p_D = \rho \phi_D + H \tag{25}$$

and

$$\rho_D = \frac{dp}{dr} \frac{\phi_D}{g} + \frac{1}{g}\left[\rho F_r - \frac{\partial H}{\partial r}\right] \tag{26}$$

where

$$H = r \int \rho F_\theta d\theta. \tag{27}$$

In Eq. (23), we use the differential form of the angular velocity specified in Eq. (2). In principle, this rotation law contributes to $a_2, a_4, \ldots, a_{10}$ coefficients, but only the a_2 and a_4 terms have been observed. To determine ϕ_D, we first expand it into its Legendre components,

$$\phi_D = \sum_k \phi_{D,k} P_{2k}(\cos \theta). \tag{28}$$

Then we determine $\phi_{D,k}$ from the Clairaut equation

$$\frac{d^2\phi_{D,k}}{dr^2} + \frac{2}{r}\frac{d\phi_{D,k}}{dr} - \frac{1}{r^2}\left[\frac{4\pi G}{g}\frac{d\rho}{dr} + 2k(2k + 1)\right]\phi_{D,k} \tag{29}$$
$$= \frac{4\pi G}{g}\left[\rho F_r - \frac{\partial H}{\partial r}\right]_k$$

where we have expanded the force into its Legendre coefficients, as well.

An order of magnitude estimate of the contribution of distortion to a_2, for 5-min period p-modes, is given by

$$a_2 \sim \frac{-\epsilon\nu}{l} \tag{30}$$

where ϵ is the solar oblateness and $\nu = \omega/2\pi$.

To calculate the effect of distortion, we use the mean rotation law determined by Dziembowski et al. (1989) from Libbrecht's (1989) data,

$$\Omega(r,\theta) = 460.2 \pm 0.2 - (58.3 \pm 1.8)\mu^2 - (73.1 \pm 2.6)\mu^4 \text{ nHz}. \tag{31}$$

In Fig. 1, the calculated values of a_2 and a_4 are compared to those in the data of Libbrecht, averaged over n and grouped in bins five l-values wide. The gradual decline in the calculated values of a_2 with decreasing l is best represented by a power law between $l^{-0.8}$ and $l^{-0.9}$ rather than by l^{-1} as in the order of magnitude estimate of Eq. (30). The slower than l^{-1} variation reflects the relatively larger distortion in the outer layers. The calculated coefficients are consistent with those in the data of Libbrecht (1989) which are small and negative, and show the same general trend to decline with decreasing l. We note that the variance in the data is too large to allow us to distinguish between a decline represented as $l^{-0.85}$ from one given by l^{-1}.

The effect of inertia from Eq. (20) is described by

$$J_I = \omega_1^2 + \int \left[m\Omega^2\xi^* \cdot (m - 2\hat{K}) \xi + \xi_s^* \xi \cdot \nabla \Omega^2\right]\rho d^3r \tag{32}$$

where $\hat{K}$ is defined in Eq. (4), subscript s denotes components parallel to the centrifugal force and ω_1 equals 2π times the frequency splitting of Eq. (3). For the case of rigid rotation, the a_2 coefficients are roughly given by

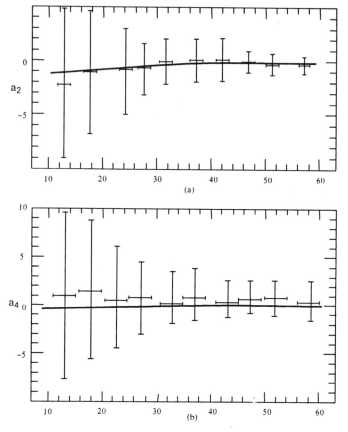

Fig. 1. Weighted averages of Libbrecht's a_2 and a_4 coefficients (nHz) vs l. The solid lines represent the calculated effect of distortion.

$$a_2 \sim l \, \frac{\left(\frac{C\Omega}{2\pi}\right)^2}{\nu} \tag{33}$$

where C is the Ledoux constant which appears quadratically because of the cancellation occurring in J_l. This cancellation may be easily seen by inserting $\hat{K}$ into the expression for J_l. The a_2 coefficient is determined after recognizing that ω_2 can be expanded in terms of the M integrals discussed in Sec. II in the context of the linear effect of rotation. Since $C \sim 10^{-3}$, the inertial term for rigid rotation is negligible for solar p-modes in the 5-min period band. Even if rotation were specified by a θ-dependent rotation like that of Eq. (31), the J_l term would still be negligible. Negligible means < 0.1 nHz, a value chosen because it is considerably smaller than the accuracy observers report in their data.

The effect of the toroidal modes J_T is easy to calculate and is negligible. The poloidal modes also make a negligible contribution J_p but this result is difficult to calculate. The poloidal mode problem has been solved by Saio (1981) for the case of uniform rotation. His approach was to derive and solve an equation for the perturbed eigenfunctions that has the same form that the perturbed hydrostatic equation has for the unperturbed oscillation. For this case, Dziembowski and Goode (1991b) showed that

$$a_2 \sim \frac{l\left(C\frac{\Omega}{2\pi}\right)^2}{\Delta \nu} \sim 10^{-6}l \tag{34}$$

where, for rigid rotation, $\Delta \nu$ is the frequency separation between successive n states for a particular l. For θ-dependent differential rotation, matters are more complicated. There, the presence of the $\cos^4 \theta$ term in Ω_2 connects modes such that an accidental degeneracy in $\Delta \nu$ is possible. In particular, internal resonances occur in pairs for modes which differ in degree by four, where

$$\nu_{n+1,l-4} \approx \nu_{nl} \tag{35}$$

if $n = l - 4$. For the 5-min-period band, these accidental degeneracies occur for l between 7 and 24. Dziembowski and Goode (1991a) have used degenerate perturbation theory to show that the poloidal mode a_2- and a_4-coefficients should still be negligible. Their calculated values of $\Delta \nu$ are as small as 0.4 μHz. However, models may not be trustworthy on that level. To be on the safe side, one should throw out the pairs of affected modes from data sets that are not averaged over n. For Libbrecht's (1989) data, we would eliminate 22 of the 678 modes. It turns out that the calculated rotation law is insensitive to whether these modes are included or excluded.

Distortion is the primary second-order effect of rotation. Because the mean rotation law from Libbrecht's data is quite similar to those from the data of Brown and Morrow (1987a) and Duvall et al. (1986), the calculated a_2- and a_4-coefficient due to centrifugal distortion would be much the same from each set. We note, however, that the data sets have markedly different average values of a_2 and a_4. This suggests the possibility of another perturbation that could be time dependent, like a magnetic field.

IV. THE MAGNETIC FIELD

We now discuss the effect on oscillation frequencies which would arise from certain magnetic fields. We focus on fields at two specific locations with two simple field geometries—a toroidal field near the base of the convection zone and relic toroidal and poloidal fields near the center of the Sun. We consider a toroidal field near the base of the convection zone because differential rotation present there, could sheer a small poloidal field into a large toroidal field. Furthermore, such a field would have the geometry and location expected for the solar dynamo. Finally, the oscillation data of Libbrecht (1989) have been inverted by Dziembowski and Goode (1989) to provide evidence of such a field of megagauss strength. In the deep interior, a large, relic field could be present because ohmic diffusivity is the only obvious decay mechanism and its time scale is comparable to the age of the Sun. We consider simple, global fields because they have their strongest signature in the reported a_2 and a_4 coefficients. Such fields have been considered by Gough and Thompson (1990) and Dziembowski and Goode (1991a, and references contained therein).

To calculate the magnetic field's perturbation of oscillation frequencies, we must include the effect of distortion due to the field, $J_D/2\omega$ and the term, $J_M/2\omega$ which is due to the perturbation of the Lorentz force by the oscillation, i.e.

$$\omega_2 = \frac{J_D + J_M}{2\omega}. \tag{36}$$

The J_M term is defined by

$$J_M = \frac{1}{4\pi} \int \xi^* \cdot [\mathbf{B} \times (\nabla \times \mathbf{B}') + \mathbf{B}' \times (\nabla \times \mathbf{B})] \mathrm{d}^3 r \tag{37}$$

where the Eulerian perturbation of the field is

$$\mathbf{B}' = \nabla \times (\xi \times \mathbf{B}). \tag{38}$$

After several integrations by parts in which it is assumed that $B = 0$ on the solar surface, we determine that

$$J_M = \frac{1}{4\pi} \sum_{i,j} \int \left\{ T_{ij} + \frac{1}{2} X_{ij} \nabla \cdot \xi + \xi \cdot \nabla X_{ij} \right\} B_i B_j \mathrm{d}^3 r \tag{39}$$

where i and j represent appropriate pairs of the spherical coordinates r, θ and ϕ. We safely assume $B = 0$ on the surface because, if the internal field is not much larger than the true surface field, the interior field will have no detectable manifestation in the spectrum of p-modes. The T and X are matrices which depend on components of the oscillations. The important remark here is that the integrand for J_M depends on terms like $B_i B_j$ multiplied by a magnetic field kernel. This implies that we may be able to determine the magnetic field from oscillation data by solving an inverse problem. However, an inverse problem is possible only if the effect of distortion in Eq. (34) can be expressed in the same form.

We remark that the (rr), $(r\theta)$, $(\theta\theta)$ and $(\phi\phi)$ components of T and X are real and imply frequency splitting. The $(r\phi)$ and $(\theta\phi)$ components of T and X are imaginary and vanish if there is no magnetic torque. If there is a magnetic torque and it is uncompensated by Reynolds stresses, for instance, then the equilibrium approach used here is not valid. We emphasize that the poloidal and toroidal components are, therefore, not coupled. This means that we can treat the two types of fields separately. In fact, this is true for each term in Eq. (36). The development of the treatment of the effect of distortion is closely analogous to that for centrifugal distortion with Eq. (23) for the centrifugal force replaced by the Lorentz force—either due to the poloidal field or the toroidal field.

A. Toroidal Field Near the Base of the Convection Zone

For an axisymmetric toroidal magnetic field, the effect of distortion on the frequency splitting can be usefully expressed as an inverse problem. For this field the distorting Lorentz force, to replace Eq. (23), is given by

$$
\mathbf{F} = \frac{1}{4\pi} (\mathbf{\nabla} \times \mathbf{B}) \times \mathbf{B} = -\frac{1}{4\pi}\left[\hat{r}\left(\frac{1}{2}\frac{\partial B_\phi^2}{\partial r}\right.\right.
$$
$$
\left.\left. + \frac{B_\phi^2}{r}\right) + \hat{\theta}\left(\frac{1}{2}\frac{\partial B_{\hat{\theta}}^2}{\partial\hat{\theta}} + B_\phi^2 ctn\hat{\theta}\right)\right] \quad (40)
$$

where $\mathbf{B} = B_\phi(r,\theta)\hat{\phi}$. For use in formulating the toroidal field kernel, we write

$$
\frac{B_\phi^2}{4\pi p} = (1 - \mu^2)\sum_{k\geq 1}\beta_k(r)\mu^{2(k-1)} \quad (41)
$$

where p is the local gas pressure. The expansion in Eq. (41) is general except that we include only even powers in μ because odd powers, if present, do not contribute to splitting at the lowest order in the distortion.

As detailed by Dziembowski and Goode (1989), the inverse problem for a toroidal field may be expressed in terms of the symmetric a coefficients as

$$
a_{2j,d} = \sum_{k\geq 1}\int\left(r\frac{d\beta_k}{dr}D_{kj,d} + \beta_k E_{kj,d}\right)dr, \quad (42)
$$

where d is the mode counter. The D_s and E_s contain roughly comparable contributions from J_D and J_M in Eq. (36). We concentrate on a quadrupole toroidal field because the available a_2 and a_4 data best sample near the base of the convection zone where we have learned that there is differential rotation that could generate, by dynamo action, a large quadrupole field from a small, simple poloidal field. In Fig. 2, we show the D and E coefficients for the quadrupole toroidal field for $l = 20$. For $n = 14$ and $l = 20$, there is considerable structure with a strong tendency for a single sign. Averaging the $n = 7 - 21$ terms together, for $l = 20$ (an average over the 5-min period band) gives a much smoother result. Integrating Eq. (42) by parts we obtain the inverse problem

$$a_{2j,d} = \sum_{k \geq 1} \int \beta_k \left[E_{kj,d} - \frac{d}{dr} (rD_{kj,d}) \right] dr. \tag{43}$$

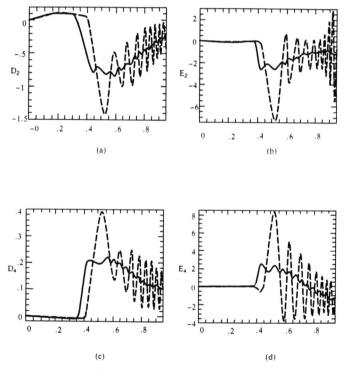

(a)

(b)

(c)

(d)

Fig. 2. Plots for a quadrupole toroidal field with $k = 2$ and the (a) D_2, (b) E_2, (c) D_4 and (d)E_4 kernels vs the fractional radius. The subscripts here are a product of k and j. The kernels are for $l = 20$ and are given in μHz. The dashed line is for $n = 14$ and the solid line is for the average over n in the 5-min period oscillation band.

The magnetic kernel in Eq. (43) is a rapidly varying function of radius. Therefore, the difference in sampling between various l and n kernels is increased with respect to those in Eq. (42). This increase improves the chances for the success of the inversion. The inverse problem posed in Eq. (43) is identical in form to that for rotation as given in the hierarchical Eq. (7)–(9). Pressing this analogy, the a_4 data could be inverted to determine β_2, then the a_2 data and β_2 would be used to calculate β_1.

In the application of Eq. (43) by Dziembowski and Goode (1989), they inverted both a_2 and a_4 data to doubly determine β_2 because both inversions gave the same results to within the errors. Before inverting Libbrecht's (1989) data, they removed the calculated effect of centrifugal distortion from the a_2 and a_4 data. As well, they eliminated the pairs of modes subject to an accidental degeneracy (see Sec. III).

Dziembowski and Goode (1989) inverted Eq. (43) using the Backus-Gilbert (1970) method of optimal averaging and the method of regularization; these methods are reviewed in the chapter by Gough and Thompson. The significant result is that there is evidence for a 2 ± 1 megagauss quadrupole toroidal field centered just beneath the base of the convection zone. Furthermore, limits can be placed on deeper-lying quadrupole toroidal fields; at $0.55\ R$ this field is less than 4 megagauss.

The 2 ± 1 megagauss field is not inconsistent with anything we know. The ratio of the magnetic pressure to the gas pressure is $\sim 10^{-3}$. This field would not cause an observable change in the speed of sound determined by Christensen-Dalsgaard et al. (1985b). Furthermore, the resulting quadrupole moment is comparable to that for a rigidly rotating Sun. The megagauss field has the proper symmetry and location to be associated with the dynamo that drives the activity cycle. However, a megagauss field is ~ 2 orders of magnitude more intense than expected for the dynamo; dynamo theory is reviewed in the chapter by Gilman and DeLuca. Magnetic buoyancy arguments would imply that a megagauss field must be confined beneath the convection zone, so such a field could be a reservoir for the dynamo.

We point out that only Libbrecht's data for the symmetric splitting coefficients are not averaged over n. Averaging his data over n sufficiently reduces the differential sampling of the resulting kernels in Eq. (43) so that no useful inversion can be performed to determine the field.

B. Relic Field in the Deep Interior

There could be a sizeable toroidal and/or poloidal magnetic field buried in the radiative interior with the toriodal component driven by dynamo action on the poloidal part. If so, what is the field's signature in the spectrum on solar oscillations? This question has been considered in various contexts by Dicke (1982), Dziembowski and Goode (1984), Gough and Taylor (1984) and Gough and Thompson (1990) among others.

Only low-degree oscillations penetrate this region so that the plausible approach to the problem is to guess simple field configurations and predict the consequences for solar oscillations. This method does, at least, provide crude limits on the core field since the requisite low-degree data are unavailable.

For a quadrupole toroidal field, we follow Gough and Thompson's (1990) definition of the field

$$\mathbf{B}_\phi(r,\theta) = a(r)\frac{dP_k}{d\theta} \tag{44}$$

where $k = 2$ and

$$a(r) = (1 + \sigma)(1 + \sigma^{-1})^\sigma B_0\left(\frac{r}{r_0}\right)^2\left[1 - \left(\frac{r}{r_0}\right)^2\right]^\sigma \qquad r < r_0 \tag{45}$$

with $\sigma = 10\,r_0 + 1$ and $a = 0$ elsewhere. The maximum value of $a(r)$ occurs at $(1 + \sigma)^{-1/2}r_0$. They chose $r_0 = 0.7R$ and $B_0 = 10^7G$. This selection was made to mimick the field chosen by Dicke (1982) in his effort to account for the 1966 Princeton oblateness observations. Such a field would cause a large quadrupole moment and a marginally observable perturbation to the a_2 and a_4 coefficients for $l > 10$. It is difficult to be more precise about comparing calculated and observed a_2 and a_4 coefficients because the dominant feature in the symmetric splitting data is a sizeable variation of the mean from data set to data set. We emphasize that these sets cover basically the same l-range. Thus, the 10^7G quadrupole toroidal field predicts a_2 and a_4 coefficients that are within the scatter of those reported by the observer after the removal of centrifugal distortion. It is possible that the scatter actually represents a trend (see Sec.IV.C).

For the poloidal field, we again follow Gough and Thompson (1990) and write

$$\mathbf{B}(r,\theta) = \left[k(k + 1)\frac{b(r)}{r^2}P_k(\cos\theta), \frac{1}{r}\frac{db}{dr}\frac{dP_k(\cos\theta)}{d\theta}, 0\right] \tag{46}$$

where the detailed form is to satisfy $\nabla \cdot \mathbf{B} = 0$. Gough and Thompson (1990) assumed, for $k = 1$ and 2, $b(r) = (r/2)a(r)$. The basic results are comparable to those for the a_2 and a_4 coefficients following from the toroidal field. We emphasize that knowledge of symmetric splittings for l from 1 to 10 would enable us to detect or, at least, sharply constrain relic fields.

It could be that there is a relic field and it is inclined to the axis of rotation. In this case, Dicke (1982) and Dziembowski and Goode (1985) have shown that the oscillation spectrum would be much more complicated. In particular, since the field's perturbation is unsteady, even if the field co-rotates with the Sun, observers will report at least $(2l + 1)^2$ frequencies for each (nl) multiplet. Without detailed splitting data for l from 1 to 10, we cannot usefully address the role, if any, for even a sizeable inclined relic field.

C. Do the Coefficients a_2 and a_4 Depend on Time?

Kuhn (1988a) has argued that the a_2 and a_4 coefficients gradually change through the solar cycle. This systematic change is reflected in a trend in the significant variations in these coefficients from data set to data set. He further argues that these changes are directly correlated with the apparent time dependence in the trend in the midlatitude surface temperature bands discovered by Kuhn et al. (1988).

According to Kuhn (1988a), the origin of these time-dependent correlated phenomena is a nonradial part that he attributes to the speed of sound

$$c^2(r,\theta) = c_0^2(r)\left[1 + \sum_{i \geq 1} f_{2i}(\cos \theta)\right] \qquad (47)$$

where $c_0(r)$ is the usual speed of sound. He says that the physical origin of the nonradial temperature perturbation implicit in Eq. (47) is a thermal shadow of the time-dependent dynamo field (Parker 1987b). In Parker's picture of a thermal shadow, the magnetic buoyancy of the dynamo field is compensated by the cool shadow above it that the field induces.

Kuhn (1988a) argues that the thermal shadow persists to the solar surface and has been observed by Kuhn et al. (1988b). Furthermore, the nonradial temperature perturbation induces the observed a_2 and a_4 coefficients. Kuhn's physical picture is simple and interesting, but what is the force which causes the perturbation?

We may write the temperature perturbation as

$$\frac{T'}{T} = \frac{\partial \ln T}{\partial \ln \rho}\bigg|_p \frac{\rho'}{\rho} + \frac{\partial \ln T}{\partial \ln p}\bigg|_\rho \frac{p'}{p}. \qquad (48)$$

From Eqs. (25) and (26), it is clear that ρ' and p' are caused by a force. Parker's picture is difficult, requiring a re-arranged flux at the bottom of the convection zone where convection is efficient. This re-arranged flux, in turn, somehow forces a nonradial perturbation to the speed of sound throughout the convection zone.

An alternative description to Kuhn's would be that the more intense toroidal field just beneath the base of the convection zone, as calculated by Dziembowski and Goode (1989), changes through the cycle and forces symmetric fine structure in the oscillation data. In either picture, the field somehow causes the global, surface-temperature dips discovered by Kuhn et al. (1988b). The energy density in the field calculated by Dziembowski and

Goode (1989) is several times the rotational energy density in the same region. In addition, the field's total energy is comparable to that from the solar luminosity over the activity cycle. Since the luminosity is so constant, such a field would have to exchange its energy adiabatically if it varies. Presumably, torques would cause the exchange with some combination of rotation and small-scale magnetic fields not manifest in the a_2 or a_4 data. Because the rotation rate in the convection zone has not changed much over the current cycle, one could speculate that there has been a sizeable change in the interior's rotation during the same time.

This and other interesting questions concerning the Sun's internal rotation and magnetism remain to be answered from oscillation data. For instance, what is the rotation rate and magnetic field in the Sun's deep interior? There is every reason to be optimistic about our ability to answer these question in the near future because greatly improved oscillation data (from groundbased networks and satellites) are expected in the next few years.

Acknowledgment.   P. R. G. is supported in part by the Air Force Office of Scientific Research.

THE INVERSION PROBLEM

D. O. GOUGH
University of Cambridge

and

M. J. THOMPSON
High Altitude Observatory

We discuss some fundamental problems associated with inverting helioseismic data to infer properties of the solar interior, and review various methods used to invert frequencies of normal modes of oscillation. After discussing linear methods, we present and compare different asymptotic methods that have been used to infer the internal solar sound speed. We also discuss numerical inversions for solar structure, and address the issue of inverting for more than one function. Methods of inverting for the nonspherically symmetric structure of the Sun are presented.

The observable properties of the Sun's normal modes of oscillation depend on the structure and dynamics of the solar interior and atmosphere. The inversion problem is to determine how to use these observables to make inferences about the solar structure and dynamics. Inversion methods have been developed extensively for remote sensing problems in a number of other disciplines: for example, in the geophysical and atmospheric sciences. There are numerous books on the subject (see, e.g., Menke 1984; Craig and Brown 1986; Tarantola 1987) as well as review articles (see, e.g., Parker 1977; Twomey 1977; Gough 1985a). In addition, there are many papers of importance to our discussion which are referenced below.

A prerequisite for solving the inverse problem is to be able to solve the forward problem, that is, for any given underlying model to be able to calculate the values that the observables would take. In this chapter, we shall confine our discussion to the inversion of one class of observable data, namely, the mode frequencies. The forward problem of calculating the frequencies is discussed in the chapter by Christensen-Dalsgaard and Berthomieu. Many of the techniques and considerations of this chapter may also be applicable to the inversion of other observable data, such as mode amplitudes, phases and linewidths, but at the present time, it is unresolved how one should solve the forward problem for these observables. (It should be noted that the solution to the forward problem may be only approximate. This might be out of necessity, as for any model of realistic complexity, or out of choice, as, for example, in the asymptotic sound-speed inversions described below. Such approximation is liable to introduce error into the solution of the inverse problem.)

I. LINEAR PROBLEMS AND LINEAR METHODS

We begin by considering linear problems, as this will enable us to discuss many of the essential elements of the inversion problem. Linear problems are also important in their own right. First-order rotational splitting is an example of such a problem: the frequency splitting is a linear functional of the rotation rate (chapter by Libbrecht and Morrow). Also, nonlinear problems may often usefully be posed by linearizing about some model (see Sec. II). For simplicity we consider initially the case where the measured data d_i are functionals of a single function of radius $f(r)$:

$$d_i = \int_0^R K_i(r)f(r)\mathrm{d}r + \epsilon_i \qquad i \in \mathcal{M}. \tag{1}$$

Here R is the radius of the Sun and r is the radial variable. This is a linear problem, because the data would be linear functionals of the underlying function $f(r)$ in the absence of errors ϵ_i. (We shall refer to *methods* as being linear if the solution is a linear functional of the data d_i.) The index i typically represents a particular mode out of a set $\mathcal{M}$, and the mode kernels K_i define the functional dependence of the data on the underlying function f. The values of the errors ϵ_i are unknown (otherwise the data d_i could be corrected for them), although one may have knowledge of their statistical properties. An example of such a linear problem is first-order frequency splitting by a latitudinally independent angular velocity $\Omega(r)$, for which the cyclic frequencies ν_{nlm} satisfy

$$\nu_{nlm} = \nu_{nl0} + m\delta\nu_{nl} \tag{2}$$
$$\delta\nu_{nl} = \int_0^R K_{nl}(r)\Omega(r)\mathrm{d}r.$$

Thus in this example i represents (n,l), the data are the splittings $\delta\nu_{nl}$ and the underlying function is $\Omega(r)$. In principle, the set $\mathcal{M}$ might contain more than one observation of the same observable, but for simplicity we shall assume that that is not the case.

The aim of an inversion might naively be stated to be to infer the function $f(r)$: that $f(r)$ *cannot* in fact be determined uniquely immediately highlights two sources of uncertainty in any solution of even so simple an inversion problem as that posed by Eq. (1). First, and most obvious, the unknown errors ϵ_i prevent f from being determined with certainty, for the values $(d_i - \epsilon_i)$ of the functionals $\int K_i(r)f(r)dr$ are not known precisely. Secondly, the function $f(r)$ cannot be uniquely determined from the data alone because the set $\mathcal{M}$ of data points is necessarily finite: even if the data were free of errors, to any solution $\bar{f}(r)$ of Eq. (1) could be added any function $g(r)$ satisfying

$$\int_0^R K_i(r)g(r)dr = 0 \text{ for all } i \in \mathcal{M} \tag{3}$$

to obtain another solution $\bar{f} + g$. (For some problems this could be so even if $\mathcal{M}$ were infinite.) The set of functions g satisfying Eq. (3) is known as the annihilator $\mathcal{A}$. Thus the inverse problem of inferring $f(r)$ from the data is underdetermined. The data can at best determine $f^{\parallel}(r)$ and contain no information whatsoever about $f^{\perp}(r)$, where

$$f(r) = f^{\parallel}(r) + f^{\perp}(r) \tag{4}$$

is the (unique) decomposition of the underlying function f into a linear combination $(f^{\parallel})$ of the mode kernels and a member of the annihilator $(f^{\perp})$.

Granted this fundamental nonuniqueness, there are different approaches that can be taken. There is the spectral expansion method, for example, which explicitly seeks to determine not f, but rather $f^{\parallel}$ (see, e.g., Parker 1977; Gough 1985a). Another approach is to introduce assumptions about the function f (such as demanding that f or its derivatives satisfy certain bounds) and to then seek a solution $\bar{f}$ to Eq. (1) that is consistent with these assumptions. It should always be borne in mind, however, that one could add any member of the annihilator $\mathcal{A}$ to one's solution and obtain another solution that satisfies the data d_i equally well. The annihilator contains functions of high spatial frequency, but it may in addition include functions with large-scale structure. Thus, to give an artificial example, if the set of kernels consisted of $\sin n\pi x$ $(0 \le x \equiv r/R \le 1)$ for $n = 2, 3, \ldots, N$, then the annihilator would contain not only the functions $\sin n\pi x (n > N)$ but also $\sin \pi x$. In practical applications, it is most likely that the annihilator will contain functions with large-scale structure only when the number of modes is small. Of course, if the mode set contains only shallowly penetrating modes, functions that have substantial amplitude only at great depth could be members of the annihilator.

A further difficulty encountered with the helioseismological inversion problem is that it is usually ill conditioned. Provided the kernels K_i are not actually linearly dependent (and generally they will not be unless the data set contains more than one observation of the same observable), one can find a solution $\bar{f}$ that satisfies the data exactly:

$$d_i = \int_0^R K_i(r)\bar{f}(r)dr \qquad i \in \mathcal{M}. \tag{5}$$

As the data contain errors, $\bar{f}$ will not be equal to f even up to an additive member of the annihilator. But worse, in helioseismological problems, many kernels tend to be very similar so that the set of kernels is almost linearly dependent. Then small changes in d_i can lead to large changes in $\bar{f}$: the problem is ill conditioned. One can get some feel for how this arises in the following way. For simplicity, let the kernels be normalized thus:

$$\int_0^R K_i^2(r)dr = 1 \qquad i = 1, \ldots, N. \tag{6}$$

Again for simplicity, suppose that the errors ϵ_i are independent and are each Gaussian distributed with zero mean and variance σ^2. If the kernels are orthogonal (in the sense that $\int K_i K_j dr = \delta_{ij}$), then a representation of one function that satisfies the data exactly is

$$\bar{f}(r) = \sum_i d_i K_i(r). \tag{7}$$

From Eqs. (1) and (5), at any point r the error in $\bar{f}$ that arises from the data errors ϵ_i is $\sum \epsilon_i K_i(r)$ (summing over i); and the variance of this error is $\sigma^2 \sum (K_i(r))^2$. On the other hand, suppose that $\int_0^R K_1 K_2 dr = \alpha \approx 1$ and that $d_1 \approx d_2$. Then one might form an orthonormal set of functions by the Gram-Schmidt orthogonalization algorithm:

$$\psi_1 = K_1, \; \psi_2 = (K_2 - \alpha K_1)/(1 - \alpha^2)^{1/2}, \ldots \tag{8}$$

so that $\bar{f}(r) = \sum \tilde{d}_i \psi_i(r)$ with

$$\tilde{d}_1 \equiv d_1 = \int \psi_1(r)f(r)dr, \quad \tilde{d}_2 \equiv \frac{d_2 - \alpha d_1}{(1 - \alpha^2)^{1/2}} = \int \psi_2 f(r)dr \tag{9}$$

etc. The error in $\tilde{d}_2$ is $(\epsilon_2 - \alpha \epsilon_1)/(1 - \alpha^2)^{1/2}$, and this has variance $\sigma^2(1 + \alpha^2)/(1 - \alpha^2)$. Thus small errors in d_1 and d_2 can lead to a large error in $\tilde{d}_2$

and hence also in $\bar{f}$. (A similar argument, by analogy with vectors, is given by Gough [1985a]. For a fuller discussion of the ill-conditioned inversion problem see, e.g., the book by Craig and Brown [1986].)

Even if the data were error free, numerical errors arising when computing the solution might similarly cause large changes in $\bar{f}$. It is evident that solving the Eq. (5) exactly is not a good strategy. What is required is a solution that is relatively insensitive to errors in the data. Most practical inversion methods have one or more parameters that control the sensitivity of the solution to random errors in the data.

Trying to infer $f(r)$ directly is not the only approach to the inversion problem. An alternative is to use the data d_i, which are approximately the values of the functionals $\int K_i(r)f(r)dr$, to estimate the values of new functionals. Of particular interest is the case when the new functionals are chosen to be of the form

$$h = \int_0^R H(r)f(r)dr \tag{10}$$

where the weighting function (or kernel) $H(r)$ is a linear combination of the mode kernels $K_i(r)$. For then h is independent of $f^\perp(r)$, and precise statements can be made about h if the statistics of the errors ϵ_i are known. Whether this is a useful procedure depends in this case on whether the new functionals are intrinsically of greater interest or are more easily interpretable than the original weighted averages $\int K_i(r)f(r)dr$ of f, and on how large the uncertainties are in the values of the new functionals. Inversion by optimally localized averages is an example of this technique.

Inversion by Optimally Localized Averages

The procedure depends on an idea by Backus and Gilbert (1968). The aim is, for any given radius r_0, to choose a linear combination of the mode kernels that "resembles" a delta function $\delta(r - r_0)$, so that the corresponding functional is an average of $f(r)$ that is highly localized about $r = r_0$. Thus one chooses coefficients $c_i(r_0)$, $i \in \mathcal{M}$, to make the averaging kernel (or resolution kernel)

$$\mathcal{K}(r;r_0) \equiv \sum_{i \in \mathcal{M}} c_i(r_0)K_i(r) \tag{11}$$

have certain desired properties. (The averaging kernel corresponds to $H(r)$ in the previous paragraph.) Clearly from Eq. (1)

$$\sum_{i \in \mathcal{M}} c_i(r_0)d_i = \int_0^R \mathcal{K}(r;r_0)f(r) \, dr + \sum_{i \in \mathcal{M}} c_i(r_0)\epsilon_i. \tag{12}$$

The left-hand side is an estimate of the functional $\int \mathcal{K}(r;r_0)f(r)dr$. Let us now suppose that the statistics of the errors ϵ_i are Gaussian with zero mean, described by a known covariance matrix E_{ij}. Then the error term in Eq. (12) is Gaussian distributed with zero mean and variance $\sum_{i,j} E_{ij} c_i(r_0) c_j(r_0)$.

The coefficients $c_i(r_0)$ must be chosen not only to give $\mathcal{K}(r;r_0)$ the desired properties but also to keep the variance of the error term small, if the estimate of the functional is to be useful. To date, the most commonly adopted choice of coefficients in helioseismological applications is such as to minimize

$$12 \cos \theta \int_0^R (r - r_0)^2 (\mathcal{K}(r;r_0))^2 dr \; + \; \mu \sin \theta \sum_{i,j} E_{ij} c_i(r_0) c_j(r_0) \qquad (13)$$

subject to

$$\int_0^R \mathcal{K}(r;r_0) dr \equiv \sum_{i \in \mathcal{M}} c_i(r_0) \int_0^R K_i(r) dr = 1 \qquad (14)$$

(Backus and Gilbert 1970; also see, e.g., Gough 1985a). The parameter θ determines the trade-off between the opposing aims of minimizing the first term in expression (13) and minimizing the error-covariance term. The parameter μ is included for convenience so that at fixed θ, the coefficients c_i are roughly independent of the magnitude of the errors in the data: a common choice is

$$\mu = (\text{trace}(E_{ij})/M)^{-1} \qquad (15)$$

(where M is the number of data points in set $\mathcal{M}$). Differentiating Eq. (13) with respect to each c_i in turn (for fixed r_0) and setting these derivatives equal to zero yields a matrix equation for the coefficients $c_i(r_0)$. The solution to this equation can then be normalized to satisfy constraint (Eq. 14).

Other definitions of "delta-ness" are possible. The function $12(r - r_0)^2$ may more generally be any function $J(r - r_0)$ that increases monotonically as its argument departs further from zero. Alternatively, expression (13) might be replaced by

$$12 \cos \theta \int_0^R \left(\int_0^r \mathcal{K}(r'; r_0) dr' - \Theta(r - r_0) \right)^2 dr$$
$$+ \; \mu \sin \theta \sum_{i,j} E_{ij} c_i(r_0) c_j(r_0) \qquad (16)$$

where $\Theta(r - r_0)$ is the Heavyside function (Parker 1977). This formulation requires only one matrix to be inverted: changing r_0 merely modifies the right-hand side of the matrix equation that determines the coefficients c_i. Other

possibilities are given by Backus and Gilbert (1968). Naturally, these formulations will generally yield different averaging kernels.

Examples of averaging kernels constructed by minimizing expression
(13) are shown in Fig. 1, taken from Christensen-Dalsgaard et al. (1990).
These are for the problem of inverting the splitting due to latitudinally independent rotation, and E_{ij} is given by $\sigma^2 \delta_{ij}$ for some σ.

If one has succeeded in constructing well-localized kernels $\mathcal{K}(r,r_0)$, then
one can identify the left-hand side of Eq. (12) as $\bar{f}(r_0)$, an estimate of the
function f at $r = r_0$.

Least-Squares Inversion Methods

A second class of methods, which may be referred to as least-squares
methods, seek a solution $\bar{f}(r)$ expressed as a linear combination of a chosen
set of base functions $\phi_j(r):(j = 1, \ldots , N)$:

$$\bar{f}(r) = \sum_{j=1}^{N} \bar{f}_j \phi_j(r) \qquad (17)$$

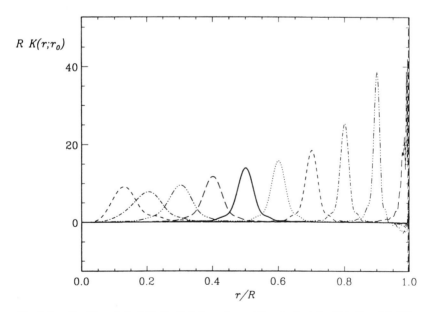

Fig. 1. Localized kernels for latitudinally independent rotation at selected radii ($r_0/R = 0.1, 0.2$,
$\ldots$, 1.0), for the method of optimally localized averages ($\theta = 10^{-2}$). The mode set used
consisted of 834 p modes with frequencies in the range 2 to 4 mHz and degrees between 1 and
200 (figure from Christensen-Dalsgaard et al. 1990).

where $\bar{f}_j$ are constants to be determined. For example, by choosing a dissection

$$0 = \bar{r}_0 < \bar{r}_1 < \ldots < \bar{r}_N = R \qquad (18)$$

and defining

$$\phi_j(r) = \begin{cases} 1 & \bar{r}_{j-1} < r < \bar{r}_j \\ 0 & \text{elsewhere} \end{cases} \qquad (19)$$

a piecewise constant function $\bar{f}$ is constructed. A variant is to construct $\bar{f}$ to be continuous and linear on each interval $[\bar{r}_{j-1}, \bar{r}_j]$. Alternatively, one could choose the ϕ_j to be some other set of orthogonal functions.

Basically one would like to choose the constants $\bar{f}_j$ to minimize

$$\sum_{i \in \mathcal{M}} \left(d_i - \int_0^R K_i(r) \bar{f}(r) dr \right)^2 / \sigma_i^2 \qquad (20)$$

where we now assume for simplicity that the errors ϵ_i are not only Gaussian but also independent, with variances σ_i^2. However, care must be taken because of the ill-conditioned nature of the problem. One could choose the number N of base functions to be very small; alternatively, in the case of a piecewise constant or continuous piecewise-linear function $\bar{f}$, one can choose to minimize

$$\sum_{i \in \mathcal{M}} \left(d_i - \int_0^R K_i(r) \bar{f}(r) dr \right)^2 / \sigma_i^2 + \mu F[\bar{f}] \qquad (21)$$

where μ is an adjustable parameter and $F[\bar{f}]$ is a "regularization" or "smoothing" term. Some choices of $F[\bar{f}]$ (and to keep the expressions simple we consider only uniform dissections) are

$$F[\bar{f}] = \frac{1}{N} R \sum_{j=1}^N \bar{f}_j^2 \qquad \approx \int_0^R \bar{f}^2 \, dr$$

$$F[\bar{f}] = \sum_{j=2}^{N-1} \frac{(\bar{f}_j - \bar{f}_{j-1})^2 + (\bar{f}_{j+1} - \bar{f}_j)^2}{(\bar{r}_{j+1} - \bar{r}_{j-1})} \approx \int_0^R \left(\frac{d\bar{f}}{dr} \right)^2 dr \qquad (22)$$

$$F[\bar{f}] = \sum_{j=2}^{N-1} \frac{8(\bar{f}_{j-1} - 2\bar{f}_j + \bar{f}_{j+1})^2}{(\bar{r}_{j+1} - \bar{r}_{j-1})} \approx \int_0^R \left(\frac{d^2\bar{f}}{dr^2} \right)^2 dr$$

(see, e.g., Phillips 1962; Tikhonov and Arsenin 1977; Twomey 1977). It is evident what kinds of functions $\bar{f}$ these regularization terms favor, but solutions that make $F[\bar{f}]$ large are not totally excluded if they are strongly demanded by the data. The second and third terms may be called first- and second-derivative smoothing, respectively.

In the case of an expansion in orthogonal functions exhibiting increasingly small-scale structure as the index j increases, one might add a regularization term

$$\sum_{j=1}^{N} \alpha_j^2 \bar{f}_j^2 \tag{23}$$

where the α_j are prescribed constants that increase with j: this would tend to inhibit the resolution of, or introduction of, small-scale structure.

Expression (21) is minimized by differentiating with respect to each $\bar{f}_j$ in turn and equating the derivatives to zero. This leads to a matrix equation of the form

$$\sum_{k=1}^{N} M_{jk}\bar{f}_k \equiv \sum_{k=1}^{N} (B^T B + \mu G^T G)_{jk}\bar{f}_k = \sum_{i \in \mathcal{M}} B_{ji}^T(d_i / \sigma_i) \tag{24}$$

where

$$B_{ij} = \int_0^R K_i(r)\phi_j(r)dr / \sigma_i \tag{25}$$

and $G^T G$ is given by the identity

$$\sum_{k=1}^{N} (G^T G)_{jk}\bar{f}_k \equiv \frac{1}{2} \frac{\partial}{\partial \bar{f}_j} F[\bar{f}]. \tag{26}$$

It follows that

$$\bar{f}(r_0) = \sum_{i \in \mathcal{M}} c_i(r_0)d_i = \int_0^R \mathcal{K}(r;r_0)f(r)dr + \sum_{i \in \mathcal{M}} c_i(r_0)\epsilon_i \tag{27}$$

where

$$c_i(r_0) = \sum_{j,k} \phi_j(r_0)M_{jk}^{-1}B_{ki}^T / \sigma_i \tag{28}$$

and $\mathcal{K}(r;r_0)$ is given by Eq. (11). Thus, just as in the Backus-Gilbert method, averaging kernels exist that show how the estimate is actually related to the unknown function $f(r)$. Note that for any given value of μ the coefficients

$c_i(r_0)$ and hence the averaging kernels are independent of the values of d_i. Examples of kernels constructed by minimizing Eq. (21) for piecewise-constant $\bar{f}$ and with $F[\bar{f}]$ given by the last of Eqs. (22) are shown in Fig. 2, for the same mode set and error properties as used for Fig. 1.

The above approach of minimizing expression (21) can be referred to as "regularized least-squares" inversion. Another approach which also takes account of the ill-conditioned nature of the problem is the "truncated singular value decomposition" (TSVD) method. As an example, we consider once again the piecewise constant approximation given by Eqs. (17), (18) and (19). We suppose that the data and kernels have been renormalized so that the data errors have unit variance. It is assumed that the number of data (M) is at least as great as the number of base functions (N), i.e., $M \geq N$. Matrix B (Eq. 25) can be decomposed as

$$B = U W V^T \qquad (29)$$

where U and V are square orthonormal matrices (of order M and N, respectively) and W is diagonal. The elements W_{kk} ($k = 1, \ldots, N$) are non-negative and are the singular values σ_k of B. It may be assumed that the σ_k's are ordered in decreasing size. The column vectors of U and V, $\mathbf{u}^{(n)}$ and $\mathbf{v}^{(n)}$

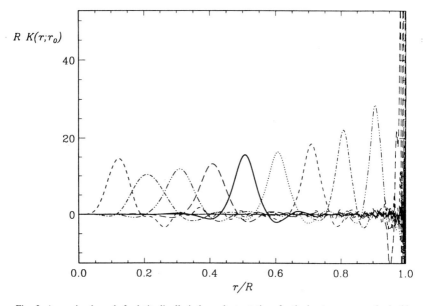

Fig. 2. Averaging kernels for latitudinally independent rotation, for the least-squares method with piecewise-constant approximation and second-derivative smoothing ($\mu = 5 \times 10^{-7}$; 100 equal intervals; mode set as for Fig. 1) at selected radii ($r_0/R = 0.1, 0.2, \ldots, 0.9$) (figure from Christensen-Dalsgaard et al. 1990).

are eigenvectors of BB^T and B^TB, respectively. Provided none of the singular elements are zero, it can be shown (see, e.g., Press et al. 1986) that

$$\bar{f}_j \equiv \sum_{i=1}^{M} (VW^{-1}U^T)_{ji}\, d_i \tag{30}$$

minimizes expression (20), where W^{-1} is defined to be the diagonal $N \times M$ matrix with diagonal elements

$$(W^{-1})_{kk} = 1/\sigma_k \qquad k = 1, \ldots, N. \tag{31}$$

In vector notation, Eq. (27) may be rewritten

$$\bar{\mathbf{f}} = \sum_{n=1}^{N} \left(\frac{\mathbf{u}^{(n)}.\mathbf{d}}{\sigma_n} \right) \mathbf{v}^{(n)}. \tag{32}$$

Clearly the data errors tend to be magnified most in those terms of this sum corresponding to the smallest singular values. If any of the singular values are zero, matrix W^{-1} as defined by Eq. (31) cannot be formed. Moreover, the rank of matrix B is then $< N$, so the vector that minimizes Eq. (20) is not unique. However, by setting to *zero* the diagonal elements of W^{-1} that correspond to zero singular values, of all vectors that minimize (Eq. 20), the solution $\bar{f}_j$ given by Eq. (30) is the one with smallest L_2 norm (see, e.g., Press et al. 1986); that is, $\bar{f}_j$ is the minimum norm solution.

Because of the tendency of small singular values to magnify errors in the data, an approach to handling the ill-conditioned nature of the problem is simply to truncate the sum in Eq. (32) for singular values below some threshold size ϵ. This is the TSVD method. It is equivalent to defining

$$(W^{-1})_{kk} = \begin{cases} 1/\sigma_k & \sigma_k > \epsilon \\ 0 & \sigma_k < \epsilon \end{cases}. \tag{33}$$

A particular case of this method, that has been used in helioseismology and geophysics, is the spectral expansion method (Parker 1977; cf. Gilbert 1971), where the base functions are chosen to be the mode kernels

$$\phi_j(r) = K_j(r) \qquad (j = 1, \ldots, N) \tag{34}$$

and so $N = M$. The solution $\bar{f}$ is therefore a linear combination of the mode kernels. Details of the method may be found in the papers by Parker (1977) and Gough (1985a). Examples of averaging kernels constructed using the method are presented by Christensen-Dalsgaard et al. (1990).

The TSVD method is equivalent to finding the minimum norm solution
to the problem

$$\text{minimize} \quad \sum_i \left(\sum_j B^*_{ij} \bar{f}_j - d_i \right)^2 \tag{35}$$

where

$$B^* = UW^*V^T \tag{36}$$

and W^* is diagonal with elements

$$(W^*)_{kk} = \begin{cases} \sigma_k & \sigma_k > \epsilon \\ 0 & \sigma_k < \epsilon. \end{cases} \tag{37}$$

Of course, provided some of the singular values are smaller than ϵ, the solution to Eq. (35) is not unique. Any linear combination of vectors $v^{(n)}$ corresponding to σ_n smaller than ϵ could be added to the solution vector $\bar{f}$. As
Sekii and Shibahashi (1988) have pointed out, choosing the minimum norm
solution is only one way of making the solution unique. One might, as they
suggest, choose instead the smoothest solution (according to whatever
smoothness criterion one selects). Sekii and Shibahashi have shown that this
can lead to more accurate recovery of simple target functions; it would of
course be of interest to examine averaging kernels for this method.

Systematic and Random Errors in the Inversion

Averaging kernels exist for any method of inverting Eq. (1) for which
the solution $\bar{f}$ is linearly related to the data d_i. Thus as Menke (1984) comments of the discrete inversion problem, there is a dualism in the theory of
linear inversions. The solution $\bar{f}(r_0)$ can be interpreted as an approximation
to $f(r)$, in which case the solution contains a *systematic error* [which depends
on the underlying $f(r)$] because the averaging kernels are not delta-functions.
Alternatively, for each value of r_0, $\bar{f}(r_0)$ can be interpreted as an estimate of
the weighted average of $f(r)$ with weighting function $\mathcal{K}(r;r_0)$.

In addition there is an error in the inversion that comes from errors in
the data. Here we shall consider mainly the effect of *random errors* in the
data, but it might be noted from Eq. (27) that the coefficients c_i determine
how *any* data errors affect the solution $\bar{f}$.

Figures 1 and 2 show averaging kernels only for particular values of the
parameters θ and μ. As these parameters are varied, the extent to which the
inversions are sensitive to random errors in the data changes. One measure
of this sensitivity is the *error magnification*. If the errors ϵ_i are independent,
normally distributed and have identical standard deviations σ, the estimate
$\bar{f}(r_0)$ will have an error $\sum_i c_i(r_0)\epsilon_i$ with standard deviation $(\sum_i (c_i(r_0))^2)^{1/2}\sigma$.

The quantity

$$\left[\sum_{i\in\mathcal{M}}(c_i(r_0))^2\right]^{1/2} \tag{38}$$

is known as the error magnification. The error magnifications for the inversions corresponding to Figs. 1 and 2 are shown in Fig. 3. It is also useful to introduce a measure of the resolution of the averaging kernels. A crude but useful measure of the width of the central maximum of the kernels is the distance between the lower quartile point, to the left of which lies one quarter of the area under the kernel, and the upper quartile point, to the right of which also lies one quarter of the area. The parameters θ and μ determine the trade-off between resolution width and error magnification. Generally, if the resolution width is decreased, the error magnification increases, and vice versa. Figure 4 shows how the resolution width and error magnification at $r_0 = 0.5\,R$ change for the inversion methods and mode set used for Figs. 1, 2 and 3, as the trade-off parameters θ and μ are varied.

If $\bar{f}$ is viewed as a pointwise estimate of f, one source of error in this estimate is the errors ϵ_i. If these are indeed random and their statistics are known, the statistics of this error in $\bar{f}$ can be calculated (either analytically or by Monte Carlo simulations) from the coefficients $c_i(r_0)$. In addition, however, there is the "systematic" error in $\bar{f}$ due to the finite resolution and nonlocal character of the averaging kernels. Thus, even if all the errors were in fact zero, $\bar{f}$ would still in general not be identical to the underlying function $f(r)$.

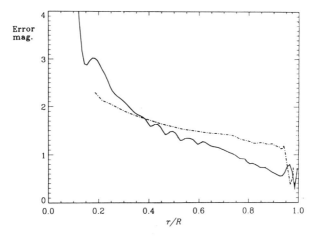

Fig. 3. The error magnification for inversions corresponding to the rotation kernels shown in Figs. 1 and 2, for the optimally localized averages (—.—.) and least squares (———————) methods (figure adapted from Christensen-Dalsgaard et al. 1990).

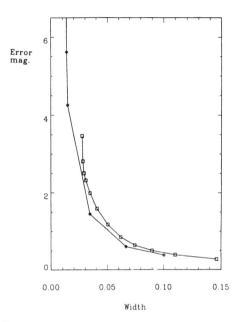

Fig. 4. The trade-off between error magnification and width of rotation averaging kernels (for the same mode set used for Fig. 1). The width is defined to be the distance between the quartile points, defined so that one quarter of the area under the averaging kernel lies to the left of the lower quartile point and one quarter to the right of the upper quartile point. Both width and error magnification are for $r_0 = 0.5\,R$. The data points for the least-squares method ($\diamond$) correspond to values of μ of 1.25×10^{-10}, 2.5×10^{-9}, 5×10^{-7}, 2.5×10^{-5} and 2.5×10^{-4} (100 equal intervals in all cases) and the points for the optimally localized averages method ($\square$) correspond to values of θ of 0.00022, 0.00046, 0.001, 0.002, 0.0046, 0.01, 0.022, 0.046, 0.1, 0.22, 0.46, and 1 (figure adapted from Christensen-Dalsgaard et al. 1990).

[We already know that this must be the case, as f can at best be determined up to an additive member of the annihilator $\mathcal{A}$. It follows that for any member g of $\mathcal{A}$, the weighted average $\int_0^R \mathcal{K}(r;r_0)g(r)\mathrm{d}r$ is zero for all r_0, as is obvious when one notes that the averaging kernels are linear combinations of the mode kernels.]

When adopting this view of $\bar{f}$, some would argue that the trade-off parameters cannot be chosen freely but must provide residuals

$$\left(d_i - \int_0^R K_i(r)\bar{f}(r)\mathrm{d}r\right) \qquad i \in \mathcal{M} \qquad (39)$$

that are consistent with the statistics of the errors. For example, in the case of independent Gaussian errors one might seek those values of the trade-off parameters that give the following equality:

$$\sum_{i \in \mathcal{M}} \left(d_i - \int_0^R K_i(r) \bar{f}(r) dr \right)^2 = \sum_{i \in \mathcal{M}} \sigma_i^2 \qquad (40)$$

(Phillips 1962). A shortcoming of this criterion is that the residuals may be systematically much greater than would be consistent with the statistics of the errors in some part of $\mathcal{M}$ (for example, shallowly penetrating modes) and systematically much less in some other part (e.g., deeply penetrating modes), so that the solution is still not actually consistent with the estimated errors in the data. Of course, even if $\bar{f}$ is consistent with the data errors one has still at best only determined "f modulo $\mathcal{A}$". [If one has a priori knowledge regarding the underlying function $f(r)$—in the form of a probability distribution, say— criterion (Eq. 40) can be modified to take this into account. This and other aspects of choosing the "best" parameter values are discussed in more detail by Craig and Brown (1986); see also the paper by Golub et al. (1979) for another approach to parameter selection.] A more pragmatic approach, motivated by the fact that the statistics of the errors of most helioseismic data are not well determined, is to adjust the parameter values to give as fine a resolution as possible without introducing what is apparently spurious small-scale structure in $\bar{f}$ (see, e.g., Gough 1985a).

Alternatively, one can view $\bar{f}(r_0)$ as an estimate of the weighted average $\int \mathcal{K}(r;r_0) f(r) dr$, in which case different inversion methods are just different prescriptions for defining the weighting. Interpreted in this way $\bar{f}(r_0)$ contains no systematic error (except for any coming from a nonrandom component of ϵ_i), and the statistics of the random error in $\bar{f}$ can be calculated as discussed above if the statistics of the data errors are known. In this view, one can choose any value of the trade-off parameter according to whether (loosely speaking) one wants broad averages with narrow limits of uncertainty or very localized averages with large uncertainty limits.

One might also consider that part of the inversion procedure is to estimate the statistics of the errors ϵ_i. Unless the mode kernels are linearly dependent, there is some function $\bar{f}$ that would satisfy Eq. (5) exactly. Thus without some a priori assumption about the true underlying function $f(r)$, one cannot use the data alone to estimate their errors. However, one might be able to exploit the substantial overlap between the mode kernels to estimate the statistics of the ϵ_i with only weak assumptions about $f(r)$, particularly if it is further assumed that the error properties vary only weakly from one mode to another. One way of doing this is by using the method of Chahine (1977), which has been applied to helioseismic data by Jeffrey (1988). This nonlinear method, which is described in the above two references, is a procedure to iterate on the solution $\bar{f}(r)$ until it violates a previously chosen assumed bound on the function f. The function $\bar{f}$ tends to be a poor solution because as much error as possible is included in the inversion subject to the bound being achieved. However, if the problem is very ill conditioned, only a very small error might cause the bounds to be violated. Then the residues (Eq. 39) at the last iteration give a broad estimate of the errors ϵ_i in the data. For very ill-conditioned problems, it is reasonable to expect that this estimate depends only weakly on the assumed bounds, provided they comfortably accommodate the true function $f(r)$ and yet are not so large that even large errors could be retained in the inversion without the bounds being violated.

There are a number of similar ways in which errors might be estimated, using bounds on the solution or on its derivatives (e.g., using the assumption that the solution must be smooth). For example, one might vary the trade-off parameter until $\bar{f}$ exhibits (what one believes to be) spurious small-scale structure, and then calculate what errors would be consistent with the residuals (cf. Gough 1985a). Alternatively, one might seek linear combinations of the mode kernels that are almost zero everywhere (this can be rigorously defined in terms of the maximum value of a weighted average of f, if bounds on f are assumed), and then use the corresponding linear combinations of the d_i to estimate the errors in the data.

II. INVERSIONS FOR SPHERICALLY SYMMETRIC STRUCTURE

Having introduced linear inversion methods and some of the concepts involved in the inversion problem, we now consider some methods more specific to helioseismology. Helioseismic inverse problems are frequently more complicated than the simple paradigm used above: for example, the data might be nonlinear functionals of the unknown function, or they might be functionals of more than one unknown function. We shall address some of these complications here. The further complication that the unknown function or functions need not depend on radius alone will be considered in Sec. III.

Asymptotic Sound-Speed Inversions

The problem considered in this subsection is that of inferring from solar p-mode frequencies the variation of sound speed within the Sun. This is a nonlinear problem. While the details of determining the annihilators and the averaging kernels of linear problems do not immediately generalize, the concept that systematic errors in the inversion procedure and errors in the data lead to uncertainties in the solution still holds. The effects of random errors on the solution can be assessed by applying the methods to artificial data in Monte Carlo simulations. Alternatively, or in addition, simple analytical expressions for the propagated error can be derived by linearization about the solution if the errors are sufficiently small. Again, even with error-free data, there is generally no unique solution to the nonlinear inverse problem given only a finite number of data (except in pathological cases). The systematic errors introduced by the inversion procedure can be investigated by performing the inversion on artificially generated data using different underlying functions.

The starting point for all the inversion methods in this subsection is the JWKB quantization for p modes. From this equation, as discussed in the Christensen-Dalsgaard and Berthomieu chapter, one can obtain Duvall's law

$$\frac{(n + \alpha(\omega))\pi}{\omega} = F\left(\frac{\omega}{L}\right) \equiv \int_{r_t}^{R}\left(1 - \frac{L^2 c^2}{\omega^2 r^2}\right)^{1/2}\frac{dr}{c} \tag{41}$$

where we adopt $L = l + \frac{1}{2}$ and r_t is the radius of the lower turning point of the mode, and the remaining notation is the same as that used by Christensen-Dalsgaard and Berthomieu. When this was first used to infer the solar sound speed, the phase α was taken to be constant, determined by finding that value of α that best collapsed the observed $(n + \alpha)/\omega$ onto a single curve when plotted against $w \equiv \omega/L$ (Christensen-Dalsgaard et al. 1985b; cf. Duvall 1982a). [Note that ω and l are directly observable, and n can be determined either by comparing frequencies with those of models or by observing sufficiently high degree modes that the f-mode ($n = 0$) ridge is seen in the five-minute band. In the latter case, the ridges can be labeled $n = 0,1,2, \ldots$ at high l, which carries back to lower-degree modes.] Thus the data determine $F(w)$, the right-hand side of Eq. (41). If the exponent in the integrand in that equation were between 0 and -1, the function F could be inverted analytically to infer the sound speed as a function of radius. One way to obtain an equation of this form is to differentiate with respect to w:

$$w^3\frac{dF}{dw} = \int_{w}^{a_s}\left(1 - \frac{a^2}{w^2}\right)^{-1/2}\frac{d\ln r}{d\ln a}da \cdot \tag{42}$$

where $a \equiv c/r$ and $\alpha_s = a(R)$ (Gough 1984b). This is an equation of the Abel type and can be inverted analytically to yield

$$r = R\exp\left[\frac{-2}{\pi}\int_{a_s}^{a}(w^{-2} - a^{-2})^{-1/2}\frac{dF}{dw}dw\right] \tag{43}$$

(Gough 1984b; Christensen-Dalsgaard et al. 1989a), which defines r as a function of a and hence implicitly defines c as a function of r. Unfortunately, $F(w)$ is not known observationally for values of w as small as a_s, as observed modes do not have lower turning points arbitrarily close to the surface. However, for small w the function $F(w)$ can be approximated by extrapolation from the observable data, by making, for example, the assumption of polytropic stratification in the surface layers (Gough 1986a). The relative errors in r thus determined by Eq. (43) become insignificant once $R - r$ is substantially greater than the depth of the layer through which the extrapolation was carried out.

The accuracy of the method can be improved by allowing α to be a function of frequency, chosen in such a way as to optimize the resemblance of $(n + \alpha(\omega))/\omega$ to a unique function of w (Gough 1986a).

A method developed by Brodsky and Vorontsov (1987,1988) also starts from Duvall's law. The frequency is considered to be a continuous function of the continuous variables n and L, and $(\partial\omega/\partial n)_L$ and $(\partial\omega/\partial L)_n$ are estimated from the data. Then from Duvall's law

$$\left.\begin{array}{rcl}\dfrac{dF}{dw} & = & \pi\dfrac{L^2}{\omega^2}\dfrac{\partial\omega}{\partial L}\Big/\dfrac{\partial\omega}{\partial n} \\[3mm] \dfrac{d(\alpha/\omega)}{d\omega} & = & \dfrac{-1}{\omega^2}\left(\omega - L\dfrac{\partial\omega}{\partial L} - n\dfrac{\partial\omega}{\partial n}\right)\Big/\dfrac{\partial\omega}{\partial n}.\end{array}\right\}\quad (44)$$

Having obtained dF/dw from the data, this can be inverted as before (Eq. 43). Also, the second Eq. (44) can be integrated to infer $\alpha(\omega)$ to within an additive linear function of ω.

Both of these methods can be extended to include the buoyancy term in Eq. (53) of the Christensen-Dalsgaard and Berthomieu chapter (Gough 1986a; Brodsky and Vorontsov 1987,1988). In principle, this could not only make the sound-speed inversion more accurate but also allow the dependence of buoyancy frequency on depth to be estimated from the data.

The starting point for both of these inversion methods is Duvall's law (Eq. 41), in which the effect of the acoustic cut-off frequency $\omega_c(r)$ and the position of the upper turning point $(r = R_t)$ are approximately incorporated into the frequency-dependent phase $\alpha(\omega)$. The inversion method of Shibahashi and Sekii (Shibahashi 1988; Sekii and Shibahashi 1989) starts not from Duvall's law but from Eq. (53) of the Christensen-Dalsgaard and Berthomieu chapter. Assuming that the buoyancy term is negligible, this may be written

$$(n + \epsilon)\pi = \int_{r_t}^{R_t}(\omega^2 - \Phi_l(r))^{1/2}\frac{dr}{c} \quad (45)$$

where

$$\Phi_l(r) = \frac{L^2c^2}{r^2} + \omega_c^2(r). \quad (46)$$

Equation (45) can be put into the Abel form by differentiating with respect to ω^2 (regarding n as a function of ω^2 and L^2):

$$2\pi\frac{\partial n}{\partial\omega^2}\Big|_{L^2} = \int_{r_t}^{R_t}(\omega^2 - \Phi_l(r))^{-1/2}\frac{dr}{c} = \int_{\tau_1(\omega,L)}^{\tau_2(\omega,L)}(\omega^2 - \Phi_l(r))^{-1/2}\,d\tau \quad (47)$$

where τ is the acoustical radius $\int_0^r c^{-1} dr$, and τ_1 and τ_2 are the acoustical radii of the lower and upper turning points, respectively. By definition, $\Phi_l = \omega^2$ at the turning points, and Shibahashi and Sekii assume further that $\Phi_l(r)$ is monotonic decreasing as r increases from r_t until, at some radius, it reaches the minimum value $\Phi_l^{\min}$, beyond which point it then increases monotonically until $r = R_t$. Thus, one can write the right-hand side of Eq. (47) as two integrals with respect to the new independent variable Φ_l, between limits $\Phi_l^{\min}$ and ω^2. As Φ_l has the dimensions of frequency squared, one can write this as

$$2\pi \frac{\partial n}{\partial \omega^2}\bigg|_{L^2} = \int_{\Phi_l^{\min}}^{\omega^2} (\omega^2 - \omega'^2)^{-1/2} \left(\frac{\partial \tau_2}{\partial \omega'^2}\bigg|_{L^2} - \frac{\partial \tau_1}{\partial \omega'^2}\bigg|_{L^2} \right) d\omega'^2. \qquad (48)$$

Here the subscripts on τ_1 and τ_2 distinguish the two branches of τ regarded as a function of Φ_l (see Fig. 5): equivalently, $\tau_1(\omega',L)$ and $\tau_2(\omega',L)$ are, respectively, acoustic radii of the lower and upper turning points of a putative mode of degree l and frequency ω'.

Equation (48) can be inverted to obtain

$$s(\omega^2,L^2) \equiv 2 \int_{\Phi_l^{\min}}^{\omega^2} (\omega^2 - \bar\omega^2)^{-1/2} \frac{\partial n}{\partial \bar\omega^2}\bigg|_{L^2} d\bar\omega^2 = \tau_2(\omega,L) - \tau_1(\omega,L). \qquad (49)$$

To a good approximation, except at very high degree, it is the case that for five-minute p modes the position of the upper turning point is independent of l while the lower turning point is insensitive to ω_c. Thus

$$\frac{\partial s}{\partial L^2}\bigg|_{\omega^2} = -\frac{\partial \tau_1}{\partial L^2}\bigg|_{\omega^2} = \left(2L^2 \frac{da}{d\ln r}\bigg|_{a=w} \right)^{-1}. \qquad (50)$$

This can be evaluated at different values of $a = w$. Hence by integration r is found as a function of a and thus c is determined implicitly.

A fourth method is to perturb Duvall's law about some reference model, whence linearization in the (presumed) small perturbations $\delta\omega/\omega, \delta c/c$ yields approximately

$$\left(\int_{r_t}^R \left(1 - \frac{L^2 c^2}{\omega^2 r^2} \right)^{-1/2} c^{-1} dr - \pi \frac{d\alpha}{d\omega} \right) \frac{\delta\omega}{\omega}$$
$$= \int_{r_t}^R \left(1 - \frac{L^2 c^2}{\omega^2 r^2} \right)^{-1/2} \frac{\delta c}{c} c^{-1} dr + \pi \frac{\delta\alpha}{\omega} \qquad (51)$$

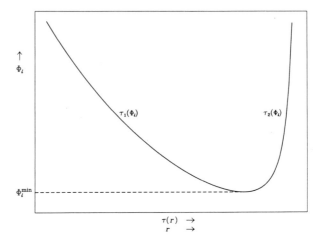

Fig. 5. A schematic drawing of the acoustic potential Φ_l as a function of the acoustic radius τ. The branches τ_1 and τ_2 of τ regarded as a function of Φ_l are labeled. This drawing is based on the case of a plane-parallel polytrope: in general, the acoustic potential may not have the simple convex form assumed in the Shibahashi and Sekii inversion method and illustrated here.

(Christensen-Dalsgaard et al. 1988c). Here all quantities refer to the reference model, except for $\delta\omega/\omega$, $\delta\alpha/\alpha$ and $\delta c/c$, which are the relative differences in frequency, phase function α (at fixed frequency) and sound speed (at fixed radius) between the unknown model (or the Sun) and the reference model. The left-hand side of Eq. (51) is determined from the data and the reference model, and the terms on the right-hand side can be separated because of their different dependences: the first is a function of w only, the second a function only of ω. The term involving $\delta c/c$ is again of the Abel type and can thus be inverted to infer $\delta c/c$ and hence the unknown sound-speed distribution (Christensen-Dalsgaard et al. 1989a). [Christensen-Dalsgaard et al. do not include the term in $d\alpha/d\omega$ on the left-hand side of Eq. (51). In experiments with artificial data, this makes little difference to the sound speed inferred throughout almost all of the star; but including the term might make a difference to the sound speed inferred in the very outer layers, because for shallowly penetrating modes the two terms multiplying $\delta\omega/\omega$ could be of similar magnitude.] We call this the differential method.

The $\delta\alpha$ term on the right-hand side of Eq. (51) might also in principle be used to make inferences about the Sun. Christensen-Dalsgaard and Pérez Hernández (1988) have investigated how $\delta\alpha$ is related to the equilibrium structure of the outer subphotospheric layers of the Sun. This work could be used as the basis for an inversion of the $\delta\alpha$ term. However, the issue is complicated by the fact that there are undoubtedly other contributions to $\delta\alpha$, from the nonadiabaticity of the modes (assuming that the theoretical frequencies do not adequately take this into account) and from additional physical phenomena such as convective motions and magnetic fields in and above the outer layers of the convective region.

In an attempt to provide a unified comparison of the four methods, we have applied all of them to a set of p-mode frequencies computed for a solar model (Model Y of Christensen-Dalsgaard et al. [1991]). This model is "standard" except for being diffusively mixed in the core by an amount intermediate between a normal unmixed model and the diffusively mixed model of Schatzman et al. (1981). The mode set used corresponds to the modes of degree 0–99 in the compilation by Duvall et al. (1988a) and the p-modes of degree $l \geq 100$ reported by Libbrecht and Kaufman (1988). In addition to inverting a set of error-free data, we have also inverted twenty sets of data to which independent realizations of artificial random errors had first been added. The errors were generated using a parametrized fit to the standard deviations quoted by Duvall et al. and Libbrecht and Kaufman. To be specific, the error added to each frequency was Gaussian distributed, with zero mean and with standard deviation σ which depended upon w and the cyclic frequency v according to (Christensen-Dalsgaard et al. 1991)

$$\sigma^2 = \sigma_w^2 + \sigma_v^2 \tag{52}$$

where

$$\sigma_w^2 = A_w^2((\frac{w*}{w})^3 + 1), \quad \sigma_v^2 = A_v^2(\frac{v}{v*})^5 \tag{53}$$

and $A_w = 0.2$ μHz, $A_v = 3$ μHz, $w*/2\pi = 120$ μHz, $v* = 5000$ μHz.

For a reference model in the differential method, we used Reference Model 3 of Christensen-Dalsgaard et al. (1991, Table 3): this differs from the "mystery" model under consideration in the equation of state and in the opacities used, and has no core mixing, and so provides quite a stringent test of the differential procedure. The relative difference in sound speed between the two models is shown in Fig. 6.

We emphasize that the results presented are only for our implementations of the methods. It is quite likely that with further "tuning" of the implementations, the results could be improved. This is particularly likely for the methods of Brodsky and Vorontsov, and Shibahashi and Sekii, which we have implemented especially for this comparison and of which we do not have much experience. As the results may in particular be sensitive to the ways in which the data were smoothed, we shall try to be quite explicit about these details.

For the inversions using Gough's method, we obtained a smooth representation of F (Eq. 41) by fitting two natural cubic splines, each with 15 knots equally spaced in log w. The dividing point between the splines was at

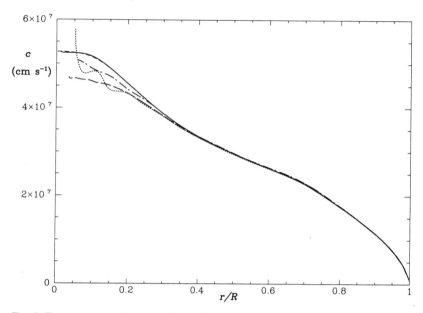

Fig. 6. Exact sound speed as a function of fractional radius in a mystery model (— — —)
together with the sound speeds inferred from the error-free frequencies of 1662 p modes using
the differential inversion method of Christensen-Dalsgaard et al. (———————)
and the methods of Gough (. . . .), Brodsky and Vorontsov (—— ——)
and Shibahashi and Sekii (—.—.).

a value of w corresponding to a mode with lower turning point at a radius of
roughly $0.73\,R$. The surface phase function $\alpha(\omega)$ was estimated in the manner
described by Gough(1986); in order not to contaminate the estimate with
effects of buoyancy and perturbations to the gravitational potential, only
modes whose lower turning points are in the convection zone were used.

For the differential method, we separated the functions of w and ω on
the right of Eq. (51) by fitting to the weighted frequency differences two
splines $\bar{H}_1(w)$ and $\bar{H}_2(\omega)$, using an unweighted least-squares fit. The term in
$d\alpha/d\omega$ on the left of Eq. (51) was neglected. Splines $\bar{H}_1$ and $\bar{H}_2$ were each
defined on twenty knots, in the former case equally spaced in log w (except
for a longer final interval at large w because of the sparsity of deeply penetrating modes; see Christensen-Dalsgaard et al. [1989a]), and in the latter
case equally spaced in ω. A few modes of low order and degree were not
used, as described by Christensen-Dalsgaard et al. The treatment of the surface layers, above the lower turning point of the most shallowly penetrating
mode, was also as described in that paper.

In the Brodsky and Vorontsov method, the derivatives $\partial\omega/\partial L$ and $\partial\omega/\partial n$
were obtained using a parabolic fit through nearest neighbors: derivatives at
end points (of l and n ranges) were not used as these were considered unreli-

able. From these, a smooth representation of dF/dw (Eq. 44) was obtained by fitting a spline with 29 knots equally spaced in log w (except for a double-length interval at large w). For values of w smaller than those corresponding to the modes used, F was approximated to be of the form $F_0 w^\nu$. Constants F_0 and ν were obtained by making a least-squares fit to log (dF/dw) in the range $3 \times 10^{-5}s^{-1} \leq w \leq 7 \times 10^{-5}s^{-1}$.

The method of Shibahashi and Sekii is apparently rather sensitive to the way in which derivatives are taken, etc. (Shibahashi, personal communication) and our limited attempts at tuning our implementation have not improved upon the results presented here. These clearly do not do full justice to the method, as can be judged from comparing our results with those presented by Shibahashi and Sekii (1988). We excluded modes with $l = 0$ and $l > 600$ in addition to f modes. The minima Φ_l^{min} of the acoustic potentials were obtained for each l individually by finding a fit of the form $\omega^2 = A_l n^2 + \Phi_l^{min}$ to the frequencies for the two lowest available values of n. Derivatives $\partial\omega^2/\partial n$ were found using a parabolic fit through nearest neighbors (and simple two-point differences at end-points). The function $s(\omega^2, L^2)$ was then computed on a uniform frequency mesh (of 500 points, though s was only calculated at every fifth point, because of the computation time required). The derivatives $\partial s/\partial L^2$ were similarly computed using parabolic fits (but derivatives at end points were not computed). A spline with 20 knots (evenly spaced in log w, except for one greater interval at large w) was then fitted to $2L^2\partial s/\partial L^2$ after points corresponding to $\nu > 3500$ µHz and certain adjacent pairs of l values had been discarded: these pairs of l values correspond to cases where the computed Φ_l^{min} varied in a nonmonotic way or jumped by a large amount, and correspond to l values ($l \geq 8$ only) where the lowest available n value changes. This was found to be necessary because otherwise the values of $2L^2\partial s/\partial L^2$ departed greatly from a single curve.

Figure 7 shows the results of applying our implementation of each of the four methods to the error-free set of frequencies for the mystery model. The final result for the differential method, which we find to be by far the most accurate in this example, was obtained by multiplying the known sound speed of the reference model by $(1 + \delta c/c)$, where $\delta c/c$ is the inferred relative sound-speed difference.

The accuracy of the inversions can be judged more precisely from Fig. 8, which shows the relative error in each of the inversions, together with the results for the 20 realizations with data errors. Of course, all the methods are in error in the surface layers, where the lack of modes that penetrate sufficiently shallowly makes the results sensitive to the extrapolation used. The Gough inversion of error-free data is accurate to better than 1% throughout most of the star, although the error is much greater than this inside $r = 0.3 R$ where the neglected effects of the perturbed gravitational potential and of buoyancy frequency become important. Indeed, for the same data, the Brodsky and Vorontsov inversion method, which is perhaps a little more accurate

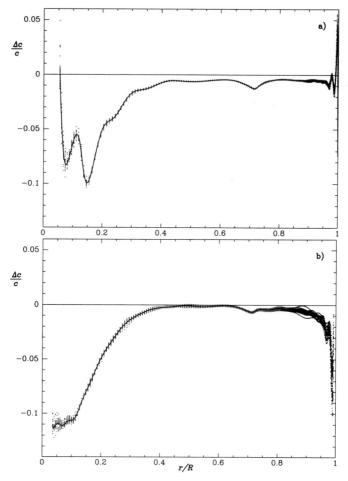

Fig. 7. (a) The relative error in the sound-speed determination for the mystery model, using Gough's inversion method. The result for error-free data is shown as a solid line; the results of 20 realizations of random errors (also discussed in the text) are indicated with dots. (b) As (a), but for the method of Brodsky and Vorontsov. (c) As (a), but for the method of Shibahashi and Sekii. (d) As (a), but for the inversion method of Christensen-Dalsgaard et al. and using the reference model discussed in the text.

in $r > 0.3 \, R$, shows a very similar growth in error from $r \approx 0.3 \, R$ down to $r \approx 0.15 \, R$. Below this point, the details may depend on the amount of smoothing: the results using Gough's method indicate that the spline has too much freedom in this region to follow departures from asymptotic behavior, whereas the spline-knot spacing has been increased at large w for our implementation of Brodsky and Vorontsov's method. Note that, for both these in-

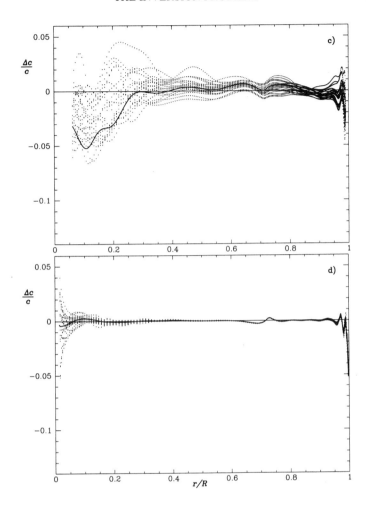

versions, the error arising from realistic data errors is small, $< 0.5\%$ almost everywhere.

The Shibahashi and Sekii inversion results are more accurate than the previous methods in the core, but are marginally worse in much of the rest of the star: the accuracy is similar to that obtained by Shibahashi and Sekii (1988) in most of the star, but they obtain an error smaller by a factor of at least 2 in the core. The effects of adding errors to the data are much greater on this inversion. The error arising from data errors is close to 1% in much of the star, greater than for the previous two methods but similar to what Shibahashi and Sekii find. (In fact, our results seem less sensitive to errors in the outer 20% of the star, although this may simply reflect differences between the precise error properties that have been assumed.) In $r \lesssim 0.4\,R$, the

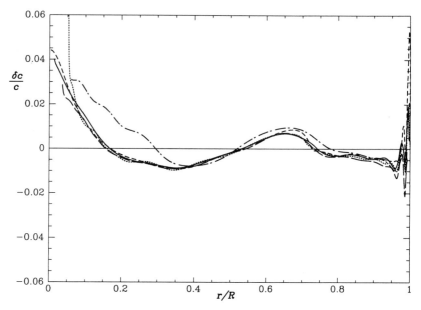

Fig. 8. The exact sound-speed difference as a function of radius (— — —) between the mystery
model and a reference model, discussed in the text, and the sound-speed differences inferred
using the inversion methods of Christensen-Dalsgaard et al. (——————),
Gough (. . . . .), Brodsky and Vorontsov (—— ——) and Shibahashi and Sekii (—.—.).

errors arising from data errors are much greater than 1%. Considering that
the spread in our 20 realizations should correspond very roughly to 2 standard
deviations, our results on the effects of data errors appear to be consistent
with what Shibahashi and Sekii found. The improved accuracy of their
method in the core may to some extent reflect a trade-off between systematic
and random errors, as discussed in Sec. I.

The differential method produces very precise results, with errors
smaller than one quarter of one percent at most radii. The sensitivity to data
errors is also small throughout most of the star. By applying the ideas of Sec.
I to this linear method, one can see that the systematic error near the base of
the convection zone, which is substantially greater than the error anywhere
else except for the inner core and the surface layers, could probably be reduced by increasing the local density of spline knots without making the
effects of random errors, which are tiny at that radius at present, unreasonably
large. In the inner core, the sensitivity to random errors appears to be similar
to that in the Brodsky and Vorontsov inversion. One can see that, although
the error-free inversion is accurate almost to the center, realistic errors in the
data render the inversion very unreliable in the inner 3% by radius. (Note that

results for the Brodsky and Vorontsov inversion, which might at first sight appear to be less sensitive to data errors in this region, are not available this close to the center because modes penetrating this deeply are excluded by our decision not to evaluate derivatives $\partial\omega/\partial l$ at end points of the l range.)

The price to be paid for the accuracy of the differential method is that the result depends on a reference model, whereas the other inversion methods are model independent. Nonetheless, it is interesting to investigate how the other methods perform when used in a differential manner (cf. Christensen-Dalsgaard et al. 1985*b*). To this end, we have used them to perform inversions for the same set of mode frequencies computed for the reference model used in the differential method. The relative difference between the sound speeds inferred for the mystery and reference models can then be computed. The hope of course is that systematic effects in the two inversions largely cancel when differences are taken. For such a cancellation to take place, the details of the implementation may be even more critical than before. (One of the reasons for the development of the differential method of Christensen-Dalsgaard et al. [1989*a*] was to avoid the need for this possibly delicate cancellation.) The results in Fig. 6 show that this procedure is not very successful for our implementation of Shibahashi and Sekii's method. The methods of Gough and of Brodsky and Vorontsov, however, recover the sound-speed difference very well: the results are comparable with those of the "straight" differential method. The inversion using Gough's method becomes unreliable in the inner core, but as mentioned above this is probably caused by having too great a density of spline knots.

We may conclude from this comparison that the methods of Gough and of Brodsky and Vorontsov can be made quite insensitive to what we suppose are realistic random errors in the data and that simultaneously it appears possible to reproduce the sound-speed distribution in much of the star with an error of $< 1\%$. The systematic errors in the core are substantial, however. It may indeed be possible to obtain more accurate results in the core using the method of Shibahashi and Sekii, but for our implementation of this method, we do find a large sensitivity to random data errors. All three inversions have the advantage that they are independent of any reference model. It does appear, however, that to reduce systematic errors in the core region by a substantial amount with mode sets similar to those currently available requires one either to go beyond the simple asymptotic description upon which these inversions have been based or else to adopt some kind of differential inversion procedure. Gough (1986), Brodsky and Vorontsov (1987,1988) and Vorontsov (1988) have shown how their methods can be extended beyond the level at which we have applied them here to take some account of higher-order asymptotic terms. Further comparisons should evidently be performed to evaluate the effects of these enhancements.

Numerical Inversions for the Solar Structure

While the problem of inferring the Sun's internal rotation is to a good approximation linear, the dependence of the mode frequencies on the solar structure is nonlinear. This is one reason why asymptotic inversion methods have generally been used to infer the solar sound speed, while "numerical" inversions have been used for inferring the rotation rate. (A second reason, that the structural inversions in general involve more than one unknown function, is addressed below.) Because the asymptotics are only approximate, however, one would expect with sufficiently accurate data that better inversions could be achieved with numerical methods.

The problem of inferring the solar structure can be made amenable to the linear inversion methods discussed above by linearizing about a reference model which has the correct radius R and mass M. Provided the model is not too far removed from reality, one hopes that the approximation introduced by linearizing will not be too severe. The method has been sketched out by Gough (1985a).

The starting point is the variational formulation of the adiabatic eigenvalue problem (see, e.g., Ledoux and Walraven 1958):

$$\omega^2 \int \rho \boldsymbol{\xi}^* \cdot \boldsymbol{\xi} dV = \int \left[\rho c^2 |\mathrm{div}\boldsymbol{\xi}|^2 + 2\mathrm{Re}((\boldsymbol{\xi}^* \cdot \nabla p)\mathrm{div}\boldsymbol{\xi}) + \rho^{-1}(\boldsymbol{\xi}^* \cdot \nabla p) \right.$$

$$(\boldsymbol{\xi} \cdot \nabla \rho) \Big] dV - G \iint |\mathbf{r} - \mathbf{r}'|^{-1} \mathrm{div}(\rho\boldsymbol{\xi}^*)\mathrm{div}'(\rho\boldsymbol{\xi}) dV' dV \qquad (54)$$

in the same notation as in the chapter by Christensen-Dalsgaard and Berthomieu. (Here primed quantities and operators are with respect to the dummy variables $\mathbf{r}'$. The integrals are over the volume of the star. The star (*) denotes complex conjugate, and Re denotes the real part of a complex quantity.) One now perturbs about the structure and eigensolutions of the reference model, which introduces perturbation quantities $\delta\rho$, δc, δp (all evaluated at constant radius r), and $\delta\omega$. However, by the variational principle the quantity $\delta\boldsymbol{\xi}$ can be dropped and the frequency perturbation $\delta\omega$ will still be accurate to first order in $\delta\rho$, δc and δp. Then linearizing in the perturbation quantities, and restricting attention to perturbations that are functions only of r, one obtains

$$2\omega^2 S \frac{\delta\omega}{\omega} = 2 \int \rho c^2 \frac{\delta c}{c} \chi^2 r^2 dr + \int [c^2 \chi^2 - \omega^2(\xi^2 + L^2\eta^2)] \frac{\delta\rho}{\rho} \rho r^2 dr$$

$$- \int \frac{d}{dr}\left(\xi^2 r^2 \frac{dp}{dr}\right) \frac{\delta\rho}{\rho} dr + \int \left(2\xi\chi + \rho^{-1}\xi^2 \frac{dp}{dr}\right) \frac{d(\delta p)}{dr} r^2 dr \qquad (55)$$

$$+ \int \delta\Phi'\left(\rho\chi + \xi\frac{dp}{dr}\right) r^2 dr + \int \Phi'\left(\chi\delta\rho + \xi\frac{d\delta\rho}{dr}\right) r^2 dr,$$

$$S = \int (\xi^2 + l(l + \delta 1)\eta^2)\rho r^2 dr \qquad (56)$$

and all the integrals are evaluated from $r = 0$ to $r = R$. Here $\chi(r)$ is given by $\text{div}\boldsymbol{\xi} = \chi(r)Y_l^m(\theta,\phi)$, Y_l^m being a surface harmonic function, and Φ' is the Eulerian perturbation to the gravitational potential (cf. chapter by Christensen-Dalsgaard and Berthomieu) and satisfies

$$\nabla^2(\Phi'Y_l^m) = 4\pi G\,\text{div}(\rho\boldsymbol{\xi}) \tag{57}$$

$\xi(r) Y_l^m(\theta,\phi)$ is the radial component of $\boldsymbol{\xi}$. All unperturbed quantities refer to the reference model. [As discussed below, some surface terms have been dropped in deriving Eq. (55).] The equation of hydrostatic support can also be perturbed to give $d(\delta p)/dr$ in terms of $\delta\rho$:

$$\frac{d(\delta p)}{dr} = -\frac{Gm}{r^2}\delta\rho - \frac{G\rho}{r^2}\int_0^r 4\pi r^2\delta\rho\,dr. \tag{58}$$

Thus the number of unknown functions in Eq. (55) can be reduced from three to two, yielding

$$\left.\frac{\delta\omega}{\omega}\right|_{nl} = \int_0^R\left[K_{c,\rho}^{(n,l)}(r)\frac{\delta c}{c}(r) + K_{\rho,c}^{(n,l)}(r)\frac{\delta\rho}{\rho}(r)\right]dr \tag{59}$$

where the kernels $K_{c,\rho}^{(n,l)}$ and $K_{\rho,c}^{(n,l)}$ are given by

$$\omega^2 S K_{c,\rho}^{(n,l)} = \rho c^2\chi^2 r^2, \tag{60}$$

$$\omega^2 S K_{\rho,c}^{(n,l)} = -\frac{1}{2}(\xi^2 + l(l+1)\eta^2)\rho\omega^2 r^2 + \frac{1}{2}\rho c^2\chi^2 r^2$$

$$- Gm\rho(\chi + \frac{1}{2}\xi\frac{d\ln\rho}{dr})\xi - 4\pi G\rho r^2\int_r^R(\chi(s) + \frac{1}{2}\xi(s)\frac{d\ln\rho}{ds})\xi(s)\rho(s)\,ds$$

$$+ Gm\rho\xi\frac{d\xi}{dr} + \frac{1}{2}G(m\frac{d\rho}{dr} + 4\pi r^2\rho^2)\xi^2 \tag{61}$$

$$- \frac{4\pi G}{2l+1}\rho\Big((l+1)r^{-l-1}(\xi - l\eta)\int_0^r(\rho\chi + \xi\frac{d\rho}{ds})s^{l+2}\,ds$$

$$- lr^{l+1}(\xi + (l+1)\eta)\int_r^R(\rho\chi + \xi\frac{d\rho}{ds})s^{-(l-1)}\,ds\Big).$$

In deriving the density kernel it is convenient to use a Green function to relate Φ' to $\text{div}(\rho\boldsymbol{\xi})$ thus:

$$\Phi'(r)Y_l^m = -\frac{4\pi G}{2l+1}\left(\int_0^r\left(\frac{s}{r}\right)^{l+1}s\,\text{div}_{(s)}(\rho\boldsymbol{\xi})\,ds + \int_r^R\left(\frac{r}{s}\right)^l s\,\text{div}_{(s)}(\rho\boldsymbol{\xi})\,ds\right). \tag{62}$$

To the integral constraints (Eq. 59) on the unknown functions, a further constraint

$$0 = \int_0^R 4\pi r^2 \rho \frac{\delta \rho}{\rho} dr \qquad (63)$$

can be added to ensure that the total mass M is conserved. Equations (59) and (63) are of the form

$$d_i = \int_0^R \left[K_{c,\rho}^{(i)} \frac{\delta c}{c} + K_{\rho,c}^{(i)} \frac{\delta \rho}{\rho} \right] dr \quad i \in \mathcal{M}. \qquad (64)$$

Examples of such kernels are shown in Fig. 9. Between the turning points, the sound-speed kernel for a higher-order p mode oscillates about roughly the value

$$c^{-1}(1 - L^2 c^2 / \omega^2 r^2)^{-1/2} / \int_0^R (1 - L^2 c^2 / \omega^2 r^2)^{-1/2} \frac{dr}{c} \qquad (65)$$

predicted by asymptotic theory, while the density kernel oscillates about zero. This is consistent with the prediction of leading-order asymptotic theory, that the frequencies are determined principally by the sound speed.

In arriving at Eq. (59) various surface terms have been omitted. At the outset, the right-hand side of Eq. (54) has additional terms unless specific boundary conditions hold at the stellar surface (cf. Christensen-Dalsgaard 1982), so that in general the variational formulation in the form Eq. (54) does not hold exactly. Further surface terms arising from integrations by parts were

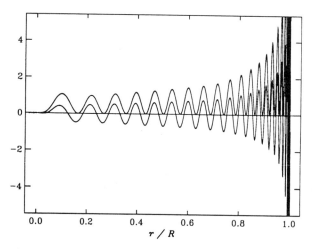

Fig. 9. Kernels $RK_{c,\rho}$ and $RK_{\rho,c}$ for an $l = 1$, $n = 21$ p mode ($\nu = 3.1$ mHz). The kernel $K_{c,\rho}$ is positive everywhere, while the kernel $K_{\rho,c}$ takes both positive and negative values.

dropped in deriving Eq. (59) from Eq. (54). A "justification" at the present time for not worrying about omitting these terms is that there are many other errors and uncertainties near the stellar surface in present calculations of normal frequencies. Nonadiabatic effects, omitted in most calculations, affect the mode dynamics in the surface layers. Also solar models omit many uncertain details near the surface, for example, convective motions, small-scale inhomogeneities and magnetic fields. These affect not only the mean effective properties of the medium but also the nature and position of the boundary of the acoustical cavity. Except at high degree, one expects from simple asymptotics that errors in the surface layers will give rise to errors in the computed frequencies (when suitably scaled) that are independent of l (see the chapter by Christensen-Dalsgaard and Berthomieu). This indeed appears to be true of the predominant discrepancy between the computed and observed frequencies of five-minute modes. Thus one should perhaps include on the right-hand side of Eq. (59) an unknown function of frequency, multiplied by S; and one's inversion technique should be made robust against the uncertainty of this function. Such a function can readily be accommodated in a least-squares inversion (Dziembowski et al. 1990; Däppen al. 1990).

Kernels for $\delta\rho/\rho$ and $\delta\gamma/\gamma$ (where γ is the adiabatic exponent $\partial\ln p/\partial\ln\rho$ at constant specific entropy) instead of for $\delta\rho/\rho$ and $\delta c/c$ can also easily be derived, by noting in Eq. (55) that

$$2\frac{\delta c}{c} = \frac{\delta\gamma}{\gamma} + \frac{\delta p}{p} - \frac{\delta\rho}{\rho} \tag{66}$$

and integrating Eq. (58) to express δp in terms of $\delta\rho$. (A boundary condition is required in order to perform this integration. In the spirit of the comments in the previous paragraph, one can for example impose $\delta p = 0$ at $r = R$.) Analogous to Eq. (64) one then has

$$d_i = \int_0^R \left[K_{\gamma,\rho}^{(i)} \frac{\delta\gamma}{\gamma} + K_{\rho,\gamma}^{(i)} \frac{\delta\rho}{\rho} \right] dr \qquad i \epsilon \mathcal{M}. \tag{67}$$

Explicit expressions for $K_{\gamma,\rho}$ and $K_{\rho,\gamma}$ are given by Gough (1991).

Both Eqs. (64) and (67) comprise a linear problem, but one that involves two functions. The question of inverting for more than one function will be addressed presently. However, if the number of unknown functions could be reduced to one, the linear methods of Sec. I could be applied directly. This can be achieved if one assumes that the equation of state of the solar material is known, as this gives a relationship between the two functions. Alternatively, one might assume that the contribution from one of the two terms in each integrand is negligible. Thus, guided by the asymptotic theory one might neglect the term $K_{\rho,c}^{(i)} \delta\rho/\rho$ in Eqs. (64). (This contribution is likewise neglected in the leading-order asymptotic inversions of Sec. II.) Similarly, Kosovichev (1968,1988a) has formulated the inversion for $\delta\rho/\rho$ using g-mode frequencies in the manner of Eq. (67) with $\delta\gamma/\gamma$ assumed to be zero, as g modes can have substantial amplitude only in the radiative interior where the solar material may reasonably be assumed to be highly ionized. [It should be noted that the formulation (Eq. 67) is not natural for p modes, as these modes are primarily sensitive to changes in the sound speed and do not readily distinguish whether such changes arise from perturbations to the adiabatic exponent or to the hydrostatic structure.]

These are just two examples of pairs of functions that can be considered to be the independent unknowns. Other examples are $\delta c/c$ and $\delta\gamma/\gamma$, and $\delta c/c$ and δN (where N is the buoyancy frequency). The kernels for such pairs can be derived in terms of the previous kernels from the equations of stellar structure, although less straightforwardly than for the cases considered thus far. As an example, consider the case of $\delta c/c$ and $\delta\gamma/\gamma$. We follow the method given by Masters (1979) in a geophysical context. Assuming that the required kernels exist, one has

$$d_i = \int_0^R \left[K_{c,\gamma}^{(i)} \frac{\delta c}{c} + K_{\gamma,c}^{(i)} \frac{\delta\gamma}{\gamma} \right] dr \qquad (68)$$

which is also equal to the right-hand side of Eq. (64). Then, using Eq. (66),

$$\int K_{c,\rho}^{(i)} \frac{\delta c}{c} + K_{\rho,c}^{(i)} \frac{\delta\rho}{\rho} \, dr = \int (K_{c,\gamma}^{(i)} + 2K_{\gamma,c}^{(i)}) \frac{\delta c}{c}$$
$$+ \int K_{\gamma,c}^{(i)} \left(\frac{\delta\rho}{\rho} - \frac{\delta p}{p} \right) dr \qquad (69)$$

whence it can be shown by expressing $\delta p/p$ in terms of $\delta\rho/\rho$ and equating the coefficients of $\delta c/c$ and $\delta\rho/\rho$ on both sides that

$$K_{\rho,c}^{(i)} = K_{\gamma,c}^{(i)} - \frac{Gm\rho}{r^2} \int_0^r \frac{K_{\gamma,c}^{(i)}(r')}{p(r')} dr'$$
$$+ 4\pi G\rho r^2 \int_0^r \frac{K_{\gamma,c}^{(i)}(r')}{p(r')} \left(\int_{r'}^r \rho(s)s^{-2}ds \right) dr' \qquad (70)$$

and

$$K_{c,\rho}^{(i)} = K_{c,\gamma}^{(i)} + 2K_{\gamma,c}^{(i)}. \qquad (71)$$

Equation (70) is an integral equation for $K_{\gamma,c}^{(i)}$ in terms of the known kernel $K_{\rho,c}^{(i)}$: this can be solved iteratively (Masters 1979). Alternatively, it can be cast in terms of a differential equation for $K_{\gamma,c}^{(i)}$ (Gough 1991). Once $K_{\gamma,c}^{(i)}$ is known, the kernel $K_{c,\gamma}^{(i)}$ can be found, using Eq. (71). An inversion for sound speed in the radiative interior of the Sun using these kernels, and assuming that the contributions from $K_{\gamma,c}^{(i)}$ $\delta\gamma/\gamma$ in this region are negligible, has been performed by Gough and Kosovichev (1988).

Inversions for More than one Function

To invert Eqs. (64), (67) or (68), for example, without further assumptions or approximation, requires methods of inverting for two or more functions simultaneously. This requirement has also arisen in the problem of inferring the two-dimensional variation of the solar rotation rate, as to date this has been approached not as a genuine two-dimensional problem but by parametrizing the rotation in terms of two or three functions of radius (cf. chapters by Libbrecht and Morrow and by Dziembowski and Goode).

Least-squares methods of inversion are in principle extendable in an obvious manner to the case of more than one function, though little work has been done on the systematic errors arising from cross-talk between the different functions through the averaging kernels. Least-squares inversions for the parametrized latitudinally dependent rotation have been performed by Dziembowski et al. (1989), Korzennik et al. (1988) and Thompson (1990) (see also the chapter by Dziembowski and Goode). A limited investigation of the systematic errors in this particular problem has been performed by Thompson (1990).

The problem of inverting for the parametrized latitudinally dependent rotation is atypical, because by choosing a suitable parametrization the inversions for the different functions of radius become decoupled in the limit of large l. This fact has been exploited by Brown et al. (1989), Christensen-Dalsgaard and Schou (1988) and Korzennik et al. (1988) to perform inversions using optimally localized averages. Brown et al. and Korzennik et al. have also used the decoupling to perform asymptotic and spectral expansion inversions, respectively.

The method of optimally localized averages can be adapted to the inversion of more than one function even when no such decoupling is possible. In the case of two functions, for example, expression (13) can be replaced by (Backus and Gilbert 1968)

$$\cos\theta\left(12 \int_0^R (r - r_0)^2 (\sum_{i\in\mathcal{M}} c_i(r_0)K_{1i}(r))^2 dr\right.$$
$$\left. + \lambda \int_0^R (\sum_{i\in\mathcal{M}} c_i(r_0)K_{2i}(r))^2 dr\right) + \mu \sin\theta \sum_{i,j} E_{ij}c_i(r_0)c_j(r_0) \tag{72}$$

with the parameter λ chosen to be positive. Here K_{1i} and K_{2i} are the mode kernels pertaining to the first and second functions. The choice of parameters θ and λ now weights the relative importance of forming a well-localized average of the first function, keeping the random error in the solution small and keeping the sensitivity to the second function small. Figure 10 shows the resulting averaging kernels when this method is used to form a localized average of $\delta c/c$ with little interference from $\delta\rho/\rho$. It is possible in this case to reduce the interference without making the averaging kernels for $\delta c/c$ (Fig. 10a) much wider than when λ is set to zero. Note that the amplitude of the resulting averaging kernel that determines the contribution from $\delta\rho/\rho$ to the determination of $\delta c/c$ has been much reduced (Fig. 10b). However, experiments with artificial data show that this may not result in a noticeably better inversion for $\delta c/c$, as the main cause of systematic error in present solar inversions for $\delta c/c$ with realistic data sets and data errors is probably the poor

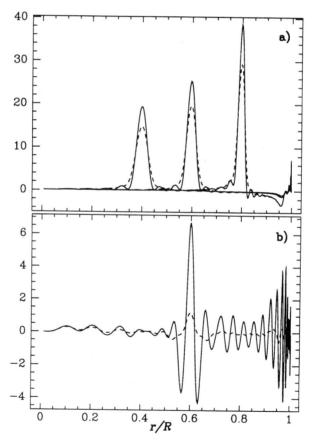

Fig. 10. (a) Optimally localized kernels for $\delta c/c$ (the other variable being $\delta\rho/\rho$) for $r_0/R = 0.4$, 0.6 and 0.8, with $\lambda = 0$ (solid curves) and with $\lambda = 0.1$ (dashed curves). (b) Corresponding to the $r_0 = 0.6\,R$ case shown in (a), the functions $\Sigma_i c_i(r_0)\,K_{p,c}^i(r)$ for $\lambda = 0$ (solid curve) and $\lambda = 0.1$ (dashed curve). [$\theta = 10^{-3}$].

resolution of the averaging kernels rather than sensitivity to contributions to the frequency from factors other than the sound speed. This balance may change in the future with extended data sets and more accurate frequency determinations.

III. ASPHERICITY INVERSIONS

Structural Inversions in More than One Spatial Dimension

The derivation of numerical kernels in the previous section can be extended to the case where the perturbations are no longer spherically symmetric. The simplest extension is to the case of an axisymmetric perturbation $\delta c/c$. If the stellar surface were not distorted and (for example) the contribution from $\delta\rho/\rho$ were negligible, Eq. (59) would generalize to

$$\frac{\delta\omega}{\omega}\bigg|_{nlm} = \int_0^R K_c^{(n,l)}(r) \left(\int_{-1}^1 (P_l^m(\cos\theta))^2 \frac{\delta c}{c} d\cos\theta \bigg/ \int_{-1}^1 (P_l^m(x))^2 dx \right) dr \quad (73)$$

where $K_c^{(n,l)}$ is simply $K_{c,\rho}^{(n,l)}$, given by Eq. (60). The angular integration follows from the three-dimensional integrals in Eq. (54). A method of allowing for a distortion of the surface has been discussed and applied by Gough and Thompson (1990), who also incorporated the density perturbation for the particular case of a perturbing axisymmetric magnetic field. If the star no longer has an axis of symmetry, the situation is complicated by the fact that the eigenfunctions no longer have the simple form assumed above (see the chapter by Christensen-Dalsgaard and Berthomieu). Nonetheless, it is straightforward to extend the perturbation theory to this case (Lavely and Ritzwoller 1991).

The least-squares inversion can again be extended to multidimensional inversions. Such inversions have been performed in geophysics (see, e.g., Giardini et al. 1987). For example, one can seek a piecewise-constant approximation to the unknown function by dividing the space into two- or three-dimensional cells, analogous to the one-dimensional dissection of Eqs. (17), (18), and (19). The method of optimally localized averages might also be adapted by modifying Eq. (13) so that r and r_0 are vectors and the integrals are over the required number of dimensions.

Just as the resolution in depth tends to become worse with increasing depth if only p-mode data are available, so too does the resolution in latitude, owing to the small number of different values of m for low-degree modes. (In the extreme case, for the very central regions penetrated only by $l = 0$ p modes, no angular resolution will be available without g-mode data.) On the other hand, near the surface the $(2l + 1)$ different modes associated with each value of n and l give very good resolution at high degree.

Asymptotic methods are also extendable to functions that vary in latitude as well as in depth. The spherically symmetric asymptotic analysis of the chapter by Christensen-Dalsgaard and Berthomieu can be generalized to admit aspherical perturbations to the sound speed. In the case of axisymmetric perturbations (so $\delta c/c$ is a function only of radius r and colatitude θ), the asymptotic limit of Eq. (73) is

$$
\int_{r_t}^{R} \left(1 - \frac{L^2 c^2}{\omega^2 r^2}\right)^{-1/2} \frac{dr}{c} \cdot \left.\frac{\delta\omega}{\omega}\right|_{nlm} \sim \int_{r_t}^{R} \left(1 - \frac{L^2 c^2}{\omega^2 r^2}\right)^{-1/2}
$$
$$
\left(\int_{-1}^{1} (P_l^m (\cos\theta))^2 \frac{\delta c}{c} d\cos\theta \,\middle/\, \int_{-1}^{1} (P_l^m(x))^2 dx\right) \frac{dr}{c}
\tag{74a}
$$

by substituting for P_l^m an asymptotic expression valid at high degree l, or from ray theory, one can also obtain the asymptotic relation

$$
\int_{r_t}^{R} \left(1 - \frac{L^2 c^2}{\omega^2 r^2}\right)^{-1/2} \frac{dr}{c} \cdot \left.\frac{\delta\omega}{\omega}\right|_{nlm} \sim \frac{1}{\pi} \int_{r_t}^{R} \left(1 - \frac{L^2 c^2}{\omega^2 r^2}\right)^{-1/2}
$$
$$
\left(\int_{-\cos\Theta}^{\cos\Theta} (\cos^2\Theta - \cos^2\theta)^{-1/2} \frac{\delta c}{c} d\cos\theta\right) \frac{dr}{c}
\tag{74b}
$$

(Gough 1990a; cf. Gough and Thompson 1990), where $\Theta = \sin^{-1}(m/L)$. Any component of $\delta c/c$ that is an odd function of $\cos\theta$ makes no contribution to this integral; therefore, for convenience, we shall suppose that $\delta c/c$ has no such component. The second equation is of the Abel type and can be written (replacing the asymptotic relation by an exact equality)

$$
d(w;\Theta) = \frac{-1}{\pi} \int_{a_s}^{w} \int_{-\cos\Theta}^{\cos\Theta} \left(1 - \frac{a^2}{w^2}\right)^{-1/2}
$$
$$
(\cos^2\Theta - \mu^2)^{-1/2} \frac{\delta c}{c} a^{-1} \frac{d\ln r}{da} d\mu \, da
\tag{75}
$$

(where $a = c/r$, $w = \omega/L$ and $\mu = \cos\theta$). Thus $d(w;\Theta)$, the (observable) left-hand side of Eqs. (74), is related by Eq. (75) to the unknown function $\delta c/c$. This can be inverted to give, for $\mu \geq 0$,

$$
\frac{\delta c(r,\mu)}{c(r)} = \frac{-2}{\pi} a(r) \frac{\partial}{\partial\ln r} \frac{\partial}{\partial\mu} \int_{a_s}^{a(r)} \left[\int_{0}^{\mu} \cos\Theta(\mu^2 - \cos^2\Theta)^{-1/2} \times \right.
$$
$$
\left. (a^2 - w^2)^{-1/2} d(w;\Theta) \, d\cos\Theta\right] dw.
\tag{76}
$$

Details on inverting an Abel integral may be found, for example, in the paper by Christensen-Dalsgaard et al. (1989). Alternatively, one may seek to determine the latitudinal dependence of $\delta c/c$ in terms of an expansion in Legendre polynomials $P_{2k}(\cos\theta)$:

$$\frac{\delta c}{c} \approx \sum_{k \geq 1} f_{2k}(r) P_{2k}(\cos\theta). \tag{77}$$

(Of course, cast in this form the problem is one of inverting for more than one function which has been discussed in the previous section.) The quantity $(\delta\omega/\omega)_{nlm}$ in Eq. (74) is the relative difference between the observed frequency ω_{nlm} and the frequency $\omega \equiv \omega_{nl}$ of a spherically symmetric reference model. This equation can be used to obtain an expression for the difference between observed frequencies due to an axisymmetric latitudinal component of the sound speed:

$$\frac{\omega_{nlm} - \omega_{nl0}}{\omega_{nl}} = S^{-1} \int_{r_t}^{R} \left[\left(1 - \frac{L^2 c^2}{\omega^2 r^2} \right)^{-1/2} \times \right.$$
$$\left. \int_{-1}^{1} \left(\frac{(P_l^m(\mu))^2}{\int_{-1}^{1}(P_l^m(x))^2 dx} - \frac{(P_l^0(\mu))^2}{\int_{-1}^{1}(P_l^0(x))^2 dx} \right) \frac{\delta c}{c} d\mu \right] \frac{dr}{c} \tag{78}$$

where

$$S = \int_{r_t}^{R} \left(1 - \frac{L^2 c^2}{\omega^2 r^2} \right)^{-1/2} \frac{dr}{c}. \tag{79}$$

The observed quantity $(\omega_{nlm} - \omega_{nl0})$ is commonly reported in terms of coefficients a_i (which strictly are functions n and l):

$$\omega_{nlm} - \omega_{nl0} \approx \sum_{i=0}^{I} La_i P_i\left(\frac{m}{L}\right) \tag{80}$$

for some I (cf. the chapter by Libbrecht and Morrow). The coefficients a_i for odd i carry information about the Sun's angular velocity distribution (see the chapters by Libbrecht and Morrow and by Dziembowski and Goode). The even-degree coefficients a_{2k} can be related to the functional coefficients $f_{2k}(r)$ by noting that

$$\int_{-1}^{1} P_{2k}(x)(P_l^m(x))^2 dx \,\bigg/\, \int_{-1}^{1} (P_l^m(x))^2 dx \sim \frac{(-1)^k (2k)!}{2^{2k} k!^2} P_{2k}(m/L) \tag{81}$$

for $l \gg k$. Thus the term $f_{2k}(r)P_{2k}(\cos\theta)$ in Eq. (77) contributes to La_{2k} an amount

$$\frac{(-1)^k (2k)!}{2^{2k} k!^2} \frac{\omega}{S} \int_{r_t}^{R} \left(1 - \frac{L^2 c^2}{\omega^2 r^2} \right)^{-1/2} f_{2k}(r) \frac{dr}{c}. \tag{82}$$

Provided the coefficients a_{2k} are first corrected for the second-order rotational effects, and assuming that there are no other significant contributions to a_{2k}, expression (82) leads to the inversion formula

$$f_{2k}(r) = -\frac{2a(r)}{\pi} \frac{d}{d\ln r} \int_{a_s}^{a(r)} (a^2(r) - w^2)^{-1/2} D_{2k} dw \qquad (83)$$

where (cf. Gough 1988a)

$$D_{2k} \equiv \frac{2^{2k}(k!)^2}{(-1)^k(2k)!} \frac{La_{2k}}{\omega_{nl}}. \qquad (84)$$

It should be emphasized that these formulae hold only when the stellar surface is undistorted and the perturbation is scalar. Moreover, the perturbation is assumed to be axisymmetric about the rotation axis of the star. The inversion in other cases, for example, in the presence of a buried magnetic field, is likely to be more complicated. A magnetic inversion using both optimally localized averages and least-squares methods has recently been performed by Dziembowski and Goode (1989), who made the problem tractable by assuming a particular simple geometry for the field.

Inversions for Latitudinally Dependent Rotation

Similar asymptotic formulae hold for latitudinally dependent rotation (Brown et al. 1989; Kosovichev and Parchevski 1988; Gough 1991). The fact that Eq. (81) holds only asymptotically means that the problems of inferring the functions f_{2k} (Eq. 83) from the coefficients a_i are coupled: the same is true for the rotational inversion, although by choosing a different expansion for the frequency splitting, the problems can be decoupled (Ritzwoller and Laveley 1990). Numerical solutions can be obtained by an iterative procedure (cf. Brown et al.) or by solving for all the unknowns simultaneously (using, for example, the techniques discussed in Sec. II).

IV. INVERSIONS OF SOLAR DATA

Rotation

Application of the inversion methods discussed in this chapter has provided, in particular, estimates of the Sun's internal rotation and sound speed. The rotation of the Sun is discussed in detail in the chapters by Libbrecht and Morrow and by Dziembowski and Goode, so we restrict ourselves here to classifying briefly the investigations of the rotation according to the inversion method used; for a discussion of the results obtained, we refer the reader to the original papers and to these other chapters.

The method of optimally localized averages has been used extensively for inferring the internal rotation. It has been used to investigate the variation of rotation rate with depth over a wide range of depths by Duvall et al. (1984), and in the outer subphotospheric layers by Hill et al. (1988*b*) and Hill et al. (1988*a*). It has been applied to the study of the rotation as a function of latitude and depth by Brown et al. (1989), Christensen-Dalsgaard and Schou (1988) and Korzennik et al. (1988).

Duvall et al. (1984) have also investigated the variation of rotation with depth using two least-squares methods (without regularization terms), one with a piecewise-constant approximation and the other using a polynomial fit. Regularized least-squares inversions to infer the latitudinal and depth dependence have been performed by Dziembowski et al. (1989) using a piecewise linear approximation and first-derivative smoothing, and by Thompson (1990) using a piecewise-constant approximation and second-derivative smoothing. Korzennik et al. (1988) have similarly performed an inversion with second-derivative smoothing; they have also presented results obtained using a spectral expansion method.

Linear asymptotic inversions for the variation of the rotation rate with depth and latitude have been performed by Kosovichev (1988*b*) and by Brown et al. (1989).

Stratification

Inversions for the variation of sound speed with radius inside the Sun using the asymptotic methods of Sec. II have been performed by Christensen-Dalsgaard et al. (1985*b*), Vorontsov (1988), Shibahashi and Sekii (1988), Christensen-Dalsgaard et al. (1988*a*) and Kosovichev (1988*c*). Gough and Kosovichev (1988), Dziembowski et al. (1990) and Däppen et al. (1990) have performed numerical inversions, the first assuming that the contribution from $\delta\gamma/\gamma$ (see Sec. II) is negligible. All these inversions have produced similar results: as an example, those of Christensen-Dalsgaard et al., in terms of the relative difference in sound speed between the Sun and Model 1 of Christensen-Dalsgaard (1982), are shown in Fig. 11. The solar sound speed is found to be higher than that of Model 1 in $0.3 \lesssim r/R \lesssim 0.5$ by $\sim 1\%$. This can be explained by the theoretical opacities being too small for temperatures corresponding to this location in the Sun: Christensen-Dalsgaard et al. (1985*b*) estimated that increasing the opacity by $\sim 20\%$ in the radiative layers immediately beneath the base of the convection zone would essentially eliminate the discrepancy.

In spite of the narrow uncertainty limits, the features in Fig. 11 above $r \approx 0.9\, R$ are possibly not significant, but may instead be due to systematic error in either the high-degree data or the asymptotic inversion: if real, however, the substantial hump in $\delta c/c$ at $r \approx 0.95\, R$ might be caused by some inadequacy of the theoretical equation of state. The features below $r/R = 0.2$ are possibly also not significant. However, the negative $\delta c/c$ at $r/R \approx 0.2$ is a recurrent feature in these inversions. It could conceivably arise from partial

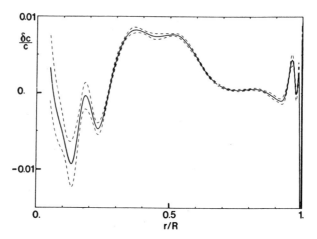

Fig. 11. The relative difference in sound speed between the Sun and Model 1 of Christensen-Dalsgaard (1982), in the sense Sun minus model, as inferred using the differential inversion method. The dashed curves indicate ± 1σ limits based on the residuals from the asymptotic fit to the scaled data (figure from Christensen-Dalsgaard et al. 1988a).

mixing of the core (cf. Fig. 23 of the Christensen-Dalsgaard and Berthomieu chapter), although this is but one possibility.

Because of the relatively abrupt change in the gradient of the square of the sound speed expected at the base of the convection zone, it has been possible from sound-speed inversions to infer that the base of the convection zone is located at $r \approx 0.71\ R$ (Christensen-Dalsgaard et al. 1985b). A careful determination using more recent data but based on the same principle has yielded an answer of $r/R = 0.713 \pm 0.003$ (Christensen-Dalsgaard et al. 1990b). These results are in agreement with the value inferred not by formal inversion techniques but by model fitting (Berthomieu et al. 1980; Shibahashi et al. 1983; Cox et al. 1989).

Inversions for sound speed and rotation are examples of primary inversions, because the frequencies and frequency splittings depend directly on these quantities. It is possible also to perform inversions for quantities upon which the data depend only indirectly but which are themselves of considerable interest. Examples are the helium abundance in the solar convection zone and the opacity of material in the Sun's radiative interior. The frequencies of p modes depend on these quantities primarily through the sound speed. The simplest way to perform such an inversion is to compute the forward problem for a number of different cases and find which (in some sense) best fits the data. This approach might be regarded as model fitting. In this way one of the earliest indirect inferences of helioseismology, using solar models evolved with different chemical compositions, was that the fractional solar helium abundance by mass (Y) is apparently 0.25 ± 0.02 (Christensen-Dalsgaard and Gough 1981; Gough 1982; Shibahashi et al. 1983).

Another method for inferring the helium abundance of the convection zone is to use the signature in the sound speed of the helium ionization zone (Däppen and Gough 1984,1986; Däppen, et al. 1988b; see also the chapter by Däppen et al.). Because the second ionization zone of helium is located in an adiabatically stratified region and the mass interior to the ionization zone is very nearly the entire solar mass, one can show that

$$\frac{r^2}{GM}\frac{dc^2}{dr} \equiv W_0(r) \approx \Theta(r) \equiv \frac{(1 - \gamma_\rho - \gamma)}{1 - \gamma_{c^2}} \tag{85}$$

in this region (Gough 1984b). (Here γ_ρ and γ_{c^2} are, respectively, the partial derivatives of log γ with respect to log ρ and log c^2 at constant c^2 and ρ.) The inference of W_0 from the p-mode frequencies is a primary inversion. Because the agreement between W_0 and Θ is likely to be very good in this region, the function Θ can be inferred, subject to whatever errors there are in the inversion for W_0. The function Θ takes the value $-2/3$ in most of the convection zone but is less negative in the He II ionization zone. Plotted against radius it exhibits a hump in that region. The height of the hump depends among other things on the helium abundance. The inference of the helium abundance Y is still however a secondary inversion, even though this determination should be independent of assumptions about the Sun's evolution, since the function $\Theta(r)$ depends not only on Y but also on the uncertain equation of state and the treatment of convection. No reliable value for the solar helium abundance has yet been produced by this method. However, if successful, the method will provide an estimate of Y that is largely independent of the uncertain details of the Sun's evolution, to which the earlier estimates may be sensitive.

One might use model fitting to investigate possible errors in the standard opacities used to construct solar models, but the number of forward calculations could be prohibitive. A more sophisticated approach along these lines is to parametrize the variations to be investigated, deduce from forward calculations the sensitivity of each mode frequency to a small change in each parameter separately, and then use the derivatives of each mode frequency with respect to each parameter to find the optimal parameter values to fit the observations. This approach has been used iteratively by Korzennik and Ulrich (1989) to perform an inversion for opacity. Depending on the precise way in which the method was applied to the solar data, they obtained a number of solutions that reduced the discrepancies between theoretical and observed frequencies by amounts that were indistinguishable amongst the different solutions. In the solution at one extreme of their solution set, they found an increase in the opacity at temperatures corresponding to $0.3 \lesssim r/R \lesssim$

0.7 in the Sun, with little change at other temperatures. This solution is consistent with the estimated opacity change presented by Christensen-Dalsgaard et al. (1985b; Cox et al. 1989). At the other extreme, the solution exhibited a substantial decrease in opacity for $r/R \lesssim 0.15$, with only minor adjustments to the standard opacities for temperatures less than 10^7 K. As Korzennik and Ulrich note, this is similar to the effect on the energy transport of introducing WIMPs (Steigman et al. 1978; Faulkner and Gilliland 1985; Spergel and Press 1985) in the solar core. The sound-speed distributions corresponding to these two solutions are similar for $r/R \gtrsim 0.3$ and are consistent with the results of direct inversions for the sound speed (Korzennik, personal communication). In the core, however, the two sound-speed distributions differ considerably: the sound speed in the WIMP-like solution is *higher* than that in the "non-WIMP" solution by $\sim 1\%$ at $r/R \approx 0.2$ and *lower* by a similar amount in $r/R \lesssim 0.1$. The non-WIMP solution agrees more closely with the results of direct sound-speed inversions. It might also be noted in this regard that the (separate) inversions for $\delta c/c$ and $\delta\rho/\rho$ performed by Gough and Kosovichev (1988) appear not to be consistent with the WIMP solar model investigated by, for example, Faulkner and Gilliland (1985).

V. CONCLUSIONS

The large data sets which it is anticipated will come from observations with groundbased networks and satellite-borne telescopes will make new demands of inversion methods. Clearly the computational expense of each method of inversion will become an important factor. The method of optimally localized averages and the spectral expansion method require the inversion of one or more matrices whose order is the number of data points, whereas the order of the matrices inverted in least-square methods is the number of base functions. Thus, the size of the data set may determine the method to be used. In this regard, the formulation (Eq. 16) of the optimally localized averages method may be preferable to the more commonly used formulation (Eq. 13), in that the former requires only one matrix inversion whereas the latter requires one inversion for each value of r_0. Clearly the relative merits of these and other formulations, in terms of their systematic errors and their robustness to data errors, need to be investigated. Some work on the computational expense of inversion methods has been carried out by Jeffrey and Rosner (1986), but more extensive assessments need to be made. In addition, pre-processing strategies may be developed to reduce the size of data sets that are used in the inversions. Such pre-processing might involve suitably combining data or reducing the modes to a subset. The question of what mode sets are required to answer specific scientific questions is also raised by the limitations of on-board computing and satellite-to-ground communication rates for observations from space.

This chapter has concentrated on inversion methods that have already been used on helioseismic problems. There are other methods whose potential needs to be explored, for example, the maximum entropy method (see, e.g., Jeffrey and Rosner 1986,1988). Also the probabilistic approach represented by Tarantola (1987), that the inversion process is one of improving on *a priori* probability distributions describing the data and the underlying structure and dynamics to produce *a posteriori* probabilities, has hardly yet been explored by helioseismologists.

New ways of presenting the helioseismic data can also lead to novel inversion methods. Ring diagrams, which are slices at constant frequency through three-dimensional power spectra (two dimensions for the horizontal wave number and one for frequency) are an example: their use for analyzing two-dimensional horizontal flows using high-degree p modes has been proposed by Brown and Morrow (Morrow 1988*a*) and by Hill (1988*a*).

Acknowledgments.   We thank J. Christensen-Dalsgaard for detailed comments on an early draft of this text, and we acknowledge useful conversations with him and with T. M. Brown and H. Shibahashi. We are grateful to S. G. Korzennik for supplying sound speeds associated with two of his inversions for opacity.

SOLAR GRAVITY MODES

H. Hill
The University of Arizona

C. Fröhlich
Physikalisch-Meteorlogisches Observatorium DAVOS

M. Gabriel
Université de Liège

and

V. A. Kotov
Crimean Astrophysical Observatory

The internal gravity modes, or g-modes, of the Sun may be important both in affecting the internal structure of the Sun and in furnishing a diagnostic probe of the solar interior. The internal structure could be altered by core mixing due to unstable g-modes or by nonlinear effects due to large-amplitude g-modes located in the core. On the other hand, small-amplitude g-modes operating in the linear regime offer the possibility of studying in detail the present state of internal solar structure such as the internal rotation rate, the Brunt-Väisälä frequency, the speed of sound and the mean molecular weight. In all of these roles, the solar g-modes may be important. The g-modes may be a contributor to the solar neutrino paradox and/or they may be a source of information leading to a more complete understanding of the physics responsible for the paradox. The observational work on solar g-modes does not exclude any of these possibilities at this time.

I. GENERAL PROPERTIES OF g-MODES

The stable eigenvalue spectrum of spherical stars is divided into two groups: p- and g-modes plus the fundamental mode which separates them (Cowling 1941). Gravity modes have an accumulation point at zero and their order $\infty > n \geq 1$ gives their position in the eigenvalue sequence of decreasing values. Such a spectrum exists for each degree ℓ associated with surface spherical harmonic functions.

While for high-order p-modes the pressure perturbation provides the restoring force, it is produced by the buoyancy in high-order g-modes. For high orders, the displacement is nearly radial for p-modes, while it is nearly horizontal for g-modes. For low-order modes there is, however, no sharp distinction between p- and g-modes. Their properties change progressively with the order of the mode. A simple way to distinguish between p- and g-modes is often to consider the oscillatory behavior of the eigenfunctions. Gabriel and Scuflaire (1979) and Gabriel (1979, 1986a) have presented discussions of the oscillatory properties of the eigenfunctions and of the characteristics of the eigenvalue spectrum in the Cowling approximation, which rests on mathematical theorems. Unfortunately, it seems impossible to extend the theory to the fourth-order problem. However, their results apply perfectly to the Sun, which shows none of the complications met for low-degree and low-order modes in models with high central condensation.

Let us define λ by

$$\lambda^2 = \frac{1}{c^2} \left(\frac{L^2}{\sigma^2} - 1 \right) (\sigma^2 - N^2) \qquad (1)$$

where c is the sound speed, $L = [\ell(\ell + 1)c^2/r^2]^{1/2}$ is the Lamb frequency, N is the Brunt-Väisälä frequency and σ is the eigenvalue. The run of L and N in the Sun is given, for instance, in Brown et al. (1986).

In regions where $\lambda^2 > 0$, eigenfunctions may have only one node, in either the radial displacement δr or in the Eulerian pressure perturbation p'. In regions where $\lambda^2 < 0$, eigenfunctions may oscillate, i.e., δr and p' may have several nodes and waves can propagate in the layers. If $\sigma^2 > L^2$ and $\sigma^2 > N^2$, pressure waves can propagate in that cavity and eigenfunctions rotate counterclockwise in the $(\delta r, p')$ plane as the radius increases. In regions where $\sigma^2 < L^2$ and $\sigma^2 < N^2$, gravity waves can propagate and eigenfunctions turn clockwise in the $(\delta r, p')$ plane as the radius increases; they are said to have a g behavior. In the Sun, low-degree and low-order internal g-modes have a g-cavity in the radiative core and a p-cavity farther out. For that reason, the amplitude of the eigenfunction of these modes does not necessarily decrease in the convective envelope. As the degree increases, the radius of the lower limit of the p-cavity increases and it disappears when the minimum of the Lamb frequency L_{min} is larger than the maximum Brunt-Väisälä frequency in the core. For high-order modes, the p-cavity disappears when $\sigma^2 < L^2_{min}$. Because for not-too-high degrees and orders, eigenfunctions may have g- and p-cavities where they rotate in opposite directions, it is necessary to assign a sign to the nodes. If we plot the eigenfunction in the $(\delta r, p')$ plane and if we count positively the nodes of δr when the solution turns counterclockwise and negatively when it turns clockwise, then the algebraic sum of the nodes is positive for p-modes and negative for g-modes. The absolute value of that sum is equal to the order of the mode. As a g-mode must have a g-cavity, the condition $\sigma^2 < N^2$ must be fulfilled somewhere in the star. Therefore, the eigenvalues of g-modes are always smaller than the maximum of the Brunt-Väisälä frequency. As, on the other hand, it can be proved (Gabriel 1986a) that $|\sigma^2|$ for a given order increases with increasing degree, the eigenvalue spectrum will become denser and denser with increasing degree. For example, in the Sun, the period separation is of the order of 1.2 min for $\ell = 30$.

The Sun has two g-cavities. One is the radiative core, where N^2 increases from zero at the center, reaches a maximum $N^2_{max} \simeq 8 \times 10^{-6}$ s^{-1} ($P_{min} \simeq$ 36 min), and then decreases to zero at the bottom of the convective envelope. The other is in the photosphere, where N^2 increases from the top of the convective envelope, reaches a maximum $N^2_{max} \simeq 1.5 10^{-3}$ s^{-2} ($P_{min} \geq 162$ s) at the temperature minimum, and then decreases in the chromosphere where its minimum is $N^2_{min} \simeq 2 \times 10^{-4}$ s^{-2} ($P_{max} \simeq 440$ s). As the two cavities are very distant, the Sun may have two kinds of g-modes. The internal g-modes are considered in this chapter and have large amplitudes in the interior. Internal g-waves are efficiently reflected in the convective envelope when there is no p-cavity or only a small one for their frequency. When there is a p-cavity, waves are reflected by the photospheric g-cavity. The other kind of g-modes have large amplitudes in the atmospheric cavity only. Their eigenvalues obey the conditions $N^2_{max} > \sigma^2 > N^2_{min}$. When $\sigma^2 < N^2_{min}$, waves are no longer reflected in the chromosphere and cannot build up standing waves. The oscillatory theorem requires that $\sigma^2 < L^2$ in a g-cavity. As $L^2 \simeq 1.4 \times 10^{-10}$ $\ell(\ell + 1)$ s^{-2} in the photosphere, these g-modes can exist only for high degrees of several thousands. To our knowledge, these photospheric modes have never been observed, probably because strong radiative damping and horizontal inhomogeneities forbid their growth.

One rather unusual property of gravity waves should also be given. The φradial components $V_\varphi = -\sigma/|\lambda|$ and V_g of the phase and group velocities, respectively, are related by (Smeyers 1984)

$$V_\varphi V_g = \sigma^2 \left(\frac{\sigma^2}{c^2} - \frac{\ell(\ell + 1)}{r^2} \frac{N^2}{\sigma^2}\right)^{-1}. \tag{2}$$

This formula shows that for g-waves, V_φ and V_g have opposite signs and both go to zero as σ^2.

A. Asymptotic Properties of g-Modes

Studies of asymptotic nonradial oscillations begin with Ledoux's (1962,1963) papers. Most of the subsequent developments can be found in Tassoul (1980) and in Smeyers and Tassoul (1987), where references to previous works are given.

For g-modes, asymptotic methods rest on several hypotheses. It is supposed that:

1. The Cowling approximation is valid;
2. Models have a zero surface temperature which implies that N^2 goes to infinity at the surface as $(R - r)^{-1}$;
3. If the outer layers are in radiative equilibrium, their structure can be represented with an effective polytropic index;
4. $L^2 >> \sigma^2$ everywhere;
5. $|N^2| >> \sigma^2$ everywhere except near the zeros of N and in the vicinity of the surface.

Condition 4 is a strong constraint for low degrees as it will be fulfilled only for very high orders. However, as long as it holds in the internal g-cavity, only the phase φ (see Eq. 3) is affected when it fails somewhere else. Condition 5 is never fulfilled in the convective envelope for any g-mode of practical interest. For this reason, the phase φ cannot be predicted accurately by asymptotic methods. Although when applicable asymptotic methods provide rigorous solutions, the algebra becomes quickly cumbersome and results are limited to the first-order terms or to the first two at best.

For models with a radiative core and a convective envelope, the first-order approximation gives for the eigenvalue σ or for the period P

$$\sigma = \frac{2[\ell(\ell + 1)]^{1/2}}{\pi(2n + \ell + \varphi)} \int_{rad} \frac{N}{r} dr \tag{3}$$

$$P_{n\ell} = \frac{P_0}{2} \frac{2n + \ell + \varphi}{[\ell(\ell + 1)]^{1/2}} \tag{4}$$

with

$$P_0 = 2\pi^2 \left[\int_{rad} \frac{N}{r} dr \right]^{-1} \tag{5}$$

where the integral extends over the radiative region. For the reasons discussed above, the phase φ also varies with the degree and the order of the modes. Equation (3) shows that condition 5 is fulfilled only when $n \gg \ell$.

Tassoul (1980) also gives the first-order formula for models with a convective core and envelope. The difference between this first-order formula and Eq. (4) is that ℓ does not appear in the numerator. This cannot be used as an argument for the nonexistence of a convective core without a more detailed discussion, because condition 5 does not hold in the convective core either.

Equation (4) predicts that high-order g-modes have periods equally spaced with a separation between two successive modes equal to $P_o/[\ell(\ell + 1)]^{1/2}$. Low-order modes have periods larger than predicted by the asymptotic formula (Berthomieu et al. 1984a; Gabriel 1984a,b).

For models with a radiative core, the eigenfunctions are given to the first order in the g-cavity and when $x \gg 1$ by

$$\delta r \simeq \left[\frac{\sigma^2}{\ell(\ell + 1)N^2 \rho^3 r^6}\right]^{1/4} \sin\left(x - \frac{\pi\ell}{2}\right) \tag{6}$$

$$\frac{1}{r}\frac{1}{\sigma^2}\frac{p'}{\rho} \simeq \left\{\sigma\left[\frac{\ell(\ell + 1)}{N^2}\right]^{1/2}\right\}^{-1}\left[\frac{\sigma^2}{\ell(\ell + 1)N^2\rho^2 r^6}\right]^{1/4} \cos\left(x - \frac{\pi\ell}{2}\right) \tag{7}$$

$$x = \left[\frac{\ell(\ell + 1)}{\sigma^2}\right]^{1/2} \int_0^r \frac{N}{r}dr \quad \cdot \tag{8}$$

As the horizontal components of the displacement are proportional to $p'/(r\sigma^2\rho)$, Eqs. (3) and (7) show that the displacement gets closer to the horizontal plane when σ decreases or when the order increases at constant ℓ, while it becomes more and more radial when ℓ increases at constant σ or constant order. Outside the g-cavity, and if $L^2 \gg \sigma^2$ everywhere, the eigenfunctions decrease essentially as $\rho^{-1/2} \exp(- \int \lambda dr)$. Consequently, in the convective envelope where $\lambda \simeq (L/c) = [\ell(\ell + 1)]^{1/2}/r$, the amplitude of high-degree modes decreases there as $\rho^{-1/2} r^{-[\ell(\ell + 1)]^{1/2}}$. As a result, only low-degree g-modes are expected to be observable in the Sun.

Gravity modes are well suited to probe the deep solar core because they have their largest amplitude there. For low-degree modes, the envelope of the curves which give the components of the relative displacement $\delta r/r$ decreases rapidly for $r \leq 0.2 R$ and then more slowly farther out.

Ellis (1984,1986) Gabriel (1986b) and Provost and Berthomieu (1986) have studied the variation for the phase φ with the degree and the order of the modes and with the depth of the convective zone. They showed that information concerning the depth of the convective zone or the behavior of N just below it can be obtained from the variation of φ, provided high overtone modes ($n \geq 40$) can be observed.

Ellis (1986) and Provost and Berthomieu (1986) also give a second-order formula for the period which can be more useful than Eq. (4) to observers who try to identify g-modes. It is

$$P_{n\ell} = \frac{P_0}{2} \frac{(2n + \ell + \varphi)}{[\ell(\ell + 1)]^{1/2}} + \frac{P_0^2}{P_{n\ell}} W_\ell \qquad (9)$$

$$W_\ell = V_1 + V_2/[\ell(\ell + 1)] \qquad (10)$$

where V_1 and V_2 are constants but φ is still a function of n and ℓ.

Table I gives the theoretical values of P_0 obtained by different groups. The various observational results reported for P_0 are given in Table VIII of Sec.III.E. The first eight entries in Table I give theoretical values for standard models. They differ, probably because of differences in the input physics. The second group of models supposes turbulent diffusion. The value of the effective Reynolds number R_e^* is given in the second column. The last of these models supposes that diffusion is inhibited by the μ gradient. All of these models, except for the last one, have P_0 values which are much too high. Therefore, either turbulent diffusion is very weak in the Sun or it is inhibited by the μ gradient. WIMPS models have P_0 values much smaller than the standard one and are compatible only with Fröhlich (1988) data. The last four models have P_0 values close to those of standard models. Two of them include gravitational settling. In Cox et al. (1989) models, the opacity has been increased by 15% below the convective envelope. In Lebreton et al. (1988) models, the axial weak-interaction coupling constant g_A has been varied from -1.26 to -1.64 and P_0 increases with $|g_A|$.

B. Excitation Mechanisms

Interest in the stability of solar g-modes first arose in connection with the neutrino problem after a suggestion by Dilke and Gough (1972) that some g-modes could be unstable. Stability analysis in the quasi-adiabatic approximation have been made by Dziembowski and Sienkiewicz (1973), Christensen-Dalsgaard et al. (1974), Shibahashi et al. (1975) and Boury et al. (1975). They found that the Sun was unstable when its age was between 2×10^8 and 3×10^9 yr. The most unstable mode was the $\ell = 1$, g_1-mode with a minimum e-folding time of the order of 2×10^5 yr. The $\ell = 1$, g_2-mode was

TABLE I
Theoretical Values of P_0

Author	Kind of Model	P_0(min)
Iben and Mahaffy (1976)	standard	36.0
Christensen-Dalsgaard et al. (1979)	"	36.2
Berthomieu et al. (1984b)	"	33.9
Gabriel (1984a,b)	"	35.5
Lebreton and Meader (1986)	"	36.5
Lebreton and Meader (1987)	"	37.0
Lebreton et al. (1988)	"	35.1
Kidman and Cox (1987)	"	~38.0
Berthomieu et al. (1984b)	$R_e^* = 100$	54.6
Berthomieu et al. (1984b)	$R_e^* = 200$	62.9
Cox and Kidman (1984)	$R_e^* = 100$	58.0
Lebreton and Meader (1987)	$R_e^* = 100$	56.3
Lebreton and Meader (1987)	$R_e^* = (8r/Hp) + R_{ec}^*$	46.6
Lebreton and Maeder (1987)	inhibited by $\nabla\mu$	37.1
Faulkner et al. (1986)	WIMPS	29.0
Däppen et al. (1986)	"	32.56
Cox et al. (1990)	Cosmions	28.8, 30.9
Gabriel (1984b)	gravitational settling	35.1
Cox et al. (1989)	" + high κ	34.5
Cox et al. (1989)	high κ	34.6
Lebreton et al. (1988)	g_A	35.1 to 40

also unstable for a shorter period of time. This instability is caused by nuclear driving.

The Lagrangian perturbation of the energy generation rate ϵ is given by Ledoux and Walraven (1958):

$$\delta\epsilon = \sum_{ij} \epsilon_{ij} \left(\mu_{ij} \frac{\delta\rho}{\rho} + \nu_{ij} \frac{\delta T}{T} + \frac{\delta X_i}{X_i} + \frac{\delta X_j}{X_j} \right) = \epsilon \left(\mu_e \frac{\delta\rho}{\rho} + \nu_e \frac{\delta T}{T} \right) \quad (11)$$

where the sum is carried over all the reactions. The perturbation of the abundance of the reagents is obtained through the perturbation of the equations of nuclear kinematics. When the lifetime of an element is much longer than the period (e.g., ^{3}He), its abundance is not perturbed ($\delta X_i = 0$). In the opposite limit (e.g., ^{2}H), the reagent keeps its equilibrium value at any time. For example, we can consider that ^{1}H(^{1}H,e^+) ^{2}H(^{1}H,γ)^{3}He is just one reaction. This leads to a value of ν_e much larger than the static value $\nu_{11} \simeq 4$ (see Boury et al. 1975). As an example, when the p-p chain always terminates through the ^{3}He-^{3}He reaction, which has $\nu_{33} \cong 16$, we find

$$\nu_e = [\epsilon(3\ ^1H \rightarrow\ ^3He)\nu_{11} + \epsilon(2\ ^3He \rightarrow\ ^4He + 2\ ^1H)\nu_{33}]/\epsilon \qquad (12)$$

or when reagents have their equilibrium value,

$$\nu_e = (13.36\nu_{11} + 12.85\nu_{33})/26.21 \approx 10 \qquad \cdot \qquad (13)$$

This high value of ν_e allows an efficient nuclear driving. In this Lagrangian interpretation, the "strong" ^{3}He gradient is not required for instability. Of course there must be a gradient in a radiative core in order for ^{3}He to reach its equilibrium abundance everywhere. Then the ^{3}He burning reactions provide a large fraction of the energy released by nuclear burning (nearly 50% in the example considered above). However, an even larger contribution of these reactions can be obtained just after mixing of, let us say, the inner 50% of the mass within a time much shorter than the ^{3}He lifetime. Then ν_e will be larger and the model more unstable.

As the Sun gets older, the surface amplitude of the eigenfunction of these unstable modes becomes larger compared to its values in the nuclear burning core. Consequently, the outer layers have a higher weight in the stability analysis and nonadiabatic computations are required for the present Sun. Christensen-Dalsgaard and Gough (1975), neglecting the Eulerian perturbation of the convective flux, find the present Sun stable. On the other hand, Saio (1980), using a theory by Gabriel et al. (1974,1975) to compute the perturbation of the convective flux, finds the $\ell = 1$, g_2-mode unstable with an e-folding time τ_e of 7×10^6 yr. Also, Cox and Kidman (1984) find the $\ell = 2$, g_1-mode unstable with $\tau_e \approx 2 \times 10^9$ yr. Kosovichev and Severny (1984a,1985) made a quasi-adiabatic calculation for two non-standard solar models and found them unstable.

Results for the present Sun are difficult to interpret because the interaction between convection and pulsation is poorly known and may not be neglected. On the other hand, the unstabilities found for a younger Sun seem to rest on safer ground because they are predicted whatever hypotheses are made concerning the influence of convection. Nevertheless, pulsations with periods equal to these of theoretically unstable modes have never been observed for either solar-type stars or less massive ones for which the same instabilities are also predicted (Noels et al. 1976). Does this imply that theoreticians have made incorrect predictions or that these pulsations have escaped detection because their amplitude is limited to small values? This problem would deserve some observational efforts, because if these instabilities turn out to be real, they can lead to a mild turbulence (Ulrich 1974a) and modify the chemical composition profile of these stars.

The only connection between the low-order g-modes and the higher-order ones observed in the Sun is, maybe, the resonant three-wave interactions of g-modes, noticed by Guenther and Demarque (1984).

Ando (1986) and Wentzel (1986,1987) have considered the possibility of exciting observed solar g-modes by resonant coupling with 5-min p-modes. Wentzel (1987) concludes that their amplitudes should be, at best, not larger than 1% of that of p-modes, while the ratio of velocities should be $< 10^{-3}$. Dziembowski et al. (1985) reach the same conclusion. Their argument is that this coupling will tend to lead to energy equipartition between g- and p-modes. If, as for a given surface amplitude, g-modes have energies ~ 10 orders of magnitude larger than p-modes, g-modes would have very small amplitudes. The same argument allows ruling out the excitation of g-modes by interaction with convection.

Dziembowski et al. (1985) suggest that the most plausible driving mechanism is the effect of a magnetic torque caused by the presence of a global magnetic field and mild turbulence in the core. For further discussion of mode excitation, reference is made to the chapter on mode excitation by Cox et al.

II. DETECTION AND CLASSIFICATION OF INTERNAL GRAVITY MODES

Are internal gravity modes of oscillation excited in the Sun to sufficient levels to be detected? A great deal of research has been performed in an effort to obtain an answer to this difficult question. For the period 1973–1981, the question could only be addressed by observations looking for the fundamental characteristic of global or normal modes of oscillations, a spatial coherence across the solar surface of the Sun and temporal coherence over long periods of time. It was apparent in the early 1970s that if these modes were excited, the amplitude of oscillations must be, relatively speaking, very small because if not, they would have already been detected. Difficult demands are placed on instrumentation and data analysis techniques because of the relatively small signals and the very complex normal mode spectrum of the Sun.

The pioneering work of the 1973–1981 period was done at SCLERA (the Santa Catalina Laboratory for Experimental Relativity by Astrometry— operated by the University of Arizona), Crimean Astrophysical Observatory, Birmingham University and Stanford University. The work of the latter three locations was directed principally toward understanding the 160-min period oscillation first announced in 1976 (Severny et al. 1976; Brookes et al. 1976). Work at SCLERA was concerned with the more general problem of g-mode detection. The first evidence of normal mode detection at SCLERA was presented in 1974 (Hill and Stebbins 1975b). For a review of this period, the reader is referred to Secs.III.A. and D.

The work of the period 1982 to the present has been concerned with the detection and classification of resolved g-modes. There have been numerous works reported on the detection and classification of g-modes. Some are based on rather extensive analyses. These works fall into several well-defined groups by the nature of the techniques used in the detection and/or mode classification process (see Table II). These groups might be ordered as follows:

1. Work on a 160-min period oscillation with primary effort placed on demonstration of temporal coherence.
2. Work on detection of g-mode signals without specific mode detections and classifications by reliance on first-order asymptotic theory predictions.
3. Work on detection of g-mode signals leading to specific mode detections and classifications by reliance on first-order asymptotic theory predictions.
4. Work on detection of g-mode signals leading to specific mode detections and classifications by extensive use of observed spatial and temporal properties of eigenfunctions.

To date, there has been very little agreement between the various findings. However, this lack of agreement may be only superficial, as it appears possible to bring some order to the current situation when consideration is given to potential problem areas. These problem areas are due in part to noise introduced by observing through the Earth's atmosphere, complications introduced by observing a rotating nonuniform solar surface, a lack of an adequate understanding of the origin of the signal being observed, a lack of spatial

TABLE II
Summary of Work on Gravity Mode Detection and Classification

Based on First-Order Asymptotic Theory	Number of Reported $\nu_{n\ell m}$
Delache and Scherrer (1983)	16
Fröhlich and Delache (1984a,b)	—
Isaak et al. (1984)	—
Pallé and Roca-Cortés (1988)	32
Fröhlich (1988)	
Based on Kotov et al. (1984)	15
Based on Pallé and Roca	
Cortés (1988)	21
Fröhlich (1988)	—
Based on SMM	
Total Irradiance	
van der Raay (1988)	32
Henning and Scherrer (1988b)	15
Not Based on Asymptotic Theory Predictions	
Hill (1985b)	—
Hill (1986a,b,c)	—
Hill and Gu (1990)	235
Rabaey and Hill (1988, 1990)	638

information, a lack of knowledge of the internal rotation, and reliance on inadequate theoretical eigenfrequency spectra. The status of the situation with regard to these various points is summarized under the separate subheadings following the more general discussions of the four groups of works defined in this section. This review is intended to serve as a resource for the researcher either entering the field or for a researcher currently working on these problems.

A. Work on the 160-min Period Oscillation: Primary Effort Placed on Demonstration of Temporal Coherence

Systematic observations of the differential Doppler velocity made in Crimea from 1974 through 1987 continue to confirm the presence of the phase-coherent, in average, solar pulsation with a period of 160 min. On the other hand, the analysis of solar flare data suggests that the 160-min pulsations in the Sun might exhibit a multiplet fine structure with a frequency spacing of the order of 3 nHz. If verified, this novel property of the pulsations can offer a new possibility for probing the Sun's interior and perhaps for the study for the 11(22)-year cycle of solar activity. But the most surprising thing is that the same 160-min period appears to be the most commensurate one for periods of close binary systems of the Galaxy and for some types of short-period variable stars. In addition, the 160-min period was recently found in the light-flux fluctuations of the cores of several Seyfert galaxies and also quasar 3C 273. The problem of the 160-min oscillations, therefore, becomes very intriguing, not only for the Sun, but also for astrophysics as a whole and especially for cosmology.

1. Historical Remarks. The idea that the Sun could be oscillating with the 160-min period ((1/9) day) was first suggested decades ago. Namely, long before the first helioseismological studies (Hill et al. 1975b; Brookes et al. 1976; Christensen-Dalsgaard and Gough 1976; Severny et al. 1976) Sevin (1946) claimed that "la période propre de vibration du Soleil, c'est-à-dire la période de son infra-son (1/9 de jour), a joué un rôle essential dans la distribution des planètes supérieures." This strange speculation was based primarily on the results of numerical distribution analysis of planetary orbits. (One can note, e.g., that the light-travel time from the Sun to Saturn and Uranus is equal to $\approx$80 and $\approx$160 min, respectively.) But the actual discovery of the 160-min pulsation of the Sun was made only in the 1970s when Severny et al. (1976) and Brookes et al. (1976) announced the finding of a significant 160-min periodicity in their respective Doppler shift measurements of Fraunhofer spectral lines. The results of these observations have been further described by Brookes et al. (1978a) and Kotov et al. (1978,1983a,b).

The Crimean observers (Kotov et al. 1978) measured the relative Doppler velocity between a central circular area with a radius of 0.66 R on the solar disk of radius R and the remaining annular area of the solar disk. The

Birmingham University group (Brookes et al. 1978c) used a resonant optical scattering method to measure Doppler shifts between the Sun and laboratory lines for the integrated light from the entire solar disk.

The first results obtained by the two groups agreed well in the value of the period, amplitude and phase. The discovery stimulated considerable discussion in the literature about the true nature of the 160-min oscillation, because (a) the period is too long to be prescribed to the fundamental mode of solar pulsations, (b) it appears to be very close to the 9th harmonic of the terrestrial day, and (c) the period happened to be also close to the value $P = 167$ min, which follows from the Ritter's formula

$$P = 2\pi /[(3\gamma - 4)GM/R^3]^{1/2} \qquad (14)$$

for radial pulsations of a star being homogeneous in density and having solar mass M (γ is the adiabatic index equal to $5/3$ for the Sun, and G is the gravitational constant).

2. The 160-min Oscillation: Pro and Con. The systematic Doppler observations undertaken in Crimea over a 7-yr period, from 1974 to 1980 (in all 325 days of observations, with $N = 22147$ individual velocity measures for 5-min integration time interval), showed the presence of significant periodicity $P_{ob} = 160.010(\pm 2)$ min (Kotov et al. 1983b) with the mean harmonic amplitude $A_h = 0.54$ m s^{-1}. It is important to note that the period differs significantly, statistically speaking, from (1/9) day (160.000 min).

Quite similar measurements were performed at Stanford University, and it was subsequently reported by Scherrer et al. (1979,1980a) and Scherrer and Wilcox (1983) that the Stanford observations, made from 1976 through 1980, tended to confirm the existence of such an oscillation.

Later, Grec et al. (1980) examined full-disk Doppler measurements made in 1979 and 1980 at the geographic South Pole for evidence of the oscillation. The participants of the expedition believed that all terrestrial effects suspected of giving rise to a spurious occurrence of the 160-min period signal at middle geographic latitudes would be eliminated in Antarctica during an austrial summer. The expedition succeeded in obtaining a 5-day interval of continuous, without day-night alternations, string of observations. The power spectrum of these data showed that one of the four major peaks for the range of periods $P < 170$ min, corresponded to a period of $161(\pm 1)$min. But the most surprising result was the superposed epoch plot for the 160.010-min period: the South Pole average velocity curve matched fairly well the two other curves obtained earlier in Crimea and Stanford (see Fig. 1). It seemed that the solar origin of the oscillation could be regarded as well established.

Nevertheless, the observations of 160-min oscillations have been questioned by many people over the last decade. The reasoning of opponents was based mostly on the fact that the period is almost exactly 1/9th of a day.

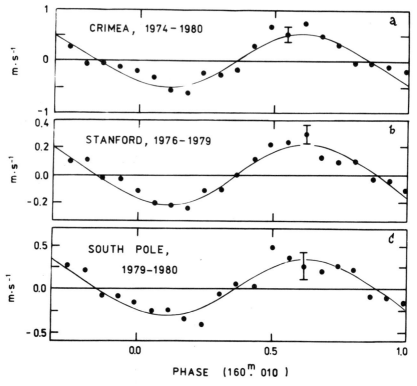

Fig. 1. Average velocity curves for the 160.010-min period according to observations made in Crimea (a), Stanford (b), and the South Pole (c). Sinusoids are fitted by the least-squares method; vertical bars represent standard errors ($\pm \sigma$). Zero phase is taken at UT 00h 00m on 1 January 1974. Positive velocity corresponds to a blue shift of the spectral line for the central portion of the solar disk.

Despite the number of strong independent findings in favor of a solar origin (Kotov et al. 1983*b*; Kotov 1985; Koutchmy et al. 1980), some authors (see, e.g., Dittmer 1977; Grec and Fossat 1979) put forward analyses which attributed the periodic signal to effects of observing through the Earth's atmosphere and to artifacts of the observing and data-reduction procedures.

The authenticity of the solar signal at the 160-min period was also disputed by Forbush et al. (1983): they claimed that one always expects to obtain some signal from the superposed-epoch analysis, and that the statistical significance of the result might be strongly overestimated due to the so-called "quasi-persistency" effect present in the data. However, the detailed consideration of this problem (Rachkovsky 1985) showed that this effect appears to be negligible for the Crimean data.

The most convincing point in favor of a solar nature of the 160-min period is a small difference between $(1/P_{ob}) = (1/160.010$ min) and a 9th daily harmonic which causes a steady phase shift by $\sim$33 min per year relative to that obtained for a signal with exactly a (1/9) day period. This phase shift has been observed consistently over 9 yr in Crimea (1974–1982) and 5 yr at Stanford (1976–1980) (see Kotov et al. 1984a; Kotov 1985; Scherrer et al. 1979).

The 160-min period was also detected in the apparent solar diameter and differential radius observations obtained at SCLERA (Hill 1985b, 1986a; Hill et al. 1986a), observations which are sensitive to both changes in the solar radius and the photospheric temperature profile, i.e., changes in the limb darkening function. These data also favor the mode classification $\ell = 2$, $m = 2$ g-mode for the oscillation based on spatial and temporal properties of the signals. The authors have also advanced the hypothesis that actually the 160-min oscillations, which are detected by Doppler shift techniques, involve temperature perturbations in conjunction with the surface rotation of the Sun and not physical displacement of the solar photosphere. This apparent-velocity hypothesis may account for some of the reported discrepancies on the 160-min oscillation and also have important implications for g-mode classification and for future instrument designs. One should note in this regard that no evidence of a 160-min period solar oscillation was found for a change in radius by Kuhn et al. (1986) from solar ellipticity fluctuations detected by the Princeton Solar Distortion Telescope in 1983 and 1984. In addition to an explanation based on the apparent velocity hypothesis, a second possible explanation of these negative observations could be the remarkable decrease in the amplitude of the 160-min oscillation in the 1980s as indicated by the Crimean measurements (see the following section).

However, note that the question of the real (solar) origin of the 160-min oscillation remains quite unclear still for a number of authors in view of the recent papers by Stanford and Birmingham-Izana authors. Indeed, Henning and Scherrer (1986) and Elsworth et al. (1987) recently cast doubts on the long-time phase stability of the P_{ob} period. It was reported that the phase stability is no longer present and that the oscillation itself could be an artifact produced by the atmospheric transparency fluctuations. It is noted, however, that this latter conclusion (Elsworth et al. 1989) is in opposition with the previous detailed analysis by Kotov et al. (1983) of the potential influence of the differential atmospheric extinction on the Doppler shift measurements.

Moreover, the Crimean observations performed 1974 to 1987 do continue to confirm the presence of the phase-coherent, in the average, solar pulsation with a period of 160.01 min. Another strong argument in favor of the solar nature of the period was obtained recently from analysis of onset times of solar flares. To study this 160-min effect in flares, Kotov and Levitskij (1987) analyzed the time series of beginnings of 18877 solar flares of importance $B \geq 1$ observed on the Sun during the 1947 to 1980 time interval according to the world network of solar observatories. This time series was subjected to a modified "power spectrum" analysis based on the superposed-epoch routine and computation of χ^2-criterion. The "power spectrum" (Fig. 2) showed that the major peak in the 160-min range studied corresponds to a period of 160.0101(± 1) min, in excellent agreement with the result of Doppler shift observations. It is also important to note that the flare data (onset times of flares) cannot be influenced by the Earth's atmospheric transparency fluctuations; the "(1/9) day" argument if applied to the flare result is completely untenable because of the well-documented solar origin for the flares.

3. Phase-Diagram and Power Spectrum.   There are additional works on the Crimean observations that should be noted. First, Kotov and Tsap (1988) reduced recently an additional 677 hr of the 1983–1984 data set which had been rejected from earlier analysis (Kotov et al. 1984b). As a result, they have increased the total amount of the 1983–1984 data set by a factor of nearly 4. Second, the authors have added new observations carried out over the period 1985 to 1987 (1984 hr of new data). In all, Kotov and Tsap (1988) analyzed 5612 hr of observations (vs 2786 hr analyzed in Kotov et al. 1984b) made during 902 days.

To establish the difference between 160.00 (1/9 day) and the 160.010-min P_{ob} period of the oscillation, Kotov and Tsap (1988) have subjected to special analysis the phases of maximum "expansion" velocity for each year separately (excluding the 1984-year data set which resulted in an almost null average harmonic amplitude: $A_h = 0.04 \pm 0.14$ m s^{-1}). The phases of the signal at 160.000-min period were computed for each year separately in the

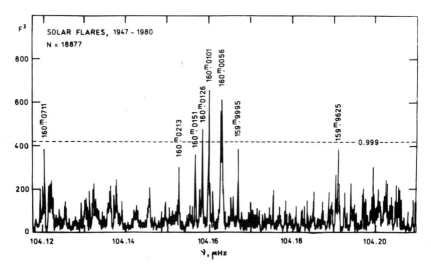

Fig. 2. χ^2-"power spectrum" for 18877 solar flares observed on the Sun during 1947–1980 time interval ($F^2 = (\chi^2)^2$ by definition).

way described by Scherrer et al. (1979) and Kotov et al. (1983*a*). In Fig. 3, all 13 Crimean phases (1974–1987) are plotted three times, with the vertical separation of exactly 160.000 min. Then using the least-squares method of Kotov (1985), the authors have found the best-fitting straight line which corresponds to a period of 160.0093(± 15) min with ≃3.3 σ-level of confidence. The slope of the straight line implies that the observed period differs slightly, by ≃0.01 min, from the 9th daily harmonic. The same procedure has also been applied to 18 yearly phases now available from three other observatories: Stanford, South Pole and Izana for 1976–1986. It was found that there exists a 2.2 σ-significant period 160.008(± 2) min in these independent data. The latter value agrees well, within the limits of error, with the Crimean period 160.009 min. It is important to note also that the average Crimean phase, 0.51 ± 0.07, agrees nicely with that deduced from the data of the three other observatories: 0.55 ± 0.07 (here the phases are determined for the period value 160.0101 min and zero epoch UT 00h 00m, 1 January 1974).

The power spectrum of the Crimean 1974–1986 Doppler velocity observations computed with very high resolution (the increment in frequency was ≃0.1 nHz) is shown in Fig. 4 (here $N = 60257$ is the number of individual velocity measurements made with a 5-min time of integration). The strongest peak corresponds to a period of 160.010(± 1) min, which has ≃5.7 σ-confidence level. One can see also 2 yearly sidelobes corresponding to 160.058- and 159.963-min periods.

It is therefore concluded (see Figs. 3 and 4) that the total set of the

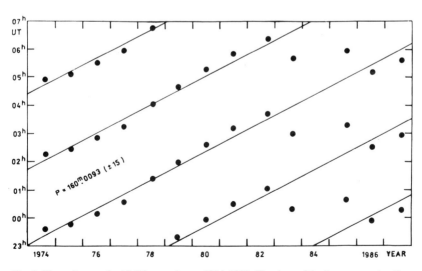

Fig. 3. Phase diagram for 13 Crimean phases, 1974–1987. The slope of the linear regression line corresponds to the best value of period $P = 160.0093(± 15)$ min for solar global pulsation. Zero phase is taken at UT 00h 00m on 1 January 1974.

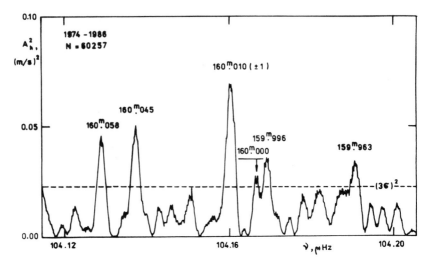

Fig. 4. Power spectrum of solar oscillations near 160-min period; Crimean observations 1974–
1986.

Crimean 1974–1987 observations confirms the existence of solar pulsation
with a period of 160.010 min, which appears to keep constant (in average)
initial phase. It should be noted simultaneously, however, that late Crimean
observations carried out in 1983–1987 seem to support the conclusions of
Henning and Scherrer (1989)and Elsworth et al. (1989) about large scattering
of the phase during several late years.

Over a 14-yr span of observations in Crimea, the amplitude of 160-min
oscillation exhibited remarkable variations during an observational season
and also from year to year (see also Scherrer et al. 1979; Kotov et al.
1983a,b). In addition, Crimean data also show that the average amplitude has
decreased by a factor of 4 between 1974–1975 and 1984–1985 time intervals.
The new 1986–1987 measurements appear to indicate a slight enhancement
of A_h again.

The amplitude changes are thought to be caused primarily by larger scatter of phases of the 160-min signal as observed on individual records. This is
quite consistent with the Stanford and Izana results about significant changes
of the phase over late years.

The average velocity curve for the 160.0101-min period deviates
strongly from a sinusoid (see Fig. 5). With the phase shift of $\simeq 0.25$ ($P_{ob}/4$),
it well resembles the relative curve for the solar flare modulation (δ in Fig.
5b is the relative amplitude of the modulation of a number of flares in each
of 8 phase bins). The nonharmonic form of the velocity curve may be ascribed to a nonlinear mechanism of excitation of the 160-min oscillation or

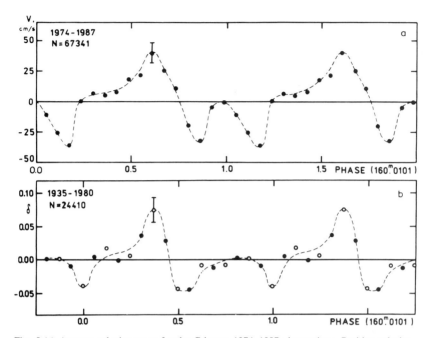

Fig. 5.(a) Average velocity curve for the Crimean 1974–1987 observations. Positive velocity corresponds to a blue shift of the spectral line for the central portion of the solar disk. (b) Average curve of solar flare modulation with the 160.0101-min period (the total number of flares N = 24410 observed on the Sun over 1935–1980 time interval. · and o −mean values of δ for each of 8 phase bins. Phase zero corresponds to UT moment 00h 00m, 1 January 1974.

to the interference of the fine-structure features of the period which plausibly has a multiplet structure.

Indeed, the flare data (Fig. 2) show that within the 160-min range, the pattern of significant peaks originate as a multiplet structure. But the most remarkable is that the spacing between the two major peaks (160.0101 and 160.0056 min) corresponds perfectly well to the average length of the solar 11-yr cycle. This gives credit to a tentative supposition (Kotov and Levitskij 1987) that some physical relation between the fine structure of the 160-min period and the solar cycle might exist.

4. So Why 160 Minutes?   The interpretation of the 160.01-min period in terms of normal *g*-modes suffers from a lack of information about phase and spatial distribution of an amplitude across the Sun's surface, and also about the true mechanism responsible for the excitation. It is quite difficult also to explain the dominant character of this oscillation in the power spectra (see power spectra computed for extended range of periods in Scherrer and Wilcox [1983]): the theory of eigenmode oscillations of the Sun predicts very large numbers of modes within the frequency domain studied (cf. Sec.I).

The difficulty in understanding the pulsations theoretically has led some authors to advance several unusual (sometimes quite "extravagant") hypotheses about the real nature of the 160-min periodicity. A discussion of this subject can be found in Kotov (1985). One unusual hypothesis noted there is a g-mode excited by the close encounter of the Sun with a passing star in the distant past. Another is the oscillation induced by gravitational waves from an external source.

It was also noticed by Kotov (1985) that there is a puzzling coincidence: the most commensurate period for periods of revolution or pulsation of different celestial bodies (asteroids, close binary systems, RR Lyr stars in globular cluster, δ Sct stars) appears to be exactly 160 min (see, e.g., Fig. 6). Finally, Kotov and Lyuty (1988) recently found traces of the 160.010-min periodicity in the variations of optical and X-ray emission of the cores of several Seyfert galaxies and also quasar 3C 273 (see Fig. 7). Consequently, it was hypothesized that the 160-min oscillation may have a cosmological nature.

But we are at a loss to imagine the exact nature of all these observational evidences and 160-min period coincidences. In any case, all discussions

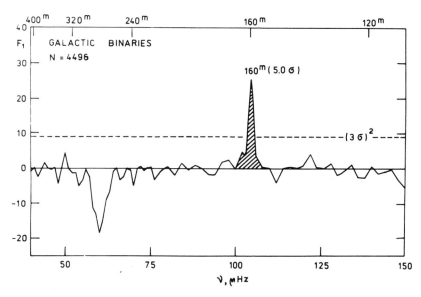

Fig. 6. Commensurability spectrum for 4496 galactic binaries with orbital periods < 4 day (according to Kotov 1985).

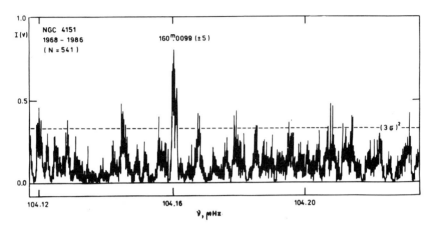

Fig. 7. Power spectrum of brightness variations of the core of the Seyfert galaxy NGC 4151 (1968–1986; the number of separate measures of the light flux N = 541; see Kotov and Lyuty 1988).

about 160-min oscillation seem to be quite interesting and stimulating for astrophysics.

B. Work on Identifying Asymptotic Properties of g-Modes Without Obtaining Specific Mode Detections and Classifications

One method that has been employed to look for g-mode signals is based on cross correlations between a theoretical eigenfrequency spectrum and observed power spectra. To date, applications of this technique have relied exclusively on use of the equal-spaced-period predictions of first-order asymptotic theory for g-modes and on power spectra of total irradiance observations (Fröhlich and Delache 1984a,b; Fröhlich 1987,1988). The results of these analyses are generally characterized by low values of P_0, $\varphi = -\frac{1}{4}$ (Eq. 9); $V_1 = V_2 = 0$ (Eq. 10) and a rotational splitting of ≈ 1 µHz. However, as no particular modes are detected and/or classified in this type of analysis, it is not possible to make detailed comparisons of eigenfrequencies with other findings. The results of these analyses with respect to inferred asymptotic properties are included in Sec. III.E along with results obtained by other techniques. For the implications of low values of P_0, reference is made to Sec.I.A and for implications of $V_1 = V_2 = 0$ and $\varphi = -\frac{1}{4}$, the reader is referred to Sec.III.D.

The following is an example of the method employed in the search for g-mode signals in the time series of the solar total irradiance from the experiment ACRIM (Active Cavity Radiometer for Irradiance Monitoring) (Willson 1984) on board the Solar Maximum Mission (SMM) spacecraft. The data used are the following: for 1980 a complete set of measurements is available

from day 49 through day 325 when the pointing system of the SMM spacecraft failed. After the repair of SMM in spring 1984, good data are again
available after day 125, 1984. In the present discussion the data acquired up
to the end of 1984 are used. The power spectra of the orbital means of both
time series, the database for the search for g-modes, are shown in Fig. 8. The
natural resolution of the power spectra is slightly below 0.05 µHz, corresponding to the length of the series of $\approx$250 days. The Nyquist frequency is
$\approx$80 µHz, corresponding to the orbital period of 94 min. A typical feature of
the spectra is the $1/\nu^2$ increase of the power towards low frequencies and the
$1/\nu$ behavior at higher frequencies. Whereas at very low frequencies, the different strength of the solar activity in the 2 yr is well recognized, the frequency spectra above $\sim$10 µHz (1-day period) are very similar (Kroll et al.
[1988a,b] report that the power level in 1980 is a factor of 2 higher than that
of 1984 for $80 \lesssim \nu \lesssim 135$ µHz).

Periods of g-modes with $n >> \ell$ can be described by asymptotic theory
(see Sec. I.A). The importance of the second order term in the asymptotic
approximation decreases with the increasing n, crossing the 1% level at
$n = 25$ for $\ell = 1$ modes (Provost and Berthomieu 1986). The fact that g-
modes are essentially equidistant in period means that their density is increasing with decreasing frequency. The spectrum is further complicated by rotational splitting. For g-modes the splitting $\delta\nu'$ due to rotation can be written
to a good approximation as

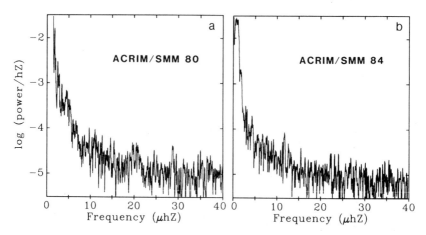

Fig. 8.(a) Power spectra of ACRIM total solar irradiance data for 1980. The spectrum is
    smoothed and the 90% confidence limits correspond to about 2/3 of a decade on the plot (factor
    of 4.5 between upper and lower limit). (b) Power spectrum of ACRIM total solar irradiance
    data for 1984. The spectrum is smoothed and the 90% confidence limits correspond to $\sim$ 2/3
    of a decade on the plot (factor of 4.5 between upper and lower limit).

$$\delta v \;=\; m \left[1 \;-\; \frac{1}{\ell(\ell + 1)} \right] (\Omega/2\pi) \qquad\qquad (15)$$

where Ω is the appropriate weighted average of the internal angular rotation (cf. the chapter by Libbrecht and Morrow on rotation).

The intrinsic resolution of the ACRIM power spectra (0.05 μHz) allows resolving the rotational splitting, if the rotational period is faster than ≈ 250 days, which is quite expected. The limit for resolving individual g-modes with $\ell = 1$ (2), on the other hand, is achieved at $n = 120$ (160) for a P_0 of 33 min, corresponding to a frequency of about 6 (8) μHz. Calculations of the visibility of g-modes have been done in some detail by Provost and Berthomieu (1988) and Berthomieu and Provost (1990), assuming energy equipartition of the different modes (see Sec. IV and Fig. 17 below; see also Cox et al. 1989). Two main findings are of relevance to the search for g-modes:

1. The predicted visibility of $\ell = 1$ and 2 modes is zero at 60 and 80 μHz;
2. The ratio of the irradiance amplitude of $\ell = 1$ and 2 at low frequencies equals 1.5 at 35 μHz and more than 2 at 20 μHz.

From this, and in view of the Nyquist frequency of the ACRIM data, the best choice for the search is the low-frequency part of the spectrum. Moreover, low frequencies correspond to high orders and thus allow the use of first-order asymptotic theory for the calculation of the g-mode frequencies. A limiting factor is, however, the solar noise due to granulation, super-granulation and other large-scale disturbances on the solar surface, which may or may not be related to the activity. Estimates indicate that the power seen in the ACRIM spectra is indeed of the order of the solar noise expected. This rather complicated situation makes it impossible to see individual peaks due to g-modes in the power spectrum in the low-frequency range, where otherwise g-modes would most likely have been seen. This is supported by the results of a simple cross correlation of the power spectra of 1980 and 1984 that do not show a significant peak at zero lag (Fossat et al. 1988). Moreover, the auto-correlation function of the irradiance spectra is dominated by peaks at lags corresponding to the modulation of the solar features by the solar rotation and its harmonics.

From these discussions, it becomes clear that the method to be used must be able to extract the g-mode signals from spectra that are dominated by solar noise. The method consists of comparing computed g-mode spectra with the real ones and searching for the one that best fits the real one in the analyzed spectral range (Fröhlich and Delache 1984a,b; Fröhlich 1987). The calculation of the frequencies of the g-modes is based on first-order asymptotic theory with a given P_0 and rotational rate $\Omega/2\pi$. Within the frequency range considered (15 to 35 μHz), the orders n range from 20 to 48 for $\ell = 1$ and

$P_0 = 33$ min, thus yielding a maximum deviation of 1.5% by using only first-order asymptotic theory, which is acceptable at the level of accuracy searched for. Thus, g-mode frequencies $v_{n\ell m}$ for a given P_0 and $\bar{\Omega}/2\pi$ are calculated using the n- and ℓ-dependent frequency shifts given by Eq. (15) and first-order asymptotic theory. As the expected amplitudes in the range from 15 to 35 μHz of the $\ell = 2$ modes are less than half of the ones of $\ell = 1$, only the latter are taken into account. The search for P_0 and $\bar{\Omega}/2\pi$ is done using the following steps:

1. Calculate for a given P_0 and $\bar{\Omega}/2\pi$ and for $\ell = 1$ and $m = \pm 1$ the $v_{n\ell m}$ of g-modes in the range 15 to 35 μHz;
2. Produce a g-mode power spectrum with a natural resolution of 0.05 μHz and with a frequency-dependent amplitude according to Provost and Berthomieu (1988);
3. Calculate the coherence c_{ij} between the power spectra f_i (ACRIM) and f_j (synthetic g-modes) as a function of frequency where $c_{ij}^2 = f_{ij}^2/(f_i f_j)$ with f_{ij} being the complex cross spectrum. In order to get reasonably unbiased estimates for f_i, f_j and f_{ij} running means, over 5 natural frequency bins are calculated before the coherence is evaluated. The $\langle c_{ij}^2 \rangle$ are summed over the range under consideration, and the mean $[\overline{c_{ij}(P_0,\bar{\Omega}/2\pi)}]^2$ is calculated;
4. Steps 1, 2 and 3 are repeated for other P_0 (25 to 40 min with 0.1-min increments) and $\bar{\Omega}/2\pi$ (0.4–2.0 μHz with 0.04 μHz increments) yielding a 151×41 matrix of $[\overline{[c_{ij}(P_0\bar{\Omega}/2\pi)]}]$ values. Before plotting this matrix as a gray tone map, a two dimensional smoothing is performed (running mean over 2 elements in the P_0 direction and 3 elements in the $\bar{\Omega}/2\pi$ direction).

In order to illustrate the behavior of these plots, simulations are performed: a synthetic g-mode spectrum with P_0 and $\bar{\Omega}/2\pi$ of 29.8 min and 1.04 μHz, respectively, is analyzed and the corresponding map is displayed in Fig. 9a. The choice of P_0 and $\bar{\Omega}/2\pi$ is based on the results of a preliminary analysis (Fröhlich 1988). The maximum at the chosen P_0 and $\bar{\Omega}/2\pi$ clearly stands out, as well as some structure which is due to the fact that the g-mode spectrum can also be approximately fitted by other P_0 and/or $\bar{\Omega}/2\pi$. In order to check this structure in the further analysis, isolines are constructed which can be laid over the gray tone maps. To study the influence of noise, the following test is performed: a noise spectrum with a normal distribution in power and a frequency dependence of $1/v$ is added to the synthetic g-mode spectrum. For the case presented in Fig. 9b, the mean rms signal-to-noise ratio in the frequency range concerned (15 to 35 μHz) is set at 1 to 4 (4 times more noise than signal). On the map (Fig. 9b), the peak at 29.8 min and 1.04 μHz is still clearly seen. The noise adds mainly power to a band crossing the map diagonally, but the general structure of Fig. 9a can still be recognized. This

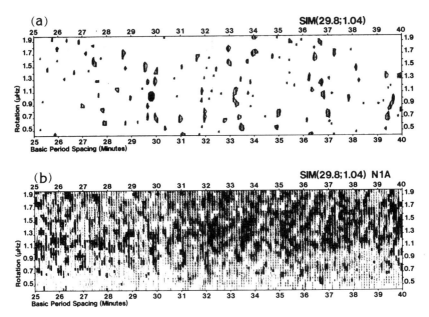

Fig. 9. Maps of the mean coherence between g-mode power spectra with P_0 between 25 and 40 min and f_R between 0.4 and 2.0 μHz and simulated power spectra. (a) Map for a simulated g-mode spectrum with $P_0 = 29.8$ min and $f_R = 1.04$ μHz; (b) same g-mode spectrum, but with noise added (*rms* signal-to-noise ratio 1:4). For both figures, the darker the point, the higher the level of coherence.

means that a g-mode spectrum can indeed be retrieved in the presence of quite high noise levels.

In Fig. 10a,b,c the maps of the ACRIM data of 1980, 1984 and the combination 80/84 are displayed. The latter is produced by calculating the geometrical mean at each index of the two matrices before smoothing the result. The following properties are noted:

1. The noise in the 1980 plot seems more pronounced than in 1984, which is probably due to the increased solar activity in 1980.
2. The diagonal noise band seems to be weaker than that generated in the maps from simulated data; this may indicate that the *rms* signal-to-noise is higher in the real data than in the simulated ones.
3. There are least two other "peaks" which could be interpreted as g-mode maxima at 36.2 min, 1.60 μHz and 34.5 min, 1.45 μHz.
4. The overlaid isoplot has to be moved slightly in order to best fit the combined data 80/84 and their general features.

Attempts to fit other simulated data, e.g., $P_0 = 36.2$ min peak, do not fit the outside features as well as the overlaid one shifted to $P_0 =$

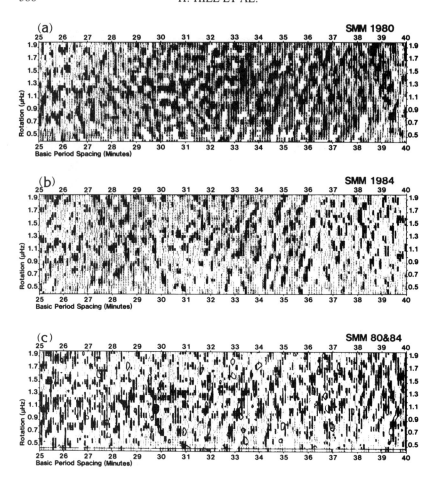

Fig. 10. Maps of the mean coherence between g-mode power spectra with P_0 between 25 and 40 min and f_R between 0.4 and 2.0 μHz and ACRIM power spectra of data from (a) 1980, (b) 1984, and (c) combined 1980/84. For these figures, the darker the point, the higher the level of coherence.

29.75min and $\Omega/2\pi = 1.08$ μHz. The use of only the first term in the asymptotic approximation means that the peak has to be smeared out and some of the farther away features have to be shifted, as the second-order term tends to stretch the scale of P_0 (see Sec.III.C). The smearing out of the main peak can be estimated from the results of Provost and Berthomieu (1986). At the $n = 20$, the error in P_0 due to the omission of the second-order term amounts to 16 s or $\simeq 3$ units on the map, at $n = 48$ the omitted term leads to an enlargement of $\simeq 2$ units. Thus, e.g., the 4 maxima of the isoplot between 36 and 37 min should be moved to the left and would fit on top of the line of peaks, thus improving the overall coincidence.

The result for the basic g-mode period spacing of $P_0 = 29.7$ min seems to best fit the ACRIM irradiance data and supports qualitatively the findings of the frequency separation $(\nu_{n,\ell} - \nu_{n-1,\ell+2})$ of low-degree p-modes (Grec et al. 1983; Claverie et al. 1983). Moreover, it agrees closely with the prediction for a solar model with increased energy transport in the core, e.g., by WIMPS.

From the splitting, a rotational period of the core where the g-modes are concentrated amounts to 10.7 day (1.08 μHz) which is 2.5 times the surface rate. This value is a plausible continuation of the rotation curve from the surface to 0.15 solar radius of Duval et al. (1984).

Although the statistical significance of the result of the g-mode search is very difficult to assess, it seems that the method is quite powerful. A comparison of the results of 1980 and 1984 indicates that the noise in the latter is smaller; hence, it looks promising to repeat the analysis for the years 1985 to 1987 during which the solar activity was very low. Such an analysis is planned for the near future.

C. Work on Detection of g-Mode Signals Leading to Specific Mode Detection and Classification by Reliance on First-Order Asymptotic Theory Predictions

Section II.B describes how the search for evidence of g-modes has been made using a statistical approach based on the equally spaced period feature of asymptotic theory predictions. The statistical approach was indicated in that work because of the anticipated relatively poor signal-to-noise ratio. For those situations where the signal-to-noise ratio is not so unfavorable, the statistical approach may not be required in lieu of an explicit identification of a series of peaks in the power spectra which exhibit the equal-spaced period pattern.

The work of Delache and Scherrer (1983) is an example of this type of search for evidence of g-modes. They first examine the observations (Stanford differential Doppler shift observations of 1979) for a measure of the signal-to-noise ratio with encouraging results. They then proceed to analyze the more prominent peaks in the power spectrum for evidence of the equal-spaced period pattern.

In work of this type, a search is typically made over a specific range in P_0 and $\Omega/2\pi$ with $\varphi = -\frac{1}{4}$. In addition to the work of Delache and Scherrer (1983), there are similar analyses by Isaak et al. (1984), Fröhlich and Delache (1984a,b), Severny et al. (1984), Pallé and Roca-Cortés (1988), Fröhlich (1988), van der Raay (1988), and Henning and Scherrer (1988b). The results of these analyses are used in Sec.III.A in a comparison of reported g-mode eigenfrequency spectra and in Sec.III.E, which summarizes reported values of asymptotic parameters.

D. Work on Detection of g-Mode Signals Leading to Specific Mode Detections and Classifications by Extensive Use of Observed Spatial and Temporal Properties of Eigenfunctions

In the 1970s, an observational technique for the study of solar oscillations with periods longer than 5 min was developed at SCLERA. The technique yields information about the temperature eigenfunction of an oscillation at the extreme solar limb (Hill et al. 1975a). The main objective of the data analysis program for 1982–1988 was to develop analysis techniques which would lead to mode classifications based on the SCLERA differential radius observations and on the combination of the differential radius observations with differential velocity observations.

The results of these programs appear in a series of works (Hill 1984a,b, 1985a,b,1986b; Rabaey et al. 1988; Rabaey and Hill 1988; Hill and Gu 1990; Rabaey 1989) and have led to the publication of a sizeable amount of data, the development of the analysis programs for the classification of modes, the classification of ≈ 1209 resolved modes belonging to 166 multiplets, including acoustic, f- and g-modes, tests of mode classification programs, inferred information on the internal rotation of the Sun, and evidence of mode coupling which may be relevant to the solar neutrino paradox. These results relevant to g-mode classifications are reviewed in the following subsections.

1. Publication Schedule of SCLERA Observational Data. Data from SCLERA observations have been presented in a series of works for the 1978, 1979 and 1985 observations. The average of 13 daily power spectra with a 30 μHz resolution is shown in Fig. 1 of Caudell et al. (1980) and also in Fig. 10 of Caudell (1980) for the 1978 observations. Also, for the 1978 observations, the diagrams of phase vs time for 12 peaks in the power spectrum are shown in Figs. 3, 4 and 5 of Caudell et al. (1980) and also in Figs. 15, 16 and 17 of Caudell (1980). For the 1979 observations, the list of works containing data is more extensive and is given in Table III. The data in these works, much of it available as early as 1982, permits certain of the results to be independently tested such as was done for the 1978 observations. In that test, Gough (1980) independently examined the phase diagrams for the 1978 observations and concurred in the conclusions of Caudell et al. (1980). The first data from the 1985 disk center intensity observations appears in the dissertation of Oglesby (1987c).

2. The SCLERA Mode Classification Program. The objective of the mode classification program developed at SCLERA is the classification of low-to-intermediate degree p-, f- and g-modes with low-to-intermediate radial orders. This program is based primarily on the differential radius observations from SCLERA supplemented with various types of Doppler shift and total irradiance observations from other observatories. These observations provide

TABLE III
SCLERA Publications of Data for the 1979 Observations

Source	Fig. Number	Resolution (in mHz)	Type of Power Spectra[c]	Frequency Range (in μHz)
Bos (1982)	4.1	30	P_1	0–3.2
	4.2	"	P_2	"
	4.3	"	P_4	"
	4.4	"	P_5	"
	4.5	0.28	P_1	240–272
	4.6	"	P_2	"
	4.7	"	P_1	420–452
	4.8	"	P_2	"
	4.9	"	P_4	"
	4.10	"	P_3	"
	4.11	"	P_1	550–582
	4.12	"	P_2	"
	4.13	"	P_1	420–452
	6.3	"	P_1	235–285
Hill et al. (1982)	1	0.28	P_1	235–285
Bos & Hill (1983)	3[a]	0.28	P_1	450–482
Hill (1984a)	1[b]	0.28	P_1	235–285
Hill (1985a)	1	0.28	P_1	2492–2502
	2	"	P_2	3227–3237
	3	"	P_2	3909–3919
	4	"	P_2	3209–3225
	5	"	P_2	3616–3632
	6	"	P_2	3890–3906
Hill et al. (1986)	1	0.28	P_1	100–110
Hill & Czarnowski (1986)	3	0.28	P_1	0–20
	4	"	P_2	"

[a] Same as Fig. 6.3 of Bos (1982) and Fig. 1 of Hill (1984a).
[b] Same as Fig. 6.3 of Bos (1982) and Fig. 1 of Hill et al. (1982).
[c] Definition of P_i nomenclature same as used by Hill (1984b).

the following categories of information for use in mode classification: (1) the multiplet fine structure of the eigenfrequency spectrum expected for a slowly rotating axisymmetric system; (2) the symmetry properties of the eigenfunction; (3) the parities of ℓ and m; (4) the magnitude of ℓ; (5) the (exp $im\phi$) dependence of the eigenfunction; and (6) the θ-dependence of the eigenfunction given by Y_ℓ^m. In addition, a theoretical eigenfrequency spectrum from a standard solar model is used in the classification of the radial order.

The subsections below review how the above information is obtained and/or applied to mode classification. It is apparent that this rather extensive list may permit not only the classification of modes but also the implementation of a number of tests of the accuracy of a given set of mode classifications. The availability of a set of independent tests has become a hallmark of the SCLERA mode classification program. These tests are in addition to those concerned with the reproducibility of results from one year to another.

Symmetry Properties of Eigenfunctions.   The determination of the symmetry properties of the eigenfunctions has been one of the most important exercises of the last few years. First, if the Sun were not an axisymmetric system as seen by the normal modes, the complexity of the multiplet fine structure could considerably complicate the mode classification process. Thus the information obtained on the eigenfunction symmetry properties was first used to determine the type of multiplet fine structure to be expected (see Bos and Hill 1983).

With the determination of the symmetry properties of the system, two analysis techniques based on the inferred symmetry properties of the eigenfunctions have been implemented. The first technique uses the symmetry properties to generate power spectra which contain essentially the signals for modes with only one combination of the parities of ℓ and m. The 4 different combinations of the parities of ℓ and m are (even, even), (even, odd), (odd, even) and (odd, odd). This technique leads to an important reduction in the density of peaks in the complex power spectra and permits the immediate classification of the associated parities of ℓ and m with a relatively high probability of being correct. Examples of the use of this technique are found in the works of Hill et al. (1982), Bos and Hill (1983), Hill (1984b, 1985a,b,1986b), Rabaey et al. (1988), Rabaey and Hill (1988), and Hill and Gu (1990).

The second technique uses the symmetry properties to test for the correctness of a particular mode classification. If the mode classifications for multiplets are correct, the signals in the complex power spectra due to every other member of a multiplet should be confined to one power spectrum while the signals due to the remaining set should be confined to a second power spectrum. Examples of this technique are found in the same works as listed in the previous paragraph. It should be noted that the second technique permits the determination of what fraction of the time incorrect peak classifications are made due to coincidental alignment of the frequency of one mode with that of an alias of a second mode (cf. Sec. IVg of Hill 1984b and Sec. IVe of Hill 1985b).

Frequency Patterns of Multiplet Fine Structure.   The determination of the symmetry properties of the eigenfunctions as those arising from an axisymmetric system furnished an extremely important foundation for defining the basic properties of the multiplet fine structure. In particular, a slowly rotating axisymmetric Sun should lead to a multiplet fine structure that to a first approximation is uniformly spaced in frequency. This frequency pattern is often referred to as a Zeeman pattern, and its simplicity has played an important role in the SCLERA mode classification program.

Azimuthal Dependence of Eigenfunctions. The azimuthal dependence $(\exp im\phi)$ of an eigenfunction expected for an axisymmetric system is the basis of an important technique developed by Hill (1984b) and extended in the work of Hill et al. (1985). The information on the m-dependence is contained in the observable D_{ij} which is a functional of $(\exp im\phi)$. Formally, D_{ij} is essentially the ratio of the finite Fourier transform of the sin $m\phi$ component of $(\exp im\phi)$ to the finite Fourier transform of the cos $m\phi$ component of $(\exp im\phi)$. The general properties of D_{ij} are that it is a quasi-periodic function of m and is antisymmetric for $m \rightarrow -m$. An example of the D_{ij} is shown in Fig. 11, which was obtained in the work on the intermediate-degree g-modes (Rabaey and Hill 1990; Rabaey 1989).

The functional D_{ij} has been important in the determination of the $m = 0$ member of a multiplet because of its symmetry properties about $m = 0$. It has also been a valuable part of the SCLERA mode classification program because it furnishes a highly statistically significant test of a particular set of mode classifications. Examples of its application are found in Hill (1984b,1985b), Rabaey et al. (1988), Rabaey and Hill (1988), and Rabaey (1989).

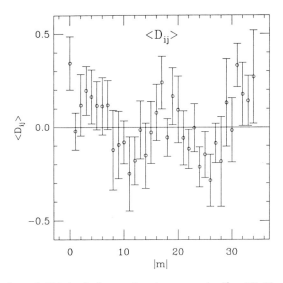

Fig. 11. The observed $\langle D_{ij} \rangle$ for the intermediate-degree g-modes ($\ell \approx 30$). The error bars represent the standard deviations obtained from the scatter in D_{ij} at each $|m|$ value.

Zeros of $\overline{Y_\ell^m}$.   A second technique has been introduced to determine the $m = 0$ mode and the ℓ value of a multiplet for modes with $|m| \gtrsim 20$. This technique is based on the properties of $\overline{Y_\ell^m}$ where $\overline{Y_\ell^m}$ is the average of Y_ℓ^m over the detector geometry used by Bos (1982; see Bos and Hill 1983). The relationship between ℓ and the values of m, m_0, for which $\overline{Y_\ell^m} = 0$, is shown in Table IV for even intermediate values of ℓ.

The first use of these properties with respect to zeros was in the work on phase-locked modes (cf. Sec. VII). For such modes, the zeros of $\overline{Y_\ell^m}$ introduce 180° phase shifts in the plots of phases of the modes vs m.

The Ratio of Amplitudes from Differential Velocity and Differential Radius Observations.   The g-mode spectrum is much more complex than the low-order, low-degree acoustic mode spectrum. Evidence for the detection of g-modes was presented by Hill and Caudell (1979) and by Bos and Hill (1983), but the classification of the modes has been hampered by the complexity of the g-mode spectrum. Hill and Caudell (1979) were able to estimate the magnitude of ℓ only. Hill et al. (1982) identified subsets of two g-mode multiplets, but were only able to place a lower limit on ℓ. However, the complexity can be managed as a consequence of the results obtained by Hill (1985b) in a test of the Hill et al. (1986a) apparent-velocity hypothesis (cf. Sec. III.B).

The development of the apparent-velocity hypothesis described in Sec.III.B arose out of a comparison of results obtained with different observing techniques. This hypothesis interprets the velocity signal obtained in differential Doppler shift studies of long-period oscillations as an apparent velocity due to the combined effect of the surface rotation of the Sun and a perturbation in the radiation intensity produced by a normal mode of oscillation. The ratio of this velocity signal to the observed amplitude of differential radius amplitude was found by Hill et al. (1986a) to be only weakly dependent on m for $2 \leq |m| \leq 10$ so that the ratio is to a good approximation only a function of ℓ. Thus, information is available on the magnitude of ℓ by combining the results of differential velocity and differential radius observations, i.e., the observed constraints on ℓ and m are decoupled. An example

TABLE IV
Values of m, m_0, for $\overline{Y_\ell^m} = 0$

ℓ	m_0
26	9
28	13
30	17
32	21
34	25

of the application of this to mode classification is found in the work of Hill (1985*b*).

III. CLASSIFIED *g*-MODE EIGENFREQUENCIES

A number of *g*-mode classifications have been made by several different groups (Delache and Scherrer 1983; Severny et al. 1984; Fröhlich 1988; Henning and Scherrer 1988*b*; Hill and Gu 1990; van der Raay 1988); this is in addition to the work on the 160-min period oscillation (cf. Table II). Unfortunately, the results of the different groups on *g*-mode detection and classification may be characterized as orthogonal with regard to reported eigenfrequencies. In particular, there are essentially no members of a given eigenfrequency spectrum that coincide with members of a second classified spectrum. This is true regardless of whether the comparisons are made with or without regard to ℓ assignments. The results of a formal comparison are presented in Sec. III.A. As for the work on the 160-min oscillation, there are widely divergent positions held regarding the possible significance of this particular work (cf. Sec. II.A).

Part of the problem encountered in working on the 160-min period signal may be a lack of understanding of what is actually being detected. In particular, are the Doppler shift observations detecting the displacement eigenfunction or the temperature eigenfunction? The displacement eigenfunction produces a velocity signal in the Doppler shift of a spectral line; the temperature eigenfunction can lead to a velocity signal if a rotating Sun is observed with poor spatial resolution. The latter situation is referred to by Hill et al. (1986*a*) as an apparent velocity produced by a normal mode of oscillation. It is important to appreciate the significance of this possibility because it may be relevant to the resolution of several problems. It may also be an important consideration in detector design. The properties of this apparent velocity and its significance are discussed in more detail in Sec. III.B below.

The lack of agreement between the reported classified *g*-mode spectra seems to indicate that at least in some cases the observed signals are not due to *g*-modes. This supposition is based on a predicted lifetime for low-degree *g*-modes of $\gtrsim 10^5$ yr (Boury et al. 1975; Shibahashi et al. 1975; Saio 1980; Kosovichev and Severny 1984*b*). Two additional important observations can be made which indicate the presence of major weaknesses in some of the mode detection and classification programs. One concerns the use of first-order asymptotic theory predictions in the mode detection and classification programs: the manner in which predictions of asymptotic theory are being applied essentially removes the possibility of identifying the *g*-mode signals. The second observation concerns the presence of signals which are produced when observing a rotating nonuniform solar surface. It has been possible to demonstrate in several cases that the reported *g*-mode signals may well be produced in part by this effect.

Because of the obvious importance of these latter two general observations about mode classification programs, they are reviewed in more detail in Secs. III.C and D, respectively. These results can also be important in indicating which reported asymptotic theory parameters, such as P_0 and $\varphi_{q\ell}$, may be based on real g-mode signals. It is for this reason that the presentation and discussion of reported asymptotic parameters in Sec. III.E are not included in Sec. III.A but deferred until after Sec. III.D.

Although it has not yet been possible to obtain agreement between independent g-mode classifications, it has been possible in another series of works to take a given set of g-mode classifications and then find evidence of these signals in a completely independent set of observations (obtained by different groups using different observing techniques during different years). This is a very important development in that g-mode signals should at least be common to all detector systems, albeit with different levels of sensitivity. For this reason these tests are reviewed in Sec. IV.

A. Comparisons of Reported g-Mode Eigenfrequency Spectra

There are six works which report evidence of g-mode detection along with eigenfrequency values: Delache and Scherrer (1983), Severny et al. (1984), Fröhlich (1988), van der Raay (1988), Henning and Scherrer (1988b) and Hill and Gu (1990). A detailed comparison of the classified spectra of these works shows that there are very few coincidences where a coincidence is operationally defined to be when two frequencies are within $\pm$ 0.065 μHz. The window of ± 0.065 μHz is typical of that used in the SCLERA mode classification program. No frequencies are common to three or more of the reported spectra.

The comparison can be made in several ways. One is to compare only eigenfrequencies with the same assigned values of ℓ. However, if g-mode signals are actually being detected but incorrectly classified according to ℓ, then a more informative type of comparison would be to compare without consideration of ℓ assignments. The results of a comparison of the latter type for $\ell \leq 3$ are summarized in Table V. This table lists, for each comparison, the common frequency window d, the number of frequencies a and b in the two referenced spectra that fall in the common frequency window and the number of coincidences c. From the Hill and Gu (1990) results, only the $\ell = 1, 2$ and 3 modes have been included in this comparison as it is expected that the remaining works as a group will have a reduced sensitivity for the higher-degree values. A comparison involving the Severny et al. (1984) classifications is also not included in the table because the assigned values of ℓ are 4. This omission does not affect the general conclusion. For example, a comparison of the $\ell = 4$ mode classifications of Severny et al. (1984) and Hill and Gu (1990) shows no common frequency.

The number of coincidences found is quite consistent with the result to be expected if the spectra are unrelated. This is demonstrated in Fig. 12,

TABLE V
Numbers of Coincidences Between Different Classified *g*-Mode Spectra

KEY a b c d	Fröhlich (1988) (based on Pallé and Roca Cortés 1988)	Fröhlich (1988) (based on Kotov et al. 1984)	van der Raay (1988)	Henning and Scherrer* (1988*b*)	Hill & Gu (1990)
Delache & Scherrer (1983)	16 14 0 45–100 μHz	4 9 0 80–100 μHz	11 32 4 50–75 μHz	11 15 0 50–82 μHz	4 40 1 75–100 μHz
Fröhlich (1988) (based on Pallé and Roca-Cortés 1988)		6 12 0 80–120 μHz	7 32 1 50–75 μHz	11 15 0 50–82 μHz	9 61 2 75–120 μHz
Fröhlich (1988) (based on Kotov et al. 1984)			— — — 0	— — — 0	15 73 3 80–140 μHz
van der Raay (1988)				20 13 2 50–75 μHz	16 13 2 60–75 μHz
Henning & Scherrer (1988*b*)					15 25 3 60–82 μHz

*The $\nu_{15,2}$ = 1/237.9 min was used as the correct frequency of 70.06 μHz, not 70.6 μ Hz.

Legend
  a: the number of eigenfrequencies in the common frequency window from the reference at left;
  b: the number of eigenfrequencies in the common frequency window from the reference at the top of column;
  c: the number of the a and b eigenfrequencies that are within ± 0.065 μHz of each other;
  d: common frequency window.

where the observed number of coincidences is plotted against the number predicted for uncorrelated spectra. The solid line has a slope of unity depicting what would be expected for uncorrelated spectra.

The statement in Sec. III that the classified spectra are essentially orthogonal is based on the results in Fig. 12 and the fact that there are no frequencies which are common to three or more of the reported spectra. This is certainly not the expected result, especially in light of the expected long lifetime of the *g*-modes. There is an obvious conclusion suggested by these findings: either (1) *g*-modes have been detected but the detection efficiency is <<1; or (2) *g*-modes simply have not been detected in at least five of the referenced works. The results of the analyses in Secs. III C, D and E do give

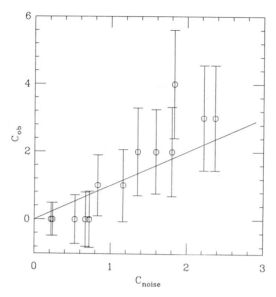

Fig. 12. The number of observed coincidences plotted vs the number predicted for uncorrelated spectra. The solid line has a slope of unity depicting what one would expect from uncorrelated spectra.

some indication of where problems may lie and, as a consequence, which of the classified spectra are more likely to be based on *g*-mode signals.

B. Properties of the Apparent Velocity Signal

Solar oscillations have been observed in the Doppler shifts of spectral lines by several groups (Brookes et al. 1976; Scherrer et al. 1980*a*; Kotov et al. 1983*a,b*). The results obtained appear to depend heavily on the details of the observing technique. In particular, the velocity amplitudes obtained by the Crimean and Stanford groups for the 160-min oscillation differ by more than 60%, while the Birmingham group has not reported evidence of this oscillation since their first results in 1976 (Brookes et al. 1978*a*). These discrepant results have led to a re-examination of how velocity fields on the solar disk are manifested in Doppler shift measurements. It has been found that solar rotation in conjunction with a normal mode of oscillation can create apparent velocity signals in the data which are not directly related to line-of-sight velocities. This apparent velocity hypothesis can account for the difference in velocity amplitudes observed by the Crimean and Stanford groups (Hill et al. 1986*a*).

Optical resonant-scattering spectrometers have been successfully used in the study of 5-min solar oscillations (Brookes et al. 1978*c*; Claverie et al. 1981*a*; Grec et al. 1983). However, similar success has not been achieved when searching for evidence of the 160-min period oscillation. It is noted that this apparent discrepancy between resonant-scattering velocity spectrometers and grating-based velocity spectrometers (such as used at Crimean and Stanford) may also be understood in terms of the work of Hill et al. (1986a). An LTE calculation by Germain and Hill (1988) shows that the resonant-scattering detection sensitivity using the K 7699 line relative to that of the Crimean spectrometer for the Fe I 5124 line (Kotov et al. 1983a,b) will be 0.24 and 0.33 for $\ell = 1$ and 2 modes, respectively. Germain and Hill (1988) also note that the corresponding ratios for a NaD line spectrometer relative to the same Crimean spectrometer for the Fe I 5124 line are 0.31 and 0.43. This leads to a predicted reduction of an order of magnitude in the signal-to-noise ratio in the resulting power spectra and could account in part for the apparent difficulty encountered in detecting g-mode signals with resonant-scattering spectrometers.

The results for grating-based and optical resonant-scattering spectrometers are summarized in Table VI. These results may be important not only for understanding the differences reported in searches for g-modes (and also, in particular, the 160-min period signal), but they may also be important in the design of future detectors for programs whose particular goal is the detection of g-mode signals.

It should be noted that the apparent velocity signal is essentially inversely proportional to the azimuthal spatial dimensions of a given detector. Thus, the sensitivity to this type of signal will be diminished for detectors put on-line with improved spatial resolutions, a situation which could lead to a lower probability for the detection of g-mode signals.

It has been observed by Hill et al. (1986a) that the properties of the apparent velocity signal are proportional to $Y_\ell^m(\theta,0)$ times a function of ℓ. This feature has been exploited by Hill (1985b) to obtain decoupled observational constraints on ℓ and m. This decoupling feature has been integrated into the SCLERA g-mode classification program (cf. Sec.II.D).

TABLE VI
Relative Sensitivities of Velocity Spectrometers to the Apparent Velocity Signal

	ℓ	
	1	2
Crimean Spectrometer		
Fe I 5124	1.00	1.00
Stanford Spectrometer		
Fe I 5124	0.31	0.48
Optical Resonant-scattering Spectrometer		
NaD	0.31	0.43
K 7699	0.24	0.33

C. Linear Representations of Asymptotic Theory Predictions

Considerable research on the detection and classification of g-modes has been based on the theoretical predictions of first-order asymptotic theory (Delache and Scherrer 1983; Fröhlich and Delache 1984a,b; Isaak et al. 1984; Severny et al. 1984; Fröhlich 1988; Pallé and Roca-Cortés 1988; van der Raay 1988). The first-order asymptotic relation is given by either Eq. (4) or (9) when the V_1 and V_2 terms are negligible and φ equal to a constant which is discussed later in this section. There have been several sets of guidelines suggested to prescribe the domain of applicability of the first-order asymptotic relation (cf. Berthomieu et al. 1984a,b; Kidman and Cox 1987). These guidelines are based on theoretical analyses of standard solar models. The criteria for establishing the guidelines have generally been based on the objectives of recovering the proper value of P_0 given by Eq. (5) and/or a linear dependence of $P_{n\ell}$ on n.

It has been tacitly assumed in applications relying on these guidelines that the intercept $P_{0\ell}$ for the liner representation of $P_{n\ell}$ remains $P_0(\ell + \varphi)/[\delta\ell(\ell + 1)]^{1/2}$ and that the quadratic dependence of $P_{n\ell}$ on n is not observable or is unimportant. Unfortunately, these assumptions appear not to be valid with serious implications for the results of these applications.

There appears to be primarily two problems arising in applications relying on first-order asymptotic theory predictions. Based on the properties of standard solar models, the V_1 and V_2 terms in Eq. (10) are not negligible for the works of Delache and Scherrer (1983), Fröhlich and Delache (1984a,b), Isaak et al. (1984), Severny et al. (1984), Fröhlich (1988), Pallé and Roca-Cortés (1988), and van der Raay (1988). For this same set of works, the phase factor φ itself may be quite far from its asymptotic value (Provost and Berthomieu 1986). These two potential problem areas are discussed further below.

1. Manifestation of V_1 *and* V_2. The least-squares fit about $n = q$ of $P_{n\ell}$ given by Eq. (9) to the linear equation

$$P_{n\ell} = A_{q\ell} + B_{q\ell}\, n \qquad (16)$$

yields

$$A_{q\ell} = \frac{P_0}{[\ell(\ell + 1)]^{1/2}} \left(\frac{\ell}{2} + \varphi_{q\ell} + \varphi'\right) \qquad (17)$$

$$B_{q\ell} = \frac{P_0\,(1 + \partial\varphi/\partial n)_{q\ell}}{[\ell(\ell + 1)]^{1/2}} \frac{1}{1 + K_{q\ell}} = \frac{P_0'}{[\ell(\ell + 1)]^{1/2}} \qquad (18)$$

where

$$\varphi' = \frac{P_0}{P_{q\ell}}[\ell(\ell + 1)]^{1/2} W_\ell + \frac{qK_{q\ell}}{1 + K_{q\ell}} \qquad (19)$$

$$K_{q\ell} = \left(\frac{P_0}{P_{q\ell}}\right)^2 W_\ell \qquad . \qquad (20)$$

Only the lowest-order correction terms involving $\partial\varphi/\partial n$ and W_ℓ have been retained. The corresponding expression for the second derivative $[\partial^2 P_{n\ell}/\partial n^2]_{n=q}$ of $P_{n\ell}$ is given by

$$\left(\frac{1}{P_{n\ell}}\frac{\partial^2 P_{n\ell}}{\partial n^2}\right)_{n=q} = \frac{2\ell(\ell + 1)K_{q\ell}^2}{W_\ell(1 + K_{q\ell})^3} \qquad . \qquad (21)$$

The values of φ' and (P_0'/P_0) appropriate for the standard solar model of Schatzman and Maeder (1981) are given in Figs. 13 and 14. These figures were generated based on the values of $P_0 = 34.3$ min, $V_1 = 0.38$ and $V_2 = 4.7$ obtained by Berthomieu et al. (1984b) for the Schatzman and Maeder (1981) model with no mixing. Similar values for V_1 and V_2 are obtained for the standard solar model of Saio (personal communication). The range of values for q used in Figs. 13 and 14 were chosen to cover the different tests for g-modes and sets of classified g-modes that have been reported.

There are three important observations that can be made about g-mode classifications that have been based on linearized asymptotic theory predictions. First, we see from Fig. 13 that for the reported classified g-mode spectra with $\nu_{q\ell} \gtrsim 20$ μHz, $\varphi' \gtrsim 1/4$. Second, we see from Fig. 14 that the P_0' for classified g-mode spectra should be observably ℓ dependent for $\nu_{q\ell} \gtrsim 30$ μHz. And finally, the values for the second derivative of $P_{n\ell}$ expressed by Eq. (21) indicate that this term should also have been observable in the classified spectra in question.

Thus the tests for g-modes and the classified spectra which are based on equally spaced $P_{n\ell}$ values with $\varphi_{q\ell} + \varphi' = -1/4$ fail on three tests which check their validity: (1) the effective ($\varphi_{q\ell} + \varphi'$) should have a non-negligible positive value for φ'; (2) the effective P_0' should have an observable ℓ dependence; and (3) there should be an observable quadratic deviation from an equally spaced period pattern. The observed signals may indeed be due to g-modes, but the single argument in support of their identification and classification as g-modes which is based on linearized asymptotic theory appears to be badly flawed.

In another application of linearized asymptotic theory predictions, the

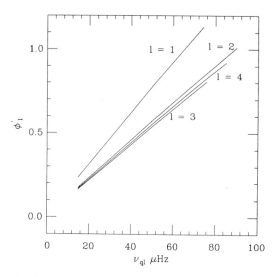

Fig. 13. The functional dependence of φ' vs frequency for the four cases of $\ell = 1,2,3,4$. These theoretical curves were obtained from the standard solar model (with no mixing) of Schatzman and Maeder (1981). Note that for $\nu_{q\ell} > 20$ µHz, φ' must be $> 1/4$.

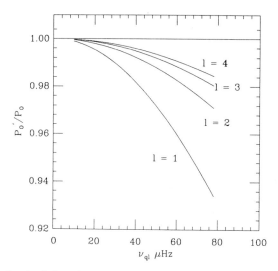

Fig. 14. The functional dependence of P_0'/P_0 vs frequency for the four cases of $\ell = 1,2,3,4$. These theoretical curves were obtained from the standard solar model (with no mixing) of Schatzman and Maeder (1981). Note that P_0' is a function of ℓ.

quantity $P_{n+1,\ell} - P_{n,\ell}$ is examined to obtain a measure of $P_0/[\ell(\ell + 1)]^{1/2}$. However, this procedure does not sidestep the issue raised with regard to $\varphi_{q\ell} + \varphi'$: in this particular application, it is necessary to use the classified $P_{n,\ell}$ and determine if the $\varphi_{q\ell} + \varphi'$ is within an acceptable range. The $\ell = 4$ classifications of Severny et al. (1984) (cf. Sec. III.E) are a case in point.

2. *Properties of the Phase Term φ.*   The lack of a detailed understanding of the properties of the phase parameter φ in Eq. 9 currently places the most serious limitation on the use of the theoretical asymptotic formalism to detect and classify g-modes and to invert the g-mode spectrum for internal properties of the Sun. An appreciation of this limitation by the solar physics community has been slow to develop.

In the work of Tassoul (1980), a value for φ of

$$\varphi = -1/4 \tag{22}$$

was obtained under the assumption that every physical quantity is continuous as well as its derivatives at the base of the convection zone. The result has been widely used by various groups to look for evidence of g-modes. However, these assumptions do not always apply for the case of real stars where, even though N^2 may vanish at the boundaries of a convection zone, its first derivative is not necessarily continuous. For example, mixing-length theories lead to an N^2 with a discontinuity in its gradient at the convective interface. Tassoul finds for such a case that

$$\varphi = -5/12 \quad . \tag{23}$$

It was also observed by Gabriel (1986b) and Provost and Berthomieu (1986) that in the derivations of the Tassoul asymptotic relations, it was assumed that $|N^2 / \sigma^2| \gg 1$ everywhere in the star except in the vicinity of the transition point at r_1, the inner radius of the convection zone. It was pointed out that this condition will be fulfilled in the convective envelope only if $\sigma^2 < 10^{-4}$ or if $P > 16,000$ min. Both Gabriel (1986b) and Provost and Berthomieu (1986) proceeded to derive expressions for the eigenfrequencies using the asymptotic expressions for the eigenfunctions given by Tassoul (1980) in the inner radiative zone and by a numerical integration in the convection zone. In the more extensive analysis of Provost and Berthomieu (1986), it was found that φ depends significantly on the frequency. This result was not widely anticipated (cf. Ellis 1986).

Ellis (1984) found that the limiting value ϕ_0 of ϕ can be expressed as

$$\lim_{\sigma \to 0} \varphi = -\frac{1}{4} - \frac{1}{2}\left(\frac{1}{2 + \beta}\right) \tag{24}$$

where β is a stratification index just below the convection zone. Explicitly, β is the exponent of the depth below the convection zone with which the polytropic index and N^2 both pull away from their respective adiabatic values in the convection zone. Gabriel (1986b) found for a convective envelope where $N^2 \neq 0$ that there is an additional relatively small correction term to Eq. (24).

There exists yet another problem which remains to be treated. In theoretical work such as found in Provost and Berthomieu (1986), their second-order asymptotic formula is based on the formalism presented in Tassoul (1980) which assumes certain stratification properties just below the convection zone. In particular, it is assumed that in the region just below the convection zone that $\beta = 1$ so that

$$N^2 \propto (r_1 - r). \tag{25}$$

The choice of $\beta = 1$ and $N^2 \propto (r_1 - r)$ in such work is not based on a physical model. For the type of transition point encountered near $r = r_1$, uniform approximations to the asymptotic solutions of second-order differential equations have been found specifically for the case of $\beta = 1$ (see Sec. 14 of Olver 1974). In particular, other values of β are not permitted for these asymptotic solutions near $r = r_1$.

Thus, we are left at the present time without an asymptotic theory which is applicable to a real star with $\beta \neq 1$ near $r = r_1$. This indeed poses a serious road block for the g-mode classification and inversion programs which rely on asymptotic theory predictions.

D. Properties of the Spectrum Produced when Observing a Rotating Nonuniform Solar Surface

The manifestations of observing a rotating nonuniform solar surface have been seen in total irradiance observations (Hill and Kroll 1986; Kroll et al. 1988a,b), in whole-disk Doppler shift observations (Anderson and Maltby 1983; Edmonds and Gough 1983; Jefferies et al. 1988) and in differential velocity and differential radius observations (Hill and Czarnowski 1986). There is also evidence that the classified g-mode spectra are in part based on a misinterpretation of spectra produced by observing a rotating nonuniform solar surface. The first such evidence was presented by Hill and Czarnowski (1986) for the classified spectrum of Delache and Scherrer (1983).

The Fourier transform of the signal produced by a long-lived perturbation on the solar surface will be a convolution of the Fourier transform of the observing window function with the Fourier transform of a signal such as that shown in Fig. 15 where $T = 1/\nu_0$ is the rotation period of the Sun at the location of the perturbed region. The function shown in Fig. 15 is an example of what might be obtained with a whole-disk velocity spectrometer for a perturbed region which had a lifetime of two solar rotations. Because the Fourier transform of the observing window function is convoluted with the Fourier transform of a function such as in Fig. 15, it is apparent that the resulting power spectrum will be characterized by Zeeman-like frequency patterns centered on harmonics of $1/T_d$ where T_d is the local solar day (see Fig. 16). For a satellite observation, T_d will be the orbital period of the satellite. For groundbased observations, T_d will, in general, be observably different from the mean solar day of 24 hr. The frequency spacing of the Zeeman patterns will be ν_0.

The classified g-mode spectra of Delache and Scherrer (1983), Fröhlich (1988) (based on the power spectra of Kotov et al. [1984a] and Pallé and Roca-Cortés [1988]), and van der Raay (1988) have been examined for this characteristic frequency pattern. In general, it has been possible to obtain better fits to the reported spectra using the model of Fig. 16 rather than the equally spaced period model. In this analysis, two dominant values for ν_0 were found. These are the surface rotation rate of the active region centered around a latitude of $\approx 20°$ and a rotation frequency of ≈ 0.45 μHz. The results of the analysis are summarized in Table VII. The results of Dicke (1983) are also included in Table VII along with the results of Dicke (1981) based on solar oblateness observations.

It is important to note that negative results were obtained by Hill and Czarnowski (1986) when testing the Hill (1985b) classified g-mode spectrum for the signature of observing a rotating nonuniform surface. As the Hill

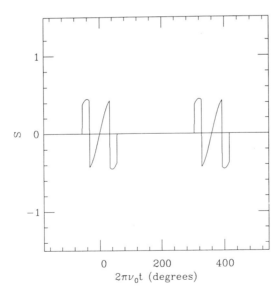

Fig. 15. An example of what one may obtain with a whole-disk spectrometer from a perturbed region on the solar surface which has a lifetime of two solar rotations. The convolution of the Fourier transform of this signal with the Fourier transform of the observing window function gives rise to Zeeman-like frequency patterns such as that shown in Fig. 16.

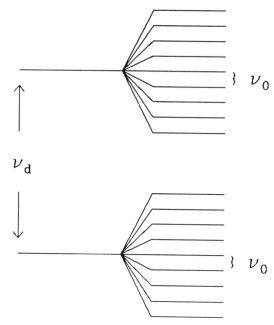

Fig. 16. An example of the Zeeman-like frequency patterns which may result from a perturbed region on the solar surfuace when observed by a whole-disk spectrometer. The frequency $\nu_d \approx 1/\text{day}$ and ν_0 is the frequency of solar rotation.

(1985*b*) classified *g*-mode spectrum is essentially a subset of the Hill and Gu (1990) *g*-mode spectrum, it is inferred that the Hill and Gu (1990) classified spectrum likewise is not derived from a spectrum produced by a rotating nonuniform surface.

E. Reported Values of Asymptotic Parameters

The reported values of asymptotic parameters generally fall into two categories: one category consists of work based on a linear representation of asymptotic properties and a second category consists of work based on second-order asymptotic theory. The primary distinction between these two categories is that second-order terms from asymptotic theory are assumed to be negligible in the linear representation. As noted in Sec. III.C, this may not be a good approximation and model-dependent corrections can be made to the linear representative findings to obtain an estimate for $\varphi_{q\ell}$.

The various results reported for asymptotic parameters are given in Table VIII. The $\varphi_{q\ell} + \varphi'$ results have been obtained from the reported values of the phase in the referenced works. Model-based values for φ' given in Sec. III.C are also included in the table. For the work of Severny et al. (1984),

TABLE VII
Inferred Values with ν_0

	ν_0 μHz
Fröhlich (1988)[a]	0.4258 ± 0.0067
Fröhlich (1988)[b]	0.4151 ± 0.0064
van der Raay (1988)[c]	0.4213 ± 0.0055
Dicke (1981)	0.4518 ± 0.0036
Claverie et al. (1982)[d]	0.442 ± 0.007
Delache and Scherrer (1983)[e]	0.4447 ± 0.0004
SCLERA (1979)[f]	0.4447 ± 0.0020
Kotov et al. (1983)[g]	0.4452 ± 0.0005
Beardsley (1987)	0.44
van der Raay (1988)[c]	0.4553 ± 0.0057

[a] Analysis results reported here for Fröhlich (1988) classified spectrum based on the work of Kotov et al. (1984).

[b] Analysis results reported here for Fröhlich (1988) classified spectrum based on the work of Pallé and Roca-Cortés (1988).

[c] Analysis results reported here for van der Raay (1988) classified spectrum.

[d] Analysis results of Dicke (1983) based on work of Claverie et al. (1982).

[e] Analysis results of Hill and Czarnowski (1986) for classified spectrum of Delache and Scherrer (1983).

[f] Analysis results of Hill and Czarnowski (1986) based on work of Bos and Hill (1983).

[g] Analysis results of Hill and Czarnowski (1986) based on work of Kotov et al. (1983).

where no value for the phase was given, the reported $P_{n\ell}$ were fit to obtain the associated value of $\varphi_{q\ell} + \varphi'$.

One of the most striking features of the results in Table VIII is the large variation in P_0. This is not surprising considering the negative results found in Sec. III.A with regard to the low commonality of reported classified g-mode spectra and the positive results discussed in Sec. III.D with regard to the presence of surface rotation effects. The negative results of Sec. III.A indicate that as many as five of the six reported classified spectra are not due to g-modes. The positive results of Sec. III.D indicate that the signals identified as due to g-modes are in fact strong candidates for signals produced by observing a rotating nonuniform surface.

A second and more insightful observation about the results in Table VIII concerns the first category (which is based on linearized asymptotic theory predictions). Comparing the model based values for φ' with the reported values for the phase $\varphi_{q\ell} + \varphi'$, it is noted that there is a very small range of values for $\varphi_{q\ell} + \varphi'$ while there is a relatively large range of values for φ'. This suggests, for example, nonphysical values for W_ℓ with $\varphi_{q\ell} \approx -1/4$ or

TABLE VIII
Observational Results on Asymptotic Parameters

	P_0' (min)	$\varphi_{q\ell} + \varphi'$	$(\varphi')^a$
Linearized Asymptotic Results			
Delache and Scherrer (1983)	38.60 ± 0.50	-0.2 ± 0.1	0.79
Isaak et al. (1984)	41.32 ± 0.12	-0.25	0.65
Fröhlich and Delache (1984a,b)	38.90	-0.25	0.29
Severny et al. (1984)	37.40 ± 2.70	2.04	1.20
Pallé and Roca-Cortés (1988)			
1981 observations	41.4	-0.25	0.79
1982 observations	42.6	-0.25	0.79
1984 observations	35.4	-0.25	0.79
	42.4	-0.25	0.79
1985 observations	37.4	-0.25	0.79
Fröhlich (1988)			
Based on Kotov et al.			
(1984)	29.85	-0.25	0.12
Based on Pallé			
& Roca-Cortés (1988)	29.85	-0.25	0.80
Based on ACRIM	29.85	-0.25	0.29
van der Raay (1988)	41.00	-0.25	0.66
Henning and Scherrer (1988b)			
$\ell = 1$	37.12	-0.05	0.96
$\ell = 2$	37.23	-0.35	0.80

Second Order Asymptotic Results			
	P_0	$\varphi_q \ell$	V_1	V_2
Gu et al. (1988)	36.31 ± 0.12	-0.43 ± 0.13	0.35	4.76

[a] Values taken from Fig. 13 based on standard solar model for the respective lowest values of ℓ used in the reported findings unless ℓ value specifically noted.

physically reasonable values for W_ℓ accompanied with nonphysical values for $\varphi_{q\ell}$ and hence β, the stratification index.

The possible nonphysical findings for $\varphi_{q\ell} + \varphi'$ reported for the first category could furnish a strong discriminant in addressing the problem of a low commonality of the reported classified g-mode spectra. Certainly, the associated values for P_0 should be used with caution in inferring properties of the solar interior from P_0. The results of Gu et al. (1988) do not impose a similar nonphysical constraint on $\varphi_{q\ell}$ or constrain W_ℓ to low values in the context of standard solar models. This latter statement does not necessarily imply that the SCLERA results are based on g-modes but does indicate that these results do satisfy a necessary condition for them to be based on g-mode signals.

IV. PROPERTIES OF SOLAR g-MODE EIGENFUNCTIONS
IN THE PHOTOSPHERE

The identification of photospheric properties of the g-mode eigenfunctions have been the objective of a number of different studies. Such work has been done extensively at SCLERA and has been concerned with the general symmetry properties for reflection about the equator and about the axis of rotation, about the detail dependence on the spherical coordinates θ and ϕ, and the ℓ, m frequency dependence. References to all but the frequency dependence work can be found in Sec. II.D. The work on the frequency dependence is particularly interesting and is reviewed in this section.

Predictions have been made by Provost and Berthomieu (1988) and Berthomieu and Provost (1990) for the distribution of observed amplitudes of the total irradiance oscillations produced by low-degree g-modes. These predictions are that the total irradiance amplitude should exhibit a minimum for eigenfrequencies corresponding to frequencies of oscillations for which the divergence of the displacement eigenfunction is zero at the surface; that is, they exhibit f-mode behavior at the surface. Their predictions are shown in Fig. 17 where energy equipartition has been assumed (10^{37} erg per mode).

The minimum in the total irradiance amplitude is associated with a min-

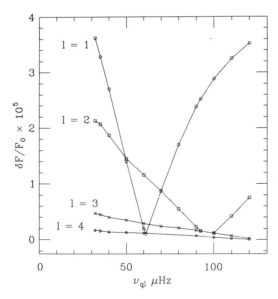

Fig. 17. The functional dependence of the amplitudes of the total irradiance perturbations for low-degree g-modes vs frequency. This figure was adapted from the work of Provost and Berthomieu (1988).

imum in the photospheric values for $\delta T'/T$ where $(\delta T'/T)Y_\ell^m$ is the Lagrangian temperature eigenfunction. Thus, as noted by Hill et al. (1988a,1989), the work of Provost and Berthomieu (1988) implies that there should be corresponding minima in the power spectra of differential radius observations. The frequencies $\sigma_\ell = 2\pi\nu_\ell$ for the adiabatic divergence-free displacement eigenfunctions at the surface are given by

$$\sigma_\ell^2 = \frac{\ell(\ell+1)}{\sigma_\ell^2}\left(\frac{g}{R}\right)^2 - 4\frac{g}{R} \qquad (26)$$

where g is surface value of gravitational field and R the radius of the Sun (Hill et al. 1989). The predicted ν_ℓ for $\ell = 1, \ldots, 5$ divergence-free oscillations are given in Table IX.

The amplitudes of the even $(\ell + m)$ g-modes classified by Hill and Gu (1990) have been examined by Hill et al. (1988a) for this predicted property. Their results for $\ell = 3$ are shown in Fig. 18 where the amplitudes are taken from Hill (1988). The results in Fig. 18 have a well-defined minimum at $\nu \approx 148$ μHz while plots of $\ell = 2$ and 4 classified modes have changes in slopes at $\nu \approx 115$ μHz and 180 μHz, respectively. The presence of these 3 observationally obtained frequency features near the theoretically predicted locations by Provost and Berthomieu (1988) and Berthomieu and Provost (1990) give support to the interpretation that the observed signals are due to oscillations in the solar atmosphere and that the ℓ classifications of the signals are also correct.

V. CROSS-CORRELATION-TYPE STUDIES OF CLASSIFIED g-MODE SPECTRUM WITH POWER SPECTRA OF DIFFERENT TYPES AND SETS OF OBSERVATIONS

Because of the relatively small signals ascribed to g-modes and because of the complexity of the theoretically predicted spectrum, a g-mode detection and classification program places rather severe demands on the quality of a

TABLE IX
Frequencies of Adiabatic Divergence-Free
Displacement Eigenfunctions at the Surface

ℓ	ν_ℓ (μHz)
1	67
2	108
3	141
4	170
5	196

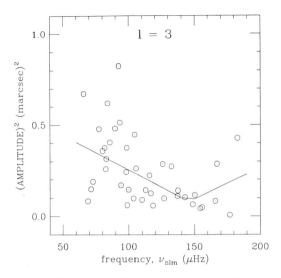

Fig. 18. The observed amplitudes for the ($\ell + m$) even g-modes classified as $\ell = 3$ by Hill and Gu (1990). Notice the well-defined minimum at $\nu \approx 148$ μHz.

given set of observations if that set is to be the only source of observational information in the program. Thus, the situation may arise where most of the available observations are not usable in such a g-mode detection and classification program. However, these same rejected observations may be quite valuable in furnishing a test for a given set of independently derived g-mode classifications.

Extensive use of this latter type of test has been made at SCLERA in p-, f- and g-mode work. Unlike the situation encountered with p- and f-modes, a classified g-mode should be observable with no detectable change in amplitude from one year to the next and by different types of detectors. If confirmation is obtained by two different types of detectors, then the relative signal strengths can be used to test the m and/or ℓ classifications. The following cross correlations have been made using the g-mode classifications of Delache and Scherrer (1983), Fröhlich (1988), van der Raay (1988) and Hill and Gu (1990):

Hill (1986):   Correlation studies of differential velocity power spectra by Kotov et al. (1983a) with 1979 differential radius power spectra by Bos and Hill (1983); used to make ℓ classifications.

Hill (1988):   Correlation studies of the 1978 diameter power spectrum by Caudell et al. (1980) with 1979 differential radius power spectra of Bos and Hill (1983) at frequencies of classified g-modes by Hill and Gu (1990); tested temporal and spatial symmetry properties.

Yi and Hill (1988):   Correlation studies of the 1983 differential radius power spectra by Yi and Czarnowski (1987) with 1979 differential radius power spectra of Bos and Hill (1983) at frequencies of classified g-modes by Hill and Gu (1990); tested temporal and spatial symmetry properties.

Oglesby (1987*b*):   Correlation studies of 1985 disk center intensity power spectra by Oglesby (1987*c*) with 1979 differential radius power spectra by Bos and Hill (1983) at frequencies of classified g-modes by Hill and Gu (1990); tested temporal properties and *m* classifications on a statistical basis.

Kroll et al. (1988*a,b*): Correlation studies of 1980 and 1984 total irradiance power spectra obtained by ACRIM (see Sec. II.B) with 1979 differential radius power spectra by Bos and Hill (1983) at frequencies of classified g-modes by Hill and Gu (1990); tested temporal properties on a statistical basis.

Hill and Czarnowski   Correlation studies of the classified g-mode spectrum
(1986):   by Delache and Scherrer (1983) based on 1979 differential velocity observations; tested temporal properties and for non-g-mode sources of classified signals.

Secs. III.A and D: Correlation studies of the classified g-mode spectra by Fröhlich (1988) and van der Raay (1988); tested temporal properties and for non-g-mode sources of classified signals.

Recognizing that there are generally no free parameters in this series of correlation studies, the positive results that have been reported are very encouraging. In this regard, it is important to note that one feature of the positive results is not model dependent: the observationally demonstrated common features of power spectra cannot be removed by altering n, ℓ and/or m classifications. This must certainly be the observational base on which g-mode detection and classifications must stand.

VI. DEVIATIONS FROM ASYMPTOTIC THEORY PREDICTIONS PRODUCED BY LOCALIZED CHANGES IN THE BACKGROUND STATE

A fine structure may be present in an eigenfrequency spectrum for a number of different reasons. The more widely known example from solar seismology is the fine structure generated by rotation of the Sun. For a given set of normal modes, a fine structure may also be caused by a localized change in the background state of the Sun when the change occurs over a distance shorter than the local length characterizing radial changes in the respective eigenfunctions. Examples of this are found in the works of Hill and Rosenwald (1986a), Gavryusev and Gavryuseva (1988), Berthomieu and Provost (1988) and Kidman and Cox (1987). Analogous effects have received much attention in the study of the normal modes of the Earth. For example, Lapwood and Usami (1981) have called this fine structure effect the "solo-tone" effect.

Evidence of such a structurally induced fine structure on the eigenfrequencies of low-degree 5-min oscillations has been reported by Hill and Rosenwald (1986a). This fine structure has been interpreted as the result of changes in the background state that occur over a relatively short distance at the base of the convection zone. A second example of fine structure which is presumably induced by significant localized changes in the background state has also been found in the study of low-degree solar g-modes (Rosenwald et al. 1987; Gu et al. 1988).

The reported g-mode fine structure effects are characterized as quasi-periodic deviations of the classified g-mode spectrum from the spectrum predicted by second-order asymptotic theory described in Sec. I.B. The deviations are periodic in radial order n for a fixed ℓ and not dependent on m. An example of the observed deviations for $\ell = 4$ is shown in Fig. 19. For more specifics concerning these deviations, reference is made to Gu et al. (1988).

The location in radius of a significant fractional change in the background state determines the period in n of the periodic deviation and the magnitude of the fractional change in the background state determines the amplitude of the periodic deviation. It should thus be possible, by using observational results and sensitivity analysis results based on a given standard solar model, to determine both the location and magnitude of the change in the background state primarily responsible for observed quasi-periodic deviations.

An inversion of the reported g-mode fine structure effects has been performed by Hill et al. (1988b). It was found that the basic features of the fine structure present in the classified low-degree g-mode spectrum reported by Gu et al. (1988) and Hill and Gu (1988) are physically plausible and can be reproduced by a positive perturbation in $A^* \simeq rN^2/g$ at $\bar{r}/R = 0.235$. The effective radial width of this perturbation is $\Delta r/R \leq 0.015$. Furthermore, it

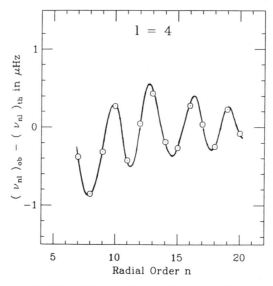

Fig. 19. The observed deviation of the classified g-mode spectrum ($\ell = 4$) from the spectrum predicted by second-order asymptotic theory. Notice the quasi-periodic nature of the deviations.

was noted that this perturbation in A^* could be produced by a fractional decrease in the mean molecular weight μ of $|\Delta\mu/\mu| = 0.009$ within the effective radial width of $\Delta r/R \leq 0.015$ at $r/R = 0.235$.

The inferred $\Delta\mu/\mu$ could be the result of a partial mixing of the core in the past. However, it was noted by Hill et al. (1988) that the mixing required to yield the inferred $\Delta\mu/\mu$ is sufficiently small so as not to be very important in resolving the solar neutrino paradox. The inferred fractional change is also sufficiently small so as not to be inconsistent with the observational based value of P_0 obtained by Gu et al. (1988) and the 5-min oscillations which indicate that the little mixing in the core has occurred.

VII. EVIDENCE OF GRAVITY-MODE COUPLING AND POTENTIAL IMPACT ON THE INFERRED INTERNAL STRUCTURE OF THE SUN AND THE SOLAR NEUTRINO PARADOX

The first evidence of solar g-modes which may have relatively large temperature perturbations in the core was obtained in the 1970s by Hill and Caudell (1979). They found evidence which indicated the excitation of a number of g-mode oscillations with frequencies near 250 μHz and 370 μHz and $20 \lesssim \ell \lesssim 40$. The implications of this discovery were potentially quite important to several areas of solar physics: if the values of ℓ were of this magnitude, then the amplitudes of the associated oscillations may be sufficiently large in the core to affect the Sun's internal properties in a significant way.

One of the more important effects of the presence of such intermediate-degree g-modes would be to lower the effective central temperature of the Sun and thus to alter the production rates of solar neutrinos. This possibility was noted in the 1970s by Hill (cf. Barabanov et al. 1978; Thomsen 1978). The reduced core temperature would be, in part, a consequence of the additional energy transport mechanism furnished by the large-amplitude g-modes.

The lack of a detailed knowledge of the internal structure of the Sun and the complexity presented by the boundary conditions (cf. Hill and Logan 1984) make it difficult to infer with some confidence the amplitudes of these oscillations in the core from knowledge of their properties at the solar surface. Thus, the work of the 1970s left us with this impasse: on one hand, we needed to know the equilibrium properties of the core and the boundary conditions at the surface to calculate the properties of the large-amplitude oscillations, and, on the other hand, if the oscillation amplitudes were indeed large, we needed to know their properties to infer the equilibrium properties of the core.

If the amplitudes of oscillation are in fact sufficiently large to alter the effective temperatures of the various physical processes in the core, then second- and possibly higher-order processes are important. If higher-order processes are important, it is possible that the oscillations with large amplitudes in the core are no longer independent of each other, i.e., they are coupled. Coupling may be manifested in different ways. One of the more widely recognized ways is that an oscillation may couple to excite harmonics of its own frequency. A second is that two or more modes may be phase-locked. In this case, a subset of the modes of a rotationally split multiplet may have their frequencies and phases nearly uniformly spaced (amplitudes $\propto \exp i[2\pi(\nu_{n,\ell} + m\nu'_{n,\ell})t + m\varphi_{n,\ell}]$, where $\nu_{n,\ell}$, $\nu'_{n,\ell}$ and $\varphi_{n,\ell}$ are constants). It should be noted that the set of phase-locked modes need not include the entire set of $(2\ell + 1)$ modes of a multiplet.

The demonstration of the two primary observable features of mode coupling of the second class, (i.e., phase-locking would exhibit uniform spacing in phase and frequency) would side-step the impasse described above and indicate that the amplitudes of the particular g-modes are sufficiently large to have higher-order processes important in the core of the Sun.

Evidence of these two features of mode coupling has been reported by Hill (1986), Rabaey and Hill (1988,1990), and Rabaey (1989) in work on the detection and classification of 20 intermediate-degree g-mode multiplets (cf. Sec. II). The frequency patterns of the multiplet fine structure and the phases of the oscillations were found to consist of sections linear in m. The phase-vs-m diagram and $\nu_{n\ell m}$-vs-m diagram for the multiplet, classified as $g_{15,30}$, are shown in Figs. 20 and 21, respectively.

We see in the figures evidence that the classified $g_{15,30}$ multiplet consists

of three groups of phase-locked modes. In particular, the results shown in
Figs. 20 and 21 represent highly statistically significant support for the phase-
locking hypothesis. Note that each of the phase-locked groups in the phase-
vs-m diagram coincide with a group in the frequency-vs-m diagram. This last
result represents an important self-consistency test. Similar results were obtained for the other 19 classified multiplets.

Evidence of a single section of coupled g-modes may lead to a modification of the neutrino production rates as predicted by a non-oscillating standard solar model. The neutrino production rates can be effected both by the
relatively large temperature eigenfunctions of the modes in the mode-coupled
section and also by the associated energy transport of these modes in the
region of $0.05 \lesssim r/R \lesssim 0.45$. In both cases, the change in the neutrino production rate is a consequence of a modification of the effective temperature
profile as defined for a given process in the Sun. Of these two mechanisms,
it is anticipated that the former will be the most important process for the
modification of the neutrino production rates (for example, the ⁸B neutrino).

Evidence of two or more sections of phase-locked g-modes may lead to
a modification of the neutrino production rates as predicted by the non-
oscillating standard solar model as well as to a periodic modulation in time
of the neutrino production rates (cf. Rabaey and Hill 1988). The results of a

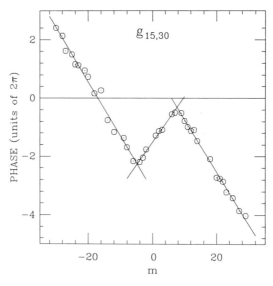

Fig. 20. The observed mode phases vs the azimuthal eigenvalue m for the $n = 15$, $\ell = 30$
gravity mode multiplet. Note the three distinct linear sections. This signature is a highly statistically significant feature.

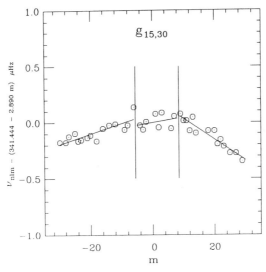

Fig. 21. The observed mode frequencies vs the azimuthal eigenvalue m for the $n = 15$, $\ell = 30$ gravity mode multiplet. To show sufficient detail in the figure, an offset value (in μHz) of $v(m) = 341.444 - 2.890\, m$ was subtracted from each mode frequency.

preliminary g-mode spectral analysis suggest a typical period for this periodicity of ≈ 2 yr (Rabaey 1989).

The ability to predict both a change in the mean rate of solar neutrino production and a temporal modulation of the solar neutrino production places solar seismology in a unique position. Unlike most if not all of the other hypotheses put forward to resolve the solar neutrino paradox, the ability to predict both a change in mean production rates and a temporal modulation of the production rates is a natural consequence of mode coupling. Furthermore, this hypothesis can be tested directly, for example, by correlation-type studies of the properties of the coupled g-modes with periodicities found in the neutrino capture rates (cf. Haubold and Gerth 1985) along with mean capture rates such as observed by the ^{8}B neutrino detector at the Homestake Gold Mine (cf. Rowley et al. 1985).

VIII. OSCILLATING QUADRUPOLE MOMENTS OF THE SUN AND POTENTIAL RELEVANCE TO SOLAR SEISMOLOGY AND GRAVITATIONAL RADIATION DETECTORS

A proposal was made in 1978 (Douglass 1978) to use a satellite to measure the oscillating gravitational multipole moments of the Sun associated with the normal modes of oscillation. This proposal was made in light of evidence for such modes which had been presented only a few years earlier at SCLERA, the Crimean Astrophysical Observatory and Birmingham University. The importance of direct measurements of the associated gravitational moments was noted in offering another observational technique for use in seismological studies of the Sun. Estimates of the magnitude of these moments were subsequently made by Johnson et al. (1980) using both an approximate solar model and a Cowling polytropic model. They reported for both models that

$$\frac{\langle J'_2 \rangle}{y_1(R)} \sim 10^{-3} \tag{27}$$

where $J'_2(GM/r)(R/r)^2 P_2$ is the oscillatory amplitude of the external quadrupole gravitational field produced by a given mode of oscillation, G is the gravitational constant, P_2 the Legendre polynomial of degree 2, $y_1 = \xi_r/r$, and $\xi_r Y_\ell^m$ is the radial component of the displacement eigenfunction. It is also important to recognize that should it be possible to detect the normal modes of oscillation by measuring perturbations in the Sun's gravitational field, then these same perturbations may present a problem as a background signal in the more sensitive gravitational radiation detectors that are being considered (Hill and Rosenwald 1986b).

The periodic variations of the Sun's gravitational field due to the $\ell = 2$ g-modes classified by Hill (1985b) have been estimated by Hill and Rosenwald (1986b). The projected values of $J'_{n,2,m}$ for the classified $\ell = 2$ g-modes of Hill (1985b) were obtained from the observed intensity amplitudes of $(T'/t)\, Y_\ell^M(\pi/2,0)$ by using a theoretically derived relationship between T'/T and Φ'/g where Φ is the gravitational potential. The typical value reported by Hill and Rosenwald (1986b) for $J'_{n,2,m}$ is of the order of 10^{-11}.

The accuracy of the reported values of $(T'/t)Y_2^m$ was estimated to be $\approx \pm 35\%$. However, the accuracy of the inferred values of $J'_{n,2,m}$ will include, in addition to this uncertainty, a systematic error arising from the theoretically derived relationship between T'/t and Φ'/g used in inferring the $J'_{n,2,m}$ results. This systematic error may be of the order of a factor of 3 or more. For this reason, the results were presented in a way which lends themselves to easy adaptation to other $T'/t - \Phi'/g$ relationships that may be used in future analyses.

There are at the present time major groundbased projects in a number of countries to detect gravitational radiation. To date, strain sensitivities of $10^{-18}/\mathrm{Hz}^{1/2}$ have been achieved in the frequency range above 800 Hz utilizing 30 and 40 m systems.

One of the objectives for future developments in that field is to lower the frequency range of low-strain sensitivity. For example, this is seen in the study of Faller et al. (1985) where the properties of a large interferometer in space are examined. Their projected sensitivity is $10^{-19}/(\mathrm{Hz})^{1/2}$ over a frequency range of 100 μHz to 0.1 Hz. The frequency range identified in the Faller et al. (1985) analysis includes the low-degree g-modes that have been classified by Hill (1985b) and Hill and Gu (1990).

Using the inferred values for $J'_{n,\ell,m}$ and the associated eigenfrequencies, the strain amplitudes $h_{n,\ell,m}$ produced by the low-degree g-modes are projected to yield a typical value for the strain $h_{n,\ell,m}$ of 1.5×10^{-20}. The phases of the strains are also projected relative to the phase of the temperature eigenfunction at the surface.

The strain amplitudes produced by the very low-frequency perturbations of the intermediate-degree g-modes reported by Hill (1986c), Rabaey and Hill (1988,1990) and Rabaey (1989) (see Sec. VII) have also been inferred. Hill and Rosenwald (1986b) find a typical value for $h_{2,m} \sim 10^{-19}$.

Although the effective $J'_{2,m}$ for the harmonics produced by the phase-locked modes with frequencies ≈ 700 μHz were not estimated by Hill and Rosenwald (1986a), they reported that a preliminary analysis suggests that the $h_{2,m}$ for these harmonics may be of the order of 1×10^{-22} for $|m| = 2$ and 1 terms.

In summary, the values of the strain produced by classified g-modes are projected to be approximately a factor of 15 larger than the strain sensitivity of 10^{-21} projected for a 10^4 s measurement with the system studied by Faller et al. (1985). For the very low-frequency perturbations with periods of 2 to 4 day produced by the second-order effects associated with the phase-locked g-modes, the values estimated for the strains are found to be as large as 10^3 times the design sensitivity of 10^{-21} for a 10^4 s measurement. In the 700 μHz range, the contributions of the second-order terms may be 10 times smaller than the design sensitivity of 10^{-21}.

These findings indicate that there may be in the future interesting opportunities and applications for solar seismology with gravitational radiation detectors. The observed values of the near-field perturbations in the Sun's gravitational field would permit a quite different examination of the solar interior in a solar seismology project. The results of the analyses also indicate the potential value of current solar seismology programs in identifying the narrow-line spectrum that may be present in the signals from gravitational radiation detectors that are g-mode produced near-field perturbations in the Sun's gravitational field.

OSCILLATION MODE EXCITATION

ARTHUR N. COX
Los Alamos National Laboratory

SHASHIKUMAR M. CHITRE
Tata Institute of Fundamental Research

SOREN FRANDSEN
Aarhus Universitet

and

PAWAN KUMAR
High Altitude Observatory

The excitation of the oscillation modes in the Sun is very different from that for the previously known variable stars. A review of the normal pulsation mechanisms seen in many classes of variable stars shows that they actually are operating in the Sun. However, most, but not all, studies of the solar mode excitation predict that radiative damping,and damping by convective processes also, overwhelm the driving to stabilize all radial and nonradial modes. This is in accord with the observations that frequently show measurable widths of the lines in the p-mode oscillation spectrum. These line widths indicate mode lifetimes of days for the p-modes. Most calculations predict that solar g-modes are stable, leading to the question of how they then can ever be observed. However, there is a possibility that low-degree and low-order g-modes could be just slightly unstable. Improvements to the predictions of lowest-order p-mode excitation by the inclusion of better radiative intensity formulations at the top of the convection zone and in the photosphere indicate even more mode stability. Calculations that show how convection may drive solar p-modes are presented.

[618]

Arguments about three-mode couplings that are not strongly damping lead to the conclusion that the solar p-*modes are probably stochastically driven by coupling with convection. Current disagreements among authors are discussed.*

I. BACKGROUND

Even though helioseismology has made rapid advances in the 1980s, the basic mechanism responsible for the excitation of solar oscillations still remains a debatable issue. The modes are either intrinsically overstable, or they could be excited by nonlinear interactions with other motions. Whatever may be the process responsible for excitation of these modes, it is essential for our understanding of the phenomenon and for any quantitative estimate of the amplitudes to have a knowledge of their intrinsic growth rates.

Clearly, solar oscillations are not at all like those for classical yellow-giant variable stars. The Sun is to the red of the instability strip red edge, and its strong convection would be expected to damp low-order radial modes as it does for Cepheids, δ Scuti variables and the RR Lyrae variables when they evolve to the red edge of the instability strip. However, the radial and non-radial high-degree, high-order p-modes in the Sun, surprisingly, seem to be excited by the convection, as we shall describe later.

The linear growth rates of 5-min oscillations for realistic model solar envelopes have been estimated in the literature. Ulrich (1970), using a highly simplified approximation to radiative transfer, studied the stability of high-degree modes. Classical effects known so well for the yellow giants were studied by Ando and Osaki (1975) who used the Paczynski (1969) evolution program to construct models that were used in pulsation studies. They found that many modes were pulsationally unstable. If such modes were unstable, like those for classical variable stars, then they should perhaps be observed at similar amplitudes. Yet the solar oscillations are very small indeed. This linear theory work was confirmed by Goldreich and Keeley (1977a) for periods longer than 6 min. However, the line widths of the peaks in the spectrum of the oscillations, at least for the higher-frequency modes ($\nu \geq 3000$ μHz), indicated lifetimes, depending on the mode, of days, not millions of years as expected for the slowly evolving classical variable stars. The finite line widths suggested that the modes were not constant in amplitude.

If a time series of observations extends longer than the lifetime of a mode, the line shape of that mode will be given by the Lorentz formula

$$P(\nu)d\nu \sim \frac{\gamma d\nu}{(\nu - \nu_o)^2 + \gamma^2} \tag{1}$$

where $P(\nu)$ is the relative power in the frequency band $d\nu$ centered around the line center ν_o, and γ is the frequency increment from the line center where the power is one half its peak. The lifetime of a mode is then given by the formula

$$\Delta t = \frac{1}{2\pi\gamma} \tag{2}$$

with Δt the time for the power to decrease by a factor of e. If the observing time is short compared to the mode lifetime, or if the time series of the observations has gaps in it, then the line shape will depend on the observing history, the so-called window function.

Recent observations (Jefferies et al. 1988) suggest that a given p-mode can be re-excited giving an amplitude and phase change while decaying, and the time-scale of the re-excitation also affects the measurement of γ. Thus, the interpretation of the measured line widths of the p-modes is complicated and depends on the observing details, the mode re-excitation, and possibly the interaction with other modes, as well as the basic mode lifetime.

Three recent publications give data about the widths of the p-mode spectral lines. Duvall et al. (1988b) give data for spherical harmonic degree l-values in the intervals 20–39, 40–59, 60–79 and 80–98. They get line widths to increase with frequency and slightly with l. Libbrecht (1988d) obtains line widths for $l = 19$–24 for various frequencies. He gets a more complicated behavior of line width with frequency, but very large widths for very high frequency. Finally, Elsworth et al. (1990a) present data for the radial ($l = 0$) and $l = 1$ modes in the 5-min band. Again, the higher-frequency modes have considerably larger line widths. The higher frequencies, for fixed l, are for high radial-order modes, and they have steeper gradients of temperature fluctuations in their eigensolutions with correspondingly more possibility of radiative damping. The higher l-modes have a greater possibility of horizontal heat flow making their damping more likely. Thus, from the viewpoint of radiative pulsation driving and damping mechanisms, the observations seem reasonable.

Arguments between authors over whether these classical mechanisms cause the Sun to be a variable star depend critically on the opacities. With some more modern mixture tables that allow for many more photospheric absorption lines and the formation of molecules, the classical κ effect does seem to cause pulsation driving. With others, the nonadiabatic eigensolutions show mode stability. A good deal of the disagreement between authors about the stability of solar oscillation modes centers around the use of accurate opacities both for the model construction and for the stability calculations.

II. SURFACE-RADIATIVE PULSATION MECHANISMS

We describe here six mechanisms that are known to operate in the solar envelope, just as they operate in classical variable stars. The most important feature of pulsation driving is whether it can overwhelm the radiative and turbulent damping that exist in stars and cause the stars to have growing amplitudes with time. We assume, as seems to be correct for the many classes of variable stars, that to exhibit light and velocity variations, the stars must be pulsationally driven in the very small-amplitude (linear) regime. More details of stellar models and pulsation theory can be found in Cox and Giuli (1968), Cox (1980) and Christensen-Dalsgaard (1986).

The first and most important mechanism is the κ effect, caused by the stellar opacity increasing when the star experiences a local compressive perturbation. Almost always, the opacity of the stellar material increases with density, as electrons are often forced back onto atoms to make them more absorptive or as free-free absorption increases with density. However, an increase in temperature usually decreases the abundance of absorbing material like partially ionized atoms, and the opacity usually decreases. However, in the hydrogen and helium ionization zones there is a temperature range where an increase in temperature produces an opacity increase. This is mostly caused by the energy of the photons in the field increasing and moving into the photon energy range that hydrogen and helium can absorb effectively. Anyway, an increase of opacity with compression and the natural temperature increase that goes with the compression, makes the outward flow of luminosity in a star decrease. A cyclical perturbation in density then dams up the luminosity at maximum compression, only to have the luminosity subsequently increase during the re-expansion part of the cycle. This lagged flow of luminosity produces a lagged pressure history and the conversion of radiative luminosity to mechanical motions.

Figure 1 shows the logarithmic derivative of the opacity with respect to the temperature in the outer 9% of the solar radius. This is less than the outer one-third of the convection zone. At the surface, one can see the effects of the negative hydrogen ion and a few molecules as they absorb photons in the photospheric layers. Deeper, the absorption by partially ionized hydrogen gives a typical value of this derivative as about 11. Deep in the convection zone, one can also see the bound-free absorption by helium undergoing ionization of its second electron that just barely gives a positive exponent for this solar envelope structure. The "Stellingwerf bump" at about 120,000 K, when the mean photon energy coincides with the helium second-ionization photoelectric absorption edge, can also be seen, but it is not a significant pulsation destabilizing effect in stars.

A parallel mechanism that also operates in the hydrogen and helium ionization zones is the γ effect. With energy going into ionization instead of into the kinetic energy of the particles, compression does not increase the temperature of the matter as much as when there is no ionization sink. Another way of saying the same thing is that with the low γ in an ionization zone, the temperature excursions are less than with a γ of completely ionized matter as 5/3. Similarly as above, a cyclical perturbation will produce cyclical luminosity variations, but with the luminosity lagged a bit because it is partially hidden during the compression stages.

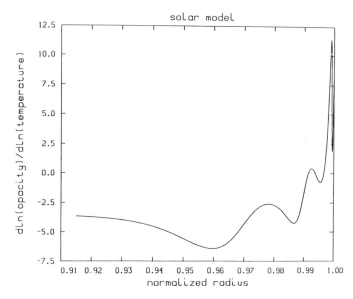

Fig. 1. The logarithmic derivative of the opacity with respect to temperature is plotted for the outer 9% of the solar radius. In this case with the Stellingwerf fit multiplied by 3, the opacity derivative rises to over 11. Different opacity tables with slightly different compositions including or not including the very small effect of molecules give both different model structures and different derivatives of this plotted function.

Figure 2 plots the $\Gamma_3 - 1$ vs radius for a solar model. The rapid change of this variable occurs at the outer edge of the hydrogen ionization zone. Actually the first helium electron is also ionized at about the same temperature, and for hydrogen-atom-dominating compositions, the separate He I ionization usually is not visible. The dip at about 0.98 of the radius is due to the final ionization stages of helium at 120,000 K.

A third, simple geometric pulsation mechanism was discussed by Baker (1966). This radius effect locally causes driving, because at maximum compression, a mass shell frequently has a smaller radiating area than its mean value. Thus, luminosity is impeded to be lagged to a later expansion phase where the resulting pressure increment can produce mechanical motions.

In the equation for the radiative luminosity another important factor is the temperature gradient. The amplitude of the cyclical variation of the temperature varies from level to level in the solar envelope. The solution of the pulsation equations, including nonadiabatic effects, results in the relative amplitude of the sinusoidal variations being smaller at the surface than deeper at the top of the convection zone. Thus, at this important level where enough of the solar mass is involved, the temperature variations are considerably larger than at the surface. This gives an increase in the temperature gradient at compression, promoting radiation leaking and pulsation damping.

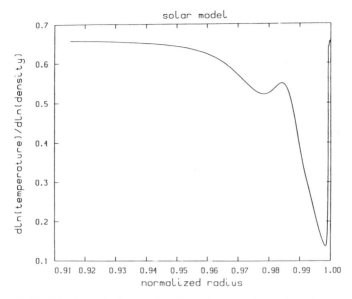

Fig. 2. The $\Gamma_3 - 1$ is plotted for the outer 9% of the solar radius. The main minimum is caused by the ionization of hydrogen and the first and even second ionizations of helium. The rapid change of the second ionization of helium with temperature at about 120,000 K gives the small dip deeper into the model. The bottom of the model part plotted here is only at 0.5×10^6 K, only the upper third of the convection zone.

In addition, there are the Cowling mechanisms. The original Cowling (1957) mechanism, for nonradial motions only, involved a strong magnetic field in the presence of a superadiabatic gradient. A rapid adiabatic displacement, say, upward in a convection zone would cool the displaced material, but it still would be hotter than its surroundings. The strong magnetic field would resist the displacement and force the material back to its original position. But, when the material returns to its original position before the small perturbation, it will be cooler than its original surroundings because of the heat it lost on the upward excursion. With rapid pressure equilibration, the cooler element would be heavier and continue to sink. Then, on its downward excursion, it would be cooler than its surroundings there, and gain heat. The magnetic field again forces the element back to its original position, but now it is too hot for its original surroundings. Analysis shows that looping in the P-V diagram would be clockwise, just as it is for the κ and γ effects, and the energy in the material would be converted to motions.

Even without the magnetic field, there are Cowling mechanisms. The Kato (1966) mechanism in semiconvection zones in evolved stars has the composition gradient as the restoring force. But the radiative and convective Cowling or δ (diffusion) mechanisms (Moore and Spiegel 1966; also see

Unno et al. [1979] for more details) that occur in the Sun need no restoring force. A cyclical displacement in a convection zone would have this sidewise heat flow either by radiation or even by convection. Since often numerical calculations do not allow any change in the convection flux with time, the convective Cowling mechanism is not allowed to occur in them. The radiative Cowling mechanism does operate in current calculations, but usually the flux is so small, for moderate l values, as shown by Ando and Osaki (1975), that no effect can be recognized.

To be complete, there are other envelope pulsation-driving mechanisms and several that occur in the deep interior. Very strong magnetic fields as seen in the Ap star pulsators can act as a restoring force to select high-order, low-degree nonradial modes. In the case of very strong convection with the bottom of the convection zone at the proper mass depth, the blocking of radiative luminosity can cause pulsations as observed in the white-dwarf variables. Another relevant deep mechanism for the Sun, with no convective core or semiconvective zone, is the Eddington ε mechanism involving the nuclear burning of H and especially ^{3}He. Cox et al. (1989) find that these last two mechanisms do not have enough driving for any solar mode, even the g-modes.

One possible g-mode excitation mechanism has been named convection blocking by Cox et al. (1987) and Pesnell (1987) in their discussions of white-dwarf star pulsations. If the time scale for the convection is long compared to the pulsation period, convection will continue to carry its mean outward luminosity without regard to the input luminosity and configuration variations. At a time during the pulsation cycle when the luminosity entering the bottom of the convection zone is higher than its mean over the cycle, the convection zone will not respond to this increased input; thus, luminosity will be blocked. If this is a time of compression of the material (as it usually is), the convection blocking will cause driving of any small perturbation. This is because, as the luminosity into the bottom of the slowly adapting convection zone is dammed up, it increases the pressure during expansion over that which would obtain for a purely adiabatic excursion. When the luminosity is less than its mean, the convection zone will still carry its mean luminosity, drawing radiation out from the radiative region below. For these expanded-material phases then, the pressure will be lower than for adiabatic motion, and there will be less resistance to a collapse. Convection blocking then takes energy out of the environment and puts it into pulsation motions.

It is also possible that there could be a convective Cowling mechanism operating inside but near the bottom of the convection zone. Here the sidewise flow of convective luminosity might be destabilizing as the g-mode motions decay in the evanescent region of the convection zone. This effect seems to be very small in unpublished calculations by Cox.

III. CALCULATIONS OF RADIATION-DIFFUSION STABILITY

Even simple photon diffusion using opacities from Cox and Tabor (1976) can allow reasonably accurate estimates for the decay rates of solar p-modes. Figure 3 gives the observed mode decay rates as inferred from the line widths (discussed in Sec. I) measured by Elsworth et al. (1990a) and by Libbrecht (1988d), and also the theoretical $l = 0$ (radial modes) decay rates calculated by Kidman and Cox (1984). Figures similar to Fig. 3 are found in the recent papers by Christensen-Dalsgaard et al. (1989) and Balmforth and Gough (1989) where results for models that include coupling between connection and pulsations are compared with the same data. Not all the surface damping is at deep enough levels below the photosphere to have the diffusion approximation accurate, but most of it is. Thus, agreement between theory and observation is not totally unexpected.

The Kidman and Cox radiative decay rates were based on a formulation of the linear nonadiabatic, nonradial pulsation equations developed by Saio

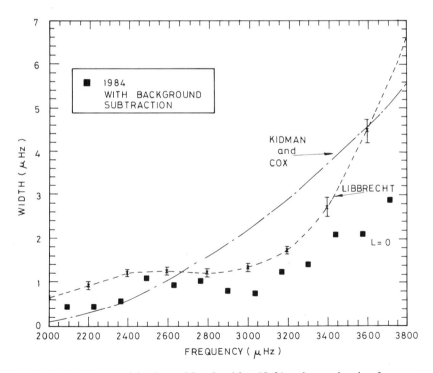

Fig. 3. The line widths of the observed $l = 0$ and $l = 19$–24 modes are plotted vs frequency from Elsworth et al. (1990a) and from Libbrecht (1988d), respectively. Also plotted are the Kidman and Cox (1984) theoretical line widths for the radial modes.

and Cox (1980). Their Eulerian technique has now been superceded by a Lagrangian method described by Pesnell (1990). In this method, each Lagrangian mass shell obeys four equations, two of which are second order: a radial momentum equation, a horizontal momentum equation, the Poisson equation for the gravitational effects that are not necessarily radial, and an energy equation. These equations have four unknowns, the radial perturbation δr, the horizontal displacement δh, the gravitational perturbation, and the entropy times temperature variation $T\delta s$, all as a function of Lagrangian mass shell. All these variations are Lagrangian, meaning that the variations refer to a fixed element of mass. A matrix solution similar to the radial case, as developed by Castor (1971) yields the complex eigenvector as well as the real and imaginary parts of the eigenvalue. The ratio of the imaginary to real parts of the eigenvalue is proportional to the growth rate per cycle of the pulsation.

The basic energy equation used in constructing the equilibrium model is

$$T\frac{ds}{dt} = \varepsilon - \frac{1}{\rho}\nabla \cdot \mathbf{F} \tag{3}$$

with ε the energy generation rate in units of erg s^{-1}, and part of the total flux vector $\mathbf{F}$ is due to the energy transport by radiation and part by convection. The radiation diffusion flux is

$$\mathbf{F}^r = -\frac{ac}{3\kappa\rho}\nabla T^4 \tag{4}$$

and if a more accurate energy transport formulation is required, the gradient is taken on the mean intensity J instead of the Planck function B used in this formula. Perturbing the energy equation, allowing the oscillations to be non-adiabatic, gives

$$T\frac{d\delta s}{dt} = \delta\varepsilon + \frac{\delta\rho}{\rho^2}\nabla \cdot \mathbf{F} - \frac{\delta(\nabla \cdot \mathbf{F})}{\rho} \tag{5}$$

and the usual assumption is that the convective part of the flux $\mathbf{F}^c$ is frozen in. For very high l, where the Cowling mechanisms operate, and for high radial order, where the convection luminosity is forced to have high gradients, convection is important in determining whether the mode is stable or unstable. However, for some solar oscillation modes, frozen-in convection may be a suitable approximation, especially when the convection time scale is long compared to the oscillation period.

Unfortunately, there are different opinions among researchers as to how the convective part of the luminosity is held fixed. There are two common views about the variation for the second term on the right-hand side of Eq. (3); either that the entire term for convection be neglected (advocated by J. P. Cox; personal communication), or that the development continues to Eq. (5), and the convection part of the last term of that right-hand side only be neglected (advocated by H. Saio and discussed briefly in Saio and Cox [1980]). Pesnell (personal communication) has followed the derivation even further to find that retaining all terms leads to some cancellations. His favored frozen-in convection prescription neglects the variations of the convective luminosity in both the radial and horizontal directions, but contains, among others, a term $k_l^2 \delta h(dL_{total}/dm - L_{conv}/(4\pi r^3 \rho))$ compared to the Cox method that has only the term $k_l^2 \delta h dL_{rad}/dm$. The comparable Saio terms are $k_l^2 \delta h(dL_{rad}/dm - 1/r^2 d/dr(r^2\delta r)dL_{conv}/dm)$. Here k_l^2 is $l(l + 1)/r^2$ and δh is the coefficient which multiplies the gradient of the spherical harmonic function for the horizontal part of the displacement $\delta \mathbf{r}$.

These procedures are rather different from those used by Ando and Osaki (1975). They assumed that the radial (and horizontal, though not discussed) variations of the convection luminosity were zero. The procedure is very much like the method Cox preferred.

A problem with this procedure is that in regions with the convective (and therefore radiative) flux varying with mass depth, there are spurious driving or damping contributions that are really not allowed for properly when the variation of the convective flux is set to zero. Blocking of a time-varying luminosity at the bottom of the convection zone when the radiative luminosity input is above its mean value acts very much like the blocking of luminosity due to the κ effect. The Cox method with its simple radiative term produces some pulsation damping as the model (equilibrium) radiative luminosity decreases, moving outward into the bottom of the solar convection zone, if δh for a particular mode is positive there. The Saio method with its additional term produces additional damping at the bottom of the convection zone for a mode with the gradient of the radial displacement positive there. The Pesnell method produces a different, smaller, amount of damping, now dependent not on the radiation and convection luminosity gradient (the gradient of the total luminosity is usually zero at the bottom of the convection zone) but on the convection luminosity itself. At the surface of the convection zone, the luminosity gradients are of the opposite sign, and the effects are in the opposite sense. The surface driving and damping is in a small amount of mass, and, depending on the model structure and the importance of the dominant κ effect in the same region, it may be negligible.

There is an important boundary condition for the highest-frequency solar p-modes. If the pressure and temperature gradients are not sharp relative to the vertical scale of the mode, the surface boundary condition of complete reflection will not obtain. Then there can be a loss of energy (damping) of these global modes as energy leaks out into the solar atmosphere. The possibility of a chromospheric mode (Unno et al. 1979), of acoustic character, is often discussed, but it may not exist because of this mechanical energy leaking even farther out into the corona.

Stability predictions are not very firm, because both the surface-layer structure and the opacity and its density and temperature derivatives are not well understood. The Kidman and Cox values, that seem to agree with low-degree observations within a factor of 2, were calculated using the Stellingwerf (1975a,b) opacity fit for both the radiation-diffusion model structure and radiation-diffusion effects on the modal stability. A better representation for the opacity might be to triple the Stellingwerf-fit opacity in the temperature range between 4000 and 6000 K to allow for the presence of the many photospheric lines and a few molecules that affect both the equation of state and the opacity. When this is done, the logarithmic derivative of the opacity is actually that given in Fig. 1. However, then the modes are pulsationally unstable with an exponential growth time of approximately 820 cycles for the p_9-mode with $l = 60$ at 3034 μHz. Using an equation of state and opacity table (Ross-Aller 1) which includes many more photospheric lines than used by Cox and Tabor (1976) and the few important molecules, for a He mass fraction $Y = 0.28$ and the heavier elements with a mass fraction of $Z = 0.02$, gives a similar result with a growth time of 750 cycles.

Yet the stable modes discussed by Kidman and Cox for very low l decay typically in hundreds of cycles and seem to agree with the observations. Figures 4 and 5 show the work per zone to cause driving for the radiation-diffusion case for the p_9 $l = 60$ mode using Stellingwerf fit (using Cox and Tabor [1976] opacities) and the case for the Ross-Aller 1 table, which includes improved absorption lines and some important molecules. The first of these calculations (the same as used by Kidman and Cox) gives stable modes while the use of a "good" table with more physical effects gives unstable modes unlike those observed. The driving seems to be caused by the κ,γ, radius and radiative Cowling effects, and the surface subphotospheric damping is caused by the gradient effect. Radiative-transfer effects reduce the pulsation driving by changes in the model structure and in the radiation flow. It appears that the use of the Stellingwerf fit to a mixture like King IVa (Cox and Tabor 1976) without the modern improvements of the Los Alamos Opacity Library (Huebner et al. 1977) compensates for the intricate radiative effects discussed in Sec. VI. The decay times for these $l = 60$ modes using the Stellingwerf fit without multiplying it by anything to allow for modern effects is about 1100 cycles, and it is about twice as long for models that use the Cox and Tabor opacity tables. At the frequency for the first case of 3040 μHz or a period of 329 s, the lifetime would be 4.19 days. The line half width for this lifetime is then $1/(86400 \times 4.2 \times 2\pi) = 2.8/2\pi = 0.4$ μHz, considerably less than the value measured by Duvall et al. (1988b), though their minimum measurable line width is 6 times this value. It is, however, within a factor of 2 of the Libbrecht half widths for lower l.

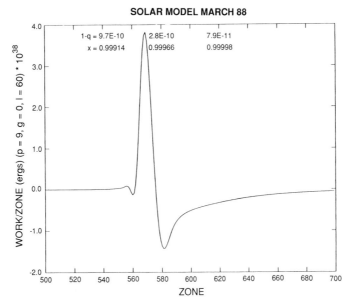

Fig. 4. The work per mass shell is plotted vs mass shell for the model using the Stellingwerf equation of state and opacity fit in both the structure and stability calculations. This work plot is plotted vs zone number to allow one to see accurately the relative contributions. The mass coordinate $1-q$ where q is the internal mass fraction, and the radius coordinate x where x is the fractional radius are also given.

Since both the equilibrium model structure and the pulsation analysis depend critically on these opacities and the treatment of the radiation flow (diffusion or detailed transport), precise lifetime predictions will not be possible until these theoretical problems are settled. We note that the Christensen-Dalsgaard and Frandsen (1983) result for radial modes with allowance for radiative-transfer effects gives decay rates in the kinetic energy of about 0.001 per cycle for modes near 5-min periods, which is 4 times slower than the Kidman and Cox (1984) rates with radiation diffusion.

The criticism of Mihalas (1984) that the radiation-transport calculations of Ando and Osaki (1975) using the Eddington approximation developed by Unno and Spiegel (1966) are incorrect does not seem to apply very strongly to solar oscillations. The terms in the energy equation that should contain radiation contributions, but were neglected by Ando and Osaki when using the Paczynski (1969) evolution program, are only slightly changed by inclusion of radiation; that is because the radiation pressure is always $< 0.1\%$ of the total pressure throughout the entire Sun. Thus, the use of the Eddington approximation used by Christensen-Dalsgaard and Frandsen (1983) should be essentially unaffected by the Mihalas discussion.

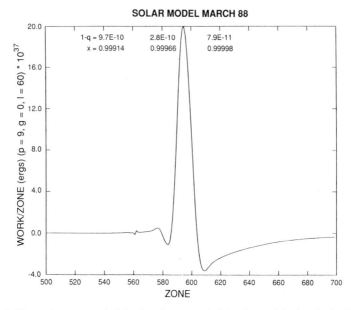

Fig. 5. The work per mass shell is plotted vs mass shell for the model using the Stellingwerf equation-of-state fit in both the structure and stability calculations. The opacity used is from the Ross-Aller 1 table which includes the large number of photospheric lines and a few important molecules. This work plot is plotted vs zone number to allow one to see accurately the relative contributions. The mass coordinate $1 - q$ where q is the internal mass fraction, and the radius coordinate x where x is the fractional radius are also given. With the larger Ross-Aller 1 opacities, driving occurs at larger zone numbers, i.e., higher in the model mass. The peak driving then occurs at a temperature of 9000 K in zone 596, which is only 1.7×10^{-10} of the mass into the Sun and 130 km below the photosphere at an optical depth of about 50. Pulsation driving is limited to cooler temperatures than for classical pulsating stars because convection becomes so strong at 9000 K, carrying there about 99% of the solar luminosity.

IV. LOW-DEGREE, LOW-ORDER g-MODE STABILITY

The stability of the solar g-modes is also uncertain, because there is the possibility that convective eddies could interact with a mode to excite it. There are regions at the top of the convection zone with time scales similar to that for the low-degree, low-order g-modes, but they seem to be in very adiabatic layers at temperatures above 30,000 K.

Recent calculations by Cox show that g-mode overstability can be caused by a combination of convection blocking at the bottom of the slowly adapting solar convection zone and the surface κ effect. He gets this overstability only for the lowest-order p- and g-modes and only for $l = 1$. The strong surface damping that is seen for the 5-min p-modes is not present for these low-degree, low-order modes because the motion in the damping region just at the photosphere is more horizontal than radial. How these modes correspond to those many higher-order g-modes observed and discussed by Hill and Gu (1990) requires further consideration.

Figure 6 gives an illustration of the work per mass shell in a completely nonadiabatic calculation for the 61.8-min (270 μHz) g_1-mode with $l = 1$. For this calculation, convection is completely frozen in, but that is not a bad approximation for modes with periods much less than the convection time scale at the bottom of the convection zone of $\sim 2 \times 10^6$ s. In the very deepest shells, there is no significant driving by the ε mechanism. Convection blocking is seen for zones 1234 to over 1300 at the bottom of the convection zone. The deep damping minimum is due to radiation flow mostly in the radial direction by the gradient effect, which (each cycle) radiates a very small amount of energy out of compressed regions between radial and angular node lines faster than these regions can expand.

Antia et al. (in preparation) have constructed a nonlocal model for the solar interior adopting a framework based on an element-conservation equa-

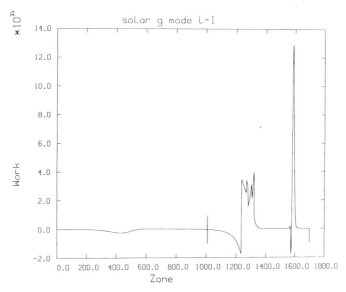

Fig. 6. The work over a pulsational cycle is plotted for each zone for the g_1-mode with degree $l = 1$. Pulsational driving occurs at the bottom of the convection zone where the periodic flow of radiative luminosity is blocked by relatively slowly changing convection as found in white-dwarf stars. The κ effect in the surface layers, similar to that in Figs. 4 and 5, is also a major source of instability. No discernible ε effect is seen at the model center. The small damping dip between zones 400 and 500 is due to gradient-effect radiative luminosity in the radial direction. The switch from the Iben (1965) procedure to complete ionization is seen at zone 1008. The two dips in the convection blocking effect are due to inaccuracies in the calculation of the relative amounts of radiation and convection luminosity caused by regime changes in the Iben (1975) opacity fit used for the model construction. Zone number, mass and radius scales are given as in Fig. 5.

tion in phase space (Spiegel 1963). The equation for the nonlocal convective flux was approximated by a second-order differential equation (Gough 1976). The global stability of such a nonlocal model was investigated to find that many gravity modes in the frequency range 66 to 257 μHz could be rendered overstable by convective processes. They get pulsation driving due to the convective flux fluctuations. More details about their use of time-dependent convection is presented in Sec. VII. Discontinuities in the plot are caused by the use of the Iben material property procedures.

V. HIGH-DEGREE f-MODE STABILITY

There is a special case for the f-modes of high degree. Shine et al. (1987) have taken very high-resolution frames of the solar granulation from space, and report that they can see evidence for the f-modes on the granulation spatial scale. An issue to determine is whether these modes are self-excited or whether they, like the p-modes on larger spatial scales, are merely excited by stochastic coupling with the convection. The latter idea seems preferable because, if the modes are self excited, modes of many l values would not necessarily be pulsating in the same phase, and the granulation would not be easily visible. The best interpretation is the traditional one, i.e., that the granulation is the top of the convective cells (the dynamically unstable g-minus-modes) and the f-modes are merely induced by this activity.

For the f-modes, there is not much variation in the density over a cycle, and therefore it has been thought that the κ effect could not be very effective. Nevertheless, there are some density and temperature changes during the cycle in the nonadiabatic calculations, and the effects described earlier can operate.

The neighboring p-modes have similar stability because they also do not have much density variation during a pulsation cycle. Just as for the low-degree p-modes, the model structure and the opacity variations are not known accurately enough to predict any definite f-mode instability. The enhancement due to the convection depends (Balmforth and Gough 1990b) on the turbulent pressure in the mixing-length formalism used. It is also probably not strong enough to overcome the radiative damping for most modes.

VI. RADIATIVE-TRANSFER EFFECTS

In the deep interior of the convective envelope of the Sun one can assume that the acoustic waves are adiabatic. As the distance to the surface diminishes, the nonadiabatic effects increase in significance. The first departures from adiabaticity due to the transport of radiation can be described as a diffusion of radiation. The radiation is still dependent on the local physical parameters because the mean free path of the photons is small compared with the scale height of temperature, density and pressure. At each point, the radiation will have either a stabilizing or driving effect essentially determined by the opacity derivatives.

Close to the surface, diffusion is no longer a good description. The photons start to escape from the Sun and to couple layers with different physical conditions, i.e., the radiation changes to a nonlocal quantity. This complicates the problem considerably.

At the top of the solar atmosphere, where the density is low, the situation is aggravated by deviations from local thermodynamic equilibrium (LTE). Even particle densities are now dependent on nonlocal conditions. The absorption becomes nongray, letting photons escape easily in some spectral regions and blocking the flux in other regions. In the chromosphere, various uncertain mechanical-flux or magnetic-diffusion terms contribute to the energy balance. At the point where the radiation increases its influence on the waves as the surface is approached, the horizontal homogeneity of the Sun disappears. The upper layers are visibly inhomogeneous as shown by the granulation, and the chromosphere changes with time and position on the Sun, greatly influenced by solar activity.

Dynamical effects of the radiation on global pulsations are small. The Sun is a high-Q oscillator where the growth or damping of waves due to radiative loss or gain takes place slowly. Also the radiation-energy density and the radiation pressure are low compared with the energy density and the pressure of the matter.

Mihalas (1984) has derived the radiation hydrodynamic equations, to first order in the ratio of fluid velocity to speed of light, v/c. The total radiative flux can be found from the radiation-energy density E as

$$\mathbf{F}^r = -\frac{c}{3\chi}\nabla E \qquad (6)$$

where χ is the sum of the thermal absorption coefficient κ_ν and the assumed isotropic coherent scattering coefficient, s_ν (both cm^{-1}) and E is in cgs units of erg cm^{-3}. Deep down, the mean intensity $J = B$, the Planck function, and the radiation-energy density $E = 4\pi J/c$ becomes only a function of the temperature

$$E = aT^4. \qquad (7)$$

This leads to the usual diffusion formula for the flux.

The radiative flux is not constant in the convection region. Therefore, even below the surface $J \neq B$. The nonlinear equations, from which the linearized equations describing the oscillations can be derived, are Eqs. (8) and (10) given in Mihalas (1984), leaving out the nuclear energy generation. The matter energy only and the total energy equations are

$$\rho\left[\frac{De}{Dt} + p\frac{D}{Dt}\left(\frac{1}{\rho}\right)\right] = \kappa\left(cE - 4\pi B\right) \tag{8}$$

and

$$\rho\left[\frac{D}{Dt}\left(e + E/\rho\right) + (p + E/3)\frac{D}{Dt}(1/\rho)\right]$$
$$= -\nabla \cdot \mathbf{F} = -c/3\nabla \cdot (1/\chi\nabla E). \tag{9}$$

The variables are the density ρ, the internal energy per unit mass e, the gas pressure p and the absorption and scattering coefficients sum χ. In the Sun, the radiation-energy density E and the radiation pressure $E/3$ have usually been neglected (Ando and Osaki 1975; Christensen-Dalsgaard and Frandsen 1983) but they are included in Kidman and Cox (1984). The radiation still affects the fluid through E on the right-hand side of these two equations. To find E one must solve the radiative-transfer equations for the optically thin regions.

The equations describing the solar oscillations are linearized equations for small perturbations from equilibrium. With proper boundary conditions one obtains an eigenvalue problem where the eigenvalue is the complex frequency ω. In the nongray case, for radial oscillations, ignoring the convective terms, the linearized right-hand side of the matter energy equation can be written

$$\delta Q = 4\pi \int_0^\infty [\delta\kappa_\nu(J_\nu - B_\nu) + \kappa_\nu(\delta J_\nu - \delta B_\nu)]d\nu. \tag{10}$$

When the pulsation-convection coupling is ignored, differences between results mainly come from the different ways the δQ term is treated. Ando and Osaki (1975,1977) ignore the first term in the parenthesis assuming that $J = B$ in a gray calculation. But if $J = B$, then one should also assume $\delta J = \delta B$ to be consistent. As mentioned earlier, this is a mistake due to the nonvanishing divergence of the radiative flux caused by the changes in the convective flux. Christensen-Dalsgaard (1988a) shows that the $J - B$ term contributes to the damping, which explains why Ando and Osaki (1975) find the p-modes to be more unstable than Christensen-Dalsgaard and Frandsen (1983) and Kidman and Cox (1984) do.

The absorption coefficient and its derivatives vary considerably through the solar envelope. Clearly, use of different opacity tables can change the amount of damping one calculates for any given mode.

It is possible to treat the δQ term in quite some detail, if it is needed. Christensen-Dalsgaard and Frandsen (1984) show that, for the determination of the complex frequencies, it is sufficient to treat the radiation as gray and use the Eddington approximation. Nevertheless, there are observations that can only be understood when accurate eigenfunctions exist. Then it is necessary to look at the linearized transfer equations (here written in plane parallel coordinates for radial oscillations)

$$\mu \frac{d\delta I_\nu}{dr} = \rho[\delta s_\nu J_\nu + \delta \kappa_\nu B_\nu - (\delta s_\nu + \delta \kappa_\nu) I_\nu \\ + s_\nu \delta J_\nu + \kappa_\nu \delta B_\nu - (s_\nu + \kappa_\nu)\delta I_\nu] \tag{11}$$

for the Lagrangian perturbation of the specific intensity I_ν in the direction given by $\mu = cos\theta$. By integration of this transfer equation over all angles or by multiplication by μ and another integration, two moment equations result:

$$\frac{d\delta H_\nu}{dr} = \rho[\delta \kappa_\nu(B_\nu - J_\nu) + \kappa_\nu(\delta B_\nu - \delta J_\nu)], \tag{12}$$

and

$$\frac{d\delta K_\nu}{dr} = -\rho[\delta \chi_\nu H_\nu + \chi_\nu \delta H_\nu] \tag{13}$$

where $\chi_\nu = \kappa_\nu + s_\nu$. Similarly to the gray case, one can assume $\delta K_\nu = \delta J_\nu/3$ (the Eddington approximation) and obtain a closed set of equations.

As often done for the static equations, the solution to the transfer equations can be written by introducing a Λ-operator. To avoid mistaking the operator describing the solution to Eqs. (8) and (9) with the static operator, we define

$$\delta J_\nu = \Lambda_{\nu,osc}(\delta B_\nu). \tag{14}$$

The operator is quite analogous to the usual Λ-operator defined by

$$J_\nu = \Lambda_\nu(B_\nu) \tag{15}$$

used to solve the static problem. Formulating the problem this way, it is possible to take over the very efficient techniques first introduced by Cannon (1973) and later refined by Scharmer (1981) and Werner and Husfeld (1985). The solution of the transfer equations are obtained by a perturbation analysis, where two numerical representations of the Λ-operator are used, one very fast and one very accurate and detailed. The solution reached corresponds to the complicated operator, but the computing time is only a fraction of the time used if a solution were attempted directly for the complicated operator.

The radiative-transfer equations must be solved under the restrictions of some additional hydrodynamic equations. This is similar to the atmospheric-model computations, where flux conservation and hydrostatic equilibrium are imposed. One generally has to work a while to find suitable parameters to solve for and to find an elimination scheme for the total set of equations. The exact procedure depends a great deal on the physics included. The important point is that the radiative damping of the pulsations can be calculated in the same detail as for the model atmospheres, taking into account line blanketing, spherical geometry and non-LTE effects.

It is feasible to explore the effects on the frequencies and the excitation of various radiative processes. For low-frequency modes, where most of the oscillation energy is concentrated in deep layers due to the evanescence in the atmosphere, the effects are small (Christensen-Dalsgaard and Frandsen 1984; Balmforth and Gough 1988,1990b), but for high-frequency modes, and certainly for chromospheric modes, measurable changes are found between different assumptions for the radiation.

The discrepancy between observed and calculated frequencies for standard solar models has existed until recently over the whole range of 5 min p-modes. As the low-frequency modes are not affected much by radiative damping, more effort has been spent on improving the equation of state and the opacities. These efforts seem to have succeeded in removing most of the discrepancies according to Christensen-Dalsgaard et al. (1988a) and Christensen-Dalsgaard (1988c), after the appearance of the MHD equations of state (Hummer and Mihalas 1988; Mihalas et al. 1988a; Däppen et al. 1988a).

There are still discrepancies remaining at high frequencies that are due to nonadiabatic radiative effects (Cox et al. 1989). Before trying to calculate the radiative terms more precisely, there are at least two subjects that should be investigated. First, the convection and the turbulent pressure must be included consistently in the static as well as in the linearized pulsation calculations. Doing so, Balmforth and Gough (1988,1990b) show that nearly all solar p-modes are marginably stable, but the balance is so delicate that small neglected terms might upset the stability. The impression is that the mechanical coupling terms are less well defined and carry more uncertainty into the damping rates than the radiative terms. More work needs to be done. Radiation enters the energy equation through the term $J - B$, which can become very small. This easily leads to instability when more degrees of freedom are introduced. The problem is well known and proper measures must be taken as described by Nordlund (1982) to ensure numerical stability. The difference $J - B$ should be used as a variable, and the Λ-operator modified to compute $J - B$ instead of J.

The second and even more difficult issue is the assumption, generally used, of a static, horizontally homogeneous model. Stein et al. (1988,1989) and Steffen (1988) find oscillations in numerical simulations of the solar granulations. The simulated volume is so small that it is difficult to scale the result to the real Sun. However, a comparison of the pulsations occurring in the simulations with oscillations, computed for a similar horizontally homogeneous slab, indicates a decrease of the frequencies for the simulations in the direction needed to remove the discrepancy among the observed and the calculated high-frequency modes. It is impossible to simulate a region large enough to match the cavities of the solar p-modes. A scaling from the small box with a few granules to a part of the solar convection zone is unavoidable. It remains to be seen whether the results of hydrodynamical simulations can be understood as simple changes of sound speed, simple forcing terms on an oscillator or other easily implemented frameworks. The problem of transmission of waves through an inhomogeneous medium has been addressed by Brown (1984b), Durney (1984) and Delache and Fossat (1988). The discussions raise more problems than provide answers. More work must be done to find a useful interpretation. Does a plane wave become scattered and delayed by the inhomogeneities?

Considering the two problems mentioned, radiation theory does not constitute the largest problem for the calculation of the frequencies and the excitation rates at present. The situation is different when the observations involved are of amplitudes and phases of the measured flux from the Sun. Observing a choice of spectral frequencies in the continuum or in spectral lines, information about the eigenfunctions in the atmosphere can be extracted. Each spectral point provides a signal dependent on the eigenfunction and the physical parameters in the Sun over a limited depth interval. It is straightforward to compute the spectral flux variations once the eigenfunctions are known. However, the result is very sensitive to the shape of the eigenfunction and the solar model.

Observations show that the evanescent waves become slightly progressive in an upward direction due to the nonadiabatic effects. They can transport energy up into the chromosphere, the more so the higher the frequency. The radiative damping time decreases as the waves travel upward leading to an increasing loss of energy from the modes. The progressive character manifests itself as a small phase change in the velocity signal measured in lines formed at different altitudes in the solar atmosphere. At the same time the adiabatic 90° phase difference, between the intensity change and the velocity, is shifted to larger values. Both effects are seen in Figs. 7 and 8.

The small observed phase changes in velocity at different altitudes is in good agreement with the values calculated up to around the temperature minimum. Also the more-than-90° intensity-velocity phase shift is predicted by the calculations. Most phase and amplitude data so far come from studies of line profiles obtained by placing a narrow slit near the center of the solar disk. In the center and the near wings of the spectral lines the light comes from a region of the Sun, where the eigenfunctions for most of the modes are well behaved (no nodes). We would be surprised if the theory and the observations did not agree. On the contrary, in the far-line wings and in the contin-

A. N. COX ET AL.

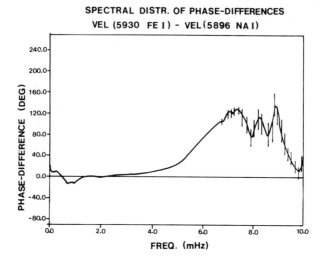

Fig. 7. The phase difference between an Fe I line and the Na I line vs frequency. In the *p*-mode range only small values and small changes are seen. Above the cutoff frequency, large phase changes occur (figure courtesy of Staiger 1987).

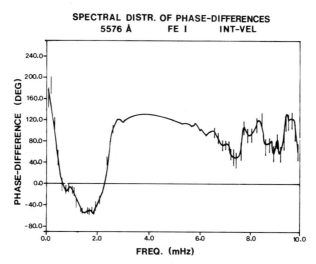

Fig. 8. The Fe I line has an intensity-minus-velocity phase difference above 90° in the whole *p*-mode frequency region. Other Fe I lines formed deeper down in the photosphere have smaller values (figure courtesy of Staiger 1987).

uum, large discrepancies have been common. Slit observations are not very useful for layers near the continuum-forming depths, because the granulation pulls the $I - \upsilon$ (intensity minus velocity) phase towards zero. The influence of the granulation is difficult to disentangle.

Lately, measurements of the full solar disk, where the granulation (and the seeing) has much less influence, have been made. Frölich and van der Raay (1984), Jiménez et al. (1988b,d) and Jiménez (1988) have published combined intensity and velocity data. The error bars have become so small that a variation of $I - \upsilon$ phase with pulsational frequency is visible (Fig. 9). An equally interesting behavior is seen when the ratio of the flux variations and the velocity are plotted against frequency (Fig. 10). Both the $I-\upsilon$ phase and the amplitude ratio show nontrivial bumps and dips.One would naively expect the phase to converge towards 90° as observations probed deeper into the Sun. Various calculations (Schmieder 1977,1979; Frandsen 1988) give conflicting results, that also do not fit the observations very well. Even after the latest results by Balmforth and Gough (1990b) where convective coupling is included, the large $I - \upsilon$ phase discrepancy remains.

Looking further into the problem, precisely the $I - \upsilon$ phase turns out to be a very sensitive indicator of the physics. The phase of the continuum intensity causes the difficulty. It is found by integrating the equation

$$\frac{d\delta I_\nu}{d\tau_\nu} = \delta I_\nu - \delta S_\nu \qquad (16)$$

where the perturbed source function is

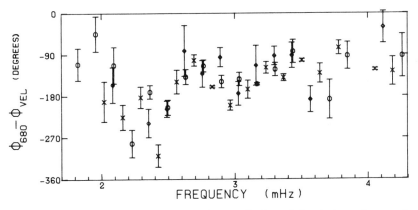

Fig. 9. Full disk results for the phase difference between the flux variations in the continuum and the velocity in the K I line is plotted vs frequency. The results differ from the measurements done by Staiger (1987) using a narrow slit on the Sun (figure from Jiménez et al. 1988d).

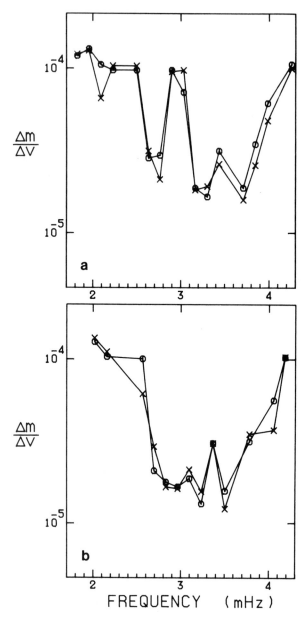

Fig. 10. The amplitude of the flux oscillations divided by the velocity for two different spectral wavelengths: 680 nm = circles, 770 nm = crosses vs frequency. The frequency dependence is interesting, showing an oscillatory behavior suggesting an interplay between the vertical wavelength of a *p*-mode and the scale heights in the photosphere (figure from Jiménez et al. 1988*d*). Plot (a) is for *l* = 0; plot (b) is for *l* = 1.

$$\delta S_{\nu} = \delta B_{\nu} - \frac{\delta \chi_{\nu}}{\chi_{\nu}}(I_{\nu} - B_{\nu}). \tag{17}$$

In the continuum-forming region, nodes are found in the eigenfunctions at many frequencies. At a node a fast phase shift of the temperature takes place. The contribution from the first and the second term of δS vary and can produce a sign change of the source function corresponding to a $180°$ phase jump. The emerging intensity fluctuation at the solar surface is therefore an average over terms pulling in all directions in phase. The balance can be quite delicate producing different results for different eigenfunctions and different physics.

Instead of seeing this as a problem, it might be thought of as a tool for studying the top of the convection region (the overshoot region) where the coupling to the pulsations surely produces phases different from phases predicted when the coupling is ignored. Necessary ingredients of a theoretical cocktail to explain Figs. 9 and 10 include a technique to handle the coupling between pulsation and convection, good upper boundary conditions and a careful handling of all major radiative effects. Progress is foreseen in the immediate future. The granulation models of Stein et al. (1988,1989) can be used to study the phase differences between the intensities emerging from the slab and the phase differences obtained from a calculation of the modes in a horizontal homogeneous model obtained by averaging, in space and time, the time series of granulation. Hopefully, an understanding of the interplay between granulation and oscillations can be reached in this way. In addition, the difference between the mean and the time-varying model can tell us whether results from a horizontal homogeneous model can be trusted.

VII. A FORMULATION FOR TIME-DEPENDENT CONVECTION

Ando and Osaki (1975), who used the Eddington approximation to model radiative transfer, but ignored the modulation of the heat and momentum fluxes due to turbulence, found that a large number of acoustic modes were overstable in the region where a considerable power in solar oscillations had been observed. Later, Gough (1980) took account of the modulation of the convective heat flux and Reynolds stresses by adopting a mixing-length approach (Baker and Gough 1979) and found the radial acoustic modes to be unstable. However, Berthomieu et al. (1980), using a similar prescription, found the nonradial acoustic modes of high degree ($l \gtrsim 200$) to be stable. This was attributed by Antia et al. (1982) to the unity value of the turbulent Prandtl number that was apparently inherent in the formulation of Berthomieu et al. For such a value of the turbulent Prandtl number, Antia et al. (1982) also found that the high-degree nonradial p-modes were stable; but, with smaller values chosen to reproduce the linearized modes of convective instability that best agreed with observed motions on the solar surface, the convection was actually found to destabilize the p modes. The oscillatory modes were computed by Antia et al. (1982) for realistic solar models by employing the Eddington approximation to radiative transfer; furthermore, the fluctuations in the convective heat flux were incorporated in the calculations in a quite rudimentary way. This situation was remedied in a later calculation by Antia et al. (1988) who improved the treatment of the dynamics of convection perturbed by solar pulsation to find that convection tends to enhance the excitation of the p-modes.

The thrust of this calculation is to determine the linear nonadiabatic eigenmodes of nonradial solar pulsation. The governing equations in an Eulerian coordinate system, for a fluid of density ρ at pressure p moving with a large-scale pulsating velocity $\mathbf{v}$ are

$$\frac{\partial \rho}{\partial t} + \boldsymbol{\nabla} \cdot (\rho \mathbf{v}) = 0 \tag{18}$$

$$\rho \frac{\partial \mathbf{v}}{\partial t} + \rho(\mathbf{v} \cdot \boldsymbol{\nabla})\mathbf{v} = -\rho \boldsymbol{\nabla} \psi - \boldsymbol{\nabla} p - \boldsymbol{\nabla} \cdot \mathbf{P} \tag{19}$$

$$\rho T \left(\frac{\partial s}{\partial t} + \mathbf{v} \cdot \boldsymbol{\nabla} s \right) = \rho \varepsilon - \boldsymbol{\nabla} \cdot \left(\mathbf{F}^r + \mathbf{F}^c \right) + Q \tag{20}$$

$$\nabla^2 \psi = 4\pi G \rho. \tag{21}$$

Here ψ is the gravitational potential, s is the specific entropy, ε is the nuclear energy generation rate per unit mass, $\mathbf{F}^r$ and $\mathbf{F}^c$ are, respectively, the radiative and convective heat fluxes, $\mathbf{P}$ is the viscous stress tensor, assumed to result solely from the turbulent momentum transport, and Q is the rate of generation of heat by viscous dissipation. This set of equations is supplemented by an equation of state, for which the medium is treated as a mixture of radiation and a perfect ionizing gas.

The radiative heat flux, $\mathbf{F}^r$ is computed using a modification of the Eddington approximation

$$\mathbf{F}^r = -\frac{4}{3\kappa\rho} \boldsymbol{\nabla}(fJ) \tag{22}$$

in which the mean intensity of radiation J is determined by the radiative equation

$$J = B - \frac{1}{4\kappa\rho} \nabla \cdot \mathbf{F}^r. \tag{23}$$

Here $B = (ac/4)T^4$ is the Planck function, T is the temperature, κ is the opacity, which was computed by interpolating the tables of Cox and Tabor (1976). We have introduced the space-dependent Eddington factor f, which is assumed not to be perturbed by the pulsation, and which is defined such that $f = 1/3$ corresponds to the usual Eddington approximation. This prescription, recommended by Christensen-Dalsgaard and Frandsen (1983), provides a solution that hopefully is closer to the solution of the full-transfer equation than is the unmodified Eddington approximation.

We have adopted a simple mixing-length prescription for computing the convective heat flux, $\mathbf{F}^c$ based on the concept of diffusive mixing. In a sense, it combines the ideas embodied in the discussions of Unno (1967,1977; see also Unno et al. 1979), Gough (1977a) and Antia et al. (1982).

In a spherically symmetrical (nonpulsating) star the convective heat flux, $\mathbf{F}^c$ is in the radial direction, and according to local mixing-length prescriptions of convection, in a chemically homogeneous fluid its magnitude, F^c can be written as

$$F^c = K_t\beta \tag{24}$$

where β is the magnitude of the superadiabatic temperature gradient. $\beta = -(T/C_p)\nabla s, C_p,$ being the specific heat at constant pressure. The turbulent conductivity K_t is itself a function of β, and of course, the local mean state of the fluid. We assume that this equation can be generalized to a good approximation to the vector equation

$$\mathbf{F}^c = K_t(\beta)\boldsymbol{\beta} \tag{25}$$

in which $K_t(\beta)$ is the same function as that applicable to a spherically symmetrical star. In general, we could expect the turbulent conductivity to be a second-rank tensor that we assume to be dominated by an isotropic component. This provides a convenient means of generalizing to nonradial pulsations any theory of time-dependent convection originally formulated for radial pulsations.

We have adopted a hybrid of the approaches of Unno (1967) and Gough (1977a) to describe the temporal modulation of the convective heat flux by the large-scale pulsation. For this purpose, we simply write

$$F^c = \rho \, c_p W \Theta \qquad\qquad K_t = F^c/\beta \qquad\qquad (26)$$

which is valid for radial pulsations and where W and Θ are the amplitudes of the vertical component of velocity and temperature fluctuation in a typical heat-transporting convective eddy. These are computed from the amplitude equations given by Gough (1977a)

$$\left[\frac{d}{dt} - \frac{1}{\Phi}\frac{d}{dt}\ln(r^2\rho)\right]W + \frac{2W^2}{\ell} - \frac{g\delta}{\Phi T}\Theta = 0 \qquad (27)$$

$$\left[\frac{d}{dt} - (C_{PT} - \delta)\frac{d\ln T}{dt} - \frac{\delta_T d\ln p}{C\ dt} + q\right]\Theta + \frac{2W\Theta}{\ell} - \beta W = 0 \quad (28)$$

where $d/dt = \partial/\partial t + \upsilon_r\,\partial/\partial r, \upsilon_r$ being the radial pulsation velocity, $g = \partial\psi/\partial r$ is the acceleration due to gravity, and

$$\delta = -\left(\frac{\partial\ln\rho}{\partial\ln T}\right)_p \qquad C = \frac{C_p\rho T}{p\delta} \qquad C_{PT} = \left(\frac{\partial\ln C_p}{\partial\ln T}\right)_p \qquad (29)$$

$$\delta_T = \left(\frac{\partial\ln\delta}{\partial\ln T}\right)_p \qquad \Phi = \frac{k^2}{k_h^2} = 1 + \frac{k_r^2}{k_h^2} \qquad k_r = \frac{\pi}{\ell} \qquad (30)$$

$$q = \frac{4acT^3k^2\chi}{3\kappa\rho^2C_p} \qquad \chi = \left[1 + \frac{k^2}{3(\kappa\rho)^2}\right]^{-1}. \qquad (31)$$

In this prescription the size of an eddy is characterized by a wave number k with horizontal and vertical components k_h and k_r, respectively. The vertical component is related to the mixing-length ℓ, and Φ is a geometrical shape factor, which takes into account the inertia of horizontally moving convective motion. In the expression for χ we have ignored the departure from radiative equilibrium. The steady amplitudes W_o and Θ_o are obtained by solving the steady versions of these equations to yield

$$W_o = \frac{1}{4}g_o\ell_o\left[\left(1 + \frac{4g_o\delta_o\beta_o}{\Phi_o T_o q_o^2}\right)^{1/2} - 1\right] \qquad (32)$$

$$\Theta_o = \frac{2\Phi_o T_o}{\ell_o g_o \delta_o} W_o^2. \qquad (33)$$

We regard the vector equation $\mathbf{F}^c = \kappa_t(\beta)\boldsymbol{\beta}$ to hold even when spherical symmetry is broken. In particular, we assume the approximation to be valid for pulsating stars too, and set $\mathbf{F}^c = \mathbf{F}_o^c + \mathbf{F}_1^c$, where the suffix o refers to the static state and the suffix 1 to the Lagrangian perturbation. We can express $\mathbf{F}_1^c$ as

$$\mathbf{F}_1^c = K_{t1}\,\boldsymbol{\beta}_o + K_{to}\boldsymbol{\beta}_1. \qquad (34)$$

In this equation, the perturbed turbulent conductivity K_{t1} is a scalar and it may be determined by identifying the vertical component of $\mathbf{F}_1^c$ with the expression obtained from the convection prescription applied to radially pulsating stars. For radial pulsations, we can write

$$\frac{F_1^c}{F_o^c} = \frac{\rho_1}{\rho_o} + \frac{C_{p1}}{C_{po}} + \frac{W_1}{W_o} + \frac{\Theta_1}{\Theta_o} \equiv \frac{K_{t1}}{K_{to}} + \frac{\beta_1}{\beta_o}. \qquad (35)$$

The perturbation of the amplitude equations is reasonably straightforward except for the mixing-length ℓ and the geometrical shape factor Φ, and these perturbed quantities can be expressed (cf. Antia et al. 1988) as

$$\frac{\ell_1}{\ell_o} = \frac{1}{1 - i\sigma}\left[\frac{\delta H}{H_o} + i\sigma\left(2\frac{\delta r}{r} + \frac{\delta\rho}{\rho_o}\right)\right] \qquad (36)$$

where $\sigma = \omega_R \ell_o/W_o$, ω_R being the pulsation frequency, H is the pressure scale height, r is the Lagrangian radial coordinate and the operator δ denotes a Lagrangian perturbation amplitude with respect to the radial component of the pulsation velocity.

We have followed the usual procedure for linearizing the governing equations, and we define the complex eigenfrequency ω with real and imaginary parts ω_R and ω_I for convenient introduction of a stability coefficient $\eta = \omega_I/\omega_R$, taken positive for overstable modes. The system of linear-pulsation equations is solved to obtain the complex eigenvalues for two representative values of degree: $l = 1$ and $l = 100$, adopting the boundary conditions and the numerical method described by Antia et al. (1982). Guided by the form of the transport coefficient implied by the analyses of Gough (1977a) and Goldreich and Keeley (1977b), the form of the turbulent viscosity coefficient is taken to be $\mu_t = \rho_o W_o \ell_o/(1 + \sigma^2)$. The turbulent viscous stress tensor can be calculated in the same manner as the convective heat flux, but we have not carried that through in the calculations reported here.

VIII. NUMERICAL RESULTS FOR ACOUSTIC MODES

The results of our computation are presented in Table I that shows the (real part of the) complex eigenfrequency and the stability coefficient for a sample of acoustic modes of degree $l = 1$ and 100.

Case A. Inviscid fluid with no convective flux perturbations, and with turbulent stresses P and turbulent energy dissipation Q ignored. This corresponds to the computations reported by Christensen-Dalsgaard and Frandsen (1983). All the modes are stable, as were found by Christensen-Dalsgaard and Frandsen. The values of the stability coefficients from the two calculations do not agree precisely, but they are comparable in magnitude, with both calculations predicting a similar increase in stability as frequency rises.

Case B. Inviscid fluid with no departure from radiative equilibrium ($J_o = B_o$) and instantaneous adjustment of the convective eddies to the changing environment.

Case C. Inviscid fluid with departure from radiative equilibrium and instantaneous adjustment of convective eddies.

Case D. Inviscid fluid with departure from radiative equilibrium and with the modulation of heat convective flux by pulsation included.

Case E. Viscous fluid with departure from radiative equilibrium and with the modulation of convective heat flux by pulsation included.

It is evident from the numerical results that the perturbation to the convective heat flux destabilizes the pulsations. Remarkably, the departure from radiative equilibrium also tends to destabilize the pulsations when convective-flux perturbations are included. This is at variance with the impression gained from the computations of Christensen-Dalsgaard and Frandsen (1983), in which convective perturbations were ignored. This may be attributed to the global effect of the term involving $(J - B)$ which influences the structure of the eigenfunction and thus contributes to the destabilization of the eigenmodes. The inclusion of turbulent viscosity in the simple manner we have adopted is clearly stabilizing; when it is included the f-mode is stabilized, but a large number of p-modes with frequencies < 5 mHz are still unstable. The difference in growth rates between modes with degree $l = 1$ and $l = 100$ can be largely ascribed to differences in the effective inertia of the fluid undergoing pulsation.

The principal conclusion of this work is that in the presence of convective heat-flux perturbations, the departure from radiative equilibrium tends to destabilize acoustic modes. Inclusion of time-dependent terms in the convective-flux perturbation destabilizes low frequency ($\nu \lesssim 2.2$ mHz) modes, but leads to a stabilization of the higher-order modes. The predominant destabilization almost certainly arises from the perturbations of the convective heat flux.

A recent investigation by Balmforth and Gough (1988) demonstrates how delicately the stability characteristics of solar acoustic modes depend on

TABLE I
Frequency and Stability Coefficients for a Sample of Acoustic Modes

Mode	Frequency (μHz)	Period (s)	Stability Coefficient ($\eta = \omega_I/\omega_R$)				
			(A)	(B)	(C)	(D)	(E)
$l=1$							
p_2	448.4	2230.4	-2.59E-7	-4.54E-8	-3.28E-8	8.58E-9	6.48E-9
p_{11}	1738.7	575.2	-1.19E-4	4.28E-6	1.28E-5	1.88E-5	1.06E-5
p_{17}	2551.5	391.9	-6.25E-4	1.99E-4	2.44E-4	1.85E-4	1.13E-4
p_{23}	3368.4	296.9	-7.34E-4	5.25E-4	5.74E-4	3.29E-4	2.00E-4
p_{29}	4196.9	238.3	-7.02E-4	5.57E-4	5.83E-4	3.16E-4	1.91E-4
p_{34}	4892.8	204.4	-1.48E-3	2.01E-4	2.10E-4	-6.13E-5	-1.66E-4
$l=100$							
f	1014.5	985.7	-4.50E-6	3.35E-9	1.84E-7	1.27E-7	-3.87E-6
p_2	1829.1	546.7	-3.48E-4	1.59E-5	4.10E-5	5.75E-5	2.07E-5
p_5	2678.6	373.3	-1.33E-3	4.46E-4	5.43E-4	4.19E-4	2.50E-4
p_8	3409.4	293.3	-1.30E-3	8.70E-4	9.59E-4	5.83E-4	3.61E-4
p_{11}	4082.7	244.9	-1.12E-3	8.73E-4	9.26E-4	5.40E-4	3.32E-4
p_{14}	4719.0	211.9	-1.67E-3	4.65E-4	4.89E-4	1.79E-4	1.69E-5

the detailed processes involving radiative transfer and convection. In their calculations, the stability of solar radial p-modes is studied with the Eddington approximation including the departure from radiative equilibrium in the expression for χ, and also using the convection theory of Gough (1977a). The computations indicate that, with the improved treatment of radiative transfer in the atmosphere, there is an increased radiative damping of radial p-modes. Thus, p-modes, which are overstable in the framework of the diffusion approximation, tend to become stabilized in the Eddington approximation. But, when convection-pulsation coupling is incorporated, even in an approximate way, in the calculations, at least some of the solar radial p-modes turn out to be overstable.

Hopefully, the results reported here will shed some light on the physics of mode excitation. It must, however, be recognized that the numerical results of stability coefficients, which depend on a delicate balance of several different mechanisms, are still rather fragile (Balmforth and Gough 1988). It is therefore not possible to draw a definitive conclusion about the viability of a self-excitation mechanism for driving these solar oscillations. One cannot help the feeling that almost every conceivable hydrodynamical effect will have to be carefully examined before finally deciding the issue whether the solar acoustic modes are self-excited or are stochastically driven.

IX. ARE SOLAR p-MODES OVERSTABLE?

The results of theoretical investigations for the linear stability of solar p-modes are found to depend sensitively on the treatment of the interaction of pulsation with convection and radiation (Ando and Osaki 1975; Goldreich and Keeley 1977a; Christensen-Dalsgaard and Frandsen 1983; Antia et al. 1988); this is particularly so in the transition region between high and low optical depths where most of the driving occurs. Therefore, it is unlikely that a linear stability calculation will provide a conclusive answer to whether or not the solar 5-min modes are excited due to overstability without a fundamental advance in modeling the interaction of turbulent convection with pulsation. Moreover, a knowledge of the linear growth rate is insufficient to calculate the amplitudes of overstable modes, and a nonlinear calculation must be carried out for this purpose.

Another approach to understand mode excitation which circumvents our ignorance about the coupling of convection with pulsation was taken by Kumar and Goldreich (1989). These authors assumed that the modes are either overstable (discussed below), or stable and driven by acoustic emission from turbulent convection (Sec. X). They then explored the implications of each of these assumptions, and compared them with observations to narrow down the possible excitation mechanisms. At first sight, it might appear that since excitation involves the poorly understood interaction of modes with convection, the results for this process must be as uncertain as the linear stability calculations. Fortunately, this is not so. In order to calculate acoustic emission, only the gross features of turbulent flow, for instance, the magnitude of velocity and its correlation lengths are required (see Sec. X) and these can be estimated by scaling arguments such as in the mixing-length theory.

If we assume that some of the solar p-modes are overstable, then we must look for a nonlinear mechanism to transfer energy from these modes to modes that are linearly damped, thereby limiting the exponential energy growth of the overstable modes. Given the small amplitude of solar p-modes (approximately 20 cm s^{-1} at the photosphere and decreasing inwards), the most efficient mechanism to accomplish this job involves couplings of 3 near-resonant modes, i.e., modes which satisfy horizontal wave-vector and frequency-matching conditions, described below. In order to serve as an amplitude-limiting process, these 3-mode couplings must be able to drain energy from overstable modes on their e-folding time scales at a rate proportional to E^{1+q}, where E is the energy in the mode and q is a positive number.

In the remainder of this section, we consider modes in a plane-parallel stratified atmosphere consisting of an adiabatic and an isothermal part. The parameters of this atmosphere are chosen to resemble closely the solar convection zone and the atmosphere. The acoustic modes in this atmosphere are uniquely specified by their frequency ω and horizontal wave vector $\mathbf{k}_h$. These modes can be identified with solar modes of degree, $l \approx k_h R_\odot$, and $m \approx k_h R_\odot \cos \phi$, where ϕ is an angle between $\mathbf{k}_h$ and some arbitrarily chosen horizontal direction.

The average rate of change of energy in a mode α, due to couplings with modes β and γ (Kumar and Goldreich 1989) is

$$\left\langle \frac{dE_\alpha}{dt} \right\rangle = \frac{9\omega_\alpha}{2\pi} \sum_{n\beta} \int d\omega_\gamma \int dk_{h\beta} \, k_{h\beta} \int_0^\pi d\theta_{\alpha\beta} |K_{\alpha\beta\gamma}|^2 \delta(\omega_\alpha + s_1\omega_\beta + s_2\omega_\gamma)$$
$$\times \left\{ \omega_\alpha E_\beta E_\gamma + s_1\omega_\beta E_\alpha E_\gamma + s_2\omega_\gamma E_\alpha E_\beta \right\}. \tag{37}$$

The terms in the above integral are restricted by the conservation of horizontal momentum expressed through the condition on the horizontal wave vectors

$$s_\alpha \mathbf{k}_{h\alpha} + s_\beta \mathbf{k}_{h\beta} + s_\gamma \mathbf{k}_{h\gamma} = 0 \tag{38}$$

and the frequency matching condition expressed by the delta function in the integrand. In the above equations, $\mathbf{k}_{h\alpha}$, $\mathbf{k}_{h\beta}$, $\mathbf{k}_{h\gamma}$ are horizontal wave vectors, ω_α, ω_β, ω_γ are frequencies and E_α, E_β, E_γ are energies for modes α, β and γ, respectively. The symbols s_α, s_β, s_γ take on the values $+1$ or -1, $s_1 = s_\beta/s_\alpha$, $s_2 = s_\gamma/s_\alpha$, and $\theta_{\alpha\beta}$ is the angle between $\mathbf{k}_{h\alpha}$ and $\mathbf{k}_{h\beta}$. The coefficient $K_{\alpha\beta\gamma}$ appearing in Eq. (37) is the overlap integral of the product of the gradients of the displacement eigenfunctions of modes α, β and γ (ξ_α, ξ_β and ξ_γ, respectively), and is defined by

$$
K_{\alpha\beta\gamma} = - \int dz \frac{p}{6} \Big[(\Gamma - 1)^2 (\nabla \cdot \xi\alpha_{s\alpha})(\nabla \cdot \xi\beta_{s\beta})(\nabla \cdot \xi_{\gamma s\gamma})
$$
$$
+ \xi^{i,j}_{\alpha s\alpha} \xi^{j,k}_{\beta s\beta} \xi^{k,i}_{\gamma s\gamma} + \xi^{i,j}_{\beta s\beta} \xi^{j,k}_{\alpha s\alpha 0} \xi^{k,i}_{\gamma s\gamma} + (\Gamma - 1) \{ (\nabla \cdot \xi\alpha_{s\alpha}) \xi^{i,j}_{\beta s\beta} \xi^{j,i}_{\gamma s\gamma} \qquad (39)
$$
$$
+ (\nabla \cdot \xi\beta_{s\beta}) \xi^{i,j}_{\gamma s\gamma} \xi^{j,i}_{\alpha s\alpha} + (\nabla \cdot \xi\gamma_{s\gamma}) \xi^{i,j}_{\alpha s\alpha} \xi^{j,i}_{\beta s\beta} \} \Big]
$$

where p is the pressure, Γ the adiabatic index, and z is vertical distance from the top of adiabatic atmosphere increasing in the direction of gravitational acceleration. Note that $K_{\alpha\beta\gamma}$ is symmetrical under the interchange of indices α, β, and γ. Equation (37), which is derived for a plane-parallel atmosphere, is applicable to large $l (l \geq 50)$ solar p-modes. These modes typically have on the order of 500 resonant couplings that result in their rate of energy transfer being independent of their line widths (Kumar and Goldreich 1989).

Nonlinear mode interactions are expected to be strongest in regions where the appropriate nonlinearity parameter, the acoustic Mach number, is large. We use simple scaling rules for the pressure, density and displacement eigenfunction to determine the part of the atmosphere that contributes most to mode couplings. For an adiabatic atmosphere of $\Gamma = 5/3$, the pressure $p \propto z^{5/2}$, the density $p \propto z^{3/2}$ and the displacement eigenfunction $\xi \propto z^{-1}$. The scaling for ξ follows from the conservation of energy flux $\rho\xi^2 c$ (c being the sound speed). In the region of the adiabatic atmosphere where a wave propagates $|\nabla D\xi| \approx \omega\xi/c$, therefore, $|\nabla\xi| \propto z^{-3/2}$. Making use of these scaling in Eq. (39) we find that the local 3-mode coupling strength decreases at least as fast as z^{-2}. Thus, the contribution to the 3-mode coupling coefficient from the adiabatic layer is concentrated close to its upper boundary. In the isothermal atmosphere, the pressure decreases, whereas ξ increases exponentially with height. However, for most triplets, the pressure decrease overwhelms the increase in ξ, resulting in the decrease of coupling strength in the isothermal atmosphere. Therefore, the dominant contribution to the 3-mode coupling comes from the upper part of the convection zone and the lower part of the isothermal layer.

For those triplets, that have a mode with frequency greater than the acoustic cutoff frequency, and another very close to it, the coupling strength actually increases with height in the isothermal atmosphere and formally diverges. In such cases, the coupling coefficients are calculated to a height at which the acoustic Mach number reaches unity. Unfortunately, the validity of the perturbation expansion becomes questionable at this point, and higher-order couplings have comparable strengths to the 3-mode couplings there.

The damping rate due to mode couplings depends on the energy spectrum and the thickness of the isothermal atmosphere (cf. Eq. 37). The ener-

gies of low *l* modes are known observationally, but there is considerable uncertainty in mode energy at high *l*. Kumar and Goldreich (1989), assuming
an *l*-independent energy spectrum (which is probably an upper limit [cf.
Kaufman 1988] and has the effect of overestimating the importance of mode
couplings), and a 2.5 scale-height thick isothermal atmosphere, find that the
damping time for modes in the 5-min band is smaller by a factor of about 4
compared to the value deduced from observation (cf. Fig. 11). In other
words, the energy decay rates due to mode couplings are somewhat smaller
compared to the energy input rates. The latter quantity is estimated observationally by taking the product of mode energy and line width. This suggests,
given the significant residual uncertainties, both in the observations and the
theoretical model, that overstable modes might be stabilized by mode couplings. However, in order to limit the exponential growth of overstable
modes, the energy decay exponent *q* must be greater than zero, as previously
mentioned.

Numerical calculations carried out by Kumar and Goldreich show that
d ln E_α/d*t* due to mode couplings is almost independent of E_α, implying that
q is very nearly equal to zero. Therefore, 3-mode couplings are incapable of
quenching the growth of overstable modes. This result follows from the geometry of the *k* − ω diagram and the observed mode energies. It can be

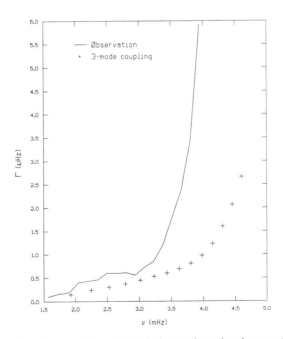

Fig. 11. Comparison of observed line width to the inverse decay time due to mode coplings.

understood qualitatively by identifying the most important mode couplings. From the horizontal momentum conservation condition (cf. Eq. 38), we find that for $k_{h\alpha} \ll k_{h\beta}$, $k_{h\beta} \sim k_h\gamma$, and since $|\nabla\xi| \sim k\xi \sim k_h\xi$, it follows from Eq. (39) that $K_{\alpha\beta\gamma} \propto k_{h\beta}k_{h\gamma} \sim k_{h\beta}^2$. Of course, there is an upper limit to $k_{h\beta}$, beyond which the contribution to mode coupling declines rapidly. This turnover arises because the mode energy spectrum falls rapidly at high frequencies (Libbrecht 1988d). So a mode couples most strongly to modes of the largest possible l just below the acoustic cutoff frequency. Using the frequency-matching condition, we conclude that the third mode, whose frequency is the sum of the frequencies of modes 1 and 2, must be propagating in the isothermal atmosphere. Couplings of these kind are a drain on the energy of trapped modes.

There is another class of couplings involving only trapped modes, but the above argument shows that they are in general significantly weaker. From the approximate dispersion relation $\omega^2 \approx gk_h(n + 3/4)$, we note that the maximum value for k_h at a fixed frequency for ridge n goes as $k \approx (n + 3/4)^{-1}$. Hence $K_{\alpha\beta\gamma}$ decreases with increasing n as $(n + 3/4)^{-2}$, and the available phase space also decreases as $(n + 3/4)^{-2}$. Finally, d ln $E_\alpha/dt \propto K_{\alpha\beta\gamma}^2 E_\beta$, which implies that d ln $E_\alpha/dt \propto (n_\beta + 3/4)^{-6}E_\beta$.

Thus, the contribution to energy decay rate due to mode couplings involving a ridge n in the $k - w$ plane decreases very rapidly with increasing n, and essentially all the contribution to d ln E_α/dt comes from the fundamental-mode ridge. These arguments show that the inverse dissipation time of an overstable mode of high order (at low ℓ the observed n range is 5 to 35) depends mainly on the energy of f-modes and to a much lesser extent on the energies of modes with $n > 0$. For instance, the contribution to d ln E_α/dt from the $n = 5$ ridge is smaller by a factor of about 10^5 compared to the $n = 0$ ridge. Therefore, we conclude that even for the lowest-order overstable mode, the dissipation rate gives a value for q very nearly equal to zero.

If we have seriously overestimated the energies of the f-modes by assuming the spectrum to be l independent, the most important couplings might involve the next higher available ridge. This would not affect our conclusion that $q \approx 0$. This result shows that the 3-mode couplings, previously considered the best candidate for limiting the energies of overstable solar p-modes, are not strong enough for this purpose, and this in turn casts severe doubt on any proposal for overstable p-modes in the Sun.

X. MODE EXCITATION DUE TO ACOUSTIC EMISSION FROM TURBULENT CONVECTION

A typical solar p-mode has an energy of $\sim 3 \times 10^{27}$ erg (Libbrecht 1988d), and there are about 10^7 modes, so the total energy in solar acoustic waves is $\sim 10^{34}$erg. The lifetime of a 5-min, low l-mode is about 5 days (4 $\times$ 10^5s), and shorter for high-frequency or high l modes. Therefore, we need of the order of 10^{28} erg s^{-1} or $10^{-6}L_\odot$ to excite solar p-modes. This large energy requirement constrains the possible mechanisms for exciting these modes. For instance, the total magnetic energy in the Sun is 10^{32} erg (perhaps an underestimate if most of the magnetic field is locked in thin flux tubes), which will have to be regenerated in just a few hours if it were to supply energy to the acoustic modes; this is highly improbable. The rotational energy of $\sim 10^{41}$ erg is sufficient to support acoustic modes for about 10 Myr if efficiency of order unity can be achieved. But the time scales for rotation (25 days) and pulsation (5 min) are so different that this does not seem possible.

Transient energetic phenomenon on the solar surface (e.g., flares and active regions) might meet the energy requirement. However, the observations carried out over the last 6 yr indicate that the average energies of p-modes have not changed significantly during this period (Deubner and Gough 1984; Libbrecht and Zirin 1986; Libbrecht 1988d), whereas the flares and active regions undergo a strong 11-yr cyclic variation. Therefore, we conclude that these transients are not viable means of supporting the observed global oscillations. Acoustic modes with $l > 15$ cannot tap the nuclear energy source in the solar core because they have negligible amplitude in the nuclear-burning region.

Of course, there is another large energy source in the form of solar luminosity, a millionth of which would be sufficient to sustain the acoustic oscillations. Indeed, the pulsation in the classical variable stars is caused by tapping radiative luminosity by varying the opacity in the H or He ionization zone. So it is not unnatural to suggest that the solar oscillations might also be excited in a similar way. About 99.9% of the luminosity in the outer 30% of the radius of the Sun is carried by convection, and the remainder by radiation. Thus, either of these two are, in principle, capable of supplying the required energy to excite the p-modes to the observed levels. Unfortunately, all the mechanisms proposed so far to extract energy from radiative or convective luminosity involve overstable modes, which we argued in the last section would inevitably lead to runaway mode amplitudes in the solar case. Therefore, we must look elsewhere.

There is yet another energy reservoir consisting of the mechanical energy of convection, i.e., the kinetic energy of bulk motion, which is assumed to be in a state of fully developed turbulent motion. The total kinetic energy in the convective motions is 5×10^{38} erg, of which approximately 5×10^{34} erg resides in the top scale height of the convection zone and has a characteristic recycling time of about 5 min. It is well known that turbulent flow is a source of sound waves of frequency bandwidth equal to the inverse energy cascade time.

The most familiar example of this process is perhaps jet noise. The theory for acoustic emission from homogeneous turbulence was first worked out by Lighthill (1952). This work has been applied by many authors to the problem of generation of acoustic and MHD waves for heating of the solar atmosphere. A little more than a decade ago, Goldreich and Keeley (1977*b*) suggested that sound radiated from turbulent convection could excite global solar oscillations. In the remainder of this section we give a simple presentation of Lighthill's theory, Goldreich and Keeley's energy equipartition result, and some recent results of Goldreich and Kumar (1988) on the interaction of acoustic waves with forced turbulence. An excellent review by Crighton (1975) on aerodynamic noise generation, and Lighthill's book (1978) on waves in fluids should be consulted for detailed and formal discussion of some of the material presented in this section.

Sound production generally involves one of the following 3 basic mechanisms:

1. A time-dependent mass flux, e.g., popping a balloon, cracking a finger, or a radially pulsating sphere;
2. A time-varying force (momentum flux), e.g., violin string, tuning fork, vibrations in general;
3. Internal stresses with no associated mass or momentum flux, e.g., free turbulence.

At distances much greater than the typical wavelength of radiation and source of linear dimension smaller than the wavelength, the angular distribution of the wave amplitude is, for the above cases: (1) spherically symmetric; (2) dipolar; and (3) quadrupolar, respectively (Crighton 1975; Lighthill 1978). This simple relation between angular pattern of radiation and the physical nature of source is only applicable to sound emission in a homogeneous medium.

To keep the discussion simple, we calculate acoustic emission due to a point mass source. Emission from dipole and higher-order multipole are obtained by considering an appropriate superposition of monopole sources. The propagation of waves in a homogeneous medium is governed by the following equation

$$\frac{\partial^2 \phi}{\partial t^2} - c^2 \nabla^2 \phi = 0 \tag{40}$$

where ϕ is the velocity potential ($\nabla \phi$ is the fluid velocity) and c is the sound speed.

The general, spherically symmetric solution of this equation corresponding to an outward propagating wave is readily obtained

$$\phi(\mathbf{x},t) = g(r - ct)/r \tag{41}$$

where $r = |\mathbf{x}|$, and $g(r - ct)$ is a function proportional to the total mass flux in the near field (distance from the source much smaller than a wavelength). The fluid velocity $\mathbf{v}$, for the above solution is given by

$$\mathbf{v} \equiv \nabla\phi = \frac{(r\dot{g} - g)\mathbf{x}}{r^3} \qquad (42)$$

where $\dot{g}$ is the derivative of g with respect to r. It follows from Eq. (42) that the total momentum of fluid enclosed inside a sphere of arbitrary radius centered at the origin is zero. Thus, there is no external force acting on the fluid. However, the total mass flux through the surface of this sphere is nonzero, and can be easily seen to be $4\pi\rho(r\dot{g} - g)$. For a periodic source of temporal frequency ω, $\dot{g} = -\omega g/c$, and so the mass flux through the surface of a sphere of radius r is $-4\pi\rho(1 + r\omega/c)g$. In the near field this reduces to $-4\pi\rho g$. Therefore, the function g is the mass input rate divided by $4\pi\rho$.

The acoustic luminosity $\mathcal{L}$, defined to be the total acoustic energy radiated per second is equal to $4\pi\rho r^2|\mathbf{v}|^2 c = 4\pi\rho\dot{g}^2 c$. For a harmonic source of linear dimension r_0, and mass flux $4\pi\rho\omega r_0^3$, the velocity in the radiation zone ($r \gg \lambda$) is $\omega^2 r_0^3/cr$, and therefore $\mathcal{L} = 4\pi\rho\omega^4 r_0^6/c$. The acoustic efficiency η, defined as the ratio of energy radiated by the source over a wave period ($\mathcal{L}/\omega$) to the total energy in the near field, is equal to the Mach number $\mathcal{M} = \omega r_0/c$, of the source.

A dipole source is obtained by the superposition of two monopole sources of opposite phase separated by a distance $\mathbf{a}$, where $|\mathbf{a}| \ll \lambda$. Obviously, in this case the total mass flux is zero. The velocity potential for the dipole follows from Eq. (41), $\phi \approx (\mathbf{a} \cdot \mathbf{x})(g + r\dot{g})/r^3$. Therefore, the momentum of fluid enclosed by a sphere of radius r is $4\pi\rho(g + r\dot{g})\mathbf{a}/3$, which changes with time as a result of the force the dipole exerts on the surrounding fluid. From the above dipole potential, we find that the fluid velocity in the far field of the dipole source is $\mathbf{x}(\mathbf{a} \cdot \mathbf{x})\ddot{g}/r^3$. Therefore, $\mathcal{L} \approx \rho a^2\ddot{g}^2 c$, and for a harmonic source the acoustic efficiency η is $\mathcal{M}^3$.

A quadrupole source is obtained by placing two dipole sources in opposite phase. In this case both the total mass and momentum transfer to the fluid are zero. The acoustic efficiency for a quadrupole source can be easily verified to be $\mathcal{M}^5$. Thus, in general for a multipole source of order n, $\eta \sim \mathcal{M}^{(2n+1)}$.

In the following discussion, we adopt a simple picture of turbulence as consisting of critically damped eddies. Energy is supplied on a spatial scale H, and time τ_H, which cascades down to a scale ε where it is dissipated by molecular viscosity. In between scales H and ε (called the inertial range), the energy dissipation is negligible. Thus, the energy flow per unit volume for eddies of size h in this range should be constant i.e., $\rho V_h^2/\tau_h = \rho V_H^2/\tau_H$, or $V_h = V_H(\tau_h/\tau_H)^{1/2}$. Using $V_h\tau_h = h$ we obtain $V_h = V_H(h/H)^{1/3}$, which implies that the energy density per unit wavenumber $k = 1/h$, is proportional to $k^{-5/3}$. This is the well known Kolmogoroff spectrum.

We now have all the machinery needed to estimate acoustic emission from turbulent flows. In the absence of external forces the lowest order acoustic emission from turbulent flow is quadrupolar, which has an efficiency of $\mathcal{M}^5$. Thus, the total acoustic energy flux $\mathcal{F}_h$ from an eddy of size h and velocity V_h is equal to $\rho V_h^2 \mathcal{M}^5 h^3/\tau_h = \rho c^3 h^2 \mathcal{M}^8$, which is uniformly spread over frequency range of 0 to τ_h^{-1}. This result was first derived by Lighthill (1952).

To estimate the mean energy E_q, in a normal mode q (frequency ω_q and wavenumber k_q) of a box containing some turbulent fluid, we need the mean energy input rate A_q and the dissipation rate η_q. The mean mode energy follows from the condition $A_q = \eta_q$. In keeping with the spirit of simplicity over rigor, we estimate $A_q^{(h)}$, $\eta_q^{(h)}$ and $E_q^{(h)}$ due to a single eddy of size h, and then sum over the available eddy spectrum to obtain E_q. Since an eddy of turnover time τ_h, emits radiation at all frequencies smaller than τ_h^{-1}, therefore, for modes of frequency $\leq \tau_h^{-1}$, $A_q^{(h)} \approx \mathcal{F}_h/N_q$, where $\mathcal{F}_h$ is the total energy flux from an eddy of size h estimated above, and N_q is the number of normal modes of the box that have frequency smaller than ω_q. For a box of volume V, $N_q \sim V k_q^3$ (where $k_q = \omega_q/c$), thus, $A_q^{(h)} \sim \rho h^5 c^3 \mathcal{M}_h^5/V(\omega_q \tau_h)^3$. For high Reynolds-number fluids, the dissipation of waves due to molecular viscosity is insignificant compared to the dissipation resulting from the interaction with turbulence. This latter process, in analogy with molecular viscosity, is modeled by an effective viscosity, $\nu_h = h V_h$, associated with an eddy of size h. For a mode q of velocity amplitude $\mathbf{V}_q$, the energy dissipation rate due to viscosity ν is $\int d^3x\, \rho\nu|\nabla\mathbf{V}_q|^2$. Thus $\eta_q^{(h)}$ due to an eddy of size h $\ll 1/k_q$, is $\rho\nu_h k_q^2 V_q^2 h^3 = E_q^{(h)} \nu_h k_q^2 h^3/V$. Equating energy gain and loss rate, we obtain mean mode energy due to a single eddy of size h, $E_q^{(h)} \sim M_h V_q^2/(\omega_q \tau_h)^5$, where $M_h = \rho h^3$ is the eddy mass. The total energy in the mode is obtained by integrating $E_q^{(h)}$ over the allowed eddy size, $E_q \sim \int_0^{h_q} dh\, E_q^{(h)}/h$, where h_q is the size of an eddy with turnover time equal to ω_q^{-1}, i.e., $h_q \sim H(\tau_h \omega_q)^{-3/2}$. Thus $E_q \sim M_q V_{hq}^2$, where M_q and V_{hq} are, respectively, the mass and velocity of the resonant eddy, i.e., the eddy with a turnover time equal to the mode period. So, we find that the energy in a mode is equal to the kinetic energy of a resonant eddy.

For modes of period longer than the turnover time of the largest eddy, the energy is $M_H V_H^3$, giving rise to a Rayleigh-Jeans blackbody spectrum. This result suggests a simple thermodynamical interpretation. Acoustic radiation can be thought of as being in thermal equilibrium with resonant eddies at acoustic temperature equal to $M_h V_h^2/k_b$, where k_b is Boltzmann's constant. It is straightforward to check that, if the fractional volume occupied by turbulent fluid increases, the energy per mode remains unchanged, and the mode line width increases linearly.

The equipartition between modal energy and the kinetic energy of a resonant eddy was derived by Goldreich and Keeley (1977b) for turbulence in a stratified medium by considering quadrupole emission and absorption due to Reynolds stress. Later on, we discuss how this result is modified when there are external forces present, but for the present, we use it to estimate energy in the solar acoustic modes. The equipartition result of Goldreich and

Keeley implies that the energy in a typical 5-min p-mode should be equal to the kinetic energy of a resonant eddy, which can be identified as the granules observed on the solar surface. Therefore, the expected p-mode energy is approximately $\rho_p H L^2 V_H^2$, where ρ_p, H, L and V_H are the mass density, scale height, granule size and convective velocity at the top of convection zone, respectively. With $\rho_p = 3 \times 10^{-7}$ g cm^{-3}, $H = 1.4 \times 10^7$ cm, $L = 10^8$ cm and $V_H = 2 \times 10^5$ cm s^{-1}, we obtain energy per mode of 1.7×10^{27} erg, which is within a factor of 5 of the observed value (Libbrecht 1988d). Given the crude nature of this estimate, the agreement is encouraging.

An intriguing feature of the p-mode energy spectrum has been recently reported by Libbrecht (1988a). He finds that the product of the observed mode energy and the line width (which is a rough measure of the energy input rate) for low-degree modes is proportional to ω^8. This dependence follows naturally from the above stochastic excitation picture, if most of the energy comes from the evanescent region near the top of the convection zone (Goldreich and Kumar, in preparation).

So far we have discussed the idealized case of free turbulent flow, which is, strictly speaking, not applicable to realistic astrophysical systems. Turbulence is an inherently dissipative process, requiring a continuous supply of energy to maintain it. The mechanism which excites turbulent flow can itself cause acoustic emission and absorption. As a result, two different turbulent fluids with identical properties of the turbulent flow may give rise to entirely different acoustic spectra; and so to estimate energy in modes, we need to take into consideration the process of generation of turbulence. We illustrate this point by considering 2 different examples of homogeneous isotropic turbulent flows analyzed by Goldreich and Kumar (1988), and the resulting modification to the equipartition result.

The first example of turbulent flow we consider is maintained by stirring the fluid with spoons of size H. The second example consists of fluid in a uniform force field acting on nondiffusive scalar contaminant, which is added and removed from the fluid randomly on spatial scale H and time τ_H. To ensure isotropy, the direction of the field is assumed to vary randomly on time τ_H. The second example is related to turbulent convection, where entropy plays the role of the scalar contaminant and gravity is the external force field. However, there is an important difference between our second example and true turbulent convection. In turbulent convection, the entropy of a fluid parcel changes only by mixing with the surrounding fluid, whereas in our second example, the scalar contaminant is added and removed from the fluid at random. To differentiate between the two cases, Goldreich and Kumar refer to the second example of turbulent flow as pseudoconvection.

The force due to spoons or the external field on the fluid causes dipole acoustic emission, which as we have seen before is more efficient by a factor of $\mathcal{M}^{-2}$ compared to the quadrupole or free turbulence case. If the wave dissipation were also to be correspondingly enhanced by the external force, then the energy per mode would remain unchanged, although the mode lifetime would decrease. However, Goldreich and Kumar (1988) have shown that the wave absorption is enhanced for the turbulence excited by spoons but not for the pseudoconvection. Thus, the energy per mode in the case of turbulent pseudoconvection is equal to $M_q c^2$, which is greater by a factor of $\mathcal{M}^{-2}$ compared with the corresponding free turbulence.

To understand wave dissipation qualitatively, we consider the motion of a fluid particle under the combined influence of the wave, an external force and other particles. Acting alone on a test particle, an acoustic wave of pressure variation $P_w \cos(\omega t - \mathbf{k} \cdot \mathbf{x})$, causes periodic velocity of amplitude $\mathbf{v}_w = kP_w/\rho(\omega - \mathbf{k} \cdot \mathbf{v})$, where $\mathbf{v}$ is the test particle's mean velocity, ω, $\mathbf{k}$ and ρ are the frequency, wave vector and ambient density, respectively.

Let us consider the absorption of a wave when fluid is pushed around by spoons of size H, corresponding to the largest eddy size. For simplicity, we analyze this in the rest frame of spoon, so that all the changes in the test particle's energy are attributable to a wave. Furthermore, let us assume that the interaction of the particle with a spoon results in the reversal of particle's instantaneous velocity. Therefore, in the presence of a wave, the particle's mean velocity changes by $\sim 2kP_w \cos(\omega t_o - \mathbf{k} \cdot \mathbf{x}_o)/\rho\omega$, due to encounter with a spoon. Here t_o is the time of encounter and $\mathbf{x}_o$ is particle's position at t_o. Multiple encounters with spoons lead to a random walk in the particle's mean velocity, which results in the increase of its mean kinetic energy. The energy gained by the particle is the energy lost by the wave. Thus, the average wave energy dissipated during a single encounter ΔE is $\sim P_w^2 H^3/\rho c^2$.

Next consider the case of pseudoconvection. The external force acting on scalar contaminant changes the velocity of energy-bearing eddy by V_H, independent of its position or velocity. Due to the change in its velocity, the particle feels the wave at a Doppler-shifted frequency $\omega - \mathbf{k} \cdot \mathbf{v}$. If $\omega\tau_H \gg 1$, then the particle oscillation responds adiabatically to the changing Doppler-shifted frequency, and there is no net acoustic energy absorbed from the wave. However, for $\omega\tau_H \lesssim 1$, the random variation of the Doppler shift results in a random walk of the particle's velocity of magnitude $P_w V_H/\rho c^2$. This amounts to an absorption of average energy $\Delta E \sim P_w^2 V_H^2 H^3/\rho c^4$ per step from the wave. Thus, in this case the absorption is smaller by a factor of $\mathcal{M}^2$ compared to our first example, or the same as the free-turbulence case. Therefore, the energy per mode in turbulent pseudoconvection is $M_q c^2$. The difference in absorption in these two examples can be traced back to the dependence and independence of the external force on fluid velocity.

How does the above result affect our earlier estimate of energy in solar p-modes? Emission associated with buoyancy force is indeed more efficient compared to Reynolds stresses by $\mathcal{M}^{-2}$, and there is a monopole emission when eddies lose heat by radiation at the solar surface. The magnitude of monopole emission, however, is comparable to buoyancy-induced dipole emission because the fractional change in volume is $\mathcal{M}^2$. Moreover, due to a subtle reason, it turns out that, for energy-bearing eddies, the dipole and the monopole radiation cancel each other leaving a residual of magnitude the same as the quadrupolar Reynolds-stress emission (Goldreich and Kumar, in preparation). Therefore, those p-modes, that are resonant with energy-bearing eddies near the top of convection zone, have energies given by the Goldreich-Keeley equipartition result. But modes that have significant interaction with subenergy-bearing eddies will have an enhanced energy of $\mathcal{M}c^2$.

XI. CONCLUSIONS

A careful reading of this chapter will reveal that there are many uncertainties and disagreements in the prediction of the stability or overstability of the observed solar oscillations. The simplest calculations using only radiation flow by diffusion, ignoring the error of this approximation near the solar surface, give stability of most modes. This assumes, however, that we know the opacity and its derivatives, and small errors in these values could result in predictions of overstable modes.

When the model structure in the top 1000 km or less, below the photosphere, is modified to include radiative-transport effects, and the variations of the transport are calculated by linear pulsation theory, stability seems to be enhanced.

The inclusion of the convective Cowling mechanism gives driving just below the photosphere that appears marginally to overcome damping effects for p-modes according to the Antia et al. (1982) calculations. They recently propose that low-order g-modes could also be excited by this near-surface convective Cowling mechanism.

We then note that if the p-modes are really overstable, some very effective amplitude-limiting mechanism would have to be present because the observed modes have extremely small amplitudes unlike those for classical variable stars like the Cepheids. Three-mode coupling, the most efficient mechanism that seems available for nonlinear amplitude limitation, is not good enough to suppress these p-modes. Therefore, it seems from that conclusion that the solar p-modes may not be self excited.

Finally we come to a view that the forcing of the p-modes by coupling to convection-induced acoustic noise is an adequate mechanism for their excitation. This concept is accepted by most workers, but the convective Cowling mechanism needs further consideration.

The prediction of low-degree, low-order g-mode instability needs further investigation, but unlike the p-modes, these modes may well be at a large amplitude with amplitude limitations due to nonlinear effects. The difficulty in seeing these modes at the surface may only be because they tunnel through the convection zone leaving only a small surface amplitude. Any large deep amplitudes can greatly affect neutrino production rates, if these modes actually occur in the Sun. However, this does not seem to be a problem, because those g-modes that are now predicted to be possibly overstable need comparable amplitudes in the surface layers as in the deep interior to invoke the radiative and convective processes operating in the subsurface layers of the Sun. The observed small surface amplitudes suggest small deep amplitudes that would not materially affect the nuclear-burning regions near the center. These modes are probably also stochastically excited by coupling with convection, just as for the p-modes, and their large radiative damping explains their very low amplitudes. These many unresolved problems we must leave for the future.

Acknowledgments.   This work was supported in part by a Smithsonian fellowship to P. K., and a grant from the National Science Foundation. The work was performed while P. K. was visiting Harvard-Smithsonian Center for Astrophysics. He is grateful to W. Press and the Smithsonian Institution for financial support. S.M.C. is also affiliated with NASA Goddard Space Flight Center.

PART III
What Surface Radiation Reveals About the Inside

THE SOLAR SPECTRUM

ROBERT L. KURUCZ
Harvard-Smithsonian Center for Astrophysics

This chapter contains a discussion of what is known about the solar spectrum as a whole from the ultraviolet to the infrared. The data that are available are described and information is provided on how the reader can obtain paper or magnetic tape copies for particular applications, be it in atomic or molecular spectroscopy, or in solar, atmospheric, planetary, cometary or stellar physics.

I. FLUX SPECTRA

The solar flux spectrum is important for its effects on objects in the solar system, including ourselves. It is less important for solar physics because in the flux spectrum much of the spatial and Doppler information about the solar atmosphere has been integrated away leaving a spectrum broadened and blended by the 2 km s^{-1} solar rotation. The flux spectrum is quite important for stellar physics, however, because the Sun serves as the "standard star." We can determine its properties much better than those of any other star. Solar flux spectra are required for planning and interpreting stellar and planetary observations. Solar spectra have the resolution and signal-to-noise ratio to show what is actually being observed. Because of inadequate resolution, blending and a low signal-to-noise ratio, most stellar observations cannot be interpreted without such additional, *a priori,* information.

Observations made from groundbased observatories include the atmospheric transmission spectrum, so it is necessary to be aware of blending and blocking by terrestrial lines and to have resolution high enough to resolve their profiles. A solar-flux spectrum observed from the ground is a useful guide for these purposes. The spectrum should have a resolving power $> 10^6$

and a signal-to-noise ratio $> 10^4$. For atmospheric chemistry of planetary and cometary atmospheres, and for space-based stellar observations, however, the true flux spectrum above the atmosphere is required.

The flux spectrum has been poorly observed. An atlas exists for the groundbased spectrum from Kitt Peak for wavelengths 300 to 1300 nm (Kurucz et al. 1984). It has very high quality by astronomical standards but still leaves considerable room for improvement. There are no high-resolution flux spectra covering other wavelength regions or above the atmosphere and no improvement is expected in the near future. In the meantime, there are three approaches to approximating the flux spectrum. The first is to model the atmospheric transmission and then to divide the groundbased spectrum by it. This should work quite well as long as the signal-to-noise ratio is very high and the transmission is not near zero. The second method is semi-empirical: fitting an intensity spectrum computed from a solar model to an observed intensity spectrum and then using the derived line parameters to generate the flux spectrum. The problem is that a significant fraction of the lines in the spectrum has not been identified so that their properties would have to be assumed. The third method is to compute a purely theoretical flux spectrum from the existing line data. At present, computing a fully realistic spectrum is beyond the state of the art, but work on all three approaches is in progress.

II. INTENSITY SPECTRA

For studying the solar atmosphere, or for atomic or molecular spectroscopy, intensity spectra are required. The spectrum provides a tremendous amount of information about the structure of the solar atmosphere. This information increases as the spatial resolution increases. Spectra-spectroheliograms show the spectrum at each resolution element, but they give almost too much information because they emphasize the instantaneous velocity field. The existing intensity atlases are spatial and temporal averages over a small area on the disk, either at the center or near the limb, such that the signal-to-noise ratio is high and the spectrum is reproducible, except in the ultraviolet where the spectrum actually varies with solar activity. Because there is no rotational broadening in intensity spectra, the lines are narrower and less blended than in flux spectra and so are better suited for spectroscopic analysis. There are still net depth-dependent Doppler shifts in the average spectra that must be considered for precise work (< 1 km s^{-1}). These can shift the cores of lines relative to the wings, or shift lines that are formed in cooler regions of the atmosphere relative to those formed in warmer regions. There is also the 636 m s^{-1} gravitational red shift and the differential rotation velocity to be considered. Despite these complications, the Sun has advantages as a spectroscopic source over laboratory sources because it shows line spectra to high energy, or to molecular levels high V and high J, that are difficult or impossible to obtain in the laboratory. For example, Melen et al. (1989) analyzed the 4-3 band of CH using lines in the solar spectrum that have not been observed in the laboratory.

The solar intensity spectrum has been poorly observed except in the visible. In the ultraviolet, the spectrum has never been resolved and the signal-to-noise ratio is inadequate—the available results are very poor. The limit on the signal-to-noise ratio should be the point where there is no more information to be gained, where no additional weak lines become measurable; and in addition, complete wavelength coverage is required. For example, a strong test of solar models is consistent predictions between the ultraviolet electronic bands and the infrared vibration-rotation bands of a molecule, but high-quality observations are not available.

III. TERRESTRIAL ATMOSPHERIC SPECTRA

Solar spectra taken through the atmosphere show absorption lines from molecules in the Earth's atmosphere. They show lines of H_2O, CO_2, O_3, N_2O, CO, CH_4, O_2 and many trace molecules. The spectra provide both a spectroscopic source for studying these molecules and a means of determining the composition of the atmosphere. Fortunately, the Air Force has funded work on atmospheric transmission and the Air Force Geophysics Laboratory publishes a line list on magnetic tape called the HITRAN data base (Rothman et al. 1987) which is continually being updated. The 1990 edition should cover the whole visible and infrared spectrum.

Kitt Peak is a good site for studying the terrestrial water-vapor spectrum. The Kitt Peak solar-flux atlas in the visible (Kurucz et al. 1984) and the central intensity atlas in the infrared (Delbouille et al. 1981) are good sources. Brault has also taken central intensity spectra that have high signal-to-noise contrast and sunset spectra that have long path lengths. These FTS (Fourier Transform Spectrograph) spectra in the visible are being reduced (Kurucz, work in progress). Water-vapor lines are present throughout the visible on these high signal-to-noise spectra. Camry-Peyret et al. (1985) and Mandin et al. (1986) have produced line lists with many classifications for the visible.

In contrast, Jungfraujoch is a good site for observing the Sun because of its high altitude (3580 m) and low water vapor. By comparing the lowest panel in Figs. 1 and 2 for Kitt Peak and for Jungfraujoch, one can easily see the difference in the infrared.

The ATMOS FTS spectrograph (Farmer 1987) flown on the Space Shuttle has taken high-quality infrared spectra for a range of heights in the atmosphere, and also of the Sun directly as discussed below. Similar experiments should be flown to investigate the whole spectrum from the ultraviolet through the infrared.

IV. ATLASES

The *Solar Flux Atlas from 296 to 1300 nm* by Kurucz et al. (1984) plots residual flux and also gives a table for converting to the absolute irradiance calibration by Neckel and Labs (1984). It is shown in Fig. 1. [Paper copies of this atlas are available from the National Solar Observatory at Sunspot, New Mexico for $13. Magnetic tape copies can be obtained at no charge from the author.] The spectrum was observed at Kitt Peak using the Fourier Transform Spectrograph on the McMath telescope. The resolving power is 522×10^3 in the red and infrared, and 348×10^3 in the ultraviolet. The resolution is not high enough to resolve the terrestrial lines so there is some ringing. The atlas was fitted together from 8 overlapping scans that have reasonable signal-to-noise contrast at the center but fall off in the region of overlap. In the final spectrum the signal-to-noise ratio varies from 2×10^3 to 9×10^3. Ideally an atlas should be made from many more overlapping scans so that only portions near maximum need be used. The wavelength scale was set from a terrestrial O_2 line. the continuum level was estimated from high points and is uncertain, especially in the ultraviolet. There are also problems caused by broad structures in the atmospheric transmission produced by ozone and O_2 "dimer." There is an error that apparently affects the low-frequency components of the Fourier transform so that in wavelength space the zero point can vary by a fraction of a percent. Thus, the depths of the strongest lines are reliable only to a fraction of a percent of the continuum.

The best central intensity spectrum in the visible is the Jungfraujoch atlas, *Spectrophotometric Atlas of the Solar Spectrum from λ3000 to λ10000* by Delbouille et al. (1973). The atlas is shown in Fig. 2. [The paper or magnetic tape copies of the atlas can be purchased from G. Roland at the Institut d'Astrophysique in Liège.] The spectrum was observed in small-wavelength sections using a rapid-scanning double-pass monochromator. The wavelengths were set from Pierce and Breckinridge's (1973) line list but only up to 900 nm. From 900 to 1000 nm, the wavelengths are not accurate because of poor wavelength standards. Brault's FTS spectra that will provide good wavelengths in this region is being reduced (Kurucz, work in progress). The spectrum was partially smoothed so it is difficult to estimate the signal-to-noise ratio. Each section was "straightened" to make a continuous spectrum but the "straightening" introduced ripples on the 0.2% level in the red (Kurucz's estimate) and as large as 5% in the blue and ultraviolet (Rutten 1988). There is also a small, unknown amount of scattered light.

The *Kitt Peak Preliminary Photoelectric Atlas* by Brault and Testerman (1972) has both center (294.2 to 1080.0 nm) and limb (367.6 to 973.9 nm) spectra in the visible. [It is available on magnetic tape from Kurucz.] The central intensity spectrum is now obsolete but the limb spectrum remains the best that is now available. Limb spectra are useful for testing damping con-

Figure 1.

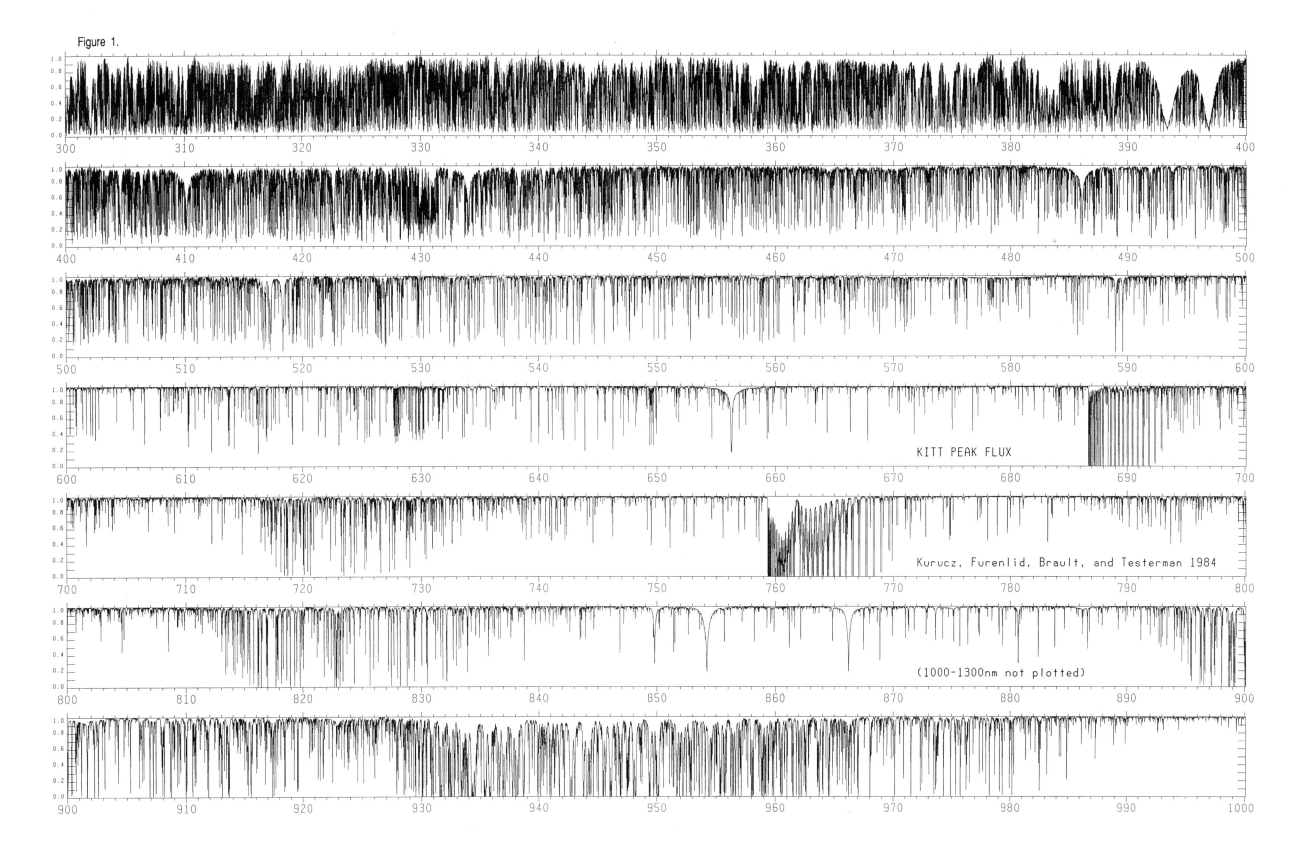

KITT PEAK FLUX

Kurucz, Furenlid, Brault, and Testerman 1984

(1000-1300nm not plotted)

Figure 2.

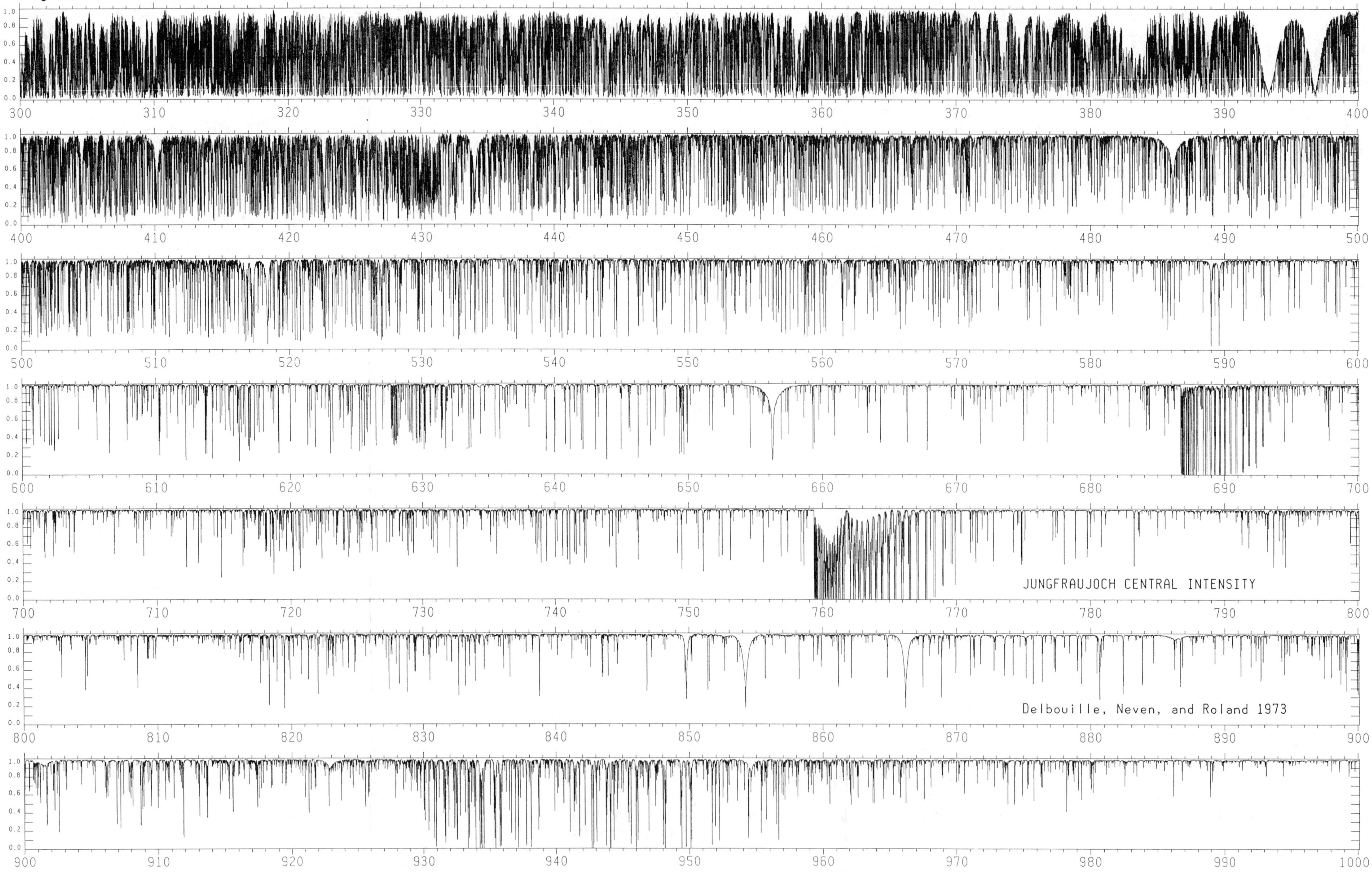

JUNGFRAUJOCH CENTRAL INTENSITY

Delbouille, Neven, and Roland 1973

Figure 3.

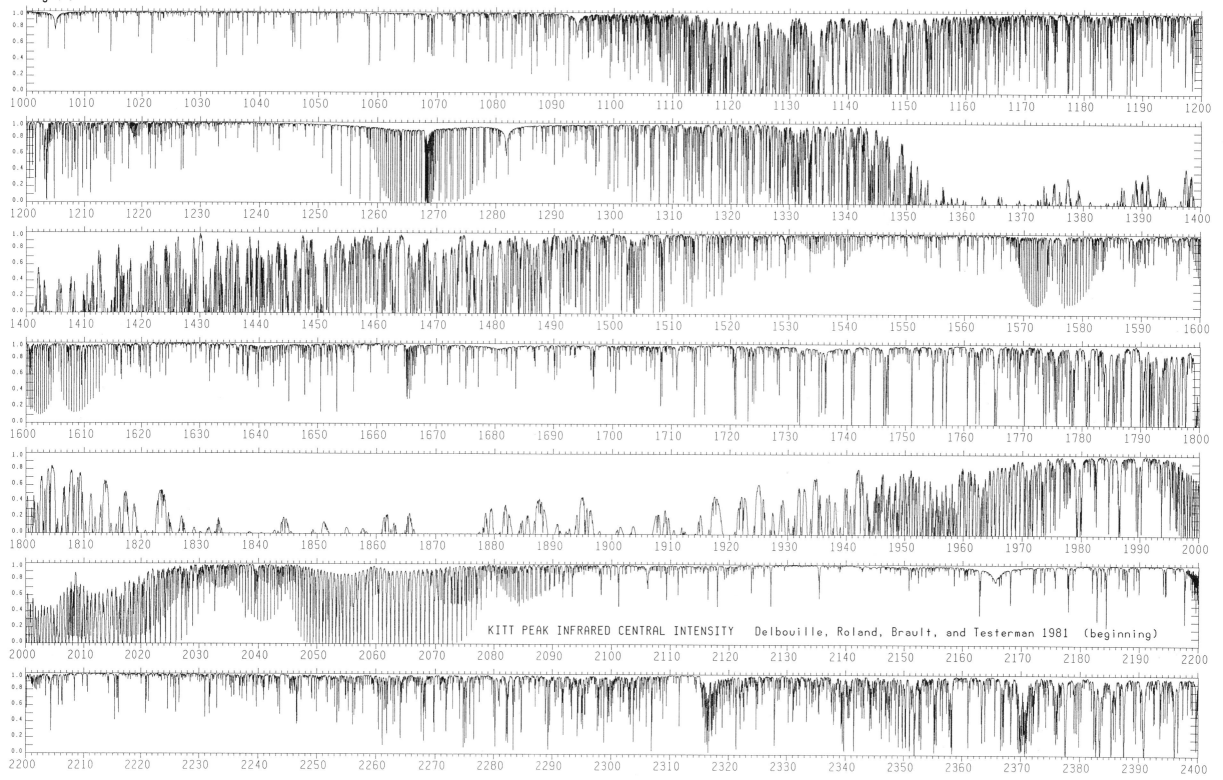

KITT PEAK INFRARED CENTRAL INTENSITY   Delbouille, Roland, Brault, and Testerman 1981 (beginning)

Figure 4.

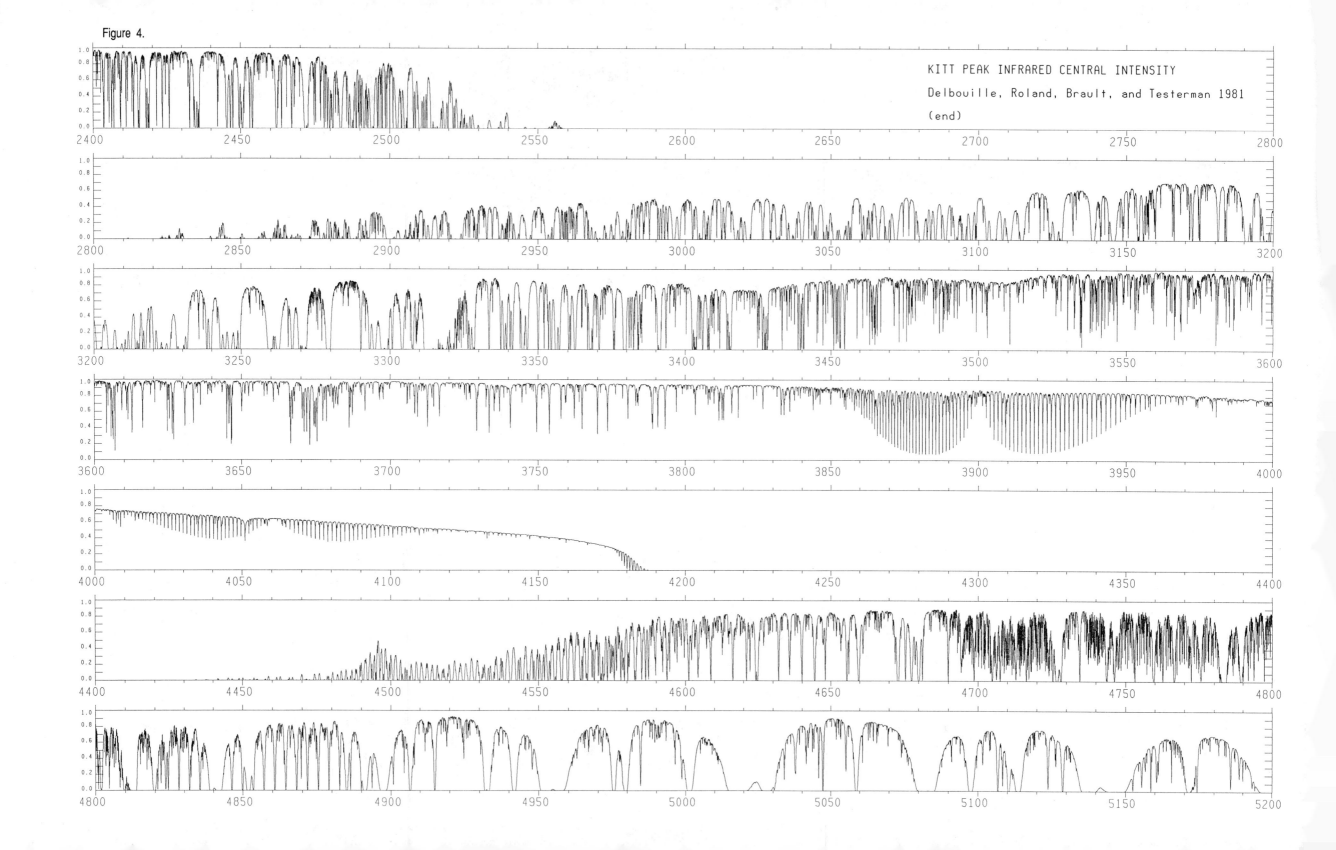

KITT PEAK INFRARED CENTRAL INTENSITY

Delbouille, Roland, Brault, and Testerman 1981

(end)

Figure 5.

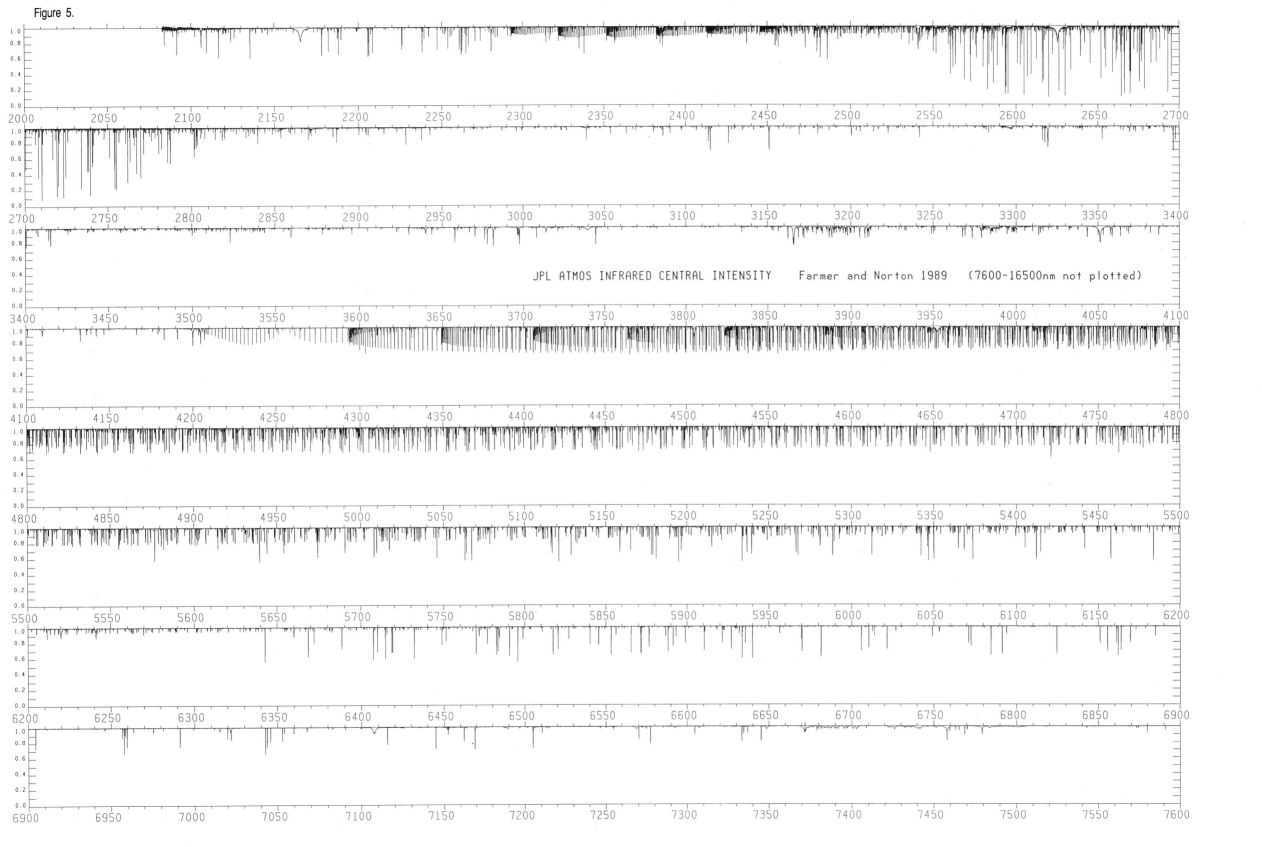

JPL ATMOS INFRARED CENTRAL INTENSITY   Farmer and Norton 1989   (7600-16500nm not plotted)

Figure 6.

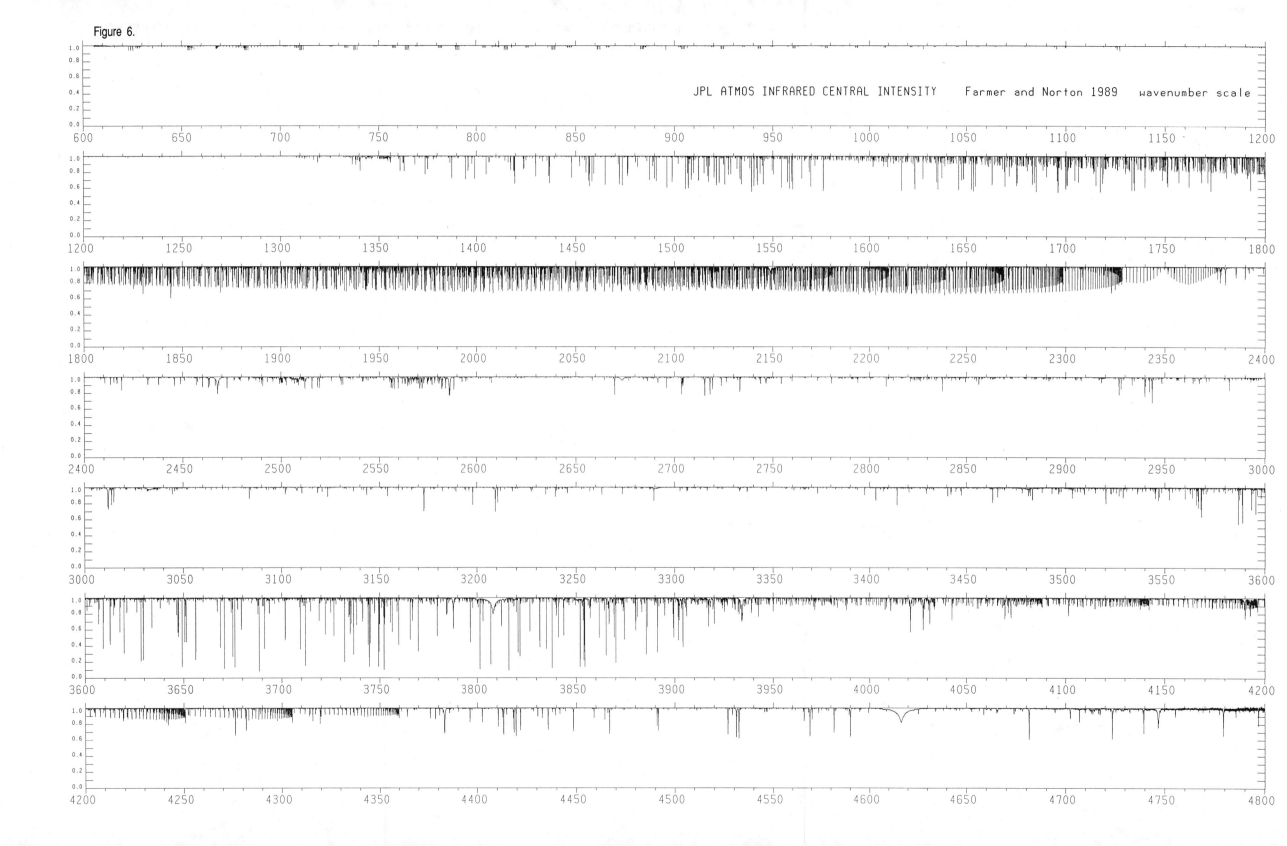

JPL ATMOS INFRARED CENTRAL INTENSITY   Farmer and Norton 1989   wavenumber scale

Figure 1.

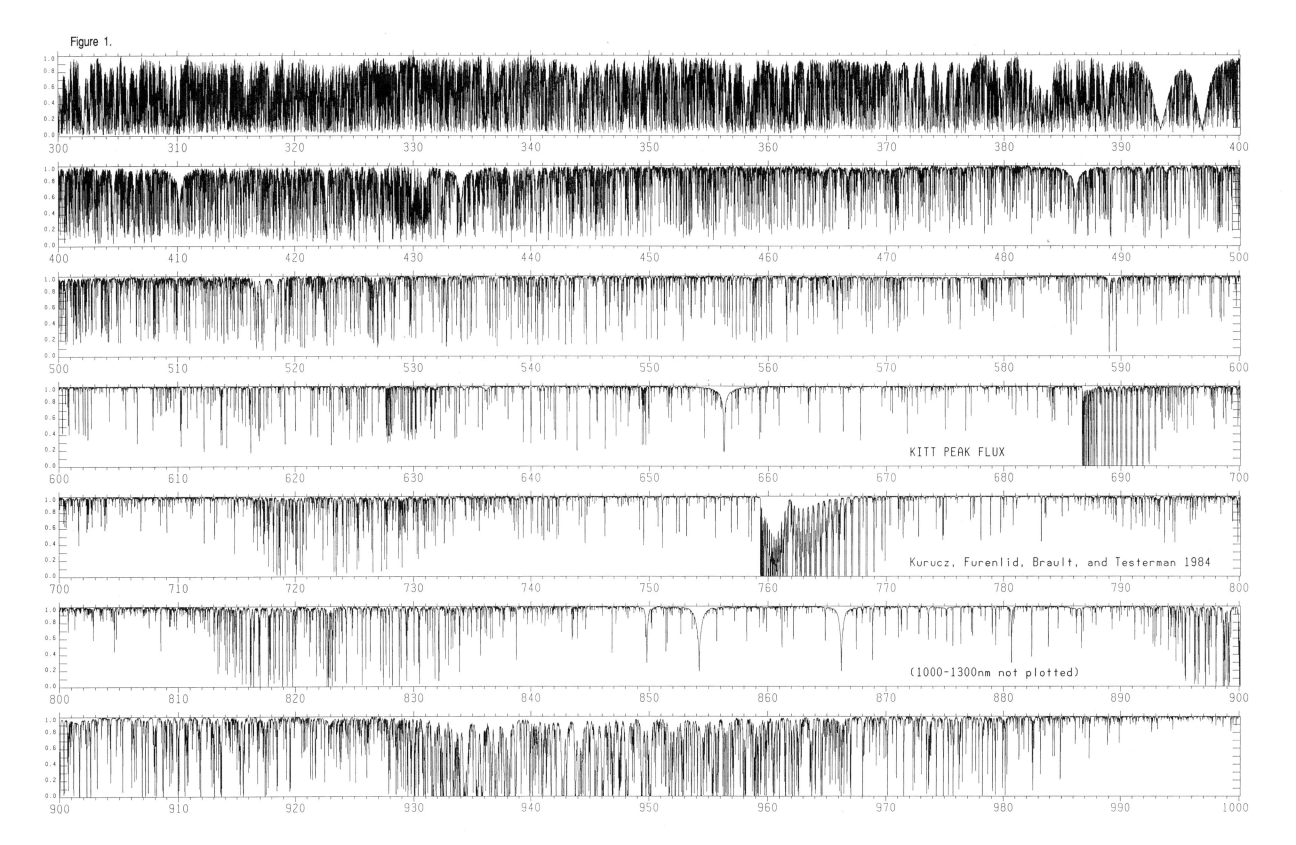

KITT PEAK FLUX

Kurucz, Furenlid, Brault, and Testerman 1984

(1000-1300nm not plotted)

Figure 2.

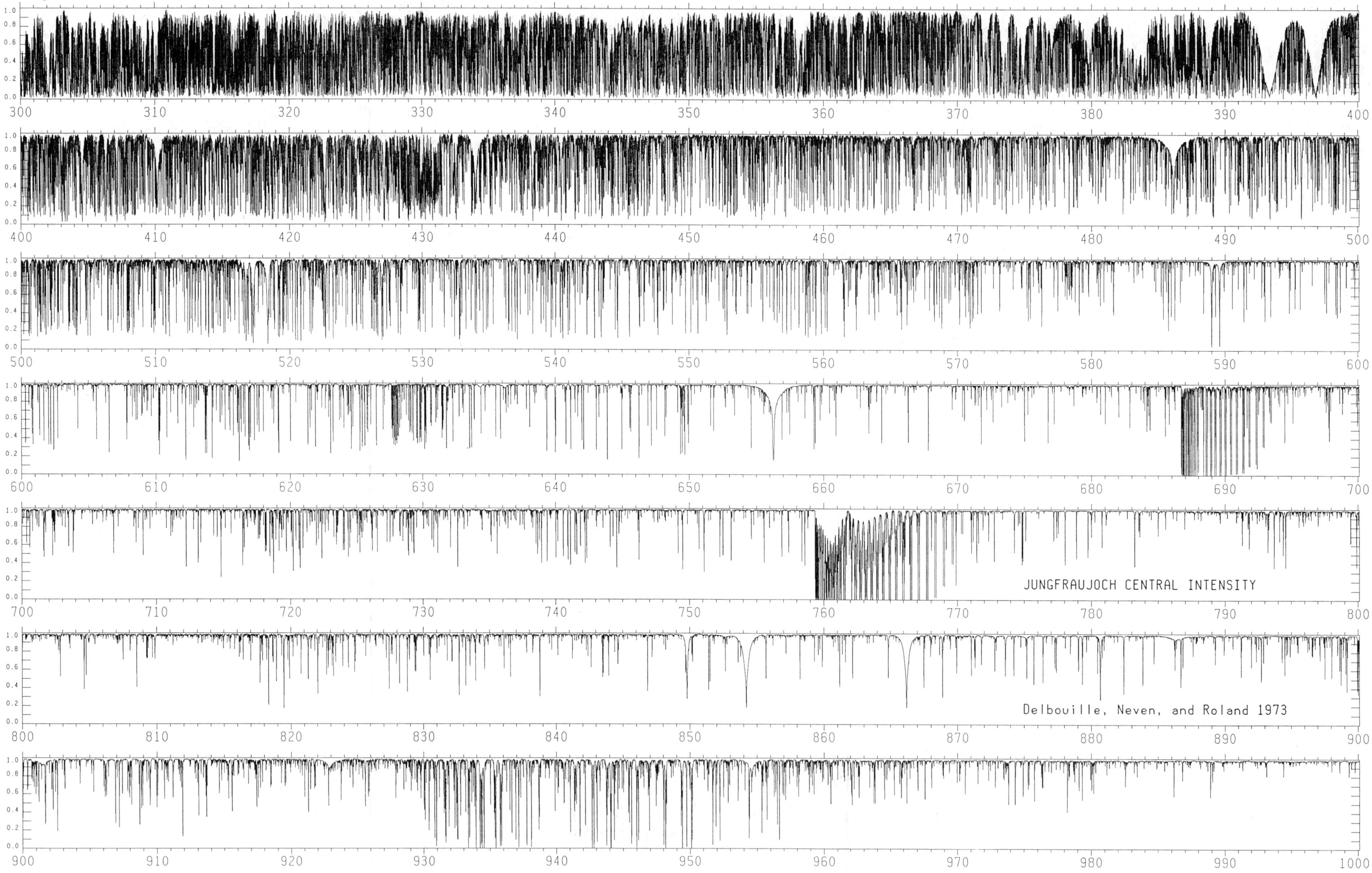

JUNGFRAUJOCH CENTRAL INTENSITY

Delbouille, Neven, and Roland 1973

Figure 3.

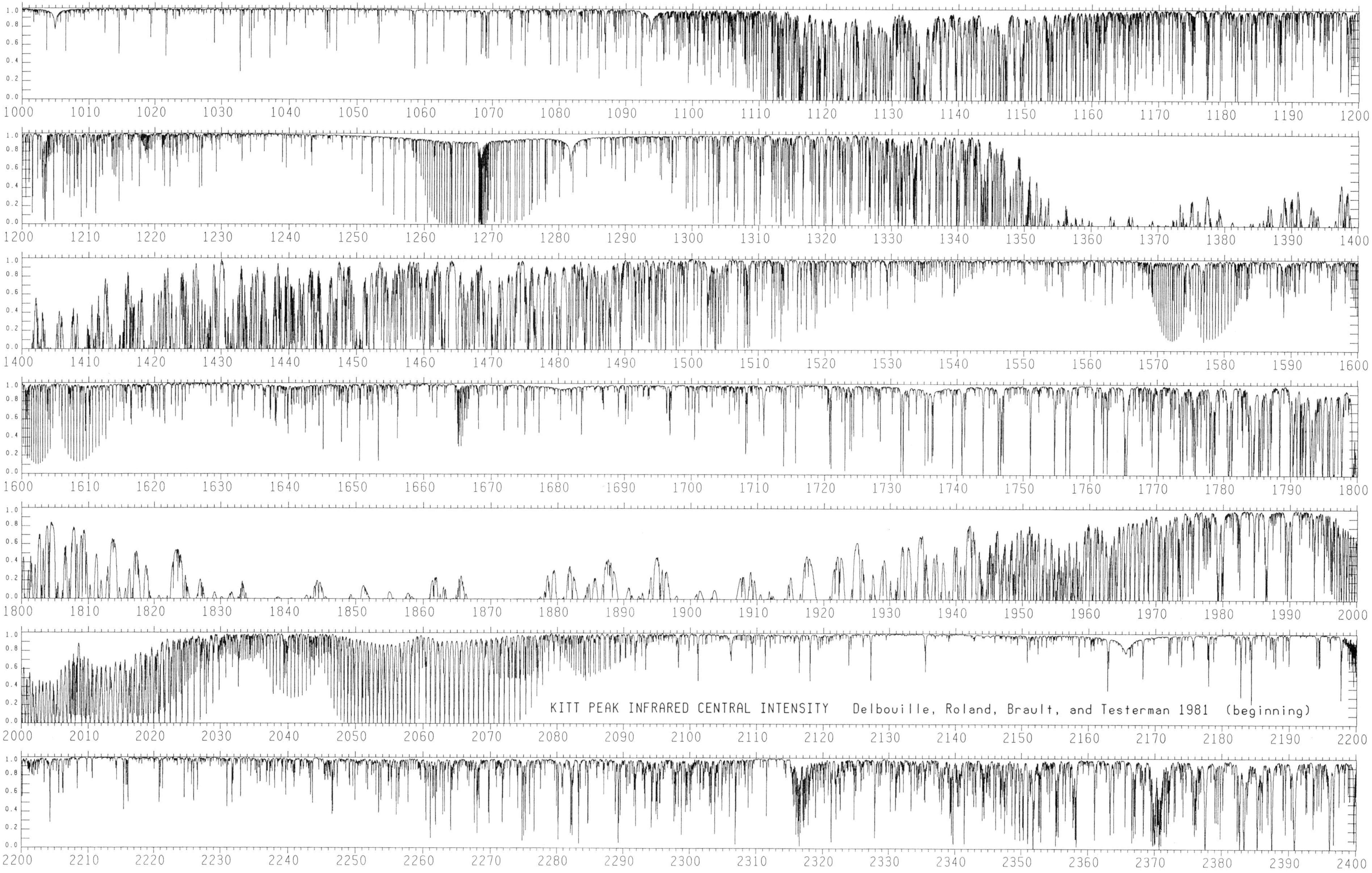

KITT PEAK INFRARED CENTRAL INTENSITY   Delbouille, Roland, Brault, and Testerman 1981   (beginning)

Figure 4.

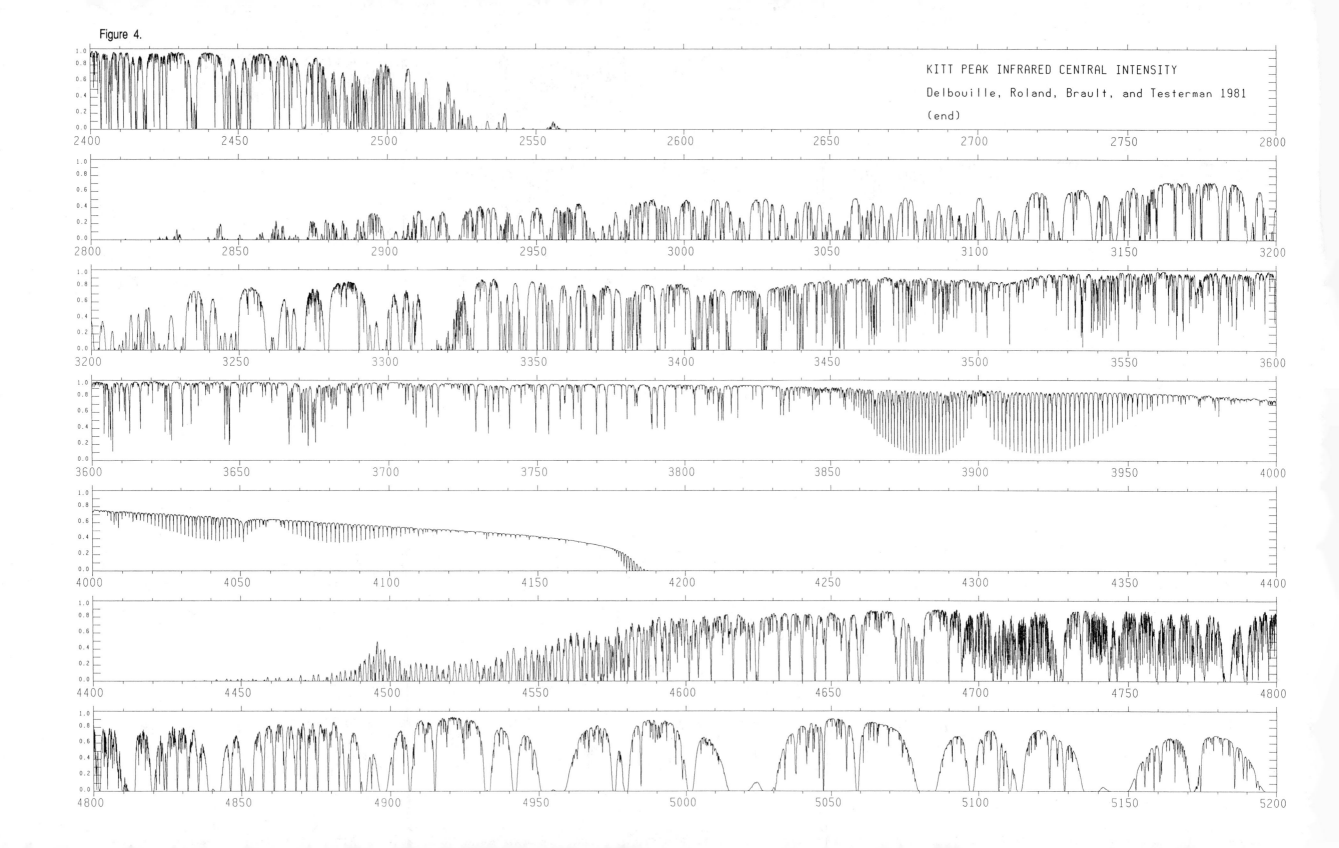

KITT PEAK INFRARED CENTRAL INTENSITY

Delbouille, Roland, Brault, and Testerman 1981

(end)

Figure 9.

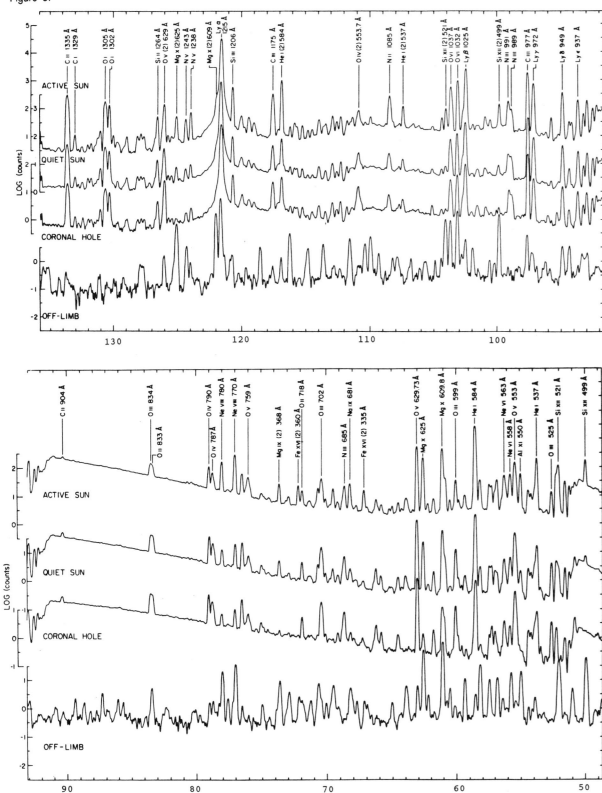

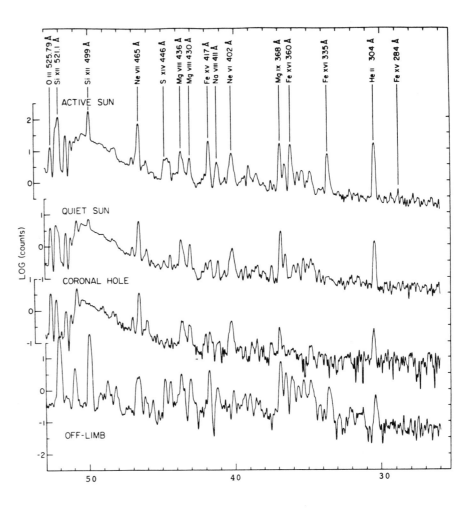

Composite solar spectra in the far ultraviolet
from 26 to 136 nm determined by Vernazza and Reeves (1978)
from Spacelab data. While the plots show only log counts, the
paper lists integrated intensities for individual features.

stants under different physical conditions because the wings of strong lines are formed higher in the atmosphere. Molecular lines are stronger in the limb spectrum.

The Kitt Peak infrared central intensity atlas by Delbouille et al. (1981), *Photometric Atlas of the Solar Spectrum from 1,850 to 10,000 cm⁻¹*, is the best available spectrum in the infrared. It is a residual intensity atlas illustrated in Figs. 3 and 4. [A paper copy can be obtained from the National Solar Observatory in Tucson. Instructions for purchasing a magnetic tape copy are given in the printed atlas.] It is the combination of 9 FTS scans on the McMath telescope with resolving power about 4×10^5 at 1 μm decreasing to about 13×10^4 at 5 μm. The signal-to-noise ratio varies from 3200 to 5200. There is a zero point error of up to 0.5% as in the FTS flux atlas. Delbouille and Roland are redoing the atlas from Jungfraujoch to improve the resolution and the signal-to-noise ratio and especially to reduce the water-vapor absorption. They will also produce a limb spectrum.

The JPL ATMOS experiment was flown on the Space Shuttle to obtain infrared FTS spectra of the atmosphere at sunset from which to measure trace molecules. Before sunset, solar intensity spectra were recorded. Farmer and Norton (1989) have produced an atlas of averaged solar and atmospheric scans, *A High-Resolution Atlas of the Infrared Spectrum of the Sun and the Earth Atmosphere from Space*. Wavelength coverage is 2 to 16 μm, resolution is 0.0147 cm⁻¹, and signal-to-noise contrast varies from 1000 to 3000. Additional problems are contamination lines produced by outgassing and background modulation by bandpass-limiting filters. Figures 5 and 6 show the atlas plotted both against wavelength and against wavenumber. Comparison with Fig. 4 shows the value of space observation. [Paper copies are available from the Government Printing Office. A magnetic tape or CD-ROM is available from the National Space Science Data Center at Goddard Space Flight Center. The original spectra are also available from NSSDC on 23 tapes.] Eventually a version with the contamination lines removed will be produced by the author. (See the chapter on the infrared by Deming in this book for a discussion on the longer-wavelength observations.)

The central intensity spectra taken by the Ultraviolet Spectrometer and Polarimeter on the Solar Maximum Mission spacecraft (SMM/UVSP) are the best in the ultraviolet but will not be ready for publication until sometime in 1990. Figure 7 shows the region from 129 to 177 nm that has been reduced by Shine et al. (personal communication, 1989) at Lockheed. [I plan to have paper and magnetic tape copies for the region from 180 to 300 nm, part of which is shown in Figure 8, at a later date.] The instrument failed before the spectrum was completely scanned so there is a gap between the two sections. Limb spectra were also taken. The resolution and signal-to-noise ratio are low by visible standards but are still much better than previous work in the ultraviolet. Better spectra are not likely to be produced in the near future.

The plan in order to produce a continuous spectrum is to fill in gaps in the SMM spectra using Harvard data (Kohl et al. 1975), and data from NRL (Moe et al. 1976) and OSO-8 (Chipman, personal communication, 1981).

There are three atlases available now in the 220 and 300 nm region. One is the Harvard rocket atlas, *Center and Limb Solar Spectrum in High Spectral Resolution: 225.2 to 319.6 nm* (Kohl et al. 1978). [This atlas can be obtained on paper or magnetic tape from Kurucz.] It is very much noisier than the SMM spectra. It has line identifications from a computed spectrum by Kurucz and Avrett (1981). The NRL atlas, *An Atlas of the Solar Ultraviolet Spectrum between 2226 and 2992 Angstroms* by Tousey et al. (1974) is similar in resolution to the Harvard spectrum but is photographic. There is no magnetic tape. The third atlas is the *High Resolution Atlas of the Solar Spectrum 2678–2931 Å* by Allen et al. (1978). [It is available from the Institute for Astronomy at Hawaii.]

Figure 9 from Vernazza and Reeves (1978) shows the spectrum at shorter wavelengths from 136 nm to 26 nm taken from Skylab. The resolution is only 0.16 nm FWHM so the lines are far from being resolved (cf. Fig. 7) and weak features cannot be seen. In this wavelength interval the spectrum depends to some extent on the level of solar activity so detailed features may vary with time.

Other spectra not discussed here are observations at high spatial resolution from the NRL High Resolution Telescope and Spectrograph (HRTS) (Brueckner 1980a) and observations of features such as flares (Cohen 1981) and sunspots (Hall 1974).

V. LINE IDENTIFICATIONS

One half of the lines in the solar spectrum are not identified. The best source for line identifications in the visible is still the Rowland Table, *The Solar Spectrum 2935Å to 8770Å,* by Moore et al. (1966). Improved identifications in the 3000 Å region have been made by Mitchell and Mohler (1969). In the ultraviolet, Moore et al. (1982) have extended the identifications down to 2095Å using NRL photographic spectra (Tousey et al. 1974). Line identifications in the near infrared are given by Swensson et al. (1970), *The Solar Spectrum from λ7498 to λ12016.* There are a number of papers by Biemont et al. (1985), for example, where he identifies infrared lines of iron group atoms.

The Sac Peak flux atlas (Beckers et al. 1976) was the predecessor to the Kitt Peak flux atlas. The Kitt Peak atlas has higher resolution and higher signal-to-noise ratio and greater wavelength coverage, but the printed copy of the Sac Peak atlas does have the advantage of line labels from the Rowland Table.

The author plans to publish or republish atlases with the lines labeled, including terrestrial lines from the AFGL HITRAN line list (Rothman et al.

1987). Each spectrum is being synthesized and one should be able eventually to deconvolve the blends and to deconvolve the atmospheric transmission where it is not near zero.

Acknowledgment.   This work was supported in part by a grant from the National Aeronautics and Space Administration.

THE PHOTOSPHERE AS A RADIATIVE BOUNDARY

LAWRENCE S. ANDERSON
University of Toledo

and

EUGENE H. AVRETT
Center for Astrophysics

We review the role of the photosphere as a radiative boundary for the solar interior, concentrating on semi-empirical and a priori models for the structure and emergent radiation. For concreteness, we define the photosphere as the region between a lower boundary where the optical depth at the 1.6 μm opacity minimum equals 3 and an upper boundary at the temperature minimum. We stress that any discussion of structures within the photosphere having sizes of roughly 300 km or less is strongly compromised by the radiant integration over these scale lengths. We show that plane-parallel models are leading to the conclusion that Boltzmann-Saha statistics are reasonably valid throughout the photosphere, except within about 100 km of the temperature minimum. Further, the semi-empirical models depart from radiative equilibrium by being too cold at the temperature minimum and (possibly) too hot in the middle photosphere. This excess temperature gradient is consistent with an intensity-weighted averaging over convective elements. We discuss the observational evidence for horizontal inhomogeneities and their characteristics, including granular fluctuations, fluctuations in the brightness of the temperature minimum region, and the recent demonstration of the presence of relatively cold, carbon monoxide absorbing regions. We conclude by relating these inhomogenities and the radiative imbalance revealed by the plane-parallel studies to the dynamic models of Nordlund.

I. INTRODUCTION

A. Emphasis

In this review, we concentrate on recent advances and lingering problems in the understanding of physical conditions in the photosphere. We further concentrate on the quiet photosphere as the equilibrium radiative surface of the Sun, which is only slightly perturbed by active regions that have their origins in mechanisms only indirectly related to this surface. Nevertheless, when appropriate we will occasionally touch on active regions and the low chromosphere in order to provide connections to other chapters in this book. Velocity fields are fundamentally related to both the thermodynamic equilibrium and the interpretation of diagnostics in the photosphere and chromosphere. However, we will only discuss them incidentally (see the chapter by Howard et al.).

An important issue in the understanding of the quiet photosphere and chromosphere is the application of what we learn from this spatially resolvable laboratory to the unresolved stars. Stars are very effective at erasing the histories of their origins through a variety of dissipation mechanisms. All that remains are the integral properties of mass, angular momentum and the evolving chemical composition. Ultimately even the "hair" on stars, i.e., chromospheres, coronae and surface activity, must arise from these properties. While the emergence of a particular local feature may be determined by chance and chaos, one might hope to learn how the properties of the ensemble are generated by the star. To help us in this understanding, we need to know how chromospheres, surface activity and even convection itself are structured on other stars. These phenomena imprint spectroscopic, time varying signatures on the star's radiation output. By observing the Sun and constructing detailed models, we can relate the signatures to the phenomena. Not only do we have to identify the signatures of activity, we also must improve our ability to determine the fundamental stellar parameters to which we wish to relate the activity. Understanding the radiation output from the quiet photosphere contributes to our knowledge of both the fundamental parameters and of the spectrum on which the signatures of activity are imprinted.

Besides the standard journal literature, there are at this time a number of useful review chapters in books and seminal papers in workshop proceedings. We will refer to some of these works at appropriate places in the body of our chapter, and here only mention ones which touch broadly on many aspects. The chapter "Radiation Output" by Athay (1986b) is an outstanding guide to the understanding of photospheric radiation (effectively a mini-course in stellar atmospheres). The paper "The NLTE Formation of Iron Lines in the Solar Photosphere" by Rutten (1986) is a fine review of the role (and hidden pitfalls therein) played by assumed structural models in the interpretation of emergent spectra. The paper "The Fine Structure of the Quiet Sun" by Muller

(1985) is a good review of the relations between granulation, network bright points, line profiles and models.

We organize the rest of this review as follows. First we present the average physical conditions found in the photosphere and describe the characteristics of various atmospheric components. Next we discuss plane-parallel models and their interpretation, including both empirical and *a priori* models (Sec. II). In Sec. III, we review the observed departures from plane-parallel structure and in Sec. IV, we compare and contrast results from dynamical convection models with the plane-parallel work. Finally, in Sec. V we return to the question of the validity of plane-parallel models and the diagnostics of nonradiative and nonplanar structures in stars.

B. Role of the Photosphere as a Radiative Surface

Average Characteristics.   The quiet photosphere is the radiative surface from which the Sun loses nearly all of its energy. The chromosphere ordinarily radiates about 2×10^{-4} of the solar luminosity (Anderson and Athay 1989b; see the chapter by Kalkofen), and even at the maximum of the spot cycle, active regions affect no more than a few percent of the luminosity. The corona radiates a far smaller fraction than the chromosphere. As our knowledge of the structure of the solar atmosphere improves, it becomes increasingly difficult to define the boundaries of the photosphere. Perhaps it is best to keep the definition intuitive. For the purposes of this chapter, we will set the upper boundary at the temperature minimum and the lower boundary at the point where the vertical optical depth in the 1.6 μm opacity minimum is 3. In the real Sun, the surfaces defined by these parameters are far from level. However, plane-parallel models provide rough estimates of the range of physical conditions found between these two surfaces. In Table I we

TABLE I^a
Average Photospheric Properties

Quantity	At $\tau_{1.6\mu} = 3$	At $\tau_{0.5\mu} = 1$	At T_{min}	Units
Column mass	6.0	4.4	0.06	g cm^{-2}
Height above $\tau_{0.5\mu} = 1$	-53	0	520	km
Optical depth at 0.5μ	4.7	1.0	3×10^{-4}	
Temperature	8000	6520	4400	K
Density	3.1×10^{-7}	2.8×10^{-7}	5.0×10^{-9}	g cm^{-3}
Pressure	1.6×10^5	1.2×10^5	1.5×10^3	dyn cm^{-2}
Pressure scale height	190	150	110	km
Electron density	8×10^{14}	8×10^{13}	3×10^{11}	cm^{-3}
Electrons per atom	6×10^{-3}	6×10^{-4}	1×10^{-4}	
d ln T/d ln P	0.54	0.48	0.00	

^{a}Table from model by Maltby et al. (1986).

list some of these conditions (Maltby et al. 1986); in Sec. II we present more detail.

The emergent intensity on a ray path s is given by the formal solution of the equation of transfer: $I_\nu^{em}(s) = \int_0^\infty S_\nu \exp(-\tau_{s,\nu}) d\tau_{s,\nu}$. Here S_ν is the source function at optical depth $\tau_{s,\nu}$ backwards along the ray and $d\tau_{s,\nu} = k_\nu dx$, where k_ν is the extinction (per cm) at the geometrical distance x along the ray. In the Eddington-Barbier approximation (S_ν, a linear function of $\tau_{s,\nu}$), I_ν^{em} is equal to the source function at the position x where $\tau_{s,\nu} = 1$. Roughly 75% of I_ν^{em} comes from optical depths $0.1 < \tau_{s,\nu} < 2.0$ if the source function is not strongly varying with position (cf. Edmonds 1962). Then, if the extinction varies exponentially along the ray (a good approximation in a plane-parallel atmosphere) with a scale length L so that $k_\nu = k_0 \exp[(x - x_0)/L]$, the optical depth range 0.1 to 2.0 covers about $3L$. In a plane-parallel atmosphere with constant opacity per gram of material, L is the density scale height H_ρ divided by the zenith angle cosine μ of the ray. Thus, the region of 75% contribution extends over the same number of *vertical* scale heights at all zenith angles.

The Eddington-Barbier approximation is reasonably accurate at most frequencies (the Hα line is an exception). The principles outlined in the previous paragraph have been used by many astronomers over the years to establish empirical models for the structure of the solar atmosphere. They were used by Münch (1945) to show that H$^-$ produces the visible continuum in the photosphere. More recently, Scharmer (1981,1984) has based an entire operator pertubation scheme for the solution of line-transfer problems on them. We wish to emphasize here that these simple principles define crucial lengths of the order of 300 km in the empirical determination of local *source functions* in the solar atmosphere. Thus, without sophisticated extraction techniques, we will never be able to resolve structural elements smaller than this size. It is also inappropriate to model smaller structures as composed of independent components with various surface-filling factors. Finally, if we desire to associate the source function with real physical parameters to be measured (e.g., temperature, density, magnetic field), we are limited further by scattering (cf. Kneer 1980). We will return to this problem in Sec. V.

Structural Elements. In the quiet photosphere, there are three well-documented horizontal structures of rather different scale and contrast. These structures, in order of decreasing size, are supergranulation, granulation and magnetic structures. Supergranulation, mainly visible through horizontal flow patterns of order 500 m s^{-1} (and the calcium network in the chromosphere), has a scale size of $\sim$30,000 km, lifetime of order one day, and very low contrast in brightness (cf. Beckers 1968; Worden 1975). Granulation, mainly visible through its brightness contrast of $\sim$20%, has a scale size of $\sim$1400 km (but see Roudier and Muller 1986), lifetime of about 15 min (but correlation lengths and times may be longer in the case of granule 'families' [Kawaguchi 1980]), and tyical vertical velocities of $\sim$1 km s^{-1} and horizontal velocities of ~ 2 km s^{-1}. The velocities are strongly correlated with the brightness. Magnetic structures outside of sunspots appear to be composed of differing concentrations of rather similar magnetic flux tubes or filaments (cf. Stenflo and Harvey 1985). The similarity is presumably due to the convective-collapse mechanism of their origin (Parker 1978; Spruit and Zweibel 1979a) which leads to a pressure equilibrium between the field and the external gas. Typical field strengths are ~ 1500 G. The horizontal scale size of these flux tubes is probably ~ 150 km; however, the determination of tube parameters is very model dependent. The properties of granulation and supergranulation are reviewed by Bray et al. (1984); see also the chapter by Topka and Title. The nature of photospheric magnetic structures is reviewed by Solanki (1987) and Schüssler (1987b); see also Nordlund (1985d) and the chapter by Spruit et al. We will not discuss magnetic structures further in this chapter.

Recent high-resolution observations and image processing techniques have revealed another scale of convection, intermediate between granulation and supergranulation. Called mesogranulation, it is most easily seen by following "test particles" set to move with the granulation pattern. It has mean sizes of ~ 7000 km, lifetimes of several hours, and horizontal velocities of ~ 50 m s^{-1}.

Convection remains the outstanding unsolved problem in photospheric physics, although much progress has been made in recent years. Below our lower boundary surface, convection carries most of the solar flux. In the first scale height above this surface, convection is replaced by radiation as photons are able to reach the optical boundary and escape to black space. The conversion of the flux carrier is conceptually straightforward. In the absence of magnetic fields, viscosity and conductivity, one may write the average vertical luminosity of energy across surface element A at depth z as $AF = \Delta(av_z 1/2\rho v/2) + \Delta(av_z \rho h) + AF_{\mathrm{rad}}$, where the Δ represents the difference between mean upward moving areas a_u and downward moving areas a_d (and $a_u + a_d = A$), $1/2\rho v/2$ is the nonthermal kinetic energy density, and ρh is the enthalpy density. Mass conservation leads to $\Delta(av_z \rho) = 0$. The second term, advection of enthalpy, carries most of the convective flux. In a diffusive environment, flux will be transported from higher to lower temperatures in proportion to the temperature gradient. If the mean free path of photons is short, then the enthalpy will rise until convection can carry the flux. But, near the optical surface of a star, the mean free path of photons rises toward infinity, and any desired flux can and will be carried by radiation. Then the convective components of the flux are free to vanish, even if the atmospheric gradient $d\ln T/d\ln P$ remains superadiabatic. The enthalpy density ρh will decrease upward exponentially with a scale height set by the characteristic temperature of the radiation.

The conversion of convective flux to radiative flux, while conceptually simple, is computationally difficult. At what actual value of the temperature and density this conversion takes place is a function of the flux, the surface gravity, the adiabatic index, the opacity and the structure of the convection itself. We believe that for a given composition, the first two quantities determine the last three. However, no one has shown convincingly the physical source of any of the scales of convection. Thus, for a given flux, the photospheric boundary condition on a convective envelope (the location of the flux conversion), which determines the stellar radius, is nontrivial, unlike the radiative zero condition on radiative envelopes. In Sec. IV we discuss dynamical models of convection and their relation to plane-parallel models of the photosphere.

The photosphere plays the additional roles of source, sink and reflecting boundary for a variety of acoustic, magneto-acoustic and buoyancy mode oscillations. These oscillations are reviewed in other chapters of this book. Here, we only state that the structural and radiative properties of the background photosphere are integrally related to these modes, on scales that range from the microscopic to several thousand kilometers.

II. PLANE-PARALLEL MODELS

In this section we review the status of plane-parallel models for the solar photosphere. There are two basic approaches. The first is the so-called *semi-empirical* approach, wherein one uses the spectral information from a real object, coupled with radiative transfer theory, to derive the depth dependence of physical parameters independent of any global energy equilibrium. The second is the so-called *a priori* approach, wherein one attempts to include enough physics to be able to calculate from first principles an emergent spectrum which matches observations. Historically, the former has been applied most frequently to the Sun, where center-to-limb variations and horizontal inhomogeneities are measurable, and where observations show that our knowledge of physics is (or was) woefully inadequate to account for structural details from first principles. Of course, the solar example has not deterred astronomers from using the *a priori* approach to analyze stellar spectra, where such confusing detail is hidden in the integration over the stellar disk. Integrated sunlight and a comparison of semi-empirical and *a priori* models for the Sun provide the connecting link, and show that the stellar astronomers are often justified in extracting gross information from *a priori* models. In recent times, an interesting reversal is developing. With better detectors, telescopes and satellite observations, we are able to see the failings of simple *a priori* models in stellar spectra, while improved physics and computers are leading toward realistically detailed *a priori* models for solar phenomena. It is likely that over the next decade or two we will see a grand synthesis of solar and stellar modeling. Avrett (1989) provides further discussion of the detailed differences between various solar models.

A. Equations

The equations governing a plane-parallel atmosphere derive from the conservation of mass, momentum and energy. We choose as the independent depth variable the column mass $m(z) = m(0) + \int_0^z \rho dz'$ grams cm^{-2}, where z is the geometrical depth measured inward in the normal direction from the boundary at $z = 0$ and ρ is the mass density. In the static plane-parallel case conservation of mass and momentum leads to the hydrostatic equilibrium equation

$$\frac{d}{dm} P = g \tag{1}$$

where g is the acceleration of gravity and P is the total pressure. We can write $P = \rho kT/\bar{\mu} + \rho v_t^2/2$, where $\bar{\mu}$ is the mean particle mass and the last term is the contribution to the pressure from small-scale mass motions (microturbulence).

The conservation of energy is usually expressed as

$$\frac{d}{dm} F = \frac{d}{dm} (F_{rad} + F_{conv} + F_{MHD}) = 0. \tag{2}$$

We have ordered the fluxes by decreasing value (although not decreasing divergence; what is lost in one must be gained in the others) and decreasing understanding. The radiative flux dominates all but the deepest layers of the photosphere. Its divergence is given by

$$\frac{d}{dm} F_{rad} = \int_0^\infty \int_{4\pi} \kappa_\nu (I_\nu - S_\nu) d\Omega d\nu. \tag{3}$$

Here $\kappa_\nu = k_\nu/\rho$ is the opacity in cm^{2}g^{-1} and I_ν is the specific intensity determined from the equations of transfer

$$\mu \frac{d}{dm} I_\nu(\mu) = \kappa_\nu (I_\nu - S_\nu). \tag{4}$$

Both κ_ν and S_ν are functions of $I_{\nu'}$, T and P. Together with statistical equilibrium equations for the population fractions of atoms in their various energy states, Eqs. (1–4) form a complete set for the physical state of the atmosphere as a function of depth.

B. Empirical Models

Empirical modeling of the solar atmosphere has enjoyed a long history. Equations (1) and (2) are global equations which relate the dependent variables $P(m)$, $T(m)$, and $I_\nu(\mu, m)$. In particular, they relate the intensities at one frequency to the intensities at all other frequencies. The numerical enormity of that problem, plus the difficulty of finding plane-parallel expressions for F_{conv} and F_{MHD}, has led workers to sidestep the energy equilibrium equation in favor of an empirical determination of the material energy content, as measured by $T(m)$. The procedure, then, is to start with a simple model (e.g., radiative equilibrium, Boltzmann-Saha statistics and the diffusion approximation $T^4(m) = 0.75T^4_{eff}[\bar{\tau}(m) + 2/3]$ where $d\bar{\tau} = \bar{\kappa}dm$, and $\bar{\kappa}$ is a suitably chosen average opacity) and adjust the temperature structure until $I^{em}_\nu(\mu)$ matches the observations at as many frequencies and zenith angle cosines as possible. The uniqueness and smoothness of the solution are assured by the Eddington-Barbier approximation discussed above. In order to determine the emergent intensities, one still needs to solve equations of transfer and statistical equilibrium. However, one only needs to solve the transfer equations at the frequencies to be used in the empirical comparisons, and the statistical equilibrium equations for the atoms which contribute to the opacity and source function at those frequencies. In general, these frequencies form much reduced and independent sets.

Three frequency domains are particularly useful for empirical modeling. The first is the H$^-$ continuum from its peak opacity at 800 nm to the total opacity minimum at 1.6 μm, and the H$^-$ free-free continuum at wavelengths longer than 1.6 μm. In principle, this region can sample temperatures throughout the layers we have chosen to call the photosphere. The free-free (bremsstrahlung) continuum eventually reaches unit optical depth at the temperature minimum at wavelengths of about 150 μm. Highly thermalized (Maxwellian) electrons are the source of the free-free continuum, so S_ν is the Planck function B_ν and directly measures T. Until recently, however, detectors of such longwave radiation have not been sufficiently sensitive to measure brightness temperatures to the required accuracy. Figure 1 shows continuum brightness temperatures for the average quiet photosphere in the near infrared at disk center compared with the values calculated from a recent empirical model from Maltby et al. (1986). The far infrared is discussed by Mankin (1977), Vernazza et al. (1976), Degiacomi et al. (1985) and Avrett (1985).

The second frequency range commonly used for empirical modeling is over the 4s ^{2}S $-$ 4p ^{2}P resonance multiplet of ionized calcium. This feature samples depths from high in the chromosphere down to within about 50 km of our lower boundary, as one moves out on the wings of the absorption profile. Throughout the photosphere (in plane-parallel models where the closest optical boundary is in the vertical direction), the transition is thermalized; that is, line photons are sufficiently trapped by the high opacity to prevent escape before collisional processes destroy them. Again, $S_\nu = B_\nu$ so that the intensity directly measures T. Near the temperature minimum, however, line photons *are* capable of scattering with redistribution from the core to the transparent wings in sufficient numbers to alter the source function. Care must

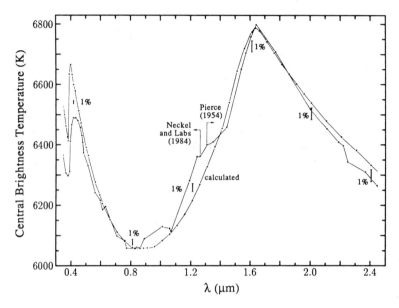

Fig. 1. Continuum brightness temperature distribution at disk center from the Maltby et al. (1986) reference model compared with the highest observed maxima from Neckel and Labs (1984) and Pierce (1954); see Maltby et al. (1986) for details.

be used to treat this redistribution accurately, particularly in the interpretation of the variation of I_ν^{em} with μ (cf. Shine et al. 1975).

The third frequency range which probes the photosphere is the ultraviolet pseudocontinuum. Disk center brightness temperatures vary from a maximum ~6500 K near 300 nm to a minimum ~4400 K near 160 nm. Shortward of 162 nm, the Mg I and later Fe, Si and C I ionization continua reach optical depth unity above the temperature minimum. The Harvard-Smithsonian Reference Atmosphere (Gingerich et al. 1971) was based on this frequency range. The ultraviolet is difficult to interpret for two reasons. First, photons scatter more with decreasing wavelength; the collisional quenching is roughly proportional to ν^{-3} (Anderson 1989). Second, the continuum is dominated by the wings of countless weak and strong lines. The lines in turn have two effects. They contribute directly to the emergent spectrum, and line radiation dominates photoionization rates and so sets the ionization balance. We are coming to grips with both problems. R. Kurucz has recently finished calculating atomic data for 1.9×10^7 lines in the first three spectra of Ca-Ni, and Anderson (1989) has devised a scheme for solving the associated statistical equilibrium equations. Avrett et al. (1984) presented new models for the temperature minimum region based on Kurucz' new data. The presence of the additional line opacity effectively traps the continuum photons and ensures their thermalization to a much higher degree than was the case in previous models (see e.g., Vernazza et al. 1981). Detailed comparisons of these new models with observations are given by Avrett (1985).

Thus, it now appears that in the photosphere, metal continua (and thus ionization) are driven toward Boltzmann-Saha statistics by the action of line opacities, and the photospheric wings of strong lines are driven to Boltzmann-Saha statistics by their own opacity. Quoting Rutten (1986): "With enormous computational sophistication, plane-parallel modeling of the solar photosphere is back again to simple LTE!" (Local thermodynamic equilibrium, or LTE, is the assumption of Boltzmann-Saha statistics; an LTE empirical reference atmosphere is that of Holweger and Müller [1974].) Problems still remain to be addressed. In the Avrett et al. (1984) models and the reference photosphere of Maltby et al. (1986), all opacities for wavelengths shorter than 386 nm have been augmented by a factor of 1.2 to compensate for lines that were estimated to be missing from the Kurucz line list available in 1984. As this chapter is being written, Kurucz is completing work on a more comprehensive list that appears to match the opacity throughout the ultraviolet. The ultraviolet line haze (a word used by Rutten to describe the background line opacity not explicitly included in the Harvard group's statistical equilibrium calculations) is presently calculated with Saha-Boltzmann statistics and arbitrarily given a depth-dependent scattering albedo. Finally, it is not clear to what extent a plane-parallel model derived from surface-averaged light represents true averages of T and P over surfaces of constant m. We will return to these problems below.

C. A Priori Models

The main reason why *a priori* models are preferred by stellar astronomers is one of expedience. There are too many stars to model each one empirically. It is much easier to create an atlas of stellar spectra from first principles, and locate observed stars in that atlas. Plane-parallel *a priori* models of the Sun include those of Bell et al. (1976), Kurucz (1979) and more recently Anderson (1989) and Anderson and Athay (1989b). The first two works employed local mixing-length theory, wherein F_{conv} at depth m is a function of the superadiabaticity of d $\ln T$/d $\ln P$ at m (for a review of convection theory, see Bray et al. [1984]), and Boltzmann-Saha statistics for κ_ν, S_ν and the electron density. A variety of nonlocal plane-parallel convection theories have been proposed over the years, but all suffer from an *ad hoc* treatment of the nonlinear processes responsible for photospheric convection. A true *a priori* understanding of the conversion of F_{conv} to F_{rad} must incorporate horizontal inhomogeneity.

Anderson (1989) chose to ignore convection in the photosphere on the grounds that its effect on the average temperature structure is smaller than or equivalent to the possible error in the plane-parallel approximation itself. Instead, he concentrated on treating the radiative transfer and statistical equilibrium as completely and accurately as possible. He mapped the millions of transitions calculated by Kurucz onto model atoms consisting of a small number of "diffuse" excitation states, and treated them as transition multiplets with constant profiles and source-function equality. Thus, at least statistically, electron cascades and Raman scattering are accounted for, both in the source functions and in the opacities. This radiative equilibrium solution once again recovers Boltzmann-Saha statistics throughout most of the photosphere ($m > 0.03$ g cm^{-2}), for the same reasons the recent empirical models do. S_ν departs from B_ν by at most 0.3 B_ν, for Fe I transitions at depths corresponding to the temperature minimum. While Anderson has not yet explored the departures for high excitation lines, this result begins to validate the Harvard group's use of LTE opacities for the iron line haze.

In the vicinity of the photospheric temperature minimum, Anderson's radiative statistical model is about 100 K warmer than the equivalent model calculated with Boltzmann-Saha statistics. There are many competing reasons for this difference, but the dominant one is that green Fe I lines, arising from the 4s a^{5}F configuration, have sufficiently reduced source functions to alter the thermal equilibrium. In addition, the slight excess of ultraviolet radiation does perturb the ionization in favor of the ionized metal and H$^-$ heating. The temperature rise is augmented by the concomitant destruction of carbon monoxide and the loss of its cooling. The overall result is that the empirical reference model for the average photosphere of Maltby et al. (1986) is some 200 K cooler than radiative equilibrium at the temperature minimum. Both models appear in Fig. 2a. Anderson and Athay (1989b) have shown that equilibrium temperatures at the temperature minimum are not strongly affected by the presence of the average chromosphere. Thus, the *a priori* model confirms the need for some process of nonradiative refrigeration in the high photosphere as suggested by Avrett (1985). Carbon monoxide by itself cannot provide this cooling without a reduced F_{rad}. Nordlund (1985b) suggests that the cooling is simply a result of granulation hydrodynamics. The surface averaged F_{rad} is dominated by contributions from hot, rising granules overshooting into the stable upper photosphere. These granules must expand as they rise in the stratified atmosphere, and in expanding attempt to cool adiabatically (i.e., d$[\Delta(av_2\rho h)]$d$m < 0$) more than the radiation field allows.

The Maltby et al. (1986) reference photosphere is consistently *hotter* than Anderson's (1989) radiative equilibrium model in the middle photosphere, $0.3 < m < 5.7$ g cm^{-2}. It is possible that this difference is again due to granulation advection. The average gradient d ln T/d ln P is strongly superadiabatic at depths $m > 4$ g cm^{-2}, and granules rising into the low photosphere have a lot of residual heat recently extracted from the recombination of hydrogen. This interpretation is consistent with the net radiative losses shown by Avrett (1985). However, small differences in the mean opacity between the models can also account for the temperature differences by shifting the optical depth scale relative to the column mass scale. The empirical model is able to "see" the conversion of F_{conv} into F_{rad} at $m \sim 6$ g cm^{-2}; beyond this

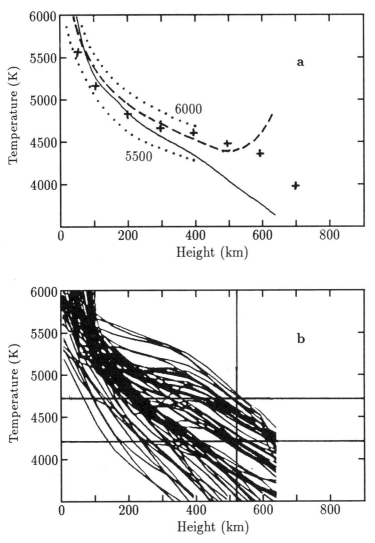

Fig. 2. (a) A variety of average-temperature profiles for the solar photosphere. *Solid line:* horizontal average of numerical convection (Nordlund 1985*b*); *dashed line:* empirical quiet Sun (Maltby et al. 1986); *dotted lines:* LTE radiative equilibrium at the labeled effective temperatures (Kurucz 1979); *plus symbols:* non-LTE radiative equilibrium (Anderson 1989). (b) A selection of $T_{x,y}(z)$ from the numerical convection of Nordlund (1985*c*). Also shown are the temperature limits of T_{min} deduced from the ultraviolet and Ca II, and the height of T_{min} deduced from the empirical plane-parallel model in plot (a).

depth, the radiative equilibrium temperature gradient rises steeply with the opacity.

III. DEPARTURES FROM PLANE-PARALLEL STRUCTURE

Granulation is the most obvious manifestation of horizontal inhomogeneity in the solar photosphere, and granular cells probably carry most of the solar luminosity into the low photosphere. Thus, this phenomenon is the most important to understand *vis-à-vis* stellar atmospheres and the interior boundary condition. We also wish to understand the relations between supergranulation, mesogranulation, granulation, magnetic structures and chromospheric heating. In this section we review some of the observations that are inconsistent with the average plane-parallel model results.

A. Granulation Studies

The horizontal scale size of granulation is of order 10 to 15 pressure scale heights, so it should be reasonable to use plane-parallel elements to construct empirical models of $S_\nu(m)$, using ray paths anywhere except near the solar limb. The average plane-parallel models described in the previous section also suggest that $S_\nu = B_\nu$ over most of the photosphere, so one can translate S_ν into $T(m)$. Unfortunately, it is very difficult in practice to get spectrally complete (e.g., over a range of $\sim 10^4$ in κ_ν), temporally resolved (~ 10 s), and spatially resolved (~ 300 km) intensity measures. It is even difficult to get spatially resolved, simultaneous I_ν^{em} over one line profile out to the continuum. Even if were possible, one cannot (without many spacecraft scattered about the Earth's orbit) measures I_ν^{em} as a function of μ for the same granule.

Often observers sidestep these difficulties by measuring not the intensity as a function of frequency at one location, but the intensity as a function of location at several frequencies. The distinction is important, because the spatial correlation between frequencies is either lost or not made use of, except to simply identify the scale lengths and coherence of fluctuations emerging from one depth $m(\tau_\omega = 1)$ with another $m'(\tau_{\nu'} = 1)$. One is left with only the (spatial) spectrum of intensity fluctuations, which one tries to interpret as temperature fluctuations. However, $m(\tau_\nu = 1)$ is itself model dependent and sensitive to temperature, so it is not clear how accurately one can assign a depth to an intensity fluctuation, particularly if the temperature fluctuations are uncorrelated with depth. Nevertheless, it is reasonably certain that the *granular* temperature fluctuations show a very steep gradient in the low photosphere, and are virtually nonexistent above roughly one-half scale height above the continuum-forming layers (cf. Kneer et al. 1980; Durrant and Nesis 1981). We stress "granular" because the intensity fluctuations remain roughly constant as a function of depth; what happens is that they rapidly lose coherence with increasing height. Why the granular fluctuation does not persist into the middle photosphere, and why there are uncorrelated fluctuations there is not obvious. Layers above both granules and intergranule regions should be roughly in radiative equilibrium with the fluxes moving through them, which differ in characteristic temperature by ~500 K. Figure 2a shows radiative equilibrium, LTE solutions for atmospheres with $T_{eff} = 5500$ and 6000 K and $\log g = 4.5$ from Kurucz (1979). We have increased the height scale by 15% to account for the solar gravity. Not until one reaches heights of several hundred km will the radiative-flux fluctuations be smoothed out through horizontal radiative transfer. The height of flux homogeneity may be reduced if the surfaces of constant optical depth slope in toward the intergranule region sufficiently. More importantly, divergent velocity fields associated with granulation or buoyancy modes contribute heating or cooling terms to the local flux divergence through the associated enthalpy divergence $\nabla \cdot (\rho \mathbf{v} h)$. We will return to this idea in Sec. IV.

B. The Temperature Minimum

A relatively wide variety of brightnesses is observed in the radiation originating from near the temperature minimum (cf. the Ca II H line data of Cram and Damé [1983], and the 160 nm data of Cook et al. [1983] and Foing and Bonnet [1984]). The ultraviolet fluctuations appear to be composed of granular bright points about 1 to 2 Mm across and separated by similar distances. The brightest points are coincident with the calcium network outlining the supergranulation cells and are presumably magnetic in origin. The cell-interior bright points are fainter and are organized in larger structures about 5 to 10 Mm across. The time-resolved calcium spectra of Cram and Damé (1983) show similar bright points in the cell interiors, and indicate that the larger scale structure is roughly oscillatory with periods consistent with chromospheric p-modes (3 to 5 min). Again, it is probably all right to model this emission with planar geometry. Cram and Damé (1983) divide the calcium data into equal-area bins based on intensity; the brightness range is relatively evenly spread over these bins, suggesting that each decile in brightness is reasonably ubiquitous. Avrett (1985) presents a series of models (labeled A'-F') which span the observed brightness range. These models have temperature minima ranging from 4225 to 4700 K. However, the models were constructed primarily to measure the temperature at the level of the temperature minimum and above. They converge to the same temperature structure at depths $m > 2$ g cm^{-2}($\tau_{0.5\mu} > 0.2$), and do not attempt to represent fluctuations in the middle and low photosphere.

C. Indirect Evidence

In a series of papers (cf. Ayres et al. 1986), Ayres and co-workers have argued that the upper atmosphere of the Sun, at heights normally considered chromospheric, must contain large fractional areas of cooler material in order to account for the low central intensities in the fundamental and first overtone vibration bands of carbon monoxide. They match average Ca II K line profiles and the CO data with a simple model having two independent components, one having chromospheric temperatures higher than the Maltby et al. (1986) reference atmosphere, and one having temperatures much closer to the radiative equilibrium solution of Anderson (1989). In view of the ubiquitousness of at least some Ca II emission in the data of Cram and Damé (1983), having spatial resolution ~500 km, the CO observations present a very perplexing problem. If the two-component model is correct, then the surface fragments must be smaller than 500 km, but with some of both components in any larger areas over nearly all of the solar surface. A model made of plane-parallel, independent components is inconsistent with such small scales, particularly in interpreting the center-to-limb variation. This problem is more directly related to chromospheric heating (chapter by Kalkofen), but its explanation may be found in photospheric temperature fluctuations.

Horizontal inhomogeneity also may be revealed by inconsistencies between center-to-limb variations $I_\nu^{em}(\mu)$ and frequency variations $I_\mu^{em}(\nu)$ near disk center (cf. Lefèvre and Pecker 1961). Recently Pecker et al. (1988) have shown that even a slight roughness (characteristic slopes of 20 to 30°) can have profound effects, whether the structures are resolved or not.

IV. DYNAMICAL CONVECTION MODELS

We do not intend to review the history of convection modeling (see instead Nordlund 1985b). Here, we wish to make a few comments about some of the results that are apparent in the three-dimensional modeling of granulation by Nordlund (see the above reference except where noted) and their relation to the empirical and a priori plane-parallel models. Nordlund is the first to couple a reasonable treatment of radiative transport with three-dimensional hydrodynamic numerical simulations of granular convection.

Nordlund's simulations produce a wide range of temperatures at all depths. Curves in Fig. 2, taken from Nordlund (1985c), show both a horizontally averaged temperature $\bar{T}(z)$ and a selection of temperatures $T_{x,y}(z)$ which contribute to the average. He argues that the enthalpy divergence contributes significantly to the thermodynamic equilibrium. The temperature structure in the middle and high photosphere shows very little correlation with the granulation intensity pattern, just as in the observations discussed in Secs. III.A and III.B (cf. Nordlund 1985a). At first glance, the wild temperature fluctuations in Fig. 2b look far larger than the observed rms 250 K. However, the curves are plotted on a common geometrical height scale. If they were plotted on a common optical-depth scale, they would lie much closer together. Part of the shift is due to density fluctuations over the granules, but most is due to the strong temperature sensitivity of the opacity. The figure is a clear demonstration of the invalidity of assigning a depth to a temperature fluctuation derived from the spatial fluctuations of the emergent intensity at one or more frequencies. Rather, one should "do" the observation on the model, and compare directly the intensity fluctuations.

While many of the $T_{x,y}(z)$ curves in Fig. 2b even show temperature inversions, most of them follow the general trend of the radiative equilibrium curves in Fig. 2a. Nearly all of them lie between the two radiative equilibrium limits in the height range 0 to 300 km. In Fig. 2a we have included both the Maltby et al. (1986) empirical model and Anderson's (1989) radiative equilibrium model for comparison. If Nordlund is accounting properly for the true thermodynamics of the middle to upper photosphere, it is not clear why his average temperature profile has a steeper gradient than the empirical model. The sharpness of the chromospheric temperature rise suggests there is not enough $\nabla \cdot F_{\mathrm{MHD}}$ below a height of 500 km to account for the difference— unless the empiricists are interpreting a distribution of $z(T_{\min})$ as an excess temperature. It may be just that Nordlund's average temperature is only geometrically weighted, whereas the empirical model is based on intensity-weighted averages made over an optical surface that is not a level plane. It may also be an effect of radiation statistics. Nordlund (1985a) shows that line source functions are enhanced above the average B_v by the differential velocity field. The atoms are exposed to more continuum radiation than normal through relative Doppler shifts (cf. Shine 1975). The same effect will alter the thermodynamic equilibrium of the plasma; thus, both the empirical and the hydrodynamic models may be wrong. This enhancement of the source function occurs primarily for lines that are not thermalized in the sense described in Sec. II.B.

We also show in Fig. 2b the range of minimum temperatures deduced from the 160 nm and Ca II data. Again, one must be wary of assuming that the brightness temperature at a particular frequency measures $T_{\min}$; however, in the calcium data one has a clear measure of $T_{\min}$ at the H_1 minimum in the profile. At face value, then, the range in minimum temperatures corresponds to a strongly corrugated surface with peak-to-valley separations of a few hundred km. However, the Cram and Damé (1983) observations would not fully resolve structures in Nordlund's models. There are about 36 observing pixels over Nordlund's horizontal plane. We remind readers that the geometrical limits set by the intensity contribution function are not much smaller.

The chromospheric radiative dissipation suggests that saturated acoustic waves are responsible for its heating (Anderson and Athay 1989a). One expects these waves to become nonlinear above a surface of constant density, and $T_{\min}$ should be just below this surface. Such surfaces are not as corrugated as constant temperature surfaces, and together with the Nordlund simulation imply a wide range for $T_{\min}$. If the $T_{\min}$ surface is near the 500-km level derived from the empirical plane-parallel models, Fig. 2b shows a range of 3000 to 5000 K. Ca II and ultraviolet emission is heavily weighted toward the upper end of this range, and its observers will not detect the low end if there is any high-temperature contamination. The contamination can be instrumental (lack of resolution) or intrinsic (contributions to the local "cold" source function from scattered photons produced in nearby warmer regions). CO absorption, on the other hand, is weighted toward the low end, and its observers will not detect the high end unless the line of sight misses any cold regions. Thus, it is possible that the CO absorption occurs in *photospheric* clouds just under $z(T_{min})$, and not at chromospheric heights (cf. Nordlund 1985c).

V. CONCLUSIONS

We conclude with a discussion of some of the problems facing us today, and suggest ways in which they might be addressed. Most of the problems are a direct consequence of structural inhomogeneity in the photosphere. Both theorists and observers must confront this issue.

In plane-parallel models, $S_\nu = B_\nu$ is a reasonably accurate approximation at photospheric depths. In an inhomogeneous but nearly static atmosphere, surfaces of constant optical "depth" (measured along paths parallel to the radiative flux) must have radii of curvature smaller than a few density scale heights to alter this conclusion. Consider the thermalization depth of any one spectral line. In a plane-parallel atmosphere, the line is thermalized because a particular column mass of material is between this depth and the escape surface. If there is no corresponding escape surface within a similar horizontal column mass (reached in one density scale height H_ρ at constant density), the line remains thermalized. Similarly, in an atmosphere with velocity fields, the nearest escape surface may be altered by differential Doppler shifts. The spatial variation of v, dv/dx, must be larger than $v_{thermal}/H_\rho$ at the thermalization depth to alter the thermalization. For metal lines, $v_{thermal}/H_\rho \approx 0.001$ s^{-1}. In all but the most pathological cases, Nordlund's structures are not sufficiently inhomogeneous to alter the thermalization depths of lines from their plane-parallel values.

We need from the observers absolute $I^{em}(\nu,x,y,t)$ over entire line profiles of several well chosen lines in the ultraviolet, visible and near infrared (cf. Elste [1986] for a simple example). They must be simultaneous measurements; temperature inhomogeneities on scales smaller than the observer's resolution will only show up as inconsistencies between lines at very different wavelengths and/or with different dk_ν/dT. From these observations, we can construct various empirical $T(\bar\tau)$ distributions, where $\bar\tau$ is a common optical depth scale for all frequencies in one line. Such temperature profiles are still relatively model independent. They, or the emergent profiles themselves, can be compared to results obtained from models such as Nordlund's. It is also possible to extract an estimate of the line-of-sight velocity as a function of $\bar\tau$ from the behavior of the so-called line bisector.

The next level of processing is to map the $\bar\tau$ scales onto a common m scale. Geometrical height is only useful as a relative scale; we have no way of determining an absolute height for a particular m. The mapping of $\bar\tau$ on m

involves solving the hydrostatic equilibrium equation in order to determine $d\bar{\tau}/dm = \bar{\kappa}(P,T)$. By choosing lines which are relatively insensitive to pressure, one can neglect hydrodynamic contributions to the pressure equilibrium. At this point, the empirical models will reveal temperature fluctuations and optical depth contours on a common depth scale (albeit not flat geometrical surfaces). They will also provide "environment" models for theoretical studies of, e.g., thin magnetic flux tubes. In order to calculate empirical radiative flux divergences, we must place adjacent x,y emergent columns together on the same geometrical depth scale and do the full multidimensional radiative transfer. Unless hydrodynamic models suggest that the geometrical surface of $\tau = 1$ in the continuum is relatively flat compared to H_ρ, this common placement is empirically impossible without triangulation from remote sensors.

There are many questions to be addressed with regard to Nordlund's hydrodynamic simulations, especially concerning the interpretation of surface-integrated fluxes. We have touched on several of these questions above. The most important is to determine whether plane-parallel models really are meaningful first approximations, and whether we can deduce granular contrasts and sizes from integrated light. The light we receive from a star with a convective photosphere comes from a corrugated, moving surface. A plane-parallel model, be it empirical or *a priori*, averages over this surface in a highly nonlinear fashion. To what degree are the quantities that we derive from such models meaningful? Consider the Sun as a star with known parallax and luminosity. An *rms* intensity fluctuation of 30% at 0.5 μm translates into an *rms* temperature fluctuation of 350 K. If we try to deduce the *effective temperature* spectroscopically in the visible, we will be in error by $(\Delta I_{rms}/I)\Delta T \approx 100$ K, and we will underestimate the solar radius by 3.6%. That is not a large error compared to likely errors in the luminosity, but, in different stars with higher granulation contrasts, the error could be much larger. Detailed comparisons between spectra determined from the hydrodynamical simulations and observations with high spatial and temporal resolution should answer many of these questions.

HIGH SPATIAL RESOLUTION TECHNIQUES

O. VON DER LÜHE
Institut für Astronomie, ETH Zürich

Observations of the solar surface from the ground are severely affected by turbulence of the Earth's atmosphere. Beside selecting observatory sites for excellent atmospheric quality, the careful design of the observing equipment, and the application of modern methods that undo the effects of the atmosphere or recover degraded data provide a wealth of information on small-scale processes. The effects of atmospheric turbulence on solar observations are reviewed, and methods that achieve high angular resolution on the Sun, such as adaptive optics and interferometry, are discussed. Some experiments that provide solar observations of high spatial resolution are presented.

I. INTRODUCTION

Under ideal circumstances, the angular resolution r of a solar telescope with a circular, unobstructed entrance opening of diameter D is

$$r = 2.44 \times 10^5 \frac{\lambda}{D} \qquad (1)$$

measured in arcsec where λ is the wavelength of light. As a rule of thumb, a 10-cm telescope resolves 1 second of arc; a 1-m telescope resolves one tenth of an arcsec, and so on, at $\lambda = 500$ nm. Consequently, small-scale structure on the Sun with dimensions of some 100 km could be resolved theoretically with today's meter-class solar telescopes. However, resolution of that quality is rarely achieved in practice.

There are two main reasons for the degradation of the resolution of a solar telescope. First, telescope optics are not perfect. Aberrations of optical components may be caused by bad figuring and thus are stationary, or they may be caused by mechanical and thermal stresses through gravity and heat input, and thus may vary slowly in time. Although high-quality optics are an absolute imperative for attaining high spatial resolution, their fabrication and testing is fundamentally similar to the methods used for larger nighttime telescopes, and will not be discussed here.

The major portion of degradation in resolution of a groundbased instrument is caused by rapid, random fluctuations of the optical path between the Sun and the detector. Those fluctuations are due to variations of the index of refraction in the Earth's atmosphere, dominated by thermal turbulence. This effect is called "seeing." At a given site, seeing limits the attainable average resolution of every large-aperture instrument to more or less the same value. The most obvious remedy for improving the resolution caused by poor seeing is putting the instrument into space. This would also allow one to cover a much wider spectral range. However, a space-borne solar telescope is at least 1 order of magnitude more expensive than a groundbased instrument of comparable size, and space projects usually take a long time to be realized. Thus, there is a strong case for improving groundbased observations.

Section II addresses the properties of a turbulent atmosphere and the consequences for the image of an astronomical source. Section III discusses how to minimize seeing caused by the instrument. It is a common experience that, when seeing conditions are generally good, there may be short moments of little or no atmospheric disturbances. Only small corrections must be applied to data obtained under these conditions. Section IV addresses methods of image-enhancement. A much more ambitious approach to overcome the atmosphere is active compensation of wavefront deformations, i.e., actively undoing the path fluctuations caused by atmospheric turbulence with a suitable servo control system (Sec. V). Solar interferometry (Sec. VI) attempts to restore information on small-scale structure *post factum*, i.e., after recording of the data. High-resolution projects in space are discussed in Sec. VII.

II. OPTICAL TRANSFER THROUGH THE ATMOSPHERE

The effects of random fluctuations of the optical path on imaging over large distances has received a lot of attention since the 1950s. Tatarskii's (1971) book on wave propagation is the classical reference that covers in detail the work done at that time. In the 1960s observers were mostly concerned with passive, direct imaging; a classical example is the often referred to paper by Fried (1966). During the 1970s, when speckle interferometry and adaptive optics were under development, the previous theories and methods were very successfully applied to explain and to analyze the performance of those new methods. Bertolotti et al. (1979) and Roddier (1981) review the state of the art in the late 1970s and early 1980s. Since then, major progress

has been made in analyzing the performance of adaptive optical systems
(Roddier 1987). A few concepts that form the basis of discussions in the next
sections are presented here.

Turbulence in the atmosphere affects the propagation of electromagnetic
fields through variations in the index of refraction n. The refractive index is
a function of air temperature, pressure and humidity, where temperature fluctuations are most important. The second-order statistics of temperature fluctuations are governed by the velocity distribution of the air mass, that is
usually assumed to show fully developed turbulence. This is the case when
the Reynolds number Re of the medium is large; for the free atmosphere, Re
$\approx 10^5$ to 10^6. However, orographics at a particular site or nearby artificial
structures may generate turbulence that is not fully developed.

The propagation of a wave through a volume with refractive index variations results in both amplitude and phase fluctuations of the outgoing wave.
The first effect accounts for scintillation, the second effect accounts for blurring and atmospheric image wander. Amplitude fluctuations are usually neglected when astronomical light sources are considered, because the "phase
screen" is a much more efficient scatterer than the "amplitude screen" associated with the fluctuations.

Let $\bar{x}$ represent a two-dimensional coordinate in the plane of the entrance
pupil of the instrument. The optical phase $\phi(\bar{x}) = 2\pi l(\bar{x})/\lambda$ [optical path
length $l(\bar{x})$] of an initially plane wave that has crossed the turbulent volume
has the following properties:

1. First-order statistics is gaussian;
2. Second-order statistics (i.e., the two-point correlation of the phase) can be
   expressed with the phase structure function

$$\langle |\phi(\bar{x}) - \phi(\bar{x} + \bar{\xi})|\rangle^2 = 6.88 \left(\frac{|\bar{\xi}|}{r_o}\right)^{5/3} \tag{2}$$

where the so-called Fried parameter r_o is defined as

$$r_o = 0.18 \cdot \lambda^{6/5} (\cos z)^{3/5} \cdot \left[\int_0^\infty C_n^2(h)dh\right]^{-3/5}. \tag{3}$$

The quantity $C_n^2(h)$ (dimension: $[m^{-1/3}]$) is the structure constant that describes the fluctuations of the index of refraction n in the atmosphere. h
represents vertical height, and z is the zenith angle.

r_o is a measure of the integral of fluctuations of the index of refraction
along the line of sight. Its dimension is a length and can be regarded as the
diameter of a telescope that has the same diffraction-limited resolution as the
average seeing. r_o is associated with the correlation length of the complex

amplitude of the disturbed electromagnetic field (i.e., the wave is essentially flat over distances of the order of r_o), but not with the correlation length of the optical path; the latter is associated with the much larger outer scale of turbulence L_o. It is also important to note that Eq. (2) only holds within the inertial range between the inner scale l_o (some 10 mm) and outer scale of turbulence (some 10 m), i.e., $l_o < |\zeta| < L_o$. Most astronomical telescopes have diameters smaller than L_o, justifying Eq. (2).

   r_o is the most important atmospheric parameter. It determines the average resolution of long-exposure observations (Fried 1966), the probability of obtaining a "lucky" short exposure with very little seeing disturbances (Fried 1978), the number of independently steerable elements of a diffraction-limited adaptive optical system (Hudgin 1977), and the transfer function associated with speckle image recovery techniques (Korff 1973). One attempts to find a site that minimizes the integral of the refractive index structure coefficient along the line of sight for a solar observatory. Good solar sites usually have $r_o = 8$ cm to 20 cm. Of course, r_o at a particular site is highly variable. It has been found that r_o is log normally distributed; extrema can be as large as 40 cm during daytime conditions (Borgnino and Brandt 1982; Brandt et al. 1987).

   Another important parameter is the atmospheric correlation time. Fluctuations in the optical path vary rapidly, and so does the point-spread function that is associated with the deformed wavefront. In a simple picture, one may consider turbulence as concentrated in a single layer above the telescope; this may well be the case occasionally at a good site. If the layer moves with constant speed u and one assumes "frozen turbulence" (the turbulence pattern is fixed and moves as an entity), the atmospheric correlation time τ_o can be estimated with $\tau_o = r_o/u$. A typical value for τ_o is 10 ms. Exposures for frame selection and interferometric methods have to be shorter than τ_o, and the required bandwidth of an adaptive optical system is also determined by the atmospheric correlation time.

   Imaging through a turbulent medium is essentially nonlinear, i.e., the effects of the atmosphere are different for different lines of sight. This effect is easily understood if one considers a single turbulent layer that is some distance away from the telescope. Different lines of sight will eventually pass through different regions in the turbulent layer, hence the wavefront deformations will be different. The angle over which wavefront deformations are correlated defines the "isoplanatic angle" θ_o. An estimate of the isoplanatic angle for the single turbulent layer model is given with $\theta_o = r_o/h$, where h is the distance from the telescope to the point of intersection of the line of sight with the turbulent layer. The size of the isoplanatic patch is generally only a few seconds of arc, which poses serious constraints on most imaging methods. Turbulent layers that are farthest away from the telescope have the strongest anisoplanatic effect.

   We now present the equations that govern the transfer of information through the atmosphere and the optical system. We are interested in the Fourier transform of the observed focal-plane intensity. Using a result of Fourier optics, we can write the Fourier transform $F(\bar{s})$ of the intensity pattern generated by a point source at position angle θ as the autocorrelation of the complex wave in the entrance pupil. $\bar{s}$ is a two-dimensional spatial frequency normalized to 1 at the telescope's diffraction limit D/λ. Letting $W(\bar{r})$ denote the transmission of the entrance pupil (that is 1 inside the pupil and 0 outside), we can write

$$F(\bar{s})_{(\text{point source})} = I_o \exp[-jkD\bar{\theta}\bar{s}] \times$$
$$\iint W(\bar{r})W(\bar{r} - D\bar{s})\exp[j(\phi(\bar{\theta},\bar{r}) - \phi(\bar{\theta},\bar{r} - D\bar{s}))]d\bar{r}. \quad (4)$$

I_o is the point source intensity, j is $\sqrt{-1}$, and k is $2\pi/\lambda$. ϕ is the optical phase; the dependence of ϕ on the angular coordinate θ indicates that waves propagating along different paths may experience different aberrations. The exponential before the integral assures that the image appears at the proper location in the focal plane.

If the source is extended and incoherent, i.e., a function $I_o(\bar{\theta})$, we obtain the Fourier transform of its image by integration Eq. (4) over $\bar{\theta}$

$$F(\bar{s})_{(\text{ext., anisopl.})} = \iint I_o(\bar{\theta})\exp[-jkD\bar{\theta}\bar{s}] \times$$
$$\iint W(\bar{r})W(\bar{r} - D\bar{s})\exp[j(\phi(\bar{\theta},\bar{r}) - \phi(\bar{\theta},\bar{r} - D\bar{s}))]d\bar{r}\,d\bar{\theta}. \quad (5)$$

This is the anisoplanatic imaging equation. There will be no simple relationship between object and image intensity distributions under anisoplanatic conditions. If the object is smaller than the isoplanatic patch, we can drop the dependency of ϕ on θ. This results in the separation of $\bar{r}$ and θ integrals

$$F(\bar{s})_{(\text{ext., anisopl.})} = \iint I_o(\bar{\theta})\exp[-jkD\bar{\theta}\bar{s}]d\bar{\theta} \times$$
$$\iint W(\bar{r})W(\bar{r} - D\bar{s})\exp[j(\phi(\bar{r}) - \phi(\bar{r} - D\bar{s}))]d\bar{r}. \quad (6)$$

The first integral is the Fourier transform $F_o(\bar{s})$ of the source intensity distribution. The second integral depends entirely on the instrument and on the atmosphere; this is the instantaneous system optical transfer function (OTF) $S_{\text{sys}}(\bar{s})$. This function is generally random because the optical phase ϕ is random. The imaging equation now takes the simple form

$$F(\bar{s}) = F_o(\bar{s}) \times S_{\text{sys}}(\bar{s}) . \qquad (7)$$

In many cases, one assumes linearity of the imaging process and therefore applies Eq. (7) in order to correct for atmosphere and telescope. One should bear in mind that for an extended source such as the Sun, Eq. (7) is generally not valid for scales smaller than the seeing disk. This is particularly true when average quantities are derived from a number of pictures that show the same object under different atmospheric conditions.

III. EVACUATED AND HELIUM-FILLED TELESCOPES

A significant amount of thermal turbulence in the light path originates from the telescope structure, the optics and the enclosure within the telescope and in its immediate vicinity. This is called internal seeing and dome seeing. Internal seeing is a more prominent problem with a solar telescope than with a nighttime telescope because of the large influx of radiant energy that heats up all surfaces in direct sunlight. Most solar telescopes are designed in such a way as to reduce these effects, typically by evacuating as much of the optical path as possible. Dome seeing can be reduced by minimizing or eliminating the dome structure. Some direct-pointing, compound telescopes designs such as THEMIS (Mein and Rayrole 1989) and LEST (Andersen et al. 1984) use a minimal protective enclosure that tracks the telescope. Others have coelostats protected by fully retractable domes. Examples are the National Solar Observatory (NSO) vacuum tower on Kitt Peak, and the German vacuum tower telescope on Teneriffe (Soltau 1989). There is free air flow through the coelostats when the instruments are in use, carrying away the turbulence. The wish to have the coelostat optics inside the vacuum with the other optics led to the turret designs of the NSO vacuum tower telescope at Sacramento Peak (Dunn 1985) and the Swedish solar tower on La Palma (Scharmer 1987).

A vacuum of a few mbar eliminates internal seeing, but requires a sealing window at the telescope entrance. This adds an additional large optical element to the design, which must be of as excellent quality as the primary optics. In particular, it must be of constant thickness. The vacuum load causes stress which the window must withstand, leading to a diameter-to-thickness ratio of approximately 15:1 for a flat glass window (Dunn 1985). The thermal load on the window originates from the direct sunlight, uneven cooling by the air outside, and heat conduction to or from the support. The loads result in variations of the optical path within the window, causing aberrations. Protection of the window cell from direct sunlight usually helps.

If the telescope is to be used for polarimetric observations, great care must be taken to minimize stress birefringence in the window. The stress caused by the supports is most important. Polarization caused by the window becomes smaller, the thinner the window is (Bernet 1979).

If the telescope uses a lens as the objective, the lens may serve as the entrance window. The Swedish solar tower on La Palma uses this approach with excellent results (Scharmer 1987).

Helium filling of a sealed telescope becomes a viable alternative to evacuation. A window will still be necessary, but it may be considerably thinner if the pressure of the helium comes close to the ambient pressure. This reduces stress in the glass and hence stress birefringence. Helium has an index of refraction lower than air and a higher thermal conductivity, and is also more viscous. Table I compares physical parameters of helium and air.

Helium will cause significantly less variations of the optical path than air for the same temperature fluctuations. The larger thermal conductivity will distribute heat more easily, thus reduce temperature fluctuations. The larger viscosity of helium results in a smaller Reynolds number, permitting larger velocities before the onset of turbulence.

Some experiments have been carried out by filling vacuum telescopes with helium, which show clearly the advantage of helium compared to air (Engvold et al. 1981, 1983). Temperature fluctuations are evidently smaller, and onset of turbulence occurs at significantly larger pressures in helium than in air. Recent experiments indicate that temperature fluctuations can be reduced even more by forcing the helium into a determined flow pattern with blowers and ducts. The Large Earthbound Solar Telescope (LEST) with a 2.4 m aperture is currently designed as a helium-filled telescope (Andersen et al. 1984).

IV. DATA PROCESSING AND IMAGE ENHANCEMENT METHODS

Most high-resolution data from a groundbased telescope result from inspection of a sometimes huge data base and selection of the very few frames or spectra with the highest quality. Based upon his experience, the analyst makes educated guesses as to whether or not his very best sets are free of seeing and to what extent they are affected by telescopic aberrations. The amount of correction applied to the data is sometimes inferred by considering the diffraction of the instrument alone, allowing for small telescopic aberrations at most (Schmidt et al. 1981; Muller and Keil 1983). In other cases, a measured correction function for the telescope can be used (Koutchmy 1977a). Results from this type of work tend to be conservative, in the sense that overcorrection is unlikely. As a consequence, the photometric accuracy is rather limited. The problem stems from the virtual lack of information on

TABLE I
Physical Parameters of Air and Helium

	Air	Helium
Index of refraction $(n-1)$	$2.93\ 10^{-4}$	$3.60\ 10^{-5}$
Density ρ [g cm^{-1}]	$1.29\ 10^{-3}$	$1.79\ 10^{-4}$
Specific heat C_p (cal g^{-1} °C)	0.236	1.24
Thermal conductance λ_t [cal s^{-1} cm^{-2} (°C / cm)$^{-1}$]	$5.8\ 10^{-5}$	$3.4\ 10^{-4}$
Kinematic viscosity ν (cm^2 s^{-1})	0.132	1.042

the quality of the imaging process present in a few, excellent pictures or spectra. In rare exceptions only, an independent estimate of imaging quality can be obtained, e.g., by simultaneous observation of the lunar limb during a partial eclipse and by deriving the correction from the lunar limb profile (Levy 1971; Deubner and Mattig 1975). Another possibility would be the simultaneous use of a wavefront sensor, measuring wavefront errors of atmosphere and telescope, from which the correction functions can be calculated (see Sec. V).

A. Single Picture Restoration

The analysis of a small number of data sets usually involves a simple Fourier inversion. One often assumes linearity of the imaging process and therefore can separate instrumental terms from the observed object. The Fourier transform $F(\bar{s})$ of the observed intensity distribution is calculated. For spectra obtained with a grating spectrograph, the transform is performed in the spatial direction along the slit only. The effect of the spectrograph slit on the image requires careful consideration.

The transform of the observed data set relates to the transform $F_o(\bar{s})$ of the object intensity via Eq. (7), if one assumes isoplanatic imaging conditions. If the optical transfer function of the system S_{sys} was known, one could recover the object transform for those frequencies for which the transfer function does not vanish, by applying

$$F_o(\bar{s}) = \frac{F(\bar{s})}{S_{sys}(\bar{s})} . \qquad (8)$$

The object intensity is then restored with an inverse Fourier transform.

The optical transfer function is generally a complex quantity, and in most cases, only its modulus MTF_{sys} (modulation transfer function) can be measured or inferred. Strictly speaking, this makes the recovery of the object intensity impossible when S_{sys} is not purely real. However, the power spectral density $P_o(\bar{s}) = |F_o(\bar{s})|^2$ of the object can still be recovered, which may suffice in some cases:

$$P_o(\bar{s}) = \frac{P(\bar{s})}{MTF^2_{sys}(\bar{s})} = \frac{|F(\bar{s})|^2}{|S_{sys}(\bar{s})|^2} . \qquad (9)$$

For a single picture or spectrum, the system transfer function can be separated into components originating from diffraction S_D, system aberrations S_A, and from seeing S_S

$$S_{sys}(\bar{s}) = S_D(\bar{s}) \times S_A(\bar{s}) \times S_S(\bar{s}) . \qquad (10)$$

If one assumes that seeing is negligible and aberrations are small, the main component is the diffraction term, for an unobstructed, circular entrance pupil with diameter D, at wavelength λ

$$S_D(\bar{s}) = \frac{2}{\pi}\left[\arccos(\lambda s/D) - \lambda s/D\sqrt{1 - (\lambda s/D)^2}\right]. \tag{11}$$

The Fourier coordinate s is expressed in units of line pairs per radian; division by a factor 206264.8 scales this coordinate into line pairs per arcsec. The product $S_D \times S_A$ can be measured interferometrically (Dunn 1985; Koutchmy 1977a; Smartt 1989).

The data usually contain noise which becomes amplified in the correction process. This can be avoided by applying a suitable filter function, such as Brault and White's (1971) optimum filter OF:

$$OF(\bar{s}) = \frac{P(\bar{s})}{P(\bar{s}) + N(\bar{s})}. \tag{12}$$

$N(\bar{s})$ represents the noise power spectrum, which can be measured directly or can be inferred from an inspection of the data spectrum and by making assumptions on the form of the noise spectrum. The filtered result, corrected for the system transfer, is finally obtained with

$$\tilde{F}_o(\bar{s}) = F_o(\bar{s}) \times OF(\bar{s}) = \frac{F(\bar{s}) \times OF(\bar{s})}{S_{sys}(\bar{s})}. \tag{13}$$

B. Frame Selection and Time Series Analysis

Modern computer equipment can perform a great deal of the selection and analysis process. This is particularly important when the data bases are very large, as is the case when time series of pictures are taken in multiple wavelengths. Using fast image analysis hardware, the "best frame selection" can be made while the data are taken, thereby greatly reducing the amount of data that is actually recorded. For example, the image acquisition system that is used with the Swedish solar tower records images at video rates (30 frames/ s) and selects the best picture within a given period (Scharmer 1987). The selection criterion is normally the image contrast. The best pictures are kept along with a time tag. Time series of pictures of exceptional quality and uniformity can be taken this way.

Further analysis steps removes image distortion from a time series. When seeing is very good, only image motion and distortion (anisoplanatic image motion) shows. The picture appears as if the solar surface is viewed through a slightly disturbed, rippled layer of water. Distortion can be measured by calculating a running average from the series over a period that covers as many frames as possible, but is short compared to the correlation time of the observed structure. The average serves as a reference. Each frame is dissected into small segments, and the displacement of a segment with respect to the reference is calculated using a correlation algorithm. The result is a vector map that represents the distortion present in the frame. The distortion is undone by shifting segments appropriately and their subsequent combination (November 1986; Topka et al. 1986).

Interpolation between pictures can then be performed that turns the existing time series into a series where frames are equally spaced in time. Three-dimensional Fourier analysis then permits the analysis of intensity oscillations, or allows one to remove them from the data. Data sets processed in this way provide an excellent basis for the study of the morphology and the dynamics of granulation, penumbral fine structure and active regions (Brandt et al. 1988).

V. ACTIVE WAVEFRONT COMPENSATION

The most ambitious approach for improving groundbased resolution is through active compensation of atmospherically produced wavefront deformations. This is currently done by placing a steerable or deformable mirror in a pupil plane image and by connecting the device with a wavefront analyzing system. Wavefront deformations are sensed by the analyzer, the signals thus generated are used to minimize the aberrations with the active device in a servo-loop configuration. This can be done to various degrees of perfection. One may expand wavefront deformations in a set of basis functions (e.g., Zernike polynomials) and correct only a few, lowest orders. Since a constant offset of the wavefront phase does not affect the image, the lowest order to be corrected would be a linear wavefront error that corresponds to image displacement. A simple, steerable (agile) mirror in a pupil image is sufficient to remove image motion. Higher orders of deformation require deformable mirrors, and the complexity of the adaptive system increases rapidly. This chapter therefore discusses image motion compensation and higher-order adaptive optics separately.

Review articles on the topic of adaptive optics can be found in Hardy (1978, 1981a). The LEST Foundation Technical Report No. 28 (Merkle et al. 1987) is solely concerned with the application of adaptive optics to solar observations and presents in great detail the current state of the art.

A. Image Motion Compensation

Figure 1 shows a schematic layout of an image motion compensation system used at the National Solar Observatory. An image of the solar surface is located at the primary focus PF. The light is deflected by a flat mirror M_1, by two imaging mirrors M_2 and M_3, and by the agile mirror AM. An image

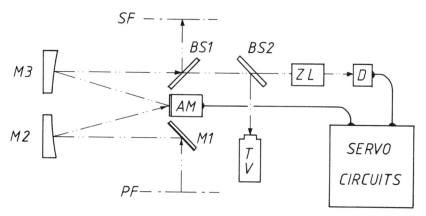

Fig. 1. National Solar Observatory image motion compensation system layout. See Sec. V.A for
a description of the components.

of the pupil is located at the agile mirror. The primary focus is re-imaged at
the secondary focus *SF* at a 1:1 scale. A beam splitter BS_1 extracts a fraction
of the light, part of which is fed to a video camera *TV;* the remainder is used
for the image motion detector *D*. A zoom lens *ZL* re-images the solar surface
onto the detector with a variable scale. The video camera is used to monitor
the region of interest on the Sun and the performance of the guider. The
separation of primary and secondary foci is roughly 60 cm. The whole system, except for servo electronics, fits into a box about 1 m × 0.5 m in size
that can be conveniently placed in front of various post-focus instruments.
Another possibility is to include the image-motion control system into the
optical path of the telescope, as was done with the Vacuum-Newton telescope
at the Observatorio del Teide, Izaña, Teneriffe.

Although only a first-order correction is made with image-motion control, the gain in resolution may be quite substantial. Theoretically, about half
of the degradation of resolution in a long-exposure observation is due to atmospheric image motion (Fried 1965). Practically, instrumental shake and
guider errors add to that. The atmospheric component of image motion is
< 1 arcsec *rms* under fair-to-good seeing conditions. For a meter-class tele-
scope, the bandwidth is relatively low; there is only little power left above 10
Hz (von der Lühe 1988). Figure 2 presents power spectra of image motion.
The corrected area depends on the size of the isoplanatic patch. Figure 3
shows measurements of residual image motion as a function of distance from
a stabilized spot. It is seen that an area of 1 to 2 arcmin was successfully
stabilized. The bandwidth of image motion increases when the telescope diameter becomes smaller; at the same time, the isoplanatic area decreases.

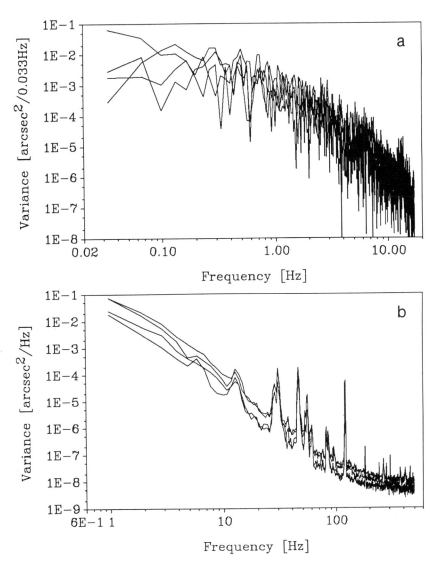

Fig. 2. Temporal power spectra of image motion, measured with the 76-cm vacuum tower telescope at National Solar Observatory - Sacramento Peak (see von der Lühe 1988). Four sample
spectra are shown for the low-frequency and high-frequency ranges.

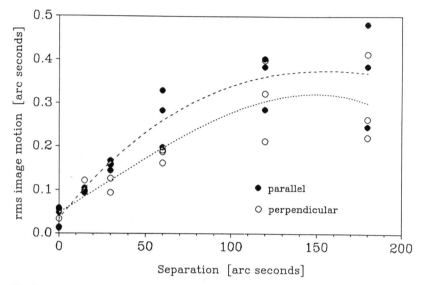

Fig. 3. Anisoplanatic image motion (distortion) shown as a function of separation from a reference area (see von der Lühe 1988). The *rms* values are plotted for the directions parallel and perpendicular to the line connecting the reference and target areas.

Agile Mirrors.   Agile mirrors are made with electrodynamic or piezo-electric actuators. Electrodynamic actuators have a large dynamic range, but they are usually soft, which complicates the dynamical response and increases the complexity of the servo circuitry.

Piezoelectric actuators are fast and powerful, and the servo electronics are much simpler. High voltages are necessary to drive piezos; modern actuators require a few hundred volts. The bandwidth is usually limited by the capacitance of the piezo (10 to 100 nF) and the internal resistance of the driving HV amplifier; a bandwidth of 100 Hz is easily achieved. The stroke of piezos and thus the dynamic range of the agile mirror is rather limited; a 10 μm to 40 μm stroke is common and the corresponding tilt range of a mirror can be several minutes of arc. The angular dynamical range Δ_F in the final focus is the product of the demagnification m of the pupil image on the agile mirror and the mirror tilt range Δ_M: $\Delta_F = m \cdot \Delta_M$. Dynamical ranges Δ_F of 10 to 40 arcsec are typical; this is ample for correcting seeing-induced image motion, but may not be sufficient to compensate for slow telescope drifts, solar rotation, etc. over long periods of time. Figure 4 presents the design of a piezo-driven agile mirror and its dynamical response in the form of amplitude and phase Bode diagrams (von der Lühe 1988).

A different approach to compensating image motion is to use a tiltable plane-parallel glass plate close to a solar image. Those devices tend to be slow because of the large masses and angles involved, but they can be useful

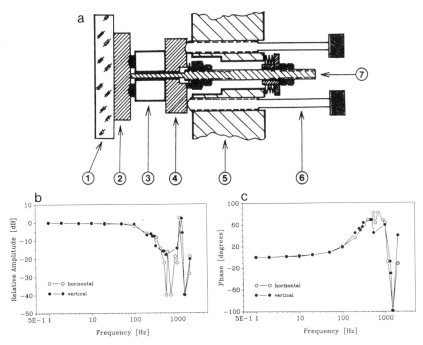

Fig. 4. (a) Design of an agile mirror. 1: mirror, 2: base plate, 3: piezo actuators (3), 4: mount plate, 5: support plate, 6: adjustment screws (3), 7: steel rod, connecting the parts using stacks of Belleville washers. (b and c) Dynamical response log amplitude and phase of the mirror above, measured for two directions.

if a pupil image is not easily accessible and there is not enough space for a separate image-motion control system (Wiehr et al. 1989).

Spot trackers.   The simplest detector for image motion consists of a quadrant cell. The difference of the photo currents of opposite quadrants is amplified and used to drive the active element. Those devices can establish lock on confined targets, which appear dark against a bright background or bright against a darker background (e.g., sunspots, pores, faculae near the limb). Under good seeing conditions, granules can be tracked for short periods. The range of objects on which spot trackers can establish lock is surprisingly large (Tarbell and Smithson 1981; von der Lühe 1988).

Correlation trackers.   More sophisticated motion detectors are required if no structure suitable for a spot tracker is in the vicinity of the area of interest. Correlation trackers permit one to lock on arbitrary structure, e.g., solar granulation. This is done by means of an array detector that is continuously scanned, a reference picture of the area and a correlation algorithm that compares the "live" picture from the detector with the reference.

O. von der LÜHE

The agile mirror is driven so as to maximize the correlation of live and reference images. Evolutionary changes of the appearance of the structure require updating of the reference picture every 1 to 2 min.

A number of correlation tracker systems have been proposed, are under development, and have been successfully tested (Edwards et al. 1987; Rayrole 1987; von der Lühe et al. 1989). They are distinguished by the detector pattern, the correlation algorithm and the correlation strategy. The detector pattern refers to the configuration of detector pixels, linear or two-dimensional arrays being most usual. A straight-forward cross covariance algorithm is normally used. Let $I_L(\bar{x})$ and $I_R(\bar{x})$ denote the intensity distribution in the live and reference pictures, respectively. The cross covariance $C(\bar{\Delta})$ is given with

$$C(\bar{\Delta}) = \iint_{\text{field}} I_L(\bar{x}) \cdot I_R(\bar{x} + \bar{\Delta}) d\bar{x} . \tag{14}$$

$C(\bar{\Delta})$ has a maximum for spatial lag $(\bar{\Delta})_{\text{max}}$, where I_L and I_R match optimally in the least-squares sense. The closely related mean-square residual function $MSRF(\Delta)$ may be alternatively used:

$$MSRF(\bar{\Delta}) = \iint_{\text{field}} \left[I_L(\bar{x}) - I_R(\bar{x} + \bar{\Delta}) \right]^2 d\bar{x} . \tag{15}$$

The $MSRF$ has a minimum at the optimal lag $(\bar{\Delta})_{\text{max}}$. The absolute value of the integrand can be taken instead of the square, which eliminates multiply operations.

The cross covariance evidently contains much more information than necessary for finding a simple displacement. Simplifications of Eq.(14) have therefore been suggested for computational efficiency. For example, a displacement error signal δ can be obtained with an algorithm proposed by Mertz (1965).

$$\delta = \iint_{\text{field}} I_L(\bar{x}) \cdot \left[I_R(\bar{x} - \bar{\Delta}_o) - I_R(\bar{x} + \bar{\Delta}_o) \right] d\bar{x} . \tag{16}$$

$(\bar{\Delta})_o$ is an adjustable, but otherwise fixed, parameter. Comparison with Eq.(15) reveals that δ is merely the difference of the cross covariance at lags Δ_o and $-\Delta_o$, so a finite difference of the cross covariance function is effectively computed. For a relative displacement between live and reference images smaller than the correlation length of the target structure, the error signal is zero if the lag Δ is zero (Fig. 5). At both sides from zero lag, the error signal has opposite sign. This method of locating a correlation peak is called a "derivative strategy"; the error signal as a function of lag resembles the

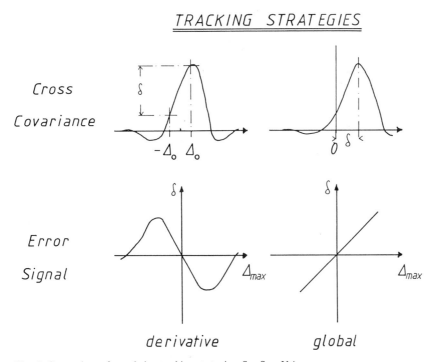

Fig. 5. Comparison of correlation tracking strategies. See Sec. V.A.

derivative of the cross covariance function and is not linear. A "global strategy" locates the cross covariance maximum in the entire available field; the resulting error signal is a direct representation of the relative displacement and thus a linear function of lag. A derivative strategy is computationally more efficient, but the target acquisition range is determined by the correlation length of the structure rather than by the detector field size.

Two correlation tracker systems have been built and successfully tested on the Sun to date. One system was designed and built by the Lockheed Palo Alto Research Laboratory (LPARL) for the use in the Coordinated Instrument Package of the Orbiting Solar Laboratory (see Sec. VII; Edwards et al. 1987), the other one was built in cooperation by the National Solar Observatory (NSO) and the Kiepenheuer-Institut (KIS, Freiburg, West Germany) for use on groundbased instruments (von der Lühe et al. 1989). Both systems use an areal tracker pattern. The LPARL system applies a derivative strategy (Eq. 16) while the NSO/KIS system applies a global strategy. Both systems have been successful in establishing lock on solar granulation. Figure 6 shows two long-exposure pictures, stabilized and not stabilized, taken with the NSO/KIS system at the NSO-Sacramento Peak vacuum tower telescope.

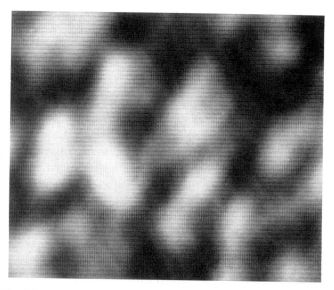

Fig. 6. Unstabilized (top) and stabilized (bottom) pictures showing a 12 × 12 arcsec field of solar granulation. The effective exposure is 2 s.

B. Adaptive Optics

Full wavefront deformation control using an adaptive optical system has been applied to solar observations with various degrees of success (Hardy 1981*b*; Smithson et al. 1984; Acton 1989). The systems used were the ITEK Corporation, Real-Time Atmospheric Compensation system (Hardy 1987), and the 19-element segmented mirror system designed by Smithson et al. (1984) for the use on the Sun. All observations to date were made during engineering and verification operations of the systems; there is currently no adaptive optical system available for routine use at a solar telescope. Dunn et al. (1989) report on the development of an adaptive optical system for the NSO - Sacramento Peak vacuum tower. The current state of the art is demonstrated in Fig. 7, taken from Acton (1989), which shows uncompensated and compensated pictures side by side. It is fair to state at this point that adaptive optics have shown their capability of consistently improving the solar image and thus have demonstrated their viability, but much work remains to be done before these systems can be commissioned for routine use.

Figure 8 shows the main components of an adaptive optical system. They include a wavefront control system (deformable mirror), a wavefront sensor, and a servo system that combines the two to a servo loop.

An important characteristic of an adaptive optical system is the number of servo channels. They directly determine the number of degrees of freedom N available to match a given wavefront deformation. As a rule of thumb, $N \geq (D/r_o)^2$ in order to achieve diffraction-limited resolution. The error in the wave leaving the control system that is caused by the limited number of de-

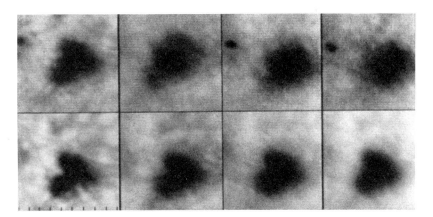

Fig. 7. Uncompensated and compensated short exposure pictures produced with an adaptive optical system. The squares are 9 arcsec on the side. The system locks on the sunspot near the center of the frame. Variability of seeing still results in variations of image quality, but the compensated pictures are consistently and significantly sharper than the uncompensated pictures (figure courtesy Lockheed Palo Alto Research Laboratory).

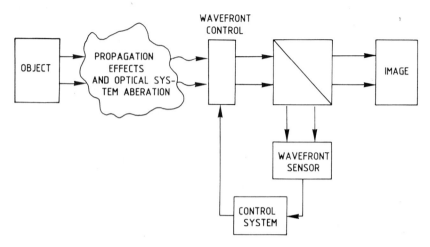

Fig. 8. Main components of an adaptive optical system (figure courtesy F. Merkle, European Southern Observatory).

grees of freedom is called the "wavefront fitting error" (Hudgin 1977). Another characteristic is the system bandwidth, which must be large enough to resolve the atmospheric correlation time. Bandwidths of 500 to 1000 Hz are considered sufficient; insufficient bandwidth produces a bandwidth error. A third characteristic is the "dynamic range" of the control element. Most systems have distinct corrector elements for image motion control and higher-order corrections; a few μm dynamic range is sufficient for the deformable element under these conditions. Another source of error is the uncertainty with which the wavefront error is estimated. All these errors are system dependent and can be reduced with an appropriate increase in effort.

Other error sources are more fundamental (Roddier 1987). Amplitude fluctuations are not removed by a deformable mirror, and chromatic errors may be produced if the observation wavelength is different from that at which the wavefront is sensed. These two errors are rather small; much more important is the isoplanatic error. An adaptive optic will improve the image only within an isoplanatic patch. In order to do so, the size of the field from which wavefront error measurements are made must not be larger than the isoplanatic patch. Hardy (1987) points out that this is possible only if $\theta_o r_o/\lambda \gg 1$. Therefore, sites should be selected which maximize this product.

Adaptive optics may be distinguished by their control type. One can correct wavefront errors at each point in the pupil independently from others. This is called zonal control and is realized in most existing systems. Alternatively, one can expand the measured wavefront distortion into Zernike polynomials and correct each mode independently; this is called modal control. Modal control seems more complex, but it allows one to concentrate on those modes that contribute most of the wavefront error, which is most efficient for a partial compensation.

Corrector elements.   There are numerous designs for deformable mirrors (Fig. 9; Merkle 1987), continuous face-plate and tip-tilt segmented mirrors being now the most common ones. Actuators are mostly piezo-ceramics, but electrodynamic, magnetostrictive and hydraulic types also exist. Recently, electrostrictive actuators were designed. Mirrors have usually one degree of freedom per actuator; the segmented piston-and-tilt type has two degrees of freedom per segment. Segmented mirrors have all the advantages of modularity; they can be expanded and segments can be replaced. However, the segments must be carefully co-phased in order for the mirror to operate with incoherent light; this requires highly linear actuators. Continuous face plates do not have this problem, but replacement of faulty actuators is often impossible. A segmented piston-and-tilt mirror is used in Smithson's solar adaptive optical system, a monolithic mirror was used by Hardy in his solar experiments (Fig. 10).

Wavefront sensors.   There are three broad types of wavefront sensors for solar adaptive optics: Hartmann-Shack sensors, image-plane correlators, and image-sharpening sensors (Fig. 11). Hartmann-Shack sensors are most common in astronomy. An array of lenslets is placed at an image of the entrance pupil. Each lenslet produces an image of the object which is shifted from some reference position according to the tilt of the wavefront averaged

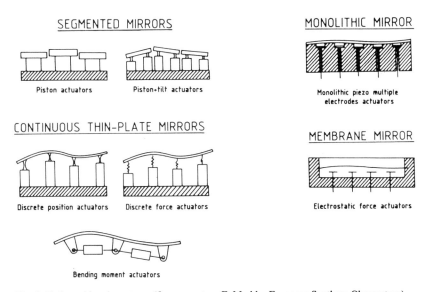

Fig. 9. Deformable mirror types (figure courtesy F. Merkle, European Southern Observatory).

Fig. 10. Deformable mirrors used for adaptive optics that were applied to solar observations. (a) A 21-actuator monolithic mirror produced by Litton - Itek Corporation. (b) The 19-segment piston-and-tilt mirror used by Smithson et al. 1984 (figure 10b courtesy Lockheed Palo Alto Research Laboratory).

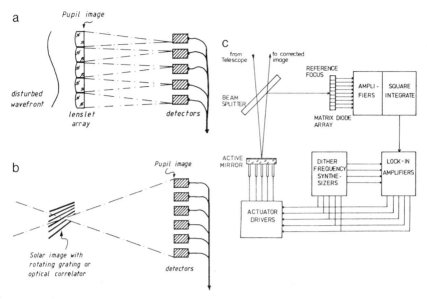

Fig. 11. Wavefront sensor principles. (a) Hartmann-Shack-type wavefront sensor. (b) Image plane correlator. (c) Image sharpening. (See Sec. V.B.)

over the lenslet area. An array of position-sensitive detectors at the image plane measures the wavefront tilts. An integrating network reconstructs the wavefront from the tilts and applies corresponding drive signals to the mirror. The position-sensitive sensors may consist of an array of spot trackers or correlation trackers; the latter approach may be prohibitively complex. An array of quadrant cells is used in Smithson's system (Smithson et al. 1984).

Image-plane correlators consist of an image plane modifier and of an array of detectors in a pupil image that sense local wavefront tilts in the form of intensity fluctuations. The best known device of this kind is the AC lateral shearing interferometer used in Itek's Real-Time Atmospheric Compensation system (Hardy et al. 1977). It consists of a rapidly rotating radial grating whose rulings intercept the light at a solar image. The spatial frequency corresponding to the grating period is continuously "chopped" by the grating. The position of that particular spatial frequency fringe in the image can then be measured from the temporal phase of the corresponding intensity variation in the pupil image. Comparison of the phases at all positions in the pupil image yields the local wavefront tilts. A diaphragm at the grating restricts the field of view of the sensor. The dimension of the diaphragm in the direction perpendicular to the rulings has to match an integer multiple of the grating period to high precision in order to avoid lock on the field stop (Hardy 1987). A more general correlator that makes more efficient use of the information present at the focus was recently proposed (von der Lühe 1988).

A wavefront sensor operating with an image sharpness criterion (Muller and Buffington 1974) is quite different from the approaches discussed previously. Small, deliberate errors are applied to the wavefront to be measured. The corresponding changes are monitored with some criterion of image sharpness, usually the *rms* contrast. Changes that improve the sharpness criterion are amplified until the image sharpness is at an optimum. This method is independent from the shape of the structure under observation, but it can be shown that, for solar small-scale structure, the signals are quite small (von der Lühe 1987*c*). It seems feasible to apply this method to a few lowest-order error terms in a modal control system (Wenhan et al. 1988).

VI. INTERFEROMETRIC METHODS

Since the early 1920s, interferometry was applied to astronomical sources in order to overcome atmospheric and instrumental resolution limits. The well-known Michelson stellar interferometer was used to measure the separation of double stars, and the diameters and center-to-limb variation of supergiants (Michelson 1927). Interferometry was applied to small structures of the Sun only recently (Harvey 1972). Introductions to interferometric methods in astronomy can be found, e.g., in Dainty (1984), and in Roddier (1987).

One can distinguish between multiple-aperture and single-aperture interferometry. In the first case, the light of several, distinct entrance apertures is combined in a single focus where a fringe pattern is produced. This may be done by using a number of co-pointing telescopes or a mask in the front of a large telescope. A special case of a multiple-aperture interferometer is a nonredundant array in which each combination of two elements has a unique spacing. In the second case, a short exposure, seeing-distorted picture is viewed as an interferogram which contains information about the source up to the diffraction limit of the aperture. This information is extracted from a number of those pictures by speckle-interferometric methods. The following two sections discuss these two approaches.

A. Michelson Interferometry and Nonredundant Arrays

A Michelson stellar interferometer consists in principle of a mask with two circular openings with diameters D and separation d, placed at the entrance pupil of a large telescope (Fig. 12). Each opening produces an Airy pattern in the telescope focus; the superposition of the two patterns generates a fringe pattern perpendicular to the direction of separation of the holes. The angular frequency of the fringes is $s_i = 2 \times 10^5 \cdot \lambda/d$ in line pairs per arcsec. The transfer function, i.e., the autocorrelation of the pupil function, of the interferometer is also shown in Fig. 12. It is seen that the interferometer effectively acts as a spatial filter that transmits information at low spatial frequencies ($s \leq s_T = 2 \times 10^5 \lambda/D$) and around the interferometer frequency

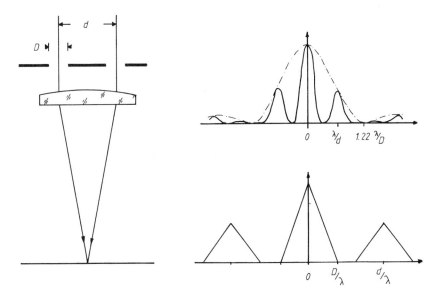

Fig. 12. Left: principle of a Michelson stellar interferometer. Top right: intensity profile in the focal plane, traced in the direction of the separation of the two apertures. The profile is a composite of an Airy profile (first minimum at $1.22 \lambda/D$ in angular coordinates) and the cosine pattern produced by the combination of the two apertures (periodicity of λ/d). Bottom right:corresponding modulus of the system transfer function.

s_I, where $s_I - s_T \leq s \leq s_I + s_T$. Fringes can be detected in the focal plane if the object distribution actually contributes to these angular frequencies.

The contrast of the fringes (the "visibility") is unaffected by the atmosphere and is directly related to the object contrast, provided that the diameter D of each element is smaller than r_o. In this case, atmospheric distortions are essentially constant across each aperture. However, the difference in optical path length between the two elements determines the position of the fringes, and the object phase cannot be determined from a single measurement. This can be done, in principle, by tracking the fringes over long periods of time and by determining their average position. A Michelson interferometer is best suited for determining only object amplitudes (Aime et al. 1977).

A nonredundant array (NRA) can be viewed as a collection of Michelson interferometers with a common focus that are all sensitive of different spatial frequencies. Figure 13 presents an often-used linear configuration with 4 elements configured like an ideal Golomb ruler, and the associated transfer function. It is essential that the vector separation (the "baseline") between any combination of two holes is unique in order to obtain seeing-independent amplitudes. If there were two pairs of holes that have the same baseline, then the resulting fringe patterns would be indistinguishable, and their relative phases would be random. This results in random reduction of fringe contrast.

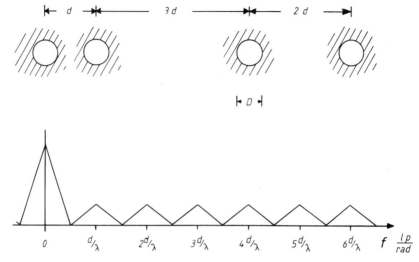

Fig. 13. Top: a linear, nonredundant array consisting of 4 apertures arranged in an ideal Golomb ruler configuration. Bottom: modulus of the associated transfer function. The transfer function can cover almost all angular frequencies up to the largest possible baseline when the aperture diameters D and the unit separation d are properly selected.

NRA's can be used for imaging the object, i.e., for determining both object amplitudes and phases in Fourier space. Let ϕ_i denote the deviation of the optical phase of the ith array element caused by the instrument and the atmosphere. The measured phase ψ_{ik} at the spatial frequency that is transmitted by holes number i and k is given with:

$$\psi_{ik} = \psi_{o,ik} + \phi_i - \phi_k . \tag{17}$$

$\psi_{o,ik}$ is the desired object phase. For each combination of three holes, a "phase closure" relation can be established that depends only on the object terms

$$\psi_{ik} + \psi_{kl} + \psi_{li} = \psi_{o,ik} + \psi_{o,kl} + \psi_{o,li} . \tag{18}$$

Unfortunately, there are not sufficient closure relations for arrays with a small number of elements to determine all object phases uniquely. Also, it is usually necessary to detect fringes for all baselines, which can be a problem if the structure contrast is low. More information can be recovered by using multiple arrays (Zirker and Brown 1986). The diameter of array elements can be increased, and the interferometer can become more efficient, if wavefront tilts across element apertures are accounted for (Zirker 1987).

B. Speckle Interferometry

Speckle interferometry is applied to data taken with a large, filled aperture telescope. The name has its origin from the appearance of star images taken with short exposures (specklegrams). The light is scattered over an area with an angular diameter of λ/r_o (seeing disk) and consists of a random pattern of bright points (speckles) that have a typical angular diameter of λ/D. The contrast of the pattern decreases if the object is theoretically resolved by the telescope, but the speckles are always present so long as the object contains structure that is not resolved by the atmosphere, i.e., structure that appears at an angle smaller than λ/r_o. Direct averaging of many specklegrams results in a blurred picture equivalent to a long exposure. The information on small-scale structures can be retrieved by computing "intelligent" averages.

The Labeyrie method.   Labeyrie (1970) proposed a method of calculating the average autocorrelation of a set of specklegrams, reasoning as follows. Consider a double star that is theoretically resolvable by the instrument, but is not resolved by the atmosphere. Both components cause identical speckle clouds which are shifted according to the separation of the stars and which do not interfere. This means that each speckle occurs twice, with the same direction and separation as the double star. Averaging the autocorrelation of many specklegrams would eventually reveal the "true" double-star autocorrelation on top of a broad, smooth background.

Similar arguments hold for a random small-scale structure, such as solar granulation. Speckle-interferometric methods were applied to solar observations as early as 1973 (Harvey and Breckinridge 1973). Much work was done on the recovery of power spectra of solar small-scale structure (Harvey and Breckinridge 1973; Harvey and Schwarzschild 1975; Aime et al. 1978b; Aime and Ricort 1979; von der Lühe and Dunn 1987). An analysis of power spectra in spectral lines was performed (Aime et al. 1985), which allows one to study the coherence between intensity and velocity fluctuations of granulation.

Speckle techniques are most suitably presented with Fourier domain quantities. Let $F_i(\bar{s})$ be the Fourier transform of the ith in a series of N specklegrams, and let $F_o(\bar{s})$ denote the Fourier transform of the intensity distribution of the object as seen with an aberration-free, nondiffracting telescope without seeing (object distribution). For the time being we may assume that imaging is a linear process and that the effects of the telescope and the atmosphere can be expressed with an instantaneous optical transfer function $S_{sys, i}(\bar{s})$ (see Eq. 7). F_i is then related to F_o via

$$F_i(\bar{s}) = F_o(\bar{s}) \cdot S_{sys, i}(\bar{s}) . \tag{19}$$

The fluctuations of the optical path in the atmosphere continuously modify $S_{sys,\,i}$, which behaves randomly within an envelope given by the diffration-limited transfer function of the telescope (Fig. 14a). The fluctuations of $S_{sys,\,i}$ are correlated over angular frequency scales of r_o/λ. This scale is called the "seeing limit." The direct average (Eq. 20) is essentially zero for frequencies larger than the seeing limit (Fig. 14b).

$$\frac{1}{N}\sum_{i=1}^{N} S_{sys,\,i}(\bar{s}) \qquad (20)$$

The Fourier transform of the autocorrelation of a function is the power spectrum (squared modulus) of the function's Fourier transform. Averaging the power spectra of the specklegrams is therefore equivalent to averaging autocorrelation functions. With the Labeyrie method, we can obtain an estimate of the Fourier amplitude of the object distribution from an average of specklegram power spectra:

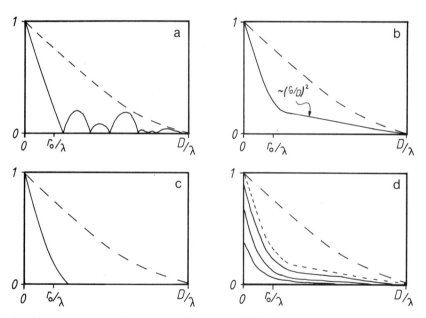

Fig. 14. Illustration of transfer functions relevant to speckle interferometry. The wide-dashed line in all panels represents the diffraction-limited transfer function of a telescope with a circular entrance aperture of diameter D (Eq. 11). (a) Modulus of a sample instantaneous optical transfer function. (b) Direct average optical transfer function. (c) Speckle transfer function (Eqs. 21 and 22). (d) Knox-Thompson cross spectrum transfer functions, for various magnitudes of Δ. The short-dashed curve shows the corresponding speckle transfer function. The solid cross spectrum transfer function curves decrease for increasing Δ.

$$\frac{1}{N} \sum_{i=1}^{N} |F_i(\bar{s})|^2 = |F_o(\bar{s})|^2 \left[\frac{1}{N} \sum_{i=1}^{N} |S_{\mathrm{sys},\, i}(\bar{s})|^2 \right] . \qquad (21)$$

This requires that the sum in the right-hand side of Eq. 21 is known. The bracketed term on the right-hand side of the equation is called the speckle transfer function (STF). There are two components (Fig 14c): a low-frequency term, that is essentially the average optical transfer function (Fig. 14b) squared, and a high-frequency tail that extends to the diffraction limit of the telescope. It is this term that permits retrieval of small-scale information.

Calibration. The correction for the STF is one of the most prominent problems in speckle interferometry. Nighttime observers usually make observations of a nearby bright star which is unresolved, interspersed with observations of the program source, and use its average power spectrum as an estimate for the STF. This approach has its problems, because seeing conditions might be slightly different while observing program and reference objects (Christou et al. 1985). Solar observers have no stable, known reference source at hand and thus have to resort to indirect methods. One of these methods (Bruck and Sodin 1980) relies on the observation that the STF can be approximated with

$$\langle |S_{\mathrm{sys},\, i}(\bar{s})|^2 \rangle \cong |\langle S_{\mathrm{sys},\, i}(\bar{s}) \rangle|^2 + \beta \left(\frac{r_o}{D} \right)^2 S_D(\bar{s}) . \qquad (22)$$

The first term on the right-hand side is the long-exposure transfer function. $S_D(\bar{s})$ is the diffraction-limited transfer function of the telescope (Eq. 11), and $\beta = 0.341$ is a constant. Object amplitudes can be recovered by subtracting a properly scaled fraction of the long-exposure power spectrum, i.e., $|\sum_{i=1}^{N} F_i(\bar{s})|^2$, and by rescaling the result. This requires the knowledge of r_o. It is often more precise to apply models of the STF (Korff 1973) rather than to use the method just described. This method is particularly imprecise at angular frequencies around r_o/λ where there may be a large amount of object information.

One can estimate r_o from the data set by one of two methods. Aime et al. (1978a) proposed analyzing the ratio of the average power spectra of two different data sets. Signatures in the ratio permit one to derive r_o for both sets. It is not necessary that the two sets are taken at the same time or even show the same structure as long as the statistics of the objects are the same (e.g., if solar granulation at the Sun's center in a quiet region is observed both times). Von der Lühe (1984) proposed a different technique that estimates r_o from a single data set by analyzing the ratio of the squared modulus of the average Fourier transform and the average power spectrum. The result is essentially the quotient of the long-exposure transfer function (Fig. 14b) and the STF (Fig. 14c), which is approximately unity up to the seeing limit and then quickly falls off. An estimate of r_o can be obtained from this decay.

Another important issue is calibration for noise. The noise present in the specklegram transforms produces a bias in the average power spectrum. Most of the bias is deterministic and can be subtracted. For nighttime sources, that have high-contrast, photon noise is correlated with image contents. This produces object-dependent cross terms in the average power spectrum. Since solar structure has mostly low contrast, cross terms are not all that important. Noise power spectra can be obtained from flat-field data (von der Lühe and Dunn 1987).

Speckle Imaging.   Images of the observed object at very high resolution can be obtained if one recovers Fourier phases along with Fourier amplitudes. There are mainly two extensions of the Labeyrie method that work successfully on the Sun, the Knox-Thompson method (Knox and Thompson 1974), and the speckle-masking method (Weigelt and Wirnitzer 1983). The application of the Knox-Thompson method to solar observations has been intensively investigated by a group at the Center for Astrophysics (Stachnik et al. 1977, 1983) and by the author (von der Lühe 1985, 1987a,b). Work of the application of speckle masking to solar data is in progress (Pehlemann and von der Lühe 1989).

With the Knox-Thompson method, averages of so-called cross spectra are analyzed:

$$\frac{1}{N}\sum_{i=1}^{N} F_i(\bar{s}) \cdot F_i^*(\bar{s} - \bar{\Delta}) = F_o(\bar{s})F_o^*(\bar{s} - \bar{\Delta})$$
$$\times \frac{1}{N}\sum_{i=1}^{N} S_{\text{sys}, i}(\bar{s}) \cdot S_{\text{sys}, i}^*(\bar{s} - \bar{\Delta}). \quad (23)$$

$\bar{\Delta}$ is a small, fixed shift in Fourier space. The Labeyrie method is a special case of the Knox-Thompson method for $\bar{\Delta} = 0$. A cross spectrum is generally a complex quantity. The important point is that the cross spectrum transfer function, i.e., the average on the right-hand side of Eq. (23), is approximately a *real* quantity, provided that $\bar{\Delta}$ is smaller than r_o/λ (it quickly tends to zero if $\bar{\Delta}$ is larger [Fig. 8d]). The explanation is that $S_{\text{sys}, i}(\bar{s})$ is correlated over scales of the order of r_o/λ, and the imaginary part of the product $S_{\text{sys}, i}(\bar{s}) \cdot S_{\text{sys}, i}^*(\bar{s} - \bar{\Delta})$ is much smaller than its real part. The phase of the average observed cross spectrum is approximately the phase of the cross spectrum of the object distribution. A few calculations show that the cross-spectrum phases represent a matrix of differences of object phases at frequencies that are separated by $\bar{\Delta}$, and summing techniques can be used to recover the object phase. Two cross spectra with linearly independent $\bar{\Delta}_{1,2}$ are required to recover the phase of two-dimensional objects. The situation is quite analogous to the recovery of a deformed wavefront from a set of wavefront-tilt measurements (Sec. V.B). More spacings $\bar{\Delta}_i$ can be added with an appropriate increase in effort; this method is called the extended Knox-Thompson method (Beletic 1988). Once object phases are reconstructed, the result can be combined with amplitudes recovered with the Labeyrie method. An inverse Fourier transform yields a reconstructed image. Fourier amplitudes could in principle be recovered from cross spectra, but the Labeyrie method is used because of signal-to-noise considerations.

A more sophisticated method is the speckle masking technique (Weigelt and Wirnitzer 1983)—other terms used are bispectrum or triple-correlation analysis. A bispectrum $F^3(\bar{s},\bar{t})$ of a Fourier transform $F(\bar{s})$ is the triple product:

$$F^3(\bar{s},\bar{t}) = F(\bar{s}) \cdot F(\bar{t}) \cdot F^*(\bar{s} + t) . \qquad (24)$$

Here $\bar{s}$ and $\bar{t}$ are independent frequencies so that, for two-dimensional Fourier transforms, the bispectrum has four dimensions. The essence of the method is to average the bispectra of the set of specklegrams

$$\frac{1}{N} \sum_{i=1}^{N} F_i^3(\bar{s},\bar{t}) = F_o^3(\bar{s},\bar{t}) \times \frac{1}{N} \sum_{i=1}^{N} S_i^3(\bar{s},\bar{t}) \qquad (25)$$

where F_i^3, F_o^3, S_i^3 denote bispectra of the observed specklegrams, the object and the instantaneous optical transfer function, respectively. The sum on the right-hand side is another transfer function, the bispectral transfer function. As with the Knox-Thompson technique, the bispectral transfer function is purely real, so the phase of the observed bispectrum is that of the object bispectrum. The information contained in a bispectrum is highly redundant and object phases can be recovered with high precision.

It has been shown (Roddier 1986; Cornwell 1987) that speckle masking is related to phase closure. A phase closure relationship is accordingly used to recover object phases:

$$\psi(\bar{s} + \bar{t}) = \phi(\bar{s},\bar{t}) - \psi(\bar{s}) - \psi(\bar{t}) . \qquad (26)$$

Here, ψ and ϕ are phases of the object and the bispectrum, respectively. Many combinations of $\bar{s}$ and $\bar{t}$ lead to the same sum $\bar{s} + \bar{t}$; this explains the high redundancy of the technique. Object phases have to be iteratively reconstructed. Three initial phase values need to be supplied, one at the origin (which is zero) and two that determine the position of the reconstruction.

Speckle masking can be viewed as extending the Knox-Thompson method to variable Δ, but useful information is still present when both $\bar{s}$ and $\bar{t}$ are larger than r_o/λ. The required computer resources are substantial, however. Memory usage of a complete bispectrum is prohibitive for reasonable sizes of the field of view, even if all symmetry relations are exploited in order to reduce the bispectrum size. One has to resort to using only a fraction of the available information by suitably truncating the bispectrum (Pehlemann and von der Lühe 1989).

Speckle Interferometry and Anisoplanatism.   When imaging conditions are not isoplanatic, the simple imaging Eq. (19) is no longer valid. Equation (7) has to be used to calculate the average power spectrum, the average cross spectrum, and the average bispectrum, respectively. This was done for the average power spectrum (Roddier et al. 1982); the result is for a single, turbulent layer

$$\langle |F_i(\bar{s})|^2 \rangle = |F_o(\bar{s})|^2 \cdot |\langle S_{\text{sys,}} \bar{s} \rangle|^2 + \left(\frac{r_o}{D}\right)^2 \iint CC_o(\bar{\theta}) \, A(\bar{s}, \bar{\theta}h/\lambda) \quad (27)$$
$$\exp[-2\pi j \bar{\theta}\bar{s}] \, d\bar{\theta} \, .$$

$CC_o(\bar{\theta})$ is the cross correlation function of the object intensity distribution. A is a term that describes the anisoplanatic effect. Its width determines the isoplanatic angle which is also a function of spatial frequency. The integral represents a Fourier transform. Comparison with Eq. (22) shows that the result is the same for frequencies smaller than the seeing limit. Imaging of low frequencies is always isoplanatic—this is what one would expect.

The relevance of Eq. (27) is easier to understand if the Fourier transform is carried out:

$$\langle |F_i(\bar{s})|^2 \rangle \simeq |F_o(\bar{s})|^2 \cdot |\langle S_{\text{sys,}} \,_i(\bar{s}) \rangle|^2 + \left(\frac{r_o}{D}\right)^2 |F_o(\bar{s})|^2 \otimes B(\bar{s}, \bar{s}') \, . \quad (28)$$

The effect of anisoplanatism at large frequencies is effectively a convolution $\otimes$. B is the Fourier transform of A, so its width is reciprocal to the size of the isoplanatic patch. The corresponding results for the Knox-Thompson and speckle-masking techniques are not much different from Eq.(28).

Anisoplanatism does not affect the results of the Labeyrie method much if statistical properties of the observed structure is studied, and its spectrum is smooth. An example would be the spectrum of solar granulation (Aime and Ricort 1979). Speckle imaging techniques are seriously affected. With the Knox-Thompson method, observed Fourier phases at closeby frequencies are no longer independent because of the convolution operation in Fourier space. The estimates of object phase differences are therefore no longer good. This can be overcome by decreasing the resolution in Fourier space or, equivalently, by reducing the size of the reconstructed field to less than the isoplanatic area. Larger fields can be reconstructed by dissection into appropriately sized subfields, which are independently reconstructed and recombined (von der Lühe 1987*b*). This approach works only so long as the subfield size remains larger than the seeing disk, which means that the Knox-Thompson method cannot be applied to an extended source when the isoplanatic patch is smaller than the seeing disk. Figure 15 shows an example where the Knox-Thompson technique was repeatedly applied to the same set of data, but with different subfield sizes.

Speckle masking does not suffer that much from anisoplanatism. Phase closure relations can be established over distances in Fourier space that are

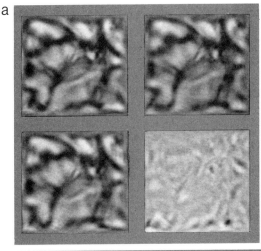

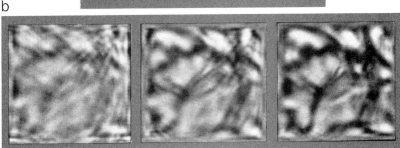

Fig. 15. Demonstration of the Knox-Thompson image reconstruction of solar features. The size of the field of view is 5.5 arcsec in all cases shown. (a) Reproducibility of the reconstruction process. The upper two frames present reconstructions from the even-numbered and odd-numbered 50 pictures out of a set of 100, the lower-left frame shows the reconstruction using the complete set. The lower-right panel shows the difference of the upper two frames. (b) Reconstructions using different subfield sizes. The picture to the left was obtained by reconstructing the full field; the center and right pictures were obtained by reconstructing fields that correspond to 1/4 and 1/16, respectively, of the total field size, and recombining the subfield reconstructions.

larger than the width of B in Eq. (28) and results in better stability of the reconstructed phase. However, the distance over which the relations can be established depends on the size of the bispectrum if it was truncated. This is demonstrated in Fig. 16 which shows examples of speckle masking reconstructions with various sizes of the truncated bispectrum.

VII. HIGH-RESOLUTION OBSERVATIONS FROM SPACE

Telescopes that operate outside the Earth's atmosphere are, of course, not affected by seeing and are not limited by the atmospheric transmissivity that restricts groundbased observations to a few, relatively narrow frequency bands. In particular, the near and vacuum ultraviolet bands as well as the full infrared regime can only be accessed from space, potentially free from any terrestrial interference.

However, space experiments are expensive; a telescope in space costs at least one order of magnitude more than a groundbased telescope of comparable size, which limits the aperture size and thus the resolution of the instrument. Space experiments need several years of development time and the design is usually frozen years before observations can be carried out. Requirements of strict product assurance result in the implementation of well-understood, standard technology, so space technology actually tends to be conservative. On-board data storage or telemetry limit the amount of data that can be taken. If an experiment operates on a free-flying satellite rather than as a shuttle payload, it may become functionally restricted or even lost if anything breaks after launch. It is almost impossible to access and upgrade instrumentation when the instruments are in space.

Some recent and planned space missions that are related to high-resolution observations are briefly described in the next sections.

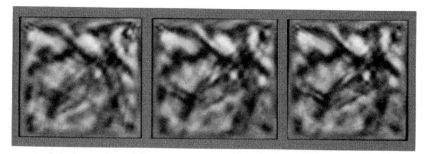

Fig. 16. Speckle masking reconstructions from the same data set that was used for Fig. 15, for different sizes of the bispectrum.

A. Solar Optical Universal Polarimeter

The Solar Optical Universal Polarimeter experiment was the first orbital space experiment dedicated to high-resolution observations of the Sun in visible light. The experiment was designed and built by the Solar and Optical Physics Laboratory of the Lockheed Palo Alto Research Laboratories for multiple flights on the Space Shuttle. Its first (and only) flight was on the August 1985 Spacelab II mission. A sketch of the experiment is shown in Fig. 17. The experiment consists of a 30 cm Cassegrain telescope, a universal tunable filter with a set of prefilters, and electronic and film cameras that record filtergrams and the photosphere in white light. The universal filter covers the 5100 Å to 6600 Å range with a spectral resolution of 55 mÅ full width at half maximum. A 244 × 248 pixel video CID camera was used to record the filtergraph data. Polarization optics allowed for generation of magnetic field maps.

During the flight, the experiment was troubled with a variety of problems that limited the observing time to two days (Acton and Pehanich 1987). After the flight, it turned out that only the data from the white-light film camera could be analyzed. Although the amount of data was far less than what was expected, the analysis of long time series of data with consistently high quality resulted in much deeper insight into the dynamics of granulation, and the relation of granular convection with magnetic fields in active regions. Fig. 18 shows a picture taken with the white-light camera of the Solar Optical Universal Polarimeter.

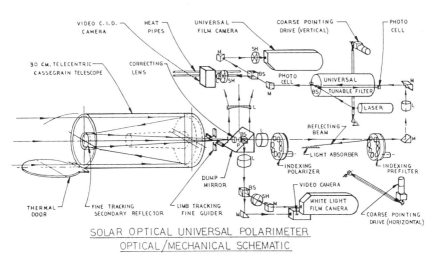

Fig. 17. An optical cartoon of the Spacelab II Solar Optical Universal Polarimeter experiment (figure courtesy of Lockheed Palo Alto Research Laboratory).

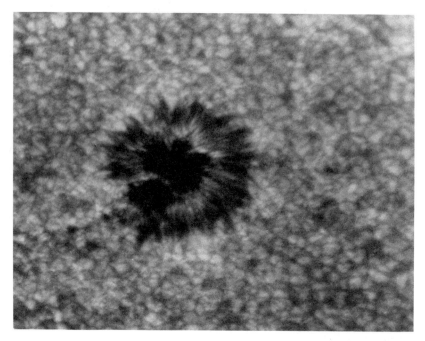

Fig. 18. Picture of a sunspot obtained with the white-light film camera of the Solar Optical
Universal Polarimeter (figure courtesy of Lockheed Palo Alto Research Laboratory).

B. Orbiting Solar Laboratory

The Orbiting Solar Laboratory, expected to be launched in 1995, will be
a free-flying complement of scientific instruments to observe the surface and
upper atmosphere of the Sun over a spectral range from the X ray to the near
infrared. The main instrument will be a 1-m Gregorian telescope that feeds a
narrowband tunable filter, a set of fixed broadband filters and a spectrograph.
These three instruments are mounted together in a common structure to form
a coordinated instrument package. An ultraviolet spectrograph and an XUV/
X-ray imager are desired co-pointing instruments that have their own independent optical systems. The coordinated instrument package and the co-
pointing systems operate with overlapping fields of view.

Figure 19 shows an artist's conception of the Orbiting Solar Laboratory
pointing at the Sun. The large central cylinder contains the 1-m aperture,
$f/24$ telescope with 0.13 arcsec resolution at 500 nm. The field of view is 3.9
arcmin. The wavelength coverage ranges from 200 nm to 1200 nm. The two
smaller cylinders attached on the sides indicate the ultraviolet spectrograph
and the XUV/X-ray imager. The coordinated instrument package is seen on

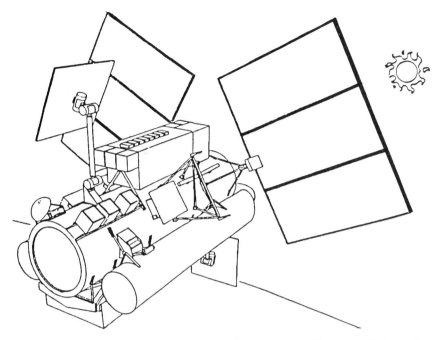

Fig. 19. Sketch of the Orbiting Solar Laboratory (figure courtesy Lockheed Palo Alto Research
Laboratory).

the top of the telescope. The large square panels are solar arrays and the small
ones are phased array TDRSS antennae.

The coordinated instrument package consists of a photometric filtergraph, a tunable filtergraph and an Echelle spectrograph. An optical cartoon
is seen in Fig. 20. The photometric filtergraph provides CCD images with a
resolution as high as 0.075 arcsec/pixel at 15 wavelengths in the range between 2100 Å and 9950 Å, with bandwidths between 400 mÅ and 100 Å.
The tunable filtergraph has a wavelength range from 4600 Å to 6600 Å, the
spectral resolution can be set to 50 mÅ or 100 mÅ. The spectrograph covers
several bands in the 2700 Å to 10,000 Å regime using 2 interchangeable
Echelle gratings, 4 CCD detectors, and 4 sets of blocking filters for each
camera.

A version of the High-Resolution Telescope and Spectrograph is studied
as the ultraviolet spectrograph experiment. The XUV/X-ray imager will probably be an array of normal-incidence telescopes with multilayer coatings.

The satellite will be launched on a Delta II-class vehicle. The design
lifetime is 3 yr, but it is anticipated that the mission will last as long as 8 yr.
A circular orbit at 510 km with 97°4 inclination provides continuous observation of the Sun for 260 days per year.

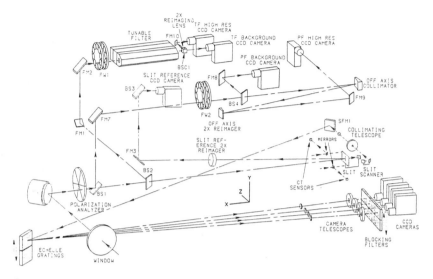

Fig. 20. An optical cartoon of the coordinated instrument package of the Orbiting Solar Laboratory (figure courtesy Lockheed Palo Alto Research Laboratory).

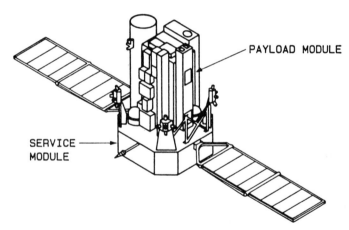

Fig. 21. Sketch of the Solar and Heliospheric Observatory spacecraft (figure courtesy European Space Agency).

C. Solar and Heliospheric Observatory

Although not strictly a high spatial resolution mission, the Solar and Heliospheric Observatory (SOHO) features a number of experiments that provide unprecedented angular resolution in the ultraviolet. The SOHO mission is a part of ESA's Solar/Terrestrial Physics Program to be carried out in collaboration with NASA, the first of 4 cornerstones of ESA's long-term pro-

TABLE II

Solar and Heliospheric Observatory Experiments and Spatial Resolution

Experiment[a]	Description	Spectral Range (nm)	Spatial Resolution (arcsec/pixel)
MDI/SOI	Velocity oscillations, high degree modes	676.8	0.7 (2.0)
SUMER	Plasma flow characteristics, chromosphere through corona	50–160	1.0
CDS	Temperatures and density, transition region and corona	15–80	2.0
EIT	Evolution of chromospheric and coronal structures	17.1, 19.5, 28.4, 30.4	3.0

[a]MDI/SOI: Michelson Doppler Imager/Solar Oscillation Imager; SUMER: Solar Ultraviolet Measurements of Emitted Radiation; CDS: Coronal Diagnostic Spectrometer; EIT: Extreme (UV) Imaging Telescope.

gram. The main objectives of SOHO are the study and understanding of solar coronal phenomena and the study of the solar structure and interior dynamics. Coronal and solar-wind studies will be carried out by remote sensing of the solar atmosphere and *in situ* measurements of the solar wind, structure, and interior dynamics will be studied by helioseismological methods.

The spacecraft (Fig. 21) carries 12 experiments, 3 for helioseismology studies, 6 for remote sensing of the solar atmosphere, and 3 for *in situ* solar-wind measurements. The satellite will be located in a halo orbit at the L_1 Lagrangian point approximately 1.5×10^6 km towards the Sun. This location has the advantage of a stable thermal environment, unobstructed sunlight and lack of interference from the Earth's aureole. Table II lists some SOHO experiments and their spatial resolution.

HIGH-RESOLUTION OBSERVATIONS OF THE SOLAR GRANULATION

K. P. TOPKA and A. M. TITLE

Lockheed Palo Alto Research Laboratory

The opportunity to observe the solar surface from space and from excellent groundbased sites, combined with enhanced digital image recording and computer technology, has led to improved high-resolution observations of solar granulation in the last decade. Today, granulation movies over 1 hr long are available with spatial resolution of nearly $1/3''$ (= 240 km on the solar surface). These new data have modified our conception of solar granulation by showing it to be a more complicated combination of different phenomena than was previously appreciated. For example, granules were formerly thought to be rising thermal plumes due to convection. While this is probably true for the larger granules, it is probably not true for many if not most of the smaller granules. Instead, they may be the result of the fragmentation of larger granules due to turbulent small-scale flows. The new data have been used to detect large-scale horizontal flows on the Sun's surface, by the direct measurement of the proper motion of granules during their lifetimes. These motions can have a substantial impact on the interpretation of granule evolution, because few if any granules remain at rest during their lifetimes. These new results may also have important implications for the heating of the solar chromosphere and corona. Also, the observed intensity pattern on the surface of the Sun is not caused by granulation alone, but is due to the superposition of different types of waves as well. Furthermore, all of these phenomena are modified in the presence of magnetic fields. The granules change appearance in magnetic areas, becoming smaller and lower in contrast. The vertical velocities of the granules, horizontal flows and waves due to the 5-min oscillation are all suppressed.

Solar granulation is the fundamental structure visible in the photosphere when the Sun is examined outside of sunspots and pores in broadband visible light or in narrowband continuum. It resembles a bright quasi-polygonal cellular structure, $\sim$ 1″ to 2″ large (1″ = 720 km on the solar surface), and separated by dark intergranular lanes narrower than the cells. By about 1980, a self-consistent picture had emerged in which granulation was thought to be the observational manifestation of solar convection, due to the convective overshoot of rising thermal plumes. The basic properties of the granules, including size and lifetime, were considered well measured, although the nature of the convective processes and how the thermal plumes rise into the photosphere was not well understood.

Our understanding of convective processes in the Sun comes from optical observations of the photosphere, helioseismological observations of the convection zone below the surface, and from theories and simulations of compressible convection. Significant progress has been made in all of these fields in recent years. In this chapter, we summarize the new, more complicated, picture of solar convection that has emerged as a result of improved high-resolution observations. For a textbook on granulation, see Bray et al. (1984). Other recent reviews include Wittman (1979), Beckers (1981), Nordlund (1985b) and Spruit et al. (1990). For recent workshop and conference proceedings see Schmidt (1985), Schroter et al. (1987), Rutten and Severino (1989), von der Lühe (1989a) and Stenflo (1990).

I. OBSERVATIONAL TECHNIQUES

While individual images of solar granulation with very high resolution have been produced for many years from balloon-borne telescopes and from the ground, only recently have high-quality time sequences (movies) been produced. This was due to guiding errors, focus shifts, and thermal bending of the optics in earlier experiments (Schwarzschild 1959; Mehltretter 1978). Recent improvements include spaceflight opportunities, the establishment of new solar observatories at sites with excellent seeing, active optics, and (perhaps most important) improved digital image recording and selection techniques at new and existing observatories.

The Solar Optical Universal Polarimeter (SOUP) instrument was built by the Lockheed Palo Alto Research Laboratory. It consists of a 30-cm Cassegrain telescope, with an active secondary mirror. Image stabilization was accomplished by driving the secondary mirror using 3 piezoelectric transducers whose control signals were generated by a set of 4 solar limb sensors located in the primary focal plane. The focal-plane instrument package contained a birefringent filter tunable over 5100 to 6600 Å, with a 0.05 Å bandpass, a set of 8 prefilters on a rotating wheel, a 35-mm film and a 244 × 248 pixel CID camera behind the filter, and a separate white-light system with 35-mm film and TV cameras (Title et al. 1986; see von der Lühe chapter, Fig. 17).

SOUP flew on Space Shuttle Challenger as part of Spacelab 2 from 29 July to 6 August, 1985. Pointing was provided by the European Space

Agency Instrument Pointing System. The image stabilization and white-light systems worked very well. Image stability of 0."003 *rms* on each axis was achieved. Some 6000 solar images were obtained from the white-light frames on Kodak SO 253 film. The field of view was 166" × 250", and images were taken starting in orbit 100 (1985 August 5, 04:01:22 UT), and continuing until orbit 111 (1985 August 5, 21:08:24 UT). These data are diffraction limited at 0."5 resolution and are free of distortions caused by atmospheric turbulence or image motion.

In addition, time sequences of very high quality have now been produced from groundbased sites that have excellent seeing. The Swedish Solar Telescope at the Observatorio del Roque de Los Muchachos on La Palma in the Canary Islands is located at 2400 m elevation at a site which often has exceptional seeing (0."3 or better). The alt-azimuth telescope has an aperture of 50 cm and is of nearly perfect optical quality (Scharmer et al. 1985; Scharmer 1987). The observatory has been in operation since 1987 and has produced some outstanding time sequences of continuum images of sunspots and granulation by using a CCD video camera coupled with real-time image selection. The best frames have a resolution of about 0."25, although some "rubber sheet" distortion due to atmospheric seeing is still present (Scharmer 1989). Image sequences of similar quality have also been produced at Pic du Midi by digitizing the best images from film taken in "burst mode." High-quality granulation movies have also been produced at Big Bear Solar Observatory (BBSO) using a CCD video camera and high-quality 0.75 inch video tape to record all images. The best video frame present in each 10-s interval was then selected from the recorded tape and used to produce a final digital movie. (Zirin and Wang 1989).

New methods for analyzing time sequences of granulation images are now possible due to the availability of digital image processing. Improvements in computer speed and on-line memory capability now make processing of large sets of digital images possible. These methods include: (a) digital registration and destretching to remove telescope tracking errors and some atmospheric seeing distortion; (b) space-time filtering (3-dimensional Fourier filtering) to separate convective processes from the 5-min oscillations; and (c) local correlation tracking for measuring granule proper motions (horizontal flows).

II. INTENSITY FLUCTUATIONS

Any theoretical model of solar convection must be able to predict the correct temperature variation due to convection, which is usually measured by the *rms* contrast in the continuum. Typical high-quality granulation pictures show *rms* contrasts of about 5% in uncorrected images (Wittman 1979). Previous measurements, summarized by Bray et al. (1984, p. 69), range from 6.6 to 8.8%. To date the highest-contrast raw images have been obtained at the Swedish Solar Telescope on La Palma. They measure *rms* contrasts as high as 10 to 11% in a 25 Å band centered at 4600 Å. The measured values need to be corrected for the telescope and atmospheric seeing modulation transfer functions (MTF), but the detailed theory needed for the instantaneous atmospheric MTF has not yet been developed. Estimates of the corrected *rms* contrast from previous observers range from 7 to 17.9%. Karpinsky (1990) reported 22% *rms* at 5000 Å based on observations from the Soviet Stratospheric Solar Observatory. The correction value is large and sensitive to the details of the telescope and atmospheric MTF, so the final value at 4600 Å could be 25% or higher.

Part of the observed *rms* contrast in the continuum is due to the solar 5-min oscillation. This is because these waves are pressure driven, causing successive compressions and rarefactions in the gas in the photosphere. Because the characteristic cooling time of the gas at this layer is short compared to 5 min, these pressure differences manifest themselves as intensity fluctuations. Fully 35% of the *rms* continuum contrast present in the SOUP data was due to 5-min oscillations (Title et al. 1989).

III. SIZE AND EVOLUTION

Continuum movies of the solar photosphere actually show a superposition of phenomena due to convection, 5-min oscillation, waves, horizontal flows and magnetic fields. Therefore, an accurate description of granulation evolution requires that the effects of the other phenomena be identified and isolated as much as possible. Fortunately, the effects of magnetic fields and 5-min oscillation can be separated from that due to convective processes.

The magnetic field causes substantial changes in solar convection. By observing well away from active regions and by taking magnetograms simultaneous with granulation movies and checking for the presence of network fields, one can restrict study to only quiet Sun regions. It is possible to remove the 5-min oscillation from a movie by space-time filtering. The oscillations occupy a well-defined region of Fourier ($\kappa_x - \kappa_y - \omega$) space, well separated from the solar convective processes, which are slower and smaller in scale. By Fourier transforming a granulation movie into ($\kappa_x - \kappa_y - \omega$) space, applying a filter which removes all Fourier components in the domain of the oscillations (by setting them equal to zero), and then inverse Fourier transforming back to real space, one can produce a granulation time sequence free of 5-min oscillation. We call this process space-time filtering, consistent with the work of Spruit et al. (1990); it was first applied to SOUP movies. Of course, one can also apply the opposite filter to produce a movie containing only the 5-min oscillation and free of convective processes. Such movies are useful for measuring, for example, how the amplitude of the 5-min oscillation changes in magnetic areas as compared to quiet Sun.

Other wave processes exist in the solar photosphere, but unlike the 5-

min oscillation, these cannot be separated from convection by space-time filtering. This is because they have similar spatial and temporal scales to convection and therefore occupy much of the same region as convection does in Fourier space. They are visible as traveling perturbations in the appearance of individual granules. The granules appear to fragment and often, but not always, reform. Furthermore, horizontal flows exist on the surface of the Sun (discussed below) that also cannot be easily removed.

The temporal autocorrelation (AC) function is one method that has been used to measure the lifetime of granules. It calculates the correlation in the intensity pattern at some arbitrarily chosen initial time with the intensity pattern for the same location at a later time. This method avoids the difficulty of identifying individual granules and tracking them forward and backward in time, but implicitly assumes that all of the intensity changes on the solar surface are due to granule evolution. We now know that this assumption is false. Previous measurements showed an AC lifetime of $\sim$ 6 min (Bahng and Schwarzschild 1961; Mehltretter 1978; Wittman 1979), and the SOUP raw movie results are 5 min in quiet Sun and 7 min in magnetic regions (Title et al. 1988,1990b). But after space-time filtering to remove the 5-min oscillation these correlation times increased to 7 min and 15 min, respectively. Thus, intensity fluctuations due to the p-modes as they move across the field of view give rise to substantial intensity decorrelation, causing all previous AC lifetime measurements to underestimate granule lifetimes. Furthermore, space-time filtering does not remove the effects of slower waves and horizontal flows, both of which can also cause substantial decorrelation. Therefore, even the SOUP space-time filtered AC lifetimes are only lower limits to granulation lifetimes.

A second method of measuring granule lifetimes is tracking them from birth to death on the movie. This means individual granules must be identified, either visually or by computer algorithm, and then followed forwards and backwards in time. Visual identification suffers from the subjective bias of the observer. As a result, published estimates of the number density of granules using visual identification has differed by a factor of 2 (Namba and Diemel 1969; Bray et al. 1984, Table 2.4).

The best computer algorithms used to identify granules are based on the fact that each granule is a quasi-polygonal bright structure that is surrounded by a darker intergranular lane, analogous to a mountain range where all mountains are separated from each other by a continuous valley. Leighton (1963) argued that this structure is evidence that granules originate from convective plumes. Turbulence, on the other hand, would lead to an intensity pattern that is more symmetric with respect to bright and dark features, and the size distribution of granules would then form a Kolmogorov spectrum. The highest-quality modern photographs of granulation ($\sim$ 0$\farcs$25 resolution) continue to show the same structure as described by Leighton.

The lane-finding algorithm identifies granules by examining every image pixel to identify those belonging to intergranular lanes. Typically, the criterion used is the existence of a local minimum in at least one direction. The result is a map of all lane pixels, in which a second algorithm fills in any missing pixels to form a continuous network. Areas completely surrounded by these lanes are defined to be granules. This method is considerably more accurate than simply finding local maxima corresponding to granule centers because of image noise and granule fragmentation.

Granules identified by this technique using SOUP and La Palma data have measured lifetimes on the order of 5 min, even after the 5-min oscillation is removed by space-time filtering (Title et al. 1989). Granule lifetimes based on visual identification are typically around 15 min (Mehltretter 1978; Wittman 1979). The lane-finding algorithm also measures granule size (by counting the number of pixels inside each closed boundary). The size distribution of granules shows no cutoff, or even turn over, at the small end (Roudier and Muller 1986; Title et al. 1989). The number density of granules increases monotonically with decreasing size until the resolution limit is reached. Figure 1 shows an image of quiet Sun granulation taken at the Pic du Midi Observatory. The limiting resolution on this frame is about $0''.25$. Note the number of small features present at or near the resolution limit. This is consistent with a Kolmogorov spectra expected if the smaller granules originate from turbulence. However, there does appear to be a cutoff in number density at the larger end, corresponding to a size of $3''$. Exploding granules (see Sec. IV) are known to expand to a size larger than $3''$, but they always fragment before reaching this size. Although the number density of granules increases towards smaller sizes, the total area covered by small granules does not. This is because the number density is increasing at a rate that is less than the square of the inverse of the average size. The area covered by granules as a function of their size has a broad peak between $0''.9$ and $1''.3$. The exact location of this peak is algorithm dependent, because different algorithms can vary by more than a factor of 2 in the number of small granules they find. The observed distribution of granule separation also shows a well-defined peak that varies from $1''.3$ to $1''.7$, depending on algorithm.

The fact that the observed size distribution of granules shows no cutoff so far at the small end naturally raises the question of whether a minimum granule size exists. Nelson and Musman (1978) have shown that convective plumes of arbitrarily small size probably cannot exist on the Sun because at diameters smaller than ~ 200 km, the plumes are small enough to exchange radiation efficiently *horizontally,* which would quickly eliminate all temperature fluctuations. Until a space-based high-resolution telescope such as NASA's Orbiting Solar Laboratory (Title 1989) can be launched, the best hope for resolving the finest-scale photospheric features lies with the techniques of solar speckle interferometry (von der Lühe and Zirker 1988; von der Lühe 1989b) and adaptive optics (Dunn et al. 1989; Acton 1989). The results to date tend to support a lower limit of ~ 200 km for the size of

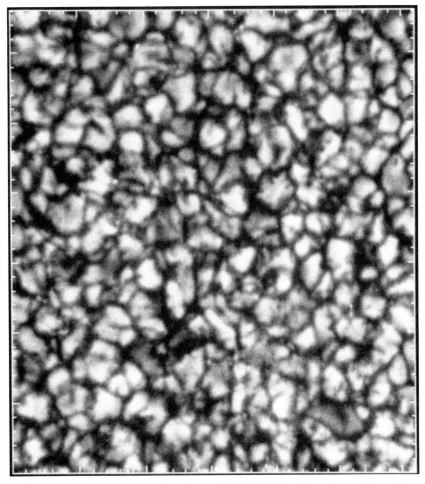

Fig. 1. Image of the solar granulation taken at Pic du Midi. The tick marks are 1″ apart. The smallest discernable features are ~ 0″.25 across.

granules (von der Lühe and Dunn 1987). This does not mean that even smaller features cannot exist in the photosphere, due, for example, to turbulent breakup of granules as they expand horizontally.

The characteristic size of granules can also be measured using the spatial AC function, which is the correlation between an intensity image and itself after a small displacement in space. The typical size of granules is then determined from the full width at half maximum (FWHM) of the spatial AC function. These results, combined with the those of the granule-finding algorithms above, allow us to say that typically granules have a size of 1″.2 to 1″.4, and a nearest neighbor distance of 1″.6 to 1″.9 (Wittman 1979; Bray et al. 1984; Title et al. 1989).

As seen in the SOUP movies the shortest-lived granules have, on average, the smallest mean sizes. This is true for lifetimes < 5 min. There is no variation of size for granules with lifetimes > 5 min. The shortest-lived granules also have the highest random velocities. The random velocities are measured by tracking the location of granule centers during their lifetime. The mean is 0.7 km s^{-1}, varying from 1.1 km s^{-1} for lifetimes < 200 s, to 0.4 km s^{-1} for lifetimes > 20 min. These results, along with the size-distribution measurements made by Roudier and Muller (1986), and Title et al. (1989), provide evidence that the smallest granules originate from turbulent eddies, rather than convective plumes.

Numerical simulations have been used to calculate the morphology and evolution of granulation (cf. Stein and Nordlund 1989). Good agreement is found between synthetic granulation and the geometrical properties of observed solar granulation (Wohl and Nordlund 1985). For further discussion on numerical simulations, see the chapter on solar convection theory by Chan et al.

IV. EXPLODING GRANULES

Granules that expand radially at 1 to 2 km s^{-1} to a size of $\sim$ 4″, while darkening in the center and eventually fragmenting were discovered by Rosch and his collaborators (Carlier et al. 1968) and called "exploding granules." Figure 2 shows the time history of a proto-typical exploding granule. Before SOUP, they were thought to be relatively rare (Mehltretter 1978; Namba and van Rijsbergen 1977), but the space-time filtered SOUP movies in particular showed them to be quite common. The birth rate is 7.7×10^{-11} km^{-2} s^{-1}, and their mean diameter at maximum expansion is 4″.2 (Title et al. 1989). This means that they would cover the entire surface of the Sun in 30 min if uniformly distributed. However, they are not uniformly distributed. Instead, they tend to occur inside mesogranule cells found in the horizontal flow field (see Sec. VI). Inside these cells, exploding granules cover the entire area every 15 min.

During the expansion phase, all neighboring granules either disappear (it appears that the expanding front of the exploder overrides them), or are displaced. The exploding granule eventually fragments into smaller pieces, some of which can persist for several minutes. Sometimes a new exploding granule will form at or near the same location as a previous exploder. The common occurrence of exploding granules and their preferred location within cells means that there is at least 3, and possibly 4, distinct modes of granule evolution. The first mode is the exploding granule process itself. The second mode is normal evolution of nonexploding granules. The third mode is interrupted evolution of nonexploding granules due to the presence of an exploder nearby. A possible fourth mode is abnormal granule evolution in magnetic areas, as discussed in Sec. VII.

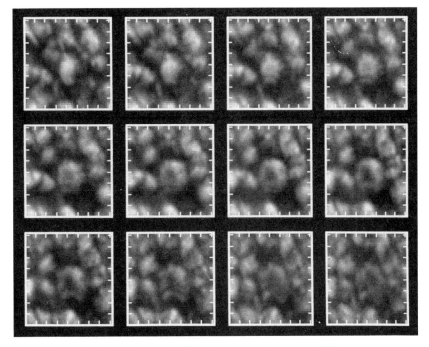

Fig. 2. Proto-typical exploding granule. These images were taken by the SOUP instrument during Spacelab 2. The images are separated by 1 min in time, and the ticks are each 1″.

Movies of solar granulation are not time invariant. They do not look the same played in reverse as when played forward because granules tend to expand during their lifetimes. In fact, all granules are observed to expand horizontally during their lifetimes unless by chance they are compressed by the expansion of a more powerful neighbor. Granules are accelerated and heated as they rise through the region of large superadiabatic temperature gradient, below the photosphere. Some of the kinetic energy present in the vertical velocity field causes a horizontal pressure gradient to form that drives the horizontal expansion. The larger the convective plume, the larger the pressure fluctuations required to drive the horizontal flow, because the horizontal velocity is roughly proprotional to the horizontal scale. When a certain critical size is reached, the energy present in the vertical velocity field becomes insufficient to set up the pressure fluctuations necessary to drive the horizontal flow, and the continuity equation cannot be satisfied. The plume cools and eventually collapses, rather than rising and expanding horizontally. This phenomenon is called buoyancy braking, and determines the maximum size of granules on the Sun ($\sim$ 4 Mm or 5″; Nelson and Musman 1978). The largest observed granules are smaller, $\sim$ 3″ across, so other factors must play a role in determining maximum granule size. Indeed, Stein and Nordlund

(1989) have shown that the turbulent breakup of large granules also plays an important role in determining their maximum size.

The large pressure fluctuations present in the largest granules explains the exploding granule phenomenon (Nelson and Musman 1978; Spruit et al. 1990). For large granules, the large pressure fluctuations causes the opacity to change, increasing in hot regions and decreasing in cool ones. The overall effect is to decrease the large-scale granule contrast. The opacity fluctuations can cause the center of granules to look darker than the expanding annulus, even though the temperature in the center is higher. This is the observational characteristic of exploding granules. The observations are consistent with exploding granules being among the largest granules present (Title et al. 1989).

V. VERTICAL VELOCITIES

The vertical velocity and intensity (or temperature) in the lower photosphere should be well correlated on a fine scale because the convective plumes overshoot the convection zone and penetrate into the convectively stable layers above. When one obtains spectra without enough resolution to resolve the granulation, the result is the well-known convective blue shift of photospheric lines, due to the correlation of velocity and intensity fluctuations (Beckers and Nelson 1978; Dravins et al. 1981). The width of the lines is a measure of the *rms* velocity. Also, when one constructs bisectors from detailed profiles of these same lines, one obtains the characteristic C-shape that is due to velocity, density and temperature fluctuations (Dravins et al. 1981). Nordlund (1985b) has compared calculations of average line profiles based on the numerical simulation of granulation to the observations averaged over many granules. This type of observation is the easiest to obtain with high precision, because very high spatial resolution and the application of large, uncertain correction factors are not needed. He finds good agreement between theory and observation with the convective blue shift, line width and C-shape of photospheric lines.

A long-standing goal of observers has been to measure the vertical velocity field after fully resolving the granulation, and then to deduce how this velocity field varies with height and magnetic field strength. Magnetic fields change the structure of granulation which in turn causes changes in the profiles of photospheric lines and wavelength shifts (Livingston 1982; Kaisig and Schroter 1983; Miller et al. 1984; Righini et al. 1984; Cavallini et al. 1985a,1987b). The interpretation of the data is made difficult because of 3 complications. First, a substantial correction needs to be applied because of finite resolution and because of image distortion due to atmospheric seeing. Even the best data may require a correction of order 2 (cf. Lites et al. 1989). Second, the 5-min oscillation contributes very significantly to the velocity field in this part of the solar atmosphere. It is difficult to separate the contribution due to granulation from that due to oscillation. Third, the measurements are made using Doppler shifts of lines that do not originate from one unique height in the atmosphere, but rather each line is formed from contributions over a range of heights.

Keil et al. (1989) report substantial changes in photospheric line asymmetry as a function of magnetic field strength. These results are consistent with the notion that magnetic fields suppress the amplitude of the vertical velocity but permit the convective plumes to penetrate higher into the atmosphere. Komm and Mattig (1989) observed the Sun simultaneously in 4 spectral ranges during a partial solar eclipse. The sharp edge of the Moon was used to correct the data for instrumental effects and for atmospheric seeing. They measured an *rms* vertical velocity due to granulation of 1.2 km s^{-1} near the photosphere, dropping rapidly to 400 m s^{-1} at a height of 200 km, and above. Nesis et al. (1989) report that the velocity field above 200 km shows no correlation with the continuum intensity due to granulation nor with the velocity field below 200 km. These results suggest that the granulation affects the vertical velocity field up to a height of $\sim$ 200 km, but not the layers higher up. Lites et al. (1989) compared very high-resolution FTS (Fourier Transform Spectrometer) observations from La Palma with synthetic granulation images from fully compressable numerical simulations. The width and depth of the simulated Fe I 6302.5 Å line agreed well with the observations, indicating that the simulations predict very nearly the correct *rms* vertical velocity. Also, the ratio of the simulated intensity *rms* contrast (16.6%) to the velocity *rms* fluctuations (1.1 m s^{-1}) agrees with the observations (8.0% and 0.52 m s^{-1}). The actual values of the intensity and velocity *rms* is larger in the simulations than in the observations because of finite telescope resolution and atmospheric seeing, the exact MTF of which is unknown. If the simulated velocity *rms* is smeared with the telescope MTF and a model atmospheric MTF made to agree with the observations (at 0.52 m s^{-1}), and the same smearing is applied to the simulated intensity *rms*, the result is 8.1%, in good agreement with the observations. Finally, the observations contain more fine-scale features in the granulation than is reproduced in the simulations.

Future work should include improved observations at highter spatial resolution, refinements in the correction applied for atmospheric seeing, and the development of dynamical models to explain the detailed shape of the bisectors at high resolution, both inside and outside of magnetic areas.

VI. HORIZONTAL VELOCITIES

The high-quality SOUP granulation movies taken in space allow for the determination of large-scale surface flows on the Sun by the direct measurement of displacements normal to the line of sight in the local intensity pattern. The measurements are made by using local correlation tracking (LCT) methods (November et al. 1987), in which a window (usually a truncated Gaussian) on a grid is applied to a registered movie. The flow is measured by determining the relative displacement that maximizes the correlation between the intensity patterns in the same window at different times. By repeating the measurement in the same window for many pairs of frames, one can determine the noise in the measurements, and the rate at which the flow changes. As granules typically live 5 to 15 min, the measurements must be made in a time short compared to this lifetime. The result has limited spatial resolution determined by the size of window, as LCT will average over this window. This technique is sensitive to the phase velocities of any traveling intensity waves present and to the intensity fluctuations due to the 5-min oscillation. The effects of the 5-min oscillation can be removed by applying space-time filtering (discussed in Sec. III). November and Simon (1988) and Brandt et al. (1988) have shown that horizontal flows can be measured using LCT and groundbased data if the seeing is sufficiently good. Bogart et al. (1988) showed that good results on large-scale surface flows using LCT can be obtained from space with spatial resolution as low as 1″.5, implying that apertures as small as 10 to 14 cm can be used.

The average flow field in quiet Sun shows a cellular pattern ranging from 4″ to 12″ in diameter in which horizontal outflow occurs. This size is considerably smaller than supergranules, but much larger than granules. These cells are probably related to the mesogranulation-scale structures detected in the vertical velocity field (November et al. 1981). Other observations reporting variations on a mesogranular scale include those by Deubner (1974), Kawaguchi (1980), Oda (1984), Deubner (1989b), Brandt et al. (1989a) and Muller et al. (1990). Mesogranules observed from the horizontal flow field span nearly the entire range in size from the largest granules to the smallest supergranules. As the size of the measuring window is changed from 0″.7 to 1″.5, the average flow speed decreases from 800 m s^{-1} to 400 m s^{-1}, but the basic flow pattern remains unchanged.

The velocity deduced from LCT is purely horizontal and can be expressed as $u_H = (u_x, u_y, 0)$. The divergence of the horizontal flow, defined as

$$\nabla \cdot u_H = \frac{\partial u_x}{\partial x} + \frac{\partial u_y}{\partial y} \tag{1}$$

is roughly proportional to the vertical component of the large-scale velocity field (November 1989). This is because of mass conservation; an upflow of new material must exist to replace that which is flowing away in areas of positive divergence (horizontal outflow), and vice versa for negative divergence areas. As discussed in Sec. V, exploding granules almost always occur in regions of positive divergence, which defines the interiors of the mesogranulation cells.

The horizontal flow field can be visualized better by calculating where an initially uniform field of hypothetical free particles ("corks") would move with time. Cork movies can be generated by moving each cork at each time step in the movie according to the local flow field at that cork's location. Divergence maps, flow-vector maps and cork movies can all be overlayed on the original granulation movies to visualize horizontal flows in mesogranules and supergranules, and to compare granule evolution inside the mesogranule cells with that occurring at the boundaries. Figure 3 shows examples of 4 frames from a cork movie illustrating how the corks move with the horizontal flow after 5 min, 15 min, 30 min and 45 min (Fig. 3 A-D). The corks start with a uniform distribution.

Typically, it takes 30 min for the corks to move from their initially uniform distribution to the mesogranule cell boundaries. The cellular pattern thus formed agrees well with that revealed by the divergence maps. The cork movies and divergence maps show that supergranules are covered by a closely packed pattern of mesogranules, which are themselves covered by a closely packed pattern of granules. Furthermore, some of the latest and as yet unpublished results reveal that the mesogranules located inside the supergranules are themselves advected towards the supergranule boundaries. Therefore, the

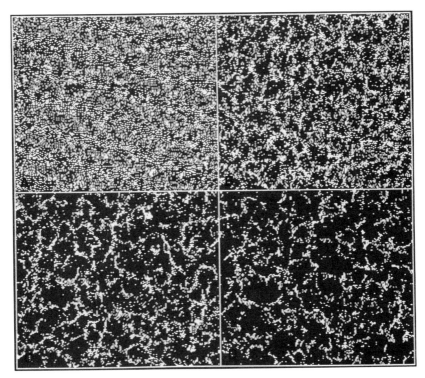

Fig. 3. Four frames from a cork movie, showing the motion of the corks after 5 min (A), 15 min (B), 30 min (C) and 45 min (D). The original granulation movie covers 58″8 × 48″1 of quiet Sun, and was taken at Pic du Midi. The corks move to the edges of the mesogranular cells, with the pattern becoming recognizable after ~ 30 min.

lifetime of mesogranules inside supergranules depends upon their location. Those formed near the center will typically live longer than those located closer to the boundary, simply because of the longer transit time to the network. Cork movies reveal that mesogranules evolve on the time scale of 1 hr. Upon reaching the network boundaries the corks do not stop moving. Instead, they continue moving along the boundary, albeit at reduced velocity. This result may be an important feature in the evolution of the magnetic field pattern.

Simon and Weiss (1989a) have shown that a simple kinematic model of convection is consistent with the observed flow fields. Their model consists of a superposition of convective plumes on the surface of the Sun, each axisymmetric with radial outward flow. Besides the flow field itself, this model can be used to predict the evolution of the magnetic field, because the magnetic network, as observed in magnetograms and in Ca II emission, is the result of the interaction between the plumes. The model does predict the existence of the network, but further predicts that the network would collapse into a relatively small number of points ("sinks") on a time scale of order 10 hr. This is not observed, which implies that the individual flux tubes offer resistence to the flow, unlike the corks which are passive test particles.

Much of the quiet Sun appears to be covered with a hierarchy of densely packed cells of all sizes ranging from the smallest granules to the largest supergranules, as many solar astronomers have always assumed. But not all of the quiet Sun is so structured, and much of the Sun in active regions lacks such well-defined cellular structure as well. The horizontal flow maps reveal the existence of several streams, some of which are 50 to 100 Mm in length and 5 to 10 Mm in width (Simon et al. 1988). In these regions, no cells larger than granules exist. Also, outflowing "moats" surrounding sunspots exist; they extend $\sim 5''$ to $8''$ from the penumbral boundary into the photosphere. In this region, the granules are advected radially outward from the sunspot at a typical velocity of 200 to 500 m s^{-1}.

Brandt et al. (1988) measured the horizontal flow field in a 79-min-long granulation movie of very high quality taken at the Swedish Solar Observatory on La Palma. They discovered a vortex that lasts for the entire movie without noticeable change. It had a diameter of ~ 5000 km, an average circulation of 4000 km s^{-1}, and a central vorticity of 1400 s^{-1}. Corks placed on a circle 2000 km in radius centered on the vortex typically require 1 to 2 hr to travel to the center. Such fluid motions can twist magnetic flux tubes and induce electric currents. This could represent a possible mechanism for heating the solar corona, especially if vortices are common on the Sun.

VII. MAGNETIC FIELDS AND GRANULATION

The structure and location of magnetic fields on the Sun can be measured directly using a magnetograph, which is sensitive to the longitudinal (line-of-sight) component of the magnetic field. The presence of a magnetic field changes the atomic emission process via the Zeeman effect, which results in measurable polarization effects in spectral lines. Today magnetograms are available with spatial resolution of $\sim$ 0″.5 (360 km). At this resolution the plage in active regions resembles "Swiss cheese," being completely filled with holes nearly devoid of field and typically ranging in size from 2″ to 5″ (Ramsey et al. 1977). When very high-resolution magnetograms are taken simultaneously with continuum images, a comparison of the structures present on both becomes possible if the alignment between these images is accurate enough. The alignment is most difficult for groundbased images, that have image distortion even at the highest resolution. Furthermore, filtergram images for the magnetogram are typically taken in the blue wing of a spectral line where the Zeeman effect is at maximum, and where there is little or no structure in common with continuum images. Very accurate image alignment can still be obtained by taking magnetograms in both the blue and red wings, and then aligning these magnetograms (Title et al. 1987c), or by taking a series of images spaced evenly throughout the spectral line and aligning successive pairs (Title et al. 1990c).

The appearance of the photosphere depends to a large extent on the size and density of magnetic elements present. When the magnetic elements are small and are spaced widely apart, corresponding to a low filling factor in the magnetogram, the granulation appears normal. The filling factor α is the fraction of the resolution element covered by line-of-sight magnetic fields of strength B. The remaining fraction 1-α is assumed to be field free. When the filling factor is very low, as in quiet Sun or weak plage, the magnetic elements are often coincident with small, bright continuum structures. These continuum structures have been called filigree, facular points and network bright points, and are seen in the continuum at very high resolution (Muller 1983a; Muller and Keil 1983; Muller et al. 1989; Muller 1990). Earlier observations showed a very likely correspondence (but not yet a definitely proved association) between the network bright points and the magnetic elements (Schussler and Solanki 1988). The latest high-resolution observations show that the magnetic elements and the continuum bright points are coincident within a small fraction of an arcsec, leaving no doubt that these structures are intimately associated (Title et al. 1990a,c). Interestingly, the relation between these bright points and the magnetic field is not one to one. There are many small, bright structures in the continuum that look identical to the network bright points but have no corresponding magnetic feature present in the magnetogram. Also, there are small magnetic features present in the magnetogram that have no continuum bright points coincident with them. The bright points are usually very small in size (0″.5 or less), form in the intergranular lane between 3 or more granules, and remain in an intergranular lane during their lifetimes, which average 18 min (Muller 1983a). These bright points occur in both quiet sun and magnetic regions.

In areas of moderate filling factor, of order 10 to 30% and corresponding to moderate-to-dense plage in active regions, the entire structure of the granulation changes. Dunn and Zirker (1973) first noticed this change of structure, and called it "abnormal granulation." The granules are smaller and the contrast between granule centers and intergranular lanes is lower (Muller et al. 1989; Title et al. 1989,1990a,c). The *rms* contrast in the best continuum frames at 6800 Å taken on La Palma by the SOUP tunable filter and CCD camera is 6.0% for quiet Sun, and 5.0% for the same frame in magnetic areas. This corresponds to ~ 9% and 7%, respectively, at 4600 Å. Figure 4 shows a photograph of granulation in quiet vs magnetic areas taken by the Lockeed Palo Alto Research Laboratory group at La Palma using an evaluation model of the tunable filter section of NASA's Orbiting Solar Laboratory. Magnetograms were taken simultaneously with the same instrument. A precisely registered magnetogram (Fig. 4A) was used as a mask (Fig. 4B) to reveal the granulation structure visible in quiet Sun (Fig. 4C), and in magnetic areas (Fig. 4D). In moderate-to-strong plage many magnetic features are still coincident with bright continuum structures but the situation is more complicated than for quiet Sun. In addition to the small facular points, there are larger bright features in the continuum, called facular knots or facular granules. Magnetic features present in moderate-to-dense plage are often > 1″, sometimes larger than a few arcsec, and their relation to the lower-contrast features present in the abnormal granulation is often less clear.

At still larger filling factor, exceeding 30% and corresponding to very dense plage, the field resides in areas immediately surrounding pores, which may be a resolution effect, and in continuum areas considerably darker on average than quiet Sun, but not yet recognizable as pores. The surrounding granulation appears abnormal. Not all of the intergranular lanes contain field structures nor are the ones that do necessarily the darkest. Furthermore, the field structures are considerably wider than the intergranular lanes, which means that they overlap many neighboring granules. Pores form when the flux per feature grows to ~ 2.5×10^{19} Mx, and finally sunspots form, constituting the largest magnetic features on the Sun, with flux larger than 5×10^{20} Mx.

Besides appearance, other properties of granulation change in magnetic areas that we can measure. Title et al. (1990c) found that the continuum in plage is brighter than in quiet Sun in areas where their magnetogram signal ranges from 200 to 600 G. This is true when averaging over an entire granulation movie, and it is true for all individual images as well. Their measurements from a single CCD frame reveal a maximum enhancement in continuum intensity near disk center of 0.8% at 6800 Å. These results are shown in Fig. 5. Notice that for very weak fields, the continuum is slightly fainter than for quiet Sun. Title et al (1990c) have discovered that the areas with the weakest magnetogram signals tend to lie between the dense plage and quiet Sun, in the same location as many of the dark intergranular lanes that sur-

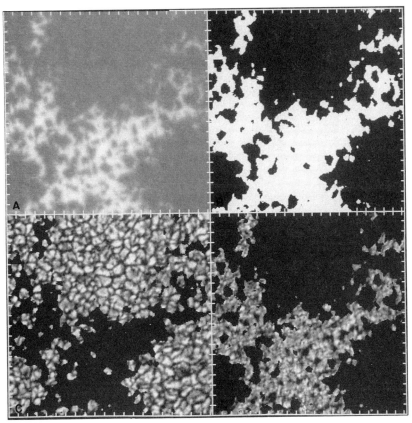

Fig. 4. Image of the Sun showing normal vs abnormal granulation: (A) magnetogram; (B) mask formed from magnetogram, threshold 150 G; (C) CCD image of granulation in a quiet Sun region; (D) same CCD image showing granulation in a magnetic region, using same gray scale as (C).

round the continuum bright points. In other words, the dip seen at 150 G in Fig. 5 is just a result of the fact that in the continuum, magnetic features appear as bright structures surrounded by dark intergranular lanes. For denser plage, corresponding to magnetogram signals > 600 G, the mean brightness of the photosphere is lower than quiet Sun and drops as the field strength increases. This drop is not due to the formation of pores, as all spots and pores were masked out of the images before the calculation was performed. When the continuum brightness is integrated over all values of the magnetogram signal strength, no net brightness enhancement is found when compared to quiet Sun. However, this apparent balance is due to the presence of a relatively small number of large magnetic structures ("magnetic knots") that are dark in the continuum but not so dark as to be classified as small pores.

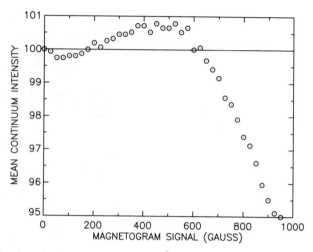

Fig. 5. Mean intensity in the continuum at 6768 Å vs magnetogram signal.

The weaker plage, corresponding to smaller flux elements, contributes a continuum brightness enhancement of $\sim 0.6\%$.

The larger flux elements corresponding to magnetic knots are much less common in the quiet Sun than in the active region. Also, it has been suggested that the facular points are brighter in quiet Sun than in magnetic regions. These results suggest that small flux tubes may be the source of the third component contributing to solar luminosity fluctuations (see Oster et al. 1982; Chapman et al. 1984; Foukal and Lean 1988b). For a more detailed discussion of this topic see the chapter by Livingston et al.

A measure of the "vigor" of the convection can be obtained from the temporal *rms* or from power spectra. The temporal *rms* starts from a single image pixel that is extracted at the same location from each image in the movie. The *rms* obtained this way is proportional to the power contained in the intensity fluctuations in the image at that point. By applying the calculation to all pixels in the image, we can construct an *rms* map that can be compared to a carefully registered magnetogram taken simultaneously. This calculation has been applied to granulation movies that were space-time filtered (Sec. III) to remove the effects of the 5-min oscillation. The result is that the *rms* power in the continuum drops by $\sim 25\%$ in plage areas when compared to quiet Sun. This means that the vigor of convection is suppressed in magnetic regions.

The horizontal flow field, discussed in Sec. VI above, is dramatically changed by the presence of magnetic fields. First, the average flow speed slows down by nearly a factor of 3. Second, the temporal *rms* of the horizontal flow field, as determined from instantaneous LCT measurements, is lower in magnetic regions than in quiet Sun. Furthermore, the stronger the magnetic field, the greater the reduction in *rms*. Also note that this remains true for a wide variety of sizes of the Gaussian mask used for the LCT calculations. For a 1″-Gaussian mask (FWHM), the results are 1.38 km s^{-1} in quiet Sun, 1.05 km s^{-1} in regions with medium-strength plage and 0.75 km s^{-1} for regions with strong plage (Title et al. 1989). When Doppler measurements are made at the limb instead of disk center, the results are a measure of the horizontal velocity, rather than vertical. In this way, Komm and Mattig (1989) obtained a corrected measure of the horizontal *rms* velocity fluctuation of 1.3 km s^{-1}, in good agreement with the results above. This is further evidence that the vigor of convection is suppressed in magnetic regions.

Simon et al. (1988) have shown that the horizontal flow field measured from the SOUP granulation movies is an excellent predictor for the motion and location of the magnetic field. These results were obtained by comparing the SOUP flow field to carefully co-aligned magnetogram movies of the same region obtained at Big Bear Solar Observatory. Vectors which define the granulation flow field point towards the magnetic field. Cells present in the flow field where the flow vectors point radially outward are coincident with cell-like regions of the magnetogram, which are regions devoid of magnetic flux in the center but are totally or partially surrounded by magnetic flux at the boundaries. Other large-scale flows occurred in the SOUP data near the sunspot and pores including a moat flow surrounding the sunspot (Sheeley 1969) in which granules stream radially outward for a distance of ~ 5″ (Title et al. 1986), and a well-organized stream flowing towards the pores. In both cases, magnetic features are seen in the BBSO magnetogram movies that move along the same flow vectors.

Cork movies, discussed in Sec. VII, can be used to predict where free-floating particles in the flow field would congregate. These projected areas are coincident with the location of the magnetic field as seen on the BBSO movies. Interestingly, the horizontal flow field was derived from SOUP movies 28 min long, while the BBSO magnetogram movies span over 9 hr, including at least 4 hr before and after SOUP. Furthermore, the cork movies show that ~ 8 hr are needed before all the corks assemble into a network-like pattern that resembles the observed magnetic structure. This suggests that the flow field observed during the 28 min SOUP observation actually has a lifetime > 8 hr, which is what we would expect for large-scale flows associated with supergranules.

The SOUP observations, as well as others, show that the flow is largest within cell interiors, in sunspot moats, and in other areas where large-scale flow exists. The flow slows down appreciably at the network boundaries where the magnetic field congregates, but it does not stop. The cork movies show motion along the boundaries until sink points are reached, which are usually vertices in the network. The BBSO movies show that the magnetic field elements move in the same way as the corks. This can have important implications for the buildup for magnetic stresses and for coronal heating, because the flow along the network boundaries can mix and twist the fields on a small scale. The effect can be enhanced by the presence of granules which bump against the field as they expand and evolve.

VIII. SUMMARY

Significant progress in our understanding of convective processes in the Sun has been made in recent years. New observations are now available with improved spatial resolution and image stabilization over longer periods of time. Increasing the spatial resolution has resulted in the detection of increasing numbers of tiny granules at or near the resolution limit, so far without limit. It is probable that the larger granules are rising convective plumes that overshoot the convective zone into the convectively stable layers above (as most researchers currently believe). The smallest observable granules, on the other hand, could be fragments that result from the turbulent breakup of large granules after they reach their maximum height and begin to expand horizontally. The data also clearly show the presence of intensity fluctuations in the continuum due to 5-min oscillations.

Exploding granules were found to play a more important role in granule evolution than previously thought, because new higher-resolution data showed them to be quite common, especially after the 5-min oscillation was removed by space-time filtering. Even more significant has been the development of new techniques to measure horizontal surface flows. With sufficiently high resolution, this can be done using granulation movies, by directly observing the displacements of granules during their lifetime. The horizontal flow field in quiet Sun is characterized by a cellular structure with a mean diameter of $\sim 7''$. Outflowing moats surrounding sunspots, large-scale streams and vortices were also detected.

The presence of magnetic fields on the solar surface causes considerable changes in the atmosphere, and in granulation. The actual appearance of granulation changes in magnetic areas, becoming smaller with lower-contrast intergranular lanes. The vertical velocity associated with the rising convective plumes is suppressed. The horizontal velocity in magnetic areas is suppressed. The data also show that small flux tubes are able to increase the solar luminosity to a small but measurable extent. The amplitude of the 5-min oscillations is suppressed in magnetic regions. There is considerable agreement between the observed flow fields derived from granulation movies and carefully registered magnetograms. The flow field predicts the structure of the magnetic network fairly well, and magnetic elements are always observed to move along flow-field stream lines. The flow fields already observed may be essential to mechanisms for heating the solar chromosphere and corona.

These new data have revolutionized our concept of time-dependent phenomena on the Sun's surface by showing them to be a more complicated combination of rising thermal plumes, chaotic small-scale flows, directed large-scale flows, waves, and the interaction of all of these phenomena with magnetic fields, than was previously appreciated. Also, mathematical models have been developed independently that predict intensity and velocity patterns that agree well with the observations. Together these results have led to an improved physical understanding of solar convection. Furthermore, the models predict the properties of the convective flow both below and above the surface. These predictions can be compared with future observations using the techniques of helioseismology to check convective flows below the surface, and high-resolution one- and two-dimensional spectroscopy to compare with theory above the surface.

LARGE-SCALE VELOCITY FIELDS

ROBERT F. HOWARD
National Solar Observatory

L. L. KICHATINOV
*Siberian Institute of Terrestrial Magnetism, Ionosphere and
Radio Wave Propagation*

RICHARD S. BOGART
Stanford University

and

ELIZABETH RIBES
Observatoire de Paris

Results in the area of solar rotation over the past few years are reviewed in this chapter. Considerable effort has gone into rotation determinations from tracers—especially sunspots—in recent years. The Greenwich data set has been extensively used by several groups to examine in detail the differential rotation characteristics of sunspot groups. The Mount Wilson data set has been used principally for studies of the rotation rate of individual spots. Efforts in Doppler determinations of rotation have been centered on studies of systematic instrumental or scattered-light phenomena and their effects on rotation measurements. The present status of our knowledge of solar supergranulation, mesogranulation and giant-scale convection is briefly reviewed. The information on meso- and supergranulation is supplied mainly by observational studies. Only a qualitative theory of these types of convection has been developed. On the other hand, our knowledge of giant-scale convection is predominantly of theoretical origin. Mixing-length theories and nonaxisymmetric computer simula-

tions strongly suggest the existence on the Sun of convective motions of mega-meter scales. However, direct Doppler measurements do not reveal these motions. Theoretical suggestions are discussed which may serve to reconcile theory and observations of giant-scale solar convection. Global meridional cir-culation in the photosphere is suggested by models of the solar rotation, which usually require pole-to-equator flows of a few m s^{-1}, and by the observed mi-gration of magnetic activity over the solar cycle. Measurements of the average meridional component of sunspot motions generally show a pattern of diver-gence from the center of the activity zone toward both the equator and the poles with an amplitude of 1 to 2 m s^{-1}, but there is considerable disagreement over the statistical significance of the results. Doppler measurements generally show a poleward flow with an amplitude more than 10 times larger, but probably reflect a nonisotropic limb shift of the spectral lines rather than large-scale motion. The solar rotation shows a latitude and cycle dependence. The effect is clearly seen in the magnetic tracers (sunspot data) as well as the spectroscopic observations, although the latter are affected by systematic errors which cannot easily be removed from a real solar signal. This variability can be understood in terms of a time-dependent convective toroidal roll pattern. Counter-rotating rolls form at intervals of 2 to 3 yr, and they are associated with the emergence of active regions. By transporting angular momentum either towards the equa-tor or towards the poles, they modulate the surface rotation in such a way as to produce alternately latitude bands rotating faster and slower than average (the torsional oscillation). Their effect is small, however, and the net equatorial acceleration is shown to be maintained by nonaxisymmetric motions. The con-nection of the rolls with rigidly rotating layers favors a solar dynamo located below the convective zone, possibly in the radiative interior which rotates like a solid body. Such a new picture of the rotation and convection has to be con-firmed. Undoubtedly helioseismology will further add to our knowledge of the solar rotation in the near future.

I. ROTATION

Several reviews of solar rotation have appeared in recent years (Howard 1984; Schröter 1985; Ribes 1986*a*; Bogart 1987; Harvey 1988). The topic is well covered in these reviews, and so only recent developments will be treated here. In the chapter by Libbrecht and Morrow, rotation is discussed with special emphasis on inferences from helioseismology, so this topic will not be covered here.

In general there are two methods for determining rotation on the Sun. One can measure the positions of tracers such as sunspots (and including oscillation wave patterns), and one can measure a line-of-sight velocity using the Doppler effect. Most of the Doppler data refer to the photosphere. Both techniques have advantages and disadvantages. Doppler measures, for ex-ample, can suffer from the effects of scattered light from other parts of the image and from interference fringes caused by various optical elements in the spectograph. Results from following tracers can suffer from problems of iden-tifying the same portion of an evolving feature and from systematic condi-tions affecting position determinations such as the Wilson effect in sunspots.

Theory has generally lagged observations in this field. Numerical models have reached a fairly advanced state and do a good job of representing the surface rotation observations. Other theoretical approaches also can represent the surface results, but agreement between the various theoretical modeling of subsurface rotation is not good. This discrepancy will have to be resolved with subsurface rotation rates determined from extensive oscillation observations.

Observations

Tracers. Sunspots are relatively good tracers of rotation and other motions at the solar surface because they are small, relatively well defined and unchanging, and often live for several days or more. On rare occasions sunspots live long enough to cross the central meridian more than once. A number of studies over the years have determined the solar rotation and its variation with latitude (differential rotation) from sunspots over long intervals of time (Newton and Nunn 1951; Ward 1966; Godoli and Mazzucconi 1979; Balthasar and Wöhl 1980; Arévalo et al. 1982; Lustig and Dvorak 1984; Balthasar et al. 1984; Howard et al. 1984a; Tuominen and Virtanen 1987). Only the Howard et al. (1984a) reference includes extensive results from individual sunspots; the others refer generally to sunspot groups. The Greenwich data in digitized form have been used extensively in nearly all of these observational studies.

Younger sunspots tend to show a rotation rate that is faster than that derived from older sunspots (Balthasar et al. 1982; Ribes 1986b). This may be the explanation for the observation that small spots rotate faster than large spots (Howard et al. 1984a), because large spots are likely to be older than small spots. Ribes (1986b) also found that young spots show a more rigid rotation profile than do older spots. Recently, a comprehensive study of the rotation of spot groups from the Greenwich data set has been completed (Balthasar et al. 1986a). The dependence of rotation rate on spot classification, and therefore age, has been confirmed. The oldest (H and J type) spot groups have the slowest rotation rate, and the youngest (C type) groups have the fastest rotation rate. Since, except for the large, individual spots, the Greenwich results refer to groups, it is not clear whether the group rates refer to the rates for the individual spots within the groups. It could be that as the groups develop and expand, to some extent leading spots appear in the average to give a relatively fast rotation rate to the groups as a whole. It is true, however, that there is deceleration observed during the lifetimes of groups containing only a single spot—a result found also from a different data set by Lustig and Dvorak (1984). This whole issue is complicated by the fact that the Greenwich positions are eye estimates.

Further results of the comprehensive work of Balthasar et al. (1986a) are that the rotation rate of sunspot groups appears to have decreased slightly early in this century. The authors point out that it is not possible now to determine whether or not this is an observational effect. They also confirm the result of Balthasar and Wöhl (1980) for Greenwich data and Gilman and Howard (1984b) for Mount Wilson data that the sunspot rotation rate shows a clear maximum near the minimum phase of the activity cycle. In addition, the Mount Wilson data show some slight evidence for a weak secondary maximum near the maximum phase of the cycle. They also confirm the correlation of rotation rate and meridional motion found by Gilman and Howard (1984a). This is an effect discovered first by Ward (1965) from the Greenwich data, and is evidence of angular momentum transport toward the solar equator.

The same Greenwich data set has been analyzed by Tuominen and Virtanen (1987) with quite similar results. They point out that an apparent decrease in the rotation rate during this century has been accompanied by an increase in the level of solar activity. Howard and Harvey (1970) have reported a slight increase in rotation rate of the Sun during this century from an examination of various spectroscopic results. Tuominen and Virtanen (1987) find that the cycle dependence mentioned above is seen principally at higher latitudes, thus the latitude differential rotation shows a cycle dependence. This same dependence of latitude differential rotation on phase in the cycle was found by Vitinsky and Ikhsanov (1985) from the Greenwich data from 1875 to 1954, and is undoubtedly related to the one-per-hemisphere torsional oscillation found from Doppler data by LaBonte and Howard (1982b).

Antalova (1986), using Greenwich data from 1921 to 1971, has reported that zones where there are many sunspots show a lower velocity than the long-term zonal average.

Recent determinations of the rotation elements (i and Ω) from the analysis of large quantities of sunspot position data (Balthasar et al. 1986b; 1987b) indicate that i is slightly smaller and Ω slightly larger than the classic Carrington results. Recurrent single sunspot results are the same as those from spot groups.

Magnetic Fields and Ca II K-line Plages.   In a series of papers, Ternullo (1986, 1987a, 1987b), using Ca spectroheliograms, showed that the differential rotation profile is steeper in the first four days of the lifetime of a plage than it is later in the lifetime of the plage. Plages decelerate with age in some latitude zones and accelerate with age in other latitude zones. These zones coincide with the fast and slow zones of the torsional oscillations (Howard and LaBonte 1980). The differential rotation profile of plages is found to be steepest at the maximum phase of the solar activity cycle. This profile flattens suddenly in the first years of declining activity. The later years of a cycle are characterized by an intermediate steepness.

A related result by Singh and Prabhu (1985) shows that synodic periods of Ca plage rotation vary in different latitude zones over time scales ranging from the cycle length down to ~ 2 yr. Variations in adjacent belts are in phase, but generally the results from the two hemispheres do not show a consistent phase relationship.

Studies of the differential rotation of magnetic-field patterns (Snodgrass 1983; Bumba and Hejna 1987a,b; Hejna 1986, 1987) show that the rotation profile is close to that of sunspots or plage magnetic fields. In this data, torsional oscillations with wave number one per hemisphere (LaBonte and Howard 1982b) could be seen. This effect was later confirmed though viewed in a different way by Snodgrass (1985) and Snodgrass and Howard (1985). Using Gegenbauer polynomials, Snodgrass (1984) was able to eliminate the cross talk between the coefficients of the latitude expansion that had been a characteristic of earlier analyses. Hejna found stationary torsional oscillation modes with wave numbers 0, 1/2 and 1 per hemisphere, thus confirming earlier results. In a similar study Shelke and Pande (1986) looked at the differential rotation of magnetic patterns using the same latitude-strip method introduced by Bumba and Howard (1969). They found that the rotation rates of background field patterns, which are not associated with coronal holes, are different from those patterns which are associated with coronal holes; the sign of the difference is opposite in the north and south hemispheres. They find evidence for differential rotation of coronal holes. Shelke and Pande (1986) also found evidence for the differential rotation of coronal holes (in disagreement with earlier results) from a study of Kitt Peak He I 10830 Å maps.

Several studies of solar rotation rates determined from integrated light measurements have appeared recently. Dreschen et al. (1984) have investigated the rotation modulation of the emission measure of the Ca II K line in the solar spectrum. They attempted to find variations in the derived rotation period with phase in the cycle but were unable to find any such effect, confirming the results of LaBonte (1982). A similar result was published by Singh and Livingston (1987), who proposed that some variations observed in the rotation rate by this means were due to circumstances of sampling and the chance birth and decay of active regions. Kotov (1987) has analyzed both the mean magnetic field of the Sun from 1968 to 1984 and the interplanetary magnetic field (in part inferred) from 1926 to 1983. Power spectra of both these data sets show peaks at 26.94 day and 28.20 day. The former is the dominant period of magnetic features at low solar latitudes. The latter is the highest peak in the interplanetary magnetic field spectrum, and is conjectured to represent a deep rooted magnetic field pattern which penetrates the radiative zone. Similar periods were found in an earlier analysis of the same data by Svalgaard and Wilcox (1975).

Studies of rotation in the polar regions of the Sun (Durney et al. 1985; Makarova and Solonskij 1987) give evidence for rotation rates that are not significantly different from the results of extrapolation to high latitudes of Doppler results.

Gopasyuk and Demkina (1987) have examined the structure of small magnetic-field elements in the photosphere using the Fe I 5250.2 Å line. These elements are stretched in the east-west direction and apparently compressed on the west side, suggesting that they rotate faster than the surrounding atmosphere. They derive a rotation difference of 2.5 to 3% from this effect. A similar result has been found in this difference from sunspots by Koch (1984) who compared spectroscopic and tracer measurements. This is a result found earlier by Foukal (1979) with a higher amplitude. Lower amplitudes of the same effect were found by Beckers (1977) and Adam (1979).

The apparent rigid rotation of magnetic-field patterns has found an explanation in the work of Sheeley et al. (1987) who have modeled the motion of magnetic fields on the solar surface and also used analytical calculations to demonstrate that a combination of meridional motion and the shearing effects of differential rotation can give rise to rigid patterns. The meridional component in their work is provided by diffusion due to supergranular motions and perhaps a small component of meridional flow. The appearance of new sources of flux retard this process and prevent the rigid-pattern rotation from developing until well into the declining phase of the activity cycle.

The rotation rate of the supergranular cell pattern has been determined by Duvall (1980). He finds a rate that is 3% faster than Doppler rates, which is consistent with a model of the supergranular convection in which angular momentum is conserved by rising and falling elements of the photosphere (Foukal 1977).

The Corona.   Recent studies of coronal rotation by Parker (1986, 1987) have used the High Altitude Observatory K-coronameter data from 1964 to 1976. Parker finds that the height gradient of rotation is of opposite sign at high and low latitudes such that there is a slow decrease of differential rotation with height and a very low differential rotation near the solar-wind source. Parker suggests that the dynamics of coronal loops may account for much of the height and latitude dependence of coronal rotation. Fisher and Sime (1984) also find a lower level of differential rotation in the corona.

Hoeksema and Scherrer (1987) have modeled the coronal magnetic field by extrapolating the photospheric magnetic field measured at the Wilcox Solar Observatory at Stanford University with a potential field approximation. The correlation of their modeled coronal field over one rotation shows much less differential rotation than in the photosphere, in agreement with the observational results of Parker discussed above. Examination of correlation peaks at multiple rotation lags shows a more complex rotational structure and different rates in the north and south, with the northern hemisphere showing a significantly faster rotation rate than the southern hemisphere.

Doppler Measures.   The importance of scattered light to photospheric Doppler measurements has been known for some years (DeLury 1939; Scherrer et al. 1980*b;* LaBonte and Howard 1981). Recently, Albregtsen and Andersen (1985) have developed a mathematical method to correct Doppler measures for this effect. They have applied this method to the Mount Wilson measures and achieved a closer correspondence of the Doppler results to the sunspot rotation rate. Andersen (1985) has applied this and a further modification of this technique to the spectroscopic results of Pierce and LoPresto (1984). This correction brings the Pierce and LoPresto rotation rate in good agreement with that of Scherrer et al. (1980b).

Livingston and Wallace (1985) have shown that the spectrum line often used for spectroscopic determinations of solar rotation—Fe I 5250.2 Å—is weakly blended by telluric water vapor lines. This blend is quite weak, but a slight error in the measured line position could result from this blend, and this error will vary with the water vapor content of the atmosphere at the observatory site.

Two recent studies (Balthasar 1984; Gadun et al. 1987) have examined the depth dependence of solar photospheric rotation using measurements from many spectrum lines. Both studies agree that little or no depth dependence can be found. These are only the latest contributions to a long series of conflicting results that started in the early years of this century. The problem is very complicated because of the difficulties of interpreting line shifts in spectrum lines which differ in atomic properties. This recent result is in agreement with the most recent oscillation results as reviewed by Harvey (1988).

Pierce and LoPresto (1984) have examined solar rotation from a number of spectrum lines and find good agreement with previous results. They also find that the small day-to-day variations in rotation rate that are seen are confined to sunspot latitudes and cannot be detected outside the sunspot zone. This may indicate that such variations are due to velocity fields associated with active regions—the Evershed effect and apparent downward motion over plages.

Theory

Numerical Models.   The nonlinear action of rotation and convection in a spherical, compressible gas have been modeled (Gilman and Miller 1986; Glatzmaier 1984, 1985a,b). Recently Glatzmaier (1987) has reviewed this aspect of the theoretical approach. With the proper choice of parameters, a good agreement with observed surface differential rotation may be achieved with such models. In these models, the subsurface rotation rate decreases with depth such that the rate is approximately constant on cylinders centered on the rotation axis. Since sunspots and other magnetic-field related features tend to rotate slightly faster than the photospheric gas, it has often been assumed that the layers immediately below the photosphere must rotate faster than the surface. This discrepancy has yet to be resolved, but one should keep in mind that the models involve a number of approximations and cannot be expected to represent all details of the rotation profile. Results from p-mode frequency measures, which are still very preliminary, suggest that the rotation rate may increase slightly with depth close to the photosphere, then decrease with depth at lower layers (Harvey 1988). It could be imagined that the faster-moving subsurface magnetic fields are rooted in the faster zone (see also Schüssler 1987a).

Analytical Methods.   The problem of the internal dynamics of the Sun is a complicated one, treated either numerically or analytically, even ignoring magnetic fields, which may well play a role. Durney (1987) has approached the problem analytically by means of a generalization of the mixing-length theory applied to rotating, compressible convection zones. The arbitrary parameters are the dimensions of the turbulent eddies. Differential rotation results from the balance of the Coriolis forces, pressure gradients and buoyancy forces. Constant rotation along cylinders centered on the rotation axis is ruled out by this approach, in conflict with the model results mentioned above. In more recent work (Durney 1989), the rotation near the lower part of the solar convection zone is studied. Here the Reynolds and viscous stresses are important, whereas in the upper layers of the convection zone Coriolis forces, pressure gradients and buoyancy forces are important. The assumption of isotropic viscosity is not a good one. Kichatinov (1987) has carried out a similar study also with success in generating the differential rotation from compressible convective motions in a rotating fluid.

Pidatella et al. (1986) have also considered a mixing-length formalism, and they find that anisotropy of viscosity is a more important effect in generating the differential rotation than is latitude-dependent heat transport modeled through perturbations in the mixing length.

Rüdiger et al. (1986) have proposed that the main observational properties of differential rotation and meridional flow can be produced by a dynamo model which is in accord with the observed mean magnetic field.

Torsional Oscillations.   Earlier explanations of the torsional oscillations of the Sun (Howard and LaBonte 1980) pointed at subsurface, dynamo-generated Lorentz force waves as the cause (Yoshimura 1981; Schüssler 1981). More recently, however, Dolginov and Muslimov (1985) have suggested that these torsional oscillations are primary eigenmode oscillations of surface layers excited by convection. These authors (and others, e.g., Snodgrass and Wilson [1987]) propose that the activity cycle results from the modulation by the torsional oscillations of the field-generating process. In a similar vein, Kliorin and Ruzmajkin (1984) have suggested that the torsional oscillation is produced by a dynamo wave of large-scale magnetic fields.

II. CONVECTION

Scales from mesogranulation and larger will be referred to as large scales in this section. This range includes the recently discovered mesogranulation, supergranulation and giant convection. There are a number of substantial differences in both the physical nature and our understanding of the meso- and supergranulation and of the giant scale convection. Whereas information on the first two is supplied mainly by observations, evidence for the very existence of giant convection rests almost entirely upon theoretical foundations. The relation to the solar rotation and magnetic field is also different. A typical lifetime of giant cells is believed to equal several periods of solar rotation with a typical Rossby number smaller than unity. Such motions must be influenced by global rotation. The back reaction distorts the rotation, thus making it differential. It is also believed that giant cells play an important role in generating large-scale magnetic fields on the Sun. On the contrary, Rossby numbers for meso- and supergranulation motions are large and the mutual influence of the rotation and magnetic fields, on the one hand, and these kinds of convection, on the other, can have only a modifying, rather than a decisive, effect. Therefore, it seems reasonable to discuss the meso- and supergranulation separately from the convection of giant cells. Recent reviews on large-scale solar convection may be found in papers by Gilman (1986) and Nordlund (1985).

Supergranulation and Mesogranulation

Observations.   Hart (1956) was, possibly, the first to observe supergranulation, which was later described in detail by Leighton et al. (1962) and Simon and Leighton (1964). These last authors observed in the photosphere a cellular pattern with an average horizontal cell size of 32 megameters (Mm), with material flowing from the center to the edges of a cell. The velocity of horizontal flows was 300 to 500 m s^{-1}. The central region of a cell shows an upward mass flow, with a typical rise velocity of $\sim$ 100 m s^{-1}. On supergranule boundaries the fluid goes down with a velocity of $\sim$ 200 m s^{-1} (Worden and Simon 1976). The average lifetime of supergranules varies, according to different estimates, from about 20 to 36 hr (Worden and Simon 1976). Some supergranules are able to persist as long as several days (Worden and Simon 1976).

Supergranulation motions penetrate high into the chromosphere, reaching heights $\sim$ 10 scale heights above the photosphere. The velocity distribution there is nearly isotropic, with a typical mean value of $\sim$ 3 km s^{-1} (November et al. 1979).

Almost at the very beginning of supergranulation studies, a spatial correspondence between the chromospheric network and supergranule boundaries was revealed (Simon and Leighton 1964). Magnetic fields are carried away by the plasma flow to the supergranule boundaries where an increased density of magnetic flux is observed (Giovanelli and Slaughter 1978).

Recently November et al. (1981, 1982) discovered a cellular pattern with a typical scale intermediate between the granulation and supergranulation which they called the mesogranulation. On the average, the relevant cells are 5 to 10 Mm in diameter. A typical velocity amplitude in mesogranules is significantly smaller than that in supergranules and is about 60 m s^{-1}; the lifetime of a mesogranule being not less than 2 hr. No relation of the mesogranulation pattern to the chromospheric network has yet been found.

Theory.   At present there is a consistent quantitative theory only for the smallest-scale cellular pattern of convection observed on the surface of the Sun, i.e. for granulation (Nordlund 1985*b*). No such theory is available for supergranules, to say nothing of mesogranules. There are only some qualitative arguments which might explain the presence in the upper layers of the convective envelope of the Sun of the three singled-out scales of convection.

Simon and Leighton (1964) were, perhaps, the first to hypothesize that granulation and supergranulation are accounted for by ionization of hydrogen and helium, respectively, in subphotospheric layers of the Sun. It is believed that the cellular patterns observed in the photosphere are manifestations of convective motions generated by convective instability of the outer layers of the Sun which are stratified superadiabatically. It seems likely that partial ionization must enhance the convection. Indeed, descending, for example, fluid elements undergo an increase in pressure of the surrounding gas, which must lead to their being compressed and heated. However, the increase of the degree of ionization will suppress the rise of temperature. This will engender an additional compression and will increase the density difference between the descending element and the surroundings. On the contrary, the release of the latent heat of ionization in the ascending fluid element must enhance the density deficit in it as compared with ambient material, thereby enhancing its buoyancy. Thus, partial ionization is able to intensify the convection. Note that an important role may also be played by an opacity increase as well as a change of the mean molecular weight in regions of partial ionization.

After the detection of mesogranulation, November et al. (1981, 1982) supposed that it was due to a first-order ionization of helium at a depth of about 7 Mm below the photosphere. The hydrogen ionization and the second-order ionization of helium give granulation and supergranulation, respectively. The above arguments are well known and appear to be reasonable ones. The depths of H$^+$, He$^+$ and He^{++} ion production roughly equal the granulation, meso- and supergranulation scales, respectively (see Gilman 1986, Fig. 1). Nevertheless, for meso- and supergranulation this reasoning remains justified only qualitatively. Difficulties of developing a quantitative theory of supergranulation are due to the complicated nature of its interaction with granulation as well as other factors mentioned by Nordlund (1985*b*). Nevertheless, it might be expected that a quantitative theory of supergranulation and mesogranulation will be constructed in the near future.

Giant convection

Theory.   As pointed out above, our knowledge of so-called giant cells is largely of a theoretical character. Therefore, it seems appropriate to begin with the theoretical aspects of the issue. Initial knowledge about the structure of the convective envelope of the Sun was based mainly on predictions of the mixing-length theory. Within the framework of this theory, heat transport in a convection zone is modeled by the random walk of fluid elements. The elements move, on the average, a distance ℓ before mixing and exchanging heat with surrounding material. The mixing length is usually taken equal, in order of magnitude, to the pressure scale height H. Initially, the mixing-length theories predicted for the depth d of the solar convection zone values of 20 to 30% of the solar radius and typical (depth-dependent) velocities of convective motions of several tens of m s^{-1} (Baker and Temesvary 1966; Spruit 1974; Gough and Weiss 1976). In these models the greater part of the convection zone has a stratification which, though superadiabatic, is close to the adiabatic one. In most of the convection zone the relative difference of the temperature gradient from the adiabatic gradient is $\sim 10^{-4}$. A substantial deviation from adiabaticity of stratification exists only at depths of ≤ 1000 km below the photosphere.

The pressure scale height in the near adiabatically stratified convection zone grows with depth, remaining comparable, in order of magnitude, with this depth. Even this factor induces us to suggest a giant (~ 100 Mm) convection in the middle and deep layers of the convection zones.

Estimates of Reynolds numbers for giant convection yield very large values ($\sim 10^{13}$). Hence the giant cells must be dynamically unstable and produce turbulent, smaller-scale motions. In other words, the convection zone is likely to be a chaotic dynamical system and what is called "giant convection" here is the largest-scale representative of turbulent motions with a broad spectrum of scales.

Data on the 5-min oscillations made it recently possible to refine estimates of the depth and other parameters of the convection zone. Lubov et al. (1980) and many others more recently showed that the best agreement with these and other data is achieved for $d = 0.3R_\odot$ and $\ell = 1.65\,H$.

There are a significant number of so-called nonlocal mixing-length models that take account of the fact that a typical spatial scale of the convective motions is not small compared to the size d of the convection zone (see, e.g., Unno et al. 1985).

Despite the fact that certain parameters of giant solar convection (the convection-zone depth, for example) are known only from mixing-length theory, this theory is subject to criticism in two aspects, at least. First, it is constructed not as a result of straightforward transformations of fundamental equations but relies on qualitative foundations. Therefore, the mixing-length theory is, *per se,* an approximate one, and estimates of its accuracy are not easily available. Second, this theory does not take account of some effects which are important for solar convection. Convection on the Sun occurs in a rotating fluid and is influenced by Coriolis and centrifugal forces. A measure of this influence is the Coriolis number $\omega = 2\tau\Omega$ (Ω being the angular velocity of global rotation; τ is the lifetime of a giant cell that can be estimated as $\tau = \ell/u$, where u is the characteristic velocity of convective motions). For giant cells, the parameter $\omega \sim 6$ (Durney and Latour 1978), i.e., the influence of rotation is essential (see, e.g., Tayler 1973). Besides, on the Sun we are dealing with magnetoconvection. The effects of magnetic fields, though substantial, seem to be of less importance than the rotational influence. The strongest magnetic fields in the convection zone of the Sun seem to have a value $B \simeq 2000$ G. A measure of disturbance of the convection of a highly conducting fluid by a magnetic field is the energy ratio $\epsilon = B^2/(\mu\rho u^2)$. For middle and lower layers of the convection zone we have an estimate $\epsilon \lesssim 1$.

Qualitatively, the influence of rotation (and of the magnetic field) can be understood from linear theory (Chandrasekhar 1961; Eltayeb 1972). Both the rotation and magnetic field suppress the convection. This affects least those modes with the largest size along the rotation axis (along the magnetic field). Therefore, in a rotating medium, convective cells elongate along the rotation axis, assuming the form of rolls. Spherical geometry of the convective envelope leads to a bending of large-scale rolls (banana-shaped structure).

A straightforward description of the essentially nonlinear solar convection beyond the scope of the mixing-length approximation seems hardly possible with the analytic approach. Therefore, numerical simulations are usually applied for this purpose. There exists an extensive literature on numerical simulations of Boussinesq and compressible convection of nonrotating fluids. It is impossible and, probably, pointless to review it here. Note only that the most important difference of these models from predictions of the mixing-length theories is probably that the largest-scale modes extend not to a single scale height but cover the entire thickness of an unstable fluid, even if it contains a large number of scale heights (see e.g., Graham and Moore 1978). Let us, then, consider the recent numerical models of solar giant convection which include compressibility and rotation.

Such numerical simulations were carried out independently by Glatzmaier (1984, 1985a,b) and Gilman and Miller (1986) with almost identical results, with the exception that Glatzmaier was considering the magnetic field as well. The starting equations used were those of (magneto-) gas dynamics written in a particular variant of the anelastic approximation (Gilman and Glatzmaier 1981). This approximation is linear in convection-induced deviations of thermodynamic quantities from a certain reference (adiabatic) profile but includes the velocity and magnetic field in a nonlinear way. It differs from the more commonly used Boussinesq approximation by taking account of the density inhomogeneity but does not allow for acoustic disturbances. Only the giant convection was described explicitly. The small-scale convective motions were parameterized by introducing effective viscosities and conductivities.

Glatzmaier (1984,1985a,b) and Gilman and Miller (1986) obtained giant convection with a well-defined roll structure at low latitudes. The axes of the rolls are distorted due to spherical geometry and differential rotation. A more cellular convection pattern was dominant in the polar regions. Convective motions did not show any well-defined scale and were characterized by a broad spectrum. A typical vertical scale of the motions coincides with the thickness of the convectively unstable layer. The rolls normally have a latitudinal extent of the order of one solar radius. There is a broad spectrum of longitudinal wave numbers m with a maximum in the region $m \simeq 10$ to 12. A typical velocity of convective motions is ~ 100 m s^{-1}. The *rms* radial velocity is smaller than the latitudinal and longitudinal velocities. A typical turnover time of a convection element is of the order of the rotation period of the Sun. Gilman and Miller (1986) note that not only the rotation, *per se*, but also its differentiality has a substantial influence upon the convection structure. On the upper boundary (lying in the models at 7 to 10% of the solar radius below the photosphere), the velocities of the horizontal motions exceed by a factor of 2 to 3 the upper bound imposed by Doppler measurements. The results differ little from earlier models (Gilman 1980b) that neglect the density stratification.

It appears likely that the above models of giant convection are presently the most developed numerical experiments on giant convection and include a majority of the relevant effects. Nevertheless, it may be confidently argued that they missed something substantial because, by taking account of magnetic fields, Glatzmaier (1985a) obtained a poleward drift of the toroidal field as opposed to the observed migration of solar activity towards the equator. It might be expected that this discrepancy will be eliminated in the near future. Despite the above-mentioned disagreement with the observations, the models under consideration have clearly indicated that convection in a rotating star is, indeed, able to generate differential rotation and magnetic fields. Probably, these models reproduce adequately the general features of the giant solar convection.

Observations.   There is no decisive observational evidence for the existence of giant convection in the Sun. Such evidence could be provided by direct Doppler observations of the velocity field in the photosphere. However, these measurements do not unambiguously show any giant cells. They give only upper bounds for velocity amplitudes. Thus, LaBonte et al. (1981) obtained upper bounds for amplitudes of longitudinal harmonics of giant convection. These bounds vary from $\simeq 12$ m s^{-1} for the longitudinal wave number $m = 1$ to $\simeq 3$ m s^{-1} for $m > 20$. Snodgrass and Howard (1984) applied a correlation analysis, rather than traditionally using synoptic maps, and showed that (a) if convective motions have a single-scale spectrum, then the velocity amplitude cannot exceed $\simeq 1$ m s^{-1}, and (b) if there is a broad spectrum, then the integral power should not be greater than $\simeq 10$ m^2 s^{-2}. Cram et al. (1983) and Durney et al. (1985) tried to find long-lived giant structures in the velocity field of the polar regions of the Sun. They obtained some indications of a giant (~ 100 Mm) structure with an amplitude $\simeq 3$ m s^{-1} near the resolution limit and found the upper amplitude bound to be $\simeq 5$ m s^{-1}. Schröter et al. (1978) studied the motions of the calcium network in order to identify giant cells. They obtained some evidence for the presence of such structures but failed to find any correlation with Doppler velocity measurements. Thus, it may be argued that if the giant convection does, indeed, penetrate into the photospheric level, the amplitudes of these motions must not exceed a few m s^{-1}.

At the same time, there is certain indirect evidence for the presence of global convection in deeper layers of the Sun. One such indication might be the presence of large-scale structures of magnetic fields. Bumba (1987) discovered structures with a scale of the order of the solar radius and a lifetime of 5 to 8 solar rotations. He also established a relation of these structures with active longitudes which may also serve as an indication of the giant convection.

If giant convection does, indeed, generate differential rotation, which is accepted by most present-day theories, then the zonal motions in giant cells must correlate with radial and meridional motions (Rüdiger 1989). Such a correlation of latitudinal and longitudinal velocities was found in motions of sunspots (Ward 1965; Gilman and Howard 1984a). It is difficult to find some other explanation for this correlation, except for the relationship of sunspots with giant convection. Sunspots seem to be anchored deep in subphotospheric layers and reflect the motions of these layers, even if these motions do not penetrate into the photosphere.

There have been attempts to discover giant cells directly from sunspot motions. So-called kinematic elements with the size of 50 to 100 Mm were identified, within which the sunspot trajectories were sufficiently alike (see Vitinsky et al. 1986, and references therein). The random component of velocities of kinematic elements has a non-Maxwellian distribution, and the *rms.* velocity is about 38 m s^{-1}. It was established that in most cases kinematic elements coincide with regions of background magnetic fields with the same polarity (Vitinsky et al. 1986), which reminds us of the findings by Bumba (1987). Note, however, that background field structures discussed by Bumba were significantly larger than kinematic elements.

How to Reconcile Theory and Observations.   The fact that giant convection, whose presence on the Sun is rather soundly suggested theoretically, is not identified through direct Doppler observations, is, perhaps, most natural to interpret by assuming that this convection does not penetrate into the photospheric level.

Masaguer and Zahn (1980), neglecting partial ionization, but taking compressibility into account, showed that near the convective-envelope surface there is a layer with negative buoyancy. This gives rise to horizontal pressure gradients which convert vertical motions to horizontal motions before the surface is reached. Such motions decrease rapidly (exponentially) towards the outer surface. Latour et al. (1981) also found this effect. Van Ballegooijen (1986) investigated the role of density stratification and found that giant convection could be hidden in subphotospheric layers owing to stratification. A certain contribution (a 2- to 3-fold decrease towards the surface) can also be made by the effect of turbulent screening (Stix 1981a). At this same time, convective motions of giant cells are able, without reaching the photosphere, to manifest themselves in the observed motions of sunspots as well as to generate large-scale structures of background magnetic fields. The observations of such phenomena have been mentioned above. Such a picture appears to be quite a probable one. However, additional theoretical and observational investigations are required in order to discuss the giant solar convection with more confidence.

III. MERIDIONAL CIRCULATION

Models and Predictions

The existence of large-scale steady meridional flows at the surface of the Sun is suggested both by the phenomenology of the solar activity cycle and by the maintenance of the differential rotation. The most obvious example is the equatorial migration of the centroid of the zone of new activity over the course of the solar cycle, equivalent to an average displacement of ~ 1 m s^{-1}. There is also an observed poleward migration of unipolar magnetic field regions and polar filament bands over the course of a cycle that can be explained in terms of advection of magnetic flux by a poleward meridional flow of ~ 10 m s^{-1} (Topka et al. 1982).

The theoretical predictions of meridional flows transporting angular momentum sufficient to sustain the observed differential rotation have been well reviewed by Gilman (1974), and will only be summarized here. There are three classes of models: axisymmetric models in which there is an anisotropic eddy viscosity; axisymmetric models in which the transport coefficients depend on latitude due to the interaction of rotation with convection; and nonaxisymmetric models of cellular convection. The first two both generally predict meridional flows from equator to pole of a few m s^{-1} (Gilman 1974; Durney 1974). Surface meridional flows are predicted by the nonaxisymmetric models as well, but their magnitude and structure tends to depend on the detailed assumptions in the model (see, e.g., Glatzmeier and Gilman 1982).

Observations

Tracers. The most easily observed photospheric tracers are sunspots, and the most useful and accessible data are the long relatively stable time series of photographic observations made by the Greenwich Royal Observatory and cooperating sites through 1976 and by the Mount Wilson Observatory. These data have been exhaustively analyzed for evidence of meridional motions, with results of marginal significance. The first problem to be confronted is that the spectrum of observed individual velocity measurements, based on pairs of observations separated by one day typically, is at least 1 to 2 orders of magnitude larger than any of the reported mean drifts (Ward 1973). This may of course reflect real variance of the surface flow velocities on a small scale. A more severe problem is related to the identification of features. Most of the published data since 1916 refer to the positions of spot *groups,* the area-weighted means of all spots known or supposed to be common to a single active region moving as a coherent feature. With large spot groups and isolated long-lived spots this is not a bad assumption, and such positions are in fact relatively stable. But this is a small subset of the data, with velocities that are perhaps uncharacteristic of broader photospheric fluid motions. Furthermore, because of the narrow latitudinal distribution of sunspots, it is tempting to include as many spots as possible at the extremes of the range to gain information about the latitudinal extent of any motions. Attempts to expand the data base are beset by their own peculiar difficulties. One obvious source of spurious motions is the identification of two distinct small spots appearing on successive days at different locations within the same group; another is the well-established evolution of the spot groups themselves as leading and following spots move relative to the center and follow different growth curves as well. Data corrected for these and similar effects run the risk of being compromised by selection effects in the inferred motions. Even analyses based on measurements of individual spot positions, such as that of Howard and Gilman (1986), must deal with the essentially similar problem of possible misidentification of spots, since the useful original photographs were typically separated by at least one day. Howard and Gilman investigated this effect and found that it becomes significant when identified spots separated by as much as $0°5$ day^{-1} (~ 70 m s^{-1} are included); this represents about one-third of the data and shows the effect of the wide velocity distribution.

Observations of possible consistent latitudinal drifts of sunspots go back to the mid-19th century. References to much of the earlier work can be found in Bogart (1987). Broadly, there have been two lines of attack, one focusing on the highly restricted subset of stable long-lived spots and the other using data on as many spots as possible. The first has been the subject of a series of papers by Tuominen and others (see Tuominen et al. [1983] for references and recent results). The long-lived spot groups appear to diverge from about latitudes $\pm 15°$ and converge toward the equator and poles, with velocities of order 1 to 2 m s^{-1}. There is evidence that the lines of divergence propagate to lower latitudes in the course of the 11-year solar cycle, from $\sim \pm 20°$ to $\sim \pm 10°$ (Touminen 1976). A complementary analysis restricted to short-lived groups was also made by Tuominen (1961), with results that were interpreted as being consistent with the divergent flows of long-lived groups, but which are rather ambiguous due to the inevitably large error bars.

Most sunspot motion studies have followed the second approach of using essentially all available data, throwing out only such measurements as would obviously skew the results due to possible misidentifications or mismeasurements (e.g., groups near the limb). The Greenwich data have been examined in this way by Ward (1973), Arévalo et al. (1982) and by Balthasar et al. (1986a). The larger error bars have led the authors to conclude that the mean meridional flows are not significant, except perhaps near the equator, but the results are consistent in latitude profile and magnitude with those of Tuominen (1976). Balthasar et al. (1986a) also investigated the effect of varying the cutoffs in day-to-day position changes to guard against misidentifications, confirming the conclusion of Howard and Gilman (1986) that the effect becomes important when the cutoff is $\leq 0°5$ in latitude. Howard and Gilman (1986) obtained essentially the same results using spot positions obtained at Mount Wilson over a 67-year period, with only the motions near the equator exceeding the error bars. Analysis of spot positions measured at Kanzelhöhe for 37 yr (Hanslmeier and Lustig 1986; Lustig and Hanslmeier 1987) generally shows the same effect, with equatorward motions at low latitudes and poleward motions at high latitude, except that the zone of equatorward motion appears to be mostly suppressed in the southern hemisphere. There is large scatter among the different cycles. The same qualitative description of the mean meridional flows diverging from midlatitude zones has been obtained by Ribes et al. (1985a) using observations made at Stanford. Coffey and Gilman (1969) analyzed three years' data from Sacramento Peak Observatory, with results that showed the same trend (divergence from midlatitudes, convergence toward the equator), even though the magnitudes of the motions and the error bars were both an order of magnitude larger and the results not significantly different from zero except perhaps near the equator. This rather curious result suggests a possible systematic error in all the determinations, but has not been pursued to our knowledge.

In principle, it should be possible to use other features such as Hα filaments to measure meridional motions directly, but the requisite long series of observations are generally lacking. The analysis of such filaments by Topka et al. (1982) was based on positions obtained from synoptic charts, so in effect was another measurement of poleward field-structure advection, although they argued on theoretical grounds that this must be due to meridional flow rather than diffusion. In any case, such determinations would still be tied to the evolution of large-scale magnetic structures which may or may not be simply advected by the ambient gas.

Doppler Measurements.   Given the limitations of tracers as a diagnostic for large-scale motions, it is natural to try to complement the observations with direct Doppler observations of the photosphere. Except for the impossibility of directly observing any kind of transverse flows near disk center and the severe difficulty of observing meridional flows near the equator with B_0 ≤ 7°.2, the measurements are not otherwise restricted to any location or time in the solar cycle and can be made both globally and more or less instantaneously. Only a few data sets are of high enough quality to detect meridional flows of ~ 10 m s^{-1}. Analyses have suggested mean poleward flows of 15 to 40 m s^{-1} (Beckers 1978; Duvall 1979; LaBonte and Howard 1982*a*; Snodgrass 1984), but the most recent results indicate much smaller values—typically < 5 m s^{-1} (Ulrich et al. 1988*b*). The contrary result of Pérez-Garde et al. (1981) that the flow is equatorward is based on only a small amount of data, but it should be noted that occasional directional reversals at midlatitudes are shown in the Mount Wilson data of Ulrich et al. (1988*b*). Although not inconsistent with the tracer results, these measurements do not suggest that a comparable steady small component of the motion has been identified.

Why do the Doppler data not confirm the tracer results? Unfortunately, the measurement of Doppler velocities to a precision of 1 m s^{-1} is not only exceedingly challenging from an instrumental standpoint, but it is complicated by several physical effects which are still being sorted out. One of these is scattered light; the proper removal of the effects of scattered light at this level is possible (Andersen 1986), but requires accurate knowledge of the differential rotation, which may be changing in time. A more serious problem is the limb shift, the observed variation of both line center and line profile from disk center to limb, presumably the result of projection of convective velocities and the correlation of upward velocities with brightness of the material. The limb effect is not the same along polar and equatorial diameters, possibly because of latitude dependence of convective parameters rather than meridional motion (Beckers and Taylor 1980; Brandt and Schröter 1982; Andersen 1984,1987; Cavallini et al. 1985*b*,1986). Cavallini et al. (1986) and Andersen (1987) have shown that the variations in the limb shift of various lines are equivalent to the effect of meridional flows of about 20 to 50 m s^{-1} in the equatorward direction, suggesting real poleward flows in the absence of an "observed" flow, or that the reported poleward flows were substantially underestimated. Andersen (1984) reported the intriguing finding that the limb shift would mimic the flow corresponding to convergence on ±45° latitude, opposite in sign and an order of magnitude larger than the apparent tracer motions. This would suggest that the limb shift may depend on latitude in a more complicated way, due to the effect of strong fields as well as rotation. This is to be expected in terms of the well-known "magnetic red-shift" (the association of excess observed red-shift with regions of strong magnetic flux), but serves to complicate an already difficult situation. Ulrich et al. (1988*b*)

have discussed all these effects in substantial detail in their recent analysis of
the Mount Wilson Doppler data.

So far no analyses have been attempted for nonaxisymmetric flows. Although the Doppler data seem scarcely able even to confirm the tracer results
for the mean flow at present, there would seem to be more hope for finding
meridional flows associated with a small number of banana-type rolls. In this
case at least the problem of the anisotropic limb shift could be made to disappear, although magnetic effects and scattered light problems would still
exist. These, however, can be dealt with in a fairly straightforward way.

Root Mean Square Fluctuations. An interesting suggestion has been
made by Gilman (1977*b*), who points out that if angular momentum is being
advected to lower latitudes, then the *rms* fluctuations of the Doppler signal
should be different at the same longitudes west and east of the meridian. This
is due to the nonvanishing correlation term $\langle vv \rangle$ between zonal and meridional
velocities representing the Reynolds stress. The sign of the difference between the east and west hemispheres would certainly indicate the direction of
a steady meridional component in any case, whether or not the magnitude
could be estimated. This observation has not yet been done. Although tracers
so far provide velocity data of substantially better precision, the systematic
errors associated with Doppler observations would largely cancel for such
measurements, as they would for direct measurements of nonaxisymmetric
modes.

IV. CYCLIC MOTIONS

Variability of the Mean Differential Rotation

Torsional Oscillations. There is some evidence for temporal variation
in the solar rotation rate over intervals of months and years. Using the magnetically active line λ 5250.216 Å of neutral iron to measure the line-of-sight
velocity at the photospheric level, the surface rotation has been obtained at
Mount Wilson for almost 22 yr (Fig. 1). A modulation of the surface rotation
has been found, with a periodicity of 11 yr. A faster than average latitude
band emerges near the pole, at the time of sunspot maximum and sweeps
toward the equator, at solar minimum, allowing the next polar wave to begin
(Howard and La Bonte 1980). The amplitude of the phenomenon is a few m
s^{-1}. A different analysis of the data has been proposed by Snodgrass (1985)
Snodgrass and Howard (1985) who suggest a less wave-like phenomenon.
Both representations, the 2-per-hemisphere wave pattern and the net pattern,
are related to the magnetic surface activity and follow the butterfly diagram.

The net torsional pattern has been corroborated independently from the
motions of magnetic tracers. Using sunspot groups tabulated in the Green-

NET TORSIONAL PATTERN

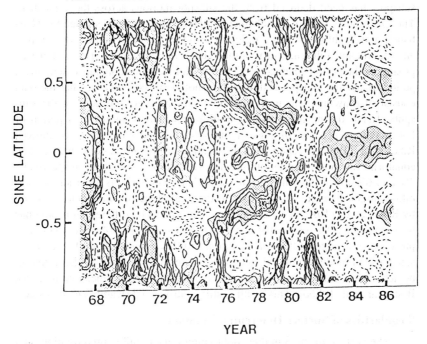

Fig. 1. Net torsional oscillation pattern derived from the Mount Wilson solar Doppler data, for cycles 20 and 21. The data are analyzed by subtracting from the daily data the total time average of daily fits separately for each latitude zone, and then subtracting the rigid body component of the solar rotation for each hemisphere. Contour intervals are ± 2, ± 4, . . . ± 12 m s^{-1}.

wich time-series, Godoli and Mazzucconi (1983) and Tuominen et al. (1983) found a similar latitudinal shear in the velocity maps.

Shorter Time-Scale Motions. Many attempts to study cycle dependence of the solar rotation have been made (see Howard [1984] and Schröter [1985] for reviews). The most extensive studies are based upon long time series of sunspots and sunspot groups covering several solar cycles. A faster rotation rate ($\Delta\omega$ up to $0°2$ day^{-1}) near the equator around sunspot minimum, is found at the 3-σ level (Tuominen and Kyrölainen 1982; Arevalo et al. 1982; Lustig 1983; Gilman and Howard 1984b; Balthasar et al. 1986a). This cycle dependence does not seem to depend upon the age and complexity of sunspots or sunspot groups (Balthasar et al. 1986a). The oscillations found in the rotation of sunspots are not migratory, in contrast with the torsional oscillation present in the Doppler data. Moreover, the amplitude ($0°5$ day^{-1}) is much larger than the Doppler signal present in the torsional oscillation.

On a long time scale (longer than the solar cycle), the rotation rate of sunspots has been deduced from the sunspot motions across the solar disk. The 17th century records of Scheiner and Hevelius have been studied by Abarbanell and Wöhl (1981) and Yallop et al. (1982). These authors do not find any significant difference in rotation rate in modern times. This is not surprising as the period spanned by Scheiner and Hevelius contains essentially a period of normal sunspot activity. One should stress, however, that the data consist of drawings which do not show very accurate positions of the sunspots. Later, during the Maunder minimum which is known as a period of low magnetic activity, timings of the solar limb and the sunspots allowed Ribes et al. (1987) to reconstruct the rotation profile of the solar surface. The equatorial rate and the rotation rate at a latitude of $\pm 15°$ were, respectively, 2% and 6% lower than the current ones. This peculiarity is an order of magnitude larger than one would expect if the larger horizontal diameter reported by Ribes et al. (1987) were erroneous. This enhanced differential profile has been confirmed by returned sunspots when crossing the solar meridian. The presence of a larger differential rotation profile coincided with the dearth of sunspots. In modern times, the equatorial rate, A coefficient, is also found to be decreasing as the level of solar activity decreases (Hanslmeier et al. 1986; Balthasar et al. 1986a). This result is still controversial (Kambry et al. 1988).

Singularities of Surface Differential Rotation

The surface of the Sun does not exhibit a mean differential rotation on a daily basis. From Doppler observations, Deubner and Vasquez (1976) have shown a strong longitude dependence of the surface rotation. The local departures may attain 4 to 5%. The rotation of the magnetic field measured in Stanford is also far from being homogeneous (Wilcox et al. 1970). At the location of coronal holes, the rotation can be rigid (Antonucci et al. 1979; Simon 1979), while the surrounding atmosphere exhibits the mean differential rotation. Such peculiarities of the surface rotation are well shown by the rotation of magnetic features (Glackin 1974). More recently, using Hα filaments as tracers of the fluid motions, Escaut-Soru et al. (1984) have followed the motions of the filaments during the course of their life. In the Carrington frame of reference, filaments are usually swept in the differentially rotating layers, according to the rotation law proposed by M. and L. d'Azambuja (1948). At some stage, however, they may pivot instead, the rotation being rigid and corresponding to the Carrington rate (Escaut-Soru et al. 1984). The pivot points are the places of sunspot-forming regions (McIntosh and Wilson 1985; Martres et al. 1986), new emerging fluxes (Mouradian et al. 1986) and flaring activity (Soru-Escaut et al. 1985; Bumba and Gezstelyi 1987).

Variability of the Meridional Circulation

As shown in Sec. III, the meridional circulation is an indicator of the angular momentum processes at work building up the surface differential rotation. The possible existence of a meridional circulation and its variability through the cycle would provide valuable information on the dynamics controlling the differential rotation, and it would be responsible for the dynamo mechanism. Here again, one should consider various methods of detection.

Detection from Spectroscopic Observations. Spectroscopic observations do not reveal any clear equatorwards trend in the meridional motions. It is possible, however, to interpret the large dispersion, in sign and amplitude, of the meridional circulation reported by various authors in terms of a latitude and cycle-dependent phenomenon (Ribes 1986a), keeping in mind that large systematic errors affecting the radial velocities are partly responsible for the observed variability.

More recently, Snodgrass (1986) determined the meridional flow from the latitude-by-latitude variations in the line-of-sight components of the velocity fields. The residuals, averaged over 273-day periods, are shown in Fig. 2. There is clearly a latitude and cycle dependence of the meridional flow, although it is of a very weak amplitude ($0\overset{\circ}{.}03$ day^{-1}).

Detection from Tracers and Magnetic Fields. Average drifts of sunspots and sunspot groups in heliographic latitude have been extensively studied. Using the Greenwich data set, a latitude dependence of the meridional circulation of sunspots has been reported by Becker (1954). Low-latitude sunspot groups moved predominantly equatorward while high-latitude spots moved polewards near sunspot maximum. Such a pattern was not seen near sunspot minimum (Becker 1954; Tuominen 1961). In addition to this cycle effect, a 22-year oscillation of the drift, with an amplitude of $\pm 0\overset{\circ}{.}0043$ day^{-1}, was reported by Tuominen (1952), and Richardson and Schwarzschild (1953). The 22-year dependence shows up as an equatorwards mean meridional circulation of sunspot groups for even cycles, and a polewards mean meridional circulation for odd cycles. The effect was found to be independent of the age of the sunspot groups. More recently, Balthasar et al. 1986a) could not confirm the 22-year cyclic effect nor the 11-year dependence over the 9 solar cycles covered by the Greenwich data). The use of sunspot groups to study temporal variations of rotation has been questioned, however (Howard et al. 1984a), as the unequal development of one polarity with respect to the other may lead to fictitious drifts. For that reason, Howard et al. (1984a) have designed an automatic procedure to determine the center of gravity of each sunspot detected on the Mount Wilson white-light pictures from 1920 onwards. The detection of sunspots is based upon a pattern recognition procedure and allows them to detect motions of ± 18 m s^{-1} on each sunspot. Their method, however, restricts the analysis to the class of sunspots which do not exhibit daily motions $> \pm 0\overset{\circ}{.}5$ day^{-1}. Gilman and Howard (1984b) did not detect a net axisymmetric meridional drift of sunspots. They found, instead, some indication of angular momentum transport, via nonaxisymmetric mo-

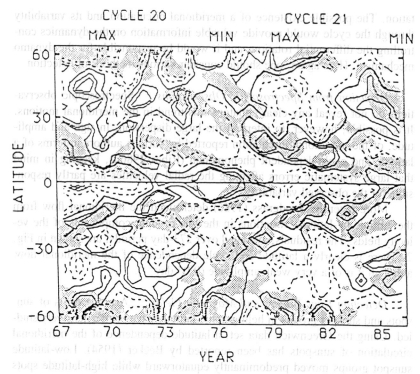

Fig. 2. Spectroscopic evidence for a moving pattern of convective rolls on the Sun for cycle 20 and 21. Solid contours represent motion away from the observer and dashed contours represent motions towards the observer. The contour intervals are at ± 1, ± 2.5 and ± 4 m s^{-1}. The shaded regions are the zones of shear enhancement.

tions. Within a latitude band, sunspots rotating faster than average move equatorwards, while sunspots rotating slower than average move polewards. The amplitude of the phenomenon is latitude dependent. As the effect is larger for high-latitude sunspots that occur mainly at the ascending phase of the solar cycle, Gilman and Howard's results can be interpreted as a cycle dependence. The magnitude of the covariance is sufficient to account for the observed equatorial acceleration (Gilman and Howard 1984b). The effect had been reported previously by Ward (1965) and confirmed later by Balthasar et al. (1986a) from the Greenwich data, and by Belvedere et al. (1976) using chromospheric faculae.

Using a digital analysis of the Meudon spectroheliograms based upon complex image processing to correct the solar image for photometric and geometric distortions (Mein and Ribes 1989), individual sunspot motion can be obtained with an accuracy of a few m s^{-1} for the best seeing conditions available at Meudon. Ribes (1986a) did not choose an automatic procedure to follow the sunspot from one day to another because a stringent vicinity test rejects a large number of sunspots (in particular the young ones), and a less-stringent vicinity test leads to an incorrect identification of sunspots and fictitious drifts and rotation rates. Instead, Ribes identified each sunspot so that the result does not suffer from an identification bias. Their sample is small, however, because it was restricted to young sunspots visible on the summer sequences of spectroheliograms. It was found that young sunspots (Ribes and Mein 1985; Ribes 1986a) and possibly old sunspots (Ribes and Bonnefond 1989) do trace a zonal meridional circulation in the form of four latitude bands, with the direction of drift changing from one band to the next. The amplitude of the meridional circulation is about 15 m s^{-1}, and can reach 100 m s^{-1}. The pattern is also time dependent and exhibits a 22-yr periodicity. Wöhl (1988) has questioned the validity of Ribes's approach. His argument is based upon the probability that a zonal pattern established with a small number of sunspots might well occur by chance. If the repeatability of pattern persists throughout the six solar cycles available at Meudon, the probability of it being by chance becomes negligible. Moreover, if old sunspots do follow the same meridional circulation pattern, as suggested recently (Ribes and Bonnefond 1989), the statistics will then become significant. The meridional circulation signal is disturbed by the rapidly growing magnetic polarities nearby: this is probably the reason why the meridional circulation signal disappears when no selection of sunspots is made. An accurate determination of the velocity field together with a careful identification of the tracer used are needed.

The detection of the zonal meridional circulation pattern, showing a cycle dependence by Ribes et al. (1985a) throws some light on earlier results (Tuominen 1952; Richardson and Schwarszchild 1953). It is likely, however, that the maintenance of the equatorial acceleration is assured by other types of motions, for example, the nonaxisymmetric cells.

Variability of the Large-Scale Magnetic Field

Variability from Magnetic Tracers.   Using the Hα filaments seen on the Kodaikanal data set as tracers of the inversion lines of the longitudinal magnetic field, Makarov (1984) designed a method to reconstruct the large-scale magnetic field. The mean latitudes corresponding to filament bands covering a substantial longitude range is time dependent, and their drift draws the trajectory of the magnetic boundaries. Ribes (1986a) used a somewhat similar method with the Stanford data, and results are shown in Fig. 3. The magnetic pattern, which is axisymmetric by construction, consists of several tori during the active phase of the solar cycle. New tori occur at intervals of 2 to 3 yr. The large-scale organization of the magnetic field seems to disappear near sunspot minimum. Each torus moves poleward, the earlier ones reaching the pole shortly after the sunspot maximum, explaining the reversal of the polar magnetic field. Using the high-latitude Hα filaments, Topka et al. (1982) observed a similar poleward drift of these features with respect to

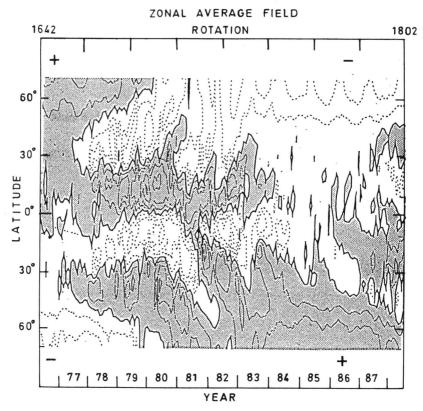

Fig. 3. Zonal magnetic-field pattern as measured at the Wilcox Observatory from 1976 onwards. Dashed and solid lines correspond to negative and positive polarity, respectively. The contour levels correspond to ±50, ±100, ±150 μTeslas.

the surrounding magnetic field. The amplitude of the drift is 10 m s⁻¹. The criterion for defining a filament band is subjective. The definition of boundaries is empirical, with an accuracy of ±5° at most. Nevertheless, in the absence of direct measurements of the magnetic field, one can obtain the large-scale magnetic pattern for the last six solar cycles.

It should be stressed that the Hα filaments do show an orientation that is close to the meridian axis, especially at low latitudes (below 50°) and at the early stage of their lives (McIntosh 1981). Such a magnetic orientation could indeed suggest that there are also convective motions in the form of short-lived meridional cells. The latter are the modes of convection that are expected to occur in a compressible rotating shell dominated by rotation (Glatzmaier 1985a; Gilman and Miller 1986). One important and relevant issue is the co-existence of these two patterns and their respective roles in the dynamo mechanism (cf. Sec. II on the solar dynamo).

Variability from Direct Measurements of the Longitudinal Field. Magnetic maps of the large-scale fields are now available (Howard and LaBonte 1981; Hoeksema 1984). Removing the contribution of active regions, Hoeksema has obtained the zonal magnetic field for the last solar cycle (Fig. 3). The comparison with the large-scale magnetic pattern deduced from the Hα filaments is quite satisfactory. The mean toroidal field strength is variable through the solar cycle and ranges from 50 to 300 μTesla. The boundaries of the magnetic polarities are defined with an accuracy of ± 5°.

Rolls. Ribes et al. (1985*a*) noticed the coincidence between the magnetic pattern and the zonal meridional circulation pattern. This coincidence occurs only in the sunspot-forming latitudes, as the tracers used for the meridional flow are confined to latitudes ranging between ± 40°. Assuming that the tracer motions reflect the fluid motions, they interpreted this phenomenon as evidence of giant azimuthal rolls. Each roll is characterized by a magnetic polarity and a direction of rotation. While the pattern of counter-rotating rolls moves poleward with time, the occurrence of new rolls drifts equatorwards. The properties of the rolls are as follows:

1. Their width is about 20° and their lifetime is several years.
2. Their velocity (up to 100 m s^{-1}) is comparable to the convective velocities predicted for giant cells.
3. The poleward motion of the rolls can explain the observed reversal of the polar magnetic field. The disappearance of rolls at a high latitude and the occurrence of new rolls closer to the equator resembles the bifurcation in convective patterns. Ribes et al. (1985*a*) suggested that *the azimuthal roll pattern could well be the long-searched-for giant convection.*
4. The meridional circulation pattern is highly latitude and time dependent. The rolls slowly drift poleward with time, as is shown by the drift of the magnetic pattern. A zonal meridional circulation is associated with the rotation of each roll. However, the main meridional circulation (averaged over a solar hemisphere) shows a privileged direction, flowing mainly equatorward for even cycles and poleward for odd cycles, with a period of 22 yr. This is because sunspot activity occurs mostly in a torus of the same polarity in a given hemisphere and for a given cycle. The presence of a latitude dependence of the meridional circulation will create alternately flattening and enhancement of the differential rotation, through the action of the Coriolis forces (Ribes et al. 1985*a;* Ribes 1986*a,b;* Snodgrass and Wilson 1987). As a result, latitude bands of faster- and slower-than-average rotation will form and move equatorward, following the appearance and disappearance of rolls. That the direction of propagation of the torsional oscillations is opposite to the poleward drift of the rolls is shown in Fig. 4: solar activity (and torsional oscillations) follow the direction of the new roll's occurrence. Thus, the rolls provide a plausible explanation

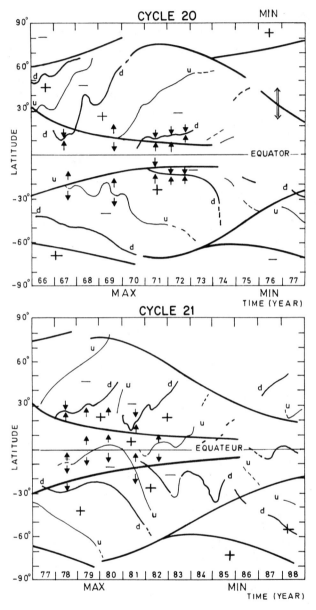

Fig. 4. Toroidal roll pattern derived from the meridional circulation motions of young sunspots (arrows) and the trajectory of the Hα filaments, for cycles 20 and 21. The solar activity is schematically represented by the green line coronal activity. Note that the solar activity frames the toroidal roll pattern.

for the modulation of the surface rotation. Moreover, the torsional oscillation signal as well as other activity indices (see Pecker [1988] for a summary) frame the roll pattern rather than being superimposed by it. This fact explains the existence of the two branches of activity, at high and at low latitudes visible on the solar surface at a particular time (Ribes and Laclare 1988).

5. Sunspot activity concentrates in the regions of enhanced rotational shear (Howard and LaBonte 1981; Snodgrass 1986), as shown in Fig. 3. An analysis of the latitude distribution of active regions with respect to the distance to the nearest roll's boundary (Ribes 1988) confirms that sunspot groups cluster near the converging rolls. Moreover, active regions occur mainly in tori having the same polarity as the dominant polarity defined by the Hale rule (Fig. 5). A natural explanation is that observed tori could be a reservoir of toroidal field as suggested by Parker (1988). In fact, this would imply that the percentage of active regions though small which occurs beyond the dominant torus would not follow the Hale polarity rule. This has never been observed; so, it is likely that the magnetic activity originates below the rolls, the latter being not the source of active regions but rather the selective amplifier. The unbalanced magnetic flux of active regions (Semel 1967) could be reconnected to the weak azimuthal field of the adjacent torus.

6. Peculiarities in the rotation rate, though not all of them, can be explained by a meridional flow associated with a time-dependent roll pattern. The

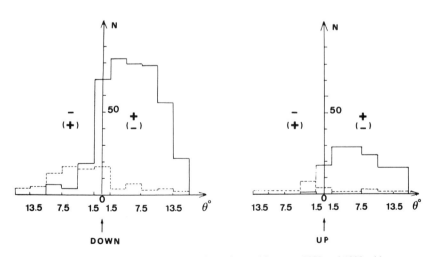

Fig. 5. Latitude distribution of the active regions observed between 1978 and 1982 with respect to their nearest roll boundary. The active regions cluster preferentially at the border of converging rolls and within the torus of the dominant polarity. "Up" and "down" refer to roll boundaries characterized by upward and downward motions.

sudden increase of the equatorial rotation rate between 1981 and 1982 from the Doppler velocity signal (Küveler and Wöhl 1989) could be due to the new roll appearing near the equator at the end of 1981 (see Fig. 5). The resulting rotation will be necessarily variable, and one should expect different rotation profiles for even and odd cycles as the sunspots move in streams of opposite directions.

7. The rolls might perturb the convection locally and cause temperature perturbations at a detectable level (Kuhn et al. 1987). A solar-limb temperature variation of 1° K has been found at a latitude of about 50°, during the summers 1983 to 1985. Their result can be interpreted as the signature of a "2 roll per hemisphere" structure inferred from the sunspot drift analysis.

Cyclic Effects on Small Scales of Convection

Granulation and Supergranulation.   The small-scale convection varies over the solar cycle. High-resolution pictures of solar granulation obtained at Pic du Midi shows the change of morphology throughout the cycle. The number of granules (or their mean size) decreases with increasing activity (Macris 1951; Macris et al. 1984; Muller and Roudier 1984). The granule-intergranular lane intensity ratio at 5200 Å is maximum (1.30) at sunspot maximum and decreases down to 1.10 at sunspot minimum. This variability also shows up in the weakenings of the photospheric lines measured at Kitt Peak National Observatory. From sunspot minimum (1976) to the next maximum (1980), Livingston and Holweger (1982) reported a decrease in the equivalent width down to 2.3%. A perturbation in the small scale of convection in response to an increase in the magnetic field could explain the shape variation of the line asymmetries. The curvature of the bisectors of the strong iron lines in the full-disk Fraunhofer spectrum, measured with the Fourier Transform Spectrometer at Kitt Peak, is reduced as the activity cycle proceeds from minimum to maximum (Livingston and Holweger 1982; Livingston 1983).

A latitudinal dependence of the granulation has been suggested by Snodgrass (1984) to account for some changes in the limb shift. Differences in the limb-shift curves along the polar and equatorial diameters can be interpreted in part as a convective modulation under the influence of cyclic magnetic fields (Andersen 1985; Cavallini et al. 1985b).

The diameters of the calcium network features are shown to be smaller by 5% at solar maximum than at solar minimum. The change in size of the network is interpreted in terms of changes in the supergranular cell (Simon and Leighton 1964; Singh and Bappu 1981).

Dynamo Mechanism

Problems Raised by Torsional Oscillations, Azimuthal Rolls and Other Peculiarities of Rotation.   The counter-rotating azimuthal rolls described above seemed to be the principal large-scale motions lying beneath the surface. They would sweep the toroidal field into this zone, collecting, amplifying or dispersing it. The problem remains as to whether or not the east-west oriented rolls are compatible with the α-ω dynamo (Parker 1955b). Clearly, the preferred convective pattern expected from numerical simulations (Glatzmaier 1985a; Gilman and Miller 1986) and fluid experiments (Hart et al. 1986) remains as meridional cells. The extent to which the Taylor-Proudman constraint is upheld depends heavily on the following observational constraints (Durney 1987): (1) Which is dominant, the Laplace forces or the Coriolis Forces? (2) Is there any radial gradient in the angular velocity? (3) Where is the dynamo situated, within, at the base or below the convective zone? The existence of toroidal rolls shows that the preferred mode of convection is controlled by the Laplace forces, at least during the active phase of the solar cycle (Geiger 1982; Jones and Galloway 1987). If the solar cycle is a delicate balance between two large-scale convective modes, one would expect to detect meridional cells near sunspot minimum (Ribes 1986b). The recent results on the rotational splitting of acoustic oscillations show that the velocity in the convective layers is not constant on cylinders (Libbrecht 1988c), in contrast to the rotation profile predicted by meridional cells (Gilman and Miller 1986). On the other hand, the existence of a strong radial shear in the angular velocity with depth is also ruled out by helioseismology observations, which indicate a rather constant rotation equatorial rate with depth, in the convective zone (Duvall et al. 1987). Lastly, as the solar activity seems to emerge with a constant rate, it is tempting to assign the location of the dynamo in the rigidly rotating layers, that is, the radiative interior (Brown and Morrow 1987a). The suggestion has already been made (Dicke 1979; Bracewell 1988), and offers an explanation independent of the conflicting predictions of a dynamo driven by compressible rotating shells (Glatzmaier 1985a,b; Gilman and Miller 1986). It does not solve the problem, however, of deciding upon the type of motions that can generate the fields in a weakly turbulent medium.

The azimuthal rolls can explain a number of existing observations. In particular, the torsional oscillation pattern and some peculiarities of the rotation could be the surface manifestation of the rolls on the rotation, through the action of the Coriolis forces. Although there is some connection between the solar activity, the azimuthal rolls and the modulation of the surface rotation, torsional oscillation or net pattern, there is no compelling evidence that the torsional wave generates the toroidal field. Yet, there is no consistent theoretical modeling showing that the rolls can produce the characteristics of the solar cycle, even though an azimuthal roll pattern clearly is the convective response of a magnetically dominated magnetic fluid. The amplitude of the radial shear in the angular velocity observed below the convective zone is not well determined. So, it is still unclear whether or not the α-ω dynamo mechanism placed at the interface between the radiative interior and the convective zone (Galloway and Weiss 1981) can work (Parker 1988d).

Observational Aspects.   There are a number of observational approaches which could throw some light on the nature of the dynamo mechanism responsible for the solar cycle. One deals with the nature of the torsional oscillation pattern. It seems likely that the net pattern can result from the meridional transport of surface momentum by deep convective rolls. On the other hand, it has been stressed that sunspots, which have been used for detecting the rolls, might not represent well the plasma motions but rather the Laplace-Lorentz forces acting at the location of discrete flux tubes. An explanation of the torsional oscillation pattern has also been proposed along this line of thought (Yoshimura 1981; Schüssler 1981). It seems difficult to interpret the amplitudes of the motions of sunspots within such a framework, even though the torsional oscillation signal could well be a surface phenomenon. A careful analysis of sunspot motion at a different phase of evolution may help to discriminate between the meridional flow carrying the magnetic regions and the magnetohydrodynamic forces. The existence of active longitudes (Trellis 1971a) and other properties of the surface activity such as the asymmetry in the production of active regions and their magnetic flux can be understood by means of tracers. On the other hand, the depth of the rolls, the co-existence of meridional cells as well as an accurate rotation profile of the solar interior are now accessible with helioseismology.

Acknowledgments.   The National Solar Observatory is a division of the National Optical Astronomy Observatories, which is operated by the Association of Universities for Research in Astronomy, Inc., under cooperative agreement with the National Science Foundation.

PART IV

Expanding Magnetic Fields and the Atmosphere

THE SOLAR ACTIVITY CYCLE

DOUGLAS M. RABIN
National Solar Observatory

C. RICHARD DeVORE, NEIL R. SHEELEY, JR.
Naval Research Laboratory

KAREN L. HARVEY
Solar Physics Research Corporation

and

J. TODD HOEKSEMA
Stanford University

Study of the solar cycle is entering a new era dominated by objective, precise measurements of magnetic, velocity and radiation fields over the surface of the Sun. This review emphasizes observations of photospheric magnetic flux during cycle 21 (1976–1986) and how these measurements have been used to model the cyclic variability of the heliospheric magnetic field. Indices of solar activity are discussed in terms of their potential to figure in theoretical or empirical models. Other recent data, such as measurements of large-scale surface flows and information on the Sun's internal rotation from helioseismology, as well as the magnetic flux observations, are considered in the context of Babcock's phenomenological model of the solar cycle: can this model still serve? Is there anything better to replace it?

I. INTRODUCTION

The solar activity cycle has fascinated scientists and amateurs alike for over a century, but its mystery remains, and even deepens, as we collect new data that reveal its full complexity. Solar activity demonstrably affects terrestrial phenomena from radio communications to the isotopic composition of tree rings, and has been alleged (with varying degrees of logic) to influence many of the vicissitudes of life. Because of this broad appeal, the solar cycle continues to inspire a vast amount of research worldwide. No survey of reasonable size can aspire to comprehensive coverage of even recent work. The present treatment has a more limited objective.

Over the last two decades, sensitive measurements of the distribution of magnetic flux over the solar disk have been made on a regular basis at several observatories. These archives constitute a new tool for analyzing the behavior of surface magnetic fields during the solar cycle and comparing the results with simulations based on physical models. We focus on these data as an example of the quantitative approach that will increasingly dominate analysis of historical data in the study of solar activity. Thus, the discussion of activity indices touches fairly briefly on classical topics such as sunspot numbers. Indices derived from spatially resolved observations of the solar disk are emphasized over integrated-disk measurements (chapter by Livingston et al.). Variations in solar activity that span many 22-yr cycles are typically studied by historical and statistical methods that cannot be discussed adequately here. Despite their historical importance in solar cycle research and their continuing use for solar-terrestrial prediction, magnetospheric and ionospheric phenomena are neglected here because of their indirect relationship to the primary manifestations of solar activity. Finally, we restrict attention throughout to *cyclic* aspects of solar activity, with reference as necessary to other chapters that consider the properties of particular types of activity, such as active regions and flares, for their own sake.

In contrast to the quantitative emphasis in our treatment of observations, our approach to the physical basis of solar activity is frankly phenomenological. In part, this is because the chapter by DeLuca and Gilman is devoted to mathematical models and simulations of the most popular candidate mechanism for the solar cycle: a regenerative dynamo seated within or just below the convection zone. However, there are two other reasons to emphasize empirical models. First, for over 25 yr the point of reference for relating the mechanism of solar activity to observations has been the empirical model of Babcock (1961). If only for lack of persuasive alternatives, the Babcock model has played a central role that cannot be ignored. Second, as DeLuca and Gilman's chapter makes clear, the daunting complexity of self-consistent dynamo models has thus far limited their role to achieving consistency with basic features of the activity cycle rather than making predictions at the detailed level of modern observations.

Because of the restrictions we have imposed, the reader could usefully consult other treatments that help to convey the wide range of phenomena associated with the solar cycle. Kiepenheuer (1953) and Zwaan (1981,1987) give useful reviews of the solar cycle in the general context of solar activity and magnetism. Geophysical effects of solar activity and solar-terrestrial prediction are considered extensively in the compendia edited by Jursa (1985) and Simon et al. (1986). *The Sun in Time,* edited by Sonett et al. (1991), is devoted to evolutionary changes in solar behavior, including possible variations in the character of the activity cycle.

II. INDICES OF SOLAR ACTIVITY

The usefulness of an activity index may be judged differently according to the goal in view: to *mark* or *predict* the course of the solar cycle, or to *understand* its physical basis. Naturally, these aims are not exclusive; but, as the solar physicist primarily seeks understanding, this discussion will emphasize indices that are most likely to figure in theoretical or empirical models.

For over a century, the number and distribution of sunspots on the Sun was the primary measure of solar activity. Although the length and continuity of the sunspot record ensures that it will continue to play an important role in characterizing the solar cycle, there are now a variety of quantitative measures that span a full cycle or more. We stress indices that make use of objective, spatially resolved observations of the solar disk, such as photospheric magnetograms. Sunspots too have much to say beyond the crude summaries embodied in conventional sunspot numbers. In several cases, our comments about an activity index are more an indication of how it could be made more quantitative or useful for interpretation than a summary of its measurements.

A. Radio and Ultraviolet Fluxes

Measurements of electromagnetic fluxes from the integrated solar disk, including cyclic variations in bolometric luminosity, are discussed in the chapter by Livingston et al. Particle fluxes will not be addressed in general. The possibility of a cyclic modulation of the solar neutrino flux is discussed in the chapter by Davis.

The integrated 10.7-cm (2.8 GHz) radio flux F_{10} is a valuable quantitative index of solar activity that now spans more than four complete cycles (since 1947). F_{10}—corrected by removing bursts and adjusted to 1 AU—correlates so strongly with the International sunspot number R_I that current literature often treats the 10-cm flux as a proxy or quantitative replacement for sunspot number; the limitations of that assumption are discussed in Sec. II.B below.

Progress in interpreting the 10-cm record will likely come by better understanding the relative contributions of various kinds of solar activity to the integrated flux. There is now a substantial body of aperture synthesis observations, mostly in the wavelength range 2 to 20 cm, that reveals the radio structure of active regions with high angular resolution, typically 1 to 10 arcsec (Kundu and Lang 1985). Nonflaring microwave emission is predominantly thermal, either bremsstrahlung or gyroresonant radiation. The observed brightness temperature usually increases with wavelength, reflecting local electron temperatures that range from $\sim 10^5$ K at 2 cm to $\sim 10^6$ K at 20 cm. The observed polarization of the microwave radiation is a powerful diagnostic that can be used in conjunction with photospheric magnetograms and images at optical, ultraviolet and X-ray wavelengths to provide a comprehensive view of the magnetic structure of active regions from the photosphere through the low corona.

For one active region, Felli et al. (1981) studied the distribution of the observed microwave flux between thermal bremsstrahlung and gyroresonant radiation and between plage areas and areas associated with sunspots. However, as Donnelly (1987) has emphasized, little is known about how such relative contributions vary from region to region, and even less is known about the contribution from quiet and enhanced network. Between the 10-cm record, aperture synthesis studies of individual regions, and whole-Sun multifrequency patrol observations (Barron et al. 1985), much of the data needed for such a study already exists. Tapping (1987) and Oster (1990) have made a start by using a part of the 10.7-cm record that is usually neglected: the daily east-west fanbeam scans with 1.5-arcmin resolution. Oster reaches the following conclusions. When a 10.7-cm emission peak can be identified with a single active region, the strength of the peak correlates strongly with the total sunspot area and with reported plage area in that region; the product-moment correlation coefficient of a bilinear fit to sunspot area and plage area is $r^2 = 0.92$. He finds no evidence for an "active network" component in addition to the contributions of individual active regions and a quiet-Sun background that does not vary during the solar cycle. However, Tapping and DeTracey (1990) find that 60% or more of the 10.7-cm flux above the quiet-Sun background is "widely distributed" in the sense that it appears as an enhancement of the background level between peaks in the east-west scans. Because the coronal field strength well outside sunspots is too weak to support significant gyroresonance emission, they attribute the widely distributed emission to bremsstrahlung. The conclusions of these two studies are not necessarily inconsistent. Here, as in other studies (Secs. II.B.2, II.C, III.D), the observation that magnetic activity tends to occur and recur in large-scale complexes blurs the meanings of "individual active regions" and "background activity." Quantitative distinctions are needed.

The flux from a centimetric microwave burst may outshine the total background (nonflaring) emission in the same waveband by 2 orders of magnitude. One of the cyclic properties of microwave bursts is discussed in Sec. II.D under the heading of flares.

In the case of integrated ultraviolet, extreme-ultraviolet, and soft X-ray fluxes, the key to incorporating their observed time variations into models of the solar cycle would again seem to be a better understanding how different spatially resolved features contribute to the integrated fluxes. The archives of observations from Skylab and the Solar Maximum Mission (SMM) could provide a start in this direction.

B. Sunspot-Related Indices

1. International Sunspot Number.   Also known as the Zürich number, the Wolf number, or simply "the sunspot number," the International sunspot number is the best known and most analyzed of all measures of solar activity. The present chapter could easily be filled by reviewing research on this famous time series during the last 5 yr alone. However, consistent with the intentions expressed above, the discussion will be limited to a few aspects of the sunspot number that bear on its use in quantitative models and how it relates to other indices. Historical perspectives on the limitations of the sunspot record, particularly for years before 1850, may be found in reviews by Eddy (1980) and Schove (1983). A guide to the literature of solar prediction using the sunspot series may be found in the compendium mentioned above (Simon et al. 1986) and the earlier monograph by Vitinskii (1965).

It is widely recognized that the measurement of the sunspot number is partly subjective. The definition is

$$R_I = k(10g + f) \tag{1}$$

where f is the number of individual spots, g is the number of sunspot groups, and k is a correction factor that depends on the telescope, observer and observing conditions. In principle, k is chosen to maintain the calibration of R_I by reducing counts made at a modern observatory to what Wolf would have counted when he introduced the relative sunspot number in 1864 (the telescope he used still exists).

As it is hardly clear that the scaling factor k can compensate for differences in a subjective judgment, such as what constitutes a sunspot group, it is tempting to dismiss R_I as a quantitative index. That judgment would be too harsh for the simple reason that, as has long been known, the sunspot number during recent cycles correlates quite strongly with objective measures such as 10-cm radio flux. It is interesting, however, that the composite nature of R_I can be seen in a detailed comparison between R_I and F_{10}. The two time series are shown in Fig. 1, smoothed by a conventional 13-month running mean (Waldmeier 1961). The overall product-moment correlation coefficient between the two series is $r^2 = 0.99$; but it is clear that the epochs of maximum have not always coincided.

Figure 2 shows a broader comparison between the number of spots, the number of groups and umbral area, as well as sunspot number and radio flux

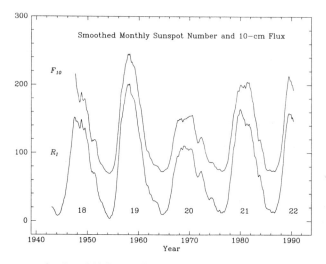

Fig. 1. Sunspot number R_I and 10.7-cm radio flux F_{10} for the period 1947 to 1990. The monthly mean values have been smoothed with a 13-month running mean.

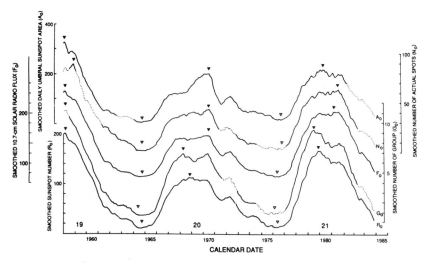

Fig. 2. Evolution of sunspot number and radio flux compared to number of sunspot groups, number of sunspots, and daily average sunspot umbral area in millionths of a hemisphere A_0 for the interval 1958 to 1984. All curves are 13-month running means. Note the similarities between R_0 and G_0 and between F_0 and N_0 (figure from Wilson et al. 1987).

(Wilson et al. 1987). Near maximum, the radio flux closely tracks the number of spots, while the sunspot number behaves more like the number of groups. Do these trends reflect a greater average magnetic complexity for groups appearing after the maximum in R_I, with correspondingly greater levels of activity (including radio flux)? Or, does the character of the groups and complexes change in such a way that the subjective assignment of groups varies during the cycle or between cycles? To what degree are the results affected by the dominance of long-lived complexes of activity near solar maximum? These questions suggest that classical activity indices can still raise interesting physical issues, which improved versions of the indices may help to resolve. For example, it will be interesting to apply objective criteria for the enumeration of sunspot groups to quantitative measurements of the white-light photographs from Mount Wilson (Howard et al. 1984) or Kodaikanal. In conjunction with the radio studies of active regions mentioned above, and objective characterizations of groups and complexes from synoptic magnetograms such as those discussed in Sec. III below, sunspot-related indices have the potential to tell us much more about the operation of the activity cycle than simply, "When was maximum?"

2. Heliographic Distribution of Sunspots. Obvious characteristics missing from the sunspot number are properties of individual spots and their heliographic distribution. Spot properties are considered below.

A graph of the range in latitude of sunspots as a function of cycle phase is the familiar butterfly or Maunder diagram. The Greenwich photoheliographic observations provide the longest series (1874–1976); the data are plotted by Yallop and Hohenkerk (1980). A large range of properties can be extracted from such data, from simple summaries such as north-south asymmetry to somewhat fanciful intimations of fine structure within the butterfly diagram. Summaries of early work on the latitude distribution of spots may be found in Kiepenheuer (1953) and Kopecký (1967,1984). Even crude separations by latitude can improve the discriminative properties of the usual sunspot indices. For example, as shown by Giovanelli (1966) and confirmed with later data, a graph of the number of old-cycle (low-latitude) and new-cycle (high-latitude) groups as a function of time is a sensitive marker of the onset of a new cycle, with potential application to near-term prediction. It is our impression that the quantitative aspects of butterfly diagrams have been underutilized in the past, perhaps because the volume of data was difficult to handle without computers.

Aspects of the sunspot butterfly diagram, and its relationship to similar diagrams constructed from properties such as coronal emission or photospheric velocity shear, is discussed further in Sec. V. Proper motions of sunspots are discussed in the chapter by Howard et al.

The distribution of sunspots in Carrington longitude is usually studied by means of a longitude-time diagram (similar to a Bartels display) for spots in a given range of latitude. As long ago as 1899, Wolfer noticed the tendency for spots to be concentrated at certain longitudes for many rotations, longer than the lifetimes of individual spots. It was realized early on that the so-called active longitudes were usually different in the northern and southern hemispheres (Kiepenheuer 1953). Recently, magnetic flux and magnetic polarity have been studied in a similar way by stacking latitude strips from synoptic charts in a Bartels-like display (Gaizauskas et al. 1983; McIntosh and Wilson 1985). The long-lived patterns shown by such displays prompted a fresh analysis by Castenmiller et al. (1986) of the Greenwich photoheliographic observations. They describe the preferred areas of sunspot occurrence as sunspot "nests" and compare the properties of nests to the properties of recurrent magnetic patterns studied by Gaizauskas et al. and the complexes of activity described by Bumba and Howard (1965).

3. Properties of Individual Sunspots.   Let $N(A)$ be the number of spots with umbral area greater than A existing over the whole solar surface at one time; A is usually measured in millionths of the area of a hemisphere. Kopecký (1967) determined the differential distribution dN/dA during cycles 15 to 18 and illustrated the distinct curvature in a log-log plot of dN/dA against A. Bogdan et al. (1988) determined dN/dA from Mount Wilson white-light plates covering 1917 to 1982. They found a similar shape and fitted it with a log-normal distribution, as shown in Fig. 3. They also found that the *same* distribution, differing only in normalization, applied to all phases of the solar cycle and all the cycles covered by their observations. This result may be compared with the conclusion in Sec. III below that the shape of the cumulative distribution of active-region areas determined from photospheric magnetograms is also independent of cycle phase. However, Kopecký claimed that the shape of $N(A)$ varies among cycles, and that a variation is particularly clear in the frequency distribution of spot *groups* according to maximum area. There is clearly more work to be done on this topic, which once again points to the importance of constructing objective (if arbitrary) definitions of sunspot groups and active regions.

Umbral area is only the simplest photometric characteristic of sunspots that might be followed through the solar cycle. For example, Albregtsen and Maltby (1981*b*) measured the umbra-to-photosphere and penumbra-to-photosphere intensity ratios of 15 large sunspots during the period 1968 to 1979. Pinhole photometer measurements were made at 10 wavelengths covering the spectral range 0.4 to 2.4 μm. Albregtsen and Maltby found significant variations in umbral contrast between different spots that correlated strongly with cycle phase but not with any measurable properties of the individual spots (e.g., magnetic field strength, size, age, or Zürich classification). There was also a somewhat weaker correlation between umbral intensity and latitude, but this was difficult to separate from the temporal effect because of the migration of the sunspot belts during the cycle. The photo-

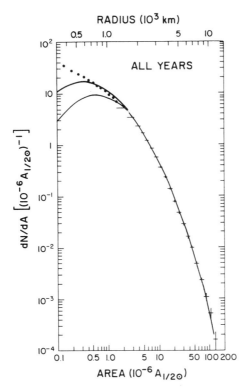

Fig. 3. Differential size spectrum for a sample of 24,615 sunspots measured on Mount Wilson white-light photographs taken during several solar cycles (1917–1982). The horizontal error bars give the bins widths; the vertical error bars show plus/minus the square root of the count in each bin. The filled circles represent small spots that are seriously affected by seeing and measurement errors. Two log-normal fits are shown which nearly coincide for areas $\geq 5 \times 10^{-6}$ hemisphere (figure from Bogdan et al. 1988).

metric properties of sunspots through the cycle should be investigated with a larger sample, particularly in the infrared where corrections for stray light are small.

It would also be of interest to follow through the cycle the average magnetic field strength within sunspots, controlled for effects such as sunspot size. Most Babcock-type magnetographs are not suitable for such a study because their magnetic signal saturates at umbral field strengths.

   4. Properties of Sunspot Groups.   Knoška and Křivský (1984) have tabulated the distribution of sunspot groups among the Zürich classes for each year of cycle 20; similar data for cycle 19 are given by Kopecký and Křivský (1966). The value of such data would be enhanced by comparing them with measures of region size and complexity based on magnetograms. Data of the latter kind are now becoming available (Sec. III below; Howard 1989).

C. Flares and Other Energetic Events

Solar flares and their terrestrial effects are monitored worldwide. Their frequency of occurrence can be used to mark the general progress of the cycle, particularly in the years around maximum. For example, there is some evidence that the frequency of large flares tends to remain high (in episodes) past the time of sunspot number maximum and then fall sharply; if confirmed by a larger sample, this property could be of use in establishing the phase of the declining cycle (Křivský et al. 1986). Butterfly diagrams and the like can be constructed for flares (Knoška and Křivský 1978).

From the viewpoint of understanding the cycle, however, the use of flare frequencies is problematic. The plethora of flare-related effects means that there are many ways to construct a sample and many ways to judge the importance of an event. For many types of events, there is no worldwide standard of quality and uniformity comparable to that established for the sunspot number. Finally, we understand flares as little as we do the solar cycle. We know in a general way that flares are associated with magnetic complexity in the photosphere and low corona, but little is known quantitatively about which kinds of complexity preferentially give rise to which kinds of energetic events, or how all but a few types of magnetic complexity arise in the first place.

Progress is being made in the study of coronal mass ejections. Observations from the SMM Coronagraph/Polarimeter (1980, 1984–1989), the P78-1 Solwind coronagraph (1979–1985), and from groundbased K coronameters now constitute a data set of sufficient quality, uniformity and temporal span that it is possible to make statistically sound studies of the properties of coronal mass ejections through the solar cycle as well as to understand the physical basis of these events (Sheeley et al. 1986b; Sime 1989; Harrison et al. 1990).

Interesting information about medium-term cyclic activity came from an unexpected direction when Rieger et al. (1984) analyzed gamma-ray events observed by the SMM satellite. A power spectrum covering 3.5 yr revealed a significant peak at a recurrence period of about 154 days. A similar peak was found in temporal power spectra of hard X-ray bursts observed by SMM (Kiplinger et al. 1984) and soft X-ray events recorded by the GOES satellites (Rieger et al. 1984). It is perhaps significant that this discovery emerged first from satellite observations, which offer a combination of characteristics rare in groundbased flare data: well-defined samples of fully quantitative measurements from closely monitored instruments operating over known duty cycles (in the case of the GOES satellites, nearly 100%). Subsequently, a periodicity of about 152 days was found in microwave events observed from the ground during 1966 to 1983 (Bogart and Bai 1985) and was claimed to be coherent between cycles 20 and 21. Lean and Brueckner (1989) detected a 155-day periodicity in other groundbased observations (sunspot blocking function, 10.7-cm radio flux and sunspot number) obtained during cycles 19 to 21. Analyzing the incidence of major X-ray flares, Brueckner and Cook (1990) find evidence that the 155-day period has maintained phase coherence between cycles 21 and 22. Bai (1987b) detected a 51-day period in the occurrence of major flares during cycle 19 selected according to the Comprehensive Flare Index (Dodson and Hedeman 1981), but this period is not significant in the data analyzed by Lean and Brueckner (1989).

In the case of the sunspot series, studies that retain spatial information have been more informative about the nature of the solar cycle than simple time-series analyses. Bai (1987a, 1988) has taken this approach to flares by analyzing their heliographic distribution during 1955 to 1985. The data base is a combination of major flares chosen according to the Comprehensive Flare Index and the list of hard X-ray bursts from SMM. Bai found that a small number of "superactive" regions, amounting to only about 1% of all active regions, accounted for almost half of all the flares in the sample. Superactive regions appeared more frequently in certain co-rotating areas of the Sun, which we shall call "active domains." There were typically two active domains in each hemisphere, separated by about 180° in longitude, but the domains did not generally line up between hemispheres to form active longitudes. Bai further claimed that some domains persisted for 2 or even 3 solar cycles without significant disruption by differential rotation. The acceptance of magnetic patterns that cross not only 11-yr sunspot cycles but 22-yr magnetic cycles would clearly impact phenomenological theories of the solar cycle (Sec. V). The joint relationship between the 154-day periodicity, active domains, sunspot nests, and magnetic complexes of activity is more than a confusion of terminology (although it is that) and deserves study.

D. Chromospheric and Coronal Indices

1. Ca K Emission.   It has long been recognized that emission in the cores of the Ca II H and K lines is a sensitive indicator of solar active regions and the chromospheric network. From the earliest magnetograms, Babcock and Babcock (1955) also recognized that there is a more quantitative relationship between calcium plage and photospheric magnetic field strength.

In recent years, much of the quantitative work on solar Ca II emission has been stimulated by its use in monitoring stellar chomospheric activity of presumed magnetic origin (chapter by Livingston et al.). The relationship between emission strength in the central 1 Å of the K line and photospheric magnetic flux density has been investigated for both quiet network (Skumanich et al. 1975) and active regions (Sivaraman et al. 1987; Schrijver et al. 1989a). Such relationships could be applied to full-disk K-line photometric images of the Sun to yield absolute magnetic flux maps. This is not entirely perverse: if the center-to-limb behavior of the relationship between calcium emission and magnetic flux were known well enough, the calcium-derived magnetic flux map might allow a better estimate of polar field strengths than is possible using line-of-sight magnetograms.

The calcium plage areas and intensities published in several places are qualitative estimates which, although useful (as, for example, in modeling solar irradiance variations, are of lesser value for the purpose of understanding the solar cycle. The National Solar Observatory has obtained photographic K-line spectroheliograms on a daily basis for some years. These data can greatly extend the range of calcium activity studies when they are photometrically calibrated.

2. Helium I λ10830. Daily full-disk spectroheliograms in the He I λ10830 line have been taken by the National Solar Observatory since early 1974 using the 512-channel magnetograph at the Vacuum Telescope on Kitt Peak. These data have provided a daily look at coronal holes and the equivalent width of the λ10830 line for a period spanning cycle 21 and the rise of cycle 22 (Harvey 1981,1984,1990a). The importance of these observations for solar cycle studies lies in the study of coronal holes (see, e.g., Sheeley and Harvey 1981; Harvey et al. 1982) and the correlation of the He I λ10830 equivalent width with extreme-ultraviolet (EUV) and ultraviolet (UV) solar spectral irradiance (Donnelly et al. 1985, 1986; Lean 1987). The λ10830 line is formed in the chromosphere and has no photospheric contribution, making it an extremely useful line for studying purely chromospheric phenomena. In addition, it has some indirect coronal sensitivity because one process of exciting the transitions in the triplet levels of the He I atom that result in the λ10830 and D_3 lines is photoionization by coronal radiation ($\lambda < 500$ Å) followed by recombination (Goldberg 1939). Such an excitation process explains the ability to detect coronal holes in λ10830 spectroheliograms.

The He I λ10830 equivalent width shows a variation that is similar to that observed in total magnetic flux, in 10.7-cm radio flux, and several EUV and UV spectral irradiance measures. During cycle 21, however, the increase in the λ10830 equivalent width from minimum to maximum, which occurs in mid to late 1981, is only about a factor of 2 compared to the more than a factor of 5 in total magnetic flux (Sec. III.C). The solar cycle variation of λ10830 more closely follows full-disk measurements of chromospheric emission, such as the Ca K-index (chapter by Livingston et al.), and shows a behavior that indicates important contributions from active regions as well as from the network originating in both the quiet Sun and the stronger, diffused magnetic regions.

3. Coronal Tracers. Sime (1985) has reviewed variations of the corona and the interplanetary medium during the solar cycle. Coronal mass ejections were briefly addressed in Sec. II.C; Sec. V.D.3 discusses high-latitude coronal tracers that have been interpreted in conjunction with other high-latitude activity as evidence for an "extended" solar cycle.

Cyclic variations in the shape and brightness of the K-corona (Koutchmy and Loucif 1984) primarily reflect cyclic changes in the large-scale organization of the coronal magnetic field. Both the K-corona and the corona seen in Fe XIV (5303 Å) rotate more rigidly on average than does the photospheric gas (Sime et al. 1989). Although the magnetic transport models described in Sec. IV account in a straightforward way for the relatively rigid rotation of the K-corona (typically measured at 1.3 to 1.5 $R_\odot$), it is less evident that the rotation of the Fe XIV corona (measured at 1.15 $R_\odot$) is adequately explained.

E. Proxy Indicators of Historical Solar Activity

As mentioned earlier, we shall mostly omit this topic, potentially of great significance but fraught with special difficulties, in favor of monographs (Pepin, et al. 1980; Sonett et al. 1991). However, one proposed proxy should be mentioned, if only as a caution, because of the extraordinary duration and precision of the information it appeared to offer about ancient solar cycles.

Williams (1981) hypothesized that cyclically laminated rocks found in the South Australian Elatina Formation ($\sim$ 680 Myr old) were periglacial varves (annual sedimentary deposits) that recorded modulations of the local climate by the solar activity cycle. The individual laminae tend to occur in groups of 10 to 14, separated by thinner dark bands. In turn, the groups commonly alternate between thick and thin cycles. These two properties are reminiscent of the typical (but variable) number of years in recent sunspot cycles and the sometime tendency for high and low cycles to alternate (as in the period 1845 to 1945). Williams (1985) collected 3 drill cores from the Elatina Formation that together provided a continuous record of some 19,000 laminae, more than 1500 cycles. This may be compared to the modern sunspot record of $\sim$25 cycles, of which only the last 12 to 13 reflect consistent, controlled observations. The Earth's climate today is modulated at most weakly by the solar cycle, and no modern varves show a clear solar signal.

It now appears that the Elatina varves are not of solar origin. According to Williams (1988), two other Precambrian formations in South Australia typically show cycles of 14 to 15 laminae, with many laminae split into semi-laminae. Comparing the structure of these cycles to modern tidal data, Williams concludes that the South Australian laminae are more easily understood as effects of the lunar tide. Thus, the sunspot series, with all its known defects, is still the best continuous record of solar activity. Some of the analysis techniques prompted by the Elatina series may yet find application to solar time series.

III. MAGNETIC FLUX MEASUREMENTS OF THE SUN

Several observational programs have been instituted to observe, systematically and over long periods, the magnetic flux passing through the solar photosphere. Observatories involved include the Mount Wilson Observatory (since 1957), the U.S. National Solar Observatory (since 1973), the Wilcox Solar Observatory (since 1976) and the Crimean Astrophysical Observatory (since 1968).

The full-disk magnetograms from Mount Wilson are made in the Fe I λ5250 line; the spatial resolution initially was 23 arcsec but has improved twice: in 1967 to 17.5 arcsec and in 1975 to 12.5 arcsec. During the Skylab period in 1973, full-disk magnetograms were obtained by the National Solar Observatory using the Fe I λ5233 line; the spatial resolution of these data was 2.5 arcsec. In 1974, a new magnetograph was installed providing 1-arcsec spatial resolution in the Fe I λ8688 line. The Wilcox Solar Observatory and Crimean Astrophysical Observatory observations are made in the Fe I λ5250 line with a spatial resolution of 3 arcmin. These lower-resolution measurements have primarily been used in investigations of the connection between large-scale photospheric magnetic fields and the interplanetary magnetic field and sector structure (see, e.g., Hoeksema 1984; Pflug and Grigoryev 1986) and rotational periods of solar activity (see, e.g., Kotov 1987). As this section concerns the cyclic behavior of the Sun's magnetic fields, the discussion will be restricted to National Solar Observatory data (Harvey 1990a) and Mount Wilson data.

A. Mount Wilson Observations

Howard (1974) and Howard and LaBonte (1981) have used the Mount Wilson data to study the variations and spatial distribution of magnetic flux on the Sun during 2 solar cycles. Their data cover a period of 13.5 yr from 1967 in the rise of cycle 20 to 1981 just after the sunspot maximum of cycle 21. The Mount Wilson data showed that the total magnetic flux increased from sunspot minimum to maximum by a factor of 2 during solar cycle 20 and by a factor of 3 during solar cycle 21. Howard and LaBonte find that the weak fields show essentially the same pattern as the stronger fields and conclude that the large-scale patterns result only from emerging active-region magnetic fields, which diffuse and move poleward. The polar fields are reversed and strengthened during episodes of rapid poleward migration of remnant magnetic flux of the active regions, predominantly of following polarity. Howard and LaBonte suggest that the poleward drift of the fields cannot be explained by diffusion alone but must result from meridional flows that early in the cycle are 5 to 10 m s^{-1} and reach 15 to 20 m s^{-1} during the cycle decline. The magnetic flux at the poles appears to be about 1% of the total flux on the Sun, indicating that most of the flux emerging in active regions must disappear well before reaching the poles.

B. National Solar Observatory Observations

The National Solar Observatory (NSO) magnetograph data have been used to study the variation of the magnetic flux on the Sun from March 1975 through January 1987. This investigation confirms and extends the results of the Mount Wilson observations. The principal objective is to separate the Sun's magnetic field into two components—active-region fields and fields outside active regions—to investigate their respective contributions to the Sun's total field, their variation as a function of the solar cycle, and their north-south asymmetry.

Conventional solar magnetographs, including the NSO instrument, measure magnetic flux rather than magnetic field strength. Measurements reported in units of gauss must therefore be interpreted as averages over the spatial resolution element. Solar magnetic elements outside of sunspots are usually spatially unresolved (chapter by Spruit et al.).

1. The Synoptic Magnetic Maps. Since early 1974, the line-of-sight component of the Sun's photospheric magnetic field has been measured on a nearly daily basis by the 512-channel magnetograph on Kitt Peak (Livingston et al. 1976) using the λ8688 line of Fe I. The full-disk magnetograms have an instrumental spatial resolution of 1 arcsec and a noise level of 7 G (gauss). The daily magnetograms are used to construct a synoptic magnetic map of the Sun as follows: the magnetic field data on each full-disk, high-resolution observation are summed within equal area bins of 1° in longitude and 0.011 in sine latitude yielding a reduced resolution magnetogram of 180×180 elements. The 180×180 daily magnetograms belonging to a given Carrington rotation are merged, weighting the data by a $\cos^4$ function in central meridian distance. The resulting synoptic magnetic map depicts an average of the magnetic fields at each Carrington longitude during its central meridian passage. Examples of the synoptic magnetic maps are shown for a quiet period (April 1975, Carrington rotation 1628) and for an active period (August 1981, Carrington rotation 1712) in Fig. 4.

For the analysis described below, the magnetic fields are assumed to be radial and have been corrected by $\cos^{-1}$ (latitude). Only data within the latitude range N 60° to S 60° have been included in the analysis due to the sparse sampling at higher latitudes. For each synoptic magnetic map, the following parameters of the magnetic field have been determined: the total magnetic flux (the sum of the absolute values of the positive and negative flux: $F_T = |F_+| + |F_-|$), the net magnetic flux ($F = F_+ + F_-$), and mean absolute flux density (F_T/area). By summing over the entire synoptic magnetic map, variations due to solar rotation, as defined by the Carrington rate, are eliminated.

2. Noise and Errors in the Magnetic Field Data. There are several sources of noise and error in the magnetic field observations used in this

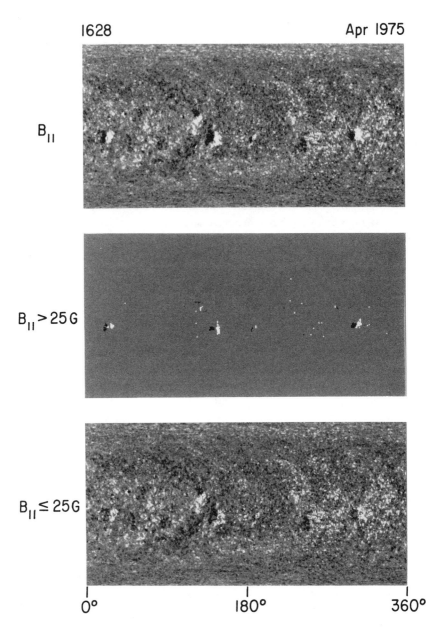

Fig. 4. Synoptic magnetic rotation maps from the National Solar Observatory. Each map shows photospheric magnetic flux for a complete Carrington rotation of the Sun (1628 or 1712). The left-hand maps were constructed from full-disk magnetograms taken near solar minimum;

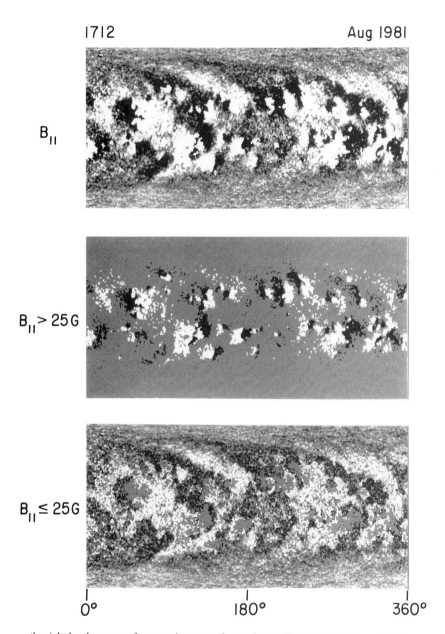

1712   Aug 1981

$B_{\parallel}$

$B_{\parallel} > 25\,G$

$B_{\parallel} \leq 25\,G$

0°   180°   360°

the right-hand maps are for a rotation near solar maximum. The top panels show all the flux, the middle panels show flux above the threshold chosen to isolate active areas (25 G), and the bottom panels show flux below that threshold (quiet areas). North is at the top; east is to the left.

study. Although these uncertainties are significant, they do not change the basic results. The influence of uncertainties will be discussed as appropriate.

The summing of the higher-resolution observations of the magnetic field within the larger-resolution elements of the synoptic maps yields a reduced noise level of 0.3 G. Due to instrumental problems, the noise level from 1975 to early 1977 was higher, leading to an artificially higher value of the total magnetic flux by $\sim$ 20% during this period. The reduced resolution of the synoptic maps also results in a decrease in the apparent magnetic flux if the original magnetogram contains opposite polarity structures within the resolution element of the synoptic map. It is estimated that about 5×10^{22} Mx is lost in the conversion to the synoptic magnetic field maps regardless of cycle phase (J. Harvey, personal communication).

Inherent in most methods currently used to measure magnetic fields on the Sun is the inability to determine sunspot magnetic fields accurately. Typically, in the NSO full-disk observations, the magnetic field within a sunspot umbra is underestimated by about 1000 G per pixel. The missing magnetic flux, which varies with the solar cycle, can be estimated from measurements of the umbral area during cycle 21 (see, e.g., Howard et al. 1984) and compared with the determination of the total magnetic flux from the NSO synoptic maps over the solar cycle. It is found that during maximum, the Sun's total magnetic flux is underestimated by at least 10 to 15%.

C. Magnetic Flux Variations during Solar Cycle 21

1. Net and Total Magnetic Flux.   The Sun's total magnetic flux and net magnetic flux are shown as functions of time in Fig. 5. The net magnetic flux varies around zero for much of the cycle, although during some intervals (such as 1975–1976) the net flux appears to be systematically shifted. This probably reflects an incorrect instrumental zero level in the magnetic field data. The offset does not substantially affect the values of the total magnetic flux during this period. The variations in the net flux appear to bear no relation to the larger short-term changes in the total magnetic flux.

The total magnetic flux during the solar cycle shows variations on several time scales. Over the long term, from minimum in 1975 to a maximum in August 1981 (1.7 yr after sunspot maximum), the total magnetic flux in the photosphere increased by a factor of $\sim$5. This is larger than the three-fold increase found by Howard and LaBonte (1981) from 1976 to sunspot maximum in 1979, and is a lower limit to the actual increase in the magnetic flux during solar cycle 21. Because the total magnetic flux during maximum is underestimated, the actual increase may be as much as 6 to 7.

On a shorter time scale there appear to be quasi-periodic pulses or episodes of enhanced magnetic activity. The duration of these pulses appears to be longer ($\sim$ 9 rotations) in the rise and decline of the cycle and shorter ($\sim$ 5 rotations) during the years of maximum. The 154-day periodicity found in flare activity (Sec. II.C) may be a consequence of activity pulses. Com-

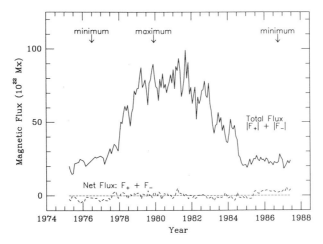

Fig. 5. Total and net magnetic fluxes during cycle 21. The resolution of the plot is one Carrington rotation. The plot of the total flux shows clearly the pulses of activity superimposed on the solar cycle variation of the magnetic flux. Maximum of the Sun's total flux occurs in August 1981, almost 2 yr after sunspot maximum based on sunspot numbers.

parisons with the magnetic field maps suggests that these pulses of activity correspond to the occurrence of active-region complexes (Gaizauskas et al. 1983) or sunspot nests (Castenmiller et al. 1986). For example, the first major pulse of activity in solar cycle 21, peaking in the first half of 1978, can be traced to the activity complex studied by Gaizauskas et al. (1983); the pulse in 1982 is associated with a long-lived activity complex that included the unusually large sunspot regions observed in June and July 1982. Although some pulses or episodes of activity can be identified with specific complexes of activity, the synoptic map for Carrington rotation 1712 (Fig. 4) indicates that the magnetic flux contributing to the large peak in August 1981 is more globally distributed and not confined to one or two active region complexes as is the case in 1982.

2. The Active Region and Quiet Sun Components. The magnetic fields in active regions were separated from those outside by setting a threshold. Magnetic fields above the threshold were defined as the strong-flux component and comprised both active regions and the active network. Magnetic fields equal to or below the threshold were defined as the weak-flux component, which included quiet Sun and network fields. After some experimentation, a threshold of 25 G was chosen as giving visually a good separation of the strong and weak components. Figure 4 shows the pictorial result for 2 Carrington rotations. Although active regions are effectively isolated by a threshold, it is clear from the figure that some fields in active regions are included as weak fields, while some network, not active, is included in the strong-flux regions.

A visual comparison of the weak and strong components indicates that, at both low and high levels of activity, most of the magnetic flux in what has been defined as the weak fields comes from dispersing active regions. What is surprising is that the measured increase of the weak-field component is no more than a factor of 2. As shown in Fig. 6, this is a significantly smaller increase than the factor of 5 observed for the Sun as a whole. In Fig. 7, the total flux of the weak- and strong-field components is compared. The increase observed in the strong-field component is at least a factor of 15 in contrast to the small increase in the weak-field component. If the increase of the weak fields over the cycle can be attributed to the dispersal of active-region fields, then only a fraction of the flux that emerges in active regions actually leaves the active region. These data suggest that this fraction is ≤30% and that ≳70% of the flux that emerges in active regions disappears *in situ*. This supports the results of Gaizauskas et al. (1983) that a significant fraction of the magnetic flux in active regions disappears within the boundaries of the active complex. The short time scale over which this takes place as well as the significant levels of flux disappearance in active regions is also indicated in Fig. 7 by the rapid increases and decreases of magnetic flux levels during the pulses of activity in the strong-field component that are not evident in the weak fields. The variation of the weak fields during the cycle is nearly in phase with that of the stronger fields, suggesting that flux disappearance con-

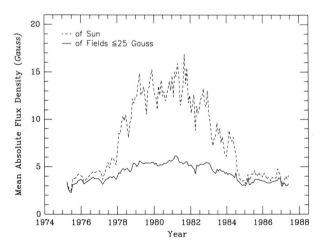

Fig. 6. Mean absolute flux density during cycle 21. Plotted are the mean absolute flux density (total flux divided by area) of all fields (dotted line) and of the weak component (solid line). The variation of the weak component is no more than a factor of 2, compared with the factor of 4 seen for all fields.

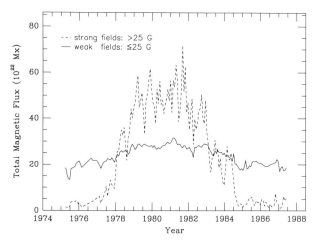

Fig. 7. Total magnetic flux divided into strong- and weak-field components. During the early and late parts of the cycle, the weak component dominates; from 1978 to 1983 the strong component substantially exceeds the level of flux observed in the weak fields. The pulses of activity evident in the strong component are not seen in weak component. The strong component varies by at least a factor of 15 from minimum to maximum while the weak component varies by a factor of 2 or less.

tinues in the weak-field component, although at a slower rate than is observed in active regions.

Until early 1978 and after 1983, except for 2 short periods, the magnetic flux in the weak-field component significantly exceeds that observed in the strong-field component. The magnetic flux in the weak component during periods of low activity remains at a high level, raising the question of what makes up the magnetic fields in the quiet Sun. Three possibilities are:

1. Dispersing active-region fields belonging to the outgoing old cycle and the incoming new cycle. This contribution is small, as the number of active regions emerging during this time is relatively low (Fig. 7). The similar cycle-variation of the strong and weak components of the magnetic field also indicates that the dispersed active-regions fields (during minimum as well) are not accumulating, but are disappearing over a time scale of months.
2. Ephemeral regions. These small-scale bipoles are below the resolution limit of synoptic magnetic field maps and, therefore, do not contribute to the level of magnetic flux observed during cycle minimum. If they were observed, their total flux would contribute $\leq 8\%$ to the magnetic flux levels at minimum. This is determined from the result of Harvey (1984,1990b) that 300 to 1000 ephemeral regions are present on the Sun at any given

time. With an average total flux of 3×10^{19} Mx (Harvey and Harvey 1976a), their total flux contribution is $\leq 1.5 \times 10^{22}$ Mx.

3. An ever-present background field that may be continuously produced. Some evidence for this has been suggested by the observations of intranetwork fields by Livingston and Harvey (1975) and Martin (1988).

3. North-South Asymmetries   An imbalance in the level and timing of activity between the northern and southern hemisphere has been recognized as a property of many recent solar cycles. To investigate the north-south asymmetry of magnetic activity, the total magnetic flux was determined separately for the northern and southern hemisphere for both the strong and weak components. In Fig. 8, the total flux measured in strong fields in the north and in the south is shown as a function of time. Evident early in the cycle is a lag of several rotations in the onset of activity in the southern relative to the northern hemisphere. At the end of 1982, the active-region flux in the northern hemisphere drops precipitously, while the southern hemisphere remains active, though decreasing. The pulses or episodes of activity seen in the whole-Sun measures also can be seen in these data. The pulses are narrower, and seem to show little relation between the hemispheres during much of the cycle until the beginning of 1983. At this point in the cycle, the pulses of activity observed in the northern and southern hemispheres appear to synchronize, suggesting that the sources of flux for the two hemispheres are

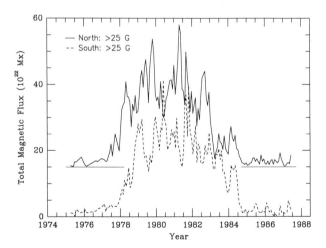

Fig. 8. Total magnetic flux in the strong component for northern and southern hemispheres separately. There the general level of activity in the southern hemisphere lags the northern hemisphere by ~ 9 rotations. Pulses of activity appear to be uncorrelated between the hemispheres until 1983, when they appear to synchronize.

interacting. This is occurring at the time of the cycle when almost all of the active regions belonging to cycle 21 are erupting within 15° of the equator.

In Fig. 9, the magnitude of the north-south asymmetries of the strong-field component is shown during the course of the solar cycle and is compared with the north-south asymmetry of the weak-field component. The very strong asymmetry seen in the active-region magnetic fields is not reflected in the quiet Sun fields. The asymmetry of the magnetic fields outside active regions, while in the same sense as for active regions, is considerably smaller, again arguing for substantial flux loss in active regions over a relatively short time scale (1–2 rotations).

D. Summary of Magnetic Flux Variations

1. The Sun's total magnetic flux increased during solar cycle 21 by at least a factor of 5, peaking in August 1981, 1.7 yr after sunspot maximum. Dividing the magnetic fields into strong and weak components by thresholding, it is found that the magnetic flux contributed by the active-region fields (>25 G) varies by at least a factor of 15 from cycle minimum to maximum, while the quiet Sun fields (≤25 G) increase by no more than a

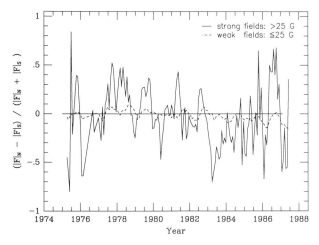

Fig. 9. Asymmetry index (ratio of the difference between northern and southern total fluxes to whole-Sun total flux) for strong and weak fields during cycle 21. Asymmetry in activity between the hemispheres occurs primarily in the strong component. The asymmetry in the weak component is smaller and may be delayed relative to the asymmetry in the strong component. This suggests that much of the magnetic flux that emerges in active regions disappears before it can diffuse into the weaker background fields. The asymmetry in the strong component shows the dominance of the northern hemisphere early in cycle 21 and the dominance of the southern hemisphere late in the cycle. The large asymmetries noted in 1975, 1985 and 1986 result from low activity during solar minimum.

factor of 2 during the same interval. It is concluded that most of the flux emerging in active regions disappears before it is able to disperse. It is estimated that more than 70% of the magnetic flux in active regions disappears *in situ* with only a small fraction (<30%) dispersing into the background.

2. The decrease of the weak fields during the decline of the cycle coincides with the decrease of the stronger fields, suggesting that flux disappearance continues in the weak-field component but at a slower rate than is observed in active regions.

3. There are strong north-south asymmetries of the active-region magnetic flux during the solar cycle. The asymmetry of the magnetic fields outside active regions are in the same sense as for active regions but of smaller magnitude, again suggesting that much of the magnetic flux in active regions disappears before it is able to disperse.

4. During cycle 21, there appear to be quasi-periodic pulses or episodes of activity of longer duration ($\sim$ 9 rotations) during the rise and decline of the cycle and of shorter duration ($\sim$ 5 rotations) during maximum. These pulses of activity appear to be a result of the occurrence of active-region complexes, and may explain the periodicity observed in flare data for this cycle. Pulses of activity in the northern and southern hemisphere appear to be unrelated, or perhaps anticorrelated, until late in the cycle, when they synchronize.

E. Modal Structure of the Magnetic Flux

Stenflo and Güdel (1988) have analyzed a 25-yr time series of synoptic maps (combining Mount Wilson and Kitt Peak data) by decomposing each map into spherical harmonic components and deriving power spectra of the harmonic coefficients. For the zonal ($m = 0$) modes, they find that odd-l modes (antisymmetric about the equator) are dominated by the 22-yr cycle period; the power in the 11-yr overtone is only about 10% as large as the power in the fundamental. The even-l (symmetric) modes behave quite differently. Their maximum amplitudes are 5 to 10 times smaller than the maximum amplitudes of odd modes of comparable degree, and they show almost no evidence of the 22-yr fundamental. Rather, the l-ν diagram for even modes appears to show a ridge of power from lower to higher wavenumber and frequency. Stenflo and Vogel (1986) discuss a speculative interpretation of this structure.

F. Solar Cycle Variations in the Size Spectrum of Active Regions

The total magnetic flux of the strong-field component shows a change of at least a factor of 15 from cycle minimum to maximum. How does this relate to the number of active regions observed in the magnetograms? This can be answered in part by a study of the count and the properties of active regions

(Harvey 1990c). Active regions that emerged on the disk were identified on the NSO full-disk magnetograms during 29 selected rotations from the period 1975 to 1986. The timing of the 29 rotations studied is shown in Fig. 10 in relation to the 27-day running mean of the 10-cm radio emission.

A total of 978 new active regions with an area > 2.5 deg² after correcting for foreshortening were identified on the total of 739 magnetograms obtained during the total interval of 844 days in the 29 time intervals. The adopted minimum area of 2.5 deg² corresponds to 370 Mm² or 125 millionths of the visible solar hemisphere. Bipolar active regions are counted only once. Although some of the measurements of the regions were made at first observation, for example (position and orientation of the region's magnetic poles) some of the intrinsic properties of the regions, such as their area, were determined at the time of the *maximum development*—that is, the moment when the emergence of magnetic flux in the region ceases. The area was estimated by a rectangle or rectangles that approximate the extent of the region. The area estimated using this method compares within ±15% with measurements of the area defined by an isogauss contour. The random error in the adopted technique of area measurement was estimated to be ~10% from remeasuring the area for 120 regions covering the entire size range of the data set.

Two corrections were applied to the data: (1) for data gaps that result in missing the emergence of short-lived active regions; and (2) for the loss of visibility of active regions with increasing radial distance. As the visibility

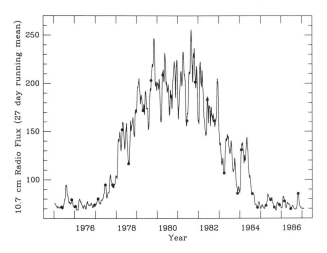

Fig. 10. Times (•) of the Carrington rotations selected for the study of active-region sizes. The times of the 29 intervals (each one rotation in duration) are compared with a 27-day running mean of the daily values of the 10.7-cm radio flux.

function depends on region size, it was determined empirically for several
size ranges by examining the distribution in radial distance of regions in equal
area bins in a narrow latitude strip at the equator. Harvey (1990a) gives the
details of this procedure.

1. Variation over Solar Cycle as a Function of Region Size.   The variation in the number of regions during the solar cycle is considered for 4 area
ranges, as shown in Table I. All region sizes show a similar variation during
the solar cycle, increasing by a factor of 5.5 to 6.0 from minimum to maximum (Fig. 11). Except for the regions with areas > 24.5 deg^2, maximum is
reached in 1981, almost 2 yr following sunspot maximum in late 1979.

2. Area Distribution of Active Regions.   Let $N(A)$ be the number of
active regions with area greater than A on the visible solar disk. The data
shown in Fig. 12 (top panel) are differential distributions, dN/dA, normalized
to 1 day and area bins of 1 deg^2. The log-log plots show a roughly linear
increase in the number of regions with decreasing area, of the form $dN/dA
= C A^{-2.49}$; A is the area in square degrees, corrected to disk center. However,
the integral of this power law differs noticeably from $N(A)$. The differential
of a cubic polynomial in log $N(A)$ and log A (solid line) provides a better fit,
both to dN/dA and $N(A)$. An extrapolation of the two fits to smaller areas
brackets the observed number of ephemeral regions, indicating a substantially
closer fit than the factor of 1000 difference suggested by the Tang et al. (1984)
study of active regions identified on the Mount Wilson magnetograms. The
sizes of regions in this study fall in the area range Tang et al. felt were undersampled in their data.

In the bottom panel of Fig. 12, area distributions are shown for active
regions identified during maximum, considered here to be the years 1979 to
1982, and during minimum, including 1975 to 1976 and 1985 to 1986. The
distributions have essentially the same form for both periods during the cycle,

Table I
Area Bins for Active Regions

Bin	deg^2	Range in Area 10^8 km^2	10^{-6} hemisphere
1	2.5–5.5	3.7–8.2	125–275
2	5.5–10.5	8.2–15.7	275–525
3	10.5–24.5	15.7–36.6	525–1830
4	> 24.5	> 36.6	> 1830

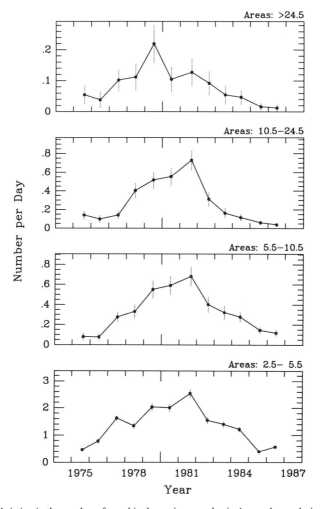

Fig. 11. Variation in the number of new bipolar regions per day in 4 area classes during cycle 21 (see Table I). The smaller regions all show a maximum in 1981, ~ 2 yr after sunspot maximum, but the largest regions peak in 1979 with a smaller peak in 1981. The vertical lines indicate the square root of the observed number of regions per day.

but the distribution at minimum is lower by a factor of 5 to 6. Thus, the distribution of active-region sizes has a characteristic form, independent of cycle phase; the normalization of the distribution shifts up and down with the rise and decline of the cycle.

Although active-region areas and magnetic flux are both good markers of cycle phase, the contribution by active regions based on area indicates only

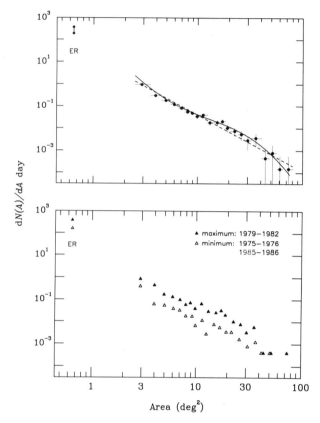

Fig. 12. (Top) Distribution of active-region areas for all regions during cycle 21. Plotted are the counts of active regions, d$N(A)$/dA, normalized to one day. The size distribution is fitted with a differential of a cubic polynomial in log $N(A)$ and log A (solid line) and a power law (dashed line). The number of ephemeral regions, estimated from daily NSO full-disk magnetograms, assumes an average lifetime of one day. The vertical lines indicate the square root of the observed number of regions in each bin; the horizontal lines indicate the width of the bin. (Bottom) Distribution of active-region areas during the minimum and the maximum phases of cycle 21. The distributions are closely similar in shape but differ in scale by 5.5 to 6.0.

a factor of 5 to 6 increase from cycle minimum to maximum, substantially smaller than the increase by a factor of 15 or more in the magnetic flux of active regions. More work is needed to understand the relation between active-region area and magnetic flux. The difference in the magnitude of their cyclic variation suggests that there may not be a simple relation between them. There is no evidence that the size spectrum of active regions varies with cycle phase (so as, for example, to favor larger active regions near solar maximum).

IV. MODELING THE EVOLUTION OF
SOLAR MAGNETIC FIELDS

The activity of magnetic fields in the solar atmosphere gives rise to a rich variety of dynamic phenomena occurring on a broad range of spatial and temporal scales. It is widely accepted that dynamo action in the Sun's convective zone is responsible for the cyclic variability in the expulsion of magnetic flux from the solar interior, but a fully consistent, quantitative description of the process has been elusive. In part, this is because the workings of the dynamo are hidden beneath the surface; prior to the advent of helioseismology, the pattern of erupting flux provided the only guide to the dynamo's operation. In contrast, the evolution of magnetic flux at the photosphere and beyond is amenable to detailed observational study. Consequently, models that attempt to describe aspects of this evolution are constrained by and can be quantitatively tested against a considerable body of data.

This section reviews models for the evolution of the solar magnetic field at the photosphere and in the corona and interplanetary medium. The discussion is restricted to long–term, large–scale phenomena relevant to the Sun's cyclic variability. Some implications of those models for two outstanding problems of solar physics, the rotation of magnetic fields and the existence of a meridional circulation, are then considered.

A. Models for Photospheric and Coronal Magnetic Fields

The first accurate measurements of relatively weak photospheric magnetic fields were reported by Babcock and Babcock (1955). Those observations established that magnetic flux erupts through the solar surface in the form of compact, bipolar regions (the largest of which almost invariably contain sunspots) and that these bipolar regions spread out and become increasingly sheared as they age. The shearing was readily understood as resulting from the Sun's nonuniform rotation, but the cause of the spreading remained obscure for several years. Soon after the discovery of the supergranulation, Parker (1963) demonstrated that such convection cells should sweep any embedded magnetic flux to their boundaries and concentrate it there (cf. Simon et al. 1988). Leighton (1964) then noted that the constant shifting of the supergranulation pattern would lead to a random walk of individual elements of flux and thus to an effective diffusion of the large-scale magnetic field. He was thereby able to account in a natural way for the observed dispersal of the photospheric flux.

During the last decade, measurements of the motions of photospheric plasma and magnetic tracers have yielded evidence for a meridional component of circulation in the solar convective zone (chapter by Howard et al.). The reported flows would add a slow poleward drift to the more rapid rotational shearing and convective dispersal of the magnetic field.

The combined effects on the photospheric magnetic field of differential rotation, supergranular diffusion and meridional flow, together with the expulsion of new flux from the convective zone, have been calculated from the transport equation (Leighton 1964; DeVore et al. 1984):

$$\frac{\partial B_p}{\partial t} + \frac{1}{R_\odot \sin\theta} \frac{\partial}{\partial \theta}(v B_p \sin\theta) + \omega \frac{\partial B_p}{\partial \phi} =$$

$$= \frac{\kappa}{R_\odot^2} \left[\frac{1}{\sin\theta} \frac{\partial}{\partial \theta}\left(\sin\theta \frac{\partial B_p}{\partial \theta}\right) + \frac{1}{\sin^2\theta} \frac{\partial^2 B_p}{\partial \phi^2} \right] + S. \quad (2)$$

Here, $B_p(\theta,\phi,t)$ is the radial component of the photospheric field, θ and ϕ are the solar colatitude and longitude, respectively, $R_\odot$ is the radius of the Sun, $v(\theta)$ is the meridional flow velocity, $\omega(\theta)$ is the angular rotation rate, κ is the diffusion coefficient, and $S(\theta,\phi,t)$ is a source term representing the eruption of new flux. This equation seeks to describe the evolution of the field on scales larger than the lifetime and size of the supergranules ($\sim$ 1 day and 3×10^4 km, respectively). Wilson et al. (1990) have criticized the transport Eq. (2) because of the physical limitations of the diffusive transport ansatz.

Given an initial distribution of flux, values for the transport parameters, and a record of emerging flux, the transport equation can be solved numerically to yield the instantaneous flux distribution at any later time. For example, Fig. 13 compares synoptic maps of the observed and simulated magnetic fields for Carrington rotation 1728 (November 1982). The calculation (Wang et al. 1989) was started in August 1976, near the beginning of sunspot cycle 21, and run for the ensuing 6 years with the only input being the deposition of newly erupting bipolar flux. A diffusion coefficient of 600 km² s⁻¹ and a poleward meridional flow with a peak speed of 10 m s⁻¹ were assumed. The qualitative features of the observed and simulated large-scale fields, particularly the elongated patterns curving towards the poles from mid and low latitudes, are clearly very similar. The quantitative agreement between the two maps, in the detailed distribution of the flux among positive- and negative-polarity regions, also is quite good.

The evolution of the mean and gross solar magnetic fields during sunspot cycle 21 is shown in Fig. 14. The mean field is the average line-of-sight field of the Sun viewed as an unresolved star, and is a measure of the net or unbalanced flux on the visible disk. It is highly correlated with both coronal holes and the interplanetary field (Sheeley et al. 1985). The gross field is the absolute radial field averaged over the visible disk, and is a measure of the total flux on the Sun. It is highly correlated with sunspot number and the intensity of chromospheric Ca II emission (Sheeley et al. 1986a). These fields are displayed in Bartels' (1934) format, in which daily values are coded and arranged in 27-day rows. The simulations reproduce fairly well the rather complex evolution of the mean field observed during cycle 21. The agreement between the simulated and observed gross fields is not quite as impressive,

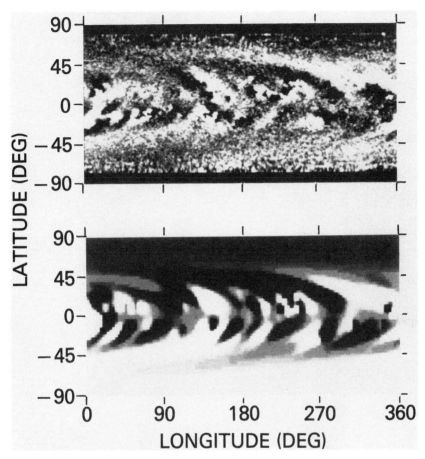

Fig. 13. Synoptic maps of the observed (upper panel) and simulated photospheric magnetic field of the Sun for Carrington rotation 1728, during the declining phase of sunspot cycle 21. The observed field has been spatially smoothed to bring out the large-scale structure, and strips have been blacked out at high latitudes to indicate the absence of reliable flux measurements near the poles. Both the observed and simulated fields are displayed nonlinearly, to emphasize the weak-field patterns. The observations were made at the National Solar Observatory at Kitt Peak; the simulations were performed at the Naval Research Laboratory (figure from Wang et al. 1989).

even though the evolution of the gross field appears to be considerably simpler. This may just reflect the sensitivity of the gross field to the spatial resolution at which it is computed.

The transport model describes the evolution of the radial component of the large-scale magnetic field at the photosphere. Of course, the expulsion of magnetic flux from the solar interior which forms a compact bipolar region at the surface also creates an associated loop structure in the corona. As the

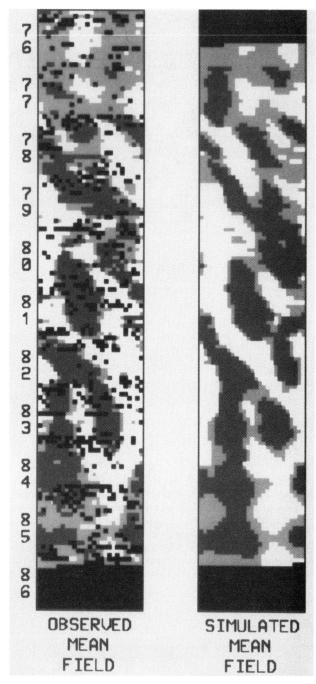

Fig. 14. Bartels displays of the observed and simulated daily values of the Sun's mean and gross magnetic fields during sunspot cycle 21. For the mean field, light and dark shading indicate positive- and negative-polarity fields with strengths greater than 0.25 G, respectively, while neutral shading refers to fields of either polarity with strengths below 0.25 G. For the gross

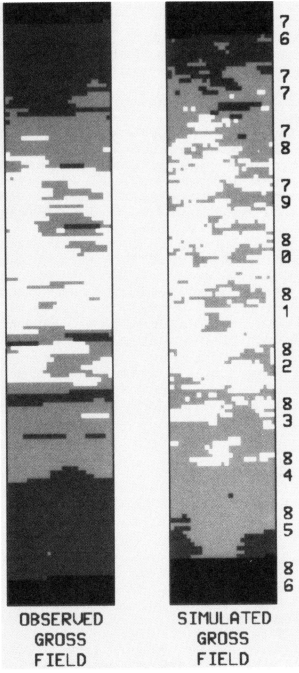

OBSERVED
GROSS
FIELD

SIMULATED
GROSS
FIELD

field, light, neutral, or dark shading indicates field strengths greater than 5.0 G, between 2.5 and 5.0 G, or below 2.5 G, respectively. Very dark entries represent missing or unreduced data. The observations were made at the Wilcox Solar Observatory; the simulations were performed at the Naval Research Laboratory.

supergranular convection disperses the flux at the photosphere, the coronal field responds by expanding both horizontally and vertically. The expansion eventually can raise the field to a sufficient height that it is carried out into the interplanetary medium by the solar wind. In this way, the evolution from compact bipolar regions to extended unipolar regions in the photosphere drives the transition from closed loops to open magnetic configurations in the corona. A key factor in this coordinated evolution is that the dynamics of the inner corona is field-dominated, i.e., the magnetic pressure greatly exceeds the plasma pressure, in general. Consequently, the magnetic field there is approximately force-free, and the plasma must move to accommodate it. In contrast, the dynamics is plasma-dominated in the photosphere and outer corona, so that the convection sweeps the magnetic field to the supergranule boundaries and draws it outward into interplanetary space.

The simplest model used to calculate the coronal and interplanetary magnetic fields from the photospheric flux distribution is the source-surface model developed by Schatten et al. (1969) and Altschuler and Newkirk (1969). In this model, the field is assumed to be current-free from the photosphere out to a spherical "source" surface of radius R_s, where it is drawn out radially by the solar wind. The field thus can be calculated from a scalar potential $\psi(r,\theta,\phi,t)$,

$$\mathbf{B} = -\nabla\psi \tag{3}$$

where ψ satisfies Laplace's equation, $\nabla^2\psi = 0$, subject to the boundary conditions

$$\frac{\partial\psi}{\partial r}\bigg|_{(R_\odot,\theta,\phi,t)} = -B_p(\theta,\phi,t)$$
$$\frac{\partial\psi}{\partial\theta}\bigg|_{(R_S,\theta,\phi,t)} = \frac{\partial\psi}{\partial\phi}\bigg|_{(R_S,\theta,\phi,t)} = 0. \tag{4}$$

The first condition matches the radial component of the extrapolated coronal field to the photospheric field, and the second conditions ensure that the extrapolated field becomes radial at the source surface. Beyond the source surface, the field strength decreases as r^{-2}. The radial component of the interplanetary field at the Earth $B_e(t)$ thus can be calculated from (see, e.g., Hoeksema 1984)

$$B_e(t) = \left(\frac{R_s}{R_e}\right)^2 B_s[\theta_e(t-t'),\phi_e(t-t'),t-t'] \tag{5}$$

where t' is the Sun-Earth transit time of the solar wind and its entrained magnetic field, and (R_e,θ_e,ϕ_e) is the heliographic position of the Earth.

The simulated and observed interplanetary magnetic fields during sunspot cycle 21 are displayed in Bartels format in Fig. 15. The calculations were made using Altschuler and Newkirk's (1969) value of 2.5 $R_\odot$ for the source-surface radius R_s, and Wilcox Solar Observatory synoptic maps (Hoeksema and Scherrer 1986), corrected for the line-of-sight projection of radial fields, for the photospheric magnetic flux distribution $B_p(\theta,\phi,t)$. The agreement between the simulated and observed fields again is imperfect but quite good overall.

In the two subsections which follow, the application of the transport and source-surface models for the solar magnetic field to problems of solar rotation and meridional convection are discussed.

B. Rotation and Magnetic Fields

The transport model provides a basis for understanding several heretofore puzzling aspects of solar rotation. It has been known for many years that long-lived patterns of the photospheric magnetic field tend to rotate more rigidly than would be expected from the rotation exhibited by small-scale tracers such as sunspots (Bumba and Howard 1965,1969; Howard 1967; Wilcox et al. 1970; Wilcox and Howard 1970; Stenflo 1974,1977). Using a model that included supergranular diffusion, differential rotation and observed sources of flux during 1959 to 1960, Schatten et al. (1972) obtained magnetic field patterns that rotated at the same rate as the observed field and more rigidly than the rate used as input to their simulations. However, they apparently regarded this result to be an artifact of their "rotation-by-rotation method of calculations," and 15 yr elapsed before it was recognized that these rigidly rotating patterns were caused by the transport of flux across latitudes (Sheeley et al. 1987). A north-south component of motion combines with the east-west component due to differential rotation to give slanted patterns that rotate rigidly, while the individual elements of flux still undergo the differential motions of the latitudes they are crossing.

In the simulations of Schatten et al., the transport across latitudes was caused by the meridional component of supergranular diffusion used in their calculations. However, Sheeley et al. found better agreement with the observed field during sunspot cycle 21 when a 10 m s^{-1} poleward meridional flow was included with diffusion at a nominal 600 km^2 s^{-1} rate. Flow reinforced the poleward diffusion above the sunspot belts to give a more rigid rotation profile, and it opposed the equatorward diffusion between the sunspot belts to prevent the accumulation of large regions of flux near the equator during the rising phase of the cycle. A comparison between the observed and simulated rotation profiles over 2 intervals during cycle 21 is shown in Fig. 16.

Within the sunspot belts themselves, the field is dominated by the flux in newly erupting active regions, which exhibit the differential rotation rate characteristic of small-scale tracers. However, in the absence of ongoing

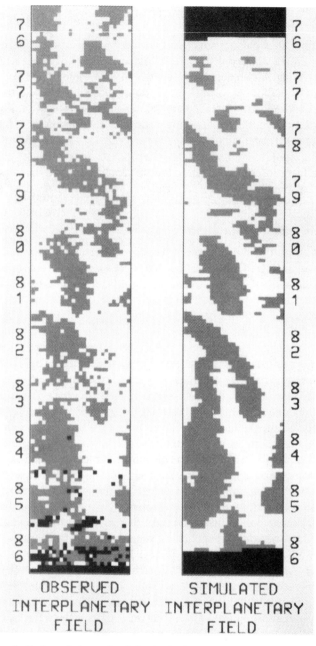

Fig. 15. Bartels displays of observed and simulated daily values of the radial component of the interplanetary magnetic field during sunspot cycle 21. Light and dark shading refer to positive- and negative-polarity fields with strengths greater than 2.5 μG, respectively, while neutral shading refers to fields of either polarity with strengths less than 2.5 μG. Very dark entries represent missing or unreduced data. The interplanetary-field data were collected by various spacecraft in near-Earth orbit; the simulations were performed at the Naval Research Laboratory.

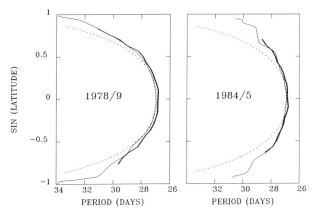

Fig. 16. Average rotation profiles of the photospheric magnetic field, obtained by cross-correlating successive synoptic maps, for two intervals during sunspot cycle 21. Thick curves show the observed profiles; thin curves show the simulated profiles. For comparison, the dashed curves show the rotation rate of individual flux elements, used as input in the calculations. The observations were made at the Wilcox Solar Observatory; the simulations were performed at the Naval Research Laboratory (figure from Wang et al. 1989).

sources, the field must evolve toward a state of rigid rotation, as DeVore (1987) deduced directly from the flux-transport equation. In his analysis, similar to that performed by Leighton (1964) for the axisymmetric field, DeVore found that the eigenfunctions of the nonaxisymmetric field are rigidly rotating, uniformly decaying patterns of flux similar to the stripes of a barberpole.

A second puzzle has been the tendency for the outer corona to rotate much more rigidly than the photosphere (Hansen et al. 1969; Parker et al. 1982; Fisher and Sime 1984; Hoeksema 1984). Wang et al. (1988) showed that this behavior could be understood in terms of the source-surface model of the coronal field. In this case, the harmonic amplitudes of the coronal field fall off with radial distance as $r^{-(l+2)}$, so that the field at 2.5 $R_\odot$ depends on only a few, relatively low-order components. Without high-order components, the outer-coronal field cannot follow the photospheric field as it winds up into increasingly narrow stripes, and must therefore rotate rigidly. Its rotation rate corresponds to the latitude of the large-scale, nonaxisymmetric photospheric flux that it can "see."

Using the transport model, Wang et al. showed that as the sunspot number declines, differential rotation and supergranular diffusion combine to wipe out most of the nonaxisymmetric flux at midlatitudes, so that the corona rotates rigidly at the 27-day period of the surviving flux near the equator. On the other hand, near the time of sunspot maximum, new sunspot groups provide an ongoing supply of flux that is spread over a wide range of latitudes by supergranular diffusion and carried poleward by meridional flow. Consequently, at this time in the cycle, the coronal field rotates at the 28- to 29-day periods characteristic of the broad distribution of photospheric flux at midlatitudes.

This is essentially the same explanation that Sheeley and DeVore (1986) had found earlier in their search for the origin of the puzzling 28- to 29-day recurrent patterns of the Sun's mean line-of-sight field (Svalgaard and Wilcox 1975). Like the field in the outer corona, the mean field is a weighted average of the photospheric field, and therefore it also sees only the large-scale components of the photospheric field. Consequently, near the time of sunspot minimum, its polarity patterns recur with the 27-day period of the residual nonaxisymmetric flux near the equator, and near the time of sunspot maximum, they recur with periods in the range 28 to 29 days, corresponding to the flux that is continually being spread at midlatitudes.

During cycle 21, the two dominant periods seen in the interplanetary magnetic field pattern originated in different solar hemispheres. The large-scale photospheric field pattern in the northern hemisphere rotated every 26.9 days while the southern field pattern rotated every 28 days (Antonucci et al. 1990). The asymmetry persisted throughout the cycle, although the signals were weaker near solar minimum, and the frequency was less constant in the south. The result was present in both National Solar Observatory and Wilcox Solar Observatory data. The same asymmetry may have been present during cycle 20, but that result is not certain. The same periods have been observed in the interplanetary magnetic field and in geomagnetic indices for the last 6 solar cycles. It will be interesting to see to what extent the transport model can account for persistent north-south rotational asymmetries, which were first noticed in potential-field extrapolations of photospheric magnetic fields (Hoeksema and Scherrer 1987).

The transport model predicts rotational properties for the gross radial field (Sheeley et al. 1986a) that differ markedly from the properties of the mean field. Constructed to provide a proxy for the Sun's chromospheric intensity, the gross field is a weighted average of the photospheric field's absolute value and is therefore dominated by flux in newly erupted bipolar regions. Its decay is principally via supergranular diffusion, which is responsible for the rapid *in situ* annihilation of flux in bipolar regions. In contrast, the mean field originates in large-scale fields, which are relatively insensitive to diffusion and decay principally by being wound up into smaller-scale stripes by differential rotation. Consequently, long-lived recurrent patterns of mean field reflect persistent distributions of large-scale flux on the Sun, whereas long-lived patterns of gross field must indicate the ongoing eruption of new flux in active domains.

Finally, the transport model provides an explanation for the rotational properties of coronal holes, including the behavior of Skylab Coronal Hole 1, which maintained its meridional structure for ~ 7 months before becoming appreciably sheared by differential rotation (Timothy et al. 1975; Bohlin

1977*b*). Regarding coronal holes to be the footpoints of open field lines in a largely current-free corona, Nash et al. (1988) noted that the current-free condition would ultimately cause the footpoints to co-rotate with their outer-coronal extensions. Otherwise, the field lines would become "curled" and volume currents would develop. Having already shown that the outer-coronal field rotates rigidly (Wang et al. 1988), they concluded that the co-rotating footpoints must continually change their connections to the differentially rotating photospheric flux distribution. Thus, a coronal hole is a kind of "shadow" that drifts uniformly across its photospheric polarity pattern as a wave of reconnecting field lines.

Pursuing these ideas further, Nash et al. found that a uniformly drifting hole must eventually encounter the neutral line at the edge of its photospheric polarity pattern. When this happens, the boundary of the hole conforms to the slanted shape of the evolving neutral line and is forced slowly toward the equator. In this way, the edges of the hole rotate differentially with the photospheric field while the center of the hole remains longitudinally co-aligned with the rigidly rotating field in the outer corona. Supergranular diffusion and meridional flow eventually offset the equatorward motion so that even the photospheric field rotates rigidly, as discussed earlier. However, the asymptotic configuration consists of purely axisymmetric polar coronal holes together with dwindling remnants of nonaxisymmetric holes near the equator, where the shearing rate is small.

The field's axisymmetric component provides the key to the puzzle of why Coronal Hole 1 was able to maintain its meridional shape for such a long time. A strong axisymmetric component delays the inexorable encounter with the neutral line by forcing it to low latitudes, where the rate of shearing is small. This was the situation during the Skylab mission in 1973, when the polar field was strengthening and new eruptions of flux were confined increasingly close to the equator. Although the axisymmetric component was responsible for maintaining the hole's meridional shape, the nonaxisymmetric component still determined the hole's rotation period, which in this case was ~ 27.5 days. A simulation of the evolution of an initial configuration similar to that of Coronal Hole 1 is shown in Fig. 17.

C. Meridional Convection and Magnetic Fields

In his original study of flux transport on the Sun, Leighton (1964) showed that in the absence of new bipolar sources, supergranular diffusion would cause the large-scale magnetic field to relax to an axisymmetric dipole configuration. This result was at least qualitatively consistent with the observed distribution of flux around the time of minimum activity, and was one of the early successes of his model. An increasing body of evidence collected over the past decade suggests, however, that a pattern of meridional convection, directed poleward at the photosphere, may be present on the Sun and affect the long-term evolution of the magnetic field (chapter by Howard et

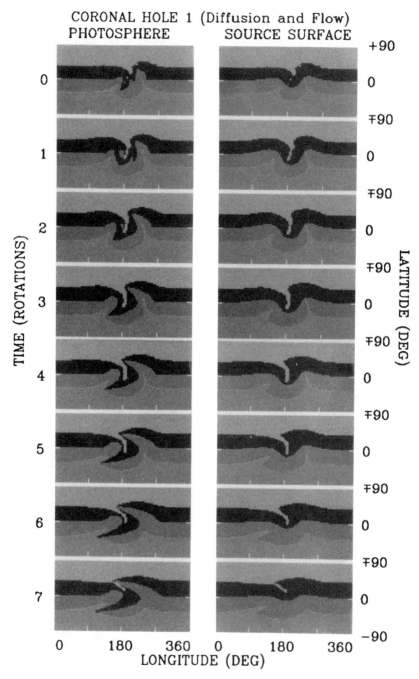

Fig. 17. Evolution of an initial configuration of 5 bipolar regions and a 1-G axisymmetric dipole field, as a model for the behavior of Skylab's Coronal Hole 1. A diffusion coefficient of 600 km^2 s^{-1} and a poleward meridional flow of 10 m s^{-1} were assumed. For $\sim$ 6 Carrington rotations, the northern-hemisphere hole (light shading) maintains a nearly undistorted meridional shape similar to that of Coronal Hole 1. The simulations were performed at the Naval Research Laboratory (figure from Nash et al. 1988).

al.). It has already been noted that the rotation profiles of the simulated and observed photospheric fields agree better when a poleward flow of the order of 10 m s^{-1} is included. Other observed features of the structure and evolution of the magnetic field can be explained by the transport and source-surface models only if a meridional flow is assumed to be present.

The distortion of the axisymmetric field near the time of sunspot minimum, away from the simple dipole calculated by Leighton, was considered by DeVore et al. (1984). They showed that a 10 m s^{-1} meridional flow could be expected to concentrate the magnetic flux near the poles until the field gradient is sufficiently steep that the equatorward diffusion and poleward convection of flux balance each other. The details of the slowly decaying distribution of flux produced by this balance depend upon the relative rates of flow and diffusion and the profile of the flow at high latitudes. The basic result— that the flux is quickly swept into a rather well-defined cap at each pole, with little flux left at lower latitudes—follows for virtually any smooth flow profile with a peak speed substantially greater than the effective global diffusion speed, $\kappa/R_\odot \sim 1$ m s^{-1}.

Several lines of evidence suggest that the solar magnetic field around the time of sunspot minimum is distributed in just such a "topknot" configuration. Svalgaard et al. (1978) assumed a model profile for the flux in the polar caps during 1976 to 1977, and calibrated it to reproduce the annual modulation of the strength of the polar fields as observed at Earth. They deduced that the flux was highly concentrated toward the poles, following a $\cos^{8\pm1}\theta$ dependence and having a peak strength of $\sim$ 12 G. Wang and Sheeley (1988) obtained essentially the same result in their analysis of Wilcox Solar Observatory data from 1984 to 1985. During the earlier interval, Duvall et al. (1979) measured the variation of the axisymmetric magnetic field with latitude. They too found a strong skewing of the flux distribution toward high latitudes, with low latitudes being relatively field free. More recently, the interaction between newly erupting, high-latitude bipolar regions and the polar coronal holes was studied by Sheeley et al. (1989). They concluded that the polar flux must be concentrated into narrow caps at the poles in order for the models to reproduce (1) the observed high-latitude (60°) boundary of the undisturbed polar holes, and (2) the observed extent and latitudinal separation of network enhancements of absorption in the He I λ10830 line, which occur at the equatorward boundary of the hole and at the trailing, poleward boundary of the newly erupted bipolar flux. Their results are summarized in Fig. 18.

All of these findings suggest that the Sun's axisymmetric flux is highly concentrated toward the poles around the time of sunspot minimum. This in turn requires the presence of a poleward meridional flow with a peak speed of the order of 10 m s^{-1}, in good general agreement with observational studies.

The polarity reversal of the Sun's polar magnetic fields near the time of

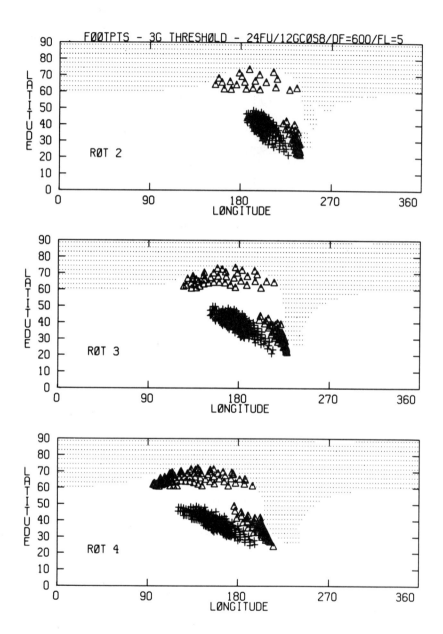

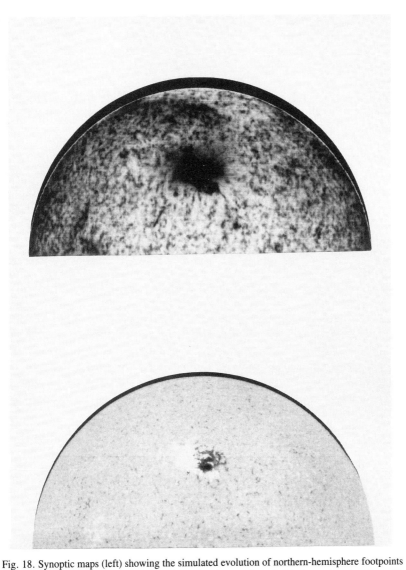

Fig. 18. Synoptic maps (left) showing the simulated evolution of northern-hemisphere footpoints of magnetic field lines for several rotations after the eruption of a bipolar region, together with the He I λ10830 image and magnetogram (upper and lower right, respectively) at the initial rotation. Dotted areas are open-field regions; triangles are footpoints of strong, negative-polarity fields connected to the bipole's positive pole; plus signs are footpoints of strong, positive-polarity fields connected to either the bipole's negative pole or the north polar field. Areas of open field correspond to coronal holes (bright areas in the He image); areas of strong, closed field (triangles and plus signs) correspond to the enhanced helium network (dark areas in the He image). The crucial features are the high-latitude boundary of the polar hole and the distinct latitudinal separation of the two regions of enhanced network. The simulations were performed at the Naval Research Laboratory; the observations were made at the National Solar Observatory at Kitt Peak.

maximum sunspot activity was first observed during cycle 19 (Babcock 1959). In addition to providing an explanation for the structure of the polar fields during the quiet interval around minimum, Leighton (1964) described its subsequent evolution. The majority of the large bipolar regions that erupt during a given cycle possess a meridional component of dipole moment which opposes the initial, global dipole moment of the Sun (Joy's law). As supergranular diffusion spreads the flux in these regions over the solar surface, the global dipole weakens and then reverses its polarity. A key ingredient in this reversal is the diffusive transport of leading polarity flux across the equator, which is required for the net flux in the two hemispheres to change. A poleward meridional flow inhibits but does not prevent this equatorial crossing, unless the flow is too fast. For a nominal rate of diffusion, simulations with the transport model suggest that the flow speed is limited to around 10 m s^{-1} (DeVore and Sheeley 1987).

Evidence for a poleward flow of this order of magnitude was reported by Howard and LaBonte (1981) from their analysis of Mount Wilson Observatory data from sunspot cycle 20. They constructed a map of the magnetic field as a function of latitude and time during the cycle, and discovered episodic transports of following-polarity flux from mid to high latitudes accompanying the polar reversals. Similar but more numerous episodic transports, of both leading and following polarity, were evident during cycle 21 (R. K. Ulrich, personal communication) and gave rise to short-term oscillations in the strength of the polar fields. An outstanding question has been whether these episodes are due solely to the intermittent eruption of bipolar flux into the photosphere or require transient increases in the rate of meridional convection. Preliminary studies with the transport model indicate that bursts in both the flux-eruption rate and in the meridional-flow speed are needed to reproduce the observations. Thus, a quantitative analysis and comparison of the models and the observations not only strongly supports the view that meridional flow is present on the Sun, but progresses further to ask more subtle questions about the latitudinal dependence of the flow pattern and about its variability during the sunspot cycle.

V. EMPIRICAL MODELS OF THE SOLAR ACTIVITY CYCLE

An empirical model aims to organize heterogeneous observations into a coherent picture of a complex physical system. The presumption is that the main physical processes can be identified but that, because the equations are incomplete or too difficult to solve, a complete mathematical description is not yet possible. By considering in some detail the empirical model of the solar activity cycle proposed by Babcock (1961), and the family of models that have elaborated or modified the original proposal, this review accepts that empirical models can be useful: that they can embody sound physical ideas, be specific enough to test, and stimulate the development of a complete theory. As good textbooks demonstrate, a few equations, approximate solutions, and sketches can often capture the essence of well-understood physical phenomena. An empirical model tries to infer and present that essence in advance of full understanding.

An empirical model is a two-edged sword. It is incomplete and therefore not unique or fully prescriptive. The simplicity and visual character that makes it attractive and easy to recall can also prolong it beyond its useful life. It assumes that the observations it incorporates are key parts of the physical systems it seeks to explain, not incidental or superficial effects. This caveat is uncomfortably relevant to the Sun: all models place essential elements of the activity cycle in the interior, but, until recently, observations sampled only the radiative skin.

In 1961, solar magnetographs had been operating for less than half a magnetic cycle; nothing was known about the internal rotation of the Sun; numerical simulations of all but the most idealized dynamos were impractical. Yet, for most astronomers and even for many solar physicists, "the Babcock model" still serves as a paradigm for understanding activity cycles on the Sun and solar-type stars. Does some form of the model deserve this primacy? If not, is there a better empirical picture to replace it? It is first necessary to distinguish the model that Babcock proposed from the plethora of revised, extended, or merely corrupted, versions that have collectively assumed its name.

A. The Babcock Model

Babcock (1961) strove to explain four principal characteristics of the solar magnetic cycle:

1. The reversal of the north-south dipolar field at intervals of ~ 11 yr;
2. Spörer's law, according to which the sunspot belt moves to lower latitudes as the cycle progresses (Maunder's butterfly diagram);
3. Hale's polarity law, according to which the dominant polarity of leading sunspots in each hemisphere matches the polarity of the polar cap in that hemisphere at the preceding solar minimum;
4. The observation that bipolar magnetic regions disappear by expanding.

Figure 19 illustrates the four principal stages of Babcock's model (the fifth stage is the same as the first stage with reversed polarity). Elements of the model had been discussed previously (references in Babcock 1961). Babcock's contribution was the attempt to synthesize these elements into a full picture of the activity cycle, with particular reference to magnetic observations.

In Stage 1, the large-scale magnetic field in the inner heliosphere is idealized as a north-south dipole. Under the surface, field lines pass along meridians in the outer part of the convective zone. The differential rotation in this layer is roughly constant on cylinders, with angular velocity increasing

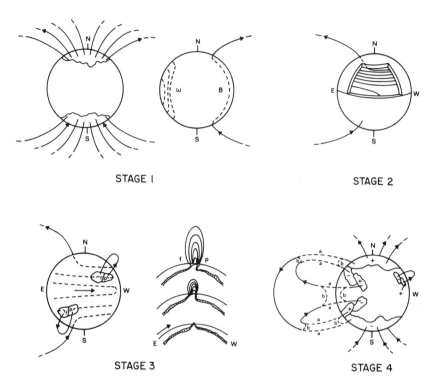

Fig. 19. Cartoons from Babcock (1961) illustrating how the solar magnetic field evolves through one half of a 22-yr activity cycle. Meridional field lines (Stage 1) are distorted and amplified by differential rotation (Stage 2) until they erupt through the photosphere as bipolar active regions (Stage 3). Remnant surface fields disperse, cancel and reconnect (Stage 4), eventually reversing the polar fields, at which point Stage 1 recommences with reversed polarity.

toward the surface and toward the equator. In Stage 2, the magnetic field lines are drawn out into spirals by the differential rotation and the field strength is amplified. The angle between a field line and the meridian is computed from the surface form of the differential rotation. When the winding is steep, the field strength in the spiral is proportional to $t \sin\phi$, where ϕ is latitude and t is the time since winding began.

Stage 3 is marked by the eruption of bipolar magnetic regions (BMR's). Convective irregularities have concentrated the subsurface field into irregular, twisted and relatively discrete flux ropes. When the field in a flux rope is strong enough, kink-like instabilities and magnetic buoyancy bring an Ω-shaped segment of magnetic field to the surface, creating a BMR that obeys Hale's polarity law (the term "flux loop" will be reserved for magnetic field lines that close on themselves). Whatever the critical field strength necessary for eruption, it will be reached first at higher latitudes. Multiple eruptions at a given latitude lead to fragmentation and weakening of the subsurface flux ropes until field amplification and eruption ceases there. Thus, three effects combine to give Spörer's law: a subsurface field strength that increases with latitude and time; a minimum field strength for the emergence of BMR's; and the ability of repeated flux eruptions to stop field amplification.

Stage 4 is the key to the cycle. The following-polarity (f) part of a BMR tends to expand or drift toward the nearer pole, while the preceding-polarity (p) part tends to shift toward the equator. This observed characteristic of the late evolution of active regions agrees in sense with the observation that the p spot of a simple BMR tends to lie closer to the equator than the f spot. Most of the flux in BMR's does not contribute to the reversal of the large-scale dipole. Rather, the expanding p area of one BMR impinges on the f area of the next BMR to the west and reconnects with it, producing a flux loop that eventually escapes outward into the corona. Only ~ 1% of the flux need evade this local neutralization to account for the reversal of the polar fields. This is accomplished by reconnecting high-latitude f flux with the polar cap while low-latitude p flux in one hemisphere reconnects with opposite-polarity p flux across the equator. The trans-equatorial field lines (marked b in Fig. 19) are required to submerge below the photosphere, where they oppose and, after ~ 11 yr, supplant the initial dipole (presumably the polar loops marked b also submerge).

B. Criticisms of the Babcock Model

Because the model is empirical, criticism should focus on internal consistency, completeness and the gross magnitudes of the physical effects.

Stage 1 envisions that, at least in the sunspot latitudes, the magnetic field lines are confined to the outer part of the convection zone and the angular velocity of rotation increases outward. These assumptions were not well justified and are probably false, as discussed below. Nevertheless, the desired physical effect (Stage 2) is simply toroidal winding such that the equatorward parts of field lines lead the poleward parts and the field strength reaches a given strength at progressively lower latitudes as time advances. The effect is crucial, not the details of how it is achieved.

The emergence of BMR's in Stage 3 involves several assumptions that are loosely justified. It is empirically required that sunspots first appear some 3 yr into the cycle at about 30° latitude. The subsurface field is therefore assumed to reach whatever strength is required for instabilities and buoyancy to bring it to the surface at that latitude and time. Fluid shear and convection are heuristically invoked to concentrate a uniform subsurface sheath of strength 250 to 300 G into flux ropes of 1000 to 2000 G so that buoyancy may be effective. However, because the assumed form of the differential rotation produces a subsurface field that strengthens monotonically with latitude, the question of why sunspots first emerge at relatively low latitudes is unanswered. One might assume that the initial subsurface field weakens as it merges with the polar caps (Babcock's assumption) or that the high-latitude field is not as effectively concentrated into ropes. Behind these arguments is the tacit assumption that the main problem is to strengthen the subsurface field until it is buoyant enough to emerge. As Parker (1984b) has stressed, the problem may instead be how to *suppress* buoyancy until the field amplifies enough to supply the observed flux in active regions and complexes.

Babcock assumes that the emergence of BMR's at a given latitude will gradually suppress amplification of the subsurface field "owing to fragmentation of the flux ropes." Even if a spate of emergence can disorganize the subsurface field for a time, it is not obvious that, after a year or two without emergence, under the continuing influence of differential rotation and convection, the field will not again be strong enough to produce spots. This is one aspect of a larger question poorly addressed by the model: how does the toroidal field disappear to make way for the much smaller poloidal field of the next cycle?

Stage 4, the most critical step for producing an alternating cycle, is unsatisfactory in several respects. First, the evolving topology of the magnetic field below the surface is not clearly specified. Although there are limitations to the field-line picture (reconnection is essential to the operation of a dynamo), the frozen-in approximation is excellent almost everywhere in the solar interior, thus, a consistent description of the magnetic topology should be part of a complete empirical model. A second questionable area is the disposition of magnetic flux loops that thread the photosphere. Loops are taken to submerge or escape as needed. No physical evidence for submergence is adduced. A flux loop cannot escape until it frees itself from the mass weighing it down, but, even in the photospheric layers of highest resistivity, the naive diffusion time scale is too long. Parker (1984b) concluded that the solar surface should be considered impervious to magnetic flux on the time scale of the activity cycle. Finally, the model does not account for one of the characteristics it sought to explain, the disappearance of active regions by expansion. Rather, as with differential rotation, this observation is incorporated as a given part of the model.

C. Leighton's Semi-Empirical Model

The important ideas introduced by Leighton (1964,1969) have occasioned references to a "Babcock-Leighton model," a label that obscures significant differences between the models.

Leighton (1964) first addressed the question of why active regions disappear by expansion, proposing that supergranular convection would in effect disperse magnetic flux according to a diffusion equation. The recent results discussed in this chapter show that, with the addition of meridional flow (also mentioned by Leighton), the transport model can account for a surprisingly wide range of characteristics of the photospheric and heliospheric magnetic field, including properties such as the quasi-rigid rotation of large-scale magnetic features that have sometimes been taken to require subsurface flows or patterns outside the scope of the transport model. The observed tendency of bipolar p flux to lie closer to the equator than associated f flux is still taken as given. Also, more recent observations indicate that a substantial fraction of the flux in active regions and complexes disappears *in situ* (Wallenhorst and Howard 1982; Gaizauskas et al. 1983), suggesting active retraction by subsurface forces (Rabin et al. 1984; Parker 1984*a*). The transport model nevertheless gives a good account of the evolution of the large-scale magnetic flux.

Leighton also advanced a pictorial explanation of how the strong toroidal field that underlies active regions can disappear to make way for the reversed field of the next cycle. Bipolar magnetic regions emerge with a meridional component opposite in sign to the large-scale dipolar component of the field at the preceding solar minimum. If this reversed component is assumed to affect the subsurface field as well, the meridional component of the subsurface field will eventually reverse, giving it a backward tilt. At that point, differential rotation will start to *unwind* the toroidal field, decreasing rather than increasing its strength. How does the characteristic surface tilt of BMR's impress itself on the subsurface field? This question was addressed by Adams (1977), who devised an idealized model of the field reversal that made explicit the connection between changes in the magnetic topology above and below the surface. The proposed process involved surface reconnection of p flux with the f flux of a different BMR at a lower latitude, followed by submergence of the reconnected arch; but the important point was that magnetic field topology could be treated consistently in a Babcock-like model.

Leighton incorporated diffusive surface transport of magnetic fields into a "magneto-kinematic" model of the activity cycle (Leighton 1969). This model was successful in its stated purpose: to occupy a middle ground between Babcock's qualitative treatment and formal dynamo models. Although the individual equations embody heuristic and parameterized elements, such as the critical field strength for magnetic eruption, the assumed average tilt of BMR's, and the diffusion ansatz, the system of equations is closed and can be numerically solved. From the solutions, emerge several general conclusions. First, a radial gradient of angular velocity maintains dynamo action more effectively than a latitudinal gradient of the same magnitude. Second, the butterfly diagram is not the purely kinematic effect it is in the Babcock model, where the latitude profile of toroidal field strength changes in normalization but not in shape as the cycle progresses. In the Leighton model, the zone of erupting flux represents a true dynamo wave, which propagates toward the equator (as observed) only if the interior rotates faster than the surface. Third, once initial conditions have been forgotten, the field is never purely poloidal as Stage 1 of the Babcock model would have it; the time at which the local field becomes meridional varies with latitude (the model averages over longitude and thus carries no information about longitudinal variations). Fourth, some of the observed randomness of the cycle, such as north-south asymmetries and fluctuations in period, can be simulated simply by allowing random variations in the rate at which a field erupts.

Babcock stressed the role of flux ropes in his model, but, apart from their role in flux emergence, they are not crucial to the scheme for reversing the dipolar field. Leighton noted that his model nowhere incorporates the concept of flux ropes and only briefly discussed what their influence might be.

Despite their considerable differences, the models of Babcock and Leighton share a central concept: they are $\alpha\omega$ dynamos in which the magnetic flux observed at the surface plays an active role. The surface flux is not just the detritus of a deeply buried machine, irrelevant to its operation or important only as a loss term. The fate of the surface flux directly affects the evolution of the subsurface field. Below, this concept will be termed "the Babcock picture." It should be understood to represent the core ideas of the Babcock and Leighton models and their potential for development, not the details of a particular model.

D. Recent Observations and Theory Bearing on Models of the Activity Cycle

A catalogue of relevant data would include at least the following categories: new information about large-scale velocity fields (including rotation) at the surface and in the interior; new forms of cyclic activity; new information about the cyclic properties of known forms of activity; and theoretical ideas concerning subsurface magnetic fields, magnetic buoyancy and the fate of emerged flux (excluding dynamo theory as such).

In considering the impact of new data on the Babcock picture, one should ask whether the data are inconsistent with the picture or simply outside its existing scope. In the former case, the aim should be to define the contradiction as precisely and narrowly as possible; in the latter case, the question is whether the scope of the picture can be suitably enlarged without sacrificing its basic premises. This process will be meaningless unless it is possible to falsify the Babcock picture in the general form in which we have deliberately presented it. Thus, we list some types of data that are taken to be inconsistent with the Babcock picture.

a. *Precisely periodic cyclicity,* because there is no underlying clock in the Babcock dynamo. Dicke (1978) applied statistical tests to the epochs of sunspot maximum and to the deuterium/hydrogen ratio in bristlecone pine samples (which he accepted as being modulated by the solar cycle) and concluded that the large phase excursions of the solar cycle are more like independent fluctuations around a precise period than a random walk in phase. However, as is so often the case, the historical record alone is too short and heterogeneous to permit a decisive conclusion. Independent data, perhaps helioseismic, could revive this hypothesis.

b. *Discrete phenomena,* such as complexes of activity, that maintain their identity across several cycles, because there is no obvious way for the regenerative dynamo to carry a precise "memory" of conditions in previous cycles, although gross relationships of amplitude and phase are expected and found in dynamo models (Yoshimura 1978*b*,1979).

c. *Activity that starts "too soon."* In the Babcock picture, erupted toroidal flux creates the reversed poloidal field, which is in turn wound up by differential rotation to create the toroidal field and active regions of the next cycle. New-cycle activity is not expected until a substantial amount of flux from the current cycle has erupted and dispersed. As will become apparent, this requirement must be formulated and applied with some care.

d. *Irrelevance of surface magnetic flux to the workings of the cycle.* The causal involvement of the magnetic flux we observe sets the Babcock picture apart from the most general $\alpha\omega$ dynamo. This point is more cautionary than practical at present, but helioseismic measurements may eventually characterize internal magnetic fields precisely enough to decide the numerical importance of surface flux.

1. Internal Rotation. The chapter by Libbrecht and Morrow discusses what helioseismology tells us, and with what confidence, about the internal rotation of the Sun. A schematic description consistent with existing data can be given as follows. The differential rotation observed at the surface persists through the convection zone, $R \gtrsim 0.7\ R_{\odot}$ (that is, the isotachs are directed approximately radially). The character of the rotation below the convection zone is only roughly known. The radiative core, $R \lesssim 0.5\ R_{\odot}$, appears to rotate approximately as a solid body with an angular velocity equal to the surface angular velocity at about 30° latitude. In a transitional zone, $0.5 \lesssim R \lesssim 0.7\ R_{\odot}$, the angular velocity increases outward at latitudes below 30° and decreases outward at latitudes above 30°.

This picture of internal rotation is clearly at odds with Babcock's assumption that the rotation is approximately constant on cylinders. His assumption that the magnetic field participating in the dynamo is confined to the upper part of the convection zone cannot be tested as yet by helioseismology, but the strong magnetic buoyancy that such fields would experience suggests to the contrary that the seat of the dynamo lies near the base of the convection zone (Parker 1984*b*). The same conclusion is suggested by the absence of a radial gradient in angular velocity in the convection zone, as a radial gradient is more effective than a latitudinal gradient in producing dynamo action. Gilman et al. (1989) have sketched a qualitative picture of how a dynamo might operate in a thin layer of overshooting convection between the convection zone and the radiative interior. They argue that fluid motions have positive helicity ($\alpha < 0$) near the top of the convection zone and negative helicity ($\alpha > 0$) near the base. Positive helicity would combine with a positive radial gradient in angular velocity below 30° latitude to drive a dynamo wave toward the equator. Above 30° latitude, $\alpha < 0$ and $d\omega/dR < 0$

would produce a poleward wave. Note that the equatorward migration of fields at low latitudes results from reversing the sign of both α and $d\omega/dR$ from the models of Leighton (1969) and others.

Two observational issues are pertinent to this scenario. First, there is evidence (discussed in the next two sections) that equatorward migration of activity begins at least as high as 45° latitude. Second, apart from doubtful indications in the torsional shear signal, activity that migrates toward a pole is most straightforwardly explained as a surface effect associated with the mixing of magnetic polarities along the retreating boundary of the polar cap.

Stenflo (1990) has determined the phase velocity of the photospheric magnetic field pattern from an autocorrelation and power-spectrum analysis of the Mount Wilson and Kitt Peak synoptic maps discussed in Sec. III. Patterns that persist for a few days show differential rotation with latitude in agreement with other analyses (Snodgrass 1983), but patterns that recur after one or more rotations exhibit quasi-rigid rotation that is faster at all latitudes than the Snodgrass rate. The rotation law for pattern recurrence appears to vary by at most a few percent over the solar cycle. Stenflo argues that the quasi-rigid, time-invariant rotation of recurring patterns is not an instrumental artifact and is not adequately explained by the flux transport model. He takes the quasi-rigid rotation law to represent the rotation of magnetic flux stored near the bottom of the convection zone, which replenishes the observed surface field on time scales shorter than 27 days (so that fresh emergences mark the internal angular velocity from one rotation to the next) but longer than a few days (so that short-lived patterns follow the angular velocity of the surface plasma). Stenflo's interpretation agrees qualitatively but disagrees quantitatively with the current heleioseismological picture of internal rotation: the angular velocity at the base of the convection zone inferred from helioseismology matches the surface rate for midlatitudes, whereas Stenflo's inferred rate is slightly faster than the surface equatorial rate. The occurrence of north-south asymmetries in the magnetic rotation rate (Sec. IV.B) also raises the question of whether (in Stenflo's picture) the field sources must rotate at different rates in the two hemispheres. In view of the success of the transport model in producing other types of quasi-rigid rotation, and the lack of a predictive theory to connect observations of surface flux with flux concentrations at the bottom of the convection zone, the interpretation of photospheric patterns that recur over long time scales remains an open question.

2. Torsional Oscillations and other Surface Flows.   The torsional oscillations are fast or slow streams (zones of latitude) superposed on the time-averaged latitude profile of differential rotation (Howard and LaBonte 1980; chapter by Howard et al.). There are typically 2 fast and 2 slow streams in each hemisphere. The lower-latitude streams migrate toward the equator at a variable rate, 0.5 to 3 m s^{-1}, which corresponds to traveling from pole to equator in about 20 yr. The migratory behavior of the high-latitude streams is less certain. Snodgrass and Howard (1985) and Snodgrass (1987a) have shown that, with a different method of analysis, the torsional pattern can appear as a polar spin-up near solar maximum, followed by (but not clearly continuous with) a single stream migrating from midlatitudes toward the equator.

Torsional oscillations are small: about ± 3 m s^{-1} compared to a typical rotation speed of 1900 m s^{-1}. This accounts for much of the difficulty in analyzing them and illustrates that they are not energetically important as such. However, they are a type of large-scale organization that was not anticipated in the Babcock picture. Their importance for empirical models centers around 2 questions: are torsional oscillations the surface signature of large-scale flows that are energetically significant and may cause solar activity? Is the evolutionary pattern of the oscillations consistent with an $\alpha\omega$ dynamo?

The shear zones between the fast and slow streams are clearly *associated* with solar activity at both high and low latitudes (LaBonte and Howard 1982b; Wilson et al. 1988). Yoshimura (1981) and Schüssler (1981) showed that the Lorentz forces associated with dynamo waves are strong enough to drive the torsional oscillations. Although LaBonte and Howard (1982b) pointed out that the evolutionary pattern of the Lorentz force in these models did not agree in detail with the pattern of the torsional oscillation, a later model (Tuominen et al. 1984) suggests that the induced velocity field may agree better. Thus, there are theoretical grounds for expecting that the torsional oscillations are an effect of cyclic magnetic activity. As yet, there is no straightforward evidence for the alternative hypothesis that the oscillations catalyze magnetic activity or are associated with larger forces that drive activity. The question of the timing of the torsional pattern is discussed below in connection with notions of an "extended" cycle.

Of the other forms of large-scale velocity fields discussed in the chapter by Howard et al., none (except supergranulation and differential rotation) has been unambiguously and consistently measured. It appears likely that there is an equator-to-pole meridional flow in the photosphere of 5 to 20 m s^{-1} at middle and high latitudes. As shown in Sec. IV, a steady flow of this magnitude improves the agreement of surface transport models with magnetic observations, although it is not clear that a steady flow can account for episodes of enhanced poleward flux transport such as those reported by Howard and LaBonte (1981). Mass conservation requires a return flow, the speed of which will depend on how deep and thick the return layer is. A return flow of 3 m s^{-1} would travel from pole to equator in $\sim$ 11 yr, so subsurface meridional circulation might play a role in the equatorward migration of active regions.

Giant convective cells or rolls are outside the scope of the Babcock picture. Until they are shown to exist and their basic properties are established, their impact on empirical models is a matter of speculation.

3. An "Extended" Solar Cycle?   Recently, Wilson et al. (1988, and references therein) have combined data on sunspots, torsional shear, ephemeral active regions and activity in the green-line corona to suggest that successive solar cycles overlap by 7 to 11 yr (Fig. 20) rather than the 2 to 3 yr seen in the butterfly diagram of sunspots alone. They assert that this degree of overlap is inconsistent with the Babcock picture. It is worth examining carefully in what respects this is true.

There is no reason in the Babcock picture that surface magnetic flux should not persist at low latitudes well beyond solar minimum into the following cycle. Surface interactions will determine how long it takes to "clean up" the remnant flux. Nor is it surprising that mild forms of activity such as ephemeral active regions should precede the onset of sunspot activity and extend the butterfly diagram to higher latitudes (Babcock's model assumes that the toroidal field amplifies for ~ 3 yr before it is strong enough to cause sunspots). However, a central feature of the Babcock model is that the dis-

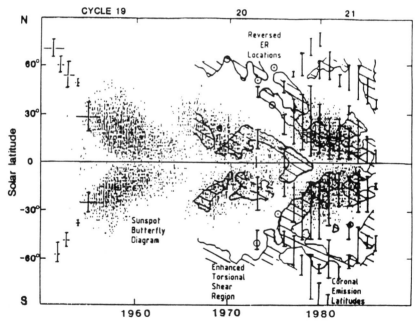

Fig. 20. Generalized butterfly diagram for cycles 19, 20 and 21. Zones of enhanced shear in the torsional oscillation pattern are shown as hatched regions within contours. The vertical lines delimit latitudes of significant emission in the coronal green line during a given year (the data before 1960 show only high-latitude maxima). The circles show typical locations of reversed-polarity ephemeral active regions during cycle 20. The generalized diagram illustrates that solar cycles overlap more than would be inferred from sunspot activity alone (figure from Wilson et al. 1988).

persal or transport of active-region fields during the currrent cycle causes the polar reversal and prepares the large-scale poloidal field for the next cycle. If there is a component of high-latitude activity that begins to migrate equatorward *before* the polar reversal and unambiguously connects with sunspot-belt activity during the following cycle, this would fall outside the scope of the Babcock model.

In each hemisphere, sunspot maximum and the polar reversal generally coincide to within about a year. In Fig. 20, reversed-polarity ephemeral regions are not common until 2 or 3 yr after sunspot maximum in 1969; similarly, Wilson et al. could not establish the presence of a high-latitude component of ephemeral regions prior to 1983, 3 yr after maximum. High-latitude torsional shear and the high-latitude maximum of green-line emission do extend back to solar maximum and perhaps 2 or 3 yr earlier. However, it is interesting that the small amount of high-latitude activity that *precedes* polar reversal appears to occur at the same (or somewhat lower) latitudes as activity at the time of reversal. The interpretation of this weak activity is not clear. Empirically, some activity is expected whenever surface fields of opposite polarity meet or intermingle. How much of the activity seen near the time of polar reversals is associated with the retreat of the polar caps (evident in the zonally averaged photospheric field and in the rush to the poles by polar crown filaments) and the subsequent period of indefinite but mixed polarity near the poles? The challenge is to separate activity incidentally associated with flux transport from activity driven by deep-seated, organized magnetic fields.

Within the broader context of $\alpha\omega$ dynamo models, the observed overlapping of cycles is not surprising. Figure 21 illustrates that the same dynamo action that strengthens a toroidal flux concentration on its low-latitude side and weakens its high-latitude side (thus propagating the toroid equatorward) also induces a weak toroidal field of the opposing polarity at higher latitudes. In this figure, the weak toroid that appears in year 5 (with strength 0.1 times the maximum toroidal field over the cycle) can be traced for 18 more years until it disappears at the same level of weakness.

4. Phenomena that Span Solar Cycles.   Section II.C discussed two types of observations in this category: a periodicity of 152 to 155 days in several types of solar activity that has been claimed to maintain phase coherence across cycle boundaries; and the possible spatial persistence of active domains for longer than one cycle. If confirmed, these phenomena should be closely examined to determine how much of the phase coherence and persistence can be associated with the overlapping of solar cycles discussed above. Phenomena that can be identified in 3 successive cycles would clearly challenge the Babcock picture. More work is needed to establish that the apparent persistence of activity across cycle boundaries is a physical connection rather than a statistical happenstance.

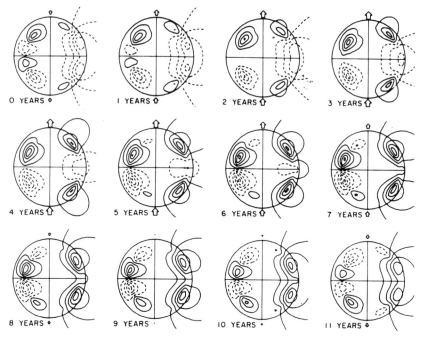

Fig. 21. Numerical solutions of an antisymmetric, oscillatory $\alpha\omega$ dynamo model with radial velocity shear ($d\omega/dR < 0$) and $\alpha \sim \cos\theta$. In each meridional cross section, contours of constant toroidal field strength are on the left, and poloidal lines of force are on the right. The levels of the curves are ± 0.1, ± 0.3, ... ± 0.9 times the maximum values (over the whole cycle) of the toroidal field or the poloidal flux function. Positive values (solid curves) indicate toroidal fields pointing out of the figure or clockwise poloidal field lines; negative values (dashed curves) indicate the opposite. The direction and magnitude of the field at the poles is indicated by vertical arrows. The negative toroidal field system that appears at high latitudes in year 5 can be followed for 18 yr as it migrates toward the equator (to follow its progress beyond year 11, resume at year 1 with reversed polarity) (figure from Stix 1976*b*).

Sunspot activity was severely depressed during the Maunder minimum (1645 to 1715), although the historical record is not adequate to establish that sunspots or other forms of activity were literally absent (Eddy 1980). Yoshimura (1987*b*) has shown how a kinematic dynamo with nonlinear feedback can produce long-term modulations in the envelope of the magnetic cycle, including prolonged periods in which there may be little or no erupted flux. The dynamo need not "decay" or "restart" to achieve this behavior. Periods such as the Maunder minimum may appear to pose a more fundamental difficulty for the Babcock picture, in which the migration and reconnection of surface fields is an active element. In this connection, it should be remembered that sunspots account for only about 40% of the surface magnetic flux during a cycle (K. Harvey, unpublished work); sunspots visible to the unaided eye probably carry at most 10% of the flux.

5. Theoretical Ideas. This section is not a résumé of recent theory; the purpose is to point to a connected group of ideas developed by Parker (1979a, and references therein, 1987abc, 1988abc) that bear on empirical models of the activity cycle.

From observations of a complex of activity (Gaizauskas et al. 1983), it is possible to estimate the strength of the subsurface azimuthal field necessary to maintain large complexes. Even if the subsurface field is recycled in episodes of emergence and retraction, azimuthal flux equivalent to a field strength of at least 3000 G over a depth of 10^5 km is required. Submerged magnetic flux experiences a buoyant force that, unopposed, will cause the flux bundle to rise through the convection zone at about the Alfvén speed, $v_A = B / (4\pi\rho)^{1/2}$. The Alfvén speed for $B = 100$ G in the middle of the convection zone (at a depth of 10^5 km) is about 2 m s^{-1}; flux rising at this rate would reach the surface in 2 yr. The problem, therefore, is how to keep the azimuthal flux submerged long enough for the ω effect to amplify it to the required strength. For $B = 3000$ G the Alfvénic rise time is too short even at the bottom of the convection zone; moreover, if the flux is concentrated in a thin layer, the field must be even stronger to supply the observed surface flux.

A magnetic field with an energy density at least as large as the kinetic energy density of the convective fluid in which it is embedded significantly reduces convective heat transport. A field $B \gtrsim 3000$ G satisfies this condition in the solar convection zone. Heat is trapped beneath the band of azimuthal field and a cool "thermal shadow" forms above the field and presses down on it. Parker has explored the complex dynamics which result from the contending forces in this scenario: the weight of the thermal shadow, the convective downdraft associated with it, the pile-up of heat beneath the azimuthal band, the expected Rayleigh-Taylor instabilities on both sides of the band, and the interaction with the surrounding convection. He shows how the accumulation of heat below the band can drive a thermal relaxation oscillation in which plumes of hot gas (carrying azimuthal field) force their way to the surface where they produce bipolar magnetic regions. The characteristic period of the oscillation is estimated to be a week or two. Once the thermal blockage is relieved, the convective downdraft retracts the flux in a characteristic time of a few months.

Parker also shows that flux bundles with diameters comparable to the local pressure scale height are subject to horizontal forces that may drive motions as large as 10 m s^{-1}. The motions arise when a flux bundle is "squeezed" asymmetrically between the thermal shadow above and the hot layer below; the motion is amplified by the counterflow that is produced in the surrounding fluid ("convective propulsion").

Although these ideas are subject to many uncertainties (not surprising in light of the complex physical situation), they have important implications for empirical models of the activity cycle. They offer a consistent scenario in which the seat of the dynamo is of necessity located near the bottom of the convection zone, as helioseismic determinations of the internal rotation also suggest, but in which the submerged reservoir of azimuthal field can communicate with the surface in a rapid and coherent way. However, if the predicted horizontal motions of the flux bundles are systematic, such motions are large enough to compete with dynamo waves and meridional circulation in determining the observed equatorward migration of active regions during the cycle.

E. The Status and Future of Empirical Models

We now attempt to answer the questions posed earlier: is the Babcock picture today a viable empirical model of the solar activity cycle, and, if not, is there a better model to replace it? Our short answers are "no" and "no."

If a good empirical model is one that could reasonably be expected to appear in an elementary textbook after the phenomenon it describes is quantitatively understood, the Babcock picture is no longer tenable. It gives an incorrect picture of the solar internal rotation; its assumption that the dynamo field is located in the outer convection zone is almost certainly wrong; it does not adequately link the reconnection of surface fields to the alteration of the magnetic topology below the surface; and its single-track evolution, in which a globally dipolar phase (allowing for remnants of low-latitude activity) is followed by the development of a single azimuthal flux system, strains to accommodate the degree of overlapping between cycles that is evident in the torsional oscillations and other indicators. However, one key aspect of the Babcock picture should continue to be investigated, because the answer will affect our interpretation of any empirical model: do the magnetic fields observed at the surface play an active role in the solar cycle, or are they just tracers? If they are tracers, with what fidelity do they reflect the evolution of subsurface fields? The most that can be said at this point is that there is no good evidence that the surface fields are actively involved. Parker's ideas suggest that young bipolar regions may reflect subsurface conditions with only a few days delay, even if the repository of azimuthal flux lies near the base of the convection zone. Most of the flux in active regions may disappear by retraction, indicating continuing subsurface control. The fact that the evolution of diffuse surface fields is well described by the transport model indicates that they are not as strictly controlled from below. How, then, are surface phenomena such as the polar polarity reversals, which may be multiple at each pole, connected to the evolution of the poloidal dynamo field? There is still much to be learned observationally and theoretically about surface fields and their subsurface connections.

Kinematic $\alpha\omega$ dynamo models (reviewed by Parker 1979a) can accommodate many observed features of the solar cycle, including torsional oscillations, highly overlapped cycles, an irregular period, north-south asymmetries and long-period modulations of the activity envelope. No well-observed active phenomenon appears to rule out the paradigm of a regenerative dynamo driven by nonuniform rotation and cyclonic turbulence. Can we then advance this paradigm as an empirical model in place of the more restricted Babcock picture? The problem is that the $\alpha\omega$ paradigm is overbroad. Models that differ significantly, such as those of Yoshimura (1978a) and Stix (1976b), exhibit many of the same essential features, including qualitative agreement with the observed butterfly diagram; yet neither model is consistent with what is now known about the internal rotation of the Sun. Within 10 yr, the behavior of the internal rotation from the surface well into the radiative core should be known with some confidence, and it is likely that a kinematic dynamo consistent with that behavior can be devised to operate near the base of the convection zone. Even then, such a model could not be labeled "empirical" until more is known about how deep-seated changes in the dynamo field manifest themselves at the surface. For now, it is well to remember that even the zeroth-order achievement of $\alpha\omega$ dynamo models—their ability to reproduce the equatorward migration of active latitudes—is in doubt. The simplest way to produce an equatorward wave from a dynamo operating in the convection zone is to assume that angular velocity increases with depth; but helioseismic measurements indicate that the radial gradient of angular velocity is too small for this purpose throughout the convection zone. Beyond that, there is no compelling reason to believe that the observed migration reflects a dynamo wave rather than subsurface meridional circulation, dynamic effects on flux bundles, or the net result of several effects.

There are few well-developed empirical models that stand clearly outside the Babcock picture. Models based on torsional oscillations of a large-scale, nonregenerative (but not necessarily "primordial") magnetic field have appeared several times (Walén 1947; Piddington 1976; Layzer et al. 1979). Although the latter references (Piddington and Layzer et al.) contain pertinent critiques of dynamo theory (dwelling particularly on the essential but questionable assumption of turbulent magnetic diffusion), the models themselves are qualitative and incomplete even by the standards of the Babcock picture. Nothing presently rules out the presence of a large-scale field in the solar core with the average strength ($B \sim 100$ G) needed to produce a torsional oscillation of the right period. Boyer and Levy (1984) have, however, shown that a large-scale, nonregenerative field stronger than a few gauss is not compatible with a conventional dynamo because the nonregenerative field introduces a strong even-odd effect in the amplitude of successive 11-yr cycles. One clear difficulty faced by torsional-oscillation models is the cyclical reversal of the polar fields seen by magnetographs. It is necessary to assume that the observed clumpy fields do not reflect a deep-seated poloidal component.

Wilson and Snodgrass (Wilson 1986,1987,1988; Snodgrass 1987ab; Snodgrass and Wilson 1987) have advanced a new schematic model of the solar cycle. They hypothesize a system of latitude- and time-dependent toroidal convective rolls in the lower part of the convection zone (Fig. 22). The rolls originate at the poles, where buoyancy is unimpeded by angular mo-

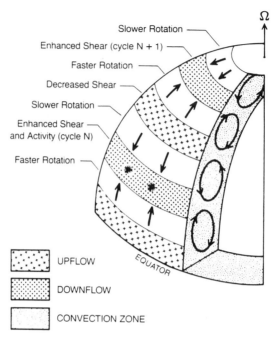

Fig. 22. Cartoon of the convective roll system proposed by Wilson and Snodgrass. Arrows on the surface show the direction of meridional flow; arrows on the ovals show the direction of circulation in the convection zone. Convective rolls develop at the poles every 11 yr and migrate toward the equator over the next 18 to 22 yr. Enhanced magnetic activity and shear in the torsional oscillations occur in latitude zones where flows converge. The convective rolls may break up into cells with longitudinal structure at lower latitudes (figure from Snodgrass 1987*b*).

mentum effects. Plasma diverges along meridians from latitude zones of rising flow and converges along meridians toward zones of downflow. The local effect of Coriolis acceleration on these flows produces latitude bands in which the rotation is alternately faster or slower than the time-averaged differential rotation at that latitude. This modulation of the differential rotation by convective rolls is identified with the observed torsional oscillation signal. At lower latitudes, the rolls may break up into giant cells with longitudinal structure.

To be consistent with observations of the fast and slow bands, a roll must originate near the pole every 11 yr and migrate toward the equator over the next 18 to 22 yr. Wilson (1988) has discussed how the overturning motions in a roll might combine with cyclonic convection to drive a dynamo wave that grows and progresses toward the equator. Where a toroidal roll dominates the convective motions, $d\omega/dR < 0$ locally because of conservation of angular momentum. The same conservation law acts to produce negative helicity (in the northern hemisphere) in rising and expanding parcels of gas. Wilson writes local dynamo equations for this system and shows that waves generated from a poloidal field that is not inclined to the local vertical at an angle greater than the co-latitude will propagate toward the equator with roughly the correct phase speed.

As the dynamo-generated toroidal field migrates toward the equator, its thermal shadow (Sec. V.D.5) accompanies it. The cool gas in the shadow above the toroid becomes a preferred location for convective downflow, so the convective roll follows the magnetic toroid.

Snodgrass and Wilson (1987) discuss two sets of data that may provide support for the existence of toroidal convective rolls. After rotation is removed, the zonal line-of-sight velocity signal in the Mount Wilson observations should be sensitive to upflow and downflow associated with rolls. A synoptic chart of velocity residuals does suggest zones of upflow and downflow that migrate toward the equator along with the fast and slow rotational bands.

Kuhn et al. (1987) have reported measurements of the latitude dependence of limb brightness temperature over the period 1983 to 1985. The pattern of small ($\lesssim 1$ K) temperature excesses and deficits in 1983 broadly corresponds with zones of upflow and downflow inferred from the convective roll picture, although the data at latitudes below $35°$ include an uncertain residual contribution from faculae. Kuhn et al. conclude that the phenomenon producing limb-temperature variations is of large scale and predominantly zonal in character, which is consistent with the toroidal roll picture.

The ideas advanced by Wilson and Snodgrass are an interesting attempt to view the solar cycle from a fresh perspective, emphasizing torsional oscillations and highly overlapped cycles as basic features. It is also instructive to consider the weaknesses of their picture, if only to illustrate the challenges an empirical model must face.

The most obvious difficulty is that there is no evidence for toroidal convective rolls that is as solid as, for example, the evidence for the torsional oscillation signal. The Mount Wilson line-of-sight velocity residuals are noisy and show a large, unexplained signal at the equator. Moreover, the association of enhanced velocity shear, downflow, and magnetic activity may be at least partly instrumental in origin. It is known that the presence of magnetic flux causes asymmetries in the Stokes V profile of Zeeman-sensitive lines that can be interpreted as a net red shift (downflow) if the line is not fully resolved (Stenflo and Harvey 1985). Also, granular convection is weaker in magnetic regions, which reduces the convective blue-shift effect (a velocity bias introduced because upwelling material is brighter on average) and causes a systematic red shift with respect to regions of lower magnetic activity.

The limb brightness temperature measurements cannot be reliably associated with large-scale patterns at low latitudes because of residual contributions from faculae. If the temperature deficit near 50° is connected with a convective roll and velocity stream, it should migrate toward the equator at the same rate as those features. Although Snodgrass and Wilson claim that the deficit does move equatorward, the inferred rate is twice as slow as expected. Kuhn et al. (1987) state only that "any latitude shift in the temperature minima near 50° (between 1983 and 1985) is small."

The roll system shown in Fig. 22 requires meridional flows that alternate in sign across at least 4 latitude zones, but the model does not specify the direction or amplitude of these motions at the surface. The observational evidence on meridional motions is confused (chapter by Howard et al.). Ribes et al. (1985a) claimed to detect toroidal convective rolls using atmospheric tracers, but their proposed *poleward*-propagating roll system does not even qualitatively match the Wilson-Snodgrass picture.

The dynamo action proposed by Wilson (1988) relies on an inward increase in angular velocity produced locally by the convective rolls. As discussed above, current helioseismic measurements do not support a significant gradient in angular velocity within the convection zone, however produced. More generally, several elements of the Wilson-Snodgrass picture derive from the local effects of Coriolis acceleration, which is only one among many dynamical effects in the convection zone. As Snodgrass and Wilson (1987) note, the most naive application of angular momentum conservation would predict that rotational angular velocity increases with latitude, opposite to what is observed.

Conceptually, the Wilson-Snodgrass picture cannot yet be considered an empirical model of the solar *cycle* because it does not address key cyclical elements. How, physically, is the toroidal field generated? Why does it amplify? How does the poloidal field reverse? Why do the two hemispheres maintain approximate but not exact synchronization? Although $\alpha\omega$ dynamo models face many difficulties, as discussed earlier, they can at least demonstrate feasibility in these regards through closed-form numerical experiments. The Wilson-Snodgrass picture has not been quantified to the extent that it can be numerically investigated.

What is the future of empirical models of the solar activity cycle? As our physical understanding of convection and dynamos becomes more sophisticated, the need for such models decreases. However, at this point there is no self-sufficient body of theory that explains the cycle and eliminates the need for a phenomenological model. The Babcock picture is inadequate, but no clearly superior alternative has been proposed. We will probably continue to try to understand new observations and ideas within the framework of $\alpha\omega$ dynamos unless the cumulative weight of disparities and incompleteness makes it impossible to continue. Improved observations of surface and (through helioseismology) subsurface magnetic and velocity fields, together with increasingly realistic numerical simulations of dynamos and magnetic transport, should set the stage for a decisive confrontation within the next decade.

Acknowledgments. DeVore and Sheeley acknowledge valuable discussions with and assistance from Y.-M. Wang and A. G. Nash, and financial support from the Office of Naval Research. Rabin is grateful to R. L. Moore for a discussion about magnetic topology in empirical models of the activity cycle. The National Solar Observatory of the National Optical Astronomy Observatories is operated by the Association of Universities for Research in Astronomy, Inc., under cooperative agreement with the National Science Foundation.

ACTIVE REGIONS, SUNSPOTS AND THEIR MAGNETIC FIELDS

MEIR SEMEL, ZADIG MOURADIAN, IRINA SORU-ESCAUT
Meudon Observatory

PER MALTBY
University of Oslo

DAVID REES
University of Sydney

MITSUGU MAKITA
Kwasan and Hida Observatories

and

TAKASHI SAKURAI
National Astronomical Observatory of Japan

Surface magnetism is the progenitor of active regions, sunspots, and all related phenomena. This cause and effect is reversible so that, using well-established empirical laws, the presence and morphology of photospheric magnetic fields can be deduced from active-region light emission structure. In the (simplifying) case of sunspots, MHD and thermodynamic theory find some success in the interpretation of the interaction of magnetic fields and solar plasma. Coronal magnetic fields also appear to be predictable by extrapolation techniques starting from the photospheric conditions. Alternatively, surface magnetism can be observed "directly" by means of the spectroscopic Zeeman effect and Stokes polarimetry. Eventually these empirical, theoretical and direct-measurement techniques must converge to identical results as we better understand the physics of active regions.

I. ACTIVE REGIONS AND THEIR MAGNETIC STRUCTURE AS DEDUCED BY MONOCHROMATIC IMAGING

The magnetic field in an active region (AR) locally modifies the energy deposition and causes different types of structures to form. Although an AR generally evolves slowly (over a few weeks), it may give rise to fast phenomena such as flares (see the chapter by Zirin et al.). Phenomena having a morphology similar to that of some of the AR structures appear in the so-called quiet Sun: polar filaments (prominences), the chromospheric network and facular mottles. These, however, will not be discussed here. There has been renewed interest in active regions over the past few years, not only as an aid to understanding stellar activity, but also as a basis for explaining solar-terrestrial relationships. The AR atmosphere can be considered as one type of stellar atmosphere, basically structured by the magnetic field.

We attempt here to relate the different structures that make up the active regions, and show how they interact. As the subject is vast, only certain aspects will be discussed in depth. Several surveys of active regions have been made in the past (see Martres and Bruzek [1977]; Zwann [1985,1987]; review papers of the Proceeding of the Skylab Active Region Workshop [Orrall 1981] and those of the Tenth European Regional Meeting [Hejna and Sobotka 1987]). In the following, we shall define the AR, and presume that magnetic flux emerges from the deep layers of the Sun and evolves.

A. What is an Active Region?

Structure. An AR is a combination of plasma and magnetic field that is constantly undergoing structural changes. It consists of a wide variety of features such as faculae, sunspots, filaments (prominences), coronal arches, flares, etc. All these can be considered as "epiphenomena" responding to the emergence of magnetic tubes of force. They are all in fact different aspects of the AR, depending on how it is observed: spectral bands from X rays to radio ranges, center of absorption lines, etc. Figure 1 shows four spotted ARs observed (a) in the violet wing of the Ca II K_1 line (photosphere and low chromosphere), (b) at the center of the Hα line (chromosphere) and (c) at the center of the Ca II K_3 line (high chromosphere). In Hα line, we also observe plage or quiescent filaments as well as vortex structures around facular plages. A simple AR is sketched in Fig. 2. Table I indicates the various structures in an AR, along with the spectral regions in which they are observed. The wavelengths indicated are those most commonly used.

An AR consists of tubes of magnetic field, filled with matter at relatively low pressure. Some tubes are open, some are closed (arches). The magnetic flux of an AR is of the order of 10^{22} to 10^{23} Mx (Maxwell). The variety of structures bears witness to the diversity both of the magnetic energy storage, and of the energy transport and release mechanisms. It should be noted that many tubes are thermally inhomogeneous, i.e., cold toward the bottom (Hα line) and hot toward the top (EUV or X-ray emission) (Rosner et al. 1978).

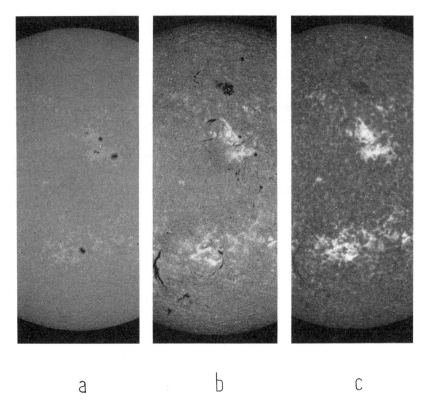

a   b   c

Fig. 1. Spectroheliograms of the Paris-Meudon Observatory (14 August 1988) showing active regions. (a) Blue wing of Ca II K line (K_{1v}): low chromosphere and photosphere, spots and plages. (b) Center of H 6563 Å line (Hα): high chromosphere, filaments, plages and vortex structures. (c) Center of Ca II K line (K_3): high chromosphere, filaments and plages.

One special case of active regions is that of the ephemeral regions. These are small ARs having no spot or flare association and a magnetic flux of about 10^{20} Mx. (Harvey et al. 1975; Martin and Harvey 1979).

Motions.   The plasma inside closed structures is often in a state of flow from one foot of an arch to the other, even when the arch is stable. Fibrils and surges are dynamic structures in which the motion of the plasma is guided by magnetic field tubes. Sprays, on the other hand, are purely dynamic structures in which the plasma is not guided. The main problem with studying the motion of ejected matter is determining what triggers them.

Evolution.   The birth of the active region is not fully understood. The source of the magnetic flux is probably a local dynamo (see the chapter by

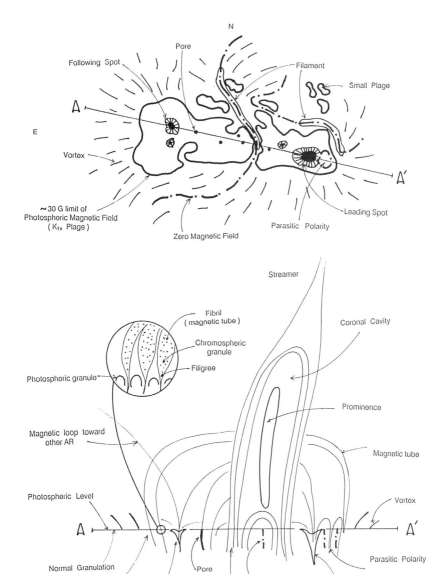

Fig. 2. Sketch of a simple, well-developed active region. Top: composite drawing of the photospheric and chromospheric structures. Bottom: cross section AA′ illustrating the height variations of the active region features.

TABLE I

The Usual Ways to Observe the Components of an Active Region

Layer	Feature	Wavelength	Remarks
Photosphere	sunspot	white light	
	lower plage	white light	visible only close to the limb.
Chromosphere	sunspot	K1v	the most intense
	plage	K3, EUV	region of K3
		Hα	plage.
Corona	vortices	Hα	
	flares	Hα	usually Hα
	filament (prominence)	Hα, K3	patrol
	hot filaments	EUV, X	at the limb only
	arches	Hα, EUV, X Fe XIV 5303	
	flares	EUV, X, γ	
	surges, sprays	Hα, EUV	

Rabin et al.) co-existing with the general dynamo that is responsible for the chromospheric network of the undisturbed Sun. But in essence, in the current state of our knowledge, the subphotospheric structure of ARs is unknown, so our report will be limited to the properties of the active solar atmosphere *after* the emergence of magnetic flux.

Švestka (1976), analyzing the X-ray bright points, suggests that each AR can be identified with an X-ray point at the beginning, and then it develops by itself. At their origin, the ARs have a closed-loop magnetic field associated with sunspots in a small plage; then the spots move away each from the other and the magnetic field opens. In the subsequent phase of evolution a filament may appear between the opposite polarities (Švestka et al. 1977). This basic evolution scheme for an AR is similar to that proposed by Kiepenheuer (1953).

The current limits on the spatial resolution of the observations (approximately 1 arcsec) hinder our understanding of the physics of the active regions. An increase in the spatial resolution, however desirable, will not necessarily lead to a direct detection of certain phenomena because the structures in question lack optical depth and may disappear because of decreased contrast.

B. What is a Complex of Activity?

An active region appears either alone or in a complex of activity (CA), i.e., together with other ARs. The very existence of a CA proves that a common source of magnetic flux exists for several ARs (Gaizauskas et al. 1983).

Several contemporary ARs may occupy a restricted area, that of a CA. The example of this is given in Fig. 3 which shows four ARs that are nearly simultaneous. The first appeared when the CA was in the invisible hemisphere, while the three others occurred when it was on the visible disk. In certain cases, filaments separate the ARs, as for example in Fig. 3. From the magnetic point of view, such a region is complex. The AR in Fig. 3 has 6 main sunspots, 3 of which are of north polarity, and the other 3 south. In most cases, the sunspots in a new AR are located in such a way that opposite polarities are positioned approximately on the same meridian. These configurations, by their magnetic complexity, stimulate the productivity of flares. On the other hand, eruptivity is weak if polarities of the same sign are located at about the same Carrington longitude. Bappu et al. (1968), Bumba and Suda (1984) and Mouradian et al. (1988) showed that new ARs tend to appear at the edge of old ones, within a given CA.

A CA will have a longer lifetime than any individual AR. Several ARs may appear successively at about the same heliographic coordinates (sequence of ARs) assembled into a CA. The example given in Fig. 4 shows 3 ARs succeeding each other during an interval of 5 rotations. This type of CA is magnetically less complex than the previous one illustrated in Fig. 3.

It should be noted that in many cases the time-space development of a CA, and the ARs making it up, is a combination of the two types mentioned above. Gaizauskas et al. (1983) presented several general properties of CAs, such as the repeated emergence of magnetic flux and its relative maintenance, the *in situ* disappearance of the magnetic flux, the correct equilibrium between the positive and negative fluxes and the fact that CAs rotate at the Carrington rate (see Sec. V), which is not the case for its individual constitutive ARs.

Long-term studies of ARs have shown that there exist "active" Carrington longitudes. Castenmiller et al. (1986) showed the existence of "sunspot nests" that are also a manifestation of the active longitudes. These active longitudes were analyzed for long periods of time by Trellis (1971b), who studied the location of the sunspot in heliographic longitude for eight cycles (12 through 19). Figure 5 shows the distribution of sunspot areas as a function of longitude. Trellis states that successive cycles can conserve the longitude of maximum frequency even over four cycles, and conserve the character of the distribution. We point out that in Fig. 5 the sunspot area distributions show that the longitudinal difference between the maxima is almost a multiple of $\pi/2$. This suggest the existence of a quadripolar component of the general dynamo of the Sun.

"Parasitic polarity" is the appearance of a small AR inside another AR that already existed (Martres et al. 1966; Rust 1968). One of the magnetic bipolar structures is identified with one of the existing AR polarities; the other one becomes a "parasitic island". Its presence transforms the AR into a small CA. Another form of association among distant ARs is the interconnection

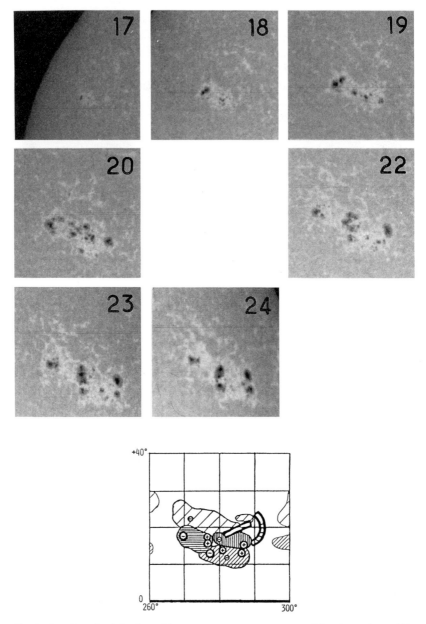

Fig. 3. Complex of activity formed by contemporaneous emergence of 4 active regions within four days (17–20 June 1982, K_{1v} spectroheliograms). Bottom: Meudon synoptic map of the complex activity (CA) with sunspot polarities.

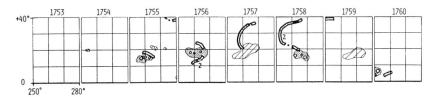

Fig. 4. Complex of activity formed by 3 active regions emerging successively within four rotations (1755–1758), as seen in the synoptic maps.

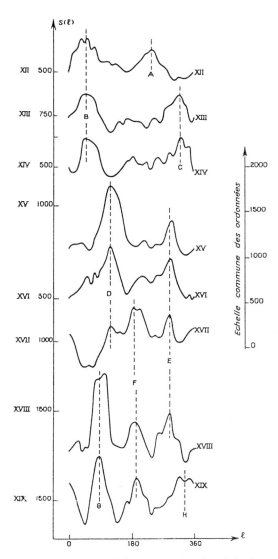

Fig. 5. Long-lasting "active longitudes." Sunspot area function of Carrington longitude. Note that the activity may last even 4 to 11 years (Trellis 1971).

by EUV or X-ray arches, which characterizes this as a surface phenomenon. A statistical analysis was made by Chase et al. (1976), who suggest that magnetic reconnection may explain this phenomenon. A visible manifestation of this reconnection might be the fast-moving Moreton waves observed in Hα (Moreton 1965) (see chapter by Zirin et al.). By their number and temporary distribution, active regions and complexes of activity define the cycle of solar activity. This is discussed in the chapter by Rabin et al.

C. The Structure of a Plage and the Active Corona

Although sunspots were historically known before plages, the plage is the basic element of the active region as it is the first structure that appears and one of the last to disappear. Even if absorbing features are seen in the Hα line before the plage appears (Martres and Soru-Escaut 1971), they are only the tips of the emerging magnetic arches, which actually form this plage. From the magnetic point of view, the plage, with or without spots, has a bipolar character. The growth of an AR is evidenced, among other things, by the increase of the plage, while its decreasing is seen by the dislocation of the plage.

The magnetic field of the plage is not homogeneous; the "filling factor" can vary over a large range. Livingston and Harvey (1969) showed that the magnetic field of the plage is structured in tubes with a mean flux of 2.8 $\times$ 10^{18} Mx. Tarbell and Title (1977) find that 85% of the tubes which have a magnetic field of more than 125 G have a field that, in fact, exceeds 1000 G. Also, Howard and Stenflo (1972) show that $<$ 10% of the total magnetic field is located outside the flux tubes. Solanki (1985), studying the properties of the magnetic flux tubes, finds values of 1500 G. On the other hand, the data of Semel (1985) and Zirin and Popp (1988) yield lesser values for the magnetic field, respectively, 600 to 1600 G and 150 to 600 G. Probably a range of field strengths are involved.

The appearance of the plage changes with altitude. In the photosphere, we observe an "abnormal granulation" (Fig. 6a), which is like the normal granulation but with the intergranular space filled with small structures (filigrees) of $<$300 km (Muller 1985; Fig. 6a) which gives to the region a low-contrast appearance. These small structures are the feet of the magnetic tubes and have a flux of the order of a few 10^{17} Mx. As they rise up into the chromosphere, the magnetic tubes reach their maximum diameter of the order of 10^3 km. Spruit and Zwaan (1981) studied the photometric properties and the dimensions of the magnetic tubes in young ARs. At the level of the chromosphere, the plage appears as small grains (facular grains) (Wilson 1981; Fang et al. 1984; Kitai and Muller 1984) that are the cold part of the tubes. Higher up, the temperature of the tubes increases and they become emissive in EUV lines. The tube structure remains at temperatures up to 7 to 8 $\times$ 10^5 K. Any tubes that rise higher, exceed 10^6 K, and group together into bundles that give them their characteristic appearance in coronal X-ray

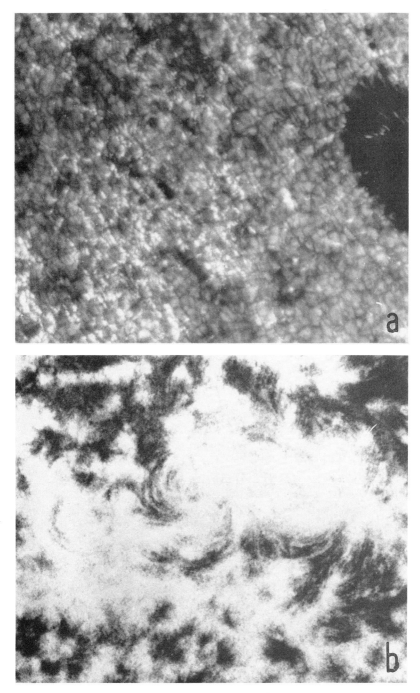

Fig. 6. The active region structure at two levels. (a) Photospheric abnormal granulation (45″ × 55″) (figure courtesy of R. Muller). (b) High chromospheric arches in Lyman α (36″ × 445″) (figure courtesy of LPSP).

observations. The X-ray corona over the ARs is covered by a "helmet streamer," which is structured by the open magnetic field and extends the AR into interplanetary space. Finally, we note that the filigrees, the plage (photospheric and chromospheric), the fibrils, the EUV and coronal arches and the streamers are different aspects of the same magnetic tube system.

The coronal arches having $T > 10^6$ K are stable, while the cooler arches exhibit major variations in visibility and may even disappear (Habbal et al. 1985). It should be pointed out that, aside from the variation in height of the arch temperatures, there also exists a variation transverse to the tube axis, with the central part of the tube cooler than the peripheral part (Levin and Withbroe 1977; Krishna 1983).

Inspection of the extensions of the plages as observed in either K_{1v} or K_3 (Fig. 1a and 1c), indicates a divergence of the magnetic field between these two heights. The brightest part of the plage observed in K_3 coincides with the plage observed in Hα (Fang and Martres 1986). The arch structures of ARs plage are shown by the Lyman α observations (Bonnet et al. 1980; Fig. 6b). The matter contained in the plage flux tubes is in motion. In the photosphere, the upward and downward velocities are of the order of 200 to 300 m s^{-1} (Harvey and Harvey 1976b). In the chromosphere and transition zone ($\leq T \leq 10^5$ K), the velocities increase up to the order of 25 km s^{-1} and then level off (Athay 1981; Mouradian et al. 1983b). The thermodynamic properties of the plages vary with altitude, but the diversity of the plages is such that the existing models sometimes differ greatly from each other. Lemaire (1988) and Jakimiec (1988) have reviewed the current models of plages. Pecker et al. (1988) have shown the effect of departures from sphericity (roughness) on diagnosis, hence upon the models of chromospheric and coronal layers.

In the Hα photograph of Fig. 1b, we see clearly that an obvious vortex structure *around* the plage extends it. This structure is made of tilted spicule-like features, the orientation of which indicates the inclination of the magnetic field in the first 10,000 km altitude. We may say that the AR concept applies beyond the plage and encompasses a lot of what is usually considered to be the "undisturbed" Sun.

D. The Sunspot as a Component of the Active Region Formation and Location

Observations of the formation of sunspots have shown that they originate by coalescence of magnetic tubes of like polarity (Zwaan 1985,1987). This is confirmed by the observed rates at which sunspots move to their future positions: the convergent motion of the photospheric plasma causes the magnetic field tubes to concentrate. The plasma motions are observed to have the same concentrative effects in the formation of the chromospheric network in the undisturbed Sun. An example of very rapid magnetic field concentration is shown in Fig. 7 where, in 31 hr 25 min., a group of sunspots multiplied

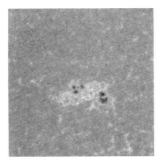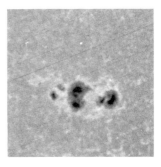

Fig. 7. Unusually quick evolution of sunspots from May 10 to May 11, 1983. Time elapsed 31 hr 25 min.

its surface area by a factor of 9. Garcia de la Rosa (1986) remarks that, at the time of their appearance, the preceding spots of the large ARs (having magnetic fluxes $>5 \times 10^{21}$ Mx) have lower latitudes than the following spots, while the small ARs have a random orientation. The author explains the difference between these two types of AR by the different depth at which they form: shallow for the small ARs and deeper for the others.

One important feature of the evolution of the sunspots is their proper motion with respect to the plasma affected by differential rotation. In recent years the Debrecen group has tried to relate this motion to the triggering of flares (Csepura and Nagy 1985). As far as the location of the sunspots in the AR is concerned, Mouradian et al. (1988) have shown that, at their maximum, the sunspots occupy preferential positions, i.e., they locate either at the edge of a former facular plage (Fig. 8a,b), or at the tip, or in the gap of a former filament (Figs. 8c,d).

E. Characteristics of the Filament as a Component of the Active Region

There are two kinds of plage filaments: (1) those that separate two plages of opposite polarity within the same AR (Fig. 1b, bottom), and which appear above all during the AR decreasing phase; and (2) those filaments that partially surround an AR, and hence point to it, or that separate two ARs within a CA (figure 1b, top). The lifetime of a plage filament may be longer than that of the plage itself. After the plage disappears, the filament then appears as a quiescent filament. Hirayama (1985) reviews the current state of our knowledge on plage filaments.

The study of plage filaments has not made as much progress as that of quiescent filaments, but we do know that plage filaments have the height of some 1.5 to 2×10^4 km (they are lower than the quiescent filaments) and, when seen at the edge of the disk as prominences, they appear more compact and brighter. Their magnetic field is of the order of 20 G, with an azimuth of $20°$ with respect to the body of the filament. The orientation of the magnetic

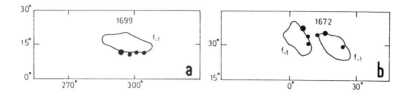

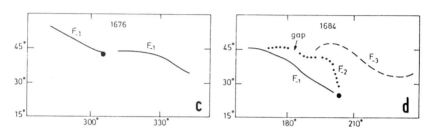

Fig. 8. Relationship between the position of sunspots (●) at their maximum development and that of previous filaments (F) and/or the facular plages (f). Sunspot located on the borders of plages of previous rotations (a, b). Sunspot located in the gap (c) or on the tip (d) of filaments of previous rotations (figure from Mouradian et al. 1988).

field is the same as that of the facula surrounding it, i.e., it is "potential-like" (Leroy et al. 1984). The local topology of the magnetic field, according to Hirayama (1985), conditions the narrow transition zone between the AR filament and the surrounding corona. The formation of a plage filament was observed by Rompolt and Bogdan (1986); they suggest that photospheric movements are responsible for this formation.

Plage filaments may disappear before a double ribbon flare starts. This "disparition brusque" (DB) may be either final or temporary (Mouradian et al. 1981; Malherbe et al. 1983). The activation of the plage filaments before a flare was examined by Martin (1980) and Gaizauskas (1989). Simon et al. (1986b) found, for partial DBs, that the heating and emergence of the magnetic flux contribute to the disappearance of the filament. In a series of articles, Molodenskij and Filippov (1987–1988) studied the expansive motion and disappearance of the filaments of the ARs in the framework of potential magnetic field theory. It should be pointed out that often any absorbing, elongated structure of the AR that can be seen on Hα images is wrongly called a filament. Filament-prominences should be distinguished from other absorbing features, with definitions as follows:

> *Spray,* an unguided plasmoid ejection of matter by the magnetic field, in a fantail motion;
> *Fibrils,* chromospheric arches in the active regions having plasma veloc-

ities and visibility variations due to the temperature and density variations (Fig. 6b, 9), (Alissandrakis et al. 1990);

Post-flare loops, arches that are cooling down after flare event;

Surges, a partial aspect of surging arches (Mouradian et al. 1983*a*);

Arch filament systems, arches specific to young ARs the sunspots of
which appear and develop.

Summarizing this section, it should be noted that the last structure of importance is the flare. The reader is referred to the chapter by Zirin et al. for
details about this phenomenon. Table II summarizes the basic average characteristics of the main elements of active regions. Despite the substantial
advances in our knowledge of active regions over the past few years, many
questions remain without a satisfactory answer. In closing, we mention a few:

What is the cause of the formation of active regions? What is the
mechanism that drives their emergence and evolution? Does local differential rotation play a role in the magnetic field decrease phase? What
degree of inhomogeneity is imposed inside the Sun by the existence of
the ARs?

Why do ARs appear in the zone between ± 40° latitude, while the
phenomena of the solar activity cycle begin to appear at the poles? If
solar activity decreases in latitude as the cycle progresses, why does the
sunspot maximum occur in one phase of this migration rather than another?

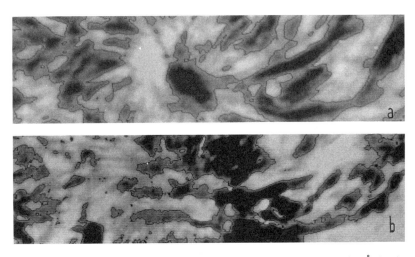

Fig. 9. Mass motion in an AR chromosphere. (a) The MSDP intensity map at 0.4 Å showing
dark as absorbing features and white as emissive features. (b) The MSDP velocity map at 0.4
Å showing dark as downward motion and white as upward motion (Mein 1977).

TABLE II

Physical Characteristics of the Components of an Active Region

Structure	Layer	Height (km)	Dimension[a] (km)	T (K)	n (cm^{-3})	v (km s^{-1})	B (G)	Lifetime (min/h/d)
Spot	photosphere	0	$\emptyset_{u+p} = 5 \times 10^4$	3700		6–7	2–3000	5–6 d
Pore		0	$\emptyset_u = 1\text{–}3000$				1500	few h
E F R	chromosphere	5000	$l = 3 \times 10^4$					
Filigree	photosphere	0	$\emptyset = 200$			≥ 10		3–4 d
Plage granul.	chromosphere	2000	$\emptyset = 700$	10^4				
Fibril	high chromo.	3000	$l = 10^4; \emptyset = 1500;$ $l = 10^4 - 10^5; \emptyset = 5000$	10^4			(300)	2 h
Loop	TZ and corona	$10^4 - 10^5$		$10^5 - 3 \times 10^6$	$5 \times 10^8 - 5 \times 10^9$	20		70 min
Vortices	chromosphere	1.1×10^4	$\emptyset = 700$	16000	10^{11}			
A F S	low corona	5000	$l = 3 \times 10^4; \emptyset = 3000$			30 50		5 min 20 min
Filament	corona	2×10^4	$l = 5 \times 10^4 - 5 \times 10^5$	7000	10^{11}	5	20–70	140 d

[a] $\emptyset$ = diameter; l = length; u = umbra; p = penumbra.

Why does the magnetic field of the plage filaments have a potential configuration while the quiescent filaments have an inverse configuration?

How does the plasma circulate between the spot and the plage (global Evershed effect in the AR)?

What causes the proper motion of spots?

To attempt to answer these questions, certain improvements are needed in our methods of observation, for example: coordinated multifrequency ground- and space-based observations; systematic observations in both wings of the Hα line with a standardization of format to facilitate data exchanges; systematic observations of the magnetic field and of the velocity field at the level of the chromosphere and of the photosphere.

Very often the various AR structures are studied individually, without taking the context into account, whereas any in-depth study demands a consideration of the environment and interactions. The main problem we have in constructing a homogenous review of this subject resides in the fact that there is as yet no global theory encompassing the structures that make up an AR. The models that do exist relate to different structures taken separately.

We can, however, conceive of an overall image to guide our thinking: deeply localized magnetic tubes of force are brought close to the surface by convection, and then emerge in response to instabilities. This emergence is possibly related to the solid rotation of the interior where the flux tubes originate. The magnetic field then opens more or less into a bipolar structure, arches develop and these guide the motions. Complexity leads to instabilities and flaring, and such epiphenomena develop until the active region finally dissipates. But under the surface there remains unstable magnetic structure and this leads to the appearance of a new AR at nearly the same place.

II. SUNSPOT ATMOSPHERES

In the sunspot atmosphere, various types of phenomena occur, ranging from slowly evolving configuration changes to rapid changes causing acceleration of charged particles to high energies (see the chapter by Zirin et al.). In this section, the discussion concentrates on processes that are slow enough to permit a magnetohydrodynamical description (see Cowling 1976; Priest 1981).

In spite of the fact that our empirical knowledge of sunspots is considerable (see, e.g., Bray and Loughhead 1964; Cram and Thomas 1981), Parker (1979) has described a sunspot as a phenomenon lacking scientific explanation. However, during recent years theoretical discussions as well as observations have given new insight into the physics of sunspots (see, e.g., Cram and Thomas 1981; Schmidt 1985; Athay and Spicer 1987; Schröter et al. 1987). Our present knowledge forms a basis for a general understanding of the sunspot atmosphere, but important properties like the filament structure and the mode of energy transport still lack convincing explanations.

A. Atmosphere of Large Sunspots

Numerous papers have been published that relate one measured parameter as a function of sunspot size. Several of these studies are valuable, but investigations dealing with the determination of physical parameters like the sunspot brightness are usually very uncertain because stray light tends to reduce the quality of the measurements (see, e.g., Zwaan 1965). Not all observers agree with this statement and claim to have detected a size dependence or, for instance, a difference between growing and stable sunspots (see, e.g., Chou 1987). We note that extremely good observing conditions and space observations have not been used frequently enough to determine with accuracy the intensity and other physical parameters of small and medium size sunspots. Thus, the following discussion will be based on our existing, somewhat wanting knowledge of large sunspots.

Several aspects of the atmosphere of individual sunspots may be discussed by regarding the sunspot as consisting of a single magnetic flux tube with the magnetic lines of force fanning out close to the solar surface (see, e.g., Cowling 1976; Meyer et al. 1977). The importance of the interchange instability for the stability of sunspots has been discussed vividly (see, e.g., Piddington 1978; Parker 1979a). In an untwisted, vertical flux tube, the buoyancy tends to stabilize the sunspot atmosphere if the fanning out of the field lines is large enough (Meyer et al. 1977; see also Priest 1984). A possible twisting of the flux tube will not increase the stability of the sunspot, but introduce dynamical effects and possible nonequilibrium (see, e.g., Parker 1979b).

B. Gas Pressure and Magnetic Field

The sunspot magnetic field influences the energy transport processes and accordingly the sunspot temperature distribution. Magnetic fields also alter the gas pressure distribution; the gas is approximately in hydrostatic balance along the field lines. The magnetic field distribution is coupled to the gas pressure distribution. The importance of this relation was first pointed out by Alfvén (1943), who found that static equilibrium in a vertical magnetic field is only possible if the temperature in the field is lower than in the environment. This argument may be extended to show that the darkness/brightness of a disturbed region on the solar surface is determined by the combined actions of the velocity field and the magnetic field in the observable layers (Dicke 1970). It should be noted, however, that the sunspot darkness/brightness is primarily determined by the energy transport processes, but the energy equation is mathematically coupled to the momentum equation.

Slowly varying changes at heights close to the photosphere suggest that

the sunspot atmosphere may be regarded as going through a series of magnetohydrodynamical equilibrium states, each without significant contributions from viscosity and ohmic dissipation. Assuming axial symmetry and
neglecting the azimuthal components of both the velocity field and the magnetic field, the momentum equation in cylindrical coordinates is considerably
simplified. Integration of the momentum equation in the horizontal plane
from the center of the sunspot ($r = 0$) to the photosphere ($r = a$) gives the
pressure difference (see, e.g., Maltby 1977),

$$P(r = a) - P(r = 0) = \frac{1}{2\mu} B_z^2(r = 0) + \frac{1}{\mu} \int_0^a B_z \frac{\partial B_r}{\partial z} dr$$
$$- \int_0^a \frac{1}{2}\rho \left(\frac{\partial v_r^2}{\partial r} + 2v_z \frac{\partial v_r}{\partial z}\right) dr \quad (1)$$

where B is the magnetic induction, μ the permeability, ρ the density and v
the velocity. Alternative formulations are possible, for example in terms of
the field-line curvature (see, e.g., Jakimiec, 1965). The largest contribution
to the lateral pressure difference comes from the magnetic field that acts both
through the magnetic pressure $B^2/2\mu$ and through the magnetic tension. Using
semi-empirical models one finds that the lateral pressure difference reaches a
maximum that exceeds the magnetic pressure at heights relatively close to an
optical depth of unity in the umbra. The exact value depends on the adopted
value for the Wilson depression.

Although analytical magnetostatic models are of interest (Schlüter and
Temesváry 1958; Osherovich et al. 1983), the focus has shifted to numerical
solutions (see, e.g., Pizzo 1987). An approach that involves numerical solutions of a free boundary value problem has been introduced (Schmidt and
Wegmann 1983; Jahn 1989). For more complicated configurations a methodology has been worked out with the aim to obtain three-dimensional magnetostatic solutions (Low 1985).

The pressure distribution in the vertical direction is usually found by
noting that for axial symmetry the momentum equation at $r = 0$ simplifies
to

$$\frac{\partial P(r = 0)}{\partial z} = -\rho(r = 0) \left[g + \frac{1}{2}\frac{\partial v_z^2(r = 0)}{\partial z}\right]. \quad (2)$$

Empirical sunspot model atmospheres are based on this equation, often without taking the contribution from the velocity term into account. In other sunspot models, the velocity term is neglected at photospheric heights, but included at heights where the observed nonthermal (turbulent) velocity is
comparable to the sound speed.

C. Velocity Field

The wavelength displacement of spectral lines in the spectra of sunspots is known as the Evershed effect. Although other interpretations are possible, the wavelength shift is usually interpreted as material motion and referred to as the Evershed flow (for a review see Sivaraman 1984). At photospheric heights the flow is nearly horizontal, directed from the umbra towards the photosphere with the highest speeds occurring in dark penumbral filaments and with the flow most probably restricted to the penumbra (see, e.g., Wiehr et al. 1986). The flow in the moat surrounding sunspots is derived from studies of moving magnetic features and has a different nature (see Brickhouse and LaBonte 1988).

The chromospheric counterpart to the Evershed flow is characterized by an inflow of material from the chromosphere into the spot along flow channels/superpenumbral filaments. The fact that the electrical conductivity is high strongly suggests that the motion is along magnetic field lines. The flow appears to have the character of a siphon flow (Meyer and Schmidt 1968). The properties of the siphon flow may be illustrated by considering the simple case of an isothermal gas in a strong magnetic flux tube with cross section A. Combining the equations of continuity and momentum the velocity v may be expressed as

$$\left(\frac{v^2}{c^2} - 1\right)\frac{\partial \ln v}{\partial s} = \frac{g_\parallel}{c^2} + \frac{\partial \ln A}{\partial s} \qquad (3)$$

where s and $g_\parallel$ are directed along the field line and c is the sound speed. A more complete discussion involves the temperature gradient and the energy equation, but certain conclusions are apparent. Subsonic and supersonic velocities will behave differently both in upflows and downflows; transitions between sub- and supersonic speeds are interesting, but require a careful discussion (see, e.g., Priest 1981). Meyer and Schmidt (1968) suggested that the flow is caused by a pressure difference, and reaches supersonic speeds in the downflow with a transition to subsonic flow through a stationary shock. A series of papers have been published with suggested application to the flow in sunspots including the siphon subsonic solution (Alissandrakis et al. 1988) and flow in a flux tube surrounded by field-free gas (Thomas 1988). An interesting discussion of the problems involved in understanding the Evershed flow is given by Spruit (1981d).

The observations show that the characteristics of the Evershed flow are maintained for most of the sunspot lifetime. Measurements in the chromosphere give a hint to how this is possible (Maltby 1975). Individual flow channels change on a time scale that is comparable to the time it takes the flow to pass from one end of the channel to the other. Persistent flows have also been observed in the sunspot transition region between the chromosphere and the corona. Whereas moderate spatial resolution observations have given conflicting results, recent high-resolution observations show both supersonic downflows in relatively small regions (4 to 6 arcsec) and subsonic upflows extending over larger regions (Kjeldeth-Moe et al. 1988). These measurements show gas nearly at rest spatially co-existing with strong downflows, suggesting a fine structure with flow channel width <1 arcsec. Although we have no direct proof, it seems very likely that these flows are aligned with the magnetic field and that flow speeds change markedly with position.

D. Semi-Empirical Models

In semi-empirical umbral models, the atmosphere is assumed to be either in hydrostatic or in hydrodynamical equilibrium. The umbral model in the deepest observable layers should account for the following continuum observations:

1. Spectral energy distribution;
2. Center-limb variations;
3. Solar-cycle variation.

The last item may be taken care of by considering a range in temperature from dark and bright umbrae. The presence of umbral dots have initiated investigations of two-component models (see, e.g., Adjabshirzadeh and Koutchmy 1983; Obridko and Staude 1988). As observations of the dot intensity are difficult, the two-component models are relatively uncertain. Another approach is to calculate a one-component model, search for possible inconsistencies in the model and evaluate the need for a two-component model. In passing, we note that in the penumbra the filament structure is so evident that a two-component model is unavoidable.

Most umbral models have been constructed by comparing theoretical umbra/photosphere continuum intensity ratios with the corresponding ratios derived from observations. Alternative approaches are possible, including a direct comparison of observed intensities on an absolute scale with calculated spectra. This direct approach is interesting, but requires an extensive compilation of atomic and molecular data applicable for calculations of synthetic high-resolution umbral spectra, modified by the absorption in the Earth's atmosphere.

A comparison of recently published umbral models shows relatively large differences between the models. Only part of these differences are due to real differences between different umbrae. A comparison between continuum observations and models by Staude et al. (1983), van Ballegooijen (1984) and Maltby et al. (1986) shows that the two latter models agree with the measurements for a medium dark umbra within the accuracy of the observations. This comparison suggests that the uncertainty in the temperature determination for the deepest observable layers is approximately 150 K. This estimate compares reasonably with the uncertainty in semi-empirical photospheric models. The semi-empirical umbral and photospheric models give the stratification with respect to the optical depth equal to unity level in the umbra and the photosphere, respectively. The height difference between these two levels is called the Wilson depression. In order to make a comparison of theoretical and empirical sunspot models meaningful, a value for the Wilson depression is needed.

For layers higher in the umbral atmosphere observations of spectral lines in the visible, infrared as well as the ultraviolet have been used (see, e.g., Cram and Thomas 1981). The non-LTE effects involved are similar to those of the undisturbed atmosphere (see the chapter by Anderson et al.), but complicated by the fact that the sunspot atmosphere interchanges radiation laterally with the surrounding atmosphere (see, e.g., Lites et al. 1987).

E. Energy Considerations

The energy transport in sunspots appears to be one area where our present knowledge still is rather limited. A better understanding of this would also be of value for the interpretation of the solar luminosity variations (see chapter by Livingston et al.; Hudson 1988). The presence of strong magnetic fields in sunspots prevent violent overturning convective motions (Biermann 1941). The problem is accordingly not to find the reason for the umbral darkness, but the source for the umbral brightness.

A sunspot model consisting of a cluster of thin flux tubes in subphotospheric layers, held together by a postulated flow, has been suggested (Parker 1979a). In this picture, the umbral dots represent intrusions of nonmagnetic plasma from subphotospheric layers. This model has not been generally accepted, because there is no observational indication of the postulated flow. The observed intensity difference among large umbrae shows a spectral distribution that is difficult to explain by a difference in the numbers of umbral dots contributing to the intensity (Albregtsen and Maltby 1981a). Spruit (1981a) has argued that the thin flux tubes merge into a single flux tube close to the photosphere and that the energy is transferred laterally by radiation from the nonmagnetic to the magnetic regions in the umbra. We note that the filament structure is found at all heights and appears to be a general characteristic of the sunspot atmosphere.

Alternative ways of explaining the umbral brightness, including sound waves and Alfvén waves, have been suggested but have not been generally accepted. Oscillations in the sunspot umbra have been observed both in the 3-min (see, e.g., Giovanelli 1974) and 5-min range (see, e.g., Balthasar et al. 1987a) and interpreted as resonance frequencies in the chromosphere and the subphotospheric layers (see, e.g., Zhugzhda et al. 1987). Whereas these oscillations appear to be without influence on the umbral brightness, another type, i.e., the overstable oscillations (see, e.g., Syrovatskii and Zhugzhda 1968; Knobloch and Wiess 1984) may be important. In overstable oscillations, heat is transferred by motion along the vertical field lines and laterally by radiation. Although the theory has not been worked out in detail, nonlinear calculations have been carried out (Cattaneo 1984; Hurlburt 1987). It is, however, too early to claim that we know the mode of energy transport. This remark applies also to the penumbra (see Schmidt et al. 1986).

The energy and momentum balance of the transition region between the chromosphere and the corona above sunspots include similar problems to those of the quiet Sun and other solar-like stars (see the chapter by Cook et al.), the main difference being that the energy balance is influenced by larger material flows in the sunspot. The importance of material motion on the ionization balance should also be kept in mind (Hansteen 1988).

III. STOKES POLARIMETRY OF SUNSPOTS

Magnetic fields in the solar atmosphere are most commonly inferred from measurements of the polarization in spectral lines split by the Zeeman effect. Correct interpretation of such data rests on an understanding of the theory of spectral line formation in the presence of a magnetic field. Central to this theory are the equations of transfer of polarized radiation. The focus of this section is on the application of this theory to observations in sunspots. The analysis of spectral polarization observed in the spatially unresolved photospheric network is reviewed by Stenflo (1989).

The state of polarization of radiation may be described by the Stokes parameters which are the components of the Stokes vector $\mathbf{I} = (I,Q,U,V)^\dagger$, ($\dagger$ = transpose). The transfer equations for the Stokes parameters were formulated originally by Unno (1956) for a Zeeman triplet assuming local thermodynamic equilibrium (LTE) and neglecting magneto-optical effects (hereafter referred to as MO) such as Faraday rotation. Rachkovsky (1962) extended the theory to include MO (see also Jefferies et al. 1989). Today, the Unno-Rachkovsky theory forms the basis for most polarization data analysis. For some time it was thought that MO could be safely neglected as a minor perturbation. Now it is recognized that MO are of fundamental importance, especially in the determination of vector magnetic fields. The crucial role of MO in sunspot magnetic field diagnostics is a recurring theme in this section.

In Sec.III.A the relevant line formation theory, including non-LTE effects, is summarized. Under certain conditions the transfer equations admit an analytic solution which is applicable to strong lines formed in the chromosphere. This solution as well as recent developments in numerical integration of the transfer equations are discussed in Sec.III.B.

Optical polarimetry of solar active regions can be divided into three classes depending on the spectral resolution $\delta\lambda$ of the observations (Landi Degl'Innocenti 1985a): (a) broadband observations ($\delta\lambda \approx 50$ to 100 Å); (b) magnetograph observations ($\delta\lambda \approx 100$ mÅ); (c) high-resolution observations of the full Stokes profiles ($\delta\lambda \approx 30$ mÅ). Recent applications of transfer theory to magnetic field diagnostics in these respective areas are summarized in Secs.III.C through E.

A. Line Formation Theory

The transfer equations are set up in the *xyz* coordinate frame (see Fig. 10), the *z*-axis being towards the observer. The magnetic field vector **B** has magnitude $|\mathbf{B}|$, inclination γ to the line of sight, and azimuth χ. In this frame, the Stokes parameters may be defined as follows (Shurcliff 1962):

$$
\begin{aligned}
I &= \text{total intensity} \\
Q &= I_{\text{lin}}(\delta = 0) - I_{\text{lin}}(\delta = \pi/2) \\
U &= I_{\text{lin}}(\delta = \pi/4) - I_{\text{lin}}(\delta = 3\pi/4) \\
V &= I_{\text{right}} - I_{\text{left}}
\end{aligned}
\tag{4}
$$

where $I_{\text{lin}}(\delta)$ is the intensity of linearly polarized light that would be transmitted by an analyzer with its axis at an angle δ measured counterclockwise from the *x*-axis, while I_{right} and I_{left} are the intensities transmitted by right- and left-hand circular analyzers. Right hand means a clockwise rotation of the electric vector of the light as seen by the observer receiving the radiation, and left hand means counterclockwise.

The non-LTE line formation theory presented here reduces in the LTE limit to the Unno-Rachkovsky theory. It is based on the following assumptions (Landi Degl'Innocenti 1976; Rees et al. 1989): electric dipole transition; *LS*-coupling; no quantum interference between Zeeman states; complete de-

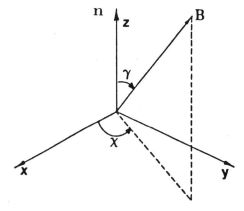

Fig. 10. Coordinate system for definition of the magnetic field and the Stokes parameters.

alignment of atomic levels (i.e., the populations of Zeeman states in any atomic level are equal); complete redistribution on scattering; stimulated emission treated as negative absorption; and steady-state atmosphere. Although more general non-LTE formulations exist that include quantum interference and partial redistribution effects (see review by Rees 1987), the complexity of the equations precludes any practical application to the analysis of sunspot spectra at this stage.

As in standard non-LTE theory in the absence of Zeeman splitting one defines a line center opacity κ_0 (for zero damping and corrected for stimulated emission) and a line source function S_l which are computed from the total populations of the upper and lower levels of the line transition (see, e.g., Mihalas 1978). In LTE, these populations are obtained from the Saha-Boltzmann equations, so that $S_l = B_\nu(T_e)$, the Planck function at the local electron temperature T_e. In non-LTE, one should strictly derive the populations by self-consistent solution of the equations of transfer and statistical equilibrium in a multilevel atomic model. Auer et al. (1977a) have shown that, to a good approximation, one can neglect the effect of Zeeman splitting and compute the populations using a nonmagnetic, non-LTE code. This is the *field-free* method suggested by Rees (1969). Implementation of van Ballegooijen's (1987) new method for computing multilevel non-LTE populations accounting for Zeeman splitting should clarify the limitations of this approximation. Finally continuum polarization is neglected and continuum processes are characterized by an opacity κ_c and source function $S_c = B_\nu(T_e)$.

The transfer equation for the Stokes vector can be written as (Lites et al. 1988)

$$\mu \frac{d\mathbf{I}}{d\tau_0} = (r_0 \mathbf{1} + \boldsymbol{\Phi})\mathbf{I} - r_0 S_c \mathbf{e_0} - S_l \boldsymbol{\Phi} \mathbf{e_0} \qquad (5)$$

where $r_0 = \kappa_c / \kappa_0$, τ_0 is the line center optical depth defined by $d\tau_0 = -\kappa_0 dz$ and $\cos^{-1}\mu$ is the heliocentric angle. This assumes a plane-parallel atmosphere, but horizontal inhomogeneities can be accommodated by using a line-of-sight optical depth instead. Here $\mathbf{1}$ is the 4×4 unit matrix, $\mathbf{e_0} = (1,0,0,0)^\dagger$ and $\boldsymbol{\Phi}$ is the line absorption matrix,

$$\boldsymbol{\Phi} = \begin{pmatrix} \phi_I & \phi_Q & \phi_U & \phi_V \\ \phi_Q & \phi_I & \phi'_V & -\phi'_U \\ \phi_U & -\phi'_V & \phi_I & \phi'_Q \\ \phi_V & \phi'_U & -\phi'_Q & \phi_I \end{pmatrix} \qquad (6)$$

where

$$\phi_I = \tfrac{1}{2}\phi_p \sin^2 \gamma + \tfrac{1}{4}(\phi_r + \phi_b)(1 + \cos^2 \gamma)$$

$$\phi_Q = \tfrac{1}{2}(\phi_p - \tfrac{1}{2}(\phi_r + \phi_b))\sin^2 \gamma \cos 2\chi$$

$$\phi_U = \tfrac{1}{2}(\phi_p - \tfrac{1}{2}(\phi_r + \phi_b))\sin^2 \gamma \sin 2\chi$$

$$\phi_V = \tfrac{1}{2}(\phi_r - \phi_b)\cos \gamma \tag{7}$$

$$\phi'_Q = \tfrac{1}{2}(\phi'_p - \tfrac{1}{2}(\phi'_r + \phi'_b))\sin^2 \gamma \cos 2\chi$$

$$\phi'_U = \tfrac{1}{2}(\phi'_p - \tfrac{1}{2}(\phi'_r + \phi'_b))\sin^2 \gamma \sin 2\chi$$

$$\phi'_V = \tfrac{1}{2}(\phi'_r - \phi'_b)\cos \gamma.$$

Since $\Phi(\chi) = \Phi(\chi \pm \pi)$, there is an intrinsic ambiguity in the specification of the azimuth. This ambiguity may be resolved in data analysis by appealing to nonpolarimetric information such as the morphology of chromospheric structures observed in Hα, or symmetry considerations (Lites and Skumanich 1989).

For an arbitrary Zeeman multiplet, the absorption and anomalous dispersion profiles $\phi_{p,b,r}$ and $\phi'_{p,b,r}$ are suitably weighted sums of the shifted profiles of the individual Zeeman components (see, e.g., Rees et al. 1989). Note that the $\phi'_{Q,U,V}$ terms give rise to MO. The subscripts p,b,r denote, respectively, components with $\Delta M = M_{up} - M_{lo} = 0, +1, -1$ where $M_{up,lo}$ are the magnetic quantum numbers of Zeeman states in the upper and lower energy levels of the line transition. For a Zeeman triplet these profiles simplify to

$$\phi_p = H(a, v + v + v_{los}), \qquad \phi_{b,r} = H(a, v \pm v_B + v_{los}) \tag{8}$$
$$\phi'_p = 2F(a, v + v_{los}), \qquad \phi'_{b,r} = 2F(a, v \pm v_B + v_{los})$$

where $H(a,v)$ and $F(a,v)$ are the Voigt and Faraday-Voigt functions:

$$H(a,v) = \frac{a}{\pi} \int_{-\infty}^{\infty} \frac{e^{-y^2}}{(v - y)^2 + a^2} \, dy$$
$$F(a,v) = \frac{1}{2\pi} \int_{-\infty}^{\infty} \frac{(v - y)e^{-y^2}}{(v - y)^2 + a^2} \, dy. \tag{9}$$

The parameters in these equations are defined as follows: $a = \Gamma\lambda_0^2/4\pi c\Delta\lambda_D$ where Γ is the line damping, $\Delta\lambda_D$ is the Doppler width, c is the velocity of light and λ_0 is the laboratory line center wavelength; $v = (\lambda - \lambda_0)/\Delta\lambda_D$ where λ is the wavelength in the line; $v_B = g_L e\lambda_0^2|\mathbf{B}|/4\pi mc^2\Delta\lambda_D$ = the Zeeman shift of the σ components, g_L being the Landé factor, and e and m the electron charge and mass; and $v_{los} = \lambda_0 V_{los}/c\Delta\lambda_D$ = the Doppler shift due to the line-of-sight macroscopic velocity field V_{los} (positive towards the observer).

B. Solutions of the Transfer Equations

The profile-fitting method described in Sec.III.E uses an analytic solution of the transfer equations in which r_0 and Φ are treated as constants, i.e., κ_c, κ_0, $\Delta\lambda_D$, Γ, $\mathbf{B}$ and V_{los} are constants over the region of line formation. The continuum source function is assumed to be linear in τ_0,

$$S_c = B_0 + B_1\tau_0 \tag{10}$$

while the line source function is written as

$$S_l = B_0 + B_1\tau_0 - \sum_i A_i e^{-\epsilon_i\tau_0}. \tag{11}$$

With an appropriate number of exponential terms this representation can mimic the nonlinear variation of $S_l(\tau_0)$ typical of strong lines formed in non-LTE in the chromosphere. Integrating Eq. (5) one obtains the emergent Stokes vector (Lites et al. 1988),

$$\mathbf{I}(0,\mu) = B_0\,\mathbf{e_0} + \mu B_1\,[r_0\mathbf{1} + \Phi]^{-1}\,\mathbf{e_0} - \sum_i A_i\mathbf{e_0}$$
$$+ \sum_i A_i(r_0 + \epsilon_i\mu)[(r_0 + \epsilon_i\mu)\mathbf{1} + \Phi]^{-1}\mathbf{e_0}. \tag{12}$$

If the exponential terms are omitted this is equivalent to the Milne-Eddington model solution commonly used in LTE data analysis. Let $\eta_{I,Q,U,V} = \eta_0\phi_{I,Q,U,V}$ and $\rho_{Q,U,V} = \eta_0\phi'_{Q,U,V}$ where $\eta_0 = 1/r_0$ and $\tilde{B}_1 = r_0B_1$. Then from Eq. (12) one obtains the familiar expressions for the emergent Stokes parameters (Landolfi and Landi Degl'Innocenti 1982):

$$I = B_0 + \mu\tilde{B}_1[(1 + \eta_I)((1 + \eta_I)^2 + \rho_Q^2 + \rho_U^2 + \rho_V^2)]/\Delta$$
$$Q = -\mu\tilde{B}_1[(1 + \eta_I)^2\eta_Q + (1 + \eta_I)(\eta_V\rho_U - \eta_U\rho_V) + \rho_Q W]/\Delta \tag{13}$$
$$U = -\mu\tilde{B}_1[(1 + \eta_I)^2\eta_U + (1 + \eta_I)(\eta_Q\rho_V - \eta_V\rho_Q) + \rho_U W]/\Delta$$
$$V = -\mu\tilde{B}_1[(1 + \eta_I)^2\eta_V + \rho_V W]/\Delta$$

where

$$W = \eta_Q\rho_Q + \eta_U\rho_U + \eta_V\rho_V \tag{14}$$

and

$$\Delta = (1 + \eta_I)^2[(1 + \eta_I)^2 - \eta_Q^2 - \eta_U^2$$
$$- \eta_V^2 + \rho_Q^2 + \rho_U^2 + \rho_V^2] - W^2. \tag{15}$$

The advantage of the exponential representation for the non-LTE case is that the last two terms in Eq. (12) have a similar form to the first two, and thus can be expressed similarly to Eq. (13) (see Sec.III.E).

In general, for an atmosphere with depth dependence in all variables, an analytic solution is not possible, and one must integrate the transfer equations numerically. Runge-Kutta integration is the most popular method, but this is unsuitable if the optical path length is large as is the case in strong chromospheric lines. A more efficient finite-difference method was introduced by Auer et al. (1977a). However, recent numerical tests indicate that the most rapid and accurate method is one based on an integral equation formulation of Eq. (5). This is called the DELO method since it involves a numerical representation of the integral lambda operator defined on the optical-depth scale associated with the diagonal element $r_0 + \phi_I$ of the matrix $r_0 \mathbf{1} + \mathbf{\Phi}$ (Rees et al. 1989; Murphy 1989). Another integral operator technique has been proposed by Landi Degl'Innocenti (1987), but its efficiency relative to the DELO method has not yet been tested.

Figure 11 shows the Stokes profiles of the Ca II K line computed as follows: non-LTE populations, neglecting Zeeman splitting, were calculated with the Kitt Peak code (Auer et al. 1972) in the HSRA solar model (Gingerich et al. 1971); then the transfer equations were integrated by the DELO method assuming a constant magnetic vector with $|\mathbf{B}| = 3000$ G, $\gamma = \pi/4$ and $\chi = 0$, and $V_{\text{los}} = 0$. This is the model studied by Auer et al. (1977a), who neglected MO. Comparison of the profiles with and without MO clearly shows the influence of MO, especially on the profiles of Q and U. The V profile has an interesting sign reversal which follows the changing slope—$dI/d(\Delta\lambda)$ of the intensity profile. The sign change occurs at the K_2 emission peak. The I emission and V reversal are mappings of the changes in slope of the line source function. Thus they provide sensitive complementary diagnostic information about the thermal structure of the atmosphere. Such features have been observed in Ca II infrared triplet Stokes profiles (Rees et al. 1991).

A recent advance in polarized transfer has been the development of methods to compute the contribution functions of the Stokes parameters. Van Ballegooijen (1985a) used a complex coherency matrix formulation of the transfer equations to compute the so-called emission contribution functions by a Runge-Kutta method. Grossmann-Doerth et al. (1988a) adapted his method to compute contribution functions for the Stokes line depression profiles $(1 - I/I_c, -Q/I_c, -U/I_c, -V/I_c)^\dagger$ where I_c is the continuum intensity. The line depression contribution functions give a better estimate of the height of formation of Zeeman-split lines. They also computed the corresponding Stokes response functions. Both types of contribution functions can be computed more efficiently using the DELO method (Rees et al. 1989; Murphy 1989).

Stokes contribution and response function analysis promises to be an

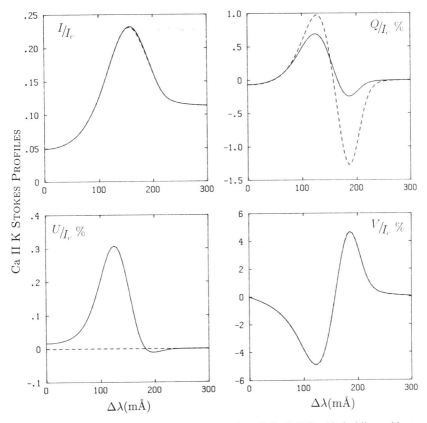

Fig. 11. Synthetic Stokes profiles for the anomalously split Ca II K line (dashed line, without MO; solid line, with MO) (figure from Rees et al. 1989, Fig. 8).

important tool for modeling magnetic structures. So far, it has been applied to polarimetry of unresolved flux tubes in the photosphere, but it should be invaluable also in the construction of sunspot models.

C. Broadband Observations

The broadband linear polarization observed in sunspots (see, e.g., Illing et al. 1974a,b,1975) can be explained in terms of a combination of magnetic intensification and MO in the spectral lines within the bandpass (see, e.g., Landi Degl'Innocenti and Calamai 1982). However, such data have limited diagnostic value. Of more interest for the study of sunspot structure is broadband circular polarization (Illing et al. 1974a,b,1975; Kemp and Henson 1983; Henson and Kemp 1984; Makita 1986b; Makita and Ohki 1986).

Indirect evidence indicates that net circular polarization (NCP) in sunspots observed with broadband filters is due to Zeeman sensitive lines. For example, Illing et al. (1975) found that polarization in the yellow band λ5824 − 5844 Å was significantly weaker than in the green band λ5253 − 5353 Å which contains more strong lines. Kemp and Henson (1983) also found a similar spectral dependence, the NCP decreasing at longer wavelengths where there are fewer lines within the bandpass. More direct evidence that NCP is essentially the integrated polarization of the lines was provided by Makita who used a spectrograph instead of a filter. He found that masking of the strong lines in the observed spectral range significantly reduced the NCP. The line spectral origin of NCP observed away from sunspots (Kemp et al. 1987) has been confirmed by integrating FTS Stokes V spectra (Mürset et al. 1988).

Illing et al. showed that for a purely longitudinal magnetic field ($\gamma = 0$) a combination of magnetic and velocity field gradients is required to produce NCP in a Zeeman sensitive line. This mechanism has been studied by Ribes et al. (1985b) and Solanki and Pahlke (1988) in attempting to explain Stokes V asymmetries in unresolved magnetic elements. NCP can also occur when there is a velocity gradient and a homogeneous field that is inclined to the line of sight (Auer and Heasley 1978). Auer and Heasley neglected MO in their analysis.

However, these simple models are too restrictive to account for the spatial distribution of NCP in sunspots. For example, Henson and Kemp (1984) found that the NCP reverses sign or diminishes to near zero at the center of the umbra relative to the outer penumbra. Moreover, the NCP shows a marked asymmetry, the penumbral region on the limbward side showing a stronger NCP than the disk-center side of the spot.

Landman and Finn (1979) computed the spatial variation of NCP in a spectral line formed in the Schlüter-Temesvàry spot model. However, sign errors in their treatment of the MO terms in the transfer equations prompted Skumanich and Lites (1987b) to repeat and extend their analysis. Figure 12 shows the NCP computed for normal emergence as a function of radius r from the center of the umbra. The thermodynamics are simplified to a Milne-Eddington model with constant opacity ratio $\eta_0 = 20$ and linear source functions $S_l = S_c = 1 + 9\tau$ (τ is the continuum optical depth). The velocity field is an accelerating upflow along the lines of force.

The solid curves are the signatures when gradients in $|\mathbf{B}|$ and γ are neglected (mean values over $\tau = 0.005$ to $\tau = 0.5$ are used at each value of r). The solid curve with the open circles neglects MO (this is the Auer-Heasley mechanism); the curve labeled MO shows the dramatic effect of MO, NCP being positive at all r. The other two curves are the results obtained when the full radial and depth variations of $|\mathbf{B}|$ and γ are included. The curve labeled B' neglects MO. Note the symmetry breaking effect of the magnetic field gradient which produces a nonzero NCP at $r = 0$ where $\gamma = 0$. This is the Illing et al. mechanism. Finally the curve labeled MO $+ B'$ shows the NCP when magnetic field gradients and MO are included. The signature is then in

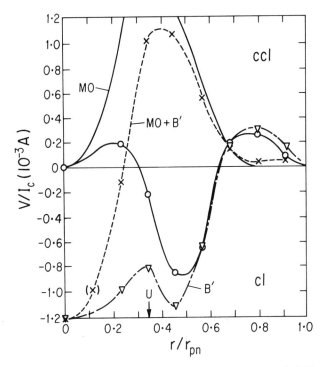

Fig. 12. Radial distribution of NCP for Landman and Finn sunspot magnetic field model with accelerating upflow along field lines: open circles, no magnetic gradients or MO, curve labeled MO, no gradients with MO; curve labeled B', gradients without MO, curve labeded MO + B', gradients with MO (figure from Skumanich and Lites 1987b, Fig. 2a).

qualitative agreement with the observations of Kemp and Henson. Calculations with more realistic thermodynamics produce similar results. Using reasonable estimates for the spectral density of lines in the broadband windows, Skumanich and Lites find good quantitative agreement with the observations. In this model, the magnetic field azimuth χ is constant along the line of sight. An alternative mechanism for producing NCP has been proposed by Makita (1986b). From a comparison of linear and circular polarization maps, he infers that the azimuth is variable, i.e., the magnetic field twists as it emerges through the sunspot. His radiative transfer calculations produce a NCP compatible with his spectrograph data and the filter data of Illing et al., but his choice of magnetic-vector variation is rather *ad hoc,* compared with the more realistic Schlüter-Temesvàry model. Skumanich and Lites note that High Altitude Observatory Stokes Polarimeter observations indicate that about as many sunspots appear not to show any measurable twist as do. The relative importance of twist in producing the NCP signature needs to be investigated further through radiative transfer computations for nonnormal lines of sight not only in the Schlüter-Temesvàry model, but also in its generalizations, such as the return flux model (Flå et al. 1982).

Kemp et al. (1984) and Landi Degl'Innocenti (1985b) suggest another mechanism that could produce spectral line NCP, namely atomic orientation. The combination of a velocity gradient and an anisotropic radiation field can induce inequalities in the Zeeman sublevel populations which give rise to line asymmetries. Radiative transfer calculations including this non-LTE phenomenon are not yet available, so the importance of atomic orientation is an open question at this time.

Although broadband polarimetry can be used to measure low levels of polarization very precisely, it is of limited use in the study of vector magnetic fields. For this purpose one requires more detailed spectral information, as provided by magnetographs and high-resolution Stokes polarimeters. Thus broadband polarimetry is unlikely to play a significant role in the future of sunspot magnetometry.

D. Magnetograph Observations

A magnetograph is a device which detects the polarization in a restricted wavelength range in the wings of a spectral line (for reviews on different instruments and operating principles see Baur [1981] and Harvey [1985b]). The standard longitudinal magnetograph which measures only circular polarization (Stokes V) provides information about the line-of-sight or longitudinal magnetic field component in the region of line formation. A vector magnetograph measures the linear polarization (Stokes Q and U) as well and, in principle, can be used to infer the vector magnetic field.

Precise calibration of a vector magnetograph is very difficult. In the region of line formation, there may be gradients in the vector magnetic field coupled with velocity gradients that distort the profiles. Moreover, the thermal structure of the atmosphere varies from point to point. All this information is mapped into the Stokes spectral profiles, but only a small portion is accessed because of the smearing effect of integration across the bandwidth and the limited wavelength coverage in the line (this last problem can be alleviated somewhat by tuning across the line). For these reasons, the inverse problem of extracting the vector magnetic field is very ill-posed.

A careful analysis of this problem was given by West and Hagyard (1983) in relation to the calibration of the Marshall Space Flight Center vector magnetograph. This instrument samples the Stokes profiles of the Fe I $\lambda 5250$ Å Zeeman triplet with a tunable 1/8 Å filter. The measured Stokes vector is

$$\tilde{\mathbf{I}}(\Delta\lambda_0) = \int \mathbf{I}(\Delta\lambda) f(\Delta\lambda - \Delta\lambda_0) d(\Delta\lambda) \qquad (16)$$

where $f(\Delta\lambda)$ is the filter profile (approximately gaussian) and $\Delta\lambda_0$ is the filter offset from line center.

It is assumed that the vector magnetic field is constant, i.e., some sort of mean magnetic field is inferred from the data. The field azimuth χ is estimated from the azimuth $\phi = 1/2 \tan^{-1}(\bar{U}/\bar{Q})$ of the plane of maximum linear polarization. As a consequence of MO, one finds that ϕ is rotated relative to χ (if there were no MO, then they would coincide). Radiative transfer calculations in a sunspot model for several field strengths and inclinations by West and Hagyard indicate that for large offsets, $|\Delta\lambda_0| \geq 120$ mÅ, one can make the approximation $\phi \approx \chi$ with an error of the order of $10°$. This well-known discrepancy between ϕ and χ was studied first by Rachkovsky (1962) and lately has been the focus of much theoretical and observational work (Kawakami 1983; Makita 1986a; Ye Shi-Hui and Jin Jie-Hai 1987). It is generally agreed that MO can be quite strong, especially in sunspot umbrae, and should not be ignored in measuring field azimuth.

Referring the Stokes parameters to the frame in which the estimated field azimuth is zero, West and Hagyard derive the field strength $|\mathbf{B}|$ and inclination γ using the formulae

$$|\mathbf{B}| \cos \gamma = C_1(\Delta\lambda_0)k(\Delta\lambda_0)(PV(\Delta\lambda_0) - PV_0) \tag{17}$$
$$|\mathbf{B}| \sin \gamma = C_2(\Delta\lambda_0)k^{1/2}(\Delta\lambda_0)(PQ(\Delta\lambda_0) - PQ_0)^{1/2}$$

where $PV = \bar{V}/\bar{I}$ and $PQ = (\bar{Q}^2 + \bar{U}^2)^{1/2}/\bar{I}$ (PV_0 and PQ_0 are background signals). The $C_{1,2}$ are calibration curves based on Kjeldseth Moe's (1968) solutions of the radiative transfer equations (ignoring MO) for a penumbral model atmosphere (Kjeldseth Moe and Maltby 1969). The parameter k is an empirically determined scale factor. This calibration procedure, in common with all other magnetograph calibrations, is therefore strongly model dependent. The issue of just how accurate are magnetograph vector field maps can only be settled by detailed comparison with more elaborate data analysis procedures of the type described below in Sec.III.E.

One of the most striking images in sunspot polarimetry is the pinwheel appearance of Marshall Space Flight Center maps of the linearly polarized intensity (Hagyard et al. 1977; West and Hagyard 1983). The sense of the spiral correlates with sunspot polarity thus: clockwise/negative; counterclockwise/positive (Landi Degl'Innocenti 1979). Neglecting MO, Hagyard et al. interpreted this phenomenon as signifying a spiralling magnetic field, but later it was shown to be due to MO (Landi Degl'Innocenti 1979; West and Hagyard 1983; Ye Shi-Hui and Jin Jie-Hai 1986).

Figure 13 compares observations with model computations for a cylindrically symmetric sunspot with a purely transverse magnetic field distributed radially from spot center. Let $\bar{I}$, $\bar{Q}$ and $\bar{U}$ be model Stokes parameters for the radial field with zero azimuth. Then, by symmetry the fractional linear polarization for the radial field with azimuth χ is $(\bar{Q} \cos 2\chi - \bar{U} \sin 2\chi)/\bar{I}$. If MO are neglected, then $\bar{U} = 0$ and changes in the fractional linear polarization simply reflect the $\cos 2\chi$ modulation (middle row of Fig. 13). With MO

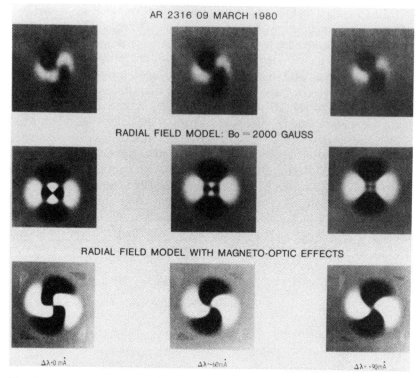

Fig. 13. Linear polarization intensity maps of a sunspot at three offset positions of 1/8 Å bandpass filter of the MSFC magnetograph: top row, observations; middle row, calculations with radial transverse field and no MO; bottom row, calculations with MO (figure from West and Hagyard 1983, Fig. 1).

included, $\tilde{U} \neq 0$ and good agreement with the observed pinwheel pattern is achieved (bottom row of Fig. 13). Thus the classical radial field configuration seems to be favored, at least for isolated circular sunspots.

Magnetographs will continue to provide important data on the configuration of vector magnetic fields in active regions, but further careful attention needs to be given to the problem of calibration.

E. Stokes Profile Observations

High spectral resolution profiles of the four Stokes parameters measured with a Stokes polarimeter (see Baur 1981; Harvey 1985b) provide the best opportunity of recovering the magnetic structure in the region of line formation. There are two approaches to the analysis of Stokes profiles:

(a) *Spectral synthesis* in which a model of the thermodynamics and velocity and magnetic fields of the active region is adjusted until the computed and observed polarization spectra agree within acceptable limits (see, e.g.,

Wittmann 1974,1977). This process is difficult to automate, and would be inappropriate for the routine analysis of large data sets.

(b) *Spectral inversion* in which one tries to extract only a small set of parameters that typify the physical conditions in an active region. Of the various methods that have been devised for this purpose (see review by Landi Degl'Innocenti 1985a), the so-called Unno-fit technique initiated by Auer et al. (1977b) (hereafter denoted by AHH) appears to be the most satisfactory. This technique involves nonlinear least-squares fitting of the Stokes profiles to an analytic solution of the Unno-Rachkovsky transfer equations (see Sec.III.B). Over the last few years considerable effort has been expended to assess the accuracy and the applicability of this method for routine data analysis. The Unno-fit method and its extensions provide a benchmark against which other Stokes profile analysis methods may be judged. For example, Ronan et al. (1987) have proposed a method using integrals of the Stokes parameters over wavelength. They compare this method with the Unno-fit method and show that the latter is superior.

AHH's formulation, which was based on the analytic solutions in Eqs. (13), with MO terms omitted, was used by Gurman and House (1981) to map the vector magnetic field in a sunspot observed in the Zeeman triplet Fe I $\lambda6302.5$ Å with the High Altitude Observatory Stokes I polarimeter (Baur et al. 1980). However, Landolfi and Landi Degl'Innocenti (1982) showed that it was essential to include MO in the fitting procedure especially to ensure a reliable estimation of the magnetic field azimuth. This and other improvements were incorporated in an up-dated version of AHH's computer code by Lites and Skumanich (1985,1989) and Skumanich and Lites (1987a).

The Lites-Skumanich inversion procedure involves up to 8 least-squares parameters associated with the magnetic region: the magnetic field parameters $|\mathbf{B}|$, γ, χ; $\lambda_c = \lambda_0(1 - V_{los}/c)$, the wavelength of the line center from which one can deduce an estimate of the line-of-sight velocity V_{los}; the line strength η_0 and broadening parameters Γ and $\Delta\lambda_D$; the source function slope $\tilde{B}_1$. In addition, they include a fraction α of unpolarized quiet Sun "stray light" (due to incomplete spatial resolution and/or scattered light in the telescope optics) with intensity I^{QS} and wavelength offset λ^{QS}. Instead of the intensity, the line depth,

$$I' = (I_c - I) + \alpha(I_c^{QS} - I^{QS}) \tag{18}$$

is fitted, where I is the true intensity of the magnetic region; I_c and I_c^{QS} are the continuum intensities for the magnetic region and the stray light component. Since $I_c = B_0 + \mu\tilde{B}_1$, the parameter B_0 is eliminated from the fit and one has from Eq. (13)

$$I' = \mu\tilde{B}_1[(1 + \eta_I)((1 + \eta_I)^2 + \rho_Q^2 \\ + \rho_U^2 + \rho_V^2)]/\Delta + \alpha(I_c^{QS} - I^{QS}). \tag{19}$$

A function is fitted to the weighted sum of squares of the residuals,

$$\sum_{\lambda} [\omega_{I'}(I'_{obs} - I')^2 + \omega_Q(Q_{obs} - Q)^2$$

$$+ \omega_U(U_{obs} - U)^2 + \omega_V(V_{obs} - V)^2] \quad (20)$$

where the sum is over the observed wavelengths in the line. An $IQUV$-fit involves all four Stokes parameters. If the intensity profile is strongly contaminated by scattered light, it can be excluded by setting $\omega_{I'} = 0$. This is called a QUV-fit.

It is evident that as presently formulated this method sacrifices the information on magnetic and velocity field gradients contained in the profiles. Even so, tests on synthetic data generated by solving the transfer equations with realistic thermodynamics and field gradients indicate that the mean field parameters recovered by the method are physically reasonable (Auer et al. 1977b; Landolfi et al. 1984; Skumanich et al. 1985). When the model magnetic field vector is constant, the $|\mathbf{B}|$ and γ are recovered to within a few percent, and χ to within 1 to $2°$ except when the field is nearly longitudinal (then large errors $\geq 10°$ in χ may occur). Landolfi (1987) describes a diagnostic method for detecting field gradients. This method, which incorporates the concept of Stokes profile response functions in the least-squares fitting procedure, has yet to be applied to real data.

Figure 14 shows the significant improvement in the fit achieved by the Lites-Skumanich method (here using a QUV-fit). The data are the Stokes parameters in the Fe I $\lambda6173$ Å line observed with the High Altitude Observatory Stokes II Polarimeter (Baur et al. 1981). The wavelength scale is measured relative to the Doppler shifted line center λ_c. Note in particular the reversal in Stokes V near line center. This cannot be fitted by the AHH method since it is due to MO. Moreover the introduction of MO greatly affects the inferred line strength: $\eta_0 = 3.0$ and 45 for AHH and Lites-Skumanich, respectively. The very broad AHH I profile is due to the fact that AHH set the damping $a = 0$ so that magnetic broadening is mistaken for thermal broadening, giving an artificially large Doppler width $\Delta\lambda_D = 76$ mÅ compared with the Lites-Skumanich value, $\Delta\lambda_D = 20$ mÅ which is compensated for by a damping value $a = 0.21$. As a result, AHH seriously underestimates $|\mathbf{B}|$. The magnetic field parameters ($|\mathbf{B}|,\gamma,\chi$) obtained are: (1798 G, $143°.7$, $7°.04$) for AHH; (2253 G, $150°.7$, $-2°.91$) for Lites-Skumanich.

In the first major application of their method, Lites and Skumanich (1989) obtained vector field maps of 4 large sunspots observed in the Fe I $\lambda6302.5$ Å line with the High Altitude Observatory Stokes II Polarimeter. They found stray light correction to be an essential element of the data analysis, especially outside the umbra, i.e., the QUV-fitting strategy is unsatisfactory because I' contains thermodynamic parameter information that is critical to the recovery of the vector magnetic field.

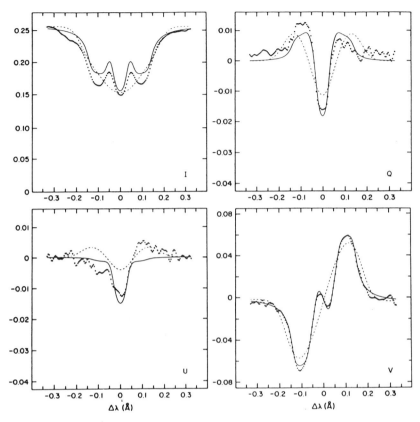

Fig. 14. Fe I λ6173 Å Stokes profile data (dots) fitted using the AHH method (dashed line) and the Lites-Skumanich method (solid line). Inclusion of MO enables one to fit the V reversal near line center (figure from Skumanich and Lites 1987a, Fig. 1).

An important conclusion is that in the penumbra the magnetic field is filamentary, the field strength at the outer edge being about 800 G with an inclination to the vertical of about 70°, in contradiction to the horizontal field predicted by the return-flux model (Flå et al. 1982). Note that, using Makita's (1979) method, which fits profiles outside one Doppler width of line center and neglects MO, Kawakami (1983) derived a similar penumbral field inclination of about 75°. In symmetric spots, the radial variation of the poloidal field strength and inclination is well represented by the potential field of a buried dipole, and some measurements show a significant azimuthal twist that suggests the presence of electric currents aligned with the poloidal field.

Lites et al. (1987,1988) combined the inversion and synthesis techniques to analyze High Altitude Observatory Stokes I profiles of the forbidden intercombination line Mg I λ4571 Å and the Mg I b lines. The intensity profiles

of the Mg I b lines, that are formed in the low chromosphere above the
temperature minimum, have deep narrow Doppler cores and extended damping wings. This shape is due to the nonlinear variation of the line source
function, the deep core being a symptom of departures from LTE. To account
for this, the inversion code was generalized to include the exponential terms
in the source function discussed in Sec.III.B. The Mg I b lines are far more
difficult to analyze than weaker photospheric lines because they are formed
over a large range of optical depths and there is substantial variation of thermodynamic parameters over that range. To invert the Mg I b lines, the fit was
restricted to the line core, and two members of the multiplet (λ5172.7 Å and
λ5183.6 Å) with different Zeeman splitting patterns were fitted simultaneously. The use of multiplets to help constrain the inferred field was pioneered by Stenflo (1973) for photospheric lines. However, this is the first
application to full Stokes profiles as well as to chromospheric lines. Figure
15 shows typical fits obtained to umbral and penumbral profiles of the two
lines.

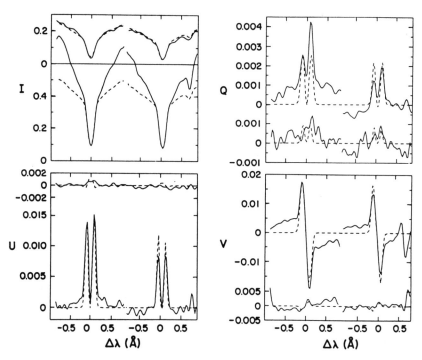

Fig. 15. Simultaneous least-squares fits (dashed line) to Stokes profile data (solid line) in the Mg
I b $\lambda\lambda$5172.7 and 5183.6 Å lines (left and right profiles, respectively). The top profiles are
from a sunspot umbra, and the bottom profiles are from the penumbra (figure from Lites et al.
1988, Fig. 11).

The Mg I b line $\lambda5172.7$ Å has two moderate-strength photospheric lines in its wings: Fc I $\lambda5171.61$ Å and Ti I $\lambda5173.75$ Å. Murphy (1989) has fitted the Stokes profiles of the three lines simultaneously in order to study vertical gradients in the vector magnetic field.

Despite the success in fitting the Mg I b lines, chromospheric line inversion needs more extensive testing on synthetic data before it becomes operational for large data sets. For example, the accuracy of the method in the presence of noise, spatial gradients in the field of view, line-of-sight velocity and magnetic field gradients, and different atmospheric models must be assessed. Model dependence is a critical issue, especially in the inversion of chromospheric lines. For this reason, alternative approaches to chromospheric vector field measurement also should be pursued. Of particular interest is the method of Deming et al. (1988b) who use the infrared emission line Mg I $\lambda12.32$ μm to study sunspot fields. Zeeman splitting is proportional to λ^2, so the Zeeman components are well separated. Also the line is almost optically thin so radiative-transfer effects such as MO are less important.

Several new polarimeters are now under construction (Harvey 1985b). These will provide data with high spatial and spectral resolution. It is expected that application of algorithms of the type described in this section will reveal far more precise details of the vector magnetic fields in active regions than have hitherto been available.

IV. OBSERVATIONS OF ACTIVE REGIONS WITH VECTOR MAGNETOGRAPHS

Vector magnetographs can be classified by the following criterion:

1. *Method of the polarization modulation.* The analyzer of the polarization is composed of wave plates and a linear polarizer. The modulation of the polarization is performed either by changing the retardation of the wave plates (say by an electro-optic effect) or by physically rotating the plates.
2. *Treatment of the instrumental polarization.* Instrumental polarization is mainly caused by reflections at mirrors or retardence in windows. The effect is minimized by putting the analyzer before as many of the optical elements as possible and by avoiding oblique reflections. Otherwise, it is necessary to use polarization compensators or to correct for instrumental polarization during data analysis by calculation (in a matrix inversion).
3. *Instrument to isolate the spectral line.* The spectral line to be measured is isolated by a filter or spectrograph. Generally, filters are smaller and lighter than spectrographs and may be mounted on-axis, in accord with criterion 2.

Vector magnetographs now in operation are listed in Table III.

TABLE III
Vector Magnetographs in Operation

Observatory	(1) Modulation	(2) Instrum. pol.	(3) Spectral Isolation	Reference
Sayan Obs. SibIZMIR, USSR	retardation	matrix inversion	spectrograph	Grigoryev et al. 1985.
Central Inst. Astrophys. Potsdam, DDR	"	"	"	Bachmann et al. 1975.
MSFC, NASA Huntsville, USA	"	on-axis	filter	Hagyard et al. 1982.
Huairou Station Beijing Astron. Obs., China	"	"	"	"
Univ. Hawaii Maui, USA	rotation	"	spectrograph	Mickey 1985.
Okayama Astrophys. Obs., Okayama Japan	"	compensator	"	Makita et al. 1985.

A. Observational Results Obtained with the Vector Magnetograph

The unique feature of a vector magnetograph is the measurement of the transverse component of magnetic fields as well as the longitudinal one. Transformation of the observed degree of linear polarization to the transverse component of the magnetic field involves some uncertainties just as in the case of the longitudinal component (see, e.g., West and Hagyard 1983; Sakurai 1987). The direction of the linear polarization as an indicator of the azimuth of magnetic fields may require that magneto-optical effects be taken into account.

Distribution of transverse magnetic fields and the magneto-optical effect. The observed distribution of the transverse magnetic field looks plausible near sunspots. However, the magneto-optical effect may not be negligible and the observed azimuth will deviate from the direction of the true transverse field (Kawakami 1983; West and Hagyard 1983; Makita 1986). This deviation varies along the line profile and makes it difficult to detect depth variation of the transverse magnetic field, i.e., twisted magnetic structures (Beckers 1969).

Magnetic Shear and Activity. In the active parts of active regions, magnetic shear (which is the azimuthal deviation of the transverse magnetic field from the expected azimuth given by the potential field assumption) is larger (see e.g., Gary et al. 1987). The correlation of the shear with the flare occurrence has been examined. Moore et al. (1987) recently studied this in several active regions. However, the results apparently are not yet conclusive.

Magnetic Field Structure and Activity. Kawakami et al. (1989) examined the correlation of the magnetic field azimuth with the elongated Hα fine structures in three active regions. Coincidence of the two azimuths decreases from center to limb. This decrease is steeper in the weaker active region. By assuming horizontal Hα structures, they conclude that there is a dominance of the vertical magnetic fields in the weaker active region. In other words, the stronger active region has a flatter magnetic field structure than the weaker active region.

Electric Current. The magnetic shear, i.e., the deviation from the potential field (Gary et al. 1987), implies the existence of electric currents. Some investigations have been made into the relation of shear with activity (see e.g., Lin and Gaizauskas 1987); however, as in the magnetic shear problem no definite conclusion has been reached yet. A possible characteristic of the electric current calculated from the transverse magnetic field is that its horizontal scales are smaller than those of the magnetic field. In the Okayama magnetograms, opposite electric currents are often seen in a uniform magnetic polarity region (see, e.g., Sakurai 1987). This was also found in the other observations (Hofmann and Staude 1987). Although the calculation accuracy of the electric current (Hagyard 1988) should be investigated carefully, such fragmental electric current distribution can theoretically be expected to arise from a shearing motion in the atmosphere (Nakagawa 1985, personal communication).

V. THE EXTRAPOLATION OF PHOTOSPHERIC MAGNETIC FIELDS INTO THE CORONA

A. Introduction

Present magnetographs are only able to measure the magnetic fields at the photospheric level. Fields in the upper atmosphere are therefore usually inferred from computational extrapolations, based on the photospheric measurements. The extrapolations are carried out by solving a basic equation for the magnetic field.

The equation for the static equilibrium of a magnetic field is written as

$$-\nabla p + \frac{1}{4\pi} \operatorname{curl} \mathbf{B} \times \mathbf{B} + \rho g = 0 \qquad (21)$$

where p is the pressure, $\mathbf{B}$ is the magnetic field, ρ is the density and g is the gravitational acceleration. Various assumptions are introduced further to simplify this equation. Generally the plasma in the corona is tenuous and the pressure and gravitational forces are smaller than the magnetic force. Therefore we obtain

$$\text{curl } \mathbf{B} \times \mathbf{B} = 0. \tag{22}$$

This equation describes the so-called force-free magnetic fields.

B. Current-Free Fields

The simplest case among force-free fields is that the electric current itself vanishes in the corona, namely,

$$\text{curl } \mathbf{B} = 0. \tag{23}$$

The magnetic field $\mathbf{B}$ is then derived from a potential ϕ via

$$\mathbf{B} = -\nabla\phi. \tag{24}$$

The divergence-free condition for $\mathbf{B}$ ($\nabla\cdot\mathbf{B} = 0$) requires that the potential ϕ should satisfy the Laplace equation,

$$\Delta\phi = 0. \tag{25}$$

This equation together with a proper boundary condition leads to the magnetic field solution $\mathbf{B}$, which is called potential or current-free.

First, we consider the case in which the region to be studied is close to the disk center and its extent is small so that one can neglect the curvature of the solar surface in that area. In such a case, we adopt a Cartesian coordinate system where the xy-plane ($z = 0$) represents the photosphere and the z-axis points upward. The observed longitudinal magnetic fields correspond to the normal component (B_n) of the field on the boundary. Therefore the normal derivative of the potential on the boundary is prescribed as

$$-\partial\phi/\partial n \ (z = 0) = B_n \tag{26}$$

and the problem reduces to the Neumann boundary-value problem of the Laplace equation.

Two kinds of methods are currently used to solve this boundary-value problem. One is the Green's function method (Schmidt 1964), and the other is the Fourier expansion method (Nakagawa and Raadu 1972; Teuber et al. 1977). Comparison between the two methods can be found in Levine (1975) and in Seehafer (1982).

When the region to be studied is away from the disk center, the observed longitudinal component of the field B_ℓ is related to the potential ϕ by

$$- \partial \phi / \partial l \ (z = 0) = B_l. \tag{27}$$

This boundary condition is different from the standard Neumann problem. Methods of solution are available both in the Green's function approach (Semel 1967) and in the Fourier approach (Wellck and Nakagawa 1973).

When the line-of-sight component of the field is specified on the whole visible surface of the Sun, Sakurai (1982) developed a Green's function approach applicable to such cases. In his approach, the distribution of the field in the invisible hemisphere of the Sun is assumed to be either symmetric or antisymmetric reflection of the field in the visible hemisphere. If the line-of-sight fields are measured over a whole rotation of the Sun, a global modeling of the magnetic field according to the method of Altschuler and Newkirk (1969) is possible. In their model, the effect of the solar wind is taken into account by requiring that the field becomes radial at an outer radius $r = r_s$. This outer surface $(r = r_s)$ is called the source surface, and the value of the potential ϕ is specified there (Dirichlet boundary condition).

C. Constant-α Force-Free Fields

The force-free field is described by Eq. (22) or equivalently by

$$\mathrm{curl}\ \mathbf{B} = \alpha\, \mathbf{B} \tag{28}$$

where α should satisfy the relation

$$\mathbf{B} \cdot \nabla \alpha = 0 \tag{29}$$

due to the condition $\nabla \cdot \mathbf{B} = 0$. The scalar α is constant along a given magnetic field line but may take different values on different field lines. The particular case in which α is constant everywhere is the simplest and has been most widely studied. Such magnetic fields are called constant-α force-free fields.

The constant-α force-free field solutions can be obtained either by the Green's function method (Chiu and Hilton 1977) or by the Fourier expansion method (Nakagawa and Raadu 1972; Seehafer 1978).

D. Nonlinear Force-Free Fields

If α is not constant but varies from one field line to another, the equation to be solved is nonlinear. Several different approaches have been developed to obtain the numerical solution to this nonlinear force-free field problem.

The magnetic field $\mathbf{B}$ can be expressed as

$$\mathbf{B} = \nabla u \times \nabla v \tag{30}$$

where u and v are called the Euler potentials for **B**. This expression satisfies the divergence-free condition for **B**. Equation (30) implies that u and v are constant along the magnetic lines of force. Sakurai (1979) showed that the force-free field can be determined by specifying the values of u and v on the boundary.

Yang et al. (1986) noted that the force-free field is the final state of the evolving magnetic field which follows the equation

$$\frac{\partial u}{\partial t} = -\nu^{-1} F_B \cdot \nabla u, \qquad \frac{\partial v}{\partial t} = -\nu^{-1} \mathbf{F}_B \cdot \nabla v \qquad (31)$$

where $\mathbf{F}_B$ stands for the Lorentz force and ν is a fictitious coefficient of friction. Starting from an initial state, the system will relax under the action of frictional force toward the static equilibrium, which corresponds to $\mathbf{F}_B = 0$ viz., the force-free field. Craig and Sneyd (1986) developed a similar frictional method by adopting a different representation for the magnetic field.

If the magnetic vector **B** is measured on the boundary (i.e., on the photosphere), the distribution of α is obtained observationally as

$$\alpha = \left[\frac{\partial B_y}{\partial x} - \frac{\partial B_x}{\partial y} \right] / B_n. \qquad (32)$$

Sakurai (1981) developed a scheme which utilizes the obtained distribution of α. The required boundary conditions are the values of B_n on the whole boundary and the values of α in the positive (or negative) polarity portion on the boundary. Since α is constant along the lines of force, which are not known until the solution is obtained, the specification of α on the whole boundary plane generally leads to a discrepancy.

Pridmore-Brown (1981) developed a scheme based on a least-square minimization of residual Lorentz force. The boundary conditions to be supplied are the values of B_n and B_y/B_x.

Wu et al. (1985) tried to solve the force-free field equation as the Cauchy problem, by integrating the equation from $z = 0$ upwards. The force-free equation is "ill-posed" when solved as the Cauchy problem, and the solution will lose accuracy quickly as it is integrated upwards.

E. Comparison with Observations

Current-free extrapolations generally show that coronal structures follow the current-free field lines in a broad sense (Poletto et al. 1975; Sakurai and Uchida 1977). Berton and Sakurai (1985) reconstructed the shape of coronal loops stereoscopically by utilizing two X-ray pictures taken on different days, and compared it with the current-free field lines (Fig. 16).

The coronal field strength inferred from radio observations, by assuming

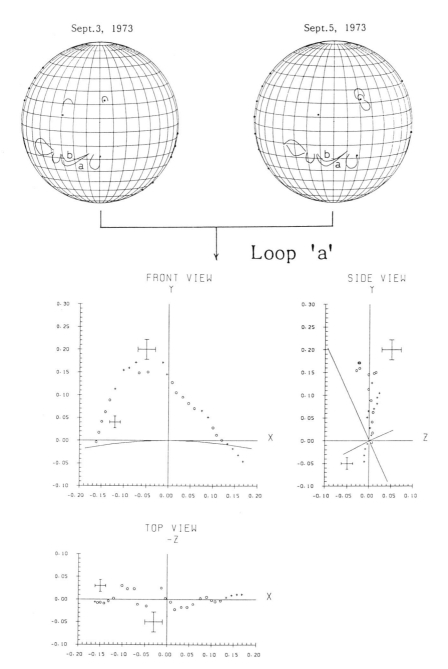

Fig. 16. Three-dimensional shape of a coronal loop, derived by using two X-ray pictures taken two days apart (figure from Berton and Sakurai 1985).

a   b

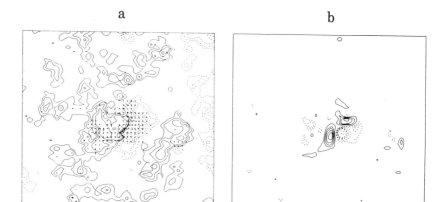

magnetic field vector   electric current

Fig. 17. (a) Observed magnetic field vector. Solid and dotted contours show positive and negative
longitudinal fields, respectively. Arrows indicate the transverse field vector. (b) Distribution of
electric currents derived from (a). Field lines have the same footpoints as Fig. 16 (figure from
Sakurai et al. 1985).

a   b

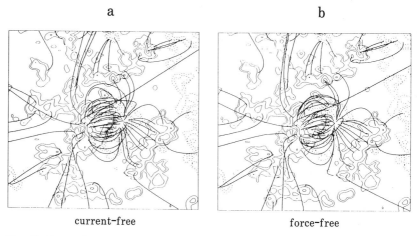

current-free   force-free

Fig. 18. Current-free field lines (a) and force-free field lines (b). Current-free field lines are
computed from the observed longitudinal magnetic field, while force-free field lines are computed by using the distribution of electric currents shown in Fig. 17b. Field lines have the same
footpoints as in Fig. 16 (figure from Sakurai et al. 1985).

the gyro resonance as the opacity source, is found to be larger than the calculated current-free field strength (Alissandrakis et al. 1980; Pallavicini et al. 1981). This may indicate a large deviation of the coronal magnetic field from the current-free state. However, it is also likely that the possible existence of filamentary structures in the radio sources could lead to an over-estimated field strength.

Flare activity is now believed to come from the magnetic energy stored in the corona. Therefore it is expected that the magnetic field would deviate from the current-free field as the energy is stored in the coronal magnetic field. The magnetic vector measurements described in Sec. IV indicated that flare-productive regions are characterized by a large amount of shear in the magnetic field. Attempts have been made to reproduce the observed field structures within the framework of the constant-α modeling. The fitting procedure, however, only indicated the sign of α at most, and a globally successful fitting has seldom been achieved (Levine 1976; Gary et al. 1987). Nonlinear force-free field calculations reviewed here are not fully developed yet and only a limited number of cases of application have been reported (see Figs. 17 and 18; Sakurai et al. 1985). The development of reliable computational schemes that can utilize the vector magnetograms is urgently needed. Quantitative studies of the flare energy buildup can only be carried out with the development of such computational tools.

FILIGREE AND FLUX TUBE PHYSICS

H. C. SPRUIT
Max-Planck Institut für Physik und Astrophysik

M. SCHÜSSLER
Kiepenheuer—Institut für Sonnenphysik

and

S. K. SOLANKI
University of St. Andrews

This review covers the properties of the small-scale (outside sunspots) magnetic field from observational and theoretical points of view. Special emphasis is put on the physics of small isolated magnetic concentrations (flux tubes). Topics discussed include the basic observational properties, the origin and disappearance of the small-scale field, the properties of individual magnetic elements, their influence on the solar irradiance and their interaction with solar oscillations.

I. BASIC OBSERVATIONS

Even relatively low-resolution magnetograms show the solar magnetic field to be highly structured. Increasing spatial resolution has revealed an ever smaller scale of magnetic structure, with discrete patches of magnetic field separated by stretches of nonmagnetic atmosphere. So far the best direct upper limit on the sizes of the smallest magnetic-flux concentrations is 200 to 350 km set by Ramsey et al. (1977). Smaller scales cannot be ruled out, but being below the best presently available spatial resolution, they cannot be studied by direct means. We shall call the small concentrations of magnetic flux simply "magnetic elements." Fortunately, a range of mostly indirect techniques for obtaining information on the structure and evolution of magnetic elements is available and has been applied to derive their properties. However, only very rudimentary information can be obtained without the help of some kind of model, and to gain proper insight into the underlying physics, both analytical and numerical studies are required. The present theoretical ideas of the formation, evolution, structure and dynamics of magnetic elements and their relation to the observations are presented in Secs. II, III and IV.

A. Field Strengths

One of the most simple and widely used models in *empirical* studies is the two-component model. It consists of a component describing the magnetic elements covering a fraction α of the surface in the resolution element, with magnetic field strength B_1, temperature structure T_1, etc. and a component describing their surroundings covering $(1 - \alpha)$ of the surface, with field strength B_2, temperature T_2, etc. Generally, the second component is taken to be identical to the quiet Sun, i.e., $B_2 = 0$, $T_2 = T_{\text{quiet}}$, etc., although evidence is mounting that the presence of magnetic fields also affects the nonmagnetic component of the atmosphere in its vicinity (Tarbell et al. 1988; Muller et al. 1989; see also Spruit et al 1990).

The basic observation underlying the two-component model structure is the measurement of the intrinsic field strength of magnetic elements. If the average magnetic field outside of sunspots and pores is determined from the amplitudes of Stokes V profiles using a magnetograph with a spatial resolution $< 1''$, then values ranging up to a few 100 Gauss are obtained. Stokes V is the difference between the spectra in left and right circularly polarized light (see the chapter by Semel et al.). Unlike the unpolarized spectrum (Stokes I), it is formed only within the magnetic elements to first order. The intrinsic field strength within the magnetic elements can be derived from more detailed measurements of V-profiles. A variety of techniques give values between 1000 G and 2000 G (see, e.g., Stenflo 1973; Harvey 1977; Tarbell and Title 1977; Wiehr 1978; Frazier and Stenflo 1978; Koutchmy and Stellmacher 1978; Robinson et al. 1980; Solanki and Stenflo 1984; Stenflo et al. 1987*b*; Solanki et al. 1987), so that in the lower and middle photosphere, where the lines used to determine the field are formed, the magnetic field strength lies between these values. The two different observations are reconciled by introducing the filling factor of the field, $\alpha < 1$, so that the average field is αB.

Despite the increasing sophistication of the analysis techniques, there are at least two basic quantities which still cannot be determined with sufficient accuracy. It is not possible to determine the exact filling factor with the data and techniques available so far, as has been pointed out by Grossmann-Doerth et al. (1987) and Schüssler and Solanki (1988). Also, it is still uncertain which fraction of the total magnetic flux is in strong-field (1 to 2 kG) form. Stokes V is sensitive only to the line of sight component of the field and changes sign when the polarity of the field changes. Therefore, magnetograms only show the *net* flux in the resolution element, i.e., the excess flux of the dominant polarity, after the flux of the other polarity has been subtracted. More than 90% of this net flux is in strong-field form (see, e.g., Howard and Stenflo 1972). If a part of the field is in a "tangled" state, i.e., both polarities intermingled on a spatial scale below the resolution of the observations, traditional methods of determining the flux and field strength, based on Stokes V fail to provide any information on it. Attempts have been made using either high spatial resolution measurements (Tarbell et al. 1979), or methods based on Stokes I, Q and U (unpolarized and linearly polarized light, sensitive to the field in a different manner from Stokes V; cf. Stenflo 1988) to set limits on the amount of flux in such a hidden or tangled form. It follows from these measurements that the polarities of any "hidden" field must be mixed on a scale smaller than 0."5 the spatially averaged strength must lie between 10 G and 100 G (the lower limit is relatively uncertain) and the field nearly isotropic. The "inner network fields" observed by Livingston and Harvey (1975), Sivaraman and Livingston (1982), Martin et al. (1985), Livi et al. (1985) and called intranetwork fields by the latter authors are perhaps such tangled fields. Their visibility does not seem to depend on disk position. This is in contrast to the network fields, which become invisible in circular polarization near the limb—an indication that they consist of mainly vertical fields. The more uniform visibility of the intranetwork fields indicates that they do not have such a preferred direction. This implies that they are sufficiently weak that the magnetic buoyancy forces are not able to maintain a preferred vertical orientation. These fields must therefore be significantly weaker than the equipartition strength with the granulation flow; probably they are less than a few 100 Gauss.

B. Distribution and Correlation with Brightness Structures

In contrast to the much rarer sunspots, magnetic elements are found at all solar latitudes. They are concentrated in a network near the boundaries of supergranular cells (see, e.g., Simon and Leighton 1964; Frazier 1970; Simon et al. 1988) and in active regions. On a scale of a few arcsec, magnetic elements are well correlated with strong Ca II H and K core emission, i.e., with Ca plages (Skumanich et al. 1975; LaBonte 1986), strengthened emission in ultraviolet lines (see, e.g., Zirin 1985) and a brightening in white light near the limb, i.e., faculae (see, e.g., Muller 1975; Libbrecht and Kuhn 1985; Wang and Zirin 1987). At higher spatial resolution they are correlated with facular points seen in white-light images, or in the wings of strong lines (see, e.g., Mehltretter 1974; Title et al. 1989), with dark intergranular lanes and the associated granular downflow (see, e.g., Title et al. 1987c).

Evidence that magnetic elements are bright in the continuum near disk center (at least 1.4 times as bright as the quiet Sun) also comes from spectral observations (Schüssler and Solanki 1988). This agrees well with very high-resolution filter measurements of the brightness contrast of facular points (Koutchmy 1977*a*; Muller and Keil 1983; Von der Lühe 1989), suggesting at least a partial equivalence of the two.

C. Sizes and Shapes

If we assume facular points to be the white-light signature of magnetic elements a few further properties can be elicited. A clear preference to be located in dark intergranular lanes is found (Dunn and Zirker 1973; Mehltretter 1974; Muller 1983*b*; Title et al. 1987*b*). Facular points are usually observed to be smaller than 250 km (Mehltretter 1974; Spruit and Zwaan 1981; Muller and Keil 1983) and many may lie undetected under the best spatial resolution of 150 to 200 km. In an active region, Von der Lühe (1987) finds a structure with a reconstructed width < 100 km (in at least one direction) from speckle interferometry, while Spruit and Zwaan (1981) also infer the presence of such small elements from estimates of the effects of seeing. These size determinations of facular points augment other measurements, including the direct upper limit on the smallest magnetic elements set by Ramsey et al. (1977) of approximately 200 km. A more indirect limit can be set by converting the smallest amount of flux seen with the instrument into a cross section by assuming an intrinsic field strength of the order of 1 kG. With this technique, Wiehr (1979) finds diameters between 390 km and 550 km, while Wang et al. (1985) find diameters between 35 km and 130 km. Until now, the lowest measured flux is restricted by the sensitivity of the instrument, i.e., the derived diameters are upper limits. On the other hand, the method fails if an intrinsically weak field component is present. The profiles of very strongly split lines, as found in the infrared, also contain information on the size of magnetic elements. An analysis based on them suggests that the widths of the magnetic elements in an active region lie between 60 km and 300 km (Zayer et al. 1989). This analysis also suggests that within such a region the diameters of most magnetic elements are relatively similar. All these indirect size determinations must be treated with caution, although such techniques do show promise for the future.

What shape do magnetic elements have? Since they are either near or beyond the limit of spatial resolution, little can be said about their horizontal outline. Facular points appear round (implying flux tubes?), but the smallest structures reconstructed from speckle interferometry are quite elongated (implying flux slabs? [Von der Lühe 1987*a*]). However, it is possible to obtain an idea of how strongly the cross-sectional area covered by the field varies with height. The variation in size with height can be derived directly by comparing magnetograms obtained in lines formed at two different heights. Since magnetic elements are too small to be individually resolved, this technique will in general only show the expansion of whole groups of magnetic elements. Investigations using this technique show that in the chromosphere the field covers a much larger area than in the photosphere, leading to the suggestion that much of the chromosphere is filled with fields forming the so-called magnetic canopy (Giovanelli 1980; Giovanelli and Jones 1982; Jones and Giovanelli 1983). Another approach is to determine the vertical gradient of the field strength and derive the change in cross section through the requirement of flux conservation. The field strength is found to decrease rapidly with height, being approximately 1500 to 1600 G in the lower photosphere (as derived from the strongly split infrared line Fe I λ15648.5; Stenflo et al. 1987b; Zayer et al. 1989), 1100 to 1200 G in the central photosphere (as derived from the line ratio of Fe I λ5250.2 and λ5247.1; see, e.g., Solanki et al. 1987) and 250 to 500 G near the temperature minimum (where the 12 μm lines are formed; Brault and Noyes 1983; Deming et al. 1988b). The detailed analysis of Zayer et al. (1989) actually suggests that at least in the lower and middle photosphere, the field-strength decrease with height is compatible with the thin-tube approximation (see, e.g., Parker 1979a). That is, the field strength is determined by pressure balance with the surroundings, and magnetic-tension forces are unimportant. This approximation is valid for tubes with diameters small compared with the pressure scale height of the atmosphere. Due to the decreasing external gas pressure, the cross-sectional area of magnetic elements increases with height approximately exponentially. The inferred field strength at the $\tau_{5000} = 1$ level in the magnetic element is approximately 2000 G.

II. ORIGIN AND DISAPPEARANCE

A. The Field-Eruption Process

The currently favored picture for the origin of the observed magnetic flux in the solar photosphere is the destabilization of a dynamo-generated horizontal magnetic-flux system located near the bottom of the solar convection zone which leads to buoyant rise of a bundle of magnetic flux tubes towards the surface (Parker 1955a; Spruit and van Ballegooijen 1982; Moreno-Insertis 1986; Chou and Fisher 1989).

One should expect that the magnetic flux before eruption is not in the same state (entropy, field strength) as the flux tubes that are subsequently observed in an active region. A transformation would then have to take place during the eruption process. The observational evidence for this is not clear. For example, there is no clear evidence that the field strength in very young erupted flux would be smaller (see Brants 1985). This means either that the transformation takes place very quickly, or that the field arrives at the surface already in strong form. The proposed mechanisms for the formation of a flux tube in concentrated form (Sec. II. B) operate on short time scales, of the order of minutes, so that it may not be surprising that only strong fields are seen in an emerging flux region. Observations of emerging flux regions, however, indicate that not only is the local field strength high, but also the *average* over a significant area. Zwaan (1978) argues that the amount of flux erupting in a small area can be so large that it becomes difficult to accommodate average pre-eruption field strengths less than a few 100 gauss. In a horizontal flux tube floating up through the convection zone, the mass per unit of magnetic flux ρ/B is conserved, so that a tube starting with 10^4 G at the base of the convection zone ($\rho = 0.2$ g cm^{-3}) would arrive at the surface ($\rho \propto 10^{-6}$ g cm^{-3}) with $B = 0.05$ G, much less than observed. More realistically, the tube does not float up horizontally but as a loop along which mass drains downward. The field near the top therefore does not decrease as much as in the case of a purely horizontal loop. Calculations of this process have been made by Moreno-Insertis (1986) and Chou and Fisher (1989). The results show that the field still arrives at the surface in a very weak form. The reason for this is that if the tube starts at the equipartition field strength of 10^4 G, it soon drops below the equipartition value due to the expansion. The time scale for draining of fluid along the tube then becomes long compared with the (convective) time scale on which the tube rises, so that the field strength decreases in much the same way as in a horizontal tube. Evidently, there is another process at work that maintains a higher field strength during the tube's rise. This may well be magnetic flux expulsion (Parker 1963; also see Proctor and Weiss 1982 for a review) by local convective flows that are not included in the calculations mentioned. This process probably contributes to maintaining intermittent magnetic structures in the convection zone with typical intrinsic field strengths determined by equipartition between the magnetic-energy density and the kinetic-energy density of the convective flows. In spite of this, calculations like those of Moreno-Insertis, which ignore convective concentration processes, reproduce some properties of active regions, such as their sizes and the time scales for their formation.

The actual process of eruption of magnetic flux through the surface, producing (part of) a new active region, is an interesting event for which new high-resolution observations, as well as theoretical calculations have recently become available. Observations of emerging flux regions always show the appearance of dipoles of magnetic flux. Both polarities are initially close together and drift apart with time at velocities ranging from 0.2 to 1.0 km s^{-1} (see, e.g., Chou and Wang 1987). A transient upflow has been observed prior to the appearance of the magnetic dipole (Bruzek 1967; Frazier 1972; Brants 1985; Tarbell et al. 1989) and is interpreted as the rise of a horizontal flux rope, which then forms a loop-like structure visible in Hα (see, e.g., Zwaan 1978). Later, a considerably longer downflow phase (≈ 1 hr), with peak velocities reaching 1 km s^{-1}, is observed (Tarbell et al. 1989). The observations cannot accurately determine whether these velocities are concentrated purely or even mainly within the magnetic features. It seems possible that the observed downdrafts represent the draining of mass out of the loop after it has crossed the surface (Frazier 1972; Brants 1985). Brants (1985) sees downflows mostly in Stokes I, suggesting that the downflows are outside the magnetic elements. He also presents indirect evidence for highly inclined fields associated with emerging flux.

On the theoretical side, we note that the field in a loop crossing the surface must transform from a state with high gas density on the field lines (high β, where β is the ratio of gas pressure to magnetic pressure) to a nearly potential (low β) field with very little mass on the field lines. The process therefore necessarily involves strong flows of mass along the rising field. The first calculations of this process were done by Shibata (1980), recent numerical simulations were done by Shibata et al. (1989). The mass at the top of the loop is pushed up through the photosphere at a substantial velocity, a few 100 m s^{-1} at the height where medium-strong line cores are formed, sufficient to explain the observed blue shifts.

B. Flux-Tube Formation

For the photospheric granulation, the equipartition field strength relevant for flux expulsion amounts to a few 100 gauss. To account for the observed kG fields, another process is held responsible, the "convective collapse" (Parker 1978b; Webb and Roberts 1978; Spruit 1979; Spruit and Zweibel 1979; Unno and Ando 1979). The following is a somewhat simplified sketch of the process. Starting with a weak initial field, the first step is concentration by convective flows (Galloway and Weiss 1981). In this stage, the field is too weak to influence the flow, and becomes concentrated into the downdrafts of the convective pattern. As the field strength increases, it starts interfering with the flow when it reaches equipartition with the kinetic energy density of the flow. In the absence of further effects, this would limit the field strength to the equipartition value, a few 100 gauss near the surface. By interfering with the flow, the convective transport of heat into a magnetic structure is reduced, so that it thermally insulates itself. Because radiative cooling at the surface is not reduced, the structure cools, from the surface downward. This reduces the pressure scale height in the structure, partially evacuating it and increasing the field strength.

In addition to these effects, a field of a few 100 gauss is still not strong enough to suppress convective instability completely. The critical field strength needed for stability to (adiabatic) motions was determined by Spruit and Zweibel (1979b). They studied the case of a thin magnetic flux tube under the assumption that $\beta = 8\pi P_i/B^2$ (with P_i tube interior gas pressure) is independent of depth, and found stability for $\beta \leq 1.8$. This corresponds to a minimal field strength of 1350 G at the solar surface. Spruit (1979) found that the collapsed state possesses a field strength in the range between 1280 and 1650 G at the surface, which is in excellent agreement with the observations. Some of this agreement may be fortuitous however, because it hinges upon the assumed constancy of β. Also, the field strength of the end state depends somewhat on the strength of the initial field (with $\beta < 1.8$); a weaker initial field gives rise to a stronger final field. Spruit also suggested that the end state may be one of overstable oscillations (see, e.g., Cowling 1976) due to radiative energy exchange between the flux tube and its superadiabatic surroundings. Nonlinear calculations have been carried out by Venkatakrishnan (1983,1985) and Hasan (1984,1985). Hasan's calculations, which include lateral radiative exchange, achieve a final state with a time-averaged field strength of approximately 1250 G at a level 50 km below $\tau = 1$ in the external medium (initial field strength $= 800$ G). A fixed field strength cannot be defined, since the flux tube oscillates overstably, with temperature and magnetic field oscillating along with the velocity. Such an oscillatory state is not particularly consistent with the present observational evidence. The calculations are still somewhat idealized, however, with respect to the treatment of radiative transport and of the conditions imposed at the lower boundary.

The formation of magnetic flux tubes has also been studied by three-dimensional simulations by Nordlund (1983,1986). These calculations follow the field evolution starting with the convective concentration phase, through the collapse process (which is not quite distinct from the first phase in these calculations). However, due to the limited horizontal resolution of the grid (corresponding to 190 km), he cannot follow the collapse to the fully concentrated state and the field strengths do not correspond to the observed ones when integrating over a realistic resolution element.

C. Relation to Sunspots

In its growing phase a spot is assembled from many smaller flux concentrations (see the chapters by Rabin et al. and Semel et al.). These concentrations can be as large as pores, but large amounts of flux are usually added to the spot in the form of very small flux tubes, whose individual motions are hard to follow. In white light, this gives the impression as if a spot is growing *in situ*. On magnetograms, the process can be followed in detail only with time sequences of very high spatial resolution, such as those obtained by Tarbell et al. (1989). The motion of the concentrations toward the spot is very systematic, and independent of the large-scale flows in the area. This implies that spots are not formed by advection of flux elements by a large-scale flow (see, e.g., Meyer et al. 1974). Instead the flux elements move by their own forces, in a way that is described well by the "rising tree" cartoon (Fig. 1; see also Zwaan 1978). Such independent behavior is understandable theoretically if the flux tubes in question have a field strength that is high enough to give them a significant buoyancy. This implies kG field strengths, which are indeed observed for the tubes in growing active regions. For calculations of the motion of small flux tubes near spots see Meyer et al. (1979): see also the discussion on field eruption in Sec. II.A.

From time to time it has been suggested that a sunspot not only originates as an ensemble of small elements but that below the surface it continues

Fig. 1. Interpretation of an emerging flux region in terms of a fragmented flux rope rising through the convection zone. Time increases from left to right.

to be a cluster of small tubes separated by field-free gaps that communicate directly with the surrounding convection zone (Severny 1965; Gokhale and Zwaan 1972). This idea has significant theoretical appeal (Parker 1979b; see also Spruit and Roberts 1983; Moreno-Insertis and Spruit 1989).

During the slow decay of spots, they are often surrounded by a small-scale field of mixed polarity that flows away from the spot (moving magnetic features; Harvey and Harvey 1973; Brickhouse and Labonte 1988). It is believed that this phenomenon is physically related to the decay of the spot; Harvey and Harvey for instance find that the spot's polarity dominates in the moving magnetic features, though only by a small amount. The motion of the magnetic features is consistent with the assumption that they are carried passively by the moat cell surrounding the spot (Brickhouse and Labonte 1988).

D. Disappearance and Destruction Processes

Due to their small size it is very difficult to determine observationally the lifetime of individual magnetic elements. Small magnetic elements are in constant motion because of their reaction to the local flows (granulation). Because of this constant buffeting, it is difficult to recognize the "theoretical flux tube." Rather, the appearance is like that of floating debris on the surface of a liquid in motion. Sivaraman and Livingston (1982) find all the elements of intranetwork flux in their field of view to survive their observing span of an hour. From video magnetograms Zirin (1985) derives a lifetime of 50 to 100 hr for network clumps (i.e., groups of magnetic elements). The relation of this time to the lifetime of individual magnetic elements is not clear, since the latter may live much longer than a network clump can hold on to its identity, or, on the other hand, the network clump may well outlive its individual constituting magnetic elements. Muller (1983) found from a time series of approximately 1 hr that individual facular points live only 18 min on average. However, it is dangerous to consider this as a measure of the life span of the (possibly) underlying magnetic structure. A fading may have a purely thermodynamic origin (e.g., overstable oscillation) and must not necessarily mean disappearance of the magnetic structure. In fact, if magnetic elements are stabilized by whirl flows (Schüssler 1984), their lifetimes are not necessarily limited by those of the granular flows. The stabilizing flow is partly due to the very existence of the flux concentration which creates a thermal circulation that feeds the whirl and influences the properties of the convection pattern in its environment (Title et al. 1987*b;* Muller et al. 1989).

An important, and probably the dominant, way in which flux is observed to disappear at the surface, both in the quiet Sun and in active regions is cancellation, described as magnetic features of opposite polarity approaching each other, merging and disappearing (see, Martin et al. 1985; Livi et al. 1985). These observations allow relatively straightforward interpretations like the submergence of a flux rope, or rather of a bundle of smaller flux ropes (Fig. 2b), or the rising of a rope through the surface (Fig. 2c). The interpretation of other observations is more puzzling. Wilson and Simon (1983), Simon and Wilson (1985) and Topka et al. (1986) find that the magnetic flux apparently changes in some small, relatively isolated unipolar magnetic features without seeming to cancel with fields of opposite polarity. One should keep in mind that it is not the magnetic flux per se which is observed, but Stokes *V,* which is influenced by many other parameters of the atmosphere inside the magnetic elements. Indeed the above authors discuss various possibilities to explain their observations. One of the mechanisms, namely changes in the temperature and brightness of magnetic elements, has been shown to be a strong source of spurious flux changes (Grossmann-Doerth et al. 1987). Also, Martin et al. (1985) and Livi et al. (1985) do not observe the decrease or increase in flux of one polarity only, they always see a cancellation. They suggest that the disappearance of a single polarity may be an artifact of insufficient sensitivity to the magnetic flux.

Cancellation events may well imply reconnection of field lines but it is important to recognize the difference between reconnection and cancellation

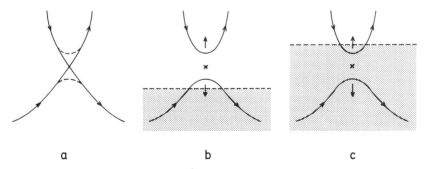

a   b   c

Fig. 2. Relation between reconnection of magnetic fields and cancellation as seen at the surface. (a) Two field lines before reconnection. (b) The reconnection point (X) lies above the surface; cancellation of the fields at the surface (dashed line) occurs if the lower field loop pulls back below the surface. (c) The reconnection point lies below the surface; cancellation is the consequence of the upper loop moving up through the surface.

as seen at the photosphere. In particular, flux cancellation does *not* imply a release of energy at the solar surface. The reconnection point can lie above the surface or below (see Fig. 2; Zwaan 1978). In the first case, the energy release will occur at some height in the atmosphere, and is in principle observable. In the second case, the energy release occurs in the dense layers below the surface and is not expected to have obvious observational consequences. The evolution after the reconnection depends on the height of the reconnection point. By reconnection above the surface a magnetic connection is formed between the footpoints at the surface, but this is in general not enough for the field to pull back below the surface. The magnetic force that brings the foot points together is small until their distance is of the same order as their diameter at the surface (Parker 1979a). This rather strong requirement is not likely to be satisfied very frequently although it may occur in regions with strong field of mixed polarity, where the chances of magnetic elements meeting is large. This scenario would then be consistent with the observations of flux cancellation on scales of arcsec or less. Alternatively, the tubes may have reconnected *below* the surface long before a cancellation event is seen at the surface. By such reconnection a U-loop is formed (Spruit et al. 1987b), in which magnetic tension and buoyancy make the loop float upward (Fig. 2b). A large amount of mass is trapped in this loop, which, because of the high conductivity of the solar plasma, can separate from the loop only by small-scale processes (Parker 1984; Spruit et al. 1987). If the reconnection took place at a significant depth below the surface, the trapped mass makes the loop expand as it floats towards the surface, and the field strength becomes low. Observations of such an event would show flux concentrations of opposite polarity separated by a significant distance, each diffusing into a patch of weak field, followed by a simultaneous decrease in strength of these two patches. From these predicted characteristics, it is clear that flux cancellation resulting from reconnection below the surface would be much harder to recognize observationally.

Magnetic field detectors have a finite sensitivity, so that magnetic flux concentrations may also disappear from view by spreading into a diffuse form. Which physical processes in magnetic tubes could lead to such diffusion? The most immediate one, magnetic diffusion by finite ohmic resistance of the plasma is also the least important. With a diffusivity of 3×10^6 cm^2 s^{-1} (valid at a depth of 300 km) spreading over 0.1 arcsec would take 4 months—much longer than the times over which individual magnetic concentrations can be followed. At the surface, the diffusivity is a factor 30 higher due to the lower degree of ionization, but this does not yield a more rapid dispersal of the tube, as the following argument shows. Horizontal pressure balance between the tube and its surroundings is established on the time a fast magnetosonic wave takes to cross the tube. For a tube of 250 km radius, this is on the order 1 min only. A decrease in field strength at the surface due to ohmic diffusion therefore would imply an increasing gas pressure inside the tube. As long as the flows along the tube are well below the sound speed, the pressure also has to be roughly in hydrostatic balance *along* the tube, which is not compatible with a local pressure increase at some level. Mass entering the tube therefore implies flows along it. The end result is a horizontal inflow at the surface level accompanied by downflow along the tube (Giovanelli 1977; Schüssler 1986). This flow pattern brings outward diffusing field lines back to the tube and thereby keeps it concentrated. The amount of downflowing gas is small compared with the mass in the deeper layers and therefore has little effect on the lifetime of the tube.

This shows that dispersal of the field into a weak form by ohmic diffusion can become effective only after the tube has fragmented to very small scales (of the order 1 km) by other processes. Candidates for such processes are fluting instability and shredding by shear flows (Kelvin-Helmholtz instability). The fragments would be dispersed by granular flows. At the same time, the granulation flow also systematically expels them into the intergranular lanes, where they would be pressed into larger structures again. A more effective way of transforming a flux tube into a weak dispersed form would be an upward flow along the tube from below, such as would happen in a large-amplitude overstable oscillation (see Sec. II.B). Another would be the formation of U-loops discussed above.

III. PROPERTIES OF MAGNETIC ELEMENTS

Existing model calculations that aim at describing the detailed structure of magnetic elements assume that the observed magnetic elements are in a quasi-equilibrium state (Spruit 1976, 1977; Deinzer et al. 1984a,b; Pneuman et al. 1986; Steiner et al. 1986; Knölker et al. 1988; Grossmann-Doerth et al. 1988; Steiner and Pizzo 1989). Since the small-scale field is variable on the time scale of granulation, one may wonder if it always stays in concentrated form. Given that 90% of the total net magnetic flux is in magnetic elements (Howard and Stenflo 1972), the strong-field phase of the evolution of a flux concentration must last much longer than the formation and destruction phases. As discussed in the preceding chapter (Semel et al.), convective collapse leads to a stable configuration which may exhibit oscillations. The observation of nearly unshifted Stokes V zero crossings (see discussion below) excludes stationary flows *within* the magnetic field. It also constrains the possibilities for time-dependent flows, since they will produce a wavelength shift due to intensity-velocity correlation. Consequently, significant convective energy transport, or large amplitudes of overstable oscillations in the interior of flux concentrations are difficult to reconcile with these observations. Being situated in intergranular downdraft regions, magnetic elements experience a converging downflow in their surroundings which probably comes into rapid swirling motion (Nordlund 1983). Such a flow contributes to stabilizing the concentrated magnetic field against fluting, i.e., the splitting of the tube into smaller flux bundles (Schüssler 1984). Moreover, the flux concentration itself drives a thermal downflow in its environment (Deinzer et al. 1984b) which enhances the normal intergranular downdrafts and stabilizes the surrounding flow pattern. This is supported by observations of a larger lifetime of granules in regions with large amounts of magnetic flux (Title et al. 1987a) and of the deformation of granules around bright points (Roudier 1986; Muller et al. 1989).

For many years magnetic elements have been considered to be the seat of strong internal downflows of 0.5 to 2.0 km s^{-1} (see, e.g., Harvey 1977; Giovanelli and Slaughter 1978; Wiehr 1985). As mentioned above, this notion is not supported by most of the recent observations (see, e.g., Stenflo and Harvey 1985; Solanki 1986; Stenflo et al. 1987a). In particular, Solanki (1986) measured the wavelength shift of the zero crossing of the Stokes V-profile for approximately 200 spectral lines and set a limit of 250 m s^{-1} on any stationary upflow or downflow within the magnetic elements. The main difference between observations showing downflows and those not showing any is the lower spectral resolution of the former. Since Stokes V-profiles of most photospheric lines when observed near disk center are distinctly asymmetric, with their blue wings stronger than the red wings (see, e.g., Stenflo et al. 1984; Solanki and Stenflo 1984), spectral smearing gives rise to a fictitious red shift. With this asymmetry given, previous observations of downflows in magnetic elements can be quantitatively explained as an effect of their lower spectral resolution (Solanki and Stenflo 1986).

On the other hand, the presence of strong downflows outside but in the immediate vicinity of magnetic elements can be inferred from the observed Stokes V asymmetries. Apart from atomic orientation (Kemp et al. 1984) which has not been worked out in detail, a combination of magnetic field and velocity gradients along the line of sight appears to be the only reasonable explanation for the observed amplitude and area asymmetries of the V-profile (Illing et al. 1975). By considering a variety of spectral lines and physically plausible magnetic field and flow profiles, Solanki and Pahlke (1988) could show that in one-dimensional (thin flux tube) geometry, it is impossible to reproduce the observed asymmetries without producing shifts of the zero crossing of Stokes V which are much larger than the limits set by observation (Solanki 1986). On the other hand, van Ballegooijen (1985b) has pointed out that an asymmetric Stokes V-profile will originate in a two-dimensional structure if the line of sight traverses both a static magnetic field and a flow field in the surrounding, nonmagnetic atmosphere. Since the magnetic field decreases strongly with height, flux concentrations flare out and, near solar disk center, lines of sight at the periphery traverse both static magnetic (upper part of the atmosphere) and downflowing nonmagnetic (lower part of the atmosphere) regions. If the transition between the two regions is as sharp as indicated by theoretical arguments (Schüssler 1986), model calculations (Knölker et al. 1988) and observations (Zayer et al. 1988), this leads to a situation in which for a significant part of the area covered by a magnetic-flux concentration, the magnetic field and velocity field are spatially separated along the same line of sight. It has been demonstrated quite generally by Grossmann-Doerth et al. (1988*b*, 1989*b*) that such a configuration leads to assymmetric *V*- profiles with unshifted zero crossings.

Near the limb, a reversal of the Stokes *V* asymmetry is seen (Stenflo et al. 1987*a;* Pantellini et al. 1988), which may be interpreted as the result of a horizontal flow towards the flux tubes feeding the strong downflows.

The thermal properties of magnetic elements prove interesting in observation and theory. Interaction with the surrounding medium and suppression of convective energy transport within the magnetic elements have been included in numerical simulations (see, e.g., Spruit 1977; Deinzer et al. 1984*a,b;* Knölker et al. 1988; Grossman-Doerth et al. 1989*a*). The models show a significant cooling of the deeper layers of the magnetic structure with respect to the surrounding atmosphere at the same geometrical height. The resulting horizontal temperature gradient drives an energy flux into the magnetic element. It is carried by convection and radiation outside the magnetic field and solely by radiation within. Partial evacuation due to cooling and convective collapse and the temperature dependence of H^- opacity increase the transparency of the flux concentration. As a result, the horizontal influx of energy is facilitated and, at the same time, the vertical energy flux is increased. It turns out that models of flux concentrations of the size of a magnetic element (100 to 200 km diameter) exhibit large continuum intensities (if the line of sight is along the vertical) in accordance with values or limits derived from observations. The results are practically independent of the assumed amount of vertical convective transport within the magnetic field: more (less) convective flux leads to less (more) horizontal heating by radiation flux such that the total vertical flux stays nearly constant and only a slight geometrical change of the optical-depth scale ensues, even if convection is totally suppressed (Deinzer et al. 1984*b*). This behavior is in contrast to the case of larger flux concentrations and sunspots which are not able to compensate the deficit of vertical convective transport by lateral radiative influx of energy because of their large optical depth in the horizontal direction (see, e.g., Knölker and Schüssler 1988).

As far as detailed temperature profiles are concerned, practically all observations point towards a higher temperature than in the quiet Sun at equal optical depth in the photosphere (see, e.g., Chapman and Sheeley 1968; Harvey and Livingston 1969; Stenflo 1975; Chapman 1977; Hirayama 1978; Koutchmy and Stellmacher 1978; Stellmacher and Wiehr 1979; Solanki 1986; Walton 1987; Keller 1989). However, this appears to be the only bit of common ground between the various models of the magnetic-element temperature structure. In almost all other respects they differ, sometimes widely, from each other. This partly reflects the different data the various models are based on, and partly on the different analysis techniques used to derive them, which vary considerably in sophistication (cf. Solanki 1989).

The observations cannot definitively decide whether magnetic elements are hotter or cooler than the quiet Sun at equal geometrical depth, although the latest empirical model, based on a least-squares fit to Stokes V-profile data (Keller 1989) suggests that deeper in the atmosphere, the temperature is lower in the magnetic elements, while higher up the opposite is true. Recent investigations suggest that the nonmagnetic part of the atmosphere in an active region is cooler than the quiet Sun (Schüssler and Solanki 1988; Immerschitt and Schröter 1989; Brandt and Solanki, in preparation). Observations and simple model calculations of line bisectors suggest that this is a result of inhibited convection in the nonmagnetic atmosphere surrounding the magnetic elements (see, e.g., Livingston 1982; Cavallini et al. 1987). The bisector changes may be the spectral signature of abnormal granulation (see, e.g., Dunn and Zirker 1973).

Detailed comparison with model calculations for the line-forming upper layers of the atmosphere within a magnetic element are presently not of much value since the treatment of radiation in the models still has to be improved. In particular, the importance of including the frequency dependence of the opacity for flux-tube models has been demonstrated by Steiner and Stenflo (1990). However, the qualitative features agree, i.e., the magnetic element is found to be significantly hotter than the mean atmosphere at the same optical depth. The progress from the diffusion approximation to a full radiative transfer has revealed an important effect: the illumination of the upper layers of a magnetic element by its hot bottom and walls leads to a significant (≈ 400 K) temperature enhancement with respect to the surrounding atmosphere at the same geometrical height (Grossmann-Doerth et al. 1988*a;* Steiner and Stenflo 1990). This radiative illumination effect, possibly together with NLTE effects (Solanki and Steenbock 1988) contributes to the observed line weakenings (Spruit 1976) and may well render unnecessary the assumption of mechanical heating of the photospheric layers of magnetic elements. This is in contrast to the upper atmosphere, where mechanical heating inside the flux tubes is clearly indicated.

The precise value of the emerging continuum intensity in theoretical two-dimensional flux-tube models depends on the amount of evacuation, which is a free parameter in the present models. The value of this parameter is fixed independently by comparison of model results with the observed magnetic field strength as a function of height (see, e.g., Zayer et al. 1989). If the evacuation is too large the field strength in the observable layers becomes too large (and vice versa). The models presented by Knölker et al. (1988) and Grossmann-Doerth et al. (1989*a*) demonstrate that the amount of evacuation that is adequate to reproduce the observed field strength at the same time leads to continuum intensities that are consistent with observations near the center of the disk. Theoretical models for a single magnetic element have been unable, however, to reproduce the center-to-limb variation of the continuum contrast of faculae. Small elements that are bright at disk center have their maximum contrast much closer to the disk center than the observations (Chapman and Klabunde 1982) show. Large structures are bright close to the limb, but dark at disk center (Spruit 1976,1977; Knölker and Schüssler 1988). An active plage, however, contains a large range of sizes, from pores to structures below the resolution limit (see, eg., Spruit and Zwaan 1981). Since each size has its own center-to-limb variation, a meaningful comparison with observations must take such a spectrum of sizes into account. Spruit (1976) showed that an appropriate mixture of tube sizes could produce the observed center-to-limb variation. The predictive power of this explanation of the center-to-limb variation of facular brightness is limited because it depends critically on the, as yet insufficiently known, spectrum of flux-tube sizes. An alternative explanation is that a facular element is a cluster of small magnetic elements: if their hot upper layers become optically thick by overlapping along the line of sight, the observed limb brightening of faculae could be reproduced (Schüssler 1987; Steiner and Stenflo 1990). The radiative illumination effect mentioned above leads to a temperature enhancement of about 400 K with respect to the same height in the mean atmosphere. This is sufficient to reproduce the observed contrast values.

How strongly do individual magnetic elements differ from each other? First consider the field strength. Stenflo and Harvey (1985) applied the Stokes V line-ratio technique to data obtained in regions with widely different filling factors and found that the intrinsic field strength of the magnetic elements is quite similar at the height at which the lines used are formed. There is a slight tendency for the field strength to increase as the filling factor increases. From a comparison of line profiles in the infrared and the visible, Zayer et al. (1989) conclude that the field strength, at the height of formation of the strongly split Fe I λ15648.5 line, must be similar for most magnetic elements within the 5″ field of view. Therefore, as far as the field strength is concerned, a two-component model is justified as a first approximation.

In contrast to this homogeneity in the magnetic field strength, the temperature does appear to be quite different for the magnetic elements in the quiet network (low filling factor) and active region plages (high filling factor), with the plage magnetic elements being cooler (Solanki and Stenflo 1984; Solanki 1986; Pantellini et al. 1988; Keller 1989). This is particularly evident on images taken at 1.6 μm (Foukal et al. 1989,1990), the wavelength that has the greatest depth of formation in the solar spectrum. There are two plausible explanations for the lower temperature of magnetic elements in plages (cf. Schüssler 1987). Since theoretical models predict a strong increase of the tube temperature (at the continuum-forming level) with decreasing size, it could be that tubes in the quiet network are on average smaller than in active regions. This agrees with the absence of larger elements (pores) in the network, a view proposed by Spruit and Zwaan (1981). Alternatively, it could be caused by the close packing of magnetic elements in plages interfering with convective energy transport, and thus leading to a net cooling of the nonmagnetic atmosphere between the elements.

IV. INTERACTION WITH THE SURROUNDING ATMOSPHERE

A. Wave Modes and Energy Transport

The results quoted in this subsection are mostly obtained in the so-called thin-tube approximation. This approximation assumes that the diameter of the tube is small compared with the other scales of interest, such as the pressure scale height of the atmosphere and the wavelength of the waves considered. Usually, it is also assumed that the time scales considered are long compared to the fast magnetosonic wave that travels across the width of the tube. The tube is then in pressure equilibrium with its surroundings.

The wave modes of a thin tube have been described by Ryutov and Ryutova (1976), Defouw (1976), Roberts and Webb (1978), Wilson (1980) and Spruit (1981,1982*a,b*). There are three basic modes. The *transverse tube wave* or kink mode (Ryutov and Ryutova 1976) consists of lateral displacements without changes in internal gas pressure. The *longitudinal tube wave* or sausage mode (Defouw 1976) changes the cross section of the tube and is compressive. The third wave is a *torsional Alfvén wave*. Further modes, in which the flow has more structure across the tube (Edwin and Roberts 1983) are conceivably important but are neglected in the thin-tube approximation. The three modes differ in the nature and degree of their interaction with the surroundings. By interacting with the surroundings, the wave modes are excited or damped. For interaction to occur, the velocity component perpendicular to the surface of the tube must be nonzero. For the torsional Alfvén wave, this is not the case and one might therefore expect this mode to be unimportant. However, the surface of a real tube is likely to be rough (it will, in fact, bear little resemblance to a circular cylinder), so that vortical flows around the tube are likely to set up motions in the tube that are equivalent to the excitation of a torsional Alfvén wave. Because vortices occur at the same kind of place as flux tubes, namely in convective downdrafts, the excitation of torsional waves might be significant (Schüssler 1984).

The second most important aspect determining the interaction is the *phase speed* of the tube wave, i.e., the frequency of the wave times its wavelength along the tube. If this is larger than the sound speed in the surrounding medium (the Cherenkov condition; Ryutov and Ryutova 1976), the tube will emit sound waves into the surroundings, causing the tube wave to be damped (Spruit 1980). Conversely, sound waves in the external medium can resonantly excite tube waves if this condition is satisfied (Ryutov and Ryutova 1976; Bogdan 1990). For photospheric conditions, all three tube waves typically have phase speeds less than the external sound speed so that resonant excitation of tube waves by external sound waves is unlikely. Weaker nonresonant excitation is still possible. Strong excitation is expected by interaction with slower processes, namely with convection. Spruit (1981c) has argued that the time scales of granulation are ideally suited for exciting transverse tube waves with periods in the range 800 s and higher. Transverse and torsional waves both can easily travel into the chromosphere; the transverse wave because the dominant exciting frequencies are below its cutoff frequency, the torsional wave because it propagates at all frequencies. Longitudinal tube waves are excited by the pressure fluctuations in the convective flow. The amount of longitudinal tube waves entering the chromosphere in this way is likely to be quite small. In the first place, the amplitude of the pressure fluctuations is rather low even near the surface; secondly, the cutoff frequency of longitudinal waves (of the order of the acoustic cutoff in a nonmagnetic atmosphere) is higher than the exciting frequencies in the granulation. The energy flux into the chromosphere and corona in the form of transverse or torsional waves has been estimated by Hollweg (1981,1984) and Spruit (1984a). Hollweg finds numbers that would suffice to heat the corona in active regions; Spruit finds an upper limit of about 5×10^5 erg cm^{-2} s^{-1}, sufficient only to explain the lowest observed level of coronal emission. In the chromosphere, both analyses yield energy fluxes that are sufficient to explain the observed emission. Hollweg et al. (1981a; Hollweg 1982) show that even the spicule phenomenon may be explained as a consequence of torsional or transverse tube waves excited by the granulation.

B. Influence on Solar Luminosity

The solar irradiance (bolometric flux integrated over the solar disk) at 1 AU is known to vary by about 0.2% over the solar cycle, with a maximum near sunspot maximum. For a more extensive discussion of this and related observations see the chapter by Livingston et al. This variation has several distinct components (cf. Hudson and Willson 1988; Foukal and Lean 1988a). Sunspots produce a reduction of the irradiance which is modeled well by the instantaneous (projected) area and brightness of the spot surface. Facular areas contribute positively by an amount that is similar in magnitude but opposite in sign to the sunspot effect (Oster et al. 1982; Chapman et al. 1984; Foukal and Lean 1988a). The two do not correlate on a day to day basis because faculae contribute mostly near the limb whereas the spot contribution peaks at the disk center. Foukal and Lean (1988a) show that, in addition to these two components, there is a third component which has a maximum near spot maximum. It makes a strong contribution but does not correlate well with either spots or faculae. Foukal and Lean identify it with the network and the enhanced network that results from decaying active regions.

Since most of the field in faculae and the network is in concentrated (flux tube) form, it follows that small flux tubes apparently are able to enhance the solar output to a measurable extent. Such an effect was predicted theoretically by Spruit (1977) who found that "small tubes such as are found in the quiet network act as leaks in the solar surface through which an excess heat flux escapes from the convection zone." This prediction was based on calculations of perturbations in the solar energy flux caused by magnetic flux tubes near the top of the convection zone (for more general theoretical discussions of the effect of magnetic perturbations on the solar luminosity and radius see Spruit [1988,1990] and Gilliland [1988]). The physics of this theory is sketched below. For alternative interpretations of irradiance variations associated with spots and faculae see Chapman et al. 1984; Schatten and Mayr 1985; Schatten et al. 1986.

The magnetic field of a vertical flux tube locally reduces the gas pressure, and thereby distorts the surface of constant optical depth (optical depth measured perpendicular to the solar surface). The energy transport by convection and radiation is therefore perturbed near the tube. It is convenient to separate the perturbation conceptually into a "local" and a "global" effect. The global effect is the net effect of the tube on the solar luminosity (or the irradiance); the local effect encompasses everything else. Let us first look at the local effects. Consider the hypothetical case of a flux tube created (somehow) instantaneously at the solar surface, such that initially only the gas pressure is perturbed (lower inside). The lower optical depth in the tube means higher temperatures at a given optical depth and hence a greater radiative flux than in the unperturbed surroundings (see also Sec. III). This cools the interior of the tube (compared with the surroundings at the same geometric depth) so that an energy flux is set up horizontally from the surroundings into the tube. This implies a cooling of the surroundings relative to their initial temperature. Thus, we predict the formation of a region of lower than normal brightness immediately surrounding the tube. The time scale for the initial cooling of the tube is on the order of seconds in the visible layers (the thermal time scale in these layers); the formation of the cooler surface area around the tube takes on the order of a granule turnover time (assuming that the horizontal energy flux toward the tube is carried by convection). How large is the net excess energy flux integrated over the tube and its surroundings? This depends on from where the horizontally inflowing energy derives. It turns out that only part of the influx derives from the immediate surroundings of the tube (giving rise to the cool ring), a part derives from the bulk of the convection zone. The reason for this is the strong decrease of the thermal resistance (the temperature gradient needed to carry a given energy flux) with depth in the convection zone (roughly as ρ^{-1} where ρ is the density). This means that a thermal perturbation that has penetrated the first few pressure scale heights of the convection zone no longer feels thermal resistance and spreads over the entire convection zone. Since the horizontal influx into the tube takes place mainly over the first two pressure scale heights (for the observed field strengths), only a fraction of the influx derives from the deeper layers in this way (cf., Spruit 1977, his Fig. 16). The fraction that *does* however, also represents a global effect; it produces a net excess flux that would disappear only on the thermal time scale of the convection zone (200,000 yr).

Recent two-dimensional models (cf. Sec. III) show a smaller net-flux excess than that originally predicted by Spruit. These models include an important effect previously neglected, namely the creation by the tube of a thermal downdraft surrounding it. The walls of the optically thin part of the tube are kept at a temperature near 6500 K by the balance between radiative loss and advection of heat by convection. This is significantly less than the normal temperatures in the convection zone at the same depth (10^4 K). By horizontal pressure balance, this implies a higher density and a strong downward acceleration. This is likely to be the explanation for the strong downdrafts outside the tube inferred from the line profiles, as discussed in Sec. III. The energy advection by this systematic flow is important for the energy balance and hence for the net flux excess of the tube. In the current models, the downflows at the tube boundary are present but probably reduced below their real speed by (numerical) viscous coupling to the tube, so that the models would underestimate the excess energy flux from the tube.

Note that the relatively low temperatures of the tube walls imply that the energy-flux excess is not really due to a hot-wall effect; we prefer the name bright-wall effect. The net excess manifests itself mainly through emission at large angles to the vertical. At these angles, the tube walls are bright because they are seen face-on, compared with the surrounding atmosphere which is strongly limb darkened.

In order to compare theoretical models of the energy-flux excess of the small flux tubes at the surface with observations of the solar luminosity, some improvement is needed in the theoretical models, but especially in our knowledge of the size spectrum of magnetic elements. The main contribution is expected from the smallest tubes (consistent with the observed importance of the network for the irradiance variation) whereas tubes larger than $1''$ have a negative contribution, like pores and spots.

C. Interaction with *p*-Mode Waves

Observations show that small magnetic flux tubes, as well as sunspots, have a surprisingly strong interaction with the 5-min oscillation. Giovanelli et al. (1978) found a substantial reduction of the *p*-mode power in active regions. Recent high-resolution observations of this phenomenon were made by Tarbell et al. (1988). Braun et al. (1987) show that sunspots absorb as much as 50% of the *p*-mode flux incident on them, at high horizontal wave numbers. Only preliminary interpretations of these observations have been made (cf., Abdelatif and Thomas 1987; Bogdan and Knölker 1988; Lou 1988). It is clear, however, that scattering of *p*-modes by flux tubes cannot be responsible for these observations. The sunspot observations by Braun et al.

rule this out because they measure directly the difference between the wave fluxes toward and away from the spot. Since the 5-min oscillations form a statistical ensemble of modes in which all propagation directions are equally represented, any scattering object is invisible in the wave field except by close inspection of the phases of the waves (just like a scattering surface is invisible in a blackbody radiation field). Reduced excitation of the waves due to a modification of the properties of granulation in active regions (which is observed; cf., Title et al. 1989) does not seem likely either. The observed sharpness of the p-mode ridges implies that the interaction of the modes with the convection (by excitation and damping) is weak, and hence their horizontal damping length large in contrast to the observed sharpness of the boundary of areas of reduced p-mode amplitude (Tarbell et al. 1988). The conclusion seems unavoidable that an additional, considerably stronger, damping mechanism operates in magnetic regions.

The magnetic field in the convection zone influences the frequencies of p-modes, because of the increased stiffness of magnetized gas (see Bogdan and Cattaneo 1989, and references therein). This might lead to an observable modulation of p-mode frequencies with the solar cycle. The effect depends not only on the amount of magnetic flux but also on the intrinsic field strength. For a given amount of flux, the effect is proportional to the intrinsic field strength (Bogdan and Zweibel 1987b; Roberts and Campbell 1988).

THE HEATING OF THE SOLAR CHROMOSPHERE

WOLFGANG KALKOFEN
Harvard-Smithsonian Center for Astrophysics

The chromospheric layers in the quiet Sun show three distinct regions: (1) Magnetic elements of relatively low field strength in the interior of supergranulation cells; these intranetwork elements are associated with bright points, which are heated by large-amplitude compressive waves with periods near the acoustic cutoff period of approximately three min. (2) Magnetic elements of high field strength in the network on the boundary of the cells; they are heated by waves with still longer periods, up to ~ 10 min. (3) Magnetic field-free regions, which appear to be heated by waves with periods of the order of one tenth the acoustic cutoff period. The atmosphere associated with magnetic fields is much brighter in the characteristic chromospheric emitters than the field-free atmosphere. Thus the chromosphere can be largely identified with the atmosphere inside magnetic elements (flux tubes). This chapter discusses the heating mainly of the bright points that are associated with the intranetwork magnetic fields. It reviews the relevant observations in line and continuum radiation and concludes that the energy dissipated by the 3-min waves is probably sufficient to heat the low and middle chromosphere in the bright points to the observed temperatures.

I. INTRODUCTION

Ever since Biermann (1946,1948) and Schwarzschild (1948) proposed that the solar chromosphere is heated by acoustic waves that are generated in the hydrogen convection zone and dissipated in shocks in the chromosphere, the problem of chromospheric heating by mechanical waves has been a theoretical solution in search of observational confirmation. Chromospheric heating therefore has been discussed and reviewed in many papers over the years (see, e.g., Schatzman 1949; Osterbrock 1961; Kuperus 1969; Ulmschneider 1970,1979,1981,1986,1989; Jordan 1981; Stein 1981, 1985; Ulmschneider and Stein 1982; Athay 1985; Kalkofen 1989; Narain and Ulmschneider 1989) and in many books (see, e.g., Thomas and Athay 1961; Bray and Loughhead 1974; Athay 1976; Zirin 1988); for extensive references to papers and reviews, cf. Narain and Ulmschneider (1989). With few exceptions, the reviews have stressed heating by the theoretically expected short-period acoustic waves, with periods of approximately one tenth the acoustic cutoff period (Ulmschneider 1970,1974), or about 0.5 min. Yet, the observational evidence in support of the proposition of heating by such waves is indirect, showing no evidence of shocks, and suggests that this heating may account only for the relatively low background level of a basal chromosphere (Schrijver 1987). Nevertheless, short-period waves are widely regarded as the only acoustic waves capable of heating the chromosphere. Accordingly, the lack of direct evidence appears to imply that the solar chromosphere is not being heated by acoustic waves of any period. Cram (1987) expressed this sentiment most forcefully in stating that "spectroscopic investigations from ground-based and space-based facilities have consistently failed to find evidence of shock-wave heating in the solar atmosphere." This sentiment is widely shared, and it has directed research to other heating mechanisms (see, e.g., Stein 1985; Cram 1987; Narain and Ulmschneider 1989). However, the observations clearly show strong heating in magnetic regions, but by waves that are not expected theoretically since their periods (repetition times) are near and longward of the acoustic cutoff period. According to the theory of low-amplitude waves propagating in a gravitationally stratified atmosphere (cf. Schatzman and Souffrin 1967), acoustic waves with periods longer than the acoustic cutoff period (approximately 3 min at the temperature minimum), are evanescent or nonpropagating. Thus, waves with periods near P_{ac} should carry little or no energy and hence are generally considered to be unimportant. However, the observations show long-period waves to be the main, if not the sole, heating agent of the low and middle chromosphere in magnetic regions of the quiet solar atmosphere.

This chapter reviews the observational evidence for the heating of the quiet solar chromosphere in the bright points by long-period waves with typically a 3-min period. Section II discusses the association of wave heating with magnetic elements; Sec. III gives evidence for wave heating from observations of Hα, the H and K lines of ionized calcium, and of continuum radiation at far infrared and radio wavelengths. Section IV draws conclusions on heating from chromospheric models and from weak-shock theory. Speculation on wave generation is presented in Sec. V; the observed long wave periods are discussed in Sec. VI; implications for semiempirical models are given in Sec. VII; and observational consequences are listed in Sec. VIII. In Sec. IX, we draw conclusions.

II. ASSOCIATION OF HEATING AND MAGNETIC FIELDS

The quiet solar chromosphere shows three distinct structures, namely, the network on the boundaries of supergranulation cells, the bright points in the interior of the cells, and the truly quiet chromosphere.

In the network, elements of intense vertical magnetic field coincide with bright emission in the cores of the Hα line as well as the H and K lines of Ca$^+$, indicating nonradiative heating (Athay 1976). The correlation between magnetic field and excess emission extends throughout the chromosphere and into the transition region between the chromosphere and the corona. And it is twofold: all regions of increased vertical magnetic field are bright in the core of Hα and in the K line, and all regions that are bright correspond to peaks in the vertical magnetic field strength (Zirin 1988).

In the bright points, the intranetwork magnetic field appears to be weaker (Stenflo 1982,1989; Zwaan 1987), but the observations show even here a one-to-one relation between magnetic elements and nonradiative heating. This is indicated by the small size of the wave-heated area, of the order of 2″ or less (Liu 1974; Cram and Damé 1983), and by the repeated occurrence of heating in the same location (Damé 1983,1984); it is also shown by the coincidence of bright points with magnetic elements observed by Sivaraman and Livingston (1982). They note that when the magnetic field moves (e.g., because of horizontal flows), the associated bright point moves with it; that a new bright point is born at the location of a previous bright point; that a bright point is never observed without an associated magnetic element; and when a magnetic element is found without a bright point, this can be understood in terms of the finite lifetime of the bright point.

The gas in the magnetic field-free medium in the cell interior is much less bright in the H and K lines and constitutes the truly quiet chromosphere. At least part of this gas is cool enough (well below 4000 K) to emit strongly in the vibration-rotation bands of the CO molecule (Ayres 1981; Ayres and Testerman 1981; Ayres et al. 1986). This medium therefore appears to receive a much lower mechanical energy flux than the atmosphere in the magnetic regions; it may be heated only at the level of a basal chromosphere (Schrijver 1987), i.e., at a heating rate that is much lower than that inside the magnetic elements, and also significantly lower than that predicted by the acoustic heating theory (cf. Cram 1987).

The observed one-to-one correspondence between significant excess emission and magnetic fields in both the network and the bright points suggests that strong nonradiative heating occurs only in association with the magnetic field. The bright chromosphere may therefore be identified with the atmosphere inside magnetic elements.

Most of the magnetic field in the quiet Sun occurs in the network in the form of concentrations of a strong magnetic field. In these regions, the pressure exerted by the magnetic field may be larger than the gas pressure. Such field concentrations are referred to as magnetic flux tubes (cf. Spruit 1981*b*). They grow from a horizontal size of a few hundred kilometers or less in the photosphere, with a filling factor of less than 1% (Stenflo 1989), until they fill all space somewhere in the chromosphere, in the so-called magnetic canopy. The intranetwork magnetic field is much weaker than the network field (Stenflo 1982) so that the gas pressure is probably larger than the magnetic pressure. Nevertheless, the appearance of the waves heating the bright points, and the small size of the bright points themselves, suggests that this field, too, is able to channel and contain the waves. It seems legitimate, therefore, to refer also to the intranetwork magnetic regions as flux tubes. Thus, the outer solar atmosphere consists of essentially two phases: (1) magnetic flux tubes, in which the ratio of gas pressure to magnetic pressure is probably less than unity in the network on the cell boundary, and larger than unity in the cell interior; and (2) a cool intertube medium, in which the magnetic field is so weak that it plays no role. The main features of chromospheric heating can be described on the basis of this two-phase model.

This chapter considers mainly the chromosphere inside intranetwork magnetic flux tubes, and in particular, the lower layers of the chromosphere that extend from the temperature minimum to the end of the temperature plateau at about 8000 K (cf. Fig. 1), i.e., layers traditionally referred to as the low and middle chromosphere. The restriction of the discussion of heating to these largely neutral layers inside magnetic elements may also entail a restriction to a particular heating mechanism.

III. OBSERVATIONAL EVIDENCE OF WAVE HEATING

A. Line Radiation: Hα and the K Line

Observations of intensity and velocity variations of a spectral feature may not allow an unambiguous interpretation of a wave phenomenon since the phase relation between intensity and velocity maxima may not clearly separate running from standing waves, or dissipation from adiabatic conditions. Thus, the observations by Jensen and Orrall (1963) and Orrall (1966) identifying the oscillatory nature of K line variability were merely suggestive of chromospheric heating.

Orrall found quasi-periodic wavelength displacements and intensity fluctuations in the K line (3933 Å). The intensity at the bottom of the line core K_3 varied with a typical period of 3 min. A particular period depended on the strength of the emission of the K_2 feature: in regions where K_2 was *faint*, defined by relatively weak emission in the violet peak, K_{2v} of the line, the oscillations had a period of about 180 s, with a tail extending to 400 s; and in regions where K_2 was *bright*, defined by strong K_{2v} emission, the long-period tail extended to 600 s. Orrall surmised that the significant positive

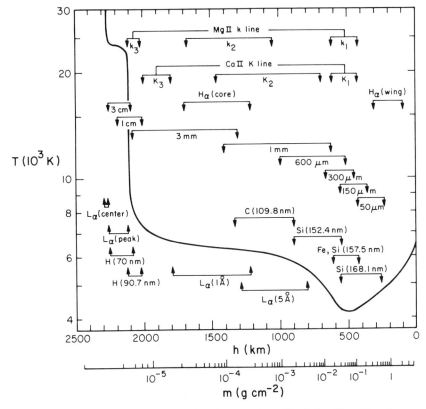

Fig. 1. The temperature structure of the quiet solar atmosphere. The depth of formation of prominent features is indicated. The "chromosphere" of this chapter is located between the heights of 500 and 2000 km (figure from Vernazza et al. 1981).

correlation between period and K_2 intensity also implied a correlation with the strength of the vertical magnetic field.

The maximum enhancement of the intensity minimum, K_1, which is formed at the temperature minimum between photosphere and chromosphere (cf. Fig. 1), occurred about one-quarter period before the maximum of the upward velocity, inferred from spectral lines formed in the same layers. From this phase relation, Giovanelli (1974) concluded that the waves were standing acoustic waves in an adiabatic gas. This agreed with the conclusion drawn by Jensen and Orrall (1963) on the basis of similar observations. On the other hand, Jensen and Orrall had also noted a small phase shift in the intensity of the K line wings indicating an outward-propagating feature.

Bhatnagar and Tanaka (1972) observed intensity variations of the Hα line in the center of the Hα-supergranulation network, in rosette centers, and

in plage granules, seeing conspicuous intensity oscillations of a fairly periodic nature. The difficulty with their observations, taken both at line center
and in the blue wing of Hα, was that they could not attribute the observed
intensity variations unambiguously to oscillations of either the velocity field
or the intrinsic brightness. On the whole, they confirmed Orrall's K line observations regarding the trend of the oscillation periods with the strength of
the violet peak K_{2v}: in the network the oscillation period in Hα was 170 ± 44s,
in the rosette centers it was 312 ± 56 s, and in plage granules, 289 ± 49 s.
Since the height of formation of the Hα profile at 0.5 Å blueward of the line
center should be the same in all these regions, they too attributed the difference in oscillation period to differences in magnetic field strength. They concluded that their observations of the intensity of Hα were consistent with
Orrall's observations of the velocity of the K line.

A major advance in observations of chromospheric heating was made by
Liu (1974), and also by Cram (1974a,b), again in the K line. Liu observed
the K line with high spatial and temporal resolution in bright points of a quiet
region at the center of the solar disk. He found intensity perturbations that
propagated from the far line wings toward the line core, with a time lag
between the intensity perturbation maxima at K_2 and K_1 that he interpreted in
terms of an upward-propagating disturbance causing local heating in the chromosphere. The phenomenon tended to repeat with a period of ~ 180 s and
occurred over 90% of the area covered by the telescope slit.

Some of the details of the observations are highly suggestive of running
waves with dissipation in shocks. Thus, the intensity enhancement first appeared symmetrically in the far wings of the line, but when it reached the line
core it resulted in a strong enhancement of only the violet peak K_{2v} (cf. Fig.
2); red peak (K_{2r}) enhancement was seen only occasionally. The disturbance
propagated faster at later phases of the process, and the intensity amplitude
increased as the wave propagated upward.

The range of periods observed by Liu was 180 ± 57 s (cf. Fig. 3), and
the intensity K_{2v} increased typically by a factor of 3. He concluded that he
had found direct evidence for local heating in the chromosphere by the passage of a disturbance. Since the region he had observed was typical of the
quiet Sun, he concluded that 90% of the Sun shows these coherent oscillations. The size of the region covered by the oscillations varied from 2000 to
4000 km, but the characteristic size of the heated region at the height of
formation of the K_{2v} peak was only 1500 km. Interpreted in light of Defouw's
(1976) study of wave propagation along a magnetic tube, this difference in
size may imply that the diameter of flux tubes at the height of formation of
the K_2 feature is of the order of 1500 km, or 2″, and that the atmosphere
outside a tube oscillates in phase with the inside gas. Defouw's study of an
infinite train of low-amplitude waves with a definite period showed no energy
loss associated with the oscillation of the gas external to the tube. However,
Liu's observations were of wave trains consisting of typically 3 pulses only,

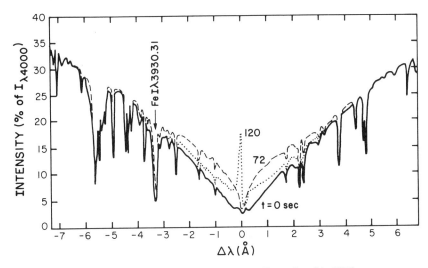

Fig. 2. The intensity of the K line at three different phases (figure from Liu 1974).

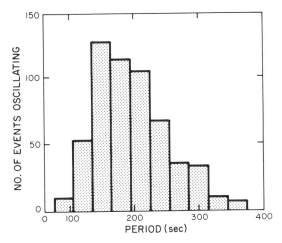

Fig. 3. Histogram of wave periods observed at the K_{2v} emission feature in the K line (figure from Liu 1974).

with one exceptional wave train containing 9. In addition, the disturbances had large amplitude. We might therefore expect that the oscillation of the external medium represents a loss of energy for the upward-propagating wave to the intertube medium. It is not known whether or not the intertube gas is heated by this sideways radiation of acoustic waves.

The relaxation of the profile after the enhancement of the K_2 feature to a state that Liu considered to be its relaxed state was much faster than was consistent with the estimate of 471 s he quoted from Ulmschneider (1971) for the radiative relaxation time of the gas. Hence Liu concluded that pure radiative cooling of the gas might not be adequate to account for his observations. Another interpretation might be that the estimated radiative cooling time is too long because the actual relaxation process differs from the one envisaged by Ulmschneider. We will return to this point in the discussion of the observations in the far infrared continuum by Lindsey and Roellig (see Sec. III.C).

Liu noted that very soon after the strong enhancement of the K_{2v} peak had occurred, this feature became invisible, and he suggested that the gas simply expanded away. Another possibility might be that Ca^+ becomes ionized to Ca^{++} by the strong energy pulse of the shock. Such a process might leave an imprint in recombination lines in calcium or in Lyman radiation (cf. Sec. VIII).

Liu estimated the energy involved in the phenomenon, both the mechanical energy in the wave that heated the gas as well as the radiative energy in the extra emission in the K line. For the energy flux in the mechanical wave he took a particular event and he assumed that the phenomenon was an acoustic wave, which allowed him to convert an observed phase speed into a group speed. He compared the mechanical energy flux of the wave with the excess radiative flux seen in the K line for the same event. Even though the inferred mechanical flux was nearly twice as large as the radiative flux, he concluded that the energy input, based on the acoustic wave assumption, was probably not sufficient to balance the energy needs of the chromosphere for all emitters. However, as Cram and Damé (1983) pointed out, Liu's estimate for the energy radiated in the K line was based on a very bright, atypical structure and may have been too large.

An empirical interpretation of the time evolution of the Ca II K line was offered by Liu and Skumanich (1974; cf. also Athay 1970; Cram 1972; Grossmann-Doerth et al. 1974) who modeled the asymmetric line profile with a downward rather than an upward-propagating pulse, as might have been suggested by Liu's observations. With a velocity amplitude of the pulse of 10 km s^{-1}, which exceeds the sound speed of the layer where the line core is formed, they were able to model the line shape. It may be noted that the velocity profile is ambiguous, and that a downflow is kinematically equivalent to an upward velocity since only the relative velocity of the layers where the line is formed matters.

In an earlier paper, Liu (1973) had reported K line observations near a small sunspot. These showed that the typical size of the emission features was again 1500 km and that the intensity enhancement was by a factor of 3 over that of the quiet background, which is comparable to that of network fragments of quiet regions of the atmosphere. In this paper, he was still puzzled as to the nature of the phenomenon, and he did not comment on its time scale.

Cram (1974*a,b*) reported similar results from observations in a time series of small-scale Ca^+ emission features located outside the network in the quiet chromosphere, again for the K line. He concluded that his observations implied unambiguously that the evolutionary behavior of the line profile was due to an outward-propagating velocity pulse; that the inferred flow parameters suggested that the wave lost mechanical energy in traversing the chromosphere; and hence, that the bright Ca II features, called K grains, were the manifestation of local heating in the chromosphere, possibly by shock waves. In the discussion following the short paper by Cram (see the discussion following Cram [1974*b*]), Tanaka commented that the grains showed a very clear oscillation in Hα, with a period of 180 s, similar to what Liu had found for the K_{2v} cell points, and Cram speculated that K grains inside the cells might be shock waves heating the chromosphere, that their diameter was about 1000 to 2000 km and their separation was about 6000 km.

B. Line Radiation: The H Line

Significant progress in the study of chromospheric heating was made by Cram and Damé (1983) with a time series of high spatial and temporal resolution of the H line (3968 Å) of Ca II (cf. also Damé [1983,1984] for observations of both the H and the K line). They suggested that the extra radiative loss they saw from the fine structures, when averaged over space and time, matched closely the radiative flux divergence implied by the semiempirical model *C* of Vernazza et al. (1981; henceforth VAL).

The H line was produced by two kinds of structures in the chromosphere: network fragments, with a typical lifetime of 10 min, and cell points, also called bright points or grains, with a lifetime of < 1 min. These H line observations are reminiscent of those made by Liu of the K line in the grains: at first a brightening appears symmetrically in both wings of the H line, and this intensity enhancement propagates towards the line core; when it reaches the emission feature H_2 (cf. Fig. 4), the line becomes highly asymmetric, with only a blue peak enhancement (H_{2v}). They suggested that the temperature fluctuations were spatially coherent between the upper photosphere and the chromosphere, with only upward-propagating features. Unlike Liu (1974) who reported observing an occasional downward-propagating feature, Cram and Damé saw none.

The time scale of the H line variations was the same as that of the corresponding K line variations, with periods in the range of 3 to 5 min. The spatial scale was of the order of 2000 to 3000 km. Again, the apparent spatial scale of the oscillation was larger than that of the brightest cell points, i.e., the heated region. The appearance of the line core H_3 suggested the breakup of the points into Doppler-shifted elements with sizes < 1000 km.

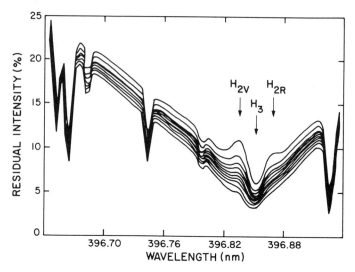

Fig. 4. The intensity of the H line in the quiet Sun ordered by the H index (figure from Cram and Damé 1983).

For the data analysis, Cram and Damé put the line profiles into an ordered sequence according to the strength of the line, measured by the H index, and placed them into 10 bins (cf. Fig 4). Averages of the profiles in the 10 bins then represented equal weightings in time and space of the H line profile. For the average profile of the whole set they estimated the excess radiative emission and compared this with the emission in the H line of VAL model C, finding 80% of the energy flux required for chromospheric heating according to the semiempirical model.

The estimate of the excess radiative flux was obtained by comparing the average profile of the whole set of data with the average profile in the lowest decile. This would give the excess energy supplied by the wave if the lowest bin contained lines only from gas in the completely relaxed state and its average profile represented the same weighting of flux tubes as the average profile of the whole set. Now, some of the profiles presumably showed no evidence of mechanical heating (cf. Bappu and Sivaraman 1971); they could thus be the background component postulated by Wilson (1970) and Cram (1972), in which the source function of the H line decreases monotonically with height. But the average profile of the lowest decile shows a weak enhancement at the frequency of the H_{2v} feature, suggesting that at least some of the profiles represent flux tubes with waves near the height of formation of the line core (a further complication is that these data include some emission from the network in addition to bright points). Thus, since the reference profile in the lowest decile does not describe only the completely relaxed state of the gas inside flux tubes, their estimate of the mechanical energy input is likely to be only a lower limit to the true value. But it is remarkable that the energy dissipation rate implied by the Cram and Damé observations is of the same order of magnitude as that required by the empirical model C, which is based on different data (cf. Sec. VII).

C. Continuum Radiation

Observations of continuum radiation at far infrared and radio wavelengths have some advantages over observations at optical wavelengths. Since electron collisions are sufficiently frequent, except where the scale length of the temperature variation is shorter than the collision mean free path, the source function for free-free transitions of H^- is given by the Planck function, which is well approximated by the Rayleigh-Jeans function. Thus, intensity data can be interpreted directly in terms of temperatures without invoking line-formation theory and departures from local thermodynamic equilibrium. The disadvantages are that there are no Doppler data from which to obtain velocities, and since the resolution of radio telescopes is much lower than that of optical telescopes, many oscillating elements contribute simultaneously to the signal.

The long-wavelength radiation that is formed in the chromosphere between the temperature minimum and the steep temperature rise at 8000 K covers approximately two decades in wavelength, from 0.3 to 30 mm (cf. Fig. 1). The observations reported here were made at 0.35, 0.8, 3.3, 3.5 and 33 mm. Here we describe them in their historical order.

Yudin (1968) measured "quasi-periodic radio-wave fluctuations" at a wavelength of 3.3 cm. The histogram of the periods showed power between 100 and 400 s, with most power between 200 and 300s. For the temperature amplitude of the phenomenon, he quoted only an upper limit.

Simon and Shimabukuro (1971) measured the intensity at 3.3 and 3.5 mm, using two different telescopes. This radiation is formed in about the same layers as the core of the K line. They observed a broad power spectrum with the strongest component at a period of 180 s which they found to be in general agreement with the K line observations of Orrall (1966).

In order to extract an energy flux from the data, they assumed that wave propagation was adiabatic. This assumption was consistent with the estimates of radiative cooling times by Noyes and Leighton (1963) from Spiegel's (1957) formula (cf. also Mihalas and Mihalas 1984,§101), which were based on the supposition that cooling occurred via H^- radiation. These thermal relaxation times, like Ulmschneider's quoted earlier, were much longer than the wave periods. With a temperature fluctuation of only 2 K, they implied a wave energy flux that was entirely negligible as a heat source for the chromosphere.

Lindsey and Roellig (1987) measured the continuum brightness variation of the quiet Sun at 350 and 800 μm. The periods in their time series ranged from 3 to 5 min and the phase lag between the intensity fluctuations at the 2 wavelengths was 25 to 35°. They speculated that the phase delay of the intensity at the longer wavelength might be attributed to the nonadiabatic response of the medium to an outward-propagating wave. The temperature fluctuations had an amplitude of only 5 K, not much different from those of Simon and Shimabukuro. However, Lindsey and Roellig based their arguments about the importance of their observations for wave heating of the chromosphere on cooling times estimated by Lindsey (1981), which assumed that thermal relaxation of those layers occurred via ionization relaxation, in which an initially partially ionized hydrogen gas becomes strongly ionized by a wave and then recombines. The resulting cooling times ranged from 14 to 60 s in the layers in which the submillimeter radiation is formed. These times were short compared to the wave period, implying that cooling by radiation was an important process in the response of the medium to a dissipating wave. Since much of the dissipated energy was put into ionizing the gas, the amplitude of the measured temperature fluctuations was only modest. Lindsey and Roellig concluded that, in spite of the low amplitude, they may have observed an important mechanism for the dissipation of mechanical energy.

IV. EVIDENCE OF WAVE HEATING FROM EMPIRICAL MODELS

Information on wave heating of the chromosphere can be obtained not only from observations of high spatial, temporal and spectral resolution, but also from semiempirical models. What is surprising is that these models contain information not only on the rate of energy dissipation but also on the nature of the waves and on their period, even though the models are *averages* of temporal and spectral data, and the spatial data are of uncertain parentage. VAL model C', for example, is based at the foot of the chromosphere on an average of flux tubes with presumably different magnetic field strengths that are heated by waves with different energy fluxes and measured at different phases.

Two routes can be taken to extract the energy dissipation rate from an empirical model. One can analyze either the cooling function, which measures the local heating rate directly, or one can try to reproduce the temperature structure and thereby infer the heating rate indirectly.

Using the cooling function has the drawback that one depends on both the empirical as well as the theoretical model having the complete set of contributing opacity sources. This is the route, however, that is usually taken. It is the one recommended by Stein (1985) who suggests that the task of any theory of chromospheric heating must be to explain the radiative cooling function of VAL model C. The temperature structure, on the other hand, depends not only on the heating rate but also on the cooling time. However, it is not necessary that the empirical model have the complete set of absorbers; it is sufficient that the theoretical model be complete in this respect.

Anderson and Athay (1989a) chose to model the temperature function of the VAL model C. They modified a radiative equilibrium program by including nonradiative heating in the energy equation. Then they adjusted the heat input until the "theoretical" temperature agreed with the temperature of the semiempirical model. With this match of the temperature profiles, the cooling rates of the theoretical and empirical models should have agreed. They did not; they differed by more than a factor of 2 in the region of the temperature plateau at 7000 K, even though both treated line formation in statistical equilibrium. The reason for this is the difference in absorbers, the list in the model of Anderson and Athay being more extensive, especially as regards the important lines of Fe^+.

The heating rate found by Anderson and Athay rises strongly from the base of the chromosphere to the plateau between 6000 and 8000 K, where it is constant, with a flux divergence per unit mass of $4.5 \times 10^9 erg\ g^{-1}\ s^{-1}$. At the steep temperature rise into the high chromosphere, the dissipation rate rises again sharply.

Anderson and Athay speculated on the origin of the heat input. Assuming that it is entirely due to acoustic waves, they inferred from the dissipation rate in the plateau region that the velocity amplitude is equal to about half the sound speed. This was consistent with the assumption that the heating is due to sound waves and suggested shock dissipation. As a further point in support of their hypothesis, they cited the value of the microturbulent velocity, required in the modeling of line profiles, which is also of the order of the sound speed in the plateau region (cf. Avrett 1985, Fig. 9). Reasoning in a similar manner, Solanki (1988) suggested that the large value for the micro- and macro-turbulent velocity, obtained by Solanki (1986) from Stokes V profiles of Fe I and II lines originating in magnetic flux tubes, could be explained by longitudinal waves.

With the result for the cooling rate of the atmosphere in the region of the chromospheric temperature plateau, we can check whether the waves that heat the solar chromosphere have low amplitude, as assumed in the derivation of the cutoff condition, for example. We find that each pulse of a 200-s wave dissipates an energy of about 1 eV per particle. This exceeds the thermal energy of the chromospheric gas, which is approximately 0.5 eV per particle. Thus, these waves have very large amplitude.

The constancy of the heating rate per unit mass suggested to Anderson and Athay a saturation effect, in which the growth of the velocity amplitude of waves propagating in the outward direction in a gravitationally stratified atmosphere is counterbalanced by dissipation. This balance is addressed by weak-shock theory (cf. Osterbrock 1961; Ulmschneider 1970,1989; Bray and Loughhead 1974; Priest 1981), which gives an analytic expression for the dissipation rate of shock waves in terms of their Mach number and period. Even though this theory is developed for plane, monochromatic, acoustic waves and therefore not strictly applicable to longitudinal waves in a magnetic flux tube, we may use the theory for rough estimates of the Mach number and the wave period; the assumption of the theory that the waves are monochromatic is consistent with the chromospheric observations in both the calcium lines and in the radio continuum.

The decay law for the mechanical flux, πF_{mech}, of a weak, saw-toothed shape shock wave with the period P and traveling in an atmosphere with the sound speed a is given by (cf. Ulmschneider 1970; Bray and Loughhead 1974; Priest 1981)

$$\frac{d\pi F_{mech}}{\rho dh} = -\frac{(\gamma + 1)}{12} a^2 \eta^3 / P \tag{1}$$

where h is distance measured in the outward direction and where η is the shock strength, defined in terms of the preshock and postshock densities ρ_1 and ρ_2, respectively, by the equation

$$\eta = \rho_2/\rho_1 - 1. \tag{2}$$

The relation between the shock strength $\bar{\eta}$ and the Mach number M of the preshock gas is

$$\eta = 2\frac{M^2 - 1}{(\gamma - 1)M^2 + 2} \cdot \tag{3}$$

If the degree of ionization and the sound speed of the gas are constant, i.e., if $|d \ln \gamma/dh| \ll H^{-1}$ and $|d \ln a/dh| \ll (3H)^{-1}$, and the shock wave has reached its limiting shock strength, where wave amplitude growth and dissipation are in balance and the shock strength $\bar{\eta}$ is constant, then $\bar{\eta}$ and the wave period P are related by

$$P = (\gamma + 1)(H/a)\eta \tag{4}$$

where H is the scale height of the atmosphere. The equation obeyed by the wave flux is then identical to the equation for hydrostatic pressure

$$\frac{d \ln F_{mech}}{dh} = -1/H. \tag{5}$$

Assuming that the ratio of specified heats is $\gamma = 5/3$ and using the dissipation rate found by Anderson and Athay, we estimate the wave period to be of the order of 1 min and the Mach number to be nearly $M = 2$. Even though such waves violate the condition of the theory that they be weak, and there are uncertainties as far as the nature of the waves, the geometry of the medium in which the waves propagate and the variation in the value of γ, we may conclude that, based on the heating rate implied by the empirical model, weak-shock theory agrees broadly with the inferences on the wave properties drawn from the chromospheric observations, i.e., that the chromosphere is heated by waves with periods near the acoustic cutoff period of 3 min and that all the energy required for the support of the chromosphere may be supplied by these waves.

The main uncertainties in this conclusion come from the empirical chromospheric model, which averages over different regions with unknown filling factors, and from measurements of phase delays (Lites and Chapman 1979; Lites et al. 1982; Deubner 1988) that show that long-period acoustic waves are evanescent and that all acoustic waves carry little mechanical energy flux beyond 800 km in the chromosphere. The measurements are made with the Ca II infrared triplet lines. These lines show no emission peaks or self-reversals (Deubner and Fleck 1989); they may be formed mainly in the field-free intertube medium that dominates in the signal because of its filling factor. Therefore, the phase-delay observations implying an insignificant energy flux of acoustic waves in the outer solar atmosphere are probably consistent with observations of a high energy flux inside the magnetic tubes.

V. WAVE GENERATION

To understand the history of chromospheric heating theory and the failure to accept the observational evidence of heating by compressional waves with periods near 3 min, one must start with the central tenet of that field of research, that waves with periods longer than the acoustic cutoff period are evanescent and hence cannot transport energy to the chromosphere. Thus, if the chromosphere is heated by acoustic waves, their periods must be short. Consequently, the waves that have been considered have periods of the order of one tenth the acoustic cutoff period (except for Leibacher et al. [1982] who have studied 3 and 5 min waves and a single pulse). This is a plausible value that corresponds to the sound travel time over a scale height. Such waves (with enough energy to heat the chromosphere) have the required property of forming the first shock at the base of the chromosphere (cf. Ulmschneider 1970,1974), i.e., if the waves emerge from the convection zone with sinusoidal shape, the combined effects of the steepening of the profile of a finite-amplitude wave and the amplitude growth for outward propagation in a gravitationally stratified medium (Stein and Schwartz 1972,1973) produce a shock at the foot of the chromosphere. In both VAL models C and C' as well as their predecessor, the *HSRA* (Harvard Smithsonian Reference Atmosphere) (Gingerich et al. 1971), this point is located at a height of 500 km above τ_{5000} = 1. If the footpoint of the chromosphere were moved to 300 km above τ_{5000}

= 1, as might be suggested by flux-tube models, still shorter periods would be required. Also, such waves might need uncomfortably high initial energy fluxes to compensate for the stronger radiation damping in the dense layers of the photosphere (cf. Ulmschneider et al. 1978).

The modeling of heating by short-period waves received material support from estimates of the power and spectrum of acoustic waves generated in the (magnetic field-free) convection zone (Stein 1967,1968; Stein and Leibacher 1974,1981). These estimates are based on the Lighthill (1952) and Proudman (1952) mechanism of producing sound waves in a turbulent medium. However, the limitations of the theory were not fully appreciated, namely, that it is based on assumed turbulence spectra, that it does not allow for the interaction of the waves with the turbulent motions that generate them, and that it employs an expansion that is valid only for periods that are short compared to the acoustic cutoff period.

In magnetic tubes, where observations show the energy flux of longitudinal waves to be very high, theory thus far has failed to demonstrate how the large fluxes are generated. Musielak and Rosner (1987,1988) and Musielak et al. (1987,1989)—for a history of recent research, cf. Musielak et al. (1989)—have attempted to generate longitudinal waves in a medium threaded by a magnetic field; but the energy flux calculated is well below that required for chromospheric heating. Work by Ulmschneider and Zähringer (1987) and Zähringer and Ulmschneider (1987) on the generation of transverse waves in tubes by means of footpoint shaking and the coupling and transfer of energy from transverse to longitudinal modes is highly suggestive, but the energy flux they obtained is inadequate. Current work by Goldreich and Kumar (personal communication; cf. also Stein 1967) on gravity waves may provide an indirect way of generating the required longitudinal waves since their energy flux in gravity waves exceeds the emission of the chromosphere by 2 orders of magnitude, and so a relatively low efficiency of conversion of gravity waves into transverse and longitudinal tube waves would be sufficient. The process might work in a manner similar to the one investigated by Ulmschneider and Zähringer, except that instead of shaking just the footpoint, the gravity waves could be imagined to shake the whole flux tube, generating perhaps copious amounts of transverse tube waves. Such waves would be directly interesting for coronal heating. However, the waves which are observed to arise deep in the photosphere, where the far wings of the H and K lines are formed, are compressional. Showing how they are generated indirectly via transverse waves or directly from the gravity waves remains a challenge.

An aspect of the waves that is likely to be critical for their generation is that the excitation of bright points in a cell appears to follow a pattern, in which the excitation sweeps across a cell, exciting waves in the bright points in some order, and the excitation of the following wave tends to follow the same order (Damé 1983). Furthermore, the pulses that make up a wave train are not identical. In a typical case, the first pulse is particularly strong, and the two following pulses are weaker; the next pulse is again strong (cf. Damé 1983).

VI. LONG-PERIOD WAVES

According to theory, acoustic waves will not propagate vertically outward in a gravitationally stratified atmosphere if their period is longer than the acoustic cutoff period. How, then, can we understand the observational evidence, which implies that all the heating in the magnetic regions of the chromosphere occurs by means of waves with periods near and above the acoustic cutoff period? Several explanations of their importance in chromospheric heating might be advanced:

1. The flux tubes are inclined and therefore the effective scale height in the atmosphere is increased.—For an inclination angle of 30° against the vertical direction, the projection increases the cutoff period by ~ 0.5 min. This is insufficient for the longest and, perhaps, strongest waves; these might require implausibly large inclinations.
2. The observed period is not the true hydrodynamic period. Because of the outward motion of the postshock gas, which starts with nearly the speed of sound after the first wave, the succeeding shock travels into a medium that is moving outward. Therefore the new wave must travel farther and needs longer to catch up with gas accelerated by the preceding wave, thereby lengthening the observed period.—While this explanation might work for waves of shorter period, it may not provide enough lengthening of the period to solve the problem of the long-period waves. For these, the gas would have to move a significant distance relative to the thickness of the chromosphere; this is especially true for later wave trains. In addition, the atmosphere tends to oscillate at the cutoff period, and so an outward-moving wave may meet inward-moving gas. Thus this effect would probably not account for the long wave periods, but it could explain the frequent observation of an inward-moving K_3-forming layer.
3. The initial wave flux is sufficiently large that enough energy enters the chromosphere in spite of the exponential damping such waves suffer.— There is a tendency for the atmosphere to reduce excess energy flux beyond a certain limit (which depends on wave period) by radiative damping in the dense layers of the photosphere (cf. Ulmschneider and Kalkofen 1977). While the density and, hence, the radiative emission inside a flux tube are reduced, the magnetic field strength in the bright points may be too low to produce a significant reduction of the radiation damping. Furthermore, 3-min waves have very long wavelengths and therefore may not suffer much radiation damping.

The basic assumption of the theory in which the cutoff period is derived is that the wave phenomenon can be separated into a background atmosphere and a wave that runs on the background without disturbing it significantly. This assumption is incompatible with the waves that are observed to heat the solar chromosphere. These waves have high amplitude and hence effect profound changes on the structure of the medium, and for them the usual treatment of small-amplitude disturbances is invalid. Another assumption made in the derivation of the cutoff condition is that the waves are pure sinusoidal oscillations, and thus have no transients. However, the waves in Liu's observations consist of only a few pulses. In addition, successive pulses tend to have different strengths. Similar waves have been found to propagate in the chromosphere (in numerical modeling) and to dissipate even when they have a fairly small amplitude (cf. Leibacher et al. 1982).

When the magnetic field is taken into account in the derivation of the cutoff period, its value is hardly changed (Spruit and Roberts 1983). But when the ratio of specific heats drops much below $\gamma = 5/3$, the value of the cutoff period may be significantly increased. This is of interest mainly in the plateau region, where γ may approach unity during part of the time because of hydrogen ionization; but the problem of passing through the temperature minimum region remains. However, for waves with periods only slightly longer than the acoustic cutoff period, the decay length may be much longer than the pressure scale height (cf. Leibacher et al.), allowing a significant amount of energy to reach the chromosphere.

Why is the period predominantly near 3 min, and why does it appear to increase with the magnetic field strength? Leibacher and Stein (1981) have identified the 3-min period with a resonance of the "cavity" formed by the layers between the steep temperature rise in the photosphere and the sharp transition to the corona. However, the strong dissipation of these flux-tube waves, as described, for example, by weak-shock theory (where the damping of the energy flux becomes exponential; cf. Eq. 5), as well as the rare occurrence of emission in the K_{2r} peak, suggest that there is little communication between the two boundaries and, hence, this resonance is probably unimportant for the generation of the waves that heat the chromosphere. But a different resonance, the eigenperiod of the gravitationally stratified atmosphere, which is also near 3 min, may be important both for the generation of the 3-min waves and for their dissipation. (This resonance is found even in an isothermal atmosphere. Thus, it is different from the resonance associated with the chromospheric cavity.) This period is identical to the acoustic cutoff period (cf. Lamb 1908).

VII. SEMIEMPIRICAL MODELS

Empirical models play a central role in the modeling of chromospheric heating. How trustworthy are they, and are they truly chromospheric models?

There are three separate height ranges to be considered: (1) the photosphere up to the base of the chromosphere; (2) the layers where the temperature rises into the chromosphere and where most of the cores of the calcium lines are formed; and (3) the temperature plateau.

In the photosphere, where the filling factor of flux tubes is very small, model C (which differs from model C' only at the base of the chromosphere) is mainly a model of the nearly field-free intertube medium. Only towards the temperature minimum, where emission in the calcium lines favors flux tubes and leads to empirical temperatures markedly higher than those observed in the infrared continuum, is there a strong contribution by flux tubes.

At the foot of the chromosphere, where model C' is based on the Cram and Damé H line observations, which have high enough spatial resolution to exclude contamination by emission from the intertube medium, model C' is essentially a flux-tube model. But in the layers of the steep temperature rise above the base of the chromosphere, where the model is based on Skylab spectroheliograms of radiation between 400 and 1400 Å, flux tubes are not resolved. The empirical model is therefore only partly a flux-tube model. To convert it into a true flux-tube model, the height-dependent filling factor would have to be known.

In the temperature plateau, especially near the end where the temperature rises from the 8000 K level to the transition zone and the corona, the magnetic field presumably fills most of the space (forming the magnetic canopy). Therefore a model based on observations of radiation originating in these layers should be a flux-tube model.

How trustworthy is the empirical model C as a physical model of the chromosphere? In the region of the temperature plateau, where much of the radio continuum is formed, the result of the constant dissipation rate per unit mass and our understanding of that result in terms of a simple theory (namely, weak-shock theory) suggests that the empirical model is physically reasonable (but note that the solar chromosphere is time-dependent, whereas the model is static). But at the base of the chromosphere, model C' suffers from the averaging process of the Cram and Damé data, in which profiles are placed into bins on the basis of the H index, i.e., the strength of the line in the vicinity of the H_3 absorption feature. In the average profile, which is the one used for model C', this includes waves for which the maximal H index reached by a wave just qualifies for membership in the bin. It also includes much stronger waves that reach the same value of the H index during phases either preceding or following the phase of maximal H index (it also includes some emission from the network). And waves of different shock strength might arise preferentially in flux tubes of different magnetic field strength (cf. Sec. III.A). Thus, the model averages over different flux tubes with waves of considerably different energy fluxes and perhaps different magnetic field strength. In addition, the medium is strongly time dependent, which may be one of the reasons for the poor match between observed and computed profiles (cf. Avrett 1985, Fig. 18). Thus, in the layers of the steep temperature rise into the chromosphere, the empirical model should be taken only as a very rough and approximate guide to the true physical conditions in the medium.

No observations constrain the height scale of the VAL models. Instead, height is based on the solution of the hydrostatic equilibrium equation. In addition to the thermal gas pressure, it includes a contribution by turbulent pressure, which is calculated from the microturbulent velocity required for matching line profiles. This contribution is negligible in the vicinity of the temperature minimum but becomes comparable to the thermal contribution in the temperature plateau where the microturbulent velocity rises to the sound speed (cf. Avrett 1985, Fig. 9). If model C' is to be understood as a flux-tube model, which it is essentially everywhere except in the layers of the photosphere, then it must include the effect of the magnetic field on the gas density. This contribution cannot be estimated without knowing the ratio of the gas and magnetic pressure contributions. But on the basis of the theory of thin flux tubes (cf. Spruit 1981b; Kalkofen et al. 1986), which is likely to be applicable only in the layers of the photosphere, one would expect the requirement of pressure equilibrium in horizontal layers to lead to a substantial density reduction inside flux tubes relative to the outside gas at the same geometrical height. A given column mass required for a particular spectral feature would then be reached deeper in the atmosphere than in the absence of the magnetic pressure support. One might therefore suppose that the effect of the magnetic field would be a downward shift of the height scale of the gas inside flux tubes relative to the outside gas, perhaps by 1 or 2 scale heights (at least in the flux tubes of the network; for the bright points, this shift might be negligible). Thus, the footpoint of the chromosphere, which is defined by the outward temperature rise observed in the H and K lines, and hence inside flux tubes, might actually lie at a distance of 300 km above $\tau_{5000} = 1$ rather than the canonical value of 500 km, and it might be different for different flux tubes.

VIII. OBSERVATIONAL CONSEQUENCES

The inference drawn by Anderson and Athay (1989) for microturbulence from the energy dissipation rate in the temperature plateau, and the connection made here between the atmosphere inside magnetic flux tubes and the solar chromosphere, imply that the microturbulent velocity should be highly anisotropic, with a strong emphasis on the vertical direction. This anisotropy should be most pronounced where the magnetic field is mainly vertical, i.e., in the layers where the field fills all available space, which may be towards the end of the temperature plateau at 8000 K, as well as in the deeper layers where the radial component of the magnetic field in the flux tubes is small compared to the vertical component.

Acoustic heating of the CO-emitting, field-free gas should result in a noticeable microturbulent velocity. If there is such heating, this could be established by modeling of the line formation of CO on the basis of an intertube model atmosphere, not on the basis of the chromospheric model C'.

If shock waves ionize the chromospheric gas, as is suggested by the rapid disappearance of the K_{2v} spectral feature and by the large amount of energy dissipated by the waves, one should expect that during certain phases of the waves a spectrum typical of a recombining gas is emitted. Prominent observable features might be transitions from Ca^{++} to Ca^{+} as well as Lyman transitions in hydrogen.

The enhancement of the red peak K_{2r} of the K line, reported by Liu, might be caused by longitudinal waves traveling in closed flux loops of bipolar regions. Such a process would entail a time correlation of neighboring heating events.

IX. CONCLUSIONS

The apparent one-to-one correspondence between regions of enhanced vertical magnetic field and regions that are bright in Hα and the H and K lines suggests that the solar chromosphere is the atmosphere inside magnetic flux tubes. These tubes are concentrations of largely vertical magnetic field surrounded by nearly field-free atmosphere. If the shock-heated grains outline the intranetwork flux tubes, then the size of the grains and their separation imply that at the height of formation of the K_2 emission feature the flux tubes have expanded to fill approximately one tenth of the area.

The conjecture that the chromosphere can be identified with the atmosphere inside magnetic flux tubes is supported not only by the close correspondence between chromospheric emission and the magnetic field but also by the spatial structure of the emission, i.e., the small size of the heated area, and by the strong tendency of the heating to occur and recur in preferred locations.

Waves, largely compressive in nature, are observed to propagate in the outward direction. Inward propagating features are rarely observed. The waves first become visible in the H and K lines when they are still deep in the photosphere. They presumably form shocks in the chromosphere and are seen in intensity fluctuations and line shifts in Hα and the H and K lines as well as in oscillations in continuum radiation between 300 μm and 3 cm.

The period of the compressive waves heating the bright points is mainly between 2 and 4 min. A typical wave train contains three pulses; the first one of them tends to be the strongest. It is not known whether the corresponding layers in the network on the boundaries of the supergranulation cells are heated the same way. The heating mechanism in the high chromosphere, where hydrogen is completely ionized, may be different from that in the layers of the low and middle chromosphere discussed in this chapter.

Each wave pulse of a 3-min wave dissipates an amount of energy that is comparable to the thermal energy of the chromospheric gas. Hence the cutoff condition that limits propagation of low-amplitude disturbances to waves with periods shorter than the cutoff period may not apply to these large-amplitude waves.

Estimates of the energy flux carried by the long-period waves are consistent with heating by only these waves, and the dissipation rate of the heating mechanism in the layers of the temperature plateau extending from 6000 to 7000 K is consistent with heating by acoustic waves alone. In addition, weak-shock theory can account for the properties of the waves in an approximate way. The weight of this evidence implies that the chromosphere is heated only by these long-period waves and that no other heating mechanism makes a significant contribution to the energy budget of the chromosphere. Uncertainties in this conclusion come from the empirical chromospheric model, which averages over different regions with unknown filling factors, and from measurements of phase delays in chromospheric layers that imply that long-period acoustic waves are evanescent and that beyond 800 km all acoustic waves carry little mechanical energy.

Acknowledgments. I have benefited from discussions of the subject matter of this chapter from comments on the manuscript by many colleagues. In particular, I wish to thank E. H. Avrett, T. R. Ayres, A. van Ballegooijen, F.-L. Deubner, P. Goldreich, P. Kumar, W. Livingston, R. F. Stein, J. Stenflo, C. A. Whitney and H. Zirin. I owe a debt of gratitude to P. Ulmschneider with whom I have collaborated over many years and who has contributed materially to our understanding of the dynamical nature of chromospheric heating. This research has been supported by NASA.

PHYSICS OF THE INFRARED SPECTRUM

DRAKE DEMING, DONALD E. JENNINGS
Goddard Space Flight Center

JOHN JEFFERIES
National Optical Astronomy Observatories

and

CHARLES LINDSEY
University of Hawaii, Manoa

We describe the diagnostic value and principal results derived from solar studies at wavelengths exceeding 1.6 µm. Since the ratio of Zeeman splitting to line width is proportional to wavelength, the infrared is a favorable region to conduct studies of the solar magnetic field. The high-n emission lines in the 12-µm spectrum are of special interest. The large Zeeman splitting, and optical thinness of the emission, makes 12-µm observations very well suited to the derivation of vector magnetic fields. Moreover, the 12-µm lines are formed several scale heights above the white-light photosphere, in a region where vector magnetic-field information is difficult to obtain by other means. However, the LTE or NLTE nature of the lines, and the mechanism of their excitation, remain poorly understood. Observations in the 12-µm lines have been facilitated by the development of cooled spectral isolators for Fourier transform spectrometers. Stokes profiles measurements are now possible at 12 µm using single detectors, and implementation of array detectors seems imminent. Polarimetry at 12 µm is facilitated by relatively low levels of telescope and instrumental polarization. The infrared continuum and many infrared spectral lines are formed in LTE, and are especially useful in studies of the temperature structure of the solar atmosphere, from the deepest observable photospheric layers at 1.6 µm, to chromospheric altitudes at wavelengths exceeding 150 μm. Far-infrared studies have been hindered by water-vapor absorption, and the relatively coarse diffraction-limited resolution of existing telescopes. Observations from aircraft and balloon altitudes, especially during total eclipses, have alleviated some of these problems. The far-infrared continuum is an excellent thermometer for the upper photosphere and chromosphere, allowing study of the average thermal state and the compressional effects of wave motions. Observations of limb brightening at far-infrared wavelengths have shown that the structure of the chromosphere is spatially inhomogenous, even at the lowest chromospheric altitudes. Time-series observations in the far-infrared show that the chromosphere exhibits a substantial thermal response to the 5-min oscillations. Further progress in far-infrared studies will result from the new generation of large-aperture submillimeter telescopes, and from the development of the theory of radiative transfer in inhomogenous media.

I. INTRODUCTION

Most studies of the solar spectrum have concentrated on the visible spectral region. Visible-region studies are facilitated by a large photon flux, good transparency of the terrestrial atmosphere and the availability of sensitive detectors. However, investigations at infrared wavelengths are also possible, and have a long and rich history. One of the earliest studies of the solar spectrum was the demonstration by Herschel (1800) that radiation existed beyond the red limit of human vision. Pioneering investigations of the infrared spectrum were made by Langley and Abbot (1900) at the Smithsonian, and subsequently by investigators at the McMath-Hulbert Observatory (McMath and Goldberg 1949; Pierce 1949; Pierce et al. 1950). The recent advent of sensitive infrared detectors in array format has led to renewed interest in solar studies at long wavelengths. This chapter is concerned primarily with observations at the longest infrared wavelengths, concentrating on the thermal infrared region near 12 μm, and the far-infrared region at still longer wavelengths. A general review of this subject was given by deJager (1975), but many new results have been reported in more recent years.

Observations of the infrared spectral region must contend with absorption by the terrestrial atmosphere, which renders much of the spectrum inaccessible from the ground. Figure 1 shows atmospheric transmission for $1 < \lambda < 1000$ μm, based on work by Traub and Stier (1976). Initial observations of the solar spectrum in these windows were made by Mohler et al. (1950), Migeotte et al. (1956) and Hall (1974). Recently, high spectral resolution observations have been made with Fourier transform spectrometers (Goldman et al. 1980; Delbouille et al. 1981). The ATMOS Fourier transform spectrometer on Spacelab 3 recently recorded the solar spectrum from 2.2 to 16 μm with 0.01 cm^{-1} resolution (Farmer and Norton 1989), and these data should reveal the infrared spectrum without the hindrance of absorption in the terrestrial atmosphere.

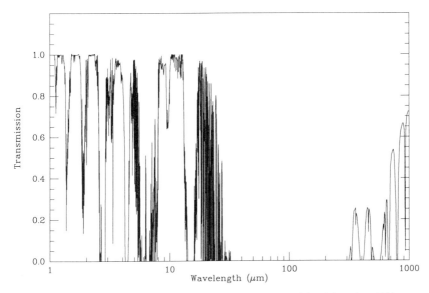

Fig. 1. Transmission of the terrestrial atmosphere in the wavelength band from 1 to 1000 μm. We have condensed the plots of Traub and Stier (1976) to represent the vertical transmission of the Earth's atmosphere from the summit of Mauna Kea, with 1 mm of precipitable water vapor along the line of sight.

II. DIAGNOSTIC VALUE OF THE INFRARED SPECTRUM

The earliest investigations of the solar infrared spectrum were motivated by the realization that the principal source of continuous opacity in the infrared is primarily free-free absorption by H^- (Chandrasekhar and Breen 1946; Geltman 1965), with some contribution due to H free-free absorption (Vernazza et al. 1976). The opacity is simple, well understood, and is in local thermodynamic equilibrium (LTE) with the electron density. At shorter wavelengths an H^- bound-free contribution must be considered, but this is zero longward of the dissociation threshold at 1.65 μm. The opacity minimum of the solar atmosphere occurs at 1.65 μm, with the H^- free-free opacity proportional to λ^2 at longer wavelengths. This means that observations at increasingly long infrared wavelengths sample the solar atmosphere at increasing heights, and the infrared continuum for $\lambda \gtrsim 150$ μm is formed above the temperature minimum in the low chromosphere (Vernazza et al. 1973,1976,1981). This is illustrated in Fig. 2, which shows weighting functions for several infrared continuum wavelengths. In addition to the simple LTE nature of the opacity, the Planck function at long wavelength is well approximated by a Rayleigh-Jeans law, in which intensity is linear in temperature. Visible and ultraviolet observations are in the regime where the Planck

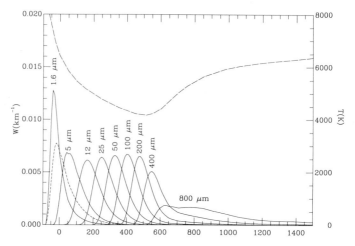

Fig. 2. Weighting functions for infrared wavelengths from 1.6 to 800 μm, shown in comparison to the solar atmospheric temperature profile (dashed line). The dot-dash line is the weighting function for the visible-region continuum. x-axis units measured in km.

function responds exponentially with temperature, and are consequently more sensitive to hot regions, while the infrared continuum gives a nearly linear average over all temperature fluctuations. For these reasons, the infrared continuum has always been regarded as an excellent diagnostic for the thermal structure of the solar atmosphere, especially for chromospheric altitudes where other indicators may no longer be in LTE.

A second way in which the infrared spectrum is diagnostic of the solar atmosphere concerns the solar magnetic field, through the wavelength dependence of Zeeman splitting. The Zeeman effect on spectral lines is the principal tool by which the solar magnetic field is investigated. However, Zeeman splitting of lines in the visible spectrum is not a large effect. For example, even a 1000 gauss field strength produces a Zeeman splitting in λ5250 which is less than the line width. This leads to complications in deriving the strength and vector geometry of the magnetic field. Babcock magnetographs (Babcock 1953) can efficiently measure the longitudinal component of the magnetic flux, but these instruments do not measure the true field strength. In the case when the Zeeman splitting is less than the line width, the derivation of vector magnetic fields is extremely difficult observationally (Harvey 1985b), due primarily to the small linear polarization from transverse field components. Moreover, the interpretation is model dependent (Stenflo 1985) in such a case. The problem is greatly simplified when the Zeeman splitting is completely resolved, as frequently occurs in the infrared region. Zeeman splitting in energy units (e.g., frequency, or wavenumbers) is constant, but Doppler line widths decrease in proportion to frequency. Consequently, the ratio of

Zeeman splitting to line width, for Doppler broadened lines, is proportional to wavelength. In addition, instrumental and telescope polarization, which potentially interfere with vector field measurements, are greatly reduced at longer wavelengths.

A limitation of solar observations in the infrared has been the limit on spatial resolution set by diffraction. The diffraction limit of the McMath solar telescope at 12 μm ($1.22\lambda/D$) is 2 arcsec; the NASA 3-m Infrared Telescope Facility (IRTF) has a 30 arcsec diffraction-limited beam at 350 μm. This means that high-resolution imaging is not currently possible at the longest infrared wavelengths, and high spatial resolution observations obtained in the visible will greatly complement infrared studies. For example, Zeeman studies at 12 μm can benefit from simultaneous high-resolution magnetograms made in the visible or near-infrared. At the longest infrared wavelengths, occultation of the solar disk by the Moon during total eclipses has been used to extract higher spatial resolution at the solar limb (see Sec. III.C).

III. DISCUSSION BY SPECTRAL REGION

Many new insights into the physics of the solar atmosphere have come from observations at infrared wavelengths. We organize and discuss these advances by spectral region. We briefly discuss the near-infrared regions shortward of the 10 μm window, and we then give more extended discussions of the 10 to 12 μm spectrum, and of the far-infrared region at still longer wavelength.

A. The Near-Infrared Region ($\lambda < 5$ μm)

An important window in the terrestrial atmospheric spectrum occurs near 1.6 μm (Fig. 1), where the solar atmospheric opacity reaches a minimum, and observations are possible to deeper layers of the photosphere than is the case in the visible. Imaging in the continuum at these wavelengths can yield new insights into the thermal structure of various components of the solar atmosphere. For example, Foukal et al. (1989) have recently imaged sunspots and faculae at 1.6 μm using an infrared CCD camera, finding that faculae are often darker than the surrounding photosphere at this wavelength. They conclude that their observations contradict the prevailing view that faculae require a source of heating by waves, electric currents or similar mechanisms. Since magnetic flux tubes play a central role in the dynamics of solar active regions, it is obviously of interest to conduct Zeeman studies as well as imaging at 1.6 μm. The first such observations were made in the very Zeeman-sensitive 1.56 μm Fe I lines by Harvey and Hall (1975), with subsequent studies by Stenflo et al. (1987b). Very recently, Rabin and Graves (1989) have measured sunspot magnetic fields in these lines, using a facility infrared array from the National Optical Astronomy Observatories.

The spectral region near 2.3 μm contains the $\Delta V = 2$ overtone bands of

CO, with the $\Delta V = 1$ fundamental bands lying near 4.6 μm. These bands were first detected in the solar spectrum by Goldberg et al. (1952), and have subsequently been investigated at high resolution using Fourier transform spectrometers (see e.g., Ayres and Testerman 1981). The $\Delta V = 1$ CO lines in the Sun are widely believed to be formed in LTE (Carbon et al. 1976; Ayres and Wiedemann 1989), and since the cores of the strongest lines are formed at $\sim$ 500 km altitude, they represent an excellent temperature diagnostic for the upper photosphere and temperature minimum region. The CO lines can be used to investigate both the average temperature structure, and its time variability. Noyes and Hall (1972) showed that the cores of the strong CO lines exhibit large brightness temperature fluctuations in response to the 5-min oscillations, and very recent time-series observations in these lines by Ayres and Brault (personal communication) should give valuable information on the thermodynamics of the oscillations. Noyes and Hall also noted that, at the solar limb, the CO line-core brightness temperatures fell to values below the minimum temperature in solar atmospheric models. Ayres and his co-workers (Ayres 1981; Ayres and Testerman 1981; Ayres et al. 1986) have developed the idea that a thermal bifurcation exists on very small spatial scales in the solar atmosphere, with substantial amounts of cool material co-existing with hotter regions. It is natural to identify the hotter regions as being heated by wave motions in small-scale magnetic flux tubes. In this view, the upper solar atmosphere only appears uniform to the extent that the hot and cool components are spatially unresolved. Ayres (1981) suggested that CO itself was largely responsible for the low temperatures in the cool regions, via radiative cooling in the strong bands. This idea is supported by some calculations (Muchmore and Ulmschneider 1985; Anderson 1989), although not by the results of Mauas et al. (1990). The concept of thermal bifurcation of the upper solar atmosphere may be related to the inhomogeneities which are inferred from chromospheric observations in the far-infrared continuum (see Sec. III.C).

B. The Region Between 10 and 13 μm

Continuum Observations.   Another very important region of the solar spectrum is accessible to groundbased observation in a window between 9 and 13 μm. Limb darkening and other photometric measurements (Pierce et al. 1950; Saiedy 1960; Lena 1968, 1970) have long been made in this spectral region. Raster-scan imaging of sunspots at 10 μm was reported by Turon and Lena (1970), who found their 10 μm images to be very similar to visible region images. Lindsey and Heasley (1981) observed faculae at 10 μm, finding them to have a relatively low contrast, with an excess emission only a few percent above the quiet Sun. They concluded that their observations strongly supported models in which the faculae are confined to narrow flux tubes in the low photosphere.

The Line Spectrum.   It is only recently that attention has been drawn to the very significant solar line spectrum which occurs in the 10-μm spectral region. Regarding the line spectrum, deJager (1975) remarked that "in this field no progress has been made in the last ten years," and that "no solar lines have with certainty been identified for $\lambda > 5$ μm." This situation prevailed until as late as 1981, when Murcray et al. (1981) noted unidentified emission features near 7.4 and 12.4 μm, using Fourier transform spectrometer (FTS) data from the ground and balloon altitudes. Goldman et al. (1981) identified pure rotation lines of the OH radical in the photospheric spectrum. In addition to the photospheric spectrum, the spectra of sunspots are very rich in molecular lines in the 10 μm region (Glenar et al. 1983).

OH Rotational Lines.   The OH pure rotation lines in the solar spectrum were used by Goldman et al. (1983) to derive the photospheric oxygen abundance. An improved analysis, using higher resolution data, was given by Sauval et al. (1984) who obtained a value of 8.91 ± 0.01, on the usual logarithmic scale where the hydrogen abundance is 12.00. A quartet of v = 0 OH lines in the photospheric spectrum is shown in Fig. 3. The nearly linear dependence of the Planck function on temperature at this frequency makes the lines appear deceptively weak. Actually, these are strong lines whose cores are formed near $\tau_{5000} \sim 0.005$ ($\sim$ 300 km altitude). The OH rotational lines are prime indicators of the photospheric oxygen abundance, for several

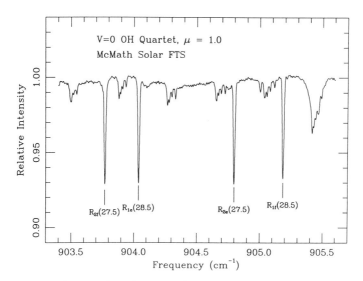

Fig. 3. McMath FTS observations of a quartet of rotational lines of OH in the v = 0 state.

reasons. They have been observed in vibrational states up to v = 3, where the lines are completely unsaturated, with consistent results. Being pure rotation transitions, deviations from LTE are unlikely. The f-values are well determined from the dipole moment function (Werner et al. 1983), and Sauval et al. (1984) found that the solar data could be used to constrain simultaneously the dipole moment function as well as the solar oxygen abundance. The dissociation energy is also well known (Carlone and Dalby 1969). Both Goldman et al. (1983) and Sauval et al. (1984) found that the photospheric model of Holweger and Muller (1974) gave excellent agreement with the equivalent widths as a function of lower state energy. However, the center-to-limb behavior of the lines is not in accord with any one-dimensional model of the upper photosphere. Deming et al. (1984) observed the OH lines with a laser heterodyne spectrometer, finding that the lines strengthen at the limb, the line core brightness temperature falling as low as 4000 K. This behavior was confirmed by Ottusch (1986), who observed additional OH lines with a tunable-diode laser heterodyne spectrometer. The behavior of these OH lines is similar to the CO lines, and it supports the idea of thermal bifurcation in the upper photosphere.

Like the CO lines, the OH lines are valuable diagnostics of the thermodynamics of the 5-min oscillations. Deming et al. (1986) performed laser heterodyne time-series observations of an OH line at solar disk center. The laser heterodyne technique (Kostiuk and Mumma 1983; Glenar et al. 1986) is limited in the sense of being able to observe at most a few spatial positions simultaneously. However, the stability and spectral purity obtainable using this technique make it particularly well suited to determine the effect of the oscillations on line profiles, e.g., on the amplitude and phase relations between temperature and velocity. Figure 4 shows the single-point power spectra obtained by Deming et al. (1986) at solar disk center. The envelope of power in the line-intensity spectrum, indicative of temperature fluctuations, is shifted to a significantly higher frequency than in the velocity spectrum; this may occur via radiative damping and other effects. The line-intensity spectrum appears to peak rather sharply at 4.3 mHz, which is close to the frequency of the chromospheric resonance as calculated by Ando and Osaki (1977), leading Deming et al. (1986) to claim detection of this resonance. It has not yet been possible to obtain a two-dimensional (k, ω) power spectrum, which is needed to identify such a resonance unequivocally. This could potentially be achieved by imaging in the OH lines cores, using a high-resolution Fabry-Perot etalon, and the new array detectors now available in this spectral region. Similar studies using the cores of the CO lines would be of great interest.

The 12-μm Emission Lines.   Perhaps the most interesting development in solar spectroscopy in recent years has centered around the emission lines observed in the 12-μm region. The 12-μm lines were first noted by Murcray

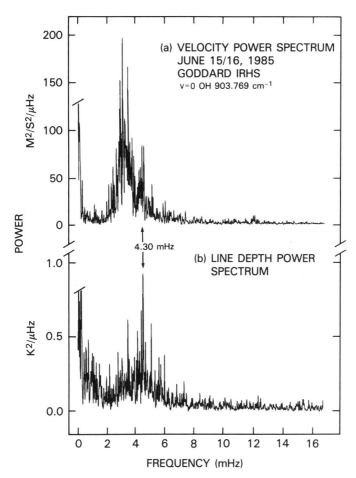

Fig. 4. Power spectra of the velocity and line depth for a v = 0 OH rotational line, measured at a single point at solar disk center by Deming et al. (1986).

et al. (1981), and were independently discovered, and extensively characterized, by Brault and Noyes (1983, hereafter BN). The lines were subsequently identified as nearly hydrogenic high-n transitions in neutral magnesium, aluminum and silicon (Chang and Noyes 1983; Chang 1984). More than fifty lines are seen in emission in this region, although most are quite weak. The two strongest lines, near 811 and 818 cm^{-1}, are illustrated in Fig. 5, where it can be seen that at disk center they are accompanied by broad, shallow absorption wings. BN demonstrated that the emission lines are present in the quiet Sun, plages and sunspot penumbrae, although not in large sunspot umbrae. The absorption component weakens and disappears for $\mu \lesssim 0.5$. The lines show very great Zeeman splitting into three components, and are the

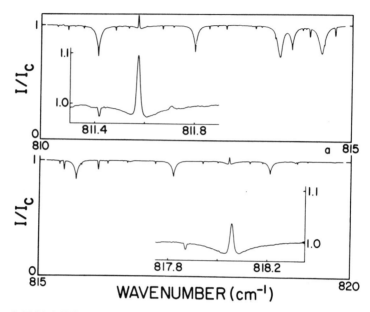

Fig. 5. McMath FTS observations of the Mg I emission lines near 811 and 818 cm⁻¹ at solar disk center (figure from Brault and Noyes 1983).

most Zeeman-sensitive lines that are currently observable in the solar spectrum. The appearance of the 811 and 818 cm⁻¹ line in sunspots and plage regions is shown in Fig. 6. Further discussions of the line properties have been given by Goldberg (1982), Noyes and Avrett (1987), Deming (1987), Deming et al. (1988b), Zirin (1988) and Zirin and Popp (1989). Jennings et al. (1986) detected the lines, in absorption, in the spectra of a few bright red giants.

The 12-μm emission lines present both a challenge and an opportunity. The challenge arises in that we cannot firmly answer the most obvious question that comes to mind: why are the lines present in emission instead of absorption? The opportunity arises in that the great Zeeman sensitivity of these lines gives us an excellent tool for very sensitive vector magnetic-field measurements, at an altitude several scale heights higher than most Zeeman measurements made in photospheric lines.

There are many clues that may enable us to determine the emission mechanism and altitude of formation for the 12-μm lines. BN regarded the lines as formed in the low chromosphere. They showed that the lines are not coronal, since their widths are less than the coronal Doppler width. There are good indications that the lines are optically thin: the intensity increases towards the limb closely proportional to 1/μ, and the Zeeman splitting in a sunspot penumbra is the same in the 811 and 818 cm⁻¹ lines (Fig. 6), al-

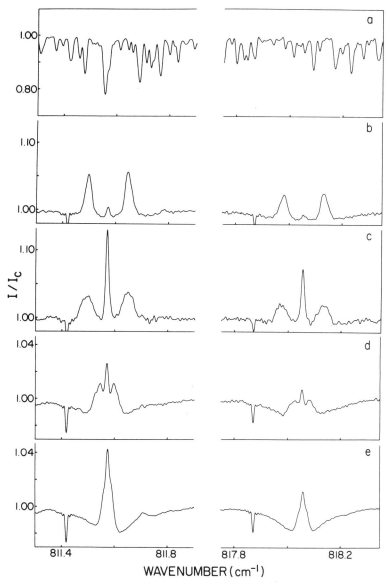

Fig. 6. McMath FTS observations of the 811 and 818 cm^{-1} emission lines in the spectra of sunspots and plages, from Brault and Noyes (1983): (a) in a large sunspot umbra; (b) on the side of the penumbra toward disk center (nearly longitudinal magnetic field); (c) on the side of the penumbra toward the limb (nearly transverse magnetic field); (d) and (e) two different regions in a plage near disk center.

though the line strength differs by a factor of 1.7. We point out that if the lines were not optically thin, then their depth of formation would vary with line strength, and the vertical magnetic-field gradient above the penumbra would lead to an $\sim 3\%$ difference in Zeeman splitting. However, Brault and Noyes (1983) found the splittings to be the same to 1%.

An obvious idea is that these lines are in emission simply because they sample the chromospheric temperature rise. It has long been realized that, at a sufficiently long wavelength, the solar spectrum should turn from a predominately absorption spectrum to a spectrum of emission lines as the increase in continuous opacity veils much of the photospheric continuum, enhancing the visibility of chromospheric emission. However, this should not happen at 12 μm if standard solar atmospheric models apply (Vernazza et al. 1981). Using standard models, the temperature is still decreasing at the level where 12-μm radiation is formed. Moreover, the 12-μm lines should not have sufficient opacity in the chromosphere to produce a noticeable emission (Chang and Noyes 1983). Including deviations from LTE (Lemke and Holweger 1987) has not helped. The difficulty of producing the requisite column density in these levels led Lemke and Holweger (1987) to suggest that the lines were formed out of LTE, but in the upper photosphere where the greater column density makes a significant emission feature easier to obtain. Zirin and Popp (1989) have recently argued that the lines are formed in LTE at relatively low altitudes, near the classical temperature-minimum region, but they argue that the temperature is rising at these altitudes, not falling. This would require that the solar atmospheric models be modified in a rather drastic manner. Even if the emission cores are formed out of LTE, we would expect LTE to hold at the depth where the line wings are formed. The fact that the absorption wings disappear for $\mu < 0.5$ is a strong argument in favor of the Zirin and Popp view. However, a temperature increase just above the 12-μm continuum-forming region is contradicted by the existence and behavior of the OH rotational lines discussed above. The Zirin and Popp view of the emission lines would force us to adopt a thermally bifurcated atmospheric model, and this is interesting, since Deming et al. (1984) found similar indications from the OH lines themselves. Such a thermal bifurcation would have to occur on a scale well below the 12-μm diffraction limit of the McMath telescope (2 arcsec), since neither the OH lines (Deming et al. 1984) nor the emission lines (Jefferies et al. personal communication, 1988) show evidence of thermal bifurcation on resolvable scales. Zirin and Popp (1989) have stressed that the $1/\mu$ limb brightening of the emission lines is typical of formation in a homogenous layer, so any spatial inhomogeneities that contribute to the emission lines must mimic a homogenous layer when spatially averaged.

Additional recent observations have given more information relevant to the altitude and mechanism of emission-line formation. Deming et al. (1988b) observed the 5-min oscillation in the 811 cm^{-1} line, and measured the Evershed flow in a sunspot penumbra. The 5-min oscillation does not show the "high frequency tail" (Evans et al. 1963) which is characteristic of chromospheric lines. The Evershed flow is an inflow, as in Hα, but of smaller amplitude. They concluded that these properties indicated an altitude of formation near the temperature minimum. No intensity oscillations were detected in the line, from which they concluded that the line was not formed in LTE. Their upper limit on intensity corresponds to a kinetic temperature oscillation of < 50 K, below what is seen in CO features (Noyes and Hall 1972), but this could perhaps be reconciled with LTE line formation under some conditions. Glenar et al. (1988b) recently detected Mg I *absorption* lines in the solar spectrum near 9 and 12 μm. The strongest of the absorption lines are from the $n = 6$ level of Mg I; they are ∼ 3 times stronger than LTE lines formed in the Holweger and Muller (1974) model, but show no emission cores. They concluded that the emission process must be confined to altitudes well above the $\tau_{5000} = 10^{-2}$ level.

The difficulty in producing the emission lines in standard solar atmospheric models gives us two choices. We must either change the models (as Zirin and Popp [1989] imply), or change the level populations within the models. In the latter instance, we must look for a process that: (A) increases all of the high-n level populations, so that an appreciable column density can be achieved at chromospheric altitudes, forming an emission line thermally; or (B) changes the relative level populations, so that the line-source function can be raised above the Planck function at upper photospheric altitudes (Lemke and Holweger 1987). It is easier to find candidate mechanisms for (A); e.g., dielectronic recombination immediately comes to mind. Dielectronic recombination is often a much more efficient process than radiative recombination (Goldberg and Dupree 1967), but it usually applies to levels considerably higher than $n = 7$. Another possibility is the suggestion by Chang (1987) that near-resonant charge exchange with hydrogen may be effective in increasing the Mg I population, e.g.,

$$H(n=7) + Mg^+ \leftrightarrow H^+ + Mg(n=7). \tag{1}$$

Rate calculations for this process in solar atmospheric models are needed. Chang (1987) also stresses the possibility of high cross sections for l-changing collisions.

The absorption wings which are present on the 811 and 818 cm^{-1} emission lines are very broad (0.17 cm^{-1}), and suggest the presence of pressure broadening. Glenar et al. (1988b) found that the wings of the Mg I absorption lines from $n = 6$ states also required pressure broadening, with a cross section of order 10^{-14} cm^2. High-n atomic lines are expected to have large cross sections for line broadening, because the electron eigenfunctions peak at large distances from the nuclear core. Deming (1987) suggested a pressure shift as the cause of the ∼ 0.01 cm^{-1} differences between the frequencies of the emission core and the absorption wings in the 811 cm^{-1} line, an effect which is not noticeable at 818 cm^{-1} (see Fig. 5). Laboratory measurements (Lemoine et al. 1987b) show that the emission core is at the rest frequency, the absorption component being shifted to higher frequency. The lower state for the 811 cm^{-1} line has the maximum allowed angular momentum. The electron eigenfunctions for the highest angular-momentum states do not penetrate to the nucleus, and are plausibly more subject to effects such as pressure shifts. Such effects are, like Zeeman splitting, larger in comparison to Doppler widths at longer wavelengths. Quantum mechanical calculations and laboratory measurements of broadening and level-shift effects via collisions with ions, electrons and neutrals are needed. Comparison of the observed line profiles with theoretical profiles based on such studies is a potentially very powerful method of determining the pressure level where the absorption wings are formed. Initial work along these lines was reported by Hoang-Binh et al. (1987), who calculated collisional broadening of hydrogenic transitions by ions and electrons.

Finally, there is another property of the emission lines which is not understood at all, and it involves an apparent interaction with the OH rotational lines. Sauval et al. (1984) noted that a few of the OH rotational lines are anomalously weak, and that this occurs whenever an OH line falls close to, but not coincident with, an emission line. This is *not* due to a simple filling in of the OH absorption by overlapping emission. Instead, they suggested that the emission-line wings produce opacity which mimics continuous opacity, thereby weakening the OH lines. However, adding opacity to an OH absorption line should decrease both the line core and continuum specific intensity, as long as the source function is decreasing with altitude. The OH line will weaken because the line core will be less affected than the continuum. However, in one instance of this interaction noted by Deming et al. (1984), the OH line core intensity *increases* substantially, as if the OH line source function were increasing with altitude. Compared to the other three lines in this quartet (see Fig. 7), the perturbation to the OH line is seen to be substantial, in spite of the fact that the perturbing emission line is so weak that it is only visible at the extreme solar limb.

Although most attention has been concentrated on the heavier element emission lines, hydrogen emission is also seen in this spectral region. Zirker (1985b) discussed observations of 10 to 20 μm region hydrogen lines in prominence spectra, showing that the results are consistent with previous work at shorter wavelengths.

Potential for Zeeman Studies Using the Emission Lines.   When the 12-μm emission lines were first described by BN, it was immediately recognized that they were a potentially powerful tool for studies of the solar magnetic field. They give information on relatively high altitudes, where other lines are unsuitable. Chang (1987) has shown that the Zeeman patterns are virtu-

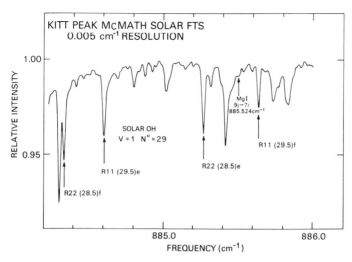

Fig. 7. Example of the perturbation of an OH rotational line by the Mg I $9j - 7i$ emission line, which is itself too weak to appear in this disk-center spectrum. The line-core intensity of the R11 (29.5)f line has been substantially increased by the presence of the emission line, as can be seen by comparison with the other v = 1 lines. In the unperturbed case the four OH line profiles should be identical.

ally identical to normal Zeeman splitting, i.e., splitting into three components. This occurs because the level fine structure, that ordinarily produces the anomalous Zeeman effect, is 2 orders of magnitude smaller than the Zeeman splitting. Under such conditions, the anomalous Zeeman pattern reverts to the normal Zeeman pattern, which is referred to as the Paschen-Back effect. Fine structure is believed to be present in the Zeeman pattern of the Mg I lines at only the 0.001 cm^{-1} level, which is well below the Doppler width of the lines. Chang's (1987) calculation of a Lande g-factor of unity for the strongest lines was confirmed by Lemoine et al. (1988) by direct laboratory measurement. Other properties of the lines are also well suited for magnetic-field studies. The optical thinness of the lines means that their polarization state is not altered by radiative transfer in the overlying atmosphere. Moreover, instrumental and telescope polarization effects, which plague measurements in the visible, are much reduced at 12 μm. For example, oblique reflections at aluminum mirror surfaces induce relatively little 12-μm polarization, because the real and imaginary coefficients of the index of refraction for aluminum are both very large at 12 μm.

Although the potential utility of these lines in solar magnetic-field studies was immediately recognized by BN, it was believed that polarization (Stokes profile) observations at 12 μm would be difficult to achieve (Harvey 1985b). It was thought that thermal background noise would severely limit the sensitivity of the observations. Even if a narrowband cryogenic filter were used to eliminate all thermal radiation outside the region of the lines, the intrinsic detector noise might be too high to take advantage of the reduced background. Moreover, no method was available to perform polarization analysis at 12 μm. Fortunately, this situation was already changing at the time of the BN observations. Doped silicon detectors were becoming available with intrinsic noise well below the level needed for use with narrowband cryogenic filters. In particular, the blocked-impurity-band (BIB) detectors manufactured by Rockwell International for use in the Cosmic Background Explorer spacecraft were extremely low-noise, stable devices. At the same time, liquid-He cooled monochromators were being built for use as ultra-narrow bandwidth filters (Wiedemann et al. 1989). The combination of a BIB detector and cryogenic monochromator was first used as a "postdisperser" on the McMath FTS in late 1986. By eliminating noise due to out-of-band radiation, 0.005 cm⁻¹ resolution spectra were recorded in 90 s, with spatial resolution close to the telescope diffraction limit (Deming et al. 1988b). Examples of these spectra taken in a sunspot penumbra are illustrated in Fig. 8. Although the Deming et al. study was made without polarization measurements,

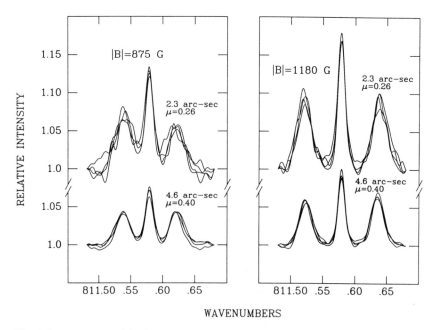

Fig. 8. Sunspot spectra of the 811 cm⁻¹ line taken by Deming et al. (1988b), using a cryogenic grating postdisperser to increase the sensitivity of the McMath FTS. Spectra taken in 90-s intervals at two values of magnetic-field strength, and two disk positions, are illustrated.

several new results concerning the magnetic structure of sunspots were obtained. From the width of the σ components in the penumbra, Deming et al. deduced that small-scale variations in field strength of order ± 235 gauss were present, possibly related to the penumbral filamentary structure which is apparent in high-resolution visible images. They were also able to confirm an effect first noted by BN, who found that the central π component becomes very narrow when the observations are made looking across the field lines, showing that nonthermal motions were suppressed by the field (e.g., see Fig. 6c). The plage surrounding the sunspot observed by Deming et al. was found to have an average field strength of 400 gauss, consistent with the early plage observations by BN, who noted that a field strength of this magnitude was consistent with the divergence of flux tubes having much greater field strengths at lower atmospheric levels. Zirin and Popp (1989) found similar values for plage field strengths, but they argue that the field is not diverging, and that their values can be extrapolated down to photospheric levels. Polarization measurements are needed to evaluate the divergence properties of plage fields at the level of 12-μm line formation, and such measurements were recently made by Hewagama et al. (1989).

Hewagama et al. (1989) successfully measured the Stokes profiles of the 811 cm^{-1} line using the McMath FTS. Figure 9 shows an example of their profiles in a penumbral region where the field was primarily transverse to the line of sight, showing large linear (Stokes U,Q) and small circular (Stokes V) polarization. Polarization analysis is performed using a ZnSe Fresnel rhomb, or quarter-wave plate, and photoresistive thin-film linear polarizers.

These developments have now made it possible to derive vector magnetic-field information from McMath FTS observations with good sensitivity. However, the single-point spatial coverage of FTS observations remains a significant limitation. Clearly it is desirable to implement array detectors, thereby observing all spatial positions within a large area simultaneously. At the same time that cryogenic postdispersers were being constructed, sensitive infrared array detectors were becoming available. The most sensitive of these are HgCdTe arrays for $\lambda < 3.5$ μm, InSb arrays operating out to 5.5 μm, Si:Ga devices operating to 17 μm, and Si:As arrays sensitive to wavelengths as long as 24 μm. The characteristics of the current generation (circa 1989) of infrared array detectors is summarized in Table I. There is great potential for application of these arrays to solar studies at many wavelengths, and both the technology and its solar applications can be expected to develop rapidly in the next few years. The current generation of arrays for 12-μm application (Si:Ga and Si:As) have more than adequate sensitivities for Stokes profile measurements of the 12-μm emission lines at high spectral and spatial resolution. A NASA-funded 12-μm imaging Stokes polarimeter, utilizing an infrared array detector and a cooled Fabry-Perot etalon, is under development at the Goddard Space Flight Center.

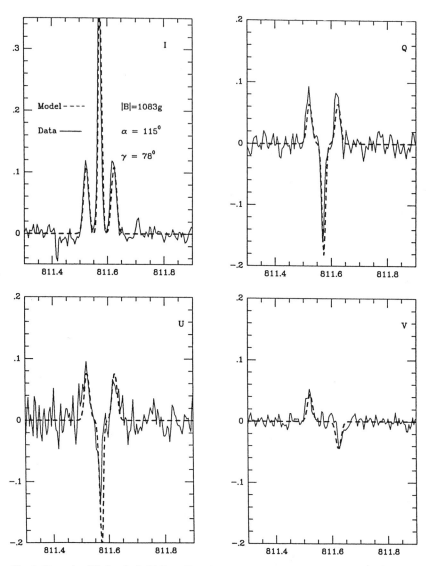

Fig. 9. Example of Stokes *I, Q, U, V* profiles of the 811-cm^{-1} emission line, obtained by Hewagama et al. (1989) in a sunspot penumbra, using a ZnSe Fresnel rhomb and photoresistive thin-film linear polarizers on the McMath FTS. The large amplitude in linear polarization (*U, Q*), and small circular polarization (*V*), indicates a magnetic field nearly transverse to the line of sight.

TABLE I
Infrared Detector Arrays

Material	Format	λ-Range	Type	Multiplexer	Manufacturer
PtSi	640 × 488	1–5.4 μm	PE[a]	DRO/CCD	Kodak, Hughes, RCA, etc.
Ge diodes	32 × 32	1–1.6 μm	PV[b]	JFET	Ford Aerospace
HgCdTe	128 × 128	1–2.5 μm	PV	CCD	Rockwell
InSb	58 × 62	1–5.5 μm	PV	CCD	SBRC[c]
Si:In	32 × 32	1–7.5 μm	PC[d]	CCD	Rockwell
Si:Ga	58 × 62	5–17.5 μm	PC	DRO[e]	SBRC
Si:As	10 × 50	5–24 μm	BIB[f]	SWIFET[g]	Rockwell

[a]Photoemissive
[b]Photovoltaic
[c]Santa Barbara Reserch Center
[d]Photoconductive

[e]Direct Readout
[f]Blocked Impurity Band
[g]Switched MOSFET

C. The Far-Infrared Region (λ > 13 μm)

The Line Spectrum.   Very few spectral lines have been observed in the solar spectrum longward of the 12-μm window. Since these generally involve transitions between high-lying levels, they are readily pressure broadened. Hoang-Binh (1982) predicted the observability of far-infrared hydrogen lines on this basis, and Boreiko and Clark (1986) observed the $n = 16$ to 15 and $n = 14$ to 13 hydrogen emission lines longward of 100 μm. These very difficult observations were made using a balloon-borne cryogenic FTS at 0.015 cm^{-1} resolution, and also showed evidence for similar hydrogenic transitions in Mg I. In contrast to the case near 12 μm, where the Mg I lines dominate the spectrum, at the longer wavelengths the hydrogen lines are stronger. This change in relative line strength has been discussed qualitatively by Zirin and Popp (1989), and it may represent a clue to the formation of the 12-μm lines.

Continuum Measurements in the Far Infrared.   The far-infrared solar continuum is an excellent thermometer for the upper solar atmosphere. A comparison between a far-infrared intensity map and an image of the Sun in any strong visible chromospheric line illustrates the quality of the far-infrared continuum as a thermometer on a very basic level. Figure 10 shows the Sun in a strong chromospheric line, the K-line of singly ionized calcium, which forms well above the temperature minimum. If this line were an excellent thermometer, the image would show limb brightening, since the temperature of the chromosphere *increases* with height. However, the excitation temperature of calcium ions *decreases* with height, as the collisional excitation rate drops with the number density, and the transition becomes radiatively coupled

Calcium K

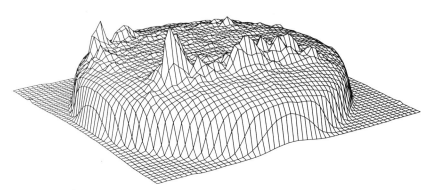

Fig. 10. Intensity map of the full solar disk in light of singly ionized calcium, showing strong limb darkening.

to cold space. All visible-region chromospheric lines show similar strong limb darkening, and are relatively poor thermometers. Figure 11 shows the Sun in 800-μm continuum radiation. This intensity map was made at the NASA IRTF on Mauna Kea, using Thijs de Graauw's heterodyne receiver (Lindsey et al. 1984*a*). It shows clear limb brightening, consistent with an overall increase of the free-electron kinetic temperature with increasing height. The far-infrared continuum is, in fact, the best known thermometer throughout most of the photosphere and chromosphere. For this reason, it has played an important role in all of the most thorough atmospheric models of the last two decades, especially for the low chromosphere and temperature minimum region. The same considerations make the far-infrared continuum equally important in regard to *dynamics* in the solar atmosphere.

The potential of far-infrared continuum photometry in solar work was well known by the late nineteen sixties (Noyes et al. 1966). Efforts to determine the far-infrared continuum solar flux include observations by Mankin (1969), Gay (1970), Eddy et al. (1969) and Gezari et al. (1973). The best measurements to date are those of Rast et al. (1978). These observations, properly assimilated, clearly show the reversal in the vertical temperature gradient in the low chromosphere. Infrared measurements in the late sixties did not satisfactorily fit solar atmospheric models of that time. The measurements of Eddy et al. argued for a cooler temperature minimum than that of the Bildeberg Continuum Atmosphere (BCA; Gingerich and deJager 1968). The Harvard-Smithsonian Reference Atmosphere (HSRA; Gingerich et al. 1971) improved on the BCA by satisfying the infrared continuum observa-

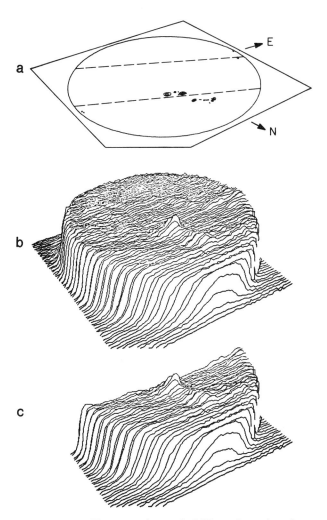

Fig. 11. The Sun in the submillimeter continuum. A visible-continuum intensity map of the Sun (a) is compared with projected relief maps of solar intensity in the 820-μm continuum (b and c). The map in (c) is cut across the middle to show the limb brightening more clearly. The 820-μm observations were made at the IRTF with Thijs de Graauw's heterodyne receiver (Lindsey et al. 1984*a*).

tions. The most careful and thorough atmospheric modelers since that time (see e.g., Vernazza et al. 1976,1981) have taken care to construct low chromospheres which satisfy the far-infrared observations.

Limb Brightening.   Besides absolute photometry, a subject of considerable interest has long been the radial brightness profile of the quiet solar disk. Simple representations of the solar atmosphere as consisting of smooth or plane-parallel layers of gas predict limb darkening, with an absorption line spectrum, for wavelengths that form below the temperature minimum, and limb brightening, with an emission-line spectrum, for wavelengths that come from above it. Investigators who encountered the technical difficulties of directly measuring the absolute far-infrared solar flux recognized the possibility of using the limb-brightness profile to determine the vertical temperature gradient. If determined over a broad range of wavelengths, this could supplement direct photometry in a very difficult region of the spectrum. This concept failed in practical application, particularly for infrared wavelengths longward of 100 μm and for radio wavelengths, where strong limb brightening was expected. Attempts to use this method of determining the overall vertical temperature gradient of the chromosphere gave results sufficiently at odds with direct photometry (see, e.g., Mankin 1969) that something had to be wrong. The structure of the chromosphere is far from smooth, apparently even at the lowest levels, and because of this, limb-brightening profiles cannot be used as a quantitative measure of the vertical temperature gradient. Far-infrared and radio observations were eventually shown to exhibit weak limb brightening (Lindsey and Hudson 1976; Lindsey et al. 1981; Horne et al. 1980), but it was also clear that the plane-parallel assumption simply would not work at any chromospheric level. Much of the insight into this problem came from models designed to explain radio observations, and a very clear and concise review was given by Simon and Zirin (1969). Limb-brightness profiles have since become of renewed importance to solar atmospheric modeling. In a complex chromosphere, their utility is no longer limited to a supplementary role to direct photometry. With the convenience of smooth, plane-parallel structure removed, constraints from limb observations become fundamental to atmospheric modeling.

Groundbased Observations.   In addition to the problem posed by the limited transparency of the terrestrial atmosphere, and the limit on spatial resolution due to diffraction (see Sec. II), a further problem arises at the longest infrared wavelengths. As the Rayleigh-Jeans limit is approached, the Planck function becomes linear in temperature, and the thermal background emission from the instrument, telescope and terrestrial atmosphere become significant relative to the solar signal. The problem is exacerbated by the substantial emissivity of the terrestrial atmosphere at wavelengths which are only partially transparent. For these reasons, high-quality solar observations at the longest wavelengths require that the background be subtracted. Noyes et al. (1968) pioneered far-infrared solar astronomy from the ground, with their two-beam scans across the solar disk. They borrowed a technique called two-beam subtraction from stellar infrared astronomy. A chopping secondary mirror alternates between two beams separated on the sky by up to several arcmin. The difference signal between the two beams is synchronously detected by phase-sensitive electronics. For solar observations the same technique works to cancel the undesirable effects of drifting sky transparency. While transparency variations at a single point on the sky can be quite severe, they are highly uniform over several arcmin, and it is thereby possible to compare brightness variations between two points on the quiet Sun to better than 0.01%.

Noyes et al. (1968) used two-beam chopping to look for brightening at the extreme limb, which was expected at 22.5 μm; they saw no evidence of limb brightening at this wavelength. They used both single- and two-beam scans across the disk to look for limb brightening at 1.2 mm. They failed to see brightening directly, because of poor angular resolution, but they deduced correctly that the extreme limb was brightened, based on the asymmetry of the two-beam limb excursion profile. Others (Beckman and Clark 1973; Beckman et al. 1973; Righini and Simon 1976) subsequently found that submillimeter limb brightening was far less consistent than with a smooth lower chromosphere. Lindsey and Hudson (1976) detected significant limb brightening using a two-beam chopping technique, but with a much greater beam throw than used by Noyes et al. (1968). With modern facilities, this limb brightening has become relatively easy to detect from 800 μm to wavelengths exceeding 1 mm (Lindsey et al. 1981;1984a; Horne et al. 1980), but it remains only a fraction of the limb brightening predicted by smooth model chromospheres.

Generally, shortward of 350 μm, limb brightening decreases and is confined to the extreme limb. Airborne observations shortward of 200-μm wavelengths normally fail to resolve it (Lindsey et al. 1984b), but eclipse observations show the extreme limb clearly brighter down to 100 μm.

Eclipse Observations.   A powerful method of surmounting the diffraction limit to obtain subarcsecond resolution of the extreme solar-limb profile is by lunar occultation in eclipse. With the exception of the 1991 eclipse over Mauna Kea (Espenak 1989), eclipse observations normally require mobile observatories, and aircraft are ideally suited for eclipse observation in the far-infrared. Beckman et al. (1975) pioneered this technique in the Mach 2 Concord for 400, 800 and 1200 μm radiation, in the total solar eclipse of 30 June 1973. They used an FTS to observe the eclipse in integrated lunar and solar radiation, while the Concord followed the path of the lunar shadow. They found the submillimeter and millimeter solar limbs to be extended $\sim$ 5 arcsec beyond the visible limb at these wavelengths, a result that has been roughly confirmed by subsequent observations. They also found the extreme limb to be brightened in a narrow spike $\sim$ 10 arcsec across.

Clark and Boreiko (1982) observed the total solar eclipse of 26 February 1979 from the NASA Lear Jet Observatory in 400-μm radiation. They found a somewhat brightened limb, but only a fraction of the brightness excess found by Beckman et al. (1975), and once again considerably below the excess expected from a smooth chromosphere.

Becklin led an observational program in the total solar eclipse of 31 July 1981 from the NASA Kuiper Airborne Observatory (KAO) (see Lindsey et al. 1986). Their observation capitalized on the more narrow (2 arcmin) beam of the 1-m KAO telescope. This allowed the detector to be concentrated on just the extreme limb crescent at the point of contact, and provided sufficient signal to allow subarcsecond discrimination of the extreme limb-brightness profile. Figure 12 shows the limb profiles obtained in 30, 50, 100 and 200 μm radiation. The peak limb brightnesses found are reasonably consistent with the models of Vernazza et al. (1973) (dashed curves). However, for wavelengths longward of 30 μm, the data show emission extending several hundred kilometers above the heights predicted by these models. On this basis, they proposed that gravitational-hydrostatic equilibrium fails very low in the chromosphere, not far above the temperature minimum. For departures from hydrostatic equilibrium to coincide with the clear manifestation of non-radiative heating suggests a relationship between the two. This idea had been previously considered by others, notably Giovanelli (1980), on the basis of chromospheric magnetograms.

Becklin's team recently observed the eclipse of 18 March 1988 from the KAO at 200, 400 and 800 μm. These observations show the solar limb extended up to 3500 km above the base of the photosphere. These, and longer

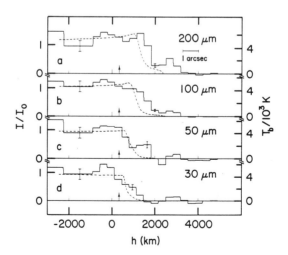

Fig. 12. Submillimeter limb-brightness profiles determined from airborne limb-occultation observations in the total solar eclipse of 31 July 1981 (Lindsey et al. 1986). The dashed curves show the limb brightness profiles expected from the model of Vernazza et al. (1973). The arrow indicates the visible limb.

wavelength data, provide an important basis for modeling inhomogenous structure in the chromosphere far from gravitational-hydrostatic equilibrium, such as spicules. This approach promises to open some entirely new concepts in solar modeling. Hermans and Lindsey (1986) explain the application to smooth models out of gravitational-hydrostatic equilibrium. Lindsey (1987), Braun and Lindsey (1987) and Jefferies and Lindsey (1988) begin to explore new approaches to LTE radiative transport in inhomogenous structure. Important and practical work in this area goes back many decades, both in the radio (Athay 1959; Ahmad and Kundu 1981) and infrared spectrum (Beckman et al. 1975). Much of our early approach to understanding the effect of rough structure in the chromosphere comes from the visible-line work of Redman (1943), Evans (1947) and Pecker (1949,1951). Modern chromospheric modeling now demands a formal theory of radiative transfer in *inhomogenous* media. Far-infrared continuum observations represent a significant area of application for such a theory. The Jefferies and Lindsey (1988) theory of radiative transfer in inhomogenous media is outlined in Sec. IV.

Sunspots and Plage Regions.   Although a sunspot is $\sim$ 2000 K cooler than the quiet photosphere, it retains approximately two thirds of the quiet photospheric intensity in the far-infrared continuum. In contrast, the visible continuum retains only a small fraction of the intensity of the quiet photosphere, and this small fraction can be severely contaminated by scattered light from the bright surrounding photosphere. The far-infrared continuum therefore has promise for sunspot radiometry, following on the work of Turon and Lena (1970) who imaged sunspots in the 10-μm continuum. Sunspot radiometry at much longer wavelengths is of great interest for temperature-minimum diagnostics, but no sunspot has yet been spatially resolved at submillimeter wavelengths. The new large submillimeter telescopes on Mauna Kea (e.g., the 15-m Maxwell Telescope) will be the first facilities able to resolve structure on the scale of normal sunspots.

Chromospheric Oscillations.   The thermal diagnostics offered by the far-infrared continuum in static atmospheric structure apply equally well to dynamics. The infrared continuum has already found application in the dynamics of the 5-min oscillation. It offers further potential in the dynamics of convection, magnetic regions, and flares.

When we look for oscillations in the infrared continuum, we are looking almost entirely at temperature variations in the atmospheric medium associated with them. These temperature variations are expected in response to work done on and by the medium over the compression cycle of the oscillation. They can be expected from the middle photosphere upward through the entire chromosphere, for variations up to several minutes in period. They are not expected to be very strong in the low photosphere, where the radiative relaxation time drops to a few seconds, damping any thermal response to compression. In the upper photosphere and chromosphere, however, the oscillations do cause a thermal response, and the infrared continuum is an ideal diagnostic for them.

Hudson and Lindsey (1974) and Lindsey (1977) observed infrared continuum brightness oscillations in the 10- to 25-μm continuum. At 25 μm they found evidence of the 5-min oscillation in their power spectrum at 3 mHz. In the 10-μm continuum, no 3-mHz signal was detected, presumably damped by rapid radiative relaxation in the photosphere. Their results are qualitatively consistent with the occurrence of prominent brightness-temperature oscillations in the cores of CO and OH lines at 5- and 10-μm, respectively (Noyes and Hall 1972; Deming et al. 1986), because the cores of these lines are formed at altitudes similar to the continuum at longer wavelengths.

Lindsey and Kaminski (1984) first observed intensity oscillations in 800-μm radiation at the IRTF. Lindsey and Roellig (1987) have recently followed this with simultaneous observations at 350 and 800 μm. They found a strong correlation between these two wavelengths, but with the 350-μm oscillations leading the 800 μm by ~ 30° (illustrated in Fig. 13). Lindsey and his co-workers have recently observed oscillations at several wavelengths simultaneously, including 50, 100 and 200 μm, from the KAO, while co-spatial Doppler observations were made from the Mees Solar Observatory in the sodium D_1 line, and from the Wilcox Observatory in the iron λ5250 line. A

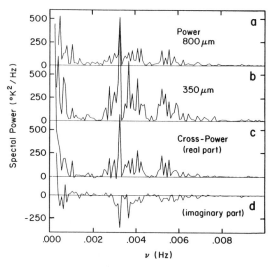

Fig. 13. Cross-power spectral analysis of brightness oscillations measured simultaneously in 350- and 800-μm radiation. Plots (a) and (b) show the power spectra of 800 and 350 μm variations, respectively, plotted linearly from 0 to 10 mHz. Plots (c) and (d) show, respectively, the real and imaginary parts of the cross-power spectrum. Note the consistent negative tendency of the imaginary part, indicating a delay of the 800-μm signal behind the 350-μm signal.

sample of their data is illustrated in Fig. 14. The infrared intensity fluctuations in this plot are all strongly correlated with velocity, but with intensity lagging significantly behind velocity. This phase shift is best shown by cross-power spectral analysis in Fig. 15. Maximum intensity follows maximum redshift by ~ 45 to ~ 70°. The phase shift at longer wavelength is larger, but the uncertainty is also much greater. There is little difference in phase between 50 and 100 μm intensity.

In the ideal case where the 5-min oscillation is an adiabatic evanescent wave in the upper photosphere and chromosphere, the phase shift between temperature and velocity is expected to be 90°, and the phase of each variable should be independent of height. The studies described above seem to indicate that this is not the case, but more work is needed to clarify the role of thermal relaxation, energy propagation and other effects (see, e.g., Kopp et al. 1988). The relation between the amplitudes and phases seen in the far-infrared continuum, and the intensities and velocities seen in the cores of CO and OH lines at shorter wavelengths, also need to be studied. More thorough infrared observations, including simultaneous Doppler measurements in chromospheric lines, promise to contribute considerably to our understanding of wave dynamics in the chromosphere.

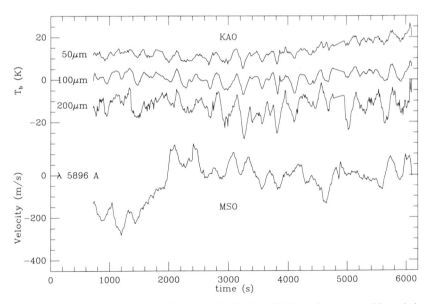

Fig. 14. Chromospheric Doppler oscillations in the D_1 line of Na I are shown over a 6-hr period (lower plot). These observations were made by Lindsey using the Stokes Polarimeter at the Mees Solar Observatory, in collaboration with D. L. Mickey, who built the instrument. The velocity is compared to simultaneous brightness-temperature oscillations in the 50, 100 and 200 μm continuum.

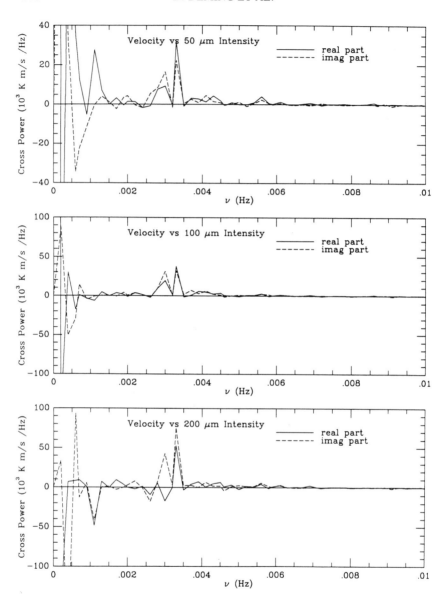

Fig. 15. Cross-power spectra of infrared brightness variations and Doppler velocity (from Fig. 14). The real parts are plotted as solid curves, and the imaginary as dashed curves. The positive imaginary values indicate phase leads of maximum redshift ahead of maximum brightness.

IV. MULTICOMPONENT MODELING BASED ON INFRARED OBSERVATIONS

The observations described in Sec. III give several reasons to believe that thermal inhomogeneities in the solar atmosphere have a profound impact on the solar infrared spectrum. Prominent examples include the bifurcation phenomenon seen in CO and OH lines, and the rough chromospheric structure seen in the far-infrared continuum. While it is natural to begin modeling infrared observations using homogenous models, this approach has not been successful, and the observational evidence for thermal inhomogeneities demands that a more general radiative-transfer theory be developed and applied. In this section we describe the initial development of a statistical theory of radiative transfer which allows for the random occurrence of inhomogeneities. We follow the more comprehensive description given by Jefferies and Lindsey (1988). We summarize a formalism for obtaining the mean and variance of the intensity of radiation emerging from a medium in which the emission and absorption coefficients at any point in the gas are described statistically. While this development has been stimulated by attempts to understand the solar infrared data, it is in no sense limited to that case. It is, however, restricted to LTE excitation, or at least to the case where the emission and absorption coefficients are independent of the local radiation intensity.

Consider a region comprised of two components: structural elements (of a single type) immersed in an ambient medium. A generalization to an arbitrary number of components is straightforward, but will not be considered here. Both the ambient medium and structures absorb and emit radiation. Let the probability that a given component occurs at any point in the medium be specified, as well as the emission and absorption coefficients for each component. We subdivide the medium into cells whose size is chosen to be the characteristic dimension of a structure. The discrete division into cells can be removed to obtain a differential formalism, though at a cost in complexity. We further specify that if a structural element falls in a cell, it falls only and completely in that cell, so that the probability that a given cell is occupied by a particular component is independent of the occupancy state of adjacent cells.

If we were to assign, *a priori,* values of the absorption coefficients and source functions at all points along a ray, then we could immediately integrate the transfer equation to obtain the emergent intensity whether the medium were homogenous or not. However, in our problem these variables are specified only statistically, i.e., they are *random* variables and a calculus of probabilities is needed to work with them. Our approach here is to derive a recurrence relationship for the random variable $I^{(m)}$—representing the statistical description of the intensity emerging from the mth cell—in terms of $I^{(m-1)}$ and the sources and sinks of the radiation.

Thus, consider the statistic $I^{(1)}$ for the radiation emerging from the first cell. At the point where the ray enters, this cell will contain a structural element or ambient material with probabilities p_1 and $1-p_1$, respectively. In either case, the specific intensity of the radiation emerging from the cell may be written

$$I^{(1)} = I_0 \exp(-\tau(1)) + \int_0^{\tau^{(1)}} S(t) \exp(t-\tau(1)) \, dt \qquad (2)$$

with I_0 the intensity incident externally on the first cell of the composite atmosphere, and S is the source function. By an obvious extension we find

$$I^{(m)} = I^{(m-1)}\exp(-\tau(m)) + \int_0^{\tau^{(m)}} S^{(m)}(t) \exp(t-\tau(m)) \, dt. \qquad (3)$$

Equation (3) is the analogue of the usual radiative transfer equation for a medium of specified structure. It is straightforward to show from Eq. (3) that the evolution of the expected value of the intensity emerging from a cell is given by the relationship

$$\langle I^{(m)} \rangle = \alpha_m \langle I^{(m-1)} \rangle + (1 - \alpha_m) S_m \qquad (4)$$

where

$$\alpha_m = \langle \exp(-\tau(m)) \rangle \qquad (5)$$

$$S_m = \langle S_m(1-\exp[-\tau(m)]) \rangle / \langle 1 - \exp[-\tau(m)] \rangle \qquad (6)$$

and the brackets $\langle \rangle$ indicate an expected value. For simplicity we have supposed that the source functions are constant across a cell. If we define the change in the expectation value of I in passing through the nth cell by

$$\langle I^{(m)} \rangle = \langle I^{(m-1)} \rangle + \Delta \langle I^{(m-1)} \rangle \qquad (7)$$

then Eq. (4) can be written

$$\Delta \langle I^{(m-1)} \rangle = - (1 - \alpha_m) [\langle I^{(m-1)} \rangle - S_m] \qquad (8)$$

which brings out clearly the analogy with the transfer equation for a smooth atmosphere, to which Eq. (8) reduces in the homogeneous case.

It is easily shown that the variance of the intensity can be expressed in a form similar to Eq. (4)

$$\sigma_m^2 = a_m \, \sigma_{m-1}^2 + b_m \qquad (9)$$

where

$$a_m = \langle \, \exp \, (-2\tau(m)) \, \rangle \qquad (10)$$

and b_m is a more complicated expression set out explicitly by Jefferies and Lindsey (1988).

It must be emphasized that the approach outlined here is only a beginning. A natural extension is to remove the discrete division into cells having a common geometry. In order to allow for more complex and realistic geometries, a differential formalism is required. Such a generalization has been developed, but is too complex to consider here.

V. FUTURE DIRECTIONS

The advance of technology is perhaps only now beginning to make infrared observations possible for solar physicists to the same degree as observations in the visible. The spectacular new advances in infrared array-detector technology promise to revolutionize solar imaging, and especially solar magnetic-field studies. We can look forward to images of the solar surface in radiation from the deepest observable layers of the photosphere and from much higher layers seen in the cores of the strong CO lines. The prospect exists for making vector magnetic-field measurements, with high sensitivity, simultaneously in the deep photosphere and the base of the chromosphere, by utilizing the very Zeeman-sensitive lines at 1.56 and 12 μm. At the longest infrared wavelengths, new large telescopes will give much improved angular resolution. The James Clerk Maxwell Telescope on Mauna Kea is the first instrument capable of resolving the thermal structure of the chromospheric supergranular network, or of sunspots, in the 350-μm to 1-mm continuum. This will be an important advance for studying the structure and dynamics of the low chromospheres and temperature minima of these features.

SOLAR FLARES

HAROLD ZIRIN
California Institute of Technology

ALEC MacKINNON
University of Glasgow

and

SUSAN M. P. MCKENNA-LAWLOR
St Patrick's College

Solar flares are discussed from three complementary viewpoints. The first is the radiative aspect which includes optical and radio observations, in addition to the magnetic circumstances of their origin. It is seen that flares are produced in regions of magnetic shear, a condition that arises mainly from sunspot motions. In this regard recipes for the prediction of flares are given. The second view point considers that flares produce great streams of energetic particles which carry away much of the flare energy. Gamma and X-ray studies of the spectra of these particles provide information on the flare's transient high-temperature plasma. A third consideration, satellite samples of flare-particle streams, provide diagnostic information on acceleration mechanisms pertinent to different kinds of flare event, while shedding complementary light on solar atmospheric composition at particular flare sites and on physical processes operating in the corona.

I. FLARE STRUCTURE AND PREDICTION

Solar flares are abrupt releases of magnetic energy that produce a panoply of high-energy effects, particles and photons. They almost invariably occur at sharp inversions of sign of the longitudinal field, in places where the magnetic-field gradient is so steep that only a force-free field, one with strong currents flowing along the field lines, may obtain. The surface migration of magnetic elements (in particular sunspots) builds up these currents, and through the flare event, the field can jump to a much lower energy potential field, usually by means of magnetic reconnection.

A. Optical Appearance: Magnetic Patterns

The location of flare occurrence is powerful confirmation of these commonly accepted ideas. Flares in active regions invariably occur at sites of magnetic neutral lines of nonpotential form (Severny 1957). Such a neutral line delineates regions of opposite polarity. Often the locale is an inclusion where a sharp polarity gradient has built up from sunspot motions. Flares never occur in unipolar regions. Often the neutral line flare will contain a filament, which usually, but not always, is disrupted by the flare. Active regions without strong gradients or anomalous magnetic structure do not produce flares, no matter how large they are. Often enormous simple spots appear on the Sun and are monitored in vain for possible flares, while smaller spots with steep gradients flare frequently on another part of the disk.

Despite this agreement between model and observation, changes in longitudinal field occurring at the time of the moderate flares have been hard to observe, but McKenna-Lawlor (1981) found evidence for the disappearance of a north polar umbra of field strength > 2000 gauss in association with the occurrence of an east limb flare associated with the production of a polar cap absorption event; Moore et al. (1984) found evidence for field disappearance after a sizeable flare, and Livi et al. (1989) and Wang and Zirin (1989) observed the disappearance of opposite small-scale fields shortly following small flares. Also Zirin (Wang et al. 1989) found a before-after longitudinal field change in a moderate flare. However, in many other cases the longitudinal magnetic field simply does not change at the moment of the flare. This has prompted the view that the transverse field lines connecting magnetic poles are what changes, an idea supported by the strong sheared fields found along neutral lines in the case of great flares. A "vector magnetograph" at the Marshall Space Flight Center has given unequivocal evidence for the occurrence of flares at such shear lines, even though the observations were not frequent enough to test for before-after changes. Now a vector videomagnetograph at Big Bear Solar Observatory (BBSO) is being employed daily, when appropriate, to study the role of transverse fields.

Only the very largest of flares are visible in the continuum (i.e., the level of the photosphere). This is but part of a wide range of evidence supporting the idea that flare energy release takes place above the surface, and chromospheric brightenings appear at the magnetic footpoints, i.e., intersections of the lines of force with the surface. In major flares, one often finds a string of impulsive footpoint brightenings along the magnetic neutral line, followed by a main-phase brightening of ribbons parallel to the neutral line. Occasionally, less energetic flares with widely separated brightenings are observed, indicating energy release quite high in the corona. Current evidence suggests that the major part of the flare energy is imparted to high-energy electrons and protons. These deposit their energy in the chromosphere to produce hot plasma which, in turn, radiates the optical emissions that we see. An enormous flux of energetic particles is produced, with electrons up to 10 MeV and nucleons to hundreds of MeV. The particles rapidly thermalize to a very hot (about 40×10^6 deg) plasma. During this thermal phase we see two bright ribbons of emission that rapidly elongate (100 km s^{-1}) on either side of the magnetic neutral line and separate at 5 to 20 km s^{-1}, while loop prominences connecting the opposite magnetic polarities rise higher and higher in the corona.

Partisans of coronal mass ejections argue that their phenomenon is the major energy sink in the flare, but the fact that many transients occur without flares, and vice versa, makes this point of view hard to accept. Similarly the close relation of flare occurrence to surface magnetic-field structure rules out a source in the high corona but not a trigger, as is still occasionally suggested.

Figure 1 shows a flare near the limb observed in the blue wing of Hα. This sequence is useful in demonstrating the three-dimensional structure of flares. The early phase has bright emission at footpoints in the penumbra; then bright loops appear, apparently condensing out of the soft X-ray coronal condensation. The initial emission is usually at the loop tops, where density is highest. How the density gets to be higher at the loop tops has never been explained, but it is thought that some kind of compression associated with thermal instability of the coronal cloud occurs there. The loops steadily expand and new loops form higher up. Although no emission is seen in the last frame, there still is footpoint emission along the center line. Note the shift in the loops; at first the lowest loops connect the nearby ribbons in the spot (a spot having polarity inverted from the Hale-Nicholson Law; see the chapter by Semel et al.); later they connect the more distant spots. Optical emission closely tracks the X-ray rise in time, but declines more slowly, depending on the balance of heat loss and the energetic inputs. Thus the continuum emission, arising in lower layers of high emissivity, falls off with the hard X-ray emission, while the chromospheric Hα comes from a region that cools slowly and lasts much longer. The location of bright optical sources at the footpoints of Hα arches (Fig. 1) confirms this picture. While X-ray and radio spikes may be quite short, the optical emission rarely changes in less than a few seconds, and the growth time to maximum of great flares may be 4 min or more.

Often the filament rises tens of minutes before the flare; it may get exceptionally dark, blue-shifted or broadened in Hα. This is the only definite flare precursor. Then the flare breaks out with brilliant Hα emission and the filament blows away. This is called the flash phase; hard X-ray, Hα and microwave flux all increase sharply at once (Zirin 1978).

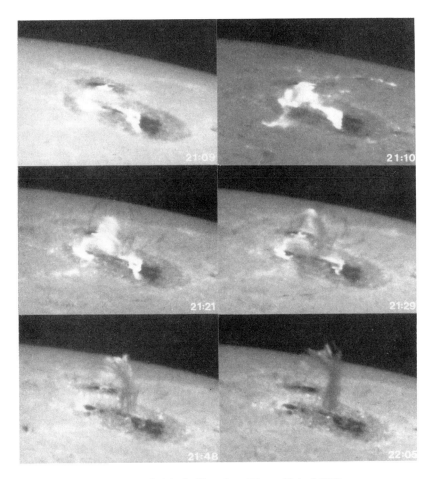

Fig. 1. Flare observed near the limb in the blue wing of Hα on 29 April 1973.

The Hα emission consists of bright, broad-profile kernels, an extensive area of narrower (∼1 Å) emission and bright post-flare loops connecting the two ribbons. The total Hα emission from each of the first two is about the same, 2×10^{30} erg s^{-1} in a big flare (Zirin and Tanaka 1973); the higher Balmer lines amount to 3 or 4 times that amount, and the Lyman α is equal to Hα. Flare ejecta include sprays, surges, streams of electrons and nuclei, and shock waves as well as coronal mass ejections.

Flares are ranked in importance by optical, X-ray or radio flux. Recently flare size classification by the 1 to 8 Å soft X-ray flux monitored by the GOES spacecraft has become popular. The flares are designated by Cn, Mn or Xn, where the first letter is determined by whether the flux at 1AU is 10^{-6}, 10^{-5} or 10^{-4} w m^{-2}, respectively, and the integer n gives the flux for each power of 10; thus M3 means a flux of 3×10^{-5} w m^{-2} at the Earth. The older classification (Zirin 1988) of class 1, 2, 3 used the flare area as seen in Hα, with suffix F, B or N according to brightness faint, bright or normal. An M-class flare is roughly similar to class 1 on the area system, and classes 2 and 3 both fall in the X class. The X-ray measurement is probably more repeatable and accurate than the area, which is subject to the vagaries of weather and poor equipment (although the X-ray satellites cost more than all the flare patrols since the beginning of time). However, a soft X-ray measurement still refers to only the soft X ray; it undervalues impulsive events and overvalues slow thermal events.

Flares are log periodic in distribution; that is, C-class flares are roughly 10 times less energetic and 10 times more frequent then M-class flares. Thus, there is about as much energy in each class all together. The reader must remember that almost all published discussions of flare energetics refer to the big flares, but the relative distribution of energy is probably the same for all flares, although little is known about energetic particles in smaller flares other than electrons.

Flares show different time and energy profiles. Tanaka (1987) defines three classes:

1. Hot-thermal, with $T = 3$ to 4×10^7 deg, but limited and hard X-ray emission from a compact source and little radio emission.
2. Impulsive, with spiky hard X-ray and microwave emission from footpoints and low corona.
3. Gradual-hard, a long-enduring (> 30 min) large event, with gradual peaks, a hard spectrum and a strong X-ray microwave source high in the corona.

Solar energetic particles reaching the Earth are preferentially produced by long-duration, multiple-peak events, or those accompanied by coronal mass ejection and Moreton waves (described later); most gamma-ray line flares, on the other hand, direct their particles inward and do not produce proton storms at the Earth. Large flares produce almost all these effects.

Observations of flares in different lines with narrowband filters reveal different aspects according to the line strength. In Hα, a wide range of chromosphere emissions, as well as elevated bright clouds associated with the flare kernel are observed; in He I D3, only the hot spots appear in emission, although ejecta are typically dark against the disk. When we observe a line in emission against the disk, the temperature must be higher than the photosphere, and the density great enough so that collisions dominate over radiative scattering (Zirin 1988). This occurs for densities above 10^{12} particles cm^{-3}, and is particularly strong in He D3, where there is little background structure. For the λ4686 $n = 4 \rightarrow 3$ transition in ionized He, the highest excitation line in the visible spectrum, Zirin and Hirayama (1985) found a temperature of 30,000 to 50,000 deg and density 3×10^{12} particles cm^{-3}. A similar density in the electron acceleration region is deduced by Ramaty et al. (1983) from 10 MeV photon emission in flares.

The temperature of flares covers a great range. The power-law electrons, of course, correspond to no thermal description, while the hot thermal soft X-ray source cools from 40×10^6 down to 3 or 4×10^6 deg. The Hα source ranges from 500,000 deg in the 10 Å wide kernels to around 10,000 deg in the secondary emission regions. The last value can be deduced from the fact that Hγ is generally the brightest Balmer line.

The greatest flares produce emission in the continuum which shows a marked blue excess (Zirin and Neidig 1981). No one knows how the flare energy reaches this deep level; electrons are too easily stopped, while there are not enough protons. The blue emission is probably a height effect; although the H⁻ opacity is lower in the blue, the appearance of images in the near ultraviolet is more like the chromosphere than the photosphere, so we see the greater heating of the higher layers. There also may be a simple Wien's law effect: a temperature of 9000 deg will produce an emission peak in the near ultraviolet. Emission may be in short-lived small kernels, or, in the case of great flares, a slowly moving bright wave (Machado and Rust 1974; Zirin and Tanaka 1973) coincident with the brightest Hα double ribbons.

B. Flare Electrons

Flare electrons carry much of the energy and produce the following easily observed effects:

1. Microwave emission (by synchrotron and free-free emission);
2. Meter-wave emission (by plasma oscillations Type III bursts);
3. Distant Hα brightenings (associated with Type III-RS bursts);
4. Direct detection at the Earth.

Type III radio bursts are produced by streams of electrons of around 40 keV, moving out from the Sun at 1/3 the speed of light; the RS (reverse slope) bursts are headed downward, probably along loops (LaBonte 1976). It is entirely possible that the electrons carry most of the energy initially imparted to the flare plasma, even that which later appears in macroscopic motion (something has to drive the ejected material).

In the decimeter range, a picturesque zoo (Slottje 1981) of zebra, fiber and tadpole bursts are recorded by spectroscopy of radio-noise storms. These are probably excited by electrons, but the phenomena are so complex, and those studying them so few, that little progress has been made in understanding the rich observational material. Four components of the flare electrons have been recognized.

1. Nonthermal electrons with a power-law distribution occurring in bursts or spikes in the impulsive phase. They can be recognized from the spiky profile of X-ray or radio emission that they produce. Their spectral distribution may range from E^{-2} to E^{-9} and they are effective microwave emitters. One can show that in most impulsive events these represent the main energy input.

2. *A hard nonthermal cloud of electrons occurring late in the flare high above the surface.* Tanaka (1986) finds these in flares without filament eruption. The X-ray spectrum of this component cannot be fitted by a thermal distribution (Ohki et al. 1982) and intense microwave emission is observed. The power-law exponent is 2.5-4.

3. *A hot thermal cloud of electrons at 25 to 50 $\times$ 10^6 deg.* This was discovered by Feldman et al. (1980) in X-ray line spectra of Fe XXV and later detected by Lin et al. (1981) from high-resolution X-ray continuum data. This kind of hot cloud appears at loop tops in the Skylab images slightly after the impulsive event, producing the most intense chromospheric emission. This cloud appears to be the result of collisional thermalization of component (1).

4. *A post-flare thermal cloud with T dropping from 20 to 3 $\times$ 10^6 deg during the cooling of component (3).*

At the onset of an impulsive flare, X ray, microwave, Hα and the continuum rise abruptly as electrons are accelerated. Even if there are several spikes, the X-ray and microwave fluxes track each other. But the decay phase varies according to the cooling rate of the source; power-law electrons disappear immediately, while the thermal coronal cloud provides a source of heating to the chromosphere, and Hα is emitted so long as the cloud is present. Some flares have thermal phases lasting many hours. This can only occur if there is continued energy input to the thermal cloud by unknown sources.

It has been argued by some that the hot thermal source is chromospheric material heated by the impulsive component which rises into the corona (the evaporation model). Support for this model is twofold: first, strong blue-shifts of Fe XXV lines corresponding to upward flows of up to 600 km s^{-1} have been observed (Acton et al. 1982; Antonucci et al. 1982; Tanaka and Zirin 1985); second, there is much more coronal material in the thermal plasma than was present before the flare. But in present models (Fisher 1987a,b), it is difficult to produce a 30 $\times$ 10^6 K plasma by heating the photosphere with particles, particularly electrons. Further, the energy of the nonthermal electrons barely equals that of the thermal plasma, so there is nothing left to compensate for other losses, such as the optical emission and the heat conducted away. All the measurements of blue-shifted Fe XXV show it occurring during the flash phase, the period of Hα expansion. This is usually before the maximum of the white-light flare, the maximum of lower-atmosphere heating when the "evaporation" would be expected to peak. On the other hand, there is plenty of chromospheric material that could be evaporated before the energy reaches the photosphere.

C. Waves, Ejections and Meter-Wave Bursts

When an energy release as large as a flare occurs, it is bound to make waves, and this is exactly what happens. Both the gas and magnetic pressure in the event produce substantial ejecta, involving the flare material itself and nearby structures. These are among the most spectacular solar phenomena and have the most profound terrestrial effects. We see ejected material, shock waves and streams of high-energy particles.

The primary source of ejected material is the flare filament, which rises gradually 20 min or so before the flare onset, whereupon it is violently expelled from the surface. Because the filament is still cool, the lifting is a magnetic phenomenon either breaking the bonds that hold the filament down or expanding fields involved in the flare. Velocities up to 2000 km s^{-1} have been observed. These events are usually followed by a shock wave and coronal mass ejection. If the ejecta are irregular and disorganized it is termed a spray. Its trajectory will be roughly straight with a weak spiraling around the invisible field lines. When filaments in the quiet Sun also erupt, the difference is that few or no energetic particles are produced at the Sun and little Hα emission occurs, although a coronal mass ejection may result. We estimate the energy involved in a prominence eruption at 500 km s^{-1} as 10^{31} erg, similar to the total radiative output. All of the energy must come from magnetic buoyancy. In most cases, the velocity is much less, and only part of the material is expelled.

The results of Widing et al. (1986) on an eruption observed in the ultraviolet show that the temperature of the ejected material is not high. The ejecta may even be cooled by adiabatic expansion as they rise. The same is true of surges, discussed below. The eruption is a purely magnetic phenomenon.

The major characteristic of the flare spray is the overwhelming of neighboring fields by the tremendous explosive power of the flare. By contrast, surges follow the existing magnetic patterns. The initial temperature is greater than the chromosphere, or else we would not see emission. The surge goes from emission to absorption as it moves out, but this may be either a density or cooling effect. The fact that surge spectra are generally of low excitation argues for cooling.

A typical surge can quite reasonably be explained by conversion of all the internal energy of a small flare into kinetic energy via a magnetic nozzle. A surge velocity of 100 km s^{-1} corresponds to thermal energy at 500,000 deg; if the surge volume is 10^{22} m^3, then a volume increase of 10^4 times takes place if the source flare has a volume of 10^{18} m^3, which is a typical value. If $T \sim 10^7$ K, there is plenty of energy. Some of these parameters have been modeled by Noci (1979).

Despite the attraction of the nozzle model, a few ejections are threaded by field lines and have a more complex behavior. Hurford and Tang (personal communication) found a surge which absorbed the emission from a radio source in an active region. The absorption was due to invisible material moving ahead of the visible surge, which also later appeared in absorption. A surge is not always simple.

Moreton waves (Moreton and Ramsey 1960; Athay and Morton 1961) are shocks moving 1000 km s^{-1} produced by large impulsive flares (microwave flux > 700 SFU (solar flux unit) where 1 SFU $= 10^4$ Jansky). They appear as a bright front in the center of Hα or a dark front (corresponding to a wavelength shift) in the Hα wing. They propagate with constant velocity in an arc about the source, following the curve of the surface, sometimes for more than a solar radius. The wave normally propagates until it reaches a magnetic boundary of some sort (such as a filament or active region). Subtraction of images in red and blue wings of Hα show that a downward shift followed by an upward recovery produces the optical effect. The center-line brightening may be due to the same effect, because a shift in the absorption line increases the intensity of light passing through the Lyot filter, or it may be excitation by particles trapped in the front. The up and down motion produced by the wave passage causes a dramatic winking of filaments as they shift in and out of the bandpass. Moreton waves are invariably accompanied by Type II bursts.

Uchida (1974) explained the Moreton wave as a weak MHD shock wave propagating spherically from the flare site; the optical phenomenon is the intersection of this expanding sphere with the surface. He fitted actual trajectories of waves to the magnetic circumstances.

The magnetic nature of the Moreton wave is shown by its disappearance at magnetic boundaries and propagation along the curve of the Sun's surface. How it is initiated is unknown. The wave becomes visible only after it travels some distance and the arc it forms becomes visible against an undisturbed chromosphere.

Flare waves and material ejection are detectable in the characteristic radio frequencies emitted by the corona as expected. In their early work Wild and his co-workers (Wild 1950; Wild and McCready 1950) classified meterwave bursts by their distinctive spectral properties:

Type I circularly polarized narrowband (5 MHz) bursts. These last a second or less with complex spectral properties superposed on a continuous background: they make up this Type I noise storm.

Type II burst of great intensity following large flares. A slow drift to lower frequencies of 1 MHz s^{-1} indicates that they are moving outward at 1000 to 1500 km s^{-1}. Harmonics at twice the frequency are often observed. The emission is thought to be produced by plasma waves excited by the Moreton wave as it travels outward.

Type III the most common flare-associated bursts, observed from 5 to 600 MHz. They are characterized by a fast drift from low to high frequency, around 20 MHz s^{-1} at 100 MHz. Connected with streams of electrons around

40 keV, the frequency drift corresponds to such a stream moving 100,000 km s^{-1} and successively exciting lower local-plasma frequencies.

Type IV broadband continuum radiation lasting for hours after a flare. The source moves outward with velocities from several hundred to 1000 km s^{-1}. Originally thought to be a noise storm initiated by the flare, Type IV was shown by Boischot (1957) to be a different phenomenon. Both moving and stationary Type IV bursts are now recognized. The moving Type IV's last around half an hour, and are partly circularly polarized; the stationary Type IV's last a day or more, and may in fact be identical to the continuum portion of noise storms. The broadband nature of the Type IV emission suggests synchrotron emission as its source. In this case, the energetic electrons would have to be trapped for a fairly long period of time within magnetic bottle of some sort.

Type V broadband continuum radiation following a Type III burst. Wild concluded that the downward drift of frequency in Types II and III was due to the outward movement of a source through the corona, exciting emission at the local plasma frequency. Interferometer measurements and other data confirmed this idea. The sources travel all the way to the Earth, emitting very long (kilometric) wave emission at the interplanetary plasma frequency. At the Earth, bursts of 40 keV electrons are detected by a spacecraft; both Type II and Type III bursts are detected (Malitson et al. 1973).

In large flares (Tang and Moore 1982), distant points brighten in Hα within a few seconds of the flash phase. Since the distance may be 100,000 km or more, the agent must be fast electrons; Tang and Moore in fact found that Type III-RS (reverse slope) bursts were emitted in these events, the reverse slope indicating that electrons were moving downward. Coronal X-ray pictures have shown long loops connecting the main flare to distant points. The velocities determined from films are lower limits because of the time resolution; the velocity is probably 100,000 km s^{-1}.

Type II and IV bursts are associated with the largest flares. Wild's conclusion that the Type II bursts were traveling outward through the corona at 100 km s^{-1} was confirmed when the Culgoora radioheliograph formed images from meter waves of the traveling disturbances. The discovery of the Moreton waves revealed the source. Finally, Dulk et al. (1976) showed that the Type II burst from a given level closely followed the white-light transient or coronal mass ejection, beginning just after the leading edge passed and ending after the transient had gone by.

D. Flare Spectra

Visible-light flare spectra are characterized by their high spatial resolution and their sensitivity to Doppler shifts in the photosphere or chromosphere, while ultraviolet spectroscopy illuminates the high-energy parts of the flare. The main observational limitation to flare spectra is the problem of pointing the spectrograph at the flare at the right time and place. This is particularly difficult in the visible, where the flare emission is weak compared to the photospheric background. In the ultraviolet, the flare emission is so strong that it is all we see in most of the lines below 400 Å, but one still must point to the right place. Most high-excitation lines are in the ultraviolet where spatial resolution is low, and the coronal lines are only emitted in the thermal post-flare regimes. The 4686 Å line of ionized helium, which as mentioned is the highest excitation flare emission observable in the visible spectrum, and has been observed in two intense flares (Zirin and Hirayama 1985).

Flare images in lines of highly ionized iron have been obtained by the Naval Research Laboratory "overlappograph" (Widing 1975; Cheng 1977). The high-ionization lines appear as tiny cores at the loop tops, while the low-ionization lines resemble Hα, showing that the height scale is not great. The behavior of these high-ionization lines tells us little about the impulsive stage of the flare; as Kane and Donnelly (1971) showed, the extreme ultraviolet and visible flare lines in the chromosphere brighten in the flash phase, and the coronal lines belong to the thermal phase. These early phases are domi-nated by chromospheric lines and those lines from the hottest plasma. The former must be excited by precipitating electrons and the latter are associated with the hot thermal plasma. Later in the flare, we see extended images in the permitted transitions of coronal ions such as Fe XIV, XV, XVI, reflecting the cooling of coronal loops. These images persist long after the higher He II lines and continuum have faded. The Fe XXIII and XXIV cores sit on the loop tops, evidence that this is the soft X-ray source, and that maybe the energy release takes place along the sheared neutral line. These ions radiate mainly in the X-ray range, but there also are screening doublets of the types $2s \rightarrow 2p$ with relatively long wavelength. In fact, Feldman et al. (1974) have identified a forbidden line (analogous to the coronal lines in the visible) of Fe XXI at 1354 Å. The continuum emission in these spectra has not been ana-lyzed but probably is due to a combination of H and He at the loop tops.

Beautiful spectra of the same flare in the 1100 to 1600 Å range were obtained by the Naval Research Laboratory slit spectrograph on Skylab. They show (Brueckner 1975) very broad Lyman α emission, broad chromospheric lines of higher ionization and many narrow lines from the low chromosphere. We also find the forbidden line of Fe XXI at 1354 Å and, late in the decay phase, the forbidden line of Fe XI, as well as a few unidentified lines of similar nature. No continuum appears in these spectra, probably because the slit was set across the neutral line, missing the footpoints. All the resonance lines are greatly enhanced. The He II 1640 (Hα-like) line is weak.

Soft X-ray line emission permits spectroscopic analysis of plasmas up to 50×10^6 deg. Ionization equilibrium data yield temperatures, and density diagnostics exist if we are lucky and the plasma has the right density for permitted and forbidden lines to change their ratios.

Each important ion has a cluster of satellite lines produced in the presence of outer-shell electrons which changes slightly the energy of the transition. The spectrum in the 1.75 to 1.85 Å region consists of the Lyman α line of Fe XXVI and dielectronic recombination satellites, the K line due to $n = 3 \rightarrow 1$ jumps in neutral Fe after K-shell photoionization, and a range of lines due to Fe XX-XXVI. The electron temperature is determined by fitting to the theoretical curves (Tanaka 1986). We never see the ionization increasing at the flare onset; Fe XXVI just appears and cools. Further, it is not known if the ions are high-speed Fe nuclei (which would give lines about 0.004 Å wide for 50 keV) or field Fe ions stripped by collisions, which would give narrower lines.

Flare spectra in the visible range have been studied for a long time. They are still useful because our new knowledge of flares enables us to ask the right questions, and many of the spectra have never been properly analyzed. The high spatial and spectroscopic resolution available in the visible is unique.

The properties of the spectrum vary considerably depending on where in the flare we look. Big flares are bright enough to show the weaker lines and last long enough to be photographed, so there is a tendency for these events to be studied. Only in the most intense kernels of emission are representative spectra of the real footpoints recorded, with broad Hα lines, He and continuum emission; outside the kernels the Hα lines are fairly narrow, as can be seen from the limited extent of flare emission in pictures made in the wing of Hα. For many purposes the echelle spectra of a type 3B flare obtained by Jefferies et al. (1959) are the best compilation. Higher-resolution spectra have been obtained by Hiei (Dinh 1980), by Neidig (1983) and by Donati-Falchi et al. (1985). None of these show the strong continuum emission observed in two-dimensional observations. Neidig notes that his slit was 3 arcsec away from the brightest kernel. His spectrum consists of intense broad H, He and resonance lines (such as Ca II H and K, Na D) which represent the heated chromosphere, and many narrow lines in the cores of Fraunhofer lines. Neidig's spectrogram is useful because it shows not only the blue continuum but also the narrow low-excitation metallic lines. These lines are less than 0.1 Å wide; there is no motion at the low heights where they are formed because the excitation is either by fluorescence of flare emission or conduction from hot thermal plasma above.

The Hα lines can be very broad; Zirin and Tanaka (1973) found Hα half widths of 10 Å in a great flare. Neidig (1983) found 5 Å in H 15. These widths correspond to thermal velocities of 500 km s^{-1} or energies of one keV per nucleon. This is occasionally referred to as "turbulence," but one way or another it is the energy of the H atoms, the dominant species, and therefore may be thought of as equivalent to a temperature of millions of degrees. Most of the Balmer emitting region has much narrower profiles resulting from simple heating (Zirin and Tanaka 1973). The kernels of broad emission are associated with the elevated bright regions as well as in footpoints. Bombard-ment by flare particles produces an extraordinary hot, dense plasma in the chromosphere.

One interesting facet of the Hα emission is the frequent presence of red shifts. They indicate that material is moving downward rapidly in the Hα emitting region. This suggests that the extremely broad Hα profiles come from footpoints where energetic heating is taking place and a shock wave is moving downwards, ionizing and heating the material below.

E. How do Flares Occur?

The understanding of why flares occur has come from the study of many Hα and magnetic observations of active regions; the critical problem now is how they take place. A successful long series of synoptic high-resolution observations of flares, with accompaning magnetograms, has made it possible to evaluate the circumstances under which they occur. While large flares are generally associated with large spots, and almost all large sunspots spawn some flares, certain spots have many, and some large ones have few. The pattern is simple: spot groups with widely separated poles of proceeding p and following f polarity are characterized by generally potential fields; since this is the minimum energy state, flares do not occur. In intermediate cases, there is some intrusion of one polarity in the opposite; if the longitudinal fields reverse in a short distance, the field lines cannot simply go up and down, they must be sheared along the neutral line, and energy is stored. In some cases great shears are produced by the close approach of large spots to one another; the presence of two umbrae of opposite polarity in a single penumbra, called a δ configuration, is the typical example.

Figure 2 shows a magnetogram of a large sunspot that crossed the Sun in July 1988 (Wang et al. 1989). While this was a very large spot, positive (white) and negative polarities were well separated and the number of flares was not great. All of the flares in this region occurred when elements of one magnetic polarity immersed in areas of the opposite. The importance of shear was emphasized by the fact that there were no flares along the main magnetic inversion line.

Figure 3 shows a magnetogram of the great active region of March 1989. This region appeared at the limb as a great p umbra surrounded by f polarity. Steady flow of this f polarity relative to the p produced great shear and nu-merous great flares. Typically, such large regions do not return on the follow-ing solar rotation passage.

While observers knew that bright Hα, which is a mark of emerging flux, was connected with high activity, the idea that it was flux emergence was first recognized by Martres et al. (1968), who termed the growing field "structure magnetique evolutive." As we learned more about emerging flux regions, and obtained better observations, the picture became clearer. Liggett and Zirin

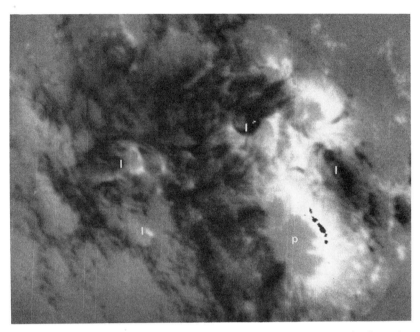

Fig. 2. A high-resolution videomagnetogram of a large bipolar region. The large leading umbra
p is situated at the right. All the umbrae appear as circular gray areas. This region was generally well separated into *p* and *f* polarity, and all the flares occurred at polarity inclusions
(marked I). Flares were most frequent following the flux emergence that created these inclusions.

(1985) have shown that there is a strong tendency for magnetic flux to erupt
in existing active regions. The new flux must either cancel the existing fields
or push them aside. Since diffusion is quite slow, only a little reconnection
takes place (marked by bright Hα), but as the old flux is pushed aside, strong
gradients are established at the edges of the emerging field. As the dipole
emerges, the proceeding spot usually moves forward at a rapid rate, squeezing the flux ahead of it. The new dipole expands in other directions, too,
replacing the old field.

The following list (Zirin and Liggett 1987) of circumstances leading to
big flares may be useful: δ spots present; umbrae obscured by Hα emission;
bright Hα, which marks flux emergence; new flux erupting on the leading
side of the penumbra of a dominant spot; a filament crossing a spot. All these
effects are associated with greatly sheared magnetic configurations.

If one follows the points made in this chapter, it is easy to determine
when a high probability of a solar flare exists. This has been demonstrated

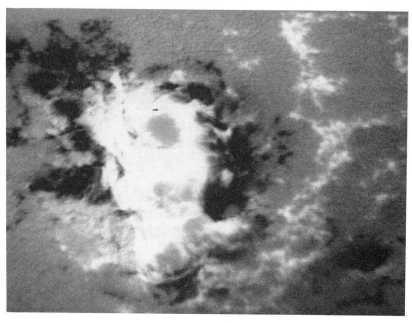

Fig. 3. Magnetogram of the March 1989 region, one of the most active sunspot groups in history, showing a giant *p* spot surrounded by *f* (black) polarity.

by the series of BEARALERTS issued by Big Bear Solar Observatory (Marquette and Zirin 1990). If there have been many flares, persistency will work; there probably will be more. If we have a big spot with umbrae covered by Hα emission, there will be flares; if the Hα emission dies down, there will be none. If a filament on a sheared neutral line shows a big blue shift, there will be flares. If a new EFR comes up rapidly in the middle of an old active region, and its proceeding spot is pushing into the following spot of the old group, there will be flares. When (the exact time) is not so easy to tell, although a study of past experience could yield probabilities per day. Being forced to predict flares should help us understand them. However, although we can confidently predict flare occurrence in certain regions in the near future (even several days ahead) we cannot state precisely when they will occur, and except for the occasional rising filament just before the flare, the vast majority of flares occur without identifiable precursors. However, it should be noted that only a small group of people actually study flares; solar radio observatories, which might give some clues, are almost nonexistent. Perhaps further study will uncover new possibilities for flare prediction.

II. ENERGETIC PARTICLES IN FLARES

A. Observations Relevant to Fast Particles

These observations fall into three classes: (1) radiation produced by the fast particles themselves (hard X rays, microwaves, continuum γ rays); (2) emissions produced by other particles after suffering collisions with energetic particles; (3) *in situ* detection of flare-associated fast particles in the interplanetary medium (see Sec. III). Obviously, the third class of observation is the most immediately informative, but of course one does not know whether the particles which escape are representative. They may even be accelerated remotely from the flare site by flare radio radiation (Sprangle and Vlahos 1983), or by flare-associated traveling interplanetary shocks (see, e.g., Cane and Stone 1984). Indeed, they seem to be much less numerous than the particles inferred at the flare site, both in the case of protons (Murphy and Ramaty 1984), and electrons (Lin and Hudson 1976).

The abundance information from such observations is unique. None of the flare radiative signatures give any clue about abundances in the fast-ion population (γ-ray lines from heavier ions are broad and not individually resolvable; see, e.g., Ramaty 1986). Cook et al. (1984) give recent, good abundance data. Abundance anomalies are common; these may reflect anomalies in the flare site (for which there is evidence from other sources; see Murphy et al. 1985; Sylwester et al. 1984), or an acceleration mechanism which singles out particular Z/A ratios, or ionization states. See Forman et al. (1986) for more details of solar flare energetic particles. At the last solar maximum, the first satellite detections of energetic neutrons at Earth were made (Chupp et al. 1982,1987). These are a consequence of nuclear interactions of accelerated ions, corroborating the interpretation of γ-ray line emission (see Sec. III).

We turn now to the flare radiative signatures. Aside from delayed γ-ray lines, extended hard X-ray (which are rarely observed) and microwave emission, and some sorts of metric-radio bursts, we are restricted to the impulsive phase of the flare. In the impulsive phase, hard X rays, microwaves, γ rays (prompt lines and continuum), extreme ultraviolet and Hα all exhibit similar time profiles, at least on time scales of $\simeq 1$ s (see, e.g., Kane 1974). Delays and noncorrelations between different wavelength bands, and between different wavelengths in the same band, may become evident on subsecond time scales (Bai et al. 1983; Costa et al. 1984; Schwartz 1984; Uralov and Nefedev 1976; Kaufmann et al. 1982). The impulsive-phase time profile of all these emissions exhibits one or more spikes of about 1 to 10 s duration.

Hard X-ray emission is optically thin, and does not suffer from propagation effects. In principle, therefore, it may be regarded as the most direct signature of the fast electrons that produce it. The hard X rays are usually regarded as being the photons with energies ≥ 10 to 20 keV, (at lower energies, the emission is dominated by atomic lines from the thermal flare plasma). The hard X-ray time profile shows the impulsive behavior described above, and occasionally an extended component, which rises and falls gradually over 10^2 to 10^3 s, and whose spectrum hardens monotonically throughout its duration (see e.g., Hoyng et al. 1976; Vilmer et al. 1982).

At least below $\simeq 100$ keV, individual spikes in the impulsive hard X-ray burst display spectral hardening until burst peak, softening again thereafter (see, e.g., Hoyng et al. 1976; Kane et al. 1980). At higher energies, however, there seems to be some evidence that spectral hardening continues through the decline of the spike (Dennis 1985). Such observations yield information on the acceleration and transport of the fast electrons, at least in principle.

Since the last solar maximum, we have obtained some information on the spatial distribution of hard X-ray emission from the HXIS instrument on the SMM satellite (van Beek et al. 1980), from the SXT instrument on the Hinotori satellite (Makishima 1982; Takakura et al. 1983), and from stereoscopic observations from two widely separated spacecraft. The results from the imaging instruments conflicted to a certain extent (see Duijveman and Hoyng 1983). HXIS detected distinct bright points, co-spatial with Hα bright points, in some events (Duijveman et al. 1982). Sources initially displaying multiple bright points evolved into single sources as the events progressed (see Machado et al. 1985). The SXT, on the other hand, found hard X-ray sources stationary in the corona (Takakura et al. 1983; Tsuneta et al. 1984). The multi-spacecraft observations of Kane et al. (1979; Kane 1983) indicate that emission at higher energies (several 10^2 keV) is concentrated at lower altitudes (< 2000 km above the photosphere). So a variety of flare morphologies seem to be observed, depending perhaps on flare size, photon energy and stage of the solar cycle (see Duijveman and Hoyng 1983).

Tanaka (1987) has proposed a flare classification scheme, based partly on hard X-ray spatial morphology and temporal behavior, and on the physical interpretation of these (see Sec. I). While such schemes are certainly not without their uses (as metric-radio-burst classification shows; see, e.g., Kundu 1965), it does seem that the small size of the existing data set must render this classification scheme at best provisional, especially if one or two of the events begin to appear intermediate between two classes (e.g., the 21 May 1980 flare; de Jager and Svestka 1985).

Also relevant is the observation (Kiplinger et al. 1982) of fast fluctuations in the hard X-ray emission on time scales $\simeq 10^{-2}$ s. These must have important implications for fast-electron acceleration and transport (Emslie 1983; Kiplinger et al. 1984). It should be noted that these fluctuations are isolated transients rather than persistent variations, and the question of their reality seems at present not clearly formulated from a statistical viewpoint (Brown et al. 1984). There is, however, no such question mark over the reality of the very rapid, intense spikes observed at radio wavelengths (see, e.g., Slottje 1978; at centimetric wavelengths); these demand coherent radiation for their interpretation and (by their very nature) have implications for electron transport and acceleration (see, e.g., Kuijpers et al. [1981] for interpretation and observations of a decimetric radio burst).

As mentioned above, the microwave emission shows a time profile similar to the hard X rays. Delays between different frequencies, and between microwaves and hard X rays are apparent on subsecond time scales (Kaufmann et al. 1983; Cornell et al. 1984). When these delays are observed, microwaves never peak first (Costa et al. 1984). The significance of this fact is not yet clear. Until recently, the spectrum of the microwave emission was described as having a single maximum, usually somewhere between 1 and 10 GHz. Guidice and Castelli (1975) and Schöchlin and Magun (1979) discuss the statistical distribution of the parameters characterizing the observed spectra. Interpretation of this spectral form, using the theory of gyrosynchrotron emission by fast electrons, yields reasonable results in terms of consistency with ideas about the hard X-ray emission (See Dulk and Marsh 1981; Dulk and Dennis 1982; Klein et al. 1986). However, more recent investigations, using the frequency-agile spectrometer at the Owens Valley Radio Observatory (Hurford et al. 1984), reveal a more complex situation. Spectra sometimes have multiple maxima, and even when a single maximum occurs, the spectrum can be much steeper on the low-frequency side than had previously been appreciated. Moreover, the frequency of peak emission remains remarkably constant during the duration of most bursts (Stähli et al. 1989).

We regard the term γ rays as embracing all photons of energies $>$ 500 keV (roughly the rest mass energy of an electron). There are two major spectral components to this emission. First, a power-law continuum, the continuation upwards in photon energy of the hard X-ray continuum, apparently extends to energies of several tens of MeV in at least some bursts (Rieger et al. 1983). The implication is that individual electrons attain comparable energies. There is also some evidence of a flattening of the spectrum (or even a bump) around 70 MeV (Forrest et al. 1985). This is particularly interesting, as it would be the signature of π° decay, with the π° production in proton–proton collisions (Murphy et al. 1987). Of particular interest is the observation of limb-brightening of the continuum, both at energies of a few 100 keV (Vestrand et al. 1987), and at > 10 MeV (Rieger et al. 1983).

From 0.511 MeV up to $\sim$ 7 to 8 MeV, many distinct, narrow lines are superimposed on the continuum. These are mostly from de-excitation of nuclei excited by fast protons or α particles. They are of particular interest as they are the only direct, radiative signatures of accelerated flare ions. Since the lifetimes of the excited states are $\ll 1$ s, the de-excitation lines are "prompt"; there is no observable delay between excitation and emission of a photon. The observed time profile of the prompt lines mirrors the other impulsive phase emissions, again with delays sometimes being observed (Chupp 1984; Rieger 1982).

The lines at 0.511 MeV, from positron annihilation, and at 2.223 MeV from neutron capture, are not prompt. Their temporal evolutions are governed, respectively, by: the lifetimes of the positron-emitting, radioactive nuclei (see Kozlovsky et al. 1987); and the lifetimes in the photosphere of the fast neutrons (Hua and Lingenfelter 1987). They are observed over a period $\simeq 10^2$ to 10^3 s after the impulsive phase (see Chupp 1984).

The quality of observations in the γ-ray regime improved dramatically at the last solar maximum, with the results from the instruments on the SMM and Hinotori satellites; see the review by Chupp (1984), Murphy (1985), Ramaty (1986), Ramaty and Murphy (1987) and Vlahos et al. (1986).

The flare emissions in the extreme ultraviolet, optical continuum and chromospheric spectral lines (the most studied of which is Hα) also display an impulsive phase component, with rapid variations in flux evolving similarly to the other impulsive phase emissions. Such observations are reviewed in Canfield et al. (1986). As mentioned above, the radiation in these wavelength ranges is a less direct indicator of nonthermal particles, in the sense that the radiation is not emitted by the particles themselves, but by other particles excited by them. In the case of the white-light emission, the essential role of radiative backwarming of the deep atmosphere (Aboudarham and Henoux 1987) makes the connection still less direct.

Many interesting and suggestive results regarding the spatial and temporal behavior of these emissions, in conjunction with radiation at other wavelengths, were obtained at the last solar maximum, in particular with the instrumentation aboard SMM. For example, Cheng et al. (1981) showed that extreme ultraviolet brightenings at different locations coincided with different peaks in the hard X-ray time profile, indicating that successive injections of fast particles took place at different spatial locations; again, see Canfield et al. (1986) for a review.

Finally, for completeness, we mention the discovery (Doyle and Bentley 1986) that even the soft X-ray emission may exhibit (comparatively) rapid fluctuations, in this instance in line profile shapes. Presumably this indicates the hydrodynamic response of the bulk of the flare plasma to energy injection.

B. Interpretation of the Observations

We present in this section a synopsis of the most popular broad picture regarding the origin of the various flare emissions. Electrons are accelerated in the corona, by some unknown mechanism, to energies between $\simeq 10$ to 20 keV, and > 10 MeV. They produce hard X-ray and γ-ray continua by electron-ion (and electron-electron above 500 keV) bremsstrahlung (Korchak 1967,1971; Bai 1977). The power-law photon energy spectrum observed implies that the electron energy distribution is also a power law (Brown 1971). Such a distribution is a natural consequence of many particle acceleration mechanisms such as stochastic acceleration in wave-particle interactions (see, e.g., Benz 1977; Benz and Smith 1987) or in the vicinity of shocks (see e.g., Blandford and Ostriker 1978; Forman et al. 1986). Moreover, the total electron energy content approaches (Hoyng et al. 1976) or even exceeds (Tanaka and Zirin 1985) the energy otherwise manifested in the flare; thus it is reasonable to suppose that the electrons play a major role in redistributing the energy released in the flare through the atmosphere. They produce most of their hard X-ray emission and deposit most of their energy in the upper chromosphere, thus driving the (impulsive) extreme ultraviolet and Hα emission. Continuum emission from deeper layers is produced via radiative backwarming (Aboudarham and Henoux 1987). When chromospheric material is heated past the peak in the radiative-loss curve, the resulting pressure imbalance drives it into the corona, where it constitutes the source of the soft X-ray emission (see, e.g., Brown 1973; Antiochos and Sturrock 1978; Acton et al. 1982).

The same electrons that produce the hard X-ray emission and drive the visible-light flare are responsible for the microwave emission, via gyrosynchrotron radiation (Holt and Ramaty 1969). If they can escape onto open field lines, they may produce Type III metric bursts (alternatively, if they cannot escape they may generate intense, coherent, bursts of microwave emission that may remotely accelerate the electrons which produce the Type III bursts; Sprangle and Vlahos 1983). In any case, the conditions for X-ray producing electrons to incite a Type III burst are evidently not always satisfied, since only $\simeq 30\%$ of hard X-ray bursts have associated Type III's (Kane 1972; see also Vlahos et al. 1986). Other types of metric bursts display an association with hard X-ray emission, so we may presume that these emissions also are produced by the same electrons, or at least that the electrons variously manifested all have a common origin (Klein et al. 1983).

In large, extended flares, various phenomena occur that indicate an extended ($\simeq 100$ s) energy release, involving particle acceleration in the high corona (Cliver et al. 1986). The fact that the photon flux sometimes maximizes later with increasing photon energy (with delays occurring of several tens of seconds) has been taken as evidence of extended, "second-stage" acceleration (Bai and Ramaty 1979; Bai et al. 1983; Bai and Dennis 1985), although such an interpretation is not undisputed; the energy dependence of fast electron collisional losses will produce qualitatively the same behavior (Vilmer et al. 1982; Bespalov et al. 1987; see also Vlahos et al. 1986). In any case, the photon spectrum hardens progressively during the burst (indicating either extended acceleration, or electron evolution in a low-density trap, or both), and the associated microwave emission is intense, but has a low peak frequency (indicating a low magnetic field, and thus an origin in the high corona; Cliver et al. 1986). Type II metric bursts, associated with the coronal shock, which can accelerate particles, often accompany such an extended phase.

We now list some of the successes and failures of the "thick-target" model of the impulsive-phase phenomena outlined above. The observations which support the above picture are largely those which led to its development: simultaneity of hard X rays, microwave, extreme ultraviolet, Hα and white-light emission in the impulsive phase. Successful joint interpretation of the hard X-ray and microwave emission lends quantitative consistency to this picture (Klein et al. 1986). The spatially separated bright points observed by HXIS (Hoyng et al. 1981; Duijveman et al. 1982) coincide spatially with Hα bright points and may result from fast electrons arriving at the footpoints of magnetic loops. The evolution of image morphology also may be consistent with the picture of electron beams heating the chromosphere (Machado et al. 1985). The SMM observations also show close simultaneity of hard X-ray and extreme ultraviolet emission, both line and continuum (Poland et al. 1984; Woodgate et al. 1983), and the details of the extreme ultraviolet line time profiles are consistent with electron-beam energy deposition (Emslie and Nagai 1985). The hydrodynamic consequences of such energy deposition have been studied by many authors (Nagai and Emslie 1984; MacNeice et al. 1984; Fisher et al. 1985a,b,c; Pallavicini et al. 1983; see also Doschek et al. 1986), and the results for the temporal evolution of the soft X-ray emission, in some aspects, broadly confirm the above picture (see, however, Kopp et al. [1986] for a discussion of the numerical difficulties and comparison of results from the various groups).

The co-temporality of ultraviolet continuum and hard X rays seemed problematical, because of the difficulty of supplying enough energy in electrons to the depths from which the continuum emission originates. But here again, the role of radiative backwarming is crucial (Machado et al. 1986). However, there are really more difficulties than positive successes, and probably none of the evidence unambiguously or uniquely supports the thick-target model. First, the Hα and extreme ultraviolet emissions do not require the energy input into the chromosphere implied by the hard X rays. In the case of the Hα emission, this is because the depth of formation of this emission is such that only higher energy electrons (≥ 100 keV) are important in driving it; these electrons are energetically insignificant compared to those at lower energies. The case of the extreme ultraviolet line emission is more serious, however (Emslie et al. 1978; Donnelly and Kane 1978). Apparently this emission demands only a small fraction of the energy supposedly dumped in the high chromosphere by the fast electrons. The problem has been recently re-examined by McClymont and Canfield (1986), and possible resolutions, involving trapping in the corona of the fast electrons, have been indicated (MacKinnon 1988; Larosa and Emslie 1988; Spicer and Emslie 1988).

We consider the hard X-ray images more closely. At first sight, footpoint observations certainly look like evidence for thick-target electron beams, but there are difficulties (with total photon fluxes, and with the spatial distribution of spectral indices), and statistical uncertainties (photon shot noise limits the extent to which timing considerations constrain the energy transport mechanism) (MacKinnon et al. 1985). Worse still for the thick-target model are the Hinotori coronal sources mentioned above (Takakura et al. 1983; Tsuneta et al. 1984). Trapping of electrons (MacKinnon 1988), a thermal origin (Takakura 1984) and collisional stopping of the (rather low energy) electrons close to the acceleration region (Tsuneta et al. 1984) are three possible explanations of these source morphologies. The third of these implies the interesting possibility that these observations might indicate the actual spatial location and extent of the acceleration region. But in any case, these observations do not resemble much the expectations of the thick-target model.

The evidence for chromospheric evaporation driven by electron energy deposition is at best ambiguous (Doschek et al. 1986). The absence from the data of the large blue shifts implied by this picture must be considered a real difficulty (Emslie and Alexander 1987; McClements and Alexander 1989). One can only conclude that the large number of energetic electrons, if they are present at all, are prevented from depositing much of their energy in the high chromosphere. There are various ways in which this might happen (Larosa and Emslie 1988; Bespalov et al. 1988; Spicer and Emslie 1988).

Perhaps the greatest difficulty with the thick-target model is not directly observational. Deposition of a large fraction of the total flare energy in nonthermal, fast electrons is very difficult to explain, and seems highly unlikely from the point of view of statistical thermodynamics. How seriously one takes this problem must depend on how well determined one considers the "total" flare energy to be (see the relevant chapters of Sturrock [1980] and Wu et al. [1986]). The total energy content of mass motions, for example, may be rather uncertain.

Further, serious problems crop up if one supposes that the fast electrons ever form a single, direct current. A neutralizing "return current" is set up, whose maintenance constitutes an additional energy loss from the beam (Knight and Sturrock 1977; Emslie 1980; Brown and Bingham 1984; Spicer and Sudan 1984; van den Oord 1990). Velocity space instability of the drifting background electron distribution in this case may further complicate matters (Hoyng et al. 1978; Rowland and Vlahos 1985; Cromwell et al. 1988).

These various difficulties continue to prompt examination of alternative scenarios for the hard X-ray productions, since only the interpretation of the hard X rays seems to demand the high-acceleration efficiency. As long as we persist in considering sources in which nonthermal electrons slow down collisionally, the thick-target model (Brown, 1971,1975) gives a strict lower limit to the total flux of electrons required (MacKinnon and Brown 1989). While doubtless essential for correct modeling of the spatial structure and temporal evolution of hard X rays, inclusion of collisionless processes (see, e.g., Hoyng et al. 1979; Emslie and Smith 1984; Vlahos and Rowland 1984; McClements 1987) serves only to worsen the efficiency problem (Brown 1975; Brown and MacKinnon 1985). "Exotic" radiation mechanisms, such as inverse Compton scattering, need extreme source parameters (Kaufmann et al. 1986; McClements and Brown 1986).

The only way of making the source more efficient is to decrease the rate at which the electrons slow down; the way to do this is to give all electrons in the source comparable energies. This leads to the so-called "thermal" models (Brown et al. 1979; Smith and Lilliequist 1979; Batchelor et al. 1985; Spicer and Emslie 1988). The controlling factor for the efficiency of these models is the effectiveness with which electrons can be contained. None of them has achieved any great degree of popularity; but on the other hand, they are hard to rule out from either spectral (Brown 1974) or existing spatial (MacKinnon et al. 1985) observations, and the potential gain in efficiency (in principle, up to 10^5) is such that it seems unwise to discard them completely at present.

We briefly discuss the interpretation of the γ-ray line emission, the only source of information on fast ions. As reviewed by Chupp (1984), Ramaty (1986) and Ramaty and Murphy (1987), it yields information on the number and spectrum of fast ions, with energies > 10 MeV per nucleon. One finds that up to 10^{33} protons are accelerated above 30 MeV. Their energy distribution falls off more steeply than a single power law (Murphy and Ramaty 1984), for example, as the modified Bessel function K_2 in rigidity found in the Ramaty-Lee model of stochastic acceleration (Forman et al. 1986). The number of lower-energy protons is uncertain, and their possible importance in the flare process debated (Simnett 1986; Brown et al. 1990). Gamma-ray lines have other diagnostic possibilities. Line shapes may yield information on the anisotropy of fast protons (Murphy et al. 1988). The width of the 0.511 MeV positron annihilation line indicates the temperature of the interaction region, a question of some interest for source models and transport (Ramaty 1986). The temporal evolution of γ-ray emissions also has diagnostic potential (Ramaty et al. 1988; Hulot et al. 1988).

C. Acceleration Mechanisms

A successful model for the acceleration of fast particles in flares must be able to accelerate 10^{34} to 10^{36} electrons s^{-1}, for up to 100 s, to energies from $\simeq 20$ keV to possibly 100 MeV, with a distribution in energy of (possibly broken) power-law form. The total energy in these electrons is 10^{30} to 10^{32} erg (Hoyng et al. 1976; Brown and Smith 1980). This acceleration must occur simultaneously at all energies to within 2 s (at least in some events) (see, e.g., Chupp 1984). Up to 10^{33} protons have to be accelerated, to energies between 10 MeV and possibly 1 GeV (Murphy et al. 1987), again simultaneously at all energies, to within 2 s. Their distribution in energy must fall off more steeply than a single power law (Murphy and Ramaty 1984). The requirement at lower proton energies remains a matter for speculation (but see Canfield and Chang 1985; MacKinnon 1989).

As we saw above, some of these requirements may be open to revision, because of model-dependent physics. Also, they depend at least partly on solution of an inverse problem, a procedure always fraught with danger (Craig and Brown 1986). One can only look with envy at the sort of observations which exist for the Earth's bow shock (see, e.g., Paschmann et al. 1981).

All acceleration of particles is achieved by electric fields, whether these are the rapidly oscillating electric fields associated with waves, the (d.c.) electric field produced inductively in the reconnection process, the electric field experienced by an electron in its rest frame as it encounters an MHD shock, or the self-consistent field produced by charge separation in a double layer. Reviews are given, e.g., by Heyvaerts (1981), Forman et al. (1986) and Vlahos et al. (1986). Acceleration in the vicinity of shock waves has been widely studied in recent years, in various branches of astrophysics (see, e.g., Drury 1983). In the flare context, it appears a likely candidate for accelerating ions to energies of 10 to 100 MeV (Decker and Vlahos 1986; Forman et al. 1986) inside the 2-s time resolution of the SMM GRS instrument. The shock formation time for collisionless shocks is a few ion gyroperiods (Cargill et al. 1988), so this is not a problem. However, the absence of any soft X-ray signature of the shocked material may constitute a problem (Smith and Brecht 1988).

The energy gain of a particle per shock crossing is proportional to the particle mass (see, e.g., Kantrowitz and Petschek 1966); shock acceleration is consequently much less effective for electrons, although it may additionally accelerate them to relativistic energies in a "second step" of acceleration (see, e.g., Bai et al. 1983). If shock acceleration were to be seriously considered for the acceleration of the thick-target inferred electron population, nonlinear investigation (including the effect of the fast electrons on the shock dynamics) would become necessary (see, e.g., Drury and Völk 1981). The same comment applies if we want shock-accelerated ions to be energetically important, as proposed by Simnett (1986).

Two major ingredients of a complete theory of ion acceleration in shocks may still need elaboration. The first concerns the preheating mechanism. Many scenarios for shock acceleration (see, e.g., Decker and Vlahos 1986) require particles already moderately accelerated (see, e.g., Forman et al. 1986) and some other mechanism must exist which is able to do this. The second ingredient is that efficient scattering mechanisms must exist both upstream and downstream of the shock (see, e.g., Forman et al. 1986). Downstream, the shock itself may excite turbulence, but upstream generation of turbulence is more problematic. The accelerated particles themselves may eventually generate waves (Bell 1978), but the initiation of this process is not yet clear. Chiueh (1988) suggests one possible way in which particles could initially be accelerated, which will work for shocks propagating quasi-perpendicular to the field.

If a nonthermal distribution of wavevectors is present in a particular mode, particles able to resonate with the waves may gain energy stochastically from them. As long as the perturbation of the particle orbits by the waves is unimportant, quasi-linear theory is appropriate. A particular, hypothetical spectrum of waves may be employed to calculate a diffusion coefficient, and acceleration times, spectra, etc., calculated (see, e.g., Kulsrud and Ferrari 1971; Kaplan et al. 1975; Melrose 1980; Achterberg 1981: Forman et al. 1986). Hydromagnetic turbulence acts too slowly to accelerate electrons in the impulsive phase (although, again, it may do so in the gradual phase, in a "second step" of acceleration), but it is a good candidate for ion acceleration (Smith and Brecht 1986,1988). Heyvaerts (1981) discusses these ideas in some detail.

If quasi-linear theory is inadequate, there are still other possible approaches. First, one might attempt to make use of a more nonlinear statistical theory, e.g., a "resonance broadening" theory, suggested by Heyvaerts (1981). Alternatively, integration of test particle orbits in the presence of large-amplitude turbulence may be attempted. In the latter case, the interesting possibility of chaotic electron orbits arises (de la Beaujardiére and Zweibel 1989).

Electron acceleration by Langmuir waves has in the past been extensively investigated, but the generation of the Langmuir waves constitutes a substantial difficulty, especially since the efficiency of the "turbulent bremsstrahlung" process has been shown to be identically zero (Melrose and Kuijpers 1987). Lower-hybrid waves have recently received attention, and appear promising in that specific mechanisms can be given for their generation, and that they will produce electrons of high enough energies fast enough (Benz and Smith 1987, and references therein).

A difficulty with all models involving stochastic acceleration by plasma waves concerns efficiency. Acceleration by a particular sort of plasma wave may hypothetically be 100% efficient, in the sense that the waves are damped completely by accelerating particles. But we still need to explain how up to half of the flare energy finds its way into the relevant wave mode. For example, the model of Benz (1977) invoked Langmuir-wave acceleration, and addressed the anti-correlation of flux and spectral index in hard X-ray data (see also Brown and Loran 1985). Benz inferred $W/nkT \simeq 10^{-5}$, where W is the total wave-energy density. Using a density, temperature and volume typical of the flaring corona, one finds that a typical intense electron beam carrying off 4×10^{29} erg s^{-1} will remove the wave energy of the entire volume in $\leq 10^{-5}$ s. This time may easily be shorter than either the time to form the wave spectrum, or the electron acceleration time (using the results of Kaplan et al. [1975]), so that a consistent treatment of the acceleration must address simultaneously the generation of the wave spectrum and the particle acceleration.

Direct electric-field acceleration has been discussed, in various forms, by various authors (see, e.g., Smith 1980; Heyvaerts 1981; Kuijpers et al. 1981; Holman 1985; Martens 1988; see also Vlahos et al. 1986). The fact that no return current can flow in the acceleration region (Smith 1980; Heyvaerts 1981; Spicer 1983) places severe restrictions on how many electrons can be accelerated in this way. Holman (1985) finds that $\simeq 10^4$ separate current channels, each with its own return current, are needed. The acceptability of such a situation is at present largely a matter of personal choice. Holman (1985) prefers a thermal interpretation of the hard X-ray emission, with joule heating in a single current sheet able to provide the necessary hot material. He neglects all loss processes, so the possible temperatures which he derives may be overestimated, perhaps significantly.

Martens (1988) attempts to circumvent the difficulty of needing lots of separate current sheets, essentially by an argument based on the geometry of the two-ribbon flare (see Kaastra 1985). Acceleration of protons and electrons to equal velocities (so that most of the energy is in fast protons) takes place as in the geomagnetic tail (cf. Speiser 1965). Hard X-ray production is not correctly addressed in this work, however; electrons are not produced in sufficient numbers, and the protons that are produced are mostly of too low an energy to produce hard X-ray photons directly (contrary to Martens [1988], Heristchi [1986] does not reconcile proton production of hard X rays with the theory of γ-ray line production). Developments of this idea might yet prove fruitful.

III. SPACECRAFT OBSERVATIONS OF SOLAR EJECTA

A. Introduction

During the past several years, it has become possible, through making observations of suprathermal and energetic particles in space, to use these particles as diagnostic tools for the remote probing of solar processes. It is, for example, now recognized that impulsive flares, coronal mass ejections, disappearing-filament events and interplanetary shock waves each have as distinctive a signature in the timing, composition and spectra of their associated accelerated particles as they do in their radio, optical, X-ray and γ-ray emissions. This new perspective has emerged from extensive comparisons of particle observations with coronagraphic and spectroscopic data. These show unexpected new correlations that can be exploited to study the physics of underlying solar processes.

Simultaneous observations of various particle populations and correlative studies with electromagnetic data provide valuable information concerning the nature, location, the time of triggering and the energy dependence of the acceleration mechanism. Further, since solar energetic particles constitute direct samples of solar material, they also carry information about such parameters as the composition, temperature and density of the local region from which they originate as well as information concerning the thickness of the atmospheric layer they have traversed.

B. Particle Acceleration in the Solar Atmosphere

Acceleration processes occurring in the solar atmosphere, or close to the Sun, display wide variations in their scales, atmospheric height, location, topological structure of associated magnetic fields and in the overall size of the acceleration region. On the basis of these variations, four distinct acceleration processes are currently recognized:

1. Short time scale (impulsive) acceleration related to the flash phase of flares;
2. Second-order Fermi (stochastic) acceleration in turbulent regions generated after a flare;
3. Low coronal shock acceleration generated immediately after the impulsive flare (sometimes operating in closed magnetic loops);
4. High coronal shocks associated with the largest flare events and with coronal mass ejections.

These different processes operate in different coronal sites and energize different particle populations to produce suprathermal and energetic particle signatures. Impulsive particle energization is likely to be due to fast magnetic-field reconnection. Thus, an understanding of this process is central to unraveling the role of field-line merging in the corona and at other sites. Acceleration by shock waves is important not only because of its central role in energizing solar particles but also because, in the case of coronal shocks, much of the shock energy is eventually converted into thermal energy and thus may be a significant contributor to coronal heating.

Two Classes of Impulsive Events.   Impulsive solar-particle events are associated with short-duration X-ray events (< 1 hr) which originate in compact regions low in the corona. During solar maximum conditions, these events occur about once per day, and generally produce relatively small fluxes of energetic particles and low maximum particle energies. The events are thought to begin with the release of energy via magnetic-field annihilation over an active region that impulsively energizes solar coronal electrons to 10 to 100 keV. This is often called "first-phase" or impulsive-phase acceleration. A substantial fraction of the total energy released in the annihilation is imparted to these electrons. As they move through the corona and into interplanetary space, where they are directly observed, the accelerated electrons produce Type III, and occasionally Type V, radio bursts (see Sec. I.C). Protons are also energized, but with a low abundance. Shock-wave acceleration, at least on a large scale, appears to be absent.

Kocharov and Kocharov (1984) have shown that the particles associated with some impulsive events are enriched in ^{3}He, while Klecker et al. (1984) have demonstrated, based on ion charge states, that the particles come from hotter coronal regions than do large-particle event fluxes. Studies by Reames et al. (1985) indicate that, in these impulsive events, electrons are accelerated to 10 to 100 keV simultaneously with the ^{3}He population.

The ^{3}He-rich events are of special interest because (a) they require a plasma resonance process to heat preferentially the ^{3}He in the corona before the occurrence of an impulsive event; (b) they take place in unusually hot coronal sites; and (c) they provide the most nearly "pure" signature of the impulsive acceleration process since shocks do not appear to be involved. It is interesting to note that one could not have anticipated the presence of a plasma resonance phenomenon that would produce a thousand-fold enhancement in the rare isotope ^{3}He on the basis of the electromagnetic spectrum alone.

A second class of impulsive-particle events is associated with the observation of nuclear γ-ray continuum and line emission. Chupp (1984) has shown, using γ-ray observations, that ion and electron acceleration during these events occurs virtually simultaneously and on a time scale of seconds or less. Only a small fraction of the number of nuclei required to explain the observed γ emission escape into interplanetary space and, for a given γ flux, a large variation may be observed in the number of protons and electrons detected by spacecraft instrumentation (Kallenrode et al. 1987).

The energetic-particle characteristics of these events include hard-electron spectra and a particularly high abundance of MeV electrons (see Evenson et al. 1984). Also, there is a correlation between the electron-to-proton ratio and the helium-to-proton ratio (Kallenrode et al. 1987). Unlike the ^{3}He-rich events, described above, γ-ray flares are associated with coronal shock waves.

Gamma-ray line flares are distinguished from other flares by a different acceleration mechanism (Bai and Dennis 1985). Protons are accelerated within closed magnetic loops and produce γ rays by interacting with the solar atmosphere (Bai 1986). An additional proton component is then accelerated by a shock wave in the high corona and escapes into interplanetary space. This partly explains the poor correlation between γ-ray fluences and interplanetary proton fluxes.

Long-Duration Events. Solar energetic-particle events associated with long-duration ($>$ 1 hr) soft X-ray emission, are characterized by the appearance of large fluxes of relativistic electrons, energetic protons, helium and heavy nuclei in interplanetary space. They occur roughly once per month, as observed at 1 AU. According to a generally accepted interpretation, these events begin with first-phase acceleration of electrons whose subsequent interactions with the solar atmosphere through, for example, explosive joule heating, produce a variety of other phenomena including Hα brightening. The explosively heated solar atmosphere produces a shock wave which propagates outward through the corona generating Type II radio bursts. Closely associated with these processes is the expulsion of material in a coronal mass ejection (Kahler et al. 1984). As this mass ejection moves away from the Sun, it triggers further magnetic field annihilation at high altitudes, producing long-duration soft X-ray emission.

The high coronal shock characterizing the second phase is likely to be the primary accelerating mechanism in these events. Models of diffusive shock acceleration (Ellison and Ramaty 1985) are able to produce energy spectra for electrons, protons and helium nuclei which are in acceptable agreement with the observations. A break in the electron spectrum at around 200 keV may indicate that the shock acceleration works at higher energies while a different (impulsive) process accelerates the electrons up to about 100 keV.

An important property of large-particle events is that energetic particles reach the Earth from flares occurring over the entire visible disk of the Sun. Since particles are bound closely to magnetic-field lines in interplanetary space, this implies the existence of some mechanism operating close to the Sun that permits transport over heliographic ranges of up to about 150 deg. If the energetic particles originate at, or near, the flare site, then some special field configuration might be responsible, or, alternatively, a process of continuous field-line merging throughout the corona might permit particle transport over large distances (see Newkirk and Wentzel 1978). Alternatively, the longitudinal extent of a particle event might be due to the large-scale size of the original acceleration process. In this scenario, sizeable coronal shocks in the outer corona spread over a large fraction of one solar hemisphere and directly produce particles on open field lines where they are immediately released into interplanetary space (Manson et al. 1984).

Filament Eruptions.   These eruptions also produce particle signatures in interplanetary space. On the disk, these events are associated with some brightenings in Hα that are insufficient to allow the event to be called a flare. Filament eruptions occur outside of active regions and have no radio or hard X-ray bursts associated with them. This implies that strong impulsive phenomena are not essential for the production of their associated energetic protons. The important connection between these phenomena and long-duration flare events is the presence of a coronal mass ejection. The acceleration mechanism involved is almost certainly a high coronal shock, with no admixture of an impulsively energized particle population. In these events, it is likely that large-scale shocks traversing the solar corona accelerate material from rather high coronal locations, perhaps as far out as several solar radii.

C. Solar Atmosphere Diagnostics

Since solar ejecta carry a sample of the solar atmosphere into interplanetary space, *in situ* measurements can provide unique information about the solar atmospheric composition and about physical processes operating in the corona. Fractionation processes operate in energetic-particle acceleration, thereby biasing the energetic-particle population compared with the ambient source material in the solar atmosphere. These effects are systematic and appear to be well ordered by the particle mass-to-ionization state (A/Z^*) ratio, making it possible to correct for this biasing if one observes a large number of different particle events over a broad energy range.

As discussed above, different kinds of solar-energetic-particle flares sample different coronal sites. While large-particle events appear to sweep up large volumes of coronal material, small ^{3}He-rich events sample compact hot sites in the corona. By comparing the composition at flare sites with solar wind and photospheric composition, it is then possible to gain insights into the physical mechanisms that transport heavy ions into the corona. Up until the present time, composition instruments for the solar wind have not been flown in conjunction with suprathermal and energetic-particle composition instruments.

At high particle energies, inelastic nuclear interactions are possible in the solar atmosphere. Secondary particles such as hydrogen isotopes, fragmented light nuclei and neutrons provide signatures of such interactions. Since the cross sections of these processes are well known, the fluxes of the secondaries can be used to infer the amount and density of solar material traversed. Composition measurements of solar-energetic-particle events can accordingly provide relevant information. Also, as shown by Evenson et al. (1985), neutron emissions can be studied by detecting medium energy protons resulting from neutron decay and these data can provide further information concerning atmospheric traversal.

Elemental Abundances. The average solar-energetic-particle abundances observed in large flares differ from photospheric and universal abundances in a manner ordered by the first ionization potential (FIP) (see McGuire et al. 1986; also Fig 4). Solar-energetic-particle abundances of elements with FIP $<$ 10 are overabundant relative to the photosphere by a factor of 3 to 4 as compared with elements with FIP $>$ 10. This same trend is found in the composition of the solar corona and of the solar wind. A similar effect has been found in galactic cosmic-ray data. The FIP bias detected in solar energetic particles appears to imply a preferential rise of ionized heavy elements relative to neutral ones from the underlying cool chromosphere (see Meyer 1985). This suggests that heavy elements with low FIP are easily ionized and a selection then takes place between ionized and neutral heavy elements in which ionized material is transported into the corona with a 3- to 4-fold increase in efficiency. This tentative interpretation requires further testing by space observations.

Isotopic Abundances. High-resolution particle measurements obtained by Mewaldt et al. (1984) using the ISEE-3 instrument indicate that, during a solar-energetic-particle event, the ^{20}Ne/^{22}Ne ratio was significantly

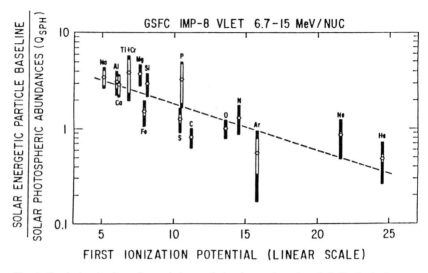

Fig. 4. Graph showing how observed elemental abundances depend on their first ionization potential.

different from that found in the solar wind. This result suggests that a different mechanism for the injection/acceleration of solar energetic particles and solar wind neon may be operative. Thus, a systematic survey of these particle isotopic abundances designed to cover a broad energy range in major and minor species should be carried out using sensitive, high-resolution instrumentation and compared with complementary solar-wind data.

³He-Rich Solar Flares. Recent studies have shown that ³He-rich solar flares occur more often than once per month and are associated with impulsive kilovolt-electron events and therefore also with Type III events (Reames et al. 1985). Since kilovolt-electron events are observed with a frequency of > 1 day, it is probable that advanced instrumentation with relatively low-energy thresholds will shortly detect ³He-rich events with a similar frequency. These events often have ³He/⁴He ratios of the order of unity, which is about 1000 to 10,000 times the abundance observed in the photosphere or in the solar wind.

The source regions of these small events have been studied by a number of authors. In particular, Kahler et al. (1985*a,b*) and Lin (1987) report that the acceleration mechanism in these events is different from the one operating in large solar-particle events, and indeed appears to be identical with the mechanism that impulsively accelerates the kilovolt electrons. The shape of the electron spectrum that typically extends down to 2 keV, indicates that the source of the electrons is probably high in the corona.

Fisk (1978) and Varvoglis and Papadopoulos (1983) interpret the ³He enrichments to be due to a resonant plasma-heating mechanism which preferentially heats the ^{3}He before some additional mechanism energizes it to MeV energies. These events are also associated with substantial heavy-nuclei enrichments of 10 to 20 times normal solar-energetic-particles composition, although the heavy-nuclei enrichment does not correlate well with FIP for these events. While the heavy-nuclei enrichments might be caused by the same plasma-heating mechanism that effects the ^{3}He, other evidence indicates that these heavy-ion enrichments are due to thermal diffusion (which would be expected to be important at high-temperature coronal sites). The conclusion is that ^{3}He flares sample a distinctly different location in the corona from that pertaining to large-particle events.

Acknowledgments.   One of us (S.McK-L) acknowledges the important source of material provided by the literature survey carried out by the COS-TEP team (of which she is Co-I) in the course of preparing their proposal to fly a suprathermal and energetic particle analyser on ESA's Solar and Heliospheric Observatory.

FINE STRUCTURE OF THE SOLAR TRANSITION REGION: OBSERVATIONS AND INTERPRETATION

J. W. COOK AND G. E. BRUECKNER
Naval Research Laboratory

We review recent observations of the solar transition region and the temperature minimum obtained with high spatial resolution, primarily ultraviolet spectra and spectroheliograms from sounding rocket and Spacelab 2 flights of the High Resolution Telescope and Spectrograph. This instrument covers the 1175 to 1710 Å wavelength range with 0.05 Å spectral resolution and arcsec spatial resolution, and samples emissions arising from the temperature-minimum region (continuum over 1525 to 1680 Å), chromosphere (lines of C I, Fe I, Si I), transition region (lines of Si IV, C IV, N V), corona (lines of Fe XII, Fe XIII, Si VIII), and a high-temperature flare line from Fe XXI. Information on intensities, velocities, emission measures and electron densities can be determined. We discuss observational results obtained from the High Resolution Telescope and Spectrograph flights, and the implications of observed fine structure for our ideas of the structure and physics of the solar atmosphere. Inclusion of fine structure in our ideas about the transition region affects the plausibility of one-dimensional average models of the atmosphere, the determination of tempera-ture gradients, possible mechanisms for nonradiative heating, and the compar-ison of transition region structures with corresponding high-resolution obser-vations from the photosphere and corona.

I. INTRODUCTION

In this chapter we discuss recent high spatial resolution observations in the extreme ultraviolet of the solar atmosphere, and illustrate why these ob-servations are important for our understanding of the basic properties of the transition region, and indeed, of the solar atmosphere in general. The chapter is not a general review of the transition region (TR); such general reviews might be found in the discussions by Withbroe and Noyes (1977), Athay (1976, 1981, 1986*a,b*), Pneuman and Orrall (1986) and Mariska (1986). We do not attempt to develop standard concepts or methods of analysis found in these reviews, such as the emission measure, model atmospheres derived from the emission measure, electron density diagnostics from line intensity ratios, etc. In addition, the observational material discussed is heavily weighted toward data obtained by the High Resolution Telescope and Spectrograph (HRTS) experiment of the Naval Research Laboratory, which since 1975 has had six flights as a sounding rocket payload and one flight on the Space Shuttle, the Spacelab 2 mission. Other experiments, in particular the Ultraviolet Spectrometer and Polarimeter (UVSP) experiment aboard the Solar Maximum Mission (SMM) satellite, and the sounding rocket flights of the Transition Region Camera of the French LPSP (Laboratoire de Physique Stellaire et Planétaire)—Lockheed collaboration, have also obtained important results. But in its combination of spectral coverage and high spatial resolution, the HRTS is so far unique, and has obtained observations pertinent to a large range of problems in solar physics. Some earlier general reviews of HRTS observations have been given by Brueckner et al. (1978), and Brueckner (1980*a, b, c,* 1981). Reviews of the Spacelab 2 general results can be found in Cook (1985), Brueckner et al. (1986), and Dere et al. (1987*a.*).

In this chapter we concentrate on observations of atmospheric fine structure, and how the interpretation of these data affects our basic understanding of the solar atmosphere. We focus on the most common, general solar features: quiet regions, active regions and sunspots, and discuss the relationship of TR structures with those observed both in the photosphere and chromosphere, using such observations as magnetograms, Hα, and temperature-minimum images, and in the corona, for example, using soft X-ray data. This is an area of obvious importance and general interest, but it is difficult to obtain simultaneous observations of a range of temperature regimes.

Because of the fortunate coincidence that the ultraviolet continuum emission arising from the temperature-minimum region occurs within the spectral range of the HRTS instrument, we obtain observations of the coolest region of the solar atmosphere with HRTS. This data is discussed for its own inherent interest, and also because it is one of the best examples of comparing the TR structure with other regions of the atmosphere. Finally, we mention briefly several upcoming space programs which are aimed at even better observations of fine structure in the future.

The TR does not generally receive as much attention in solar physics texts as the photosphere or chromosphere (extensively studied from the ground) or the corona. The TR, as suggested by its name, is often thought of only as an interface between chromosphere and corona as temperature rapidly climbs to coronal values. It represents a small amount of material compared to the chromosphere or corona (see Sec. VIII). It is actually more interesting than this, predominantly because it is the ideal place to observe the effects of small-scale heating in its most violent and easily observed form. In the lower TR (say 10^5 K), some quiet Sun plasmas may have negligible magnetic fields with an energy balance where conduction is not important compared to radiation, it is an ideal place to obtain highly turbulent effects because small-scale heating is not as readily redistributed by conduction along field lines but must be dissipated locally. The TR may also be the best place observationally to look for diagnostic signatures of possible wave heating of the corona, and for the small-scale sources of the solar wind.

The most important diagnostics of the TR are spectral emission lines in the ultraviolet formed over the approximate range 2×10^4 to 10^6 K. In a general way, we can think of the TR starting at the approximate temperature of formation of Lyman α 1215 Å, while by 10^6 K we are dealing with coronal emissions. In the wavelength range between 1175 and 1710 Å covered by the HRTS instrument there are over 3000 emission lines (see Sandlin et al. 1986). We can observe a range of both chromospheric (lines of C I, Fe I, Fe II, Si I, etc.) and TR (lines of C II - C IV, Si III - Si IV, N IV - N V, etc.) emissions. The major resonance lines of many of the abundant metals from ionization stages within this temperature range occur in this region of the solar spectrum. There are also a limited number of coronal lines (Fe XII, Fe XIII, Si VIII) present, although they are not strong on the disk. We even have one excellent diagnostic of the high temperatures encountered in solar flares, the 11×10^6 K Fe XXI line at 1354 Å. Finally, we also can observe continuum emissions from the very coolest layer of the Sun, the temperature-minimum region, over the 1525 to 1680 Å range. There is excellent coverage up to approximately 2.5×10^5 K (O V) for the lower TR, but for higher temperatures we need to go to a shorter-wavelength range, and so we miss the upper TR observing with HRTS. This particular region of the solar atmosphere may be one of the most interesting areas for future experiments (see Secs. XII, XIV and XVI).

II. TRANSITION REGION FINE STRUCTURE

Important results were obtained from many earlier lower-resolution observations of the solar TR from such experiments as the Orbiting Solar Observatory series, culminating in OSO-8, from sounding rocket programs, and especially from the Skylab instruments flown on the Apollo Telescope Mount in 1973–74. A traditional interpretation of the TR evolved: the observed differential emission measure (DEM) distribution yielded a model atmosphere with a very steep temperature gradient and small scale height. It was discovered that TR emission line profiles were turbulently broadened, but attempts to interpret line broadening, or Doppler velocities from line wavelength shifts, as evidence of acoustic wave heating by dissipation of an amount of energy sufficient to account for the TR, were negative (see, e.g. Athay and White 1979). This work was not yet at high spatial resolution, which we will consider as 5 arcsec or better resolution, which roughly begins to resolve spatially individual network elements in the quiet Sun.

What is interesting about the fine-structure observations of the TR more recently obtained? First, of course, we naturally want to know what is really there. The discovery of high-velocity TR explosive events at arcsec spatial scale (see Sec. X), for example, was completely unexpected. But the reason for interest which we emphasize here is the modifications which the fine structure may impose on earlier interpretations of lower-resolution observations, that treated the atmosphere at some spatial scale as a plane parallel layer (an "average" model). The physical importance of the fine structure arises from the nonlinearity in scaling of many physical quantities when their fine-structure values are compared to average model values: the average over the true fine structure is not the same as the average model value. As an example, taken from HRTS temperature-minimum observations, we will see (in Sec. XV) that the average brightness temperature of the temperature-minimum region observed with 1-arcsec spatial resolution is some 25 K cooler than the brightness temperature corresponding to the average intensity observed at low spatial resolution (which is averaged over the fine-structure elements). At the temperature minimum, a 25 K difference represents a non-negligible difference in the heating requirement because of the great efficiency of H^- as a radiator. In the TR itself, taking into account actual fine structure reduces the temperature gradient determined from the observed differential emission measure (DEM) distribution, and reduces the value obtained for the average nonthermal broadening of emission lines.

III. THE HIGH RESOLUTION TELESCOPE AND SPECTROGRAPH

The High Resolution Telescope and Spectrograph (HRTS) experiment was designed from the beginning as a high spatial resolution probe of the solar chromosphere and TR. The solar physics group at the Naval Research Laboratory (NRL), already experienced in building spectrographs in the far ultraviolet, was responsible for the SO82B spectrograph which flew aboard Skylab in 1973–74. That instrument covered a nominal wavelength range from 970 to 1960 Å, but had no spatial resolution along its 2 × 60 arcsec slit. It was felt that higher spatial resolution would be the key to further advances, and a new fully stigmatic design with a novel tandem Wadsworth grating mount was constructed (see Bartoe and Brueckner 1975, 1978). This HRTS instrument has since flown six times as a sounding rocket payload, and was flown aboard the Spacelab 2 mission from 29 July to 6 August 1985.

HRTS consists of a 30-cm cassegrain (for sounding rockets) or gregorian (on Spacelab 2) telescope; a broadband spectroheliograph tuned to a wavelength region around 1600 Å on some flights to view the temperature-minimum region on the disk, or to 1550 Å to view structures at the limb in

C IV 1548 Å and 1550 Å; a slit spectrograph which can cover a wavelength range from 1175 to 1710 Å with 0.05 Å spectral resolution (although in individual flights it has been configured in several different ways); and an Hα system, using a temperature controlled Fabry-Perot filter, that could both display images using a TV camera and record them on film. Both the spectroheliograph and Hα systems used reflected images from the spectrograph slit jaw mirrors. Slit spectra and spectroheliograph images were recorded by film exposures.

The spectrograph slit was 920 arcsec, or approximately a solar radius in length. Its theoretical spatial resolution is sub-arcsec. In sounding rocket flights spatial resolution of up to 0.8 arcsec along the slit was achieved on HRTS I. In later sounding rocket flights, the resolution has sometimes been poorer, possibly because of the difficulty in establishing thermal equilibrium within the structure just before launch, and from the small deformation during launch acceleration. For HRTS II, the spatial resolution was 2 arcsec; for HRTS III, 1.2 arcsec; and for HRTS IV, 3 arcsec. With HRTS V, a mechanism to focus the telescope during flight based on the Hα image was successfully used, and the resolution was again near 1 arcsec. During the Spacelab 2 mission, jitter in the Spacelab 2 pointing system typically limited resolution to 1 to 2 arcsec in practice.

Figure 1 illustrates a HRTS spectrum from Spacelab 2. On a circle rep-

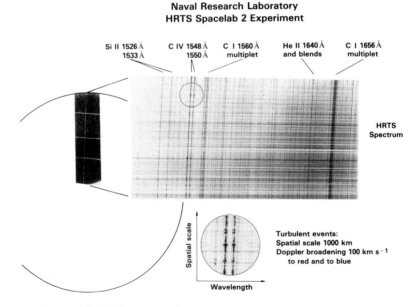

Fig. 1. Format of the HRTS spectrograph.

resenting the solar disk is shown to scale a HRTS Hα image, where the spectrograph slit and three fiducial wires crossing the slit can be seen. A section of a HRTS spectrum that includes the C IV lines near 1550 Å is illustrated, and a region along the slit showing high-velocity C IV explosive events is circled and shown enlarged at bottom (see Sec. X). This is meant to give a sense of the format and scale of the HRTS observations.

IV. OBSERVATIONS POSSIBLE WITH HRTS

The strong point of HRTS is the simultaneous coverage potentially available over a broad wavelength range at high spatial and spectral resolution. We can observe emission lines and measure directly their intensities, bulk flow velocities from the shift of the line profile as a whole from its rest wavelength, and nonthermal or turbulent velocities from the broadening of the line profile beyond its thermal width for the temperature of formation of the line. In addition, we can often obtain an estimate of the electron density of the emitting plasma using density-sensitive line-intensity ratios that vary with density over some appropriate range (see the review by Dere and Mason 1981). Within the wavelength range of the HRTS spectrograph we have useful line ratios involving Si III, O III, O IV, O V, Mg V and N IV, giving good coverage of the range of electron densities from 10^9 to 10^{12} cm^{-3} of interest for the Sun. We can also estimate the differential emission measure (DEM) distribution from the intensities of lines formed over a broad range of temperatures, together with the necessary atomic data for the individual lines. The DEM distribution (n_e^2 (d$\log T$/dx)$^{-1}$ vs log T in a one-dimensional geometry) is a very useful summary of the temperature-density information implied by the line intensities over a broad temperature range, and a good quantity for comparison with predicted values from a solar model; we can even extract the temperature gradient alone from the observed DEM distribution with appropriate assumptions on the geometry. Finally, we can try to obtain these sorts of observations for the range of known solar features, and look for morphological relationships with corresponding structures in the lower atmosphere, from such observations as photospheric magnetograms, He I 10830 Å images, Hα images, and T_{min} images, and in the corona from such observations as X-ray or radio data.

There are also of course some drawbacks to the HRTS data. In contrast to the spectroheliograph and Hα systems, the spectrograph does not directly obtain a two-dimensional image. It can be difficult to tell exactly what the slit is viewing, or to co-align it with other data. An image can be constructed afterward on the computer if the spectrograph was run in a raster pattern, but this is not always done, and if it is, it introduces compromises of spatial coverage and temporal resolution. The French TRC instrument has obtained beautiful images of transition region fine structure on the disk in Lyman α

(Bonnet et al. 1980); presumably we could directly image similar structure in C IV at 10^5 K with the proper instrument.

The temporal coverage possible during a 5-min rocket flight also limits the kinds of problems which HRTS has attacked. The discovery of $\sim$ 150-s oscillations in C IV intensities and velocities in sunspot umbrae by the UVSP experiment on the Solar Maximum Mission (Gurman et al. 1982; see Sec. XIII), and the C IV intensity flashes around a sunspot (Shine and Schrijver 1988; see Sec. XIII) seen during the rotation of an active region across the disk, need an orbiting instrument for their study. It is also perfectly possible that some properties of fine structures may change over the course of the solar cycle.

Finally, the wavelength coverage of the HRTS spectrograph means that we miss the TR above 2.5×10^5 K. This is actually one of the most interesting areas for future work. We do not know what happens to the high-velocity TR explosive events in the upper TR, and we do not know much about the morphological transformation from the supergranular network to coronal loop structures that occurs somewhere above 5×10^5 K (see Secs. X and XII).

V. BASIC QUESTIONS ABOUT THE
TRANSITION REGION

What are the primary observational facts we want to understand and the interpretive questions about the TR for which we want answers? We suggest here a short list of major questions influenced by the types of data which we have worked on (other workers would obviously have additional questions). We have in mind quiet and active regions predominantly, although we could ask the same questions of all the variety of observed solar phenomena. We would like to know:

1. What is the morphological structure of the solar atmosphere in the TR and its relationship to structures hotter and cooler?
2. Why are TR observed structures mostly low-density "empty" space? That is, why is the fill factor of TR structures found to be so small (of order 1%), even down to spatial resolution of 1 arcsec?
3. As a useful summary of the information implied by intensities of spectral lines formed over a range of temperatures, how do we explain the general shape of the DEM distribution? We can understand the higher-temperature rise up to 10^6 K as a consequence of an energy balance where the conductive heat flux from the corona, going as $T^{5/2}dT/dx$, is nearly constant, giving a DEM distribution ξ with slope $d(\log\xi)/d(\log T) = 3/2$ with the very simplest assumptions. But how do we explain the minimum of ξ at 10^5 K and its rise as T^{-3} to lower temperatures?
4. What are the actual temperature gradients in the TR? Are they really the

large gradients derived from one-dimensional "average" modeling of the DEM distribution?

5. Most basic of all, what is the heating mechanism that keeps this system in the balance we observe? It is widely agreed that this mechanism must vitally involve the magnetic field, but exactly how?

In the following sections we discuss observations that are pertinent to these questions, and suggest some possible interpretations of the observations. However, the reader will not find a complete answer to any of these basic questions.

VI. MORPHOLOGICAL STRUCTURE

What is the basic morphological structure of the TR? As discussed in the next section, we are not yet even close to observing the most basic structural elements of the TR. The gross morphological structures which we *can* resolve at arcsec spatial resolution have only $\sim 1\%$ of their apparent volume actually filled with emitting plasma. What are these gross morphological structures?

It was already clear from the limited resolution of the OSO-4 atlas (Reeves and Parkinson 1970) that active regions are quite similar in location and area over a large range of TR temperatures, and are the same plage areas observed in Ca II K line images. With more spatial resolution, the quiet Sun supergranular pattern is found to be very similar in morphology over a range of temperatures through the chromosphere, from the temperature-minimum region and through a large part of the TR, up to at least 5×10^5 K (see Sec. XII). In the temperature-minimum region (see Sec. XV), the individual network elements are composed of clumped arcsec-scale bright points. These individual bright points are quickly lost as we go to hotter atmospheric regions and the network elements thicken into more continuous patches of emission with the spreading of magnetic field lines with decreasing pressure. Dere et al. (1987b) have studied the size scale of individual network element features in C IV quiet Sun spectra from Spacelab 2, and find that the average dimension is 2400 km, or about 3 to 4 arcsec. The basic quiet Sun structure in the TR up to at least 5×10^5 K is closely related to structure at lower temperatures.

On Spacelab 2 the HRTS spectroheliograph was tuned to 1550 Å and adapted for viewing structure in C IV (10^5 K) at the limb. Coatings of different reflectivity were used on the two slit jaw mirrors, so that the less reflectant side could be positioned on the limb to decrease scattered light, while the more reflectant side was used for viewing above the limb. Individual exposures have a noticeable jump in exposure level on the limb side of the slit.

The HRTS Spacelab 2 observations at the limb are discussed by Brueckner et al. (1986) and Dere et al. (1989). Figure 2 illustrates representative

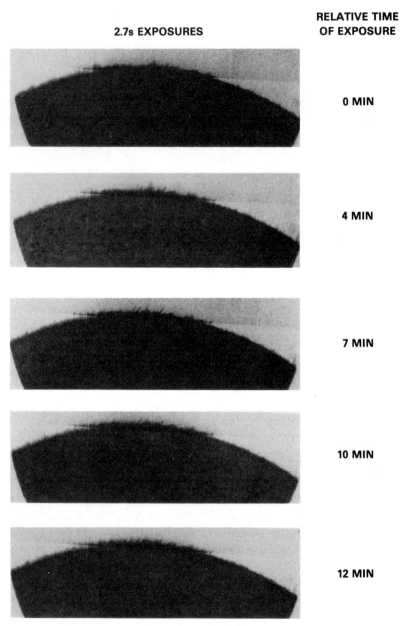

0 MIN

4 MIN

7 MIN

10 MIN

12 MIN

Fig. 2. C IV ultraviolet spicules and superspicules observed at the limb during the Spacelab 2 mission by the HRTS spectroheliograph.

images over 12 min from a 15-min series of exposures with good pointing stability, with the slit placed tangentially to the limb. The sharp boundary in exposure level across the top third of each image marks the slit position. Three fiducial wires cross the slit at left, center and right. On the disk, continuum and line emission are mixed in the instrument passband, but above the limb the spicular structures seen arise from line emission, primarily from C IV. Limb structures appear to have two components. Ordinarily ultraviolet spicules, extending up to a height of around 10 arcsec, occur everywhere. It is difficult to estimate a lifetime from this short a time series, but it appears consistent within the order of 10 to 15 min. These ultraviolet spicules are also probably appearing as finger-like mottles extending from network elements seen on the disk close to the limb. They appear to be the ultraviolet analog of Hα spicules.

Also present are more transient, fainter (3 to 10 times fainter than the ultraviolet spicules) superspicules, which extend up to a height of 20 arcsec, and which can form, develop and disappear on a shorter time scale, as short as 3 min. They are much less common than the ultraviolet spicules, and perhaps have a connection with the high-velocity TR events discussed in Sec. X.

Hα observations during the same exposure sequence show Hα spicules. As shown in Fig. 3, the Hα and ultraviolet spicules occur at similar locations and appear to have some degree of correspondence. The ultraviolet spicules are seen to the left and right on the upper image (shown as a white positive), where the solar limb is below the slit. The central limb lying above the slit, on the highly reflectant side of the slit, is well exposed to show the superspicules but is overexposed to see the individual ultraviolet spicules forming a solid base in this area. The Hα image is shown displaced below, as a black negative image.

There is some evidence for possible rotation of these ultraviolet spicules. Cook et al. (1984a,b) have discussed limb observations from the HRTS IV flight where the slit covered a spicular structure toward one edge of the field of view. Although observationally marginal, the spectrograph seems to show a tilted spectral signature that could be interpreted as evidence of a rotational velocity of ~ 50 km s^{-1}. If confirmed, this result may have a bearing on the possible heating of spicules by torsional Alfvén waves.

VII. FILL FACTOR

A remarkable property of the TR structures that have been observed so far down to arcsec spatial scales with HRTS, is that they are actually composed of even smaller structures occupying only the order of 1% or less of the gross volume of the observable structure. This was discovered by Doschek et al. (1977a) in flare observations a decade ago, but has been repeatedly verified with other features wherever the necessary data is available.

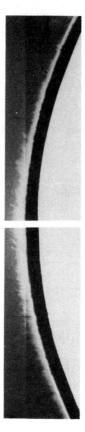

Fig. 3. Comparison of ultraviolet spicules from the C IV images and the corresponding HRTS
  Hα image. The C IV image is shown above as a white positive, while the Hα image is shown
  displaced below as a black negative.

This conclusion is reached from a comparison of the total absolute intensity of emission in a spectral line from a structure of gross observed volume
V_G, to the electron density in the region of line emission found from an
appropriate electron-density diagnostic line-intensity ratio. The total observed emission is expressed as the emission measure $n_e^2 \Delta V$, which can be
calculated from the line intensity knowing the appropriate atomic data; here
ΔV is the volume of the actual region of line emission. Dividing this by the
electron density squared determined from the density diagnostic ratio, we can
then directly determine $\Delta V/V_G$, the fill factor of the observed structure. This
fill factor is typically 1% or less. For example, in the study (Dere et al.
1987b) of C IV observations from the Spacelab 2 flight mentioned earlier, the
emission measures of the individual structures observed along the slit (having
an average spatial scale of 2400 km) were determined and their densities

estimated using a density-sensitive ratio from lines of O IV. Even here, at arcsec-scale spatial resolution, the fill factor was still estimated at 1% or less.

The most common idea about the geometry of the unresolved fine structure is of long, thin, filamentary tubes, probably aligned along the magnetic field lines. There must be many of these within any resolved observable structure in order to fill and define the gross volume; the observed scale heights of emission at the limb and other evidence for larger spatial scales of TR emission suggest a longer length in one dimension. A filamentary geometry with a 1% fill factor at 1 arcsec resolution would require structures with a diameter much less than 50 km.

VIII. DIFFERENTIAL EMISSION MEASURE DISTRIBUTION

The differential emission measure (DEM) distribution is one of the most useful ways to summarize the temperature and density information contained in observations of emission lines formed over a range of temperatures from a multithermal plasma. The remarkable property that it exhibits is the constancy of its shape for a range of solar features. The DEM distribution ($\xi = n_e^2 (\mathrm{d}\log T/\mathrm{d}x)^{-1}$ vs log T for a one-dimensional geometry) has a negative slope of approximately T^{-3} for temperatures below 10^5 K, reaches a minimum around 2×10^5 K, and then increases toward 10^6 K with a slope of approximately $T^{3/2}$. All the DEM directly tells us is that there is relatively little material at TR temperatures, compared to the chromosphere or corona. To interpret the DEM, for example, to extract the temperature-height run of the atmosphere, we need an idea of the actual geometry and structure found from other evidence.

How do we account for its remarkable constancy of shape, basically increasing by a multiplicative factor as we go from quiet to active regions? We can understand its rise up to 10^6 K as a consequence of the energy-balance equation between heating, radiation and conduction for models of hot coronal loops (see, e.g., Rosner et al. 1978). Qualitatively, at these temperatures conduction is so efficient that the conductive flux $F_{cond} = 10^{-6} T^{5/2} \mathrm{d}T/\mathrm{d}x$ is essentially constant down the loop legs, thus $\mathrm{d}\log T/\mathrm{d}x \propto T^{-7/2}$, and so $\xi = n_e^2 (\mathrm{d}\log T/\mathrm{d}x)^{-1} \propto T^{3/2}$, where we use the fact that the pressure scale height at these temperatures is so large that the electron pressure $P_e = n_e T$ is nearly constant. But these hot-loop models will never have a DEM which rises again below 10^5 K, as is observed. The hot-loop models match fairly well the observed coronal temperature loops, but they also result in a TR of very small scale height, with a steep temperature gradient.

Several ideas have been tried out in the effort to explain the DEM. It is clear that the observed DEM is based on emission line intensities that arise from all the emitting plasma structures within the field of view. One might try to explain the DEM as arising from an ensemble of loops of different temperatures. This will not work for an ensemble of loops of the hot-loop variety, but will only modify the positive slope toward 10^6 K.

An interesting recent attempt to explain the rise again below 10^5 K comes out of the cool-loop model developed by Antiochos and Noci (1986). It is clear that TR structures are more extended, with less steep temperature gradients, than one-dimensional modeling suggests, and that TR plasmas form network elements, spicules and other structures that are very different from the coronal-temperature loops considered by the hot-loop models. Antiochos and Noci (1986) pointed out the existence of another class of loop solutions, which balance gravity, radiation and heating, and which do not rise to coronal temperatures, making the conductive term in the energy-balance equation unimportant in magnitude. We will call these cool-loop solutions, and Antiochos and Noci (1986) suggest that they are the proper models of the atmosphere below 2×10^5 K. With a mixed ensemble of these cool loops together with the hot loops for the atmosphere above 2×10^5 K, they were able to mimic the observed behavior of the DEM.

Without accepting all the claims of the cool-loop model, it does have several attractive features. The scale height of loops is the gravitational scale height, giving more extended structures. The temperature gradient seems more like that suggested by observation (see the following section). The basic idea, of a different model for lower TR structures than an essentially coronal model, is surely correct.

IX. TEMPERATURE GRADIENT

Ideas on the spatial scales and temperature gradients in the TR have been heavily influenced by interpretations of the observed DEM distribution. Knowing the electron density at one or several temperatures, a model was typically calculated assuming a one-dimensional or "average" atmosphere by integrating $n_e^2(\mathrm{d}\log T/\mathrm{d}x)^{-1}$. This procedure yielded large temperature gradients and small scale heights for the TR; in fact, in analogy with gradients from semi-empirical model atmospheres, the brighter the feature, the smaller the scale heights and the larger the temperature gradients.

We have already discussed the small value of the fill factor in the TR, and this might immediately alert us to possible difficulties with an "average" model temperature gradient. Nicolas et al. (1979) discussed the effect of fill factors on determination of a temperature gradient for a very simple assumed geometry, and the inclusion of a fill factor can greatly reduce the estimated temperature gradient and increase scale heights. There is an abundance of evidence (see, e.g., from Skylab observations: Doschek et al. 1977b; Kjeldseth-Moe and Nicolas 1977) for structures in the TR at much greater heights than predicted from either "average" semi-empirical model atmospheres such as VAL (Vernazza et al. 1976,1981) or from the temperature-height variation of "average" model atmospheres determined from the observed DEM.

The HRTS data allows another approach to finding a temperature gradient. Dere et al. (1983) have compared images of HRTS III spectrograph features in the continuum, C I, Fe II and C IV, and examined their offsets in spatial position in an attempt to determine their temperature structure. This can now be carried further. During the Spacelab 2 mission a spectrum was obtained, illustrated in Fig. 4, which covers more completely the range of temperature from the temperature minimum continuum at 4400 K to O V at

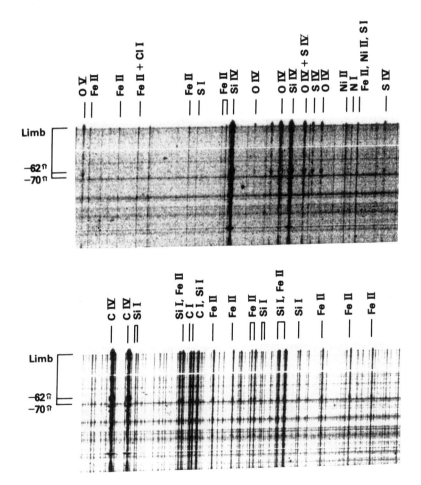

Fig. 4. A solar feature (large spicule?) observed 60 to 70 arcsec inside the solar limb during the Spacelab 2 mission. The feature is systematically displaced in distance from the limb as a function of the temperature of formation of the line in which it is seen.

2.5×10^5 K. This spectrum includes a small event which is located only 60 to 70 arcsec inside the limb, and so allows a good viewing angle for resolving the vertical structure. This feature is probably a brighter network element or ultraviolet spicule/mottle; we see it here because of an unusually long exposure time (60 s) and its fortunate position relative to the limb and the slit. In Fig. 5, we have plotted the height of the center of emission of the feature in various lines (or its appearance as a continuum) against the temperature of emission of the line (or of the continuum). The relative height is determined from the difference in displacement of each image from the continuum image taken as reference, with a correction to give a radial height instead of the line-of-sight observed displacement. We also show as a dashed line the gravitational scale height $H = 2kT/mg_\odot$. A representative straight line through the data gives a logarithmic temperature gradient d$\log T$/dx of 2×10^{-9} cm^{-1}, orders of magnitude less than "average" values. We recall that the cool-loop models of Antiochos and Noci (1986) have a temperature scale height of the order of the gravitational scale height. Without having to accept all features

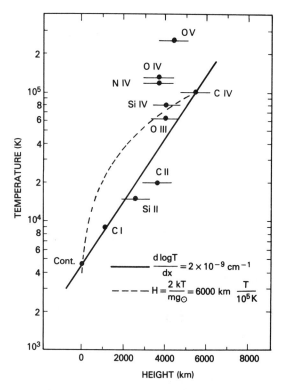

Fig. 5. Relative height of the spectral feature from Fig. 4, seen in a range of emission lines, plotted as a function of temperature of formation of the line.

of such a cool-loop model, it is interesting that it at least begins to approach the scale heights that the observations suggest. In addition, we can compare the radiative loss term $n_e^2 \Lambda(T)$ with the conductive loss term $\nabla \cdot \mathbf{F}_{cond} =$ $d/dx \, (10^{-6} T^{5/2}/dT/dx)$ for $d\log T/dx = 2 \times 10^{-9} \, cm^{-1}$. Using $n_e \sim 10^{10} \, cm^{-3}$ and $\Lambda(T = 10^5 \, K) \sim 10^{-22} \, erg \, cm^3 s^{-1}$, the radiative loss term is $10^{-2} \, erg$ $cm^{-3} \, s^{-1}$ and $\nabla \cdot \mathbf{F}_{cond} \sim 8 \times 10^{-7} \, erg \, cm^{-3} \, s^{-1}$. The energy balance is essentially between radiation and heating alone.

X. HIGH-VELOCITY EXPLOSIVE EVENTS IN THE TRANSITION REGION

One of the most interesting discoveries of the HRTS program has been the spectrographic observation of small (several arcsec) explosive TR events, seen best in the strong C IV 1548 Å and 1550 Å lines but observable in other lines from $2 \times 10^4 \, K$ (Lyman α, C II) to $2.5 \times 10^5 \, K$ (O V), that have broadened line profiles of order $100 \, km \, s^{-1}$ to the blue, to the red or both. These events are common on the Sun in quiet areas, and will be seen anywhere with observations of sufficient spatial resolution. What contribution might these events make in mass, momentum and energy to the upper solar atmosphere? The statistical properties of these events are difficult to establish from the limited observations of a sounding rocket flight.

The first analysis of these events, from HRTS I and HRTS II observations, was performed by Brueckner and Bartoe (1983; see also Dere et al. [1989] for a later analysis of HRTS III data on these events), who estimated the mass and kinetic energy fluxes contained in energetic events. We can see now that their answer, that these events can directly supply the mechanical energy to heat the TR and corona through dissipation of the kinetic energy flux, is too extreme. This is because the value used for the average event velocity, $400 \, km \, s^{-1}$, was too large, and the kinetic energy flux goes as the cube of this velocity. As became clearer after examining more data, the HRTS II C IV events are the most dramatic and energetic yet observed, and not typical. It was not until the Spacelab 2 mission that a large data base could be obtained to study the statistics of these events.

We briefly mention the problem of nomenclature. These events have loosely been called by a number of names. Brueckner and Bartoe (1983) attempted to establish that blue-shifted events were "jets" or "bullets," while both red- and blue-shifted events were "turbulent" or "explosive" events; they did not observe enough red-shifted events to assign any name. In practice these names have become very confused. With more observations, and consideration of them, it is clearer that individual events can evolve in time from one category to another, that all are highly variable, and that these probably all arise from the same mechanism. We prefer to call these high-velocity TR events simply explosive events. We discuss later in this section their connection with the idea of "microflares" or "nanoflares" as a heating mechanism.

For Spacelab 2, a survey program was designed and executed which sampled approximately 25% of the solar disk in 13 individual raster sequences over 4 consecutive orbits; each raster consisted of 60 steps of the 920-arcsec length slit, with a 1-arcsec step width. An analysis of these observations has been given by Cook et al. (1987).

Cook et al. (1987) looked for small spatial scale (typically 1 to 8 arcsec) events in C IV showing velocity signatures in the profile of at least 50 km s^{-1} to the blue, to the red or both. Figure 6 shows part of a raster from these observations, showing events with a range of velocities. The distribution of events was very roughly a third for each: 35% blue shifted; 25% red shifted; 40% both red and blue shifted.

The histogram distribution of the leading edge velocity of the line profiles is shown in Fig. 7. It is roughly gaussian, with all three classes of events peaking in number near a leading-edge velocity of 80 km s^{-1}. In this data set the highest velocity encountered is 200 km s^{-1}, while in the HRTS II data examples reaching 400 km s^{-1} are present. Extrapolating to the whole Sun, there would be approximately 4000 events present at one time.

The size distribution (Fig. 7) was also measured along the slit direction, and was found to peak at 2 arcsec; this was also consistent with the spatial scale over which events are visible in the raster direction.

The distribution of event lifetimes was also studied (from another observation where an area was repeatedly rastered with 1-min temporal resolution). The average lifetime (Fig. 7) was 90 s, although this is really only an upper

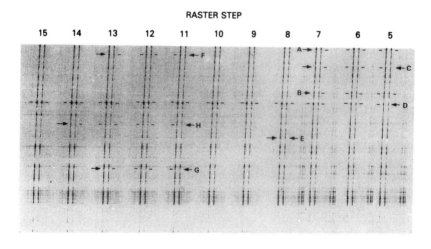

HRTS SURVEY, EXECUTION 7, RASTER STEPS 5-15

Fig. 6. Example of a C IV raster from the Spacelab 2 disk survey. Raster steps were 1 arcsec apart, and a complete raster consisted of 60 slit positions. Individual high-velocity C IV explosive events are lettered A–F.

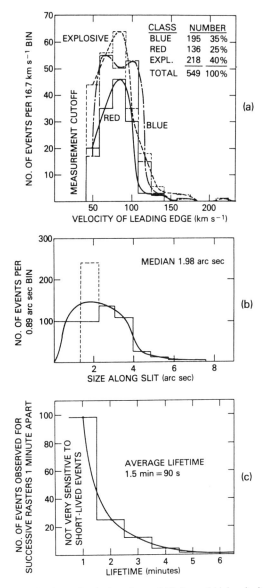

Fig. 7. Histogram distributions of velocity, size and lifetime of high-velocity C IV explosive events. (a) Distribution of the leading-edge velocity. Number of events is binned in 16.7 km s^{-1} intervals down to a cut-off just below 50 km s^{-1}. Blueshifted ("blue"), redshifted ("red") and explosive (both red and blueshifted: "expl") are separately treated. (b) Size distribution of high-velocity events along the spectrograph slit (all three classes combined). (c) Lifetime distribution of events (all three classes combined).

bound because of the insensitivity to shorter lifetimes imposed by the available temporal resolution; a peak in the lifetime distribution was not turned. The average global birth rate of these events is then $4000 / 90$ s $= 44$ s^{-1}.

Can these events provide the energy required to support the quiet solar TR or corona? By this we mean, if their direct mechanical energy were dissipated *in situ,* would they provide the energy required to heat the TR or corona? This required energy is typically estimated from summing up the total observed radiative output at various levels, or from the energy input required to maintain a semi-empirical model atmosphere. Typical energy flux estimates are around 5×10^6 erg cm^{-2} s^{-1} for the quiet chromosphere, 3×10^5 erg cm^{-2} s^{-1} for the quiet corona, and up to 10^7 erg cm^{-2} s^{-1} to heat the corona in an active region (see, e.g., Withbroe and Noyes 1977).

Cook et al. (1987) estimated the average energy fluxes from the explosive events arising from kinetic energy and enthalpy flow, and the radiative losses, using an average velocity of 80 km s^{-1}, a size of 2 arcsec, electron density $= 10^{10}$ cm^{-3}, and $T = 10^5$ K. For a global total of 4000 events, and dividing by the Sun's surface area, they estimated a generous possible heating from the global energetic events of around 2.5×10^4 erg cm^{-2} s^{-1}, at least a factor of 10 or more too low to permit these events to heat directly the upper atmosphere.

We have recently studied another aspect of the TR explosive events, their location relative to the network (Moses et al. 1990). On 11 December 1987 the HRTS V rocket was launched, together with an American Science and Engineering (AS&E) soft X-ray instrument launched a half hour earlier. In a collaborative "bright point" campaign, photospheric magnetograms, He I 10830 Å images, and Hα images were obtained at Kitt Peak and Big Bear. We co-aligned HRTS C IV spectra and temperature minimum spectroheliograms, a Kitt Peak magnetogram, a Kitt Peak He I 10830 Å image, and an X-ray image from the AS&E exposures, and measured the location of TR explosive events observed in C IV. We found that these events appear to be located at the edges of strong field network elements, or even within network elements of weakest magnetic field strength. The C IV events are not X-ray bright points, which instead occur at sites of magnetic dipoles whose corresponding features in the HRTS C IV data are larger and more intense network elements. The TR explosive events in the observed quiet area preferentially occur within an X-ray dark lane which covers a part of the field of view, and tend to avoid areas of hazy, faint X-ray emission possibly corresponding to faint loop systems. A greater number of explosive events per unit area were found than would be suggested by the earlier Spacelab 2 survey. This probably occurred because of the sampling of a particularly rich area associated with the X-ray dark lane, and from an extended detection threshold because of superior spatial resolution in the HRTS V data over the Spacelab 2 resolution.

Although we have concluded that the explosive events do not directly heat the upper atmosphere through their mechanical energy, they still may be important markers of sites of heating. Cook et al. (1987) argue that the explosive events are the by-products of extremely turbulent energy dissipation from a nonthermal primary heating mechanism. They suggest that the average explosive event velocity of 80 km s^{-1}, interpreted as an Alfvén speed, is consistent with a rough equipartition of kinetic, turbulent and magnetic gas pressures, a sign of an extremely turbulent plasma. The actual primary heating mechanism is presumably magnetic reconnection. Brueckner et al (1988) studied C IV spectra of a larger area showing nonthermal broadening of line profiles (~ 100 km s^{-1}) which was connected with a well-observed pattern of emerging magnetic flux. This example may be a scaled-up version of the process producing the smaller explosive events.

Recently the idea of "microflares" or "nanoflares" as a TR or coronal heating mechanism has become popular. These terms generally suggest a small spatial scale, rapid release of energy which manifests itself observationally as brightening, plasma motions and plasma turbulence. The C IV explosive events are one type of observed phenomenon which could be a candidate for such an event. Their signature has been highly broadened line profiles. Another class of events, small, short time-scale brightenings in TR lines in the network, have been studied in UVSP data from SMM by Porter and Moore (1988). From the statistical properties of ultraviolet brightenings, they suggested that these events could provide enough energy to heat the corona.

The signature of the SMM events is a quick brightening, while the signature of TR explosive events is a greatly broadened line profile, not necessarily representing a bright feature. The relationship of these two is not clear. In addition, the estimate of the energy connected with the ultraviolet brightenings used by Porter and Moore (1988) is an extrapolation of a mechanism proposed to estimate the magnetic energy release in solar flares, scaled down to an assumed TR field strength in the ultraviolet brightenings, not an estimate of the direct mechanical energy of the events. There is no direct contradiction in the analyses of the ultraviolet brightenings vs the explosive events.

We finish this section by giving an example of an event which we would truly call a "microflare." The observational signature of a flare is supercoronal temperatures. As mentioned earlier, the HRTS wavelength coverage includes one true flare line, the 11×10^6 K Fe XXI line at 1354 Å. HRTS has not so far observed a solar flare, but we show in Fig. 8 a spectrum from the HRTS II flight which shows on the disk a small area where Fe XXI emission is occurring. It is unmistakable from the broad, fuzzy line profile, which this time arises not from turbulence but from the simple thermal broadening of an 11×10^6 K plasma. The other emission lines at this position are not enhanced, and we would otherwise not know of the presence of this flare plasma. This is so far the only example of such an event in the HRTS data.

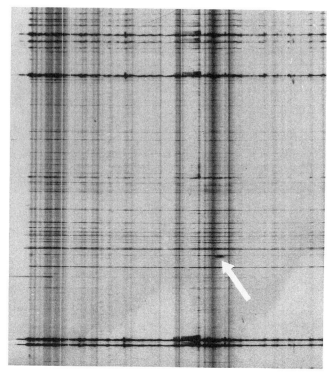

Fig. 8. HRTS II observations of a high-velocity downflow in a sunspot. A "microflare" observed in the 11×10^6 K line of Fe XXI 1354 Å can be seen below the region of the sunspot flow, and is indicated by an arrow.

XI. TURBULENCE IN THE TRANSITION REGION

The emission lines in the TR generally show line profiles that are significantly broadened beyond their thermal width. This excess broadening can be expressed as a nonthermal broadening V_t in km s^{-1}, where the line profile full width at half maximum W in Å is given by $W = [W_I^2 + 4 \ln 2(\lambda/c)^2 (2kT_i/M_i + V_t^2)]^{1/2}$. Here W_I is the instrumental width (0.05 Å or 0.10 Å for the HRTS spectrograph, depending upon the flight), λ is the line's wavelength in Å, and T_i and M_i are the ion temperature (taken equal to the electron temperature at the temperature of maximum abundance of the ion) and ion mass in grams. This assumes that the line profile and all broadening mechanisms are gaussian, but will still give a definable parameter V_t for actual observed profiles.

This nonthermal broadening has often been interpreted as a turbulent broadening, that is, as a nonthermal random mass motion. Most directly however, it is just the extra line broadening in an observed profile beyond the thermal broadening. Conventionally, we will call it the turbulent velocity.

The distribution of turbulent velocity with temperature has been extensively studied using lower-resolution data, such as spectra from the NRL SO82B spectrograph on Skylab (Kjeldseth Moe and Nicolas 1977; Doschek et al. 1977b). Its value is typically 20 to 30 km s^{-1} near 10^5 K. Turbulence broadening is important as a possible diagnostic of the dissipation of energy, such as from acoustic or Alfvén waves, in the TR. Interpretations along this line have been common.

We have seen that the TR is composed of discrete gross structures, with further unresolvable fine structure. Both these gross (3 to 4 arcsec) and unresolved fine structures can affect the amount of turbulent broadening determined from lower-resolution observations, by convolving in the total observed emission-line profile the actual velocity patterns, both bulk flow and genuine nonthermal broadening, of all the plasma structures within the unresolved volume. Dere et al. (1987b) have studied the line widths in C IV of the quiet region structures discussed earlier, and found an average nonthermal velocity of 16 to 19 km s^{-1}, much less than found from lower-resolution data. This means that some of the line broadening in earlier data was from unresolved separate gas parcels with different bulk flow velocities. We do not know what would be found if we could resolve the bulk flow of the unresolved (1% fill factor) fine structure and determine the true average nonthermal width.

XII. RANGE IN THE ATMOSPHERE OF NETWORK STRUCTURE

The observed structure of the solar photosphere is a granular pattern with individual granules of order 1 to 2 arcsec that rise and fall vertically, have sideways motions of order 1 km s^{-1}, and last around 8 min. But already in the temperature-minimum region (see Sec. XV), the observed structure in quiet regions is the supergranular network, where small network elements composed of clumped arcsec-scale bright points occur at edges of the supergranulation. This is coincident with stronger field regions in photospheric magnetograms, while the cell centers are filled with the order of 30 individual bright points evolving on a 1-min time scale. This supergranular network structure of quiet regions, but with network elements thickened into continuous patches of emission, can be observed through the chromosphere into the TR, with a fair degree of correspondence to location of network elements in images from a range of temperatures.

This network structure persists up to at least 5×10^5 K, as shown in Fig. 9 of spectroheliograms in Ne VII obtained by the NRL SO82A instrument flown on Skylab in 1973–74. The Ne VII image (5×10^5 K) shows

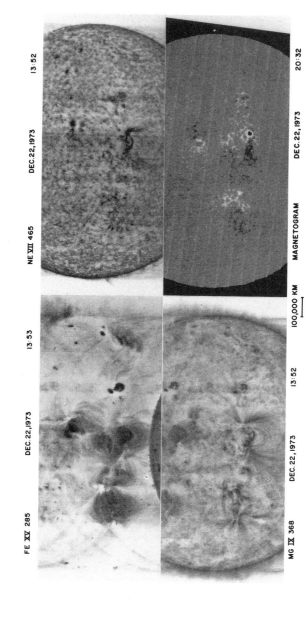

Fig. 9. Skylab SO 82A spectroheliograph images and corresponding Kitt Peak magnetogram (from Sheeley et al. 1975).

that this temperature represents something of a transitional range for solar morphology. The quiet disk is still organized in the network pattern, but loop structures, sometimes just disjointed loop legs, are beginning to appear in active regions. At temperatures of 10^6 K (Mg IX) and above (Fe XV), the disk and limb show large complete loop systems, typically connecting separated active regions, and the supergranular pattern has disappeared.

At what temperature does the network really begin to disappear? What is the reason for this, and does it tell us something fundamental about the solar atmosphere and fine structure? As we have noted earlier, this temperature range cannot be studied with the HRTS instrument. Below we mention two ideas that may have a bearing on this subject.

What we have called hot-loop models of coronal temperature loops, as opposed to cool-loop models where the maximum temperature at the loop top never reaches coronal values, have a scaling law relating loop length L, base pressure p, and maximum temperature T_{max}, which is a direct consequence of an energy balance between the heating rate, conduction and radiation for a static loop. In the formulation of Rosner et al. (1978), this scaling law is $T_{max} = 1.4 \times 10^3 \, (pL)^{1/3}$. For solar base pressures of the order 0.1 (quiet region) to 1.0 (active region) dyne cm^{-2}, hot loops are first possible, for lengths of the order of 15 arcsec or greater, only at coronal temperatures. A shorter, stable hot loop would require unrealistically high pressures. The basic point is that it is only at coronal temperatures that a stable hot loop can span the distances necessary to connect adjacent active regions.

The second idea is the possible dependence of the stability of cool loops, whose maximum temperature only reaches TR values or lower, and which are the presumed structural elements of the lower TR, on the shape of the radiative loss function $\Lambda (T)$ of optically thin plasmas, determining radiative losses in the energy-balance equation as $n_e^2 \Lambda(T)$ erg cm^{-3} s^{-1}. The cool-loop model of Antiochos and Noci (1986) becomes unstable above 2×10^5 K because of the negative slope of the conventional $\Lambda(T)$ function, so that radiation can no longer balance additional heating. In Sec. XIV we discuss how a consequence of the possible difference in elemental abundances in the corona from the traditional photospheric values is a change in the shape of $\Lambda(T)$, which develops an extended plateau up to coronal temperatures before falling off again (see Fig. 10). Cook et al. (1989) have looked at the consequences of the new radiative-loss function on the cool-loop models, and suggest that these models may extend beyond 2×10^5 K to temperatures closer to 10^6 K, in better agreement with the observations discussed earlier showing no break in the quiet Sun supergranular structure up to at least the temperature of Ne VII, 5×10^5 K. The change from the supergranular pattern to coronal loops is then the change from cool loops to hot loops as the basic structural feature.

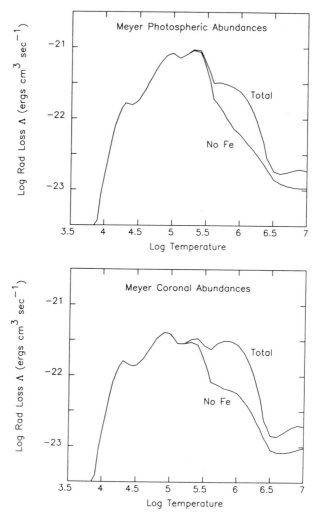

Fig. 10. The radiative loss function $\Lambda(T)$ vs log T, for the photospheric (a) and coronal (b) elemental abundances of Meyer (1985).

XIII. SUNSPOT FLOWS

The very first flight of HRTS in 1975 observed a sunspot. An analysis of these observations was given by Nicolas et al. (1982), where model atmospheres were constructed from the sunspot differential emission measure distribution. In future flights, sunspots were also observed by the HRTS II sounding rocket and during the Spacelab 2 mission. Here we discuss only one aspect of the HRTS sunspot observations, the presence of complex transition-region velocity flows, both blue and redshifted, with structure at small spatial scales.

On all three flights sunspots showed complex velocity patterns in TR emission. In HRTS I and II, the primary pattern is a red shift, of order 100 km s^{-1} in HRTS I and, much more dramatically, 200 km s^{-1} in HRTS II. The Spacelab 2 observations were more complex, with redshifted flows of 80 km s^{-1} and a smaller blueshifted flow of approximately 5 to 20 km s^{-1} (Kjeldseth-Moe et al. 1988). The sunspot observations from HRTS II are illustrated in Fig. 8. The C II doublet near 1334 Å is seen at the left of the dispersion, and the complex of Si IV and O IV lines near 1400 Å are seen at the right. The sunspot flow pattern seen in the small area along the slit center covers some 10 arcsec, but ends abruptly and sharply at what we know from the 1600-Å spectroheliograms and Hα images is the edge of a light bridge through the sunspot. This light bridge is seen as a band of continuum emission at the base of the maximum velocity flow pattern in the individual TR emission lines. A similar connection of the flow pattern with a light bridge is seen in the HRTS I data, and the sunspot viewed during Spacelab 2 also showed a light bridge. This would appear fundamental to the physical interpretation of what is occurring; the magnetic field geometry implied by an extension of lower field-strength regions into the high field-strength sunspot suggests the picture of a funnel into the light bridge. At least in HRTS II, the extention of the flow pattern to the surrounding active region suggests a source for the flow.

We also mention two other remarkable sunspot observations, from the UVSP experiment on the Solar Maximum Mission. The first is the discovery of ~ 150-s oscillations in the TR of sunspot umbrae (Gurman et al. 1982) observed in both intensity and velocity in C IV 1548 Å. The significance of this discovery is in its diagnostic value for the sunspot resonant cavity, once the proper wave type of the oscillation is identified and modeled. We also point out the study by Shine and Schrijver (1988) of synoptic observations of an active region over approximately 5 days as it crossed the solar disk. Throughout this period the UVSP C IV images show a series of bright flashes in the active region surrounding the sunspot. These types of observations require the temporal coverage available from an orbiting instrument.

XIV. UNCONVENTIONAL PHYSICAL ASSUMPTIONS

Most analyses of the TR use a number of conventional physical assumptions. We have in mind such common assumptions as ionization equilibrium, equilibrium electron velocity distributions, conventional transport coefficients such as the Spitzer value for thermal conductivity, and the standard photospheric elemental abundances. In this section we briefly discuss whether we have any reasons to suspect some of the common assumptions generally made in analyses of TR observations, i.e., are there *observations* that discredit such assumptions?

We begin with the question of elemental abundances. For many years, theoretical arguments have been made that the steep changes in temperature, density and degree of ionization in the solar atmosphere might produce elemental abundance anomalies in the solar chromosphere, TR and corona from diffusive processes driven by these steep gradients. Some observations seemed to support this idea, for example, a suggested coronal oxygen depletion (Parkinson 1977). Recently Meyer (1985) reviewed elemental abundance determinations from a variety of observations, and suggests depletions in the coronal abundances of a number of elements relative to the photosphere. Recent work by Widing and Feldman (1989) strongly supports such abundance anomalies, and suggests their connection with open vs closed magnetic field regions. Meyer (1985) and Vauclair and Meyer (1985) suggest that these anomalies arise from a diffusive process anomalously separating low from high first-ionization-potential elements (high FIP elements such as C, N and O are depleted in the corona while abundances of low FIP elements such as Mg, Si and Fe are essentially unchanged), and that this process occurs in the middle chromosphere.

The use of these coronal abundances would clearly affect a range of problems in TR analysis. A few examples are: the calculation of a differential emission measure (DEM) distribution from observed line intensities; model solar atmospheres determined from an observed DEM distribution; spectroscopic line-intenstiy ratio diagnostics of electron density using two lines from different elements; the numerical value and qualitative shape of the radiative loss function $\Lambda(T)$; and the onset of thermal instabilities dependent on the slope of $\Lambda(T)$ as a function of temperature.

We illustrate in Fig. 10 a recomputation of the loss function by Cook et al. (1989) for the coronal and photospheric abundance sets of Meyer (1985). The new curve has a gentle plateau to higher temperatures in contrast to the conventional curve, which strongly peaks at 2×10^5 K. We have discussed (Sec. XII) the significance of the new coronal curve for the TR cool-loop models of Antiochos and Noci (1986), with the motivation of suggesting a reason for the disappearance of the supergranular network pattern somewhere in the 5×10^5 to 10^6 K range.

Next, we mention the question of non-Maxwellian electron velocity distributions in the TR. It has been suggested that steep temperature gradients in the TR may allow energetic electrons from hotter regions to leak through to lower temperatures as a high-energy tail to the velocity distribution. An example of this idea is the numerical solution of the Fokker-Planck equation by Shoub (1983) for a constant pressure quiet Sun atmosphere. A good diagnostic of this effect would be a density-sensitive diagnostic line ratio involving a line with high excitation energy, where a non-Maxwellian tail can significantly increase the collisional excitation rate to the upper level. Keenan et al. (1989) have analyzed HRTS Spacelab 2 observations of line intensity ratios of Si III, including the ratio R (1301 Å/1312 Å), where Si III 1312 Å is a high excitation line that can be affected by a non-Maxwellian tail. They found an anomalous value of this ratio alone in a sunspot; the other line intensity ratios of Si III analyzed gave consistent densities in quiet and active areas and in the sunspot. The difference in the affected ratio was in the direction suggested by a calculation of the ratio for the non-Maxwellian velocity distribution of Shoub (1983). This may represent observational evidence for nonequilibrium velocity distributions in the TR, although other explanations of the observed anomalous diagnostic value are possible.

Finally, we briefly mention the question of nonequilibrium ionization balances. It is perhaps a little surprising that so far we have not encountered firm observational evidence for transient or nonequilibrium effects. Two phenomena in the HRTS data are possible candidates for such effects. The TR explosive events evolve on a 20-s time scale; Dere et al. (1981) calculated the cooling of a transiently ionized plasma for comparison with the TR explosive events, obtaining a comparable lifetime. Lastly, the high-velocity downflows in sunspots associated with light bridges are a good place to look for the presence of nonequilibrium ionization balances from reactive flows. So far this has not been modeled.

XV. THE TEMPERATURE MINIMUM REGION

The HRTS spectroheliograph produces a broadband image of tunable central wavelength that covers approximately 7.5 × 15 arcmin, with the slit of the spectrograph running down the middle of the long dimension. The spectroheliograph has been tuned to 1600 Å on the HRTS II and HRTS V flights, and to 1550 Å to cover the C IV doublet at 1548 Å and 1550 Å, in order to view TR structure at the limb on the HRTS III, HRTS IV and Spacelab 2 flights. The limb observations in C IV have been discussed in Sec. VI.

The 1600-Å disk observations cover a wavelength band where the main flux contributor is continuum emission arising from the solar temperature-minimum region. Mixed with this is some line emission from chromospheric and TR emissions, but the continuum component is approximately 70% in quiet regions.

An analysis of the HRTS II 1600-Å observations was given by Cook et al. (1983). Quiet regions were composed of bright-point emission features of 1 to 2 arcsec spatial scale. These bright points occurred clumped together in network elements, where they varied only slowly in brightness, lasting the flight duration, while in cell centers the order of 30 bright points were highly variable on a 1-min time scale. Bright points were also observed packed in plage areas, with perhaps a general background emission also present. The bright-point pattern and emission remained constant in these plage areas over the 4 min of observations. Clearly the temporal variability varies inversely as the magnetic field strength.

Cook et al. (1983) demonstrated the close correspondence in the T_{min} image of network elements in quiet regions, and of the structure of plage areas, with morphologically similar structures observed in Ca II K line images and in photospheric magnetograms, illustrated in Fig. 11. One of the most interesting unanswered questions is whether the cell-center bright points also are correlated with the small-scale inner network fields.

Similar data on the 1600-Å region has been obtained by the TRC experiment of the French LPSP group in sounding rocket flights in collaboration with the Lockheed group (Bonnet et al. 1982; Foing and Bonnet 1984).

The 1600-Å data is closely comparable with Ca II K_3 data, which arises from the same region of the atmosphere (Vernazza et al. 1976). The advantage of the 1600-Å data is their more direct interpretation. The Si I continuum, the main opacity source at 1600 Å, is not far from LTE, and observed brightness temperatures give us a rough guide to the actual temperatures.

The HRTS V flight obtained 1600-Å spectroheliograph observations, and in addition a Kitt Peak magnetogram is also available. Cook and Ewing (1990) studied a 450 arcsec square area at position angle $\mu = 0.84$, reasonably close to Sun center. They determined the distribution of intensities, expressed as brightness temperatures, and the relationship of brightness temperature with magnetic field strength in the magnetogram.

A histogram distribution of brightness temperature in this area is illus-

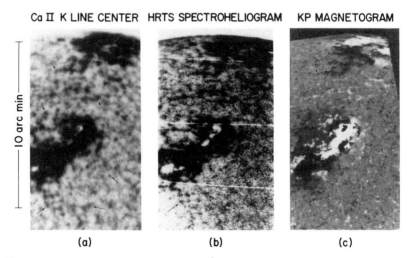

Fig. 11. The correspondence of the HRTS 1600-Å temperature minimum image (b), the Ca II K image from Sac Peak (a), and the Kitt Peak magnetogram (c), obtained during the HRTS II flight.

trated in Fig. 12; the average brightness temperature is 4377 K. However, the brightness temperature which corresponds to the average intensity of the distribution is 4401 K, approximately 25 K hotter. This fact was used in Sec. II as an example of how nonlinear scaling of physical parameters over fine structure leads to inaccurate "average" values based on low-resolution observations.

Cook and Ewing (1990) determined a quantitative relationship between magnetic field strength from photospheric magnetograph observations and the brightness temperature of the fine-structure elements observed at 1600 Å. They then used this function to produce a pseudo T_{min} image from the magnetogram, which is compared with the actual observation. The functional relationship is shown in Fig. 13 between absolute value of magnetic field strength in the Kitt Peak magnetogram and brightness temperature in the T_{min} image. It is essentially linear from 10 to 200 gauss (they studied only a quiet region, and the relationship cannot be scaled up to kilogauss sunspot field strengths). A comparison of the actual 1600-Å image, the magnetogram, and the pseudo 1600-Å image constructed from the magnetogram using the functional relationship of Fig. 13, is shown in Fig. 14.

These results are consistent with a nonmagnetic basal heating which occurs generally, while bright points are produced in magnetic regions with temperature enhancements proportional to [**B**]. Excess heating ΔT above a general basal background value T requires an excess heating flux ΔE which varies as ΔT in an LTE approximation, with $\Delta E = 16\sigma T^3 \Delta T \tau_0$ (Athay 1976), where σ is the Stefan-Boltzmann constant and τ_0 is the optical depth at 5000

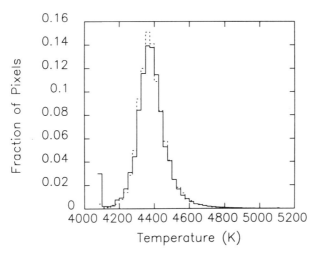

Fig. 12. The brightness temperature histogram distribution from a quiet area of a HRTS V 1600-Å spectroheliogram.

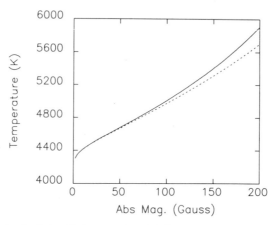

Fig. 13. Functional relationship between the 1600-Å brightness temperature and the absolute value of magnetic field strength from a Kitt Peak magnetograph.

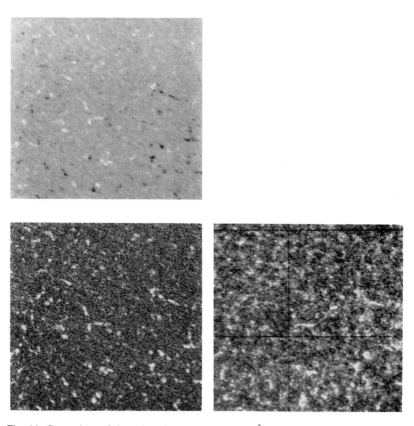

Fig. 14. Comparison of the quiet area seen in the 1600-Å HRTS image (bottom right), the original Kitt Peak magnetogram (top left), and the pseudo 1600 Å image constructed from the magnetogram using the relationship of Fig. 13 (bottom left).

Å of the region (of order $\tau_0 = 3 \times 10^{-4}$ in the T_{min}). With ΔT observed to vary linearly with B, the excess heating ΔE must also vary as B. This is at least consistent with heating by dissipation of a fraction of the energy of Alfvén waves, whose flux is linear in B. The Alfvén wave flux is $\rho v^2 V_A$, where $V_A = B/(4\pi\rho)^{1/2}$. Alfvén waves produced in the photosphere ($\rho \sim 2.7 \times 10^{-7}$ g cm^{-3}) from granular buffeting of order 1 km s^{-1} in magnetic regions of field strength B would produce an Alfvén wave flux of $1.5 \times 10^6\,B$ erg cm^{-2} s^{-1}. The excess heating in bright points in the T_{min} is $1.4 \times 10^5\,B$ erg cm^{-2} s^{-1}, using the observed linear relation of B with T in the previous equation for ΔE. If we can dissipate approximately 10% of this hypothetical Alfvén wave flux in the T_{min}, we could explain the heating of the bright points.

Finally, we can integrate the excess energy equation over the brightness temperature distribution of the bright points and estimate the average heating requirement of the bright points. This gives 7.5×10^6 erg cm^{-2} s^{-1}, which is comparable to the ordinary background flux in nonmagnetic areas calculated by Anderson and Athay (1990) as necessary to produce the general background T_{min}. That is, the extra heating per unit area required to produce the bright points is as much as the "average" heating to produce the background.

XVI. FUTURE EXPERIMENTS

Where do we go from here? We have discussed some of the currently available observations of the TR fine structure, but have still not answered any of our list of basic questions about the TR (Sec. V).

Even better observations should be obtained in the next decade. Two promising future missions are the Solar and Heliospheric Observatory (SOHO) and the Orbiting Solar Laboratory (OSL). SOHO is an approved mission currently under development by the European Space Agency and NASA, and is scheduled for launch in 1995. OSL is not as far advanced, but it is hoped that OSL will be approved by NASA as a 1991 new start. With OSL, temporal coverage can be extended to the longer periods impossible with a 5-min rocket flight. In addition, we will almost automatically obtain the range of observational comparisons with magnetograms and photospheric observations that is so highly desirable. We can also hope for high-resolution images of the TR, as opposed to the one-dimensional look with a slit which has been available so far from the HRTS spectrograph. The HRTS spectroheliograph images of ultraviolet spicules at the limb in C IV, UVSP time series showing successive transient brightenings around sunspots and intensity and velocity oscillations in sunspot umbrae, and TRC images of fine structure in Lyman α all show how important imaged data can be.

It does not appear that we will obtain a high spatial resolution look at the 500 to 1100 Å range in the near future, which would show us what is happening in the upper TR between 2.5×10^5 and 10^6 K. We do not know what happens to the high-velocity C IV explosive events at hotter temperatures (we do not see them at 10^6 K); how they finally dissipate is a major question. Just this range of temperatures might show if they are a source of material for the solar wind. We would like to understand how and why the supergranular network, still visible in Ne VII at 5×10^5 K, has largely disappeared by 10^6 K, where the basic structural element has become coronal loops. The upper transition region may also be the best place to observe clearly the small-scale sources of the solar wind.

Acknowledgments.   The authors acknowledge support by the National Aeronautics and Space Administration and by the Office of Naval Research.

MODEL FOR THE RELATIONSHIP OF GRANULATION
AND SUPERGRANULATION

V. KRISHAN
Indian Institute of Astrophysics

*In this chapter a mechanism is proposed for producing the observed solar su-
pergranulation from the photospheric granulation by a dissipative decay of two
dimensional turbulence, which leads to concentration of the energy spectrum to
the longest wavelengths. This concentration of convective eddies by selective
dissipation to the scale with the maximum available spatial dimension and with
a much longer time scale is verified by mode-mode coupling seen in computer
simulations as well as in laboratory experiments. Theoretical predictions for
these granulation scales and magnetic structures can be tested by high quality
observations of the solar surface.*

I. INTRODUCTION

Radiation and convection are the two main energy transport processes in
the solar interior. The convective transport becomes operative where the temperature and density gradients are such that a fluid element, when displaced
from its equilibrium position, keeps moving away from it. This stratification,
through unstable convection, produces turbulence in the medium. The fluid
eddies of varying sizes then carry energy as they propagate and dissipate.
The cellular patterns observed on the solar surface are believed to be the
manifestations of convective phenomena occurring in the subphotospheric
layers. The cellular velocity fields are seen prominently on two scales: the
granulation and the supergranulation, though mesogranulation and giant cells
are also suspected to be present. The formation of granules with an average
size of 1000 km and a life time of a few minutes can be understood either

from the mixing length (Schwarzschild 1975) or from the linear instability (Bogart et al. 1980) description of the convection in the hydrogen ionization zone of the subphotospheric medium. The supergranules, with an average size of 30,000 km and a life time of 20 hr, do not have an unambiguous association with a subphotospheric region. Attempts have been made to identify this region and to seek an explanation for the energy concentration at the supergranular scale. Simon and Leighton (1964) suggested helium ionization to be responsible for accumulation of energy at supergranular scales. Convective modes with dominant growth rates at the two scales have been favored by Simon and Weiss (1968), Bogart et al. (1980) and Antia et al. (1984).

Here, a new mechanism of making supergranules is presented. It is based on the very special redistribution of the energy associated with the granules in a turbulent medium. Before discussing the particular processes that facilitate the formation of supergranules from granules, a few comments on the general properties of a turbulent medium are in order.

Formation of ordered structures in a turbulent medium relates to the concept of self-organization, which occurs when a system has two or more invariants in the absence of dissipation. The invariants suffer selective decay in the presence of dissipation. One conserved quantity has a higher decay rate than the others. The cascading process is such that the slowly decaying quantity transfers towards smaller wavenumber and thus appears in the form of a large-scale organized pattern. The system can be described using a variational principle where the fast decaying quantity is minimized keeping the slowly decaying quantity constant. Kraichnan (1967) found that in a two-dimensional hydrodynamic turbulence, the energy cascades toward large spatial scales and enstrophy, which is the total squared vorticity, and towards small spatial scales where it suffers heavy dissipation. It is this property of selective decay that facilitates the formation of large structures, whose dimensions are determined from the ratio of energy and enstrophy.

The condition of two dimensionality needs to be clarified. It is shown in the following sections that a velocity field $V = (V_x(x,y), V_y(x,y), V_z)$ with a constant vertical V_z component, and with the other two components varying only horizontally, satisfies the requirements of two-dimensional hydrodynamic turbulence. We shall call this the generalized "2-D" situation. The observed nearly two-dimensional velocity field associated with supergranules encourages us to investigate the role of 2-D hydrodynamic turbulence in the formation of supergranules. The inertial range of the turbulent spectrum is derived in Sec. II. Section III deals with the inverse cascade through mode-mode interaction. The concept of self-organization in 2-D turbulence is discussed in Sec. IV and a model of supergranulation is proposed in Sec. V. The inverse cascade in 3-D, and how a 3-D situation develops into a quasi 2-D one, are discussed in Sec. VI and finally the role of the magnetic field is addressed in Sec. VII.

II. THE INERTIAL RANGE OF THE TURBULENT SPECTRUM

The hydrodynamic equations describing the motion of an element in an incompressible fluid are:

$$\frac{d\mathbf{V}}{dt} = \frac{\partial \mathbf{V}}{\partial t} + (\mathbf{V} \cdot \nabla)\mathbf{V} = -\nabla T + \nu \nabla^2 \mathbf{V} \tag{1}$$

$$\nabla \cdot \mathbf{V} = 0. \tag{2}$$

Here $\mathbf{V}$ is the velocity, $T = p/\rho$ is a normalized temperature, p is pressure, ρ is density and ν is the kinematic viscosity. The equation for the vorticity vector Ω can be derived from Eqs. (1) and (2) as:

$$\frac{d\Omega}{dt} = \frac{\partial \Omega}{\partial t} + (\mathbf{V} \cdot \nabla)\Omega = \nu \nabla^2 \Omega \tag{3}$$

$$\Omega = \nabla \times \mathbf{V}. \tag{4}$$

In generalized 2-D, $\mathbf{V}$ and Ω may be expressed by a scalar stream function ψ as:

$$\mathbf{V} = -\nabla \psi \times \hat{z} + V_z \hat{z} \tag{5}$$

$$\Omega = \nabla^2 \psi \, \hat{z}. \tag{6}$$

Here $\hat{z}$ is a unit vector and V_z is the constant z-component of velocity $\mathbf{V}$. Equation (3) can be rewritten as:

$$\frac{\partial}{\partial t} \nabla^2 \psi \hat{z} + \{-\nabla \psi \times \hat{z} + V_z \hat{z}\} \cdot \nabla \{\nabla^2 \psi \hat{z}\} - \nu \nabla^4 \psi = 0 \tag{7}$$

where

$$\nabla \equiv \frac{\partial}{\partial x} \hat{x} + \frac{\partial}{\partial y} \hat{y}. \tag{8}$$

One notes that the term $(V_z \hat{z}) \cdot \nabla(\nabla^2 \psi \, \hat{z})$ vanishes for the generalized 2-D system. If the viscosity is small, i.e., the Reynolds number is large, the time evolution of the velocity field is determined by the second term in Eq. (7), which represents the coupling of various spatial fourier components in a turbulent state. The equation for mode coupling is obtained from Eq. (7) by expressing ψ as:

$$\psi = \tfrac{1}{2} \left[\sum_k \psi_K(t) \exp\left(i\,\mathbf{K} \cdot \mathbf{X}\right) + C.C. \right] \tag{9}$$

where $\mathbf{K}$ is a two-dimensional wave vector and ψ_K is the fourier amplitude. Equation (6) can be rewritten as:

$$\frac{d\psi_K}{dt} + K^2 \nu\, \psi_K = \tfrac{1}{2} \sum_{K=K'+K''} \Lambda^K_{K'K''} \, \psi_{K'} \, \psi_{K''} \tag{10}$$

where

$$\Lambda^K_{K'K''} = \frac{1}{K^2}\,(\mathbf{K}' \times \mathbf{K}'') \cdot \hat{z}\,(K''^2 - K'^2). \tag{11}$$

The total energy W and the enstrophy U are defined as:

$$W = \tfrac{1}{2} \int V^2 d^3r, \qquad U = \int \frac{\Omega^2}{2}\, d^3r. \tag{12}$$

The conservation of energy and enstrophy in the absence of dissipation ($\nu = 0$) can be easily proved if the fluid is surrounded by either a periodic boundary or a rigid boundary, so that the normal component of the velocity vanishes on the boundary. Enstrophy conservation remains valid as long as the vorticity Ω is along $\hat{z}$ and ∇ is in the (x, y) plane. Since there are two invariants, two types of inertial ranges are expected, one for energy and the other for enstrophy. These can be derived by using Kolmogorov arguments. The inertial range for energy is the well-known Kolmogorov law:

$$W(K) = C\left(\frac{\epsilon}{\rho}\right)^{2/3} K^{-5/3} \tag{13}$$

where ϵ is the dissipation rate of energy at a sink, C is a universal dimensionless constant, and $\int W(K)dK$ gives the total energy. The enstrophy density is given by $K^2 V_K^2$, and the inertial range for enstrophy requires that $(\rho K^2 V_K^2)(KV_K) = \epsilon' = $ constant. The energy spectrum in this range is given by:

$$W(K) = C'\left(\frac{\epsilon'}{\rho}\right)^{2/3} K^{-3}. \tag{14}$$

The range of validity of the two inertial ranges (Eqs. 13 and 14) can be established by investigating the cascading process (Sec. III).

III. INVERSE CASCADE THROUGH MODE-MODE COUPLING

Let there be a source at $K = K_S$ with energy W_S. Through mode-mode coupling, this would decay to two modes with wavenumbers K_1 and K_2. Since energy W and enstrophy U are conserved, one can calculate the energies W_1 and W_2 and enstrophies U_1 and U_2 of the modes K_1 and K_2 as (Hasegawa 1985):

$$W_S = W_1 + W_2, \qquad U_S = U_1 + U_2 \tag{15}$$
$$K_S^2 W_S = K_1^2 W_1 + K_2^2 W_2.$$

From Eq. (15) one finds

$$W_1 = \frac{K_2^2 - K_S^2}{K_2^2 - K_1^2} W_S, \qquad U_1 = K_1^2 W_1 \tag{16}$$
$$W_2 = \frac{K_S^2 - K_1^2}{K_2^2 - K_1^2} W_S, \qquad U_2 = K_2^2 W_2.$$

For $W_1, W_2 > 0$, $K_2^2 > K_S^2 > K_1^2$ should be satisfied. Thus K_S decays to two modes, one with wavenumber $K_1 < K_S$, and to another mode with $K_2 > K_S$. Hasegawa and Kodama (1978) have shown that the decay rate is maximum when:

$$p = \frac{K_1^2}{K_S^2} = (\sqrt{2} - 1) \tag{17}$$
$$K_2^2 = K_S^2 + K_1^2.$$

Then

$$W_1 = p W_S \qquad W_2 = (1-p)W_S, \tag{18}$$
$$U_1 = p^2 U_S, \qquad U_2 = (1-p^2)U_S.$$

In the next step of the cascade, the mode at K_1^2 decays to modes at $pK_1^2 = p^2 K_S^2$ and $(1+p)K_1^2 = p(1+p)K_S^2$. The mode at K_2 decays to modes at $pK_2^2 = p(1+p)K_S^2$ and $(1+p)K_2^2 = (1+p)^2 K_S^2$. The corresponding energy partitions are $p^2 W_S$, $2p(1-p)W_S$, and $(1-p)^2 W_S$ for wavenumbers at $p^2 K_S^2$, $p(1+p)K_S^2$, and $(1+p)K_S^2$, respectively. Continuing to the nth step, the energy distribution is given by a binomial distribution for a parameter (r/n) such that

$$W(K^2 = p^{n-r} (1+p)^r K_S^2) = {}^n C_r \, p^{n-r}(1-p)^r \, W_S. \tag{19}$$

Equation (19) gives the energy spectrum which results from a series of cascades at a fixed ratio $(K_I^2/K_S^2) = p$ at each step, where $K_I^2 + K_S^2 = K_2^2$. It can easily be shown that the energy spectrum condenses at $K \to 0$ as $n \to \infty$. Hence an inverse cascade and condensation of the spectrum at $K \to 0$ is expected from this model. Inverse cascade obtained this way is a consequence of conservation of energy and enstrophy.

IV. SELF-ORGANIZATION IN TWO-DIMENSIONAL TURBULENCE

Kraichnan's hypothesis of inverse cascade and inertial range spectra (Kraichnan 1967) has been tested by solving Eq. (10) numerically (Batchelor 1969; Lilly 1969; Fornberg 1977) as shown in Fig. 1. The creation of large-scale structures in the stream function in two-dimensional fluids has also been observed in laboratory experiments. The condensation of energy at the longest wavelengths permitted, due either to the finite size of the container or to the periodic boundary condition, has been reproduced in computer simulations (Hossain et al. 1983).

From modal transfer, it is clear that enstrophy cascades towards the shortest length scales and then suffers viscous dissipation. Thus, if the enstrophy $\int(\Omega^2/2) \, d^3r$ vanishes during normal cascade, the total energy W attains constancy even in the presence of viscosity. This, together with the experimental evidence for the inverse cascade, indicates that the system will

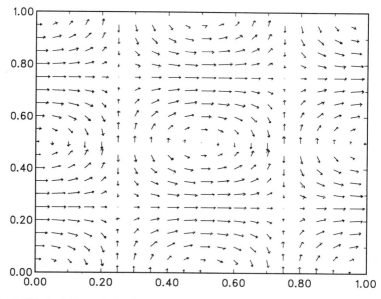

Fig. 1. Velocity field as calculated from Eq. (10).

evolve to a state of minimum enstrophy with constant energy. Such a dissipative process is called selective dissipation (Kraichnan and Montgomery 1980). Thus the large-scale structure appears as a result of minimization of enstrophy with the constraint of constant energy. This is expressed as

$$\delta U - \lambda \, \delta W = 0. \qquad (20)$$

For periodic boundary conditions or for a viscous boundary such that $\Omega = 0$ at the boundary, one finds:

$$\nabla \times (\nabla \times \mathbf{V}) - \lambda \, \mathbf{V} = 0 \qquad (21)$$

which can be solved by using the stream function that is determined by

$$\nabla^2 \psi + \lambda \psi = 0. \qquad (22)$$

Since λ gives the ratio of enstrophy to energy, Eq. (22) should be solved for the minimum eigenvalue λ. If the fluid has a periodic boundary condition with the periods a and b in the x and y directions, then

$$\psi = \psi_0 \cos \frac{2\pi x}{a} \cos \frac{2\pi y}{b}. \qquad (23)$$

The self-organized state obtained here is also a stationary solution of the dynamical Eq. (1). Substituting Eq. (22) into Eq. (1) and setting $\partial V/\partial t = 0$ and $\nu = 0$, one gets

$$\nabla \left(\frac{V^2}{2} + T + \frac{\Omega^2}{2\lambda} \right) = 0. \qquad (24)$$

This gives the temperature profile $T(x,y)$ shown in Fig. 2.

V. APPLICATION TO SOLAR SUPERGRANULATION

The observed nearly two-dimensional nature of the velocity field in the supergranules permits us to use the results of Secs. II, III and IV. Based on this, we would like to propose and test the following model for formation of supergranular cells on the solar surface:

1. The supergranulation is produced as a result of redistribution of energy associated with granulation.

2. The redistribution of energy takes place in a region with predominantly horizontal velocity fields, i.e., between the middle chromosphere and photosphere, below which the velocity field becomes three dimensional and isotropic.

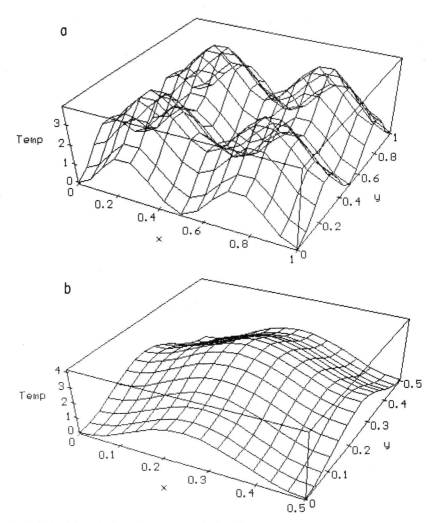

Fig. 2. Spatial distribution of temperature calculated from Eq. (23) in a region $(x/a, y/a) = (0,0)$ to $(1,1)$ plot (a) and $(0,0)$ to $(0.5,0.5)$ plot (b).

3. The redistribution of energy responsible for supergranulation occurs through the inverse cascade of energy towards larger scales, a consequence of the mode-mode interaction in a two-dimensional system with two invariants, the energy and the enstrophy.

4. The largest spatial scale is determined from the ratio of energy and enstrophy. From Eqs. (22) and (23), we find

$$\lambda = \left(\frac{2\pi}{a}\right)^2 + \left(\frac{2\pi}{b}\right)^2 \qquad (25)$$

where a and b are the dimensions of the organized structures, here the super-granular cell, and λ is the ratio of enstrophy U to energy W. Therefore, for $a \sim b \sim L$, the size of the cell, one finds

$$L = \left(\frac{8\pi^2}{\lambda}\right)^{1/2} = \left(\frac{8\pi^2 W}{U}\right)^{1/2}. \qquad (26)$$

For horizontal velocities $V \sim 0.5$ km s^{-1}, the energy per unit density and unit volume is $(1/2)(0.5 \times 10^5)^2$ cm^2 s^{-2}. Therefore, to get $L \sim 30{,}000$ km, the value of enstrophy per unit density and unit volume is required to be 10^{-8} s^{-2}. It is instructive to compare this with the square of the average velocity gradient in the supergranular cell, i.e., with $(V/L)^2 \sim (0.5 \times 10^5/3 \times 10^9)^2 \sim 0.028 \times 10^{-8}$ s^{-2}. Thus, the required value of the enstrophy corresponds to a stronger velocity circulation.

5. The spatial variation of temperature within the supergranular cell is given by Eq. (24). For $a = b$, one finds $\partial T/\partial x$ is maximum at $(x = (2n+1)a/8,$ $y)$ and $\partial T/\partial y$ is maximum at $(x, y = (2n+1)a/8)$.

6. The rate of modal transfer is given by the nonlinear term, and the associated time scale is $\sim(KV_K)^{-1}$. For the two inertial ranges one finds:

$$KV_K = K^{3/2}[W(K)]^{1/2} \propto K^{2/3} \text{ for } K < K_S \qquad (27)$$
$$\propto K^0 \text{ for } K > K_S.$$

Therefore, the characteristic time increases with the increase in the spatial scale, which means that the larger cells will have larger time scales.

7. The velocity field (V_x, V_y) given by Eq. (5) is plotted in Fig. 3 for the case $a = b$. The circulation pattern is clearly visible. This attains special significance in view of the recent observations of vortex formation in the granules (Brandt et al. 1988).

8. The distributions in energy and enstrophy would give a range of spatial scales, the largest of which may correspond to the giant cells.

Proposed tests of the model are the following:

A. If the energy input for supergranulation is at the granular scale K_S, then the energy spectrum should show a break at K_S: the spectrum should go as K^{-3} for $K > K_S$, and as $K^{-5/3}$ for $K < K_S$. Duvall (1987) has proposed two experiments to check the spectral behavior: (i) Doppler shift measurements, which have the advantage of providing a high-precision map of motions over the surface. The disadvantage is that one gets only one component of the horizontal motion, as the Doppler effect gives only the line-of-sight compo-

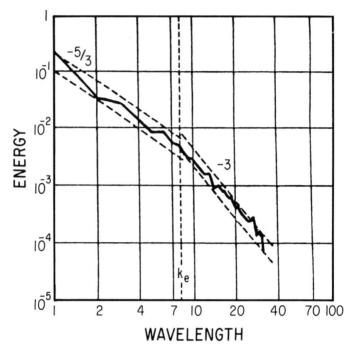

Fig. 3. An omnidirectional energy of two-dimensional Navier-Stokes turbulence obtained numerically (Lilly 1969). The initial spectrum dominated by the source spectrum at the source
wave K_e, is shown to relax to the inertial range spectra for enstrophy at $K > K_e$ and energy at
$K < K_e$.

nent; (ii) tracer measurements in which small magnetic elements can be followed and both horizontal components measured. The disadvantage is that
one does not obtain a very dense grid of tracers, and this would yield a noisy
measurement. Under the assumption that the two components of the horizontal motion are approximately equal, the Doppler method looks quite promising.

    B. The constancy of energy and enstrophy can be verified with detailed
observations of velocity fields.

    C. The observed spatial variation of temperature when compared with
the prediction of Eq. (23) will provide another test of this model.

VI. SOLAR GRANULATION AND 3-D HYDRODYNAMIC TURBULENCE

    It has been shown in the previous sections that in a 2-D situation, the
inverse cascade of energy can lead to the formation of large coherent struc-

tures, which in the case of the Sun may be the supergranular cells. But how good is the assumption of 2-D for the Sun? Levich (1985, and references therein) has shown that the inverse cascade occurs even in 3-D hydrodynamic turbulence. A qualitative description of this phenomenon, and investigation of the very important question of how a 3-D situation develops into a 2-D or a quasi 2-D one, are briefly attempted here. In analogy with the Earth's atmosphere (where the energy released is the latent heat of vaporization), the energy injection into the solar atmosphere occurs by convective upward motions, and the energy associated with the latent heat of ionization is released. It is estimated that 2 to 3% of the thermal energy is transferred into kinetic energy of random motions. This begs the question whether the excitation of random small-scale motions can lead to large organized structures that are observed in the form of granules, supergranules and even giant cells. Here, a picture that emphasizes the role of large helicity fluctuations in the cascading process is presented, as developed by Levich (1985) and co-workers for the formation of large cloud clusters, cyclones, and other related structures in the Earth's atmosphere. The helicity density, a measure of the knottedness of the vorticity field, is given by $\gamma = \mathbf{V} \cdot (\nabla \times \mathbf{V})$. A turbulent medium exhibits large fluctuations in helicity even though the mean helicity $< \mathbf{V} \cdot (\nabla \times \mathbf{V}) >$ $= 0$. The reason for this is that in a nonequilibrium system, any quantity is expected to fluctuate strongly if there are no special restrictions. The fluctuating topology of the vorticity field in turbulent flows with $<\gamma> = 0$ is characterized by a statistical helicity invariant I, which represents the conserved mean square helicity per unit volume:

$$I = \lim_{V \to \infty} \frac{1}{V} \left\langle \left(\int \mathbf{V} \cdot \mathbf{\Omega} d^3 r \right)^2 \right\rangle \propto \int [W(K)]^2 dK \qquad (28)$$

which is a constant for a nondissipative system. One recalls that the nonlinear term in the Navier-Stokes equation is $\nabla \times (\mathbf{V} \times \mathbf{\Omega})$. Thus if in some volume I is large, i.e., $(\mathbf{V} \cdot \mathbf{\Omega})$ is large, then $(\mathbf{V} \times \mathbf{\Omega})$ is very small. In other words, if $\mathbf{V}$ and $\mathbf{\Omega}$ are strongly aligned, the nonlinear interaction term is vanishingly small. Therefore the energy cascade to small scales is inhibited in this volume. Using Kolmogorov arguments (see Sec. II), one can determine the inertial range of the I invariant. If one substitutes $W(K) \propto K^{-5/3}$, which is the inertial range for energy in 3-D, into Eq. (28), one finds total energy $E = \int W(K) dK \propto L^{2/3}$ and $I \propto L^{7/3}$. Therefore, as was argued in the 2-D case, it is not possible to have both E and I conserved in the identical inertial range. Thus energy (like enstrophy in 2-D with larger K dependence) cascades to smaller scales and I to larger scales. It is more appropriate to say that the correlation length of helicity fluctuations increases, without carrying much energy with it. In the case of highly anisotropic flow, with the vertical scale $L_z \ll L_H$, the horizontal scale, as well as $V_z \ll V_x, V_y$, one gets

$$\frac{D}{Dt} V_z = \frac{\partial V_z}{\partial t} + \left(V_x \frac{\partial}{\partial x} + V_y \frac{\partial}{\partial y}\right) V_z = 0 \qquad (29)$$

i.e., V_z is convected by the horizontal velocity field. Therefore $\langle V_z^2 \rangle$ is $\sim$ constant and independent of (x, y), which was previously called "generalized 2-D." The helicity density can be approximated as $V_x \Omega_x \sim V_y \Omega_y \ll V_z \Omega_z$. Then

$$I = \int \langle (V_z \Omega_z)^2 \rangle \, dxdydz \propto z_0 \int \langle (V_z \Omega_z)^2 \rangle \, dxdy \qquad (30)$$
$$= z_0 \langle V_z^2 \rangle \int \langle \Omega_z^2 \rangle \, dxdy \propto L^{2/3}$$

and from $\int I(K) dK = I$, $I(K) \propto K^{-5/3}$, where L is the largest characteristic scale in the (x, y) plane. Thus the $I(K)$ spectrum coincides with the energy spectrum of 2-D turbulence, $W(K) \propto K^{-5/3}$, corresponding to the inverse cascade. Thus, the cascade of the I invariant to large scales is indistinguishable from that of the energy, in contrast to the fully 3-D case where a small energy flux accompanies the cascade of the I invariant. Intermediate situations correspond to various degrees of anisotropy. As anisotropy increases, the fraction of energy transferred to larger scales also increases. These conclusions can be summarized as:

(a) An anisotropic situation can develop from a completely isotropic one if the growth in the vertical direction is restricted due to some condition in the atmosphere. For example, in the solar photosphere this maximum vertical extent may be limited by the size of the region in which the temperature gradient remains superadiabatic.

(b) In the initial stages when isotropy dominates, most of the energy cascades to small scales where it suffers viscous dissipation. The cascade of the I invariant results in an increase of the correlation length of helicity fluctuations.

(c) When this correlation length becomes equal to the vertical scale imposed by requiring superadiabaticity, for example, the correlation can grow only in the horizontal plane. This gives rise to anisotropy.

(d) The anisotropic fluctuations act as a new anisotropic stirring force accompanied by increasing amounts of energy transfer to large scales.

(e) As the anisotropy grows, it facilitates the accumulation of energy at larger scales and thus the formation of large structures like supergranular cells.

(f) The growth of large structures in the case of anisotropic turbulence can again be interrupted as a result of symmetry braking for example, caused by the Coriolis force. At the length scale L_c where the nonlinear term of the Navier-Stokes equation becomes comparable to the Coriolis force, the inverse cascade is inhibited. In the quasi 2-D situation that exists, the Coriolis force, together with a lack of reflectional symmetry with respect to the horizontal plane, favors helical structures with a definite sign of helicity. It is found that in quasi 2-D, the Coriolis force favors cyclonic circulation, and the sign of helicity corresponding to the updraft cyclonic motion can be fixed. If downward motion is present, it must be anticyclonic to retain the same sign of helicity. It is also found that much greater energy is needed to maintain coherent structures at the scale L_c.

There are several related questions of the energetics, the life times, and the temporal evolution of these structures which need a detailed investigation (keeping in view the available observations of solar granulation), and which may give direction to what more needs to be measured about solar granulation.

VII. ROLE OF MAGNETIC FIELD

The enhancement of magnetic field at the supergranular cell boundaries due to horizontal motion has been discussed by Simon and Leighton (1964), where the maximum magnetic field builds up to the equipartition value. In the chromosphere the ratio of magnetic to kinetic energy is found to be greater than unity. The nature of magnetohydrodynamic turbulence in the presence of a strong magnetic field changes significantly (Montgomery and Turner 1981). The turbulent spectrum results from the interplay of three effects: (1) the driving mechanism; (2) dissipative mechanisms; and (3) the modal transfer due to nonlinear interaction between various spatial modes. From laboratory studies, it is found that in the presence of a strong magnetic field $\mathbf{B}$, anisotropy is set up and the turbulent spectrum splits into two parts: (1) two-dimensional magnetohydrodynamic fluctuations carrying most of the energy in a plane perpendicular to the mean field $\mathbf{B}_0$; and (2) a more isotropic spectrum which is identified with Alfvén waves. In two-dimensional incompressible MHD, the square of the vector potential A^2 and the magnetic field $\mathbf{B}$ take the roles of V and Ω in two-dimensional Navier-Stokes turbulence. Thus an inverse cascade in the spectrum of A^2 can lead to an organized state such that

$$\delta \int (\nabla \times \mathbf{A})^2 \, dV - \lambda \delta \int A^2 dV = 0 \qquad (31)$$

which gives

$$\nabla^2 \mathbf{A} + \lambda \mathbf{A} = 0. \qquad (32)$$

The constant-A contours correspond to the magnetic lines of force, and the expected self-organized state is a pair of long wavelength circular magnetic fields. The cascade of A^2 to small wavenumbers and of magnetic field energy to large wavenumbers has been demonstrated by Pouquet (1978). The time scales of the 2-D MHD fluctuations are governed by nonlinear terms, whereas for the isotropic part these are the Alfvénic time scales. The Alfvénic part of the turbulent spectrum can be associated with the spicules, which Osterbrock (1961) described as slow-mode disturbances carrying chromospheric material up along the magnetic lines of force into the corona. The organized two-dimensional MHD turbulence could be an explanation for the formation of the magnetic network.

The predictions of the above model can be tested by measuring the correlation lengths along the mean field $\mathbf{B}_0$ and perpendicular to it. One expects that the correlation lengths along $\mathbf{B}_0$ are much longer than those transverse to $\mathbf{B}_0$. This is true for velocity and magnetic field fluctuations. The root-mean-square values of the transverse magnetic fluctuations are much larger than the longitudinal (along $\mathbf{B}_0$) fluctuations. The single-point frequency measurements for both magnetic and velocity field fluctuations are expected to show steep power-law frequency spectra that are negligibly small for frequencies below either the ion gyrofrequency or the ratio of the Alfvén speed to the correlation length, thus indicating the weakness of the higher-frequency Alfvénic spectrum.

The organizational properties of three-dimensional magnetohydrodynamic turbulence have been successfully used to delineate the structure of coronal loops (Krishan 1985a,b). The lowest-energy state emerges as a force-free state which reproduces the observed spatial variations of pressure in the coronal loops. In 3-D MHD, the new invariant, magnetic helicity, turns out to be a useful indicator of the pre-flare configuration of a flaring loop, as it is conjectured that during a flare, a high-energy, nearly force-free state decays to the lowest-energy state with the release of magnetic energy, while the magnetic helicity (due to inverse cascade) remains nearly constant (Krishan 1986).

Thus, these studies indicate the relevance and the resourcefulness of the concept of self-organization in magnetohydrodynamic turbulence in the solar context.

VIII. CONCLUSIONS

The inverse cascade of energy in two-dimensional hydrodynamic turbulence favors the formation of large organized structures. Application of this idea to the production of supergranulation seems to account for the observed spatial scale of the cellular motion. With inclusion of a strong magnetic field, for example, for the conditions obtained in the chromosphere, the turbulent spectrum consists of two parts: (1) anisotropic two-dimensional MHD fluctuations in the transverse direction; and (2) an Alfvénic spectrum along $\mathbf{B}_0$. The predictions about correlation lengths in the two directions along and perpendicular to $\mathbf{B}_0$ need to be tested by obtaining high-quality observations.

Acknowledgments.   The author is grateful to D. Ter Haar for introducing her to the works of E. Levich and collaborators. Figure 3 has been reproduced from Hasegawa (1985).

CORONAL ACTIVITY

SERGE KOUTCHMY, JACK B. ZIRKER
National Solar Observatory, Sac Peak

RICHARD S. STEINOLFSON
University of Texas, Austin

and

JOSEPH D. ZHUGZDA
Izmiran-Moscow

This chapter considers the rather extended, very inhomogeneous and dynamical part of the solar atmosphere called the corona. Coronal structures (Sec. I) are first described from the point of view of quasi-static structure in order to consider their thermodynamical gross properties, including heating and magnetic fields. The solar-cycle-related variations of coronal structures are reviewed as well. Solar prominences (Sec. II), are a well-observed ingredient of the corona; they form its cool component. Their properties, including magnetic field, thermal and velocity field structures are reviewed before considering the problem of their formation. Coronal mass ejections (Sec. III), are the most dramatic phenomena occurring in a short time in the corona and are almost inevitably related to a prominence ejection. Their observations are described with reference to several related solar phenomena in order to understand the driving mechanism. Models of coronal mass ejections are described based on numerical simulations including consideration of waves. Finally, the very important and still open question of the waves in the solar corona (Sec. IV) is briefly discussed from a theoretical point of view, emphasizing the possible role of hydromagnetic waves.

I. CORONAL STRUCTURES

It is generally accepted that the analysis of coronal structures of the Sun (including small-scale structures) gives an important insight into the knowledge of the solar magnetic fields, their evolution and their origin and, accordingly, of the solar activity (see, e.g., Pneuman and Kopp 1971; Kahler et al. 1981). Indeed, coronal structures behave like a tracer of the magnetic lines of force, even when a dynamical corona is considered. Eclipse photographs show that structures overlying the coronal hole regions are open, as are the magnetic field lines of force and, conversely, arch-like and loop-like structures are observed around active regions and filaments, corresponding to closed magnetic field lines, which are eventually "disrupted" during a coronal mass ejection (CME). The origin, dynamics and physics of coronal structures are reviewed by Zirker (1985a), Rosner (1986) and Altrock (1988); their main examples are large streamers, enhancements and holes (see Koutchmy [1977b] for a detailed illustrated description). First, we briefly consider what observations show, without attention to the cool part of the corona (prominences) which is treated further in Sec. II.

A. Structures Revealed by White-Light Eclipse Photographs, X-ray Pictures, Coronagraphic Images and Radio Maps

White-light eclipse photographs do not properly record the numerous coronal features unless special filters are used. The most useful method is to have a radial neutral filter to compensate the large gradient of light, especially over the inner corona. However, the *overlapping* of the details superposed along the line of sight and, also, the large differences often observed between equatorial and polar regions, still make the photographs different from what would be "seen" by an experienced observer. This is illustrated in Fig. 1 which shows, with the example of the well-studied eclipse photograph of 30 June 1973, a processed picture that leads to a result not far from the interpretation given when a drawing is produced from a naked eye observation. Such a white-light picture shows best the distribution of densities in the corona, because the amount of observed light is linearly related to the amount of free electrons and, consequently, to the plasma densities, integrated along the line of sight. It is quite evident, from the inspection of this figure, that the plasma is largely confined to small-scale structures, i.e., thread-like filaments and loops. This impression is largely confirmed by the inspection of the X-ray picture superposed in the same Fig. 1, which shows the inner corona observed simultaneously on the solar disk. Although the light observed in X rays consists of several line emissions, the picture is a good representation of the distribution of plasma densities in loops and bright points and, additionally, it shows precisely the contours of coronal holes.

Recent pictures (maps) constructed from high spatial resolution radio-

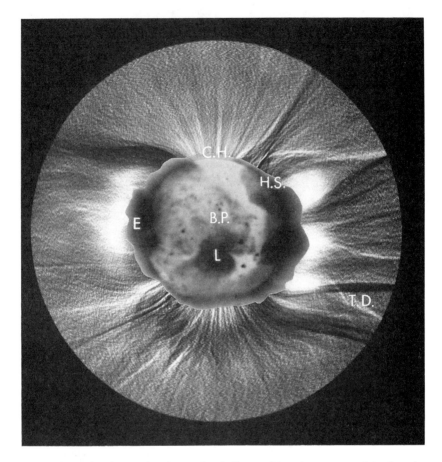

Fig. 1. Composite of a negative picture taken in X rays of the solar corona overlying the solar disk (courtesy of D. Webb, A.S.E.) and of a processed picture taken simultaneously in white light to show the extensions of the coronal structures outside the solar disk (30 June 1973 solar total eclipse—CNRS France). Prominent regions of the corona are: CH: corona hole; HS: helmet streamer; L: loop system; E: enhancements; BP: bright points. Note the fine thread-like structures and discontinuities at the edges of streamers viewed in white light on both sides of the plane of the sky; note also the north-pole coronal hole (at the top) going down on the disk and crossing the equatorial region.

emission measurements at the VLA (Lang et al. 1988) confirm the conclusions drawn from X-ray pictures, showing bright points (Fu et al. 1987), holes, the enhancements above active regions and even the filament channels with faint loops. Their interpretation is, however, more difficult, because the sensitivity of radio emission to plasma density and temperature is not trivial. However, full disk images can be made at any time unlike eclipse pictures. The same is true with X-ray pictures; EUV pictures yield a better discrimination among the lines. Plasma densities are far better deduced from white-light observations and temperature structures can also be more easily ana-lyzed with groundbased coronagraphs (Altrock 1988). Many observations of coronal structures seen in optical emission lines have been accumulated, pri-marily in the popular Fe XIV 530.3 nm line, and even a complete movie of coronal structures observed during more than one sunspot cycle has been assembled at Sacramento Peak Observatory. Both the richness and the com-plexity of structures are evident, suggesting permanent dynamical phenom-ena (flows) occurring essentially at very small scales and sudden instabilities occurring at large scales. Although all scales are largely revealed in these numerous observations, an even better spatial resolution is required to under-stand fully the physics of these structures.

B. Coronal Magnetic Fields

The best way to study coronal structures would be to measure the co-ronal magnetic field directly; unfortunately, no direct measurement exists, although some attempts have been made. The prominence magnetic field has been measured by several groups using both the Zeeman effect and the Hanle effect (see Sec. II on prominences), giving at least some feeling of what could be the average field in the cooler plasma. Radio astronomers have extensively discussed their measurements in terms of coronal magnetic fields (see Dulk and McLean 1978), but their interpretations are largely model dependent, so their conclusions are not essentially different from what is deduced from gen-eral considerations of pressure equilibrium of the coronal plasma. Briefly, assuming an isothermal corona, the gas pressure P in the structures can bal-ance the magnetic field pressure, at least in the middle corona where $P \sim r^{-6}$, because the strength B of the large-scale magnetic field (low-order harmonic components of the photospheric field) decreases as r^{-3} there (r being the radial distance from the Sun center, in units of solar radii). Using a potential field approximation and some practical boundary layer conditions outside, numerical methods have been largely applied (see, e.g., Levine et al. 1977) to extrapolate the magnetic field measured at photospheric levels. The deduced global morphology mimics quite well the observed density structures, so it is tempting to believe in the obtained amplitudes. At very large scales, the approximation is probably correct. Unfortunately, the corona is filled with very small scale structures like jets and threads, and even the beta ($\beta = 8\pi P/B^2$) of the plasma is not known so the magnetic field can only be conjectured. This is possible in the case of coronal discontinuities, for example. Therefore, a detailed knowledge of the plasma density distribution is required to calculate the magnetic field needed to balance the lateral gas pressure.

Loops are a special case as a dynamical equilibrium should be consid-ered and the field is likely to be nonpotential; there, temperature inhomoge-neities also play a hidden role as the so-called scaling law relating the temperatures, the pressure and the loop length, prevails (see Rosner 1986). It seems doubtful that much progress can now be made in the knowledge of the coronal magnetic fields, and consequently, of the coronal physics, *without attacking the fundamental problem of their direct measurements,* a problem which requires the use of a large aperture coronagraph.

C. Coronal Holes and the Active Corona

In a very crude approximation, the corona can be divided into open parts, the coronal holes producing the fast solar winds, and the more closed parts rising above active regions which, eventually, are extended or stretched out into the interplanetary medium in a heliospheric sheet surrounding the Sun near the equatorial plane (see Fig. 2). Even this very simple picture raises fundamental questions such as: what is the source of the slow wind or what

Fig. 2. Composite of white-light eclipse pictures showing the extension of coronal structures into the heliosphere (observations and processing by the Los Alamos group and the Paris CNRS group of the 30 June 1973 eclipse). At this phase of the sunspot cycle, the corona can be crudely considered to consist of two parts: open coronal holes arising above the polar regions and the more closed equatorial regions extending in streamers and stretched out toward the heliosphere equatorial sheet.

is the source of energy maintaining the high temperature of the dense closed regions?

If we consider the physics of coronal holes, without entering in to details that are hidden at very small scales, it is impossible to explain the origin of the fast wind from both the point of views of mass balance and of energy balance (see Lallement et al. 1986). Lately, some confusion has appeared at this point. Indeed, a few observations have been added to what was known after the SkyLab Workshop (see Zirker 1977), and this is enough to make the first models proposed at that time (see also Orrall 1981) completely obsolete. It is quite obvious that models should now take into account the highly structured state of the plasma in both these morphologically opposite regions (open and closed). Even the physics of the large streamers should take into account the existence of tiny elongated structures like filaments or threads stretching out in quasi-parallel "flowing lines" to explain their slow evolution and their merging in the heliospheric sheet. Finally, in the most active parts of the corona such as sporadic condensations and loop systems, we are faced with the fundamental problem of the source of heating. Temperatures rise by more than 2 orders of magnitude between the high chromosphere and the ambient coronal structures; there again the plasma should be considered to be in highly organized and compact structures to allow electric current dissipation or wave generation and propagation. Indeed the best coronagraphic observations suggest that this is true (see Smartt and Zhang 1987).

D. Thermodynamical Description of Coronal Structures

1. Density Determinations and Results.   As the intensity of the white-light corona, which is due to Thomson scattering, is directly related to the free electron densities, eclipse observations provide direct measurements of coronal densities. However, a good absolute intensity calibration is required to derive precise values. This is especially needed when *average* densities are derived, i.e., when the coronal atmosphere is considered homogeneous and radially stratified. In this case only a radial variation is then derived. This approach has been extensively developed in the past to describe low-resolution observations, (Saito 1972), and especially, to derive coronal models (Osherovich et al. 1984, 1985). Starting from the classical Parker (1957) model of the solar wind, all models describing the radial expansion of the coronal plasma use homogeneous distributions, including those trying to model coronal holes; a recent paper considering the problem is by Withbroe (1988). Although average density distributions are useful for comparing the corona at different phases of the sunspot cycle, they are far from the realistic coronal structures. The most conservative estimate of the structures' filling factor in the corona gives a value of 1% or less, so average values correspond to a vastly oversimplified picture of the corona, even without considering really small-scale structures at scales less than 10 arcsec.

Let us consider the surface brightness of the corona $B = B(\rho)$ with ρ

being the projected (on the plane of the sky) radial distance (in solar radii R). It has been found convenient to represent $B(\rho)$ by a sum of terms of the form ρ^{-m}, with m taking different predominant values (from 17 to 3.5), depending on the value of the radial distance. Between the surface and few solar radii ($r < 3$ or 4), the derived radial density distribution with $m_1 = 17$ and $m_2 = 7$, follows nicely a curve described by a hydrostatic law with a constant temperature of 1 to 2 × 10⁶ K (see Fig. 3). This seems to indicate that the departure from hydrostatic equilibrium due to the hydrodynamic flow (solar wind) is small in this region. Assuming μ the mean molecular weight of particles in the corona is known from consideration of abundances and ionization degrees (H and He are fully ionized in first approximation, so

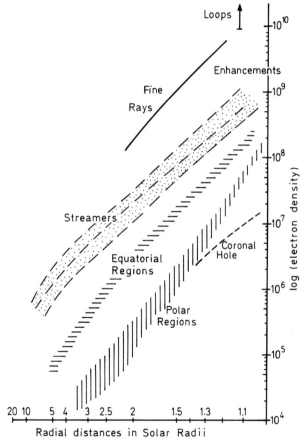

Fig. 3. Radial variation of electron densities deduced from the analysis of eclipse results assuming distribution with different degrees of inhomogeneity. Note the large difference of densities in holes, fine rays and loops.

$\mu \simeq 0.6$), M_H the mass of a proton and g the constant of solar gravity at the surface, we can write:

$$N_e(r) = N_e (R_\odot) \exp \left[-(\mu M_H g R_\odot / kT)(1/R_\odot - 1/r) \right] \qquad (1)$$

with $N_e (R_\odot)$ the density at the surface, and $R_\odot$ the solar radius.

Using this approximation, Newkirk (1961) and recently Badalyan and Livshits (1986) and Badalyan (1986) found it more convenient to describe the distribution of the inner electron densities and, consequently, of the inner coronal brightnesses, as an exponential function of r instead of a power of r. However, in the very inner corona an even better fit is obtained in describing the radial variation of densities as an exponential function of the height above the limb, instead of r.

Farther out in the corona, we should return to interpretation of observed white-light intensity modulations, which, in its simplest form, can be written as:

$$\Delta B = \sigma_e W_\lambda B_\odot N_e H \qquad (2)$$

where σ_e is the Thomson scattering coefficient (considered here isotropic), W_λ is the dilution factor $\sim r^{-2}$, $B_\odot$ the average solar brightness, N_e the local electron density and H the *equivalent* thickness of the plasma producing the scattering along the line of sight. Considering the high degree of inhomogeneity of the corona (Fig. 3), the approximation (Eq. 2) can be conveniently used to interpret modulations of white-light intensities observed on a coronal eclipse photograph. We then consider the integrated intensities as produced by a collection of structures overlapping along the line of sight. Finally, we notice that the filling factor (ff) or the fraction of volume occupied by structures with local average densities, is very small and depends on the spatial resolution. With the available observations, it is already possible to state that $ff \leq 0.01$, a value deduced from the comparison of the densities deduced from homogeneous models and of the local densities deduced using Eq. (2).

Temperature and Velocity Determinations.   Up to now groundbased coronagraphic observations (see, e.g., Tsubaki 1975) have attained the best simultaneous spatial and spectral resolution in coronal studies (see Fig. 4) although only above the limb where overlapping effects are important. X-EUV coronal observations on the disk with sometimes excellent spatial resolution, are not yet capable of providing full line-profile information with enough spatial resolution to resolve individual structures above the transition region.

Besides the hydrostatic temperature T_h, which has been deduced from

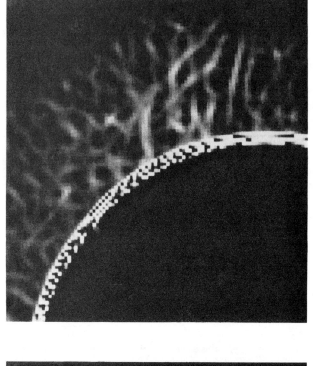

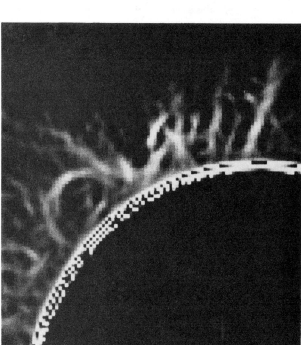

Fig. 4. Illustrations of the temperature inhomogeneity of the corona evidenced from line emissions at different regime: 637.4 nm emissions of Fe X near 1×10^6 K and 530.3 nm emission of the Fe XIV near 2×10^6 K, observed simultaneously at the west limb of the eclipse corona 31 July 1981 (original observations by J. Sykora).

Eq. (1) using electron densities calculated for different models of structure, the best determined temperatures T_i are those deduced from ionization equilibrium calculations. In the inner corona, several observations have shown essentially four regimes of coronal temperatures:

1. The most probable temperature, close to 2×10^6 K, evidenced by the emission of ions like Fe XIV at 530.3 nm, that match density structures quite well;

2. A regime of temperature close to 1×10^6 K, revealing a plasma in probably a more transitory state, as in coronal holes or in regions that have not reached a relative equilibrium; the emission of the ion Fe X at 637.4 nm is representative of this state;

3. Extreme temperatures are also met in special cases. In sporadic condensations above flaring regions high temperatures are recorded; emission of the Ca XV line at 569.6 nm is typical of this region situated above 3×10^6 K;

4. Finally, a cool component of coronal emissions is often observed in slowly erupting prominences, for example, or above active regions; emission of low excitation lines like Hα is typical of this region.

Figure 4 is a good illustration of the temperature inhomogeneity of the corona deduced from line emissions.

The line profile analysis of ionic emissions permits the determination of the so-called line-width Doppler temperatures T_D. Considering a gaussian line profile described with the function $\exp(-\Delta\lambda^2/\sigma^2)$ of full width at half maximum W with $\sigma = W/2 \, (\ln 2)^{1/2}$ and λ the wavelength, we have:

$$\sigma = (\lambda/c)(2kT_D/M_i)^{1/2} \qquad (3)$$

or, taking into account the value of the ionization temperature of the line:

$$\sigma = (\lambda/c)(2kT_i/M_i + V_t^2)^{1/2} \qquad (4)$$

where V_t is the so-called nonthermal turbulent velocity, with a typical value of 20 to 30 km s^{-1} and M_i the mass of the ion considered. A more detailed analysis of the line profiles seems to show slightly shifted multicomponent line profiles, so a Doppler velocity can be deduced. Typical values cover the range of a few km s^{-1} to several tens of km s^{-1} (in rare occasions). Oscillations have also been discovered, with ± 2 km s^{-1} amplitude and rather short periods. However, large Doppler velocities have not been detected in the inner corona and in the more external parts, because intensities are too low to permit a precise measurement with the present instruments. In coronal holes, a recent line profile analysis performed at Sacramento Peak over a well-developed polar coronal hole of sunspot minimum, definitely favors rather large velocities, e.g., 80 km s^{-1}, as deduced from the emissions of the 637.4 nm line of Fe X in the very inner parts of the hole.

Models of Streamers: Geometric Parameters and Density Distribution.
With coarse resolution, eclipse photographs show (Fig. 2) the prominent streamers with presumably a quasi-gaussian shaped cross section. These streamers should not be confused with the inner coronal enhancements that are well observed with groundbased coronagraphs or in the meter range of radio emissions. Models of average density distributions inside large streamers have been deduced (see Fig. 3). At greater heights, the coronal streamers seem to open out (Koutchmy 1977*b*), and it is reasonable to consider streamers as the coronal counterpart of the heliosheet, at least in equatorial regions of the corona near sunspot minimum activity. At maximum activity, by contrast, streamers are observed stretching out beyond the equatorial plane of symmetry. With a high-resolution photograph, it is evident that streamers are made of a collection of fine rays or sheets that follow the same dominant direction, like a "flowing curve." Often a tangential discontinuity is observed edge on. However, at their base, elements of arches, or curved detached structures can be identified, rising sometimes above dark voids with the same morphology. When a prominence is viewed edge-on at the limb, these voids form a kind of cavity.

Models of Coronal Holes: Present Difficulties.   Besides streamers, coronal holes are by far the most important regions to understand, in connection with the physics of the solar wind (see Zirker 1977). Models have been proposed with a homogeneous distribution of densities across the section of the hole, raising the question of the constriction of streamers (Kopp and Holzer 1976). It is clear that the acceleration of the coronal plasma occurs in a region of open magnetic field lines; however, no such large-scale constriction is observed in eclipse pictures (Fig. 1). More difficulties are apparent when the energy balance is considered (Lallement et al. 1986). Considering more carefully the density distribution inside a coronal hole, we see that conspicuous polar rays or plumes are distributed over coronal holes. With high spatial resolution, these rays show a very small cross section, that exemplifies plasma densities several orders of magnitude higher than those derived from homogeneous models. Proper motions have been reported also, suggesting that intermittent effects are important. Clearly, more observations, including line-profile analysis, are needed before the present difficulties can be resolved.

Models of Loops: Coronal Condensations and Crossing Phenomena.
Loop systems are observed both on the disk, with radio telescopes like the VLA, with the X-EUV filtergrams, and above the limb, with groundbased coronagraphs. These active-region coronal structures need magnetic field

measurements with good spatial and temporal resolution to be understood, especially for the short-lived, post-flare loops. They have been extensively studied (see, e.g., Orrall 1981; Zirker 1985a). The loops of coronal condensations are of considerable interest in relation to flares, but this relation is rather complex, because the magnetic fields are not potential and because of the great temperature inhomogeneity and the dynamical state of the plasma in these regions. The question of heating the coronal loop is not yet resolved and probably needs higher-resolution observations. Smartt and Zhang (1987) have described interesting cases of crossing phenomena in loops occurring in different situations, including post-flare loops. These crossings could correspond to reconnections occurring high in the corona showing the importance of interactions of loops.

Fine Structures and Discontinuities.   When a high-resolution coronal white-light photograph is examined, many quasi-radial coronal sheets can be identified. Almost invariably, they are parts of a large coronal streamer. Indeed they represent the fine-scale structure of the streamers (Fig. 1); they are more visible on an unprocessed picture showing the very characteristic signature of a coronal tangential discontinuity, a large and definite jump of electron density. A "void" or "dark lane," often follows the discontinuity (D). It has not yet been interpreted; it could correspond to a local peculiarity of the orientation of the magnetic field, which is enhanced near the jump. Discontinuities are quasi-radial so that, considering the pressure equilibrium on either side of D, we simply write:

$$\Delta \left[P_i + B_i^2/8\pi \right] = 0 \tag{5}$$

assuming the magnetic field parallel to D. Typical values for a discontinuity observed on eclipse pictures are given in a review by Altrock (1988 p. 208). The connection of small-scale discontinuities with photospheric structures still seems open and needs very high-resolution observations in order to provide a firm observational basis for the model of coronal heating suggested by Parker (1988e); he assumes tangential discontinuity "shaking" processes. Finally, fine structures also show evidence of a twist, but the ratio between the pitch of the twist and cross-sectional diameter of the structure is rather large and even variable along the same filament. Loops also show evidence of twist, which is probably connected with their dynamics.

E. Solar Cycle Variations of Coronal Structure

White-Light Variations.   Variations of the white-light structure of the eclipse corona have been known for a century. Roughly the gross shape of the corona is related to the sunspot cycle: near the minimum of activity, the corona is considerably flattened toward the equator and near maximum, structures are stretched out in all directions outside the equatorial plane. A more careful analysis of eclipse data shows considerable details; for example, the extension of polar holes and the deviation of the axis of large streamers with respect to the radial direction behave systematically during the cycle. Figure 5 gives the cyclic behavior of the best observed global photometric parameter, the (due to Luddendorf; see Koutchmy and Nitschelm 1984) flattening index; a definite phase shift is observed, suggesting that the behavior of the corona is not just a reflection of the sunspot cycle.

Monochromatic Emission Variations (Fe XIV, Fe X lines) and Overlapping Cycles.   A considerable amount of data, covering more than four sunspot cycles, have been collected on the 530.3 nm emission of Fe XIV and also with other coronal monochromatic emissions, thanks to the use of groundbased Lyot coronagraphs (see Leroy and Trellis 1974; Altrock 1988). To look at the latitude distribution of the coronal activity as a function of the sunspot cycle, we took the Pic du Midi data covering the period from 1945 to 1975 of absolute 530.3 nm brightnesses and displayed them after super-

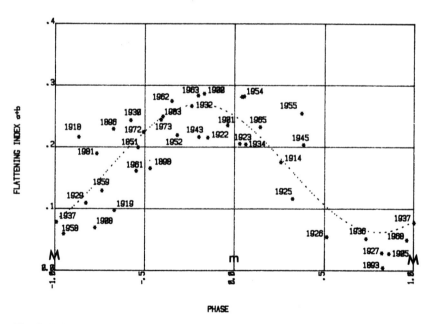

Fig. 5. Variations of the coronal flattening index as a function of the phase of the sunspot cycle (Luddendorf's definition; Koutchmy and Nitschelm 1984). *m* and *M* correspond to the epoch of minimum and maximum sunspot activity. All cycles where the eclipse corona was observed have been superposed to increase the visibility of the graph. Note the definite shift of phase of the best-fitted curve.

posing 3 cycles of activity (see Fig. 6). This figure shows very well two important features of the coronal cycle: (a) the double maxima of activity in the equatorial belt, and (b) the north-south asymmetry of activity.

Leroy and Noëns (1983) looked at these data and used relative brightnesses in order to enhance the activity variability. They confirmed the importance of polar region activity (see also Altrock 1988) to define the coronal cycles. Overlapping cycles seem to emerge when the whole corona is considered, strongly suggesting that the magnetic field cycle of the Sun cannot be confined to the sunspot field alone as low-order harmonic spatial components are more significant to explain the coronal cycle.

The Heliosheet and Coronal Hole Solar-Cycle Variations.   As already shown in Fig. 2, large coronal structures of the inner corona extend far into the interplanetary medium, forming, at the time of reduced sunspot activity, a heliosheet surrounding the equatorial plane of the Sun. This sheet is bent and seems to be influenced by the activity of polar coronal holes rotating together with the Sun. At sunspot minimum, coronal holes can cover a large part of the polar regions; the process starts after the polar reversal of magnetic polarity which occurs asynchronously on both hemispheres, creating a north-south asymmetry in polar regions. The polar reversal is preceded by the so-called "rush to the pole" of the polar crown filaments, that lie under a system of helmet streamers. The relation of these streamers to the heliosheet is unclear, as well as the appearance of the heliosheet near sunspot maximum, when coronal mass ejection activity is high. Although the coronal holes have been regarded as a fundamental component of coronal activity, (see Zirker 1977), today their apparent rigid rotation has been strongly questioned and new ideas about solar magnetism seems to give more importance to the evolution of the closed inner coronal structures and to the large-scale dynamical phenomena. However, coronal holes produce the fast solar wind and there are even suggestions that besides the coronal mass ejections, all the solar wind is coming from coronal holes. Therefore we are inclined to admit that coronal holes and the heliosheet are different parts of the same phenomena. Coronal-hole cycle variations are now studied with 1083.0 nm spectroheliograms of He I line and 530.3 nm routine coronal scans of Fe XIV (see Altrock 1988).

F. Heating-Related Problems and the Source of the Fast Wind

Observations.   The heating of the coronal plasma, connected with the question of its origin, is an old problem still not resolved. A part of this problem may be related to the activity of the corona at very small scales, especially in loop systems and tangential discontinuities (see Parker 1988e). In such regions, MHD waves could be generated; they would dissipate and maintain the plasma at coronal temperature. Dissipation of electric currents have also been suggested (see Ionson 1984) but theoretical considerations seem to favor very small scales which cannot be directly observed. The mi-

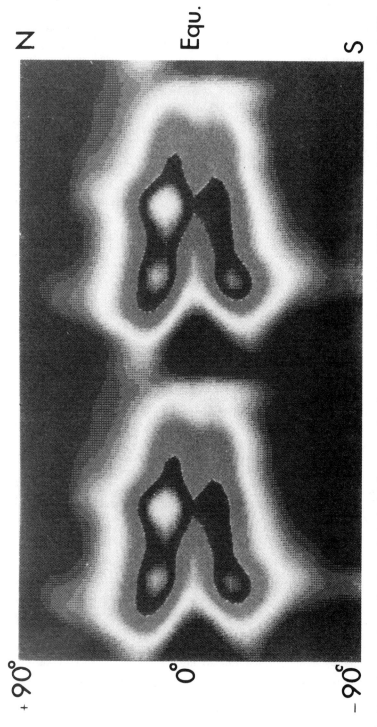

Fig. 6. Superposed coronal cycles as obtained using 3 sunspot cycles of measurements performed during 1945–75 at the Pic du Midi Observatory. Note: the NS asymmetries; the powerful secondary maximum of coronal activity in the north-hemisphere; the "rush" to the poles occurring before or during the first maximum of activity.

croturbulence of the plasma, well observed by its effect on the profile of ion line emissions, could be the key to understanding the heating mechanism(s). Precise observations of line profiles are still of limited spatial resolution and we know that the structures are small. It is tempting to attribute a part of the line profile broadening to the overlapping effect of dynamically active small-scale features. A few observations have already shown that the line profiles are not systematically gaussian; more work is in progress.

Small-scale Impulsive Events and Related Phenomena.   Observation of the solar disk with high time and spatial resolution in visible, UV, X rays and radio waves show that the transition-region emissions from small magnetic bipoles are intermittent and impulsive. Parker (1988*e*) has proposed calling the basic unit of impulsive energy release a "nanoflare"; then the active inner corona could be understood as a swarm of nanoflares. The best observations of impulsive events on the disk are from the NRL experiment (HRTS) flown three times on board sounding rockets during 1975–78 (Brueckner and Bartoe 1983). Although only a few minutes of observations of transition-region line emission have been collected, this experiment gave the first direct evidence of impulsive events starting at the base of the corona and showing a large Doppler effect, up to 400 km s^{-1}, of blue-shifted isolated emission. These events can be interpreted as very small exploding loops called "jets." Turbulent events are also observed, including red-shifted events. In jets, the acceleration is of the order of 5 km s^{-2}, with a lifetime of 80 s, so these events should contribute to the corona. Analyzing high-resolution eclipse photographs, Koutchmy and Stellmacher (1976) found white-light spikes over a polar coronal hole; these are related to chromospheric emissions and strongly suggest striation and intermittence in coronal holes. They could be a special case of the jet phenomena, where the coronal magnetic field is open. The deduced coronal densities in spikes are high, indeed comparable to those of loops or fine rays (Fig. 3). Because coronal holes are filled with fine rays, at times matching the distribution in the chromospheric network (polar plumes, polar rays), it is tempting to identify the source of the fast wind with jets and spikes. Coronal observations in the inner and intermediate corona are needed to confirm this assumption. X-EUV observations of bright points on the disk with a better resolution (both temporal and spatial) could also help to understand the energetics of the processes which seem to emerge from the high chromosphere in the form of bubble-like or loop-like tiny structures. A careful consideration of very small-scale phenomena occurring at the prominence-corona interface seems to show similar phenomena; a common physical mechanism should be looked for.

G. Importance of the Study of Coronal Structures

We have tried to show how a careful consideration of the fine structures in the corona could help to attack old problems of the physics of coronal activity from a new point of view. Both coronal-hole activity and streamers need high spatial resolution analysis; loops and active-regions structures are probably connected with even smaller-scale phenomena but there the magnetic field is also of higher amplitude and needs special attention. Models built on the basis of homogeneous distribution of densities have failed to explain the most fundamental properties of the corona, its heating and the source of the fast wind. The first step is to introduce in models the filling factor although a more realistic approach is to consider the actual structures. This raises the question of interpretation of fine measurements like the analysis of "instantaneous" coronal line profiles. We should go further than just assume a turbulent field of velocities leading to a gaussian symmetric profile. It is likely that coronal waves may also be detected with improved resolution, individual structures producing an easily measurable modulation of the line profile.

II. THE PHYSICS OF PROMINENCES

Solar prominences are among the largest, most persistent structures in the solar atmosphere. They have attracted a vast amount of attention from solar physicists in the past, and even more so lately. Many regularities are recognized now in prominence behavior and fresh observations, particularly of their vector magnetic fields, have stimulated new models. And yet, with all this work, we are still unable to answer definitively some of the essential questions pertaining to prominences; for example, what is the configuration of the magnetic field, within and around a prominence? Does a prominence form by condensation from the corona, or injection from the chromosphere, or both? How is the energy of prominences supplied? Several good reviews are available on quiescent prominences (see, e.g., Priest 1988; Ballester and Priest 1988; Poland 1986; Zirker 1989).

A. Magnetic Field

A quiescent prominence is about a hundred times denser than the corona, and would fall into the photosphere without some mechanical support. Magnetic fields in and around the prominence are commonly thought to provide this support. The magnetic field in the quiet corona is too weak to be measured with conventional optical methods. Radio techniques (see, e.g., Holman and Kundu 1985) provide some estimates in special, simple cases. Thus, in general, the field configuration surrounding a prominence must be deduced from indirect evidence, and we still lack a definitive knowledge of the magnetic configuration within and around a prominence.

Eclipse photographs of the white-light corona (e.g., Fig. 1) show that a quiescent prominence is often embedded within a helmet streamer. The streamer, seen edge-on, is a large, fan-like structure that extends above the neutral line of the large-scale photospheric longitudinal magnetic field, with its footpoints in regions of opposite magnetic polarity. The prominence also lies along the magnetic neutral line. A region of low coronal density (the cavity) surrounds the prominence. White-light coronal arches are occasionally visible above the cavity (Saito and Tandberg-Hanssen 1972). On the disk such arches (or arcades of arches) have been observed occasionally in X-ray and EUV images (Vaiana et al. 1973; Schmahl. et al. 1982). Thus the prominence is part of a much larger, more complex structure, which presumably outlines the weak coronal magnetic field. Pneuman (1972) has stressed this connection, and offers some qualitative suggestions for prominence formation. However, the magnetic environment of a prominence is still uncertain.

Movies of quiescent prominences filmed in the $H\alpha$ line at subarcsec resolution (Dunn 1960) show that a prominence consists of many fine threads and knots. In hedgerow prominences, the threads are predominantly horizontal and axial at low heights and connect the feet, while at greater heights the threads are approximately vertical. Figure 7 shows simultaneous prominence images in $H\alpha$ and He D_3 ($\lambda5876$ Å) (Cram and Smartt, personal communication). Note the difference in structure as seen in these two lines. If, as we have good reason to think, the magnetic field is frozen into the plasma, the threads should outline the field lines. This complicated picture is somewhat at variance with the vector field observations which we consider next.

Polarization Measurements. During the past decade three experimental groups have explored the internal magnetic field of prominences. The French and U.S. groups (Leroy et al. 1984; Athay et al. 1983) have interpreted the observed depolarization of optical emission lines (the Hanle effect) in order to derive the vector field, while the Soviet-French group (Nikolsky et al. 1986) have relied on the Zeeman effect to find the longitudinal field component. Leroy (1987) has summarized the experimental results as follows:

1. The field strength in quiescent prominences ranges from about 2 to 20 gauss;
2. The field vector is nearly horizontal (within 30°) and lies at a small angle (~25°) to the long axis of the prominence;
3. The field strength and azimuth angle change only slightly with increasing height;
4. The field seems remarkably uniform over the surface of the prominence sheet, *within a resolution element of about 5 arcsec.* The dispersion of field strength, for example, is smaller than the dispersion of spectrum line intensities;
5. The field parameters vary only slightly in tens of minutes.

Thus, the polarization measurements imply a prominence field that is reasonably uniform, horizontal and slowly varying, while the emission line movies imply a field that is nonuniform, vertical (in places) and rapidly vary-

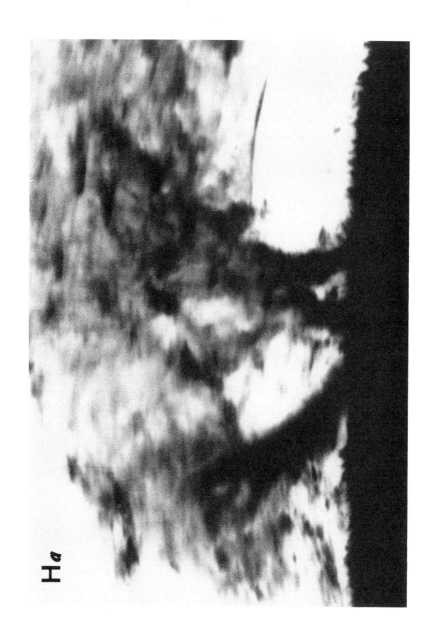

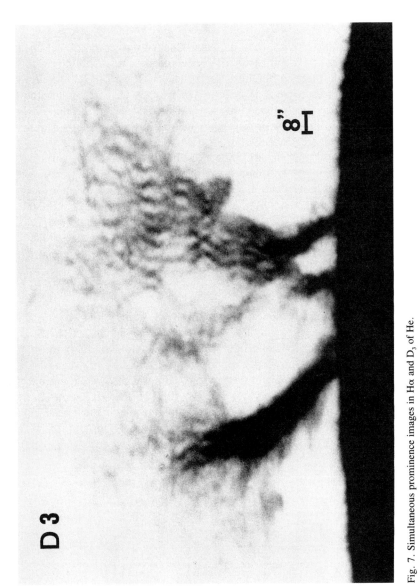

Fig. 7. Simultaneous prominence images in Hα and D₃ of He.

ing (i.e., within a few minutes). The discrepancy between these sets of results may reflect only the lower spatial and temporal resolution of the polarization measurements; or, they may imply a more fundamental effect—a drift of the plasma across field lines.

Magnetostatic Models   As the large-scale coronal magnetic field near prominences has not been observed directly, theorists have only a few constraints to guide them in building MHD models. These are (a) the vector magnetic field measurements described previously, (b) the observed thickness of the prominence sheet (about 5000 km in Hα), and (c) the location of the sheet along the photospheric magnetic neutral line, suggesting a quasipotential field at large distances.

Two classical models are shown in Fig. 8. In the original (K-S) Kippenhahn-Schluter (1957) model, the plasma is supported by the Lorentz force in a current sheet at the top of a coronal arch. In later variations (see, e.g., Malherbe and Priest 1983), a loop may develop at the top of the arch. In the (K-R) Raadu and Kuperus (1973,1974) model the plasma is supported by the repulsion of a line current in the prominence and a line current below the photosphere. Anzer (1985) has argued that the K-R model, especially in its "figure-8" form, is self-pinching, hence unstable. Another K-R form, with field lines opening outward to infinity, would require a strong solar wind to maintain it.

Note that a prime distinction between the K-S and K-R models is the direction of the magnetic field through the plane of the prominence. In the K-S model the sense of direction follows that in the photosphere, in the K-R model it is reversed. The French group (Leroy 1987) have investigated this relationship in a large sample of prominences. Unfortunately, the Hanle effect yields ambiguous results (i.e., two solutions) for the field direction, but by using supplementary observations, Leroy et al. were able to draw some conclusions. They found that for high prominences at high latitudes, the field direction is anti-parallel, as in the K-R model, while for low-lying promi-

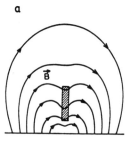
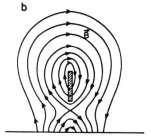

Fig. 8. Classical models of solar prominences.

nences at lower latitudes, it is parallel. The U.S. group is more cautious, stating only that if the prominence and photospheric fields are parallel, the prominence field is nearly normal to the prominence plane, and if anti-parallel, then parallel to the plane.

Low (1975,1981,1982a,b) has found a number of elegant analytic K-S prominence models that demonstrate mechanical equilibrium. However, their density distributions are in general too broad to agree with observations.

Recently, Priest et al. (1989) have proposed a twisted flux-tube model. In their picture, the prominence forms in a large flux tube that lies parallel to the final long axis of the prominence. The footpoints of the tube are assumed to be twisted by photospheric motions. A critical amount of twist can produce a dip at the underside of the tube, providing a stable pocket in which cool plasma can accumulate. As the twist increases, the field component normal to the long axis increases and the prominence lengthens. With sufficient twist the prominence may erupt. This model has many interesting and desirable features. In particular, the current generated by the condensation process has the correct direction to support the plasma. We shall return to this model later.

B. Thermal Structure

The radiative cooling time for a typical quiescent prominence ($\sim 10^2$ s) is much shorter than its observed lifetime ($< 10^4$ s). Thus energy must be continuously supplied to maintain the prominence. The source of this heating is unknown, as in the corona itself. Because a steady state exists, however, the energy input must balance the radiative losses. These can be estimated if the thermal structure of the prominence is determined. Optical and EUV spectra provide the best clues we have to the thermal structure.

A simple isothermal, isobaric slab model suffices to explain much of the optical spectrum with considerable accuracy. Heasley and Milkey (1976,1978) calculated the emergent radiation in several strong optical lines (e.g., Ca II 8542, Hβ, D_3, etc.) from such slabs. They were successful in reproducing the line *ratios* observed photoelectrically by Landman and Illing (1976,1977), if they assumed a slab temperature of 7500 K, a pressure of 0.01 dyne cm^{-2} and adjustable mass column densities. Their models fail, however, to predict the absolute line fluxes unless the slabs are thicker than Hα observations permit.

Engvold (1980) pointed out that Landman's observations did not resolve the subarcsec emission threads that appear in high-quality Hα images (Dunn 1960), so that his data (and the slab models) refer to spatially averaged spectra. Engvold measured the Hα brightness of threads on Dunn's films and used the Heasley-Milkey calculations to show that the pressure in threads was closer to 0.1 than 0.01 dyne cm^{-2}. No satisfactory resolution of this discrepancy has yet been reached.

The fine structure of prominences is evidently of prime importance in any discussion of thermal equilibrium. Increasing attention has been paid in recent research to incorporating small-scale structures in thermal models. Unfortunately, with rare exceptions, this fine structure is not resolved in observations, and the models must employ free parameters (e.g., a filing factor) to cope with it.

Hirayama's (1986) recent study of prominence hydrogen lines yields valuable limits on the diameters and spatial density of unresolved threads. He has determined the electron density in many positions in five hedgerow prominences from the Stark broadening of Hα. The line emission measure (N_e^2L) and the electron density (N_e) combine to yield the effective plasma thickness L of a 10-arcsec slice through the prominence. If Hirayama assumes a thread diameter of 300 km (Dunn 1960), he requires 8 unresolved threads in his 10 arcsec sampling aperture to account for L. Some derived values of L are so small (e.g., 2 km) that they imply thread diameters less than 150 km. If the overall prominence thickness were 5000 km (as seen in Hα filaments, for example), Hirayama's results would imply that threads occupy only 0.5% of the prominence volume. This is an incredibly small filling factor.

EUV spectra contain information on prominence plasma at transition-zone temperatures ($10^{4.5} - 10^{5.5}$K). Monochromatic EUV images from Skylab (Orrall and Schmahl 1976) show that all this material is co-spatial, within the 5 arcsec resolution of the observations. Schmahl and Orrall (1972) tested two models for the fine structure: unresolved cylinders or resolved thin slabs, both with cool cores and hot sheaths. Their test relied on the observed absorption (in the hydrogen Lyman continuum) of emission lines with wavelengths shorter that 912 Å. They found that if the unresolved cylinders were perfectly aligned in the line of sight (admittedly, an unlikely situation), five or more would have to be present in each 5 × 5 arcsec sampling aperture.

Later Schmahl and Orrall (1986) allowed for a random spatial distribution of unresolved vertical threads, and also considered two more structural types: isothermal threads with different temperatures, and threads with temperature gradients parallel to the magnetic field. The latter model is attractive because it would allow heat conduction to balance radiative losses. Randomly distributed cylinders with cool cores and hot sheaths can only explain the observed temperature variation of the line emission if their gas pressure increases radially outward at an improbably large rate. Isothermal threads utterly fail to predict the observed temperature variation of the Lyman continuum absorption. Finally, the third geometry (with temperature gradient parallel to field) requires several threads or sheets along the line of sight to account for the Lyman continuum absorption. If these lie on the same magnetic field line, however, the conductive flux will be insufficient to supply the radiative losses.

Thus, no model of slabs or threads has been found that satisfies all constraints by the EUV spectra. Rabin (1986) has offered a suggestion, pointing out that heat conduction across the magnetic field lines (which is commonly neglected) may play an important role if the fine structure is sufficiently thin. If the cross-field area of a thin sheet were 10^4 times the longitudinal area, heat conduction from the corona could account for the observed differential measures of EUV lines in the complete temperature range $10^4 < T < 10^6$ K. Thus, Rabin conjectures that the fine structure consists of thin ribbons; this idea needs further development.

C. Velocity Field

Prominence movies taken at the limb in Hα show predominantly downward or lateral motions of the fine structures (Engvold 1976) with speeds ranging from 5 km s^{-1} in the coarse knots near the top edges, to 15 to 30 km s^{-1} in the narrow vertical threads. If these motions were real, and not some sort of excitation wave, they could imply substantial mass fluxes. Athay (1976) estimated that the corona would be drained in about 20 days by 10 large prominences having such real motions.

Since these motions may only be apparent, however, most observers prefer to measure Doppler velocities in filaments, on the solar disk. The problem then is that the thread structure may be transparent and undetectable against the disk and these Doppler velocities in filaments may then refer to the lower, denser structures within the prominence sheet. The French group (Malherbe, Schmieder and associates) has published a series of papers on this topic. Their main results are as follows:

1. Doppler velocities in filaments are small (1 to 3 km s^{-1}) and usually directed upward. At the feet of hedgerow prominences, the velocity may reach 10 km s^{-1}, either upwards or downwards (Malherbe et al. 1981,1983a,b);
2. No evidence of narrow vertical channels (i.e., the threads) appears in Doppler velocity maps of filaments. The velocity pattern is fairly coarse (5 to 10 arcsec), although the claimed resolution may be as good as a few arcsec (Schmieder et al. 1985).
3. Simultaneous observations in Hα and C IV (Schmieder et al. 1985; Simon et al. 1986a; Malherbe et al. 1987) indicate larger speeds and dispersions in C IV, but with little correlation to speeds in Hα.

Malherbe and Priest (1983) have interpreted these rising motions in terms of the photospheric motions of the footpoints of the prominence's magnetic field. Motions that converge toward (diverge from) the central plane of the prominence produce rising motions in a K-R (K-S) model.

D. Prominence Formation

Although the bulk of a quiescent prominence resides in the low corona, we have little information on the process of formation there. Instead, the most complete observations available refer to the chromosphere and to the lower portions of a filament. Martin (1973,1986) has given detailed descriptions on the evolution of active and quiescent filaments, as seen in Hα filtergrams and in video magnetograms. Some of the salient features observed are:

1. A quiescent filament forms only when large photospheric regions of opposite magnetic polarity come into contact by some unknown process;
2. Additionally, the filament appears only after the Hα chromospheric fibrils at the boundaries of each polarity repel each other to form a narrow channel where the fibrils are aligned nearly end to end. Short strands of dark Hα material, which do not originate in the fibrils but are aligned with them, begin to form in the channel;
3. Patches of opposite magnetic polarity drift toward each other into the channel and merge or cancel.

When observed inside the limb in Hα, a typical hedgerow prominence seems to form by first growing a foot (a dense column of plasma rooted in the chromosphere and extending into the low corona). This foot is gradually connected to the existing prominence by horizontal strands to form the characteristic bridge-like structure. The process takes a few hours.

As mentioned previously, we have virtually no observations of prominence formation in the corona. In contrast, most theoretical work on prominence formation is based on the idea of condensation and cooling of hot coronal plasma, an idea first suggested by Kiepenheuer (1953). The physical basis for this idea is simple: a plasma of solar composition is radiatively unstable at temperatures between 0.1 and 1×10^6 K. If its temperature drops slightly, the plasma radiates more efficiently, cools, contracts and radiates still more efficiently (Field 1965; Parker 1957; Cox and Tucker 1969). Thermal conduction tends to stabilize this instability, but can fail if the conduction path exceeds a critical length. The presence of low-density cavities around mature prominences suggests that the prominence mass is drawn primarily from the corona, but recent estimates of the initial mass in cavities are too low, so that additional mass injection from the chromosphere seems necessary. In fact, we do not really know if a prominence forms by condensation from the corona, or injection from the chromosphere, or by both ways.

Kuperus and Tandberg-Hanssen (1967) introduced the idea that the condensation process occurs preferentially at a current sheet that separates coronal magnetic fields of opposite polarity. The current sheet they proposed would be unstable to the tearing mode (Furth et al. 1963) and would tend to form closed loops or helices that thermally insulate the proto-prominence from the hot corona. A great deal of theoretical work and modeling has been done recently to fill in the details of this picture.

For example, Malherbe et al. (1983a,b) explored the thermal instability of a pre-existing Kuperus-Raadu magnetic field model. Their calculations are intended only to find the dominant time and distance scales, but do offer some insights into the physics. They found four destabilizing effects:

1. The scale length for thermal conduction may exceed a critical value;
2. The K-R field may be sheared beyond a critical angle, again reducing the effective conductivity;
3. The initial compression at the current sheet, which is required to balance the lateral pressure, may exceed a critical value;
4. Coronal heating may be reduced near the prominence. If the heating is produced by MHD waves, these will follow the magnetic field lines, avoiding the neutral sheet.

These calculations are based essentially on linear stability arguments. To follow the condensation process through its nonlinear development, numerical simulations are required. Several authors have attempted these recently with varying degrees of realism. Van Hoven et al. (1987) investigated nonlinear radiative condensation in a two-dimensional, sheared field. The sheared field suppresses parallel heat conduction near the axial plane, and condensation begins there, as expected. In their model the axial temperature falls from $10^{6.3}$ to 10^4 K in 730 Alfvénic times, or with reasonable estimates for initial conditions, in about 9 hr. Because reconnection of field lines is not included in the physics, however, the density rises only by 50%, not the factor of 10^2 that is observed.

Malherbe et al. (1984) ignored gravity, but included the effects of reconnection in their simulation. Their model develops *both* a K-S and K-R magnetic configuration, starting from a uniform field with a neutral sheet. The K-R portion is unstable, however, and erupts.

Priest et al. (1989) propose a twisted flux model (Sec. II. above; see also An 1984,1985,1986) as an alternative to the usual two-dimensional planar geometry. They show, with analytic examples, how increasing the twist in a long tube can generate dips in the field lines, which then can collect cool dense gas. The gas may either condense from the corona or be injected from the chromosphere. This schematic model has several attractive features. It is essentially three-dimensional in concept and takes into account the strong axial magnetic field component present in prominences. It sketches physical mechanisms for twisting the footpoints of the flux tube, although the time scale for at least one seems rather long—35 days. It offers a reason why prominences erupt. However, the Hα observations suggest that cool gas first appears at the footpoints, not at the center of the span between footpoints. Moreover, it is not obvious that this model results in a flat sheet of cool gas.

E. Prospect for the Future

Many basic aspects of quiescent prominence structure and evolution remain to be clarified in a physically consistent and complete manner. The problems involved are quite complicated, but the basic processes are reasonably well understood, and are likely to yield to further numerical simulations.

The subarcsec threads in prominences seem to control their thermal structure. It would be surprising if the magnetic field did not also possess a fine structure, comparable in scale and strongly associated with the fine threads. In order to guide the theorists, we need better spectroscopic observations, particularly in the EUV, and if possible, polarimetric observations with higher spatial resolution. The prospects for such improvements are reasonably good, with the Orbiting Solar Laboratory, Project THEMIS, the Large Earth-Based Telescope (LEST) and adaptive optics, all under development.

III. CORONAL MASS EJECTIONS

A. Introduction

Coronal mass ejections (CMEs) are now recognized as an important component of the large-scale evolution of the solar corona. They have been observed routinely, although not continuously, since the early 1970s by white-light coronagraphs, which measure the Thomson scattering of photospheric light by coronal ejections; relatively bright regions contain excess mass while comparatively dark regions have less mass. Distinguishing characteristics of CMEs are the appearance of new, bright, coherent features, comparable in size to the solar diameter, in the coronagraph field of view and temporal changes on time scales of minutes to hours (Hundhausen et al. 1984*a*). They originate near the coronal base, have a predominantly outward motion, and involve the addition of both mass and energy to the corona. They subsequently pass out of the coronagraph upper field of view and continue into interplanetary space. A considerably longer period (on the order of a day) is required for the corona to return to near its pre-event state.

CMEs were first observed about 15 yr ago by the Orbiting Solar Observatory (OSO-7) white-light coronagraph (Tousey 1973). Since then an appreciable data base has been accumulated with results from three subsequent orbiting instruments. The Skylab coronagraph operated during 1973–1974 and recorded 77 events (Munro et al. 1979). This was followed by the Solwind coronagraph on the P-78 satellite, which obtained in excess of 1200 CME observations during 1979–1985 (Sheeley et al. 1980). The coronagraph polarimeter on the Solar Maximum Mission (SMM) satellite recorded approximately 70 CMEs over a time period of a few months until it failed in 1980 but has been operational since its repair in 1984. The orbiting coronagraphs individually observed over different portions of the corona varying from a minimum of 1.6 solar radii (coronagraph polarimeter on SMM) out to a maximum of 10 solar radii (Solwind). This data set was supplemented by the groundbased High Altitude Observatory K-coronameter at Mauna Loa, Hawaii (Fisher and Poland 1981), which observed the corona nearer the Sun (from 1.2 to 2.0 solar radii), and by the zodiacal light photometers on the

Helios spacecraft (Jackson and Leinart 1985), which detected interplanetary transients.

Among the more recent general reviews of CMEs are those by Hundhausen et al. (1984*b*), Kahler (1987) and Hundhausen (1988). In addition, Kahler (1988) concentrated on observations and Steinolfson (1988) reviewed observations and theory relating to possible driving mechanisms. Some statistical properties of CMEs derived from large data sets have been discussed by MacQueen (1985), Howard et al. (1984) and Wagner (1984).

We begin in the following section by reviewing some representative observations of CMEs, with emphasis on more recent results, subsequently, focusing on recent observations and theory as they relate to the following aspects of CMEs: (1) the role of the waves in determining the white-light signature, and (2) the mechanism by which the CME is driven (or launched) into the corona.

B. Observations

CMEs are observed to occur in a wide variety of sizes and shapes and at various latitudes (Munro et al. 1979; Howard et al. 1984). We are primarily concerned with the subset in which the definitive leading bright signature has the appearance of a radially expanding loop. It is not unusual for such CMEs to be associated with the equatorial streamer belt, which is the streamer "blowout" classification used by Howard et al. (1985). Our motivation for emphasizing the loop-shaped CME is that (a) they are a significant fraction of the total observed (20% in the Skylab data set [Sime et al. 1984] and at least 2/3 of those in the SMM data in 1980 [Hundhausen 1988]), and (b) they are geometrically simple and, hence, are most easily studied theoretically.

A loop-like CME detected by the SMM coronagraph-polarimeter instrument is shown in Fig. 9 (Hundhausen et al. 1984*a*). CMEs are often observed to have the three-part structure displayed in this figure, that is, a preceding bright loop followed by a dark cavity containing an additional outward moving bright structure, which in this case has the appearance of a second expanding bright loop. The large dark circular shape is part of a Sun-centered occulting disk that blocks out the Sun. The CME had an outward velocity of about 260 km s^{-1}, which is near the low end of the observed speed range (from tens to thousands of km s^{-1}). The leading bright loop and following dark cavity are the necessary components of CMEs we are primarily concerned with. The second bright structure may or may not be present or detectable in individual events. During 1980, it could be detected in up to 3/4 of the loop-like CMEs seen by SMM (Hundhausen 1988); its presence and shape may be due to a solar-cycle effect.

The loop-like CMEs often appear to originate near the base of, and propagate outward through pre-existing, well-formed helmets or coronal streamers, that may be part of the equatorial streamer belt as mentioned above. The 4 loop-like Skylab CMEs studied by Sime et al. (1984) and all but 1 of the

Fig. 9. A representative loop-like CME observed by the SMM coronagraph. This image was obtained on 14 April 1980 and shows the characteristic three-part structure of a bright leading loop followed by a dark cavity containing a second bright structure (figure from Hundhausen 1988).

16 used in CME onset program (Harrison et al. 1990), originated near the base of coronal streamers. Since different images are used to identify CMEs in the Solwind data, it is difficult to determine the pre-event corona in this data set.

Coronal streamers form over polarity inversion (neutral) lines in the radial magnetic field on the solar surface. Streamers consist of low-lying closed field lines surrounded by field lines open to the interplanetary medium. The corona flows outward along these open field lines and is trapped within the closed region. Prominences tend to form within the closed-field region and also lie over local neutral lines that may be coincident with the streamer neutral line. This general picture is supported by observations indicating that CMEs can often be associated with eruptive prominences (Munro et al. 1979;

Webb and Hundhausen 1987). A schematic drawing of the initial state and the CME is shown in Fig. 10. In this simplified sketch the prominence overlies the same inversion line as the streamer. Recent observations, however, using approximately simultaneous images from the SMM coronagraph polarimeter, the MLSO Mark III K coronameter and the MLSO prominence monitor show the eruptive prominence often to be distinctly asymmetric with respect to the CME bright loop (Hundhausen 1988). This would indicate that the prominence may form over a local inversion line in the active region and

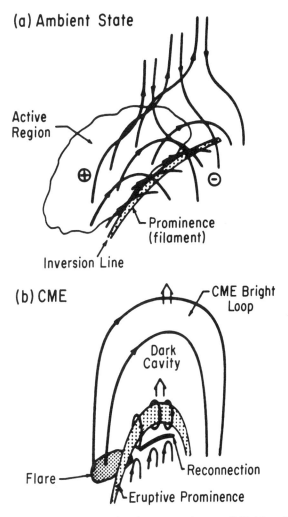

Fig. 10. A schematic drawing illustrating the corona prior to a CME (a), and the relation of various phenomena during the ejection (b).

preferentially on one side of the streamer inversion line. In addition, if a flare occurs in association with a CME, it almost invariably occurs under one leg of the expanding bright CME loop (Harrison 1986; Simnett and Harrison 1984) and generally under the loop leg closer to the equator. More will be said about the temporal spatial relation of flares, prominences, and CMEs in Sec.III.D on driving mechanisms. We note here that the picture that seems to evolve from various data sets is one in which the streamer forms in a large-scale, global magnetic structure while the prominence and flare form in smaller magnetic structures imbedded in local active regions within the global field.

The characteristic three-part structure of the CME mentioned above is indicated in the schematic drawing in Fig. 10b. The observed bright inner structure, as seen in Fig. 9, has been interpreted as the eruptive prominence, and the dark cavity as the cavity originally around the prominence. Whether or not the eruptive prominence material appears loop-like may depend partially on its orientation in the plane of the sky. Waves extended beyond the bright CME loop, particularly to the sides, may also be a portion of the entire phenomenon, as suggested in the model by Steinolfson (1985). If the waves steepen sufficiently, they may contribute to the bright signature. Simulations show that shock compression may be responsible for part of the increased brightness near the radial leading edge of the loop. Away from the bright loop, the orientation of the wave normal to the ambient magnetic field in streamers is such that the wave produces a very small compression of the corona and probably would not be detected in coronagraphs.

C. The Role of Waves in CMEs

During the early analyses of CME observations, the general consensus appeared to be that the CME (in particular, the bright loop) was magnetically controlled and driven much as in the conceptual models of Kopp and Pneuman (1976) and Mouschovias and Poland (1978). The bright loop was a result of dense material carried outward from the lower corona by the expanding magnetic field. The first self-consistent MHD simulations of CMEs, on the other hand, suggested the interpretation that the bright loop resulted from shock compression of ambient coronal plasma by a fast-mode MHD shock (Dryer et al. 1979). The shock model has subsequently been shown not to be applicable to loop-like CMEs for several reasons, but primarily because of the observed restricted lateral motion of the loop legs that directly contradicts the model (Sime et al. 1984). The expanding magnetic loop model may be conceptually correct, but this remains to be demonstrated with a quantitative self-consistent study.

Although the bright CME loop may not be entirely a result of shock compression, waves must be involved in CMEs. The corona is a compressible medium and the movement of a disturbance through it at typical observed CME velocities must generate a wave response. The type of wave response is governed to a large extent by the magnitude of the CME speed relative to characteristic wave velocities in the corona (or the velocities at which information can be transmitted through the corona). As mentioned above, CME speeds cover a wide range, but histograms (Gosling et al. 1976; Howard et al. 1985) show that typical velocities are a few 100 km s^{-1}, say 200 to 500 km s^{-1}. For a coronal temperature of 10^6 K, the sound speed is about 170 km s^{-1}. The magnitude of the coronal magnetic field is not well known, but extrapolation of interplanetary observations gives a value of about 0.2 gauss at a heliocentric distance of 3 solar radii. Using a typical number density at this location implies that the Alfvén speed is about 620 km s^{-1}, or that the plasma β is about 0.1. Typical CME speeds, then, are supersonic but sub-Alfvénic.

The relevant wave speeds are the slow, intermediate and fast speeds in the direction of the wave normal. The presence of the magnetic field, of course, introduces (in addition to multiple characteristic speeds) the further complication of making the corona anisotropic to wave propagation. For outward propagation along a radial magnetic field and for $\beta < 1$, the slow speed becomes the sound velocity and the fast and intermediate speeds become the Alfvén velocity. Hence, it is reasonable to consider the general coronal wave response in terms of the CME speed relative to the sound and Alfvén speeds.

If the CME speed exceeds the fast-mode (Alfvén) speed, one would expect formation of fast-mode MHD shocks. This is the situation studied in Dryer's simulations (Dryer et al. 1979) although the CME loop would generally not be associated with the shock-compressed plasma. Sime and Hundhausen (1987) have shown that in most cases there is no noticeable disturbance of adjacent structures prior to arrival of the bright shock. However, there are readily observable disturbances of ambient corona structures outside the bright CME loop—particularly latitudinally displaced from the loop, although not necessarily ahead of it. This behavior can be understood from the model and simulations of Steinolfson (1985) and Steinolfson and Hundhausen (1988). They show that when the CME velocity exceeds the fast-mode speed, a fast MHD shock does form and is coincident with the outermost part of the bright CME loop. At the flanks, however, the shock propagates away from the loop and produces the observable displacements of nearby ambient structures. Except near the top of the CME loop, the shock is a fast MHD shock propagating at small angles to the ambient (predominantly radial) magnetic field. The shock thus produces a very small density rise with its primary effects being to turn the ambient field and accelerate the corona parallel to the shock front. Consequently, the shock easily disturbs ambient structures without producing a detectable brightness increase.

When the CME speed is supersonic and sub-Alfvénic, fast MHD shocks do not form, but slow shocks may develop. By considering the magnetic field change across slow shocks and the expected field configuration of a CME loop, Hundhausen et al. (1987) suggested that such slow shocks would be

concave upward as shown in Fig. 11b. The situation for a preceding fast
shock is shown schematically in Fig. 11a. The authors reasoned that such a
concave shape for the slow shock would explain the flat tops of some mass
ejections, as well as the deflection of adjacent structures.

Steinolfson and Hundhausen (1988, 1989, 1990) used time-dependent
numerical solutions of the MHD equations in two-dimensions to generate
wave systems propagating outward through simplified static coronas with various prescribed field geometries, uniform thermodynamics, and no gravity.
For an initially radial field and disturbance speed larger than the sound and
less than the Alfvén velocities, a slow shock is formed as shown in Fig. 12,
but it is concave downward and is preceded by a fast-mode expansion. The
solution is symmetric about the left edge, and the driver was initiated at the

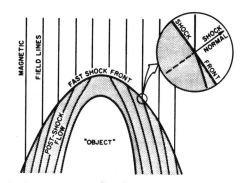

a. Spatial Configuration of a "Fast" Bow Shock

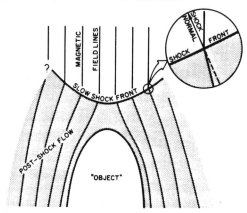

b. Proposed Spatial Configuration of a "Slow" Bow Shock

Fig. 11. The speculated geometry of the fast (a) and slow (b) MHD shocks generated as part of
the CME (figure from Hundhausen et al. 1987).

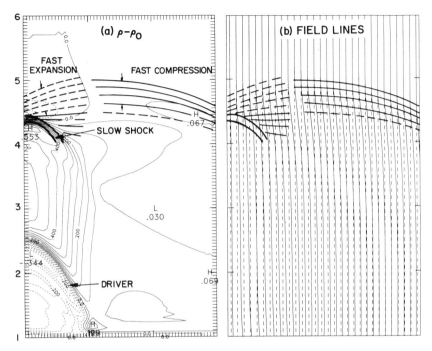

Fig. 12. Slow-shock geometry produced in MHD simulations using a vertical magnetic field and a thermal driver (figure from Steinolfson and Hundhausen 1989); (a) density structure and (b) magnetic field lines.

bottom which is taken to represent the base of the corona. The authors suggested that the slow-shock geometry may be a result of the fact that the field lines in this particular study are constrained to be essentially vertical behind the slow shock. Recent unpublished results by these authors show that a concave-upward shock is produced in the more realistic case of a magnetic driver inside a closed-field region in which case the field lines are bent away from the center symmetry axis. If the disturbance speed is increased so it slightly exceeds the Alfvén velocity, intermediate shocks are formed near the symmetry axis that merge continuously with fast shocks away from the symmetry axis. Results from the MHD Rankine-Hugoniot equations have been used to confirm the numerical results and demonstrate how intermediate shocks must be present in the flow field. The major result of this study is to demonstrate the importance of considering the global configuration and, in particular, cross-flow (perpendicular to the velocity) interactions in studying wave propagation. A multi-dimensional analysis is essential in order to allow different regions of the flow field to communicate and interact with each other.

D. Driving Mechanisms

Observations. A recently derived result is that the flare impulsive phase follows the CME onset by a relatively long time period. The time delay is long enough that any energy release in the flare cannot be solely responsible for driving the CME outward. This was convincingly demonstrated by combining data from the hard X-ray imaging spectrometer (HXIS) on SMM with coronagraph-polarimeter data (Simnett and Harrison 1984; Harrison 1986). Some results from this study are shown in the schematic drawing in Fig. 13, where the line labeled CME locates the leading edge of the bright CME loop. Extrapolation of the CME trajectory back to the surface with no acceleration shows that the CME onset coincides with a weak precursor some tens of minutes prior to the flare impulsive phase. In addition to providing the time sequence of events, the imaging capability of the X-ray instrument locates the flare site in one of the footpoints of the large magnetic arch that

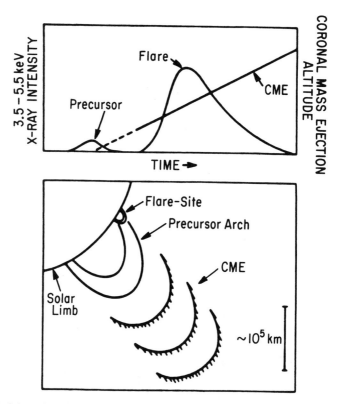

Fig. 13. Schematic drawing of the temporal and spatial relation between the precursor, flare impulsive phase, and the CME bright loop (figure from Harrison 1986).

brightens in X rays as the precursor. This asymmetry also argues against the flare as the driving mechanism.

Other simultaneous data sets have been used to determine the possible role of the eruptive prominence (filament) in the CME phenomena. Kahler et al. (1988) used Hα data from the Big Bear Solar Observatory to locate the filament and hard X-ray emission from instruments on the ISEE spacecraft for the flare impulsive phase. They were able to show that the filament eruption begins several minutes before the impulsive phase and that there is no appreciable change in the filament motion during the flare. They concluded that the filament eruption was not driven as a result of any pressure pulse associated with the impulsive flare, thereby corroborating the relatively passive role as found by Harrison and others of the flare in driving CMEs.

Additional studies combining coronagraph and Hα data have examined the relative outward motion of CMEs and erupting filaments (Wagner 1983; Illing and Hundhausen 1986). The main results are that the CME and prominence begin moving outward at approximately the same time although the CME (bright loop) velocity exceeds that of the erupting prominence.

Observations do not support either the flare impulsive phase or the eruptive prominence as likely candidates for propelling CMEs outward. On the other hand, observations do suggest that the initiating agent may be a loss of equilibrium in the large-scale (global) magnetic field configuration. That is, the global field may slowly be stressed to the extent that it can no longer remain in equilibrium. In this picture, the CME, flare-impulsive phase and eruptive prominence are all secondary effects resulting from the nonlinear evolution as the corona adjusts to a new global equilibrium.

Some support for the gradual buildup to a loss of equilibrium hypothesis can be seen in synoptic maps derived from SMM coronagraph data. Daily averaged observations at a fixed height above the limb are shown in Fig. 14. Noting that time runs from right to left in this presentation, one can see several examples of a gradual broadening of the bright region over several days (e.g., from day 310 to day 317 on the east limb) followed by an abrupt reduction in brightness. The broadening signal is due to a gradual expansion of a coronal streamer, and a CME is observed to occur on the day in which there is a sudden decrease in brightness, the interpretation being that the streamer configuration is slowly stressed to the extent that it loses equilibrium and generates a CME. The sudden disappearance of the bright signature following the CME suggests that the magnetic fields have reconnected.

Models: Theory and Simulation.   The observations have established the spatial and temporal relation of various individual structures identified by their unique emission characteristics during a mass ejection. The different emission characteristics occur for quite different physical (primarily thermodynamic) conditions. Although the magnetic field cannot be directly mea-

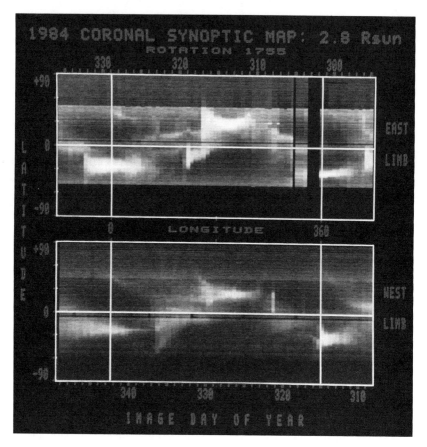

Fig. 14. Synoptic map produced from the SMM coronagraph observations showing daily averaged brightness at 2.8 solar radii and at both limbs (figure from unpublished results supplied
by A. J. Hundhausen).

sured in the corona, it is generally accepted (and supposed by indirect inference) to be an important, if not dominant, contribution to the CME initiation
and propagation. Some of the analytic and numerical models that have been
used to try to understand the physical processes in these phenomena are discussed briefly below. A common, and necessary ingredient in all models is
the magnetic field.

The first numerical simulations of CMEs used a local increase in thermal
pressure as the driving mechanism. Such a driver was based on the previous
view that energy released during the flare impulsive phase may become available to create the CME. Although such a driver is not now supported by the
observation, studies using thermal drivers have been useful in studying the

coronal response to an outward propagating disturbance. They have demonstrated the significant influence of the physical conditions (magnetic configuration, flow velocity, thermodynamics) in the ambient corona on the CME. In fact, all of the observed characteristics of loop-like CMEs discussed by Sime et al. (1984) have been reproduced by simulations of a CME initiated by a thermal driver at the base of a coronal streamer in a heated corona (Steinolfson and Hundhausen 1988). The initial state of the corona is probably the primary factor in determining the brightness signature rather than the details of the driver.

The only studies using magnetic, rather than thermal, forces to drive CMEs are limited to a single radial dimension and neglect interactions with the surrounding atmosphere (see, e.g., Mouschovias and Poland 1978). Such models demonstrate that radially unbalanced magnetic pressure and force can, for not unreasonably parametric values, produce an outward motion comparable to that observed in some CMEs.

However, observations suggest that the entire phenomena is initiated by a loss of equilibrium in the global magnetic field following a comparatively slow evolution to a highly stressed state. One way in which such a nonpotential stressed state could arise is as a result of slow evolution of the coronal field due to photospheric motion of the footpoints. The response of the coronal field to photospheric shear motion has been studied analytically for many years (see, e.g., Low 1982a,b; Birn and Schindler 1981). These simplified models generally consider only force-free magnetic field evolution and neglect the interaction of the field with the coronal motion and thermodynamics. The usual approach is to find sequences of equilibrium solutions for a given form of a generating function relating the sheared field to a parameter α. As α is changed monotonically, a value is reached beyond which there is no equilibrium solution, and this point is then identified with the onset of a more rapid evolution. The major problem with these studies is that the solutions do not uniquely relate to footpoint motion. A multiplicity of solutions for given α led Jockers (1978) to suggest that in an actual situation the critical value of α may never be achieved. This latter view is supported by recent numerical simulations (Klimchuk and Mariska 1988; Biskamp and Welter 1988). The study by Klimchuk et al. also assumed force-free evolution as in the analytic studies. It should be pointed out that their results only demonstrate that at least one equilibrium solution can be found beyond the critical shear. A more complete study would consider all possible sheared states beyond the critical value. Biskamp and Welter include more physics in their model (magnetic field-flow interaction) and, therefore, the analytic results for force-free evolution may not apply. In addition to showing the continued slow evolution of a single magnetic arcade beyond analytically predicted stable limits, Biskamp and Welter showed that shearing of at least 3 adjacent arcades was necessary to give rise to an ejection.

E. Discussion

The potential role of waves in the CME has been clarified to some extent by numerical simulations. More convincing evidence of their presence and contributions would be provided if brightness signatures attributed solely to the waves could be identified in the observations. We have already mentioned that the disturbance of adjacent structures seen in this data is probably due to a fast-mode MHD wave, that produces an undetectable density rise. Of more current interest, then, would be to clarify their effect nearer and perhaps ahead of the bright CME loop. For most typical CME velocities, it was shown how an expansion wave precedes the slow (or intermediate) MHD shock near the outer edge of the central portion of the loop. Such an expansion would produce a density depression and reduction in brightness ahead of the bright shock compression. In addition, if the compression near the outer leading edge is produced by a concave-outward slow shock or by an intermediate shock, the outer central portion of the loop would be depressed. Such flat tops have been seen in some CMEs. However, a more quantitative and thorough comparison with the data is needed to clearly establish the role of the MHD shocks.

Most of the available observational evidence implies that a loss of equilibrium in the global magnetic field initiates the complex phenomena collectively referred to as a CME. The precise nature of the driving mechanism and, more specifically, how an unstable, stressed magnetic configuration nonlinearly evolves to a stable configuration at lower energy is poorly understood. Intimately involved with this issue is the question of how the field loses equilibrium, and whether it occurs primarily in the open-field region of a streamer or in the underlying closed field, that tends to be more highly stressed. Simplified analytic studies will continue to provide useful guides in determining equilibrium magnetic field configurations. Unfortunately, the limited physics that can be included in the analytic models raises the question of how well they apply to a realistic corona. Additional driving mechanisms, such as emerging flux due to magnetic buoyancy (the nonlinear Parker instability), and energy conversion by field reconnection should also receive continued theoretical and numerical study.

IV. WAVES IN THE SOLAR CORONA

The words "waves" and "coronal heating" have been tightly connected for 40 yr. Bierman (1948), Schwarzschild (1948) and Schatzman (1949) suggested that small-amplitude acoustic waves generated in the convective zone would transport energy from those sources to the upper hot layers of the solar atmosphere. Although it is now known that the 5-min oscillations are acoustic waves or, more strictly, p-modes, plane acoustic waves cannot penetrate into the solar corona. Coronal heating is impossible without dynamical processes in the solar atmosphere, and the magnetic field plays a key role, because the heating of the chromosphere-corona takes place only over magnetically enhanced regions.

A. Acoustic Wave Propagation and Reflection

Wave propagation in a stratified atmosphere has some peculiarities. The first is the existence of a cut-off frequency, which is manifested as a wave reflection after approaching a definite layer in the atmosphere. The main barrier for acoustic waves in the solar atmosphere is the temperature minimum, which is transparent only for the waves with periods shorter than 2 min. Strong reflection of acoustic waves from the chromosphere-corona transition layer is also present. Additionally, the acoustic waves transform into shocks on their path through the chromosphere and, as a result, dissipate rapidly, decreasing the energy flux transmitted to the corona.

Many attempts to calculate the acoustic flux into the corona depend on uncertain boundary conditions on the wave spectrum upgoing from the convective zone (Ulmschneider 1979). The problem was solved when an acoustic flux estimation, based on observations in the upper chromosphere, were made possible. The measurements of wave amplitudes and phase shift have shown that acoustic heating of the solar corona is impossible (Athay and White 1978).

Let us estimate the energy flux in the corona with a simple formula:

$$I = \rho v^2 C \tag{6}$$

where ρ is density, v is wave amplitude velocity and C is sound velocity. The mean turbulent velocities deduced from coronal lines do not exceed 25 km s^{-1}. Let us suppose that the broadening of spectral lines is entirely due to waves. With this assumption and taking for the particle density (in the lower corona) 10^9 cm^{-3}, the energy flux deduced from Eq. (6) is about 10^3 erg cm^{-2} s^{-1}. That is 10^3 times smaller than the flux needed for coronal heating. It follows from this that magnetic fields must be taken in account and hydromagnetic waves be considered because the sound velocity into Eq. (6) can be exchanged for Alfvén velocity.

B. Hydromagnetic Waves: Propagation and Transformation

The solar corona is governed by magnetic field lines. Consequently, any mechanism for coronal heating must include energy channeling along field lines. Moreover, the empirical relation between the maximum loop temperature, loop length and pressure inside the loop derived by Rosner et al. (1978) must be explained by a coronal heating mechanism. The channeling of the energy along field lines can be made by Alfvén waves, but it should be kept in mind that pure Alfvén waves are practically impossible in the solar atmosphere. The magneto-hydrodynamic waves in a stratified conducting atmosphere permeated by the magnetic field are named magneto-atmospheric waves. These waves originate from three restoring forces: pressure, magnetic forces and gravity. Their main property is the linear interaction of the slow and fast modes of acoustic and gravity waves (Zhugzda and Dzhalilov 1982). In many cases, the magneto-atmospheric waves are an indivisible mixture of modes.

The acoustic waves generated in the convective zone can be transformed into fast and Alfvén modes on their path to the chromosphere and the corona due to linear interactions in a stratified atmosphere. This interaction is not dependent on the waves' amplitudes and arises instantly. There is a strong reflection of waves on the path between the photosphere and the corona due to differences in Alfvén velocities. The ratio of Alfvén velocities in the corona and the photosphere is about 10^5 to 10^4 due to the decrease of density. The wave flux of Alfvén waves must decrease by this factor. This is a very serious restriction because we need a wave flux in the photosphere of the same order of magnitude as the radiation flux through the photosphere. This is impossible. This restriction concerns not only pure Alfvén waves. All modes of magneto-atmospheric waves with the exception of slow modes are subjected to strong reflection of the same order because their phase velocity is tightly connected with Alfvén velocities. This conclusion is true for waves in flux tubes including surface waves (Ionson 1978).

Zhugzda and Locans (1982) and Hollweg (1981) independently discovered a phenomenon that possibly is an effective mechanism for pumping waves toward the corona, i.e., the propagation of Alfvén waves along a coronal magnetic arch. Such an arch, which has its legs in the chromosphere, acts as an interference filter for Alfvén waves. The interference of waves reflected from different ends of the coronal arch significantly changes the coefficient of wave transmission into the corona for the resonance frequencies of the arch. The boundary between the corona and the chromosphere serves as a mirror with a high reflection coefficient. In an Alfvén filter, the space between mirrors is not a plane parallel slab with constant "refractive index" as for an optical filter.

The central frequencies of the interference filter correspond to the case when the wavelength is a multiple of the double distance between the mirrors. In this case, reflected waves interfere and cancel each other, while the amplitude of transmitted waves increases. The first calculations of Zhugzda, Locans and Hollweg demonstrate that 100% transmission of waves from the photosphere into the corona is possible. However, this is inconsistent with the restrictions on wave amplitudes in the corona, because high transmission corresponds to large-amplitude resonance oscillations of the arch. Moreover, in this case wave flux is only transmitted through the corona back to the chromosphere, because absorption is absent. An Alfvén coronal filter with absorption was considered by Zhugzda and Locans (1982). The calculations demonstrate that the effective wave absorption in the filter is possible when a definite relation between the reflection coefficient of the mirrors and the damping decrement of waves is considered. As a rule, the damping of Alfvén waves is negligible due to the linearity of the process, and most of the wave energy must pass through the Alfvén coronal filter back to chromosphere and photosphere.

A coronal interference filter can arise for other waves, when the conditions of wave channeling along a coronal arch and of strong reflection at its legs are fulfilled. In a strong magnetic field, the slow waves propagate along field lines with acoustic velocity. If there is an Alfvén velocity discontinuity at the boundary of a coronal arch, the channeling of surface waves is possible. The second condition of a strong wave reflection is caused by the temperature and density jumps on the chromosphere-corona boundary. Thus, many wave modes can propagate in the flux tubes of a coronal arch.

Interference filters start to be effective after a short time. The transient process may be qualitatively described by a multiple wave reflection in an arch. Upon the second reflection from both ends of the arch the wave again interferes with the one reflected from the near end and reduces the total reflected wave flow, thereby the transmission coefficient increases. In this manner, each subsequent reflection increases the transmission of the waves to the corona. The transient time is about $Q \cdot P$, where Q is the quality factor and P is the period of oscillations; for Alfvén filters in the corona, it may be several hours or longer. The transient time is therefore comparable with the evolution time of physical parameters of the arch and the coherence time of incident oscillations. Consequently, the resonance oscillations of arches are constantly destroyed by arch evolutionary variations along with phase variations of upgoing waves. Transient processes are possible because of the poor efficiency of resonance phenomena in the corona.

C. Waves in the Corona: Observations and Interpretations

Resonance oscillations can be excited by 5-min oscillations. In a strong magnetic field, acoustic waves are channeled along field lines. The length of the resonant arch must be about $(n + 0.5) Pc$, where P is the period, c is the acoustic velocity and $n = 0, 1, 2. \ldots$ For 5-min oscillations in the corona Pc is about 30,000 km. Consequently, an acoustic resonance must arise, but a shock forms as a result of effective damping and destroying of the phase coherence. Thus, acoustic resonance in a coronal arch is questionable, and special investigations are needed.

The 5-min oscillations in the solar corona were observed by Tsubaki (1975, 1977) and Koutchmy (1981). These observations indicate the absence of oscillations in intensity fluctuations and their presence in velocity fluctuations. They contradict the model of acoustic resonance in arches because acoustic waves must generate intensity fluctuations and cannot generate displacements perpendicular to the plane of the arch as observed on the limb.

The 5-min oscillations in a coronal arch cannot be resonance Alfvén oscillations of the arch because it is difficult to construct an arch model with such a
low resonance frequency (Zhugzda and Locans 1982).

Koutchmy (1981) observed arch oscillations with periods of 300, 80 and
43 ± 2 s. The period of 43 s is the 7th harmonic of 5-min oscillations.
Koutchmy et al. (1983) assumed that the observed coronal arch resonates on
Alfvén oscillations with periods of 80 and 43 s. The oscillations were considered as pure Alfvén waves because the influence of gravity can be neglected
for horizontal displacements of the oscillating elements.

The height of the coronal arch above the photosphere is equal to 5×10^4 km, the distance between feet and the photospheric level is 3.3×10^4
km. They assumed that the magnetic field in footpoints is 1500 gauss and 4.4
gauss at the top, the temperature being 1.8×10^6 K and the concentration at
the lowest coronal level being 2.5×10^8 cm^{-3}. The resonance periods of
oscillation for this arch are near the observed ones of $P = 84.5$ s; $P_n = 86/n$, $n = 2, 3. \ldots$ The most striking effect is that nonlinear interactions between 5-min and Alfvén oscillations take place with a period of 43 s. The
estimation of the heating by Alfvén waves is difficult due to uncertainty in
the value of the damping coefficient. Observations at two heights are needed.

D. Discussion

The small value of the observed coronal velocities contradicts the idea
of a dynamical heating of the solar corona. Observation of coronal dynamics
is an extremely difficult problem. Available information about coronal waves
is rather poor. Fine structure of the corona can influence the wave propagation
and may hide dynamical phenomena from the present observations. Finally,
we notice that other mechanisms have been proposed (Parker 1988e) to heat
the corona, based on the shaking and twisting by photospheric motions of
magnetic field lines and dissipation at the location of tangential discontinuities. Obviously, high spatial resolution observations are needed to resolve this
problem.

Acknowledgments.   S. Koutchmy wishes to thank the AFGL and NSO/
SP for support during most of the period in which this chapter was written.
The work of K. Lang is greatly appreciated for improving the manuscript.
R. S. Steinolfson thanks the High Altitude Observatory for its hospitality.

THE SOLAR WIND AND ITS CORONAL ORIGINS

GEORGE L. WITHBROE
Harvard-Smithsonian Center for Astrophysics

WILLIAM C. FELDMAN
Los Alamos National Laboratory

and

HARJIT S. AHLUWALIA
University of New Mexico

Coronal holes are the most well-established coronal source of steady-state solar wind, the high-speed solar wind streams which typically have asymptotic flow speeds of 700 km s^{-1}. Coronal mass ejections associated with flares and/or eruptive prominences are another clearly identified source of solar wind, a transient component that accounts for approximately 5% of the total solar-wind mass loss. The role of other coronal structures in the generation of the solar wind is less clear. Streamers and the interfaces between streamers and other coronal regions are likely sources of low-speed wind ($\bar{V} \approx 330$ km s^{-1}), a hypothesis consistent with existing observations, but measurements of mass flows in these features are needed to confirm this. These measurements can be obtained by new instruments under development. Small-scale dynamical phenomena observed at the base of the corona (spicules, macrospicules and high-speed jets) and small-scale structures observed in polar coronal holes (polar plumes) may or may not play a significant role in supplying mass, momentum and energy to the solar wind. Improved measurements are required to determine the role, if any, of these small-scale structures in the generation of the solar wind.

I. INTRODUCTION

The plasma in the outer solar corona is sufficiently hot that it is not bound by the Sun's gravitational field, but flows outward into interplanetary space forming the solar wind. In the low corona the ratio of the gas and magnetic pressures β is less than unity and, as a result, most of the surface, ~80%, appears to be covered with closed magnetic fields. The remainder of the surface has open magnetic configurations where only one end of each magnetic field line is attached to the surface. Several solar radii above the solar surface, the magnetically open regions expand to occupy the entire volume and the solar-wind flow becomes predominately radial. Figure 1 shows a schematic view illustrating this. The most prominent magnetically open regions are large unipolar magnetic regions known as coronal holes. These regions are sources of steady-state high-speed solar wind streams with characteristic speeds of 700 km s^{-1}. The other primary component of the solar wind is the low-speed wind, which typically has velocities of about 330 km s^{-1}. The source of the low-speed wind has not been unambiguously identified, but is likely to originate in open magnetic field regions associated with streamers or the interfaces between streamers and other coronal regions. Coronal mass ejections provide a third type of solar wind, which has a wide range of speeds and physical conditions and is often associated with interplanetary shocks.

Section II summarizes the physical conditions in the various components

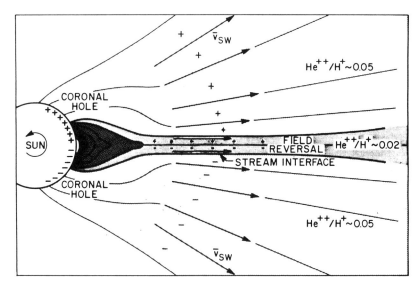

Fig. 1. One possible simple phenomenological model for a coronal hole, streamer and streamer/coronal hole interface (figure from Gosling et al. 1981).

of the solar wind. Sections III, IV and V describe, respectively, coronal mass ejections, coronal holes and streamers as sources of solar wind. Section VI discusses the possible role of small-scale structures in the generation of the solar wind; Sec. VII discusses heating and acceleration of the solar wind; and Sec. VIII discusses heliospheric structure. Finally, Sec. IX, summarizes our knowledge of the origin of the solar wind.

II. PHYSICAL CONDITIONS IN SOLAR-WIND FLOWS

In order to connect specific solar-wind flows with their coronal origins, attention must be confined to radial distances inside of about 1 AU. Beyond 1 AU, stream-stream interactions induced by solar rotation strongly perturb ambient conditions leading to a new flow state that has lost all memory of its coronal origins (see the review by Burlaga 1988, and references therein).

Solar-wind flows inside of 1 AU can be separated into one of three categories: the high-speed, slow-speed and transient solar wind. This categorization was suggested on the basis of lengthy statistical and time-sequence studies of all measurable flow parameters at, and inside of 1 AU. In particular, those flows having bulk speeds $\gtrsim 600$ km s^{-1} were singled out as representing a relatively uniform, structure-free state (Bame et al. 1977; Feldman et al. 1986). Flows having speed $\lesssim 400$ km s^{-1} are characterized by a distinct, albeit heterogeneous flow state (see, e.g., Hundhausen 1972, and references therein), that is most clearly recognizable as a distinct class from a statistical survey of flow speeds in the inner solar wind (Rosenbauer et al. 1977). This classification is reinforced by the fact that interfaces between slow and fast streams steepen with decreasing heliocentric distances.

Lastly, transient flows resulting from solar flares (and, generally, coronal mass ejections) were first recognized by the characteristic sequence of time-dependent flow states observed to follow interplanetary shock waves at 1 AU (see, e.g., Hundhausen 1972, and references therein). This identification was later generalized to include a host of unusual solar-wind parameter excursions found first to occur within solar-wind shock disturbances, but later also found when shocks are not observed. Salient features of these special conditions include (1) a high helium abundance; (2) an unusually low electron and proton temperature; (3) bistreaming suprathermal electron populations; (4) unusually high ionization states of heavy-ion populations; and (5) magnetic clouds (Hirshberg et al. 1972; Montgomery et al. 1974; Bame et al. 1979,1981; Klein and Burlaga 1982; Zwickl et al. 1983; Nuegebauer 1988). A new addition to this list may include observation of slow-mode shocks (see, e.g., Hundhausen 1988) although such shocks have been observed only rarely beyond 0.3 AU (Richter 1988).

Bulk flow parameters characterizing low- and high-speed solar wind have been tabulated in Feldman et al. (1977), Feldman (1981) and Schwenn (1983). Rather than devote space here to an enumeration of all these charac-

teristics, we refer the reader to the foregoing reviews and include Table I from
Feldman et al. (1977) for completeness here. We concentrate in the remainder
of this section on those characteristics that specifically help identify the suite
of coronal conditions and associated physical mechanisms that shape the
solar-wind outflow.

III. CORONAL MASS EJECTIONS

Identification of arbitrary solar-wind volumes in interplanetary space is
difficult because many of the flow characteristics measured far from the Sun
have changed significantly from their corresponding values low in the corona.
The first successful identification coupled isolated, large-amplitude shock-
wave disturbances with major, isolated solar flares (see, e.g., Hundhausen
1972; Schwenn 1983). Although many pairs of flare-shock events were obviously coupled by virtue of their spatial and temporal alignments, many
defied a unique pairing because the Sun was too active. Even so, many shock-
wave disturbances were known to be generated in high-temperature solar transients, because they contained anomalously high ionization states of iron
(Bame et al. 1979; Fenimore 1980; Galvin et al. 1984; Ipavich et al. 1986).

The shock-wave disturbances in the solar wind which have been associated with solar flares are a subset of phenomena caused by coronal mass
ejections (CME's), which appear to be the primary transient coronal source
of solar wind. CME's are often associated with solar flares and/or eruptive
prominences. Figure 2 shows photographs of CME and eruptive prominence
phenomena. These phenomena appear to be caused by changes in the global
magnetic field (see the review by Hundhausen 1988). During CME's, sub-

TABLE I
Characteristics of Various Types of Solar-Wind Flows at 1 AU[a]

Parameter	Average		Low Speed		High Speed	
	Mean	σ	Mean	σ	Mean	σ
N (cm^{-3})	8.7	6.6	11.9	4.5	3.9	0.6
V (km s^{-1})	468	116	327	15	702	32
NV (10^8 cm^{-2} s^{-1})	3.8	2.4	3.9	1.5	2.7	0.4
Total Energy Flux (erg cm^{-2} s^{-1})	2.05	0.9	1.68	0.5	2.15	0.2
T_p (10^5 K)	1.2	0.9	0.34	0.15	2.3	0.3
T_e (10^5 K)	1.4	0.4	1.3	0.3	1.0	0.1
T_α (10^5 K)	5.8	5.0	1.1	0.8	14.2	3.0
$\langle \delta V^2 \rangle^{1/2}$ (km s^{-1})	20.5	12.1	9.6	2.9	34.9	6.2
$(N_\alpha V_\alpha)/(N_p V_p)$	0.047	0.019	0.038	0.018	0.048	0.005
$B(\gamma)$	6.2	2.9	6.2	2.9	6.2	2.9

[a]Data from Feldman et al. 1977.

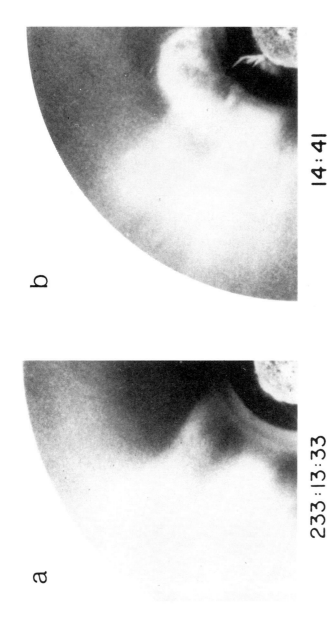

233 : 13 : 33   14 : 41

Fig. 2. (a) Eruptive prominence (near solar limb in He II λ304 image) and (b) large coronal mass ejection (outer white light coronagraph image) observed by Skylab telescopes (figure courtesy Naval Research Laboratory and High Altitude Observatory).

stantial amounts of coronal material, typically a few times 10^{15} g, are ejected from the Sun. CME's appear to occur most frequently at solar maximum at a rate of approximately 2 per day on the average. At solar minimum, the rate appears to be a factor of about 6 smaller (Sheeley et al. 1986b; Howard et al. 1986). The speeds typically range between 200 and 1000 km s^{-1}. Comparisons of coronagraphic observations by the P-78 spacecraft and *in situ* measurements of the solar wind by Helios indicate that coronal mass ejections are often associated with interplanetary shocks (see the review by Schwenn 1986). Although coronal mass ejections are clearly an important source of the variable component of the solar wind, they contribute only about 5% of the solar-wind mass loss from the Sun. For more information on these dynamic coronal phenomena, see reviews by MacQueen (1980), Dryer (1982), Wagner (1984), Sheeley et al. (1986b), Hundhausen (1988) and Kahler (1988).

IV. CORONAL HOLES

Coronal holes were the first coronal regions unambiguously identified as sources of steady-state solar wind. An example of a large coronal hole is shown in the X-ray photograph presented in Fig. 3. Coronal holes are regions with open magnetic field configurations where the field lines extend from the

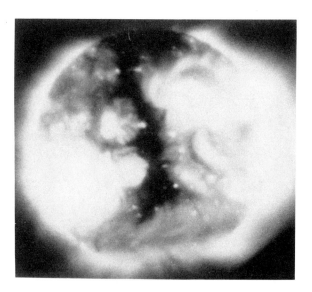

MAY 31, 1332 UT

Fig. 3. Skylab soft X-ray photograph showing a large coronal hole extending from the north pole into equatorial regions (figure courtesy American Science and Engineering and Harvard College Observatory).

solar surface out into interplanetary space. Since reliable measurements of coronal magnetic fields in coronal holes do not exist, it has been necessary to infer the magnetic structure from observations of coronal structures in, and adjacent to, coronal holes (see, e.g., Munro and Jackson 1977) and via calculations of coronal fields from measurements of magnetic fields at the photospheric base of the corona (cf., Levine 1977,1982). Figure 4 illustrates the large-scale configuration of the calculated coronal magnetic field. Only magnetic field lines extending into interplanetary space are plotted. We see that

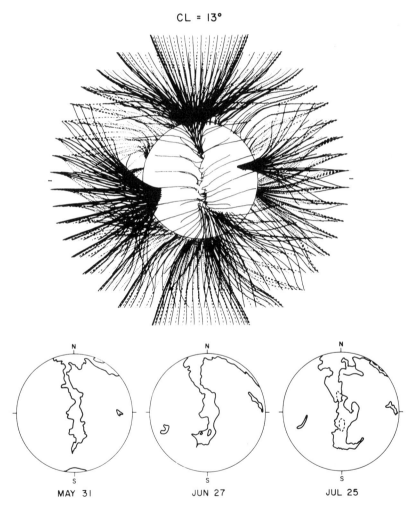

Fig. 4. Comparison of calculated magnetic field configuration for field lines swept out by solar wind (upper part of figure) and contour maps (lower part of figure) giving locations of coronal holes determined from X-ray photographs (figure from Levine 1977).

the footpoints coincide with the locations of coronal holes determined from X-ray data. It must be emphasized that such calculations *assume* a potential field (no electric currents in the corona) and thus are only approximations to the actual fields. The calculated field configurations also depend upon the assumed location of the source surface (where the plasma outflow and field lines become radial). Hence, they may not indicate the origins of all the open field lines.

In 1973, Krieger et al. traced a recurrent high-speed solar wind stream back to the Sun assuming constant velocity and found that it mapped into the location of an equatorial coronal hole. Subsequently, the connection between recurrent high-speed streams and corona holes was solidified using *in situ* and radio measurements of solar-wind speeds far from the Sun and optical observations of coronal holes from OSO-7, Skylab and the ground (cf., reviews by Zirker 1977, 1981; Rickett and Coles 1983; Coles and Rickett 1986; and references cited therein). These studies made use of the fact that many coronal holes, especially during the declining phase of the solar cycle, are long-lived, lasting many solar rotations, and thus can be clearly related to recurrent high-speed solar-wind streams. Plots of solar-wind speed measured *in situ* which are time-shifted to correct for transit from the Sun to 1 AU assuming constant velocity, showed a one-to-one overlay of high-speed wind with the central meridian passage of equatorial-crossing coronal holes. Since then many lines of evidence have reinforced this association. Most convincing is the coincidence between global maps of high-speed flows using radio scintillation observation of compact radio sources, and those of the size and shape of polar coronal holes (Coles et al. 1980; Kojima and Kakinuma 1987). Reinforcement also came from identification of (1) the large unipolar magnetic sectors that fill individual high-speed flows with similar unipolar sectors that fill coronal holes (see, e.g., Hundhausen 1977; Sheeley and Harvey 1981); and (2) of the electron temperature in the upper coronal region inferred from observed properties of the solar wind beyond 0.3 AU (Feldman 1981; Galvin et al. 1984; Ipavich et al. 1986; Pilipp et al. 1987) with the electron temperature observed within coronal holes in the low solar corona (Munro and Withbroe 1972; Vaiana et al. 1976; Mariska 1978).

Current knowledge concerning relationships between coronal holes and the solar wind can be summarized as follows: There is strong evidence that both equatorial and polar coronal holes are sources of high-speed solar wind. However, they may not be the only nontransient source, as there is limited evidence that other regions can be sources of high-speed wind (see, e.g., Nolte et al. 1977; Burlaga et al. 1978; Sheeley and Harvey 1981). Our knowledge concerning the details of the coronal magnetic structure inside and around coronal holes is limited, particularly concerning the locations of the footpoints of the open field lines. Finally, we still do not know how the coronal plasma in these regions is heated and accelerated to form the solar wind; the coronal heating mechanism is unknown and there are uncertainties as to the role of wave-particle interactions in accelerating the solar wind (see Sec. VII).

V. STREAMERS

Association of low-speed solar-wind flows with their coronal footpoints has proven most difficult. However, there are several independent types of measurements which strongly support an origin in or near streamers. Figure 1 shows a schematic drawing of a streamer. Streamers are the prominent bright features in photographs of electron-scattered white-light emissions from the corona (e.g., Fig. 5). The magnetic structure of streamers has been inferred from (1) comparisons between observed coronal structures and magnetic field configurations calculated from measurements of photospheric fields; and (2) the finding that streamers overlie neutral lines where the large-scale photospheric field reverses sign. The inner regions within about 1.5 solar radii are believed to have closed magnetic fields, while the surrounding outer envelope and the nearly radial extensions are believed to be regions with open magnetic fields where the solar wind carries the field lines into interplanetary space (see Fig. 1).

Over the years a number of individuals have suggested that streamers and the interface areas between streamers and other regions are a likely source of the low-speed solar wind. The strongest evidence for this is that magnetic

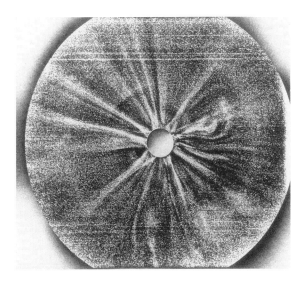

Fig. 5. Photograph of white light corona obtained at 1980 solar eclipse (figure courtesy Los Alamos Scientific Laboratory).

polarity reversals detected in the solar wind trace back to the locations of streamers (Gosling et al. 1981). In addition, polarity reversals are often associated with maxima in the proton (and hence electron) density as would be expected. Another type of evidence supporting an association between low-speed wind and streamers is the coincidence between measured ionization state temperature of oxygen and iron in the slow-speed wind with the measured scale-height temperature of coronal streamers (Feldman et al. 1981; Newkirk 1967). Additional strong evidence is the demonstrated long-term overlap of the constant velocity projection coronal maps of slow-speed wind using interplanetary scintillation measurements with maps of the solar magnetic neutral line between 1973 and 1985 (Kajima and Kakinuma 1987; see also Rickett and Coles 1983; Coles and Rickett 1986). The results of the last two methods also show that some of the observed low-speed flows could have originated in the magnetically open parts of the predominantly closed quiet corona.

Simple phenomenological models for streamers and the streamer/coronal-hole interface (e.g., Fig. 1) account for many of the observed features of the solar wind, including wind speed (high in coronal holes, low in streamers and the interface region), proton density (low in coronal holes and high-speed wind, high in streamers and low-speed wind) and magnetic field configuration (unipolar in coronal holes, opposite polarities separated by a neutral sheet in streamers). Gosling et al. (1981) also suggest that the low-speed wind originating from streamers has low helium densities.

Although streamers are a likely source of the low-speed solar wind, there is much we do not know. We do not know the relative contributions of streamers and the interfaces between streamers and other regions. Nor do we know with any certainty the surface origins of the open field lines associated with streamers and these interfaces. Do these field lines sometimes terminate in active regions as suggested by some calculations of coronal magnetic fields (see, e.g., Levine 1977,1982), observations of the location of Type III radio sources (Dulk et al. 1986), and the appearance of magnetic field configurations inferred from X-ray photographs of active regions (see, e.g., Svestka et al. 1977)? Do they terminate primarily within the boundaries of coronal holes or in "quiet" regions of the corona (regions other than coronal holes and active regions)? What lies at the footprint of the interface between two streamers? Are there regions where open and closed magnetic fields are mixed (cf., Kahler et al. 1983)? In short, there is much to be learned about the origins of the low-speed solar wind and the role that streamers have in its generation. Answers to some of the above questions can be provided by remote sensing instruments, such as those under development for the Solar Heliospheric Observatory (SOHO), which is scheduled for launch in late 1995, and by *in situ* observations of the solar wind close to the Sun, such as will be afforded by the proposed Solar Probe.

VI. FINE-SCALE PHENOMENA AND STRUCTURES

Up to this point we have been concerned with large-scale coronal structures such as coronal holes and streamers. It is possible that smaller-scale structures may play a significant role in the generation of the solar wind. Until the role, if any, of these structures is understood, our understanding of the physics of solar-wind generation will be incomplete, perhaps incorrect. Figure 6 illustrates some of these features in schematic fashion. The dark shaded area represents the base of a coronal hole which is a large-scale magnetically open structure. Within the coronal hole are small magnetically closed regions with overlying regions of higher-than-average density known as polar plumes (see, e.g., Saito 1965; Bohlin 1977a) which most likely have open magnetic configurations similar to streamers (see Fig. 1). (The plumes and loops are drawn several times wider than scale in the illustration.) Plumes can contain 10 to 20% of the coronal mass in a polar coronal hole (Ahmad and Withbroe 1977). Do they contribute a comparable amount of mass to the solar wind? To be more specific, what is the relative contribution to the mass outflow of

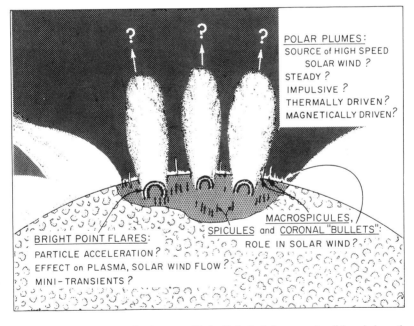

Fig. 6. Schematic drawing of polar coronal hole (dark shaded area on the disk and above the limb) with loops in small bipolar regions (arches) and overlying polar plumes which extend far above the surface. (The loops and the plumes are several times wider than actual scale for this illustration.) Also illustrated as short vertical lines are various small-scale jets known as spicules, macrospicules and coronal bullets or high-energy jets (see text).

the plumes and interplume regions? Plumes often have coronal bright points at their bases, features that usually mark the location of small-scale bipolar magnetic regions (cf., Golub et al. 1974; Ahmad and Withbroe 1977; Ahmad and Webb 1978). Coronal bright points found in coronal holes appear to be heated impulsively by a stochastic process, perhaps by rapid magnetic reconnection (Habbal and Withbroe 1981; Habbal et al. 1990). Are the overlying polar plumes a source of varying solar wind? Is this wind driven by thermal or magnetic forces (Mullan and Ahmad 1982)? Coronal bright points often flare (exhibiting rapid flare-like phenomena where the X-ray brightness increases by a factor of 10 for several minutes). Since ordinary active-region flares often are associated with coronal transients, are bright-point flares associated with mini-transients?

A second class of small-scale features which may or may not play a role in the generation of the solar wind are the short-lived (minutes) jets observed at the base of the corona. When observed with high spatial resolution, the atmosphere at the base of the corona is found to be dynamic with a large fraction of the surface covered by chromospheric and transition-region material moving up and down in structures with characteristic sizes of ~ 1000 km. Three types of "jets" are observed: (1) the ubiquitous spicules with typical velocities of 25 km s^{-1}; (2) macrospicules, large spicules best observed in coronal holes with velocities up to 150 km s^{-1}; and (3) high-energy jets with velocities of 100 to 400 km s^{-1} (see Beckers 1972; Bohlin 1977a; Withbroe et al. 1976; Brueckner and Bartoe 1983; Dere et al. 1986,1989). The role, if any, played by these features in the generation of the solar wind is unknown. Spicules may be the source of the mass lost in the solar-wind outflow; the upward mass flux provided by spicules is approximately 100 times that lost in the solar-wind outflow. Hence, most of this mass, if heated to coronal temperatures, must cool and fall back into the chromosphere, perhaps causing the systematic red shifts observed in spectral lines formed in the chromospheric transition region (cf., Athay and Holzer 1982), with only a small fraction being carried out in the solar wind. However, it is not known what fraction of the mass observed in spicules gets heated to coronal temperatures (10^6 K). There is no direct empirical evidence that significant amounts of spicular material are heated to coronal temperatures, and some limited evidence that most of the mass in spicules remains at lower temperatures (Withbroe 1983).

The high-energy jets detected by the HRTS (High Resolution Telescope Spectrograph) rocket experiment are perhaps a more interesting phenomena (Brueckner and Bartoe 1983). These small-scale, energetic events (also known as explosive events) have been observed in spectral lines formed at transition-region temperatures ($T \sim 10^5$ K) and have characteristic velocities of 100 to 400 km s^{-1} (see the chapter by Cook and Brueckner). Given their observed birth rates, lifetimes, masses and energies, it appears that they could provide sufficient mass to supply the solar wind. However, they appear to contain insufficient energy to account for the energy flux of the solar wind (Dere et al. 1989).

In summary, the role of small-scale phenomena (spicules, high-speed jets, polar plumes, coronal bright-point flares) in the generation of the solar wind is unknown at the present time. They may or may not have a significant role in supplying mass, momentum and/or energy to the solar wind. Improved observations, particularly by instruments or groups of instruments, which can simultaneously observe mass and energy flows in the chromosphere, transition region and corona with high spatial and temporal resolutions are needed. This requires a new generation of high-resolution coronal instruments with appropriate spectroscopic diagnostic capabilities, such as the coronal instruments under development for the SOHO mission.

VII. HEATING AND ACCELERATION OF THE SOLAR WIND

As indicated in the introduction, the plasma in the outer solar corona is sufficiently hot, of the order of 10^6 K, that it is not bound by the Sun's gravitation field, but flows outward into interplanetary space. Parker (1958,1965) demonstrated that a thermally driven wind will attain an asymptotic flow speed of several 100 km s^{-1} for a coronal plasma with temperatures comparable to those observed. At the present time the mechanism responsible for heating the corona to these temperatures is unknown.

It has proven difficult to infer coronal heating mechanisms from measurements of solar-wind properties beyond 0.3 AU. The first partial success in this regard demonstrated that the nearly pure, large-amplitude outward-propagating Alfvén waves which mark high-speed solar-wind flows at 1 AU must originate inside the Alfvén critical point (Belcher and Davis 1971). Subsequent studies have reinforced this result by revealing increasing Alfvén amplitudes and mode purity with decreasing radial distances (Bavasano et al. 1982; Denskat and Neubauer 1982; Roberts et al. 1987; Luttrell and Richter 1988). However, although there are signatures of nonthermal phenomena which could be caused by Alfvén waves in the corona (cf., Withbroe et al. 1985; Hollweg 1986; Withbroe 1988), Alfvén waves have not yet been uniquely identified there. Furthermore, the Alfvén wave flux at 0.3 AU does not have sufficient amplitude to be energetically important. Nevertheless, models of the high-speed solar wind indicate that Alfvén waves could play an important role in boosting the flow speed of the wind beyond the sonic point (but inside of the Alfvén critical point) (cf., Hollweg 1986; review by Leer 1988, and references therein). This boost is needed, because a purely thermally driven wind, such as initially proposed by Parker (1958,1965) from coronal regions with the conditions found in coronal holes, will attain an asymptotic flow speed of approximately half the 700 km s^{-1} speed measured in high-speed solar-wind streams.

This acceleration is illustrated in Fig. 7 which compares empirical

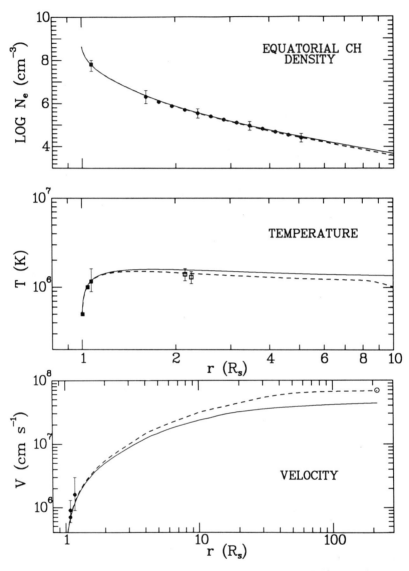

Fig. 7. Comparison of empirical densities, temperatures and flow velocities (points) with those calculated with a single fluid solar-wind model (curves). The solid curves are for a purely thermally driven wind; the dashed curves show the effect of adding sufficient energy in the form of Alfvén waves to raise the asymptotic flow speed to that observed (figure from Withbroe 1988).

(points) and calculated (solid curves) for radial variation of the temperature, density and flow speed in a typical equatorial coronal hole (e.g., Fig. 2). In order to accelerate the wind to the observed speeds, an additional mechanism besides the thermal pressure gradient is needed. One attractive mechanism is wave-particle interactions between Alfvén waves and the solar-wind plasma, where Alfvén waves from the lower solar atmosphere pass through the dense lower layers of the corona and deliver momentum and energy to the solar-wind plasma in the outer corona and beyond, and accelerate it to high speeds (see reviews by Holzer 1988; Leer 1988). The dashed lines in Fig. 7 illustrate the effects of introducing sufficient Alfvén wave flux to bring the calculated wind speed in agreement with the measurements. The required Alfvén wave velocity amplitudes are consistent with measurements of nonthermal broadening of spectral lines observed in the low corona (see Withbroe 1988). Clearly, new experiments to probe conditions within coronal holes inside of the Alfvén critical point are needed to determine the role of Alfvén waves in the physics of the solar-wind flow from these regions. It is interesting to note, though, that large-amplitude Alfvén waves are generally not present in the low-speed solar wind where the extra boost in flow speed has not occurred. However, present observations do not preclude the possibility that large-amplitude Alfvén waves and a super acceleration of the flow to form the high-speed solar wind are both caused by some, as yet unidentified, mechanism.

Other intriguing observations which point to specific heating mechanisms that support a hot corona and its outward expansion, stem from the internal state of ions in the high-speed solar wind. These observations pertain only to the high-speed wind, because only there are Coulomb collisions sufficiently negligible to allow the signature of collisionless processes to survive to 1 AU (Feldman et al. 1974a; Neugebauer and Feldman 1979; Lin et al. 1986). Ubiquitous in the high-speed solar wind are the presence of relatively convecting distinct proton beams (Feldman et al. 1976a; Marsh et al. 1982a), a bulk convection of alpha particle and all heavy ions relative to the center of mass proton convection (Asbridge et al. 1976; Mitchell et al. 1981; Marsch et al. 1982b; Bochsler et al. 1985). The point of interest here is that all ion relative streaming velocities stay close to the local Alfvén speed and therefore increase with decreasing heliocentric distance. In consequence, the alpha-particle bulk convection at 0.3 AU alone amounts to more than a third of the proton bulk convection flux. How close to the Sun does this growth in relative streaming of a secondary proton beam and all heavy ion beams persist? Note that the bulk convection energy flux is proportional to ρV^3 so that the ratio of alpha to proton kinetic flux is $(\rho_\alpha V_\alpha{}^3)/(\rho_P V_P{}^3) \approx 0.2(V_\alpha/V_P)^2$ in the high-speed solar wind (Feldman et al. 1977). If $V_\alpha - V_P$ continues to grow with the Alfvén speed inside of 0.3 AU, then at the Alfvén critical point near about 15 $R_\odot$ (Pizzo et al. 1983; Marsch and Richter 1984), $V_\alpha/V_P = 2$ and $(\rho_\alpha V_\alpha{}^3)/(\rho_P V_\alpha{}^3) = 0.8$. Clearly, these secondary ion beams may dominate the solar-wind expansion in the lower corona. New measurements of flow conditions inside of 15 $R_\odot$, such as proposed for a Solar Probe, will be needed to resolve this issue.

At this point it is useful to assemble many seemingly disparate observations of the outer solar atmosphere to form a coherent (yet highly speculative and nonunique) model of at least some of the processes that work in concert to heat the open corona and accelerate the solar wind. (1) Analysis of Skylab EUV data revealed ubiquitous, small-scale transient brightenings in transition-zone lines covering the entire disk of the Sun (see, e.g., Mariska 1986, and references therein). (2) Analyses of HRTS data indicate a fine-scale, filamentary structure to the chromosphere/transition zone. This pattern is interrupted randomly, yet spaced uniformly over the solar disk, by explosive-type events resulting in upward-traveling jets (Brueckner and Bartoe 1983; Dere et al. 1986,1989). (3) Additionally, microflares yielding X-ray fluxes above $\sim 7 \times 10^{-3}$ (cm^2 s keV)$^{-1}$ at 20 keV have a recurrence rate of about once every 5 min integrated over the full solar disk (Lin et al. 1984). These recurrence rates seem to be increasing with decreasing threshold without any hint of saturation at 20 keV. If this rate continues to increase with decreasing energy below 20 keV, then the total energy liberated in hot electrons alone would be a significant fraction of that required to maintain the active corona. Although these X-ray measurements were interpreted to reflect active-region phenomena by Lin et al. (1984), the instrumentation had no imaging capability. It is therefore possible that at least some of these microflares may have come from quiet regions, as evidenced by the ultraviolet events observed by HRTS (item 2 above) and ultraviolet microflares observed by SMM (Solar Maximum Mission) (Porter et al. 1987). The ultraviolet observations suggest that microflares are a common, nearly continuous phenomenon in the quiet Sun. (4) Another strong indicator of the ubiquitous occurrence rate of fine-scale filamentary jets in the open corona comes from interpretations of interplanetary scintillation measurements of solar-occulted compact radio sources (Eckers and Little 1971; Armstrong et al. 1986). Here observations of the radial speeds of density fluctuation patterns integrated along the line of sight, show variations of outflow velocities that change by up to an order of magnitude in a matter of hours at about 4 $R_\odot$ (Armstrong et al. 1986). Furthermore, the *rms* speed variations within each measurement are comparable to the bulk outflow velocity at this radius as well.

If indeed fine-scale jets are ubiquitous in the open solar corona, the internal nonequilibrium states of ions measured in the high-speed solar wind can be readily understood. Consider the model of magnetic flux emergence proposed by Heyvaerts et al. (1977) to explain some classes of solar flares and modified here in Fig. 8. If such emergence occurs uniformly throughout coronal holes, as well as in active regions, then current sheets leading to magnetic reconnection will form as illustrated in the figure. Using the terrestrial magnetosphere as a guide, such reconnection will most likely proceed routinely but in a somewhat patchy fashion (Russell and Elphic 1979) in

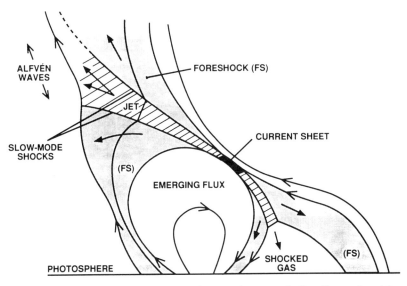

Fig. 8. Schematic drawing showing model for emerging magnetic flux (figure adapted from Heyvaerts et al. 1977). See text.

accordance with the Petschek mechanism (Petschek 1964). Detailed *in situ* measurements in the deep geomagnetic tail reveals the formation of slow-mode shocks leading to ion jets (see, e.g., Feldman 1988). The main virtue of the application of this mechanism to the suite of solar observations just summarized is that it provides a very efficient way of transforming magnetic energy in the form of an upward-moving Poynting flux into convected ion enthalpy. Indeed, in the distant geomagnetic tail, the ions carry the overwhelming fraction of the magnetic input energy both downstream of the shock (denoted in Fig. 8 by the curved hatched area) as well as upstream in the foreshock (denoted in Fig. 8 by the four stippled areas). In the present geometry the left-hand member of the shocked flow is decelerated by neighboring open field lines and the upward facing foreshock contains a strong ion stream. If the energy-conversion process proceeds in a patchy fashion, the leftward-going decelerated shocked flow will generate both upward and downward-going Alfvén waves and the upward-going component of the foreshock ion flow will appear as a jet. Since all this occurs well below the Alfvén critical point, only the upward-going component of the Alfvén waves will escape into the solar wind and the sequence of time-dependent jets will eventually overrun each other at higher altitudes to form the relatively streaming ion velocity distributions which are routinely observed in the high-speed solar wind. This situation is schematically illustrated in Fig. 9 (Feldman et al. 1974*b*). If true, then the ultimate source of energy that heats the corona and

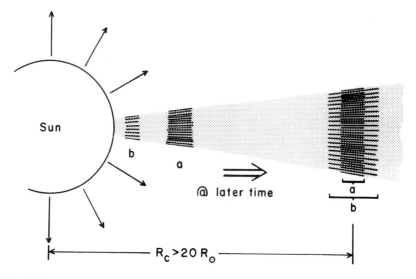

Fig. 9. Schematic drawing showing fast plasma (b) overtaking slower plasma (a) in solar-wind flux tube.

accelerates the high-speed solar wind originates in the buoyant emergence of magnetic flux through the solar photosphere. The jets eventually merge with overlying material to form a uniform outflow that may suffer further acceleration at higher altitudes through direction momentum transfer from, and perhaps heating by, the Alfvén waves (Esser et al. 1986; Hollweg 1986; Hollweg and Johnson 1988).

VIII. THE HELIOSPHERE

The outflow of the solar wind from the Sun has carved out a cavity in the interstellar medium where the Sun controls the local environment. This region is the heliosphere. The awareness of the existence of the heliosphere and its rudimentary structure evolved from the early, systematic studies of a variety of observed temporal changes in cosmic-ray intensity, over time periods ranging from hours to several years. These investigations, conducted over decades (Forbush 1966, and references therein), led to the realization that the Earth may be immersed in a region of tangled interplanetary magnetic field (IMF) of solar origin (Morrison 1956), probably confined within a cavity of a mean radius of about 200 AU (Davis 1955; Meyer et al. 1956). Biermann (1953) proposed that the flow of plasma from the Sun could explain the behavior of comet tails. Shortly thereafter, Parker (1958) developed a quantitative theory for the solar wind, containing an IMF drawn into Archimedes spirals. Axford et al. (1963) showed that the solar wind and IMF may be confined in a cavity of mean radius ~50 AU. Since then spacecraft have flown to the outer limits of the solar system and demonstrated that the heliosphere extends out to at least 44 AU (cf., Lee 1988).

Because measurements by spacecraft have been limited to near the ecliptic, other types of measurements are needed to infer the three-dimensional structure of the heliosphere. Measurements of interplanetary scintillations of remote astronomical radio sources have provided information about the three-dimensional variation of the solar-wind speed within 1 AU. These data have shown that near solar maximum, when the polar coronal holes are small or nonexistent, most of the interplanetary space is filled with low-speed wind. However, away from solar maximum, when the polar coronal holes are large, high-speed wind dominates at high latitudes and low-speed wind is confined to low latitudes (see Rickett and Coles 1983; Coles and Rickett 1986; Kojima and Kakinuma 1987). Away from solar maximum, recurrent high-speed streams from low-latitude coronal holes, and occasionally from polar holes, produce high-speed streams at low latitudes. Measurements of Lyman-α $\lambda1216$ radiation scattered by the interplanetary medium show that the solar-wind mass flux near solar minimum, when there are large polar holes, is smaller at high latitudes than at low latitudes (cf., Lallement et al. 1985). Finally, the observed modulation of cosmic rays in the heliosphere can be used to probe the structure of the heliosphere (see Ahluwalia and Dessler 1962; Krymskiy 1964; Parker 1968, and references therein). A highly successful diffusion-convection model has been developed to account for a variety of observed cosmic-ray modulation effects (Forman and Gleeson 1975, and references therein). Cosmic rays, being charged particles, are sensitive to electromagnetic conditions in the heliosphere, particularly the large-scale structure of the IMF. Therefore, one can use cosmic rays of appropriate energies to probe the electromagnetic states of heliospheric regions far from the Earth and obtain estimates of parameters that play important roles in the transport of energetic particles in the heliosphere from the local interstellar medium (cf., Kota and Jokipii 1983; Riker and Ahluwalia 1987). One may then make predictions that can be tested when *in situ* measurements are actually made at off-ecliptic sites by the out-of-the-ecliptic Ulysses mission.

It appears the heliosphere may have two natural states (Ahluwalia 1988). They correspond to the large-scale restructuring of IMF that follows epochs when the direction of the solar magnetic dipole moment reverses (Ahluwalia 1980). However, there are several unanswered questions and details that need to be worked out. What is the size of the heliosphere and how does it change with solar activity? How does the heliosphere connect to the local interstellar medium? How does the transition from one natural state to the other affect the mode of propagation of the solar wind? Are the coronal holes instrumental in triggering the transition or are they themselves the end product of the transition? *In situ* measurements being made in the outer heliosphere by spacecraft have already produced some surprises (see McKibben 1988). Nevertheless, one expects the interplay between theory and measurements will continue to broaden our horizons for the rest of the century.

IX. SUMMARY

Coronal holes are the most well-established coronal source of steady-state solar wind, the high-speed solar wind streams that typically have asymptotic flow speeds of 700 km s^{-1}. Coronal mass ejections associated with flares and/or eruptive prominences are another clearly identified source of solar wind, a transient component that accounts for approximately 5% of the total solar mass loss. The role of other coronal structures in the generation of the solar wind is less clear. Streamers and the interfaces between streamers and other coronal regions are potential sources of low-speed wind ($V \approx 330$ km s^{-1}), a hypothesis consistent with existing observations, but measurements of mass flows in these features are needed to confirm this. These measurements can be obtained by new instruments under development. Small-scale dynamical phenomena observed at the base of the corona (spicules, macrospicules and high-speed jets) and small-scale structures observed in polar coronal holes (polar plumes) may or may not play a significant role in supplying mass, momentum and energy to the solar wind. New, improved measurements are required to determine the role, if any, of small-scale structures in the generation of the solar wind.

PART V
Our Sun, Other Suns, Other Stars

SUN-AS-A-STAR SPECTRUM VARIABILITY

WILLIAM LIVINGSTON
National Solar Observatory, NOAO

RICHARD F. DONNELLY
National Oceanic and Atmospheric Administration

VICTOR GRIGORYEV, M. L. DEMIDOV
SibIZMIR

JUDITH LEAN
Naval Research Laboratory

MATTHIAS STEFFEN
Kiel University

ORAN R. WHITE
Lazy FW Ranch

and

RICHARD L. WILLSON
Jet Propulsion Laboratory

The Sun is observed as a star in order to determine luminosity change, detect minute variability in average granulation and facular signals, and to use as a standard against which other stars may be compared. In this regard, topics discussed include: total irradiance variability as measured from space by the

Activity Cavity Radiometer Irradiance Monitor and Earth Radiation Budget radiometers; Fraunhofer line heights of formation and examples of their variability in visible wavelengths; ultraviolet and extreme ultraviolet irradiance variability as observed from space; the magnetic origin of irradiance change; and the observed mean magnetic field of the Sun.

I. INTRODUCTION

If we could observe the Sun from the location of a neighboring star, and assuming powerful instruments but no resolution of its disk, what spectrum variability could be deduced? This is the same as asking: what is the irradiance spectrum variability of the Sun? Observations to answer this question have only been acquired over the last decade and are displayed here in the form of time series of various spectral features. In the case of the total irradiance and the visible spectrum, the observed variability is small but significantly above the noise. In the extreme ultraviolet (EUV), the signals become large in terms of fractional change. A summary of our present knowledge on irradiance varibility is found in Table I.

The reason why variability increases to shorter wavelengths follows from the dependence of opacity on wavelength and hence the height of formation of spectral features. Opacity is at a minimum at 1600 nm. At shorter and longer wavelengths, opacity increases and we see higher and higher in the atmosphere. In the visible, we sense photospheric layers; in the extreme ultraviolet or far infrared, and in the center of strong lines like Hα and Ca H and K, we sense the chromosphere; X rays, gamma rays and long radio waves originate in the corona. Magnetic fields associated with solar activity have more and more effect on these outer, less-dense regions, and their presence modulates the Sun's output accordingly.

Because we *can* observe the resolved solar disk, we are able to go a step further and inquire as to the causes of variability. We know that magnetic fields clump in ways that produce sunspots, faculae and coronal structure. The Sun's 27-day rotation causes these phenomena to transit the disk, inducing rotational modulation of magnetic-related signals. The 11-yr activity cycle creates an envelope of such modulation. Spectrum variability is thus interpreted in terms of solar magnetic fields, the rotational modulation of these fields and the activity cycle.

Why bother to observe the Sun as a star when its disk is revealed in such enticing detail? A practical reason is that it is the energy output summed over the entire solar disk that influences terrestrial temperatures. Small but persistent changes in total irradiance of the order of 1% could invoke climate change in the long term. This is discussed in Sec. II. In the short term, activity-related flare-ups in extreme ultraviolet irradiance cause expansion of our outer atmosphere (the thermosphere), and this induces the drag which brings down low-Earth orbit satellites (Sec. IV).

Another reason for the study of integrated sunlight has to do with better

TABLE I.
Observed Variability of Solar Irradiance Spectrum Features

Technique	Parameter[a]	Value at Solar Min.	Solar-Cycle Modulation	Secular	Rot. Modulation at Solar Max.
ACRIM	total irrad.	1367.10 Wm^{-2}	0.08%	?	0.2%
Fe 525.0 nm	magnetic field	0.0 gauss	$\sim \pm 1$ gauss	—	$\sim \pm 1$ gauss
C 538.0 nm	W_λ	0.00222 nm	none	0.03%yr^{-1}	none?
Fe 537.9 nm	W_λ	0.00625 nm	none?	0.01%yr^{-1}	0.8%
Fe 539.3 nm	W_λ	0.01387 nm	none	none	0.7%
O 777.4 nm	W_λ	0.00638 nm	1.0%	—	0.3%
Mn 539.4 nm	W_λ	0.00799 nm	2%	—	1.3%
CN 388.3 nm	0.3 Å index	0.00580 nm	4%	—	1.8%
H 656.2 nm	central depth	0.841	1.2%	—	?
Ca K 393.3 nm	1 Å index	0.0085 nm	20%	—	13%
	central intensity	0.058	46%	—	21%
	WB Effect	1.545	2.6%	—	1%
	V/R	1.30	12%	—	6%
	λK_3	393.3680 nm	0.1%	—	?
He 1083.0 nm	W_λ	0.00400 nm	200%	—	50%
Mg 280.0 nm	Mg index intensity	2.63	10%	—	4%
Lyα 121.5 nm	line intensity	2.3×10^{11}	200%	—	30%
10.7 cm	flux	70×10^{-32} W m^{-2} Hz^{-1}	>400%	—	~200%

[a] W_λ = equivalent width; 0.3 Å index = $1 - W_\lambda$ taken over 0.3 Å to blue of CN bandhead; central depth = depth of line center relative to the continuum (see Fig. 5); 1 Å index = $1 - W_\lambda$ of Ca K ± 0.5 Å from K_3; central intensity = $1 -$ central depth; WB Effect = Wilson-Bappu Effect; V/R = K2V/K2R with K2V and K2R taken relative to K_3 intensity; λK_3 is wavelength of K_3 relative to nearby photospheric lines.

understanding of the Sun itself. For example, do surface changes in magnetic flux affect solar granulation? If we examine the spectrum from point to point over the solar disk, we encounter tremendous variety in line shifts and strengths. The finer the image detail, the higher the dispersion in deduced velocities, temperatures and densities. If we ask how these observables change with time, we need their average values. One way to average spatially is to defocus the solar image. The presence of evolving granulation patterns and the 5-min oscillations set limits on this averaging method. Only by averaging over the entire solar disk can these disturbances be adequately suppressed. Then we can inquire if granular convection changes over the activity cycle (Livingston 1987). The answer, which is not yet known, will provide a basis for magnetohydrodynamic (MHD) models of solar convection.

Finally, the Sun is a standard against which other solar-type stars may be compared. Can we hope to detect planets around stars similar to the Sun from measures of periodic, minute Doppler shifts? Or do activity-related spectral disturbances mask such signals (Wallace et al. 1988)? What can we infer about the Sun from observations of younger or older stars of like spectral type? These and other questions can best be answered through the study of irradiance spectrum variability.

II. TOTAL IRRADIANCE VARIABILITY FROM 1980 TO 1988

Sustained efforts to detect variability in the total solar irradiance (or solar constant) began in the late 19th century. Experiments from groundbased observatories, aircraft and balloons during the first three quarters of the 20th century were unable to detect intrinsic total solar-irradiance variation. The principal unsolved problems for the early experiments were uncertainties in instrument performance and atmospheric attenuation (Fröhlich 1977). In this section, we discuss the solution to both problems and the implementation of long-term space-based monitoring of the total solar irradiance during the last decade.

Notice that one cannot deduce that the solar luminosity is, or is not, variable from Earth-based total irradiance measurements. Variability discussed herein is derived from instruments observing the Sun in its equatorial plane, and so inferences are restricted to that perspective. Whether or not any brightness increase in the equatorial region is compensated by a corresponding decrease in the polar regions is not known at this time.

The instrumentation problem was solved by the development of a new generation of electrically self-calibrating, cavity pyroheliometers at several laboratories around the world during the 1960s, 70s, and 80s. These new cavity detectors provided a higher precision and accuracy than earlier sensors and several were capable of automatic, remote operation. The solution to the atmospheric attenuation problem was provided by access to space-flight platforms with extended observation opportunities during the late 1970s facilitating operation of solar monitoring experiments above the Earth's atmosphere.

The ACRIM I Instrument

The most accurate technology for measuring total solar irradiance is a specialized form of radiometry referred to as pyrheliometry in which the heat produced by the absorption of solar radiation by a cavity detector is compared with the heat produced on the same detector by the dissipation of a known amount of electrical power. Pyrheliometric detectors are the most accurate means of defining the radiation scale for solar total irradiance.

The Active Cavity Radiometer Irradiance Monitor (ACRIM I) instrument employs the dual cavity Active Cavity Radiometer (ACR) type IV sensor (Willson 1979). It operates in a differential (shuttered), full-time electrical self-calibration mode (see Fig. 1). The shutters on one or more of ACRIM I's three nominally identical sensors open or close every 65.536 s, providing solar and internal reference data. Up to 3×10^4 independent samples of total solar irradiance are acquired per day by each sensor in normal Solar Maximum Mission (SMM) spacecraft operations.

The shutters over two of the ACRIM I sensors (channels B and C) are closed most of the time, protecting them from the solar ultraviolet and particulate fluxes that slowly degrade the continuously monitoring sensor (channel A). Periodic comparisons of the solar observations of 2 or 3 sensors provides an effective long-term relative calibration of the optical degradation of channel A. This calibration shows the sensitivity of channel A to have decreased by 400 parts-per-million (ppm) of the average total irradiance during the first 8 yr of the mission. ACRIM I results are corrected for this degradation with an uncertainty of 30 ppm. The ACRIM I instrument is shown in Fig. 2.

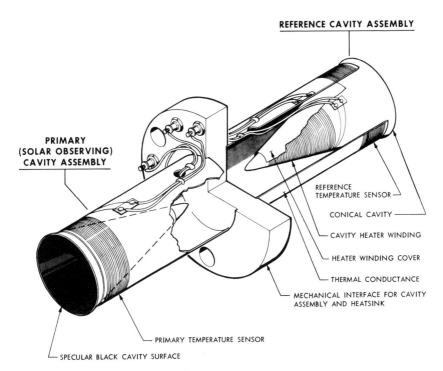

REFERENCE CAVITY ASSEMBLY

PRIMARY
(SOLAR OBSERVING)
CAVITY ASSEMBLY

REFERENCE
TEMPERATURE SENSOR

CONICAL CAVITY

CAVITY HEATER WINDING

HEATER WINDING COVER

THERMAL CONDUCTANCE

MECHANICAL INTERFACE FOR CAVITY
ASSEMBLY AND HEATSINK

PRIMARY TEMPERATURE SENSOR

SPECULAR BLACK CAVITY SURFACE

Fig. 1. The Active Cavity Radiometer Sensor.

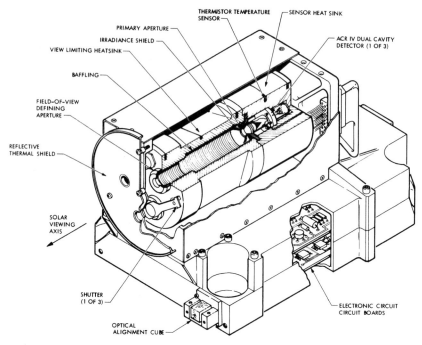

THERMISTOR TEMPERATURE SENSOR
SENSOR HEAT SINK
PRIMARY APERTURE
IRRADIANCE SHIELD
ACR IV DUAL CAVITY DETECTOR (1 OF 3)
VIEW LIMITING HEATSINK
BAFFLING
FIELD-OF-VIEW DEFINING APERTURE
REFLECTIVE THERMAL SHIELD
SOLAR VIEWING AXIS
SHUTTER (1 OF 3)
ELECTRONIC CIRCUIT CIRCUIT BOARDS
OPTICAL ALIGNMENT CUBE

Fig. 2. The Active Cavity Radiometer Irradiance Monitor Instrument on the Solar Maximum Mission.

Solar-Irradiance Monitoring in Space

The first full implementation of electrically, self-calibrating cavity (ESSC) solar pyrheliometry in a space-flight experiment occurred in 1976, with the flight of a JPL Active Cavity Radiometer (ACR) on the first of a series of flights of NASA's solar-irradiance rocket program. This series of brief space observations was designed to calibrate satellite solar-irradiance monitoring observations.

The results from satellite space-flight solar pyrheliometry experiments, extending over many solar cycles will ultimately be required for monitoring the variability of total solar irradiance on time scales of interest to climatology. The first satellite pyroheliometric experiment utilizing a cavity sensor with an electrical self-calibration capability was the Earth Radiation Budget (ERB) experiment on the Nimbus 7 spacecraft. The solar data base from this experiment, launched in late 1978 and continuing into the present, is the longest currently available.

The Nimbus 7/ERB Hickey-Friedan (HF) solar sensor is operated in a self-calibrating mode about every 14 days. An empirical algorithm derived from these ESCC operations is used to approximate sensor performance in between. The principal objective of the HF sensor was to provide the ERB solar insolation observations with ±0.5% accuracy. The results lacked the precision required for recognition of intrinsic solar variability until after it was discovered in the results of the SMM/ACRIM I experiment in early 1980 (Hickey et al. 1980).

The first dedicated total solar-irradiance monitoring experiment was the NASA SMM/ACRIM I launched in February 1980 (Willson et al. 1981). Unequivocal evidence of solar variability was soon discovered in the ACRIM I results. Intrinsic solar variability has since been detected on virtually every observable time scale, from ACRIM I's 1.024 s sampling interval, to the duration of the mission. While variations on all time scales are of interest for solar physics, the inertia of the Earth's climate system apparently obscures measurable responses to small solar variations with shorter than multiyear periodicities. The possibility of shorter-term solar influences on weather patterns is an open subject at present.

Sensors similar to those in the ACRIM I experiment were included on the ERB experiments launched on the ERB satellite in October, 1984 and the National Oceanic and Atmospheric Administration NOAA-9 satellite in December, 1984. Although the ERB solar sensors operate in the full ESSC mode, the priority given to the nadir-looking ERB experiments, together with the lack of solar pointing, impose limitations on their performances. The sensors operate bi-weekly for short intervals at orbital sunrise as the Sun transits their fields of view. ERB solar observations have a precision of better than 0.1%, which is adequate for the requirements of the ERB solar insolation observations, yet the observation frequency and precision are not adequate for providing the long-term solar-irradiance monitoring data base required for climatology and solar physics.

Results of Monitoring the Total Irradiance

Results of the SMM/ACRIM I and NIMBUS 7/ERB observations are shown in Figs. 3 and 4. The variability on several time scales is discussed below.

Variability on Solar-Cycle Time Scales (Years). A systematic long-term decrease in total irradiance was detected between their launches and mid-1985 in both the ACRIM I and Nimbus 7/ERB results. Linear least-square fits to these two data bases yield nearly identical values of 0.02% yr^{-1} for the rate of irradiance decrease between 1980 and 1985. The total decrease of total irradiance over this period is just under 0.1%. Results from the two ERB experiments detected a similar trend in the 1984–85 period of their operation.

Recent results from the ACRIM I experiment show that the irradiance remained nearly constant from mid-1985 to early 1987, and then began to

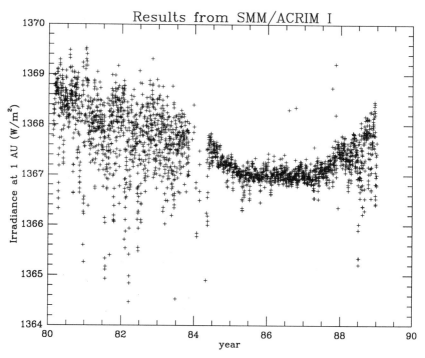

Fig. 3. Summary of solar total-irradiance observations from the SMM/ACRIM I, space-flight experiment. The large negative excursions (up to −0.25%) in the ACRIM I results are due to transit of large sunspot groups across the solar disk. Smaller (up to + 0.1%) increases, significant relative to the noise of only the ACRIM I results are due to large facular areas in solar-active regions. Quieter results between 1984 and 1987 reflect the lower solar-activity level during that period and provide a reference for the intrinsic noise in each data base.

increase. The relatively constant portion of the irradiance is centered near the magnetic solar minimum (September, 1986). A strong upward trend continued throughout 1987, regaining nearly half the deficit incurred over the period 1980 to 1985. Assuming the upward trend continues into the activity maximum of solar-cycle 22, a direct solar-activity cycle/luminosity relationship will have been observed for the first time. The direct correlation of total irradiance and solar activity is a major discovery from the ACRIM I experiment. It agrees in sense with that predicted from the coincidence of the historic "Little Ice Age" climate anomaly and the "Maunder Minimum" of solar activity during the 16th and 17th centuries. For long-term variations fewer sunspots implies a lower irradiance.

ACRIM I results through mid-1989 show a continuation of the systematic irradiance increase, amounting to nearly 0.06% since early 1987. This was followed by a transition to the high-variance irradiance mode, like that

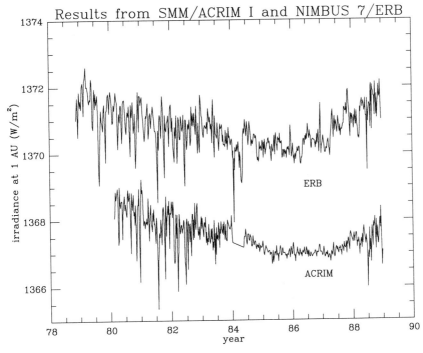

Fig. 4. Results from SMM/ACRIM I and Nimbus 7/ERB compared. A 6-day running mean has been passed through the data.

observed during the maximum period of solar-cycle 21. On a time scale of 10 days, this mode is characterized by irradiance deficits and excesses caused by the development of sunspot and facular areas, respectively.

Variability on Solar-Activity Time Scales (Days-Months). An inverse relationship between sunspot area and total solar irradiance was discovered from the ACRIM I data during the Solar Maximum Mission. As active regions containing sunspots rotate onto, or form and grow on the visible side of the solar disk, a corresponding decrease in total irradiance was detected. These decreases last for the duration of the transit of the sunspots and have been detected with amplitudes of up to −0.25% of the total irradiance. Confirmation of this result was obtained by the subsequent identification of similar features in the Nimbus 7/ERB results for the larger sunspot events.

A more subtle, direct relationship between active-region faculae and features in the ACRIM I irradiance record has been found. A coincidence of irradiance peaks and the transit of active regions containing a large facular area has been frequently observed. On other occasions, a facular area has decreased the net deficit produced by the dark sunspots, relative to deficits produced by a comparable sunspot area when fewer faculae were present.

The nature of energy flow and balance in active regions has been investigated using variability results from these solar-active regions. Initial irradiance deficits caused by sunspots appear to be compensated for by excess persistent facular emission in the same active region on time scales of months.

Short-Term Variability (Solar Global Oscillations).   Solar global oscillations have been detected in the ACRIM I total-irradiance data, including pressure modes and possible gravity modes. Pressure-mode oscillations of low degree ($l = 0, 1, 2$) and orders 15 to 25 have been derived by a careful statistical analysis of individual ACRIM I shutter-cycle observations (averages over 32 s spaced every 131.072 s). Peak amplitudes of these 5-min oscillations are 3 ppm of the total-irradiance signal. Inversion of p-mode measurements provides unique insights into the physical properties of the convection zone and, to some extent, the outer solar core. The most significant ACRIM I p-mode results are:

1. Support for the relativistic interpretation of Mercury perihelion observations. The value of the solar gravitational quadrupole moment estimated from the ACRIM observations of the p-mode oscillation is consistent with Einstein's general relativistic calculation of the precession of Mercury;
2. An upper limit for radial variation of solar rotation of 2.2 times the photospheric rate (see the chapter by F. Hill et al.);
3. Mode coherence times of 1 to 2 days;
4. A shift in frequency of the modes between the 1980 (high solar activity) and 1984–86 (solar quiet) periods. Apparently this is another facet of the solar activity/luminosity trend.

In addition to the p-mode oscillations, there is evidence for the presence of detectable gravity mode (g-mode) oscillation effects in the ACRIM I data. Although results are preliminary, at least one group claims to have isolated g-mode signals in the 10 to 80 μHz frequency range (period range of one day to several hours). The great interest in the g-modes derives from their likely ability to provide information on the physics of the solar core.

III. FRAUNHOFER LINE VARIABILITY OBSERVED
FROM THE GROUND

Changes in Fraunhofer line strength depend in a sensitive way on changes in temperature and pressure in the solar atmosphere. In exceptional cases (like He 1083.0 nm), non-LTE radiative pumping by coronal emission plays a role in determining line absorption.

Observations

Measurements of line strength which are described in this section have been made mainly with the 13.5-m double-pass spectrometer of the McMath Telescope on Kitt Peak (Livingston et al. 1977). Unfocused sunlight is incident on a rather wide (0.4 mm) entrance slit which projects a pinhole image of the Sun on the grating. Selected wavelength intervals of about 0.3 to 1.0 nm are rapidly scanned to generate precision spectra whose noise level is approximately 0.05% after optimal filtering. In the reduction process the above spectra are normalized to a fixed continuum level, and line parameters are then measured.

Line Parameters Defined

Referring to Fig. 5, the central depth of a line is defined as the normalized depth at line center. The central intensity is then equal to one minus line depth. Because the central depth is very sensitive to instrument resolution, and thus to focus and internal spectrometer seeing, the preferred parameter of line strength is the equivalent width W_λ. In the hypothetical absence of blends, W_λ is independent of spectrometer resolution. For real spectra, which always contain blends to some degree, W_λ is dependent on resolution (but less so than central intensity).

There are broad lines such as Mn 539.4 nm for which the central intensity is useful, or Ca K 393.4 nm for which central depth is a reliable measurement. Line indices are employed for the measurement of strong lines with complex profiles (see White and Livingston 1981). For Ca K, we define the "1A index" as $1 - W_\lambda$ measured between $\pm$ 0.5 Å, or 0.05 nm, from line center K_3. For the CN bandhead, we define a "CN index" to be measured as $1 - W_\lambda$ between the band edge at 388.32903 nm and blueward 0.02845 nm. The foregoing line parameters are all critically dependent on the position of zero intensity. To determine accurately this zero, a double-pass spectrometer is essential. Periodically, the saturated telluric lines of O_2 near 761.5 nm are observed to verify that the zero level is correct. In the case of Ca K, there are

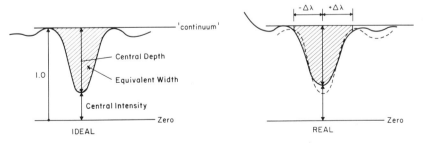

Fig 5. Measured line parameters in the ideal case (left) and, somewhat exaggerated, the real situation (right).

profile measurements useful as activity indicators that are independent of zero point. For example, there is the Wilson-Bappu parameter, which here is defined as the log of the wavelength difference between the mid-position of K2V - K1V and K2R - K1R. V/R is simply the ratio (K2V - K1V)/(K2R - K1R). Another zero-point independent measurement is the wavelength of K_3 relative to nearby photospheric lines. Certain of these parameters may be valuable when equipment must be used that has an inexact zero level.

Heights of Formation of Spectral Lines

To get an idea in what atmospheric layers the continuum intensity and the absorption lines are formed, appropriate contribution functions and heights of formation for representative spectral features have been computed. Results for C I 538.03 nm (7.68 eV), Fe I 537.96 nm (3.63 eV), Mn I 539.4 nm (0.00 eV) and the CN bandhead near 388.3 nm are reported here. First the contribution functions for intensity and for relative line depression are discussed. Results obtained by using the solar model of Holweger and Mueller (1974), assuming local thermodynamic equilibrium (LTE), are presented.

The formal solution of the equation of radiative transfer immediately gives an expression for the contribution to the emergent *intensity* (Magain 1986, Eq. 5)

$$C_I\,(x,\mu,\lambda) \; = \; \frac{\ln 10}{\mu}\,\tau_{ref}\,\frac{\kappa}{\kappa_{ref}}\,S_\lambda\,e^{-\tau/\mu} \tag{1}$$

from which the emergent intensity is obtained as

$$I\,(\mu,\lambda) \; = \; \int_{-\infty}^{\infty} C_I\,(x,\mu,\lambda)\,dx. \tag{2}$$

Here τ and τ_{ref} are the optical depths at λ and λ_{ref}, respectively, $x = \log \tau_{ref}$, κ and κ_{ref} are the absorption coefficients at λ and λ_{ref}, respectively, S_λ is the source function and $\mu = \cos\theta$. To obtain the contribution function for the flux spectrum (Sun as a star), we assume plane-parallel conditions. In this case, we simply have to integrate the contribution functions over μ with weight μ at each depth

$$C_{I,F}\,(x,\lambda) \; = \; \int_0^1 \cdot 2\mu\,C_I(x,\mu,\lambda)\,d\mu. \tag{3}$$

For this purpose, it was sufficient to take 10 equidistant μ values. The full-disk intensity, $F(\lambda)$, is then given by

$$F(\lambda) = \int_{-\infty}^{\infty} C_{I,F}(x,\lambda)\,dx. \tag{4}$$

In the examples of this section, we arbitrarily use the optical depth at 500 nm as the reference scale ($x = \log \tau_{500}$) and display $C_{I,F}(x,\lambda)$ for the continuum intensity. In the figures, $C_{I,F}$ is normalized to $\int_{-\infty}^{\infty} C_{I,F}(x,\lambda)\,dx = 100$. The mean height of formation of the continuum intensity we then define as the center of gravity of the corresponding contribution function (on the $\log \tau_{500}$ scale) and indicate it by a vertical bar in the figures.

The intensity contribution function discussed so far is only used for the continuum intensity. It could be used in the same way to obtain the contribution function for the intensity at any position along a line profile. However, we are not so much interested in where the line intensity originates, but rather would like to know where the line absorption is formed. The intensity in a weak line, or in the wings of any line, always will be found to come from near the continuum-forming layers. In contrast, this is not necessarily true for the line absorption. Even a very weak line may be formed in very different layers than the continuum intensity (an extreme case is a telluric line). To compute the contribution function for the line absorption, we follow Magain (1986). His approach correctly describes the formation of the relative line depression. Even in the extreme case of a telluric line, the resulting contribution function would peak in the Earth's atmosphere.

To begin with, Magain derived a transfer equation for the relative line depression D formed in a stellar atmosphere. Note that $D = 1 - I_l/I_c$, with I_l the emergent intensity in the line and I_c the emergent continuum intensity. D is the appropriate variable for our problem because it is the relative line strength that is measured when analyzing the variability of the solar line spectrum. From this transfer equation, an expression is obtained for the contribution function to the relative line depression, $C_D(x,\mu,\lambda)$, in a way that is perfectly analogous to the derivation of the intensity contribution function (Magain 1986, Eq. 21). At least in LTE, this expression can be considered as the only rigorous contribution function to the relative line depression. As above, we obtain the line-depression contribution function for the flux spectrum, $C_{D,F}(x,\lambda)$ by integration over μ, as

$$C_{D,F}(x,\lambda) = \int_{0}^{1} 2\,\mu\,C_D(x,\mu,\lambda)\,d\mu. \tag{5}$$

In terms of $C_{D,F}$, the relative line depression is given by

$$D(\lambda) = \int_{-\infty}^{\infty} C_{D,F}(x,\lambda) \, dx. \tag{6}$$

The calculated line depression depends on the assumed value of macroturbulence ζ. If $C_{D,F}(x,\lambda)$ denotes the contribution function for vanishing macroturbulence, we obtain the corresponding contribution function that takes macroturbulence into account $C^*_{D,F}(x,\lambda,\zeta)$ by convolution with the (gaussian) macroturbulence distribution function

$$C^*_{D,F}(x,\lambda,\zeta) = \int_{-\infty}^{\infty} C_{D,F}(x-z,\lambda) \, e^{-[z/\zeta]^2} \, dz. \tag{7}$$

On the other hand, by integrating $C_{D,F}(x, \lambda)$ over the absorption line, we obtain the contribution function to equivalent width in the full-disk spectrum,

$$C_{W,F}(x) = \int_{blue}^{red} C_{D,F}(x,\lambda) \, d\lambda \tag{8}$$

which is independent of macroturbulence. The equivalent width in the full-disk spectrum W_F, is then given by $W_F = \int_{-\infty}^{+\infty} C_{W,F}(x) dx$. In the figures, the resulting contribution functions are normalized to

$$\int_{-\infty}^{+\infty} C^*_{D,F}(x,\lambda,\zeta) dx = 100 \, D \tag{9}$$

where $C^*_{D,F}$ was computed with $\zeta = 1.6$ km s^{-1}, and

$$\int_{-\infty}^{+\infty} C_{W,F}(x) dx = 100. \tag{10}$$

Again, we calculate the mean heights of formation as the centers of gravity of the corresponding contribution functions on the log (τ_{500}) scale (see Magain 1986, Eq. 22).

From Fig. 6 we see that the mean height of formation of the full-disk continuum intensity is about $x = -0.2$ at 538.0 nm, while the continuum

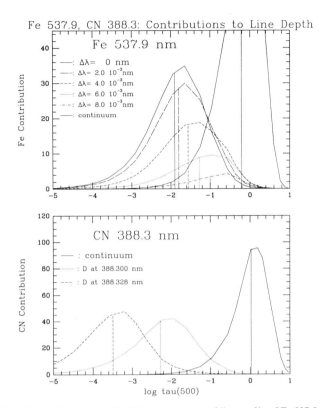

Fig. 6. Contribution function for full-disk measurements of line profile of Fe 537.9 nm and CN 388.3 nm, and the continuum at these wavelengths. Vertical lines indicate mean heights of formation. Note: 1Å = 0.1 nm; 1 mÅ = 10^{-4}nm.

radiation comes from somewhat deeper layers in the ultraviolet at 388.3 nm ($x = 0.0$). Figure 6a shows that the line core of Fe I 537.9 nm is formed at about $x = 1.9$. Proceeding from the line center into the wings, we find the absorption to be formed in increasingly deeper layers. Approximately 8.0 nm into the line wing, the main contribution to absorption still comes from significantly higher layers ($x = -1.0$) than the continuum intensity ($x = -0.2$). Even in the far line wings the absorption comes from a mean height not deeper than $x = -0.65$. Figure 6b demonstrates that the strong absorption at the CN bandhead at 388.33 nm is formed in very high layers ($x = -3.5$). Since CN is believed to be formed in LTE (Mount et al. 1973), this result is probably realistic. Figure 7 shows the contribution functions to equivalent width in the full-disk spectrum. We see from Fig. 7a that the C I line at 538.03 nm originates from a relatively narrow range of height, the average height of formation being only slightly higher ($x = -0.4$) than for

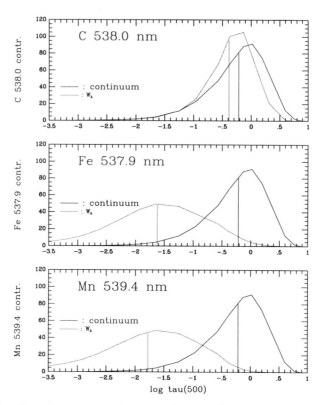

Fig. 7. Contribution functions for full-disk equivalent-width W_λ measurements for C 538.0 nm, Fe 537.9 nm and Mn 539.4 nm.

the continuum intensity ($x = -0.2$). The stronger lines, Fe I 537.96 nm and Mn I 539.47 nm, are formed in a broad range of height centered at significantly higher layers than in the case of C I 538.03 nm. The mean height of formation of the full-disk equivalent width is $x = -1.6$ for Fe I 537.9 nm, Fig. 7b and $x = -1.8$ for Mn 539.4 nm, Fig. 7c.

It must be noted that while the contribution functions presented here give an idea in what levels of the solar atmosphere the process of absorption takes place, it is not possible to conclude from this information how a particular line responds to a given perturbation of the atmosphere (e.g., a change in the temperature structure). For this purpose it is necessary to calculate corresponding response functions, a topic which is, however, beyond the scope of this chapter.

Examples of Line-Variability Time Series

In this section, we compare the behavior of a number of photospheric lines with simultaneous (same-day) observations of the Ca K index. The latter reports on chromospheric emission and therefore the state of the activity cycle. In all instances, the data have been smoothed by a 4-point running mean. Corrections have been made for water-vapor blends where appropriate. One should bear in mind that these are not daily samples but 2- to 4-day sequences at roughly monthly intervals.

Lines Not Modulated by the 11-yr Activity Cycle.   The equivalent widths of the carbon line, C 538.0 nm, and a relatively weak line of iron, Fe 537.9 nm, appear unaffected by the activity cycle (Fig. 8). Although regular spectral-irradiance observations have been made since late in 1975, improvements in the instrument and alignment techniques produced data early on that are irreconcilable with material obtained since 1978. Over the past decade, however, instrument and observing technique have been immutable. Within this epoch, both C 538.0 nm and Fe 537.9 nm have increased in strength at the rate of 0.03 % yr^{-1} and 0.01 % yr^{-1}, respectively. We have, at present, no explanation for this secular rise. Perhaps it is an instrumental artifact, although Fe 539.3 nm (see Fig. 9) appears immutable over this same time span. The moderately strong line Fe 539.3 nm is constant in strength with time within errors of measurement. Occasional short-term changes are noted that are common to other lines; for instance both Fe 539.4 nm and C 538.0 nm display a significant peak at the beginning of 1985 (coincident with the onset of minimum activity in Ca K). These short-term correlations appear to be the exception rather than the rule. Behaving somewhat differently than Fe 539.3 nm, the strong line Si 1082.7 nm rose steadily in strength up to 1985, after which it has returned to nearly its initial value. (The data are too noisy to be certain about this final decline; additional observations are required.)

Lines Modulated by the 11-yr Activity Cycle.   Both the CN index and Mn 539.4 nm (see Fig. 10) mimic Ca K. It is reasonable that CN index should track Ca K. Formed at the temperature minimum, CN is weakened by the same mechanism that weakens Ca K: magnetic heating. But contribution calculations place Mn 539.4 nm low in the atmosphere, well below where faculae form in the diverging surface magnetic fields, so we do not understand this weakening of Mn 539.4 nm. It is noteworthy that Mn 539.4 nm recorded solar minimum coincident in time with Ca K. On the other hand, activity minimum was reported by the CN index 18 months later.

Unlike any of the other lines we observe, both the oxygen triplet (O 777.2 nm, O 777.4 nm and O 777.5 nm) and He 1083.0 nm (see Figs. 10 and 11) strengthen with solar activity. All other lines weaken in the visible wavelengths. (Recall that an index may strengthen, but this is really equal to $1 - W_\lambda$.) Since the three components of the triplet act in unison, we average their equivalent widths and call this mean $\bar{W}_\lambda$ O 777.4 nm. He 1083.0 nm is

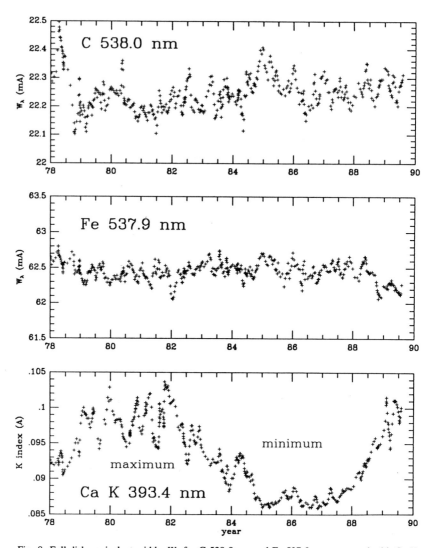

Fig. 8. Full-disk equivalent widths W_λ for C 538.0 nm and Fe 537.9 nm compared with Ca K index as representative of the activity cycle.

of special interest because daily full-disk maps of this line are made at Kitt Peak (to study coronal holes) and these maps can be integrated to yield the irradiance values. Time series for this line based on daily observations are available back as far as 1974. Foukal and Lean (1988*a*) have shown that there is an excellent correlation between He 1083.0 nm W_λ and ACRIM total irradiance (after correction for sunspot blocking; see Sec. V).

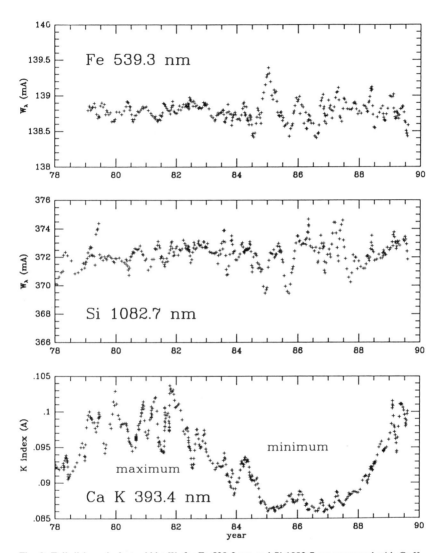

Fig. 9. Full-disk equivalent widths W_λ for Fe 539.3 nm and Si 1082.7 nm compared with Ca K index. These lines have displayed little or no variation with the activity cycle.

IV. SOLAR EXTREME ULTRAVIOLET, ULTRAVIOLET AND LYMAN α FROM SPACE

State of EUV and UV Irradiance Measurements

The EUV-UV spectrum of the Sun starts at 400 nm and extends to the hard X-ray region. Figure 12 shows the 160–400 nm irradiance spectrum measured with the Solar Backscatter Ultraviolet photometer on 7 November

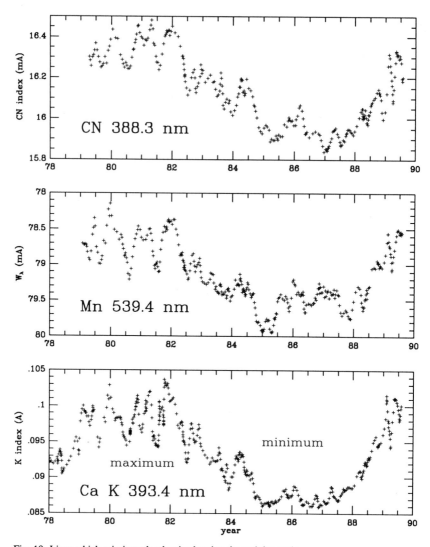

Fig. 10. Lines which mimic each other in showing the activity cycle.

1978 (Heath 1980). The intensity declines steadily because of the decrease of the Planck function at shorter wavelengths and because of increased opacity from the crowding by metallic lines. Important chromospheric features are the Ca II H and K lines (396.8 and 393.4 nm) and the companion doublet of Mg II at 280 nm. Al I and Mg I absorption edges cause the sharp intensity drops below 210 and 254 nm, respectively. Chromospheric emission lines begin to appear at 182 nm as the photospheric continuum weakens. Below

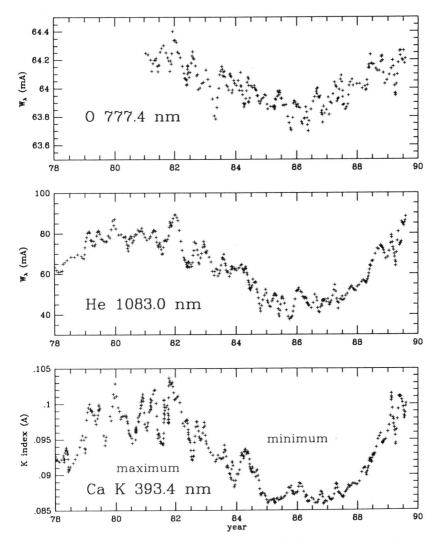

Fig. 11. Other lines which display the activity cycle. Probably O 777.4 nm is a photospheric line that is subject to non-LTE effects.

about 160 nm the spectrum becomes an emission spectrum of ions abundant in the chromosphere, chromosphere-corona transition region and corona.

Figure 13 indicates the transition from a Fraunhofer-type spectrum to an emission spectrum at shorter EUV wavelengths. The use of a logarithmic intensity scale allows one to see the emission lines superposed on the weak background as well as the free-bound continua of neutral hydrogen, helium and carbon.

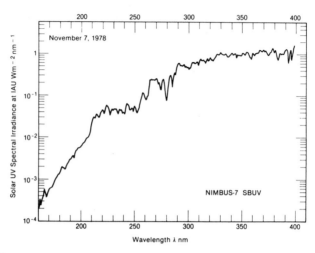

Fig. 12. The solar ultraviolet spectrum as measured by the Solar Backscatter Ultraviolet Photometer (Heath 1980) on 7 November 1978 from Nimbus 7. The spectral resolution is 1.1 nm.

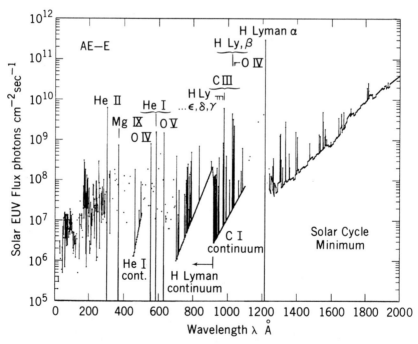

Fig. 13. A plot of Hinteregger's EUV-UV reference spectrum for solar minimum (July 1976). Emission lines and free-bound continua are identified (figure from Tobiska 1989).

The sources of measurements needed for a study of the variability in the EUV and UV spectrum are listed in Table II. For the EUV (0.1–121 nm), continuous satellite observations only exist from July 1977 to December 1980, which is the ascending phase of Cycle 21. Continuous full-disk measurements of the longer-wavelength band from 120 to 400 nm did not begin until the launch of Nimbus 7 in 1978 and Solar Mesosphere Explorer in 1981. The descending phase of Cycle 21 and the beginning of Cycle 22 are now covered by continuous measurements from the SME, NOAA-9 and NOAA-11 satellites. Both the Solar Maximum Mission and the Solar Mesosphere Explorer will regrettably re-enter the Earth's atmosphere in 1989–90, thus ending our current observing programs for the total irradiance and spectral irradiances above 120 nm. Resumption of these measurements with the Upper Atmosphere Research Satellite is anticipated in 1991.

From the view of solar physics, current measurements describe variability of the irradiance originating in the upper photosphere, chromosphere and chromosphere-corona transition region. No strong coronal lines are being measured by experiments in progress today. Since irradiances in strong lines are direct measures of radiative energy losses in the outer solar atmosphere, their variation and correspondence to surface features on the Sun are fundamental physical properties of the solar cycle.

Temporal variations of the solar EUV-UV fluxes are caused by the evolution of structures associated with solar magnetic fields (sunspots, plages, faculae, bright network, ephemeral regions, flares). In addition to these evolutionary changes, the fluxes display a 27-day modulation due to rotation of

TABLE II
Major Experiments to Measure Solar EUV and UV

Spacecraft	Photometer	Group	Wavelengths, nm	Duration
AE-E	EUV	AFGL	0–200 nm	Jul 1977– 1980
NIMBUS 7	SBUV	GSFC	160–400	Nov 1978– 1986
SME	FDP	LASP	120–300	Oct 1981– 1989
NOAA9	SBUV2	NOAA	160–400	Apr 1985– to date
SL1	SOLSPEC	ESA	115–410	Dec 1983
SL2	SUSIM	NRL	115–410	Aug 1987
UARS	SUSIM	NRL	115–410	1991– 1996
UARS	SOLSTICE	LASP	120–450	1991– 1996
SOHO	SUMER	Planck	EUV	1995

brightness structures across the solar disk as seen from the Earth. The degree of fluctuation in the irradiance spectrum decreases from the X-ray region to the visible spectrum because of the growing contribution of the photosphere toward longer wavelengths. The details of time variations at different wavelengths depend critically on emission mechanisms and the geometry of the emitting structure.

We do not discuss solar-flare influences on solar irradiance variations here since they are covered in the chapter by Rabin et al. Flares are no longer being studied or monitored in full-disk EUV and UV measurements for upper atmospheric research. An exception is the full-disk soft X-ray flux (e.g., 1 to 8 Å) which is routinely monitored to detect flares and estimate their strengths.

Solar-cycle Variations

Until solar-cycle 21, the single source of daily irradiance data has been the 10.7-cm radio flux (F 10.7). This situation changed in 1975 when several observing programs began at the National Solar Observatory for the express purpose of constructing a data base of daily measurements of solar irradiance spectra in the visible spectrum. More significantly, during the period from 1976 to 1981, four satellite experiments (AE-E, Nimbus 7, SMM and SME) became operational. As a result, we have the spectral irradiance data coverage for solar-cycle 21 shown in Figs. 14 and 15 as well as the total irradiance measurement from ACRIM on SMM and ERB on Nimbus 7 (see Sec. II above). The companion Fig. 16 shows a continuation of the longer-wavelength measurements into solar-cycle 22. When the total irradiance record from SMM is also included, we have, for the first time, an empirical picture of solar radiative output variations over all phases of a solar cycle. The exception is in the EUV, where measurements below 120 nm exist only for the ascending phase of Cycle 21 from 1977 to 1980.

Figure 14 illustrates the variability of specific EUV emission lines formed in the chromosphere-corona transition region and corona. The Fe XVI line originates in a low-density plasma with a temperature of 3×10^6 K, while H I Lyman β comes from higher-density regions with temperatures in the 20,000 to 50,000 K range. As the figure shows, the range of variability in coronal lines is 2 to 100 times, while the lower-ionization lines vary by factors of 2 or so. This behavior illustrates the high sensitivity of coronal emissions to the increased temperatures and densities due to solar activity. Solar EUV radiation is a major contributor to heating of the Earth's upper atmosphere which, as a consequence, expands and contracts in synchronism with the activity cycle.

Figure 15 shows the variability of Lyman α of neutral hydrogen, the resonance lines of Mg II and Ca II and the 10.7-cm radio flux for Cycle 21. These data show clearly the transition to solar minimum conditions in 1985 and how the minimum epoch differs between F 10.7 and the spectrum lines. Pulses at 27-day intervals are rotational modulation due to bright, active re-

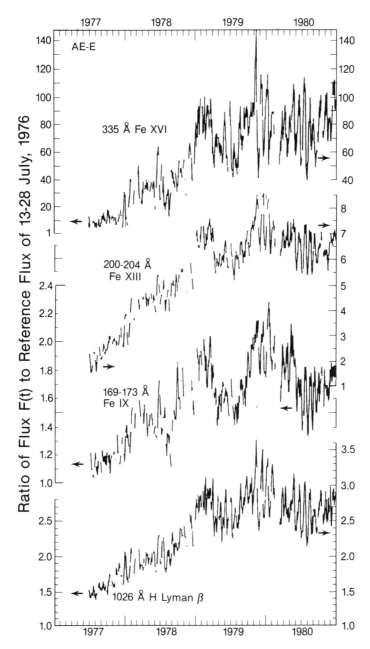

Fig. 14. Four EUV line irradiances from AE-E in the ascending phase and maximum of Cycle 21. The intensity scale is relative to the solar minimum spectrum (Fig. 13) and shows the high-temperature line of Fe XVI increasing by a factor of 100 from minimum to maximum, Fe XIII increasing by 8, Fe IX increasing by 2 and Lyman β by 3 (figure based on measurements from Hinteregger et al. 1981).

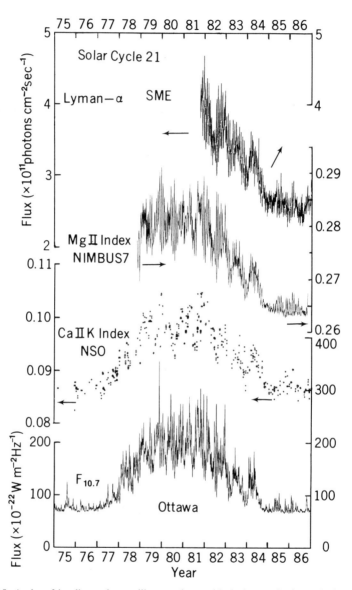

Fig. 15. A plot of irradiance data to illustrate the empirical picture of solar-cycle 21 (1976–1986). The Lyman α and 10.7-cm data are absolute irradiances while the data for Ca II and Mg II are relative intensity indices sensitive to chromospheric activity. Sources: SME measurements from Rottman (1988); Nimbus 7 measurements from Heath and Schlesinger (1986 and personal communication 1989); Ca K line measurements from White et al. (1987) and 10.7-cm radio-flux measurements from Algonquin Radio Observatory, Canada.

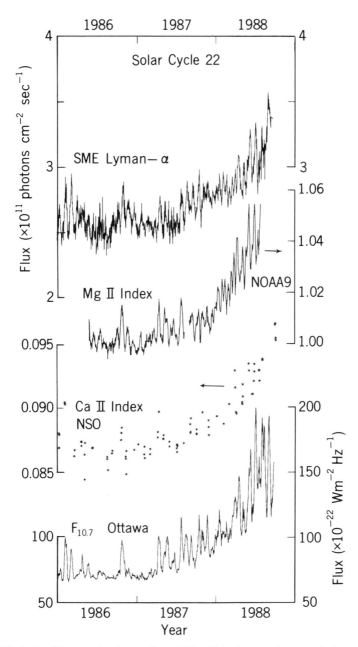

Fig. 16. A plot of the same irradiances shown in Fig. 15 for the ascending part of solar-cycle 22 (1986–1988). Note that this new data for the Mg II, that are normalized to the monthly mean value for the September 1986 index, now come from the improved SBUV2 spectrometer on board the NOAA 9 satellite (Donnelly 1988).

gions moving across the visible solar hemisphere. The range of variability from solar maximum to minimum is a factor of 2 for Lyman α, 6% for Mg II 280, 18% for Ca II K, and a factor of 3 for the 10.7-cm flux.

We have not shown the AE-E data for Lyman α (Fukui 1988) because of a controversy over the absolute calibration. The AE-E data near solar maximum in 1979 appears to be about a factor of 3 too large even though the agreement at solar minimum is to within 15% for Vidal-Madjar's (1977) result. Broadband EUV measurements by Brace et al. (1988) suggest that the Lyman α flux is fairly level at the beginning and end of 1981, and this indicates that the AE-E data overestimate the Lyman α irradiance by a factor of 3; see Lean (1987) for a discussion of the difficulty.

In Fig. 16, we see the beginning of Cycle 22 as displayed by Lyman α, Mg II 280 nm, Ca II K and F 10.7. We emphasize that these data represent new ground since neither Lyman α nor Mg II 280 nm were measured during the ascending phase of the previous cycle. Episodes of 27-day modulation are present in these lines, but the durations are quite short (3 to 4 rotations) relative to lifetimes observed near solar maximum.

Figure 17 summarizes variability of the EUV-UV spectrum over Cycle

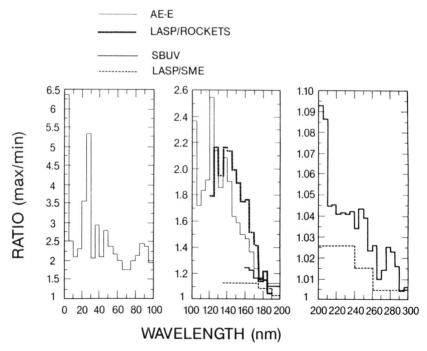

Fig. 17. The wavelength dependence of the ratio of solar EUV-UV flux at solar maximum to solar minimum (figure from Lean 1987).

21 from all data available up until 1986. The full range of variability from the X-ray region to 300 nm is a factor of 5 to 0.01 for the wavelength intervals studied by Lean (1987). This again emphasizes the solar-cycle dependence of the short-wavelength coronal emissions. Figure 18 is another estimate of solar-cycle variability of the range from 120 nm to 300 nm from current SME data. The expected changes in variability at the Al I absorption edge (210 nm) and Mg II doublet (280 nm) do not show in Fig. 18. Solar-cycle variability at wavelengths above 160 nm appears to be undetectable below the 5% level by the SME photometer because of the difficulty of long-term absolute calibration.

Rotational Modulation

The most obvious variations in spectral irradiance time series are the long-term variations due to the solar cycle itself and the 27-day modulation of this slow variation by bright solar regions crossing the solar disk. Such modulation by coronal-chromospheric activity is seen in other stars as well, and is used to estimate the strength of their chromospheric activity and stellar rotation periods. Figure 19 shows the modulation curves for 8 wavelength intervals during a particularly active period from Oct 1979 to Feb 1980. These examples illustrate the variation in 27-day modulation for emissions originating in different regions of the solar atmosphere. Clearly the 1 to 8 Å coronal X-ray band has a strong variation relative to the other examples, but the chromosphere-corona transition region emissions of He II 30.4 nm, H I Lyman β and H I Lyman α fall into a group with a more or less common waveform. The F 10.7 and Fe XV 28.4-nm fluxes form another group be-

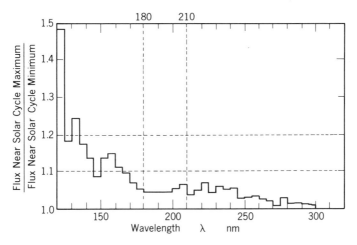

Fig. 18. The wavelength dependence of the ratio of solar ultraviolet flux at solar maximum to solar minimum as determined from SME data (figure from Simon 1989).

cause of their contributions from coronal regions. Curves for 200–205 nm are representative of photospheric time variations.

The strength of the modulation varies with ionic species and wavelength. For example, the ratio of peak-to-minimum amplitudes in Fig. 19 is 1.06 for the ultraviolet flux at 200 nm from the upper photosphere and lower chromosphere; 1.6 for He II 30.4 nm from the upper chromosphere and lower transition region; 2.8 for Fe XV 28.4 nm from the corona; and 25 for the 1 to 8 Å X-ray flux. This order of increasing ratios is caused by the increasing role of active-region emission relative to the quiet Sun flux rather than in total energy emitted. Figure 20 summarizes the variation of the strength of the 27-day modulation over the spectrum from 160 to 300 nm. One conclusion is that the modulation amplitude increases with line strength, or roughly with height, since it is seen clearly in the strong lines (C IV 155 nm, Si II 181 nm and Mg II 280 nm) and the Al edge at 210 nm. This increase arises from the greater contrast of active regions observed for EUV and UV emission lines. In addition, lines of highly ionized coronal species have large modulation amplitude even though they are not optically thick.

All modulation curves and the sunspot-number series have power at a period of 27 days, but certain irradiances display a periodicity at 13.5 days. Figure 21 shows a sample of four such power spectra in which three show a 13.5-day peak. Notice that the 13.5-day periodicity is missing in the F 10.7 radio-flux curve. The peak at half the rotation period arises from two sources: (1) recurrence of solar active regions 180 deg apart in longitude, and (2) the shape of the visibility curve. In the case when two bright, active regions lie 180 deg apart, the strength of the peak at 13.5 days depends strongly on the shape of the limb-darkening function at the very edge of the solar disk. The extreme limb is not sharp for coronal emissions such as for F 10.7; consequently, emission from regions at the east and west limbs combine to reduce the 13.5-day periodicity.

Activity Complexes and Episodes of 27-Day Modulation

The amplitude of the rotational modulation component waxes and wanes throughout a solar cycle indicating that sources of bright emission emerge, dominate for several solar rotations and then fade. The periods from October 1979 to March 1980 and May 1982 to March 1983 in Figs. 14, 15, 19 and 22 are examples of the occurrence of episodes of rotational modulation. These intermediate-term variations reflect the evolution of not only single active regions, but larger aggregates of active regions occurring in preferred longitudinal bands. Such aggregates are identified as "activity complexes" by Bumba and Howard (1965) and described further by Gaizauskas et al. (1983).

An activity complex is a large-scale magnetic phenomenon involving emerging flux regions, sunspots, plage and flux dissipation or dispersion. The link between the magnetic activity and changes in irradiance is the occurrence of bright regions at positions where solar magnetic flux tubes occur (see Sec.

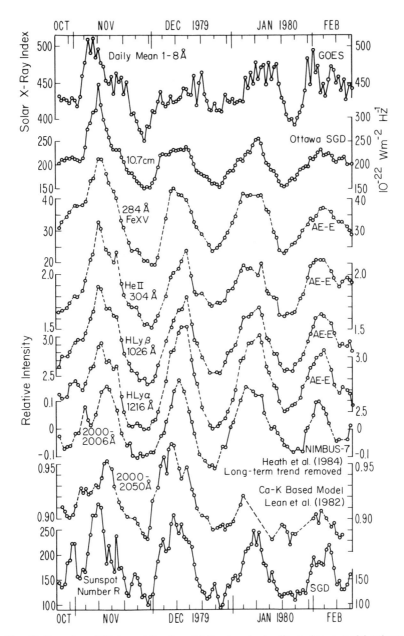

Fig. 19. Examples of 27-day rotational modulation in solar irradiance data, a model estimate, and the Zurich sunspot number for solar-maximum conditions, Oct 1979 to Mar 1980 (figure from Donnelly et al. 1986).

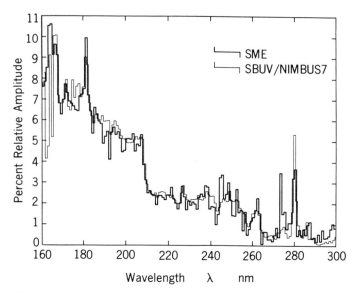

Fig. 20. The wavelength variation of the 27-day rotational modulation in the ultraviolet under solar maximum conditions during August 1982 (figure from Simon 1989).

V below). However, here we are seeing an aggregate effect of many magnetic structures, all of which are occurring in a limited region on the Sun and which are recurring on the visible hemisphere of the Sun every 27 days.

These slower variations persist for 4 to 8 solar rotations and display a more uniform amplitude for the upper photospheric UV and chromospheric EUV and UV fluxes than for X rays, coronal EUV fluxes, sunspot number and the 10.7-cm solar radio flux (Donnelly et al. 1986). Earlier observational results are consistent with this empirical picture: chromospheric plages and photospheric faculae last longer than sunspots (Kiepenheurer 1953). Implied is that the hot coronal emissions rise, peak and decay more rapidly than the associated chromospheric and photospheric plage emission. Neupert (1967) showed that the Fe XIII ($T \sim 2 \times 10^6$ K) emission of a major active region evolves more slowly and persists longer than the Fe XVI ($T \sim 3 \times 10^6$ K) emission. Donnelly (1987) found that the persistence of the 27-day modulation produced by groups of bright active regions was very high for ultraviolet fluxes from the upper photosphere, high for Fe XIII and chromospheric fluxes, low for hot ($T \geq 3 \times 10^6$ K) coronal fluxes (Fe XV 28.4 nm and Fe XVI 33.5 nm) and very low for soft X rays and transition-region fluxes (Fe IX).

To describe these episodes better, we have analyzed the full Lyman α data set from SME, using the analytic signal technique (Bracewell 1965; Simon et al. 1988). The results are shown in Fig. 22. This method isolates the

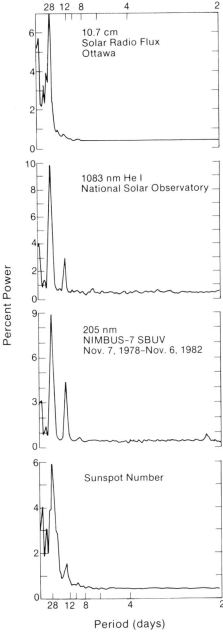

Fig. 21. Power spectra computed from the time series of the sunspot number, 10.7-cm radio flux, 205-nm irradiance and the strength of the He I 1083 nm line (figure from Donnelly et al. 1985).

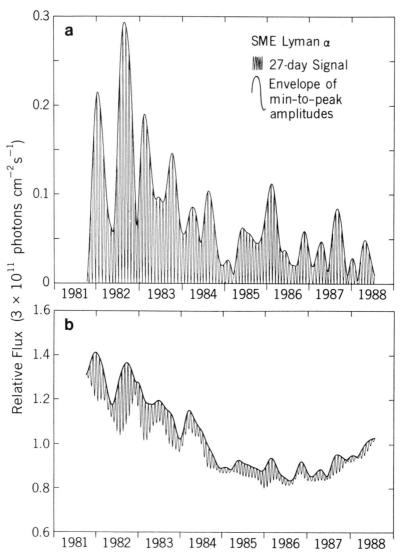

Fig. 22. The solar-cycle variation of 27-day modulation in H I Lyman α from 1981 to summer 1988. (a) The narrowband 27-day signal and its envelope are combined with the long-period components of the slower solar-cycle variation. (b) Total maximum amplitude (max-min) of the 27-day component. Data are from Rottman's (1988) SME measurements.

27-day component numerically with a passband filter and then yields the amplitude of the resulting narrowband signal. Figure 22a shows this 27-day component combined with the lower-frequency components that give both the intermediate term and longer solar-cycle variations in the period from 1982 to 1988. The upper envelope of the modulation shows the intermediate-term variability due to activity complexes. Figure 22b shows a further isolation of the 27-day component. This plot is obtained by subtracting the lower envelope from the data in panel (a). The resulting 27-day modulation curve is what one would see if the pulses of emission occurred every 27 days on a constant intensity background. One can count 13 episodes in this 7-yr interval, each one lasting from 3 to 4 rotations (51 to 189 days). These durations are the same as quoted for the activity complexes by Gaizauskas et al. (1983). The strength of the episodes shows that the variability of the 27-day modulation decreases by approximately a factor of 3 from solar maximum to minimum.

Research Issues in EUV-UV Measurement and Analysis

Programs to measure the total and spectral irradiances in solar-cycle 21 have yielded valuable and perhaps surprising results. We now know that the total irradiance varies in phase with the solar cycle, as do spectral irradiances in the EUV and UV. Large sunspots do, indeed, cause temporary decreases in the total irradiance, but these decreases are superimposed on a slow rise and decline that follows the overall level of solar activity.

What solar structures cause the increase in irradiance? Lean (1987) and collaborators continue to emphasize that plage emission falls short by a factor of 2 in producing the observed spectral irradiance increases in lines sensitive to chromospheric activity. One likely source of increased radiative output is the so-called active chromospheric network, assumed to arise from the dissolution of the bright active regions. Since the active network is a pattern of small-scale features limited in latitude, what is the possibility of even smaller structures uniformly distributed on the Sun brightening to cause the observed changes on a global scale? Observations of the Ca K line at the center of the solar disk by Livingston and White (personal communication, 1989) indicate little change in the spectrum of the quiet chromosphere over Cycle 21. Kuhn et al. (1988*b*), however, present new results showing latitudinal bands of increased brightness that appear to follow predominant sunspot belts in Cycle 21. The experiment allows separation of the facular and quiet Sun component; thus, these results suggest a change in the quiet photosphere that produces the observed residual changes in total output.

Confidence in the reliability of the irradiance time series has grown because of the improved accuracies (0.01% for total irradiance, 0.1% for Ca K and 2–5% for UV lines) and because of the similarities in the various time series reported by Livingston et al. (1988) and Lean (1988). The intermediate-term variations (periods >100 days) in the total irradiance (corrected for sunspot effects) and spectral irradiances showing chromospheric activity (Ca K, Lyman α, 205-nm band) are very similar. Accepting the reality of the solar-cycle variation of the Sun's total radiative output, we are led further to an interesting scenario on the relationship between the solar cycle and the Earth's climate. Eddy (1976) discusses the Maunder and Spörer Minima of solar activity and their relationship to a colder terrestrial climate lasting for decades. The obvious conjecture is that the Earth's atmosphere cools in the absence of solar activity. Climatologists argue that the amplitude of the total-irradiance variation is insufficient by a factor of 10 to produce meaningful changes in the mean global temperature. What then is the mechanism by which solar activity or its absence may alter the Earth's climate as suggested by Eddy's research? (See Sonett et al. 1991.)

One approach to understanding irradiance variations throughout a solar cycle is to use models of known surface structure as *post facto* estimators of the variation (see Skumanich et al. 1984; Lean et al. 1982; Lean 1988). These models use average photometric data for sunspots, plages, faculae and network. There is only one observational program today for daily systematic photometry of these solar structures (Chapman 1987*a*). A more comprehensive measurement of evolution of active regions and related fine structure is required for a better understanding of irradiance changes.

Analysis of the time series discussed here shows the need for continuous, daily measurements of solar irradiances. This requirement can be met by multiple sites for groundbased observations and the use of space platforms for continuous remote operation. EUV measurements need to be resumed during solar cycle 22. Progress on the problem of the source of the irradiance variations throughout a solar cycle will, likewise, come through continuation of groundbased programs to record quantitatively the properties of active regions, faculae and other solar surface structures. Such complete data bases will allow both observers and theorists to assist in the interpretation of irradiance measurements. Space operations, in particular, are sensitive to the effects of solar activity on spacecraft and the Earth's atmosphere. One has to allow for thermospheric heating and the resulting increased satellite drag in planning space missions near times of maximum solar activity. All civilian and military satellites in Earth orbits are subject to radiation, high-energy particles and plasma from the Sun; consequently, knowledge of fluxes and their durations is important in instrument design as well as in the management of the system in orbit.

V. SOLAR IRRADIANCE MODULATION BY
MAGNETIC ACTIVITY

This section describes the connection between spatially resolved observations of the Sun's disk, which identify and characterize magnetic active regions, and Sun-as-star observations, which integrate the Sun's emission from both active and quiet regions over the entire disk.

Solar Radiance and Irradiance

When there are no active regions on the solar disk, and the Sun is observed as a star, the measured quiet Sun spectral irradiance $F_Q(\lambda)$ at wavelength λ can be expressed in terms of the radiance I_Q (λ, μ) at heliocentric location μ by $F_Q(\lambda) = 2\pi \int I_Q(\lambda,\mu)\mu d\mu$. The total (i.e., spectrally integrated) irradiance of the quiet Sun is $S_Q = \int F_Q(\lambda)d\lambda$. As the Sun's magnetic activity waxes and wanes throughout the solar cycle, active regions appear, evolve and decay on the disk. Since the Sun's emission in active regions differs from its emission in quiet regions, the occurrence of active regions on the solar disk modulates the quiet Sun irradiance by an amount that can be estimated from

$$F_A(\lambda,t) = 2\pi I_Q(\lambda,1)\sum_i A_i(t)\mu_i R(\lambda,\mu_i)[C_i(\lambda,\mu_i) - 1] \qquad (11)$$

where A_i is the area (in units of the solar hemisphere) and C_i the (resolution consistent) contrast of the ith active region located at μ_i and $R(\lambda,\mu) = I(\lambda,\mu)/I(\lambda,1)$ is the center-to-limb variation.

Active region phenomena of particular importance for solar irradiance variations are sunspots, where the emission is diminished relative to that of the surrounding background Sun, and faculae and plage, where the emission is enhanced. Figure 23 illustrates the contrasts of sunspots and faculae/plages as functions of wavelength. Figure 24 shows the fraction of the disk covered by (a) sunspots, $\Sigma A_s\mu$, and (b) Ca K plages, $\Sigma A_p\mu$, during 1980 (near the maximum of magnetic activity in solar cycle 21), as recorded by the World Data Center of the National Oceanic and Atmospheric Administration (WDC/NOAA). This figure demonstrates that the total area on the solar disk covered by sunspots is typically an order of magnitude less than that covered by plages, and that these two different active-region phenomena are not simply linearly related, the ratio $\Sigma A_p\mu/\Sigma A_s\mu$ varying from 210 (DOY 77) to 5 (DOY 100). Data similar to those in Figs. 23 and 24 can be used to estimate the irradiance modulation associated with both sunspots and plages on a particular day.

When calculating solar spectral irradiance variations at ultraviolet wavelengths, the modulation by magnetic activity is assumed to be dominated by bright plage emission alone (Cook et al. 1980; Lean et al. 1982; Lean and Skumanich 1983; Lean 1984). At 250 nm, for example, it can be seen in Fig. 23 that $| C_P - 1 | / | C_S - 1 | \approx 1$ so that when $\Sigma A_p\mu/\Sigma A_s\mu = 10$, the magnitude of the ultraviolet irradiance modulation by sunspots is an order of magnitude less than that by plages. Solar ultraviolet irradiance variations corresponding to the WDC/NOAA Ca K plage data have been estimated by

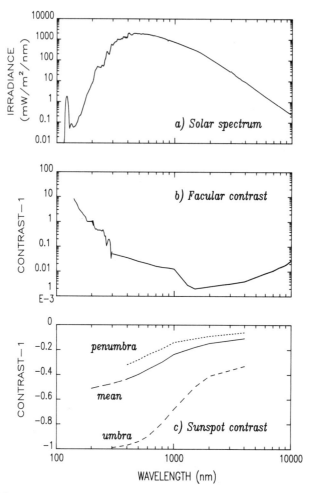

Fig. 23. Solar spectrum and the contrast of faculae (λ >300 nm)/plage (λ <300 nm) and sun-spots. The sunspot data are from Allen (1973), with the mean sunspot contrast determined from the umbra and penumbra data assuming a constant ratio of 0.2 for the umbral to penumbral areas. The plage contrasts are from Cook et al. (1980) at λ <210 nm, from Lean (1984) at 210 <λ <300 nm and from Herse (1979) at λ >300 nm, scaled to a contrast of ~2% at 500 nm and adjusted to merge smoothly with the shorter-wavelength data.

$$F(\lambda<250,t) = F_Q + 2\pi I_Q(\lambda,1)[C_P(\lambda)-1]\sum_i A_{P_i}(t)\mu_i w_i R(\lambda,\mu_i) \quad (12)$$

where A_{P_i} are the plage areas, w_i are observed relative brightness estimates and C_P is an average plage contrast (Lean et al. 1982).

Both sunspots and faculae modulate the Sun's total irradiance, about half of which is radiated at visible wavelengths. At 500 nm, for example, $|C_P - 1|$

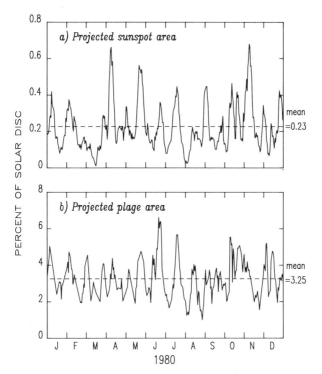

Fig. 24. Time series of daily projected sunspot areas $\Sigma A_s \mu$ and projected plage areas $\Sigma A_p \mu$ in 1980. Data are from the WDC/NOAA. Variations in the solar B_0 angle, that modify slightly the values of μ, have been omitted. The dashed lines are the mean values of the daily data for the entire year.

$/|C_S - 1| \simeq 0.05$ so that when $\Sigma A_P \mu / \Sigma A_S \mu = 10$, the magnitude of the reduced emission from sunspots is twice that of the enhanced emission from the Ca K WDC/NOAA plages. Total irradiance variations associated with magnetic activity are typically calculated by

$$S(t)/S_Q = 1 + \sum_i [\bar{C}_P(\mu_i) - 1] A_{Pi} \mu_i (3\mu_i + 2)/2$$

$$+ [\bar{C}_S - 1] \sum_i A_{Si} \mu_i (3\mu_i + 2)/2 \quad (13)$$

where $(3\mu + 2)/5$ is the photospheric limb-darkening function, and $\bar{C}_S$ and $\bar{C}_P$ the bolometric sunspot and plage contrasts, respectively (Foukal 1981; Hoyt and Eddy 1982; Hudson et al. 1982; Schatten et al. 1982; Schatten 1988; Willson and Hudson 1988).

In the following sections, calculations of the modulation of the solar irradiance by magnetic active regions are compared with observations of the Sun's total irradiance, over time scales of the solar rotation ($\sim$ 27 days), active region evolution ($\sim$ 3 to 9 months) and the solar cycle ($\sim$ 11 yr). It will be seen that while the most conspicuous magnetic active disk features (i.e., the sunspots and plages recorded by the WDC/NOAA) can account for the observed rotation modulation, calculations of the WDC/NOAA Ca K plage emission underestimate, by about a factor of 2, the solar-cycle variations observed in those full-disk emissions whose shorter-term variations are dominated by enhanced emission from bright faculae or plages. A simple but fundamental question still to be answered is whether this additional solar-cycle variability observed in the Sun as a star can be attributed to the magnetic modulation of the quiet Sun's irradiance by active regions other than those recorded by the WDC/NOAA (Foukal and Lean 1988a; Schatten 1988), or must global changes such as a thermal pulsation (Willson and Hudson 1988) or thermal variations (Kuhn et al 1988b) be invoked?

Rotation of Magnetically Active Features

Active regions are typically distributed asymmetrically in heliocentric longitude, and thus appear to move across the face of the Sun seen at the Earth as the Sun rotates. Because of this, total solar irradiance observations exhibit an approximate 27-day periodicity. A comparison of the measured rotation modulation of full disk emissions with calculations of active region emission allow a reliable calibration of the solar rotation tracers important for that emission.

Figure 25 illustrates that the observed rotation modulation of the Sun's ultraviolet irradiance at Lyman α and 205 nm can be reasonably well described by calculations of the enhanced emission from bright plages (Lean and Skumanich 1983; Lean 1984), with the sunspot contribution neglected. When calculating the modulation of the Sun's total irradiance, however, both sunspot darkening and facular brightening must be considered. This is demonstrated in Fig. 26; shown in panels (a) and (b) are the daily values of the total irradiance in 1980, measured by ACRIM, and the sunspot blocking function, calculated from (Hoyt and Eddy 1982)

$$P_s = S_Q \left\{ 1 + \sum_i (0.36 + 0.84\mu_i \right.$$
$$\left. - 0.2\mu_i^2) \left[(C_U - 1)A_{Ui}\mu_i + (C_P - 1)A_{Pi}\mu_i \right] \right\} \quad (14)$$

with $S_Q = 1366.8$ Wm^{-2}. In this evaluation of the sunspot modulation of the total irradiance, the projected sunspot umbral and penumbral areas, $A_U\mu$ and $A_P\mu$, with contrasts $C_U = 0.25$ and $C_P = 0.75$, are evaluated specifically. Panels (c) and (d) in Fig. 26 demonstrate that when P_s is subtracted from the

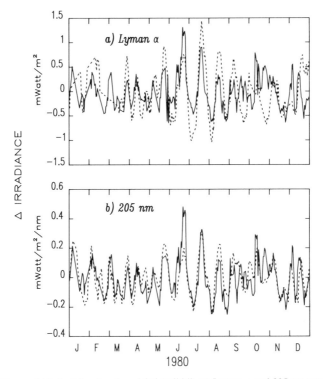

Fig. 25. Observed (dashed lines) and modeled (solid lines) Lyman α and 205-nm rotation mod-ulation in 1980. The Lyman α and 205-nm data are from the AE-E (Hinteregger et al. 1981) and Nimbus 7 SBUV (Heath et al. 1984) satellite experiments, respectively. The calculations at Lyman α are from Lean and Skumanich (1983) and at 205 nm from Lean (1984). Both the AE-E Lyman α data and the calculated Lyman α plage emission have been detrended by removing a quadratic fit to the daily data. The observed and calculated 205-nm irradiances have been detrended by removing a linear fit to the daily data.

total irradiance S, the residual times series, $S - P_s$ is well correlated with the solar spectral irradiance variations at 205 nm, which were shown in Fig. 25 to be caused by the excess emission from the bright Ca K plage areas recorded by the WDC/NOAA.

Growth, Evolution and Decay of Magnetically Active Features Throughout the 11-yr Solar Cycle

The totally quiet Sun (a condition never realized at present) is character-ized by an absence of sunspots, faculae and plages. As solar magnetic activity increases, the fractional disk areas of both sunspots and faculae/plages in-crease. Individual active regions tend to emerge at preferred heliocentric lon-gitudes, in groups called active complexes. Within these complexes, en-hanced magnetic activity is sustained by new flux injections for a number of

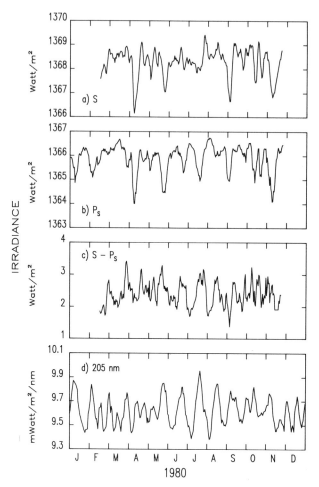

Fig. 26. Daily irradiance variations in 1980 (adapted from Foukal and Lean 1986). In (a) are the total-irradiance measurements made by ACRIM. The calculated modulation of the total irradiance by sunspot blocking alone, with $S_Q = 1366.8$ W m^{-2}, is given in (b). The residual time series $S - P_s$ is shown in (c), and in (d) are the daily solar irradiances at 205 nm, measured by the SBUV instrument. Each of the four time series has been detrended by subtracting from the data a linear fit to the daily data during the entire year, then adding to the data the yearly mean.

solar rotations (see, e.g., Gaizaiskas et al. 1983). The growth, evolution and decay of active regions within complexes modulates the amplitude and phase of the 27-day irradiance variations associated with solar rotation, and generates additional irradiance variations over time scales of months and years.

Shown in Fig. 27 are the solar Lyman α and 205-nm irradiances, and the spectrally integrated facular emission (as determined from $S - P_s$), averaged over approximately 3 solar rotations (81 days) during solar-cycle 21.

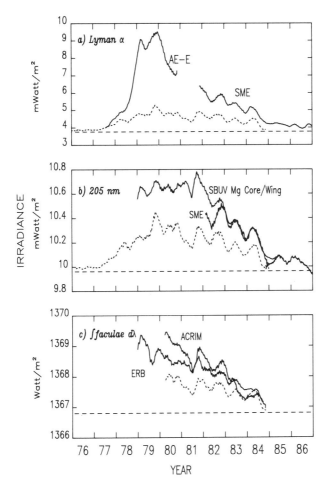

Fig. 27. Comparison of 81-day running means of the daily measured total solar-irradiance variations during solar-cycle 21 (solid lines) with the calculated variations due to WDC Ca K plage emission (dotted lines) at (a) Lyman α, (b) 205 nm, and (c) all wavelengths, as determined by $S - P_S$. The dashed lines are the adopted irradiances of the quiet Sun. The AE-E data have been normalized to the plage emission in 1977. The $S - P_S$ faculae data have been normalized to the adopted quiet Sun total irradiance.

Although some discrepancies between observations made by different instruments are evident, there is general agreement that both the intermediate-term variations and the overall long-term downward trend during the descending phase on the cycle, are indeed real solar variations.

For each of the time series in Fig. 27 it is the modulation by the disk-integrated emission from bright active regions, rather than by dark sunspots, which controls the variations, at least over solar-rotation time scales. Also shown in Fig. 27 are calculations of the changes in the bright emission from the WDC/NOAA Ca K plage. It is evident that the intermediate-term, low-amplitude irradiance variations correlate well with plage emission variations, but the overall long-term downward trend present in the total solar-irradiance observations is not present in the Ca K plage index. This is true for each of the three spectral regions, Lyman α from the upper chromosphere, 205 nm from the middle photosphere and the spectrally integrated emission dominated by the emissions near $\tau = 1$. It has also been shown to be true for the full-disk Ca K variations (Skumanich et al. 1984) that originate in the middle chromosphere. These results suggests that a real solar component of the total-irradiance variations is absent in the WDC/NOAA plage data, which reports only on compact plage areas.

Figure 28 compares the ACRIM-P_s residuals with the full-disk helium 1083-nm equivalent-width variations, as well as with the plage index. As in Fig. 27, the total solar-irradiance data exhibit similar long-term downward trends but the plage data do not. Because of this, the helium data provide a more uniformly distributed regression with $S - P_s$ than do the Ca K plages. By using the linear correlation between the ACRIM-P_s and helium data, Foukal and Lean (1988a) have reconstructed the facular irradiance contribution to the Sun's total irradiance since 1974, which allows the total-irradiance variations caused by both dark spots and bright faculae then to be estimated throughout Cycle 21; these calculations are shown in Fig. 29.

Observational information about the Sun's total- and ultraviolet-irradiance variations during epochs prior to Cycle 21 is sparse. In the absence of daily, continuous total solar-irradiance data prior to Cycle 21, other than at 10.7 cm, Lean and Foukal (1988) have estimated historical variations in the total solar irradiance during Cycles 19, 20 and 21 by using a linear correlation between the monthly $S - P_s$ residuals and the 10.7-cm flux during the period 1981–84. These results, shown in Fig. 30, illustrate how the magnetic modulation of the total irradiance depends on the magnitude of the competing effects of sunspots and faculae. Surprisingly, in Fig. 30, Cycle 19, the most active cycle in the 20th century, does not exhibit as large a total irradiance variation as does Cycle 21.

Discussion

Modulation of the Sun's radiative output by magnetic activity is a primary cause of total solar irradiance variability. A simple construction of both the total and ultraviolet irradiance variations derived from groundbased observations of sunspots and plages, and their average contrasts, reproduces the observed irradiance variations sufficiently well to establish that active-region modulation is almost completely responsible for irradiance variations over time scales of solar rotation and active region evolution, as well as for at least half of the solar-cycle variations. It remains to be determined whether the additional solar-cycle variations are caused by global solar changes or by

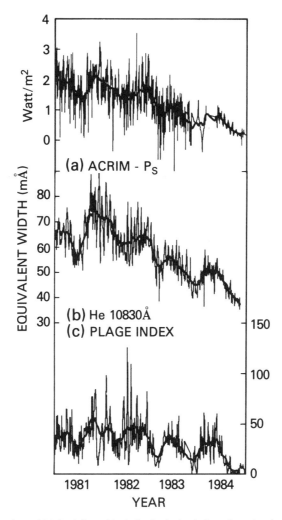

Fig. 28. Comparison of (a) the daily residuals $S - P_S$ obtained after subtracting the sunspot blocking function from the ACRIM total irradiance data with daily values of (b) the He 1083-nm
equivalent line widths and (c) the Ca K plage index. Smooth curves are 81-day running means
(figure from Foukal and Lean 1988a).

simply the enhanced emission from bright regions of a smaller spatial scale
than those recorded by the WDC/NOAA.

There are a number of assumptions implicit in the expressions which are
currently used to calculate $F(\lambda,t)$ and $S(t)$. F_Q is considered invariant from
one solar cycle to the next. C_A is not a function of time nor is A a function of
wavelength (i.e., height in the solar atmosphere). Representative values of C_S

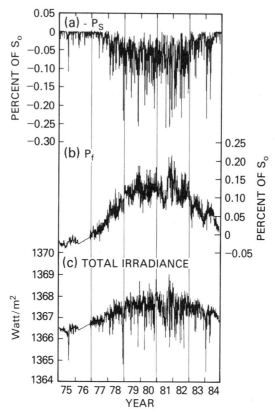

Fig. 29. Solar-cycle 21 behavior of (a) the daily sunspot blocking function P_S; (b) the daily facular irradiance excess calculated from the He 1083-nm data; and (c) the reconstructed total solar irradiance (figure from Foukal and Lean 1988a).

and C_p are used for all sunspots and plages, respectively. The center-to-limb dependence of ultraviolet plage contrast is unknown. Thus far, calculations of $S(t)$ have employed only bolometric sunspot and faculae contrasts while calculations of emission variations at λ <250 nm neglect the sunspot blocking effect. Additional data are needed to provide a complete data base by which the total solar irradiance observations can be constructed from active-region emissions. Of particular value would be solar cycle observations of the Ca K emission from selected high-latitude regions on the solar disk, such as have been measured at the disk center during Cycle 21 (e.g., Sec. III). These data might help to resolve the dilemma of global change vs magnetic modulation as the origin of solar-cycle total irradiance variations. Note that Schatten (1988) has postulated an important role for polar faculae as a source of magnetic modulation of the total irradiance, in addition to that by bright regions associated with the WDC/NOAA Ca K plages.

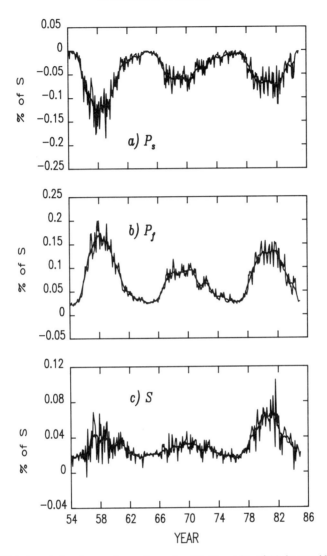

Fig. 30. Variations during the past three solar cycles, 19, 20 and 21, of (a) the monthly sunspot blocking function P_s; (b) the monthly facular irradiance excess calculated from a correlation between monthly $S - P_s$ residuals and monthly 10.7-cm radio fluxes, and (c) the reconstructed total solar irradiance (figure from Lean and Foukal 1988a).

VI. MEAN MAGNETIC-FIELD VARIABILITY

The line-of-sight mean magnetic field (MMF) of the Sun as a star is deduced from the Zeeman splitting of one or more Fraunhofer lines as measured with a Babcock-type magnetograph. Traditionally, the line of Fe I 525.02 nm has been used. As most field regions on the Sun are bipolar, the

MMF is small, ranging between about ± 1 gauss depending on the time in the activity cycle (Fig. 31). Chance imbalance of flux, as active regions rotate onto and off the disk, are part of the source of a MMF. Brightness weighting by sunspot umbrae, penumbrae and faculae may be another source. More interesting is the apparent development of large-scale unipolar regions that have a significant influence on the MMF and, as it turns out, on the interplanetary magnetic field. Long-term measurements of the MMF have been carried out at the Crimean, Mt. Wilson and Stanford Observatories. Recently, the Sayan Observatory has joined in this effort.

Observational Methods

In common with the measurement of total irradiance Fraunhofer line variability (Sec. III), the optical system of the telescope and spectrograph must provide for illumination of the diffraction grating so that all portions of the solar disk participate with equal weight. Minor defects arising from mirror blemishes introduce proportional errors in the sampling of the solar disk. Limb darkening, solar rotation and the convective limb red shift lead to a

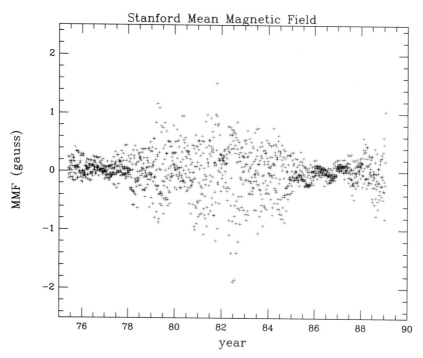

Fig. 31. The mean magnetic field as measured at the Wilcox Solar Observatory (Stanford). A 4-day running mean has been applied to the data.

preferential weighting of disk center (Scherrer 1973). Details of the optical schemes for MMF measurements have been published by Severny (1969), Scherrer (1973), Grigoryev et al. (1983), and Grigoryev and Demidov (1987).

Crucial to MMF measurements is the determination of the instrumental zero. At the Sayan Observatory, continuous zero control is accomplished as follows. Following each magnetic measurement, another recording is repeated with a half-wavelength plate introduced into the light beam in front of the coelostat. The half-wavelength plate reverses the directions of polarization of the σ-components of Zeeman splitting. This introduces a change of sign of the magnetograph signal, but only for that part of it that is due to the magnetic field; the component of the zero signal caused by the instrumental polarization and electronics remains unaltered. The half-difference of the two measurements then gives a result that is free from zero point error.

Imbalance of Magnetic Fields and Solar-activity Cycle

Evidence for an imbalance of magnetic fields on the Sun when opposite-polarity fluxes do not fully cancel each other, and the Sun then behaves as a "magnetic monopole" during a time interval longer than one rotation, was noticed from the very beginning of magnetographic measurements (Babcock 1959; Babcock and Babcock 1955; Bumba and Howard 1965; Severny 1966,1968). Recently Grigoryev and Demidov (1987) have analyzed Stanford and Sayan MMF strength for the interval 1982–84. They find conclusive evidence for long-duration imbalances of the MMF. The data from the two observatories agree that the MMF was consistently negative during this interval.

Results presented in Table III show that the magnetic monopole of the Sun was negative during 1968–1970 and positive from 1970 to 1981 (except for the years 1974 and 1978, which may well be a real fluctuation), negative from 1982 to 1986 and positive in 1987, the first year of a new cycle of activity. Data on a global imbalance of solar magnetic fields for the time interval 1957–1968 may be found in a number of papers (Babcock 1959; Bumba et al. 1967; Severny 1971; Howard 1974). Thus, early observational data, together with the data listed in the table, cover a >30-yr time interval and affords a unique possibility of analyzing the asymmetry of magnetic fields. A change of sign of the observed solar monopole, plotted from these data, is shown in Fig. 32. It is evident that changes of sign in Cycles 19–21 occurred at epochs of maxima activity, with the sign remaining unchanged in between. But in 1987, a change of sign of the monopole occurred at the epoch of minimum activity. If data of the subsequent years happen to confirm conservation of sign, then it is important to continue MMF observations for this purpose. Figure 32 also shows schematically that magnetic field polarity at the solar poles displayed an alteration. Comparison shows that the MMF of

TABLE III
The Yearly Mean Values $\bar{H}$ (μT) of the Solar MMF Strength as Measured at Three Observatories

Year	Crimea $\bar{H}$	Sayan $\bar{H}$	Stanford $\bar{H}$	$\bar{H}'^{a}$	$\tilde{g}_0^{b}$
1968	−28				
1969	−39				
1970	−2				
1971	8				
1972	34				
1973	14				
1974	−24				
1975	−28		4		
1976	−15		4		
1977			4	3	2
1978			−2	−1	−2
1979			6	3	4
1980			6	4	2
1981			5	+0	6
1982		−13	−2	−3	−0
1983		−8	−3	−4	−4
1984		−22	−8	−6	−6
1985			−4	−4	−4
1986			−1	−1	
1987			3	2	

[a] $\bar{H}'$ represent yearly mean values obtained by comparing yearly series with a period of 27 days.
[b] $\tilde{g}_0$ is a yearly mean value of the zero harmonic in the expansion of synoptic maps for background magnetic fields.

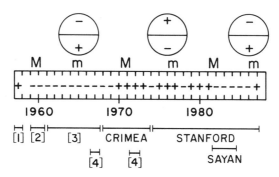

Fig. 32. The polarity of the solar magnetic monopole during the period 1957–1987. M and m denote epochs of maxima and minima of solar activity. At the top, one can see a schematic drawing of the magnetic field polarity in the polar regions of the Sun.

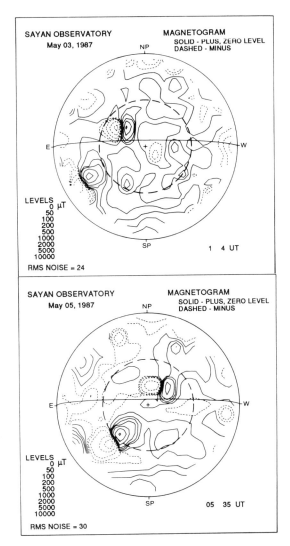

Fig. 33. Magnetograms of background magnetic fields from the Sayan Observatory for May 3 (plot a) and 5 (plot b) 1987. A dashed circle shows the central area 16′ in diameter.

the Sun coincides in sign with the field polarity at the north pole during 1957–1986 and with the polarity at the south pole in 1987. Because activity was higher in the northern and southern hemispheres in Cycles 19–21 and in the year 1987, respectively, it might be anticipated that the polarity of the observed monopole is determined by the magnetic-field polarity at the pole of that hemisphere of the Sun in which the activity is higher.

Short-Period Variations of the Mean Magnetic Field

The existence of rapid variations of the MMF was pointed out as early as two decades ago by Severny (1971). Analyses of long-term continuous MMF observations at the Crimean Observatory and at Mount Wilson Observatory has provided evidence for the existence of MMF oscillations with a period of 160 min associated, perhaps, with the well-known solar velocity oscillations (Kotov et al. 1976). Possibly, MMF variability occurs also on a shorter time scale. Ioshpa et al. (1973) reported the detection of 5-min oscillations of the mean magnetic field of the Sun. Because of the insufficient amount of observational data, the question as to whether the short-period variations of the MMF do exist remains an open question that begs to be answered.

Recently Demidov et al. (1988) returned to the analysis of MMF observations from the Crimea and Mount Wilson Observatories, combined with new observational data from the Sayan Observatory. All data were examined to determine the reality of the MMF variations in the range of periods from 300 to 5 min. Evidence was found for weak power at the 160-min period. On the other hand, Scherrer (1989) recently reported no evidence of a 160-min period in the Stanford data. So the issue remains undecided.

From Doppler measurements, it is known that changing from center-disk, line-of-sight velocities to full disk causes a decrease of observed oscillation amplitude. Instead of an amplitude of ~ 300 m s^{-1} in center disk for 5-min velocity oscillations, full-disk measurements of the Sun yield velocities 10^2 times smaller (Claverie et al. 1981d). A similar situation should be expected for MMF oscillations as well. If the local oscillations of the magnetic field have an amplitude of 200 to 300 μT (Tanenbaum et al. 1971), then in the MMF the oscillations amplitude must be of order of a few μT.

However, the situation may be more complicated than indicated above. Figure 33 shows the magnetograms of large-scale distribution of the magnetic field on the resolved disk, where a dashed circle indicates a region of spatial averaging of the magnetograph signal in which the mean magnetic field was measured locally. When the magnetograph aperture is occupied largely by a unipolar field (May 3, 1987, for example), then the oscillation amplitude is greater than when the same-size area is located on a bipolar region (May 5, 1987). Thus, the character of the 5-min mean magnetic field oscillation can be expected to vary with time.

WHAT CAN OTHER STARS TELL US ABOUT THE SUN?

ROBERT W. NOYES, S. L. BALIUNAS,
Smithsonian Astrophysical Observatory

and

E. F. GUINAN
Villanova University

We focus on those aspects of the solar-stellar connection where observations of other stars give information about the Sun which could not be obtained from study of the Sun alone. Solar-like phenomena on other stars with a range of values for key parameters (e.g., rotation) yield the dependence on those parameters and hence better physical insight into the governing mechanisms. A key presumption is that the Sun is a normal star, so that the same mechanisms govern its behavior. Also if the Sun is normal, then study of stars with similar masses but different ages gives information on the evolution of solar properties, such as structure, internal dynamics, activity, etc. First, we discuss stellar observations which yield information on the internal structure of solar-like stars, including central density, and helium abundance. Such results support the standard value for the solar helium abundance, as well as standard stellar structure theory. We note how stellar seismology can, in principle, determine the stellar radius, as well as the degree of mixing in stellar cores, as a function of age. However, to date we can only anticipate how such results might help us refine the solar picture. Next, we discuss information on the internal dynamics of the Sun, and its evolution, as inferred from the study of the time history of surface rotation in solar-type stars. We use the rotation-activity-age connection to infer how the Sun's rotation and activity level has changed with age, the latter both in terms of chromospheric activity and photometric modulation (spots). We discuss what the dependence of activity on mass and rotation can tell us about the nature and location of the solar dynamo. Finally, we discuss the solar activity cycle and its likely change over the lifetime of the Sun, inferred from observations of other stars.

I. INTRODUCTION

The solar-stellar connection is a well-mapped highway to understanding stars similar to the Sun. Traffic on this highway moves mostly in one direction: detailed observations of the Sun give meaning to much sketchier data on other stars. But it is not a one-way road, for other stars can contribute to our understanding of the Sun itself. Here we restrict our discussion to the return route, noting some key ways in which study of other stars illuminate the workings of our own Sun. Also, to keep this review to manageable length, we concentrate on slowly varying aspects of the Sun—its internal structure and evolution, and its magnetic activity when viewed on time scales of lifetimes of active regions or longer.

Underlying the use of other lower-main-sequence stars to understand the workings of the Sun is the assumption that the Sun is a normal star, with properties similar to those on other normal stars. In studying the Sun by itself we can investigate only one point in a multi-dimensional space of stellar properties, defined (primarily) by mass, age, rotation and chemical composition. Other stars allow us to explore this multi-dimensional space, and even the relatively crude picture that results may help us to understand better how our own Sun works.

Of course, there are no *a priori* reasons why the Sun should be normal, that is, with behavior not very different from that of other stars with similar mass, age, rotation rate and chemical composition. Observations, where direct comparisons are possible, provide some reassurance on this score. First, a single theory of stellar structure and evolution predicts extremely well the basic observable properties of both the Sun and similar stars (i.e., luminosity and radius as a function of mass, age and chemical composition). Second, the nonspherically symmetric features of the Sun, evidently associated with its magnetic fields, appear grossly similar to those of other stars. For example, the Sun obeys the relation between rotation and magnetic activity (e.g., chromospheric and coronal emission) defined by other lower-main-sequence stars, has activity cycles like those of other stars of similar mass and rotation rate, and shows spot-associated photometric modulation appropriately scaled down from more rapid rotators of similar mass.

Beyond the assumption that the Sun is normal, we further assume that the evolution of the Sun's structure and magnetic fields has been the same as that of other stars with the same mass and chemical composition. This requires that all Sun-like stars follow an evolutionary timeline depending only on mass and chemical composition, and independent of initial conditions such as initial angular momentum. For this assumption to be valid, we must restrict the term "Sun-like" to lower-main-sequence stars that are single, or at least stars whose companions are so small or so distant from the primary that they do not affect its rotation. (Binaries where there is significant tidal interaction, such as RS CVn stars, show significant departures from the activity-age connection displayed by single stars. Apparently the angular momentum reservoir in the orbital motion of the binary keeps the tidally locked stellar rotation from decaying as it does in single stars, and the rotationally induced activity remains high as well.)

The assumption of independence on initial conditions greatly simplifies the interpretation of observational data, because then important stellar properties, such as rotation and magnetic activity, depend mainly on mass and age. For example, we may hope to infer the age of a star from its measured mass and rotation rate (Soderblom et al. 1991; Barry 1988). The assumption also allows us to study stars with the same mass as the Sun, but different ages, and infer the behavior of the Sun at a younger or older age than at present. This is one of the key ways in which other stars tell us about the Sun. Finally, however, we must keep in mind that our two assumptions—that the Sun is typical of 1 $M_\odot$ stars, and that the rotation and activity of such stars depends only on their mass and age—are not really established facts, and retain a healthy skepticism about the evolutionary scenario they imply for the Sun.

II. WHAT OTHER STARS CAN TELL US ABOUT THE STRUCTURE AND EVOLUTION OF THE SOLAR INTERIOR

Helioseismology and solar neutrino observations are two recently developed tools for probing the solar interior, While it is unlikely that we can record the stellar analogs of solar neutrinos in the foreseeable future, asteroseismology should reveal enough about the internal structure of other Sun-like stars to provide useful information on the solar interior, both the structure of the deep interior and convection in the envelope. Additional information on solar convection can be supplied by inferences on stellar granulation from line profiles. Finally, information on stellar interior structure is provided by eclipsing binary systems, and this too has implications for solar structure.

A. Asteroseismology and Solar Interior Structure

Asteroseismology will soon become an important tool for directly probing stellar interiors. But what can it tell us about the Sun? Determination of the spectrum of low-degree p-modes will shed light on the interiors of a few Sun-like stars (Gough 1987) and thereby, indirectly at least, on the solar interior. This is particularly true for stars that are components of binary systems, such as α Cen A (G2V), or Procyon (F5IV-V), for which there exist good mass determinations. A possible result should be further confirmation of the standard theory of evolution of the solar interior, and perhaps refinement of some detailed aspects of that theory (e.g., the mixing-length formalism for the convection zone) to produce optimum accord with data for stars of different mass, rotation rate and age. For example, we might hope to learn whether the mixing-length to scale-height ratio has the same value for stars of different masses and convective zone properties; such information could help us understand solar convection better. For single stars, however, uncertainties in their mass determination may seriously compromise the structural information available from p-mode frequencies (Gough 1987).

A longer-term and more difficult goal would be to obtain data on splitting of low-degree modes, at least for rapidly rotating (and hence younger) Sun-like stars. This, combined with surface rotation measurements, could yield information on the depth gradient of rotation within these stars. Doppler-imaging observations might facilitate the separation of prograde and retrograde modes. The results could lead to better understanding of the evolution of the Sun's interior rotation, which presently seems surprisingly small compared to dynamical evolution calculations (see, e.g., Pinsonneault et al. 1989).

In addition to the spectrum of p-modes, asteroseismology yields the envelope of excitation of these modes. For example, Brown et al. (1990) have found that the p-mode spectrum for Procyon appears to peak at about an 18 min period. This provides information on the mechanism of excitation of stellar acoustic modes. If acoustic modes are excited in stars by turbulent convection, as is commonly thought, then the envelope of these modes for other stars, analogous to the solar p-mode envelope peaking at about 5 min, should peak at a period corresponding to the convective overturn time at the top of the convection zone, modified by the acoustic cutoff frequency for the stellar photosphere (Christensen-Dalsgaard and Frandsen 1983; Libbrecht 1988a). This convective turnover time depends on the velocity and scale length for convection near the surface (granulation), both of which can be modeled by mixing-length theory. (Independent information on stellar granulation velocities comes from spectral line profile observations; see below.) Also, the energy in the stellar p-mode oscillations should scale with convective velocity. But if the excitation is, say, through the kappa mechanism, neither the energy nor the envelope of the oscillation spectrum should show such relations. Therefore, stellar oscillation data may provide either confirmation or refutation of the current leading candidate for excitation of stellar and hence solar p-modes. In addition, measurements of the lifetimes of stellar p-mode oscillations may give information on the damping mechanism of p-mode oscillations.

B. Information on Solar Convection and Granulation from Stellar Spectra

Exciting progress has been made in recent years in both observation and modeling of stellar granulation (see, e.g., Dravins 1989). The observed C-shape of line profile bisectors in the Sun is thought to be caused by the statistical mix of hot and bright upward-moving granular elements and cold downward-moving ones. A similar C shape is found in other Sun-like stars such as α Cen A and Procyon. Nordlund and Dravins (1990) have carried out detailed numerical simulations, in which the line profile from the stellar disk is synthesized from a mix of up- and down-moving elements whose properties are determined by numerical hydrodynamics and radiative transfer. They have succeeded in making a detailed match with observed profiles, and in the process deduced many properties of granulation on these and other stars, such as size of granules, brightness contrast, center-to-limb appearance and velocity spectrum.

What does this tell us about solar granulation? First, the impressive match for other stars with differing gravities, opacities and effective temperatures gives confidence in the numerical hydrodynamic codes, which also reproduce the main characteristics of the solar granulation, *without* recourse to the concepts of mixing length, or micro- or macroturbulence. Second, the observed profiles from other G stars give information on the evolution of convection in the Sun. Nordlund and Dravins (1990; see also Dravins 1989) have found that strong photospheric lines in α Cen A are more asymmetric than the corresponding solar lines; they conclude that the lower surface gravity of α Cen A (by a factor of 2 relative to the Sun, due to its increased radius as it evolves toward the subgiant stage) permits a greater degree of convective overshoot, so that higher velocities are seen in the strong lines. The numerical simulations of Nordlund and Dravins reproduce this effect, and in addition indicate that the granules are greater in size on α Cen A, in proportion to its greater photospheric scale height. These results give a reasonable prediction of the nature of solar granulation some 3 Gyr hence.

C. Internal Structure of Sun-like Stars from Eclipsing Binaries

If a star is a member of an eclipsing binary system with an eccentric orbit, it is sometimes possible to infer its internal mass distribution by measuring the apsidal motion of its orbit. The apsidal motion is measured from the time rate of change of longitude of periastron ω of the orbit, as determined from observations of the times of primary and secondary eclipses over a few decades of time.

In an isolated binary system, apsidal motion arises from the classical quadrupole-moment produced by the tidal and rotational distortions in the shapes of the stars. (An additional contribution due to general relativity must be subtracted.) Expressions for the classical and general relativistic apsidal motion are given in Guinan and Maloney (1985). They include internal density concentration parameters k_2 for the two components, which are the only quantities in the expressions which cannot be directly measured from observations; therefore, measurements of the apsidal motion allows the determination of the mean value of k_2 for the two stars. The best determination of k_2 for stars of a given mass are obtained from binaries containing a pair of similar stars; fortunately many close binary systems have components of nearly equal masses and spectral types.

The k_2 parameter is related to the ratio of the central to mean density.

This ratio depends on the mass, evolutionary age and chemical composition of the star (see Jeffery 1984; Hejlesen 1987). Thus measurement of apsidal motion of an eclipsing binary with well-determined physical and orbital properties and with components of equal mass allows us to determine the internal mass distribution of the stars and compare it with modern stellar structure calculations.

A number of such studies have been carried out over the past 20 yr (Jeffery 1984; Gimenez and Garcia-Pelayo 1982), with generally satisfactory agreement between observations and theory. Unfortunately, there are very few suitable eccentric eclipsing binaries with solar-type components. One is V1143 Cyg (HD 185912; HR7484). This consists of two nearly identical F5V stars in a highly eccentric orbit ($e = 0.54$) and orbital period of 7.64 d. It has deep, narrow eclipses, permitting accurate timing of eclipses and hence determinations of apsidal motion (see, e.g., Anderson et al. 1987). The most recent timings yield an apsidal motion of 3.26 ± 0.15 deg/100 yr (Burns and Guinan, in preparation). Subtracting a relativistic contribution of 1.86 deg/100 yr leaves a classical quadrupole-moment effect of 1.40 deg/100 yr, implying $k_2 = 0.0041$. In turn this yields a ratio of central to mean density $\rho_c/\rho = 210$, or $\rho_c = 170$ g cm^{-3}, using a mean density $\rho = 0.81$ g cm^{-3} and the observationally determined mass and radius of the two stars.

Comparison with models indicates very good agreement between observations and theory, providing the helium abundance is appropriately chosen. For hydrogen, helium and metal abundance $X = 0.70$, $Y = 0.28$, and $Z = 0.02$, Hejlesen (1987) models give $k_2 = 0.0043$, close to the value from observations. A value of $Y = 0.18$ gives a substantially poorer fit. This implies a helium abundance close to $Y = 0.28$ for these stars, in good agreement with the helium abundance $Y_\odot = 0.27$ for standard solar models (see, e.g., Bahcall and Ulrich 1988).

Thus the apsidal motion study of V1143 Cyg provides important information, difficult to obtain any other way, on the internal structure of young solar-type stars; the results support the predictions of standard stellar structure theory. Moreover, the inferred helium abundance is in good accord with the helium abundance of standard solar models, and provides some measure of confidence in such models, at least for stars near the Sun's mass.

A few other eclipsing binaries containing Sun-like stars with well-determined physical properties also offer an opportunity for determining the helium abundance Y of Sun-like stars. A recent detailed study of an eclipsing binary with solar type stars, AI Phoenices (F7V:1.20 M$_\odot$ + KO IV:1.24 M$_\odot$, $P = 24.6$ d) by Andersen et al. (1988) yields a helium abundance of $Y = 0.27 \pm 0.02$. This value is well determined and is essentially identical to the inferred solar helium abundance. AI Phe is the only known eclipsing binary which contains stars similar to the Sun in mass and age ($\tau = 4.1 \pm 0.4$ Gyr). With one component near the main sequence and its slightly more massive partner already on the early giant branch, this is an ideal binary pair for providing an empirical check of stellar evolution models. Because the masses, radii and colors of both stars are known with great precision ($\leq$ 1%), their age and helium abundance can be derived with some confidence.

Furthermore, the values derived for the helium abundances of V1143 Cyg and AI Phe are in excellent agreement with the value of $Y = 0.27 \pm 0.03$ indicated by analysis of the Hyades eclipsing binary HD 27130 (Popper and Ulrich 1986). The results for these three binaries strongly support modern theoretical models with helium abundances for solar-type stars and in particular the Sun. The properties of the stars (masses, spectral types, metal abundance and age) along with the helium abundances are given in Table I.

III. WHAT OBSERVATIONS OF STELLAR ROTATION AND ACTIVITY CAN TELL US ABOUT THE SUN

A. History of Solar Rotation and Activity

1. The Evolution of Solar Rotation. In recent years observations of stars in clusters have given us a good picture of how the surface rotation of Sun-like stars varies with time, and by extension, the history of solar surface rotation. This, together with theory on the coupling of interior and surface rotation, suggests how the solar interior rotation may have evolved. For reviews of recent work, see Hartmann and Noyes (1987) and Stauffer and Hartmann (1986).

Briefly, pre-main-sequence (T Tauri) stars of 1 $M_\odot$ are found to have rather small rotational velocities (about 10 to 30 km s^{-1}) compared to breakup velocities of about 200 km s^{-1}, so that they probably lost substantial angular momentum during star formation or early in their T Tauri phase. But data on the very young (main-sequence age $\sim$ 50 Myr) α Persei cluster (Stauffer et

TABLE I
Solar-Type Stars with Well-Determined Helium Abundances

Star	Sp,Mass	Sp,Mass	[Fe/H]	Age(Gyr)	Y	Ref[a]
V1143 Cyg	F5 V 1.39	F5 V 1.35	+0.08	0.6	.28 ± .03	1
HD 27130	G8V 1.06	K2 V 0.765	+0.12	0.6	.27 ± .03	2
AI Phe	F7 V 1.196	K0 IV 1.236	−0.14	4.1	.27 ± .02	3
Sun	G2 V 1.00		0.00	4.6	.27 ± .02	4

[a] 1: Guinan and Burns 1989; 2: Popper and Ulrich 1986; 3: Anderson et al. 1988; 4: Bahcall and Ulrich 1988.

al. 1985), imply that at least some T Tauri stars experience surface spin-up during their final contraction toward the main sequence. This can be explained (Hartmann et al. 1986) by the decrease of moment of intertia as the star contracts about threefold in radius, assuming little or no angular momentum loss during this time.

By the age of the Pleiades of about 70 Myr (Maeder and Mermilliod 1981)—that is, slightly older than α Persei, the G dwarfs appear to have slowed substantially, whereas the K and M dwarfs have not. A plausible explanation is that only the surface convection zones are slowing, and that the relatively shallow G star convection zones, having less moment of inertia than those of K or M dwarfs, spin down more quickly. This requires that the radiative interior of lower-main-sequence stars continues to rotate rapidly. By the age of the Hyades (~ 0.6 Gyr; Maeder and Mermilliod 1981), both G and K dwarfs have slowed to a rate consonant with the well-known Skumanich (1972) law, $v_{rot} \sim age^{1/2}$. Furthermore, the spread in rotation among stars of the same mass has declined to about 20%, in contrast to the much greater spread among similar Pleiades stars. This suggests that a mass-dependent "feedback" is operative, such that faster or slower rotators experience greater or lesser spindown, respectively. Presumably faster rotators have greater activity, more powerful winds, and hence greater spindown, with opposite trends for slower rotators.

Applied directly to the Sun, the above results suggest that the surface rotation period of the Sun was only about 3 d at the age of the Pleiades, and slowed to about 10 d by the age of the Hyades (Fig. 1). This behavior is in rough agreement with a model of Endal and Sofia (1981), except that their model predicts much too high a surface rotation for the present-day Sun. More serious is the disagreement in their model between the interior rotation of the Sun and the interior rotation implied by helioseismology data. Endal and Sofia predicted that a rapidly rotating interior should be found in the Sun today, extending out to the base of the convection zone. As they noted at the time, this disagrees with upper limits to the solar gravitational quadrupole moment. Also, recent helioseismic data (Libbrecht 1989) show that a rapidly rotating interior, if it exists, must be confined to the deep interior ($R/R_\odot \lesssim 0.2$ (see Demarque and Guenther 1988). Pinsonneault et al. (1989) have calculated evolutionary solar models including solar-wind angular-momentum loss and internal redistribution of angular momentum; they find that the angular velocity of the present-day Sun should be significantly higher than indicated by helioseismic data for $R < 0.4R_\odot$. Thus there is a serious conflict between theory and observation.

Furthermore, the best model of Pinsonneault et al (1989), when evolved onto the subgiant branch, implies a rotation period of order 50 d, whereas if the Sun is rotating essentially uniformly today, it should have a rotation period of > 150 d when it becomes a subgiant. The sparse data for rotation of subgiants of solar mass indicates that they have much shorter rotation periods,

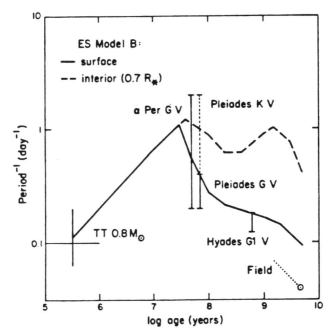

Fig. 1. Measured surface rotation periods of solar-type stars as a function of age, and calculated rotation periods at the surface (solid line) and near the base of the convection zone (dashed), according to Model B of Endal and Sofia (1981). Dotted line is the Skumanich law fit to the Sun (figure from Hartmann and Noyes 1987).

in the range 40 to 50 d; five such cases are given in the tabulation of Noyes et al. (1984b). (Here the rotation periods were not directly measured, because the amplitude of rotational modulation was undetectably low, but rather predicted by the star's chromospheric emission, assuming the same scaling between rotation and activity found for lower-main-sequence stars. Clearly, directly measured rotation periods are needed for verification.) Unless the inferred rotation periods are gross underestimates, the combined subgiant and helioseismology data imply a region of fast rotation within the inner 0.2 $R_\odot$ of the Sun today (see also Demarque and Guenther 1988).

Summarizing, from study of rotation in lower-main-sequence stars of different mass and age, we learn about the Sun the following:

(a) It probably experienced spin-up as it contracted to the ZAMS, where it had a rotation period of only a few days. Large angular momentum loss through magnetic stellar winds, coupled with the relatively small moment of inertia of its convection zone, produced rapid spindown. A feedback mechanism, presently not well understood, caused it later (certainly by the age of the Hyades) to join the equilibrium evolutionary track of surface velocity with age generally associated with the Skumanich law.

(b) The relation between early spindown and convection zone depth suggests that the Sun should have (at least initially) retained much of the angular momentum of its radiative interior. Helioseismology results imply that the present-day rotation curve is flat to about $R/R_\odot = 0.2$, a result that requires effective angular redistribution mechanisms beyond those contained in the theoretical models of Pinsonneault et al. Such a flat rotation curve would also be inconsistent with the sparse existing data on rotation of 1 $M_\odot$ subgiants, and further observations (both helioseismology and stellar rotation measurements) and theory are needed to resolve this conflict.

2. *Evolution of the Mean Level of Solar Activity.*   It is now well accepted that chromospheric and coronal emission is an excellent index of stellar magnetic activity (see, e.g., Skumanich et al. 1975; Linsky 1980). Much of what we know about the evolution of stellar, and hence solar magnetic activity, is based upon measured chromospheric emission from other stars, because chromospheric emission may be easily measured from groundbased telescopes (principally in the Ca II H and K lines), and therefore there is a very large data base from such observations.

Recent work has established a quantitative relation between Ca II H and K emission and magnetic flux density from point to point on the spatially resolved Sun (Schrivjer et al. 1989b). A minimum level of K-line core emission occurs even in nonmagnetic areas, probably through acoustic heating; the amount is about the same as seen in the disk-averaged flux of the least active solar-type stars, which therefore may have negligible surface fields. In solar magnetic regions there is an excess K-line core emission which increases nonlinearly with increasing magnetic flux density, saturating at a level similar to that in the most active solar-type stars; this suggests that such stars may be characterized by atmospheric structure similar to that above solar plages, but covering essentially the entire stars.

The variation of mean levels of stellar H and K emission with age in Sun-like stars therefore should tell us considerably about the evolution of magnetic activity on the Sun. It is well known that the long-term averaged chromospheric flux level is correlated both with the star's rotation and its main-sequence age (Kraft 1967; Skumanich 1972). (Here "long-term" means over times longer than the star's characteristic activity cycle time and perhaps much longer, if the solar Maunder minimum is any guide; this imposes obvious uncertainties given the relatively short data base on stellar chromospheric activity.) The long-term averaged chromospheric Ca II K emission flux decays smoothly for solar-mass stars, ranging in age from that of the Pleiades (about 70 Myr) to the Sun (4.6 Gyr). The surface rotation declines roughly in step with surface activity, and it appears likely that a decrease in stellar rotation rate, rather than the aging process itself, is the cause of the decreasing overall level of chromospheric activity. The best evidence that rotation rather than age is the controlling factor is provided by the binary RS CVn and BY Dra stars, that maintain their rapid rotation rates while aging (they draw upon the angular momentum of the orbital motion of the binary components), and also maintain their high levels of chromospheric magnetic activity.

In recent years, an extensive data base has been acquired on both rotation and magnetic activity in older, slowly rotating stars like the Sun. Rotation is difficult to measure by traditional line broadening analyses for slowly rotating stars because the Doppler broadening is subtle. However, nightly monitoring of the brightness of emission in a chromospheric line such as Ca II H and K (hereafter labeled HK), which reflects the magnetic flux in active regions on the facing hemisphere of the star, reveals quasi-periodic modulation at the rotation period. In addition, the same data reveal the mean level of magnetic activity through long-term averages of the nightly data (Vaughan et al. 1981; Baliunas et al. 1983; Noyes et al. 1984a).

This time-serial technique can also be applied to the *photometric* modulation of chromospherically active stars whose surface coverage by dark inhomogeneities is significant (several percent) and therefore measureable. Rotation modulation in brightness has been detected, for example, in Hyades Sun-like dwarf stars and other active stars (Lockwood et al. 1984; Radick et al. 1987). HK measurements and photometric measurements also provide independent information about the distribution of activity over the surface of the star, and in combination they give information about the relative size and distribution of bright active areas and dark starspots on the stellar surface.

Recent HK data from Mount Wilson and elsewhere (see, e.g. Noyes et al. 1984b; Catalano and Marilli 1983; Soderblom 1985; Barry 1988; Soderblom et al. 1991) have given strong support to earlier findings (Skumanich 1972) that both mean HK emission and rotation decline with age, at least for main-sequence stars older than the 70-Myr age of the Pleiades. The precise functional form of the decay of chromospheric emission with age is uncertain, due both to scatter of observed emission fluxes for stars of like ages, and to uncertainties in ages. A recent, thorough analysis (Soderblom et al. 1991) suggests that a power-law flux $\sim T^{-n}$ where $n \sim \frac{1}{2}$ to $\frac{3}{4}$,, may be acceptable, but it would imply a recent burst of local star formation at an age $t \sim 1$ Gyr and younger. A constant star-formation rate can also be accomodated by the data, with a slightly modified functional fit. The exact function describing the decrease of emission with age may be uncertain, but the decrease itself is not.

Thus, G dwarfs in the Hyades moving group or in the Hyades cluster itself, with age ~ 0.6 Gyr, have mean levels of HK emission about 2 to 3 times that of the Sun. If the surface brightness of the HK emission in plages on these stars is similar to that of the Sun, then the fractional surface coverage is correspondingly higher. For stars of the age of the Pleiades or younger, the fractional surface coverage can approach unity, and indeed the HK flux seems to saturate for these stars at a level corresponding to a mean surface flux some 3 to 5 times that of the Sun (Simon 1988), perhaps because the fractional surface flux is approaching unity.

In the distribution of HK fluxes with mass, or equivalently color index
B-V, there is an apparent gap, or paucity of stars, at activity levels corresponding to an age of order 1 Gyr, slightly older than the Hyades, as was first noted
by Vaughan and Preston (1980). Perhaps this is simply a statistical artifact
(Hartmann et al. 1984), but if real, it could imply that the ratio between mean
activity and rotation suffers a sharp decline at an age of order 1 Gyr. A possible mechanism for such a decline would be a changing topology of magnetic field lines, such that there are fewer closed loop structures and/or a
larger separation of footpoints (with consequently less magnetic heating)
when the rotation and mean magnetic field decline below a critical value. If
so, the Sun presumably would have experienced the same history in the past.
However, further data are needed to determine whether the gap is due to a
real physical effect.

Photometric modulation data show that the surface fractional coverage
by cool starspots decreases with age, just as does the chromospheric plage
coverage. The fractional area covered by spots in the 4.6-Gyr-old Sun ranges
up to about 0.001, while for 0.6-Gyr-old Hyades G dwarfs it ranges up to
about 0.003 (Radick et al. 1982); the Hyades moving group star HD1835 also
shows $\sim$ 1% modulation (Chugainov 1980). At an age of 0.3 Gyr, the Ursa
Major cluster G0-1 dwarfs χ^1 Ori and π^1 UMa have been recently discovered
to have $\sim$ 5-d light variations of 2 to 3% arising from cool starspots (Guinan
and McCook, in preparation). For the 70-Myr-old dwarfs in the Pleiades, the
corresponding figure is a few percent (Radick et al. 1983), and a nearby G0
V member of the Pleiades moving group, HD 129333, has a 4 to 5% light
variation due to spot modulation at a 2.7 d rotation period (Dorren and
Guinan 1990). Available data do not yet indicate whether the increased coverage for younger G dwarfs is due to larger spots or more numerous spots of
solar size. If the latter, however, they would have to be bunched in "active
longitudes" to produce the observed amplitude of rotational modulation.

To summarize, mean levels of both chromospheric activity and spot coverage have decreased over the lifetime of the Sun, as its surface rotation
slowed. Table II shows rotation and activity parameters inferred from the
Sun-like stars mentioned above.

3. The Evolution of the Solar Cycle.   The magnetohydrodynamic
theory behind the solar cycle is complex, and a successful model requires
detailed knowledge about rotation and convection within the Sun. While such
knowledge is now beginning to become available, it is limited at best. In
addition, of course, we are unable to experiment with the effects of changing
rotation and convection properties within the Sun. The study of stellar activity cycles can add a new perspective by elucidating the dependence of cycle
properties upon mass (which dictates convective zone properties), surface
rotation (which is related to mean interior rotation and differential rotation
within the interior) or chromospheric activity levels (related to photospheric

TABLE II
Evolution of Solar Activity Inferred from Other Sun-like Stars

Age (Myr)	Star	Association	Sp	P_{rot}	f_{spot}	$R'_{HK} / R'_{HK} \odot *$
70	HD129333	Pleiades MG	G0 V	2.7	0.04–0.05	5.5
600	HD1835	Hyades MG	G2V	7.6	0.02	3.3
4500	Sun		G2V	25.4	0.001	1.0

*$R'_{HK}\odot$ is the ratio of HK emission flux to stellar bolometric flux after correcting for photospheric emission within the instrument bandpass (see Noyes et al. 1984a).

magnetic field strengths). Cycle measurements have the great appeal of being relatively unambiguous: a measured cycle time scale may be identified precisely with the dynamo period. In contrast, HK emission is related to the physics of stellar dynamos in a complicated way, depending on such poorly understood processes as magnetic flux amplification, convective structuring of surface fields, magnetohydrodynamic energy transport into the atmosphere, nonthermal energy deposition, and non-LTE emission—all of which may depend on spectral type. A difficulty with cycle measurements is the long observation time needed; a valid determination of stellar cycle properties requires many cycle periods, with each cycle lasting from several to many years (or an ensemble of observations of only a cycle or two, but in *many* stars with similar mass and rotation rate.)

4. Information on the Solar Cycle from Stellar HK Data.   For over two decades, research at Mount Wilson Observatory has been providing time-serial chromospheric fluxes in nearly 100 lower-main-sequence stars. With twelve years of data, Wilson (1978) discovered that Sun-like cycles of chromospheric emission were common rather than rare among lower-main-sequence stars, and from the same data Vaughan (1980) noted that Sun-like cycles (that is, rather smoothly varying long-term emission with amplitude, period and shape similar to the solar cycle) were largely confined to old, and presumably slowly rotating stars. Vaughan et al. (1981) confirmed that the rotation periods of the stars with Sun-like cycles were in excess of ~ 20 d.

As already noted, the true nature of stellar activity cycles can be determined only with longer time series than presently are available. The sunspot record itself illustrates the problem: the value of 11 yr for the sunspot cycle time scale is an average over the past 2 centuries, during which the cycle-to-cycle lengths have been as short as 7 or as long as 15 yr. Thus, periodicities in stellar cycles must be verified by averages of numerous cycle lengths or by a statistically large sample of stars. In addition, with only a 20-yr timespan, we can verify periods only if they are on the order of 10 yr or less, no matter how many stars are observed.

The variety of long-term behavior in the Mount Wilson sample of lower-main-sequence stars is shown in Fig. 2. The values of the Ca II H and K emission strengths S are plotted as a function of time between 1966 and 1985. After 1980, measurements were scheduled nightly and only monthly averages of those points are shown.

In Fig. 2, the star HD 10700 (τ Cet, G8V, $P_{rot} \sim 40$ d) has no significant long-term variation and is labeled N. The time series for HD 9562 (G2, $P_{rot} \sim 27$ d) displays a long-term trend L which, if periodic, would imply a period of at least several decades. Some stars, such as HD 26913 (G3, $P_{rot} \sim 2$ d) vary significantly but with no clear periodicity, and are labeled V. The next four panels display activity variations that are probably periodic C, and the apparent period is listed in parentheses. The star HD 17925 (K0, $P_{rot} \sim 6.6$ d) appears to have a cycle longer than 18 yr but possibly near 20 yr. The

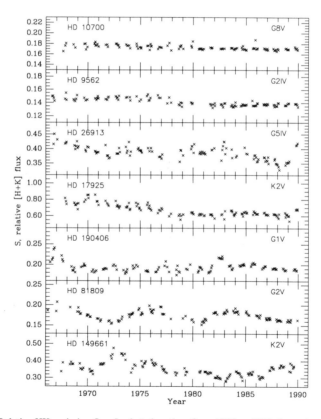

Fig. 2. Relative HK emission flux S, plotted vs time from 1966 to 1985, for various dwarf and subgiant stars as labeled. These seven typical examples show no significant long-term variability; long-term trend (longer than the data sample); variation but no apparent cycle; and cycle, respectively. See text for details.

period in HD 190406 (G1, $P_{rot} \sim 13.5$ d) is a short 2.6 yr. In the Sun-like star HD 81809 (G2, $P_{rot} \sim 22$ d) the average period is 8.3 yr. The star HD 149661 (K0, $P_{rot} \sim 21$ d) shows two significant peaks in its power spectrum, corresponding to periods of 12.4 and 4.4 yr. However, the data shown make it evident that there are real uncertainties in using short time series for accurate period determinations.

Even within the limitations of the relatively short (< 25 yr) and small (< 100 stars) present database, the data yield interesting hints about the activity cycle in our own Sun. A few are discussed below.

(a) Magnetic activity cycles are common in lower-main-sequence stars. About 2/3 of all the stars observed, with a range of masses and ages, show either definite cycles or possibly cyclic behavior. Hence there is no reason to doubt that the present solar-cycle behavior reflects the Sun's normal magnetic variability.

(b) However, there are some stars like the Sun which have some chromospheric emission but do not show observable long-term variability of any sort. The best example is τ Cet (Fig. 2, top). This star has chromospheric HK emission comparable to the Sun, and rotation period of ~ 40 d (v sin $i = 0.9$ km s^{-1}; Gray 1984). For this spectral type and rotation rate, dynamo activity might be expected to produce magnetic variability comparable to that of the Sun; the increase of activity expected with advancing spectral type (see below) should at least partially compensate the decrease of activity expected with decreasing rotation. Why is no variability seen? It is possible that this star is in a Maunder minimum state. Such a conjecture is supported by the observation that about 1/4 to 1/3 of old, Sun-like stars show no significant variability; this conforms statistically to solar behavior inferred from tree-ring measurements spanning several thousand years, which imply that the present Sun spends about 1/3 of its time in a minimum state (Baliunas 1988; Baliunas and Jastrow 1990). While such a conclusion is rather speculative given the present data sample, it is clear that, in principle, observations of enough G dwarfs of solar age could reveal the statistics of long-term solar activity minima like the Maunder minimum. Furthermore such data could tell us more of what the Sun was really like during the Maunder minimum—for example, to what level of HK flux the chromospheric emission declined at that time.

(c) With a few exceptions, stars with clean cycles have rotation periods $\gtrsim 20$ d (Vaughan et al. 1981). If this is a general property of stellar activity cycles, it implies for the Sun that at age $\lesssim 3$ Gyr (when $P_{rot} \sim 20$ d for the Sun, according to Skumanich law), there was no clear cycle visible, but larger-amplitude episodic activity.

If there are regular cycles in young Sun-like stars, they are apparently buried beneath large-amplitude episodic activity, perhaps rapidly evolving active regions or possibly even shorter transients such as very large flares. Support for large amplitude transient activity in young Sun-like stars is provided by a large X-ray flare discovered with EXOSAT on the rapidly rotating young G1V star π^1 UMa (Landini et al. 1986). π^1 UMa is a chromospheri-cally active star with rotation period $\sim$ 5 d and, as a member of the UMa moving group, has an age $\sim$ 0.3 Gyr. The energy emitted by the X-ray flare was more than 10 times the total energy release in the largest solar flares. More recently, Lockwood et al. (in preparation) have detected a "white light" flare on the very young (70 Myr) G0 V star HD 129333. These results for proxies of the young Sun indicate that flaring probably occurred on a much larger scale in the Sun in the past than it does today.

Because of the strong episodic activity in young rapid rotators, we can-not yet say whether, buried beneath the irregular activity, these stars have a cycle as regular and as large in amplitude as do old stars like the Sun. An alternative would be that regular cycle activity starts only when the rotation rate has slowed to a certain value (perhaps 20 d), at a time ($\sim$ 3 Gyr) when the irregular activity has coincidentally decreased to a level such that the newly emerged cycle is not masked. More data are needed to distinguish between these alternatives. It would be useful to know this as some theories of the dynamo suggest that the dynamo process itself is irregular at rapid rotation rates and more regular at slower ones (Parker 1971; see also Durney et al. 1981).

It is possible that activity cycles in younger stars, while masked in HK by short-term chromospheric variability, are detectable from the photospheric brightness variations produced by starspots; as in the Sun, these variations might be relatively uncontaminated by flares or other short-term effects. Sup-port for this possibility comes from observations by Dorren and Guinan (1990) that there is an apparent 7-yr spot cycle for the Pleiades moving group members HD 129333, during which the star's luminosity changes by 5%. If this is really evidence that a clean cycle underlies the star's complex chro-mospheric variations, it suggests that Sun-like dynamo activity extends to rather rapid rotators (in this case rotation period of 2.7 d; see Table II).

(d) Stars with clean cycles with a few exceptions have cycle time scales in the neighborhood of 10 yr (see, e.g., Vaughan 1980; Noyes et al. 1984a). This suggests that the period of the solar cycle has not changed greatly in the past 1 Gyr or so.

Two notable exceptions in the HK data to the trends noted above are: HD 190406 (Fig. 2), which shows a clean 2.6-yr period, and HD 152391(G8) which shows a period near 10 yr but is a more rapid rotator ($P_{rot} \sim$ 11 d), has greater activity, and is presumably younger than the typical older stars with regular cycles. However, putting these exceptions aside, Noyes et al. (1984a) found that for a sample of 13 slow rotators including the Sun, stars of similar mass have P(cyc) proportional to P(rot); if verified by further data, this would imply that the solar cycle period has been gradually lengthening from about 9 yr to the present 11 yr over the past 1.5 Gyr, as its rotation period increased from 20 d to the present 25 d.

We have already noted that the Pleiades moving group star HD 129333 appears to have a spot cycle of length about 7 yr. If this is true and is indicative of the behavior of the Sun at age 70 Myr, then the young Sun may have had a significantly shorter and more vigorous activity cycle than today, with luminosity variations as great as 5%, i.e., some 60 times the solar luminosity variation during its most recent activity cycle (see below).

   5. Further Information on the Solar Cycle from Stellar Photometric Variability.   Observations of long-term stellar photometric variability add to our understanding of the solar cycle, beyond what can be learned from HK data. A major step ahead in their interpretation comes from recent solar data from the ACRIM and ERB experiments aboard the SMM and NIMBUS-7 satellites. These instruments recorded correlated irradiance variations that signaled a drop of about 0.08% in solar irradiance during the declining phase of solar cycle 21, followed by a rise toward the maximum of cycle 22. Foukal and Lean (1988a) modeled the irradiance with negative (blocked light) contributions from sunspot area coverage and positive contributions from magnetic faculae and plage areas (Chapman 1987b). Irradiance blocked by sunspots is more than offset by plage contributions, so that maximum irradiance occurs at sunspot maximum. The physical mechanism responsible for such solar irradiance imbalances appears to be the impediment of heat flow within convecting flux tubes and a resulting change in density and temperature structure (Spruit 1988).

   The solar photometric variation with the activity cycle is positively correlated with the solar HK variation; i.e., the maximum in both solar irradiance and chromospheric HK flux occurs together, at sunspot maximum. It is now possible to examine the relationship between magnetic and luminosity variations on the late-type stars, as an aid to understanding the mechanisms involved in the solar case, and also to explore the past history of the Sun. Two relevant sets of time series are measurements of the chromospheric HK fluxes from Mount Wilson Observatory and extremely precise (0.003 mag) photometric differential magnitudes, primarily from Lowell Observatory. Approximately a 4-yr overlap of the time series exists for about three dozen lower-main-sequence stars, spanning a range in ages of < 1 Gyr to at least 4 Gyr, and a range in spectral type of mid F to late G. The precise photometry, in the Stromgren *b* and *y* passbands, shows rotational modulation due to the presence of dark spots, and small but significant longer-term changes that are probably caused by variations in overall magnetic activity level (cyclic or otherwise).

   Cross analysis of the two time series (Baliunas 1988; Radick et al. 1990) suggests that on short time scales, the diminution in visible light caused by the rotation of starspots onto the face of the star is accompanied by an increase in the HK flux, for both very active and relatively inactive stars, just as for the Sun (cf. Foukal and Lean 1986,1988a). An example of this effect in an active, evolved star δ CrB (G3.5 IV-III) is shown in Fig. 3. The 0.1-

mag variation in broadband light (*V* filter) implies a surface coverage by dark spots of about 10%, very large compared to sunspot coverage. Both the photometric data and the HK data yield a rotation modulation period of 57 d but the variations are inversely correlated: maximum brightness occurs at times of minimum HK flux. The (anti-) correlation coefficient is essentially perfect considering the measurements' uncertainties, suggesting that the spots and their associated plages are temporally and spatially coincident, as in the Sun. Because we have no information about the contrast or the spatial distribution of the HK emission in the active areas relative to the quiescent atmosphere, and because likewise the distribution of starspots is indeterminate (due to the uncertainties in key parameters such as the inclination of the stellar rotation axis to our line of sight), we cannot compare the contributions by spots and plages or magnetic faculae in quantitative detail as Foukal and Lean (1988*a*) did for the solar data. However, we can at least conclude that both for relatively inactive stars like the Sun and very active stars like δ CrB, photo-

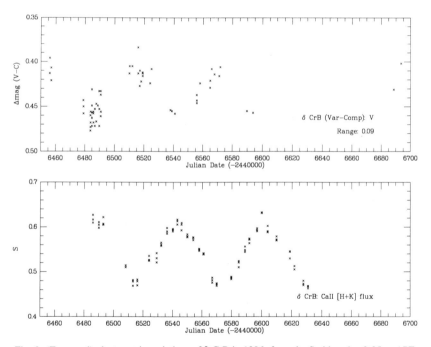

Fig. 3. (Top panel) photometric variations of δ CrB in 1986, from the Smithsonian 0.25-m APT (differential *V*-band magnitude expressed in terms of magnitude difference between variable and comparison star [*V-C*]; higher-lying points with smaller values of *V-C* are brighter). (Bottom panel) HK emission index *S* of δ CrB for the same time interval, from Mt. Wilson. The anti-correlation between photometric and HK flux rotational modulation is evident (figure from Baliunas et al. 1987).

spheric irradiance decreases and chromospheric emission increases are positively correlated on rotation time scales.

On longer time scales, however, the situation is different. Figure 4 shows the seasonal averages of the Lowell photometry and the HK fluxes for an old, relatively inactive star (Fig. 4a) and a young, active star (Fig. 4b). Long-term trends over 4 seasons are evident. The star HD 76572 (F3V) is very weak in chromospheric activity; the chromospheric radiative loss normalized to the stellar bolometric luminosity (Noyes et al. 1984*b*) is less than in the Sun. Like the Sun, HD 76572 is a comparatively old star. While the photometric (Lowell) and chromospheric (Mount Wilson) variations over 4 seasons are small, there is a very significant positive correlation between magnetic activity and luminosity, as is the case for the Sun when viewed on cycle time scales. In contrast, Fig. 4b shows the corresponding behavior for VB 64, a relatively young Hyades star, with age of ~ 0.7 Gyr. This young, active star has a behavior *opposite* to that of the Sun: the photometric light is dark when chromospheric activity increases.

Such behavior is supported by data from a larger sample of stars (Radick et al. 1990), but a longer time span covering entire cycles is necessary before the results can be considered definitive. However, it is reasonable to conjecture that in younger, more active stars, the spots are so large that their flux deficit is able to overcome the contribution of photospheric faculae and produce a net luminosity deficit at times of high activity, whereas spots in older stars like the Sun are too small relative to the photospheric faculae to do so.

Regarding the Sun, this idea implies that when the Sun was younger and more active, the relation between solar irradiance and the overall envelope of

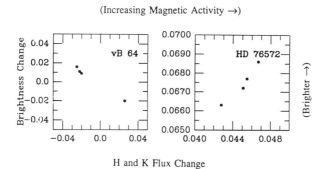

Fig. 4. Annual mean differential magnitudes Δmag (*b* and *y* filter combined) over 4 seasons for two stars, plotted vs seasonal average HK emission index variations Δ*S* (offset from an arbitrary level). The right plot represents the old, weak emission-line star HD 76572, that shows a positive correlation of photospheric brightness and magnetic activity, like the Sun's behavior. The left plot represents the young Hyades dwarf vB 64, that shows an opposite correlation; this suggests that the correlation between photospheric brightness and magnetic activity in the Sun may have been of opposite sign when the Sun was younger.

activity was in the opposite sense from what it is today (Radick et al. 1990). We have already suggested from comparison with the Hyades G2V dwarf HD 1835 that when the Sun was at the age of the Hyades stars, its spot coverage was perhaps 10 to 30 times larger than at present, whereas the HK emission was only about 3 times larger. If (as for the present Sun) there were a relatively good correlation between photospheric facular brightness and HK brightness, it is in fact quite reasonable that the irradiance decrease due to spots would dominate over any increase due to faculae. Hence, the young Sun's irradiance may have varied by a percent or so during its activity cycle, a variation which might have had effects on the terrestrial climate.

B. The Physics of Solar Magnetism

1. The Origins of Solar Magnetic Activity. It is commonly thought that solar magnetism is created through a dynamo process, through the combined action of differential rotation and convection. In the standard mean-field dynamo picture (Parker 1979a), differential rotation stretches poloidal magnetic fields in the toroidal direction, strengthening them in the process, and cyclonic convection then brings the buoyant magnetic regions to the surface. Coriolis forces acting on the rising toroidal fields give them again a poloidal component, but in the opposite sense from the original poloidal fields. This process gives rise to the 22-yr alternating cycle of magnetic activity. The "period" of the cycle is governed by the characteristic time for magnetic diffusion from active-region latitudes toward the poles, as the newly created field diffuses into the region occupied by the original polar field and overwhelms it, creating a new polar field of opposite polarity.

A basic parameter of mean-field dynamo theory is the dynamo number N_D, given by

$$N_D = \alpha \Omega' d^4 / \eta^2 \qquad (1)$$

where α is the product of the mean helicity of convection $<\mathbf{v} \cdot (\nabla \times \mathbf{v})>$ and the convective overturn time τ_c, Ω' is the radial gradient of angular velocity, d is the characteristic length scale for the convection, and η is the turbulent magnetic diffusivity. If $v\tau_c$ scales like d, $(\nabla \times \mathbf{v})$ scales like Ω, η scales like d^2/τ_c, and Ω scales like Ω/d, then

$$N_D \sim (\Omega\tau_c)^2 \sim 1/Ro^2 \qquad (2)$$

where Ro, the Rossby number, is the ratio of the rotation period P_{rot} to the convective overturn time τ_c in the region of dynamo generation of magnetic fields (see, e.g., Durney and Latour 1978). τ_c increases with decreasing mass, or advancing spectral type. According to one set of convective zone models (see Noyes et al. 1984a), it is about 12 d for the Sun and about 24 d for a K5

dwarf such as 61 Cyg A. For a star of given mass, it is to first order independent of age, and hence of P_{rot}.

For stars of the same mass, and hence roughly the same convection zone
properties, the mean level of magnetic activity decreases with increasing P_{rot},
and thus the "rotation-activity connection," in which the level of stellar chromospheric and coronal emission declines with decreasing rotation rate, is in
qualitative agreement with dynamo theory. One would also expect from dynamo theory that magnetic activity depends on stellar mass through the variation of τ_c. A number of authors (see, e.g., Noyes et al. 1984a; Simon et al.
1985) have concluded that the mean level of chromospheric and coronal emission does seem to depend on mass as well as rotation, and, in fact, scales
approximately inversely with the Rossby number. If this could be quantitatively verified, it would be strong support for the dynamo theory as an explanation for the origins of solar magnetism.

A difficulty with quantitative verification of dynamo theory from chromospheric and coronal indicators is the complex physics connecting magnetic
fields in the interior with the observed energy emission from the atmosphere.
The connection depends on nonmagnetic properties (especially the amplitude
of turbulent convection) whose variation with mass along the lower main
sequence could mask mass-dependent terms like τ_c that enter dynamo theory.
Thus, although the ratio of chromospheric or coronal flux to stellar bolometric flux scales well with the inverse Rossby number, the flux by itself scales
better with the rotation velocity alone, (see, e.g., Rutten and Schrijver 1987;
Schrijver et al. 1989), contrary to expectations from dynamo theory.

However, other measures of stellar magnetic activity independently imply greater magnetic activity both for more rapid rotation and for longer convective turnover times. One is the area covered by starspots. Sunspots, with
a surface coverage of $\leq 0.1\%$ of the solar disk, are not visible in integrated
sunlight except from space. However, G dwarfs do have observable spots if
they are rotating fast enough; examples are the G dwarfs in the Pleiades or
Hyades with rotation periods of several days and spot coverage of up to several percent (Radick et al. 1983,1987). And spots are also observable for slow
rotators if their mass is small enough (i.e., τ_c is large enough). An example
is 61 CygA (K5V, with $\tau_c \sim 24$ d, as noted above), whose 40-d rotation
period is 60% longer than the Sun's, but which nevertheless has spot coverage
of $> 1\%$ (Dorren and Guinan 1981). Finally, stars which are both of low
mass *and* rapidly rotating can have spot coverage of 10% or more; for example, EQ Vir, with the same K5V spectral type as 61 CygA but rotation
period of only 4 d, has about 10% spot coverage.

Figure 5 illustrates these trends; in the top panel are plotted the fractional
spot coverage for G0 to G5 dwarfs (filled circles) and K0 to K5 dwarfs (open
circles), vs inverse rotation period (i.e., angular velocity). We see that spot
coverage increases with angular velocity for both G and K dwarfs, and also
that for a given angular velocity, K dwarfs have greater spot coverage than G

dwarfs. In the lower panel, the same data are plotted vs the inverse Rossby number τ_c/P_{rot}, where τ_c is determined from the spectral type using the relations discussed by Noyes et al. (1984b). We note that, as in the case of the chromospheric flux ratio R_{HK}, the fractional spot coverage shows a dependence on inverse Rossby number, with rather small scatter. This gives further support to the idea that magnetic activity in G and K dwarfs (and hence in the Sun) owes its origin to a process explainable in outline, if not in detail, by standard dynamo theory.

In addition, direct observations of photospheric magnetic fields indicate that the total magnetic flux (i.e., the quantity Bf where B is the photospheric field strength and f the fractional surface coverage) increases both with increasing rotation rate and decreasing mass (see, e.g., Marcy and Basri 1989), again in accord with predictions of dynamo theory.

Finally, dynamo theory suggests that there should be a relation between magnetic activity cycle periods and Rossby number. As we have discussed,

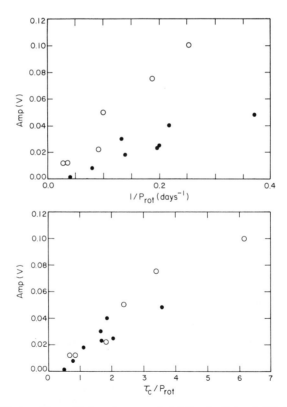

Fig. 5. (top) Photometric amplitude vs inverse period, $1/P_{rot}$, for G dwarfs (filled circles) and K dwarfs (open circles); (bottom) the same data plotted vs inverse Rossby number τ_c/P_{rot} (see text).

the measured timing of a cycle is not subject to complexities of interpretation such as are encountered, say, in relating chromospheric or coronal emission to basic dynamo processes, so this would seem to be a good testing ground for dynamo theories. Noyes et al. (1984b), in a search for regularity in cycle periods among 13 stars (including the Sun) with rotation rates > 20 d and clearly visible cycles, found that their cycle periods varied approximately as the Rossby number. If this is true, not only does it lend support to solar dynamo theory, but it allows comparison of specific solar dynamo mechanisms. For example, the linear relation found is consonant with a dynamo in which saturation occurs by escape of buoyant magnetic flux through the surface, but not with one in which saturation occurs by magnetic freezing of the differential rotation in the region of field amplification.

The stars give us another useful lesson for thinking about the solar dynamo. They warn us that we should not try to force nature to fit our preconceptions of a stellar dynamo. Already in the solar case we are finding it necessary to modify our concepts in the light of new data. We find, for example, that the expected depth gradient of rotation through the convection zone, earlier thought necessary to drive the dynamo, may not exist. Other stars compound this problem. W UMa stars, which are binaries so close that they share a common envelope, almost certainly have negligible differential rotation, and yet they show very strong magnetic activity, including large starspots. RS CVn stars are also ones where the gradient of rotation with depth is less than (or at least different from) single stars, because tidal forces have forced equality of rotation and orbital angular velocities. Yet these stars show large spots, and indeed obey the same rotation-activity relations seen in single stars (Basri 1987; Rutten and Schrijver 1987). It would thus seem to be an open question whether in the Sun a depth gradient of rotation must exist in the region of dynamo formation. Finally, there seems to be a wide range of surface magnetic effects in other stars varying greatly from the familiar solar ones. In some stars, spot latitudes apparently migrate, not toward the equator as in the Sun, but toward the poles (cf. Vogt 1983; Catalano and Marilli 1983). The lesson seems to be that magnetic dynamos are common in any star with both significant rotation and convection, regardless of the detailed rotation profile with depth and latitude. Solar dynamo theory probably should not be so tightly constrained that it does not admit generalization to stars with rather different interior dynamics.

2. What Determines the Strength of Solar Magnetic Fields?   A widely accepted explanation of the concentration of solar surface fields (outside sunspots) into flux knots with field strength in the 1 to 2 kG range is convective collapse of field in convective downdrafts, to a concentration level where the magnetic pressure is a fraction $f \sim 0.6$ of the surrounding photospheric gas pressure (Spruit and Zweibel 1979a). If so, one would expect a 50% increase in photospheric magnetic field strength with advancing spectral type along the lower main sequence (again outside spots), as the photospheric gas pressure increases from about 7×10^4 cgs (at G2) to about 1.6×10^5 cgs (at M0). Unfortunately, current techniques do not permit a sufficiently accurate separation of the product Bf of magnetic field strength and fractional surface-area coverage into its component factors that we may distinguish such a slight trend; all that is known at present is that field strengths in the range 1000 to 1600 Gauss are characteristic of stellar photospheric fields spanning the range from G through late K (Marcy and Basri 1989), and the data are consistent with field strength determined by the photospheric gas pressure. On the other hand, the fractional areas f covered by strong fields appear to increase significantly with decreasing Rossby number. (This is by no means unexpected, given the increase in spot area with decreasing Rossby number discussed above.)

3. Where is the Seat of the Solar Dynamo? Current pictures of the solar dynamo favor the field generation occurring at or slightly beneath the bottom of the convection zone; a dynamo distributed throughout the convection zone would produce fields that rise too rapidly to the surface to generate a cycle of the observed duration (Parker 1975a). Angular velocity gradients associated with the transition between the convective exterior and the radiative interior could provide the needed angular velocity shear to induce dynamo activity. A test of this would be to search for a cutoff of magnetic activity around M5, where stars become fully convective and hence no longer have such transition layers (Bookbinder 1985). However, strong X-ray emission, including X-ray flaring, now appears to prevail in stars as late as M7 (see, e.g., Pallavicini 1989); this is an almost sure sign of magnetic activity in some stars that are fully convective.

The picture is not quite settled, however, for another indicator of magnetic activity, starspots, may not continue to increase in size as we progress into the fully convective region of the lower main sequence. Spots in early M dwarfs can be very large, much larger than those seen on late M dwarfs. Examples of large spots on early M dwarfs include AU Mic (M0 Ve), whose photometric variations imply $> 30\%$ spot coverage, and BY Dra (K6Ve), with spot coverage of $\sim 25\%$ (Torres and Ferraz-Mello 1973). (This is significantly larger than the spots illustrated in Fig. 5. If these stars were to fit on the Rossby relation shown there, they would have to have convective overturn time τ_c of order 60 d, rather longer than might be expected from simple extrapolation of calculated values into the very late K and early M dwarf region.) On the other hand, late M dwarfs do not generally have large apparent spot coverage; for example, CM Dra (M5 V + M5 V), $P_{rot} \sim 1.1$ d), shows only 0.04 mag photometric variation. Perhaps in fully convective stars, magnetic active regions (as revealed by X-ray emission) exists, but photospheric spots are for some reason less prominent. Or, there could be substantial spot coverage, but more uniformly distributed over the surface in fully convective stars, such that the modulation amplitude is less. Or, the small sample of late M stars with measured photometric modulation is anomalous. More photometric data are clearly needed.

In summary, while present data cannot provide definitive information about the location of stellar dynamo activity, there is some evidence that an interface between a convective zone and a radiative interior region is *not* necessary for a dynamo to exist. It appears that dynamo activity can occur in any star with both rotation and convection.

IV. CONCLUSION

This survey of some ways that other stars can tell us about the Sun is quite selective; for example, it does not deal with transient activity, with stellar winds, or with X-ray emission and temperature structure of other stellar coronae. Even within the areas we do treat, we have not attempted to provide an exhaustive discussion of current literature, but rather to focus on a few specific issues, such as the indications of how rotation and mass together govern the nature and evolution of solar and stellar magnetic activity. As we have indicated, these issues are presently by no means settled. However, as we hope that the discussion illustrates, the questions are most likely to be answered by more and better observations, in many cases over time scales typical of stellar activity cycles.

Data on stellar activity, by showing how activity depends on properties like rotation, mass and age, can reveal aspects of the underlying physics quite inaccessible from observations of the Sun alone. Key observables include stellar chromospheric and coronal emission (not just in H and K lines emphasized here, but at other visible, ultraviolet and X-ray wavelengths known to respond to stellar magnetic activity), photometric data over both rotation and cycle time scales, direct magnetic flux measurements, and stellar p-mode oscillations. These are needed for a broad range of Sun-like stars, i.e., stars on the lower main sequence at various ages, and also evolved stars that have lower-main-sequence progenitors. Particularly important, of course, are stars of ~ 1 $M_\odot$ solar mass, which provide direct information on the evolutionary history of the Sun. Although much progress has been made in recent years, we may have just begun to decipher the clues that other stars can provide us about the Sun.

Acknowledgments. We are grateful to J. Christensen-Dalsgaard for useful discussions, and to D. Duncan and C. Zwaan for a careful reading of the manuscript. We would also like to express our appreciation for the devoted efforts of the staff of Mount Wilson Observatory. This research is supported by the Scholarly Studies Program, the Langley-Abbot Program and other funds of the Smithsonian Institution; the National Science Foundation; the Richard Lounsberry Foundation; the George C. Marshall Institute and the Mobil Foundation, Inc.

POST-MAIN-SEQUENCE SOLAR EVOLUTION

PIERRE DEMARQUE

and

DAVID B. GUENTHER
Yale University

In this chapter our objective is to discuss some current research frontiers in solar-stellar physics. First, we describe and update the standard evolution of a 1-$M_\odot$ star from the zero-age main sequence to the helium flash. Then, we focus on two areas of solar-stellar research which are presently in a state of rapid development. These two areas are: (1) the theory of solar and stellar nonradial acoustic (p-mode) oscillations, which can be used to test the pressure stratification in the stellar interior; and (2) the theory of the evolution of rotating stars, which predicts the evolution of surface rotational velocities and the mixing of chemical elements to the surface. We note in addition that oscillations and rotation are closely interrelated, because each individual p-mode is also split in frequency by internal rotation in the region through which it propagates. Observations of p-mode splittings thus can provide information on the state of rotation of the solar and stellar interiors.

I. INTRODUCTION

The close connection between solar physics and the study of stellar evolution was recognized as early as 1908 by George Ellery Hale (*The Study of Stellar Evolution*, p. 7). In his monograph he states:

"In entering upon our consideration of the study of stellar development, we may think of the subject in either one of two ways. Some will prefer to regard it as the general problem of stellar evolution,

[1186]

in its broad application to the universe at large. But others will find it easier to conceive of the question as an investigation of the Sun, tracing it, through analogies afforded by stars in earlier stages of growth, from its origin in a nebula to those final chapters which, though not yet written for the Sun itself, may be read in the life-histories of red stars. Viewed from whichever standpoint, the task of the investigator remains the same, since in either case it is concerned with stellar origin, development, and decay."

Our objective here, in the same spirit as Hale's monograph, is to discuss some current research frontiers in solar-stellar physics.

A. The Sun as an Evolving Star

The internal structure of the present Sun can only be understood properly within the context of stellar evolution theory. In the chapter by Sofia et al. they have discussed the pre-main-sequence evolutionary phases, and how the study of rotational velocities and Li and Be surface abundances in young star clusters can be used, together with theoretical models, to understand the spin-down due to stellar winds and the early rotational history of the interior of the Sun and solar-type stars. In the same way, we will see that more advanced phases of evolution, beyond the phase of the present Sun, can provide information about the deep interior of the Sun that cannot be derived in any other way. And because the Sun's life expectancy is ~ 12 Gyr (see Sec. III), which is approximately the age of the oldest stars in the galactic disk, we are in the fortunate position that analogs of all phases of solar evolution are accessible to direct observation in our own galactic neighborhood.

B. Brief History of Solar Interior Modeling

The history of solar modeling is the history of stellar structure, and follows in step the history of physics. It seems fair to credit Lane (1869) with constructing the first solar model. He wrote down a set of equations describing a gas sphere in hydrostatic equilibrium. His purpose was to determine the Sun's surface temperature and density, which he did getting $T_{surface} = 30,000$ K and $\rho_{surface} = 0.0004$ g cm^{-3}. These are very good results when one realizes, as noted by Chandrasekhar (1939), that the Stefan-Boltzman law had not yet been published. The solar model was improved by Eddington who included a description of the transport of energy by radiation (1926). This was made possible by the development of atomic physics and the first calculations of absorption coefficients. Eddington's solar and stellar models were static models, i.e., they did not include the evolutionary changes brought about by the central nuclear reactions. At the time, the source of the Sun's luminosity was not known. Regardless of this missing piece of information, by assuming that the structure of the star could be approximated by a polytrope, Eddington was able to use his stellar models to deduce the observed mass-luminosity relationship of stars. He also noted that in order for the Sun to satisfy the mass-luminosity relationship it had to be primarily composed of ionized hydrogen.

With nuclear physics came the concept of thermonuclear reactions. In the early 1930s, it was realized that the central temperature and densities in stars are high enough to allow nuclear reactions of hydrogen to take place. The Sun and other stars derive their principal energy from the transmutation of four hydrogen atoms into one helium atom, a source of energy which is capable of sustaining the star's luminosity for billions of years. But not until the calculation of interior opacities had improved sufficiently, did it become possible to distinguish between the CN cycle, originally proposed by Bethe (1939), and the proton-proton chain as the main source of the solar luminosity (Oke 1950). Together with the discovery of the importance of the proton-proton chain came the realization that the core of the Sun must be in radiative equilibrium, rather than convective (Cowling 1935), as would be expected if the CN cycle dominated the nuclear energy production.

From this point on, models of the Sun (no longer static models) were based on a sequence of models evolved from a zero-age model (Schwarzschild et al. 1957). The evolutionary sequence of models include the effects of the nuclear reactions which ultimately leads to changes in the models' structure. Today's computers permit solar model calculations to include as much physics as is reasonable to calculate. The current limitations appear in the approximately determined constitutive quantities such as the opacity, nuclear cross sections, and equation of state and the neglect of "second-order" effects such as internal dynamics and magnetic fields.

In the next section, we begin by describing the standard solar model, the sources of numerical errors in constructing a standard model and uncertainties in the physical assumptions. Section III follows with a discussion of the standard evolution of a $1\text{-}M_\odot$ star from the zero-age main sequence (ZAMS) to the helium flash. This section includes a description of the evolution of surface abundances and of the properties of p-mode oscillations along the evolutionary track. Section IV focuses on the importance of rotation on solar post-main-sequence evolution. Diagnostics for measuring the rate of internal rotation in the Sun are briefly reviewed. Tests from other rotating stars, and the implications of these tests for understanding the history of the distribution of angular momentum in stars are also considered and compared to the solar data.

II. THE STANDARD SOLAR MODEL

A. Introduction

Nearly all stellar evolutionary calculations are calibrated with respect to the standard solar model. The standard solar model is derived from the conservation laws and energy transport equations of physics, applied to a spherically symmetric gas (plasma) sphere and constrained by the luminosity, radius, age and composition of the Sun. The standard solar model is used as a test case for the stellar evolution calculation because the luminosity, radius, age and composition of the Sun are well determined. In fact, the only way to determine the two free parameters of the stellar evolution calculation, *viz.*, the helium abundance and the mixing-length parameter, is to adjust the standard solar model to "fit" the observed Sun.

To obtain the standard solar model, a 1 $M_\odot$, zero-age stellar model (a model with a homogeneous distribution of chemical elements which is just beginning to derive most of its luminosity from nuclear reactions) is evolved to the age of the Sun. The structural changes in the Sun, as it evolves, are caused by the nuclear reactions occurring in the central regions of the Sun. The transmutation of four hydrogen atoms into one helium reduces the number density of particles in the central regions which decreases the pressure. The pressure decrease does not actually occur because the surrounding layers respond to the force imbalance by contracting in the central regions. Half of the gravitational energy released from the contraction goes to raising the temperature of the central regions (the other half, according to the Virial Theorem, is radiated away). The increased temperature, by the ideal gas law, increases the pressure of this region and restores the balance between the pressure and gravitational forces. The larger mean molecular weight increases the luminosity of the star, and the rate of nuclear reactions. Also, while the central layers contract, the outer regions expand, in a sense, compensating for the steepening temperature gradients in the central regions. Therefore, as the Sun evolves from the zero age both its luminosity and radius increase.

The definition of the standard model depends critically on fitting the model to some of the Sun's observable characteristics. The 1 $M_\odot$ model must have the observed radius and luminosity of the Sun at the Sun's inferred age. In addition, the model's surface composition of hydrogen to heavy elements ratio must match the observed solar abundances. Even though both the Sun's neutrino flux and the p-mode oscillation spectrum have been measured, they are not normally used to constrain the definition of the standard solar model. In fact, as is well known, the standard solar model produces significantly more neutrinos than are observed by the Davis (1978) experiment. Also, the p-mode frequencies of the standard solar model do not precisely match (the error is of the order of $\pm 0.3\%$) the observed frequencies.

The standard solar model serves two purposes: (1) it provides estimates for the helium abundance and mixing-length parameter α in the Sun by forcing the solar model to have the correct luminosity and radius at the Sun's age; and (2) it provides a benchmark to compare "improved" solar models which have additional physics, such as rotation, magnetic fields, and diffusion or *ad hoc* improvements, such as turbulent diffusion, overshooting, and metal-rich cores. The former role comes about from the fact that neither the helium abundance nor α can be precisely determined from observation or laboratory experiment. As a consequence, many stellar evolutionary calculations of Population I stars assume a solar mixing-length parameter and a solar helium abundance. Below, we describe the recipe followed by stellar evolutionists to construct a standard solar model.

B. Recipe for Constructing a Standard Solar Model

Conservation Laws.   The most firmly grounded laws used in modeling the evolution of the Sun from zero age to its current age are the conservation laws of physics, *viz.*, the conservation of mass, momentum and energy. These three laws along with the energy transport equation and the equations describing the nuclear reaction network constitute the complete set of equations which describe the evolution of a star. They are discussed at length in all books on stellar structure and evolution. Some classic books on the theory of stellar structure and evolution, in addition to the landmark books by Eddington and Chandrasekhar mentioned in the introduction, are Schwarzschild's *Structure and Evolution of the Stars* (1965), Clayton's *Principles of Stellar Evolution and Nucleosynthesis,* (1968), and Cox and Giuli's *Principles of Stellar Structure* (1968).

In the standard solar model the physics is simplified by assuming that the Sun is spherically symmetric. The symmetry breaking physics of magnetic fields and rotation are neglected because they increase the complexity of the problem enormously and are generally not believed to affect the structure and evolution of the model by very much. We note that the energy arguments used to prove the unimportance of magnetic fields and rotation are based on the surface values of these quantities. It is possible that magnetic fields and rotation in the interior are orders of magnitude stronger.

The amount of mass converted into energy over the entire lifetime of the star is less than 0.007 the total mass of the nuclear burning regions of the star and thus can be ignored. Also, in current standard solar models, mass loss from winds is usually not included, again, because the loss is considered negligible (10^{-14} $M_\odot$ yr^{-1}; Cassinelli and MacGregor 1986) when compared to the total mass. The total mass and the mass in any given layer is, therefore, assumed to be constant during the evolution of the Sun. Substantial mass loss, of the order of several tenths of a solar mass, occurs much later in the life of the solar-type star.

The dynamical time scale of the Sun is approximately one hour, hence, during the Sun's 4.5 Gyr evolution it has had enough time to establish hydrostatic equilibrium. For any volume element within the Sun, the weight of the element is assumed to be exactly balanced by the sum of all pressure forces acting on the element. The total pressure includes both the gas pressure and the radiation pressure. In the Sun, the radiation pressure at the center accounts for only 0.05% of the total pressure.

The energy balance of any shell in the Sun is obtained by equating the sum of the energy entering the bottom of the shell and the energy produced by nuclear reactions within the shell to the sum of the energy lost through the top of the shell plus the work energy done by the shell on its surroundings. This balance is represented mathematically by: ˙

$$\frac{dL}{dr} = 4\pi r^2 \rho \left(\varepsilon + T \frac{dS}{dt} \right) \tag{1}$$

where L is the total luminosity (energy per unit time per unit area) passing outwards through the shell at radius r, ε is the nuclear energy generation rate per unit mass, and S is the entropy per unit mass. The $T\,dS/dt$ term ("entropy term") is the sole term in the basic stellar evolution equations that includes time explicitly. It describes the internal energy gains and losses associated with the structural changes that occur as the star evolves. The structural changes, as mentioned earlier, are due to the continually occurring nuclear processes. If the entropy term were dropped then the equations would describe a static non-evolving model.

Energy Transport. Three forms of energy transport are modeled in stellar evolutionary calculations: convection, conduction and radiation. Only convective and radiative transport are important in the Sun. It is the efficiency of the energy transport (determined by the opacity and temperature gradients in the star) which determines the overall luminosity of the star.

The relation between the temperature gradient and the power flux of photons is described by the radiative transport equation. Even though energy is clearly flowing out of the Sun, the temperature gradients required to account for the Sun's luminosity are shallow enough that locally the Sun's interior can be considered to be in thermodynamic equilibrium. This permits the equations describing the radiative transport of energy to be linearized. In the Sun, this linearization is accurate to one part in 10^{10}, i.e., the neglected second-order terms are 10^{10} times smaller than the linear terms. The radiative transport equation is:

$$L = -4\pi r^2 \frac{4ac}{3} \frac{T^3}{\kappa\rho} \frac{dT}{dr} \tag{2}$$

where a is Stefan's constant, c is the speed of light, and κ is the Rosseland mean opacity. As the temperature gradient steepens or the opacity decreases, the rate of energy flux passing through a shell increases. The opacity, a function of temperature, density, and composition, is probably the least accurately determined term in the equations of stellar evolution, having an error of up to $\pm 20\%$.

When the temperature gradient required to transport all the energy by radiation exceeds the adiabatic temperature gradient, the layer is unstable to convection and the material physically transports the energy outwards. Because convection is very efficient at transporting energy, the temperature gradient in the convective region needs to be only slightly superadiabatic in order for convection to carry nearly all the energy. This is not true near the surface where the density is low enough that radiative transport must be taken into account. The real temperature gradients in this thin outer layer are significantly steeper than the adiabatic temperature gradient. Except for this outermost layer, the temperature gradient in a convective region is taken to be the adiabatic temperature gradient. In the outer convection zone, the process is modeled by the mixing-length theory.

In analogy with Prandtl's (1952) fluid dynamical picture, the mixing length can be defined as the length traveled by a convective element before it loses its identity. In the stellar context, where there are steep pressure gradients in the convection zone, it is convenient to define a mixing-length parameter α, which is the ratio of the mixing length to the local pressure scale height (Böhm-Vitense 1958). The original Prandtl meaning has been partially lost in stellar evolutionary calculations where α is used to control the radius of the model; therefore, it effectively characterizes the efficiency of energy transport by controlling the steepness of the temperature gradient in the very outer superadiabatic layer of the Sun. This gradient is adjusted, via α, until the radius of the model matches the observed radius of the Sun. The temperature gradient throughout the entire convective envelope is, therefore, characterized by one number. And because the flow of energy through this layer is controlled by the temperature gradient, ultimately it is the mixing-length parameter which controls this flow. As a consequence, the actual temperature stratification in the convective envelope of the Sun may be different, especially in the very outer superadiabatic region, from that predicted by the standard solar model using the mixing-length theory.

The Equation of State.   The four equations in five variables: P, T, ρ, M_r and L are solved with the help of the equation of state which provides a relation between P, T and ρ. The manner in which the equation of state is implemented by different stellar evolution codes is not fixed. In the interior of the Sun, where hydrogen and helium are fully ionized, the simple ideal gas law provides an adequate equation of state. This is not adequate in the outer layers where minimally the equation of state must account for the hydrogen and helium ionization regions. For the models calculated here using the Yale Stellar Evolution code, the single ionization of H, the first ionization of the metals, and both ionizations of He are taken into account, via the Saha equation, for temperatures below 10^6 K. Above 10^6 K, the metals, H and He are assumed to be fully ionized.

The equation of state is currently the subject of investigation by several groups and is probably an area where the standard solar model will be improved. Preliminary comparisons (Christensen-Dalsgaard et al. (1988*b*) between the detailed MHD (Däppen et al. 1988*a*) tables of the equation of state for a solar mixture and the simple Eggleton formulation (used in many stellar evolution programs) indicate that the differences are not large with the largest differences occurring in the interior and at the base of the convection zone. Models constructed using the MHD tables have *p*-mode frequencies which are 5 to 10 μHz closer to the observed frequencies than models employing the Eggleton formulation and 1 to 5 μHz closer than models employing the Yale Stellar Evolution code formulation (Y.-C. Kim et al. 1991).

Nuclear Reactions.   The nuclear reaction network followed in most stellar evolution programs includes the hydrogen to helium burning reactions, i.e., *p-p* I, *p-p* II, *p-p* III chains and the CNO bi-cycle. Other reactions such as helium burning are typically included in stellar evolution calculations but are not important for the Sun. Light-element burning of, for example, Li, does not significantly affect the total energy output, hence, the corresponding nuclear reaction chains are not generally included in the stellar evolution reaction network. The reaction network includes all the hydrogen-burning reaction chains summarized in Table I.

The cross sections or reaction rates, with the exception of the *p-p* I chain, are determined in the laboratory and extrapolated to the temperature density regime encountered in the Sun. Unless a resonance exists, which can significantly increase the cross section, the cross sections are believed to be well determined with an error of $\pm 1\%$ (Parker 1986).

The reaction equations serve two purposes: (1) they determine the energy output per unit time of a given shell; and (2) they determine the evolution of the abundances of the elements involved in the nuclear reactions, such as hydrogen and helium. The former is used in the energy-balance equation, and the latter is used to specify the evolution of the mean molecular weight in a given shell.

The reaction network equations are kept distinct from the four partial differential equations describing the physics of the model. Nuclear reaction rates are calculated for each shell based on the shell's temperature and density. The new composition is then calculated. The energy output of the shell and its "evolved" composition are then input as constants into the four stellar evolution equations.

Pinsonneault (personal communication, 1990) reports that comparisons made between our solar models and those calculated by Bahcall and Ulrich (1988), both constructed using the latest nuclear cross sections, have predicted neutrino fluxes and physical variables which agree to better than 1 part in 10^3 in the nuclear burning regions. Note that the latest reaction cross sections are not used in the model calculations presented here.

TABLE I
Hydrogen Nuclear Burning Chains

	p-p Chains			
	$^1H + {}^1H$	$\rightarrow$	$^2D + e^+ + \nu$	
	$^1H + e^- + {}^1H$	$\rightarrow$	$^2D + \nu$	*(pep)*
p-p I	$^2D + {}^1H$	$\rightarrow$	$^3He + \gamma$	
	$^3He + {}^3He$	$\rightarrow$	$^4He + {}^1H + {}^1H$	
	$^3He + {}^1H$	$\rightarrow$	$^4He + e^+ + \nu$	
	$^3He + {}^4He$	$\rightarrow$	$^7Be + \gamma$	
p-p II	$^7Be + e^-$	$\rightarrow$	$^7Li + \nu$	
	$^7Li + {}^1H$	$\rightarrow$	$^4He + {}^4He$	
	$^7Be + {}^1H$	$\rightarrow$	$^8B + \gamma$	
p-p III	^8B	$\rightarrow$	$^8Be^* + e^+ + \nu$	
	$^8Be^*$	$\rightarrow$	$^4He + {}^4He$	
	CNO Chains			
	$^{12}C + {}^1H$	$\rightarrow$	$^{13}N + \gamma$	
	^{13}N	$\rightarrow$	$^{13}C + e^+ + \nu$	
	$^{13}C + {}^1H$	$\rightarrow$	$^{14}N + \gamma$	
	$^{14}N + {}^1H$	$\rightarrow$	$^{15}O + \gamma$	
	^{15}O	$\rightarrow$	$^{15}N + e^+ + \nu$	
	$^{15}N + {}^1H$	$\rightarrow$	$^{12}C + {}^4He$	
or				
	$^{15}N + {}^1H$	$\rightarrow$	$^{16}O + \gamma$	
	$^{16}O + {}^1H$	$\rightarrow$	$^{17}F + \gamma$	
	^{17}F	$\rightarrow$	$^{17}O + e^+ + \nu$	
	$^{17}O + {}^1H$	$\rightarrow$	$^{14}N + {}^4He$	

Opacities. The tables of Rosseland mean opacities, a function of density, temperature and composition, used in stellar evolution, are obtained from detailed numerical calculations. Until recently, the opacity tables of Cox and Stewart (1969, 1970*b*) and Cox and Tabor (1976) have been used extensively in stellar evolution calculations. Two groups, one at Los Alamos and the other at Lawrence Livermore, currently provide opacity tables. A third group consisting of Däppen et al. (1988*a*) are currently working on a new calculation of the equation of state and opacity. The results of comparisons made among the three groups have not yet been published at the time of this writing but early reports (A. N. Cox, personal communication) indicate that the Los Alamos opacities are lower, by as much as 20%, than the Lawrence Livermore tables in the temperature density regime found near the base of the convection zone. In addition, Cox et al. (1989) find that *p*-mode spectrum of their solar model shows better agreement with the observed *p*-mode oscillation spectrum if they increase their LAOL opacities at the base of the convection by 20%. Due to the complexities inherent to the construction of opacities, the opacities potentially represent the most significant source of error in the construction of the standard solar model. The errors in the bound-free opacities have been estimated to be of the order of 20%, although, they may only have an error of 10% in the solar interior (Demarque et al. 1988a).

The group responsible for the Los Alamos Opacity Library (LAOL) (Huebner et al. 1977) provides a set of tapes containing extinction coefficients, a function of photon frequency, density and temperature, for the twenty most abundant elements in the Sun. The extinction coefficients of the individual elements are combined in the specific abundance ratios required and then averaged over photon frequency to obtain the Rosseland mean opacities. The Lawrence Livermore opacities are provided in completed tables (as well as the MHD tables) as their calculation method does not permit the separation of the extinction coefficients into individual elements. The details of the two calculations are described in the chapter by Däppen et al. Because of our need to try a variety of heavy element mixtures, we have used the LAOL opacities which enable us to construct opacity tables for different mixtures.

The opacities provided by the LAOL do not cover the entire astrophysical temperature and density regime. For example, the tables do not include opacities below 10,000 K because absorption due to molecules are not included in the calculation. This is a serious drawback because it prevents the structure of the critical outermost layers of the Sun from being determined accurately. Los Alamos does provide a set of opacity tables which do include the effects of molecules and extend to lower temperatures but they are available only for the Ross and Aller (1976) mixture, hence, can only be used in the modeling of solar-type stars. They have not been incorporated in the models presented here.

Another problem associated with the current form of the opacities, but one that could easily be remedied, is the coarseness of the temperature, density grid at which the extinction coefficients are calculated. The resolution of the opacity grid in density is such that only one or two opacities are calculated for each decade change in density. The opacity can change by as much as 100% between neighboring grid points. This limits the accuracy of interpolation in the tables. This is demonstrated by comparisons made of interpolated tables constructed by Bahcall and Ulrich (1988) and ourselves. When the interpolation schemes and grid resolutions match the opacities at specified densities and temperatures, opacities were found to agree to within 1% (Bahcall and Ulrich 1988; Y.-C. Kim, personal communication, 1990).

Constraints.   The standard solar model is distinguished from models of other stars by the constraints imposed on the model. The standard solar model must have the Sun's luminosity and radius at the Sun's age.

The basic constraints of mass, radius and luminosity are directly measurable. The accuracy of the mass determination depends directly on the determination of G. The mass of the Sun is 1.9891×10^{33} g with a relative uncertainty of $\pm 0.02\%$ (Cohen and Taylor 1986). The radius of the Sun for stellar structure calculations is defined at an optical depth of $\tau = 2/3$, and is known, through transit and eclipse measurements to be 6.96×10^{10}cm at $\tau = 0.001$ (Ulrich and Rhodes 1983) with an error of $\pm 0.01\%$. Here, as noted by Ulrich and Rhodes, it is important to translate the glancing angle measurements at the Sun's limb to determine the Sun's radius to a $\tau = 2/3$ optical depth measured perpendicular to the Sun's surface. The luminosity is determined from solar constant measurements from space on both the Nimbus 7 and the SMM satellites: ERB-Nimbus 7 measures 1371.0 ± 0.765 W m^{-2} and SMM/ACRIM measures 1367.7 ± 0.802 W m^{-2} (Hickey and Alton 1983). The luminosity of the Sun is 3.846×10^{33} erg s^{-1} from SMM/ACRIM and 3.857×10^{33} erg s^{-1} from ERB-Nimbus 7. Taking the average of the two and setting the error equal to the difference between them, one obtains $L_\odot = 3.8515 \pm 0.011 \times 10^{33}$ erg s^{-1}.

The age of the Sun is inferred from the ages of the oldest meteorites. Although, the age is commonly quoted as being 4.6 Gyr or 4.7 Gyr, Guenther (1989) has recently rederived the age using more current information and obtains 4.49 ± 0.04 Gyr. Guenther notes that the latest determinations of the ages of the oldest meteorites (see discussion by Tilton [1988]) sets the age of meteoritic condensation at 4.56 Gyr which revises his earlier age estimate. The new best estimate of the Sun's age is now 4.52 ± 0.04 Gyr.

The abundances of most of the elements can be directly determined from the photospheric spectrum of the Sun. Although the abundance measurements represent the surface abundances of the Sun and not the interior abundances, it is assumed that these abundances are identical to the abundance of the elements in the zero-age model (with the exception of Li and Be which are affected by nuclear burning and diffusion). The inert gases helium, neon and argon are not visible in the photospheric spectrum hence their abundances must be inferred. Neon and argon abundances are adopted from their measured abundances in the solar corona, solar winds, nebula and hot stars (Meyer 1979). Helium, as the second most abundant element in the Sun, is left as a free parameter of the standard solar model. Because the luminosity of a stellar model is very sensitive to the mean molecular weight ($L \propto \mu^{7.5}$) the abundance of helium is adjusted to produce a solar model with the Sun's luminosity.

It turns out, however, that the abundance of helium so determined, is sensitive to the mixture of other elements and to the specific opacity tables used in the model's construction. Guenther et al. (1989) show that standard solar models constructed using the opacity mixture and tables of Cox and Stewart (1969,1970b) will typically have helium mass fractions of $Y = 0.24$. Models constructed using the more recently determined Grevesse mixture (Grevesse 1984b) and opacities of the LAOL have $Y = 0.28$. See Table II for a comparison of these models.

TABLE II
Standard Solar Model—Mixture Dependence

No.[a]	Opacity Source	Mixture	Z	Y	α	$\log(L/L_\odot)$	$\log(R/R_\odot)$
1	Cox and Stewart	Cox and Stewart	0.0200	0.239	1.36	−1.24E-4	−0.6E-4
2	LAOL	Cox and Stewart	0.0200	0.28	1.24		
3	LAOL	Ross and Aller	0.0169	0.278	1.24	−0.2E-5	−0.1E-5
4	LAOL	Grevesse	0.0194	0.281	1.26	−0.1E-5	−0.1E-5

[a]Models 1 and 2 have 1500 shells and were evolved in 10 time steps from the ZAMS to 4.5 Gyr. Models 3 and 4 have 2500 shells and were evolved from the ZAMS to 4.5 Gyr in 80 time steps.

TABLE III
Mixture-Number Fraction Abundance

Element	Abundance Cox and Stewart (1970b)	Abundance Ross-Aller (1976)	Abundance Grevesse (1984b)
C	0.19874	0.30279	0.29661
N	0.05606	0.06326	0.05918
O	0.44504	0.50249	0.49226
Ne	0.25071	0.02699	0.06056
Na	0.00094	0.00138	0.00129
Mg	0.01249	0.02892	0.02302
Al	0.00084	0.00241	0.00179
Si	0.01596	0.03244	0.02149
P		0.00023	0.00017
S		0.01151	0.00982
Cl		0.00023	0.00019
Ar	0.01669	0.00073	0.00230
Ca		0.00163	0.00139
Ti		0.00008	0.00006
Cr		0.00037	0.00028
Mn		0.00019	0.00017
Fe	0.00252	0.02297	0.02833
Ni		0.00138	0.00108

Table III and Fig. 1 compare the heavy element abundance determinations of Cox and Stewart, Ross and Aller (1976), and Grevesse. Based on the better agreement between the more recent determinations of Ross and Aller and Grevesse than to those of Cox and Stewart, it appears that the abundance determinations are improving. It is no surprise that a large disagreement remains among the abundance determinations for Ne and Ar. Although the abundance of Ar is very low in the Sun the abundance of Ne is not, being

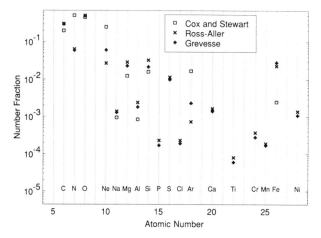

Fig. 1. The number fraction abundance is plotted against the atomic number for each element included in the mixtures of Cox and Stewart, Ross and Aller, and Grevesse. The Ross-Aller and Grevesse mixtures are more recent determinations and are in good agreement with each other for all elements except Ne and Ar which cannot be directly observed in the photospheric spectrum of the Sun.

comparable to abundance of N. Ne will certainly play a role in future standard solar models as a tuning parameter (Guenther et al. 1989).

It should be noted that the published solar abundance measurements include a helium mass fraction Y and an overall heavy element mass fraction Z (for Grevesse's mixture $Z = 0.0194$). Neither the total mass fraction of heavy elements nor the mass fraction of helium can be determined directly from observation but are based on values required to produce a standard solar model. Therefore, the list of abundance determinations should be interpreted as a list of relative abundances which are to be appropriately scaled to produce an accurate solar model.

Modeling Procedure. The standard solar model is constructed through an iterative procedure. An initial zero-age model with a guessed mixing-length parameter and helium abundance is evolved to the current age of the Sun. The model's luminosity and radius are then compared to the Sun's. Based on the discrepancy between the model's and the Sun's radius and luminosity, corrections to the mixing-length parameter and helium abundance, respectively, are calculated.

The accuracy to which the standard solar model is required to match the observed values of luminosity and radius is determined by the specific application. Historically, the standard solar model was used primarily to calibrate the unknown helium abundance, and mixing-length parameter was determined. Because the accuracy of stellar observations rarely exceeds $\pm 1\%$ in the $\log(T_{eff})$ − $\log(L/L_{\odot})$ plane, no attempt was made to construct a standard solar model (or stellar model) to a greater precision than $\pm 1\%$.

Recently the spectrum of p-modes have been used to test and constrain the standard solar model. The frequencies are measured to a remarkable accuracy of $\pm 0.01\%$. Where 300 shells were enough to achieve the desired reproducible accuracy required for stellar evolution applications, 1500 shells are now required for solar pulsation work. Not only have the number of shells been increased but also their distribution. Most stellar evolution programs distribute shells based on what is happening in a particular region. If, for example, hydrogen is near exhaustion in the core, then the program will increase the number of shells in this region and conversely, the program will reduce the number of shells in the convective envelope. The requirements of solar pulsation are different. Because most p-mode oscillations are confined to the surface layers and because the frequency of their oscillation is primarily determined by the sound speed in these layers more shells are required in the outer layers.

Conclusions. From the description presented here of the standard solar model it is clear that a unique standard solar model is not defined. There still remain several degrees of freedom which, due to complexity of calculation, have not been fully explored. Consider for example, the atmosphere calculation which is used to determine the surface boundary conditions. Although most stellar evolution programs model the atmosphere with an Eddington gray atmosphere, some static solar models of the Sun are based upon more accurate solar atmosphere models such as the Harvard Smithsonian Reference atmospheres (Vernazza et al. 1981) and the Krishna Swamy (1966) empirical fit to the Sun's T-τ relation. With regard to reproducing the observed highest frequencies of the solar p-mode oscillation spectrum, the differences appear to be significant (Guenther 1991). The treatment of surface boundary conditions also have significant implications for the shape of evolutionary tracks (see VandenBerg and Bell [1985] and the recent discussion by Demarque et al. [1988]).

Overall, the standard solar model does reflect the level of our knowledge with respect to the established physical processes involved in the evolution of a solar-type star. The model can be improved by attacking the weak points of the model, *viz.*, the opacities, the equation of state, the nuclear reaction cross sections, the convective modeling, the element abundance determinations, and the atmospheric structure. The model can also be improved by adding new physics to the description including the effects of rotation, element diffusion and magnetic fields.

The standard solar model has evolved significantly from the time of Lane's first model and so has our understanding of the Sun's interior. The diagnostic ability of both the solar p-mode oscillation spectrum and the spectrum of solar neutrino flux observations cannot help but to continue raising the standard of the standard solar model.

III. THE EVOLUTION OF THE SUN WITHOUT ROTATION

In this section we describe the evolution of the Sun from the zero-age main sequence, through its current age, on through hydrogen core exhaustion, and up to the tip of the giant branch. This pattern of evolution, including the giant branch, was first explained by Hoyle and Schwarzschild (1955) using a sequence of time independent static models. The first evolutionary calculation is due to Haselgrove and Hoyle (1956). Many authors have calculated giant branch evolution since then (see, e.g., the reviews by Iben [1974] and by Renzini [1977]).

A. How the Models Were Calculated

The evolutionary sequence presented in this section was calculated using the Yale Stellar Evolution code, YREC (Yale Rotating Evolution Code) in its nonrotating configuration (Pinsonneault et al. 1989, hereafter PKSD). The initial zero-age main-sequence model is a one solar mass stellar model with a homogeneous distribution of elements whose relative abundances are determined by the Grevesse mixture. We chose $Z = 0.0194$ as determined by Grevesse. The helium mass fraction Y and the mixing length α were chosen so that the model when evolved to an age of 4.5 Gyr will have the Sun's luminosity and radius. After several iterations, we determined $Y = 0.28$ and $\alpha = 1.25$.

YREC interpolates (cubic spline) opacity tables, based on the Grevesse mixture, constructed using the LAOL table of extinction coefficients. In the models presented here, we use Cox and Stewart (1969) opacities below 10,000 K (molecular opacities were not used). The equation of state is calculated assuming that hydrogen and helium are totally ionized in all regions except the outer envelope and atmosphere. The atmosphere is a gray atmosphere calculated in the Eddington approximation. The number of shells in the interior, envelope and atmosphere is approximately 500, with the actual number increasing as the solar model evolved off of the main sequence.

B. Basic Evolution

Figure 2 shows the evolution of the solar model from the zero-age main sequence to shortward of the tip of the giant branch (due to the large number of models that would be required, the approach to helium flash was not calculated) in the theoretical HR diagram. Approximately ten models are required to get to the Sun's age, 70 models to get to the base of the giant branch, and 1450 models to get to the tip of the giant branch. The evolutionary track is typical for a 1 $M_\odot$ Population I star.

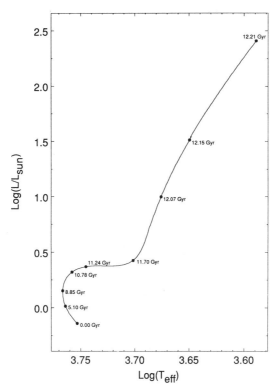

Fig. 2. The stellar evolutionary track for the Sun from zero age to just before helium flash is shown in a theoretical HR diagram. The age, measured from the zero-age main sequence, is noted at various points along the evolutionary track, in Gyr. Note that the track is terminated shortward of helium flash.

The solar model begins its evolution with hydrogen burning in the core. The transmutation of hydrogen into helium increases the mean molecular weight and as a result the layers surrounding the nuclear burning core slowly contract. The contraction of the central regions is compensated by the expansion of the outer layers. Because the contraction raises the central temperatures, the nuclear burning rates also increase. As the Sun evolves along the main sequence the solar model increases in both luminosity and radius. The p-p I chain provides most of the Sun's luminosity.

Hydrogen burning continues in the core until it is exhausted at about 10.5 Gyr. Hydrogen continues burning in a shell around the core. The shell thins as the solar model climbs the giant branch. This is due to the increasing mass of the inert helium core which acquires material from the hydrogen shell as the hydrogen shell consumes hydrogen as it moves outward. The isothermal helium core will remain inert until the solar model reaches the tip of the giant branch. At this point the temperatures ($\sim 10^8$ K) and densities ($\sim 10^6$ g cm^{-3}) are high enough to permit helium burning by the triple-alpha reaction (see Sec. III.C).

Figure 3 shows the rise in central temperature and density as the Sun evolves. Just before hydrogen exhaustion in the core at 10.5 Gyr the central temperature slows its increase as the available fuel of hydrogen becomes the important limiting factor in determining the reaction rates. At hydrogen exhaustion, the region surrounding the core contracts, heating the core. The core is nearly isothermal from this point on.

The jog in the curve just before 12 Gyr coincides with the passage of the hydrogen burning shell through the helium discontinuity left behind by the convective envelope at its deepest mass penetration (Thomas 1967). This jog corresponds to a momentary decrease in the star's rate of luminosity evolution along the giant branch, and is responsible for a bump in the luminosity function of star clusters. The observed luminosity of this bump is a function of chemical composition. Observations of this bump also yield information about the precise depth of convective mixing on the giant branch, which is not well known because of uncertainties in opacities and the efficiency of convective overshoot (King et al. 1985). The growth of the mass in the convective envelope is shown in Fig. 4. Figure 5 shows the temperature and density at the base of the convection zone for our standard models.

The relative importance of the basic hydrogen burning chains (see Table I) to contributing to the total luminosity is shown in Fig. 6. The *p-p* I chain initially dominates. With the buildup of ^{3}He by the *p-p* I chain and the rising central temperatures, the *p-p* II chain increases in importance. The larger Coulomb barrier of the carbon nucleus requires higher temperatures; hence CNO burning does not begin until later in the evolution. After core hydrogen

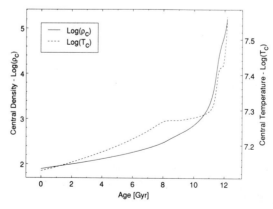

Fig. 3. The logarithm of the solar model's central density and temperature plotted opposite the model's age.

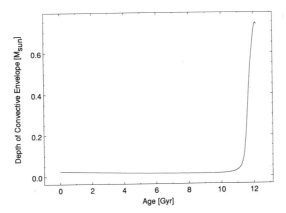

Fig. 4. The depth of the convective envelope, in solar mass units plotted against the age. The mass of the convective envelope initially decreases then increases rapidly at the point where the model reaches the base of the giant branch.

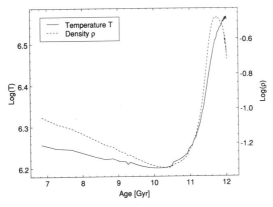

Fig. 5. The logarithm of the temperature and density at the base of the convective envelope plotted opposite the evolutionary age of the model. Noise in the curves is due to shell resolution which demands that the position of the base fall on a shell edge (the shell edge is not repositioned to match the precise location of the convective envelope base).

exhaustion the CNO cycle dominates the energy production of the star in a hydrogen burning shell. The hydrogen burning shell is initially very thick (Fig. 7) but as the solar model evolves up the giant branch it thins as the temperature rises. The hydrogen burning shell also moves outwards in mass toward the surface due to the ever increasing He core.

The deepening of the convective envelope is visible in the surface abundances of the isotopes ^{12}C, ^{13}C, ^{14}N and ^{3}He shown in Fig. 8. When the

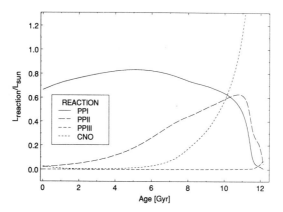

Fig. 6. The luminosity of the *p-p* I, *p-p* II, *p-p* III, and CNO nuclear reaction chains, in units of $L_\odot$ plotted against age. The luminosity of helium burning (not plotted) is entirely negligible during the main-sequence and giant-branch evolution of the Sun.

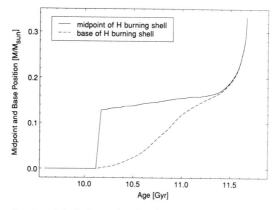

Fig. 7. The mass location of the hydrogen burning shell represented here by its base position and its midpoint position. The midpoint is defined as the mass shell where the maximum nuclear energy output is located.

convective envelope dips into regions previously partially processed by nu-clear reactions, the primordial surface mixture is altered. Initially we see a rise in the surface abundance of ^{3}He as the convective envelope deepens. Later its abundance is diluted by further mixing. As the convection zone deepens, ^{13}C and later ^{14}N are mixed to the surface. It is the ratio of ^{12}C to ^{13}C that is observed. The decline in this ratio as the solar model climbs the giant branch is rapid when the base of the convection zone reaches its maxi-mum depth (in mass). The ratio then remains constant (see Fig. 9). The same is true for the ^{14}N surface abundances.

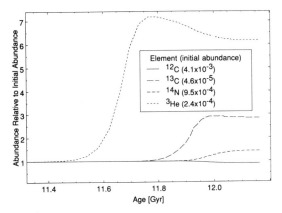

Fig. 8. The surface abundance of ^{12}C, ^{13}C, ^{14}N and ^{3}He plotted as a function of evolutionary time. The abundances are normalized relative to their initial abundance (zero-age main-sequence abundance). The surface abundances do not change until the convective envelope dips into regions which have been processed by nuclear reactions.

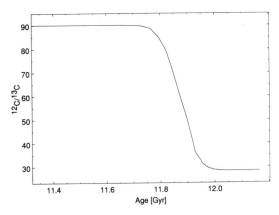

Fig. 9. The surface of abundance of ^{12}C divided by the surface abundance of ^{13}C plotted opposite the model's age.

During the last few years, extensive spectroscopic studies have been made on giant stars in the field and in star clusters that reveal the presence of internal mixing well beyond the predictions of the standard stellar evolution theory, and into layers that are believed to be extremely stable against convection. Mixing mechanisms that have been suggested include: (1) thermal instabilities in the hydrogen burning shell, similar in character to the helium shell flashes that take place on the asymptotic giant branch (AGB) (Dearborn et al. 1975a); (2) mass loss on the giant branch (Dearborn et al. 1975b); and

(3) rotational mixing (Sweigart and Mengel 1979). More detailed studies have shown the first mechanism to be unlikely. The importance of mass loss on early giant-branch evolution is severely constrained by observations of surface lithium abundances. Internal rotation remains as the most promising mechanism to explain the mixing of CNO-processed elements to the surface (see Sec. IV.D). But some mass loss takes place on the giant branch in a quiescent way. The Reimers' (1975) formula for mass loss, derived empirically, shows that most of the mass loss occurs at the highest luminosities on the giant branch, and amounts to about 0.20 $M_\odot$ for the Sun (see also Renzini 1977). By the time the tip of the giant is reached, the Sun's mass will then be ~ 0.80 $M_\odot$.

In Fig. 10 we show the basic hydrogen burning chains responsible for producing the solar neutrino rate in the Davis (1978) experiment. It is primarily sensitive to the neutrinos produced by the ^8B reaction (see Table I) which is clearly the dominant contributor at 5 Gyr. With the CNO cycle increasing in importance with respect to its contribution to the total luminosity as the model evolves, so do the ^{13}N and ^{15}O neutrinos increase in importance.

Figure 11 shows the sensitivity of the various solar neutrino detectors to the neutrino flux of the evolving solar model using the cross sections of Bahcall (1987). Note that only the ^{71}Ga detector is able to detect neutrinos from the *p-p* reaction, and that the ^7Li detector is particularly sensitive to the *pep* neutrinos.

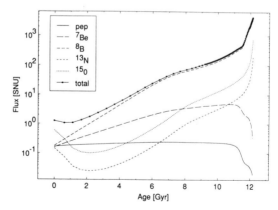

Fig. 10. The flux of neutrinos, in SNU, produced by the individual hydrogen burning nuclear reactions which produce neutrinos (see Table I) plotted against the model's age. The total neutrino flux is also plotted. The flux is determined for the Davis detector which at the age of the Sun is very sensitive to the ^8B reaction's neutrino.

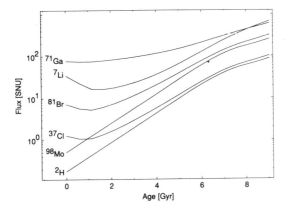

Fig. 11. The neutrino fluxes, in SNU, for various detectors. The differences are due to the fact that the detectors have different neutrino energy thresholds, hence are sensitive to different neutrino reactions occurring in the Sun.

C. Helium Burning and Advanced Evolutionary Phases

Core Helium Flash.   For stars of low mass like the Sun, the evolution up the giant branch is terminated by the runaway burning of helium in the core (the helium flash). Because of neutrino losses near the center, helium burning begins off-center. The initial burning of helium heats the central region; the reaction rates then increase, and because the degenerate electron gas dominates, the temperature rises still further. This is unlike the situation near the main sequence, where an increase in temperature results in an increase in pressure which restores hydrostatic equilibrium by a cooling expansion. Most of the energy produced in the helium flash goes into internal energy, i.e., into removing the degeneracy of the electron gas, and does not affect the luminosity of the star outside the helium core (Schwarzschild and Härm 1962; Demarque and Mengel 1971).

Because the development of the helium flash takes place on a relatively short time scale (years), the possibility of dynamic events must be considered. Much interest has been focused on the possibility of complete or partial mixing of the star (Larson 1965; Rood 1970; Petersen 1972; Smith and Demarque 1980), and of mass ejection. The current view is that no major mass loss takes place as a result of the core helium flash (Cole and Deupree 1980,1981). No mixing of the star leading to major structural effects is believed to occur at this point, although a small amount of mixing of fresh hydrogen into the helium core cannot be ruled out if the flash is far off-center or particularly violent (Paczynski and Tremaine 1977; Deupree and Cole 1983). It is also possible that even a small departure from spherical symmetry due to rotation, which is likely to be present (see Sec. IV.D), could lead to some mixing. This mixing could have important consequences for nucleosynthesis (Cowan et al. 1982), and could slightly modify subsequent evolutionary time scales (Seidel et al. 1987).

Core Helium Burning.   The flash ends when the degeneracy of the electron gas has been removed. The star settles into a double energy source phase, with quiescent helium burning in the convective inner part of the helium core, and hydrogen burning in a shell surrounding the helium core. A 0.8 $M_\odot$ star with the Sun's chemical composition occupies a position at the red end of the horizontal branch in the HR diagram, right against the giant branch, at about 60 $L_\odot$. The red clump observed in old disk star clusters represents this phase of evolution (Faulkner and Cannon 1973; Seidel et al. 1987), which for the Sun will last ~ 0.1 Gyr.

Helium Shell Flashes, Planetary Nuclei and White Dwarf Phases.   When the inner portion of the helium core has exhausted its fuel supply at the end of the red clump phases, the star derives its luminosity from two thin energy producing shells: the helium-burning shell, just outside the carbon-oxygen core produced by helium burning, and the hydrogen burning shell, just outside the helium zone, at the base of the hydrogen-rich envelope. Because of the high temperature sensitivity of helium burning, the helium shell is subject to thermal instabilities during which the helium luminosity undergoes rapid rises (Schwarzschild and Härm 1965). These runaway increases in helium luminosity are checked by the development of convection in the intershell region of the star. As a result, the temperature in the hydrogen shell drops, and the hydrogen luminosity vanishes.

As in the case of the core helium flash, the possibility has repeatedly been raised whether the convection induced by the flash results in the mixing of protons from the envelope into the hot helium zone. For stars below 1 $M_\odot$, the standard models quite clearly do not favor this mixing (Sweigart 1974). And yet there seems to be s-process enrichment taking place in these low-mass AGB stars (Sanders 1967). Nonstandard processes such as plumes have been suggested as a mode of mixing (Scalo and Ulrich 1973). The effects of rotational distortion will be discussed in Sec. IV.D.

We note in passing that while substantially more massive stars (say, with $M > 5$ $M_\odot$) describe blueward loops in the HR diagram through the cepheid instability strip during helium shell flash, stars like the Sun in mass and composition remain near the giant branch during most of their shell flashes (Sweigart 1974). Only the final flash before or during the blueward crossing that follow planetary nebula (PN) ejection is expected to result in a "cepheid loop" in the HR diagram (Mengel 1973; Gingold 1974). The field star FG Sag may be in this rare evolutionary phase (Langer et al. 1974b; Ulrich 1974b). The PN nucleus then embarks in its final phase of evolution on a cooling curve on its way to become a white dwarf. We will see in Sec. IV that even stars in these most advanced phases of evolution retain some memory of their main-sequence properties and thus provide valuable constraints on models for the current Sun.

D. Interior Structure of Selected Models

We have selected models near 0, 4, 8, 10 and 11 Gyr and plotted the density, pressure, temperature and luminosity as a function of radius fraction (see Figs. 12 through 15). As the model evolves off of the main sequence, the effects of hydrogen burning are apparent in the resultant structural changes. Figures 12 and 13 show the density and pressure evolution (note: the ordinate is logarithmic). As hydrogen is replaced by helium in the core, the central regions contract and the outer layers expand. The stationary shell, i.e., the radius fraction at which the radius and pressure do not change, is at the edge of the hydrogen burning core. Between 10 Gyr and 11 Gyr the solar model exhausts its central hydrogen and the rate at which the central regions contract increases dramatically. This is seen in the run of pressure and density for the 11-Gyr model which shows a much more concentrated core than the 10-Gyr model.

The temperature gradient in the model follows along with the pressure and density (Fig. 14). As the model evolves, the central region heats up because of contraction and the outer layers cool down because of the expansion. At 10 Gyr, a very small isothermal core develops. The amount of hydrogen fuel is so low that the nuclear energy production in this region is negligible hence there is no energy left to induce a temperature gradient. When hydrogen is completely exhausted in the core, the subsequent contraction of the core pumps enough energy into the core to raise its temperature.

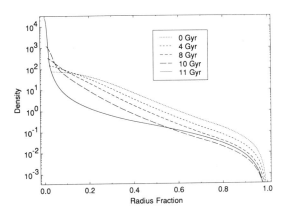

Fig. 12. For models at 0, 4, 8, 10 and 11 Gyr, the density plotted as a function of radius fraction ($X \equiv r/R_{total}$). The plot shows that as the Sun evolves, the density in the central region increases and the density of the outer layers decreases.

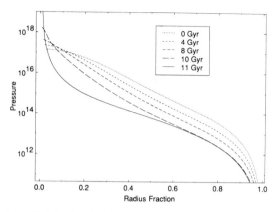

Fig. 13. For models at 0, 4, 8, 10 and 11 Gyr, the pressure plotted as a function of radius fraction. The plot shows that as the Sun evolves, the pressure in the central region increases and the pressure of the outer layers decreases.

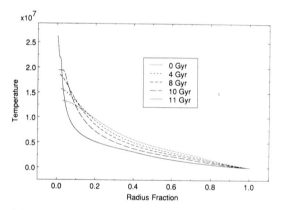

Fig. 14. For models at 0, 4, 8, 10 and 11 Gyr, the temperature plotted as a function of radius fraction. The plot shows that as the Sun evolves, the temperature in the central region increases and the temperature of the outer layers decreases. Note the flattened temperature profile for $X < 0.02$ for the 10-Gyr model where hydrogen burning is no longer important.

The luminosity is plotted for the central burning regions in Fig. 15. As the temperature increases in the core, so do the nuclear reaction rates, hence, so does the luminosity. At 10 Gyr, the model has depleted its central hydrogen fuel to the point where the nuclear reactions, in the core, do not provide very much energy. At 11 Gyr, after central hydrogen exhaustion, the small helium core contracts rapidly, releasing enough gravitational energy to increase the central temperatures. The luminosity associated with this contraction, i.e.,

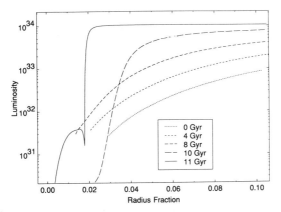

Fig. 15. For models at 0, 4, 8, 10 and 11 Gyr, the luminosity plotted as a function of radius fraction in the central regions. The plot shows that as the Sun evolves, the luminosity increases except in the inert helium core regions of the 10- and 11-Gyr models. Gravitation provides the small amount of luminosity in the inert helium core of the 11-Gyr model.

one half the gravitation energy lost during the contraction, is $< 1\%$ of the total luminosity generated by the hydrogen burning shell.

E. Evolution of the p-Mode Oscillations

Evolution of Frequencies. The Sun is currently the only solar-type star for which a large number of p-modes have been observed. Regardless, it is interesting to follow the evolution of the p-mode frequencies as the Sun evolves from zero age to core hydrogen exhaustion. After hydrogen is exhausted in the core, the isothermal helium core provides a cavity in which g-modes are trapped. This directly affects the p-mode oscillation modes.

We have calculated the p-mode frequencies of $l = 0$, 1, 2 and 3 and $n = 20$, 21, 22, 23, 24 and 25 for models near 0, 4, 8, 10 and 11 Gyr. Figures 16 and 17 show the evolution of the frequencies of these modes. Clearly, as the solar model evolves, the frequencies decrease, a direct consequence of the rarefaction of the outer layers and the increasing radius of the models.

The thickness of the bundle of lines in Figs. 16 and 17 also decreases as the solar model evolves. The thickness of the bundle is related to the characteristic frequency separation of the oscillations. In Fig. 18 we plot the frequency separation of adjacent in n modes of degree $l = 1$. This separation $\Delta\nu$, as is well known, is inversely proportional to the radius of the star. In Fig. 19 we have plotted: a cubic spline fit to $\Delta\nu$ vs the age of the model; the radius fraction X (scale on opposite Y axis) vs the age; and the product of these two quantities $X\,\Delta\nu$ vs the age. Not surprisingly, as the solar model evolves from the main sequence the product decreases slowly (see Christen-

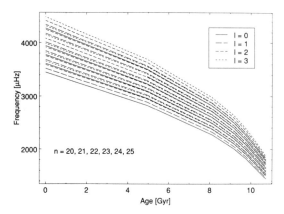

Fig. 16. The frequency of the p-modes $(n; l) = (0, 1, 2, 3; 20, 21, 22, 23, 24, 25)$ plotted as a function of model age. See Fig. 17.

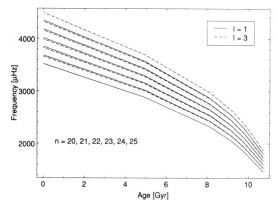

Fig. 17. Similar to Fig. 16, the frequency of the p-modes $(l; n) = (1, 3; 20, 21, 22, 23, 24, 25)$ plotted as a function of model age. Note the well-known overlap of mode frequencies.

sen-Dalsgaard 1986). The frequency separation between adjacent modes depends not only on the radius of the model but also on its structure. As the model evolves, the outer regions become more rarified which along with the increasing radius decreases the frequency spacing between the $n = 21$, $l = 1$ mode and the $n = 20$, $l = 1$ mode.

The second-order frequency separation parameter $\delta v \equiv v_{l,n} - v_{l+2,n-1}$ is sensitive to the interior structure. Specifically, as the density increases in the interior δv decreases. This is shown in Fig. 20. Notice, however, that δv begins to increase near hydrogen core exhaustion. This is because, as shown below, the selected p-modes become evanescent in the central regions, as

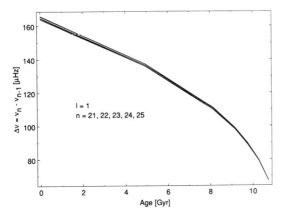

Fig. 18. Related to the asymptotic frequency spacing, the frequency separation between neighboring-in-n modes $\Delta\nu$ plotted opposite the model's age.

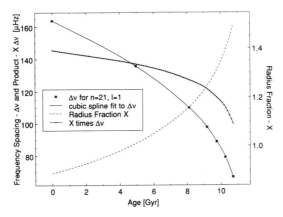

Fig. 19. The frequency spacing between neighboring-in-n modes $\Delta\nu$ plotted opposite the model's age as in Fig. 18. A cubic spline fit is drawn through these points. The radius fraction $X = R/R_\odot$ and the product $\Delta\nu X$ is also plotted. For most of the main-sequence evolution, the product is relatively constant.

hydrogen nears exhaustion, effectively reducing the cavity size to the layers above this "core."

Evolution of p-*Mode Amplitude Displacements.*   The amplitude of the radial displacement δr of a p-mode increases quite rapidly with increasing radius with the largest amplitude of the mode being confined to layers near the surface. Higher-degree modes are confined to a narrower layer near the surface than lower-degree modes. Even the amplitude of the low-degree mode

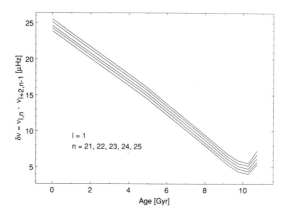

Fig. 20. The second-order spacing $\delta\nu$ plotted against the age. See text for discussion.

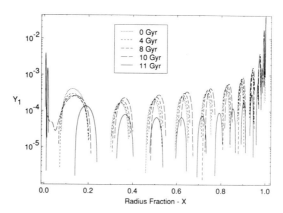

Fig. 21. The logarithm of the radial displacement $Y_1 \equiv \delta r/r$ plotted opposite radius fraction for 0, 4, 8, 10 and 11 Gyr models. The logarithmic ordinate permits the large amplitude differences between the interior and the surface to be shown, although the negative sweep of the radial displacement cannot be shown. Note the large peaks near the center of the 11-Gyr model.

discussed here ($l = 1$, $n = 20$) decreases rapidly with increasing depth. To better represent δr, we depart from tradition and plot the logarithm of $Y_1 \equiv \delta r/r$ vs radius (see Fig. 21). Although this representation does not permit the negative sweep of the amplitude from being drawn, it does permit a clear representation of all the peaks from the center to the surface.

In Fig. 21 we have plotted Y_1 vs radius fraction of the ($l = 1$, $n = 20$) mode for the 0, 4, 8, 10, and 11 Gyr models. The shape of Y_1 does not change very much until hydrogen exhaustion in the core occurs (between 10 and 11 Gyr). After hydrogen exhaustion, the peaks in Y_1 are shifted in position and their amplitudes are reduced by a factor of one-half to one-eighth. Figure 22

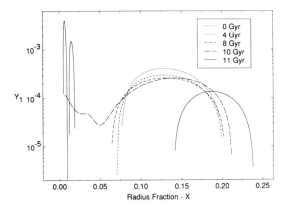

Fig. 22. Expanded horizontal scale view of Fig. 21 showing the two centrally located g-mode peaks of the 11-Gyr model. See text for discussion.

expands the radius fraction scale to reveal more detail in the central regions. Referring to Fig. 7, we note that the midpoint of the hydrogen burning shell between 10 to 11 Gyr is located near mass fraction 0.15. This corresponds to a radius fraction of 0.025 (see Fig. 15). Figure 22 shows that there are two peaks of considerable amplitude trapped inside this hydrogen burning shell.

By examining the phase relation between Y_1 and $Y_2 \equiv (p'/\rho + \Phi')/(gr)$ it is possible to determine whether the mode is a gravity mode or an acoustic mode. In evolved models it is common to have modes of mixed character. The mode shown in Fig. 22 for the 11-Gyr model is such a mode. The principal restoring force of the two peaks in the center is gravity or buoyancy and the principal restoring force for the rest of the peaks outside this region is pressure. The $l = 1$, $n = 20$ mode, therefore, changes character as the Sun evolves. Up to hydrogen exhaustion, the mode is a pure acoustic mode. Thereafter, it is a combined gravity and acoustic mode with the hydrogen burning shell completely isolating the two flavors of the mode.

The reason for the isolation is shown in Fig. 23 which plots the square of the Brunt Väisälä frequency N^2 vs radius fraction. As the star evolves, a large peak develops outside the core, where the mean molecular weight suddenly changes. The peak effectively separates oscillations on one side from oscillations on the other. Recall that a p-mode of frequency ν is evanescent in a region where N is $> \nu$.

The isolated nature of the dual flavored mode renders useless the usual identification scheme where the order n of the mode is defined as the number of p-mode amplitude peaks minus the number of g-mode peaks. For a given l this scheme provides a one to one correspondence between n and the oscillation modes of the star. Table IV shows how this one-to-one correspondence breaks down for the trapped modes found in the 11-Gyr model.

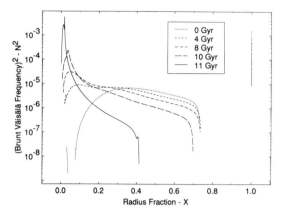

Fig. 23. The square of the Brunt-Väisälä frequency plotted against radius fraction for the 0, 4, 8, 10 and 11 Gyr models. A large spike develops at the edge of the helium core where a large μ gradient exists. The spike effectively traps g-modes inside this region. The negative Brunt-Väisälä frequency near the center of the 0 Gyr model is due to a small convective core near the main sequence.

The phenomenon of mode trapping, introduced here for the Sun, is not new to stellar oscillation theory and has been thoroughly studied for more massive stars (see Unno et al. 1979, and references therein). Further investigations of mode trapping in solar-type stars are in progress (Guenther 1991).

IV. EVOLUTION WITH ROTATION

A. Introduction

Although standard solar models do describe the internal mass and temperature distribution of the Sun remarkably well, there still remain small, but significant and persistent discrepancies with observation. One is the neutrino discrepancy illustrated in the previous section. Another is the failure to predict precisely the observed p-mode oscillation spectrum. Still another is the low observed solar Li surface abundance. The first two of these discrepancies are still unexplained, but progress has been made recently in modeling the evolution of rotating stars and the associated internal mixing which is responsible for Li depletion. We will see that there is ample evidence that rotation is present in the interiors of solar-type stars in all phases of their evolution, and in particular, plays a significant role in the mixing of chemical elements, which can then serve as tracers of rotational history. The study of advanced evolutionary phases thus provide us with the unique opportunity to probe the state of rotation in the deepest layers of the Sun. Rotation also controls the strength of magnetic activity and the properties of stellar activity cycles (Noyes et al. 1984b; Rutten 1987; Simon and Fekel 1987).

TABLE IV
Mode Character of p-Mode ($l=1$) for the 11-Gyr Model

Frequency µHz	n	No. of g-mode nodes	No. of p-mode nodes
1001.588	16	5	21
1033.396	19	3	22
1070.953	20	3	23
1110.753	19	5	24
1171.780	21	4	24
1203.847	24	2	26
1243.008	25	2	27
1281.302	26	2	28
1315.941	27	2	29
1350.039	26	4	30
1387.473	27	4	31

B. Models of an Evolving Sun

The problem is to understand the efficiency of the solar wind torque in slowing down the solar interior during its evolution (Schatzman 1962), and to determine the amount and distribution of the angular momentum which has remained trapped in the solar interior, primarily through the effect of stabilizing mean molecular weight barriers. Following up on the pioneering effort of Endal and Sofia (1981), PKSD (Pinsonneault et al. 1989) have calculated the evolution of the Sun from the Hayashi pre-main-sequence phase to the present. Their work includes a revised treatment of the solar wind torque by Kawaler (1968a), and of the effect of rotationally induced instabilities on angular momentum transport and chemical mixing.

Another new feature in the PKSD study is its reliance on observations of rotational velocities and of Li and Be abundances in young star clusters to set constraints on the parameterization of the angular momentum transfer. This approach offers the advantage of providing a theoretical tool to explore the efficiency of otherwise poorly known hydrodynamic processes, and a framework which predicts a solar rotation curve which can be understood in terms of the efficiency of distinct and precisely defined physical processes. Results are shown in the chapter by Sofia et al.

C. Direct Diagnostics of Internal Rotation in the Sun

There are two kinds of observations of the Sun which test directly its rate of internal rotation. They are the oblateness of the solar disk, and the rotational splitting of p-modes.

Oblateness of the Solar Disk. The solar oblateness is a measure of the quadrupole moment of the Sun. We currently have an upper limit to solar oblateness, set by the measurements of Hill and Stebbins (1975a). The Hill-Stebbins constraint is easily satisfied by the PKSD models, even though these models exhibit rapid rotation in their innermost part. Solar oblateness is on the other hand very sensitive to the state of rotation of the radiative layers just below the convection zone (Endal and Sofia 1981). It is hoped that the observations to be made from space with the Solar Disk Sextant (Sofia et al. 1984), will yield sufficiently precise data to replace the Hill-Stebbins upper limit by an actual measurement of the solar oblateness, permitting a direct comparison with the predicted oblateness of PKSD's rotating models, and helping differentiate between possible rotation curves near the solar center.

The Rotational Splitting of Solar p-*Modes.*   Observations of the splitting of solar *p*-modes provide in principle a strong test of the solar internal rotation curve. However, at this time, only modes with degree $l>8$ have splitting data of sufficient quality to compare with theoretical models (as of this writing new observations are being reported which surpass those referenced here). Modes with $l>8$ are primarily sensitive to the propagation of sound waves outside 0.4 $R_\odot$, and are not able to detect the presence or absence of rapid rotation inside 0.4 $R_\odot$, where most of the mass of the Sun is located, as shown in Figs. 24 and 25 due to Demarque and Guenther (1988).

Figure 24 shows three test rotation profiles that were used to calculate *p*-mode splittings. All curves are flat from the surface to 0.4 $R_\odot$. At 0.4 $R_\odot$ one curve rises by a factor of 5 over its surface rotation rate, the second remains flat and the third is reduced by a factor of 5. In Fig. 25, the *p*-mode splittings associated with these rotation curves are plotted along with some low *l* solar *p*-mode splitting data, kindly provided by K. Libbrecht (personal communication 1988). The splitting coefficients were calculated for our solar model using formulae contained in Gough (1981). Clearly, within the noise of the data it is not easy to distinguish the various rotation curves by their frequency splittings. In fact, the frequency splittings of similar test rotation profiles bent at 0.2 $R_\odot$ are virtually indistinguishable from solid body rotation on the scale of Fig. 25.

For lower *l*-values, which are needed to probe the deeper layers, the current data are still too noisy for reliable splitting estimates. It is true that a number of published rotation curves derived by inversion techniques extend to within 0.2 $R_\odot$. However, Fig. 25 shows that it is not possible to derive meaningful rotation curves so close to the center of the Sun from the available data.

Nonetheless, if indeed the observed splittings derived from observations of low *l* modes reflect the state of rotation in the solar interior, they can be used to put quite severe constraints on the rotation rate between 0.4 and 0.6 $R_\odot$, where the sensitivity to rotation is higher. In particular, the available data would restrict solar models to a flatter rotation curve between 0.4 and

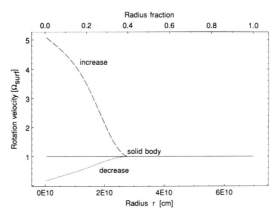

Fig. 24. Test rotation curves used to calculate the rotational *p*-mode splittings shown in Fig. 25.

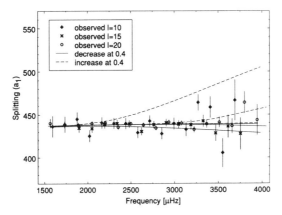

Fig. 25. The low-*l* frequency splittings from Libbrecht drawn with quoted errors along with the frequency splittings calculated for the test rotation curves shown in Fig. 24. Note that the frequency splittings of rotation curves which are similar to those shown in Fig. 24 but bent at 0.2 $R_\odot$ rather than 0.4 $R_\odot$ would appear as horizontal straight lines at $a_1 = 438$ on the scale of this figure. The curves with the largest deflection are the $l = 10$ modes. The nearly horizontal straight curves are $l = 20$ modes.

0.6 $R_\odot$ than the model labeled 'best model' by PKSD. This would provide a useful guide to test the efficiency of angular momentum transport by the secular shear instability, which controls the slope of the rotation profile in this region of the Sun within the framework of the PKSD models. Precise quantitative estimates will require more detailed sensitivity studies, and cannot be made from simple inspection of the splitting data.

The question of whether or not the Sun has a rapidly rotating interior below 0.4 $R_\odot$ remains unanswered. The PKSD models suggests that angular momentum has remained in the solar core, trapped by the increasing mean molecular weight gradients due to the nuclear transformation of hydrogen into helium. It is possible that an important mechanism for angular momentum transfer is missing in the PKSD models. On the other hand, we will see below that other lines of evidence indicate that the interior of the Sun probably rotates faster than its surface. It is hoped that observations of low-l modes ($l = 1$ and 2) with much improved noise levels, obtained by networks of stations observing the Sun as a star by the Birmingham and Nice groups, will provide in the near future the critical data to answer this question. (We note in passing that already in 1981, Claverie et al. (1981a) conjectured the presence of a rapidly rotating core in the Sun on the basis of very preliminary splitting data of low-l p-modes.)

D. Subgiants, Giants and More Evolved Stars

There are at present no detailed models of the evolution of the internal angular momentum in solar-type stars beyond the subgiant phase. But additional clues to the state of internal rotation of the Sun can be obtained from observations of solar-type stars that are more evolved than the Sun. The two most important clues come from rotational velocities of subgiant stars and from the $^{12}C/^{13}C$ isotopic ratio and CN-band strength on the giant branch. Strong additional constraints can also be derived from observations of still more evolved stars.

Rotational Velocities of Subgiant Stars.   We have seen in the previous sections that as solar-type stars exhaust their central hydrogen and evolve toward the giant branch, their envelope expands and their convection zone deepens. The expansion of the radius leads to an increase in the moment of inertia, which in the case of rigid rotation, is followed by a decrease in surface rotational velocity. The increase in the envelope mass means that layers that used to be below the convection zone are now included in the convection zone, within which exchanges of angular momentum are very effective. It follows that any excess angular momentum that might have been present in the formerly radiative layers is redistributed in the convection zone, leading to an increase in surface rotational velocity.

It is an easy matter to calculate the expected change in rotation period P during subgiant evolution under the assumption of rigid rotation for the present Sun, at least in the layers of the Sun which become incorporated in the convection zone. The expected sharp increase in P is illustrated in Fig. 26. On the other hand, evolutionary solar models with rotation rates increasing toward the center (such as the PKSD models) should show little or no increase in P, as dredge-up angular momentum compensates for the increase in radius. Figure 26 shows the available rotation data from the work of Noyes et al. (1984b), which lends support to the dredge-up hypothesis and the PKSD

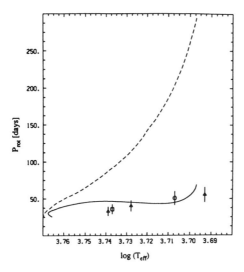

Fig. 26. Observed surface rotation periods of subgiant stars as a function of effective temperature compared with the rotation period the Sun would have as it evolved from the main sequence to the giant branch if it rotated as a solid body (dashed line), or as predicted by one of PKSD's models with a rotating interior (solid line). The data points are analogs of the Sun whose rotation periods are inferred from their chromospheric activity (Noyes et al. 1984*b*). Solar-mass giants are denoted by triangle points. The circle and square data points represent stars slightly more and less massive than the Sun, respectively.

models. Clearly more data are needed to confirm these results. If confirmed, these observations cannot be understood without the existence of a reservoir of angular momentum below the convection zone in stars with the same evolutionary status as the Sun.

Z_o *Mixing of CNO Processed Elements in Giants.*   We have seen in Sec. III that the standard evolution of a 1-$M_\odot$ star predicts a rapid variation of the $^{12}C/^{13}C$ isotopic ratio, from 90 to 27, on the giant branch after about 12 Gyr (Fig. 9). The predicted enhancement in ^{14}N surface abundance at the expense of ^{12}C is only 30%, and is therefore not easily measured. (Differences in nitrogen abundance are reflected in variations in the strength of the CN band in the spectrum.)

Internal rotation leads to enhanced ^{13}C and ^{14}N abundances at the surface compared to standard evolution. Its effects are two-fold: slow diffusive mixing spreads out the tail of the ^{13}C distribution toward the surface, which results in earlier and less abrupt convective dredge-up of CNO processed material along the giant branch until maximum dredge-up is reached: contrast with standard evolution, CNO processing might continue to decrease the $^{12}C/^{13}C$ ratio and ^{14}N enrichment beyond maximum dredge-up due to rotationally induced mixing below the convection zone, in a way similar to the models of Sweigart and Mengel (1979), so that lower $^{12}C/^{13}C$ ratios, approaching the CNO equilibrium value of 4, could be explained.

Detailed comparison with observations of the $^{12}C/^{13}C$ ratio along the giant branch of old star clusters would help constrain the shape of the ^{13}C distribution before dredge-up, and the efficiency of mixing beyond dredge-up. At the same time, the variations in CN-band strength observed in star clusters would be naturally explained by rotational mixing (Freeman and Norris 1981; Smith 1987), as all stars would not be expected to rotate at the same rate. We note that the CNO mixing diagnostic tests the state of rotation in deeper layers than surface rotational velocities of subgiants, and would in particular tell us about the efficiency of mean molecular weight gradients to stabilize the Goldreich-Schubert instability (Goldreich and Schubert 1967; Fricke 1968) and the mixing efficiency of meridional circulation.

Mixing on the Asymptotic Giant Branch.   On the horizontal branch and on the AGB, the rotational velocities involved are too small to modify the internal structure in a significant way. However, internal rotation can induce mixing due to shear instabilities and meridional circulation. In the case of AGB stars, with thin helium burning shells subject to thermal instabilities, the additional effects of nonspherical perturbations on the shell could result in the slow mixing of protons from the envelope into the helium zone needed for s-processing (Sanders 1967). In a linear-mode analysis of the response of a helium shell flash to nonradial perturbations, Richstone (1974) has shown that instabilities can take place with growth rates of the order of magnitude of convective lifetimes of intershell convection. This suggestive result deserves further investigation as an attractive mechanism for s-processing in AGB stars of low mass.

Angular Momentum Constraints from Evolved Stars.   A lower limit on internal angular momentum near the main sequence can also be placed from observations of the rotation rates of core helium burning stars, nuclei of planetary nebulae, and white dwarfs. In core helium burning stars (horizontal-branch stars), Peterson (1985) has measured substantial rotational velocities, which show that at least in some stars, angular momentum compatible in quantity with the Kraft (1970) curve estimate is preserved in the interior beyond giant branch evolution. Still beyond this stage of evolution, nuclei of planetary nebulae are known to rotate. Indeed, the rotational splitting of g-modes has been observed in these objects, which have been the subject of elegant seismological research (Kawaler et al. 1986). Finally, we note the observations of rotational velocities of white dwarfs well in excess of what would be expected if these stars had rotated as solid bodies on the main sequence (see Pilachowski and Milkey [1987] and the discussion of their data in PKSD).

Acknowledgments.   We gratefully acknowledge partial support for this research from grants of the Solar-Terrestrial Theory Program and the Astrophysics Theory Program of the National Aeronautics and Space Administration.

PART VI
Basic Data

APPENDIX A:
SOLAR ELEMENT ABUNDANCES

NICOLAS GREVESSE
Université de Liège

and

EDWARD ANDERS
University of Chicago

I. INTRODUCTION

Solar elemental abundances can be derived from spectroscopic studies of the photosphere, sunspots, chromosphere and corona as well as from particle measurements in the solar wind (SW) and solar energetic particles (SEP), and from gamma-ray spectroscopy of solar flares.

For many reasons (summarized by Meyer [1985*a*], by Grevesse [1984*a*] and by Aller [1987]), photospheric results provide undoubtably the most reliable and most accurate set of solar abundances not only because the photosphere is well mixed by convection and turbulence but also because the photospheric spectra are of very high quality and the physical conditions and physical processes are better known there than anywhere else in the Sun. Even when the solar wind and solar energetic particle results themselves are accurate, fractionation processes have altered the abundance pattern in the corona (Sec. III).

II. ABUNDANCE RESULTS

The key role played by atomic and molecular data, and mainly by transition probabilities, in solar abundance studies has been reviewed by Biémont and Grevesse (1977), Grevesse (1984b) and Grevesse (1989). Close collaborations between atomic physicists and astronomers have led to major improvements in the accuracy of the gf-values and, as a direct consequence, in the accuracy of the abundance results. In many cases, the uncertainties of the photospheric abundance results reflect the quality of the oscillator strengths used.

We present in Table I an updated version of Grevesse's (1984a) critical review of solar photospheric abundances (on the usual astronomical scale relative to hydrogen, $A_{el} = \log N_{el}/N_H + 12$). It is beyond the scope of this appendix to give details and references concerning all the results. Those may be found in Grevesse (1984a) and in Anders and Grevesse (1989) who give a detailed review of meteoritic as well as photospheric and coronal abundances.

Special comments are made below concerning a few important elements, i.e., the rare gases (He, Ne, Ar), C, N, O and Fe.

He

Because of the low accuracy of the prominence results and of the possible fractionation of He in the solar wind and the solar energetic particles, the solar He content must be derived from recent standard solar models. The results agree with data for Uranus and for extrasolar sources (H II regions, hot stars; Anders and Grevesse 1989; Meyer 1989).

Ne and Ar

These elements are fractionated relative to low first-ionization-potential (FIP) metals in the corona (see Sec. III). For this reason, we shall again rely partly upon local galactic values (H II regions, H I gas, stars) as reviewed by Meyer (1989). These values agree with the "photospheric" values derived from SW and SEP corrected by a constant fractionation factor (see Fig. 1; Anders and Grevesse 1989).

C, N and O

The new photospheric values for C, N and O are based on a careful analysis of vibration-rotation and pure rotation lines of CO, CH, NH and OH that are present in the solar infrared spectrum, as obtained from space by the ATMOS-SL3 (Atmospheric Trace Molecule Spectroscopy Experiment on board Spacelab 3) (Farmer et al. 1987). The new analysis made by A. J. Sauval (Bruxelles), C. B. Farmer and R. H. Norton (Jet Propulsion Laboratory) and N. Grevesse (Liège) with the help of E. Van Dishoeck (Caltech) and D. L. Lambert (Austin) leads to C and N results that differ somewhat from those previously adopted (Grevesse 1984a).

TABLE I
Abundance in the Solar Photosphere (log N_H = 12.00)

Element	Photosphere[a]		Meteorites[b]		Phot.–Met.[a]
1 H	12.00		[12.00]		—
2 He	[10.99	±0.035]	[10.99]		—
3 Li	1.16	±0.1	3.31	±0.04	−2.15
4 Be	1.15	±0.10	1.42	±0.04	−0.27
5 B	(2.6	±0.3)	2.88	±0.04	(−0.28)
6 C	8.60	±0.05	[8.60]		—
7 N	8.00	±0.05	[8.00]		—
8 O	8.93	±0.035	[8.93]		—
9 F	4.56	±0.3	4.48	±0.06	+0.08
10 Ne	[8.09	±0.10]	[8.09	±0.10]	—
11 Na	6.33	±0.03	6.31	±0.03	+0.02
12 Mg	7.58	±0.05	7.58	±0.02	0.00
13 Al	6.47	±0.07	6.48	±0.02	−0.01
14 Si	7.55	±0.05	7.55	±0.02	0.00
15 P	5.45	±(0.04)	5.57	±0.04	−0.12
16 S	7.21	±0.06	7.27	±0.05	−0.06
17 Cl	5.5	±0.3	5.27	±0.06	+0.23
18 Ar	[6.56	±0.10]	[6.56	±0.10]	—
19 K	5.12	±0.13	5.13	±0.03	−0.01
20 Ca	6.36	±0.02	6.34	±0.03	+0.02
21 Sc	3.10	±(0.09)	3.09	±0.04	+0.01
22 Ti	4.99	±0.02	4.93	±0.02	+0.06
23 V	4.00	±0.02	4.02	±0.02	−0.02
24 Cr	5.67	±0.03	5.68	±0.03	−0.01
25 Mn	5.39	±0.03	5.53	±0.04	−0.14
26 Fe	7.67	±0.03[c]	7.51	±0.01	+0.16[c]
27 Co	4.92	±0.04	4.91	±0.03	+0.01
28 Ni	6.25	±0.04	6.25	±0.02	0.00
29 Cu	4.21	±0.04	4.27	±0.05	−0.06
30 Zn	4.60	±0.08	4.65	±0.02	−0.05
31 Ga	2.88	±(0.10)	3.13	±0.03	−0.25
32 Ge	3.41	±0.14	3.63	±0.04	−0.22
33 As	—		2.37	±0.05	—
34 Se	—		3.35	±0.03	—
35 Br	—		2.63	±0.08	—
36 Kr	—		3.23	±0.07	—
37 Rb	2.60	±(0.15)	2.40	±0.03	+0.20
38 Sr	2.90	±0.06	2.93	±0.03	−0.03
39 Y	2.24	±0.03	2.22	±0.02	+0.02
40 Zr	2.60	±0.03	2.61	±0.03	−0.01
41 Nb	1.42	±0.06	1.40	±0.01	+0.02
42 Mo	1.92	±0.05	1.96	±0.02	−0.04
44 Ru	1.84	±0.07	1.82	±0.02	+0.02
45 Rh	1.12	±0.12	1.09	±0.03	+0.03
46 Pd	1.69	±0.04	1.70	±0.03	−0.01
47 Ag	(0.94	±0.25)	1.24	±0.01	(−0.30)
48 Cd	1.86	±0.15	1.76	±0.03	+0.10
49 In	(1.66	±0.15)	0.82	±0.03	(+0.84)

TABLE I (*continued*)

Element	Photosphere[a]		Meteorites[b]		Phot.–Met.[a]
50 Sn	2.0	±(0.3)	2.14	±0.04	−0.14
51 Sb	1.0	±(0.3)	1.04	±0.07	−0.04
52 Te	—		2.24	±0.04	—
53 I	—		1.51	±0.08	—
54 Xe	—		2.23	±0.08	—
55 Cs	—		1.12	±0.02	—
56 Ba	2.13	±0.05	2.21	±0.03	−0.08
57 La	1.22	±0.09	1.20	±0.01	+0.02
58 Ce	1.55	±0.20	1.61	±0.01	−0.06
59 Pr	0.71	±0.08	0.78	±0.01	−0.07
60 Nd	1.50	±0.06	1.47	±0.01	+0.03
62 Sm	1.00	±0.08	0.97	±0.01	−0.03
63 Eu	0.51	±0.08	0.54	±0.01	−0.03
64 Gd	1.12	±0.04	1.07	±0.01	+0.05
65 Tb	(−0.1	±0.3)	0.33	±0.01	(−0.43)
66 Dy	1.1	±0.15	1.15	±0.01	−0.05
67 Ho	(0.26	±0.16)	0.50	±0.01	(−0.24)
68 Er	0.93	±0.06	0.95	±0.01	−0.02
69 Tm	(0.00	±0.15)	0.13	±0.01	(−0.13)
70 Yb	1.08	±(0.15)	0.95	±0.01	+0.13
71 Lu	(0.76	±0.30)	0.12	±0.01	(+0.64)
72 Hf	0.88	±(0.08)	0.73	±0.01	+0.15
73 Ta	—		0.13	±0.01	—
74 W	(1.11	±0.15)	0.68	±0.02	(+0.43)
75 Re	—		0.27	±0.04	—
76 Os	1.45	±0.10	1.38	±0.03	+0.07
77 Ir	1.35	±(0.10)	1.37	±0.03	−0.02
78 Pt	1.8	±0.3	1.68	±0.03	+0.12
79 Au	(1.01	±0.15)	0.83	±0.06	(+0.18)
80 Hg	—		1.09	±0.05	—
81 Tl	(0.9	±0.2)	0.82	±0.04	(+0.08)
82 Pb	1.85	±0.05	2.05	±0.03	−0.20
83 Bi	—		0.71	±0.03	—
90 Th	0.12	±(0.06)	0.08	±0.02	+0.04
92 U	(<−0.47)		−0.49	±0.04	—

[a]Values in parentheses are uncertain.
[b]Values in brackets are based on solar or other astronomical data.
[c]See text.

Fe

The photospheric result reported in Table I (7.67) is obtained by Black-well et al. (1984,1986) through study of Fe I lines of rather low excitation with accurately known oscillator strengths. This value is 45% larger than that in meteorites (7.51). These low-excitation Fe I lines could possibly be affected by departures from LTE. A substantially lower value of 7.56 ± 0.08

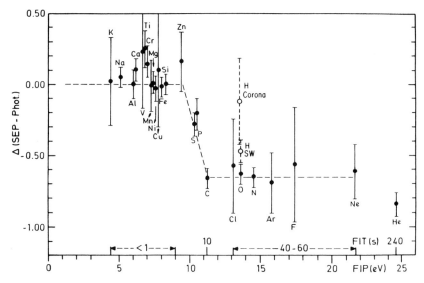

Fig. 1. Abundances of solar energetic particles (SEP) relative to the solar photosphere (Table II). Elements K to Zn, of first ionization potential (FIP) $\leq$ 10 eV, are present in normal abundance relative to Si, but elements C to Ne, of higher FIP, are depleted by an approximately constant factor. Apparently these elements are depleted in the corona by a process depending on FIP (or first ionization time, FIT). The points for P and S would fall closer to the dashed trend line if they were referenced to meteoritic rather than photospheric abundances. As H cannot be reliably determined in SEP, we plotted representative spectroscopic and solar-wind data from Table II.

was obtained by O'Mara (personal communication) using new transition probabilities of a few high-excitation lines (Milford et al. 1989). Results from Fe II (by far the dominant species) should allow the question to be settled. Unfortunately, accurate transition probabilities for good solar Fe II lines are very rarely available. Pauls et al. (1990) recently obtained the *gf*-values for 3 Fe II lines, which lead to $A_{Fe} = 7.66 \pm 0.06$. Details can be found in Anders and Grevesse (1989); the question of the solar iron abundance remains in a state of flux, but there is some hope that full agreement will soon be reached by joint efforts of atomic physicists and astronomers.

III. DISCUSSION

In Table I, we compare the solar results with the best meteoritic values as recently reviewed by Anders and Grevesse (1989), converted to the astronomical scale by adding 1.554 to the log of the meteoritic abundances on the Si = 10^6 scale. Meteoritic analyses have now converged to the point where most elements are known to better than 10 %.

TABLE II

Abundance in the Solar Corona[a]

Element	Spectro-scopic[b]	Solar Wind[c]	SEP[d]	Corona Adopted	Photosphere[e]	Corona–Phot.
1 H	11.88 ±0.30	11.53 ±0.08	—	—	≡12.00	—
2 He	(10.88 ±0.48)	10.13 ±0.10	10.14 ±0.06	10.14 ±0.06	10.99 ±0.035	-0.85
6 C	8.33 ±0.48	7.88 ±0.02	7.92 ±0.04	7.90 ±0.06	8.60 ±0.05	-0.70
7 N	7.55 ±0.23	7.42 ±0.15	7.40 ±0.03	7.40 ±0.06	8.00 ±0.05	-0.60
8 O	8.25 ±0.20	8.25 ±0.15	8.30 ±0.03	8.30 ±0.06	8.93 ±0.035	-0.63
9 F	—	—	(4.00 ±0.30)	(4.00 ±0.30)	4.56 ±0.30	(-0.56)
10 Ne	7.50 ±0.20	7.48 ±0.05	7.44 ±0.04	7.46 ±0.06	8.07 ±0.18	-0.61
11 Na	6.40 ±0.23	—	6.38 ±0.04	6.38 ±0.06	6.33 ±0.03	+0.05
12 Mg	7.53 ±0.11	—	7.59 ±0.03	7.59 ±0.06	7.58 ±0.05	+0.01
13 Al	6.40 ±0.23	—	6.47 ±0.03	6.47 ±0.06	6.47 ±0.07	0.00
14 Si	≡7.55 ±0.11	≡7.55 ±0.13	≡7.55 ±0.03	≡7.55 ±0.05	7.55 ±0.05	0.00
15 P	—	—	5.24 ±0.06	5.24 ±0.08	5.45 ±0.04	-0.21
16 S	6.89 ±0.23	—	6.93 ±0.02	6.93 ±0.05	7.21 ±0.06	-0.28
17 Cl	—	—	4.93 ±0.14	4.93 ±0.14	5.5 ±0.3	-0.57
18 Ar	6.28 ±0.26	5.85 ±0.10	5.93 ±0.06	5.89 ±0.10	6.58 ±0.18	-0.69
19 K	—	—	5.14 ±0.17	5.14 ±0.17	5.12 ±0.13	+0.02
20 Ca	6.43 ±0.20	—	6.46 ±0.06	6.46 ±0.08	6.36 ±0.02	+0.10

21 Sc	—		(4.04 ±0.40)	(4.04 ±0.40)	3.10 ±0.09	(+0.96)
22 Ti	—		5.24 ±0.12	5.24 ±0.13	4.99 ±0.02	+0.25
23 V	—		(4.23 ±0.40)	(4.23 ±0.40)	4.00 ±0.02	(+0.23)
24 Cr	—		5.81 ±0.08	5.81 ±0.09	5.67 ±0.03	+0.14
25 Mn	—		5.38 ±0.17	5.38 ±0.18	5.39 ±0.03	−0.01
26 Fe	7.55 ±0.18	7.53 ±0.27	7.65 ±0.04	7.65 ±0.06	7.67 ±0.03	−0.02
28 Ni	6.29 ±0.23		6.22 ±0.06	6.22 ±0.08	6.25 ±0.04	−0.03
29 Cu	—		(4.31 ±0.40)	(4.31 ±0.40)	4.21 ±0.04	(+0.10)
30 Zn	—		4.76 ±0.18	4.76 ±0.19	4.60 ±0.08	+0.16

[a] All logarithmic abundances have been normalized to the photospheric scale, with $\log A_{Si} \equiv$ 7.55. See triple equal signs. Parenthesized values are very uncertain.

[b] Coronal spectroscopic results apply variously to the ordinary quiet corona, active regions, coronal holes or prominences; they are taken from a review by Meyer (1985b). The hydrogen value has been slightly modified from Meyer's value, on the basis of the Si/H and Ca/H ratios derived by Veck and Parkinson (1981). The helium value is taken from prominence results.

[c] From a review by Bochsler (1987), except for Fe (Schmidt et al. 1988).

[d] From Breneman and Stone (1985), except for He (Cook et al. 1984; McGuire et al. 1986). We quote here the solar energetic particles (SEP) values corrected for the Q/M-dependent fractionation (SEP-derived corona in Table I of Breneman and Stone), which, however, depend on the assumed Fe/Si ratio.

[e] From Table I except for Ne and Ar, which are the local galactic values of Meyer (1985a, 1988b, 1989).

The agreement between meteoritic and solar abundances is now remarkably good. Past discrepancies have vanished as the solar values have become more accurate, thanks mainly to improved transition probabilities. Among the elements which disagree by > 0.10 to 0.15 dex, twenty elements (B, F, P, Cl, Ga, Rb, Ag, In, Sn, Sb, Ce, Tb, Ho, Tm, Yb, Lu, Hf, Pt, Au, Tl) are without doubt poorly determined in the solar photosphere due either to severe blending of their few available lines or to lack of accurate gf-values, or even both. If we retain only 29 accurately known elements, having a sufficient number of good lines with accurate transition probabilities, then the solar and meteoritic results agree to within $\pm$ 9 %. Further details concerning this comparison as well as a discussion of the discordant elements (Mn, Fe, Ge, Pb) may be found in Anders and Grevesse (1989).

Coronal abundances can be derived either by spectroscopy or by measurement of particles originating in the corona such as SW and SEP. The most recent values are given in Table II. The SEP and SW values are in most cases quite accurate; they generally agree with the spectroscopic values within the rather large error limits of these data. The "corona-adopted" values in Table II are based on SEP-SW values. Differences between these coronal and the photospheric values are given in the last column of Table II and are shown in Fig. 1. Although it is well known that elements with high first ionization potentials (FIP) are depleted relative to elements of lower FIP, the improved photospheric data for CNO now show, more clearly than before, that this depletion is constant at $-$ 0.65 dex (factor of 4.5) for CNO and presumably for most elements of FIP $\geq$ 11 eV (except for H and He). This confirms the two-plateau pattern first suggested by Cook et al. (1979) and suggests that separation processes at relatively low temperature fractionate the gas supplied to the corona. Elements with high FIP's, that are neutral at this temperature, are depleted relative to elements of lower FIP, that are ionized. Further details are given by Anders and Grevesse (1989, and references therein).

The abundances of Table I have been used to compute the mass fractions of H, He and heavier elements (X, Y, Z where $X + Y + Z = 1$). The values (in percent) are given below, with an alternative set for the solar Fe value (7.67) in parentheses: $X = 70.671$ (70.631), $Y = 27.426$ (27.411), $Z = 1.903$ (1.958). We therefore adopt the following best values: $X = 70.651$ ($\pm$ 2.5%), $Y = 27.418$ ($\pm$ 6 %), $Z = 1.930$ ($\pm$ 8.5 %). The uncertainties are conservative estimates, calculated by varying the individual elements within their error limits.

APPENDIX B:
SOLAR INTERIOR MODELS

J. A. GUZIK
Los Alamos National Laboratory

and

Y. LEBRETON
Observatoire de Paris

We present below the interior properties of four solar models from the recent astrophysical literature. The first three are standard solar models, each with somewhat different input physics. The fourth is an unconventional model designed to address the solar neutrino problem, with reduced central opacity to simulate the effects of weakly interacting massive particles.

The standard solar model of Lebreton and Dappen (1988) has metallicity $Z = 0.02$, initial He mass fraction $Y = 0.278$, and mixing-length/pressure-scale-height ratio $\alpha = 2.161$. The model is calculated using the latest equation of state developed by Mihalas (1988a; see also Hummer and Mihalas 1988; Däppen et al. 1988a). This equation of state is based on the free-energy minimization method which implies thermodynamical consistency, and contains a large number of atomic and ionic species, with detailed internal partition functions containing weighted occupation probabilities. The opacities used are from the Los Alamos Opacity Library (Huebner et al. 1977), supplemented by molecular absorption coefficients at low temperature (Cox, personal communication, 1981). Nuclear reaction rates have been updated for the *p-p* chain (Parker 1987) and the CNO cycle (Harris et al. 1983; Rolfs 1985). Abundances of elements heavier than H and He are from Grevesse's (1984) data. Further details regarding the input physics and numerical methods involved in calculating this model can be found in Lebreton and Maeder (1987) and Lebreton (1988).

The standard model of Bahcall and Ulrich (1988) has metallicity $Z = 0.0196$ and initial He mass fraction $Y = 0.271$. Instead of α, the initial value of S, an entropy-like variable which depends on the mixing length and defines the adiabat of the convection zone, is chosen, along with the initial He abundance, to calibrate the final solar luminosity and radius (Bahcall et al. 1982). Mixing-length theory is used only for temperatures $\lesssim 3 \times 10^5$ K. The equation of state, developed by Ulrich (1982), includes radiation pressure, Debye-Huckel screening, and electron degeneracy corrections, and accounts for scattering states using the methods of Larkin (1960), Ebeling et al. (1977) and Ebeling and Sandig (1973). Opacity tables for the solar mixture of Grevesse (1984) were specially prepared with the codes of the Los Alamos Opacity Library (Huebner et al. 1977). See Bahcall and Ulrich (1988) and Bahcall et al. (1982) for a detailed bibliography of sources for nuclear reaction rates.

The standard model of Cox et al. (1989) has metallicity $Z = 0.02$, initial He mass fraction $Y = 0.291$, and mixing-length/pressure-scale-height ratio $\alpha = 1.894$. In the Cosmion model of Cox et al. (1990), weakly interacting massive particles (cosmions) effectively reduce the opacity by a factor of approximately 10^{-3} in the central 1/10 of the solar radius, which produces a lower-temperature isothermal core, and thereby decreases the number of ^{8}B neutrino captures predicted for the ^{37}Cl experiment. This model also has $Z = 0.02$, and requires a reduced initial He mass fraction ($Y = 0.277$) and increased mixing-length/pressure-scale-height ratio $\alpha = 2.015$ to attain 1 $L_\odot$ and 1 $R_\odot$ at age 4.6 Gyr. Both models were evolved using the Iben (1965) evolution code, with several improvements to the equation of state and opacity. The equation of state is calculated using the Iben (1963*b*) procedure for ionization regions, in which only the ionization of H and He electrons, plus a single electron from a representative heavy element are considered, and the procedure of Eggleton et al. (1973) for high-temperature regions where a small amount of electron degeneracy occurs. Coulomb pressure-reducing effects (Clayton 1968) were incorporated by Cox et al. (1989,1990), due to their importance for accurately calculating *p*-mode oscillation frequencies. The Iben (1975) analytical method for calculating opacities was adjusted to match the Los Alamos Opacity Library values (Huebner et al. 1977). At temperatures below 1.5×10^6 K, the Iben procedure was replaced by the analytical fit of Stellingwerf (1975*a,b*), and altered to account for molecular absorption. In addition, opacites below the convection zone, between 2 and 7×10^6 K, were increased within theoretical uncertainties (15–20%) to improve agreement between observed and calculated *p*-mode frequencies.

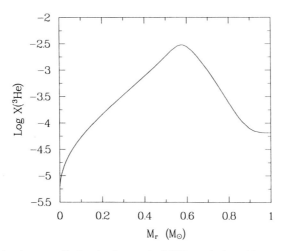

Fig. 1. ³He abundance profile for the Cox et al. (1989) standard model at age 4.6 Gyr. The abundances for mass fraction $\lesssim 0.6$ $M_\odot$ below the peak represent equilibrium $p\text{-}p$ chain abundances attained at high interior temperatures.

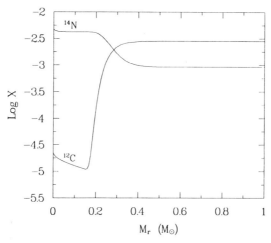

Fig. 2. ¹²C and ¹⁴N abundance profile for the Cox et al. (1989) standard model at age 4.6 Gyr. The abundances for mass fraction $\lesssim 0.3$ $M_\odot$ are altered by partial CN-cycle processing.

Nuclear-reaction rates are from Fowler et al. (1975). Abundances of elements heavier than H and He are from Ross and Aller (1976).

Tables I and II summarize the evolution of the standard models of Lebreton and Däppen (1988) and Cox et al. (1989). Tables III through VI present the interior mass fraction, and the corresponding radius, luminosity, pressure, temperature, density, opacity and abundances of H, ³He, ¹²C and

^{14}N of the four models at the present solar age. Figure 1 shows the ^{3}He profile, while Fig. 2 shows the ^{12}C and ^{14}N profiles of the Cox et al. (1989) standard model at age 4.6 Gyr. Table VII gives the predicted neutrino fluxes from the model of Lebreton and Dappen (1988), while Table VIII compares the predicted neutrino capture rates for the ^{37}Cl detector for the four models.

TABLE I
Evolution of the Lebreton and Däppen (1988) Standard Model[a]

Age(Gyr)	Log L (L$_\odot$)	Log T$_{eff}$ (K)	R(R$_\odot$)	T$_c$ (10^6 K)	ρ$_c$ (g cm^{-3})	X$_c$ (H)	M$_{ce}$ (M$_\odot$)	T$_{ce}$ (10^6 K)	R$_{ce}$ (R$_\odot$)
0.01	−0.1815	3.747	0.8705	13.66	74.55	0.701	0.974	2.29	0.642
0.58	−0.1209	3.753	0.9072	13.72	85.33	0.668	0.976	2.24	0.663
1.09	−0.1079	3.754	0.9164	13.87	90.35	0.632	0.978	2.18	0.675
1.61	−0.0943	3.755	0.9259	14.05	96.00	0.593	0.978	2.16	0.682
2.12	−0.0799	3.756	0.9364	14.26	102.3	0.553	0.979	2.14	0.690
2.63	−0.0647	3.758	0.9476	14.49	109.4	0.513	0.979	2.12	0.698
3.14	−0.0488	3.759	0.9595	14.74	117.4	0.473	0.980	2.10	0.706
3.40	−0.0406	3.760	0.9657	14.87	121.9	0.452	0.980	2.08	0.711
3.91	−0.0239	3.761	0.9797	15.14	131.9	0.410	0.981	2.05	0.721
4.37	−0.0081	3.762	0.9930	15.40	142.3	0.372	0.981	2.03	0.730
4.60	0.0001	3.762	0.9999	15.54	148.0	0.352	0.982	2.02	0.735

[a]Subscripts c refer to central values, while subscripts ce refer to values at the base of the envelope conveection zone.

TABLE II
Evolution of the Cox, Guzik and Kidman (1989) Standard Model[a]

Age(Gyr)	$\log L\ (L_\odot)$	$\log T_{\rm eff}$ (K)	$R(R_\odot)$	T_c (10^6 K)	ρ_c (g cm^{-3})	X_c (H)	M_{ce} ($M_\odot$)	T_{ce} (10^6 K)	R_{ce} ($R_\odot$)
0.00	-0.1522	3.7520	0.8755	13.69	81.44	0.688	0.970	2.52	0.631
0.50	-0.1252	3.7542	0.8939	13.75	90.19	0.652	0.972	2.44	0.646
1.00	-0.1110	3.7553	0.9050	13.92	95.55	0.615	0.973	2.42	0.653
1.50	-0.0969	3.7562	0.9153	14.12	101.5	0.577	0.973	2.41	0.660
2.00	-0.0829	3.7572	0.9268	14.32	108.3	0.538	0.974	2.39	0.667
2.50	-0.0675	3.7581	0.9386	14.54	115.6	0.500	0.974	2.37	0.674
3.00	-0.0520	3.7590	0.9516	14.78	124.3	0.460	0.975	2.35	0.682
3.50	-0.0369	3.7597	0.9651	15.04	133.8	0.420	0.975	2.34	0.691
4.00	-0.0205	3.7605	0.9801	15.32	145.2	0.378	0.975	2.32	0.700
4.64	-0.0001	3.7612	1.0000	15.71	162.2	0.324	0.976	2.30	0.712

[a]Subscripts c refer to central values, while subscripts ce refer to values at the base of the envelope convection zone.

TABLE III
Interior Structure of the Lebreton and Dappen (1988) Standard Model at Age 4.6 Gyr

M_r ($M_\odot$)	r($R_\odot$)	L_r ($L_\odot$)	P (dynes cm⁻²)	T_r (10⁶ K)	ρ_r (g cm⁻³)	κ (cm² g⁻¹)	X(H)	X(³He)	X(¹²C)	X(¹⁴N)
0.0000	0.0000	0.0000	2.278 + 17	15.54	148.0	1.24	0.352	8.61 − 6	1.59 − 5	5.91 − 3
0.0005	0.0164	0.0040	2.239 + 17	15.46	145.3	1.24	0.359	9.17 − 6	1.57 − 5	5.91 − 3
0.0009	0.0205	0.0076	2.217 + 17	15.41	143.7	1.25	0.363	9.53 − 6	1.56 − 5	5.91 − 3
0.0039	0.0340	0.0322	2.116 + 17	15.18	136.6	1.28	0.382	1.13 − 5	1.51 − 5	5.91 − 3
0.0099	0.0470	0.0781	1.986 + 17	14.87	127.6	1.31	0.407	1.42 − 5	1.45 − 5	5.91 − 3
0.0199	0.0604	0.1482	1.830 + 17	14.49	117.1	1.37	0.439	1.86 − 5	1.37 − 5	5.91 − 3
0.0400	0.0783	0.2708	1.604 + 17	13.88	102.6	1.45	0.485	2.76 − 5	1.25 − 5	5.91 − 3
0.0604	0.0919	0.3760	1.431 + 17	13.38	92.08	1.52	0.520	3.80 − 5	1.16 − 5	5.92 − 3
0.1028	0.1143	0.5497	1.160 + 17	12.51	76.37	1.63	0.572	6.49 − 5	1.00 − 5	5.92 − 3
0.1430	0.1319	0.6707	9.680 + 16	11.83	65.56	1.73	0.606	9.96 − 5	1.04 − 5	5.92 − 3
0.1992	0.1537	0.7898	7.616 + 16	11.00	54.04	1.86	0.638	1.69 − 4	4.81 − 4	5.15 − 3
0.2506	0.1720	0.8632	6.140 + 16	10.35	45.64	1.97	0.659	2.64 − 4	2.27 − 3	2.46 − 3
0.3006	0.1892	0.9116	4.979 + 16	9.778	38.77	2.08	0.672	3.99 − 4	3.52 − 3	1.19 − 3
0.3497	0.2057	0.9438	4.036 + 16	9.262	32.96	2.23	0.681	5.91 − 4	4.00 − 3	9.76 − 4
0.4026	0.2237	0.9668	3.193 + 16	8.742	27.52	2.41	0.688	9.06 − 4	4.16 − 3	9.48 − 4
0.4502	0.2401	0.9804	2.563 + 16	8.296	23.22	2.61	0.693	1.35 − 3	4.20 − 3	9.45 − 4
0.4966	0.2567	0.9892	2.045 + 16	7.920	20.01	2.85	0.695	2.04 − 3	4.21 − 3	9.44 − 4
0.5685	0.2841	0.9968	1.399 + 16	7.209	14.55	3.33	0.698	3.17 − 3	4.21 − 3	9.44 − 4
0.6091	0.3010	0.9985	1.105 + 16	6.830	12.13	3.54	0.699	2.59 − 3	4.21 − 3	9.44 − 4
0.6584	0.3231	0.9994	8.082 + 15	6.376	9.501	3.79	0.700	1.54 − 3	4.21 − 3	9.44 − 4
0.7039	0.3459	0.9998	5.853 + 15	5.957	7.365	4.09	0.701	8.65 − 4	4.21 − 3	9.44 − 4
0.7539	0.3748	1.0000	3.902 + 15	5.487	5.332	4.57	0.702	4.25 − 4	4.21 − 3	9.44 − 4
0.8050	0.4104	1.0000	2.381 + 15	4.977	3.588	5.31	0.702	1.91 − 4	4.21 − 3	9.44 − 4
0.8532	0.4531	1.0000	1.335 + 15	4.449	2.251	6.29	0.702	9.10 − 5	4.21 − 3	9.44 − 4
0.9000	0.5099	1.0000	6.340 + 14	3.853	1.234	8.03	0.702	5.27 − 5	4.21 − 3	9.44 − 4
0.9497	0.6077	1.0000	1.864 + 14	3.005	0.4654	11.4	0.702	4.24 − 5	4.21 − 3	9.44 − 4
0.9700	0.6763	1.0000	8.102 + 13	2.499	0.2433	14.7	0.702	4.16 − 5	4.21 − 3	9.44 − 4
0.9914	0.8054	1.0000	1.429 + 13	1.356	0.0794	42.0	0.702	4.15 − 5	4.21 − 3	9.44 − 4
0.9993	0.9280	1.0000	7.499 + 11	0.4233	0.0135	359.	0.702	4.15 − 5	4.21 − 3	9.44 − 4
1.0000	1.0000	1.0000	6.839 + 04	0.0058	0.0000	0.652	0.702	4.15 − 5	4.21 − 3	9.44 − 4

TABLE IV
Interior Structure of the Bahcall and Ulrich (1988) Standard Model at Age 4.6 Gyr

M_r ($M_\odot$)	r($R_\odot$)	L_r ($L_\odot$)	P (dynes cm^{-2})	T_r (10^6 K)	ρ_r (g cm^{-3})	X(H)	$X(^3$He)	$X(^{12}$C)	$X(^{14}$N)
0.0000	0.0000	0.000	2.29 + 17	15.6	148.	0.341	7.74 − 6	2.61 − 5	6.34 − 3
0.0004	0.0158	0.003	2.26 + 17	15.6	145.	0.349	8.29 − 6	2.56 − 5	6.29 − 3
0.0014	0.0237	0.012	2.21 + 17	15.5	142.	0.359	9.06 − 6	2.50 − 5	6.22 − 3
0.0021	0.0277	0.018	2.18 + 17	15.4	140.	0.365	9.58 − 6	2.47 − 5	6.18 − 3
0.0032	0.0317	0.027	2.15 + 17	15.3	137.	0.372	1.02 − 5	2.43 − 5	6.14 − 3
0.0046	0.0358	0.038	2.12 + 17	15.2	135.	0.380	1.09 − 5	2.38 − 5	6.10 − 3
0.0062	0.0400	0.051	2.08 + 17	15.1	132.	0.389	1.18 − 5	2.34 − 5	6.06 − 3
0.0083	0.0442	0.067	2.03 + 17	15.0	129.	0.398	1.28 − 5	2.29 − 5	6.02 − 3
0.0108	0.0484	0.085	1.99 + 17	14.9	126.	0.408	1.39 − 5	2.24 − 5	5.98 − 3
0.0211	0.0616	0.157	1.83 + 17	14.5	116.	0.442	1.85 − 5	2.08 − 5	5.88 − 3
0.0429	0.0804	0.287	1.60 + 17	13.8	101.	0.492	2.85 − 5	1.87 − 5	5.81 − 3
0.0577	0.0902	0.363	1.47 + 17	13.5	93.3	0.518	3.60 − 5	1.76 − 5	5.79 − 3
0.0808	0.1033	0.466	1.31 + 17	13.0	84.0	0.551	4.93 − 5	1.62 − 5	5.78 − 3
0.1038	0.1147	0.553	1.18 + 17	12.5	76.4	0.577	6.51 − 5	1.50 − 5	5.77 − 3
0.1500	0.1346	0.688	9.60 + 16	11.7	64.5	0.615	1.07 − 4	1.68 − 5	5.77 − 3
0.2040	0.1551	0.798	7.66 + 16	10.9	54.0	0.646	1.79 − 4	5.28 − 4	5.17 − 3
0.2520	0.1719	0.865	6.29 + 16	10.3	46.4	0.666	2.73 − 4	2.18 − 3	3.24 − 3

0.3000	0.1881	0.912	5.16+16	9.74	39.9	0.679	4.10−4	3.42−3	1.80−3
0.3500	0.2047	0.945	4.18+16	9.20	34.0	0.689	6.20−4	3.92−3	1.21−3
0.4000	0.2212	0.966	3.36+16	8.70	28.8	0.696	9.38−3	4.08−3	1.03−4
0.4500	0.2381	0.981	2.68+16	8.22	24.2	0.700	1.44−3	4.12−3	9.80−4
0.5000	0.2555	0.990	2.10+16	7.76	20.1	0.703	2.26−3	4.13−3	9.67−4
0.5500	0.2739	0.996	1.62+16	7.32	16.5	0.705	3.22−3	4.13−3	9.64−4
0.6200	0.3020	0.999	1.09+16	6.70	12.0	0.707	2.49−3	4.14−3	9.63−4
0.6550	0.3176	1.000	8.69+15	6.39	10.1	0.708	1.74−3	4.14−3	9.63−4
0.6900	0.3344	1.000	6.82+15	6.08	8.34	0.709	1.15−3	4.14−3	9.63−4
0.7600	0.3737	1.000	3.88+15	5.44	5.32	0.709	4.67−4	4.14−3	9.63−4
0.7950	0.3975	1.001	2.77+15	5.09	4.06	0.710	2.95−4	4.14−3	9.63−4
0.8650	0.4597	1.001	1.18+15	4.33	2.03	0.710	1.38−4	4.14−3	9.63−4
0.9000	0.5038	1.001	6.59+14	3.88	1.27	0.710	1.12−4	4.14−3	9.63−4
0.9556	0.6195	1.000	1.55+14	2.91	0.396	0.710	1.01−4	4.14−3	9.63−4
0.9662	0.6559	1.000	9.94+13	2.64	0.281	0.710	1.00−4	4.14−3	9.63−4
0.9742	0.6906	1.000	6.53+13	2.38	0.205	0.710	1.00−4	4.14−3	9.63−4
0.9803	0.7230	1.000	4.37+13	2.11	0.154	0.710	1.00−4	4.14−3	9.63−4
0.9850	0.7523	1.000	2.98+13	1.82	0.122	0.710	1.00−4	4.14−3	9.63−4
0.9913	0.8015	1.000	1.44+13	1.36	0.0792	0.710	1.00−4	4.14−3	9.63−4
0.9949	0.8406	1.000	7.19+12	1.03	0.0522	0.710	1.00−4	4.14−3	9.63−4
0.9990	0.9197	1.000	8.95+11	0.454	0.0015	0.710	1.00−4	4.14−3	9.63−4

TABLE V

Interior Structure of the Cox, Guzik and Kidman (1989) Standard Model at Age 4.6 Gyr

M_r ($M_\odot$)	r($R_\odot$)	L_r ($L_\odot$)	P (dynes cm^{-2})	T_r (10^6 K)	ρ_r (g cm^{-3})	κ (cm^2 g^{-1})	X(H)	X(^{3}He)	X(^{12}C)	X(^{14}N)
0.0000	0.0000	0.0000	2.477+17	15.71	162.2	1.16	0.324	6.28−6	2.27−5	4.76−3
0.0005	0.0166	0.0047	2.428+17	15.61	162.2	1.16	0.332	6.85−6	2.21−5	4.70−3
0.0010	0.0208	0.0089	2.401+17	15.56	157.2	1.16	0.338	7.22−6	2.18−5	4.66−3
0.0020	0.0263	0.0174	2.357+17	15.47	153.9	1.17	0.347	7.83−6	2.13−5	4.61−3
0.0029	0.0298	0.0249	2.325+17	15.40	151.5	1.18	0.353	8.30−6	2.09−5	4.58−3
0.0040	0.0333	0.0338	2.291+17	15.33	149.0	1.19	0.360	8.83−6	2.05−5	4.54−3
0.0061	0.0383	0.0499	2.236+17	15.21	145.2	1.20	0.371	9.73−6	2.00−5	4.49−3
0.0080	0.0423	0.0652	2.189+17	15.11	141.6	1.21	0.380	1.06−5	1.96−5	4.45−3
0.0103	0.0462	0.0819	2.144+17	15.01	138.4	1.22	0.389	1.15−5	1.92−5	4.42−3
0.0209	0.0597	0.1568	1.959+17	14.59	126.2	1.26	0.424	1.56−5	1.77−5	4.32−3
0.0406	0.0766	0.2766	1.716+17	14.00	110.6	1.33	0.470	2.34−5	1.60−5	4.24−3
0.0599	0.0894	0.3768	1.534+17	13.52	99.09	1.40	0.504	3.19−5	1.47−5	4.22−3
0.0803	0.1007	0.4673	1.377+17	13.08	89.86	1.46	0.532	4.21−5	1.37−5	4.20−3
0.1026	0.1116	0.5514	1.234+17	12.65	81.64	1.52	0.557	5.51−5	1.27−5	4.20−3
0.1514	0.1326	0.6937	9.857+16	11.83	67.66	1.65	0.597	9.19−5	1.09−5	4.20−3
0.2023	0.1520	0.7969	7.889+16	11.10	56.62	1.79	0.626	1.48−4	1.82−4	4.00−3
0.2453	0.1673	0.8583	6.561+16	10.55	49.06	1.91	0.643	2.14−4	1.11−3	2.92−3
0.3036	0.1873	0.9150	5.108+16	9.864	40.34	2.08	0.659	3.48−4	2.26−3	1.58−3

0.3496	0.2028	0.9447	4.176 + 16	9.367	34.51	2.23	0.668	5.08 − 4	2.63 − 3	1.14 − 3
0.4051	0.2216	0.9683	3.248 + 16	8.803	28.42	2.43	0.676	8.03 − 4	2.78 − 3	9.72 − 4
0.4476	0.2365	0.9802	2.656 + 16	8.391	24.32	2.59	0.679	1.15 − 3	2.81 − 3	9.37 − 4
0.4981	0.2546	0.9896	2.066 + 16	7.920	20.01	2.81	0.683	1.81 − 3	2.82 − 3	9.24 − 4
0.5547	0.2763	0.9959	1.525 + 16	7.405	15.79	3.10	0.685	2.86 − 3	2.82 − 3	9.21 − 4
0.5985	0.2943	0.9983	1.182 + 16	7.012	12.93	3.36	0.686	2.82 − 3	2.82 − 3	9.20 − 4
0.6470	0.3161	0.9994	8.686 + 15	6.577	10.13	3.70	0.687	1.85 − 3	2.82 − 3	9.20 − 4
0.6979	0.3419	0.9999	6.048 + 15	6.114	7.584	4.16	0.688	1.01 − 3	2.82 − 3	9.20 − 4
0.7493	0.3721	1.0001	3.975 + 15	5.632	5.418	4.75	0.689	5.04 − 4	2.82 − 3	9.20 − 4
0.7991	0.4073	1.0002	2.460 + 15	5.138	3.679	5.55	0.689	2.46 − 4	2.82 − 3	9.20 − 4
0.8492	0.4524	1.0002	1.356 + 15	4.594	2.268	6.75	0.689	1.25 − 4	2.82 − 3	9.20 − 4
0.8989	0.5129	1.0002	6.198 + 14	3.965	1.231	8.80	0.689	7.88 − 5	2.82 − 3	9.20 − 4
0.9504	0.6177	1.0001	1.731 + 14	3.082	0.4386	13.6	0.689	6.63 − 5	2.82 − 3	9.20 − 4
0.9598	0.6474	1.0001	1.223 + 14	2.854	0.3339	15.6	0.689	6.58 − 5	2.82 − 3	9.20 − 4
0.9700	0.6864	1.0001	7.745 + 13	2.539	0.2370	19.4	0.689	6.55 − 5	2.82 − 3	9.20 − 4
0.9755	0.7118	1.0001	5.721 + 13	2.295	0.1899	24.5	0.689	6.53 − 5	2.82 − 3	9.20 − 4
0.9803	0.7363	1.0001	4.203 + 13	2.031	0.1604	31.5	0.689	6.53 − 5	2.82 − 3	9.20 − 4
0.9850	0.7643	1.0001	2.882 + 13	1.748	0.1278	43.1	0.689	6.53 − 5	2.82 − 3	9.20 − 4
0.9900	0.7998	1.0001	1.696 + 13	1.416	0.0930	67.6	0.689	6.53 − 5	2.82 − 3	9.20 − 4
0.9951	0.8489	1.0001	7.059 + 12	1.000	0.0549	99.6	0.689	6.53 − 5	2.82 − 3	9.20 − 4
0.9990	0.9171	1.0001	1.177 + 12	0.4927	0.0190	581.	0.689	6.53 − 5	2.82 − 3	9.20 − 4
0.9999	0.9634	1.0001	9.941 + 10	0.1896	0.0043	2000.	0.689	6.53 − 5	2.82 − 3	9.20 − 4

TABLE VI
Interior Structure of the Cox, Guzik and Raby (1990) Cosmion Model at Age 4.6 Gyr[a]

M_r ($M_\odot$)	r ($R_\odot$)	L_r ($L_\odot$)	P (dynes cm^{-2})	T_r (10^6 K)	ρ_r (g cm^{-3})	κ (cm^2 g^{-1})	X(H)	X(^{3}He)	X(^{12}C)	X(^{14}N)
0.0000	0.0000	0.0000	2.970 + 17	13.00	209.0	1.50[a]	0.431	3.35 − 5	1.36 − 5	4.22 − 3
0.0005	0.0149	0.0032	2.904 + 17	13.00	204.4	2.06 − 3	0.434	3.38 − 5	1.36 − 5	4.22 − 3
0.0010	0.0193	0.0067	2.862 + 17	13.00	197.6	1.93 − 3	0.436	3.40 − 5	1.36 − 5	4.22 − 3
0.0020	0.0242	0.0130	2.803 + 17	13.00	196.4	1.95 − 3	0.440	3.44 − 5	1.36 − 5	4.22 − 3
0.0030	0.0277	0.0191	2.755 + 17	13.00	192.6	2.05 − 3	0.442	3.47 − 5	1.36 − 5	4.22 − 3
0.0040	0.0307	0.0256	2.711 + 17	13.00	189.2	2.20 − 3	0.445	3.49 − 5	1.36 − 5	4.22 − 3
0.0060	0.0352	0.0375	2.638 + 17	12.99	183.6	2.55 − 3	0.449	3.54 − 5	1.36 − 5	4.22 − 3
0.0082	0.0394	0.0512	2.564 + 17	12.99	178.2	3.02 − 3	0.454	3.58 − 5	1.36 − 5	4.22 − 3
0.0104	0.0428	0.0642	2.502 + 17	12.99	173.5	3.56 − 3	0.458	3.62 − 5	1.36 − 5	4.22 − 3
0.0204	0.0549	0.1222	2.267 + 17	12.99	155.3	7.49 − 3	0.473	3.78 − 5	1.35 − 5	4.22 − 3
0.0405	0.0716	0.2305	1.935 + 17	12.98	130.4	2.97 − 2	0.498	4.06 − 5	1.35 − 5	4.21 − 3
0.0599	0.0841	0.3264	1.699 + 17	12.95	113.0	1.02 − 1	0.520	4.34 − 5	1.33 − 5	4.21 − 3
0.0802	0.0955	0.4191	1.502 + 17	12.87	98.89	3.24 − 1	0.543	4.75 − 5	1.31 − 5	4.20 − 3
0.1023	0.1065	0.5096	1.329 + 17	12.68	87.20	0.80 − 1	0.566	5.50 − 5	1.27 − 5	4.20 − 3
0.1509	0.1281	0.6692	1.040 + 17	11.97	69.97	1.56	0.608	8.69 − 5	1.12 − 5	4.20 − 3
0.2017	0.1482	0.7838	8.193 + 16	11.20	57.69	1.78	0.638	1.42 − 4	1.70 − 4	4.01 − 3
0.2445	0.1639	0.8508	6.747 + 16	10.62	49.56	1.90	0.656	2.08 − 4	1.11 − 3	2.91 − 3
0.3011	0.1839	0.9105	5.232 + 16	9.931	40.64	2.07	0.673	3.39 − 4	2.27 − 3	1.56 − 3
0.3473	0.2000	0.9425	4.241 + 16	9.412	34.53	2.22	0.682	5.02 − 4	2.64 − 3	1.13 − 3

0.4030	0.2194	0.9676	$3.270 + 16$	8.828	28.23	2.42	0.690	$8.06 - 4$	$2.76 - 3$	$9.70 - 4$
0.4457	0.2347	0.9801	$2.656 + 16$	8.403	24.04	2.59	0.693	$1.17 - 3$	$2.78 - 3$	$9.36 - 4$
0.4966	0.2535	0.9898	$2.049 + 16$	7.916	19.66	2.81	0.697	$1.88 - 3$	$2.81 - 3$	$9.24 - 4$
0.5535	0.2759	0.9963	$1.500 + 16$	7.388	15.41	3.11	0.699	$2.99 - 3$	$2.82 - 3$	$9.21 - 4$
0.5976	0.2946	0.9987	$1.155 + 16$	6.987	12.54	3.38	0.700	$2.80 - 3$	$2.82 - 3$	$9.20 - 4$
0.6465	0.3172	0.9997	$8.422 + 15$	6.544	9.767	3.73	0.701	$1.80 - 3$	$2.82 - 3$	$9.20 - 4$
0.6981	0.3441	1.0002	$5.802 + 15$	6.071	7.258	4.20	0.702	$9.61 - 4$	$2.82 - 3$	$9.20 - 4$
0.7500	0.3756	1.0004	$3.774 + 15$	5.140	5.037	4.83	0.703	$4.76 - 4$	$2.82 - 3$	$9.20 - 4$
0.7999	0.4122	1.0005	$2.317 + 15$	5.083	3.464	5.66	0.703	$2.34 - 4$	$2.82 - 3$	$9.20 - 4$
0.8504	0.4594	1.0005	$1.262 + 15$	4.534	2.118	6.90	0.703	$1.21 - 4$	$2.82 - 3$	$9.20 - 4$
0.9002	0.5224	1.0004	$5.697 + 14$	3.900	1.138	9.06	0.703	$7.76 - 5$	$2.82 - 3$	$9.20 - 4$
0.9497	0.6258	1.0004	$1.677 + 14$	3.044	0.4254	$1.40 + 1$	0.703	$6.65 - 5$	$2.82 - 3$	$9.20 - 4$
0.9594	0.6561	1.0004	$1.187 + 14$	2.804	0.3259	$1.63 + 1$	0.703	$6.60 - 5$	$2.82 - 3$	$9.20 - 4$
0.9697	0.6952	1.0004	$6.887 + 13$	2.454	0.2360	$2.14 + 1$	0.703	$6.57 - 5$	$2.82 - 3$	$9.20 - 4$
0.9750	0.7184	1.0004	$5.715 + 13$	2.200	0.1993	$2.74 + 1$	0.703	$6.57 - 5$	$2.82 - 3$	$9.20 - 4$
0.9801	0.7439	1.0004	$4.121 + 13$	1.932	0.1637	$3.65 + 1$	0.703	$6.57 - 5$	$2.82 - 3$	$9.20 - 4$
0.9849	0.7711	1.0004	$2.834 + 13$	1.665	0.1307	$4.93 + 1$	0.703	$6.57 - 5$	$2.82 - 3$	$9.20 - 4$
0.9899	0.8056	1.0004	$1.674 + 13$	1.351	0.0953	$7.48 + 1$	0.703	$6.57 - 5$	$2.82 - 3$	$9.20 - 4$
0.9951	0.8534	1.0004	$6.988 + 12$	0.9556	0.0564	$1.41 + 1$	0.703	$6.57 - 5$	$2.82 - 3$	$9.20 - 4$
0.9990	0.9194	1.0004	$1.176 + 12$	0.4722	0.0197	$6.99 + 2$	0.703	$6.57 - 5$	$2.82 - 3$	$9.20 - 4$
0.9999	0.9642	1.0004	$1.001 + 11$	0.1831	0.0045	$2.49 + 4$	0.703	$6.57 - 5$	$2.82 - 3$	$9.20 - 4$

[a]In the cosmion opacity expression used by Cox et al. (1990), the opacity reverts to the normal opacity as r approaches 0.

J. GUZIK AND Y. LEBRETON

TABLE VII

Predicted Neutrino Fluxes from the Lebreton and Däppen (1988)
Standard Model

	Flux $(10^{10}$ cm^{-2} s$^{-1})$	Flux (SNU[a]) ^{37}Cl	Flux (SNU[a]) ^{71}Ga
p-p	6.0	0.000	70.51
pep	0.014	0.220	2.95
^{7}Be	0.46	1.099	33.81
^{8}B	0.00057	6.026	13.82
^{13}N	0.038	0.064	2.36
^{15}O	0.029	0.193	3.39
^{17}F	0.00048	0.003	0.06
Total		7.605	126.90

[a]Neutrino absorption cross sections from Bahcall and Ulrich (1988); 1 SNU = 10^{-36} absorptions per target atom per second.

TABLE VIII

Predicted Neutrino Fluxes for the ^{37}Cl Detector

	Flux (SNU[a])			
	Lebreton and Däppen (1988)	Bahcall and Ulrich (1988)	Cox, Guzik and Kidman (1989)	Cox, Guzik and Raby (1990)
pep	0.22	0.22	0.25	0.31
^{7}Be	1.10	1.12	1.52	0.82
^{8}B	6.03	6.11	9.21	1.00
^{13}N	0.06	0.10	0.11	0.02
^{15}O	0.19	0.35	0.32	0.06
Total	7.60	7.90	11.41	2.21

[a]Neutrino absorption cross sections from Bahcall and Ulrich (1988); 1 SNU = 10^{-36} absorptions per target atom per second.

APPENDIX C:
SOLAR p-MODE FREQUENCIES

PERE L. PALLÉ

Instituto de Astrofísica de Canarias

The aim of this appendix is to present some mathematical formulae which have been derived from observed frequencies and which allow the reader, not only to reproduce the observed frequencies accurately, but also to extrapolate them to limits beyond present observations. It must be remembered that, until now, more than 1500 solar p-mode frequencies have been measured for different values of l *and* n. *The listing of all of them would take some 15 pages in this book, which is more space than could be allotted.*

Since 1979, low-degree ($l \leq 3$) p-modes have been measured continuously by the Tenerife-Birmingham Group. Sporadic measurements have also been made by others groups (Nice, SMM, etc.); the agreement among all of these measurements, within observational errors, makes the frequency values very secure and precise. On the other hand, high-degree solar p-modes ($4 \leq l \leq 100$) have been measured in the last years by different groups: NSO in Tucson, Caltech in California, HAO in Boulder and WO at Stanford. A complete list of all measured p-mode frequencies can be found in Duvall et al. (1988a) and Libbrecht et al. (1990).

To derive the relations for computing the frequencies of solar p-modes, we separate them into two groups: high l and low l. The main reason for this is the existence, for the second group of modes, of assymptotic relations first derived by Tassoul in 1980, which are only valid for modes with $n \gg l$. We present the different relations together with the achieved accuracies in each case.

I. LOW-DEGREE p-MODES ($l \leq 3$)

Three different relationships have been used, each one giving different accuracies for this set of frequencies. In all cases, the frequencies found by Jiménez et al. (1988e) were used, because they are a mean value over 8 yr of observations. The covered range is $11 \leq n \leq 33$.

Tassoul first-order assympotic relation

The frequencies are given by the relation:

$$\nu_{n,l} = \nu_0 \left[n + \frac{l}{2} + \delta\right] = \nu_0 x_{n,l} + \delta' \qquad (1)$$

with $x_{n,l} = n + l/2$. Then, fitting a straight line for different l values, we obtain the coefficients:

l	$\nu_0(\mu Hz)$	δ	$\sigma(\mu Hz)$
0	135.4 ± .1	1.43 ± .02	3.5
1	135.7 ± .1	1.36 ± .02	4.1
2	135.4 ± .1	1.36 ± .02	3.4
3	135.7 ± .1	1.24 ± .01	2.8

with σ the standard deviation of the linear fit.

The frequencies calculated from Eq. (1) using the above coefficients, show differences with the observed ones of 2 to 5 μHz at the edges of the frequency band ($n < 15$ and $n > 25$) and < 2 μHz in the central part ($15 < n < 25$). In order to reduce the number of coefficients needed, one can try using mean values of ν_0 and δ for different degrees ; these values are $\nu_0 = 135.5$ μHz and $\delta = 1.35$ for $l = 0,1,2$ and 3. In this case the differences become bigger: up to 15 μHz at the extreme regions of the spectrum and 2 to 4 μHz in the central one.

Tassoul Second-Order Asymptotic Relation

In this approximation, the frequencies are given by:

$$\nu_{n,l} = \nu_0 \left[n + \frac{l}{2} + \delta - \frac{l(l+1)\alpha - \beta}{n + \frac{l}{2} + \delta}\right]. \qquad (2)$$

To obtain the values of the four coefficients (ν_0, α, β, δ), a nonlinear fitting procedure was used in order to minimize the difference between observed and predicted frequencies. After a trial-and-error procedure, the best set of coefficients, together with the standard deviation σ of the fit, is found to be:

l	ν_0	α	β	δ	$\sigma(\mu Hz)$
0	137.0		5.6	0.90	1.8
1	137.9	0.20	7.8	0.62	2.0
2	137.4	0.15	7.8	0.70	1.6
3	137.0	0.20	7.4	0.80	1.4

with errors of the order of 10 to 15%.

Using these coefficients and comparing them with observed frequencies, the agreement is now much improved; discrepancies now being of ~ 2 μHz in the extreme regions and < 1 μHz in the center. Mean values of coefficients for $l \leq 3$ are given by $\nu_0 = 137.3$ μHz, $\alpha = 0.20$, $\beta = 7.1$ and $\delta = 0.75$. Using this set of values the discrepancies increase up to 8 μHz and ~ 2 μHz, respectively.

Polynomial Relation

The asymptotic relations for p-mode frequencies $\nu_{n,l}$ motivate a polynomial fit of the form (Scherrer et al. 1983; Christensen-Dalsgaard 1984):

$$\nu_{n,l} = \Delta\nu_l + \bar{\nu}_l x + \gamma_l x^2 + \ldots \tag{3}$$

with $x = n + l/2 - n_0$; $n_0 = 22$ is a suitable chosen reference order and $\Delta\nu_l$ and $\bar{\nu}_l$ are linear functions of $l(l + 1)$ such as:

$$\Delta\nu_l = \Delta\nu_0 - l(l + 1)D_0$$
$$\bar{\nu}_l = \nu_0 + l(l + 1)d_0. \tag{4}$$

Fitting the observed frequencies, we obtain a new set of coefficients:

l	$\Delta\nu_l$	$\bar{\nu}_l$	γ_l	$\sigma(\mu Hz)$
0	3169.4 $\pm$ .3	135.31 $\pm$ .03	.090 $\pm$ .006	1.0
1	3166.2 $\pm$ .3	135.52 $\pm$ .04	.105 $\pm$ .006	1.2
2	3160.5 $\pm$ .3	135.35 $\pm$ .03	.085 $\pm$ .006	1.1
3	3150.8 $\pm$ .3	135.52 $\pm$ .04	.070 $\pm$ .006	1.1

and by linear fit to Eq. (4):

$$\Delta \nu_0 = 3169.4 \pm .2\mu Hz \quad D_0 = 1.54 \pm .03\mu Hz \quad \sigma = 0.9\mu Hz$$
$$\nu_0 = 135.35 \pm .02\mu Hz \quad d_0 = .012 \pm .004\mu Hz \quad \sigma = 3.2\mu Hz.$$

As can be seen by simple inspection of the standard deviations (σ) of these fits, this third method, polynomial relation, is the one that gives the highest accuracy for reproducing the observed frequencies. Differences are smaller than 1.5 μHz and 0.5 μHz in the extreme and central regions, respectively. However, if values of the coefficients are averaged for different l, the discrepancies are much higher than before due to the variation of $\Delta \nu_l$ with the degree l.

II. HIGH-DEGREE p-MODES ($4 \leq l \leq 100$)

To fit the observed high-degree p-modes, we have used the frequencies measured by the NSO group at the South Pole and by the Caltech group at Big Bear Observatory. The total number of measured frequencies is almost 1200 and they are restricted to the range $3 \leq n \leq 100$. Two different methods have been used; one is based on parabolic fits and the other is a relationship taken from Hathaway (1989). Also, the Duvall dispersion law has been used in order to show only its application to infer the properties of the interior of the Sun and not to reproduce accurate values of the frequencies.

Parabolic Fits

When observed high l frequencies are plotted as a function of degree (l) for a given n value (Fig. 1), the well-known $k - \omega$ diagram is obtained. The figure shows a parabolic shape of the constant n ridges. The extreme values of n, ($n < 6$ and $n > 20$), correspond to very high (~ 4.1 mHz) and very low (~ 1.5 mHz) frequencies. It is in these regions where peak-frequency determinations are less confident because of the high noise level (worse S/N ratio).

Therefore, a second-order polynomial equation is fitted to each ridge of $n =$ const. as a function of l, and from $n = 3$ to $n = 24$:

$$\nu_{n,l} = a_0(n) + a_1(n)l + a_2(n)l^2 \tag{5}$$

with $4 \leq l \leq 100$.

The coefficients a_i have an n dependence and therefore they are fitted to second-order polynomials in n. The expression in matrix form is:

$$\begin{pmatrix} a_0 \\ a_1 \\ a_2 \end{pmatrix} = \begin{pmatrix} b_{00} & b_{01} & b_{02} \\ b_{10} & b_{11} & b_{12} \\ b_{20} & b_{21} & b_{22} \end{pmatrix} \cdot \begin{pmatrix} 1 \\ n \\ n^2 \end{pmatrix} \tag{6}$$

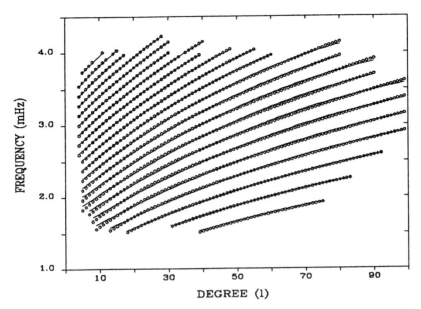

FREQUENCY (mHz)

DEGREE (1)

Fig. 1. Diagram of the observed high *l* solar *p*-modes, ($4 \leq l \leq 100$). The circles represent the observed values and the continuous line the parabolic fits, Eq. (5) in the text. The lower ridge corresponds to $n = 3$ and the one on top to $n = 24$.

and the best set of coefficients for the matrix b_{ij} is found to be:

$$B = \begin{pmatrix} 643 & 101.3 & 0.71 \\ 8.6 & 2.9 & -0.047 \\ -0.025 & -0.008 & -0.0002 \end{pmatrix}.$$

Then, for a given n value, parameters a_0, a_1 and a_2 can be obtained from Eq. (6) and substituted in (5) to get all frequencies from $l = 4$ to $l = 100$.

The general relations (Eqs. 5 and 6) will produce discrepancies with respect to the observed frequencies, but they are a close approximation for general purposes. Duvall et al. (1988*a*) and Libbrecht et al. (1990) give a complete list of all frequencies with their errors, if required.

The accuracy achieved with the above relations depends both on n and *l*:

a. High *n*: ($n > 18$). At the extreme values of *l*, differences are < 6 μHz and < 1 in the intervening part.

b. Intermediate *n*: ($8 < n < 18$). These intervals correspond to frequencies that are measured from $l \sim 4$ to $l \sim 100$. A parabolic fit is then quite complicated due to the poor knowledge of the observed frequencies at the extremes of the acoustical power spectrum. As shown in the graph, the fit is very unsatisfactory there but rather good in the intervening region, for which the discrepancies are < 4 μHz.

c. Low n: ($n < 8$). All these lines lie at the low end of the frequency interval (< 2 mHz), where S/N ratio is very poor. Discrepancies as large as 10 μHz are found at the edges of the spectrum and ~ 2 μHz in the intervening regions.

Hathaway's Relation

Alternatively, Hathaway (1989) derived a relation to calculate high l-mode frequencies for desired n values. This relation is given by:

$$\nu(n,l) = 2354.2(n + 1.57) \exp\left[0.2053\{[(\ln x - 14.523)^2 \right. \quad (7)$$
$$\left. + 4.1175]^{1/2} - \ln x\}\right]$$

in units of μHz with $x = (n + 1.57)\pi R_\odot [l(l + 1)]^{-1/2}$ being $R_\odot$ the solar radius in km. For $3 \leq n \leq 13$, this relationship gives discrepancies between 2 and 2.5 times lower than parabolic fits, while for $14 \leq n \leq 24$, parabolic fits gave discrepancies 1.5 to 2.8 times lower than by using Eq. (7).

Duvall Dispersion Law

A second approach to fit the high-degree solar p-mode frequencies is based on an idea of Duvall (1982) for reducing the p-mode frequency data to a more manageable form. As explained in a previous chapter (see the chapter by Christensen-Dalsgaard and Berthomieu) the idea is to collapse the $\kappa - \omega$ diagram to a single ridge, which means that a single function essentially determines the frequencies of all p-modes. Therefore, in a plot of the quantity ω/κ_h, [κ_h being the horizontal wavenumber: $\kappa_h = \sqrt{l(l + 1)}/R_\odot$ and ω the angular frequency: $\omega = 2\pi\nu$] vs $(n + \alpha)\pi/\omega$, the ridge structure should collapse onto a single curve for a given value of α. To determine the best value of α, the standard deviation of fifth-order polynomial fits to the resulting points was computed as a function of α. The minimum value of the standard deviation, 0.5% of the mean value of $(n + \alpha)\pi/\omega$, was found for $\alpha = 1.67 \pm 0.01$. In Fig. 2 are shown the same p-mode frequencies as in Fig. 1 but following Duvall's dispersion law, for the value of α that produces the tightest curve. Although this method is not very convenient for reproducing high l solar p-mode frequencies (errors > 30 μHz), the dispersion law is of great interest to infer the interior properties of the Sun.

Finally, we point out that modes with $n < 3$ have not been included in

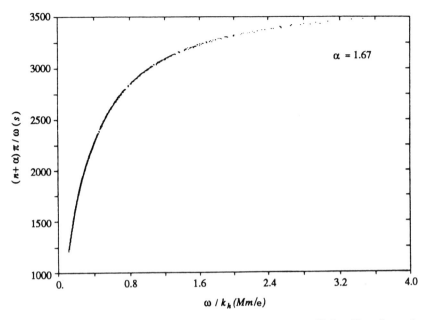

Fig. 2. Diagram for the observed high *l* solar *p*-modes according to Duvall's law. The collapse of the ridges shown in Fig. 1 onto the single curve shown here, reflects the fact that a single function, the sound speed as a function of depth inside the Sun, essentially determines the frequencies of all high *l* *p*-modes. The value of $\alpha = 1.67$ is the one that produces the tightest curve.

this section. The main reason for this is that, unlike the observed frequencies presented up to now that have been measured by different authors using different techniques, the modes with $n < 3$, (which have periods up to 60 min) have frequencies measured by only one group (University of Arizona) and their values are not yet well established.

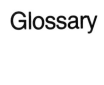

Glossary

GLOSSARY*

Compiled by Mary Guerrieri

acoustic cutoff period
: the period that separates the regimes of propagating and evanescent (nonpropagating) acoustic waves.

active mirror
: a concave mirror whose focal length and figure is continually adjusted to achieve optimum performance.

active region
: an area on the solar surface characterized by the presence of sunspots and plages.

adaptive optics
: a system which attempts to rapidly correct for wavefront distortion introduced by the Earth's atmosphere.

adiabatic
: a thermodynamic process that occurs without any gain or loss of energy for the parcel of matter being considered.

α-ω dynamo mechanism
: a way of generating the solar magnetic field by the interplay of cyclonic motions (α) and rotation (ω) shearing.

Alfvén waves
: a type of magnetohydrodynamic wave in which the wave amplitude is transverse to the direction of the magnetic field lines.

AU (astronomical unit)
: the mean distance of the Earth from the Sun, equal to 1.496×10^{13} cm.

asymptotic theory
: the mathematical approximation to global solar oscillation modes for those which have a high radial order compared to the number of node lines on the Sun.

---

*We have used some definitions from *Glossary of Astronomy and Astrophysics* by J. Hopkins (by permission of the University of Chicago Press, copyright 1980 by the University of Chicago), from *Astrophysical Quantities* by C. W. Allen (London: Athlone Press, 1973), from *Glossary of Geology,* edited by M. Gary, R. McAfee and C. L. Wolff (Washington, D.C.: American Geological Institute, 1972), and from *The Planetary System* by David Morrison and Tobias Owen (Reading, Mass.: Addison-Wesley Publishing Co., 1988). We also acknowledge definitions and helpful comments from various chapter authors.

azmuthal order   a description of an oscillation mode that gives the number of surface node lines that are perpendicular to the equator.

azimuthal rolls   an observed velocity pattern on the solar surface characterized by meridional motions, constant in longitude, suggesting a large-scale circulatory pattern.

barn   a unit of area equal to 10^{-24} cm^2 used in measuring cross sections.

basal chromosphere   the low-level chromospheric emission from regions free of magnetic field.

Bohr magneton   (μ_o or μ_B) magnetic moment of an electron in the first Bohr orbit ($\mu_o = eh/(4\pi mc)$).

Boltzmann-Saha statistics   a theory describing the distribution of atoms of partially ionized matter over possible excitation and ionization states, in the limit of low density.

Boussinesq approximation   convection in a system where buoyancy depends only on temperature with the density structure unchanging.

bright points   emitting regions prominent in Hα and the H and K lines of ionized calcium, located on the face of supergranulation cells, and having typical sizes of the order of 2 arcsec.

Brunt-Väisälä frequency   the oscillation frequency of a small matter parcel undergoing adiabatic motions. The maximum in the Sun or a model is the high-frequency limit for gravity modes.

butterfly diagram   a plot of the heliographic latitude of individual sunspots vs time in years or solar cycles.

calcium network   a pattern of emission features seen in the ionized Ca line.

canopy (magnetic)   the observed spreading out of magnetic fields as they emerge from the photosphere.

CCD   charge-coupled device. A solid state detector used for low light level imaging.

central meridian
: the meridian on the solar surface which at any one time passes through disk center.

Cerenkov light
: the light emitted when a particle has a velocity that exceeds the velocity of light in a medium. This light is used to detect neutrinos by the light from a very high-velocity scattered electron in water detectors.

chromosphere
: the part of the solar atmosphere between the photosphere and the corona in which there occurs an outward rise in temperature (which indicates nonradiative energy input, e.g., by mechanical dissipation, into the atmosphere).

chromospheric mode
: a mode of oscillation trapped in the chromosphere.

CNO cycle
: the series of nuclear reactions that produce a small part of the solar luminosity using carbon, nitrogen, and oxygen a catalysts for converting hydrogen to helium.

convection zone
: a surface layer with a thickness of about 30% of the solar radius where the opacity is so large that energy is carried by turbulent motions instead of photon diffusion.

contribution function
: an estimation of the effective height of origin of solar radiation in a wavelength interval.

core
: the deepest layers of the Sun or of solar models where the nuclear energy is generated and where hydrogen is depleted by its conversion to helium.

Coriolis force
: a pseudo force that appears in the dynamical equations when set up for a rotating coordinate system. In the Earth's northern hemisphere, it tends to turn all velocity vectors to the right.

Coriolis number
: twice the angular rotation velocity times the lifetime of a convective cell as estimated by the ratio of the mixing length divided by a typical eddy velocity.

chlorine experiment
: the experiment located in the Homestake mine when ^{37}Cl absorbs solar neutrinos to produce radioactive ^{37}Ar.

convective   a description of solar layers that have energy transport
                       mostly by mass motions instead of by photon diffusion.

cork movie   computer simulation which predicts horizontal surface
                       flow patterns starting from an observed initial condition.

corona   the outermost atmosphere of the Sun immediately above
                       the chromosphere, consisting of hot, low-density gas
                       ($T = 2 \times 10^6$ K, $n_e = 10^9$ cm^{-3}, $B = 10^5$ gauss) that
                       extends for millions of kilometers from the Sun's sur-
                       face. The E corona is the emission-line corona, as op-
                       posed to the dust scattered light of the F corona, display-
                       ing Fraunhofer lines. The K corona or electron corona
                       or inner corona is produced by electron scattering with
                       no spectrum lines seen because of Doppler smearing.

coronagraph   a low scattered light telescope for coronal observations
                       featuring an occulting disk to cover the solar image.

coronal bright   small (typically about 10,000 km across) region with
    point   strong magnetic fields, usually bipolar, which is charac-
                       terized by bright soft X-ray emission.

coronal bullets   another name for high-energy jets in the corona.

coronal hole   low-density region of the corona which is much less
                       bright in coronal emission lines and soft X-rays than the
                       surrounding quiet and active regions. This is a source
                       region for some of the solar wind.

coronal mass   a coronal transient phenomena where large amounts of
    ejection   coronal material (typically a few times 10^{15} g) are ejected
                       from the Sun, most likely as a result of instabilities in
                       the coronal magnetic field.

coronal streamers   high-density, bright, coronal structures which have ray-
                       like forms extending approximately radially outward to
                       large distances from the Sun.

cosmic rays   particles that are accelerated to relativistic velocities on
                       galactic space scales, but are influenced by the time-
                       changing magnetic fields of the Sun.

Coulomb barriers	the electric field repulsion experienced when two charged nuclear particles approach. When the barrier is overcome, nuclear reactions that produce the solar luminosity can occur.
Cowling approximation	the mathematical assumption that the gravitation force is always pointed directly to the solar center even for complicated nonradial oscillation motions.
cross section	the effective area of interaction (measured usually in barns) for nuclear reactions.
cryogenic spectrometer	a spectrometer for the infrared in which the grating and other optics are cooled to near-liquid nitrogen temperature.
degree	generally used to represent the angular order of the oscillation mode, or the number of node lines on the Sun.
diagnostic diagram	a two-dimensional power spectrum of solar oscillations; also known as the k, ω or l, ν diagram.
differential rotation	the variation of angular rotation rate with solar latitude. The equatorial regions rotate faster than higher latitude ones.
diffusion	the process where photons or nuclear particles slowly move through the solar material by a very large number of scatterings in random directions.
Doppler shift	the red or blue shift of photons caused by either the recession or approach of a source relative to the observer.
double-pass spectrometer	the effective coupling of two spectrometers in series to reduce instrumental scattered light.
Duvall's law	an approximate formula for solar oscillation frequencies that has a theoretical basis in asymptotic theory.
dynamo	the generation process of magnetic fields from convective and global rotational motions.

echelle diagram   a rearrangement of the spectrum of low-degree global solar oscillations in which the spectrum is cut into strips and stacked.

eigenfrequency   the theoretical frequency of a material body as derived from general properties of mass, momentum and energy conservation and its detailed structure.

entropy   a thermodynamic property of matter. The entropy per gram is almost constant throughout the convection zone, but it decreases toward the solar center in the subadiabatic radiative layers.

ephemeral region   a short-lived tiny active region.

equation of state   the theoretical relation between pressure and energy as dependent variables, as a function of the independent variables of temperature and density as used in solar structure modeling.

evershed flow   an outward, mainly horizontal flow of matter in sunspot penumbrae. The amplitude of the effect is a few km s^{-1}.

faculae   bright regions of the photosphere seen in white light and visible only near the limb.

fermion   a matter particle which obeys the Fermi statistics.

filament   a dark, thread-like structure seen in a chromospheric line (such as Hα) which represents a cloud of relatively cool material at the level of the low corona. Beyond the limb of the Sun, a filament is seen in emission and is called a prominence.

filigree   tiny bright points of sub-arcsec size seen in off-band Hα, typically in the vicinity of active regions.

filtergram   a picture taken in monochromatic light, often set at the wavelength of a chromospheric line such as Hα.

five-minute oscillations   a ubiquitous oscillation pattern seen in the Doppler effect in solar absorption lines or in light variations at nearly all locations on the solar disk. These represent global oscillation modes, some of which extend to the solar center.

f-mode   the fundamental nonradial oscillation mode of the Sun.

Fourier transform   a spectrometer which operates by obtaining an optical
  spectrometer   autocorrelation of the spectrum, using beam-splitting
  (FTS)   and recombining optics. The original spectrum is then
                             obtained by performing a mathematical Fourier trans-
                             form of this so-called interferogram.

Fraunhofer line   an absorption feature in the solar spectrum caused by
                             photon absorption by atoms cooler than the background.

GALLEX   the gallium experiment in Italy that is measuring solar
                             neutrinos of low energy from the proton-proton reac-
                             tions.

gauss   unit of magnetic field strength. 10^4 gauss = 1 tesla.

giant-scale   the largest scale of solar convection cells, predicted by
  convection   theory, but not yet confirmed by observation.

GONG   the global oscillation network group effort to use 6 sta-
                             tions on Earth to continuously monitor the solar oscilla-
                             tions.

g-mode   global gravity oscillation mode with a specific spatial
                             pattern and temporal frequency.

granulation   fine white-light cellular pattern seen on the solar surface.
                             Convective in origin, granules are the order of 1000 km
                             in size and have life times of a few minutes.

gravitational   energy released as the stellar mass shrinks into its poten-
  energy   tial well. This energy contributes very little to the solar
                             luminosity.

gravitational   the gravitational waves that are emitted into space at very
  radiation   low amplitude from *g*-mode solar oscillations.

gravity waves   long-period oscillatory mass motions in the solar interior
                             that may play a role in mixing layers of different com-
                             position.

Hayashi track — a nearly vertical track, in the Hertzsprung-Russell diagram, of stellar evolution toward the main sequence during phases when the star is largely or completely in convective equilibrium.

Hale's law — the polarity of the proceeding spot of a bi-polar group is reversed in sense to the same spot configuration in the opposite hemisphere. Each cycle the spots' polarities reverse.

HE (hydrostatic equilibrium) — layers in the Sun in or a model where the pressure gradient exactly equals the opposite force of gravity.

helioseismology — the probing of the interior of the Sun through measurements of its oscillations.

heliosphere — the cavity carved out in the interstellar medium by the solar wind. The heliosphere, where the Sun controls the local environment, is believed to extend on the order of 100 AU from the Sun.

heterodyne spectrometer — a spectrometer which operates by mixing source radiation with the output of a coherent local oscillator, usually a laser. The result of this mixing is a difference frequency spectrum, which represents a small portion of the source spectrum shifted to lower frequencies and analyzed electronically.

high-energy jets — small-scale explosive features observed in ultraviolet spectral lines formed at temperatures of $\sim 10^5$ K which exhibit nonthermal velocities of $100-400$ km s^{-1}.

high-speed solar wind — solar wind with flow speeds greater than about 600 km s^{-1}.

Homestake — the South Dakota gold mine where the ^{37}Cl neutrino detector is located.

index of solar activity — observable parameter, such as sunspot number, which is an indicator of global activity.

interplanetary magnetic field — extension of the Sun's surface fields as drawn into space by the solar wind.

inversion   the mathematical process that takes global solar oscillation frequencies and derives solar data such as the internal rotation structure, internal magnetic field, or the convection zone helium content.

irradiance   amount of power per square meter from the Sun at the Earth. Total irradiance is the solar constant.

Kamiokande II   the electronic detector for neutrinos located in Japan.

Kelvin-Helmoltz   the contraction of a star contemplated by Kelvin and Helmholtz as a consequence of a star's radiating its thermal energy in the absence of thermonuclear energy sources.

K line features   the absorption feature K1 in the line wing is formed in the upper photosphere and at the temperature minimum; the feature K2 in the emission core is formed in the low to middle chromosphere; the central absorption feature K3 is formed in the middle chromosphere.

Lamb frequency   the frequency obtained by dividing the horizontal wavelength of a nonradial solar oscillation into the local sound velocity. For pressure to act as a restoring force, a mode should have a frequency larger than the Lamb frequency.

Laplace-Lorentz force   the force on a charged particle moving in electric and magnetic fields, equal to the particle's charge times the sum of the electric field and the cross product of the particle's velocity with the magnetic flux density.

lepton   any fermion that does not participate in the strong interactions. Leptons include the electron family and the muon family.

limb shift   any systematic wavelength shift as a function of distance from disk center. Most lines show strong red shifts near the solar limb.

linear perturbation theory   the theory that perturbations are so small that they can be approximated by only the linear term in a power expansion. Such a theory is used in many applications such as for solar oscillation solutions.

low and middle chromosphere	the chromosphere extending from the base near the temperature minimum at about 4000 K to about 8000 K. In these layers, which have a thickness of approximately 1500 km, the gas is mainly neutral. Beyond the 8000 K level, the temperature rises steeply to the high chromosphere, the transition region, and the corona.
low-speed solar wind	solar wind with flow speeds of several hundred km s^{-1}, typically 300 to 400 km s^{-1}.
LTE (local thermodynamic equilibrium)	the assumption that all distribution functions characterizing the material and its interaction with the radiation field (but not that of the radiation itself) at a point in the star are given by thermodynamic equilibrium relations at local values of the temperature and density.
macrospicules	jet-like, cool features best observed in ultraviolet lines from at a few times 10^4 to a few times 10^5 K in polar coronal holes. Lifetimes are of the order of 10 minutes, velocities up to about 150 km s^{-1}, with heights up to about 25,000 km above the solar surface.
magnetic flux tubes	magnetic elements with a typical size of at most a few hundred kilometers in the photosphere. Flux tubes spread out with height to merge with the magnetic canopy in the middle or upper chromosphere.
magnetic braking	the interactions of the solar rotation with open magnetic field lines that transfer angular momentum from the star to matter out in space.
magnetic moment	the magnetism that a fundamental particle has that can interact with external magnetic fields.
magnetic reconnection	changes in the topology of the magnetic field where the field lines reorient themselves via new connections. This may be an important heating mechanism of the solar corona.
magnetogram	a computer drawn picture or map of solar magnetic flux based on Zeeman-effect measurements.
maxwell	unit of magnetic flux equal to one gauss cm^2.

Maxwell-Boltzmann distribution	the distribution function that any species of particle will have if it is in thermodynamic equilibrium.
mean magnetic field	magnetic field of the Sun as it would be seen from a distance like the other stars.
meridional motion	motion in the north-south direction on the solar surface.
mesogranulation	a convection scale with dimensions of a few thousand kilometers, roughly midway between granulation and supergranulation.
meson	a nuclear particle with a mass intermediate between that of a proton and that of an electron, which is believed to be responsible for the strong nuclear force. Their charge can be positive, negative, or zero.
MHD (magneto-hydrodynamics)	sometimes called hydromagnetics, the study of the collective motions of charged particles in a magnetic field. This name has also been given to the equation of state procedures of Mihalas, Hummer, and Däppen.
microflares	ubiquitous small brightenings lasting a few minutes which can be observed in ultraviolet, extreme ultraviolet and X-ray emissions from chromospheric-coronal spectral lines from quiet regions, coronal holes, and active regions of the Sun.
mixing angle	the angle which when used as the cosine of that angle gives the relative abundance of two fundamental particle or neutrino species.
mixing-length theory	a phenomenological theory of the luminosity due to convection that is widely used in constructing solar models.
moat (of sunspot)	area immediately around a sunspot that appears free of plage.
Morton wave	a circular Hα brightening centered on a flare that expands radially as a disturbance across the solar disk; probably a shock wave.

M region	a magnetized region on the solar surface.
MSW effect	the interaction of electron neutrinos with electrons in the solar core that converts them to invisible neutrinos for electron neutrino detectors such as at Homestake, SAGE, or GALLEX.
muon (μ)	an elementary particle, formerly called a mu-meson but now classified with the leptons because it seems to be identical with the electron except for its much greater mass (207 times that of an electron). The muon family includes the muons and their neutrinos (and their antiparticles). Muons may have a positive or a negative charge.
network	a set of observing instruments distributed on the Earth to obtain continuous solar observations (see also calcium network).
neutrino	a massless or almost massless fundamental particle.
neutrino oscillations	the changing of one flavor of neutrino to another in matter or in vacuum.
neutron	a nuclear particle with zero charge and with a mass slightly greater than that of a proton (mass of neutron 1.008665 amu $= 1.6748 \times 10^{-24}$ g).
NLTE	nonlocal thermodynamic equilibrium where the usual equilibrium equations for the abundances of atoms in ionization or excitation stages cannot be used.
nonradial modes	oscillation modes for the Sun that are not spherical at all times during their cycle.
nuclear energy	the energy obtained by nuclear reactions at the solar center that is ultimately converted to the solar luminosity.
opacity	the mean absorption of photons that is used with a diffusion equation to calculate the photon luminosity flow in the Sun.
order	used to indicate the number of radial nodes in the global solar oscillation mode.

overshooting | the phenomenon that allows convective eddies to move beyond the formal boundary of the convection zone.

partition function | the total number of separate ways an ionization stage of an atom can exist. This internal partition function is used to obtain the relative number of atoms in each ionization stage.

phase diagram | a representation of the relative phases as a function of frequency and/or wavelength of the oscillations in the Sun's atmosphere.

photosphere | the region of a star which gives rise to its continuum radiation. The visible surface of the Sun (temperature about 6000 K), just below the chromosphere and just above the convective zone.

pion (also called π-meson) | an unstable nuclear particle of mass intermediate between that of a proton and an electron (π^+ and π^-: 273 m_e; π°: 264 m_e). The pions are believed to be the particles exchanged by nucleons, resulting in the strong nuclear force; they play a role in the strong interactions analogous to that of the photons in electromagnetic interactions. A charged pion usually decays into a muon and a neutrino; a neutral pion, into two γ-rays.

plage | (or chromospheric faculae). This is a region of emission in a strong chromospheric line. Plages always accompany and outlive sunspot groups. Plages represent the presence of magnetic fields in the photosphere below.

Planck function | the energy distribution of blackbody radiation under conditions of thermal equilibrium at a temperature T: $B_\nu = (2h\nu^3/c^2) [\exp(h\nu/kT) - 1]^{-1}$, where h is Planck's constant and ν is the frequency.

p-mode | an acoustic mode of oscillation of the Sun.

polar plume | bright, ray-like structures often observed extending from the polar regions of the Sun to heights of several hundred thousand km above the surface.

poloidal | pertaining to structures that are aligned with the poles of the Sun.

power spectrum	a representation of the power as a function of frequency, and/or the three components of wavelength of solar oscillations.
p-p chain	three reaction networks that produce nuclear energy by the fusion of hydrogen to helium in the solar core.
positron (also called antielectron)	a particle with the mass of an electron but with an equal positive charge. It is the antiparticle with respect to the electron.
pre-main sequence	the stage of solar evolution before nuclear reactions began. In this early stage the Sun was completely convective.
radiative	pertaining to those regions of the Sun where all the luminosity is carried by photons.
radiative transfer	the detailed theory where the angle and energy dependence of photons needs to be considered. Regions at and above the photosphere need this more complicated theory.
Raman scattering	scattering of radiation by an atom or molecule accompanied by excitation of internal degrees of freedom.
RE (radiative equilibrium)	equal energy flow into and out of a layer in the Sun or a model.
Reynolds number (*Re*)	a dimensionless number which is proportional to the ratio of inertial force to viscous force in a fluid flow and which is, therefore, a measure of the relative importance of viscosity.
Reynolds stress	the correlated eddy momentum components (e.g. [u' v'], where u' and v' denote the eastward and northward eddy velocities) associated with the small-scale motions which act as a stress field on the large-scale flow.
ring diagram	a slice of a three-dimensional power spectrum of solar oscillations.
Rossby number (*Ro*)	the nondimensional ratio of inertial to Coriolis acceleration, given by $Ro = U/fL$, where U is the characteristic

horizontal velocity, $f = 2\Omega\sin\lambda$ is the vorticity and L is the characteristic horizontal distance scale.

rotational modulation

the fluctuation with time of solar indices arising from their disk passage from solar rotation.

rotational splitting

the apparent change in solar oscillation frequencies caused by the rotation of the Sun.

SAGE

the Soviet-American gallium experiment at Baksan that can measure the low-energy neutrinos from the proton-proton reactions in the Sun.

Saha equation

the ionization equilibrium equation that relates the number of atoms in adjacent ionization stages.

screening

the shielding of electric charge by intervening particles to reduce the electrostatic effect.

seeing

fluctuation in an image due to refractive inhomogenieties in the Earth's atmosphere.

semi-empirical solar models

models that are constructed by adjustment of the temperature so as to match observed line profiles and continua.

slow-speed solar wind

another name for low-speed wind.

SNU (solar neutrino unit)

$1\ SNU = 10^{-36}$ solar-neutrino captures per second per target atom.

SOHO

the joint NASA and ESA solar and heliospheric observatory.

solar evolution

the calculation of a large series of models that follow the depletion of hydrogen in the central regions over time and the changes of the solar radius and luminosity.

solar cycle

the approximately 11-year cycle representing the growth and decay of solar activity. Taking Hale's law of magnetic polarity into account leads to a 22-year cycle.

solar energetic particles	nuclei of elements ejected from the Sun by flares and detected by spacecraft and on the Earth.
solar flare	a sudden brightening in an active region observed in chromospheric and coronal emissions which typically lasts tens of minutes. The increase in X-ray emissions is particularly strong, because the temperature of the hot flare plasma typically reaches several times 10^7 K.
solar flux spectrum	irradiance or the spectrum of the Sun as seen from a distance as the other stars.
solar wind	plasma which flows outward from the Sun with speeds of several hundred to over 1000 km s^{-1}.
spatial filter	a mathematical pattern used in helioseismology to isolate a single spherical harmonic.
specific heat	the derivative of the internal forms of energy with respect to temperature. The specific heats with volume or with pressure held constant are frequently used for solar modeling.
speckle	pattern of bright points seen in a short exposure stellar image. Speckle is a consequence of atmospheric seeing.
speckle imaging	technique of reconstructing an image based on a series of speckle patterns.
spherical harmonic	a mathematical function pattern of vibration, or mode of oscillation, on a sphere.
spicules	small, jet-like features extending a few thousand kilometers above the solar surface; typical lifetimes are 5 to 10 minutes with velocities about 25 km s^{-1}.
stability	the property of theoretical solutions which indicates whether a perturbation will grow or decay with time. Solar oscillations are mostly stable, but they are frequently initiated by interaction with the solar convection.
standard solar model	a recent term to signify a solar model with the most simple approximations. Nonstandard models invoke un-

usual and perhaps unlikely features that are frequently assumed to reduce the solar neutrino output to closer to that observed.

Stokes profiles spectral profiles of atomic and molecular lines measured as Stokes parameters, i.e., as differences in orthogonal polarizations. There are four Stokes parameters, one (Stokes V) measures the difference in right and left circular polarization, two (Stokes Q and U) measure differences in linear polarization, and one (Stokes I) is measured without polarization preference.

sunspot belts the heliographic latitude zones where sunspots are found.

sunspot number a measure of the total number of spots plus groups on the disk weighted by observational quality.

sunspots dark regions on the solar disk associated with large concentrations of magnetic fields. A typical spot has a central umbra surrounded by a penumbra, although either feature can exist without the other.

supergranulation a convection pattern seen at the solar surface with dimensions of about 10,000 to 20,000 km and circulatory velocity amplitudes of about 1 km s^{-1}. This pattern covers the entire solar surface except in plages and sunspots.

thermal bifurcation the term used to denote the presence of material in the solar atmosphere at two widely differing temperatures over small spatial scales, and also to denote the creation of this condition by radiative instabilities.

toroidal rolls an observed velocity pattern on the solar surface characterized by latitudinal motions suggesting a large-scale circulatory pattern.

torsional oscillation a low-amplitude, large-scale modulation of the solar rotation which varies with time.

T Tauri stars eruptive variable subgiant stars associated with interstellar matter and believed to be in the process of gravitational contraction.

turbulence | the chaotic mass motions associated with convection and usually observed by broadened spectral lines with a range of Doppler shifts.

vector magnetic field | magnetic field defined by the magnitude and the direction of the field, e.g., by giving its components in heliographic coordinates.

viscosity | the property of matter that converts bulk motions into molecular motions or heat.

white-light flare | a flare seen in white light. For this to happen it must be an extraordinarily strong event.

Wilson effect | an apparent variation in the shape of sunspot penumbrae as a function of distance from the disk center, most clearly seen near the limb. This effect results from the depressed, bowl-like shape of a sunspot.

WIMPs | weakly interacting massive particles that have been assumed to orbit the solar core with just enough interaction with matter to efficiently conduct heat rapidly away from the center to cool it.

X | the mass fraction of hydrogen in a parcel of solar matter.

Y | the mass fraction of helium in a parcel of solar matter.

Z | the mass fraction of elements heavier than hydrogen and helium in a parcel of solar matter.

Zeeman effect | the splitting of spectral lines in a magnetic field. In the simplest case, denoted as normal Zeeman splitting, a line splits into three components. One component (the π component) is not displaced, and is linearly polarized. Two components (the σ components) are shifted by equal amounts to lower and higher frequencies, the magnitude of the shift being proportional to magnetic field strength. In the general case, the σ components are both circularly and linearly polarized.

Bibliography

BIBLIOGRAPHY

Compiled by Mary Guerrieri

Aardsma, G., Allen, R. C., Anglin, J. D., Bercovitch, M., Carter, A. L., Chen, H. H., Davidson, W. F., Doe, P. J., Earle, E. D., Evans, H. C., Ewan, G. T., Hallman, E. D., Hargrove, C. K., Jagam, P., Kessler, D., Lee, H. W., Leslie, J. R., MacArthur, J. D., Mak, H.-B., McDonald, A. B., McLatchie, W., Robertson, B. C., Simpson, J. J., Sinclair, D., Skensved, P., and Storey, R. S. 1987. A heavy water detector to resolve the solar neutrino problem. *Phys. Lett.* B194:321–325.

Abarbanell, C., and Wöhl, H. 1981. Solar rotation velocity as determined from sunspot drawings of J. Hevelius in the 17th century. *Solar Phys.* 70:197–203.

Abdallah, J., Clark, R. E. H., and Cowan, R. D. 1988. Los Alamos Natl. Lab Publ. LA-11436-M.

Abdelatif, T. E., Lites, B. W., and Thomas, J. H. 1986. The interaction of solar *p* modes with a sunspot. I. Observations. *Astrophys. J.* 311:1015–1024.

Abdelatif, T. E., and Thomas, J. H. 1987. The interaction of solar *p*-modes with a sunspot. II. Simple theoretical models. *Astrophys. J.* 320:884–892.

Abetti, G. 1929. Solar physics. In *Hanbuch der Astrophysik,* vol. 4 (Berlin: Springer-Verlag), pp. 57–230

Aboudarham, J., and Henoux, J.-C. 1987. Non-thermal excitation and ionization of hydrogen in solar flares. II. Effects on the temperature minimum region: Energy balance and white light flares. *Astron. Astrophys.* 174:270–274.

Achterberg, A. 1981. On the nature of small amplitude Fermi acceleration. *Astron. Astrophys.* 97:259–264.

Acker, A., Pakvasa, S., and Panteleone, J. 1990. The solar neutrino problem: Some old solutions revisited. Univ. of Hawaii Preprint UH-511–720–90 (to be published).

Acton, D. S. 1989. Results from the Lockheed solar adaptive optics system. In *High Spatial Resolution Solar Observations,* ed. O. von der Lühe (Sunspot, N.M.: Sacramento Peak Obs.), pp. 71–89.

Acton, L. W., and Pehanich, S. J. 1987. Solar science from on high—the flight of Spacelab 2. Lockheed Horizons, Lockheed Missiles & Space Company, Inc., PO Box 3504, Sunnyvale, CA 94088–3504.

Acton, L. W., Canfield, R. C., Gunkler, T. A., Hudson, H. S., Kiplinger, A. L., and Leibacher, J. W. 1982. Chromospheric evaporation in a well-observed compact flare. *Astrophys. J.* 263:409–422.

Adam, M. G. 1979. A determination of solar rotation using sunspot spectra. *Mon. Not. Roy. Astron. Soc.* 188:819–827.

Adams, W. M. 1977. *The Solar Cycle Field Reversal,* Big Bear Solar Obs. Publ. No. 163.

Adelberger, E. G., and Haxton, W. C. 1990. The ^{37}Cl solar neutrino capture cross section. *Phys. Rev.* C36:879–882.

Adjabshirzadeh, A., and Koutchmy, S. 1983. Photometric analysis of sunspot umbral dots. *Astron. Astrophys.* 122:1–8.

Aglietta, M., and Members of the Mt. Blanc Collaboration. 1989. An Analysis of the Data Recorded by the Mt. Blanc Neutrino Detector and by the Maryland and Rome Gravitational Wave Detectors during SN 1987A. *Nuovo Cim.* 12C:75–103.

Ahluwalia, H. S. 1980. Solar polar field reversals and secular variation of cosmic ray intensity. In *Solar and Interplanetary Dynamics,* eds. M. Dryer and E. Tandberg-Hanssen (Dordrecht: D. Reidel), pp. 79–86.

Ahluwalia, H. S. 1988. The regimes of the east-west and the radial anisotropies of cosmic rays in the heliosphere. *Planet. Space Sci.* 36:1451–1459.

Ahluwalia, H. S., and Dessler, A. J. 1962. Diurnal variation of cosmic radiation intensity produced by a solar wind. *Planet. Space Sci.* 9:195–210.

Ahmad, I. A., and Kundu, M. R. 1981. Microwave solar limb brightening. *Solar Phys.* 69:273–287.

Ahmad, I. A., and Webb, D. 1978. X-ray analysis of polar plume. *Solar Phys.* 58:323–336.

Ahmad, I. A., and Withbroe, G. L. 1977. EIV analyses of polar plumes. *Solar Phys.* 53:397–408.

Aime, C., and Ricort, G. 1979. Solar seeing and the statistical properties of the photospheric granulation. III. Solar speckle interferometry. *Astron. Astrophys.* 76:324–335.

Aime, C., Ricort, G., and Grec, G. 1977. Solar seeing and the statistical properties of the photospheric solar granulation. II. Power spectrum calibration via Michelson stellar interferometry. *Astron. Astrophys.* 54:505–516.

Aime, C., Ricort, G., Roddier, C., and Lago, G. 1978a. Changes in the atmospheric-lens modulation transfer function used for calibration in solar speckle interferometry. *J. Opt. Soc. Amer.* 68:1063–1066.

Aime, C., Ricort, G., and Harvey, J. W. 1978b. One-dimensional speckle interferometry of the solar granulation. *Astrophys. J.* 221:362–367.

Aime, C., Borgnino, P., Druesne, P., Harvey, J. W., Martin, F., and Ricort, G. 1985. Speckle interferometry technique applied to the study of granular velocities. In *High Resolution in Solar Physics,* ed. R. Muller (Berlin: Springer-Verlag), pp. 103–107.

Aindow, A., Elsworth, Y. P., Isaak, G. R., McLeod, C. P., New, R., and van der Raay, H. B. 1988. The current status of the Birmingham solar seismology network. In *Seismology of the Sun and Sun-Like Stars,* ed. E. J. Rolfe, ESA SP-286, pp. 157–160.

Aizenman, M. L., and Smeyers, P. 1977. An analysis of the linear adiabatic oscillations of a star in terms of potential fields. *Astrophys. Space Sci.* 48:123–136.

Aizenman, M. L., Smeyers, P., and Weigert, A. 1977. Avoided crossing of modes of non-radial stellar oscillations. *Astron. Astrophys.* 58:41–46.

Ajzenberg-Selove, F. 1988. Energy levels of light nuclei A = 5 − 10. *Nucl. Phys.* A490:1–225.

Akhmedov, E. Kh. 1988a. Resonant amplification of neutrino spin rotation in matter and the solar neutrino problem. *Phys. Lett.* B213:64–68.

Akhmedov, E. Kh. 1988b. Resonant enhancement of neutrino spin precession. In *Matter and the Solar Neutrino Problem. Soviet J. Nucl. Phys.* 48:382–383.

Akhmedov, E. Kh. 1989. Interplay of resonant spin-flavor precession and resonant oscillations of neutrinos. *Soviet Phys. JETP* 68:690.

Akhmedov, E. Kh., and Bychuk, O. V. 1989. Resonant flavor-changing neutrino spin rotation and the solar neutrino problem. *Zh. Eksp. Teor. Fiz.* 95:442.

Akhmedov, E. Kh., and Khlopov, M. Yu. 1988. Resonance enhancement of neutrino-oscillations in a longitudinal magnetic field. *Soviet J. Nucl. Phys.* 47:689–691.

Alamanni, N., Bertello, L., Cavallini, F., Ceppatelli, G., and Righini, A. 1990a. Depth dependence of the intensity-velocity phase difference in the solar "5-min" oscillations. *Astron. Astrophys.* 231:518–524.

Alamanni, N., Cavallini, F., Ceppatelli, G., and Righini, A. 1990b. Ca I 6162 Å oscillations observed in a solar active region. *Astron. Astrophys.* 228:517–521.

Albregtsen, F., and Andersen, B. N. 1985. The effect of spatial smearing on solar Doppler measurements. *Solar Phys.* 95:239–249.

Albregtsen, F., and Maltby, P. 1981a. Constraints on umbral core models as derived from broadband observations. In *The Physics of Sunspots,* eds. L. Cram and J. H. Thomas (Sunspot, N. M.: Sacramento Peak Obs.), pp. 127–139.

Albregtsen, F., and Maltby, P. 1981b. Solar cycle variation of sunspot intensity. *Solar Phys.* 71:269–283.

Alecian, G., and Vauclair, S. 1981. Element stratification in the atmospheres of main sequence stars: The silicon accumulation. *Astron. Astrophys.* 101:16–25.

Alexander, T. K., Ball, G. C., Lennard, W. N., Geissel, H., and Mak, H.-B. 1984. Measurement of the absolute cross section of the ^{3}He(^{4}He,γ)^{7}Be reaction at $E_{cm} = 525$ keV. *Nucl. Phys.* A427:526–544.

Alexeyev, E. N., Alexeyeva, L. N., Chudakov, A. E., and Krivosheina, I. V. 1987. Are there neutrinos of a few tens MeV in solar flares? In *Proc. 20th Intl. Cosmic Ray Conf.*, vol. 4, pp. 351–354.

Alfvén, H. 1943. On the effect of a vertical magnetic field in a conducting atmosphere. *Arkiv Mat. Astr. Fysik* 29(11):1–12.

Alfvén, H. 1950. *Cosmical Electrodynamics* (Oxford: Clarendon Press).

Alissandrakis, C. E., Kundu, M. R., and Lantos, P. 1980. A model for sunspot associated emission at 6 cm wavelength. *Astron. Astrophys.* 82:30–40.

Alissandrakis, C. E., Dialetis, D., Mein, P., Schmider, B., and Simon, G. 1988. The Evershed flow in the solar photosphere, chromosphere and chromosphere-corona transition region. *Astron. Astrophys.* 201:339–349.

Alissandrakis, C. E., Tsiropoula, G., and Mein, P. 1990. Physical parameters of solar Hα absorption features derived with the cloud model. *Astron. Astrophys.* 230:200–212.

Allen, C. W. 1973. *Astrophysical Quantities* (London: Athlone Press).

Allen, M. S., McAllister, H. C., and Jeffries, J. T. 1978. *High Resolution Atlas of the Solar Spectrum 2678–2931 Å* (Honolulu: Univ. of Hawaii Inst. for Astronomy).

Aller, L. H. 1987. Chemical abundances. In *Spectroscopy of Astrophysical Plasmas*, eds. A. Dalgarno and D. Layzer (Cambridge: Cambridge Univ. Press), pp. 90–124.

Aller, L. H., and Chapman, S. 1960. Diffusion in the Sun. *Astrophys. J.* 132:461–472.

Altrock, R. C., ed. 1988. *Solar and Stellar Coronal Structure and Dynamics: Proc. of the 9th Sac Peak Summer Symp.* (Sunspot, N.M.: Sacramento Peak Obs.).

Altrock, R. C., and Musman, S. 1976. Physical conditions in granulation. *Astrophys. J.* 203:533–540.

Altrock, R. C., Musman, S., and Cook, M. C. 1984. The evolution of an average solar granule. In *Small-Scale Dynamical Processes in Quiet Stellar Atmospheres*, ed. S. L. Keil (Sunspot, N. M.: Sacramento Peak Obs.), pp. 130–140.

Altschuler, M. D., and Newkirk, G., Jr. 1969. Magnetic fields and the structure of the solar corona. *Solar Phys.* 9:131–149.

Alvarez, L. A. 1949. Proposed Experimental Test of the Neutrino Theory. Univ. California Radiation Laboratory Report UCRL-328.

Alvarez, L. W. 1973. A Signal Generator for Ray Davis' Neutrino Detector. Univ. California Radiation Laboratory Physics Notes, Memo 767.

An, C. 1984. Condensation modes in magnetized cylindrical plasmas. *Astrophys. J.* 276:755–760.

An, C. 1985. Formation of prominences by condensation modes in magnetized cylindrical plasmas. *Astrophys. J.* 298:409–413.

An, C. 1986. Condensation modes in sheared magnetic fields. *Astrophys. J.* 304:532–541.

Anders, E., and Grevesse, N. 1989. Abundances of the elements: Meteoritic and solar. *Geochim. Cosmochim. Acta* 53:197–214.

Andersen, B. N. 1984. Limb effect of solar absorption lines. *Solar Phys.* 94:49–56.

Andersen, B. N. 1985. Straylight correction to Doppler rotation measurements. *Solar Phys.* 98:173–176.

Andersen, B. N. 1986. The effect of spatial smearing on solar Doppler measurements. II. General formulation and application to limb effect and large-scale velocity fields. *Solar Phys.* 107:27–38.

Andersen, B. N. 1987. Solar limb effect and meridional flows. Results for Fe I 512.4, 543.4, and 709.0 nm. *Solar Phys.* 114:207–222.

Andersen, B. N. 1989. Helioseismology and plasma physics. In *Plasma Astrophysics*, vol. 1, eds. T. D. Guyenne and J. J. Hunt, ESA SP-285, pp. 83–93.

Andersen, B. N., and Maltby, P. 1983. Has rapid solar core rotation been observed? *Nature* 302:808–810.

Andersen, B. M., Domingo, V., Jones, A. R., Korzennik, S., Jiménez, A., Pallé, P., Regulo, C., and Roca-Cortés, T. 1988. Solar luminosity oscillation telescope (SLOT). In *Seismology of the Sun and the Sun-Like Stars*, ed. E. J. Rolfe, ESA SP-286, pp. 175–176 (abstract).

Andersen, J., Garcia, J. M., Giménez, A., and Nordstrom, B. 1987. Absolute dimensions of eclipsing binaries. *Astron. Astrophys.* 174:107–115.

Andersen, J., Clausen, J. V., Gustafsson, B., Nordström B., and VandenBerg, D. A. 1988. Absolute dimensions of eclipsing binaries. XIII. AI Phoenicis: A case study in stellar evolution. *Astron. Astrophys.* 196:128–140.

Andersen, T. E., Dunn, R. B., and Engvold, O. 1984. *LEST Design Study,* LEST Foundation Tech. Rept. No. 7 (Oslo: Univ. of Oslo).

Anderson, E. R., Duvall, T. L., Jr., and Jefferies, S. M. 1990. Modeling of solar oscillation power spectra. *Astrophys. J.* 364:699–705.

Anderson, L. S., 1989. Line blanketing without local thermodynamic equilibrium. II. A solar-type model in radiative equilibrium. *Astrophys. J.* 339:558–578.

Anderson, L. S., and Athay, R. G. 1989*a.* Chromospheric and coronal heating. *Astrophys. J.* 336:1089–1091.

Anderson, L. S., and Athay, R. G. 1989*b.* Model solar chromosphere with prescribed heating. *Astrophys. J.* 346:1010–1018.

Ando, H. 1985. Examination of wave behaviors in the differentially rotating systems. *Publ. Astron. Soc. Japan* 37:47–68.

Ando, H. 1986. Resonant excitation of the solar g-modes through coupling of 5-min oscillations. *Astrophys. Space Sci.* 118:177–181.

Ando, H., and Osaki, Y. 1975. Nonadiabatic nonradial oscillations: An application to the five minute oscillations of the Sun. *Publ. Astron. Soc. Japan* 27:581–603.

Ando, H., and Osaki, Y. 1977. The influence of the chromosphere and corona on the solar atmospheric oscillations. *Publ. Astron. Soc. Japan* 29:221–223.

Andreason, G. K. 1988. Stellar consequences of enhanced metal opacity. I. An alternative solution of the Cepheid period ratio discrepancies. *Astron. Astrophys.* 201:72–79.

Anglin, J. D. 1975. The relative abundances and energy spectra of solar-flare-accelerated deuterium, tritium, and helium-3. *Astrophys. J.* 198:733–753.

Anguera Gubau, M., Pallé, P. L., Perez Hernandez, and Roca-Cortés, T. 1990. An attempt to identify low ℓ and low n solar acoustic models. *Solar Phys.* 128:79–90.

Antalova, A. 1986. The latitude distribution of sunspot areas and variations of the differential rotation of the Sun during the period 1921–1971. *Contrib. Astron. Obs. Skalnate Pleso* 14:121–144.

Antia, H. M., Chitre, S. M., and Narashima, D. 1982. Overstability of acoustic modes and the solar five minute oscillations. *Solar Phys.* 77:303–327.

Antia, H. M., Chitre, S. M., and Narashima, D. 1983. Influence of turbulent pressure on solar convective modes. *Mon. Not. Roy. Astron. Soc.* 204:865–881.

Antia, H. M., Chitre, S. M., and Narashima, D. 1984. Convection in the envelope of red giants. *Astrophys. J.* 282:574–583.

Antia, H. M., Chitre, S. M., and Gough, D. O. 1988. On the excitation of solar five minute oscillations. In *Advances in Helio- and Asteroseismology: Proc. IAU Symp. 123,* eds. J. Christensen-Dalsgaard and S. Frandsen (Dordrecht: D. Reidel), pp. 371–374.

Antiochos, S. K., and Noci, G. 1986. Structure of the static corona and transition region. *Astrophys. J.* 310:440–447.

Antiochos, S. K., and Sturrock, P. A. 1978. Evaporative cooling of flare plasma. *Astrophys. J.* 220:1137–1143.

Antonucci, E., Azzarelli, L., Casalini, P., and Denoth, F. 1979. Chromospheric rotation. II. Dependence on the size of chromospheric features. *Solar Phys.* 63:17–30.

Antonucci, E., Gabriel, A. H., Acton, L. W., Culhane, J. L., Doyle, J. G., Leibacher, J. W., Machado, M. E., Orwig, L. E., and Rapley, C. G. 1982. Impulsive phase of flares in soft X-ray emission. *Solar Phys.* 78:107–123.

Anzer, U. 1985. The global structure of magnetic fields which support quiescent prominences. In *Measurement of Solar Vector Magnetic Fields,* ed. M. J. Hagyard, NASA CP-2374, pp. 101–106.

Appenzeller, I., and Tscharnuter, W. 1975. On the luminosity of spherical protostars. *Astron. Astrophys.* 40:397–399.

Appourchaux, T. 1987. The Faraday effect in a magneto-optical filter. *Solar Phys.* 109:393–397.

Appourchaux, T. 1988. Solar oscillations instrumentation and measurement theory. In *Seismology of the Sun and Sun-Like Stars,* ed. E. J. Rolfe, ESA SP-286, pp. 595–600.

Appourchaux, T. 1989. Optimization of parameters for helioseismology experiments measuring solar radial velocities. *Astron. Astrophys.* 222:361–366.

Arévalo, M. J., Gomez, R., Vázquez, M., Balthasar, M., and Wöhl, H. 1982. Differential rotation and meridional motions of sunspots from 1874 to 1902. *Astron. Astrophys.* 111:266–271.

Armstrong, J. W., Coles, W. A., Kojima, M., and Rickett, B. J. 1986. Solar wind observations near the Sun. In *The Sun and the Heliosphere in Three Dimensions*, ed. B. G. Marsden (Dordrecht: D. Reidel), pp. 59–64.

Arnett, W. D., and Truran, J. W. 1969. Carbon-burning nucleosynthesis at constant temperature. *Astrophys. J.* 157:339–365.

Asbridge, J. R., Bame, S. J., Feldman, W. C., and Montgomery, M. D. 1976. Helium and hydrogen velocity differences in the solar wind. *J. Geophys. Res.* 81:2719–2727.

Assenbaum, H. J., Langanke, K., and Rolfs, C. 1987. Effects of electron screening on low-energy fusion cross section. *Z. Phys.* A327:461–468.

Athay, R. G. 1959. A model of the chromosphere from radio and optical data. In *Paris Symp. on Radio Astronomy*, ed. R. N. Bracewell (Stanford: Stanford Univ. Press), pp. 98–104.

Athay, R. G. 1966. Radiative energy loss from the solar chromosphere and corona. *Astrophys. J.* 146:223–240.

Athay, R. G. 1970. Emission cores in H and K lines. *Solar Phys.* 11:347–354.

Athay, R. G. 1976. *The Solar Chromosphere and Corona: Quiet Sun* (Dordrecht: D. Reidel).

Athay, R. G. 1981. The chromosphere and transition region. In *The Sun as a Star*, ed. S. Jordan, NASA SP-450, pp. 85–133.

Athay, R. G. 1985. Fine structure and dynamics of the chromosphere and transition region and future directions of models. In *Theoretical Problems in High Resolution Solar Physics*, ed. H. U. Schmidt (München: Max-Planck-Institut), pp. 205–231.

Athay, R. G. 1986*a*. Chromospheric fine structure. In *Physics of the Sun*, vol. 1, eds. P. A. Sturrock, T. E. Holzer, D. M. Mihalas and R. K. Ulrich (Dordrecht: D. Reidel), pp. 51–69.

Athay, R. G. 1986*b*. Radiation output. In *Physics of the Sun*, vol. 2, eds. P. A. Sturrock, T. E. Holzer, D. M. Mihalas and R. K. Ulrich (Dordrecht: D. Reidel), pp. 1–50.

Athay, R. G., and Holzer, T. E. 1982. The role of spicules in heating the solar atmosphere. *Astrophys. J.* 255:743–752.

Athay, R. G., and Morton, G. E. 1961. Impulsive phenomena of the solar atmosphere. I. Some optical events associated with flares showing explosive phase. *Astrophys. J.* 133:935–945.

Athay, R. G., and Spicer, D. S., eds. 1987. *Theoretical Problems in High Resolution Solar Physics II*, NASA CP-2483.

Athay, R. G., and White, S. 1978. Chromospheric and coronal heating by sound waves. *Astrophys. J.* 226:1135–1139.

Athay, R. G., and White, O. R. 1979. Chromospheric oscillations observed with OSO 8. IV. Power and phase spectra for C IV. *Astrophys. J.* 229:1147–1162.

Athay, R. G., Querfeld, C. W., Smartt, R. N., Landi, E., and Bommier, V. 1983. Vector magnetic fields in prominences. *Solar Phys.* 89:3–35.

Atkinson, R. d'E., and Houtermans, F. G. 1929. Zur Frage der Aufbaumoglichkeit der Elemente in Sterne. *Z. Phys.* 54:656–665.

Attolini, M. R., Cecchini, S., Castagnoli, G. C., and Galli, M. 1988. On a biennial variation of the solar neutrino flux. *Astron. Lett. and Comm.* 27:55–61.

Auer, L. H., and Heasley, J. N. 1978. The origin of broad-band circular polarization in sunspots. *Astron. Astrophys.* 64:67–71.

Auer, L. H., Heasley, J. N., and Milkey, R. W. 1972. *A Computational Program for the Solution of Non-LTE Transfer Problems by the Complete Linearization Method*. KPNO Contrib. 555 (Tucson: Kitt Peak National Obs.).

Auer, L. H., Heasley, J. N., and House, L. L. 1977*a*. Non-LTE line formation in the presence of magnetic fields. *Astrophys. J.* 216:531–539.

Auer, L. H., Heasley, J. N., and House, L. L. 1977*b*. The determination of vector magnetic fields from Stokes profiles. *Solar Phys.* 55:47–61.

Avignone, F. T., and Brodzinski, R. L. 1988. A review of recent developments in double beta decay. In *Progress in Particle and Nuclear Physics*, vol. 21, ed. A. Faessler (London: Pergamon), pp. 99–181.

Avrett, E. H. 1985. Recent thermal models of the chromosphere. In *Chromospheric Diagnostics and Modelling*, ed. B. W. Lites (Sunspot, N.M.: Natl. Solar Obs.), pp. 67–127.

Avrett, E. H., Kurucz, R. L., and Loeser, R. 1984. New models of the solar temperature minimum region and low chromosphere. *Bull. Amer. Astron. Soc.* 16: 450 (abstract).

Axford, W. I., Dessler, A. J., and Gottlieb, B. 1963. Termination of solar wind and solar magnetic field. *Astrophys. J.* 137:1268–1278.

Ayres, T. R. 1981. Thermal bifurcation in the solar outer atmosphere. *Astrophys. J.* 244:1064–1071.

Ayres, T. R., and Testerman, L. 1981. Fourier transform spectrometer observations of solar carbon monoxide. I. The fundamental and first overtone bands in the quiet Sun. *Astrophys. J.* 245:1124–1140.

Ayres, T. R., and Wiedemann, G. 1989. Non-LTE CO, revisited. *Astrophys. J.* 338:1033–1046.

Ayres, T. R., Testerman, L., and Brault, J. W. 1986. Fourier transform spectrometer observations of solar carbon monoxide. II. Simultaneous cospatial measurements of the fundamental and first-overtone bands, and Ca II K, in quiet and active regions. *Astrophys. J.* 304:542–559.

Babcock, H. W. 1953. The solar magnetograph. *Astrophys. J.* 118:387–396.

Babcock, H. W. 1959. The Sun's polar magnetic field. *Astrophys. J.* 130:364–365.

Babcock, H. W. 1961. The topology of the Sun's magnetic field and the 22-year cycle. *Astrophys. J.* 133:572–587.

Babcock, H. W., and Babcock, H. D. 1955. The Sun's magnetic field, 1952–1954. *Astrophys. J.* 121:349–366.

Babu, K. S., and Mohapatra, R. N. 1989. Model for large transition moment of an electron neutrino. *Phys. Rev. Lett.* 63:228–231.

Bacher, A. D., and Tombrello, T. A. 1965. p – ^{4}He final-state interaction in ^{3}He(^{3}He,2p)^{4}He. *Rev. Mod. Phys.* 37:433–437.

Backus, G., and Gilbert, F. 1968. The resolving power of gross Earth data. *Geophys. J. Roy. Astron. Soc.* 16:169–205.

Backus, G., and Gilbert, F. 1970. Uniqueness in the inversion of inaccurate gross Earth data. *Phil. Trans. Roy. Soc. London* A266:123–192.

Badalyan, O. B. 1986. Polarization of white-light coronal under hydrostatic density distribution. *Astron. Astrophys.* 169:305–312.

Badalyan, O. B., and Livshits, M. A. 1986. The K-corona under hydrostatic density distribution relevance to solar wind. *Solar Phys.* 103:385–392.

Baglin, A. 1972. Short period variable stars. IX. Rotation and mixing in the outer layers of A stars. Turbulent mixing due to the meridional circulation velocity field. *Astron. Astrophys.* 19:45–50.

Baglin, A. 1975. Rotation mixing and variability in A stars. In *Multiple Periodic Variable Stars: Proc. IAU Coll. 29*, ed. W. S. Fitch (Budapest: Hungarian Academy of Sciences), pp. 233–246.

Baglin, A., and Lebreton, Y. 1990. Surface abundance of light elements as diagnostic of transport process in the Sun and solar-type stars. In *Inside the Sun: Proc. IAU Coll. 121*, eds. G. Berthomeiu and M. Cribier (Dordrecht: Kluwer), pp. 437–448.

Baglin, A., and Morel, P. 1989. Observational constraints on the turbulent diffusion coefficient in late type stars. In *Rotation and Mixing in Stellar Interiors*, eds. M.-J. Goupil and J.-P. Zahn (Berlin: Springer-Verlag), pp. 167–172.

Baglin, A., Morel, P., and Schatzman, E. 1985. Stellar evolution with turbulent diffusion mixing. V. Lithium abundance on the lower main sequence. *Astron. Astrophys.* 149:309–314.

Bahcall, J. N. 1964. Solar neutrino cross sections and nuclear beta decay. *Phys. Rev.* B135:137–146.

Bahcall, J. N. 1966. Non-resonant nuclear reactions at stellar temperatures. *Astrophys. J.* 143:259–261.

Bahcall, J. N. 1978. Solar neutrino experiments. *Rev. Mod. Phys.* 50:881–903.

Bahcall, J. N. 1979. Solar neutrinos: Theory versus observation. *Space Sci. Rev.* 24:227–251.

Bahcall, J. N. 1980. Solar neutrinos. *Prog. Part. Nucl. Phys.* 6:111–123.

Bahcall, J. N. 1987. Neutrino-electron scattering and solar neutrino experiments. *Rev. Mod. Phys.* 59:505–521.

Bahcall, J. N. 1989. *Neutrino Astrophysics* (Cambridge: Cambridge Univ. Press).

Bahcall, J. N., and Bethe, H. A. 1990. A solution of the solar neutrino problem. *Phys. Rev.* 65:2233–2235.

Bahcall, J. N., and Davis, R. 1976. Solar neutrinos: A scientific puzzle. *Science* 191:264–267.

Bahcall, J. N., and Davis, R. 1982. An account of the development of the solar neutrino problem. In *Essays in Nuclear Astrophysics*, eds. C. A. Barnes, D. D. Clayton and D. N. Schramm (Cambridge: Cambridge Univ. Press), p. 243.

Bahcall, J. N., and Haxton, W. C. 1989. Matter-enhanced neutrino oscillations in the standard solar model. *Phys. Rev.* D40:931–941.

Bahcall, J. N., and Holstein, R. R. 1986. Solar neutrinos from decay of ^{8}B. *Phys. Rev.* C33:2121–2127.

Bahcall, J. N., and Loeb, A. 1990. Element diffusion in stellar interiors. *Astrophys. J.* 360:267–274.

Bahcall, J. N., and May, R. M. 1969. The rate of the proton-proton reaction and some related reactions. *Astrophys. J.* 155:501–510.

Bahcall, J. N., and Moeller, C. P. 1969. The ^{7}Be electron-capture rate. *Astrophys. J.* 155:511–514.

Bahcall, J. N., and Pinsonneault, M. H. 1991. Standard solar models with diffusion. *Astrophys. J.* submitted.

Bahcall, J. N., and Press, W. H. 1991. Solar cycle modulation of the event rates in the chlorine solar neutrino experiment. *Astrophys. J.*, 370:730–742.

Bahcall, J. N., and Primakov, H. 1978. Neutrino-antineutrino oscillations. *Phys. Rev.* D18:3463–3466.

Bahcall, J. N., and Sears, R. L. 1972. Solar neutrinos. *Ann. Rev. Astron. Astrophys.* 10:25–44.

Bahcall, J. N., and Ulrich, R. K. 1988. Solar models, neutrino experiments, and helioseismology. *Rev. Mod. Phys.* 60:297–372.

Bahcall, J. N., Fowler, W. A., Iben, K., Jr., and Sears, R. L. 1963. Solar neutrino flux. *Astrophys. J.* 137:344–345.

Bahcall, J. N., Bahcall, N. A., and Shaviv, G. 1968. Present status of the theoretical predictions for the ^{37}Cl solar-neutrino experiment. *Phys. Rev. Lett.* 20:1209–1213.

Bahcall, J. N., Bahcall, N. A., and Ulrich, R. K. 1969. Sensitivity of the solar-neutrino fluxes. *Astrophys. J.* 156:559–568.

Bahcall, J. N., Cabibbo, N., and Yahil, A. 1972. Are neutrinos stable particles? *Phys. Rev. Lett.* 28:316–318.

Bahcall, J. N., Cleveland, B. T., Davis, R., Dostrovsky, I., Evans, J., Frati, W., Friedlander, G., Lande, K., Rowley, J. K., Stoenner, R. W., and Weneser, J. 1978. Proposed solar neutrino experiment using ^{71}Ga. *Phys. Rev. Lett.* 40:1351–1354.

Bahcall, J. N., Huebner, W. F., Lubow, S. H., Parker, P. D., and Ulrich, R. K. 1982. Standard solar models and the uncertainties in predicted capture rates of solar neutrinos. *Rev. Mod. Phys.* 54:767–801.

Bahcall, J. N., Field, G. B., and Press, W. H. 1987a Is solar neutrino capture rate correlated with sunspot number? *Astrophys. J.* 320:L69–73.

Bahcall, J. N., Gelb, J. M., and Rosen, S. P. 1987b. Mikheyev-Smirnov-Wolfenstein effect in electron-neutrino scattering experiments. *Phys. Rev.* D35:2976–2982.

Bahcall, J. N., Davis, R., and Wolfenstein, L. 1988. Solar neutrinos: A field in transition. *Nature* 334:487–493.

Bahcall, J. N., Kubodera, K., and Nozawa, S. 1989. Neutral current reactions of solar and supernova neutrinos on deuterium. *Phys. Rev.* D38:1030–1039.

Bahng, J., and Schwarzschild, M. 1961. Lifetime of solar granules. *Astrophys. J.* 134:312–322.

Bai, T. 1977. Studies of Solar Hard X-Rays and Gamma-Rays: Compton Backscatter, Anistropy, Polarization, and Evidence for Two Phases of Acceleration. Ph.D. Thesis, Univ. of Maryland.

Bai, T. 1986. Two classes of gamma-ray/proton flares: Impulsive and gradual. *Astrophys. J.* 308:912–928.

Bai, T. 1987a. Distribution of flares on the Sun: Superactive regions and active zones of 1980–1985. *Astrophys. J.* 314:795–807.

Bai, T. 1987b. Periodicities of the flare occurrence rate in solar cycle 19. *Astrophys. J.* 318:L85–L91.

Bai, T. 1988. Distribution of flares on the Sun during 1955–1985: "Hot spots" (active zones) lasting for 30 years. *Astrophys. J.* 328:860–878.

Bai, T., and Dennis, B. R. 1985. Characteristics of gamma-ray line flares. *Astrophys. J.* 292:699–715.

Bai, T., and Ramaty, R. 1979. Hard X-ray time profiles and acceleration processes in large solar flares. *Astrophys. J.* 227:1072–1081.

Bai, T., Hudson, H. S., Pelling, R. M., Lin, R. P., Schwartz, R. A., and von Rosenvinge, T. T.

1983. First-order Fermi-acceleration in solar flares as a mechanism for the second-step acceleration of prompt protons and relativistic electrons. *Astrophys. J.* 267:433–441.

Bailey, G. M., Griffiths, G. M., Olivo, M. A., and Helmer, R. L. 1970. Gamma-ray yields from the reaction D(p,γ)^{3}He at low energies. *Canadian J. Phys.* 48:3059–3061.

Baker, N. H. 1966. Simplified models for Cepheid instability. In *Stellar Evolution,* eds. R. F. Stein and A. G. W. Cameron (New York: Plenum Press), pp. 333–346.

Baker, N. H., and Gough, D. O. 1979. Pulsations of model RR Lyrae stars. *Astrophys. J.* 234:232–244.

Baker, N. H., and Temesvary, S. 1966. *Tables of Convective Stellar Envelope Models,* 2nd. ed. (New York: Goddard Inst. for Space Studies).

Balachandran, S. 1988. Ph.D. Thesis, Univ. of Texas at Austin.

Balachandran, S. 1990. Lithium depletion and rotation in main sequence stars. *Astrophys. J.* 354:310–332.

Balachandran, S., Lambert, D. L., and Stauffer, J. R. 1988. Lithium in the main sequence stars of the alpha Persei cluster. *Astrophys. J.* 333:267–276.

Balandin, A. L., Grigoryev, V. M., and Demidov, M. L. 1987. On spatial filtering of low-degree global oscillations of the Sun. *Solar Phys.* 112:197–209.

Baliunas, S. L. 1988. Luminosity and magnetic activity variations on cool stars. In *Solar Radiative Output Variations,* ed. P. Foukal (Cambridge, Mass.: Cambridge Research and Instrumentation), pp. 230–240.

Baliunas, S. L., and Jastrow, R. 1990. Evidence for long-term brightness changes of solar-type stars. *Nature* 348:520.

Baliunas, S. L., Donahue, R. A., Loeser, J. G., Guinan, E. F., Genet, R. M., and Boyd, L. J. Broadband photometry of bright stars: The first year of APTS at the F. L. Whipple Observatory. In *New Generation Small Telescopes,* eds. D. D. Hayes, D. R. Genet and R. M. Genet (Mesa, Ariz.: Fairborn Press), pp. 97–116.

Ballester, J. L., and Priest, E. R. 1988. A 2D prominence model. In *Dynamics and Structure of Solar Prominences,* eds. J. L. Ballester and E. R. Priest (Majorca: Palma), pp. 137–143.

Balmforth, N. J., and Gough, D. O. 1988. On radiative and convective influences on stellar pulsational stability. In *Seismology of the Sun and Sun-Like Stars,* ed. E. J. Rolfe, ESA SP–286, pp. 47–52.

Balmforth, N., and Gough, D. O. 1990*a*. High frequency solar *p* modes. *Astrophys. J.,* in press.

Balmforth, N. J., and Gough, D. O. 1990*b*. Mixing-length theory and the excitation of solar acoustic oscillations. *Solar Phys.* 128:161–193.

Balthasar, H. 1984. Asymmetries and wavelengths of solar spectral lines and the solar rotation determined from Fourier transform spectra. *Solar Phys.* 93:219–241.

Balthasar, H., and Wöhl, H. 1980. Differential rotation and meridional motion of sunspots in the years 1940–1968. *Astron. Astrophys.* 92:111–116.

Balthasar, H., Schussler, M., and Wöhl, H. 1982. On changes of the rotation velocities of stable, recurrent sunspots and their interpretation with a flux tube model. *Solar Phys.* 76:21–28.

Balthasar, H., Lustig, G., and Wöhl, H. 1984. On the determination of heliographic positions and rotation velocities of sunspots. III. Effects caused by wrong solar image radii and their corrections. *Solar Phys.* 91:55–59.

Balthasar, H., Vázquez, M., and Wöhl, H. 1986*a*. Differential rotation of sunspot groups in the period from 1874 through 1976 and changes of the rotation velocity within the solar cycle. *Astron. Astrophys.* 160:277–279.

Balthasar, H., Lustig, G., Stark, D., and Wöhl, H. 1986*b*. The solar rotation elements ι and ω derived from sunspot groups. *Astron. Astrophys.* 160:277–279.

Balthasar, H., Küveler, G., and Wiehr, E. 1987*a*. A comparison of the oscillations in sunspot umbrae, penumbrae, and the surrounding photosphere. *Solar Phys.* 112:37–48.

Balthasar, H., Stark, D., and Wöhl, H. 1987*b*. The solar rotation elements ι and ω derived from recurrent single sunspots. *Astron. Astrophys.* 174:359–360.

Balthasar, H., Grosser, H., Schroter, E. H., and Wiehr, E. 1990. Intensity correlations in granular spectra. *Astron. Astrophys.* 235:437–440.

Baltz, A. J., and Weneser, J. 1987. Effect of transmission through the Earth on neutrino oscillations. *Phys. Rev.* D35:528–535.

Baltz, A. J., and Weneser, J. 1988. Matter oscillations: Neutrino transformation in the Sun and regeneration in the Earth. *Phys. Rev.* D37:3364–3367.

Baltz, A. J., and Weneser, J. 1991. Comment on "A Solution of the solar neutrino problem." *Phys. Rev. Lett.* 66:520.

Baltz, A. J., Rowley, J. K., and Weneser, J. *On Maximum Likelihood Statistics and the Empirical* ^{37}Cl *Rate Determinations.* BNL Rept. 43625 (Upton, N.Y.: Brookhaven National Lab).

Bame, S. J., Hundhausen, A. J., Asbridge, J. R., and Strong, I. B. 1968. Solar wind ion composition. *Phys. Rev. Lett.* 20:393–395.

Bame, S. J., Asbridge, J. R., Feldman, W. C., and Montgomery, M. D. 1975. Solar wind heavy ion abundances. *Solar Phys.* 43:463–473.

Bame, S. J., Asbridge, J. R., Feldman, W. C., and Gosling, J. T. 1977. Evidence for a structure-free state at high solar wind speeds. *J. Geophys. Res.* 82:1487–1492.

Bame, S. J., Asbridge, J. R., Feldman, W. C., Fenimore, E. E., and Gosling, J. T. 1979. Solar wind heavy ions from flare heated coronal plasma. *Solar Phys.* 62:179–201.

Bame, S. J., Asbridge, J. R., Feldman, W. C., Gosling, J. T., and Zwickl, R. D. 1981. Bidirectional streaming of solar wind electrons >80ev: ISEE evidence for a closed-field structure within the driver gas of an interplanetary shock. *Geophys. Res. Lett.* 8:173–176.

Bappu, M. K. V., and Sivaraman, K. R. 1971. K emission-line widths and the solar chromosphere. *Solar Phys.* 17:316–330.

Bappu, M. K. V., Grigoriev, V. M., and Stepanov, V. E. 1968. On the development of magnetic fields in active regions. *Solar Phys.* 4:409–421.

Barabonov, I. R., Egorov, A. I., Garvin, V. N., Kopysov, Yu. S., and Zatsepin, G. T. 1978. Present state and outlook for development of the solar neutrino astronomy. In *Neutrino '77* (Moscow: Acad. of Sciences of the USSR), pp. 20–41.

Barabanov, I. R., Veretenkin, E. P., Gavrin, V. N., Danshin, D. N., Eroshkina, L. A., Zatsepin, G. T., Zakharov, Yu. I., Klimova, S. A., Klimov, Yu. B., Knodel, T. V., Kopylov, A. V., Orekhov, I. V., Tikhonov, A. A., and Churmaeva, M. I. 1985. Pilot installation of the gallium-germanium solar neutrino telescope. In *Solar Neutrinos and Neutrino Astronomy,* eds. M. L. Cherry, K. Lande and W. A. Fowler (New York: American Inst. of Physics), pp. 175–184.

Barbieri, R., and Mohapatra, R. N. 1989. A neutrino with a large magnetic moment and a naturally small mass. *Phys. Lett.* B218:225–229.

Barger, V., Whisnant, K., Pakvasa, S., and Phillips, R. J. N. 1980. Matter effects on three-neutrino oscillations. *Phys. Rev.* 22D:2718–2726.

Barger, V., Philips, R. J. N., and Whisnant, K. 1986. Level crossings in solar-neutrino oscillations. *Phys. Rev.* D34:980–983.

Barger, V., Phillips, R. N. J., and Whisnant, K. 1990. Long-wavelength vacuum neutrino oscillations and the solar-neutrino deficit. *Phys. Rev. Lett.* 65:3084–3087.

Barger, V., Phillips, R. N. J., and Whisnant, K. 1991. Re-examination of neutrino oscillation solutions of the solar neutrino problem. *Phys. Rev.* D43:1110–1128.

Barnes, C. A. 1971. Nucleosynthesis by charged-particle reactions. In *Advances in Nuclear Physics,* vol. 4, eds. M. Baranger and E. Vogt (New York: Plenum Press), pp. 133–204.

Barr, S. M., Freier, E. M., and Zee, A. 1990. Mechanism for large neutrino magnetic moments. *Phys. Rev. Lett.* 65:2626–2629.

Barron, R. W., Cliver, E. W., Cronin, J. P., and Guidice, D. A. 1985. Solar radio emission. In *Handbook of Geophysics and the Space Environment,* ed. A. S. Jursa (Washington, D.C.: U.S. Air Force), pp. 11–1–11–15.

Barry, D. C. 1988. The chromospheric age dependence of the birthrate, composition, motions, and rotation of late F and G dwarfs within 25 parsecs of the Sun. *Astrophys. J.* 334:436–448.

Bartels, J. 1934. Twenty-seven day recurrences in terrestrial-magnetic and solar activity 1923–1933. *Terrest. Magnet.* 39:201–202.

Bartoe, J.-D. F., and Brueckner, G. E. 1975. New stigmatic, coma-free, concave-grating spectrograph. *J. Opt. Soc. Amer.* 65:13–21.

Bartoe, J.-D. F., and Brueckner, G. E. 1978. Applications of a new high spatial and spectral resolution spectrography design principle to solar and stellar ultraviolet spectroscopy. In *New Instrumentation for Space Astronomy,* eds. K. Van Der Hucht and G. S. Vaiana (Oxford: Pergamon), pp. 81–84.

Basri, G. S. 1987. Stellar activity in synchronized binaries. II. A correlation analysis with single stars. *Astrophys. J.* 316:377–388.

Basu, D. 1982. Solar neutrinos and solar particles. *Solar Phys.* 81:363–365.

Batchelor, D. A., Crannell, C. J., Wiehl, H. J., and Magun, A. 1985. Evidence for collisionless conduction fronts in impulsive solar flares. *Astrophys. J.* 295:258–274.

Batchelor, G. K. 1967. *An Introduction to Fluid Dynamics* (Cambridge: Cambridge Univ. Press).

Batchelor, G. K. 1969. Computation of the energy spectrum in homogeneous two-dimensional turbulence. II. *Phys. Fluids* 12:233–239.

Baur, T. G. 1981. Optical polarimeters for solar research. *Opt. Eng.* 20:2–13.

Baur, T. G., House, L. L., and Hull, H. K. 1980. A spectrum scanning polarimeter. *Solar Phys.* 65:111–146.

Baur, T. G., Elmore, D. E., Lee, R. H., Querfeld, C. W., and Rogers, S. R. 1981. Stokes II— A new polarimeter for solar observations. *Solar Phys.* 70:395–410.

Bavassano, B., Dobrowolny, M., Mariani, F., and Ness, N. F. 1982. Radial evolution of power spectra of interplanetary Alfvén turbulence. *J. Geophys. Res.* 87:3617–3662.

Bazilevskaya, G. A., Mukhamedzhanov, A. M.-A. Nikol'skii, S. I., Stozhkov, Yu. I., and Charakhch'yan, T. N. 1984. Cosmic rays and the neutrino flux in the Davis Experiment. *Soviet J. Nucl. Phys.* 39:543–550.

Beardsley, B. J. 1987. The Visual Shape and Multipole Moments of the Sun. Ph.D. Thesis, Univ. of Arizona.

Becker, U. 1954. Die eigenbewegung der sonnenflecken in breite. *Z. Astrophys.* 21:129–136.

Beckers, J. C. 1968. Photospheric brightness differences associated with the solar granulation. *Solar Phys.* 5:309–322.

Beckers, J. C. 1977. Material motions in sunspot umbrae. *Astrophys. J.* 213:900–905.

Beckers, J. M. 1969. The profiles of Fraunhofer lines in the presence of Zeeman splitting. *Solar Phys.* 9:372–386.

Beckers, J. M. 1972. Solar spicules. *Ann. Rev. Astron. Astrophys.* 10:73–100.

Beckers, J. M. 1981. Dynamics of the solar photosphere. In *The Sun as a Star,* ed. S. Jordan, NASA SP-450, pp. 11–64.

Beckers, J. M., and Nelson, G. D. 1978. Some comments on the limb shift of solar lines. II. The effect of granular motions. *Solar Phys.* 58:243–261.

Beckers, J. M., and Taylor, W. R. 1980. Some comments on the limb shift of solar lines. III. Variation of limb shift with solar latitude, across plages, and across supergranules. *Solar Phys.* 68:41–47.

Beckers, J. M., Bridges, C. A., and Gilliam, L. B. 1976. *A High Resolution Spectral Atlas of the Solar Irradiance from 380 to 700 nm.* AFGL-TR-76-0126.

Beckman, J. E., and Clark, C. D. 1973. Studies of the solar chromosphere from millimetre and submillimetre observations. I. Isophotometric mapping. *Solar Phys.* 29:25–39.

Beckman, J. E., Clark, C. D., and Ross, J. 1973. Studies of the solar chromosphere from millimetre and submillimetre observations. II. Simple models of the lower chromosphere. *Solar Phys.* 31:319–338.

Beckman, J. E., Lesurf, J. C. G., and Ross, J. 1975. Submillimetre brightness spike at the solar limb. *Nature* 254:38–39.

Beichman, C. A., Myers, P. C., Emerson, J. P., Harris, S., Mathieu, R., Benson, P. J., and Jennings, R. E. 1986. Candidate solar-type protostars in nearby molecular cloud cores. *Astrophys. J.* 307:337–349.

Belcher, J. W., and Davis, L., Jr. 1971. Large-amplitude Alfvén waves in the interplanetary medium. 2. *J. Geophys. Res.* 76:3534–3563

Beletic, J. W. 1988. Comparison of Know-Thompson and bispectrum algorithms for reconstructing phase of complex extended objects. In *High Resolution Imaging by Interferometry,* ed. F. Merkle, part I (Garching: European Southern Obs.), pp. 357–371.

Bell, A. R. 1978. The acceleration of cosmic rays in shock fronts. I. *Mon. Not. Roy. Astron. Soc.* 182:147–156.

Bell, R. A., Eriksson, K., Gustafsson, B., and Nordlund, Å. 1976. A grid of model atmospheres for metal-deficient giant stars. II. *Astron. Astrophys. Suppl.* 23:37–95.

Belmonte, J. A., Elsworth, Y., Isaak, G. R., New, R., Pallé, P. L., and Roca-Cortés, T. 1988. Effect of atmospheric extinction on solar radial velocity measurements. In *Seismology of the Sun and Sun-Like Stars,* ed. E. J. Rolfe, ESA SP-286, pp. 177–179.

Belvedere, G., and Paterno, L. 1977. Convection in a rotating deep compressible spherical shell: Application to the Sun. *Solar Phys.* 54:289–312.

Belvedere, G., Godoli, G., Motta, S., Patterno, L., and Zappala, R. A. 1976. Surface Reynolds stresses determined from the analysis of facular motions and the maintenance of the Sun's differential rotation. *Solar Phys.* 46:23–28.

Benz, A. O. 1977. Spectral features in solar hard X-ray and radio events and particle acceleration. *Astrophys. J.* 211:270–280.

Benz, A. O., and Smith, D. F. 1987. Stochastic acceleration of electrons in solar flares. *Solar Phys.* 107:299–309.

Berger, A., Imbrie, J., Hays, J., Kukla, G., and Saltzman, B., eds. 1984. *Milankovitch and Climate: Understanding the Response to Astronomical Forcing* (Dordrecht: D. Reidel).

Berger, Ch., Fröhlich, M., Mönch, H., Nisius, R., Raupach, F., Blum, D., Bourdarios, C., Dudelzak, B., Eschstruth, P., Jullian, S., Lalanne, D., Laplanche, F., Longvemare, C., Paulot, C., Perdereau, O., Roy, Ph., Szklarz, G., Behr, L., Degrange, B., Nguyen-Khac, U., Tisserant, S., Arpesella, C., Bareyre, P., Barlovtaud, R., Borg, A., Chardin, G., Ernwein, J., Glicenstein, J. F., Mosca, L., Moscoso, L., Becker, J., Becker, K. H., Daum, H. J., Jacobi, B., Kuznik, B., Meyer, H., Möller, R., Schubnell, M., and Wei, Y. 1990. A study of atmospheric neutrino oscillations in the Frejus Experiment. *Phys. Lett.* B245: 305–310.

Bernet, G. 1979. Polarization of glass disks under mechanical and thermal stresses. *JOSO Annual Rept.*, ed. A. V. Alvensleben.

Bertello, L., and Caccin, B. 1990. Non-adiabatic modelling of "5-min" oscillations: Effects on photospheric line profiles. *Astron. Astrophys.* 231:509–517.

Berthomieu, G., and Provost, J. 1987. Sismologie stellaire: Probleme direct. In *Méthodes Mathématiques Pour l'Astrophysique*, eds. M. Auvergne and A. Baglin (Paris: Soc. Française des Spécialistes d'Astronomie), pp. 525–549.

Berthomieu, G., and Provost, J. 1988. Asymptotic properties of low degree gravity modes. In *Advances in Helio- and Asteroseismology: Proc. IAU Symp. 123*, eds. J. Christensen-Dalsgaard and S. Frandsen (Dordrecht: D. Reidel), pp. 121–125.

Berthomieu, G., and Provost, J. 1990. Light and velocity visibility of solar *g*-mode oscillations. *Astron. Astrophys.* 227:563–576.

Berthomieu, G., Cooper, A. J., Gough, D. O., Osaki, Y., Provost, J., and Rocca, A. 1980. Sensitivity of five minute eigenfrequencies to the structure of the Sun. In *Nonradial and Nonlinear Stellar Pulsation*, eds. H. A. Hill and W. A. Dziembowski (Heidelberg: Springer-Verlag), pp. 307–312.

Berthomieu, G., Provost, J., and Schatzman, E. 1984*a*. Influence of turbulent diffusion on asymptotic low degree solar gravity modes. *Mem. Soc. Astron. Italiana* 55:107–114.

Berthomieu, G., Provost, J., and Schatzman, E. 1984*b*. Solar gravity models as a text of turbulent diffusion mixing. *Nature* 308:254–257.

Bertolotti, M., Carnevale, M., Consortini, A., and Ronchi, L. 1979. Optical propagation: Problems and trends. *Optica Acta* 26:507–529.

Berton, R., and Sakurai, T. 1985. Stereoscopic determination of the three-dimensional geometry of coronal magnetic loops. *Solar Phys.* 96:93–111.

Bertout, C. 1989. T Tauri stars: Wild as dust. *Ann. Rev. Astron. Astrophys.* 27:351–395.

Bertout, C., Basri, G., and Bouvier, J. 1988. Accretion disks around T Tauri stars. *Astrophys. J.* 330:350–373.

Bespalov, P. A., Zaitsev, V. V., and Stepanov, A. V. 1987. On the origin of time delays in hard X-ray and gamma-ray emission of solar flares. *Solar Phys.* 114:127–140.

Bethe, H. A. 1939. Energy production in stars. *Phys. Rev.* 55:434–456.

Bethe, H. A. 1986. Possible explanation of the solar-neutrino puzzle. *Phys. Rev. Lett.* 56:1305–1308.

Bethe, H. A. 1989. Solar neutrino experiments. *Phys. Rev. Lett.* 63:837–839.

Bethe, H. A., and Critchfield, C. L. 1938. The formation of deuterons by proton combination. *Phys. Rev.* 54:248–254.

Bhatnagar, A., and Tanaka, K. 1972. Intensity oscillation in Hα-fine structure. *Solar Phys.* 24:87–97.

Bieber, J. W., Seckel, D., Stanev, T., and Steigman, G. 1990. Variation of the solar neutrino flux with the Sun's activity. *Nature* 348:403–407.

Biémont, E., and Grevesse, N. 1977. f-values and abundances of the elements in the Sun and stars. *Phys. Scripta* 16:39–47.

Biémont, E., Brault, J., Delbouille, L., and Roland, G. 1985. An investigation of iron in the infrared solar spectrum based on FTS laboratory measurements. *Astron. Astrophys. Suppl.* 61:107–125.

Bienaymé, O., Schatzman, E., and Maeder, A. 1984. Stellar evolution with turbulent diffusion mixing. IV. Intermediate and low mass stars and $^{12}C/^{13}C$ ratio in giants of the first ascending branch. *Astron. Astrophys.* 131:316–326.

Biermann, L. 1932. Untersuchungen uberden inneren Aufbau der Sterne. IV. Konvektionszonen im Innern der Sterne. *Z. Astrophys.* 5:117–139.

Biermann, L. 1937. *Astron. Nach.* 263:185–199.

Biermann, L. 1941. Der gegenwärtige Stand der Theorie konvektiver Sonnenmodelle. *Vierteliahrsschrift Astron. Gesselschaft* 76:194–200.

Biermann, L. 1946. Zur Deutung der chromosphärischen Turbulenz und des Exzesses der UV-Strahlung der Donne. *Naturwiss.* 33:118–119.

Biermann, L. 1948. Über die Ursache der chromosphärischen Turbulenz und des UV-Exzesses der Sonnenstrahlung. *Z. Astrophysik* 25:161–177.

Biermann, L. 1953. Physical processes in comet tails and their relationship to solar activity. *Extrait des Mem. Soc. Roy. Liège Collection* 13:271–302.

Bilenky, S. M., and Petcov, S. T. 1987. Massive neutrinos and neutrino oscillations. *Rev. Mod. Phys.* 59:671–754.

Bionta, A. M., Blewitt, G., Bratton, C. B., Casper, D., Ciocio, A., Claus, R., Cortez, B., Crouch, M., Dye, S. T., Errede, S., Foster, G. W., Gajewski, W., Ganezer, K. S., Goldhaber, M., Haines, T. J., Jones, T. W., Kielczewska, D., Kropp, W. R., Learned, J. G., LoSecco, J. M., Matthews, J., Miller, R., Mudan, M. S., Park, H. S., Price, L. R., Reines, F., Schulz, J., Seidel, S., Shumard, E., Sinclair, D., Sobel, H. W., Stone, J. L., Sulak, L. R., Svoboda, R., Thorton, G., van der Velde, J. C., and Wuest, C. 1987. Observation of a neutrino burst in coincidence with supernova 1987A in the Large Magellanic Cloud. *Phys. Rev. Lett.* 58:1494–1496.

Birn, J., and Schindler, K. 1981. Two ribbon flares: Magnetostatic equilibria. In *Solar Flare Magnetohydrodynamics,* ed. E. R. Priest (New York: Gordon and Breach), pp. 337–377.

Biskamp, D., and Welter, H. 1988. Magnetic arcade evolution and instability. *Solar Phys.,* submitted.

Blackwell, D. E., Booth, A. J., and Petford, A. D. 1984. Is there an abundance anomaly for the 2.2 eV FeI lines in the solar spectrum? *Astron. Astrophys.* 132:236–239.

Blackwell, D. E., Booth, A. J., Haddock, D. J., Petford, A. D., and Leggett, S. K. 1986. Measurement of the oscillator strengths of very weak 1 eV FeI lines. *Mon. Not. Roy. Astron. Soc.* 220:549–553.

Blandford, R. R., and Ostriker, J. P. 1978. Particle acceleration by astrophysical shocks. *Astrophys. J.* 221:L29–L32.

Bocshler, P. 1987. Solar wind ion composition. *Phys. Scripta* T18:55–60.

Bochsler, P., and Geiss, J. 1975. Solar abundance of light nuclei and mixing of the Sun. *Solar Phys.* 32:3–91.

Bochsler, P., Geiss, J., and Joos, R. 1985. Kinetic temperatures of heavy ions in the solar wind. *J. Geophys. Res.* 90:10779–10789.

Bochsler, P., Geiss, J., and Kunz, S. 1986. Abundances of carbon, oxygen, and neon in the solar wind during the period from August 1978 to June 1982. *Solar Phys.* 103:177–201.

Bodenheimer, P. 1965. Studies in stellar evolution. II. Lithium depletion during the pre-main-sequence contraction. *Astrophys. J.* 142:451–461.

Bodenheimer, P. 1972. Stellar evolution toward the main sequence. *Rept. Prog. Phys.* 35:1–54.

Bodenheimer, P., and Sweigart, A. 1968. Dynamic collapse of the isothermal sphere. *Astrophys. J.* 152:515–522.

Bodenheimer, P., Forbes, J. E., Gould, N. L., and Henyey, L. G. 1965. Studies in stellar evolution. I. The influence of initial CNO abundances in a star of mass 2.3. *Astrophys. J.* 141:1019–1042.

Bodenheimer, P., Tohline, J. E., and Black, D. C. 1980. Criteria for fragmentation in a collapsing rotating cloud. *Astrophys. J.* 242:209–218.

Boercker, D. B. 1987. Collective effects on Thomson scattering in the solar interior. *Astrophys. J.* 316:L95–L98.

Boesgaard, A. M. 1987. Lithium in the coma star cluster. *Astrophys. J.* 321:967–974.

Boesgaard, A. M., and Tripico, M. J. 1986*a*. Lithium in the Hyades cluster. *Astrophys. J.* 302:L49–L53.

Boesgaard, A. M., and Tripico, M. J. 1986*b*. Lithium in early F dwarfs. *Astrophys. J.* 303: 724–739.

Boesgaard, A. M., Budge, K. G., and Burch, E. E. 1988*a*. Lithium and metallicity in Ursae Major group. *Astrophys. J.* 325:749–758.

Boesgaard, A. M., Budge, K. G., and Ramsey, M. E. 1988*b*. Lithium in the Pleiades and in alpha Persei cluster. *Astrophys. J.* 327:389–398.

Bogart, R. S. 1987. Large-scale motions on the Sun: An overview. *Solar Phys.* 110:23–24.

Bogart, R. S., and Bai, T. 1985. Confirmation of a 152 day periodicity in the occurrence of solar flares inferred from microwave data. *Astrophys. J.* 299:L51–L55.

Bogart, R. S., Gierasch, P. J., and MacAuslan, J. M. 1980. Linear modes of convection in the solar envelope. *Astrophys. J.* 236:285–293.

Bogart, R. S., Ferguson, S. H., Scherrer, P. H., Tarbell, T. D., and Title, A. M. 1988. On the feasibility of correlation tracking at moderate resolution. *Solar Phys.* 116:205–214.

Bogdan, T. J. 1987*a*. Propagation of compressive waves through fibril magnetic fields. II. Scattering from a slab of magnetic flux tubes. *Astrophys. J.* 318:888–895.

Bogdan, T. J. 1987*b*. Propagation of compressive waves through fibril magnetic fields. III. Waves that propagate along the magnetic field. *Astrophys. J.* 318:896–903.

Bogdan, T. J. 1990. On the resonance scattering of sound by slender magnetic flux tubes. *Astrophys. J.* submitted.

Bogdan, T. J., and Cattaneo, F. 1989. The normal modes of a resonant cavity containing discrete inhomogeneities: The influence of fibril magnetic fields on the solar acoustic oscillations. *Astrophys. J.* 342:545–557.

Bogdan, T. J., and Lerche, I. 1985. Dynamical evolution of large-scale two-dimensional fibril magnetic fields. *Astrophys. J.* 296:719–738.

Bogdan, T. J., and Zweibel, E. G. 1985. Effect of a fibril magnetic field on solar *p* modes. *Astrophys. J.* 298:867–875.

Bogdan, T. J., and Zweibel, E. G. 1987*a*. Propagation of compressive waves through fibril magnetic fields. *Astrophys. J.* 312:444–456.

Bogdan, T. J., and Zweibel, E. G. 1987*b*. Propagation of compressive waves through fibril magnetic fields. II. Scattering from a slab of magnetic flux tubes. *Astrophys. J.* 318:888–895.

Bogdan, T. J., Gilman, P. A., Lerche, I., and Howard, R. 1988. Distribution of sunspot umbral areas: 1917–1982. *Astrophys. J.* 327:451–456.

Bohlin, J. D. 1977*a*. An observational definition of coronal holes. In *Coronal Holes and High Speed Solar Wind Streams,* ed. J. B. Zirker (Boulder: Colorado Assoc. Univ. Press), pp. 27–69.

Bohlin, J. D. 1977*b*. Extreme-ultraviolet observations of coronal holes. *Solar Phys.* 51:377–398.

Bohm, K. H. 1963. The mixing of matter in the layer below the outer solar convection zone. *Astrophys. J.* 138:297–301.

Böhm-Vitesse, E. 1958. Über die Wasserstoffkonvektionszone in Sternen verschiedener Effektivtemperaturen und Leuchtkräfte. *Z. Astrophys.* 46:108–143.

Bohugas, J., Carrasco, L., Torres, C. A. O., and Quast, G. R. 1986. Rotational braking of late-type main sequence stars. *Astron. Astrophys.* 157:278–292.

Boischot, A. 1957. Caractères d'un type d'émission hertzienne associé à certaines éruptions chromosphériques. *Compt. Rend.* 244:1326.

Bondarenko, L. N., Kurguzov, V. V., Prokofev, Yu. A., Rogov, E. V., and Spivak, P. E. 1978. Measurement of the neutron half-life. *JETP Lett.* 28:303–307.

Bonet, J. A., Márquez, I., Vázquez, M., and Wöhl, H. 1988. Temporal and center-to-limb variations of the K I 769.9 nm line profiles in quiet and active solar regions. *Astron. Astrophys.* 198:322–330.

Bonnet, R. M. 1990. Future prospects of helioseismology from space. In *Inside the Sun: Proc. IAU Coll. 121,* eds. G. Berthomieu and M. Cribier (Dordrecht: Kluwer), pp. 289–304.

Bonnet, R. M., Bruner, E. C., Jr., Acton, L. W., Brown, W. A., and Decaudin, M. 1980. High-resolution Lyman-alpha filtergrams of the Sun. *Astrophys. J.* 237:L47–L50.

Bonnet, R. M., Bruner, M., Acton, L. W., Brown, W. A., Decaudin, M., and Foing, B. 1982.

Rocket photographs of fine structure and wave patterns in the solar temperature minimum. *Astron. Astrophys.* 111:125–129.

Bookbinder, J. A. 1985. Observations of Non-Thermal Radiation from Late-Type Stars. Ph.D. Thesis, Harvard Univ.

Bopp, P., Dubbers, D., Hornig, L., Klemt, E., Last, J., Schutze, H., Freedman, S. J., and Scharpf, O. 1986. Beta-decay asymmetry of the neutron and g_A/g_V. *Phys. Rev. Lett.* 56: 919–922.

Boreiko, R. T., and Clark, T. A. 1986. Far-IR solar emission lines from high-n states of hydrogen. *Astron. Astrophys.* 157:353–356.

Borgnino, J., and Brandt, P. N. 1982. Daytime and nighttime ρ_0 measurements at la Palma in June 1982. In *JOSO Annual Rept.*, ed. A. V. Alvensleben, pp. 9–17.

Borrini, G., Gosling, J. T., Bame, S. J., Feldman, W. C., and Wilcox, J. M. 1981. Solar wind helium and hydrogen structure near the heliospheric current sheet: A signal of coronal streamers at 1 AU. *J. Geophys. Res.* 86:4565–4573.

Borrini, G., Gosling, J. T., Bame, S. J., and Feldman, W. C. 1982a. An analysis of shock wave disturbances observed at 1 AU from 1971 through 1978. *J. Geophys. Res.* 87:4365–4373.

Borrini, G., Gosling, J. T., Bame, S. J., and Feldman, W. C. 1982b. Helium abundance enhancements in the solar wind. *J. Geophys. Res.* 87:7370–7378.

Bos, R. J. 1982. Observations of Individual Solar Eigenmodes: Their Properties and Implications. Ph.D. Thesis, Univ. of Arizona.

Bos, R. J., and Hill, H. A. 1983. Detection of individual normal modes of oscillation of the Sun in the period range from 2 hr to 10 min in solar diameter studies. *Solar Phys.* 82:89–102.

Boss, A. P. 1987. Protostellar formation in rotating interstellar clouds. VI. Nonuniform initial conditions. *Astrophys. J.* 319:149–161.

Boss, A. P. 1988. Binary stars: Formation by fragmentation. *Comments Astrophys.* 12:169–190.

Bouchez, J., Cribier, M., Hampel, W., Rich, J., Spiro, M., and Vignaud, D. 1986. Matter effects for solar neutrino oscillations. *Zeit. Phys.* 32C:499–511.

Boury, A., Gabriel, M., Noëls, A., Scuflaire, R., and Ledoux, P. 1975. Vibrational instability of a 1 $M_\odot$ star towards non-radial oscillations. *Astron. Astrophys.* 41:279–285.

Boussinesq, J. 1903. *Théorie Analytique de la Chaleur, Tome II* (Paris: Gauthier-Villars).

Bouvier, J. 1990. Stellar rotation prior to the main sequence. In *Rotation and Mixing in the Stellar Interior*, eds. M.-J. Goupil and J.-P. Zahn (Berlin: Springer-Verlag), pp. 47–68.

Bouvier, J., Bertout, C., Benz, W., and Mayor, M. 1986. Rotation in T Tauri stars. I. Observations and immediate analysis. *Astron. Astrophys.* 165:110–119.

Boyer, D. W., and Levy, E. H. 1984. Oscillating dynamo magnetic field in the presence of an external nondynamo field. The influence of a solar primordial field. *Astrophys. J.* 277: 848–861.

Brace, L. H., Hoegy, W. R., and Theis, R. F. 1988. Solar EUV measurements at Venus based on photoelectron emission from the Pioneer Venus Langmuir probe. *J. Geophys. Res.* 93:7282–7296.

Bracewell, R. N. 1965. *The Fourier Transform and Its Applications* (New York: McGraw-Hill).

Bracewell, R. N. 1988. Varves and solar physics. *Quart. J. Roy. Astron. Soc.* 29:119–128.

Brandenburg, A., Moss, D., Rudiger, G., and Tuominen, I. 1990. The nonlinear solar dynamo and differential rotation: A Taylor number puzzle? *Solar Phys.* 128:243–251.

Brandt, P. N., and Schröter, E. H. 1982. On the centre-to-limb variations and latitude dependence of the asymmetry and wavelength shift of the solar line $\lambda5576$. *Solar Phys.* 79:3–18.

Brandt, P. N., Mauter, H. A., and Smartt, R. N. 1987. Day-time seeing statistics at Sacramento Peak Observatory. *Astron. Astrophys.* 188:163–168.

Brandt, P. N., Scharmer, G. B., Ferguson, S., Shine, R. A., Tarbell, T. D., and Title, A. M. 1988. Vortex flow in the solar photosphere. *Nature* 335:238–240.

Brandt, P. N., Ferguson, S., Scharmer, G. B., Shine, R. A., Tarbell, T. D., and Title, A. M. 1989a. Variation of granulation properties on a meso-granular scale. In *High Spatial Resolution Solar Observations*, ed. O. von der Lühe (Sunspot, N.M.: Sacramento Peak Obs.), pp. 437–488.

Brandt, P. N., Scharmer, G. B., Ferguson, S. H., Shine, R. A., Tarbell, T. D., and Title, A. 1989b. Vortex motion of the solar granulation. In *Solar and Stellar Granulation*, eds. R. J. Rutten and G. Severino (Dordrecht: Kluwer), pp. 305–310.

Brants, J. J. 1985. High-resolution spectroscopy of active regions. 3. Relations between the intensity, velocity, and magnetic structure in an emerging flux region. *Solar Phys.* 98: 197–217.

Brault, J. W., and Noyes, R. W. 1983. Solar emission lines near 12 microns. *Astrophys. J.* 269:L61–L66.

Brault, J. W., and Testerman, L. 1972. *Preliminary Kitt Peak Photoelectric Atlas* (Tucson: Kitt Peak Natl. Obs.).

Brault, J. W., and White, O. R. 1971. The analysis and restoration of astronomical data via the fast Fourier transform. *Astron. Astrophys.* 13:169–189.

Braun, D. C., and Lindsey, C. 1987. A solar chromosphere and spicule model based on far-infrared observations. *Astrophys. J.* 320:898–903.

Braun, D. C., Duvall, T. J., Jr., and LaBonte, B. J. 1987. Acoustic absorption by sunspots. *Astrophys. J.* 319:L27–L31.

Braun, D. C., Duvall, T. L., Jr., and LaBonte, B. J. 1988. The absorption of high-degree p-mode oscillations in and around sunspots. *Astrophys. J.* 335:1015–1025.

Braun, D. C., Labonte, B. J., and Duvall, T. L., Jr. 1990. The spatial distribution of p-mode absorption in active regions. *Astrophys. J.* 335:1015–1025.

Bray, R. J., and Loughhead, R. D. 1964. *Sunspots* (London: Chapman and Hall).

Bray, R. J., and Loughhead, R. E. 1974. *The Solar Chromosphere* (London: Chapman and Hall).

Bray, R. J., Loughhead, R. E., and Durrant, C. J. 1984. *The Solar Granulation*, 2nd ed. (Cambridge: Cambridge Univ. Press).

Breneman, H. H., and Stone, E. C. 1985. Solar coronal and photospheric abundances from solar energetic particles. *Astrophys. J.* 299:L57–61.

Brickhouse, N. S., and LaBonte, B. J. 1988. Mass and energy flow near sunspots. I. Observations of moat properties. *Solar Phys.* 115:43–60.

Brodsky, M. A., and Vorontsov, S. V. 1987. An asymptotic technique for solving the inverse problem of helioseismology. *Soviet Astron. Lett.* 13:179–181.

Brodsky, M. A., and Vorontsov, S. V. 1988. On the technique of the inversion of helioseismological data. In *Advances in Helio- and Asteroseismology: Proc. IAU Symp. 123,* eds. J. Christensen-Dalsgaard and S. Frandsen (Dordrecht: D. Reidel), pp. 137–140.

Brookes, J. R., Isaak, G. R., and van der Raay, H. B. 1976. Observations of free oscillations of the Sun. *Nature* 259:92–95.

Brookes, J. R., Isaak, G. R., and van der Raay, H. G. 1978a. A resonant scattering solar spectrometer. *Mon. Not. Roy. Astron. Soc.* 185:1–17.

Brookes, J. R., Isaak, G. R., and van der Raay, H. B. 1978b. The observation of a rotating body using high-resolution spectroscopy. *Mon. Not. Roy. Astron. Soc.* 185:19–22.

Brookes, J. R., Isaak, G. R., McLeod, C. P., van der Raay, H. G., and Roca-Cortés, T. 1978c. The search for solar oscillations. *Mon. Not. Roy. Astron. Soc.* 184:759–767.

Brown, J. C. 1971. The deduction of energy spectra on non-thermal electrons in flares from the observed dynamic spectra of hard X-ray bursts. *Solar Phys.* 18:489–502.

Brown, J. C. 1973. Chromospheric flares heated by non-thermal electrons. *Solar Phys.* 31: 143–169.

Brown, J. C. 1974. On the thermal interpretation of hard X-ray bursts from solar flares. In *Coronal Disturbances: Proc. IAU Symp. 57,* ed. G. Newkirk (Dordrecht: D. Reidel), pp. 395–412.

Brown, J. C. 1975. The interpretation of spectra, polarization, and directivity of solar hard x-rays. In *Solar Gamma-, X- and EUV Radiation: Proc. IAU Symp. 68,* ed. S. R. Kane (Dordrecht: D. Reidel), pp. 245–282.

Brown, J. C., and Bingham, R. 1984. Electrodynamic effects in beam/return current systems and their implications for solar impulsive bursts. *Astron. Astrophys.* 131:L11–L14.

Brown, J. C., and Loran, J. 1985. Possible evidence for stochastic acceleration of electrons in solar hard X-ray bursts observed by SMM. *Mon. Not. Roy. Astron. Soc.* 212:245–255.

Brown, J. C., and MacKinnon, A. L. 1985. Bremsstrahlung spectra from thick-target electron beams with non-collisional energy losses. *Astrophys. J.* 292:L31–L34.

Brown, J. C., and Smith, D. F. 1980. Solar flares. *Rept. Prog. Phys.* 43:125–197.

Brown, J. C., Melrose, D. B., and Spicer, D. S. 1979. Production of a collisionless conduction front by rapid coronal heating and its role in solar hard X-ray bursts. *Astrophys. J.* 228: 592–597.

Brown, J. C., Loran, J. M., and MacKinnon, A. L. 1984. The shortest time scales present in solar hard X-ray bursts. *Astron. Astrophys.* 147:L10–L12.

Brown, J. C., Karlicky, M., MacKinnon, A. L., and van den Oord, G. H. J. 1990. Beam heating in solar flares: Electrons or protons. *Astrophys. J. Suppl.* 73:343–348.

Brown, J. M., Mihalas, B. W., and Rhodes, E. J., Jr. 1986. Solar waves and oscillations. In *Physics of the Sun,* vol. 1, eds. P. A. Sturrock, T. E. Holzer, D. M. Mihalas and R. K. Ulrich (Dordrecht: D. Reidel), pp. 177–247.

Brown, T. M. 1982. Seeing-independent definitions of the solar limb position. *Astron. Astrophys.* 116:260–264.

Brown, T. M. 1984*a*. The Fourier Tachometer II—an instrument for measuring global solar velocity fields. In *Solar Seismology from Space,* NASA JPL Publ. 84-84, pp. 157–163.

Brown, T. M. 1984*b*. Solar *p*-mode eigenfrequencies are decreased by turbulent convection. *Science* 226:687–689.

Brown, T. M. 1985. Solar rotation as a function of depth and latitude. *Nature* 317:591–594.

Brown, T. M. 1986. Measuring the Sun's internal rotation using solar *p*-mode oscillations. In *Seismology of the Sun and the Distant Stars,* ed. D. O. Gough (Dordrecht: D. Reidel), pp. 199–214.

Brown, T. M. 1988*a*. Automated *p*-mode identification using Bayes' theorem. In *Advances in Helio- and Asteroseismology: Proc. IAU Symp. 123,* eds. J. Christensen-Dalsgaard and S. Frandsen (Dordrecht: D. Reidel), pp. 491–494.

Brown, T. M. 1988*b*. Techniques for observing solar oscillations. In *Advances in Helio- and Asteroseismology: Proc. IAU Symp. 123,* eds. J. Christensen-Dalsgaard and S. Frandsen (Dordrecht: D. Reidel), pp. 453–465.

Brown, T. M. 1988*c*. Solar diameter measurements. *Bull. Amer. Astron. Soc.* 20:376 (abstract).

Brown, T. M. 1990*a*. An inverse method for *p*-mode scattering measurements. *Solar Phys.* 128:133–141.

Brown, T. M. 1990*b*. The source of solar high-frequency acoustic modes: Theoretical expectations. *Astrophys. J.* 371:396–401.

Brown, T. M., and Christensen-Dalsgaard, J. 1990. A technique for estimating complicated power spectra from time series with gaps. *Astrophys. J.* 349:667–674.

Brown, T. M., and Harrison, R. L. 1980. Evidence for trapped gravity waves in the solar atmosphere. *Astrophys. J.* 236:L169–L173.

Brown, T. M., and Morrow, C. A. 1987*a*. Depth and latitude dependence of solar rotation. *Astrophys. J.* 314:L21–L26.

Brown, T. M., and Morrow, C. A. 1987*b*. Observations of solar *p*-mode rotational splittings. In *The Internal Solar Angular Velocity,* eds. B. R. Durney and S. Sofia (Dordrecht: D. Reidel), pp. 7–17.

Brown, T. M., Mihalas, B. W., and Rhodes, E. J., Jr. 1986. Solar waves and oscillations. In *Physics of the Sun: The Solar Interior,* eds. P. A. Sturrock, T. E. Holzer, D. M. Mihalas and R. K. Ulrich (Dordrecht: D. Reidel), pp. 177–247.

Brown, T. M., Christensen-Dalsgaard, J., Dziembowski, W. A., Goode, P., Gough, D. O., and Morrow, C. A. 1989. Inferring the Sun's internal angular velocity from observed *p*-mode frequency splittings. *Astrophys. J.* 343:526–546.

Brown, T. M., Gilliland, R., Noyes, R. W., and Ramsey, L. 1990. Detection of possible *p*-mode oscillations on Procyon. *Astrophys. J.* 368:599–609.

Bruck, Y. M., and Sodin, L. G. 1980. A method for processing speckle images requiring no reference point source. *Astron. Astrophys.* 87:188–191.

Brueckner, G. E. 1975. Flare-like ultraviolet spectra of active regions. *Solar Gamma-, X-, and EUV Radiation: Proc. IAU Symp. 68,* ed. S. R. Kane (Dordrecht: D. Reidel), pp. 105–107.

Brueckner, G. E. 1980*a*. A high resolution view of the solar chromosphere and corona. In *Highlights in Astronomy,* vol. 5, ed. P. A. Wayman (Dordrecht: D. Reidel), pp. 557–569.

Brueckner, G. E. 1980*b*. The dynamics of active regions. In *Solar Active Regions,* ed. F. Q. Orrall (Boulder: Colorado Associated Univ. Press), pp. 113–127.

Brueckner, G. E. 1980*c*. Solar physics in the EUV: The importance of high resolution observations. *Appl. Opt.* 19:3994–4001.

Brueckner, G. E. 1981. High resolution ultraviolet solar observations from sounding rockets and Spacelab. *Space Sci. Rev.* 29:407–418.

Brueckner, G. E., and Bartoe, J.-D. F. 1983. Observations of high-energy jets in the corona above the quiet Sun, the heating of the solar corona, and the acceleration of the solar wind. *Astrophys. J.* 272:329–348.

Brueckner, G. E., and Cook, J. W. 1990. Return of the hard solar flares' 154 day periodicity in solar cycle 22 and evidence for a phase relationship of this periodicity between cycles 21 and 22. *Astrophys. J.* 350:L21–L24.

Brueckner, G. E., Bartoe, J.-D. F., and VanHoosier, M. E. 1978. High spatial resolution observations of the solar EUV spectrum. In *Proc. Nov. 7–10, 1977 OSO-8 Workshop,* (Boulder: Lab. of Atmospheric and Space Physics), pp. 380–418.

Brueckner, G. E., Bartoe, J.-D. F., Cook, J. W., Dere, K. P., and Socker, D. G. 1986. HRST results from Spacelab 2. *Adv. Space Res.* 6:263–272.

Brueckner, G. E., Bartoe, J.-D. F., Cook, J. W., Dere, K. P., Socker, D. G., Kurokawa, H., and McCabe, M. 1988. Plasma motions in an emerging flux region. *Astrophys. J.* 335: 986–995.

Bruzek, A. 1967. On arch filament systems in spotgroups. *Solar Phys.* 2:451–461.

Bumba, V. 1987. Does the large-scale solar magnetic field distribution really reflect the convective velocity field? *Solar Phys.* 110:51–57.

Bumba, V., and Gesztelyi, L. 1987. Formation of the white-light flare region of April 1984 (NOAA 4474) within the 21st cycle of solar activity. *Bull. Astron. Inst. Czech.* 39:86–92.

Bumba, V., and Hejna, L. 1987a. Observation of solar differential rotation with the aid of magnetic tracers. *Solar Phys.* 110:109–113.

Bumba, V., and Hejna, L. 1987b. Solar differential rotation reflected in the distribution of background magnetic fields. *Bull. Astron. Inst. Czech.* 38:29–36.

Bumba, V., and Howard, R. 1965. Large-scale distribution of solar magnetic fields. *Astrophys. J.* 141:1502–1512.

Bumba, V., and Howard, R. 1969. Solar activity and recurrences in magnetic-field distribution. *Solar Phys.* 7:28–38.

Bumba, V., and Suda, J. 1984. Processes observable in the photosphere during the formation of an active region. II. Development of a normal active region: Growth of a sunspot penumbra. *Bull. Astron. Inst. Czech.* 35:28–39.

Bumba, V., Howard, R., and Smith, S. 1967. Magnetic and related stars. In *Magnetic and Related Stars* (Baltimore: Mono), pp. 131–136.

Burbridge, E. M., Burbridge, G. R., Fowler, W. A., and Hoyle, F. 1957. Synthesis of the elements in stars. *Rev. Mod. Phys.* 29:547–651.

Burgers, J. M. 1960. Statistical plasma dynamics. In *Plasma Dynamics,* ed. F. H. Clauser (Reading, Mass.: Addison-Wesley), pp. 119–186.

Burgers, J. M. 1969. *Flow Equations for Composite Gases* (New York: Academic Press).

Burgi, A., and Geiss, J. 1986. Helium and minor ions in the corona and solar wind: Dynamics and charge states. *Solar Phys.* 103:347–383.

Burlaga, L. F. 1988. Interaction regions in the distant solar wind. In *Proc. Sixth Intl. Solar Wind Conf.,* eds. V. J. Pizzo, T. E. Holzer and D. G. Sime, NCAR TN-306 (Boulder: Natl. Center for Atmos. Res.), pp. 547–562.

Burlaga, L. F., Behannon, K. W., Hansen, S. F., Pneuman, G. W., and Feldman, W. C. 1978. Sources of magnetic fields in recurrent interplanetary streams. *J. Geophys. Res.* 83:4177–4185.

Burrows, A. 1984. On detecting stellar collapse with neutrinos. *Astrophys. J.* 283:848–852.

Busse, F. H. 1970. Differential rotation in stellar convection zones. *Astrophys. J.* 159:629–639.

Butler, R. P., Cohen, R. D., Duncan, D. K., and Marcy, G. W. 1987. The Pleiades rapid rotators: Evidence for an evolutionary sequence. *Astrophys. J.* 319:L19–L22.

Byrne, J., Morse, J., Smith, K. F., Shaikh, F., Green, K., and Greene, F. L. 1980. A new measurement of the neutron lifetime. *Phys. Lett.* B92:274–278.

Cacciani, A., and Rhodes, E. J., Jr. 1984. The magneto-optical filter, working principles and recent progress. In *Solar Seismology from Space,* NASA JPL Publ. 84–84, pp. 115–123.

Cacciani, A., Ricci, D., Rosati, P., Rhodes, E. J., Jr., Smith, E., Tomczyk, S., and Ulrich, R. K. 1988. Acquisition and reduction procedures for MOF Doppler-magnetograms. In *Seismology of the Sun and Sun-Like Stars,* ed. E. J. Rolfe, ESA SP-286, pp. 185–188.

Cacciani, A., Paverani, E., Ricci, D., Rosati, P., and Tomczyk, S. 1990. An experiment to measure the solar $\ell = 1$ rotational frequency splitting. In *Oji Intl. Seminar on Progress of*

Seismology of the Sun and Stars, eds. Y. Osaki and H. Shibahashi (Berlin: Springer-Verlag), in press.

Caccin, B., Fofi, M., Smaldone, L. A., and Torelli, M. 1990. Anisotropies in solar oscillations. *Astron. Astrophys.* 232:516–519.

Cameron, A. G. W. 1958. Modifications of the proton-proton chain. *Bull. Phys. Soc.* 3:227.

Campbell, W. R., and Roberts, B. 1989. The influence of a chromospheric magnetic field on the solar *p* and *f* modes. *Astrophys. J.* 338:538–556.

Camry-Peyret, C., Flaud, J.-M., Mandin, J.-Y., Chevillard, J.-P., Brault, J., Ramsay, D. A., Verloet, M., and Chauville, J. 1985. The high resolution spectrum of water vapor between 16500 and 25250 cm^{-1}. *J. Molec. Spect.* 113:208–228.

Cane, H. V., and Stone, R. G. 1984. Type II solar radio bursts, interplanetary shocks, and energetic particle events. *Astrophys. J.* 282:339–344.

Canfield, R. C., and Chang, E. S. 1985. Lyα and Hα emission by superthermal proton beams. *Astrophys. J.* 292:275–284.

Canfield, R. C., and Mehltretter, J. P. 1973. Fluctuations of brightness and vertical velocity at various heights in the photosphere. *Solar Phys.* 33:33–48.

Canfield, R. C., Gely-Dubau, E., Brown, J. C., Dulk, G. A., Emslie, A. G., Enome, S., Gabriel, A. H., Kundu, M. R., Melrose, D., Neidig, D. F., Ohki, K., Petrosian, V., Poland, A., Rieger, E., Tanaka, K., and Zirin, H. 1986. Type II Cepheids: A comparison of theory with observations. In *Energetic Phenomena on the Sun,* eds. M. R. Kundu and B. E. Woodgate, NASA CP-2439, pp. 3–1—3–46.

Cannon, C. J. 1973. Frequency-quadrature perturbations in radiative transfer theory. *Astrophys. J.* 185:621–630.

Carbon, D. F., Milkey, R. W., and Heasley, J. N. 1976. Departures from LTE in the fundamental bands of cool stars. *Astrophys. J.* 207:253–262.

Cargill, P. J., Goodrich, C. C., and Vlahos, L. 1988. Collisionless shock formation and the prompt acceleration of solar flare ions. *Astron. Astrophys.* 189:254–262.

Carlier, A., Chauveau, R., Hogon, M., and Rosch, J. 1968. High resolution spatial resolution cinematography of photospheric granulation. *Circ. Roy. Acad. Sci. Paris* 226:199–201.

Carlone, C., and Dalby, F. W. 1969. Spectrum of the hydroxyl radical. *Canadian J. Phys.* 47:1945–1957.

Carlson, E. D. 1986. Terrestrially enhanced neutrino oscillations. *Phys. Rev.* 34D:1454–1459.

Carroro, C., Schafer, A., and Koonin, S. E. 1988. Dynamic screening of thermonuclear reactions. *Astrophys. J.* 331:565–571.

Carson, T. R., Stothers, R., and Vemury, S. K. 1981. Type II Cepheids: A comparison with observations. *Astrophys. J.* 244:230–241.

Cash, J. R., and Moore, D. R. 1980. A high order method for the numerical solution of two-point boundary value problems. *BIT* 20:44–52.

Cassidy, G. L. 1973. The solar neutrino background and the photonuclear cross section. In *Proc. 13th Intl. Cosmic Ray Conf.,* vol. 3, pp. 1958–1961.

Cassinelli, J. P., and MacGregor, K. B. 1986. Stellar chromospheres, coronae, and winds. In *Physics of the Sun,* vol. 3, eds. P. A. Sturrock, T. E. Holzer, D. M. Mihalas and R. K. Ulrich (Dordrecht: D. Reidel), pp. 47–123.

Castenmiller, M. J. M., Zwaan, C., and van der Zalm, E. B. J. 1986. Sunspot nests. Manifestations of sequences in magnetic activity. *Solar Phys.* 105:237–255.

Castor, J. I. 1971. On the calculation of linear nonadiabatic pulsations of stellar models. *Astrophys. J.* 166:109–129.

Catalano, S., and Marilli, E. 1983. Ca II chromospheric emission and rotation of main sequence stars. *Astron. Astrophys.* 121:190–197.

Cattaneo, F. 1984. Oscillatory convection in sunspots. In *The Hydromagnetics of the Sun,* eds. T. D. Guyenne and J. J. Hunt, ESA SP-220, pp. 47–50.

Cattaneo, F., and Hughes, D. W. 1986. Magnetic fields in the overshoot zone: The great escape. In *Theoretical Problems in High Resolution Solar Physics II,* eds. G. Athay and D. Spicer, NASA CP-2483, pp. 101–104.

Cattaneo, F., and Hughes, D. W. 1988. The nonlinear breakup of a magnetic layer: Instability of interchange modes. *J. Fluid Mech.* 196:323–344.

Cattaneo, F., Hurlburt, N. E., and Toomre, J. 1990. Supersonic convection. *Astrophys. J.* 349:L6–L66.

Caudell, T. P. 1980. Long Period Solar Oscillations: A Seismological and Intercomparative Study. Ph.D. Thesis, Univ. of Arizona.

Caudell, T. P., Knapp, J., Hill, H. A., and Logan, J. D. 1980. Recent observations of solar oscillations at SCLERA. In *Nonradial and Nonlinear Stellar Pulsation,* eds. H. A. Hill and W. A. Dziembowski (Berlin: Springer-Verlag), pp. 206–218.

Cavallini, F., Ceppatelli, B., and Righini, A. 1985*a*. Asymmetry and shift of three Fe I photospheric lines in solar active regions. *Astron. Astrophys.* 143:161–121.

Cavallini, F., Ceppatelli, B., and Righini, A. 1985*b*. Meridional and equatorial center-to-limb variation of the asymmetry and shift of three Fe I solar photospheric lines around 6300 Å. *Astron. Astrophys.* 150:256–265.

Cavallini, F., Ceppatelli, G., and Righini, A. 1986. Solar limb effect and meridional flow: Results on the Fe I lines at 5569.6 Å and 5576.1 Å. *Astron. Astrophys.* 163:219–228.

Cavallini, F., Ceppatelli, G., Righini, A., and Alamanni, N. 1987*a*. Five-minute oscillations in the wings and bisectors of solar photospheric Fe I lines. *Astron. Astrophys.* 173:161–166.

Cavallini, F., Ceppatelli, G., and Righini, A. 1987*b*. Interpretation of shifts and symmetries of Fe I lines in solar facular areas. *Astron. Astrophys.* 173:155–160.

Cayrel, R., Cayrel de Strobel, G., Campbell, B., and Däppen, W. 1984. The lithium abundance of Hyades main-sequence stars. *Astrophys. J.* 283:205–208.

Champagne, A. E., Dodge, G. E., Kovzcs, R. T., Lowry, M. M., McDonald, A. B., and Roberson, M. W. 1988*a*. Gamma decays of isobaric analog states relevant to neutrino detection. *Phys. Rev.* C38:900–904.

Champagne, A. E., Kovzcs, R. T., Lowry, M. M., and McDonald, A. B. 1988*b*. Decay of the first isobaric analog state in ^{59}Ge, *Phys. Rev* C38:2430–2431.

Chahine, M. T. 1977. Generalization of the relaxation method for the inverse solution of nonlinear and linear transfer equations. In *Inversion Methods in Atmospheric Remote Sounding,* ed. A. Deepak (New York: Academic Press), pp. 67–111.

Chan, K. L. 1983. Application of the alternating direction implicit method to the computation of time-dependent compressible convection. In *Numerical Methods in Laminar and Turbulent Flow,* eds. C. Taylor, J. A. Johnson and W. R. Smith (Swansea, U.K.: Pineridge Press), pp. 898–907.

Chan, K. L., and Mayr, H. G. 1989. Differential rotation in and around the solar convection zone. *Eos: Trans. AGU* 70:417.

Chan, K. L., and Sofia, S. 1986. Turbulent compressible convection in a deep atmosphere. III. Tests on the validity and limitation of the numerical approach. *Astrophys. J.* 307:222–241.

Chan, K. L., and Sofia, S. 1987. Validity tests of the mixing-length theory of deep convection. *Science* 235:465–467.

Chan, K. L., and Sofia, S. 1988. The mixing-length ratio, eddy diffusivity, and acoustic waves. In *Atmospheric Diagnostics of Stellar Evolution: Chemical Peculiarity, Mass Loss, and Explosion,* ed. K. Nomoto (New York: Springer-Verlag), pp. 191–192.

Chan, K. L., and Sofia, S. 1989. Turbulent compressible convection in a deep atmosphere. IV. Results of three-dimensional computations. *Astrophys. J.* 336:1022–1040.

Chan, K. L., and Wolff, C. L. 1982. ADI on staggered mesh—a method for the calculation of compressible convection. *J. Comput. Phys.* 47:109–129.

Chan, K. L., Wolff, C. L., and Sofia, S. 1981. A variable mixing-length ratio for convection theory. *Astrophys. J.* 244:582–586.

Chan, K. L., Sofia, S., and Wolff, C. L. 1982. Turbulent compressible convection in a deep atmosphere. I. Preliminary two-dimensional results. *Astrophys. J.* 263:935–943.

Chan, K. L., Sofia, S., and Mayr, H. G. 1987. Mixing-length, shears, and differential rotation. In *The Internal Solar Angular Velocity,* eds. B. R. Durney and S. Sofia (Dordrecht: D. Reidel), pp. 347–360.

Chandrasekhar, S. 1934*a*. The solar chromosphere. *Mon. Not. Roy. Astron. Soc.* 94:14–35.

Chandrasekhar, S. 1934*b*. The solar chromosphere. II. *Mon. Not. Roy. Astron. Soc.* 94:726–737.

Chandrasekhar, S. 1939. *An Introduction to the Study of Stellar Structure* (New York: Dover).

Chandrasekhar, S. 1961. *Hydrodynamic and Hydromagnetic Stability* (Oxford: Clarendon Press).

Chandrasekhar, S. 1964. A general variational principle governing the radial and the non-radial oscillations of gaseous masses. *Astrophys. J.* 139:664–674.

Chandrasekhar, S., and Breen, F. H. 1946. On the continuous absorption coefficient of the negative hydrogen ion. III. *Astrophys. J.* 104:430–445.

Chang, E. S. 1984. Non-penetrating Rydberg states of silicon from solar data. *J. Phys.* B17: L11–L17.

Chang, E. S. 1987. Solar emission lines revisited: Extended study of magnesium. *Phys. Scripta* 35:792–797.

Chang, E. S., and Noyes, R. W. 1983. Identification of the solar emission lines near 12 microns. *Astrophys. J.* 275:L11–L13.

Chapman, G. A. 1977. Facular line profiles and facular models. *Astrophys. J. Suppl.* 33:35–54.

Chapman, G. A. 1987*a*. Solar variability due to sunspots and faculae. *J. Geophys. Res.* 92: 809–812.

Chapman, G. A. 1987*b*. Variations of solar irradiance due to magnetic activity. *Ann. Rev. Astron. Astrophys.* 25:633–667.

Chapman, G. A., and Klabunde, D. P. 1982. Measurements of the limb darkening of faculae near the solar limb. *Astrophys. J.* 261:387–395.

Chapman, G. A., and Sheeley, N. R., Jr. 1968. The photospheric network. *Solar Phys.* 5: 442–461.

Chapman, G. A., Herzog, A. D., Lawrence, J. K., and Shelton, J. C. 1984. Solar luminosity fluctuations and active region photometry. *Astrophys. J.* 282:L99–L101.

Chapman, S. 1917. Thermal diffusion and the stars. *Mon. Not. Roy. Astron. Soc.* 77:539–540.

Chapman, S. 1958. Thermal diffusion in ionized gases. *Proc. Phys. Soc.* 72:353–362.

Chapman, S., and Cowling, T. G. 1970. *The Mathematical Theory of Non-Uniform Gases,* 3rd ed. (Cambridge: Cambridge Univ. Press).

Charbonneau, P., and Michaud, G. 1990. Turbulence and the Li abundance in main sequence stars. *Astrophys. J.* 352:681–688.

Charbonneau, P., Michaud, G., and Proffitt, C. R. 1989. Lithium in cluster giants: Constraints on meridional circulation transport on the main sequence. *Astrophys. J.* 347:821–834.

Chase, R. C., Krieger, A. S., Svestka, Z., and Vaiana, G. S. 1976. Skylab observations of X-ray loops connecting separate active regions. *Space Res.* XVI:917–922.

Chen, C. H., Kramer, S. D., Allman, S. L., and Hurst, G. S. 1984. Selective counting of krypton atoms using resonance ionization spectroscopy. *Appl. Phys. Lett.* 44:640–642.

Chen, H. H. 1985. Direct approach to resolve the solar neutrino problem. *Phys. Rev. Lett.* 55:1534–1536.

Chen, H. H. 1988. The Sudbury Neutrino Observatory. *Nucl. Inst. Meth. Phys. Res.* A264: 48–54.

Cheng, C.-C. 1977. Evolution of the high-temperature plasma in the 15 June 1973 flare. *Solar Phys.* 55:413–429.

Cheng, C.-C., Tanderg-Hanssen, E., Bruner, E. C., Orwig, L. E., Frost, K. J., Woodgate, B. E., and Shine, R. A. 1981. Spatial and temporal structures of impulsive bursts from solar flares observed in UV and hard X-rays. *Astrophys. J.* 248:L39–L43.

Cherry, M. L., and Lande, K. 1988. Prospects for a day-night ^{37}Cl solar-neutrino oscillation experiment. *Phys. Rev.* D36:3571–3574.

Cherry, M. L., Fowler, W. A., and Lande. K., eds. 1985. *Solar Neutrinos and Neutrino Astronomy* (New York: American Inst. of Physics).

Childress, S. 1979. Alpha-effect in flux ropes and sheets. *Phys. Earth Planet. Int.* 20:172–180.

Chiosi, C. 1986. Advancements in the stellar evolution theory: The role of convective overshooting all across the HR diagram. In *Spectral Evolution of Galaxies,* eds. C. Chiosi and A. Renzini (Dordrecht: D. Reidel), pp. 237–262.

Chiu, Y. T., and Hilton, H. H. 1977. Exact Green's function method of solar force-free magnetic-field computations with constant α. I. Theory and basic test cases. *Astrophys. J.* 212:873–885.

Chiueh, T. 1988. Multiple-encounter shock-drift acceleration in nearly perpendicular shocks. *Astrophys. J.* 333:366–385.

Chou, D.-Y. 1987. The cooling time scales of growing sunspots. *Astrophys. J.* 312:955–962.

Chou, D.-Y., and Fisher, G. H. 1989. Dynamics of anchored flux tubes in the convection zone. I. Details of the model. *Astrophys. J.* 341:533–548.

Chou, D.-Y., and Wang, H. 1987. The separation velocity of emerging flux. *Solar Phys.* 110: 81–99.

Choudhuri, A. R., and Gilman, P. A. 1987. The influence of the coriolis force on flux tubes rising through the solar convection zone. *Astrophys. J.* 316:788–800.

Christensen, C. J., Nielsen, A., Bahnsen, A., Brown, W. K., and Rustad, B. M. 1972. Free-neutron beta-decay half-life. *Phys. Rev.* D5:1628–1640.

Christensen-Dalsgaard, B. L. 1982. On the use of the finite-element method with Lagrange multipliers in scattering theory. *J. Phys.* A15: 2711–2722.

Christensen-Dalsgaard, J. 1976. On isolated stars in non-radial oscillation. *Mon. Not. Roy. Astron. Soc.* 174:87–89.

Christensen-Dalsgaard, J. 1980. On adiabatic non-radial oscillations with moderate or large ℓ. *Mon. Not. Roy. Astron. Soc.* 190:765–791.

Christensen-Dalsgaard, J. 1981. The effect of non-adiabaticity on avoided crossings of non-radial stellar oscillations. *Mon. Not. Roy. Astron. Soc.* 194:229–250.

Christensen-Dalsgaard, J. 1982. On solar models and their periods of oscillation. *Mon. Not. Roy. Astron. Soc.* 199:735–761.

Christensen-Dalsgaard, J. 1984. What will asteroseismology teach us? In *Space Research Prospects in Stellar Activity and Variability,* eds. A. Mangeney and F. Praderie, pp. 11–46.

Christensen-Dalsgaard, J. 1984a. Solar oscillations. In *Theoretical Problems in Stellar Stability and Oscillations* (Liège: Inst. d'Astrophysique), pp. 155–207.

Christensen-Dalsgaard, J. 1984b. Solar oscillations. In *The Hydromagnetics of the Sun,* eds. T. D. Guyenne and J. J. Hunt, ESA SP-220, pp. 3–12.

Christensen-Dalsgaard, J. 1986. Theoretical aspects of helio- and asteroseismology. In *Seismology of the Sun and the Distant Stars,* ed. D. O. Gough (Dordrecht: D. Reidel), pp. 23–53.

Christensen-Dalsgaard, J. 1987. In *Méthodes mathématiques pour l'astrophysique,* eds. M. Auvergne and A. Baglin (Paris: Société Français des Spécialistes d'Astronomie), pp. 479–524.

Christensen-Dalsgaard, J. 1988a. Excitation of solar oscillations: Theoretical possibilities and observational consequences. In *Multimode Stellar Pulsations,* eds. G. Kovacs, L. Szabados and B. Szeidl (Budapest: Kultura Press and Konkoly Obs.), pp. 153–180.

Christensen-Dalsgaard, J. 1988b. A Hertzsprung-Russell diagram for stellar oscillations. In *Advances in Helio- and Asteroseismology: Proc. IAU Symp. 123,* eds. J. Christensen-Dalsgaard and S. Frandsen (Dordrecht: D. Reidel), pp. 295–298.

Christensen-Dalsgaard, J. 1988c. Study of solar structure based on *p*-mode helioseismology. In *Seismology of the Sun and Sun-Like Stars,* ed. E. J. Rolfe, ESA SP-286, pp. 431–450.

Christensen-Dalsgaard, J. 1989. The effect of rotation on whole-disk Doppler observations of solar oscillations. *Mon. Not. Roy. Astron. Soc.* 239:977–994.

Christensen-Dalsgaard, J. 1990a. Helioseismic investigation of solar internal structure. In *Inside the Sun,* eds. G. Berthomieu and M. Cribier (Dordrecht: Kluwer), pp. 305–326.

Christensen-Dalsgaard, J. 1990b. Helioseismic measurements of the solar internal rotation. In *Accretion and Winds,* ed. G. Klare (Berlin: Springer-Verlag), pp. 313–349.

Christensen-Dalsgaard, J. 1990c. Interior structure of the Sun. In *Basic Plasma Processes on the Sun: Proc. IAU Symp. 142,* eds. E. R. Priest and V. Krishnan (Dordrecht: Kluwer), pp. 23–32.

Christensen-Dalsgaard, J., and Frandsen, S. 1983. Radiative transfer and solar oscillations. *Solar Phys.* 82:165–204.

Christensen-Dalsgaard, J., and Frandsen, S. 1984. Non-grey radiative transfer in solar oscillations. *Mem. Soc. Astron. Italiana* 55:285–291.

Christensen-Dalsgaard, J., and Gough, D. O. 1975. Non-adiabatic non-radial oscillation of a solar model. *Mem. Soc. Roy. Liège* 8:309–316.

Christensen-Dalsgaard, J., and Gough, D. O. 1976. Towards a heliological inverse problem. *Nature* 259:89–92.

Christensen-Dalsgaard, J., and Gough, D. O. 1980. Is the Sun helium-deficient? *Nature* 288:544–547.

Christensen-Dalsgaard, J., and Gough, D. O. 1981. Comparison of observed solar whole-disk oscillation frequencies with the predictions of a sequence of solar models. *Astron. Astrophys.* 104:173–176.

Christensen-Dalsgaard, J., and Gough, D. O. 1982. On the interpretation of five-minute oscillations in solar spectrum line shifts. *Mon. Not. Roy. Astron. Soc.* 198:141–171.

Christensen-Dalsgaard, J., and Gough, D. O. 1984. Implications of observed frequencies of solar *p* modes. In *Solar Seismology from Space,* NASA JPL Publ. 84–84, pp. 199–204.

Christensen-Dalsgaard, J., and Gough, D. O. 1989. Rotational enhancement of Doppler measurements of solar and stellar hexadecapole oscillations. *Solar Phys.* 119:5–18.

Christensen-Dalsgaard, J., and Perez-Hernandez, F. 1988. On the relation of the Duvall phase function to the upper layers of the convection zone. In *Seismology of the Sun and Sun-Like Stars,* ed. E. J. Rolfe, ESA SP-286, pp. 499–503.

Christensen-Dalsgaard, J., and Schou, J. 1988. Differential rotation in the solar interior. In *Seismology of the Sun and Sun-Like Stars,* ed. E. J. Rolfe, ESA SP-286, pp.149–153.

Christensen-Dalsgaard, J., Dilke F. W., and Gough, D. O. 1974. The stability of a solar model to non-radial oscillations. *Mon. Not. Roy. Astron. Soc.* 169:429–445.

Christensen-Dalsgaard, J., Gough, D. O., and Morgan, J. G. 1979. Dirty solar models. *Astron. Astrophys.* 73:121–128 and 79:260 (erratum).

Christensen-Dalsgaard, J., Dziembowski, W., and Gough, D. O. 1980. How deep is the solar convection zone? In *Nonradial and Nonlinear Stellar Pulsation,* eds. H. A. Hill and W. A. Dziembowski (Berlin: Springer-Verlag), pp. 313–341.

Christensen-Dalsgaard, J., Duvall, T. L., Gough, D. O., Harvey, J. W., and Rhodes, E. J. 1984. Speed of sound in the solar interior. *Nature* 315:378–382.

Christensen-Dalsgaard, J., Gough, D. O., and Toomre, J. 1985a. Seismology of the Sun. *Science* 229:923–931.

Christensen-Dalsgaard, J., Duvall, T. L., Jr., Gough, D. O., Harvey, J. W., and Rhodes, E. J., Jr. 1985b. Speed of sound in the solar interior. *Nature* 315:378–382.

Christensen-Dalsgaard, J., Gough, D. O., and Thompson, M. J. 1988a. Determination of the solar internal sound speed by means of a differential asymptotic inversion. In *Seismology of the Sun and Sun-Like Stars,* ed. E. J. Rolfe, ESA SP-286, pp. 493–497.

Christensen-Dalsgaard, J., Däppen, W., and Lebreton, Y. 1988b. Solar oscillation frequencies and the equation of state. *Nature* 336:634–638.

Christensen-Dalsgaard, J., Gough, D. O., and Pérez Hernàndez, F. 1988c. Stellar disharmony. *Mon. Not. Roy. Astron. Soc.* 235:875–880.

Christensen-Dalsgaard, J., Gough, D. O., and Thompson, M. J. 1988a. Differential asymptotic soundspeed inversions. *Mon. Not. Roy. Astron. Soc.* 238:481–502.

Christensen-Dalsgaard, J., Gough, D. O., and Libbrecht, K. G. 1989b. Seismology of solar oscillation line widths. *Astrophys. J.* 341:L103–L106.

Christensen-Dalsgaard, J., Schou, J., and Thompson, M. J. 1990. A comparison of methods for inverting helioseismic data. *Mon. Not. Roy. Astron. Soc.* 242:353–369.

Christensen-Dalsgaard, J., Gough, D. O., and Thompson, M. J. 1991. The depth of the solar convection zone. *Astrophys. J.,* submitted.

Christou, J. C., Cheng, A. Y. S., Hege, E. K., and Roddier, C. 1985. Seeing calibration of optical astronomical speckle interferometric data. *Astron. J.* 90:264–265.

Chugainov, P. F. 1980. *Isv. Krymsk. Ap. Obs.* 61:124.

Chupp, E. L. 1984. High-energy neutral radiations from the Sun. *Ann. Rev. Astron. Astrophys.* 22:359–387.

Chupp, E. L., Forrest, D. J., Ryan, J. M., Heslin, J., Reppin, C., Pinkau, K., Kanbach, G., Rieger, E., and Share, G. H. 1982. A direct observation of solar neutrons following the 0118 UT flare on 1980 June 21. *Astrophys. J.* 263:L95–L99.

Chupp, E. L., Debrunner, H., Flückiger, E., Forrest, D. J., Golliez, F., Kanbach, G., Vestrand, W. T., Cooper, J., and Share, G. 1987. Solar neutron emissivity during the large flare in 1982 June 3. *Astrophys. J.* 318:913–925.

Cisneros, A. 1971. Effect of a neutrino magnetic moment on solar neutrino observations. *Astrophys. Space Sci.* 10:87–92.

Clark, A., and Johnson, A. C. 1967. Magnetic-field accumulation in supergranules. *Solar Phys.* 2:433–440.

Clark, T. A., and Boreiko, R. T. 1982. Airborne total eclipse observation of the extreme solar limb at 400 μm. *Solar Phys.* 76:117–128.

Clarke, D. 1978. Brightness oscillations of daytime sky. *Nature* 274:670–671.

Claverie, A., Isaak, G. R., McLeod, C. P., van der Raay, H. B., and Roca-Cortés, T. 1979. Solar structure from global studies of the five-minute oscillation. *Nature* 282:591–594.

Claverie, A., Isaak, G. R., McLeod, C. P., and van der Raay, H. B. 1981a. Rapid rotation of solar interior. *Nature* 293:443–445.

Claverie, A., Isaak, B. R., McLeod, C. P., van der Raay, H. B., and Roca-Cortés, T. 1981b. Rotational splitting of solar five-minute oscillations of low degree. *Solar Phys.* 82:233–234.

Claverie, A., Isaak, G. R., and McLeod, C. P. 1981c. Short-period intensity fluctuations of integral sunlight. *Solar Phys.* 74:73–78.

Claverie, A., Isaak, G. R., McLeod, C. P., van der Raay, H. B. and Roca-Cortés, T. 1981d. Structure in the 5 minute oscillations of integral sunlight. *Astron. Astrophys.* 91:L9–L10.

Claverie, A., Isaak, G. R., McLeod, C. P., van der Raay, H. B., Pallé, P. L., and Roca-Cortés, T. 1982. Solar core rotation. *Nature* 299:704–707.

Claverie, A., Isaak, G. R., McLeod, C. P., van der Raay, H. B., and Roca-Cortés, T. 1983. Rapid rotation of the solar interior. *Nature* 293:443–445.

Claverie, A., Isaak, G. R., McLeod, C. P., van der Raay, H. B., Pallé, P. L., and Roca-Cortés, T. 1984. Continuous observation of solar oscillations from two suitably spaced ground stations. *Mem. Soc. Astron. Italiana* 55:63–66.

Clayton, D. D. 1968. *Principles of Stellar Evolution and Nuclear Synthesis* (New York: McGraw-Hill).

Cleveland, B. T. 1983. The analysis of radioactive decay with a small number of counts by the method of maximum likelihood. *Nucl. Instr. Meth.* 214:451–458.

Cliver, E. W., Dennis, B. R., Kiplinger, A. L., Kane, S. R., Neidig, D. F., Sheeley, N. R., and Koomen, M. J. 1986. Solar gradual hard X-ray bursts and associated phenomena. *Astrophys. J.* 305:920–935.

Cloutman, L. D. 1979. A physical model of the solar granulation. *Astrophys. J.* 227:614–628.

Cloutman, L. D. 1987. A new estimate of the mixing length and convective overshooting in massive stars. *Astrophys. J.* 313:699–710.

Cloutman, L. D., and Whitaker, R. 1980. On convective and semiconvective mixing in massive stars. *Astrophys. J.* 237:900–902.

Coffey, H. E., and Gilman, P. A. 1969. Sunspot motion statistics for 1965–67. *Solar Phys.* 9:423–426.

Cogan, B. C. 1975. Convective overshooting in main-sequence models. *Astrophys. J.* 201:637–640.

Cohen, E. R., and Taylor, B. N. 1986. *The 1986 Adjustments of the Fundamental Physical Constants,* Codata Bulletin No. 63 (New York: Pergamon).

Cohen, L. 1981. *An Atlas of Solar Spectra Between 1175 and 1950 Angstroms Recorded on Skylab with the NRL's Apollo Telescope Mount Experiment,* NASA RP-1069.

Cohen, M., and Kuhi, L. V. 1979. Observational studies in pre-main-sequence evolution. *Astrophys. J.* 41:743–843.

Cole, P. W., and Deupree, R. G. 1980. The core helium flash with two-dimensional convection. *Astrophys. J.* 239:284–291.

Cole, P. W., and Deupree, R. G. 1981. The violent phase of the core helium flash. *Astrophys. J.* 247:607–613.

Colella, P., and Woodward, P. R. 1984. The piecewise parabolic method (PPM) for gasdynamical simulations. *J. Comput. Phys.* 54:174–201.

Coles, W. A., and Rickett, B. J. 1986. Interplanetary scintillation observations of the solar wind at high latitudes. In *The Sun and the Heliosphere in Three Dimensions,* ed. R. G. Marsden (Dordrecht: D. Reidel), pp. 143–151.

Coles, W. A., Rickett, B. J., Rumsey, V. H., Kaufman, J. J., Turley, D. J., Ananthakrishnan, S., Armstrong, J. W., Harmons, J. K., Scott, S. L., and Sime, D. G. 1980. Solar cycle changes in the polar solar wind. *Nature* 286:239–241.

Connes, J., and Connes, P. 1984. A numerical solution of the data gaps problem. In *Space Research Prospects in Stellar Activity and Variability,* eds. A. Mangeney and F. Praderie (Meudon: Obs. de Paris), pp. 135–149.

Cook, J. W. 1985. HRTS results from Spacelab 2. In *Theoretical Problems in High Resolution Solar Physics,* ed. H. U. Schmidt (München: MPI für Physics and Astrophysics), pp. 308–310.

Cook, J. W., and Ewing, J. A. 1990. Relationship of magnetic field strength and brightness of fine structure elements in the solar temperature minimum region. *Astrophys. J.* 355:719–725.

Cook, J. W., Brueckner, G. E., and VanHoosier, M. E. 1980. Variability of the solar flux in the far ultraviolet 1175–2100 Å. *J. Geophys. Res.* 85:2257–2268.

Cook, J. W., Brueckner, G. E., and Bartoe, J.-D. F. 1983. High resolution telescope and spectrograph observations of solar fine structure in the 1600 Å region. *Astrophys. J.* 270:L89–L93.

Cook, J. W., Brueckner, G. E., Bartoe, J.-D. F., and Socker, D. G. 1984*a*. HRST observations of spicular emission at transition region temperature above the solar limb. *Adv. Space Sci.* 4:59–62.

Cook, J. W., Brueckner, G. E., Bartoe, J.-D. F., and Socker, D. G. 1984*b*. HRST evidence for rotation of transition region temperature spicules. In *Proc. Eighth Intl. Coll. Ultraviolet and X-Ray Spectroscopy of Astrophysical and Laboratory Plasmas: Proc. IAU Coll. 86*, ed. G. A. Doschek (Washington, D.C.: Naval Research Lab), pp. 32–35.

Cook, J. W., Lund, P. A., Bartoe, J.-D. F., Brueckner, G. E., and Socker, D. G. 1987. Statistical properties of small high-velocity events in the solar transition region. In *Fifth Cambridge Workshop on Cool Stars, Stellar Systems and the Sun,* eds. J. L. Linsky and R. E. Stencel (New York: Springer-Verlag), pp. 150–153.

Cook, J. W., Cheng, C.-C., Jacobs, V. L., and Antiochos, S. K. 1989. Effect of coronal elemental abundances on the radiative loss function. *Astrophys. J.* 338:1176–1183.

Cook, W. R., Stone, E. C., Vogt, R. E., Trainor, J. H., and Webber, W. R. 1979. Elemental composition of solar energetic particles in 1977 and 1978. In *Proc. 16th Intl. Cosmic Ray Conf.,* vol. 12 (Kyoto: Univ. of Tokyo), pp. 265–270.

Cook, W. R., Stone, E. C., and Vogt, R. E. 1984. Elemental composition of solar energetic particles. *Astrophys. J.* 279:827–838.

Coplan, M. A., Ogilvie, K. W., Bochsler, P., and Geiss, J. 1984. Interpretation of ^{3}He abundance variations in the solar wind. *Solar Phys.* 93:415–434.

Cornell, M. E., Hurford, G. J., Kiplinger, A. L., and Dennis, B. R. 1984. The relative timing of microwaves and hard X-rays in solar flares. *Astrophys. J.* 279:875–881.

Cornwell, T. J. 1987. Radio-interferometric imaging of weak objects in conditions of poor phase stability: The relationship between speckle masking and phase closure methods. *Astron. Astrophys.* 180:269–274.

Costa, J. E. R., Kaufmann, P., and Takakura, T. 1984. Timing analysis of hard X-ray emission and 22 GHz flux and polarization in a solar burst. *Solar Phys.* 94:369–378.

Cowan, G. A., and Haxton, W. C. 1982. Solar neutrino production of technicium-97 and technicium-98 *Science* 216:51–54.

Cowan, R. D. 1981. *The Theory of Atomic Structure and Spectra* (Berkeley: Univ. of California Press).

Cowan, J. J., Cameron, A. G. W., and Truran, J. W. 1982. The thermal runaway r-process. *Astrophys. J.* 252:348–355.

Cowling, T. G. 1934. The magnetic field of sunspots. *Mon. Not. Roy. Astron. Soc.* 94:39–48.

Cowling, T. G. 1935. The stability of gaseous stars. *Mon. Not. Roy. Astron. Soc.* 96:42–60.

Cowling, T. G. 1941. The non-radial oscillations of polytropic stars. *Mon. Not. Roy. Astron. Soc.* 101:367–375.

Cowling, T. G. 1957. *Magnetohydrodynamics* (New York: Interscience Tract).

Cowling, T. G. 1976. *Magnetohydrodynamics* (Bristol: A. Hilger).

Cowling, T. G., and Newing, R. A. 1949. The oscillations of a rotating star. *Astrophys. J.* 109:149–158.

Cox, A. N. 1965. Stellar absorbtion coefficients and opacities. In *Stars and Stellar Systems,* vol. 8, eds. I. H. Aller and D. R. McLaughlin (Chicago: Univ. of Chicago Press), pp. 195–267.

Cox, A. N. 1990*a*. Periods and stability of solar *g*-modes. *Solar Phys.* 128:123–131.

Cox, A. N. 1990*b*. Solar opacities constrained by solar neutrinos and solar oscillations. In *Inside the Sun: Proc. IAU Coll. 121,* eds. G. Berthomieu and M. Cribier (Dordrecht: Kluwer), pp. 61–80.

Cox, A. N., and Giuli, R. T. 1968. *Principles of Stellar Structure* (New York: Gordon and Breach).

Cox, A. N., and Kidman, R. B. 1984. Effects of core mixing on solar oscillation frequencies. In *Theoretical Problems in Stellar Stability and Oscillations,* eds. A. Noëls and M. Gabriel (Liège: Inst. d'Astrophysique), pp. 259–262.

Cox, A. N., and Stewart, J. N. 1969. *Sci. Inform. Astron. Council USSR* 15:1.

Cox, A. N., and Stewart, J. N. 1969. Radiative and conductive opacities for twenty-three stellar mixtures. *Nauch. Info.* 15:3–103.

Cox, A. N., and Stewart, J. N. 1970*a*. Rosseland opacity tables for population I compositions. *Astrophys. J. Suppl.* 19:243–259.

Cox, A. N., and Stewart, J. N. 1970*b*. Rosseland opacity tables for population II compositions. *Astrophys. J. Suppl.* 19:261–279.

Cox, A. N., and Tabor, J. E. 1976. Radiative opacity tables for 40 stellar mixtures. *Astrophys. J. Suppl.* 31:271–312.

Cox, A. N., Starrfield, S. G., Kidman, R. B., and Pesnell, W. D. 1987. Pulsations of white, dwarf stars with thick hydrogen or helium surface layers. *Astrophys. J.* 317:303–324.

Cox, A. N., Guzik, J. A., and Kidman, R. G. 1989. Oscillations of solar models with internal element diffusion. *Astrophys. J.* 342:1187–1206.

Cox, A. N., Guzik, J. A., and Raby, S. 1990. Oscillations of condensed-out iron and cosmion solar models. *Astrophys. J.* 353:698–711.

Cox, D. P., and Tucker, W. H. 1969. Ionization equilibrium and radiative cooling of a low density plasma. *Astrophys. J.* 157:1157–1164.

Cox, J. P. 1980. *Theory of Stellar Pulsation* (Princeton, N.J.: Princeton Univ. Press).

Cox, J. P., and Giuli, R. T. 1968. *Principles of Stellar Structure* (New York: Gordon and Breach).

Craig, I. J. D., and Brown, J. C. 1986. *Inverse Problems in Astronomy: A Guide to Inversion Strategies for Remotely Sensed Data* (Bristol: A. Hilger).

Craig, I. J. D., and Sneyd, A. D. 1986. A dynamic relaxation technique for determining the structure and stability of coronal magnetic fields. *Astrophys. J.* 311:451–459.

Cram, L. E. 1972. Multi-component models for the formation of the chromospheric Ca II K line. *Solar Phys.* 22:375–386.

Cram, L. E. 1974*a*. High resolution spectroscopy of the disk chromosphere. *Solar Phys.* 37: 75–83.

Cram, L. E. 1974*b*. High resolution spectroscopy of the disk chromosphere. In *Chromospheric Fine Structure: Proc. IAU Symp. 56*, ed. R. G. Athay (Dordrecht: D. Reidel), pp. 51–53.

Cram, L. E. 1978. High-resolution spectroscopy of the disk chromosphere. VI. Power, phase and coherence spectra of atmospheric oscillations. *Astron. Astrophys.* 70:345–354.

Cram, L. E. 1987. Heating of chromospheres and coronae: Present status of theory. In *Cool Stars, Stellar Systems and the Sun*, eds. J. L. Linsky and R. E. Stencel (New York: Springer-Verlag), pp. 123–134.

Cram, L. E., and Damé, L. 1983. High spatial and temporal resolution observations of the solar Ca II H line. *Astrophys. J.* 272:355–361.

Cram, L. E., and Thomas, J. H., eds. 1981. *The Physics of Sunspots* (Sunspot, N.M.: Sacramento Peak Obs.).

Cram, L. E., Durney, B. R., and Guenther, D. B. 1983. Preliminary observations of velocity fields at the solar poles. *Astrophys. J.* 267:442–454.

Cribier, M., Pichard, B., Rich, J., Spiro, M., Vignaud, D., Besson, A. Bevilacqua, A., Caperan, F., Dupont, G., Sire, P., Gorry, J., Hampel, W., and Kirsten, T. 1988. Study of a high intensity 746 KeV neutrino source for the calibration of solar neutrino detectors. *Nucl. Inst. Meth.* A265:574–586.

Crighton, D. G. 1975. Basic principles of aerodynamic noise generation. *Prog. Aerospace Sci.* 16:31–96.

Cromwell, D., McQuillan, P., and Brown, J. C. 1988. Beam-driven return current instability and anomalous plasma heating in solar flares. *Solar Phys.* 115:289–312.

Csepura, G., and Nagy, I. 1985. Sunspot Motions Before Large Flares. Heliophys. Obs. Hungarian Acad. Sci. Preprint No. 4.

Cuperman, S., and Metzler, N. 1975. Solution of three-fluid model equations with anomalous transport coefficients for the quiet solar wind. *Astrophys. J.* 196:205–219.

Cuypers, J. 1980. On the calculation of the frequency splitting of adiabatic nonradial stellar oscillations by slow differential rotation. *Astron. Astrophys.* 89:207–208.

Dainty, J. C. 1984. Stellar speckle interferometry. In *Laser Speckle and Related Phenomena*, 2nd ed., ed. J. C. Dainty (Berlin: Springer-Verlag), pp. 255–279.

Damé, L. 1983. Eruptions solaires en lumière blanche et structures fines, oscillations et chauffage de la chromosphère solaire calme. Thèse, Univ. de Paris VII.

Damé, L. 1984. Small-scale dynamical processes in the solar atmosphere. In *Small-Scale Dynamical Processes in Quiet Stellar Atmospheres*, ed. S. L. Keil (Sunspot, N.M.: Sacramento Peak Obs.), pp. 54–64.

Damé, L. 1985. Meso-scale structures: An oscillatory phenomenon? In *Theoretical Problems in High Resolution Solar Physics*, ed. H. U. Schmidt (Munich: MPI für Astrophysik), pp. 244–250.

Damé, L. 1988. The GOLF helioseismometer on board SOHO. In *Seismology of the Sun and Sun-Like Stars*, ed. E. J. Rolfe, ESA SP-286, pp. 367–370.

Damé, L., Gouttebroze, P., and Malherbe, J.-M. 1984. Observation and analysis of intensity oscillations in the solar K line. *Astron. Astrophys.* 130:331–340.

D'Antonia, F., and Mazzitelli, I. 1984. Lithium depletion in stars. Pre-main sequence burning and extra mixing. *Astron. Astrophys.* 138:431–442.

Däppen, W. 1980. An analytical version of the free-energy minimization method for the equation of state of stellar plasmas. *Astron. Astrophys.* 91:212–220.

Däppen, W. 1987. Solar oscillations and the equation of state. In *Strongly Coupled Plasma Physics*, eds. F. Rogers and H. DeWitt (New York: Plenum Press), pp. 179–182.

Däppen, W., and Gough, D. O. 1984. On the determination of the helium abundance of the solar convection zone. In *Theoretical Problems in Stellar Stability and Oscillations*, eds. M. Gabriel and A. Noëls (Liège: Inst. d'Astrophysique), pp. 264–269.

Däppen, W., and Gough, D. O. 1986. Progress report on helium abundance determination. In *Seismology of the Sun and the Distant Stars*, ed. D. O. Gough (Dordrecht: D. Reidel), pp. 275–280.

Däppen, W., Gilliland, R. L., and Christensen-Dalsgaard, J. 1986. Weakly interacting massive particles, solar neutrinos and solar oscillations. *Nature* 321:229–231.

Däppen, W., Anderson, L. S., and Mihalas, D. 1987. Statistical mechanics of partially ionized plasmas: The Planck-Larkin partition function, polarization shifts, and simulation of optical spectra. *Astrophys. J.* 319:195–206.

Däppen, W., Mihalas, D., Hummer, D. G., and Mihalas, B. W. 1988*a*. The equation of state for stellar envelopes. III. Thermodynamic quantities. *Astrophys. J.* 332:261–270.

Däppen, W., Gough, D. O., and Thompson, M. J. 1988*b*. Further progress on the helium abundance determination. In *Seismology of the Sun and Sun-Like Stars*, ed. E. J. Rolfe, ESA SP-286, pp. 505–510.

Däppen, W., Gough, D. O., Kosovichev, A. G., and Thompson, M. J. 1990. In *Challenges to Theories of the Structure of Moderate-Mass Stars*, eds. D. O. Gough and J. Toomre (Heidelberg: Springer-Verlag), in press.

Dar, A., Mann, A., Melina, Y., and Zajman, D. 1987. Neutrino oscillations and the solar-neutrino problem. *Phys. Rev.* 35D:3607–3620.

Davids, C. N., Wang, T. F., Almad, I., Holzmann, R., and Janssens, R. V. F. 1987. Electron capture branching ratio of ^{81}Kr^m. *Phys. Rev.* C35:1114–1118.

Davis, L. 1955. Interplanetary magnetic fields and cosmic rays. *Phys. Rev.* 100:1440–1444.

Davis, R. 1952. Nuclear recoil following neutrino emission from beryllium-7. *Phys. Rev.* 86:976–985.

Davis, R. 1955. Attempt to detect the antineutrinos from a nuclear reactor by th ^{37}Cl$(v,e^-)^{37}$ Ar. *Phys. Rev.* 97:766–769.

Davis, R. 1956. Nuclear emulsions; general nuclear physics. *Bull. Amer. Phys. Soc. II* 1: 219–220.

Davis, R. 1957. An attempt to observe the capture of reactor neutrinos in ^{37} Cl. In *Radioisotopes in Scientific Research*, vol. 1, ed. R. C. Extermann (New York: Pergamon), pp. 728–735.

Davis, R. 1964. Solar neutrinos. II. Experimental. *Phys. Rev. Lett.* 12:303–305.

Davis, R. 1978. The solar neutrino experiment. In *Proc. Informal Conference on Status and Future of Solar Neutrino Research*, vol. 1, ed. G. Friedlander, BNL Rept. 50879 (Upton, N.Y.: Brookhaven National Lab), pp. 1–33.

Davis, R. 1986. Report on status of solar neutrino experiments. In *ICOBAN Conference*, Toyama, Japan, ed. J. Arafune, pp. 237–276.

Davis, R. 1989. Fred Reines and solar neutrinos. In *Neutrino '88: Proc. 13th Intl. Conf. on Neutrino Physics and Astrophysics*, eds. J. Schneps, T. Kafka, W. A. Mann and P. Nath (Singapore: World Scientific), p. 502.

Davis, R., and Evans, J. C. 1973. Experimental limits on extraterrestrial neutrino flux. In *Proc. 13th Intl. Cosmic Ray Conf.*, vol. 3, pp. 2001–2003.

Davis, R., and Harmer, D. S. 1959. Attempt to detect the neutrinos from a nuclear reactor by the ^{37}Cl$(v,e^-)^{37}$Ar reaction. *Bull. Amer. Phys. Soc.* 4:217 (abstract).

Davis, R., Harmer, D. S., and Hoffman, K. C. 1968. Search for neutrinos from the Sun. *Phys. Rev. Lett.* 20:1205–1209.

Davis, R., Evans, J. C., Rogers, L. C., and Radejka, V. 1972. Report on the Brookhaven solar neutrino experiment. In *Neutrino '72*, vol. 1, eds. A. Frenkel, G. Marx and E. L. F. Tarsvlat, pp. 5–27.

Davis, R., Evans, J. C., Fowler, E. C., and Meyer, S. 1973. *Study of the neutrino capture cross section in ^{37}Cl with μ^+ decay neutrinos.* Proposal 53 for the LAMPF Accelerator Beam Stop.

Davis, R., Cleveland, B. T., Rowley, J. K., Katcoff, S., Remsberg, L. P., Friedlander, G., Weneser, J., Kirsten, T., Hampel, W., Scholtz, R., and Dostrovsky, I. 1981. *Proposal for a Fundamental Test of the Theory of Nuclear Fusion in the Sun with a Gallium Solar Neutrino Detector* (Upton, N.Y.: Brookhaven National Lab).

Davis, R., Cleveland, B. T., and Rowley, J. D. 1987. Variations in the solar neutrino flux. In *Proc. 20th Intl. Cosmic Ray Conf.*, vol. 4, pp. 328–331.

Davis, R., Cleveland, B. T., and Rowley, J. K. 1988. Report on the Chlorine Solar Neutrino Experiment. In *Underground Physics Conf.*, eds. G. V. Domogatskii et al., pp. 1–14.

Davis, R., Lande, K., Cleveland, B. T., Ullman, J., and Rowley, J. K. 1989a. Report on the Chlorine Solar Neutrino Experiment. In *Neutrino '88: Proc. 13th Intl. Conf. on Neutrino Physics and Astrophysics*, eds. J. Schneps, T. Kafka, W. A. Mann and P. Nath (Singapore: World Scientific), pp. 518–525.

Davis, R., Mann, A. K., and Wolfenstein. 1989b. Solar neutrinos. *Ann. Rev. Nucl. Part. Sci.* 39:467–506.

Davis, R., Lande, K., Lee, C. K., Wildenhain, P., Weinberger, A., Dailey, T., Cleveland, B. T., and Ullman, J. 1990. The time dependence of the solar neutrino flux observed at Homestake. In *Proc. 21st Intl. Cosmic Ray Conf.*, vol. 12, ed. R. J. Protheroe (Adelaide: Univ. of Adelaide), pp. 143–151.

d'Azambuja, L., and d'Azambuja, M. 1948. Etude d'ensemble des protubérances solaires et de leur évolution. *Ann. Obs. Paris* 6:1–278.

Dearborn, D. S. P., and Fuller, G. M. 1989. Neutrino oscillations and uncertainty in the solar model. *Phys. Rev.* D39:3543–3548.

Dearborn, D. S. P., Lambert, D. L., and Tomkin, J. 1975a. The $^{12}C/^{13}C$ ratio in stellar atmospheres. V. Twelve K giants and subgiants. *Astrophys. J.* 200:675.

Dearborn, D. S. P., Bolton, A. J. C., and Eggleton, P. P. 1975b. The effect on the $^{12}C/^{13}C$ ration of repeated deep mixing to the hydrogen burning shell in a red giant. *Mon. Not. Roy. Astron. Soc.* 170:7p–10p.

Deardorff, J. W. 1970. A numerical study of three-dimensional turbulent channel flow at large Reynolds numbers. *J. Fluid Mech.* 41:453–480.

Deardorff, J. W. 1971. On the magnitude of the subgrid scale eddy coefficient. *J. Comput. Phys.* 7:120–133.

Decker, R. B., and Vlahos, L. 1986. Numerical studies of particle acceleration at turbulent, oblique shocks with an astrophysical application to prompt ion action during solar flares. *Astrophys. J.* 306:710–729.

Defouw, R. J. 1976. Wave propagation along a magnetic tube. *Astrophys. J.* 209:266–269.

Degiacomi, C. G., Kneubühl, F. K., and Huguenin, D. 1985. Far-infrared solar imaging from a balloon-borne platform. *Astrophys. J.* 298:918–933.

Deinzer, W., Hensler, G., Schüssler, M., and Weisshaar, E. 1984a. Model calculations of magnetic flux tubes. I. Equations and method. *Astron. Astrophys.* 139:426–434.

Deinzer, W., Hensler, G., Schüssler, M., and Weisshaar, E. 1984b. Model calculations of magnetic flux tubes. II. Stationary results for solar magnetic elements. *Astron. Astrophys.* 139:435–449.

de Jager, C. 1975. The Sun in the far-infrared and submillimeter region. *Space Sci. Rev.* 17: 645–654.

de Jager, C., and Svestka, Z. 1985. 21 May flare review. *Solar Phys.* 100:435–463.

de la Beaujardière, J.-F., and Zweibel, E. G. 1989. Magnetohydrodynamic waves and particle acceleration in the solar corona. *Astrophys. J.* 336:1059–1072.

Delache, P. 1965. *Compt. Rend.* 261:643.

Delache, P. 1967. Contribution à l'étude de la zone de transition chromosphère-couronne. *Ann. Astrophys.* 30:827–860.

Delache, P., and Fossat, E. 1988. Effects of convective velocities on solar pressure mode frequencies. In *Seismology of the Sun and Sun-Like Stars,* ed. E. J. Rolfe, ESA SP-286, pp. 671–672.

Delache, P., and Scherrer, P. H. 1983. Detection of solar gravity mode oscillations. *Nature* 306:651–653.

Delache, P., Laclare, F., and Sadsaoud, H. 1985. Long period oscillations in solar diameter measurements. *Nature* 317:416–418.

de la Zerda-Lerner, A., and O'Brien, K. 1987. Atmospheric radioactivity and variations in the solar neutrino flux. *Nature* 310:353–354.

Delbouille, L., Roland, G., and Neven, L. 1973. *Spectrophotometric Atlas of the Solar Spectrum from λ3000 to λ10000* (Liège: Inst. d'Astrophysique).

Delbouille, L., Roland, G., Brault, J. W., and Testerman, L. 1981. *Photometric Atlas of the Solar Spectrum from 1850 to 10000 cm⁻¹* (Tucson: Kitt Peak Natl. Obs.).

Delcroix, A., and Grevesse, N. 1968. Diffusion des éléments dans le soleil. *Compt. Rend.* 266:356–358.

Deliyannis, C. P., Demarque, P., and Kawaler, S. D. 1990. Lithium in halo stars from standard stellar evolution. *Astrophys. J. Suppl.* 73:21–65.

DeLuca, E. E. 1986. Dynamo Theory for the Interface Between a Convection Zone and the Radiative Interior of a Star. Ph.D. Thesis, Univ. of Colorado.

DeLuca, E. E., and Gilman, P. A. 1986. Dynamo theory for the interface between the convection zone and the radiative interior of a star. Part I. Model equations and exact solutions. *Geophys. Astrophys. Fluid Dyn.* 37:85–127.

DeLuca, E. E., and Gilman, P. A. 1988. Dynamo theory for the interface between the convection zone and the radiative interior of a star. Part II. Numerical solutions of the nonlinear equations. *Geophys. Astrophys. Fluid Dyn.* 43:119–148.

DeLury, R. E. 1939. The law of solar rotation. *J. Roy. Astron. Soc. Canada* 33:345–378.

Demarque, P., and Guenther, D. B. 1988. Is the Sun really a rigid rotator? In *Seismology of the Sun and Sun-Like Stars,* ed. E. J. Rolfe, ESA SP-286, pp. 99–102.

Demarque, P., and Mengel, J. G. 1971. Advanced evolution of Population II stars. I. Red giants and the helium flash. *Astrophys. J.* 164:317–330.

Demarque, P., and Percy, J. R. 1964. A series of solar models. *Astrophys. J.* 140:541–543.

Demarque, P., King, C. R., and Diaz, A. 1982. The globular cluster metallicity scale: Evidence from stellar models. *Astrophys. J.* 259:154–158.

Demarque, P., Christensen-Dalsgaard, J., and Guenther, D. B. 1988a. GONG Solar Model Workshop, Yale Center for Solar and Space Research, May 27–29, 1987. *Comm. Astrophys.* 12:157–163.

Demarque, P., Guenther, D. B., King, C. R., and Green, E. M. 1988b. Isochrone comparisons, stellar physics and implications for stellar ages. In *Calibration of Stellar Ages,* ed. A. G. D. Philip (Schenectady, N.Y.: L. Davis Press), pp. 101–106.

Demidov, M. L., Kotov, V. A., and Grigoryev, V. M. 1988. On the short-period variations of the global magnetic field of the Sun. *Izv. Krymsk. Astrofiz. Obs.* 82, in press.

Deming, D. 1987. Infrared spectroscopy of the Sun and stars. In *Proc. Fifth Cambridge Workshop on Cool Stars, Stellar Systems and the Sun,* eds. J. L. Linsky and R. E. Stencel (New York: Springer-Verlag), pp. 361–372.

Deming, D., Hillman, J. J., Kostiuk, T., Mumma, M. J., and Zipoy, D. M. 1984. Thermal bifurcation in the upper photosphere inferred from heterodyne spectroscopy of OH rotational lines. *Solar Phys.* 94:57–74.

Deming, D., Glenar, D. A., Kaufl, H. U., Hill, A. A., and Espenak, F. 1986. Infrared helioseismology: Detection of the chromospheric mode. *Nature* 322:232–234.

Deming, D., Glenar, D. A., Käufl, H. U., and Espenak, F. 1988a. Infrared helioseismology: Detection of the chromospheric mode. In *Advances in Helio- and Asteroseismology: Proc. IAU Symp. 123,* eds. J. Christensen-Dalsgaard and S. Frandsen (Dordrecht: D. Reidel), pp. 425–428.

Deming, D., Boyle, R. J., Jennings, D. E., and Wiedemann, G. 1988b. Solar magnetic field studies using the 12-micron emission lines. I. Quiet Sun time series and sunspot slices. *Astrophys. J.* 333:978–995.

Dennis, B. R. 1985. Solar hard-X-ray bursts. *Solar Phys.* 100:465–490.

Denskat, K. R., and Neubauer, F. M. 1982. Statistical properties of low-frequency magnetic field fluctuations in the solar wind from 0.29 to 1.0 AU during solar minimum conditions: Helios 1 and Helios 2. *J. Geophys. Res.* 87:2215–2223.

Dere, K. P., and Mason, H. E. 1981. Spectroscopic diagnostics of the active region: Transition zone and corona. In *Solar Active Regions*, ed. F. Q. Orrall (Boulder: Colorado Assoc. Univ. Press), pp. 129–164.

Dere, K. P., Bartoe, J.-D. F., Brueckner, G. E., Dykton, M. D., and VanHoosier, M. E. 1981. Transient plasmas in the solar transition zone. *Astrophys. J.* 249:333–339.

Dere, K. P., Bartoe, J.-D. F., and Brueckner, G. E. 1983. High-resolution telescope and spectrograph observations of the quiet solar chromosphere and transition zone. *Astrophys. J.* 281:870–873.

Dere, K. P., Bartoe, J.-D. F., Brueckner, G. E., Cook, J. W., and Socker, D. G. 1987a. Ultraviolet observations of solar fine structure. *Science* 238:1267–1269.

Dere, K. P., Bartoe, J.-D. F., Brueckner, G. E., Cook, J. W., and Socker, D. G. 1987b. Discrete subresolution structures in the solar transition zone. *Solar Phys.* 114:223–237.

Dere, K. P., Bartoe, J.-D. F., and Brueckner, G. E. 1989. Explosive events in the solar transition zone. *Solar Phys.* 123:41–68.

Deubner, F.-L. 1972. Some properties of velocity fields in the solar photosphere. IV. Long periods, five-minute oscillations, and the supergranulation at lower layers. *Solar Phys.* 22: 263–275.

Deubner, F.-L. 1974a. On the energy distribution in wavenumber spectra of the granular velocity field. *Solar Phys.* 36:299–301.

Duebner, F.-L. 1974b. Some properties of velocity fields in the solar photosphere. V. Spatiotemporal analysis of high-resolution spectra. *Solar Phys.* 39:31–48.

Deubner, F.-L. 1975. Observations of low-wave number nonradial eigenmodes of the Sun. *Astron. Astrophys.* 44:371–375.

Deubner, F.-L. 1976a. Observations of low-wavenumber nonradial eigenmodes of the Sun. In *Physique des Mouvements dans les Atmosphères Stellaires*, eds. R. Cayrel and M. Steinberg (Paris: CNRS), pp. 259–262.

Deubner, F.-L. 1976b. Observations of short period acoustic waves bearing on the interpretation of "microturbulence." *Astron. Astrophys.* 51:189–194.

Deubner, F.-L. 1981. Detection of low-order *p*-modes in brightness fluctuations of the Sun. *Nature* 290:682–693.

Deubner, F.-L. 1985. Oscillations as a diagnostic of chromospheric structure. In *Chromospheric Diagnostics and Modelling*, ed. B. W. Lites (Sunspot, N.M.: Sacramento Peak Obs.), pp. 279–281.

Deubner, F.-L. 1986a. Observations of gravity waves in the solar atmosphere. In *Seismology of the Sun and the Distant Stars*, ed. D. O. Gough (Dordrecht: D. Reidel), pp. 81–82.

Deubner, F.-L. 1986b. Is there an anisotropy of *p* modes? In *Seismology of the Sun and the Distant Stars*, ed. D. O. Gough (Dordrecht: D. Reidel), pp. 83–84.

Deubner, F.-L. 1988. Observations of solar oscillations. In *Pulsation and Mass Loss in Stars*, eds. R. Stalio and L. A. Willson (Dordrecht: Kluwer), pp. 163–179.

Deubner, F.-L. 1989a. Mesogranulation: A convective phenomenon. *Astron. Astrophys.* 216:259–264.

Deubner, F.-L. 1989b. Mesogranulation—a convective phenomenon. In *High Spatial Resolution Solar Observations*, ed. O. von der Lühe (Sunspot, N.M.: Sacramento Peak Obs.), pp. 489–490.

Deubner, F.-L. 1990. Waves and oscillations in the non-magnetic photosphere. In *Solar Photosphere: Structure, Convection and Magnetic Fields: Proc. IAU Symp. 138*, ed. J. O. Stenflo (Dordrecht: Kluwer), pp. 217–228.

Deubner, F.-L., and Fleck, B. 1989. Dynamics of the solar atmosphere. III. Cell-network distinctions of chromospheric oscillations. Preprint.

Deubner, F.-L., and Gough, D. O. 1984. Helioseismology: Oscillations as a diagnostic of the solar interior. *Ann. Rev. Astron. Astrophys.* 22:593–619.

Deubner, F.-L., and Mattig, W. 1975. New observations of the granular intensity fluctuations. *Astron. Astrophys.* 45:167–171.

Deubner, F.-L., and Vasquez, M. 1976. Differential rotation and the sunspot zones. *Solar Phys.* 43:87–90.

Deubner, F.-L., and Fleck, B. 1989. Dynamics of the solar atmosphere. I. Spatio-temporal analysis of waves in the quiet solar atmosphere. *Astron. Astrophys.* 213:423–428.

Deubner, F.-L., and Fleck, B. 1990. Dynamics of the solar atmosphere. III. Cell-network distinctions of chromospheric oscillations. *Astron. Astrophys.* 228:506–512.

Deubner, F.-L., and Gough, D. O. 1984. Helioseismology: Oscillations as a diagnostic of the solar interior. *Ann. Rev. Astron. Astrophys.* 22:593–619.

Deubner, F.-L., and Mattig, W. 1975. *Astron. Astrophys.* 14:167ff.

Deubner, F.-L., Ulrich, R. K., and Rhodes, E. J., Jr. 1979. Solar *p*-mode oscillations as a tracer of radial differential rotation. *Astron. Astrophys.* 72:177–185.

Deubner, F.-L., Endler, F., and Staiger, J. 1984. Phase relations of high-degree modes revisited. *Mem. Soc. Astron. Italiana* 55:135–146.

Deubner, F.-L., Reichling, M., and Langhanki, R. 1988. On the energy of short period waves in the solar atmosphere. In *Advances in Helio- and Asteroseismology: Proc. IAU Symp. 123*, eds. J. Christensen-Dalsgaard and S. Frandsen (Dordrecht: D. Reidel), pp. 439–442.

Deubner, F.-L., Fleck, B., Marmolino, C., and Severino, G. 1990. Dynamics of the solar atmosphere. IV. Evanescent waves of small amplitude. *Astron. Astrophys.* 236:509–514.

Deupree, R. G. 1984a. Two- and three-dimensional numerical simulations of the core helium flash. *Astrophys. J.* 282:274–286.

Deupree, R. G. 1984b. Two- and three-dimensional numerical simulations of the hydrodynamic phase of the core helium flash. *Astrophys. J.* 287:268–281.

Deupree, R. G. 1977. The theoretical red edge of the RR Lyrae gap. *Astrophys. J.* 211:509–526.

Deupree, R. G. 1979. Another fundamental, but correctable, inconsistency of the local mixing length theory. *Astrophys. J.* 234:228–231.

Deupree, R. G., and Cole, P. W. 1983. A survey of core helium flash with dynamic convection. *Astrophys. J.* 269:676–685.

Deupree, R. G., and Varner, T. M. 1980. The value of the mixing length. *Astrophys. J.* 237: 558–562.

DeVore, C. R. 1987. The decay of the large-scale solar magnetic field. *Solar Phys.* 112: 17–35.

DeVore, C. R., and Sheeley, N. R., Jr. 1987. Simulations of the Sun's polar magnetic fields during sunspot cycle 21. *Solar Phys.* 108:47–59.

DeVore, C. R., Sheeley, N. R., Jr., and Boris, J. P. 1984. The concentration of the large-scale solar magnetic field by a meridional surface flow. *Solar Phys.* 91:1–14.

Dewar, R. L. 1970. Interaction between hydromagnetic waves and a time dependent inhomogeneous medium. *Phys. Fluids* 13:2710–2720.

Dicke, R. H. 1979. Is there a chronometer hidden deep in the Sun? *Nature* 276:676–680.

Dicke, R. H. 1981. Seismology and geodesy of the Sun: Solar geodesy. *Proc. Natl. Acad. Sci. USA* 78:1309–1312.

Dicke, R. H. 1982. A magnetic core in the Sun? The solar rotator. *Solar Phys.* 78:3–16.

Dicke, R. H. 1983. Evidence for a distorted solar core rotating with a 12.4-day period. *Nature* 303:292–295.

Didkovskii, L. V. 1989. Internal differential rotation of the Sun: *p*-mode frequency splitting obtained from measurements of the brightness oscillations. *Soviet Astron. Lett.* 15: 484–487.

Didkovskii, L. V., Kotov, V. A., and Tarasova, T. N. 1988. Five-minute oscillations of the solar brightness: Observations with a photodiode array. *Bull. Crimean Astrophys. Obs.* 79: 174–178.

Diesendorf, M. O. 1970. Electron correlations and solar neutrino counts. *Nature* 227:266–267.

Diesendorf, M. O., and Ninham, B. W. 1969. The effect of quantum correlations on electron-scattering opacities. *Astrophys. J.* 156:1069–1073.

Dilke, F. W., and Gough, D. O. 1972. The solar spoon. *Nature* 240:262–264, 293–294.

Dilts, G. A. 1985. Computation of spherical harmonic expansion coefficients via FFTs. *J. Comput. Phys.* 57:439–453.

Dinh, N. V. 1980. Spectral analysis of solar flares. I. Observational results. *Publ. Astron. Soc. Japan* 32:495–514.

Dittmer, P. H. 1977. Large-Scale Periodic Solar Velocities: An Observational Study. Stanford Univ. IPR Rept. No. 686.

Dittmer, P. H., Scherrer, P. H., and Wilcox, J. M. 1978. An observational search for large-scale organization of the five-minute oscillations on the Sun. *Solar Phys.* 57:3–11.

Dodson, H. W., and Hedeman, E. R. 1981. Experimental comprehensive solar flare indices for "major" and certain lesser flares 1975–1979. Report UAG-80, World Data Center A for Solar-Terrestrial Physics (Boulder: NOAA).

Doi, M., Kotani, T., and Takasugi, E. 1985. Double beta decay and Majorana neutrino. *Prog. Theoret. Phys. (Kyoto)* Suppl. 83.

Dolginov, A. Z., and Muslimov, A. G. 1985. Torsional 11-year oscillations of the Sun. *Soviet Astron. Lett.* 11:112–115.

Domingo, V. 1988a. Helioseismology from space, the SOHO project. In *Advances in Helio- and Asteroseismology: Proc. IAU Symp. 123*, ed. J. Christensen-Dalsgaard and S. Frandsen (Dordrecht: D. Reidel), pp. 545–548.

Domingo, V. 1988b. SOHO status. In *Seismology of the Sun and Sun-Like Stars*, ed. E. J. Rolfe, ESA SP-286, pp. 363–366.

Domingo, V., and Poland, A. 1988. The SOHO project: Helioseismology investigations. *Adv. Space Res.* 11(11):109–115.

Domogatskii, G. V., and Eramzhyan, R. A. 1977. Estimation of the ^{37}Ar production rate due to cosmic-ray neutrinos in tests for solar neutrinos. *Bull. Acad. Sci. USSR: Phys. Ser.* 41: 169–171.

Donati-Falchi, A., Falciani, R., and Smaldone, L. A. 1985. Analysis of the optical spectra of solar flares. IV. The "blue" continuum of white light flares. *Astron. Astrophys.* 152:165–169.

Donnelly, R. F. 1987. Temporal trends of solar EUV and UV full-disk fluxes. *Solar Phys.* 109:37–58.

Donnelly, R. F. 1988. The solar UV Mg II core-to-wing ratio from the NOAA9 satellite during the rise of solar cycle 22. *Adv. Space Res.* 8:77–80.

Donnelly, R. F., and Kane, S. R. 1978. Impulsive extreme-ultraviolet and hard X-ray emission during solar flares. *Astrophys. J.* 222:1043–1053.

Donnelly, R. F., Harvey, J. W., Heath, D. F., and Repoff, T. P. 1985. Temporal characteristics of the solar UV flux and He I line at 1083 nm. *J. Geophys. Res.* 90:6267–6273.

Donnelly, R. F., Hinteregger, H. E., and Heath, D. F. 1986. Temporal variations of solar EUV, UV, and 10830Å radiations. *J. Geophys. Res.* 91:5567–5578.

Dorman, L. I., and Wolfendale, A. W. 1991. The correlation of the solar neutrino rate with solar activity. *J. Phys. G,* in press.

Dorren, J. D., and Guinan, E. F. 1981. Photometric variability of single stars with activity cycles. In *Second Cambridge Workshop on Cool Stars, Stellar Systems and the Sun*, vol. 2, eds. M. S. Giampapa and L. Golub, SAO Special Rept. No. 392, pp. 49–64.

Dorren, J. D., and Guinan, E. F. 1990. HD 129333—the Sun in its infancy. *Astrophys. J.,* submitted.

Doschek, G. A., Feldman, U., and Rosenberg, F. D. 1977a. EUV spectra of the 1973 June 15 solar flare observed from Skylab. I. Allowed transitions in chromospheric and transition zone ions. *Astrophys. J.* 215:329–344.

Doschek, G. A., Feldman, U., VanHoosier, M. E., and Bartoe, J.-D. F. 1977b. The emission-line spectrum above the limb of the quiet Sun: 1175–1940 Å. *Astrophys. J. Suppl.* 31: 417–443.

Doschek, G. A., Antiochos, S. K., Antonucci, E., Cheng, C. C., Culhane, J. L., Fisher, G. H., Jordan, C., Leibacher, J. W., MacNiece, P., McWhirter, R. W. P., Moore, R. L., Rabin, D. M., Rust, D. M., and Shine, R. A. 1986. Chromospheric explosions. In *Energetic Phenomena on the Sun*, eds. M. R. Kundu and B. E. Woodgate, NASA CP-2439, pp. 4–1—4–42.

Dostrovsky, I. 1978. In *Proc. Informal Conf. on Status and Future of Solar Neutrino Research*, ed. G. Friedlander, BNL 50879 (Upton, N.Y.: Brookhaven National Lab), p. 231.

Douglass, D. H. 1978. A close-up of the Sun. In *JPL Publications*, pp. 78–80.

Doyle, J. G., and Bentley, R. 1986. Broadening of soft X-ray lines during the impulsive phase of solar flares: Random or directed mass motions? *Astron. Astrophys.* 155:278–282.

Dravins, D. 1989. Stellar granulation: Modeling of stellar surfaces and photospheric line asymmetries. In *Solar and Stellar Granulation*, eds. R. J. Rutten and G. Severino (Dordrecht: Kluwer), pp. 493–519.

Dravins, D., Lindegren, L., and Nordlund, Å. 1981. Solar granulation: Influence of convection on spectral line asymmetries and wavelength shifts. *Astron. Astrophys.* 96:345–364.

Dravins, D., Larsson, B., and Nordlund, Å. 1986. Solar Fe II line asymmetries and wavelength shifts. *Astron. Astrophys.* 158:83–88.

Dravins, D., Lindegren, L., and Nordlund, Å. 1981. Solar granulation: Influence of convection on spectral line asymmetries and wavelength shifts. *Astron. Astrophys.* 96:345–364.

Dreschen, T., Wöhl, H., and Küveler, G. 1984. On the determination of the solar rotation and indications of the solar differential rotation from an analysis of solar integrated light. In *The Hydromagnetics of the Sun: Proc. Fourth European Meeting on Solar Physics,* eds. T. D. Guyenne and J. J. Hunt, ESA SP-220, pp. 29–32.

Drury, L. O'C. 1983. An introduction to the theory of diffusive shock acceleration of energetic particles in tenuous plasmas. *Rept. Prog. Phys.* 46:973–1027.

Drury, L. O'C., and Völk, H. J. 1981. Hydromagnetic shock structure in the presence of cosmic rays. *Astrophys. J.* 248:344–351.

Dryer, M. 1982. Coronal transient phenomena. *Space Sci. Rev.* 33:233–275.

Dryer, M., Wu, S. T., Steinolfson, R. S., and Wilson, R. M. 1979. Magnetohydrodynamic models of coronal transients in the meridional plane. II. Simulation of the coronal transient of 1973 August 21. *Astrophys. J.* 227:1059–1071.

Duebner, F.-L., and Gough, D. O. 1984. Helioseismology: Oscillations as a diagnostic of the solar interior. *Ann. Rev. Astron. Astrophys.* 22:593–619.

Duijveman, A., and Hoyng, P., 1982. Imaging of three flares during the impulsive phase. *Solar Phys.* 86:279–288.

Dulk, G. A., and Dennis, B. R. 1982. Microwaves and hard X-rays from solar flares: Multithermal and non-thermal interpretations. *Astrophys. J.* 260:875–884.

Dulk, G. A., and Marsh, K. A. 1981. Simplified expressions for the geosynchrotron radiation from mildly relativistic, nonthermal and thermal electrons. *Astrophys. J.* 259:350–358.

Dulk, G. A., and McLean, D. J. 1978. Coronal magnetic fields. *Solar Phys.* 57:279–295.

Dulk, G. A., Smerd, S. F., MacQueen, R. M., Gosling, J. T., Magun, A., Stewart, R. T., Sheridan, K. V., Robinson, R. D., and Jacques, S. 1976. White light and radio studies of the coronal transient of 14–15 September, 1973. I. Material motions and magnetic field. *Solar Phys.* 49:369–394.

Dulk, G. A., Steinberg, J. L., Hoang, S., and Lecacheux, A. 1986. Latitude distribution of interplanetary magnetic field lines rooted in active regions. In *The Sun and the Heliosphere in Three Dimensions,* ed. R. G. Marsden (Dordrecht: D. Reidel), pp. 229–233.

Dunn, R. B. 1960. Ph.D. Thesis, Harvard Univ.

Dunn, R. B. 1985. High resolution solar telescopes. *Solar Phys.* 100:1–20.

Dunn, R. B., and Zirker, J. B. 1973. The solar filigree. *Solar Phys.* 33:281–304.

Dunn, R. B., Streander, G., and von der Lühe, O. 1989. Adaptive optical system at Sac Peak: A progress update. In *High Spatial Resolution Solar Observations* ed. O. von der Lühe (Sunspot, N.M.: Sacramento Peak Obs.), pp. 53–70.

Dupree, A. K. 1972. Analysis of the extreme-ultraviolet quiet solar spectrum. *Astrophys. J.* 178:527–541.

Durney, B. R. 1970. Nonaxysymmetric convection in rotating spherical shell. *Astrophys. J.* 161:1115–1127.

Durney, B. R. 1974. On the Sun's differential rotation: Its maintenance by large-scale meridional motions in the convection zone. *Astrophys. J.* 190:211–221.

Durney, B. R. 1976. On the constancy along cylinders of the angular velocity in the solar convection zone. *Astrophys. J.* 204:596–598.

Durney, B. R. 1984. On the influence of turbulent motions on non-radial oscillations. In *Solar Seismology from Space,* NASA JPL Publ. 84–84, pp. 325–334.

Durney, B. R. 1987. The generalization of mixing length theory to rotating convection zones and applications to the Sun. In *The Internal Solar Angular Velocity,* eds. B. R. Durney and S. Sofia (Dordrecht: D. Reidel), pp. 235–262.

Durney, B. R. 1989. On the behavior of the angular velocity in the lower part of the solar convection zone. *Astrophys. J.* 338:509–527.

Durney, B. R. 1990. On the numerical calculation of the solar rotational splitting coefficients. *Astrophys. J.* 351:682–686.

Durney, B. R., and Latour, J. 1978. On the angular momentum loss of late type stars. *Geophys. Astrophys. Fluid Dyn.* 9:241–255.

Durney, B. R., and Roxburgh, I. W. 1971. Inhomogeneous convection and the equatorial acceleration of the Sun. *Solar Phys.* 16:3–20.

Durney, B. R., and Spruit, H. C. 1979. On the dynamics of stellar convection zones: The effect of rotation on the turbulent viscosity and conductivity. *Astrophys. J.* 234:1067–1078.

Durney, B. R., Gilman, P. A., and Stix, M. 1976. Summary of final discussion. In *Basic Mechanisms of Solar Activity: Proc. IAU Symp. 71*, eds. V. Bumba and J. Kleczek (Dordrecht: D. Reidel), pp. 479–484.

Durney, B. R., Mihalas, D., and Robinson, R. D. 1981. A preliminary interpretation of stellar chromospheric Ca II emission variations within the framework of stellar dynamo theory. *Publ. Astron. Soc. Pacific* 93:537–543.

Durney, B. R., Cram, L. E., Guenther, D. B., Keil, S. L., and Lytle, D. M. 1985. A search for long-lived velocity fields at the solar poles. *Astrophys. J.* 292:752–762.

Durney, B. R., Hill, F., and Goode, P. 1988. On the expansion of the rotational eigenfrequencies in Legendre polynomials. *Astrophys. J.* 326:486–489.

Durrant, C. J., and Nesis, A. 1981. Vertical structure of the solar photosphere. *Astron. Astrophys.* 95:221–228.

Durrant, C. J., and Schröter, E. H. 1983. Solar global velocity oscillations and active region rotation. *Nature* 301:589–591.

Duvall, T. L., Jr. 1979. Large-scale solar velocity fields. *Solar Phys.* 63:3–15.

Duvall, T. L., Jr. 1980. The equatorial rotation rate of the supergranular cells. *Solar Phys.* 66:213–221.

Duvall, T. L., Jr. 1982*a*. A dispersion law for solar oscillations. *Nature* 300:242–243.

Duvall, T. L., Jr. 1982*b*. The equatorial photospheric rotation rate. *Solar Phys.* 76:137–143.

Duvall, T. L., Jr. 1990. A review of observational helioseismology. In *Inside the Sun: Proc. IAU Coll. 121*, eds. G. Berthomieu and M. Cribier (Dordrecht: Kluwer), pp. 253–264.

Duvall, T. L., Jr., and Harvey, J. W. 1983. Observations of solar oscillations of low and intermediate degree. *Nature* 302:24–27.

Duvall, T. L., Jr., and Harvey, J. W. 1984. Rotational frequency splitting of solar oscillations. *Nature* 310:19–22.

Duvall, T. L., Jr., and Harvey, J. W. 1986. Solar Doppler shifts: Sources of continuous spectra. In *Seismology of the Sun and the Distant Stars*, ed. D. O. Gough (Dordrecht: D. Reidel), pp. 105–116.

Duvall, T. L., Jr., Scherrer, P. H., Svalgaard, L., and Wilcox, J. M. 1979. Average photospheric poloidal and toroidal magnetic field components near solar minimum. *Solar Phys.* 61:233–245.

Duvall, T. L., Jr., Jones, H. P., and Harvey, J. W. 1983. Solar oscillations with a 13-day period. *Nature* 304:517–518.

Duvall, T. L., Jr., Dziembowski, W. A., Goode, P. R., Gough, D. O., Harvey, J. W., and Leibacher, J. W. 1984. Internal rotation of the Sun. *Nature* 310:22–25.

Duvall, T. L., Jr., Harvey, J. W., and Pomerantz, M. A. 1986. Latitude and depth variation of solar rotation. *Nature* 321:500–501.

Duvall, T. L., Jr., Harvey, J. W., and Pomerantz, M. A. 1987. Latitude and depth variation of solar rotation. In *The Internal Solar Angular Velocity*, eds. B. R. Durney and S. Sofia (Dordrecht: D. Reidel), pp. 19–22.

Duvall, T. L., Jr., Harvey, J. W., Libbrecht, K. G., Popp, B. D., and Pomerantz, M. A. 1988*a*. Frequencies of solar *p*-mode oscillations. *Astrophys. J.* 324:1158–1171.

Duvall, T. L., Jr., Harvey, J. W., and Pomerantz, M. A. 1988*b*. Intermediate degree solar oscillations. In *Advances in Helio- and Asteroseismology: Proc. IAU Symp. 123*, eds. J. Christensen-Dalsgaard and S. Frandsen (Dordrecht: D. Reidel), pp. 37–40.

Dwarakanth, M. R. 1974. ^{3}He(^{3}He,$2p$)^{4}He and the termination of the proton-proton chain. *Phys. Rev.* C9:805–809.

Dwarakanath, M. R., and Winkler, H. 1971. ^{3}He(^{3}He,$2p$)^{4}He total cross-section measurements below the Coulomb Barrier. *Phys. Rev.* C4:1532–1541.

Dyachkov, L. G., Kobzev, G. A., and Pankratov, P. M. 1988. The hydrogen Balmer spectrum in the near-threshold region: Unified semiclassical calculation of line and continuum contributions. *J. Phys.* B21:1939–1949.

Dyson, J., and Schutz, B. F. 1979. Perturbations and stability of rotating stars. I. Completeness of normal modes. *Proc. Roy. Soc. London* A368:389–410.

Dziembowski, W. A., and Goode, P. R. 1984. Simple asymptotic estimates of the fine structure in the spectrum of solar oscillations due to rotation and magnetism. *Mem. Soc. Astron. Italiana* 55:185–213.

Dziembowski, W. A., and Goode, P. R. 1985. Frequency splitting in Ap stars. *Astrophys. J.* 296:L27–L30.

Dziembowski, W. A., and Goode, P. R. 1988. The magnetic field inside the Sun. In *Advances in Helio- and Asteroseismology: Proc. IAU Symp. 123,* eds. J. Christensen-Dalsgaard and S. Frandsen (Dordrecht: D. Reidel), pp. 171–174.

Dziembowski, W. A., and Goode, P. R. 1989. The toroidal magnetic field inside the Sun. *Astrophys. J.* 347:540–550.

Dziembowski, W. A., and Goode, P. R. 1991a. The Sun's relict magnetic field? In preparation.

Dziembowski, W. A., and Goode, P. R. 1991b. The second order effect of rotation in stars. In preparation.

Dziembowski, W. A., and Sienkiewicz, R. 1973. Vibrational stability of a 1 $M_\odot$ star in the phase of central hydrogen burning. *Acta Astron.* 23:273–281.

Dziembowski, W. A., Paterno, L., and Ventura, R. 1985. Excitation of solar oscillation gravity modes by magnetic torque. *Astron. Astrophys.* 151:47–51.

Dziembowski, W. A., Goode, P. R., and Libbrecht, K. G. 1989. The radial gradient in the Sun's rotation. *Astrophys. J.* 337:L53–L57.

Dziembowski, W. A., Pamyatnykh, A. A., and Sienkiewicz, R. 1990. Solar model from helioseismology and the neutrino flux problem. *Mon. Not. Roy. Astron. Soc.* 244:542–550.

Ebeling, W., and Sandig, R. 1973. *Ann. Phys.(Leipzig)* 28:289.

Ebeling, W., Kraeft, W. D., and Kremp, D. 1976. *Theory of Bound States and Ionization Equilibrium in Plasmas and Solids* (Berlin: Akademie Verlag).

Ebeling, W., Kraeft, W. D., Kremp, D., and Röpke, G. 1985. Energy levels in hydrogen plasmas and the Planck-Larkin partition function: A comment. *Astrophys. J.* 290:24–27.

Eckart, C. 1960. *Hydrodynamics of Ocean and Atmospheres* (New York: Pergamon).

Eckers, R. D., and Little, L. T. 1971. The motions of the solar wind close to the Sun. *Astron. Astrophys.* 10:310–316.

Eddington, A. S. 1920. The internal constitution of the stars. *Nature* 106:14–20.

Eddington, A. S. 1926. *The Internal Constitution of the Stars* (Cambridge: Cambridge Univ. Press), esp. pp. 289–320.

Eddy, J. A. 1976. The Maunder minimum. *Science* 192:1189–1202.

Eddy, J. A. 1980. The historical record of solar activity. In *The Ancient Sun: Fossil Record in the Earth, Moon and Meteorites,* eds. R. O. Pepin, J. A. Eddy and R. B. Merrill (New York: Pergamon), pp. 119–134.

Eddy, J. A., Léna, P. J., and McQueen, R. M. 1969. Far infrared measurement of the solar minimum temperature. *Solar Phys.* 10:330–341.

Edmonds, F. N. 1962. A coherence analysis of Fraunhofer line fine structure and continuum brightness fluctuations near the center of the solar disk. *Astrophys. J.* 136:507–533.

Edmunds, M. G., and Gough, D. O. 1983. Solar atmospheric temperature inhomogeneities induce a 13-day oscillation in full disk Doppler measurements. *Nature* 302:810–812.

Edwards, C. G., Levay, M., Gilbreth, C. W., Tarbell, T. D., Title, A. M., Wolfson, C. J., and Torgerson, D. D. 1987. The correlation tracker image stabilization system for HRSO. *Bull. Amer. Astron. Soc.* 19:929 (abstract).

Edwin, P. M., and Roberts, B. 1983. Wave propagation in a magnetic cylinder. *Solar Phys.* 88:179–191.

Eggleton, P. P. 1971. The evolution of low mass stars. *Mon. Not. Roy. Astron. Soc.* 151:351–364.

Eggleton, P. P., Faulkner, J., and Flannery, B. P. 1973. An approximate equation of state for stellar material. *Astron. Astrophys.* 23:325–330.

Ehgamberdiev, S., Khalikov, S., and Fossat, E. 1990. General presentation of a single IRIS raw data analysis problem. *Solar Phys.,* in press.

Ehrlich, R. 1982. Possible time variations in ^{37}Cl solar neutrino data. *Phys. Rev.* D25:2282–2286.

Eisenfeld, J. 1969. A completeness theorem for an integro-differential operator. *J. Math. Anal. Appl.* 26:357–375.

Eliezer, S., Ghatak, A., and Hora, H. 1986. *An Introduction to Equations of State: Theory and Applications* (Cambridge: Cambridge Univ. Press).

Ellis, A. N. 1984. The base of the solar convection zone. In *Theoretical Problems in Stellar Stability and Oscillations,* eds. A. Noëls and M. Gabriel (Liège: Inst. d'Astrophysique), pp. 290–293.

Ellis, A. N. 1986. An improved asymptotic formula for solar gravity-mode periods. In *Seismology of the Sun and the Distant Stars,* ed. D. O. Gough (Dordrecht: D. Reidel), pp. 173–175.

Ellis, A. N. 1987. Seismological Traits of the Sun's Interior Structure. Ph.D. Thesis, Univ. of Cambridge.

Ellis, A. N. 1988. Inversion of asymptotic gravity-mode frequencies and its application to the Sun. In *Advances in Helio- and Asteroseismology: Proc. IAU Symp. 123,* eds. J. Christensen-Dalsgaard and S. Frandsen (Dordrecht: D. Reidel), pp. 147–150.

Ellison, D. C., and Ramaty, R. 1984. Shock acceleration of electrons and ions in solar flares. *Astrophys. J.* 298:400–408.

Elowitz, M., Hill, F., and Duvall, T. L., Jr. 1989. A test of a modified algorithm for computing spherical harmonic coefficients using an FFT. *J. Comput. Phys.* 80:506–511.

Elste, G. 1986. Manganese and carbon lines as temperature indicators. *Solar Phys.* 107:47–56.

Elsworth, Y. P., Isaak, G. R., Jefferies, S. M., McLeod, C. P., New, R., van der Raay, H. B., Pallé, P. L., Régulo, C., and Roca-Cortés, T. 1988. Experience in operating a limited global network of stations measuring the full-disc oscillations of the Sun. In *Advances in Helio- and Asteroseismology: Proc. IAU Symp. 123,* eds. J. Christensen-Dalsgaard and S. Frandsen (Dordrecht: D. Reidel), pp. 535–539.

Elsworth, Y. P., Jeffries, S. M., McLeod, C. P., New, R., Pallé, P. L., van der Raay, H. B., Régulo, C., and Roca-Cortés, T. 1989. The 160 minute solar oscillation: An artifact? *Astrophys. J.* 338:557–562.

Elsworth, Y. P., Isaak, G. R., Jefferies, S. M., McLeod, C. P., New, R., Pallé, P. L., Régulo, C., and Roca-Cortés, T. 1990a. Linewidth of low-degree acoustic modes of the Sun. *Mon. Not. Roy. Astron. Soc.* 242:135–140.

Elsworth, Y. P., Howe, R., Isaak, G. R., McLeod, C. P., and New, R. 1990b. Variation of low-order acoustic solar oscillations over the solar cycle. *Nature* 345:322–324.

Eltayeb, I. A. 1972. Hydromagnetic convection in a rapidly rotating fluid layer. *Proc. Roy. Soc. London* 326A:229–254.

Elwyn, A. J., Holland, R. E., Davids, C. N., and Ray, W. 1982. $^7Li(d,p)^8Li$ reaction cross section near 0.78 MeV. *Phys. Rev.* C25:2168–2173.

Emden, V. R. 1907. *Gaskugeln* (Leipzig: G. B. Teubner).

Emslie, A. G. 1980. The effect of reverse currents on the dynamics of non-thermal electron beams in solar flares and on their emitted X-ray bremsstrahlung. *Astrophys. J.* 235:1055–1065.

Emslie, A. G. 1983. Thick-target bremsstrahlung interpretation of short time-scale solar hard X-ray features. *Astrophys. J.* 271:367–375.

Emslie, A. G., and Alexander, D. 1987. Line profiles in the impulsive phase of electron-heated solar flares. *Solar Phys.* 110:295–303.

Emslie, A. G., and Nagai, F. 1985. Gas dynamics in the impulsive phase of solar flares. II. The structure of the transition region—A diagnostic of energy transport processes. *Astrophys. J.* 288:779–788.

Emslie, A. G., and Smith, D. F. 1984. Microwave signature of thick-target electron beams in solar flares. *Astrophys. J.* 279:882–895.

Emslie, A. G., Brown, J. C., and Donnelly, R. F. 1978. The inter-relationship of hard X-ray and EUV bursts during solar flares. *Solar Phys.* 57:175–190.

Endal, A. S., and Sofia, S. 1978. The evolution of rotating stars. II. Calculations with time-dependent redistribution of angular momentum from 7 and 10 $M_\odot$ stars. *Astrophys. J.* 220:279–290.

Endal, A. S., and Sofia, S. 1981. Rotation of solar-type stars. I. Evolutionary models for the spin-down of the Sun. *Astrophys. J.* 243:625–640.

Endler, F., and Deubner, F.-L. 1983. The influence of seeing on the observation of short period fluctuations in the solar atmosphere. *Astron. Astrophys.* 121:291–296.

Engstler, S., Krauss, A., Neldner, K., Rolfs, C., Schroder, U., and Langanke, K. 1988. Effects of electron screening on the ^{3}He$(d,p)^4$He low-energy cross sections. *Phys. Lett.* B202:179–184.

Engvold, O. 1976. The fine structure of prominences. *Solar Phys.* 49:283–297.

Engvold, O. 1980. Thermodynamic models and fine structure prominences. *Solar Phys.* 67: 351–355.

Engvold, O., Brahde, R., and Fossum, B. 1981. Solar telescopes filled with helium? *Publ. Astron. Soc. Pacific* 93:526–527.

Engvold, O., Dunn, R. B., Livingston, W. C., and Smartt, R. N. 1983. Tests of vacuum vs helium in a solar telescope. *Appl. Opt.* 22:10–12.

Erozolimskii, B. G., Frank, A. I., Mostovoi, Yu. A., Arzumanov, S. S., and Voitzik, L. R. 1979. Measurement of the spin-electron correlation coefficient in the decay of polarized neutrons and determination of the G_A/G_V ratio. *Soviet J. Nucl. Phys.* 30:356–361.

Escaut-Soru, I., Martres, M.-J., and Mouradian, Z. 1984. Points singuliers de la rotation solaire. *Circ. Acad. Sci. Paris II* 299:545–548.

Espenak, F. 1989. Predictions for the total solar eclipse of 1991. *J. Roy. Astron. Soc. Canada* 83:157–178.

Esser, R., Leer, E., Habbal, S. R., and Withbroe, G. L. 1986. A two fluid solar wind model with Alfvén waves: Parameter study and application to observations. *J. Geophys. Res.* 91:2950–2960.

Evans, D. J., and Roberts, B. 1990. The influence of a chromospheric magnetic field on the solar *p*- and *f*-modes. II. Uniform chromospheric field. *Astrophys. J.* 356:704–719.

Evans, D. S. 1947. An analysis of solar atmospheric structure. *Mon. Not. Roy. Astron. Soc.* 107:433–451.

Evans, J. C., Davis, R., and Bahcall, J. N. 1974. Brookhaven solar neutrino detector and collapsing stars. *Nature* 251:486–488.

Evans, J. W., and Catalano, C. P. 1972. Observed oddities in the lines H, K, b, and H_β. *Solar Phys.* 27:299–301.

Evans, J. W., and Michard, R. 1962. Observational study of the macroscopic inhomogeneities in the solar atmosphere. III. Vertical oscillatory motions in the solar photosphere. *Astrophys. J.* 136:493–506.

Evans, J. W., Michard, R., and Servajean, R. 1963. Observational study of macroscopic inhomogeneities in the solar atmosphere. V. Statistical study of the time variations of solar inhomogeneities. *Ann. Astrophys.* 26:368–382.

Evenson, P., Meyer, P., Yanagita, S., and Forrest, D. J. 1984. Electron rich particle events and the production of gamma-rays by solar flares. *Astrophys. J.* 283:439–449.

Evenson, P., Koreger, R., and Meyer, P. 1985. Further observations of protons resulting from the decay of neutrons ejected by solar flares. In *Proc. 19th Intl. Cosmic Ray Conf.*, vol. 4, eds. F. C. Jones, J. Adams and G. M. Mason, NASA CP-2376, pp. 130–133.

Ewan, G. T., Evans, H. C., Lee, H. W., Leslie, J. R., Mak, H.-B., McLatchie, W., Robertson, B. C., Skiensved, P., Allen, R. C., Bühler, G., Chen, H. H., Doe, P. J., Sinclair, D., Tanner, N. W., Anglin, J. D., Bercovitch, M., Davidson, W. F., Hargrove, C. K., Mes, H., Storey, R. S., Earle, E. D., Milton, G. M., Jagam, P., Simpson, J. J., McDonald, A. B., Hallman, E. D., Carter, A. L., and Kessler, D. 1987. *Sudbury Neutrino Observatory Proposal* (Ottawa: SNO-87-12).

Fahlman, G. G., and Ulrych, T. J. 1982. A new method for estimating the power spectrum of gapped data. *Mon. Not. Roy. Astron. Soc.* 199:53–65.

Faller, J. E., Bender, P. L., Hall, J. L., Hils, D., and Vincent, M. A. 1985. Space antenna for gravitational wave astronomy. In *Proc. Coll. on Kilometric Optical Arrays in Space*, eds. N. Longdon and O. Melita, ESA SP-226, pp. 157–163.

Fan, C. Y., Gloeckler, G., and Hovestadt, D. 1984. The composition of heavy ions in solar energetic particle events. *Space Sci. Rev.* 38:143–178.

Fang, C., and Martres, M. J. 1986. Chromospheric "active region loops." *Solar Phys.* 105: 51–65.

Fang, C., Mouradian, Z., Bannos, G., Dumont, S., and Pecker, J. C. 1984. Structure and physics of solar faculae. VI. Chromospheric granular structure. *Solar Phys.* 91:61–70.

Farmer, C. B. 1987. High resolution infrared spectroscopy of the Sun and the Earth's atmosphere from space. *Mikrochim. Acta* III:189–214.

Farmer, C. B., and Norton, R. H., eds. 1989. *A High-Resolution Atlas of the Infrared Sectrum of the Sun and Earth Atmosphere from Space*, vol. 1, *The Sun*, NASA RP-1224.

Farmer, C. B., Raper, O. F., and O'Callaghan, F. G. 1987. *Final Report on the First Flight of the ATMOS Instrument during the Spacelab 3 Missions, April 29 through May 6, 1985.* NASA JPL Publ. 87-32, pp. 1–45.

Faulkner, D. J., and Cannon, R. D. 1973. Models of Population I clump giants. *Astrophys. J.* 180:435–446.

Faulkner, J., and Gilliland, R. L. 1985. Weakly interacting, massive particles and the solar neutrino flux. *Astrophys. J.* 299:994–1000.

Faulkner, J., Gough, D. O., and Vahia, M. N. 1986. Weakly interacting massive particles and solar oscillations. *Nature* 321:226–229.

Feldman, W. C. 1981. Electron characteristics in the high speed solar wind. In *Proc. Fourth Intl. Solar Wind Conf.*, ed. H. Rosenbauer, MPAE-W-100–81–31 (Katlenburg-Lindau: MPI für Aeronomie), pp. 217–225.

Feldman, W. C. 1988. Slow-mode shocks in the Earth's magnetosphere. In *Proc. Sixth Intl. Solar Wind Conf.*, eds. V. J. Pizzo, T. E. Holzer and D. G. Sime, NCAR TN-306 (Boulder: Natl. Center for Atmos. Res.), pp. 447–462.

Feldman, W. C., Asbridge, J. R., and Bame, S. J. 1974a. The solar wind He^{2+} to H^+ temperature ratio. *J. Geophys. Res.* 79:2319–2323.

Feldman, W. C., Asbridge, J. R., Bame, S. J., and Montgomery, M. D. 1974b. Interpenetrating solar wind streams. *Rev. Geophys. Space Phys.* 12:715–723.

Feldman, W. C., Abraham-Shrauner, B., Asbridge, J. R., and Bame, S. J. 1976a. The internal plasma state of the high speed solar wind at 1 AU. In *Physics of Solar Planetary Environments*, ed. D. J. Williams (Washington, D.C.: American Geophysical Union), pp. 413–427.

Feldman, W. C., Asbridge, J. R., Bame, S. J., and Gosling, J. T. 1976b. High-speed solar wind flow parameters at 1 AU. *J. Geophys. Res.* 81:5054–5060.

Feldman, W. C., Asbridge, J. R., Bame, S. J., and Gosling, J. T. 1977. Plasma and magnetic fields from the Sun. In *The Solar Output and Its Variations*, ed. O. R. White (Boulder: Colorado Assoc. Univ. Press), pp. 351–382.

Feldman, W. C., Asbridge, J. R., Bame, S. J., and Gosling, J. T. 1978. Long-term variations of selected solar wind properties: IMP 6, 7, and 8 results. *J. Geophys. Res.* 83:2177–2189.

Feldman, W. C., Doschek, G. A., Kpeplin, R. W., and Mariska, J. T. 1980. High-resolution X-ray spectra of solar flares. IV. General spectral properties of M type flares. *Astrophys. J.* 241:1175–1185.

Feldman, W. C., Asbridge, J. R., Bame, S. J., Fenimore, E. E., and Gosling, J. T. 1981. The solar origins of solar wind interstream flows: Near equatorial coronal streamers. *J. Geophys. Res.* 86:5408–5416.

Felli, M., Lang, K. R., and Willson, R. F. 1981. VLA observations of solar active regions. *Astrophys. J.* 247:325–337.

Fenimore, E. E. 1980. Solar wind flows associated with hot heavy ions. *Astrophys. J.* 235:247–257.

Ferziger, J. G., and Kaper, H. G. 1972. *Mathematical Theory of Transport Processes in Gases* (Amsterdam: North-Holland Press).

Fetisov, V. N., and Kopysov, Y. S. 1972. Are the solar-neutrino experiments suggestive of the existence of a resonance in the $^3He + ^3He$ system? *Phys. Lett.* B40:602–604.

Fetisov, V. N., and Kopysov, Y. S. 1975. Solar neutrinos and experiments to search for the hypothetical level in ^6Be. *Nucl. Phys.* A239:511–529.

Field, G. B. 1965. Thermal instability. *Astrophys. J.* 142:531–567.

Filippone, B. W., and Vogel, P. 1990. Statistical analysis of the time dependence of the solar neutrino capture rate. 1990. *Phys. Rev. Lett.* B246:546–550.

Filippone, B. W., Elwyn, A. J., Ray, W., and Koetke, D. D. 1982. Absolute cross section for $^7Li(d,p)^8Li$ and solar neutrino capture rates. *Phys. Rev.* C24:2174–2181.

Filippone, B. W., Elwyn, A. J., Davids, C. N., and Koetke, D. D. 1983. Proton capture cross section of ^7Be and the flux of high energy solar neutrinos. *Phys. Rev.* C28:2222–2230.

Finn, M. J., and Ott, E. 1988. Chaotic flows and fast magnetic dynamos. *Phys. Fluids* 31(10):2992–3011.

Fiorentini, G., and Mezzorani. 1991. Solar neutrinos, sunspot numbers, and the magnetic field in the convection zone. *Phys. Lett.* B253:181–184.

Fireman, E. L., Cleveland, B. T., Davis, R., and Rowley, J. K. 1985. Cosmic ray depth studies with ^{39}K-^{37}Ar detectors. In *Solar Neutrinos and Neutrino Astronmy*, eds. M. L. Cherry, K. Lande and W. A. Fowler (New York: American Inst. of Physics), pp. 22–31.

Fisher, G. H. 1987*a*. Chromospheric downflow velocities as a diagnostic in solar flares. *Solar Phys.* 113:307–311.

Fisher, G. H. 1987*b*. Explosive evaporation in solar flares. *Astrophys. J.* 317:502–513.

Fisher, G. H., Canfield, R. C., and McClymont, A. N. 1985*a*. Flare loop radiative hydrodynamics. IV. Chromospheric evaporation due to heating by non-thermal electrons. *Astrophys. J.* 289:425–433.

Fisher, G. H., Canfield, R. C., and McClymont, A. N. 1985*b*. Flare loop radiative hydrodynamics. V. Response to thick-target heating. *Astrophys. J.* 289:414–424.

Fisher, G. H., Canfield, R. C., and McClymont, A. N. 1985*c*. Flare loop radiative hydrodynamics. VII. Dynamics of the thick-target heated chromosphere. *Astrophys. J.* 289:434–441.

Fisher, R. R., and Sime, D. G. 1984. Rotation characteristics of the white-light solar corona. *Astrophys. J.* 287:959–968.

Fisher, R. R., and Poland, A. I. 1981. Coronal activity below 2 $R_\odot$: 1980 February 15–17. *Astrophys. J.* 246:1004–1009.

Fisk, L. A. 1978. ^{3}He-rich flares: A possible explanation. *Astrophys. J.* 224:1048–1055.

Fleck, B., and Deubner, F.-L. 1989. Dynamics of the solar atmosphere. II. Standing waves in the solar chromosphere. *Astron. Astrophys.* 224:245–252.

Flao, T., Osherovich, V., and Skumanich, A. 1982. On the magnetic and thermodynamic consequences of the return-flux sunspot model. *Astrophys. J.* 261:700–709.

Focas, J. H., and Banos, C. J. 1964. Photometric study of the atmospheric activity on the planet Jupiter and peculiar activity in its equatorial area. *Ann. Astrophys.* 27:36–45.

Foing, B., and Bonnet, R. M. 1984. On the origin of the discrete character of the solar brightness in the 160 nanometer continuum. *Astrophys. J.* 279:848–856.

Foley, H. M. 1946. The pressure broadening of spectral lines. *Phys. Rev.* 69:616–628.

Fontaine, G., Graboske, H. C., and van Horn, H. M. 1977. Equations of state for stellar partial ionization zones. *Astrophys. J. Suppl.* 35:293–358.

Forbush, S. E. 1966. Time-variations of cosmic rays. In *Encyclopedia of Physics*, vol. 1, ed. S. Flügge (New York: Springer-Verlag), pp. 159–247.

Forbush, S. E., Pomerantz, M. A., Duggal, S. P., and Tsao, C. H. 1983. Statistical considerations in the analysis of solar oscillation data by the superposed epoch method. *Solar Phys.* 82:113–122.

Forman, M. A., and Gleeson, L. J. 1975. Cosmic-ray streaming and anisotropies. *Astrophys. Space Sci.* 32:77–94.

Forman, M. A., Ramaty, R., and Zweibel, E. G. 1986. The acceleration and propagation of solar flare energetic particles. In *The Physics of the Sun*, vol. 2, eds. P. A. Sturrock, T. E. Holzer, D. M. Mihalas and R. K. Ulrich (Dordrecht: D. Reidel), pp. 249–289.

Fornberg, B. 1977. A numerical study of 2-D turbulence. *J. Comput. Phys.* 25:1–31.

Fossat, E. 1981. Solar and stellar oscillations. In *Solar Phenomena in Stars and Stellar Systems*, eds. R. M. Bonnet and A. K. Dupree (Dordrecht: D. Reidel), pp. 75–98.

Fossat, E. 1984. Atmospheric limitations in stellar seismology. Should one measure radial velocity or brightness fluctuations? In *Proc. Workshop on Improvements to Photometry*, eds. W. J. Borucki and A. Young, NASA CP-2350, pp. 68–78.

Fossat, E. 1985. Global oscillations. In *Future Missions in Solar, Heliospheric and Space Plasma Physics*, eds. E. Rolfe and B. Battrick, ESA SP-235, pp. 209–217.

Fossat, E. 1988*a*. IRIS: A network for full disk helioseismology. In *Seismology of the Sun and Sun-Like Stars*, ed. E. J. Rolfe, ESA SP-286, pp. 161–162.

Fossat, E. 1988*b*. Multi-year variations of solar oscillations. *Adv. Space Res.* 8(7):107–114.

Fossat, E. 1990. The IRIS network for full disk helioseismology. Present status of the program. *Solar Phys.*, in press.

Fossat, E., and Ricort, G. 1973. Contribution to the observation of photospheric oscillations. *Solar Phys.* 28:311–317.

Fossat, E., and Ricort, G. 1975. Photospheric oscillations. I. Large scale observations by optical resonance method. *Astron. Astrophys.* 43:243–252.

Fossat, E., and Roddier, F. 1971. A sodium experiment for photospheric velocity field observations. *Solar Phys.* 18:204–210.

Fossat, E., Harvey, J. W., Hausman, M., and Slaughter, C. 1977. Apparent solar oscillations and fluctuations in the Earth's atmosphere. *Astron. Astrophys.* 59:279–282.

Fossat, E., Grec, G., and Harvey, J. W. 1981. Power spectrum of differential refraction and comparison with solar diameter fluctuation measurements. *Astron. Astrophys.* 94:95–99.

Fossat, E., Gelly, B., Grec, G., and Pomerantz, M. A. 1987. Search for solar *p*-mode frequency changes between 1980 and 1985. *Astron. Astrophys.* 177:L47–L48.

Fossat, E., Gelly, B., Grec, G., and Schmider, F.-X. 1989. Full-disk helioseismology in the Antarctic. In *Astrophysics in Antarctica,* eds. D. J. Mullan, M. A. Pomerantz and T. Stanev (New York: American Inst. of Physics), pp. 231–233.

Forrest, D. J., Vestrand, W. T., Chupp, E. L., Rieger, E., Cooper, J., and Share, G. 1985. Neutral pion production in solar flares. In *Proc. 19th Intl. Cosmic Ray Conf.,* vol. 4, eds. F. C. Jones, J. Adams and G. M. Mason, NASA CP-2376, pp. 146–149.

Fossat, E., Gavrjuseva, E., Garjusev, W., and Grec, G. 1988. Searching for solar *g*-mode tests of a statistical method. In *Seismology of the Sun and Sun-Like Stars,* ed. E. J. Rolfe, ESA SP-286, pp. 393–398.

Foukal, P. V. 1977. Supergranulation and the dynamics of gas and magnetic field below the solar photosphere. *Astrophys. J.* 218:539–546.

Foukal, P. V. 1979. A Doppler measurement with low scattered light of the higher rotation rate of sunspot magnetic fields at the photosphere. *Astrophys. J.* 234:716–722.

Foukal, P. V. 1981. Sunspots and changes in the global output of the Sun. In *The Physics of Sunspots,* eds. L. E. Cram and J. H. Thomas (Sunspot, N.M.: Sacramento Peak Obs.), pp. 391–423.

Foukal, P. V., and Jokipii, J. R. 1975. On the rotation of gas and magnetic fields at the photosphere. *Astrophys. J.* 199:L71–L73.

Foukal, P. V., and Lean, J. 1986. The influence of faculae on total solar irradiance and luminosity. *Astrophys. J.* 302:826–835.

Foukal, P. V., and Lean, J. 1988*a*. Magnetic modulation by photospheric activity. *Astrophys. J.* 328:347–357.

Foukal, P. V., and Lean, J. 1988*b*. Magnetic modulation of solar luminosity by photospheric activity. *Astrophys. J.* 336:L33–L55.

Foukal, P. V., Little, R., and Mooney, J. 1989. Infrared imaging of sunspots and faculae at the photospheric opacity minimum. *Astrophys. J.* 336:L33–L35.

Foukal, P. V., Little, R., Graves, J., Rabin, D., and Lynch, D. 1990. Infrared imaging of faculae at the deepest photospheric layers. *Astrophys. J.* 353:712–715.

Fowler, W. A. 1958. Completion of the proton-proton reaction chain and the possibility of energetic neutrino emission by hot stars. *Astrophys. J.* 127:551–556.

Fowler, W. A. 1972. What cooks with solar neutrinos? *Nature* 238:24–26.

Fowler, W. A., Caughlin, G. R., and Zimmerman, B. A. 1967. Thermonuclear reaction rates. *Ann. Rev. Astron. Astrophys.* 5:525–570.

Fowler, W. A., Caughlin, G. R., and Zimmerman, B. A. 1975. Thermonuclear reaction rates. II. *Ann. Rev. Astron. Astrophys.* 13:69–112.

Fox, P. 1985. Compressible Convection in the Sun. Ph.D. Thesis, Monash Univ., Australia.

Fox, P., and Van der Borght, R. 1985. Solar granulation: The influence of viscosity laws on theoretical models. *Proc. Astron. Soc. Australia* 6:60–62.

Frandsen, S. 1986. Phase studies of solar 5 min oscillations. In *Seismology of the Sun and the Distant Stars,* ed. D. O. Gough (Dordrecht: D. Reidel), pp. 73–80.

Frandsen, S. 1988. Diagnosis of the solar atmosphere using solar oscillations. In *Advances in Helio- and Asteroseismology: Proc. IAU Symp. 123,* eds. J. Christensen-Dalsgaard and S. Frandsen (Dordrecht: D. Reidel), pp. 405–419.

Frazier, E. N. 1968. A spatio-temporal analysis of velocity fields in the solar photosphere. *Zeit. Astrophys.* 68:345–356.

Frazier, E. N. 1970. Multichannel magnetograph observations. II. Supergranulation. *Solar Phys.* 14:89–111.

Frazier, E. N. 1972. The magnetic structure of arch filament systems. *Solar Phys.* 26:130–141.

Frazier, E. N., and Stenflo, J. O. 1978. Magnetic velocity and brightness structures in solar faculae. *Astron. Astrophys.* 70:789–799.

Freeman, K. C., and Norris, J. E. 1981. The chemical composition, structure, and dynamics of globular clusters. *Ann. Rev. Astron. Astrophys.* 19:319–356.

Fricke, K. 1968. Instabilität Stationären Rotations in Sternen. *Zeit. Astrophys.* 68:317–344.

Fried, D. L. 1965. Statistics of a geometric representation of wavefront distortion. *J. Opt. Soc. Amer.* 55:1427–1435.

Fried, D. L. 1966. Optical resolution through a randomly inhomogenous medium for very long and very short exposures. *J. Opt. Soc. Amer.* 56:1372–1379.

Fried, D. L. 1978. Probability of getting a lucky short-exposure image through turbulence. *J. Opt. Soc. Amer.* 68:1651–1658.

Friedlander, G. 1978. Report on an informal conference on the status and future of solar-neutrino research. *Comments Astrophys.* 8:47–54.

Friedlander, G., and Weneser, J. 1987. Solar neutrinos: Experimental approaches. *Science* 235:760–765.

Fröhlich, C. 1977. Contemporary measures of the solar constant. In *The Solar Output and Its Variation,* ed. O. R. White (Boulder: Colorado Assoc. Univ. Press), pp. 93–109.

Fröhlich, C. 1984. Wavelength dependence of solar luminosity fluctuations in the five-minute range. *Mem. Soc. Astron. Italiana* 55:237–243.

Fröhlich, C. 1987. Solar oscillations and helioseismology from ACRIM/SMM irradiance data. In *New and Exotic Phenomena,* eds. O. Fackler and J. T. T. Vân (Gif sur Yvette: Editions Frontières), pp. 397–404.

Fröhlich, C. 1988. Solar gravity modes from ACRIM/SMM irradiance data. In *Advances in Helio- and Asteroseismology: Proc. IAU Symp. 123,* eds. J. Christensen-Dalsgaard and S. Frandsen (Dordrecht: D. Reidel), pp. 83–86.

Fröhlich, C., and Delache, P. 1984a. Solar gravity modes for ACRIM/SMM irradiance data. *Mem. Soc. Astron. Italiana* 55:99–105.

Fröhlich, C., and Delache, P. 1984b. Solar gravity modes from ACRIM/SMM irradiance data. In *Solar Seismology from Space,* NASA JPL Publ. 84–84, pp. 183–193.

Fröhlich, C., and van der Raay, H. B. 1984. Global solar oscillations in irradiance and velocity: A comparison. In *The Hydromagnetics of the Sun,* eds. T. D. Guyenne and J. J. Hunt, ESA SP-220, pp. 17–20.

Fröhlich, C., Andersen, B. N., Berthomieu, G., Crommelynck, D., Delache, P., Domingo, V., Jiménez, A., Jones, A. R., Roca-Cortés, T., and Wehrli, C. 1988a. VIRGO: The solar monitor experiment on SOHO. In *Seismology of the Sun and Sun-Like Stars,* ed. E. J. Rolfe, ESA SP-286, pp. 371–374.

Fröhlich, C., Bonnet, R. M., Bruns, A. V., Delaboudinière, J. P., Domingo, V., Kotov, V. A., Kollath, Z., Rashkovsky, D. N., Toutain, T., Vial, J. C., and Wehrli, C. 1988b. IPHIR: The helioseismology experiment on the PHOBOS mission. In *Seismology of the Sun and Sun-Like Stars,* ed. E. J. Rolfe, ESA SP-286, pp. 359–362.

Fröhlich, C., Toutain, T., Bonnet, R. M., Bruns, A. V., Delaboudinière, J. P., Domingo, V., Kotov, V. A., Kollath, Z., Rashkovsky, D. N., Vial, J. C., and Wehrli, C. 1990. Recent helioseismological results from space: Solar *p*-mode oscillations from IPHIR on the PHOBOS mission. In *Inside the Sun: Proc. IAU Coll. 121,* eds. G. Berthomieu and M. Cribier (Dordrecht: Kluwer), pp. 279–288.

Fu, Q., Kundu, M. R., and Schmahl, E. J. 1987. Coronal bright points at 6 cm wavelength. *Solar Phys.* 108:99–111.

Fujikawa, K., and Shrock, R. E. 1980. Magnetic moment of a massive neutrino and neutrino-spin rotation. *Phys. Rev. Lett.* 45:963–966.

Fukugita, M., and Yanagida, T. 1987. Particle-physics model for the Voloshin-Vysotskii-Okun solution to the solar neutrino problem. *Phys. Rev. Lett.* 58:1807–1809.

Fukugita, M., Kohyama, Y., Kubodera, K., and Kuromoto, T. 1989. Reaction cross sections for "solar flare neutrinos" with ^{37}Cl and ^{16}O targets. *Astrophys. J.* 337:L59–L63.

Fukui, K. 1988. Unique increase of solar Lyman alpha flux observed by Ae-E satellite during cycle 21. *Eos: Trans. AGU* 69:417.

Fuller, G. A., and Myers, P. C. 1987. Dense cores in dark clouds. In *Physical Processes in Interstellar Clouds,* eds. G. E. Morfill and M. Scholer (Dordrecht: D. Reidel), pp. 137–160.

Furth, H. P., Killeen, J., and Rosenbluth, M. N. 1963. Finite-resistivity instabilities of a sheet pinch. *Phys. Fluids* 6:959–965.

Gabriel, A. H., Bocchia, R., Bonnet, R. M., Cesarky, C., Christensen-Dalsgaard, J., Damé, L., Delache, P., Deubner, F.-L., Foing, B., Fossat, E., Fröhlich, C., Gorisse, M., Gough, D. O., Grec, G., Hoyng, P., Pallé, P. L., Paul, J., Robillot, J. P., Roca-Cortés, T., Stenflo,

J., Ulrich, R. K., and van der Raay, H. B. 1989. GOLF—global oscillations at low frequencies for SOHO mission. In *The SOHO Mission*, ESA SP-1104, pp. 13–17.

Gabriel, M. 1979. Properties of non-radial oscillations of spherical stars. *Bul. Acad. Roy. Belgique* 64:530–538.

Gabriel, M. 1980. On the multiplicity of the eigenvalues of nonradial oscillations. *Astron. Astrophys.* 82:8–13.

Gabriel, M. 1984a. Solar g modes. Comparison between theoretical and observed periods. *Astron. Astrophys.* 134:387–389.

Gabriel, M. 1984b. Solar g modes: Comparison between observations and theoretical periods for 3 solar modes. In *Theoretical Problems in Stellar Stability and Oscillations*, eds. A. Noëls and M. Gabriel (Liège: Inst. d'Astrophysique), pp. 284–289.

Gabriel, M. 1986a. A theorem concerning nonradial oscillations. *Astron. Astrophys.* 158:395.

Gabriel, M. 1986b. Solar g-modes: A method to find depth of the convective envelope. In *Seismology of the Sun and the Distant Stars*, ed. D. O. Gough (Dordrecht: D. Reidel), pp. 177–186.

Gabriel, M. 1989. The D_{nl} values and the structure of the solar core. *Astron. Astrophys.* 226: 278–283.

Gabriel, M., and Noëls, A. 1976. Stability of a 30 $M_\odot$ star towards g^+ modes of high spherical harmonic values. *Astron. Astrophys.* 53:149–157.

Gabriel, M., and Scuflaire, R. 1979. Properties of non-radial stellar oscillations. *Acta Astron.* 29:135–149.

Gabriel, M., Scuflaire, R., Noëls, A., and Boury, A. 1974. *Bul. Acad. Roy. Belgique* 40: 866–887.

Gabriel, M., Scuflaire, R., Noëls, A., and Boury, A. 1975. Influence of convection on the vibrational stability of stars towards non-radial oscillations. *Astron. Astrophys.* 40:33–39.

Gabriel, M., Scuflaire, R., and Noëls, A. 1982. The solar structure and the low ℓ five-minute oscillation. I. *Astron. Astrophys.* 110:50–53.

Gadun, A. S., Kostyk, R. I., and Sheminova, V. A. 1987. On the depth dependence of the solar rotation velocity. *Astr. Zh.* 64:1066–1071 (in Russian).

Gaizauskas, V. 1989. Preflare activity. In *Solar and Stellar Flares: Proc. IAU Coll. 104*, eds. B. M. Haisch and M. Rodonò (Dordrecht: Kluwer), pp. 135–152.

Gaizauskas, V., Harvey, K. L., Harvey, J. W., and Zwaan, C. 1983. Large-scale patterns formed by solar active regions during the ascending phase of cycle 21. *Astrophys. J.* 265:1056–1065.

Galloway, D. J., and Moore, D. R. 1979. Axysymmetric convection in the presence of a magnetic field. *Geophys. Astrophys. Fluid Dyn.* 12:73–105.

Galloway, D. J., and Weiss, N. O. 1981. Convection and magnetic fields in stars. *Astrophys. J.* 243:945–953.

Galvin, A. B., Ipavich, F. M., Gloeckler, G., Hovestadt, D., Klecker, B., and Scholer, M. 1984. Solar wind ionization temperatures inferred from charge state composition of diffuse particle events. *J. Geophys. Res.* 89:2655–2671.

Gamow, G. 1928. Zur Quantentheorie der Atomzertrümmerung. *Z. Phys.* 52:510–515.

Garcia de la Rosa, J. I. 1986. On the initial orientation of emerging active regions. *Solar Phys.* 103:249–257.

Garcia-Lopez, R., and Spruit, H. 1990. *Astrophys. J.*, submitted.

Gari, M. 1978. The solar p + p reaction—A well known process? In *Proc. Informal Conf. on Status and Future of Solar Neutrino Research*, vol. 1, ed. G. Friedlander, BNL Report 50879 (Upton, N.Y.: Brookhaven National Lab), pp. 137–156.

Gary, G. A., Moore, R. L., Hagyard, M. J., and Haisch, B. M. 1987. Nonpotential features observed in the magnetic field of an active region. *Astrophys. J.* 314:782–794.

Gavrin, V. N. 1991. In *Neutrino '90: Proc. 14th Intl. Conf. on Neutrino Physics and Astrophysics*, ed. J. Panman (Amsterdam: North-Holland Press), in press.

Gavrin, V. N., and Kopylov, A. V. 1984. Does the Davis Neutrino Experiment show a variable [37]Ar production rate? *Soviet Astron. Lett.* 10:62–64.

Gavrin, V. N., Kopysov, Yu. S., and Makeev. 1990. Study of Sun's neutrino brightness curve with the help of a chlorine-argon neutrino detector. *JETP Lett.* 35:608–611.

Gavryusev, V. G., and Gavryuseva, E. A. 1988. The role of nuclear reactions in the excitation of stellar oscillations. In *Advances in Helio- and Asteroseismology: Proc. IAU Symp. 123*, eds. J. Christensen-Dalsgaard and S. Frandsen (Dordrecht: D. Reidel), pp. 363–370.

Gavryuseva, E. A., Gavryusev, V. G., and Rosljakov, A. G. 1990. *Statistical analysis of neutrino capture rate in* ^{37}Cl *experiment.* Inst. Nuclear Research Report 0672.

Gay, J. 1970. Balloon observations in the submillimeter region: An absolute measurement of the brightness of the Sun and the transparency of the stratosphere above 25 km. *Astron. Astrophys.* 6:327–348.

Gazdag, J. 1976. Time-differencing schemes and transform methods. *J. Comput. Phys.* 20: 196–207.

Gear, G. W. 1971. *Numerical Initial Value Problems in Ordinary Differential Equations* (Englewood Cliffs, N.J.: Prentice Hall), pp. 104–111, 129–131.

Geiger, G. 1982. Patterns of thermal convection in slowly rotating weakly magnetic shells. *Phys. Earth Planet. Int.* 28:185–190.

Geiss, J. 1973a. Noble gas isotopes and deuterium in the solar system. In *Origin of the Solar System,* ed. F. Reeves (Paris: CNRS), pp. 217–236.

Geiss, J. 1973b. Solar wind composition and implications about the history of the solar system. In *Proc. 13th Intl. Cosmic Ray Conf.,* vol. 5, pp. 3375–3398.

Geiss, J. 1982. Processes affecting abundances in the solar wind. *Space Sci. Rev.* 33:201–217.

Geiss, J., and Bochsler, P. 1985. Ion composition in the solar wind in relation to solar abundances. In *Isotopic Ratios in the Solar System* (Paris: Editions CNES), pp. 213–229.

Geiss, J., and Bochsler, P. 1986. Solar wind composition and what we expect to learn from out-of-ecliptic measurements. In *The Sun and the Heliosphere in Three Dimension,* ed. B. G. Mardsen (Dordrecht: D. Reidel), pp. 173–186.

Geiss, J., and Burgi, A. 1986a. Diffusion and thermal diffusion in partially ionized gases in the atmospheres of the Sun and planets. *Astron. Astrophys.* 159:1–15.

Geiss, J., and Burgi, A. 1986b. Diffusion and thermal diffusion in partially ionized gases in the atmospheres of the Sun and planets (erratum). *Astron. Astrophys.* 166:398.

Geiss, J., and Burgi, A. 1986c. Thermal diffusion in partially ionized gases: The case of unequal temperatures. *Astron. Astrophys.* 178:286–291.

Geiss, J., Hirt, P., and Leutwyler, H. 1970. On acceleration and motion of ions in corona and solar wind. *Solar Phys.* 12:458–483.

Geiss, J., Buehler, F., Cerruti, H., Eberhardt, P., and Filleaux, Ch. 1972. Solar wind composition experiment. In *Apollo 16 Preliminary Science Report,* NASA SP-315, pp.14–1— 14–10.

Gelly, B. 1990. The IRIS data management. *Solar Phys.,* in press.

Gelly, B., Fossat, E., and Grec, G. 1988a. New analysis of the solar *p*-mode frequency change from 1980 to 1986. *Astron. Astrophys.* 200:L29–L31.

Gelly, B., Fossat, E., Grec, G., and Schmider, F.-X. 1988b. Solar calibration of asteroseismology. In *Seismology of the Sun and the Sun-Like Stars,* ed. E. J. Rolfe, ESA SP-286, pp. 579–582.

Gelly, B., Fossat, E., Grec, G., and Schmider, F.-X. 1988c. Solar calibration of asteroseismology. *Astron. Astrophys.* 200:207–212.

Geltman, S. 1965. Continuum states of H^- and the free-free absorption coefficient. *Astrophys. J.* 141:376–394.

Germain, M. E., and Hill, H. A. 1988. Gravity mode detection sensitivity of resonant-scattering spectrometers. *Bull. Amer. Astron. Soc.* 19:1120 (abstract).

Gezari, D. Y., Joyce, R. R., and Simon, M. 1973. Measurement of the solar brightness temperature at 345μm, 450μm and 1000μm. *Astron. Astrophys.* 26:409–411.

Giardini, D., Li, X.-D., and Woodhouse, J. H. 1987. Three-dimensional structure of the Earth from splitting in free-oscillation spectra. *Nature* 325:405–411.

Gilbert, A., Frish, U., and Pouquet, A. 1988. Helicity is unnecessary for alpha effect dynamos, but it helps. *Geophys. Astrophys. Fluid Dyn.* 42:151–161.

Gilbert, F. 1971. Ranking and winnowing gross Earth data for inversion and resolution. *Geophys. J. Roy. Astron. Soc.* 23:125–128.

Gilliland, R. L. 1981. Solar radius variations over the past 265 years. *Astrophys. J.* 248:1144–1155.

Gilliland, R. L. 1988. Mechanisms of slow change in solar luminosity. In *Solar Radiative Output Variations,* ed. P. V. Foukal (Cambridge, Mass.: Cambridge Research and Instrumentation), pp. 289–300.

Gilliland, R. L., and Däppen, W. 1988. Oscillations in solar models with weakly interacting massive particles. *Astrophys. J.* 324:1153–1157.

Gilliland, R. L., Faulkner, J., Press, W. H., and Spergel, D. N. 1986. Solar models with energy transport by weakly interacting particles. *Astrophys. J.* 306:703–709.

Gilman, P. A. 1972. Nonlinear Boussinesq convection roll for large scale solar circulation. *Solar Phys.* 27:3–26.

Gilman, P. A. 1974. Solar rotation. *Ann. Rev. Astron. Astrophys.* 12:47–70.

Gilman, P. A. 1977*a*. Nonlinear dynamics of Boussinesq convection in a deep rotating spherical shell. I. *Geophys. Astrophys. Fluid Dyn.* 8:93–135.

Gilman, P. A. 1977*b*. A note on estimating the latitudinal angular momentum transport in the solar photosphere from Doppler velocities. *Astron. Astrophys.* 58:315–318.

Gilman, P. A. 1980*a*. Differential rotation in stars with convection zones. In *Stellar Turbulence: Proc. IAU Coll. 51*, eds. D. F. Gray and J. L. Linsky (Berlin: Springer-Verlag), pp. 19–37.

Gilman, P. A. 1980*b*. Differential rotation in stars with convection zones. *Lecture Notes in Phys.* 114:19–37.

Gilman, P. A. 1983. Dynamically consistent nonlinear dynamos driven by convection in a rotating spherical shell. II. Dynamos with strong cycles and feedbacks. *Astrophys. J. Suppl.* 53:243–268.

Gilman, P. A. 1986. The solar dynamo: Observations and theories of solar convection, global circulation, and magnetic fields. In *Physics of the Sun*, vol. 1, eds. P. A. Sturrock, T. E. Holzer, D. M. Mihalas and R. K. Ulrich (Dordrecht: D. Reidel), pp. 95–160.

Gilman, P. A. 1989. In *Systematic Observations of the Sun, IAU Joint Discussion Commissions 10 and 12*, in press.

Gilman, P. A., and Glatzmaier, G. A. 1981. Compressible convection in a rotating spherical shell. I. Anelastic equations. *Astrophys. J. Suppl.* 45:335–349.

Gilman, P. A., and Howard, R. 1984*a*. On the correlation of longitudinal and latitudinal motions of sunspots. *Solar Phys.* 93:171–175.

Gilman, P. A., and Howard, R. 1984*b*. Variations in solar rotation with the sunspot cycle. *Astrophys. J.* 283:385–391.

Gilman, P. A., and Howard, R. 1985. Rotation rates of leader and follower sunspots. *Astrophys. J.* 295:233–240.

Gilman, P. A., and Howard, R. 1986. Rotation expansion within sunspot groups. *Astrophys. J.* 303:480–485.

Gilman, P. A., and Miller, J. 1981. Dynamically consistent convection in a rotating spherical shell. *Astrophys. J. Suppl.* 46:211–238.

Gilman, P. A., and Miller, J. 1986. Nonlinear convection of compressible fluid in a rotating spherical shell. *Astrophys. J. Suppl.* 61:585–608.

Gilman, P. A., Morrow, C. A., and DeLuca, E. E. 1989. Angular momentum transport and dynamo action in the Sun: Implications of recent oscillation measurements. *Astrophys. J.* 338:528–537.

Giménez, A., and Garcia-Pelayo, J. M. 1982. On the internal structure of main-sequence stars. In *Binary and Multiple Stars as Tracers of Stellar Evolution: Proc. IAU Coll. 69* (Dordrecht: D. Reidel), pp. 37–46.

Gingerich, O., and de Jager, C. 1968. The Bilderberg model of the photosphere and low chromosphere. *Solar Phys.* 3:5–15.

Gingerich, O., Noyes, R. W., Kalkofen, W., and Cuny, Y. 1971. The Harvard-Smithsonian reference atmosphere. *Solar Phys.* 18:347–365.

Gingold, R. A. 1974. Asymptotic giant branch evolution of a $0.6M_\odot$ star. *Astrophys. J.* 193:177–185.

Giovanelli, R. G. 1966. Sunspot minima. In *Proc. Meeting on Sunspots*, ed. G. Barbèra (Florence: Galileo Quadricentennial Committee), pp. 57–58.

Giovanelli, R. G. 1974. Waves and oscillations in the chromosphere in active and quiet regions. In *Chromospheric Fine Structure: Proc. IAU Symp. 56*, ed. R. G. Athay (Dordrecht: D. Reidel), pp. 137–151.

Giovanelli, R. G. 1977. Gas entry into non-spot magnetic fields. *Solar Phys.* 52:315–325.

Giovanelli, R. G. 1980. An exploratory two-dimensional study of the coarse structure of network magnetic fields. *Solar Phys.* 68:49–69.

Giovanelli, R. G., and Jones, H. P. 1982. The three dimensional structure of atmospheric magnetic fields in two active regions. *Solar Phys.* 79:267–278.

Giovanelli, R. G., and Slaughter, C. 1978. Motions in solar magnetic tubes. I. The downflow. *Solar Phys.* 57:255–260.

Glackin, D. L. 1974. Differential rotation of solar filaments. *Solar Phys.* 36:51–60.

Glashow, S. L., and Krauss, L. M. 1987. "Just so" neutrino oscillations. *Phys. Lett.* 190B: 199–207.

Glatzmaier, G. A. 1984. Numerical simulations of stellar convective dynamos. I. The model and method. *J. Comput. Phys.* 55:461–484.

Glatzmaier, G. A. 1985a. Numerical simulations of stellar convective dynamos. II. Field propagation in the convection zone. *Astrophys. J.* 291:300–307.

Glatzmaier, G. A. 1985b. Numerical simulation of stellar convective dynamos. III. At the base of the convection zone. *Geophys. Astrophys. Fluid Dyn.* 31:137–150.

Glatzmaier, G. A. 1987. A review of what numerical simulations tell us about the internal rotation of the Sun. In *The Internal Solar Angular Velocity: Theory, Observations and Relationship to Solar Magnetic Fields,* eds. B. R. Durney and S. Sofia (Dordrecht: D. Reidel), pp. 263–274.

Glatzmaier, G. A., and Gilman, P. A. 1981. Compressible convection in a rotating spherical shell. II. A linear anelastic model. *Astrophys. J. Suppl.* 45:351–380.

Glatzmeier, G. A., and Gilman, P. A. 1982. Compressible convection in a rotating spherical shell. V. Induced differential rotation and meridional circulation. *Astrophys. J.* 256:316–330.

Glenar, D. A., Deming, D., Jennings, D. E., Kostiuk, T., and Mumma, M. J. 1983. Diode laser heterodyne observations of silicon monoxide in sunspots: A test of three sunspot models. *Astrophys. J.* 269:309–318.

Glenar, D. A., Deming, D., Espenak, F., Kostiuk, T., and Mumma, M. J. 1986. A laser heterodyne spectrometer for helioseismology. *Appl. Opt.* 25:58–62.

Glenar, D. A., Käufl, H. U., Deming, D., Kostiuk, T., and Mumma, M. J. 1988a. Infrared heterodyne spectroscopy: A tool for helioseismology. In *Advances in Helio- and Asteroseismology: Proc. IAU Symp. 123,* eds. J. Christensen-Dalsgaard and S. Frandsen (Dordrecht: D. Reidel), pp. 481–484.

Glenar, D. A., Reuter, D. C., Deming, D., and Chang, E. S. 1988b. Mg I absorption features in the solar spectrum near 9 and 12 microns. *Astrophys. J.* 335:L35–L38.

Global Oscillation Network Group. 1987. *Report Number 5: The 1987 Artificial Data Workshop* (Tucson: Natl. Optical Astron. Obs.).

Godoli, G., and Mazzucconi, F. 1979. On the rotation rates of sunspot groups. *Solar Phys.* 64:247–254.

Godoli, G., and Mazzucconi, F. 1983. Relationships between photospheric plasma, angular velocity and solar activity. *Solar Phys.* 83:339–347.

Gokhale, M. H., and Zwaan, C. 1972. The structure of sunspots. I. Observational constraints, current sheet models. *Solar Phys.* 26:52–75.

Goldberg, L. 1939. The temperature of the solar chromosphere. *Astrophys. J.* 89:673–678.

Goldberg, L. 1982. Possible origins for the 12 μm emission lines in the solar spectrum. In *Activity in Red Dwarf Stars: Proc. IAU Symp. 71,* eds. P. B. Byrne and M. Rodonò (Dordrecht: D. Reidel), pp. 327–330.

Goldberg, L., and Dupree, A. K. 1967. Population of atomic levels by dielectronic recombination. *Nature* 215:41–43.

Goldberg, L., McMath, R. R., Mohler, O. C., and Pierce, A. K. 1952. Identification of CO in the solar atmosphere. *Phys. Rev.* 85:140.

Goldman, A., Blatherwick, R. D., Murcray, F. H., Van Allen, J. W., Bradford, C. M., Cook, G. R., and Murcray, D. G. 1980. *New Atlas of IR Solar Spectra: Volume 1—Line Positions* and *Identifications* and *Volume 2—The Spectra* (Denver: Univ. of Denver).

Goldman, A., Murcray, F. J., Gillis, J. R., and Murcray, D. G. 1981. Identifications of new solar OH lines in the 10–12 micron region. *Astrophys. J.* 248:L133–L135.

Goldman, A., Murcray, D. G., Lambert, D. L., and Dominy, J. F. 1983. The pure rotation spectrum of the hydroxyl radical and the solar oxygen abundance. *Mon. Not. Roy. Astron. Soc.* 203:767–776.

Goldreich, P., and Keeley, D. A. 1977a. Solar seismology. I. The stability of the solar *p*-modes. *Astrophys. J.* 211:934–942.

Goldreich, P., and Keeley, D. A., 1977*b*. Solar seismology. II. The stochastic excitation of the solar *p*-modes by turbulent convection. *Astrophys. J.* 212:243–251.

Goldreich, P., and Kumar, P. 1988. The interaction of acoustic radiation with turbulence. *Astrophys. J.* 326:462–478.

Goldreich, P., and Nicholson, P. 1989*a*. Tides in rotating fluids. *Astrophys. J.* 342:1075–1078.

Goldreich, P., and Nicholson, P. 1989*b*. Tidal friction in early-type stars. *Astrophys. J.* 342:1079–1084.

Goldreich, P., and Schubert, G. 1967. Differential rotation in stars. *Astrophys. J.* 150:571–587.

Golub, L., Krieger, A. S., Silk, J. K., Timothy, A. F., and Vaiana, G. S. 1974. Solar X-ray bright points. *Astrophys. J.* 189:L93–L97.

Golub, G. H., Heath, M., and Wahba, G. 1979. Generalized cross-validation as a method for choosing a good ridge parameter. *Technometrics* 21:215–223.

Golub, L., Rosner, R., Vaiana, G. S., and Weiss, N. O. 1981. Solar magnetic fields: The generation of emerging flux. *Astrophys. J.* 243:309–316.

Gomez, M. T., Marmolino, C., Roberti, G., and Severino, G. 1987. Temporal variations of solar spectral line profiles induced by the five-minute photospheric oscillation. *Astron. Astrophys.* 188:169–177.

Gonczi, G., and Osaki, Y. 1980. On local theories of time-dependent convection in the stellar pulsation problem. *Astron. Astrophys.* 84:304–310.

Goode, P. R., and Kuhn, J. R. 1990. Consistency in trends in helioseismic data and photospheric temperature data through the solar cycle. *Astrophys. J.* 356:310–314.

Goode, P. R., Dziembowski, W. A., Rhodes, E. J., Jr., and Korzennik, S. G. 1990. Has the Sun's internal rotation changed through this activity cycle? In *Oji Intl. Seminar on Progress of Seismology of the Sun and Stars*, eds. H. Shibahashi and Y. Osaki (Berlin: Springer-Verlag), in press.

Goode, P. R., Dziembowski, W. A., Korzennik, S. G., and Rhodes, E. J., Jr. 1991. What we know about the Sun's internal rotation from solar oscillations. *Astrophys. J.*, in press.

Goodman, C. D. 1985. Can (p,n) reactions help us determine neutrino cross sections? In *Solar Neutrinos and Neutrino Astronomy*, eds. M. L. Cherry, K. Lande and W. A. Fowler (New York: American Inst. of Physics), pp. 109–120.

Goossens, M. 1976. Nonradial oscillations and stability of a polytrope n = 3 with a toroidal magnetic field. *Astrophys. Space Sci.* 44:397–404.

Gopasyuk, S. I., and Demkina, L. B. 1987. Rotation of the Sun and structure of magnetic elements. *Soviet Astron.* 31:205–207.

Gosling, J. T., Hildner, E., MacQueen, R. M., Munro, R. H., Poland, A. I., and Ross, C. L. 1976. The speeds of coronal mass ejection events. *Solar Phys.* 48:389–398.

Gosling, J. T., Asbridge, J. R., Bame, S. J., Feldman, W. C., and Zwickl, R. D. 1980. Observations of large fluxes of He$^+$ in the solar wind following an interplanetary shock. *J. Geophys. Res.* 85:3431.

Gosling, J. T., Borrini, G., Asbridge, J. R., Bame, S. J., Feldman, W. C., and Hansen, R. T. 1981. Coronal streamers in the solar wind at 1 AU. *J. Geophys. Res.* 86:5438–5448.

Gough, D. O. 1969. The anelastic approximation for thermal convection. *J. Atmos. Sci.* 26:448–456.

Gough, D. O. 1976. The current state of mixing-length theory. In *Problems of Stellar Convection: Proc. IAU Coll. 38*, eds. E. A. Spiegel and J. P. Zahn (Berlin: Springer-Verlag), pp. 15–56.

Gough, D. O. 1977*a*. Mixing length theory for pulsating stars. *Astrophys. J.* 214:196–213.

Gough, D. O. 1977*b*. Random remarks on solar hydrodynamics. In *The Energy Balance and Hydrodynamics of the Solar Chromosphere and Corona: Proc. IAU Coll. 36*, eds. R. M. Bonnet, P. Delache and G. de Bussac (Clermont-Ferrand: G. de Bussac), pp. 3–36.

Gough, D. O. 1978. The significance of solar oscillations. In *Pleins Feux sur la Physique Solaire*, ed. J. Rösch (Paris: CNRS), pp. 81–103.

Gough, D. O. 1980. Some theoretical remarks on solar oscillations. In *Nonradial and Nonlinear Stellar Pulsation*, eds. H. A. Hill and W. Z. Dziembowski (Berlin: Springer-Verlag), pp. 273–299.

Gough, D. O. 1981. A new measure of the solar rotation. *Mon. Not. Roy. Astron. Soc.* 196:731–735.

Gough, D. O. 1982. Diagnostics of the solar interior. *Europhys. News* 13:3–5.

Gough, D. O. 1984a. Immediate and long-term prospects for helioseismology. *Adv. Space Res.* 4:85–102.

Gough, D. O. 1984b. On the rotation of the Sun. *Phil. Trans. Roy. Soc. London* A313:27–38.

Gough, D. O. 1984c. Towards a solar model. *Mem. Soc. Astron. Italiana* 55:13–35.

Gough, D. O. 1985a. Inverting helioseismic data. *Solar Phys.* 100:65–99.

Gough, D. O. 1985b. Theory of solar oscillations. In *Future Missions in Solar, Heliospheric and Space Plasma Physics*, eds. E. Rolfe and B. Battrick, ESA SP-235, pp. 183–197.

Gough, D. O. 1986a. Asymptotic sound-speed inversion. In *Seismology of the Sun and the Distant Stars*, ed. D. O. Gough (Dordrecht: D. Reidel), pp. 125–140.

Gough, D. O. 1986b. EBK quantization of stellar waves. In *Hydrodynamic and Magnetohydrodynamic Problems in the Sun and Stars*, ed. Y. Osaki (Tokyo: Univ. of Tokyo Press), pp. 117–143.

Gough, D. O. 1987. Seismological measurement of stellar ages. *Nature* 326:257–259.

Gough, D. O. 1988a. On taking observers seriously. In *Seismology of the Sun and Sun-Like Stars*, ed. E. J. Rolfe, ESA SP-286, pp. 679–684.

Gough, D. O. 1988b. Prediction of solar oscillation frequencies. *Nature* 336:720.

Gough, D. O. 1990. Open questions. In *Inside the Sun: Proc. IAU Coll. 121*, eds. G. Berthomieu and M. Cribier (Dordrecht: Kluwer), pp. 451–475.

Gough, D. O. 1990. Shaky clues to solar activity. *Nature* 345:768–769.

Gough, D. O. 1991. Linear adiabatic stellar pulsation. In *Astrophysical Fluid Dynamics*, eds. J.-P. Zahn and J. Zinn-Justin (Amsterdam: North Holland Press), in press.

Gough, D. O., and Kosovichev, A. G. 1988. An attempt to understand the Stanford *p*-mode data. In *Seismology of the Sun and Sun-Like Stars*, ed. E. J. Rolfe, ESA SP-286, pp. 195–201.

Gough, D. O., and Taylor, P. P. 1984. Influence of radiation and magnetic fields on stellar oscillation eigenfrequencies. *Mem. Soc. Astron. Italiana* 55:215–226.

Gough, D. O., and Thompson, M. J. 1988a. Magnetic perturbations to stellar oscillation eigenfrequencies. In *Advances in Helio- and Asteroseismology: Proc. IAU Symp. 123*, eds. J. Christensen-Dalsgaard and S. Frandsen (Dordrecht: D. Reidel), pp. 155–160.

Gough, D. O., and Thompson, M. J. 1988b. On the implications of the symmetric component of the frequency splitting reported by Duvall, Harvey, and Pomerantz. In *Advances in Helio-and Asteroseismology: Proc. IAU Symp. 123*, eds. J. Christensen-Dalsgaard and S. Frandsen (Dordrecht: D. Reidel), pp. 175–180.

Gough, D. O., and Thompson, M. J. 1990. The effect of rotation and a buried magnetic field on stellar oscillations. *Mon. Not. Roy. Astron. Soc.* 242:25–55.

Gough, D. O., and Weiss, N. O. 1976. The calibration of stellar convection theories. *Mon. Not. Roy. Astron. Soc.* 176:589–607.

Gough, D. O., Spiegel, E. A., and Toomre, J. 1975a. Highly stretched meshes as functionals of solutions. In *Lecture Notes in Physics*, vol. 35, ed. R. D. Richtmyer (Heidelberg: Springer-Verlag), pp. 191–196.

Gough, D. O., Spiegel, E. A., and Toomre, J. 1975b. Model equations for cellular convection. *J. Fluid Mech.* 68:695–719.

Graboske, H. C., Harwood, D. J., and Rogers, F. M. 1969. Thermodynamic properties of non-ideal gases. I. Free energy minimization method. *Phys. Rev.* 186:210–225.

Graham, E. 1975. Numerical simulation of two-dimensional compressible convection. *J. Fluid Mech.* 70:698–703.

Graham, E. 1977. Compressible convection. In *Problems of Stellar Convection*, eds. E. A. Spiegel and J.-P. Zahn (New York: Springer-Verlag), pp. 151–155.

Graham, E., and Moore, D. R. 1978. The onset of compressible convection. *Mon. Not. Roy. Astron. Soc.* 183:617–632.

Gray, D. F. 1984. Measurements of rotation and turbulence in F, G, and K dwarfs. *Astrophys. J.* 281:719–722.

Grec, G., and Fossat, E. 1977. Photospheric oscillations. III. Search for long period modes. *Astron. Astrophys.* 55:411–420.

Grec, G., and Fossat, E. 1979. Calculation of pseudo solar narrow band oscillations produced by atmospheric differential extinction. *Astron. Astrophys.* 77:351–353.

Grec, G., Fossat, E., and Pomerantz, M. 1980. Solar oscillations: Full disk observations from the geographic south pole. *Nature* 288:541–544.

Grec, G., Fossat, E., and Pomerantz, M. 1983. Full-disk observations of solar oscillations from the geographic south pole: Latest results. *Solar Phys.* 82:55–66.

Grec, G., Fossat, E., Gelly, B., and Schmider, F.-X. 1990. The IRIS sodium cell instrument. *Solar Phys.*, in press.

Green, J. 1960. Rand Corp. Rept. RM-2580-AEC.

Grevesse, N. 1984*a*. Abundances of the elements in the Sun. In *Frontiers of Astronomy and Astrophysics,* ed. R. Pallavicini (Florence: Italian Astron. Soc.), pp. 71–82.

Grevesse, N. 1984*b*. Accurate atomic data and solar photospheric spectroscopy. *Phys. Scripta* T8:49–58.

Grevesse, N. 1989. The abundances of matter in the Sun. In *Cosmic Abundances of Matter,* ed. C. J. Waddington (New York: American Inst. of Physics), in press.

Griem, H. 1974. *Spectral Line Broadening by Plasmas* (New York: Academic Press).

Griem, H. R., Baranger, M., Kolb, A. C., and Oertel, G. 1962. Double injection in insulators. *Phys. Rev.* 125:126–141.

Griffiths, G. M., Lal, M., and Scarfe, C. D. 1963. The reaction of ^{2}D$(p,\gamma)^3$He below 50 keV. *Canadian J. Phys.* 41:724–736.

Grigoryev, V. M., and Demidov, M. L. 1987. Observations of the mean magnetic field at Sayan Observatory during 1982–1984. *Solar Phys.* 114:147–163.

Grigoryev, V. M., Demidov, M. L., and Osak, B. F. 1983. The measurement of the mean solar magnetic field at the Sayan Observatory: Methods and preliminary results. *Issled. Geomagn. Aeron. Fiz. Solntsa* 65:13–22.

Grimshaw, R. 1984. Wave action and wave-mean flow interaction with applications to stratified shear flows. *Ann. Rev. Fluid Mech.* 16:11–44.

Grotz, K., Klapdor, H. V., and Metzinger, J. 1986. Microscopic calculation of the neutrino capture rates in ^{69,71}Ga and the detection of solar and galactic neutrinos. *Phys. Rev.* C33:1263–1269.

Grossmann-Doerth, U., Kneer, F., and von Uexküll, M. 1974. Properties of the solar Ca II K line in high spatial resolution. *Solar Phys.* 37:85–97.

Grossmann-Doerth, U., Pahlke, K.-D., and Schüssler, M. 1987. Spurious variation of photospheric magnetic flux. *Astron. Astrophys.* 176:139–145.

Grossmann-Doerth, U., Larsson, B., and Solanki, S. K. 1988*a*. Contribution and response functions for Stokes line profiles formed in a magnetic field. *Astron. Astrophys.* 204:266–274.

Grossmann-Doerth, U., Schüssler, M., and Solanki, S. K. 1988*b*. Unshifted, asymmetric Stokes V-profiles: Possible solution of a riddle. *Astron. Astrophys.* 206:L37–L39.

Grossmann-Doerth, U., Knölker, M., Schüssler, M., and Weisshaar, E. 1989*a*. Models of magnetic flux sheets. In *Solar and Stellar Granulation,* eds. R. Rutten and G. Severino (Dordrecht: D. Reidel), pp. 481–490.

Grossmann-Doerth, U., Schüssler, M., and Solanki, S. K. 1989*b*. Stokes V asymmetry and shift of spectral lines. *Astron. Astrophys.* 221:338–341.

Gu, Y., Hill, H. A., and Rosenwald, R. D. 1988. Observed asymptotic properties of low-degree solar gravity-mode eigenfrequencies. In *Seismology of the Sun and Sun-Like Stars,* ESA SP-286, ed. E. J. Rolfe pp. 399–401.

Guenther, D. B. 1989. Age of the Sun. *Astrophys. J.* 339:1156–1159.

Guenther, D. B. 1991. High sensitivity of *p*-modes near the acoustic cut-off frequency to solar model parameters. *Astrophys. J.* 369:247–254.

Guenther, D. B., and Demarque, P. 1984. Resonant three-wave interaction of solar *g*-modes. *Astrophys. J.* 277:L17–L19.

Guenther, D. B., and Sarajedini, A. 1988. Insensitivity of solar *p*-mode frequencies to changes in the helium abundance. *Astrophys. J.* 327:993–997.

Guenther, D. B., Jaffe, A., and Demarque, P. 1989. The standard solar model: Composition, opacities, and seismology. *Astrophys. J.* 345:1022–1033.

Guidice, D. A., and Castelli, J. P. 1975. Spectral distributions of microwave bursts. *Solar Phys.* 44:155–172.

Guinan, E. F., and Maloney, F. P. 1985. The apsidal motion of the eccentric eclipsing binary DI Herculis—An apparent discrepancy with general relativity. *Astron. J.* 90:1519–1528.

Gurm, H. S., and Wentzel, D. G. 1967. On the mixing in stellar interior caused by magnetic fields. *Astrophys. J.* 149:139–144.

Gurman, J. B., and House, L. L. 1981. Vector magnetic fields in sunspots. I. Weak line observations. *Solar Phys.* 71:5–20.

Gurman, J. B., Leibacher, J. W., Shine, R. A., Woodgate, B. E., and Henze, W. 1982. Transition region oscillations in sunspots. *Astrophys. J.* 253:939–948.

Gustafsson, B. 1973. *Uppsala Astron. Obs. Ann.* 5(6).

Gustafsson, B., Bell, R. A., Eriksson, K., and Nordlund, A. 1975. A grid of model atmospheres for metal-deficient giant stars. I. *Astron. Astrophys.* 42:407–432.

Guzik, J. A., and Cox, A. N. 1991. Effects of opacity and equation of state on solar structure and oscillations. *Astrophys. J.,* in press.

Guzik, J. A., Willson, L. A., and Brunish, W. M. 1987. A comparison between mass-losing and standard solar models. *Astrophys. J.* 319:957–965.

Habbal, S. R., and Withbroe, G. L. 1981. Spatial and temporal variations of EUV coronal bright points. *Solar Phys.* 69:177–197.

Habbal, S. R., Ronan, R., and Withbroe, G. L. 1985. Spatial and temporal variations of solar coronal loops. *Solar Phys.* 98:323–340.

Habbal, S. R., Dowdy, J. F., Jr., and Withbroe, G. L. 1990. A comparison between bright points in a coronal hole and a quiet Sun region. *Astrophys. J.* 352:333–342.

Haber, D. A., and Hill, F. 1990. The effect of magnetic fields on intermediate degree solar oscillations. In *Challenges to Theories of Moderate Mass Stars,* eds. D. Gough and J. Toomre (Berlin: Springer-Verlag), submitted.

Haber, D. A., Toomre, J., Hill, F., and Gough, D. O. 1988a. Local effects of a major flare on solar five-minute oscillations. In *Seismology of the Sun and the Sun-Like Stars,* ed. E. J. Rolfe, ESA SP-286, pp. 301–304.

Haber, D. A., Toomre, J., and Hill, F. 1988b. Response of the solar five-minute oscillations to a major flare. In *Advances in Helio- and Asteroseismology: Proc. IAU Symp. 123,* eds. J. Christensen-Dalsgaard and S. Frandsen (Dordrecht: D. Reidel), pp. 59–62.

Haber, D. A., Hill, F., and Toomre, J. 1990. The role of the f mode in rotational inversions. In *Challenges to Theories of Moderate Mass Stars,* eds. D. Gough and J. Toomre (Berlin: Springer-Verlag), submitted.

Haber, D. A., Hill, F., and Toomre, J. 1991. Spatial filtering effects on high ℓ power spectra and their inversion. In *Challenges to Theories of Moderate Mass Stars,* eds. D. Gough and J. Toomre (Berlin: Springer-Verlag), submitted.

Hagyard, M. J. 1988. Observed nonpotential magnetic fields and the inferred flow of electric currents at a location of related flaring. *Solar Phys.* 115:107–124.

Hagyard, M. J., West, E., and Cumings, N. P. 1977. The spiral configuration of sunspot magnetic fields. *Solar Phys.* 53:3–13.

Hagyard, M. J., Cumings, N. P., West, E. A., and Smith, J. E. 1982. The MSFC vector magnetograph. *Solar Phys.* 80:33–51.

Hale, G. E. 1908. Preliminary note on the rotation of the Sun as determined from the motion of the hydrogen flocculi. *Astrophys. J.* 27:219–229.

Hale, G. E. 1908. *The Study of Stellar Evolution* (Chicago: Univ. of Chicago Press).

Hall, D. N. B. 1974. *An Atlas of the Solar Photosphere and of Sunspot Umbrae in the Spectral Intervals 4040–5095, 5550–6700, 7400–8790 cm^{-1}* (Tucson: Kitt Peak Natl. Obs.).

Hall, D. N. B. 1974. *An Atlas of Infrared Spectra of the Solar Photosphere and of Sunspot Umbrae, in the Spectral Intervals 4040 cm^{-1}–5095 cm^{-1}, 5550 cm^{-1}–6700 cm^{-1}, 7400 cm^{-1}–8790 cm^{-1}* (Tucson: Kitt Peak Natl. Obs.).

Hall, D. N. B. 1975. Spectroscopic detection of solar ^{3}He. *Astrophys. J.* 197:509–512.

Halprin, A. 1986. Neutrino oscillations in nonuniform matter. *Phys. Rev.* 34D:3462–3466.

Hansen, C. J., Cox, J. P., and Van Horn, H. 1977. The effects of differential rotation on the splitting of nonradial modes of stellar oscillation. *Astrophys. J.* 217:151–159.

Hansen, J. P. 1973. Statistical mechanics of dense ionized matter. I. Equilibrium properties of the classical one-component plasma. *Phys. Rev.* A8:3096–3109.

Hansen, R. T., Hansen, S. F., and Loomis, H. G. 1969. Differential rotation of the solar electron corona. *Solar Phys.* 10:135–149.

Hanslmeier, A., Lustig, A., and Wöhl, H. 1986. Rotation und meridionale strömungen auf der sonne. *Mitt. Astron. Ges.* 67:285.

Hansteen, V. 1988. *Emissionsmal-analyse og overgangssonen.* Thesis, Univ. of Oslo.

Hardie, R. E., Filippone, B. W., Elwyn, A. J., Wiescher, M., and Segel, R. E. 1984. Resonant alpha capture by ^{7}Be and ^{7}Li. *Phys. Rev.* C29:1199–1206.

Hardy, J. W. 1978. Active optics: A new technology for the control of light. *Proc. IEEE* 66: 651–697.

Hardy, J. W. 1981a. Active optics in astronomy. In *Scientific Importance of High Angular Resolution at Infrared and Optical Wavelengths,* eds. M. H. Ulrich and K. Kjär (Garching: European Southern Obs.), pp. 25–39.

Hardy, J. W. 1981b. Solar isoplanatic patch measurements. In *Solar Instrumentation: What's Next?,* ed. R. B. Dunn (Sunspot, N.M.: Sacramento Peak Obs.), pp. 421–431.

Hardy, J. W. 1987. Adaptive optics for solar telescopes. In *Adaptive Optics in Solar Astronomy,* eds. F. Merkle, O. Engvold and R. Falomo, LEST Foundation Tech. Rept. No. 28 (Oslo: Univ. of Oslo).

Hardy, J. W., Lefebvre, J. E., and Koliopoulos, C. L. 1977. Real-time atmospheric compensation. *J. Opt. Soc. Amer.* 67:360–369.

Harris, M. J., Fowler, W. A., Caughlan, G. R., and Zimmerman, B. A. 1983. Thermonuclear reaction rates. III. *Ann. Rev. Astron. Astrophys.* 21:165–176.

Harrison, R. A. 1986. Solar coronal mass ejections and flares. *Astron. Astrophys.* 162:283–291.

Harrison, R. A., Hildner, E., Hundhausen, A. J., Sime, D. G., and Simnett, G. M. 1990. The launch of coronal mass ejections: Results from the coronal mass ejection onset program. *J. Geophys. Res.* 95:917–937.

Hart, A. B. 1956. Motions in the Sun at the photospheric level. VI. Large scale motions in the equatorial region. *Mon. Not. Roy. Astron. Soc.* 116:38–55.

Hart, J. E., Glatzmaier, G. A., and Toomre, J. 1986. Space laboratory and numerical simulations of thermal convection in a rotating hemisphere shell with radial gravity. *J. Fluid Mech.* 173:519–544.

Hart, M. H. 1973. Linear convective modes and the energy transport in stellar convection zones. *Astrophys. J.* 184:587–603.

Hartmann, L. W., and Kenyon, S. 1988. Accretion disks around young stars. In *Formation and Evolution of Low Mass Stars,* eds. A. K. Dupree and M. T. V. T. Lago (Dordrecht: Kluwer), pp. 163–179.

Hartmann, L. W., and Noyes, R. W. 1987. Rotation and magnetic activity in main-sequence stars. *Ann. Rev. Astron. Astrophys.* 25:271–301.

Hartmann, L. W., and Stauffer, J. R. 1989. Additional measurements of pre-main sequence stellar rotation. *Astrophys. J.* 97:873–880.

Hartmann, L. W., Soderblom, D., Noyes, R., Burnham, N., and Vaughan, A. 1984. An analysis of the Vaughan-Preston Survey of chromospheric emission. *Astrophys. J.* 276:254–265.

Hartmann, L. W., Hewett, R., Stahler, S., and Mathieu, R. D. 1986. Rotational and radial velocities of T Tauri stars. *Astrophys. J.* 309:275–293.

Harvey, J. W. 1972. Interferometry applied to visible solar features. *Nature Phys. Sci.* 235: 90–91.

Harvey, J. W. 1977. Observations of small-scale photospheric magnetic fields. In *Highlights of Astronomy,* vol. 4, ed. E. A. Müller, pp. 223–239.

Harvey, J. W. 1981. Variation of the solar He I 10830 Å line: 1977–1980. In *Variations of the Solar Constant,* ed. S. Sofia, NASA CP-2391, pp. 265–272.

Harvey, J. W. 1984. He I 10830 Å irradiance: 1975–1983. In *Workshop on Solar Variability on Active Region Time-Scales,* eds. B. LaBonte, G. Chapman and H. Hudson, NASA CP-2310, pp. 197–211.

Harvey, J. W. 1985a. High-resolution helioseismology. In *Future Missions in Solar, Heliospheric, and Space Plasma Physics,* eds. E. Rolfe and B. Battrick, ESA SP-235, pp. 199–208.

Harvey, J. W. 1985b. Trends in measurement of solar vector magnetic fields using the Zeeman effect. In *Measurements of Solar Vector Magnetic Fields,* ed. M. Hagyard, NASA CP-2374, pp. 109–120.

Harvey, J. W. 1988. Solar internal rotation from helioseismology. In *Seismology of the Sun and Sun-Like Stars,* ed. E. J. Rolfe, ESA SP-286, pp. 55–66.

Harvey, J. W. 1989. Solar observing conditions at the South Pole. In *Astrophysics in Antarctica,* eds. D. J. Mullan, M. A. Pomerantz and T. Stanev (New York: American Inst. of Physics), pp. 227–230.

Harvey, J. W. 1990a. He I 10830 Å irradiance: 1975–1990. In preparation.

Harvey, J. W. 1990b. Trends in helioseismology observations and data reduction. In *Oji Intl. Seminar on Progress of Seismology of the Sun and Stars,* eds. Y. Osaki and H. Shibahashi (Berlin: Springer-Verlag), in press.

Harvey, J. W., and Breckinridge, J. B. 1973. Photoelectric speckle interferometry of the solar granulation. *Astrophys. J.* 182:L137–L139.

Harvey, J. W., and Duvall, T. L., Jr. 1984. Observations of intermediate-degree solar oscillations. In *Solar Seismology from Space,* NASA JPL Publ. 84–84, pp. 165–172.

Harvey, J. W., and Hall, D. N. B. 1975. Magnetic field observations with Fe I λ15648 Å. *Bull. Amer. Astron. Soc.* 7:459 (abstract).

Harvey, J. W., and Livingston, W. 1969. Magnetograph measurements with temperature-sensitive lines. *Solar Phys.* 10:283–293.

Harvey, J. W., and Schwarzschild, M. 1975. Photoelectric speckle interferometry of the solar granulation. *Astrophys. J.* 196:221–226.

Harvey, J. W., Pomerantz, M. A., and Duvall, T. L., Jr. 1982. Astronomy on ice. *Sky and Teles.* 64:520–523.

Harvey, J. W., Kennedy, J. R., and Leibacher, J. W. 1987. The Global Oscillation Network Group Project. *Sky & Teles.* 74:470–476.

Harvey, J. W., and the GONG Instrument Development Team. 1988a. The GONG instrument. In *Seismology of the Sun and Sun-Like Stars,* ed. E. J. Rolfe, ESA SP-286, pp. 203–208.

Harvey, J. W., Hill, F., Kennedy, J. R., Leibacher, J. W., and Livingston, W. C. 1988b. The Global Oscillation Network Group (GONG). *Adv. Space Res.* 8(11):117–120.

Harvey, K. L. 1984. Solar cycle variation of ephemeral active regions. In *The Hydromagnetics of the Sun,* eds. T. D. Guyenne and J. J. Hunt, ESA SP-220, pp. 235–236.

Harvey, K. L. 1990a. Magnetic fields on the Sun: Variations during cycles 21 and 22. In preparation.

Harvey, K. L. 1990b. The global properties of ephermeral regions. In preparation.

Harvey, K. L. 1990c. Small active regions. In preparation.

Harvey, K. L., and Harvey, J. W. 1973. Observations of moving magnetic features near sunspots. *Solar Phys.* 28:61–71.

Harvey, K. L., and Harvey, J. W. 1976a. *Characteristics of Individual Ephemeral Regions.* Air Force Geophysics Lab Rept. AFGL-TR-76–0255, Part II.

Harvey, K. L., and Harvey, J. W. 1976b. A study of the magnetic and velocity fields in an active region. *Solar Phys.* 47:233–246.

Harvey, K. L., Harvey, J. W., and Martin, S. F. 1975. Ephemeral active regions in 1970 and 1973. *Solar Phys.* 40:87–102.

Harvey, K. L., Sheeley, N. R., Jr., and Harvey, J. W. 1982. Magnetic measurements of coronal holes during 1975–1980. *Solar Phys.* 79:149–160.

Hasan, S. S. 1984. Convective instability in a solar flux tube. I. Nonlinear calculations for an adiabatic inviscid fluid. *Astrophys. J.* 285:851–857.

Hasan, S. S. 1985. Convective instability in a solar flux tube. II. Nonlinear calculations with horizontal radiative heat transport and finite viscosity. *Astron. Astrophys.* 143:39–45.

Hasegawa, A. 1985. Self organization processes in continuous media. *Adv. Phys.* 34:1–42.

Hasegawa, A., and Kodama, Y. 1978. Spectrum cascades by mode coupling in drift-wave turbulence. *Phys. Rev. Lett.* 41:1470–1473.

Hasselgrove, C. B., and Hoyle, F. 1956. A mathematical discussion of the problems of stellar evolution, with reference to the use of an automatic digital computer. *Mon. Not. Roy. Astron. Soc.* 116:515–526.

Hathaway, D. H. 1989. *Solar Phys.,* submitted.

Hathaway, D. H. 1988. Simulating photospheric Doppler velocity fields. *Solar Phys.* 117:329–341.

Hathaway, D. H., and Wilson, R. M. 1990. Solar rotation and the sunspot cycle. *Astrophys. J.* 357:271–274.

Haubold, H. J., and Gerthe, E. 1985. The search for possible time variations in Davis' measurements of the argon production rate in the solar neutrino experiment. *Astron. Nachr.* 306(4):203–211.

Haubold, H. J., and Gerthe, E. 1990. On the Fourier spectrum analysis of the solar neutrino data. *Solar Phys.* 127:347–356.

Haxton, W. C. 1986. Adiabatic conversion of solar neutrinos. *Phys. Rev. Lett.* 57:1271–1274.

Haxton, W. C. 1987. Analytic treatments of matter enhanced solar-neutrino oscillations. *Phys. Rev.* 35D:2352–2364.

Haxton, W. C. 1988a. Radiochemical neutrino detection via $^{127}I(\nu e, e^-)^{127}Xe$. *Phys. Rev. Lett.* 60:768–771.

Haxton, W. C. 1988b. ^{37}Ar as a calibration source for solar neutrino detectors. *Phys. Rev.* C38:2474–2477.

Haxton, W. C., and Donnelly, W. T. 1977. Solar neutrino induced reactions. *Phys. Lett.* B66:123–126.

Haxton, W. C., and Johnson, C. W. 1988. Geochemical integrations of the neutrino flux from stellar collapses. *Nature* 333:325–329.

Haxton, W. C., and Stephenson, G. J., Jr. 1984. Double beta decay. In *Progress in Particle and Nuclear Physics,* vol. 12, ed. D. Wilkinson (London: Pergamon), pp. 409–479.

Hayashi, C. 1961. Stellar evolution in early phases of gravitational contraction. *Publ. Astron. Soc. Japan* 13:450–452.

Hayashi, C., and Nakano, T. 1965. Thermal and dynamical properties of a protostar and its contraction to the stage of quasi-static equilibrium. *Prog. Theor. Phys.* 34:754–775.

Hayashi, C., Hoshi, R., and Sugimoto, D. 1962. Evolution of the stars. *Prog. Theor. Phys. Suppl.* No. 22.

Heasley, J., and Milkey, R. 1976. Structure and spectrum of quiescent prominences. II. Hydrogen and helium spectra. *Astrophys. J.* 210:827–835.

Heasley, J., and Milkey, R. 1978. Structure and spectrum of quiescent prominences. III. Application of theoretical models in helium abundance determination. *Astrophys. J.* 221:677–688.

Heath, D. F. 1980. A review of observational evidence for short and long term ultraviolet flux variability of the Sun. In *Sun and Climate* (Toulouse: CNES), pp. 447–471.

Heath, D. F., and Schlesinger, B. M. 1986. The Mg 280-nm doublet as a monitor of changes in solar ultraviolet irradiance. *J. Geophys. Res.* 91:8672–8682.

Heath, D. F., Repoff, T. P., and Donnelly, R. F. 1984. NIMBUS 7 observations of solar UV spectral irradiance variations caused by solar rotation and active-region evolution for the period November 7, 1978-November 1, 1980. NOAA Tech. Memo ERL ARL-129 (Boulder: Air Resources Lab).

Hejlesen, P. M. 1987. Studies in stellar evolution. III. The internal structure constants. *Astron. Astrophys. Suppl.* 69:251–262.

Hejna, L. 1986. On the differential rotation of solar background magnetic fields. 1. On the possible existence of torsional waves with the wave number $k = 1$/hemisphere. *Bull. Astron. Inst. Czech.* 37:175–179.

Hejna, L. 1987. On the differential rotation of solar background magnetic fields. 3. Use of Gegenbauer polynomials and low modes of stationary torsional waves. *Bull. Astron. Inst. Czech.* 38:184–189.

Hejna, L., and Sobotka, M., eds. 1987. *Proc. 10th European Regional Astronomy Meeting of the IAU, Vol. 1: The Sun,* Publ. Astron. Inst. of Czech. Acad. Sci. no. 66 (Ondrejov: Astronomical Inst. of Czechoslovak Academy of Sciences).

Henyey, L. G., Wilets, L., Böhm, K.-H., Lelevier, R., and Levee, R. D. 1959. Method for automatic computation of stellar evolution. *Astrophys. J.* 129:628–636.

Henning, H., and Scherrer, P. H. 1988a. Helioseismic observations at Stanford, 1977–1986. In *Advances in Helio- and Asteroseismology, IAU Symp. 123,* eds. J. Christensen-Dalsgaard and S. Frandsen (Dordrecht D. Reidel), pp. 29–32.

Henning, H. M., and Scherrer, P. H. 1988b. The search for solar gravity modes. In *Proc. Symp. on Seismology of the Sun and Sun-Like Stars,* ed. E. J. Rolfe, ESA SP-286, pp. 419–423.

Henson, G. D., and Kemp, J. C. 1984. Broad-band circular polarimetry of sunspots, 0.4–1.7 microns: Spatial scans with a 3.4 arc sec diameter aperture. *Solar Phys.* 93:289–299.

Henyey, L. G., Forbes, J. E., and Gould, N. L. 1964. A new method of automatic computation of stellar evolution. *Astrophys. J.* 139:306–317.

Herbig, G. 1977. Eruptive phenomena in early stellar evolution. *Astrophys. J.* 217:693–715.

Herbig, G. H. 1989. FU Orionis eruptions. In *ESO Workshop on Low Mass Star Formation and Pre-Main Sequence Objects,* ed. B. Reipurth (Garching: European Southern Obs.), pp. 233–246.

Heristchi, D. 1986. Hard X-ray and gamma-ray bremsstrahlung production by high-energy protons in solar flares. *Astrophys. J.* 311:474–484.

Hermans, L., and Lindsey, C. 1986. Solar chromospheric modeling based on submillimeter limb brightness profile. *Astrophys. J.* 310:907–911.

Herschel, W. 1800. *Phil. Trans. Roy. Soc.* 90:284.

Herse, M. 1979. High resolution photographs of the Sun near 200 nm. *Solar Phys.* 63:35–60.

Hewagama, T., Jennings, D., Deming, D., Boyle, R., and Zipoy, D. M. 1989. The magnetic field structure in sunspots and plages: Stokes profile observations of the Mg I emission at 12.32 μm. *Bull. Amer. Astron. Soc.* 21:839 (abstract).

Heyer, M. H. 1988. The magnetic evolution of the Taurus molecular clouds. II. A reduced role of the magnetic field in dense core regions. *Astrophys. J.* 324:311–320.

Heyvaerts, J. 1981. Particle acceleration in solar flares. In *Solar Flare Magnetohydrodynamics,* ed. E. R. Priest (London: Gordon and Breach), pp. 429–455.

Heyvaerts, J., Priest, E. R., and Rust, D. M. 1977. An emerging flux model for the solar flare phenomenon. *Astrophys. J.* 216:123–137.

Hickey, J. R., and Alton, B. M. 1983. Status of solar measurements and data reduction for ERB-Nimbus 7. In *Solar Irradiance Variations of Active Region Time Scales,* eds. B. J. LaBonte, G. A. Chapman, H. S. Hudson and R. C. Wilson, NASA CP-2310, pp. 43–58.

Hickey, J. R., Stowe, L. L., Jacbowitz, H., Pellegrino, P., Maschoff, R. H., House, F., and Von der Haar, T. H. 1980. Initial determinations from Nimbus 7 cavity radiometer measurements. *Science* 208:281–283.

Hilgemeier, M., Becker, H. W., Rolfs, C., Trautvetter, H. P., and Hammer, J. W. 1988. Absolute cross section of the ³He(α,γ)⁷Be reaction. *Z. Phys.* A329:243–254.

Hill, F. 1984a. The effects of a nearly 100% duty cycle on observations of solar oscillations. In *Solar Seismology from Space,* NASA JPL Publ. 84–84, pp. 271–277.

Hill, F. 1984b. The effects of image motion on the ℓ-v diagram. In *Solar Seismology from Space,* NASA JPL Publ. 84–84, pp. 255–262.

Hill, F. 1987a. Numerical simulation of the effects of seeing on the GONG data. In *GONG Report Number 5: The 1987 Artificial Data Workshop* (Tucson: NOAO), pp. 23–25.

Hill, F. 1987b. The equatorial rotation rate in the solar convection zone. In *The Internal Solar Angular Velocity,* eds. B. R. Durney and S. Sofia (Dordrecht: D. Reidel), pp. 45–50.

Hill, F. 1988a. Rings and trumpets—three dimensional power spectra of solar oscillations. *Astrophys. J.* 333:996–1013.

Hill, F. 1988b. Oscillation ring diagrams and the thermodynamics of the outer solar convection zone. In *Seismology of the Sun and Sun-Like Stars,* ed. E. J. Rolfe, ESA SP-286, pp. 103–108.

Hill, F. 1989a. *GONG Report Number 6: A Selected Bibliography on Helio- and Asteroseismology* (Tucson: NOAO).

Hill, F. 1989b. Solar oscillation ring diagrams and large-scale flows. *Astrophys. J.* 343: L69–L71.

Hill, F. 1990a. Networks for helioseismic observations. In *Inside the Sun: Proc. IAU Coll. 121,* eds. G. Berthomieu and M. Cribier (Dordrecht: Kluwer), pp. 265–278.

Hill, F. 1990b. A map of the horizontal flows in the solar convection zone. *Solar Phys.,* in press.

Hill, F., and Leibacher, J. W. 1991. Ground-based helioseismology networks. *Adv. Space Res.,* in press.

Hill, F., and Newkirk, G., Jr. 1985. On the expected performance of a solar oscillation network. *Solar Phys.* 95:201–219.

Hill, F., Gough, D. O., and Toomre, J. 1984a. Attempt to measure the solar subsurface velocity. *Mem. Soc. Astron. Italiana* 55:153–161.

Hill, F., Toomre, J., Merryfield, W., and Gough, D. 1984b. Numerical simulation of the impact of atmospheric turbulence. In *Probing the Depths of a Star: The Study of Solar Oscillations from Space,* eds. R. W. Noyes and E. J. Rhodes, Jr., NASA JPL Publ. 400–237, pp. 37–42.

Hill, F., Toomre, J., November, L. J., and Gebbie, K. B. 1984c. On the determination of the lifetime of vertical velocity patterns in mesogranulation and supergranulation. In *Small-Scale Dynamical Processes in Quiet Stellar Atmospheres,* ed. S. L. Keil (Sunspot, N.M.: Sacramento Peak Obs.) pp. 160–172.

Hill, F., Gough, D. O., and Toomre, J. 1984d. Sensitivity of inferred subphotospheric velocity field to mode selection, analysis technique, and noise. In *Solar Seismology from Space*, NASA JPL Publ. 84–84, pp. 95–111.

Hill, F., Haber, D. A., Toomre, J., and November, L. J. 1986. Influence of spatial filtering on possible anisotropies in solar oscillations. In *Seismology of the Sun and the Distant Stars*, ed. D. O. Gough (Dordrecht: D. Reidel), pp. 85–92.

Hill, F., and the GONG Site Survey Team. 1988*a*. The GONG site survey. In *Seismology of the Sun and Sun-Like Stars*, ed. E. J. Rolfe, ESA SP-286, pp. 209–215.

Hill, F., Rust, D. M., and Appourchaux, T. 1988*b*. Rotation in the solar convection zone inferred from Fabry-Perot observations of the 5-min oscillations. In *Advances in Helio- and Asteroseismology*, eds. J. Christensen-Dalsgaard and S. Frandsen (Dordrecht: D. Reidel), pp. 49–52.

Hill, F., Gough, D. O., Toomre, J., and Haber, D. A. 1988*c*. Solar equatorial rotation rate from inversion of frequency splitting of high-degree modes. In *Advances in Helio- and Asteroseismology*, eds. J. Christensen-Dalsgaard and S. Frandsen (Dordrecht: D. Reidel), pp. 45–48.

Hill, F., Rhodes, E. J., Jr., Brown, T. M., Korzennik, S. G., and Cacciani, A. 1990*b*. Convection zone flow maps from Mt. Wilson full-disk Doppler images. In *Challenges to Theories of Moderate Mass Stars*, eds. D. Gough and J. Toomre (Berlin: Springer-Verlag), submitted.

Hill, F., Gough, D. O., Merryfield, W. J., and Toomre, J. 1991. Simulation of effects of atmospheric seeing on the observation of high-degree solar oscillations. *Astrophys. J.* 369: 237–246.

Hill, H. A. 1978. Seismic sounding of the Sun. In *The New Solar Physics*, ed. J. A. Eddy (Boulder: Westview Press), pp. 135–214.

Hill, H. A. 1984*a*. Rotational splitting of global solar oscillations and its relevance to tests of general relativity. *Intl. J. Theoret. Phys.* 23:683–699.

Hill, H. A. 1984*b*. The observed low-order, low-degree acoustic mode eigenfrequency spectrum and its properties. *SCLERA Mono. Series in Astrophys.* No. 1.

Hill, H. A. 1985*a*. Detection and classification of resolved multiplet members of the solar five minute oscillations through solar diameter observations. *Astrophys. J.* 290:765–781.

Hill, H. A. 1985*b*. Multiplet classifications for the 160 min and similar long period solar oscillations and the rotational splitting of these multiplets. *SCLERA Mono. Series in Astrophys.* No. 3.

Hill, H. A. 1986*a*. Comparison of differential infrared and differential radius observations of the solar 160 minute period oscillation. *Astrophys. J.* 303:489–494.

Hill, H. A. 1986*b*. Test of a uniformly rotating convection zone model and detection of intermediate-degree *f*-modes of the Sun. *SCLERA Mono. Series in Astrophys.* No. 4.

Hill, H. A. 1986*c*. Evidence of gravity-mode coupling and its relevance to solar neutrino production rates. In *Neutrino '86: The 12th Intl. Conf. on Neutrino Physics and Astrophysics*, eds. T. Kitagaki and H. Yuta (Singapore: Kim Hup Lee), pp. 221–230.

Hill, H. A. 1988. Tests of the detection and mode classifications of low-degree solar gravity modes with 1978 solar diameter observations. *Astrophys. J. Suppl.*, submitted.

Hill, H. A., and Caudell, T. P. 1979. Global oscillations of the Sun: Observed as oscillations in the apparent solar limb darkening function. *Mon. Not. Roy. Astron. Soc.* 186:327–342.

Hill, H. A., and Caudell, T. P. 1985. Confirmation of the detection and classification of low-order, low-degree acoustic modes with the 1978 solar diameter observations. *Astrophys. J.* 299:517–525.

Hill, H. A., and Czarnowski, W. M. 1986. Comparison of differential velocity and differential radius observations of solar oscillations for evidence of gravity modes. *SCLERA Mono. Series in Astrophys.* No. 6.

Hill, H. A., and Gu, Y. 1990. Extension of the range in radial order in the detection and mode classification of solar low-degree gravity modes. *Science in China* A33:67–79.

Hill, H. A., and Kroll, R. 1986. A comparison of total irradiance observations with differential velocity and differential radius observations of long-period solar oscillations. *SCLERA Mono. Series in Astrophys.* No. 5.

Hill, H. A., and Logan, J. D. 1984. The effects of the nonlocal character of the mean intensity, deviation from radiative equilibrium and a nongray atmosphere on oscillations in stellar envelopes. *Astrophys. J.* 285:386–399.

Hill, H. A., and Rosenwald, R. D. 1986*a*. Deviations from the normal mode spectrum of asymptotic theory. I. Identification of quasi-periodic departures in the low-degree spectrum of the solar 5-min oscillations. *Astrophys. Space Sci.* 126:335–356.

Hill, H. A., and Rosenwald, R. D. 1986*b*. Solar oscillations, gravitational multipole field of the Sun and the solar neutrino paradox. *SCLERA Mono. Series in Astrophys.* No. 7.

Hill, H. A., and Stebbins, R. T. 1975*a*. The intrinsic visual oblateness of the Sun. *Astrophys. J.* 200:471–483.

Hill, H. A., and Stebbins, R. T. 1975*b*. Recent solar oblateness observations: Data, interpretation and significance for earlier work. In *Annals New York Academy Sci.*, vol. 262, eds. P. G. Bergmann, E. J. Fenyves and L. Motz, pp. 472–480.

Hill, H. A., Stebbins, R. T., and Oleson, J. R. 1975*a*. The finite Fourier transform definition of an edge on the solar disk. *Astrophys. J.* 200:484–498.

Hill, H. A., Stebbins, R. T., and Brown, T. M. 1975*b*. Recent solar oblateness observations: Data, interpretation and significance for earlier work. In *Atomic Masses and Fundamental Constants 5*, eds. J. H. Sanders and A. H. Wapstra (New York: Plenum), pp. 622–628.

Hill, H. A., Bos, R. J., and Goode, P. R. 1982. Preliminary determination of the gravitational quadrupole moment of the Sun from rotational splitting of global oscillations and its relevance to tests of general relativity. *Phys. Rev. Lett.* 49:1794–1797.

Hill, H. A., Bos, R. J., and Caudell, T. P. 1983. On the origin of oscillations in a solar diameter observed through the Earth's atmosphere: A terrestrial atmospheric or a solar phenomenon. *Solar Phys.* 82:129–138.

Hill, H. A., Yakowitz, D. S., Rosenwald, R. D., and Campbell, W. 1984. The internal rotation of the Sun inferred from the rotational splitting of acoustic and gravity mode multiplets. In *The Hydromagnetics of the Sun*, eds. T. D. Guyenne and J. J. Hunt, ESA SP-220, pp. 187–188.

Hill, H. A., Alexander, J. C., and Caudell, T. P. 1985. The horizontal spatial properties of the eulerian perturbation of radiation intensity produced by the low-order, low-degree acoustic modes of the Sun. *SCLERA Mono. Series in Astrophys.* No. 2.

Hill, H. A., Tash, J., and Padin, C. 1986*a*. Interpretation and implications of diameter differential radius observations of the 160 minute period solar oscillations. *Astrophys. J.* 304:560–578.

Hill, H. A., Rabaey, G. F., Yakowitz, D. S., and Rosenwald, R. D. 1986*b*. The radial dependence of the $P_2(\cos \tau)$ component and the scale height of the $P_4(\cos \tau)$ component of the internal rotation of the Sun. *Astrophys. J.* 310:444–455.

Hill, H. A., Cornuelle, C. S., and Germain, M. E. 1988*a*. Properties of solar gravity-mode eigenfunctions in the photosphere. *Bull. Amer. Astron. Soc.* 20:1012 (abstract).

Hill, H. A., Gao, Q., and Rosenwald, R. D. 1988*b*. Inversion of quasi-periodic deviations between low-degree solar gravity mode eigenfrequencies and asymptotic theory eigenfrequencies. In *Proc. Symp. Seismology of the Sun and Sun-Like Stars*, ed. E. J. Rolfe, ESA SP–286, pp. 403–406.

Hill, T. L. 1960. *Statistical Thermodynamics* (Addison-Wesley), chpt. 15.

Hinteregger, H. E., Fukui, K., and Gilson, B. R. 1981. Observational, reference and model data on solar EUV, from measurements on AE-E. *Geophys. Res. Lett.* 8:1147–1150.

Hirata, K., et al. 1987*a*. Univ. of Tokyo ICEPP Report #UT-87–04.

Hirata, K. S., Kajita, T., Koshiba, M., Nakahata, M., Oyama, Y., Sato, N., Suzuki, A., Takta, M., Totsuka, Y., Kifune, T., Suda, T., Takahoshi, K., Tanimori, T., Miyano, K., Yamada, M., Beier, E. W., Feldscher, L. R., Kim, S. B., Mann, A. K., Newcomer, F. M., Van Berg, R., and Zhang, W. 1987*b*. Observation of a neutrino burst from supernova SN 1987A. *Phys. Rev. Lett.* 58:1490–1493.

Hirata, K. S., and the Kamiokande Collaboration. 1988*a*. Search for correlation of neutrino events with solar flares in Kamiokande. *Phys. Rev. Lett.* 61:2653–2656.

Hirata, K. S., and the Kamiokande Collaboration. 1988*b*. Observation in the Kamiokande II Detector of the neutrino burst from supernova SN 1987A. *Phys. Rev.* D38:448–458.

Hirata, K. S., Kajita, T., Kifune, K., Kirhara, K., Nakahata, M., Nakamura, K., Ohara, S., Oyama, Y., Sato, N., Takita, M., Totsuka, Y., Yaginuma, Y., Mori, M., Suzuki, A., Takahashi, K., Tanimori, T., Yamada, M., Koshiba, M., Suda, T., Miyano, K., Miyata, H., Takei, H., Kaneyuki, K., Nagashima, Y., Suzuki, Y., Beier, E. W., Feldscher, L. R., Frank, E. D., Frati, W., Kim, S. B., Mann, A. K., Newcomer, F. M., VanBerg, R., and Zhang,

W. 1989. Observation of ^{8}B solar neutrinos in the Kamiodande—II detector. *Phys. Rev. Lett.* 63:16–19.

Hirata, K. S., et al. 1990*a*. Constraints on neutrino-oscillation parameters from the Kamiokande—II solar neutrino data. *Phys. Rev. Lett.* 65:1301–1304.

Hirata, K. S., and the Kamiokande Collaboration. 1990*b*. Results from one thousand days of real-time, directional solar-neutrino data. *Phys. Rev. Lett.* 65:1297–1300.

Hirata, K. S., and the Kamiokande Collaboration. 1991. Real time directional measurement of the ^{8}B solar neutrinos in the Kamiokande Detector. *Phys. Rev.*, in press.

Hirayama, T. 1978. A model of solar faculae and their lifetime. *Publ. Astron. Soc. Japan* 30:337–352.

Hirayama, S. 1986. The density and thickness of quiescent prominences. In *Coronal and Prominence Plasmas,* ed. A. I. Poland, NASA CP-2442, pp. 149–156.

Hirayama, T. 1985. Modern observations of solar prominences. *Solar Phys.* 100:415–434.

Hirshberg, J., Bame, S. J., and Robbins, D. E. 1972. Solar flares and solar wind He enrichments: July 1965-July 1967. *Solar Phys.* 23:467–486.

Hoang-Binh, D. 1982. On solar hydrogen lines in the far-infrared and submillimeter spectrum. *Astron. Astrophys.* 112:L3–L5.

Hoang-Binh, D., Brault, P., Picart, J., Tran-Minh, N., and Valle, O. 1987. Ion-collision broadening of solar lines in far-infrared and submillimeter spectrum. *Astron. Astrophys.* 181: 134–137.

Hobbs, L. M., and Pilachowski, C. 1986*a*. Lithium in NGC 752. *Astrophys. J.* 309:L17–L21.

Hobbs, L. M., and Pilachowski, C. 1986*b*. Lithium in M67. *Astrophys. J.* 311:L37–L40.

Hobbs, L. M., and Pilachowski, C. 1988. Lithium in old open clusters: NGC 188. *Astrophys. J.* 334:734–745.

Hofmann, A., and Staude, J. 1987. Electric current density in the sunspot photosphere derived from vector magnetograms. In *Proc. 10th European Regional Astronomy Meeting of the IAU, Vol. 1: The Sun,* eds. L. Hejna and M. Sobotka, Publ. Astron. Inst. of Czech. Acad. Sci. no. 66 (Ondrejov: Astronomical Inst. of Czechoslovak Academy of Sciences), pp. 105–107.

Hoeksema, J. T. 1984. Structure and Evolution of the Large Scale Solar and Heliospheric Magnetic Fields. Ph.D. Thesis, Stanford Univ.

Hoeksema, J. T., and Scherrer, P. H. 1986. The solar magnetic field—1976 through 1985. Report UAG-94, World Data Center A for Solar-Terrestrial Physics (Boulder: NOAA).

Hoeksema, J. T., and Scherrer, P. H. 1987. Rotation of the coronal magnetic field. *Astrophys. J.* 318:428–436.

Hoeksema, J. T., Scherrer, P. H., Title, A. M., and Tarbell, T. D. 1988. The Michelson Doppler Imager for the Solar Oscillations Imager program on SOHO. In *Seismology of the Sun and Sun-Like Stars,* ed. E. J. Rolfe, ESA SP-286, pp. 407–412.

Hollweg, J. V. 1981*a*. Alfvén waves in the solar atmosphere. II. Open and closed magnetic flux tubes. *Solar Phys.* 70:25–66.

Hollweg, J. V. 1981*b*. Minor ions in the low corona. *J. Geophys. Res.* 86:8899–8905.

Hollweg, J. V. 1982. On the origin of spicules. *Astrophys. J.* 257:345–353.

Hollweg, J. V. 1984. Resonances of coronal loopes. *Astrophys. J.* 277:392–403.

Hollweg, J. V. 1986. Transition region, corona, and solar wind in coronal holes. *J. Geophys. Res.* 91:4111–4125.

Hollweg, J. V., and Johnson, W. 1988. Transition region, corona, and solar wind in coronal holes: Some two fluid models. *J. Geophys. Res.* 93:9547–9554.

Hollweg, J. V., Jackson, S., and Galloway, D. 1981. Alfvén waves in the solar atmosphere. III. Nonlinear waves in open flux tubes. *Solar Phys.* 75:35–61.

Holman, G. D. 1985. Acceleration of runaway electrons and Joule heating in solar flares. *Astrophys. J.* 293:584–594.

Holman, G. D., and Kundu, M. 1985. The microwave structure of hot coronal loops. *Astrophys. J.* 292:291–296.

Holt, S. S., and Ramaty, R. 1969. Microwave and hard X-ray bursts from solar flares. *Solar Phys.* 8:119–141.

Holweg, J. V. 1984. Resonances of coronal loops. *Astrophys. J.* 277:392–408.

Holweger, H., and Kneer, F. 1989. Spatially resolved spectra of solar granules. In *Solar and Stellar Granulation,* eds. R. Rutten and G. Severino (Dordrecht: Kluwer), pp. 173–186.

Holweger, H., and Müller, E. A. 1974. The photospheric barium spectrum: Solar abundance and collision broadening of Ba II lines by hydrogen. *Solar Phys.* 39:19–30.

Holzer, T. E. 1988. Acceleration of stellar winds. In *Proc. Sixth Intl. Solar Wind Conf.*, eds. V. J. Pizzo, T. E. Holzer and D. G. Sime, NCAR TN-306 (Boulder: Natl. Center for Atmos. Res.), pp. 3–22.

Hopfinger, E. J., Browand, F. K., and Gagne, Y. 1982. Turbulence and waves in a rotating tank. *J. Fluid Mech.* 125:505–534.

Horne, K., Hurford, G. J., Zirin, H., and de Graauw, T. 1980. Solar limb brightening at 1.3 millimeters. *Astrophys. J.* 244:340–344.

Hossain, M., Matthaeus, W. H., and Montgomery, D. 1983. Long-time states of inverse cascades in the presence of a maximum length scale. *J. Plasma Phys.* 30:479–493.

Howard, R. A. 1967. Magnetic field of the Sun (observational). *Ann. Rev. Astron. Astrophys.* 5:1–24.

Howard, R. A. 1974. Studies of solar magnetic fields. I. The average field strengths. *Solar Phys.* 38:59–67.

Howard, R. A. 1984. Solar rotation. *Ann. Rev. Astron. Astrophys.* 22:131–155.

Howard, R. A. 1989. The magnetic fields of active regions. I. Data and first results. *Solar Phys.* 123:271–284.

Howard, R. A., and Gilman, P. A. 1986. Meridional motions of sunspots and sunspot groups. *Astrophys. J.* 307:389–394.

Howard, R. A., and Harvey, J. 1970. Spectroscopic determinations of the solar rotation. *Solar Phys.* 12:23–51.

Howard, R. A., and LaBonte, B. J. 1980. The Sun is observed to be a torsional oscillator with a period of 11 years. *Astrophys. J.* 239:L33–L36.

Howard, R. A., and LaBonte, B. J. 1981. Surface magnetic fields during the solar activity cycle. *Solar Phys.* 74:131–145.

Howard, R. A., and LaBonte, B. J. 1983. The observed relationships between some solar rotation parameters and the activity cycle. In *Solar and Stellar Magnetic Fields*, ed. J. O. Stenflo (Dordrecht: D. Reidel), pp. 101–111.

Howard, R. W., and Stenflo, J. W. 1972. On the filamentary nature of solar magnetic fields. *Solar Phys.* 22:402–417.

Howard, R. A., Gilman, P. A., and Gilman, P. I. 1984a. Rotation of the Sun measured from Mount Wilson white-light images. *Astrophys. J.* 283:373–384.

Howard, R. A., Sheeley, N. R., Jr., Koomen, M. J., and Michels, D. J. 1984b. The statistical properties of coronal mass ejections during 1979–1981. *Adv. Space Res.* 4:307–317.

Howard, R. A., Sheeley, N. R., Jr., Koomen, M. J., and Michels, D. J. 1985. Coronal mass ejections: 1979–1981. *J. Geophys. Res.* 90:8173–8178.

Howard, R. A., Sheeley, N. R., Jr., Michels, D. J., and Kooman, M. J. 1986. The solar cycle dependence of coronal mass ejections. In *The Sun and the Heliosphere in Three Dimensions*, ed. R. G. Marsden (Dordrecht: D. Reidel), pp. 107–111.

Howard, R., Gilman, P. A., and Gilman, P. I. 1984. Rotation of the Sun measured from Mount Wilson white-light images. *Astrophys. J.* 283:373–384.

Hoyle, F., and Schwarzschild, M. 1955. The evolution of type II stars. *Astrophys. J. Suppl.* 2: 1–40.

Hoyng, P. 1989. On the sensitivity of resonant scattering spectrometers for whole-disk solar velocity oscillation measurements. *Astrophys. J.* 345:1088–1103.

Hoyng, P., Brown, J. C., and van Geek, H. F. 1976. High time resolution analysis of solar hard X-ray flares observed on board the ESRO TD-1A satellite. *Solar Phys.* 48:197–254.

Hoyng, P., Knight, J. W., and Spicer, D. S. 1978. Diagnostics of solar flare hard X-ray sources. *Solar Phys.* 58:139–148.

Hoyng, P., Melrose, D. B., and Adams, J. C. 1979. Relaxation and bremsstrahlung of thick-target electron streams: A simple astrophysical application of the Legendre expansion method. *Astrophys. J.* 230:950–960.

Hoyng, P., Duijveman, A., Machado, M. E., Rust, D. M., Svestka, D. M., Beoelee, A., de Jager, C., Frost, K. J., Lafleur, H., Simnett, G. M., van Beek, H. F., and Woodgate, B. F. 1981. Origin and location of the hard X-ray emission in a two-ribbon flare. *Astrophys. J.* 246:L155–L159.

Hoyt, D. V., and Eddy, J. A. 1982. An atlas of variations in the solar constant caused by sunspot blocking and facular emissions from 1874 to 1981. NCAR TN-194 + str (Boulder: Natl. Center for Atmos. Res.).

Hua, X.-M., and Lingenfelter, R. E. 1987. A determination of the He/H ratio in the solar photosphere from flare gamma-ray line observations. *Astrophys. J.* 319:555–566.

Hudgin, R. 1977. Wave-front compensation error due to finite correction-element size. *J. Opt. Soc. Amer.* 67:393–395.

Hudson, H. S. 1988. Observed variability of the solar luminosity. *Ann. Rev. Astron. Astrophys.* 26:473–507.

Hudson, H. S., and Willson, R. 1988. The nature of the solar cycle variation of total irradiance. In *Solar Radiative Output Variations,* ed. P. V. Foukal (Cambridge, Mass.: Cambridge Research and Instrumentation), pp. 318–322.

Hudson, H. S., and Lindsey, C. A. 1974. Direct observation of temperature amplitude of solar 300-second oscillations. *Astrophys. J.* 187:L35–L36.

Hudson, H. S., Silva, S., Woodard, M., and Willson, R. C. 1982. The effects of sunspots on solar irradiance. *Solar Phys.* 76:211–219.

Huebner, W. F. 1986. Atomic and radiative processes in the solar interior. In *Physics of the Sun,* vol. 1, eds. P. A. Sturrock, T. E. Holzer, D. M. Mihalas and R. K. Ulrich (Dordrecht: D. Reidel), pp. 33–75.

Huebner, W. F., Merts, A. L., Magee, N. H., and Argo, M. F. 1977. *Astrophysical Opacity Library,* Los Alamos Scientific Lab Manual No. LA-6760-M.

Huebner, W. F., Argo, M. F., and Olsen, L. D. 1978. Photoelectric cross sections for ions scaled from their neutral atoms. *J.Q.S.R.T.* 19:93–97.

Hui, A. K., Armstrong, B. H., and Wray, A. A. 1978. Rapid computation of the Voigt and complex error functions. *J.Q.S.R.T.* 19:509–516.

Hulot, E., Vilmer, N., and Trottet, G. 1988. Solar prompt gamma-ray lines and hard X-ray emission expected from a trap plus precipitation model for electrons and protons. *Astron. Astrophys.* 213:383–396.

Hummer, D. G., and Mihalas, D. 1988. The equation of state for stellar envelopes. I. An occupation probability formalism for the truncation of internal partition function. *Astrophys. J.* 331:794–814.

Hundhausen, A. J. 1972. *Coronal Expansion and Solar Wind* (Berlin: Springer-Verlag).

Hundhausen, A. J. 1977. An interplanetary view of coronal holes. In *Coronal Holes and High Speed Wind Streams,* ed. J. Zirker (Boulder: Colorado Assoc. Univ. Press), pp. 225–329.

Hundhausen, A. J. 1988. The origin and propagation of coronal mass ejections. In *Proc. Sixth Intl. Solar Wind Conf.,* eds. V. J. Pizzo, T. E. Holzer and D. G. Sime, NCAR TN-306 (Boulder: Natl. Center for Atmos. Res.), p. 181.

Hundhausen, A. J., Sawyer, C. B., House, L., Illing, R. M. E., and Wagner, W. J. 1984*a*. Coronal mass ejections observed during the solar maximum mission: Latitude distribution and rate of occurrence. *J. Geophys. Res.* 89:2639–2646.

Hundhausen, A. J., Burlaga, L. F., Feldman, W. C., Gosling, J. T., Hildner, E., House, L. L., Howard, R. A., Krieger, A. S., Kundu, M. R., Low, B. C., Sheeley, N. R., Jr., Steinolfson, R. S., Stewart, R. T., Stone, R. G., and Wu, S. T. 1984*b*. Coronal transients and their interplanetary effects. In *Solar Terrestrial Physics: Present and Future,* eds. D. M. Butler and K. Papadopoulos, NASA RP-1120, pp. 6.1–6.32.

Hundhausen, A. J., Holzer, T. E., and Low, B. C. 1987. Do slow shocks precede some coronal mass ejections? *J. Geophys. Res.* 92:173–178.

Hurford, G. J., Mewaldt, R. A., Stone, E. C., and Vogt, R. E. 1975. Enrichment of heavy nuclei in ³He-rich flares. *Astrophys. J.* 201:L95–L97.

Hurford, G. J., Read, R. B., and Zirin, H. 1984. Frequency-agile interferometer for solar microwave spectroscopy. *Solar Phys.* 94:413–426.

Hurlburt, N. E. 1983. Compressible Convection. Ph.D. Thesis, Univ. of Colorado, Boulder.

Hurlburt, N. E. 1987. Nonlinear compressible convection in regions of intense magnetic fields. In *The Role of Fine-Scale Magnetic Fields on the Structure of the Solar Atmosphere,* eds. E.-H. Schröter, M. Vázquez and A. A. Wyller (Cambridge: Cambridge Univ. Press), pp. 210–213.

Hurlburt, N. E., and Toomre, J. 1988. Magnetic fields interacting with nonlinear compressible convection. *Astrophys. J.* 327:920–932.

Hurlburt, N. E., Toomre, J., and Massaguer, J. M. 1984. Two-dimensional compressible convection extending over multiple scale heights. *Astrophys. J.* 282:557–577.

Hurlburt, N. E., Toomre, J., and Massaguer, J. M. 1986. Nonlinear compressible convection penetrating into stable layers and producing internal gravity waves. *Astrophys. J.* 311: 563–577.

Hurst, G. S., Payne, M. G., Kramer, S. D., and Young, J. P. 1979. Resonance ionization spectroscopy and one atom detection. *Rev. Mod. Phys.* 51:767–819.

Hurst, G. S., Payne, M. G., Kramer, S. D., and Chen, E. H. 1980. Counting atoms. *Physics Today* 9:24–29.

Hurst, G. S., Chen, C. H., Kramer, S. D., Cleveland, B. T., Davis, R., Rowley, J. K., Gabbard, F., and Schima, F. J. 1984. Feasibility of the $^{81}Kr(\nu,e^-)^{81}Kr$ solar neutrino experiment. *Phys. Rev. Lett.* 53:1116–1119.

Iben, I. 1963a. A comparison between homogeneous stellar models and observations. *Astrophys. J.* 138:452–470.

Iben, I., Jr. 1963b. Massive stars in quasi-static equilibrium. *Astrophys. J.* 138:1090–1096.

Iben, I., Jr. 1965. Stellar evolution. I. The approach to the main sequence. *Astrophys. J.* 141:993–1018.

Iben, I., Jr. 1969. The ^{37}Cl solar neutrino experiment and the solar helium abundance. *Ann. Phys. New York* 54:164–203.

Iben, I., Jr. 1974. Post main sequence evolution of single stars. *Ann. Rev. Astron. Astrophys.* 12:215–256.

Iben, I., Jr. 1975. Thermal pulses; *p*-capture, α-capture, *s*-process nucleo synthesis and convective in a star of intermediate mass. *Astrophys. J.* 196:525–547.

Iben, I., Jr., and MacDonald, J. 1985. The effects of diffusion due to gravity and due to composition gradients on the rate of hydrogen burning in a cooling degenerate dwarf. I. The case of a thick helium buffer layer. *Astrophys. J.* 296:540–553.

Iben, I., Jr., and Mahaffy, J. 1976. On the Sun's acoustical spectrum. *Astrophys. J.* 209: L39–L43.

Iben, I., Jr., Kalata, K., and Schwarz, J. 1967. The effect of ^7Be K-capture on the solar neutrino flux. *Astrophys. J.* 150:1001–1004.

Iglesias, C. A., Rogers, F. J., and Wilson, B. G. 1987. Reexamination of the metal contribution to astrophysical opacity. *Astrophys. J.* 322:L45–L48.

Illing, R. M. E., and Hundhausen, A. J. 1986. Disruption of a coronal streamer by an eruptive prominence and coronal mass ejection. *J. Geophys. Res.* 91:10951–10960.

Illing, R. M. E., Landman, D. A., and Mickey, D. L. 1974a. Broad-band circular and linear polarization in sunspots: Centre-to-limb variation. *Astron. Astrophys.* 35:327–331.

Illing, R. M. E., Landman, D. A., and Mickey, D. L. 1974b. Observations of broad-band circular polarization in sunspots: Magnetic field correspondence. *Astron. Astrophys.* 37:97–99.

Illing, R. M. E., Landman, D. A., and Mickey, D. L. 1975. Broad-band circular polarization of sunspots: Spectral dependence and theory. *Astron. Astrophys.* 41:183–185.

Immerschitt, S., and Schröter, E. H. 1989. On the behaviour of asymmetry and other profile parameters of the Fe I λ5576.1 Å line in solar regions of varying magnetic flux. *Astron. Astrophys.* 208:307–313.

Ionson, J. A. 1978. Resonant absorption of Alfvénic surface waves and the heating of solar coronal loops. *Astrophys. J.* 226:650–673.

Ionson, J. A. 1984. A unified theory of electrodynamic coupling in coronal magnetic loops: The coronal heating problem. *Astrophys. J.* 276:357–368.

Iospha, B. A., Obridko, V. N., and Shelting, B. D. 1973. Short-period oscillations of the magnetic field of the Sun as a star. *Solar Phys.* 29:385–392.

Ipavich, F. M., Galvin, A. B., Gloeckler, G., Hovestadt, D., Bame, S. J., Klecker, B., Scholer, M., Fisk, L. A., and Fan, C. Y. 1986. Solar wind Fe and CNO measurements in high-speed flows. *J. Geophys. Res.* 91:4133–4141.

Isaak, G. R. 1961. An atomic beam spectrophotometer. *Nature* 189:373–374.

Isaak, G. R. 1980. Solar oscillations, stellar oscillations and cosmology. *Nature* 283:644–645.

Isaak, G. R. 1982. Is the Sun an oblique magnetic rotator? *Nature* 296:130–131.

Isaak, G. R. 1986. Linewidth and rotational splitting of low-degree modes in the five-minute region. In *Seismology of the Sun and the Distant Stars*, ed. D. O. Gough (Dordrecht: D. Reidel), pp. 223–228.

Isaak, G. R., and Jones, A. R. 1988. Continuous magnetic calibration of (velocity) sensitivity of oscillation spectrometers. In *Advances in Helio- and Asteroseismology: Proc. IAU Symp. 123*, eds. J. Christensen-Dalsgaard and S. Frandsen (Dordrecht: D. Reidel), pp. 467–469.

Isaak, G. R., van der Raay, H. B., Pallé, P. L., Roca-Cortés, T., and Delache, P. 1984. Solar *g* modes. *Mem. Soc. Astron. Italiana* 55:91–97.

Isaak, G. R., Jefferies, S. M., McLeod, C. P., New, R., van der Raay, H. B., Pallé, P. L., Régulo, C., and Roca-Cortés, T. 1988. Solar cycle dependence of solar *p* modes. In *Advances in Helio- and Asteroseismology: Proc. IAU Symp. 123*, eds. J. Christensen-Dalsgaard and S. Frandsen (Dordrecht: D. Reidel), pp. 201–204.

Isaak, G. R., McLeod, C. P., van der Raay, H. B., Pallé, P. L., and Roca-Cortés, T. 1989. Solar oscillations as seen in the Na I and II absorption lines. *Astron. Astrophys.* 208:297–302.

Isaak, G. R., Elsworth, Y. P., Howe, R., McLeod, C. P., and New, R. 1990. Precision measurements of low ℓ and *p* modes of the Sun. In *Oji Intl. Seminar on Progress of Seismology of the Sun and Stars*, eds. Y. Osaki and H. Shibahashi (Berlin: Springer-Verlag), in press.

Itoh, N., and Kohyama, Y. 1978. Statistical theory of nuclear neutrino capture (II). Inclusion of first-forbidden transitions. *Nucl. Phys.* A306:527–535.

Itoh, N., Kohyama, Y., and Fujii, A. 1977. A statistical theory of nuclear neutrino capture. *Nucl. Phys.* A287:501–505.

Jackson, B. V., and Leinert, C. 1985. Helios images of solar mass ejections. *J. Geophys. Res.* 90:10759–10765.

Jackson, J. D. 1962. *Classical Electrodynamics* (New York: Wiley).

Jakimiec, J. 1965. Magnetohydrostatic models of sunspots. I. *Acta Astron.* 15:145–176.

Jakimiec, J. 1988. Investigation of the coronal part of solar active regions from X-ray observations. In *The Sun: Proc. 10th European Regional Meeting of the IAU*, eds. L. Hejna and M. Sobotka, pp. 219–227.

Jefferies, J. T., and Lindsey, C. A. 1974. Direct observation of temperature amplitude of solar 300-second oscillations. *Astrophys. J.* 187:L35–L36.

Jefferies, J. T., Smith, E. P., and Smith, H. J. 1959. The flare of September 18, 1957. *Astrophys. J.* 129:146–163.

Jefferies, J. T., Lites, B. W., and Skumanich, A. 1989. Transfer of line radiation in a magnetic field. *Astrophys. J.* 343:920–935.

Jefferies, S. M., Pallé, P. L., van der Raay, H. B., Régulo, C., and Roca-Cortés, T. 1988. Frequency stability of solar oscillations. *Nature,* in press.

Jefferies, S. M., McLeod, C. P., van der Raay, H. B., Pallé, P. L., and Roca-Cortés, T. 1988. Splitting of the low λ solar *p* modes. In *Advances in Helio- and Asteroseismology*, eds. J. Christensen-Dalsgaard and S. Frandsen (Dordrecht: D. Reidel), pp. 25–28.

Jefferies, J. T., and Lindsey, C. 1988. Radiative transfer in inhomogeneous atmospheres: A statistical approach. *Astrophys. J.* 335:372–382.

Jefferies, S. M., McLeod, C. P., van der Raay, H. B., Pallé, P. L., and Roca-Cortés, T. 1988*a*. Splitting of the low ℓ solar *p* modes. In *Advances in Helio- and Asteroseismology: Proc. IAU Symp. 123*, eds. J. Christensen-Dalsgaard and S. Frandsen (Dordrecht: D. Reidel), pp. 25–28.

Jefferies, S. M., Pallé, P. L., van der Raay, H. B., Régulo, C., and Roca-Cortés, T. 1988*b*. Frequency stability of solar oscillations. *Nature* 333:646–649.

Jefferies, S. M., Pomerantz, M. A., Duvall, T. L., Jr., Harvey, J. W., and Jaksha, D. B. 1988*c*. Helioseismology from the South Pole: Comparison of 1987 and 1981 results. In *Seismology of the Sun and the Sun-Like Stars*, ed. E. J. Rolfe, ESA SP-286, pp. 279–284.

Jefferies, S. M., Duvall, T. L., Jr., Harvey, J. W., Osaki, Y., and Pomerantz, M. A. 1990*a*. Characteristics of intermediate-degree solar *p*-mode line widths. *Astrophys. J.,* submitted.

Jefferies, S. M., Pomerantz, M. A., Duvall, T. L., Jr., and Harvey, J. W. 1990*b*. Helioseismology from the South Pole: Results from the 1987 campaign. In *Oji Intl. Seminar on Progress of Seismology of the Sun and Stars*, eds. H. Shibahashi and Y. Osaki (Berlin: Springer-Verlag), in press.

Jeffery, C. S. 1984. Apsidal motion in main-sequence binary stars. *Mon. Not. Roy. Astron. Soc.* 207:323–337.

Jeffrey, W. A. 1988. Variational Problems in Astrophysics. Ph.D. Thesis, Harvard Univ.

Jeffrey, W. A., and Rosner, R. 1986. Optimization algorithms: Simulated annealing and neural network processing. *Astrophys. J.* 310:473–481.

Jeffrey, W. A., and Rosner, R. 1988. An optimal approach to the inverse problem. In *Advances in Helio- and Asteroseismology,* eds. J. Christensen-Dalsgaard and S. Frandsen (Dordrecht: D. Reidel), pp. 129–132.

Jennings, D. E., Deming, D., Wiedemann, G. R., and Keady, J. J. 1986. Detection of 12 micron Mg I and OH lines in stellar spectra. *Astrophys. J.* 310:L39–L43.

Jensen, E., and Orrall, F. Q. 1963. Observational study of macroscopic inhomogeneities in the solar atmosphere. IV. Velocity and intensity fluctuations observed in the K line. *Astrophys. J.* 138:252–270.

Jiménez, A., Pallé, P. L., Roca-Cortés, T., Domingo, V., and Korzennik, S. 1987. Ground-based measurements of solar intensity oscillations. *Astron. Astrophys.* 172:323–326.

Jiménez, A., Pallé, P. L., Roca-Cortés, T., and Domingo, V. 1988a. Correlation between velocity and luminosity measurements of solar oscillations. *Astron. Astrophys.* 193:298–302.

Jiménez, A., Pallé, P. L., Pérez Hernandez, F., Régulo, C., and Roca-Cortés, T. 1988b. The observed background solar velocity noise. *Astron. Astrophys.* 192:L7–L9.

Jiménez, A., Pallé, P. L., Andersen, N. B., Domingo, V., Alvarez, M., Ledezma, E., and Roca-Cortés, T. 1988d. Solar luminosity oscillations from two stations and correlation with velocity measurements. In *Seismology of the Sun and Sun-Like Stars,* ed. E. J. Rolfe, ESA SP-286, pp. 163–168.

Jiménez, A., Pallé, P., Pérez, J. C., Régulo, C., Roca-Cortés, T., Isaak, G. R., McLeod, C. P., and van der Raay, H. 1988e. The solar oscillations spectrum and the solar cycle. In *Advances in Helio- and Asteroseismology: Proc. IAU Symp. 123,* eds. J. Christensen-Dalsgaard and S. Frandsen (Dordrecht: D. Reidel), pp. 205–209.

Jiménez, A., Alvarez, M., Andersen, N. B., Domingo, V., Jones, A., Pallé, P. L., and Roca-Cortés, T. 1990. Phase differences between luminosity and velocity measurements of the acoustic modes. *Solar Phys.* 126:1–19.

Jockers, K. 1978. Bifurcation of force-free solar magnetic fields: A numerical approach. *Solar Phys.* 56:37–54.

Johnson, W. W., Wingent, D. E., Douglass, D. H., and Van Horn, H. M. 1980. Time-varying gravitational multipole moments. In *Nonradial and Nonlinear Stellar Pulsation,* eds. H. Hill and W. Dziembowski (Berlin: Springer-Verlag), pp. 357–368.

Jokipii, J. R. 1965. Ph.D. Thesis, California Inst. of Technology.

Jokipii, J. R. 1966. In *The Solar Wind,* eds. R. J. Makin and M. Neugebauer (New York: Pergamon), p. 215.

Jones, H. P. 1989. Formation of Fourier phase shifts in the solar Ni I 6768 Å line. *Solar Phys.* 120:211–234.

Jones, H. P., and Giovanelli, R. G. 1983. Magnetic canopies in unipolar regions. *Solar Phys.* 87:37–42.

Jordan, S. D. 1981. Chromospheric heating. In *The Sun as a Star,* ed. S. D. Jordan, NASA SP-450, pp. 301–319.

Joselyn, J., and Holzer, T. E. 1978. A steady three-fluid coronal expansion for non-spherical geometries. *J. Geophys. Res.* 83:1019–1026.

Jursa, A. S., ed. 1985. *Handbook of Geophysics and the Space Environment* (Bedford, Mass.: Hanscom Air Force Base).

Kaastra, J. 1985. Ph.D. Thesis, Univ. of Utrecht.

Kahler, S. 1987. Coronal mass ejections. *Rev. Geophys.* 25:663–675.

Kahler, S. W. 1988. Observations of coronal mass ejections near the Sun. In *Proc. Sixth Intl. Solar Wind Conf.,* eds. V. J. Pizzo, T. E. Holzer, and D. G. Sime, NCAR TN-306 (Boulder: Natl. Center for Atmos. Res.), pp. 181–214.

Kahler, S. W., Webb, D. F., and Moore, R. L. 1981. X-ray and Hα observations of a filament-disappearance flare: An empirical analysis of the magnetic field configuration. *Solar Phys.* 70:335–350.

Kahler, S. W., Davis, J. M., and Harvey, J. W. 1983. Comparison of coronal holes observed in soft x-rays and HE I 10830 angstrom spectroheligrams. *Solar Phys.* 87:47–56.

Kahler, S. W., Sheeley, N. R., Jr., Howard, R. A., Koomen, M. J., Michels, D. J., McGuire, R. E., von Rosenvinge, T. T., and Reames, D. 1984. Associations between coronal mass ejections and solar energetic proton events. *J. Geophys. Res.* 89:9683–9693.

Kahler, S. W., Lin, R. P., Reames, D. V., Stone, R. G., and Liggett, M. 1985a. Solar source regions of the ³He rich solar particle events. In *Proc. 19th Intl. Cosmic Ray Conf.,* eds. F. C. Jones, J. Adams and G. M. Mason, NASA CP-2376, p. 269.

Kahler, S. W., Reames, D. V., Sheeley, N. R., Jr., Howard, R. A., Koomen, M. J., and Michels, D. H. 1985*b*. A comparison of solar ³He rich events with Type II bursts and coronal mass ejections. *Astrophys. J.* 290:742–747.

Kahler, S. W., Lin, R. P., Reames, D. V., Stone, R. G., and Ligget, M. 1987. Characteristics of solar coronal source regions producing ³He-rich particle events. *Solar Phys.* 107:385–394.

Kahler, S. W., Moore, R. L., Kane, S. R., and Zirin, H. 1988. Filament eruptions and the impulsive phase of solar flares. *Astrophys. J.* 328:824–829.

Kaisig, M., and Schroter, E. H. 1983. The asymmetry of photospheric absorption lines. II. The asymmetry of medium-strong Fe I lines in quiet and active regions of the Sun. *Astron. Astrophys.* 117:305–313.

Kajino, T., and Arima, A. 1984. Resonating-group calculations of radiative capture reactions $\alpha(^3He,\gamma)^7Be$ and $\alpha(t,\gamma)^7Li$. *Phys. Rev. Lett.* 52:739–743.

Kajita, T. 1989. In *Workshop on Elementary Particle Picture of the Universe,* Fuji-Yoshida, Japan, Univ. of Tokyo Inst. Cosmic Ray Research (185–89–2).

Kalkofen, W. 1989. Chromospheric heating. *Astrophys. J.* 346:L37–L40.

Kalkofen, W., Rosner, R., Ferrari, A., and Massaglia, S. 1986. The equilibrium structure of thin magnetic flux tubes. II. *Astrophys. J.* 304:519–525.

Kallenrode, M. B., Rieger, E., Wibberenz, G., and Forrest, D. J. 1987. Energetic charged particles resulting from solar flares with gamma-ray emission. In *Proc. 20th Intl. Cosmic Ray Conf.,* Moscow, August.

Kambry, M. A., Nishikawa, J., Sakurai, T., Ichimoto, K., and Kiei, E. 1988. Solar differential rotation from sunspot observations. *Proc. Astron. Soc. Japan,* in press.

Kane, S. R. 1972. Evidence for a common origin of the electrons responsible for the impulsive X-ray and Type III radio bursts. *Solar Phys.* 27:174–181.

Kane, S. R. 1974. In *Coronal Disturbances: Proc. IAU Symp. 57,* ed. G. Newkirk (Dordrecht: D. Reidel), pp. 105–141.

Kane, S. R. 1983. Spatial structure of high energy photon sources. *Solar Phys.* 86:355–365.

Kane, S. R., and Donnelly, R. F. 1971. Impulsive hard X-ray and ultraviolet emission during solar flares. *Astrophys. J.* 164:151–163.

Kane, S. R., Anderson, K. A., Evans, W. D., Klebesadel, R. W.; and Laros, J. 1979. Observation of an impulsive solar X-ray burst from a coronal source. *Astrophys. J.* 233:L151–L155.

Kane, S. R., et al. 1980. In *Solar Flares,* ed. P. A. Sturrock (Boulder: Univ. of Colorado), pp. 187–229.

Kantrowitz, A., and Petschek, H. E. 1966. In *Plasma Physics in Theory and Astrophysical Application,* ed. W. B. Kunkel (New York: McGraw-Hill), pp. 148–155.

Kaplan, S. A., Pikel'ner, S. B., and Tsytovitch, V. N. 1975. Plasma physics of the solar atmosphere. *Phys. Repts.* 15:1–82.

Karpinsky, V. N. 1990. Properties of the solar granulation. In *Solar Photosphere: Structure, Convection, and Magnetic Fields: Proc. IAU Symp. 138,* ed. J. O. Stenflo (Dordrecht: Kluwer), pp. 67–79.

Karzas, W. J., and Latter, R. 1958. The Rand Corp. RM-2091-AEC.

Karzas, W. J., and Latter, R. 1961. Electron radiative transition in a Coulomb field. *Astrophys. J. Suppl.* 6:167–212.

Kato, S. 1966. Overstable convection in a medium stratified in mean molecular weight. *Publ. Astron. Soc. Japan* 18:374–383.

Kaufman, J. M. 1988. Measurements of high degree solar oscillation amplitudes. In *Seismology of the Sun and Sun-Like Stars,* ed. E. J. Rolfe, ESA SP-286, pp. 31–35.

Kaufmann, P., Strauss, F. M., and Costa, J. E. R. 1982. Time delays in solar bursts measured in the mm-cm range of wavelengths. *Solar Phys.* 81:159–172.

Kaufmann, P., Strauss, F. M., Costa, J. E. R., Dennis, B. R., Kiplinger, A. L., Frost, K. J., and Orwig, L. E. 1983. Microwave and hard X-ray observations of a solar flare with a time resolution better than 100 ms. *Solar Phys.* 84:311–319.

Kaufmann, P., Correia, E., Costa, J. E. R., and Zodi Vaz, A. M. 1986. A synchrotron/inverse Compton interpretation of a solar burst producing fast pulses at $\lambda < 3$ mm and hard X-rays. *Solar Phys.* 157:159–172.

Kavanagh, R. W. 1960. Proton capture in ⁷Be. *Nucl. Phys.* 15:411–421.

Kavanagh, R. W. 1972. Reaction rates in the proton-proton chain. In *Cosmology, Fusion, and Other Matters*, ed. F. Reines (Boulder: Colorado Assoc. Univ. Press), p. 169–185.

Kavanagh, R. W. 1982. Solar power. In *Essays in Nuclear Astrophysics*, eds. C. A. Barnes, D. D. Clayton and D. N. Schramm (Cambridge: Cambridge Univ. Press), pp. 159–170.

Kavanagh, R. W., Tombrello, T. A., Moser, J. M., and Goosman, D. R. 1969. The $^7\mathrm{Be}(p,\gamma)^8\mathrm{B}$ cross section. *Bull. Amer. Phys. Soc.* 14:1209 (abstract).

Kawaguchi, I. 1980. Morphological study of the solar granulation. II. The fragmentation of granules. *Solar Phys.* 65:207–220.

Kawakami, H. 1983. Polarimetric study of unipolar sunspots. *Publ. Astron. Soc. Japan* 35: 459–489.

Kawakami, S., Makita, M., and Kurokawa, H. 1989. Detailed comparison of transverse magnetic fields of the Sun with H-alpha fine structures. *Publ. Astron. Soc. Japan* 41:175–195.

Kawaler, S. D. 1987. Angular momentum in stars: The Kraft curve revisited. *Publ. Astron. Soc. Pacific* 99:1322–1328.

Kawaler, S. D. 1988a. Angular momentum loss in low-mass stars. *Astrophys. J.* 333:236–247.

Kawaler, S. D. 1988b. The stellar seismology of the hot white dwarf star PG1159–035. In *Advances in Helio- and Asteroseismology: Proc. IAU Symp. 123*, eds. J. Christensen-Dalsgaard and S. Frandsen (Dordrecht: D. Reidel), pp. 329–332.

Kawaler, S. D., Winget, D. E., Hansen, C. J., and Iben, I., Jr. 1986. The helium shell game: Nonradial *g*-mode instabilities in hydrogen deficient planetary nebula nuclei. *Astrophys. J.* 306:L41–L44.

Keeley, D. A. 1977. Linear stability analysis of stellar models by the inverse iteration method. *Astrophys. J.* 211:926–933.

Keenan, F. P., Cook, J. W., Dufton, P. L., and Kingston, A. E. 1989. Solar Si III line ratios from the high-resolution telescope and spectrograph on board Spacelab 2: The effects of non-maxwellian electron distribution functions. *Astrophys. J.* 340:1135–1139.

Keil, S. 1980. The structure of solar granulation. I. Observations of the spatial and temporal behavior of vertical motions. *Astrophys. J.* 237:1024–1034.

Keil, S., and Mossman, A. 1989. Observations of high-frequency waves in the solar atmosphere. In *Solar and Stellar Granulation*, eds. R. J. Rutten and G. Severino (Dordrecht: Kluwer), pp. 333–345.

Keil, S. L., and Canfield, R. C. 1978. The height variation of velocity and temperature fluctuations in the solar photosphere. *Astron. Astrophys.* 70:169–179.

Keil, S. L., Roudier, T., Cambell, E., Koo, B. C., and Marmolino, C. 1989. Observation and interpretation of photospheric line asymmetry changes near active regions. In *Solar and Stellar Granulation*, eds. R. J. Rutten and G. Severino (Dordrecht: Kluwer), pp. 273–281.

Keller, C. U. 1989. Empirical photospheric fluxtube models fom inversion of Stokes V data. In *Solar Photosphere: Structure, Convection and Magnetic Fields: Proc. IAU Symp. 138*, ed. J. O. Stenflo (Dordrecht: Kluwer), pp. 121–124.

Kemble, E. C. 1937. *The Fundamental Principles of Quantum Mechanics* (New York: McGraw-Hill).

Kemp, J. C., and Henson, G. D. 1983. The broad-band circular polarization of sunspots, 0.37–4.5 microns. *Astrophys. J.* 266:L69–L72.

Kemp, J. C., Macek, J. H., and Nehring, F. W. 1984. Induced atomic orientation, an efficient mechanism for magnetic circular polarization. *Astrophys. J.* 278:863–873.

Kemp, J. C., Henson, G. D., Steiner, C. T., and Powell, E. R. 1987. The optical polarization of the Sun measured at a sensitivity of parts in ten million. *Nature* 326:270–273.

Kennedy, J. R., and Pintar, J. A. 1988. The Global Oscillation Network Group (GONG) helioseismic data reduction and analysis system. In *Astronomy from Large Databases*, eds. F. Murtagh and A. Heck (Garching: European Southern Obs.), pp. 367–372.

Kenyon, S., and Hartmann, L. 1988. The FU Orionis variables: Accretion and mass loss. In *Pulsation and Mass Loss in Stars*, eds. R. Stalio and L. A. Willson (Dordrecht: Kluwer), pp. 133–154.

Kichatinov, L. L. 1987. A mechanism for differential rotation based on angular momentum transport by compressible convection. *Geophys. Astrophys. Fluid Dyn.* 38:273–292.

Kidman, R. B., and Cox, A. N. 1984. The stability of the low degree five minute solar oscillations. In *Solar Seismology from Space,* eds. R. K. Ulrich, J. Harvey, E. J. Rhodes and J. Toomre, NASA Publ. 84–84, pp. 335–343.

Kidman, R. B., and Cox, A. N. 1987. Nonadiabatic, nonradial solar oscillations. In *Lecture Notes in Physics* No. 274, eds. A. N. Cox, W. M. Sparks and S. G. Starrfield (Berlin: Springer-Verlag), pp. 326–329.

Kiepenheuer, K. O. 1953. Solar activity. In *The Sun,* ed. G. P. Kuiper (Chicago: Univ. of Chicago Press), pp. 322–465.

Kim, C. W., Nussinov, S., and Sze, W. K. 1987. Non-adiabatic resonant conversion of solar neutrinos in three generations. *Phys. Lett.* 184B:403–409.

Kim, Y.-C., Demarque, P., and Guenther, D. B. 1991. The effect of the Mihalas-Hummer-Däppen equation of state and the molecular opacity on the standard solar model. *Astrophys. J., in press.

King, C. R., Da Costa, G. S., and Demarque, P. 1985. The luminosity function of the subgiant branch of 47 Tucane: A comparison of observation and theory. *Astrophys. J.* 299:674–682.

Kiplinger, A. L., Dennis, B. R., Emslie, A. G., Frost, K. J., and Orwig, L. W. 1982. Millisecond time variations in hard X-ray solar flares. *Astrophys. J.* 265:L99–L104.

Kiplinger, A. L., Dennis, B. R., Frost, K. J., and Orwig, L. E. 1984. Fast variations in high-energy X-rays from solar flares and their constraints on nonthermal models. *Astrophys. J.* 287:L105–L108.

Kiplinger, A. L., Dennis, B. R., and Orwig, L. E. 1984. Detection of a 158 day periodicity in the solar hard X-ray flare rate. *Bull. Amer. Astron. Soc.* 16:891 (abstract).

Kippenhahn, R. 1963. Differential rotation in stars with convective envelopes. *Astrophys. J.* 137:664–678.

Kippenhahn, R., and Schulter, A. 1957. A theory of solar filaments. *Z. Ap.* 43:63.

Kippenhahn, R., and Thomas, H. C. 1981. Rotation and stellar evolution. In *Fundamental Problems in Stellar Evolution Theory,* eds. D. Sugimoto, D. Q. Lamb and D. W. Schramm (Dordrecht: D. Reidel), pp. 237–256.

Kirsten, T. 1990. The GALLEX project. In *Inside the Sun: Proc. IAU Coll. 121,* eds. G. Berthomieu and M. Cribier (Dordrecht: Kluwer), pp. 187–199.

Kitai, R., and Muller, R. 1984. On the relation between chromospheric and photospheric fine structure in an active region. *Solar Phys.* 90:303–314.

Kjeldseth-Moe, O. 1968. A generalized theory for line formation in a homogeneous magnetic field. *Solar Phys.* 4:267–285.

Kjeldseth-Moe, O., and Maltby, P. 1969. A model for the penumbra of sunspots. *Solar Phys.* 8:275–283.

Kjeldseth-Moe, O., and Nicolas, K. R. 1977. Emission measures, electron densities, and nonthermal velocities from optically thin UV lines near a quiet solar limb. *Astrophys. J.* 211:579–586.

Kjeldseth-Moe, O., Brynildsen, N., Brekke, P., Engvold, O., Maltby, P., Bartoe, J.-D. F., Brueckner, G. E., Cook, J. W., Dere, K. P., and Socker, D. 1988. Gas flows in the transition region above sunspots. *Astrophys. J.* 334:1066–1075.

Klecker, B., Hovestadt, D., Gloeckler, G., Ipavich, F. M., Scholer, M., Fan, C. V., and Fisk, L. A. 1984. Direct determination of the ionic charge distribution of helium and iron in ³He-rich solar energetic particle events. *Astrophys. J.* 281:458–462.

Klein, K.-L., Anderson, K. A., Pick, M., Trottet, G., and Vilmer, N. 1983. Association between gradual hard X-ray emission and metric continua during large flares. *Solar Phys.* 84:295–310.

Klein, K.-L., Trottet, G., and Magun, A. 1986. Microwave diagnostics of energetic electrons in flares. *Solar Phys.* 104:243–252.

Klein, L. W., and Burlaga, L. F. 1982. Interplanetary magnetic clouds at 1 AU. *J. Geophys. Res.* 87:613–624.

Klimchuk, J. A., and Mariska, J. T. 1988. Heating-related flows in cool solar loops. *Astrophys. J.* 328:334–343.

Kliorin, N. I., and Ruzmajkin, A. A. 1984. On the nature of the 11-year torsional oscillation of the Sun. *Soviet Astron. Lett.* 10:390–392.

Kneer, F. J. 1980. Multidimensional radiative transfer in stratified atmospheres. III. Non-LTE line formation. *Astron. Astrophys.* 93:387–394.

Kneer, F. J., and von Uexküll, M. 1985. Oscillations of the Sun's chromosphere. II. Hα line center and wing filtergram time sequences. *Astron. Astrophys.* 144:443–451.

Kneer, F. J., Mattig, W., Nesis, A., and Werner, W. 1980. Coherence analysis of granular intensity. *Solar Phys.* 68:31–39.

Kneer, F. J., Newkirk, G., Jr., and von Uexküll, M. 1982. New features of the oscillation spectrum of the Sun. *Astron. Astrophys.* 113:129–134.

Knight, J. W., and Sturrock, P. A. 1977. Reverse current in solar flares. *Astrophys. J.* 218: 306–310.

Knobloch, E., and Weiss, N. O. 1984. Convection in sunspots and the origin of umbral dots. *Mon. Not. Roy. Astron. Soc.* 207:203–214.

Knölker, M., and Schüssler, M. 1988. Model calculations of magnetic flux tubes IV. Convective energy transport and the nature of intermediate size flux concentrations.

Knölker, M., and Stix, M. 1983. A convenient method to obtain stellar eigenfrequencies. *Solar Phys.* 82:331–341.

Knölker, M., Schüssler, M., and Weisshaar, E. 1988. Model calculations of magnetic flux tubes. III. Properties of solar magnetic elements. *Astron. Astrophys.* 194:257–267.

Knosvka, S., and Kříský, L. 1978. Time-latitude occurrence of flares in solar cycle no 20 (1965–1976). *Bull. Astron. Inst. Czech.* 29:352–354.

Knosvka, S., and Krvïvský, L. 1984. Types of sunspot groups occurring in the course of solar cycle no. 20. *Bull. Astron. Inst. Czech.* 35:271–272.

Knox, K. T., and Thompson, B. J. 1974. Recovery of images from atmospherically degraded short-exposure photographs. *Astrophys. J.* 193:L45–L48.

Koch, A. 1984. Plasma motion in umbrae and the surrounding photosphere derived from spectroscopic Doppler measurements and tracer measurements of spots. *Solar Phys.* 93:53–72.

Kocharov, L. G., and Kocharov, G. E. 1984. ³He-rich solar flares. *Space Sci. Rev.* 38:89–141.

Kohl, J. L., Parkinson, W. H., and Kurucz, R. L. 1978. *Center and Limb Solar Spectrum in High Spectral Resolution: 225.2 to 319.6 nm* (Cambridge, Mass.: Harvard-Smithsonian Center for Astrophysics).

Kohler, H. 1970. Differential rotation caused by anisotropic turbulent viscosity. *Solar Phys.* 13:3–18.

Kohler, H. 1973. The solar dynamo and estimates of the magnetic diffusivity and the α-effect. *Astron. Astrophys.* 25:467–419.

Kojima, M., and Kakinuma, T. 1987. Solar cycle evolution of solar wind stream structure between 1973 and 1985 observed with the interplanetary scintillation method. *J. Geophys. Res.* 92:7269–7279.

Kolb, E. W., Turner, M., and Walker, T. P. 1986. Yet another possible explanation of the solar-neutrino puzzle. *Phys. Lett.* 175B:478–484.

Komm, R., and Mattig, W. 1989. Results on the height dependence of granular velocity fluctuations. In *High Spatial Resolution Solar Observations,* ed. O. von der Lühe (Sunspot, N.M.: Sacramento Peak Obs.), pp. 330–338.

Kopecký, M. 1967. The periodicity of the sunspot groups. *Adv. Astron. Astrophys.* 5:189–266.

Kopecký, M. 1984. Sunspot activity according to Greenwich observations. *Solar Phys.* 93: 181–187.

Kopecký, M., and Křïvský, L. 1966. Proton flares and types of spot groups in the 11-year cycle. *Bull. Astron. Inst. Czech.* 17:360–365.

Kopp, R. A,. and Holzer, T. E. 1976. Dynamics of coronal hole regions. Steady polytropic flows with multiple critical points. *Solar Phys.* 49:43–56.

Kopp, R. A., and Pneuman, G. W. 1976. Magnetic reconnection in the corona and the loop prominence phenomenon. *Solar Phys.* 50:85–98.

Kopp, G., Lindsey, C., Becklin, E., Orrall, F., Roellig, T., and Werner, M. 1988. Chromospheric dynamics based on infrared observations. Solar Interior and Atmosphere Conference, 15–18 November, Tucson, Arizona, Abstract Booklet, p. 45.

Kopp, R., et al. 1986. In *Energetic Phenomena on the Sun,* eds. M. R. Kundu and B. E. Woodgate, NASA CP-2439, pp. 7-1--7-9.

Korchak, A. A. 1967. Possible mechanisms for generating hard X-rays in solar flares. *Soviet Astron. AJ* 11:258–263.

Korchak, A. A. 1971. On the origin of solar flare X-rays. *Solar Phys.* 18:284–304.

Korff, D. L. 1973. Analysis of a method for obtaining near-diffraction-limited information in the presence of atmospheric turbulence. *J. Opt. Soc. Amer.* 63:971–980.

Korzennik, S. G., and Ulrich, R. K. 1989. Seismic analysis of the solar interior. I. Can opacity changes improve the theoretical frequencies? *Astrophys. J.* 339:1144–1155.

Korzennik, S. G., Cacciani, A., Rhodes, E. J., Jr., Tomczyk, S., and Ulrich, R. K. 1988. Inversion of the solar rotation rate versus depth and latitude. In *Seismology of the Sun and Sun-Like Stars,* ed. E. J. Rolfe, ESA SP-286, pp. 117–124.

Kosovichev, A. G. 1986. Solution of an inverse helioseismological problem from observations on solar gravitational oscillations. *Bull. Crimean Astrophys. Obs.* 75:36–42.

Kosovichev, A. G. 1988*a*. An attempt to determine the structure of the solar core from observed *g*-mode frequencies. In *Advances in Helio- and Asteroseismology,* eds. J. Christensen-Dalsgaard and S. Frandsen (Dordrecht: D. Reidel), pp. 141–146.

Kosovichev, A. G. 1988*b*. The internal rotation of the Sun from helioseismological data. *Soviet Astron. Lett.* 14:145–149.

Kosovichev, A. G. 1988*c*. Determination of the solar sound speed by an asymptotic inversion technique. In *Seismology of the Sun and Sun-Like Stars,* ed. E. J. Rolfe, ESA SP-286, pp. 533–537.

Kosovichev, A. G., and Parchevskii, K. V. 1988. An asymptotic solution of the inverse problem of helioseismology for the internal differential rotation of the Sun. *Soviet Astron. Lett.* 14:201–204.

Kosovichev, A. G., and Severny, A. B. 1984*a*. The stability of solar gravity-mode oscillations and the structure of the Sun. *Soviet Astron. Lett.* 10:284–286.

Kosovichev, A. G, and Severny, A. B. 1984*b*. Instability of non-radial *g*-mode oscillations of the Sun with low z interior. *Mem. Soc. Astron. Italiana* 55:129–134.

Kosovichev, A. G., and Severny, A. B. 1985. Chemical effects on the stability of the Sun's natural gravitational oscillations. *Bull. Crimean Astrophys. Obs.* 72:162–170.

Kostiuk, T., and Mumma, M. J. 1983. Remote sensing by IR heterodyne spectroscopy. *Appl. Opt.* 22:2644–2654.

Kota, J., and Jokipii, J. R. 1983. Effects of drifts on the transport of cosmic rays. *Astrophys. J.* 265:573–581.

Kotov, V. A. 1985. The 160 minute oscillation. *Solar Phys.* 100:101–113.

Kotov, V. A. 1987. Rotation of the Sun and rotation of its general magnetic field. *Publ. Crimean Astrophys. Obs.* 77:39–50 (in Russian).

Kotov, V. A., and Levitsij, L. S. 1987. The 160 minutes period, internal rotation and 11-year cycle of the Sun: Evidence for a relationship? *Izv. Krymsk. Astrofiz. Obs.* 77:51–71.

Kotov, V. A., and Lyuty, V. M. 1988*a*. The 160.01 minutes oscillations of compact extragalactic objects and plausible cosmological consequences. *Izv. Krymsk. Astrofiz. Obs.* 79:139–157.

Kotov, V. A., and Lyuty, V. M. 1988*b*. Photometry of the Seyfert galaxy NGC 4151 and quasar 3C 273: The 160.010 minutes period. *Izv. Krymsk. Astrofiz. Obs.* 78:89–111.

Kotov, V. A., and Lyuty, V. M. 1989. *Compt. Rend.,* in press.

Kotov, V. A., and Tsap, T. T. 1988. *Izv. Krymsk. Astrofiz. Obs.* 82, in preparation.

Kotov, V. A., Severny, A. B., and Tsap, T. T. 1976. Time variations of the mean solar magnetic field. Some properties of the general solar magnetic field. In *The Problems of Magnetic Fields in the Cosmos,* Proc. Intl. Symp., Crimea, pt. 2, pp. 38–49.

Kotov, V. A., Severny, A. B., and Tsap, T. T. 1978. Observations of oscillations of the Sun. *Mon. Not. Roy. Astron. Soc.* 183:61–78.

Kotov, V. A., Koutchmy, S., Severny, A. B., and Tsap, T. T. 1980. The effect of the equation of time on the 160-min solar oscillation. *Soviet Astron. Lett.* 6:233–235.

Kotov, V. A., Severny, A. B., and Tsap, T. T. 1982. An investigation of the global oscillations of the Sun. I. Method and instruments. *Bull. Crimean Astrophys. Obs.* 65:1–30.

Kotov, V. A., Severny, A. B., Tsap, T. T., Moiseev, I. G., Efanov, V. A., and Nesterov, N. S. 1983*a*. Manifestation of the 160-min solar oscillations in velocity and brightness (optical and radio observations). *Solar Phys.* 82:9.

Kotov, V. A., Severny, A. B., and Tsap, T. T. 1983*b*. A study of solar global oscillations. II. Results of observations in 1974–1980, their analysis, and some conclusions. *Izv. Krymsk. Astrofiz. Obs.* 66:1–61. In English.

Kotov, V. A., Severny, A.B., and Tsap, T. 1984*a*. 160-minute and other long-period oscillations of the Sun: Analysis and interpretation of 9-year observations. *Mem. Soc. Astron. Italiana* 55:117–122.

Kotov, V. A., Severny, A. B., and Tsap, T. T. 1984b. *The Hydromagnetics of the Sun*, eds. T. D. Guyenne and J. J. Hunt, ESA SP-220, pp. 189–190.

Koutchmy, S. 1977a. Photospheric faculae: The contrasts at the center of the solar disk using filigree pictures. *Astron. Astrophys.* 61:397–404.

Koutchmy, S. 1977b. Solar corona. In *Illustrated Glossary for Solar and Solar-Terrestrial Physics*, eds. A. Bruzek and C. J. Durrand (Dordrecht: D. Reidel), pp. 39–52.

Koutchmy, S. 1981. A search of short-period coronal waves. *Space Sci. Rev.* 29:375–376.

Koutchmy, S., and Lebecq, D. 1986. The solar granulation. II. Photographic and photoelectric analysis of photospheric intensity fluctuations at the meso-granulation scale. *Astron. Astrophys.* 169:323–328.

Koutchmy, S., and Loucif, M. 1984. Empirical model of the solar corona using solar-cycle related parameters. In *The Hydromagnetics of the Sun*, eds. T. D. Guyenne and J. J. Hunt, ESA SP-220, pp. 265–267.

Koutchmy, S., and Nitschelm, C. 1984. Photometric analysis of the June 11, 1983 solar corona. *Astron. Astrophys.* 138:161–163.

Koutchmy, S., and Stellmacher, G. 1976. Photometric study of chromospheric and coronal spikes observed during the total solar eclipse of 30 June, 1973. *Solar Phys.* 49:253–265.

Koutchmy, S., and Stellmacher, G. 1978. Photospheric faculae. II. Line profiles and magnetic field in the bright network of the quiet Sun. *Astron. Astrophys.* 67:93–102.

Koutchmy, S., Koutchmy, O., and Kotov, V. A. 1980. Detection of the 160 min solar intensity variations: Sampling effect. *Astron. Astrophys.* 90:372–376.

Koutchmy, S., Zhugzhda, Y. D., and Locans, V. 1983. Short period coronal oscillations: Observation and interpretation. *Astron. Astrophys.* 120:185–191.

Kozlovsky, B., Lingenfelter, R. E., and Ramaty, R. 1987. Positrons from accelerated particle interactions. *Astrophys. J.* 316:801–818.

Kraeft, W. D., Kremp, D., Ebeling, W., and Röpke, G. 1986. *Quantum Statistics of Charged Particle Systems* (New York: Plenum Press).

Kraft, R. P. 1965. Studies in stellar rotation. I. Comparison of rotational velocities in the Hyades and Coma cluster. *Astrophys. J.* 142:681–702.

Kraft, R. P. 1967. Studies of stellar rotation. V. The dependence of rotation on age among solar-type stars. *Astrophys. J.* 150:551–570.

Kraft, R. 1969. Stellar rotation. In *Stellar Astronomy*, eds. H.-Y. Chiu, R. L. Warisala and J. I. Remo (New York: Gordon and Breach), pp. 317–367.

Kraft, R. P. 1970. Stellar rotation. In *Spectroscopic Astrophysics*, ed. G. H. Herbig (Berkeley: Univ. of California Press), pp. 385–422.

Kraichnan, R. H. 1976. Inertial ranges in two-dimensional turbulence. *Phys. Fluids* 10:1417–1423.

Kraichnan, R. H., and Montgomery, D. 1980. Two-dimensional turbulence. *Rept. Progr. Phys.* 43:547–619.

Kramer, S. D., Hurst, G. S., Young, J. P., and Payne, M. G. 1979. Resonance ionization spectroscopy of lithium. *Optics Comm.* 30:47–50.

Krasnikov, Yu. G. 1977. *Soviet Phys. JETP* 46(2):170.

Krastev, P. I., and Petcov, S. T. 1988. Resonance amplification and T-violation effects in three-neutrino oscillations in the Earth. *Phys. Lett.* 205B:84–92 and 214E:660–661.

Krauss, A., Becker, H. W., Trautvetter, H. P., and Rolfs, C. 1987. Astrophysical S(E) factor of ^{3}He(^{3}He,2p)^{4}He at solar energies. *Nucl. Phys.* A467:273–290.

Krauss, L. M. 1990. Correlation of solar neutrino modulation with solar cycle variation in p-mode acoustic spectra. *Nature* 348:403–407.

Krauss, L. M., Glashow, S. L., and Schramm, D. N. 1984. Antineutrino astronomy and geophysics. *Nature* 310:191–198.

Kräwinkel, H., Becker, H. W., Buchmann, L., Görres, J., Kettner, K. U., Keiser, W. E., Santo, R., Schmalbrock, P., Trautvetter, H. P., Vlieks, A., Rolks, C., Hammer, J. W., Azuma, R. E., and Rodney, W. S. 1982. The ^{3}He(α,γ)^{7}Be reaction and the solar neutrino problem. *Z. Phys.* A304:307–332.

Krieger, A. S., Timothy, A. F., and Roelof, E. C. 1973. A coronal hole as the source of a high velocity solar wind stream. *Solar Phys.* 29:505–525.

Krishan, V. 1985a. Two-dimensional pressure structure of a coronal loop. *Solar Phys.* 97: 183–189.

Krishan, V. 1985b. Statistical mechanics of velocity and magnetic fields in solar active regions. *Solar Phys.* 95:269–280.

Krishan, V. 1986. A probable initial configuration of a flaring loop. *Plasma Phys. Controlled Fusion* 28:509–510.

Krishna-Swamy, K. S. 1966. Profiles of strong lines in K dwarfs. *Astrophys. J.* 145:174–194.

Krvîvský, L., Krüger, A., and Ruzvicvková-Topolová, B. 1986. The sudden decrease of the large flare occurrence and its manifestation in interplanetary space after the maximum of the solar cycle. *Bull. Astron. Inst. Czech.* 37:111–114.

Krofcheck, D., Sugarbaker, E., Wagner, A. J., Rapaport, J., Wang, D., Bahcall, J. N., Byrd, R. C., Foster C. C., Goodman, C. D., Gaarde, C., Horen, D. J., Cary, T., and Taddeucci, T. N. 1987. Gamow-Teller strength distribution in ^{81}Kr and consequences for a ^{81}Br solar neutrino detector. *Phys. Rev. Lett.* 189:299–303.

Krohn, V. E., and Ringo, G. R. 1974. G_A/G_V measured in the decay of polarized neutrons. *Phys. Lett.* 55B:175–182.

Kroll, R. J., Chen, J., and Hill, H. A. 1988a. Properties of solar gravity mode signals in total irradiance observations. In *Seismology of the Sun and Sun-Like Stars,* ed. E. J. Rolfe, ESA SP-286, pp. 415–418.

Kroll, R. J., Hill, H. A., and Chen, J. 1988b. Test for the presence of solar gravity mode signals in the total irradiance observations. *Adv. Space Res.* 88:115–142.

Krymskiy, G. F. 1964. Diffusion mechanism of diurnal cosmic ray variations. *Geomag. Astron.* 4:763–769.

Kuchowitz, B. 1976. Neutrinos from the Sun. *Repts. Prog. Phys.* 39:291–343.

Kuhfub, R. 1986. A model for time-dependent turbulent convection. *Astron. Astrophys.* 160:116–120.

Kuhn, J. R. 1982. Recovering spectral information from unevenly sampled data: Two machine-efficient solutions. *Astron. J.* 87:196–202.

Kuhn, J. R. 1984. Spectral information from gapped data: A comparison of techniques. In *Solar Seismology from Space,* NASA JPL Publ. 84–84, pp. 293–303.

Kuhn, J. R. 1988a. Helioseismological splitting measurements and the non-spherical solar temperature structure. *Astrophys. J.* 331:L131-L134.

Kuhn, J. R. 1988b. Radial and temporal structure of the internal solar asphericity. In *Seismology of the Sun and Sun-Like Stars,* ed. E. J. Rolfe, ESA SP-286, pp. 87–90.

Kuhn, J. R. 1989a. Helioseismic observations of the solar cycle. *Astrophys. J.* 339:L45-L47.

Kuhn, J. R. 1989b. Calculating the internal solar asphericity from frequency splitting measurements. *Solar Phys.* 123:1–5.

Kuhn, J. R., and Gough, D. O. 1989. Solar oscillation. *Nature* 338:384.

Kuhn, J. R., and O'Hanlon, M. 1983. Low-σ1,5-min oscillation observations. *Solar Phys.* 87:207–219.

Kuhn, J. R. Libbrecht, K. G., and Dicke, R. H. 1986. Solar ellipicity fluctuations yield no evidence of *g*-modes. *Nature* 319:128–131.

Kuhn, J. R., Libbrecht, K. G., and Dicke, R. H. 1987. Evidence of global circulation currents from solar-limb temperature variations. *Nature* 328:326–327.

Kuhn, J. R., O'Neill, C. M., and Gilliam, L. B. 1988a. Solar 5-min oscillation amplitude anisotropy and Doppler velocity systematics. In *Advances in Helio- and Asteroseismology: Proc. IAU Symp. 123,* eds. J. Christensen-Dalsgaard and S. Frandsen (Dordrecht: D. Reidel), pp. 63–65.

Kuhn, J. R., Libbrecht, K. G., and Dicke, R. H. 1988b. The surface temperature of the Sun and changes in the solar constant. *Science* 242:908–911.

Kuijpers, J., van der Post, P., and Slottje, C. 1981. Runaway acceleration in a radio flare. *Astron. Astrophys.* 103:331–338.

Kulsrud, R. M., and Ferrari, A. 1971. The relativistic quasilinear theory of particle acceleration by hydromagnetic turbulence. *Astrophys. Space Sci.* 12:302–318.

Kumar, P., and Goldreich, P. 1989. Nonlinear interactions among solar acoustic modes. *Astrophys. J.* 343:558–575.

Kumar, P., Franklin, J., and Goldreich, P. 1988. Distribution functions for the time averaged energies of stochastically excited solar *p* modes. *Astrophys. J.* 328:879–887.

Kumar, P., Duvall, T. L., Jr., Harvey, J. W., Jefferies, S. M., Pomerantz, M. A., and Thompson, M. J. 1990. What are the observed high frequency solar acoustic modes? In *Oji Intl. Seminar on Progress of Seismology of the Sun and Stars,* eds. Y. Osaki and H. Shibahashi (Berlin: Springer-Verlag), in press.

Kundu, M. R. 1965. *Solar Radio Astronomy* (New York: Wiley).

Kundu, M. R., and Lang, K. R. 1985. The Sun and nearby stars: Microwave observations at high resolution. *Science* 228:9–15.

Kuo, T. K., and Pantaleone, J. 1987. Three-neutrino oscillations and the solar-neutrino experiments. *Phys. Rev.* 35D:3432–3446.

Kuo, T. K., and Pantaleone, J. 1989. Neutrino oscillations in matter. *Rev. Modern Phys.* 61: 937–979.

Kuperus, M. 1969. Chromospheric heating. *Space Sci. Rev.* 9:713–739.

Kuperus, M., and Tandberg-Hanssen, E. 1967. The nature of quiescent solar prominences. *Solar Phys.* 2:39–69.

Kurucz, R. L. 1979. Model atmospheres for G, F, A, B, and O stars. *Astrophys. J. Suppl.* 40: 1–34.

Kurucz, R. L., and Avrett, E. H. 1981. *Solar Spectrum Synthesis. I. A Sample Atlas from 224 to 300 nm.* Smithsonian Astrophys. Obs. Spec. Rept. No. 391.

Kurucz, R. L., Furenlid, I., Brault, J., and Testerman, L. 1984. *Solar Flux Atlas from 296 to 1300 nm* (Sunspot, N.M.: National Solar Obs.).

Küveler, G., and Wöhl, H. 1989. Daily variations of the photospheric equatorial rotation rate velocity of the Sun and its absolute values in 1981 and 1982 as determined from measurements using a two-dimensional photodiode array. *Solar Phys.,* in press.

Kuzmin, V. A. 1966. Detection of solar neutrinos by means of the ^{71}Ga$(\nu,e^-)^{71}$Ge reaction. *Soviet Phys. JETP* 22:1051–1052.

Labeyrie, A. 1970. Attainment of diffraction limited resolution in large telescopes by Fourier analysing speckle patterns in stare images. *Astron. Astrophys.* 6:85–87.

LaBonte, B. J. 1982. Solar calibration of stellar rotation tracers. *Astrophys. J.* 260:647–654.

LaBonte, B. J. 1986. Spectra of plages on the Sun and stars. I. CA II H and K lines. *Astrophys. J. Suppl.* 62:229–239.

LaBonte, B. J., and Howard, R. 1981. Solar rotation measurements at Mount Wilson. II. Systematic instrumental effects and the absolute rotation rate. *Solar Phys.* 73:3–12.

LaBonte, B. J., and Howard, R. 1982a. Solar rotation measurements at Mount Wilson. III. Meridional flow and limbshift. *Solar Phys.* 80:361–372.

LaBonte, B. J., and Howard, R. 1982b. Torsional waves on the Sun and the activity cycle. *Solar Phys.* 75:161–178.

LaBonte, B. J., Howard, R., and Gilman, P. A. 1981. An improved search for large-scale convection cells in the solar atmosphere. *Astrophys. J.* 250:796–798.

La Clare, F. 1983. Mesures du diamètre solaire à l'astrolabe. *Astron. Astrophys.* 125:200–203.

Lallement, R., Bertaux, J. L., and Kurt, V. G. 1985. Solar wind decrease at high heliographic latitudes detected from Prognoz interplanetary Lyman alpha mapping. *J. Geophys. Res.* 90:1413–1423.

Lallement, R., Holzer, T. E., and Munro, R. H. 1986. Solar wind expansion in a polar coronal hole: Inferences from coronal WL and interplanetary Lyman alpha observations. *J. Geophys. Res.* 91:6751–6759.

Lamb, H. 1909. On the theory of waves propagated vertically in the atmosphere. *Proc. London Math. Soc.* (2) 7:122–141.

Lamb, H. 1932. *Hydrodynamics,* 6th ed. (Cambridge: Cambridge Univ. Press).

Lande, K., Bozoki, G., Frati, W., Lee, C. K., Fenyves, E., and Saavendra, O. 1974. Possible anti-neutrino pulse of extragalactic origin. *Nature* 251:485–486.

Landi Degl'Innocenti, E. 1976. MALIP—a programme to calculate the Stokes parameters profiles of magnetoactive Fraunhofer lines. *Astron. Astrophys. Suppl.* 25:379–390.

Landi Degli'Innocenti, E. 1985a. Generation and transfer of polarized radiation in the solar atmosphere: Physical mechanisms and magnetic field diagnostics. In *Measurements of Solar Vector Magnetic Fields,* ed. M. J. Hagyard, NASA CP-2374, pp. 279–299.

Landi Degl'Innocenti, E. 1985b. Radiative transfer in small-scale magnetic structures. In *Theoretical Problems in High Resolution Solar Physics,* ed. H. U. Schmidt (Munich: MPI für Astrophysik), pp. 162–166.

Landi Degl'Innocenti, E. 1987. Transfer of polarized radiation, using 4 × 4 matrices. In *Numerical Radiative Transfer,* ed. W. Kalkofen (Cambridge: Cambridge Univ. Press), pp. 265–278.

Landi Degl'Innocenti, E., and Calamai, G. 1982. Broadband linear polarization from magnetized stellar atmospheres. Numerical tables for the magnetic intensification mechanism. *Astron. Astrophys. Suppl.* 49:677–685.

Landman, D. A., and Finn, G. D. 1979. Radiative transfer through a model sunspot. *Solar Phys.* 63:221–235.

Landman, D. A., and Illing, R. 1976. Further observations of helium and hydrogen emission in quiescent prominences. *Astron. Astrophys.* 49:277–283.

Landman, D. A., and Illing, R. 1977. Measurements of line emission in quiescent prominences: Ca^+, H and He. *Astron. Astrophys.* 55:103–109.

Landolfi, M. 1987. A possible diagnostic method for magnetic field and velocity gradients in sunspots. *Solar Phys.* 109:287–306.

Landolfi, M., and Landi Degl'Innocenti, E. 1982. Magneto-optical effects and the determination of vector magnetic fields from Stokes profiles. *Solar Phys.* 78:355–364.

Landolfi, M., Landi Degl'Innocenti, E., and Arena, P. 1984. On the diagnostic of magnetic fields in sunspots through the interpretation of Stokes parameters profiles. *Solar Phys.* 93:269–287.

Lane, J. H. 1869. On the theoretical temperature of the Sun under the hypothesis of a gaseous mass, maintaining its volume by its internal heat, and depending on the laws of gases as known to terrestrial experiment. *Amer. J. Sci. Arts* 2nd Ser. 50:57–74.

Lanford, W. A., and Wildenthal, B. H. 1972. Some comments on the cross section of ^{37}Cl for solar neutrino absorption. *Phys. Rev. Lett.* 29:606–608.

Lang, K. R., Willson, R. F., and Trottet, G. 1988. High-resolution VLA maps of the quiescent corona at 90 cm wavelength. *Astron. Astrophys.* 199:325–328.

Langer, G. E., Kraft, R. P., and Anderson, K. S. 1974. FG Sagittae: The *s*-process episode. *Astrophys. J.* 189:509–521.

Langley, S. P., and Abbot, C. G. 1900. The absorption lines in the infrared spectrum of the Sun. *Ann. Astrophys. Obs. Smithsonian Inst.* 1:5–21.

Lanzerotti, L. J., and Raghavan, R. S. 1981. Solar activity and solar neutrino flux. *Nature* 293:122–124.

Lapwood, E. R., and Usami, T. 1981. The solotone effect. In *Free Oscillations of the Earth* (Cambridge: Cambridge Univ. Press), pp. 14–15, 125–157.

Larkin, A. I. Thermodynamic functions of a low-temperature plasma. *Soviet Phys. JETP* 11:1363–1364.

LaRosa, T. N., and Emslie, A. G. 1988. A self-consistent interpretation of the solar flare extreme-ultraviolet to hard X-ray ratio in large events. *Astrophys. J.* 326:997–1001.

Larson, R. B. 1965. An attempted explanation of the horizontal branch. *Publ. Astron. Soc. Pacific* 77:452–455.

Larson, R. B. 1969a. Numerical calculations of the dynamics of a collapsing proto-star. *Mon. Not. Roy. Astron. Soc.* 145:271–295.

Larson, R. B. 1969b. The emitted spectrum of a proto-star. *Mon. Not. Roy. Astron. Soc.* 145:297–308.

Larson, R. B. 1972a. The collapse of a rotating cloud. *Mon. Not. Roy. Astron. Soc.* 156: 437–458.

Larson, R. B. 1972b. The evolution of spherical protostars with masses 0.25 $M_\odot$ to 10 $M_\odot$ *Mon. Not. Roy. Astron. Soc.* 157:121–145.

Larson, R. B. 1973. The evolution of protostars—Theory. *Fund. Cosmic Phys.* 1:1–70.

Larson, R. B. 1978. Calculations of three-dimensional collapse and fragmentation. *Mon. Not. Roy. Astron. Soc.* 184:69–85.

Larson, R. B. 1983. Angular momentum and protostellar disks. *Rev. Mexicana Astron. Astrof.* 7:219–227.

Larson, R. B. 1984. Gravitational torques and star formation. *Mon. Not. Roy. Astron. Soc.* 206:197–207.

Larson, R. B. 1985. Cloud fragmentation and stellar masses. *Mon. Not. Roy. Astron. Soc.* 214: 379–398.

Larson, R. B. 1989. The evolution of protostellar disks. In *Formation and Evolution of Planetary*

Systems, eds. H. A. Weaver, F. Paresce and L. Danly (Cambridge: Cambridge Univ. Press), pp. 33–54.

Last, J., Arnold, M., Dohner, J., Dubbers, D., and Freedman, S. J. 1988. Pulsed-beam neutron-lifetime measurement. *Phys. Rev. Lett.* 60:995–998.

Latour, J., Spiegel, E. A., Toomre, J., and Zahn, J.-P. 1976. Stellar convection theory. I. The anelastic modal equations. *Astrophys. J.* 207:233–243.

Latour, J., Toomre, J., and Zahn, J.-P. 1981. Stellar convection theory. III. Dynamical coupling of the two convection zones in A-type stars by penetrative motions. *Astrophys. J.* 248:1081–1098.

Latour, J., Toomre, J., and Zahn, J.-P. 1983. Nonlinear anelastic modal theory for solar convection. *Solar Phys.* 82:387–400.

Lattimer, J. M., and Cooperstein, J. 1988. Limits on the neutrino magnetic moment from SN1987A. *Phys. Rev. Lett.* 61:23–26.

Lavely, E. M., and Ritzwoller, M. H. 1991. *Phil. Trans. Roy. Soc.,* in press.

Layzer, D., Rosner, R., and Doyle, H. T. 1979. On the origin of solar magnetic fields. *Astrophys. J.* 229:1126–1137.

Lean, J. 1984. Estimating the variability of the solar flux between 200 and 300 nm. *J. Geophys. Res.* 89:1–9.

Lean, J. L. 1987. Solar ultraviolet irradiance variations: A review. *J. Geophys. Res.* 92:839–868.

Lean, J. L. 1988. Solar ultraviolet emission and total irradiance variations in solar cycle 21. *Science,* submitted.

Lean, J. L., and Brueckner, G. E. 1989. Intermediate-term solar periodicities: 100–500 days. *Astrophys. J.* 337:568–578.

Lean, J. L., and Foukal, P. 1988. A model of solar luminosity modulation by magnetic activity between 1954–1984. *Science* 240:906–908.

Lean, J. L., and Skumanich, A. 1983. Variability of the Lyman alpha flux with solar activity. *J. Geophys. Res.* 88:5751–5759.

Lean, J. L., White, O. R., Livingston, W. C., Heath, D. F., Donnelly, R. F., and Skumanich, A. 1982. A three-component model of the variability of the solar ultraviolet: 145–200 nm. *J. Geophys. Res.* 87:10307–10317.

Lebreton, Y., and Däppen, W. 1988. The influence of the equation of state on the ZAMS and the Sun. In *Seismology of the Sun and Sun-Like Stars,* ed. E. J. Rolfe, ESA SP-286, pp. 661–664.

Lebreton, Y., and Maeder, A. 1986. The evolution and helium content of the Sun. *Astron. Astrophys.* 161:119–124.

Lebreton, Y., and Maeder, A. 1987. Stellar evolution with turbulent diffusion mixing. VI. The solar model, surface ^{7}Li and ^{3}He abundances, solar neutrinos and oscillations. *Astron. Astrophys.* 175:99–112.

Lebreton, Y., Berthomieu, G., Provost, J., and Schatzman, E. 1988. Influence of the axial-vector coupling constant on solar models: Solar neutrino fluxes, helium content and oscillations. *Astron. Astrophys.* 200:L5–L8.

Ledoux, P. 1951. The nonradial oscillations of gaseous stars and the problem of Beta Canis Majoris. *Astrophys. J.* 114:373–384.

Ledoux, P. 1962. Asymptotic form of the adiabatic radial pulsations of a star. I. *Bul. Acad. Roy. Belgique* 48:240–254.

Ledoux, P. 1963. Asymptotic form of the adiabatic radial pulsations of a star. II. Asymptotic behavior of the amplitudes. *Bul. Acad. Roy. Belgique* 49:286–302.

Ledoux, P., and Walraven, H. 1958. Variable stars. In *Hanbüch der Physik,* vol. 51, ed. S. Flügge (Berlin: Springer-Verlag), pp. 353–604.

Lee, M. A. 1988. The solar wind termination shock and the heliosphere beyond. In *Proc. Sixth Intl. Solar Wind Conf.,* eds. V. J. Pizzo, T. E. Holzer and D. G. Sime, NCAR TN-306 (Boulder: Natl. Center for Atmos. Res.), pp. 635–650.

Leer, E. 1988. Wave acceleration mechanisms for the solar wind. In *Proc. Sixth Intl. Solar Wind Conf.,* eds. V. J. Pizzo, T. E. Holzer and D. G. Sime, NCAR TN-306 (Boulder: Natl. Center for Atmos. Res.), pp. 3–22.

Lefèvre, J., and Pecker, J.-C. 1961. Ecarts à l'équilibre et abondances dans les photosphères solaires et stellaires. VI. La variation des intensités centrales des rais métalliques entre le bord du soleil *Ann. Astrophys.* 24:238–250.

Legait, A. 1986. Radiative convection in a stratified atmosphere. *Astron. Astrophys.* 168: 173–183.

Legrand, and Simon, P. A. 1981. Ten cycles of solar and geomagnetic activity. *Solar Phys.* 70:173–195.

Leibacher, J. W., and Stein, R. F. 1971. A new description of the solar five-minute oscillation. *Astrophys. Lett.* 7:191–192.

Leibacher, J. W., and Stein, R. F. 1981. Oscillations and pulsations. In *The Sun as a Star,* ed. S. Jordan, NASA SP-450, pp. 263–287.

Leibacher, J. W., Gouttebroze, P., and Stein, R. F. 1982. Solar atmospheric dynamics. II. Non-linear models of the photospheric and chromospheric oscillations. *Astrophys. J.* 258: 393–403.

Leibacher, J. W., Noyes, R. W., Toomre, J., and Ulrich, R. K. 1985. Helioseismology. *Sci. Amer.* 253(3):48–57.

Leifsen, T., and Maltby, P. 1988. Solar infrared intensity oscillations. In *Seismology of the Sun and Sun-Like Stars,* ed. E. J. Rolfe, ESA SP-286, pp. 169–173.

Leifsen, T., and Maltby, P. 1990. New light on solar infrared intensity oscillations. *Solar Phys.* 125:241–249.

Leighton, R. B. 1961. Untitled comments on observations of supergranulation and five-minute oscillations. In *Aerodynamic Phenomena in Stellar Atmospheres: Proc. IAU Symp. 12,* ed. R. N. Thomas. *Nuovo Cimento Suppl. Ser. 10* 22:321–327.

Leighton, R. B. 1963. The solar granulation. *Ann. Rev. Astron. Astrophys.* 1:19–40.

Leighton, R. B. 1964. Transport of magnetic fields on the Sun. *Astrophys. J.* 140:1547–1562.

Leighton, R. B. 1969. A magneto-kinematic model of the solar cycle. *Astrophys. J.* 156:1–26.

Leighton, R. B., Noyes, R. W., and Simon, G. W. 1962. Velocity fields in the solar atmosphere. I. Preliminary report. *Astrophys. J.* 135:474–499.

Lemaire, P. 1988. Physical structure and diagnostics of solar active regions deduced from optical, visible and UV observations. In *Proc. 10th European Regional Astronomy Meeting of the IAU, Vol. 1: The Sun,* eds. L. Hejna and M. Sobotka, Publ. Astron. Inst. of Czech. Acad. Sci. no. 66 (Ondrejov: Astronomical Inst. of Czechoslovak Academy of Sciences) pp. 185–194.

Lemke, M., and Holweger, H. 1987. A non-LTE study of the solar emission lines near 12 microns. *Astron. Astrophys.* 173:375–382.

Lemoine, B., Demuynck, C., and Destombes, J. L. 1988. Rydberg transitions of neutral magnesium in the infrared: Frequency measurements and Zeeman effect. *Astron. Astrophys.* 191: L4–L6.

Léna, P. 1968. Observations of the center-to-limb variation of the solar brightness in the far infrared (10 to 25 microns). *Solar Phys.* 3:28–35.

Léna, P. 1970. Le rayonnement continu infrarrouge de la photosphère solaire. *Astron. Astrophys.* 4:202–219.

Leroy, J.-L. 1987. Observations of prominence magnetic field. *Astrophys. Space Sci.* 150: 77–111.

Leroy, J.-L., and Noëns, J.-C. 1983. Does the solar activity cycle extend over more than an 11-year period? *Astron. Astrophys.* 120:L1–L2.

Leroy, J.-L., and Trellis, M. 1974. Three activity cycles on the lower corona. *Astron. Astrophys.* 35:283–288.

Leroy, J.-L., Bommier, V., and Sahal-Brechot, S. 1984. New data on the magnetic structure of quiescent prominences. *Astron. Astrophys.* 131:33–44.

Levich, D. K. 1969. Numerical simulation of two dimensional turbulence. II. *Phys. Fluids* 12:240–249.

Levine, R. H. 1975. The representation of magnetic field lines from magnetograph data. *Solar Phys.* 44:365–370.

Levine, R. H. 1976. Evidence for opposed currents in active region loops. *Solar Phys.* 46: 159–170.

Levine, R. H. 1977. Large scale solar magnetic fields and coronal holes. In *Coronal Holes and High Speed Wind Streams,* ed. J. B. Zirker (Boulder: Colorado Assoc. Univ. Press), pp. 103–143.

Levine, R. H. 1982. Open magnetic fields and the solar cycle I: Photospheric origins of open magnetic flux. *Solar Phys.* 79:203–230.

Levine, R. H., and Withbroe, G. L. 1977. Physics of an active region loop system. *Solar Phys.* 51:83–101.

Levine, R. H., Altschuler, M. D., Harvey, J. W., and Jackson, B. V. 1977. Open magnetic structures on the Sun. *Astrophys. J.* 215:636–651.

Levy, M. 1971. Analyse photométrique statistique de la granulation corrigée de l'influence de l'atmosphère terrestre et de l'instrument. *Astron. Astrophys.* 14:15–23.

Libbrecht, K. G. 1986. Is there an unusual solar core? *Nature* 319:753–755.

Libbrecht, K. G. 1988a. The excitation and damping for solar *p*-modes. In *Seismology of the Sun and Sun-Like Stars*, ed. E. J. Rolfe, ESA SP-286, pp. 3–10.

Libbrecht, K. G. 1988b. Solar and stellar seismology. *Space Sci. Rev.* 47:275–301.

Libbrecht, K. G. 1988c. Solar *p*-mode frequency splittings. In *Seismology of the Sun and Sun-Like Stars*, ed. E. J. Rolfe, ESA SP-286, pp. 131–136.

Libbrecht, K. G. 1988d. Solar *p*-mode phenomenology. *Astrophys. J.* 334:510–516.

Libbrecht, K. G. 1989. Solar *p*-mode frequency splittings. *Astrophys. J.* 336:1092–1097.

Libbrecht, K. G. 1990. Comparison of solar *p*-mode oscillations in surface brightness and velocity. *Astrophys. J.* 359:232–234.

Libbrecht, K. G., and Kaufman, J. M. 1988. Frequencies of high-degree solar oscillations. *Astrophys. J.* 324:1172–1183.

Libbrecht, K. G., and Kuhn, J. R. 1985. On the facular contrast near the solar limb. *Astrophys. J.* 299:1047–1050.

Libbrecht, K. G., and Woodard, M. F. 1990a. Solar-cycle effects on solar oscillation frequencies. *Nature* 345:779–782.

Libbrecht, K. G., and Woodard, M. F. 1990b. Observations of solar cycle variations in solar *p*-mode frequencies and splittings. In *Oji Intl. Seminar on Progress of Seismology of the Sun and Stars*, eds. Y Osaki and H. Shibahashi (Berlin: Springer-Verlag), in press.

Libbrecht, K. G., and Zirin, H. 1986. Properties of intermediate-degree solar oscillation modes. *Astrophys. J.* 308:413–423.

Libbrecht, K. G., Woodard, M. F., and Kaufman, J. M. 1990. Frequencies of solar oscillations. *Astrophys. J. Suppl.* 74:1129–1149.

Liggett, M. A., and Zirin, H. 1985. Emerging flux in active regions. *Solar Phys.* 97:51.

Lighthill, M. J. 1952. On sound generated aerodynamically. I. General theory. *Proc. Roy. Soc. London* A211:564–587.

Lighthill, M. J. 1978. *Waves in Fluids* (Cambridge: Cambridge Univ. Press).

Lilly, D. K. 1969. Numerical simulation of two dimensional turbulence II. *Phys. Fluids* 12: 240–249.

Lim, C.-S., and Marciano, W. J. 1988. Resonant spin-flavor prescession of solar and supernova neutrinos. *Phys. Rev.* D37:1368–1373.

Lin, R. P. 1987. Solar particle acceleration and propagation. *Rev. Geophys. Space Phys.* 25:676–684.

Lin, R. P., and Hudson, H. S. 1976. Non-thermal processes in large solar flares. *Solar Phys.* 50:153–178.

Lin, R. P., Schwartz, R. A., Pelling, R. M., and Hurley, K. C. 1981. A new component of hard X-rays in solar flares. *Astrophys. J.* 251:L109–L114.

Lin, R. P., Schwartz, R. A., Kane, S. R., Pelling, R. M., and Hurley, K. C. 1984. Solar hard X-ray microflares. *Astrophys. J.* 283:421–425.

Lin, Y.-Z., and Gaizauskas, V. 1987. Coincidence between H-alpha flare kernels and peaks of observed longitudinal electric current densities. *Solar Phys.* 109:81–90.

Lindholm, E. 1946. *Ark. Mat. Astron. Fys.* 32a(17).

Lindsey, C. A. 1977. Infrared continuum observations of five-minute oscillations. *Solar Phys.* 52:263–281.

Lindsey, C. A. 1981. Heating of the solar chromosphere by ionization pumping. *Astrophys. J.* 244:659–677.

Lindsey, C. A. 1987. LTE modeling of inhomogenous chromospheric structure using high-resolution limb observations. *Astrophys. J.* 320:893–897.

Lindsey, C. A., and Braun, D. C. 1989. Helioseismic imaging of sunspots at their antipodes. *Solar Phys.* 126:101–115.

Lindsey, C. A., and Heasley, J. N. 1981. Far-infrared continuum observations of solar faculae. *Astrophys. J.* 247:348–353.

Lindsey, C. A., and Hudson, H. S. 1976. Solar limb brightening in submillimeter wavelengths. *Astrophys. J.* 203:753–759.

Lindsey, C. A., and Kaminski, C. 1984. Temporal variations in the solar submillimeter continuum. *Astrophys. J.* 282:L103–L106.

Lindsey, C. A., and Roellig, T. 1987. Submillimeter diagnostics of the response of the solar chromosphere to compressional waves. *Astrophys. J.* 313:877–892.

Lindsey, C. A., Hildebrand, R. H., Keene, J., and Whitcomb, S. E. 1981. Solar limb brightening at 350 microns. *Astrophys. J.* 248:830–835.

Lindsey, C. A., de Graauw, T., de Vries, C., and Lindholm, S. 1984*a*. Solar limb brightening at 820 microns. *Astrophys. J.* 277:424–428.

Lindsey, C. A., Becklin, E. E., Jefferies, J. T., Orall, F. Q., Werner, M. W., and Gatley, I. I. 1984*b*. Observations of the brightness profile of the Sun in the 30–200 micron continuum. *Astrophys. J.* 281:862–869.

Lindsey, C., Becklin, E. E., Jefferies, J. T., Orrall, F. Q., Werner, M. W., and Gatley, I. I. 1986. Extreme limb profiles of the Sun at far-infrared and submillimeter wavelengths. *Astrophys. J.* 308:448–458.

Lindsey, C. A., Kopp, G., Becklin, E. E., Roellig, T., Werner, M. W., Jefferies, J. T., Orrall, F. Q., Braun, D. C., and Mickey, D. L. 1990. Far infrared intensity variations caused by five-minute oscillations. *Astrophys. J.* 350:475–479.

Linsky, J. L. 1980. Stellar chromospheres. *Ann. Rev. Astron. Astrophys.* 18:439–488.

Linsky, J. L. 1985. Evidence for nonradiative activity in stars with $\tau_{eff} < 10,000$ K. In *The Origin of Nonradiative Heating/Momentum in Hot Stars,* NASA CP-2358, pp. 24–46.

Lites, B. W., and Chipman, E. G. 1979. The vertical propagation of waves in the solar atmosphere. I. Observations of phase delays. *Astrophys. J.* 231:570–588.

Lites, B. W., and Skumanich, A. 1985. The inference of vector magnetic fields from polarization measurements with limited spectral resolution. In *Measurements of Solar Vector Magnetic Fields,* ed. M. J. Hagyard, NASA CP-2374, pp. 342–367.

Lites, B. W., and Skumanich, A. 1989. Stokes profile analysis and vector magnetic fields. V. The magnetic field structure of large sunspots observed with Stokes II. *Astrophys. J.,* in press.

Lites, B. W., Chipman, E. G., and White, O. R. 1982. The vertical propagation of waves in the solar atmosphere. II. Phase delays in the quiet chromosphere and cell-network distinctions. *Astrophys. J.* 253:367–385.

Lites, B. W., Skumanich, A., Rees, D. E., Murphy, G. A., and Carlsson, M. 1987. Stokes profile analysis and vector magnetic fields. III. Extended temperature minima of sunspot umbrae as inferred from Stokes profiles of Mg I λ4571. *Astrophys. J.* 318:930–939.

Lites, B. W., Skumanich, A., Rees, D. E., and Murphy, G. A. 1988. Stokes profile analysis and vector magnetic fields. IV. Synthesis and inversion of chromospheric Mg I b lines. *Astrophys. J.* 330:493–512.

Lites, B. W., Nordlund, Å., and Scharmer, G. B. 1989. Constraints imposed by very high resolution spectra and images on theoretical simulations of granular convection. In *Solar and Stellar Granulation,* eds. R. J. Rutten and G. Severino (Dordrecht: Kluwer), pp. 349–357.

Litherland, T., Fireman, E. L., and Rowley, J. K. 1987. Are AMS [7]Be measurements of a lithium solar neutrino detector practical. *Nucl. Inst. Meth.* B29:387–388.

Liu, S.-Y. 1973. Outflow of chromospheric emission features from the rim of a sunspot. *Solar Phys.* 31:127–129.

Liu, S.-Y. 1974. Direct observational evidence for the propagation and dissipation of energy in the chromosphere. *Astrophys. J.* 189:359–365.

Liu, S.-Y., and Skumanich, A. 1974. An empirical interpretation for the time evolution of the Ca II K line. *Solar Phys.* 38:109–115.

Livi, S. H. B., Martin, S., Wang, H., and Guoxiang, A. 1989. The association of flares to cancelling magnetic features on the Sun. *Solar Phys.* 121:197–214.

Livi, S. H. B., Wang, J., and Martin, S. F. 1985. The cancellation of magnetic flux. I. On the quiet Sun. *Australian J. Phys.* 38:855–873.

Livingston, W. C. 1982. Magnetic fields, convection and solar luminosity variability. *Nature* 297:208–210.

Livingston, W. C. 1983. Solar magnetic fields. Origins and coronal effects. In *Solar and Stellar Magnetic Fields: Origins and Coronal Effects: Proc. IAU Symp. 102,* ed. J. O. Stenflo (Dordrecht: D. Reidel) pp. 149–153.

Livingston, W. C. 1987. Line asymmetry and the activity cycle. In *The Role of Fine-Scale Magnetic Fields on Structure of the Solar Atmosphere*, eds. E.-H. Schröter, M. Vázquez and A. A. Wyller (Cambridge: Cambridge Univ. Press).

Livingston, W. C., and Duvall, T. L., Jr. 1979. Solar rotation, 1966–1978. *Solar Phys.* 61: 219–231.

Livingston, W. C., and Harvey, J. W. 1969. Observational evidence of quantization in photospheric magnetic flux. *Solar Phys.* 10:294–296.

Livingston, W. C., and Harvey, J. W. 1975. A new component of solar magnetism—The inner network fields. *Bull. Amer. Astron. Soc.* 7:346 (abstract).

Livingston, W. C., and Holweger, H. 1982. Solar luminosity variation. IV. The photospheric lines. *Astrophys. J.* 252:375–385.

Livingston, W. C., and Wallace, L. 1985. Water vapor and 5250.2. *Solar Phys.* 95:251–252.

Livingston, W. C., Harvey, J., Slaughter, C., and Trumbo, D. 1976. Solar magnetograph employing integrated diode arrays. *Appl. Opt.* 15:40–52.

Livingston, W. C., Milkey, R., and Slaughter, D. 1977. Solar luminosity variation. I. C I 5380 as a temperature indicator and a search for global oscillations. *Astrophys. J.* 211:281–287.

Livingston, W. C., Wallace, L., and White, O. R. 1988. Spectrum line intensity as a surrogate for solar irradiance variations. *Science* 240:1765–1767.

Lockwood, G. W., Thompson, D. T., Radick, R. R., Osborn, W. H., Baggett, W. E., Duncan, D. K., and Hartmann, L. W. 1984. The photometric variability of solar-type stars. IV. Detection of rotational modulation among Hyades stars. *Publ. Astron. Soc. Pacific* 96:714–722.

Losecco, J. M. 1985. Comment on the possible explanation of the solar neutrino puzzle. *Phys. Rev. Lett.* 54:2299–2301.

Lossecco, J. M. 1986. Comment on the possible explanation of the solar neutrino puzzle. *Phys. Rev. Lett.* 57:652.

Lou, Y.-Q. 1988. Viscous magnetohydrodynamic modes and *p*-mode absorption by sunspots. In *Seismology of the Sun and Sun-Like Stars*, ed. E. J. Rolfe, ESA SP-286, pp. 305–310.

Lou, Y.-Q. 1990. Viscous magnetohydrodynamic modes and *p*-mode absorption by sunspots. *Astrophys. J.* 350:452–462.

Low, B. C. 1975. Nonisothermal magnetostatic equilibria in a uniform gravity field. II. Sheet models of quiescent prominences. *Astrophys. J.* 198:211–214.

Low, B. C. 1981. The field and plasma configuration of a filament overlying a solar bipolar magnetic region. *Astrophys. J.* 246:538–548.

Low, B. C. 1982a. The vertical filamentary structures of quiescent prominences. *Solar Phys.* 75: 119–131.

Low, B. C. 1982b. Nonlinear force-free magnetic fields. *Rev. Geophys. Space Phys.* 20: 145–159.

Low, B. C. 1985. Three-dimensional structures of magnetostatic atmospheres. I. Theory. *Astrophys. J.* 293:31–43.

Lowry, M. M., Kouzes, R. T., Loeser, F., McDonald, A. B., and Naumann, R. A. 1987. Electron capture decay of $^{81}Kr^m$. *Phys. Rev.* C35:1950–1953.

Lubov, S. H., Rhodes, E. J., and Ulrich, R. K. 1980. Five minute oscillations as a probe of the solar interior. *Lecture Notes in Physics* 125:300–306.

Lustig, G. 1983. Solar rotation 1947–1981 determined from sunspot data. *Astron. Astrophys.* 125:335–358.

Lustig, G., and Dvorak, R. 1984. Solar rotation rate from stable recurrent sunspot tracings. *Astron. Astrophys.* 141:105–107.

Luttrell, A. H., and Richter, A. K. 1988. The role of Alfvénic fluctuations in MHD turbulence evolution between 0.3 and 1.0 AU. In *Proc. Sixth Intl. Solar Wind Conf.*, eds. V. J. Pizzo, T. E. Holzer and D. G. Sime, NCAR TN-306 (Boulder: Natl. Center for Atmos. Res.), pp. 335–339.

Machado, M. E., and Rust, D. M. 1974. Analysis of the August 7, 1972 white-light flare: Its spectrum and vertical structure. *Solar Phys.* 38:499–516.

Machado, M. E., Emslie, A. G., and Mauas, P. J. 1986. A mechanism for deep chromospheric heating during solar flares. *Astron. Astrophys.* 159:33–39.

Machado, M. E., Rovira, M. G., and Sneibrun, C. 1985. Hard X-ray imaging evidence of non-thermal and thermal burst components. *Solar Phys.* 99:189–217.

MacKinnon, A. L. 1988. Coulomb collisional precipitation of fast electrons in solar flares. *Astron. Astrophys.* 194:279–287.

MacKinnon, A. L. 1989. A potential diagnostic for low energy, nonthermal protons in solar flares. *Astron. Astrophys.*, in press.

MacKinnon, A. L., and Brown, J. C. 1989. On the bremsstrahlung efficiency of nonthermal hard X-rays source models. *Solar Phys.* 122:303–311.

MacKinnon, A L., and Brown, J. C., and Hayward, J. 1985. Quantitative analysis of hard X-ray "footprint" flares observed by the solar maximum mission. *Solar Phys.* 99:231–261.

MacNeice, P., McWhirter, R. W. P., Spicer, D. S., and Burgess, A. 1984. A numerical model of a solar flare based on electron beam heating of the chromosphere. *Solar Phys.* 90: 357–382.

MacQueen, R.M. 1980. Coronal transients. *Phil. Trans. Roy. Soc.* 279:605–620.

MacQueen, R. M. 1985. Coronal mass ejections: Acceleration and surface associations. *Solar Phys.* 95:359–361.

Macris, C. J. 1951. The possible variation of photospheric granules. *Observatory* 75:122–123.

Macris, C. J., Muller, R., Rosch, J., and Roudier, Th. 1984. Variation of the mesh of the granular network along the solar cycle. In *Small-Scale Dynamical Processes in Quiet Stellar Atmospheres,* ed. S. Keil (Sunspot, N. M.: Sacramento Peak Obs.), pp. 265–275.

Maeder, A. 1975. Stellar evolution. III. The overshooting from convective cores. *Astron. Astrophys.* 40:303–310.

Maeder, A. 1983. Evolution of chemical abundances in massive stars. *Astron. Astrophys.* 120:113–129.

Maeder, A. 1990. A look on non-standard solar models. In *Inside the Sun,* eds. G. Berthomieu and M. Cribier (Dordrecht: Kluwer), pp. 133–144.

Maeder, A., and Mermilliod, J. C. 1981. The extent of mixing in stellar interiors: Evolutionary models and tests based on the HR diagrams of 34 open clusters. *Astron. Astrophys.* 93: 136–149.

Magain, P. 1986. Contribution functions and the depth of formation of spectrum lines. *Astron. Astrophys.* 163:135–139.

Magee, N. H., Merts, A. L., and Huebner, W. F. 1975. Improved opacity calculations. *Astrophys. J.* 196:617–620.

Makarov, V. I. 1984. Do prominences migrate equatorwards? *Solar Phys.* 93:59–79.

Makarova, V. V., and Solonskij, Yu. A. 1987. Rotation of solar high-latitude regions in 1971–1978. *Soln. Dannye, Byull.* 11:56–63 (in Russian).

Maki, Z., Nakagawa, M., and Sakata, S. 1962. Remarks on the unified model of elementary particles. *Prog. Theoret. Phys. (Kyoto)* 28:870–880.

Makishima, K. 1982. Hard X-ray imaging by SXT: Compact sources. In *Proc. Hinotori Symp. on Solar Flares* (Tokyo: Inst. of Space and Astronautical Science), pp. 120–129.

Makita, M. 1979. Determination of the magnetic field from the Zeeman line profile. *Publ. Astron. Soc. Japan* 31:575–584.

Makita, M. 1986a. A detection of the Faraday rotation by the solar vector magnetograph. *Solar Phys.* 103:1–10.

Makita, M. 1986b. An interpretation of the broad-band circular polarization of sunspots. *Solar Phys.* 106:269–286.

Makita, M., and Ohki, Y. 1986. The broad-band polarization of sunspots observed from 8 February to 30 August 1982. *Ann. Tokyo Astron. Obs.* 21:1–30.

Malherbe, J.-M., and Priest, E. R. 1983. Current sheet models for solar prominences. I. Magnetohydrostatics of support and evolution through quasi-static models. *Astron. Astrophys.* 123:80–88.

Malherbe, J.-M., Schmieder, B., and Mein, P. 1981. Dynamics in the filaments. I. Oscillations in a quiescent filament. *Astron. Astrophys.* 102:124–128.

Malherbe, J. M., Simon, G., Mein, P., Mein, N., Schmieder, B., and Vial, J. C. 1983. Preflare heating of filaments. *Adv. Space Res.* 2(11):53–56.

Malherbe, J.-M., Forbes, T. G., and Priest, E. R. 1984. A numerical simulation of the formation of solar prominences. In *Hydromagnetics of the Sun,* eds. T. D. Guyenne and J. J. Hunt, ESA SP-220, pp. 119–122.

Malherbe, J.-M., Schmieder, B., Mein, P., and Tandberg-Hanssen, E. 1987. Dynamics of solar filaments. V. Oscillations in the Hα and 1548Å CIV lines. *Astron. Astrophys.* 119:197–206.

Malitson, H. H., Fainberg, J., and Stone, R. G. 1973. Observation of a type II solar radio burst to 37 R$_\odot$. *Astrophys. Lett.* 14:111–114.

Maltby, P. 1975. The chromospheric Evershed flow. *Solar Phys.* 43:91–105.

Maltby, P. 1977. On the difference in darkness between sunspots. *Solar Phys.* 55:335–346.

Maltby, P., Avrett, E. H., Carlsson, M., Kjeldseth-Moe, O., Kurucz, R. L., and Loeser, R. 1986. A new sunspot umbral model and its variation with the solar cycle. *Astrophys. J.* 306:284–303.

Mampe, W., Ageron, P., Bates, C., Pendlebury, J. M., and Steyerl, A. 1989. Neutron lifetime measured with stored ultracold neutrons. *Phys. Rev. Lett.* 63:593–596.

Mandin, J.-Y., Chevillard, J.-P., Camy-Peyret, C., Elaud, J.-M., and Brault, J. 1986. The high resolution spectrum of water vapor between 13200 and 165000 cm^{-1}. *J. Molec. Spect.* 116:167–190.

Mankin, W. G. 1969. *Solar Limb Darkening in the Far Infrared and Its Relation to the Temperature of the Chromosphere*, NASA Report N69–30814.

Mankin, W. G. 1977. The solar spectrum between 10 and 10000 µm. In *The Solar Output and Its Variation*, ed. O. R. White (Boulder: Colorado Assoc. Univ. Press), pp. 151–168.

Mannheim, P. 1988. Derivation of the formalism for neutrino matter oscillations from the neutrino relativistic field equations. *Phys. Rev.* 37D:1935–1941.

Marcus, P. S. 1979. Stellar convection. I. Modal equations in sphere and spherical shells. *Astrophys. J.* 231:176–192.

Marcus, P. S. 1980. Stellar convection. III. Convection at large Rayleigh numbers. *Astrophys. J.* 240:203–217.

Marcy, G., and Basri, G. 1989. Physical realism in the analysis of stellar magnetic fields. II. K dwarfs. *Astrophys. J.* 345:480–488.

Mariska, J. T. 1978. Analysis of extreme-ultraviolet observations of a polar coronal hole. *Astrophys. J.* 225:252–258.

Mariska, J. T. 1986. The quiet solar transition region. *Ann. Rev. Astron. Astrophys.* 24:23–48.

Markov, M. A. 1977. Introductory talk. In *Neutrino '77*, ed. M. A. Markov (Moscow: Nauk), pp. 10–12.

Marmolino, C., and Severino, G. 1990. On the five-minute photospheric oscillation and its modelling. In *Solar Photosphere: Structure, Convection and Magnetic Fields: Proc. IAU Symp. 138*, ed. J. O. Stenflo (Dordrecht: Kluwer), pp. 251–254.

Marmolino, C., and Stebbins, R. T. 1989. Wave behavior in the solar photosphere: A comparison of theory and observation. *Solar Phys.* 124:23–26.

Marquette, W., and Zirin, H. 1990. BEARALERTS: A new program of flare prediction. *Bull. Amer. Astron. Soc.* 21:836 (abstract).

Marsch, E., and Richter, A. 1984. Distribution of solar wind angular momentum between particles and magnetic field: Inferences about the Alfvén critical point from Helios observations. *J. Geophys. Res.* 89:5386–5394.

Martens, P. C. H. 1988. The generation of proton beams in two-ribbon flares. *Astrophys. J.* 330:L131–L133.

Martin, S. F. 1973. The evolution of prominences and their relationship to active centers. *Solar Phys.* 31:3–21.

Martin, S. F. 1980. Preflare conditions, changes and events. *Solar Phys.* 68:217–236.

Martin, S. F. 1986. Recent observations of the formation of filaments. In *Coronal and Prominence Plasmas*, ed. A. I. Poland, NASA CP-2442, pp. 73–80.

Martin, S. F. 1988. The identification and interpretation of network, intranetwork, and ephemeral region magnetic fields. *Solar Phys.* 117:243–259.

Martin, S. F., and Harvey, K. L. 1979. Ephemeral active regions during solar minimum. *Solar Phys.* 64:93–108.

Martin, S. F., Livi, S. H. B., and Wang, J. 1985. The cancellation of magnetic flux. II. In a decaying active region. *Australian J. Phys.* 38:929–959.

Martres, M.-J., and Bruzek, A. 1977. Active regions. In *Illustrated Glossary for Solar-Terrestrial Physics*, eds. A. Bruzek and J. C. Durrant (Dordrecht: D. Reidel), pp. 53–70.

Martres, M.-J., and Soru-Escaut, I. 1971. Chromospheric absorbing features promising the appearance and the development of an active centre. *Solar Phys.* 21:137–145.

Martres, M.-J., Michard, R., and Soru-Iscovici, I. 1966. Etude morphologique de la structure magnétique des régions actives en relation avec les phénomènes chromosphériques et les éruptions solaires. *Ann. Astrophys.* 29:245–253.

Martres, M.-J., Michard, R., Soru-Iscovici, I., and Tsap, T. T. 1968. Etude de la localisation des éruptions dans la structure magnétique évolutive des régions actives solaires. *Solar Phys.* 5:187–206.

Martres, M.-J., Mouradian, Z., and Soru-Escaut, I. 1986. Effect of local rigid rotation on sunspots. *Astron. Astrophys.* 161:376–380.

Mason, G. M., Gloeckler, G., and Hovestadt, D. 1984. Temporal variations of nucleonic abundances in solar flare energetic particle events. II. Evidence for large-scale shock acceleration. *Astrophys. J.* 280:902–916.

Massaguer, J. M., and Zahn, J.-P. 1980. Cellular convection in a stratified atmosphere. *Astron. Astrophys.* 87:315–327.

Massaguer, J. M., Latour, J., Toomre, J., and Zahn, J. P. 1984. Penetrative cellular convection in a stratified atmosphere. *Astron. Astrophys.* 140:1–16.

Masters, G. 1979. Observational constraints on the chemical and thermal structure of the Earth's deep interior. *Geophys. J. Roy. Astron. Soc.* 57:507–534.

Mathews, G. J., Bloom, S. D., Fuller, G. M., and Bahcall, J. N. 1985. Shell model calculation for the $^{71}Ga(\nu,e)^{71}Ge$ solar neutrino detector. *Phys. Rev.* C32:796–804.

Mattig, W., Hanslmeier, A., and Nesis, A. 1989. Granulation line asymmetries. In *Proc. Capri Workshop in Solar and Stellar Granulation*, eds. R. Rutten and G. Severino (Dordrecht: Kluwer), pp. 187–193.

Mauas, P. J., Avrett, E. H., and Loeser, R. 1989. On carbon monoxide cooling in the solar atmosphere. *Astrophys. J.* 345:1104–1113.

May, R., and Clayton, D. D. 1968. Neutron tunneling in $^3He(^3He,2p)^4He$. *Astrophys. J.* 153:855–863.

Mayer, J. E. 1950. The theory of ionic solutions. *J. Chem. Phys.* 18:1426–1436.

McClements, K. G. 1987. The quasi-linear relaxation and bremsstrahlung of thick target electron beams in solar flares. *Astron. Astrophys.* 175:255–262.

McClements, K. G., and Alexander, D. 1989. Observations of the Ca XIX resonance line during the impulsive phase of solar flares. *Solar Phys.* 123:161–176.

McClements, K. G., and Brown, J. C. 1986. The inverse Compton interpretation of fast-time structures in solar microwave and hard X-ray bursts. *Astron. Astrophys.* 165:235–243.

McClymont, A. N., and Canfield, R. C. 1986. The solar flare extreme ultraviolet to hard X-ray ratio. *Astrophys. J.* 305:936–946.

McCrea, W. H. 1929. The hydrogen chromosphere. *Mon. Not. Roy. Astron. Soc.* 89:483–497.

McCrea, W. H. 1934. Theories of the solar chromosphere. *Mon. Not. Roy. Astron. Soc.* 95: 80–84.

McDonald, A. B., Alexander, T. K., Beene, J. E., and Mak, H. B. 1977. The low solar neutrino flux: A search for a resonance in ^6Be. *Nucl. Phys.* A288:529–532.

McGuire, R. E., von Rosenvinge, T. T., and McDonald, F. B. 1986. The composition of solar energetic particles. *Astrophys. J.* 301:938–961.

McIntosh, P. S. 1981. The birth and evolution of sunspots: Observations. In *The Physics of Sunspots*, eds. L. E. Cram and J. H. Thomas (Sunspot, N.M.: Assoc. of Univ. for Research in Astronomy), pp. 7–54.

McIntosh, P. S., and Wilson, P. R. 1985. A new model for flux emergence and the evolution of sunspots and the large-scale fields. *Solar Phys.* 97:59–79.

McKibben, R. B. 1988. Cosmic ray modulation. In *Proc. Sixth Intl. Solar Wind Conf.*, eds. V. J. Pizzo, T. E. Holzer and D. G. Sime, NCAR TN-306 (Boulder: Natl. Center for Atmos. Res.), pp. 615–633.

McLeod, D. B., and Isaak, G. R. 1988. A detector of small gradients of transparency of the terrestrial atmosphere. In *Seismology of the Sun and Sun-Like Stars*, ed. E. J. Rolfe, ESA SP-286, pp. 223–225.

McMath, R. R., and Goldberg, L. 1949. Recent exploration of the infrared solar spectrum at the McMath-Hulbert Observatory. *Proc. Amer. Phil. Soc.* 93:362–372.

Mehltretter, J. P. 1974. Observations of photospheric faculae at the center of the solar disk. *Solar Phys.* 38:43–57.

Mehltretter, J. P. 1978. Balloon-borne imagery of the solar granulation. *Astrophys. J.* 62: 311–316.

Mein, N. 1977. Wave propagation in the quiet solar chromosphere. *Solar Phys.* 52:283–292.

Mein, N., and Schmieder, B. 1981. Mechanical flux in the solar chromosphere. III. Variation of the mechanical flux. *Astron. Astrophys.* 97:310–316.

Mein, P. 1966. Champ macroscopique des vitesses dans l'atmosphère solaire d'après les mesures de déplacement des raies de Fraunhofer. *Ann. Astrophys.* 29:153–191.

Mein, P. 1977. Multi-channel subtractive spectrograph and filament observations. *Solar Phys.* 54:45–51.

Mein, P., and Mein, N. 1976. Velocity waves in the quiet solar chomosphere. *Solar Phys.* 49:231–248.

Mein, P., and Rayrole, J. 1989. High resolution observations with THEMIS: Prospects in magnetic field measurements. In *High Spatial Resolution Solar Observations,* ed. O. von der Lühe (Sunspot, N.M.: Sacramento Peak Obs.), pp. 12–26.

Mein, P., and Ribes, E. 1989. A method of detection for magnetic tracers based on image-processing of digitized spectroheliograms. *Astron. Astrophys.,* in press.

Mein, P., Simon, G., Vial, J. C., and Shine, R. A. 1982. Mass motions in the solar chromosphere and transition zone. *Astron. Astrophys.* 111:136–139.

Mein, P., Malherbe, J. M., Schmieder, B., Simon, G., and Tanberg-Hanssen, E. 1985. Mass fluxes and magnetic structures in the chromosphere and transition region; canopies. In *Chromospheric Diagnostics and Modelling,* ed. B. W. Lites (Sunspot, N.M.: National Solar Obs.).

Melen, F., Grevesse, N., Sauval, A. J., Farmer, C. B., Norton, R. H., Bredohl, H., and Dubois, I. 1989. A new analysis of the vibration-rotation spectrum of CH from solar spectra. *J. Molec. Spec.* 134:305–313.

Melrose, D. B. 1980. *Plasma Astrophysics* (London: Gordon and Breach).

Melrose, D. B., and Kuijpers, J. 1987. On the controversy concerning turbulent bremsstrahlung. *Astrophys. J.* 323:338–345.

Mendoza, V. E. E. 1968. Infrared excess in T Tauri stars and related objects. *Astrophys. J.* 151:977–989.

Mengel, J. G. 1973. The evolutionary status of Population II Cepheids. In *Proc. IAU Coll. 21,* ed. J. D. Fernie (Dordrecht: D. Reidel), pp. 214–220.

Menke, W. 1984. *Geophysical Data Analysis: Discrete Inverse Theory.* (Orlando: Academic Press).

Mercer-Smith, J. A., Cameron, A. G. W., and Epstein, R. I. 1984. On the formation of stars from disk accretion. *Astrophys. J.* 279: 363–366.

Merkle, F. 1987. Mirror developments for adaptive optics. In *Adaptive Optics in Solar Astronomy,* LEST Foundation Tech. Rept. No. 28, eds. F. Merkle, O. Engvold and R. Falomo (Oslo: Univ. of Oslo), pp. 35–54.

Merkle, F., Engvold, O., and Falomo, eds. 1987. *Adaptive Optics in Solar Astronomy,* LEST Foundation Tech. Rept. No. 28 (Oslo: Univ. of Oslo).

Merryfield, W. S., Gough, D. O., and Toomre, J. 1991. Nonlinear behaviour of solar gravity modes driven by ^{3}He in the core. II. Numerical simulations. *Astrophys. J.* 367:658–665.

Mertz, L. 1965. *Transformation in Optics* (New York: Wiley).

Messiah, A. 1986. Treatment of neutrino oscillations in solar matter: The MSW effect. In *Proc. 1986 Moriond Workshop on Massive Neutrinos,* eds. O. Fackler and J. T. T. Vân (Gif-sur-Yvette: Editions Frontières), pp. 373–389.

Mestel, L. 1965. Meridional circulation in stars. In *Stars and Stellar Systems v. 8: Stellar Structure,* eds. L. H. Aller and D. B. McLaughlin (Chicago: Univ. of Chicago Press), pp. 465–497.

Mestel, L. 1983. Angular momentum loss during pre-main sequence contraction. In *3rd Cambridge Workshop on Cool Stars, Stellar Systems and the Sun,* eds. J. Linsky and R. Stencel (Berlin: Springer-Verlag), pp. 49–59.

Mestel, L., and Spruit, H. 1987. On magnetic braking of late-type stars. *Mon. Not. Roy. Astron. Soc.* 226:57–66.

Mestel, L., and Weiss, N. 1987. Magnetic fields and nonuniform rotation in stellar radiative zones. *Mon. Not. Roy. Astron. Soc.* 226:123–135.

Mewaldt, R. A., Spalding, J. D., and Stone, E. C. 1984. A high resolution study of the isotopes of solar flare nuclei. *Astrophys. J.* 280:892–901.

Meyer, F., and Schmidt, H. U. 1968. Magnetisch ausgerichtete Strömungen zwischen Sonnen-flecken. *Zeit. Angew. Math. Mech.* 48:218–221.

Meyer, F., Schmidt, H. U., Weiss, N. O., and Wilson, P. R. 1974. The growth and decay of sunspots. *Mon. Not. Roy. Astron. Soc.* 169:35–57.

Meyer, F., Schmidt, H. U., and Weiss, N. O. 1977. The stability of sunspots. *Mon. Not. Roy. Astron. Soc.* 179:741–761.

Meyer, F., Schmidt, H. U., Simon, G. W., and Weiss, N. O. 1979. Buoyant magnetic flux tubes in supergranules. *Astron. Astrophys.* 76:35–45.

Meyer, J.-P. 1979. The argon and neon abundances in the solar neighborhood. In *Les Eléments et leurs Isotopes dans l'Univers* (Liège: Inst. d'Astrophysique), pp. 477–489.

Meyer, J.-P. 1985a. The baseline composition of solar energetic particles. *Astrophys. J. Suppl.* 57:151–171.

Meyer, J.-P. 1985b. Solar-stellar outer atmospheres and energetic particles, and galactic cosmic rays. *Astrophys. J. Suppl.* 57:173–204.

Meyer, J.-P. 1988a. Cosmic rays: Material from coronae of ordinary stars and from He-burning zones. In *Symp. on the Origin and Distribution of the Elements*, ed. G. J. Mathews (Singapore: World Scientific), pp. 310–336.

Meyer, J.-P. 1988b. Everything you always wanted to ask about local galactic abundances, but were afraid to know. In *Symp. on the Origin and Distribution of the Elements*, ed. G. J. Mathews (Singapore: World Scientific), pp. 337–348.

Meyer, J.-P. 1989. The abundances of matter in local interstellar medium. In *Cosmic Abundances of Matter*, ed. C. J. Waddington (New York: American Inst. Physics), in press.

Meyer, P., Parker, E. N., and Simson, J. A. 1956. Solar cosmic rays of February, 1956 and their propagation through interplanetary space. *Phys. Rev.* 104:768–783.

Michaud, G. 1970. Diffusion processes in peculiar A stars. *Astrophys. J.* 160:641–658.

Michaud, G. 1977. Diffusion time scales and accretion in the Sun. *Nature* 266:433–434.

Michaud, G. 1985. Particle transport in solar type stars. In *Solar Neutrinos and Neutrino Astronomy*, eds. M. L. Cherry, K. Lande and W. A. Fowler (New York: American Inst. of Physics), pp. 75–87.

Michaud, G. 1986. The lithium abundance gap in the Hyades F stars: The signature of diffusion. *Astrophys. J.* 302:650–655.

Michaud, G., Charland, Y., Vauclair, S., and Vauclair, G. 1976. Diffusion in main sequence stars: Radiation forces, time scales, anomalies. *Astrophys. J.* 210:447–465.

Michaud, G., Dupuis, J., Fontaine, G., and Montmerle, T. 1987. Selective mass loss, abundance anomalies and helium-rich stars. *Astrophys. J.* 322:302–314.

Michelson, A. A. 1927. Application of interference to astronomical investigation. In *Studies in Optics* (Chicago: Univ. of Chicago Press).

Migeotte, M., Neven, L., and Swensson, J. 1956. The solar spectrum from 2.8 to 23.7 microns. *Mem. Roy. Soc. Liège* special vol. no. 1.

Mihalas, B. W. 1984. Self-consistent radiation-hydrodynamic equations for stellar oscillations. I. Nonlinear form. *Astrophys. J.* 284:299–302.

Mihalas, B. W., and Toomre, J. 1981. Internal gravity waves in the solar atmosphere. I. Adiabatic waves in the chromosphere. *Astrophys. J.* 249:349–371.

Mihalas, D. 1978. *Stellar Atmospheres*, 2nd ed. (San Francisco: Freeman).

Mihalas, D., and Mihalas, B. W. 1984. *Foundations of Radiation Hydrodynamics* (New York: Oxford Univ. Press).

Mihalas, D., Däppen, W., and Hummer, D. G. 1988a. The equation of state for stellar envelopes. II. Algorithm and selected results. *Astrophys. J.* 331:815–825.

Mihalas, D., Hummer, D. G., Mihalas, B. W., and Däppen, W. 1988b. The equation of state for stellar envelopes. IV. Thermodynamic quantities and selected ionization fractions for six elemental mixes. *Astrophys. J.*, submitted.

Mikheyev, S., and Smirnov, A. Yu. 1985. Resonance enhancement of oscillations in matter and solar neutrino spectroscopy. *Soviet J. Nucl. Phys.* 42:913–917.

Mikheyev, S., and Smirnov, A. Yu. 1986. Resonant amplification of neutrino oscillations in matter and solar-neutrino spectroscopy. *Nuovo Cimento* 9C:17–26.

Mikheyev, S., and Smirnov, A. Yu. 1987. Resonance oscillations of neutrino in matter. *Soviet Phys. Uspekhi* 30:759–790.

Mikheyev, S., and Smirnov, A. Yu. 1989. Resonant neutrino oscillations in matter. *Prog. Particle Nucl. Phys.* 23:41–76.

Milford, P. N., O'Mara, B. J., and Ross, J. E. 1988. Measurements of relative intensities of FeI lines of astrophysical interest. *J. Quant. Spect. Rad. Trans.*, in press.

Miller, P., Foukal, P., and Keil, S. L. 1984. On the interpretation of Fraunhofer line shifts at supergranule boundaries. *Solar Phys.* 92:33–46.

Milne, E. A. 1924. The equilibrium of the calcium chromosphere. *Mon. Not. Roy. Astron. Soc.* 85:111–141.

Milne, E. A. 1925. The equilibrium of the calcium chromosphere. II. *Mon. Not. Roy. Astron. Soc.* 86:8–28.

Minakata, H., and Nunokawa, H. 1989. Hybrid solution of the solar neutrino problem in anti-correlation with sunspot activity. *Phys. Rev. Lett.* 63:121–124.

Mitchell, D. G., Roelof, E. C., Feldman, W. C., Bame, S. J., and Williams, D. J. 1981. Thermal iron ions in high speed solar wind streams. 2. Temperatures and bulk velocities. *Geophys. Res. Lett.* 8:827–830.

Mitchell, D. G., Roelof, E. C., and Bame, S.J. 1983. Solar wind iron abundance variations at speeds > 600 km s^{-1}, 1972–1976. *J. Geophys. Res.* 88:9059–9068.

Mitchell, W. E., and Mohler, O. C. 1969. Revision of the ultraviolet solar spectrum in the range 3650–3000 Å. *Astrophys. J. Suppl.* 18:379–427.

Moe, O. K., van Hoosier, M. E., Bartoe, J.-D. F., and Brueckner, G. E. 1976. *A Spectral Atlas of the Sun Between 1175 and 2100 Angstroms,* Naval Research Lab Rept. 8057.

Moffatt, H. K. 1978. *Magnetic Field Generation in Electrically Conducting Fluids* (Cambridge: Cambridge Univ. Press).

Mohler, O. C., Pierce, A. K., McMath, R. R., and Goldberg, L. 1950. *Photometric Atlas of the Near Infrared Solar Spectrum from λ8465 to λ25242* (Ann Arbor: Univ. of Michigan Press).

Molodenskij, M. M., and Filippov, B. P. 1978. Rapid motions of filaments in solar active regions. *Astron. Z.* 64:825–834, 1079–1087, and 65:396–402, 1047–1057.

Monchick, L., and Mason, E. A. 1985. A reconsideration of thermal diffusion in ionized gases: Quantal and dynamic shielding effects. *Phys. Fluids* 28:3341–3348.

Montgomery, D., and Turner, L. 1981. Anisotropic magnetohydrodynamic turbulence in a strong external magnetic field. *Phys. Fluids* 24:825–831.

Montgomery, M. D., Asbridge, J. R., Bame, S. J., and Feldman, W. C. 1974. Solar wind electron temperature depressions following some interplanetary shock waves: Evidence for magnetic merging? *J. Geophys. Res.* 79:3103–3110.

Montmerle, T., and Michaud, G. 1976. Diffusion in stars: Ionization and abundance effects. *Astrophys. J. Suppl.* 31:489–515.

Moore, C. E. 1976. *Selected Tables of Atomic Spectra,* NSRDBS-NBS 3, Sect. 7 (Oxygen I).

Moore, C. E., Minnaert, M. G. J., and Houtgast, J. 1966. *The Solar Spectrum 2935 Å to 8770 Å,* NBS Mono. 61.

Moore, C. E., Tousey, R., and Brown, C. M. 1982. *The Solar Spectrum 3069 Å to 2095 Å,* NRL Rept. 8653.

Moore, D. W., and Spiegel, E. A. 1966. A thermally excited non-linear oscillator. *Astrophys. J.* 143:871–887.

Moore, R. L., Hurford, G. J., Jones, H. P., and Kane, S. R. 1984. Magnetic changes observed in a solar flare. *Astrophys. J.* 276:379–390.

Moore, R. L., Hagyard, M. J., and David, J. M. 1987. Flare research with the NASA/MSFC vector magnetograph: Observed characteristics of sheared magnetic fields that produce flares. *Solar Phys.* 113:347–352.

Moreno-Insertis, F. 1986. Nonlinear evolution of kink-unstable magnetic flux tubes in the convection zone of the Sun. *Astron. Astrophys.* 166:291–305.

Moreno-Insertis, F., and Spruit, H. C. 1989. Stability of sunspots to convective motions. I. Adiabatic instability. *Astrophys. J.* 342:1158–1171.

Moreton, G. E. 1965. Flare-associated filament changes. In *Stellar and Solar Magnetic Fields: Proc. IAU Symp. 22,* ed. R. Lüst (Amsterdam: North-Holland Press), pp. 371–373.

Moreton, G. E., and Ramsey, H. E. 1960. Recent observations of dynamical phenomena associated with solar flares. *Publ. Astron. Soc. Pacific* 72:357–358 (abstract).

Morrison, P. 1956. Solar origin of cosmic-ray time variations. *Phys. Rev.* 101:1397–1404.

Morrow, C. A. 1988a. A New Picture for the Internal Rotation of the Sun. Ph.D. Thesis, Univ. of Colorado and NCAR (Cooperative Thesis No. 116).

Morrow, C. A. 1988b. Solar rotation and the a_1, a_3, and a_5 splitting coefficients for solar acoustic oscillations. In *Seismology of the Sun and Sun-Like Stars*, ed. E. J. Rolfe, ESA SP-286, pp. 91–98.

Morrow, C. A. 1988c. A way to examine rotational splittings without explicit forward or inverse computations. In *Seismology of the Sun and Sun-Like Stars*, ed. E. J. Rolfe, ESA SP-286, pp. 137–140.

Morrow, C. A., and Brown, T. M. 1988. A Bayesian approach to ridge fitting in the ω-k diagram of the solar five-minute oscillations. In *Advances in Helio- and Asteroseismology: Proc. IAU Symp. 123*, eds. J. Christensen-Dalsgaard and S. Frandsen (Dordrecht: D. Reidel), pp. 485–489.

Morrow, C. A., Gilman, P. A., and DeLuca, E. E. 1988. In *Seismology of the Sun and Sun-Like Stars*, ed. E. J. Rolfe, ESA SP-286, pp. 109–115.

Moses, J. D., Cook, J. W., Bartoe, J.-D. F., Brueckner, G. E., Dere, K. P., Webb, D. F., Davis, J. M., Recely, F., Martin, S. F., and Zirin, H. 1990. Correspondence between solar fine-scale structures in the corona, transition region, and lower atmosphere from collaborative observations. In preparation.

Mount, G. H., Linsky, J. L., and Shine, R. A. 1973. One- and multi-component models of the photosphere based on molecular spectra. I. The violet system of CN (0,0). *Solar Phys.* 32:13–30.

Mouradian, Z., Martres, M. J., and Soru-Escaut, I. 1981. Is the "Disparition Brusque" phenomenon always an effective dissappearance? In *Proc. Japan-France Seminar on Solar Physics*, eds. F. Moriyama and J. C. Henoux, pp. 195–198.

Mouradian, Z., Martres, M. J., and Soru-Escaut, I. 1983a. The emerging magnetic flux and the elementary eruptive phenomenon. *Solar Phys.* 87:309–328.

Mouradian, Z., Dumont, S., Pecker, J. C., Chipman, E., Artzner, G. E., and Vial, J. C. 1983b. Structure and physics of solar facuale. II. The non-thermal velocity field above faculae. *Solar Phys.* 78:83–100.

Mouradian, Z., Martres, M.-J., Soru-Escaut, I., and Gesztelyi, L. 1986. Local rigid rotation and the emergence of active centers. *Astron. Astrophys.* 199:318–324.

Mouradian, Z., Martres, M.-J., and Soru-Escaut, I. 1988. Localisation of sunspots at their maximum development. *Astron. Astrophys.* 199:318–324.

Mouschovias, T. Ch., and Poland, A. I. 1978. Expansion and broadening of coronal loop transients: A theoretical explanation. *Astrophys. J.* 220:675–682.

Muchmore, D., and Ulmschneider, P. 1985. Effects of CO molecules on the outer solar atmosphere: A time-dependent approach. *Astron. Astrophys.* 142:393–400.

Mullan, D. J. 1974. Is magnetic convection important in the Sun? *Solar Phys.* 38:9–13.

Mullan, D. J. 1989. *g*-mode pulsations in polytropes: High-precision eigenvalues and the approach to asymptotic behavior. *Astrophys. J.* 337:1017–1022.

Mullan, D. J., and Ahmad, I. A. 1982. Coronal holes: Mass loss driven by magnetic reconnection. *Solar Phys.* 75:347–350.

Mullan, D. J., and Ulrich, R. K. 1988. Radial and non radial pulsations of polytropes: High-precision eigenvalues and the approach of p-modes to asymptotic behavior. *Astrophys. J.* 331:1013–1028.

Müller, E. A., Petremann, E., and de la Reza, R. 1975. The solar lithium abundance. II. Synthetic analysis of the solar lithium feature at $\lambda 6707.8$ Å. *Solar Phys.* 41:53–65.

Muller, R. A. 1983a. The dynamical behavior of facular points in the quiet photosphere. *Solar Phys.* 85:113–121.

Muller, R. A. 1983b. A model of photospheric faculae deduced from white light high resolution pictures. *Solar Phys.* 45:105–114.

Muller, R. A. 1985. The fine structure of the quiet Sun. *Solar Phys.* 100:237–255.

Muller, R. A. 1990. Fine structure of the photospheric faculae. In *Solar Photosphere: Structure, Convection, and Magnetic Fields: Proc. IAU Symp. 138*, ed. J. O. Stenflo (Dordrecht: Kluwer), pp. 85–96.

Muller, R. A., and Buffington, A. 1974. Real-time correction of atmospherically degraded telescope images through image sharpening. *J. Opt. Soc. Amer.* 64:1200–1210.

Muller, R. A., and Keil, S. L. 1983. The characteristic size and brightness of facular points in the quiet photosphere. *Solar Phys.* 87:243–250.

Muller, R. A., Roudier, Th., and Hulot, J. C. 1989. Perturbation of the granulation pattern by the presence of magnetic flux tubes. *Solar Phys.* 119:229–243.

Muller, R. A., Roudier, Th., and Vigneau, J. 1990. The large-scale pattern formed by the spatial distribution of granules. *Solar Phys.* 126:53–67.

Münch, G. 1945. A theoretical discussion of the continuous spectrum of the Sun. *Astrophys. J.* 102:385–394.

Munro, R. H., and Jackson, B. V. 1977. Physical properties of a polar coronal hole from 2 to 5 $R_\odot$. *Astrophys. J.* 213:874–886.

Munro, R. H., and Withbroe, G. L. 1972. Properties of a coronal "hole" derived from EUV observations. *Astrophys. J.* 176:511–520.

Munro, R. H., Gosling, J. T., Hildner, E., MacQueen, R. M., Poland, A. I., and Ross, C. L. 1979. The association of coronal mass ejection transients with other forms of solar activity. *Solar Phys.* 61:201–215.

Murcray, F. J., Goldman, A., Murcray, F. H., Bradford, C. M., Murcray, D. G., Coffey, M. T., and Mankin, W. G. 1981. Observations of new emission lines in the infrared solar spectrum near 12.33, 12.22, and 7.38 microns. *Astrophys. J.* 247:L97–L99.

Murphy, G. A. 1989. The Synthesis and Inversion of Stokes Spectral Profiles. Ph.D. Thesis, Univ. of Sydney.

Murphy, R. J. 1985. Gamma Rays and Neutrons from Solar Flares. Ph.D. Thesis, Univ. of Maryland.

Murphy, R. J., and Ramaty, R. 1984. Solar-flare neutrons and gamma rays. *Adv. Space Res.* 4:127–136.

Murphy, R. J., Ramaty, R., Forrest, D. J., and Kozlovsky, B. 1985. Abundances from solar-flare gamma-ray line spectroscopy. In *Proc. 19th Intl. Cosmic Ray Conf.*, vol. 4, eds. F. C. Jones, J. Adams and G. M. Mason, NASA CP-2376, pp. 249–252.

Murphy, R. J., Dermer, C. D, and Ramaty, R. 1987. High-energy processes in solar flares. *Astrophys. J. Suppl.* 63:721–748.

Murphy, R. J., Kozlovsky, B., and Ramaty, R. 1988. Gamma-ray spectroscopic tests for the anisotropy of accelerated particles in solar flares. *Astrophys. J.* 331:1029–1035.

Mürset, U., Solanki, S. K., and Stenflo, J. O. 1988. Interpretation of broad band circular polarization measurements using Stokes V spectra. *Astron. Astrophys.* 204:279–285.

Musielak, Z. E., and Rosner, R. 1987. On the generation of magnetohydrodynamic waves in a stratified and magnetized fluid. I. Vertical propagation. *Astrophys. J.* 315:371–384.

Musielak, Z. E., and Rosner, R. 1988. On the generation of magnetohydrodynamic waves in a stratified and magnetized fluid. II. Magnetohydrodynamic energy fluxes for late-type stars. *Astrophys. J.* 329:376–383.

Musielak, Z. E., Rosner, R., and Ulmschneider, P. 1987. Magnetic flux tubes as sources of wave generation. In *Cool Stars, Stellar Systems and the Sun,* eds. J. L. Linsky and R. E. Stencel (Berlin: Springer-Verlag), pp. 66–68.

Musielak, Z. E., Rosner, R., and Ulmschneider, P. 1989. On the generation of flux tube waves in stellar convection zones. I. Longitudinal tube waves driven by external turbulence. *Astrophys. J.* 337:470–484.

Musman, S., and Nelson, G. D. 1976. The energy balance of granulation. *Astrophys. J.* 207:981–988.

Myers, P. C. 1983. Dense cores in dark clouds. III. Subsonic turbulence. *Astrophys. J.* 270:105–118.

Myers, P. C. 1987. Dense cores and young stars in dark clouds. In *Star Forming Regions: Proc. IAU Symp. 115*, eds. M. Peimbert and J. Jugaku (Dordrecht: D. Reidel), pp. 33–43.

Myers, P. C., and Goodman, A. A. 1988. Magnetic molecular clouds: Indirect evidence for magnetic support and ambipolar diffusion. *Astrophys. J.* 329:392–405.

Myers, P. C., Fuller, G. A., Mathieu, R. D., Beichman, C. A., Benson, P. J., Schild, R. E., and Emerson, J. P. 1987. Near-infrared and optical observations of IRAS sources in and near dense cores. *Astrophys. J.* 319:340–357.

Nagai, F., and Emslie, A. G. 1984. Gas dynamics in the impulsive phase of solar flares. I. Thick-target heating by nonthermal electrons. *Astrophys. J.* 279:896–908.

Nagatani, K., Dwarakanath, M. R., and Ashery, D. 1969. The ^{3}He$(\alpha,\gamma)^7$Be reaction at very low energy. *Nucl. Phys.* A128:325–332.

Nakada, M. P. 1969. A study of the composition of the lower solar corona. *Solar Phys.* 7: 302–320.

Nakada, M. P. 1970. A study of the composition of the solar corona and solar wind. *Solar Phys.* 14:475–479.

Nakagawa, M., Kohyama, Y., and Itoh, N. 1987. Relativistic free-free Gaunt factor of the dense high-temperature stellar plasma. *Astrophys. J. Suppl.* 63:661–684.

Nakagawa, Y., and Raadu, M. A. 1972. On practical representation of magnetic field. *Solar Phys.* 25:127–135.

Namba, O., and Diemel, W. E. 1969. A morphological study of the solar granulation. *Solar Phys.* 7:167–177.

Namba, O., and van Rijsbergen, R. 1977. Evolution pattern of the exploding granules. In *Problems of Solar and Stellar Convection: Proc. IAU Coll. No. 38*, eds. E. A. Spiegal and J. P. Zahn (Berlin: Springer-Verlag), pp. 119–125.

Narain, U., and Ulmschneider, P. 1989. Chromospheric and coronal heating mechanisms. *Space Sci. Rev.*, submitted.

Narashima, D., and Antia, H. M. 1982. Consistency of the mixing length theory. *Astrophys. J.* 262:358–368.

Nash, A. G., Sheeley, N. R., Jr., and Wang, Y.-M. 1988. Mechanisms for the rigid rotation of coronal holes. *Solar Phys.* 117:359–389.

Neckel, H., and Labs, D. 1984. The solar radiation between 3300 and 12500 Å. *Solar Phys.* 90:205–258.

Neidig, D. F. 1980. Solar rotation studies using sunspot data (1967–1974). *Solar Phys.* 66: 205–211.

Neidig, D. F. 1983. Spectral analysis of the optical continuum in the 24 April 1981 flare. *Solar Phys.* 85:285–302.

Neldner, K. 1988. Diplomarbeit. Univ. Münster.

Nelson, G. D., and Musman, S. 1977. A dynamical model of solar granulation. *Astrophys. J.* 214:912–916.

Nelson, G. D., and Musman, S. 1977. The scale of solar granulation. *Astrophys. J.* 222: L69–L72.

Neng-Ming, W., Novatrskii, V. N., Osetinskii, G. M., Nai-Kung, C., and Chepurchenko, I. A. 1966. Investigation of the reaction ^{3}He + ^{3}He. *Soviet J. Nucl. Phys.* 3:777–781.

Nesis, A., Durrant, C. J., and Mattig, W. 1988. Overshoot of horizontal and vertical velocities in the deep solar photosphere. *Astron. Astrophys.* 201:153–160.

Nesis, A., Fleig, K.-H., and Mattig, W. 1989. New results on the hydrodynamics of the overshoot layers in "active regions." In *High Spatial Resolution Solar Observations*, ed. O. von der Lühe (Sunspot, N.M.: Sacramento Peak Obs.), pp. 321–329.

Neugebauer, M. M. 1988. The problem of associating interplanetary and transient solar events. In *Proc. Sixth Intl. Solar Wind Conf.*, eds. V. J. Pizzo, T. E. Holzer and D. G. Sime, NCAR TN-306 (Boulder: Natl. Center for Atmos. Res.), pp. 243–259.

Neugebauer, M. M., and Feldman, W. C. 1979. Relation between superheating and superacceleration of helium in the solar wind. *Solar Phys.* 63:201–205.

Neupert, W. M. 1967. The solar corona above active regions: A comparison of extreme ultraviolet line emission with radio emission. *Solar Phys.* 2:294–315.

New, R. 1990. Watching the wobbling Sun. *New Scientist* 125(1702):54–56.

Newkirk, G. R., Jr. 1961. The solar corona in active regions and the thermal origin of the slowly varying component of solar radio radiation. *Astrophys. J.* 133:983–1013.

Newkirk, G. R., Jr. 1967. Structure of the solar corona. *Ann. Rev. Astron. Astrophys.* 5: 213–266.

Newkirk, G. R., Jr., and Wentzel, D. G. 1978. Rigidity-independent propagation of cosmic rays in the solar corona. *J. Geophys. Res.* 83:2009–2015.

Newman, M. J. 1986. The solar neutrino problem: Gadfly for solar evolution theory. In *Physics of the Sun*, vol. 3, eds. P. A. Sturrock, T. E. Holzer, D. M. Mihalas and R. K. Ulrich (Dordrecht: D. Reidel), pp. 33–45.

Newman, M. J., and Winkler, K.-H. 1980. The pre-main sequence Sun: A dynamical approach.

In *The Ancient Sun: Fossil Record in the Earth, Moon and Meteorites,* eds. R. O. Pepin, J. A. Eddy and R. B. Merrill (New York: Pergamon), pp. 551–558.

Newton, H. W., and Nunn, M. L. 1951. The Sun's rotation derived from sunspots 1934–1944 and additional results. *Mon. Not. Roy. Astron. Soc.* 11:413–421.

Nicolas, K. R., Bartoe, J.-D. F., Brueckner, G. E., and VanHooser, M. E. 1979. The energy balance and pressure in the solar transition zone for network and active region features. *Astrophys. J.* 233:741–755.

Nicolas, K. R., Kjeldseth-Moe, O., Bartoe, J.-D. F., and Brueckner, G. E. 1982. High resolution EUV structure of the chromosphere-corona transition region above a sunspot. *Solar Phys.* 81:253–280.

Nikolsky, G., Kim, I., Koutchmy, S., Stepanov, A., and Stellmacher, G. 1986. Measurement of magnetic fields in solar prominences. *Soviet Astron.* 29:669–675.

Nishikawa, J., Hamana, S., Mizugaki, K., and Hirayama, T. 1986. Detection of solar five-minute oscillations through white-light intensity. *Publ. Astron. Soc. Japan* 38:277–283.

Noci, G. 1979. A model of surge. In *Proc. IAU Symp. 91,* eds. M. Dryer and E. Tandberg-Hanssen, p. 307.

Noëls, A., Boury, A., Gabriel, M., and Scuflaire, R. 1976. Vibrational stability towards nonradial oscillations during central hydrogen burning. *Astron. Astrophys.* 49:103–106.

Noëls, A., Scuflaire, R., and Gabriel, M. 1984. Influence of the equation of state on the solar five-minute oscillations. *Astron. Astrophys.* 130:389–396.

Noerdlinger, P. D. 1977. Diffusion of helium in the Sun. *Astron. Astrophys.* 57:407–415.

Noerdlinger, P. D. 1978. Automatic computation of diffusion rates. *Astrophys. J. Suppl.* 36: 259–273.

Nolte, J. T., Davis, J. M., Gerassimenko, M., Laxaras, A. T., and Sullivan, J. D. 1977. A comparison of solar wind streams and coronal structures near solar minimum. *Geophys. Res. Lett.* 4:291–294.

Nordlund, Å. 1976. A two-component representation of stellar atmospheres with convection. *Astron. Astrophys.* 50:23–29.

Nordlund, Å. 1980. Numerical simulation of granular convection: Effects on photospheric spectral line profiles. In *Stellar Turbulence,* eds. D. F. Gray and J. L. Linsky (Berlin: Springer-Verlag), pp. 213–224.

Nordlund, Å. 1982. Numerical simulations of the solar granulation. I. Basic equations and methods. *Astron. Astrophys.* 107:1–10.

Nordlund, Å. 1983. Numerical 3-D simulations of the collapse of photospheric flux tubes. In *Solar and Stellar Magnetic Fields: Origins and Coronal Effects: Proc. IAU Symp. 102,* ed. J. O. Stenflo (Dordrecht: D. Reidel), pp. 79–83.

Nordlund, Å 1985*a.* NLTE spectral line formation in a three-dimensional atmosphere with velocity fields. In *Progress in Stellar Spectral Line Formation Theory,* eds. J. E. Beckman and L. Crivellari (Dordrecht: D. Reidel), pp. 215–224.

Nordlund, Å 1985*b.* Solar convection. *Solar Phys.* 100:209–235.

Nordlund, Å. 1985*c.* The dynamics of granulation, and its interaction with the radiation field. In *Theoretical Problems in High-Resolution Solar Physics,* ed. H. U. Schmidt (München: MPI für Astrophysik), pp. 1–24.

Nordlund, Å. 1985*d.* The 3-D structure of the magnetic field, and its interaction with granulation. In *Theoretical Problems in High-Resolution Solar Physics,* ed. H. U. Schmidt (München: MPI für Astrophysik), pp. 101–119.

Nordlund, Å. 1986. 3-D model calculation. In *Small Scale Magnetic Flux Concentrations in the Solar Photosphere,* eds. W. Deinzer, M. Knölker and H.-H. Voigt (Göttingen: Vandenhoeck & Ruprecht), pp. 83–102.

Nordlund, Å., and Dravins, D. 1990. Stellar granulation. III. Hydrodynamic model atmospheres. *Astron. Astrophys.* 228:155–183.

Nordlund, Å., and Stein, R. F. 1989. Simulating magnetoconvection. In *Solar and Stellar Granulation,* eds. R. J. Rutten and G. Severino (Dordrecht: Kluwer), pp. 453–470.

Nordlund, Å., and Stein, R. F. 1990. 3-D simulation of solar and stellar convection and magnetoconvection. *Computer Phys. Comm.,* in press.

Nötzold, D. 1987. Exact analytic solutions for Mikheyev-Smirnov-Wolfenstein level crossings. *Phys. Rev.* 36D:1625–1633.

Nötzold, D. 1988. New bounds on neutrino magnetic moment from stellar collapse. *Phys. Rev.* D38:1658–1668.

November, L. J. 1986. Measurement of geometric distortion in a turbulent atmosphere. *Appl. Opt.* 25:392–397.

November, L. J. 1989. The vertical component of the supergranular convection. *Astrophys. J.* 344:494–503.

November, L. J., and Simon, G. W. 1988. Precise proper-motion measurement of solar granulation. *Astrophys. J.* 333:427–442.

November, L. J., Toomre, J., Gebbie, K. B., and Simon, G. W. 1979. The height variation of supergranular velocity fields determined from simultaneous OSO 8 satellite and ground-based observations. *Astrophys. J.* 227:600–613.

November, L. J., Toomre, J., Gebbie, K. B., and Simon, G. W. 1981. The detection of mesogranulation on the Sun. *Astrophys. J.* 245:L123–L126.

November, L. J., Toomre, J., Gebbie, K. B., and Simon, G. W. 1982. Vertical flows of supergranular and mesogranular scale observed on the Sun with OSO 8. *Astrophys. J.* 258: 846–859.

November, L. J., Simon, G. W., Tarbell, T. D., Title, A. M., and Ferguson, S. H. 1987. Large-scale horizontal flows from SOUP observations of solar granulation. In *Theoretical Problems in High Resolution Solar Physics II*, eds. R.G. Athay and D. S. Spicer, NASA CP-2483, pp. 121–127.

Noyes, R. W. 1967. Observational studies of velocity fields in the solar photosphere and chromosphere. In *Aerodynamic Phenomena in Stellar Atmospheres: Proc. IAU Symp. 28*, ed. R. N. Thomas (New York: Academic Press), pp. 293–345.

Noyes, R. W., and Avrett, E. H. 1987. The solar chromosphere. In *Spectroscopy of Astrophysical Plasmas*, eds. A. Dalgarno and D. Layzer (London: Cambridge Univ. Press), p. 125.

Noyes, R. W., and Hall, D. N. B. 1972. Thermal oscillations in the high solar photosphere. *Astrophys. J.* 176:L89-L92.

Noyes, R. W., and Leighton, R. B. 1963. Velocity fields in the solar atmosphere. II. The oscillatory field. *Astrophys. J.* 138:631–647.

Noyes, R. W., and Rhodes, E. J., Jr., eds. 1984. *Probing the Depths of a Star: The Study of Solar Oscillations from Space*, NASA JPL Publ. 400-237.

Noyes, R. W., Gingerich, O., and Goldberg, L. 1966. On the infrared continuum of the Sun and stars. *Astrophys. J.* 145:344–347.

Noyes, R. W., Beckers, J. M., and Low, F. J. 1968. Observational studies of the solar intensity profile in the far-infrared and millimeter region. *Solar Phys.* 3:36–46.

Noyes, R. W., Weiss, N. O., and Vaughan, A. H. 1984a. The relation between stellar rotation rate and activity cycle periods. *Astrophys. J.* 287:769–773.

Noyes, R. W., Jr., Hartmann, L. W., Baliunas, S. L., Duncan, D. K., and Vaughan, A. H. 1984b. Rotation, convection, and magnetic activity in lower main-sequence stars. *Astrophys. J.* 279:763–777.

Nunokawa, H., and Minakata, H. A. 1991. A statistical analysis of the chlorine solar neutrino experiment. Report TMUP-HEL-9008. *Intl. J. Mod. Phys.*, submitted.

Obridko, V. N., and Staude, J. 1988. A two-component working model for the atmosphere of a large umbra. *Astron. Astrophys.* 189:232–242.

Oda, N. 1984. Morphological study of the solar granulation. III. The mesogranulation. *Solar Phys.* 93:243–255.

Ogilvie, K. W., and Hirshberg, J. 1974. The solar cycle variations of the solar wind helium abundance. *J. Geophys. Res.* 79:4595–4602.

Oglesby, P. H. 1987a. Confirmation of detection and classification of low-order, low-degree acoustic modes with 1985 observations. In *Stellar Pulsation*, eds. A. N. Cox, W. M. Sparks and S. G. Starrfield (Berlin: Springer-Verlag), pp. 314–317.

Oglesby, P. H. 1987b. Evidence of solar gravity modes detected in 1985 radiation intensity measurements. *Bull. Amer. Astron. Soc.* 19:1119 (abstract).

Oglesby, P. H. 1987a. Global Solar Oscillations Observed in the Visible to Near Infrared Continuum. Ph.D. Thesis, Univ. of Arizona.

Ogura, Y., and Phillips, N. A. 1962. Scale analysis of deep and shallow convection in the atmosphere. *J. Atmos. Sci.* 19:173–179.

Ohki, K., Nitta, N., Tsuneta, S., Takakura, T., Makishima, K., Murakami, T., Ogawara, Y., and Oda, M. 1982. Results from the hard X-ray spectrometer (HXM). In *Proc. Hinotori Symp. on Solar Flares* (Tokyo: Inst. of Space and Astronautical Science), pp. 69–88.

Oke, J. B. 1950. A theoretical Hertzprung-Russell diagram for red dwarf stars. *J. Roy. Astron. Soc. Canada* 44:135–148.

Oleson, J. R., Zanoni, C. A., Hill, H. A., Healy, A. W., Clayton, P. D., and Patz, D. L. 1974. SCLERA: An astrometric telescope for experimental relativity. *Appl. Opt.* 13:206–211.

Olver, F. W. J. 1956. The asymptotic solution of linear differential equations of the second order in a domain containing one transition point. *Phil. Trans.* A249:65–97.

Olver, F. W. J. 1974. In *Asymptotics and Special Functions*, ed. W. Rheinbolt (New York: Academic Press).

Orall, F. Q. 1966. Observational study of macroscopic inhomogeneities in the solar atmosphere. VIII. Vertical chromospheric oscillations measured in K_3. *Astrophys. J.* 143:917–927.

Orall, F. Q., ed. 1981. *Solar Active Regions* (Boulder: Colorado Assoc. Univ. Press).

Orall, F. Q., and Schmahl, E. J. 1976. The prominence-corona interface compared with the chromosphere-corona transition region. *Solar Phys.* 50:365–381.

Orrall, F. W. 1966. Observational study of macroscopic inhomogeneities in the solar atmosphere. VIII. Vertical chromospheric oscillations measured in K_3. *Astrophys. J.* 143:917–927.

Osaki, Y. 1975. Nonradial oscillations of a 10 solar mass star in the main-sequence stare. *Publ. Astron. Soc. Japan* 27:237–258.

Osaki, Y., and Hansen, C. J. 1973. Nonradial oscillations of cooling white dwarfs. *Astrophys. J.* 185:277–292.

Osborne, J.L., Barnes, C.A., Kavanagh, R.W., Kremer, R.M., Mathews, G.J., Zyskind, J.L., Parker, P.D., and Howard, A.J. 1982. Low energy ^{3}He$(\alpha,\gamma)^7$Be cross section measurements. *Phys. Rev. Lett.* 48:1664–1667.

Osborne, J.L., Barnes, C.A., Kavanagh, R.W., Kremer, R.M., Mathews, G.J., Zyskind, J.L., Parker, P.D., and Howard, A.J. 1984. Low-energy behavior of the ^{3}He$(\alpha,\gamma)^7$ cross section. *Nucl. Phys.* A419:115–132.

Osherovich, V.A., Flan, T., and Chapman, G.A., 1983. Magnetohydrostatic model of solar faculae. *Astrophys. J.* 268:412–419.

Osherovich, V.A., Tzur, I., and Gliner, E.B. 1984–1985. Theoretical model of the solar corona during sunspot minimum. *Astrophys. J.* 284:412–421 and 288:396–400.

Oster, L. 1990. Reconstructing the emission at 10.7 cm from individual active regions in terms of optical data. *J. Geophys. Res.,* in press.

Oster, L., Schatten, K. H., and Sofia, S. 1982. Solar irradiance variations due to active regions. *Astrophys. J.* 256:768–773.

Osterbrock, D. E. 1961. The heating of the solar chromosphere, plages, and corona by magnetohydrodynamic waves. *Astrophys. J.* 134:347–388.

Ottusch, J. J. 1986. Ph.D Thesis, Univ. of California, Berkeley.

Paczynski, B. 1969. Envelopes of red supergiants. *Acta Astron.* 19:1–22.

Paczynski, B., and Tremaine, S. 1977. Core helium flash and the origin of CH and carbon stars. *Astrophys. J.* 216:57–60.

Pallavicini, R. 1989. X-ray emission from stellar coronae. *Astron. Astrophys. Rev.* 1:177–207.

Pallavicini, R., Sakurai, T., and Vaiana, G. S. 1981. X-ray, EUV, and centimetric observations of solar active regions: An empirical model for bright radio sources. *Astron. Astrophys.* 98: 316–327.

Pallavicini, R., Peres, G., Serio, S., Vaiana, G., Acton, L., Leibacher, J., and Rosner, R. 1983. Closed coronal structures. V. Gasdynamic models of flaring loops and comparison with SMM observations. *Astrophys. J.* 270:270–287.

Pallé, P. L., and Roca-Cortés, T. 1988. Search for solar g-modes from 1981 to 1985. In *Advances in Helio- and Asteroseismology*, eds. J. Christensen-Dalsgaard and S. Frandsen (Dordrecht: D. Reidel), pp. 79–82.

Pallé, P. L., Pérez, J. C., Régulo, C., Roca-Cortés, T., Isaak, G. R., McLeod, C. P., and van der Raay, H. B. 1986a. The global oscillation spectrum of the Sun. I. Analysis of daily power spectra of velocity measurements. *Astron. Astrophys.* 169:313–318.

Pallé, P. L., Pérez, J. C., Régulo, C., Roca-Cortés, Isaak, G. R., McLeod, C. P., and van der Raay, H. B. 1986b. The global oscillation spectrum of the Sun. II. the observed low-ℓ high-n solar p-mode spectrum. *Astron. Astrophys.* 170:114–119.

Pallé, P. L., Régulo, C., and Roca-Cortés,T. 1988a. Frequency shift of solar *p* modes as seen by crosscorrelation analysis. In *Seismology of the Sun and Sun-Like Stars,* ed. E. J. Rolfe, ESA SP-286, pp. 285–289.

Pallé, P. L., Pérez Hernández, F., and Roca-Cortés,T. 1988b. Further implications of solar *p* modes as measured in Na and K resonance lines. In *Seismology of the Sun and the Sun-Like Stars,* ed. E. J. Rolfe, ESA SP-286, pp. 513–515.

Pallé, P. L., Hernández, F. P., Régulo, C., and Roca-Cortés, T. 1988c. Rotational splitting of low ℓ solar *p*-modes. In *Seismology of the Sun and Sun-Like Stars,* ed. E. J. Rolfe, ESA SP-286, pp. 125–130.

Pallé, P. L., Régulo, C., and Roca-Cortés, T. 1989. Solar cycle induced variations of the low ℓ solar acoustic spectrum. *Astron. Astrophys.* 224:253–258.

Pallé, P. L., Régulo, C., and Roca-Cortés, T. 1990a. Variations of the low ℓ solar acoustic spectrum correlated with the activity cycle. In *Inside the Sun: Proc. IAU Coll. 121,* eds. G. Berthomieu and M. Cribier (Dordrecht: Kluwer), pp. 349–355.

Pallé, P. L., Régulo, C., and Roca-Cortés, T. 1990b. The spectrum of solar *p*-modes and the solar activity cycle. In *Oji Intl. Seminar on Progress of Seismology of the Sun and Stars,* eds. Y. Osaki and H. Shibahashi (Berlin: Springer-Verlag), in press.

Pantellini, F. G. E., Solanki, S. K., and Stenflo, J. O. 1988. Velocity and temperature in solar magnetic fluxtubes from a statistical centre-to-limb analysis. *Astron. Astrophys.* 189:263–276.

Papaloizou, J., and Pringle, J. E. 1978. Non-radial oscillations of rotating stars and their relevance to the short-period oscillations of cataclysmic variables. *Mon. Not. Roy. Astron. Soc.* 182:423–442.

Paquette, C., Pelletier, C., Fontaine, G., and Michaud, G. 1986a. Diffusion coefficients in stellar plasmas. *Astrophys. J. Suppl.* 61:177–195.

Paquette, C., Pelletier, C., Fontaine, G., and Michaud, G. 1986b. Diffusion in white dwarfs: New results and comparative study. *Astrophys. J. Suppl.* 61:197–217.

Parke, S. J. 1986. Non-adiabatic level crossing in resonant neutrino oscillations. *Phys. Rev. Lett.* 57:1275–1278.

Parke, S. J., and Walker, T. P. 1986. Resonant-solar-neutrino-oscillation experiments. *Phys. Rev. Lett.* 57:2322–2325.

Parker, E. N. 1955a. The formation of sunspots from the solar toroidal field. *Astrophys. J.* 121:491–507.

Parker, E. N. 1955b. Hydromagnetic dynamo models. *Astrophys. J.* 122:293–314.

Parker, E.N. 1957. Instability of thermal fields. *Astrophys. J.* 117:431–436.

Parker, E.N. 1958. Dynamics of the interplanetary gas and magnetic fields. *Astrophys. J.* 128:664–675.

Parker, E.N. 1963. Kinematical hydromagnetic theory and its application to the low solar photosphere. *Astrophys. J.* 138:552–575.

Parker, E.N. 1965. Dynamical theory of the solar wind. *Space Sci. Rev.* 4:666–708.

Parker, E.N. 1968. *Interplanetary Dynamical Processes,* Interscience Mono. and Texts in Physics and Astronomy, vol. 8 (New York: Wiley).

Parker, E. N. 1975. The generation of magnetic fields in astrophysical bodies. X. A solar dynamo based on horizontal shear. *Astrophys. J.* 198:205–209.

Parker, E. N. 1978. Hydraulic concentration of fields in the solar photosphere. VI. Adiabatic cooling and concentration in downdrafts. *Astrophys. J.* 221:368–377.

Parker, E. N. 1979a. *Cosmical Magnetic Fields: Their Origin and Their Activity* (Oxford: Clarendon Press).

Parker, E. N. 1979b. Sunspots and the physics of magnetic flux tubes. I. The general nature of sunspots. *Astrophys. J.* 230:905–913.

Parker, E. N. 1982a. The dynamics of fibril magnetic fields. I. Effect of flux tubes on convection. *Astrophys. J.* 256:292–301.

Parker, E. N. 1982b. The dynamics of fibril magnetic fields. II. The mean field equations. *Astrophys. J.* 256:302–315.

Parker, E. N. 1984a. Depth of origin of solar active regions. *Astrophys. J.* 280:423–427.

Parker, E. N. 1984b. Magnetic buoyancy and the escape of magnetic fields from stars. *Astrophys. J.* 281:839–845.

Parker, E. N. 1987a. The dynamic oscillation and propulsion of magnetic fields in the convection zone of a star. I. General considerations. *Astrophys. J.* 312:868–879.

Parker, E. N. 1987b. The dynamical oscillation and propulsion of magnetic fields in the convective zone of a star. II. Thermal shadows. *Astrophys. J.* 321:984–1008.

Parker, E. N. 1987c. The dynamic oscillation and propulsion of magnetic fields in the convection zone of a star. III. Accumulation of heat and the onset of the Rayleigh-Taylor instability. *Astrophys. J.* 321:1009–1030.

Parker, E. N. 1988a. The dynamic oscillation and propulsion of magnetic fields in the convection zone of a star. IV. Eruption to the surface. *Astrophys. J.* 325:880–890.

Parker, E. N. 1988b. The dynamic oscillation and propulsion of magnetic fields in the convection zone of a star. V. Instability and propulsion of flux bundles. *Astrophys. J.* 326:395–406.

Parker, E. N. 1988c. The dynamic oscillation and propulsion of magnetic fields in the convection zone of a star. VI. Small flux bundles, network fields, and ephemeral active regions. *Astrophys. J.* 326:407–411.

Parker, E. N. 1988d. The dynamo dilemma. *Solar Phys.* 110:11–21.

Parker, E. N. 1988e. Nanoflares and the solar x-ray corona. *Astrophys. J.* 330:474–479.

Parker, G. D. 1986. Lifetime, age, and rotation of coronal magnetic fields. *Solar Phys.* 104:333–345.

Parker, G. D. 1987. Radial variation of differential rotation in the solar electron corona. *Solar Phys.* 108:77–87.

Parker, G. D., Hansen, R. T., and Hansen, S. F. 1982. Coronal rotation during solar cycle 20. *Solar Phys.* 80:185–198.

Parker, P. D. 1966. ^{7}Be(p,γ)^{8}B reaction. *Phys. Rev.* 150:851–857.

Parker, P. D. 1968. Reanalysis of the ^{7}Be(p,γ)^{8}B cross section data. *Astrophys. J.* 153:L85-L86.

Parker, P. D 1972. Comments on the destruction of ^{7}Be in the solar interior. *Astrophys. J.* 175:261–264.

Parker, P.D. 1986. Thermonuclear reactions in the solar interior. In *Physics of the Sun,* vol. 1, eds. P. A. Sturrock, T. E. Holzer, D. M. Mihalas and R. K. Ulrich (Dordrecht: D. Reidel), pp. 15–32.

Parker, P. D. 1987. Unpublished communication at the GONG Workshop on Solar Models, Yale University, May 1987.

Parker, P. D., and Kavanagh, R. W. 1963. ^{3}He(α,γ)^{7}Be reaction. *Phys. Rev.* 131:2578–2582.

Parker, P. D., Bahcall, J. N., and Fowler, W. A. 1964. Termination of the proton-proton chain in stellar interiors. *Astrophys. J.* 139:602–621.

Parker, R. L. 1977. Understanding inverse theory. *Ann. Rev. Earth Planet. Sci.* 5:35–64.

Parkinson, J. H. 1977. Solar abundances from X-ray observations. *Astron. Astrophys.* 57:185–191.

Paschmann, G., Sckopke, N., Papamastorakis, I., Asbridge, J. R., Bame, S. J., and Gosling, J. T. 1981. Characteristics of reflected and diffuse ions upstream from the Earth's bow shock. *J. Geophys. Res.* 86:4355–4364.

Peach, G. 1965. A general formula for the calculation of absorption cross sections for free-free transitions in the field of positive ions. *Mon. Not. Roy. Astron. Soc.* 130:361–367.

Peach, G. 1967a. Free-free absorption coefficients for non-hydrogenic atoms. *Mem. Roy. Astron. Soc.* 71:1–11.

Peach, G. 1967b. A revised general formula for the calculation of atomic photoionization cross sections. *Mem. Roy. Astron. Soc.* 71:13–27.

Peach, G. 1967c. Total continuous absorption coefficients for complex atoms. *Mem. Roy. Astron. Soc.* 71:29–45.

Peach, G. 1970. Continuous absorption coefficients for non-hydrogenic atoms. *Mem. Roy. Astron. Soc.* 73:1–123.

Pecker, J.-C. 1949. Variation des raies sur le disque solaire. *Ann. Astrophys.* 12:9–20.

Pecker, J.-C. 1951. Contribution à la théorie du type spectral. V. Le "blanketing effect" et la structure de la photosphère solaire. *Ann. Astrophys.* 14:152–175.

Pecker, J.-C. 1982. Caractères météorologiques du climat et activité solaire: Quelques remarques. In *Compendium in Astronomy,* eds. E. G. Mariolopoulos, P. S. Theocars and L. N. Mavridis (Dordrecht: D. Reidel), pp. 151–160.

Pecker, J.-C. 1988. New vistas on the solar activity cycle. *Irish Astron. J.* 18:133–146.

Pecker, J.-C., and Roberts, W. O. 1955. Solar corpuscles responsible for geomagnetic distur-bances. *J. Geophys. Res.* 60:33–44.

Pecker, J.-C., Dumont, S., and Mouradian, Z. 1988. The center-to-limb variation on solar chro-mospheric and transition lines, and the effect of roughness of the emitting layers. *Astron. Astrophys.* 196:269–276.

Pehlemann, E., and von der Lühe, O. 1989. Technical aspects of the speckle masking phase reconstruction algorithm. *Astron. Astrophys.* 216:337–346.

Pelletier, C., Fontaine, G., Wesemael, F., Michaud, G., and Wegner, G. 1986. Carbon pollution in helium-rich white dwarf atmospheres: Time-dependent calculations of the dredge-up pro-cess. *Astrophys. J.* 307:242–252.

Pepin, R. O., Eddy, J. A., and Merrill, R. B., eds. 1980. *The Ancient Sun: Fossil Record in the Earth, Moon and Meteorites* (New York: Pergamon).

Pérez Garde, M., Vézquez, M., Schwan, H., and Wöhl, H. 1981. Large-scale solar motions as determined by Doppler shift measurements using a linear photodiode array. *Astron. Astro-phys.* 93:67–70.

Pesnell, W. D. 1987. A new driving mechanism for stellar pulsations. *Astrophys. J.* 314: 598–604.

Pesnell, W. D. 1990. Nonradial, nonadiabatic stellar pulsations. *Astrophys. J.* 363:227–233.

Petcov, S. T. 1988. Exact analytic description of two-neutrino oscillations in matter with expo-nentially varying density. *Phys. Lett.* 200B:373–379.

Petersen, J. O. 1972. Mixed models for blue horizontal branch stars. *Astron. Astrophys.* 19: 197–199.

Peterson, R. C. 1985. The rotation of horizontal-branch stars. IV. Members of the globular clus-ter NGC288. *Astrophys. J.* 294:L35-L37.

Petschek, H. E. 1964. Magnetic field annihilation. In *AAS-NASA Symp. on Solar Flares,* NASA SP-50, pp. 425–439.

Pflug, F., and Grigoryev, V. M. 1986. The chain of solar magnetographs and its results. *Skalnate Pleso Astrophys. Obs. Contrib.* 15:453–468.

Phillips, D. L. 1962. A technique for the numerical solution of certain integral equations of the first kind. *J. Assoc. Comp. Mech.* 9:84–97.

Pidatella, R. M., and Stix, M. 1986. Convection overshoot at the base of the Sun's convection zone. *Astron. Astrophys.* 157:338–340.

Pidatella, R. M., Stix, M., Belvedere, G., and Paterno, L. 1986. The role of inhomogen-eous heat transport and anisotropic momentum exchange in the dynamics of stellar con-vection zones: Applications to models of the Sun's differential rotation. *Astron. Astrophys.* 156:22–32.

Piddington, J. H. 1975. Solar magnetic fields and convection. III. Recent development in dy-namo and related theories. *Astrophys. Space Sci.* 38:157–166.

Piddington, J. H. 1976. Solar magnetic fields and convection. VII. A review of the primordial field theory. In *Basic Mechanisms of Solar Activity: Proc. IAU Symp. 71,* eds. V. Bumba and J. Kleczek (Dordrecht: D. Reidel), pp. 389–407.

Piddington, J. H. 1978. The flux-rope-fibre theory of solar magnetic fields. *Astrophys. Space Sci.* 55:401–425.

Pierce, A. K. 1949. Solar observations with a Perkin-Elmer infrared spectrometer. *Publ. Astron. Soc. Pac.* 61:217–219.

Pierce, A. K. 1954. Solar limb darkening in the region λλ7793–24388. *Astrophys. J.* 120: 221–232.

Pierce, A. K., and Breckinridge, J. B. 1973. *The Kitt Peak Table of Photographic Solar Spec-trum Wavelengths.* Kitt Peak Natl. Obs. Contrib. No. 559.

Pierce, A. K., and Lo Presto, J. C. 1984. Solar rotation from a number of Fraunhofer lines. *Solar Phys.* 93:155–170.

Pierce, A. K., McMath, R. R., Goldberg, L., and Mohler, O. C. 1950. Observations of solar limb darkening between 0.5 and 10.2 μm. *Astrophys. J.* 112:289–298.

Pilachowski, C. A., and Milkey, R. W. 1987. The rotational velocity of white dwarfs. *Publ. Astron. Soc. Pacific* 99:836–838.

Pilipp, W. G., Miggenrieder, H., Montgomery, M. D., Mühlhäuser, K.-H., Rosenbauer, H., and Schwenn, R. 1987. Characteristics of electron velocity distribution functions in the solar wind derived from the Helios plasma experiment. *J. Geophys. Res.* 92:1075–1092.

Pinch, T. 1986. *Confronting Nature, The Sociology of Solar Neutrino Detection* (Dordrecht: D. Reidel).

Pinsonneault, M., Kawaler, S., and Demarque, P. 1989*a*. Rotation of low mass stars: A new probe of stellar evolution. *Astrophys. J. Suppl.*, in press.

Pinsonneault, M. H., Kawaler, S. D., Sofia, S., and Demarque, P. 1989*b*. Evolutionary models of the rotating Sun. *Astrophys. J.* 338:424–452.

Pintar, J. A., and the GONG Data Team. 1988. The GONG data reduction and analysis system. In *Seismology of the Sun and Sun-Like Stars*, ed. E. J. Rolfe, ESA SP-286, pp. 217–221.

Pizzo, V. 1987. Magnetic field morphology in the upper layers. In *Theoretical Problems in High Resolution Solar Physics II*, eds. G. Athay and D. S. Spicer, NASA CP-2483, pp. 1–13.

Pizzo, V., Schwenn, R., Marsch, E., Rosenbauer, H., Mühlhäuser, K.-H., and Neubauer, F. M. 1983. Determination of the solar wind angular momentum flux from the Helios data—an observational test of the Weber and Davis theory. *Astrophys. J.* 271:335–354.

Pizzochero, P. 1987. Non-adiabatic level crossings in neutrino oscillations for an exponential solar density profile. *Phys. Rev.* 36D:2293–2296.

Pneuman, G. 1972. Temperature-density structure in coronal helmets: The quiescent prominence and coronal cavity. *Astrophys. J.* 177:793–805.

Pneuman, G. W., and Kopp, R. A. 1971. Gas-magnetic field interactions in the solar corona. *Solar Phys.* 18:258–270.

Pneuman, G. W., and Orrall, F. Q. 1986. Structure, dynamics, and heating of the solar atmosphere. In *Physics of the Sun*, vol. 2, eds. P. A. Sturrock, T. E., Holzer, D. M. Mihalas and R. K. Ulrich (Dordrecht: D. Reidel), pp. 71–134.

Pneuman, G. W., Solanki, S. K., and Stenflo, J. O. 1986. Structure and merging of solar magnetic fluxtubes. *Astron. Astrophys.* 154:231–242.

Pochoda, P., and Reeves, H. 1964. A revised solar model with a solar neutrino spectrum. *Planet. Space Sci.* 12:119–126.

Poland, A. I., ed. 1986. In *Coronal and Prominence Plasmas*, NASA CP-2442.

Poland, A. I., Orwig, L. E., Mariska, J. T., Nakatsuka, R., and Auer, L. E. 1984. The energy relation between hard X-ray and O v emission in solar flares. *Astrophys. J.* 280:457–463.

Poletto, G., Vaiana, G. S., Zombeck, M. V., Krieger, A. X., and Timothy, A. F. 1975. A comparison of coronal X-ray structures of active regions with magnetic fields computed from photospheric observations. *Solar Phys.* 44:83–99.

Pomerantz, M. A. 1986. Astronomy on ice. *Proc. Astron. Soc. Australia* 6:403–415.

Pontecorvo, B. 1946. Inverse β processes (a lecture). Chalk River Lab. Rept. PD-205.

Pontecorvo, B. 1958*a*. Mesonium and antimesonium. *Soviet Phys. JETP* 6:429–431.

Pontecorvo, B. 1958*b*. Inverse beta processes and nonconservation of lepton charge. *Soviet Phys. JETP* 7:172–173.

Pontecorvo, B. 1968*a*. Neutrino experiments and the problem of conservation of leptonic charge. *Soviet Phys. JETP* 26:984–988.

Pontecorvo, B. 1968*b*. Superweak interactions and double beta decay. *Phys. Lett.* 26B:630–632.

Pontecorvo, B. 1983. Pages in the development of neutrino physics. *Soviet Phys. Usp.* 26:1087–1108.

Popper, D., and Ulrich, R. 1986. Can binary stars test solar models? *Astrophys. J.* 307:L61-L64.

Porter, J. G., and Moore, R. L. 1988. Coronal heating by microflares. In *Solar and Stellar Coronal Structure and Dynamics*, ed. R. C. Altrock (Sunspot, N.M.: National Solar Obs.), pp. 125–129.

Porter, J. G., Moore, R. L., and Reichmann, E. J. 1987. Microflares in the solar magnetic network. *Astrophys. J.* 323:380–390.

Pouquet, A. 1978. On two-dimensional magnetohydrodynamic turbulence. *J. Fluid Mech.* 88:1–16.

Prandtl, L. 1952. *Essentials of Fluid Dynamics* (New York: Hafmen).

Press, W. H. 1981. Radiative and other effects from internal waves in solar and stellar interiors. *Astrophys. J.* 245:286–303.

Press, W. H., and Rybicki, G. B. 1981. Enhancement of radiative diffusion and suppression of heat flux in a fluid with time-varying shear. *Astrophys. J.* 248:751–766.

Press, W. H., Flannery, B. P., Teukolsky, S. A., and Vetterling, W. T. 1986. *Numerical Recipes* (Cambridge: Cambridge Univ. Press).

Pridmore-Brown, D.C. 1981. The computation of solar magnetic fields from observational data. Aerospace Corporation Rept. ATR-81(7813)-1 (El-Segundo, Calif.: The Aerospace Corp.).

Priest, E. R., ed. 1981. *Solar Flare Magnetohydrodynamics* (London: Gordon and Breach).

Priest, E. R. 1988. *Dynamics and Structure of Quiescent Prominences* (Dordrecht: Kluwer).

Priest, E. R., Hood, A. W., and Anzer, U. 1989. A twisted flux-tube model for solar prominences. I. General properties. *Astrophys. J.* 344:1010–1025.

Primakoff, H. 1978. Neutrino oscillations and the solar neutrino problem. In *The Status and Future of Solar Neutrino Research,* vol. 2, ed. G. Friedlander, pp. 211–232.

Proctor, M. R. E., and Weiss, N. O. 1982. Magnetoconvection. *Rept. Progr. Phys.* 45:1317–1379.

Proffitt, C. R., and Michaud, G. 1989. Pre-main sequence depletion of ^{6}Li and ^{7}Li. *Astrophys. J.* 346:976–982.

Proffitt, C. R., and Michaud, G. 1991. Gravitational settling in solar models. *Astrophys. J.*, in press.

Proudman, I. 1952. The generation of noise by isotropic turbulence. *Proc. Roy. Soc. London* 214A:119.

Provost, J. 1984. Solar constraints. In *Observational Tests of Stellar Evolution Theory: Proc. IAU Symp. 105,* eds. A. Maeder and A. Renzini (Dordrecht: D. Reidel), pp. 47–65.

Provost, J., and Berthomieu, G. 1986. Asymptotic properties of low degree solar gravity modes. *Astron. Astrophys.* 165:218–226.

Provost, J., and Berthomieu, G. 1988. Integrated light and velocity of solar *g*-mode oscillations. In *Seismology of the Sun and Sun-Like Stars,* ed. E. J. Rolfe, ESA SP-286, pp. 387–391.

Provost, J., Berthomieu, G., and Rocca, A. 1981. Low frequency oscillations of a slowly rotating star: Quasi-toroidal modes. *Astron. Astrophys.* 94:126–133.

Pulido, J. 1990. Solar neutrinos: The magnetic transition. *Phys. Lett.* B244:88–94.

Raadu, M. A., and Kuperus, M. 1973. Thermal instability of coronal neutral sheets and the formation of quiescent prominences. *Solar Phys.* 28:77–94.

Raadu, M. A., and Kuperus, M. 1974. The support of prominences formed in neutral sheets. *Astron. Astrophys.* 31:189–142.

Rabaey, G. F. 1989. The Observed Properties of the Intermediate-Degree Gravity Modes and Their Relevance to the Solar Neutrino Paradox. Ph.D. Thesis, Univ. of Arizona.

Rabaey, G. F., and Hill, H. A. 1988. The intermediate-degree *g*-mode spectrum of the Sun, its properties and their implications to the solar neutrino paradox. *Ann. New York Acad. Sci.* 571:594–600.

Rabaey, G. F., and Hill, H. A. 1990. The observed properties of the intermediate-degree gravity modes. *Astrophys. J.* 362:734–744.

Rabaey, G. F., Hill, H. A., and Barry, C. T. 1988. The observed intermediate-degree *f*-mode eigenfrequency spectrum of the Sun and tests of the mode classification proficiency. *Astrophys. Space Sci.* 143:81–97.

Rabin, D. M. 1986. The prominence-corona interface and its relationship to the chromosphere-corona transition. In *Coronal and Prominence Plasmas,* ed. A. I. Poland, NASA CP-2442, pp. 135–142.

Rabin, D. M., and Graves, J. E. 1989. Measuring sunspot magnetic fields with the infrared line Fe I λ15649. *Bull. Amer. Astron. Soc.* 21:854 (abstract).

Rabin, D. M., Moore, R., and Hagyard, M. J. 1984. A case for submergence of magnetic flux in a solar active region. *Astrophys. J.* 287:404–411.

Rachkovsky, D. N. 1962. Magnetic rotation effects in spectral lines. *Izv. Krymsk. Astrofiz. Obs.* 28:259–270.

Rachkovsky, D. N. 1985. Investigation of magnetic fields by the line-ratio method outside of active regions on the Sun. *Izv. Krymsk. AstroFiz. Obs.* 71:79–87.

Radick, R. R., Hartmann, L., Mihalas, D., Worden, S. P., Africano, J. L., Klimke, A., and Tyson, E. T. 1982. The photometric variability of solar-type stars. I. Preliminary results for the Pleiades, Hyades, and the Malmquist field. *Publ. Astron. Soc. Pacific* 94:934–944.

Radick, R. R., Lockwood, G. W., Thompson, D. T., Warnock, A., Hartmann, L. W., Mihalas, D., Worden, S. P., Henry, G. W., and Sherlin, J. M. 1983. The photometric variability of solar-type stars. III. Results from 1981–82, including parallel observations of thirty-six Hyades stars. *Publ. Astron. Soc. Pacific* 95:621–634.

Radick, R. R., Thompson, D. T., Lockwood, G. W., Duncan, D. K., and Baggett, W. E. 1987. The activity, variability, and rotation of lower main-sequence Hyades stars. *Astrophys. J.* 321:459–472.

Radick, R. R., Lockwood, G. W., and Baliunas, S. L. 1990. Stellar activity and brightness variations: A glimpse at the Sun's history. *Science* 247:39–44.

Raghavan, R. S., and Pakvasa, S. 1988. Probing the nature of the neutrino: The boron solar-neutrino experiment. *Phys. Rev.* 37D:849–857.

Ramaty, R. 1986. Nuclear processes in solar flares. In *Physics of the Sun,* eds. P. A. Sturrock, T. E. Holzer, D. M. Mihalas and R. K. Ulrich (Dordrecht: D. Reidel), pp. 291–323.

Ramaty, R., and Murphy, R. J. 1987. Nuclear processes and accelerated particles in solar flares. *Space Sci. Rev.* 45:213–268.

Ramaty, R., Murphy, R. J., and Kozlovsky, B. 1983. Implications of high-energy neutron observations from solar flares. *Astrophys. J.* 273:L41-L45.

Ramaty, R., Miller, J. A., Hua, X.-M., and Lingenfelter, R. E. 1988. In *Nuclear Spectroscopy in Astrophysics,* eds. G. H. Share and N. Gehrels (New York: American Inst. of Physics), pp. 628–632.

Ramsey, H. E., Schoolman, S. A., and Title, A. M. 1977. On the size, structure, and strength of the small-scale solar magnetic field. *Astrophys. J.* 215:L41-L42.

Rapaport, J., Taddeucci, T., Welch, P., Gaarde, C., Larsen, J., Goodman, C., Foster, C. C., Goulding, C. A., Horen, D., Sugarbaker, E., and Masterson, T. 1981. Empirical evaluation of Gamow-Teller strength function for ^{37}Cl $\rightarrow$ ^{37}Ar and its implication in the cross section for solar neutrino absorption by ^{37}Cl. *Phys. Rev. Lett.* 47:1518–1521.

Rapaport, J., Welch, P., Bahcall, J., Sugarbaker, E., Taddeucci, T. N., Goodman, C. D., Foster, C. F., Horen, D., Gaard, C., Larsen, J., and Masterson, T. 1985. Solar-neutrino detection: Experimental determination of Gamow-Teller strengths via the ^{98}M$_\odot$ and ^{115}In (e,n) reactions. *Phys. Rev. Lett.* 54:2325–2328.

Rast, J., Kneubuhl, F. K., and Muller, E. A. 1978. Measurement of the solar brightness temperature near its minimum with a balloon-borne Lamellar-grating interferometer. *Astron. Astrophys.* 68:229–238.

Raychaudhuri, P. 1986. Solar neutrino flux, cosmic rays, and the solar activity cycle. *Solar Phys.* 104:415–424.

Raychaudhuri, P. 1989. Solar neutrino flux variation and its implications. *Phys. Essays* 2: 118–123.

Rayrole, J. 1987. The French polarization-free telescope THEMIS. In *The Role of Fine-Scale Magnetic Fields on the Structure of the Solar Atmosphere,* eds. E.-H. Schröter, M. Vázquez and A. A. Wyller (Cambridge: Cambridge Univ. Press), pp. 367–369.

Reames, D. V., and Lin, R. P. 1985. In *Proc. 19th Intl. Cosmic Ray Conf.,* vol. 4, eds. F. C. Jones, J. Adams and G. M. Mason, NASA CP-2376, p. 273.

Reames, D. V., von Rosevinge, T. T., and Lin, R. P. 1985. Solar ^{3}He-rich events and nonrelativistic electron events: A new association. *Astrophys. J.* 292:716–724.

Redman, R. O. 1943. Spectrographic observations at the total solar eclipse of 1940 October 1. III. The spectrum of the extreme limb of the Sun. *Mon. Not. Roy. Astron. Soc.* 103:173–190.

Reed, M., and Simon, B. 1975. *Methods of Modern Mathematical Physics. II. Fourier Analysis, Self-Adjointness* (New York: Academic Press).

Rees, D. E. 1969. Line formation in a magnetic field. *Solar Phys.* 10:268–282.

Rees, D. E. 1987. A gentle introduction to polarized radiative transfer. In *Numerical Radiative Transfer,* ed. W. Kalkofen (Cambridge: Cambridge Univ. Press), pp. 213–239.

Rees, D. E., Murphy, G. A., and Durrant, C. J. 1989. Stokes profile analysis and vector magnetic fields. II. Formal numerical solutions of the Stokes transfer equations. *Astrophys. J.* 339:1093–1106.

Rees, D. E., Lites, B. W., and Skumanich, A. 1991. Stokes profile analysis and vector magnetic field. VIII. Synthesis and inversion of the Ca II infrared triplet lines. In preparation.

Reeves, E. M., and Parkinson, W. H. 1970. An atlas of extreme-ultraviolet spectroheliograms from OSO-IV. *Astrophys. J. Suppl.* 21:1–30.

Reeves, H., and Meyer, S. P. 1978. Cosmic-ray nucleosynthesis and the infall rate of extragalactic material in the solar neighborhood. *Astrophys. J.* 226:613–631.

Reimers, D. 1975. Circumstellar envelopes and mass loss of red giant stars. In *Problems of*

Stellar Atmospheres and Envelopes, eds. B. Baschek, W. H. Kegel and G. Traving (New York: Springer-Verlag), pp. 229–256.

Reines, R., and Trimble, V. 1973*a*. In *Proc. Solar Neutrino Conf.,* Univ. of California, Feb. 1972.

Reines, R., and Trimble, V. 1973*b*. The solar neutrino problem—A progress report. *Rev. Mod. Phys.* 45:1–5.

Renzini, A. 1977. The evolution of Population II stars and mass loss and stellar evolution. In *Advanced Stages of Stellar Evolution,* eds. P. Bouvier and A. Maeder (Geneva: Geneva Obs.), pp. 151–284.

Rhodes, E. J., Jr. 1977. Non-Radial *p*-mode Oscillations as a Seismic Probe of Solar Structure. Ph.D. Thesis, Univ. of California at Los Angeles.

Rhodes, E. J., Jr., Ulrich, R. K., and Simon, G. W. 1977. Observations of nonradial *p*-mode oscillations on the Sun. *Astrophys. J.* 218:901–919.

Rhodes, E. J., Cacciani, A., Woodard, M., Tomczyk, S., Korennik, S., and Ulrich, R. K. 1987. Estimates of the solar internal angular velocity obtained with the Mt. Wilson 6-foot solar tower. In *The Internal Solar Angular Velocity,* eds. B. R. Durney and S. Sofia (Dordrecht: D. Reidel), pp. 75–82.

Rhodes, E. J., Jr., Cacciani, A., and Korzennik, S. G. 1988*a*. Initial high-degree *p*-mode frequency splittings from the 1988 Mt. Wilson 60-foot Tower solar oscillation program. In *Seismology of the Sun and Sun-Like Stars,* ed. E. J. Rolfe, ESA SP-286, pp. 81–86.

Rhodes, E. J., Jr., Woodard, M. F., Cacciani, A., Tomczyk, S., Korzennik, S. G., and Ulrich, R. K. 1988*b*. On the constancy of intermediate-degree *p*-mode frequencies during the declining phase of solar cycle 21. *Astrophys. J.* 326:479–485.

Rhodes, E. J., Jr., Cacciani, A., Korzennik, S. G., Tomczyk, S., Ulrich, R. K., and Woodard, M. F. 1988*c*. Radial and latitudinal gradients in the solar internal angular velocity. In *Seismology of the Sun and Sun-Like Stars,* ed. E. J. Rolfe, ESA SP-286, pp. 73–80.

Rhodes, E. J., Jr., Cacciani, A., Korzennik, S. G., Tomczyk, S., Ulrich, R. K., and Woodard, M. F. 1990. Depth and latitude dependence of the solar internal angular velocity. *Astrophys. J.* 351:687–700.

Ribes, E. 1986*a*. The large-scale solar variability through the cycle. *Adv. Space Res.* 6:221–228.

Ribes, E. 1986*b*. Etude de la dynamique de la zone convective solaire et ses conséquences sur le cycle d'activité. *Circ. Acad. Sci. Paris II* 302:871–876.

Ribes, E. 1988. The influence of the toroidal roll pattern on the sunspot activity. In *Seismology of the Sun and Sun-Like Stars,* ed. E. J. Rolfe, ESA SP-286, pp. 291–293.

Ribes, E., and Bonnefond, F. 1989. Magnetic tracers, a probe of the convective layers. In preparation.

Ribes, E., and Laclare, F. 1988. Toroidal convective rolls: A challenge to theory. *Geophys. Astrophys. Fluid Dyn.* 41:171–180.

Ribes, E., and Mein, P. 1984. Search for giant convective cells from the analysis of Meudon spectroheliograms. In *Proc. European Astron. Meeting,* pp. 283–288.

Ribes, E., Mein, P., and Mangeney, A. 1985*a*. A large-scale circulation in the convective zone. *Nature* 318:170–171.

Ribes, E., Rees, D. E., and Fang, Ch. 1985*b*. Observational diagnostics for models of magnetic flux tubes. *Astrophys. J.* 298:268–277.

Ribes, E., Ribes, J.-C., and Barthalot, R. 1987. Evidence for a larger Sun with a slower rotation during the seventeenth century. *Nature* 326:52–55.

Richardson, R. S., and Schwarzschild, M. 1953. On the possibility of a 22-year oscillation of the Sun. *Acad. Naz. Linc. Fond. Ales. Volta. Atti Conv.* 11:228–247.

Richstone, D. O. 1974. The occurrence of a nonspherical instability in red giant stars. *Astrophys. J.* 188:327–333.

Richter, A. K. 1988. Interplanetary slow shocks: A review. In *Proc. Sixth Intl. Solar Wind Conf.,* eds. V. J. Pizzo, T. E. Holzer and D. G. Sime, NCAR TN-306 (Boulder: Natl. Center for Atmos. Res.), pp. 411–420.

Richtmyer, R. D., and Morton, K. W. 1968. *Difference Method for Initial Value Problems* (New York: Interscience).

Rickett, B. J.,and Coles, W. A. 1983. Solar cycle evolution of the solar wind in three dimensions. In *Proc. Fifth Intl. Solar Wind Conf.,* ed. M. Neugebauer, NASA CP-2280, pp. 315–321.

Rieger, E. 1982. Gamma ray measurements during solar flares with the gamma ray detector on the SMM—An overview. In *Proc. Hinotori Symp. on Solar Flares* (Tokyo: Inst. of Space and Astronautical Science), pp. 246–262.

Rieger, E., Reppin, C., Kanbach, G., Forrest, D. J., Chupp, E. L., and Share, G. H. 1983. In *Proc. 18th Intl. Cosmic Ray Conf.,* vol. 10, eds. N. Durgaprasad, S. Ramadurai, P. V. Raman Murthy, M. V. S. Rao and K. Sivaprasad (Bombay: Tata Inst. of Fundamental Research), pp. 338–341.

Rieger, E., Share, G. H., Forrest, D. J., Kanbach, G., Reppin, C., and Chupp, E. L. 1984. A 154-day periodicity in the occurrence of hard solar flares? *Nature* 312:623–625.

Righini, G., and Simon, M. 1976. Solar brightness temperature distribution at 350 and 450 microns. *Astrophys. J.* 203:L95-L97.

Righini, A., Gavallini, F., and Geppatelli, G. 1984. Shifts and asymmetries of photospheric lines. In *Small-Scale Dynamical Processes in Quiet Stellar Atmospheres,* ed. S. Keil (Sunspot, N.M.: Sacramento Peak Obs), pp. 300–305.

Riker, J.F., and Ahluwalia, H. S. 1987. A survey of cosmic ray diurnal variation during 1973–79. 2. Application of diffusion convection model to diurnal anisotropy data. *Planet. Space Sci.* 35:1117–1122.

Ritter, A. 1878. Untersuchungen über die Höhe der Atmosphäre und die Constitution gasförmiger Weltkörper. VI. *Ann. Physik Chemie neue Folge* 5:543–558.

Ritter, A. 1882. Untersuchungen über die Höhe der Atmosphäre und die Constitution gasförmiger Weltkörper. VIII. *Ann. Physik Chemie neue Folge* 16:166–192.

Ritzwoller, M., and Laveley, E. 1990. *Astrophys. J.,* submitted.

Robe, H. 1968. Les oscillations non radiales des polytropes. *Ann. Astrophys.* 31:475–482.

Roberts, B., and Campbell, W. R. 1988. The influence of a chromospheric magnetic field on *p*- and *f*-modes. In *Seismology of the Sun and Sun-Like Stars,* ed. E. J. Rolfe, ESA SP-286, pp. 311–314.

Roberts, B., and Webb, A. R. 1978. Vertical motions in an intense magnetic flux tube. *Solar Phys.* 56:5–35.

Roberts, D. A., Goldstein, M. L., Klein, L. W., and Matthaeus, W. H. 1987. Origin and evolution of fluctuations in the solar wind: Helios observations and Helios-voyager comparisons. *J. Geophys. Res.* 92:12023–12035.

Roberts, P. H. 1967. *An Introduction to Magnetohydrodynamics* (New York: Elsevier).

Roberts, P. H. 1971. Dynamo theory. In *Mathematical Problems in the Geophysical Sciences, Vol 2,* ed. W. H. Reid, pp. 129–206.

Roberts, P. H. 1972. Kinetic dynamo models. *Phil. Trans. Roy. Soc. London* A272:663–698.

Roberts, P. H., and Stix, M. 1971. *The Turbulent Dynamo* NCAR-Tech. Note/IA-60 (Boulder: Natl. Center for Atmos. Res.).

Roberts, W. O. 1945. A preliminary report on chromospheric spicules of extremely short lifetimes. *Astrophys. J.* 101:136–140.

Robertson, R. G. H. 1990. Neutrino mass: Recent results. In *Fundamental Symmetries in Nuclei and Particles,* eds. H. Hendrikson and P. Vogel (Singapore: World Scientific), pp. 86–100.

Robertson, R. G. H., Dyer, P., Bowles, T. J., Brown, R. E., Jarmie, N., Maggiore, C. J., and Austin, S. M. 1983. Cross section of the capture reaction $^3He(\alpha,\gamma)^7Be$. *Phys. Rev.* C27: 11–18.

Robillot, J. M., Bocchia, R., Fossat, E., and Grec, G. 1984. Solar large scale velocity structures from optical resonance method. *Astron. Astrophys.* 137:43–50.

Robinson, R. D., Worden, S. P., and Harvey, J. W. 1980. Observations of magnetic fields on two late-type dwarf stars. *Astrophys. J.* 236:L155-L158.

Roddier, F. 1981. The effects of atmospheric turbulence in optical astronomy. In *Progress in Optics,* vol. XIX, ed. E. Wolff (New York: Elsevier).

Roddier, F. 1986. *Opt. Comm.* 60:145–148.

Roddier, F. 1987. Seeing and atmospheric turbulence: Parameters relevant to adaptive optics. In *Adaptive Optics in Solar Astronomy,* eds. F. Merkle, O. Engvold and R. Falomo, LEST Foundation Tech. Rept. No. 28 (Oslo: Univ. of Oslo).

Roddier, F. 1987. Signal-to-noise ratios and beam combination. Interferometric imaging in astronomy. *High Resolution Imaging,* ed. J. W. Goad, pp. 135–138.

Roddier, F., Gilli, J. M., and Vernin, J. 1982. On the isoplanatic patch size in stellar speckle interferometry. *J. Optics (Paris)* 13:63–72.

Rogers, F. J. 1977. On the compensation of bound and scattering state contributions to the partition function. *Phys. Lett.* 61A:358–360.

Rogers, F. J. 1981. Equation of state of dense, partially degenerate, reacting plasmas. *Phys. Rev.* A24:1531–1543.

Rogers, F. J. 1986. Occupation numbers for reacting plasmas: The role of the Planck-Larkin partition function. *Astrophys. J.* 310:723–728.

Rogers, F. J., Wilson, B. G., and Iglesias, C. A. 1988. Parametric potential method for generating atomic data. *Phys. Rev.* A38:5007–5020.

Rolfs, C. 1973. Spectroscopic factors from radiative capture reactions. *Nucl. Phys.* A217: 29–70.

Rolfs, C. 1985. In *Proc. NATO Advanced Research Workshop (5th Moriond Astrophysics Meeting) on Nucleosynthesis and its Implications on Nuclear and Particle Physics* (Dordrecht: D. Reidel), p. 431.

Rolfs, C., and Azuma, R. E. 1974. Interference effects in $^{12}C(p,\gamma)^{14}N$ and direct capture to unbound states. *Nucl. Phys.* A227:291–308.

Rolfs, C., and Kavanagh, R. W. 1986. The $^{7}Li(p,\alpha)^{4}He$ cross section at low energies. *Nucl. Phys.* A455:179–188.

Rolfs, C., and Rodney, W. 1974. Proton capture by ^{15}N at stellar energies. *Nucl. Phys.* A235:450–459.

Rolfs, C., and Rodney, W. S. 1988. *Cauldrons in the Cosmos* (Chicago: Univ. of Chicago Press).

Rompold, B., and Bogdan, T. 1986. On the formation of active region prominences (H alpha filaments). In *Coronal and Prominence Plasmas,* ed. A. I. Poland, NASA CP-2442, pp. 81–87.

Ronan, R. S., Mickey, D. L., and Orrall, F. Q. 1987. The derivation of vector magnetic fields from Stokes profiles: Integral versus least squares fitting techniques. *Solar Phys.* 113: 353–359.

Ronan, R. S., Harvey, J. W., and Duvall, T. L., Jr. 1990. Wavelength variation of p mode intensity fluctuations. *Astrophys. J.,* submitted.

Rood, R. T. 1970. Models for partially mixed stars. *Astrophys. J.* 162:939–946.

Rood, R. T. 1978. Review of non-standard solar models. In *Proc. Informal Conf. on Status and Future of Solar Neutrino Research,* vol. 1, ed. G. Friedlander, BNL 50789 (Upton, N.Y.: Brookhaven National Lab), pp. 175–206.

Rood, R. T. 1989. In *Proc. Third ESO CERN Symp. on Astronomy, Cosmology, and Fundamental Physics,* eds. M. Caffo et al. (Dordrecht: Kluwer).

Rosen, S. P. 1989. Double beta decay: A theoretical overview. In *Neutrino '88: Proc. 13th Intl. Conf. on Neutrino Physics and Astrophysics,* eds. J. Schneps, T. Kafka, W. A. Mann and P. Nath (Singapore: World Scientific), pp. 78–99.

Rosen, S. P. 1990. Neutrino oscillations: An essay in honor of Felix Böhm. In *Fundamental Symmetries in Nuclei and Particles,* eds. H. Hendrikson and P. Vogel (Singapore: World Scientific), pp. 101–115.

Rosen, S. P., and Gelb, J. M. 1986a. Mikheyev-Smirnov-Wolfenstein enhancement of oscillations as a possible solution to the solar-neutrino problem. *Phys. Rev.* 34D:969–979.

Rosen, S. P., and Gelb, J. M. 1986b. Matter oscillations and solar neutrinos: A review of the MSW effect. In *Proc. of the XXIII Intl. Conf. on High Energy Physics,* vol. 2, ed. S. C. Loken (Singapore: World Scientific), pp. 909–920.

Rosen, S. P., and Gelb, J. M. 1989. Neutrino-electron scattering and the choice between different Mikheyev-Smirnov-Wolfenstein solutions of the solar-neutrino problem. *Phys. Rev.* 39D:3190–3193.

Rosenbauer, H., Schwenn, R., Marsch, E., Meyer, B., Miggenrieder, H., Montgomery, M., Mühlhäuser, K.-H., Pilipp, W., Voges, W., and Zink, S. K. 1977. A survey on initial results of the Helios plasma experiment. *J. Geophys.* 42:561–580.

Rosenwald, R. D., Hill, H. A., and Gu, Y. 1987. Observed quasi-periodic deviations from second order asymptotic theory of low-degree gravity mode eigenfrequencies. *Bull. Amer. Astron. Soc.* 19:1120 (abstract).

Rosner, R. 1986. On the origins and dynamics of spatial structure in the outer solar atmosphere. In *Hydrodynamic and Magnetohydrodynamic Problems in the Sun and Stars,* ed. Y. Osaka (Tokyo: Univ. of Tokyo), pp. 37–51.

Rosner, R., and Weiss, N. O. 1985. Differential rotation and magnetic torques in the interior of the Sun. *Nature* 317: 790–792.

Rosner, R., Tucker, W. H., and Vaiana, G. S. 1978. Dynamics of the quiescent solar corona. *Astrophys. J.* 220:643–665.

Ross, J. E., and Aller, L. H. 1974. The solar abundance of beryllium. *Solar Phys.* 36:11–19.

Ross, J. E., and Aller, L. H. 1976. The chemical composition of the Sun. *Science* 191:1223–1229.

Roth, M. L., and Weigert, A. 1979. More on avoided level crossing of non-radial stellar oscillations. *Astron. Astrophys.* 80:48–52.

Rothman, L. S., Gamache, R. R., Goldman, A., Brown, L. R., Toth, R. A., Pickett, H. M., Poynter, R. L., Elaud, J.-M., Camry-Peyret, C., Barge, A., Husson, N., Rinsland, C. P., and Smith, M. A. H. 1987. The HITRAN database: 1986 edition. *Appl. Opt.* 26:4058–4097.

Rottman, G. J. 1988. Results from space measurements of solar UV and EUV flux. In *Solar Radiative Output Variations*, ed. P. V. Foukal (Cambridge, Mass.: Cambridge Research and Instrumentation), pp. 71–86.

Roudier, T. 1986. Thesis, Univ. of Toulouse.

Roudier, T., and Muller, R. 1986. Structure of the solar granulation. *Solar Phys.* 107:11–26.

Rouse, C. A. 1983. Comment on the Planck-Larkin partition function. *Astrophys. J.* 272:377–379.

Rouse, C. A. 1986. Evidence for a small, high-Z, iron-like solar core. II. Agreements with observed frequencies of oscillation in the five-minute band. *Solar Phys.* 106:205–216.

Roussel-Dupré, R. 1981. Computations of ion diffusion coefficients from the Boltzmann-Fokker-Planck equation. *Astrophys. J.* 243:329–343.

Roussel-Dupré, R. 1982. Diffusion and viscosity coefficients from helium. *Astrophys. J.* 252:393–401.

Roussel-Dupré, R., and Beerman, C. 1981. Effects of diffusion and mass flows on C IV and Si IV lines formed in the solar atmosphere. *Astrophys. J.* 250:408–423.

Rowland, H. L., and Vlahos, L. 1985. Return currents in solar flares: Collisionless effects. *Astron. Astrophys.* 142:219–224.

Rowley, J. K. 1978. The ^{7}Li-^{7}Be experiment. In *Proc. Informal Conf. on Status and Future of Solar Neutrino Research*, vol. 1, ed. G. Friedlander, BNL 50879 (Upton, N.Y.: Brookhaven National Lab), pp. 265–291.

Rowley, J. K., Cleveland, B. T., Davis, R., Hampel, W., and Kirsten, T. 1980. The present and past luminosity of the Sun. In *The Ancient Sun: Fossil Record in the Earth, Moon and Meteorites*, eds. R. O. Pepin, J. A. Eddy and R. B. Merrill (New York: Pergamon), pp. 45–62.

Rowley, J. K., Cleveland, B. T., and Davis,R.,Jr. 1985. The chlorine solar neutrino experiment. In *Solar Neutrinos and Neutrino Astronomy*, eds. M. L. Cherry, K. Lande and W. A. Fowler (New York: American Inst. of Physics), pp. 1–21.

Roxburgh, I. 1978. Convection and stellar structure. *Astron. Astrophys.* 65:281–285.

Roxburgh, I. W. 1983. Stellar winds and spin-down in solar-type stars. In *Solar and Stellar Magnetic Fields*, ed. J. O. Stenflo (Dordrecht: D. Reidel), pp. 449–460.

Roxburgh, I. 1985. Present problems of the solar interior. *Solar Phys.* 100:21–51.

Roxburgh, I. W. 1989. Integral constraints on convective overshooting. *Astron. Astrophys.* 211:361–364.

Rudiger, G. 1980. Reynolds stresses and differential rotation. I. Recent calculation of zonal fluxes in slowly rotating stars. *Geophys. Astrophys. Fluid Dyn.* 16:239–261.

Rüdiger, G. 1989. *Differential Rotation and Stellar Convection: Sun and Solar-Type Stars* (New York: Gordan and Breach).

Rüdiger, G., Tuominen, I., Krause, F., and Virtanen, H. 1986. Dynamo-generated flows in the Sun. I. Foundations and first results. *Astron. Astrophys.* 166:306–318.

Ruppel, H. M., and Norton, J. L. 1975. *Theoretical Simulation of the Gas Explosive Simulation Technique (GEST) Experiments*, LA-6154-MS (Los Alamos, N.M.: Los Alamos Sci. Lab).

Russell, C. T., and Elphic, R. C. 1979. ISEE observations of flux transfer events at the dayside magnetopause. *Geophys. Res. Lett.* 6:33–36.

Russell, H. N. 1929. On the composition of the Sun's atmosphere. *Astrophys. J.* 70:11–82.

Rust, D. M. 1968. Chromospheric explosions and satellite sunspots. In *Structure and Development of Solar Active Regions: Proc. IAU Symp. 35*, ed. K. O. Kiepenheuer (Dordrecht: D. Reidel), pp. 77–84.

Rust, D. M., and Appourchaux, T. 1988. The stable solar analyzer. In *Seismology of the Sun and Sun-Like Stars,* ed. E. J. Rolfe, ESA SP-286, pp. 227–233.

Rutten, R. J. 1986. The NLTE formation of iron lines in the solar photosphere. In *Physics of Formation of Fe II Lines Outside LTE: Proc. IAU Coll. 94,* ed. R. Viotti (Dordrecht: D. Reidel), pp. 185–210.

Rutten, R. G. M. 1987. Magnetic structure in cool stars. *Astron. Astrophys.* 177:131–142.

Rutten, R. J. 1988. Oscillator strengths from the high S/N solar spectrum. In *The Impact of Very High S/N Spectroscopy on Stellar Physics: Proc. IAU Symp. 132,* eds. G. Cayrel de Strobel and M. Spite (Dordrecht: Kluwer), pp. 367–371.

Rutten, R. G. M., and Schrivjer, C. J. 1987. Magnetic structure in cool stars. XIII. Appropriate unities for the rotation-activity relation. *Astron. Astrophys.* 177:155–162.

Rutten, R. J., and Severino, G., eds. 1989. *Solar and Stellar Granulation* (Dordrecht: Kluwer).

Rutten, R. J., Bruls, J. H. M. J., Gomez, M. T., and Severino, G. 1988. The granulation sensitivity of helioseismology lines. In *Seismology of the Sun and Sun-Like Stars,* ed. E. J. Rolfe, ESA SP-286, pp. 251–255.

Ruzvdjak, V., and Vujnović, V. 1977. Statistically extended recombination continuum and line dissolution in an analysis of the Balmer spectrum at the line merging region. *Astron. Astrophys.* 54:751–755.

Ryutov, D. D., and Ryutova, M. P. 1976. Sound oscillations in a plasma with "magnetic filaments." *Soviet Phys. JETP* 43:491–497.

Ryzhikova, N. N. 1988*a*. On the nature of the solar brightness oscillation. I. *Soviet Astron. Lett.* 14:269–271.

Ryzhikova, N. N. 1988*b*. On the nature of the solar brightness oscillation. II. *Soviet Astron. Lett.* 14:271–274.

Sackman, I.-J., Boothroyd, A., and Fowler, W. A. 1990. Our Sun. I. The standard model success and failures. *Astrophys. J.* 360:727–736.

Saiedy, F. 1960. Solar intensity and limb darkening between 8.6 and 13u. *Mon. Not. Roy. Astron. Soc.* 121:483–495.

Saio, H. 1980. Stability of nonradial g^+-mode pulsations in 1 $M_{\odot}$ models. *Astrophys. J.* 240:685–692.

Saio, H. 1981. Rotational and tidal perturbations of nonradial oscillations in a polytrope star. *Astrophys. J.* 244:299–315.

Saio, H., and Cox, J. P. 1980. Linear nonadiabatic analysis of nonradial oscillations of massive near main sequence stars. *Astrophys. J.* 236:549–559.

Saito, K. 1965. Polar rays of the solar corona. II. *Publ. Astron. Soc. Japan* 17:1–26.

Saito, K. 1972. A non-spherical axisymmetric model of the solar K-corona of the minimum type. *Annals of the Tokyo Astron. Obs.* XII:53–120.

Saito, K., and Tandberg-Hansen, E. 1972. The arch systems, cavities and prominences in the helmet streamer observed at the solar eclipse, Nov 12, 1966. *Solar Phys.* 31:105–121.

Sakurai, K. A. 1990. Possible chaotic process in the solar interior as inferred from the observed time variation of the neutrino flux from the Sun. In *Proc. 21st Intl. Cosmic Ray Conf.,* vol 7, ed. R. J. Protheroe (Adelaide: Univ. of Adelaide), pp. 172–175.

Sakurai, T. 1979. A new approach to the force-free field and its application to the magnetic field of solar active regions. *Publ. Astron. Soc. Japan* 31:209–230.

Sakurai, T. 1982. Green's function methods for potential magnetic fields. *Solar Phys.* 76: 301–321.

Sakurai, T. 1987. A study of magnetic energy build-up based on vector magnetograms. *Solar Phys.* 113:137–142.

Sakurai, T. 1990. Helioseismology observations by SOLAR-A satellite. In *Oji Inst. Seminar on Progress of Seismology of the Sun and Stars,* eds. Y. Osaki and H. Shibahashi (Berlin: Springer-Verlag), in press.

Sakurai, T., and Uchida, Y. 1977. Magnetic field and current sheets in the corona above active regions. *Solar Phys.* 52:397–416.

Sakurai, T., Makita, M., and Shibasaki, K. 1985. Observation of magnetic field vector in solar active regions. In *Theoretical Problems in High Resolution Solar Physics,* ed. H. U. Schmidt (MPI: MPA-212), pp. 312–315.

Salpeter, E. E. 1954. Electron screening and thermonuclear reactions. *Australian J. Phys.* 7: 373–388.

Salpeter, E. E. 1968. Neutrinos from the Sun. *Comments Nucl. Part. Phys.* 2:97–102.

Sampson, D. H. 1959. The opacity at high temperatures due to Compton scattering. *Astrophys. J.* 129:734–751.

Sanders, R. H. 1967. s-process nucleosynthesis in thermal relaxation cycles. *Astrophys. J.* 150:971–977.

Sandlin, G. D., Bartoe, J.-D. F., Brueckner, G. E., Tousey, R., and VanHoosier, M. E. 1986. The high resolution solar spectrum, 1175–1710Å. *Astrophys. J. Suppl.* 61:801–898.

Saslaw, W. G., and Schwarzschild, M. 1965. Overshooting from stellar convective cores. *Astrophys. J.* 142:1468–1480.

Sauval, A. J., Grevesse, N., Brault, J. W., Stokes, G. M., and Zander, R. 1984. The pure rotation spectrum of OH and the solar oxygen abundance. *Astrophys. J.* 282:330–338.

Scharmer, G. B. 1987. The Swedish 50 cm vacuum solar telescope: Concepts and auxiliary instrumentation. In *The Role of Fine-Scale Magnetic Fields on the Structure of the Solar Atmosphere,* eds. E.-H. Schröter, M. Vázquez and A. A. Wyller (Cambridge: Cambridge Univ. Press), pp. 349–353.

Scharmer, G. B. 1989. High resolution granulation observations from La Palma: Techniques and first results. In *Solar and Stellar Granulation,* eds. R. J. Rutten and G. Severino (Dordrecht: Kluwer), pp. 161–167.

Scharmer, G. B., Brown, D. S., Pettersson, L., and Rehn, J. 1985. Concepts for the Swedish 50 cm vacuum solar telescope. *J. Appl. Opt.* 24:2558–2564.

Scharmer, R. 1981. Solutions to radiative transfer problems using approximate lambda operators. *Astrophys. J.* 248:720–730.

Scharmer, R. 1984. Accurate solutions to non-LTE problems using approximate lambda operators. In *Methods in Radiative Transfer,* ed. W. Kalkofen (Cambridge: Cambridge Univ. Press), pp. 173–210.

Schatten, K. H. 1988. A model for solar constant secular change. *Geophys. Res. Lett.* 15: 121–124.

Schatten, K. H., and Mayr, H. G. 1985. On the maintenance of sunspots: An ion hurricane mechanism. *Astrophys. J.* 299:1051–1062.

Schatten, K. H., and Sofia, S. 1987. Forecast of an exceptionally large even-numbered solar cycle. *Geophys. Res. Lett.* 14:632–635.

Schatten, K. H., Wilcox, J. M., and Ness, N. F. 1969. A model of interplanetary and coronal magnetic fields. *Solar Phys.* 6:442–455.

Schatten, K. H., Leighton, R. B., Howard, R., and Wilcox, J. M. 1972. Large-scale photospheric magnetic field: The diffusion of active region fields. *Solar Phys.* 26:283–289.

Schatten, K. H., Miller, N., Sofia, S., and Oster, L. 1982. Solar irradiance modulation by active regions from 1969 through 1980. *Geophys. Res. Lett.* 9:49–51.

Schatten, K. H., Mayr, H. G., Omidvar, K., and Maier, E. 1986. A hillock and cloud model for faculae. *Astrophys. J.* 311:460–473.

Schatzman, E. 1949. The heating of the solar corona and chromosphere. *Ann. Astrophys.* 12:203–228.

Schatzman, E. 1951. The ³He isotope in stars, application to the theory of supernovae and white dwarfs. *Compt. Rend.* 232:1740–1751.

Schatzman, E. 1962. A theory of the role of magnetic activity during star formation. *Ann. Astrophys.* 25:18–29.

Schatzman, E. 1981. In *Turbulent Diffusion and the Solar Neutrino Problem.* CERN, p. 81.

Schatzman, E. 1984. Physical mechanisms of mixing in stellar interiors. In *Observational Tests of Stellar Evolution Theory: Proc. IAU Symp. 105,* eds. A. Maeder and A. Renzini (Dordrecht: D. Reidel), pp. 491–512.

Schatzman, E. 1987. Solar rotation and age. In *The Internal Solar Angular Velocity,* eds. B. R. Durney and S. Sofia (Dordrecht: D. Reidel), pp. 159–171.

Schatzman, E. 1989. Stellar rotation, dynamo, electromagnetic braking, age and lithium burning. In *Turbulence and Non-Linear Dynamics in Magneto-Hydrodynamic Flows,* eds. M. Menmeguzzi, A. Pouquet and P. L. Sulem (Amsterdam: North Holland Press), pp. 1–18.

Schatzman, E. 1990*a.* Inside the Sun: Unsolved problems. In *Inside the Sun,* eds. G. Berthomieu and M. Cribier (Dordrecht: Kluwer), pp. 5–17.

Schatzman, E. 1990*b.* Rotation, lithium and mixing. In *Rotation and Mixing in Stellar Interiors,* eds. M. J. Goupil and J.-P. Zahn (Berlin: Springer-Verlag), pp. 3–26.

Schatzman, E., and Baglin, A. 1990. On the physics of lithium depletion. *Astron. Astrophys.*, submitted.

Schatzman, E., and Maeder, A. 1981. Stellar evolution with turbulent diffusion mixing. III. The solar model and the neutrino problem. *Astron. Astrophys.* 96:L1-L16.

Schatzman, E., and Ribes, E. 1987. The solar neutrino problem: Structure, variability of the Sun (an astrophysicist's point of view). In *New and Exotic Phenomena*, eds. O. Gackler and J. T. T. Vân (Gif-sur-Yvette: Editions Frontières), pp. 367–386.

Schatzman, E., and Souffrin, P. 1967. Waves in the solar atmosphere. *Ann. Rev. Astron. Astrophys.* 5:67–84.

Schatzman, E., Maeder, A., Angrand, F., and Glowinski, R. 1981. Stellar evolution with turbulent diffusion mixing. III. The solar model and the neutrino problem. *Astron. Astrophys.* 96:1–16.

Scherrer, P. H. 1973. A Study of the Mean Solar Magnetic Field. Ph.D. Thesis, Stanford Univ.

Sherrer, P. H. 1984. Detection of solar gravity mode oscillations. In *Solar Seismology from Space*, NASA JPL Publ. 84–84, pp. 173–182.

Scherrer, P. H. 1986. Comments on techniques for spectral deconvolution. In *Seismology of the Sun and the Distant Stars*, ed. D. O. Gough (Dordrecht: D. Reidel), pp. 117–120.

Scherrer, P. H., and Wilcox, J. M. 1983. Structure of the solar oscillations with period near 160 min. *Solar Phys.* 82:37–42.

Scherrer, P. H., Wilcox, J. M., Kotov, V. A., Severny, A. B., and Tsap, T. T. 1979. Observations of solar oscillations with a period of 160 minutes. *Nature* 277:635–637.

Scherrer, P. H., Wilcox, J. M., Severny, A. B., Kotov, V. A., and Tsap, T. T. 1980*a*. Further evidence of solar oscillations with a period of 160 minutes. *Astrophys. J.* 237:L97-L98.

Scherrer, P. H., Wilcox, J. M., and Svalgaard, L. 1980*b*. The rotation of the Sun: Observations at Stanford. *Astrophys. J.* 241:811–819.

Scherrer, P. H., Wilcox, J. M., Christensen-Dalsgaard, J., and Gough, D. O. 1982. Observation of additional low-degree five-minute modes of solar oscillation. *Nature* 297:312–313.

Scherrer, P. H., Wilcox, J. M., Christensen-Dalsgaard, J., and Gough, D. O. 1983. Detection of solar five-minute oscillations of low degree. *Solar Phys.* 82:75–87.

Scherrer, P. H., Hoeksema, J. T., Bogart, R. S., and the SOI Co-Investigator Team. 1988. The solar oscillations imager for SOHO. In *Seismology of the Sun and Sun-Like Stars*, ed. E. J. Rolfe, ESA SP-286, pp. 375–379.

Schiff, L. I. 1949. *Quantum Mechanics* (New York: McGraw Hill).

Schlüter, A., and Temesváry, S. 1958. The internal constitution of sunspots. In *Electromagnetic Phenomena in Cosmical Physics: Proc. IAU Symp. 6*, ed. B. Lehnert (Cambridge: Cambridge Univ. Press), pp. 263–275.

Schmahl, E. J., and Orrall, F. O. 1979. Evidence for continuum absorption above the quiet sun transition zone. *Astrophys. J.* 231:L41-L44.

Schmahl, E. J., and Orrall, F. O. 1986. Interpretation of the prominence differential emission measure for three geometries. In *Coronal and Prominence Plasmas*, ed. A. I. Poland, NASA CP-2442, pp. 127–136.

Schmahl, E. J., Mouradian, F., Martres, M. J., and Soru-Escaut, I. 1982. EUV arcades: Signatures of filament instability. *Solar Phys.* 81:91–105.

Schmid, J., Bochsler, P., and Geiss, J. 1987. Velocity of iron ions in the solar wind. *J. Geophys. Res.* 92:9901–9906.

Schmid, J., Bochsler, P., and Geiss, J. 1988. Abundance of iron ions in the solar wind. *Astrophys. J.* 329:956–966.

Schmidt, H. U. 1964. On the observable effects of magnetic energy storage and release connected with solar flares. In *The Physics of Solar Flares*, ed. W. N. Hess, NASA SP-50, pp. 107–114.

Schmidt, H. U., ed. 1985. *Theoretical Problems in High Resolution Solar Physics* (Garching: MPI für Physik und Astrophysik).

Schmidt, H. U., and Wegmann, R. 1983. A free boundary value problem for sunspots. In *Dynamical Problems in Mathematical Physics*, eds. B. Brosowski and E. Martensen (Frankfurt: P. Lang), pp. 137–150.

Schmidt, H. U., Spruit, H. C., and Weiss, N. O. 1986. Energy transport in sunspot penumbrae. *Astron. Astrophys.* 158:351–360.

Schmidt, W., Deubner, F.-L., Mattig, W., and Mehltretter, J. P. 1979. On the center to limb variation of the granular brightness fluctuations. *Astron. Astrophys.* 75:223–227.

Schmidt, W., Knölker, M., and Schröter, E. H. 1981. RMS-value and power spectrum of the photospheric intensity fluctuations. *Solar Phys.* 73:217–231.

Schmieder, B. 1977. Linear hydrodynamic equations coupled with radiative transfer in a non-isothermal atmosphere. I. Method. *Solar Phys.* 54:269–288.

Schmieder, B. 1979. Waves in the low solar chromosphere. *Astron. Astrophys.* 74:273–279.

Schmidt-Kaler, T., and Winkler, C. 1984. Ground-based observations of solar intensity oscillations. *Astron. Astrophys.* 136:299–305.

Schmieder, B. 1976. Wave propagation in the photosphere. *Solar Phys.* 47:435–460.

Schmieder, B. 1977. Linear hydrodynamical equations coupled with radiative transfer in a non-isothermal atmosphere. I. Method. *Solar Phys.* 54:269–288.

Schmieder, B. 1978. Linear hydrodynamical equations coupled with radiative transfer in a non-isothermal atmosphere. II. Application to solar photospheric observations. *Solar Phys.* 57:245–253.

Schmieder, B., Malherbe, J.-M., Poland, A. I., and Simon, G. 1985. Dynamics of filaments. IV. Structure and mass flow of an active region filament. *Astron. Astrophys.* 153:64–70.

Schmitt, D. 1987. An alpha-omega dynamo with an alpha effect due to magnetostrophic waves. *Astron. Astrophys.* 174:281–287.

Schmitt, J. H. M. M., Rosner, R., and Gohn, H. U. 1984. The overshoot region at the base of the convection zone. *Astrophys. J.* 282:316–329.

Schöchlin, W., and Magun, A. 1979. A statistical investigation of microwave burst spectra for the determination of source inhomogeneities. *Solar Phys.* 64:349–357.

Schove, D. J., ed. 1983. *Sunspot Cycles* (Stroudsburg, Penn.: Hutchinson Ross).

Schramm, D. N. 1989. *Comm. Nucl. Part. Phys.* 17:5239.

Schrijver, C. J. 1987. Heating of stellar chromospheres and coronae: Evidence for non-magnetic heating. In *Cool Stars, Stellar Systems and the Sun,* eds. J. L. Linsky and R. E. Stencel (Berlin: Springer-Verlag), pp. 135–144.

Schrijver, C. J., Coté, J., Zwaan, C., and Saar, S. H. 1989a. The photospheric magnetic field and the emission from the outer atmospheres of cool stars. I. The solar Ca II K line core emission. *Astrophys. J.* 337:964–976.

Schrijver, C. J., Coté, J., Zwaan, C., and Saar, S. H. 1989b. Relations between the photospheric magnetic field and the emission from the outer atmospheres of cool stars. I. The solar Ca II K line core emission. *Astrophys. J.* 337:964–975.

Schröder, U., Becker, H. W., Bogaert, G., Görres, J. Rolfs, C., Trautvetter, H. P., Azuma, R. E., Campbell, C., King, J. D., and Vise, J. 1987. Stellar reaction rate of $^{14}N(p,\gamma)^{15}O$ and hydrogen burning in massive stars. *Nucl. Phys.* A467:240–260.

Schröter, E. H. 1957. Zur Deutung der Rotverschiebung und der Mitte-Rand-Variation der Fraunhoferlinien bei Berücksichtigung der Temperaturschwankungen der Sonnenatmosphäre. *Z. Astrophys.* 41:141–181.

Schröter, E. H. 1985. The solar differential rotation: Present status of observations. *Solar Phys.* 100:141–169.

Schröter, E. H., Wöhl, H., Soltau, D., and Vázques, M. 1978. An attempt to compare the differential rotation of the Ca^+-network with that of the photospheric plasma. *Solar Phys.* 60:181–201.

Schröter, E.-H., Vázquez, M., and Wyller, A. A., eds. 1987. *The Role of Fine-Scale Magnetic Fields on the Structure of the Solar Atmosphere* (Cambridge: Cambridge Univ. Press).

Schüssler, M. 1979. Magnetic buoyancy revisited: Analytical and numerical results for rising flux tubes. *Astron. Astrophys.* 71:79–91.

Schüssler, M. 1980. Flux tube dynamo approach to the solar cycle. *Nature* 288:150–152.

Schüssler, M. 1980. Flows along magnetic flux tubes. I. Equilibrium and buoyancy of slender magnetic loops in the interior of a star. *Astron. Astrophys.* 89:26–32.

Schüssler, M. 1981. The solar torsional oscillator and dynamo models of the solar cycle. *Astron. Astrophys.* 94:L17-L18.

Schüssler, M. 1984. The interchange instability of small flux tubes. *Astron. Astrophys.* 140:453–458.

Schüssler, M. 1986. MHD models of solar photospheric magnetic flux concentrations. In *Small*

Scale Magnetic Flux Concentrations in the Solar Photosphere, eds. W. Deinzer, M. Knölker and H.-H. Voigt (Göttingen: Vandenhoeck & Ruprecht), pp. 103–120.

Schüssler, M. 1987*a*. Magnetic fields and the rotation of the solar convection zone. In *The Internal Solar Angular Velocity: Theory, Observations and Relationship to Solar Magnetic Fields,* eds. B. R. Durney and S. Sofia (Dordrecht: D. Reidel), pp. 303–320.

Schüssler, M. 1987*b*. Structure and dynamics of small magnetic flux concentrations: Observation versus theory. In *The Role of Fine-Scale Magnetic Fields on the Structure of the Solar Atmosphere,* eds. E.-H. Schröter, M. Vázquez and A. A. Wyller (Cambridge: Cambridge Univ. Press), pp. 223–242.

Schüssler, M., and Solanki, S. K. 1988. Continuum intensity of magnetic flux concentrations: Are magnetic elements bright points? *Astron. Astrophys.* 192:338–342.

Schwarschild, M. 1948. On noise arising from solar granulation. *Astrophys. J.* 107:1–5.

Schwartz, R. A. 1984. High Resolution and Hard X-Ray Spectra of Solar and Cosmic Sources. Ph.D. Thesis, Univ. of California, Berkeley.

Schwarzschild, K. 1906. Uber das gleichgewicht der sonnen atmosphere. In *Göttinger Nachrichten Math. Physs Kl.,* 1–13.

Schwarzschild, M. 1946. On the helium content of the Sun. *Astrophys. J.* 104:203–207.

Schwarzschild, M. 1948. On noise arising from the solar granulation. *Astrophys. J.* 107:1–5.

Schwarzschild, M. 1958. *Structure and Evolution of Stars* (Princeton: Princeton Univ. Press).

Schwarzschild, M. 1961. Convection in stars. *Astrophys. J.* 134:1–8.

Schwarzschild, M. 1965. *Structure and Evolution of the Stars* (New York: Dover).

Schwarzschild, M. 1959. Photographs of the solar granulation taken from the stratosphere. *Astrophys. J.* 130:345–363.

Schwarzschild, M. 1975. On the scale of photospheric convection in red giants and supergiants. *Astrophys. J.* 195:137–144.

Schwarzschild, M., and Härm, R. 1962. Red giants of Population II. *Astrophys. J.* 136:152–157.

Schwarzschild, M., and Härm, R. 1965. Thermal instabilities in non-degenerate stars. *Astrophys. J.* 142:855–867.

Schwarzschild, M., Howard, R., and Härm, R. 1956. A solar model with convective envelope and inhomogeneous interior. *Astrophys. J.* 125:233–241.

Schwarzschild, M., Howard, R., and Härm, R. 1957. Inhomogeneous stellar models. V. A solar model with convective envelope and inhomogeneous interior. *Astrophys. J.* 125:233–259.

Schwenn, R. 1983. The "average" solar wind in the inner heliosphere: Structures and slow variations. In *Proc. Fifth Intl. Solar Wind Conf.,* ed. M. Neugebauer, NASA CP-2280, pp. 489–508.

Schwenn, R. 1983. Direct correlation between coronal transients and interplanetary disturbances. *Space Sci. Rev.* 34:85–99.

Schwenn, R. 1986. Relationship of coronal transients to interplanetary shocks: 3D aspects. In *The Sun and the Heliosphere in Three Dimensions,* ed. B. G. Marsden (Dordrecht: D. Reidel), pp. 119–121.

Scott, R. 1976. Possibility of using [81]Kr to detect solar neutrinos. *Nature* 264:729–730.

Scuflaire, R. 1974. The non radial oscillations of condensed polytropes. *Astron. Astrophys.* 36:107–111.

Scuflaire, R., Gabriel, M., Noëls, A., and Boury, A. 1975. Oscillatory periods in the Sun and theoretical models with or without mixing. *Astron. Astrophys.* 45:15–18.

Sears, R. L. 1964. Helium content and neutrino fluxes in solar models. *Astrophys. J.* 140:477–484.

Seaton, M. 1987. Atomic data for opacity calculations. I. General description. *J. Phys. B: Atom. Molec. Phys.* 20:6363–6378.

Seehafer, N. 1978. Determination of constant α force-free solar magnetic fields from magnetograph data. *Solar Phys.* 58:215–223.

Seehafer, N. 1982. A comparison of different solar magnetic field extrapolation procedures. *Solar Phys.* 81:69–80.

Seidel, E., Demarque, P., and Weinberg, D. 1987. Evolution of red clump stars: Theoretical sequences. *Astrophys. J. Suppl.* 63:917–945.

Sekii, T., and Shibahashi, H. 1988. An inversion method based on the Moore-Penrose generalized inverse matrix. In *Seismology of the Sun and Sun-Like Stars,* ed. E. J. Rolfe, ESA SP-286, pp. 521–523.

Seikii, T., and Shibahashi, H. 1989. An asymptotic inversion method of inferring the sound velocity distribution in the Sun from p-mode oscillations spectrum. *Publ. Astron. Soc. Japan* 41:311–331.

Semel, M. 1967. Contributions à l'étude des champs magnétiques dans les régions actives solaires. *Ann. Astrophys.* 30:1–39.

Semel, M. 1985. Determination of magnetic fields in unresolved features. In *High Resolution in Solar Physics,* ed. R. Müller (Berlin: Springer-Verlag), pp. 178–197.

Severny, A. B. 1957. Some results of investigations of nonstationary processes on the Sun. *Astron. Z.* 34:684–693.

Severny, A. B. 1965. The sunspot magnetic field. *Soviet Astron. AJ* 9:171–182.

Severny, A. B. 1966. An investigation of the general magnetic field of the Sun. *Izv. Krymsk. Astrofiz. Obs.* 35:97–138.

Severny, A. B. 1968. Magnetic asymmetry and variations of the general magnetic field of the Sun. *Izv. Krymsk. Astrofiz. Obs.* 38:3–51.

Severny, A. B. 1969. Is the Sun a magnetic rotator? *Nature* 224:53–54.

Severny, A. B. 1971. The polar fields and time fluctuations of the general field of the Sun. In *Solar Magnetic Fields: Proc. IAU Symp. 43,* ed. R. Howard (Dordrecht: D. Reidel), pp. 675–695.

Severny, A.B., Kotov, V.A., and Tsap, T.T. 1976. Observations of solar pulsations. *Nature* 259:87–89.

Severny, A.B., Kotov, V.A., and Tsap, T.T. 1984. Power spectrum of long period solar oscillations and 160-min pulsations during 1974–1982. *Nature* 307:247–249.

Sevin, E. 1946. Astronomie—Sur la structure du système solaire. *Compt. Rend.* 222:220–221.

Share, G.H., Matz, S.M., Messina, D.E., Nolan, P.L., Chupp, E.L., Forrest, D.J., and Cooper, J.F. 1986. SMM observation of a cosmic gamma-ray burst from 20 keV to 100 MeV. *Adv. Space Res.* 6:15–18.

Shaviv, G., and Salpeter, E.E. 1973. Convective overshooting in stellar interior models. *Astrophys. J.* 184:191–200

Sheeley, N. R. 1969. The evolution of the photospheric network. *Solar Phys.* 9:347–357.

Sheeley, N. R., Jr., and DeVore, C. R. 1986. The origin of the 28- to 29-day recurrent patterns of the solar magnetic field. *Solar Phys.* 104:425–429.

Sheeley, N. R., Jr., and Harvey, J. 1981. Coronal holes, solar wind streams, and geomagnetic disturbances during 1978 and 1979. *Solar Phys.* 70:237–249.

Sheeley, N. R., Jr., Howard, R. A., Michels, D. J., and Koomen, M. J. 1980. Solar observations with a new Earth-orbiting coronagraph. In *Solar and Interplanetary Dynamics,* eds. M. Dryer and E. Tandberg-Hanssen (Hingham, Mass.: D. Reidel), pp. 55–59.

Sheeley, N. R., Jr., DeVore, C. R., and Boris, J. P. 1985. Simulations of the mean solar magnetic field during sunspot cycle 21. *Solar Phys.* 98:219–239.

Sheeley, N. R., Jr., DeVore, C. R., and Shampine, L. R. 1986a. Simulations of the gross solar magnetic field during sunspot cycle 21. *Solar Phys.* 106:251–268.

Sheeley, N. R., Jr., Howard, R. A., Koomen, M. J., and Michels, D. J. 1986b. Solwind observations of coronal mass ejections during 1975–1985. In *Solar Flares and Coronal Physics Using P/OF as a Research Tool,* eds. E. Tandberg-Hanssen, R. Wilson and H. Hudson, NASA CP-2421, pp. 241–256.

Sheeley, N. R., Jr., Nash, A. G., and Wang, Y.-M. 1987. The origin of rigidly rotating magnetic field patterns on the Sun. *Astrophys. J.* 319:481–502.

Sheeley, N. R., Jr., Wang, Y.-M., and Harvey, J. W. 1989. The effect of newly erupting flux on the polar coronal holes. *Solar Phys.* 119:323–340.

Sheldon, W. R. 1969. Possible relation of a null solar neutrino flux to the 11 year solar cycle. *Nature* 221:650–651.

Shelke, R. N., and Pande, M. C. 1986. Semi regularities in the photospheric magnetic fields. *Solar Phys.* 105:257–263.

Shibahashi, H. 1979. Modal analysis of stellar nonradial oscillations by an asymptotic method. *Publ. Astron. Soc. Japan* 31:87–104.

Shibahashi, H. 1988. Inverse problem: Acoustic potential vs acoustic length. In *Advances in Helio- and Asteroseismology,* eds. J. Christensen-Dalsgaard and S. Frandsen (Dordrecht: D. Reidel), pp. 133–136.

Shibahashi, H., and Osaki, Y. 1976. Overstability of gravity modes in massive stars with the semi-convective zone. *Publ. Astron. Soc. Japan* 28:199–214.

Shibahashi, H., and Osaki, Y. 1981. Theoretical eigenfrequencies of solar oscillations of low harmonic degree ℓ in five-minute range. *Publ. Astron. Soc. Japan* 33:713–719.

Shibahashi, H., and Sekii, T. 1988. Sound velocity distribution in the Sun inferred from asymptotic inversion of p-mode spectrum. In *Seismology of the Sun and Sun-Like Stars*, ed. E. J. Rolfe, ESA SP-286, pp. 471–474.

Shibahashi, H., Osaki, Y., and Unno, W. 1975. Nonradial g-mode oscillations and the stability of the Sun. *Publ. Astron. Soc. Japan* 27:401–410.

Shibahashi, H., Noëls, A., and Gabriel, M. 1983. Influence of the equation of state and of the Z value on the solar five-minute oscillation. *Astron. Astrophys.* 123:283–288.

Shibata, K. 1980. On the origin of strong downdrafts associated with the birth of sunspots. *Solar Phys.* 66:61–70.

Shibata, K., Tajima, T., Matsumoto, R., Horiuchi, T., Hanawa, T., Rosner, R., and Uchida, Y. 1989. Nonlinear Parker instability of isolated magnetic flux in a plasma. *Astrophys. J.* 388:471–492.

Shine, R. A. 1975. The effect of intermediate-scale motions on line formation. *Astrophys. J.* 202:543–550.

Shine, R. A., and Schrijver, C. 1988. Active region evolution in the chromosphere and transition region. In *Max '91: Flare Research at the Next Solar Maximum, Workshop #1: Scientific Objectives*, eds. R. C. Canfield and B. R. Dennis, pp. 29–32.

Shine, R.A., Milkey, R. W., and Mihalas, D. 1975. Resonance line transfer with partial redistribution. IV. A generalized formulation for lines with common upper states. *Astrophys. J.* 199:718–723.

Shine, R. A., Title, A. M., Tarbell, T. D., and Topka, K. P. 1987. White light sunspot observations from the solar optical universal polarimeter on Spacelab 2. *Science* 238:1264–1267.

Shoub, E. C. 1983. Invalidity of local thermodynamic equilibrium for electrons in the solar transition region. I. Fokker-Planck results. *Astrophys. J.* 266:339–369.

Shu, F. H. 1977. Self-similar collapse of isothermal spheres and star formation. *Astrophys. J.* 214:488–497.

Shu, F. H., Adams, F. C., and Lizano, S. 1987. Star formation in molecular clouds: Observation and theory. *Ann. Rev. Astron. Astrophys.* 25:23–81.

Shurcliff, W. A. 1962. *Polarized Light* (Cambridge, Mass.: Harvard Univ. Press).

Sienkiewicz, R., Paczynski, B., and Ratcliff, S. J. 1988. Neutrino emission from solar models with a metal-depleted core. *Astrophys. J.* 326:392–394.

Sienkiewicz, R., Bahcall, J. N., and Paczynski, B. 1990. Mixing and the solar neutrino problem. *Astrophys. J.* 349:641–646.

Sime, D. G. 1985. The corona and interplanetary medium during the solar cycle. In *Future Missions in Solar, Heliospheric and Space Plasma Physics*, eds. E. Rolfe and B. Battrick, pp. 23–26.

Sime, D. G. 1989. Coronal mass ejections and the evolution of the large-scale K-coronal density distribution. *J. Geophys. Res.* 94:151–158.

Sime, D. G., and Hundhausen, A. J. 1987. The coronal mass ejection of July 6, 1980: A candidate for interpretation as a coronal shock wave. *J. Geophys. Res* 92:1049–1056.

Sime, D. G., MacQueen, R. M., and Hundhausen, A. J. 1984. Density distribution in loop-like coronal transients: A comparison of observations and a theoretical model. *J. Geophys. Res.* 89:2113–2121.

Sime, D. G., Fisher, R. R., and Altrock, R. C. 1989. Rotation characteristics of the Fe XIV (5303Å) solar corona. *Astrophys. J.* 336:454–467.

Simnett, G. M. 1986. A dominant role for protons at the onset of solar flares. *Solar Phys.* 106:165–183.

Simnett, G. M., and Harrison, R. A. 1984. The relationship between coronal mass ejections and solar flares. *Adv. Space Res.* 4:279–285.

Simon, G. W., and Leighton, R. B. 1964. Velocity fields in the solar atmosphere. III. Large-scale motions, the chromospheric network, and magnetic fields. *Astrophys. J.* 140:1120–1147.

Simon, G. W., and Weiss, N. O. 1968. Supergranules and the hydrogen convection zone. *Zeit. Astrophysik* 69:435–450.

Simon, G. W., and Weiss, N. O. 1989a. Simulation of large-scale flows at the solar surface. *Astrophys. J.* 345:1060–1078.

Simon, G. W., and Weiss, N. O. 1989b. Simulating plumes and sinks observed at the solar surface. In *High Spatial Resolution Solar Observations,* ed. O. von der Lühe (Sunspot, N.M.: Sacramento Peak Obs.), pp. 529–539.

Simon, G. W., and Wilson, P. R. 1985. Flux changes in small magnetic regions. II. Further observations and analysis. *Astrophys. J.* 295:241–257.

Simon, G. W., Schmieder, B., Demoulin, P., and Poland, A. I. 1986a. Dynamics of filaments. VI. Center-to-limb study of Hα and CIV velocities in a quiescent filament. *Astron. Astrophys.* 166:319–325.

Simon, G. W., Gesztelyi, L., Schmieder, B., and Mein, N. 1986b. Filament eruption connected to photospheric activity. In *Coronal and Prominence Plasmas,* ed. A. I. Poland, NASA CP-2442, pp. 229–233.

Simon, G. W., Title, A. M., Tokpa, K. P., Tarbell, T. D., Shine, R. A., Ferguson, S. H., Zirin, H., and the SOUP Team. 1988. On the relation between photospheric flow fields and the magnetic field distribution on the solar surface. *Astrophys. J.* 327:964–967.

Simon, M., and Zirin, H. 1969. The coarse structure of the solar atmosphere. *Solar Phys.* 9: 317–327.

Simon, M., and Shimabukuro, F. I. 1971. Observations of the solar oscillatory component at a wavelength of 3 millimeters. *Astrophys. J.* 168:525–529.

Simon, P. A. 1979. Polar coronal holes and solar cycles. *Solar Phys.* 63:339–410.

Simon, P. A., Heckman, G., and Shea, M. A., eds. 1986. *Solar-Terrestrial Predictions* (Boulder: Natl. Oceanic Atmos. Admin.).

Simon, P. C. 1989. Solar spectral irradiance for middle atmosphere studies. COSPAR XXVII. *Advances in Space Research,* in press.

Simon, P. C., Rottmann, G. J., White, O. R., and Knapp, B. K. 1988. Short term variability between 120 and 300 nm from SME observations. In *Solar Radiative Output Variations,* ed. P. V. Foukal (Cambridge, Mass.: Cambridge Research and Instrumentation), pp. 125–129.

Simon, T. 1988. In *A Decade of UV Astronomy with the IUE Satellite,* vol. 1, eds. N. Longdon and E. J. Rolfe, ESA SP-281, p. 279.

Simon, T., and Fekel, F. C. 1987. The dependence of ultraviolet chromospheric emission upon rotation among late-type stars. *Astrophys. J.* 316:434–448.

Simon, T., Herbig, G., and Boesgaard, A. M. 1985. The evolution of chromospheric activity and the spin-down of solar-type stars. *Astrophys. J.* 293:551–574.

Singh, J., and Bappu, M. K. V. 1981. A dependence on solar cycle of the size of the CaII network. *Solar Phys.* 71:161–187.

Singh, J., and Livingston, W. C. 1987. Sun as a star: Rotation rates from the calcium K index. *Solar Phys.* 109:387–391.

Singh, J., and Prabhu, T. P. 1985. Variations in the solar rotation rate derived from Ca$^+$K plage areas. *Solar Phys.* 97:203–212.

Sivaraman, K. R. 1984. Observations of the Evershed effect—a past and present. *Kodaikanal Obs. Bull.* 4:11–17.

Sivaraman, K. R., and Livingston, W. C. 1982. CA II K$_{2v}$ spectral features and their relation to small-scale photospheric magnetic fields. *Solar Phys.* 80:227–231.

Sivaraman, K. R., Bagare, S. P., Gupta, S. S., and Kariyappa, R. 1987. Calibration on the Sun for stellar magnetic fields. In *Cool Stars, Stellar Systems and the Sun,* eds. J. L. Linsky and R. E. Stencel (Berlin: Springer-Verlag), pp. 47–50.

Skumanich, A. 1972. Time scales for CA II emission decay, rotational braking, and lithium depletion. *Astrophys. J.* 171:565–567.

Skumanich, A., and Lites, B. W. 1987a. Stokes profile analysis and vector magnetic fields. I. Inversion of photospheric lines. *Astrophys. J.* 322:473–482.

Skumanich, A., and Lites, B. W. 1987b. The polarization properties of model sunspots: The broad-band polarization signature of the Schlüter-Temesvàry representation. *Astrophys. J.* 322:483–493.

Skumanich, A., Smythe, C., and Frazier, E. N. 1975. On the statistical description of inhomogeneities in the quiet solar atmosphere. I. Linear regression analysis and absolute calibration of multichannel observations of the Ca$^+$ emission network. *Astrophys. J.* 200:747–764.

Skumanich, A., Lean, J. L., White, O. R., and Livingston, W. C. 1984. The Sun as a star:

Three component analysis of chromospheric variability in the CA II K line *Astrophys. J.* 282:776–783.

Skumanich, A., Rees, D. E., and Lites, B. W. 1985. Least squares inversion of Stokes profiles in the presence of velocity gradients. In *Measurements of Solar Vector Magnetic Fields,* ed. M. J. Hagyard, NASA CP-2374, pp. 306–321.

Slansky, R. C. 1989. Comment on ^{37}Ar as a calibration source for solar neutrino dectors. *Phys. Rev.* C39:2080.

Slottje, C. 1978. Millisecond microwave spikes in a solar flare. *Nature* 275:520–521.

Slottje, C. 1981. Atlas of dynamic fine structure. Dwingeloo, NFRA.

Smagorinsky, J. S. 1963. General circulation experiments with the primitive equations. I. The basic experiment. *Mon. Weather Rev.* 91:99–165.

Smartt, R. N. 1989. Measurement of telescope system aberrations. In *High Spatial Resolution Solar Observations,* ed. O. von der Lühe (Sunspot, N.M.: Sacramento Peak Obs.), pp. 232–238.

Smartt, R. N., and Zhang, Z. 1987. Loop interaction in the visible emission corona—morphological details. In *High Resolution Solar Physics,* eds. R. G. Athay and D. S. Spicer, NASA CP-2483, pp. 129–133.

Smeyers, P. 1984. Introductory report: Non-radial oscillations. In *Theoretical Problems in Stellar Stability and Oscillations,* eds. A. Noëls and M. Gabriel (Liège: Inst. d'Astrophysique), pp. 68–91.

Smeyers, P., and Tassoul, M. 1987. Asymptotic approximations for higher order nonradial oscillations of a spherically symmetric star. *Astrophys. J. Suppl.* 65:429–449.

Smith, D. F. 1980. First phase acceleration mechanisms and implications for hard X-ray burst models in solar flares. *Solar Phys.* 66:135–148.

Smith, D. F., and Brecht, S. H. 1986. Acceleration, storage, and release of solar flare protons: Erratum. *Astrophys. J.* 306:317–322.

Smith, D. F., and Brecht, S. H. 1988. Shock waves versus stochastic acceleration of impulsive solar flare protons. *Solar Phys.* 115:133–148.

Smith, D. F., and Lilliequist, C. G. 1979. Confinement of hot, hard X-ray producing electrons in solar flares. *Astrophys. J.* 232:582–589.

Smith, G. H. 1987. The chemical inhomogeneity of globular clusters. *Publ. Astron. Soc. Pacific* 99:67–90.

Smith, J. A., and Demarque, P. 1980. CH-subgiants and the mixing hypothesis. *Astron. Astrophys.* 92:163–166.

Smithson, R. C., Marshall, N. K., and Sharbaugh, R. J. 1984. A 57 actuator active mirror for solar astronomy. In *Small-Scale Dynamical Processes in Quiet Stellar Atmospheres,* ed. S. L. Keil (Sunspot, N.M.: Sacramento Peak Obs.), pp. 66–73.

Snodgrass, H. B. 1983. Magnetic rotation of the solar photosphere. *Astrophys. J.* 270:288–299.

Snodgrass, H. B. 1984. Separation of large-scale photospheric Doppler patterns. *Solar Phys.* 94:13–31.

Snodgrass, H. B. 1985. Solar torsional oscillations: A net pattern with wavenumber 2 as artifact. *Astrophys. J.* 291:339–343.

Snodgrass, H. B. 1986. Spectroscopic evidence for a moving pattern of azimuthal convective rolls on the Sun. *Astrophys. J.* 316:L13–L31.

Snodgrass, H. B. 1987*a*. Spectroscopic evidence for a moving pattern of azimuthal rolls on the Sun. *Astrophys. J.* 316:L91–L94.

Snodgrass, H. B. 1987*b*. Torsional oscillations and the solar cycle. *Solar Phys.* 110:35.

Snodgrass, H. B., and Howard, R. 1984. Limits on photospheric Doppler signatures for solar giant cells. *Astrophys. J.* 284:848–855.

Snodgrass, H. B., and Howard, R. 1985. Torsional oscillations of the Sun. *Science* 228: 945–952.

Snodgrass, H. B., and Wilson, P. R. 1987. Solar torsional oscillations as a signature of giant cells. *Nature* 328:696–699.

Soderblom, D. R. 1983. Rotational studies of late-type stars. II. Ages of solar type stars and the rotational history of the Sun. *Astrophys. J. Suppl.* 53:1–15.

Soderblom, D. R. 1985. A survey of chromospheric emission and rotation among solar-type stars in the solar neighborhood. *Astron. J.* 90:2103–2115.

Soderblom, D. R., Jones, B. F., and Walker, B. F. 1983. Rapid rotation among Pleiades K dwarfs. *Astrophys. J.* 274:L37–L41.

Soderblom, D. R., Duncan, D. K., and Johnson, D. R. H. 1989. *Astrophys. J.,* in press.

Sofia, S., and Chan, K. L. 1984. Turbulent compressible convection in a deep atmosphere. II. Two-dimensional results for main-sequence *Astrophys. J.* 282:550–556.

Sofia, S., Chiu, H.-Y., Maier, E., Schatten, K. H., Minott, P., and Endal, A. S. 1984. Solar disk sextant. *Appl. Opt.* 23:1235–1237.

Solanki, S. K. 1985. High spectral resolution and properties of small magnetic flux tubes. In *Theoretical Problems in High Resolution Solar Physics,* ed. H. U. Schmidt, pp. 172–175.

Solanki, S. K. 1986. Velocities in solar magnetic fluxtubes. *Astron. Astrophys.* 168:311–329.

Solanki, S. K. 1987. Structure of magnetic flux tubes as derived from observations with moderate spatial resolution. In *The Role of Fine-Scale Magnetic Fields on the Structure of the Solar Atmosphere,* eds. E.-H. Schröter, M., Vázquez and A. A. Wyller (Cambridge: Cambridge Univ. Press), pp. 67–81.

Solanki, S. K. 1988. Magnetic fields: Observations and theory. In *The Sun: 10th European Regional Meeting of the IAU,* Publ. No. 66 of Astron. Inst. of Czech. Acad. Sci., eds. L. Hejna and M. Sobotka (Ondrejov: Astronomical Inst. Czechoslovak Academy of Sciences), pp. 95–102.

Solanki, S. K., and Pahlke, K. D. 1988. Can stationary velocity fields explain the Stokes V asymmetry observed in solar magnetic elements? *Astron. Astrophys.* 201:143–152.

Solanki, S. K., and Steenbock, W. 1988. NLTE effects in solar magnetic fluxtubes. *Astron. Astrophys.* 189:243–253.

Solanki, S. K., and Stenflo, J. O. 1984. Properties of solar magnetic fluxtubes as revealed by Fe I lines. *Astron. Astrophys.* 140:185–198.

Solanki, S. K., Keller, C., and Stenflo, J. O. 1987. Properties of solar magnetic fluxtubes from only two spectral lines. *Astron. Astrophys.* 188:183–197.

Soltau, D. 1989. High resolution observations with the new German VTT on Teneriffe. In *High Spatial Resolution Solar Observations,* ed. O. von der Lühe (Sacramento Peak, N.M.: Sacramento Peak Obs.), pp. 3–11.

Sonett, C. P., and Williams, G. E. 1987. Frequency modulation and stochastic variability of the Elatina varve record: A proxy for solar cyclicity? *Solar Phys.* 110:397–410.

Sonett, C. P., Giampapa, M. S., and Matthews, M. S., eds. 1991. *The Sun In Time* (Tucson: Univ. of Arizona Press).

Soru-Escaut, I., Martres, M.-J., and Mouradian, Z. 1985. Singularity of solar rotation and flare productivity. *Astron. Astrophys.* 145:19–24.

Soward, A. M. 1987. Fast dynamo action in steady flow. *J. Fluid Mech.* 180:267–295.

Speiser, T. W. 1965. Particle trajectories in model current sheets. *J. Geophys. Res.* 70:4219–4226.

Spergel, D. N. 1990. Solar cosmions. In *Inside the Sun: Proc. IAU Coll. 121,* eds. G. Berthomieu and M. Cribier (Dordrecht: Kluwer), pp. 145–152.

Spergel, D. N., and Faulkner, J. 1988. Weakly interacting massive particles in horizontal-branch stars. *Astrophys. J.* 331:L21–L24.

Spergel, D. N., and Press, W. H. 1985. Effect of hypothetical, weakly interacting, massive particles on energy transport in the solar interior. *Astrophys. J.* 294:663–673.

Spicer, D. S. 1983. Magnetic energy storage and thermal versus non-thermal hard X-ray hypothesis. *Adv. Space Res.* 2:135–137.

Spicer, D. S., and Emslie, A. G. 1988. A new quasi-thermal trap model for solar flare hard X-ray bursts: An electrostatic trap model. *Astrophys. J.* 330:997–1007.

Spicer, D. S., and Sudan, R. N. 1984. Beam-return current systems in solar flares. *Astrophys. J.* 280:448–456.

Spiegel, E. A. 1957. The smoothing of temperature fluctuations by radiative transfer. *Astrophys. J.* 126:202–207.

Spiegel, E. A. 1963. A generalization of the mixing length theory of turbulent convection. *Astrophys. J.* 138:216–225.

Spiegel, E. A., and Unno, W. 1962. On convective growth-rates in a polytropic atmosphere. *Publ. Astron. Soc. Japan* 14:28–32.

Spiegel, E. A., and Veronis, G. 1960. On the Boussinesq approximation for a compressible fluid. *Astrophys. J.* 31:442–447.

Spinka, H., Tombrello, T. A., and Winkler, H. 1971. Low-energy cross sections for $^7Li(p,\alpha)^4He$ and $^6Li(p,\alpha)^3He$. *Nucl. Phys.* A164:1–10.

Spitzer, L. 1962. *Physics of Fully Ionized Gases* (New York: Interscience).

Sprangle, R., and Vlahos, L. 1983. Electron cyclotron wave acceleration outside a flaring loop. *Astrophys. J.* 273:L95–L99.

Spruit, H. C. 1974. A model of the solar convection zone. *Solar Phys.* 34:277–290.

Spruit, H. C. 1976. Pressure equilibrium and energy balance of small photospheric fluxtubes. *Solar Phys.* 50:269–295.

Spruit, H. C. 1977. Heat flow near obstacles in the solar convection zone. *Solar Phys.* 55:3–34.

Spruit, H. C. 1979. Convective collapse of flux tubes. *Solar Phys.* 61:363–378.

Spruit, H. C. 1981*a*. A cluster model for sunspots. In *The Physics of Sunspots*, eds. L. Cram and J. H. Thomas (Sunspot, N.M.: Sacramento Peak Obs.), pp. 98–103.

Spruit, H. C. 1981*b*. Magnetic flux tubes. In *The Sun as a Star*, ed. S. Jordan, NASA SP-450, pp. 385–412.

Spruit, H. C. 1981*c*. Motion of magnetic flux tubes in the convection zone and chromosphere of the Sun. *Solar Phys.* 98:155–160.

Spruit, H. C. 1981*d*. Small scale phenomena in umbras and penumbras. In *The Physics of Sunspots*, eds. L. Cram and J. H. Thomas (Sunspot, N.M.: Sacramento Peak Obs.), pp. 359–368.

Spruit, H. C. 1984*a*. The interaction of flux tubes with convection. In *Small-Scale Dynamical Processes in Quiet Stellar Atmospheres*, ed. S. L. Keil (Sunspot, N.M.: Sacramento Peak Obs.), p. 249–259.

Spruit, H. C. 1984*b*. Mixing in the solar interior. In *The Hydromagnetics of the Sun*, eds. T. D. Guyenne and J. J. Hunt, ESA SP-220, pp. 21–27.

Spruit, H. C. 1987. Angular momentum transport in the radiative interior of the Sun. In *The Internal Solar Angular Velocity*, eds. B. R. Durney and S. Sofia (Dordrecht: D. Reidel), pp. 185–200.

Spruit, H. C. 1988. Influence of magnetic activity on the solar radius and luminosity. In *Solar Radiative Output Variations*, ed. P. V. Foukal (Cambridge, Mass.: Cambridge Research and Instrumentation), pp. 254–288.

Spruit, H. C. 1990. Angular momentum and transport and magnetic fields in the solar interior. In *Inside the Sun: Proc. IAU Coll. 121*, eds. G. Berthomieu and M. Cribier (Dordrecht: Kluwer), pp. 415–423.

Spruit, H. C., and Roberts, B. 1983. Magnetic flux tubes on the Sun. *Nature* 304:401–406.

Spruit, H. C., and van Ballegooijen, A. A. 1982. Stability of toriodal flux tubes in stars. *Astron. Astrophys.* 106:58–66.

Spruit, H. C., and Zwaan, C. 1981. The size dependence of contrasts and numbers of small magnetic flux tubes in an active region. *Solar Phys.* 70:207–228.

Spruit, H. C., and Zweibel, E. G. 1979*a*. Convective instability of thin flux tubes. *Solar Phys.* 62:15–22.

Spruit, H. C., and Zweibel, E. G. 1979*b*. The size dependence of contrasts and numbers of small magnetic flux tubes in an active region. *Solar Phys.* 70:207–228.

Spruit, H. C., van Ballegooijen, A. A., and Title, A. M. 1987. Is there a weak mixed polarity background field? Theoretical arguments. *Solar Phys.* 110:115–127.

Spruit, H. C., Nordlund, Å., and Title, A. M. 1990. Solar convection. *Ann. Rev. Astron. Astrophys.* 28:263–301.

Stachnik, R. V., Nisenson, P., Ehn, D. C., Hudgin, R. H., and Schirf, V. E. 1977. Speckle image reconstruction of solar features. *Nature* 266:149–151.

Stachnik, R. V., Nisenson, P., and Noyes, R. W. 1983. Speckle image reconstruction of solar features. *Astrophys. J.* 271:L37–L40.

Stahler, S. W. 1983. The birthline for low-mass stars. *Astrophys. J.* 274:822–829.

Stahler, S. W. 1988*a*. Deuterium and the stellar birthline. *Astrophys. J.* 332:804–825.

Stahler, S. W. 1988*b*. Understanding young stars: A history. *Publ. Astron. Soc. Pacific* 100:1474–1485.

Stahler, S. W., Shu, F. H., and Taam, R. E. 1980*a*. The evolution of protostars. I. Global formulation and results. *Astrophys. J.* 241:637–654.

Stahler, S. W., Shu, F. H., and Taam, R. E. 1980*b*. The evolution of protostars. II. The hydrostatic core. *Astrophys. J.* 242:226–241.

Stahler, S. W., Shu, F. H., and Taam, R. E. 1981. The evolution of protostars. III. The accretion envelope. *Astrophys. J.* 248:727–737.

Stähli, M., Gary, D. E., and Hurford, G. J. 1989. High-resolution microwave spectra of solar bursts. *Solar Phys.* 120:351–368.

Staiger, J. 1985. Wellenausbreitung in der Sonnenatmosphaere. Ein Beispiel fuer den Einsatz von Halbleiterdetektoren in der Sonnenphysic. Thesis, Freiburg.

Staiger, J. 1987. Observations of oscillatory phase shifts with diode arrays. *Astron. Astrophys.* 175:263–270.

Staiger, J., Schmieder, B., Deubner, F.-L., and Mattig, W. 1984. Phase spectra of solar oscillations. *Mem. Soc. Astron. Italiana* 55:147–152.

Staude, J., Fürstenberg, F., Hildebrandt, J., Krüger, A., Jakimiec, J., Obridko, V. N., Siarkowski, M., Sylwester, B., and Sylwester, J. 1983. A working model of sunspot structure in photosphere, chromosphere and corona, derived from X-ray, EUV, optical and radio observations. *Acta Astron.* 33:441–460.

Stauffer, J. R. 1987. Rotational velocity evolution on and prior to the main sequence. In *Cool Stars, Stellar Systems and the Sun: Proc. 5th Cambridge Workshop,* eds. J. L. Linsky and R. E. Stencel, pp. 182–191.

Stauffer, J. R., and Hartmann, L. W. 1986. Rotational studies of lower main-sequence stars. *Publ. Astron. Soc. Pacific* 91:737–745.

Stauffer, J. R., and Hartmann, L. W. 1986. The rotational velocities of low-mass stars. *Publ. Astron. Soc. Pacific* 96:1233–1251.

Stauffer, J. R., Hartmann, L., Soderblom, D. R., and Burnham, N. 1984. Rotational velocities of low mass stars in the Pleiades. *Astrophys. J.* 280:202–212.

Stauffer, J. R., Hartmann, L., Burnham, N., and Jones, B. 1985. Evolution of low-mass stars in the Alpha Persei Cluster. *Astrophys. J.* 289:247–261.

Stebbins, R. T., and Goode, P. R. 1987. Waves in the solar photosphere. *Solar Phys.* 110: 237–253.

Stebbins, R. T., and Wilson, C. 1983. The measurement of long period oscillations at Sacramento Peak and the South Pole. *Solar Phys.* 82:43–54.

Steenbeck, M., and Krause, F. 1969a. Zur Dynamotheorie stellarer und planetarer Magnetfelder. I. Berechnung sonnenahnlicher Wechselfeldgeneratoren. *Astron. Nachr.* 291:49–84.

Steenbeck, M., and Krause, F. 1969b. Zur Dynamotheorie stellarer und planetarer Magnetfelder. II. Berechnung planetenahnlicher Gleichfeldgeneratoren. *Astron. Nachr.* 291:271–286.

Steenbeck, M., Krause, F., and Radler, K.-H. 1966. Berechnung der Mittleren Lorentz-Feldstarke $V \times B$ für ein elktrisch leitendes Medium in turbulenter, durch Coriolis-Krafte beeinflubter Bewegung. *Z. Naturforschung* 21a:369–376.

Stefanik, R. P., Ulmschneider, P., Hammer, R., and Durrant, C. J. 1984. Nonlinear two-dimensional dynamics of stellar atmospheres. I. A computational code. *Astron. Astrophys.* 134:77–86.

Steffen, M. 1987. In *The Role of Fine-Scale Magnetic Fields on the Structure of the Solar Atmosphere,* eds. E.-H. Schröter, M. Vázquez and A. .A. Wyller (Cambridge: Cambridge Univ. Press), pp. 47.

Steffen, M. 1988. Interactions of convection and oscillations in the solar atmosphere: Numerical results. In *Advances in Helio- and Asteroseismology: Proc. IAU Symp. 123,* eds. J. Christensen-Dalsgaard and S. Frandsen (Dordrecht: D. Reidel), pp. 379–382.

Steffen, M., and Muchmore, D. 1988. Can granular fluctuations in the solar photosphere produce temperature inhomogeneities at the height of the temperature minimum? *Astron. Astrophys.* 193:281–290.

Steffen, M., Ludwig, H.-G., and Kruess, A. 1989. A numerical simulation study of granular convection in cells of different horizontal dimension. *Astron. Astrophys.* 193:281–290.

Steigman, G., Sarazin, C. L., Quintana, H., and Faulkner, J. 1978. Dynamical interactions and astrophysical effects of stable heavy neutrinos. *Astron. J.* 83:1050–1061.

Stein, R. F. 1967. Generation of acoustic and gravity waves by turbulence in an isothermal stratified atmosphere. *Solar Phys.* 2:385–432.

Stein, R. F. 1968. Waves in the solar atmosphere. I. The acoustic energy flux. *Astrophys. J.* 154:297–306.

Stein, R. F. 1981. Stellar chromospheric and coronal heating by magnetohydrodynamic waves. *Astrophys. J.* 246:966–971.

Stein, R. F. 1985. Mechanisms for chromospheric heating. In *Chromospheric Diagnostics and Modelling,* ed. B. W. Lites (Sunspot, N.M.: National Solar Obs.), pp. 213–227.

Stein, R. F., and Leibacher, J. 1974. Waves in the solar atmosphere. *Ann. Rev. Astron. Astrophys.* 12:407–435.

Stein, R. F., and Leibacher, J. 1981. Wave generation. In *The Sun as a Star,* ed. S. D. Jordan, NASA SP-450, pp. 289–300.

Stein, R. F., and Nordlund, Å. 1989. Topology of convection beneath the solar surface. *Astrophys. J.* 342:L95-L98.

Stein, R. F., and Schwartz, R. A. 1972. Waves in the solar atmosphere. II. Large-amplitude acoustic pulse propagation. *Astrophys. J.* 177:807–828.

Stein, R. F., and Schwartz, R. A. 1973. Waves in the solar atmosphere. III. The propagation of periodic wave trains in a gravitational atmosphere. *Astrophys. J.* 186:1083–1089.

Stein, R. F., Nordlund, Å, and Kuhn, J. R. 1988. Convection and p-mode oscillations. In *Seismology of the Sun and Sun-Like Stars,* ed. E. J. Rolfe, ESA SP-286, pp. 529–532.

Stein, R. F., Nordlund, Å., and Kuhn, J. R. 1989. Convection and waves. In *Solar and Stellar Granulation,* eds. R. J. Rutten and G. Severino (Dordrecht: Kluwer), pp. 381–399.

Steiner, O., and Pizzo, V. J. 1989. A parametric survey of model solar fluxtubes. *Astron. Astrophys.* 211:447–462.

Steiner, O., and Stenflo, J. O. 1990. Model calculations of the photospheric layers of solar magnetic fluxtubes. In *Solar Photosphere: Structure, Convection and Magnetic Fields: Proc. IAU Symp. 138,* ed. J. O. Stenflo (Dordrecht: Kluwer).

Steinolfson, R. S. 1985. Theories of shock formation in the solar atmosphere. In *Collisionless Shocks in the Heliosphere: Reviews of Current Research,* eds. B. T. Tsurutani and R. G. Stone (Washington, D.C.: American Geophysical Union), pp. 1–12.

Steinolfson, R. S. 1988. Driving mechanisms for coronal mass ejections. In *Outstanding Problems in Solar System Plasma Physics: Theory and Instrumentation,* eds. J. Birch and J. Hunter, in press.

Steinolfson, R. S., and Hundhausen, A. J. 1988. Density and white-light brightness in loop-like coronal mass ejections: Temporal evolution. *J. Geophys. Res.,* in press.

Steinolfson, R. S., and Hundhausen, A. J. 1989. Waves in low-beta plasmas: Slow shocks. *J. Geophys. Res.* 94:1222–1234.

Steinolfson, R. S., and Hundhausen, A. J. 1990. MHD intermediate shocks in coronal mass ejections. *J. Geophys. Res.* 95:6389–6401.

Stellingwerf, R. F. 1975a. Modal stability of RR Lyrae stars. *Astrophys. J.* 195:441–446.

Stellingwerf, R. F. 1975b. Nonlinear effects in double mode Cepheids. *Astrophys. J.* 199:705–709.

Stellingwerf, R. F. 1982. Convection in pulsating stars. I. Nonlinear hydrodynamics. *Astrophys. J.* 262:330–338.

Stellmacher, G., and Wieher, E. 1979. A common model for solar filigree and faculae. *Astron. Astrophys.* 75:263–267.

Stenflo, J. O. 1973. Magnetic-field structure of the photospheric network. *Solar Phys.* 32:41–63.

Stenflo, J. O. 1974. Differential rotation and sector structure of solar magnetic fields. *Solar Phys.* 36:495–515.

Stenflo, J. O. 1975. A model of the supergranulation network and of active-region plages. *Solar Phys.* 42:79–105.

Stenflo, J. O. 1977. Solar-cycle variations in the differential rotation of solar magnetic fields. *Astron. Astrophys.* 61:797–804.

Stenflo, J. O. 1982. The Hanle effect and the diagnostics of turbulent magnetic fields in the solar atmosphere. *Solar Phys.* 80:209–226.

Stenflo, J. O. 1985. Diagnostics of vector magnetic fields. In *Measurements of Solar Vector Magnetic Fields,* ed. M. Hagyard (Huntsville, Ala.: NASA/Marshall Space Flight Center), pp. 263–278.

Stenflo, J. O. 1988. Observational constraints on a "hidden" turbulent magnetic field on the Sun. *Solar Phys.* 114:1–19.

Stenflo, J. O. 1989. Small-scale magnetic structures on the Sun. *Astron. Astrophys. Rev.* 1:3–48.

Stenflo, J. O., ed. 1990. *Solar Photosphere: Structure, Convection, and Magnetic Fields: Proc. IAU Symp. 138* (Dordrecht: Kluwer).

Stenflo, J. O. 1990. Time invariance of the Sun's rotation rate. *Astron. Astrophys.* 233:220–228.

Stenflo, J. O., and Güdel, M. 1988. Evolution of solar magnetic fields: Modal structure. *Astron. Astrophys.* 191:137–148.

Stenflo, J. O., and Harvey, J. W. 1985. Dependence of the properties of magnetic fluxtubes on area factor or amount of flux. *Solar Phys.* 95:99–118.

Stenflo, J. O., and Vogel, M. 1986. Global resonances in the evolution of solar magnetic fields. *Nature* 319:285–290.

Stenflo, J. O., Harvey, J. W., Brault, J. W., and Solanki, S. K. 1984. Diagnostics of solar magnetic fluxtubes using a Fourier transform spectrometer. *Astron. Astrophys.* 131:333–346.

Stenflo, J. O., Solanki, S. K., and Harvey, J. W. 1987a. Center-to-limb variation of Stokes profiles and the diagnostics of solar magnetic fluxtubes. *Astron. Astrophys.* 171:305–316.

Stenflo, J. O., Solanki, S. K., and Harvey, J. W. 1987b. Diagnostics of solar magnetic fluxtubes with the infrared line FE I λ15648.54 Å. *Astron. Astrophys.* 173:167–179.

Stevenson, D. J., and Salpeter, E. E. 1977. The phase diagram and transport properties for hydrogen-helium fluid planets. *Astrophys. J.* 35:221–237.

Stix, M. 1973. Spherical $\alpha\omega$-dynamos by a variational method. *Astron. Astrophys.* 47:243–254.

Stix, M. 1976a. Differential rotation and the solar dynamo. *Astron. Astrophys.* 47:243–254.

Stix, M. 1976b. Dynamo theory and the solar cycle. In *Basic Mechanisms of Solar Activity: Proc. IAU Symp. 71*, eds. V. Bumba and J. Kleczek (Dordrecht: D. Reidel), pp. 367–388.

Stix, M. 1981a. Screening effect in the solar convection zone. *Astron. Astrophys.* 93:339–340.

Stix, M. 1981b. The theory of the solar cycle. *Solar Phys.* 74:79–101.

Stix, M. 1987a. Models for a differentially rotating solar convection zone. In *The Internal Solar Angular Velocity*, eds. B. R. Durney and S. Sofia (Dordrecht: D. Reidel), pp. 329–342.

Stix, M. 1987b. On the origin of stellar magnetic fields. In *Lecture Notes in Physics, vol. 292: Solar*, eds. E.-H. Schröter and M. Schüssler (Berlin: Springer-Verlag), pp. 15–38.

Stix, M., and Knölker, M. 1987. On the frequency of solar oscillations. In *Physical Processes in Comets, Stars, and Active Galaxies*, eds. W. Hillebrandt, E. Meyer-Hofmeister and H.-C. Thomas (Berlin: Springer-Verlag), pp. 67–77.

Stix, M., and Wühl, H. 1974. The center-to-limb variation of the photospheric wave spectrum. *Solar Phys.* 37:63–74.

Strang, G., and Fix, G. J. 1973. *An Analysis of the Finite Element Method* (New York: Prentice-Hall).

Stratawa, C., Dobrozemsky, R., and Weinzierl, P. 1978. Ratio G_A/G_V derived from the proton spectrum in free-neutron decay. *Phys. Rev.* D18:3970–3979.

Strauss, H. R. 1986. Resonant fast dynamos. *Phys. Rev.* L57:2231–2233.

Straus, J. M., Blake, J. B., and Schramm, D. N. 1976. Effects of convective overshoot on lithium depletion in main-sequence stars. *Astrophys. J.* 204:481–487.

Strebel, H., and Thüring, B. 1932. Untersuchungen zu einer photometrischen statistik der granulation der sonnenoberfläche. *Zeit. Astrophys.* 5:348–385.

Stringfellow, G. S., Swenson, F. J., and Faulkner, J. 1987. Is there a classical lithium classical Hyades lithium problem? *Bull. Amer. Astron. Soc.* 19:1020 (abstract).

Strom, S. E., Strom, K. M., and Edwards, S. 1988. Energetic winds and circumstellar disks associated with low mass young stellar objects. In *Galactic and Extragalactic Star Formation*, eds. R. E. Pudritz and M. Fich (Dordrecht: Kluwer), pp. 53–88.

Sturrock, P. A. 1980. *Solar Flares* (Boulder: Univ. of Colorado Press).

Sturrock, P. A., Holzer, T. E., Mihalas, D. M., and Ulrich, R. K., eds. 1986. *Physics of the Sun* (Dordrecht: D. Reidel).

Subramanian, A., and Lal, S. 1987. On the statistical significance of possible variations in the solar neutrino flux. *Astron. Nachr.* 308:127–134.

Subramanian, A., Lal, S., and Yvas, P. 1988. On the origin of transient effects in the solar neutrino experiment. *Astron. Nachr.* 309:363–371.

Svalgaard, L., and Wilcox, J. M. 1975. Long term evolution of solar sector structure. *Solar Phys.* 41:461–475.

Svalgaard, L., Duvall, T. L., Jr., and Scherrer, P. H. 1978. The strength of the Sun's polar fields. *Solar Phys.* 58:225–240.

Svestka, A., Solodyna, C. V., Howard, R., and Levine, R. H. 1977. Open magnetic fields in active regions. *Solar Phys.* 55:359–369.

Svestka, Z. 1976. Development of solar active regions. In *Proc. Intl. Symp. on Solar Terrestrial Physics,* June 7–16, Boulder, Colo., ed. D. J. Williams, pp. 129–143.

Svestka, Z., Solodyna, C. V., Howard, R., and Levine, R. H. 1977. Open magnetic fields in active regions. *Solar Phys.* 55:359–369.

Sweigart, A. V. 1974. Do helium shell flashes cause extensive mixing in low mass stars? *Astrophys. J.* 189:289–291.

Sweigart, A. V., and Mengel, J. G. 1979. Meridional circulation and CNO anomalies in red giant stars. *Astrophys. J.* 229:624–641.

Swenson, F. J., Stringfellow, G. S., and Faulkner, J. 1990. Is there a classical Hyades lithium problem? *Astrophys. J.* 348:L33-L36.

Swensson, J. W., Benedict, W. S., Delbouille, L., and Roland, G. 1970. The solar spectrum from λ7498 to λ12016, a table of measures and identifications. *Mem. Soc. Roy. Sci. Liège,* special vol. no. 5.

Sylvester, J., Lemen, J. R., and Mewe, R. 1984. Variation in observed coronal calcium abundance of X-ray flare plasma. *Nature* 310:665–666.

Syrovatskii, S. I., and Zhugzhda, Y. D. 1968. Oscillatory convection of a conducting gas in a strong magnetic field. *Soviet Astron.* 11:945–952.

Takakura, T. 1984. Steady models for the hard X-ray loops in the solar corona. *Solar Phys.* 91:311–324.

Takakura, T., Ohki, K., Nitta, N., Makishima, K., Murakami, T., Ogawara, Y., Oda, M., and Miyamoto, S. 1983. Hard X-ray imaging of a solar limb flare with the X-ray telescope aboard the Hinotori Satellite. *Astrophys. J.* 270:L83-L87.

Tanaka,K. 1986. Solar flare X-ray spectra of Fe XXVI and Fe XXV from the Hinotori satellite. *Publ. Astron. Soc. Japan* 38:225–249.

Tanaka, K. 1987. Impact of X-ray observations from the Hinotori satellite on solar flare research. *Publ. Astron. Soc. Japan* 39:1–45.

Tanaka, K., and Zirin, H. 1985. The great flare of 1982 June 6. *Astrophys. J.* 299:1036–1046.

Tandberg-Hanssen, E., and Emslie, A. G. 1988. *The Physics of Solar Flares* (Cambridge: Cambridge Univ. Press).

Tanenbaum, A. S., Wilcox, J. M., Frazier, E. N., and Howard, R. 1969. Solar velocity fields: Five-minute oscillations and supergranulation. *Solar Phys.* 9:328–342.

Tanenbaum, A. S., Wilcox, J. M., and Howard, R. 1971. Five-minute oscillations in the solar magnetic field. In *Solar Magnetic Fields: Proc. IAU Symp. 43,* ed. R. Howard (Dordrecht: D. Reidel), pp. 348–355.

Tang, F., and Moore, R. L. 1982. Remote flare brightenings and Type III reverse slope bursts. *Solar Phys.*77:263–276.

Tang, F., Howard, R., and Adkins, J. M. 1984. A statistical study of active regions 1967–1981. *Solar Phys.* 91:75–86.

Tapping, K. F. 1987. Recent solar radio astronomy at centimeter wavelengths: The temporal variability of the 10.7-cm radio flux. *J. Geophys. Res.* 92:829–838.

Tarantola, A. 1987. *Inverse Problem Theory: Methods for Data Fitting and Model Parameter Estimations* (Amsterdam: Elsevier).

Tarbell, T. D., and Smithson, R. C. 1981. A simple, image motion compensation system for solar observations. In *Solar Instrumentation: What's Next?,* ed. R. B. Dunn (Sunspot, N.M.: Sacramento Peak Obs.), pp. 491–501.

Tarbell, T. D., and Title, A. M. 1977. Measurements of magnetic fluxes and field strengths in the photospheric network. *Solar Phys.* 52:13–25.

Tarbell, T. D., Title, A. M., and Schoolman, S. A. 1979. Weak and strong magnetic fields in the solar photosphere. *Astrophys. J.* 229:387–392.

Tarbell, T. D., Peri, M., Frank, Z., Shine, R., and Title, A. 1988. Observations of f- and p-mode oscillations of high degree ($500 < \ell < 2500$) in quiet and active Sun. In *Seismology of the Sun and Sun-Like Stars,* ed. E. J. Rolfe, ESA SP-286, pp. 315–319.

Tassoul, J.-L. 1978. *Theory of Rotating Stars* (Princeton, N.J.: Princeton Univ. Press).

Tassoul, J.-L., and Tassoul, M. 1989. The internal rotation of the Sun. *Astron. Astrophys.* 213:397–401.

Tassoul, M. 1980. Asymptotic approximations for stellar nonradial pulsations. *Astrophys. J. Suppl.* 43:469–490.

Tassoul, M. 1990. Second-order asymptotic approximations for stellar nonradial acoustic modes. *Astrophys. J.* 358:313–327.

Tassoul, M., and Tassoul, J.-L. 1984*a*. Meridional circulation in rotating stars. VII. The effects of chemical inhomogeneities. *Astrophys. J.* 279:384–393.

Tassoul, M., and Tassoul, J.-L. 1984*b*. Meridional circulation in rotating stars. VIII. The solar spin-down problem. *Astrophys. J.* 286:350–358.

Tatarskii, V. I. 1971. *The Effect of the Turbulent Atmosphere on Wave Propagation,* Israel Program for Scientific Translations.

Tayler, R. J. 1973. Convection in rotating stars. *Mon. Not. Roy. Astron. Soc.* 165:39–52.

Tegner, P. E., and Bargholtz, C. 1983. The rate of the $^3He(p,e^+v)^4He$ reaction. *Astrophys. J.* 272:311–316.

Terebey, S., Shu, F. H., and Cassen, P. 1984. The collapse of the cores of slowly rotating isothermal clouds. *Astrophys. J.* 286:529–551.

Teuber, D., Tandberg-Hanssen, E., and Hagyard, M. J. 1977. Computer solutions for studying correlations between solar magnetic fields and Skylab X-ray observations. *Solar Phys.* 53: 97–110.

Thomas, H.-C. 1967. Der Helium-Flash bei einem Stern von 1.3 Sonnen massen. *Z. Astrophys.* 67:420–455.

Thomas, J. H. 1988. Siphon flows in isolated magnetic flux tubes. *Astrophys. J.* 333:407–419.

Thomas, J. H., Cram, L. E., and Nye, A. H. 1982. Five-minute oscillations as a subsurface probe of sunspot structure. *Nature* 297:485–487.

Thomas, R. N., and Athay, R. G. 1961. *Physics of the Solar Chromosphere* (New York: Interscience).

Thompson, M. J. 1990. A new inversion of solar rotational splitting data. *Solar Phys.* 125:1–12.

Thomsen, D. E. 1978. Sun shakes. *Sci. News* 113:253.

Thompson, M. F. 1990. A new inversion of solar rotational splitting data. *Solar Phys.* 125: 1–12.

Tikhonov, A. N., and Arsenin, V. Y. 1977. *Solutions of Ill-Posed Problems* (Washington, D.C.: Winston).

Tilton, G. R. 1988. Age of the solar system. In *Meteorites and the Early Solar System,* eds. J. F. Kerridge and M. S. Mattews (Tucson: Univ. of Arizona Press), pp. 259–275.

Timothy, A. F., Krieger, A. S., and Vaiana, G. S. 1975. The structure and evolution of coronal holes. *Solar Phys.* 42:135–156.

Title, A. M. 1989. An overview of the Orbiting Solar Laboratory. In *High Spatial Resolution Solar Observations,* ed. O. von der Lühe (Sunspot, N.M.: Sacramento Peak Obs.), pp. 147–165.

Title, A. M., Tarbell, T. D., Simon, G. W., and the SOUP Team. 1986. White-light movies of the solar photosphere from the SOUP instrument on Spacelab 2. *Adv. Space Res.* 6:253–262.

Title, A., Tarbell, T., and the SOUP Team. 1987*a*. First results on quiet and magnetic granulation from SOUP. In *Proc. of the Second Workshop on Theoretical Problems in High Resolution Solar Physics,* eds. G. Athay and D. S. Spicer, NASA CP-2483, pp. 55–77.

Title, A. M., Tarbell, T. D., Topka, K. P., Shine, R. A., Simon, G. W., Zirin, H., and the SOUP Team. 1987*b*. New ideas about granulation based on data from the solar optical universal polarimeter instrument on Spacelab 2 and magnetic data from Big Bear Solar Observatory. In *Solar and Stellar Physics,* eds. E.-H. Schröter and M. Schüssler (Berlin: Springer-Verlag), pp. 173–186.

Title, A. M., Tarbell, T. D., and Topka, K. P. 1987*c*. On the relation between magnetic field structures and the granulation. *Astrophys. J.* 317:892–899.

Title, A M., Tarbell, T., Topka, K., Acton, L., Duncan, D., Ferguson, S., Finch, M., Frank, Z., Kelley, G., Lindgren, R., Morrill, M., Pope, T., Reeves, R., Rehse, R., Shine, R., Simon, G., Harvey, J., Leibacher, J., Livingston, W., November, L., and Zirker, J. 1988. Correlation lifetimes of quiet and magnetic granulation from the SOUP instrument on Spacelab 2. *Astro. Lett. and Commun.* 27:141–149.

Title, A. M., Tarbell, T. D., Topka, K. P., Ferguson, S. H., Shine, R. A., and the SOUP team. 1989. Statistical properties of solar granulation derived from the SOUP instrument on Spacelab 2. *Astrophys. J.* 336:475–494.

Title, A. M., Tarbell, T., Topka, K., Cauffman, D., Balke, C., and Scharmer, G. 1990a. Magnetic flux tubes and their relation to continuum and photospheric features. In *Physics of Magnetic Flux Ropes,* eds. C. T. Russel, E. R. Priest and L. C. Lee (Washington, D.C.: American Geophys. Union), pp. 171–179.

Title, A. M., Shine, R. A., Tarbell, T. D., Topka, K. P., and Scharmer, G. B. 1990b. High resolution observations of the photosphere. In *Solar Photosphere: Structure, Convection, and Magnetic Fields: Proc. IAU Symp. 138,* ed. J. O. Stenflo (Dordrecht: Kluwer), pp. 49–66.

Title, A. M., Topka, K., Tarbell, T., Schmidt, W., Balke, C., and Scharmer, G. 1990c. On the differences between plage and quiet Sun in the solar photosphere. *Astrophys. J.,* submitted.

Tobiska, W. K. 1988. A Solar Extreme Ultraviolet Flux Model. Ph.D. Thesis, Univ. of Colorado.

Tombrello, T. A. 1965. The capture of protons by ^{7}Be. *Nucl. Phys.* 71:459–464.

Tombrello, T. A. 1967. Astrophysical problems. In *Nuclear Research with Low Energy Accelerators,* eds. J. B. Marion and D. M. Van Patter (Academic Press), pp. 195–212.

Tombrello, T. A., and Parker, P. D. 1963. Direct-capture model for the ^{3}He(α,γ)^{7}Li reactions. *Phys. Rev.* 131: 2582–2590.

Tomczyk, S. 1988. A Measurement of the Rotational Frequency Splitting of the Solar Five-Minute Oscillations. Ph.D. Thesis, Univ. of California, Los Angeles.

Tomczyk, S., Cacciani, A., Korzennik, S. G., Rhodes, E. J., Jr., and Ulrich, R. K. 1988. Measurement of the rotational frequency splitting of the solar five-minute oscillations from magneto-optical filter observations. In *Seismology of the Sun and Sun-Like Stars,* ed. E. J. Rolfe, ESA SP-286, pp. 141–147.

Toomre, J. 1986. Properties of solar oscillations. In *Seismology of the Sun and the Distant Stars,* ed. D. O. Gough (Dordrecht: D. Reidel), pp. 1–22.

Toomre, J., Zahn, J.-P., Latour, J., and Speigel, E. 1976. Stellar convection theory. II. Single-mode study of the second convection zone in a A-type star. *Astrophys. J.* 207:545–563.

Topka, K., Moore, R., LaBonte, B. J., and Howard, R. 1982. Evidence for a poleward meridional flow on the Sun. *Solar Phys.* 79:231–245.

Topka, K. P., Tarbell, T. D., and Title, A. M. 1986. High-resolution observations of changing magnetic features on the Sun. *Astrophys. J.* 306:304–316.

Torres, C. A. O., and Ferraz-Mello, S. 1973. On variable dMe stars. *Astron. Astrophys.* 27: 231–236.

Torres-Peimbert, S., Ulrich, R. K., and Simpson, E. 1969. Studies in stellar evolution. VII. Solar models. *Astrophys. J.* 155:957–964.

Toshev, S. 1987a. Exact analytical solution of the two-neutrino evolution equation in matter with exponentially varying density. *Phys. Lett.* 196B:170–174.

Toshev, S. 1987b. Non-adiabatic neutrino transitions in matter with linearly varying density. *Phys. Lett.* 198B:551–555.

Tousey, R. 1973. The solar corona. In *Space Research XIII,* eds. M. J. Rycroft and S. K. Runcorn (Berlin: Akademie-Verlag), pp. 713–730.

Tousey, R., Milone, E. F., Purcell, J.D., Schneider, W. P., and Tilford, S. G. 1974. *An Atlas of the Solar Ultraviolet Spectrum Between 2226 and 2992 Angstroms,* Naval Research Lab Rept. 778.

Toutain, T., and Gouttebroze, P. 1988. Visibility of nonradial pulsation modes in solar continuum intensity measurements. In *Seismology of the Sun and Sun-Like Stars,* ed. E. J. Rolfe, ESA SP-286, pp. 241–246.

Townsend, A. A. 1958. The effect of radiative transfer on turbulent flow in a stratified fluid. *J. Fluid Mech.* 4:361–375.

Traub, W., and Stier, M.T. 1976. Theoretical atmospheric transmission in the mid- and far-infrared at four altitudes. *Appl. Opt.* 15:364–377.

Travis, L. D., and Matsushima, S. 1973. The role of convection in stellar atmospheres. I. Observable effects of convection in the solar atmosphere. *Astrophys. J.* 180:975–985.

Trellis, M. 1966a. Sur une relation possible entre l'aire des taches solaires et la position des planètes. *Compt. Rend.* 262:312–315.

Trellis, M. 1966b. Influence de la configuration du système solaire sur la naissance des centres d'activité. *Compt. Rend.* 262:376–377.

Trellis, M. 1971a. Fréquence des groupes de taches solaires en fonction de leur aire moyenne. *Circ. Acad. Sci. Paris* 272:1026–1028.

Trellis, M. 1971*b*. Persistance dans le temps des zones favorables à l'apparition des grands groupes des taches solaires. *Circ. Roy. Acad. Sci. Paris* B272:549–552.

Tsubaki, T. 1975. Line profile analysis of a coronal formation observed near a quiescent prominence: Intensities, temperatures, and velocity fields. *Solar Phys.* 43:147–175.

Tsubaki, T. 1977. Periodic oscillations found in coronal velocity fields. *Solar Phys.* 51:121–130.

Tsuneta, S., Takakura, T., Nitta, N., Ohki, K., Tanaka, K., Makishima, K., Murakami, T., Oda, M., Ogawara, Y., and Kondo, I. 1984. Hard X-ray imaging of the solar flare on 1982 May 13 with the Hinotori Spacecraft. *Astrophys. J.* 280:887–891.

Tuominen, I., and Kyröläinen, J. 1982. On the latitude drift of sunspot groups and solar rotation. *Solar Phys.* 79:161–172.

Tuominen, I., and Rudiger, G. 1989. Solar differential rotation as a multiparameter turbulence problem. *Astron. Astrophys.* 217:217–228.

Tuominen, I., and Virtanen, H. 1987. Solar rotation variations from sunspot group statistics. In *The Internal Solar Angular Velocity,* eds. B. R. Durney and S. Sofia (Dordrecht: D. Reidel), pp. 83–88.

Tuominen, J. 1952. On the dependence of the systematic drift of sunspots in heliographic latitude on phase in the 22-year cycle of the Sun. *Z. Astrophys.* 30:261–274.

Tuominen, J. 1961. On the latitude drift of sunspot groups. *Z. Astrophys.* 51:91–94.

Tuominen, J. 1976. 22-year cycle or 11-year cycle in the latitude drift of sunspot groups? *Solar Phys.* 47:541–550.

Tuominen, J., Tuominen, I., and Kyröläinen, J. 1983. Eleven year cycle in solar rotation and meridional motions as derived from the position of sunspot groups. *Mon. Not. Roy. Astron. Soc.* 205:691–704.

Tuominen, I., Virtanen, I., Krause, F., and Rüdiger, G. 1984. Cyclic flow in the solar convection zone due to the dynamo reaction. In *The Hydromagnetics of the Sun,* eds. T. D. Guyenne and J. J. Hunt, ESA SP-220, pp. 225–226.

Turk-Chièze, S. 1990. On the accuracy of solar modelling. In *Inside the Sun: Proc. IAU Coll. 121,* eds. G. Berthomieu and M. Cribier (Paris: Kluwer), pp. 125–132.

Turck-Chièze, S., Cahen, S., Cassé, M., and Doom, C. 1988. Revisiting the standard solar model. *Astrophys. J.* 335:415–424.

Turnullo, M. 1986. The rotation of calcium plages in the years 1967–1970. *Solar Phys.* 105: 197–204.

Turnullo, M. 1987*a*. The impact of aging on the rotation rate of calcium plages in the years 1967–1977. *Solar Phys.* 112:153–163.

Turnullo, M. 1987*b*. The old calcium plages differential rotation latitudinal profile: Its gross and fine structure evolution with the solar cycle. *Solar Phys.* 112:143–151.

Turon, P. J., and Léna, P. J. 1970. High resolution solar images at 10 microns: Sunspot details and photometry. *Solar Phys.* 14:112–124.

Twomey, S. 1977. Some aspects of the inversion problem. In *Inversion Methods in Atmospheric Remote Sounding,* ed. A. Deepak (New York: Academic Press), pp. 41–62.

Uchida, Y. 1974. Behavior of the flare-produced coronal MHD wavefront and the occurrence of Type II radio bursts. *Solar Phys.* 39:431–449.

Ulmschneider, P. 1970. On the frequency and strength of shock waves in the solar atmosphere. *Solar Phys.* 12:403–415.

Ulmschneider, P. 1971. On the propagation of a spectrum of acoustic waves in the solar atmosphere. *Astron. Astrophys.* 14:275–282.

Ulmschneider, P. 1974. Radiation loss and mechanical heating in the solar chromosphere. *Solar Phys.* 39:327–336.

Ulmschneider, P. 1979. Stellar chromospheres. *Space Sci. Rev.* 24:71–100.

Ulmschneider, P. 1981. Theories of heating of solar and stellar chromospheres. In *Solar Phenomena in Stars and Stellar Systems,* eds. R. M. Bonnet and A. K. Dupree (Dordrecht: D. Reidel), pp. 239–263.

Ulmschneider, P. 1986. The present state of wave heating theories of stellar chromospheres. *Adv. Space Res.* 6(8):39–46.

Ulmschneider, P. 1989. On the chromospheric emission from acoustically heated stellar atmospheres. *Astron. Astrophys.* 222:171–178.

Ulmschneider, P., and Kalkofen, W. 1977. Acoustic waves in the solar atmosphere. III. A theoretical temperature minimum. *Astron. Astrophys.* 57:199–209.

Ulmschneider, P., and Stein, R. F. 1982. Heating of stellar chromospheres when magnetic fields are present. *Astron. Astrophys.* 106:9–13.

Ulmschneider, P., and Zähringer, K. 1987. The dynamics of solar magnetic flux tubes subjected to resonant foot point shaking. In *Cool Stars, Stellar Systems and the Sun,* eds. J. L. Linsky and R. E. Stencel (Berlin: Springer-Verlag), pp. 63–65.

Ulmschneider, P., Schmitz, F., Kalkofen, W., and Bohn, H. U. 1978. Acoustic waves in the solar atmosphere. V. On the chromospheric temperature rise. *Astron. Astrophys.* 70:487–500.

Ulrich, R. K. 1970. The five minute oscillation on the solar surface. *Astrophys. J.* 162:993–1002.

Ulrich, R. K. 1974a. Solar models with low neutrino fluxes. *Astrophys. J.* 188:369–378.

Ulrich, R. K. 1974b. Studies of evolved stars. III. Models of SG Sagittae consistent with s-process nucleosynthesis. *Astrophys. J.* 192:507–516.

Ulrich, R. K. 1982. The influence of partial ionization and scattering states on the solar interior structure. *Astrophys. J.* 258:404–413.

Ulrich, R. K. 1986. Determination of stellar ages from asteroseismology. *Astrophys. J.* 306: L37–L40.

Ulrich, R. K., and Rhodes, E. J., Jr. 1983. Testing solar models with global solar oscillations in the five-minute band. *Astrophys. J.* 265:551–563.

Ulrich, R. K., and Rhodes, E. J. 1984. The sensitivity of solar eigenfrequencies to the treatment of the equation of state. In *Solar Seismology from Space,* NASA JPL Publ. 84-84, pp. 371–377.

Ulrich, R. K., Rhodes, E. J., Jr., Tomczyk, S., Dumont, P. J., and Brunish, W. M. 1983. The analysis of solar models-neutrinos and oscillations. In *Science Underground, AIP Conf. Proc. 96,* eds. M. M. Nieto et al. (New York: American Inst. of Physics), pp. 66–79.

Ulrich, R. K., Rhodes, E. J., Jr., Cacciani, A., and Tomczyk, S. 1984. The effects of seeing on noise. In *Solar Seismology from Space,* NASA JPL Publ. 84-84, pp. 263–270.

Ulrich, R. K., Boyden, J. E., Webster, L., and Shieber, T. 1988a. Magnetically induced spectral line redshifts. Full disk measurements. In *Seismology of the Sun and Sun-Like Stars,* ed. E. J. Rolfe, ESA SP-286, pp. 325–332.

Ulrich, R. K., Boyden, J. E., Webster, L., Snodgrass, H. B., Padilla, S. P., Gilman, P., and Shieber, T. 1988b. Solar rotation measurements at Mount Wilson. V. Reanalysis of 21 years of data. *Solar Phys.* 117:291–328.

Unno, W. 1956. Line formation of a normal Zeeman triplet. *Publ. Astron. Soc. Japan* 8: 108–125.

Unno, W. 1957. Anisotropy of solar convection. *Astrophys. J.* 126:259–265.

Unno, W. 1967. The stellar radial pulsation coupled with convection. *Publ. Astron. Soc. Japan* 19:140–153.

Unno, W. 1969. Theoretical studies on stellar stability. II. Undisturbed convective nongrey atmospheres. *Publ. Astron. Soc. Japan* 21:240–262.

Unno, W. 1977. Wave generation and pulsation in stars with convective zones. In *Problems of Stellar Convection,* eds. E. A. Spiegel and J.-P. Zahn (Berlin: Springer-Verlag), pp. 315–324.

Unno, W., and Ando, H. 1979. Instability of a thin magnetic flux tube in the solar atmosphere. *Geophys. Astrophys. Fluid Dyn.* 12:107–115.

Unno, W., and Spiegel, E. A. 1966. The Eddington approximation in the radiative heat equation. *Publ. Astron. Soc. Japan* 18:85–95.

Unno, W., Osaki, Y., Ando, H., and Shibahashi, H. 1979. *Nonradial Oscillations of Stars* (Tokyo: Univ. of Tokyo Press).

Unno, W., Kondo, M., and Da-Run, X. 1985. Solar convection zone given by nonlocal mixing length theory. *Publ. Astron. Soc. Japan* 37:235–244.

Unno, W., Osaki, Y., Ando, H., Saio, H., and Shibahashi, H. 1989. *Nonradial Oscillations of Stars,* 2nd ed. (Tokyo: Univ. of Tokyo Press).

Uralov, A., and Nefedev, V. 1976. Pulsed solar microwave bursts with a quasi thermal spectrum. *Soviet Astron. AJ* 20:590–592.

Vaiana, G. S., Davis, J. M., Giacconi, R., Krieger, A. S., Silk, J. K., Timothy, A. F., and Zombeck, M. 1973. X-ray observations of the solar corona: Preliminary results from Skylab. *Astrophys. J.* 185:L47–L51.

Vaiana, G. S., Krieger, A. S., Timothy, A. F., and Zombeck, M. 1976. ATM observations, X-ray results. *Astrophys. Space Sci.* 39:75–101.

van Ballegooijen, A. A. 1982. The overshoot region at the base of the solar convection zone and the problem of magnetic flux storage. *Astron. Astrophys.* 113:99–112.

van Ballegooijen, A. A. 1984. On the temperature structure of sunspot umbrae. *Solar Phys.* 91:195–217.

van Ballegooijen, A. A. 1985a. Contribution functions for Zeeman-split lines, and line formation in photospheric faculae. In *Measurements of Solar Vector Magnetic Fields,* ed. M. J. Hagyard, NASA CP-2374, pp. 322–334.

van Ballegooijen, A. A. 1985b. Cascade of magnetic energy as a mechanism of coronal heating. In *Theoretical Problems in High Resolution Solar Physics,* ed. H. U. Schmidt (München: Max-Planck-Institut für Astrophysik), pp. 268–272.

van Ballegooijen, A. A. 1986. On the surface response of solar giant cells. *Astrophys. J.* 304:828–837.

van Ballegooijen, A. A. 1987. Radiative transfer in the presence of strong magnetic fields. In *Numerical Radiative Transfer,* ed. W. Kalkofen (Cambridge: Cambridge Univ. Press), pp. 279–304.

van Beek, H. F., Hoyng, P., Lafleur, B., and Simnett, G. M. 1980. The hard X-ray imaging spectrometer (HXIS). *Solar Phys.* 65:39–52.

Vandakurov, Yu. V. 1967. On the frequency distribution of stellar oscillations. *Soviet Astron. AJ* 11:630–638.

VandenBerg, D. A., and Bell, R. A. 1985. Theoretical isochrones for globular clusters with predicted BVRI and Strömgren photometry. *Astrophys. J. Suppl.* 58:561–621.

van den Oord, G. H. J. 1990. The electrodynamics of beam/return current systems in the solar corona. *Astron. Astrophys.* 234:496–518.

Van der Borght, R. 1975. Finite-amplitude convection in a compressible medium and its application to solar granulation. *Mon. Not. Roy. Astron. Soc.* 173:85–95.

Van der Borght, R., and Fox, P. 1983. A convective model of solar granulation. *Proc. Astron. Soc. Australia* 5:166–168.

van der Raay, H. B. 1984. Solar oscillations: Excitation, decay and rotational splitting. In *Theoretical Problems in Stellar Stability and Oscillations,* eds. A. Noëls and M. Gabriel (Liège: Inst. d'Astrophysique), pp. 215–219.

van der Raay, H. B. 1988. Long period solar oscillations. In *Seismology of the Sun and Sun-Like Stars,* ed. E. J. Rolfe, ESA SP-286, pp. 339–351.

van der Raay, H. B., Pallé, P. L., and Roca-Cortés, T. 1986. Rotational splitting of $\ell = 1$ p modes. In *Seismology of the Sun and the Distant Stars,* ed. D. O. Gough (Dordrecht: D. Reidel), pp. 215–221.

Van Hoven, G. S., Sparks, L., and Schnack, D. 1987. Nonlinear radiative condensation in a sheared magnetic field. *Astrophys. J.* 317:L91–L94.

Van Leer, B. 1977. Towards the ultimate conservative difference scheme. III. Upstream-centered finite-difference schemes for ideal compressible flow. *J. Comput. Phys.* 23:276–275.

van Leeuwen, F., and Alphenaar, P. 1982. Variable K-type stars in the Pleiades cluster. *ESO Messenger* 28:15–18.

van Leeuwen, F., Alphenaar, P., and Meyrs, J. J. M. 1987. VBLUW observations of Pleiades G and K dwarfs. *Astron. Astrophys. Suppl.* 67:483–506.

Vauclair, S. 1975. An explanation for helium-rich and helium-variable stars: Diffusion in a radial mass-loss flux. *Astron. Astrophys.* 45:233–235.

Vauclair, S., and Meyer, J. P. 1985. Diffusion in the chromosphere, and the composition of the solar corona and energetic particles. In *Proc. 19th Intl. Cosmic Ray Conf.,* vol. 4, eds. F. C. Jones, J. Adams and G. M. Mason, NASA CP-2376, pp. 233–236.

Vauclair, S., and Vauclair, G. 1982. Element segregation in stellar outer layers. *Ann. Rev. Astron. Astrophys.* 20:37–60.

Vauclair, S., Vauclair, G., Schatzman, E., and Michaud, G. 1978. Hydrodynamical instabilities in the envelopes of main-sequence stars: Constraints implied by the lithium, beryllium, and boron observations. *Astrophys. J.* 223:567–582.

Vauclair, S., Hardorp, J., and Peterson, D. A. 1979. Silicon levitation in chemically peculiar stars and the oblique rotator model. *Astrophys. J.* 227:526–533.

Vaughan, A. H. 1980. Comparison of activity cycles in old and young main-sequence stars. *Publ. Astron. Soc. Pacific* 92:392–396.

Vaughan, A. H. 1984. The magnetic activity of sunlike stars. *Science* 225:793–800.

Vaughan, A. H., and Preston, G. W. 1980. Flux measurements of Ca II H and K emission. *Publ. Astron. Soc. Pacific* 90:267–274.

Vaughan, A. H., Baliunas, S. L., Middelkoop, F., Hartmann, L. W., Mihalas, D., Noyes, R. W., and Preston, G. W. 1981. Stellar rotation in lower main sequence stars measured from time variations in H and K emission-line fluxes. I. Initial results. *Astrophys. J.* 250:276–283.

Vaughn, F. J., Chalmers, R. A., Kohler, D., and Chase, L. R., Jr. 1970. Cross sections for the $^7Be(p,\gamma)^8Be$ reaction. *Phys. Rev.* C2:1657–1665.

Veck, N. J., and Parkinson, J. H. 1981. Solar abundances from X-ray flare observations. *Mon. Not. Roy. Astron. Soc.* 197:41–55.

Venkatakrishnan, P. 1983. Nonlinear development of convective instability within slender flux tubes. I. Adiabatic flow. *J. Astrophys. Astron.* 4:135–149.

Venkatakrishnan, P. 1985. Nonlinear development of convective instability within slender flux tubes. II. The effect of radiative heat transport. *J. Astrophys. Astron.* 6:21–34.

Veretenkin, E. P., Gavrin, V. N., and Yanovich, E. A. 1985. Use of metallic lithium for detecting solar neutrinos. *Soviet Atomic Energy* 58:82–83.

Vernazza, J. E., and Reeves, E. M. 1978. Extreme ultraviolet composite spectra of representative solar features. *Astrophys. J. Suppl.* 27:485–513.

Vernazza, J. E., Avrett, E. H., and Loeser, R. 1973. Structure of the solar chromosphere. I. Basic computations and summary of the results. *Astrophys. J.* 46:604–631.

Vernazza, J. E., Avrett, E. H., and Loeser, R. 1976. Structure of the solar chromosphere. II. The underlying photosphere and temperature-minimum region. *Astrophys. J. Suppl.* 30:1–60.

Vernazza, J. E., Avrett, E. H., and Loeser, R. 1981. Structure of the solar chromosphere. III. Models of the EUV brightness components of the quiet Sun. *Astrophys. J. Suppl.* 45:635–725.

Veronis, G. 1963. Penetrative convection. *Astrophys. J.* 137:641–663.

Veselov, A. I., Vysotskii, M. I., and Yurov, V. P. 1987. Half-year variations in the solar-neutrino flux given by the Davis data of 1979–1982. *Soviet J. Nucl. Phys.* 45:865–871.

Vestrand, W. T., Forrest, D. J., Chupp, E. L., Rieger, E., and Share, G. H. 1987. The directivity of high-energy emission from solar flares. *Astrophys. J.* 322:1010–1027.

Vidal-Madjar, A. 1977. The solar spectrum at Lyman-alpha 1216Å. In *The Solar Output and Its Variations,* ed. O. R. White (Boulder: Colorado Assoc. Univ. Press), pp. 213–234.

Vignaud, D. 1991. Report on the Gallex Experiment. In *Proc. XI Moriond Workshop, Tests of Fundamental Laws of Physics,* Les Arcs.

Vigneron, C., Schatzman, E., Catala, C., and Schwatzman, E. 1990. Angular momentum transport in pre-main sequence stars of intermediate mass. *Solar Phys.* 128:287–298.

Vilmer, N., Kane, S. R., and Trottet, G. 1982. Impulsive and gradual hard X-ray sources in a solar flare. *Astron. Astrophys.* 108:306–313.

Vitense, E. 1953. Die Wasserstoffkonvektionszone der Sonne. *Z. Astrophys.* 32:135–164.

Vitinskii, Yu. I. 1965. *Solar-Activity Forecasting* (Jerusalem: Israel Prog. Sci. Trans.).

Vitinsky, Yu. I., and Ikhsanov, R. N. 1985. On time variations of characteristics of the solar differential rotation determined from sunspot groups with an average area greater than 300 m.s.h. *Soln. Dannye, Byull.* 10:77–83 (in Russian).

Vitinsky, Yu. I., Kopecky, M., and Kuklin, G. V. 1986. *Statistics of Sunspot Activity* (Moscow: Nauka Press), in Russian.

Vlahos, L., and Rowland, H. L. 1984. Electron precipitation in solar flares: Collisionless effects. *Astron. Astrophys.* 139:263–270.

Vlahos, L., Machado, M. E., Ramaty, R., Murphy, R. J., Alissandrakis, C., Bai, T., Batchelor, D., Benz, A. O., Chupp, E., Ellison, D., Evenson, P., Forrest, D. J., Holman, G., Kane, S. R., Kaufmann, P., Kundu, M. R., Lin, R. P., MacKinnon, A., Nakajima, H., Pesses, M., Pick, M., Ryan, J., Schwartz, R. A., Smith, D. F., Trottet, G., Tsuneta, S., and VanHoven, G. 1986. Particle acceleration. In *Energetic Phenomena on the Sun,* eds. M. R. Kundu and B. E. Woodgate, NASA CP-2439, pp. 127–224.

Vogel, S. N., and Kuhi, L. V. 1981. Rotational velocities of pre-main sequence stars. *Astrophys. J.* 245:960–976.

Vogt, S. S. 1983. Spots, stop-cycles, and magnetic fields of late-type dwarfs. In *Activity in Red Dwarf Stars,* eds. P. B. Byrne and M. Rodonò (Dordrecht: D. Reidel), pp. 137–156.

Voight, H. H. 1956. "Drei-Strom-Modell" der Sonnenphotosphare und Asymmetrie der Linien des infraroten Sauerstoff-Tripletts. *Z. Astrophys.* 40:157–190.

Volk, H., Kräwinkel, H., Santo, R., and Walleck, L. 1983. Activation measurement of the ^{3}He(^{4}He,γ)^{7}Be reaction. *Z. Phys.* A310:91–94.

Voloshin, M. B., Vysotskii, M. I., and Okun, L. B. 1986*a.* Electrodynamics and possible consequences for solar neutrinos. *Soviet Phys. JETP* 64:446–452.

Voloshin, M. B., Vysotskii, M. I., and Okuń, L. B. 1986*b.* Properties of the neutrino and possible semiannual variations in the solar neutrino flux. *Soviet J. Nucl. Phys.* 44:440–441.

von der Lühe, O., 1984. Estimating Fried's parameter from a time series of an arbitrary resolved object imaged through atmospheric turbulence. *J. Opt. Soc. Amer.* 1A:510–519.

von der Lühe, O. 1985. High resolution speckle imaging of solar small scale structure: The influence of anisoplanatism. In *High Resolution in Solar Physics,* ed. R. Muller (Berlin: Springer-Verlag), pp. 96–102.

von der Lühe, O. 1987*a.* Photospheric fine structure close to a sunspot. In *The Role of Fine-Scale Magnetic Fields on the Structure of the Solar Atmosphere,* eds. E.-H. Schröter, M. Vázquez and A. A. Wyller (Cambridge: Cambridge Univ. Press), pp. 156–161.

von der Lühe, O. 1987*b.* Application of the Knox-Thompson method to solar observations. In *Interferometric Imaging in Astronomy,* ed. J. Goad (Tucson: Natl. Optical Astron. Obs.), pp. 37–40.

von der Lühe, O. 1987*c.* Photon noise analysis for LEST multidither adaptive optical system. In *Adaptive Optics in Solar Astronomy,* eds. F. Merkle, O. Engvold and R. Falomo, LEST Tech. Rept. No. 28 (Oslo: Univ. of Oslo), pp. 255–262.

von der Lühe, O. 1988. Measurements of characteristics of image motion with a solar image stabilizing device. *Astron. Astrophys.* 205:354–360.

von der Lühe, O. 1988. Wavefront error measurement technique using extended, incoherent light sources. *Opt. Eng.* 27:1078–1087.

von der Lühe, O., ed. 1989*a.* *High Spatial Resolution Solar Observations* (Sunspot, N.M.: Sacramento Peak Obs.).

von der Lühe, O. 1989*b.* Solar speckle imaging. In *High Spatial Resolution Solar Observations,* ed. O. von der Lühe (Sunspot, N.M.: Sacramento Peak Obs.), pp. 147–165.

von der Lühe, O., and Dunn, R. B. 1987. Solar granulation power spectra from speckle interferometry. *Astron. Astrophys.* 177:265–276.

von der Lühe, O., and Zirker, J. B. 1988. Scientific goals for solar interferometry. In *High Resolution Imaging by Interferometry Part I,* ed. R. Merkle (Garching: European Southern Obs.), pp. 77–94.

von der Lühe, O., Widener, A. L., Rimmele, Th., Spence, G., Dunn, R. B., and Wilborg, P. 1989. A solar feature correlation tracker for ground-based telescopes. *Astron. Astrophys.,* in press.

von Neuman, J., and Wigner, E. 1929. Über merkwürdige diskrete Eigenwerte. Über das Verhalten von Eigenwerten bei adiabatischen Prozessen. *Phys. Z.* 30:467–470.

von Uexküll, M., Kneer, F., Malherbe, J. M., and Mein, P. 1989. Oscillations of the Sun's chromosphere. V. Importance of network dynamics for chromospheric heating. *Astron. Astrophys.* 208:290–296.

von Weizsäcker, C. F. 1937. Ü ber Elementumwandlungen im Innern derSterne I. *Physik. Z.* 38:176–191.

von Weizsäcker, C. F. 1938. Über Elementumwandlungen im Innern der Sterne II. *Physik. Z.* 39:663–666.

Vorontsov, S. V. 1988. Helioseismological inverse problem: Sound speed in the solar interior. In *Seismology of the Sun and Sun-Like Stars,* ed. E. J. Rolfe, ESA SP-286, pp. 475–480.

Vorontsov, S. V., and Zharkov, V. N. 1989. Helioseismology. *Soviet Sci. Rev. A: Astrophys. Space Sci. Rev.* 7(1):1–103.

Vuilleumier, J. L. 1990. Neutrino physics at reactors. In *Fundamental Symmetries in Nuclei and Particles,* eds. H. Hendrikson and P. Vogel (Singapore: World Scientific), pp. 116–130.

Wagner, W. J. 1983. SERF studies of mass motions arising in flares. *Adv. Space Res.* 2:203–213.

Wagner, W. J. 1984. Coronal mass ejections. *Ann. Rev.Astron. Astrophys.* 22:267–284.

Wagoner, R. V. 1969. Synthesis f the elements within obs exploding from very high temperatures. *Astrophys. J. Suppl.* 18:247–295.

Waldmeier, M. 1961. *The Sunspot-Activity in the Years 1610–1960* (Zürich:Schulthess).

Walén, C. 1947. On the distribution of the solar general magnetic field and remarks concerning geomagnetism and solar rotation. *Ark. Mat. Astr. Fys.* 33A, no. 18.

Wallace, L., Huang, Y. R., and Livingston, W. 1988. The Sun as a star: On wavelength stability. *Astrophys. J.* 327:399–404.

Wallenhorst, S. G., and Howard, R. 1982. On the dissolution of sunspot groups. *Solar Phys.* 76:203–209.

Walton, S. R. 1987. Flux tube models of solar plages. *Astrophys. J.* 312:909–929.

Wambsganss, J. 1988. Hydrogen-helium-diffusion in solar models. *Astron. Astrophys.* 205: 125–128.

Wang, H., and Zirin, H. 189. Flows, flares and formation of umbrae and light bridges in BBSO No. 1167. *Solar Phys.*, submitted.

Wang, J., Zirin, H., and Shi, Z. 1985. The smallest observable elements of magnetic flux. *Solar Phys.* 98:241–252.

Wang, H., Zirin, H., Patterson, A., Ai, G., and Zhang, H. 1989. Seventy-five hours of coordinated videomagnetograph observations. *Astrophys. J.* 343:489–493.

Wang, Y.-M., and Sheeley, N. R., Jr. 1988. The solar origin of long-term variations of the interplanetary magnetic field strength. *J. Geophys. Res.* 93:11227–11236.

Wang, Y.-M., Sheeley, N. R., Jr., Nash, A. G., and Shampine, L. R. 1988. The quasi-rigid rotation of coronal magnetic fields. *Astrophys. J.* 327:427–450.

Wang, Y.-M., Nash, A. G., and Sheeley, N. R., Jr. 1989. Magnetic flux transport on the Sun. *Science* 245:712–718.

Ward, F. 1965. The general circulation of the solar atmosphere and the maintenance of the equatorial acceleration. *Astrophys. J.* 141:534–547.

Ward, F. 1966. Determination of the solar rotation rate from the motion of identifiable features. *Astrophys. J.* 145:416–425.

Ward, F. 1973. The latitudinal motion of sunspots and solar meridional circulations. *Solar Phys.* 30:527–537.

Wasiutynski, J. 1946. Studies in hydrodynamics and structure of stars and planets. *Astrophys. Norvegica* vol. 4.

Wasiutynski, J. 1958. The dynamic constitution of the outer layer of the Sun. *Ann. Astrophys.* 21:137–150.

Watson, W. D. 1969. The effect of auto-ionization lines on the opacity of stellar interiors. *Astrophys. J.* 157:375–387.

Webb, A. R., and Roberts, B. 1978. Vertical motions in an intense magnetic flux tube. II. Convective instability. *Solar Phys.* 59:249–274.

Webb, D. F., and Hundhausen, A. J. 1987. Activity associated with the solar origin of coronal mass ejections. *Solar Phys.* 108:383–401.

Weigelt, G. P., and Wirnitzer, B. 1983. Image reconstruction by the speckle-masking method. *Opt. Lett.* 8:389–391.

Weiss, N. O. 1964. Convection in the presence of restraints. *Phil. Trans. Roy. Soc. London* A256:99–147.

Weiss, N. O. 1965. Convection and the differential rotation of the Sun. *Observatory* 85:37–39.

Weiss, N. O. 1966. The expulsion of magnetic flux by eddies. *Proc. Roy. Soc.* A293:310–328.

Weiss, N. O. 1981a. Convection in an imposed magnetic field. I. The development of nonlinear convection. *J. Fluid Mech.* 108:247–272.

Weiss N. O. 1981b. Convection in an imposed magnetic field. II. The dynamical regime. *J. Fluid Mech.* 108:273–289.

Wellck, R. E., and Nakagawa, Y. 1973. Force-free magnetic field computation. I. NCAE Tech. Note STR-87 (Boulder: Natl. Center for Atmos. Res).

Weneser, J., and Friedlander, G. F. 1987. Solar neutrinos: Questions and hypotheses. *Science* 235:755–769.

Wenhan, J., Yueai, L., Zhang, L., Fang, S., and Goumao, T. 1988. Study of a modal multidither image sharpening adaptive optics system for solar applications. In *LEST Foundation Tech. Rept.* No. 30, eds. O. Engvold and O. Hauge (Oslo: Univ. of Oslo).

Wentzel, D. G. 1976. Coronal heating by Alfvén waves. II. *Solar Phys.* 50:343–360.

Wentzel, D. G. 1986. Solar activity and the coupling of g-mode oscillations. *Astrophys. J.* 300:824–829.

Wentzel, D. G. 1987. Solar oscillations: Generation of a g-mode by two p-modes. *Astrophys. J.* 319:966–970.

Werner, H. J., Rosmus, P., and Reinsch, E. A. 1983. Molecular properties from MCSCF-SCEP wave functions. I. Accurate dipole moment functions of OH, OH$^-$ and OH$^+$. *J. Chem. Phys.* 79:905–916.

Werner, K., and Husfeld, D. 1985. Multilevel non-LTE line formation calculations using approximate λ-operators. *Astron. Astrophys.* 148:417–422.

Werntz, C., and Brennan, J. G. 1973. Solar neutrinos from the ^{3}He$(p,e^+v)^4$He reaction. *Phys. Rev.* C8:1545–1546.

West, E. A., and Hagyard, M. J. 1983. Interpretation of vector magnetograph data including magneto-optic effects. *Solar Phys.* 88:51–64.

Weymann, R., and Sears, R. L. 1965. The depth of the convective envelope on the lower main sequence and the depletion of lithium. *Astrophys. J.* 142:174–181.

Whitaker, W. A. 1963. Heating of the solar corona by gravity waves. *Astrophys. J.* 137:914–930.

White, O. R., and Livingston, W. C. 1981. Solar luminosity variation. III. Calcium K variation from solar minimum to maximum in cycle 21. *Astrophys. J.* 249:798–816.

Widing, K. G. 1975. Fe XXVI emission in solar flares observed with the NRL/ATM XUV slitless spectrograph. In *Solar Gamma-, X-, and EUV Radiation: Proc. IAU Symp. 68*, ed. S. R. Kane (Dordrecht: D. Reidel), pp. 153–163.

Widing, K. G., and Feldman, U. 1989. Abundance variations in the outer solar atmosphere observed in Skylab spectroheliograms. *Astrophys. J.* 344:1045–1050.

Widing, K. G., Feldman, U., and Bhatia, A. K. 1986. The extreme-ultraviolet spectrum (300–630Å) of an erupting prominence observed from Skylab. *Astrophys. J.* 308:982–992.

Wiedemann, G., Jennings, D. E., Moseley, H., Lamb, G., Hanel, R., Kunde, V., Stapelbrock, M. G., and Petroff, M. D. 1989. Postdispersion system for astronomical observations with Rourier transform spectrometers in the thermal infrared. *Appl. Opt.* 28:139–145.

Wiehr, E. 1978. A unique magnetic field range for non-spot solar magnetic regions. *Astron. Astrophys.* 69:279–284.

Wiehr, E. 1979. Evidence for a lower limit of solar magnetic field strengths. *Astron. Astrophys.* 73:L19–L20.

Wiehr, E. 1985. Spatial and temporal variation of circular Zeeman profiles in isolated solar Ca$^+$K structures. *Astron. Astrophys.* 149:217–220.

Wiehr, E., and Kneer, F. 1988. Spectroscopy of the solar photosphere with high spatial resolution. *Astron. Astrophys.* 195:310–314.

Wiehr, E., Stellmacher, G., Knolker, M., and Grosser, H. 1986. The sharp decrease of Evershed effect and magnetic field at the outer sunspot border. *Astron. Astrophys.* 155:402–406.

Wiehr, E., Pahlke, K.-D., and Koch, A. 1989. Compensation of image motion and discrimination of blurring for spectroscopy. In *High Spatial Resolution Solar Observations*, ed. O. von der Lühe (Sunspot, N.M.: Sacramento Peak Obs.), pp. 100–103.

Wiese, W. L., Kelleher, D. E., and Paquette, D. R. 1972. Detailed study of the Stark broadening of Balmer lines in a high density plasma. *Phys. Rev.* A6:1132–1153.

Wiesmeier, A., and Durrant, C. J. 1981. The analysis of solar limb observations. *Astron. Astrophys.* 104:207–210.

Wiezorek, C., Kräwinkel, H., Santo, R., and Wallek, L. 1977. Study of the ^{7}Be$(p,\gamma)^8$B reaction. *Z. Phys.* A282:121–123.

Wilcox, J. M., and Howard, R. 1970. Differential rotation of the photospheric magnetic field. *Solar Phys.* 13:251–260.

Wilcox, J. M., Schatten, K. H., Tanenbaum, A. S., and Howard, R. 1970. Photospheric magnetic field rotation: Rigid and differential. *Solar Phys.* 13:251–260.

Wild, J. P. 1950. Observations of the spectrum of high-intensity solar radiation at metre wavelengths. III. Isolated bursts. *Australian J. Sci.* A3:541–557.

Wild, J. P., and McCready, L. L. 1950. Observations of the spectrum of high-intensity solar radiation at metre wavelengths. I. The apparatus and spectral types of solar burst observed. *Australian J. Sci.* A3:387–398.

Wildt, R. 1939. Negative ions of hydrogen and the opacity of stellar atmospheres. *Astrophys. J.* 89:617–620.

Wilkinson, D. H. 1982. Analysis of neutron B-decay. *Nucl. Phys.* A377:474–504.

Williams, G. E. 1981. Sunspot periods in the late Precambrian glacial climate and solar-planetary relations. *Nature* 291:624–628.

Williams, G. E. 1985. Solar affinity of sedimentary cycles in the late Precambrian Elatina formation. *Australian J. Phys.* 38:1027–1043.

Williams, G. E. 1988. Cyclicity in the late precambrian Elatina formation, South Australia: Solar or tidal signature? *Clim. Change* 13:117–128.

Williams, R. D., and Koonin, S. E. 1981. Direct-capture cross sections at low energy. *Phys. Rev.* C23:2773–2774.

Willson, R. C. 1979. Active cavity radiometer type IV. *J. Appl. Opt.* 18:179–188.

Willson, R. C. 1981. Solar total irradiance observations by active cavity radiometers. *Solar Phys.* 74:217–229.

Willson, R. C. 1984. Measurement of solar total irradiance and its variability. *Space Sci. Rev.* 38:203–242.

Willson, R. C., and Hudson, H. S. 1988. Solar luminosity variations in solar cycle 21. *Nature* 332:810–813.

Willson, R. C., Gulkis, S., Janssen, M., Hudson, H. S., and Chapman, G. A. 1981. Observations of solar irradiance variability. *Science* 211:700–702.

Wilson, O. C. 1978. Chromospheric variations in main-sequence stars. *Astrophys. J.* 226:379–396.

Wilson, P. R. 1970. A three-component model for the formation of the chromospheric Ca II K line. *Solar Phys.* 15:139–147.

Wilson, P. R. 1980. The interaction of acoustic waves with flux tubes. *Astrophys. J.* 237:1008–1014.

Wilson, P. R. 1981. Faculae, filigree and calcium bright points. *Solar Phys.* 69:9–14.

Wilson, P. R. 1986. The generation of magnetic fields in photospheric layers. *Solar Phys.* 106:1–28.

Wilson, P. R. 1987. Solar rotation and the giant cells. *Solar Phys.* 110:59–71.

Wilson, P. R. 1988. The solar dynamo and the convective rolls. *Solar Phys.* 117:217–226.

Wilson, P. R., and Simon, G. W. 1983. Flux changes in small magnetic regions. *Astrophys. J.* 273:805–821.

Wilson, P. R., Altrock, R. C., Harvey, K. L., Martin, S. F., and Snodgrass, H. B. 1988. The extended solar activity cycle. *Nature* 333:748–750.

Wilson, R. M. 1987. On the proposed associations of solar neutrino flux with solar particles, cosmic rays, and the solar activity cycle. *Solar Phys.* 112:1–5.

Wilson, R. M., Rabin, D., and Moore, R. L. 1987. 10.7-cm solar radio flux and the magnetic complexity of active regions. *Solar Phys.* 111:279–285.

Wilson, P. R., McIntosh, P. S., and Snodgrass, H. B. 1990. The reversal of the solar polar magnetic fields. I. The surface transport of magnetic flux. *Solar Phys.* 127:1–9.

Winkler, K.-H. A., and Newman, M. J. 1980a. Formation of solar-type stars in spherical symmetry. I. The key role of the accretion shock. *Astrophys. J.* 236:201–211.

Winkler, K.-H. A., and Newman, M. J. 1980b. Formation of solar-type stars in spherical symmetry. II. Effects of detailed constitutive relations. *Astrophys. J.* 238:311–325.

Withbroe, G. L. 1983. Role of spicules in heating the solar atmosphere. *Astrophys. J.* 267:825–836.

Withbroe, G. L. 1988. The temperature structure, mass, and energy flow in the corona and inner solar wind. *Astrophys. J.* 267:442–467.

Withbroe, G. L., and Noyes, R. W. 1977. Mass and energy flow in the solar chromosphere and corona. *Ann. Rev. Astron. Astrophys.* 15:363–387.

Withbroe, G. L., et al. 1976. EUV transients observed at the solar pole. *Astrophys. J.* 203:528–532.

Withbroe, G. L., Kohl, J. L., Weiser, H., and Munro, R. H. 1985. Coronal temperatures heating and energy flow in a polar region of the Sun at solar maximum. *Astrophys. J.* 297:324–337.

Wittmann, A. 1974. Computation and observation of Zeeman multiplet polarization in Fraunhofer lines. II. Computation of Stokes parameter profiles. *Solar Phys.* 35:11–29.

Wittmann, A. 1977. Spectral synthesis in a magnetic field. *Astron. Astrophys.* 54:175–181.

Wittmann, A. 1979. Observations of solar granulation—A review. In *Small Scale Motions on the Sun*, Mitteilungen aus dem Kipenheuer-Institut, No. 179, pp. 29–54.

Wöhl, H. 1988. On the possible detection of large-scale meridional motions by analyzing A-type spots from the Greenwich photoheliographic results. *Solar Phys.*, in press.

Wohl, H., and Nordlund, Å. 1985. A comparison of artificial solar granules with real solar granules. *Solar Phys.* 97:213–221.

Wolfendale, A. W., Young, E. C. M., and Davis, R. 1972. Indirect determination of the photonuclear cross section above 20 GeV. *Nature Phys. Sci.* 238:1300–1301.

Wolfenstein, L. 1978*a*. Neutrino oscillations in matter. *Phys. Rev.* 17D:2369–2374.

Wolfenstein, L. 1978*b*. Oscillations among three neutrino types and CP violation. *Phys. Rev.* 18D:958–960.

Wolff, C. L. 1972. The five-minute oscillations as nonradial pulsations of the entire Sun. *Astrophys. J.* 177:L87–L92.

Wolfs, F. L. H., Freedman, S. J., Nelson, J. E,. Dewey, M. S., and Greene, G. L. 1989. Measurement of the ^{3}He$(n,\gamma)^4$He cross section at thermal neutron energies. *Phys. Rev. Lett.* 63:2721–2724.

Wolfsberg, K., and Kocharov, G. E. 1991. Solar neutrinos and the history of the Sun. In *The Sun in Time*, eds. C. P. Sonett, M. S. Giampapa and M. S. Matthews (Tucson: Univ. of Arizona Press), pp. 288–310.

Woodgate, B. E., Shine, R. A., Poland, A. I., and Orwig, L. E. 1983. Simultaneous ultraviolet and hard X-ray bursts in the impulsive phase of solar flares. *Astrophys. J.* 265:530–534.

Woodard, M. F. 1984*a*. Short Period Oscillations in the Total Solar Irradiance. Ph.D. Thesis, Univ. of California, San Diego.

Woodard, M. F. 1984*b*. Upper limit on solar interior rotation. *Nature* 309:530–532.

Woodard, M. F. 1989. Distortion of high-degree solar p-mode eigenfunctions by latitudinal differential rotation. *Astrophys. J.* 347:1176–1182.

Woodard, M. F., and Hudson, H. S. 1983. Frequencies, amplitudes, and linewidths of solar oscillations from total irradiance observations. *Nature* 305:589–593.

Woodard, M. F., and Libbrecht, K. G. 1988. On the measurement of solar rotation using high-degree p-mode oscillations. In *Seismology of the Sun and Sun-Like Stars*, ed. E. J. Rolfe, ESA SP-286, pp. 67–71.

Woodard, M. F., and Noyes, R. W. 1985. Change of solar oscillation eigenfrequencies with the solar cycle. *Nature* 318:449–450.

Woodard, M. F., and Noyes, R. W. 1986. Change of solar oscillation eigenfrequencies with the solar cycle. In *Seismology of the Sun and the Distant Stars*, ed. D. O. Gough (Dordrecht: D. Reidel), pp. 303–304.

Woodard, M. F., and Noyes, R. W. 1988. Time variations of the frequencies of low-degree solar p modes. In *Advances in Helio- and Asteroseismology: Proc. IAU Symp. 123*, eds. J. Christensen-Dalsgaard and S. Frandsen (Dordrecht: D. Reidel), pp. 197–200.

Woods, D. T., and Cram, L. E. 1981. High resolution spectroscopy of the disk chromosphere. VII. Oscillations in plage and quiet Sun regions. *Solar Phys.* 69:233–238.

Woodward, P. R., and Porter, D. H. 1987. Simulations of compressible convection in stars. *Bull. Amer. Astron. Soc.* 19:1023. (abstract).

Worden, S. P. 1975. Infrared observations of supergranule temperature structure. *Solar Phys.* 45:521–532.

Worden, S. P., and Simon, G. W. 1976. A study of supergranulation using a diode array magnetograph. *Solar Phys.* 46:73–91.

Wu, S. T., Chang, H. M., and Hagyard, M. J. 1985. On the numerical computation of nonlinear force-free magnetic fields. In *Measurements of Solar Vector Magnetic Fields*, ed. M. J. Hagyard, NASA CP-237, pp. 17–48.

Wu, S. T., Jager, C. de, Dennis, B. R., Hudson, H. S., Simnett, G. M., Strong, K. T., Bentley, R. D., Bornmann, P. L., Bruner, M. E, Cargill, P. J., Crannell, C. J., Doyle, J. G., Hyder, C. L., Kopp, R. N., Lemer, J. R., Martin, S. F., Pallavicini, R., Peres, G., Serio, S., Sylwester, J., and Veck, N. J. 1986. Flare energetics. In *Energetic Phenomena on the Sun*, eds. M. R. Kundu and B. E. Woodgate, NASA CP-2439, pp. 377–492.

Xiong, D. R. 1981. Statistical theory of non-local convection in chemically inhomogeneous stars. *Scientia Sinica* 24:1406–1417.

Yallop, B. D., and Hohenkerk, C. Y. 1980. Distribution of sunspots 1874–1976. *Solar Phys.* 68:303–305.

Yallop, B. D., Hohenkerk, C., Murdin, L., and Clark, D. H. 1982. Solar rotation from 17th century records. *Quart. J. Roy. Astron. Soc.* 23:213–222.

Yamagushi, Sh. 1984. Compressible convection at large initial superadiabatic temperature gradient. *Publ. Astron. Soc. Japan* 36:613–632.

Yamagushi, Sh. 1985. Compressible convection with fixed heat flux. *Publ. Astron. Soc. Japan* 37:735–746.

Yang, W. H., Sturrock, P. A., and Antiochos, S. K. 1986. Force-free magnetic fields: The magneto-frictional method. *Astrophys. J.* 309:383–391.

Ye, S.-H., and Jin, J.-H. 1986. Monochromatic images in Stokes parameters and the structure of magnetic fields in sunspots. *Solar Phys.* 104:273–285.

Ye, S.-H., and Jin, J.-H. 1987. The Faraday rotation of sunspots. *Solar Phys.* 112:305–312.

Yeh, T. 1970. A three-fluid model of solar winds. *Planet. Space Sci.* 18:199–215.

Yerle, R. 1986. Limb darkening fluctuations: The discrepancy with Doppler data and the 160-minute oscillation. *Astron. Astrophys.* 161:L5–L8.

Yi, L., and Hill, H. A. 1988. Evidence of gravity mode signals in 1983 differential radius observations of th Sun. *Bull. Amer. Astron. Soc.* 20:1011 (abstract).

Yi, L., and Czarnowski, W. M. 1987. Comparison of 1983 and 1979 SCLERA observations. In *Stellar Pulsation* (Berlin: Springer-Verlag), pp. 311–313.

Yoshimura, H. 1971. Complexes of activity of the solar cycle and very large scale convection. *Solar Phys.* 18:417–433.

Yoshimura, H. 1972. On the dynamo action of the global convection in the solar convection zone. *Astrophys. J.* 178:863–886.

Yoshimura, H. 1975*a*. Solar-cycle dynamo wave propagation. *Astrophys. J.* 201:740–748.

Yoshimura, H. 1975*b*. A model of the solar cycle driven by the dynamo action of the global convection in the solar convection zone. *Astrophys. J. Suppl.* 29:467–494.

Yoshimura, H. 1978*a*. Nonlinear astrophysical dynamos: The solar cycle as a nonlinear oscillation of the general magnetic field driven by the nonlineary dynamo and the associated modulation of the differential-rotation-global-convection system. *Astrophys. J.* 220:692–711.

Yoshimura, H. 1978*b*. Nonlinear astrophysical dynamos: Multiple-period dynamo wave oscillations and long-term modulations of the 22 year solar cycle. *Astrophys. J.* 226:706–719.

Yoshimura, H. 1979. The solar-cycle period-amplitude relation as evidence of hysteresis of the solar-cycle nonlinear magnetic oscillation and the long-term (55 year) cyclic modulation. *Astrophys. J.* 227:1047–1058.

Yoshimura, Y. 1981. Solar cycle Lorentz force waves and the torsional oscillations of the Sun. *Astrophys. J.* 247:1102–1112.

Young, R. E. 1974. Finite amplitude thermal convection in a spherical shell. *J. Fluid Mech.* 63:695–721.

Yudin, O. I. 1968. Quasiperiodic low-frequency solar radiowave fluctuations. *Soviet Phys.—Doklady* 13:503–505.

Zahkarov, Yu. 1977. Internal contamination backgrounds in some radiochemical solar neutrino detectors. *Bull. Acad. Sci. USSR. Phys. Ser.* 41:172–174.

Zahn, J.-P. 1970. Forced oscillations in close binaries. The adiabatic approximation. *Astron. Astrophys.* 4:452–461.

Zahn, J.-P. 1974. Rotational instabilities and stellar evolution. In *Stellar Instability and Evolution: Proc. IAU Symp. 129,* eds. P. Ledoux, A. Noëls and A. W. Rodgers (Dordrecht: D. Reidel), pp. 185–195.

Zahn, J.-P. 1983. Instability and mixing processes in upper main sequence stars. In *Astrophysical Processes in Upper Main Sequence Stars,* eds. A. N. Cox, S. Vauclair and J.-P. Zahn (Geneva: Geneva Obs), pp. 253–329.

Zahn, J.-P. 1987. Turbulent transport in the radiative zone of a rotating star. In *The Internal Solar Angular Velocity,* eds. B. Durney and S. Sofia (Dordrecht: D. Reidel), pp. 201–212.

Zahn, J.-P. 1987. In *Astrophysical Fluid Dynamics,* eds. J. P. Zahn and J. Zinn-Justin (New York: Elsevier), in press.

Zahn, J.-P. 1988. Preprint. Lectures given at the Summer School of Les Houches.

Zahn, J.-P., Toomre, J., and Latour, J. 1982. Nonlinear model analysis of penetrative convection. *Geophys. Astrophys. Fluid Dyn.* 22:159.

Zähringer, K., and Ulmschneider, P. 1987. Adiabatic longitudinal-transverse magnetohydrodynamic tube waves. In *The Role of Fine-Scale Magnetic Fields on the Structure of the Solar Atmosphere*, eds. E.-H. Schröter, M. Vázquez and A. A. Wyller (Cambridge: Cambridge Univ. Press), pp. 243–247.

Zatsepin, G. T. 1982. In *Proc. 8th Intl. Workshop on Weak Interactions and Neutrinos*, ed. A. Morales, p. 754.

Zatsepin, G. T., and Kuzmin, V. A. 1964. *Vest. Akad. Nauk SSR* 34:50–55.

Zatsepin, G. T., Kopolov, A. V., and Shirokova, E. K. 1981. Investigation of the background processes from cosmic ray muons for the chlorine solar neutrino detector. *Soviet J. Nucl. Phys.* 33:200–205.

Zayer, I., Solanki, S. K., and Stenflo, J. O. 1988. The internal magnetic field distribution and the diameters of solar magnetic elements. *Astron. Astrophys.* 211:463–475.

Zeldovich, Ya. B., Ruzmaikin, A. A., and Sokoloff, D. D. 1983. *Magnetic Fields in Astrophysics* (London: Gordon and Breach).

Zhugzhda, Y. D., and Dzhalilov, N. S. 1982. Transformation of magnetogravitational waves in the solar atmosphere. *Astron. Astrophys.* 112:16–23.

Zhugzhda, Yu. D., and Locans, V. 1979. In *Leningrad Seminar on Cosmic Physics*, ed. G. Kocharov, p. 28.

Zhugzhda, Yu. D., and Locans, V. 1982. Tunneling and interference of Alfvén waves. *Solar Phys.* 76:77–108.

Zhugzhuda, Yu. D., Locans, V., and Staude, J. 1987. The interpretation of oscillations in sunspot umbrae. *Astr. Nachr.* 308:257–269.

Zirin, H. 1978. Studies of solar flares using optical, X-ray and radio data. *Solar Phys.* 58:95–120.

Zirin, H. 1985. Evolution of weak solar magnetic fields. *Austrian J. Phys.* 38:961–969.

Zirin, H. 1988. *Astrophysics of the Sun* (Cambridge: Cambridge Univ. Press).

Zirin, H., and Hirayama, T. 1985. HeII emission from solar flares. *Astrophys. J.* 299:536–541.

Zirin, H., and Liggett, M. A. 1987. Delta spots and great flares. *Solar Phys.* 113:267–283.

Zirin, H., and Neidig, D. F. 1981. Continuum emission in the 1980 July 1 solar flare. *Astrophys. J.* 248:L45–L48.

Zirin, H., and Popp, B. 1988. Observations of the 12 micron Mg I lines in various solar features. *Astrophys. J.* 340:571–578.

Zirin, H., and Tanaka, K. 1973. The flares of August 1972. *Solar Phys.* 32:173–207.

Zirin, H., and Wang, H. 1989. Video image selection studies of granules, pores, and penumbral flows near a large sunspot. *Solar Phys.* 119:245–255.

Zirker, J. B., ed. 1977. *Coronal Holes and High Speed Solar Wind Streams* (Boulder: Colorado Assoc. Univ. Press).

Zirker, J. B. 1981. The solar corona and the solar wind. In *The Sun as a Star*, ed. S. Jordan, NASA SP-450, pp. 135–162.

Zirker, J. B. 1985a. Progress in coronal physics. *Solar Phys.* 100:281–287.

Zirker, J. B. 1985b. Prominence hydrogen lines at 10–20 microns. *Solar Phys.* 102:33–40.

Zirker, J. B. 1987. Interferometric imaging: A numerical simulation. *Solar Phys.* 111:235–242.

Zirker, J. B. 1989. Quiescent prominences. *Solar Phys.* 119:341–357.

Zirker, J. B., and Brown, T. M. 1986. Phase recovery with dual nonredundant arrays. *J. Opt. Soc. Amer.* 3:2077–2081.

Zwaan, C. 1965. Sunspot models. A study of sunspot spectra. *Rech. Astr. Obs. Utrecht* 17(4):1–182.

Zwaan, C. 1978. On the appearance of magnetic flux in the solar photosphere. *Solar Phys.* 60:213–240.

Zwaan, C. 1981. Solar magnetic structure and the solar activity cycle, review of observational data. In *The Sun as a Star*, ed. S. Jordan, NASA SP-450, pp. 163–179.

Zwaan, C. 1985. The emergence of magnetic flux. *Solar Phys.* 100:397–414.

Zwaan, C. 1987. Elements and patterns in the solar magnetic field. *Ann. Rev. Astron. Astrophys.* 25:83–111.

Zweibel, E. G., and Bogdan, T. J. 1986. Effects of fibril magnetic fields on solar p modes. II. Calculation of mode frequency shifts. *Astrophys. J.* 308:401–412.

Zweibel, E. G., and Däppen, W. 1989. Effects of magnetic fibrils on solar oscillation frequencies: Mean field theory. *Astrophys. J.* 343:994–1003.

Zwickl, R. D., Asbridge, J. R., Bame, S. J., Feldman, W. C., Gosling, J. T., and Smith, E. J. 1983. Plasma properties of driver gas following interplanetary shocks observed by ISEE-3. In *Proc. Fifth Intl. Solar Wind Conf.*, ed. M. Neugebauer, NASA CP-2280, pp. 711–717.

Zyskind, J. L., and Parker, P. D. 1979. Remeasurement of the low energy cross section for the $^{15}N(p,\alpha_0)^{12}C$ reaction. *Nucl. Phys.* A320:404–412.

Acknowledgments

ACKNOWLEDGMENTS

The editors acknowledge National Aeronautics and Space Administration Grant NASW-4389, the Los Alamos National Laboratory and the University of Arizona for support of the preparation of this book. They wish to thank M. Magisos, who organized the "Solar Interior and Atmospheres" conference. The editors would also like to thank J. E. Frecker, who volunteered as one of the proofreaders of this book. The following authors wish to acknowledge specific funds involved in supporting the preparation of this book.

Anderson, L. S.: NASA Grant NAGW-99-8200
Avrett, E. H.: NASA Grant NSG-7054
Bogart, R. S.: NSF Grant ATM9022249, ONR Grant N00014-89-J-1024 *and* NASA Grant NGR-559
Chan, K. L.: NSF Grant AST-88-15457
Cook, J. W.: NASA Grant DPRW-14, 541
Davis, R.: NSF Grants AST-88-12418 *and* AST-86-11924
Demarque, P.: NASA Solar-Terrestrial Theory Program Grant NAGW-777 *and* NASA Astrophysics Theory Program Grant NAGW-778
Deming, D.: RTOP Grant 170-38-53-10
Goode, P. R.: Air Force Office of Scientific Research Grant AF-89-0048
Guenther, D. B.: NASA Solar-Terrestrial Theory Program Grant NAGW-777 *and* NASA Astrophysics Theory Program Grant NAGW-778
Hill, H.: U.S. Dept. of Energy Grant DE-FG02-87ER13670 *and* Air Force Office of Scientific Research AFOSR 87-0105
Hoeksema, J. R.: NSF Grant ATM9022249, ONR Grant N00014-89-J-1024 *and* NASA Grant NGR-559
Kalkofen, W.: NASA Grant NAGW-1568
Kawaler, S.: NASA Grant NAGW-778
Kurucz, R. L.: NASA Grant NSG-7054
Nordlund, A.: NSF Grant AST-83-16231
Noyes, R. W.: NSF Grant AST-86-16545
Steffen, M.: Deutsche Forschungsgemeinschaft Grant Ho 596/21-2
Stein, R. F.: NSF Grant AST-83-16231
Title, A. M.: NASA Contracts NAS-8-32805-SOUP *and* NAS-5-26813-OSL
Ulrich, R.: NSF Grant AST-90-150815108 and NASA Grant NAGW-472

Index

INDEX*